BIOCHEMISTRY

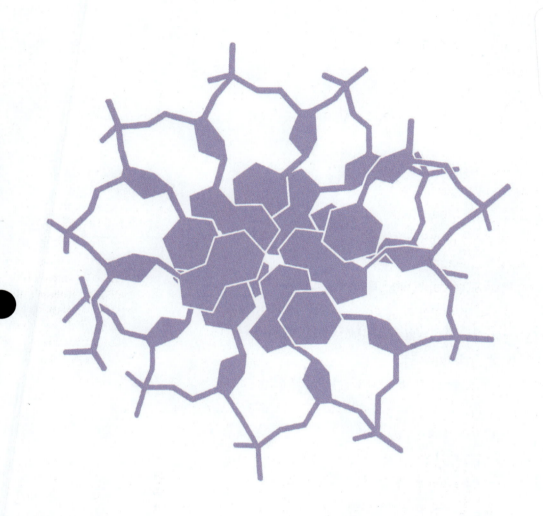

TENTH EDITION
BIOCHEMISTRY

Jeremy M. Berg

Gregory J. Gatto, Jr.

Justin K. Hines

Jutta Beneken Heller

John L. Tymoczko

Lubert Stryer

Austin • Boston • New York • Plymouth

Senior Vice President, STEM: Daryl Fox
Executive Program Director: Sandy Lindelof
Senior Program Manager: Liz Simmons
Marketing Manager: Maureen Rachford
Director of Development: Barbara Yien
Senior Development Editor: Jennifer Angel
Editorial Project Manager: Jennifer Hart
Director of Content: Heather Southerland
Executive Media Editor: Amy Thorne
Senior Media Editors: Cassandra Korsvik, Kelsey Hughes
Senior Lead Content Developer: Lily Huang
Assistant Editor: Angelica Hernandez
Senior Director of Content Management Enhancement: Tracey Kuehn
Senior Managing Editor: Lisa Kinne
Senior Content Project Manager: Peter Jacoby
Workflow Project Manager: Paul Rohloff
Director of Design, Content Management: Diana Blume
Senior Design Services Manager: Natasha A. S. Wolfe
Senior Cover Design Manager: John Callahan
Text Designer: Maureen McCutcheon
Art Manager: Matthew McAdams
Illustrations: Jeremy Berg with Network Graphics, Gregory J. Gatto,
    Jr., Adam Steinberg, Lumina Datamatics, Inc.
Senior Director of Digital Production: Keri deManigold
Senior Media Production Manager: Elton Carter
Senior Media Permissions Manager: Christine Buese
Photo Researchers: Richard Fox and Krystyna Borgen,
    Lumina Datamatics, Inc.
Composition: Lumina Datamatics, Inc.
Printing and Binding: Lakeside Book Company

Library of Congress Control Number: 2022943738

Gregory J. Gatto, Jr., is an employee of GlaxoSmithKline (GSK), which
has not supported or funded this work in any way. Any views
expressed herein do not necessarily represent the views of GSK.

Student Edition Paperback:
ISBN-13: 978-1-319-33362-1
ISBN-10: 1-319-33362-1

Student Edition Loose-leaf:
ISBN-13: 978-1-319-49840-5
ISBN-10: 1-319-49840-5

International Edition:
ISBN-13: 978-1-319-49850-4
ISBN-10: 1-319-49850-7

Printed in the United States of America

1 2 3 4 5 6   28 27 26 25 24 23

Macmillan Learning
120 Broadway
New York, NY 10271
www.macmillanlearning.com

w.h.freeman
Macmillan Learning

In 1946, William Freeman founded W. H. Freeman and Company and published Linus Pauling's *General Chemistry*, which revolutionized the chemistry curriculum and established the prototype for a Freeman text. W. H. Freeman quickly became a publishing house where leading researchers can make significant contributions to mathematics and science. In 1996, W. H. Freeman joined Macmillan and we have since proudly continued the legacy of providing revolutionary, quality educational tools for teaching and learning in STEM.

To our teachers and our students,
including our late coauthor John Tymoczko,
whose devotion to his students was an inspiration

BRIEF CONTENTS

1 Biochemistry in Space and Time 1

2 Protein Composition and Structure 33

3 Binding and Molecular Recognition 68

4 Protein Methods 100

5 Enzymes: Core Concepts and Kinetics 141

6 Enzyme Catalytic Strategies 179

7 Enzyme Regulatory Strategies 210

8 DNA, RNA, and the Flow of Genetic Information 236

9 Nucleic Acid Methods 264

10 Exploring Evolution and Bioinformatics 301

11 Carbohydrates and Glycoproteins 324

12 Lipids and Biological Membranes 352

13 Membrane Channels and Pumps 381

14 Signal-Transduction Pathways 412

15 Metabolism: Basic Concepts and Themes 446

16 Glycolysis and Gluconeogenesis 472

17 Pyruvate Dehydrogenase and the Citric Acid Cycle 515

18 Oxidative Phosphorylation 542

19 Phototrophy and the Light Reactions of Photosynthesis 585

20 The Calvin–Benson Cycle and the Pentose Phosphate Pathway 610

21 Glycogen Metabolism 640

22 Fatty Acid and Triacylglycerol Metabolism 665

23 Protein Turnover and Amino Acid Catabolism 701

24 Integration of Energy Metabolism 732

25 Biosynthesis of Amino Acids 763

26 Nucleotide Biosynthesis 791

27 Biosynthesis of Membrane Lipids and Steroids 814

28 DNA Replication, Repair, and Recombination 845

29 RNA Functions, Biosynthesis, and Processing 879

30 Protein Biosynthesis 916

31 Control of Gene Expression 949

32 Principles of Drug Discovery and Development 977

CONTENTS

PREFACE xxiv

ACKNOWLEDGMENTS xxxiv

ABOUT THE AUTHORS xxxvii

CHAPTER 1

Biochemistry in Space and Time 1

1.1 Biochemical Unity Underlies Biological Diversity 2

1.2 DNA Illustrates the Interplay Between Form and Function 4

DNA is constructed from four building blocks 4

Two single strands of DNA combine to form a double helix 5

DNA structure explains heredity and the storage of information 6

1.3 Concepts from Chemistry Explain the Properties of Biological Molecules 6

The formation of the DNA double helix is a key example 6

The double helix can form from its component strands 6

Atoms and molecules undergo random motions that help define the timescales for biochemical interactions 7

Covalent bonds and noncovalent interactions are important for the structure and stability of biological molecules 7

The formation of DNA's double helix is an expression of the rules of chemistry 11

The laws of thermodynamics govern the behavior of biochemical systems 12

By releasing heat, the formation of the double helix obeys the Second Law of Thermodynamics 14

Double-helix formation can be monitored one molecule at a time 15

Acid–base reactions are central in many biochemical processes 17

Acid–base reactions can disrupt the double helix 17

Buffers regulate pH in organisms and in the laboratory 19

EXAMPLE Applying the Henderson–Hasselbalch Equation 20

1.4 DNA Sequencing Is Transforming Biochemistry, Medicine, and Other Fields 21

Genome sequencing has become remarkably fast and inexpensive 21

Characterization of genetic variation between individuals is powerful for many applications 22

The most important function of genomic sequences is to encode proteins 25

Comparing genomes offers great insights into evolution 25

1.5 Biochemistry Is an Interconnected Human Endeavor 27

CHAPTER 2

Protein Composition and Structure   33

2.1 Several Properties of Protein Structure Are Key to Their Functional Versatility   34

2.2 Proteins Are Built from a Repertoire of 20 Amino Acids   35

The diversity of amino acids arises from the different side chains   36

Biochemists postulate several reasons why this set of amino acids is conserved across all species   40

2.3 Primary Structure: Amino Acids Are Linked by Peptide Bonds to Form Polypeptide Chains   41

Proteins have unique amino acid sequences specified by genes   43

Polypeptide chains are flexible yet conformationally restricted   44

2.4 Secondary Structure: Polypeptide Chains Can Fold into Regular Structures   46

The alpha helix is a coiled structure stabilized by intrachain hydrogen bonds   46

SCIENTIST PROFILE Herman Branson   46

Beta sheets are stabilized by hydrogen bonding between polypeptide strands   48

Polypeptide chains can change direction by making reverse turns and loops   50

2.5 Tertiary Structure: Proteins Can Fold into Globular or Fibrous Structures   50

Globular proteins form tightly packed structures   51

Fibrous proteins form extended structures that provide support for cells and tissues   53

2.6 Quaternary Structure: Polypeptide Chains Can Assemble into Multisubunit Structures   55

2.7 The Amino Acid Sequence of a Protein Determines Its Three-Dimensional Structure   55

Amino acids have different propensities for forming α helices, β sheets, and turns   57

Protein folding is a highly cooperative process   59

Proteins fold by progressive stabilization of intermediates rather than by random search   59

Prediction of three-dimensional structure from sequence remains a great challenge   61

Protein misfolding and aggregation are associated with some neurological diseases   62

Posttranslational modifications confer new capabilities to proteins   63

CHAPTER 3

Binding and Molecular Recognition   68

3.1 Binding Is a Fundamental Process in Biochemistry   69

Binding depends on the concentrations of the binding partners   69

Proteins can selectively bind certain small molecules   70

Binding is a dynamic process involving association and dissociation   72

3.2 Myoglobin Binds and Stores Oxygen   73

More myoglobin binds oxygen as the oxygen partial pressure is increased   73

A bond is formed between oxygen and iron in heme   73

The structure of myoglobin prevents the release of reactive oxygen species   74

Compared with model compounds, myoglobin discriminates between oxygen and carbon monoxide   74

3.3 Hemoglobin Is an Efficient Oxygen Carrier   75

Human hemoglobin is an assembly of four myoglobin-like subunits   75

Hemoglobin binds oxygen cooperatively   76

Oxygen binding markedly changes the quaternary structure of hemoglobin   77

Hemoglobin cooperativity can be potentially explained by several models   78

Structural changes at the heme groups are transmitted to the $\alpha_1\beta_1-\alpha_2\beta_2$ interface   79

2,3-Bisphosphoglycerate in red blood cells is crucial in determining the oxygen affinity of hemoglobin   80

Hydrogen ions and carbon dioxide promote the release of oxygen   81

Mutations in genes encoding hemoglobin subunits can result in disease   83

3.4 The Immune System Depends on Key Binding Proteins   84

The innate immune system recognizes molecules characteristic of pathogens   84

Antibodies bind specific molecules through specific hypervariable loops   86

Antibodies possess distinct antigen-binding and effector units   88

Recombination events equip the adaptive immune system with millions of unique antibodies   89

Major-histocompatibility-complex proteins present peptide antigens on cell surfaces for recognition by T-cell receptors   89

3.5 Quantitative Terms Can Describe Binding Propensity   91

Dissociation constants are useful in describing binding reactions quantitatively   91

SCIENTIST PROFILE Pamela Bjorkman   91

EXAMPLE Determining the Fraction of Bound Receptors   92

Specificity can be quantified by comparing dissociation constants   93

Kinetic parameters can also describe binding processes   94

CHAPTER 4

Protein Methods   100

4.1 The Purification of Proteins Is an Essential First Step in Understanding Their Function   101

The assay: How do we recognize the protein we are looking for?   101

Proteins must be released from the cell to be purified   101

Proteins can be purified according to solubility, size, charge, and binding affinity 102

Proteins can be separated by gel electrophoresis and displayed 105

A protein purification scheme can be quantitatively evaluated 109

EXAMPLE Calculating the Effectiveness of Protein Purification 110

Ultracentrifugation is valuable for separating biomolecules and determining their masses 111

Recombinant DNA technology can make protein purification easier 112

4.2 Immunology Provides Important Techniques for Investigating Proteins 112

Antibodies to specific proteins can be generated 112

Monoclonal antibodies with virtually any desired specificity can be readily prepared 113

Proteins can be detected and quantified by using an enzyme-linked immunosorbent assay 114

Western blotting permits the detection of proteins separated by gel electrophoresis 116

Co-immunoprecipitation enables the identification of binding partners of a protein 116

Fluorescent markers make the visualization of proteins in the cell possible 117

4.3 Mass Spectrometry Is a Powerful Technique for the Identification of Peptides and Proteins 118

Peptides can be sequenced by mass spectrometry 120

Proteins can be specifically cleaved into small peptides to facilitate analysis 121

Genomic and proteomic methods are complementary approaches to deducing protein structure and function 123

The amino acid sequence of a protein provides valuable information 123

Individual proteins can be identified by mass spectrometry 124

The proteome is the functional representation of the genome 125

4.4 Peptides Can Be Synthesized by Automated Solid-Phase Methods 126

4.5 Three-Dimensional Protein Structures Can Be Determined Experimentally 129

X-ray crystallography reveals three-dimensional structure in atomic detail 129

SCIENTIST PROFILE Rosalind Franklin 129

Nuclear magnetic resonance spectroscopy can reveal the structures of proteins in solution 132

Cryo-electron microscopy can be used to determine the structures of large proteins and macromolecular complexes 135

CHAPTER 5

Enzymes: Core Concepts and Kinetics 141

5.1 Enzymes Are Powerful and Highly Specific Catalysts 142

Most enzymes are classified by the types of reactions they catalyze 143

Many enzymes require cofactors for activity 144

Enzymes can transform energy from one form into another 144

5.2 Gibbs Free Energy Is a Useful Thermodynamic Function for Understanding Enzymes 145

The free-energy change provides information about the spontaneity but not the rate of a reaction 145

The standard free-energy change of a reaction is related to the equilibrium constant 146

EXAMPLE Calculating and Comparing $\Delta G°'$ and ΔG 147

Enzymes alter only the reaction rate and not the reaction equilibrium 148

5.3 Enzymes Accelerate Reactions by Facilitating the Formation of the Transition State 148

The formation of an enzyme–substrate complex is the first step in enzymatic catalysis 149

The active sites of enzymes have some common features 150

The binding energy between enzyme and substrate is important for catalysis 151

Because the transition state collapses randomly, the activation energies determine the accumulation of either product or substrate 151

5.4 The Michaelis–Menten Model Accounts for the Kinetic Properties of Many Enzymes 151

Kinetics is the study of reaction rates 152

The steady-state assumption aids a description of enzyme kinetics 152

The Michaelis–Menten model explains many observations of enzyme kinetics 153

The Michaelis–Menten equation describes the relationship between initial velocity and substrate concentration 153

SCIENTIST PROFILE Maud Menten 154

EXAMPLE Applying the Michaelis–Menten Equation 155

Variations in K_M can have physiological consequences 156

K_M and V_{max} values can be determined by several means 156

K_M and k_{cat} values are important enzyme characteristics 156

k_{cat}/K_M is a measure of catalytic efficiency 158

Most biochemical reactions include multiple substrates 159

Allosteric enzymes often do not obey Michaelis–Menten kinetics 161

Temperature affects enzymatic activity 161

5.5 Enzymes Can Be Studied One Molecule at a Time 162

Single-molecule kinetics confirm results obtained from ensemble studies 163

Single-molecule studies continue to reveal new information about enzyme molecular dynamics 165

5.6 Enzymes Can Be Inhibited by Specific Molecules 165

The different types of reversible inhibitors are kinetically distinguishable 166

Transition-state analogs are potent competitive inhibitors 169

Irreversible inhibitors can be used to map the active site 169

EXAMPLE Determining Inhibitor Type from Data 171

Penicillin irreversibly inactivates a key enzyme in bacterial cell-wall synthesis 172

CHAPTER 6

Enzyme Catalytic Strategies   179

6.1   Enzymes Use a Core Set of Catalytic Strategies   180

6.2   Proteases Facilitate a Fundamentally Difficult Reaction   180

Chymotrypsin possesses a highly reactive serine residue   181

Chymotrypsin action proceeds in two steps linked by a covalently bound intermediate   182

Serine is part of a catalytic triad that also includes histidine and aspartate   183

Catalytic triads are found in other hydrolytic enzymes   186

Scientists have dissected the catalytic triad using site-directed mutagenesis   187

Some proteases cleave peptides at other locations besides serine residues   188

Protease inhibitors are important drugs   189

6.3   Carbonic Anhydrases Make a Fast Reaction Faster   190

Carbonic anhydrase contains a bound zinc ion essential for catalytic activity   190

Catalysis involves zinc activation of a water molecule   191

Rapid regeneration of the active form of carbonic anhydrase depends on proton availability   192

6.4   Restriction Enzymes Catalyze Highly Specific DNA-Cleavage Reactions   194

Cleavage is by direct displacement of 3'-oxygen from phosphorus by magnesium-activated water   194

Restriction enzymes require magnesium for catalytic activity   196

The complete catalytic apparatus is assembled only within complexes of cognate DNA molecules, ensuring specificity   197

Host-cell DNA is protected by the addition of methyl groups to specific bases   200

6.5   Molecular Motor Proteins Harness Changes in Enzyme Conformation to Couple ATP Hydrolysis to Mechanical Work   200

ATP hydrolysis proceeds by the attack of water on the gamma phosphoryl group   201

Formation of the transition state for ATP hydrolysis is associated with a substantial conformational change   202

The altered conformation of myosin persists for a substantial period of time   203

Actin forms filaments along which myosin can move   204

CHAPTER 7

Enzyme Regulatory Strategies   210

7.1   Allosteric Regulation Enables Control of Metabolic Pathways   211

Many allosterically regulated enzymes do not follow Michaelis–Menten kinetics   212

ATCase consists of separable catalytic and regulatory subunits   212

Allosteric interactions in ATCase are mediated by large changes in quaternary structure   212

Allosteric regulators modulate the T-to-R equilibrium   216

7.2   Isozymes Provide a Means of Regulation Specific to Distinct Tissues and Developmental Stages   217

7.3   Covalent Modification Is a Means of Regulating Enzyme Activity   218

Kinases and phosphatases control the extent of protein phosphorylation   219

Phosphorylation is a highly effective means of regulating the activities of target proteins   221

Cyclic AMP activates protein kinase A by altering the quaternary structure   221

Mutations in protein kinase A can cause Cushing's syndrome   222

The phosphorylation states of the proteome can be measured   223

7.4   Many Enzymes Are Activated by Specific Proteolytic Cleavage   223

Chymotrypsinogen is activated by specific cleavage of a single peptide bond   224

Proteolytic activation of chymotrypsinogen leads to the formation of a substrate-binding site   225

The generation of trypsin from trypsinogen leads to the activation of other zymogens   226

Some proteolytic enzymes have specific inhibitors   226

7.5   Enzymatic Cascades Allow Rapid Responses Such as Blood Clotting   228

Prothrombin must bind to Ca^{2+} to be converted to thrombin   229

Fibrinogen is converted by thrombin into a fibrin clot   229

Vitamin K is required for the formation of γ-carboxyglutamate   231

The clotting process must be precisely regulated   231

CHAPTER 8

DNA, RNA, and the Flow of Genetic Information   236

8.1   A Nucleic Acid Consists of Four Kinds of Bases Linked to a Sugar–Phosphate Backbone   237

RNA and DNA differ in the sugar component and one of the bases   237

Nucleotides are the monomeric units of nucleic acids   238

DNA molecules are very long and have directionality   239

8.2   A Pair of Nucleic Acid Strands with Complementary Sequences Can Form a Double-Helical Structure   240

The double helix is stabilized by hydrogen bonds and van der Waals interactions   240

DNA can assume a variety of structural forms   242

The major and minor grooves are lined by sequence-specific hydrogen-bonding groups   243

Some DNA molecules are circular and supercoiled   244

Single-stranded nucleic acids can adopt elaborate structures   244

8.3 The Double Helix Facilitates the Accurate Transmission of Hereditary Information   246

Differences in DNA density established the validity of the semiconservative replication hypothesis   246

The double helix can be reversibly melted   247

8.4 DNA Is Replicated by Polymerases That Take Instructions from Templates   248

DNA polymerase catalyzes phosphodiester-bridge formation   248

The genes of some viruses are made of RNA   249

8.5 Gene Expression Is the Transformation of DNA Information into Functional Molecules   250

Several kinds of RNA play key roles in gene expression   250

All cellular RNA is synthesized by RNA polymerases   251

RNA polymerases take instructions from DNA templates   252

Transcription begins near promoter sites and ends at terminator sites   253

Transfer RNAs are the adaptor molecules in protein synthesis   254

8.6 Amino Acids Are Encoded by Groups of Three Bases Starting from a Fixed Point   255

Major features of the genetic code   255

SCIENTIST PROFILE Har Gobind Khorana   256

Messenger RNA contains start and stop signals for protein synthesis   257

The genetic code is nearly universal   258

8.7 Most Eukaryotic Genes Are Mosaics of Introns and Exons   258

RNA processing generates mature RNA   258

Many exons encode protein domains   259

CHAPTER 9

Nucleic Acid Methods   264

9.1 The Exploration of Genes Relies on Key Tools   265

Restriction enzymes split DNA into specific fragments   265

Restriction fragments can be separated by gel electrophoresis and visualized   266

DNA can be sequenced by controlled termination of replication   267

DNA probes and genes can be synthesized by automated solid-phase methods   268

Selected DNA sequences can be greatly amplified by the polymerase chain reaction   269

PCR is a powerful technique in medical diagnostics, forensics, and studies of molecular evolution   271

The tools for recombinant DNA technology have been used to identify disease-causing mutations   271

9.2 Recombinant DNA Technology Has Revolutionized All Aspects of Biology   272

Restriction enzymes and DNA ligase are key tools in forming recombinant DNA molecules   272

Plasmids and λ phage are choice vectors for DNA cloning in bacteria   274

Specific genes can be cloned from digests of genomic DNA   277

Complementary DNA prepared from mRNA can be expressed in host cells   278

Proteins with new functions can be created through directed changes in DNA   279

9.3 Complete Genomes Have Been Sequenced and Analyzed   282

The genomes of organisms ranging from bacteria to multicellular eukaryotes have been sequenced   282

The sequence of the human genome has been completed   283

Next-generation sequencing methods enable the rapid determination of a complete genome sequence   284

Comparative genomics is a powerful research tool   286

9.4 Eukaryotic Genes Can Be Quantitated and Manipulated with Considerable Precision   287

Gene-expression levels can be comprehensively examined   287

New genes inserted into eukaryotic cells can be efficiently expressed   289

Transgenic animals harbor and express genes introduced into their germ lines   290

Gene disruption and genome editing provide clues to gene function and opportunities for new therapies   290

RNA interference enables disruption of gene expression and presents new therapeutic opportunities   294

SCIENTIST PROFILE Emmanuelle Charpentier and Jennifer Doudna   294

Foreign DNA can be introduced into plants   295

CHAPTER 10

Exploring Evolution and Bioinformatics   301

10.1 Homologs Are Descended from a Common Ancestor and Can Be Detected by Sequence Alignments   302

Orthologs and paralogs are two different classes of homologous proteins   302

Statistical analysis of sequence alignments can detect homology   302

The statistical significance of alignments can be estimated by shuffling   305

Distant evolutionary relationships can be detected through the use of substitution matrices   305

Databases can be searched to identify homologous sequences   308

10.2 Examination of Three-Dimensional Structure Enhances Our Understanding of Evolutionary Relationships   310

Tertiary structure is more conserved than primary structure   310

Knowledge of three-dimensional structures can aid in the evaluation of sequence alignments   311

Repeated motifs can be detected by aligning sequences with themselves   311

Convergent evolution illustrates common solutions to biochemical challenges   313

Comparison of RNA sequences can be a source of insight into RNA secondary structures   314

EXAMPLE Interpreting an RNA Alignment   315

10.3 Evolutionary Trees Can Be Constructed on the Basis of Sequence Information   316

Evolutionary trees can be calibrated using fossil record data   316

Horizontal gene transfer events may explain unexpected branches of the evolutionary tree   317

SCIENTIST PROFILE Russell Doolittle   318

10.4 Modern Techniques Make the Experimental Exploration of Evolution Possible   318

Ancient DNA can sometimes be amplified and sequenced   318

Molecular evolution can be examined experimentally   319

CHAPTER 11

Carbohydrates and Glycoproteins   324

11.1 Monosaccharides Are the Simplest Carbohydrates   325

There are many monosaccharides but they are structurally similar   325

Most monosaccharides exist as interchanging cyclic forms   327

Pyranose and furanose rings can assume different conformations   329

D-Glucose is an important fuel for most organisms   330

Glucose is a reducing sugar and reacts nonenzymatically with hemoglobin   330

Monosaccharides are joined to alcohols and amines through glycosidic linkages by specific enzymes   331

Phosphorylated sugars are key intermediates in metabolism   332

11.2 Monosaccharides Are Linked to Form Complex Carbohydrates   333

Sucrose, lactose, and maltose are common disaccharides   333

Maltase inhibitors can help to maintain blood glucose homeostasis   334

Human milk oligosaccharides protect newborns from infection   335

Glycogen and starch are storage polysaccharides of glucose   335

Cellulose is the main structural polysaccharide of plants   335

Chitin is the main structural polysaccharide of fungi and arthropods   336

Chitin can be processed to a molecule with a variety of uses   337

11.3 Carbohydrates Can Be Linked to Proteins to Form Glycoproteins   337

Carbohydrates can be linked to proteins through asparagine (N-linked) or through serine or threonine (O-linked) residues   338

The glycoprotein erythropoietin is a vital hormone   339

Glycosylation functions in nutrient sensing   339

Proteoglycans have important structural roles   339

Proteoglycans are important components of cartilage   340

Mucins are glycoprotein components of mucus   341

Protein glycosylation takes place in the lumen of the endoplasmic reticulum and in the Golgi complex   341

Specific enzymes are responsible for oligosaccharide assembly   342

Blood groups are based on protein glycosylation patterns   343

Errors in glycosylation can result in pathological conditions   344

Biochemists use several techniques to analyze the oligosaccharide components of glycoproteins   345

11.4 Lectins Are Specific Carbohydrate-Binding Proteins   346

Lectins promote interactions between cells and within cells   346

Lectins are organized into two large classes   346

Influenza virus binds to sialic acid residues   347

SCIENTIST PROFILE Carolyn Bertozzi   347

CHAPTER 12

Lipids and Biological Membranes   352

12.1 Fatty Acids Are Key Constituents of Lipids   353

Fatty acid names are based on their parent hydrocarbons   353

Chain length and degree of unsaturation affect fatty acid properties   354

12.2 Biological Membranes Are Composed of Three Common Types of Membrane Lipids   355

Phospholipids are the major class of membrane lipids   355

Glycolipids include carbohydrate moieties   357

Cholesterol is a lipid based on a steroid nucleus   357

SCIENTIST PROFILE Marie M. Daly   357

Archaeal membranes are built from ether lipids with branched chains   358

A membrane lipid is an amphipathic molecule containing a hydrophilic and a hydrophobic moiety   358

12.3 Phospholipids and Glycolipids Readily Form Bimolecular Sheets in Aqueous Media   359

Lipid vesicles can be formed from phospholipids   360

Lipid bilayers are highly impermeable to ions and most polar molecules   361

12.4 Proteins Carry Out Most Membrane Processes 362

Proteins associate with the lipid bilayer in a variety of ways 363

Proteins interact with membranes in a variety of ways 363

Some proteins associate with membranes through covalently attached hydrophobic groups 367

Transmembrane helices can be accurately predicted from amino acid sequences 367

12.5 Lipids and Many Membrane Proteins Diffuse Rapidly in the Plane of the Membrane 369

The fluid mosaic model allows lateral movement but not rotation through the membrane 370

Membrane fluidity is controlled by fatty acid composition and cholesterol content 370

Lipid rafts are highly dynamic complexes formed between cholesterol and specific lipids 372

All biological membranes are asymmetric 373

12.6 Prokaryotes and Eukaryotes Differ in Their Use of Biological Membranes 373

Eukaryotic cells contain compartments bounded by internal membranes 374

Membrane budding and fusion are highly controlled processes 375

CHAPTER 13

Membrane Channels and Pumps 381

13.1 The Transport of Molecules Across a Membrane May Be Active or Passive 382

Many molecules require protein transporters to cross membranes 382

Free energy stored in concentration gradients can be quantified 382

EXAMPLE Calculating the Energetic Cost of Ion Transport 383

13.2 Two Families of Membrane Proteins Use ATP Hydrolysis to Actively Transport Ions and Molecules Across Membranes 384

P-type ATPases couple phosphorylation and conformational changes to pump calcium ions across membranes 384

Digoxin specifically inhibits the Na^+–K^+ pump by blocking its dephosphorylation 387

P-type ATPases are evolutionarily conserved and play a wide range of roles 387

Multidrug resistance highlights a family of membrane pumps with ATP-binding cassette domains 388

13.3 Lactose Permease Is an Archetype of Secondary Transporters That Use One Concentration Gradient to Power the Formation of Another 390

13.4 Specific Channels Can Rapidly Transport Ions Across Membranes 392

Action potentials are mediated by transient changes in Na^+ and K^+ permeability 392

Patch-clamp conductance measurements reveal the activities of single channels 393

The structure of a potassium ion channel is an archetype for many ion-channel structures 394

SCIENTIST PROFILE Baldomero Olivera 395

The structure of the potassium ion channel reveals the basis of ion specificity 395

The structure of the potassium ion channel explains its rapid rate of transport 398

Voltage gating requires substantial conformational changes in specific ion-channel domains 398

A channel can be inactivated by occlusion of the pore: the ball-and-chain model 399

The acetylcholine receptor is an archetype for ligand-gated ion channels 400

Action potentials integrate the activities of several ion channels working in concert 402

EXAMPLE Calculating Equilibrium Potentials 403

Disruption of ion channels by mutations or chemicals can be potentially life-threatening 404

Hyperpolarization-activated ion channels enable pacemaker activity in the heart 405

13.5 Gap Junctions Allow Ions and Small Molecules to Flow Between Communicating Cells 406

13.6 Specific Channels Increase the Permeability of Some Membranes to Water 407

CHAPTER 14

Signal-Transduction Pathways 412

14.1 Many Signal-Transduction Pathways Share Common Themes 413

Signal transduction depends on molecular circuits 413

14.2 Epinephrine Signaling: Heterotrimeric G Proteins Transmit Signals and Reset Themselves 414

Ligand binding to 7TM receptors leads to the activation of heterotrimeric G proteins 415

Activated G proteins transmit signals by binding to other proteins 418

Cyclic AMP stimulates the phosphorylation of many target proteins by activating protein kinase A 418

G proteins spontaneously reset themselves through GTP hydrolysis 419

Some 7TM receptors activate the phosphoinositide cascade 420

Calcium ion is a widely used second messenger 421

Calcium ion often activates the regulatory protein calmodulin 423

Some receptors signal through G proteins that inhibit rather than stimulate adenylate cyclase 423

G-protein βγ-dimers can also directly participate in signaling 424

7TM receptors trigger signaling through G proteins in many other cell types 424

SCIENTIST PROFILE Eva Neer 425

14.3 Insulin Signaling: Phosphorylation Cascades Are Central to Many Signal-Transduction Processes   425

The insulin receptor is a protein kinase that is autoinhibited prior to insulin binding   426

Insulin binding results in the cross-phosphorylation and activation of the insulin receptor   427

The activated insulin-receptor kinase initiates a kinase cascade   428

Insulin signaling is terminated by the action of phosphatases   430

14.4 Epidermal Growth Factor: Receptor Dimerization Can Drive Signaling   431

The EGF receptor undergoes phosphorylation of its carboxyl-terminal tail   432

EGF signaling leads to the activation of Ras, a small G protein   432

Activated Ras initiates a protein kinase cascade   432

EGF signaling is terminated by protein phosphatases and the intrinsic GTPase activity of Ras   433

14.5 Defects in Signal-Transduction Pathways Can Lead to Cancer and Other Diseases   433

Monoclonal antibodies can be used to inhibit signal-transduction pathways activated in tumors   434

Protein kinase inhibitors can be effective anticancer drugs   435

14.6 Sensory Systems Are Based on Specialized Signal-Transduction Pathways   435

A huge family of 7TM receptors detect a wide variety of organic compounds   436

Vision relies on a specialized 7TM receptor to signal in response to absorbed light   437

Light absorption induces a specific isomerization of bound 11-*cis*-retinal   439

Color vision is mediated by three cone receptors that are homologs of rhodopsin   440

Hearing depends on hair cells that use mechanosensitive ion channels to detect tiny motions   440

Comparison of different organisms yields insights into sensory system evolution   441

CHAPTER 15

Metabolism: Basic Concepts and Themes   446

15.1 Metabolism Is Composed of Many Interconnected Reactions   447

Metabolism consists of destructive and constructive reactions that typically yield or require energy   448

A thermodynamically unfavorable reaction can be driven by a favorable reaction   449

15.2 ATP Is the Universal Currency of Free Energy in Biological Systems   449

ATP hydrolysis is exergonic   450

ATP hydrolysis drives metabolism by shifting the equilibrium of coupled reactions   451

The high phosphoryl potential of ATP results from structural differences between ATP and its hydrolysis products   452

Phosphoryl-transfer potential is an important form of cellular energy transformation   453

EXAMPLE Calculating ΔG for a Coupled Reaction Under Real Conditions   455

15.3 The Oxidation of Carbon Fuels Is an Important Source of Cellular Energy   456

Compounds with high phosphoryl-transfer potential can couple carbon oxidation to ATP synthesis   457

Ion gradients across membranes provide an important form of cellular energy that can be coupled to ATP synthesis   458

Phosphates play a prominent role in biochemical processes   458

Energy from food is extracted in three stages   459

15.4 Metabolic Pathways Contain Many Recurring Motifs   460

Activated carriers exemplify the modular structure and economy of metabolism   460

Many activated carriers are derived from vitamins   463

Key reactions are reiterated throughout metabolism   464

Metabolic processes are regulated in three principal ways   467

CHAPTER 16

Glycolysis and Gluconeogenesis   472

16.1 Glycolysis Is an Energy-Conversion Pathway in Most Organisms   473

Glucose is generated from dietary carbohydrates   473

A family of transporters enables glucose to enter and leave animal cells   474

16.2 Glycolysis Can Be Divided into Two Parts   474

Stage 1 begins: Hexokinase traps glucose in the cell and begins glycolysis   476

Fructose 1,6-bisphosphate is generated from glucose 6-phosphate   477

The six-carbon sugar is cleaved into two three-carbon fragments   478

Mechanism: Triose phosphate isomerase salvages a three-carbon fragment   478

Stage 2 begins: The oxidation of an aldehyde powers the formation of a compound with high phosphoryl-transfer potential   480

Mechanism: Phosphorylation is coupled to the oxidation of glyceraldehyde 3-phosphate by a thioester intermediate   482

ATP is formed by phosphoryl transfer from 1,3-bisphosphoglycerate   482

Additional ATP is generated with the formation of pyruvate   484

Two ATP molecules are formed in the conversion of glucose into pyruvate   485

NAD^+ is regenerated from the metabolism of pyruvate   486

Fermentations provide usable energy in the absence of oxygen   488

Fructose is converted into glycolytic intermediates by fructokinase   489

Galactose is converted into glucose 6-phosphate   490

⚕ Galactose can be highly toxic with a defective metabolic pathway   491

⚕ Many adults worldwide are intolerant of milk because they are deficient in lactase   492

16.3 The Glycolytic Pathway Is Tightly Controlled   492

Glycolysis in muscle is regulated to meet the need for ATP   493

The regulation of glycolysis in the liver illustrates the biochemical versatility of the liver   494

The enzymes of glycolysis are physically associated with one another   497

⚕ Aerobic glycolysis is a property of tumor cells and other rapidly growing cells   497

⚕ Cancer and endurance training affect glycolysis in a similar fashion   498

16.4 Glucose Can Be Synthesized from Noncarbohydrate Precursors   499

Gluconeogenesis is not a reversal of glycolysis   499

The conversion of pyruvate into phosphoenolpyruvate begins with the formation of oxaloacetate   501

Oxaloacetate is shuttled into the cytoplasm and converted into phosphoenolpyruvate   502

The conversion of fructose 1,6-bisphosphate into fructose 6-phosphate and orthophosphate is an irreversible step   503

The generation of free glucose occurs only in some tissues and is an important control point   503

Six high-transfer-potential phosphoryl groups are spent in synthesizing glucose from pyruvate   504

16.5 Gluconeogenesis and Glycolysis Are Reciprocally Regulated   505

Glycolysis and gluconeogenesis are regulated by adenosine nucleotides and other metabolic intermediates   505

In mammals, glycolysis and gluconeogenesis in the liver are controlled by hormones sensitive to blood-glucose concentration   506

Substrate cycles amplify metabolic signals and produce heat   508

Lactate and alanine formed by contracting muscle and peripheral tissues are used by other organs   508

⚕ Deficiencies in glycolytic or gluconeogenic enzymes are rare genetic disorders   509

SCIENTIST PROFILE Gerty Cori   509

Glycolysis and gluconeogenesis are evolutionarily intertwined   511

CHAPTER 17

Pyruvate Dehydrogenase and the Citric Acid Cycle   515

17.1 The Citric Acid Cycle Harvests High-Energy Electrons   516

17.2 The Pyruvate Dehydrogenase Complex Links Glycolysis to the Citric Acid Cycle   517

Mechanism: The synthesis of acetyl coenzyme A from pyruvate requires three enzymes and five coenzymes   518

Flexible linkages allow lipoamide to move between different active sites   520

17.3 The Citric Acid Cycle Oxidizes Two-Carbon Units   522

Citrate synthase forms citrate from oxaloacetate and the acetyl group from acetyl coenzyme A   522

Mechanism: The mechanism of citrate synthase prevents undesirable reactions   523

Citrate is isomerized into isocitrate   524

Isocitrate is oxidized and decarboxylated to alpha-ketoglutarate   525

Succinyl coenzyme A is formed by the oxidative decarboxylation of alpha-ketoglutarate   526

A compound with high phosphoryl-transfer potential is generated from succinyl coenzyme A   526

Mechanism: Succinyl coenzyme A synthetase transforms types of biochemical energy   527

Oxaloacetate is regenerated by the oxidation of succinate   528

SCIENTIST PROFILE Hans Krebs   528

The citric acid cycle produces high-transfer-potential electrons, ATP, and CO_2 529

17.4 Entry to the Citric Acid Cycle and Metabolism Through It Are Controlled   531

The pyruvate dehydrogenase complex is regulated allosterically and by reversible phosphorylation   531

⚕ Diabetic neuropathy may be due to inhibition of the pyruvate dehydrogenase complex   532

The citric acid cycle is regulated at several points   533

17.5 The Citric Acid Cycle Is a Source of Biosynthetic Precursors   534

The citric acid cycle must be capable of being rapidly replenished   534

⚕ The disruption of pyruvate metabolism is the cause of beriberi and poisoning by mercury and arsenic   535

The citric acid cycle likely evolved from preexisting pathways   536

17.6 The Glyoxylate Cycle Enables Plants and Bacteria to Grow on Acetate   536

⚕ Blocking the glyoxylate cycle may lead to new treatments for tuberculosis   538

CHAPTER 18

Oxidative Phosphorylation   542

18.1 Cellular Respiration Drives ATP Formation by Transferring Electrons to Molecular Oxygen   543

Eukaryotic oxidative phosphorylation takes place in mitochondria   543

Mitochondria are the result of an endosymbiotic event   544

SCIENTIST PROFILE Lynn Margulis   545

18.2 Oxidative Phosphorylation Depends on Electron Transfer   545

The electron-transfer potential of an electron is measured as redox potential   545

EXAMPLE Calculating the Standard Free Energy of a Reaction from Reduction Potentials   547

Electron flow from NADH to molecular oxygen powers the formation of a proton gradient   548

18.3 The Respiratory Chain Consists of Four Complexes: Three Proton Pumps and a Physical Link to the Citric Acid Cycle   549

Iron–sulfur clusters are common components of the electron-transport chain   551

The high-potential energy electrons of NADH enter the respiratory chain at NADH-Q oxidoreductase   552

Ubiquinol is the entry point for electrons from $FADH_2$ of flavoproteins   553

Electrons flow from ubiquinol to cytochrome c through Q-cytochrome c oxidoreductase   554

The Q cycle funnels electrons from a two-electron carrier to a one-electron carrier while pumping protons   555

Cytochrome c oxidase catalyzes the reduction of molecular oxygen to water   556

Most of the electron-transport chain is organized into a larger complex called the respirasome   559

Toxic derivatives of molecular oxygen such as superoxide radicals are scavenged by protective enzymes   560

Electrons can be transferred between groups that are not in contact   561

18.4 A Proton Gradient Powers the Synthesis of ATP   562

The chemiosmotic hypothesis suggested that ATP formation is powered by a proton gradient   562

ATP synthase is composed of a proton-conducting unit and a catalytic unit   564

Proton flow through ATP synthase leads to the release of tightly bound ATP via the binding-change mechanism   565

Rotational catalysis is the world's smallest molecular motor   567

Proton flow around the c ring powers ATP synthesis   567

ATP synthase and G proteins have several common features   569

18.5 Many Shuttles Allow Movement Across Mitochondrial Membranes   570

Electrons from cytoplasmic NADH enter mitochondria by shuttles   570

The entry of ADP into mitochondria is coupled to the exit of ATP by ATP-ADP translocase   572

18.6 The Regulation of Cellular Respiration Is Governed Primarily by the Need for ATP   573

The complete oxidation of glucose yields about 30 molecules of ATP   573

The rate of oxidative phosphorylation is determined by the need for ATP   574

ATP synthase can be regulated   575

Regulated uncoupling leads to the generation of heat   575

Reintroduction of UCP-1 into pigs may be economically valuable   577

Oxidative phosphorylation can be inhibited at many stages   577

New mitochondrial diseases are constantly being discovered   578

Mitochondria play a key role in apoptosis   579

18.7 Proton Gradients Generated by Respiratory Chains Drive Many Biochemical Processes   579

Proton flow through a rotary motor allows bacteria to swim   579

Power transmission by proton gradients is a central motif of bioenergetics   580

CHAPTER 19

Phototrophy and the Light Reactions of Photosynthesis   585

19.1 Phototrophy Converts Light Energy into Chemical Energy   586

Photosynthesis comprises light reactions and dark reactions   587

The same biochemical principles govern both respiration and photosynthesis   587

Two kinds of light reactions take place in the green plants   588

19.2 In Eukaryotes, Photosynthesis Takes Place in Chloroplasts   588

The primary events of photosynthesis take place in thylakoid membranes   588

Chloroplasts arose from an endosymbiotic event   589

19.3 Light Absorption by Chlorophyll Molecules Induces Electron Transfer   589

Transferring electrons allows energy to be captured instead of lost as heat   590

A "special pair" of chlorophylls initiate charge separation   590

A proton gradient across the membrane is established   592

Cyclic electron flow reduces the cytochrome of the reaction center   592

19.4 Two Photosystems Generate a Proton Gradient and Reducing Power in Cyanobacteria and Photosynthetic Eukaryotes   593

Photosystem II transfers electrons from water to plastoquinone and generates a proton gradient   593

Photosystem II is comparable to the purple bacterial reaction center   594

Cytochrome bf links photosystem II to photosystem I   596

Photosystem I uses light energy to generate reduced ferredoxin, a powerful reductant   597

Ferredoxin–NADP$^+$ reductase converts NADP$^+$ into NADPH ... 598

SCIENTIST PROFILE Peter Mitchell and André Jagendorf ... 599

19.5 A Proton Gradient across the Thylakoid Membrane Drives ATP Synthesis ... 599

The ATP synthase of chloroplasts closely resembles those of mitochondria and prokaryotes ... 600

The activity of chloroplast ATP synthase is regulated ... 601

Cyclic electron flow through photosystem I leads to the production of ATP instead of NADPH ... 601

The absorption of eight photons yields one O_2, two NADPH, and three ATP molecules ... 602

19.6 Accessory Pigments Funnel Energy into Reaction Centers ... 602

Resonance energy transfer allows energy to move from the site of initial absorbance to the reaction center ... 603

Accessory pigments also protect plants from reactive oxygen ... 604

Increasing the efficiency of photosynthesis will increase crop yields ... 604

The components of photosynthesis are highly organized ... 605

Many herbicides inhibit the light reactions of photosynthesis ... 605

19.7 The Ability to Convert Light into Chemical Energy Is Ancient ... 606

Artificial photosynthetic systems may provide clean, renewable energy ... 606

Photosensitive proteins are transforming other fields ... 607

CHAPTER 20

The Calvin–Benson Cycle and the Pentose Phosphate Pathway ... **610**

20.1 The Calvin–Benson Cycle Synthesizes Hexoses from Carbon Dioxide and Water ... 611

Stage 1: Carbon dioxide reacts with ribulose 1,5-bisphosphate to form two molecules of 3-phosphoglycerate ... 611

SCIENTIST PROFILE Andrew Benson ... 611

Rubisco activity depends on magnesium and carbamate ... 612

Rubisco also catalyzes a wasteful oxygenase reaction ... 614

Stage 2: Hexose phosphates are made from phosphoglycerate ... 615

Stage 3: Ribulose 1,5-bisphosphate is regenerated ... 615

18 ATP and 12 NADPH molecules are used to bring six carbon dioxides to the level of a hexose ... 618

Starch and sucrose are the major carbohydrate stores in plants ... 618

Inspired by the Calvin–Benson cycle, scientists are developing new methods for fixing carbon dioxide ... 619

20.2 The Activity of the Calvin–Benson Cycle Depends on Environmental Conditions ... 620

Rubisco is activated by light-driven changes in proton and magnesium ion concentrations ... 621

Thioredoxin plays a key role in regulating the Calvin–Benson cycle ... 621

The C_4 pathway of tropical plants and grasses accelerates photosynthesis by concentrating carbon dioxide ... 622

Crassulacean acid metabolism permits growth in arid ecosystems ... 624

20.3 The Pentose Phosphate Pathway Generates NADPH and Synthesizes Pentoses ... 624

Two molecules of NADPH are generated in the conversion of glucose 6-phosphate into ribulose 5-phosphate ... 626

The pentose phosphate pathway and glycolysis are linked by transketolase and transaldolase ... 626

Transketolase and transaldolase stabilize carbanionic intermediates by different mechanisms ... 628

20.4 The Metabolism of Glucose 6-Phosphate by the Pentose Phosphate Pathway Is Coordinated with Glycolysis ... 631

The rate of the oxidative phase of the pentose phosphate pathway is controlled by the level of NADP$^+$ ... 631

The flow of glucose 6-phosphate depends on the need for NADPH, ribose 5-phosphate, and ATP ... 631

The pentose phosphate pathway is required for rapid cell growth ... 633

The Calvin–Benson cycle and the pentose phosphate pathway are essentially mirror images of one another ... 633

20.5 Glucose 6-Phosphate Dehydrogenase Plays a Key Role in Protection Against Reactive Oxygen Species ... 634

Glucose 6-phosphate dehydrogenase deficiency causes a drug-induced hemolytic anemia ... 634

A deficiency of glucose 6-phosphate dehydrogenase can be protective against malaria ... 635

CHAPTER 21

Glycogen Metabolism ... **640**

21.1 Glycogen Metabolism Is the Regulated Release and Storage of Glucose in Multiple Tissues ... 641

21.2 Glycogen Breakdown Requires the Interplay of Several Enzymes ... 642

Phosphorylase catalyzes the phosphorolytic cleavage of glycogen to release glucose 1-phosphate ... 642

Mechanism: Pyridoxal phosphate participates in the phosphorolytic cleavage of glycogen ... 643

Debranching enzyme also is needed for the breakdown of glycogen ... 645

Phosphoglucomutase converts glucose 1-phosphate into glucose 6-phosphate ... 646

The liver contains glucose 6-phosphatase, a hydrolytic enzyme absent from muscle 646

21.3 Phosphorylase Is Regulated by Allosteric Interactions and Controlled by Reversible Phosphorylation 647

Liver phosphorylase produces glucose for use by other tissues 647

Muscle phosphorylase is regulated by changes in AMP and ATP concentrations 648

Biochemical characteristics of muscle fiber types differ 649

Phosphorylation promotes the conversion of phosphorylase *b* to phosphorylase *a* 650

Phosphorylase kinase is activated by phosphorylation and calcium ions 650

21.4 Glucagon and Epinephrine Signal the Need for Glycogen Breakdown 651

G proteins transmit the signal for the initiation of glycogen breakdown 651

Glycogen breakdown must be rapidly turned off when necessary 653

21.5 Glycogen Synthesis Requires Several Enzymes and Uridine Diphosphate Glucose 653

UDP-glucose is an activated form of glucose 653

Glycogen synthase catalyzes the transfer of glucose from UDP-glucose to growing chains 654

A branching enzyme forms α-1,6 linkages 655

Glycogen synthase is the key regulatory enzyme in glycogen synthesis 655

Glycogen is an efficient storage form of glucose 656

21.6 Glycogen Breakdown and Synthesis Are Reciprocally Controlled by Hormones 656

Protein phosphatase 1 reverses the effects of kinases on glycogen metabolism 656

Insulin stimulates glycogen synthesis by inactivating glycogen synthase kinase 658

Glycogen metabolism in the liver regulates the blood-glucose concentration 659

Biochemists have uncovered the biochemical basis of multiple glycogen-storage diseases 660

CHAPTER 22

Fatty Acid and Triacylglycerol Metabolism 665

22.1 Triacylglycerols Are Highly Concentrated Energy Stores 666

Dietary lipids are digested by pancreatic lipases 667

Dietary lipids are transported in chylomicrons 667

22.2 The Use of Fatty Acids as Fuel Requires Three Stages of Processing 668

Mobilization: Triacylglycerols are hydrolyzed by hormone-stimulated lipases 668

Mobilization continues: Free fatty acids and glycerol are released into the blood 669

Activation: Fatty acids are linked to coenzyme A before they are oxidized 670

Transport: Carnitine carries long-chain activated fatty acids into the mitochondrial matrix 671

Breakdown: Acetyl CoA, NADH, and $FADH_2$ are generated in each round of fatty acid oxidation 672

The complete oxidation of palmitate yields 106 molecules of ATP 673

22.3 Unsaturated and Odd-Chain Fatty Acids Require Additional Steps for Degradation 674

An isomerase and a reductase are required for the oxidation of unsaturated fatty acids 674

Odd-chain fatty acids yield propionyl CoA in the final thiolysis step 675

Vitamin B_{12} contains a corrin ring and a cobalt atom 676

Mechanism: Methylmalonyl CoA mutase catalyzes a rearrangement to form succinyl CoA 677

SCIENTIST PROFILE Dorothy Hodgkin 677

Fatty acids are also oxidized in peroxisomes 678

Some fatty acids contribute to the development of pathological conditions 679

22.4 Ketone Bodies Are a Fuel Source Derived from Fats 679

Ketone bodies are a major fuel in some tissues 680

Diabetic ketoacidosis is a dangerous pathological condition caused by excessive ketone body formation 682

Animals cannot convert fatty acids into glucose 683

22.5 Fatty Acids Are Synthesized by Fatty Acid Synthase 683

Fatty acid degradation and synthesis mirror each other in their chemical reactions 684

Fatty acids are synthesized and degraded by different pathways 685

The formation of malonyl CoA is the committed step in fatty acid synthesis 685

Intermediates in fatty acid synthesis are attached to an acyl carrier protein 685

Fatty acid synthesis consists of a series of condensation, reduction, dehydration, and reduction reactions 686

Fatty acids are synthesized by a multifunctional enzyme complex in animals 687

The synthesis of palmitate requires 8 molecules of acetyl CoA, 14 molecules of NADPH, and 7 molecules of ATP 690

Citrate carries acetyl groups from mitochondria to the cytoplasm for fatty acid synthesis 690

Several sources supply NADPH for fatty acid synthesis 691

Fatty acid metabolism is altered in tumor cells 691

Triacylglycerols may become an important renewable energy source 692

22.6 The Elongation and Unsaturation of Fatty Acids Are Accomplished by Accessory Enzyme Systems 692

Membrane-bound enzymes generate unsaturated fatty acids 692

Eicosanoid hormones are derived from polyunsaturated fatty acids 693

22.7 Acetyl CoA Carboxylase Plays a Key Role in Controlling Fatty Acid Metabolism   694

Acetyl CoA carboxylase is regulated by conditions in the cell   694

Acetyl CoA carboxylase is controlled by a variety of hormones   695

AMP-activated protein kinase is a key regulator of metabolism   696

CHAPTER 23

Protein Turnover and Amino Acid Catabolism   701

23.1 Proteins Are Degraded to Amino Acids   702

The digestion of dietary proteins begins in the stomach and is completed in the intestine   702

Cellular proteins are degraded at different rates   703

23.2 Protein Turnover Is Tightly Regulated   703

Ubiquitin tags proteins for destruction   703

The proteasome digests the ubiquitin-tagged proteins   705

The ubiquitin pathway and the proteasome have prokaryotic counterparts   707

Protein degradation can be used to regulate biological function   707

23.3 The First Step in Amino Acid Degradation Is the Removal of Nitrogen   708

Alpha-amino groups are converted into ammonium ions by the oxidative deamination of glutamate in the liver   708

SCIENTIST PROFILE Cecile Pickart   708

Mechanism: Pyridoxal phosphate forms Schiff-base intermediates in aminotransferases   709

Aspartate aminotransferase is an archetypal pyridoxal-dependent transaminase   711

Blood levels of aminotransferases serve a diagnostic function   711

Pyridoxal phosphate enzymes catalyze a wide array of reactions   711

Serine and threonine can be directly deaminated   712

Peripheral tissues transport nitrogen to the liver   712

23.4 Ammonium Ions Are Converted into Urea in Most Terrestrial Vertebrates   713

The urea cycle begins with the formation of carbamoyl phosphate   714

Carbamoyl phosphate synthetase I is the key regulatory enzyme for urea synthesis   714

Carbamoyl phosphate reacts with ornithine to begin the urea cycle   715

The urea cycle is linked to gluconeogenesis   716

Inherited defects of the urea cycle cause hyperammonemia and can lead to brain damage   717

Urea is not the only means of disposing of excess nitrogen   718

23.5 Carbon Atoms of Degraded Amino Acids Emerge as Major Metabolic Intermediates   718

Pyruvate is an entry point into metabolism for a number of amino acids   719

Oxaloacetate is an entry point into metabolism for aspartate and asparagine   719

Alpha-ketoglutarate is an entry point into metabolism for amino acids with five-carbon chains   720

Succinyl coenzyme A is a point of entry for several amino acids   721

Methionine degradation requires the formation of a key methyl donor, S-adenosylmethionine   721

Threonine deaminase initiates the degradation of threonine   721

The branched-chain amino acids yield acetyl CoA, acetoacetate, or propionyl CoA   722

Oxygenases are required for the degradation of aromatic amino acids   723

Protein metabolism helps to power the flight of migratory birds   725

EXAMPLE Determining Metabolic Products of Amino Acid Degradation   725

23.6 Inborn Errors of Metabolism Can Disrupt Amino Acid Degradation   726

Branched-chain ketoaciduria is a serious disorder of branched-chain amino acid degradation   727

Phenylketonuria is one of the most common metabolic disorders   727

CHAPTER 24

Integration of Energy Metabolism   732

24.1 Caloric Homeostasis Is a Means of Regulating Body Weight   733

The brain plays a key role in caloric homeostasis   734

Short-term signals from the gastrointestinal tract induce feelings of satiety   735

Leptin and insulin regulate long-term control over caloric homeostasis   736

Leptin is one of several hormones secreted by adipose tissue   736

Leptin resistance may be a contributing factor to obesity   737

24.2 The Fasted–Fed Cycle Is a Response to Eating and Sleeping Behaviors   738

The postprandial state follows a meal   739

The postabsorptive state occurs at the beginning of a fast   739

The refed state occurs at the end of a long fast   741

24.3 Diabetes Is a Common Metabolic Disease Often Resulting from Obesity   741

Insulin initiates a complex signal-transduction pathway in muscle   742

Metabolic syndrome often precedes type 2 diabetes   743

Excess fatty acids in muscle modify metabolism   743

Insulin resistance in muscle facilitates pancreatic failure   744

Metabolic alterations in type 1 diabetes result from insulin insufficiency and glucagon excess   746

24.4 Exercise Beneficially Alters the Biochemistry of Cells   746

Fuel choice during exercise is determined by the intensity and duration of activity   746

The perplexing symptoms of McArdle disease result from the distinct ways skeletal muscle produces ATP   749

Mitochondrial biogenesis is stimulated by muscular activity   749

Exercise alters muscle and whole-body metabolism   750

EXAMPLE Measuring the Impact of a Single Athletic Activity on Caloric Homeostasis   752

24.5 Starvation Induces Protein Wasting and Ketone Body Formation   752

The first priority during starvation is the maintenance of blood-glucose concentration   752

Metabolic adaptations in prolonged starvation minimize protein degradation   753

24.6 Ethanol Alters Energy Metabolism in the Liver   755

Ethanol metabolism leads to an excess of NADH   755

Ethanol metabolites cause liver damage   755

Excess ethanol consumption disrupts vitamin metabolism   756

Ethanol and defects in central energy metabolism contribute to the development of cancer   758

CHAPTER 25

Biosynthesis of Amino Acids   763

25.1 Nitrogen Fixation: Microorganisms Use ATP and a Powerful Reductant to Reduce Atmospheric Nitrogen to Ammonia   764

Biological nitrogen fixation is catalyzed by the nitrogenase complex   764

The iron–molybdenum cofactor of nitrogenase binds and reduces atmospheric nitrogen   765

Ammonium ion is assimilated into an amino acid through glutamate and glutamine   766

25.2 Amino Acids Are Made from Intermediates of the Citric Acid Cycle and Other Major Pathways   768

Human beings can synthesize some amino acids but must obtain others from their diet   768

Aspartate, alanine, and glutamate are formed by the addition of an amino group to an alpha-ketoacid   769

SCIENTIST PROFILE Beverly Guirard   770

A common step determines the chirality of all amino acids   770

The formation of asparagine from aspartate requires an adenylated intermediate   770

Glutamate is the precursor of glutamine, proline, and arginine   771

3-Phosphoglycerate is the precursor of serine, cysteine, and glycine   772

Tetrahydrofolate carries activated one-carbon units at several oxidation levels   772

S-Adenosylmethionine is the major donor of methyl groups   774

Cysteine is synthesized from serine and homocysteine   776

High homocysteine levels correlate with vascular disease   776

Shikimate and chorismate are intermediates in the biosynthesis of aromatic amino acids   776

Tryptophan synthase illustrates substrate channeling in enzymatic catalysis   779

25.3 Feedback Inhibition Regulates Amino Acid Biosynthesis   780

Branched pathways require sophisticated regulation   780

The sensitivity of glutamine synthetase to allosteric regulation is altered by covalent modification   782

25.4 Amino Acids Are Precursors of Many Biomolecules   783

Glutathione, a gamma-glutamyl peptide, serves as a sulfhydryl buffer and an antioxidant   783

Nitric oxide, a short-lived signal molecule, is formed from arginine   784

Amino acids are precursors for a number of neurotransmitters   785

Porphyrins are synthesized from glycine and succinyl coenzyme A   786

Porphyrins accumulate in some inherited disorders of porphyrin metabolism   788

CHAPTER 26

Nucleotide Biosynthesis   791

26.1 Nucleotides Can Be Synthesized by de Novo or Salvage Pathways   792

26.2 The Pyrimidine Ring Is Assembled from CO_2, Ammonia, and Aspartate   793

Bicarbonate and other oxygenated carbon compounds are activated by phosphorylation   793

The side chain of glutamine can be hydrolyzed to generate ammonia   794

The pyrimidine ring is completed and coupled to ribose   794

Nucleotide mono-, di-, and triphosphates are interconvertible   796

CTP is formed by amination of UTP   796

Salvage pathways recycle pyrimidine bases   796

26.3 Purine Bases Can Be Synthesized from Glycine, Aspartate, and Other Components   797

The purine ring system is assembled on ribose phosphate   797

The purine ring is assembled by successive steps of activation by phosphorylation followed by displacement   797

AMP and GMP are formed from IMP   799

Enzymes of the purine biosynthesis pathway associate
with one another   800

Salvage pathways economize intracellular resource
consumption   800

An alternative to adenine is used by some viruses   801

**26.4 Deoxyribonucleotides Are Synthesized by
the Reduction of Ribonucleotides   801**

Ribonucleotide reduction occurs via a radical
mechanism   802

Stable radicals are present in ribonucleotide
reductases   803

SCIENTIST PROFILE JoAnne Stubbe   804

Thymidylate is formed by the methylation of
deoxyuridylate   804

Several valuable anticancer drugs block the
synthesis of thymidylate   805

**26.5 Key Steps in Nucleotide Biosynthesis Are
Regulated by Feedback Inhibition   807**

Pyrimidine biosynthesis is regulated by aspartate
transcarbamoylase   807

The synthesis of purine nucleotides is controlled by
feedback inhibition at several sites   807

The synthesis of deoxyribonucleotides is controlled
by the regulation of ribonucleotide reductase   808

**26.6 Disruptions in Nucleotide Metabolism Can
Cause Pathological Conditions   809**

The loss of adenosine deaminase activity results
in severe combined immunodeficiency   809

Gout is induced by high serum levels of urate   810

Lesch–Nyhan syndrome is a dramatic consequence
of mutations in a salvage-pathway enzyme   810

CHAPTER 27

**Biosynthesis of Membrane Lipids and
Steroids   814**

**27.1 Phosphatidate Is a Common Intermediate in
the Synthesis of Phospholipids and
Triacylglycerols   815**

The synthesis of phospholipids requires an activated
intermediate   816

Some phospholipids are synthesized from an
activated alcohol   817

Phosphatidylcholine is an abundant phospholipid   818

Base-exchange reactions can generate phospholipids   818

Sphingolipids are synthesized from ceramide   819

Tay–Sachs disease results from the disruption
of lipid metabolism   820

Phosphatidic acid phosphatase is a key regulatory
enzyme in lipid metabolism   821

**27.2 Cholesterol Is Synthesized from Acetyl
Coenzyme A in Three Stages   821**

Stage 1: The synthesis of mevalonate initiates
the synthesis of cholesterol   822

Stage 2: Squalene (C_{30}) is synthesized from six
molecules of isopentenyl pyrophosphate (C_5)   823

Stage 3: Squalene cyclizes to form cholesterol   824

**27.3 The Regulation of Cholesterol Biosynthesis
Takes Place at Several Levels   825**

Lipoproteins transport cholesterol and triacylglycerols
throughout the organism   827

Low-density lipoproteins play a central role in
cholesterol metabolism   829

The absence of the LDL receptor leads to
hypercholesterolemia and atherosclerosis   830

Mutations in the LDL receptor prevent LDL
release and result in receptor destruction   831

Cycling of the LDL receptor is regulated   832

HDL appears to protect against atherosclerosis   832

The clinical management of cholesterol levels
can be understood at the biochemical level   832

**27.4 Important Biochemicals Are Synthesized
from Cholesterol and Isoprene   833**

Steroids are hydroxylated by cytochrome P450
monooxygenases that use NADPH and O_2 834

Cytochrome P450s are widespread and perform
many functions   836

SCIENTIST PROFILE Namandjé Bumpus   836

Pregnenolone is a precursor of many other steroids   836

Vitamin D is derived from cholesterol by
the ring-splitting activity of light   839

Five-carbon units are joined to form a wide variety
of biomolecules   840

Some isoprenoids have industrial applications   841

CHAPTER 28

**DNA Replication, Repair, and
Recombination   845**

**28.1 DNA Replication Proceeds by the
Polymerization of Deoxyribonucleoside
Triphosphates Along a Template   846**

DNA polymerases require a template and a primer   846

DNA polymerases have common structural features   846

Bound metal ions participate in the polymerase
reaction   847

The specificity of replication is dictated by
complementarity of shape between bases   847

An RNA primer enables DNA synthesis to begin   848

SCIENTIST PROFILE Tsuneko and Reiji Okazaki   849

One strand of DNA is made continuously, whereas
the other strand is synthesized in fragments   849

DNA ligase seals breaks in double-stranded DNA   849

The separation of DNA strands requires specific
helicases and ATP hydrolysis   850

**28.2 DNA Unwinding and Supercoiling Are
Controlled by Topoisomerases   851**

The linking number, a topological property,
determines the degree of supercoiling   853

Topoisomerases prepare the double helix for
unwinding   853

Type I topoisomerases relax supercoiled structures   854

Type II topoisomerases introduce negative supercoils
through coupling to ATP hydrolysis   855

28.3 DNA Replication Is Highly Coordinated   857

DNA replication requires highly processive
polymerases   857

The leading and lagging strands are synthesized in a
coordinated fashion   858

DNA replication in *E. coli* begins at a unique site   860

DNA replication in eukaryotes is initiated
at multiple sites   861

The eukaryotic cell cycle ensures coordination
of DNA replication and cell division   862

Telomeres are protective structures at the ends of
linear chromosomes   863

Telomeres are replicated by telomerase, a specialized
polymerase that carries its own RNA template   863

**28.4 Many Types of DNA Damage Can Be
Repaired   864**

Errors can arise in DNA replication   864

DNA can be damaged by oxidizing agents, alkylating
agents, and light   865

DNA damage can be detected and repaired by a
variety of systems   866

The presence of thymine instead of uracil in DNA
permits the repair of deaminated cytosine   868

Some genetic diseases are caused by the
expansion of repeats of three nucleotides   869

Many cancers are initiated by the defective
repair of DNA   869

Many potential carcinogens can be detected by their
mutagenic action on bacteria   871

**28.5 DNA Recombination Plays Important
Roles in Replication, Repair, and Other
Processes   872**

RecA can initiate recombination by promoting strand
invasion   872

Some recombination reactions proceed through
Holliday-junction intermediates   873

CHAPTER 29

**RNA Functions, Biosynthesis, and
Processing   879**

**29.1 RNA Molecules Play Different Roles,
Primarily in Gene Expression   880**

RNAs play key roles in protein biosynthesis   880

Some RNAs can guide modifications of themselves
or other RNAs   880

Some viruses have RNA genomes   880

Messenger RNA vaccines provide protection
against diseases   880

29.2 RNA Polymerases Catalyze Transcription   881

RNA synthesis comprises three stages: initiation,
elongation, and termination   882

RNA polymerases catalyze the formation of a
phosphodiester bond   882

RNA chains are formed de novo and grow in the
5′-to-3′ direction   884

RNA polymerases backtrack and correct errors   885

RNA polymerase binds to promoter sites on the
DNA template in bacteria to initiate transcription   886

Sigma subunits of RNA polymerase in bacteria
recognize promoter sites   886

The template double helix must be unwound for
transcription to take place   887

Elongation takes place at transcription bubbles that
move along the DNA template   888

Sequences within the newly transcribed RNA signal
termination   888

In bacteria, the rho protein helps to terminate the
transcription of some genes   889

29.3 Transcription Is Highly Regulated   890

Alternative sigma subunits in bacteria control
transcription in response to changes in conditions   890

Some messenger RNAs directly sense metabolite
concentrations   891

Control of transcription in eukaryotes is highly
complex   892

Eukaryotic DNA is organized into chromatin   893

Three types of RNA polymerase synthesize RNA in
eukaryotic cells   894

Three common elements can be found in the RNA
polymerase II promoter region   896

Regulatory cis-acting elements are recognized by
different mechanisms   896

The TFIID protein complex initiates the assembly
of the active transcription complex in eukaryotes   896

Enhancer sequences can stimulate transcription at
start sites thousands of bases away   898

**29.4 Some RNA Transcription Products Are
Processed   898**

Precursors of transfer and ribosomal RNA are cleaved
and chemically modified after transcription   898

RNA polymerase I produces three ribosomal RNAs   899

RNA polymerase III produces transfer RNAs   900

The product of RNA polymerase II, the pre-mRNA
transcript, acquires a 5′ cap and a 3′ poly(A) tail   900

Sequences at the ends of introns specify splice sites
in mRNA precursors   901

Splicing consists of two sequential transesterification
reactions   902

Small nuclear RNAs in spliceosomes catalyze the
splicing of mRNA precursors   903

Mutations that affect pre-mRNA splicing cause
disease   906

Most human pre-mRNAs can be spliced in alternative
ways to yield different proteins   906

Transcription and mRNA processing are coupled   908

Small regulatory RNAs are cleaved from larger
precursors   908

RNA editing can lead to specific changes in mRNA   908

**29.5 The Discovery of Catalytic RNA Revealed
a Unique Splicing Mechanism   909**

Some RNAs can promote their own splicing   909

RNA enzymes can promote many reactions, including
RNA polymerization   912

SCIENTIST PROFILE Thomas Cech   912

CHAPTER 30

Protein Biosynthesis   **916**

30.1 Protein Biosynthesis Requires the Translation of Nucleotide Sequences into Amino Acid Sequences   917

The biosynthesis of long proteins requires a low error frequency   917

Transfer RNA (tRNA) molecules have a common design   918

Some transfer RNA molecules recognize more than one codon because of wobble in base-pairing   920

30.2 Aminoacyl-tRNA Synthetases Establish the Genetic Code   921

Amino acids are first activated by adenylation   921

Aminoacyl-tRNA synthetases have highly discriminating amino acid activation sites   922

Proofreading by aminoacyl-tRNA synthetases increases the fidelity of protein biosynthesis   923

Kinetic proofreading increases the fidelity of protein biosynthesis   924

Synthetases recognize various features of transfer RNA molecules   925

Aminoacyl-tRNA synthetases are divided into two classes   926

30.3 The Ribosome Is the Site of Protein Biosynthesis   926

Ribosomal RNAs (5S, 16S, and 23S rRNA) play central roles in protein biosynthesis   927

Ribosomes have three transfer RNA-binding sites that bridge the 30S and 50S subunits   928

The start signal is usually AUG preceded by several bases that pair with 16S rRNA   929

Bacterial protein biosynthesis is initiated by N-formylmethionyl-transfer RNA   930

N-Formylmethionyl-tRNA^{fMet} is placed in the P site of the ribosome in the formation of the 70S initiation complex   931

Elongation factors deliver aminoacyl-tRNAs to the ribosome   931

Peptidyl transferase catalyzes peptide-bond formation   932

GTP hydrolysis-driven translocation of tRNAs and mRNA follows peptide-bond formation   933

In bacteria, transcription and translation are coupled in space and time   935

Protein biosynthesis is terminated by release factors that read stop codons   935

Eukaryotic protein biosynthesis differs from bacterial protein biosynthesis primarily in translation initiation   936

Ribosomes selectively control gene expression   938

Scientists have manipulated protein biosynthesis pathways to incorporate unnatural amino acids in preselected positions   938

SCIENTIST PROFILE Ada Yonath   938

30.4 Ribosomes Bound to the Endoplasmic Reticulum Manufacture Secretory and Membrane Proteins   939

Protein biosynthesis begins on ribosomes that are free in the cytoplasm   939

Signal sequences mark proteins for translocation across the endoplasmic reticulum membrane   940

Transport vesicles carry cargo proteins to their final destinations   941

30.5 A Variety of Antibiotics and Toxins Inhibit Protein Biosynthesis   942

Some antibiotics inhibit protein biosynthesis   942

Diphtheria toxin blocks protein biosynthesis in eukaryotes by inhibiting translocation   943

Some toxins modify 28S ribosomal RNA   944

CHAPTER 31

Control of Gene Expression   **949**

31.1 Bacterial DNA-Binding Proteins Bind to Specific Regulatory Sites   950

Many DNA-binding proteins match the symmetry in their target DNA sequences   950

The helix-turn-helix motif is common to many bacterial DNA-binding proteins   951

31.2 In Bacteria, Genes Are Often Arranged into Clusters Under the Control of a Single Regulatory Sequence   951

An operon consists of regulatory elements and protein-encoding genes   952

The *lac* repressor protein can block transcription   953

Ligand binding can induce structural changes in regulatory proteins   954

The operon is a common regulatory unit in bacteria   955

Some DNA-binding proteins stimulate transcription   955

31.3 Regulatory Circuits Can Result in Switching Between Patterns of Gene Expression   956

The λ repressor regulates its own expression   957

A circuit based on the λ repressor and Cro forms a genetic switch   957

31.4 Regulation of Gene Expression Is More Complex in Eukaryotes   958

A range of DNA-binding motifs are employed by eukaryotic DNA-binding proteins   958

Activation domains interact with other proteins   960

Multiple transcription factors interact with eukaryotic regulatory regions   960

31.5 The Control of Gene Expression in Eukaryotes Can Require Chromatin Remodeling   961

Chromatin remodeling and DNA methylation regulate access to DNA-binding sites   961

Epigenetic modifications influence gene expression   962

SCIENTIST PROFILE Sarah Stewart   962

Enhancers stimulate transcription by recruiting activator proteins that alter chromatin structure   962

Nuclear hormone receptors are transcription factors that cause changes in chromatin structure   963

Nuclear hormone receptors regulate transcription by recruiting coactivators to the transcription complex   965

Chromatin structure is modulated through covalent modifications of histone tails   965

Transcriptional repression can be achieved through histone deacetylation and other modifications   967

31.6 Gene Expression Can Be Controlled at the Posttranscriptional Level   967

Attenuation regulates transcription in bacteria through the modulation of nascent RNA secondary structure   967

Eukaryotes use different mechanisms to control gene expression at the posttranscriptional level   969

Genes associated with iron metabolism are translationally regulated in animals   970

Small RNAs are involved in posttranscriptional gene regulation in eukaryotes   971

CHAPTER 32

Principles of Drug Discovery and Development   977

32.1 Drug Discovery Begins with Target Identification and Validation   978

Drug targets must be validated and tractable   978

Serendipitous observations can drive drug development   979

32.2 Lead Molecules Can Be Discovered in Many Ways   980

Natural products are a valuable source of lead molecules   980

High-throughput screening expands the opportunity for lead identification   981

Screening libraries can be prepared using combinatorial chemistry   982

DNA-encoded libraries provide very large compound libraries for lead identification   984

Phenotypic screening provides an alternative to the target-centered approach   985

32.3 Compounds Must Meet Stringent Criteria to Be Developed into Drugs   986

Drugs must be potent and selective   986

EXAMPLE Determining the IC_{50} for an Inhibitor   987

Drugs must have suitable properties to reach their targets   988

Toxicity can limit drug effectiveness   992

Lead molecules can be optimized on the basis of three-dimensional structural information about their targets   993

32.4 Biologics Are a Growing Family of Drugs   995

The majority of biologics are recombinant proteins   995

Monoclonal antibodies are highly specific and potent recombinant protein biologics   995

SCIENTIST PROFILE Gertrude Elion   996

32.5 The Clinical Development of Medicines Proceeds Through Several Phases   996

Clinical trials are time-consuming and expensive   996

The evolution of drug resistance can limit the utility of drugs for infectious agents and cancer   998

CHEMISTRY REVIEW APPENDIX   A1

ANSWERS TO SELF-CHECK QUESTIONS   A5

ANSWERS TO PROBLEMS   A11

INDEX   I1

Fifty years ago, the first edition of *Biochemistry* hit the market and forever changed the way biochemistry is taught. In the tenth edition, we remain true to the hallmarks of Lubert Stryer's vision while presenting molecular structure and function, metabolism, regulation, and laboratory techniques in new ways that engage students actively in the process of learning biochemistry.

THEMATIC THREADS

It has always been our goal to help students connect biochemistry to their own lives and the world around them. We use an evolutionary perspective to reveal the threads that bind organisms, clinical applications to show physiological relevance, and biotechnological and industrial applications to show how biochemistry is used to improve our lives. Pathways and processes are presented in a physiological context so students can see how biochemistry works in the body when it is at rest and during exercise, during health and disease, and under stress. Many of these examples are tagged with a caduceus symbol. (For a full list of these clinical applications, see p. xxxii.) New for this edition, we have highlighted the importance of time as a variable, linking the time scales associated with fundamental molecular processes, enzymatic reactions, physiological responses, and evolutionary change.

Several chapters provide the foundation for students' understanding of key aspects of biochemistry.

- Chapter 3 illustrates the remarkable abilities of biological molecules to interact specifically with one another, using as examples the oxygen-binding proteins myoglobin and hemoglobin and key molecules from the immune system.

- Chapters 5 through 7 explain where enzymes derive the catalytic power, specificity, and control they have to drive chemical and energetic transformations.

- Chapter 15 introduces the basic concepts and design of metabolic pathways before students dive into the chapters on specific pathways. ▼

- Chapter 24 integrates the concepts of energy metabolism by focusing on healthy caloric homeostasis and its dysfunctional extremes: obesity and starvation. The chapter also presents the biochemical basis for the many benefits of regular exercise and the pathology of diabetes.

AREAS OF FOCUS IN THE TENTH EDITION

In developing the tenth edition of *Biochemistry*, we stayed true to the integrity of Lubert Stryer's original textbook while updating our approach to reflect today's biochemistry instructors and students. Within our main text, you will find our hallmark clear writing and straightforward illustrations, which together make the language of biochemistry accessible for students learning the subject for the first time. We present a logical flow of ideas, clearly written text, and informative figures to help students grasp and retain biochemical concepts and details.

The strengths of the tenth edition textbook are enhanced with videos, interactive figures, and more in our interactive e-book. The e-book is available within our ≋ Achie√e platform and is downloadable through the VitalSource app, enabling students to read offline. Students can highlight, take notes, search by keyword, and even have the book read aloud. The ≋ Achie√e course further builds on the strengths of the tenth edition with interactive assessments, media, and abundant resources for use during class time or during study.

The tenth edition supports three organizing principles:

- Advances in biochemistry rely on diverse teams.

- Visualization is a critical teaching tool in biochemistry.

- Biochemistry requires conceptual understanding, critical thinking, effective problem solving, and practice.

Advances in Biochemistry Rely on Diverse Teams

In the tenth edition, we include short profiles of scientists, some very well-known and some less so, who have contributed to different aspects of biochemistry. In addition to introducing some fascinating people, the purpose

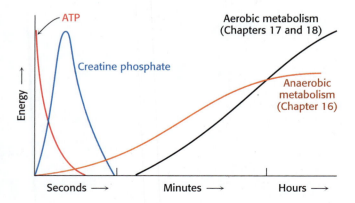

FIGURE 15.6 The sources of ATP change as exercise duration increases, even within the first few seconds. In the initial seconds of exertion, power is generated by existing high-phosphoryl-transfer compounds (ATP and creatine phosphate). Subsequently, the ATP must be regenerated by metabolic pathways.

of these profiles is to illustrate some of the wide range of training and career paths that different individuals have taken to allow them to make important contributions to biochemistry. Some profiles highlight how different life experiences have contributed to the selection of scientific problems or approaches. ▶

Through these profiles, students may be able to apply their own experiences to understanding topics in biochemistry and other sciences. A list of all the profiles is on p. xxxiii.

Visualization Is a Critical Teaching Tool in Biochemistry

The ability to visualize concepts and how they are connected helps students gain a deeper conceptual understanding. Two innovations introduced in the first edition of *Biochemistry* were an open, uncluttered layout of text and figures that was easy on the eyes (and brain) and the presentation of one concept per figure. We continue these traditions in the tenth edition, while also presenting an updated art program that enhances the power of figures to help students learn. Concise figures and ample margins are part of a layout that is well organized and easy to read.

- **One Idea at a Time.** Each figure illustrates a single concept, which helps students zoom in on the main points without the distraction of excess detail. **Figure captions** begin by identifying the concept in the figure.

- **Macromolecular Structures.** Beautifully rendered art depicting structures and structural models were prepared by authors Jeremy Berg and Gregory Gatto, based on the latest x-ray crystallographic, NMR, and cryo-electron microscopy data. For most molecular models, the **PDB number** at the end of the figure gives the student easy access to the file used in generating the structure (https://www.wwpdb.org).

JAMES TENSUAN/The New York Times/Redux

CAROLYN BERTOZZI Chemist Carolyn Bertozzi is a leader in developing the field of glycobiology and communicating its exciting possibilities. Initially planning to go to medical school, her interest in chemistry developed in college, and she went on to complete a PhD focused on synthesizing

In the tenth edition, chapter reading is enhanced with a variety of embedded multimedia tools in ≋ Achie∕e to enliven the narrative, while promoting strong visual understanding of the material. Each of these enhancements are available for stand-alone use in ≋ Achie∕e or within interactive questions and activities:

- **Interactive Figures.** Bring protein structures to life with interactive 3-D models using Jmol. Students can zoom and rotate structures and experiment with different display styles (space-filling, ball-and-stick, ribbon, and backbone) by means of a user-friendly interface. These interactive molecular structures can be launched within the e-book wherever you see the INTERACT callouts. ▼

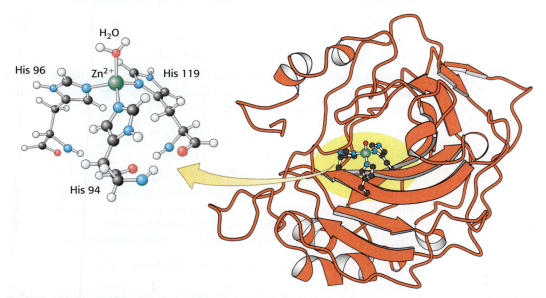

FIGURE 6.18 The active site of human carbonic anhydrase II includes a bound zinc ion. The active site includes a zinc ion that is bound to the imidazole rings of three histidine residues as well as to a water molecule in a cleft near the center of the enzyme. [Drawn from 1CA2.pdb.]

INTERACT with this model in
≋ Achie∕e

- **Animated Mechanisms.** Watch the mechanisms of iconic enzymes as short animations.

- **Animated Technique Videos.** Reinforce understanding of laboratory techniques by watching animations that include an overview of protocol and then show what is happening at the chemical level. Laboratory technique animations can be launched within the e-book wherever you see the VIEW callouts.

- **Interactive Metabolic Map.** EXPLORE callouts highlight metabolic processes that students can study in greater depth in ≋ Achieve. ▶

    With this interactive tool, students can navigate and zoom between overviews and detailed views of the most commonly taught metabolic pathways. Embedded Tours take students through the pathways step by step. Such intimate interaction helps students visualize concepts in isolation as well as understand how concepts are interconnected. Metabolic Map Pathways include glycolysis, gluconeogenesis, the citric acid cycle, oxidative phosphorylation, the pentose phosphate pathway, glycogen metabolism, fatty acid synthesis, the urea cycle, and β oxidation.

Biochemistry Requires Conceptual Understanding, Critical Thinking, Effective Problem Solving, and Practice

Every chapter of *Biochemistry*, Tenth Edition, provides numerous opportunities for students to practice problem-solving skills and apply the concepts described in the text.

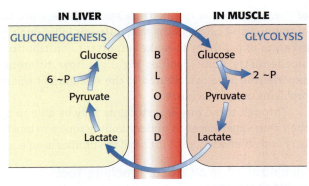

FIGURE 16.36 The Cori cycle allows the liver to better support other tissues, such as skeletal muscle. Lactate formed by active muscle is converted into glucose by the liver. This cycle shifts part of the metabolic burden of active muscle to the liver. The symbol, ~P represents nucleoside triphosphates.

EXPLORE this pathway further in the Metabolic Map in ≋ Achieve

- **NEW! Self-Check Questions.** New Self-Check questions provide periodic "speed bumps" in the chapter narrative to allow students to assess their comprehension and practice applying concepts they have just learned. ▼

❖ SELF–CHECK QUESTION

Recall our study of enzymes in previous chapters: Why do isolated F_1 subunits of ATP synthase catalyze ATP hydrolysis?

- **NEW! Examples.** The boxed examples (listed on p. xxxiii) provide point-of-use, worked-through examples of the types of quantitative and reasoning questions students will frequently encounter in biochemistry courses and in laboratory work. ▶

EXAMPLE   Determining the Fraction of Bound Receptors

PROBLEM: What fraction of receptors are occupied with ligand at a ligand concentration of 5 μM, given that half of the receptors are occupied with ligand at a total ligand concentration of 20 μM?

GETTING STARTED: Equation 2 gives the relationship between the fraction of receptors occupied with ligands as a function of the total ligand concentration and the dissociation constant:

$$[RL]/R_{total} = L_{total}/(K_d + L_{total})$$

We are interested in what happens at a total ligand concentration of 5 μM. In order to use equation 2, we need to know the dissociation constant.

SOLVE: We showed that the half of the receptors are occupied with ligands when the total ligand concentration is equal to the dissociation constant. Since half of the receptors are occupied when the ligand concentration is 20 μM, we know that $K_d = 20$ nM. Substituting this and $L_{total} = 5$ μM into equation 2, we find that

$$[RL]/R_{total} = 5\ \mu M/(20\ \mu M + 5\ \mu M) = 0.20$$

REFLECT: This result makes sense since the total ligand concentration is somewhat less than 20 μM, the ligand concentration at which half (0.50) of the receptors are occupied.

In 〰 Achieve, engagement is encouraged with adaptive quizzing, feedback-rich problem-solving practice, case study applications, and more.

- 〰 Achieve **Adaptive Quizzing.** This game-like activity motivates reading and provides individualized learning and feedback as students build towards a target score. All questions are tied back to the e-book to encourage students to explore the reading, figures, and interactive resources at hand. Adaptive quizzing includes a student dashboard with their progress and a personalized study plan. ▼

- **NEW! Every end-of-chapter problem from** *Biochemistry,* **Tenth Edition, is assignable in** 〰 Achieve. Raise comprehension and problem-solving skills with robust end-of-chapter questions, that are available as pencil-and-paper assignments or as autograded, feedback-rich online homework. The end-of-chapter problems in *Biochemistry*, Tenth Edition, vary in presentation and degree of difficulty; they include higher-level critical thinking questions that ask students to suggest or describe a chemical mechanism, draw conclusions from data taken from real research papers, and connect concepts across chapters.

Label the **cell membrane** using the terms.

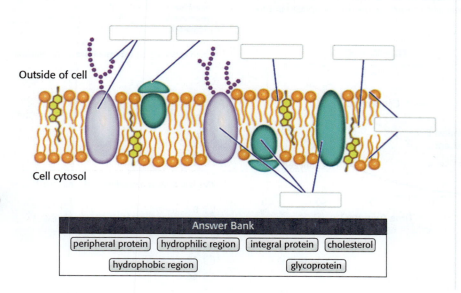

Outside of cell

Cell cytosol

Answer Bank
peripheral protein   hydrophilic region   integral protein   cholesterol
hydrophobic region   glycoprotein

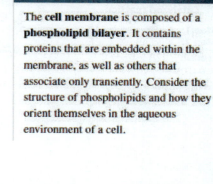

Hint

The **cell membrane** is composed of a **phospholipid bilayer**. It contains proteins that are embedded within the membrane, as well as others that associate only transiently. Consider the structure of phospholipids and how they orient themselves in the aqueous environment of a cell.

Label the **cell membrane** using the terms.

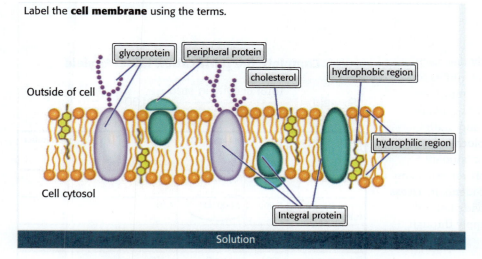

glycoprotein   peripheral protein

cholesterol

hydrophobic region

Outside of cell

hydrophilic region

Cell cytosol

Integral protein

Solution

Need more help with this question? Review your previous course materials, or use online resources to facilitate your **understanding of the functional components of the fluid mosaic model of biological membranes.**

The **phospholipid bilayer** that composes the **cell membrane** is responsible for separating the contents of the cell from the outside environment. Phospholipids consist of a hydrophilic phosphate head that associates with water and a hydrophobic fatty acid chain that repels water. As water is the predominant molecule on both sides of the membrane, phospholipids associate such that the hydrophilic heads form the inner and outer surfaces of the membrane, facing intra- and extracellular water molecules, respectively. Conversely, the hydrophobic fatty acid chains form a nonpolar membrane interior. As

Feedback

One or more terms are misplaced.

Integral proteins, such as transporters, extend into the membrane. Other proteins, called **peripheral proteins** because of their locations, have functions such as helping move substances to other parts of the cell.

- In 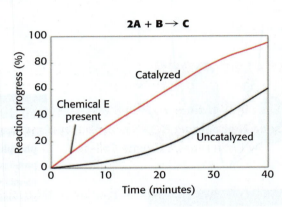 Achieve, each autograded question delivers hints and immediate, individualized feedback so students get guidance targeted to the underlying misconception of each specific wrong answer. Upon completion of each question, students can see fully worked solutions that explain how to reason through and solve the question. ▼

- Achieve **Question Bank.** In addition to end-of-chapter problems, instructors can explore and assign thousands of additional interactive, autograded questions for more problem-solving practice. Questions in the question bank are frequently customizable and algorithmic so students working together see variations of the same problems. Achieve promotes student success through practice and targeted feedback, while saving instructors time and providing meaningful insights with assessment reports.

Most chemical reactions occur in a step-wise fashion, requiring more than one **elementary step** (or **elementary reaction**) to form the final products. During a step-wise reaction process, a later elementary step consumes the **intermediate** or intermediates formed by an earlier elementary step. Therefore, like catalysts, reaction intermediates do not appear in the overall schematic of a chemical reaction.

Uncatalyzed reaction

Step 1: $A + B \longrightarrow D$
Step 2: $A + D \longrightarrow C$

Overall: $2A + B \longrightarrow C$

2A + B → C

Catalyzed reaction

Step 1: $A + B + E \longrightarrow EAB$
Step 2: $A + EAB \longrightarrow C + E$

Overall: $2A + B \longrightarrow C$

Consider the given series of **elementary reactions** and their **overall reaction**.

reaction 1:	$A + B \longrightarrow AB$
reaction 2:	$AB + C \longrightarrow AC + B$
overall:	$A + C \longrightarrow AC$

Which species is an **intermediate** in the reaction?

- ○ C
- ○ AC
- ○ B
- ○ A
- ○ AB

Which species is a **catalyst** in the reaction?

- ○ B
- ○ C
- ○ A
- ○ AB
- ○ AC

- Achieve **NEW! Skills You Need Activities.** Each chapter includes a short activity that identifies one to three skills or concepts that students may have learned in prerequisite courses—such as general chemistry, organic chemistry, biology, and math—and will need to apply to the current biochemistry chapter. Developed with guidance from student reviewers, these activities provide short refreshers on the prerequisite content then allow students to assess their mastery of the content and provide helpful feedback and search terms they can apply through their preferred methods of study. ▶

Complete the purification table: % yield

$$\% \text{ yield} = \frac{\text{total activity in step}}{\text{total activity step 1}} \times 100$$

step (procedure)	total protein (mg)	activity (units)	specific activity (units/mg)	% yield	purification factor
crude extract	20,000	4,000,000	200	100	
precipitation (salt)	5,000	3,000,000	600	75	
precipitation (pH)	4,000	1,000,000	250		
ion-exchange chromatography	200	800,000	4,000		
affinity chromatography	50	750,000	15,000		
size-exclusion chromatography	45	675,000	15,000		

- ![Achieve] **Readiness Activities.** Whereas Skills You Need activities are aligned with one chapter at a time, Readiness Activities are designed to get students ready for the course as a whole. Students can review the math, general chemistry, and organic chemistry skills they will need to use as biochemistry students, and each question provides hints, answer-specific feedback, and solutions.

- ![Achieve] **Problem-Solving Videos.** With diagrams, graphs, and narration, these videos walk students through problems on topics that typically prove difficult—such as data interpretation—helping students understand the right approach to the solution.

- ![Achieve] **Case Studies.** Each case study introduces students to a biochemical mystery and allows them to determine what investigations will solve it. Accompanying assessments ensure that students have fully completed and understood the case study, showing that they have applied their knowledge of the concepts. ▼▶

Case Study Topics

Biomolecules

- Observation Over Intuition: Thermodynamics of Protein Folding
- pH Peril: pH and Amino Acid Structure
- Profound Relevance: Hemoglobin Regulation (Bohr Effect)
- **NEW!** Mile High Stadium and Silent Killer: Hemoglobin Dysfunction
- **NEW!** Toxic Alcohols: Enzyme Function
- A Likely Story: Enzyme Inhibition

Signal Transduction/Metabolism

- **NEW!** Failure to Propagate: Signal Transduction
- **NEW!** Sudden Onset: Introduction to Metabolism
- An Unexplained Death: Carbohydrate Metabolism
- A Day at the Beach: Fatty Acid Metabolism
- **NEW!** The Narrow Window: Electron Transport Chain and Oxidative Phosphorylation
- Runner's Experiment: Integration of Energy Metabolism

Mile High Stadium and the Silent Killer
Case Study on Hemoglobin Regulation and Dysfunction
By Justin Hines, Lafayette College

Photo Credit: Zoonar /A. Kravchenko /AGE Fotostock

Act 1 - Trouble at Mile High Stadium

Gino had hemoglobin on the mind. He had been embarrassed earlier when he failed to remember how pH was related to oxygen transport in the human body.

As an undergraduate in his junior year who hoped to go to medical school, Gino's day job shadowing emergency room physician Dr. Rose Smith was otherwise going well. But the routine patient visits were interrupted when he saw a large crowd gathered just outside in the hospital lobby. Most had flowers, but some, strangely, were carrying footballs and wearing jerseys.

"Dr. Smith, what's that all about?" Gino pointed to the noisy crowd.

"Oh that... yes, we have a patient up on three recovering from a sickle cell attack."

"But why so many people?" asked Gino. "And why do some have footballs? Is it a child or something?"

"No, actually he's an NFL star... plays defensive end. Most of those people are just fans. He played a game last night here in Denver at the new Mile High Stadium. The combination of vigorous exercise from sprinting all over a 100-yard field and the fact that the game was here in Denver triggered a sickle cell crisis. He'll probably never play in Denver again."

"He has sickle cell anemia and plays in the NFL??!"

"Actually, he has sickle cell *trait*, meaning he is heterozygous for the sickle cell allele, HbS. Some people with sickle cell trait can be top athletes. But, the right conditions can still bring about a crisis."

Achieve: Your ONE-STOP-SHOP FOR INSTRUCTIONAL RESOURCES

Integrated with the tenth edition are activities, media, and resources to support in-person, virtual, and hybrid teaching. The resources found in the Achieve platform include tools for class preparation, instruction, practice and assessment, and reporting and analytics:

Class Preparation

- **Transition Guide** for navigating the changes between editions

- **Goal-Setting and Reflection Surveys** to help students identify their goals, develop strategies to reach those goals, and realign their study habits throughout the course

- Assignable chapter-by-chapter **Skills You Need** and course-level **Readiness Activities** to refresh students on content from courses frequently taken as prerequisites

- **Interactive e-book** with assignable sections and chapters

- Assignable **LearningCurve Adaptive Quizzing** tasks to ensure reading comprehension

Instruction

- Editable all-in-one **Lecture Slides** that include content, images, iClicker questions, multimedia tools, and activities

- **Standalone Slide Decks** streamlined to just include content, images, and iClicker activities

- **Problem-Solving Activities** and other **In-Class Activities** with **Instructor Activity Guides** and student **Worksheets** that follow a consistent approach for solving biochemistry problems and include recommendations for use in face-to-face, virtual, and hybrid classes

- **Cloud-based iClicker**, an in-class response system with integrated Gradebook

- Presentation-ready media such as the **Interactive Metabolic Map**, **Animated Mechanisms**, **Animated Techniques**, **Problem-Solving Videos**, **Living Figures**, and **Case Studies** integrated into **Lecture Slides**, **Problem-Solving Activities**, and other **In-Class Activities**; they are also available as stand-alone resources.

Practice and Assessment

- **Editable, curated, autograded homework assignments** that match the order and question in the text

- **Question Bank** with thousands of additional questions to create an assignment from scratch or add to a curated assignment

- **Abbreviated Solutions** and **Extended Solutions** for all end-of-chapter questions

- Assignable **Case Studies**

- **Test Banks** and accompanying software (located on Macmillan Learning's catalog page) to create tests outside of the Achieve environment

Reporting and Analytics

- **Learning Objectives** aligned with course content and resources

- **Insights** on student performance on top learning objectives and course data, such as time on task and other metrics

- **Detailed reporting** by class, by individual student, and by learning objective

- **Gradebook that syncs with iClicker** for an all-in-one gradebook

- **Integration** with most campus Learning Management Systems

RECENT ADVANCES AND DISCOVERIES INCLUDED IN THE NEW EDITION

Biochemistry is an exciting and dynamic field of study. In the tenth edition of *Biochemistry*, we have updated examples and explanations to provide the latest and most interesting discoveries in biochemistry, along with material essential for understanding biochemical concepts. Some of these new topics are:

Discussion of SARS-CoV-2 and COVID-19 in multiple contexts (Chapter 1–4, 14, 29)

Discussion and illustration of binding phenomena at the single-molecule level (Chapter 1)

Expanded discussion of the interpretation of variations in genome sequences in terms of ancestry and evolutionary timescales (Chapter 1)

New section highlighting different career paths related to biochemistry, and the importance of diverse teams of people for advancing our field of knowledge (Chapter 1)

Updated discussion of protein structure prediction, including the development of AlphaFold (Chapter 2)

New chapter focused on molecular recognition, using the examples of steroid-hormone binding, oxygen transport by hemoglobin, and key proteins from the immune system (Chapter 3)

Expanded discussion of timescale and single-molecule kinetics, including new figures and equations (Chapter 5)

Expanded discussion of molecular motor proteins, particularly myosin (Chapter 6)

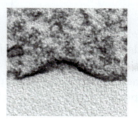
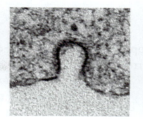
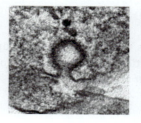
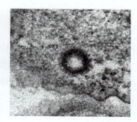

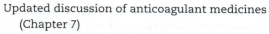

Clathrin

Clathrin-coated pit

Dynamin

FIGURE 12.38 Many cells take up molecules by receptor-mediated endocytosis. Receptor binding on the surface of the cell induces the membrane to invaginate, with the assistance of specialized intracellular proteins such as clathrin and dynamin. The process results in the formation of a vesicle within the cell. [Credit: *Top:* Republished with permission of Company of Biologists, from Membrane remodeling in clathrin-mediated endocytosis. Volker Haucke, Michael M. Kozlov, J Cell Sci (2018) 131 (17): jcs216812. Figure 1. Permission conveyed through Copyright Clearance Center, Inc.; *Bottom:* Schmid SL. Reciprocal regulation of signaling and endocytosis: Implications for the evolving cancer cell. J Cell Biol. 2017 Sep 4;216(9):2623-2632. doi: 10.1083/jcb.201705017. Epub 2017 Jul 3. PMID: 28674108; PMCID: PMC5584184. Copyright © 2017 Schmid. Illustration by Marcel Mettlen. Used with permission.]

Updated discussion of anticoagulant medicines (Chapter 7)

Discussion of the major and minor grooves of double-stranded DNA (Chapter 8)

Expanded discussion of the methods for quantitation of gene expression to include RNA-seq (Chapter 9)

Updated presentation of carbohydrate structures and the role of glycoproteins in biology (Chapter 11)

Expanded discussion of lipid rafts (Chapter 12) ▲

Updated discussion of the ion channels responsible for cardiac pacemaker activity (Chapter 13)

Illustration of signal-transduction processes in the context of the "fight-or-flight" response (Chapter 14)

Discussion of signal-transduction processes involved in sensory systems, including sight, smell, and hearing (Chapter 14)

Updated information regarding the health effects of fructose consumption (Chapter 16)

Discussion of inborn errors of metabolism (Chapter 16)

Discussion of the role of the proton motive force in the flagellar motor (Chapter 18)

New emphasis on phototrophy versus photosynthesis, and autotrophy versus heterotrophy (Chapter 19)

Discussion of optogenetics, including bacteriorhodopsin and its relationship to rhodopsin and channelrhodopsin (Chapter 19)

Updated discussion of the evolution and enzymology of C_4 plants (Chapter 20)

Updated discussion of the UPS system, the role of degrons in regulation, and the alternative functions of ubiquitin (Chapter 23)

Updated discussion of the fasted-fed cycle, diabetes, and obesity (Chapter 24)

Discussion of McArdle glycogen storage disease and its effects on the fuel choices available to muscles (Chapter 24)

Discussion of metabolism and cancer (Chapter 24)

Discussion of some viruses that utilize an alternative nucleobase (Chapter 26)

Discussion of the involvement of 3-metal ions in catalysis by DNA polymerase (Chapter 28)

Added examples of RNA function (Chapter 29)

Discussion of the role of kinetic proofreading in increasing the fidelity of protein synthesis (Chapter 30)

Discussion of the manipulation of protein-synthesis pathways to incorporate unnatural amino acids in preselected positions (Chapter 30)

Discussion of how ribosomes selectively control gene expression (Chapter 30)

Updated and greatly revised discussion on drug discovery to emphasize target identification, validation, and tractability (Chapter 32)

Expanded discussion of the role of high-throughput screening in lead identification during drug discovery (Chapter 32)

Discussion of biologics (Chapter 32)

CLINICAL APPLICATIONS

This icon signals the start of a clinical application in the text. Additional, briefer clinical correlations appear in the text as appropriate.

Osteogenesis imperfecta	54
Protein-misfolding diseases	62
2,3-BPG and fetal hemoglobin	81
Carbon monoxide poisoning	82
Sickle-cell anemia	83
Vaccines, including for SARS-CoV-2	89
Monoclonal antibodies	114
Aldehyde dehydrogenase deficiency	156
Competitive inhibitors as drugs	166
Action of penicillin	172
Toxins that modify reactive serine residues	182
Protease inhibitors as drugs	189
Protein kinase A mutations and Cushing's syndrome	222
Deficiency of α_1-antitrypsin and lung disease	227
Anticoagulant drugs	231
Identifying disease-causing mutations using recombinant DNA technology	271
Using RNAi to treat disease	295
Monitoring changes in glycosylated hemoglobin (A1C levels)	331
Using maltase inhibitors to control blood glucose	334
Human milk oligosaccharides and newborn protection	335
Erythropoietin (EPO)	339
Mucin overexpression and disease	341
Blood type O and cholera susceptibility	344
Errors in glycosylation and I-cell disease	344
Liposomes for drug delivery	361
Demyelinating diseases	362
Action of aspirin and ibuprofen	366
Gram staining to characterize bacterial infections	373
Digoxin and congestive heart failure	387
ATP-binding cassette domains and multidrug resistance	388
Drug side effects caused by interaction with the K^+ channel hERG	405
Signal-transduction pathways and cancer	433
COVID-19 effects on sense of smell	437
Vitamins	464
Effects of excessive fructose consumption	489
Galactosemia	491
Lactose intolerance	492
Glucokinase and type 2 diabetes	496
Aerobic glycolysis and cancer	497
Activation of transcription factor HIF1 by both cancer and endurance training	498
Effects of substrate cycles on health and disease	508
Deficiencies of glycolytic or gluconeogenic enzymes	509
Pyruvate dehydrogenase complex inhibition and diabetic neuropathy	532
Beriberi and mercury and arsenic poisoning	535
Blocking the glyoxylate cycle as potential tuberculosis treatment	538
Frataxin mutations and Friedreich's ataxia	552
Reactive oxygen species (ROS) and signal transduction	561
Mild uncouplers as drugs	578
Mitochondrial diseases	578
Photosensitive proteins and optogenetics	607
Glucose 6-phosphate dehydrogenase deficiency and drug-induced hemolytic anemia	634
Drugs for type 2 diabetes that target liver phosphorylase	660
Glycogen-storage diseases	660
Carnitine deficiency	672
Zellweger syndrome	679
Fatty acids and pathological conditions	679
Ketogenic diets as treatment for epilepsy	682
Diabetic ketosis	682
Fatty acid synthase inhibitors as drugs	691
Effects of aspirin on signaling pathways	694
Diseases resulting from defects in E3 proteins	705
Proteasome inhibitors and tuberculosis treatment	707
Blood levels of aminotransferases as indicators of liver damage	711
Inherited defects of the urea cycle and hyperammonemia	717
Phenylketonuria and other disorders of amino acid degradation	726
Causes of type 1 and type 2 diabetes	741
McArdle disease	749
Effects of ethanol on energy metabolism and link to cancer	758
High homocysteine levels and vascular disease	776
Inherited disorders of porphyrin metabolism	788
Herpes drug that targets viral thymidine kinase	797
Anticancer drugs that block the synthesis of thymidylate	805
Adenosine deaminase and severe combined immunodeficiency	809
Gout	810
Lesch–Nyhan syndrome	810
Cardiolipin and Parkinson's disease	817
Disruption of lipid metabolism and Tay–Sachs disease	820
Hypercholesterolemia and atherosclerosis	830
Mutations in the LDL receptor	831
Role of the protease PCSK9 in cardiovascular disease	832
Clinical management of cholesterol levels	832
Protective function of the cytochrome P450 system	836
Aromatase inhibitors in the treatment of breast and ovarian cancer	838
Rickets and vitamin D	840

Antibiotics that target DNA gyrase	857
Using DNA-repair mechanisms to study disease	868
Huntington's disease	869
mRNA vaccines	880
Burkitt lymphoma and B-cell leukemia	898
Diseases of defective RNA splicing	906
Diphtheria	943
Serendipitous discovery of penicillin	979
HMG-CoA reductase inhibitors	981
HIV protease inhibitors	994

BOXED EXAMPLE PROBLEMS

The boxed Examples provide point-of-use, worked-through examples of the types of quantitative and reasoning questions students will frequently encounter in the course.

Applying the Henderson–Hasselbalch Equation	20
Determining the Fraction of Bound Receptors	92
Calculating the Effectiveness of Protein Purification	110
Calculating and Comparing $\Delta G^{\circ\prime}$ and ΔG	147
Applying the Michaelis–Menten Equation	155
Determining Inhibitor Type from Data	171
Interpreting an RNA Alignment	315
Calculating the Energetic Cost of Ion Transport	383
Calculating Equilibrium Potentials	403
Calculating ΔG for a Coupled Reaction Under Real Conditions	455
Calculating the Standard Free Energy of a Reaction from Reduction Potentials	547
Determining Metabolic Products of Amino Acid Degradation	725
Measuring the Impact of a Single Athletic Activity on Caloric Homeostasis	752
Determining the IC_{50} for an Inhibitor	987

SCIENTIST PROFILES

Scientist profiles throughout the book illustrate some of the wide range of training and career paths that different individuals have taken to allow them to make important contributions to biochemistry.

Herman Branson	46
Pamela Bjorkman	91
Rosalind Franklin	129
Maud Menten	154
Har Gobind Khorana	256
Emmanuelle Charpentier and Jennifer Doudna	294
Russell Doolittle	318
Carolyn Bertozzi	347
Marie M. Daly	357
Baldomero Olivera	395
Eva Neer	425
Gerty Cori	509
Hans Krebs	528
Lynn Margulis	545
Peter Mitchell and André Jagendorf	599
Andrew Benson	611
Dorothy Hodgkin	677
Cecile Pickart	708
Beverly Guirard	770
JoAnne Stubbe	804
Namandjé Bumpus	836
Tsuneko and Reiji Okazaki	849
Thomas Cech	912
Ada Yonath	938
Sarah Stewart	962
Gertrude Elion	996

ACKNOWLEDGMENTS

With the publication of this tenth edition of *Biochemistry*, we are reminded of the tremendous responsibility in continuing the tradition started by Lubert Stryer in 1975. Throughout our preparation and writing of this edition, we sought to reaffirm our commitment to the guiding principles that make this book special: clear and precise writing, simple and informative illustrations, and the awareness that our work would be consumed by clever and engaged students and their dedicated and meticulous teachers. It is an honor that we do not take lightly. We are grateful to all our colleagues who supported and advised us throughout this process, and to all the pioneering scientists whose discoveries and insights provide us with the subject material that inspires our passion for biochemistry.

It has been a joy to work with the Macmillan Learning team throughout the creative process. While the undertaking was invariably arduous, they continually kept our spirits up and our eyes focused on the ultimate goal by coaxing without nagging, and critiquing while supporting us. We have many people to thank for this experience, some of whom are first timers to *Biochemistry*. We were delighted to work with Senior Program Manager Liz Simmons for the first time. She was unfailing in her enthusiasm and tremendous leadership through an enormously tumultuous time during the pandemic. Senior Development Editor Jennifer Angel provided countless thoughtful suggestions to improve the book in every facet, from the writing to the page layout. Director of Development Debbie Hardin (and Barbara Yien in the final months of the project), provided support and encouragement throughout the project. A special word of thanks to our Senior Media Editor, Cassandra Korsvik, who provided extraordinary support as we worked to synchronize our print and online assessments. Special thanks also to Assistant Editor Angelica Hernandez, who tirelessly juggled countless tasks in support of our team.

It is almost impossible to put into words the amount of coordination and effort required to shepherd a textbook project from authors' first draft to publication. We have many to thank for directing us through this complex process. Editorial Project Manager Jennifer Hart and Content Project Manager Peter Jacoby managed the flow of the entire project, from draft writing to copyediting to bound book, with amazing efficiency. Diana Blume, Director of Design, Content Management; Natasha Wolfe, Design Services Manager; and Maureen McCutcheon, Designer, produced a design and layout that makes the book uniquely attractive and modern, while still emphasizing its ties to past editions. Permissions Manager Christine Buese and Photo Researchers Richard Fox and Krystyna Borgen found many new photographs in addition to diligently tracking down classics from the published literature. Matthew McAdams, Art Manager, deftly directed the rendering of new illustrations and the conversion of legacy art for a new printing process. Paul Rohloff, Senior Workflow Project Manager, made sure that the significant difficulties of scheduling, composition, and manufacturing were smoothly overcome. Heather Southerland, Amy Thorne, Cassandra Korsvik, Kelsey Hughes, and Lily Huang did a wonderful job in their management of the media program. Maureen Rachford, Marketing Manager for the Physical Sciences, enthusiastically introduced this newest edition of *Biochemistry* to the academic world. We are deeply appreciative of Greg David and his sales staff for their support. Without their able and enthusiastic presentation of our text to the academic community, all of our efforts would be in vain.

Thanks also to our many colleagues at our own institutions as well as throughout the country who patiently answered our questions and encouraged us on our quest. Finally, we owe a debt of gratitude to our families: our spouses—Wendie Berg, Megan Williams, Catherine Hines, and Mark Heller—and our children. Without their support, comfort, and understanding, this endeavor could never have been undertaken, let alone successfully completed.

We also especially thank those who served as reviewers for this tenth edition. Their thoughtful comments, suggestions, and encouragement have been of immense help to us in maintaining the excellence of the preceding editions. These reviewers are listed below.

Balasubrahmanyam Addepalli
University of Cincinnati, Main Campus

Dolapo Adedeji
Elizabeth City State University

James Ames
University of California, Davis

Mary Anderson
Texas Woman's University

Gerald Audette
York University

Dana Baum
Saint Louis University, Main Campus

Donald Beitz
Iowa State University

Matthew Berezuk
Azusa Pacific University

Mark Berry
Memorial University

Joshua Blose
State University of New York, Brockport

Paul Bond
Shorter University

Michael Borenstein
Temple University

Adam Brummett
University of Iowa

Christopher Calderone
Carleton College

Sara C. Zapico
New Jersey Institute of Technology

Michael Cascio
Duquesne University

Nancy Castro
University of Southern California

Yongli Chen
Hawaii Pacific University, Hawaii Loa

Greg Ciesielski
Auburn University, Montgomery

Jade Clement
Texas Southern University

Robert Congdon
State University of New York, Broome Community College

Douglas Conklin
University of Albany

John Conrad
University of Nebraska, Omaha

Rebecca Corbin
Ashland University

Rikki Corniola
California North State University, College of Health Sciences

Garland Crawford
Mercer University

Tuhin Das
City University of New York, John Jay College of Criminal Justice

Madeleine De Beer
University of Dayton

Tomas Ding
North Carolina Central University

Brittney A. Dinkel
Buena Vista University

Cassidy Dobson
Truman State University

Pasha Ebrahimi
El Camino Community College District

Shawn Ellerbroek
Wartburg College

Steven R. Ellis
University of Louisville

Nuran Ercal
Missouri University of Science and Technology

Stylianos Fakas
Alabama A&M University

Kirsten Fertuck
Northeastern University

Sheree J. Finley
Alabama State University

Barbara S. Frank
Idaho State University

Linnea Freeman
Furman University

Brooke Gardner
University of California, Santa Barbara

Yulia Gerasimova
University of Central Florida

Dipak K. Ghosh
North Carolina A&T State University

Christina Goode
Western University of Health Sciences

Sarah Goomeshi Nobary
Columbia College

Neena Grover
Colorado College

Christopher S. Hamilton
Hillsdale College

Tiffany Hayden
Erskine College And Seminary

James A. Hebda
Austin College

Newton P. Hilliard Jr.
Arkansas Tech University

Cheryl Ingram-Smith
Clemson University

Lori Isom
University of Central Arkansas

Kelly E. Johanson
Xavier University of Louisiana

Brian Kalet
Colorado State University, Fort Collins

Dmitry Kolpashchikov
University of Central Florida

Ivan Korendovych
Syracuse University

Chandrika Kulatilleke
City University of New York, Bernard M Baruch College

Allison C. Lamanna
University of Michigan, Ann Arbor

Yun Li
Delaware Valley University

Jie Li
University of South Carolina, Columbia

Zhengchang Liu
University of New Orleans

Debra Martin
Saint Mary's University of Minnesota

Jonathon Mauser
Winona State University

Karen McPherson
Delaware Valley University

Michael Mendenhall
University of Kentucky

Jonathan M. Meyers
Columbus State University

Geoffrey Mitchell
Wofford College

David Mitchell
Saint John's University

Victoria Del Gaizo Moore
Elon University

Ifedayo Victor Ogungbe
Jackson State University

S. Ryan Oliver
University of Alaska, Fairbanks

Justin P'Pool
Franklin College

Siva S. Panda
Augusta University

Ann Paterson
Williams Baptist College

John D. Patton
Southwest Baptist University

Alfred S. Ponticelli
State University of New York at Buffalo Medical School

Ramin Radfar
Wofford College

Supriyo Ray
Bowie State University

Katarzyna Roberts
Rogers State University

Douglas Root
University of North Texas

Gillian E. A. Rudd
Georgia Gwinnett College

Megan E. Rudock
Wake Forest University

Usha Sankar
Fordham University

Michael Sehorn
Clemson University

Kavita Shah
Purdue University, Main Campus

Juyoung K. Shim
University of Maine, Augusta

Aaron J. Sholders
Colorado State University, Fort Collins

Gabriela M. Smeureanu
City University of New York, Hunter College

Sarah Smith
Bucknell University

Jennifer Sniegowski
Arizona State University, Downtown Phoenix

Amy L. Stockert
Ohio Northern University

Nina V. Stourman
Youngstown State University

Kent Strodtman
Columbia College

Suresh Kumar Thallapuranam
University of Arkansas

Kerstin Tiedemann
Concordia University, Loyola Campus

Peter Tieleman
University of Calgary

Candace Timpte
Georgia Gwinnett College

Yufeng Tong
University of Windsor

Andrew T. Torelli
Ithaca College

Vishwa D. Trivedi
Bethune Cookman University

Robert J. Warburton
Shepherd University

Kathrine Weeks
Centenary College of Louisiana

Rodney Weilbaecher
Southern Illinois University, Carbondale

Cindy L. White
Harding University

Vladimira Wilent
Temple University

Jacek Wower
Auburn University

Adrienne Wright
University of Alberta, Edmonton

Naciem Yousif
University of the District of Columbia

Alexander G. Zestos
American University

Joshua Zhu
Dordt College

Courtesy Jeremy Berg

JEREMY M. BERG received his BS and MS degrees in Chemistry from Stanford University (where he did research with Keith Hodgson and Lubert Stryer) and his PhD in Chemistry from Harvard with Richard Holm. He then completed a post-doctoral fellowship with Carl Pabo in Biophysics at Johns Hopkins University School of Medicine. He was an Assistant Professor in the Department of Chemistry at Johns Hopkins from 1986 to 1990. He then moved to Johns Hopkins University School of Medicine as Professor and Director of the Department of Biophysics and Biophysical Chemistry, where he remained until 2003. He then became Director of the National Institute of General Medical Sciences at the National Institutes of Health. In 2011, he moved to the University of Pittsburgh, where he is now Professor of Computational and Systems Biology and Pittsburgh Foundation Chair and Director of the Institute for Personalized Medicine. He served as President of the American Society for Biochemistry and Molecular Biology from 2011 to 2013 and as Editor-in-Chief for *Science* magazine and the *Science* family of journals from 2016 to 2019. Dr. Berg has received numerous awards for his research, teaching, and public service. He is an elected member of the National Academy of Medicine and the American Academy of Arts and Sciences. He is coauthor, with Stephen J. Lippard, of the textbook *Principles of Bioinorganic Chemistry*. He greatly enjoys sharing his life with his wife, three grown children, and grandchildren.

Karpagam Aravindhan

GREGORY J. GATTO, JR., received his AB degree in Chemistry from Princeton University, where he worked with Martin F. Semmelhack and was awarded the Everett S. Wallis Prize in Organic Chemistry. In 2003, he received his MD and PhD degrees from the Johns Hopkins University School of Medicine, where he studied the structural biology of peroxisomal targeting signal recognition with Dr. Berg and received the Michael A. Shanoff Young Investigator Research Award. He completed a postdoctoral fellowship in 2006 with Christopher T. Walsh at Harvard Medical School, where he studied the biosynthesis of the macrolide immunosuppressants. Dr. Gatto is currently a Scientific Director in the Novel Human Genetics Research Unit at GlaxoSmithKline. While he enjoys losing at board games, attempting but not completing crossword puzzles, and watching baseball games at every available opportunity, he treasures most the time he spends with his wife Megan and sons Timothy and Mark.

Clay Wegrzynowicz/Lafayette College Communications Dept

JUSTIN K. HINES is Professor of Chemistry at Lafayette College, where he teaches general chemistry and biochemistry courses and conducts education and NIH-funded laboratory research on protein misfolding with undergraduates. He received both his BS and PhD in Biochemistry from Iowa State University, where he studied the structure and regulation of the enzymes of central metabolism with Richard B. Honzatko and Herbert J. Fromm. He then completed a postdoctoral fellowship with Elizabeth A. Craig in Biochemistry at the University of Wisconsin–Madison. Professor Hines has won numerous awards for teaching and research, including being named a Cottrell Scholar by the Research Corporation for Science Advancement and a Henry-Dreyfus Teacher-Scholar

by the Camille and Henry Dreyfus Foundation. He is also the author of the case-studies series for Macmillan's three biochemistry textbooks. He enjoys running, hiking, games of any kind, and spending time with his wife and children. He would like to dedicate his work on this book to his loving mother Kathryn, who passed away in 2021.

Jutta B. Heller

JUTTA BENEKEN HELLER is a Teaching Professor at the University of Washington Tacoma. She primarily teaches upper-division undergraduate courses in the Biomedical Sciences, with a focus on molecular biology. She also acts as a faculty mentor in an NSF-funded program that supports the academic success of undergraduate students with low-income or underrepresented backgrounds. Professor Heller's top priority is the academic and personal success of her students as she strives to create a welcoming and equitable learning environment for everyone. She received her AB in Molecular Biology from Princeton University in 1995 and her PhD from Johns Hopkins University School of Medicine in 2001. Her PhD thesis work was completed in the lab of Dr. Daniel Leahy, where she determined the crystal structure of the Homer EVH1 domain in complex with a peptide ligand. She completed a postdoctoral fellowship at the University of Chicago and began her teaching career at Loyola University Chicago in 2004. She received the 2010 Edwin T. and Vivijeanne F. Sujack Award for Teaching Excellence at Loyola University. She enjoys exploring mountains, forests, and beaches with her husband Mark and their dog Ziggy.

Hai Ngo

JOHN L. TYMOCZKO was Towsley Professor of Biology Emeritus at Carleton College, where he taught from 1976 until his death in 2019. He taught a variety of courses, including Biochemistry, Biochemistry Laboratory, Oncogenes and the Molecular Biology of Cancer, and Exercise Biochemistry, and cotaught an introductory course, Energy Flow in Biological Systems. Professor Tymoczko received his BA from the University of Chicago in 1970 and his PhD in Biochemistry from the University of Chicago with Shutsung Liao at the Ben May Institute for Cancer Research. He then had a postdoctoral position with Hewson Swift of the Department of Biology at the University of Chicago. The focus of his research was on steroid receptors, ribonucleoprotein particles, and proteolytic processing enzymes.

Saul Loeb/AFP/Getty Images

LUBERT STRYER is Winzer Professor of Cell Biology, Emeritus, in the School of Medicine and Professor of Neurobiology, Emeritus, at Stanford University, where he has been on the faculty since 1976. He received his MD from Harvard Medical School. Professor Stryer has received many awards for his research on the interplay of light and life, including the Eli Lilly Award for Fundamental Research in Biological Chemistry, the Distinguished Inventors Award of the Intellectual Property Owners' Association, and election to the National Academy of Sciences and the American Philosophical Society. He was awarded the National Medal of Science in 2006. The publication of his first edition of *Biochemistry* in 1975 transformed the teaching of biochemistry.

Biochemistry in Space and Time

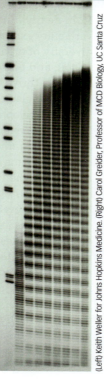

Nobel Prize winners Peter Agre, MD, and Carol Greider, PhD, celebrated on the occasion of the announcement that Dr. Greider was sharing the Nobel Prize in Medicine or Physiology in 2009. Dr. Agre had won a Nobel in 2003. Dr. Greider was honored for determining a key aspect of how chromosomes are replicated faithfully. Some data from Dr. Greider's laboratory notebook (right) shows regularly spaced signals, indicating that she had found a biochemical process suspected to exist, but not previously discovered. This work not only helped Dr. Greider earn her PhD degree but was also cited when she shared the Nobel Prize.

(Left) Keith Weller for Johns Hopkins Medicine. (Right) Carol Greider, Professor of MCD Biology, UC Santa Cruz

❖ LEARNING GOALS

By the end of this chapter, you should be able to:

1. Provide examples of the unity of biology at the biochemical level.

2. Describe the double-helical structure of DNA including its formation from two component strands in the context of important atomic interactions: covalent bonds, ionic interactions, hydrogen bonds, van der Waals interactions, and the hydrophobic effect.

3. Discuss the first and second laws of thermodynamics and the concepts of entropy and enthalpy.

4. Discuss the importance of single-molecule experiments in terms of understanding reaction rates.

5. Explain the roles of buffers in stabilizing the pH of solutions.

6. Perform calculations using the definitions of pH and pK_a and the Henderson–Hasselbalch equation.

7. Briefly describe some discoveries enabled by advances in DNA sequencing technologies, including the concept of molecular clocks.

OUTLINE

1.1 Biochemical Unity Underlies Biological Diversity

1.2 DNA Illustrates the Interplay Between Form and Function

1.3 Concepts from Chemistry Explain the Properties of Biological Molecules

1.4 DNA Sequencing Is Transforming Biochemistry, Medicine, and Other Fields

1.5 Biochemistry Is an Interconnected Human Endeavor

Biochemistry is the study of the chemistry of life processes. Scientists have explored the chemistry of life with great intensity for nearly two centuries. Many of the most fundamental mysteries of how living things function at a biochemical level have been solved. However, much remains to be investigated. As is often the case, each discovery raises at least as many new questions as it answers. We are now in an age of unprecedented opportunity for the application of this tremendous body of knowledge of biochemistry to problems in medicine, agriculture, forensics, environmental sciences, and many other fields.

1.1 Biochemical Unity Underlies Biological Diversity

The biological world is magnificently diverse. Plant and animal species range from nearly microscopic insects to elephants, whales, and giant sequoias. This variability extends further when we descend into the microscopic world. A great diversity of organisms such as protozoa, yeast, and bacteria are found in water, in soil, and both on and within larger organisms. Some organisms can survive and even thrive in seemingly hostile environments such as hot springs, deep-sea vents, and glaciers.

The development of the microscope revealed a key unifying feature that underlies this diversity: large organisms are built up of **cells**, resembling, to some extent, single-celled microscopic organisms. The construction of animals, plants, and microorganisms from cells suggested that these diverse organisms might have more in common than is apparent from their outward appearance. With the development of biochemistry, this notion has been tremendously supported and expanded. At the biochemical level, all organisms have many common features (**Figure 1.1**).

Biochemical processes entail the interplay of two different classes of molecules: large molecules such as proteins and nucleic acids, referred to as **biological macromolecules**, and low-molecular-weight molecules such as glucose and glycerol, referred to as **metabolites**, that are chemically transformed in biological processes.

Members of both these classes of molecules are common, with minor variations, to all living things. For example, **deoxyribonucleic acid (DNA)** stores genetic information in all cellular organisms. **Proteins**, macromolecules that are the key participants in most biological processes, are built from the same set of 20 building blocks in all organisms. Furthermore, proteins that play similar

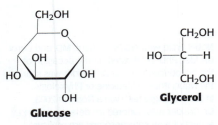

Glucose

Glycerol

Examples of metabolites

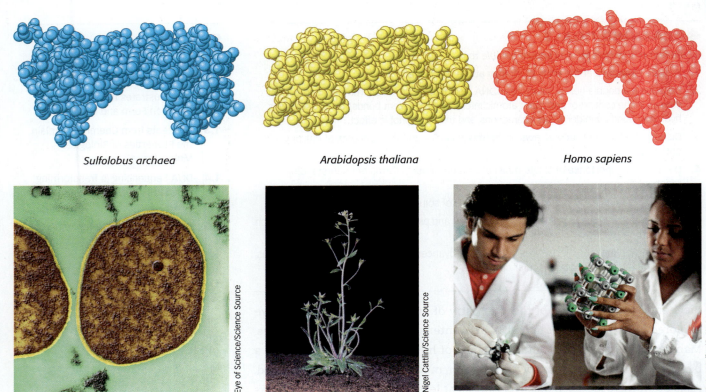

Sulfolobus archaea

Arabidopsis thaliana

Homo sapiens

Eye of Science/Science Source

Nigel Cattlin/Science Source

Peter Muller/Getty Images

FIGURE 1.1 Despite tremendous biological diversity, there is remarkable similarity in the molecules that make up living organisms. The shape of a key molecule in gene regulation (the TATA-box-binding protein) is similar in three very different organisms that are separated from one another by billions of years of evolution.

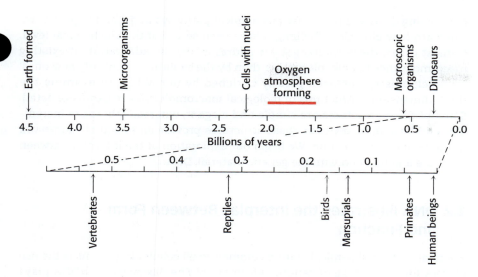

FIGURE 1.2 A possible timeline for biochemical evolution shows that life on Earth began approximately 3.5 billion years ago. Selected key events are indicated, including the relatively recent emergence of human beings. The times of emergence of classes of animals from the fossil record are shown below the main timeline.

roles in different species often have very similar, or sometimes identical, three-dimensional structures (Figure 1.1).

Key metabolic processes also are common to many organisms. For example, the set of chemical transformations that converts glucose and oxygen into carbon dioxide and water is essentially identical in simple bacteria such as *Escherichia coli* (*E. coli*) and human beings. Even processes that appear to be quite distinct often have common features at the biochemical level. Remarkably, the biochemical processes by which plants capture light energy and convert it into more-useful forms are strikingly similar to steps used in animals to capture energy released from the breakdown of glucose.

These observations overwhelmingly suggest that all living things on Earth have a **common ancestor** and that modern organisms have evolved from this ancestor into their present forms. Geological and biochemical findings support a timeline for this evolutionary path (**Figure 1.2**).

On the basis of their biochemical characteristics, the diverse organisms of the modern world can be divided into three fundamental groups called *domains*: Eukarya, Bacteria, and Archaea.

Domain **Eukarya** (also called **eukaryotes**) comprises all multicellular organisms, including all animals, plants, and human beings, as well as many microscopic unicellular organisms such as yeast. The defining characteristic of eukaryotes is the presence of a well-defined nucleus within each cell.

Unicellular organisms that lack a nucleus are referred to as **prokaryotes**. The prokaryotes were classified as two separate domains in response to Carl Woese's discovery in 1977 that certain unicellular organisms are biochemically quite distinct from previously characterized bacterial species. These organisms, now recognized as having diverged from the domain **Bacteria** early in evolution, make up the domain **Archaea**.

Evolutionary paths from a common ancestor to modern organisms can be deduced on the basis of biochemical information, with one possible such path shown in **Figure 1.3**. Much of this book will explore the chemical reactions and the associated biological macromolecules and metabolites that are found in biological processes common to all organisms. The unity of life at the biochemical level makes this approach possible. At the same time, different species have specific needs, depending on the particular biological niche in which they evolved and live. By comparing and

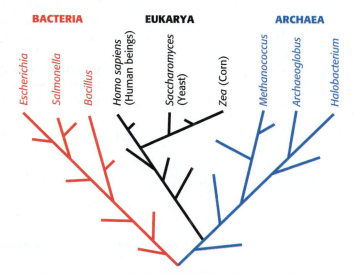

FIGURE 1.3 The relationships of all modern organisms can be displayed on a tree of life with just three major branches. A possible evolutionary path from a common ancestor is represented at the bottom of the tree, with organisms found in the modern world at the top.

contrasting details of particular biochemical pathways in different species, we can learn how biological challenges have been solved at the biochemical level. In most cases, these challenges are addressed by the adaptation of existing macromolecules to new roles rather than by the evolution of entirely new ones.

Biochemistry has been greatly enriched by our ability to examine the three-dimensional structures of biological macromolecules in exquisite detail. Some of these structures are simple and elegant, whereas others are incredibly complicated. In any case, these structures provide an essential framework for understanding function. We begin our exploration of the interplay between structure and function with the genetic material, DNA.

1.2 DNA Illustrates the Interplay Between Form and Function

A fundamental biochemical feature common to all cellular organisms is the use of DNA for the storage of genetic information. The discovery that DNA plays this central role was first made in studies of bacteria in the 1940s. A compelling proposal for the three-dimensional structure of DNA followed in 1953, setting the stage for many of the advances in biochemistry and numerous other fields, extending to the present.

The structure of DNA powerfully illustrates a basic principle common to all biological macromolecules: the intimate relationship between structure and function. The remarkable chemical properties of this substance allow it to function as an efficient and robust vehicle for storing information. We start with an examination of the covalent structure of DNA and its extension into three dimensions.

DNA is constructed from four building blocks

DNA is a linear polymer made up of four different types of monomers. It has a fixed backbone from which protrude variable substituents, referred to as **nitrogenous bases**, or simply, **bases** (**Figure 1.4**). The backbone is built of repeating sugar–phosphate units. The sugars are molecules of deoxyribose from which DNA receives its name. Each sugar is connected to two phosphate groups through different linkages. Moreover, each sugar is oriented in the same way, and so each DNA strand has directionality, with one end distinguishable from the other. This **polarity** is very important for many processes involving DNA. Joined to each deoxyribose is one of four possible nitrogenous bases: adenine (A), cytosine (C), guanine (G), and thymine (T).

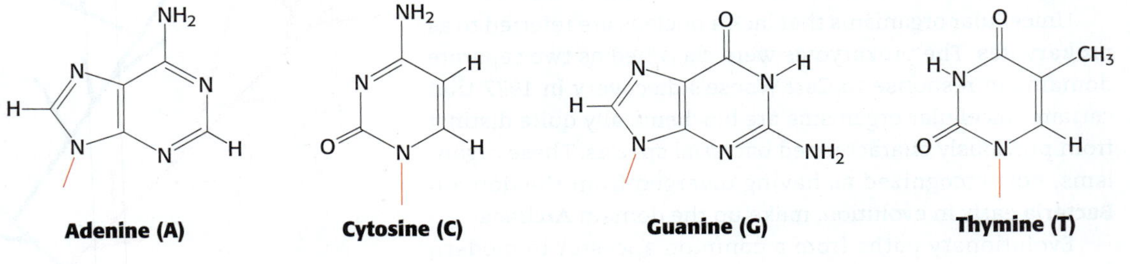

Adenine (A)   **Cytosine (C)**   **Guanine (G)**   **Thymine (T)**

These bases are connected to the sugar components in the DNA backbone through the bonds shown in black in Figure 1.4. All four bases are planar but differ significantly in other respects. Thus, each monomer of DNA, called a *nucleotide*, consists of a sugar–phosphate unit and one of four bases attached to the sugar. These bases can be arranged in any order along a strand of DNA.

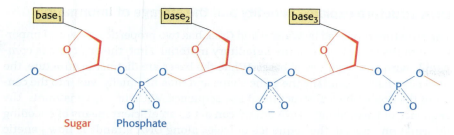

FIGURE 1.4 The covalent structure of DNA is a linear polymer of monomeric units.
Each monomer is composed of a sugar (deoxyribose), a phosphate, and a variable nitrogenous
base that protrudes from the sugar–phosphate backbone.

Two single strands of DNA combine to form a double helix

Most DNA molecules consist of not one but two strands (**Figure 1.5**). In 1953,
James Watson and Francis Crick deduced the arrangement of these strands
and proposed a three-dimensional structure for DNA molecules based, in part,
on experimental data from Rosalind Franklin. This structure is a **double helix**
composed of two intertwined strands arranged such that the sugar–phosphate
backbone lies on the outside and the bases on the inside. The key to this structure is that the bases form specific **base pairs** held together by noncovalent
interactions called *hydrogen bonds* (Section 1.3): adenine pairs with thymine
(A–T) and guanine pairs with cytosine (G–C), as shown in **Figure 1.6**. Hydrogen
bonds are much weaker than covalent bonds such as the carbon–carbon or
carbon–nitrogen bonds that define the structures of the bases themselves. Such
weak interactions are crucial to biochemical systems; they are weak enough to
be reversibly broken in biochemical processes, yet they are strong enough, particularly when many form simultaneously, to help stabilize specific structures
such as the double helix.

FIGURE 1.5 The double helix has two antiparallel strands. This representation of the
double-helical structure of DNA proposed by Watson and Crick shows the sugar–phosphate backbones
of the two chains in red and blue, and the bases in green, purple, orange, and yellow. The two strands
are antiparallel, running in opposite directions with respect to the axis of the double helix, as indicated
by the arrows.

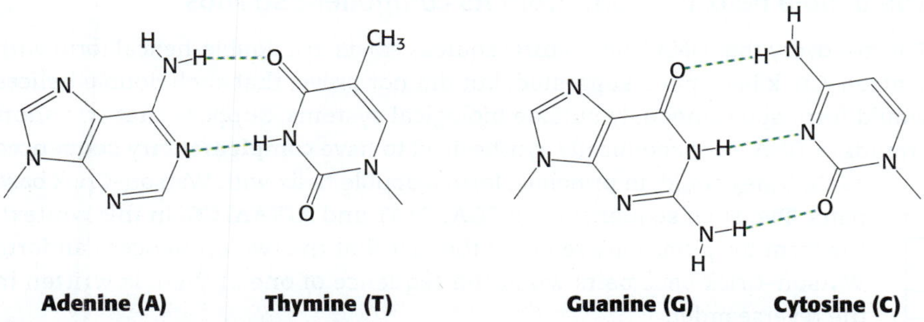

Adenine (A)   **Thymine (T)**   **Guanine (G)**   **Cytosine (C)**

FIGURE 1.6 The nitrogenous bases in DNA form specific Watson–Crick base pairs.
Adenine pairs with thymine (A–T) and guanine with cytosine (G–C). The dashed green lines
represent hydrogen bonds.

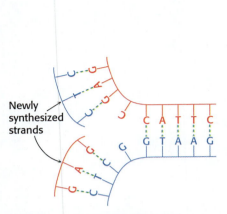

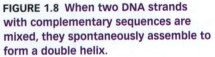

Newly synthesized strands

FIGURE 1.7 Once separated, each strand of the DNA double helix can act as a template for the generation of its partner strand.

DNA structure explains heredity and the storage of information

The structure proposed by Watson and Crick has two properties of central importance to the role of DNA as the hereditary material. First, the structure is compatible with any sequence of bases. While the bases are distinct in structure, the base pairs have essentially the same shape and thus fit equally well into the center of the double-helical structure of any sequence. Without any constraints, the sequence of bases along a DNA strand can act as an efficient means of encoding information. Indeed, the sequence of bases along DNA strands is how genetic information is stored. The DNA sequence determines the sequences of the ribonucleic acid (RNA) and protein molecules that carry out most of the activities within cells.

Second, because of base pairing, the sequence of bases along one strand completely determines the sequence along the other strand. As Watson and Crick so coyly wrote: "It has not escaped our notice that the specific pairing we have postulated immediately suggests a possible copying mechanism for the genetic material." Thus, if the DNA double helix is separated into two single strands, each strand can act as a template for the generation of its partner strand through specific base-pair formation (**Figure 1.7**). The three-dimensional structure of DNA beautifully illustrates the close connection between molecular form and function.

1.3 Concepts from Chemistry Explain the Properties of Biological Molecules

We have seen how a chemical insight into the hydrogen-bonding capabilities of the bases of DNA led to a deep understanding of a fundamental biological process. To lay the groundwork for the rest of the book, we begin our study of biochemistry by examining selected concepts from chemistry and showing how these concepts apply to biological systems. The concepts include the types of chemical interactions; the structure of water, the solvent in which most biochemical processes take place; the First and Second Laws of Thermodynamics; and the principles of acid–base chemistry.

The formation of the DNA double helix is a key example

We will use these concepts to examine an archetypical biochemical process—the formation of a DNA double helix from its two component strands. The process is but one of many examples that could have been chosen to illustrate these topics. Keep in mind that, although the specific discussion is about DNA and double-helix formation, these concepts are quite general and will apply to many other classes of molecules and processes that will be discussed in the remainder of the book.

The double helix can form from its component strands

The discovery that DNA from natural sources exists in a double-helical form with Watson–Crick base pairs suggested, but did not prove, that such double helices would form spontaneously outside biological systems. Suppose that two short strands of DNA were chemically synthesized to have **complementary sequences** so that they could, in principle, form a double helix with Watson–Crick base pairs. Two such sequences are CGATTAAT and ATTAATCG. In this context, the term *complementary* refers to the fact that the two sequences can form Watson–Crick base pairs when the sequence of one of them is written in the reverse order.

The structures of these molecules in solution can be examined through a variety of techniques. In isolation, each sequence exists almost exclusively as a single-stranded molecule. However, when the two sequences are mixed, a double helix with Watson–Crick base pairs does form (**Figure 1.8**).

FIGURE 1.8 When two DNA strands with complementary sequences are mixed, they spontaneously assemble to form a double helix.

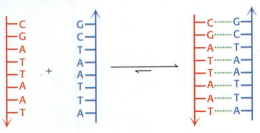

❖ SELF–CHECK QUESTION

What DNA sequence can form a double helix with 5′-AGGTCCGATA-3′?

What forces cause the two strands of DNA to bind to each other? To analyze this binding process, we must consider several factors: the types of interactions and bonds in biochemical systems, the thermal motion of molecules, the energetic favorability of the process, and the influence of the solution conditions—in particular, the consequences of acid–base reactions. Remarkably, this process proceeds nearly to completion in less than a millisecond when the single strands are mixed at appropriate concentrations. How does this process occur so rapidly?

Atoms and molecules undergo random motions that help define the timescales for biochemical interactions

The atoms and molecules within and around us are in constant thermal motion: translating, vibrating, rotating, and colliding with one another very rapidly. How fast do these processes occur? One useful estimate to consider is that the gas molecules that make up the air you are breathing now are colliding with one another about 1 billion (10^9) times per second, meaning the relevant timescale for these interactions is best measured in nanoseconds (10^{-9} s).

On a similar timescale to these gaseous molecular interactions are the biochemical interactions taking place in your cells. The timescale for biochemical interactions and processes are broadly on the order of picoseconds to microseconds (10^{-12}–10^{-6} s), which we shall refer to as the **biochemical timescale**. Consideration of this timescale can be quite illuminating in understanding biochemical phenomena. The rates of molecular interactions can be challenging to comprehend because, relative to them, human beings perceive time very slowly. Indeed, our perceptions are dependent on complex cascades of these very interactions. By carefully considering this timescale from the molecular perspective, we can improve our intuitions about the speed of chemical and biochemical events. For example, imagine a reaction that occurs once per millisecond (10^{-3} s), or 1000 times per second. Is this reaction occurring quickly or slowly? Our intuition about the passage of time might compel us to consider this a very fast reaction; this is, in fact, an imperceptibly short time for a human being. However, consider that molecules colliding one billion times per second will experience 1 million (10^6) unproductive collisions in this timeframe, on average, before a single reaction event occurs. From the perspective of a molecule undergoing these interactions, this reaction does not seem to be occurring particularly fast at all, but rather is the result of events that occur only rarely.

Many aspects of biochemistry depend on promoting specific processes and reactions so that they occur at biologically useful rates, harnessing the underlying rates of molecular interactions. In Chapters 5–7, we will focus explicitly on biochemical reaction rates and their control. In addition, we will visit the topic of biological timescale many times throughout the book as we explore molecular interactions in the context of biological systems.

Covalent bonds and noncovalent interactions are important for the structure and stability of biological molecules

Counteracting the chaos of thermal motion are attractive interactions that stabilize biological molecules. Atoms favorably interact with one another in several ways, including through the covalent bonds that define the structure of molecules, as well as a variety of noncovalent interactions that are of great importance to biochemistry. Among the strongest chemical interactions are **covalent bonds**, such as the bonds that hold the atoms together within DNA's individual bases. A covalent bond is formed by the sharing of a pair of electrons between adjacent atoms. A typical carbon–carbon (C—C) covalent bond has a bond length of 1.54 Å—where 1 Å (angstrom) is equivalent to one ten-billionth of a meter, or approximately 1/750,000th the width of a human hair—and a bond energy of 355 kJ mol^{-1} (85 kcal mol^{-1}). Because covalent bonds are so strong, considerable energy must be expended to break them. Recall that one joule (J) is the amount of energy required to move 1 meter against a force of 1 newton, and a kilojoule (kJ) is 1000 joules.

More than one electron pair can be shared between two atoms to form a multiple covalent bond. For example, three of the bases in Figure 1.6 include carbon–oxygen (C=O) double bonds. These bonds are shorter in length and even stronger than C—C single bonds, with energies near 730 kJ mol^{-1}, or 175 kilocalories (kcal) mol^{-1}.

For some molecules, more than one pattern of covalent bonding can be written; these alternative patterns are called **resonance structures**.

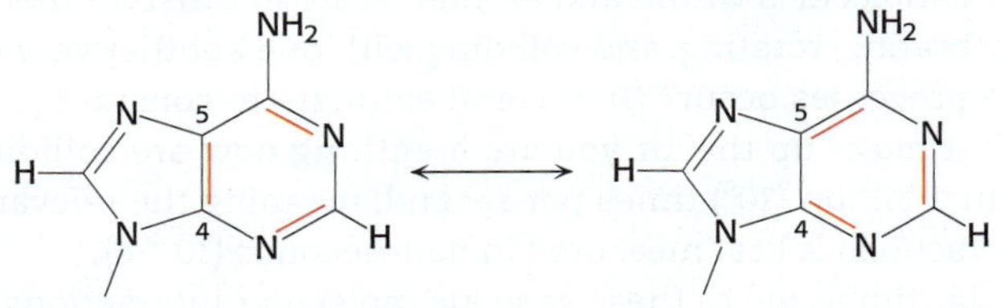

These adenine structures depict alternative arrangements of single and double bonds that are possible within the same structural framework, but the true structure of adenine is a composite of its two resonance structures. The composite structure has properties that are a blend of the two alternative states. For example, the observed length for the bond joining carbon atoms C-4 and C-5 is 1.40 Å, which is in between that expected for a C—C single bond (1.54 Å) and a C=C double bond (1.34 Å). A molecule that can be written as several resonance structures of approximately equal energies generally has greater stability than does a molecule without multiple resonance structures.

Noncovalent interactions are weaker than covalent bonds but are crucial for biochemical processes such as the formation of a double helix. Four fundamental noncovalent interaction types are ionic interactions, hydrogen bonds, van der Waals interactions, and hydrophobic interactions. They differ in geometry, strength, specificity, and physical origin. Furthermore, these interactions are affected in vastly different ways by the presence of water. Let us consider the characteristics of each type:

Ionic interactions. In an **ionic interaction**, a charged group on one molecule can attract an oppositely charged group on the same or another molecule. The energy of an ionic interaction (sometimes called an *electrostatic interaction*) is given by the Coulomb energy:

$$E = kq_1q_2/Dr$$

where E is the energy, q_1 and q_2 are the charges on the two atoms (in units of the electronic charge), r is the distance between the two atoms (in angstroms), D is the dielectric constant (which decreases the Coulomb energy depending on the intervening solvent or medium), and k is a proportionality constant ($k = 1389$ for energies in units of kJ mol^{-1}, or 332 for energies in kcal mol^{-1}).

The ionic interaction between two ions bearing single opposite charges separated by 3 Å in water (which has a dielectric constant of 80) has an energy of 5.8 kJ mol^{-1} (1.4 kcal mol^{-1}). Note the importance of the dielectric constant of the medium. For the same ions separated by 3 Å in a solvent such as hexane (which has a dielectric constant of 2), the energy of this interaction is 232 kJ mol^{-1} (55 kcal mol^{-1}).

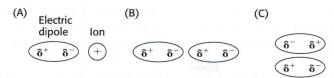

FIGURE 1.9 **Electric dipoles can form a variety of noncovalent interactions.** (A) An electric dipole can interact with an ion, with the negative end of the dipole pointing toward the positively charged ion. (B) Two dipoles can interact, with the negative end of one pointing toward to the positive end of the other. (C) Two dipoles position themselves so that both the positive and negative ends can favorably interact.

Thus far we have considered interactions between ions, atoms, or molecules that have one or more full electric charges. However, even molecules that have no overall charge can have regions where the electron distribution is uneven, such they are partially negatively or partially positively charged. This leads to **electric dipoles** (or simply **dipoles**) that can interact with ions or with other dipoles (**Figure 1.9**).

Hydrogen bonds. Hydrogen bonds are a specific type of dipole-dipole interactions that are so important in biochemistry that we single them out for a separate discussion. We already introduced hydrogen bonds in the context of their role in specific base-pair formation in the DNA double helix. In a **hydrogen bond**, an electropositive hydrogen atom is partially shared by two electronegative atoms such as nitrogen or oxygen. The **hydrogen-bond donor** is the group that includes both the atom to which the hydrogen atom is covalently bonded and the hydrogen atom itself, whereas the **hydrogen-bond acceptor** is the lone pair of electrons that is on the atom less tightly linked to the hydrogen atom (**Figure 1.10**). The electronegative atom to which the hydrogen atom is covalently bonded pulls electron density away from the hydrogen atom, which thus develops a partial positive charge (δ^+). This hydrogen atom with a partial positive charge can interact with a lone pair of electrons on an atom having a partial negative charge (δ^-) through a dipole-dipole interaction.

Hydrogen bonds are much weaker than covalent bonds. They have energies ranging from 4 to 20 kJ mol^{-1} (from 1 to 5 kcal mol^{-1}). Accordingly, hydrogen bonds are also somewhat longer than covalent bonds; their bond lengths (measured from the hydrogen atom) range from 1.5 Å to 2.6 Å. Hence, a distance ranging from 2.4 Å to 3.5 Å separates the two nonhydrogen atoms in a hydrogen bond.

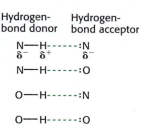

Hydrogen-   Hydrogen-
bond donor   bond acceptor

N—H------:N
δ^- δ^+ δ^-

N—H------:O

O—H------:N

O—H------:O

FIGURE 1.10 **Hydrogen bonds are weak noncovalent interactions between specific hydrogen atoms and lone pairs of electrons.** Hydrogen bonds are depicted as dashed green lines and the positions of the partial charges as δ^+ and δ^-.

Hydrogen-   Hydrogen-bond
bond donor   acceptor

0.9 Å / 2.0 Å

N—H ------:O

180°

Hydrogen bonds tend to be straight, such that the hydrogen-bond donor, the hydrogen atom, and the hydrogen-bond acceptor lie in a line. This tendency toward linearity can be important for orienting interacting molecules with respect to one another. Hydrogen-bonding interactions are responsible for many of the properties of water that make it such an important solvent, as will be described shortly.

van der Waals interactions. The basis of a **van der Waals interaction** is that the distribution of electronic charge around an atom fluctuates with time. At any instant, the charge distribution is not perfectly symmetric, leading to a dipole. This dipole acts through dipole-dipole interactions to induce a complementary dipole in its neighboring molecules, causing an attraction between the molecules. This attraction increases as the two atoms come closer to each

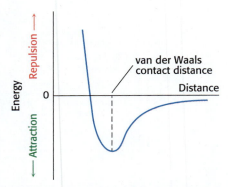

FIGURE 1.11 The energy of a van der Waals interaction varies and is most favorable at the van der Waals contact distance. Owing to electron–electron repulsion, the energy rises rapidly as the distance between the atoms becomes shorter than the contact distance.

other, until they are separated by what is termed the **van der Waals contact distance** (**Figure 1.11**). At distances shorter than the van der Waals contact distance, very strong repulsive forces become dominant because the outer electron clouds of the two atoms overlap.

Energies associated with van der Waals interactions are quite small; typical interactions contribute from 2 to 4 kJ mol^{-1} (0.5 to 1 kcal mol^{-1}) per atom pair. When the surfaces of two large molecules come together, however, a large number of atoms are in van der Waals contact; the net effect, summed over many atom pairs, can be substantial.

The hydrophobic effect. Before we define the last noncovalent interaction, it's important to remember that water is the solvent in which most biochemical reactions take place; thus its properties are essential to the formation of macromolecular structures and the progress of chemical reactions. Two properties of water are especially relevant:

- *Water is a polar molecule.* The water molecule is bent, not linear, and so the distribution of charge is asymmetric. The oxygen nucleus draws electrons away from the two hydrogen nuclei, which leaves the region around each hydrogen atom with a net positive charge. The water molecule thus has a permanent electric dipole.

The polar nature of water is responsible for its high dielectric constant of 80. Molecules in aqueous solution interact with water molecules through the formation of hydrogen bonds and through ionic interactions. These interactions make water a versatile solvent, able to readily dissolve many species, especially polar and charged compounds that can participate in these interactions.

- *Water is highly cohesive.* Water molecules interact strongly with one another through hydrogen bonds. These interactions are apparent in the structure of ice (**Figure 1.12**). Networks of hydrogen bonds hold the structure together; similar interactions link molecules in liquid water and account for many of the properties of water. In the liquid state, approximately one in four of the hydrogen bonds present in ice is broken.

The **hydrophobic effect** is a manifestation of the properties of water just described: it is the increased tendency for nonpolar molecules to aggregate in

FIGURE 1.12 The structure of ice involves extensive hydrogen bonding. Hydrogen bonds (shown as dashed green lines) are formed between water molecules, producing a highly ordered and open structure.

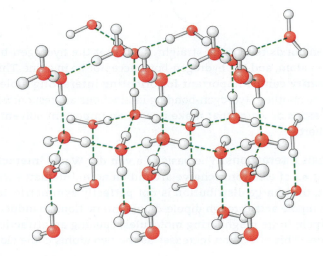

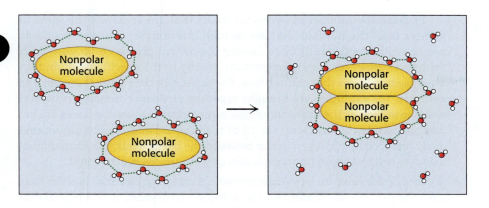

FIGURE 1.13 The hydrophobic effect is due to the behavior of water molecules. The aggregation of nonpolar groups in water leads to the release of water molecules, initially interacting with the nonpolar surface, into bulk water. The release of water molecules into solution makes the aggregation of nonpolar groups favorable.

water compared with other, less polar and less self-associating, solvents. Some molecules (termed **nonpolar molecules**) lack any dipoles and therefore cannot participate in hydrogen bonding or ionic interactions. Polar water molecules interact much more favorably with each other than they do with nonpolar molecules. In the presence of nonpolar molecules, water molecules form cage-like structures around them, becoming more ordered than water molecules free in solution. However, when two such nonpolar molecules come together, some of the water molecules that had been surrounding them are released from the "cage." The released water molecules interact freely with bulk water (**Figure 1.13**). Why the release of water from such cages is favorable will be considered shortly. The result is that nonpolar molecules show an increased tendency to associate with one another in water. This tendency is called the *hydrophobic effect*; the associated interactions are called *hydrophobic interactions*.

The formation of DNA's double helix is an expression of the rules of chemistry

Let us now explore how these noncovalent interactions work together to drive the association of two strands of DNA to form a double helix.

- *The aqueous environment surrounding DNA favors helix formation, overcoming the ionic repulsion between charged phosphate groups.* Each phosphate group in a DNA strand carries a negative charge and thus interacts unfavorably with other phosphate groups over distances. Even though the phosphate groups are far apart in the double helix, with distances greater than 10 Å, many such interactions take place (**Figure 1.14**). Thus, ionic interactions oppose the formation of the double helix. However, the strength of these repulsive ionic interactions is diminished by the high dielectric constant of water and the presence of ionic species such as Na^+ or Mg^{2+} ions in solution. These positively charged species interact with the phosphate groups and partly neutralize their negative charges.

- *Hydrogen bonds help ensure the specificity of base-pairing.* The second type of noncovalent interactions to consider in the formation of the double helix are hydrogen bonds. In single-stranded DNA, the hydrogen-bond donors and acceptors are exposed to solution and can form hydrogen bonds with water molecules. When two single strands come together, these hydrogen bonds with water are broken and new hydrogen bonds between the bases are formed.

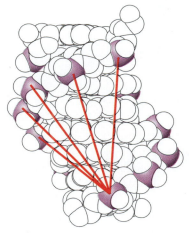

FIGURE 1.14 Ionic interactions among phosphate groups oppose the formation of the DNA double helix. Each unit within the double helix includes a phosphate group (the phosphorus atom shown in purple) that bears a negative charge. The unfavorable (repulsive) interactions of one phosphate with several others are shown by red lines.

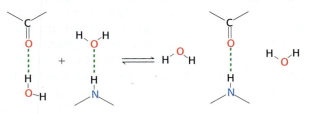

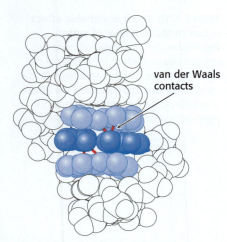

van der Waals
contacts

FIGURE 1.15 Optimal base stacking allows many weak, but favorable, van der Waals contacts that drive double-helix formation. In the DNA double helix, adjacent base pairs are stacked nearly on top of one another, and so many atoms in each base pair are separated by their van der Waals contact distance. The central base pair is shown in dark blue and the two adjacent base pairs in light blue. Several van der Waals contacts are shown in red.

Because the number of hydrogen bonds that are broken is the same or greater than the number that are formed, these new hydrogen bonds do not contribute substantially to driving the overall process of double-helix formation. However, they do contribute greatly to the specificity of binding. Suppose two bases that cannot form Watson–Crick base pairs (adenine and cytosine, for example) are brought together. Hydrogen bonds with water must be broken as the bases come into contact, but because the bases are not complementary in structure, not all of the broken bonds can be simultaneously replaced by hydrogen bonds between the bases. Thus, the formation of a double helix between noncomplementary sequences is disfavored.

- *Within a double helix, van der Waals interactions encourage base stacking.* The base pairs are parallel and stacked nearly on top of one another, with a typical separation of 3.4 Å between the planes of adjacent base pairs and approximately 3.6 Å between the most closely approaching atoms. This separation distance corresponds nicely to the van der Waals contact distance (**Figure 1.15**). Bases tend to stack even in single-stranded DNA molecules. However, the base stacking and associated van der Waals interactions are nearly optimal in a double-helical structure.

- *The hydrophobic effect also contributes to the favorability of base stacking.* More-complete base stacking moves the nonpolar surfaces of the bases away from water molecules and into contact with each other.

The principles of double-helix formation between two strands of DNA apply to many other biochemical processes. Many weak interactions contribute to the overall energetics of the process, some favorably and some unfavorably. Furthermore, **surface complementarity** is a key feature: when complementary surfaces meet, hydrogen-bond donors align with hydrogen-bond acceptors, and nonpolar surfaces come together to maximize van der Waals interactions and minimize nonpolar surface area exposed to the aqueous environment. The properties of water play a major role in determining the importance of these interactions. Interactions between complementary surfaces play central roles in a great many biochemical processes, and we will encounter them repeatedly throughout the book.

The laws of thermodynamics govern the behavior of biochemical systems

We can look at the formation of the double helix from a different perspective by examining the laws of thermodynamics, general principles that apply to all physical (and biological) processes. The laws distinguish between a system and its surroundings; a *system* refers to the matter and energy within a defined region of space, and the matter and energy in the rest of the universe is called the *surroundings*.

The laws of thermodynamics are of great importance because they determine the conditions under which specific processes can or cannot take place. We will consider these laws from a general perspective first and then apply the principles that we have developed to the formation of the double helix.

- *The First Law of Thermodynamics states that the total energy of a system and its surroundings is constant.* In other words, the energy content of the universe is constant; energy can be neither created nor destroyed. Energy can take different forms, however. Heat, for example, is one form of energy. Heat is a manifestation of the *kinetic energy* associated with the random motion of molecules. Alternatively, energy can be present as *potential energy*—energy of position that will be released on the occurrence of some process. Consider, for example, a ball held at the top of a tower. In this position the ball has considerable gravitational potential energy because, when it is released, the ball will develop kinetic energy associated with its motion as it falls. Chemical potential energy is related to the plausible ways that atoms might react with one another. For instance, a mixture of gasoline and oxygen has a large potential energy because these molecules can react to form carbon dioxide and water and release energy as heat.

Energy is always required to break chemical bonds and is released when new bonds form. The First Law requires that any energy released in the formation of chemical bonds must be used to break other bonds, released as heat or light, or stored in some other form.

- *The Second Law of Thermodynamics states that the total entropy of a system plus that of its surroundings always increases.* **Entropy** is a measure of the degree of randomness or disorder in a system. For example, the hydrophobic effect—the release of water from nonpolar surfaces—is favorable because water molecules that are free in solution are more disordered than they are when they are associated with nonpolar surfaces.

At first glance, the Second Law appears to contradict much common experience, particularly about biological systems. Many biological processes, such as the generation of a leaf from carbon dioxide gas and other nutrients, clearly increase the level of order and hence decrease entropy. Entropy may be decreased locally in the formation of such ordered structures only if the entropy of other parts of the universe is increased by an equal or greater amount. The local decrease in entropy is usually accomplished by a release of heat, which increases the entropy of the surroundings.

We can analyze this process in quantitative terms. First, consider the system. The entropy (S) of the system may change in the course of a chemical reaction by an amount ΔS_{system}. If heat flows from the system to its surroundings, then the heat content, often referred to as the **enthalpy** (H), of the system will be reduced by an amount ΔH_{system}. To apply the Second Law, we must determine the change in entropy of the surroundings. If heat flows from the system to the surroundings, then the entropy of the surroundings will increase.

The precise change in the entropy of the surroundings depends on the temperature; the change in entropy is greater when heat is added to relatively cold surroundings than when heat is added to surroundings at high temperatures; these surroundings are already in a high degree of disorder. To be even more specific, the change in the entropy of the surroundings will be proportional to the amount of heat transferred from the system and inversely proportional to the temperature (T) of the surroundings. In biological systems, T, absolute temperature in kelvins (K), is usually assumed to be constant. Thus, a change in the entropy of the surroundings is given by

$$\Delta S_{surroundings} = -\Delta H_{system}/T \tag{1}$$

The total entropy change is given by the expression

$$\Delta S_{total} = \Delta S_{system} + \Delta S_{surroundings} \tag{2}$$

Substituting equation 1 into equation 2 yields

$$\Delta S_{total} = \Delta S_{system} - \Delta H_{system}/T \tag{3}$$

Multiplying by $-T$ gives

$$-T\Delta S_{total} = \Delta H_{system} - T\Delta S_{system}$$

The function $-T\Delta S$ has units of energy and is referred to as **free energy** or **Gibbs free energy**, after Josiah Willard Gibbs, who developed this function in 1878:

$$\Delta G = \Delta H_{system} - T\Delta S_{system}$$

The free-energy change, ΔG, will be used throughout this book to describe the energetics of biochemical reactions. The Gibbs free energy can also be considered the energy available to do useful work; it is essentially an accounting tool that keeps track of both the entropy of the system (directly) and the entropy of the surroundings (in the form of heat released from the system).

Recall that the Second Law of Thermodynamics states that, for a process to take place, the entropy of the universe must increase. Examination of equation 3 shows that the total entropy will increase if and only if

$$\Delta S_{system} > \Delta H_{system}/T$$

Rearranging gives $T\Delta S_{system} > \Delta H_{system}$ or, in other words, entropy will increase if and only if

$$\Delta G = \Delta H_{system} - T\Delta S_{system} < 0$$

Thus, the free-energy change must be negative for a process to take place spontaneously. There is negative free-energy change when and only when the overall entropy of the universe is increased.

❖ SELF–CHECK QUESTION

The free energy can be calculated based on the system only. Why don't the surroundings need to be considered separately?

By releasing heat, the formation of the double helix obeys the Second Law of Thermodynamics

Let's see how the principles of thermodynamics apply to the formation of the double helix (**Figure 1.16**). Suppose solutions containing each of the two single strands are mixed. Before the double helix forms, each of the single strands is free to translate (change position) and rotate in solution, whereas each matched

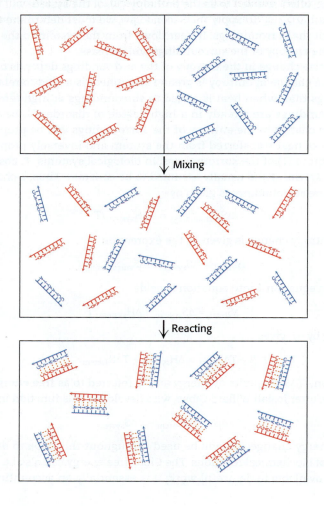

FIGURE 1.16 Double-helix formation entails a loss of entropy. When solutions containing DNA strands with complementary sequences are mixed, the strands react to form double helices. This process results in a loss of entropy from the system, indicating that heat must be released to the surroundings to prevent a violation of the Second Law of Thermodynamics.

Mixing

Reacting

pair of strands in the double helix must move together. Additionally, the free single strands exist in more conformations than possible when bound together in a double helix. Thus, the formation of a double helix from two single strands appears to result in a decrease rather than an increase in disorder, and therefore entropy, of the system.

On the basis of this analysis, we expect that the double helix cannot form without violating the Second Law of Thermodynamics unless heat is released to increase the entropy of the surroundings. Experimentally, we can measure the heat released by allowing the solutions containing the two single strands to come together within a water bath, which here corresponds to the surroundings. We then determine how much heat must be absorbed by the water bath or released from it to maintain it at a constant temperature.

This experiment establishes that a substantial amount of heat is released— namely, approximately 250 kJ mol^{-1} (60 kcal mol^{-1})—consistent with our expectation that significant heat would have to be released to the surroundings for the process not to violate the Second Law. We see in quantitative terms how order within a system can be increased by releasing sufficient heat to the surroundings to ensure that the entropy of the universe increases. We will encounter this general theme throughout this book.

Double-helix formation can be monitored one molecule at a time

One of the most exciting advances in biochemistry in recent years has been the development of methods to monitor some types of biochemical reactions one molecule at a time. To see why this is such an astounding achievement, recall that chemistry typically deals in terms of fractions of moles, and that a mole contains Avogadro's number (6.0×10^{23}) of molecules, a truly astronomical number. Being able to observe single molecules requires technological improvements to provide adequate signal to noise, as well as clever conceptions of experiments that often utilize other discoveries in biochemistry. We will discuss examples of these so-called *single-molecule methods* in some detail several times in this book. For now, we will examine the results when single molecules are monitored during the formation of double helices from single-stranded DNA molecules.

New methods allow one single-stranded DNA molecule to be exposed very abruptly to a solution containing a normal laboratory concentration, typically on the order of 1 μM (10^{-6} moles/liter) of the complementary single-stranded DNA. Once exposed to the solution, the potential double-helix formation process is allowed to occur for a fixed period of time on the millisecond (10^{-3} s) timescale. At the completion of this period, the result of the process can be probed, to see if the exposed single-stranded molecule found and bound a partner to form a double helix (**Figure 1.17**).

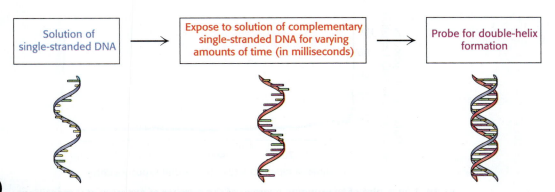

FIGURE 1.17 DNA double-helix formation can be probed at the single-molecule level. After one DNA single strand is exposed to a solution containing the complementary strand for a specific period of time, double-helix formation is probed. This experiment can be repeated many times with different individual DNA single strands.

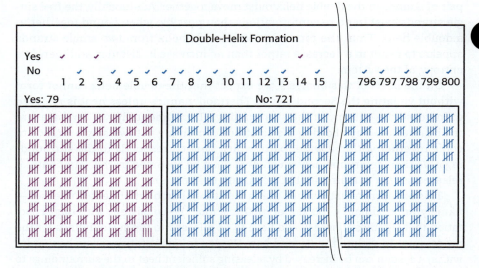

FIGURE 1.18 **A tally sheet for repeating a single-molecule experiment shows the combined results of numerous individual trials.** DNA double-helix formation was probed 800 times with an exposure time of 2.0 milliseconds, resulting in 79 times when a double helix was detected and 721 times where it was not.

Suppose that the first time we do the experiment, with an exposure time of 2.0 milliseconds, we observe that the double helix is formed. Perhaps surprisingly, however, when we repeat the identical experiment, no double helix is observed. We push on and repeat the experiment 800 times and observe that sometimes double-helix formation is observed, but usually it is not (**Figure 1.18**).

Overall, the double-helix formation was observed in 9.9% (79/800) of the experiments. This value fluctuates somewhat as the number of experiments increases. For example, after 100 experimental runs were completed, 15 (or 15.0%) showed double-helix formation, whereas after 500 runs, 51 instances (or 10.2%) of double-helix formation were seen. A plot of the percentage showing double-helix formation as a function of the number of experimental runs reveals fluctuations prior to reaching a stable value (**Figure 1.19**).

These results show that, at the single molecule level, the outcome of a particular attempt is not predictable—sometimes the double-helix forms, and sometimes it does not. Such processes are often referred to as **stochastic**. However, when averaged over hundreds of repetitions, the chance of double-helix formation converges to a well-defined probability. Under most circumstances, we observe processes involving thousands, billions, or billions of billions of

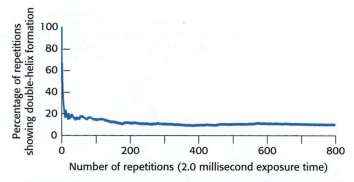

FIGURE 1.19 **A plot of the running average of the number of experimental repetitions observing double-helix formation reveals eventual convergence on a stable value.** For example, for the point at 500 repetitions, 51 had shown double-helix formation, for a percentage of 51/500 = 10.2%.

molecules, so that the variability at the single-molecule level is not apparent. However, it is best to keep in mind the reality of the variability at the level of single molecules; understanding this stochasticity is key to understanding some biochemical processes.

Acid–base reactions are central in many biochemical processes

Throughout our consideration of the formation of the double helix, we have dealt only with the noncovalent interactions that occur or are disrupted in this process. In addition, many biochemical processes entail the formation and cleavage of covalent bonds. A particularly important class of reactions prominent in biochemistry is **acid–base reactions**, in which hydrogen ions are added to molecules or removed from them. Throughout the book, we will encounter numerous processes in which the addition or removal of hydrogen ions is crucial.

A hydrogen ion, often written as H^+, corresponds to a proton. In fact, hydrogen ions exist in solution bound to water molecules, thus forming what are known as *hydronium ions*, H_3O^+. For simplicity, we will continue to write H^+, but we should keep in mind that H^+ is shorthand for the actual species (H_3O^+) that is present.

The concentration of hydrogen ions in solution is expressed as the pH Specifically, the **pH** of a solution is defined as

$$pH = -\log[H^+]$$

where $[H^+]$ is in units of molarity (M). Thus, pH 7.0 refers to a solution for which $-\log[H^+] = 7.0$, and so $\log[H^+] = -7.0$ and $[H^+] = 10^{\log[H^+]} = 10^{-7.0} = 1.0 \times 10^{-7}$ M.

The pH also indirectly expresses the concentration of hydroxide ions, $[OH^-]$, in solution. To see how, we must realize that water molecules can dissociate to form H^+ and OH^- ions in an equilibrium process:

$$H_2O \rightleftharpoons H^+ + OH^-$$

The equilibrium constant (K) for the dissociation of water is defined as

$$K = [H^+][OH^-]/[H_2O]$$

and has a value of $K = 1.8 \times 10^{-16}$. Note that an equilibrium constant does not formally have units. Nonetheless, the value of the equilibrium constant given assumes that particular units are used for concentration (sometimes referred to as standard states); in this case and in many others, units of molarity (M) are assumed.

The concentration of water, $[H_2O]$, in pure water is 55.5 M, and this concentration is constant under most conditions. Thus, we can define a new constant, K_W:

$$K_W = K[H_2O] = [H^+][OH^-]$$
$$K[H_2O] = 1.8 \times 10^{-16} \times 55.5$$
$$= 1.0 \times 10^{-14}$$

Because $K_W = [H^+][OH^-] = 1.0 \times 10^{-14}$, we can calculate

$$[OH^-] = 10^{-14}/[H^+] \text{ and } [H^+] = 10^{-14}/[OH^+]$$

With these relationships in hand, we can easily calculate the concentration of hydroxide ions in an aqueous solution, given the pH. For example, at pH = 7.0, we know that $[H^+] = 10^{-7}$ M and so $[OH^-] = 10^{-14}/10^{-7} = 10^{-7}$ M. In acidic solutions, the concentration of hydrogen ions is higher than 10^{-7} and, hence, the pH is below 7. For example, in 0.1 M HCl, $[H^+] = 10^{-1}$ M and so pH = 1.0 and $[OH^-] = 10^{-14}/10^{-1} = 10^{-13}$ M.

Acid–base reactions can disrupt the double helix

The formation process that we have been considering between two strands of DNA to form a double helix takes place readily at pH 7.0. Suppose that we take

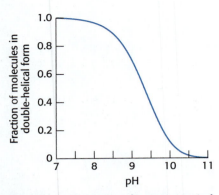

FIGURE 1.20 DNA strands are separated by the addition of a base. The addition of a base to a solution of double-helical DNA initially at pH 7 causes the double helix to separate into single strands. The process is half complete at slightly above pH 9.

the solution containing the double-helical DNA and treat it with a solution of concentrated base (i.e., with a high concentration of OH^-). As the base is added, we monitor the pH and the fraction of DNA in double-helical form (**Figure 1.20**). When the first additions of base are made, the pH rises, but the concentration of the double-helical DNA does not change significantly. However, as the pH approaches 9, the DNA double helix begins to dissociate into its component single strands. As the pH continues to rise from 9 to 10, this dissociation becomes essentially complete.

Why do the two strands dissociate as pH increases? The hydroxide ions can react with nitrogenous bases in DNA base pairs to remove certain protons. The most susceptible proton is the one bound to the N-1 nitrogen atom in a guanine base.

Guanine (G)

Proton dissociation for a substance HA (such as that bound to N-1 on guanine) has an equilibrium constant defined by the expression

$$K_a = [H^+][A^-]/[HA]$$

The susceptibility of a proton to removal by reaction with a base is often described by its **pK_a value**:

$$pK_a = -\log(K_a)$$

When the pH is equal to the pK_a, we have $pH = pK_a$

$$pH = pK_a$$

and so

$$-\log[H^+] = -\log([H^+][A^-]/[HA])$$

and

$$[H^+] = [H^+][A^-]/[HA]$$

Dividing by $[H^+]$ reveals that

$$1 = [A^-]/[HA]$$

and so

$$[A^-] = [HA]$$

So, when the pH equals the pK_a, the concentration of the deprotonated form of the group or molecule is equal to the concentration of the protonated form; the deprotonation process is halfway to completion. The pK_a for the proton on N-1 of guanine is typically 9.7. When the pH approaches this value, the proton on N-1 is lost (see Figure 1.20). Replacement of the partial positive change on the proton with a full negative charge substantially destabilizes the DNA double helix.

The DNA double helix is also destabilized by low pH. Below pH 5, some of the hydrogen-bond acceptors that participate in base-pairing become protonated. In their protonated forms, these bases can no longer form these hydrogen bonds, and the double helix separates. So, acid–base reactions that either remove or donate protons at specific positions on the DNA bases can disrupt the double helix.

Buffers regulate pH in organisms and in the laboratory

These observations about DNA reveal that a significant change in pH can disrupt molecular structure. The same is true for many other biological macromolecules; changes in pH can protonate or deprotonate key groups, potentially disrupting structures and initiating harmful reactions. Thus, biological systems have evolved to mitigate changes in pH.

Solutions that resist changes in pH are called **buffers**. Specifically, when acid is added to an unbuffered aqueous solution, the pH drops in proportion to the amount of acid added. In contrast, when acid is added to a buffered solution, the pH drops more gradually. Buffers also mitigate the pH increase caused by the addition of base and changes in pH caused by dilution.

Compare the result (shown in **Figure 1.21**) of adding a 1 M solution of the strong acid HCl drop by drop to (1) pure water, versus (2) adding it to a solution containing 100 mM of the buffer sodium acetate ($Na^+CH_3COO^-$). The process of gradually adding known amounts of reagent to a solution with which the reagent reacts while monitoring the results is called a **titration**. For pure water, the pH drops from 7 to close to 2 on the addition of the first few drops of acid. However, for the sodium acetate solution, the pH first falls rapidly from its initial value near 10, then changes more gradually until the pH reaches 3.5, and then falls more rapidly again.

Why does the pH decrease so gradually in the middle of the titration? The answer is that when hydrogen ions are added to this solution, they react with acetate ions to form acetic acid. This reaction consumes some of the added hydrogen ions so that the pH does not drop. Hydrogen ions continue reacting with acetate ions until essentially all the acetate ion is converted into acetic acid. After this point, added protons remain free in solution and the pH begins to fall sharply again.

We can analyze the effect of the buffer in quantitative terms. The equilibrium constant for the deprotonation of an acid is

$$K_a = [H^+][A^-]/[HA]$$

Taking logarithms of both sides yields

$$\log(K_a) = \log([H^+]) + \log([A^-]/[HA])$$

Recalling the definitions of pK_a and pH and rearranging gives

$$pH = pK_a + \log([A^-]/[HA])$$

This expression is referred to as the **Henderson–Hasselbalch equation**.

We can apply the equation to our titration of sodium acetate. The pK_a of acetic acid is 4.75. We can calculate the ratio of the concentration of acetate ion to the concentration of acetic acid as a function of pH by slightly rearranging the Henderson–Hasselbalch equation:

$$[\text{Acetate ion}]/[\text{Acetic acid}] = [A^-]/[HA] = 10^{pH-pK_a}$$

At pH 9, this ratio is $10^{9-4.75} = 10^{4.25} = 17{,}800$; very little acetic acid has been formed.

At pH 4.75 (when the pH equals the pK_a), the ratio is $10^{4.75-4.75} = 10^0 = 1$.

At pH 3, the ratio is $10^{3-4.75} = 10^{-1.25} = 0.02$; almost all the acetate ion has been converted into acetic acid.

We can follow the conversion of acetate ion into acetic acid over the entire titration (**Figure 1.22**). The graph shows that the region of relatively constant pH corresponds precisely to the region in which acetate ion is being protonated to form acetic acid.

From this discussion, we see that a buffer functions best close to the pK_a value of its acid component. Physiological pH is typically about 7.4.

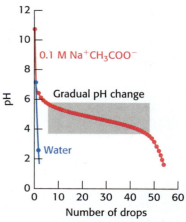

FIGURE 1.21 Buffers resist changes in pH over a narrow pH range. The addition of a strong acid, 1 M HCl, to pure water results in an immediate drop in pH to near 2. In contrast, the addition of the acid to a 0.1 M sodium acetate ($Na^+CH_3COO^-$) solution results in a much more gradual change in pH until the pH drops below 3.5.

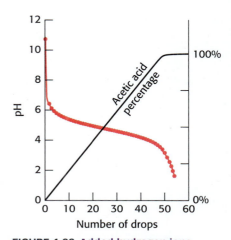

FIGURE 1.22 Added hydrogen ions protonate buffer components rather than remaining free. When acid is added to sodium acetate, the added hydrogen ions are used to convert acetate ion into acetic acid. Because the proton concentration does not increase significantly, the pH remains relatively constant until all the acetate has been converted into acetic acid.

An important buffer in biological systems is based on phosphoric acid (H_3PO_4). The acid can be deprotonated in three steps to form a phosphate ion.

$$H_3PO_4 \xrightleftharpoons[pK_a = 2.12]{H^+} H_2PO_4^- \xrightleftharpoons[pK_a = 7.21]{H^+} HPO_4^{2-} \xrightleftharpoons[pK_a = 12.67]{H^+} PO_4^{3-}$$

At about pH 7.4, inorganic phosphate exists primarily as a nearly equal mixture of $H_2PO_4^-$ and HPO_4^{2-}. Thus, phosphate solutions function as effective buffers near pH 7.4. The concentration of inorganic phosphate in blood is typically approximately 1 mM, providing a useful buffer against processes that produce either acid or base.

❖ **SELF–CHECK QUESTION**

Butyric acid ($CH_3CH_2CH_2COOH$) has a pK_a of 4.82. What is the ratio of butyrate ($CH_3CH_2CH_2COO^-$) to butyric acid at pH 6.0?

EXAMPLE **Applying the Henderson–Hasselbalch Equation**

PROBLEM: You need to prepare 2 liters of 0.100 M acetate buffer, pH 5.00 for acetic acid, sodium acetate, and water. How much acetic acid and sodium acetate do you use?

GETTING STARTED: The Henderson–Hasselbalch equation relates the pH to the ratio of ionized and unionized buffer components

$$pH = pK_a + \log([A^-]/[HA])$$

where $[A^-]$ is the concentration of acetate ion and $[HA]$ is the concentration of acetic acid. We know that the pK_a of acetic acid is 4.75 and the buffer is supposed to have a pH of 5.00 with a total concentration of acetate-containing species of 0.100 M; that is, $[CH_3COO^-] + [CH_3COOH] = 0.100$ M.

SOLVE: Substituting into the Henderson–Hasselbalch equation, we find that

$$5.00 = 4.75 + \log([CH_3COO^-]/[CH_3COOH])$$

Solving for the ratio of species, we find that

$$\log([CH_3COO^-]/[CH_3COOH]) = 5.00 - 4.75 = 0.25$$

so that $[CH_3COO^-]/[CH_3COOH] = 10^{0.25} = 1.78$.

Thus, $[CH_3COO^-] = 1.78[CH_3COOH]$.

We can then substitute this expression into the expression for the total concentration of acetate-containing species, $[CH_3COO^-] + [CH_3COOH] = 0.100$ M

That is, $1.78[CH_3COOH] + [CH_3COOH] = 0.100$ M

so that $(1.78 + 1.00)[CH_3COOH] = 2.78[CH_3COOH] = 0.100$ M.

Thus, $[CH_3COOH] = (1/2.78)(0.100 \text{ M}) = 0.036$ M and $[CH_3COO^-] = 1.78(0.036 \text{ M}) = 0.064$ M.

We need to make 2 liters of buffer, so we need (2 liters)(0.036 M) = 0.072 moles of acetic acid and (2 liters)(0.064 M) = 0.128 moles of acetate ion, provided as sodium acetate.

The molecular weights of acetic acid and sodium acetate are 60.052 g/mol and 82.034 g/mol, respectively.

Thus, we need (0.072 moles)(60.052 g/mol) = 4.32 g of acetic acid

and (0.128 moles)(82.034 g/mol) = 10.50 g of sodium acetate.

After weighing out these materials, we add water to make a total volume of 2 liters.

REFLECT: You can confirm the result by calculating the pH using the Henderson–Hasselbalch equation with the concentrations of acetic acid and acetate ion that you added.

1.4 DNA Sequencing Is Transforming Biochemistry, Medicine, and Other Fields

The discovery of the structure of DNA suggested the hypothesis that hereditary information is stored as a sequence of nitrogenous bases along long strands of DNA. This remarkable insight provided an entirely new way of thinking about biology. However, for years this hypothesis remained unconfirmed, and many features needed to be elucidated. How is the sequence information read and translated into action? What are the sequences of naturally occurring DNA molecules, and how can such sequences be experimentally determined?

Through advances in biochemistry and related sciences, we now have essentially complete answers to these questions. Indeed, in the past two decades, scientists have determined the complete genome sequences of thousands of different organisms, including simple microorganisms, plants, animals of varying degrees of complexity, and human beings. Comparisons of these genome sequences, with the use of methods discussed in Chapter 10, have provided many insights that have transformed biochemistry. In addition to its experimental and clinical aspects, biochemistry has now become an information science.

Genome sequencing has become remarkably fast and inexpensive

The sequencing of a human genome was a daunting task, because it contains approximately 3 billion (3×10^9) base pairs. We can look at a small section of the human genome to get a sense of its overall complexity. For example, the sequence

> ACATTTGCTTCTGACACAACTGTGTTCACTAGCAACCTC
> AAACAGACACCATGGTGCATCTGACTCCTG**A**GGAGAAGT
> CTGCCGTTACTGCCCTGTGGGGCAAGGTGAACGTGGA . . .

is a part of one of the genes that encodes hemoglobin, the oxygen carrier in our blood, discussed in Chapter 3. This gene is found on the end of chromosome 9 of our 24 distinct chromosomes. If we were to include the complete sequence of our entire genome, this chapter would run to more than 500,000 pages. The sequencing of our genome is truly a landmark in human history. This sequence contains a vast amount of information, some of which we can now extract and interpret, but much of which we are only beginning to understand. For example, some human diseases have been linked to particular variations in genomic sequence. Sickle-cell anemia is caused by a single base change of an A (noted in boldface red type in the preceding sequence) to a T. In this course of this book, we will encounter many other examples of diseases that have been linked to specific DNA sequence changes.

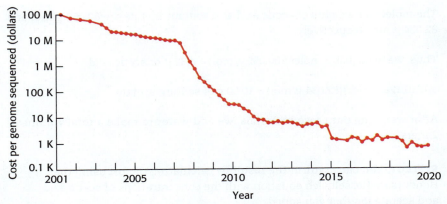

FIGURE 1.23 **The cost to sequence a human genome has decreased dramatically over the past two decades.** Through the Human Genome Project, the costs of DNA sequencing dropped steadily due to new methods and are now under $1000 for a complete human genome sequence. [National Human Genome Research Institute. https://www.genome.gov/about-genomics/fact-sheets/Sequencing-Human-Genome-cost]

Determining the first human genome sequences was a huge challenge. It required the efforts of large teams of geneticists, molecular biologists, biochemists, and computer scientists, as well as billions of dollars, because there was no previous framework for aligning the sequences of various DNA fragments: many methods had to be developed along the way. Fortunately, one human genome sequence can serve as a reference for others: the availability of reference sequences and modern techniques enables much more rapid characterization of partial or complete genomes from other individuals, as we will discuss in Chapter 9.

Methods for sequencing DNA have improved rapidly, resulting in dramatic increases in the sequencing rate and decreases in costs (**Figure 1.23**). The availability of such powerful sequencing technology is transforming many fields, including medicine, dentistry, microbiology, pharmacology, and ecology, although a great deal remains to be done to improve the interpretation of these large data sets.

Characterization of genetic variation between individuals is powerful for many applications

Each person has a unique sequence of DNA base pairs. How different are we from one another at the genomic level? Examination of the differences in our genomes reveals that, on average, each pair of individuals has a different base in one position per 200 bases; that is, the difference is approximately 0.5%. This genomic variation between individuals who are not closely related is quite substantial compared with differences in populations. The average difference between two people within one population or ethnic group is greater than the difference between the averages of two different populations or ethnic groups (**Figure 1.24**).

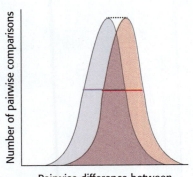

FIGURE 1.24 **DNA sequences vary more within populations than they do between populations.** The distribution of differences between pairs of individuals within two populations (from Europe and Sub-Saharan Africa) are shown in blue and red, respectively. The widths of these two distributions (shown by the blue and red lines) are larger than the average difference between the two populations (shown by the black dotted line). [Adapted from Witherspoon *et al.* Genetic differences between and within human populations. *Genetics* **176**, 351–359 (2007).]

- *Providing insights into ancestry.* While average differences in DNA sequence between different human populations are relatively small, DNA analysis can be a powerful tool for providing information about ancestry. In a landmark study, DNA from more than 1300 individuals with known ancestry from different parts of Europe was analyzed. The study looked at more than 200,000 positions spanning the genome that were known to vary across populations and visualized the differences in two dimensions. Breathtakingly, this plot reflected a remarkably accurate map of Europe (**Figure 1.25**)! More recent studies of individuals from 50 ethnolinguistic groups across Africa have revealed greater genomic variation than is seen in European populations

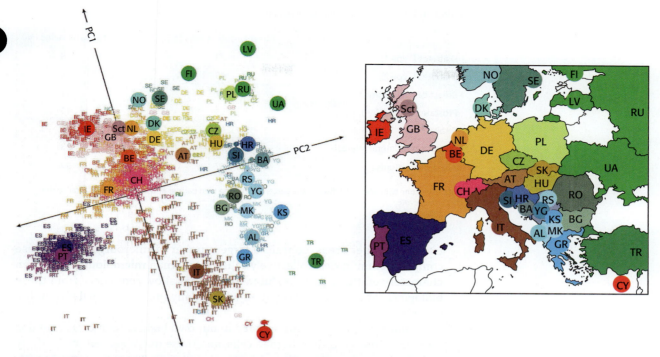

FIGURE 1.25 Genomic information can confirm locations of ancestry. Statistical analysis of the genomes of more than 1300 Europeans reproduced a map of Europe, providing a framework for inferring ancestry from genomic information. [Reprinted by permission from Macmillan Publishers Ltd: from Novembre, J., Johnson, T., Bryc, K. et al. Genes mirror geography within Europe. *Nature* 456, 98–101 (2008). Figure 1. https://doi.org/10.1038/nature07331.]

and have suggested migrations of groups of people across this continent that have occurred over the course of 2000 years.

- *Understanding the molecular basis of disease.* Scientists have identified the genetic variations associated with thousands of diseases, such as sickle-cell anemia, for which the cause can be traced to a single gene. For other diseases and traits, such as heart disease, we know that variation in many different genes contributes to disease in significant and often complex ways. Furthermore, in most cases, the presence of a particular variation or set of variations does not inevitably result in the onset of a disease but, instead, leads to a predisposition, or a heightened probability, for the development of the disease.

  The observation that genes alone do not determine disease outcomes is vividly illustrated through twin studies. Monozygotic (or "identical") twins share the same genome, whereas dizygotic (or "fraternal") twins share the same parents and typically develop in the same environment, but are not more genetically similar on average than any two siblings. By examining pairs of twins, scientists can begin to dissect the roles of genes versus environmental and other factors. For diseases like sickle-cell disease, monozygotic twins are always concordant; meaning, if one twin has the disease, so does the other. Most concordance rates, even for monozygotic twins, are well below 50%, indicating that environmental factors and chance are important in determining disease occurrence (**Table 1.1**). These often-modest levels of concordance reveal that environmental effects, and chance, can affect who gets any particular disease, revealing the complexity of biochemistry and biology that will be explored throughout the book.

- *Assessing the identity and role of our microbiome.* Our own genes are not the only ones that can contribute to our health and disease. Our bodies, including our skin, mouth, digestive tract, genitourinary tract, respiratory

TABLE 1.1 Disease concordance rates in twins

	Monozygotic twins (%)	Dizygotic twins (%)
Diabetes (Type 1)	23	5
Diabetes (Type 2)	45	19
Breast cancer (Women)	17	11
Prostate cancer (Men)	33	15
Stroke	18	4
Amyotrophic lateral sclerosis	18	< 3
Alzheimer's disease	67	22

Note: Given one twin who has a disease, the percentage whose twin partner also has the disease is called the *concordance rate.*

tract, and other areas, contain a large number of microorganisms. These complex communities, collectively called the human **microbiome**, have been characterized through powerful methods that allow sequences from these biological samples to be examined without any previous knowledge of the organisms present.

Remarkably, we are outnumbered in our own bodies! Each of us contains more microbial cells than human cells, and these microbial cells collectively include many more genes than do our own genomes. These microbiomes differ between regions of the body, from one person to another, and in the same individual over time (**Figure 1.26**). They appear to play roles in health and in diseases such as obesity and dental caries.

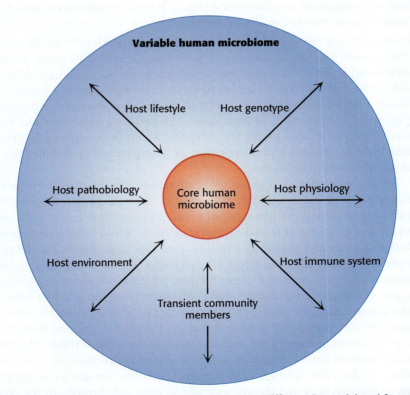

FIGURE 1.26 The human microbiome comprises many different bacterial and fungal species. The species composition varies greatly between individuals and within a given individual depending on a range of biological and environmental factors. [From Turnbaugh, P., Ley, R., Hamady, M. *et al.* The human microbiome project. *Nature* **449**, 804–810 (2007). https://doi.org/10.1038/nature06244]

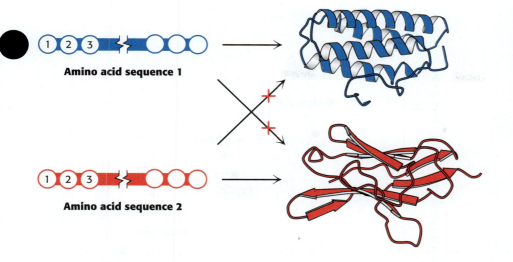

FIGURE 1.27 Protein folding is primarily dictated by the specific sequence of amino acids. Proteins are linear polymers of amino acids that fold into elaborate structures. The sequence of amino acids is the most important factor that determines the three-dimensional structure. Thus, amino acid sequence 1 gives rise only to a protein with the shape depicted in blue, not the shape depicted in red.

Amino acid sequence 1

Amino acid sequence 2

The most important function of genomic sequences is to encode proteins

What kinds of information do DNA's base sequences store, and how is it expressed? The most fundamental role of DNA is to encode the sequences of proteins. Like DNA, proteins are linear polymers. However, proteins differ from DNA in two important ways. First, proteins are built from 20 building blocks, called **amino acids**, rather than just 4, as in DNA. The chemical complexity provided by this variety of building blocks enables proteins to perform a wide range of functions. Second, proteins spontaneously fold into elaborate three-dimensional structures, determined almost exclusively by their amino acid sequences (**Figure 1.27**). This enables the transition from a largely one-dimensional world of DNA sequences into a three-dimensional world of intricate, interacting objects with diverse properties that characterize biochemistry. Proteins and protein folding will be discussed extensively in the next chapter. Interestingly, the human genome includes approximately 21,000 protein-coding genes, much smaller than earlier estimates of up to 100,000 or more.

Comparing genomes offers great insights into evolution

Studies of comparative genomics, in which one organism's genome is compared with the genomes of other organisms, are confirming the tremendous unity that exists at the level of biochemistry. Comparative genomics is also revealing key steps in the course of evolution. For example, many genes that are key to the function of the human brain and nervous system have evolutionary and functional relatives, even in the genomes of bacteria. Sequence comparisons can reveal a fascinating connection between biochemistry, evolution, and time.

The major mechanism for evolution involves the accumulation of small changes in DNA sequences. In rare cases, these changes (mutations) help the organism survive and reproduce under a particular set of conditions, leading to selection of organisms that carry these changes. More often, the changes are neutral, providing no advantage but causing no significant harm, and they accumulate because there is no reason or mechanism to remove them. Linus Pauling, a two-time Nobel Prize winner (for Chemistry and Peace) and his colleague followed this logic and suggested that the extent of sequence difference between the same protein from two organisms could be used as a measure of the time that had passed since these organisms had a common ancestor. This is called the **molecular clock hypothesis**.

FIGURE 1.28 The similarities between hemoglobin sequences from different species provide evidence for a molecular clock. A comparison of the sequence identity between a component of the protein hemoglobin from human beings and those from representation of modern species from different classes of animals is shown. The nearly linear relationship between the sequence similarity and the time of emergence of the class of animals suggests that sequence similarity of this protein between two species can be used to measure evolutionary time since these species had a common ancestor.

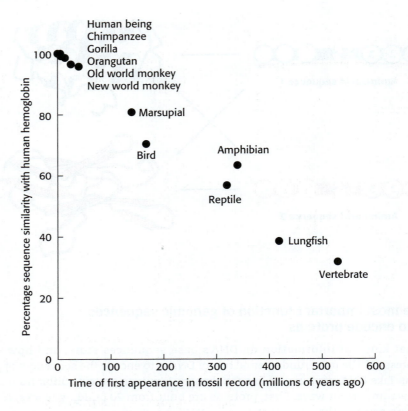

To give one example of the sorts of conclusions we can reach from applying the molecular clock hypothesis, let us examine the number of sequence differences between one component of human hemoglobin and the same component from other species. The sequences of this hemoglobin subunit from human beings and our closest living relatives, chimpanzees, are identical, while those from human beings and a slightly more distant relative, gorillas, are 99.3% identical. In contrast, the sequences between human beings and chickens are only 70.4% identical. We can plot these differences and those from representatives from the classes of animals from Figure 1.2 versus the times for the emergence of these classes from the fossil record (**Figure 1.28**).

❖ SELF–CHECK QUESTION

The amino acid sequence of hemoglobin from an unknown species is found to be 83.5% identical to that from human beings. Could this be from a primate?

Different molecular clocks are also used for entities such as viruses that evolve very rapidly. In essence, a **virus** comprises a nucleic acid genome (DNA or the related molecule ribonucleic acid, abbreviated RNA) surrounded by a protein coat. Viruses are not alive in that they cannot grow or reproduce on their own, but instead infect specific cells and take advantage of the components of these cells to produce new virus particles by copying their genomes and synthesizing their protein coats and other components encoded by the viral genome. These virus particles then escape from these cells to go on and infect other cells. The sequence of a viral genome changes over time, even slightly every time the virus replicates.

Comparison of viral sequences allows the timing of key evolutionary events to be estimated. For example, both the virus that caused Severe Acute Respiratory Syndrome (SARS) in 2002–2004, termed SARS-CoV, and the virus that causes COVID-19, SARS-CoV-2, diverged over periods of a few years from viruses found in horseshoe bats. Sequence comparison also helps reveal the origins and spread of variant viruses across the human population.

Molecular clocks provide a framework that will be used throughout the book. Specifically, we can infer evolutionary relatedness from sequence comparison, with the simple rule of thumb that nearly identical sequences had a recent common evolutionary ancestor whereas more different versions of the same sequence have had more time to diverge. Biochemical characteristics of modern organisms have much to teach us about the history of life of Earth and about recent events as well.

1.5 Biochemistry Is an Interconnected Human Endeavor

Biochemical knowledge is generated through the research efforts of human beings, individually and in teams, at many different stages along a variety of career paths, some of which are outlined in **Figure 1.29**. To give some examples of the human journeys involved in establishing this knowledge, we will introduce you to a sampling of interconnected career paths and collegial partnerships.

This chapter opened with a photograph of Carol Greider, PhD, celebrating with Peter Agre, MD, on the occasion of her award of the Nobel Prize in Medicine or Physiology in 2009 for research she started as a graduate student. Her discoveries, which were initially motivated by curiosity about fundamental biological process, have continued throughout her career, shedding light on both cancer and mechanisms of cellular aging. Her path to success was not always clear or easy, however. Although she was a strong student, she had struggled to get into college and graduate school because of relatively poor performance on standardized tests due to dyslexia.

Dr. Agre, shown with Dr. Greider in the photograph, had been awarded a Nobel Prize six years earlier. He was trained as a hematologist; his research was on Rh factor, a molecule important for blood compatibility for transfusions and

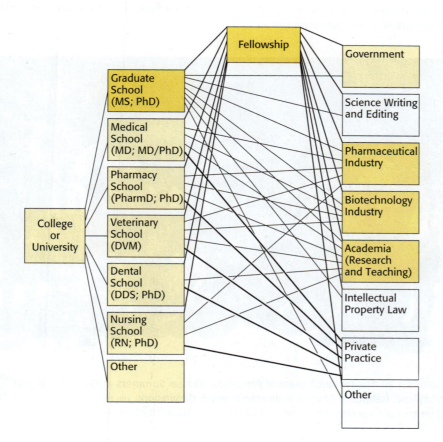

FIGURE 1.29 There are a variety of possible career paths for those involved in biochemical research. The intensity of the yellow shading corresponds to the intensity of the biochemical research activity in different positions and the lines highlight career paths that are commonly pursued.

in other contexts. In the course of this work, Dr. Agre and his coworkers discovered an abundant but uncharacterized component of red blood cells that turned out to be a long-hypothesized protein that allows red blood cells to take up and release water rapidly. For this discovery, Dr. Agre shared the Nobel in Chemistry, and he found this ironic, since he had temporarily withdrawn from high school because he received a "D" in chemistry.

One of the authors of this book, Jeremy Berg, PhD, was fortunate to have been colleagues with both Dr. Agre and Dr. Greider at Johns Hopkins University. Dr. Berg's research experience began as an undergraduate in the laboratory of Lubert Stryer, MD. Based on this and other experiences, his interests focused on the roles of metal ions in biological systems. After completing a PhD in chemistry, his research changed directions as a postdoctoral fellow, focusing on the structures of proteins that bind DNA. He was able to combine these interests with the discovery that certain DNA- and RNA-binding proteins contained bound metal ions such as zinc. During his career in academia and in government, Dr. Berg worked with undergraduates, PhD and MD/PhD students, and postdoctoral fellows with a wide range of backgrounds, interests, and future career directions.

Science often involves friendly competitions between research teams working on the same problems. For Dr. Berg, one such team was led by Michael Summers, PhD, who also began in inorganic chemistry and expanded into the structures of biological molecules. Dr. Summers and his coworkers were the first to determine the three-dimensional structure of an RNA-binding protein from Human Immunodeficiency Virus (HIV) that contains zinc and is essential for this virus to incorporate its own genome into its protein coat. Summers is a co-leader of a nationally recognized program directed to providing substantial research experiences for undergraduate students. He and an ever-changing group of undergraduates, along with some PhD students, technicians, and postdoctoral fellows (**Figure 1.30**), have continued this research over many years, elucidating more details about how this protein performs its function and expanding into other areas of HIV research.

Courtesy of Michael Summers, UMBC

FIGURE 1.30 The research team of Professor Michael Summers at the University of Maryland, Baltimore County, is diverse in many dimensions. His team consists primarily of undergraduate students, with some graduate students and postdoctoral fellows.

Among those who conducted research with Dr. Summers is Chelsea Pinnix, MD/PhD (**Figure 1.31**), who worked with Dr. Summers while she was an undergraduate student and contributed to the work that detailed the three-dimensional structure of the zinc-containing HIV protein bound to an RNA target. After graduating, she completed a combined MD/PhD program, earning her PhD in cell and molecular biology. She then completed additional medical training in radiation oncology and now uses her knowledge in all of these fields as a faculty member at an academic medical center.

Our knowledge of biochemistry is constantly expanding through the efforts of many individuals and teams working in universities and research institutes, and also in private industrial and government laboratories. One of Dr. Berg's former students and coauthor of this book is Gregory Gatto, MD/PhD, who works on developing new drugs at a pharmaceutical company. Coauthor Justin Hines, PhD, began conducting research on protein structure and function as an undergraduate and now teaches and investigates protein misfolding with a research team made up solely of undergraduates. Coauthor Jutta Heller, PhD, studied protein structure in graduate school and now teaches molecular biology to undergraduates. Many students of biochemistry, however, pursue nonacademic careers, applying their biochemical knowledge to other aspects of human endeavor, such as medicine, intellectual property law, communication, and finance.

Biochemistry enriches many aspects of modern society. Throughout this book, we will continue to highlight individuals who have made important contributions to biochemistry while pursuing various career paths. We hope these profiles help you learn about some outstanding scientists and appreciate the varied paths available to students of biochemistry.

 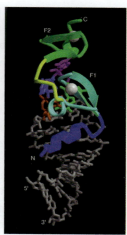

FIGURE 1.31 Chelsea Pinnix, MD/PhD, helped to determine the three-dimensional structure of an HIV zinc-containing protein bound to RNA.

[(left) Courtesy of Dr. Chelsea Pinnix; (right) Republished with permission of American Association for the Advancement of Science, from Roberto N. De Guzman, Zheng Rong Wu, Chelsea C. Stalling, Lucia Pappalardo, Philip N. Borer, Michael F. Summers, "Structure of the HIV-1 Nucleocapsid Protein Bound to the SL3 Ψ-RNA Recognition Element." Science 16 Jan 1998: Vol. 279, Issue 5349, pp. 384–388, DOI: 10.1126/science.279.5349.384, Figure 4a. Permission conveyed through Copyright Clearance Center, Inc.]

Summary

1.1 Biochemical Unity Underlies Biological Diversity

- Many biochemical processes and pathways are quite similar even in organisms that appear quite different from one another.
- Organisms can be divided into three fundamental groups called domains.
- Eukarya (eukaryotes) includes all multicellular organisms, including all animals, plants, and human beings as well as some unicellular organisms, whereas Bacteria and Archaea represent two distinct groups of microscopic microorganisms (prokaryotes).

1.2 DNA Illustrates the Interplay Between Form and Function

- Each DNA strand consists of a phosphodiester backbone with four nitrogenous bases.
- Most DNA molecules consist of two strands that wind around a common axis to form a double helix with specific base pairs in the center.

- The double-helical structure of DNA can explain the role of DNA in storing hereditary information; the information is stored in the sequence of bases, and each strand can serve as a template for the synthesis of a new copy of the opposite strand.

1.3 Concepts from Chemistry Explain the Properties of Biological Molecules

- The DNA double helix serves as a useful example to illustrate how concepts from chemistry apply to essentially all other molecules in biochemistry.
- Most biochemical interactions take place on timescales from picoseconds to microseconds (10^{-12}–10^{-6} s).
- Covalent bonds are strong bonds formed by the sharing of a pair of electrons between adjacent atoms.
- Noncovalent interactions are generally weaker than covalent bonds and include ionic

interactions, hydrogen bonds, van der Waals interactions, and interactions associated with the hydrophobic effect.

- Most biochemical processes take place in water, which has a highly cohesive structure due to water molecules (H_2O) interacting with one another through hydrogen bonds.

- The hydrophobic effect refers to the tendency of nonpolar groups, those that lack the ability to participate in ionic interactions and hydrogen bonds, to come together, minimizing exposure to aqueous solution.

- A variety of methods have been developed to observe single molecules participating in different processes, revealing the underlying randomness of chemical processes.

- The laws of thermodynamics are general principles that apply to all physical and chemical processes: The First Law states that the total energy of a system and its surroundings is constant, and The Second Law states that the total entropy of a system plus that of its surroundings always increases.

- Acid–base reactions, which involve the addition or removal of hydrogen ions from molecules, are generally characterized by the equilibrium constant $K_a = [H^+][A^-]/[HA]$.

- Buffers are solutions that resist changes in pH due to the fact that they contain both a weak acid and the associated base.

1.4 DNA Sequencing Is Transforming Biochemistry, Medicine, and Other Fields

- DNA sequencing has become rapid and inexpensive and is transforming many fields, including biochemistry and medicine.

- On average, the genome sequences of two individual human beings differ by 0.5%; this difference is generally larger than the average difference between populations.

- While variations in genome sequences can be important in contributing to traits such as disease susceptibility, environmental factors and chance are also important.

- A major role of the genome sequence is to encode the amino acid sequences of proteins, the molecules that perform most of the functions within cells.

- Molecular clocks have been developed based on analyzing variations in DNA or protein sequences between different organisms to infer how long ago these organisms shared a common ancestor.

1.5 Biochemistry Is an Interconnected Human Endeavor

- Knowledge of biochemistry advances through the efforts of scientists generally working together in small teams.

- A wide variety of career paths is accessible to individuals who study biochemistry.

Key Terms

cell (p. 2)

biological macromolecule (p. 2)

metabolite (p. 2)

deoxyribonucleic acid (DNA) (p. 2)

protein (p. 2)

common ancestor (p. 3)

Eukarya (eukaryotes) (p. 3)

prokaryotes (p. 3)

Bacteria (p. 3)

Archaea (p. 3)

nitrogenous bases (bases) (p. 4)

polarity (p. 4)

double helix (p. 5)

base pair (p. 5)

complementary sequences (p. 6)

biochemical timescale (p. 7)

covalent bond (p. 7)

resonance structure (p. 8)

ionic interaction (p. 8)

electric dipole (dipole) (p. 9)

hydrogen bond (p. 9)

hydrogen-bond donor (p. 9)

hydrogen-bond acceptor (p. 9)

van der Waals interaction (p. 9)

van der Waals contact distance (p. 10)

hydrophobic effect (p. 10)

nonpolar molecule (p. 11)

surface complementarity (p. 12)

entropy (p. 13)

enthalpy (p. 13)

free energy (Gibbs free energy) (p. 13)

stochastic (p. 16)

acid–base reactions (p. 17)

pH (p. 17)

pK_a value (p. 18)

buffer (p. 19)

titration (p. 19)

Henderson–Hasselbalch equation (p. 19)

microbiome (p. 24)

amino acid (p. 25)

molecular clock hypothesis (p. 25)

virus (p. 26)

Problems

1. Identify the hydrogen-bond donors and acceptors in each of the four bases on page 4. ❖ **2**

2. Draw a stable, uncharged resonance form for the structure shown. Be sure to include lone pairs of electrons. ❖ **2**

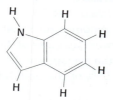

3. What types of noncovalent interactions hold together the following solids? ❖ **2**

(a) Table salt (NaCl), which contains Na^+ and Cl^- ions.

(b) Graphite (C), which consists of sheets of covalently bonded carbon atoms.

4. Given the following values for the changes in enthalpy (ΔH) and entropy (ΔS), which of the following processes can take place at 298 K without violating the Second Law of Thermodynamics? ❖ **3**

(a) $\Delta H = -84$ kJ mol^{-1} (-20 kcal mol^{-1}),
  $\Delta S = +125$ J mol^{-1} K^{-1} ($+30$ cal mol^{-1} K^{-1})

(b) $\Delta H = -84$ kJ mol^{-1} (-20 kcal mol^{-1}),
  $\Delta S = -125$ J mol^{-1} K^{-1} (-30 cal mol^{-1} K^{-1})

(c) $\Delta H = +84$ kJ mol^{-1} ($+20$ kcal mol^{-1}),
  $\Delta S = +125$ J mol^{-1} K^{-1} ($+30$ cal mol^{-1} K^{-1})

(d) $\Delta H = +84$ kJ mol^{-1} ($+20$ kcal mol^{-1}),
  $\Delta S = -125$ J mol^{-1} K^{-1} (-30 cal mol^{-1} K^{-1})

5. For double-helix formation, ΔG can be measured to be -54 kJ mol^{-1} (-13 kcal mol^{-1}) at pH 7.0 in 1 M NaCl at 25°C (298 K). The heat released indicates an enthalpy change of -251 kJ mol^{-1} (-60 kcal mol^{-1}). For this process, calculate the entropy change for the system and the entropy change for the surroundings. ❖ **3**

6. Calculate the pH of a solution that is 0.153 M HCl. ❖ **6**

7. Given a density of 1 g/ml and a molecular weight of 18 g/mol, calculate the concentration of water in water.

8. A solution is prepared by adding 0.01 M acetic acid (pK_a = 4.75) and 0.01 M ethylamine to water and adjusting the pH to 7.4. The pK_a of the ethylammonium ion (CH$_3$ CH$_2$ NH$_3^+$) is 10.70. What is the ratio of acetate to acetic acid? What is the ratio of ethylamine to ethylammonium ion? ❖ **6**

9. 100 mL of a solution of hydrochloric acid with pH 5.0 is diluted to 1 L. Similarly, 100 mL of a 0.1 mM buffer solution made from acetic acid and sodium acetate with pH 5.0 is diluted to 1 L. What are the pH values of the two diluted solutions? ❖ **5**

10. A dye that is an acid and that appears as different colors in its protonated and deprotonated forms can be used as a pH indicator. Suppose that you have a 0.001 M solution of a dye with a pK_a of 7.2. From the color, the concentration of the protonated form is found to be 0.0002 M. Assume that the remainder of the dye is in the deprotonated form. What is the pH of the solution? ❖ **6**

11. During exercise when the body lacks an adequate supply of oxygen to support energy production, a compound called pyruvate that is produced from the breakdown of glucose is converted into lactic acid.
The dissociation of lactic acid to lactate is shown in the reaction. Lactic acid has a pK_a of 3.86.
A solution containing a mixture of lactic acid and lactate was found to have a pH of 4.13. Calculate the ratio of the lactate concentration to the lactic acid concentration in this solution. ❖ **6**

12. Through the use of nuclear magnetic resonance spectroscopy, it is possible to determine the ratio between the protonated and deprotonated forms of buffers. Use the Henderson–Hasselbalch equation for a propanoic acid solution (CH$_3$CH$_2$CO$_2$H, pK_a = 4.87) to calculate the quotient [A$^-$]/[HA] at three different pH values: **(a)** pH = 4.52; **(b)** pH = 4.87; **(c)** pH = 5.08.

13. Given that phosphoric acid (H$_3$PO$_4$) can give up three protons with different pK_a values, sketch a plot of pH as a function of added drops of sodium hydroxide solution, starting with a solution of phosphoric acid at pH 1.0. ❖ **6**

14. Your laboratory is out of materials to make phosphate buffer and you are considering using sulfate to make a buffer instead. The pK_a values for the two hydrogens in H$_2$SO$_4$ are -10 and 2. **(a)** Will this approach work for making a buffer effective near pH 7? **(b)** Around what pH might a sulfate-based buffer be useful? ❖ **6**

15. You wish to prepare a buffer consisting of acetic acid and sodium acetate with a total acetic acid plus acetate concentration of 250 mM and a pH of 5.0. What concentrations of acetic acid and sodium acetate should you use? Assuming you wish to make 2 liters of this buffer, how many moles of acetic acid and sodium acetate will you need? How many grams of each will you need (molecular weights: acetic acid 60.05 g mol^{-1}, sodium acetate, 82.03 g mol^{-1})? ❖ **5**

16. Norman Good and his colleagues developed a series of compounds that are useful buffers near neutral pH. Three of these buffers are abbreviated MES (pK_a = 6.15), MOPS (pK$_a$ = 7.15), and HEPPS (pK$_a$ = 8.00). **(a)** Which of these buffers would be most effective near pH 7.0? **(b)** Near 6.0? ❖ 5

17. Would you expect the double helix in a short segment of DNA to be more stable in 100 sodium phosphate buffer at pH 7.0 or in pure water? Why? ❖ 2

18. Suppose two phosphate groups in DNA (each with a charge of –1) are separated by 12 Å. What is the energy of the ionic interaction between these two phosphates, assuming a dielectric constant of 80? Repeat the calculation, assuming a dielectric constant of 2. ❖ 2

19. Describe the noncovalent interactions that are primarily responsible for the fact that oil and water do not mix easily. ❖ 2

20. Hemoglobin from the Galapagos tortoise is 64% identical to that from human beings. In contrast, a protein called histone H2B is 98% identical in these two organisms. What level of identity would you expect for histone H2B for gorillas compared with human beings? Would histone H2B be as useful a molecular clock as hemoglobin? ❖ 7

21. Assume that 10% of the members of a population will get a particular disease over the course of their lifetime. Genomic studies reveal that 5% of the population have sequences in their genomes such that their probability of getting the disease over the course of their lifetimes is found to be 50%. What is the average lifetime risk of this disease for the remaining 95% of the population without these sequences? ❖ 7

Protein Composition and Structure

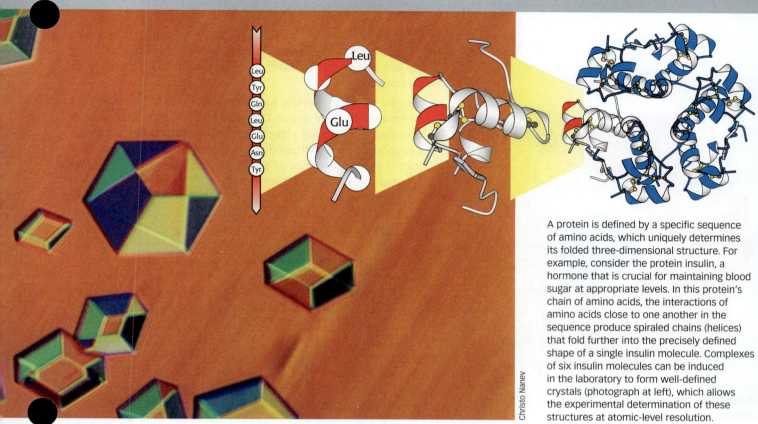

Christo Nanev

A protein is defined by a specific sequence of amino acids, which uniquely determines its folded three-dimensional structure. For example, consider the protein insulin, a hormone that is crucial for maintaining blood sugar at appropriate levels. In this protein's chain of amino acids, the interactions of amino acids close to one another in the sequence produce spiraled chains (helices) that fold further into the precisely defined shape of a single insulin molecule. Complexes of six insulin molecules can be induced in the laboratory to form well-defined crystals (photograph at left), which allows the experimental determination of these structures at atomic-level resolution.

❖ LEARNING GOALS

By the end of this chapter, you should be able to:

1. Identify the 20 amino acids and their corresponding three-letter and one-letter abbreviations, and group them according to the chemical properties of their side chains.

2. Distinguish between primary, secondary, tertiary, and quaternary structures.

3. Describe the properties of the principal types of secondary structure, including the α helix, the β sheet, and the reverse turn.

4. Explain how the hydrophobic effect serves as the primary driving force for folding of polypeptide chains into globular proteins.

5. Describe the nucleation-condensation model of protein folding, and explain why it is preferable to a random sampling model.

OUTLINE

2.1 Several Properties of Protein Structure Are Key to Their Functional Versatility

2.2 Proteins Are Built from a Repertoire of 20 Amino Acids

2.3 Primary Structure: Amino Acids Are Linked by Peptide Bonds to Form Polypeptide Chains

2.4 Secondary Structure: Polypeptide Chains Can Fold into Regular Structures

2.5 Tertiary Structure: Proteins Can Fold into Globular or Fibrous Structures

2.6 Quaternary Structure: Polypeptide Chains Can Assemble into Multisubunit Structures

2.7 The Amino Acid Sequence of a Protein Determines Its Three-Dimensional Structure

Proteins are the most versatile macromolecules in living systems and serve crucial functions in essentially all biological processes. They function as catalysts, transport and store other molecules such as oxygen, provide mechanical support and immune protection, generate movement, transmit nerve impulses, and control growth and differentiation. Indeed, much of this book will focus on understanding what proteins do and how they perform these functions. In order to fully appreciate this functional diversity, we must first understand the essential structural components shared by all proteins. For the biochemist, recognition of this architectural composition is critical and fundamental, akin to learning the alphabet of a new language.

2.1 Several Properties of Protein Structure Are Key to Their Functional Versatility

As we shall see throughout this text, proteins can adopt a breathtakingly wide variety of structures, which, in turn, enables them to participate in a broad range of functions. There are several key features of protein structure that allow for this remarkable diversity:

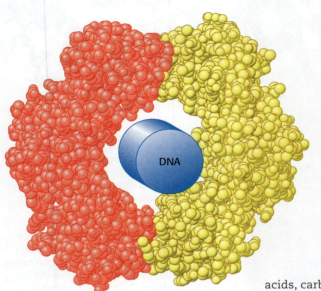

FIGURE 2.1 The structure of a protein determines its function. The protein component of the DNA replication machinery, which consists of two identical subunits (shown in red and yellow), acts as a clamp that allows large segments of DNA (depicted as a cylinder) to be copied without the replication machinery dissociating from the DNA. [Drawn from 2POL.pdb.]

INTERACT with this model in
 Achieve

- *Proteins are linear polymers built of monomer units called amino acids, which are linked end to end.* The sequence of linked amino acids is called the primary structure. Remarkably, proteins spontaneously fold up into three-dimensional structures that are determined by the sequence of amino acids in the protein polymer. Three-dimensional structure formed by hydrogen bonds between amino acids near one another is called secondary structure, whereas tertiary structure is formed by long-range interactions between amino acids. Protein function depends directly on this three-dimensional structure (**Figure 2.1**). Thus, proteins are the embodiment of the transition from the one-dimensional world of sequences to the three-dimensional world of molecules capable of diverse activities. Many proteins also display quaternary structure, in which the functional protein is composed of several distinct polypeptide chains.

- *Proteins contain a wide range of functional groups.* These functional groups include alcohols, thiols, thioethers, carboxylic acids, carboxamides, and a variety of basic groups, such as amines. Most of these groups are chemically reactive. When combined in various sequences, this array of functional groups accounts for the broad spectrum of protein function. For instance, their reactive properties are essential to the function of **enzymes**—the proteins that catalyze specific chemical reactions in biological systems (Chapters 5 through 7).

- *Proteins can interact with one another and with other biological macromolecules to form complex assemblies.* The proteins within these assemblies can act synergistically to generate capabilities that individual proteins may lack. Examples of these assemblies include macromolecular machines that replicate DNA, transmit signals within cells, and enable muscle cells to contract (**Figure 2.2**).

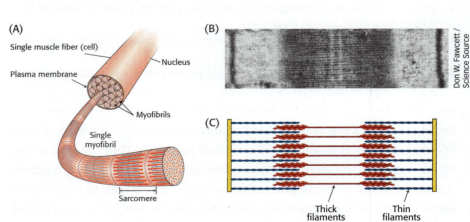

FIGURE 2.2 Some proteins form complex assemblies. (A) A single muscle cell contains multiple myofibrils, each of which is comprised of numerous repeats of a complex protein assembly known as the sarcomere. (B) The banding pattern of a sarcomere is evident by electron microscopy. (C) The banding is caused by the interdigitation of filaments made up of many individual proteins.

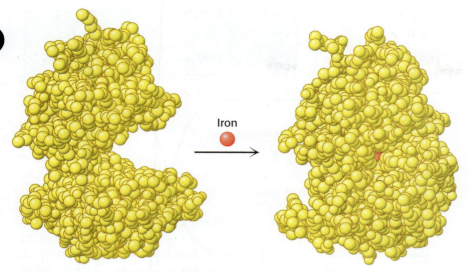

FIGURE 2.3 Conformational changes allow flexibility in protein function. On binding iron, the protein lactoferrin undergoes a substantial change in conformation that allows other molecules to distinguish between the iron-free and the iron-bound forms. [Drawn from 1LFH.pdb and 1LFG.pdb.]

INTERACT with this model in
Achieve

- *Some proteins are quite rigid, whereas others display considerable flexibility.* Rigid units can function as structural elements in the cytoskeleton (the internal scaffolding within cells) or in connective tissue. Proteins with some flexibility may act as hinges, springs, or levers. In addition, conformational changes within proteins enable the regulated assembly of larger protein complexes as well as the transmission of information within and between cells (**Figure 2.3**).

2.2 Proteins Are Built from a Repertoire of 20 Amino Acids

Amino acids are the building blocks of proteins. An α-amino acid consists of a central carbon atom, called the α **carbon**, linked to an amino group, a carboxylic acid group, a hydrogen atom, and a distinctive **R group** (also referred to as the **side chain**). With four different groups connected to the tetrahedral α-carbon atom, α-amino acids are **chiral**: they may exist in one or the other of two mirror-image forms, called the L isomer and the D isomer (**Figure 2.4**).

Only L isomers of amino acids (called L *amino acids*) are constituents of proteins. What is the basis for the preference for L amino acids? The answer has been lost to evolutionary history. It is possible that the preference for L over D amino acids was a consequence of a chance selection. However, there is evidence that L amino acids are slightly more soluble than a racemic mixture of D and L amino acids, which tend to form crystals. This small solubility difference could have been amplified over time so that the L isomer became dominant in solution.

Amino acids in solution at physiological pH (around 7.4) exist predominantly as **dipolar ions** (also called **zwitterions**). In the dipolar form, the amino group is protonated ($-NH_3^+$) and the carboxyl group is deprotonated ($-COO^-$). The ionization state of an amino acid varies with pH (**Figure 2.5**). In acid solution (e.g., pH 1), the amino group is protonated ($-NH_3^+$) and the carboxyl group is not dissociated ($-COOH$). As the pH is raised, the carboxylic acid is the first group to give up a proton, inasmuch as its pK_a is near 2. (The concept of pK_a was explained in Section 1.3.) The dipolar form persists until the pH approaches 9.5, when the protonated amino group loses a proton.

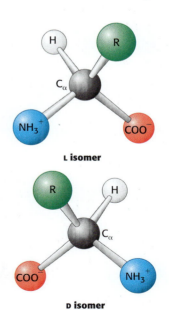

FIGURE 2.4 The L and D isomers of amino acids are mirror images of each other. The letter R refers to the side chain. Only the L amino acids are found in proteins.

FIGURE 2.5 The ionization state of amino acids is altered by changes in pH. At low pH, near the pK_a for the carboxylic acid, pK_1, the —COOH proton is lost from the fully protonated form. As the pH approaches physiological levels, the zwitterionic form predominates. At high pH, near the pK_a for the amino group, pK_2, one of the —NH$_3^+$ protons is lost to form the fully deprotonated species.

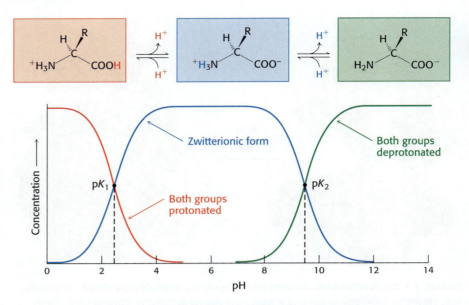

The diversity of amino acids arises from the different side chains

Twenty kinds of side chains varying in size, shape, charge, hydrogen-bonding capacity, hydrophobic character, and chemical reactivity are commonly found in proteins. Indeed, all proteins in all species—bacterial, archaeal, and eukaryotic—are constructed from the same set of 20 amino acids with only a few exceptions. This fundamental alphabet for the construction of proteins is several billion years old. The remarkable range of functions mediated by proteins results from the diversity and versatility of these 20 building blocks. Understanding how this alphabet is used to create the intricate three-dimensional structures that enable proteins to carry out so many biological processes is an exciting area of biochemistry, and one that we will return to in Section 2.7.

Although there are many ways to classify amino acids, we will sort these molecules into four groups, on the basis of the general chemical characteristics of their R groups:

1. Hydrophobic amino acids with nonpolar R groups

2. Polar amino acids with neutral R groups but the charge is not evenly distributed

3. Positively charged amino acids with R groups that have a positive charge at physiological pH

4. Negatively charged amino acids with R groups that have a negative charge at physiological pH

Hydrophobic amino acids. The simplest amino acid is glycine, which has a single hydrogen atom as its side chain. With two hydrogen atoms bonded to the α-carbon atom, glycine is unique in being achiral; that is, its mirror images are superimposable. Alanine, the next simplest amino acid, has a methyl group (—CH$_3$) as its side chain (**Figure 2.6**).

Larger hydrocarbon side chains are found in valine, leucine, and isoleucine. Methionine contains a largely aliphatic side chain that includes a thioether (—S—) group. The side chain of isoleucine includes an additional chiral center; only the isomer shown in Figure 2.6 is found in proteins. These aliphatic side chains are especially hydrophobic; that is, they tend to cluster together rather than contact water. The three-dimensional structures of water-soluble proteins are stabilized by this tendency of hydrophobic groups to come together (the hydrophobic effect, Section 1.3). The different sizes and shapes of these hydrocarbon side chains enable them to pack together to form compact structures

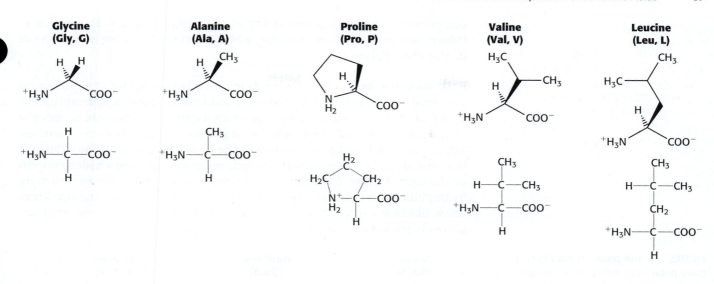

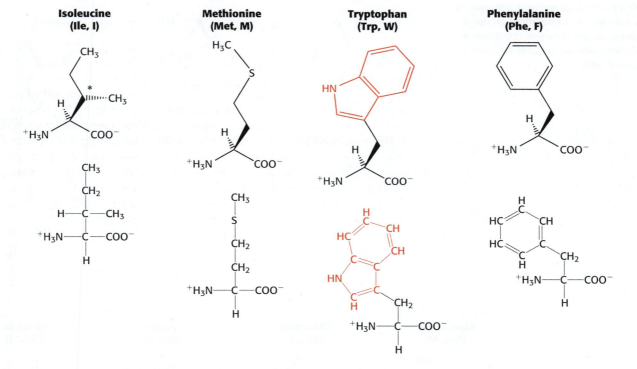

FIGURE 2.6 The hydrophobic amino acids have nonpolar R groups. For each amino acid, a stereochemically realistic formula (top) shows the geometric arrangement of bonds around atoms and a Fischer projection (bottom) shows all bonds as being perpendicular for a simplified representation. The additional chiral center in isoleucine is indicated by an asterisk. The indole group in the tryptophan side chain is shown in red.

with little empty space. Proline also has an aliphatic side chain, but it differs from other members of the set of 20 in that its side chain is bonded to both the nitrogen and the α-carbon atoms, yielding a pyrrolidine ring. Proline markedly influences protein architecture because its cyclic structure makes it more conformationally restricted than the other amino acids.

Two amino acids with relatively simple aromatic side chains are part of the fundamental repertoire. Phenylalanine, as its name indicates, contains a phenyl ring attached in place of one of the hydrogen atoms of alanine. Tryptophan has an indole group joined to a methylene (—CH_2—) group; the indole group comprises two fused rings containing an NH group (Figure 2.6, shown in red). Phenylalanine is purely hydrophobic, whereas tryptophan is less so because of its side chain NH group.

Polar amino acids. Seven amino acids are polar but uncharged at physiological pH. Three amino acids, serine, threonine, and tyrosine, contain hydroxyl groups (—OH) attached to a hydrophobic side chain (**Figure 2.7**). Serine can be thought of as a version of alanine with a hydroxyl group attached. Threonine resembles valine with a hydroxyl group in place of one of valine's methyl groups. Tyrosine is a version of phenylalanine with the hydroxyl group replacing a hydrogen atom on the aromatic ring. The hydroxyl group makes these amino acids much more hydrophilic (water loving) and reactive than their hydrophobic analogs. Threonine, like isoleucine, contains an additional asymmetric center; again, only one isomer is present in proteins.

FIGURE 2.7 The polar amino acids have polar but uncharged R groups at physiological pH. The additional chiral center in threonine is indicated by an asterisk. The carboxamide groups of asparagine and glutamine are shown in red. The imidazole group of histidine is shown in blue.

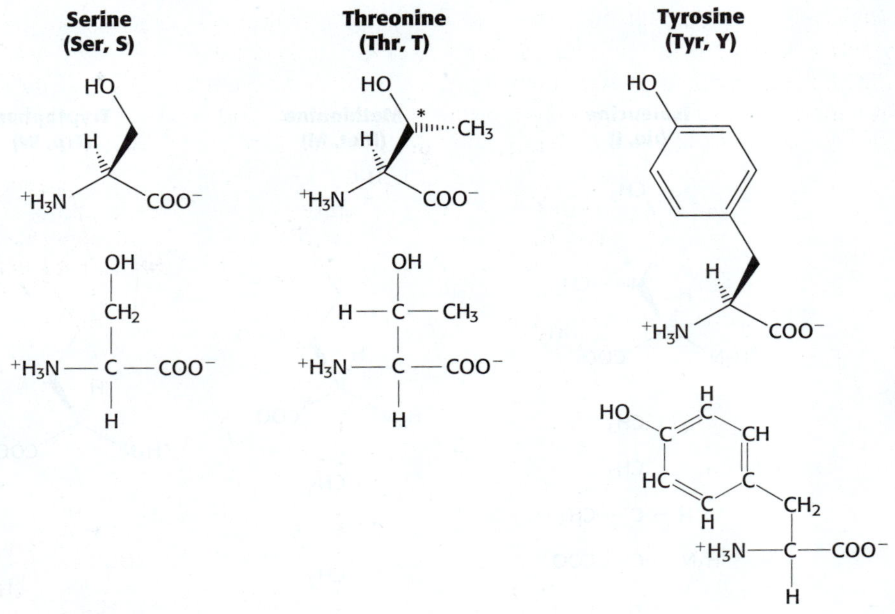

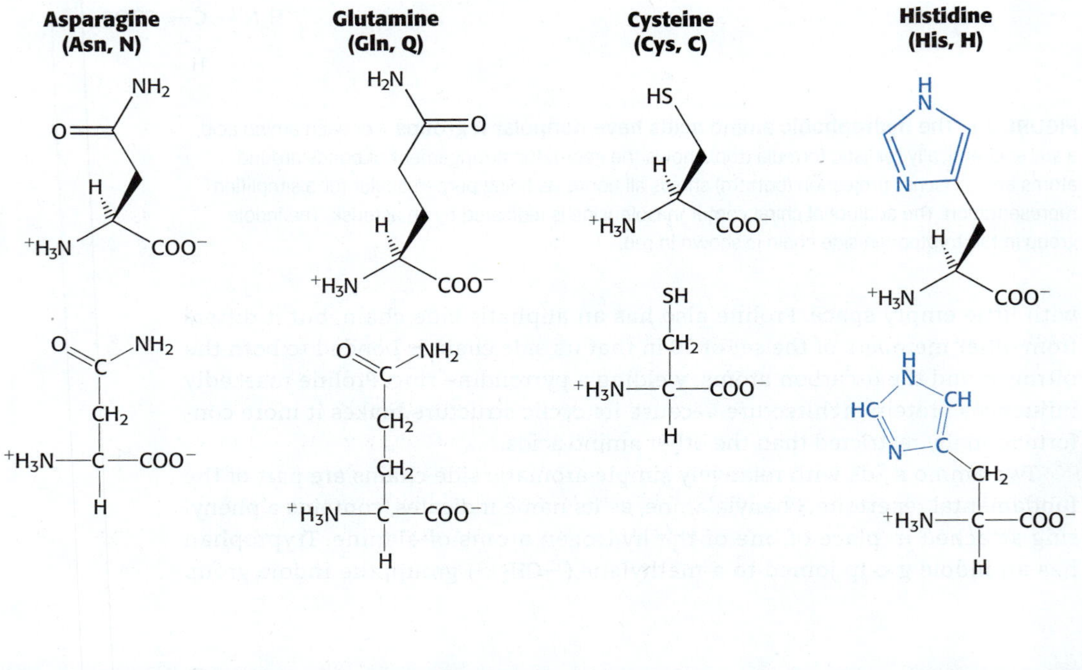

In addition, the set includes asparagine and glutamine, two amino acids that contain a terminal carboxamide, highlighted in red in Figure 2.7. The side chain of glutamine is one methylene group longer than that of asparagine.

Cysteine is structurally similar to serine but contains a sulfhydryl, or thiol (—SH), group in place of the hydroxyl group. The sulfhydryl group is much more reactive. Pairs of sulfhydryl groups may come together to form disulfide bonds, which are particularly important in stabilizing some proteins, as will be discussed shortly.

Histidine contains an imidazole group, an aromatic ring that is uncharged at pH 7 (Figure 2.7, shown in blue). However, with a pK_a value near 6, the imidazole group can be uncharged or positively charged near neutral pH, depending on its local environment (**Figure 2.8**). Histidine is often found in the active sites of enzymes, where the imidazole ring can bind and release protons in the course of enzymatic reactions.

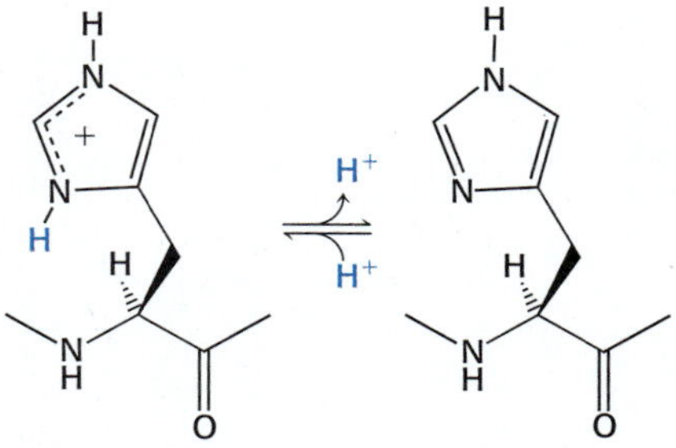

FIGURE 2.8 Histidine can bind or release protons near physiological pH.

Positively charged amino acids. We turn now to amino acids with complete positive charges that render them highly hydrophilic (**Figure 2.9**). Lysine and arginine have long side chains that terminate with groups that are positively charged at neutral pH. Lysine is capped by a primary amino group and arginine by a guanidinium group (Figure 2.9, shown in red).

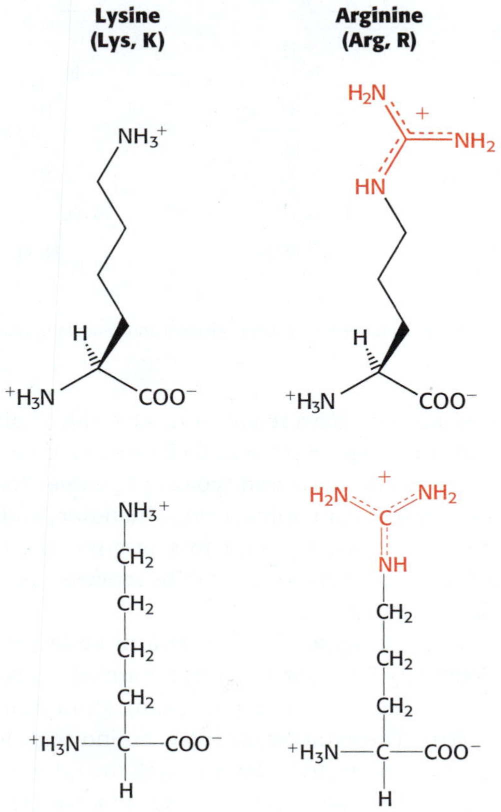

FIGURE 2.9 Two amino acids are positively charged at physiological pH. The guanidinium group in the arginine side chain is shown in red.

FIGURE 2.10 Two amino acids are negatively charged at physiological pH.

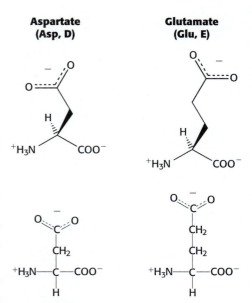

Negatively charged amino acids. This set of amino acids contains two with acidic side chains: aspartic acid and glutamic acid (**Figure 2.10**). These amino acids are charged derivatives of asparagine and glutamine (Figure 2.7), with a carboxylic acid in place of a carboxamide. Aspartic acid and glutamic acid are often called *aspartate* and *glutamate* to emphasize that, at physiological pH, their side chains usually lack a proton that is present in the acid form and hence are negatively charged. Nonetheless, these side chains can accept protons in some proteins, often with functionally important consequences.

TABLE 2.1 Typical pK_a values of ionizable groups in proteins

Group	Acid	⇌	Base	Typical pK_a^*
Terminal α-carboxyl group		⇌		3.1
Aspartic acid Glutamic acid		⇌		4.1
Histidine		⇌		6.0
Terminal α-amino group		⇌		8.0
Cysteine		⇌		8.3
Tyrosine		⇌		10.0
Lysine		⇌		10.4
Arginine		⇌		12.5

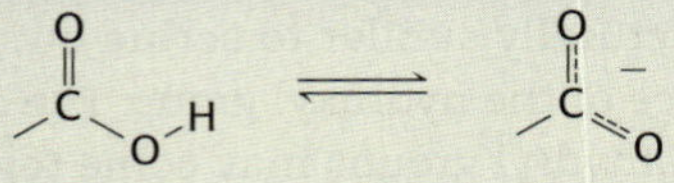

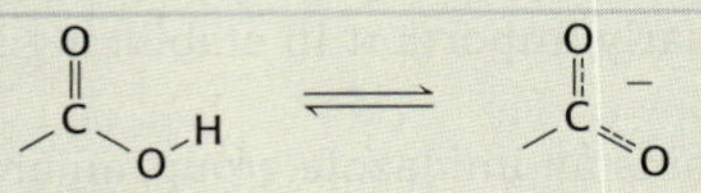

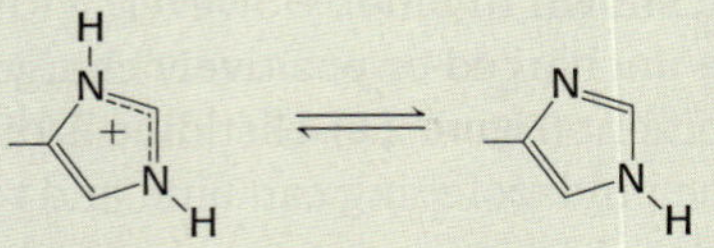

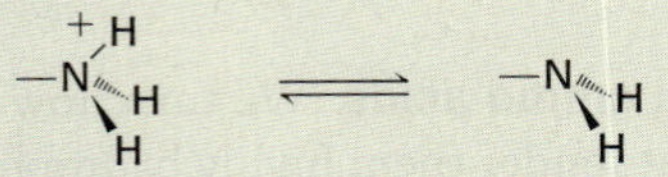

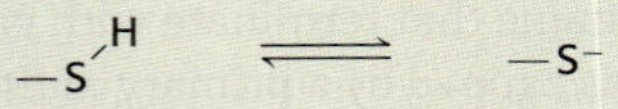

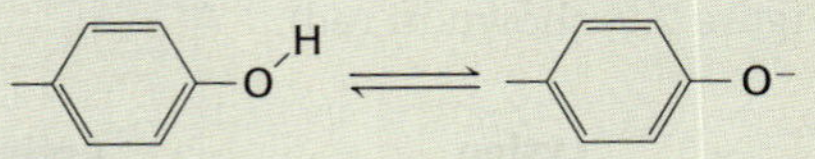

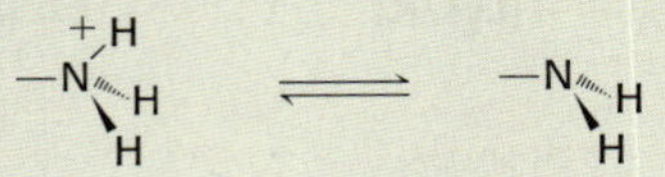

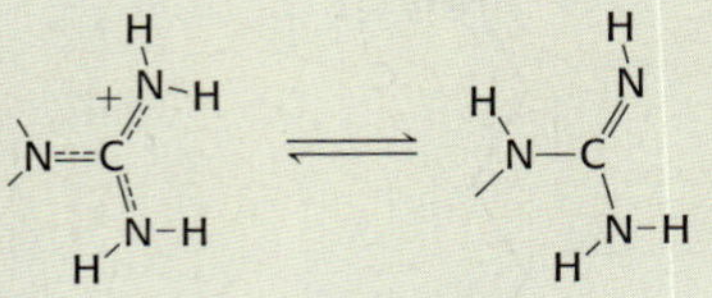

*pK_a values depend on temperature, ionic strength, and the microenvironment of the ionizable group

Seven of the 20 amino acids have readily ionizable side chains. These 7 amino acids are able to donate or accept protons to facilitate reactions as well as to form ionic bonds. **Table 2.1** gives equilibria and typical pK_a values for ionization of the side chains of tyrosine, cysteine, arginine, lysine, histidine, and aspartic and glutamic acids in proteins. Two other groups in proteins—the terminal α-amino group and the terminal α-carboxyl group—can be ionized, and typical pK_a values for these groups also are included in Table 2.1.

Amino acids are often designated by either a three-letter abbreviation or a one-letter symbol (**Table 2.2**). The abbreviations for amino acids are the first three letters of their names, except for asparagine (Asn), glutamine (Gln), isoleucine (Ile), and tryptophan (Trp). The symbols for many amino acids are the first letters of their names (e.g., G for glycine and L for leucine); the other symbols have been agreed on by convention. These abbreviations and symbols are an integral part of the vocabulary of biochemists.

Biochemists postulate several reasons why this set of amino acids is conserved across all species

How did this particular set of amino acids become the building blocks of proteins? First, as a set, they are diverse: their structural and chemical properties span a wide range, endowing proteins with the versatility to assume many functional roles. Second, many of these amino acids were probably available from prebiotic reactions; that is, from reactions that took place before

TABLE 2.2 Abbreviations for amino acids

Amino acid	Three-letter abbreviation	One-letter abbreviation	Amino acid	Three-letter abbreviation	One-letter abbreviation
Alanine	Ala	A	Methionine	Met	M
Arginine	Arg	R	Phenylalanine	Phe	F
Asparagine	Asn	N	Proline	Pro	P
Aspartic acid	Asp	D	Serine	Ser	S
Cysteine	Cys	C	Threonine	Thr	T
Glutamine	Gln	Q	Tryptophan	Trp	W
Glutamic acid	Glu	E	Tyrosine	Tyr	Y
Glycine	Gly	G	Valine	Val	V
Histidine	His	H	Asparagine or aspartic acid	Asx	B
Isoleucine	Ile	I			
Leucine	Leu	L	Glutamine or glutamic acid	Glx	Z
Lysine	Lys	K			

Homoserine

Serine

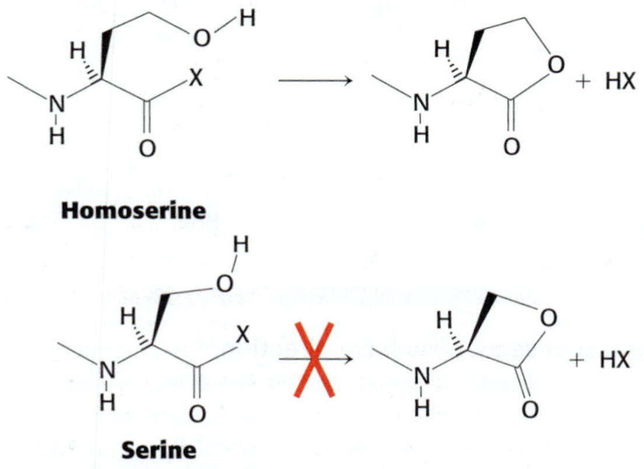

FIGURE 2.11 Some amino acids are not found in proteins because they are too reactive. Homoserine readily cyclizes to form a stable, five-membered ring, potentially resulting in peptide-bond cleavage. In contrast, serine cyclizes to a four-membered ring, which is significantly less stable.

the origin of life. Finally, other possible amino acids may have simply been too reactive. For example, amino acids such as homoserine and homocysteine tend to form five-membered cyclic forms that limit their use in proteins (**Figure 2.11**); the alternative amino acids that are found in proteins — serine and cysteine — do not readily cyclize, because the rings in their cyclic forms are too small to be stable.

❖ **SELF–CHECK QUESTION**

Which two amino acids have additional chiral centers besides their α-carbon atoms?

2.3 Primary Structure: Amino Acids Are Linked by Peptide Bonds to Form Polypeptide Chains

Proteins are linear polymers formed by linking the α-carboxyl group of one amino acid to the α-amino group of another amino acid. This type of linkage is called a **peptide bond** (or an **amide bond**). The formation of a dipeptide from two amino acids is accompanied by the loss of a water molecule (**Figure 2.12**). Under most conditions, the equilibrium of this reaction lies on

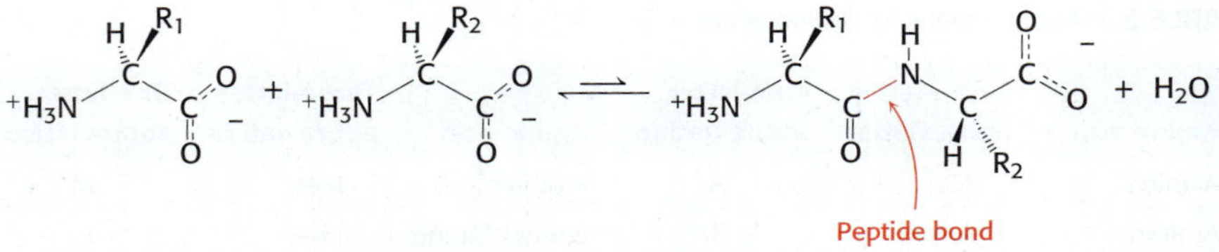

FIGURE 2.12 The formation of the peptide bond linking two amino acids is accompanied by the loss of a molecule of water.

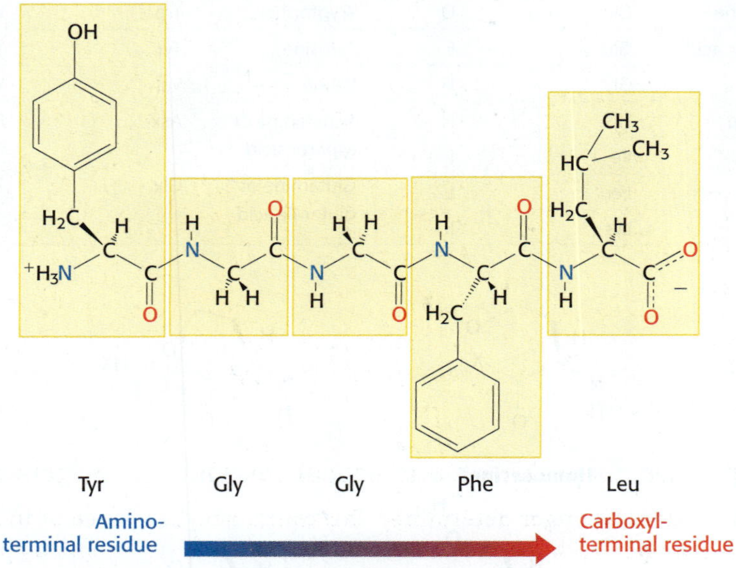

Tyr Gly Gly Phe Leu

Amino-terminal residue

Carboxyl-terminal residue

FIGURE 2.13 Amino acid sequences have direction. This illustration of the pentapeptide Tyr-Gly-Gly-Phe-Leu (YGGFL) shows the sequence from the amino terminus to the carboxyl terminus. This pentapeptide, Leu-enkephalin, is an opioid peptide that modulates the perception of pain. The reverse pentapeptide, Leu-Phe-Gly-Gly-Tyr (LFGGY), is a different molecule and has no such effects.

the side of hydrolysis rather than synthesis. Hence, the biosynthesis of peptide bonds requires an input of free energy. Nonetheless, peptide bonds are quite stable kinetically because the rate of hydrolysis is extremely slow; the lifetime of a peptide bond in aqueous solution in the absence of a catalyst approaches 1000 years.

A series of amino acids joined by peptide bonds forms a polypeptide chain, and each amino acid unit in a polypeptide is called a **residue**. A polypeptide chain has directionality because its ends are different: an α-amino group is present at one end and an α-carboxyl group at the other. The amino end is taken to be the beginning of a polypeptide chain; by convention, the sequence of amino acids in a polypeptide chain is written starting with the amino-terminal residue. Thus, in the polypeptide Tyr-Gly-Gly-Phe-Leu (YGGFL), tyrosine is the amino-terminal (N-terminal) residue and leucine is the carboxyl-terminal (C-terminal) residue (**Figure 2.13**). Leu-Phe-Gly-Gly-Tyr (LFGGY) is a different polypeptide, with different chemical properties.

A polypeptide chain consists of a regularly repeating part, called the **main chain** (or **backbone**), and a variable part, comprising the distinctive side chains (**Figure 2.14**). The polypeptide backbone is rich in hydrogen-bonding potential. Each residue contains a carbonyl group (C=O), which is a good hydrogen-bond acceptor, and, with the exception of proline, an NH group, which is a good hydrogen-bond donor. These groups interact

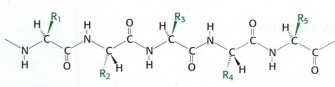

FIGURE 2.14 A polypeptide chain consists of a constant backbone and variable side chains. The side chains are shown in green.

with each other and with functional groups from side chains to stabilize particular structures, as we will discuss shortly.

Most natural polypeptide chains contain between 50 and 2000 amino acid residues and are commonly referred to as **proteins**. The largest single polypeptide known is the muscle protein titin, which consists of more than 27,000 amino acids. Polypeptide chains made of small numbers of amino acids are called **oligopeptides** (or, simply, **peptides**). The mean molecular weight of an amino acid residue is about 110 g mol^{-1}, and so the molecular weights of most proteins are between 5500 and 220,000 g mol^{-1}. We can also refer to the *molecular mass*, or the weight of a single molecule, of a protein in units of **daltons**; one dalton is equal to one atomic mass unit. A protein with a molecular weight of 50,000 g mol^{-1} has a molecular mass of 50,000 daltons, or 50 kDa (kilodaltons).

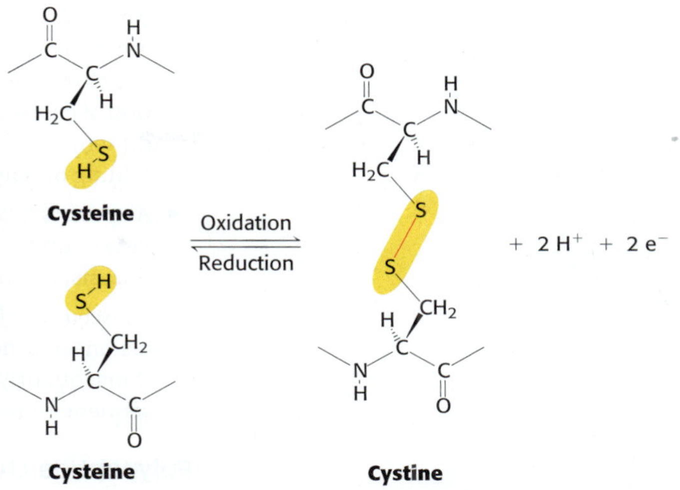

In some proteins, the linear polypeptide chain is cross-linked. The most common cross-links are **disulfide bonds**, formed by the oxidation of a pair of cysteine residues (**Figure 2.15**). The resulting unit of two linked cysteines is called *cystine*. Extracellular proteins often have several disulfide bonds, whereas intracellular proteins usually lack them. Rarely, nondisulfide cross-links derived from other side chains are present in proteins. For example, collagen fibers in connective tissue are strengthened in this way, as are fibrin blood clots.

FIGURE 2.15 The formation of a disulfide bond from two cysteine residues is an oxidation reaction.

Proteins have unique amino acid sequences specified by genes

In 1953, Frederick Sanger determined the amino acid sequence of insulin, a protein hormone (**Figure 2.16**). This work is a landmark in biochemistry because it showed for the first time that a protein has a precisely defined amino acid sequence consisting only of L amino acids linked by peptide bonds. This accomplishment inspired other scientists to carry out sequence studies of a variety of proteins. Currently, the complete amino acid sequences of millions of proteins are known. The amino acid sequence of a protein is referred to as its **primary structure**.

A series of incisive studies in the late 1950s and early 1960s revealed that the amino acid sequences of proteins are determined by the nucleotide sequences of genes. The sequence of nucleotides in DNA specifies a complementary sequence of nucleotides in RNA, which in turn specifies the amino acid sequence of a protein. In particular, each of the 20 amino acids of the repertoire is encoded by one or more specific sequences of three nucleotides (Section 8.6).

Knowing amino acid sequences is important for several reasons:

- *Knowledge of the sequence of a protein is usually essential to elucidating its function (e.g., the catalytic mechanism of an enzyme).* In fact, proteins with novel properties can be generated by varying the sequence of known proteins.

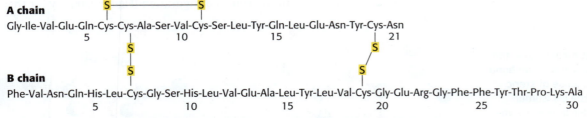

FIGURE 2.16 The primary structure of bovine insulin is its amino acid sequence. The cross links between the polypeptide chains, formed by disulfide bonds, are also shown.

- *Amino acid sequences determine the three-dimensional structures of proteins.* The amino acid sequence is the link between the genetic message in DNA and the three-dimensional structure that performs a protein's biological function. Analyses of the relationships between amino acid sequences and three-dimensional structures of proteins are uncovering the rules that govern the folding of polypeptide chains.

- *Alterations in amino acid sequence can lead to abnormal protein function and disease.* Severe and sometimes fatal diseases, such as sickle-cell anemia and cystic fibrosis, can result from a change in a single amino acid within a protein.

- *The sequence of a protein reveals much about its evolutionary history.* Proteins resemble one another in amino acid sequence only if they have a common ancestor. Consequently, molecular events in evolution can be traced from amino acid sequences; molecular paleontology is a flourishing area of research.

Polypeptide chains are flexible yet conformationally restricted

Examination of the geometry of the protein backbone reveals several important features:

- *The peptide bond is essentially planar* (**Figure 2.17**). Thus, for a pair of amino acids linked by a peptide bond, six atoms lie in the same plane: the α-carbon atom and CO group of the first amino acid and the NH group and α-carbon atom of the second amino acid. The nature of the chemical bonding within a peptide accounts for the bond's planarity.

- *The peptide bond resonates between a single bond and a double bond.* Because of this partial double-bond character, rotation about this bond is prevented, and conformation of the peptide backbone is constrained.

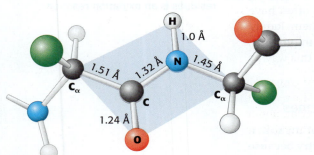

FIGURE 2.17 Peptide bonds are planar and have partial double-bond character. In a pair of linked amino acids, six atoms (C_α, C, O, N, H, and C_α) lie in a plane. Side chains are shown as green balls. The length of the peptide bond (1.32 Å) is between the expected values for a single bond and a double bond.

Peptide-bond resonance structures

The partial double-bond character is also expressed in the length of the bond between the CO and the NH groups. As shown in Figure 2.17, the C—N distance in a peptide bond is typically 1.32 Å, which is between the values expected for a C—N single bond (1.49 Å) and a C=N double bond (1.27 Å). Finally, the peptide bond is uncharged, allowing polymers of amino acids linked by peptide bonds to form tightly packed globular structures.

- *Almost all peptide bonds have a trans configuration.* In the trans configuration, the two α-carbon atoms are on opposite sides of the peptide bond. In the cis configuration, these groups are on the same side of the peptide bond. The preference for trans over cis can be explained by *steric repulsion*, the fact that two atoms will strongly oppose being brought together closer than their van der Waals contact distance (Section 1.3). Steric repulsion between groups attached to the α-carbon atoms hinders the formation of the cis configuration but does not arise in the trans configuration (**Figure 2.18**). By far the

FIGURE 2.18 For most amino acids, the trans configuration of the peptide bond is strongly favored because of steric clashes that arise in the cis isomer. Steric clashes are indicated by orange semicircles. The exception to this rule is found in the next figure.

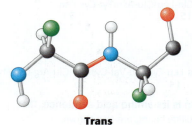

Trans

Cis

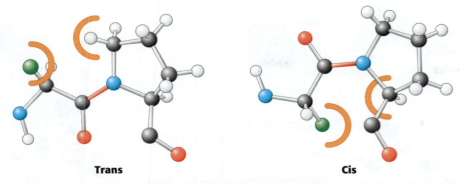

Trans   **Cis**

FIGURE 2.19 For peptide bonds between proline and its preceding residue, cis bonds are as likely as trans bonds. Neither configuration is favored because steric clashes exist for both the cis and trans forms.

most common cis peptide bonds are found between proline and its preceding amino acid. Because the nitrogen of proline is bonded to two tetrahedral carbon atoms, the steric differences between the trans and cis forms are minimal, and both forms occur (**Figure 2.19**).

- *The main chain bonds connected to the α carbon have rotational flexibility.* In contrast with the peptide bond, the bonds between the amino group and the α-carbon atom and between the α-carbon atom and the carbonyl group are pure single bonds. The two adjacent rigid peptide units can rotate about these bonds, taking on various orientations. The freedom of rotation about these two bonds of each amino acid allows proteins to fold in many different conformations.

The rotations about these bonds can be specified by **torsion angles**, or **dihedral angles** (**Figure 2.20**). The angle of rotation about the bond between the nitrogen and the α-carbon atoms is called *phi* (ϕ). The angle of rotation about the bond between the α-carbon and the carbonyl carbon atoms is called *psi* (ψ). A clockwise rotation about either bond as viewed from the nitrogen atom toward the α-carbon atom or from the α-carbon atom toward the carbonyl group corresponds to a positive value. The ϕ and ψ angles determine the path of the polypeptide chain.

Despite their rotational flexibility, not all values of ϕ and ψ are possible. In 1963, Gopalasamudram Ramachandran demonstrated that many combinations of ϕ and ψ are forbidden because of steric collisions between atoms. The allowed values can be visualized on a two-dimensional plot called a **Ramachandran plot**

(A)   (B)   (C)

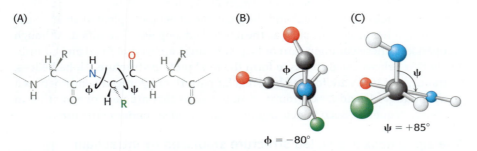

$\phi = -80°$ $\psi = +85°$

FIGURE 2.20 The structure of each amino acid in a polypeptide adjusts by rotation about two single bonds. (A) Phi (ϕ) is the angle of rotation about the bond between the nitrogen and the α-carbon atoms, whereas psi (ψ) is the angle of rotation about the bond between the α-carbon and the carbonyl carbon atoms. (B) A view down the bond between the nitrogen and the α-carbon atoms shows how ϕ is measured. (C) A view down the bond between the α-carbon and the carbonyl carbon atoms shows how ψ is measured.

FIGURE 2.21 A Ramachandran plot shows that not all φ and ψ values are possible without collisions between atoms. The most favorable regions are shown in dark blue; borderline regions are shown in light blue. The structure on the right is disfavored because of steric clashes.

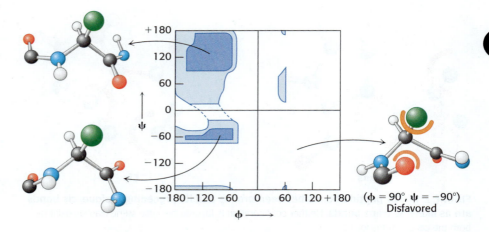

(φ = 90°, ψ = −90°)
Disfavored

(**Figure 2.21**). Three-quarters of the possible (φ, ψ) combinations are excluded simply by local steric clashes. Steric exclusion, the fact that two atoms cannot be in the same place at the same time, can be a powerful organizing principle.

The ability of biological polymers such as proteins to fold into well-defined structures is remarkable thermodynamically. An unfolded polymer exists as a random coil: each copy of an unfolded polymer will have a different conformation, yielding a mixture of many possible conformations. The favorable entropy associated with a mixture of many conformations opposes folding and must be overcome by interactions favoring the folded form. Thus, highly flexible polymers with a large number of possible conformations do not fold into unique structures. The rigidity of the peptide unit and the restricted set of allowed φ and ψ angles limits the number of structures accessible to the unfolded form sufficiently to allow protein folding to take place.

❖ SELF–CHECK QUESTION

Ramachandran plots of 18 amino acids exhibit allowed regions similar to those shown in Figure 2.21. However, the plots for two amino acids exhibit strikingly different patterns. Which two, and why?

2.4 Secondary Structure: Polypeptide Chains Can Fold into Regular Structures

Ramachandran demonstrated that many conformations of a polypeptide backbone are disallowed on the basis of steric exclusion. Do the allowed conformations permit a polypeptide chain to fold into a regularly repeating structure? In 1951, Linus Pauling, Robert Corey, and Herman Branson defined two periodic structures, the α helix (alpha helix) and the β pleated sheet (beta pleated sheet). Subsequently, other structures such as the β turn and loop were identified. Although not periodic, these common turn or loop structures are well-defined and contribute with α helices and β sheets to form the final protein structure. Alpha helices, β strands, and turns are formed by a regular pattern of hydrogen bonds between the peptide N—H and C=O groups of amino acids that are near one another in the linear sequence. Such folded segments are called **secondary structure**.

The alpha helix is a coiled structure stabilized by intrachain hydrogen bonds

In evaluating potential structures, Pauling, Corey, and Branson considered which conformations of peptides were sterically allowed and which most fully exploited the hydrogen-bonding capacity of the backbone NH and CO groups. The first of their proposed structures, the α **helix**, is a rodlike structure (**Figure 2.22**). A tightly

HERMAN BRANSON While Pauling and Corey often receive the most recognition for the discovery of the α helix, a third scientist, Herman Branson, also made key contributions to the discovery and was a coauthor on the seminal paper. Dr. Branson, a physics professor at Howard University who was a visiting scientist in Linus Pauling's lab at the California Institute of Technology in 1948, performed computational analysis using Pauling's x-ray measurements to narrow the possible structures down to the α helix and a looser structure called the gamma helix. Following this work, his scientific career continued for two more decades, after which he served as the president at two historically Black universities, Central State University from 1968 to 1970, and Lincoln University from 1970 to 1985.

(A)   (B)   (C)

(D)

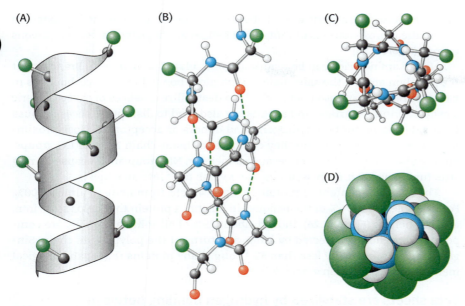

FIGURE 2.22 **Various depictions of an α helix reveal several of its key features.** (A) A ribbon depiction shows the α-carbon atoms and side chains (green). (B) A side view of a ball-and-stick version depicts the hydrogen bonds (dashed lines) between NH and CO groups. (C) An end view shows the coiled backbone as the inside of the helix and the side chains (green) projecting outward. (D) A space-filling view of part C shows the tightly packed interior core of the helix.

coiled backbone forms the inner part of the rod and the side chains extend outward in a helical array.

The α helix is stabilized by hydrogen bonds between the NH and CO groups of the main chain. In particular, the CO group of each amino acid forms a hydrogen bond with the NH group of the amino acid that is situated four residues ahead in the sequence (**Figure 2.23**). Thus, except for amino acids near the ends of an α helix, all the main-chain CO and NH groups are hydrogen bonded.

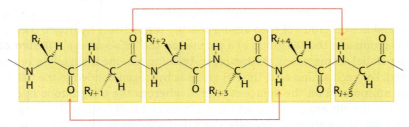

FIGURE 2.23 **In an α helix, hydrogen bonds form between the CO group of residue *i* and the NH group of residue *i* + 4.**

Each residue in an α helix is related to the next one by a *rise* (or *translation*) of 1.5 Å along the helix axis and a rotation of 100 degrees; this gives 3.6 amino acid residues per turn of the helix. Therefore, amino acids spaced three and four apart in the sequence are spatially quite close to one another in an α helix. In contrast, amino acids spaced two apart in the sequence are situated on opposite sides of the helix and so are unlikely to make contact. The *pitch* of the α helix is the length of one complete turn along the helix axis and is equal to the product of the rise (1.5 Å) and the number of residues per turn (3.6), or 5.4 Å.

The direction of rotation about the axis of an α helix (called *screw sense*) can be right-handed (clockwise) or left-handed (counterclockwise). While the Ramachandran plot reveals that both the right-handed and the left-handed helices are among allowed conformations (**Figure 2.24**), essentially all α helices found in proteins are right-handed. Right-handed helices are energetically more favorable

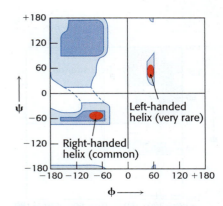

FIGURE 2.24 **Both right- and left-handed helices lie in regions of allowed conformations in the Ramachandran plot.** However, essentially all α helices in proteins are right-handed.

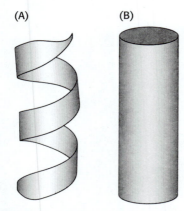

(A)　　　(B)

FIGURE 2.25 There are two common ways to draw α helices. (A) A ribbon depiction. (B) A cylindrical (rod) depiction.

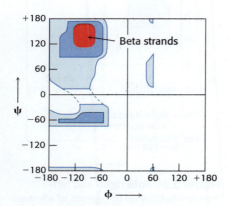

FIGURE 2.26 Ferritin, an iron-storage protein, is composed largely of α helices. [Drawn from 1AEW.pdb.]

INTERACT with this model in
 Achieve

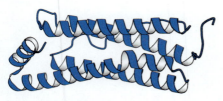

FIGURE 2.27 β strands occupy an allowed region of the Ramachandran plot.

because there is less steric clash between the side chains and the backbone. In schematic representations of proteins, α helices are depicted as twisted ribbons or rods (**Figure 2.25**).

Not all amino acids can be readily accommodated in an α helix. Branching at the β-carbon atom—the side chain carbon atom bonded to the α carbon—as in valine, threonine, and isoleucine, tends to destabilize α helices because of steric clashes. Serine, aspartate, and asparagine also tend to disrupt α helices because their side chains contain hydrogen-bond donors or acceptors in close proximity to the main chain, where they compete for main-chain NH and CO groups. Proline also is a helix breaker because it lacks an NH group and because its ring structure prevents it from assuming the φ value to fit into an α helix.

The α-helical content of proteins ranges widely, from none to almost 100%. For example, about 75% of the residues in ferritin, a protein that helps store iron, are in α helices (**Figure 2.26**). Indeed, about 25% of all soluble proteins are composed of α helices connected by loops and turns of the polypeptide chain. Single α helices are usually less than 45 Å long. Many proteins that span biological membranes also contain α helices.

Beta sheets are stabilized by hydrogen bonding between polypeptide strands

Pauling and Corey proposed another periodic structural motif, which they named the β **pleated sheet** (β because it was the second structure they elucidated, the α helix having been the first). The β pleated sheet (or, more simply, the β sheet) differs markedly from the rodlike α helix. It is composed of two or more polypeptide chains called β **strands**. A β strand is almost fully extended rather than being tightly coiled as in the α helix. A range of extended structures are sterically allowed (**Figure 2.27**).

The distance between adjacent amino acids along a β strand is approximately 3.5 Å, in contrast with a distance of 1.5 Å along an α helix. The side chains of adjacent amino acids point in opposite directions (**Figure 2.28**). A β sheet is formed by linking two or more β strands lying next to one another through hydrogen bonds. Adjacent strands in a β sheet can run in opposite directions (antiparallel β sheet) or in the same direction (parallel β sheet). In the antiparallel arrangement, the NH group and the CO group of each amino acid are respectively hydrogen bonded to the CO group and the NH group of a partner on the adjacent chain (**Figure 2.29**). In the parallel arrangement, the hydrogen-bonding scheme is slightly more complicated. For each amino acid, the NH group is hydrogen bonded to the CO group of one amino acid on the adjacent strand, whereas the CO group is hydrogen bonded to the NH group on the amino acid two residues farther along the chain (**Figure 2.30**). Many strands, typically 4 or 5 but as many as 10 or more, can come together in β sheets. Such β sheets can be purely antiparallel, purely parallel, or mixed (**Figure 2.31**).

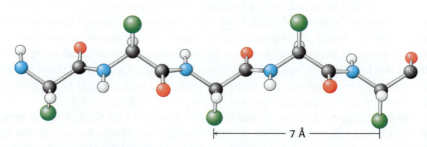

FIGURE 2.28 The side chains of a β strand project alternately above and below the plane of the strand. In this extended conformation, the side chains (shown in green) on the same face of the strand are positioned about 7 Å apart.

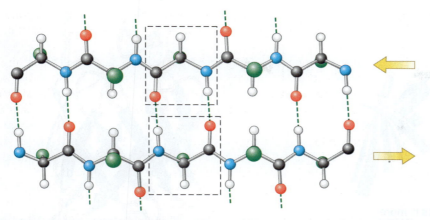

FIGURE 2.29 In an antiparallel β sheet, adjacent β strands run in opposite directions. Note that hydrogen bonds between NH and CO groups connect each amino acid to a single amino acid on an adjacent strand, stabilizing the structure.

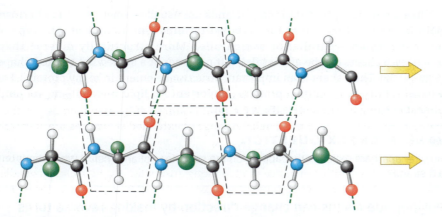

FIGURE 2.30 In a parallel β sheet, adjacent β strands run in the same direction. Note that hydrogen bonds connect each amino acid on one strand with two different amino acids on the adjacent strand.

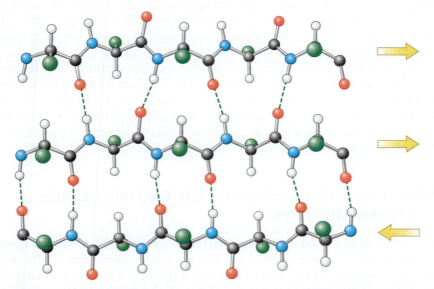

FIGURE 2.31 In a mixed β sheet, the β strands have a combination of parallel and antiparallel pairings.

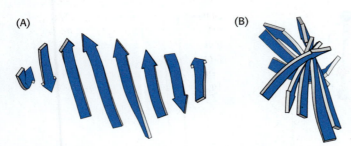

FIGURE 2.32 The twisted shape of some β sheets is more evident in a schematic depiction of β strands. (A) β strands are represented by broad arrows pointing towards their carboxy-terminal end. (B) When rotated by 90 degrees, the schematic view more clearly illustrates the twist in the β sheet.

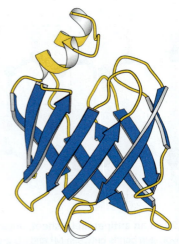

FIGURE 2.33 Fatty acid-binding protein is rich in β sheets. [Drawn from 1FTP.pdb.]

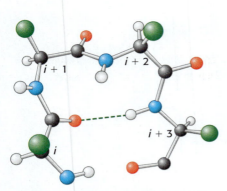

FIGURE 2.34 In a reverse turn, the CO group of residue *i* of the polypeptide chain is hydrogen bonded to the NH group of residue *i* + 3. This hydrogen bond stabilizes the turn.

In schematic representations, β strands are usually depicted by broad arrows pointing in the direction of the carboxyl-terminal end to indicate the type of β sheet formed — parallel or antiparallel. More structurally diverse than α helices, β sheets can be almost flat, but most adopt a somewhat twisted shape (**Figure 2.32**). The β sheet is an important structural element in many proteins. For example, fatty acid-binding proteins, important for lipid metabolism, are built almost entirely from β sheets (**Figure 2.33**).

❖ SELF–CHECK QUESTION

How far apart are residues that appear on the same face of an α helix? What about on a β strand?

Polypeptide chains can change direction by making reverse turns and loops

As we will explore in Section 2.5, most proteins have compact, globular shapes owing to reversals in the direction of their polypeptide chains. Many of these reversals are accomplished by a common structural element called the **reverse turn** (also known as the **β turn** or **hairpin turn**), illustrated in **Figure 2.34**. In many reverse turns, the CO group of residue i of a polypeptide is hydrogen bonded to the NH group of residue i+3. This interaction stabilizes abrupt changes in direction of the polypeptide chain.

In other cases, more-elaborate structures are responsible for chain reversals. These structures are called **loops**. Unlike α helices and β strands, loops do not have regular, periodic structures. Nonetheless, loop structures are often rigid and well-defined (**Figure 2.35**). Turns and loops invariably lie on the surfaces of proteins and often participate in interactions between proteins and other molecules.

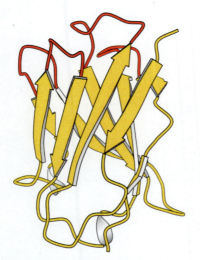

FIGURE 2.35 Loops are found on protein surfaces. A part of an antibody molecule has surface loops (shown in red) that mediate interactions with other molecules. [Drawn from 7FAB.pdb.]

2.5 Tertiary Structure: Proteins Can Fold into Globular or Fibrous Structures

Let us now examine how amino acids are grouped together in a complete protein. X-ray crystallographic, nuclear magnetic resonance (NMR), and cryo-electron microscopic (cryo-EM) studies (Section 4.5) have revealed the detailed three-dimensional structures of thousands of proteins. We begin here with an examination of myoglobin, the first protein to be seen in atomic detail,

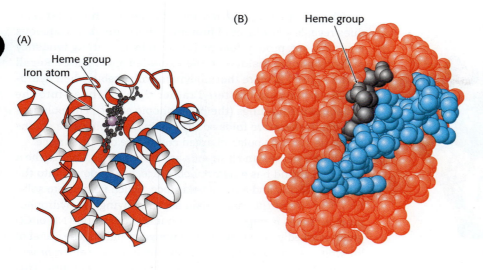

FIGURE 2.36 Myoglobin folds into a compact structure. (A) A ribbon diagram shows that the protein consists largely of α helices. (B) A space-filling model in the same orientation shows how tightly packed the folded protein is. Notice that the heme group is nestled into a crevice in the compact protein with only an edge exposed. One helix is blue to allow comparison of the two structural depictions. [Drawn from 1A6N.pdb.]

in work reported by John Kendrew and colleagues in 1958. Myoglobin, the oxygen storage protein in muscle, is a single polypeptide chain of 153 amino acids (Chapter 3) that forms an extremely compact molecule. Its overall dimensions are $45 \times 35 \times 25$ Å, an order of magnitude less than if it were fully stretched out (**Figure 2.36**). About 70% of the main chain is folded into eight α helices, and much of the rest of the chain forms turns and loops between helices.

INTERACT with the models shown in Figures 2.35–2.42 in ≋ Achie√e

The folding of the main chain of myoglobin, like that of most other proteins, is complex and devoid of symmetry. The overall course of the polypeptide chain of a protein is referred to as its **tertiary structure**. A unifying principle emerges from the distribution of side chains. Strikingly, the interior consists almost entirely of nonpolar residues such as leucine, valine, methionine, and phenylalanine (**Figure 2.37**). Charged residues such as aspartate, glutamate, lysine, and arginine are absent from the inside of myoglobin. The only polar residues inside are two histidine residues, which play critical roles in binding iron and oxygen. The outside of myoglobin, on the other hand, consists of both polar and nonpolar residues. The space-filling model shows that there is very little empty space inside.

Globular proteins form tightly packed structures

The tight packing of myoglobin into a highly compact structure, its lack of symmetry, and its solubility in water are characteristics of **globular proteins**, which perform a wide array of important functions, including regulatory, signaling, and enzymatic activities.

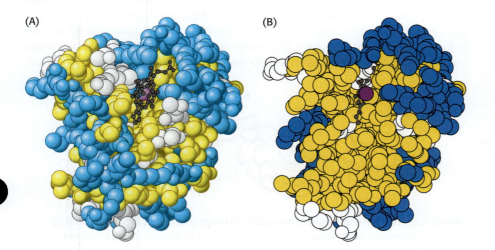

FIGURE 2.37 Hydrophobic amino acids are preferentially distributed in the interior of myoglobin. (A) A space-filling model of myoglobin with hydrophobic amino acids shown in yellow, charged amino acids shown in blue, and others shown in white. Notice that the surface of the molecule has many charged amino acids, as well as some hydrophobic amino acids. (B) In this cross-sectional view, notice that mostly hydrophobic amino acids are found on the inside of the structure, whereas the charged amino acids are found on the protein surface. [Drawn from 1MBD.pdb.]

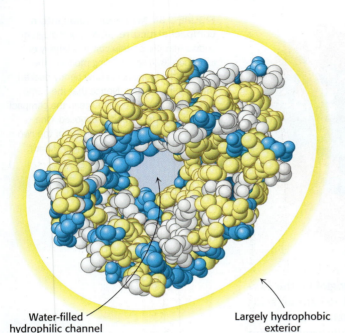

Water-filled
hydrophilic channel

Largely hydrophobic
exterior

FIGURE 2.38 Porin, found in the hydro-phobic environment of the membrane, exhibits an "inside out" amino acid distribution. The outside of porin (which contacts hydrophobic groups in membranes) is covered largely with hydrophobic residues, whereas the center includes a water-filled channel lined with charged and polar amino acids. [Drawn from 1PRN.pdb.]

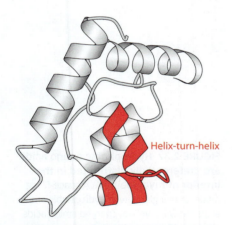

Helix-turn-helix

FIGURE 2.39 The helix-turn-helix motif is a supersecondary structural element found in many DNA-binding proteins. [Drawn from 1LMB.pdb.]

The contrasting distribution of polar and nonpolar residues reveals a key facet of protein architecture: in an aqueous environment, protein folding is driven by the strong tendency of hydrophobic residues to be excluded from water. Recall that a system is more thermodynamically stable when hydrophobic groups are clustered rather than extended into the aqueous surroundings (the hydrophobic effect). The polypeptide chain therefore folds so that its hydrophobic side chains are buried and its polar, charged chains are on the surface.

Many α helices and β strands are amphipathic; that is, the α helix or β strand has a hydrophobic face, which points into the protein interior, and a more polar face, which points into solution. The fate of the main chain accompanying the hydrophobic side chains is important, too. An unpaired peptide NH or CO group markedly prefers water to a nonpolar milieu. The secret of burying a segment of main chain in a hydrophobic environment is to pair all the NH and CO groups by hydrogen bonding. This pairing is neatly accomplished in an α helix or β sheet.

Van der Waals interactions between tightly packed hydrocarbon side chains also contribute to the stability of proteins. We can now understand why the set of 20 amino acids contains several that differ subtly in size and shape. They provide a palette from which to choose to fill the interior of a protein neatly and thereby maximize van der Waals interactions, which require intimate contact.

Some proteins that span biological membranes are "the exceptions that prove the rule" because they have the reverse distribution of hydrophobic and hydrophilic amino acids. For example, consider porins—proteins found in the outer membranes of many bacteria (**Figure 2.38**). Membranes are built largely of hydrophobic chains. Thus, porins are covered on the outside largely with hydrophobic residues that interact with the neighboring alkane chains. In contrast, the center of the protein contains many charged and polar amino acids that surround a water-filled channel running through the middle of the protein. Because porins function in hydrophobic environments, they are "inside out" relative to proteins that function in aqueous solution.

Certain combinations of secondary structure are present in many proteins and frequently exhibit similar functions. These combinations are called **motifs** (or **supersecondary structures**). For example, an α helix separated from another α helix by a turn, called a *helix-turn-helix* motif, is found in many proteins that bind DNA (**Figure 2.39**).

Some polypeptide chains fold into two or more compact regions that may be connected by a flexible segment of polypeptide chain, rather like pearls on a string. These compact globular units, called **domains**, range in size from about 30 to 400 amino acid residues. For example, the extracellular part of CD4, a protein on the surface of certain cells of the immune system, comprises four similar domains of approximately 100 amino acids each (**Figure 2.40**). Proteins may have domains in common even if their overall tertiary structures are different.

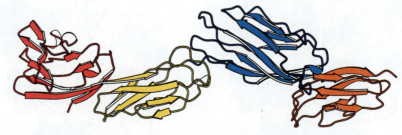

FIGURE 2.40 Domains are compact regions within proteins that are connected by flexible polypeptide segments. The cell-surface protein CD4 consists of four similar domains. [Drawn from 1WIO.pdb.]

❖ **SELF–CHECK QUESTION**

Proteins that span biological membranes often contain α helices. Given that the insides of membranes are highly hydrophobic, predict what type of amino acids would be in such an α helix. Why is an α helix particularly suited to existence in the hydrophobic environment of a membrane interior?

Fibrous proteins form extended structures that provide support for cells and tissues

In contrast to globular proteins, **fibrous proteins** form long, extended structures that feature repeated sequences. These proteins, such as α-keratin and collagen, use special types of helices that facilitate the formation of long fibers that serve a structural role.

α-Keratin, which is an essential component of wool, hair, and skin, consists of two right-handed α helices intertwined to form a type of left-handed superhelix called an *α-helical coiled coil*. α-Keratin is a member of a superfamily of proteins referred to as *coiled-coil proteins* (**Figure 2.41**). In these proteins, two or more α helices can entwine to form a very stable structure, which can have lengths of 1000 Å (100 nm, or 0.1 µm) or more. There are approximately 60 members of this family in humans, including intermediate filaments, proteins that contribute to the cell cytoskeleton (internal scaffolding in a cell), and the muscle proteins myosin and tropomyosin. Members of this family are characterized by a central region of 300 amino acids that contains imperfect repeats of a sequence of 7 amino acids called a *heptad repeat*.

(A)

(B)

FIGURE 2.41 The α-helical coiled coil adopts an extended structure. (A) A space-filling model. (B) A ribbon diagram. The two helices wind around one another to form a superhelix. Such structures are found in many proteins, including keratin in hair, quills, claws, and horns. [Drawn from 1C1G.pdb.]

The two helices in α-keratin associate with each other by weak interactions such as van der Waals forces and ionic interactions. The left-handed supercoil alters the two right-handed α helices such that there are 3.5 residues per turn instead of 3.6. As a result, the pattern of side-chain interactions can be repeated every seven residues, forming the heptad repeats. Two helices with such repeats are able to interact with one another if the repeats are complementary (**Figure 2.42**). For example, the repeating residues may be hydrophobic, allowing van der Waals interactions, or have opposite charge, allowing ionic interactions. In addition, the two helices may be linked by disulfide bonds formed by neighboring cysteine residues.

The bonding of the helices accounts for the physical properties of wool, an example of an α-keratin. Wool is extensible and can be stretched to nearly twice its length because the α helices stretch, breaking the weak interactions between neighboring helices. However, the covalent disulfide bonds resist breakage and return the fiber to its original state once the stretching force is released. The number of disulfide bond cross-links further defines the fiber's properties. Hair and wool, having fewer cross-links, are flexible. Horns, claws, and hooves, having more cross-links, are much harder.

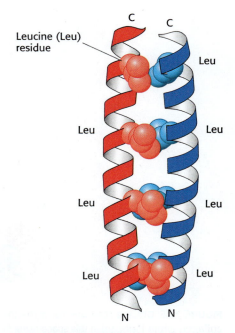

FIGURE 2.42 Heptad repeats from each helix in a coiled-coil protein interact with one another. Every seventh residue in each helix is leucine. The two helices are held together by van der Waals interactions primarily between the leucine residues. [Drawn from 2ZTA.pdb.]

13
-Gly-Pro-Met-Gly-Pro-Ser-Gly-Pro-Arg-
22
-Gly-Leu-Hyp-Gly-Pro-Hyp-Gly-Ala-Hyp-
31
-Gly-Pro-Gln-Gly-Phe-Gln-Gly-Pro-Hyp-
40
-Gly-Glu-Hyp-Gly-Glu-Hyp-Gly-Ala-Ser-
49
-Gly-Pro-Met-Gly-Pro-Arg-Gly-Pro-Hyp-
58
-Gly-Pro-Hyp-Gly-Lys-Asn-Gly-Asp-Asp-

FIGURE 2.43 The amino acid sequence of a part of a collagen chain reveals that every third residue is a glycine. Proline and hydroxyproline (Hyp) also are abundant.

A different type of helix is present in collagen, the most abundant protein of mammals. Collagen is the main fibrous component of skin, bone, tendon, cartilage, and teeth. This extracellular protein is a rod-shaped molecule, about 3000 Å long and only 15 Å in diameter. It contains three helical polypeptide chains, each nearly 1000 residues long. Glycine appears at every third residue in the amino acid sequence, and the sequence glycine-proline-hydroxyproline recurs frequently (**Figure 2.43**). Hydroxyproline is a derivative of proline that has a hydroxyl group in place of one of the hydrogen atoms on the pyrrolidine ring.

The collagen helix has properties different from those of the α helix. Hydrogen bonds within a strand are absent, and each strand adopts an extended conformation (**Figure 2.44**). The helix is stabilized by steric repulsion of the pyrrolidine rings of the proline and hydroxyproline residues. The pyrrolidine rings keep out of each other's way when the polypeptide chain assumes its helical form, which has about three residues per turn. Three strands wind around one another to form what is called a *superhelical cable*, a structure that is stabilized by hydrogen bonds between strands. The hydrogen bonds form between the peptide NH groups of glycine residues and the CO groups of residues on the other chains. The hydroxyl groups of hydroxyproline residues also participate in hydrogen bonding.

FIGURE 2.44 A single strand of a collagen triple helix adopts an extended conformation.

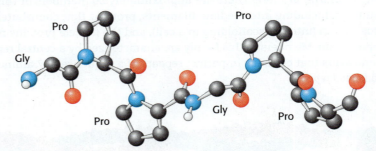

The inside of the triple-stranded helical cable is very crowded and accounts for the requirement that glycine be present at every third position on each strand (**Figure 2.45A**). The only residue that can fit in an interior position is glycine. The amino acid residue on either side of glycine is located on the outside of the cable, where there is room for the bulky rings of proline and hydroxyproline residues (**Figure 2.45B**).

The importance of the positioning of glycine inside the triple helix is illustrated in the disorder osteogenesis imperfecta, also known as brittle bone disease. In this condition, which can vary from mild to very severe, other amino acids replace the internal glycine residue. This replacement leads to a delayed and improper folding of collagen. The most serious symptom is severe bone fragility. Defective collagen in the eyes causes the whites of the eyes, or *sclera*, to have a blue tint.

(A)

(B)

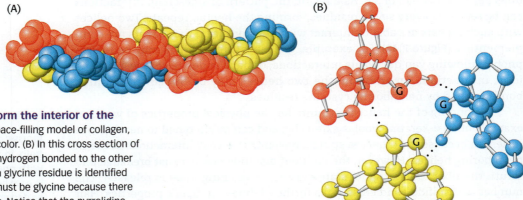

FIGURE 2.45 Glycine residues form the interior of the collagen triple helix. (A) In this space-filling model of collagen, each strand is shown in a different color. (B) In this cross section of a model of collagen, each strand is hydrogen bonded to the other two strands. The α-carbon atom of a glycine residue is identified by the letter G. Every third residue must be glycine because there is no space in the center of the helix. Notice that the pyrrolidine rings of the proline residues are on the outside.

2.6 Quaternary Structure: Polypeptide Chains Can Assemble into Multisubunit Structures

Four levels of structure are cited in discussions of protein architecture. So far, we have considered three: primary, secondary, and tertiary. We now turn to proteins containing more than one polypeptide chain. Such proteins exhibit a fourth level of structural organization. Each polypeptide chain in such a protein is called a **subunit**. The spatial arrangement of subunits and the nature of their interactions is called **quaternary structure**.

The simplest sort of quaternary structure is a **homodimer**, consisting of two identical subunits. For example, this organization is present in the DNA-binding protein Cro, found in a bacterial virus called *bacteriophage* λ (**Figure 2.46**). However, more-complicated quaternary structures are also common. More than one type of subunit can be present, often in variable numbers. For example, human hemoglobin, the oxygen-carrying protein in blood, consists of two subunits of one type (designated α) and two subunits of another type (designated β), as illustrated in **Figure 2.47**. Thus, the hemoglobin molecule exists as an $\alpha_2\beta_2$ tetramer. Subtle changes in the arrangement of subunits within the hemoglobin molecule allow it to carry oxygen from the lungs to tissues with great efficiency (Chapter 3).

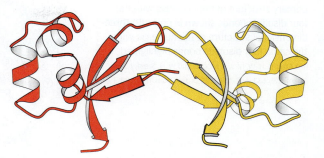

FIGURE 2.46 Quaternary structure refers to the arrangement of multiple polypeptide chains. The Cro protein of bacteriophage λ adopts the simplest type of quaternary structure, a homodimer (two identical subunits). [Drawn from 5CRO.pdb.]

INTERACT with the models shown in Figure 2.46 and Figure 2.47 in
≋ Achie√e

(A)   (B)

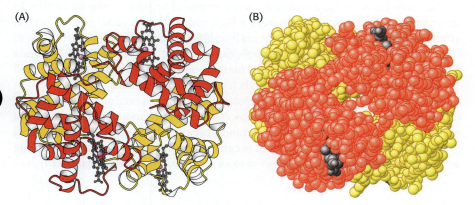

FIGURE 2.47 Human hemoglobin is composed of a $\alpha_2\beta_2$ tetramer. The structure of the two α subunits (red) is similar but not identical to that of the two β subunits (yellow). The molecule contains four heme groups (gray with the iron atom shown in purple). (A) The ribbon diagram highlights the similarity of the subunits and shows they are composed mainly of α helices. (B) The space-filling model illustrates how the heme groups occupy crevices in the protein. [Drawn from 1A3N.pdb.]

Viruses make the most of a limited amount of genetic information by forming coats that use the same kind of subunit repetitively in a symmetric array. For example, the coat of rhinovirus, the virus that causes the common cold, includes 60 copies of each of four subunits (**Figure 2.48**). The subunits come together to form a nearly spherical shell that encloses the viral genome. Although the rhinovirus coat ultimately requires 240 polypeptide chains, the symmetry achieved by its quaternary structure enables this structure to form from only four distinct sequences.

2.7 The Amino Acid Sequence of a Protein Determines Its Three-Dimensional Structure

Thus far, we have seen that proteins can adopt a wide variety of structures, ranging from highly compact to fully extended. These structures reflect the broad diversity of protein functions and represent a theme we will encounter

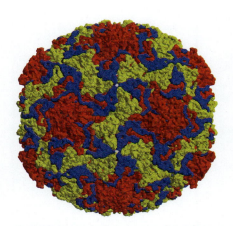

FIGURE 2.48 Viral coat proteins adopt complex quaternary structures with many copies of relatively few unique subunits. The coat of human rhinovirus, the cause of the common cold, comprises 240 individual polypeptide chains, 60 copies of each of four subunits. The three most prominent subunits are shown as different colors.

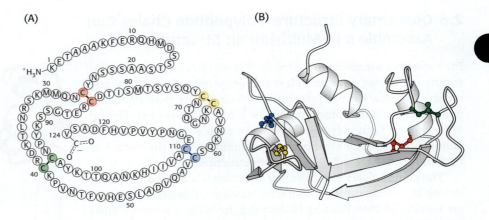

(A)   (B)

numerous times throughout this book. How polypeptide chains achieve these elaborate tertiary structures has been the focus of decades of research that still continues today.

The classic work of Christian Anfinsen in the 1950s on the enzyme ribonuclease revealed the relationship between the amino acid sequence of a protein and its conformation. Ribonuclease is a single polypeptide chain consisting of 124 amino acid residues cross-linked by four disulfide bonds (**Figure 2.49**). Anfinsen's plan was to destroy the three-dimensional structure of the enzyme and to then determine what conditions were required to restore the structure. In order to unfold ribonuclease, he used a high concentration (8 M) of urea, which effectively disrupts a protein's noncovalent interactions. Complete unfolding of ribonuclease also requires the cleavage, or reduction, of its four disulfide bonds (Figure 2.15). Using a large excess of β-mercaptoethanol, Anfinsen fully reduced the disulfides (cystines) into sulfhydryls (cysteines) (**Figure 2.50**).

Most polypeptide chains devoid of cross-links assume a random-coil conformation in 8 M urea. When ribonuclease was treated with β-mercaptoethanol in 8 M urea, the product was a fully reduced, randomly coiled polypeptide chain devoid of enzymatic activity. When a protein is converted into a randomly coiled peptide without its normal activity, it is referred to as a **denatured protein** (**Figure 2.51**).

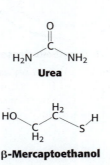

Urea

β-Mercaptoethanol

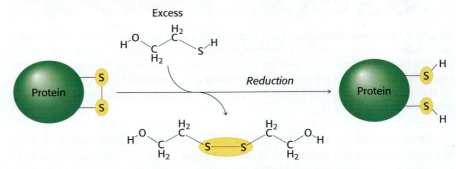

FIGURE 2.50 β-Mercaptoethanol disrupts disulfide bonds. Notice as the disulfides are disrupted (reduced), the β-mercaptoethanol is oxidized and forms dimers.

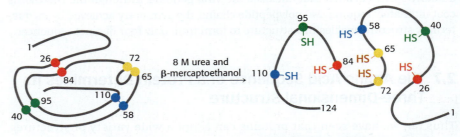

Native ribonuclease   **Denatured reduced ribonuclease**

FIGURE 2.51 Ribonuclease is denatured with urea and β-mercaptoethanol.

Anfinsen then made the critical observation that the denatured ribonuclease, introduced into a solution without urea and β-mercaptoethanol by a technique known as *dialysis* (Section 4.1), slowly regained enzymatic activity. He perceived the significance of this chance finding: the sulfhydryl groups of the denatured enzyme re-form disulfide bonds when exposed to air. Detailed studies then showed that nearly all the original enzymatic activity was regained if the disulfide bonds were oxidized under suitable conditions. All the measured physical and chemical properties of the refolded enzyme were virtually identical with those of the **native protein**; that is, the protein as it was originally isolated. These experiments showed that the information needed to specify the catalytically active structure of ribonuclease is contained in its amino acid sequence. Subsequent studies have established the generality of this central principle of biochemistry: sequence specifies conformation.

A quite different result was obtained when ribonuclease was reoxidized while it was still in 8 M urea, followed by the removal of the urea. Ribonuclease treated in this way had only 1% of the enzymatic activity of the native protein. Why were the outcomes so different when reduced ribonuclease was reoxidized in the presence and absence of urea? The reason is that the wrong disulfides formed pairs in urea. There are 105 different ways of pairing eight cysteine molecules to form four disulfides; only one of these combinations is enzymatically active. The 104 wrong pairings have been picturesquely termed "scrambled" ribonuclease. Anfinsen found that scrambled ribonuclease spontaneously converted into fully active, native ribonuclease when trace amounts of β-mercaptoethanol were added to an aqueous solution of the protein (**Figure 2.52**). The added β-mercaptoethanol catalyzed the rearrangement of disulfide pairings until the native structure was regained in about 10 hours. This process was driven by the decrease in free energy as the scrambled conformations were converted into the stable, native conformation of the enzyme. The native disulfide pairings of ribonuclease thus contribute to the stabilization of the thermodynamically preferred structure.

Similar refolding experiments have been performed on many other proteins. In many cases, the native structure can be generated under suitable conditions. For other proteins, however, refolding does not proceed efficiently. In these cases, the unfolded protein molecules usually become tangled up with one another to form aggregates. Inside cells, proteins called *chaperones* block such undesirable interactions. Additionally, it is now evident that some proteins do not assume a defined structure until they interact with molecular partners.

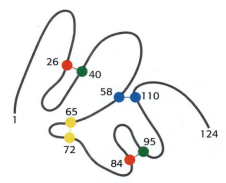

Scrambled ribonuclease

Trace of
β-mercaptoethanol

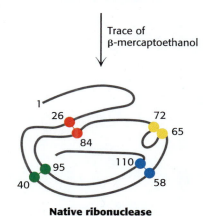

Native ribonuclease

FIGURE 2.52 **Native ribonuclease can be re-formed from scrambled ribonuclease in the presence of a trace of β-mercaptoethanol.**

❖ SELF–CHECK QUESTION

In his renaturation experiments, Anfinsen used a concentration of ribonuclease around 1 mg ml^{-1}. Inside a cell, the protein concentration is estimated to be around 100 mg ml^{-1}. Predict the outcome of Anfinsen's experiments had he used 100 mg ml^{-1} ribonuclease.

Amino acids have different propensities for forming α helices, β sheets, and turns

How does the amino acid sequence of a protein specify its three-dimensional structure? How does an unfolded polypeptide chain acquire the form of the native protein? These fundamental questions in biochemistry can be approached by first asking a simpler one: What determines whether a particular sequence in a protein forms an α helix, a β strand, or a turn? One source of insight is to examine the frequency of occurrence of particular amino acid residues in these secondary structures (**Table 2.3**). Residues such as alanine, glutamate, and leucine tend to be present in α helices, whereas valine and isoleucine tend to be present in β strands. Glycine, asparagine, and proline are more commonly observed in turns.

TABLE 2.3 Relative frequencies of amino acid residues in secondary structures

Amino acid	α helix	β sheet	Reverse turn
Glu	1.59	0.52	1.01
Ala	1.41	0.72	0.82
Leu	1.34	1.22	0.57
Met	1.30	1.14	0.52
Gln	1.27	0.98	0.84
Lys	1.23	0.69	1.07
Arg	1.21	0.84	0.90
His	1.05	0.80	0.81
Val	0.90	1.87	0.41
Ile	1.09	1.67	0.47
Tyr	0.74	1.45	0.76
Cys	0.66	1.40	0.54
Trp	1.02	1.35	0.65
Phe	1.16	1.33	0.59
Thr	0.76	1.17	0.96
Gly	0.43	0.58	1.77
Asn	0.76	0.48	1.34
Pro	0.34	0.31	1.32
Ser	0.57	0.96	1.22
Asp	0.99	0.39	1.24

Note: The amino acids are grouped according to their preference for α helices (top group), β sheets (middle group), or turns (bottom group).
Source: T. E. Creighton, *Proteins: Structures and Molecular Properties*, 2nd ed. (W. H. Freeman and Company, 1992), p. 256.

Studies of proteins and synthetic peptides have revealed some reasons for these preferences. Branching at the β-carbon atom, as in valine, threonine, and isoleucine, tends to destabilize α helices because of steric clashes. These residues are readily accommodated in β strands, where their side chains project out of the plane containing the main chain. Serine and asparagine tend to disrupt α helices because their side chains contain hydrogen-bond donors or acceptors in close proximity to the main chain, where they compete for main-chain NH and CO groups. Proline tends to disrupt both α helices and β strands because it lacks an NH group and because its ring structure restricts its ϕ value to near 60 degrees. Glycine readily fits into all structures, but its conformational flexibility renders it well-suited to reverse turns.

Can we predict the secondary structure of a protein by using this knowledge of the conformational preferences of amino acid residues? Accurate predictions of secondary structure adopted by even a short stretch of residues have proved to be difficult. Note that the conformational preferences of amino acid residues are not tipped all the way to one structure (Table 2.3). For example, glutamate, one of the strongest helix formers, prefers α helix to β strand by only a factor of three. The preference ratios of most other residues are smaller. Indeed, some penta- and hexapeptide sequences have been found to adopt one structure in one protein and an entirely different structure in another (**Figure 2.53**). Hence, some amino acid sequences do not uniquely determine secondary structure.

Tertiary interactions—interactions between residues that are far apart in the sequence—may be decisive in specifying the secondary structure of some

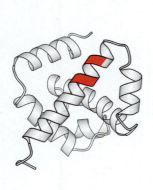

FIGURE 2.53 Many sequences can adopt alternative conformations in different proteins. Here the sequence VDLLKN shown in red assumes an α helix in one protein context (left) and a β strand in another (right). [Drawn from (left) 3WRP.pdb and (right) 2HLA.pdb.]

segments. Context is often crucial: the conformation of a protein has evolved to work in a particular environment. Nevertheless, substantial improvements in secondary structure prediction have been achieved by using families of related sequences, each of which adopts the same structure.

Protein folding is a highly cooperative process

Proteins can be denatured by any treatment that disrupts the weak bonds stabilizing tertiary structure, such as heating or chemical denaturants such as urea or guanidinium chloride. For many proteins, a comparison of the degree of unfolding as the concentration of denaturant increases reveals a sharp transition from the folded, or native, form to the unfolded, or denatured form, suggesting that only these two conformational states are present to any significant extent (**Figure 2.54**). A similar sharp transition is observed if denaturants are removed from unfolded proteins, allowing the proteins to refold.

The sharp transition seen in Figure 2.54 suggests that protein folding and unfolding is an "all or none" process that results from a **cooperative transition**. Consider this example: suppose that a protein is placed in conditions under which some part of the protein structure is thermodynamically unstable. As this part of the folded structure is disrupted, the interactions between it and the remainder of the protein will be lost. The loss of these interactions, in turn, will destabilize the remainder of the structure. Thus, conditions that lead to the disruption of any part of a protein structure are likely to unravel the protein completely. The structural properties of proteins provide a clear rationale for the cooperative transition.

The consequences of cooperative folding can be illustrated by considering the contents of a protein solution under conditions corresponding to the middle of the transition between the folded and the unfolded forms. Under these conditions, the protein is "half folded." Yet the solution will appear to have no partly folded molecules but, instead, look as if it is a 50/50 mixture of fully folded and fully unfolded molecules (**Figure 2.55**). Although the protein may appear to behave as if it exists in only two states, this simple two-state existence is an impossibility at a molecular level. Even simple reactions go through reaction intermediates, and so a complex molecule such as a protein cannot simply switch from a completely unfolded state to the native state in one step. Unstable, transient intermediate structures must exist between the native and denatured state. Determining the nature of these intermediate structures is an area of intense biochemical research.

Proteins fold by progressive stabilization of intermediates rather than by random search

How does a protein make the transition from an unfolded structure to a unique conformation in the native form? One theoretical possibility would be that all possible conformations are sampled to find the energetically most favorable one. How long would such a random search take? Consider a small protein with 100 residues. In 1969, Cyrus Levinthal calculated that, if each residue can assume three different conformations, the total number of structures would be 3^{100}, which is equal to 5×10^{47}. If it takes 10^{-13} s to convert one structure into another, the total search time would be $5 \times 10^{47} \times 10^{-13}$ s, which is equal to 5×10^{34} s, or 1.6×10^{27} years. In reality, small proteins can fold in less than a second. Clearly, it would take much too long for even a small protein to fold properly by randomly trying out all possible conformations. The enormous difference between calculated and actual folding times is called *Levinthal's paradox*. This paradox clearly reveals that proteins do not fold by trying every possible conformation; instead,

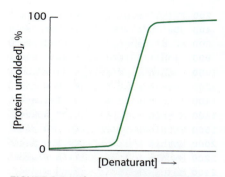

FIGURE 2.54 The transition from folded to unfolded state is abrupt. Most proteins show a sharp transition from the folded to the unfolded form on treatment with increasing concentrations of denaturants.

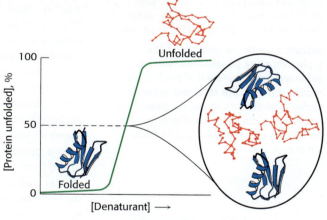

FIGURE 2.55 At the denaturation midpoint, half the molecules are fully folded and half are fully unfolded.

```
 200  ?T(\G{+s x[A.N5~,#ATxSGpn`e□@
 400  oDr'Jh7s DFR:W41'u+^v6zpJseOi
 600  e2ih'8zs n527x818d_ih=H1dseb.
 800  S#dh>}/s ]tZqC%1P%DK<|!^aseZ.
1000  V0th>nLs ut/is]1_kwojjwMasef.
1200  juth+nvs it is[lukh?SCw=ase5.
1400  Iithdn4s it isOl/ks/IxwLase~.
1600  M?thinrs it is 1Xk?T"_woasel.
1800  MSthinWs it is 1wkN7□Kw(asel.
2000  Mhthin`s it is likv,aww_asel.
2200  MMthinns it is lik+5avw1asel.
2400  MethinXs it is likydaqw)asel.
2600  Methin4s it is lik2dasweasel.
2800  MethinHs it is like□aTweasel.
2883  Methinks it is like a weasel.
```

```
 200  )z~hg)W4{{cu!kO{d6jS!N1EyUx}p
 400  "W hi\kR.<&CfA%4-Y1G!iT$6({|6
 600  .L=hinkm4(uMGP^1AWoE6k1wW=yiS
 800   AthinkaPa_vYH liR\Hb,Uo4\-"(
1000  OFthinksP)@fZO li8v] /+Eln26B
1200  6ithinksMVt -V likm+gl#K~)BFk
1400  vxthinksaEt □w like.S1Geutks.
1600  :Othinks<it MC likesN2[eaVe4.
1800  uxthinksqit 0r likeQh)weaoeW.
2000  Y/thinks it id like7a1wea)e&.
2200  Methinks it iW like a[weaWel.
2400  Methinks it is like a:weasel.
2431  Methinks it is like a weasel.
```

FIGURE 2.56 In the analogy of the typing monkey, the desired result is achieved more rapidly if individual correct keystrokes are retained. In the two computer simulations shown, the cumulative number of keystrokes required to write a line from Shakespeare's *Hamlet* is given at the left of each line.

they must follow at least a partly defined folding pathway consisting of intermediates between the fully denatured protein and its native structure.

The way out of this paradox is to recognize the power of cumulative selection. Richard Dawkins, in *The Blind Watchmaker*, asked how long it would take a monkey poking randomly at a typewriter to reproduce Hamlet's remark to Polonius, "Methinks it is like a weasel" (**Figure 2.56**). An astronomically large number of keystrokes, on the order of 10^{40}, would be required. However, suppose that we preserved each correct character and allowed the monkey to retype only the wrong ones. In this case, only a few thousand keystrokes, on average, would be needed. The crucial difference between these cases is that the first employs a completely random search, whereas, in the second, partly correct intermediates are retained.

The essence of protein folding is the tendency to retain partly correct intermediates. However, the protein-folding problem is much more difficult than the one presented to our simian Shakespeare. First, the criterion of correctness is not a residue-by-residue scrutiny of conformation by an omniscient observer but rather the total free energy of the transient species. Second, proteins are only marginally stable. The free-energy difference between the folded and the unfolded states of a typical 100-residue protein is 42 kJ mol^{-1} (10 kcal mol^{-1}), and thus each residue contributes on average only 0.42 kJ mol^{-1} (0.1 kcal mol^{-1}) of energy to maintain the folded state. This amount is less than the amount of thermal energy, which is 2.5 kJ mol^{-1} (0.6 kcal mol^{-1}) at room temperature. This meager stabilization energy means that correct intermediates, especially those formed early in folding, can be lost. The analogy is that the monkey would be somewhat free to undo its correct keystrokes. Nonetheless, the interactions that lead to cooperative folding can stabilize intermediates as structure builds up. Thus, local regions that have significant structural preference, though not necessarily stable on their own, will tend to adopt their favored structures and, as they form, can interact with one other, leading to increasing stabilization. This conceptual framework is often referred to as the *nucleation-condensation model*.

A simulation of the folding of a protein, based on the nucleation-condensation model, is shown in **Figure 2.57**. This model suggests that certain pathways may be preferred. Although Figure 2.57 suggests a discrete pathway, each of the intermediates shown represents an ensemble of similar structures, and thus a protein follows a general rather than a precise pathway in its transition from the unfolded to the native state. The energy surface for the overall process of protein folding can be visualized as a funnel (**Figure 2.58**). The wide rim of the funnel represents the wide range of structures accessible to the ensemble of denatured protein molecules. As the free energy of the population of protein molecules decreases, the proteins move down into narrower parts of the funnel, and fewer conformations are accessible. At the bottom of the funnel is the folded state with its well-defined conformation. Many paths can lead to this same energy minimum.

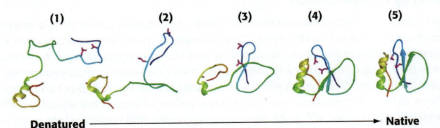

(1) (2) (3) (4) (5)

Denatured ⟶ Native

FIGURE 2.57 The proposed folding pathway of chymotrypsin inhibitor-2 demonstrates the nucleation-condensation model. Local regions with sufficient structural preference tend to adopt their favored structures initially (1). These structures come together to form a nucleus with a nativelike, but still mobile, structure (4). This structure then fully condenses to form the native, more rigid structure (5). [From A. R. Fersht and V. Daggett. *Cell* 108:573–582, 2002; with permission from Elsevier.]

Prediction of three-dimensional structure from sequence remains a great challenge

The prediction of three-dimensional structure from sequence has proved to be extremely difficult. The local sequence appears to determine only between 60% and 70% of the secondary structure; long-range interactions are required to stabilize the full secondary structure and the tertiary structure.

Investigators are exploring two fundamentally different approaches to predicting three-dimensional structure from amino acid sequence:

1. Ab initio *predictions attempt to predict the folding of an amino acid sequence without prior knowledge about similar sequences in known protein structures.* (Ab initio, from the Latin, means "from the beginning.") Computer-based calculations are employed that attempt to minimize the free energy of a structure with a given amino acid sequence or to simulate the folding process. The utility of these methods is limited by the vast number of possible conformations, the marginal stability of proteins, and the subtle energetics of weak interactions in aqueous solution.

2. *Knowledge-based methods take advantage of our growing knowledge of the three-dimensional structures of many proteins.* The amino acid sequence of unknown structure is examined for compatibility with known protein structures or fragments therefrom. If a significant match is detected, the known structure can be used as an initial model. Knowledge-based methods have been a source of many insights into the three-dimensional conformation of proteins of known sequence but unknown structure.

The prediction of protein folding from sequence information is a formidable task, one that has compelled the scientific community to explore creative methods for achieving progress. In 1994, John Moult and Krzysztof Fidelis organized the first protein-structure prediction challenge called Critical Assessment of Structure Prediction (CASP). In this competition, contestants are provided with sequences of proteins whose structures had been recently solved but not yet published, and are asked to submit their structural predictions. The submissions are then graded based on how well their predictions match the experimentally determined structures. Importantly, the entire process is *blinded*: the entrants have no prior knowledge of the experimentally solved structures, and the judges have no information about the identity of the competitors and their prediction methods.

Since its inception, CASP competitions have been conducted every two years, with participation from over 200 groups in the most recent event in 2020. Highlighting the difficulty of the task, improvements in structure prediction have been modest, with the results largely reaching a plateau in the early 2010s. However, significant improvement has been achieved in the most recent two events, largely based on the advancement of artificial intelligence (AI) methods developed by the company DeepMind. Their algorithm, AlphaFold, recently achieved an average accuracy of nearly 90%, almost equivalent to what would be expected from experimental methods. For example, their best structural prediction of the protein ORF8 from the SARS-CoV-2 virus was 89% accurate, far exceeding the competition (**Figure 2.59**). Their success suggests that the accurate computational prediction of protein folding may very well be within reach. Additionally, these results serve as evidence of the achievements made possible through collaborative research.

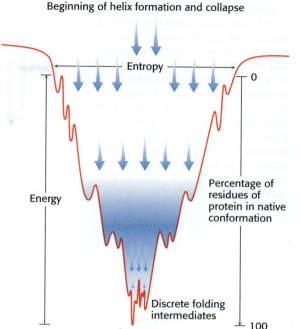

FIGURE 2.58 The folding funnel depicts the thermodynamics of protein folding. The top of the funnel represents all possible denatured conformations—that is, maximal conformational entropy. Depressions on the sides of the funnel represent semistable intermediates that can facilitate or hinder the formation of the native structure, depending on their depth. Secondary structures, such as helices, form and collapse onto one another to initiate folding. [After D. L. Nelson and M. M. Cox, *Lehninger Principles of Biochemistry*, 5th ed. (W. H. Freeman and Company, 2008), p. 143.]

FIGURE 2.59 AlphaFold represents a significant step forward in protein folding prediction. One of the targets from the most recent CASP competition was the ORF8 protein from the SARS-CoV-2 virus. (A) The AlphaFold algorithm (red bar) predicted the structure with nearly 90% accuracy (dashed line), far better than the other competitors (gray bars). (B) Comparison of the predicted structure with the experimentally determined structure shows the extensive similarity. [(A) Drawn from data obtained from predictioncenter.org/casp14/ and (B) 7JTL.pdb.]

(A)

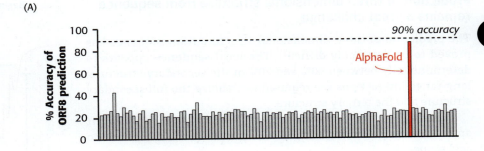

(B)

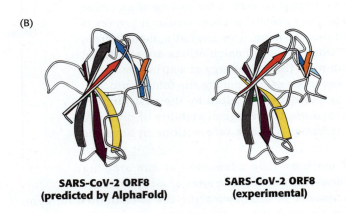

**SARS-CoV-2 ORF8
(predicted by AlphaFold)**

**SARS-CoV-2 ORF8
(experimental)**

Protein misfolding and aggregation are associated with some neurological diseases

Understanding protein folding and misfolding is of more than academic interest. A host of neurological diseases, including Alzheimer disease, Parkinson disease, and Huntington disease are associated with improperly folded proteins. All of these diseases result in the deposition of protein aggregates, called **amyloid fibrils** or **amyloid plaques**. A common feature of such diseases is that normally soluble proteins are converted into insoluble fibrils rich in β sheets. The correctly folded protein is only marginally more stable than the incorrect form. But the incorrect form aggregates, pulling more correct forms into the incorrect form. For example, the brains of patients with Alzheimer disease contain amyloid plaques that consist primarily of a single polypeptide termed Aβ. Polypeptide Aβ is prone to form insoluble aggregates. As expected, the structure is rich in β strands, which come together to form extended parallel β-sheet structures (**Figure 2.60**).

How do such aggregates lead to the death of the cells that harbor them? The answer is still controversial. One hypothesis is that the large aggregates

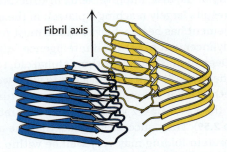

Fibril axis

FIGURE 2.60 The structure of the Aβ amyloid fibril reveals its extensive β sheet network. The Aβ amyloid fibril structure shows that protein aggregation is due to the formation of large parallel β sheets. The black arrow indicates the long axis of the fibril. [Drawn from 5OQV.pdb].

themselves are not toxic but, instead, smaller aggregates of the same proteins may be the culprits, perhaps damaging cell membranes. Nevertheless, the amount of fibrillar Aβ deposited in the brain is correlated with the severity of disease. In fact, the extent of amyloid fibril accumulation can be measured in patients using a technique called *positron emission tomography* (PET) scanning. These images can be used to anticipate the progressive worsening of neurological function (**Figure 2.61**).

Posttranslational modifications confer new capabilities to proteins

Proteins are able to perform numerous functions that rely solely on the versatility of their 20 amino acids. This array of functions can be expanded even further by the incorporation of **posttranslational modifications**, alterations in the structure of a protein after its synthesis in the cell. These modifications may include the covalent attachment of groups to amino acid side chains (**Figure 2.62**) or the cleavage or modification of the polypeptide backbone.

One example of a posttranslational modification is hydroxylation. As discussed in Section 2.5, the addition of hydroxyl groups to many proline residues stabilizes fibers of newly synthesized collagen. The biological significance of this modification is evident in the disease scurvy: a deficiency of vitamin C results in insufficient hydroxylation of collagen, and the abnormal collagen fibers that result are unable to maintain normal tissue strength. Other covalent modifications shown in Figure 2.62 will be discussed in future chapters.

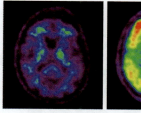 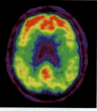

Patient A   Patient B

FIGURE 2.61 The extent of amyloid fibril aggregation in the brain can predict the progression of Alzheimer disease. With the aid of a molecule that binds Aβ fibrils, positron emission tomography scanning can measure the extent of amyloid aggregation in the brains of Alzheimer disease patients. The scan of Patient A (low) exhibited a relatively low signal, while the scan of Patient B (right) revealed a high degree of labeling throughout the brain. The mild impairment exhibited by Patient A remained stable for more than 4.5 years after the scan was taken, while Patient B progressed to dementia 3 years after the scan. [Courtesy of Professor Christopher Rowe MD, Director, Australian Dementia Network University of Melbourne.]

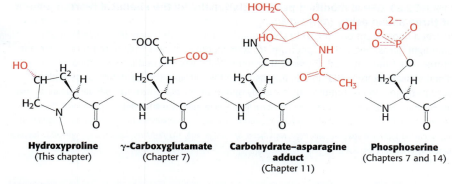

Hydroxyproline	γ-Carboxyglutamate	Carbohydrate–asparagine	Phosphoserine
(This chapter)	(Chapter 7)	adduct	(Chapters 7 and 14)
		(Chapter 11)	

FIGURE 2.62 Posttranslational modifications to proteins include covalent attachments. Examples of common and important covalent modifications of amino acid side chains are shown in red. These will be discussed in the indicated chapters.

Other special groups are generated by chemical rearrangements of side chains and, sometimes, the peptide backbone. For example, the jellyfish *Aequorea victoria* produces green fluorescent protein (GFP), which emits green light when stimulated with blue light. The source of the fluorescence is a group formed by the spontaneous rearrangement and oxidation of the sequence Ser-Tyr-Gly within the center of the protein (**Figure 2.63A**). Since the discovery of GFP, a number of mutants have been engineered which absorb and emit light across the entire visible spectrum (**Figure 2.63B**). These proteins are of great service to researchers as markers within cells (**Figure 2.63C**).

Finally, many proteins are cleaved and trimmed after synthesis. For example, digestive enzymes are synthesized as inactive precursors that can be stored safely in the pancreas. After release into the intestine, these precursors become activated by peptide-bond cleavage (Section 7.4). In blood clotting, peptide-bond cleavage converts soluble fibrinogen into insoluble fibrin. A number of polypeptide hormones, such as adrenocorticotropic hormone, arise from the splitting

(A)

(B)

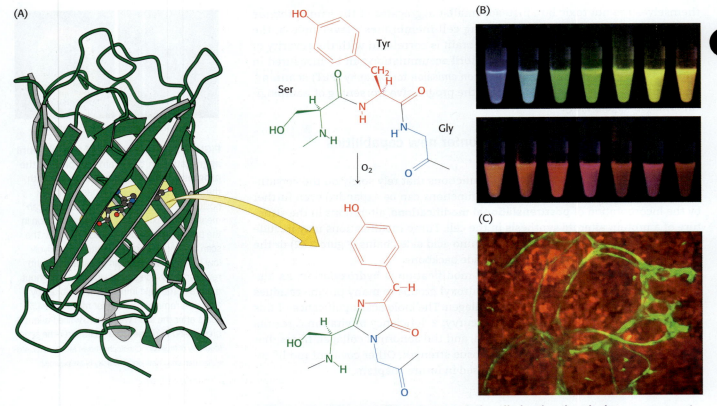

(C)

FIGURE 2.63 GFP is modified posttranslationally by the chemical rearrangement of three amino acids. (A) The rearrangement and oxidation of the sequence Ser-Tyr-Gly in green fluorescent protein (GFP) is the source of its fluorescence. (B) Mutants of GFP emit light across the visible spectrum. (C) A melanoma cell line engineered to express one of these GFP mutants, red fluorescent protein (RFP), was injected into a mouse whose blood vessels express GFP. In this fluorescence micrograph, the formation of new blood vessels (green) in the tumor (red) is readily apparent. [(A) Drawn from 1GFL.pdb; (B) Published in the Nobel lecture by Roger Y. Tsien 2008, and the Nobel lecture in its entirety is ©The Nobel Foundation © The Nobel Foundation 2008. (C) Yang, M. et al. Dual-color fluorescence imaging distinguishes tumor cells from induced host angiogenic vessels and stromal cells. *Proc. Natl. Acad. Sci. U.S.A.* 100, 14259–14262 (2003). Copyright 2003 National Academy of Sciences, U.S.A.]

of a single large precursor protein. Likewise, many viral proteins are produced by the cleavage of large polyprotein precursors. We shall encounter many more examples of modification and cleavage as essential features of protein formation and function. Indeed, these finishing touches account for much of the versatility, precision, and elegance of protein action and regulation.

Summary

2.1 Several Properties of Protein Structure Are Key to Their Functional Versatility

- Proteins are linear polymers built of amino acids.
- Proteins contain a wide range of functional groups.
- Proteins interact with each other to form assemblies.
- Proteins can vary from rigid to highly flexible.

2.2 Proteins Are Built from a Repertoire of 20 Amino Acids

- Proteins are polymers of L amino acids which can be classified on the properties of their side chains.
- Hydrophobic amino acids with nonpolar R groups include: glycine, alanine, valine, leucine, isoleucine, methionine, proline, phenylalanine, and tryptophan.

- Polar amino acids include: serine, threonine, tyrosine, asparagine, glutamine, cysteine, and histidine.
- Positively charged amino acids include: lysine and arginine.
- Negatively charged amino acids include: aspartate and glutamate.

2.3   Primary Structure: Amino Acids Are Linked by Peptide Bonds to Form Polypeptide Chains

- Amino acids are linked together into polypeptides by peptide bonds.
- The sequence of amino acids in a protein is called its primary structure.
- Polypeptide chains are directional, in that the ends of the chain are different: by convention, polypeptide sequences are written from the amino-terminal residue to the carboxyl-terminal residue.
- Each protein has its own unique primary structure.
- Peptide bonds have a planar geometry due to their partial double-bond character.
- The main chain bonds preceding and following the α carbon have rotational flexibility, which can be depicted in a Ramachandran plot.

2.4   Secondary Structure: Polypeptide Chains Can Fold into Regular Structures

- Secondary structure refers to the formation of regular structures near one another in the protein sequence.
- The α helix is a tightly packed rod where the CO group from one amino acid is hydrogen bonded to the NH group four residues further along the chain.
- The β sheet features a series of fully extended polypeptide chains (β strands) hydrogen-bonded to each other.
- Turns and loops are shorter, less-regular secondary structures that enable the polypeptide chain to change direction.

2.5   Tertiary Structure: Proteins Can Fold into Globular or Fibrous Structures

- Tertiary structure refers to the fully folded structure of a single polypeptide chain.
- The formation of tertiary structure in water-soluble proteins is largely driven by packing hydrophobic residues in the interior of the protein and exposing hydrophilic resides on the surface.
- Proteins can fold into highly compact and asymmetric globular proteins or fully extended fibrous proteins.

2.6   Quaternary Structure: Polypeptide Chains Can Assemble into Multisubunit Structures

- Quaternary structure refers to the organization of multiple polypeptide chains (subunits) into an organized complex.
- These assemblies can range from simple dimers to highly complicated viral coats with hundreds of subunits.
- Quaternary structures are held together by numerous noncovalent bonds.

2.7   The Amino Acid Sequence of a Protein Determines Its Three-Dimensional Structure

- The amino acid sequence determines the three-dimensional structure and, hence, all other properties of a protein.
- Protein folding does not occur by random searching, but by progressive stabilization of intermediate structural elements.
- Protein folding is a highly cooperative process; structural intermediates between the unfolded and folded forms do not accumulate.
- The versatility of proteins is further increased by posttranslational modifications, including the chemical modification or rearrangement of side chains and the cleavage of the peptide backbone.

Key Terms

enzymes (p. 34)

amino acid (p. 35)

α carbon (p. 35)

side chain (R group) (p. 35)

chiral (p. 35)

dipolar ion (zwitterion) (p. 35)

peptide bond (amide bond) (p. 41)

residue (p. 42)

main chain (backbone) (p. 42)

proteins (p. 43)

oligopeptides (peptides) (p. 43)

dalton (p. 43)

disulfide bond (p. 43)

primary structure (p. 43)

torsion angle (dihedral angle) (p. 45)

Ramachandran plot (p. 45)

secondary structure (p. 46)

α helix (p. 46)

β pleated sheet (β sheet) (p. 48)

β strand (p. 48)

reverse turn (β turn; hairpin turn) (p. 50)

loop (p. 50)

tertiary structure (p. 51)

globular protein (p. 51)

motif (supersecondary structure) (p. 52)

domain (p. 52)

fibrous protein (p. 53)

subunit (p. 55)

quaternary structure (p. 55)

homodimer (p. 55)

denatured protein (p. 56)

native protein (p. 57)

cooperative transition (p. 59)

amyloid fibrils (amyloid plaques) (p. 62)

posttranslational modification (p. 63)

Problems

1. Examine the following four amino acids (A–D):

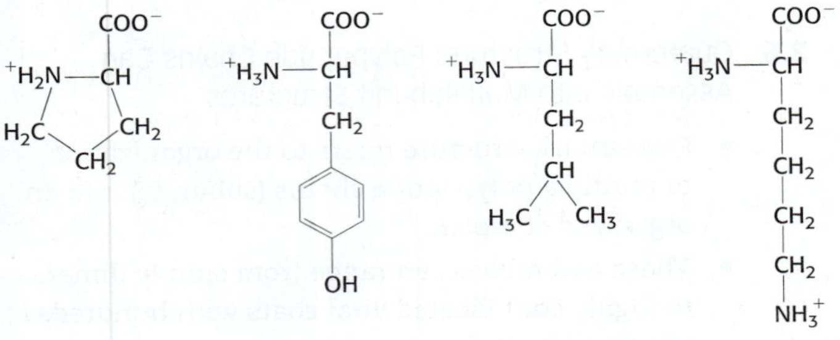

(A)   (B)   (C)   (D)

What are their names, three-letter abbreviations, and one-letter symbols? ❖ **1**

2. Sort the amino acids shown in problem 1 according to the category in which it fits. ❖ **1**

(a) Basic side chain _____

(b) Acidic side chain _____

(c) Neutral polar side chain _____

(d) Hydrophobic side chain _____

3. Which of the following amino acids have R groups that have hydrogen-bonding potential? Ala, Gly, Ser, Phe, Glu, Tyr, Ile, and Thr. ❖ **1**

4. Examine the segment of a protein shown here.

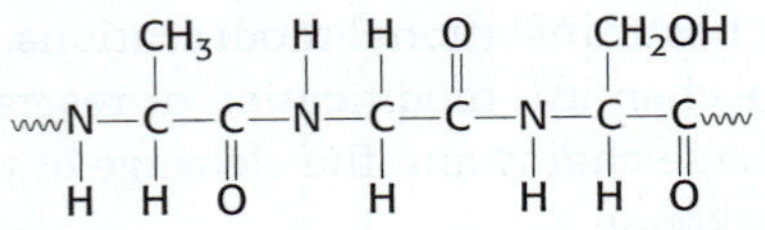

(a) What three amino acids are present?

(b) Of the three, which is the N-terminal amino acid?

(c) How many peptide bonds are explicitly shown in this peptide?

5. Draw the structure of the dipeptide Gly-His. What is the charge on the peptide at pH 5.5? pH 7.5? ❖ **1**

6. Draw the resonance structure for the peptide bond shown in the image.

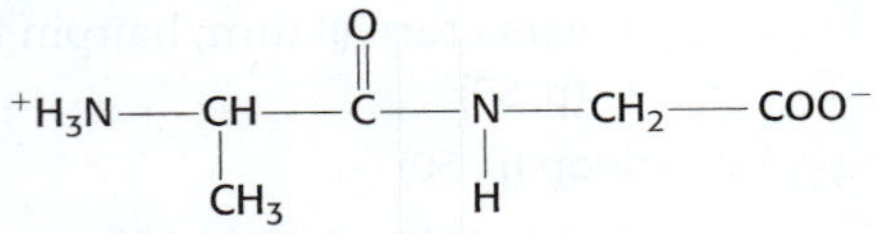

7. On Figure 2.5, indicate the pH value at which a solution of this amino acid would have no net charge.

8. All amino acids have two ionizable functional groups: an α-amino group (typical pK_a of 8.0) and an α-carboxylic acid group (typical pK_a of 3.1). Histidine also has an ionizable side chain (R group) with a pK_a of about 6.0. One of the possible ionization states of histidine is shown in the image.

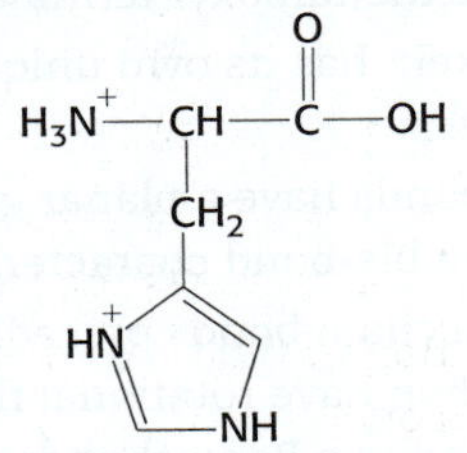

(a) At what pH would the structure be the predominant ionization state? Consider the ionization state of all three of the functional groups.

    pH 11.6   pH 8.4   pH 7.0   pH 6.0   pH 3.1   pH 1.5

(b) The protonated form of the R group of histidine is shown in the structure. The ratio of the charged (protonated) form to the deprotonated form depends on the pK_a of the R group and the pH of the solution. At which of these pH values would the charged form of the R group predominate?

    pH 3.0   pH 4.9   pH 6.0   pH 11.1

9. (a) Tropomyosin, a 70-kDa muscle protein, is a two-stranded α-helical coiled coil. Estimate the length of the molecule. **(b)** Suppose that a 40-residue segment of a protein folds into a two-stranded antiparallel β structure with a 4-residue hairpin turn. What is the longest dimension of this motif? ❖ **3**

10. Poly-L-leucine in an organic solvent such as dioxane is α helical, whereas poly-L-isoleucine is not. Why do these amino acids with the same number and kinds of atoms have different helix-forming tendencies? ❖ **3**

11. An enzyme that catalyzes disulfide–sulfhydryl exchange reactions, called protein disulfide isomerase (PDI), has been isolated. PDI rapidly converts inactive scrambled ribonuclease into enzymatically active ribonuclease. In contrast, insulin is rapidly inactivated by PDI. What does this important observation imply about the relationship between the amino acid sequence of insulin and its three-dimensional structure?

12. What is the length of a 12 kDa single-stranded α-helical protein segment? Assume a mean residue mass of 110 daltons. ❖ **3**

13. Which of these four peptides is most likely to form a β sheet?

(a) Lys-Thr-Val-Ile-Trp-Pro-Phe-Tyr-Ile-Gln-Ile-Gly

(b) Met-Leu-Lys-Ala-Ser-Ala-Leu-Glu-Lys-Leu-Ser-Glu

(c) Arg-Ser-Tyr-Glu-Gly-Leu-Lys-Arg-Ile-Ala-Glu-Ser

(d) Ala-Glu-Met-Leu-Gln-Lys-Arg-Gly-Cys-Gly-Asp-Glu

14. Consider the Ramachandran plots in Figures 2.21, 2.24, and 2.27.

(a) Which observation most likely describes part of an observed α helix? (choose one) ❖ **3**

    (a) φ = –60°; ψ = –47°; many Ala residues

    (b) φ = –140°; ψ = +130°; many Val residues

    (c) φ = –59°; ψ = +150°; many Ser residues

    (d) φ = +60°; ψ = +40°; many Gly residues

    (e) φ = –57°; ψ = –47°; many Pro residues

(b) Which observations would not likely occur in a β sheet? (may be more than one)

    (a) φ = –57°; ψ = –49°; many Ala residues

    (b) φ = +60°; ψ = +60°; many Gly residues

    (c) φ = –51°; ψ = +153°; many Gly and Pro residues

    (d) φ = –139°; ψ = +135°; many Val residues

    (e) φ = –120°; ψ = +120°; many Tyr residues

15. Classify each phrase as a description of α helices, β sheets, reverse turns, or all: ❖ **3**

(a) Contains—NH hydrogen bonded to C=O

(b) Successive R groups point in opposite directions

(c) All—NH groups point in the same direction

(d) First residue hydrogen bonded to fourth residue (i.e., residue i is H-bonded to residue i + 3)

(e) First residue hydrogen bonded to fifth residue (i.e., residue i is H-bonded to residue i + 4)

16. The α and β subunits of hemoglobin bear a remarkable structural similarity to myoglobin. However, certain residues that are hydrophilic in myoglobin are hydrophobic in the subunits of hemoglobin. Why might this be the case?

17. For an amino acid such as alanine, the major species in solution at pH 7 is the zwitterionic form. Assume a pK$_a$ value of 8 for the amino group and a pK$_a$ value of 3 for the carboxylic acid. Estimate the ratio of the concentration of the neutral amino acid species (with the carboxylic acid protonated and the amino group neutral) to that of the zwitterionic species at pH 7 (Section 1.3).

18. Translate the following amino acid sequence into one-letter code: Glu-Leu-Val-Ile-Ser-Ile-Ser-Leu-Ile-Val-Ile-Asn-Gly-Ile-Asn-Leu-Ala-Ser-Val-Glu-Gly-Ala-Ser. ❖ **1**

19. Would you expect Pro—X peptide bonds to tend to have cis conformations like those of X—Pro bonds? Why or why not?

20. For each of the amino acid derivatives shown here (A–E), find the matching set of φ and ψ values (a–e).

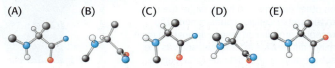

(A)   (B)   (C)   (D)   (E)

(a)	(b)	(c)	(d)	(e)
φ = 120,	φ = 180,	φ = 180,	φ = 0,	φ = –60,
ψ = 120	ψ = 0	ψ = 180	ψ = 180	ψ = –40

21. In the 1950s, Christian Anfinsen demonstrated the renaturation of the protein ribonuclease *in vitro*. After reduction and the addition of urea, the protein was in an unfolded state. After removing the urea and then the reducing agent, the protein oxidized and refolded, with greater than 90% activity. If reducing agent removal occurs before removing the urea, the protein showed less than 5% activity. Why does ribonuclease refold incorrectly if the reducing agent is removed before urea removal?

3

Binding and Molecular Recognition

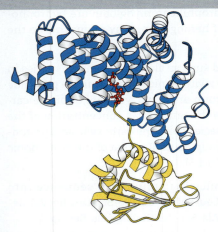

Penguins can recognize and find their mates even in a crowd of quite similar-looking and sounding birds. At the level of molecules, recognition plays a hugely important role in biochemical interactions. The complex shown above involves a protein that recognizes and binds extended amino acid sequences on other proteins and helps target them to the proper cellular compartment. The receptor protein can readily distinguish even minor differences in its target, such as the substitution of an isoleucine for a leucine.

Stefan Christmann/naturepl.com

OUTLINE

3.1 Binding Is a Fundamental Process in Biochemistry

3.2 Myoglobin Binds and Stores Oxygen

3.3 Hemoglobin Is an Efficient Oxygen Carrier

3.4 The Immune System Depends on Key Binding Proteins

3.5 Quantitative Terms Can Describe Binding Propensity

❖ LEARNING GOALS

By the end of this chapter, you should be able to:

1. Explain in qualitative terms how the interaction between a receptor and a ligand that it binds depends on the concentration of the ligand.

2. Describe the key principles for the binding of oxygen by myoglobin and hemoglobin.

3. Discuss the concept of binding cooperativity and how it applies to hemoglobin, and how cooperativity enhances the ability of hemoglobin to function as a highly effective oxygen carrier.

4. Describe some key binding proteins in the immune system, including toll-like receptors, immunoglobulins, and major histocompatibility complex proteins, and how binding reactions involving these proteins function in the immune system.

5. Calculate parameters of binding reactions using the concentrations of the reactants and the characteristics such as the dissociation constant and kinetic rate constants.

Organisms and cells are packed with different kinds of small molecules, including building blocks and metabolites, as well as macromolecules including proteins and nucleic acids, all bustling around due to thermal motion. Yet, out of this chaos, order emerges with intricately organized assemblies on a range of scales and specific biochemical pathways that are the subject of much of this book. The key process that enables this organization is specific molecular binding. The ability of many biological macromolecules to bind a specific set of molecules out of extremely complicated mixtures is very unusual compared to the capacity of non-biological molecules and can seem astounding, but it is the consequence of chemical principles and the selective power of molecular evolution.

3.1 Binding Is a Fundamental Process in Biochemistry

Sets of molecules stick together so that they spend much more time together than they would without this binding, and this occurs despite the fact that there are many other potential binding partners around them. This ability for certain molecules to bind to one another in the face of many alternatives is often referred to as **molecular recognition**. Furthermore, this binding is specific in another way: the structures of the complexes formed have particular three-dimensional shapes. This enables the complexes to have well-defined properties including participating in other binding events and, in some cases, playing key roles in processes such as the transmission of information within and between cells.

In this chapter, we will focus primarily on two examples, likely familiar in some aspects of everyday life, to illustrate these principles. These are the oxygen storage and carrier proteins myoglobin and hemoglobin, and important proteins from the immune system, including antibodies. **Myoglobin** binds oxygen and stores it for use when muscles expend energy and need to convert molecules from food into useable forms. Myoglobin, by virtue of bound **heme**, is responsible for the color of red meat. **Hemoglobin** is a related heme-containing protein found in blood that acts as oxygen carrier, transporting oxygen that it binds in the lungs to tissues throughout the body. Hemoglobin is remarkably efficient in fulfilling this role. It binds and releases oxygen in a manner that is quite sensitive to the amount of available oxygen in its environment and also to pH and the presence of carbon dioxide. These binding properties allow hemoglobin to deliver oxygen specifically to tissues where it is most needed due to ongoing metabolic activity.

Our immune systems must respond to pathogens such as viruses and bacteria, even those to which we have never previously been exposed. Remarkably, our immune systems can generate proteins called *antibodies* that bind specifically to components of these pathogens. Each antibody has a set of polypeptide loops on its surface that are selected to facilitate binding to a particular pathogen component. The ability to generate an essentially infinite set of binding proteins by combining a limited set of component loops has kept members of our and other species protected in the face of an ever-evolving set of potentially deadly pathogens.

We begin with a discussion of some of the general characteristics of binding processes. For this, we will use examples related to the estrogen receptor, a key protein that functions physiologically by binding certain steroid hormones and is also the target of some important drugs and environmental pollutants.

Binding depends on the concentrations of the binding partners

Suppose we start with a receptor R, typically a protein, and add a potential binding partner, often referred to as a **ligand**, L. These might come together to form a complex, RL.

$$R + L \rightleftharpoons RL \tag{1}$$

If R and L have a very strong tendency to bind to one another, then every molecule of L that is added will combine with a molecule of R, if one is available, to form a complex. As we add more L, this will continue until all of the R molecules are now in the RL form. On the other hand, if R and L have little tendency to combine, then, even if R and L collide, no complex will form as thermal motions will dominate. In between these two limiting cases, some of the added L will combine with R to form RL while the remainder remains free. These cases are summarized in **Figure 3.1**.

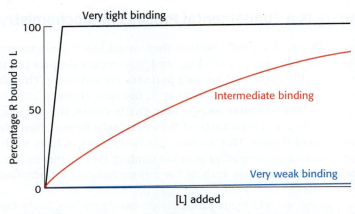

FIGURE 3.1 Comparing the percentage of receptor R bound to ligand L with the concentration of added L reveals how tightly R binds L.

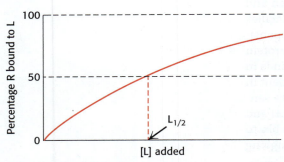

FIGURE 3.2 The binding of L to R can be characterized by $L_{1/2}$. The parameter $L_{1/2}$ is the concentration of L at which half (50%) of the total amount of R is bound to L.

Let us examine the case of intermediate binding in more detail. As more L is added, the probability that a given receptor molecule is bound to L gradually increases. A parameter that is useful for characterizing such curves is the concentration of L at which half of the receptor is bound to L and half is free. This is designated as $L_{1/2}$ in **Figure 3.2**. This parameter is useful because it reveals the concentration range of L around which the receptor makes the transition from being mostly free to mostly bound. We will explore these curves in more quantitative terms later in this chapter.

Proteins can selectively bind certain small molecules

To put the importance of binding within biochemical systems in context, let us consider the estrogen receptor and its ligands. Estrogens are steroid hormones that play primary roles in the development of the female reproductive system and secondary sex characteristics. The most prevalent estrogen is β-estradiol or, more simply, estradiol (**Figure 3.3**). The first step in estrogen exerting its biological activity is binding to estrogen receptors. These receptors are present in tissues such as mammary glands and the endometrium. Estradiol binds to these receptors with an $L_{1/2}$ of approximately 1 nM (10^{-9} M). The amount of estradiol in blood serum varies with age and time within the menstrual cycle, but on average it is around 1.1 nM, just in the range likely to trigger substantial receptor occupancy and activity.

This binding occurs despite the fact that serum contains thousands of different small molecules, many of which are present at concentrations higher than that for estradiol. For example, testosterone, a steroid hormone associated with the development of the male reproductive system and secondary sex

β-Estradiol (Estradiol)

Estrone

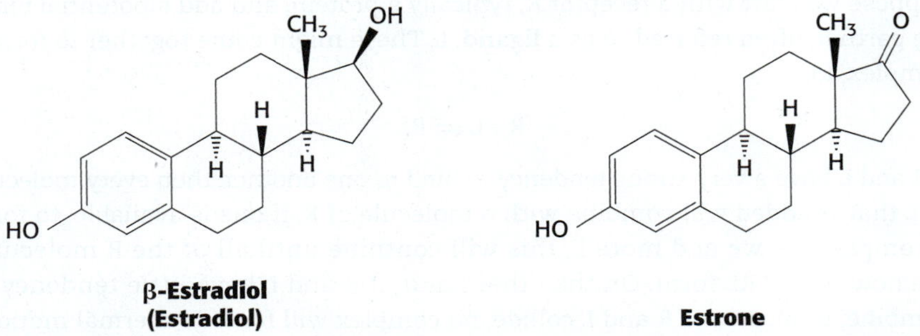

FIGURE 3.3 Estrogens are hormones that are bound by estrogen receptors. β-Estradiol is the most prevalent and potent estrogen, and estrone is one of a number of other estrogens that play somewhat different biological roles.

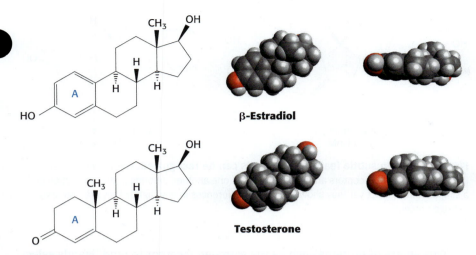

FIGURE 3.4 The steroid hormones β-estradiol and testosterone appear quite similar, yet receptors recognize the molecules as distinct. Their chemical structures are shown together with space-filling depictions from two orientations. The side views on the right reveal key differences in thickness that aid molecular recognition. The ring marked "A" is aromatic (and therefore planar) in β-estradiol, but not in testosterone.

characteristics, is present at concentrations comparable to or even higher than those for estradiol, even in women. However, the binding of testosterone to the estrogen receptor is very weak with $L_{1/2} > 1 \mu M$, more than 1000 times larger than that for estradiol.

The ability of the estrogen receptor to recognize estradiol, that is, to discriminate between estradiol and testosterone, is impressive given the similarities in their structures (**Figure 3.4**). Both structures are based on the same four-ring framework characteristic of steroids. However, one six-membered ring (the A ring shown in Figure 3.4) in estrogens such as estradiol is aromatic and, therefore, planar, whereas the corresponding ring in testosterone is not aromatic and is bonded to an additional methyl group absent in estrogens. Examination of the three-dimensional structure of the estrogen receptor bound to estradiol reveals that this protein has a narrow binding pocket lined with hydrophobic residues such as leucine and phenylalanine that fits neatly around the ligand but would likely sterically clash with thicker ligands such as testosterone. Moreover, the pocket includes groups that recognize both the oxygen and the hydrogen of the hydroxyl group in estradiol (**Figure 3.5**).

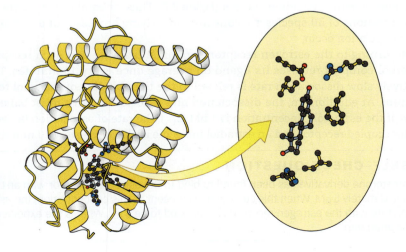

FIGURE 3.5 Proteins such as estrogen receptors include binding pockets that are complementary to the shapes of specific ligands. This enables the proteins to bind these ligands while rejecting ligands with even subtly different shapes. In the case of estrogen receptors, and some but not all proteins, some conformational changes occur upon ligand binding.

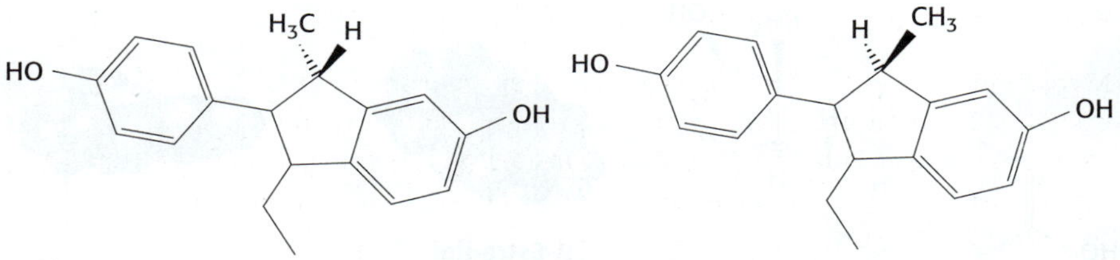

Strong binding
$L_{1/2} = 11$ nM

Weak binding
$L_{1/2} = 292$ nM

FIGURE 3.6 Even subtle features of ligands can be recognized. These two synthetic ligands for estrogen receptors are mirror images of one another, differing in the orientation of a single methyl group, yet they show substantial differences in how tightly they are bound by estrogen receptors.

This ability of proteins such as the estrogen receptor to bind ligands selectively has been probed further through the use of synthetic compounds that share some characteristics with estrogens. For example, the two compounds in **Figure 3.6** are mirror images of each other by virtue of the orientation of a single methyl group. Despite this relatively small change in structure, these compounds are bound by the estrogen receptor with $L_{1/2}$ values that differ from one another by a factor of more than 20.

This example of ligand binding by the estrogen receptor is typical of a great many proteins that we will encounter throughout this book. The ability to bind small molecules with selectivity allows proteins to pick particular molecules out of complicated mixtures to facilitate sensing of the presence of these molecules and, in some cases, to use the bound molecules for other purposes.

❖ **SELF-CHECK QUESTION**

Human growth hormone is a protein with a molecular weight of approximately 22,000 grams/mol and a concentration in the bloodstreams of children in the range of 50 nanograms/ml. What is its molar concentration? What range would you anticipate for its $L_{1/2}$ value for the growth hormone receptor?

Binding is a dynamic process involving association and dissociation

At a given ligand concentration at equilibrium, the concentrations of receptor and receptor-ligand complex are constant. For example, when the ligand concentration is $L_{1/2}$, the concentrations of the receptor and the receptor-ligand complex are equal to one another and, therefore, equal to half of the total receptor concentration. Yet, each receptor is constantly releasing one ligand molecule and, subsequently, rebinding another (**Figure 3.7**). Thus, at equilibrium where the concentrations of all species are constant, underlying processes of association and dissociation occur.

Returning to the estrogen receptor, experiments reveal that each receptor-estradiol complex releases its ligand on average once every 10 minutes. This relatively slow dissociation rate is related to the tight binding of estradiol to its receptor. At equilibrium, the dissociation and association rates must balance. Thus, if the estradiol concentration is 1 nM ($L_{1/2}$), the rate of association between an unoccupied receptor and an estradiol molecule is also once every 10 minutes.

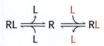

FIGURE 3.7 Dissociation and association are constantly occurring, even at equilibrium where the concentrations of all species are constant.

❖ **SELF-CHECK QUESTION**

A testosterone derivative has been found to bind to the estrogen receptor with an $L_{1/2}$ of approximately 5 μM. When this compound is present at this concentration, the association rate with the estrogen receptor is estimated to be 5 s^{-1}. What is the expected dissociation rate?

3.2 Myoglobin Binds and Stores Oxygen

Myoglobin is a fairly abundant protein representing about 0.5% by weight in our muscles, and up to nearly 10% by weight in species such as sperm whales. This protein functions to bind and store oxygen, picking it up from the bloodstream and storing in muscle for use when needed. The abundance of myoglobin in sperm whale muscles is one of the adaptations that allows these marvelous creatures to dive very deep and to stay under water for up to two hours.

More myoglobin binds oxygen as the oxygen partial pressure is increased

We can characterize the oxygen-binding properties of myoglobin by observing its oxygen-binding curve, a plot of the fractional saturation versus the concentration of oxygen. The **fractional saturation** (Y) is defined as the fraction of possible binding sites that contain bound oxygen. The value of Y can range from 0 (all sites empty) to 1 (all sites filled). The concentration of oxygen is most conveniently measured by its partial pressure, pO_2. For myoglobin, a binding curve indicating a simple chemical equilibrium is observed (**Figure 3.8**). Notice that the curve rises sharply as pO_2 increases and then levels off. Half-saturation of the binding sites, referred to as P_{50} (analogous to $L_{1/2}$ introduced previously) occurs at approximately 2 torr (mm Hg) for human myoglobin. For context, the partial pressure of oxygen at sea level is around 150 torr. Thus, a P_{50} value of 2 torr reveals that myoglobin binds oxygen with high affinity.

A bond is formed between oxygen and iron in heme

Sperm whale myoglobin was the first protein for which the three-dimensional structure was determined (**Figure 3.9**). Myoglobin consists largely of α helices that are linked to one another by turns to form a globular structure. The ability of myoglobin to bind oxygen depends on the presence of a *heme* molecule (**Figure 3.10**). The heme group consists of an organic component and a central iron atom. The organic component, called protoporphyrin IX, is made up of four linked pyrrole rings. The iron atom lies in the center of the protoporphyrin, bonded to the four pyrrole nitrogen atoms. The iron ion in myoglobin can form two additional bonds, one on each side of the heme plane. These binding sites are referred to as the fifth and sixth coordination sites. The fifth coordination site is occupied by the imidazole ring of a histidine residue from the protein, known as the **proximal histidine**.

Myoglobin can exist in an oxygen-free form called **deoxymyoglobin** or in a form with an oxygen molecule bound called **oxymyoglobin**. In deoxymyoglobin,

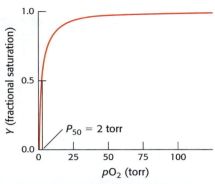

FIGURE 3.8 Half saturation of human myoglobin occurs when the partial pressure of oxygen is 2 torr. This point is revealed by plotting the fraction of myoglobin molecules that have oxygen bound (fractional saturation) as a function of the partial pressure of oxygen (pO_2).

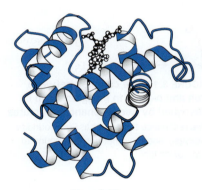

Myoglobin

FIGURE 3.9 The structure of myoglobin consists primarily of α helices surrounding a heme molecule. The heme molecule functions as the oxygen-binding site.

INTERACT with this model in
Achieve

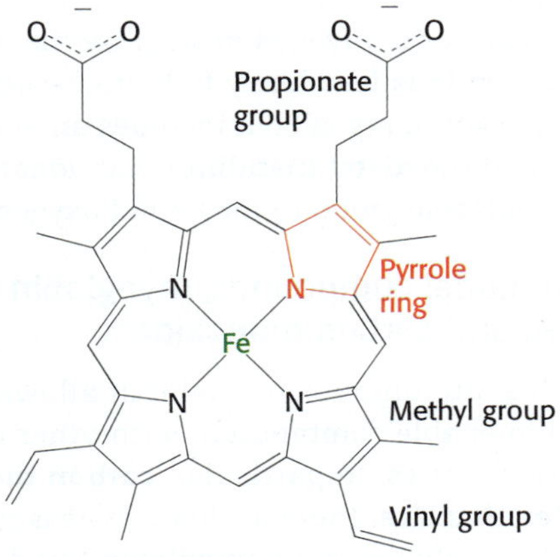

**Heme
(Fe-protoporphyrin IX)**

FIGURE 3.10 The structure of heme comprises four pyrrole rings that use their nitrogen atoms to bind to a central iron atom. Four methyl groups, two vinyl groups, and two propionate side chains are attached to the central tetrapyrrole.

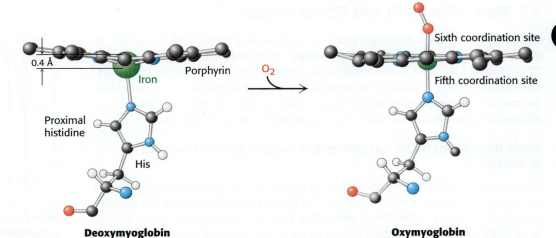

Deoxymyoglobin   **Oxymyoglobin**

FIGURE 3.11 Oxygen binding changes the position of the heme iron ion. In deoxymyoglobin, the iron ion lies slightly outside the plane of the porphyrin and moves into the plane of the porphyrin upon oxygen binding.

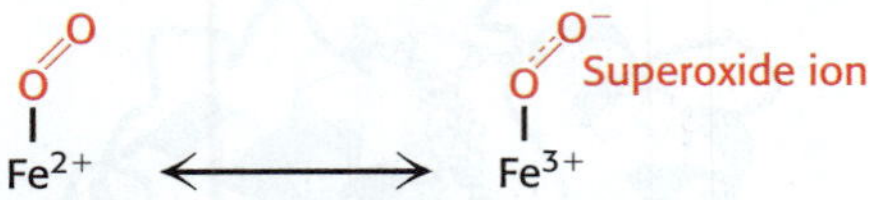

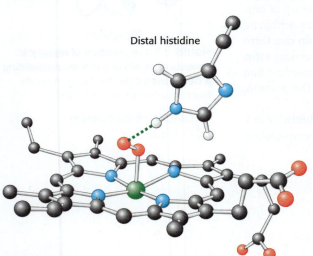

FIGURE 3.12 The complex between iron and oxygen in myoglobin can be described by two resonance structures. One resonance structure has Fe^{2+} bound to dioxygen, and the other has Fe^{3+} bound to superoxide ion.

the iron is in the ferrous (Fe^{2+}) oxidation state, and the sixth site remains unoccupied. The ferrous iron ion is slightly too large to fit into the well-defined hole within the porphyrin ring; it lies approximately 0.4 Å outside the porphyrin plane (**Figure 3.11**).

Oxymyoglobin forms when oxygen binding occurs at the sixth coordination site. This oxygen binding substantially rearranges the electrons within the iron so that the ion becomes effectively smaller, allowing it to move into the plane of the porphyrin (Figure 3.11, right). Oxygen binding to iron in heme is accompanied by the partial transfer of an electron from the ferrous ion to oxygen. In many ways, the structure is best described as a complex between ferric ion (Fe^{3+}) and superoxide anion O_2^- (**Figure 3.12**).

The structure of myoglobin prevents the release of reactive oxygen species

It is crucial that oxygen, when it is released, leaves as dioxygen rather than superoxide ion, for two important reasons. First, superoxide ion and other species generated from it are reactive oxygen species that can be damaging to many biological materials. Second, release of superoxide would leave the iron ion in the ferric state, producing a species, termed *metmyoglobin*, that does not bind oxygen. Thus, potential oxygen-storage capacity would be lost.

Features of myoglobin stabilize the oxygen complex such that superoxide is less likely to be released. In particular, the binding pocket of myoglobin includes an additional histidine residue (termed the **distal histidine**) that donates a hydrogen bond from its imidazole group to the bound oxygen molecule (**Figure 3.13**).

FIGURE 3.13 Bound oxygen is stabilized through a hydrogen bond between the distal histidine and one of the oxygen atoms. The hydrogen bond is shown as a dotted green line. The superoxide character of the bound oxygen species strengthens this interaction. Thus, the protein component of myoglobin controls the intrinsic reactivity of heme, making it more suitable for reversible oxygen binding.

Compared with model compounds, myoglobin discriminates between oxygen and carbon monoxide

While the available site on deoxymyoglobin allows oxygen binding, it also has the potential to enable combination with other molecules of similar size and binding characteristics. In particular, **carbon monoxide** (CO) is a potential competitor for this site. Indeed, this odorless gas can be deadly due to its ability to bind myoglobin and hemoglobin. Binding studies indicate that deoxymyoglobin binds CO very tightly, with a P_{50} of approximately 0.02 torr, a hundredfold tighter than oxygen (**Figure 3.14**). This is a potential problem, since oxygen cannot bind if CO is occupying the site. Environmental exposure to CO can result in blocking and inactivating both myoglobin and hemoglobin, the oxygen-carrying protein in the blood to be discussed shortly. In addition, some CO is produced by the body, ironically during the breakdown of heme, and this, too, can compete for oxygen-binding sites.

It may seem surprising that myoglobin does not discriminate better against CO given the potential risks of CO poisoning. Scientists have probed the chemical basis underlying this phenomenon using synthetic heme model compounds that mimic the essential characteristics of the myoglobin ligand-binding site. Some of these model compounds have affinities for oxygen similar to that for myoglobin, with a P_{50} of 0.6 torr. However, these model compounds bind CO with a P_{50} of 0.00002 torr; their affinity for CO is 30,000 times higher than that for oxygen. This suggests that myoglobin has evolved to show enhanced recognition of oxygen versus CO but is limited by the much higher intrinsic affinity of iron for CO.

Site-directed mutagenesis studies of sperm whale myoglobin support this conclusion and highlight an important role of the distal histidine. Removal of the imidazole group by the replacement of histidine with glycine results in a protein that binds oxygen approximately 13 times less tightly than does the wide-type protein but binds CO five times more tightly. This combination decreases the recognition of oxygen over CO by a factor of 65. Other substitutions can increase this factor further. These results reveal a major mechanism by which myoglobin has evolved to improve recognition of a key ligand.

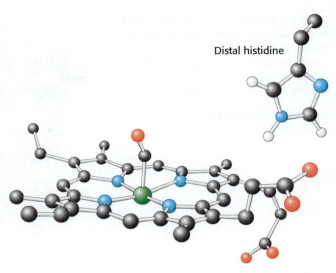

Distal histidine

FIGURE 3.14 Carbon monoxide can bind to the iron ion in myoglobin, blocking the oxygen-binding site.

3.3 Hemoglobin Is an Efficient Oxygen Carrier

As we have seen, myoglobin functions as an oxygen storage protein. Hemoglobin serves a more complicated role as an oxygen carrier. This role requires considerable sophistication, as this protein must bind oxygen in the lungs and then release as much oxygen as possible to tissues, particularly where it is most needed. We shall see that the oxygen-binding properties of hemoglobin differ from those of simpler proteins like myoglobin, and how additional ligands including carbon dioxide and hydrogen ions can influence these binding properties. Hemoglobin serves as an archetype for many other proteins with complicated properties that we will encounter throughout the book.

Human hemoglobin is an assembly of four myoglobin-like subunits

The three-dimensional structure of hemoglobin from horse heart was solved by Max Perutz and colleagues shortly after the determination of myoglobin's structure. Since then, the structures of hemoglobins from other species, including humans, have been determined. Hemoglobin consists of four polypeptide chains, two identical α chains, and two identical β chains (**Figure 3.15**). Each of the subunits consists of a set of α helices in the same arrangement as the α helices in myoglobin. The recurring structural motif is called a globin fold.

The hemoglobin tetramer, referred to as hemoglobin A (HbA), is best described as a homodimer of heterodimers, that is, a pair of identical αβ dimers ($\alpha_1\beta_1$ and $\alpha_2\beta_2$) that associate to form the tetramer. In deoxyhemoglobin, these αβ dimers are linked by an extensive interface, which includes the carboxyl

FIGURE 3.15 Hemoglobin is an assembly of four subunits, each of which is very similar to myoglobin. There are two α subunits and two β subunits that come together to form two αβ dimers. (A) A ribbon diagram. (B) A space-filling model.

INTERACT with this model in

 Achieve

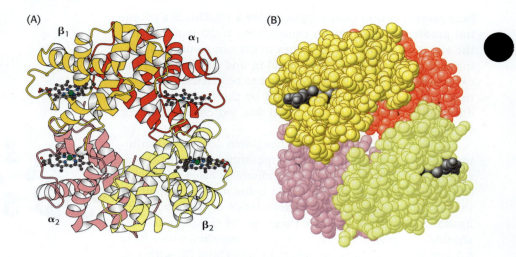

(A)   (B)

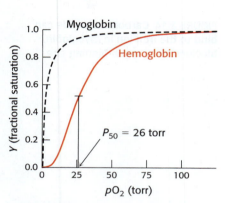

FIGURE 3.16 The oxygen-binding curve for hemoglobin, with its sigmoid (S-like) shape, is markedly different from that for myoglobin. The P_{50} value for hemoglobin is 26 torr compared with 2 torr for myoglobin.

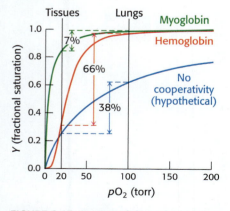

FIGURE 3.17 Cooperativity enhances oxygen delivery by hemoglobin. Because of cooperativity between oxygen-binding sites, hemoglobin delivers more oxygen to actively metabolizing tissues than would myoglobin or, indeed, any noncooperative protein, even with an optimized oxygen affinity.

terminus of each chain. The heme groups are well separated in the tetramer by iron–iron distances ranging from 24 to 40 Å.

Hemoglobin binds oxygen cooperatively

The oxygen-binding curve for hemoglobin in red blood cells shows some remarkable features (**Figure 3.16**). It does not look like a simple binding curve such as that for myoglobin; instead, it resembles an "S." Such curves are described as **sigmoid** for their S-like shape. In addition, oxygen binding for hemoglobin (P_{50} = 26 torr) is significantly weaker than that for myoglobin. Note that this binding curve is derived from hemoglobin in red blood cells.

A sigmoid binding curve indicates that a protein exhibits a particular binding behavior. For hemoglobin, the sigmoid shape suggests that the binding of oxygen at one site within the hemoglobin tetramer increases the likelihood that oxygen binds at the remaining unoccupied sites. Conversely, the unloading of oxygen at one heme facilitates the unloading of oxygen at the others. This sort of binding behavior is referred to as *cooperative*, because the binding reactions at individual sites in each hemoglobin molecule are not independent of one another. We have seen cooperativity in a different context, protein folding, where folding of one part of a protein is not independent of the folding of other parts. We will return to the mechanism of oxygen-binding cooperativity shortly.

What is the physiological significance of the cooperative binding of oxygen by hemoglobin?

- *Cooperative binding leads to efficient oxygen transport.* Oxygen must be transported in the blood from the lungs, where the partial pressure of oxygen is relatively high (approximately 100 torr), to the actively metabolizing tissues, where the partial pressure of oxygen is much lower (typically, 20 torr). Let us consider how the cooperative behavior indicated by the sigmoid curve leads to efficient oxygen transport (**Figure 3.17**). In the lungs, hemoglobin becomes nearly saturated with oxygen such that 98% of the oxygen-binding sites are occupied. When hemoglobin moves to the tissues and releases oxygen, the saturation level drops to 32%. Thus, a total of 66% (98% – 32%) of the potential oxygen-binding sites contribute to oxygen transport.

- *Cooperative release of oxygen favors more complete delivery to tissues.* The cooperative release of oxygen favors a more complete unloading of oxygen in the tissues. If myoglobin were employed for oxygen transport, it would be 98% saturated in the lungs but would remain 91% saturated in the tissues, and so only 7% of the sites would contribute to oxygen transport; myoglobin binds oxygen too tightly to be useful in oxygen transport. Nature might have solved this problem by weakening the affinity of myoglobin for oxygen to maximize the difference in saturation between 20 and 100 torr. However, for such a protein, the most oxygen that could be transported from a region in which pO_2 is 100 torr to one in which it is 20 torr is 38% (63% − 25%), as indicated by the blue curve in Figure 3.17. Thus, the cooperative binding and release of oxygen by hemoglobin enables it to deliver nearly 10 times as much oxygen as could be delivered by myoglobin and more than 1.7 times as much as could be delivered by an optimized noncooperative protein.

- *Oxygen is delivered to tissues where it is most needed.* Closer examination of oxygen concentrations in tissues at rest and during exercise underscores the effectiveness of hemoglobin as an oxygen carrier (**Figure 3.18**). Under resting conditions, the oxygen concentration in muscle is approximately 40 torr, but during exercise the concentration is reduced to 20 torr. In the decrease from 100 torr in the lungs to 40 torr in resting muscle, the oxygen saturation of hemoglobin is reduced from 98% to 77%, and so 21% of the oxygen is released over a drop of 60 torr. In a decrease from 40 torr to 20 torr, the oxygen saturation is reduced from 77% to 32%, corresponding to an oxygen release of 45% over a drop of 20 torr. Thus, because the change in oxygen concentration from rest to exercise corresponds to the steepest part of the oxygen-binding curve, oxygen is effectively delivered to tissues where it is most needed. Later, we shall examine other properties of hemoglobin that enhance its physiological responsiveness.

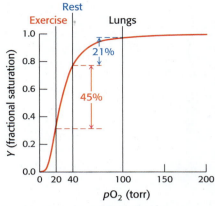

FIGURE 3.18 Oxygen delivery by hemoglobin responds to exercise. The drop in oxygen concentration from 40 torr in resting tissues to 20 torr in exercising tissues corresponds to the steepest part of the oxygen-binding curve. This allows a substantial quantity of oxygen to be released to exercising tissues.

Oxygen binding markedly changes the quaternary structure of hemoglobin

The cooperative binding of oxygen by hemoglobin requires that the binding of oxygen at one site in the hemoglobin tetramer influences the oxygen-binding properties at the other sites. Given the large separation between the iron sites, direct interactions are not possible. Thus, indirect mechanisms for coupling the sites must be at work. These mechanisms are intimately related to the quaternary structure of hemoglobin.

Hemoglobin undergoes substantial changes in quaternary structure on oxygen binding: the $\alpha_1\beta_1$ and $\alpha_2\beta_2$ dimers rotate approximately 15 degrees with respect to one another (**Figure 3.19**). The dimers themselves are relatively unchanged,

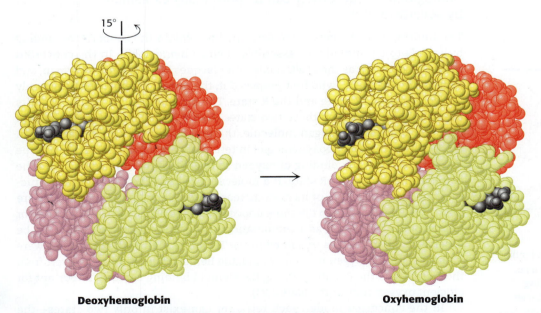

Deoxyhemoglobin   **Oxyhemoglobin**

FIGURE 3.19 Substantial quaternary structural changes occur in hemoglobin upon oxygen binding. Comparing the structures of deoxyhemoglobin and oxyhemoglobin reveals that one αβ dimer has rotated by 15 degrees relative to the other.

INTERACT with this model in
≋ Achieᴠe

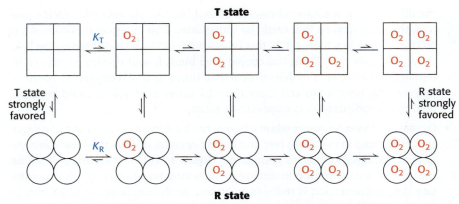

T state

R state

FIGURE 3.20 A concerted model can explain cooperative oxygen binding by hemoglobin. The hemoglobin tetramer can exist in the T state or the R state. With no oxygen bound, the equilibrium between these states strongly favors the T state. As each oxygen binds, the equilibrium shifts so that when all four sites have oxygen bound, the equilibrium strongly favors the R state. Binding sites in the R state have a higher affinity for oxygen than those in the T state.

although there are localized conformational shifts. Thus, the interface between the $\alpha_1\beta_1$ and $\alpha_2\beta_2$ dimers is most affected by this structural transition. In particular, the $\alpha_1\beta_1$ and $\alpha_2\beta_2$ dimers are more free to move with respect to one another in the oxygenated state than they are in the deoxygenated state.

The quaternary structure observed in the deoxy form of hemoglobin, **deoxyhemoglobin**, is often referred to as the **T state** (T for *tense*), because it is quite constrained by subunit–subunit interactions. In contrast, the quaternary structure of the fully oxygenated form, **oxyhemoglobin**, is referred to as the **R state** (R for *relaxed*). In light of the observation that the R form of hemoglobin is less constrained, the tense and relaxed designations seem particularly apt. Importantly, in the R state, the oxygen-binding sites are free of strain and are capable of binding oxygen with higher affinity than are the sites in the T state. By triggering the shift of the hemoglobin tetramer from the T state to the R state, the binding of oxygen to one site increases the binding affinity of other sites.

Hemoglobin cooperativity can be potentially explained by several models

Two limiting models have been developed to explain the cooperative binding of ligands to a multisubunit assembly such as hemoglobin. In the **concerted model**, also known as the MWC model after Jacques Monod, Jeffries Wyman, and Jean-Pierre Changeux, who first proposed it, the overall assembly can exist only in two forms: the T state and the R state. The binding of ligands simply shifts the equilibrium between these two states (**Figure 3.20**). Thus, as a hemoglobin tetramer binds each oxygen molecule, the probability that the tetramer is in the R state increases. Deoxyhemoglobin tetramers are almost exclusively in the T state. However, the binding of oxygen to one site in the molecule shifts the equilibrium toward the R state. If a molecule assumes the R quaternary structure, the oxygen affinity of its sites increases. Additional oxygen molecules are now more likely to bind to the three unoccupied sites. Thus, the binding curve for hemoglobin can be seen as a combination of the binding curves that would be observed if all molecules remained in the T state or if all of the molecules were in the R state. As oxygen molecules bind, the hemoglobin tetramers convert from the T state into the R state, yielding the sigmoid binding curve so important for efficient oxygen transport (**Figure 3.21**).

In the concerted model, each tetramer can exist in only two states—the T state and the R state. In an alternative model, the **sequential model**, the binding of a ligand to one site in an assembly increases the binding affinity of neighboring sites without inducing a full conversion from the T into the R state (**Figure 3.22**).

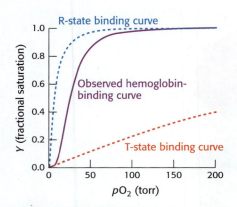

FIGURE 3.21 The sigmoid oxygen-binding curve of hemoglobin can be seen as a combination of the T state binding curve and the R state binding curve. As oxygen binds, more hemoglobin tetramers are converted from the T state to the R state, increasing the oxygen affinity of the available binding sites.

FIGURE 3.22 A sequential model can account for oxygen-binding cooperativity. As each oxygen molecule binds to one site in the hemoglobin tetramer, the affinities of the remaining sites increase, resulting in a sigmoidal binding curve.

Is the cooperative binding of oxygen by hemoglobin better described by the concerted or the sequential model? Neither model in its pure form fully accounts for all of the behavior of hemoglobin. Instead, a combined model is required. Hemoglobin behavior is concerted in that the tetramer with three sites occupied by oxygen is almost always in the quaternary structure associated with the R state. The remaining open binding site has an affinity for oxygen more than 20-fold greater than that of fully deoxygenated hemoglobin binding its first oxygen. However, the behavior is not fully concerted, because hemoglobin with oxygen bound to only one of four sites remains primarily in the T-state quaternary structure. Yet, this molecule binds oxygen three times as strongly as does fully deoxygenated hemoglobin, an observation consistent only with a sequential model. These results highlight the fact that the concerted and sequential models represent idealized limiting cases, which real systems may approach but rarely attain.

❖ SELF–CHECK QUESTION

Which is higher: (a) the affinity of a deoxyhemoglobin molecule for its first oxygen, or (b) the affinity of a hemoglobin molecule with three bound oxygens for its last oxygen?

Structural changes at the heme groups are transmitted to the $\alpha_1\beta_1$–$\alpha_2\beta_2$ interface

We will now examine how oxygen binding at one site is able to shift the equilibrium between the T and R states of the entire hemoglobin tetramer. As in myoglobin, oxygen binding causes each iron atom in hemoglobin to move from outside the plane of the porphyrin into the plane. When the iron atom moves, the proximal histidine residue moves with it. This histidine residue is part of an α helix, which also moves (**Figure 3.23**). The carboxyl terminal end of this α helix lies in the interface between the two αβ dimers. The change in position of the carboxyl terminal end of the helix favors the T-to-R transition. Consequently, the structural transition at the iron ion in one subunit is directly transmitted to the other subunits. The rearrangement of the dimer interface provides a pathway for communication between subunits, enabling the cooperative binding of oxygen.

This mechanism predicts that if the bonds connecting the proximal histidines to the α helices were to be severed, then the resulting hemoglobin molecules would bind oxygen with relatively high affinity, retain the T quaternary structure even upon oxygen binding, and show no or substantially less cooperativity. This thought experiment has actually been realized by replacing the proximal histidines with glycines and adding free imidazole, which binds to the iron ion in place of the histidine side chain (**Figure 3.24**). Such binding studies, along

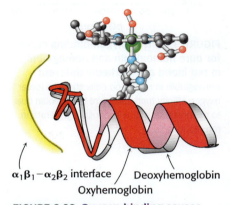

$\alpha_1\beta_1$–$\alpha_2\beta_2$ interface / Deoxyhemoglobin
Oxyhemoglobin

FIGURE 3.23 Oxygen binding causes the iron to move into the porphyrin. The motion of the iron pulls the proximal histidine and the associated helix, transmitting the structure change to the $\alpha_1\beta_1$–$\alpha_2\beta_2$ interface. The structure of oxyhemoglobin is shown in red on top of that for deoxyhemoglobin, in gray.

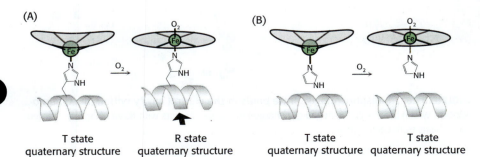

(A)   (B)

T state   R state   T state   T state
quaternary structure   quaternary structure   quaternary structure   quaternary structure

FIGURE 3.24 The movement of the distal histidine and its associated α helix upon oxygen binding drives cooperativity. (A) In normal hemoglobin, the structural changes at iron upon oxygen binding in normal hemoglobin result in motion of the α helix that includes the proximal histidine, indicated by the arrow. (B) In hemoglobin, with the proximal histidine replaced with glycine with added imidazole bound to iron, oxygen binding results in the same changes at iron but the motion of the α helix does not occur.

with structural studies, have confirmed this prediction, although some cooperativity remains, indicating that some other pathways for inter-subunit communication must be present.

❖ SELF–CHECK QUESTION

Why is the oxygen affinity of the hemoglobin variant with the proximal histidines detached (as shown in Figure 3.24) relatively high?

2,3-Bisphosphoglycerate in red blood cells is crucial in determining the oxygen affinity of hemoglobin

The oxygen-binding curves for hemoglobin shown up to this point have been for hemoglobin in red blood cells. Surprisingly, highly purified hemoglobin binds oxygen much more tightly than does hemoglobin in red blood cells (**Figure 3.25**). This dramatic difference is due to the presence within red blood cells of 2,3-bisphosphoglycerate (2,3-BPG).

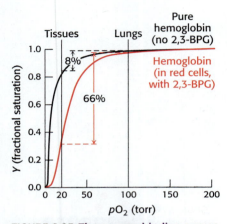

FIGURE 3.25 The oxygen-binding curves for pure hemoglobin and hemoglobin in red blood are markedly different. Hemoglobin in red blood cells shows a lower oxygen affinity than does pure hemoglobin due to the presence of 2,3-bisphosphoglycerate in red blood cells.

2,3-Bisphosphoglycerate (2,3-BPG)

This highly anionic compound is present in red blood cells at approximately the same concentration as that of hemoglobin (~2 mM). Without 2,3-BPG, hemoglobin would be an extremely inefficient oxygen transporter, releasing only 8% of its cargo in the tissues.

How does 2,3-BPG lower the oxygen affinity of hemoglobin so significantly? Examination of the crystal structure of deoxyhemoglobin in the presence of 2,3-BPG reveals that a single molecule of 2,3-BPG binds in the center of the tetramer, in a pocket present only in the T state (**Figure 3.26**). During the T-to-R

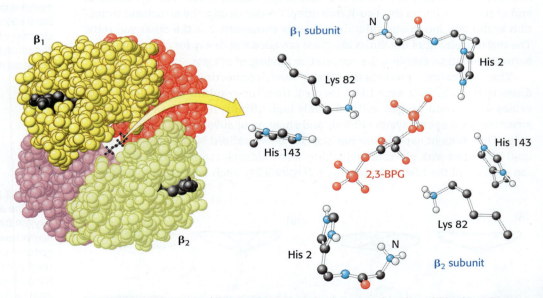

INTERACT with this model in
≋ Achie√e

FIGURE 3.26 2,3-Bisphosphoglycerate binds in the central cavity within deoxyhemoglobin. Within the cavity, the negatively charged molecule interacts with three positively charged side chains from each β chain.

transition, this pocket collapses, and 2,3-BPG is released. Thus, in order for the structural transition from T to R to take place, the interactions between hemoglobin and 2,3-BPG must be disrupted. In the presence of 2,3-BPG, more oxygen-binding sites within the hemoglobin tetramer must be occupied in order to induce the T-to-R transition, and so hemoglobin remains in the lower-affinity T state until higher oxygen concentrations are reached.

This mechanism of regulation is remarkable because 2,3-BPG does not in any way resemble oxygen, the molecule on which hemoglobin carries out its primary function. 2,3-BPG is referred to as an **allosteric effector** (from the Greek *allos*, "other," and *stereos*, "structure"), a molecule that affects the binding of another molecule from a distance. Because they bind in sites other than the active site, allosteric effector molecules can have structures that are completely different from the other molecules to which a protein binds. We will encounter allosteric effectors again when we consider enzyme regulation in Chapter 5. The binding of 2,3-BPG by deoxyhemoglobin but not by oxyhemoglobin is another illustration of the utility of molecular recognition.

The binding of 2,3-BPG to hemoglobin has other crucial physiological consequences. For example, the hemoglobin produced by human fetuses is different from that produced by adults. The differences are relatively subtle, but one of their consequences is that fetal hemoglobin has a lower affinity for 2,3-BPG than does adult hemoglobin. Consequently, the oxygen affinity of fetal red blood cells is higher than that for maternal (adult) red blood cells (**Figure 3.27**). This difference in oxygen affinity allows oxygen to be effectively transferred from maternal to fetal red blood cells, helping to ensure that the fetus receives an adequate oxygen supply.

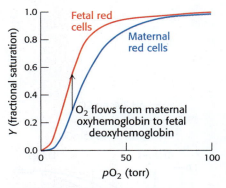

FIGURE 3.27 Fetal red blood cells obtain oxygen from maternal red blood cells by virtue of fetal hemoglobin's lower affinity for 2,3-BPG. This results in fetal hemoglobin having a higher affinity for oxygen.

Hydrogen ions and carbon dioxide promote the release of oxygen

We have seen how cooperative release of oxygen from hemoglobin helps deliver oxygen to where it is most needed: tissues exhibiting low oxygen partial pressures. This ability is enhanced by the facility of hemoglobin to respond to other cues in its physiological environment that signal the need for oxygen. For example, consider rapidly metabolizing tissues, such as contracting muscle, which generate large amounts of hydrogen ions and carbon dioxide. To release oxygen where the need is greatest, hemoglobin has evolved to respond to higher levels of these substances. Like 2,3-BPG, hydrogen ions and carbon dioxide are allosteric effectors of hemoglobin that bind to sites on the molecule that are distinct from the oxygen-binding sites. The regulation of oxygen binding by hydrogen ions and carbon dioxide is called the **Bohr effect** after Christian Bohr, who described this phenomenon in 1904.

The effect of hydrogen ions. The oxygen affinity of hemoglobin decreases as pH decreases from a value of 7.4 (**Figure 3.28**). Consequently, as hemoglobin moves into a region of lower pH, its tendency to release oxygen increases. For example, transport from the lungs, with pH 7.4 and an oxygen partial pressure of 100 torr, to active muscle, with a pH of 7.2 and an oxygen partial pressure of 20 torr, results in a release of oxygen amounting to 77% of total carrying capacity. Only 66% of the oxygen would be released in the absence of any change in pH.

Structural and chemical studies have revealed much about the chemical basis of the Bohr effect. Several chemical groups within the hemoglobin tetramer are important for sensing changes in pH; all of these have pK_a values near pH 7. Consider histidine β146, the residue at the C terminus of the β chain. In its protonated form, the histidine sidechain participates in an ionic interaction with the negatively charged aspartate β94 in the same chain. This interaction locks the terminal carboxylate group of β146 in a position where it can interact

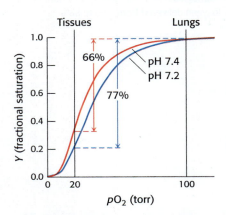

FIGURE 3.28 Lowering the pH decreases the oxygen affinity of hemoglobin and facilitates the release of oxygen into metabolically active tissues. The oxygen-binding curve at pH 7.2 (blue curve) is compared with that at pH 7.4 (red curve).

FIGURE 3.29 Ionic interactions involving the β chain C-terminal histidine are responsible for some of the pH sensitivity of hemoglobin oxygen affinity. As the pH decreases, the likelihood that the imidazole side chain of His β146 is protonated increases. This promotes an ionic interaction between the terminal carboxylate and a lysine side chain from the other αβ dimer, stabilizing the T state.

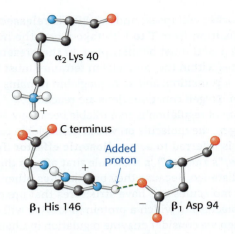

α_2 Lys 40

C terminus

Added proton

β_1 His 146 β_1 Asp 94

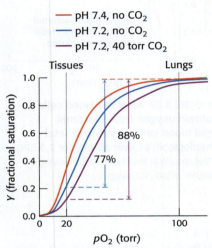

— pH 7.4, no CO_2
— pH 7.2, no CO_2
— pH 7.2, 40 torr CO_2

FIGURE 3.30 The presence of carbon dioxide decreases the oxygen affinity of hemoglobin, even beyond the effect due to a decrease in pH. At a constant pH 7.2, the presence of CO_2 at a level of 40 torr decreases the oxygen affinity of hemoglobin and increases the amount of oxygen delivered from 77% to 88%.

electrostatically with a lysine residue from the α subunit of the other αβ dimer in the T-state conformation (**Figure 3.29**).

In addition to His β146, the α-amino groups at the amino termini of the α chains and the side chain of histidine α122 also participate in ionic interactions in the T state. The formation of these ionic interactions stabilizes the T state, leading to a greater tendency for oxygen to be released.

The effect of carbon dioxide. Carbon dioxide stimulates oxygen release by two mechanisms. First, carbon dioxide reacts with water to form to form bicarbonate ion, HCO_3^- and H^+, resulting in a drop in pH that stabilizes the T state by the mechanism discussed previously. Second, a direct chemical interaction between carbon dioxide and hemoglobin stimulates oxygen release.

The effect of carbon dioxide on oxygen affinity can be seen by comparing oxygen-binding curves in the absence and in the presence of carbon dioxide at a constant pH (**Figure 3.30**). In the presence of carbon dioxide at a partial pressure of 40 torr at pH 7.2, the amount of oxygen released increases to 88% of the maximum carrying capacity.

Carbon dioxide stabilizes deoxyhemoglobin by reacting with the terminal amino groups to form carbamate groups, which are negatively charged, in contrast with the neutral or positive charges on the free amino groups.

$$\underset{H}{\overset{R}{N}}-H + \underset{O}{\overset{O}{C}} \rightleftharpoons \underset{H}{\overset{R}{N}}-\overset{O}{\underset{O}{C}}{}^- + H^+$$

Carbamate

The amino termini lie at the interface between the αβ dimers, and these negatively charged carbamate groups participate in ionic interactions that stabilize the T state, favoring the release of oxygen. The binding of hydrogen ions and carbon dioxide by hemoglobin illustrate another example of molecular recognition based on chemical reactivity.

The binding mode of carbon dioxide is very different from the binding mode for carbon monoxide. Just as for myoglobin, carbon monoxide binds to the heme irons in hemoglobin and competes with oxygen for binding. Carbon monoxide poisoning from burning materials with inadequate ventilation is a frequent cause of emergency department visits and deaths in many countries.

A variant of neuroglobin, a monomeric protein evolutionarily and structurally related to myoglobin and hemoglobin, is being developed as a potential treatment for carbon monoxide that could be administered by first responders. In the course of studying neuroglobin, scientists replaced the distal histidine

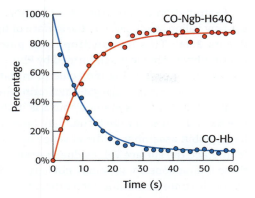

FIGURE 3.31 Carbon monoxide is rapidly transferred from carboxyhemoglobin to Ngb-H64Q. When carboxyhemoglobin (CO-Hb) is mixed with ligand-free Ngb-H64Q, CO is rapidly transferred away from hemoglobin, with exchange complete in less than one minute. [Information from I. Azarov et al., 2016, *Sci. Trans. Med.* 8:368ra173, Fig. 2c.]

with glutamine. This substitution resulted in a protein with an extremely high affinity for oxygen and, most importantly, for carbon monoxide. With three additional changes to improve the solubility, a protein termed Ngb-H64Q was generated which binds carbon monoxide 500 times more tightly that does hemoglobin. When carboxyhemoglobin was mixed with free Ngb-H64Q, CO transfer was very rapid, with almost complete conversion in less than one minute (**Figure 3.31**)! The discovery of this mutant neuroglobin revealed an exciting potential therapeutic option for the treatment of carbon monoxide poisoning, which has since been directly tested in animal studies.

Mutations in genes encoding hemoglobin subunits can result in disease

In modern times, particularly after the sequencing of the human genome, it is routine to think of genetically encoded variations in protein sequence as a factor in specific diseases. But this was not always the case. In 1949, four years prior to the proposal of the structure of the DNA double helix, the notion that diseases might be caused by molecular defects was proposed by Linus Pauling, Harvey Itano, and colleagues to explain the blood disease **sickle-cell anemia**. The name of the disorder comes from the abnormal sickle shape of red blood cells deprived of oxygen in people suffering from this disease (**Figure 3.32**). Based on experimental studies, they proposed that sickle-cell anemia might be caused by a specific variation in the amino acid sequence of one hemoglobin chain. Today, we know that this bold hypothesis is correct.

People with sickled red blood cells experience a number of dangerous symptoms. Examination of the contents of these cells reveals that the hemoglobin molecules have formed large fibrous aggregates (**Figure 3.33**). These fibers extend across the red blood cells, distorting them so that they clog small capillaries and impair blood flow. In addition, red blood cells from sickle-cell patients adhere more readily to the walls of blood vessels than those from normal individuals, prolonging the opportunity to block capillaries. The results may be painful swelling of the extremities and a higher risk of stroke or bacterial infection, due to poor circulation. The sickled red blood cells also do not remain in circulation as long as normal cells do, in part because even a single event when sickling occurs can increase the tendency of the cells to break open, or lyse, leading to anemia.

What is the molecular defect associated with sickle-cell anemia? Vernon Ingram and coworkers demonstrated in 1956 that a single amino acid substitution in the β chain of hemoglobin is responsible—namely, the replacement of a glutamate residue with valine in position 6. The mutated form is referred to as hemoglobin S (HbS). In people with sickle-cell anemia, both alleles (one from the mother and one from the father) of the hemoglobin β-chain gene are mutated. The HbS substitution substantially decreases the solubility of deoxyhemoglobin, although it does not markedly alter the properties of oxyhemoglobin.

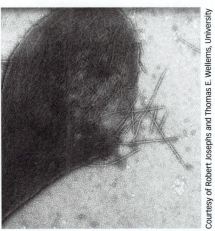

FIGURE 3.32 Red blood cells in patients with sickle-cell anemia adopt a sickle shape when oxygen levels are low. The micrograph shows a sickled red blood cell adjacent to normally shaped red blood cells

Eye of Science/Science Source

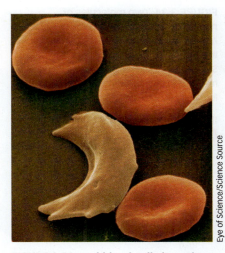

Courtesy of Robert Josephs and Thomas E. Wellems, University of Chicago.

FIGURE 3.33 Hemoglobin in individuals with sickle-cell disease forms long fibers. An electron micrograph depicting a ruptured sickled red blood cell is shown with fibers of sickle-cell hemoglobin emerging.

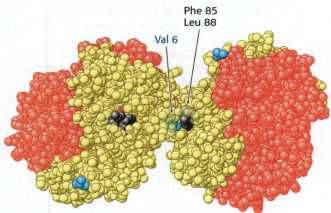

FIGURE 3.34 Two molecules of hemoglobin S interact through hydrophobic patches. These patches involve the Val β6 residue characteristic of hemoglobin S, and Phe β85 and Leu β88 on an adjacent hemoglobin tetramer. This interaction drives the formation of hemoglobin S fibers.

INTERACT with this model in
 Achieve

Examination of the structure of hemoglobin S reveals that the new valine residue lies on the surface of the T-state molecule (**Figure 3.34**). This new hydrophobic patch interacts with another hydrophobic patch formed by Phe 85 and Leu 88 of the β chain of a neighboring molecule to initiate the aggregation process. Why do these aggregates not form when hemoglobin S is oxygenated? When oxyhemoglobin S is in the R state, residues Phe 85 and Leu 88 on the β chain are largely buried inside the hemoglobin assembly. In the absence of a partner with which to interact, the surface Val residue in position 6 is benign. While these interactions are generally not very specific, hemoglobin S fibers illustrate how many such interactions can work together to form a specific oligomeric structure.

3.4 The Immune System Depends on Key Binding Proteins

The immune system faces a daunting challenge in molecular recognition, namely the ability to identify and respond to signs of invading foreign pathogens such as bacteria and viruses in the presence of vast numbers of highly similar molecules from normal cells. Several elaborate and elegant systems have evolved that address the problem of recognition of non-self in the context of self. Here, the term "non-self" can refer to pathogens such as viruses and bacteria but also to tissue from another individual in the case of organ transplantation.

The **innate immune system** is an evolutionarily ancient defense system found, at least in some form, in all multicellular plants and animals. The innate immune system represents a first line of defense against foreign pathogens. It relies on the recognition of classes of molecules common to invading organisms to identify and eliminate these threats.

Because intracellular pathogens do not leave markings on the exteriors of infected cells, vertebrates have evolved mechanisms to display components of interior contents, both self and foreign, on the cell surface. Some internal proteins are broken into peptides, which are then bound to proteins encoded by the **major histocompatibility complex (MHC)**. T cells continually scan the bound peptides to find and kill cells that display foreign motifs on their surfaces. In all of these systems, molecular recognition processes are fundamental, facilitating the detection of components of foreign pathogens and then elimination of the invaders.

The innate immune system recognizes molecules characteristic of pathogens

Perhaps the most important components of the innate immune system are members of a family of receptors that can recognize specific classes of molecules present in many pathogens but normally absent in the host. The best-understood of these receptors are the Toll-like receptors (TLRs). The name "Toll-like" is derived from a receptor known as Toll encoded in the genome of the fruit fly *Drosophila melanogaster*. The TLRs, which are present on the outer surfaces of cells, have a common C-shaped three-dimensional structure (**Figure 3.35**). Each receptor consists of a large domain that sits on the outside of cells and is built primarily from imperfectly repeated amino acid sequences termed leucine-rich repeats (LRRs). Each LRR typically contains 20–30 residues, including 6 that are usually leucine. The human TLRs have from 18 to 27 LRR repeats. In addition to the LRR domains, each TLR has additional domains that anchor the protein to the cell surface and participate in signaling processes inside the cell.

Each TLR targets a specific molecular class, often called a *pathogen-associated molecular pattern (PAMP)*, found primarily on invading organisms. Typically, a PAMP is a critical and widespread component of the pathogen. Examples include certain key molecules found on bacterial cell surfaces and certain classes of

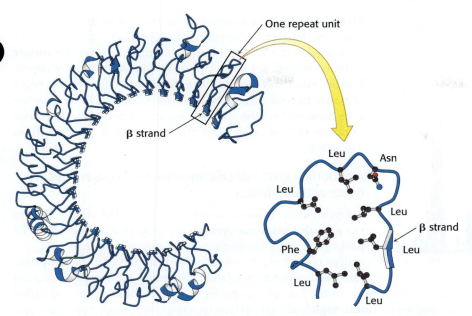

FIGURE 3.35 **The structure of the LRR domain from TLR3 reveals how many LRR units come together to form a structure with binding surfaces for potential partners.** Each LRR has a single β strand and included leucine residues that come together to form a hydrophobic core. The interior and the two sides of the C-like structure are surfaces that recognize potential partners.

nucleic acids, including double-stranded RNAs. RNA is a relative of DNA that has an additional hydroxyl group on each backbone sugar unit and has uracil instead of thymine, present in DNA. Uracil lacks the methyl group in thymine.

RNA molecules are present inside normal cells in single-stranded forms. However, many viruses utilize RNA as their genomes; RNAs outside cells are therefore markers of infections or cellular damage. The difference between uracil and thymine as well as differences in the backbone conformation due to the additional hydroxyl group present opportunities for recognition of RNA versus DNA.

How do TLRs recognize PAMPs? The hook-like structures from the extracellular domains of TLRs have surfaces that are capable of recognizing a variety of a ligands in a wide array of binding modes. For example, the PAMPs for TLR3 are double-stranded RNA molecules of appropriate lengths which fit in the cavity between two monomers (**Figure 3.36**). As might be anticipated from this structure, RNA molecules that are much shorter than 46 base pairs are bound less tightly, as they cannot interact with the two TLR3 molecules optimally at the same time. In addition, studies show that more than 1000 times the concentration of double-stranded DNA molecules are required to achieve the same level of binding as for double-stranded RNA. This can be explained by the fact that double-stranded

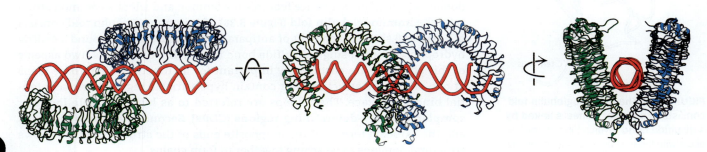

FIGURE 3.36 **The innate immunity component TLR3 recognizes its PAMP, a double-stranded RNA.** The structure of the complex between two molecules of the extracellular domain of TLR3 and a 46-base-pair double-stranded RNA molecule is shown from three perspectives showing how the receptor recognizes the structure and length of the RNA molecule.

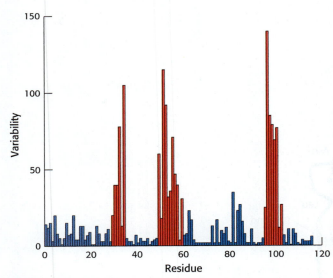

FIGURE 3.37 Immunoglobulin sequences include regions of extensive sequence variability. A plot of sequence variability in a portion of immunoglobulins as a function of position reveals three regions (red) with much greater variability than the rest of the sequence (blue). Sequence variability is measured by the number of different amino acids that were observed within a set of immunoglobulin sequences. [After R. A. Goldsby, T. J. Kindt, and B. A. Osborne, *Kuby Immunology*, 4th ed. (W. H. Freeman and Company, 2000), p. 91.]

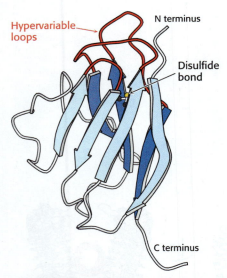

FIGURE 3.38 The immunoglobulin fold consists of a pair of β sheets linked by a disulfide bond. Hydrophobic interactions occur with three hypervariable loops at one end of the structure.

INTERACT with this model in

Achieve

RNA molecules have a somewhat different double-helical structure from those of double-stranded DNA molecules.

Because the TLRs and other components of the innate immune system are always present on the surfaces of appropriate cells, they provide the host organism with a rapid response to resist attack by pathogens. However, many pathogens have evolved the ability to escape detection by the innate immune system. How does the host defend itself in these instances?

Antibodies bind specific molecules through specific hypervariable loops

For protection against novel pathogens, the host relies on the **adaptive immune system**, which is able to target specific pathogens, even those that it has never encountered in the course of evolution. The adaptive immune system comprises two parallel but interrelated responses: humoral and cellular immune responses. In the humoral immune response, soluble proteins called **antibodies** (or **immunoglobulins**) function to recognize foreign molecules and mark them, signaling foreign invasion. In the cellular immune response, cells called **cytotoxic T lymphocytes** (also commonly called **killer T cells**) destroy cells that have been invaded by a pathogen.

How does a system capable of recognizing essentially any foreign molecule function? A clue emerged from comparison of the amino acid sequences of a large number of antibodies of a type referred to as *immunoglobulin G* (IgG). For most proteins, the amino acid sequences from many different individuals, and even across species such as human beings and mice, show only minor variations. However, for portions of these IgG molecules, especially three stretches of approximately 7 to 12 amino acids within each chain, these sequences are highly variable (**Figure 3.37**). These segments are referred to as the *hypervariable regions*.

Based on this observation, one could envisage a process involving a vast collection of potential binding proteins, each of which is constructed with a relatively constant scaffold with other regions that vary substantially from one member of the collection to another. If the collection is large and varied enough, then the odds are that one or more members of the collection have a surface that will recognize any foreign molecule. Molecules that induce immune responses are referred to as **antigens**. The challenge is to find the "needle in the haystack" for the given antigen and then produce the antibody at adequate levels to be of use. This is, indeed, the principle by which the adaptive immune system operates.

Let us look carefully at the structure of IgG to examine this adaptive ability in more detail. An IgG molecule consists of 12 structural domains. These domains have many sequence features in common and adopt a common structure, the **immunoglobulin fold** (**Figure 3.38**). The immunoglobulin fold consists of a pair of β sheets, each built of antiparallel β strands, that surround a central hydrophobic core. A single disulfide bond bridges the two sheets. Two aspects of this structure are particularly important for its function. First, three loops at one end of the structure that contain hypervariable regions form a potential binding surface. These loops are referred to as **hypervariable loops** or **complementarity-determining regions (CDRs)**. Second, the amino terminus and the carboxyl terminus are at opposite ends of the structure, which allows structural domains to be strung together to form chains.

The 12 immunoglobulin fold domains in an IgG molecule are arranged to form a Y-shaped structure (**Figure 3.39**). Each IgG molecule consists of four chains: two so-called *heavy chains*, each of which has four immunoglobulin domains;

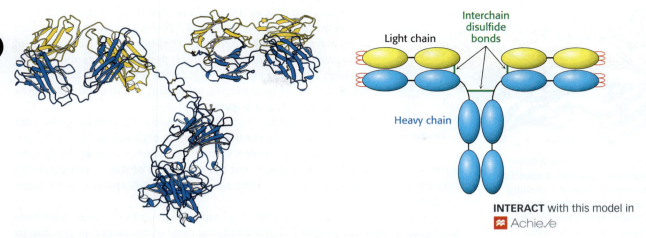

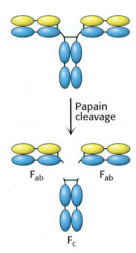

INTERACT with this model in
Achieve

FIGURE 3.39 An IgG molecule has a Y-shaped structure comprising two identical arms with hypervariable regions at the end. The heavy chains are shown in blue, the light chains in yellow, and the hypervariable loops at the end of the two arms in red. The positions of intrachain disulfide bonds are indicated.

and two *light chains*, each of which has two immunoglobulin domains. The two heavy chains are linked together by a disulfide bond, and each heavy chain is also linked to one light chain by an additional disulfide bond. The positions of the CDRs are striking. These hypervariable sequences, present in three loops of each domain, come together so that all six loops come together at the end of each arm.

Biochemists have dissected IgG using a protein called papain, isolated from papaya fruit (**Figure 3.40**). Papain catalyzes the cleavage of certain exposed peptide binds; treatment of an IgG molecule with papain results in two products in a 2:1 ratio. The first products are F_{ab} fragments, which comprise a light chain as well as a portion of the heavy chain; the second product, termed F_c, comprises two disulfide-linked portions of the heavy chain. This dissection allows the binding and structural properties of IgG molecules to be studied more easily; the three-dimensional structure of an F_{ab} fragment shows how the CDRs come together to form a binding surface (**Figure 3.41**).

Let us now examine in detail the interactions in one representative complex between an antibody F_{ab} fragment and a typical protein: lysozyme from egg whites. This antibody binds two polypeptide segments of lysozyme that are widely separated in the primary structure. All six CDRs of the antibody

FIGURE 3.40 Immunoglobulin G molecules can be cleaved into two F_{ab} fragments and one F_c fragment by treatment with papain. The F_{ab} fragments include the antigen binding sites.

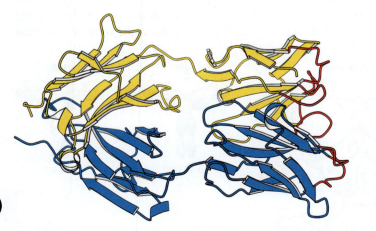

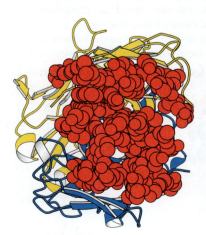

FIGURE 3.41 The structure of an F_{ab} fragment reveals how the six hypervariable loops come together to form a binding surface. (A) Side view. (B) End view looking down on the binding surface.

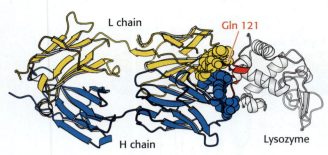

FIGURE 3.42 The structure of a complex between lysozyme and a specific F_{ab} fragment reveals how the binding surface on the F_{ab} interacts with the antigen. The hypervariable loops are shown as space-filling models. A single residue from lysozyme, glutamine 121, penetrates relatively deeply into the antibody.

INTERACT with this model in
 Achie√e

participate directly in molecular recognition. The region of contact is quite extensive (about 30×20 Å), and the apposed surfaces are rather flat. The only exception is the side chain of glutamine 121 of lysozyme, which penetrates deeply into the antibody's binding site, where it forms a hydrogen bond with a main-chain carbonyl oxygen atom and is surrounded by three aromatic side chains (**Figure 3.42**).

The formation of 12 hydrogen bonds and numerous van der Waals interactions contributes to the high affinity ($L_{1/2} = 20$ nM) of this antibody–antigen interaction. Examination of the F_{ab} molecule without bound protein reveals that the structures of the immunoglobulin domains change little on binding, although they slide 1 Å apart to allow more intimate contact with lysozyme.

Studies of thousands of large and small antigens bound to F_{ab} molecules have been sources of much insight into the structural basis of molecular recognition by antibodies. The interaction between complementary shapes results in numerous contacts between amino acids at the binding surfaces of both molecules. Many hydrogen bonds, ionic interactions, and van der Waals interactions, reinforced by hydrophobic interactions, combine to give specific and strong binding.

❖ SELF–CHECK QUESTION

How many total hypervariable loops does an immunoglobulin G molecule have? Not counting duplicates more than once, how many unique loops does it have?

Antibodies possess distinct antigen-binding and effector units

The major antibody in the serum is immunoglobulin G. As noted previously, IgG molecules have Y-shaped structures with two F_{ab} arms that participate in antigen binding and one F_c arm (Figure 3.39). The F_c arm is not merely a scaffold but is a target for molecular recognition itself, as it is recognized by specific proteins on the surfaces of certain cells. This allows the IgG molecule to act as an adaptor, linking an antigen bound by the F_{ab} arms to cells expressing the F_c receptor, initiating other steps that can help destroy the pathogen.

Why does an IgG have more than one F_{ab} arm? Many invaders such as viruses and bacteria have surfaces with repeating structures. Having two F_{ab} arms joined by a flexible hinge allows the IgG to bind to such a surface, with each F_{ab} binding to a different instance of the structural repeat. This allows the IgG to stick more avidly to the surface with a longer dissociation time. The polyvalent interactions are analogous to Velcro®, in which many hooks and loops interact relatively weakly but together can produce a strong interaction.

Recall that an IgG protein is a heterotetramer with two heavy chains and two light chains. Five different classes of antibody—IgG, IgM, IgA, IgD, and IgE—differ in their heavy chains and, hence, in their effector functions (**Figure 3.43**). By virtue

FIGURE 3.43 Five classes of immunoglobulin have the same light chain combined with a different heavy chain (γ, α, μ, δ, or ε). The light chains are shown in yellow, and disulfide bonds are indicated by green lines.

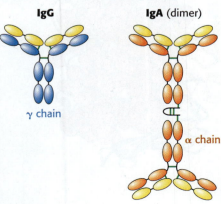
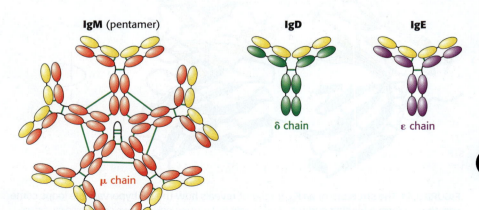

of their different effector domains and, hence, different interaction partners, each of these classes can participate in different physiological processes. For example, IgM has a noteworthy pentameric structure with ten potential antigen-binding sites. IgMs are the first class of antibody produced after an infection; in general, each IgM antigen-binding site has a relatively modest affinity for antigen, but the overall affinity is higher due to the large number of connected potential binding sites.

Recombination events equip the adaptive immune system with millions of unique antibodies

The genes that encode immunoglobulin light and heavy chains undergo fascinating recombination events that introduce different combinations of CDRs, resulting in more than 300 possible light chains and more than 8000 possible IgG heavy chains. As each light chain can come together with each heavy chain, this produces more than $300 \times 8000 = 2,400,000$ potential antibodies. This explains how the adaptive immune system can respond to unknown invaders; it is prepared with many potential molecular recognition templates, some of which will adequately match any new threat.

Through the use of vaccines, the adaptive immune system can be primed for agents known to be potential threats. Vaccines, which expose the immune system to molecules associated with a potential infection without causing serious disease, are perhaps the most powerful public health measures ever developed. Vaccines can be complete bacteria or viruses that have been rendered relatively harmless by various means, or they may be molecular components that trigger protective immune responses. The development of several vaccines against the so-called "spike" protein found on the surface of SARS-CoV-2, the virus that causes COVID-19, were hugely important advances. The vaccines can induce antibodies that bind to the spike protein and block its ability to bind to and infect cells (**Figure 3.44**).

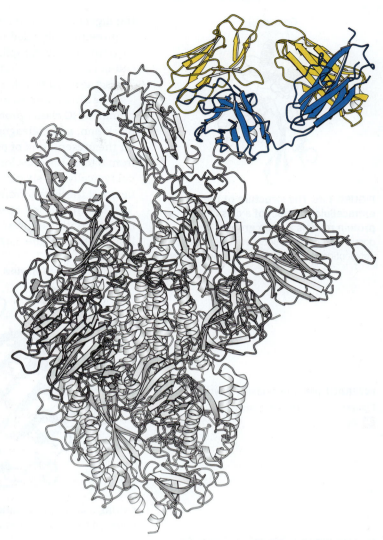

FIGURE 3.44 Antibodies elicited against the spike protein of SARS-CoV-2 can bind to this protein and block the ability of the virus to infect cells. The structure shows the trimeric spike protein targeted by an F_{ab} fragment from one specific antibody.

Major-histocompatibility-complex proteins present peptide antigens on cell surfaces for recognition by T-cell receptors

Soluble antibodies are highly effective against extracellular pathogens, but they confer little protection against microorganisms that are predominantly intracellular, such as some viruses and mycobacteria (which cause tuberculosis and leprosy). These pathogens are shielded from antibodies by the host-cell membrane. A different and more subtle strategy, **cell-mediated immunity**, evolved in response to the damaging properties of intracellular pathogens. The task is not simple; intracellular microorganisms are not so obliging as to leave telltale traces on the cell surfaces of their host without assistance. Quite the contrary, successful pathogens are masters of the art of camouflage. However, vertebrates use an elaborate mechanism—cut and display—to reveal the presence of stealthy intruders. The process starts in the cytoplasm with the cutting step; self-proteins as well as those of pathogens are cleaved into short peptides. Nearly all vertebrate cells exhibit on their surfaces a sample of peptides derived from

Groove

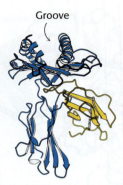

FIGURE 3.45 The structure of the extracellular domain of a class I MHC protein reveals a deep groove. The domain is a complex between two different polypeptide chains, shown in blue and yellow.

INTERACT with the models shown in Figures 3.45, 3.46, and 3.48 in
Achieve

the digestion of proteins in their cytoplasm. These peptides are then displayed by proteins embedded in the cell membrane that are encoded by the major histocompatibility complex (MHC); they are called **class I MHC proteins**. These cell-surface proteins tightly grip their bound peptides, which are then touched and scrutinized by other cells.

The three-dimensional structure of a large extracellular fragment of a human MHC class I protein termed HLA-A2, solved in 1987 by Don Wiley, Pamela Bjorkman, and colleagues, reveals a deep groove that serves as the binding site for the presentation of peptides (**Figure 3.45**). The groove can be filled by a peptide from 8 to 10 residues long in an extended conformation (**Figure 3.46**). An individual typically has six distinct MHC class I proteins with different peptide specificities, and these vary from individual to individual across the population.

The first class I MHC protein to have its structure determined binds peptides that almost always have leucine in the second position and valine in the last position (**Figure 3.47**). Side chains from the MHC molecule interact with the

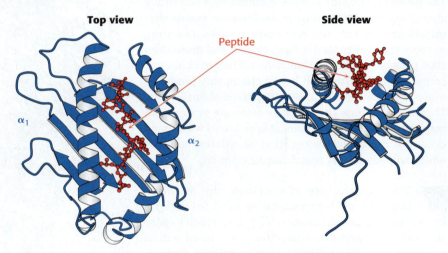

FIGURE 3.46 In class I MHC proteins, peptides can bind in the groove. The peptide (shown in red) binds in a relatively extended conformation.

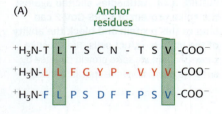

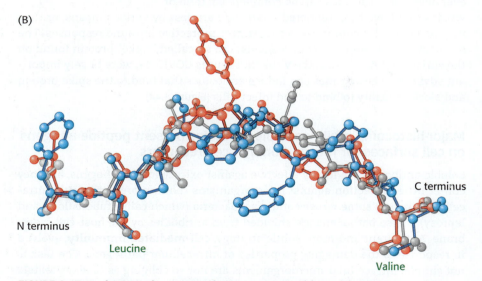

FIGURE 3.47 Each MHC class I protein presents peptides with some recurring features. (A) The amino acid sequences of three peptides that bind to the class I MHC protein HLA-A2 are shown. Each of these peptides has leucine in the second position and valine in the carboxyl-terminal position. (B) Comparison of the structures of these peptides reveals that the amino and carboxyl termini, as well as the side chains of the leucine and valine residues, are in essentially the same positions in each peptide, whereas the remainder of the structures are quite different.

amino and carboxyl termini and with amino acids in these two key positions. These two residues are often referred to as the *anchor residues*. The other residues are highly variable. Thus, many millions of different peptides can be presented by this particular class I MHC protein; the identities of only two of the nine residues are crucial for binding to the MHC protein. Each different MHC molecule recognizes a unique set of anchor residues, so a tremendous range of peptides can be presented by these molecules.

One face of the bound peptide is exposed to solution, where it can be examined by other molecules, particularly receptors on the surfaces of T cells. These T-cell receptors have a structure that is closely related to the F_{ab} portion of an antibody, comprising two chains with immunoglobulin folds and hypervariable loops. Each T-cell receptor is, in principle, capable of recognizing an MHC-peptide complex by forming a structure analogous to an antibody–antigen complex (**Figure 3.48**).

Neither the foreign peptide alone nor the MHC protein alone forms a complex with the T-cell receptor. Molecular recognition depends on the complete structure identifying fragments of an intracellular pathogen, after they are presented in a context that allows their detection and an appropriate response. An assembly consisting of the foreign peptide-MHC complex, the T-cell receptor, and some additional proteins triggers a cascade that induces death of the infected cell.

We close this section by considering an essential biological aspect of molecular recognition in the immune system; namely, the ability to distinguish between self and non-self. How is this achieved? There is no special molecular characteristic of self-proteins that pathogen proteins lack. Instead, the answer lies in developmental biology: antibody-producing cells and T cells that react to self-proteins are identified and blocked before they are released into the circulation, preventing dangerous immune responses to normal tissues. This process can fail, however, as revealed by the occurrence of autoimmune diseases such as type 1 diabetes, rheumatoid arthritis, and dozens of other conditions.

3.5 Quantitative Terms Can Describe Binding Propensity

Molecular recognition is not a "yes or no" process. The tendency of a receptor to bind a particular ligand depends on the ligand concentration. We have thus far used parameters such as $L_{1/2}$ and P_{50} to capture this concentration dependence. In order to describe binding propensity in more quantitative terms, we will now apply concepts from chemical equilibrium for ligand-binding reactions.

Dissociation constants are useful in describing binding reactions quantitatively

Recall the ligand-binding reaction (equation 1)

$$R + L \rightleftharpoons RL$$

The equilibrium constant for this reaction is

$$K = \{RL\}/\{R\}\{L\}$$

where {X} is the thermodynamic activity for component X. These activities are ratios of the concentrations of the components relative to standard states; they are unitless quantities and, therefore, so is K. The standard states are usually

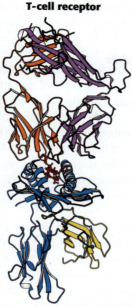

T-cell receptor

MHC class I

FIGURE 3.48 The T-cell receptor recognizes specific class I MHC complexes, contacting both the MHC protein and the bound peptide. The bound peptide is shown in red.

Lance Hayashida/Caltech

PAMELA BJORKMAN Soon after entering graduate school with a strong background in chemistry, Pamela Bjorkman was intrigued by a presentation by Assistant Professor Don Wiley, who was trying to determine the three-dimensional structures of an important viral protein. She also learned that another graduate student in Professor Jack Strominger's laboratory had painstakingly purified substantial amounts of an MHC protein. She convinced the two professors that she should try to solve the MHC structure for her PhD dissertation project. She was stymied in solving the structure by what turned out to be unexpected peptides bound in the groove, but she completed the structure as a postdoctoral fellow. The elucidation of this MHC-peptide structure represents a key milestone in immunology. Dr. Bjorkman has gone on to an extremely productive career at the California Institute of Technology, where she works with students, fellows, and staff to reveal other important structures of the immune system, with an aim to developing interventions that improve human health.

defined as concentrations of 1 M. With certain assumptions, the equilibrium constant can be written in terms of concentration as

$$K = [RL]/[R][L]$$

where [X] is the concentration of component X in molarity. Since each concentration has units of molarity, the equilibrium constant has units of M^{-1}; the appearance of units of molarity reflects the choice of standard states.

The equilibrium constant for this reaction is often written as K_a, where the subscript "a" refers to the fact this this is an association reaction. (This K_a should not be confused with K_a related to acidity and pH.) An equivalent but generally more useful quantity is the equilibrium constant for the reaction written as a dissociation process: $RL \rightleftharpoons R + L$

$$K_d = [R][L]/[RL]$$

K_d has units of concentration (in molarity with standard states of 1 M). These units are easier to understand intuitively than are units of inverse concentration.

Suppose that $[L] = L_{1/2}$. Recall that under these conditions, half of the receptor is bound and half is unbound, therefore, $[R] = [RL]$. Making these two substitutions for [L] and [R] in the equation for K_d above gives us:

$$K_d = [RL][L_{1/2}]/[RL] \text{ or } K_d = L_{1/2}$$

In many experiments, the concentration of the receptor-ligand complex, RL, is measured as the concentration of added L is increased. In general, the concentrations of free R and free L are not measured, but rather the total concentrations of R and L are known. We can use these parameters in the equilibrium constant expression:

$$R_{total} = [R] + [RL] \text{ so } [R] = R_{total} - [RL]$$

$$\text{and } L_{total} = [L] + [RL] \text{ so } [L] = L_{total} - [RL]$$

$$\text{Thus, } K_d = (R_{total} - [RL])(L_{total} - [RL])/[RL]$$

Except for ligands that are bound very tightly, the concentration of [RL] is usually much lower than the total concentration of L added, that is, $L_{total} - [RL] \approx L_{total}$.

Then, $K_d = (R_{total} - [RL])(L_{total} - [RL])/[RL] \approx (R_{total} - [RL])(L_{total})/[RL]$

$$= R_{total}L_{total}/[RL] - L_{total}$$

Rearranging, we find that $K_d + L_{total} = R_{total}L_{total}/[RL]$ so that

$$[RL]/R_{total} = L_{total}/(K_d + L_{total}) \tag{2}$$

Now, suppose that the total added ligand concentration is equal to K_d.

$$[RL]/R_{total} = L_{total}/(K_d + L_{total}) = L_{total}/(L_{total} + L_{total}) = 1/2$$

At this point, the one half of the total receptor concentration is bound to L. Therefore, the dissociation constant K_d is equivalent to $L_{1/2}$. Also notice that $[RL]/R_{total}$ is the fraction of the total receptor bound to a ligand; thus, the degree of saturation of the receptor with a ligand can be found from equation 2 once the K_d value is known (**Figure 3.49**).

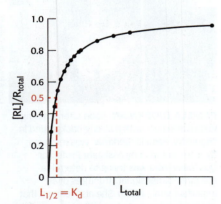

FIGURE 3.49 Binding can be monitored by measuring the fraction of receptors with ligand bound as a function of the amount of ligand added. The data curve is characterized by the value of K_d, which is the same of $L_{1/2}$.

EXAMPLE **Determining the Fraction of Bound Receptors**

PROBLEM: What fraction of receptors are occupied with ligand at a ligand concentration of 5 µM, given that half of the receptors are occupied with ligand at a total ligand concentration of 20 µM?

GETTING STARTED: Equation 2 gives the relationship between the fraction of receptors occupied with ligands as a function of the total ligand concentration and the dissociation constant:

$$[RL]/R_{total} = L_{total}/(K_d + L_{total})$$

We are interested in what happens at a total ligand concentration of 5 μM. In order to use equation 2, we need to know the dissociation constant.

SOLVE: We showed that the half of the receptors are occupied with ligands when the total ligand concentration is equal to the dissociation constant. Since half of the receptors are occupied when the ligand concentration is 20 μM, we know that $K_d = 20$ nM. Substituting this and $L_{total} = 5$ μM into equation 2, we find that

$$[RL]/R_{total} = 5\,\mu M/(20\,\mu M + 5\,\mu M) = 0.20$$

REFLECT: This result makes sense since the total ligand concentration is somewhat less than 20 μM, the ligand concentration at which half (0.50) of the receptors are occupied.

Specificity can be quantified by comparing dissociation constants

Molecular recognition is reflected not just in the affinity of a protein for its preferred ligands, but also in the affinities of the protein for other potentially competitor ligands. For example, estrogen receptors can bind ligands such as estradiol but also competitor ligands such as bisphenol A (BPA), a precursor used in the manufacture of some types of plastics, and tamoxifen, a cancer drug that is bound without receptor activation.

For estradiol, the $L_{1/2}$, which we now know is equal to the dissociation constant K_d, is 1 nM. Thus, if the estradiol concentration is 1 nM, the fraction of receptors that will be occupied by estradiol is 0.5. For BPA, $L_{1/2} = K_d = 200$ nM. If we assume that BPA is present at a concentration of 1 nM, we can apply equation 2 to find that the fraction of receptors occupied is 1 nM/(200 nM + 1 nM) = 0.005. The concentration of BPA must be increased to 200 nM in order for the fraction of receptors to be occupied with BPA to reach 0.5.

Suppose we have estrogen receptors in the presence of 1 nM estradiol and then add 1 μM BPA (**Figure 3.50**). We cannot simply use equation 2 for each ligand, because this equation was derived assuming that $R_{total} = [R] + [RL]$ without any other species present. But with two ligands, say L_1 and L_2, R_{total} must be replaced with $[R] + [RL_1] + [RL_2]$, since the receptor can be unoccupied or bound to either ligand. Using this expression, we can derive the concentrations of RL_1 and RL_2 as a function of the concentrations of L_1 and L_2, although we will not go through this derivation here. With 1 nM estradiol and 1 μM BPA, we find that the fraction of receptors occupied with estradiol is 0.14 and the fraction of receptors occupied with BPA is 0.71. Thus, the fraction of receptors bound to estradiol drops due to the addition of substantial concentrations of BPA. Conceptually, we can understand this in the following way. Initially, half of the receptors are bound to estradiol and half are unoccupied. As the concentration of BPA increases, BPA will bind to the free receptors. This will drive the equilibrium

$$\text{Receptor} + \text{estradiol} \rightleftharpoons \text{Receptor-estradiol}$$

away from the receptor-estradiol complex as the occupied receptor concentration drops.

This analysis provides a quantitative measure of how well estrogen receptors recognize estradiol compared with BPA as an alternative ligand based on their different K_d values; if the two ligands are present at the same concentration,

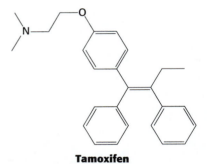

Bisphenol A (BPA)

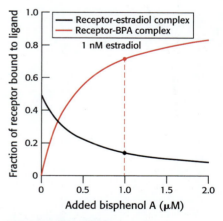

Tamoxifen

FIGURE 3.50 At high enough concentrations, ligands with lower affinity for a receptor can compete for binding with high-affinity ligands. The curves show the concentrations of an estrogen receptor-estradiol complex (black) and an estrogen receptor-BPA complex (red) as a function of increasing concentrations of BPA. At a BPA concentration of 1 μM, the fraction of estrogen receptor bound to estradiol is reduced from 0.5 to 0.14.

FIGURE 3.51 At equilibrium, the rates of the forward and reverse reactions are the same. A ligand-binding reaction is shown. In (A), no ligands are bound and the forward reaction will proceed faster than the reverse reaction to reach the equilibrium distribution in (B). In (C), all receptor sites are filled and the reverse reaction will proceed faster to reach (B). At equilibrium, the forward and reverse reactions can proceed at the same rates to interconvert the distributions in (B) and (D), which have the same fraction of receptors filled.

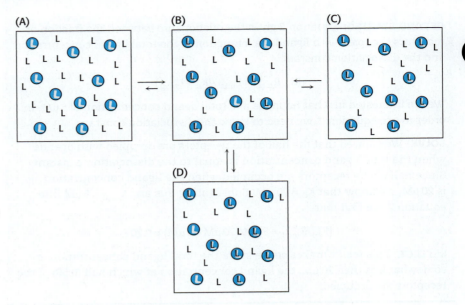

then a much larger fraction of receptors will bind estradiol; a much higher concentration of BPA is required for BPA to successfully compete with estradiol for the receptor-binding sites.

Kinetic parameters can also describe binding processes

Equilibria can also be viewed from the perspective of the rates of forward and reverse reactions.

$$R + L \underset{k_{-1}}{\overset{k_1}{\rightleftharpoons}} RL$$

The forward reaction, the association of R and L to form RL, is a second-order reaction, meaning that its rate depends on the concentrations of two species, R and L. The rate of the forward reaction is given by

$$\text{Rate}_{\text{forward}} = k_1[R][L]$$

The forward rate constant, k_1, has units $M^{-1}s^{-1}$. The reverse reaction, the dissociation of RL, is a first-order reaction, depending only on the concentration of RL. The rate of the reverse reaction is given by

$$\text{Rate}_{\text{revserse}} = k_{-1}[RL]$$

The reverse rate constant, k_{-1}, has units s^{-1}.

At equilibrium, the concentrations of all species are unchanging. The rates of the forward and reverse reactions must be the same even though both reactions are still continuing (**Figure 3.51**). $\text{Rate}_{\text{forward}} = \text{Rate}_{\text{revserse}}$, so that $k_1[R][L] = k_{-1}[RL]$; therefore, $k_{-1}/k_1 = [R][L]/[RL] = K_d$. Thus, we can also determine the dissociation constant of a complex, K_d, by examining the ratio of the reverse rate constant to the forward rate constant.

❖ SELF–CHECK QUESTION

A receptor binds a ligand L with a rate of 10^4 s^{-1} when the ligand concentration is 1 mM. The dissociation rate is 50 s^{-1}. What is the dissociation constant for the receptor-ligand complex?

Let us consider again the complex between the estrogen receptor and estradiol. The forward rate constant is found to be approximately 1.2×10^6 $M^{-1}s^{-1}$, and the reverse rate constant is found to be 1.2×10^{-3} s^{-1}, consistent with $K_d = (1.2 \times 10^{-3}$ $s^{-1})/(1.2 \times 10^6$ $M^{-1}s^{-1}) = 10^{-9}$ M = 1 nM. For comparison, for BPA, the forward rate constant is found to be approximately 1.3×10^6 $M^{-1}s^{-1}$ and the reverse rate constant is 2.7×10^{-1} s^{-1}, consistent with $K_d = (2.7 \times 10^{-1}$ s$^{-1})/(1.3 \times 10^6$ M^{-1}s$^{-1}) = 2 \times 10^{-7}$ M = 210 nM.

Notice that the forward rate constants are quite similar to one another, at approximately 1×10^6 M^{-1}s^{-1}. There is a limit for such forward rate constants, as two molecules cannot form a complex any faster than they can diffuse together and collide. The rate constants for such diffusion processes for proteins and small molecules in solution are approximately 10^9 M^{-1}s^{-1}. Of course, not all collisions result in complex formation. The ratio of the observed rate constant to the diffusion limit of 10^6 M^{-1}s$^{-1}/10^9$ M^{-1}s^{-1} = 0.001 indicates that approximately 1 in 1000 collisions is productive.

Empirically, many forward rate constants for biochemical binding reactions are found to fall in the $10^6 - 10^7$ M^{-1}s^{-1} range, providing a useful tool for estimating such rate constants in the absence of other information, although there are exceptions. For example, the drug tamoxifen binds to the estrogen receptor with a forward rate constant of only 4.5×10^3 M^{-1}s^{-1}. However, the reverse rate constant for the estrogen-tamoxifen complex is 1.0×10^{-3} s^{-1}, consistent with $K_d = (1.0 \times 10^{-3}$ s$^{-1})/(4.5 \times 10^3$ M^{-1}s$^{-1}) = 22 \times 10^{-7}$ M = 220 nM. The strikingly low forward rate constant indicates that only a small number of collusions result in complex formation, consistent with a substantial conformational change being necessary for binding.

We close the chapter with a reminder that rate constants are ensemble properties and, as such, represent the average behavior of the single molecules within the ensemble. For instance, for the estrogen receptor-tamoxifen complex the dissociation rate constant is 1.0×10^{-3} s^{-1}, but if we were to examine such complexes at the single molecule level, we would see a distribution of dissociation times (**Figure 3.52**). The histogram of these dissociation times can be fit to an exponential equation of the form

$$RL(t) = RL_0 e^{-k_{-1}t}$$

The average dissociation time in this distribution is approximately 12 minutes, a duration that is quite long in the context of the biochemical timescale. This may be important for the role of tamoxifen as a drug since, once bound, it remains bound to the receptor and blocks estradiol binding for a substantial period of time.

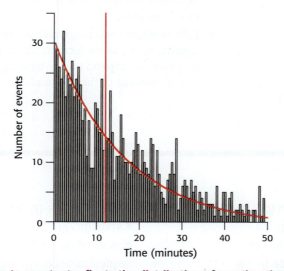

FIGURE 3.52 A rate constant reflects the distribution of reaction times at the single molecule level. The results of a simulation with 1000 events of dissociation times for an estrogen receptor–tamoxifen complex are shown, with dissociation times ranging from nearly zero to more than 50 minutes, with an average of 12 minutes. The rate constant of 1.0×10^{-3} s^{-1} characterizes the exponential decline in this distribution, shown in red.

Summary

3.1 Binding Is a Fundamental Process in Biochemistry

- Many biochemical macromolecules form specific complexes with other macromolecules or metabolites, and these molecular recognition processes facilitate the organization and specificity of binding.
- The tendency of molecules to form complexes increases as the concentration of the partners rises, with the key concentrations required for substantial complex formation varying greatly from one complex to another.
- The concentrations of many biomolecules fall in the range required to drive substantial complex formation.

3.2 Myoglobin Binds and Stores Oxygen

- The muscle protein myoglobin binds oxygen via a bound heme group with a central iron atom.
- Myoglobin binds oxygen via a simple binding reaction; myoglobin is half-saturated with oxygen at a partial pressure of 2 torr, corresponding to a high oxygen affinity.

3.3 Hemoglobin Is an Efficient Oxygen Carrier

- Human hemoglobin consists of four myoglobin-like subunits of two types, forming a homodimer of heterodimeric structures.
- Hemoglobin binds oxygen cooperatively, with relatively low affinity when no oxygen is bound, but higher affinity as each hemoglobin tetramer has already bound an increasing number of oxygen molecules.
- This cooperativity together with changes in the oxygen affinity in response to pH and the presence of carbon dioxide make hemoglobin quite efficient in delivering oxygen to tissues where it is most needed.

- Sickle-cell disease illustrates how the change of a single amino acid in a large protein can lead to disease through the formation of an unusual macromolecular complex.

3.4 The Immune System Depends on Key Binding Proteins

- The immune system includes two components: an innate system that stands ready to respond immediately to signs of pathogens; and an adaptive system that develops specific components tailored to new pathogens encountered by the host.
- Antibodies, also known as immunoglobulins, have structures comprising a conserved framework with hypervariable loops that form binding surfaces able to recognize almost any foreign antigen.
- Class I major compatibility complex proteins include a deep binding groove that presents peptides from foreign proteins on surfaces where they can be sensed by T-cell receptors, leading to an immune response.

3.5 Quantitative Terms Can Describe Binding Propensity

- Binding reactions are chemical equilibria that can be described in quantitative terms, allowing the calculation of the concentrations of various species based on a small number of parameters, including the dissociation constant and kinetic rate constants.
- Different molecules (ligands) can compete for binding to the same receptor site; the relative occupancies depend on the binding affinities and concentrations of these ligands.

Key Terms

molecular recognition (p. 69)

myoglobin (p. 69)

heme (p. 69)

hemoglobin (p. 69)

ligand (p. 69)

fractional saturation (p. 73)

proximal histidine (p. 73)

deoxymyoglobin (p. 73)

oxymyoglobin (p. 73)

distal histidine (p. 74)

carbon monoxide (p. 74)

sigmoid (p. 76)

deoxyhemoglobin (p. 78)

T state (p. 78)

oxyhemoglobin (p. 78)

R state (p. 78)

concerted model (p. 78)

sequential model (p. 78)

allosteric effector (p. 81)

Bohr effect (p. 81)

sickle-cell anemia (p. 83)

innate immune system (p. 84)

major histocompatibility complex (MHC) (p. 84)

adaptive immune system (p. 86)

antibodies (immunoglobulins) (p. 86)

cytotoxic T lymphocytes (killer T cells) (p. 86)

antigen (p. 86)

immunoglobulin fold (p. 86)

hypervariable loops (complementarity-determining regions, CDRs) (p. 86)

cell-mediated immunity (p. 89)

class I MHC proteins (p. 90)

Problems

1. What fraction of estrogen receptors is expected to be free in the presence of an estradiol concentration of 1 nM assuming no other ligands are present? ❖ **1**

2. The molecular weight of sperm whale myoglobin is 17.8 kDa. The myoglobin content of sperm whale muscle is about $80.0 \ \text{g} \cdot \text{kg}^{-1}$. In contrast, the myoglobin content of some human muscles is about $8.00 \ \text{g} \cdot \text{kg}^{-1}$.

(a) Compare the amounts of O_2 bound to myoglobin in human muscle and in sperm whale muscle. Assume that the myoglobin is saturated with O_2, and that the molecular weights of human and sperm whale myoglobin are the same. ❖ **2**

(b) The amount of oxygen dissolved in tissue water at 37°C is about 3.5×10^{-5} M. What is the ratio of myoglobin–bound oxygen to dissolved oxygen in the tissue water of sperm whale muscle?

3. Suppose myoglobin is prepared with Zn^{2+} in the place of Fe^{2+} in the porphyrin. Would you expect this modified myoglobin to bind oxygen? ❖ **2**

4. Does myoglobin exhibit a Bohr effect? Why or why not? ❖ **3**

5. Alkyl isocyanides, R-NC, are bound by myoglobin and hemoglobin.

$$R \underuni{} \overset{+}{N} \equiv \overset{-}{C}:$$

Would you expect that n-hexyl isocyanide would be bound more or less tightly than methyl isocyanide? Briefly explain. ❖ **2**

6. The pK_a of an acid depends partly on its environment. Predict the effect of each of the following environmental changes on the pK_a of a glutamic acid side chain. ❖ **3**

(a) If a lysine side chain is brought into proximity, the pK_a of the glutamic acid chain is _____.

(b) If the terminal carboxyl group of the protein is brought into proximity to the glutamic acid side chain, the pK_a of the side chain is _____.

(c) If the glutamic acid side chain is shifted from the outside of the protein to a nonpolar site inside of the protein, the pK_a is _____.

7. In 1913, Archibald Hill described an alternative formulation for cooperative processes such as the binding of oxygen by hemoglobin by considering the hypothetical equilibrium

$$Hb + nO_2 \rightleftharpoons Hb(O_2)_n$$

Analysis leads to the Hill equation:

$$\log(Y/1 - Y) = \log(pO_2{}^n/P_{50}{}^n) = n \log(pO_2) - n \log(P_{50})$$

where Y is the fractional saturation. This equation suggests that a plot of $\log(Y/1 - Y)$ versus $\log(pO_2)$ should yield a line with slope n.

Plots for myoglobin and hemoglobin are shown below:

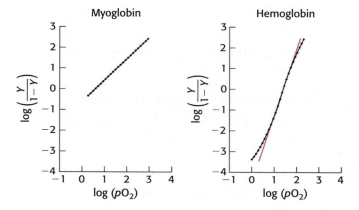

(a) Estimate the apparent slope of the plot for myoglobin and comment on the value.

(b) Estimate the apparent slope in the center of the plot for hemoglobin and comment on the value. ❖ **3**

8. Suppose Gina climbs a high mountain where the oxygen partial pressure in the air decreases to 70 torr. Assume that the pH of her tissues and lungs is 7.4 and the oxygen concentration in her tissues is 20 torr. The P_{50} of hemoglobin is 26 torr. The degree of cooperativity of hemoglobin (n) based on the Hill equation in problem 7 is 2.8.

(a) Estimate the percentage of the oxygen–carrying capacity that she utilizes. Calculate your answer to one decimal place.

After Gina spends a day at the mountaintop, where the oxygen partial pressure is 70 torr, the concentration of 2,3–bisphosphoglycerate (2,3–BPG) in her red blood cells increases.

(b) Why does increasing the concentration of 2,3–BPG in Gina's blood cells help her function well at high altitudes? ❖ **3**

9. Arthropods such as lobsters have oxygen carriers quite different from those of hemoglobin. The oxygen-binding sites do not contain heme but, instead, are based on two copper (Cu) ions. The structural changes that accompany oxygen binding are shown below.

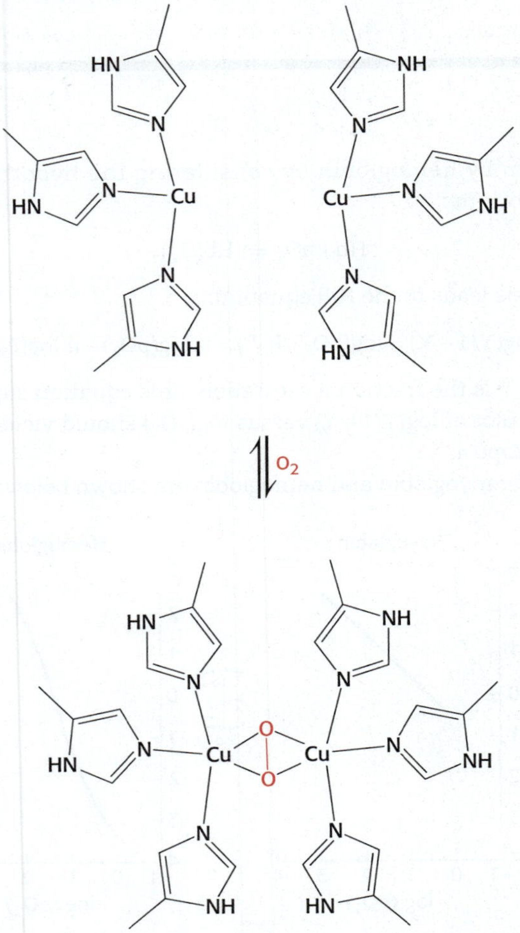

How might these changes be used to facilitate cooperative oxygen binding? ❖ 3

10. Blood cells from some birds do not contain 2,3-bisphosphoglycerate but, instead, contain one of the compounds shown below, which plays an analogous functional role. Which compound do you think is most likely to play this role? Explain briefly. ❖ 3

(a)

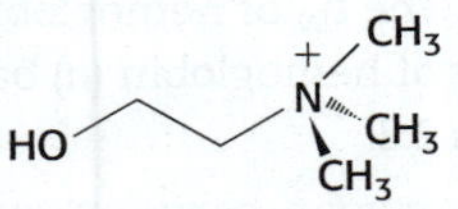

Choline

(b)

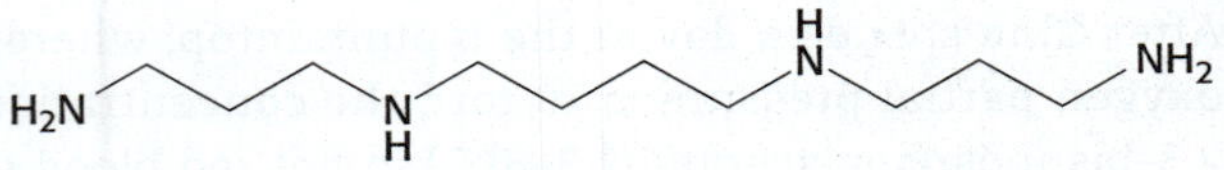

Spermine

(c)

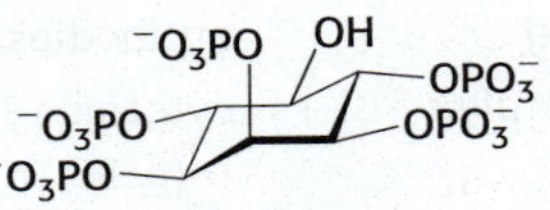

Inositol pentaphosphate

(d)

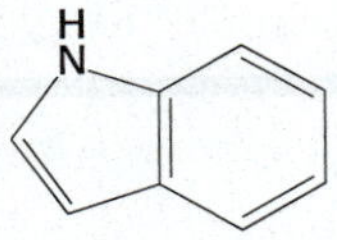

Indole

11. Some Toll-like receptors bind nucleic acids that contain uracil, but not thymine or similar bases. What is the biological rationale for this binding preference? ❖ 4

12. Some of the first vaccines against COVID-19 are based on RNA molecules that lead to the production of the SARS-CoV-2 spike protein. The RNA molecules are modified so that one base is replaced with an alternative. What base is most likely to need to be replaced? ❖ 4

13. Addition of an IgG molecule specific for hemoglobin to a solution of hemoglobin results in the formation of a red precipitate. In contrast, addition of the F_{ab} fragments from this antibody to myoglobin results in no such precipitate. Propose an explanation. ❖ 4

14. The amino acid sequence of a small protein is MSR-LASKNLIRSDHAGGLLQATYSAVSSIKNTMSFGAWSNAALNDS-RDA. Which peptide would most likely be presented by the class I MHC molecule HLA–A2? ❖ 4

(a) LLQATYSAV   (c) NLIRS

(b) MSRLASKNLIRSD   (d) SSIKNTMSF

15. A receptor-ligand complex has a dissociation constant of 5 μM. What is the ratio of the receptor-ligand complex to total receptor when the total receptor concentration is 10 nM with total added ligand concentrations of 1 μM, 10 μM, 100 μM, and 1 mM? ❖ 5

16. Recall that when we calculated that estrogen receptors with 1 nM estradiol and 1 μM BPA, the fraction of receptors occupied with estradiol was 0.14 and the fraction of receptors occupied with BPA was 0.71.

(a) Calculate the concentration of free receptor in terms of the total concentration of receptor, R_{total}.

(b) Calculate the concentration of free estradiol in terms of the total concentration of receptor, R_{total}.

(c) Calculate the concentration of free BPA in terms of the total concentration of receptor, R_{total}.

(d) Confirm the approximate values for the dissociation constants for the estrogen receptor-estradiol and estrogen receptor-BPA complexes. ❖ 5

17. A receptor-ligand complex has a dissociation constant of $K_d = 20$ nM. The rate of receptor-ligand complex formation with an added ligand concentration of $10 \, \mu M$ is $5 \times 10^3 \, s^{-1}$. What is the value of the reverse rate constant, k_{-1}? ❖ **5**

18. An F_{ab} fragment binds to lysozyme with a dissociation constant of $K_D = 10^{-11}$ M. A 1 nM (10^{-9} M) solution of lysozyme is treated with increasing concentrations of the F_{ab} fragment. At what concentration of added F_{ab} will half of the lysozyme be bound to F_{ab}? ❖ **5**

19. A different F_{ab} fragment binds to lysozyme with a dissociation constant of $K_D = 10^{-6}$ M. A 1 nM solution of lysozyme is treated with increasing concentrations of this F_{ab} fragment. At what concentration of added F_{ab} will half of the lysozyme be bound to this F_{ab}? ❖ **5**

20. Deoxyhemoglobin is treated with a 1:1 mixture of oxygen and carbon monoxide under conditions where essentially all of the iron atoms in hemoglobin are bound to a ligand. Approximately what percentage of the irons will be bound to oxygen? ❖ **2,** ❖ **5**

Protein Methods

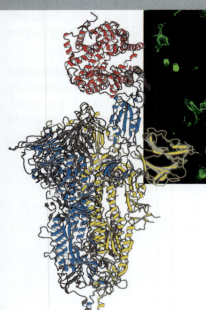

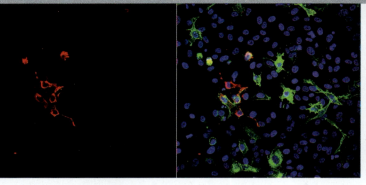

Courtesy of Dr. Peng Zhou, Principle Investigator, Bat virus infection and immunity group, Wuhan Institute of Virology, Chinese Academy of Scienc

In 2019, researchers in China identified a new strain of coronavirus that caused acute respiratory syndrome in several thousand individuals in the city of Wuhan. Using antibodies labeled with fluorescent tags, they showed that the virus (center, labeled red) preferentially infected cells that expressed angiotensin-converting enzyme 2 (ACE2, left, labeled green). Note that the signals overlap in the combined image (right; cell nuclei are stained in blue). These data suggested that ACE2 served as the target for the virus particles. Less than a year later, the structure of the SARS-CoV-2 spike protein bound to ACE2 (shown at left) was solved by cryo-electron microscopy. These experiments were critical in understanding how the virus functions and developing effective vaccines. [Drawn from 7DF4.pdb.]

OUTLINE

4.1 The Purification of Proteins Is an Essential First Step in Understanding Their Function

4.2 Immunology Provides Important Techniques for Investigating Proteins

4.3 Mass Spectrometry Is a Powerful Technique for the Identification of Peptides and Proteins

4.4 Peptides Can Be Synthesized by Automated Solid-Phase Methods

4.5 Three-Dimensional Protein Structures Can Be Determined Experimentally

❖ LEARNING GOALS

By the end of this chapter, you should be able to:

1. Explain the importance of protein purification and how it can be quantified.

2. Describe the different types of chromatography used to purify proteins.

3. Describe the different methods available for determining the mass of a protein.

4. Explain how antibodies can be used as tools for protein identification, purification, and quantitation.

5. Describe how proteins are sequenced, and explain why sequence determination is important.

6. Distinguish between the commonly used methods for protein structure determination, and describe the advantages and disadvantages of each.

A major goal of the biochemist is to understand how individual proteins adopt their unique structures and perform their functions. The first step in these experiments is usually the purification of the protein of interest so as to avoid confounding effects from contaminants. Once purified, a protein's molecular mass and amino acid sequence can be precisely determined. Many protein sequences are now available in vast sequence databases. Such a protein might be identified by matching its mass to those deduced for proteins in these databases.

After a protein has been purified and its identity confirmed, the challenge remains to determine its function and its three-dimensional structure within a physiologically relevant context. Throughout this chapter, we will discuss some of the more commonly used—and still rapidly improving—methods to achieve the purification, functional evaluation, and structural determination of a given protein. The exploration of proteins by this array of physical and chemical techniques has greatly enriched our understanding of the molecular basis of life. These techniques make it possible to tackle some of the most challenging questions of biology in molecular terms.

4.1 The Purification of Proteins Is an Essential First Step in Understanding Their Function

An adage of biochemistry is "never waste pure thoughts on an impure protein." Starting from pure proteins, we can determine amino acid sequences and investigate biochemical functions. From the amino acid sequences, we can map evolutionary relationships between proteins in diverse organisms (Chapter 10). By using crystals grown from pure protein, we can obtain x-ray data that will provide us with a picture of the protein's tertiary structure—the shape that determines function.

The assay: How do we recognize the protein we are looking for?

Purification should yield a sample containing only one type of molecule—the protein in which the biochemist is interested. This protein sample may be only a fraction of 1% of the starting material, whether that starting material consists of one type of cell in culture or a particular organ from a plant or animal. How is the biochemist able to isolate a particular protein from a complex mixture of proteins?

A protein can be purified by subjecting the impure mixture of the starting material to a series of separations based on physical properties such as size and charge. To monitor the success of this purification, the biochemist needs a test, called an **assay**, for some unique identifying property of the protein. A positive result on the assay indicates that the protein is present.

Although assay development can be a challenging task, the more specific the assay, the more effective the purification. For enzymes, the assay usually measures catalytic activity—that is, the ability of the enzyme to promote a particular chemical reaction. This activity is often measured indirectly. Consider the enzyme lactate dehydrogenase, which catalyzes the following reaction in the synthesis of glucose:

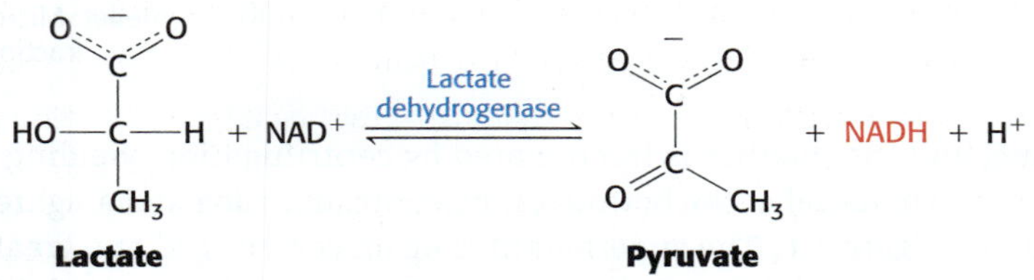

Reduced nicotinamide adenine dinucleotide (NADH, Section 15.4) absorbs light at 340 nm, whereas oxidized nicotinamide adenine dinucleotide (NAD$^+$) does not. Consequently, we can follow the progress of the reaction by examining how much light-absorbing ability is developed by a sample in a given period of time—for instance, within 1 minute after the addition of the enzyme. Our assay for enzyme activity during the purification of lactate dehydrogenase is thus the increase in the absorbance of light at 340 nm observed in 1 minute.

To analyze how our purification scheme is working, we need one additional piece of information—the amount of total protein present in the mixture being assayed. There are various rapid and reasonably accurate means of determining protein concentration. With these two experimentally determined numbers—enzyme activity and protein concentration—we then calculate the **specific activity**, the ratio of enzyme activity to the amount of protein in the mixture. Ideally, the specific activity will rise as the purification proceeds, and the protein mixture will contain the protein of interest to a greater extent. For a pure enzyme, the specific activity will have a constant value. In essence, the overall goal of the purification is to maximize the specific activity.

Proteins must be released from the cell to be purified

Having found an assay and chosen a source of protein, we now fractionate the cell into components and determine which component is enriched in the protein of interest. In the first step, a **homogenate** is formed by disrupting the cell

FIGURE 4.1 Differential centrifugation separates the contents of cells based on their density. Cells are disrupted in a homogenizer and the resulting mixture is centrifuged in a step-by-step fashion of increasing centrifugal force. Centrifugal forces are represented as multipliers of *g*, the force of gravity. The denser material will form a pellet at lower centrifugal force than will the less-dense material. The isolated fractions can be used for further purification.

[Photos credit: Courtesy of S. Fleischer and B. Fleischer.]

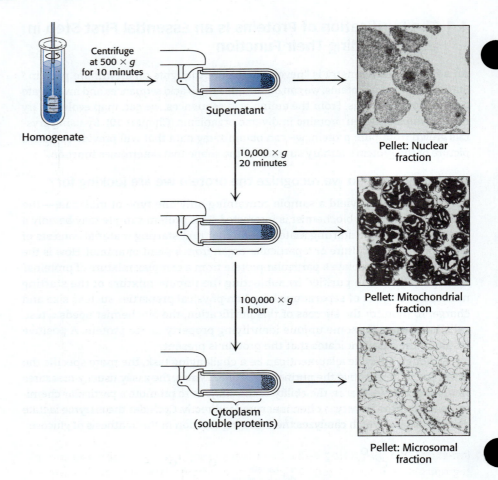

membrane, and the mixture is fractionated by centrifugation, yielding a dense pellet of heavy material at the bottom of the centrifuge tube and a lighter supernatant above (**Figure 4.1**). The supernatant is again centrifuged at a greater force to yield yet another pellet and supernatant. This procedure, called *differential centrifugation*, yields several fractions of decreasing density, each still containing hundreds of different proteins. The fractions are each separately assayed for the desired activity. Usually, one fraction will be enriched for such activity, and it then serves as the source of material to which more-discriminating purification techniques are applied.

Proteins can be purified according to solubility, size, charge, and binding affinity

Thousands of proteins have been purified in active form on the basis of such characteristics as solubility, size, charge, and binding affinity. Usually, protein mixtures are subjected to a series of separations, each based on a different property. At each step in the purification, the preparation is assayed and its specific activity is determined. A variety of purification techniques are available.

Salting out. Most proteins are less soluble at high salt concentrations, an effect called *salting out*. The salt concentration at which a protein falls out of solution, or precipitates, differs from one protein to another. Hence, salting out can be used to fractionate proteins. For example, 0.8 M ammonium sulfate precipitates fibrinogen, a blood-clotting protein, whereas a concentration of 2.4 M is needed to precipitate serum albumin. Salting out is also useful for concentrating dilute solutions of proteins, including active fractions obtained from other purification steps. Dialysis can be used to remove the salt, if necessary.

Dialysis. Proteins can be separated from small molecules such as salt by **dialysis** through a semipermeable membrane, such as a cellulose membrane with pores (**Figure 4.2**). The protein mixture is placed inside the dialysis bag, which is then submerged in a buffer solution that is devoid of the small molecules to be separated away. Molecules having dimensions significantly greater than the pore diameter remain inside the dialysis bag. Smaller molecules and ions capable of passing through the pores of the membrane diffuse down their concentration gradients and emerge in the solution outside the bag. This technique is useful for removing a salt or other small molecule from a cell fractionate, but it does not distinguish between proteins effectively.

Gel-filtration chromatography. More precise separations on the basis of size can be achieved by the technique of **gel-filtration chromatography**, also known as molecular exclusion chromatography (**Figure 4.3**). The sample is applied to the top of a column consisting of porous beads made of an insoluble but highly hydrated polymer such as dextran or agarose (carbohydrates) or polyacrylamide. Sephadex, Sepharose, and Biogel are commonly used commercial preparations of these beads, which are typically 100 µm (0.1 mm) in diameter. Small molecules can enter these beads; large ones cannot.

The result is that small molecules are distributed in the aqueous solution both inside the beads and between them, whereas large molecules are located only in the solution between the beads. Large molecules flow more rapidly through this column and emerge first because a smaller volume is accessible to them. Molecules of medium size occasionally enter the beads and will flow from the column at an intermediate position, while small molecules, which take a longer path through the beads, will exit last.

Ion-exchange chromatography. To obtain a protein of high purity, one chromatography step is usually not sufficient. For example, other proteins in the crude mixture of similar size as the desired protein will likely pass through the column together during a separation using gel filtration chromatography. Additional purity can be achieved by performing sequential separations based on distinct molecular properties.

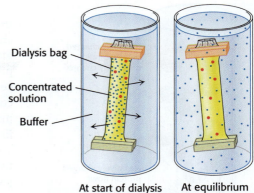

At start of dialysis   At equilibrium

FIGURE 4.2 Dialysis separates proteins from smaller molecules. Protein molecules (red) are retained within the dialysis bag, whereas small molecules (blue) diffuse down their concentration gradient into the surrounding medium.

FIGURE 4.3 Gel-filtration chromatography separates proteins based on size. A mixture of proteins in a small volume is applied to a column filled with porous beads. Because large proteins cannot enter the internal volume of the beads, they emerge sooner than do small ones.

VIEW an animation of this technique in
Achie√e

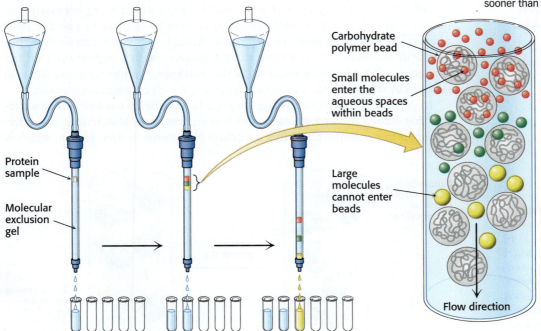

Carbohydrate polymer bead

Small molecules enter the aqueous spaces within beads

Protein sample

Molecular exclusion gel

Large molecules cannot enter beads

Flow direction

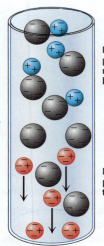

FIGURE 4.4 Ion-exchange chromatography separates proteins based on their net charge. As shown here, proteins with a net positive charge will bind to a column in which the beads are coated with negatively charged groups. Proteins with a net negative charge will pass through much more quickly. This is an example of cation exchange.

VIEW animations of the techniques shown in Figure 4.5 and Figure 4.7 in
Achieve

For example, in addition to size, proteins can be separated on the basis of their net charge by **ion-exchange chromatography**. If a protein has a net positive charge at pH 7, it will usually bind to a column of beads containing carboxylate groups, whereas a negatively charged protein will not (**Figure 4.4**). The bound protein can then be eluted, or released, from the column by increasing the concentration of sodium chloride or another salt in the eluting buffer; sodium ions compete with positively charged groups on the protein for binding to the column. Proteins that have a low density of net positive charge will tend to emerge first, followed by those having a higher charge density.

This procedure is also referred to as **cation exchange** to indicate that positively charged groups will bind to the anionic beads. Positively charged proteins (cationic proteins) can be separated by chromatography on negatively charged carboxymethylcellulose (CM-cellulose) columns. Conversely, negatively charged proteins (anionic proteins) can be separated by **anion exchange** on positively charged diethylaminoethylcellulose (DEAE-cellulose) columns.

Carboxymethyl (CM) group (ionized form)

Diethylaminoethyl (DEAE) group (protonated form)

Affinity chromatography. Another powerful means of purifying proteins that is highly selective for the protein of interest is **affinity chromatography**. This technique takes advantage of the high affinity of many proteins for specific molecules, called *ligands*. For example, the plant protein concanavalin A is a carbohydrate-binding protein that has affinity for glucose. When a crude extract is passed through a column of beads containing covalently attached glucose residues, concanavalin A binds to the beads, whereas most other proteins do not (**Figure 4.5**). The bound concanavalin A can then be eluted from the column by adding a concentrated solution of glucose. At a high concentration, the added glucose will replace the bead-bound glucose as the ligand for concanavalin A, releasing it from the column.

Affinity chromatography is a powerful means of isolating transcription factors—proteins that regulate gene expression by binding to specific DNA sequences. A protein mixture is passed through a column containing specific DNA sequences attached to a matrix; proteins with a high affinity for the sequence will bind and be retained. In this instance, the transcription factor is released by washing with a solution containing a high concentration of salt.

In general, affinity chromatography can be effectively used to isolate a protein that recognizes a particular ligand X by (1) covalently attaching X or a derivative of it to a column; (2) adding a mixture of proteins to this column, which is

FIGURE 4.5 Affinity chromatography enables the separation of proteins that can bind to a specific ligand. In this example, concanavalin A (shown in yellow), a glucose-binding protein, can be retained on a column composed of beads with covalently attached glucose residues (G). After washing away proteins that cannot bind the ligand, concanavalin can be eluted by adding a concentrated solution of glucose.

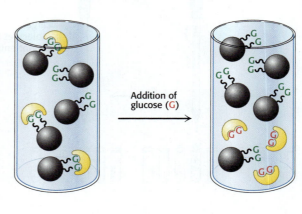

Glucose-binding protein attaches to glucose residues (G) on beads

Addition of glucose (G)

Glucose-binding proteins are released on addition of glucose

then washed with buffer to remove unbound proteins; and (3) eluting the desired protein by adding a high concentration of a soluble form of X or altering the conditions to decrease binding affinity. Affinity chromatography is most effective when the interaction between the protein and its ligand is highly specific.

Affinity chromatography can also be applied to proteins that do not have an identified ligand using the power of recombinant DNA technology (Section 9.2). Simply put, the desired protein is attached to a polypeptide that can bind to a highly specific, known ligand. For example, repeats of the amino acid histidine may be added such that the expressed protein has a string of histidine residues (called a *His tag*) on one end. The tagged proteins are then passed through a column of beads containing covalently attached, immobilized nickel(II). The His tag binds tightly to the immobilized metal ion, capturing the desired protein, while other proteins flow through the column. The tagged protein can then be eluted from the column by the addition of imidazole that binds to the metal ions and displaces the protein.

High-performance liquid chromatography. A technique called **high-performance liquid chromatography (HPLC)** is an enhanced version of the column techniques already discussed. The solid beads within the column have a much smaller diameter, and, as a consequence, possess a greater surface area that yields more interaction sites. With this increase in surface area, HPLC results in greater resolving power. Because the column is made of finer material, pressure must be applied to the column to obtain adequate flow rates. The net result is both sharper separations between proteins and a more rapid separation.

In a typical HPLC setup, the presence of protein emerging from the column is measured in real time. Since the peptide bond absorbs light at a wavelength of 220 nm, a detector that monitors light absorbance at this wavelength is placed immediately after the column. In the sample HPLC elution profile shown in **Figure 4.6**, proteins emerging from the column are readily noted by the presence of sharp peaks measured by the detector. In a short span of 10 minutes, a number of individual proteins can be readily identified.

Proteins can be separated by gel electrophoresis and displayed

How can we tell that a purification scheme is effective? One way is to ascertain that the specific activity rises with each purification step. Another is to determine that the number of different proteins in each sample declines at each step. The technique of electrophoresis makes the latter method possible.

Gel electrophoresis. A molecule with a net charge, such as a protein or a nucleic acid, will move in an electric field. This phenomenon, termed *electrophoresis*, offers a powerful means of separating mixtures of proteins or nucleic acids. Electrophoretic separations are nearly always carried out in porous gels (or on solid supports such as paper) because the gel serves as a molecular sieve that enhances separation (**Figure 4.7**). Molecules that are small compared with the

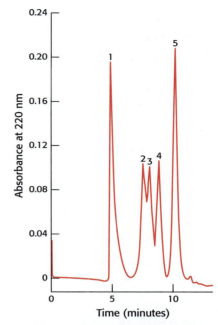

FIGURE 4.6 High-performance liquid chromatography (HPLC) can separate proteins with high resolving power over short elution times. Gel filtration by HPLC clearly defines the individual proteins because of its greater resolving power: (1) thyroglobulin (669 kDa), (2) catalase (232 kDa), (3) bovine serum albumin (67 kDa), (4) ovalbumin (43 kDa), and (5) ribonuclease (13.4 kDa). [Data from K. J. Wilson and T. D. Schlabach. In *Current Protocols in Molecular Biology*, vol. 2, suppl. 41, F. M. Ausubel, R. Brent, R. E. Kingston, D. D. Moore, J. G. Seidman, J. A. Smith, and K. Struhl, Eds. (Wiley, 1998), p. 10.14.1.]

FIGURE 4.7 Polyacrylamide gel electrophoresis can be used to separate and visualize mixtures of proteins. (A) A microliter pipette is used to place solutions of proteins in the wells of the slab. A cover is then placed over the gel chamber and voltage is applied. The negatively charged SDS (sodium dodecyl sulfate)–protein complexes migrate in the direction of the anode, at the bottom of the gel. (B) The sieving action of a porous gel separates proteins according to size, with the smallest moving most rapidly.

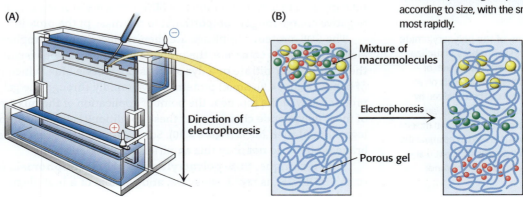

(A)

Direction of electrophoresis

(B)

Mixture of macromolecules

Electrophoresis

Porous gel

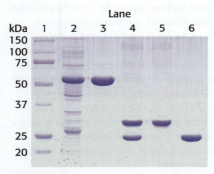

Na⁺
SO₃⁻

Sodium dodecyl sulfate (SDS)

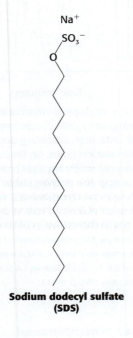

FIGURE 4.9 Staining of the polyacrylamide after electrophoresis enables visualization of the protein bands. Mixtures of proteins subjected to electrophoresis on an SDS–polyacrylamide gel can be visualized by staining with Coomassie blue. The first lane contains a mixture of proteins of known molecular weights, which can be used to estimate the sizes of the bands in the samples. [Credit: Li, et al. (2013) *PLoS One* 8(6): e65682. doi: 10.1371/journal. pone.0065682]

Acrylamide   +   **Methylenebisacrylamide**

$S_2O_8^{2-}$ (persulfate)

$2\ SO_4^{\cdot-}$ (sulfate radical, initiates polymerization)

CONH₂ CONH₂   CONH₂ CONH₂

CONH₂ CONH₂   CONH₂ CONH₂

Polyacrylamide gel

FIGURE 4.8 A polyacrylamide gel is a highly cross-linked three-dimensional mesh. The gel is formed by polymers of acrylamide with intermittently spaced cross-linker (red).

pores in the gel readily move through the gel, whereas molecules much larger than the pores are almost immobile. Intermediate-size molecules move through the gel with various degrees of facility. The electric field is applied such that proteins migrate from the negative to the positive electrodes, typically from top to bottom. **Gel electrophoresis** is performed in a thin, vertical slab of a gel composed of the polymer polyacrylamide **(Figure 4.8)**. Polyacrylamide gels are choice supporting media for electrophoresis because they are chemically inert and readily formed by the polymerization of acrylamide with a small amount of a cross-linking agent to make a three-dimensional mesh.

Proteins can be separated largely on the basis of mass by electrophoresis in a polyacrylamide gel under denaturing conditions. The mixture of proteins is first dissolved in a solution of sodium dodecyl sulfate (SDS), an anionic detergent that disrupts nearly all noncovalent interactions in native proteins. Then, β-mercaptoethanol (2-thioethanol) or dithiothreitol is added to reduce disulfide bonds. Anions of SDS bind to main chains at a ratio of about one SDS anion for every two amino acid residues. The negative charge acquired on binding SDS is usually much greater than the charge on the native protein; the contribution of the protein to the total charge of the SDS–protein complex is thus rendered insignificant. As a result, this complex of SDS with a denatured protein has a large net negative charge roughly proportional to the mass of the protein.

The SDS–protein complexes are then subjected to electrophoresis. When the electrophoresis is complete, the proteins in the gel can be visualized by staining them with silver nitrate or a dye such as Coomassie blue, which reveals a series of bands **(Figure 4.9)**. Small proteins move rapidly through the gel, whereas large proteins stay at the top, near the point of application of the mixture. The mobility of most polypeptide chains under these conditions is linearly proportional to the logarithm of their mass **(Figure 4.10)**. Some carbohydrate-rich proteins and membrane proteins do not obey this empirical relation, however.

This technique, **SDS–polyacrylamide gel electrophoresis** (often referred to as SDS-PAGE), is rapid, sensitive, and capable of a high degree of resolution.

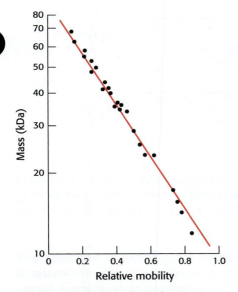

FIGURE 4.10 Electrophoresis can determine the mass of a protein. The electrophoretic mobility of many proteins in SDS–polyacrylamide gels is inversely proportional to the logarithm of their mass.

[Data from K. Weber and M. Osborn, *The Proteins*, vol. 1, 3rd ed. (Academic Press, 1975), p. 179.]

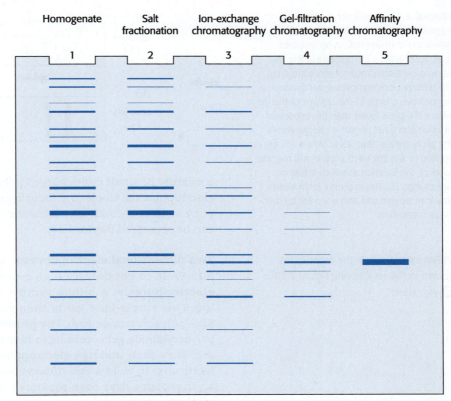

FIGURE 4.11 Electrophoretic analysis can be used to monitor the progress of protein purification. At each step of a protein purification scheme, the sample was analyzed by SDS-PAGE. Each lane contained 50 μg of sample. The effectiveness of the purification can be seen as the band for the protein of interest becomes more prominent relative to other bands.

As little as 0.1 μg (~2 pmol) of a protein gives a distinct band when stained with Coomassie blue, and even less (~0.02 μg) can be detected with a silver stain. Proteins that differ in mass by about 2% (e.g., 50 and 51 kDa, arising from a difference of about 10 amino acids) can usually be distinguished with SDS-PAGE.

We can examine the efficacy of a purification scheme by analyzing a part of each fraction by SDS-PAGE. The initial fractions will display dozens to hundreds of proteins. As the purification progresses, the number of bands will diminish, and the prominence of one of the bands should increase (**Figure 4.11**). This band should correspond to the protein of interest.

❖ SELF–CHECK QUESTION

The relative electrophoretic mobilities of a 30-kDa protein and a 92-kDa protein used as standards on an SDS–polyacrylamide gel are 0.80 and 0.41, respectively. What is the apparent mass of a protein having a mobility of 0.62 on this gel?

Isoelectric focusing. Proteins can also be separated electrophoretically on the basis of their relative contents of acidic and basic residues. The **isoelectric point** (pI) of a protein is the pH at which its net charge is zero. At this pH, its electrophoretic mobility is zero. For example, the pI of cytochrome c, a highly basic electron-transport protein found primarily in mitochondria, is 10.6, whereas that of serum albumin, an acidic protein in blood, is 4.8.

Suppose that a mixture of proteins undergoes electrophoresis in a pH gradient in a gel in the absence of SDS. Each protein will move until it reaches a position in the gel at which the pH is equal to the pI of the protein. This method of separating proteins according to their isoelectric point is called **isoelectric focusing**. The pH gradient in the gel is formed first by subjecting

FIGURE 4.12 Isoelectric focusing separates proteins based on their isoelectric point (pI). A pH gradient is established in a gel before loading the sample. (A) Each protein, represented by the different-colored circles, will possess a net positive charge in the regions of the gel where the pH is lower than its respective pI value and a net negative charge where the pH is greater than its pI. When voltage is applied to the gel, each protein will migrate to its pI, the location at which it has no net charge. (B) The proteins form bands that can be excised and used for further experimentation.

VIEW animations of the techniques shown in Figure 4.12 and Figure 4.13 in
≈ Achieve

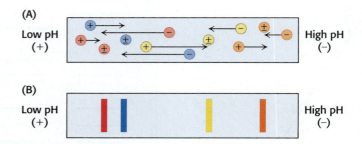

a mixture of small multicharged polymers having many different pI values to electrophoresis. Isoelectric focusing can readily resolve proteins that differ in pI by as little as 0.01, which means that proteins differing by one net charge can be separated (**Figure 4.12**).

Two-dimensional electrophoresis. Isoelectric focusing can be combined with SDS-PAGE to obtain very high-resolution separations by **two-dimensional electrophoresis**. A single sample is first subjected to isoelectric focusing. This single-lane gel is then placed horizontally across the top of an SDS–polyacrylamide slab. The proteins are thus spread across the top of the polyacrylamide gel according to how far they migrated during isoelectric focusing. They then undergo electrophoresis again in a perpendicular direction (vertically) to yield a two-dimensional pattern of spots (**Figure 4.13A**). In such a gel, proteins have been separated in the horizontal direction on the basis of isoelectric point and in the vertical direction on the basis of mass. Remarkably, more than a thousand different proteins in the bacterium *Escherichia coli* can be resolved in a single experiment by two-dimensional electrophoresis (**Figure 4.13B**).

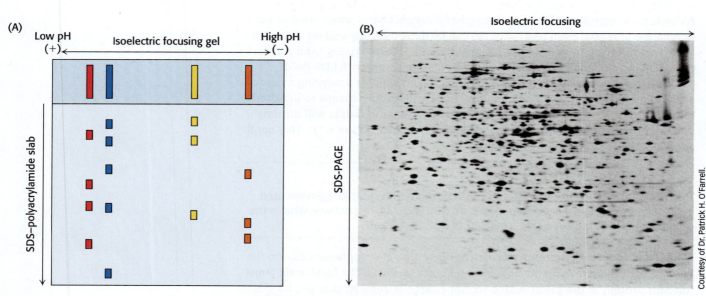

FIGURE 4.13 Two-dimensional electrophoresis uses isoelectric focusing and polyacrylamide gel electrophoresis to separate complex mixtures of proteins. (A) A protein sample is initially fractionated in one dimension by isoelectric focusing as described in Figure 4.12. The isoelectric focusing gel is then attached to an SDS–polyacrylamide gel, and electrophoresis is performed in the second dimension, perpendicular to the original separation. Proteins with the same pI are now separated on the basis of mass. (B) Proteins from *E. coli* were separated by two-dimensional gel electrophoresis, resolving more than a thousand different proteins.

Courtesy of Dr. Patrick H. O'Farrell.

(A)

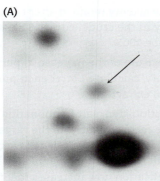

Normal colon mucosa

(B)

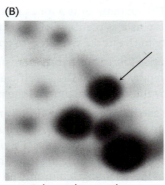

Colorectal tumor tissue

FIGURE 4.14 Alterations in protein levels under differing cellular conditions can be detected by two-dimensional gel electrophoresis. Samples of (A) normal colon mucosa and (B) colorectal tumor tissue from the same person were analyzed by two-dimensional gel electrophoresis. In the gel section shown, changes in the intensity of several spots are evident, including a dramatic increase in levels of the protein indicated by the arrow, corresponding to the enzyme glyceraldehyde-3-phosphate dehydrogenase. [Credit: Republished with permission of The American Society for Biochemistry and Molecular Biology, from Xuezhi Bi, Qingsong Lin, et al., "Proteomic analysis of colorectal cancer reveals alterations in metabolic pathways: mechanism of tumorigenesis," *Mol. Cell Proteomics.* 2006 Jun; 5(6):1119–30, Figure 2. Permission conveyed through Copyright Clearance Center, Inc.]

Two-dimensional electrophoresis can be used to resolve proteins isolated from cells under different physiological conditions. The intensities of individual spots on the gels can then be compared, with any differences indicating that the concentrations of specific proteins have changed in response to the physiological state (**Figure 4.14**). How can we discover the identity of a protein that is showing such responses? Although many proteins are displayed on a two-dimensional gel, they are not identified. We will examine how it is possible to identify proteins by coupling two-dimensional gel electrophoresis with mass spectrometric techniques (Section 4.3).

A protein purification scheme can be quantitatively evaluated

To determine the success of a protein purification scheme, we monitor each step of the procedure by determining the specific activity of the protein mixture and by subjecting it to SDS-PAGE analysis. Consider the results for the purification of a fictitious enzyme, summarized in **Table 4.1** and Figure 4.11. At each step, the following parameters are measured:

- *Total Protein.* The quantity of protein present in a fraction is obtained by determining the protein concentration of a part of each fraction and multiplying by the fraction's total volume.

- *Total Activity.* The total activity for the fraction is obtained by measuring the enzyme activity in the volume of fraction used in the assay and multiplying by the fraction's total volume.

- *Specific Activity.* This parameter is obtained by dividing total activity by total protein.

TABLE 4.1 Quantification of a purification protocol for a fictitious enzyme

Step	Total protein (mg)	Total activity (units)	Specific activity (units mg^{-1})	Yield (%)	Purification level
Homogenization	15,000	150,000	10	100	1
Salt fractionation	4600	138,000	30	92	3
Ion-exchange chromatography	1278	115,500	90	77	9
Gel-filtration chromatography	68.8	75,000	1100	50	110
Affinity chromatography	1.75	52,500	30,000	35	3000

- *Yield*. This parameter is a measure of the activity retained after each purification step as a percentage of the activity in the crude extract. The amount of activity in the initial extract is taken to be 100%.

- *Purification Level*. This parameter is a measure of the increase in purity and is obtained by dividing the specific activity, calculated after each purification step, by the specific activity of the initial extract.

As we see in Table 4.1, the first purification step, salt fractionation, leads to an increase in purity of only 3-fold, but we recover nearly all the target protein in the original extract, given that the yield is 92%. After dialysis to lower the high concentration of salt remaining from the salt fractionation, the fraction is passed through an ion-exchange column. The purification now increases to 9-fold compared with the original extract, whereas the yield falls to 77%. Gel-filtration chromatography brings the level of purification to 110-fold, but the yield is now at 50%. The final step is affinity chromatography with the use of a ligand specific for the target enzyme. This step, the most powerful of these purification procedures, results in a purification level of 3000-fold but lowers the yield to 35%. The SDS-PAGE analysis in Figure 4.11 shows that, if we load a constant amount of protein onto each lane after each step, the number of bands decreases in proportion to the level of purification, and the amount of protein of interest increases as a proportion of the total protein present.

A good purification scheme takes into account both purification levels and yield. A high degree of purification and a poor yield leave little protein with which to experiment. A high yield with low purification leaves many contaminants (proteins other than the one of interest) in the fraction and complicates the interpretation of subsequent experiments.

EXAMPLE Calculating the Effectiveness of Protein Purification

PROBLEM: In the final column chromatography step of a protein purification sequence, the total protein drops from 50 mg to 11 mg, and the total activity goes from 11,000 units to 1040 units. Was this final purification step effective?

GETTING STARTED: The problem provides us with the data for only one purification step within a scheme, the total protein and the total activity before and after the column. Recall that the goal of protein purification is to maximize specific activity. How do we calculate specific activity given the values provided?

SOLVE: The specific activity of a protein sample is simply the activity (usually in units) divided by the total amount of protein present. Before the column, the specific activity of the sample was:

$$\text{Specific activity} = \frac{11,000 \text{ units}}{50 \text{ mg}} = 220 \text{ units mg}^{-1}$$

After the purification step, the specific activity is now:

$$\text{Specific activity} = \frac{1040 \text{ units}}{11 \text{ mg}} = 94.5 \text{ units mg}^{-1}$$

Another helpful equation for assessing efficacy is the purification level. Usually, this is determined by dividing the specific activity of the current sample with the specific activity of the crude extract. The problem does not provide any data about the crude extract, but we can measure the purification level for this particular step:

$$\text{Purification level} = \frac{94.5 \text{ units mg}^{-1}}{220 \text{ units mg}^{-1}} = 0.43$$

REFLECT: What is the significance of this result? In a single purification step, the amount of total protein has decreased, but this is always the case (notice, for example, that in Table 4.1, the amount of total protein drops with each step). More importantly, however, the specific activity of the sample has *decreased* after this column step, resulting in a purification level of less than one. Clearly, this final chromatography step was not effective in the purification of our protein.

Ultracentrifugation is valuable for separating biomolecules and determining their masses

We have already seen that centrifugation is a powerful and generally applicable method for separating a crude mixture of cell components. At higher velocities, this technique, referred to as **ultracentrifugation**, is also valuable for the analysis of the physical properties of biomolecules. Using ultracentrifugation, we can determine such parameters as mass and density, learn about the shape of a molecule, and investigate the interactions between molecules.

A particle will move through a liquid medium when subjected to a centrifugal force. A convenient means of quantifying the rate of movement is the sedimentation coefficient of a particle. Sedimentation coefficients are usually expressed in Svedberg units (S). The smaller the S value, the more slowly a molecule moves in a centrifugal field. The S values for a number of biomolecules are listed in **Table 4.2**. In general, a more massive particle sediments more rapidly than does a less-massive particle of the same shape and density. Also, elongated particles sediment more slowly than do spherical ones of the same mass.

The mass of a protein can be directly determined by sedimentation equilibrium, in which a sample is centrifuged at low speed such that a concentration gradient of the sample is formed. However, this sedimentation is counterbalanced by the diffusion of the sample from regions of high to low concentration. When equilibrium has been achieved, the shape of the final gradient depends solely on the mass of the sample.

The sedimentation-equilibrium technique for determining mass is very accurate and can be applied without denaturing the protein. Thus, the native quaternary structure of multimeric proteins is preserved. In contrast, SDS–polyacrylamide gel electrophoresis provides only an estimate of the mass of dissociated polypeptide chains under denaturing conditions. Note that, if we know the mass of the dissociated components of a multimeric protein as determined by SDS–polyacrylamide analysis and the mass of the intact multimer as determined by sedimentation-equilibrium analysis, we can determine the number of copies of each polypeptide chain present in the protein complex.

TABLE 4.2 S values and molecular weights of sample proteins

Protein	S value (Svedberg units)	Molecular weight
Pancreatic trypsin inhibitor	1	6520
Cytochrome *c*	1.83	12,310
Ribonuclease A	1.78	13,690
Myoglobin	1.97	17,800
Trypsin	2.5	23,200
Carbonic anhydrase	3.23	28,800
Concanavalin A	3.8	51,260
Malate dehydrogenase	5.76	74,900
Lactate dehydrogenase	7.54	146,200

Source: T. Creighton, *Proteins*, 2nd ed. (W. H. Freeman and Company, 1993), Table 7.1.

Recombinant DNA technology can make protein purification easier

In Chapter 9, we shall consider the widespread effect of recombinant DNA technology on all areas of biochemistry and molecular biology. The application of recombinant methods to the overproduction of proteins has enabled dramatic advances in our understanding of their structure and function. Before the advent of this technology, proteins were isolated solely from their native sources, often requiring a large amount of tissue to obtain a sufficient amount of protein for analytical study. For example, the purification of bovine deoxyribonuclease in 1946 required nearly ten pounds of beef pancreas to yield one gram of protein. As a result, biochemical studies on purified material were often limited to abundant proteins.

Recombinant DNA technology affords a number of advantages in the production and purification of proteins:

- *Proteins can be expressed in large quantities.* For recombinant systems, a host organism that is amenable to genetic manipulation, such as the bacterium *E. coli* or the yeast *Pichia pastoris*, is utilized to express a protein of interest. The biochemist can exploit the short doubling times and ease of genetic manipulation of such organisms to produce large amounts of protein from manageable amounts of culture. As a result, purification can begin with a homogenate that is often highly enriched with the desired molecule. Moreover, a protein can be easily obtained regardless of its natural abundance or its species of origin.

- *Affinity tags can be fused to proteins.* As described above, affinity chromatography can be a highly selective step within a protein purification scheme. Recombinant DNA technology enables the attachment of any one of a number of possible affinity tags to a protein (such as the "His tag" mentioned earlier). Hence, the benefits of affinity chromatography can be realized even for those proteins for which a binding partner is unknown or not easily determined.

- *Proteins with modified primary structures can be readily generated.* Recombinant DNA technology gives biochemists the ability to manipulate genes to generate variants of a native protein sequence. Recall from Section 2.5 that many proteins consist of compact domains connected by flexible linker regions. With the use of recombinant DNA technology, fragments of a protein that encompass single domains can be generated, an advantageous approach when expression of the entire protein is limited by its size or solubility. Additionally, amino acid substitutions can be introduced into a protein to precisely probe the significance of these residues.

4.2 Immunology Provides Important Techniques for Investigating Proteins

The purification of a protein enables the biochemist to explore its function and structure within a precisely controlled environment. However, the isolation of a protein removes it from its native context within the cell, where its activity is most physiologically relevant. Advances in the field of immunology have enabled the use of antibodies as critical reagents for exploring the functions of proteins within the cell. The exquisite specificity of antibodies for their target proteins provides a means to tag a specific protein so that it can be isolated, quantified, or visualized.

Antibodies to specific proteins can be generated

Immunological techniques begin with the generation of antibodies to a particular protein. Recall from Section 3.4 that antibodies have specific and high affinity for the antigens—foreign proteins, polysaccharides, and nucleic acids—that stimulated their production. Small foreign molecules, such as synthetic peptides, also can elicit antibodies, provided that the small molecule is attached to a macromolecular carrier. An antibody recognizes a specific group or cluster of amino acids on the target molecule called an **antigenic determinant** or **epitope**.

Animals have a very large repertoire of antibody-producing cells, each producing an antibody that contains a unique surface for antigen recognition. When an antigen is introduced into an animal, it is recognized by a select few cells from this population, stimulating the proliferation of these cells. This process ensures that more antibodies of the appropriate specificity are produced.

Immunological techniques depend on the ability to generate antibodies to a specific antigen. To obtain antibodies that recognize a particular protein, a biochemist injects the protein into a rabbit or other lab mammal. The injected protein acts as an antigen, stimulating the reproduction of cells producing antibodies that recognize it. Blood is drawn from the immunized rabbit several weeks later and centrifuged to separate blood cells from the supernatant. This supernatant contains antibodies to all antigens to which the rabbit has been exposed, and is therefore called *antiserum*.

Only some of the antibodies in the antiserum will react with the injected protein. Moreover, antibodies that recognize a particular antigen are not a single molecular species. For instance, when 2,4-dinitrophenol (DNP) is used as an antigen to generate antibodies, the dissociation constants of individual anti-DNP antibodies range from about 0.1 nM to 1 μM. Correspondingly, a large number of bands are evident when the anti-DNP antibody mixture is subjected to isoelectric focusing.

These results indicate that cells are producing many different antibodies, each recognizing a different surface feature of the same antigen. These antibodies are termed **polyclonal antibodies**, referring to the fact that they are derived from multiple antibody-producing cell populations (**Figure 4.15**). The heterogeneity of polyclonal antibodies can be advantageous for certain applications, such as the detection of a protein of low abundance, because each protein molecule can be bound by more than one antibody at multiple distinct antigenic sites.

Monoclonal antibodies with virtually any desired specificity can be readily prepared

The discovery of a means of producing **monoclonal antibodies** of virtually any desired specificity was a major breakthrough that intensified the power of immunological approaches. As when working with impure proteins, working with a mixture of different antibodies makes it difficult to interpret data. Ideally, biochemists would isolate a clone of cells producing a single, identical antibody. The problem is that antibody-producing cells isolated from an organism have short life spans.

Immortal cell lines that produce monoclonal antibodies do exist. These cell lines are derived from a type of cancer called *multiple myeloma*, which is a malignant disorder of antibody-producing cells. In this cancer, a single transformed plasma cell (a type of white blood cell) divides uncontrollably, generating a very large number of cells of a single kind. Such a group of cells is called a *clone* because the cells are descended from the same cell and have identical properties. The identical cells of the myeloma secrete large amounts of a single immunoglobulin generation after generation. While these antibodies have proven useful for elucidating antibody structure, nothing is known about their specificity. Hence, they have little utility for the immunological methods described in the next pages.

César Milstein and Georges Köhler discovered that large amounts of antibodies of nearly any desired specificity can be obtained by fusing a short-lived antibody-producing cell with an immortal myeloma cell. An antigen is injected

Polyclonal antibodies

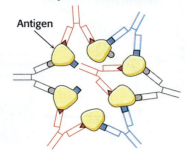

Antigen

Monoclonal antibodies

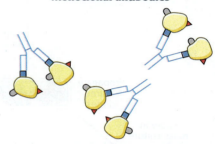

FIGURE 4.15 Polyclonal and monoclonal antibodies differ in the number of epitopes they recognize. Most antigens have several epitopes. Polyclonal antibodies are heterogeneous mixtures of antibodies, each specific for one of the various epitopes on an antigen. Monoclonal antibodies are all identical, produced by clones of a single antibody-producing cell. They recognize one specific epitope. [Information from R. A. Goldsby, T. J. Kindt, and B. A. Osborne, *Kuby Immunology*, 4th ed. (W. H. Freeman and Company, 2000), p. 154.]

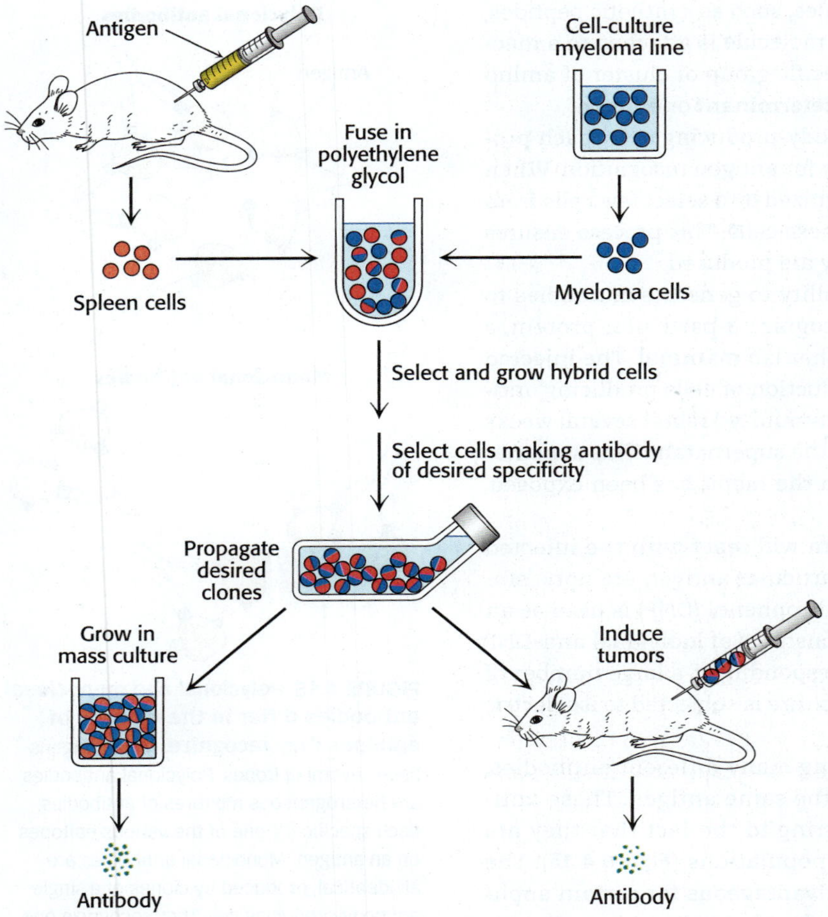

Antigen

Cell-culture myeloma line

Fuse in polyethylene glycol

Spleen cells

Myeloma cells

Select and grow hybrid cells

Select cells making antibody of desired specificity

Propagate desired clones

Grow in mass culture

Induce tumors

Antibody

Antibody

FIGURE 4.16 Monoclonal antibodies are produced by hybridoma cells formed by the fusion of antibody-producing cells and myeloma cells. The hybrid cells are allowed to proliferate by growing them in selective medium. They are then screened to determine which ones produce antibody of the desired specificity. [Information from C. Milstein. Monoclonal antibodies. Copyright © 1980 by Scientific American, Inc. All rights reserved.]

into a mouse, and its spleen is removed several weeks later (**Figure 4.16**). A mixture of plasma cells from this spleen is then fused in vitro with myeloma cells. Each of the resulting hybrid cells, called **hybridoma** cells, indefinitely produces the identical antibody specified by the parent cell from the spleen. Hybridoma cells can then be screened by a specific assay for the antigen–antibody interaction to determine which ones produce antibodies of the preferred specificity. Collections of cells shown to produce the desired antibody are subdivided and assayed again. This process is repeated until a pure cell line, a clone producing a single antibody, is isolated. These positive cells can be grown in culture medium or injected into mice to induce myelomas. Alternatively, the cells can be frozen and stored for long periods.

The hybridoma method of producing monoclonal antibodies has had a dramatic impact on biology and medicine. Biochemists can now readily prepare large amounts of identical antibodies with tailor-made specificities and use them to gather insight into the relationship between antibody structure and specificity. Monoclonal antibodies attached to solid supports can be used as affinity columns to purify scarce proteins. For example, this method has been used to purify interferon (an antiviral protein) 5000-fold from a crude mixture. Moreover, as we shall see below, biologists can use monoclonal antibodies as precise analytical and preparative reagents.

Clinical laboratories use monoclonal antibodies in many assays. For example, the detection in blood of proteins that are normally localized in the heart points to a heart attack. Blood transfusions have been made safer by antibody screening of donor blood for viruses that cause AIDS (acquired immune deficiency syndrome), hepatitis, and other infectious diseases. Monoclonal antibodies can also be used as therapeutic agents (Chapter 32). For example, trastuzumab (Herceptin) is a monoclonal antibody useful for treating some forms of breast cancer.

Proteins can be detected and quantified by using an enzyme-linked immunosorbent assay

Antibodies can be used as exquisitely specific analytic reagents to quantify the amount of a protein or other antigen present in a biological sample. The **enzyme-linked immunosorbent assay (ELISA)** makes use of an enzyme, such as horseradish peroxidase or alkaline phosphatase, that reacts with a colorless substrate to produce a colored product. The enzyme is covalently linked to a specific antibody that recognizes a target antigen. If the antigen is present, the antibody–enzyme complex will bind to it and, on addition of the substrate, the enzyme will catalyze the reaction, generating the colored product. Thus, the presence of the colored product indicates the presence of the antigen. Rapid and convenient, ELISAs can detect less than a nanogram (10^{-9} g) of a specific protein. ELISAs can be performed with either polyclonal or monoclonal antibodies, but the use of monoclonal antibodies yields more-reliable results.

Of the several types of ELISA that exist, we will further examine two types:

- *The indirect ELISA is used to detect the presence of antibodies.* This method is the basis of the test for HIV infection. The HIV test detects the presence of antibodies that recognize viral core protein antigens. Viral core proteins are coated on the bottom of a well. Antibodies from the person being tested are then added to the coated well. Only someone infected with HIV will have antibodies that bind to the antigen. Finally, enzyme-linked antibodies to human antibodies (e.g., enzyme-linked goat antibodies that recognize human antibodies) are allowed to react in the well, and unbound antibodies are removed by washing. Substrate is then applied. An enzyme reaction yielding a colored product suggests that the enzyme-linked antibodies were bound to human antibodies, which in turn implies that the patient has antibodies to the viral antigen (**Figure 4.17A**). This assay is quantitative: the rate of the color-formation reaction is proportional to the amount of antibody originally present.

- *The sandwich ELISA is used to detect an antigen rather than an antibody.* An antibody to a particular antigen—the capture antibody—is first coated on the bottom of a well. Next, solution containing the antigen (such as blood or urine, in medical diagnostic tests) is added to the well and binds to the antibody. Finally, a second antibody—the detection antibody—that also recognizes the antigen is added. Importantly, this antibody binds to the antigen in a region distinct from, and thus does not displace, the first antibody. This antibody is enzyme-linked and is processed as described for indirect ELISA. In this case, the rate of color formation is directly proportional to the amount of antigen present. Consequently, it permits the measurement of small quantities of antigen (**Figure 4.17B**).

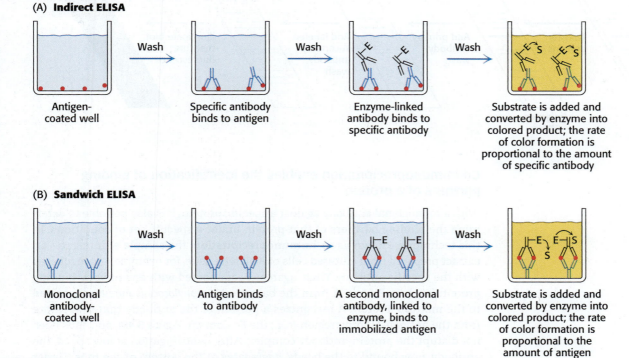

(A) Indirect ELISA

| Antigen-coated well | Specific antibody binds to antigen | Enzyme-linked antibody binds to specific antibody | Substrate is added and converted by enzyme into colored product; the rate of color formation is proportional to the amount of specific antibody |

(B) Sandwich ELISA

| Monoclonal antibody-coated well | Antigen binds to antibody | A second monoclonal antibody, linked to enzyme, binds to immobilized antigen | Substrate is added and converted by enzyme into colored product; the rate of color formation is proportional to the amount of antigen |

FIGURE 4.17 Indirect ELISA and sandwich ELISA are sensitive methods for detecting the presence of antigen or antibody in a sample. (A) In indirect ELISA, the production of color indicates the amount of an antibody to a specific antigen. (B) In sandwich ELISA, the production of color indicates the quantity of antigen. [Information from R. A. Goldsby, T. J. Kindt, and B. A. Osborne, *Kuby Immunology*, 4th ed. (W. H. Freeman and Company, 2000), p. 162.]

Western blotting permits the detection of proteins separated by gel electrophoresis

Very small quantities of a protein of interest in a cell or in body fluid can be detected by an immunoassay technique called **western blotting** (Figure 4.18). A sample is subjected to electrophoresis on an SDS–polyacrylamide gel. A polymer sheet is pressed against the gel, transferring the resolved proteins on the gel to the sheet, which makes the proteins more accessible for reaction. An antibody that is specific for the protein of interest, called the *primary antibody*, is added to the sheet and reacts with the antigen. The antibody–antigen complex on the sheet can then be detected by rinsing the sheet with a second antibody, called the *secondary antibody*, that is specific for the primary antibody (e.g., a goat antibody that recognizes mouse antibodies). Typically, the secondary antibody is fused to an enzyme that produces a chemiluminescent or colored product or contains a fluorescent tag, enabling the identification and quantitation of the protein of interest.

Western blotting makes it possible to find a protein in a complex mixture, the proverbial needle in a haystack. It is the basis for the test for infection by hepatitis C, where it is used to detect a core protein of the virus. This technique is also very useful in monitoring protein purification and in the cloning of genes.

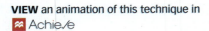

VIEW an animation of this technique in Achie/e

FIGURE 4.18 Western blotting uses antibodies to detect specific proteins after gel electrophoresis. Proteins on an SDS–polyacrylamide gel are transferred to a polymer sheet. The sheet is first treated with a primary antibody, which is specific for the protein of interest, and then washed to remove unbound antibody. Next, the sheet is treated with a secondary antibody, which recognizes the primary antibody, and washed again. Since the secondary antibody is labeled (here, with a fluorescent tag indicated by the yellow circle), the band containing the protein of interest can be identified.

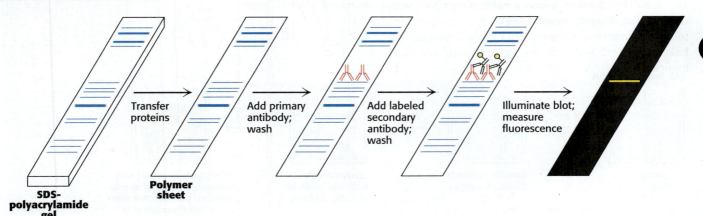

Co-immunoprecipitation enables the identification of binding partners of a protein

With a monoclonal antibody against a specific protein, it is also possible to determine the binding partners of that protein under a specific set of conditions. In this technique, known as **co-immunoprecipitation**, the sample of interest—an extract prepared from cultured cells or isolated tissue, for example—is incubated with the specific antibody. Then, agarose beads coated with an antibody-binding protein (such as Protein A from the bacterium *Staphylococcus aureus*) are added to the mixture. Protein A recognizes a portion of the antibody that is separate from the antigen-binding region (e.g., the Fc domain, Figure 3.40), and thus does not disrupt the protein-antibody complex. After centrifugation at low speed, the antibody, now bound to the beads, aggregates at the bottom of the tube. Under optimal buffer conditions, the antibody-protein complex will also precipitate any additional proteins that are bound to the original protein (Figure 4.19). Subsequent analysis of the precipitate by SDS-PAGE, followed by either western blot or mass spectrometric fingerprinting (Section 4.3), enables the identification of the binding partners.

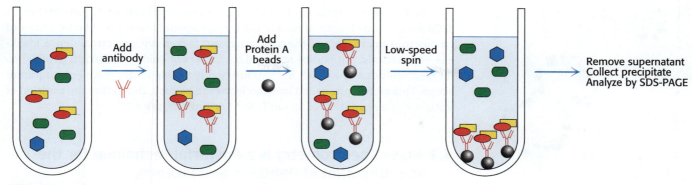

Cell or tissue extract

FIGURE 4.19 Co-immunoprecipitation can identify binding partners to a specific protein. An antibody specific for a particular protein (here, the red oval) is added to an extract isolated from cells or tissue. After an incubation period, agarose beads coated with Protein A are added, and the mixture is subjected to low-speed centrifugation. The beads bind to the antibody-protein complex and precipitate. Any additional proteins that interact with the target protein (here, the yellow rectangle) will also precipitate and can be identified by SDS-PAGE followed by western blotting or mass spectrometry.

Fluorescent markers make the visualization of proteins in the cell possible

Fluorescent markers provide a powerful means of examining proteins in their biological context. Cells can be stained with fluorescence-labeled antibodies and examined by **fluorescence microscopy** to reveal the location of a protein of interest. For example, proteins that play key roles in embryonic development have been identified with the use of monoclonal antibodies as tags (**Figure 4.20**).

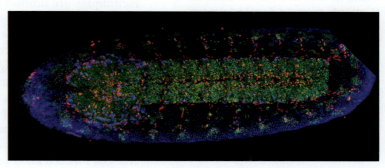

By tracking protein location, fluorescent markers also provide clues to protein function. For instance, the mineralocorticoid receptor protein binds to steroid hormones, including cortisol. The receptor was linked to a yellow variant of green fluorescent protein (GFP), a naturally fluorescent protein isolated from the jellyfish *Aequorea victoria* (Section 2.7). Fluorescence microscopy revealed that, in the absence of the hormone, the receptor is located in the cytoplasm (**Figure 4.21A**). On addition of the steroid, the receptor is translocated to the nucleus, where it binds to DNA (**Figure 4.21B**). These results indicate that the mineralocorticoid receptor protein is a transcription factor that controls gene expression.

FIGURE 4.20 Fluorescence-labeled monoclonal antibodies can detect proteins in a multicellular embryo imaged by microscopy. Here, a developing *Drosophila* embryo was stained with three different fluorescence-labeled monoclonal antibodies (red, green, and blue), each directed against a different protein important for gene regulation. [Credit: Laurina Manning, Chris Doe Lab, HHMI at University of Oregon.]

(A)   (B)

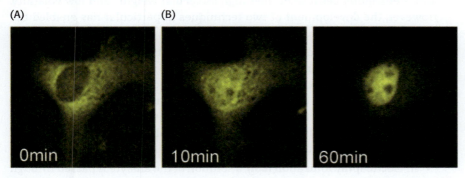

0min   10min   60min

FIGURE 4.21 Fluorescence microscopy can detect changes in protein localization within the cell. (A) The mineralocorticoid receptor, made visible by attachment to a yellow variant of GFP, is located predominantly in the cytoplasm of the cultured cell. (B) Subsequent to the addition of corticosterone (a glucocorticoid steroid that also binds to the mineralocorticoid receptor), the receptor moves into the nucleus. [Credit: Republished with permission of Society for Neuroscience, from Nishi, M. et al., "Visualization of Glucocorticoid Receptor and Mineralocorticoid Receptor Interactions in Living Cells with GFP-Based Fluorescence Resonance Energy Transfer," *The Journal of Neuroscience*, 24, 21, 2004, 4918-27, Figure 7. Permission conveyed through Copyright Clearance Center, Inc.]

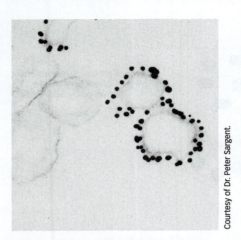

Courtesy of Dr. Peter Sargent.

**FIGURE 4.22 Immunoelectron micros-
copy can detect proteins at very high
spatial resolution.** The opaque particles
(150-Å, or 15-nm, diameter) in this electron
micrograph are clusters of gold atoms bound
to antibody molecules. A gold-labeled antibody against a channel protein (Section 13.1)
identifies membrane vesicles at the termini of
neurons that contain this protein.

The highest resolution of fluorescence microscopy is about 0.2 μm (200 nm, or 2000 Å), the wavelength of visible light. Finer spatial resolution can be achieved by electron microscopy if the antibodies are tagged with electron-dense markers. For example, antibodies attached to clusters of gold or to ferritin (which has an electron-dense core rich in iron) are highly visible under the electron microscope. This technique, called *immunoelectron microscopy*, can define the position of antigens to a resolution of 10 nm (100 Å) or finer (**Figure 4.22**).

4.3 Mass Spectrometry Is a Powerful Technique for the Identification of Peptides and Proteins

In several of the techniques described in the previous section, identification of the proteins present in the sample is often critical. For example, we may use co-immunoprecipitation to discover an unexpected protein that interacts with our sample protein. The identity of this unknown protein must be determined to advance our understanding of how our protein works. Antibody-based techniques, such as ELISA, can be very helpful toward this goal. However, these techniques are limited to the detection of proteins for which an antibody is already available.

Mass spectrometry enables the highly precise and sensitive measurement of the atomic composition of a particular molecule, or *analyte*, without prior knowledge of its identity. Originally, this method was relegated to the study of the chemical composition and molecular mass of gases or volatile liquids. However, technological advances have dramatically expanded the utility of mass spectrometry to the study of proteins, even those found at very low concentrations within highly complex mixtures, such as the contents of a particular cell type.

Mass spectrometers operate by converting analyte molecules into gaseous, charged forms (called *gas-phase ions*). Through the application of electrostatic potentials, the ratio of the mass of each ion to its charge (termed the *mass-to-charge ratio*, or *m/z*) can be measured. Although a wide variety of techniques employed by mass spectrometers are used in current practice, each of them comprises three essential components: the ion source, the mass analyzer, and the detector. Let us consider the first two in greater detail, because improvements in them have contributed most significantly to the analysis of biological samples:

1. *The ion source achieves the first critical step in mass spectrometric analysis: ionization, or the conversion of the analyte into gas-phase ions.* Proteins are difficult to ionize efficiently because of their high molecular weights and low volatility. However, the development of two techniques in particular has enabled the clearing of this significant hurdle.

   - In **matrix-assisted laser desorption/ionization (MALDI)**, the analyte is evaporated to dryness in the presence of a volatile, aromatic compound (the matrix) that can absorb light at specific wavelengths. A laser pulse tuned to one of these wavelengths excites and vaporizes the matrix, converting some of the analyte into the gas phase. Subsequent gaseous collisions enable the intermolecular transfer of charge, ionizing the analyte.

   - In **electrospray ionization (ESI)**, a solution of the analyte is passed through an electrically charged nozzle. Droplets of the analyte, now charged, emerge from the nozzle into a chamber of very low pressure, evaporating the solvent and ultimately yielding the ionized analyte.

2. *In the second step, the newly formed analyte ions then enter the mass analyzer, where they are distinguished on the basis of their mass-to-charge ratios.* There are a number of different types of mass analyzers. For this discussion, we will consider

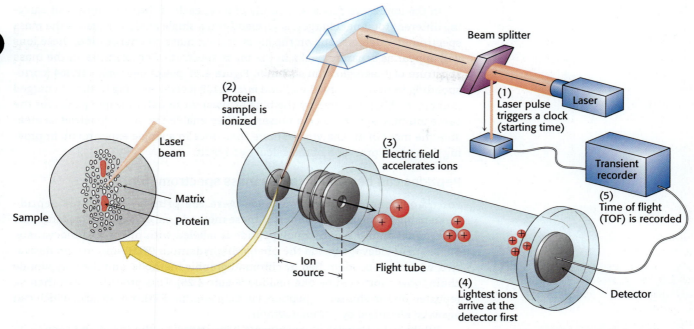

one of the simplest, the **time-of-flight (TOF) mass analyzer**, in which ions are accelerated through an elongated chamber under a fixed electrostatic potential. Given two ions of identical net charge, the smaller ion will require less time to traverse the chamber than will the larger ion. The mass of each ion can be determined by measuring the time required for each ion to pass through the chamber.

The sequential action of the ion source and the mass analyzer enables the highly sensitive measurement of the mass-to-charge ratios of very large ions, such as the charged forms of proteins. Consider an example of a MALDI ion source coupled to a TOF mass analyzer: the MALDI-TOF mass spectrometer (**Figure 4.23**). Gas-phase ions generated by the MALDI ion source pass directly into the TOF analyzer, where the mass-to-charge ratios are recorded. In **Figure 4.24**, the MALDI-TOF mass spectrum of a mixture of 5 pmol (5×10^{-12} mol) each of insulin and lactoglobulin is shown. The masses determined by MALDI-TOF are 5733.9 and 18,364, respectively. A comparison with the calculated values of 5733.5 and 18,388 reveals that MALDI-TOF is clearly an accurate means of determining protein mass.

FIGURE 4.23 The MALDI-TOF mass spectrometer measures ionized samples based on their mass-to-charge ratio. (1) A laser pulse simultaneously starts a clock and is directed toward the sample. (2) The protein sample, embedded in an appropriate matrix, is ionized by the application of the laser pulse. (3) An electric field accelerates the ions through the flight tube toward the detector. (4) The lightest ions arrive first. (5) Based on the clock reading initiated in Step 1, the time of flight (TOF) for the ions can be measured. [Information from J. T. Watson, *Introduction to Mass Spectrometry*, 3rd ed. (Lippincott-Raven, 1997), p. 279.]

VIEW an animation of this technique in Achie√e

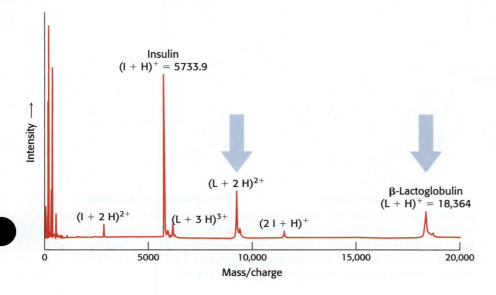

FIGURE 4.24 MALDI-TOF mass spectrometry can detect molecular masses with a high degree of sensitivity and accuracy. A mixture of 5 pmol each of insulin (I) and β-lactoglobulin (L) was ionized by MALDI, which produces predominantly singly charged molecular ions from peptides and proteins—the insulin ion $(I+H)^+$ and the lactoglobulin ion $(L+H)^+$. Peaks corresponding to β-lactoglobulin but carrying different total net charges are indicated by the blue arrows, as well as small quantities of a singly charged dimer of insulin $(2I+H)^+$ also are produced. [Data from J. T. Watson, *Introduction to Mass Spectrometry*, 3rd ed. (Lippincott-Raven, 1997), p. 282.]

In the ionization process, a family of ions, each of the same mass but carrying different total net charges, is formed from a single analyte. Because the mass spectrometer detects ions on the basis of their mass-to-charge ratio, these ions will appear as separate peaks in the mass spectrum. For example, in the mass spectrum of β-lactoglobulin shown in Figure 4.24, peaks near $m/z = 18,364$ (corresponding to the +1 charged ion) and $m/z = 9183$ (corresponding to the +2 charged ion) are visible (indicated by the blue arrows). Although multiple peaks for the same ion may appear to be a nuisance, they enable the spectrometrist to measure the mass of an analyte ion more than once in a single experiment, improving the overall precision of the calculated result.

Peptides can be sequenced by mass spectrometry

For many years, chemical methods were the primary means for peptide sequencing. In the most common of these methods, called *Edman degradation*, the N-terminal amino acid of a polypeptide is labeled with phenyl isothiocyanate. Subsequent cleavage yields the phenylthiohydantoin (PTH)-amino acid derivative, which can be identified by chromatographic methods, and the polypeptide chain, now shortened by one residue (**Figure 4.25**). This procedure can then be repeated on the shortened peptide, yielding another PTH–amino acid, which can again be identified by chromatography.

While technological advancements have improved the speed and sensitivity of the Edman degradation, these parameters have largely been surpassed by the application of mass spectrometric methods. The use of mass spectrometry for protein sequencing takes advantage of the fact that ions of proteins that have been analyzed by a mass spectrometer, called *precursor ions*, can be broken into smaller peptide chains by bombardment with atoms of an inert gas such as

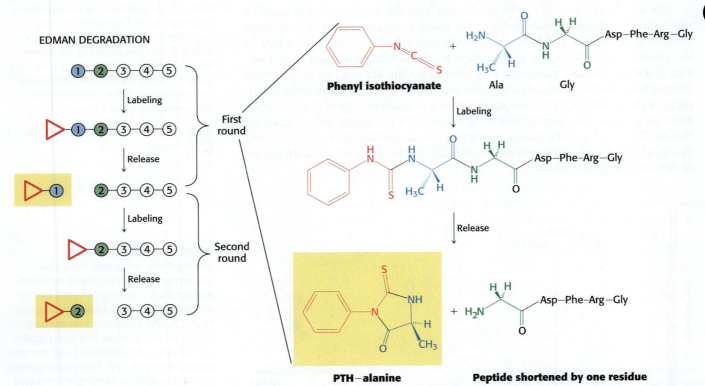

FIGURE 4.25 The Edman degradation is a chemical method for peptide sequencing. The labeled amino-terminal residue (PTH–alanine in the first round) can be released without hydrolyzing the rest of the peptide. Hence, the amino-terminal residue of the shortened peptide (Gly-Asp-Phe-Arg-Gly) can be determined in the second round. Three more rounds of the Edman degradation reveal the complete sequence of the original peptide.

(A)

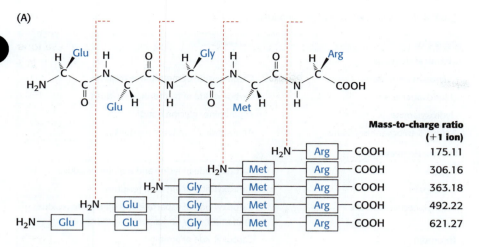

	Mass-to-charge ratio (+1 ion)
H₂N—[Arg]—COOH	175.11
H₂N—[Met]—[Arg]—COOH	306.16
H₂N—[Gly]—[Met]—[Arg]—COOH	363.18
H₂N—[Glu]—[Gly]—[Met]—[Arg]—COOH	492.22
H₂N—[Glu]—[Glu]—[Gly]—[Met]—[Arg]—COOH	621.27

(B)

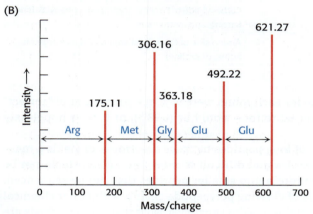

FIGURE 4.26 Peptides can be sequenced by tandem mass spectrometry. (A) Within the mass spectrometer, peptides can be fragmented by bombardment with inert gaseous ions to generate a family of product ions in which individual amino acids have been removed from one end. In this example, the carboxyl fragment of the cleaved peptide bond is ionized. (B) The product ions are detected in the second mass analyzer. The mass differences between the peaks indicate the sequence of amino acids in the precursor ion. [Data from H. Steen and M. Mann, *Nat. Rev. Mol. Cell Biol.* 5:699–711, 2004.]

VIEW an animation of this technique in
≋ Achieve

helium or argon. These new fragments, called *product ions*, can be passed through a second mass analyzer for further mass characterization.

Using two mass analyzers arranged in this manner is referred to as **tandem mass spectrometry**. Importantly, the product-ion fragments are formed in chemically predictable ways that can provide clues to the amino acid sequence of the precursor ion. For polypeptide analytes, disruption of individual peptide bonds will yield two smaller peptide ions, containing the sequences before and after the cleavage site. Hence, a family of ions can be detected; each ion represents a fragment of the original peptide with one or more amino acids removed from one end (**Figure 4.26A**). For simplicity, only the carboxyl-terminal peptide fragments are shown in Figure 4.26A. **Figure 4.26B** depicts a representative mass spectrum from a fragmented peptide. The mass differences between the peaks in this fragmentation experiment indicate the amino acid sequence of the precursor peptide ion.

❖ **SELF–CHECK QUESTION**

Which two amino acids are indistinguishable by tandem mass spectrometry as described above, and why?

Proteins can be specifically cleaved into small peptides to facilitate analysis

In principle, it should be possible to sequence an entire protein using the Edman degradation or mass spectrometric methods. In practice, the Edman degradation is limited to peptides of 50 residues, because not all peptides in the reaction mixture release the amino acid derivative at each step. For instance,

TABLE 4.3 Specific cleavage of polypeptides

Reagent	Cleavage site
Chemical cleavage	
Cyanogen bromide	Carboxyl side of methionine residues
O-Iodosobenzoate	Carboxyl side of tryptophan residues
Hydroxylamine	Asparagine–glycine bonds
2-Nitro-5-thiocyanobenzoate	Amino side of cysteine residues
Enzymatic cleavage	
Trypsin	Carboxyl side of lysine and arginine residues
Clostripain	Carboxyl side of arginine residues
Staphylococcal protease	Carboxyl side of aspartate and glutamate residues (glutamate only under certain conditions)
Thrombin	Carboxyl side of arginine
Chymotrypsin	Carboxyl side of tyrosine, tryptophan, phenylalanine, leucine, and methionine
Carboxypeptidase A	Amino side of C-terminal amino acid (not arginine, lysine, or proline)

if the efficiency of release for each round were 98%, the proportion of "correct" amino acid released after 60 rounds would be (0.98^{60}), or 0.3 — a hopelessly impure mix.

Similarly, sequencing of long peptides by mass spectrometry yields a mass spectrum that can be complex and difficult to interpret. This obstacle can be overcome by cutting the protein into smaller peptides that can be sequenced. Table 4.3 lists several ways of cleaving polypeptide chains, with either chemical reagents or proteolytic (protein-breaking) enzymes. Note that these methods are sequence specific: they disrupt the protein backbone at particular amino acid residues in a predictable manner. The peptides obtained by these methods are separated by some type of chromatography. The sequence of each purified peptide can then be more easily determined. At this point, the amino acid sequences of segments of the protein are known, but the order of these segments is not yet defined.

How can we order the peptides to obtain the primary structure of the original protein? The necessary additional information is obtained from **overlap peptides** (Figure 4.27). A second enzyme is used to split the polypeptide chain at different linkages. For example, chymotrypsin cleaves preferentially on the carboxyl side of aromatic and some other bulky nonpolar residues, while trypsin cleaves on the carboxyl side of positively charged residues (Table 4.3). Because these chymotryptic peptides overlap two or more tryptic peptides, they can be used to establish the order of the peptides. The entire amino acid sequence of the polypeptide chain is then known.

Additional steps are necessary if the initial protein sample is actually several polypeptide chains. SDS-PAGE under reducing conditions should display the number of chains. Alternatively, the number of distinct N-terminal amino acids could be determined. After a protein has been identified as being made up of two or more polypeptide chains, the chains must be separated from one another first. Denaturing agents, such as urea or guanidine hydrochloride, are used to dissociate chains held together by noncovalent interactions; polypeptide chains linked by covalent disulfide bonds are separated by reduction with thiols such as

FIGURE 4.27 Overlap peptides are used to establish the order of sequenced peptides. The peptide obtained by chymotryptic digestion overlaps two tryptic peptides, establishing their order.

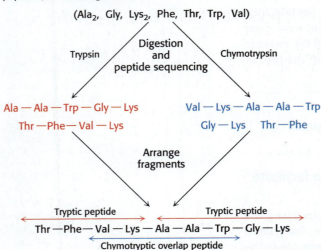

β-mercaptoethanol or dithiothreitol. To prevent the cysteine residues from recombining, they are alkylated with iodoacetate to form stable S-carboxymethyl derivatives (**Figure 4.28**).

Genomic and proteomic methods are complementary approaches to deducing protein structure and function

Despite the technological advancements in both chemical and mass spectrometric methods of peptide sequencing, heroic effort is required to elucidate the sequence of large proteins, those with more than 1000 residues. For sequencing such proteins, a complementary experimental approach based on recombinant DNA technology is often more efficient. As will be discussed in Chapter 9, long stretches of DNA can be cloned and sequenced, and the nucleotide sequence can be translated to reveal the amino acid sequence of the protein encoded by the gene. Recombinant DNA technology has generated a wealth of amino acid sequence information at a remarkable rate.

Even with the use of the DNA base sequence to determine primary structure, however, there is still a need to work with isolated proteins. The amino acid sequence deduced by reading the DNA sequence is that of the nascent protein, the direct product of the translational machinery. However, many proteins undergo posttranslational modifications after their syntheses (Section 2.7). Some have their ends trimmed, and others arise by cleavage of a larger initial polypeptide chain. Cysteine residues in some proteins are oxidized to form disulfide links, connecting either parts within a chain or separate polypeptide chains. Specific side chains of some proteins are altered. Amino acid sequences derived from DNA sequences are rich in information, but they do not disclose these modifications. Chemical analyses of proteins in their mature form are needed to delineate the nature of these changes, which are critical for the biological activities of most proteins.

The amino acid sequence of a protein provides valuable information

Regardless of the method used for its determination, the amino acid sequence of a protein provides the biochemist with a wealth of information as to the protein's structure, function, and history.

- *The sequence of a protein of interest can be compared with all other known sequences to determine whether significant similarities exist.* A search for kinship between a newly sequenced protein and the millions of previously sequenced ones takes only a few seconds on a personal computer (Chapter 10). If the newly isolated protein is a member of an established class of protein, we can begin to infer information about the protein's structure and function. For instance, chymotrypsin and trypsin are members of the serine protease family, a clan of proteolytic enzymes that have a common catalytic mechanism based on a reactive serine residue (Chapter 6). If the sequence of the newly isolated protein shows sequence similarity with trypsin or chymotrypsin, the result suggests that it, too, may be a serine protease.

- *Comparison of sequences of the same protein in different species yields information about evolutionary pathways.* Evolutionary relationships between species can be inferred from sequence differences between their proteins. If we assume that the random mutation rate of proteins over time is constant, then careful sequence comparison of related proteins between two organisms can provide an estimate for when these two evolutionary lines diverged. For example, a comparison of serum albumins found in primates indicates that human beings and African apes diverged 5 million years ago, not 30 million years ago as was once thought. Sequence analyses have opened a new perspective on the fossil record and the pathway of human evolution.

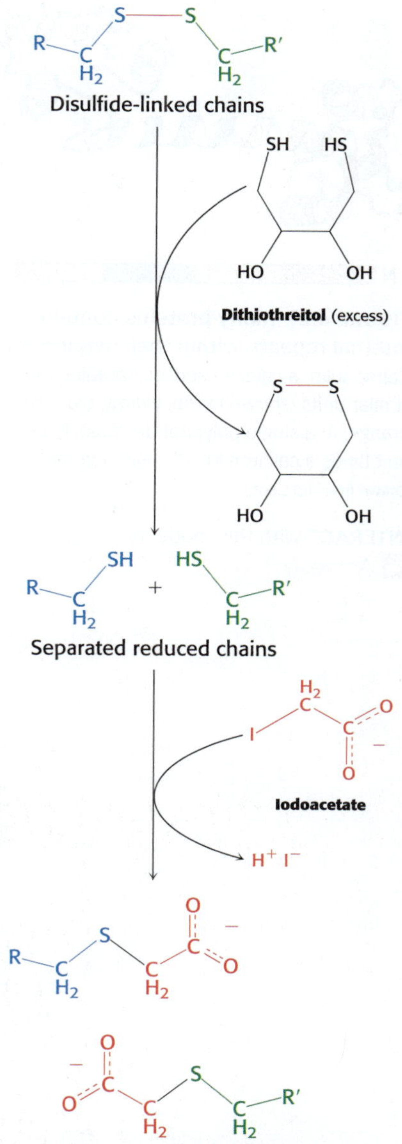

Disulfide-linked chains

Dithiothreitol (excess)

Separated reduced chains

Iodoacetate

Separated carboxymethylated chains

FIGURE 4.28 Reduction and alkylation are used to separate disulfide-linked polypeptides prior to sequencing. Polypeptides linked by disulfide bonds can be separated by reduction with dithiothreitol followed by alkylation to prevent them from re-forming.

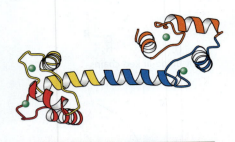

N ▮▮▮▮ C

FIGURE 4.29 Many proteins contain internal repeats within their sequences. Calmodulin, a calcium sensor, contains four similar units (shown in red, yellow, blue, and orange) in a single polypeptide chain. Each unit binds a calcium ion (shown in green). [Drawn from 1CLL.pdb.]

INTERACT with this model in
🔲 Achie√e

- *Amino acid sequences can be searched for the presence of internal repeats.* Such internal repeats can reveal the history of an individual protein itself. Many proteins apparently have arisen by duplication of primordial genes followed by their diversification. For example, calmodulin, a ubiquitous eukaryotic calcium sensor, contains four similar calcium-binding modules that arose by gene duplication (**Figure 4.29**).

- *Many proteins contain amino acid sequences that serve as signals designating their destinations or controlling their processing.* For example, a protein destined for export from a cell or for location in a membrane contains a signal sequence—a stretch of about 20 hydrophobic residues near the amino terminus that directs the protein to the appropriate membrane. Another protein may contain a stretch of amino acids that functions as a nuclear localization signal, directing the protein to the nucleus. Identification of these sequences within a protein provides clues to its function within the cell.

- *Sequence data provide a basis for preparing antibodies specific for a protein of interest.* One or more parts of the amino acid sequence of a protein will elicit an antibody when injected into a mouse or rabbit. These specific antibodies can be very useful in determining the amount of a protein present in solution or in the blood, ascertaining its distribution within a cell, or cloning its gene (Section 4.2).

- *Amino acid sequences are valuable for making DNA probes that are specific for the genes encoding the corresponding proteins.* Knowledge of a protein's primary structure means that DNA sequences corresponding to a part of the amino acid sequence can be constructed on the basis of the genetic code. These DNA sequences can be used as probes to isolate the gene encoding the protein so that the entire sequence of the protein can be determined. The gene in turn can provide valuable information about the physiological regulation of the protein. Protein sequencing is an integral part of molecular genetics, just as DNA cloning is central to the analysis of protein structure and function. We will revisit some of these topics in more detail in Chapter 9.

Individual proteins can be identified by mass spectrometry

The combination of mass spectrometry with chromatographic and peptide-cleavage techniques enables highly sensitive protein identification in complex biological mixtures. When a protein is cleaved by chemical or enzymatic methods (Table 4.3), a specific and predictable family of peptide fragments is formed. We learned in Chapter 2 that each protein has a unique, precisely defined amino acid sequence. Hence, the identity of the individual peptides formed from this cleavage reaction—and, importantly, their corresponding masses—is a distinctive signature for that particular protein. Protein cleavage, followed by chromatographic separation and mass spectrometry, enables rapid identification and quantitation of these signatures, even if they are present at very low concentrations. This technique for protein identification is referred to as **peptide mass fingerprinting**.

The speed and sensitivity of mass spectrometry has made this technology critical for the study of proteomics. Let us consider the analysis of the nuclear pore complex from yeast, which facilitates the transport of large molecules into and out of the nucleus. This huge macromolecular complex—around 100,000 kDa!—was purified from yeast cells, then fractionated by HPLC followed by gel electrophoresis. Individual bands from the gel were isolated, cleaved with trypsin, and analyzed by MALDI-TOF mass spectrometry. The fragments produced were compared with amino acid sequences deduced from the DNA sequence of the yeast genome, as shown in **Figure 4.30**. A total of 174 nuclear pore proteins were identified in this manner.

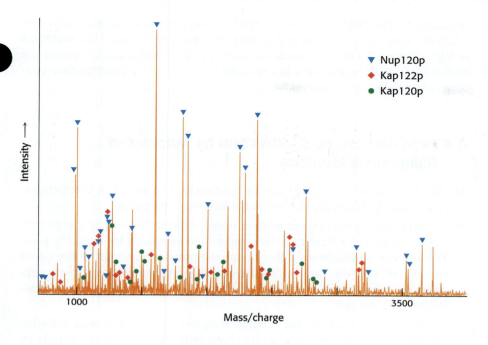

FIGURE 4.30 Peptide mass fingerprinting can identify individual proteins within complex mixtures. This mass spectrum was obtained by analyzing a trypsin-treated band in a gel derived from a yeast nuclear-pore sample. Many of the peaks were found to match the masses predicted for peptide fragments from three proteins (Nup120p, Kap122p, and Kap120p) within the yeast genome. [Data from M. P. Rout, J. D. Aitchison, A. Suprapto, K. Hjertaas, Y. Zhao, and B. T. Chait, *J. Cell Biol.* 148:635–651, 2000.]

Many of the proteins found in this experiment had not previously been identified as being associated with the nuclear pore despite years of study. Furthermore, mass spectrometric methods are sensitive enough to detect essentially all components of the pore if they are present in the samples used. Thus, a complete list of the components constituting this macromolecular complex could be obtained in a straightforward manner.

The proteome is the functional representation of the genome

Thus far, we have discussed several techniques, including two-dimensional electrophoresis and peptide mass fingerprinting, which allow for the rapid and scalable detection and quantitation of large numbers of proteins in a single experiment. These methods can be used to (1) explore the compete catalog of proteins expressed in a cell under a specific set of conditions, and (2) investigate how this inventory changes when these conditions are altered. These studies, referred to as *proteomics*, can be very powerful in understanding the mechanisms by which cells adapt to changing environments.

As will be discussed in Chapter 9, the complete DNA base sequences, or genomes, of many organisms are now available. For example, the roundworm *Caenorhabditis elegans* has a genome of 97 million bases and about 19,000 protein-encoding genes, whereas that of the fruit fly *Drosophila melanogaster* contains 180 million bases and about 14,000 genes. The completely sequenced human genome contains 3 billion bases and about 23,000 genes. However, these genomes are simply inventories of the genes that could be expressed within a cell under specific conditions. Only a subset of the proteins encoded by these genes will actually be present in a given biological context. The **proteome**—derived from proteins expressed by the genome—of an organism signifies a more complex level of information content, encompassing the types, functions, and interactions of proteins within its biological environment.

The proteome is not a fixed characteristic of the cell. Because it represents the functional expression of information, it varies with cell type, developmental stage, and environmental conditions. The proteome is much larger than the genome because almost all gene products are proteins that can be chemically modified in a variety of ways. Furthermore, these proteins do not exist in isolation; they often interact with one another to form complexes with specific functional properties. Whereas the genome is "hard wired," the proteome is highly dynamic. The techniques we have described in this section enable the understanding of the proteome by providing a means of investigating, characterizing, and cataloging proteins.

4.4 Peptides Can Be Synthesized by Automated Solid-Phase Methods

Peptides of defined sequence can be synthesized to assist in biochemical analysis. These peptides are valuable tools for several purposes.

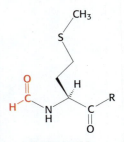

fMet peptide

- *Synthetic peptides can serve as antigens to stimulate the formation of specific antibodies.* Suppose we want to isolate the protein expressed by a specific gene. Peptides can be synthesized that match the translation of part of the gene's nucleic acid sequence, and antibodies can be generated that target these peptides. These antibodies can then be used to isolate the intact protein or localize it within the cell.

- *Synthetic peptides can be used to isolate receptors for many hormones and other signal molecules.* For example, white blood cells are attracted to bacteria by formylmethionyl (fMet) peptides released in the breakdown of bacterial proteins. Synthetic formylmethionyl peptides have been useful in identifying the cell-surface receptor for this class of peptide. Moreover, synthetic peptides can be attached to agarose beads to prepare affinity chromatography columns for the purification of receptor proteins that specifically recognize the peptides.

- *Synthetic peptides enable the study of the effects of nonproteinogenic amino acids on protein structure and function.* Recall from Section 2.2 that proteins synthesized within the cell are composed from a repertoire of 20 natural, or *proteinogenic*, amino acids. Chemical synthesis of peptides provides a convenient method for inserting *nonproteinogenic* amino acids—that is, those not directly coded into the genome—into peptides. Examples of nonproteinogenic amino acids include the D isomers of the proteinogenic amino acids (e.g., D-alanine) and amino acids that contain posttranslational modifications (Section 2.7), such as hydroxyproline. Peptides containing these amino acids are valuable tools in the study of protein structure and function.

- *Synthetic peptides can serve as drugs.* Vasopressin is a peptide hormone that stimulates the reabsorption of water in the kidney, leading to the formation of more-concentrated urine. Patients with diabetes insipidus are deficient in vasopressin (also called *antidiuretic hormone*), and so they excrete large volumes of dilute urine (more than 5 liters per day) and are continually thirsty. This defect can be treated by administering 1-desamino-8-D-arginine vasopressin (also called desmopressin), a synthetic analog of the missing hormone (**Figure 4.31**). This synthetic peptide is degraded in the body much more slowly than vasopressin and does not increase blood pressure.

- *Studying synthetic peptides can help define the rules governing the three-dimensional structure of proteins.* We can ask whether a particular sequence by itself tends to fold into an α helix, a β strand, or a hairpin turn, or behaves as a random coil (Section 2.4). The peptides created for such studies can incorporate amino acids not normally found in proteins, allowing greater variation in chemical structure than is possible with the use of only 20 amino acids.

(A)

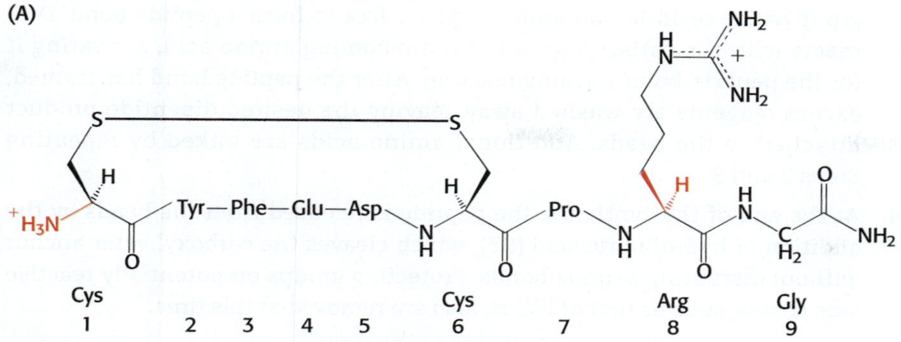

8-Arginine vasopressin
(antidiuretic hormone, ADH)

(B)

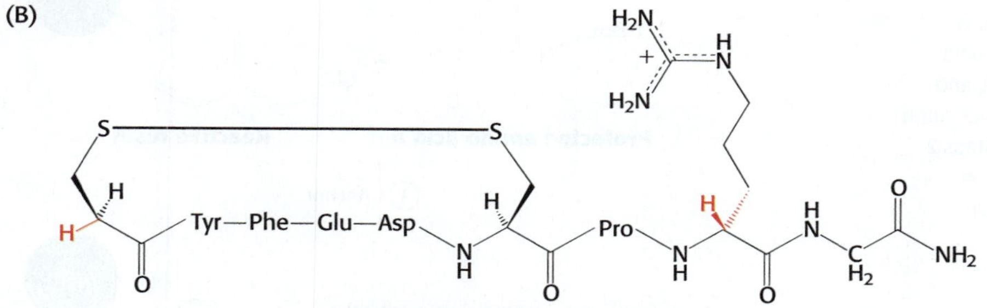

1-Desamino-8-D-arginine vasopressin
(Desmopressin)

FIGURE 4.31 Desmopressin, a synthetic vasopressin analog, can be used to treat patients deficient in vasopressin. Structural formulas of (A) vasopressin, a peptide hormone that stimulates water resorption, and (B) 1-desamino-8-D-arginine vasopressin (desmopressin), a more stable synthetic analog of this antidiuretic hormone.

How are these peptides constructed? The **solid-phase method**, first developed by R. Bruce Merrifield, provides a highly efficient approach to the synthesis of polypeptides (**Figure 4.32**). Recall that the amino group of one amino acid can link to the carboxyl group of any another. However, a unique product will form if only a single amino group and a single carboxyl group are available for reaction. To achieve this, it is necessary to block some groups and to activate others to prevent unwanted reactions:

1. First, the carboxyl-terminal amino acid (shown in red in Figure 4.32) is anchored to an insoluble resin, usually beads, by its carboxyl group, effectively protecting it from further peptide-bond-forming reactions. The α-amino group of this amino acid is blocked with a protecting group such as the *tert*-butyloxycarbonyl (*t*-Boc) or the 9-fluorenylmethyloxycarboyl (Fmoc) group. In this example, we will use the *t*-Boc protecting groups.

t-**Butyloxycarbonyl amino acid**
(*t*-Boc amino acid)

9-Fluorenylmethyloxycarbonyl amino acid
(Fmoc amino acid)

2. The *t*-Boc protecting group of this amino acid is then removed with trifluoroacetic acid, freeing the α-amino group for the next coupling reaction.

3. The next amino acid (in the protected *t*-Boc form, shown in blue in Figure 4.32) and dicyclohexylcarbodiimide (DCC) are added together. At this stage, only the carboxyl group of the incoming amino acid and the amino

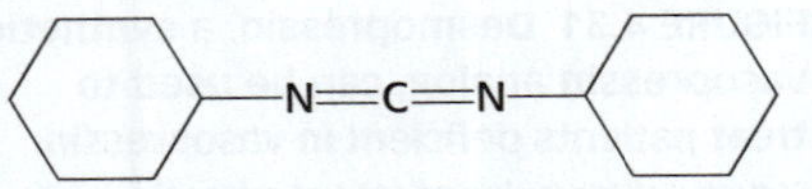

Dicyclohexylcarbodiimide (DCC)

group of the resin-bound amino acid are free to form a peptide bond. DCC reacts with the carboxyl group of the incoming amino acid, activating it for the peptide-bond-forming reaction. After the peptide bond has formed, excess reagents are washed away, leaving the desired dipeptide product attached to the beads. Additional amino acids are linked by repeating Steps 2 and 3.

4. At the end of the synthesis, the peptide is released from the beads by the addition of hydrofluoric acid (HF), which cleaves the carboxyl ester anchor without disrupting peptide bonds. Protecting groups on potentially reactive side chains, such as that of lysine, also are removed at this time.

FIGURE 4.32 Solid-phase peptide synthesis is achieved by repeated coupling and deprotection steps. The sequence of steps in solid-phase synthesis is: (1) anchoring of the *t*-Boc-protected C-terminal amino acid to a solid resin, (2) deprotection of the amino terminus, and (3) coupling of the free amino terminus with the DCC-activated carboxyl group of the next *t*-Boc-protected amino acid. Steps 2 and 3 are repeated for each added amino acid. Finally, in step 4, the completed peptide is released from the resin.

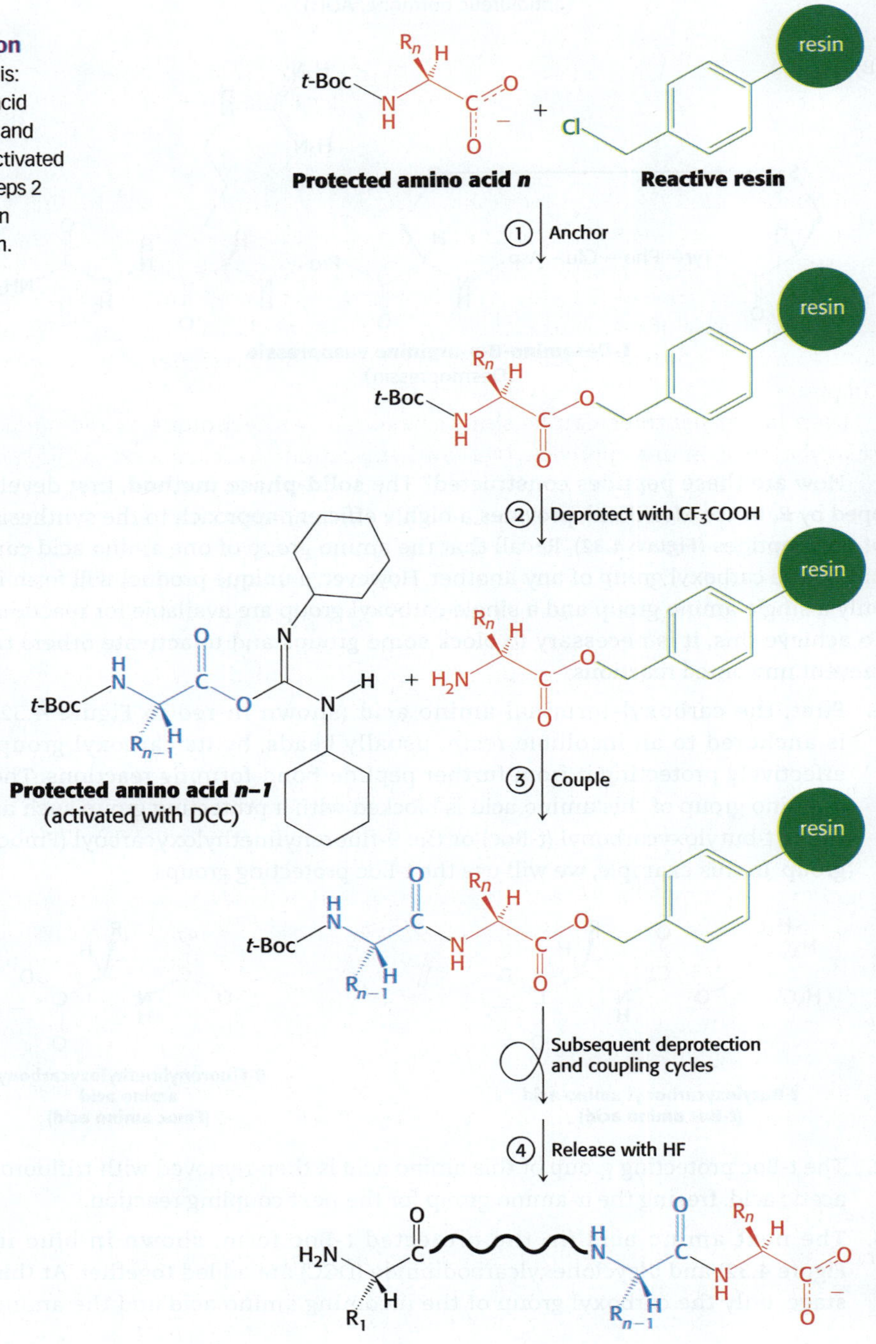

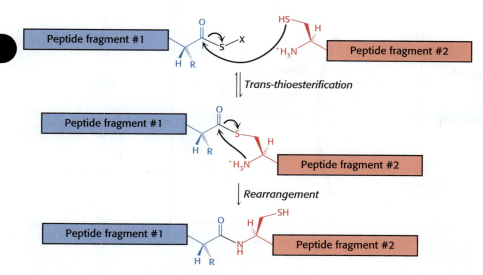

Trans-thioesterification

Rearrangement

FIGURE 4.33 Native chemical ligation can be used to link two synthetic peptides. The sulfur atom of the N-terminal cysteine of one synthetic peptide ("Peptide fragment #2") undergoes a trans-thioesterification reaction with the C-terminal thioester of the other peptide ("Peptide fragment #1"). A rapid, irreversible rearrangement leads to the formation of a peptide bond between the two fragments.

A major advantage of the solid-phase method is that the desired product at each stage is bound to beads that can be rapidly filtered and washed. Hence, there is no need to purify intermediates. Additionally, all reactions are carried out in a single vessel, eliminating losses caused by repeated transfers of products, and the cycle of reactions can be readily automated, which makes it feasible to routinely synthesize peptides containing about 50 residues in good yield and purity.

Even larger peptides can be synthesized using a technique called *native chemical ligation*. In this approach, two synthetic peptide fragments can be linked together provided the C-terminus of one peptide features a thioester and the N-terminal residue of the other peptide is a cysteine residue (**Figure 4.33**). In the first step, a trans-thioesterification reaction occurs, where the side-chain sulfur of the cysteine residue displaces the thiol group of the incoming thioester. This is followed by the rapid, irreversible rearrangement of the new thioester, yielding a new peptide bond. Native chemical ligation can be used to link together multiple peptide fragments, allowing for the synthesis of fully active proteins over 100 residues in length.

4.5 Three-Dimensional Protein Structures Can Be Determined Experimentally

Elucidating the three-dimensional structure of a protein can provide a tremendous amount of insight into its corresponding function. For example, the precise atomic arrangements within active sites and binding sites provide clues to their specificities. Knowing a protein's structure enables the biochemist to predict its mechanism of action, the effects of mutations on its function, and the desired features of drugs that may inhibit or augment its activity. X-ray crystallography, nuclear magnetic resonance spectroscopy, and cryo-electron microscopy are the most important techniques for elucidating the conformation of proteins.

X-ray crystallography reveals three-dimensional structure in atomic detail

The first method developed to determine protein structure in atomic detail was **x-ray crystallography**. This technique provides the clearest visualization of the precise three-dimensional positions of most atoms within a protein. Of all forms

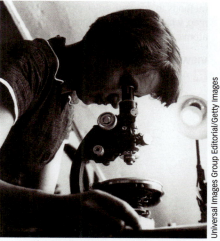

Universal Images Group Editorial/Getty Images

ROSALIND FRANKLIN Today, Rosalind Franklin is known as the x-ray crystallographer who generated clear diffraction patterns for DNA. Her data were used by James Watson and Francis Crick to develop their proposed structural model of the DNA double helix. However, the importance of her contributions to this landmark achievement were not fully appreciated or acknowledged at the time. Dr. Franklin had completed her PhD in physical chemistry, studying the density and porosity of coal, and she then learned x-ray crystallography to examine structural changes in carbon upon heating. She was a very skilled experimentalist and began to apply x-ray crystallographic techniques to the structure of DNA two years before the structural proposal by Watson and Crick. She subsequently did outstanding work on the structures of viruses. Her structural analysis of the polio virus was abruptly cut short by her untimely death from ovarian cancer at the age of 37. The subsequent publication of the polio virus structure by John Finch and Aaron Klug was dedicated to her memory.

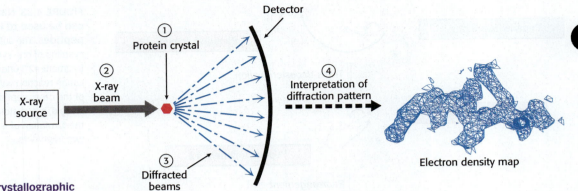

FIGURE 4.34 An x-ray crystallographic experiment is composed of several fundamental steps. (1) A crystal of the protein is prepared, then (2) placed within the path of an x-ray beam. (3) The crystal diffracts the beam, and the resulting pattern is collected on a detector. (4) The diffraction pattern is interpreted using computational methods to generate an electron density map, which enables the determination of the atomic positions of the protein.

VIEW an animation of this technique in
Achieve

of radiation, x-rays provide the best resolution for the determination of molecular structures because their wavelength approximately corresponds to the length of a covalent bond. While the details of crystallography can be technically challenging, there are several fundamental steps required in an x-ray crystallographic experiment (**Figure 4.34**):

1. *Preparing a protein crystal.* X-ray crystallography first requires the preparation of a protein or protein complex in crystal form, in which all protein molecules are oriented in a fixed, repeated arrangement with respect to one another. Slowly adding ammonium sulfate or another salt to a concentrated solution of protein to reduce its solubility favors the formation of highly ordered crystals—the process of salting out discussed in Section 4.1.

    Protein crystallization can be challenging: a concentrated solution of highly pure material is required, and it is often difficult to predict which experimental conditions will yield the most-effective crystals. Typically, hundreds of conditions must be tested to obtain crystals fully suitable for crystallographic studies. Nevertheless, increasingly large and complex proteins have been crystallized. For example, poliovirus, an 8500-kDa assembly of 240 protein subunits surrounding an RNA core, has been crystallized and its structure solved by x-ray methods. Crucially, proteins frequently crystallize in their biologically active configuration, and thus enzyme crystals may display catalytic activity if the crystals are suffused with substrate.

2. *Exposing the crystal to a source of x-rays.* After a suitably pure crystal of protein has been obtained, a source of x-rays is required. A beam of x-rays of wavelength 1.54 Å is produced by accelerating electrons against a copper target. Alternatively, x-rays can be produced by the acceleration of electrons in circular orbits at speeds close to the speed of light, a phenomenon referred to as *synchrotron radiation*. Synchrotron-generated x-ray beams are much more intense than those generated by electrons hitting copper. The higher intensity allows high-quality data to be gathered from smaller crystals over a shorter exposure time. There are around 100 facilities across the world that generate synchrotron radiation.

3. *Detecting the diffraction of the x-ray by the crystal.* When a narrow beam of x-rays is directed at the protein crystal, most of the beam passes directly through the crystal while a small part is scattered in various directions. These scattered, or diffracted, x-rays can be detected by x-ray film or by a solid-state electronic detector.

    The scattering pattern produced by the diffracted x-rays provides abundant information about protein structure. The basic physical principles underlying the technique are:

    *   *Electrons scatter x-rays.* The amplitude of the wave scattered by an atom is proportional to its number of electrons. Thus, a carbon atom scatters six times as strongly as a hydrogen atom does.

- *The scattered waves recombine.* Each diffracted beam comprises waves scattered by each atom in the crystal. The scattered waves reinforce one another at the film or detector if they are in phase—that is, they are aligned—and they cancel one another if they are out of phase.

- *The way in which the scattered waves recombine depends only on the atomic arrangement.* The protein crystal is mounted and positioned in a precise orientation with respect to the x-ray beam and the film. The crystal is rotated so that the beam can strike the crystal from many directions. This rotational motion results in an x-ray photograph consisting of a regular array of spots called *reflections.*

The intensities and positions of these reflections are the basic experimental data of an x-ray crystallographic analysis. The x-ray photograph shown in **Figure 4.35** is a two-dimensional section through a three-dimensional array of tens of thousands of reflections. Each reflection is formed from a wave with an amplitude proportional to the square root of the observed intensity of the spot. Each wave also has a phase—that is, the timing of its crests and troughs relative to those of other waves. Additional experiments or calculations must be performed to determine the phases corresponding to each reflection.

4. *Interpreting the diffraction pattern into an electron density map.* The next step is to reconstruct an image of the protein from the observed reflections. In light microscopy or electron microscopy, the diffracted beams are focused by lenses to directly form an image. However, appropriate lenses for focusing x-rays do not exist. Instead, the image is formed by applying a mathematical calculation called a *Fourier transform* to the measured amplitudes and calculated phases of every observed reflection. The image obtained is referred to as the **electron-density map**. This map is a three-dimensional graphic representation of where the electrons are most densely localized and it is used to determine the positions of the atoms in the crystallized molecule (**Figure 4.36**).

Critical to the interpretation of the electron-density map is its resolution, which is established by the number of scattered intensities used in the Fourier transform. The fidelity of the image depends on this resolution. A resolution of 6 Å reveals the course of the polypeptide chain but few other structural details. The reason is that polypeptide chains pack together so that their centers are between 5 Å and 10 Å apart. Maps at higher resolution are needed to delineate groups of atoms, which lie between 2.8 Å and 4.0 Å apart,

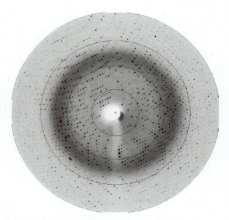

FIGURE 4.35 A protein crystal diffracts x-rays to produce a pattern of spots, or reflections, on the detector surface. The white silhouette in the center of the image is from a beam stop which protects the detector from the intense, undiffracted x-ray beam.

[Credit: Reproduced with permission of the International Union of Crystallography, from Ellis et al., Resolution improvement from 'in situ annealing' of copper nitrite reductase crystals, *Acta Cryst.* D58: 456–8, 2002, Fig.1B. © 2002 IUCr (https://journals.iucr.org/). https://doi.org/10.1107/S0907444902000136]

(A)

(B)

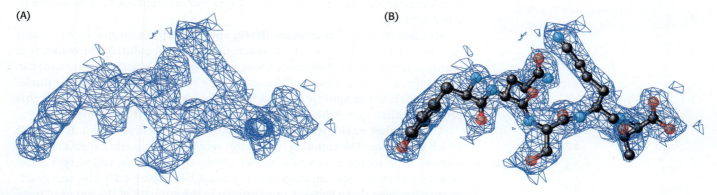

FIGURE 4.36 The electron-density map enables the determination of atomic positions of the crystallized protein. (A) A segment of an electron-density map is drawn as a three-dimensional contour plot, in which the regions inside the "cage" represent the regions of highest electron density. (B) A model of the protein is built into this map so as to maximize the placement of atoms within this density. [Drawn from 1FCH.pdb.]

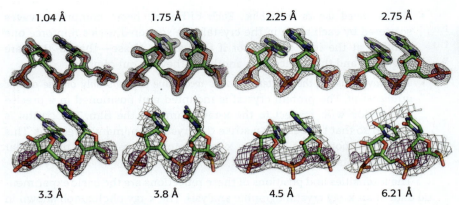

1.04 Å   1.75 Å   2.25 Å   2.75 Å

3.3 Å   3.8 Å   4.5 Å   6.21 Å

FIGURE 4.37 Resolution affects the quality of the electron density map. Electron density maps at six different resolution levels are shown. At the low resolution levels (4.5 Å and 6.21 Å), only a poorly defined area of density is visible, whereas at the higher resolutions, individual atoms are distinguishable. [Credit: Kevin S Keating, Anna Marie Pyle, "Semiautomated model building for RNA crystallography using a directed rotameric approach," Figures 1B–1I from https://www.pnas.org/content/107/18/8177 *Proc. Natl. Acad. Sci. U.S.A.* 2010 May 4; 107(18): 8177–8182. doi: 10.1073/pnas.0911888107.]

and individual atoms, which are between 1.0 Å and 1.5 Å apart (**Figure 4.37**). The ultimate resolution of an x-ray analysis is determined by the degree of perfection of the crystal. For proteins, this limiting resolution is often about 2 Å; however, in exceptional cases, resolutions of 1.0 Å have been obtained.

❖ SELF–CHECK QUESTION

Structures of proteins comprising domains separated by flexible linker regions can be quite difficult to solve by x-ray crystallographic methods. Why might this be the case?

Nuclear magnetic resonance spectroscopy can reveal the structures of proteins in solution

While x-ray crystallography is a powerful method for determining protein structures, some proteins do not readily crystallize. Furthermore, nonphysiological conditions (such as pH and added salts) are often required for successful crystallization. In addition, crystallized proteins adopt conformations that may be influenced by constraints imposed by the crystalline environment. However, since most proteins function not in crystalline form but in solution under physiological conditions, greater insight into protein function may be gained by other methods.

Nuclear magnetic resonance (NMR) spectroscopy is unique in being able to reveal the atomic structure of macromolecules in solution, provided that highly concentrated solutions (~ 1 mM, or 15 mg ml^{-1} for a 15-kDa protein) can be obtained. This technique depends on the fact that certain atomic nuclei are intrinsically magnetic. Only a limited number of isotopes display this property, called *spin*; **Table 4.4** lists those most important to biochemistry.

The simplest example of an isotope with spin is the hydrogen nucleus (^{1}H), which is a proton. The spinning of a proton generates a property called a *magnetic moment*. This moment can take either of two orientations, called *spin states* (α and β), when an external magnetic field is applied (**Figure 4.38**). The energy difference between these states is proportional to the strength of the imposed magnetic field. The α state has a slightly lower energy because it is aligned with this applied field. Hence, in a given population of nuclei, slightly more will occupy the α state (by a factor of the order of 1.00001 in a typical experiment). A spinning proton in an α state can be raised to an excited state (β state) by applying a pulse

TABLE 4.4 Biologically important nuclei giving NMR signals

Nucleus	Natural abundance (% by weight of the element)
^{1}H	99.984
^{2}H	0.016
^{13}C	1.108
^{14}N	99.635
^{15}N	0.365
^{17}O	0.037
^{23}Na	100.0
^{25}Mg	10.05
^{31}P	100.0
^{35}Cl	75.4
^{39}K	93.1

of electromagnetic radiation (a radio-frequency, or RF, pulse), provided that the frequency corresponds to the energy difference between the α and the β states. In these circumstances, the spin will change from α to β; in other words, resonance will be obtained.

These properties can be used to examine the chemical surroundings of the hydrogen nucleus. The flow of electrons around a magnetic nucleus generates a small, local magnetic field that opposes the applied field. The degree of such shielding depends on the surrounding electron density. Consequently, nuclei in different environments will change states, or resonate, at slightly different field strengths or radiation frequencies.

One-dimensional NMR. Most protons in many proteins can be resolved by using a technique called *one-dimensional NMR*, in which a resonance spectrum for a molecule is obtained by keeping the magnetic field constant and varying the frequency of the electromagnetic radiation. The nuclei of the perturbed sample absorb electromagnetic radiation at a measurable frequency. The different frequencies, termed **chemical shifts**, are expressed in fractional units δ (parts per million, or ppm) relative to the shifts of a standard compound, such as a water-soluble derivative of tetramethylsilane, that is added with the sample. For example, a —CH$_3$ proton typically exhibits a chemical shift (δ) of 1 ppm, compared with a chemical shift of 7 ppm for an aromatic proton. The chemical shifts of most protons in protein molecules fall between 0 and 9 ppm (**Figure 4.39**). With this information, we can then deduce changes to a particular chemical group under different conditions, such as the conformational change of a protein from a disordered structure to an α helix in response to a change in pH.

Two-dimensional NMR. We can garner even more information by examining how the spins on different protons affect their neighbors. Through the application of a radio-frequency pulse to briefly induce magnetization in the sample, we can alter the spin on one nucleus and examine the effect on the spin of a neighboring nucleus. Especially revealing is a two-dimensional spectrum obtained by <u>n</u>uclear <u>O</u>verhauser <u>e</u>nhancement <u>s</u>pectroscop<u>y</u> (NOESY), which graphically displays pairs of protons that are in close proximity, even if they are not close together in the primary structure. The basis for this technique is the **nuclear Overhauser effect**, an interaction between nuclei that is proportional to the

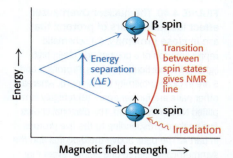

FIGURE 4.38 The transition between spin states forms the basis of NMR spectroscopy. The energies of the two orientations of an intrinsically magnetic nucleus depend on the strength of the applied magnetic field. Absorption of electromagnetic radiation of appropriate frequency induces a transition from the lower to the upper level.

VIEW an animation of NMR spectroscopy in ✻ Achie√e

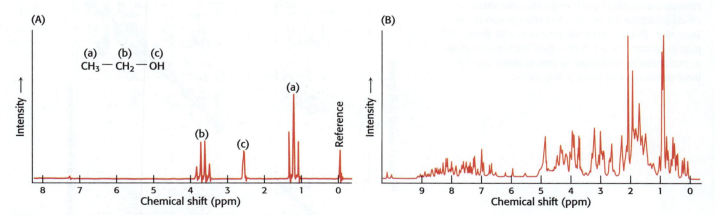

FIGURE 4.39 One-dimensional ^{1}H NMR spectra reveal the chemical shifts of protons within the sample. (A) ^{1}H-NMR spectrum of ethanol (CH$_3$CH$_2$OH) shows that the chemical shifts for the hydrogen are clearly resolved. (B) ^{1}H-NMR spectrum of a 55 amino acid fragment of a protein having a role in RNA splicing shows a greater degree of complexity. A large number of peaks are present, and many overlap. [(A) Data from C. Branden and J. Tooze, *Introduction to Protein Structure* (Garland, 1991), p. 280; (B) courtesy of Dr. Barbara Amann and Dr. Wesley McDermott.]

FIGURE 4.40 The nuclear Overhauser effect identifies pairs of protons that are in close proximity. (A) Schematic representation of a polypeptide chain highlighting five particular protons. Protons 2 and 5 are in close proximity (~4 Å apart), whereas other pairs are farther apart. (B) A highly simplified NOESY spectrum. The diagonal shows five peaks corresponding to the five protons in part A. The peak above the diagonal and the symmetrically related one below reveal that proton 2 is close to proton 5.

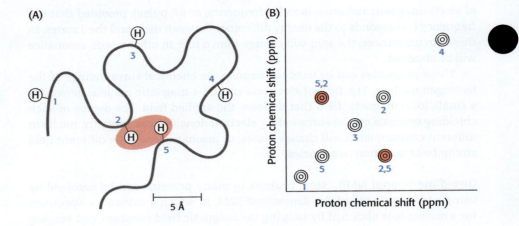

inverse sixth power of the distance between them. Magnetization is transferred from an excited nucleus to an unexcited one if the two nuclei are less than about 5 Å apart (**Figure 4.40A**).

In other words, the nuclear Overhauser effect provides a means of detecting the location of atoms relative to one another in the three-dimensional structure of the protein. The peaks that lie along the diagonal of a NOESY spectrum (shown in white in **Figure 4.40B**) correspond to those present in a one-dimensional NMR experiment. The peaks apart from the diagonal (shown in red in Figure 4.40B), referred to as *off-diagonal peaks* or *cross-peaks*, provide crucial new information: they identify pairs of protons that are less than 5 Å apart.

A two-dimensional NOESY spectrum for a protein comprising 55 amino acids is shown in **Figure 4.41**. The large number of off-diagonal peaks reveals short proton–proton distances. The three-dimensional structure of a protein can be reconstructed with the use of such proximity relationships. Structures are calculated such that proton pairs identified by NOESY cross-peaks must be separated by less than 5 Å in the three-dimensional structure (**Figure 4.42**). If a sufficient number of distance constraints are applied, the three-dimensional structure can nearly be determined uniquely.

FIGURE 4.41 The NOESY spectrum for a protein reveals many short proton–proton distances. Each off-diagonal peak corresponds to a short proton–proton separation. This spectrum, obtained from a 55-amino-acid protein domain, reveals hundreds of such short proton–proton distances, which can be used to determine the three-dimensional structure of this domain.

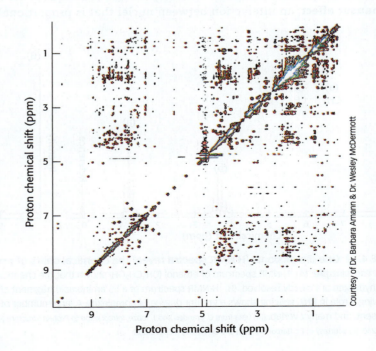

Courtesy of Dr. Barbara Amann & Dr. Wesley McDermott

(A)

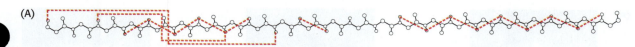

(B)

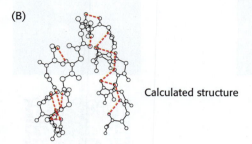

Calculated structure

FIGURE 4.42 Structures are calculated on the basis of NMR constraints. (A) NOESY observations show that protons (connected by dotted red lines) are close to one another in space. (B) A three-dimensional structure calculated with these proton pairs constrained to be close together.

In practice, a family of related structures is generated by two-dimensional NMR spectroscopy for three reasons (**Figure 4.43**). First, not enough constraints may be experimentally accessible to fully specify the structure. Second, the distances obtained from analysis of the NOESY spectrum are only approximate. Finally, the experimental observations are made not on single molecules but on a large number of molecules in solution that may have slightly different structures at any given moment. Thus, the family of structures generated from NMR structure analysis indicates the range of conformations for the protein in solution. With the ability of recombinant DNA technology to produce proteins labeled uniformly or at specific sites with ^{13}C, ^{15}N, and ^{2}H, structural determination by NMR spectroscopy can be applied to proteins up to 50 kDa. While solving the structure of proteins above this limit remains challenging, creative approaches have allowed NMR spectroscopy to address many important questions relating to very large proteins.

Cryo-electron microscopy can be used to determine the structures of large proteins and macromolecular complexes

As we shall see throughout this book, the structures determined by x-ray crystallography and NMR spectroscopy have illuminated much of our understanding of protein function. However, certain proteins, especially those that form large macromolecular complexes or are embedded within lipid membranes, represent a unique challenge. While crystallographic methods have been developed to enable structure determination of such proteins, these techniques can be quite complex and are not always reliable. An additional method for protein structure determination, **cryo-electron microscopy (cryo-EM)**, has emerged as a viable alternative.

In order to perform cryo-EM, a thin layer of the protein solution is prepared in a fine grid and then frozen very quickly, trapping the molecules in an ensemble of orientations. The sample is then placed in a transmission electron microscope under vacuum conditions and exposed to an incident electron beam. Each protein interacts with the beam to produce a two-dimensional projection on the image capture device, or detector. Many projections are detected, each capturing a molecule in a different orientation (**Figure 4.44A**). Using a process called *single-particle analysis*, these projections are assembled to build a three-dimensional representation of the protein (**Figure 4.44B**).

With improvements in image quality and sample preparation, cryo-EM is delivering structures at resolutions approaching 3 Å or better. At this level, details of atomic groups emerge. For example, cryo-EM studies were able to identify the location and structure of small molecules within the structure of TRPV1, the tetrameric protein present in the plasma membranes of pain-sensing neurons and depicted in Figure 4.44. At this resolution, cryo-EM can be used to aid the discovery and optimization of drug molecules (Chapter 32).

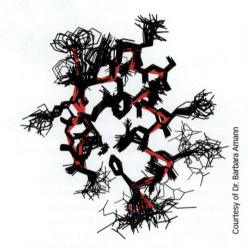

FIGURE 4.43 NMR spectroscopy generates a family of related structures. A set of 25 structures placed on top of each other for a 28-amino-acid domain from a DNA-binding protein. The red line traces the average course of the protein backbone. Each of these structures is consistent with hundreds of constraints derived from NMR experiments. The differences between the individual structures are due to a combination of imperfections in the experimental data and the dynamic nature of proteins in solution.

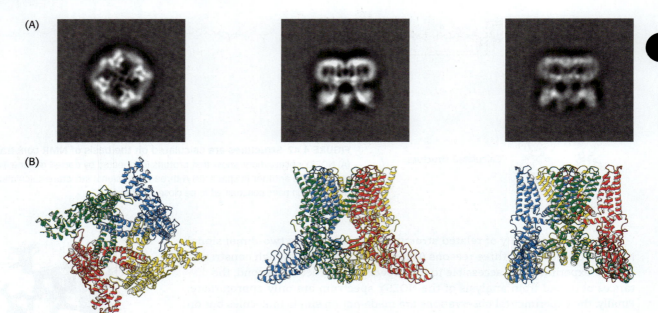

FIGURE 4.44 Cryo-EM captures numerous projections of a protein complex. The projections detected in a cryo-EM experiment represent two-dimensional images of the protein trapped in various orientations. Three such projections of the tetrameric protein TRPV1 are shown in (A). Collection of many such projections enables the reconstruction of a three-dimensional model of the protein, as depicted in (B).

[(A) Reprinted by permission from Macmillan Publishers Ltd: from M. Liao, et al. Structure of the TRPV1 ion channel determined by electron cryomicroscopy. *Nature*, 504, 107–112, 2013, Fig. 1c; Permission conveyed through Copyright Clearance Center, Inc. (B) Drawn from 3J5P.pdb.]

In addition, cryo-EM enables the visualization of very large molecular complexes. Consider the spliceosome, a complex of 4 oligonucleotide chains and over 30 protein subunits that plays an important role in the processing of eukaryotic messenger RNA (Chapter 29). While x-ray crystallographic methods have been unable to determine the structure of this massive complex, it has been solved to nearly 3.5 Å resolution by cryo-EM.

By the end of 2020, x-ray crystallography, NMR spectroscopy, and cryo-EM had revealed the structures of over 170,000 proteins. Several new structures are now determined each day. The coordinates are collected at the Protein Data Bank (www.pdb.org), and the structures can be accessed for visualization and analysis. Knowledge of the detailed molecular architecture of proteins has provided insight into how proteins recognize and bind other molecules, how they function as enzymes, how they fold, and how they have evolved. This extraordinarily rich harvest is continuing at a rapid pace and is greatly influencing the entire field of biochemistry as well as other biological and physical sciences.

Summary

4.1 The Purification of Proteins Is an Essential First Step in Understanding Their Function

- Effective protein purification relies on the development of an assay, used to determine the specific activity of a sample.

- The starting point for protein purification is the homogenate, in which the cellular membranes have been disrupted.

- Chromatography can separate proteins on the basis of size (gel-filtration chromatography), charge (ion-exchange chromatography), and specific binding to a ligand (affinity chromatography).

- Proteins can be separated and easily visualized by gel electrophoresis.

- Isoelectric focusing separates proteins on the basis of their isoelectric points.

- Two-dimensional electrophoresis combines isoelectric focusing and gel electrophoresis to separate complex mixtures of proteins.

4.2 Immunology Provides Important Techniques for Investigating Proteins

- Antibodies against a specific protein can be generated by injecting the protein, or a fragment of the protein, into an animal and collecting the antiserum.

- Monoclonal antibodies can be obtained by the hybridoma method.

- Using antibodies, proteins can be sensitively detected using enzyme-linked immunosorbent assays and western blotting.

- Labeled antibodies can be used to visualize proteins within cells and tissues.

4.3 Mass Spectrometry Is a Powerful Technique for the Identification of Peptides and Proteins

- Tandem mass spectrometry enables the rapid and highly accurate sequencing of peptides.
- Knowledge of a sequence provides valuable clues to the conformation, function, and evolutionary history of a protein.
- Peptide mass fingerprinting can be used to identify proteins, even in complex mixtures and at very low concentrations.
- Mass spectrometric techniques are central to proteomics because they make it possible to analyze the constituents of large macromolecular assemblies or other collections of proteins.

4.4 Peptides Can Be Synthesized by Automated Solid-Phase Methods

- Synthetic peptides have many important applications, including as antigens for the generation of antibodies.

- In the solid phase method, repeated cycles of peptide coupling and deprotection of a growing peptide attached to resin enables rapid synthesis without isolation of intermediates.

4.5 Three-Dimensional Protein Structures Can Be Determined Experimentally

- X-ray crystallography uses diffracted x-rays from a protein crystal to generate an electron density map, which indicates the atomic positions of the protein.
- Nuclear magnetic resonance spectroscopy reveals the structure and dynamics of proteins in solution by identifying protons in close proximity to one another.
- Cryo-electron microscopy is a rapidly developing method that can elucidate the structures of large multimeric complexes at increasingly higher resolutions.
- The three-dimensional structures of over 170,000 proteins are now known in atomic detail.

Key Terms

assay (p. 101)

specific activity (p. 101)

homogenate (p. 101)

dialysis (p. 103)

gel-filtration chromatography (p. 103)

ion-exchange chromatography (p. 104)

cation exchange (p. 104)

anion exchange (p. 104)

affinity chromatography (p. 104)

high-performance liquid chromatography (HPLC) (p. 105)

gel electrophoresis (p. 106)

SDS–polyacrylamide gel electrophoresis (SDS-PAGE) (p. 106)

isoelectric point (p. 107)

isoelectric focusing (p. 107)

two-dimensional electrophoresis (p. 108)

ultracentrifugation (p. 111)

antigenic determinant (epitope) (p. 113)

polyclonal antibodies (p. 113)

monoclonal antibody (p. 113)

hybridoma (p. 114)

enzyme-linked immunosorbent assay (ELISA) (p. 114)

western blotting (p. 116)

co-immunoprecipitation (p. 116)

fluorescence microscopy (p. 117)

mass spectrometry (p. 118)

matrix-assisted laser desorption/ionization (MALDI) (p. 118)

electrospray ionization (ESI) (p. 118)

time-of-flight (TOF) mass analyzer (p. 119)

tandem mass spectrometry (p. 121)

overlap peptides (p. 122)

peptide mass fingerprinting (p. 124)

proteome (p. 125)

solid-phase method (p. 127)

x-ray crystallography (p. 129)

electron-density map (p. 131)

nuclear magnetic resonance (NMR) spectroscopy (p. 132)

chemical shift (p. 133)

nuclear Overhauser effect (p. 133)

cryo-electron microscopy (cryo-EM) (p. 135)

Problems

1. Which of the following statements about SDS-PAGE under reducing conditions is true? Select all that apply. ❖ 3

(a) Smaller proteins migrate faster through the polyacrylamide gel.

(b) Protein-SDS complexes migrate toward the negative electrode.

(c) Proteins are separated in a polyacrylamide gel matrix.

(d) Protein-SDS complexes have similar mass-to-charge ratios; therefore, separation is by size.

(e) Proteins are visualized using a dye that binds to the gel matrix, but not to proteins.

(f) Sodium dodecyl sulfate binds proteins, resulting in protein-SDS complexes that are similar in size.

2. Complete the following table. ❖ 1

Purification procedure	Total protein (mg)	Total activity (units)	Specific activity (units mg⁻¹)	Purification level	Yield (%)
Crude extract	20,000	4,000,000		1	100
(NH₄)₂SO₄ precipitation	5000	3,000,000			
DEAE-cellulose chromatography	1500	1,000,000			
Gel-filtration chromatography	500	750,000			
Affinity chromatography	45	675,000			

3. (a) A purified protein is in a buffer solution at pH 7 with 800 mM NaCl. A 1 mL sample of the protein solution is placed in a dialysis membrane and dialyzed against 2.0 L of the same buffer with 0 mM NaCl. Small molecules and ions, such as Na, Cl, and the buffer, can diffuse across the membrane, but the protein cannot. Once the dialysis has come to equilibrium, what is the concentration of NaCl in the protein sample (assuming no change in volume inside the membrane)? **(b)** If the original 1 mL sample were dialyzed twice, successively, against 200 mL of the same buffer in 0 mM NaCl, what would be the final NaCl concentration in the sample?

4. Fluorescence-activated cell sorting (FACS) is a powerful technique for separating cells according to their content of particular molecules. For example, a fluorescence-labeled antibody specific for a cell-surface protein can be used to detect cells containing such a molecule. Suppose that you want to isolate cells that possess a receptor enabling them to detect bacterial degradation products. However, you do not yet have an antibody directed against this receptor. Which fluorescence-labeled molecule would you prepare to identify such cells?

5. (a) The octapeptide AVGWRVKS was digested with the enzyme trypsin. Which method would be most appropriate for separating the products: ion-exchange or gel-filtration chromatography? Explain. **(b)** Suppose that the peptide was digested with chymotrypsin. What would be the optimal separation technique? Explain. ❖ 2

6. Your frustrated colleague hands you a mixture of four proteins with the following properties:

	Isoelectric point (pI)	Molecular weight (in kDa)
Protein A	4.1	80
Protein B	9.0	81
Protein C	8.8	37
Protein D	3.9	172

(a) Propose a method for the isolation of Protein B from the other proteins. **(b)** If Protein B also carried a His tag at its N-terminus, how could you revise your method? ❖ 2

7. The amide hydrogen atoms of peptide bonds within proteins can exchange with protons in the solvent. In general, amide hydrogen atoms in buried regions of proteins and protein complexes exchange more slowly than those on the solvent-accessible surface do. Determination of these rates can be used to explore the protein-folding reaction, probe the tertiary structure of proteins, and identify the regions of protein–protein interfaces. These exchange reactions can be followed by studying the behavior of the protein in solvent that has been labeled with deuterium (²H), a stable isotope of hydrogen. What two methods described in this chapter could be readily applied to the study of hydrogen–deuterium exchange rates in proteins?

8. In this chapter, we described co-immunoprecipitation as a method for identifying binding partners to a protein of interest. A simpler variation of this method can also be used to isolate proteins of low abundance in a complex mixture. Arrange the steps in sequential order to use this technique for this purpose: ❖ 4

(a) The antibody-protein complexes become insoluble.

(b) Protein A beads are added to the mixture.

(c) An antibody against the protein of interest is added to the mixture.

(d) The protein mixture is centrifuged, the supernatant is removed, and the pellet is washed.

(e) The antibody binds to the protein of interest.

9. You have isolated a protein from the bacterium *E. coli* and seek to confirm its identity by trypsin digestion and mass spectrometry. Determination of the masses of several peptide fragments has enabled you to deduce the identity of the protein. However, there is a discrepancy with one of the peptide fragments, which you believe should have the sequence MLNSFK and an (M+H)⁺ value of 739.38. In your experiments, you repeatedly obtain an (M+H)⁺ value of 767.38. What is the cause of this discrepancy, and what does it tell you about the region of the protein from which this peptide is derived? ❖ 5

10. Which solution conditions would be best used for eluting a transcription factor from a DNA affinity column? ❖ 2

(a) High imidazole concentration

(b) High salt concentration

(c) High glucose concentration

(d) A mixture of free DNA nucleotides

11. A peptide has the sequence Gly-Ser-Lys-Ala-Gly-Arg-Ser-Arg.

(a) How many fragments would result from cleaving the sequence with trypsin? What is the sequence of the smallest fragment?

(b) How many fragments result from chymotrypsin cleavage? What is the sequence of the largest fragment?

12. The results of a separation using two-dimensional electrophoresis are:

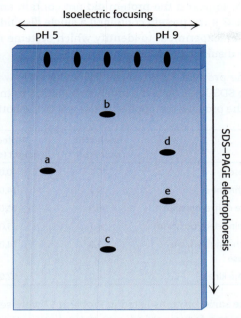

Isoelectric focusing

pH 5   pH 9

SDS–PAGE electrophoresis

(a) Which protein(s) have the highest pI value? **(b)** Which protein has the highest molecular weight? ❖ **3**

13. A protein has been isolated, but now one needs to determine its amino acid sequence. The results are shown:

• After one cycle of Edman degradation, the following phenylthiohydantoin (PTH) derivative was produced:

• Treatment with cyanogen bromide gave the fragments: KPAM, PD, LICWM, and TRM.
• One of the fragments produced after treatment with chymotrypsin was MPD.

What is the sequence of the peptide? ❖ **5**

14. Determine the sequence of decapeptide on the basis of the following data.

Trysin digestion gave two fragments with multiple residues (not in order):

  T1: Ala, Arg, Phe, Gly, Thr, Trp, Tyr

  T2: Lys, Met, Val

Chymotrypsin digestion gave four fragments with multiple residues (not in order):

  CT1: Ala, Phe

  CT2: Thr, Trp

  CT3: Lys, Met, Tyr, Val

  CT4: Arg, Gly

Treatment with cyanogen bromide yielded a single amino acid, methionine, and a nonapeptide.
What is the sequence of this decapeptide? ❖ **5**

15. Polyacrylamide gel electrophoresis experiments can be run under different sample preparation conditions, with informative results. Native PAGE does not include SDS. Reducing SDS-PAGE uses both SDS and a reducing agent, such as β-mercaptoethanol. Nonreducing SDS-PAGE uses SDS, but no reducing agent.

(a) A protein sample complex consists of two proteins, a smaller protein X, and a larger protein Y. Protein X is composed of two polypeptide chains lined by disulfide bonds. Protein Y is composed of three polypeptide chains linked by disulfide bonds.
Which type of PAGE was used to generate the each of the gels below?

(A)

1000 kDa	▬
500 kDa	▬
250 kDa	▬ ▬
100 kDa	▬
50 kDa	▬
30 kDa	▬

(B)

250 kDa	▬
150 kDa	▬ ▬
100 kDa	▬ ▬
60 kDa	▬
50 kDa	▬
30 kDa	▬

(C)

250 kDa	▬
150 kDa	▬
100 kDa	▬
60 kDa	▬ ▬
50 kDa	▬ ▬
30 kDa	▬ ▬

(b) What are the molecular weights of the protein X polypeptides?

(c) What are the molecular weights of the protein Y polypeptides? ❖ **3**

16. Performic acid cleaves the disulfide linkage of cystine and converts the sulfhydryl groups into cysteic acid residues, which are then no longer capable of disulfide-bond formation.

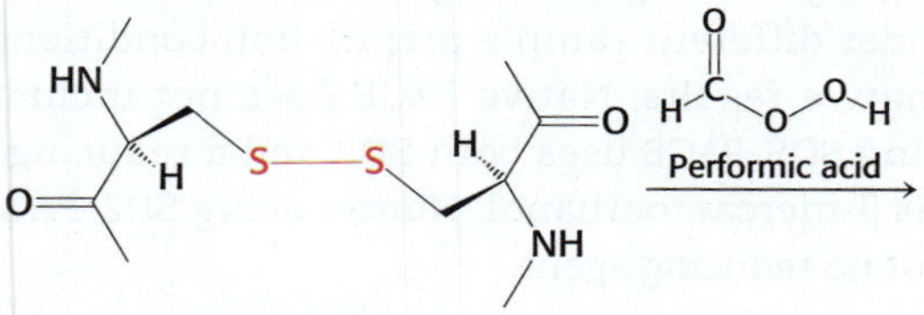

Cystine

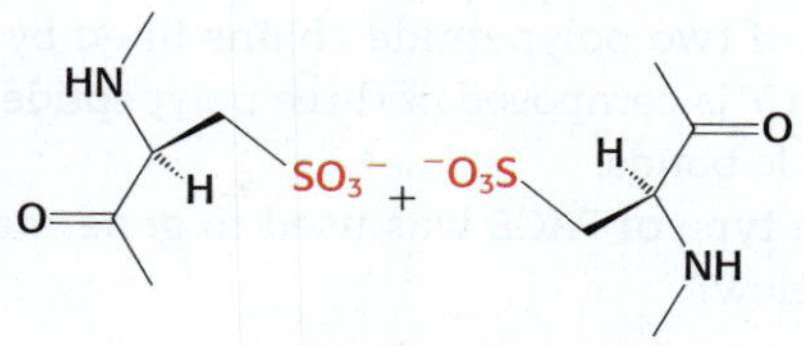

Cysteic acid

Consider the following experiment: You suspect that a protein containing three cysteine residues has a single disulfide bond. You digest the protein with trypsin and subject the mixture to electrophoresis along one end of a sheet of paper. After treating the paper with performic acid, you subject the sheet to electrophoresis in the perpendicular direction and stain it with a reagent that detects proteins. How would the paper appear if the protein did not contain any disulfide bonds? If the protein contained a single disulfide bond? Propose an experiment to identify which cysteine residues form the disulfide bond.

17. If the proteins listed in the following table were subjected to SDS-PAGE, give the order of migration from the top (where the proteins are applied to the gel) to the bottom.

Protein	Isoelectric point (pI)	Molecular weight (daltons)
(a) Aldolase	10.1	40,000
(b) Cytochrome *c*	9.2	14,000
(c) Lactate dehydrogenase	8.5	18,000
(d) Ornithine decarboxylase	4.1	75,000
(e) Phosphoenolpyruvate carboxylase	5.0	74,000
(f) Pyruvate kinase	6.0	62,000

18. If the same proteins listed in Problem 17 were separated by isoelectric focusing, indicate their distribution in the gradient from low pH to high pH.

Enzymes: Core Concepts and Kinetics

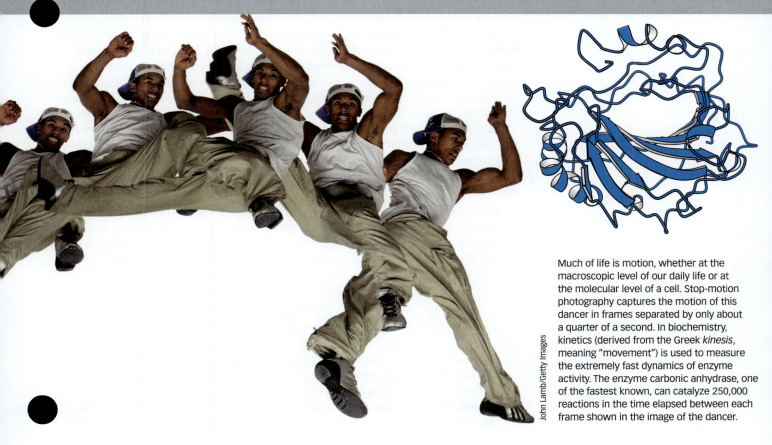

John Lamb/Getty Images

Much of life is motion, whether at the macroscopic level of our daily life or at the molecular level of a cell. Stop-motion photography captures the motion of this dancer in frames separated by only about a quarter of a second. In biochemistry, kinetics (derived from the Greek *kinesis*, meaning "movement") is used to measure the extremely fast dynamics of enzyme activity. The enzyme carbonic anhydrase, one of the fastest known, can catalyze 250,000 reactions in the time elapsed between each frame shown in the image of the dancer.

❖ LEARNING GOALS

By the end of this chapter, you should be able to:

1. Explain the thermodynamics of catalyzed and uncatalyzed reactions using Gibbs free energy diagrams.
2. Describe the central role of the formation of the transition state in enzyme catalysis.
3. Explain how reaction velocity is determined and used to characterize enzyme activity in terms of the fundamental kinetic parameters of an enzyme, K_M and k_{cat}.
4. Differentiate between types of reversible and irreversible inhibitors by mechanisms of action and using kinetics data.

OUTLINE

5.1 Enzymes Are Powerful and Highly Specific Catalysts

5.2 Gibbs Free Energy Is a Useful Thermodynamic Function for Understanding Enzymes

5.3 Enzymes Accelerate Reactions by Facilitating the Formation of the Transition State

5.4 The Michaelis–Menten Model Accounts for the Kinetic Properties of Many Enzymes

5.5 Enzymes Can Be Studied One Molecule at a Time

5.6 Enzymes Can Be Inhibited by Specific Molecules

Enzymes like carbonic anhydrase are biological catalysts that determine the patterns of chemical transformations and mediate the transformation of one form of energy into another. Enzymes are also the target of many drugs; in fact, the majority of all pharmaceuticals ever created have been molecules that reduce enzyme function. For instance, a host of drugs used to manage pain, inflammation, and fever are inhibitors of a class of enzymes called cyclooxygenases. One of the most widely used medications in the world, acetylsalicylic acid, commonly known as aspirin, is among this group. About a quarter of the genes in the human genome encode enzymes, a testament to their importance to life. In this chapter, you will learn how enzymes accelerate reactions, how the speed of their effects is measured, and how their action can be inhibited.

5.1 Enzymes Are Powerful and Highly Specific Catalysts

Agents called **catalysts** speed up chemical reactions without being consumed by them. **Enzymes** are powerful biological catalysts—made by all living organisms—that dramatically enhance and control the rates of chemical reactions. Almost all enzymes are proteins, which are highly effective catalysts for an enormous diversity of chemical reactions because of their capacity to specifically bind a very wide range of molecules. Using the full repertoire of intermolecular forces, enzymes optimally orient reactant molecules to make and break chemical bonds. However, proteins do not have an absolute monopoly on catalysis; the discovery of catalytically active RNA molecules, called **ribozymes**, provides compelling evidence that RNA was a biocatalyst early in evolution. Two remarkable properties of enzymes to consider from the start are their speed and their specificity.

- *Speed.* Enzymes can accelerate reactions by factors of a billion or more (**Table 5.1**). Indeed, most reactions in biological systems do not take place at perceptible rates in the absence of enzymes. Even a reaction as simple as the hydration of carbon dioxide is catalyzed by an enzyme—namely, carbonic anhydrase.

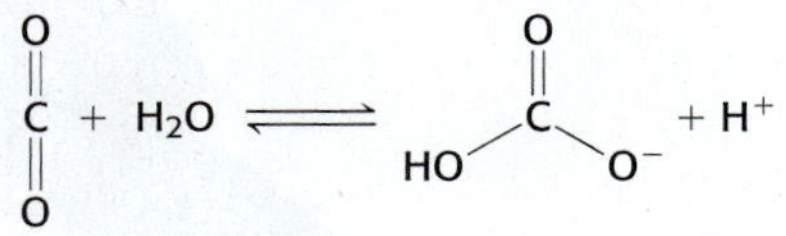

$$\underset{\substack{\| \\ O}}{\overset{\substack{O \\ \|}}{C}} + H_2O \;\rightleftharpoons\; \underset{HO}{\overset{O}{\underset{\|}{C}}}{-O^-} + H^+$$

The transfer of CO_2 from the tissues to the blood and then to the air in the alveolae of the lungs would be less complete in the absence of this enzyme. In fact, carbonic anhydrase is one of the fastest enzymes known. Each enzyme molecule can hydrate 10^6 molecules of CO_2 per second. This catalyzed reaction is 10^7 times as fast as the uncatalyzed one. We will consider the mechanism of carbonic anhydrase catalysis in Chapter 6.

TABLE 5.1 Rate enhancement by selected enzymes

Enzyme	Nonenzymatic half-life	Uncatalyzed rate (k_{un} s^{-1})	Catalyzed rate (k_{cat} s^{-1})	Rate enhancement (k_{cat} s^{-1}/k_{un} s^{-1})
OMP decarboxylase	78,000,000 years	2.8×10^{-16}	39	1.4×10^{17}
Staphylococcal nuclease	130,000 years	1.7×10^{-13}	95	5.6×10^{14}
AMP nucleosidase	69,000 years	1.0×10^{-11}	60	6.0×10^{12}
Carboxypeptidase A	7.3 years	3.0×10^{-9}	578	1.9×10^{11}
Ketosteroid isomerase	7 weeks	1.7×10^{-7}	66,000	3.9×10^{11}
Triose phosphate isomerase	1.9 days	4.3×10^{-6}	4300	1.0×10^{9}
Chorismate mutase	7.4 hours	2.6×10^{-5}	50	1.9×10^{6}
Carbonic anhydrase	5 seconds	1.3×10^{-1}	1×10^{6}	7.7×10^{6}

Abbreviations: OMP, orotidine monophosphate; AMP, adenosine monophosphate.
Source: After A. Radzicka and R. Wolfenden, *Science* 267:90–93, 1995.

- *Specificity.* Enzymes are highly specific both in the reactions that they catalyze and in the reactants they bind, which are called **substrates**. An enzyme usually catalyzes a single chemical reaction or a set of closely related reactions. For example, enzymes known as **proteases** catalyze proteolysis, the hydrolysis of a peptide bond.

Most proteolytic enzymes also catalyze a different but related reaction in vitro—namely, the hydrolysis of an ester bond. Such reactions are more easily monitored than is proteolysis and are useful in experimental investigations of these enzymes.

Ester **Acid** **Alcohol**

Proteolytic enzymes differ markedly in their degree of substrate specificity. Papain, which is found in papaya plants, is quite undiscriminating: it will cleave any peptide bond with little regard to the identity of the adjacent side chains. The digestive enzyme trypsin, on the other hand, is quite specific and catalyzes the splitting of peptide bonds only on the carboxyl side of lysine and arginine residues (**Figure 5.1A**). Thrombin, an enzyme that participates in blood clotting (Section 10.4), is even more specific than trypsin. It catalyzes the hydrolysis of Arg–Gly bonds in particular peptide sequences only (**Figure 5.1B**). The specificity of an enzyme is due to the precise interaction of the substrate with the enzyme, which is a result of the intricate three-dimensional structure of the enzyme.

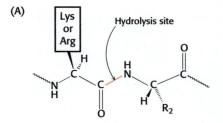

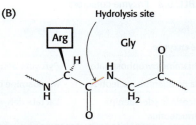

FIGURE 5.1 Enzymes are highly specific to particular substrates and chemical reactions. (A) Trypsin cleaves on the carboxyl side of arginine and lysine residues, whereas (B) thrombin cleaves Arg–Gly bonds in particular sequences only.

Most enzymes are classified by the types of reactions they catalyze

While some enzymes have common names—like papain, trypsin, and thrombin—that provide little information regarding their function, most enzymes are named for one of their substrates and for the reactions that they catalyze, with the suffix "-ase" added. Thus, a peptide hydrolase is an enzyme that hydrolyzes peptide bonds, whereas ATP synthase is an enzyme that synthesizes ATP. It is important to note, however, that enzymes catalyze chemical reactions in both the forward and reverse directions, yet only one direction of the reaction is typically denoted in the name.

To bring some consistency to enzyme nomenclature, a classification system for enzymes—developed by the International Union of Biochemistry—divides reactions into six major groups (**Table 5.2**). These groups are further subdivided

TABLE 5.2 Six major classes of enzymes

Class	Type of reaction	Example	Chapter
1. Oxidoreductases	Oxidation–reduction	Lactate dehydrogenase	16
2. Transferases	Group transfer	Nucleoside monophosphate kinase (NMP kinase)	6
3. Hydrolases	Hydrolysis reactions (transfer of functional groups to water)	Chymotrypsin	6
4. Lyases	Addition or removal of groups to form double bonds	Fumarase	17
5. Isomerases	Isomerization (intramolecular group transfer)	Triose phosphate isomerase	16
6. Ligases	Ligation of two substrates at the expense of ATP hydrolysis	Aminoacyl-tRNA synthetase	30

so that a four-number code preceded by the letters EC could precisely identify all enzymes. Consider as an example nucleoside monophosphate (NMP) kinase, an enzyme that we will examine in detail in Section 9.4. It catalyzes the following reaction:

$$ATP + NMP \rightleftharpoons ADP + NDP$$

NMP kinase transfers a phosphoryl group from ATP to any NMP to form a nucleoside diphosphate (NDP) and ADP. Consequently, it is a transferase, or member of group 2. Transferases that shift a phosphoryl group are designated 2.7. If a phosphate is the acceptor, the transferase is designated 2.7.4, and most precisely EC 2.7.4.4 if a nucleoside monophosphate is the acceptor. Although the common names are used routinely, the classification number is used when the precise identity of the enzyme is not clear from the common name alone.

Many enzymes require cofactors for activity

The catalytic activity of many enzymes depends on the presence of small molecules termed **cofactors**, although the precise role varies with the cofactor and the enzyme. Generally, these cofactors are able to execute chemical reactions that cannot be performed by the standard set of twenty amino acids. An enzyme without its cofactor is referred to as an **apoenzyme**; the complete, catalytically active enzyme is called a **holoenzyme**.

$$apoenzyme + cofactor = holoenzyme$$

Cofactors can be subdivided into two groups: (1) metals, whose importance to enzymatic activity we will explore in Chapter 6, and (2) small organic molecules called **coenzymes** (Table 5.3). Often derived from vitamins, coenzymes either can be tightly or loosely bound to the enzyme.

Tightly bound coenzymes are called **prosthetic groups**. Prosthetic groups are catalytic in that they are unchanged in the overall chemical reaction. In contrast, loosely associated coenzymes often behave more like second substrates (cosubstrates) because they bind to the enzyme, are changed by it, and then are released from it. Thus, these are also sometimes called *stoichiometric coenzymes* because they must be present in stoichiometric ratios with other substrates. The use of the same loosely associated coenzyme by a variety of enzymes sets these coenzymes apart from normal substrates, however, as does their source in vitamins (Table 5.3 and Section 15.4). Enzymes that use the same coenzyme usually perform catalysis by similar mechanisms. Throughout the book, we will see how coenzymes and their enzyme partners operate in their biochemical context.

TABLE 5.3 Enzyme cofactors

Cofactor	Enzyme
Coenzyme	
Thiamine pyrophosphate	Pyruvate dehydrogenase
Flavin adenine nucleotide	Monoamine oxidase
Nicotinamide adenine dinucleotide	Lactate dehydrogenase
Pyridoxal phosphate	Glycogen phosphorylase
Coenzyme A (CoA)	Acetyl CoA carboxylase
Biotin	Pyruvate carboxylase
5′-Deoxyadenosylcobalamin	Methylmalonyl mutase
Tetrahydrofolate	Thymidylate synthase
Metal	
Zn^{2+}	Carbonic anhydrase
Zn^{2+}	Carboxypeptidase
Mg^{2+}	*Eco*RV
Mg^{2+}	Hexokinase
Ni^{2+}	Urease
Mo	Nitrogenase
Se	Glutathione peroxidase
Mn	Superoxide dismutase
K^+	Acetyl CoA thiolase

❖ SELF–CHECK QUESTION

Which terms (cofactor, coenzyme, and/or prosthetic group) correctly describe the heme from cytochrome c oxidase, which doesn't dissociate from the protein?

Enzymes can transform energy from one form into another

A key activity in all living systems is the conversion of one form of energy into another. For example, in photosynthesis, light energy is converted into chemical potential energy. In cellular respiration, which takes place in mitochondria, the free energy contained in small molecules derived from food is converted first into the free energy of an ion gradient and then into a different currency—the free energy of adenosine triphosphate. Given their centrality to life, it should come as no surprise that enzymes play vital roles in energy transformation.

After enzymes perform fundamental roles in photosynthesis and cellular respiration, other enzymes can then use the chemical potential energy of ATP in diverse ways. For instance, the enzyme myosin converts the energy of ATP into the mechanical energy of contracting muscles (Section 6.5). Pumps in the membranes of cells and organelles, which can be thought of as enzymes that move substrates rather than chemically alter them, use the energy of ATP to transport molecules and ions across the membrane (Chapter 13). The chemical and electrical gradients resulting from the unequal distribution of these molecules and ions are themselves forms of potential energy that can be used for a variety of purposes, such as sending nerve impulses (Section 13.4).

Recent developments show that the power of enzymes may be harnessed to generate energy for entire communities, as well as reducing landfill. Unsorted municipal waste can be treated with a cocktail of enzymes that includes an array of proteases as well as carbohydrate- and lipid-degrading enzymes, turning much of the waste into a bioliquid of sugars, amino acids, and other biomolecules. The bioliquid can then be used to fuel the growth of methane-producing bacteria, and the methane harvested and burned to generate electricity. Any waste not degraded by the enzyme cocktail is recycled or incinerated to produce electricity.

5.2 Gibbs Free Energy Is a Useful Thermodynamic Function for Understanding Enzymes

Enzymes speed up the rate of chemical reactions, but the properties of the reaction—whether it can take place at all and the degree to which the enzyme accelerates the reaction—depend on energy differences between reactants (the initial state) and products (the final state). Gibbs free energy (G), which was discussed in Section 1.3, is a thermodynamic property that is a measure of useful energy, or the energy that is capable of doing work. To understand how enzymes operate, we need to consider only two thermodynamic properties of the reaction: (1) the free-energy difference (ΔG) between the products and reactants; and (2) the energy required to initiate the conversion of reactants into products. The former determines whether the reaction will take place spontaneously, whereas the latter determines the rate of the reaction. Enzymes affect only the rate.

The free-energy change provides information about the spontaneity but not the rate of a reaction

The free-energy change of a reaction (ΔG) tells us if the reaction can take place spontaneously:

- *A reaction can take place spontaneously only if ΔG is negative.* Such reactions are said to be *exergonic*.

- *A system is at equilibrium and no net change can take place if ΔG is zero.*

- *A reaction cannot take place spontaneously if ΔG is positive.* An input of free energy is required to drive such a reaction. These reactions are termed *endergonic*.

- *The ΔG of a reaction depends only on the free energy of the products minus the free energy of the reactants.* The ΔG of a reaction is independent of the molecular mechanism of the transformation. For example, the ΔG for the oxidation of glucose to CO_2 and H_2O is the same whether it takes place by combustion or by a series of enzyme-catalyzed steps in a cell.

- *The ΔG provides no information about the rate of a reaction.* A negative ΔG indicates that a reaction can take place spontaneously, but it does not signify whether it will proceed at a perceptible rate.

The standard free-energy change of a reaction is related to the equilibrium constant

As for any reaction, we need to be able to determine ΔG for an enzyme-catalyzed reaction to know whether the reaction is spontaneous or requires an input of energy, measured in kilojoules (kJ) or kilocalories (kcal); one kilojoule is equivalent to 0.239 kilocalories. To determine this important thermodynamic parameter, we need to take into account the nature of both the reactants and the products as well as their concentrations.

Consider the reaction

$$A + B \rightleftharpoons C + D$$

The ΔG of this reaction is given by

$$\Delta G = \Delta G^{\circ} + RT \ln \frac{[C][D]}{[A][B]} \tag{1}$$

in which ΔG° is the standard free-energy change, R is the gas constant, T is the absolute temperature, and [A], [B], [C], and [D] are the molar concentrations (more precisely, the activities) of the reactants. ΔG° is the free-energy change for this reaction under standard conditions — that is, when each of the reactants A, B, C, and D is present at a concentration of 1.0 M (for a gas, the standard state is usually chosen to be 1 bar, which is very close to 1 atmosphere). Thus, the ΔG of a reaction depends on the nature of the reactants (expressed in the ΔG° term of equation 1) and on their concentrations (expressed in the logarithmic term of equation 1).

A convention has been adopted to simplify free-energy calculations for biochemical reactions, in which the standard state is defined as having a pH of 7. Consequently, when H^+ is a reactant, its activity has the value 1 (corresponding to a pH of 7) in equations 1 and 2. The activity of water also is taken to be 1 in these equations. The standard free-energy change at pH 7, denoted by the symbol $\Delta G^{\circ\prime}$, will be used throughout this book. A simple way to determine $\Delta G^{\circ\prime}$ is to measure the concentrations of reactants and products when the reaction has reached equilibrium. At equilibrium, there is no net change in reactants and products; in essence, the reaction has stopped and $\Delta G = 0$. At equilibrium, equation 1 then becomes

$$0 = \Delta G^{\circ\prime} + RT \ln \frac{[C]_{eq}[D]_{eq}}{[A]_{eq}[B]_{eq}}$$

and so

$$\Delta G^{\circ\prime} = -RT \ln \frac{[C]_{eq}[D]_{eq}}{[A]_{eq}[B]_{eq}} \tag{2}$$

The equilibrium constant under standard conditions, K'_{eq}, is defined as

$$K'_{eq} = \frac{[C]_{eq}[D]_{eq}}{[A]_{eq}[B]_{eq}} \tag{3}$$

Substituting equation 3 into equation 2 gives

$$\Delta G^{\circ\prime} = -RT \ln K'_{eq} \tag{4}$$

which can be rearranged to give

$$K'_{eq} = e^{-\Delta G^{\circ\prime}/RT}$$

Substituting $R = 8.315 \times 10^{-3}$ kJ mol^{-1} K^{-1} and $T = 298$ K (corresponding to 25°C) gives

$$K'_{eq} = e^{-\Delta G^{\circ\prime}/2.48} \tag{5}$$

where $\Delta G^{\circ\prime}$ is here expressed in kilojoules per mole because of the choice of the units for R in equation 4. Thus, the standard free energy and the equilibrium constant of a reaction are related by a simple expression. For example, an equilibrium constant of 10 gives a standard free-energy change of -5.69 kJ mol^{-1} or -1.36 kcal mol^{-1} at 25°C (**Table 5.4**). Note that, for each 10-fold change in the equilibrium constant, the $\Delta G^{\circ\prime}$ changes by 5.69 kJ mol^{-1} or 1.36 kcal mol^{-1}.

It is important to stress that whether the ΔG for a reaction is larger, smaller, or the same as $\Delta G^{\circ\prime}$ depends on the concentrations of the reactants and products. The criterion of spontaneity for a reaction under real conditions is ΔG, not $\Delta G^{\circ\prime}$. This point is important because reactions that are not spontaneous based on $\Delta G^{\circ\prime}$ can be made spontaneous by adjusting the concentrations of reactants and products. This principle is commonly observed in enzyme-catalyzed reactions in metabolic pathways (Chapter 15).

TABLE 5.4 Relationship between $\Delta G^{\circ\prime}$ and K'_{eq} (at 25°C)

	$\Delta G^{\circ\prime}$	
K'_{eq}	kJ mol^{-1}	kcal mol^{-1}
10^{-5}	28.53	6.82
10^{-4}	22.84	5.46
10^{-3}	17.11	4.09
10^{-2}	11.42	2.73
10^{-1}	5.69	1.36
1	0.00	0.00
10	-5.69	-1.36
10^2	-11.42	-2.73
10^3	-17.11	-4.09
10^4	-22.84	-5.46
10^5	-28.53	-6.82

EXAMPLE Calculating and Comparing $\Delta G^{\circ\prime}$ and ΔG

PROBLEM: Calculate $\Delta G^{\circ\prime}$ and ΔG for a specific reaction that occurs during glucose metabolism:

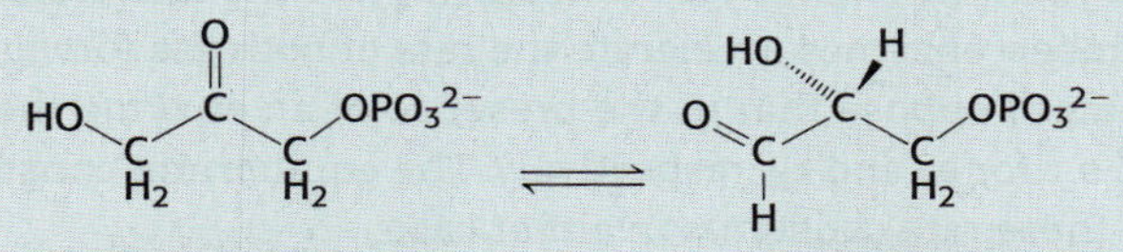

Dihydroxyacetone phosphate (DHAP) ⇌ **Glyceraldehyde 3-phosphate (GAP)**

At equilibrium, the ratio of GAP to DHAP is 0.0475 at 25°C (298 K) and pH 7; hence, $K'_{eq} = 0.0475$. Typical concentrations in the cell are 2×10^{-4} M for DHAP and 3×10^{-6} M for GAP.

GETTING STARTED: Locate the equations used to calculate $\Delta G^{\circ\prime}$ and ΔG (equations 1 and 4). Besides the values given in the problem, is there anything else we need to know to solve these equations? Recall that the gas constant, R, is 8.315×10^{-3} kJ mol^{-1} K^{-1}

CALCULATE: Let's begin by calculating $\Delta G^{\circ\prime}$, the standard free-energy change for this reaction, using equation 4 and substituting in the values for R, the temperature in Kelvin, and the value of K'_{eq}:

$$\Delta G^{\circ\prime} = -RT \ln K'_{eq}$$
$$= -8.315 \times 10^{-3} \times 298 \times \ln (0.0475)$$
$$= +7.53 \text{ kJ mol}^{-1} \text{ or } +1.80 \text{ kcal mol}^{-1}$$

What does $\Delta G^{\circ\prime}$ tell us? We can see from this result that under standard conditions the reaction is endergonic and DHAP will not spontaneously convert into GAP. However, this pertains only to standard conditions, not actual physiological conditions. For those we need to calculate ΔG, by substituting the real cellular concentrations into the equation for ΔG (equation 1):

$$\Delta G = 7.53 \text{ kJ mol}^{-1} - RT \ln \frac{3 \times 10^{-6} \text{ M}}{2 \times 10^{-4} \text{ M}}$$
$$= 7.53 \text{ kJ mol}^{-1} - (10.42 \text{ kJ mol}^{-1})$$
$$= -2.89 \text{ kJ mol}^{-1} \text{ or } -0.69 \text{ kcal mol}^{-1}$$

REFLECT: What is the significance of the difference between ΔG and $\Delta G^{\circ\prime}$? Note that ΔG for this reaction is negative, even though $\Delta G^{\circ\prime}$ is positive. The negative value for ΔG indicates that the conversion of DHAP to GAP is exergonic and can take place spontaneously when these species are present at the concentrations given.

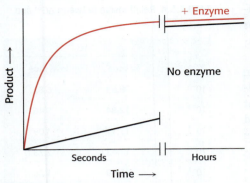

FIGURE 5.2 Enzymes do not affect thermodynamics and therefore do not change the equilibrium point. The same equilibrium point is reached much more quickly in the presence of an enzyme.

Enzymes alter only the reaction rate and not the reaction equilibrium

We have established that enzymes are excellent catalysts, but it is equally important to understand what they cannot do: an enzyme cannot alter the laws of thermodynamics and consequently cannot alter the equilibrium of a chemical reaction. Consider an enzyme-catalyzed reaction, the conversion of substrate, S, into product, P. **Figure 5.2** shows the rate of product formation with time in the presence and absence of enzyme. Note that the amount of product formed is the same whether or not the enzyme is present but, in the current example, the amount of product formed in seconds when the enzyme is present might take hours (or centuries; see Table 5.1) to form if the enzyme were absent.

Why does the rate of product formation level off with time? The reaction has reached equilibrium. Substrate S is still being converted into product P, but P is being converted into S at a rate such that the amount of P present stays the same.

Let us examine the equilibrium in a more quantitative way. Suppose that, in the absence of enzyme, the forward rate constant (k_F) for the conversion of S into P is 10^{-4} s^{-1} and the reverse rate constant (k_R) for the conversion of P into S is 10^{-6} s^{-1}. Remember: enzymes accelerate the rate of both the forward and reverse reactions. So, also suppose that in the presence of an enzyme these values are 10^5 s^{-1} and 10^3 s^{-1} for k_F and k_R, respectively. The equilibrium constant K is given by the ratio of these rate constants in either case:

$$S \underset{10^{-6}\,s^{-1}}{\overset{10^{-4}\,s^{-1}}{\rightleftharpoons}} P \qquad\qquad S \underset{10^3\,s^{-1}}{\overset{10^5\,s^{-1}}{\rightleftharpoons}} P$$

$$K_{Uncatalyzed} = \frac{[P]}{[S]} = \frac{k_F}{k_R} = \frac{10^{-4}}{10^{-6}} = 100 \qquad K_{Catalyzed} = \frac{[P]}{[S]} = \frac{k_F}{k_R} = \frac{10^5}{10^3} = 100$$

Notice that the equilibrium concentration of P is 100 times that of S, whether or not enzyme is present. However, it might take a very long time to approach this equilibrium without enzyme, whereas equilibrium would be attained rapidly in the presence of a suitable enzyme (Table 5.1).

By catalyzing both the forward and reverse directions of reactions, enzymes accelerate the attainment of equilibria but do not shift their positions. The equilibrium position is a function only of the free-energy difference between reactants and products.

5.3 Enzymes Accelerate Reactions by Facilitating the Formation of the Transition State

The free-energy difference between reactants and products accounts for the equilibrium of the reaction, but enzymes accelerate how quickly this equilibrium is attained. How can we explain the rate enhancement in terms of thermodynamics? To do so, we have to consider not the end points of the reaction, but the chemical pathway between the end points.

A chemical reaction of substrate S to form product P goes through a **transition state** X‡ that has a higher free energy than does either S or P.

$$S \longrightarrow X^\ddagger \longrightarrow P$$

The double dagger (‡) denotes the transition state. The transition state is a transitory molecular structure that is no longer the substrate but is not yet the product. The transition state is the least-stable and most-seldom-occupied species along the reaction pathway because it is the one with the highest free energy. The difference in free energy between the transition state and the substrate is called the **Gibbs free energy of activation** or simply the **activation energy**, symbolized by $\Delta G^\ddagger$ (**Figure 5.3**).

$$\Delta G^\ddagger = G_{X^\ddagger} - G_S$$

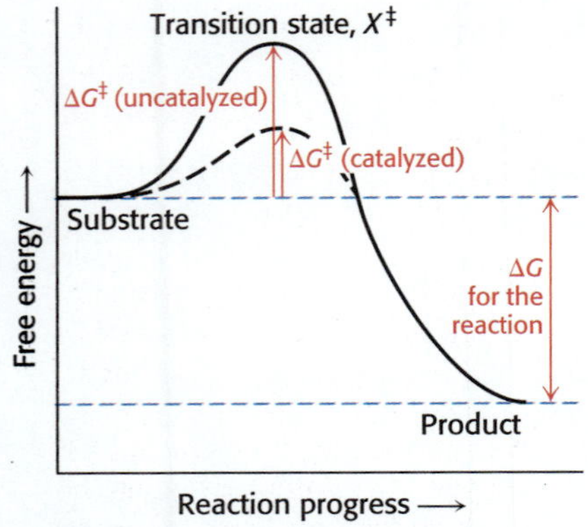

FIGURE 5.3 Enzymes accelerate reactions by decreasing $\Delta G^\ddagger$, the free energy of activation.

Note that $\Delta G^{\ddagger}$ does not enter into the final ΔG calculation for the reaction, because the energy required to generate the transition state is released when the transition state forms the product. This is also true when we consider the reaction from the reverse direction. The activation-energy barrier immediately suggests how an enzyme enhances the reaction rates in both directions without altering ΔG of the reaction: enzymes accelerate reactions by decreasing $\Delta G^{\ddagger}$, the activation energy, thus facilitating the formation of the transition state from either substrate or product.

The combination of substrate and enzyme creates a reaction pathway whose transition-state energy is lower than that of the reaction in the absence of enzyme (Figure 5.3). Because the activation energy is lower, more molecules have the energy required to reach the transition state. Decreasing the activation barrier is analogous to lowering the height of a high-jump bar; more athletes will be able to clear the bar.

The formation of an enzyme–substrate complex is the first step in enzymatic catalysis

Much of the catalytic power of enzymes comes from their binding to and then altering the structure of the substrate to promote the formation of the transition state. Thus, the first step in catalysis is the formation of an enzyme–substrate (ES) complex. Substrates bind to a specific region of the enzyme called the **active site**. Most enzymes are highly selective in the substrates that they bind. Indeed, the catalytic specificity of enzymes depends in part on the specificity of binding.

What is the evidence for the existence of an enzyme–substrate complex?

- The first clue was the observation that, at a constant concentration of enzyme, the reaction rate increases with increasing substrate concentration until a maximal velocity is reached (**Figure 5.4**), suggesting that a discrete ES complex has formed. In contrast, uncatalyzed reactions do not show this saturation effect. At a sufficiently high substrate concentration, all the catalytic sites are filled, or saturated, and so the reaction rate cannot increase. Although indirect, the ability to saturate an enzyme with substrate is the most general evidence for the existence of ES complexes.

- The spectroscopic characteristics of many enzymes and substrates change on the formation of an ES complex. These changes are particularly striking if the enzyme contains a colored prosthetic group.

- X-ray crystallography has provided high-resolution images of substrates and substrate analogs bound to the active sites of many enzymes (**Figure 5.5**). In Chapter 6, we will take a close look at several of these complexes.

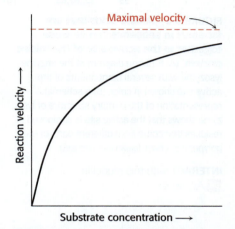

FIGURE 5.4 As substrate concentration increases, an enzyme-catalyzed reaction asymptotically approaches a maximal velocity.

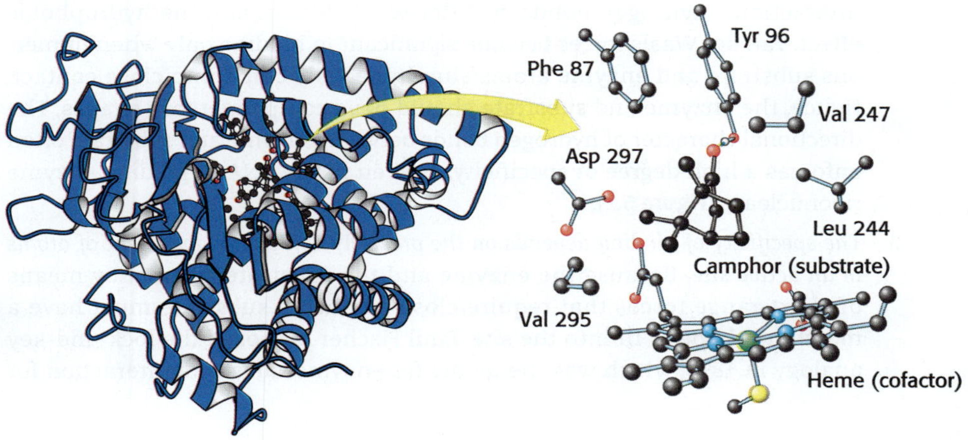

Maximal velocity

Reaction velocity ⟶

Substrate concentration ⟶

INTERACT with this model in **Achieve**

FIGURE 5.5 In the active site, substrates are surrounded by enzyme residues and sometimes cofactors. (Left) The structure of the complex between the enzyme cytochrome P450 and its substrate camphor. (Right) The substrate sits in the active site, which is a cavity formed by residues from the enzyme and the heme prosthetic group. [Drawn from 2CPP.pdb.]

Phe 87

Tyr 96

Val 247

Asp 297

Leu 244

Camphor (substrate)

Val 295

Heme (cofactor)

(A)

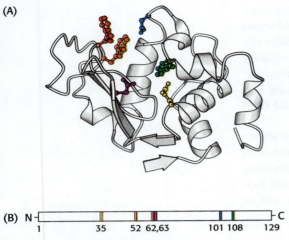

(B)

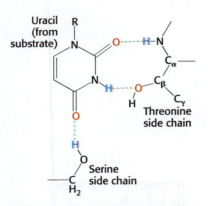

FIGURE 5.6   Amino acids that are far apart in sequence can be close together in the active site of the folded protein. (A) A ribbon diagram of the enzyme lysozyme with several components of the active site shown in color. (B) A schematic representation of the primary structure of lysozyme shows that the active site is composed of residues that come from different parts of the polypeptide chain. [Drawn from 6LYZ.pdb.]

INTERACT with this model in
≋ Achieve

FIGURE 5.7   The geometry of hydrogen bonds often creates specificity between an enzyme and its substrate. The enzyme ribonuclease forms hydrogen bonds with the uridine component of the substrate.
[Information from F. M. Richards, H. W. Wyckoff, and N. Allewell, In *The Neurosciences: Second Study Program*, F. O. Schmidt, Ed. (Rockefeller University Press, 1970), p. 970.]

The active sites of enzymes have some common features

Recall from Chapter 2 that proteins are not rigid structures, but are flexible and exist in an array of conformations. Thus, the interaction of the enzyme and substrate at the active site and the formation of the transition state is a dynamic process. Although enzymes differ widely in structure, specificity, and mode of catalysis, at least five generalizations can be made concerning their active sites:

1. *The active site is a three-dimensional cleft, or crevice, formed by groups that come from different parts of the amino acid sequence.* Indeed, residues far apart in the amino acid sequence may interact more strongly than adjacent residues in the sequence, which may be sterically constrained from interacting with one another. For example, in the enzyme lysozyme, the residues that contribute important groups to the active site are numbered 35, 52, 62, 63, 101, and 108 in the sequence of 129 amino acids (**Figure 5.6**). Lysozyme, found in a variety of organisms and tissues, including human tears, degrades the cell walls of some bacteria.

2. *The active site takes up a small part of the total volume of an enzyme.* Although residues in an enzyme are not in contact with the substrate, the cooperative motions of the entire enzyme help to correctly position the catalytic residues at the active site. Experimental attempts to reduce the size of a catalytically active enzyme show that the minimum size requires about 100 residues. In fact, nearly all enzymes are larger than this, which gives them a mass normally greater than 10 kDa and a diameter of more than 25 Å, suggesting that all residues in a protein, not just those at the active site, are ultimately required to form a functional enzyme.

3. *Active sites are unique microenvironments.* Water is usually excluded from the active site cleft unless it is a reactant. Often a nonpolar microenvironment in the cleft enhances the binding of substrates as well as catalysis. The cleft may also contain polar residues, some of which may acquire special properties essential for substrate binding or catalysis. The internal positions of these polar residues are biologically crucial exceptions to the general rule that polar residues are located on the surface of proteins, exposed to water.

4. *Substrates are bound to enzymes by multiple weak attractions.* The noncovalent interactions in ES complexes are much weaker than covalent bonds, which have energies between –210 and –460 kJ mol^{-1} (between –50 and –110 kcal mol^{-1}). In contrast, ES complexes usually have equilibrium constants that range from 10^{-2} to 10^{-8} M corresponding to free energies of interaction ranging from about –13 to –50 kJ mol^{-1} (from –3 to –12 kcal mol^{-1}). As discussed in Section 1.3, these weak reversible contacts are mediated by ionic interactions, hydrogen bonds, van der Waals forces, and the hydrophobic effect. Van der Waals forces become significant in binding only when numerous substrate and enzyme atoms simultaneously come into close contact. Hence, the enzyme and substrate should have complementary shapes. The directional character of hydrogen bonds between enzyme and substrate often enforces a high degree of specificity, as seen in the RNA-degrading enzyme ribonuclease (**Figure 5.7**).

5. *The specificity of binding depends on the precisely defined arrangement of atoms in an active site.* Because the enzyme and the substrate interact by means of short-range forces that require close contact, a substrate must have a matching shape to fit into the site. Emil Fischer proposed the lock-and-key analogy in 1890, which was the model for enzyme–substrate interaction for

several decades (**Figure 5.8A**). We now know that enzymes are flexible and that the shapes of the active sites can be markedly modified by the binding of substrate, a process of dynamic recognition called **induced fit** (**Figure 5.8B**). Moreover, the substrate may bind to only certain conformations of the enzyme, in what is called *conformation selection*. Thus, the mechanism of catalysis is dynamic, involving structural changes in both the substrates and the enzyme until the substrate reaches the transition state

The binding energy between enzyme and substrate is important for catalysis

Enzymes lower the activation energy, but where does the energy to lower the activation energy come from? Free energy is released by the formation of a large number of weak interactions between a complementary enzyme and its substrate. The free energy released on binding is called the **binding energy**. Only the correct substrate can participate in most or all of the interactions with the enzyme and thus maximize binding energy, accounting for the exquisite substrate specificity exhibited by many enzymes. Furthermore, the full complement of such interactions is formed—and the maximal binding energy is released—only when the substrate is converted into the transition state. The energy released by the interaction between the enzyme and the substrate can be thought of as lowering the activation energy.

Because the transition state collapses randomly, the activation energies determine the accumulation of either product or substrate

Molecular movements resulting in the optimal alignment of functional groups at the active site occur fleetingly. The transition state is too unstable to exist for long, collapsing randomly to either substrate or product. A consideration of the biochemical timescale, in which molecule motion is occurring on the order of microseconds to nanoseconds, is once again illuminating: a substrate may reach the transition state and collapse back to substrate again many times in a single active site within a fraction of a second before collapsing to product.

Whether substrate or product ultimately accumulates is determined only by the energy difference between the substrate and the product—that is, by the ΔG of the reaction, but why? If the direction of the collapse from the transition state is random, then the accumulation of either product or substrate can be explained by the activation energy in either direction because it dictates how frequently substrates or products reach the transition state. By lowering this activation energy, enzymes accelerate reactions in both directions by increasing the rate at which both products and substrates reach the transition state.

(A) **Lock and Key**

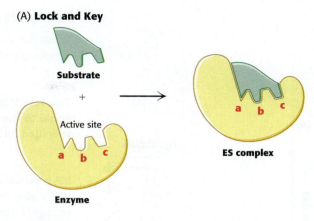

(B) **Induced Fit**

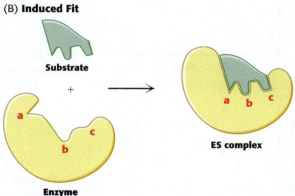

FIGURE 5.8 The old lock-and-key model of enzyme binding has been replaced by an induced-fit model. (A) In the lock-and-key model of enzyme–substrate binding, the active site of the unbound enzyme was proposed to be complementary in shape to the substrate, whereas in (B) the current induced-fit model of enzyme–substrate binding, the enzyme changes shape on substrate binding, with the active site forming a shape complementary to the substrate only after the substrate has been bound. The enzyme then continues to change shape as it induces changes in the substrate, converting it into the transition state.

5.4 The Michaelis–Menten Model Accounts for the Kinetic Properties of Many Enzymes

The study of the rates of chemical reactions is called *kinetics*, and the study of the rates of enzyme-catalyzed reactions is called *enzyme kinetics*. A kinetic description of enzyme activity will help us understand how enzymes function.

Kinetics is the study of reaction rates

What do we mean when we say the "rate" of a chemical reaction? Consider a simple reaction:

$$A \longrightarrow P$$

The rate V is the quantity of A that disappears in a specified unit of time. It is equal to the rate of the appearance of P, or the quantity of P that appears in a specified unit of time.

$$V = -d[A]/dt = d[P]/dt$$

If A is yellow and P is colorless, we can follow the decrease in the concentration of A by measuring the decrease in the intensity of yellow color with time. Consider only the change in the concentration of A for now. The rate of the reaction is directly related to the concentration of A by a proportionality constant, k, called the *rate constant*.

$$V = k[A]$$

Reactions that are directly proportional to the reactant concentration are first-order reactions. First-order rate constants have the units of s^{-1}.

Many important biochemical reactions include two reactants. For example,

$$2A \longrightarrow P$$

or

$$A + B \longrightarrow P$$

They are called *bimolecular reactions*, and the corresponding rate equations often take the form

$$V = k[A]^2$$

and

$$V = k[A][B] \tag{6}$$

The rate constants, called second-order rate constants, have the units $M^{-1} s^{-1}$.

Sometimes, second-order reactions can appear to be first-order reactions. For instance, in the reaction described by equation 6, if B is present in excess and A is present at low concentrations, the reaction rate will be first order with respect to A and will not appear to depend on the concentration of B. These reactions are called *pseudo-first-order reactions*, and we will see them a number of times in our study of biochemistry.

Interestingly, under some conditions, a reaction can be zero order. In these cases the rate is independent of reactant concentrations. Enzyme-catalyzed reactions can approximate zero-order reactions under some circumstances, as we will soon see.

The steady-state assumption aids a description of enzyme kinetics

The simplest way to investigate the reaction rate is to follow the increase in reaction product as a function of time. First, the extent of product formation is determined as a function of time for a series of substrate concentrations (**Figure 5.9A**). As expected, in each case, the amount of product formed increases with time, although eventually a time is reached when there is no net change in the concentration of S or P. The enzyme is still actively converting substrate into product and vice versa, but the reaction equilibrium has been attained.

However, enzyme kinetics is more readily comprehended if we consider only the forward reaction, which is most closely described by the initial rates of the reaction. Note that the rate of the reaction (the slope of the tangent line in Figure 5.9A) decreases over time. Additionally, only at the beginning of the reaction is the exact

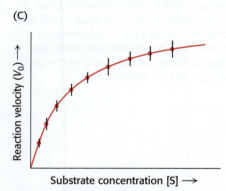

FIGURE 5.9 The relation between initial velocity and substrate concentration can be determined by a series of assays. (A) The amount of product formed at different substrate concentrations is plotted as a function of time. The initial velocity (V_0) for each substrate concentration is determined from the slope of the curve at the beginning of a reaction, when the reverse reaction is insignificant. (B) The values for initial velocity determined in part A are then plotted, with error bars, against substrate concentration. (C) The data points are connected to clearly reveal the relationship of initial velocity to substrate concentration.

concentration of the substrate known because it is being converted to product as the reaction progresses. For these two reasons, it is often most useful to focus on initial rate enzyme kinetics. We can define the rate of catalysis V_0, or the initial rate of catalysis, as the number of moles of product formed per second when the reaction is just beginning—that is, when $t \approx 0$ and $[P] \approx 0$, so the rate of the reverse reaction is negligible (Figure 5.9A). These experiments are repeated three to five times with each substrate concentration to ensure the accuracy of and assess the variability of the values attained.

Next, we plot V_0 versus the initial substrate concentration used in each experiment, [S], assuming a constant amount of enzyme and showing the data points with error bars (**Figure 5.9B**). Finally, the data points are connected, yielding the results shown in **Figure 5.9C**. The rate of catalysis rises linearly as substrate concentration increases and then begins to level off and approach a maximum at higher substrate concentrations. For convenience, we will show idealized data without error bars throughout the text, but it is important to keep in mind that in reality, all experiments are repeated multiple times.

The Michaelis–Menten model explains many observations of enzyme kinetics

In 1913, Leonor Michaelis and Maud Menten proposed a simple model to account for these kinetic characteristics. The critical feature in their treatment is that a specific ES complex is a necessary intermediate in catalysis. The model they proposed is

$$E + S \underset{k_{-1}}{\overset{k_1}{\rightleftharpoons}} ES \underset{k_{-2}}{\overset{k_2}{\rightleftharpoons}} E + P$$

An enzyme E (here representing the "free enzyme") combines with substrate S ("free substrate") to form an ES complex, with a rate constant k_1.

The ES complex has two possible fates. It can dissociate to E and S, with a rate constant k_{-1}, or it can proceed to form product P, with a rate constant k_2. The ES complex can also be reformed from E and P by the reverse reaction with a rate constant k_{-2}. However, as before, we can simplify these reactions by considering the rate of reaction at times close to zero (hence, V_0) when there is negligible product formation and thus no back reaction ($k_{-2}[E][P] \approx 0$).

$$E + S \underset{k_{-1}}{\overset{k_1}{\rightleftharpoons}} ES \overset{k_2}{\longrightarrow} E + P \tag{7}$$

Thus, for the graph in **Figure 5.10**, V_0 is determined for each substrate concentration by measuring the rate of product formation at early times before P accumulates (Figure 5.9A).

The Michaelis–Menten equation describes the relationship between initial velocity and substrate concentration

We want an expression that relates the rate of catalysis to the concentrations of substrate and enzyme and the rates of the individual steps. For any chemical reaction, the overall reaction rate is determined only by the slowest, or rate-determining step. Thus, our starting point is the assumption that catalysis, resulting from the ES complex, is slow relative to the rate of substrate binding, i.e., the formation of the ES complex. Thus, the overall reaction rate can be expressed solely as the rate of this one step, which is the product of the concentration of the ES complex and k_2.

$$V_0 = k_2[ES] \tag{8}$$

Now we need to express [ES] in terms of known quantities. The rates of formation and breakdown of ES are given by

$$\text{Rate of formation of ES} = k_1[E][S] \tag{9}$$

$$\text{Rate of breakdown of ES} = (k_{-1} + k_2)[ES] \tag{10}$$

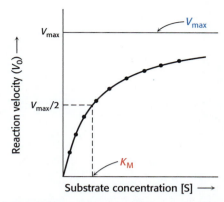

FIGURE 5.10 Michaelis–Menten kinetics are marked by a hyperbolic relationship between initial velocity and substrate concentration. A plot of the reaction velocity (V_0) as a function of the substrate concentration [S] for an enzyme that obeys Michaelis–Menten kinetics shows that the maximal velocity (V_{max}) is approached asymptotically, which is to say, V_{max} will be attained only at an infinite substrate concentration. The Michaelis constant (K_M) is the substrate concentration yielding a velocity of $V_{max}/2$.

We will use what biochemists term the *steady-state assumption* to simplify matters. In a steady-state system, the concentrations of intermediates—in this case, [ES]—stay the same even though the concentrations of starting materials and products are changing and the system is not in equilibrium. An analogy is a sink filled with water that has the tap open just enough to match the loss of water down the drain: the level of the water in the sink never changes even though water is constantly flowing from the faucet through the sink and out through the drain. This steady state is reached when the rates of formation and breakdown of the ES complex are equal. Setting the right-hand sides of equations 9 and 10 equal gives

$$k_1[E][S] = (k_{-1} + k_2)[ES] \tag{11}$$

By rearranging equation 11, we obtain

$$[E][S]/[ES] = (k_{-1} + k_2)/k_1 \tag{12}$$

Equation 12 can be simplified by defining a new constant, K_M, called the **Michaelis constant**:

$$K_M = \frac{k_{-1} + k_2}{k_1} \tag{13}$$

Note that K_M has the units of concentration and is independent of enzyme and substrate concentrations. As will be explained, K_M is an important characteristic of enzyme–substrate interactions.

Inserting equation 13 into equation 12 and solving for [ES] yields

$$[ES] = \frac{[E][S]}{K_M} \tag{14}$$

Now let us examine the numerator of equation 14. Because the substrate is usually present at a much higher concentration than that of the enzyme, the concentration of uncombined (free) substrate [S] is very nearly equal to the total substrate concentration. The concentration of uncombined (free) enzyme [E] is equal to the total enzyme concentration $[E]_T$ minus the concentration of the ES complex:

$$[E] = [E]_T - [ES]$$

Substituting this expression for [E] in equation 14 gives

$$[ES] = \frac{([E]_T - [ES])[S]}{K_M}$$

Solving the above equation for [ES] gives

$$[ES] = \frac{[E]_T[S]/K_M}{1 + [S]/K_M}$$

or

$$[ES] = [E]_T \frac{[S]}{[S] + K_M}$$

By substituting this expression for [ES] into equation 8, we obtain

$$V_0 = k_2[E]_T \frac{[S]}{[S] + K_M} \tag{15}$$

The **maximal rate**, V_{max}, is attained when the catalytic sites on the enzyme are saturated with substrate—that is, when $[ES] = [E]_T$. Thus,

$$V_{max} = k_2[E]_T \tag{16}$$

Substituting equation 16 into equation 15 yields the **Michaelis–Menten equation**:

$$V_0 = V_{max} \frac{[S]}{[S] + K_M} \tag{17}$$

Smithsonian Institution Archives, Accession 90-105, Science Service Records, Image No. SIA2008-5999

MAUD MENTEN Paving new career paths for women at the dawn of the 20th century, Maud Menten was among the first Canadian women to obtain an MD degree, in 1911; she then earned a PhD in 1916. Although she is most famous for her groundbreaking work on enzyme kinetics as a graduate student, she had a long and productive scientific career at the University of Pittsburgh and the British Columbia Medical Research Institute, teaching and publishing over 70 papers on a variety of subjects. She invented a still-used histochemical assay for alkaline phosphatase, characterized bacterial toxins and early vitamin C deficiency, and pioneered the use of electrophoresis for the study of proteins.

This equation accounts for the kinetic data given in Figure 5.10. At very low substrate concentration, when [S] is much less than K_M, $V_0 = (V_{max}/K_M)[S]$; that is, the reaction is first order with the rate directly proportional to the substrate concentration. At high substrate concentration, when [S] is much greater than K_M, $V_0 = V_{max}$; that is, the rate is maximal. The reaction is zero order, independent of substrate concentration.

Notice that because V_0 is a function of [S], then if the K_M value is known, the equation gives the velocity as a fraction of V_{max} at any concentration of substrate. Thus, the velocity can be calculated at any [S], using the Michaelis–Menten equation, and expressed either in real units (if V_{max} is known) or as a fraction of V_{max}. Likewise, if the fraction of V_{max} is known, then the relationship between [S] and K_M can be calculated.

The significance of K_M is clear when we set [S] = K_M in equation 17. When [S] = K_M, then $V_0 = V_{max}/2$. Thus, K_M is equal to the substrate concentration at which the reaction rate is half its maximal value. As we will see, K_M is an important characteristic of an enzyme-catalyzed reaction and is significant for its biological function.

EXAMPLE Applying the Michaelis–Menten Equation

PROBLEM: Assume that the initial velocity of an enzyme is 80% of V_{max}. What is the ratio of [S] to K_M under these conditions?

GETTING STARTED: The Michaelis–Menten equation relates the three terms discussed in the problem, so let's begin with it:

$$V_0 = V_{max} \frac{[S]}{[S] + K_M}$$

What do we already know? The question states the V_0 = 80% V_{max}. This is the only value we are provided, so we should be able to solve the problem with that value and the equation.

SOLVE: First, for convenience, let's convert 80% to 0.8 and substitute the 0.8 V_{max} for V_0 in the equation:

$$0.8 V_{max} = V_{max} \frac{[S]}{[S] + K_M}$$

Now we see that we can divide both sides of the equation by V_{max}, which yields

$$0.8 = \frac{[S]}{[S] + K_M}$$

Now we can solve the equation for [S]/K_M.

$$0.8[S] + 0.8K_M = [S]$$
$$0.8K_M = [S] - 0.8[S]$$
$$0.8K_M = 0.2[S]$$
$$4 = \frac{[S]}{K_M}$$

Thus, when [S] is fourfold greater than K_M, $V_0 = 0.8 V_{max} = 80\% V_{max}$.

REFLECT: You can confirm this answer by solving the Michaelis–Menten equation using [S] = 4K_M and solving for V_0 as a fraction of V_{max}.

Variations in K_M can have physiological consequences

The physiological consequence of K_M is illustrated by the sensitivity of some persons to ethanol. Such persons exhibit facial flushing and rapid heart rate (tachycardia) after ingesting even small amounts of alcohol. In the liver, alcohol dehydrogenase converts ethanol (CH_3CH_2OH) into acetaldehyde (CH_3CHO).

$$CH_3CH_2OH + NAD^+ \xrightarrow[\text{dehydrogenase}]{\text{Alcohol}} CH_3CHO + NADH + H^+$$

Normally, the acetaldehyde, which is the cause of the symptoms when present at high concentrations, is processed to acetate by aldehyde dehydrogenase.

$$CH_3CHO + NAD^+ + H_2O \xrightarrow[\text{dehydrogenase}]{\text{Aldehyde}} CH_3COO^- + NADH + 2H^+$$

Most people have two forms of the aldehyde dehydrogenase, a low K_M mitochondrial form and a high K_M cytoplasmic form. In susceptible persons, the mitochondrial enzyme is less active owing to the substitution of a single amino acid, and acetaldehyde is processed only by the cytoplasmic enzyme. Because this enzyme has a high K_M, it achieves a high rate of catalysis only at very high concentrations of acetaldehyde. Consequently, less acetaldehyde is converted into acetate; excess acetaldehyde escapes into the blood and accounts for the physiological effects.

K_M and V_{max} values can be determined by several means

K_M is equal to the substrate concentration that yields $V_{max}/2$; however, like perfection, V_{max} is approached but never attained. How, then, can we experimentally determine K_M and V_{max}, and how do these parameters enhance our understanding of enzyme-catalyzed reactions? The Michaelis constant, K_M, and the maximal rate, V_{max}, can be readily derived from rates of catalysis measured at a variety of substrate concentrations if an enzyme operates according to the simple scheme given in equation 17. The derivation of K_M and V_{max} is most commonly achieved with the use of curve-fitting programs on a computer. However, an older method, although rarely used because the data points at high and low concentrations are weighted differently and thus sensitive to errors, is a source of further insight into the meaning of K_M and V_{max}.

Before the availability of computers, the determination of K_M and V_{max} values required algebraic manipulation of the Michaelis–Menten equation. The Michaelis–Menten equation is transformed into one that gives a straight-line plot that yields values for V_{max} and K_M. Taking the reciprocal of both sides of equation 17 gives

$$\frac{1}{V_0} = \frac{K_M}{V_{max}} \times \frac{1}{[S]} + \frac{1}{V_{max}} \tag{18}$$

A plot of $1/V_0$ versus $1/[S]$, called a **Lineweaver–Burk plot** or sometimes just called a *double-reciprocal plot*, yields a straight line with a y-intercept of $1/V_{max}$ and a slope of K_M/V_{max} (**Figure 5.11**). The intercept on the x-axis is $-1/K_M$.

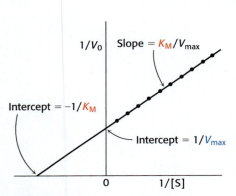

FIGURE 5.11 A Lineweaver–Burk plot shows $1/V_0$ as a function of $1/[S]$. The slope is K_M/V_{max}, the intercept on the vertical axis is $1/V_{max}$, and the intercept on the horizontal axis is $-1/K_M$.

K_M and k_{cat} values are important enzyme characteristics

The K_M values of enzymes range widely (**Table 5.5**). For most enzymes, K_M lies between 10^{-1} and 10^{-7} M. The K_M value for an enzyme depends on the particular substrate and on environmental conditions such as pH, temperature, and ionic strength. The Michaelis constant, being equal to the concentration of substrate at which half the active sites are filled, thus provides a measure of the substrate concentration required for significant catalysis to take place.

For many enzymes, experimental evidence suggests that the K_M value provides an approximation of the substrate concentration in vivo, which in turn suggests that most enzymes evolved to have a K_M approximately equal to the substrate concentration commonly available. Why might it be beneficial to have a K_M value approximately equal to the commonly available substrate concentration? If the normal concentration of substrate is near K_M, the enzyme will display significant activity, and yet the activity will be sensitive to changes in environmental conditions—that is, changes in substrate concentration. This property is known as **elasticity**. At values below K_M, enzymes are very sensitive to changes in substrate concentration—in other words, they are elastic—but display little activity. At substrate values well above K_M, enzymes have greater catalytic activity but are insensitive—or inelastic—to changes in substrate concentration. Thus, with the normal substrate concentration being approximately K_M, the enzymes have significant activity ($1/2\ V_{max}$) but are still elastic to changes in substrate concentration.

Under certain circumstances, K_M reflects the strength of the enzyme–substrate interaction. In equation 13, K_M is defined as $(k_{-1} + k_2)/k_1$. Consider a case in which k_{-1} is much greater than k_2. Under such circumstances, the ES complex dissociates to E and S much more rapidly than product is formed. Under these conditions, $(k_{-1} \gg k_2)$

$$K_M \approx \frac{k_{-1}}{k_1} \tag{19}$$

Equation 19 describes the dissociation constant, K_d (Chapter 3), of the ES complex.

$$K_{ES} = \frac{[E][S]}{[ES]} = \frac{k_{-1}}{k_1}$$

In other words, K_M is equal to the dissociation constant (K_d) of the ES complex if k_2 is much smaller than k_{-1}. When this condition is met, K_M is a measure of the strength of the ES complex: a high K_M indicates weak binding; a low K_M indicates strong binding. It must be stressed that K_M indicates the affinity of the ES complex only when k_{-1} is much greater than k_2.

The maximal rate, V_{max}, reveals the **turnover number** of an enzyme, which is the number of substrate molecules converted into product by an enzyme molecule in a unit time when the enzyme is fully saturated with substrate. For a single active site, it is equal to the rate constant k_2, which is also called k_{cat}. The maximal rate, V_{max}, reveals the turnover number of an enzyme if the concentration of active sites $[E]_T$ is known, because

$$V_{max} = k_{cat}[E]_T$$

and thus

$$k_{cat} = V_{max}/[E]_T \tag{20}$$

For example, a 10^{-6} M solution of carbonic anhydrase catalyzes the formation of 0.6 M H_2CO_3 per second when the enzyme is fully saturated with substrate. Hence, k_{cat} is 6×10^5 s^{-1}. This turnover number is one of the largest known. Each catalyzed reaction takes place in a time equal to, on average, $1/k_{cat}$, which is 1.7 μs for carbonic anhydrase. The turnover numbers of most enzymes with their physiological substrates range from 1 to 10^6 per second (**Table 5.6**).

K_M and V_{max} also permit the determination of f_{ES}, the fraction of active sites filled. This relation of f_{ES} to K_M and V_{max} is given by the following equation:

$$f_{ES} = \frac{V_0}{V_{max}} = \frac{[S]}{[S] + K_M} \tag{21}$$

TABLE 5.5 K_M values of some enzymes

Enzyme	Substrate	K_M (μM)
Chymotrypsin	Acetyl-L-tryptophanamide	5000
Lysozyme	Hexa-N-acetylglucosamine	6
β-Galactosidase	Lactose	4000
Threonine deaminase	Threonine	5000
Carbonic anhydrase	CO_2	8000
Penicillinase	Benzylpenicillin	50
Pyruvate carboxylase	Pyruvate	400
	HCO_3^-	1000
	ATP	60
Arginine-tRNA synthetase	Arginine	3
	tRNA	0.4
	ATP	300

TABLE 5.6 Turnover numbers of some enzymes

Enzyme	Turnover number (per second)
Carbonic anhydrase	600,000
3-Ketosteroid isomerase	280,000
Acetylcholinesterase	25,000
Penicillinase	2000
Lactate dehydrogenase	1000
Chymotrypsin	100
DNA polymerase I	15
Tryptophan synthetase	2
Lysozyme	0.5

❖ SELF-CHECK QUESTION

Explain why K_M is an intrinsic property of an enzyme while V_{max} is not?

k_{cat}/K_M is a measure of catalytic efficiency

Recall that when the substrate concentration is much greater than K_M, the rate of catalysis is equal to V_{max}, which is a function of k_{cat}, the turnover number. However, most enzymes are not normally saturated with substrate. Under physiological conditions, the $[S]/K_M$ ratio is typically between 0.01 and 1.0. When $[S] = K_M$, the enzymatic rate is much less than k_{cat} because most of the active sites are unoccupied.

Is there a number that characterizes the kinetics of an enzyme under these more typical cellular conditions? Indeed there is, as can be shown by combining equations 8 and 14 to give

$$V_0 = \frac{k_{cat}}{K_M}[E][S]$$

When $[S] \ll K_M$, the concentration of free enzyme $[E]$, is nearly equal to the total concentration of enzyme $[E]_T$; so

$$V_0 = \frac{k_{cat}}{K_M}[S][E]_T$$

Thus, when $[S] \ll K_M$, the enzymatic velocity depends on the values of k_{cat}/K_M, $[S]$, and $[E]_T$. Under these conditions, k_{cat}/K_M is the rate constant for the interaction of S and E.

The rate constant **k_{cat}/K_M**, called the **specificity constant**, is a measure of catalytic efficiency because it takes into account both the rate of catalysis with a particular substrate (k_{cat}) and the nature of the enzyme–substrate interaction (K_M). For instance, by using k_{cat}/K_M values, we can compare an enzyme's preference for different substrates. **Table 5.7** shows the k_{cat}, K_M, and k_{cat}/K_M values for several different substrates of chymotrypsin. Chymotrypsin clearly has a preference for cleaving next to bulky, hydrophobic side chains.

How efficient can an enzyme be? We can approach this question by determining whether there are any physical limits on the value of k_{cat}/K_M. Note that the k_{cat}/K_M ratio depends on k_1, k_{-1}, and k_{cat}, as can be shown by substituting for K_M.

$$k_{cat}/K_M = \frac{k_{cat}k_1}{k_{-1} + k_{cat}} = \left(\frac{k_{cat}}{k_{-1} + k_{cat}}\right)k_1 < k_1$$

Note that the value of k_{cat}/K_M is always less than k_1. Suppose that the rate of formation of product (k_{cat}) is much faster than the rate of dissociation of the ES complex (k_{-1}).

TABLE 5.7 Substrate preferences of chymotrypsin

Amino acid in ester	Amino acid side chain	k_{cat}/K_M ($s^{-1}M^{-1}$)
Glycine	—H	1.3×10^{-1}
Valine	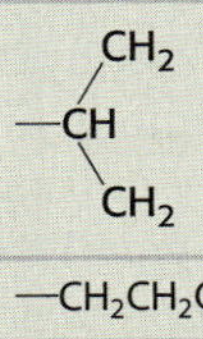 $\begin{array}{c} CH_2 \\ -CH \\ CH_2 \end{array}$	2.0
Norvaline	$-CH_2CH_2CH_3$	3.6×10^2
Norleucine	$-CH_2CH_2CH_2CH_3$	3.0×10^3
Phenylalanine	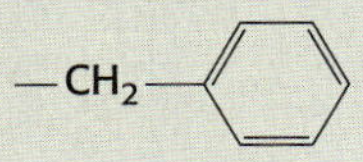 $-CH_2-\bigcirc$	1.0×10^5

Source: Information from A. Fersht, *Structure and Mechanism in Protein Science: A Guide to Enzyme Catalysis and Protein Folding* (W. H. Freeman and Company, 1999), Table 7.3.

The value of k_{cat}/K_M then approaches k_1. Thus, the ultimate limit on the value of k_{cat}/K_M is set by k_1, the rate of formation of the ES complex, which is limited by the diffusion-controlled encounter of an enzyme and its substrate. Because diffusion limits the value of k_1, it cannot be higher than between 10^8 and 10^9 s^{-1}M^{-1}.

Enzymes that have k_{cat}/K_M ratios in the range of these upper limits, such as superoxide dismutase, acetylcholinesterase, and triose phosphate isomerase, have attained what is called *kinetic perfection*. Their catalytic velocity is restricted only by the rate at which they encounter substrate in the solution (Table 5.8). Any further gain in catalytic rate can come only by decreasing the time for diffusion of the substrate into the enzyme's immediate environment.

Remember that the active site is only a small part of the total enzyme structure. Yet, for catalytically perfect enzymes, every encounter between enzyme and substrate is productive. In these cases, there may be attractive electrostatic forces on the enzyme that entice the substrate to the active site. These forces are sometimes referred to poetically as *Circe effects*, named for the figure from Greek mythology known for her powers of seduction and transformation.

The diffusion of a substrate throughout a solution can also be partly overcome by confining substrates and products in the limited volume of a multienzyme complex, such that the product of one enzyme is very rapidly found by the next enzyme. In effect, products are channeled from one enzyme to the next, much as in an assembly line.

TABLE 5.8 Enzymes with k_{cat}/K_M near the diffusion-controlled limit

Enzyme	k_{cat}/K_M (s^{-1} M^{-1})
Acetylcholinesterase	1.6×10^8
Carbonic anhydrase	8.3×10^7
Catalase	4×10^7
Crotonase	2.8×10^8
Fumarase	1.6×10^8
Triose phosphate isomerase	2.4×10^8
β-Lactamase	1×10^8
Superoxide dismutase	7×10^9

Source: Information from A. Fersht, *Structure and Mechanism in Protein Science: A Guide to Enzyme Catalysis and Protein Folding* (W. H. Freeman and Company, 1999), Table 4.5.

❖ SELF–CHECK QUESTION

If a mutation occurs in the gene encoding an enzyme which decreases k_{cat} for a specific substrate by a factor of 2 and decreases K_M for that substrate by a factor of 3, did the mutation make the enzyme a more or less efficient catalyst with respect to that substrate?

Most biochemical reactions include multiple substrates

Most reactions in biological systems are bisubstrate reactions, which start with two substrates and yield two products, represented by:

$$A + B \rightleftharpoons P + Q$$

Many bisubstrate reactions transfer a functional group, such as a phosphoryl or an ammonium group, from one substrate to the other. Those that are oxidation–reduction reactions transfer electrons between substrates. Multiple substrate reactions can be divided into two classes: sequential reactions and double-displacement reactions.

Sequential reactions. In **sequential reactions**, all substrates must bind to the enzyme before any product is released. Specifically, in a bisubstrate reaction, a ternary complex—composed of the enzyme and both substrates—forms.

- In *ordered sequential mechanisms, the substrates bind the enzyme in a defined sequence.* For example, many enzymes that have NAD$^+$ or NADH as a substrate exhibit the ordered sequential mechanism. Consider lactate dehydrogenase, an important enzyme in glucose metabolism (Section 16.1). This enzyme reduces pyruvate to lactate while oxidizing NADH to NAD$^+$.

Pyruvate   + NADH + H$^+$ ⇌ **Lactate** + NAD$^+$

In the ordered sequential mechanism, the coenzyme always binds first, and the lactate is always released first. This sequence can be represented by using a notation developed by W. Wallace Cleland:

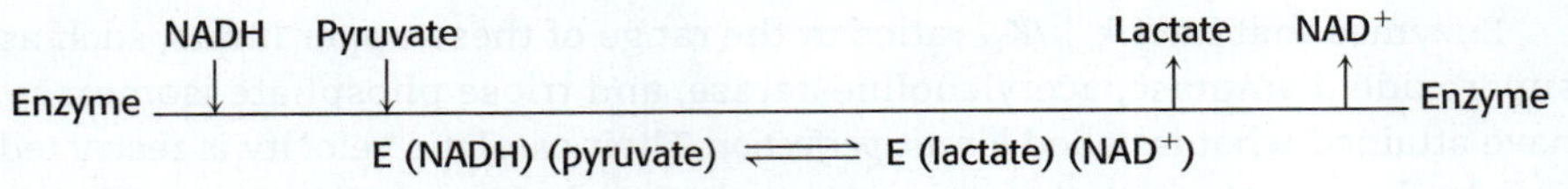

The enzyme exists as a ternary complex consisting of, first, the enzyme and substrates and, after catalysis, the enzyme and products.

- *In the random sequential mechanism, the order of the addition of substrates and the release of products is arbitrary.* An example of a random sequential reaction is the formation of phosphocreatine and ADP from creatine and ATP which is catalyzed by creatine kinase (Section 15.2).

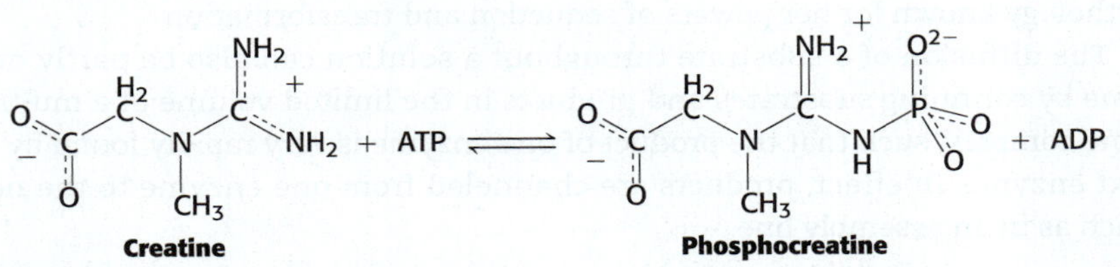

Creatine   **Phosphocreatine**

Either creatine or ATP may bind first, and either phosphocreatine or ADP may be released first. Phosphocreatine is an important energy source in muscle. Random sequential reactions also can be depicted in the Cleland notation.

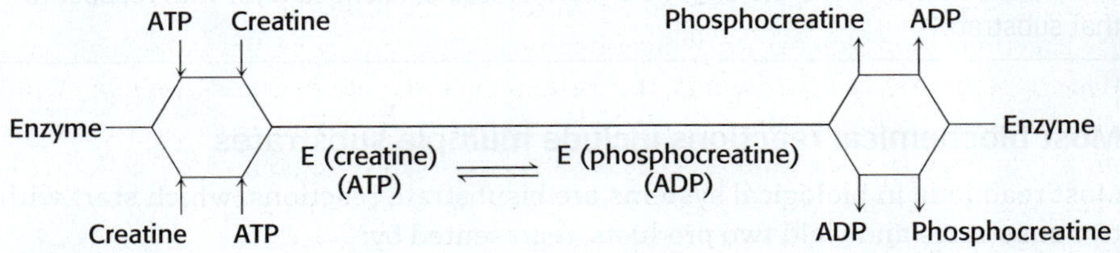

Although the order of certain events is random, the reaction still passes through the ternary complexes, including the substrates first and then the products.

Double-displacement reactions. In **double-displacement reactions** (also called **Ping-Pong reactions**), one or more products are released before all substrates bind the enzyme. The defining feature of double-displacement reactions is the existence of an enzyme intermediate in which the enzyme is temporarily modified.

Reactions that shuttle amino groups between amino acids and α-ketoacids are classic examples of double-displacement mechanisms. The enzyme aspartate aminotransferase catalyzes the transfer of an amino group from aspartate to α-ketoglutarate.

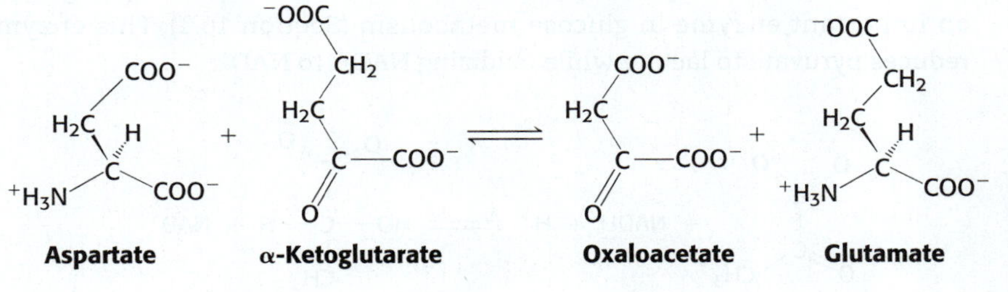

Aspartate   **α-Ketoglutarate**   **Oxaloacetate**   **Glutamate**

The sequence of events can be portrayed as the following Cleland notation:

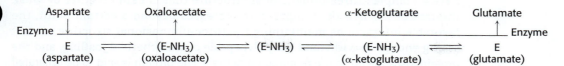

After aspartate binds to the enzyme, the enzyme accepts aspartate's amino group to form the substituted enzyme intermediate. The first product, oxaloacetate, subsequently departs. The second substrate, α-ketoglutarate, binds to the enzyme, accepts the amino group from the modified enzyme, and is then released as the final product, glutamate. In the Cleland notation, the substrates appear to bounce on and off the enzyme much as a Ping-Pong ball bounces on a table. Note that for double-displacement reactions, substrate binding order is always fixed (ordered) and never random.

Allosteric enzymes often do not obey Michaelis–Menten kinetics

The Michaelis–Menten model has greatly assisted the development of enzymology. Its virtues are simplicity and broad applicability. However, the Michaelis–Menten model cannot account for the kinetic properties of many enzymes. An important group of enzymes that often do not obey Michaelis–Menten kinetics are **allosteric enzymes**, which consist of multiple subunits and multiple active sites. The binding of substrate to one active site can alter the properties of other active sites in the same enzyme molecule (an effect known as *allostery*, Chapter 3). Allosteric enzymes often display sigmoidal plots of the reaction velocity V_0 versus substrate concentration [S] (**Figure 5.12**), rather than the hyperbolic plots predicted by the Michaelis–Menten equation (see Figure 5.10).

The interaction between an allosteric enzyme's multiple subunits can exhibit positive cooperativity. In other words, the binding of substrate to one active site facilitates the binding of substrate to the other active sites, akin to the behavior of oxygen binding to the multiple subunits of the allosteric protein hemoglobin (see Section 3.3). Such positive cooperativity results in a sigmoidal plot of V_0 versus [S]. In addition, the activity of an allosteric enzyme may be altered by regulatory molecules that reversibly bind to specific sites other than the catalytic sites. Furthermore, the enzymes themselves can often interconvert between more- or less-active quaternary states, typically denoted as "R" and "T," respectively, by the conventions established for hemoglobin.

The complexities of allosteric enzymes often require more-elaborate models than the Michaelis–Menten model to account for multiple active sites and quaternary state transitions. The catalytic properties of allosteric enzymes make them key regulators of metabolic pathways in ways we will explore more fully in Chapter 7.

Temperature affects enzymatic activity

As the temperature rises, the rate of most reactions, including enzyme-catalyzed reactions, increases. For most enzymes, there is a temperature at which the increase in catalytic activity ceases and there is a precipitous loss of activity.

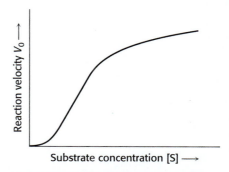

FIGURE 5.12 Allosteric enzymes often display a sigmoidal dependence of reaction velocity on substrate concentration. Shown are the kinetics for an allosteric enzyme exhibiting positive cooperativity.

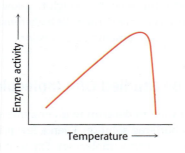

What is the basis of this loss of activity? Recall from Chapter 2 that proteins have a complex three-dimensional structure that is held together by weak interactions. When the temperature increases beyond a certain point, the favorable interactions maintaining the three-dimensional structure are not strong enough to withstand the polypeptide chain's thermal jostling, and the protein loses the structure required for activity. The protein is said to be *denatured* (Section 2.7).

In endotherms, organisms such as ourselves that maintain a constant body temperature, the effect of outside temperature on enzyme activity is minimized. However, in ectotherms, organisms that assume the temperature of their environment, temperature is an important regulator of biochemical and, indeed, biological activity. Lizards, for instance, are most active in warmer temperatures—a behavioral manifestation of biochemical activity.

Although endotherms are not as sensitive to ambient temperature as ectotherms, slight tissue temperature alterations are sometimes important and can even have dramatic effects. For example, the characteristic coloration of Siamese cats can be explained by variation in enzyme sensitivity due to temperature. Their fur color is due to the presence of the pigment melanin, the same pigment that is responsible for human skin color. The first steps in the synthesis of melanin are catalyzed by the enzyme tyrosinase:

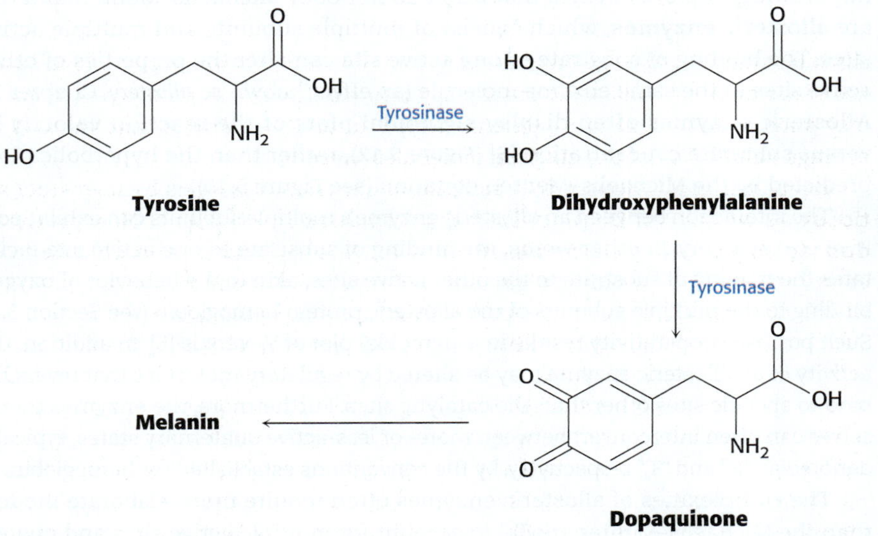

Tyrosine   →(Tyrosinase)→   **Dihydroxyphenylalanine**   →(Tyrosinase)→   **Dopaquinone**   →   **Melanin**

Most Siamese cats are born with very little coloration; sometimes they are even white. As they mature, their extremities—tips of the ears, snout, paws, and end of the tail—darken to black. A hint to understanding this phenomenon comes from the knowledge that a cat's skin temperature is coolest at the extremities. Biochemists have established that there is a mutation in the Siamese tyrosinase that results in a loss of activity above 37–39°C. The temperature of a Siamese kitten is too warm for the enzyme to be active, but as the kitten grows, the temperature at its extremities cools enough to allow the enzyme to become active and pigment to form. The core of the cat's body remains above the threshold temperature, however, and thus remains pale in color.

5.5 Enzymes Can Be Studied One Molecule at a Time

Most experiments performed to determine an enzyme's characteristics use an enzyme preparation in a buffered solution. Even a few microliters of such a solution will contain millions of enzyme molecules. Experiments on this scale, called

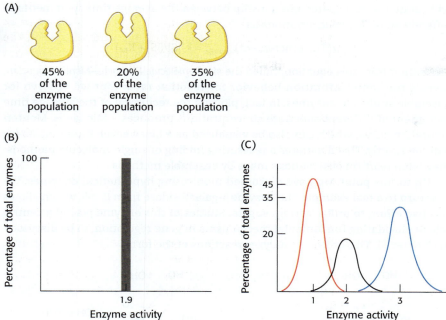

(A)

45% of the enzyme population

20% of the enzyme population

35% of the enzyme population

(B)

Percentage of total enzymes

100

1.9

Enzyme activity

(C)

Percentage of total enzymes

45
35

20

1   2   3

Enzyme activity

FIGURE 5.13 Single-molecule studies can reveal molecular heterogeneity. (A) Complex biomolecules, such as enzymes, display molecular heterogeneity. (B) When measuring an enzyme property using ensemble methods, an average value of all of the enzymes present can be obtained. (C) Single-enzyme studies reveal molecular heterogeneity, with the various forms showing different properties.

ensemble studies, make the basic assumption that all of the enzyme molecules are the same or very similar. For example, when we determine an enzymatic property such as the value of K_M in ensemble studies, we assume that value is an average value of all of the enzyme molecules present.

Keep in mind, however, that individual enzyme molecules behave stochastically, as we saw for DNA double helix formation in Chapter 1. New methods, dubbed *single-molecule experiments*, now allow one individual molecule to be examined at a time, yielding a great deal of new information but with a potential pitfall: how can we be certain that the molecule is representative and not an outlier? We can overcome this challenge by studying enough individuals to satisfy statistical analysis for validity.

Let us consider a biochemical situation. **Figure 5.13A** shows a hypothetical enzyme that displays molecular heterogeneity, with three active forms that catalyze the same reaction but at different rates. These forms have slightly different stabilities, but thermal motion is sufficient to interconvert them. Each form is present as a fraction of the total enzyme population, as indicated. If we were to perform an ensemble method experiment to determine enzyme activity under a particular set of conditions, we would get a single value, which would represent the average of the heterogeneous assembly (**Figure 5.13B**). However, were we to perform a sufficient number of single-molecule experiments, we would discover that the enzyme has three different molecular forms with very different activities (**Figure 5.13C**). Moreover, these different forms would most likely correspond to important biochemical differences.

Single-molecule kinetics confirm results obtained from ensemble studies

Now suppose that we watch an individual enzyme molecule in action, monitoring every time it produces a molecule of product. **Figure 5.14** illustrates what we would expect to see, with each line representing a single reaction event. Because individual enzyme molecules behave stochastically, the time between consecutive reaction events appears to vary randomly. However, if we watch an individual enzyme molecule turn over many thousands of times, we find a pattern in the distribution of the time intervals separating consecutive events that allows

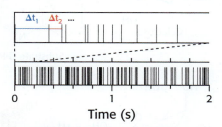

Δt_1 Δt_2 ...

0   1   2

Time (s)

FIGURE 5.14 An anticipated time course for a single-enzyme molecule reveals stochastic behavior. Each line represents an enzyme molecule turning a substrate molecule into product. These events occur stochastically, with variable time increments between consecutive events.

us to state the anticipated relationship between the average time increments at different substrate concentrations as

$$1/(\text{Time increment})_{average} = k_2[S]/([S] + K_M)$$

The output from this equation, called the *single-molecule Michaelis–Menten equation*, reveals the same saturation behavior for reaction rates that we observe for ensemble studies of enzymes. In fact, plotting the reciprocal of the average time increment at different substrate concentrations produces a Michaelis–Menten saturation curve, which can also be visualized as a Lineweaver–Burk plot, as we will see shortly. This illustrates a reassuring finding of single-molecule methods: they often reaffirm observations made by ensemble methods.

Up to this point, we have discussed monitoring hypothetical enzymes. We now turn to a real example, the enzyme β-galactosidase from E. coli, which allows this bacterium to utilize certain sugars. Studies of this enzyme played a central role in elucidating fundamental mechanisms in gene regulation, to be discussed in Chapter 31. This enzyme catalyzes reactions of the form

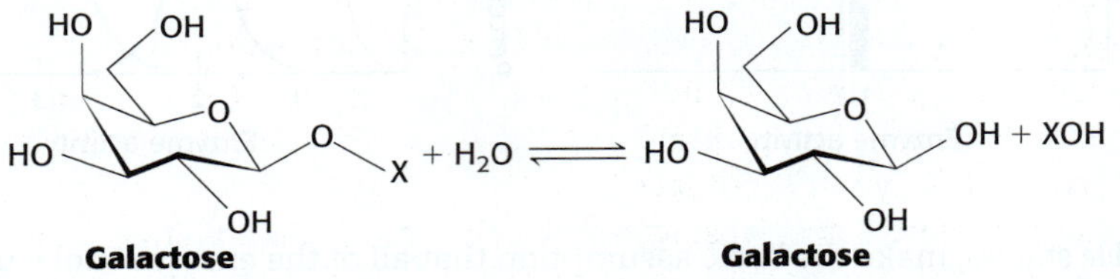

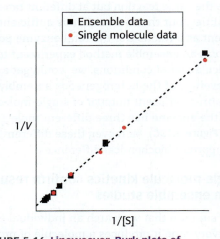

While the enzyme is quite specific for the carbohydrate galactose (to be discussed in Chapter 11), substantial variations in the other half of the substrate can be accommodated, permitting the binding of probe substrates such as Substrate G-R (**Figure 5.15**). Cleavage of this substrate by β-galactosidase releases a fluorescent product.

Using immobilized enzyme molecules and a fluorescence microscope, product release (and, hence, enzyme activity) can be monitored directly at the single-molecule level. With this system, experiments were performed at a range of substrate concentrations and the results compared with those obtained from ensemble studies using the same enzyme and substrate (**Figure 5.16**).

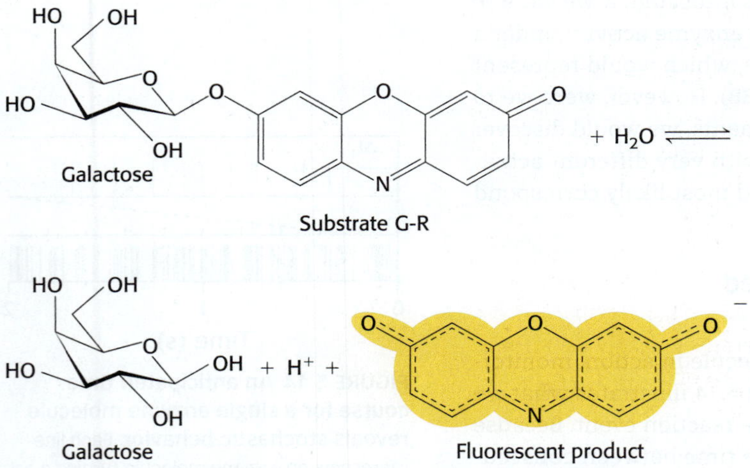

FIGURE 5.15 Upon reaction with β-galactosidase, Substrate G-R produces a highly fluorescent product. The formation of individual molecules of this product can be detected, allowing for the observation of single-enzyme turnovers to be tracked.

FIGURE 5.16 Lineweaver–Burk plots of single-molecule data compare favorably to those from ensemble studies. The results from single-molecule kinetic studies of β-galactosidase (red) are nearly indistinguishable from those from ensemble studies of the same enzyme at the same substrate concentration (black).

[Adapted from B. P. English et al., *Nature Chemical Biology* 2:87–94, 2006.]

Single-molecule studies continue to reveal new information about enzyme molecular dynamics

While it is quite reassuring to see in a real example that the results of technically challenging single-molecule studies agree so well with those of ensemble studies, the single-molecule results don't tell us anything new about enzyme behavior. However, depicting the distribution of time increments on a log plot does reveal an important new insight (**Figure 5.17**). At a low substrate concentration of 10 μM, this plot is relatively linear. However, at a higher concentration of 100 μM, this plot is decidedly nonlinear, changing from a relatively steep slope at short time increments to a more gradual slope at longer time increments.

We can explain this observation in terms of the dynamic nature of individual enzyme molecules. Each molecule can exist and convert between multiple conformational states, as we envisioned hypothetically in Figure 5.13, each conformation having distinct catalytic properties. The structural differences of these conformational states could be as simple as the orientations of particular side chains or as complex as large, interdomain movements. Further studies are required to gain such structural information, but regardless of the structural details, single-molecule experiments provide direct evidence for a critical aspect of enzymes: their dynamic nature.

Using the many powerful single-molecule techniques that now exist, we can observe events at a molecular level to reveal rare or transient structures and fleeting events in a reaction sequence, as well as to measure mechanical forces affecting or generated by an enzyme. Single-molecule studies open a new vista on the function of enzymes in particular and all large biomolecules in general. We will examine additional single-molecule studies in future chapters.

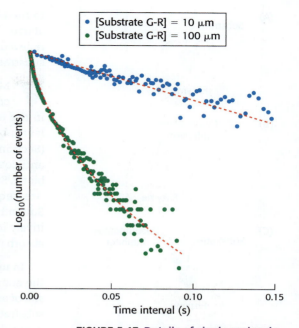

FIGURE 5.17 Details of single-molecule time increment distributions for β-galactosidase indicate heterogeneity at the molecular level. At a relatively high substrate concentration (100 μM), a plot of the logarithm of the number of events in each time interval is not linear, indicated by the distribution not following a simple exponential fall off. This observation reveals that the enzyme cannot be modeled as a single species at the single-molecule level. [Adapted from B. P. English et al., *Nature Chemical Biology* 2:87–94, 2006.]

5.6 Enzymes Can Be Inhibited by Specific Molecules

The activity of many enzymes can be inhibited by the binding of specific small molecules and ions. Enzyme inhibition serves as a major control mechanism in biological systems, especially by regulating the activity of allosteric enzymes. In addition, many drugs and toxic agents act by inhibiting enzymes (Chapter 32). These inhibitors are often designed by scientists or result from a chance discovery of an inhibitory molecule. Examining inhibition can give insight into the mechanism of enzyme action; for example, specific inhibitors are often used to identify residues critical for catalysis. Enzyme inhibition can be either irreversible or reversible.

- *An irreversible inhibitor covalently bonds to its target enzyme and thus does not dissociate at any appreciable rate.* The mode of action of some important drugs is irreversible inhibition. For example, as we will discuss shortly, penicillin acts by covalently modifying the enzyme transpeptidase, thereby preventing the synthesis of bacterial cell walls and thus killing the bacteria. Aspirin acts by covalently modifying the enzyme cyclooxygenase, reducing the synthesis of signaling molecules in inflammation.

- *Reversible inhibition is characterized by the dissociation of the enzyme–inhibitor complex, which is formed by noncovalent interactions.* Depending on their target and their effect, reversible inhibitors are classified as competitive, uncompetitive, or noncompetitive.

In **competitive inhibition**, the inhibitor competes with the substrate for binding to the enzyme and thus reduces the proportion of enzyme molecules bound to substrate. The enzyme can bind the substrate (**Figure 5.18A**) or an inhibitor (**Figure 5.18B**), but not both at the same time. Competitive inhibitors can bind anywhere on the enzyme, but they most often resemble the substrate and bind

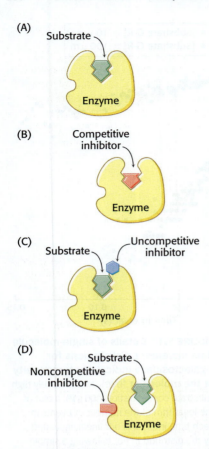

FIGURE 5.18 Reversible inhibitors can be distinguished by whether they bind to the enzyme with or without the substrate present. (A) An enzyme–substrate complex; (B) competitive inhibitors prevent the substrate from binding, usually by themselves binding at the active site; (C) an uncompetitive inhibitor binds only to the enzyme–substrate complex; (D) a pure noncompetitive inhibitor does not affect the binding of substrate.

to the active site. The substrate is thereby prevented from binding to the same active site. At any given inhibitor concentration, competitive inhibition can be relieved by increasing the substrate concentration. Under these conditions, the substrate successfully competes with the inhibitor for the enzyme.

Many competitive inhibitors are useful drugs. Drugs such as ibuprofen are competitive inhibitors of enzymes that participate in signaling pathways in the inflammatory response. Statins are drugs that reduce high cholesterol levels by competitively inhibiting a key enzyme in cholesterol biosynthesis (Section 26.3). One of the earliest examples was the use of sulfanilamide as an antibiotic. Sulfanilamide is an example of a sulfa drug, a sulfur-containing antibiotic. Structurally, sulfanilamide mimics *p*-aminobenzoic acid (PABA), a metabolite required by bacteria for the synthesis of the coenzyme folic acid. Sulfanilamide binds to the enzyme that normally metabolizes PABA and competitively inhibits it, preventing folic acid synthesis. Human beings, unlike bacteria, absorb folic acid from the diet and are thus unaffected by the sulfa drug.

In **uncompetitive inhibition**, the inhibitor binds not to the enzyme itself but to the enzyme–substrate complex; it is therefore substrate-dependent. The binding site of an uncompetitive inhibitor is created only on interaction of the enzyme and substrate (**Figure 5.18C**). Uncompetitive inhibitors inhibit catalysis, rather than prevent substrate binding, and cannot be overcome by the addition of more substrate. The herbicide glyphosate, also known as Roundup, is an uncompetitive inhibitor of an enzyme in the biosynthetic pathway for aromatic amino acids.

In **pure noncompetitive inhibition**, the inhibitor and substrate can bind simultaneously to an enzyme molecule at two different binding sites (**Figure 5.18D**). Unlike competitive or uncompetitive inhibition, a noncompetitive inhibitor can bind free enzyme or the enzyme–substrate complex, but in either case it decreases the rate catalysis.

Noncompetitive inhibitors can be either pure or mixed. A pure noncompetitive inhibitor binds equally well to the enzyme with or without substrate bound, and the effect is that only turnover number is decreased. Thus, pure noncompetitive inhibitors act by decreasing the concentration of functional enzyme rather than by altering the proportion of enzyme molecules that are bound to substrate. Noncompetitive inhibition, like uncompetitive inhibition, cannot be overcome by increasing the substrate concentration. **Mixed noncompetitive inhibition** is more complex in that the inhibitor binds preferentially to either the free enzyme or the enzyme–substrate complex, resulting in a pattern of inhibition that alters both substrate binding and turnover number simultaneously. The commonly prescribed antibiotic doxycycline, which is used to treat periodontal disease, functions at low concentrations as a noncompetitive inhibitor of a proteolytic enzyme (collagenase).

The different types of reversible inhibitors are kinetically distinguishable

How can we determine whether a reversible inhibitor acts by competitive, uncompetitive, or noncompetitive inhibition? Considering only enzymes that exhibit Michaelis–Menten kinetics, we can use measurements of the rates of catalysis at different concentrations of substrate and inhibitor and Lineweaver–Burk plots as clues to which type of reversible inhibition is occurring.

Competitive inhibition. In competitive inhibition, where the inhibitor competes with the substrate for the enzyme, the dissociation constant for the inhibitor is given by

$$K_i = [E][I]/[EI]$$

The smaller the K_i, the more potent the inhibition. The hallmark of competitive inhibition is that it can be overcome by a sufficiently high concentration of substrate (**Figure 5.19**). The effect of a competitive inhibitor is to increase the

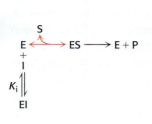

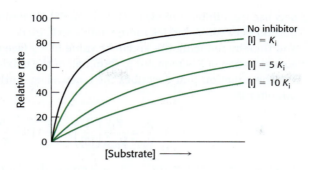

FIGURE 5.19 Michaelis–Menten plots of a competitive inhibitor reveal changes in the apparent K_M but not V_{max}. As the concentration of a competitive inhibitor increases, higher concentrations of substrate are required to attain a particular reaction velocity. The reaction pathway suggests how sufficiently high concentrations of substrate can completely relieve competitive inhibition resulting in the same V_{max} value at infinite substrate concentration.

apparent value of K_M, meaning that more substrate is needed to obtain the same reaction rate. This new apparent value of K_M, called K_M^{app}, is numerically equal to

$$K_M^{app} = K_M(1 + [I]/K_i)$$

where $[I]$ is the concentration of inhibitor and K_i is the dissociation constant for the enzyme–inhibitor complex. In the presence of a competitive inhibitor, an enzyme will have the same V_{max} as in the absence of an inhibitor. At a sufficiently high concentration, virtually all the active sites are filled with substrate, and the enzyme is fully operative.

A Lineweaver–Burk plot can also reveal competitive inhibition. In competitive inhibition, the intercept on the y-axis of the plot of $1/V_0$ versus $1/[S]$ is the same in the presence and in the absence of inhibitor, although the slope is increased (**Figure 5.20**).

The intercept is unchanged because a competitive inhibitor does not alter V_{max}. The increase in the slope of the $1/V_0$ versus $1/[S]$ plot indicates the strength of binding of a competitive inhibitor. In the presence of a competitive inhibitor, equation 18 is replaced by

$$\frac{1}{V_0} = \frac{1}{V_{max}} + \frac{K_M}{V_{max}}\left(1 + \frac{[I]}{K_i}\right)\left(\frac{1}{[S]}\right) \tag{22}$$

In other words, the slope of the plot is increased by the factor $(1 + [I]/K_i)$ in the presence of a competitive inhibitor. Consider an enzyme with a K_M of 10^{-4} M. In the absence of inhibitor, when $V_0 = V_{max}/2$, $[S] = 10^{-4}$ M. In the presence of a 2×10^{-3} M competitive inhibitor that is bound to the enzyme with a K_i of 10^{-3} M, the apparent K_M (K_M^{app}) will be equal to $K_M(1 + [I]/K_i)$, or 3×10^{-4} M. Substitution of these values into equation 22 gives $V_0 = V_{max}/4$, when $[S] = 10^{-4}$ M. Thus, the presence of the competitive inhibitor cuts the reaction rate in half at this substrate concentration.

Uncompetitive inhibition. In uncompetitive inhibition, where the inhibitor binds only to the ES complex, the resulting enzyme–substrate–inhibitor (ESI) complex does not go on to form any product. Because some unproductive ESI complex will always be present, V_{max} will be lower in the presence of inhibitor than in its absence (**Figure 5.21**). The uncompetitive inhibitor lowers the apparent value of K_M

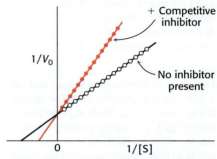

FIGURE 5.20 Competitive inhibition is easily recognized using a Lineweaver–Burk plot. A Lineweaver–Burk plot of enzyme kinetics in the presence and absence of a competitive inhibitor illustrates that the inhibitor has no effect on V_{max} but increases the apparent K_M.

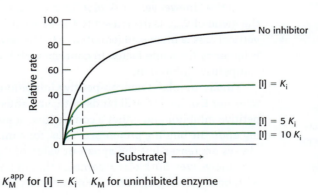

K_M^{app} for $[I] = K_i$ K_M for uninhibited enzyme

FIGURE 5.21 Michaelis–Menten plots of an uncompetitive inhibitor reveal decreases in both the apparent K_M and V_{max}. The reaction pathway shows that the inhibitor binds only to the enzyme–substrate complex. Consequently, V_{max} cannot be attained, even at high substrate concentrations. The apparent value for K_M is lowered, becoming smaller as more inhibitor is added.

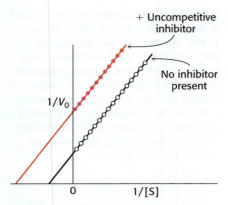

+ Uncompetitive inhibitor

No inhibitor present

$1/V_0$

0     $1/[S]$

FIGURE 5.22 Uncompetitive inhibition is easily distinguished by parallel lines in a Lineweaver–Burk plot. An uncompetitive inhibitor does not affect the slope of the Lineweaver–Burk plot. V_{max} and K_M are reduced by equivalent amounts.

FIGURE 5.23 Michaelis–Menten plots of a pure noncompetitive inhibitor reveal a decrease in V_{max}, while K_M remains unchanged. The reaction pathway shows that the inhibitor binds both to free enzyme and to an enzyme–substrate complex. Consequently, as with uncompetitive inhibition, V_{max} cannot be attained. In pure noncompetitive inhibition, K_M remains unchanged, and so the reaction rate increases more slowly at low substrate concentrations than is the case for uncompetitive competition.

because the inhibitor binds to ES to form ESI, depleting ES. To maintain the equilibrium between E and ES, more S binds to E, increasing the apparent value of k_1 and thereby reducing the apparent value of K_M (see equation 13). Thus, a lower concentration of S is required to form half of the maximal concentration of ES.

The equation that describes the Lineweaver–Burk plot for an uncompetitive inhibitor is

$$\frac{1}{V_0} = \frac{K_M}{V_{max}}\left(\frac{1}{[S]}\right) + \frac{1}{V_{max}}\left(1 + \frac{[I]}{K_i}\right)$$

The slope of the line, K_M/V_{max}, is the same as that for the uninhibited enzyme, but the intercept on the y-axis will be increased by $1 + [I]/K_i$. Consequently, the lines in Lineweaver–Burk plots will be parallel (**Figure 5.22**).

Noncompetitive inhibition. Recall that in noncompetitive inhibition, the inhibitor can bind at a different site from the substrate either the enzyme or the enzyme–substrate complex; in either case, the enzyme–inhibitor–substrate complex does not proceed to form product. In pure noncompetitive inhibition, the K_i for the inhibitor binding to E is the same as for binding to ES complex (**Figure 5.23**).

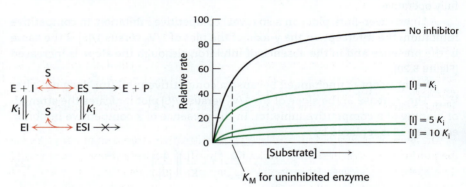

The apparent value of V_{max} is decreased to a new value called V_{max}^{app}, whereas the value of K_M is unchanged. The maximal velocity in the presence of a pure noncompetitive inhibitor, V_{max}^{app}, is given by

$$V_{max}^{app} = \frac{V_{max}}{1 + [I]/K_i} \tag{23}$$

Why is V_{max} lowered though K_M remains unchanged? In essence, the inhibitor simply lowers the concentration of functional enzyme because some quantity of enzyme is always bound to the inhibitor and inactive at any given moment, regardless of the substrate concentration. Because the inhibitor and the substrate do not affect the binding of one another, pure noncompetitive inhibition cannot be overcome by increasing the substrate concentration.

In the Lineweaver–Burk plot for pure noncompetitive inhibition (**Figure 5.24**), the value of V_{max} is decreased to the new value V_{max}^{app}, and so the intercept on the vertical axis is increased (equation 23). The new slope, which is equal to K_M/V_{max}^{app}, is larger by the same factor. In contrast with V_{max}, K_M is not affected by pure noncompetitive inhibition.

Finally, in mixed noncompetitive inhibition, the inhibitor binds preferentially to either E or ES, but still binds both, and thus behaves as a blend of a noncompetitive inhibitor and either a competitive or an uncompetitive inhibitor, depending upon the binding preference. Thus, for a mixed noncompetitive inhibitor V_{max} always decreases but the apparent value of K_M can either increase or decrease; the Lineweaver–Burk plot resembles that of a pure noncompetitive inhibitor but with lines converging either above or below the x-axis.

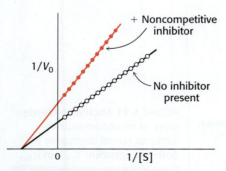

+ Noncompetitive inhibitor

$1/V_0$

No inhibitor present

0     $1/[S]$

FIGURE 5.24 Pure noncompetitive inhibition can be distinguished on a Lineweaver–Burk plot by lines converging at the x-axis. A Lineweaver–Burk plot of enzyme kinetics in the presence and absence of a pure noncompetitive inhibitor shows that K_M is unaltered and V_{max} is decreased.

Transition-state analogs are potent competitive inhibitors

We turn now to compounds that provide the most intimate views of the catalytic process itself. Linus Pauling proposed in 1948 that compounds resembling the transition state of a catalyzed reaction should be very effective inhibitors of enzymes. These mimics are called **transition-state analogs**.

The inhibition of proline racemase is an instructive example. The racemization of proline proceeds through a transition state in which the tetrahedral α-carbon atom has become trigonal (**Figure 5.25**). In the trigonal form, all three bonds are in the same plane; C_α also carries a net negative charge. This symmetric carbanion can be reprotonated on one side to give the L isomer or on the other side to give the D isomer. Pauling's hypothesis was supported by the finding that the inhibitor pyrrole 2-carboxylate binds to the racemase active site 160 times as tightly as does proline. The α-carbon atom of this inhibitor, like that of the transition state, is trigonal.

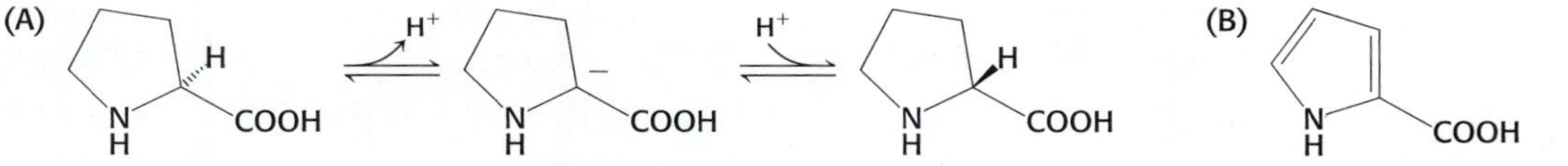

(A) L-Proline Planar transition state D-Proline **(B)** Pyrrole 2-carboxylic acid (transition-state analog)

FIGURE 5.25 Transition-state analogs are potent competitive inhibitors because they structurally resemble the transition state of a catalyzed reaction. (A) The isomerization of L-proline to D-proline by proline racemase proceeds through a planar transition state in which the α-carbon atom is trigonal rather than tetrahedral. (B) Pyrrole 2-carboxylic acid, a transition-state analog because of its trigonal geometry, is a potent competitive inhibitor of proline racemase. An analog that also carries a negative charge on C_α would be expected to bind even more tightly.

In general, highly potent and specific competitive inhibitors of enzymes can be produced by synthesizing compounds that more closely resemble the transition state than the substrate itself. Because only molecules that are structurally similar to the real transition state will bind tightly to the active site, potent competitive inhibition—or its absence—by these molecules can be used to support or refute hypotheses regarding the structure of the transition state for a given enzyme-catalyzed reaction, providing evidence for particular chemical mechanisms. The inhibitory power of transition-state analogs underscores the essence of enzyme catalysis: catalysis is accomplished by selective binding of the transition state structure by the active site.

❖ SELF–CHECK QUESTION

An inhibitor binds noncovalently to an allosteric site and induces a change in protein structure such that the enzyme can no longer bind substrate in the active site. Conversely, when the substrate is bound in the active site, the allosteric site for the inhibitor is closed such that the inhibitor cannot bind. What type of inhibition is being described here?

Irreversible inhibitors can be used to map the active site

The first step in obtaining the chemical mechanism of an enzyme is to determine what functional groups are required for enzyme activity. How can we ascertain what these functional groups are? X-ray crystallography of the enzyme bound to its substrate or substrate analog provides one approach. Irreversible inhibitors that covalently bond to the enzyme provide an alternative and often complementary approach: the inhibitors modify the functional groups, which can then be identified. Irreversible inhibitors can be divided into three categories: group-specific reagents, affinity labels, and suicide inhibitors.

- **Group-specific reagents** inhibit enzymes by reacting with specific side chains of the enzyme's amino acids. An example of a group-specific reagent is diisopropylphosphofluoridate (DIPF). DIPF modifies only 1 of the 28 serine residues in the proteolytic enzyme chymotrypsin and yet inhibits the enzyme, implying that this serine residue is especially reactive. We will see in Chapter 6 that this serine residue is indeed located at the active site. DIPF also revealed a reactive serine residue in acetylcholinesterase, an enzyme important in the transmission of nerve impulses (**Figure 5.26**). Thus, DIPF and similar compounds that bind and inactivate acetylcholinesterase are potent nerve gases. Most group-specific reagents do not display the exquisite specificity shown by DIPF. Consequently, more specific means of modifying the active site are required.

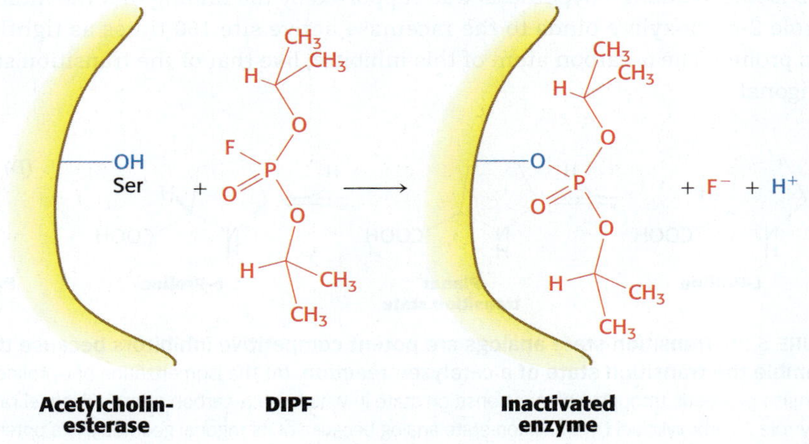

FIGURE 5.26 A group-specific reagent irreversibly inhibits enzymes by reacting with specific side chains. For example, DIPF can inhibit an enzyme by covalently modifying a crucial serine residue.

- **Affinity labels (reactive substrate analogs)** inhibit enzymes by being structurally similar to the substrate and covalently bonding to active-site residues. They are thus more specific for the enzyme's active site than are group-specific reagents. Tosyl-L-phenylalanine chloromethyl ketone (TPCK) is a substrate analog for chymotrypsin (**Figure 5.27**). TPCK binds at the active site and then reacts irreversibly with a histidine residue at that site, inhibiting the enzyme.

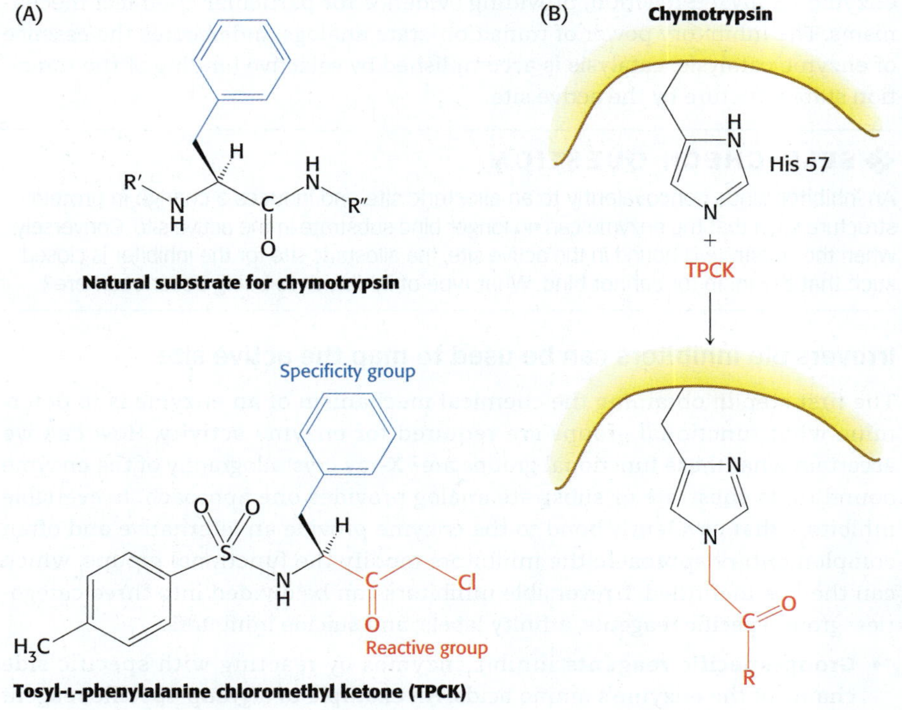

FIGURE 5.27 Affinity labels are highly specific irreversible inhibitors because they are structurally similar to the enzyme substrate. (A) Tosyl-L-phenylalanine chloromethyl ketone (TPCK) is a reactive analog of the normal substrate for the enzyme chymotrypsin. (B) TPCK binds at the active site of chymotrypsin and modifies an essential histidine residue.

- **Mechanism-based (suicide) inhibitors** are modified substrates that provide the most specific means for modifying an enzyme's active site. The inhibitor binds to the enzyme as a substrate and is initially processed by the normal catalytic mechanism. The mechanism of catalysis then generates a chemically reactive intermediate that inactivates the enzyme through covalent modification. The fact that the enzyme participates in its own irreversible inhibition strongly suggests that the covalently modified group on the enzyme is vital for catalysis.

How would a molecule like TPCK—or other types of irreversible inhibitors, for that matter—affect the kinetic parameters V_{max} and K_M that we have discussed throughout this chapter? We will address that question in the context of all the inhibitors discussed in this chapter in the example below.

EXAMPLE Determining Inhibitor Type from Data

PROBLEM: The graph shows the effect of two different concentrations of a molecule X on an enzyme. What kind of molecule is X, and how is it affecting enzyme activity?

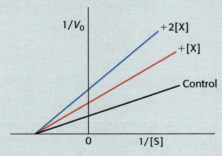

GETTING STARTED: What kind of plot is this? The first step is to examine the axes: the x-axis shows 1/[S], while the y-axis displays $1/V_0$. Note that these are two reciprocal values, so this is a double reciprocal plot, specifically a Lineweaver–Burk plot. Next, consider what we know about these plots: they can reveal how molecules affect the apparent values of the kinetic parameters K_M and V_{max} based on where the line crosses the two axes. Recall that the point where the line crosses the y-axis is $1/V_{max}$, while the intersection with the x-axis is $-1/K_M$.

ANALYZE: How is the molecule affecting the enzyme? If we first consider the x-axis, we see that all of the lines converge there. Therefore, the apparent K_M is not altered in the presence of X. However, when X is present, the lines cross the y-axis at increasing values, indicating that $1/V_{max}$ increases with increasing amounts of X. If $1/V_{max}$ increases in the presence of X, then the apparent V_{max} must be decreasing.

Let's summarize these effects and then consider the possibilities. The apparent K_M is unchanged, but the apparent V_{max} is reduced. This latter observation means that X must be some type of an inhibitor of the enzyme. Let's consider some possibilities: competitive inhibitors, including transition-state analogs, alter the apparent K_M but don't affect V_{max}. Uncompetitive inhibitors decrease both the apparent values of K_M and V_{max}. Finally, mixed noncompetitive inhibitors may raise or lower the apparent value of K_M but do not leave it unchanged. None of these descriptions fit the data. The only type of reversible inhibitor that affects an enzyme this way is a pure noncompetitive inhibitor. Recall that pure noncompetitive inhibitors lower the apparent V_{max} without altering K_M.

However, this is not the end of the problem! One more possibility exists that we have not considered: irreversible inhibitors. Because irreversible inhibitors permanently inactivate enzyme molecules, essentially removing a stoichiometric amount of them from pool of active enzyme molecules involved in the observed reaction, the effect is the same as if there were less enzyme originally added to the reaction mixture. That is, the apparent V_{max} is lowered because there are fewer active enzyme molecules, but the K_M is unaffected because those excess enzyme molecules which do not react with the inhibitor are completely unaffected by it.

REFLECT: All irreversible inhibitors give rise to the same pattern as a pure noncompetitive inhibitor with respect to the observed changes to the apparent values of K_M and V_{max}; so, in this case, additional information is required to distinguish whether molecule X is an irreversible or a pure noncompetitive inhibitor. The effects of various types of inhibitors are summarized in **Table 5.9**.

Table 5.9 Summary of effects of enzyme inhibitors on kinetic parameters

Type of inhibition	K_M	V_{max}
Competitive	Increased	Unchanged
Uncompetitive	Decreased	Decreased
Pure noncompetitive	Unchanged	Decreased
Mixed noncompetitive	Increased or decreased	Decreased
Irreversible	Unchanged	Decreased

Penicillin irreversibly inactivates a key enzyme in bacterial cell-wall synthesis

Penicillin, the first antibiotic ever discovered, is a clinically useful suicide inhibitor. Penicillin consists of a thiazolidine ring fused to a β-*lactam* ring to which a variable R group is attached by a peptide bond (**Figure 5.28A**). In benzylpenicillin, for example, R is a benzyl group (**Figure 5.28B**). The β-lactam ring is very labile due to significant strain in the four-membered ring; this instability is closely tied to the antibiotic action of penicillin.

How does penicillin inhibit bacterial growth? Let us consider *Staphylococcus aureus*, the most common cause of staph infections. Like the vast majority

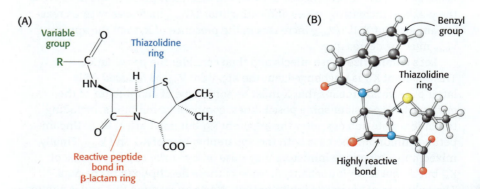

FIGURE 5.28 The reactive site of penicillin is the peptide bond of its β-lactam ring. (A) Structural formula of penicillin. (B) Representation of benzylpenicillin.

of all bacteria, the cell wall of S. *aureus* is made up of a macromolecule, called *peptidoglycan* (**Figure 5.29**), which consists of linear polysaccharide chains that are cross-linked by short peptides. The enormous peptidoglycan molecule provides a continuous, covalently linked mesh that confers mechanical support and prevents bacteria from bursting in response to their high internal osmotic pressure. Glycopeptide transpeptidase catalyzes the formation of the cross-links that make the peptidoglycan so stable. Note that bacterial cell walls are distinctive in containing D amino acids, which form cross-links by a mechanism different from that used to synthesize proteins.

Penicillin is an effective antibiotic because it interferes with the synthesis of the bacterial cell wall by irreversibly inhibiting the cross-linking transpeptidase. The transpeptidase normally forms an acyl intermediate with the penultimate D-alanine residue of the D-Ala-D-Ala peptide, which then reacts with the amino group of the terminal glycine in another peptide to form the cross-link. But penicillin can be accepted into the active site of the transpeptidase instead, because it mimics the D-Ala-D-Ala moiety of the normal substrate (**Figure 5.30**). Bound penicillin then forms a covalent bond with an active-site serine, permanently inactivating the enzyme and preventing the completion of cell-wall synthesis.

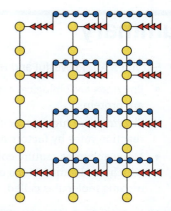

FIGURE 5.29 A schematic representation of a peptidoglycan section in S. aureus shows the crosslinking between sugar fibers. The sugars are shown in yellow, the tetrapeptides in red, and the pentaglycine bridges in blue. Only a small section is represented; the cell wall is a single, enormous, bag-shaped macromolecule because of extensive crosslinking.

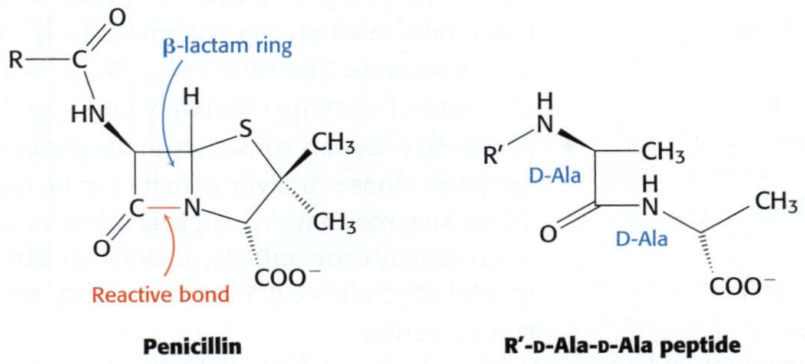

FIGURE 5.30 Penicillin resembles the transition state of the transpeptidase reaction. In the vicinity of its reactive peptide bond, the conformation of penicillin resembles the postulated conformation of the transition state of R'-D-Ala-D-Ala in the transpeptidation reaction. [Information from B. Lee, *J. Mol. Biol.* 61:463–469, 1971.]

Why is penicillin such an effective inhibitor of the transpeptidase? The highly strained, four-membered β-lactam ring of penicillin makes it especially reactive. On binding to the transpeptidase, the serine residue at the active site attacks the carbonyl carbon atom of the lactam ring to form the penicilloyl-serine derivative (**Figure 5.31**). Because the peptidase participates in its own inactivation, penicillin acts as a suicide inhibitor.

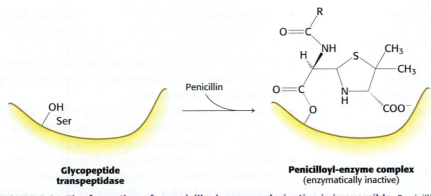

FIGURE 5.31 The formation of a penicilloyl-enzyme derivative is irreversible. Penicillin reacts irreversibly with the transpeptidase to inactivate the enzyme.

Summary

5.1 Enzymes Are Powerful and Highly Specific Catalysts

- Enzymes are biological catalysts, most of which are proteins.
- Enzymes are highly specific and can enhance reaction rates by factors as great as 10^{17}.
- Many enzymes require cofactors for activity which can be metal ions or small, vitamin-derived organic molecules called coenzymes.

5.2 Gibbs Free Energy Is a Useful Thermodynamic Function for Understanding Enzymes

- A reaction can take place spontaneously only if the change in free energy (ΔG) is negative.
- The free-energy change of a reaction under standard conditions is called the standard free-energy change ($\Delta G°$) while biochemists use $\Delta G°'$, the standard free-energy change at pH 7.
- Enzymes do not alter reaction equilibria; rather, they increase the rate at which equilibrium is attained.

5.3 Enzymes Accelerate Reactions by Facilitating the Formation of the Transition State

- Enzymes decrease the free energy of activation of a chemical reaction by providing a reaction pathway in which the transition state (the highest-energy species) has a lower free energy and hence is more rapidly formed than in the uncatalyzed reaction; the result is that the reaction is accelerated in both directions.
- The first step in catalysis is the formation of an enzyme–substrate complex in which substrates are bound at active-site clefts from which water is largely excluded.
- The specificity of enzyme–substrate interactions arises mainly from weak reversible contacts mediated by ionic interactions, hydrogen bonds, and van der Waals forces, and from the shape of the active site, which rejects molecules that do not have a sufficiently complementary shape and charge.
- Enzymes facilitate formation of the transition state by a dynamic process in which the substrate binds to specific conformations of the enzyme, accompanied by conformational changes at active sites that result in catalysis.

5.4 The Michaelis–Menten Model Accounts for the Kinetic Properties of Many Enzymes

- The kinetic properties of many enzymes are described by the Michaelis–Menten model in which an enzyme (E) combines with a substrate (S) to form an enzyme–substrate (ES) complex, which can proceed to form a product (P) or to dissociate into E and S.

$$E + S \underset{k_{-1}}{\overset{k_1}{\rightleftharpoons}} ES \overset{k_2}{\longrightarrow} E + P$$

- The rate of formation of product V_0 is given by the Michaelis–Menten equation:

$$V_0 = V_{max} \frac{[S]}{[S] + K_M}$$

in which V_{max} is the reaction rate when the enzyme is fully saturated with substrate and K_M, the Michaelis constant, is the substrate concentration at which the reaction rate is half maximal.

- The kinetic constant k_{cat}, called the turnover number, is the number of substrate molecules converted into product per unit time at a single catalytic site when the enzyme is fully saturated with substrate. The ratio of k_{cat}/K_M provides a measure of enzyme efficiency and specificity.
- Allosteric enzymes constitute an important class of enzymes whose catalytic activity can be regulated. These enzymes have multiple active sites which often display cooperativity, as evidenced by a sigmoidal dependence of reaction velocity on substrate concentration.

5.5 Enzymes Can Be Studied One Molecule at a Time

- Single-molecule methods reaffirm the key findings from ensemble studies but also reveal a distribution of enzyme characteristics rather than an average value.

5.6 Enzymes Can Be Inhibited by Specific Molecules

- Irreversible inhibitors bind covalently to enzymes and can provide a means of mapping the enzyme's active site, while reversible inhibition is characterized by a more rapid and less stable interaction between enzyme and inhibitor.
- Competitive inhibitors prevent the substrate from binding to the active site and thereby reduce the reaction velocity by diminishing the proportion of enzyme molecules that are bound to substrate. Competitive inhibition can be overcome by raising the substrate concentration.
- Uncompetitive inhibitors bind only to the enzyme–substrate complex, decreasing both the turnover number and the apparent K_M.
- Pure noncompetitive inhibitors decrease only the turnover number, while mixed noncompetitive inhibitors can have a variety of effects.
- Transition-state analogs are stable compounds that mimic key features of the transition state and are potent and specific competitive inhibitors of enzymes.

Key Terms

catalyst (p. 142)

enzyme (p. 142)

ribozyme (p. 142)

substrate (p. 142)

protease (p. 142)

cofactor (p. 144)

apoenzyme (p. 144)

holoenzyme (p. 144)

coenzyme (p. 144)

prosthetic group (p. 144)

transition state (p. 148)

Gibbs free energy of activation
(activation energy) (p. 148)

active site (p. 149)

induced fit (p. 151)

binding energy (p. 151)

K_M (Michaelis constant) (p. 154)

V_{max} (maximal rate) (p. 154)

Michaelis–Menten equation (p. 154)

Lineweaver–Burk plot (p. 156)

elasticity (p. 157)

turnover number (p. 157)

k_{cat}/K_M (specificity constant) (p. 158)

sequential reaction (p. 159)

double-displacement (Ping-Pong)
reaction (p. 160)

allosteric enzyme (p. 161)

competitive inhibition (p. 165)

uncompetitive inhibition (p. 166)

pure noncompetitive inhibition
(p. 166)

mixed noncompetitive inhibition
(p. 166)

transition-state analog (p. 169)

group-specific reagent (p. 169)

affinity label (reactive substrate
analog) (p. 170)

mechanism-based (suicide) inhibitor
(p. 171)

Problems

1. What does an apoenzyme require to become a holoenzyme?

2. Consider the reaction: $S \rightleftharpoons P$. Which of the following effects are produced by an enzyme on the general reaction? ❖ **1**, ❖ **2**

(a) The reaction equilibrium is shifted away from the products.

(b) The concentration of the products is increased.

(c) ΔG for the reaction increases.

(d) The activation energy for the reaction is lowered.

(e) The formation of the transition state is promoted.

(f) The rate constant for the forward reaction increases.

3. Why does the activation energy of a reaction not appear in the final ΔG of the reaction? ❖ **1**

4. The illustrations below show the reaction-progress curves for two different reactions. Indicate the activation energy as well as the ΔG for each reaction. Which reaction is endergonic? Exergonic? ❖ **1**

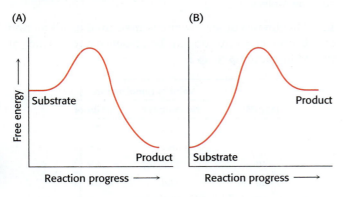

5. Suppose that, in the absence of enzyme, the forward rate constant (k_F) for the conversion of S into P is 10^{-4} s^{-1} and the

reverse rate constant (k_R) for the conversion of P into S is 10^{-6} s^{-1}. ❖ **1**, ❖ **2**

$$S \underset{10^{-6} \text{ s}^{-1}}{\overset{10^{-4} \text{ s}^{-1}}{\rightleftharpoons}} P$$

(a) What is the equilibrium for the reaction? What is the $\Delta G^{\circ\prime}$?

(b) Suppose an enzyme enhances the rate of the reaction 100-fold. What are the rate constants for the enzyme-catalyzed reaction? What is the equilibrium constant? The $\Delta G^{\circ\prime}$?

6. What would be the result of an enzyme having a greater binding energy for the substrate than for the transition state? ❖ **2**

7. Match the K'_{eq} values with the appropriate $\Delta G^{\circ\prime}$ values. ❖ **1**

K'_{eq}	$\Delta G^{\circ\prime}$ (kJ mol^{-1})
(a) 1	28.53
(b) 10^{-5}	−11.42
(c) 10^{4}	5.69
(d) 10^{2}	0
(e) 10^{-1}	−22.84

8. Assume that you have a solution of 0.1 M glucose 6-phosphate (G 6-P). To this solution, you add the enzyme phosphoglucomutase, which catalyzes the following reaction:

$$G\ 6\text{-}P \underset{}{\overset{\text{Phosphoglucomutase}}{\rightleftharpoons}} G\ 1\text{-}P$$

The $\Delta G^{\circ\prime}$ for the reaction is +7.5 kJ mol^{-1} (+1.8 kcal mol^{-1}). ❖ **1**

(a) Does the reaction proceed as written? If so, what are the final concentrations of G 6-P and G 1-P?

(b) Under what cellular conditions could you produce G 1-P at a high rate?

9. Consider the following reaction:

$$G\ 6\text{-}P \underset{}{\overset{\text{Phosphoglucomutase}}{\rightleftharpoons}} G\ 1\text{-}P$$

After reactant and product were mixed and allowed to reach equilibrium at 25°C, the concentration of each compound was measured:

$$[\text{Glucose 1-phosphate}]_{eq} = 0.01\ M$$
$$[\text{Glucose 6-phosphate}]_{eq} = 0.19\ M$$

Calculate K_{eq} and $\Delta G°'$. ❖ **1**

10. The affinity between a protein and a molecule that binds to the protein is frequently expressed in terms of a dissociation constant K_d. ❖ **3**

$$\text{Protein} + \text{small molecule} \rightleftharpoons \text{Protein} - \text{small molecule complex}$$

$$K_d = \frac{[\text{protein}][\text{small molecule}]}{[\text{protein} - \text{small molecule complex}]}$$

Does K_M measure the affinity of the enzyme complex? Under what circumstances might K_M approximately equal K_d?

11. Match the term with the description or compound. ❖ **3, ❖ 4**

(a) Competitive inhibition

_____

(b) Uncompetitive inhibition

_____

(c) Noncompetitive inhibition

_____

1. Inhibitor and substrate can bind simultaneously
2. V_{max} remains the same but the $K_M{}^{app}$ increases
3. Sulfanilamide
4. Binds to the enzyme–substrate complex only
5. Lowers V_{max} and $K_M{}^{app}$
6. Roundup
7. K_M remains unchanged but V_{max} is lower
8. Doxycycline
9. Prevents S from binding to the active site

12. Many biochemists go bananas, and justifiably, when they see a Michaelis–Menten plot like the one shown below. To see why, determine the V_0 as a fraction of V_{max} when the substrate concentration is equal to 10 K_M and 20 K_M. Please control your outrage. ❖ **3**

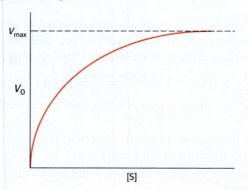

13. The hydrolysis of pyrophosphate to orthophosphate is important in driving forward biosynthetic reactions such as the synthesis of DNA. This hydrolytic reaction is catalyzed in E. coli by a pyrophosphatase that has a mass of 120 kDa and consists of six identical subunits. For this enzyme, a unit of activity is defined as the amount of enzyme that hydrolyzes 10 μmol of pyrophosphate in 15 minutes at 37°C under standard assay conditions. The purified enzyme has a V_{max} of 2800 units per milligram of enzyme. ❖ **3**

(a) How many moles of substrate are hydrolyzed per second per milligram of enzyme when the substrate concentration is much greater than K_M?

(b) How many moles of active sites are there in 1 mg of enzyme? Assume that each subunit has one active site.

(c) What is the turnover number of the enzyme? Compare this value with others mentioned in this chapter.

14. Penicillin is hydrolyzed and thereby rendered inactive by penicillinase (also known as β-lactamase), an enzyme present in some penicillin-resistant bacteria. The mass of this enzyme in *Staphylococcus aureus* is 29.6 kDa. The amount of penicillin hydrolyzed in 1 minute in a 10-ml solution containing 10^{-9} g of purified penicillinase was measured as a function of the concentration of penicillin. Assume that the concentration of penicillin does not change appreciably during the assay. ❖ **3**

[Penicillin] μM	Amount hydrolyzed (nmol)
1	0.11
3	0.25
5	0.34
10	0.45
30	0.58
50	0.61

(a) Plot V_0 versus [S] and $1/V_0$ versus $1/$[S] for these data. Does penicillinase appear to obey Michaelis–Menten kinetics? If so, what is the value of K_M?

(b) What is the value of V_{max}?

(c) What is the turnover number of penicillinase under these experimental conditions? Assume one active site per enzyme molecule.

15. The kinetics of an enzyme is measured as a function of substrate concentration in the presence and absence of 100 μM inhibitor. ❖ **3, ❖ 4**

| [S] (μM) | Velocity (μmol minute^{-1}) | |
	No inhibitor	Inhibitor
3	10.4	2.1
5	14.5	2.9
10	22.5	4.5
30	33.8	6.8
90	40.5	8.1

(a) What are the values of V_{max} and K_M in the presence of this inhibitor?

(b) What type of inhibition is it?

(c) What is the dissociation constant of this inhibitor?

(d) If [S] = 30 µM, what fraction of the enzyme molecules has a bound substrate in the presence and in the absence of 100 µM inhibitor?

16. The plot of $1/V_0$ versus $1/[S]$ is sometimes called a Lineweaver–Burk plot. Another way of expressing the kinetic data is to plot V_0 versus $V_0/[S]$, which is known as an Eadie–Hofstee plot. ❖ **3**, ❖ **4**

(a) Rearrange the Michaelis–Menten equation to give V_0 as a function of $V_0/[S]$.

(b) What is the significance of the slope, the y-intercept, and the x-intercept in a plot of V_0 versus $V_0/[S]$?

(c) Sketch a plot of V_0 versus $V_0/[S]$ in the absence of an inhibitor, in the presence of a competitive inhibitor, and in the presence of a noncompetitive inhibitor.

17. What is the defining characteristic for an enzyme catalyzing a sequential reaction? A double-displacement reaction?

18. You have isolated two versions of the same enzyme, a wild type and a mutant differing from the wild type at a single amino acid. Working carefully but expeditiously, you then establish the following kinetic characteristics of the enzymes. ❖ **1**, ❖ **3**

	Maximum velocity	K_M
Wild type	100 µmol/min	10 mM
Mutant	1 µmol/min	0.1 mM

(a) With the assumption that the reaction occurs in two steps in which k_{-1} is much larger than k_2, which enzyme has the higher affinity for substrate?

(b) What is the initial velocity of the reaction catalyzed by the wild-type enzyme when the substrate concentration is 10 mM?

(c) Which enzyme alters the equilibrium more in the direction of product?

19. For a one-substrate, enzyme-catalyzed reaction, double-reciprocal plots were determined for three different enzyme concentrations. Which of the following three families of curve would you expect to be obtained? Explain. ❖ **3**

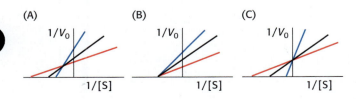

(A) $1/V_0$ ... $1/[S]$ (B) $1/V_0$ ... $1/[S]$ (C) $1/V_0$ ... $1/[S]$

20. The amino acid asparagine is required by cancer cells to proliferate. Treating patients with the enzyme asparaginase is sometimes used as a chemotherapy treatment. Asparaginase hydrolyzes asparagine to aspartate and ammonia. The below illustration shows the Michaelis–Menten curves for two asparaginases from different sources, as well as the concentration of asparagine in the environment (indicated by the arrow). Which enzyme would make a better chemotherapeutic agent? ❖ **3**

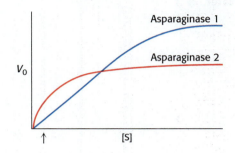

21. Picture in your mind the velocity-versus-substrate concentration curve for a typical Michaelis–Menten enzyme. Now, imagine that the experimental conditions are altered as described below. For each of the conditions described, fill in the table indicating precisely (when possible) the effect on V_{max} and K_M of the imagined Michaelis–Menten enzyme. ❖ **3**, ❖ **4**

Experimental condition	V_{max}	K_M
(a) Twice as much enzyme is used.		
(b) Half as much enzyme is used.		
(c) A competitive inhibitor is present.		
(d) An uncompetitive inhibitor is present.		
(e) A pure noncompetitive inhibitor is present.		

22. In the conversion of A into D in the following biochemical pathway, enzymes E_A, E_B, and E_C have the K_M values indicated under each enzyme. If all of the substrates and products are present at a concentration of 10^{-4} M and the enzymes have approximately the same V_{max}, which step will be rate limiting and why? ❖ **3**

$$A \underset{E_A}{\rightleftharpoons} B \underset{E_B}{\rightleftharpoons} C \underset{E_C}{\rightleftharpoons} D$$

$$K_M = \quad 10^{-2}\ M \quad 10^{-4}\ M \quad 10^{-6}\ M$$

23. What is the biochemical advantage of having a K_M approximately equal to the substrate concentration normally available to an enzyme? ❖ **3**

24. Succinylcholine is a fast-acting, short-duration muscle relaxant that is used when a tube is inserted into a patient's trachea or when a bronchoscope is used to examine the trachea and bronchi for signs of cancer. Within seconds of the administration of succinylcholine, the patient experiences muscle paralysis and is placed on a ventilator while the examination proceeds. Succinylcholine is a competitive inhibitor of acetylcholinesterase, a nervous system enzyme, and this inhibition causes paralysis. However, succinylcholine is hydrolyzed by blood-serum cholinesterase, which shows a broader substrate specificity than does the nervous system enzyme. Paralysis lasts until the succinylcholine is hydrolyzed by the serum cholinesterase, usually several minutes later. ❖ 3

(a) As a safety measure, serum cholinesterase is measured before the examination takes place. Explain why this measurement is a good idea.

(b) What would happen to the patient if the serum cholinesterase activity were only 10 units of activity per liter rather than the normal activity of about 80 units?

(c) Some patients have a mutant form of the serum cholinesterase that displays a K_M of 10 mM, rather than the normal 1.4 mM. What will be the effect of this mutation on the patient?

Enzyme Catalytic Strategies

Chess and enzymes have in common the use of strategy, consciously thought out in the game of chess and selected by evolution for the action of an enzyme. The three amino acid residues at the right, denoted by the white bonds, represent a particular chemical strategy for catalyzing a reaction that is normally very slow. This strategy relies on the properties of the different amino acids, much as chess strategy relies on different rules for moving each type of piece.

Wendie Berg

❖ LEARNING GOALS

By the end of this chapter, you should be able to:

1. Discuss four general strategies used by enzymes to accelerate particular reactions.

2. Give examples of specific chemical features of enzyme active sites that facilitate increasing the rates of specific reactions.

3. Give an example of when high absolute rate acceleration is physiologically important and how an enzyme achieves this acceleration.

4. Understand when high specificity is important for an enzyme and how this specificity is achieved.

5. Describe an example of when large conformational changes that occur during an enzymatic reaction cycle are used to drive other processes.

6. Discuss some of the experimental approaches biochemists use to elucidate enzymatic mechanisms.

OUTLINE

6.1 Enzymes Use a Core Set of Catalytic Strategies

6.2 Proteases Facilitate a Fundamentally Difficult Reaction

6.3 Carbonic Anhydrases Make a Fast Reaction Faster

6.4 Restriction Enzymes Catalyze Highly Specific DNA-Cleavage Reactions

6.5 Molecular Motor Proteins Harness Changes in Enzyme Conformation to Couple ATP Hydrolysis to Mechanical Work

What are the sources of the catalytic power and specificity of enzymes? This chapter presents the catalytic strategies used by four classes of enzymes, each of which catalyze reactions that involve the addition of water to a substrate. Scientists have revealed the mechanisms of these enzymes using incisive experimental probes, including protein structure determination, site-directed mutagenesis, detailed kinetic studies, and the use of isotope labels.

Each of the four classes of enzymes addresses a different challenge. Serine proteases promote a reaction that is almost immeasurably slow at neutral pH in the absence of a catalyst; carbonic anhydrases achieve a high absolute rate of reaction, suitable for integration with other rapid physiological processes; and restriction enzymes attain a high degree of specificity. Finally, molecular motor proteins such as myosin harness the free energy associated with the hydrolysis of ATP to drive other processes.

6.1 Enzymes Use a Core Set of Catalytic Strategies

In Chapter 5, we learned that enzymatic catalysis begins with substrate binding. The **binding energy** is the free energy released in the formation of a large number of weak interactions between the enzyme and the substrate. The use of this binding energy is the first common strategy used by enzymes. This binding energy serves two purposes: it establishes substrate specificity and increases catalytic efficiency. Often only the correct substrate can participate in most or all of the interactions with the enzyme and thus optimize binding energy, accounting for the exquisite substrate specificity exhibited by many enzymes.

Furthermore, the full complement of such interactions is formed only when the combination of enzyme and substrate is in the transition state. Thus, interactions between the enzyme and the substrate stabilize the transition state, thereby lowering the free energy of activation. The binding energy can also promote structural changes in both the enzyme and the substrate that facilitate catalysis, a process referred to as **induced fit**.

In addition to the first strategy involving binding energy, enzymes commonly employ one or more of the following four additional strategies to catalyze specific reactions:

1. *Covalent catalysis.* In **covalent catalysis**, the active site contains a reactive group that becomes temporarily covalently attached to a part of the substrate in the course of catalysis. The proteolytic enzyme chymotrypsin provides an excellent example of this strategy (Section 6.2).

2. *General acid–base catalysis.* In **general acid–base catalysis**, a molecule other than water plays the role of a proton donor or acceptor. These include histidine residues in chymotrypsin and carbonic anhydrase, an aspartate residue in EcoRV, and a phosphate group of the ATP substrate for myosin and kinesin.

3. *Catalysis by approximation.* Many reactions have two distinct substrates, including all four classes of hydrolases considered in detail in this chapter. In such cases, the reaction rate may be considerably enhanced by bringing the two substrates together along a single binding surface on an enzyme, a process called **catalysis by approximation**. For example, carbonic anhydrase binds carbon dioxide and water in adjacent sites to facilitate their reaction.

4. *Metal ion catalysis.* In **metal ion catalysis**, metal ions function catalytically in several ways. For instance, a metal ion may facilitate the formation of reactive species such as hydroxide ion by direct coordination. A zinc(II) ion serves this purpose in catalysis by carbonic anhydrase. Alternatively, a metal ion may serve to stabilize a negative charge on a reaction intermediate, a role played by magnesium(II) ion in EcoRV. Finally, a metal ion may serve as a bridge between enzyme and substrate, increasing the binding energy and holding the substrate in a conformation appropriate for catalysis. This strategy is used by myosin and kinesin and, indeed, by essentially all enzymes that use ATP as a substrate.

Although we shall not consider catalytic RNA molecules explicitly in this chapter, the principles described for protein enzymes also apply to these catalysts.

❖ SELF–CHECK QUESTION

The enzyme adenylate kinase can convert two molecules of ADP to one molecule of AMP and one molecule of ATP. Suggest at least one catalytic strategy likely used by this enzyme.

6.2 Proteases Facilitate a Fundamentally Difficult Reaction

Peptide bond hydrolysis is an important process in living systems. Proteins that have served their purpose must be degraded so that their constituent amino acids can be recycled for the synthesis of new proteins; for example, proteins ingested in the diet must be broken down into small peptides and amino acids for absorption in the gut. Furthermore, as described in detail in Chapter 7, proteolytic reactions are important in regulating the activity of certain enzymes and other proteins.

Proteases cleave proteins by a hydrolysis reaction — the addition of a molecule of water to a peptide bond:

$$R_1 - \overset{\overset{\displaystyle O}{\|}}{C} - \underset{\underset{\displaystyle H}{|}}{N} - R_2 \; + \; H_2O \; \rightleftharpoons \; R_1 - \overset{\overset{\displaystyle O}{\|}}{C} - O^- \; + \; R_2 - NH_3^+$$

Although the hydrolysis of peptide bonds is thermodynamically favorable, such reactions are extremely slow. In the absence of a catalyst, the half-life for the hydrolysis of a typical peptide at neutral pH is estimated to be between 10 and 1000 years. Yet, peptide bonds must be hydrolyzed within milliseconds in some biochemical processes.

The chemical nature of peptide bonds is responsible for their kinetic stability. Specifically, the resonance structure that accounts for the planarity of peptide bonds also makes them resistant to hydrolysis. This resonance structure endows them with partial double-bond character:

$$R_1 - \overset{\overset{\displaystyle O}{\|}}{C} - \underset{\underset{\displaystyle H}{|}}{N} - R_2 \; \longleftrightarrow \; R_1 - \overset{\overset{\displaystyle O^-}{|}}{C} = \underset{\underset{\displaystyle H}{|}}{\overset{+}{N}} - R_2$$

The carbon–nitrogen bond is strengthened by its double-bond character. More importantly, the carbonyl carbon atom is less **electrophilic** and less susceptible to **nucleophilic** attack than are the carbonyl carbon atoms in more reactive compounds such as carboxylate esters. Consequently, to promote peptide-bond cleavage, an enzyme must facilitate nucleophilic attack at a normally unreactive carbonyl group.

Chymotrypsin possesses a highly reactive serine residue

A number of proteolytic enzymes participate in the breakdown of proteins in the digestive systems of mammals and other organisms. One such enzyme, chymo-trypsin, cleaves peptide bonds selectively on the carboxyl-terminal side of the large hydrophobic amino acids such as tryptophan, tyrosine, phenylalanine, and methionine (**Figure 6.1**). Chymotrypsin is a good example of the use of covalent catalysis. The enzyme employs a powerful nucleophile to attack the unreactive carbonyl carbon atom of the substrate. This nucleophile becomes covalently attached to the substrate transiently in the course of catalysis.

What is the nucleophile that chymotrypsin employs to attack the substrate carbonyl carbon atom? A clue comes from the fact that chymotrypsin contains an extraordinarily reactive serine residue. Chymotrypsin molecules treated with organofluorophosphates such as diisopropylphosphofluoridate (DIPF) lose

FIGURE 6.1 Chymotrypsin cleaves proteins specifically on the carboxyl side of aromatic or large hydrophobic amino acids. The targeted amino acids are highlighted, and the bonds cleaved by chymotrypsin are indicated in red.

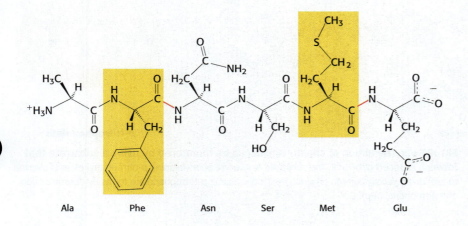

Ala   Phe   Asn   Ser   Met   Glu

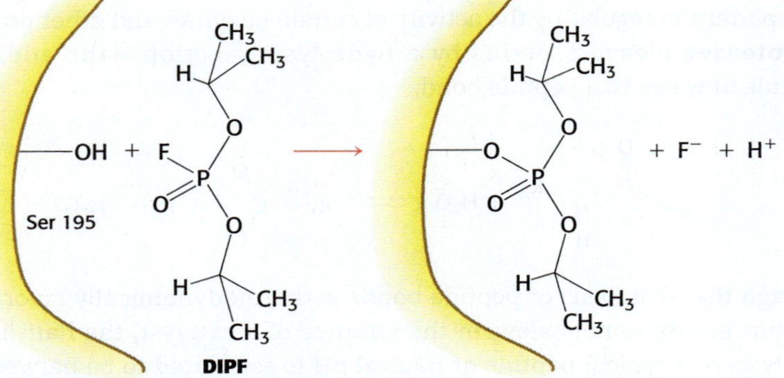

FIGURE 6.2 Chymotrypsin is inactivated by treatment with diisopropylphos-phofluoridate (DIPF). DIPF reacts only with one serine among 28 possible serine residues, revealing the unusual reactivity of serine 195.

all activity irreversibly (**Figure 6.2**). Only a single residue, serine 195, is modified. This **chemical modification reaction** suggests that this unusually reactive serine residue plays a central role in the catalytic mechanism of chymotrypsin.

Phosphorus-based agents that modify reactive serine residues can be potent toxins. Compounds such as malathion, used to kill pest insects, and nerve agents such as sarin, developed as a chemical weapon, each modify a reactive serine residue in a key enzyme in the nervous system and disrupt nerve function.

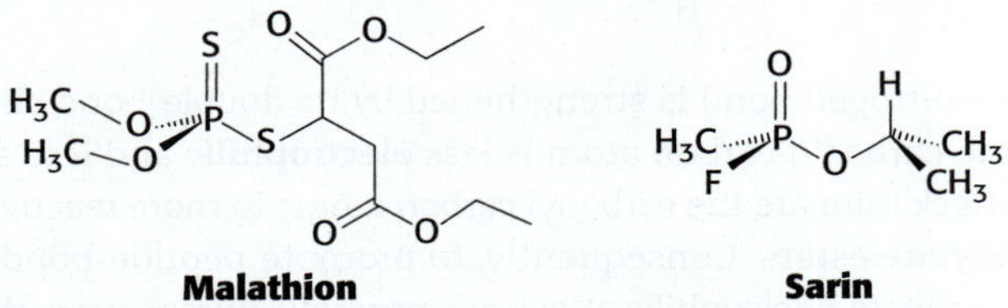

Chymotrypsin action proceeds in two steps linked by a covalently bound intermediate

A study of the kinetics of chymotrypsin provides a second clue to its catalytic mechanism. Enzyme kinetics is often monitored by having the enzyme act on a substrate analog, called a **chromogenic substrate**, that forms a colored product. For chymotrypsin, one such substrate is N-acetyl-L-phenylalanine p-nitrophenyl ester. You might notice that this substrate is an ester rather than an amide, but many proteases will also hydrolyze esters. One of the products formed cleavage of this substrate by chymotrypsin is p-nitrophenolate, which has a yellow color (**Figure 6.3**). Measurements of the absorbance of light reveal the amount of p-nitrophenolate being produced.

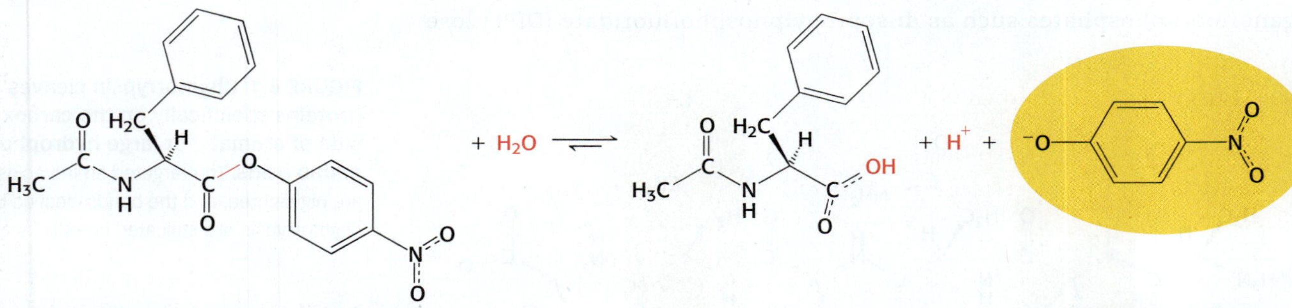

N-Acetyl-L-phenylalanine p-nitrophenyl ester

p-Nitrophenolate

FIGURE 6.3 The action of chymotrypsin can be monitored by using a substrate that forms a colored product. The substrate N-Acetyl-L-phenylalanine p-nitrophenyl ester is cleaved to produce a yellow-colored product, p-nitrophenolate. p-Nitrophenolate forms by deprotonation of p-nitrophenol at pH 7.

Under steady-state conditions, the cleavage of this substrate obeys Michaelis–Menten kinetics with K_M of 20 µM and k_{cat} of 77 s^{-1}. More insight into the mechanism can be gained by monitoring the initial phase of the reaction by using the stopped-flow method, which makes it possible to mix enzyme and substrate and monitor the results within a millisecond. This method reveals an initial rapid burst of colored product, followed by its slower formation as the reaction reached the steady state (**Figure 6.4**), confirming that hydrolysis proceeds in two phases. In the first reaction cycle that takes place on each enzyme molecule immediately after mixing, only the first phase must take place before the colored product is released. In subsequent reaction cycles, both phases must take place. Note that the burst is observed because the first phase is substantially more rapid than the second phase for this substrate.

The two phases are explained by the formation of a covalent enzyme–substrate intermediate (**Figure 6.5**). First, the acyl group of the substrate becomes covalently attached to the enzyme as p-nitrophenolate (or an amine if the substrate is an amide rather than an ester) is released. The enzyme–acyl group complex is called the acyl-enzyme intermediate. Second, the acyl-enzyme intermediate is hydrolyzed to release the carboxylic acid component of the substrate and regenerate the free enzyme. Thus, one molecule of p-nitrophenolate is produced rapidly from each enzyme molecule as the acyl-enzyme intermediate is formed. However, it takes longer for the enzyme to be reset by the hydrolysis of the acyl-enzyme intermediate, and both phases are required for enzyme turnover.

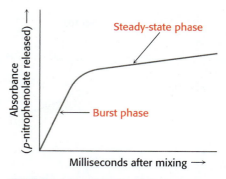

FIGURE 6.4 Two kinetic phases are evident by monitoring the initial action of chymotrypsin. When chymotrypsin is rapidly mixed with a solution of N-acetyl-L-phenylalanine p-nitrophenyl ester, a rapid burst phase of colored product is observed followed by a steady-state phase. This observation provides insight into the chymotrypsin mechanism.

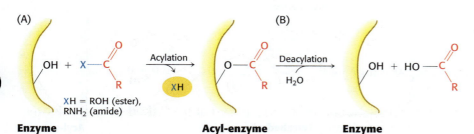

FIGURE 6.5 Hydrolysis by chymotrypsin involves covalent catalysis. In step (A), the enzyme is covalently modified to form an acyl-enzyme intermediate. In step (B), the enzyme is freed by deacylation.

Serine is part of a catalytic triad that also includes histidine and aspartate

The three-dimensional structure of chymotrypsin reveals that this enzyme is roughly spherical and comprises three polypeptide chains, linked by disulfide bonds (**Figure 6.6**). Note that the active site of chymotrypsin, marked by serine 195, lies in a cleft on the surface of the enzyme. The structure of the active site explains the special reactivity of serine 195 (**Figure 6.7**). The side chain of serine 195 is hydrogen bonded to the imidazole ring of histidine 57. The —NH group of this imidazole ring is, in turn, hydrogen bonded to the carboxylate group of aspartate 102. This constellation of residues is referred to as the **catalytic triad**.

How does this arrangement of residues lead to the high reactivity of serine 195? The histidine residue serves to position the serine side chain and to polarize its hydroxyl group so that it is poised for deprotonation. In the presence of the substrate, the histidine residue accepts the proton from the serine 195 hydroxyl group. In doing so, the histidine acts as a general base catalyst. The withdrawal of the proton from the hydroxyl group generates an alkoxide ion, which is a much more powerful nucleophile than is an alcohol. The aspartate residue helps orient the histidine residue and make it a better proton acceptor through hydrogen bonding and electrostatic effects.

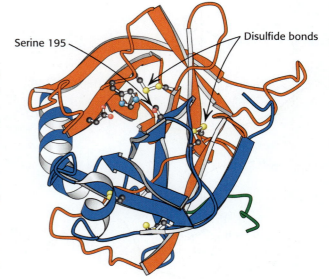

FIGURE 6.6 The active site serine 195 sits in a cleft on the surface of chymotrypsin. In the ribbon diagram shown here, chymotrypsin's three chains are indicated in orange, blue, and green. Two intrastrand and two interstrand disulfide bonds are also shown. [Drawn from 1GCT.pdb.]

INTERACT with this model in
 Achieve

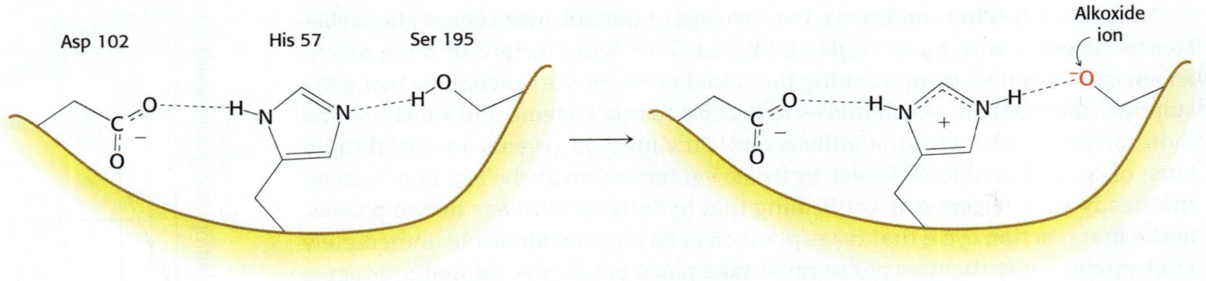

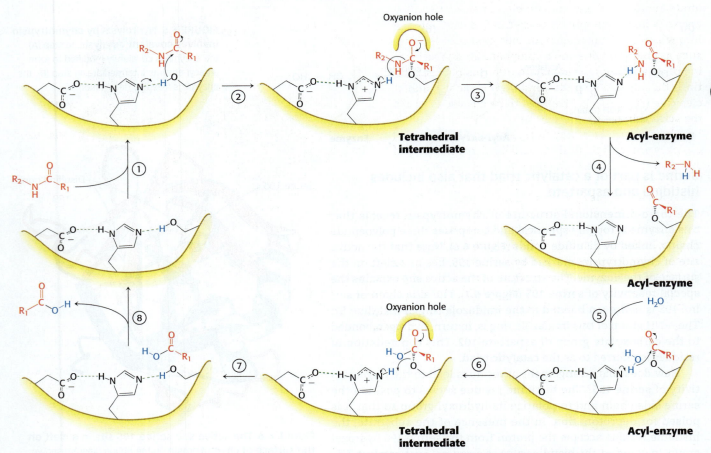

FIGURE 6.7 The catalytic triad increases the reactivity of serine 195. The catalytic triad poises this residue for formation of an alkoxide ion, a much stronger nucleophile than a hydroxyl group.

These observations suggest a mechanism for peptide hydrolysis (**Figure 6.8**). After substrate binding (step 1), the reaction begins with the oxygen atom of the side chain of serine 195 making a nucleophilic attack on the carbonyl carbon atom of the target peptide bond (step 2). There are now four atoms bonded to the carbonyl carbon, arranged as a tetrahedron, instead of three atoms in a planar arrangement. This inherently unstable tetrahedral intermediate bears a formal negative charge on the oxygen atom derived from the carbonyl group. This charge is stabilized by interactions with NH groups from the protein in a site

FIGURE 6.8 The mechanism of peptide hydrolysis illustrates the principles of covalent and acid–base catalysis. The mechanism proceeds in eight steps: (1) substrate binding, (2) nucleophilic attack of serine on the peptide carbonyl group, (3) collapse of the tetrahedral intermediate, (4) release of the amine component, (5) water binding, (6) nucleophilic attack of water on the acyl-enzyme intermediate, (7) collapse of the tetrahedral intermediate, and (8) release of the carboxylic acid component. The dashed green lines represent hydrogen bonds.

termed the **oxyanion hole** (Figure 6.9). These interactions also help stabilize the transition state that precedes the formation of the tetrahedral intermediate. This tetrahedral intermediate collapses to generate the acyl-enzyme (step 3), facilitated by the transfer of the proton being held by the positively charged histidine residue to the amino group formed by cleavage of the peptide bond. The amine component is now free to depart from the enzyme (step 4), completing the first stage of the hydrolytic reaction—acylation of the enzyme. Such acyl-enzyme intermediates have been observed using x-ray crystallography by trapping them through adjustment of conditions such as the nature of the substrate, pH, or temperature.

The next stage—deacylation—begins when a water molecule takes the place occupied earlier by the amine component of the substrate (step 5). The ester group of the acyl-enzyme is now hydrolyzed by a process that essentially repeats steps 2 through 4. Again acting as a general base catalyst, histidine 57 draws a proton away from the water molecule. The resulting OH⁻ ion attacks the carbonyl carbon atom of the acyl group, forming a tetrahedral intermediate (step 6). This structure breaks down to form the carboxylic acid product (step 7). Finally, the release of the carboxylic acid product (step 8) readies the enzyme for another round of catalysis.

This mechanism accounts for all characteristics of chymotrypsin action except the observed preference for cleaving the peptide bonds just past residues with large, hydrophobic side chains. Examination of the three-dimensional structure of chymotrypsin with substrate analogs and enzyme inhibitors reveals the presence of a deep hydrophobic pocket, called the S_1 pocket, into which the long, uncharged side chains of residues such as phenylalanine and tryptophan can fit. The binding of an appropriate side chain into this pocket positions the adjacent peptide bond into the active site for cleavage (Figure 6.10). The bond to be cleaved is called the **scissile bond**.

The specificity of chymotrypsin depends almost entirely on which amino acid is directly on the amino-terminal side of the peptide bond to be cleaved, but other proteases have more-complex specificity patterns. Such enzymes have additional pockets on their surfaces for the recognition of other residues in the substrate. Figure 6.11 shows a generic example of the nomenclature used in diagramming protease-substrate interactions. Residues on the amino-terminal side of the scissile bond are labeled P_1, P_2, P_3, and so forth, heading away from the scissile bond. Likewise, residues on the carboxyl side of the scissile bond are labeled P_1', P_2', P_3', and so forth. The corresponding sites on the enzyme are referred to as S_1, S_2 or S_1', S_2', and so forth.

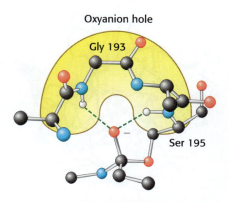

FIGURE 6.9 The oxyanion hole stabilizes the tetrahedral intermediates that occur during the chymotrypsin reaction. Note the hydrogen bonds (shown in green) that link peptide NH groups and the negatively charged oxygen atom of the intermediate.

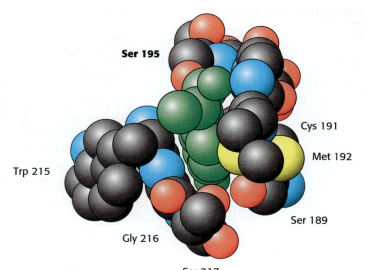

FIGURE 6.10 The specificity pocket of chymotrypsin favors the binding of residues with long hydrophobic side chains. The pocket is deep and lined with hydrophobic residues, with active-site serine residue (serine 195) positioned to cleave the peptide backbone between the residue bound in the pocket (in this case, phenylalanine, shown in green) and the next residue in the sequence.

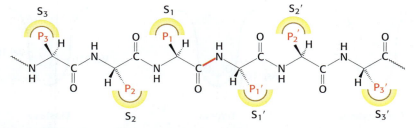

FIGURE 6.11 The specificity nomenclature for protease–substrate interactions shows the potential sites of interaction. The protease site is designated P, and the corresponding binding sites on the enzyme are designated S. The scissile bond (shown in red) is the reference point.

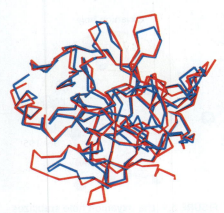

FIGURE 6.12 Chymotrypsin and trypsin are structurally similar. This similarity is revealed by overlaying the structures of chymotrypsin (red) and trypsin (blue). Only α-carbon-atom positions are shown. The mean deviation in position between corresponding α-carbon atoms is quite small, only 1.7 Å. [Drawn from 5PTP.pdb and 1GCT.pdb.]

INTERACT with this model in
 Achieve

Catalytic triads are found in other hydrolytic enzymes

Many other peptide-cleaving proteins have subsequently been found to contain catalytic triads similar to that discovered in chymotrypsin. Some, such as trypsin and elastase, are obvious homologs of chymotrypsin; that is, they clearly evolved from a common ancestor. The sequences of these proteins are approximately 40% identical with that of chymotrypsin, and their overall structures are quite similar (**Figure 6.12**). These proteins operate by mechanisms identical with that of chymotrypsin. However, the three enzymes differ markedly in substrate specificity. Chymotrypsin cleaves at the peptide bond after residues with an aromatic or long, nonpolar side chain. Trypsin cleaves at the peptide bond after residues with long, positively charged side chains—namely, arginine and lysine. Elastase cleaves at the peptide bond after amino acids with small side chains—such as alanine and serine.

Comparison of the S_1 pockets of these enzymes reveals that these different specificities are due to small structural differences. In trypsin, an aspartate residue (Asp 189) is present at the bottom of the S_1 pocket in place of a serine residue in chymotrypsin. The aspartate residue attracts and stabilizes a positively charged arginine or lysine residue in the substrate. In elastase, two residues at the top of the pocket in chymotrypsin and trypsin are replaced by much bulkier valine residues (Val 190 and Val 216). These residues close off the mouth of the pocket so that only small side chains can enter (**Figure 6.13**).

Other members of the chymotrypsin family include a collection of proteins that take part in blood clotting, to be discussed in Chapter 7, as well as the tumor marker protein prostate-specific antigen (PSA). In addition, a wide range of proteases found in bacteria, viruses, and plants belong to this clan.

Other enzymes that are not homologs of chymotrypsin have been found to contain very similar active sites to that of chymotrypsin, not because of a shared common ancestor, but because a similar active evolved independently. Subtilisin, a protease in bacteria such as *Bacillus amyloliquefaciens*, is a particularly well-characterized example. The active site of this enzyme includes both the catalytic triad and the oxyanion hole. However, one of the NH groups that forms the oxyanion hole comes from the side chain of an asparagine residue rather

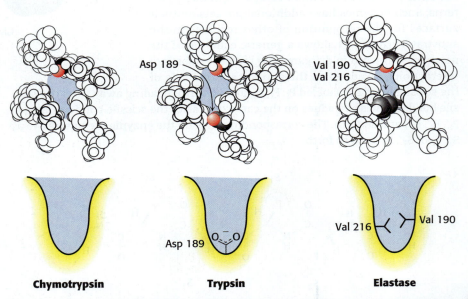

FIGURE 6.13 The S_1 pockets of chymotrypsin, trypsin, and elastase show the residues that play key roles in determining enzyme specificity. The side chains of these residues, as well as those of the active-site serine residues, are shown in color, and schematic depictions are shown below.

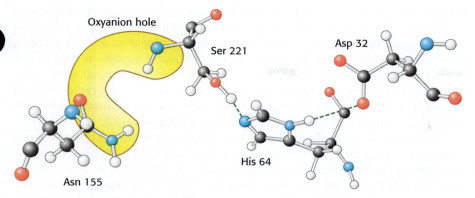

FIGURE 6.14 The catalytic triad and oxyanion hole of subtilisin includes two NH groups, one from the backbone and one from the side chain of Asn 155. The NH groups will stabilize a negative charge that develops on the peptide bond attacked by nucleophilic serine 221 of the catalytic triad.

than from the peptide backbone (**Figure 6.14**). Subtilisin is the founding member of another large family of proteases. Still other proteases have been discovered that contain an active-site serine or threonine residue that is activated not by a histidine–aspartate pair but by a primary amino group from the side chain of lysine or by the N-terminal amino group of the polypeptide chain.

The catalytic triad in proteases has clearly emerged several times in the course of evolution. We can conclude that this catalytic strategy must be an especially effective approach to the hydrolysis of peptides and related bonds.

Scientists have dissected the catalytic triad using site-directed mutagenesis

How can we test the validity of the mechanism proposed for the catalytic triad? One way is to test the contribution of individual amino acid residues to the catalytic power of a protease by using site-directed mutagenesis. Subtilisin has been extensively studied by this method. Each of the residues within the catalytic triad, consisting of aspartic acid 32, histidine 64, and serine 221, has been individually converted into alanine, and the ability of each mutant enzyme to cleave a model substrate has been examined (**Figure 6.15**).

As expected, the conversion of active-site serine 221 into alanine dramatically reduces catalytic power; the value of k_{cat} falls to less than one millionth of its value for the wild-type enzyme. The value of K_M is essentially unchanged; its increase by no more than a factor of two indicates that substrate continues to bind normally. The mutation of histidine 64 to alanine reduces catalytic power to a similar

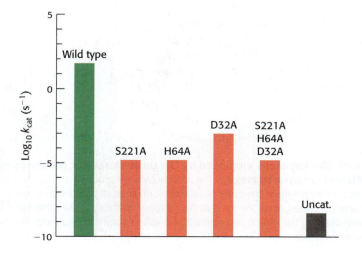

FIGURE 6.15 Mutating the residues of the catalytic triad to alanine allows the importance of each residue for catalytic activity to be quantified. Note that the activities are displayed on a logarithmic scale. The mutations are identified as follows: the first letter is the one-letter abbreviation for the amino acid being altered; the number identifies the position of the residue in the primary structure; and the second letter is the one-letter abbreviation for the amino acid replacing the original one. *Uncat.* refers to the estimated rate for the uncatalyzed reaction.

degree. The conversion of aspartate 32 into alanine reduces catalytic power by less, although the value of k_{cat} still falls to less than 0.005% of its wild-type value. The simultaneous conversion of all three residues into alanine is no more deleterious than the conversion of serine or histidine alone. These observations support the notion that the catalytic triad and, particularly, the serine–histidine pair act together to generate a nucleophile of sufficient power to attack the carbonyl carbon atom of a peptide bond. Despite the reduction in their catalytic power, the mutated enzymes still hydrolyze peptides a thousand times as fast as buffer at pH 8.6.

Site-directed mutagenesis also offers a way to probe the importance of the oxyanion hole for catalysis. The mutation of asparagine 155 to glycine eliminates the side-chain NH group from the oxyanion hole of subtilisin. The elimination of the NH group reduces the value of k_{cat} to 0.2% of its wild-type value but increases the value of K_M by only a factor of two. These observations demonstrate that the NH group of the asparagine residue plays a significant role in stabilizing the tetrahedral intermediate and the transition state leading to it.

Some proteases cleave peptides at other locations besides serine residues

Not all proteases use strategies based on activated serine residues. Classes of proteins have been discovered that employ three alternative approaches to peptide-bond hydrolysis. In each case, the strategy is to generate a nucleophile that attacks the peptide carbonyl group (**Figure 6.16**).

Cysteine proteases. The strategy used by the cysteine proteases is most similar to that used by the chymotrypsin family. In these enzymes, a cysteine residue, activated by a histidine residue, plays the role of the nucleophile that attacks the peptide bond in a manner quite analogous to that of the serine residue in serine proteases (Figure 6.16). Because the sulfur atom in cysteine is inherently a better nucleophile than is the oxygen atom in serine, cysteine proteases appear to require only this histidine residue in addition to cysteine and not the full catalytic triad.

Aspartyl proteases. The central feature of the active sites in aspartyl proteases is a pair of aspartic acid residues that act together to allow a water molecule to attack the peptide bond. One aspartic acid residue (in its deprotonated form) activates the attacking water molecule by poising it for deprotonation. The other aspartic acid residue (in its protonated form) polarizes the peptide carbonyl group so that it is more susceptible to attack (Figure 6.16).

(A) Cysteine proteases   **(B) Aspartyl proteases**   **(C) Metalloproteases**

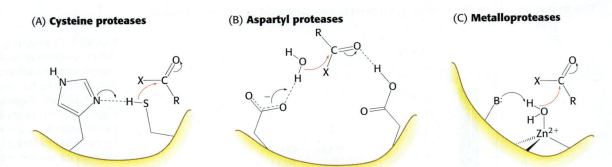

FIGURE 6.16 The key steps employed by the three classes of proteases reveal three different catalytic strategies. The peptide carbonyl group is attacked by (A) a histidine-activated cysteine in the cysteine proteases, (B) an aspartate-activated water molecule in the aspartyl proteases, and (C) a metal-activated water molecule in the metalloproteases. For the metalloproteases, the letter B represents a base (often glutamate) that helps deprotonate the metal-bound water.

Metalloproteases. The active site of a metalloprotease contains a bound metal ion, most often zinc, that activates a water molecule to act as a nucleophile to attack the peptide carbonyl group. In each of these three classes of enzymes, the active site includes features that act to (1) activate a water molecule or another nucleophile, (2) polarize the peptide carbonyl group, and (3) stabilize a tetrahedral intermediate (Figure 6.16).

❖ SELF–CHECK QUESTION

Write out a mechanism for a cysteine protease analogous with that for chymotrypsin.

Protease inhibitors are important drugs

Because of the fundamental biological roles of proteases, these enzymes are important drug targets. Drugs that block protease activity are called **protease inhibitors**. For example, captopril, a drug used to regulate blood pressure, is one of many inhibitors of the angiotensin-converting enzyme (ACE), a metalloprotease. Indinavir (Crixivan), retrovir, and many other compounds used in the treatment of AIDS are inhibitors of HIV protease, an aspartyl protease. HIV protease cleaves multidomain viral proteins into their active forms; blocking this process completely prevents the virus from being infectious. HIV protease inhibitors, in combination with inhibitors of other key HIV enzymes, have dramatically reduced deaths due to AIDS, assuming that the cost of the treatment could be covered. In many cases, these drugs have converted AIDS from a death sentence to a treatable chronic disease.

Indinavir, one of the first HIV protease inhibitors developed but not one that is widely used at present, reveals an important strategy for developing enzyme inhibitors. Indinavir was designed to resemble a peptide substrate of the enzyme. X-ray crystallographic studies revealed that indinavir adopts a confirmation that approximates the twofold symmetry of the enzyme (**Figure 6.17**). The active site of HIV protease is covered by two flexible flaps that fold down on top of the bound inhibitor. The OH group of the central alcohol interacts with the two aspartate residues of the active site and was intended to mimic a water molecule observed in crystal structures of HIV protease but generally absent in cellular aspartyl proteases. This OH group was intended to contribute to the specificity of indinavir for HIV protease. To prevent side effects, enzyme inhibitors used as drugs should be relatively specific for one enzyme, minimally inhibiting other similar enzymes within the body.

FIGURE 6.17 HIV protease has twofold symmetry. (Left) The three-dimensional structure of HIV protease is shown with the inhibitor indinavir bound in the active site. (Right) The drug has been rotated to reveal its approximately twofold symmetric conformation. [Drawn from 1HSH.pdb.]

INTERACT with this model in
🅜 Achieve

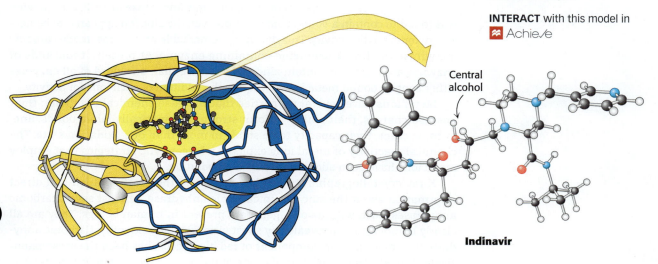

Indinavir

6.3 Carbonic Anhydrases Make a Fast Reaction Faster

Carbon dioxide is a major end product of aerobic metabolism. In mammals, this carbon dioxide is released into the blood and transported to the lungs for exhalation. While in the red blood cells, carbon dioxide reacts with water. The products of this reaction are bicarbonate ion (HCO_3^-) and a proton. These are a form of carbonic acid, H_2CO_3, a moderately strong acid, with $pK_a = 3.5$.

Bicarbonate ion   **Carbonic acid**

Even in the absence of a catalyst, this hydration reaction proceeds at a moderately fast pace. At 37°C near neutral pH, the second-order rate constant k_1 is 0.0027 M^{-1} s^{-1}. This value corresponds to an effective first-order rate constant of 0.15 s^{-1} in water ($[H_2O] = 55.5$ M). The reverse reaction, the dehydration of HCO_3^-, is even more rapid, with a rate constant of $k_{-1} = 50$ s^{-1}. These rate constants correspond to an equilibrium constant of $K_1 = 5.4 \times 10^{-5}$ and a ratio of $[CO_2]$ to $[H_2CO_3]$ of 340:1 at equilibrium.

Carbon dioxide hydration and HCO_3^- dehydration are often coupled to rapid processes, particularly transport processes. Thus, almost all organisms contain enzymes, referred to as **carbonic anhydrases**, that increase the rate of reaction beyond the already relatively high spontaneous rate. For example, carbonic anhydrases dehydrate HCO_3^- in the blood to form CO_2 for exhalation as the blood passes through the lungs. Conversely, they convert CO_2 into HCO_3^- to generate the aqueous humor of the eye and other secretions. For this reason, one of the drugs used for treating glaucoma, a condition associated with too much pressure inside the eye, acts by inhibiting carbonic anhydrase. Furthermore, both CO_2 and HCO_3^- are substrates and products for a variety of enzymes, and the rapid interconversion of these species may be necessary to ensure appropriate substrate levels.

Carbonic anhydrases accelerate CO_2 hydration dramatically. The most-active enzymes hydrate CO_2 at rates as high as $k_{cat} = 10^6$ s^{-1}, or a million times a second per enzyme molecule. Fundamental physical processes such as diffusion and proton transfer ordinarily limit the rate of hydration, and so the enzymes employ special strategies to attain such high rates.

Carbonic anhydrase contains a bound zinc ion essential for catalytic activity

Less than 10 years after the discovery of carbonic anhydrase in 1932, this enzyme was found to contain a bound zinc ion. Moreover, the zinc ion appeared to be necessary for catalytic activity. This discovery, remarkable at the time, made carbonic anhydrase the first known zinc-containing enzyme. At present, thousands of enzymes are known to contain zinc. In fact, more than one-third of all enzymes either contain bound metal ions or require the addition of such ions for activity.

Metal ions have several properties that increase chemical reactivity: their positive charges, their ability to form strong yet kinetically labile bonds, and, in some cases, their capacity to be stable in more than one oxidation state. The chemical reactivity of metal ions explains why catalytic strategies that employ metal ions have been adopted throughout evolution.

X-ray crystallographic studies have supplied the most detailed and direct information about the zinc site in carbonic anhydrase. At least seven carbonic anhydrases, each with its own gene, are present in human beings. They are all clearly homologous, as revealed by substantial sequence identity. Carbonic anhydrase II, a major protein component of red blood cells, has been the most extensively studied (**Figure 6.18**). It is also one of the most active carbonic anhydrases.

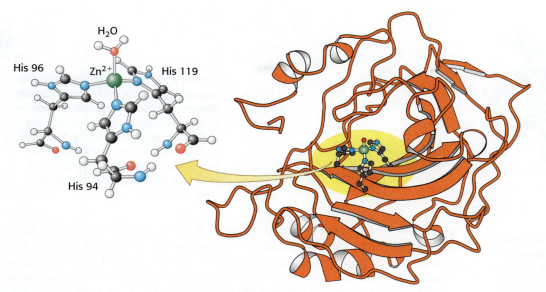

FIGURE 6.18 The active site of human carbonic anhydrase II includes a bound zinc ion. The active site includes a zinc ion that is bound to the imidazole rings of three histidine residues as well as to a water molecule in a cleft near the center of the enzyme. [Drawn from 1CA2.pdb.]

INTERACT with this model in
Achieve

Zinc is found only in the +2 state in biological systems. A zinc atom is essentially always bound to four or more specific molecules or groups, called **ligands**. In carbonic anhydrase, three coordination sites are occupied by nitrogen ligands from the imidazole rings of three histidine residues, and an additional coordination site is occupied by a water molecule (or a hydroxide ion, depending on pH). Because the ligands occupying the coordination sites are neutral, the overall charge on the $Zn(His)_3$ unit remains +2.

Catalysis involves zinc activation of a water molecule

How does this zinc complex facilitate carbon dioxide hydration? A major clue comes from the pH profile of enzymatically catalyzed carbon dioxide hydration (**Figure 6.19**). At pH 8, the reaction proceeds near its maximal rate. As the pH decreases, the rate of the reaction drops. The midpoint of this transition is near pH 7, suggesting that a group that loses a proton at pH 7 ($pK_a = 7$) plays an important role in the activity of carbonic anhydrase. Moreover, the curve suggests that the deprotonated (high pH) form of this group participates more effectively in catalysis. Although some amino acids, notably histidine, have pK_a values near 7, a variety of evidence suggests that the group responsible for this transition is not an amino acid but is the zinc-bound water molecule.

The binding of a water molecule to the positively charged zinc center reduces the pK_a of the water molecule from 15.7 to 7 (**Figure 6.20**). With the pK_a lowered, the water molecule can more easily lose a proton at neutral pH, generating a substantial concentration of hydroxide ion (bound to the zinc atom). A zinc-bound

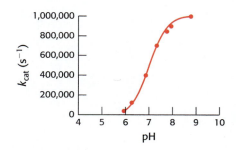

FIGURE 6.19 The activity of carbonic anhydrase varies with pH. The enzyme is maximally active at pH values near 9.

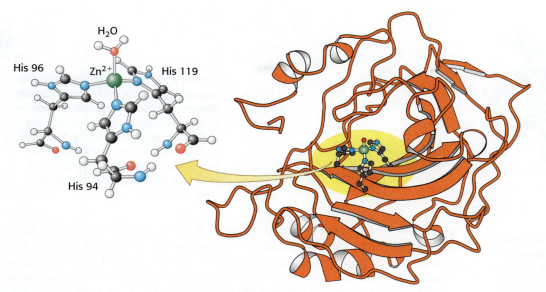

FIGURE 6.20 Upon binding to zinc in carbonic anhydrase, the pK_a of water decreases from 15.7 to approximately 7.

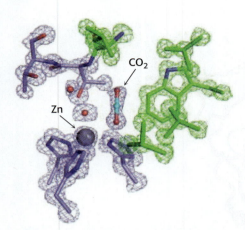

FIGURE 6.21 Carbonic anhydrase has a carbon dioxide binding site adjacent to the active site zinc ion. Crystals of carbonic anhydrase were exposed to carbon dioxide gas at high pressure and low temperature, and x-ray diffraction data were collected. The electron density (shown as cages) for carbon dioxide, clearly visible adjacent to the zinc and its bound water, reveals the carbon dioxide binding site. Hydrophobic amino acids are shaded green, and the purple ones are hydrophilic.

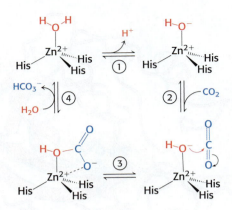

FIGURE 6.22 The zinc-bound hydroxide mechanism for the hydration of carbon dioxide proceeds in four steps. These are (1) water deprotonation, (2) carbon dioxide binding, (3) nucleophilic attack by hydroxide on carbon dioxide, and (4) displacement of bicarbonate ion by water. This mechanism provides a good example of one type of metal ion catalysis.

hydroxide ion (OH^-) is a potent nucleophile able to attack carbon dioxide much more readily than water does. Adjacent to the zinc site, carbonic anhydrase also possesses a hydrophobic patch that serves as a binding site for carbon dioxide (**Figure 6.21**). Based on these observations, a simple mechanism for carbon dioxide hydration can be proposed (**Figure 6.22**):

1. The zinc ion facilitates the release of a proton from a water molecule, which generates a hydroxide ion.

2. The carbon dioxide substrate binds to the enzyme's active site and is positioned to react with the hydroxide ion.

3. The hydroxide ion attacks the carbon dioxide, converting it into bicarbonate ion, HCO_3^-.

4. The catalytic site is regenerated with the release of HCO_3^- and the binding of another molecule of water.

Thus, the binding of a water molecule to the zinc ion favors the formation of the transition state by facilitating proton release and by positioning the water molecule to be in close proximity to the other reactant.

Rapid regeneration of the active form of carbonic anhydrase depends on proton availability

In the first step of a carbon dioxide hydration reaction, the zinc-bound water molecule must lose a proton to regenerate the active form of the enzyme (**Figure 6.23**). Let use consider this in detail:

- The equilibrium constant for this process is $K = 10^{-7}$ M (since $pK_a = 7$) and $K = k_1/k_{-1}$.

- The rate of the reverse reaction, the protonation of the zinc-bound hydroxide ion, is limited by the rate of proton diffusion. Protons diffuse very rapidly with second-order rate constants near 10^{11} M^{-1} s^{-1}.

- The forward rate constant is given by $k_1 = K \times k_{-1}$. Since $k_{-1} \leq 10^{11}$ M^{-1} s^{-1}, k_1 must be less than or equal to 10^4 s^{-1}.

As noted earlier, some carbonic anhydrases can hydrate carbon dioxide at rates as high as a million times a second (10^6 s^{-1}). This rate is so high that it seems to violate the analysis above. If carbon dioxide is hydrated at a rate of 10^6 s^{-1}, then every step in the mechanism (Figure 6.22) must take place at least this fast. How is this apparent paradox resolved?

FIGURE 6.23 The ratio of the rate constant for water deprotonation to that for hydroxide protonation must equal 10^{-7}. This is true since the pK_a of this water molecule in carbonic anhydrase is 7.

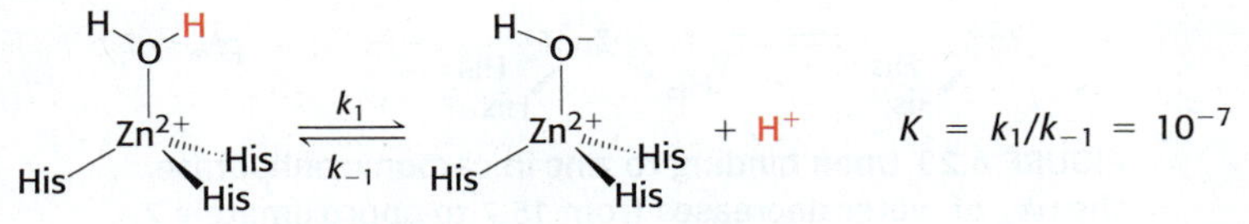

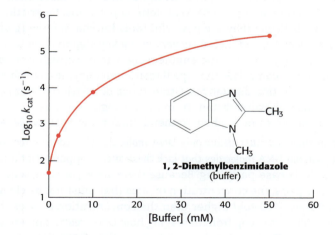

$$K = k_1'/k_{-1}' \approx 1$$

FIGURE 6.24 The deprotonation of the zinc-bound water molecule in carbonic anhydrase can be aided by buffer component B. This is optimized if the pK_a of the buffer is near 7 and if the buffer is present at a sufficiently high concentration.

The role of buffers. The answer became clear with the observation that the highest rates of carbon dioxide hydration require the presence of buffer, suggesting that the buffer components participate in the reaction. The buffer can bind or release protons. The advantage is that, whereas the concentrations of protons and hydroxide ions are limited to 10^{-7} M at neutral pH, the concentration of buffer components can be much higher, of the order of several millimolar. Let us consider the role of buffer components in more detail:

- If the buffer component BH^+ has a pK_a of 7 (matching that for the zinc-bound water molecule), then the equilibrium constant for the reaction in **Figure 6.24** is 1.
- The second-order rate constants k_1' and k_{-1}' will be limited by buffer diffusion to values less than approximately 10^9 M^{-1} s^{-1}.
- The rate of proton abstraction is given by $k_1' \times [B]$.
- If buffer concentrations greater than $[B] = 10^{-3}$ M (or 1 mM) are used, then carbon dioxide hydration rates $\geq 10^6$ M^{-1} s^{-1} may be possible because

$$k_1' \times [B] = (10^9 \text{ } M^{-1} \text{ } s^{-1}) \times (10^{-3} \text{ } M) = 10^6 \text{ } s^{-1}$$

The prediction that the rate increases with increasing buffer concentration has been confirmed experimentally (**Figure 6.25**).

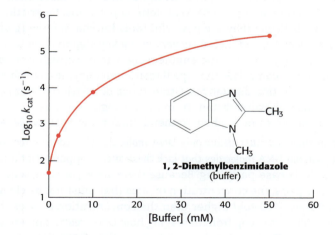

FIGURE 6.25 The rate of carbon dioxide hydration by carbonic anhydrase increases with the concentration of the buffer 1,2-dimethylbenzimidazole. If present at high enough concentrations, such buffers enable the enzyme to achieve its highest catalytic rates.

❖ SELF–CHECK QUESTION

Suppose that the rate of CO_2 hydration is measured in 1 mM buffer at pH 7.0. What is the maximum expected rate?

A "proton shuttle" for large buffers. The molecular components of many buffers are too large to reach the active site of carbonic anhydrase. Carbonic anhydrase II has evolved a **proton shuttle** to allow buffer components to participate in the reaction from solution. The primary component of this shuttle is histidine 64. This residue transfers protons from the zinc-bound water molecule to the protein surface and then to the buffer (**Figure 6.26**). Thus, catalytic function has been enhanced through the evolution of an apparatus for controlling proton transfer from and to the active site. Because protons participate in many biochemical reactions, the manipulation of the proton inventory within active sites is crucial to the function of many enzymes and explains the prominence of acid–base catalysis.

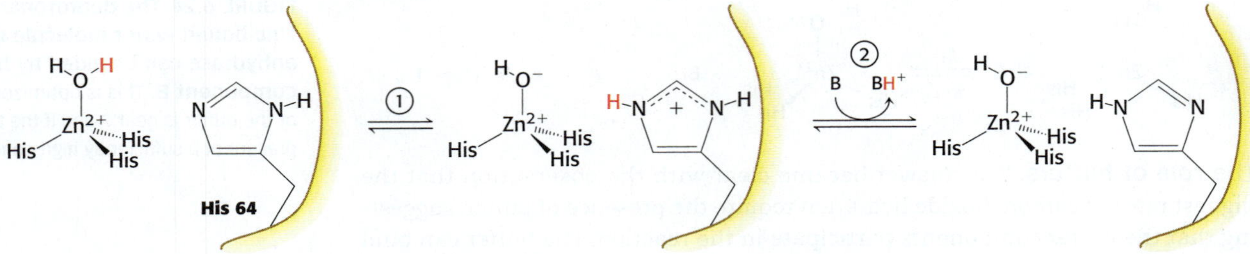

FIGURE 6.26 Histidine 64 acts as a proton shuttle. Histidine 64 (1) removes a proton from the zinc-bound water molecule and (2) transfers the proton to buffer molecules in solution.

6.4 Restriction Enzymes Catalyze Highly Specific DNA-Cleavage Reactions

We next consider a hydrolytic reaction that results in the cleavage of DNA. Bacteria and archaea have evolved mechanisms to protect themselves from viral infections. Many viruses inject their DNA genomes into cells; once inside, the viral DNA hijacks the cell's machinery to drive the production of viral proteins and, eventually, of progeny virus. Often, a viral infection results in the death of the host cell.

A major protective strategy for the host is to use **restriction enzymes** (also called **restriction endonucleases**) to degrade the viral DNA on its introduction into a cell. These enzymes recognize particular base sequences, called **recognition sequences** or recognition sites, in DNA molecules and cleave such molecules (hereafter referred to as *cognate DNA*) at defined positions. Restriction enzymes are best known because they are powerful tools for characterizing and manipulating DNA molecules, and their discovery was a key step that led to a revolution in molecular biology and genetic engineering. The most well-studied class of restriction enzymes comprises the type II restriction enzymes, which cleave DNA within their recognition sequences. Other types of restriction enzymes cleave DNA at positions somewhat distant from their recognition sites.

Restriction enzymes must show tremendous specificity at two levels:

1. *Restriction enzymes must cleave only DNA molecules that contain recognition sites, without cleaving DNA molecules that lack these sites.* Suppose that a recognition sequence is six base-pairs long. Because there are 4^6, or 4096, sequences having six base pairs, the concentration of sites that must not be cleaved will be approximately 4000-fold higher than the concentration of sites that should be cleaved. Thus, to keep from damaging host DNA, restriction enzymes must cleave cognate DNA molecules much more than 4000 times as efficiently as they cleave nonspecific sites. We shall return to the mechanism used to achieve the necessary high specificity after considering the chemistry of the cleavage process.

2. *Restriction enzymes must not degrade host DNA containing the recognition sequences.* How do these enzymes manage to degrade viral DNA while sparing their own? In *E. coli*, the restriction endonuclease EcoRV cleaves double-stranded viral DNA molecules that contain the sequence 5'-GATATC-3', but it leaves intact host DNA containing hundreds of such sequences. We shall return to the strategy by which host cells protect their own DNA at the end of this section.

Cleavage is by direct displacement of 3'-oxygen from phosphorus by magnesium-activated water

A restriction enzyme catalyzes the hydrolysis of the phosphodiester backbone of DNA. Specifically, the bond between the 3'-oxygen atom and the phosphorus atom is broken. The products of this reaction are DNA strands with a free 3'-hydroxyl

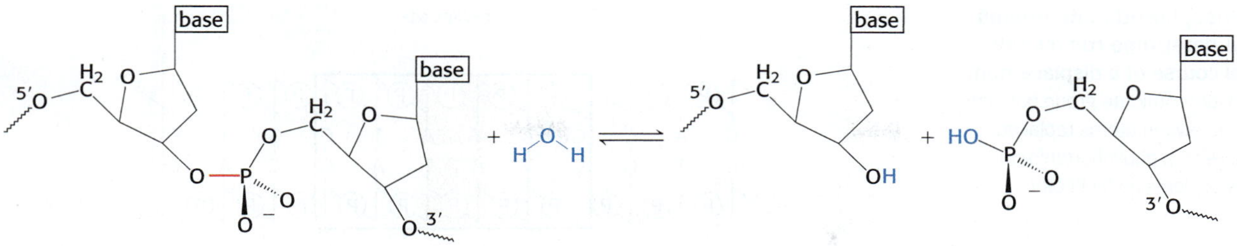

FIGURE 6.27 Restriction enzymes catalyze the hydrolysis of DNA phosphodiester bonds. The reaction leaves a hydroxyl group at the 3′ end of one product fragment and a phosphoryl group attached to the 5′ end of the other. The bond that is cleaved is shown in red.

group and a 5′-phosphoryl group at the cleavage site (**Figure 6.27**). This reaction proceeds by nucleophilic attack at the phosphorus atom. We will consider two alternative mechanisms, suggested by analogy with the proteases.

1. The restriction enzyme might cleave DNA through a covalent intermediate, employing a potent nucleophile (Nu):

Mechanism 1 (covalent intermediate)

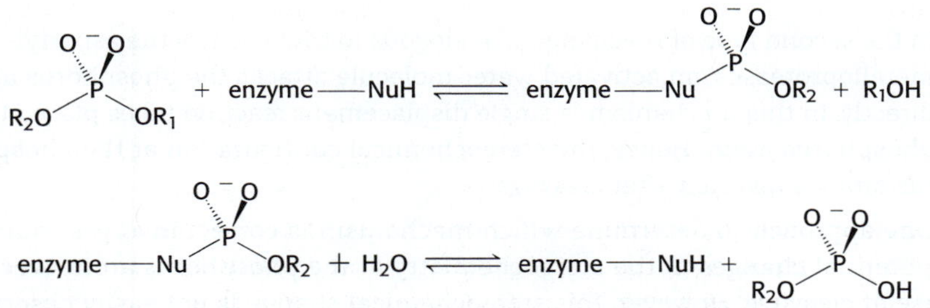

2. Alternatively, the restriction enzyme might cleave DNA through direct hydrolysis:

Mechanism 2 (direct hydrolysis)

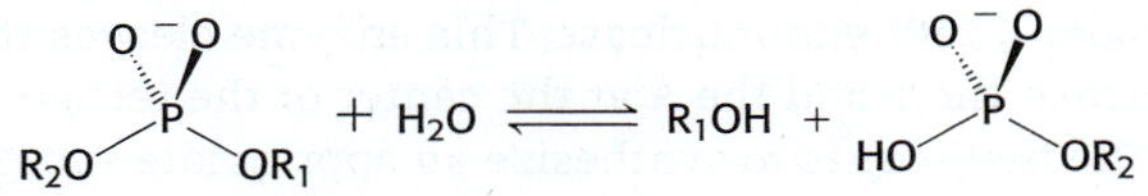

Each mechanism postulates a different nucleophile to attack the phosphorus atom. In either case, each reaction takes place by in-line displacement, in which the incoming nucleophile attacks the phosphorus atom, and a pentacoordinate transition state is formed:

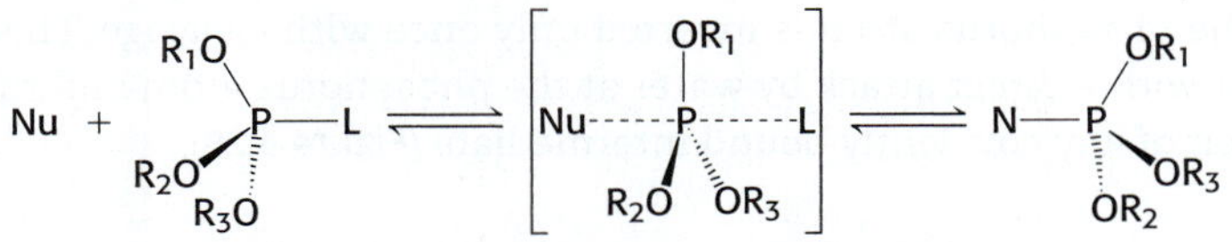

The transition state has a trigonal bipyramidal geometry centered at the phosphorus atom, with the incoming nucleophile at one apex of the two pyramids and the group that is displaced (the leaving group, L) at the other apex. Note that the displacement inverts the stereochemical conformation at the tetrahedral phosphorous atom, analogous to the interconversion of stereoisomers around a tetrahedral carbon center.

The two mechanisms differ in the number of times that the displacement takes place in the course of the reaction:

1. In the first type of mechanism, a nucleophile in the enzyme (analogous to serine 195 in chymotrypsin) attacks the phosphate group to form a covalent intermediate. Next, this intermediate is hydrolyzed to produce the final products. In this case, two displacement reactions take place at the phosphorus atom. Consequently, the stereochemical configuration at the phosphorus atom would be inverted and then inverted again, and the overall configuration would be retained.

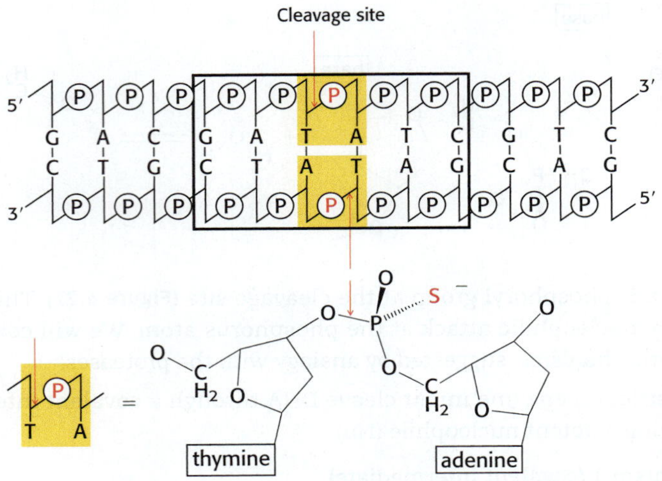

FIGURE 6.28 Phosphorothioate groups can be used to determine the overall stereochemical course of a displacement reaction. A phosphorothioate group has one of the nonbridging oxygen atoms replaced by a sulfur atom. Here, a phosphorothioate is placed at sites of cleavage by EcoRV endonuclease.

FIGURE 6.28 Phosphorothioate groups can be used to determine the overall stereochemical course of a displacement reaction. A phosphorothioate group has one of the nonbridging oxygen atoms replaced by a sulfur atom. Here, a phosphorothioate is placed at sites of cleavage by EcoRV endonuclease.

2. In the second type of mechanism, analogous to that used by the aspartyl- and metalloproteases, an activated water molecule attacks the phosphorus atom directly. In this mechanism, a single displacement reaction takes place at the phosphorus atom. Hence, the stereochemical configuration at the phosphorus atom is inverted after cleavage.

One approach to determine which mechanism is correct involves examining potential changes in the stereochemistry at the phosphorus atom over the course of cleavage. However, this stereochemical change is not easily observed, because two of the groups bound to the phosphorus atom are simply oxygen atoms, identical with each other. This challenge can be overcome by replacing one oxygen atom with sulfur (producing a species called a phosphorothioate).

Let us consider EcoRV endonuclease. This enzyme cleaves the phosphodiester bond between the T and the A at the center of the recognition sequence 5′-GATATC-3′. The first step is to synthesize an appropriate substrate for EcoRV containing phosphorothioates at the sites of cleavage (**Figure 6.28**). The reaction is then performed in water that has been greatly enriched in ^{18}O to allow the incoming oxygen atom to be tagged. The location of the ^{18}O label with respect to the sulfur atom indicates whether the reaction proceeds with inversion or retention of stereochemistry. This experiment reveals that the stereochemical configuration at the phosphorus atom is inverted only once with cleavage. This result is consistent with a direct attack by water at the phosphorus atom and rules out the formation of any covalently bound intermediate (**Figure 6.29**).

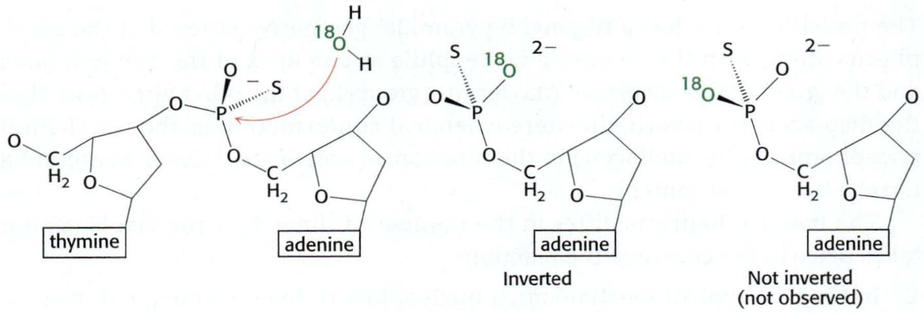

FIGURE 6.29 Cleavage of DNA by EcoRV endonuclease results in overall inversion of the stereochemical configuration at the phosphorus atom. This is indicated by the stereochemistry of the phosphorus atom bound to one bridging oxygen atom, one ^{16}O, one ^{18}O, and one sulfur atom. Two possible products are shown, only one of which is observed, indicating direct attack of water at the phosphorous atom.

Restriction enzymes require magnesium for catalytic activity

Many enzymes that act on phosphate-containing substrates require Mg^{2+} or some other similar divalent cation for activity. One or more Mg^{2+} (or similar) cations are essential to the function of restriction endonucleases. What are the functions of these metal ions?

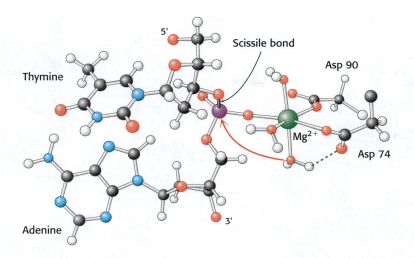

FIGURE 6.30 A magnesium ion forms a bridge between the enzyme and the DNA substrate. In this position, the ion helps activate a water molecule and position it so that it can attack the phosphorus atom at the cleavage site.

Direct visualization of the complex between EcoRV endonuclease and cognate DNA molecules in the presence of Mg^{2+} by crystallization has been challenging, because the enzyme cleaves the substrate under these circumstances. Nonetheless, metal ion complexes can be visualized through several approaches. In one approach, crystals of EcoRV endonuclease are prepared bound to oligonucleotides that contain the enzyme's recognition sequence. These crystals are grown in the absence of magnesium to prevent cleavage; after their preparation, the crystals are soaked in solutions containing the metal ion. Alternatively, crystals have been grown with the use of a mutated form of the enzyme that is less active. Finally, Mg^{2+} can be replaced by metal ions such as Ca^{2+} that bind but do not result in much catalytic activity. In all cases, no cleavage takes place, and so the locations of the metal ion-binding sites are readily determined.

As many as three metal ions per active site have been found. One ion-binding site is occupied in essentially all structures. This metal ion is coordinated to the protein through two aspartate residues and to one of the phosphate-group oxygen atoms at the site of cleavage. This metal ion binds the water molecule that attacks the phosphorus atom, helping to position and activate the water molecule in a manner similar to that for the Zn^{2+} ion of carbonic anhydrase (**Figure 6.30**).

The complete catalytic apparatus is assembled only within complexes of cognate DNA molecules, ensuring specificity

We now return to the question of cleavage site-specificity, the defining feature of restriction enzymes.

The recognition sequences for most restriction enzymes are inverted repeats, giving the three-dimensional structure of the recognition site a twofold rotational symmetry, meaning that the structure is the same if rotated by $360°/2 = 180°$ (**Figure 6.31**). The restriction enzymes display a corresponding symmetry: they are

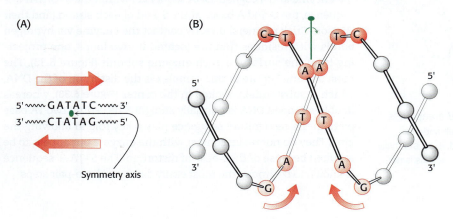

FIGURE 6.31 The recognition site for EcoRV displays twofold rotational symmetry. (A) The sequence of the recognition site, which is symmetric around the axis of rotation designated in green. (B) The inverted repeat within the recognition sequence of EcoRV (and most other restriction endonucleases) endows the DNA site with twofold rotational symmetry.

(A)

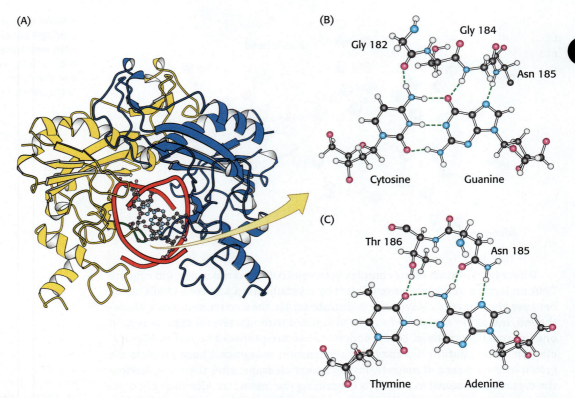

(B)

Gly 182

Gly 184

Asn 185

Cytosine   Guanine

(C)

Thr 186

Asn 185

Thymine   Adenine

INTERACT with this model in
≈ Achieve

FIGURE 6.32 The three-dimensional structure of the complex between EcoRV and a cognate DNA molecule reveals interactions responsible for the sequence specificity. (A) This view of the structure of EcoRV endonuclease bound to a cognate DNA fragment is down the helical axis of the DNA. The two protein subunits are in yellow and blue, and the DNA backbone is in red. One of the DNA-binding loops of EcoRV endonuclease is shown interacting with the base pairs of its cognate DNA-binding site. Key amino acid residues are shown hydrogen-bonding with (B) a CG base pair and (C) an AT base pair. [Drawn from 1RVB.pdb.]

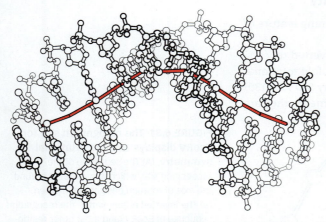

FIGURE 6.33 The cognate DNA from the EcoRV complex is substantially bent. The DNA is represented as a ball-and-stick model. The path of the DNA helical axis, shown in red, is substantially distorted on binding to the enzyme. For B form DNA, the axis is normally straight (not shown).

dimers whose two subunits are related by an appropriate 180° rotation. The matching symmetry of the recognition sequence and the enzyme facilitates the recognition of cognate DNA by the enzyme. This similarity in structure has been confirmed by the determination of the structure of the complex between EcoRV endonuclease and DNA fragments containing its recognition sequence (**Figure 6.32**). The enzyme surrounds the DNA in a tight embrace.

Surprisingly, binding studies performed in the absence of magnesium have demonstrated that the EcoRV endonuclease binds to all sequences, both cognate and noncognate, with approximately equal affinity. Why, then, does the enzyme cleave only cognate sequences? The answer lies in a unique set of interactions between the enzyme and a cognate DNA sequence. Within the 5'-GATATC-3' sequence, the G and A bases at the 5' end of each strand (and their Watson–Crick partners) directly contact the enzyme via hydrogen bonds with residues that are located in two loops, one projecting from the surface of each enzyme subunit (Figure 6.32). The most striking feature of this complex is the distortion of the DNA, which is substantially kinked in the center (**Figure 6.33**), whereas double-stranded DNA is normally straight. The central two TA base pairs in the recognition sequence play a key role in allowing the kink. They do not make contact with the enzyme but appear to be required because of their ease of distortion. The 5'-TA-3' sequence is known to be among the most easily deformed base-pair steps.

The structures of complexes formed with noncognate DNA fragments are strikingly different from those formed with cognate DNA; the noncognate DNA conformation is not substantially distorted (Figure 6.34). This lack of distortion has important consequences with regard to catalysis. No phosphate is positioned sufficiently close to the active-site aspartate residues to complete a magnesium ion-binding site (Figure 6.30). Hence, the nonspecific complexes do not bind the magnesium ions and the complete catalytic apparatus is never assembled. The distortion of the substrate and the subsequent binding of the magnesium ion account for the catalytic specificity of more than a million-fold that is observed for EcoRV endonuclease. Thus, enzyme specificity may be determined not by the specificity of substrate binding but rather by the specificity of forming a structure capable of DNA cleavage.

We can now see the role of binding energy in this strategy for attaining catalytic specificity. The distorted DNA makes additional contacts with the enzyme, increasing the binding energy, but the increase in binding energy is canceled by the energetic cost of distorting the DNA from its relaxed conformation into one that is catalytically competent, that is, into a conformation that allows DNA cleavage to proceed (Figure 6.35). Thus, for EcoRV endonuclease, there is little difference in binding affinity for cognate and nonspecific DNA fragments. However, the distortion in the cognate complex dramatically affects DNA hydrolysis by completing the magnesium ion-binding site. This example illustrates how enzymes can use available binding energy to deform substrates and position them for chemical transformation. Interactions that take place within the distorted substrate complex stabilize the transition state leading to DNA hydrolysis.

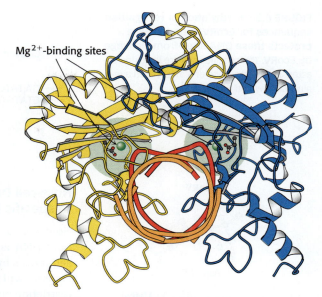

Mg^{2+}-binding sites

FIGURE 6.34 A comparison of the positions of the nonspecific and the cognate DNA within EcoRV shows the absence and presence of substantial DNA distortion. Note that, in the nonspecific complex (shown in orange compared to the specific complex in red), the DNA backbone is too far from the enzyme to complete the magnesium ion-binding sites required for DNA cleavage. [Drawn from 1RVB.pdb.]

INTERACT with this model in
♨ Achie√e

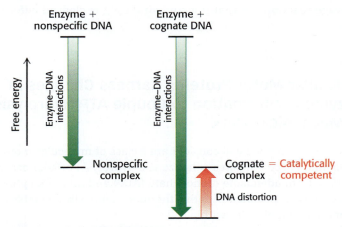

FIGURE 6.35 The binding energy drives DNA distortion. The additional interactions between EcoRV endonuclease and cognate DNA increase the binding energy, driving distortions that form a complex that is catalytically competent for DNA cleavage.

❖ SELF–CHECK QUESTION

The restriction enzyme EcoRI binds DNA fragments containing its recognition site very tightly and specifically; Changing even a single base results is a loss of approximately 1000-fold in binding affinity. How does this compare with EcoRV?

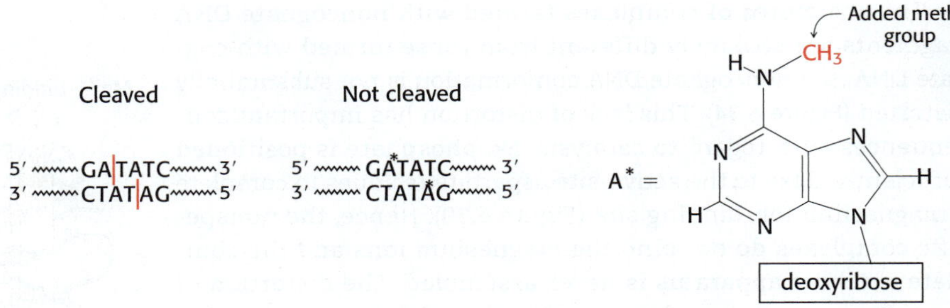

FIGURE 6.36 Methylation of recognition sequences for EcoRV endonuclease protects these DNA sites from cleavage by EcoRV. The specific sites are indicated by asterisks; the structure of a methylated adenine is shown on the right.

Cleaved   Not cleaved

5′ ⟳⟳ GA|TATC ⟳⟳ 3′   5′ ⟳⟳ GA*TATC ⟳⟳ 3′
3′ ⟳⟳ CTAT|AG ⟳⟳ 5′   3′ ⟳⟳ CTATĂG ⟳⟳ 5′

A* =

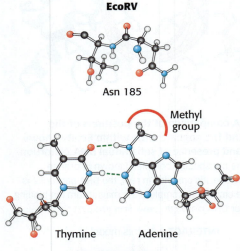

EcoRV

Asn 185

Methyl group

Thymine   Adenine

Methylated DNA

FIGURE 6.37 The methylation of adenine blocks the formation of hydrogen bonds between EcoRV endonuclease and cognate DNA molecules. This blockage prevents the hydrolysis of methylated DNA.

Host-cell DNA is protected by the addition of methyl groups to specific bases

How does a host cell harboring a restriction enzyme protect its own DNA? The host DNA is methylated on specific adenine bases within host recognition sequences by other enzymes called **DNA methylases** (**Figure 6.36**). A restriction enzyme will not cleave DNA if its recognition sequence is methylated. For each restriction enzyme, the host cell produces a corresponding methylase that marks the host DNA at the appropriate methylation site. These pairs of enzymes are referred to as **restriction-modification systems**.

The distortion of the DNA explains how methylation blocks catalysis and protects host-cell DNA. The host *E. coli* adds a methyl group to the amino group of the adenine nucleotide at the 5′ end of the recognition sequence for EcoRV endonuclease. The presence of the methyl group blocks the formation of a hydrogen bond between the amino group and the side-chain carbonyl group of asparagine 185 in the endonuclease (**Figure 6.37**). This asparagine residue is closely linked to the other amino acids that would form specific contacts with the DNA. The absence of the hydrogen bond disrupts other interactions between the enzyme and the DNA substrate, and the distortion necessary for cleavage will not take place. Thus, these added methyl groups prevent cleavage and protect the host DNA; the methylation reactions are slower than the DNA-cleavage reactions so that injected viral DNA is cleaved before it can be protected.

6.5 Molecular Motor Proteins Harness Changes in Enzyme Conformation to Couple ATP Hydrolysis to Mechanical Work

The final enzymes that we will consider are a class of molecular motor proteins called **myosins**. These enzymes catalyze the hydrolysis of adenosine triphosphate (ATP) to form adenosine diphosphate (ADP) and inorganic phosphate (P$_i$) and use the energy associated with this thermodynamically favorable reaction to drive the motion of molecules within cells.

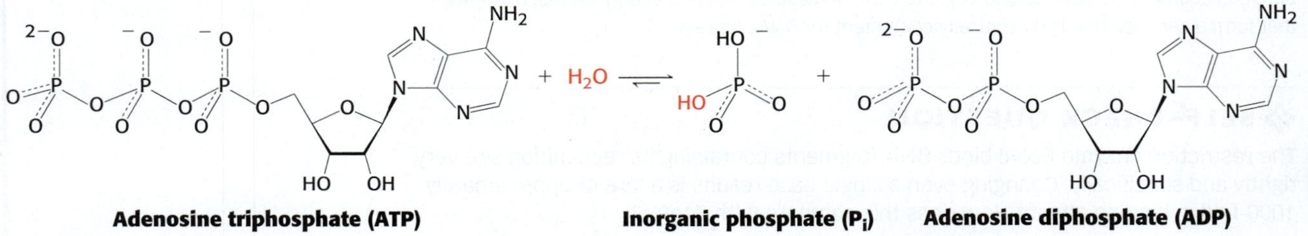

Adenosine triphosphate (ATP) + H$_2$O ⇌ **Inorganic phosphate (P$_i$)** + **Adenosine diphosphate (ADP)**

For example, when we lift a book, the energy required comes from ATP hydrolysis catalyzed by myosin in our muscles. Myosins are found in all eukaryotes, and the human genome encodes more than 40 different myosins, defined by large proteins termed myosin heavy chains. Myosins have elongated structures with globular domains that actually carry out ATP hydrolysis at one end, extended α-helical structures that promote dimer formation, and ancillary associate proteins termed light chains (**Figure 6.38**). We will focus on the globular **ATPase** domains, particularly the strategies that allow myosins to hydrolyze ATP in a controlled manner and to use the free energy associated with this reaction to promote substantial conformational changes within the myosin molecule. These conformational changes are amplified by other structures in the elongated myosin molecules to transport proteins or other cargo substantial distances within cells.

ATP is used as the major currency of energy inside cells. Many enzymes use ATP hydrolysis to drive other reactions and processes. In almost all cases, an enzyme that hydrolyzed ATP without coupling the reaction to other processes would simply drain the energy reserves of a cell without benefit.

ATP hydrolysis proceeds by the attack of water on the gamma phosphoryl group

In our examination of the mechanism of restriction enzymes, we learned that an activated water molecule performs a nucleophilic attack on phosphorus to cleave the phosphodiester backbone of DNA. The cleavage of ATP by myosins follows an analogous mechanism. To understand the myosin mechanism in more detail, we must first examine the structure of the myosin ATPase domain.

The structures of the ATPase domains of several different myosins have been examined. One such domain, that from the soil-living amoeba *Dictyostelium discoideum*, an organism that has been extremely useful for studying cell movement and molecular-motor proteins, has been studied in great detail. The crystal structure of this protein fragment in the absence of nucleotides reveals a single globular domain comprising approximately 750 amino acids. A water-filled pocket is present toward the center of the structure, suggesting a possible nucleotide-binding active site (**Figure 6.39**). However, when crystals of this protein were soaked in a solution containing ATP, the resulting structure reveals intact ATP bound in the proposed active site, but with very little change in the overall structure and without evidence of significant hydrolysis. The ATP is also bound to a Mg^{2+} ion.

Kinetic studies of myosins, as well as many other enzymes having ATP or other nucleoside triphosphates as a substrate, reveal that these enzymes are essentially inactive in the absence of divalent metal ions such as magnesium (Mg^{2+}) or manganese (Mn^{2+}) but acquire activity on the addition of these ions. In contrast with the enzymes discussed so far, the metal is not a component of the active site. Rather, nucleotides such as ATP bind these ions, and it is the metal ion–nucleotide complex that is the true substrate for the enzymes. The dissociation constant for the ATP–Mg^{2+} complex is approximately 0.1 mM, and thus, given that intracellular Mg^{2+} concentrations are typically in the millimolar range, essentially all nucleoside triphosphates (NTP) are present as NTP–Mg^{2+} complexes. Magnesium or manganese complexes of nucleoside triphosphates are the true substrates for essentially all NTP-dependent enzymes.

Since the ATP–Mg^{2+} complex was present in the crystallography study discussed above, why was there no evidence for hydrolysis? The nucleophilic attack

FIGURE 6.38 A schematic view of a myosin molecule reveals an elongated dimeric structure. The two heavy chains are shown in purple and yellow, and the light chains are shown in blue and orange.

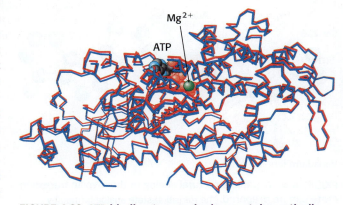

FIGURE 6.39 ATP binding to myosin does not dramatically alter its structure. An overlay of the structures of the ATPase domain from *Dictyostelium discoideum* myosin with nothing bound (blue) and the complex of this protein with ATP and magnesium bound (red) reveals that these two structures are quite similar to one another. [Drawn from 1FMV.pdb and 1FMW.pdb.]

INTERACT with this model in
Achieve

by a water molecule on the γ-phosphoryl group requires some mechanism to activate the water, such as a basic residue or a bound metal ion. Examination of the myosin–ATP complex structure shows no basic residue in an appropriate position and reveals that the bound Mg^{2+} ion is too far away from the phosphoryl group to play this role. These observations suggest why the ATP complex is relatively stable: The enzyme is not in a conformation that is suitable for promoting the ATP-hydrolysis reaction. This conclusion suggests that the domain must undergo a conformational change in order to carry out catalysis.

Formation of the transition state for ATP hydrolysis is associated with a substantial conformational change

The catalytically competent conformation of the myosin ATPase domain must bind and stabilize the transition state of the reaction. In analogy with restriction enzymes, we expect that ATP hydrolysis includes a pentacoordinate transition state.

Pentacoordinate
transition state

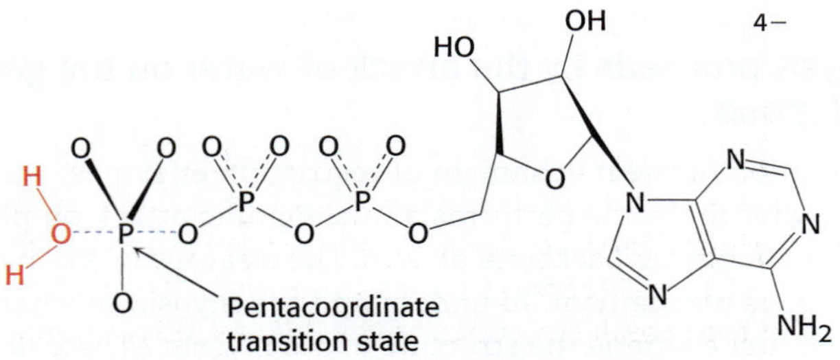

Such pentacoordinate structures based on phosphorus are too unstable to be readily observed. However, transition-state analogs in which other atoms replace phosphorus are more stable. The transition metal vanadium, in particular, forms similar structures. After crystallizing the myosin ATPase domain in the presence of ADP and vanadate, VO_4^{3-}, a complex forms that closely matches the expected transition-state structure (**Figure 6.40**).

As expected, the vanadium atom is coordinated to five oxygen atoms, including one oxygen atom from ADP diametrically opposite an oxygen atom that is analogous to the attacking water molecule in the transition state. The Mg^{2+} ion is coordinated to one oxygen atom from the vanadate, one oxygen atom from the ADP, two hydroxyl groups from the enzyme, and two water molecules. In this position, this ion does not appear to play any direct role in activating the attacking water. However, an additional residue from the enzyme, Ser 236, is well positioned to play a role in catalysis (**Figure 6.41**).

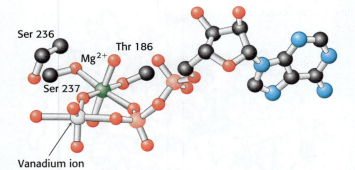

FIGURE 6.40 A stable structural analog of the myosin transition state can be prepared. The transition-state analog formed by treating the myosin ATPase domain with ADP and vanadate (VO_4^{3-}) in the presence of magnesium shows the vanadium ion (analogous to phosphorus) coordinated to five oxygen atoms including one from ADP. The positions of two residues that bind magnesium as well as Ser 236, a residue that appears to play a direct role in catalysis, are shown. [Drawn from 1VOM.pdb.]

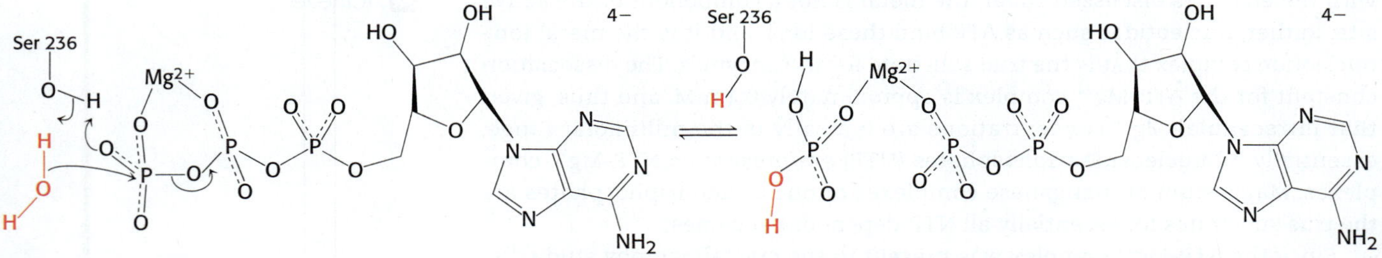

FIGURE 6.41 Water attack is facilitated by Ser 236. The attack of the water molecule on the γ-phosphoryl group of ATP is facilitated by Ser 236, which helps to deprotonate the water molecule. The serine sidechain, in turn, is deprotonated by one of the oxygen atoms of the γ-phosphoryl group forming the $H_2PO_4^-$ product.

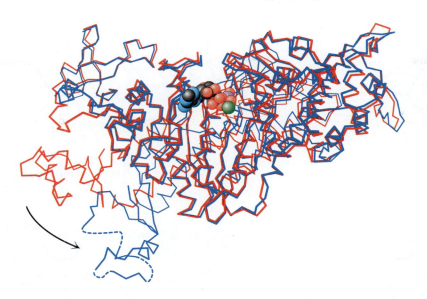

FIGURE 6.42 A substantial conformation change is associated with the formation of the myosin transition state. A comparison of the overall structures of the myosin ATPase domain with ATP bound (shown in red) and that with the transition-state analog ADP–vanadate (shown in blue) reveals dramatic conformational changes of a region at the carboxyl-terminus of the domain (see arrow), some parts of which move as much as 25 Å. [Drawn from 1FMW.pdb and 1VOM.pdb.]

In the proposed mechanism of ATP hydrolysis based on this structure, the water molecule attacks the γ-phosphoryl group, with the hydroxyl group of Ser 236 facilitating the transfer of a proton from the attacking water to the hydroxyl group of Ser 236, which, in turn, is deprotonated by one of the oxygen atoms of the γ-phosphoryl group. Thus, in effect, the ATP serves as a base to promote its own hydrolysis.

Comparison of the overall structures of the myosin ATPase domain complexed with ATP and with the ADP–vanadate reveals some remarkable differences. Relatively modest structural changes occur in and around the active site. In particular, a stretch of amino acids moves closer to the nucleotide by approximately 2 Å and interacts with the oxygen atom that corresponds to the attacking water molecule. These changes aid the hydrolysis reaction by stabilizing the transition state. However, examination of the overall structure shows even more striking changes.

A region comprising approximately 60 amino acids at the carboxyl-terminus of the domain adopts a different configuration in the ADP–vanadate complex, displaced by as much as 25 Å from its position in the ATP complex (**Figure 6.42**). This displacement tremendously amplifies the relatively subtle changes that take place in the active site, and it even affects locations beyond from the ATPase domain, because the carboxyl-terminus is connected to other atoms within the elongated structures typical of myosin molecules (Figure 6.38). Thus, the conformation that is capable of promoting the ATP hydrolysis reaction is substantially different from other conformations that are present in the course of the catalytic cycle.

The altered conformation of myosin persists for a substantial period of time

Myosins are slow enzymes, typically turning over approximately once per second. What steps limit the rate of turnover? In a particularly revealing experiment, the hydrolysis of ATP was catalyzed by the myosin ATPase domain from mammalian muscle. The reaction took place in water labeled with ^{18}O to track the incorporation of solvent oxygen into the reaction products; then the fraction of oxygen in the phosphate product was analyzed. In the simplest case, the phosphate would be expected to contain one oxygen atom derived from water and three initially present in the terminal phosphoryl group of ATP. Instead,

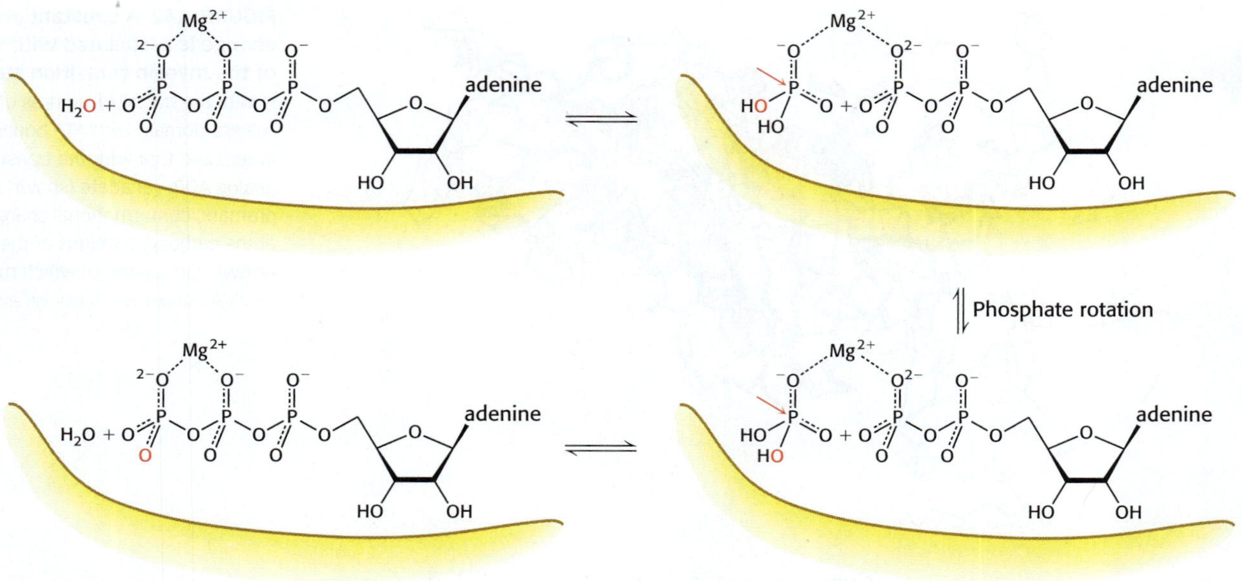

FIGURE 6.43 The hydrolysis of ATP is reversible within the active site of myosin. This is revealed by the fact that more than one atom of oxygen from water is incorporated in inorganic phosphate. The oxygen atoms are incorporated in cycles of hydrolysis of ATP to ADP and inorganic phosphate, phosphate rotation within the active site, and reformation of ATP now containing oxygen from water.

between two and three of the oxygen atoms in the phosphate were found, on average, to be derived from water. These observations indicate that the ATP hydrolysis reaction within the enzyme active site is reversible. Each molecule of ATP is cleaved to ADP and P_i and then re-formed from these products several times before the products are released from the enzyme (**Figure 6.43**).

At first glance, this observation is startling because ATP hydrolysis is a very favorable reaction with an equilibrium constant of approximately 140,000. However, this equilibrium constant applies to the molecules free in solution, not within the active site of an enzyme. Indeed, further analysis suggests that this equilibrium constant on the enzyme is approximately 10.

This result illustrates a general strategy used by enzymes: Enzymes catalyze reactions by stabilizing the transition state. The structure of this transition state is intermediate between the enzyme-bound reactants and the enzyme-bound products. Many of the interactions that stabilize the transition state will help equalize the stabilities of the reactants and the products. Thus, the equilibrium constant between enzyme-bound reactants and products usually 10 or below and is often close to 1, regardless of the equilibrium constant for the reactants and products free in solution.

These observations reveal that the hydrolysis of ATP to ADP and P_i is not the rate-limiting step for the reaction catalyzed by myosin. Instead, the release of the products, particularly P_i, from the enzyme is rate limiting. The fact that a conformation of myosin with ATP hydrolyzed but still bound to the enzyme persists for a significant period of time is critical for coupling conformational changes that take place in the course of the reaction to other processes.

❖ SELF–CHECK QUESTION

Myosin is treated with ATP in the presence of water labelled with $H_2{}^{18}O$. After a period of time, the remaining ATP is isolated and found to contain ^{18}O. Explain.

Actin forms filaments along which myosin can move

Myosin molecules use the free energy of hydrolysis of ATP to drive their own macroscopic movement along a filamentous protein termed *actin*, a 42-kDa protein that very abundant in eukaryotic cells, typically accounting for as much as 10% of the protein. Actin monomers come together to form actin filaments (**Figure 6.44**). Actin filaments have helical structures; each monomer is related to the preceding one by a translation of 27.5 Å and a rotation of 166 degrees around the helical axis.

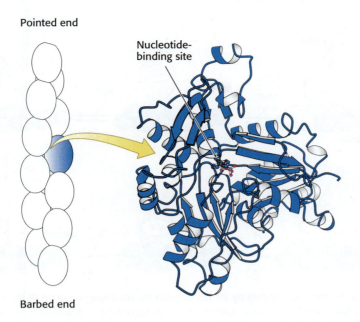

Pointed end

Nucleotide-
binding site

Barbed end

**FIGURE 6.44 An actin filament has a
polymeric structure.** In the schematic view
at left, actin monomers are depicted as ovals.
The detailed structure of one actin monomer
(blue) is shown at right. [Drawn from 1J6Z.pdb.]

INTERACT with this model in
Achieve

Because the rotation is nearly 180 degrees, F-actin resembles a two-stranded cable. Note that each actin monomer is oriented in the same direction along the actin filament, and so the structure is polar, with discernibly different ends. One end is called the barbed (plus) end, and the other is called the pointed (minus) end. Each actin monomer contains a bound nucleotide, ATP or ADP.

Using a variety of physical methods, scientists have been able to watch single myosin molecules moving along actin filaments. For example, a myosin family member termed myosin V can be labeled with fluorescent tags so that it can be localized when fixed on a surface with a precision of less than 15 Å. When this myosin in the absence of ATP is placed on a surface coated with actin filaments, each molecule remains in a fixed position. However, when ATP is added, each molecule moves along the surface. Tracking individual molecules reveals that each moves in steps of approximately 74 nm (**Figure 6.45**). The observation of steps of a fixed size as well as the determination of this step size helps reveal details of the mechanism of action of these tiny molecular machines.

How does ATP hydrolysis drive this motion? A key observation is that the addition of ATP to a complex of myosin and actin results in the dissociation of the complex. Thus, ATP binding and hydrolysis cannot be directly responsible for the power stroke. We can combine this fact with observations from

(A)

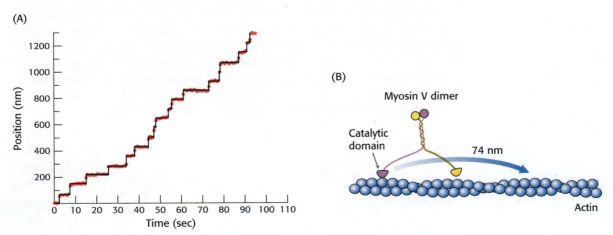

(B)

Myosin V dimer

Catalytic
domain

74 nm

Actin

FIGURE 6.45 A molecular motor such as myosin V moves in discrete steps. (A) A trace of the position of a single dimeric myosin V molecule as it moves across a surface coated with actin filaments. (B) A model of how the dimeric molecule moves in discrete steps with an average size of 74 ± 5 nm.

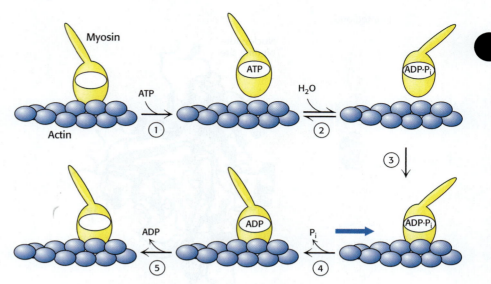

Myosin

ATP

H₂O

ADP·Pᵢ

Actin

① ATP

②

③

ADP

ADP

Pᵢ

ADP·Pᵢ

⑤

④

FIGURE 6.46 **The mechanism by which myosin moves along an actin filament involves five steps.** The binding of ATP (1) results in the release of myosin (yellow) from actin (blue). The reversible hydrolysis of ATP bound to myosin (2) can result in the reorientation of the lever arm. With ATP hydrolyzed but still bound to actin, myosin can bind actin (3). The release of P_i (4) results in the reorientation of the lever arm and the concomitant motion of actin relative to myosin. The release of ADP (5) completes the cycle.

three-dimensional structures of myosin in various forms to construct a mechanism for the motion of myosin along actin (**Figure 6.46**).

Let us begin with nucleotide-free myosin bound to actin. The binding of ATP to actin results in the dissociation of myosin from actin. With ATP bound and free of actin, the myosin domain can undergo the conformational change associated with the formation of the transition state for ATP hydrolysis. This conformational change results in the reorientation of the lever arm. In this form, the myosin head can dock onto the actin filament; phosphate is released with an accompanying motion of the lever arm. This conformational change represents the power stroke and moves the body of the myosin molecule relative to the actin filament. The release of ADP completes the cycle.

The most well-studied actin-myosin motor system is that from skeletal muscle (**Figure 6.47**). Muscle contains organized assemblies of actin molecules, referred to as thin filaments, together with multiheaded assemblies of myosin molecules that form thick filaments. Upon activation, the myosin motors move along the actin filaments to cause the combined assembly to contract.

Other cells contain elaborate meshworks of actin filaments that form an internal cellular skeleton or **cytoskeleton**, which plays a major role in determining cell shape. Two techniques for observing the cytoskeleton have traditionally been used, but each has shortcomings: Electron microscopy can produce clear images but cannot be performed on living cells, whereas light microscopy that visualizes fluorescent tags attached to actin can be performed on living cells, but the clarity of the images is limited by the relatively long wavelengths of visible light.

FIGURE 6.47 **A myosin assembly from muscle has a two-headed structure.** A schematic view shows how multiple myosin molecules come together to form the thick filament.

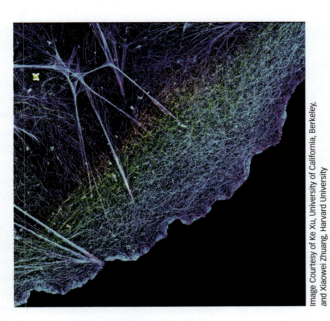

Image Courtesy of Ke Xu, University of California, Berkeley, and Xiaowei Zhuang, Harvard University

FIGURE 6.48 Actin networks can be imaged in great detail using modern fluorescence microscopy methods. A section of a live mammalian fibroblast cell with tagged actin clearly reveals individual actin filaments.

Recently, scientists have developed new microscopic methods based on the principle that the position of an isolated fluorescent molecule can be determined much more precisely than the image produced by monitoring the fluorescence. These methods have produced beautiful pictures of the intricate tracks inside living cells along which molecular motors can move (Figure 6.48). With these new tools, scientists can now ask new questions about the elaborate movements that occur within living cells driven by molecular motors.

Summary

6.1 Enzymes Use a Core Set of Catalytic Strategies

- Enzymes adopt conformations that are structurally and chemically complementary to the transition states of the reactions that they catalyze.
- Enzymes use five basic strategies to form and stabilize the transition state: the use of binding energy to promote both specificity and catalysis, covalent catalysis, general acid–base catalysis, catalysis by approximation, and metal ion catalysis.

6.2 Proteases Facilitate a Fundamentally Difficult Reaction

- The cleavage of peptide bonds by chymotrypsin is initiated by the attack by a serine residue on the peptide carbonyl group with the serine activated by a catalytic triad of Ser-His-Asp.
- The product of this initial reaction is a covalent intermediate formed by the enzyme and an acyl group derived from the bound substrate that is subsequently hydrolyzed.
- Tetrahedral intermediates that occur during these reactions have a negative charge on the peptide carbonyl oxygen atom that is stabilized by interactions with peptide NH groups in a region on the enzyme termed the oxyanion hole.

- Some other proteases employ the same catalytic strategy based on a similar catalytic triad, including homologs of chymotrypsin such as trypsin and elastase and other proteins such as subtilisin that evolved the same catalytic apparatus independently.

6.3 Carbonic Anhydrases Make a Fast Reaction Faster

- Carbonic anhydrases catalyze the reaction of water with carbon dioxide at rates as high as 1 million times per second to generate bicarbonate ion which can be protonated to form carbonic acid.
- A tightly bound zinc ion binds a water molecule, promotes its deprotonation to generate a hydroxide ion at neutral pH, and uses this nucleophile to attack carbon dioxide.
- To overcome limitations imposed by the rate of proton transfer from the zinc-bound water, the most-active carbonic anhydrases have evolved a proton shuttle to transfer protons to buffer components in solution.

6.4 Restriction Enzymes Catalyze Highly Specific DNA-Cleavage Reactions

- Restriction enzymes that cleave DNA at specific recognition sequences discriminate between molecules that contain these recognition sequences and those that do not.
- Restriction enzymes use magnesium ions to bind and activate a water molecule to attacks the phosphodiester backbone and facilitate DNA cleavage.
- EcoRV distorts DNA molecules containing the proper sequence are in a manner that allows magnesium ion binding and, hence, DNA cleavage.
- Restriction enzymes are prevented from acting on the DNA of a host cell by the methylation of key sites within its recognition sequences, blocking specific interactions between the enzymes and the DNA.

6.5 Molecular Motor Proteins Harness Changes in Enzyme Conformation to Couple ATP Hydrolysis to Mechanical Work

- Molecular motors called myosins catalyze the hydrolysis of adenosine triphosphate, using the free energy generated to drive large-scale molecular movement.
- Myosin associated with the transition state for ATP hydrolysis has a distinct structure indicating a large-scale conformational change.
- The rate of ATP hydrolysis by myosin is relatively low and is limited by the rate of product release from the enzyme.
- Myosins move along a filamentous protein called actin, binding and releasing actin filaments over the course of its catalytic cycle to facilitate movement.
- Actin filaments occur in muscle and also form a cytoskeleton inside eukaryotic cells along which myosins can move; the cytoskeleton also functions like scaffolding, giving the cell its shape.

Key Terms

binding energy (p. 180)
induced fit (p. 180)
covalent catalysis (p. 180)
general acid–base catalysis (p. 180)
catalysis by approximation (p. 180)
metal ion catalysis (p. 180)
protease (p. 181)
electrophilic (p. 181)
nucleophilic (p. 181)

chemical modification reaction (p. 182)
chromogenic substrate (p. 182)
catalytic triad (p. 183)
oxyanion hole (p. 185)
scissile bond (p. 185)
protease inhibitor (p. 189)
carbonic anhydrase (p. 190)
ligand (p. 191)

proton shuttle (p. 193)
restriction enzyme (restriction endonuclease) (p. 194)
recognition sequence (p. 194)
DNA methylase (p. 200)
restriction-modification system (p. 200)
myosin (p. 200)
ATPase (p. 201)
cytoskeleton (p. 206)

Problems

1. Examination of the cleavage of the amide substrate, A, by chymotrypsin with the use of stopped-flow kinetic methods reveals no burst. The reaction is monitored by noting the color produced by the release of the amino part of the substrate (highlighted in orange). Why is no burst observed? ❖ 1, ❖ 6

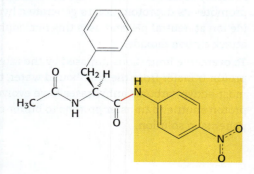

2. Consider the subtilisin substrates A and B.

Phe-Ala-Gln-Phe-X   Phe-Ala-His-Phe-X

A   B

These substrates are cleaved (between Phe and X) by native subtilisin at essentially the same rate. However, the His 64-to-Ala mutant of subtilisin cleaves substrate B more than 1000-fold as rapidly as it cleaves substrate A. Propose an explanation. ❖ 2

3. Consider the following argument. In subtilisin, mutation of Ser 221 to Ala results in a 10^6-fold decrease in activity. Mutation of His 64 to Ala results in a similar 10^6-fold decrease. Therefore, simultaneous mutation of Ser 221 to Ala and His 64 to Ala should result in a $10^6 \times 10^6 = 10^{12}$-fold reduction in activity. Is this correct? Why or why not?

4. In chymotrypsin, a mutant was constructed with Ser 189, which is in the bottom of the substrate-specificity pocket, changed to Asp. What effect would you predict for this mutation?

5. In carbonic anhydrase II, mutation of the proton-shuttle residue His 64 to Ala was expected to result in a decrease in the maximal catalytic rate. However, in buffers such as imidazole with relatively small molecular components, no rate reduction was observed. In buffers with larger molecular components, significant rate reductions were observed. Propose an explanation.

6. Restriction endonucleases are, in general, quite slow enzymes with typical turnover numbers of $1 \, s^{-1}$. Suppose that endonucleases were faster, with turnover numbers similar to those for carbonic anhydrase such that they act faster than do methylases. Would this increased rate be beneficial to host cells, assuming that the fast enzymes have similar levels of specificity?

7. Treatment of carbonic anhydrase with high concentrations of the metal chelator EDTA (ethylenediaminetetraacetic acid) results in the loss of enzyme activity. Propose an explanation. ❖ **2,** ❖ **3**

8. Elastase is specifically inhibited by an aldehyde derivative of one of its substrates:

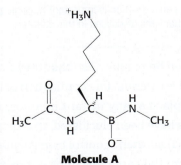

(a) Which residue in the active site of elastase is most likely to form a covalent bond with this aldehyde?

(b) What type of covalent link would be formed?

9. Use the provided structure of molecule A to complete the passage.

Molecule A contains a side chain similar to that of _____. Molecule A effectively inhibits _____ by binding to the _____ residue in the enzyme's S_1 pocket. ❖ **2**

Molecule A

10. At pH 7.0, carbonic anhydrase exhibits a k_{cat} of $600,000 \, s^{-1}$. Estimate the value expected for k_{cat} at pH 6.0.

11. To terminate a reaction in which a restriction enzyme cleaves DNA, researchers often add high concentrations of the metal chelator EDTA (ethylenediaminetetraacetic acid). Why does the addition of EDTA terminate the reaction?

12. Many patients become resistant to HIV protease inhibitors with the passage of time, owing to mutations in the HIV gene that encodes the protease. Mutations are not found in the aspartate residue characteristic of aspartyl proteases. Why not? ❖ **1,** ❖ **2**

13. Serine 236 in *Dictyostelium discoideum* myosin has been mutated to alanine. The mutated protein showed modestly reduced ATPase activity. Analysis of the crystal structure of the mutated protein revealed that a water molecule occupied the position of the hydroxyl group of the serine residue in the wild-type protein. Propose a mechanism for the ATPase activity of the mutated enzyme. ❖ **1,** ❖ **2,** ❖ **6**

14. The catalytic power of an enzyme can be defined as the ratio of the rate of the enzyme catalyzed reaction to that for the uncatalyzed reaction. Using the information in Figure 6.15 for subtilisin and in Figure 6.19 for carbonic anhydrase, calculate the catalytic powers for these two enzymes.

15. Would a reaction catalyzed by a version of subtilisin with all three residues in the catalytic triad mutated have less or more activity compared to the uncatalyzed reaction (see Figure 6.15)? Propose an explanation for any remaining catalytic power.

16. On the basis of the information provided in Figure 6.16, complete the mechanisms for peptide-bond cleavage by **(a)** a cysteine protease, **(b)** an aspartyl protease, and **(c)** a metalloprotease. ❖ **1,** ❖ **2**

17. Most enzymes are quite specific, catalyzing a particular reaction on a set of substrates that are structurally quite similar to one another. Discuss why this is advantageous from a biological perspective. Suggest why this is likely to be true from a chemical perspective. Why would it be difficult to evolve an enzyme with high catalytic activity but low specificity? ❖ **4**

18. In the absence of actin, myosin will hydrolyze ATP, but very slowly. Based on the mechanism shown in Figure 6.46, propose an explanation for this slow rate.

Enzyme Regulatory Strategies

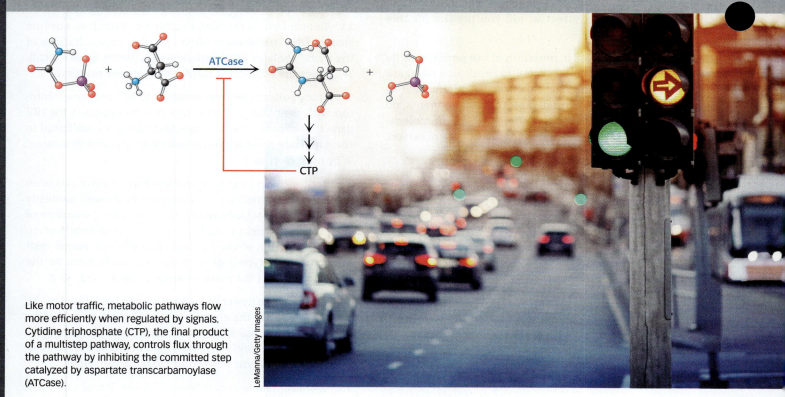

Like motor traffic, metabolic pathways flow more efficiently when regulated by signals. Cytidine triphosphate (CTP), the final product of a multistep pathway, controls flux through the pathway by inhibiting the committed step catalyzed by aspartate transcarbamoylase (ATCase).

LeManna/Getty Images

OUTLINE

7.1 Allosteric Regulation Enables Control of Metabolic Pathways

7.2 Isozymes Provide a Means of Regulation Specific to Distinct Tissues and Developmental Stages

7.3 Covalent Modification Is a Means of Regulating Enzyme Activity

7.4 Many Enzymes Are Activated by Specific Proteolytic Cleavage

7.5 Enzymatic Cascades Allow Rapid Responses Such as Blood Clotting

◈ LEARNING GOALS

By the end of this chapter, you should be able to:

1. Explain how the kinetic properties of allosteric enzymes differ from Michaelis–Menten enzymes.

2. Describe the role of feedback inhibition in the regulation of enzymes.

3. Differentiate between homotropic and heterotropic effects and explain how they modify the equilibrium between the T and R forms of an allosteric enzyme.

4. Explain the role of isozymes in the regulation of metabolism.

5. Describe the activity of protein kinases and protein phosphatases.

6. Explain why phosphorylation is an effective means of regulating the activities of target proteins.

7. Explain why the mechanism of zymogen activation is important physiologically and how it differs from the other control mechanisms discussed in this chapter.

The activity of enzymes must often be regulated so that they function at the proper time and place. This control is essential for coordination of the vast array of biochemical processes taking place at any instant in an organism. While numerous regulatory mechanisms have been elucidated, this chapter explores four commonly utilized methods: (1) allosteric control by regulatory molecules, (2) alternate enzymes that catalyze the same reactions but exhibit different catalytic and regulatory properties, (3) reversible covalent modifications, and (4) irreversible proteolytic cleavage to yield an active enzyme product. Another method, adjusting the amount of enzyme present, usually takes place at the level of gene transcription and will be discussed in Chapter 31.

7.1 Allosteric Regulation Enables Control of Metabolic Pathways

As we shall discuss in Section 8.1, pyrimidines and purines are the aromatic groups that compose the unique bases found in DNA and RNA. Aspartate transcarbamoylase (ATCase) catalyzes the first step in the biosynthesis of pyrimidines: the condensation of aspartate and carbamoyl phosphate to form N-carbamoylaspartate and orthophosphate (**Figure 7.1**). This reaction is the **committed step** in the pathway, which consists of 10 reactions that will ultimately yield the pyrimidine nucleotides uridine triphosphate (UTP) and cytidine triphosphate (CTP). The committed step means that the products of the reaction are committed to ultimate synthesis of the end products of the pathway, in this case UTP and CTP. The committed step, in any pathway, is irreversible under cellular conditions and is the reaction catalyzed by allosteric enzymes. How is ATCase regulated to generate precisely the amount of pyrimidines needed by the cell?

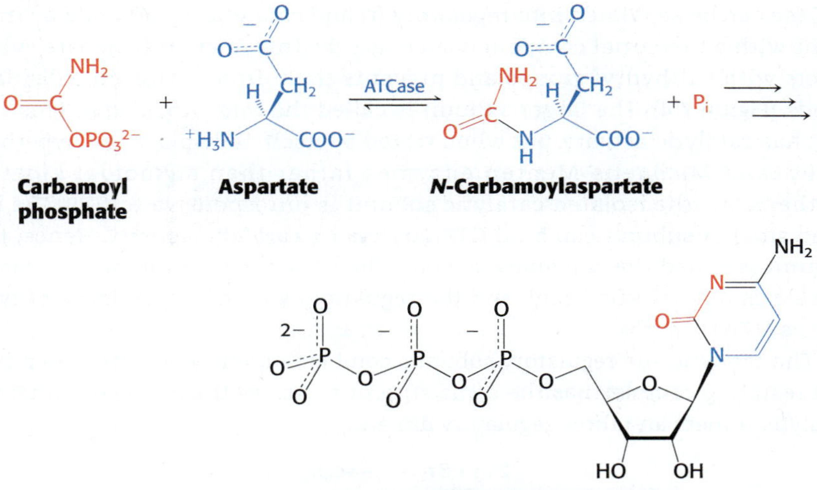

Carbamoyl phosphate   **Aspartate**   **N-Carbamoylaspartate**

Cytidine triphosphate (CTP)

FIGURE 7.1 Aspartate transcarbamoylase (ATCase) catalyzes the committed step in pyrimidine synthesis. The reaction is the condensation of aspartate and carbamoyl phosphate to form N-carbamoylaspartate.

ATCase is inhibited by CTP, the final product of the ATCase-initiated pathway. The rate of the reaction catalyzed by ATCase is fast at low concentrations of CTP but slows as CTP concentration increases (**Figure 7.2**). Thus, the pathway continues to make new pyrimidines until sufficient quantities of CTP have accumulated. The inhibition of ATCase by CTP is an example of **feedback inhibition**, the inhibition of an enzyme by the end product of the pathway. Feedback inhibition by CTP ensures that N-carbamoylaspartate and subsequent intermediates in the pathway are not needlessly formed when pyrimidines are abundant.

The inhibitory ability of CTP is remarkable because CTP is structurally quite different from the substrates of the reaction (Figure 7.1). Thus, CTP must bind to a site distinct from the active site at which substrate binds. Such sites are called **allosteric** (or **regulatory**) **sites**. CTP is an example of an allosteric inhibitor. In ATCase (but not all allosterically regulated enzymes), the catalytic sites and the regulatory sites are on separate polypeptide chains.

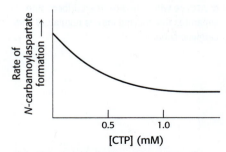

FIGURE 7.2 CTP inhibits ATCase. Cytidine triphosphate, an end product of the pyrimidine-synthesis pathway, inhibits aspartate transcarbamoylase despite having little structural similarity to reactants or products.

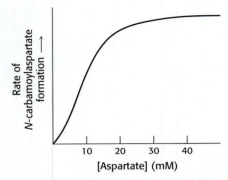

FIGURE 7.3 ATCase displays sigmoidal kinetics. A plot of product formation as a function of substrate concentration produces a sigmoidal curve because the binding of substrate to one active site increases the activity at the other active sites. Thus, the enzyme shows cooperativity.

FIGURE 7.4 *p*-Hydroxymercuribenzoate reacts with crucial cysteine residues in aspartate transcarbamoylase. Treatment of ATCase with *p*-hydroxymercuribenzoate separates the enzyme into its regulatory and catalytic subunits.

Many allosterically regulated enzymes do not follow Michaelis–Menten kinetics

Allosteric enzymes are distinguished by their response to changes in substrate concentration in addition to their susceptibility to regulation by other molecules. Let us examine the rate of product formation as a function of substrate concentration for ATCase (**Figure 7.3**). The curve differs from that expected for an enzyme that follows Michaelis–Menten kinetics. The observed curve is referred to as sigmoidal because it resembles the letter "S." The vast majority of allosteric enzymes display sigmoidal kinetics. Recall from the discussion of hemoglobin (Chapter 3) that sigmoidal curves result from cooperation between subunits: the binding of substrate to one active site in a molecule increases the likelihood that substrate will bind to other active sites. To understand the basis of sigmoidal enzyme kinetics and inhibition by CTP, we need to examine the structure of ATCase.

ATCase consists of separable catalytic and regulatory subunits

What is the evidence that ATCase has distinct regulatory and catalytic sites? ATCase can be separated into regulatory (r) and catalytic (c) subunits by treatment with a mercurial compound such as *p*-hydroxymercuribenzoate, which reacts with sulfhydryl groups and prevents them from re-forming disulfide bonds (**Figure 7.4**). The larger subunit is called the *catalytic subunit*. This subunit has catalytic activity, but when tested by itself, it displays the hyperbolic kinetics of Michaelis–Menten enzymes rather than sigmoidal kinetics. Furthermore, the isolated catalytic subunit is unresponsive to CTP. The isolated smaller subunit can bind CTP but has no catalytic activity. Hence, that subunit is called the *regulatory subunit*. The catalytic subunit (c_3) consists of three c chains (34 kDa each), and the regulatory subunit (r_2) consists of two r chains (17 kDa each).

The catalytic and regulatory subunits combine rapidly when they are mixed. The resulting complex has the same structure, c_6r_6, as the native enzyme: two catalytic trimers and three regulatory dimers.

$$2\,c_3 + 3\,r_2 \longrightarrow c_6r_6$$

Most strikingly, the reconstituted enzyme has the same allosteric and kinetic properties as those of the native enzyme. Thus, ATCase is composed of discrete catalytic and regulatory subunits, and the interaction of the subunits in the native enzyme produces its regulatory and catalytic properties. The fact that the enzyme can be separated into isolated catalytic and regulatory subunits, which can be reconstituted back to the functional enzyme, allows for a variety of experiments to characterize the allosteric properties of the enzyme.

Allosteric interactions in ATCase are mediated by large changes in quaternary structure

What are the subunit interactions that account for the properties of ATCase? Significant clues have been provided by the three-dimensional structure of various forms of ATCase (**Figure 7.5**). Two catalytic trimers are stacked one on top of the other, linked by three dimers of the regulatory chains. There are significant contacts between the catalytic and the regulatory subunits: each r chain within a regulatory dimer interacts with a c chain within a catalytic trimer. The c chain makes contact with a structural domain in the r chain that is stabilized by a zinc ion bound to four cysteine residues. The zinc ion is critical for the interaction of the r chain with the c chain. *p*-Hydroxymercuribenzoate is

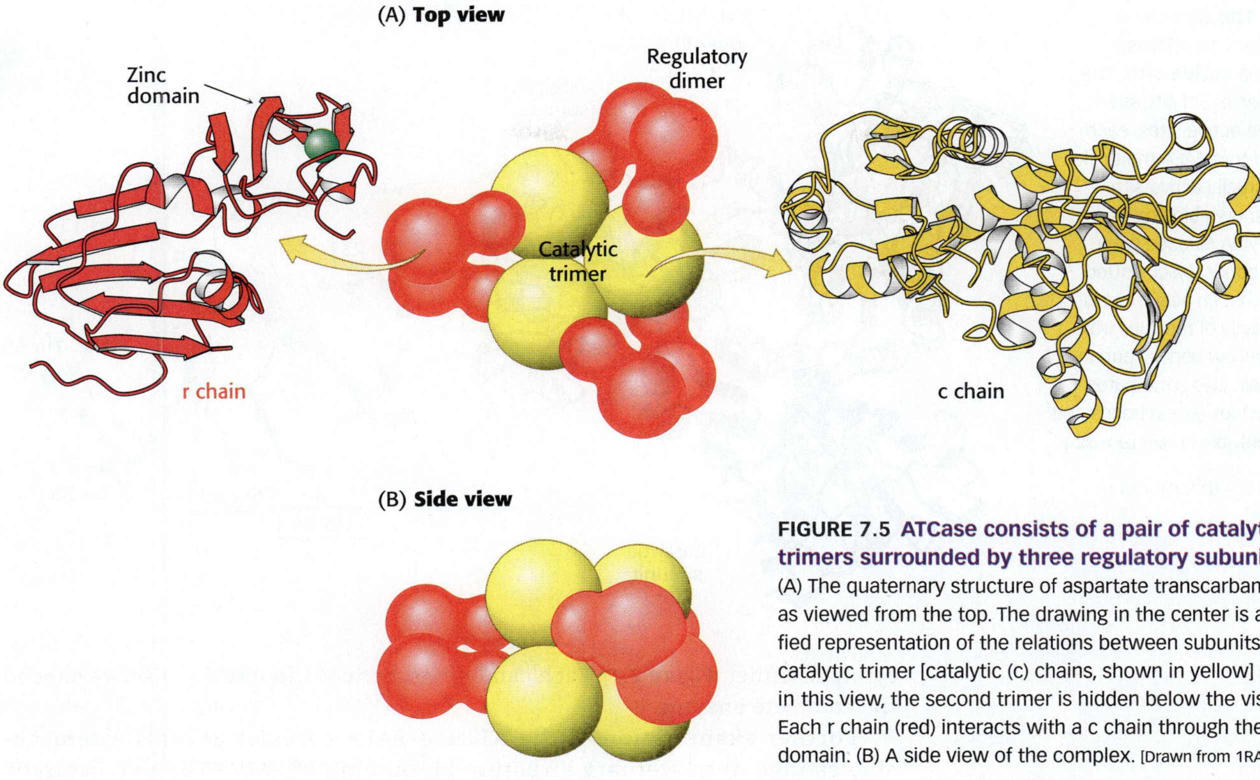

(A) **Top view**

Zinc
domain

Regulatory
dimer

Catalytic
trimer

r chain

c chain

(B) **Side view**

FIGURE 7.5 ATCase consists of a pair of catalytic subunit trimers surrounded by three regulatory subunit dimers.
(A) The quaternary structure of aspartate transcarbamoylase as viewed from the top. The drawing in the center is a simplified representation of the relations between subunits. A single catalytic trimer [catalytic (c) chains, shown in yellow] is visible; in this view, the second trimer is hidden below the visible one. Each r chain (red) interacts with a c chain through the zinc domain. (B) A side view of the complex. [Drawn from 1RAI.pdb.]

INTERACT with this model in
≋ Achieve

able to dissociate the catalytic and regulatory subunits because mercury binds strongly to the cysteine residues, displacing the zinc and preventing interaction with the c chain.

To locate the active sites, the enzyme has been crystallized in the presence of *N*-(phosphonacetyl)-L-aspartate (PALA), a *bisubstrate analog* (an analog of the two substrates) that resembles an intermediate along the pathway of catalysis (**Figure 7.6**). PALA is a potent competitive inhibitor of ATCase that binds to and blocks the active sites. The structure of the ATCase–PALA complex reveals that PALA binds at sites lying at the boundaries between pairs of c chains within a

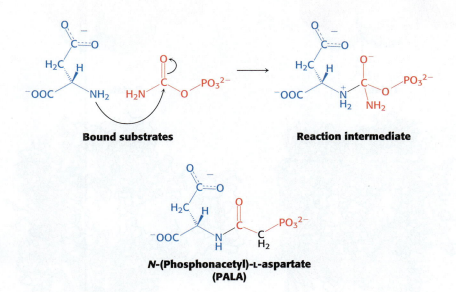

Bound substrates   **Reaction intermediate**

N-**(Phosphonacetyl)-L-aspartate
(PALA)**

FIGURE 7.6 PALA is a bisubstrate analog representing an intermediate along the ATCase catalytic pathway.
(Top) Nucleophilic attack by the amino group of aspartate on the carbonyl carbon atom of carbamoyl phosphate generates an intermediate on the pathway to the formation of *N*-carbamoylaspartate. (Bottom) *N*-(Phosphonacetyl)-L-aspartate (PALA) is an analog of the reaction intermediate and a potent competitive inhibitor of aspartate transcarbamoylase.

FIGURE 7.7 The structure of PALA bound to ATCase identifies the active site. The catalytic trimer, c_3, of ATCase contains three active sites, each shown bound to a PALA molecule. Some of the crucial active-site residues are shown binding to the inhibitor PALA (shaded gray) via hydrogen bonds (black dotted lines). Notice that an active site is composed mainly of residues from one c chain (yellow bonds), but an adjacent c chain also contributes important residues (green bonds and boxed in green). [Drawn from 1EKX.pdb.]

INTERACT with this model in
≈ Achieve

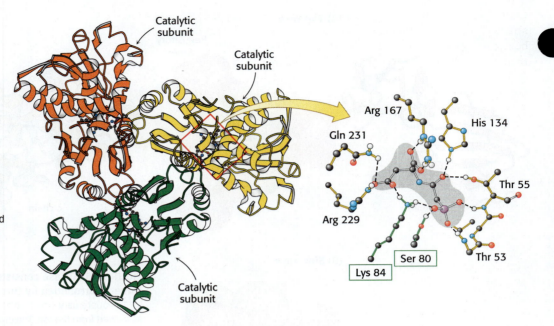

catalytic trimer (**Figure 7.7**). Each catalytic trimer contributes three active sites to the complete enzyme.

Further examination of the ATCase–PALA complex reveals a remarkable change in quaternary structure on binding of PALA. The two catalytic trimers move 12 Å farther apart and rotate approximately 10 degrees about their common threefold axis of symmetry. Moreover, the regulatory dimers rotate approximately 15 degrees to accommodate this motion (**Figure 7.8**). The enzyme literally expands on PALA binding. In essence, ATCase has two distinct

FIGURE 7.8 ATCase exists in two conformations: the tense (T) state and the relaxed (R) state. The structure of ATCase changes dramatically in the transition from the compact, relatively inactive T state to the expanded, active R state. PALA (blue) binding stabilizes the R state. [Drawn from 6AT1.pdb and 8ATC.pdb.]

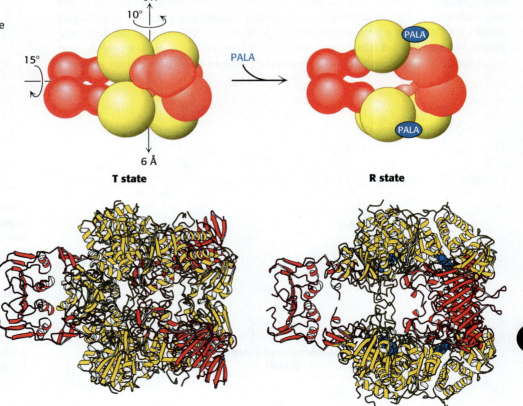

quaternary forms: one that predominates in the absence of substrate or substrate analogs, and another that predominates when substrates or analogs are bound. We call these forms the T (for tense) state and the R (for relaxed) state, respectively, as we did for the two quaternary states of hemoglobin.

How can we explain the enzyme's sigmoidal kinetics in light of the structural observations? Like hemoglobin, the enzyme exists in an equilibrium between the T state and the R state.

$$T \rightleftharpoons R$$

In the absence of substrate, almost all the enzyme molecules are in the T state because the T state is energetically more stable than the R state. The ratio of the concentration of enzyme in the T state to that in the R state is called the *allosteric coefficient* (L). For most allosteric enzymes, L is on the order of 10^2 to 10^3.

$$L = \frac{T}{R}$$

The T state has a low affinity for substrate and hence exhibits low catalytic activity. The occasional binding of a substrate molecule to one active site in an enzyme increases the likelihood that the entire enzyme shifts to the R state with its higher binding affinity for substrate.

The addition of more substrate has two effects. First, it increases the probability that each enzyme molecule will bind at least one substrate molecule. Second, it increases the average number of substrate molecules bound to each enzyme. The presence of additional substrate will increase the fraction of enzyme molecules in the more active R state because the position of the equilibrium depends on the number of active sites that are occupied by substrate. We considered this property, called *cooperativity* because the subunits cooperate with one another, when we discussed the sigmoidal oxygen-binding curve of hemoglobin in Chapter 3. The cooperative effects of substrates on allosteric enzymes are also referred to as **homotropic effects** (from the Greek *homós*, "same"), as they reflect the effects of the binding in one site on the affinity at the comparable site elsewhere in the enzyme. In the case of substrate cooperativity, these homotropic effects act on the enzyme active sites.

Recall that in Chapter 3, we discussed two limiting models for cooperativity: the *concerted* and the *sequential* models (Section 3.3). The mechanism for allosteric regulation of ATCase presented here is best described by the concerted model, as the change in the enzyme is "all or none"; the entire enzyme is converted from T into R, affecting all of the catalytic sites equally. In contrast, the sequential model assumes that the binding of ligand to one site on the complex increases the affinity of neighboring sites without causing a full T-to-R transition of all the sites. Although the concerted model explains the behavior of ATCase well, most other allosteric enzymes exhibit features of both models.

The sigmoidal curve for ATCase can be pictured as a composite of two Michaelis–Menten curves, one corresponding to the less-active T state and the other to the more-active R state. At low concentrations of substrate, the curve closely resembles that of the T state enzyme. As the substrate concentration is increased, the curve progressively shifts to resemble that of the R state enzyme (**Figure 7.9**).

What is the biochemical advantage of sigmoidal kinetics? Allosteric enzymes transition from a less active state to a more active state within a narrow range of substrate concentration. The benefit of this behavior is illustrated in **Figure 7.10**, which compares the kinetics of a Michaelis–Menten enzyme (blue curve) to that of an allosteric enzyme (red curve). In this example, the Michaelis–Menten enzyme requires an approximately 27-fold increase in substrate concentration to increase V_o from $0.1\ V_{max}$ to $0.8\ V_{max}$. In contrast, the allosteric enzyme requires only about a fourfold increase in substrate concentration to attain the same increase in velocity. Simply put, the activity of allosteric enzymes is more

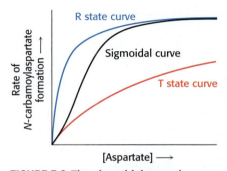

FIGURE 7.9 The sigmoidal curve in cooperative enzymes can be understood as the combination of two Michaelis–Menten enzymes. The first enzyme corresponds to the T state, with a high K_M value, while the second enzyme corresponds to the R state, with a low K_M value. As the concentration of substrate is increased, the equilibrium shifts from the T state to the R state, which results in a steep rise in activity with respect to substrate concentration.

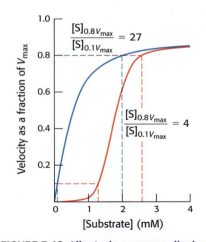

FIGURE 7.10 Allosteric enzymes display threshold effects. As the T-to-R transition occurs, the velocity increases over a narrower range of substrate concentration for an allosteric enzyme (red curve) than for a Michaelis–Menten enzyme (blue curve).

sensitive to changes in substrate concentration near K_M than are Michaelis–Menten enzymes with the same V_{max}. This sensitivity is called a *threshold effect*: below a certain substrate concentration, there is little enzyme activity. However, after the threshold has been reached, enzyme activity increases rapidly. In other words, much like an "on/off" switch, cooperativity ensures that most of the enzyme is either on (R state) or off (T state). The vast majority of allosteric enzymes display sigmoidal kinetics.

❖ SELF–CHECK QUESTION

What would be the effect of a mutation in an allosteric enzyme that resulted in an allosteric coefficient of 0?

Allosteric regulators modulate the T-to-R equilibrium

We now turn our attention to the effects of pyrimidine nucleotides on ATCase activity. As noted earlier, CTP inhibits the action of ATCase. X-ray studies of ATCase in the presence of CTP reveal (1) that the enzyme is in the T state when bound to CTP and (2) that a binding site for this nucleotide exists in each regulatory chain in a domain that does not interact with the catalytic subunit (**Figure 7.11**). Each active site is more than 50 Å from the nearest CTP-binding site. The question naturally arises: how can CTP inhibit the catalytic activity of the enzyme when it does not interact with the catalytic chain?

The quaternary structural changes observed on substrate-analog binding suggest a mechanism for inhibition by CTP (**Figure 7.12**). The binding of the inhibitor CTP to the T state shifts the T-to-R equilibrium in favor of the T state, decreasing net enzyme activity. CTP increases the allosteric coefficient from 200 in its absence to 1250 when all of the regulatory sites are occupied by CTP. Because the binding of CTP makes it more difficult for substrate binding to convert the enzyme into the R state, CTP increases the initial phase of the sigmoidal

FIGURE 7.11 CTP stabilizes the T state. The binding of CTP (green) to the regulatory subunit of ATCase stabilizes the T state. [Drawn from 6AT1.pdb and 5AT1.pdb.]

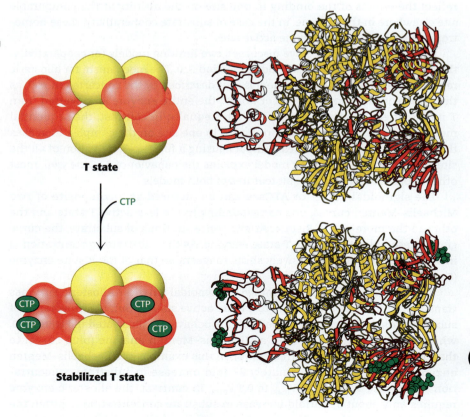

T state

CTP

Stabilized T state

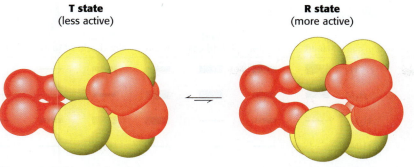

T state
(less active)

R state
(more active)

Favored by CTP binding　　　　Favored by substrate binding

FIGURE 7.12 The R state and the T state are in equilibrium. Even in the absence of any substrate or regulators, ATCase exists in equilibrium between the R and the T states. Under these conditions, the T state is favored by a factor of approximately 200.

curve (**Figure 7.13**). More substrate is required to attain a given reaction rate. UTP, the immediate precursor to CTP, also regulates ATCase. While unable to inhibit the enzyme alone, UTP synergistically inhibits ATCase in the presence of CTP.

Interestingly, ATP, too, is an allosteric effector of ATCase, binding to the same site as CTP. However, ATP binding stabilizes the R state, lowering the allosteric coefficient from 200 to 70 and thus *increasing* the reaction rate at a given aspartate concentration (Figure 7.13). Hence, ATP is an allosteric activator of ATCase, while CTP is an allosteric inhibitor. In the presence of sufficient ATP, the kinetic profile shows a less-pronounced sigmoidal behavior. Because ATP and CTP bind at the same site, high levels of ATP prevent CTP from inhibiting the enzyme. The effects of nonsubstrate molecules on allosteric enzymes (such as those of CTP and ATP on ATCase) are referred to as **heterotropic effects** (from the Greek *héteros*, "different"), since they bind at sites distinct from the sites that they affect. In the case of ATCase, CTP and ATP binding at their corresponding site acts on the enzyme active site. In summary, substrates generate the sigmoidal curve (homotropic effects), whereas regulators shift the K_M (heterotropic effects). Note, however, that both types of effects impact ATCase activity by altering the T/R ratio.

The increase in ATCase activity in response to increased ATP concentration has two potential physiological ramifications. First, high ATP concentration signals a high concentration of purine nucleotides in the cell; the increase in ATCase activity will tend to balance the purine and pyrimidine pools. As we shall see in Chapter 8, this balance is important for the structure and synthesis of DNA. Second, a high concentration of ATP indicates that energy is available for mRNA synthesis and DNA replication and leads to the synthesis of pyrimidines needed for these processes.

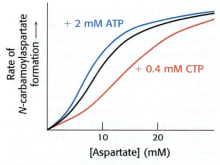

FIGURE 7.13 CTP and ATP impact ATCase kinetics differently. CTP stabilizes the T state of ATCase, making it more difficult for substrate binding to convert the enzyme into the R state. As a result, the curve is shifted to the right, as shown in red. ATP is an allosteric activator of ATCase because it stabilizes the R state, making it easier for substrate to bind. As a result, the curve is shifted to the left, as shown in blue.

7.2 Isozymes Provide a Means of Regulation Specific to Distinct Tissues and Developmental Stages

Enzymes that differ in amino acid sequence yet catalyze the same reaction are called **isozymes**, or **isoenzymes**. Typically, these enzymes display different kinetic parameters, such as K_M, or respond to different regulatory molecules. They are encoded by different genes, which usually arise through gene duplication and divergence. Isozymes can often be distinguished from one another by physical properties such as electrophoretic mobility. *Isoform* is a more generic term used when the protein in question is not an enzyme.

The existence of isozymes permits the fine-tuning of metabolism to meet the needs of a given tissue or developmental stage. Consider the example of lactate dehydrogenase (LDH), an enzyme that catalyzes a step in anaerobic glucose metabolism and glucose synthesis. Human beings have two isozymic polypeptide chains for this enzyme: the H protein is highly expressed in heart muscle, and the M protein is highly expressed in skeletal muscle. The amino acid sequences are 75% identical. Each functional enzyme is tetrameric, and many different combinations

(A)

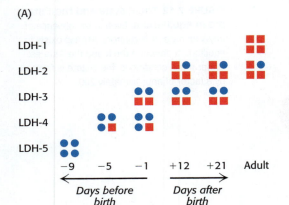

(B)

FIGURE 7.14 **The five isozymes of lactate dehydrogenase differ in their timing of expression during development and their tissue specificity.** (A) The rat heart lactate dehydrogenase (LDH) isozyme profile changes in the course of development. The H isozyme is represented by squares and the M isozyme by circles. (B) LDH isozyme content varies by tissue. The thickness of the green bars represents the relative amounts of the isozymes. [(A) Data from W.-H. Li, *Molecular Evolution* (Sinauer, 1997), p. 283; (B) Source: K. Urich, *Comparative Animal Biochemistry* (Springer Verlag, 1990), p. 542.]

of the two isozymic polypeptide chains are possible. The H_4 isozyme, found in the heart, has a higher affinity for substrates than does the M_4 isozyme. In addition, high levels of pyruvate allosterically inhibit the H_4 but not the M_4 isozyme. The other combinations, such as H_3M, have intermediate properties. We will consider these isozymes in their biological context in Chapter 16.

The M_4 isozyme functions optimally in the anaerobic environment of hardworking skeletal muscle, whereas the H_4 isozyme does so in the aerobic environment of heart muscle. Indeed, the proportions of these isozymes change throughout the development of the rat heart as the tissue switches from an anaerobic environment to an aerobic one (**Figure 7.14A**). The distribution of tissue-specific forms of lactate dehydrogenase in adult rat tissues is shown in **Figure 7.14B**. Essentially all of the enzymes that we will encounter in later chapters, including allosteric enzymes, exist in isozymic forms.

7.3 Covalent Modification Is a Means of Regulating Enzyme Activity

For many enzymes, the covalent attachment of a modifying functional group can markedly alter its catalytic activity. Most covalent modifications are reversible. Phosphorylation and dephosphorylation are common means of reversible covalent modification, which we discuss in detail below.

The attachment of acetyl groups to lysine residues by acetyltransferases and their removal by deacetylases are another example of reversible covalent modifications. Histones—proteins that are packaged with DNA into chromosomes—are extensively acetylated and deacetylated in vivo on lysine residues (Section 31.5). More heavily acetylated histones are associated with genes that are being actively transcribed. Although protein acetylation was originally discovered as a modification of histones, we now know that it is a major means of regulation, with more than 2000 different proteins in mammalian cells regulated by acetylation. Protein acetylation appears to be especially important in the regulation of metabolism. The acetyltransferase and deacetylase enzymes are themselves regulated by phosphorylation, showing that the covalent modification of a protein can be controlled by the covalent modification of the modifying enzymes.

In some cases, the covalent modification is not reversible. The irreversible attachment of a lipid group causes some proteins in signal-transduction pathways, such as Ras (a GTPase), to become affixed to the cytoplasmic face of the plasma membrane. Fixed in this location, the proteins are better able to receive and transmit information that is being passed along their signaling pathways (Section 14.4). Mutations in Ras are seen in a wide array of cancers.

Virtually all metabolic processes are regulated in part by covalent modification. Indeed, the allosteric properties of many enzymes are altered by covalent modification. **Table 7.1** lists some common covalent modifications.

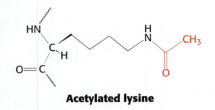

Acetylated lysine

TABLE 7.1 Common covalent modifications of protein activity

Modification	Donor molecule	Example of modified protein	Protein function
Phosphorylation	ATP	Glycogen phosphorylase	Glucose homeostasis; energy transduction
Acetylation	Acetyl CoA	Histones	DNA packing; transcription
Myristoylation	Myristoyl CoA	Src	Signal transduction
ADP ribosylation	NAD^+	RNA polymerase	Transcription
Farnesylation	Farnesyl pyrophosphate	Ras	Signal transduction
γ-Carboxylation	HCO_3^-	Thrombin	Blood clotting
Sulfation	3'-Phosphoadenosine-5'-phosphosulfate	Fibrinogen	Blood clot formation
Ubiquitination	Ubiquitin	Cyclin	Control of cell cycle

Kinases and phosphatases control the extent of protein phosphorylation

Phosphorylation is a regulatory mechanism in virtually every metabolic process in eukaryotic cells. Indeed, as much as 30% of eukaryotic proteins are phosphorylated. The enzymes catalyzing the phosphorylation of protein substrates are called **protein kinases**. These enzymes constitute one of the largest protein families known: there are more than 500 homologous protein kinases in human beings. This multiplicity of enzymes allows regulation to be fine-tuned according to a specific tissue, time, or substrate.

ATP is the most common donor of phosphoryl groups. The terminal (γ) phosphoryl group of ATP is transferred to a specific amino acid of the acceptor protein or enzyme. In eukaryotes, the acceptor residue is commonly one of the three containing a hydroxyl group in its side chain. Transfers to serine and threonine residues are handled by one class of protein kinases and transfers to tyrosine residues by another. Tyrosine kinases, which are unique to multicellular organisms, play pivotal roles in growth regulation, and mutations in these enzymes are commonly observed in cancer cells (Section 14.3).

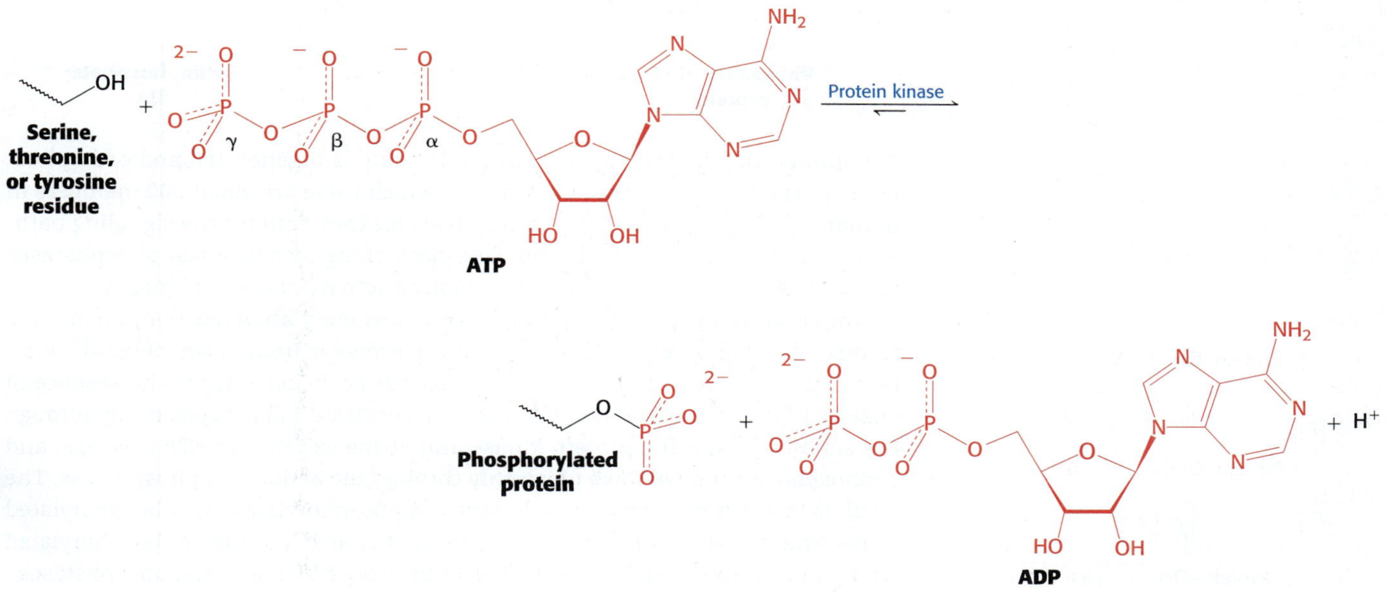

TABLE 7.2 Examples of serine and threonine kinases and their activating signals

Signal	Enzyme
Cyclic nucleotides	Cyclic AMP-dependent protein kinase Cyclic GMP-dependent protein kinase
Ca^{2+} and calmodulin	Calmodulin-dependent protein kinase Phosphorylase kinase Glycogen synthase kinase
AMP	AMP-activated kinase
Diacylglycerol	Protein kinase C
Metabolic intermediates and other "local" effectors	Many target-specific enzymes, such as pyruvate dehydrogenase kinase and branched-chain ketoacid dehydrogenase kinase

Source: Information from D. Fell, *Understanding the Control of Metabolism* (Portland Press, 1997), Table 7.2.

Table 7.2 lists a few of the known serine and threonine protein kinases. Proteins that undergo protein-phosphorylation reactions are located inside cells, where the phosphoryl-group donor ATP is abundant. Proteins that are entirely extracellular are not regulated by reversible phosphorylation.

Comparisons of amino acid sequences of many phosphorylation sites show that an individual kinase may recognize many related sequences. For example, the **consensus sequence** recognized by protein kinase A is Arg-Arg-X-*Ser*-Z or Arg-Arg-X-*Thr*-Z, in which X is a small residue, Z is a large hydrophobic one, and *Ser* or *Thr* is the site of phosphorylation. However, this sequence is not absolutely required. Lysine, for example, can substitute for one of the arginine residues but with some loss of affinity. Thus, the primary determinant of specificity is the amino acid sequence surrounding the serine or threonine phosphorylation site. However, distant residues can contribute to specificity. For instance, a change in protein conformation can open or close access to a possible phosphorylation site.

Protein phosphatases reverse the effects of kinases by catalyzing the removal of phosphoryl groups attached to proteins. The enzyme hydrolyzes the bond attaching the phosphoryl group.

Phosphorylated protein + H_2O $\xrightarrow{\text{Protein phosphatase}}$ ... + **Orthophosphate (P_i)**

The unmodified hydroxyl-containing side chain is regenerated and orthophosphate (P_i) is produced. These enzymes, of which there are about 200 members in human beings, play a vital role in cells because they turn off the signaling pathways that are activated by kinases. One class of highly conserved phosphatases called PP2A suppresses the cancer-promoting activity of certain kinases.

Importantly, the phosphorylation and dephosphorylation reactions are not the reverse of one another; each is essentially irreversible under physiological conditions. Furthermore, both reactions take place at negligible rates in the absence of enzymes. Thus, phosphorylation of a protein substrate will take place only through the action of a specific protein kinase and at the expense of ATP cleavage, and dephosphorylation will take place only through the action of a phosphatase. The result is that target proteins cycle between unphosphorylated and phosphorylated forms. The rate of cycling between the phosphorylated and the unphosphorylated states depends on the relative activities of the specific kinases and phosphatases.

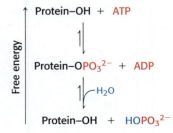

Free energy

Protein−OH + ATP

Protein−OPO_3^{2-} + ADP

+ H_2O

Protein−OH + HOPO_3^{2-}

Phosphorylation is a highly effective means of regulating the activities of target proteins

Phosphorylation is a common covalent modification of proteins in all forms of life, which leads to the question: What makes protein phosphorylation so valuable in regulating protein function that its use is ubiquitous? Phosphorylation is a highly effective means of controlling the activity of proteins for several reasons:

- *The free energy of phosphorylation is large.* Of the -50 kJ mol^{-1} (-12 kcal mol^{-1}) provided by ATP, about half is consumed in making phosphorylation irreversible; the other half is conserved in the phosphorylated protein. A free-energy change of 5.69 kJ mol^{-1} (1.36 kcal mol^{-1}) corresponds to a factor of 10 in an equilibrium constant. Hence, phosphorylation can change the conformational equilibrium between different functional states by a large factor, of the order of 10^4. In essence, the energy expenditure allows for a stark shift from one conformation to another.

- *A phosphoryl group adds two negative charges to a protein.* These new charges may disrupt electrostatic interactions present in the unmodified protein and allow new electrostatic interactions to be formed. Such structural changes can markedly alter substrate binding and catalytic activity.

- *A phosphoryl group can form three or more hydrogen bonds.* The tetrahedral geometry of a phosphoryl group makes these bonds highly directional, allowing for specific interactions with hydrogen-bond donors.

- *Phosphorylation and dephosphorylation can take place in less than a second or over a span of hours.* The kinetics can be adjusted to meet the timing needs of a particular physiological process.

- *Phosphorylation often evokes highly amplified effects.* A single activated kinase can phosphorylate hundreds of target proteins in a short interval. If the target protein is an enzyme, it can in turn transform a large number of substrate molecules.

- *ATP is the cellular energy currency.* The use of this compound as a phosphoryl-group donor links the energy status of the cell to the regulation of metabolism (Section 15.2).

❖ SELF–CHECK QUESTION

What is the energy cost associated with a single cycle of phosphorylation by a protein kinase and dephosphorylation by a protein phosphatase?

Cyclic AMP activates protein kinase A by altering the quaternary structure

Let us examine a specific protein kinase that helps animals, including humans, cope with stressful situations. As we shall see in Section 14.2, the "fight-or-flight" response—including an increase in heart rate, dilation of the airway muscles, and constriction of the blood vessels in muscle—is common to many animals presented with a dangerous or exciting situation. This response is the result of the activity of the hormone epinephrine (also called adrenaline), which triggers the formation of cyclic AMP (cAMP), an intracellular messenger formed by the cyclization of ATP. Cyclic AMP subsequently activates a key enzyme, protein kinase A (PKA). This kinase alters the activities of target proteins by phosphorylating specific serine or threonine residues. The striking finding is that most effects of cAMP in eukaryotic cells are achieved through the activation of PKA by cAMP.

Similar to ATCase, PKA provides an example of allosteric regulation of enzyme activity. In fact, the quaternary structure of PKA is reminiscent of that of ATCase in that it is comprised of distinct regulatory and catalytic subunits. In the

Cyclic adenosine monophosphate (cAMP)

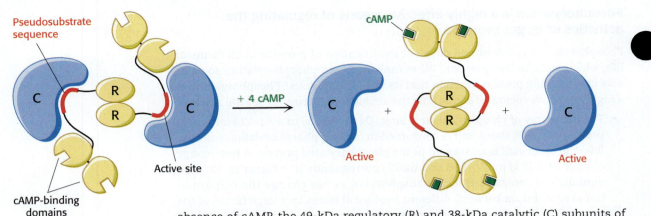

FIGURE 7.15 The binding of cAMP to the regulatory subunits of PKA activates the kinase. The binding of four molecules of cAMP activates protein kinase A by dissociating the inhibited holoenzyme (R_2C_2) into a regulatory subunit (R_2) and two catalytically active subunits (C). Each R chain includes cAMP-binding domains and a pseudosubstrate sequence.

INTERACT with this model in
≈ Achieve

absence of cAMP, the 49-kDa regulatory (R) and 38-kDa catalytic (C) subunits of PKA in muscle form an R_2C_2 complex that is enzymatically inactive (**Figure 7.15**). The binding of two molecules of cAMP to each of the regulatory subunits leads to the dissociation of R_2C_2 into an R_2 subunit and two C subunits. These free catalytic subunits are enzymatically active. Thus, the binding of cAMP to the regulatory subunit relieves its inhibition of the catalytic subunit. PKA, like most other kinases, exists in isozymic forms for fine-tuning regulation to meet the needs of a specific cell or developmental stage. In mammals, four isoforms of the R subunit and three of the C subunit are encoded in the genome.

How does the binding of cAMP activate the kinase? All four isoforms of the R chain contain an inhibitor site sequence which binds in the active site pocket of the C chains. Two of these R chains contain the sequence Arg-Arg-Gly-Ala-Ile, which matches the consensus sequence for phosphorylation except for the presence of alanine in place of serine. In the R_2C_2 complex, this **pseudosubstrate sequence** of R occupies the catalytic site of C, thereby preventing the entry of protein substrates (Figure 7.15). The remaining two R chains contain the sequence Arg-Arg-Gly-Ser-Ile. In the tetrameric complex, these sequences are packed into the catalytic site of the C chain such that ATP cannot bind, keeping these sequences unphosphorylated. In either case, the binding of cAMP to the R chains allosterically moves the inhibitor site sequences out of the catalytic domains.

Kinetic studies of the activated form of PKA indicate that cAMP enhances the dissociation rate of the C chain from the R chain by almost 700-fold. The released C chains are then free to bind and phosphorylate substrate proteins. However, even after this rate enhancement, the binding affinity between the R and C chains is still significant, and R chains are present in the cell in large molar excess. The net result is that while C chains are free to exert their cellular effects, they are probably re-captured quite rapidly by the R subunits. Hence, the activation mechanism of PKA has evolved to enable control over the active form of kinase.

Mutations in protein kinase A can cause Cushing's syndrome

As we have just seen, the allosteric regulation of kinases such as PKA enables precise timing for the activation and the inhibition of their catalytic activities. The significance of this control is emphasized by disease states in which these mechanisms fail. For example, Cushing's syndrome, a collection of diseases resulting from excess cortisol secretion by the adrenal cortex, is a metabolic disorder characterized by a variety of symptoms such as muscle weakness, thinning skin that is easily bruised, and osteoporosis. Cortisol, a steroid hormone (Section 27.4), has a number of physiological effects including stimulation of glucose synthesis, suppression of the immune response, and inhibition of bone growth.

The most common cause of Cushing's syndrome, called Cushing's disease, is a tumor of the pituitary gland that overstimulates cortisol secretion by the adrenal cortex. Mutations within the catalytic subunit of PKA have been identified in a number of patients with these tumors. These mutations disrupt the interaction between the R and C subunits. Thus, the enzyme is active even in the absence of cAMP, with the ultimate result being unregulated secretion of cortisol. In Section 14.5, we will discuss other instances where dysregulation of enzymes can lead to disease.

The phosphorylation states of the proteome can be measured

In Section 4.3, we introduced the concept of the proteome, the array of proteins expressed within a cell at a given time under specific conditions. The techniques used to study the proteome have been expanded to include the analysis of protein phosphorylation under specific conditions. The **phosphoproteome** is the name given to all proteins that are modified by phosphorylation. Through study of the phosphoproteome, termed *phosphoproteomics*, one can elucidate the pathways switched on or off under a specific set of conditions.

For example, as we will see more clearly in later chapters, exercise is fundamental to sustaining good health, in part by regulating energy metabolism and maintaining whole-body insulin sensitivity. Given the prominence of phosphorylation in biological regulation, it should come as no surprise that exercise modifies the phosphoproteome. Recent research has shown that exercise results in the phosphorylation of more than 1000 potential phosphorylation sites on almost 600 different proteins. Protein kinase A and AMP-activated kinase (AMPK, Section 22.7) as well as other kinases catalyze these modifications. These modifications alter a number of biological functions, most notably an increase in the ability to process fuels aerobically (Chapters 17 and 18). Determining the functions of the modified proteins will surely keep exercise biochemists engaged for many years.

7.4 Many Enzymes Are Activated by Specific Proteolytic Cleavage

While many enzymes acquire full enzymatic activity as they spontaneously fold into their characteristic three-dimensional forms, the folded forms of other enzymes are inactive until one or a few specific peptide bonds are cleaved. The inactive precursor is called a **zymogen** (or **proenzyme**). Since an energy source such as ATP is not needed for specific proteolytic cleavage, even extracellular proteins can be activated by this means. This contrasts with reversible enzyme regulation by phosphorylation, which requires ATP. Another noteworthy difference is that proteolytic activation, unlike allosteric control and reversible covalent modification, is irreversible, taking place just once in the life of an enzyme molecule.

Specific proteolysis is a common means of activating enzymes and other proteins in biological systems. For example:

- The digestive enzymes that hydrolyze foodstuffs are synthesized as zymogens in the stomach and pancreas (Table 7.3).

- Blood clotting is mediated by a cascade of proteolytic activations that ensures a rapid and amplified response to trauma.

TABLE 7.3 Zymogens of the stomach and pancreas

Site of synthesis	Zymogen	Active enzyme
Stomach	Pepsinogen	Pepsin
Pancreas	Chymotrypsinogen	Chymotrypsin
Pancreas	Trypsinogen	Trypsin
Pancreas	Procarboxypeptidase	Carboxypeptidase

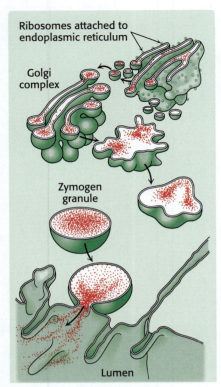

FIGURE 7.16 The acinar cells of the pancreas secrete zymogens. Zymogens are synthesized on ribosomes attached to the endoplasmic reticulum. They are subsequently processed in the Golgi apparatus and packaged into zymogen or secretory granules. With the proper signal, the granules fuse with the plasma membrane, discharging their contents into the interior (lumen) of the pancreatic ducts. Cell cytoplasm is depicted as pale green. Membranes and lumen are shown as dark green.

- Some protein hormones are synthesized as inactive precursors. For example, *insulin* is derived from *proinsulin* by proteolytic removal of a peptide.
- The fibrous protein collagen, the major constituent of skin and bone, is derived from procollagen, a soluble precursor.

Many developmental processes are controlled by the activation of zymogens. For example, in the metamorphosis of a tadpole into a frog, large amounts of collagen are resorbed from the tail in the course of a few days. Likewise, much collagen is broken down in a mammalian uterus after delivery. The conversion of procollagenase into collagenase, the active protease responsible for collagen breakdown, is precisely timed in these remodeling processes.

Programmed cell death, also called *apoptosis*, is mediated by proteolytic enzymes called *caspases*, which are synthesized in precursor form as *procaspases*. When activated by various signals, caspases function to cause cell death in most organisms, ranging from *C. elegans* to human beings. Apoptosis provides a means of sculpting the shapes of body parts in the course of development and is a means of eliminating damaged or infected cells.

Chymotrypsinogen is activated by specific cleavage of a single peptide bond

Let us examine the activation and control of zymogens, using enzymes responsible for digestion. Chymotrypsin is a member of a large family of serine proteases whose mechanism of action was described in detail in Section 6.2. Specifically, it is a digestive enzyme that cleaves peptide bonds on the carboxyl side of amino acid residues with large, hydrophobic R groups (Table 5.7). Its inactive precursor, chymotrypsinogen, is synthesized in the pancreas, as are several other zymogens and digestive enzymes. Indeed, the pancreas is one of the most active organs in synthesizing and secreting proteins. The enzymes and zymogens are synthesized in the acinar cells of the pancreas and stored inside membrane-bounded granules (**Figure 7.16**). The zymogen granules accumulate at the apex of the acinar cell; when the cell is stimulated by a hormonal signal or a nerve impulse, the contents of the granules are released into a duct leading into the duodenum, the first segment of the small intestine into which the stomach empties.

Chymotrypsinogen, a single polypeptide chain consisting of 245 amino acid residues, is virtually devoid of enzymatic activity. It is converted into a fully active enzyme when the peptide bond joining arginine 15 and isoleucine 16 is cleaved by trypsin (**Figure 7.17**). The resulting active enzyme, called π-chymotrypsin, then acts on other π-chymotrypsin molecules by removing two dipeptides to yield α-chymotrypsin, the stable form of the enzyme. The three resulting chains in

FIGURE 7.17 Proteolysis activates chymotrypsinogen. The three chains of α-chymotrypsin are linked by two interchain disulfide bonds (A to B and B to C). The approximate positions of both inter- and intrachain disulfide bonds are shown.

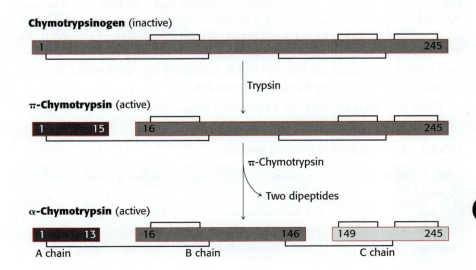

Chymotrypsinogen (inactive)

1   245

Trypsin

π-Chymotrypsin (active)

1   15   16   245

π-Chymotrypsin

Two dipeptides

α-Chymotrypsin (active)

1   13   16   146   149   245

A chain   B chain   C chain

α-chymotrypsin remain linked to one another by two interchain disulfide bonds. The striking feature of this activation process is that cleavage of a single specific peptide bond transforms the protein from a catalytically inactive form into one that is fully active.

Proteolytic activation of chymotrypsinogen leads to the formation of a substrate-binding site

How does cleavage of a single peptide bond activate the zymogen? The cleavage of the peptide bond between amino acids 15 and 16 triggers key conformational changes, which were revealed by the elucidation of the three-dimensional structure of chymotrypsinogen.

- The newly formed amino-terminal group of isoleucine 16 turns inward and forms an ionic bond with aspartate 194 in the interior of the chymotrypsin molecule (**Figure 7.18**).

- This electrostatic interaction triggers a number of conformational changes. Methionine 192 moves from a deeply buried position in the zymogen to the surface of the active enzyme, and residues 187 and 193 move farther apart from each other. These changes result in the formation of the substrate-specificity site for aromatic and bulky nonpolar groups. One side of this site is made up of residues 189 through 192 (Figure 6.10). This cavity for binding part of the substrate is not fully formed in the zymogen.

- The repositioning of aspartate 194 also rearranges the main chain amino group of glycine 195. Recall that the tetrahedral transition state generated by chymotrypsin has an oxyanion (a negatively charged carbonyl oxygen atom) that is stabilized by hydrogen bonds with two NH groups of the main chain of the enzyme (the oxyanion hole, Figure 6.9). The new position of the NH group of glycine 195 enables the proper formation of the oxyanion hole (Figure 7.18).

The conformational changes elsewhere in the molecule are very small. For example, the positions of the residues forming the catalytic triad (Ser 193, His 57,

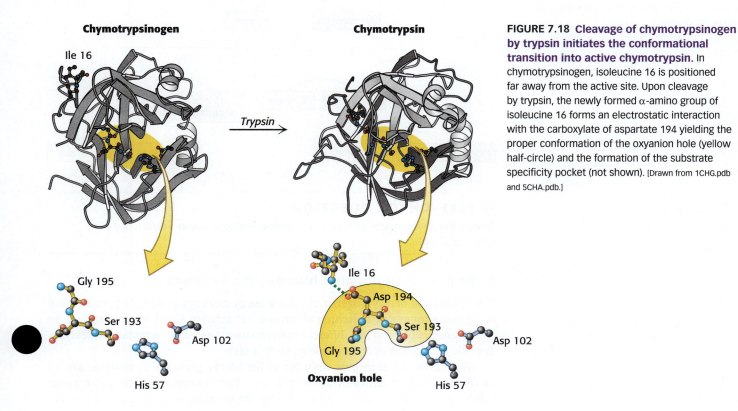

FIGURE 7.18 Cleavage of chymotrypsinogen by trypsin initiates the conformational transition into active chymotrypsin. In chymotrypsinogen, isoleucine 16 is positioned far away from the active site. Upon cleavage by trypsin, the newly formed α-amino group of isoleucine 16 forms an electrostatic interaction with the carboxylate of aspartate 194 yielding the proper conformation of the oxyanion hole (yellow half-circle) and the formation of the substrate specificity pocket (not shown). [Drawn from 1CHG.pdb and 5CHA.pdb.]

and Asp 102) are essentially the same in both structures (Figure 7.18). Thus, the switching on of enzymatic activity in a protein can be accomplished by discrete, highly localized conformational changes that are triggered by the hydrolysis of a single peptide bond.

The generation of trypsin from trypsinogen leads to the activation of other zymogens

The structural changes accompanying the activation of trypsinogen, the precursor of the proteolytic enzyme trypsin, another serine protease, are different from those in the activation of chymotrypsinogen. Four regions of the polypeptide are very flexible in the zymogen, whereas they have a well-defined conformation in trypsin. The resulting structural changes also complete the formation of the oxyanion hole.

The digestion of proteins and other molecules in the duodenum requires the concurrent action of several enzymes, because each is specific for a limited number of side chains. Thus, the zymogens must be switched on at the same time. Coordinated control is achieved by the action of trypsin as the common activator of all the pancreatic zymogens—trypsinogen, chymotrypsinogen, proelastase, procarboxypeptidase, and prolipase, the inactive precursor of a lipid-degrading enzyme.

To produce active trypsin, the cells that line the duodenum display a membrane-embedded enzyme, enteropeptidase, which hydrolyzes a unique lysine–isoleucine peptide bond in trypsinogen as the zymogen enters the duodenum from the pancreas. The small amount of trypsin produced in this way activates more trypsinogen and the other zymogens (**Figure 7.19**). Thus, the formation of trypsin by enteropeptidase is the master activation step.

FIGURE 7.19 Enteropeptidase initiates the activation of the pancreatic zymogens by activating trypsin, which then activates other zymogens. Active enzymes are shown in yellow; zymogens are shown in orange.

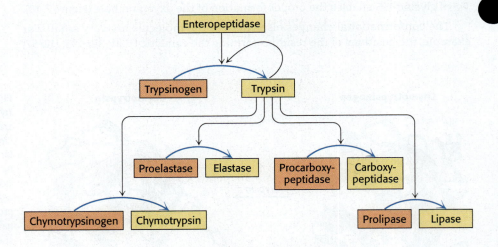

❖ SELF–CHECK QUESTION

Predict the physiological effects of a mutation that resulted in a deficiency of enteropeptidase.

Some proteolytic enzymes have specific inhibitors

The conversion of a zymogen into a protease by cleavage of a single peptide bond is a precise means of switching on enzymatic activity. However, this activation step is irreversible, and so a different mechanism is needed to terminate proteolysis. Specific protease inhibitors accomplish this task.

Inhibitors called *serpins*, shorthand for <u>ser</u>ine <u>p</u>rotease <u>in</u>hibitors, are an example of one such family of inhibitors. For example, pancreatic trypsin inhibitor, a 6-kDa serpin, inhibits trypsin by binding very tightly to its active

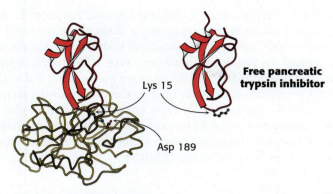

Free pancreatic
trypsin inhibitor

Lys 15

Asp 189

**Pancreatic trypsin inhibitor
bound to trypsin**

**FIGURE 7.20 Trypsin binds very tightly
to pancreatic trypsin inhibitor, a serpin.**
In the structure of a complex of trypsin
(yellow) and pancreatic trypsin inhibitor (red),
lysine 15 of the inhibitor penetrates into the
active site of the enzyme. There it forms an
ionic bond with aspartate 189 in the active
site. Notice that bound inhibitor and the free
inhibitor are almost identical in structure.
[Drawn from 2PTC.pdb and 1BPI.pdb.]

INTERACT with this model in
 Achieve

site. The dissociation constant of the complex is 0.1 pM, which corresponds to a standard free energy of binding of about -75 kJ mol^{-1} (-18 kcal mol^{-1}). In contrast with nearly all known protein assemblies, this complex is not dissociated into its constituent chains by treatment with denaturing agents such as 8 M urea (Section 2.7).

The reason for the exceptional stability of the complex is that pancreatic trypsin inhibitor is a very effective substrate analog. X-ray analyses show that the inhibitor lies in the active site of the enzyme, positioned such that the side chain of lysine 15 of this inhibitor interacts with the aspartate side chain in the specificity pocket of trypsin. In addition, there are many hydrogen bonds between the main chain of trypsin and that of its inhibitor. Furthermore, the carbonyl group of lysine 15 and the surrounding atoms of the inhibitor fit snugly in the active site of the enzyme.

Comparison of the structure of the inhibitor bound to the enzyme with that of the free inhibitor reveals that the structure is essentially unchanged on binding to the enzyme (**Figure 7.20**). Thus, the inhibitor is preorganized into a structure that is highly complementary to the enzyme's active site. Indeed, the peptide bond between lysine 15 and alanine 16 in pancreatic trypsin inhibitor is cleaved but at a very slow rate: the half-life of the trypsin–inhibitor complex is several months. In essence, the inhibitor is a substrate, but its intrinsic structure is so nicely complementary to the enzyme's active site that it binds very tightly, rarely progressing to the transition state, and is turned over slowly.

Why does trypsin inhibitor exist? Recall that trypsin activates other zymogens. Consequently, preventing even small amounts of trypsin from initiating the cascade while the zymogens are still in the pancreas or pancreatic ducts is vital. Trypsin inhibitor binds to any prematurely activated trypsin molecules in the pancreas or pancreatic ducts. This inhibition prevents severe damage to those tissues, which could lead to acute pancreatitis.

Pancreatic trypsin inhibitor is not the only important protease inhibitor. A 53-kDa plasma protein, α_1-antitrypsin (also called α_1-*antiproteinase*), protects tissues from digestion by elastase, a secretory product of neutrophils (white blood cells that engulf bacteria). Antielastase would be a more accurate name for this inhibitor, because it blocks elastase much more effectively than it blocks trypsin. Like pancreatic trypsin inhibitor, α_1-antitrypsin blocks the action of target enzymes by binding nearly irreversibly to their active sites.

Genetic disorders leading to a deficiency of α_1-antitrypsin illustrate the physiological importance of this inhibitor. For example, the substitution of lysine for glutamate at residue 53 in the type Z mutant slows the secretion of this inhibitor from liver cells. Serum levels of the inhibitor are about 15% of normal in people homozygous for this defect. The consequence is that excess elastase destroys alveolar walls in the lungs by digesting elastic fibers and other connective-tissue proteins.

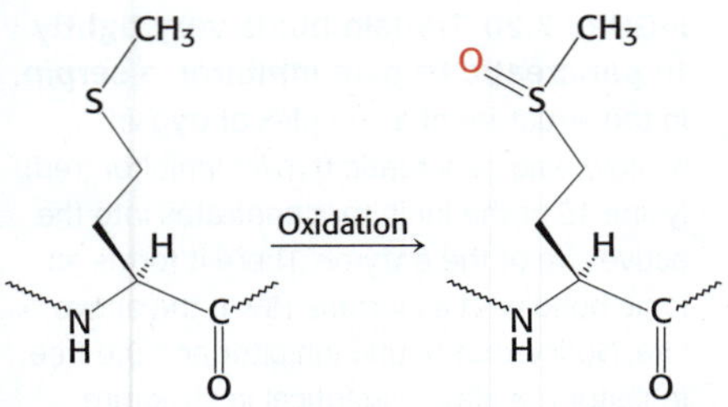

FIGURE 7.21 Cigarette smoke induces the oxidation of methionine to methionine sulfoxide. In α_1-antitrypsin, the oxidation of methionine 538 leads to reduced inhibition of elastase and subsequent progressive destruction of the alveolar walls within the lungs.

The resulting clinical condition is called *emphysema*. People with emphysema must breathe much harder than normal people to exchange the same volume of air because their alveoli are much less resilient than normal. Cigarette smoking markedly increases the likelihood that even a type Z heterozygote will develop emphysema. The reason is that smoke oxidizes methionine 358 of the inhibitor (**Figure 7.21**), a residue essential for binding elastase. Indeed, this methionine side chain is the bait that selectively traps elastase. The methionine sulfoxide oxidation product, in contrast, does not lure elastase, a striking consequence of the insertion of just one oxygen atom into a protein and a remarkable example of the effect of human behavior on biochemistry.

7.5 Enzymatic Cascades Allow Rapid Responses Such as Blood Clotting

Biochemical systems often use a series of zymogen activations, called an **enzymatic cascade**, to achieve a rapid response. In a cascade, a signal initiates a series of steps, each of which is catalyzed by an enzyme. At each step, the signal is amplified. For instance, if a signal molecule activates an enzyme that in turn activates 10 enzymes, and each of the 10 enzymes in turn activates 10 additional enzymes, after four steps the original signal will have been amplified 10,000-fold.

In the previous section, we discussed the cascade of proteolytic enzymes that accompanies digestion of a meal. Another physiological process that requires enzymatic cascades is **hemostasis**, the process of blood clot formation and dissolution. Because the activated form of one clotting factor catalyzes the activation of the next (**Figure 7.22**), a very small amount of the initial factors suffice to trigger the cascade, ensuring a rapid response to trauma.

Two means of initiating blood clotting have been elucidated. The *intrinsic pathway* is activated by exposure of anionic surfaces upon rupture of the endothelial lining of the blood vessels. The *extrinsic pathway*, which appears to be most crucial in blood clotting, is initiated when trauma exposes tissue factor (TF), an integral membrane glycoprotein. Upon exposure to the blood, tissue factor binds to factor VII to activate factor X. Both the intrinsic and extrinsic pathways lead to the activation of factor X (a serine protease), which in turn converts prothrombin into thrombin, the key protease in clotting. Thrombin then amplifies the clotting process by activating enzymes and factors that lead to the generation of yet more thrombin, an example of positive feedback. Note that the active forms of the clotting factors are designated with a subscript "a," whereas factors that are enzymes or enzyme cofactors activated by thrombin are designated with an asterisk.

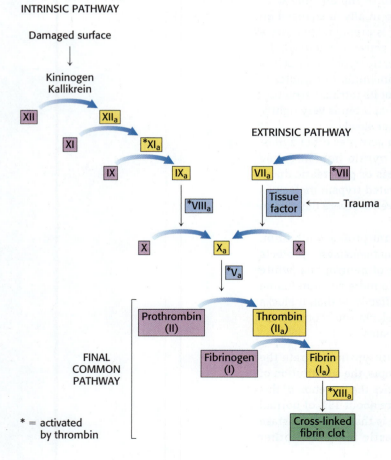

FIGURE 7.22 Blood clotting is the result of an enzymatic cascade. A fibrin clot is formed by the interplay of the intrinsic, extrinsic, and final common pathways. The intrinsic pathway begins with the activation of factor XII (Hageman factor) by contact with abnormal surfaces produced by injury. The extrinsic pathway is triggered by trauma, which releases tissue factor (TF). TF forms a complex with VII, which initiates a cascade-activating thrombin. Inactive forms of clotting factors are shown in purple; their activated counterparts (indicated by the subscript "a") are in yellow. Stimulatory proteins that are not themselves enzymes are shown in blue boxes. A striking feature of this process is that the activated form of one clotting factor catalyzes the activation of the next factor.

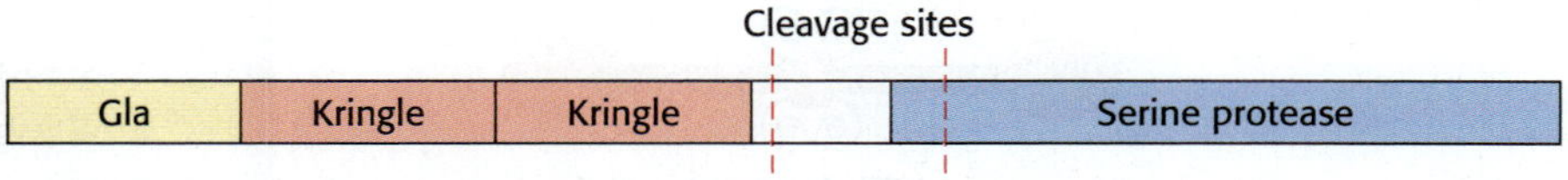

Gla	Kringle	Kringle		Serine protease

Cleavage sites

FIGURE 7.23 Prothrombin contains four distinct protein domains. Cleavage of two peptide bonds yields active thrombin. All the γ-carboxyglutamate residues are in the gla domain.

Prothrombin must bind to Ca²⁺ to be converted to thrombin

Thrombin is synthesized as a zymogen called prothrombin. The inactive molecule is comprised of four major domains, with the serine protease domain at its carboxyl terminus (**Figure 7.23**). The first domain, called the *gla domain*, is rich in γ-carboxyglutamate residues (abbreviation gla), and the second and third domains are called *kringle domains* (named after a Danish pastry that they resemble). These three domains work in concert to keep prothrombin in an inactive form. Moreover, because it is rich in γ-carboxyglutamate, the gla domain is able to bind Ca^{2+} (**Figure 7.24**). What is the effect of this binding? The binding of Ca^{2+} by prothrombin anchors the zymogen to phospholipid membranes derived from blood platelets after injury. This binding is crucial because it brings prothrombin into close proximity to two clotting proteins, factor X_a and factor V_a (a stimulatory protein), that catalyze its conversion into thrombin. Factor X_a cleaves the bond between arginine 274 and threonine 275 to release a fragment containing the first three domains. Factor X_a also cleaves the bond between arginine 323 and isoleucine 324 to yield active thrombin.

Fibrinogen is converted by thrombin into a fibrin clot

The best-characterized part of the clotting process is the final step in the cascade: the conversion of fibrinogen into fibrin by thrombin. Fibrinogen is a large glycoprotein composed of three nonidentical chains, Aα, Bβ, and γ, and is found in the blood plasma of all vertebrates. The overall composition of this 340-kDa protein is $(A\alpha)_2(B\beta)_2\gamma_2$. Three globular units are connected by two rods, and the rod regions are triple-stranded α-helical coiled coils, a recurring motif in proteins (Section 2.5; **Figure 7.25A**). Thrombin cleaves four arginine–glycine peptide bonds in the central globular region of fibrinogen (**Figure 7.25B**). On cleavage, an A peptide of 18 residues is released from each of the two Aα chains, as is a B peptide of 20 residues from each of the two Bβ chains. These A and B peptides are called

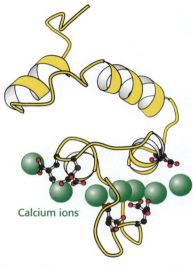

Calcium ions

FIGURE 7.24 The gla domain of pro-thrombin contains a calcium-binding region. Prothrombin binds calcium ions with the carboxyl groups of the modified amino acid γ-carboxyglutamate (red). [Drawn from 2PF2.pdb.]

INTERACT with this model in
 Achie√e

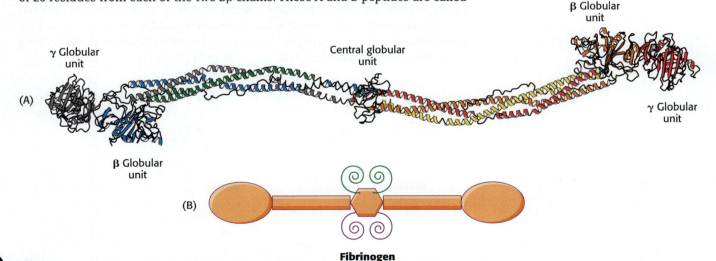

γ Globular unit

β Globular unit

Central globular unit

β Globular unit

γ Globular unit

(A)

(B)

Fibrinogen

FIGURE 7.25 Fibrinogen contains three globular regions connected by rod-shaped structures. (A) A ribbon diagram. The two rod regions are α-helical coiled coils, connected to the three globular regions. (B) A schematic representation showing the positions of the fibrinopeptides (green and purple) in the central globular region. [Part A drawn from 1M1J.pdb.]

INTERACT with this model in
 Achie√e

FIGURE 7.26 The cleavage of fibrinogen by thrombin initiates fibrin clot formation. (1) Thrombin cleaves fibrinopeptides from the central globule of fibrinogen. (2) Globular domains at the carboxyl-terminal ends of the β and γ chains interact with the peptides exposed upon thrombin cleavage.

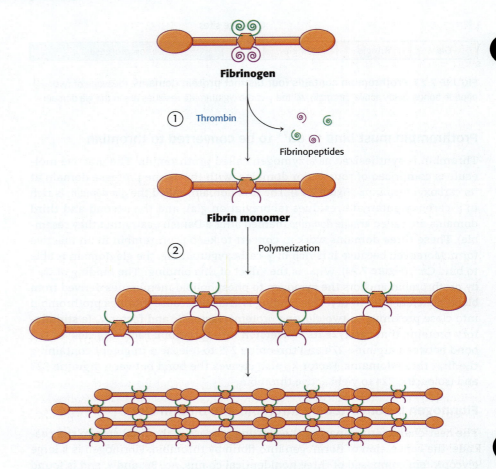

fibrinopeptides (**Figure 7.26**). A fibrinogen molecule devoid of these fibrinopeptides is called a *fibrin monomer* and has the subunit structure $(\alpha\beta\gamma)_2$.

Upon thrombin cleavage, amino acid sequences are exposed in the central globular unit that interact with the γ and β subunits of other monomers. Polymerization occurs as more fibrin monomers interact with one another (Figure 7.26). Thus, analogous to the activation of chymotrypsinogen, peptide-bond cleavage exposes new amino termini that can participate in specific interactions. The newly formed "soft clot" is stabilized by the formation of amide bonds between the side chains of lysine and glutamine residues in different monomers. This cross-linking reaction is catalyzed by transglutaminase (factor XIII$_a$), which itself is activated from its protransglutaminase form by thrombin.

Glutamine   **Lysine**

Cross-link

Vitamin K is required for the formation of γ-carboxyglutamate

Vitamin K (**Figure 7.27**) has been known for many years to be essential for the synthesis of prothrombin and several other clotting factors. Indeed, it is called vitamin K because a deficiency in this vitamin results in defective blood <u>k</u>oagulation (Scandinavian spelling). After ingestion, vitamin K is reduced to a dihydro derivative that is required by γ-glutamyl carboxylase to convert the first 10 glutamate residues in the amino-terminal region of prothrombin into γ-carboxyglutamate (**Figure 7.28**). Recall from above that γ-carboxyglutamate, a strong chelator of Ca^{2+}, is required for the activation of prothrombin.

Medicines that reduce blood clot formation, known as **anticoagulant drugs**, play an important role in the prevention of stroke, myocardial infarction (heart attack), and pulmonary embolism in patients with increased clotting risk. One most commonly prescribed anticoagulants is warfarin, a vitamin K antagonist sold under the name of coumadin (Figure 7.27). Warfarin appears to inhibit the epoxide reductase and quinone reductase that are required to regenerate the dihydro derivative of vitamin K (Figure 7.28). Newer anticoagulants have also entered the market, including inhibitors of factor X_a and thrombin.

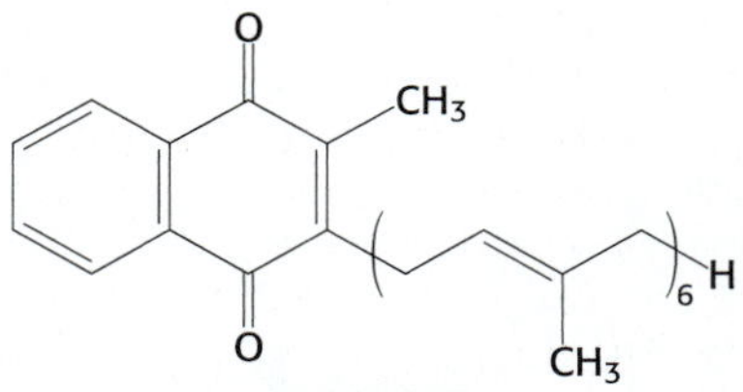

The clotting process must be precisely regulated

There is a fine line between hemorrhage and thrombosis, the formation of blood clots in blood vessels. Clots must form rapidly yet remain confined to the area of injury. What are the mechanisms that normally limit clot formation to the site of injury? The lability of clotting factors contributes significantly to the control of clotting. Activated factors are short-lived because they are diluted by blood flow, removed by the liver, and degraded by proteases. For example, the stimulatory protein factors V_a and VIII_a are digested by protein C, a protease that is switched on by the action of thrombin. Thus, thrombin has a dual function: it catalyzes the formation of fibrin, and it initiates the deactivation of the clotting cascade.

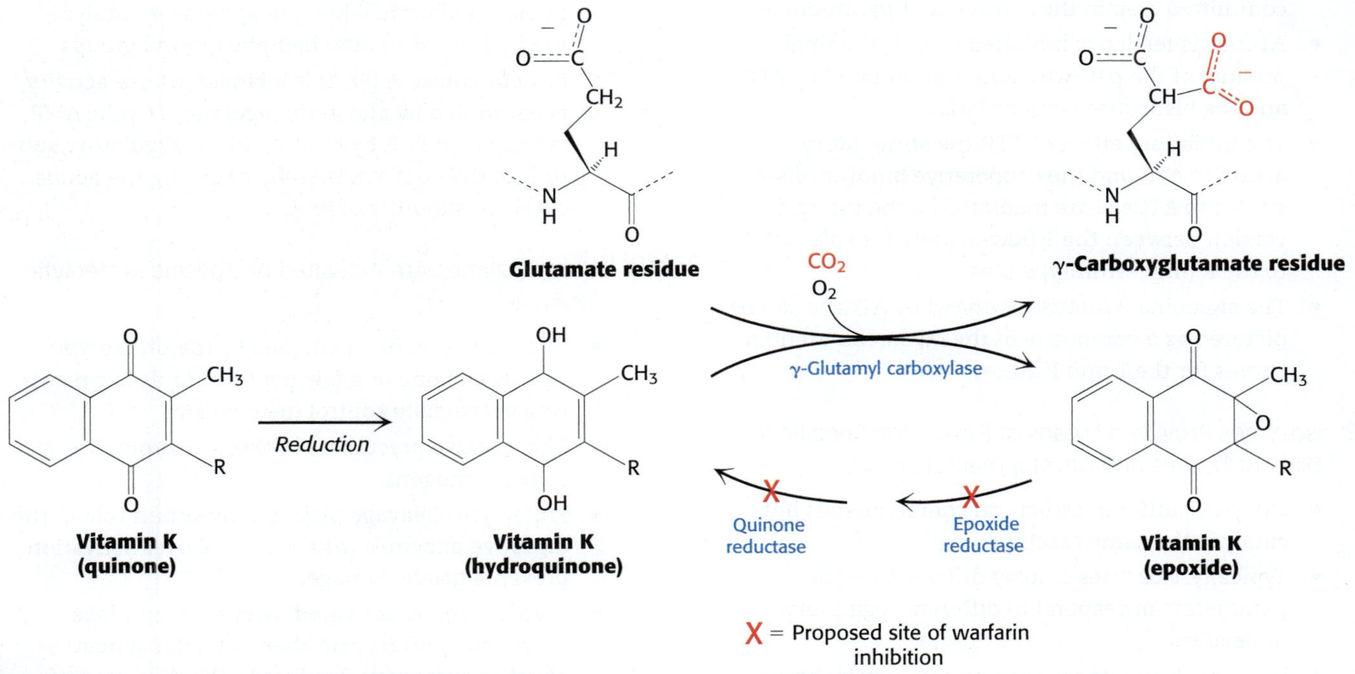

Vitamin K

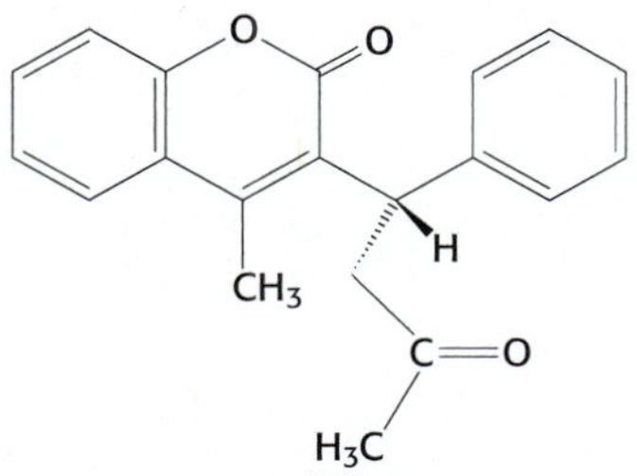

Warfarin

FIGURE 7.27 **Vitamin K and the anticoagulant warfarin share structural features.**

Glutamate residue

CO_2
O_2

γ-Glutamyl carboxylase

γ-Carboxyglutamate residue

Vitamin K (quinone)

Reduction

Vitamin K (hydroquinone)

Quinone reductase

Epoxide reductase

Vitamin K (epoxide)

X = Proposed site of warfarin inhibition

FIGURE 7.28 **Vitamin K is required for the synthesis of γ-carboxyglutamate by γ-glutamyl carboxylase.** The formation of γ-carboxyglutamate requires the hydroquinone (reduced) derivative of vitamin K, which is then regenerated from the epoxide derivative by the sequential action of epoxide reductase and quinone reductase, both of which are inhibited by warfarin.

Specific inhibitors of clotting factors are also critical in the termination of clotting. For instance, tissue factor pathway inhibitor (TFPI) inhibits the complex of $TF–VII_a–X_a$ that activates thrombin. Another key inhibitor is antithrombin III, a member of the serpin family of protease inhibitors that forms an irreversible inhibitory complex with thrombin. Antithrombin III resembles α_1-antitrypsin except that it inhibits thrombin much more strongly than it inhibits elastase (Figure 7.20). Antithrombin III also blocks other serine proteases in the clotting cascade—namely, factors XII_a, XI_a, IX_a, and X_a.

The inhibitory action of antithrombin III is enhanced by the glycosamino-glycan *heparin*, a negatively charged polysaccharide (Section 11.3) found in mast cells near the walls of blood vessels and on the surfaces of endothelial cells. Heparin acts as an anticoagulant by increasing the rate of formation of irreversible complexes between antithrombin III and the serine protease clotting factors.

❖ SELF–CHECK QUESTION

Antithrombin III forms an irreversible complex with thrombin but not with prothrombin. What is the most likely reason for this difference in reactivity?

Summary

7.1 Allosteric Regulation Enables Control of Metabolic Pathways

- Allosteric proteins constitute an important class of proteins whose biological activity can be altered by binding to distinct regulatory sites.

- Aspartate transcarbamoylase (ATCase), one of the best-understood allosteric enzymes, catalyzes the synthesis of N-carbamoylaspartate, the first and committed step in the synthesis of pyrimidines.

- ATCase is feedback inhibited by CTP, the final product of the pathway, and is stimulated by ATP and the bisubstrate analog PALA.

- The inhibitory effect of CTP, the stimulatory action of ATP, and the cooperative binding of substrates to ATCase are mediated by the interconversion between the T (low-affinity for substrate) to the R (high-affinity) states.

- The sigmoidal kinetics exhibited by ATCase can be pictured as a composite of the Michaelis–Menten curves for the T and R states.

7.2 Isozymes Provide a Means of Regulation Specific to Distinct Tissues and Developmental Stages

- Isozymes differ in structural characteristics but catalyze the same reaction.

- Typically, isozymes display different kinetic parameters or respond to different regulatory molecules.

- Isozymic forms of an enzyme can vary by tissue or developmental stage, providing a means of fine-tuning metabolism.

7.3 Covalent Modification Is a Means of Regulating Enzyme Activity

- Reversible and irreversible covalent modifications are powerful means of controlling enzyme activity.

- Phosphorylation is a common type of reversible covalent modification in all forms of life.

- Kinases are the enzymes responsible for protein phosphorylation, while phosphatases catalyze the hydrolysis of attached phosphoryl groups.

- Protein kinase A (PKA) is a kinase whose activity is controlled by allosteric regulation. Cyclic AMP switches on PKA by binding to the regulatory subunit of the enzyme, thereby releasing the active catalytic subunits of PKA.

7.4 Many Enzymes Are Activated by Specific Proteolytic Cleavage

- The activation of an enzyme by the proteolytic cleavage of one or a few peptide bonds is a recurring irreversible control mechanism.

- The inactive precursors of these enzymes are called zymogens.

- Proteolytic cleavage plays an important role in the digestive enzymes, where control over activation prevents tissue damage.

- Trypsinogen is activated by enteropeptidase or trypsin, and trypsin then activates a host of other zymogens, leading to the digestion of foodstuffs.

- Specific inhibitors of proteolytic enzymes, including the serpins, also provide an additional layer of regulatory control.

7.5 Enzymatic Cascades Allow Rapid Responses Such as Blood Clotting

- Blood clotting is an example of a process that utilizes an enzymatic cascade to achieve a rapid response.

- The activated form of one clotting factor, often a serine protease, catalyzes the activation of the next precursor.

- In the final step of clot formation, fibrinogen, a highly soluble molecule in the plasma, is converted by thrombin into fibrin by the hydrolysis of four arginine–glycine bonds.

Key Terms

committed step (p. 211)
feedback inhibition (p. 211)
allosteric (regulatory) site (p. 211)
homotropic effect (p. 215)
heterotropic effect (p. 217)

isozymes (isoenzymes) (p. 217)
protein kinase (p. 219)
consensus sequence (p. 220)
protein phosphatase (p. 220)
pseudosubstrate sequence (p. 222)

phosphoproteome (p. 223)
zymogen (proenzyme) (p. 223)
enzymatic cascade (p. 228)
hemostasis (p. 228)
anticoagulant drugs (p. 231)

Problems

1. Suppose you have a dimeric allosteric enzyme that follows the concerted model. Which of the following statements are true? ❖ **3**

(a) The equilibrium between the T state and R state favors the T state.

(b) The enzyme can exist as a RR dimer.

(c) The enzyme can exist as an RT dimer.

(d) The RR form of the enzyme is most active.

2. An allosteric enzyme that follows the concerted model has an allosteric coefficient (T/R) of 300 in the absence of substrate. Suppose that a mutation reversed the ratio. Select all of the effects this mutation will have on the relationship between the rate of the reaction (V) and substrate concentration, [S]. ❖ **3**

(a) The enzyme would be more active.

(b) The enzyme would mostly be in the T form.

(c) The enzyme would likely follow Michaelis–Menten kinetics.

(d) The plot of V versus [S] would be shaped like a hyperbola.

(e) The plot of V versus [S] would be sigmoidal.

3. Differentiate between homotropic and heterotropic effectors. ❖ **3**

4. As shown in Figure 7.2, CTP inhibits ATCase; however, the inhibition is not complete. Can you suggest another molecule that might enhance the inhibition of ATCase? Hint: See Figure 25.2. ❖ **1**

5. Classify the following phrases as describing a kinase, a phosphatase, neither, or both. ❖ **5**

(a) May use ATP as a phosphoryl group donor

(b) PKA is an example.

(c) Removes phosphoryl groups from proteins

(d) Catalyzes reactions that are the reverse of dephosphorylation reactions

(e) Regulates the activity of other proteins

(f) Catalyzes phosphorylation reactions

(g) Turns off signaling reactions triggered by kinases

(h) In eukaryotes, transfers phosphoryl groups to acidic amino acids

6. Which statements about isozymes are true? ❖ **4**

(a) Isozymes are encoded by the same gene.

(b) Isozymes may be expressed in different tissues.

(c) Isozymes catalyze different reactions.

(d) Isozymes have different amino acid sequences.

(e) Isozymes have identical regulatory properties.

(f) Lactate dehydrogenase is an example.

7. Consider the following molecules: ATP, CTP, UTP, cAMP, and PALA. With respect to ATCase, which of these molecules is ... ❖ **2**

(a) ... a feedback inhibitor?

(b) ... a bisubstrate analog?

(c) ... an allosteric activator?

8. Order the steps below for the activation of PKA by cAMP.

(a) The free catalytic subunits interact with proteins to phosphorylate Ser and Thr residues.

(b) The regulatory subunits move out of the active sites of the catalytic subunits, and the R_2C_2 complex dissociates.

(c) Cytosolic cAMP concentration increases.

(d) The cAMP molecules bind to each PKA regulatory subunit.

9. Consider the following reaction scheme:

$$W \xrightarrow{\text{enzyme 1}} X \xrightarrow{\text{enzyme 2}} Y \xrightarrow{\text{enzyme 3}} Z$$

Product Z is a feedback inhibitor of enzyme 1. Assuming Z eventually diffuses out of the cell, in the presence of a high concentration of Z: ❖ 2

(a) Would the concentration of X increase, decrease, or stay the same?

(b) Would the concentration of Y increase, decrease, or stay the same?

(c) Would the concentration of Z increase, decrease, or stay the same?

(d) What happens to the activity of enzyme 1 if the concentration of Z decreases?

10. A drug company has decided to use recombinant DNA methods to prepare a modified α_1-antitrypsin that will be more resistant to oxidation than is the naturally occurring inhibitor. Which single amino acid substitution would you recommend?

11. Consider the velocity curve of an allosteric enzyme, the thick black curve labeled "(a)" in the graph below.

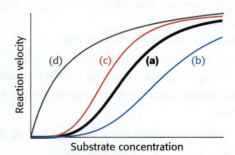

Which of the four depicted curves correctly illustrates the effect of a negative modifier? ❖ 1

12. The proteolytic enzyme trypsin is produced in the pancreas as the zymogen trypsinogen. Trypsinogen is cleaved to yield active trypsin, which, in turn, activates other pancreatic zymogens.

(a) Which of the following enzymes is used in the activation of trypsinogen: chymotrypsin, enteropeptidase, fibrin, trypsin?

(b) Which of the following zymogens is directly activated by trypsin: chymotrypsinogen, proelastase, procarboxypeptidase, prolipase?

13. Arrange the events of blood clot formation in the correct order:

(a) Fibrinogen is converted to fibrin.

(b) Trauma releases tissue factor (TF).

(c) Transglutaminase cross-links fibrin, yielding a stabilized clot.

(d) TF activates factor X to factor X_a.

(e) Fibrin threads align, forming a "soft clot."

(f) Prothrombin is converted to thrombin.

14. The following graph shows the fraction of an allosteric enzyme in the R state (f_R) and the fraction of active sites bound to substrate (Y) as a function of substrate concentration. Which model, the concerted or sequential, best explains these results? ❖ 1

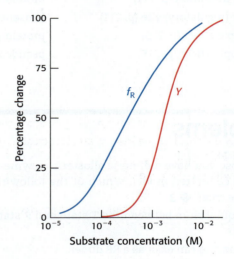

15. Examine the following metabolic pathway:

$$A \xrightarrow{e_1} B \xrightleftharpoons{e_3} C \xrightarrow{e_4} D \xrightarrow{e_5} E \xrightleftharpoons{e_6} F \xrightarrow{e_7} G$$

$$B \xrightarrow{e_2} B'$$

(a) Which enzymes catalyze irreversible reactions?

(b) Which of the enzymes, identified as "e" with a numeric subscript, is likely to be the allosteric enzyme that controls the synthesis of G?

16. Recent studies have suggested that protein kinase A may be important in establishing behaviors in many organisms, including humans. One study investigated the role of PKA in locust behavior. Certain species of locust live solitary lives until crowded, at which point they become gregarious—they prefer the crowded life.

Locusts were grouped together for one hour, and then allowed to stay with the group or move away. Prior to crowding, some insects were injected with a PKA inhibitor, a cyclic

GMP-dependent kinase inhibitor, or no inhibitor, as indicated. The results are shown here: ❖ 5

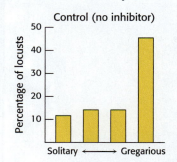

Control (no inhibitor)

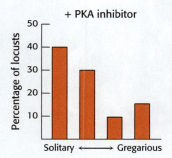

+ PKA inhibitor

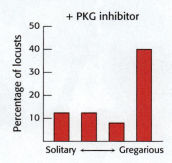

+ PKG inhibitor

(a) What is the response of the control group to crowding?

(b) What is the result if the insects are first treated with PKA inhibitor? PKG inhibitor?

(c) What was the purpose of the experiment with the PKG inhibitor?

(d) What do these results suggest about the role of PKA in the transition from a solitary to a gregarious lifestyle?

17. Recall that PALA is a potent inhibitor of ATCase because it mimics the two physiological substrates. However, in the presence of substrates, low concentrations of this unreactive bisubstrate analog *increase* the reaction velocity. On the addition of PALA, the reaction rate increases until an average of three molecules of PALA are bound per molecule of enzyme. This maximal velocity is 17-fold greater than it is in the absence of PALA. The reaction rate then decreases to nearly zero on the addition of three more molecules of PALA per molecule of enzyme. Why do low concentrations of PALA activate ATCase?

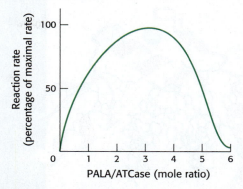

18. According to the concerted model, the relationship between the concentrations of the T and R states in the presence of substrates can be expressed as:

$$\frac{[T_i]}{[R_i]} = c^i L$$

where c is the ratio of the dissociation constants of the substrate for the R state to the T state, that is:

$$c = \frac{K_d^R}{K_d^T},$$

L is the ratio of [T] to [R] *in the absence of substrate*, and i is the number of substrate molecules bound.

Assume we have an allosteric enzyme for which a substrate binds 100 times as tightly to the R state as to its T state.

(a) By what factor does the binding of one substrate molecule per enzyme molecule alter the ratio of the concentrations of enzyme molecules in the R and T states?

(b) Suppose that L, the ratio of [T] to [R] in the absence of substrate, is 10^7 and that the enzyme contains four binding sites for substrate. What is the ratio of enzyme molecules in the R state to those in the T state in the presence of saturating amounts of substrate, assuming that the concerted model is obeyed? ❖ 3

DNA, RNA, and the Flow of Genetic Information

Family resemblance, shown here in the likeness of a puppy to its parents, results from having genes in common. Genes must be expressed to exert an effect, and proteins regulate such expression. One such regulatory protein, a zinc-finger protein (zinc ions are blue, protein is red), is shown bound to a control region of DNA (black). [Drawn from 1AAY.pdb.]

Lelusy/Shutterstock

OUTLINE

8.1 A Nucleic Acid Consists of Four Kinds of Bases Linked to a Sugar–Phosphate Backbone

8.2 A Pair of Nucleic Acid Strands with Complementary Sequences Can Form a Double-Helical Structure

8.3 The Double Helix Facilitates the Accurate Transmission of Hereditary Information

8.4 DNA Is Replicated by Polymerases That Take Instructions from Templates

8.5 Gene Expression Is the Transformation of DNA Information into Functional Molecules

8.6 Amino Acids Are Encoded by Groups of Three Bases Starting from a Fixed Point

8.7 Most Eukaryotic Genes Are Mosaics of Introns and Exons

❖ LEARNING GOALS

By the end of this chapter, you should be able to:

1. Identify the bases for DNA and RNA and distinguish between nucleosides and nucleotides.
2. Distinguish between DNA and RNA in both structure and function.
3. Explain how DNA is replicated.
4. Explain how information flows from DNA to protein.
5. Identify some key differences between bacterial and eukaryotic genes.

DNA and RNA are long linear polymers, called nucleic acids, that carry information in a form that can be passed from one generation to the next. Due to its remarkable long-term stability and double-helical structure, DNA polymers are well-suited for the storage and replication of the genetic information for all cells and many viruses. However, DNA is not the direct template for protein synthesis. Rather, a DNA strand is copied into a class of RNA molecules called messenger RNA (mRNA). The scheme that underlies information processing at the level of gene expression was first proposed by Francis Crick in 1958.

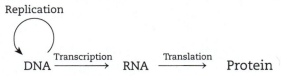

Replication

DNA →(Transcription)→ RNA →(Translation)→ Protein

Crick called this scheme the *central dogma*. The basic tenets of this dogma are true, but, as we will see later, this scheme is not as simple as depicted. This flow of information depends on the genetic code, which defines the relation between the sequence of bases in DNA and the sequence of amino acids in a protein.

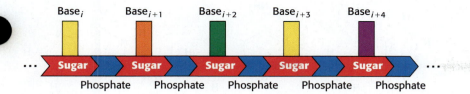

8.1 A Nucleic Acid Consists of Four Kinds of Bases Linked to a Sugar–Phosphate Backbone

The nucleic acids **deoxyribonucleic acid (DNA)** and **ribonucleic acid (RNA)** are well suited to function as the carriers of genetic information by virtue of their covalent structures. These macromolecules are linear polymers built from similar units connected end to end (**Figure 8.1**). Each monomer unit within the polymer is a **nucleotide**. A single nucleotide unit consists of three components: a sugar, a phosphate, and one of four bases. The sequence of bases in the polymer uniquely characterizes a nucleic acid and constitutes a form of linear information—information analogous to the letters that spell a person's name.

RNA and DNA differ in the sugar component and one of the bases

The sugar in DNA is **deoxyribose**. The prefix deoxy- indicates that the 2′-carbon atom of the sugar lacks the oxygen atom that is linked to the 2′-carbon atom of **ribose**, as shown in **Figure 8.2**. Note that sugar carbons are numbered with primes to differentiate them from atoms in the bases. The sugars in both nucleic acids are linked to one another by phosphodiester bridges. Specifically, the 3′-hydroxyl (3′-OH) group of the sugar moiety of one nucleotide is esterified to a phosphate group, which, in turn, is joined to the 5′-hydroxyl group of the adjacent sugar.

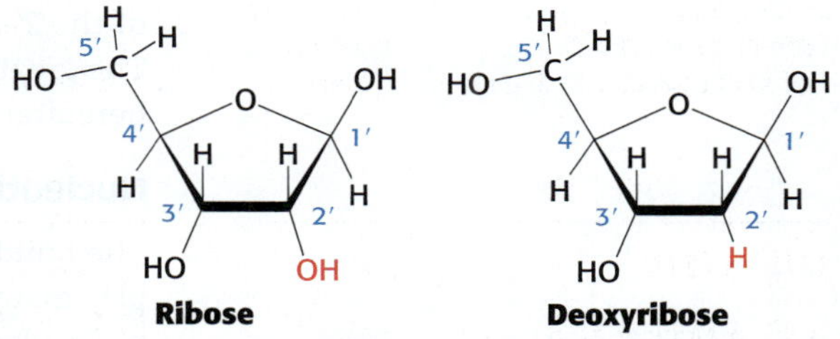

The chain of sugars linked by phosphodiester bridges is referred to as the *backbone* of the nucleic acid (**Figure 8.3**). Whereas the backbone is constant in a nucleic acid, the bases vary from one monomer to the next. Two of the bases of DNA are derivatives of **purine**—adenine (A) and guanine (G)—and two of **pyrimidine**—cytosine (C) and thymine (T), as shown in **Figure 8.4**.

RNA, like DNA, is a long, unbranched polymer consisting of nucleotides joined by 3′-to-5′ phosphodiester linkages (Figure 8.3). The covalent structure of RNA differs from that of DNA in two respects. First, the sugar units in RNA are riboses rather than deoxyriboses. Ribose contains a 2′-hydroxyl group not present in deoxyribose (Figure 8.2). Second, one of the four major bases in RNA is uracil (U) instead of thymine (T) (Figure 8.4).

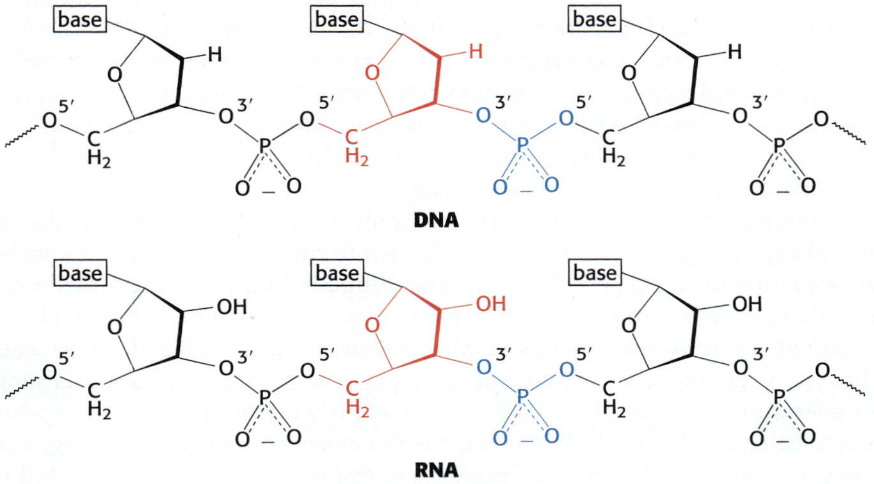

Ribose

Deoxyribose

FIGURE 8.2 Ribose and deoxyribose are the sugars found in RNA and DNA, respectively. Atoms in sugar units are numbered with primes to distinguish them from atoms in bases (see Figure 8.4).

DNA

RNA

FIGURE 8.3 The sugar–phosphate backbones of nucleic acids are formed by 3′-to-5′ phosphodiester linkages. For each nucleic acid, a sugar unit is highlighted in red and a phosphate group in blue.

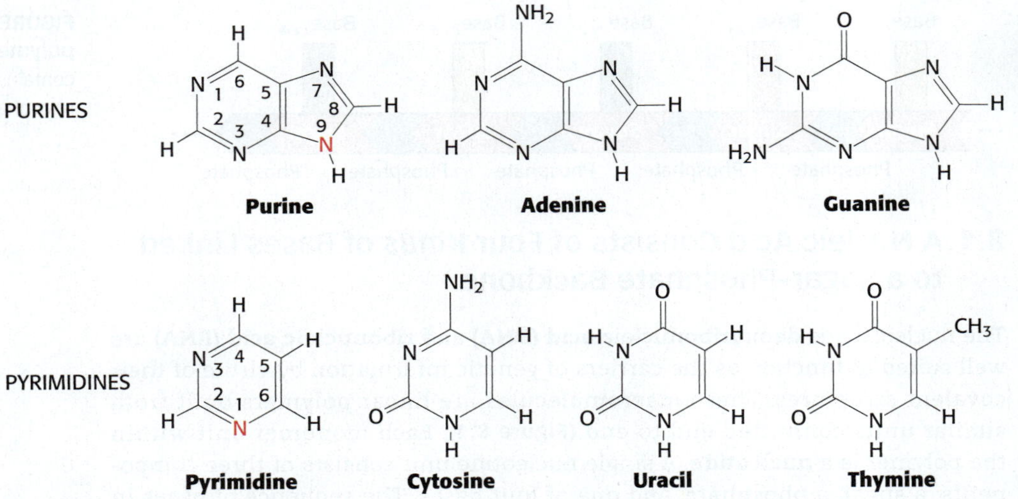

FIGURE 8.4 The bases in nucleic acids are either purines or pyrimidines. Atoms within bases are numbered without primes. Uracil is present in RNA instead of thymine. N-9 in purines and N-1 in pyrimidines form a bond with the sugar (see Figure 8.5).

PURINES

Purine   Adenine   Guanine

PYRIMIDINES

Pyrimidine   Cytosine   Uracil   Thymine

Note that each phosphodiester bridge has a negative charge (Figure 8.3). This negative charge repels nucleophilic species such as hydroxide ions, which are capable of hydrolytic attack on the phosphate backbone. This resistance is crucial for maintaining the integrity of information stored in nucleic acids. The absence of the 2′-hydroxyl group in DNA further increases its resistance to hydrolysis. The greater stability of DNA probably accounts for its use rather than RNA as the hereditary material in all modern cells and in many viruses.

Nucleotides are the monomeric units of nucleic acids

The building blocks of nucleic acids and the precursors of these building blocks play many other roles throughout the cell—for instance, as energy currency and as molecular signals. Consequently, it is important to be familiar with the nomenclature of nucleotides and their precursors. A unit consisting of a base bonded to a sugar is referred to as a **nucleoside**. The four nucleoside units in RNA are called *adenosine, guanosine, cytidine,* and *uridine,* whereas those in DNA are called *deoxyadenosine, deoxyguanosine, deoxycytidine,* and *thymidine.* (Thymidine contains deoxyribose, and by convention, the prefix deoxy- is not added because thymine-containing nucleosides are only rarely found in RNA.)

In each nucleoside, N-9 of a purine or N-1 of a pyrimidine is attached to C-1′ of the sugar by an N-β-glycosidic linkage (**Figure 8.5**). The base lies above the plane of the sugar when the structure is written in the standard orientation; that is, the configuration of the N-glycosidic linkage is β (Section 11.1).

A nucleotide is a nucleoside joined to one or more phosphoryl groups by an ester linkage (Figure 8.5). Nucleoside triphosphates, nucleotides containing three phosphoryl groups, are the precursors that form RNA and DNA. The four nucleotide units that comprise DNA are nucleoside monophosphates called *deoxyadenylate, deoxyguanylate, deoxycytidylate,* and *thymidylate.* Note that a pyrophosphate is released when the nucleotides are linked (Section 8.4). Similarly, the most common nucleotides found in RNA are nucleoside monophosphates: *adenylate, guanylate, cytidylate,* and *uridylate.*

A more precise nomenclature is commonly used to describe the number of phosphoryl groups and the site of attachment to carbon of the ribose. Consider the structure of ATP (Figure 8.5). This compound is formed by the attachment of a phosphoryl group to C-5′ of a nucleoside sugar (the most common site of phosphate esterification). In this naming system for nucleotides, the number of phosphoryl groups and the attachment site are designated. Hence, ATP is short for adenosine 5′-triphosphate. ATP is tremendously important because, in addition to being a building block for RNA, it is the most commonly used energy currency. The energy released from cleavage of the triphosphate group is used to power many cellular processes (Chapter 15).

β-Glycosidic linkage

**A nucleoside
(Adenosine)**

β-Glycosidic linkage

**A nucleotide
(Adenosine 5′-triphosphate)**

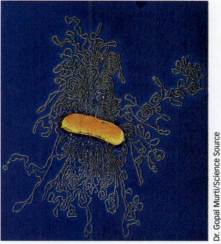

FIGURE 8.5 Nucleosides and nucleotides are distinguished by the presence of phosphoryl groups. A nucleoside contains a sugar and a base joined by an *N*-β-glycosidic linkage. A nucleotide contains a nucleoside linked to one or more phosphoryl groups.

DNA molecules are very long and have directionality

A striking characteristic of naturally occurring DNA molecules is their length. A DNA molecule must comprise many nucleotides to carry the genetic information necessary for even the simplest organisms. For example, the DNA of a virus such as polyoma, which can cause cancer in certain organisms, consists of two paired strands of DNA, each 5100 nucleotides in length. The E. coli genome is a single DNA molecule consisting of two strands of 4.6 million nucleotides each (**Figure 8.6**).

The DNA molecules of higher organisms can be much larger. The human genome comprises approximately 3 billion nucleotides in each strand of DNA, divided among 24 distinct molecules of DNA called chromosomes (22 autosomal chromosomes plus the X and Y sex chromosomes) of different sizes. One of the largest known DNA molecules is found in the Indian muntjac, an Asiatic deer; its genome is nearly as large as the human genome but is distributed on only 3 chromosomes (**Figure 8.7**). The largest of these chromosomes has two strands of more than 1 billion nucleotides each. If such a DNA molecule could be fully extended, it would stretch more than 1 foot in length. Some plants contain even larger DNA molecules.

FIGURE 8.6 The *E. coli* genome is composed of two paired nucleic acid strands, each with more than 4.5 million nucleotides. As shown by this electron micrograph, the DNA must be tightly compacted in order to fit in such a small volume.

Dr. Gopal Murti/Science Source

FIGURE 8.7 The Indian muntjac genome is about as large as a human's but is divided among three chromosome pairs. In cells from a female Indian muntjac (left), the three pairs of very large chromosomes are stained orange in the right image. The image also shows a pair of human chromosomes (stained green) for comparison. [Credits: (Left) Super Prin/Shutterstock; (Right) Reprinted by permission from Macmillan Publishers Ltd: *Nature Genetics*, J-Y Lee, M. Koi, E.J. Stanbridge, M. Oshimura, A.T. Kumamoto, & A.P. Feinberg, Simple purification of human chromosomes to homogeneity using muntjac hybrid cells, vol. 7, pp. 29–33, ©1994, Figure 2b. Permission conveyed through Copyright Clearance Center, Inc.]

FIGURE 8.8 The structure of a nucleic acid strand can be simplified to the identity of its bases. A simplified depiction of a nucleic acid (compare with Figure 8.3). The strand has a 5′ end, which is usually attached to a phosphoryl group, and a 3′ end, which is usually a free hydroxyl group.

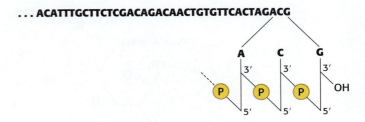

Note that, like a polypeptide (Section 2.3), a DNA chain has directionality, commonly called polarity. One end of the chain has a free 5′-OH group or a 5′-OH group attached to a phosphoryl group, and the other end has a free 3′-OH group, neither of which is linked to another nucleotide (**Figure 8.8**). When presenting very long sequences of nucleotides, only the identity of the bases are written, given that these bases represent the genetic information. However, it is important to remember that even if only the bases are presented, the base sequence is written in the 5′-to-3′ direction. Thus, ACG indicates that the unlinked 5′-OH group is on deoxyadenylate, whereas the unlinked 3′-OH group is on deoxyguanylate. Because of this polarity, ACG and GCA correspond to different compounds.

8.2 A Pair of Nucleic Acid Strands with Complementary Sequences Can Form a Double-Helical Structure

As discussed in Section 1.2, the covalent structure of nucleic acids accounts for their ability to carry information in the form of a sequence of bases along a nucleic acid strand. The bases on the two separate nucleic acid strands form specific base pairs in such a way that a helical structure is formed. The double-helical structure of DNA facilitates the replication of the genetic material—that is, the generation of two copies of a nucleic acid from one.

The double helix is stabilized by hydrogen bonds and van der Waals interactions

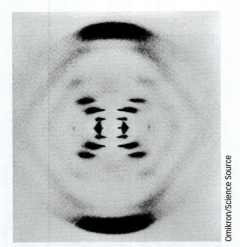

FIGURE 8.9 The x-ray diffraction photograph of a hydrated DNA fiber provided experimental evidence of DNA's three-dimensional structure. When crystals of a biomolecule are irradiated with x-rays, the x-rays are diffracted, and these diffracted x-rays are seen as a series of spots, called reflections, on a screen behind the crystal. The structure of the molecule can be determined by the pattern of the reflections (Section 4.5). In regard to DNA crystals, the central cross is diagnostic of a helical structure. The strong arcs on the meridian arise from the stack of nucleotide bases, which are 3.4 Å apart.

Omikron/Science Source

The ability of nucleic acids to form specific base pairs was discovered in the course of studies directed at determining the three-dimensional structure of DNA. Maurice Wilkins and Rosalind Franklin obtained x-ray diffraction photographs of fibers of DNA (**Figure 8.9**). The characteristics of these diffraction patterns indicated that DNA is formed of two strands that wind in a regular helical structure. From these data and others, James Watson and Francis Crick deduced a structural model for DNA that accounted for the diffraction pattern and was the source of some remarkable insights into the functional properties of nucleic acids (**Figure 8.10**).

The features of the Watson–Crick model of DNA deduced from the diffraction patterns are:

- Two helical polynucleotide strands are coiled around a common axis with a right-handed screw sense (Section 2.4). The strands are antiparallel, meaning that they have opposite directionality.

- The sugar–phosphate backbones are on the outside and the purine and pyrimidine bases lie on the inside of the helix.

- The bases are nearly perpendicular to the helix axis, and adjacent bases are separated by approximately 3.4 Å. The helical structure repeats on the order of every 34 Å, with about 10.4 bases per turn of helix. There is a rotation of nearly 36 degrees per base (360 degrees per full turn/10.4 bases per turn).

- The diameter of the helix is about 20 Å.

How is such a regular structure able to accommodate an arbitrary sequence of bases, given the different sizes and shapes of the purines and pyrimidines?

(A) Side view

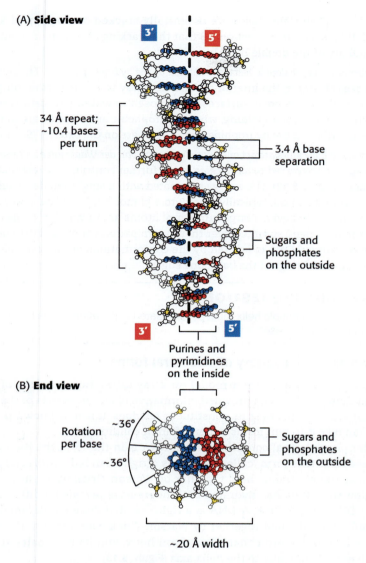

34 Å repeat; ~10.4 bases per turn

3.4 Å base separation

Sugars and phosphates on the outside

Purines and pyrimidines on the inside

(B) End view

Rotation per base

~36°

~36°

Sugars and phosphates on the outside

~20 Å width

FIGURE 8.10 The Watson–Crick model of double-helical DNA reveals critical structural features. In this figure, the bases of the individual strands are colored red and blue, while the phosphates are colored yellow. (A) The side view reveals that adjacent bases are separated by 3.4 Å. The structure repeats along the helical axis (vertical) at intervals of 34 Å, which corresponds to approximately 10 nucleotides on each chain. (B) The end view, looking down the helix axis, reveals a rotation of 36° per base and shows that the bases are stacked on top of one another. [Drawn from 3BSE.pdb.]

In attempting to answer this question, Watson and Crick discovered that guanine can be paired with cytosine and adenine with thymine to form base pairs that have essentially the same shape (**Figure 8.11**). These base pairs are held together by specific hydrogen bonds, which, although weak (4–21 kJ mol^{-1}, or 1–5 kcal mol^{-1}), stabilize the helix because of their large numbers in a DNA molecule. These base-pairing rules account for the observation originally made by biochemist Erwin Chargaff in the 1940s—and subsequently referred to as *Chargaff's rules*—that the ratios of adenine to thymine and of guanine to cytosine are nearly the same in all species studied, whereas the adenine-to-guanine ratio varies considerably (**Table 8.1**).

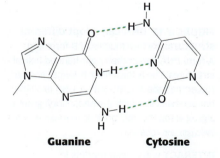

Guanine   Cytosine

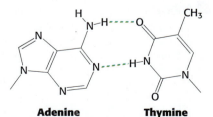

Adenine   Thymine

FIGURE 8.11 Watson and Crick proposed two base pairs. In their model, guanine paired with cytosine and adenine paired with thymine.

TABLE 8.1 Base compositions experimentally determined for a variety of organisms

Organism	A : T	G : C	A : G
Human being	1.00	1.00	1.56
Salmon	1.02	1.02	1.43
Wheat	1.00	0.97	1.22
Yeast	1.03	1.02	1.67
Escherichia coli	1.09	0.99	1.05
Serratia marcescens	0.95	0.86	0.70

Base stacking
(van der Waals interactions)

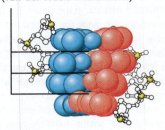

FIGURE 8.12 A side view of DNA reveals the stacking of base pairs within the double helix. The stacked bases interact via van der Waals forces. Such stacking forces help stabilize the double helix. [Drawn from 3BSE.pdb.]

Inside the helix, the bases are essentially stacked one on top of another (Figure 8.10B). Recall from Section 1.3 that the stacking of base pairs contributes to the stability of the double helix in two ways:

- *The formation of the double helix is facilitated by the hydrophobic effect.* The hydrophobic bases cluster in the interior of the helix away from the surrounding water, whereas the more polar surfaces are exposed to water. This arrangement is reminiscent of protein folding, where hydrophobic amino acids are in the protein's interior and the hydrophilic amino acids are on the exterior (Section 2.5).

- *The stacked base pairs attract one another through van der Waals forces.* These forces, appropriately referred to as *stacking forces*, further contribute to stabilization of the helix (**Figure 8.12**). The energy associated with a single van der Waals interaction is quite small, typically from 2 to 4 kJ mol^{-1} (0.5–1.0 kcal mol^{-1}). In the double helix, however, a large number of atoms are in van der Waals contact, and the net effect, summed over these atom pairs, is substantial. In addition, base stacking in DNA is favored by the conformations of the somewhat rigid five-membered rings of the backbone sugars.

❖ SELF–CHECK QUESTION

DNA in the form of a double helix must be associated with cations, usually Mg^{2+}. Why is this requirement the case?

DNA can assume a variety of structural forms

Watson and Crick based their model on x-ray diffraction patterns of highly hydrated DNA fibers, which provided information about properties of the double helix that are averaged over its constituent residues. It is now known that DNA helices can take different forms under different conditions.

Under physiological conditions, most DNA is in the form that Watson and Crick modeled. This form, called *B-DNA*, is a right-handed double helix made up of antiparallel strands held together by Watson–Crick base-pairing. X-ray diffraction studies of less-hydrated DNA fibers have revealed a different form called *A-DNA*. Like B-DNA, A-DNA is a right-handed double helix made up of antiparallel strands held together by Watson–Crick base-pairing. The A-form helix is wider and shorter than the B-form helix, and its base pairs are tilted rather than perpendicular to the helix axis (**Figure 8.13**).

FIGURE 8.13 DNA can adopt different structural forms. Models of B-form and A-form DNA depict their right-handed helical structures. The B-form helix is longer and narrower than the A-form helix. Z-DNA is a left-handed helix in which the phosphoryl groups zigzag along the backbone. [Drawn from 3BSE.pdb, 3V9D.pdb, and 4OCB.pdb.]

INTERACT with these models in
≋ Achieve

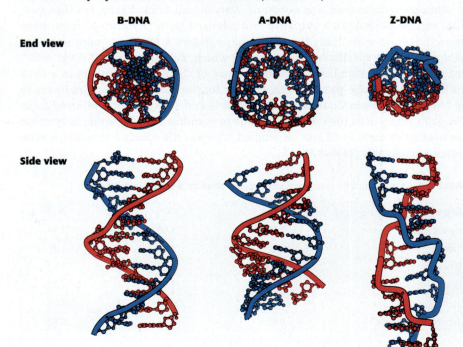

If the A-form helix were simply a property of dehydrated DNA, it would be of little significance. However, double-stranded regions of RNA and at least some RNA–DNA hybrids adopt a double-helical form very similar to that of A-DNA. What is the biochemical basis for differences between the two forms of DNA? Many of the structural differences between B-DNA and A-DNA arise from different puckerings of their ribose units (Figure 8.14). In A-DNA, C-3′ lies out of the plane (a conformation referred to as C-3′ endo) formed by the other four atoms of the ring; in B-DNA, C-2′ lies out of the plane (a conformation called C-2′ endo). The C-3′-endo puckering in A-DNA leads to an 11-degree tilting of the base pairs away from being perpendicular to the helix.

RNA helices are further induced to take the A-DNA form because of steric hindrance from the 2′-hydroxyl group: the 2′-oxygen atom would be too close to three atoms of the adjoining phosphoryl group and to one atom in the next base. In an A-form helix, in contrast, the 2′-oxygen atom projects outward, away from other atoms. The phosphoryl and other groups in the A-form helix bind fewer H_2O molecules than do those in B-DNA. Hence, dehydration favors the A form.

A third type of double helix is left-handed, in contrast with the right-handed screw sense of the A and B helices. Furthermore, the phosphoryl groups in the backbone are zigzagged; hence, this form of DNA is called Z-DNA (Figure 8.13). Although the biological role of Z-DNA is still under investigation, Z-DNA-binding proteins have been isolated, one of which is required for viral pathogenesis of poxviruses, including variola, the agent of smallpox. The existence of Z-DNA shows that DNA is a flexible, dynamic molecule whose parameters are not as fixed as depictions suggest. The properties of B-, A-, and Z-DNA are compared in Table 8.2.

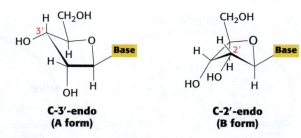

FIGURE 8.14 Sugar pucker explains many of the structural differences between the B-form and A-form of DNA. In A-form DNA, the C-3′ carbon atom lies above the approximate plane defined by the four other sugar nonhydrogen atoms (called C-3′ endo). In B-form DNA, each deoxyribose is in a C-2′-endo conformation, in which C-2′ lies out of the plane.

TABLE 8.2 Comparison of B-, A-, and Z-DNA

	B	A	Z
Shape	Intermediate	Broadest	Narrowest
Rise per base pair	3.4 Å	2.3 Å	3.8 Å
Helix diameter	~20 Å	~26 Å	~18 Å
Screw sense	Right-handed	Right-handed	Left-handed
Glycosidic bond*	*anti*	*anti*	Alternating *anti* and *syn*
Base pairs per turn of helix	10.4	11	12
Pitch per turn of helix	35.4 Å	25.3 Å	45.6 Å
Tilt of base pairs from perpendicular to helix axis	1 degree	19 degrees	9 degrees

*Syn and anti refer to the orientation of the N-glycosidic bond between the base and deoxyribose. In the *anti* orientation, the base extends away from the deoxyribose. In the *syn* orientation, the base is above the deoxyribose. Pyrimidines can be in *anti* orientations only, whereas purines can be *anti* or *syn*.

The major and minor grooves are lined by sequence-specific hydrogen-bonding groups

An examination of the DNA molecule in the B form, as shown in Figure 8.13, reveals the presence of two distinct grooves, called the major groove and the minor groove. These grooves arise because the glycosidic bonds of a base pair are not diametrically opposite each other (Figure 8.15A). In B-DNA, the major groove is wider (12 Å versus 6 Å) and deeper (8.5 Å versus 7.5 Å) than the minor groove (Figure 8.15B).

Each groove is lined by potential hydrogen-bond donor and acceptor atoms that enable interactions with proteins. These interactions are essential for replication and transcription because particular proteins bind to DNA, recognizing specific hydrogen-bond donors and acceptors on the surfaces of the grooves, to catalyze these processes (Chapters 28, 29, and 31).

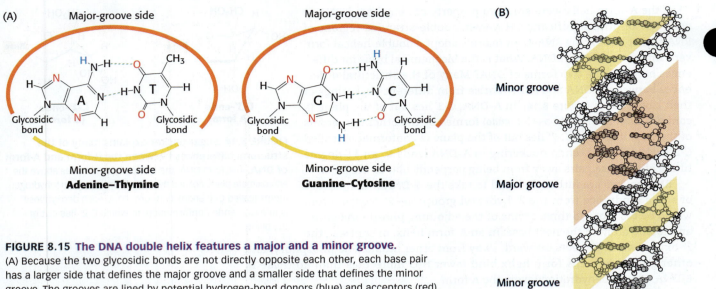

FIGURE 8.15 The DNA double helix features a major and a minor groove.
(A) Because the two glycosidic bonds are not directly opposite each other, each base pair has a larger side that defines the major groove and a smaller side that defines the minor groove. The grooves are lined by potential hydrogen-bond donors (blue) and acceptors (red). (B) In a side view of the double helix, the major (orange) and the minor (yellow) grooves alternate with each other.

In the minor groove, N-3 of adenine or guanine and O-2 of thymine or cytosine can serve as hydrogen acceptors, and the amino group attached to C-2 of guanine can be a hydrogen donor. In the major groove, N-7 of guanine or adenine is a potential acceptor, as are O-4 of thymine and O-6 of guanine. The amino groups attached to C-6 of adenine and C-4 of cytosine can serve as hydrogen donors. The methyl group of thymine also lies in the major groove. Note that the major groove displays more features that distinguish one base pair from another than does the minor groove. The larger size of the major groove in B-DNA makes it more accessible for interactions with proteins that recognize specific DNA sequences.

Some DNA molecules are circular and supercoiled

The DNA molecules in human chromosomes are linear. However, electron microscopic and other studies have shown that intact DNA molecules from bacteria and archaea are circular (**Figure 8.16A**). DNA molecules inside cells necessarily have a very compact shape. Note that the *E. coli* chromosome, fully extended, would be about 1000 times as long as the greatest diameter of the bacterium.

A closed DNA molecule has a property unique to circular DNA. The axis of the double helix can itself be twisted into a *superhelix*, where the helix is wound about itself, in a process known as *supercoiling* (**Figure 8.16B**). A circular DNA molecule without any superhelical turns is known as a *relaxed molecule*. Supercoiling is biologically important for two reasons that will be considered further in Chapter 28:

- A supercoiled DNA molecule is more compact than its relaxed counterpart.
- Supercoiling may hinder or favor the capacity of the double helix to unwind and thereby affect the interactions between DNA and other molecules.

Single-stranded nucleic acids can adopt elaborate structures

Single-stranded nucleic acids often fold back on themselves to form well-defined structures. Such structures are especially prominent in RNA and RNA-containing complexes such as the ribosome—a large complex of RNAs and proteins on which proteins are synthesized.

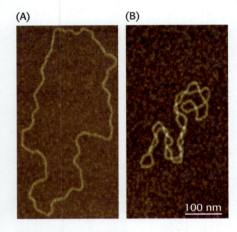

FIGURE 8.16 Atomic force microscopy of circular DNA from bacteria reveals both relaxed and supercoiled forms.
(A) Relaxed form. (B) Supercoiled form

[Credit: Republished with permission of Oxford University Press, from Witz, Guillaume; Stasiak, Andrzej, "DNA super-coiling and its role in DNA decatenation and unknotting," (2010) *Nuc. Acids Res.* 38(7) 2119-2133, Figure 1. Permission conveyed through Copyright Clearance Center, Inc.]

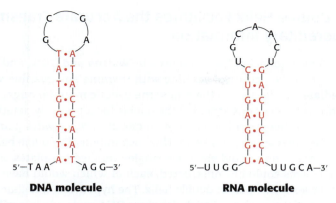

FIGURE 8.17 Stem-loop structures can be formed from single-stranded DNA and RNA molecules.

DNA molecule   **RNA molecule**

The simplest and most-common structural motif formed is a **stem-loop**, created when two complementary sequences within a single strand come together to form double-helical structures (**Figure 8.17**). In many cases, these double helices are made up entirely of Watson–Crick base pairs. In other cases, however, the structures include mismatched base pairs or unmatched bases that bulge out from the helix. Such mismatches destabilize the local structure but introduce deviations from the standard double-helical structure that can be important for higher-order folding and for function (**Figure 8.18**).

Single-stranded nucleic acids can adopt complex structures through the interaction of more widely separated bases. Often, three or more bases interact to stabilize these structures. In such cases, hydrogen-bond donors and acceptors that do not participate in Watson–Crick base pairs participate in hydrogen bonds to form nonstandard pairings (Figure 8.18B). Metal ions such as magnesium ion (Mg^{2+}) often assist in the stabilization of these more elaborate structures. These complex structures allow RNA to perform a host of functions that the double-stranded DNA molecule cannot. Indeed, the complexity of some RNA molecules rivals that of proteins, and these RNA molecules perform a number of functions that had formerly been thought the exclusive domain of proteins.

(A)

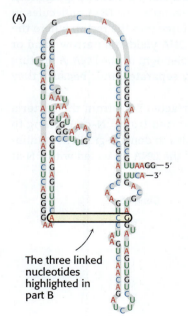

The three linked nucleotides highlighted in part B

(B)

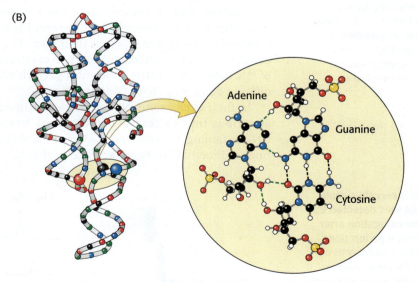

FIGURE 8.18 RNA molecules can adopt complex structures. In this example, a single-stranded RNA molecule folds back on itself. (A) The nucleotide sequence shows Watson–Crick base pairs and other nonstandard base pairings in stem-loop structures. (B) In the three-dimensional structure at the left, cytidine nucleotides are shown in blue, adenosine in red, guanosine in black, and uridine in green. The detailed projection at the right shows an important long-range interaction between three of the bases. Hydrogen bonds within the Watson–Crick base pair are shown as dashed black lines; additional hydrogen bonds are shown as dashed green lines.

8.3 The Double Helix Facilitates the Accurate Transmission of Hereditary Information

The double-helical model of DNA proposed by Watson and Crick, and in particular the specific base pairing of adenine with thymine and guanine with cytosine, immediately suggested to them how the genetic material might replicate: The sequence of bases of one strand of the double helix precisely determines the sequence of the other strand. A guanine on one strand is always paired with a cytosine on the other strand, and so on. Thus, separation of a double helix into its two component strands would yield two single-stranded templates onto which new double helices could be constructed, each of which would have the same sequence of bases as the parent double helix. The hypothesis was that as DNA is replicated, one of the strands of each daughter DNA molecule is newly synthesized, whereas the other is passed unchanged from the parent DNA molecule. This distribution of parental atoms is called **semiconservative replication**.

Differences in DNA density established the validity of the semiconservative replication hypothesis

Matthew Meselson and Franklin Stahl carried out a critical test of this hypothesis in 1958. They labeled the parent DNA with ^{15}N, a heavy isotope of nitrogen, to make it denser than ordinary DNA. The labeled DNA was generated by growing E. coli for many generations in a medium that contained $^{15}NH_4Cl$ as the sole nitrogen source. After the incorporation of heavy nitrogen was complete, the bacteria were abruptly transferred to a medium that contained only ^{14}N, the ordinary isotope of nitrogen. The question asked was: What is the distribution of ^{14}N and ^{15}N in the DNA molecules after successive rounds of replication?

The distribution of ^{14}N and ^{15}N was revealed by the technique of density-gradient equilibrium sedimentation. A small amount of the E. coli DNA was dissolved in a concentrated solution of cesium chloride. This solution was centrifuged until it was nearly at equilibrium. At that point, the opposing processes of sedimentation and diffusion created a stable density gradient in the concentration of cesium chloride across the sample. The DNA molecules in this density gradient were driven by centrifugal force into the region where the solution's density was equal to their own. The DNA yielded a narrow band or bands that were detected by absorption of ultraviolet light by the DNA. A mixture of ^{14}N DNA and ^{15}N DNA molecules gave clearly separate bands because they differ in density by about 1% (**Figure 8.19**).

In their experiment, Meselson and Stahl extracted DNA from the bacteria at various times after the bacteria were transferred from ^{15}N-containing to ^{14}N-containing medium and subjected the samples to density-gradient equilibrium sedimentation. **Figure 8.20** shows that all of the DNA is labeled with ^{15}N at

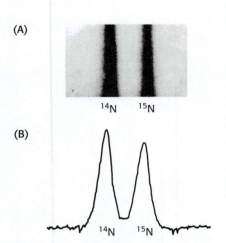

(A)

(B)

FIGURE 8.19 Meselson and Stahl resolved ^{14}N DNA and ^{15}N DNA by density-gradient centrifugation. (A) An ultraviolet-absorption photograph of a special centrifuged tube shows the two distinct bands of DNA. (B) Densitometric tracing of the absorption photograph. [Photo credit: M. Meselson and F.W. Stahl. Proc. Nat. Acad. Sci. 44 (1958):671.]

FIGURE 8.20 Semiconservative replication of *E. coli* DNA was detected by density-gradient centrifugation after incubation over several generations. The position of a band of DNA depends on its content of ^{14}N and ^{15}N. [Photo credit: From M. Meselson and F. W. Stahl. Proc. Natl. Acad. Sci. U.S.A. 44(1958):671.]

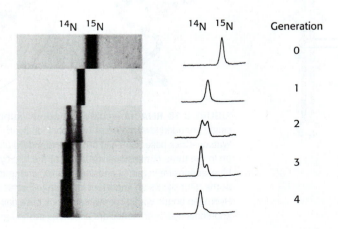

the start of the experiment (generation 0). After one generation, all of the DNA is still in a single band, but the band has shifted. The density of this band (called *hybrid DNA*) is precisely halfway between the densities of the ^{14}N DNA and ^{15}N DNA bands. After two generations, there are equal amounts of two bands of DNA. One is hybrid DNA, and the other is ^{14}N DNA. After three rounds of replication, the positions of the two bands remain unchanged, but there is three times the amount of ^{14}N DNA as there is hybrid DNA.

The absence of ^{15}N DNA indicated that parental DNA was not preserved as an intact unit after replication. The absence of ^{14}N DNA in generation 1 indicated that all of the daughter DNA derived some of their atoms from the parent DNA. Additionally, the proportion of atoms derived from the parent had to be half because the density of the hybrid DNA band was halfway between the densities of the ^{14}N DNA and ^{15}N DNA bands.

Meselson and Stahl concluded from these incisive experiments that replication was indeed semiconservative: each new double helix contains a parent strand and a newly synthesized strand. Their results agreed perfectly with the Watson–Crick model for DNA replication (**Figure 8.21**).

The double helix can be reversibly melted

During DNA replication and transcription, the two strands of the double helix must be separated from each other, at least in a local region. The two strands of a DNA helix readily come apart when the hydrogen bonds between base pairs are disrupted. In the laboratory, the double helix can be disrupted by heating a solution of DNA or by adding acid or alkali to ionize its bases.

The dissociation of the double helix is called *melting* because it occurs abruptly at a certain temperature. The **melting temperature** (T_m) of DNA is defined as the temperature at which half the helical structure is lost. Inside cells, however, the double helix is not melted by the addition of heat. Instead, proteins called *helicases* use chemical energy (from ATP) to disrupt the helix (Chapter 29).

Stacked bases in nucleic acids absorb less ultraviolet light at a wavelength of 260 nm than do unstacked bases, an effect called *hypochromism* (**Figure 8.22A**). Thus, as a sample of DNA is heated, the strands separate, increasing the proportion of single-stranded DNA, which is monitored by measuring the increase in absorption of 260 nm light (**Figure 8.22B**).

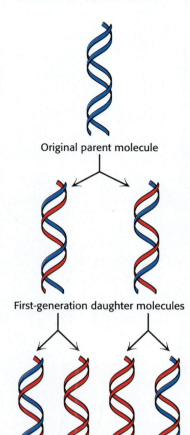

FIGURE 8.21 In semiconservative replication, each daughter DNA molecule contains one parental strand and one newly synthesized strand. Original parental DNA is shown in blue and newly synthesized DNA in red. [Information from M. Meselson and F. W. Stahl, *Proc. Natl. Acad. Sci. U.S.A.* 44: 671–682, 1958.]

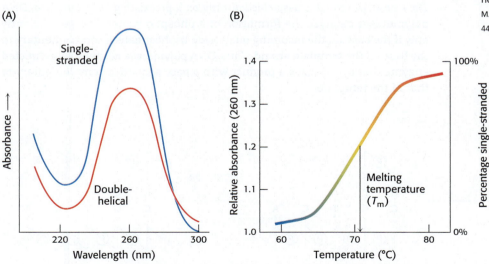

FIGURE 8.22 Single-stranded DNA absorbs light more effectively than does double-helical DNA. (A) The absorbance spectrum of similar quantities of single-stranded and double-helical DNA reveals a lower peak at around 260 nm for the double-helical sample. (B) The absorbance of a DNA solution at a wavelength of 260 nm increases when the double helix is melted into single strands.

Separated complementary strands of nucleic acids spontaneously reassociate to form a double helix when the temperature is lowered below T_m. This renaturation process is called *annealing*. The facility with which double helices can be melted and then reassociated is crucial for the biological functions of nucleic acids. The ability to melt and reanneal DNA reversibly in the laboratory provides a powerful tool for investigating sequence similarity. We will return to this important technique in Chapter 9.

❖ SELF–CHECK QUESTION

In general, the greater the percentage of GC base pairs within a region of DNA, the higher the melting temperature. **(a)** What are the chemical forces that stabilize the double helix? **(b)** Based on your answers to (a), what are some possible explanations for the relationship between GC content and T_m?

8.4 DNA Is Replicated by Polymerases That Take Instructions from Templates

We now turn to the molecular mechanism of DNA replication. The full replication machinery in a cell comprises more than 20 proteins engaged in intricate and coordinated interplay. The primary catalytic components of the replication machinery are enzymes, called *DNA polymerases*, that promote the formation of the bonds joining units of the DNA backbone. *E. coli* has a number of DNA polymerases that participate in DNA replication and repair (Chapter 28).

DNA polymerase catalyzes phosphodiester-bridge formation

DNA polymerases catalyze the step-by-step addition of deoxyribonucleotide units to a DNA strand (**Figure 8.23**). The reaction catalyzed, in its simplest form, is

$$(DNA)_n + dNTP \rightleftharpoons (DNA)_{n+1} + PP_i$$

where dNTP stands for any deoxyribonucleotide and PP_i is a pyrophosphate ion.
 DNA synthesis has the following characteristics:

- *The reaction requires all four activated precursors*—that is, the deoxynucleoside 5′-triphosphates dATP, dGTP, dCTP, and TTP—as well as Mg^{2+} ion.

- *The new DNA strand is assembled directly on a preexisting DNA template*. DNA polymerases catalyze the formation of a phosphodiester linkage efficiently only if the base on the incoming nucleoside triphosphate is complementary to the base on the **template strand**. Thus, DNA polymerase is a template-directed enzyme that synthesizes a product with a base sequence complementary to that of the template.

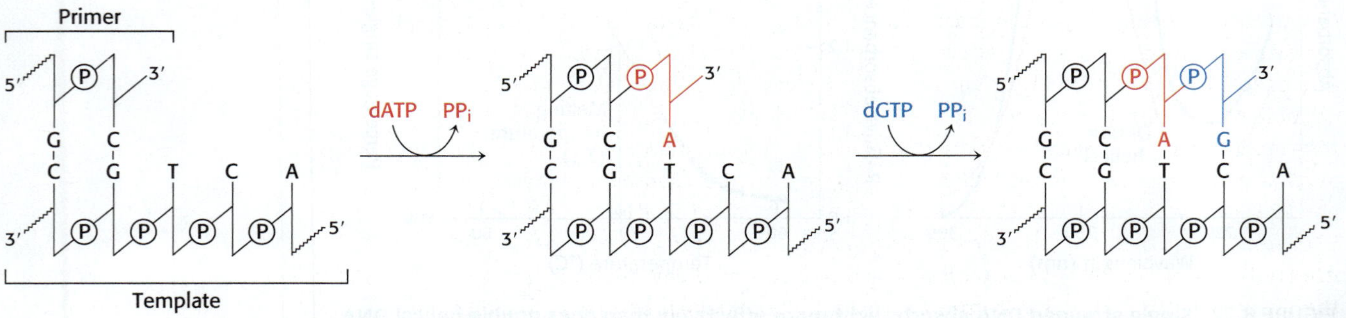

FIGURE 8.23 DNA polymerase catalyzes the polymerization reaction using nucleotide 5′-triphosphates as substrates. The reaction requires a preexisting DNA template and a primer with a free 3′-OH group.

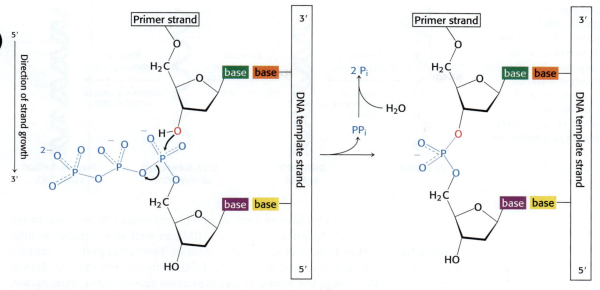

FIGURE 8.24 DNA polymerases catalyze the strand-elongation reaction in the 5′-to-3′ direction. The product of the reaction is the formation of a phosphodiester bridge. [Source: J. L. Tymoczko, J. Berg, and L. Stryer, *Biochemistry: A Short Course*, 2nd ed. (W. H. Freeman and Company, 2013), Fig. 34.2.]

- *DNA polymerases require a primer to begin synthesis.* A short single-stranded oligonucleotide having a free 3′-OH group, called a **primer**, must be already bound to the template strand. The chain-elongation reaction catalyzed by DNA polymerases is a nucleophilic attack by the 3′-OH terminus of the growing strand on the innermost phosphorus atom of the deoxynucleoside triphosphate (**Figure 8.24**). A phosphodiester bridge is formed and pyrophosphate is released.

  At this point, the reaction is readily reversible. The subsequent hydrolysis of pyrophosphate to yield two ions of orthophosphate (P_i) by pyrophosphatase, an irreversible reaction under cellular conditions, helps drive the polymerization forward. This is an example of coupled reactions, where a second reaction provides the energy to drive the first reaction forward, a common occurrence in biochemical pathways (Section 15.1).

- *Elongation of the DNA chain proceeds in the 5′-to-3′ direction.*

- *Many DNA polymerases are able to correct mistakes in DNA by removing mismatched nucleotides.* These polymerases have a distinct nuclease activity that allows them to excise incorrect bases by a separate reaction. This nuclease activity contributes to the remarkably high fidelity of DNA replication, which has an error rate of less than 10^{-8} per base pair.

The genes of some viruses are made of RNA

Genes in all cellular organisms are made of DNA. The same is true for some viruses, but for others, the genetic material is RNA. Viruses are genetic elements enclosed in protein coats that can move from one cell to another but are not capable of independent growth. A well-studied example of an RNA virus is the tobacco mosaic virus, which infects the leaves of tobacco plants. This virus consists of a single strand of RNA (6390 nucleotides) surrounded by a protein coat of 2130 identical subunits. An RNA polymerase that takes direction from an RNA template, called an *RNA-directed RNA polymerase*, copies the viral RNA. The infected cells die because of virus-instigated programmed cell death; in essence, the virus instructs the cell to commit suicide. Cell death results in discoloration in the tobacco leaf in a variegated pattern, hence the name mosaic virus.

Another important class of RNA virus comprises the **retroviruses**, so called because the genetic information flows from RNA to DNA rather than from DNA

FIGURE 8.25 In retroviruses, information flows from RNA to DNA. The single-stranded RNA genome of a retrovirus is converted into double-stranded DNA by reverse transcriptase, an enzyme brought into the cell by the infecting virus particle. Reverse transcriptase possesses several activities and catalyzes the synthesis of a complementary DNA strand, the digestion of the RNA, and the subsequent synthesis of the second DNA strand.

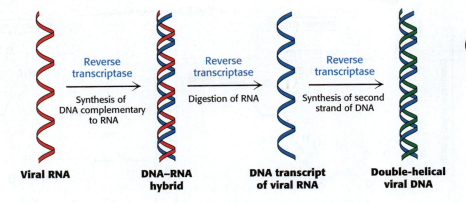

Viral RNA | Reverse transcriptase — Synthesis of DNA complementary to RNA | DNA–RNA hybrid | Reverse transcriptase — Digestion of RNA | DNA transcript of viral RNA | Reverse transcriptase — Synthesis of second strand of DNA | Double-helical viral DNA

to RNA. This class includes human immunodeficiency virus 1 (HIV-1), the cause of acquired immunodeficiency syndrome (AIDS), as well as a number of RNA viruses that produce tumors in susceptible animals. Retrovirus particles contain two copies of a single-stranded RNA molecule. On entering the cell, the RNA is copied into DNA through the action of a viral enzyme called *reverse transcriptase*, which acts as both a DNA polymerase and an RNase (**Figure 8.25**). The resulting double-helical DNA version of the viral genome can become incorporated into the chromosomal DNA of the host and is replicated along with the normal cellular DNA. At a later time, the integrated viral genome is expressed to form viral RNA and viral proteins, which assemble into new virus particles.

8.5 Gene Expression Is the Transformation of DNA Information into Functional Molecules

The information stored as DNA becomes useful when it is expressed in the production of RNA and proteins. This rich and complex topic is the subject of several chapters later in this book, but here we introduce the basics of gene expression.

DNA can be thought of as archival information, stored and manipulated in ways that minimize damage (mutations). It is expressed in two steps. First, an RNA copy is made that encodes directions for protein synthesis. This messenger RNA can be thought of as a temporary photocopy of the original information; it can be made in multiple copies, used, and then disposed of. Second, the information in messenger RNA is translated to synthesize functional proteins. Other types of RNA molecules exist to facilitate this translation.

Several kinds of RNA play key roles in gene expression

RNA is more than a passive conveyor of information during gene expression. Scientists have demonstrated that RNA plays a variety of roles, from catalysis to regulation. Cells contain three major kinds of RNA that are involved in gene expression (**Table 8.3**):

1. **Messenger RNA (mRNA)** is the template for protein synthesis. An mRNA molecule may be produced for each gene or group of genes that is to be expressed in bacteria, whereas a distinct mRNA is produced for each gene in eukaryotes. Consequently, mRNA is a heterogeneous class of molecules. The average length of an mRNA molecule is about 1.2 kilobases (kb) in bacteria and about 2.2 kb in mammals. The mRNA of all organisms has structural features, such as stem-loop structures, that regulate its efficiency of translation and lifetime. Such structural features are more prominent in eukaryotic mRNA (Chapters 29 and 31).

2. **Transfer RNA (tRNA)** carries amino acids in an activated form to the ribosome for peptide-bond formation, in a sequence dictated by the mRNA template. There is at least one kind of tRNA for each of the 20 amino acids. Transfer RNA consists of about 75 nucleotides (having a mass of about 25 kDa).

TABLE 8.3 RNA molecules in *E. coli*

Type	Relative amount (%)	Sedimentation coefficient (S)	Mass (kDa)	Number of nucleotides
Ribosomal RNA (rRNA)	80	23	1.2×10^3	3700
		16	0.55×10^3	1700
		5	3.6×10^1	120
Transfer RNA (tRNA)	15	4	2.5×10^1	75
Messenger RNA (mRNA)	5		Heterogeneous	

3. **Ribosomal RNA (rRNA)** is the major component of ribosomes (Chapter 30). In bacteria, there are three kinds of rRNA, called 23S, 16S, and 5S RNA because of their sedimentation behavior (Section 4.1). One molecule of each of these species of rRNA is present in each ribosome. As we shall discuss in Chapter 30, eukaryotic rRNAs, as well as their corresponding ribosomes, are larger. Ribosomal RNA was once thought to play only a structural role in ribosomes. We now know that rRNA is the actual catalyst for protein synthesis.

Ribosomal RNA is the most abundant of these three types of RNA. Transfer RNA comes next, followed by messenger RNA, which constitutes only 5% of the total RNA. Eukaryotic cells contain additional small RNA molecules that play a variety of roles including the regulation of gene expression, the processing of RNA, and the synthesis of proteins. We will examine these small RNAs in later chapters. In this chapter, we will consider rRNA, mRNA, and tRNA.

All cellular RNA is synthesized by RNA polymerases

The synthesis of RNA from a DNA template is called **transcription** and is catalyzed by the enzyme *RNA polymerase* (**Figure 8.26**). RNA polymerase catalyzes the initiation and elongation of RNA chains. The reaction catalyzed by this enzyme is

$$(\text{RNA})_{n \text{ residues}} + \text{ribonucleoside triphosphate} \rightleftharpoons (\text{RNA})_{n+1 \text{ residues}} + \text{PP}_i$$

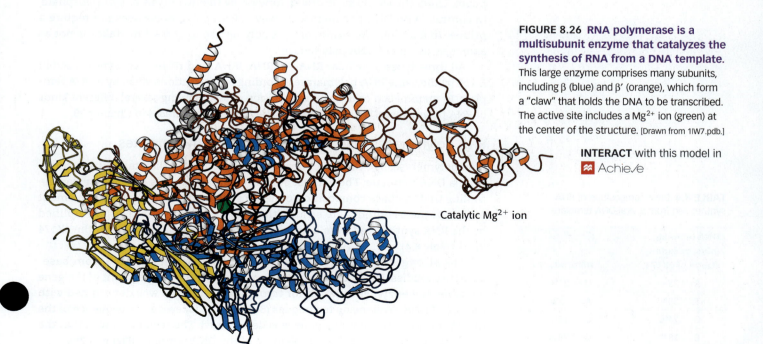

Catalytic Mg^{2+} ion

FIGURE 8.26 RNA polymerase is a multisubunit enzyme that catalyzes the synthesis of RNA from a DNA template. This large enzyme comprises many subunits, including β (blue) and β′ (orange), which form a "claw" that holds the DNA to be transcribed. The active site includes a Mg^{2+} ion (green) at the center of the structure. [Drawn from 1IW7.pdb.]

INTERACT with this model in
Achieve

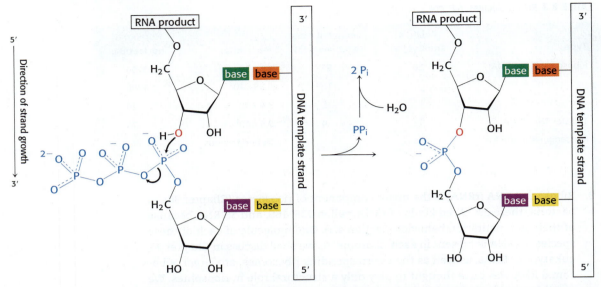

FIGURE 8.27 RNA polymerase catalyzes the strand-elongation reaction in the 5′-to-3′ direction. Note the similarity in mechanism to that of DNA polymerase shown in Figure 8.24. [Source: J. L. Tymoczko, J. Berg, and L. Stryer, *Biochemistry: A Short Course*, 2nd ed. (W. H. Freeman and Company, 2013), Fig. 36.3.]

RNA polymerase requires the following components:

1. *A template.* The preferred template is double-stranded DNA. Single-stranded DNA also can serve as a template. RNA, whether single or double stranded, is not an effective template, nor are RNA–DNA hybrids.

2. *Activated precursors.* All four ribonucleoside triphosphates—ATP, GTP, UTP, and CTP—are required.

3. *A divalent metal ion.* Either Mg^{2+} or Mn^{2+} is effective.

The synthesis of RNA is like that of DNA in several respects (**Figure 8.27**). First, the direction of synthesis is 5′ → 3′. Second, the mechanism of elongation is similar: the 3′-OH group at the terminus of the growing chain makes a nucleophilic attack on the innermost phosphoryl group of the incoming nucleoside triphosphate. Third, the synthesis is driven forward by the hydrolysis of pyrophosphate. In contrast with DNA polymerase, however, RNA polymerase does not require a primer. In addition, the ability of RNA polymerase to correct mistakes is not as extensive as that of DNA polymerase.

All three types of cellular RNA—mRNA, tRNA, and rRNA—are synthesized in *E. coli* by the same RNA polymerase according to instructions given by a DNA template. In mammalian cells, there is a division of labor among several different kinds of RNA polymerases. We shall return to these RNA polymerases in Chapter 29.

RNA polymerases take instructions from DNA templates

RNA polymerase, like the DNA polymerases described earlier, takes instructions from a DNA template. The earliest evidence supporting this conclusion was the finding that the base composition of newly synthesized RNA is the complement of that of the DNA template strand (the plus or coding strand), as exemplified by the RNA synthesized from a template of single-stranded DNA from the φX174 virus (**Table 8.4**).

The strongest evidence for the fidelity of transcription came from base-sequence studies. For instance, the nucleotide sequence of a segment of the gene encoding the enzymes required for tryptophan synthesis was determined with the use of DNA-sequencing techniques (Section 9.1). Likewise, the sequence of the mRNA for the corresponding gene was determined. The results showed that the RNA sequence is the precise complement of the DNA template (**Figure 8.28**).

TABLE 8.4 Base composition of RNA synthesized from a viral DNA template

DNA template (plus, or coding, strand of φX174)		RNA product	
A	25%	U	25%
T	33%	A	32%
G	24%	C	23%
C	18%	G	20%

5′—GCGGCGACGCGCAGUUAAUCCCACAGCCGCCAGUUCCGCUGGCGGCAU—3′   **mRNA**
3′—CGCCGCTGCGCGTCAATTAGGGTGTCGGCGGTCAAGGCGACCGCCGTA—5′   **Template strand of DNA**
5′—GCGGCGACGCGCAGTTAATCCCACAGCCGCCAGTTCCGCTGGCGGCAT—3′   **Coding strand of DNA**

FIGURE 8.28 The sequences of the template DNA strand and its corresponding mRNA are complementary. The base sequence of mRNA (red) is the complement of that of the DNA template strand (blue). The sequence shown here is from the tryptophan operon, a segment of DNA containing the genes for five enzymes that catalyze the synthesis of tryptophan. The other strand of DNA (black) is called the coding strand because it has the same sequence as the RNA transcript except for thymine (T) in place of uracil (U).

Transcription begins near promoter sites and ends at terminator sites

RNA polymerase must detect and transcribe discrete genes from within large stretches of DNA. What marks the beginning of the unit to be transcribed? DNA templates contain regions called **promoter sites** that specifically bind RNA polymerase and determine where transcription begins. While not every promoter site of a given type carries the exact same sequence, they typically vary from an idealized single sequence, or *consensus sequence*, by only one or two residues. In prokaryotes, two sequences on the 5′ (upstream) side of the first nucleotide to be transcribed function as promoter sites (**Figure 8.29A**). One of them, called the *Pribnow box*, has the consensus sequence TATAAT and is centered at –10 (10 nucleotides on the 5′ side of the first nucleotide transcribed, which is denoted by +1). The other, called the –35 *region*, has the consensus sequence TTGACA. The first nucleotide transcribed is usually a purine.

Eukaryotic genes encoding proteins have promoter sites with a TATAAA consensus sequence, called a *TATA box* or a *Hogness box*, centered at about –25 (**Figure 8.29B**). Many eukaryotic promoters also have a *CAAT box* with a GGNCAATCT consensus sequence centered at about –75. The transcription of eukaryotic genes is further stimulated by enhancer sequences, which can be quite distant (as many as several kilobases) from the start site, on either its 5′ or its 3′ side.

In *E. coli*, RNA polymerase proceeds along the DNA template, transcribing one of its strands until it synthesizes a terminator sequence. This sequence encodes a termination signal, which is a stem-loop structure on the newly synthesized RNA molecule (**Figure 8.30**). This structure is formed by base-pairing of self-complementary sequences that are rich in G and C. The newly synthesized RNA spontaneously dissociates from RNA polymerase when this stem-loop

(A) Prokaryotic promoter sites

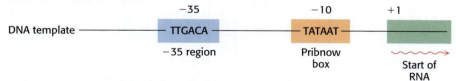

(B) Eukaryotic promoter sites

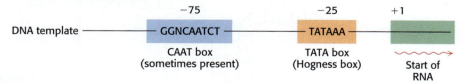

FIGURE 8.29 Promoter sites for transcription initiation differ between prokaryotes and eukaryotes. Consensus sequences are shown for (A) prokaryotic and (B) eukaryotic promoters. The first nucleotide to be transcribed is numbered +1. The adjacent nucleotide on the 5′ side is numbered –1. The sequences shown are those of the coding strand of DNA.

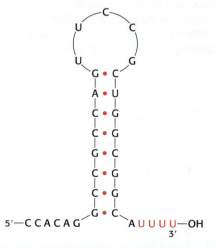

FIGURE 8.30 The base sequence of the 3′ end of an mRNA transcript in *E. coli* forms a stem-loop structure. This structure is followed by a sequence of uridine (U) residues.

FIGURE 8.31 mRNA in eukaryotes is modified after transcription. A nucleotide "cap" structure is added to the 5′ end, and a poly(A) tail is added at the 3′ end.

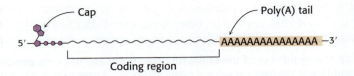

is followed by a string of U residues. Alternatively, RNA synthesis in *E.coli* can be terminated by the action of the protein rho. Less is known about the termination of transcription in eukaryotes. A more detailed discussion of the initiation and termination of transcription will be given in Chapter 30. The important point now is that discrete start and stop signals for transcription are encoded in the DNA template.

In eukaryotes, the messenger RNA is modified after transcription (**Figure 8.31**). A "cap" structure—a guanosine nucleotide with an unusual 5′-5′ triphosphate linkage—is attached to the 5′ end, and a poly(A) "tail"—a sequence of adenylates—is added to the 3′ end. These modifications will be presented in detail in Chapter 30.

Transfer RNAs are the adaptor molecules in protein synthesis

We have seen that mRNA is the template for protein synthesis. How then does it direct amino acids to become joined in the correct sequence to form a protein? Specific adaptor RNA molecules called transfer RNAs (tRNAs) bring the amino acids to the mRNA. The structure and reactions of these remarkable molecules will be considered in detail in Chapter 30. For the moment, it suffices to note that tRNAs contain an amino acid-attachment site and a template-recognition site.

Each tRNA molecule is composed of around 75 nucleotides and contains an amino acid-attachment site and a template-recognition site. Within the tRNA structure are several regions of base-paired segments in multiple stem-loops (**Figure 8.32**). The tRNA recognizes the mRNA template via three base pairs at the end of one of the stem-loops. Since the three coding bases on the mRNA template are known as the **codon**, the three complementary base pairs on the tRNA are referred to as the **anticodon**.

FIGURE 8.32 A tRNA molecule contains multiple stem-loop structures. The amino acid is attached at the 3′ end of the RNA. The anticodon (shown in red) is the template-recognition site. The tRNA has a cloverleaf structure with many hydrogen bonds (green dots) between bases.

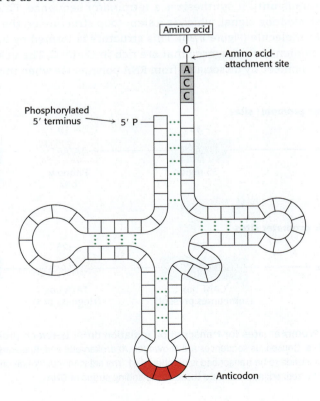

Initially, the amino acid is attached to the either the 3'- or 2'-hydroxyl group of the ribose at the 3' end of the tRNA molecule. However, the amino acid can rapidly equilibrate between the two positions; ultimately, the 3'-hydroxyl-bound form is the substrate for peptide bond formation. The sequence at the 3' end of the tRNA invariably contains two cytidylates followed by an adenylate, yielding what is referred to as the *CCA terminal region* or *CCA terminus* (**Figure 8.33**). The joining of an amino acid to a tRNA molecule to form an **aminoacyl-tRNA** is catalyzed by a specific enzyme called an *aminoacyl-tRNA synthetase*. There is at least one specific synthetase for each of the 20 amino acids.

8.6 Amino Acids Are Encoded by Groups of Three Bases Starting from a Fixed Point

The **genetic code** is the relation between the sequence of bases in DNA (or its RNA transcripts) and the sequence of amino acids in proteins. Numerous experiments established the following features of the genetic code by 1961:

- *Three nucleotides encode an amino acid.* Proteins are built from a basic set of 20 amino acids, but there are only four bases. Simple calculations show that a minimum of three bases is required to encode at least 20 amino acids. Genetic experiments showed that an amino acid is in fact encoded by a group of three bases, or codon.

- *The code is nonoverlapping.* Consider a base sequence ABCDEF. In an overlapping code, ABC specifies the first amino acid, BCD the next, CDE the next, and so on. In a nonoverlapping code, ABC designates the first amino acid, DEF the second, and so forth. Genetic experiments again established the code to be nonoverlapping.

- *The code has no "punctuation."* In principle, one base (denoted as Q) might serve as a "comma" between groups of three bases.

…QABCQDEFQGHIQJKLQ …

However, this is not the case. Rather, the sequence of bases is read sequentially from a fixed starting point, without punctuation.

- *The code has directionality.* The code is read from the 5' end of the messenger RNA to its 3' end.

- *The genetic code is degenerate, in that most amino acids are encoded by more than one codon.* There are 64 possible base triplets and only 20 amino acids; 61 of the 64 possible triplets specify particular amino acids. The remaining three triplets (called **stop codons**) designate the termination of translation.

Major features of the genetic code

All 64 codons have been deciphered (**Table 8.5**). Because the code is highly degenerate, only tryptophan and methionine are encoded by just one triplet each. Each of the other 18 amino acids is encoded by two or more. Indeed, leucine, arginine, and serine are specified by six codons each.

Codons that specify the same amino acid are called *synonyms*. For example, CAU and CAC are synonyms for histidine. Note that synonyms are not distributed haphazardly throughout the genetic code. In Table 8.5, an amino acid specified by two or more synonyms occupies a single box (unless it is specified by more than four synonyms). The amino acids in a box are specified by codons that have the same first two bases but differ in the third base, as exemplified by GUU, GUC, GUA, and GUG. Thus, most synonyms differ only in the last base of the triplet. Inspection of the code shows that XYC and XYU always encode the same amino acid, and XYG and XYA usually encode the same amino acid as well.

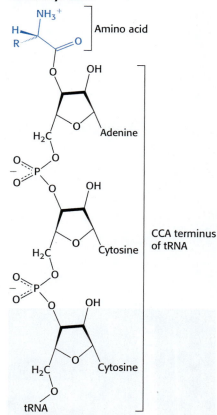

Aminoacyl-tRNA

FIGURE 8.33 An amino acid is attached to the 3' end of a tRNA molecule. The amino acid (shown in blue) is esterified to the 3'-hydroxyl group of the terminal adenylate of tRNA. The two bases preceding the 3' adenylate are cytidylates; hence, this region of the tRNA is often called the CCA terminus. [Source: J. L. Tymoczko, J. Berg, and L. Stryer, *Biochemistry: A Short Course*, 2nd ed. (W. H. Freeman and Company, 2013), Fig. 39.3.]

TABLE 8.5 The genetic code

First position	Second position				Third position
(5′ end)	U	C	A	G	(3′ end)
U	Phe	Ser	Tyr	Cys	U
	Phe	Ser	Tyr	Cys	C
	Leu	Ser	Stop	Stop	A
	Leu	Ser	Stop	Trp	G
C	Leu	Pro	His	Arg	U
	Leu	Pro	His	Arg	C
	Leu	Pro	Gln	Arg	A
	Leu	Pro	Gln	Arg	G
A	Ile	Thr	Asn	Ser	U
	Ile	Thr	Asn	Ser	C
	Ile	Thr	Lys	Arg	A
	Met	Thr	Lys	Arg	G
G	Val	Ala	Asp	Gly	U
	Val	Ala	Asp	Gly	C
	Val	Ala	Glu	Gly	A
	Val	Ala	Glu	Gly	G

Note: This table identifies the amino acid encoded by each triplet. For example, the codon 5′-AUG-3′ on mRNA specifies methionine, whereas CAU specifies histidine. UAA, UAG, and UGA are termination signals. AUG is part of the initiation signal, in addition to coding for internal methionine residues.

University of Wisconsin-Madison Archives Collections

HAR GOBIND KHORANA Born in 1922 in a small village in Punjab in British India, Har Gobind Khorana completed his master's degree in India and his PhD at the University of Liverpool. In 1952, he began his independent career at the University of British Columbia (UBC). He synthesized RNA polymers with repeating sequences of two, three, or four nucleotides and used these synthetic RNAs to guide protein synthesis. This work contributed greatly to the elucidation of the genetic code for which he shared the Nobel Prize in 1968. The ability to synthesize oligonucleotides, developed by him and his coworkers, fueled the development of PCR and many other methods. His work at UBC, the University of Wisconsin, and the Massachusetts Institute of Technology demonstrated his mastery of one of the pillars of chemistry: the powerful ability to synthesize even very complex molecules for a range of applications.

The structural basis for these equivalences of codons becomes evident when we consider the nature of the anticodons of tRNA molecules (Section 30.1).

What is the biological significance of the extensive degeneracy of the genetic code? If the code were not degenerate, 20 codons would designate amino acids and 44 would lead to chain termination. The probability of mutating to chain termination would therefore be much higher with a nondegenerate code. Chain-termination mutations usually lead to inactive proteins, whereas substitutions of one amino acid for another are usually rather harmless. Moreover, the code is constructed such that a change in any single nucleotide base of a codon results in a synonym or an amino acid with similar chemical properties. Thus, degeneracy minimizes the deleterious effects of mutations.

As we will see many times in our study of biochemistry, often things are not as straightforward as they first appear. Recent research has revealed a nonrandom use of synonymous codons in genes in different organisms, a phenomenon called *codon bias*. In other words, different organisms prefer different sets of synonymous codons. For example, there are six codons that encode arginine, but the most commonly used codons vary widely, from the bacterium *E. coli* to the yeast *S. cerevisiae* to humans (**Figure 8.34**). The biochemical benefit of codon bias is not yet clearly established, but codon bias may help regulate translation. From an experimental perspective, codon bias can have a significant impact on the efficacy of expressing a human protein within bacteria or yeast using recombinant DNA technology, as we shall discuss in Section 9.2. By adjusting the codon usage within the gene to be expressed to match the host organism, a process called *codon optimization*, one can express the protein at a higher yield without altering the protein sequence.

❖ SELF–CHECK QUESTION

The amino acid sequences of a yeast protein and a human protein having the same function are found to be 60% identical. However, the corresponding DNA sequences are only 45% identical. Account for this differing degree of identity.

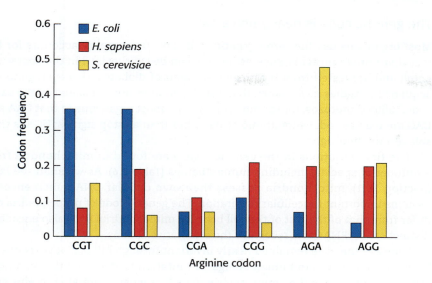

FIGURE 8.34 Different species exhibit different codon usage preferences. The relative preferences for the different arginine-encoding codons across three species—*E. coli* (blue), humans (red), and *S. cerevisiae* (yellow)—are shown. Although all of these codons encode the same amino acid, each organism exhibits a unique set of preferences, or codon bias. [Source: https://www.genscript.com/tools/codon-frequency-table.]

Messenger RNA contains start and stop signals for protein synthesis

Messenger RNA is translated into proteins on **ribosomes**—large molecular complexes assembled from proteins and ribosomal RNA. How is mRNA interpreted by the translation apparatus?

The start signal for protein synthesis in bacteria is complex. Polypeptide chains start with a modified amino acid—namely, formylmethionine (fMet). A specific tRNA, called the *initiator tRNA*, carries fMet and recognizes the codon AUG. However, AUG is also the codon for an internal methionine residue. Hence, the signal for the first amino acid in a bacterial polypeptide chain must be more complex than that for all subsequent ones. In fact, the AUG codon is only part of the initiation signal (**Figure 8.35A**). The initiating AUG codon is preceded several nucleotides away by a purine-rich sequence, called the **Shine–Dalgarno sequence** (named after John Shine and Lynn Dalgarno, who first described the sequence), that base-pairs with a complementary sequence in a ribosomal RNA molecule (Section 30.3).

In eukaryotes, the AUG closest to the 5′ end of an mRNA molecule is usually the start signal for protein synthesis (**Figure 8.35B**). This particular AUG is read by an initiator tRNA conjugated to methionine. After the initiator AUG has been located, the order of the three nonoverlapping nucleotides, or *reading frame*, is established.

As already mentioned, UAA, UAG, and UGA designate chain termination in most species. These codons are read not by tRNA molecules but rather by specific proteins called *release factors* (Section 30.3). Binding of a release factor to the ribosome releases the newly synthesized protein.

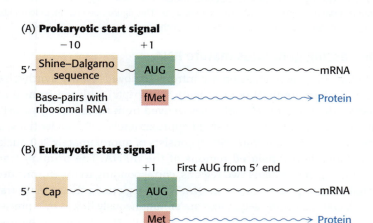

(A) Prokaryotic start signal

(B) Eukaryotic start signal

FIGURE 8.35 Initiation of protein synthesis differs between prokaryotes and eukaryotes. In prokaryotes (A), protein synthesis starts at the AUG codon about 10 bases downstream of a Shine–Dalgarno sequence, while in eukaryotes (B), the AUG closest to the 5′ end of the mRNA is usually the start signal.

TABLE 8.6 Distinctive codons of human mitochondria

Codon	Standard Code	Mitochondrial code
UGA	Stop	Trp
UGG	Trp	Trp
AUA	Ile	Met
AUG	Met	Met
AGA	Arg	Stop
AGG	Arg	Stop

The genetic code is nearly universal

Most organisms use the same genetic code. This universality accounts for the fact that human proteins, such as insulin, can be synthesized in the bacterium *E. coli* and harvested from it for the treatment of diabetes. However, genome-sequencing studies have shown that not all genomes are translated by the same code. Ciliated protozoa, for example, differ from most organisms in that UAA and UAG are read as codons for amino acids rather than as stop signals; UGA is their sole termination signal.

The first variations in the genetic code were found in mitochondria from a number of species, including human beings (**Table 8.6**). As we shall see later (Section 18.1), mitochondria possess their own circular DNA which encodes proteins important for cellular respiration. The genetic code of mitochondria can differ from that of the rest of the cell because mitochondrial DNA also encodes a distinct set of tRNAs that recognize the alternative codons.

Why has the code remained nearly invariant through billions of years of evolution, from bacteria to human beings? A mutation that altered the reading of mRNA would change the amino acid sequence of most, if not all, proteins synthesized by that particular organism. Many of these changes would undoubtedly be deleterious, and so there would be strong selection against a mutation with such pervasive consequences.

8.7 Most Eukaryotic Genes Are Mosaics of Introns and Exons

In bacteria, polypeptide chains are encoded by a continuous array of triplet codons in DNA. For many years, genes in higher organisms were assumed to be organized in the same manner. This view was unexpectedly shattered in 1977, when investigators discovered that most eukaryotic genes are *discontinuous*. For example, the gene for the β chain of hemoglobin is interrupted within its amino acid-coding sequence by a short stretch of 120 noncoding base pairs and a long one of 550 base pairs. Thus, the β-globin gene is split into three coding sequences (**Figure 8.36**). Noncoding regions are called **introns** (for <u>in</u>tervening sequences), whereas coding regions are called **exons** (for <u>ex</u>pressed sequences). The average human gene has 8 introns, and some have more than 100. Intron size ranges from 50 to 10,000 nucleotides.

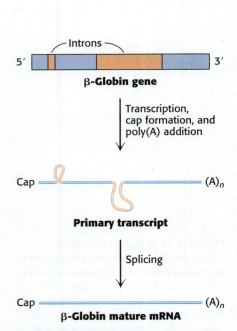

FIGURE 8.37 Transcription and processing of the β-globin gene includes an intermediate step where introns are removed. The gene is transcribed to yield the primary transcript, which is modified by cap and poly(A) addition. The introns in the primary RNA transcript are then removed to form the mature mRNA.

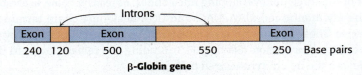

β-Globin gene

FIGURE 8.36 The β-globin gene is discontinuous. The coding regions, or exons (in blue), are interrupted by noncoding regions, or introns (in orange). This figure does not include regulatory regions (or transcribed but not translated regions) of the gene.

RNA processing generates mature RNA

At what stage in gene expression are introns removed? Newly synthesized RNA molecules (called *pre-mRNAs* or *primary transcripts*) isolated from nuclei are much larger than the mRNA molecules derived from them. In regard to β-globin RNA, the primary transcript consists of approximately 1600 nucleotides, and the processed mRNA (called *mature mRNA*) consists of approximately 900 nucleotides, which includes nontranslated regions of the mRNA. The primary transcript of the β-globin gene contains two regions, corresponding to the introns described above, that are not present in the mRNA. These regions in the primary transcript are excised, and the coding sequences are simultaneously linked by a precise splicing complex to form the mature mRNA (**Figure 8.37**). A common feature in the

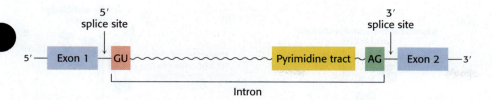

FIGURE 8.38 A consensus sequence ensures that splicing occurs at the correct locations. The introns typically begin with a GU pair and end with an AG pair that is preceded by a pyrimidine-rich tract.

expression of discontinuous, or split, genes is that their exons are ordered in the same sequence in mRNA as in DNA. Thus, the codons in split genes, like continuous genes, are in the same linear order as the amino acids in the polypeptide products.

Splicing is a complicated series of reactions carried out by complexes called *spliceosomes*, which are assemblies of proteins and small nuclear RNA molecules (snRNAs). RNA plays the catalytic role (Section 29.4). Spliceosomes recognize signals in the primary transcript that specify the splice sites. Introns nearly always begin with GU and end with an AG that is preceded by a pyrimidine-rich tract (**Figure 8.38**). This consensus sequence is part of the signal for splicing.

Many exons encode protein domains

Most genes of higher eukaryotes, such as birds and mammals, are split. Lower eukaryotes, such as yeast, have a much higher proportion of continuous genes. In bacteria, split genes are extremely rare. From an evolutionary perspective, this raises the question: have introns been inserted into genes in the evolution of higher organisms or have they been removed from genes to form the streamlined genomes of bacteria and simple eukaryotes? Comparisons of the DNA sequences of genes encoding evolutionarily conserved proteins suggest that introns were present in ancestral genes and were lost in the evolution of organisms that have become optimized for very rapid growth, such as bacteria. The positions of introns in some genes are at least 1 billion years old. Furthermore, the hypothesis that a common mechanism of splicing developed before the divergence of fungi, plants, and vertebrates is supported by the finding that mammalian cell extracts can splice yeast RNA.

What advantages might split genes confer? An attractive hypothesis is that new proteins arose in evolution by the rearrangement of exons encoding discrete structural elements, binding sites, and catalytic sites, a process called **exon shuffling**. Because it preserves functional units but allows them to interact in new ways, exon shuffling is a rapid and efficient means of generating novel genes. **Figure 8.39** shows the composition of a gene that was formed in part by

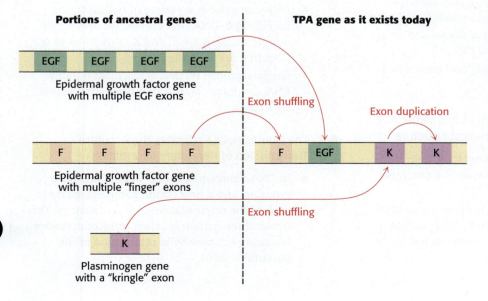

FIGURE 8.39 The tissue plasminogen activator (TPA) gene was generated by exon shuffling. The gene for TPA encodes an enzyme that functions in hemostasis (Section 7.4). This gene consists of 4 exons, one (F) derived from the fibronectin gene which encodes an extracellular matrix protein, one from the epidermal growth factor gene (EGF), and two from the plasminogen gene (K, Section 7.4), the substrate of the TPA protein. The K domain appears to have arrived by exon shuffling and then to have been duplicated to generate the TPA gene that exists today. [Information from: www.ehu.es /ehusfera/genetica/2012/10/02/demostracion-molecular -de-microevolucion/.]

FIGURE 8.40 Alternative splicing generates mRNAs that are templates for different forms of a protein. For example, (A) a membrane-bound antibody on the surface of a lymphocyte and (B) its soluble counterpart, exported from the cell, are products of alternative splicing of the same transcript. The membrane-bound antibody is anchored to the plasma membrane by a helical segment (highlighted in yellow) that is encoded by its own exon.

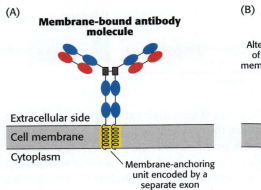

(A)
Membrane-bound antibody molecule

Extracellular side
Cell membrane
Cytoplasm

Membrane-anchoring unit encoded by a separate exon

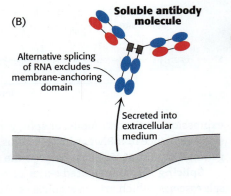

(B)
Soluble antibody molecule

Alternative splicing of RNA excludes membrane-anchoring domain

Secreted into extracellular medium

exon shuffling. DNA can break and recombine in introns with no deleterious effect on encoded proteins. In contrast, the exchange of sequences within different exons usually leads to loss of protein function.

Another advantage of split genes is the potential for generating a series of related proteins by **alternative splicing** of the primary transcript. For example, a precursor of an antibody-producing cell forms an antibody that is anchored in the cell's plasma membrane (**Figure 8.40**). The attached antibody recognizes a specific foreign antigen, an event that leads to cell differentiation and proliferation. The activated antibody-producing cells then splice their primary transcript in an alternative manner to form soluble antibody molecules that are secreted rather than retained on the cell surface. Alternative splicing is a facile means of forming a set of proteins that are variations of a basic motif without requiring a gene for each protein. Due in large part to alternative splicing, the proteome is more diverse than the genome in eukaryotes.

Summary

8.1   A Nucleic Acid Consists of Four Kinds of Bases Linked to a Sugar–Phosphate Backbone

- DNA and RNA are linear polymers of a limited number of nucleotide monomers. Each nucleotide contains a sugar, a phosphate, and a base.

- In DNA, the sugar is deoxyribose, and the bases are adenine (A), thymine (T), guanine (G), and cytosine (C). In RNA, the sugar is ribose, and thymine is replaced with uracil (U).

- Nucleic acids have directionality and are written in the 5′-to-3′ direction.

8.2   A Pair of Nucleic Acid Strands with Complementary Sequences Can Form a Double-Helical Structure

- Cellular DNA consists of two very long, helical polynucleotide strands coiled around a common axis: the double helix.

- The sugar–phosphate backbone of each strand is on the outside of the double helix, whereas the purine and pyrimidine bases are on the inside stabilized by stacking forces.

- The two strands run in opposite directions and are held together by hydrogen bonds between pairs of bases: adenine is always paired with thymine, and guanine is always paired with cytosine. Hence, one strand of a double helix is the complement of the other.

- DNA can exist in a variety of helical forms. including B-DNA (the classic Watson–Crick double helix), A-DNA, and Z-DNA. Most of the DNA in a cell is in the B-form.

- Single-stranded nucleic acids, most notably RNA, can form complicated three-dimensional structures.

8.3   The Double Helix Facilitates the Accurate Transmission of Hereditary Information

- In DNA replication, the strands of the helix separate, and a new strand complementary to each of the original strands is synthesized. This mode of replication is called semiconservative because each new helix retains one of the parental strands.

- In order for replication to take place, the strands of the double helix must be separated, or melted.
- In the laboratory, the double helix can be melted by heating the solution above its melting temperature. This process is reversible, and the strands will reassemble spontaneously below the melting temperature.

8.4 DNA Is Replicated by Polymerases That Take Instructions from Templates

- The replication of DNA is a complex process carried out by many proteins, including several DNA polymerases.
- Replication by DNA polymerase requires: all four activated precursors (deoxynucleotide 5′-triphosphates), a DNA template, and a primer with a free 3′-OH group.
- The new strand is synthesized in the $5' \rightarrow 3'$ direction.
- DNA polymerases catalyze the formation of a phosphodiester linkage only if the base on the incoming nucleotide is complementary to the base on the template strand.

8.5 Gene Expression Is the Transformation of DNA Information into Functional Molecules

- The flow of genetic information in normal cells is from DNA to RNA to protein.
- Transcription is the synthesis of RNA from a DNA template, and translation is the synthesis of a protein from an RNA template.
- Cells contain several kinds of RNA, among which are messenger RNA (mRNA), transfer RNA (tRNA), and ribosomal RNA (rRNA), which vary in size from 75 to more than 5000 nucleotides.

- As with DNA polymerases, RNA polymerases synthesize cellular RNA in the 5′-to-3′ direction according to instructions given by DNA templates. RNA polymerase differs from DNA polymerase in not requiring a primer.

8.6 Amino Acids Are Encoded by Groups of Three Bases Starting from a Fixed Point

- The genetic code is the relationship between the sequence of bases in DNA (or its RNA transcript) and the sequence of amino acids in proteins.
- Amino acids are encoded by codons—groups of three bases—starting from a fixed point known as the start codon.
- Sixty-one of the 64 possible codons specify particular amino acids, whereas the other 3 codons (UAA, UAG, and UGA) are signals for chain termination. Thus, for most amino acids, there is more than one code word.
- Natural mRNAs contain start and stop signals for translation.

8.7 Most Eukaryotic Genes Are Mosaics of Introns and Exons

- Most genes in higher eukaryotes are discontinuous: coding sequences (exons) are separated by noncoding sequences (introns).
- Introns are removed in the conversion of the primary transcript into mature mRNA by a process known as splicing.
- A striking feature of many exons is that they encode functional domains in proteins. New proteins probably arose in the course of evolution by the shuffling of exons.

Key Terms

deoxyribonucleic acid (DNA) (p. 237)
ribonucleic acid (RNA) (p. 237)
nucleotide (p. 237)
deoxyribose (p. 237)
ribose (p. 237)
purine (p. 237)
pyrimidine (p. 237)
nucleoside (p. 238)
stem-loop (p. 245)
semiconservative replication (p. 246)

melting temperature (T_m), (p. 247)
template strand (p. 248)
primer (p. 249)
retrovirus (p. 249)
messenger RNA (mRNA) (p. 250)
transfer RNA (tRNA) (p. 250)
ribosomal RNA (rRNA) (p. 251)
transcription (p. 251)
promoter site (p. 253)
codon (p. 254)

anticodon (p. 254)
aminoacyl-tRNA (p. 255)
genetic code (p. 255)
stop codon (p. 255)
ribosome (p. 257)
Shine–Dalgarno sequence (p. 257)
intron (p. 258)
exon (p. 258)
splicing (p. 259)
exon shuffling (p. 259)
alternative splicing (p. 260)

Problems

1. Classify the following descriptions as pertaining to nucleosides, nucleotides, or both nucleosides and nucleotides: ❖ **1**

(a) Contain a base, a sugar, and a phosphate group

(b) Are found in RNA and DNA

(c) Do not contain a phosphate group

(d) May contain either a purine or a pyrimidine

(e) One example is adenosine 5′-monophosphate

(f) Are the product when a base bonds at C1 of ribose or deoxyribose

(g) Are the monomers of nucleic acids

(h) Contain a base and a sugar

2. Name the following nucleosides/nucleotides: ❖ **1**

(a)

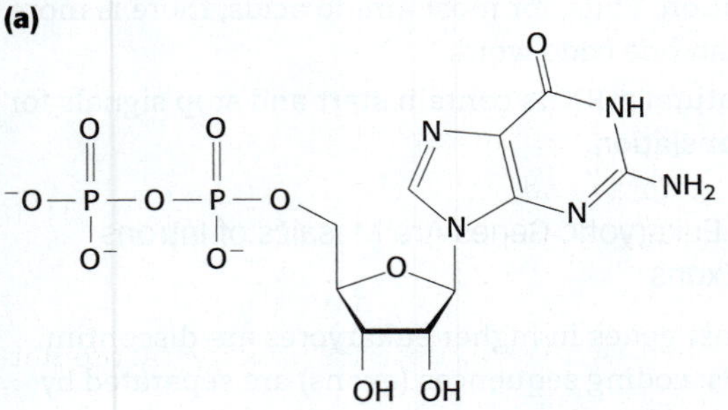

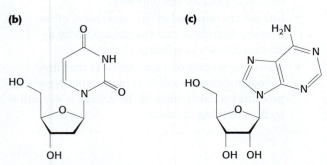

(b)   (c)

3. Biochemist Erwin Chargaff was the first to note that, in DNA, [A] = [T] and [G] = [C]. Using this rule, determine the percentages of all the bases in DNA that is 20% thymine. ❖ **1**, ❖ **2**

4. Write the complementary sequence (in the standard 5′ → 3′ notation) for

(a) GATCAA

(b) TCGAAC

(c) ACGCGT

(d) TACCAT

5. Classify the following statements as describing DNA or RNA: ❖ **2**

(a) Is usually single-stranded

(b) Is the genome for prokaryotic organisms

(c) Includes the sugar deoxyribose

(d) Includes the base uracil

(e) Can form a complex secondary structures

(f) Is usually double-stranded

(g) Can be translated into a protein

6. Classify each nucleotide or nucleoside by the type of base (purine vs. pyrimidine) and the type of sugar (ribose vs. deoxyribose) ❖ **2**

(a)   (b)

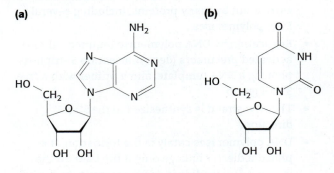

(c)

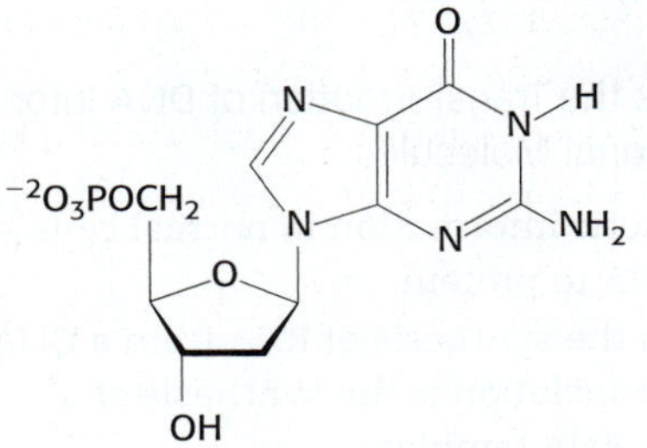

7. Classify the following statements as describing B-DNA, A-DNA, Z-DNA, or all three. ❖ **2**

(a) Widest helix diameter

(b) Has a helix diameter of ~20 Å

(c) C-3′ of deoxyribose is out of the plane of the other sugar ring atoms

(d) Left-handed

(e) Antiparallel strands

(f) Most stable under physiological conditions

(g) Narrowest helix diameter

(h) G pairs with C; A pairs with T

(i) More likely observed in dehydrated DNA fibers

8. (a) Suppose that you want to radioactively label DNA but not RNA in dividing and growing bacterial cells. Which radioactive molecule would you add to the culture medium? ❖ **3**

(b) Suppose that you want to prepare DNA in which the backbone phosphorus atoms are uniformly labeled with ^{32}P. Which precursors should be added to a solution containing DNA polymerase and primed template DNA? Specify the position of radioactive atoms in these precursors.

9. A solution contains DNA polymerase and the Mg^{2+} salts of dATP, dGTP, dCTP, and TTP. The following DNA molecules are added to aliquots of this solution. Which of them would lead to DNA synthesis? Explain your answer. ❖ **3**

(a) A single-stranded closed circle containing 1000 nucleotide units

(b) A double-stranded closed circle containing 1000 nucleotide pairs

(c) A single-stranded closed circle of 1000 nucleotides base-paired to a linear strand of 500 nucleotides with a free 3′-OH terminus

(d) A double-stranded linear molecule of 1000 nucleotide pairs with a free 3′-OH group at each end

10. In their famous experiment, Matthew Meselson and Franklin Stahl grew *E. coli* cells in a medium containing only the "heavy" isotope of nitrogen, ^{15}N. These cells were then transferred to a medium with the "light" isotope of nitrogen, ^{14}N. The results of the experiment supported the hypothesis of semiconservative replication. ❖ **3**

(a) Heavy DNA (^{15}N DNA), hybrid DNA, and light DNA (^{14}N DNA) can be separated by centrifugation. If cells containing ^{15}N DNA are transferred to a medium with only $^{14}NH_4Cl$ as a nitrogen source, what percentage of daughter molecules are composed of hybrid DNA after 2 generations?

(b) Predict what percentage of daughter molecules would be composed of hybrid DNA if DNA exhibited conservative replication (i.e., if a daughter DNA molecule were composed of newly synthesized DNA only).

11. What is the axial ratio (length:diameter) of a DNA molecule 20 μm long?

12. The DNA of a deletion mutant of λ bacteriophage has a length of 15 μm instead of the 17 μm found in the wild type phage. How many base pairs are missing from this mutant? ❖ **2**

13. Match the components in the right-hand column with the appropriate process in the left-hand column. Terms in the left-hand column may be used more than once. ❖ **4**

(a) Replication _____
(b) Transcription _____
(c) Translation _____

1. RNA polymerase
2. DNA polymerase
3. Ribosome
4. dNTP
5. tRNA
6. NTP
7. mRNA
8. Primer
9. rRNA
10. Promoter

14. **(a)** Write the sequence of the mRNA molecule synthesized from a DNA template strand having the following sequence: ❖ **4**

5′ – ATCGTACCGTTA – 3′

(b) What amino acid sequence is encoded by the following base sequence of an mRNA molecule? Assume that the reading frame starts at the 5′ end.

5′ – UUGCCUAGUGAUUGGAUG – 3′

(c) What is the sequence of the polypeptide formed on addition of poly(UUAC) to a cell-free protein-synthesizing system?

15. Compare DNA polymerase and RNA polymerase from *E. coli* in regard to each of the following features: ❖ **4**

(a) Activated precursors
(b) Direction of chain elongation
(c) Conservation of the template
(d) Need for a primer

16. What are the key characteristics of the genetic code? ❖ **4**

17. Classify the following terms on whether they are associated with tRNA, mRNA, or rRNA. ❖ **4**

(a) Codon
(b) Combines with protein to form ribosomes
(c) Transports amino acids
(d) Product of transcription
(e) Anticodon

18. Do the following phrases describe prokaryotic promoters, eukaryotic promoters, or both? ❖ **5**

(a) Includes two conserved sequences at –35 and –10
(b) Contains highly conserved DNA sequences that specify the start of transcription
(c) Includes a TATA, or Hogness, box upstream of the transcription start site
(d) Includes a Pribnow box upstream of the transcription start site
(e) May by further stimulated by enhancer sequences

19. What is the significance of the fact that human mRNA can be accurately translated in *E. coli*? ❖ **5**

20. Match the components in the right-hand column with the appropriate process in the left-hand column. ❖ **5**

(a) fMet _____
(b) Shine–Dalgarno _____
(c) Introns _____
(d) Exons _____
(e) Pre-mRNA _____
(f) mRNA _____
(g) Spliceosome _____

1. Continuous message
2. Removed during processing
3. The first of many amino acids
4. Joins exons
5. Joined to make the final message
6. Locates the start site
7. Discontinuous message

Nucleic Acid Methods

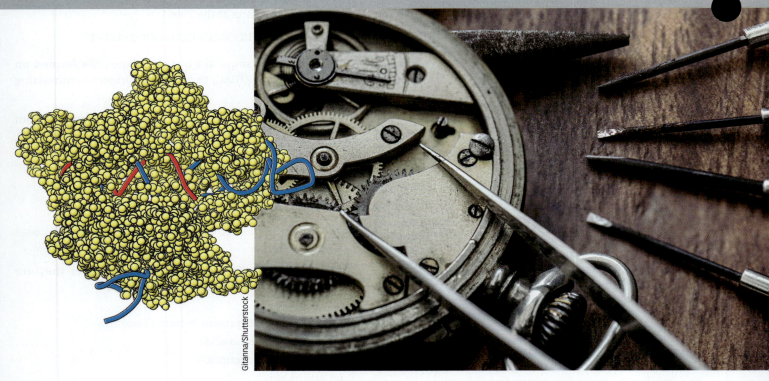

Gitanna/Shutterstock

The repair of delicate machinery, such as an old pocket watch, requires a steady hand, considerable experience, and the right tools for the job. Likewise, the advent of recombinant DNA technology has unlocked the potential to disrupt and repair DNA sequences with considerable precision. More recent developments have given us the tools such as the Cas9 endonuclease (left) to make directed changes to genomic DNA sequences with a high degree of specificity.

OUTLINE

9.1 The Exploration of Genes Relies on Key Tools

9.2 Recombinant DNA Technology Has Revolutionized All Aspects of Biology

9.3 Complete Genomes Have Been Sequenced and Analyzed

9.4 Eukaryotic Genes Can Be Quantitated and Manipulated with Considerable Precision

◈ LEARNING GOALS

By the end of this chapter, you should be able to:

1. Explain how restriction enzymes work and their importance to recombinant DNA technology.
2. Describe several methods commonly used for the sequencing of DNA.
3. List the fundamental steps of the polymerase chain reaction.
4. Describe the various types of vectors used in DNA cloning, and explain the difference between a cloning vector and an expression vector.
5. Describe some commonly used methods to introduce mutations into DNA.
6. Explain how genes can be altered in living organisms, and describe some of the applications of these techniques.

Recombinant DNA technology is the fruit of several decades of basic research on DNA, RNA, and viruses. These methods are rapidly evolving in terms of their speed, precision, and complexity. Nevertheless, even the more complex processes can be distilled to their essential biochemical principles, which we explore in this chapter.

The impact of recombinant DNA technology, generally referred to as *biotechnology*, cannot be overstated. In addition to the dramatic expansion of our understanding of human disease, these methods have enabled wide-ranging advancements across a broad spectrum of disciplines, including agriculture, environmental science, forensics, and evolutionary biology.

9.1 The Exploration of Genes Relies on Key Tools

The rapid progress in biotechnology—indeed, its very existence—is a result of a few key techniques.

- *Restriction-enzyme analysis.* Restriction enzymes are precise molecular scalpels that allow an investigator to manipulate DNA segments.
- *Blotting techniques.* Southern and northern blots are used to separate and identify DNA and RNA sequences, respectively. The western blot, which uses antibodies to characterize proteins, was described in Chapter 4.
- *DNA sequencing.* The exact nucleotide sequence of a molecule of DNA can be determined. Sequencing has yielded a wealth of information concerning gene architecture, the control of gene expression, and protein structure.
- *Solid-phase synthesis of nucleic acids.* Specific sequences of nucleic acids can be synthesized de novo and used to identify or amplify other nucleic acids.
- *The polymerase chain reaction* (PCR). The polymerase chain reaction leads to a billionfold amplification of a segment of DNA. One molecule of DNA can be amplified to quantities that permit characterization and manipulation. This powerful technique can be used to detect pathogens and genetic diseases, determine the source of a hair left at the scene of a crime, and resurrect genes from the fossils of extinct organisms.
- *Bioinformatics.* A final set of techniques relies on the computer, without which it would be impossible to catalog, access, and characterize the abundant information generated by the methods outlined above. Such uses of the computer, referred to as *bioinformatics*, will be presented in Chapter 10.

Restriction enzymes split DNA into specific fragments

As discussed in Chapter 6, **restriction enzymes**, also called **restriction endonucleases**, recognize specific base sequences in double-helical DNA and cleave both strands of that duplex at specific places. To biochemists, these exquisitely precise scalpels are marvelous gifts of nature. They are indispensable for analyzing chromosome structure, sequencing very long DNA molecules, isolating genes, and creating new DNA molecules that can be cloned.

Restriction enzymes are found in a wide variety of prokaryotes. Their biological role is to cleave foreign DNA molecules, providing the host organism with a primitive immune system. Many restriction enzymes recognize specific sequences of four to eight base pairs and hydrolyze a phosphodiester bond in each strand in this region. A striking characteristic of these cleavage sites is that they almost always possess twofold rotational symmetry. In other words, the recognized sequence is a palindrome, that is, a sequence that reads the same forwards or backwards. As a consequence of this symmetry, the cleavage sites are also symmetrically positioned. For example, the sequence recognized by a restriction enzyme from *Streptomyces achromogenes* is

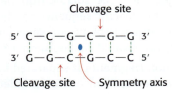

In each strand, the enzyme cleaves the C–G phosphodiester bond on the 3′ side of the symmetry axis. As we saw in Chapter 6, this symmetry corresponds to that of the structures of the restriction enzymes themselves.

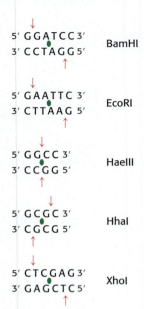

FIGURE 9.1 Restriction endonucleases vary in their sequence specificity and sites of cleavage. The sequences that are recognized by these enzymes contain a twofold axis of symmetry. The two strands in these regions are related by a 180-degree rotation about the axis marked by the green symbol. The cleavage sites are denoted by red arrows. The abbreviated name of each restriction enzyme is given at the right of the sequence that it recognizes. Note that the cuts may be staggered or even.

VIEW an animation of a restriction digest and gel electrophoresis in 🅰 Achieve

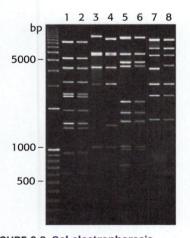

FIGURE 9.2 Gel-electrophoresis separates DNA fragments on the basis of their sizes. This gel shows the fragments produced by cleaving DNA from two viral strains (odd- vs. even-numbered lanes) with each of four restriction enzymes. These fragments were made fluorescent by staining the gel with ethidium bromide. [Credit: Carr et al., Emerging Infectious Diseases, www.cdc.gov/eid, Vol. 17, No. 8, August 2011.]

Several thousand restriction enzymes have been purified and characterized. Their names consist of a three-letter abbreviation for the host organism (e.g., Eco for _Escherichia coli_, Hin for _Haemophilus influenzae_, Hae for _Haemophilus aegyptius_) followed by a strain designation, if needed, and a roman numeral to distinguish multiple enzymes from the same strain. The specificities of several of these enzymes are shown in **Figure 9.1**.

Restriction enzymes are used to cleave DNA molecules into specific fragments that are more readily analyzed and manipulated than the entire parent molecule. For example, the 5.1-kilobase circular duplex DNA of the tumor-producing SV40 virus is cleaved at one site by EcoRI, at four sites by HpaI, and at 11 sites by HindIII. A piece of DNA, called a restriction fragment, produced by the action of one restriction enzyme can be specifically cleaved into smaller fragments by another restriction enzyme. The pattern of such fragments can serve as a fingerprint of a DNA molecule. Indeed, complex chromosomes containing hundreds of millions of base pairs can be mapped by using a series of restriction enzymes.

❖ SELF–CHECK QUESTION

The restriction enzyme AluI cleaves at the sequence 5′-AGCT-3′, and NotI cleaves at 5′-GCGGCCGC-3′. What would be the average distance between cleavage sites for each enzyme on digestion of double-stranded DNA? Assume that the DNA contains equal proportions of A, C, G, and T.

Restriction fragments can be separated by gel electrophoresis and visualized

In Chapter 4, we considered the use of gel electrophoresis to separate protein molecules (Section 4.1). Because the phosphodiester backbone of DNA is highly negatively charged, this technique is also suitable for the separation of nucleic acid fragments. Among the many applications of DNA electrophoresis are the detection of mutations that affect restriction fragment size (such as insertions and deletions) and the isolation, purification, and quantitation of a specific DNA fragment.

For most gels, the shorter the DNA fragment, the farther the migration. Poly-acrylamide gels are used to separate, by size, fragments containing as many as 1000 base pairs, whereas more-porous agarose gels are used to resolve mixtures of larger fragments (as large as 20 kilobases, kb). An important feature of these gels is their high resolving power. In certain kinds of gels, fragments differing in length by just one nucleotide out of several hundred can be distinguished. Individual bands or spots of DNA that have been labeled with radioactivity can be visualized by autoradiography after gel electrophoresis. Alternatively, a gel can be stained with a dye such as ethidium bromide, which fluoresces an intense orange under irradiation with ultraviolet light when bound to a double-helical DNA molecule (**Figure 9.2**). A band containing only 10 ng of DNA can be readily seen.

It is often necessary to determine if a particular base sequence is represented in a given DNA sample. For example, one may wish to confirm the presence of a specific mutation in genomic DNA isolated from patients known to be at risk for a particular disease. This specific sequence can be identified by hybridizing it with a labeled complementary DNA strand using a powerful technique named **Southern blotting**, for its inventor Edwin Southern (**Figure 9.3**).

1. A mixture of restriction fragments is separated by electrophoresis through an agarose gel, denatured to form single-stranded DNA, and transferred to a membrane composed of nitrocellulose or nylon. The positions of the DNA fragments in the gel are preserved during the transfer.

2. The membrane is then exposed to a ^{32}P-labeled or fluorescently tagged **DNA probe**, a short stretch of single-stranded DNA which contains a known base sequence.

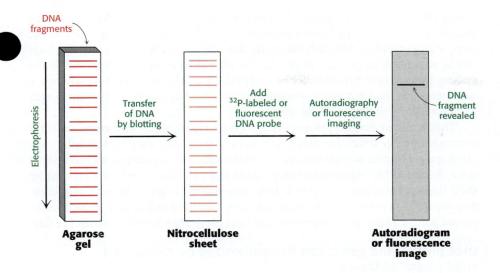

FIGURE 9.3 Southern blotting is used to detect the presence of a particular DNA sequence within a mixture of DNA molecules. A DNA fragment containing a specific sequence can be identified by separating a mixture of fragments by electrophoresis, transferring them to nitrocellulose, and hybridizing with a labeled probe complementary to the sequence. The fragment containing the sequence is then visualized by autoradiography or fluorescence imaging.

3. The probe hybridizes with a restriction fragment having a complementary sequence.

4. Autoradiography or fluorescence imaging then reveals the position of the restriction-fragment–probe duplex.

A particular fragment amid a million others can be readily identified in this way.

In a similar manner, RNA molecules of a specific sequence can also be readily identified. After separation by gel electrophoresis and transfer to nitrocellulose, specific sequences can be detected by DNA probes. This analogous technique for the analysis of RNA has been whimsically termed **northern blotting**.

DNA can be sequenced by controlled termination of replication

The analysis of DNA structure and its role in gene expression has been markedly facilitated by the development of powerful techniques for the sequencing of DNA molecules. One of the first and most widely used techniques for DNA sequencing is **controlled termination of replication**, also referred to as the **Sanger dideoxy method** for its pioneer, Frederick Sanger. The key to this approach is the generation of DNA fragments whose length is determined by the last base in the sequence (**Figure 9.4**).

In the current application of this method, a DNA polymerase is used to make the complement of a particular sequence within a single-stranded DNA molecule. The synthesis is primed by a chemically synthesized fragment that is complementary to a part of the sequence known from other studies. In addition to the four deoxyribonucleoside triphosphates, the reaction mixture contains a small amount of the 2′,3′-dideoxy analog of each nucleotide, each carrying a different fluorescent label attached to the base (e.g., a green emitter for termination at A and a red one for termination at T).

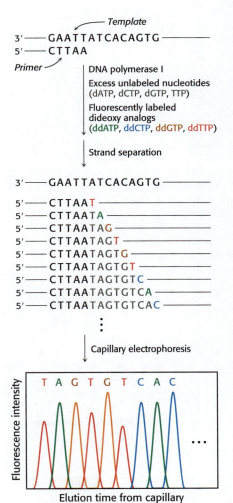

FIGURE 9.4 Controlled termination of replication uses the separation of fluorescently labeled oligonucleotide fragments to determine an overall sequence. A sequencing reaction is performed with four chain-terminating dideoxy nucleotides, each labeled with a tag that fluoresces at a different wavelength, with the color of each fragment indicating the identity of the last base in the chain. The fragments are separated by size using capillary electrophoresis, and the fluorescence at each of the four wavelengths indicates the sequence of the complement of the original DNA template.

2′, 3′-Dideoxy analog

The incorporation of the dideoxy nucleotide blocks further growth of the new chain because this analog lacks the 3′-hydroxyl terminus needed to form the next phosphodiester bond. The concentration of the dideoxy analog is low

VIEW an animation of the dideoxy sequencing method in
🔊 Achieve

enough that chain termination will take place only occasionally. The polymerase will insert the correct nucleotide sometimes and the dideoxy analog other times, stopping the reaction. For instance, if the dideoxy analog of dATP is present, fragments of various lengths are produced, but all will be terminated by the dideoxy analog. Importantly, this dideoxy analog of dATP will be inserted only where a T was located in the DNA being sequenced. Thus, the fragments of different length will correspond to the positions of T.

The resulting fragments are separated by a technique known as *capillary electrophoresis*, in which the mixture is passed through a very narrow tube containing a gel matrix at high voltage to achieve efficient separation within a short time. As the DNA fragments emerge from the capillary, they are detected by their fluorescence; the sequence of their colors directly gives the base sequence. Sequences of as many as 1000 bases can be determined in this way. Indeed, automated Sanger sequencing machines can read more than 1 million bases per day.

DNA probes and genes can be synthesized by automated solid-phase methods

DNA strands, like polypeptides (Section 4.4), can be synthesized by the sequential addition of activated monomers to a growing chain that is linked to a solid support (resin). The activated monomers are protected deoxyribonucleoside 3′-phosphoramidites. The synthesis of specific sequences proceeds by the repetition of the following cycle (**Figure 9.5**).

1. The 3′-phosphorus atom of the incoming unit is joined to the 5′-oxygen atom of the growing chain to form a phosphite triester. The 5′-OH group of the activated monomer is unreactive because it is blocked by a dimethoxytrityl (DMT) protecting group, and the 3′-phosphoryl oxygen atom is rendered unreactive by attachment of the β-cyanoethyl (βCE) group. Likewise, amino groups on the purine and pyrimidine bases are blocked. This step, called *coupling*, is carried out under anhydrous conditions because water reacts with phosphoramidites.

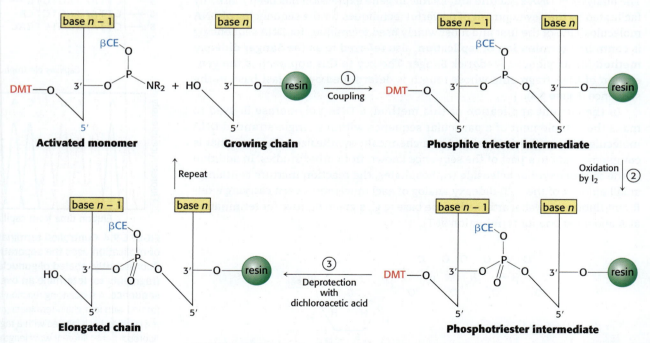

FIGURE 9.5. Oligonucleotides are synthesized in the solid-phase by the phosphite triester method. The activated monomer added to the growing chain is a deoxyribonucleoside 3′-phosphoramidite containing a dimethoxytrityl (DMT) protecting group on its 5′-oxygen atom, a β-cyanoethyl (βCE) protecting group on its 3′-phosphoryl oxygen atom, and a protecting group on the base.

2. The phosphite triester (in which P is trivalent) is oxidized by iodine to form a phosphotriester (in which P is pentavalent).

3. The DMT protecting group on the 5′-OH group of the growing chain is removed by the addition of dichloroacetic acid, which leaves other protecting groups intact. The DNA chain is now elongated by one unit and ready for another cycle of addition.

Each cycle takes only about 10 minutes and usually elongates more than 99% of the chains that are linked to the solid support resin.

This solid-phase approach is ideal for the synthesis of DNA, as it is for polypeptides, because the desired product stays on the insoluble support until the final release step. All the reactions take place in a single vessel, and excess soluble reagents can be added to drive reactions to completion. At the end of each step, soluble reagents and by-products are washed away from the resin that bears the growing chains.

At the end of the synthesis, NH_3 is added to remove all protecting groups and release the oligonucleotide from the solid support. Because elongation is never 100% complete, the new DNA chains are of diverse lengths—the desired chain is the longest one. The sample can be purified by high-performance liquid chromatography or by electrophoresis on polyacrylamide gels. DNA chains of as many as 100 nucleotides can be readily synthesized by this automated method.

The ability to rapidly synthesize DNA chains of any selected sequence opens many experimental avenues. For example, a synthesized DNA probe labeled at one end with ^{32}P or a fluorescent tag can be used to search for a complementary sequence in a very long DNA molecule or even in a genome consisting of many chromosomes. Such a probe can be used as a primer to initiate the replication of neighboring DNA by DNA polymerase.

Another application of the solid-phase approach is the synthesis of new tailor-made genes. New proteins with novel properties can now be produced in abundance by the expression of synthetic genes. Finally, the synthesis described in the steps above can be slightly modified for the solid-phase synthesis of RNA oligonucleotides, which can be very powerful reagents for the degradation of specific mRNA molecules in living cells by a technique known as RNA interference (Section 9.4).

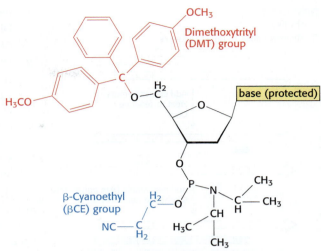

A deoxyribonucleoside 3′-phosphoramidite with DMT and βCE attached

Selected DNA sequences can be greatly amplified by the polymerase chain reaction

In 1984, Kary Mullis devised an ingenious method called the **polymerase chain reaction (PCR)** for amplifying specific DNA sequences. Consider a DNA duplex consisting of a target sequence surrounded by nontarget DNA. Millions of copies of the target sequences can be readily obtained by PCR if the sequences flanking the target are known. PCR is carried out by adding the following components to a solution containing the target sequence: (1) a pair of primers that hybridize with the flanking sequences of the target, (2) all four deoxyribonucleoside triphosphates (dNTPs), and (3) a heat-stable DNA polymerase. A PCR cycle consists of three steps (Figure 9.6).

1. *Strand separation.* The two strands of the parent DNA molecule are separated by heating the solution to 95°C for 15 s.

2. *Annealing of primers.* The solution is then abruptly cooled to 54°C to allow each primer to hybridize to a DNA strand, a process called *annealing*. One primer hybridizes to the 3′ end of the target on one strand, and the other primer

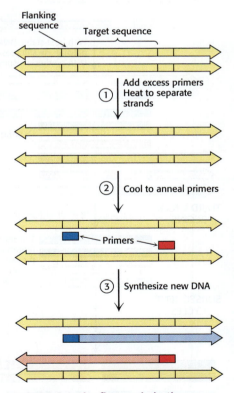

FIGURE 9.6 The first cycle in the polymerase chain reaction (PCR) generates two double-stranded DNA molecules. A cycle consists of three steps: DNA double-strand separation, the hybridization of primers, and the extension of primers by DNA synthesis.

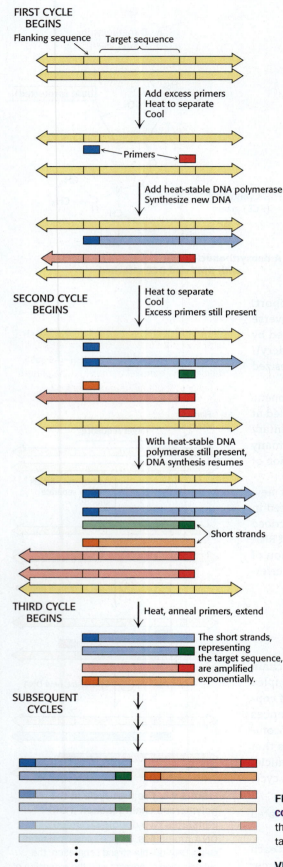

FIRST CYCLE
BEGINS

Flanking sequence — Target sequence

Add excess primers
Heat to separate
Cool

← Primers →

Add heat-stable DNA polymerase
Synthesize new DNA

SECOND CYCLE
BEGINS

Heat to separate
Cool
Excess primers still present

With heat-stable DNA
polymerase still present,
DNA synthesis resumes

Short strands

THIRD CYCLE
BEGINS

Heat, anneal primers, extend

The short strands,
representing
the target sequence,
are amplified
exponentially.

SUBSEQUENT
CYCLES

hybridizes to the 3′ end on the complementary target strand. Parent DNA duplexes do not form, because the primers are present in large excess. Primers are typically from 20 to 30 nucleotides long.

3. *DNA synthesis.* The solution is then heated to 72°C, the optimal temperature for heat-stable polymerases. One such enzyme is *Taq* DNA polymerase, which is derived from *Thermus aquaticus*, a thermophilic bacterium that lives in hot springs. The polymerase elongates both primers in the direction of the target sequence because DNA synthesis is in the 5′-to-3′ direction. DNA synthesis takes place on both strands but extends beyond the target sequence in this initial round.

These three steps — strand separation, annealing of primers, and DNA synthesis — constitute one cycle of the PCR amplification and can be carried out repetitively just by changing the temperature of the reaction mixture. The thermostability of the polymerase makes it feasible to carry out PCR in a closed container; no reagents are added after the first cycle. At the completion of the second cycle, four duplexes containing the target sequence have been generated (**Figure 9.7**). Of the eight DNA strands comprising these duplexes, two short strands constitute only the target sequence — the sequence including and bounded by the primers. Subsequent cycles will amplify the target sequence exponentially. Ideally, after n cycles, the desired sequence is amplified 2^n-fold. The amplification is a millionfold after 20 cycles and a billionfold after 30 cycles, which can be carried out in less than an hour.

Several features of this remarkable method for amplifying DNA are noteworthy. First, the sequence of the target need not be known. All that is required is knowledge of the flanking sequences so that complementary primers can be synthesized. Second, the target can be much larger than the primers. Targets larger than 10 kb have been amplified by PCR. Third, primers do not have to be perfectly matched to flanking sequences to amplify targets. With the use of primers derived from a gene of known sequence, it is possible to search for variations on the theme. In this way, families of genes are being discovered by PCR. Fourth, PCR is highly specific because of the stringency of hybridization at relatively high temperature. *Stringency* is the required closeness of the match between primer and target, which can be controlled by temperature and salt. At high temperatures, only the DNA between hybridized primers is amplified. A gene constituting less than a millionth of the total DNA of a higher organism is accessible by PCR. Fifth, PCR is exquisitely sensitive. A single DNA molecule can be amplified and detected.

❖ **SELF–CHECK QUESTION**

How would altering the stringency (for example, by changing the temperature of hybridization) affect PCR amplification? Suppose that you have a particular yeast gene A and that you wish to see if it has a counterpart in humans. How would controlling the stringency of the hybridization help you?

FIGURE 9.7 Multiple cycles of the polymerase chain reaction generate numerous copies of the amplified fragment. The two short double-stranded fragments produced at the end of the third cycle represent the target sequence. Subsequent cycles will amplify the target sequence exponentially and the parent sequence (yellow) arithmetically.

VIEW an animation of this technique in
 Achieve

PCR is a powerful technique in medical diagnostics, forensics, and studies of molecular evolution

PCR has been utilized in a wide variety of applications across many different disciplines:

- *Diagnosing certain infections caused by viruses and bacteria.* With the use of specific primers, PCR can reveal the presence of small amounts of DNA from the human immunodeficiency virus (HIV) in persons who have not yet mounted an immune response to this pathogen. In these patients, assays designed to detect antibodies against the virus would yield a false negative test result. Similarly, in patients suspected of having tuberculosis, PCR can be used to rapidly detect as few as 10 tubercle bacilli per million human cells, whereas finding *Mycobacterium tuberculosis* bacilli in tissue specimens would be slow and laborious.

- *Detecting certain cancers early and monitoring chemotherapy.* PCR can identify mutations of certain growth-control genes, such as the *ras* genes (Chapter 14). PCR is also ideal for detecting leukemias caused by chromosomal rearrangements. In addition, tests using PCR can detect when cancerous cells have been eliminated and treatment can be stopped; they can also detect a relapse and the need to immediately resume treatment.

- *Solving forensics cases.* An individual DNA profile is highly distinctive because many genetic loci are highly variable within a population. For example, each chromosome contains regions which vary in length from individual to individual. The pattern of DNA fragment sizes from PCR amplification across these regions provides a unique fingerprint for that individual. Numerous assault and rape cases rely on analyses of blood stains and semen samples by PCR (**Figure 9.8**). The root of a single shed hair found at a crime scene contains enough DNA for typing by PCR.

- *Studying molecular evolution.* DNA is a remarkably stable molecule, particularly when shielded from air, light, and water. Under such circumstances, large fragments of DNA can remain intact for thousands of years or longer. PCR provides an ideal method for amplifying such ancient DNA molecules so that they can be detected and characterized. As will be discussed in Chapter 10, sequences from these PCR products can be sources of considerable insight into evolutionary relationships between organisms.

FIGURE 9.8 PCR can be applied to forensic cases. DNA isolated from sperm obtained during the examination of a rape victim was amplified by PCR, then compared with DNA from the victim and two potential suspects using gel electrophoresis and auto-radiography. Sperm DNA matched the pattern of Suspect 1, but not that of Suspect 2.

The tools for recombinant DNA technology have been used to identify disease-causing mutations

Let us consider how the techniques just described have been used to study the molecular underpinning of human disease. Amyotrophic lateral sclerosis (ALS) was first described clinically in 1869 as a fatal neurodegenerative disease of progressive weakening and atrophy of voluntary muscles. ALS is commonly referred to as *Lou Gehrig's disease,* for the baseball legend whose career and life were prematurely cut short as a result of this devastating disease. For many years, little progress had been made in the study of the mechanisms underlying ALS.

Five percent of all patients suffering from ALS have family members who also have been diagnosed with the disease. A heritable disease pattern is indicative of a strong genetic component of disease causation. To identify these disease-causing genetic alterations, researchers identify **polymorphisms** (instances of genetic variation) within an affected family that correlate with the emergence of disease. These polymorphisms may themselves cause disease or be closely linked to the responsible genetic alteration. **Restriction-fragment-length polymorphisms (RFLPs)** are polymorphisms within restriction sites that change the sizes of DNA fragments produced by the appropriate restriction enzyme.

Using restriction digests and Southern blots of the DNA from members of ALS-affected families, researchers identified RFLPs that were found preferentially in those family members diagnosed with the disease. For some of these families, strong evidence was obtained for the disease-causing mutation within a specific region of chromosome 21.

After the probable location of one disease-causing gene had been identified, this same research group compared the locations of the ALS-associated RFLPs with the known sequence of chromosome 21. They noted that this chromosomal locus contains the *SOD1* gene, which encodes the Cu/Zn superoxide dismutase protein SOD1, an enzyme important for the protection of cells against oxidative damage (Section 18.3). PCR amplification of regions of the *SOD1* gene from the DNA of affected family members, followed by Sanger dideoxy sequencing of the targeted fragment, enabled the identification of 11 disease-causing mutations from 13 different families. This work was pivotal for focusing further inquiry into the roles that superoxide dismutase and its corresponding mutant forms play in the pathology of some forms of ALS.

9.2 Recombinant DNA Technology Has Revolutionized All Aspects of Biology

The development of recombinant DNA technology has taken biology from an exclusively analytical science to a synthetic one. New combinations of unrelated genes can be constructed in the laboratory by applying recombinant DNA techniques. These novel combinations can be cloned—amplified many-fold—by introducing them into suitable cells, where they are replicated by the DNA-synthesizing machinery of the host. The inserted genes are often transcribed and translated in their new setting. What is most striking is that the genetic endowment of the host can be permanently altered in a designed way.

Restriction enzymes and DNA ligase are key tools in forming recombinant DNA molecules

Let us begin by exploring how novel DNA molecules can be constructed in the laboratory. An essential tool for the manipulation of recombinant DNA is a **vector**, a DNA molecule that can replicate autonomously in an appropriate host organism. Vectors are designed to enable the rapid, covalent insertion of DNA fragments of interest. **Plasmids**, naturally occurring circles of DNA that act as accessory chromosomes in bacteria, and the virus **bacteriophage lambda (λ phage)** are choice vectors for cloning in *E. coli*. DNA fragments can be inserted into these vectors in several steps.

- *The vector is cut by restriction enzymes, creating sticky ends.* A vector can be prepared for accepting a new DNA fragment by cleaving it at a single specific site with a restriction enzyme. For example, the plasmid pSC101, a 9.9-kb double-helical circular DNA molecule, is split at a unique site by the EcoRI restriction enzyme. The staggered cuts made by this enzyme produce complementary single-stranded ends, which have specific affinity for each other and hence are known as **cohesive** or **sticky ends**. Any DNA fragment can be inserted into this plasmid if it has the same cohesive ends. Such a fragment can be extracted from a larger piece of DNA by using the same restriction enzyme as was used to open the plasmid DNA (**Figure 9.9**).

- *DNA ligase inserts the DNA fragment within the cut vector.* The single-stranded ends of the fragment are complementary to those of the cut plasmid. The sticky ends of the DNA fragment and the cut plasmid can base pair with each other, or *anneal*, and then joined by *DNA ligase,* which catalyzes the formation of a phosphodiester bond at a break in a DNA chain. DNA ligase

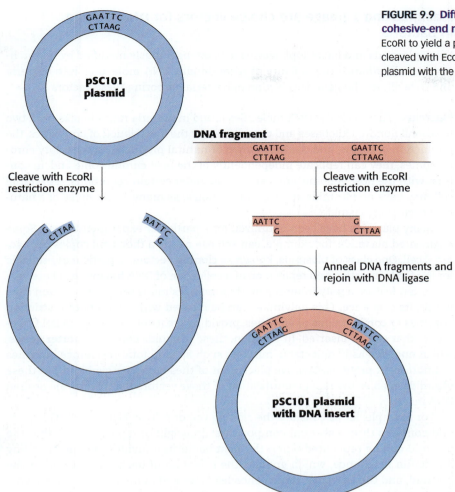

FIGURE 9.9 Different DNA molecules can be joined by the cohesive-end method. The pSC101 plasmid (blue) is cleaved once by EcoRI to yield a pair of cohesive ends. A fragment of DNA (red), also cleaved with EcoRI, can be ligated to the cut plasmid to form a new plasmid with the DNA fragment inserted.

requires a free 3′-hydroxyl group and a 5′-phosphoryl group. Furthermore, the chains joined by ligase must be in a double helix. An energy source such as ATP or NAD^+ is required for the joining reaction, as will be discussed in Chapter 28.

- *When necessary, linkers can be added to DNA fragments to yield cohesive ends.* What if the target DNA is not naturally flanked by the appropriate restriction sites? How is the fragment cut and annealed to the vector? The cohesive-end method for joining DNA molecules can still be used in these cases by adding a short, chemically synthesized DNA linker that can be cleaved by restriction enzymes. First, the linker is covalently joined to the ends of a DNA fragment. For example, the 5′ ends of a decameric linker and a DNA molecule are phosphorylated by polynucleotide kinase and then joined by the ligase from T4 phage (**Figure 9.10**). This particular ligase can form a covalent bond between blunt-ended double-helical DNA molecules. Finally, cohesive ends are produced when these terminal extensions are cut by an appropriate restriction enzyme. Thus, cohesive ends corresponding to the product of a particular restriction enzyme can be added to virtually any DNA molecule. We see here the fruits of combining enzymatic and synthetic chemical approaches in crafting new DNA molecules.

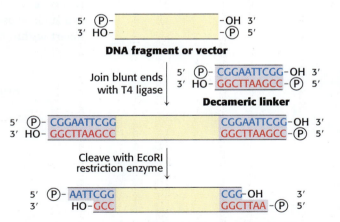

FIGURE 9.10 Cohesive ends can be added to any DNA fragment. A chemically synthesized linker containing a specific restriction site is joined to the ends of the fragment by T4 ligase. Subsequent cleavage by the restriction enzyme yields the new cohesive ends.

Plasmids and λ phage are choice vectors for DNA cloning in bacteria

Many plasmids and bacteriophages have been ingeniously modified by researchers to both enhance the delivery of recombinant DNA molecules into bacteria and to facilitate the genetic selection of bacteria harboring these vectors.

Plasmids. These circular DNA molecules found in bacteria range in size from two to several hundred kilobases and carry genes for the inactivation of antibiotics, the production of toxins, and the breakdown of natural products. As accessory chromosomes, they can replicate independently of the host chromosome and, in contrast with the host genome, are dispensable under certain conditions. A bacterial cell may have no plasmids at all, or it may house as many as 20 copies of a naturally occurring plasmid.

Many plasmids have been optimized for a particular experimental task. Some engineered plasmids, for example, can achieve nearly a thousand copies per bacterial cell. One class of plasmids, known as **cloning vectors**, is particularly suitable for the facile insertion and replication of a collection of DNA fragments. These vectors often feature a **polylinker** region that includes many unique restriction sites within its sequence. This polylinker can be cleaved with a variety of restriction enzymes or combinations of enzymes, providing great versatility in the DNA fragments that can be inserted. In addition, these plasmids contain **reporter genes**, which encode easily detectable markers such as antibiotic-resistance enzymes or fluorescent proteins. Creative placement of these reporter genes within these plasmids enables the rapid identification of those vectors that harbor the desired DNA insert.

For example, let us consider the cloning vector pUC18 (**Figure 9.11**). This plasmid contains three essential components: an origin of replication, so that the plasmid can be replicated when the host bacterium divides; a gene encoding ampicillin resistance, which enables the selection of bacteria that harbor the plasmid; and the *lacZ* gene, which encodes an essential fragment of the enzyme β-*galactosidase*, an enzyme which naturally cleaves the milk sugar lactose (Section 11.2). Insertion of a DNA fragment into the polylinker region disrupts the *lacZ* gene, an effect called *insertional inactivation*. β-Galactosidase also cleaves the synthetic substrate X-gal, releasing a blue dye. Bacterial cells containing a DNA insert within the polylinker will no longer produce the dye in the presence of

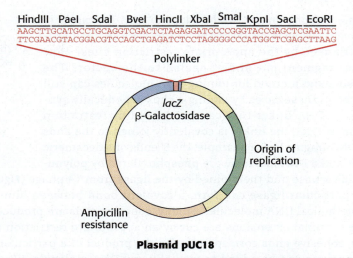

FIGURE 9.11 A polylinker contains multiple unique restriction sites. In addition to an origin of replication and a gene encoding ampicillin resistance, the plasmid pUC18 includes a polylinker within an essential fragment of the β-galactosidase gene (often called the *lacZ* gene).

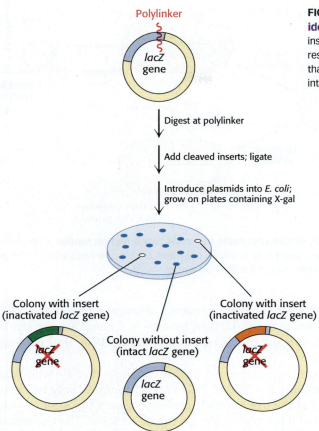

Polylinker

lacZ gene

Digest at polylinker

Add cleaved inserts; ligate

Introduce plasmids into *E. coli*; grow on plates containing X-gal

Colony with insert (inactivated *lacZ* gene)

Colony without insert (intact *lacZ* gene)

Colony with insert (inactivated *lacZ* gene)

lacZ gene

lacZ gene

lacZ gene

FIGURE 9.12 Insertional inactivation allows for the rapid identification of colonies that have new insertions. Successful insertion of DNA fragments into the polylinker region of pUC18 will result in the disruption of the β-galactosidase gene. Bacterial colonies that harbor such plasmids will no longer convert the substrate X-gal into a colored product, and will appear white on the plate.

X-gal. Hence, bacteria harboring vectors that have received an insert are readily identified by their white color (**Figure 9.12**).

Another class of plasmids has been optimized for use as **expression vectors** for the production of large amounts of protein. Many different expression vectors are currently in use. **Figure 9.13** depicts the typical elements of a vector designed to express a protein in prokaryotes. In addition to an antibiotic-resistance gene and an origin of replication, an expression vector contains promoter and terminator sequences specifically designed to drive the transcription of large amounts of a protein-coding DNA sequence. As in the cloning vectors, these plasmids also contain a polylinker region, which often contain sequences flanking the cloning site that simplify the addition of fusion tags to the protein of interest (Section 4.1), greatly facilitating purification of the overexpressed protein. Many expression vectors also possess transcriptional repressor and operator sequences (Section 31.2), which enable the precise control of when production of the protein will be turned on.

λ phage. Another widely used vector is the bacteriophage lambda, or λ phage. Bacteriophages are viruses that infect and replicate within bacteria. The λ phage enjoys a choice of lifestyles: it can destroy its host or become part of its host (**Figure 9.14**). In the lytic pathway, viral functions are fully expressed: viral DNA and proteins are quickly produced and packaged into virus particles, leading to the lysis (destruction) of the host cell and the sudden appearance of about 100 progeny virus particles, also called *virions*. In contrast, in the lysogenic pathway, the phage DNA becomes inserted into the host-cell genome and can be replicated together with host-cell DNA for many generations, remaining inactive. Certain environmental changes can trigger the expression of this dormant

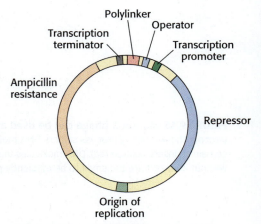

Polylinker

Operator

Transcription terminator

Transcription promoter

Transcription terminator

Ampicillin resistance

Repressor

Origin of replication

FIGURE 9.13 Expression vectors are designed to generate protein product. As with the cloning vector (Figure 9.11), the prokaryotic expression vector shown here contains an origin of replication and an antibiotic-resistance gene. In addition, these plasmids include transcription initiation (promoter) and termination sequences, required for the expression of a protein-coding gene inserted in the polylinker. Many expression plasmids also contain repressor and operator sequences which allow for the control of the timing of protein production.

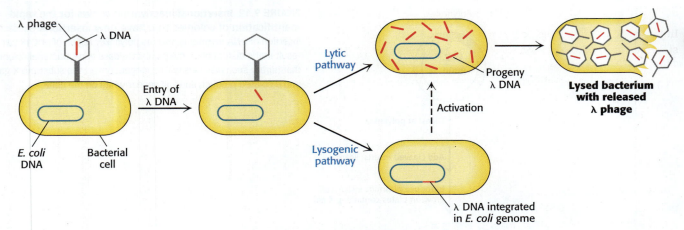

FIGURE 9.14 λ phage alternates between two infection modes. λ phage can multiply within a host and lyse it (lytic pathway), or its DNA can become integrated into the host genome (lysogenic pathway), where it is dormant until activated.

viral DNA, which leads to the formation of progeny viruses and lysis of the host. Large segments of the 48-kb DNA genome of λ phage are not essential for productive infection and can be replaced by foreign DNA, thus making λ phage an ideal vector.

Scientists have constructed mutant λ phages that are well-suited for cloning. An especially useful one called λgt-λβ contains only two EcoRI cleavage sites instead of the five normally present (**Figure 9.15**). After cleavage, the middle segment of this λ DNA molecule can be removed. The two remaining pieces of DNA (called arms) have a combined length equal to 72% of a normal genome length. This amount of DNA is too little to be packaged into a λ particle, which can take up only DNA measuring from 78% to 105% of a normal genome. However, a suitably long DNA insert (such as 10 kb) between the two ends of λ DNA enables such a recombinant DNA molecule (93% of normal length) to be packaged. Nearly all infectious λ particles formed in this way will contain an inserted piece of foreign DNA. An advantage of using these modified viruses as vectors is that they enter bacteria much more easily than do plasmids.

FIGURE 9.15 Mutant λ phage can be used as a cloning vector. The packaging process selects DNA molecules that contain an insert (colored red). DNA molecules that have resealed without an insert are too small to be efficiently packaged.

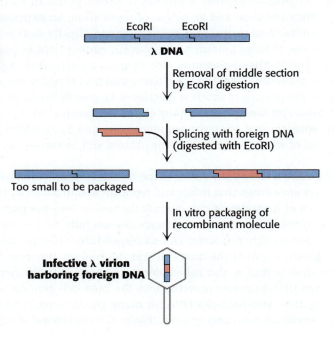

Specific genes can be cloned from digests of genomic DNA

Ingenious cloning and selection methods have made it possible to isolate small stretches of DNA in a genome containing more than 3×10^6 kb (i.e., 3×10^9 bases). The approach is to prepare a large collection, called a **genomic library**, of DNA fragments and then to identify those members of the collection that have the gene of interest. Hence, to clone a gene that is present just once in an entire genome, two critical components must be available: a specific oligonucleotide probe for the gene of interest, and a DNA library that can be screened rapidly.

How is a specific probe obtained? In one method, a probe for a gene can be prepared if a part of the amino acid sequence of the protein encoded by the gene is known. Peptide sequencing of a purified protein (Chapter 4) or knowledge of the sequence of a homologous protein from a related species (Chapter 10) are two potential sources of such information. However, a problem arises because a single peptide sequence can be encoded by a number of different oligonucleotides (**Figure 9.16**). Thus, for this purpose, peptide sequences containing tryptophan and methionine are preferred, because these amino acids are specified by a single codon, whereas other amino acid residues have between two and six codons (Table 8.5). All the possible DNA sequences (or their complements) that encode the targeted peptide sequence are synthesized by the solid-phase method and made radioactive by phosphorylating their 5′ ends with ^{32}P.

Amino acid sequence	...	Cys	Pro	Asn	Lys	Trp	Thr	His	...

Potential oligonucleotide sequences:

$$TG_{T}^{C} \quad CC_{G}^{A}_{T}^{C} \quad AA_{T}^{C} \quad AA_{G}^{A} \quad TGG \quad AC_{G}^{A}_{T}^{C} \quad CA_{T}^{C}$$

FIGURE 9.16 Oligonucleotide probes can be designed from the target protein sequence. A probe can be generated by synthesizing all possible oligonucleotides encoding a particular sequence of amino acids. Because of the degeneracy of the genetic code, 256 distinct oligonucleotides must be synthesized to ensure that the probe matching the sequence of seven amino acids in this example is present.

To prepare the DNA library, a sample containing many copies of total genomic DNA is first mechanically sheared or partly digested by restriction enzymes into large fragments (**Figure 9.17**). This process yields a nearly random population of overlapping DNA fragments. These fragments are then separated by gel electrophoresis to isolate the set of all fragments that are about 15 kb long. Synthetic linkers are attached to the ends of these fragments, cohesive ends are formed, and the fragments are then inserted into a vector, such as λ phage DNA, prepared with the same cohesive ends. E. coli bacteria are then infected with these recombinant phages. These phages replicate themselves and then lyse their bacterial hosts. The resulting lysate contains fragments of genomic DNA housed in a sufficiently large number of virus particles to ensure that nearly the entire genome is represented in the library. Phages can be propagated indefinitely such that the library can be used repeatedly over long periods.

This genomic library is then screened to find the very small number of phages harboring the gene of interest. For the human genome, a calculation shows that a 99% probability of success requires screening about 500,000 clones; hence, a very rapid and efficient screening process is essential. Rapid screening can be accomplished by DNA hybridization.

A dilute suspension of the recombinant phages is first plated on a lawn of bacteria (**Figure 9.18**). Where each phage particle has landed and infected a bacterium, a *plaque* containing identical phages develops on the plate. A replica of this master plate is then made by applying a sheet of nitrocellulose. Infected bacteria and phage DNA released from lysed cells adhere to the sheet in a pattern of spots corresponding to the plaques. Intact bacteria on this sheet are lysed with

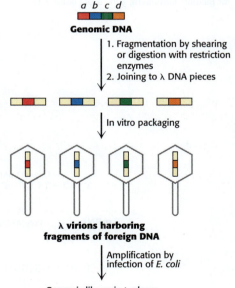

FIGURE 9.17 A genomic library can be created from a digest of a whole genome. After fragmentation of the genomic DNA into overlapping segments, the DNA is inserted into the λ phage vector (shown in yellow). Packaging into virions and amplification by infection in E. coli yields a genomic library.

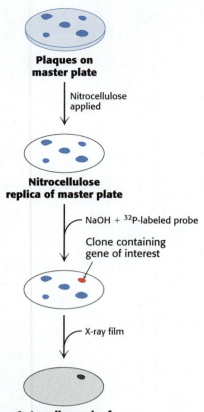

Plaques on master plate

Nitrocellulose applied

Nitrocellulose replica of master plate

NaOH + ³²P-labeled probe

Clone containing gene of interest

X-ray film

Autoradiograph of probe-labeled nitrocellulose

FIGURE 9.18 A genomic library is screened for a specific gene using a labeled DNA probe. Here, a plate is tested for plaques containing gene *a* of Figure 9.17.

sodium hydroxide, which also serves to denature the DNA so that it becomes accessible for hybridization with a ³²P-labeled probe. The presence of a specific DNA sequence in a single spot on the replica can be detected by using a radioactive complementary DNA or RNA molecule as a probe. Autoradiography then reveals the positions of spots harboring recombinant DNA. The corresponding plaques are picked out of the intact master plate and grown. A single investigator can readily screen a million clones in a day. This method makes it possible to isolate virtually any gene, provided that a probe is available.

❖ SELF–CHECK QUESTION

Suppose that a human genomic library is prepared by exhaustive digestion of human DNA with the EcoRI restriction enzyme. What is the average fragment length generated by this enzyme? Is this procedure suitable for cloning large genes? Why or why not?

Complementary DNA prepared from mRNA can be expressed in host cells

The preparation of eukaryotic genomic libraries presents unique challenges, especially if the researcher is interested primarily in the protein-coding region of a particular gene. Recall that most mammalian genes are mosaics of introns and exons. These interrupted genes cannot be expressed by bacteria, which lack the machinery to splice introns out of the primary transcript. However, this difficulty can be circumvented by causing bacteria to take up recombinant DNA that is complementary to mRNA, where the introns have been removed.

The key to forming **complementary DNA**, or **cDNA**, is the enzyme reverse transcriptase. As discussed in Section 8.4, a retrovirus uses this enzyme to form a DNA–RNA hybrid in replicating its genomic RNA. Reverse transcriptase synthesizes a DNA strand complementary to an RNA template if the transcriptase is provided with a DNA primer that is base-paired to the RNA and contains a free 3′-OH group. We can use a simple sequence of linked thymidine [oligo(T)] residues as the primer. This oligo(T) sequence pairs with the poly(A) sequence at the 3′ end of most eukaryotic mRNA molecules (Section 4.5), as shown in **Figure 9.19**.

1. The reverse transcriptase synthesizes the rest of the cDNA strand in the presence of the four deoxyribonucleoside triphosphates (dNTPs).

2. The template RNA strand of this RNA–DNA hybrid is hydrolyzed by raising the pH (also called *alkali digestion*). Unlike RNA, DNA is resistant to alkaline hydrolysis.

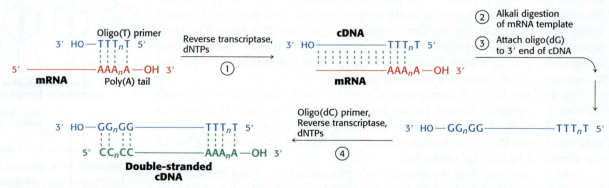

FIGURE 9.19 Double-stranded cDNA is prepared from mRNA using reverse transcriptase. A complementary DNA (cDNA) duplex is created from mRNA by (1) use of reverse transcriptase to synthesize a cDNA strand, (2) digestion of the original RNA strand, (3) addition of several G bases to the DNA by terminal transferase, and (4) synthesis of a complementary DNA strand using the newly synthesized cDNA strand as a template.

3. The enzyme terminal transferase adds nucleotides—for instance, several residues of dG—to the 3′ end of DNA.

4. Oligo(dC) can bind to dG residues and prime the synthesis of the second DNA strand. Synthetic linkers can be added to this double-helical DNA for ligation to a suitable vector.

Complementary DNA for all mRNA that a cell contains can be made, inserted into vectors, and then inserted into bacteria. Such a collection is called a **cDNA library**.

cDNA libraries can be prepared in expression vectors to enable the production of the corresponding protein of interest. Clones are then screened on the basis of their capacity to direct the synthesis of a foreign protein in bacteria, a technique referred to as *expression cloning*. A labeled antibody specific for the protein of interest can be used to identify colonies of bacteria that express the corresponding protein product (**Figure 9.20**). As described earlier, spots of bacteria on a replica plate are lysed to release proteins, which bind to an applied nitrocellulose filter. With the addition of labeled antibody specific for the protein of interest, the location of the desired colonies on the master plate can be readily identified. This immunochemical screening approach can be used whenever a protein is expressed and corresponding antibody is available.

Complementary DNA has many applications beyond the generation of genomic libraries. The overproduction and purification of most eukaryotic proteins in prokaryotic cells necessitates the insertion of cDNA into plasmid vectors. For example, proinsulin, a precursor of insulin, is synthesized by bacteria-harboring plasmids that contain DNA complementary to mRNA for proinsulin (**Figure 9.21**) to produce much of the insulin used today by millions of diabetics.

Proteins with new functions can be created through directed changes in DNA

Many insights about genes and proteins can be gained by analyzing the effects of genetic mutations on protein structure and function. In the classic genetic approach, mutations are generated randomly throughout the genome of a host organism, and those individuals exhibiting a phenotype of interest are selected. Analysis of these mutants then reveals which genes are altered, and DNA sequencing identifies the precise nature of the changes. Recombinant DNA technology makes the creation of specific mutations feasible in vitro. We can construct new genes with designed properties by making three types of directed changes: deletions, insertions, and substitutions. A variety of methods can be used to introduce these mutations.

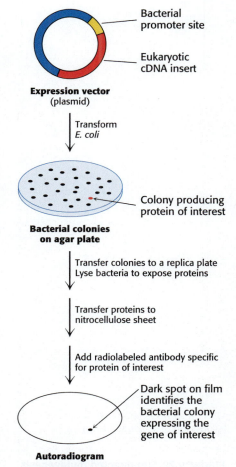

FIGURE 9.20 Expression cloning identifies colonies in a cDNA library that express the protein of interest. Protein products are identified by staining with antibody specific to the desired protein.

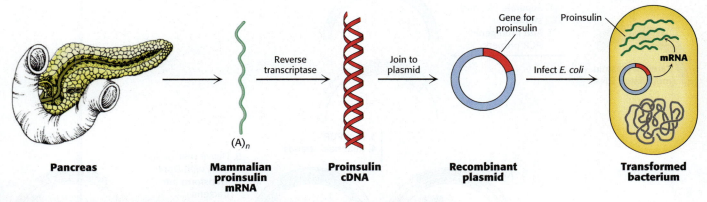

FIGURE 9.21 Bacteria infected with proinsulin cDNA can express the eukaryotic protein. Proinsulin, a precursor of insulin, can be synthesized by transformed (genetically altered) clones of *E. coli*. The clones contain the mammalian proinsulin gene.

Site-directed mutagenesis. Mutant proteins with single amino acid substitutions can be readily produced by **site-directed mutagenesis**. Suppose that we want to replace a particular serine residue with cysteine. This mutation can be made if (1) we have a plasmid containing the gene or cDNA for the protein and (2) we know the base sequence around the site to be altered. If the serine of interest is encoded by TCT, mutation of the central base from C to G yields the TGT codon, which encodes cysteine. This type of mutation is called a *point mutation* because only one base is altered. One method for introducing this mutation into our plasmid is shown in **Figure 9.22**.

1. The two strands of the plasmid are separated by raising the temperature.

2. Mutagenic PCR primers are added to the mixture. These oligonucleotides contain the desired base change but are otherwise complementary to the neighboring sequence. The single mismatch is tolerable if the annealing is carried out at an appropriate temperature.

3. Using these two primers and the original plasmid as a template, a PCR is run. With each subsequent cycle, new DNA is synthesized that is identical to the parental plasmid, except the single base change is incorporated. Because there is no DNA ligase in the reaction mixture, the product strands are linear.

4. At the completion of the PCR, the strands are reannealed. Since the newly synthesized strands are linear, they contain two single nicks at either side of the PCR primer site. Importantly, the newly mutated DNA product is present in vast excess relative to the parental plasmid.

5. As we have seen in Section 6.3, DNA in bacterial cells is methylated to protect itself from cleavage by its own restriction endonucleases. The *Diplococcus pneumoniae* restriction enzyme DpnI is unique in that it specifically cleaves methylated sequences over a 4 bp recognition site. Since the newly mutated plasmid generated by PCR is not methylated—it has never been inside a living cell—we can use DpnI to cleave any remaining parental plasmid DNA.

6. The mutated plasmid is then transformed into bacteria, where the nicks are sealed.

We will encounter many examples of the use of site-directed mutagenesis to precisely alter regulatory regions of genes and to produce proteins with tailor-made features.

FIGURE 9.22 Site-directed mutagenesis uses mismatched primers in a polymerase chain reaction. The products will be linear strands complementary to the original plasmid strands, except for the desired single base change encoded in the PCR primers. Subsequent cleavage of the methylated parental DNA, using DpnI, and transformation into bacteria will yield the altered plasmid.

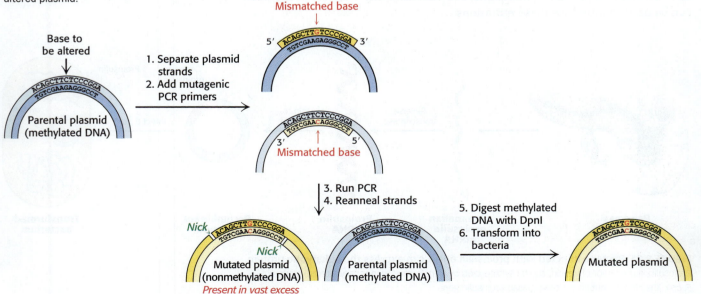

Cassette mutagenesis. In **cassette mutagenesis**, a variety of mutations, including insertions, deletions, and multiple point mutations, can be introduced into the gene of interest. A plasmid harboring the original gene is cut with a pair of restriction enzymes to remove a short segment (**Figure 9.23**). A synthetic double-stranded oligonucleotide—the *cassette*—carrying the genetic alterations of interest is prepared with cohesive ends that are complementary to the ends of the cut plasmid. Ligation of the cassette into the plasmid yields the desired mutated gene product.

Mutagenesis by PCR. A number of methods have been developed that use the creative design of PCR primers (Section 9.1) to introduce specific insertions, deletions, and substitutions into a particular sequence. One such method, called *inverse PCR*, introduces deletions into plasmid DNA (**Figure 9.24**). In this approach, primers are designed to flank the sequence to be deleted. However, these primers are oriented in the opposite direction, such that they direct the amplification of the entire plasmid, minus the region to be deleted. If each of the primers contains a 5′ phosphate group, the amplified product can be recircularized with DNA ligase, yielding the desired deletion mutation.

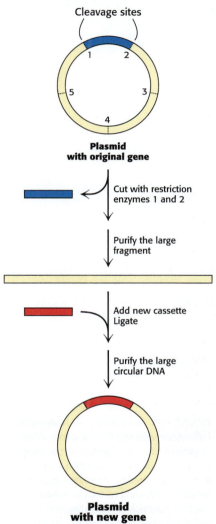

FIGURE 9.23 In cassette mutagenesis, a synthetic DNA fragment is swapped into the plasmid using compatible restriction sites. DNA is cleaved at a pair of unique restriction sites by two different restriction enzymes. A synthetic oligonucleotide with ends that are complementary to these sites (the cassette) is then ligated to the cleaved DNA. The method is highly versatile because the inserted DNA can have any desired sequence.

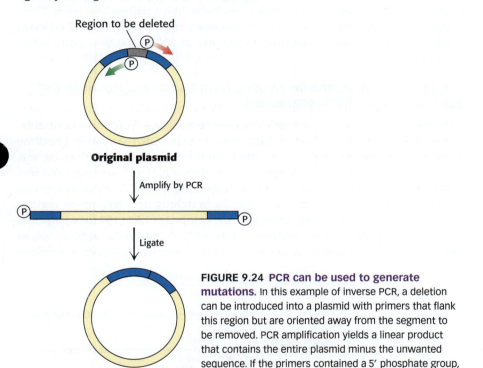

FIGURE 9.24 PCR can be used to generate mutations. In this example of inverse PCR, a deletion can be introduced into a plasmid with primers that flank this region but are oriented away from the segment to be removed. PCR amplification yields a linear product that contains the entire plasmid minus the unwanted sequence. If the primers contained a 5′ phosphate group, this product can be recircularized using DNA ligase, generating a plasmid with the desired mutation.

Designer genes. Novel proteins can also be created by splicing together gene segments that encode domains that are not associated in nature. For example, a gene for an antibody can be joined to a gene encoding a toxic protein, yielding a chimeric protein that kills any cells that are recognized by the antibody. These *immunotoxins* are being evaluated as anticancer agents. Furthermore, noninfectious coat proteins of viruses can be produced in large amounts by recombinant DNA methods. They can serve as *synthetic vaccines* that are safer than conventional vaccines prepared by inactivating pathogenic viruses. For example, a subunit of the hepatitis B virus produced in yeast is proving to be an effective vaccine against this debilitating viral disease. Finally, entirely new genes can be synthesized de novo by the solid-phase method described above. These genes can encode proteins with no known counterparts in nature.

9.3 Complete Genomes Have Been Sequenced and Analyzed

The methods just described are extremely effective for the isolation and characterization of fragments of DNA. However, the genomes of organisms ranging from viruses to human beings contain considerably longer sequences, arranged in very specific ways that are crucial for their integrated functions. Is it possible to sequence complete genomes and analyze them?

For small genomes, complete sequencing was accomplished soon after DNA-sequencing methods were developed. Sanger and his coworkers determined the complete sequence of the 5386 bases in the genome of the φX174 DNA virus in 1977, just a quarter century after Sanger's pioneering elucidation of the amino acid sequence of a protein. This tour de force was followed several years later by the determination of the sequence of human mitochondrial DNA, a double-stranded circular DNA molecule containing 16,569 base pairs that encodes 2 ribosomal RNAs, 22 transfer RNAs, and 13 proteins. Many other viral genomes were sequenced in subsequent years.

However, the genomes of free-living organisms present a much greater challenge because even the simplest comprises more than 1 million base pairs. Thus, sequencing projects require both rapid sequencing techniques and efficient methods for assembling many short stretches of 300 to 500 base pairs into a complete sequence.

The genomes of organisms ranging from bacteria to multicellular eukaryotes have been sequenced

With the development of automatic DNA sequencers based on fluorescent dideoxynucleotide chain terminators, high-volume, rapid DNA sequencing became a reality. The genome sequence of the bacterium *Haemophilus influenzae* was determined in 1995 using a "shotgun" approach in which the genomic DNA was sheared randomly into fragments that were then sequenced. Computer programs then assembled the complete sequence by matching up overlapping regions between fragments. The *H. influenzae* genome comprises 1,830,137 base pairs and encodes approximately 1740 proteins (**Figure 9.25**). Using similar approaches, as well as more advanced methods described below, investigators have determined

FIGURE 9.25 The genome of *Haemophilus influenzae* was the first of a free-living organism to be completely sequenced. The genome encodes more than 1700 proteins (indicated by the colored bars in the outer circle) and 70 RNA molecules. The likely function of approximately one-half of the proteins was determined by comparisons with sequences of proteins already characterized in other species. Gaps in the final sequence were filled by sequencing genomic DNA inserts from 300 λ phage clones (indicated by the blue bars in the inner circle), many of these overlap one another. [Credit: Republished with permission of AAAS. "Whole genome random sequencing and assembly of *Haemophilus influenzae* Rd." Fleischmann et. al., *Science* 269:496-512, 1995. Permission conveyed through Copyright Clearance Center, Inc. Scan courtesy of The Institute for Genomic Research.]

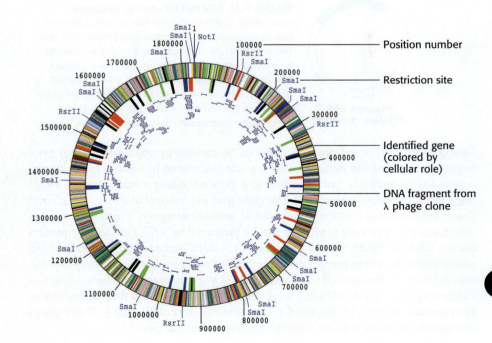

the sequences of more than 10,000 bacterial and archaeal species, including key model organisms such as E. coli, *Salmonella typhimurium*, and *Archaeoglobus fulgidus*, as well as pathogenic organisms such as *Yersinia pestis* (bubonic plague) and *Bacillus anthracis* (anthrax).

The first eukaryotic genome to be completely sequenced was that of baker's yeast, *Saccharomyces cerevisiae*, in 1996. This yeast genome comprises approximately 12 million base pairs, distributed on 16 chromosomes, and encodes more than 6000 proteins. This achievement was followed in 1998 by the first complete sequencing of the genome of a multicellular organism, the nematode *Caenorhabditis elegans*, which contains 97 million base pairs and includes more than 19,000 genes. The genomes of many organisms widely used in biological and biomedical research have now been sequenced, including those of the fruit fly *Drosophila melanogaster*, the model plant *Arabidopsis thaliana*, the mouse, the rat, and the dog. However, note that even after a genome sequence has been considered complete, some sections, such as the repetitive sequences that make up heterochromatin, may be missing because these DNA sequences are very difficult to manipulate with the use of standard techniques.

The sequence of the human genome has been completed

The ultimate goal of much of genomics research has been the sequencing and analysis of the human genome. Given that the human genome comprises approximately 3 billion base pairs of DNA distributed among 24 chromosomes (**Figure 9.26**), the challenge of producing a complete sequence was daunting. However, through an organized international effort of academic laboratories and private companies, the human genome progressed from a draft sequence first reported in 2001 to a finished sequence reported in late 2004.

The human genome is a rich source of information about many aspects of humanity, including our biochemistry and evolution. Analysis of the genome will continue for many years to come. Developing an inventory of protein-encoding genes was one of the first tasks. At the beginning of the genome-sequencing project, the number of such genes was estimated to be approximately 100,000. With the availability of the completed (but not finished) genome, this estimate was reduced to between 30,000 and 35,000. With the finished sequence, the estimate fell to between 20,000 and 25,000. We will use the estimate of 23,000 throughout this book.

The reduction in this estimate is due, in part, to the realization that there are a large number of **pseudogenes**, formerly functional genes that have accumulated mutations such that they no longer produce proteins. For example, more

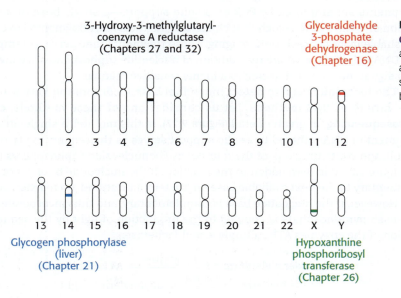

FIGURE 9.26 **The human genome has been completely sequenced.** The human genome is arrayed on 46 chromosomes—22 pairs of autosomes and the X and Y sex chromosomes. The locations of several genes associated with important pathways in biochemistry are highlighted.

than half of the genomic regions that correspond to olfactory receptors—key molecules responsible for our sense of smell—are pseudogenes (Section 14.6). In contrast, the corresponding regions in the genomes of other primates and rodents encode functional olfactory receptors.

Nonetheless, the surprisingly small number of genes belies the complexity of the human proteome. Many genes encode more than one protein through mechanisms such as alternative splicing of mRNA and posttranslational modifications of proteins. The different proteins encoded by a single gene often display important variations in functional properties.

The human genome contains a large amount of DNA that does not encode proteins. A great challenge in modern biochemistry and genetics is to elucidate the roles of this noncoding DNA. Much of this DNA is present because *mobile genetic elements*, which are related to retroviruses (Section 8.4), have inserted themselves throughout the genome over time. Most of these elements have accumulated mutations and are no longer functional. For example, more than 1 million *Alu sequences*, each approximately 300 bases in length, are present in the human genome. *Alu* sequences are examples of SINEs, or short interspersed elements. The human genome also includes nearly 1 million LINEs, or long interspersed elements, sequences that can be as long as 10 kilobases (kb). The roles of these elements as neutral genetic parasites or instruments of genome evolution are under current investigation.

Next-generation sequencing methods enable the rapid determination of a complete genome sequence

Since the introduction of the Sanger dideoxy method in the mid-1970s, significant advances have been made in DNA-sequencing technologies, enabling the readout of progressively longer sequences with higher fidelity and shorter run times. The development of **next-generation sequencing (NGS)** platforms has extended this capability to formerly unforeseen levels. By combining technological breakthroughs in the handling of very small amounts of liquid, high-resolution optics, and computing power, these methods have already made a significant impact on the ability to obtain whole genome sequences rapidly and cheaply (Section 1.4).

Next-generation sequencing refers to a family of technologies, each of which uses a unique approach for the determination of a DNA sequence. All of these methods are highly parallel: from 1 million to 1 billion DNA fragment sequences are individually acquired in a single experiment. How are NGS methods capable of determining such a high number of sequences in parallel? Individual DNA fragments are amplified by PCR on a solid support—a single bead or a small region of a glass slide—such that clusters of identical DNA fragments are distinguishable by high-resolution imaging. These fragments then serve as templates for DNA polymerase, where the addition of nucleotide triphosphates is converted to a signal that can be detected in a highly sensitive manner.

The technique used to detect individual base incorporation varies among the variety of NGS methods. Let us consider one of these methods, called **pyrosequencing**, in greater detail (**Figure 9.27**). In this method, a single-stranded fragment of DNA tethered to a solid support serves as the sequencing template. A solution containing one of the four deoxyribonucleoside triphosphates (dATP in Figure 9.27A) is then added to the reaction. If the nucleotide base is not complementary to the next available base on the template strand, no reaction occurs.

However, if the nucleotide base is complementary, and thus incorporated, the released pyrophosphate is coupled to the production of light by the sequential action of the enzymes ATP sulfurylase and luciferase:

$$\text{PP}_i + \text{adenylylsulfate} \underset{}{\overset{\text{ATP sulfurylase}}{\rightleftharpoons}} \text{ATP} + \text{sulfate}$$

$$\text{ATP} + \text{luciferin} \underset{}{\overset{\text{luciferase}}{\rightleftharpoons}} \text{oxyluciferin} + \textbf{light}$$

(A)

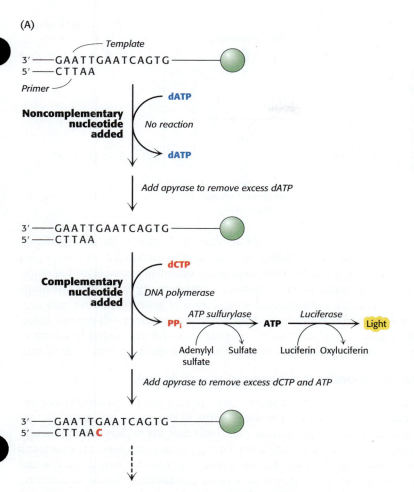

(B) **Sequence = CTTAGT**

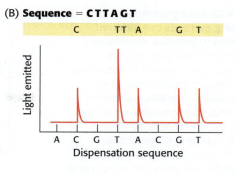

FIGURE 9.27 Pyrosequencing is an example of next-generation sequencing. (A) If a nucleotide is successfully incorporated into the growing duplex, the newly produced pyrophosphate is enzymatically converted into light. (B) Light is measured at each deoxynucleotide addition. A peak indicates the next base in the sequence, and the height of the peak indicates how many of the same deoxynucleotide were incorporated in a row.

The emitted light is measured, indicating that a reaction has occurred, thus revealing the identity of the base at that position (Figure 9.27B). If the template contains a run of one particular base, then multiple consecutive additions will take place in a single reaction; the length of this run can be determined by the intensity of the emitted light.

After each nucleotide addition, the remaining deoxynucleotide and ATP are removed by the enzyme apyrase.

$$\text{dNTP} \xrightleftharpoons{\text{apyrase}} \text{dNMP} + 2P_i$$
$$\text{ATP} \xrightleftharpoons{\text{apyrase}} \text{AMP} + 2P_i$$

The next nucleotide is added, and the steps are repeated. Nucleotides are added in a defined order, referred to as the *dispensation sequence*, and the pattern of peaks that emerge reveals the sequence of the template strand (Figure 9.27B).

Other NGS methods are also commonly used. These methods can be understood simply by considering the overall reaction of chain elongation catalyzed by DNA polymerase (**Figure 9.28**). In the reversible terminator method, the four nucleotides are added to the template DNA, with each base tagged with a unique fluorescent label and a reversibly blocked 3′ end. The blocked end assures that only one phosphodiester linkage will form. Once the nucleotide is incorporated into the growing strand, it is identified by its fluorescent tag, the blocking agent is removed, and the process is repeated. The protocol for ion semiconductor sequencing is similar to pyrosequencing, except that nucleotide incorporation is detected by sensitively measuring the very small changes in pH of the reaction mixture due to the release of a proton upon nucleotide incorporation.

Regardless of the sequencing method, the technology exists to quantify the signal produced by millions of DNA fragment templates simultaneously. However,

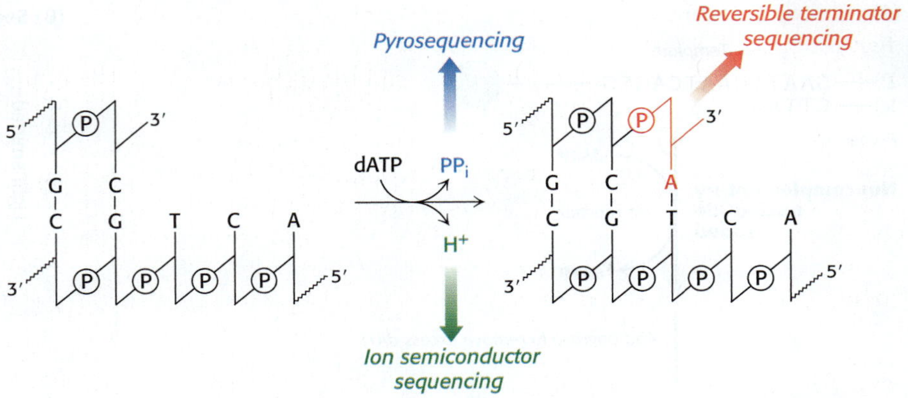

FIGURE 9.28 Next-generation sequencing methods use different approaches to measure nucleotide incorporation. Measurement of base incorporation in next-generation sequencing methods relies on the detection of the various products of the DNA polymerase reaction. Reversible terminator sequencing measures the nucleotide incorporation in a manner similar to Sanger dideoxy sequencing, while pyrosequencing and ion semiconductor sequencing detect the release of pyrophosphate and protons, respectively.

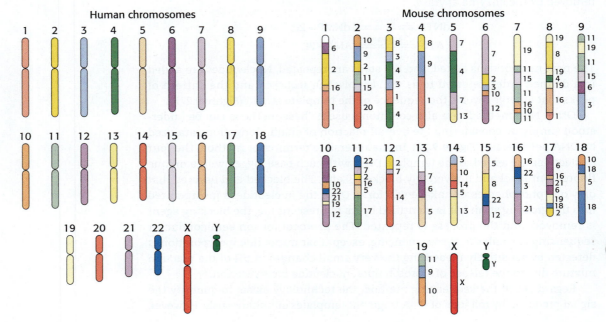

for many approaches, as few as 50 bases are read per fragment. Hence, significant computing power is required to both store the massive amounts of sequence data and perform the necessary alignments required to assemble a completed sequence.

NGS methods are being used to answer an ever-growing number of questions in many areas, including genomics, the regulation of gene expression, and evolutionary biology. Additionally, individual genome sequences will provide information about genetic variation within populations and may usher in an era of personalized medicine, when these data can be used to guide treatment decisions.

Comparative genomics is a powerful research tool

Comparisons with genomes from other organisms are sources of insight into the human genome. The sequencing of the genome of the chimpanzee, our closest living relative, as well as that of other mammals that are widely used in biological research, such as the mouse and the rat, has been completed. Comparisons reveal that an astonishing 99% of human genes have counterparts in mouse and rat genomes. However, these genes have been substantially reassorted among chromosomes in the estimated 75 million years of evolution since humans and these rodents had a common ancestor (**Figure 9.29**).

The genomes of other organisms also have been determined specifically for use in comparative genomics. For example, the genomes of two species of puffer fish, *Takifugu rubripes* and *Tetraodon nigroviridis*, have been determined. These genomes were selected because they are very small and lack much of the

FIGURE 9.29 The human and mouse genomes are closely related, although they have been rearranged relative to each other. A schematic comparison of the human genome and the mouse genome shows reassortment of large chromosomal fragments. The small numbers to the right of the mouse chromosomes indicate the human chromosome to which each region is most closely related.

noncoding DNA present in such abundance in the human genome. The puffer fish genomes include fewer than 400 megabase pairs (Mbp), one-eighth of the number in the human genome, yet the puffer fish and human genomes contain essentially the same number of genes.

Comparison of the genomes of these puffer fish species with that of humans revealed more than 1000 formerly unrecognized human genes. Furthermore, comparison of the two species of puffer fish, which had a common ancestor approximately 25 million years ago, is a source of insight into more-recent events in evolution. Comparative genomics is a powerful tool, both for interpreting the human genome and for understanding major events in the origin of genera and species (Section 10.3).

9.4 Eukaryotic Genes Can Be Quantitated and Manipulated with Considerable Precision

After a gene of interest has been identified, cloned, and sequenced, it is often desirable to understand how that gene and its corresponding protein product function in the context of a whole cell or organism. It is now possible to determine how the expression of a particular gene is regulated, how mutations in the gene affect the function of the corresponding protein product, and how the behavior of an entire cell or model organism is altered by the introduction of mutations within specific genes. In addition, levels of transcription of large families of genes within cells and tissues can be readily quantified and compared across a range of environmental conditions.

Gene-expression levels can be comprehensively examined

Most genes are present in the same quantity in every cell—namely, one copy per haploid cell or two copies per diploid cell. However, the level at which a gene is expressed, as indicated by mRNA quantities, can vary widely, ranging from no expression to hundreds of mRNA copies per cell. Gene-expression patterns vary from cell type to cell type, distinguishing, for example, a muscle cell from a nerve cell. Even within the same cell, gene-expression levels may vary as the cell responds to changes in physiological circumstances. Although mRNA levels sometimes correlate with the levels of proteins expressed, this correlation does not always hold. Thus, care must be exercised when interpreting the results of mRNA levels alone.

Quantitative PCR. The quantity of individual mRNA transcripts can be determined by **quantitative PCR (qPCR)**, or real-time PCR. RNA is first isolated from the cell or tissue of interest. With the use of reverse transcriptase, cDNA is prepared from this RNA sample. In one qPCR approach, the transcript of interest is PCR amplified with the appropriate primers in the presence of the dye SYBR Green I, which fluoresces brightly when bound to double-stranded DNA. In the initial PCR cycles, not enough duplex is present to allow a detectable fluorescence signal. However, after repeated PCR cycles, the fluorescence intensity exceeds the detection threshold and continues to rise as the number of duplexes corresponding to the transcript of interest increases (**Figure 9.30A**).

Importantly, the cycle number at which the fluorescence becomes detectable over a defined threshold (or C_T) is inversely proportional to the number of copies of the original template (**Figure 9.30B**). After the relation between the original copy number and the C_T has been established with the use of a known standard, subsequent qPCR experiments can be used to determine the number of copies of any desired transcript in the original sample, provided the appropriate primers are available.

(A)

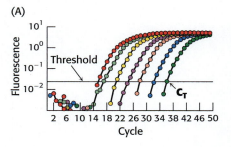

(B)

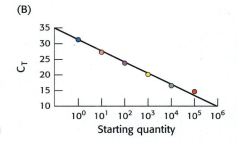

FIGURE 9.30 Quantitative PCR detects the amount of a particular transcript in a sample. (A) In qPCR, fluorescence is monitored in the course of PCR amplification to determine C_T, the cycle at which this signal exceeds a defined threshold. Each color represents a different starting quantity of cDNA. (B) C_T values are inversely proportional to the number of copies of the original cDNA template. [Data from N. J. Walker, *Science* 296:557–559, 2002.]

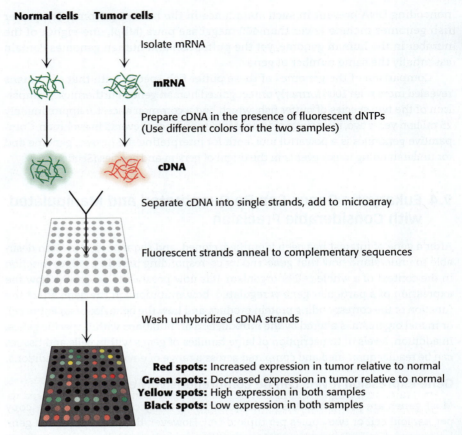

FIGURE 9.31 DNA microarrays can measure gene expression changes in a tumor. mRNA is isolated from two samples, tumor cells and a control sample. From these transcripts, cDNA is prepared in the presence of a fluorescent nucleotide, with a red label for the tumor sample and a green label for the control sample. Spots that are red indicate genes that are expressed more highly in the tumor, while the green spots indicate reduced expression relative to control. Spots that are black or yellow indicate comparable expression at either low or high levels, respectively. [Information from D. L. Nelson and M. M. Cox, *Lehninger Principles of Biochemistry*, 6th ed. (W. H. Freeman and Company, 2013).]

DNA microarrays. Although qPCR is a powerful technique for quantitation of a small number of transcripts in any given experiment, we can now use our knowledge of complete genome sequences to investigate an entire **transcriptome**, the pattern and level of expression of all genes in a particular cell or tissue. One of the most powerful methods for this purpose is based on hybridization. Single-stranded oligonucleotides whose sequences correspond to coding regions of the genome are affixed to a solid support such as a microscope slide, creating a **DNA microarray**. Importantly, the position of each sequence within the array is known. **Figure 9.31** outlines the steps in the creation of a DNA microarray.

1. mRNA is isolated from the cells of interest (a tumor, for example) as well as a control sample.

2. From this mRNA, cDNA is prepared (Section 9.2) in the presence of fluorescent nucleotides using different labels, usually green and red, for the two samples.

3. The samples are combined, separated into single strands, and added to the microarray.

4. The fluorescent strands anneal their complementary sequences.

5. Any unhybridized cDNA strands are washed off, leaving an array in which relative levels of green and red fluorescence at each spot indicate the differences in expression for each gene.

DNA microarrays have been prepared such that thousands of transcript levels can be assessed in a single experiment. Hence, over several arrays, the differences in expression of many genes across a number of different cell types or conditions can be measured (**Figure 9.32**).

RNA-seq. One disadvantage of the qPCR and microarray methods is that the transcript sequences must be known prior to the experiment, in order to design primers or microarray probes. However, the emergence of next-generation sequencing techniques, such as those described in the previous section, has enabled the detection of the presence and quantitation of virtually all the RNA molecules in a cell under specific conditions. This method, termed **RNA-seq**, consists of several essential steps:

1. *Isolation of RNA from the cells or tissue of interest.* Deoxyribonuclease (DNase), an enzyme which specifically cleaves DNA, can be used at this step to degrade any unwanted genomic DNA.

2. *RNA selection.* At this step, mRNA can be selectively enriched by filtering for sequences that contain poly(A) tails. This step allows for the removal of the abundant rRNAs from the sample.

3. *cDNA synthesis.* cDNA is prepared from the RNA sample using reverse transcriptase, as described earlier in the chapter.

4. *Sequencing and assembly.* The cDNA molecules within the sample are then sequenced using an NGS method, such as pyrosequencing. Each sequencing run is typically quite short, yielding only 50–100 bases, but millions of sequences are generated in a single experiment. Computational algorithms are then used to assemble the sequencing data into a comprehensive understanding of the transcripts present within the sample.

RNA-seq is more sensitive than microarray methods, enabling the detection of very low abundance transcripts. This sensitivity has led to the application of RNA-seq to the determination of transcript profiles of individual cells (called *single-cell RNA-seq*, or *scRNA-seq*). These single-cell experiments have broad application across a variety of disciplines, yielding insight on the variability of cancer cells within a single tumor or the changes within individual cell lineages during embryonic development.

New genes inserted into eukaryotic cells can be efficiently expressed

We have seen how bacteria can be turned into factories for the production of a wide range of prokaryotic and eukaryotic proteins. However, bacteria lack the necessary enzymes to carry out posttranslational modifications unique to these proteins, such as the specific cleavage of polypeptides and the attachment of carbohydrate units. Thus, many eukaryotic genes can be expressed correctly only in eukaryotic host cells.

Moreover, the introduction of recombinant DNA molecules into cells of higher organisms can be a source of insight into how their genes are organized and expressed. How are genes turned on and off in embryological development? How does a fertilized egg give rise to an organism with highly differentiated cells that are organized in space and time? These central questions of biology can now be fruitfully approached by expressing foreign genes in mammalian cells.

Recombinant DNA molecules can be introduced, or *transformed*, into animal cells in several ways:

• Foreign DNA molecules precipitated by calcium phosphate can be taken up by animal cells by a process known as *transfection.* A small fraction of the imported DNA becomes stably integrated into the chromosomal DNA. While the efficiency of incorporation is low, this method is useful because it is easy to apply.

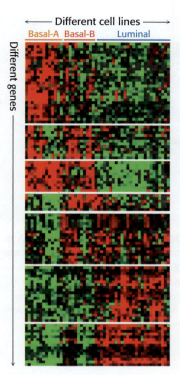

FIGURE 9.32 Microarrays detect numerous expression changes in a single experiment. The expression levels of thousands of genes can be simultaneously analyzed with DNA microarrays. Here, an analysis of 8750 genes in 50 breast cancer cell lines reveals that these samples can be divided into distinct classes (basal-A, basal-B, and luminal) based on their gene-expression patterns. In this "heat map" representation, each row represents a different gene, and each column represents a different breast cancer cell line (i.e., a separate microarray experiment). Red corresponds to gene induction, and green corresponds to gene repression. [Credit: Kao J, Salari K, Bocanegra M, Choi YL, Girard L, Gandhi J, Kwei KA, Hernandez-Boussard T, Wang P, Gazdar AF, Minna JD, Pollack JR. Molecular profiling of breast cancer cell lines defines relevant tumor models and provides a resource for cancer gene discovery. February 2009. *PLoS One* 4(7):e6146, Figure 1B. DOI:10.1371/journal. pone.0006146.]

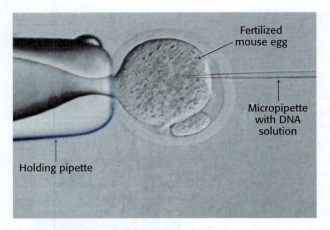

FIGURE 9.33 Cloned plasmid DNA is microinjected into a fertilized mouse egg. [Credit: Marh et al, Hyperactive self-inactivating piggyBac for transposase-enhanced pronuclear microinjection transgenesis, PNAS, Vol 109, no. 47, pp. 19184–19189. Copyright 2012 National Academy of Sciences.]

- DNA can be microinjected into cells. A fine-tipped glass micropipette containing a solution of foreign DNA is inserted into the nucleus (**Figure 9.33**). A skilled investigator can inject hundreds of cells per hour.

- Viruses can be used to introduce new genes into animal cells through a process called *transduction*. The most effective vectors are retroviruses, whose genomes are encoded by RNA and replicate through DNA intermediates. The retroviruses used for this purpose typically do not harm their hosts.

A striking feature of the life cycle of a retrovirus is that the double-helical DNA form of its genome, produced by the action of reverse transcriptase, becomes randomly incorporated into host chromosomal DNA. This DNA version of the viral genome, called *proviral DNA*, can be efficiently expressed by the host cell and replicated along with normal cellular DNA. Foreign genes have been efficiently introduced into mammalian cells by infecting them with vectors derived from the Moloney murine leukemia virus, a retrovirus which can accept inserts as long as 6 kb. Some genes introduced by this vector into the genome of a transformed host cell are efficiently expressed.

Two other vectors based on two large-genome viruses have been engineered to express DNA inserts efficiently. *Vaccinia* virus, a large DNA-containing virus, replicates in the cytoplasm of mammalian cells, where it shuts down host-cell protein synthesis. *Baculovirus* infects insect cells, which can be conveniently cultured. Insect larvae infected with this virus can serve as efficient protein factories.

Transgenic animals harbor and express genes introduced into their germ lines

As shown in Figure 9.33, plasmids harboring foreign genes can be microinjected into the male pronucleus of fertilized mouse eggs, which are then inserted into the uterus of a foster-mother mouse. A subset of the resulting embryos in this host will then harbor the foreign gene; these embryos may develop into mature animals. Southern blotting or PCR analysis of DNA isolated from the progeny can be used to determine which offspring carry the introduced gene. These **transgenic mice** are a powerful means of exploring the role of a specific gene in the development, growth, and behavior of an entire organism. Transgenic animals often serve as useful models for a particular disease process, enabling researchers to test the efficacy and safety of a newly developed therapy.

Consider the example of ALS, introduced in Section 9.1. Research groups have generated transgenic mouse lines that express forms of human superoxide dismutase that harbor mutations matching those identified in earlier genetic analyses. Many of these strains exhibit a clinical picture similar to that observed in ALS patients: progressive weakness of voluntary muscles and eventual paralysis, motor-neuron loss, and rapid progression to death. Since their first characterization in 1994, these strains continue to serve as valuable sources of information for the exploration of the mechanism and potential treatment of ALS.

Gene disruption and genome editing provide clues to gene function and opportunities for new therapies

The function of a gene can also be probed by inactivating or editing it and looking for resulting abnormalities.

Gene knockouts. Powerful methods have been developed for accomplishing such **gene disruption** (or **gene knockout**) in organisms such as yeast and mice. These methods rely on the process of **homologous recombination**

(Section 28.5), in which two DNA molecules with strong sequence similarity exchange segments. If a region of foreign DNA is flanked by sequences that have high homology to a particular region of genomic DNA, two recombination events will yield the transfer of the foreign DNA into the genome (**Figure 9.34**). In this manner, specific genes can be targeted if their flanking nucleotide sequences are known.

For example, the gene-knockout approach has been applied to the genes encoding gene-regulatory proteins (also called *transcription factors*; Section 29.3) that control the differentiation of muscle cells. When both copies of the gene for the regulatory protein myogenin are disrupted in mice, the animals lack functional skeletal muscle. The impaired formation of respiratory muscles, such as the diaphragm, leads to the inability to breathe and death shortly after birth. Microscopic inspection reveals that the tissues from which muscle normally forms contain precursor cells that have failed to differentiate fully (**Figure 9.35A**). Heterozygous mice containing one normal myogenin gene and one disrupted gene appear normal, suggesting that a reduced level of myogenin gene expression is still sufficient for normal muscle development. The generation and characterization of this knockout mouse provided strong evidence that functional myogenin is essential for proper development of skeletal muscle tissue (**Figure 9.35B**).

Analogous studies have probed the function of many other genes in the context of a living organism. Furthermore, mice harboring gene knockouts can be used as animal models for known human genetic diseases, enabling the assessment of new potential therapies for treatment. However, manipulation of genomic DNA using homologous recombination, while a powerful tool, has limitations. Recombination at the desired site can be inefficient and time-consuming. In addition, this method is generally limited to specific model organisms, such as yeast, mice, and fruit flies.

Genome editing. Over the past 10 years, new methods for the highly specific modification of genomic DNA, or **genome editing**, have emerged. These approaches rely on the introduction of double-strand breaks at precisely determined sequences within genomic DNA. The resulting cleavage site is repaired by the DNA repair machinery of the host cell in one of two ways (Section 28.4) If no template is provided, the cell will repair the break using a process known

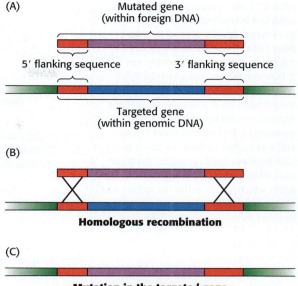

FIGURE 9.34 Foreign DNA can be inserted into the host genome by homologous recombination. (A) A mutated version (purple) of the gene to be disrupted (blue) is constructed, maintaining some regions of homology with the normal gene (red). When the foreign mutated gene is introduced into an embryonic stem cell, (B) recombination takes place at regions of homology and (C) the normal (targeted) gene is replaced, or "knocked out," by the foreign gene. The cell is inserted into embryos, and mice lacking the gene (knockout mice) are produced.

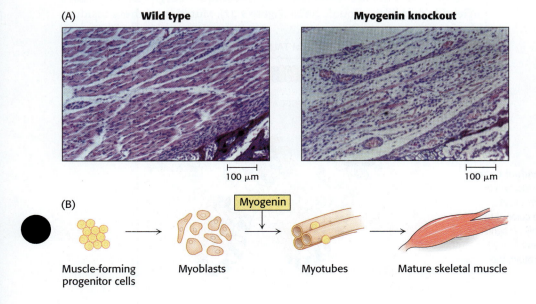

FIGURE 9.35 In knockout mice, gene disruption can have dramatic consequences. (A) Sections of muscle from normal (left) and myogenin-knockout (right) mice, as viewed under the light microscope. Muscles do not develop properly in mice having both myogenin genes disrupted. (B) Through the gene-disruption studies in (A), myogenin was identified as an essential component of the development of mature skeletal muscle. [(A) Reprinted by permission from Macmillan Publishers Ltd: *Nature,* v. 364, Hasty, P., Bradley, A., Morris, J. H., Edmondson, D. G., Venuti, J. M., Olson, E. N., Klein, W. H., Muscle deficiency and neonatal death in mice with a targeted mutation in the myogenin gene, pp. 501–506, copyright 1993. (B) Information from S. Hettmer and A. J. Wagers, *Nat. Med.* 16:171–173, 2010, Fig. 1.]

FIGURE 9.36 Genome editing starts with the introduction of precisely generated double-strand breaks in host DNA. These breaks are repaired in the cell by two possible mechanisms. One is non-homologous end joining (NHEJ), a process that is error-prone and may lead to the introduction of insertions or deletions. Such errors may introduce frameshift mutations that ultimately knock out the entire gene. Alternatively, if a donor template DNA fragment is also provided, the break can be repaired by homology directed repair (HDR), which results in the incorporation of the desired modifications (green) into the targeted gene.

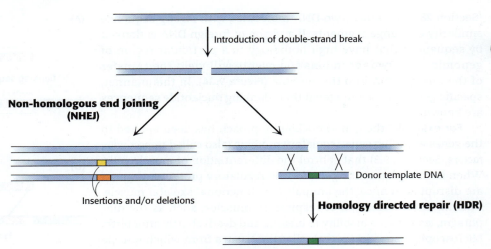

as *non-homologous end joining* (NHEJ; **Figure 9.36**). However, this process is error-prone, and a variety of insertions or deletions will be introduced into the repair site. It is likely that a subset of these modifications will introduce a premature stop codon into the targeted gene, resulting in a gene knockout. Alternatively, if a DNA template containing the desired sequence change is simultaneously introduced with the nucleases, the repair machinery will use this donor template to introduce these changes directly into the genomic sequence, in a process known as *homology directed repair* (HDR; Figure 9.36).

How are double-stranded breaks introduced specifically into the gene of interest? In one approach, a sequence-specific nuclease is engineered by fusing the nonspecific nuclease domain of the restriction enzyme FokI to a DNA-binding domain designed to bind to a particular DNA sequence. Two examples of this strategy are in common usage:

- **Zinc-finger nucleases (ZFNs)** are formed by combining the FokI nuclease domain with the DNA-binding domain that contains a series of zinc-finger domains, small zinc-binding motifs that each recognize a sequence of three base pairs. The preferred DNA-binding sequence can be altered by changing the identity of only four contact residues within each finger.

- **Transcription activator-like effector nucleases (TALENs)** are formed from the fusion of the FokI nuclear domain with an array of TALE repeats. Each repeat contains 34 amino acids and two α-helices, yet only two of these residues (at positions 12 and 13) are responsible for the unique recognition of a single nucleotide within the double helix (**Figure 9.37**). Mutation of these residues

FIGURE 9.37 TALE repeats recognize individual bases in DNA. Each TALE repeat contains 34 amino acids, two of which specify its nucleotide binding partner. In this figure, the identity of these residues is indicated by the color of the repeat. TALE proteins can be designed to uniquely recognize extended oligonucleotide sequences. In this example, a 22 base-pair sequence is bound by a single TALE protein, the bacterial effector PthXo1. [Drawn from 3UGM.pdb.]

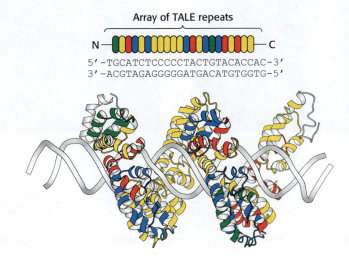

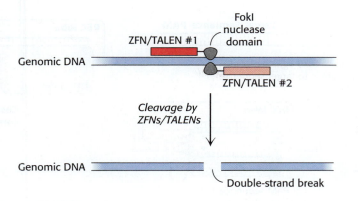

FIGURE 9.38 A pair ZFNs or TALENs introduce double-strand breaks into DNA. The nuclease domain of FokI functions as a dimer. Thus, in order to achieve a sequence-specific double-strand break, two ZFNs or TALENs are required, one for each DNA strand.

within an array of repeats enables the recognition of a vast number of possible DNA target sequences with a high degree of specificity.

Since the FokI nuclease functions as a dimer, two ZFN/TALEN constructs are necessary to generate a double-stranded break, one for each strand of DNA (**Figure 9.38**).

- The **CRISPR-Cas system** enables even greater diversity and potential for customization in the generation of precise double-strand DNA breaks. Clustered regularly interspaced palindromic repeats (or CRISPRs) were first discovered in the late 1980s in the genome of the bacterium E. coli, then subsequently identified in numerous bacterial and archaeal species. Near these repeats are clusters of genes encoding CRISPR-associated, or *Cas*, proteins, which were predicted to function as enzymes that bind to and cleave DNA. Twenty years later, it was confirmed that these loci constitute a prokaryotic immune system, enabling the cleavage of foreign DNA sequences. Since this discovery, CRISPR-Cas systems have been cleverly adapted to enable customized sequence-specific DNA cleavage.

How do CRISPR-Cas systems achieve sequence-specific double-strand breaks in DNA? Cleavage of target DNA by the CRISPR-Cas system of the bacterium *Streptococcus pyogenes* requires only a single protein, the nuclease Cas9, and an engineered, single-stranded **single guide RNA**, or **sgRNA**. At its 5′ end, the sgRNA contains approximately 20 nucleotides that can be customized to complement the desired target site, followed by multiple stem-loop structures at its 3′ end, which are required for binding to the nuclease (**Figure 9.39**). Cas9 is a large, 158 kDa protein which contains two lobes: a REC lobe, which binds to the duplex formed between the sgRNA and the target strand of DNA; and a NUC lobe, which

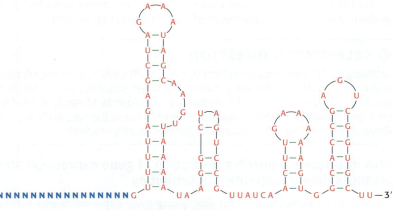

Single guide RNA (sgRNA)

FIGURE 9.39 The single guide RNA contains the sequence that will drive site-specific cleavage. A single guide RNA (sgRNA) contains 20 nucleotides at its 5′ end (blue) which specify the target sequence, followed by multiple stem-loop structures (red) which mediate the interaction with Cas9.

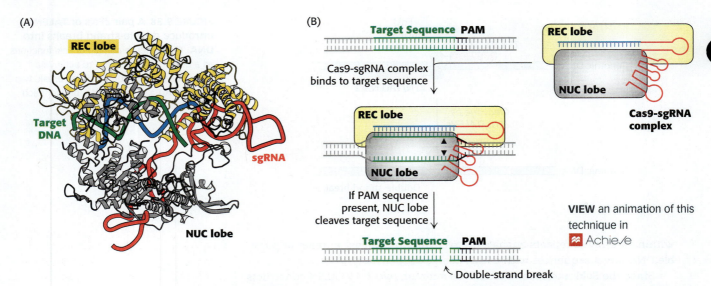

(A)

(B)

FIGURE 9.40 **DNA is cleaved by the CRISPR-Cas system.** (A) The structure of a Cas9-sgRNA complex reveals the two lobes of Cas9: the REC lobe (yellow) and the NUC lobe (gray). The REC lobe mediates the interaction with the sgRNA, shown in blue and red as in Figure 9.39. The NUC lobe contains the nuclease domain that will cleave the target DNA (green). (B) If the target DNA sequence contains an adjacent PAM sequence, the sgRNA-Cas9 complex will bind and cleave both strands of the target, at the sites indicated by the black triangles, yielding a double-strand break.

[Part (A) Drawn from 4OO8.pdb.]

EMMANUELLE CHARPENTIER AND JENNIFER DOUDNA French microbiologist and biochemist Emmanuelle Charpentier (right) has spent much of her career investigating bacterial pathogens and the factors that contribute to their virulence. In 2011, then working at Umeå University in Sweden, she and her research team published a detailed study on the CRISPR-Cas9 system of *Streptococcus pyogenes*, including characterization of the RNA components that determine the sequence specificity of the Cas9 nuclease enzyme. While presenting her work at a conference later that year, she met Jennifer Doudna (left), a biochemist at the University of California, Berkeley, and a world leader in the structure and function of RNA molecules, who had recently developed an interest in the CRISPR-Cas9 system. Their mutual and complementary interests formed the basis for a collaboration, and the following year they together published a landmark paper demonstrating how an engineered single guide RNA could direct the Cas9-mediated cleavage of almost any DNA sequence. Their work has transformed the field of genome editing and underscores the power of collaboration in the scientific endeavor. In 2020, Charpentier and Doudna shared the Nobel Prize in Chemistry for their discovery.

contains the two nuclease domains that are responsible for cleavage of the two strands of the target DNA (**Figure 9.40A**).

Cas9 also contains a region that recognizes a short (usually three- or four-nucleotide) sequence of DNA known as the *protospacer-adjacent motif* (PAM). If the target DNA sequence complementary to the sgRNA is adjacent to a PAM, the Cas9 complex cleaves both target strands using the two nuclease domains of the NUC lobe (**Figure 9.40B**). As the specificity of this nuclease is driven by a complementary sequence of RNA, the CRISPR-Cas system can be engineered to cleave virtually any DNA sequence, provided it is adjacent to a PAM, without the need for designing new protein recognition domains such as those required for ZFNs or TALENs.

Site-specific nuclease-based genome editing methods have now been applied to a variety of species, including model organisms used in the laboratory (rat, zebrafish, and fruit fly), various forms of livestock (pig, cow), and a number of plants. In addition, their use as therapeutic tools in humans is currently under investigation. For example, a ZFN that inactivates the human CCR5 gene, a coreceptor for cellular invasion of human immunodeficiency virus (HIV), is currently in clinical trials for the treatment of patients infected with HIV.

❖ SELF–CHECK QUESTION

Targeting a DNA sequence for editing by the CRISPR-Cas9 system requires that a PAM sequence be present just downstream of the target sequence. For *S. pyogenes* Cas9, this PAM is 5′-NGG-3′, where N represents any nucleotide. Mutants of this Cas9 enzyme have been developed which recognize different PAM sequences, such as 5′-NGAN-3′. Why might these mutants be useful for genome editing studies?

RNA interference enables disruption of gene expression and presents new therapeutic opportunities

An extremely powerful tool for disrupting gene expression was serendipitously discovered in the course of studies that required the introduction of RNA into a cell.

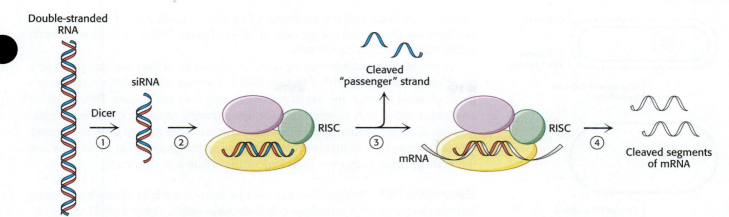

The introduction of a specific double-stranded RNA molecule into a cell was found to suppress the transcription of genes that contained sequences present in the double-stranded RNA molecule. Thus, the introduction of a specific RNA molecule can interfere with the expression of a specific gene.

The mechanism of this process, called **RNA interference (RNAi)**, has been largely established (**Figure 9.41**):

1. When a double-stranded RNA molecule is introduced into an appropriate cell, the RNA is cleaved by the enzyme Dicer into fragments approximately 21 nucleotides in length. Each fragment, termed a small interfering RNA (siRNA), consists of 19 bp of double-stranded RNA and 2 bases of unpaired RNA on each 5′ end.

2. The siRNA is loaded into an assembly of several proteins referred to as the *RNA-induced silencing complex* (RISC).

3. The RISC unwinds the RNA duplex and cleaves one of the strands, the so-called *passenger strand*. The uncleaved single-stranded RNA segment, the *guide strand*, remains incorporated into the enzyme.

4. The fully assembled RISC cleaves mRNA molecules that contain exact complements of the guide-strand sequence. Thus, levels of such mRNA molecules are dramatically reduced.

The technique of RNA interference is called *gene knockdown*, because the expression of the gene is reduced but not eliminated, as is the case with gene knockouts.

The machinery necessary for RNA interference is found in many cells. In some organisms such as *C. elegans*, RNA interference is quite efficient. Indeed, RNA interference can be induced simply by feeding *C. elegans* strains of *E. coli* that have been engineered to produce appropriate double-stranded RNA molecules. Although not as efficient in mammalian cells, RNA interference has emerged as a powerful research tool for reducing the expression of specific genes. Moreover, in 2017, an siRNA targeting the protein transthyretin achieved positive results in a Phase III clinical trial for the treatment of familial amyloid polyneuropathy.

Foreign DNA can be introduced into plants

Several methods have been developed to introduce foreign DNA into plant cells.

Ti plasmids. The common soil bacterium *Agrobacterium tumefaciens* infects plants and introduces foreign genes into plant cells, causing a lump of tumor tissue called a *crown gall* to grow at the site of infection (**Figure 9.42**). Crown galls synthesize opines, a group of amino acid derivatives that are metabolized by the infecting bacteria. In essence, the bacterium hijacks the metabolism of the plant cells to satisfy its highly distinctive appetite. **Tumor-inducing plasmids (Ti plasmids)** that are carried by *A. tumefaciens* carry instructions for the switch

FIGURE 9.41 Specific mRNA transcripts are degraded by RNA interference. A double-stranded RNA molecule is cleaved into 21-bp fragments by the enzyme Dicer to produce siRNAs. These siRNAs are incorporated into the RNA-induced silencing complex (RISC), where the single-stranded RNAs guide the cleavage of mRNAs that contain complementary sequences.

FIGURE 9.42 Crown gall, a plant tumor, is caused by the bacterium *Agrobacterium tumefaciens*. This bacterial species introduces a tumor-inducing plasmid (Ti plasmid) into the plant host. [Credit: Matthew A. Escobar, Edwin L. Civerolo, Kristin R. Summerfelt, and Abhaya M. Dandekar. RNAi-mediated oncogene silencing confers resistance to crown gall tumorigenesis. *PNAS* 2001 98 (23) 13437–13442. Copyright 2001 National Academy of Sciences, U.S.A.]

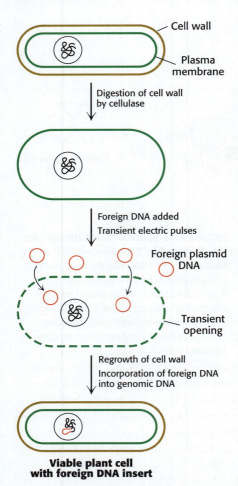

FIGURE 9.43 Foreign DNA can be introduced into plant cells by electroporation. In this technique, the application of intense electric fields makes the plant plasma membranes transiently permeable to large molecules, such as plasmid DNA.

to the tumor state and the synthesis of opines. A small part of the Ti plasmid becomes integrated into the genome of infected plant cells; this 20-kb segment is called *T-DNA*, for transferred DNA.

Ti-plasmid derivatives can be used as vectors to deliver foreign genes into plant cells. First, a segment of foreign DNA is inserted into the T-DNA region of a small plasmid through the use of restriction enzymes and ligases. This synthetic plasmid is added to A. *tumefaciens* colonies harboring naturally occurring Ti plasmids, and, by recombination, Ti plasmids containing the foreign gene are formed. These Ti vectors are valuable tools for exploring the genomes of plant cells and modifying plants to improve their agricultural value and crop yield.

Electroporation. Foreign DNA can also be introduced into plants by applying intense electric fields, a technique called **electroporation** (**Figure 9.43**). First, the cellulose wall surrounding plant cells is removed by adding cellulase; this treatment produces protoplasts, plant cells with exposed plasma membranes. Electric pulses are then applied to a suspension of protoplasts and plasmid DNA. Because high electric fields make membranes transiently permeable to large molecules, plasmid DNA molecules enter the cells. The cell wall is then allowed to re-form, and the plant cells are again viable. Moreover, the transfected cells efficiently express the plasmid DNA. Corn and carrot cells have been stably transfected in this way with the use of plasmid DNA that includes genes for resistance to herbicides. Electroporation is also an effective means of delivering foreign DNA into mammalian and bacterial cells.

Gene guns. The most effective means of transforming plant cells is through the use of so-called *gene guns*, or *bombardment-mediated transformation*. DNA is coated onto 1-μm-diameter tungsten pellets, and these microprojectiles are fired at the target cells with a velocity greater than 400 m s^{-1}. Despite its apparent crudeness, this technique is an effective way of transforming plants, especially important crop species such as soybean, corn, wheat, and rice.

The aforementioned techniques afford an opportunity to develop genetically modified organisms (GMOs) with beneficial characteristics, such as the ability to grow in poor soils, resistance to natural climatic variation, resistance to pests, and fortified nutritional content. These crops might be most useful in regions where malnutrition is prevalent. However, the use of GMOs is highly controversial, as some fear that their safety risks have not been adequately addressed.

The first GMO to come to market was a tomato characterized by delayed ripening. This was achieved by disrupting the gene for the enzyme that breaks down pectin, the polysaccharide that gives tomatoes their firmness. During ripening, pectin is destroyed, and the tomatoes soften, making shipment difficult; by disrupting production of the pectin-destroying enzyme, the tomatoes stayed firm longer. However, the tomato's poor taste hindered its commercial success. An especially successful result of the use of Ti plasmid to modify crops is golden rice. Golden rice is a variety of genetically modified rice that contains the genes for β-carotene synthesis, a required precursor for vitamin A synthesis in humans. Consumption of this rice has the potential to benefit regions of the world where rice is a dietary staple and vitamin A deficiency is common.

Summary

9.1   The Exploration of Genes Relies on Key Tools

- Restriction enzymes recognize specific base sequences in double-helical DNA and cleave both strands of the duplex, forming specific fragments of DNA.

- A DNA fragment containing a particular sequence can be identified by hybridizing it with a labeled single-stranded DNA probe (Southern blotting).

- DNA can be sequenced by controlled interruption of replication. The fragments produced are separated by capillary electrophoresis and visualized by fluorescent tags at their 5′ ends.

- DNA probes for hybridization reactions, as well as new genes, can be synthesized by the automated solid-phase method.

- The polymerase chain reaction makes it possible to greatly amplify specific segments of DNA in vitro. The region amplified is determined by the placement of a pair of primers that are added to the target DNA along with a thermostable DNA polymerase and deoxyribonucleoside triphosphates.

9.2 Recombinant DNA Technology Has Revolutionized All Aspects of Biology

- Novel DNA molecules are made by joining fragments that have complementary cohesive ends produced by the action of a restriction enzyme. DNA ligase seals breaks in DNA chains.

- Vectors for propagating the DNA include plasmids and λ phage. Two broad categories of plasmids include cloning vectors, used for easy insertion and replication of DNA fragments, and expression vectors, for generating protein product in large quantities.

- Specific genes can be cloned from a genomic library with the use of a DNA or RNA probe.

- Complementary DNA (cDNA) is produced by reverse transcription of mRNA, yielding DNA corresponding to expressed genes. Because cDNA is generated from mature mRNA, all introns have been removed.

- Mutations can be generated in vitro to engineer novel proteins using site-directed mutagenesis, cassette mutagenesis, and mutagenesis by PCR.

9.3 Complete Genomes Have Been Sequenced and Analyzed

- More than 50,000 bacterial, archaeal, and eukaryotic genomes have been sequenced, including those from key model organisms and important pathogens.

- The sequence of the human genome has now been completed with nearly full coverage and high precision.

- Only from 20,000 to 25,000 protein-encoding genes appear to be present in the human genome, a substantially smaller number than earlier estimates.

- Next-generation sequencing methods have greatly accelerated the pace at which large genomes can be sequenced and analyzed.

- Comparative genomics has become a powerful tool for analyzing individual genomes and for exploring evolution.

9.4 Eukaryotic Genes Can Be Quantitated and Manipulated with Considerable Precision

- Changes in gene expression can be readily determined in cells and tissues by such techniques as quantitative PCR, microarrays, and RNA-seq.

- Transgenic mice, in which a gene has been inserted into the host organism, and knockout mice, in which a gene has been removed, have been a source of considerable insight into disease mechanisms and possible therapies.

- Genome editing techniques, such as those enabled by the CRISPR-Cas system and ZFNs/TALENs, involve the introduction of precise double-strand breaks into the host genome. Repair of these breaks can yield insertions and deletions, often disrupting the target gene, or the template-based replacement of the target gene with a new sequence.

- Another method of disrupting the expression of a particular gene is through RNA interference, which depends on the introduction of specific double-stranded RNA molecules into eukaryotic cells.

- DNA can be transformed into plant cells by the Ti plasmids of the soil bacterium *Agrobacterium tumefaciens*, or by other techniques such as electroporation or gene guns.

Key Terms

restriction enzyme (restriction endonuclease) (p. 265)

Southern blotting (p. 266)

DNA probe (p. 266)

northern blotting (p. 267)

controlled termination of replication (Sanger dideoxy method) (p. 267)

polymerase chain reaction (PCR) (p. 269)

polymorphism (p. 271)

restriction-fragment-length polymorphism (RFLP) (p. 271)

vector (p. 272)

plasmid (p. 272)

bacteriophage lambda (λ phage) (p. 272)

cohesive (sticky) ends (p. 272)

cloning vector (p. 274)

polylinker (p. 274)

reporter gene (p. 274)

expression vector (p. 275)

genomic library (p. 277)

complementary DNA (cDNA) (p. 278)

cDNA library (p. 279)

site-directed mutagenesis (p. 280)

cassette mutagenesis (p. 281)

pseudogene (p. 283)

next-generation sequencing (NGS) (p. 284)

pyrosequencing (p. 284)

quantitative PCR (qPCR) (p. 287)

transcriptome (p. 288)

DNA microarray (p. 288)

RNA-seq (p. 289)

transgenic mouse (p. 290)

gene disruption (gene knockout) (p. 290)

homologous recombination (p. 290)

genome editing (p. 291)

zinc-finger nuclease (ZFN) (p. 292)

transcription activator-like effector nuclease (TALEN) (p. 292)

CRISPR-Cas system (p. 293)

single guide RNA (sgRNA) (p. 293)

RNA interference (RNAi) (p. 295)

tumor-inducing plasmid (Ti plasmid) (p. 295)

electroporation (p. 296)

Problems

1. Samples of a linear segment of unknown DNA are digested using the restriction enzymes EcoRI, BamHI, and a combination of EcoRI and BamHI. The digests are then run on an agarose gel in order to separate the resulting fragments by size. Use the results of the gel electrophoresis shown in the image to draw a restriction map, indicating the locations of the restriction sites and the distances between them. ❖ **1**

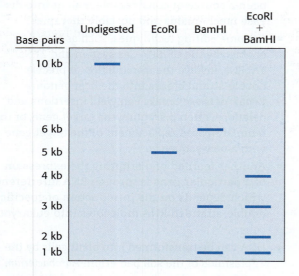

2. A series of people are found to have difficulty eliminating certain types of drugs from their bloodstreams. The problem has been linked to a gene X, which encodes an enzyme Y. Six people were tested with the use of various techniques of molecular biology. Person A is a normal control, person B is asymptomatic but some of his children have the metabolic problem, and persons C through F display the trait. Tissue samples from each person were obtained. Southern analysis was performed on the DNA after digestion with the restriction enzyme HindIII. Northern analysis of mRNA also was done. In both types of analysis, the gels were probed with labeled X cDNA. Finally, a western blot with an enzyme-linked monoclonal antibody was used to test for the presence of protein Y. The results

are shown below. Why is person B without symptoms? Suggest possible mutations in the other people. ❖ **1**

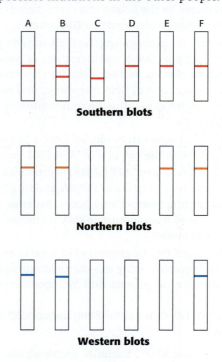

Southern blots

Northern blots

Western blots

3. Representations of sequencing chromatograms for variants of the α chain of human hemoglobin are shown here. What is the nature of the amino acid change in each of the variants? The first triplet encodes valine. ❖ **2**

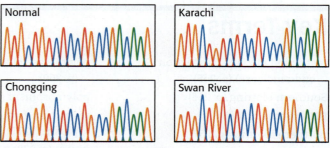

Colors: ddATP, ddCTP, ddGTP, ddTTP

4. You wish to amplify a 250 bp segment of DNA from a plasmid template by PCR with the use of the following primers: 5′-GGATCGATGCTCGCGA-3′ and 5′-AGGATCGGGTCGC-GAG-3′. Despite repeated attempts, you fail to observe a PCR product of the expected length after electrophoresis on an agarose gel. Instead, you observe a bright smear on the gel with an approximate length of 25 to 30 base pairs. Explain these results. ❖ **3**

5. A gel pattern displaying PCR products shows four strong bands. The four pieces of DNA have lengths that are approximately in the ratio of 1:2:3:4. The largest band is cut out of the gel, and PCR is repeated with the same primers. Again, a ladder of four bands is evident in the gel. What does this result reveal about the structure of the encoded protein? ❖ **3**

6. A young investigator would like to express and purify a newly discovered gene. She decides to clone the gene into an expression plasmid that contains the sequence for a hexahistidine tag. Order the steps for construction of the expression plasmid and generation of the protein. (The gene of interest will be inserted between the NotI and HindIII sites.) ❖ **4**

(a) Ligate the cut plasmid and DNA fragment.

(b) Introduce NotI and HindIII restriction sites at the ends of the gene of interest.

(c) Transfect the modified plasmid into *E. coli* for expression.

(d) Digest the plasmid and the gene insert with NotI and HindIII.

7. The map of a hypothetical plasmid, pSAP, is shown here:

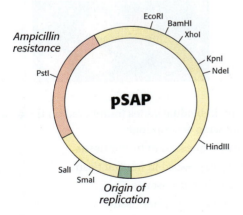

Suppose you have a DNA fragment you would like to insert into pSAP. The fragment has PstI and EcoRI restriction endonuclease sites near the 5′ end, and HindIII and SmaI restriction endonuclease sites near the 3′ end. What are the best restriction enzymes to use to digest both the DNA fragment and pSAP? ❖ **4**

8. You have identified a gene that is located on human chromosome 20 and wish to identify its location within the mouse genome. On which chromosome would you be most likely to find the mouse counterpart of this gene?

9. Which of the following amino acid sequences would yield the most optimal oligonucleotide probe?

    Ala-Met-Ser-Leu-Pro-Trp

    Gly-Trp-Asp-Met-His-Lys

    Cys-Val-Trp-Asn-Lys-Ile

    Arg-Ser-Met-Leu-Gln-Asn

How many unique oligonucleotides would be required to detect all sequence possibilities for the optimal probe?

10. Suppose a researcher previously cloned gene Y into M13 bacteriophage vector. Gene Y encodes a product called peptide Y. A region of gene Y contains the DNA sequence ATG–CGC–GAA–CTG–GTG–AAC–TAA. The researcher wishes to change a Val residue to an Ala residue in this region of peptide Y using site-directed mutagenesis. What should be the sequence of the mutant oligonucleotide primer in this region? ❖ **5**

11. Targeting a DNA sequence for editing by the CRISPR-Cas9 system requires that a PAM sequence be present just downstream of the target sequence. For *S. pyogenes* Cas9, this PAM is 5′-NGG-3′, where N represents any nucleotide. Assuming a fragment of DNA includes equal amounts of A, C, G, and T bases, how far apart would one expect to find these PAMs? ❖ **6**

12. (a) Using RNA interference, it is possible to significantly reduce homologous mRNA transcripts and thereby reduce the expression level of a specific gene of interest. What methods could be used to confirm that an RNAi knockdown experiment has been successful? **(b)** The relative ease of RNAi studies means that large-scale screens can be performed to assess a wide array of targets. What source would produce the most functional RNAi library for conducting genome-wide screens: whole-genome DNA, whole-genome proteins, or whole-genome cDNA?

13. The image below shows a time versus temperature plot for one cycle of PCR. Match the events below with the indicated regions (1, 2, and 3) on the plot. ❖ **3**

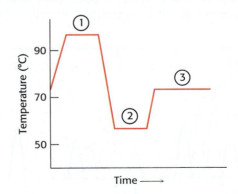

(a) DNA polymerase extends primers.

(b) DNA separates into single strands.

(c) Primers anneal to single-stranded DNA.

14. You are interested in identifying mutations in a region of DNA that, in most individuals, has 5 cleavage sites for

EcoRI. In the figure below, these sites are labeled A through E. You also have a probe that is complementary to the region between sites C and D:

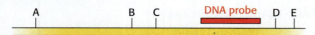

DNA samples from four individuals were cleaved with EcoRI. The DNA fragments were separated by gel electrophoresis, transferred to a membrane, and hybridized with your DNA probe. The image of the Southern blot shows the labeled DNA bands and molecular weight (MW) markers. The lane labels I, II, III, and IV correspond to individuals I, II, III, and IV. Assume that fragments such as C–D and C–E are clearly resolved in this gel system. Fragment sizes are as given: A–B is 4 kb, B–C is 1 kb, C–D is 5 kb, and D–E is 650 bp. Individual I represents the positive control, and has all five EcoRI sites intact.

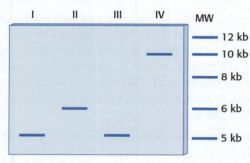

(a) Which individual has at least one point mutation that eliminates restriction site C?

(b) Which individual has mutation(s) that eliminate restriction sites B and C?

(c) Which individual has at least one point mutation that eliminates restriction site A?

15. In the course of studying a gene and its possible mutation in humans, you obtain genomic DNA samples from a collection of persons and PCR amplify a region of interest within this gene. For one of the samples, you obtain the sequencing chromatogram shown here. Provide an explanation for the appearance of these data at position 49 (indicated by the arrow):

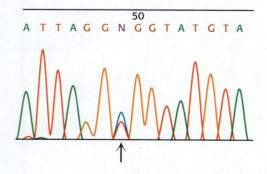

16. The restriction enzymes KpnI and Acc65I recognize and cleave the same 6-bp sequence. However, the sticky end formed from KpnI cleavage cannot be ligated directly to the sticky end formed from Acc65I cleavage. Explain why.

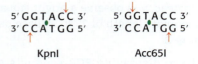

17. Suppose a single linear molecule of double-stranded DNA is amplified by PCR. How many molecules of double-stranded DNA will there be after 1 cycle? After 3 cycles? After 30 PCR cycles (a typical cycle count used in PCR experiments)? ❖ **3**

18. Which of the following reagents are needed for a typical PCR? ❖ **3**

(a) DNA ligase

(b) Four nucleotides (A, U, C, G)

(c) *Taq* DNA polymerase

(d) Template DNA

(e) Two primers

(f) *E. coli* DNA polymerase

(g) Four nucleotides (A, T, C, G)

19. In preparation for a microarray experiment, cDNA was isolated from tumor tissue and labeled with a red fluorescent probe, while the cDNA from normal tissue was labeled with a green fluorescent probe. A small segment of the resulting microarray data is shown below.

Classify the individual genes (numbered on the array) based on the following descriptions:

(a) Primarily expressed in tumor tissue

(b) Primarily expressed in normal tissue

(c) Expressed in both tissues

(d) Not expressed in either tissue.

20. As RNA-seq becomes cheaper and less computationally intensive, many researchers are replacing microarray analysis with RNA-seq as the preferred method to measure cellular transcript levels. What are some advantages of RNA-seq over microarray analysis?

Exploring Evolution and Bioinformatics

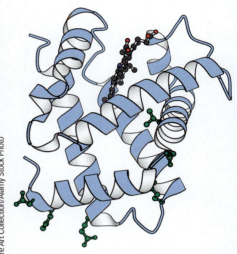

Evolutionary relationships are clear in protein sequences. The now-extinct woolly mammoth (*Mammuthus primigenius*) coexisted with early humans, as evidenced by Late Stone Age cave paintings such as those at the Rouffignac Cave in France. Using samples obtained from frozen preserved tissue, researchers used recombinant DNA technology to determine the complete sequence of the woolly mammoth genome and the structural and functional adaptations of its proteins. In the case of α-hemoglobin (right), only six residues (shown in green) are nonidentical and dissimilar between the woolly mammoth and human sequences. Amino acid sequence comparisons with living species and functional expression of these extinct proteins have provided new insights into how this species adapted to very cold environments. [Drawn from 3VRE.pdb.]

The Picture Art Collection/Alamy Stock Photo

❖ LEARNING GOALS

By the end of this chapter, you should be able to:

1. Distinguish between homologs, paralogs, and orthologs.

2. Describe how two protein sequences are aligned, and how one can determine whether that alignment is significant.

3. Explain how substitution matrices can be used to facilitate the identification of related protein sequences.

4. Distinguish between divergent and convergent evolution.

5. Describe one way the evolution of biomolecules can be demonstrated in the laboratory.

OUTLINE

10.1 Homologs Are Descended from a Common Ancestor and Can Be Detected by Sequence Alignments

10.2 Examination of Three-Dimensional Structure Enhances Our Understanding of Evolutionary Relationships

10.3 Evolutionary Trees Can Be Constructed on the Basis of Sequence Information

10.4 Modern Techniques Make the Experimental Exploration of Evolution Possible

Like members of a human family, members of molecular families often have features in common. Such family resemblance is most easily detected by comparing three-dimensional structure, the aspect of a molecule most closely linked to function. However, such structures have been determined for only a small proportion of the total number of proteins. In contrast, gene sequences and the corresponding amino acid sequences are available for a great number of proteins, largely owing to the tremendous power of DNA cloning and sequencing techniques, including complete-genome sequencing (Section 9.3). As we will discuss in this chapter, sequence databases can be probed for matches to a newly elucidated sequence to identify related molecules, providing considerable insight into the function, mechanism, and evolutionary history of the molecule encoded by that sequence.

10.1 Homologs Are Descended from a Common Ancestor and Can Be Detected by Sequence Alignments

The exploration of biochemical evolution consists largely of an attempt to determine how proteins, other molecules, and biochemical pathways have been transformed through time. Sequence comparisons can often reveal pathways of evolutionary descent and estimated dates of specific evolutionary landmarks. This information can be used to construct evolutionary trees that trace the evolution of a particular protein or nucleic acid, in many cases from Archaea and Bacteria through Eukarya, including human beings.

Orthologs and paralogs are two different classes of homologous proteins

The most fundamental relationship between two entities is *homology*; two molecules are said to be *homologous* if they are derived from a common ancestor. Homologous molecules, or **homologs**, can be divided into two classes (**Figure 10.1**):

- **Orthologs** are homologs that are present within different species and have very similar or identical functions. In Section 2.6, we discussed the structure of ribonuclease from cows (Figure 10.1; yellow). This protein is highly similar in structure to human ribonuclease (Figure 10.1; red). Both proteins share very similar functions, namely, the hydrolytic cleavage of RNA molecules.

- **Paralogs** are homologs that are present within one species. Paralogs often differ in their detailed biochemical functions. For example, angiogenin (Figure 10.1; blue), a protein that stimulates the growth of new blood vessels, is also structurally similar to ribonuclease—so similar that both angiogenin and ribonuclease are clearly members of the same protein family. Paralogs such as angiogenin and ribonuclease strongly suggest a common ancestor at some earlier stage of evolution.

Understanding the homology between molecules can reveal the evolutionary history of the molecules as well as information about their function; if a newly sequenced protein is homologous to an already characterized protein, we have a strong indication of the new protein's biochemical function.

FIGURE 10.1 Homologs can be divided into two classes. Homologs that perform identical or very similar functions in different species are called orthologs, whereas homologs that perform different functions within one species are called paralogs. [Drawn from 8RAT.pdb, 2RNF.pdb, and 2ANG.pdb.]

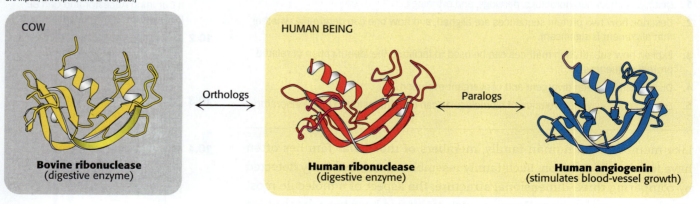

COW

Orthologs

Bovine ribonuclease
(digestive enzyme)

HUMAN BEING

Paralogs

Human ribonuclease
(digestive enzyme)

Human angiogenin
(stimulates blood-vessel growth)

Statistical analysis of sequence alignments can detect homology

How can we tell whether two human proteins are paralogs or whether a yeast protein is the ortholog of a human protein? A significant sequence similarity between two molecules implies that they are likely to have the same evolutionary origin and, therefore, similar three-dimensional structures, functions, and mechanisms. Both nucleic acid and protein sequences can be compared to detect homology. However, the possibility exists that the observed agreement

Human hemoglobin (α chain)

VLSPADKTNVKAAWGKVGAHAGEYGAEALERMFLSFPTTKTYFPHFDLSHG
SAQVKGHGKKVADALTNAVAHVDDMPNALSALSDLHAHKLRVDPVNFKLLS
HCLLVTLAAHLPAEFTPAVHASLDKFLASVSTVLTSKYR

Human myoglobin

GLSDGEWQLVLNVWGKVEADIPGHGQEVLIRLFKGHPETLEKFDKFKHLKS
EDEMKASEDLKKHGATVLTALGGILKKKGHHEAEIKPLAQSHATKHHIPVK
YLEFISECIIQVLQSKHPGDFGADAQGAMNKALELFRKDMASNYKELGFQG

FIGURE 10.2 The amino acid sequences of human hemoglobin (α chain) and human myoglobin will be used to demonstrate sequence alignments. α-Hemoglobin is composed of 141 amino acids; myoglobin consists of 153 amino acids. (One-letter abbreviations designating amino acids are used; see Table 2.2.)

between any two sequences is solely a product of chance. Because nucleic acids are composed of fewer building blocks than proteins (4 bases versus 20 amino acids), the likelihood of random agreement between two DNA or RNA sequences is significantly greater than that for protein sequences. For this reason, detection of homology between protein sequences is typically far more effective.

To illustrate sequence-comparison methods, let us consider the globins, the critical oxygen carrying proteins which we first encountered in Chapter 3. Recall that myoglobin is a monomer, while each human hemoglobin molecule is composed of four heme-containing polypeptide chains, two identical α chains and two identical β chains. Here, we shall consider only the α chain. To examine the similarity between the amino acid sequence of the human α chain of hemoglobin and that of human myoglobin (**Figure 10.2**), we apply a method, referred to as a **sequence alignment**, in which the two sequences are systematically aligned with respect to each other to identify regions of significant overlap.

How can we tell where to align the two sequences? In the course of evolution, the sequences of two proteins that have an ancestor in common will have diverged in a variety of ways. Insertions and deletions may have occurred at the ends of the proteins or within the functional domains themselves. Individual amino acids may have been mutated to other residues of varying degrees of similarity. To understand how the methods of sequence alignment take these potential sequence variations into account, let us first consider the simplest approach, where we slide one sequence past the other, one amino acid at a time, and count the number of matched residues, or sequence **identities** (**Figure 10.3**). For α-hemoglobin and myoglobin, the best alignment reveals 23 sequence identities spread throughout the central parts of the sequences.

However, careful examination of all the possible alignments and their scores suggests that important information regarding the relationship between myoglobin and hemoglobin α has been lost with this method. In particular, we see that another alignment, featuring 22 identities, is nearly as good (Figure 10.3). This alignment is shifted by six residues relative to the preceding alignment and yields identities that are concentrated toward the amino-terminal end of the sequences. By introducing a gap into one of the sequences, the identities found in both alignments will be represented (**Figure 10.4**). The insertion of gaps allows the alignment method to compensate for the insertions or deletions of nucleotides that may have taken place in the gene for one molecule but not the other in the course of evolution.

The use of gaps substantially increases the complexity of sequence alignment because a vast number of possible gaps, varying in both position and length, must be considered throughout each sequence. Moreover, the introduction of an excessive number of gaps can yield an artificially high number of identities. Nevertheless, methods have been developed for the insertion of gaps in the automatic alignment of sequences. These methods use scoring systems to compare different alignments, including penalties for gaps to prevent the insertion of an unreasonable number of them. For example, in one scoring system, each identity between aligned sequences is counted as +10 points, whereas each gap

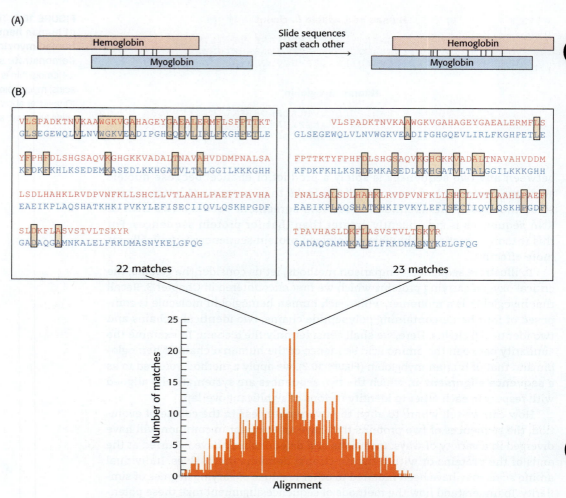

FIGURE 10.3 Amino acid sequences are compared by sliding one sequence past the other, one amino acid at a time. (A) At each step, we count the number of amino acid identities between the two proteins. Shown are representations of just two of the >100 possible alignments. (B) The two alignments with the largest number of matches are shown above the graph, which plots the matches as a function of alignment.

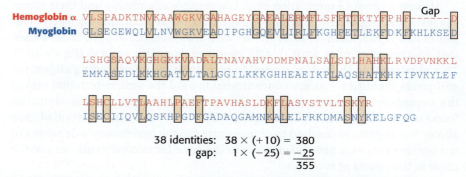

FIGURE 10.4 Addition of a gap improves the alignment between myoglobin and α-hemoglobin. In the scoring method shown in this figure, a penalty is applied for the gap introduced.

introduced, regardless of size, counts for −25 points. For the alignment shown in Figure 10.4, there are 38 identities (38 × 10 = 380) and 1 gap (1 × −25 = −25), producing a score of (380 + −25 = 355). Overall, there are 38 matched amino acids in an average length of 147 residues; thus, the sequences are 25.9% identical.

Alignments such as the one just described inform us as to the percent identity between two sequences. However, the question remains: how do we interpret this value? For example, we have determined that α-hemoglobin and myoglobin, when allowing for the insertion of gaps, are 25.9% identical. Is this value sufficiently high to suggest a relationship between these two proteins? To answer these questions, we must determine the significance of alignment scores and percent identities.

❖ SELF-CHECK QUESTION

Why is it important to assess penalties for the placement of gaps when scoring an alignment?

The statistical significance of alignments can be estimated by shuffling

The sequence similarities in Figure 10.4 appear striking, yet there remains the possibility that a grouping of sequence identities has occurred by chance alone. Because proteins are composed of the same set of 20 amino acid monomers, the alignment of any two unrelated proteins will yield some identities, particularly if we allow the introduction of gaps. Even if two proteins have identical amino acid composition, they may not be linked by evolution. It is the order of the residues within their sequences that implies a relationship between them.

We can assess the significance of our alignment by "shuffling," or randomly rearranging, one of the sequences (**Figure 10.5**), performing the sequence alignment, and determining a new alignment score. This process is repeated many times to yield a histogram showing, for each possible score, the number of shuffled sequences that received that score (**Figure 10.6**). If the original score is appreciably different from the scores from the shuffled alignments, then it is unlikely that the original alignment is merely a consequence of chance.

When this procedure is applied to the sequences of myoglobin and α-hemoglobin, the authentic alignment (indicated by the red bar in Figure 10.6) clearly stands out. Its score is far above the mean for the alignment scores based on shuffled sequences. The probability that such a deviation occurred by chance alone is approximately 1 in 10^{20}. Thus, we can comfortably conclude that the two sequences are genuinely similar; the simplest explanation for this similarity is that these sequences are homologous—that is, the two molecules have descended from a common ancestor.

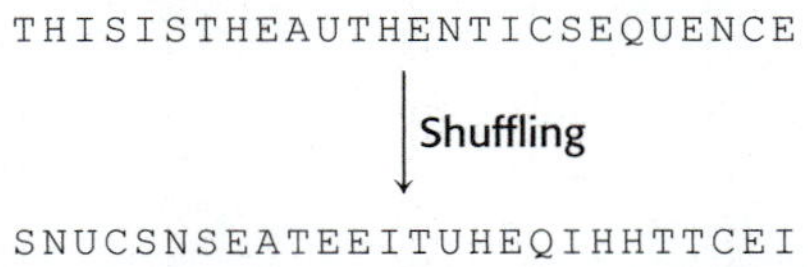

Distant evolutionary relationships can be detected through the use of substitution matrices

The scoring scheme described above assigns points only to positions occupied by identical amino acids in the two sequences. No credit is given for any pairing that is not an identity. However, as already discussed, two proteins related by evolution undergo amino acid substitutions as they diverge. A scoring system based solely on amino acid identity cannot account for these changes. To add greater sensitivity to the detection of evolutionary relationships, methods have been developed to compare two amino acids and assess their degree of *similarity*.

Not all substitutions are equivalent. For example, amino acid changes can be classified as structurally conservative or nonconservative. A **conservative substitution** replaces one amino acid with another that is similar in size and chemical properties. Conservative substitutions may have only minor effects on protein structure and often can be tolerated without compromising protein function. In contrast, in a **nonconservative substitution**, an amino acid is replaced by one that is structurally dissimilar. Amino acid changes can also be classified by the fewest number of nucleotide changes necessary to achieve the corresponding amino acid change. Some substitutions arise from the replacement of only a single nucleotide in the gene sequence, whereas others require two or three

THISISTHEAUTHENTICSEQUENCE

| Shuffling

SNUCSNSEATEEITUHEQIHHTTCEI

FIGURE 10.5 The individual amino acids are scrambled to generate a shuffled sequence.

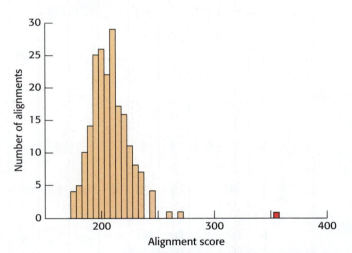

FIGURE 10.6 Comparison of alignment scores reveals the statistical significance of an unshuffled alignment. Alignment scores are calculated for many shuffled sequences, and the number of sequences generating a particular score is plotted against the score. The resulting plot is a distribution of alignment scores occurring by chance. The alignment score for unshuffled α-hemoglobin and myoglobin (shown in red) is substantially greater than any of these scores, strongly suggesting that the sequence similarity is significant.

replacements. Conservative and single-nucleotide substitutions are likely to be more common than are substitutions with more radical effects.

How can we account for the type of substitution when comparing sequences? We can approach this problem by first examining the substitutions that have been observed in proteins known to be evolutionarily related. From an examination of appropriately aligned sequences, substitution matrices have been deduced. A **substitution matrix** describes a scoring system for the replacement of any amino acid with each of the other 19 amino acids. In these matrices, a large positive score corresponds to a substitution that occurs relatively frequently, whereas a large negative score corresponds to a substitution that occurs only rarely.

One commonly used substitution matrix, the Blosum-62 (for <u>Bl</u>ocks of amino acid <u>s</u>ubstitution <u>m</u>atrix), is illustrated in **Figure 10.7**. In this depiction, each column in this matrix represents 1 of the 20 amino acids, whereas the position of the single-letter codes within each column specifies the score for the

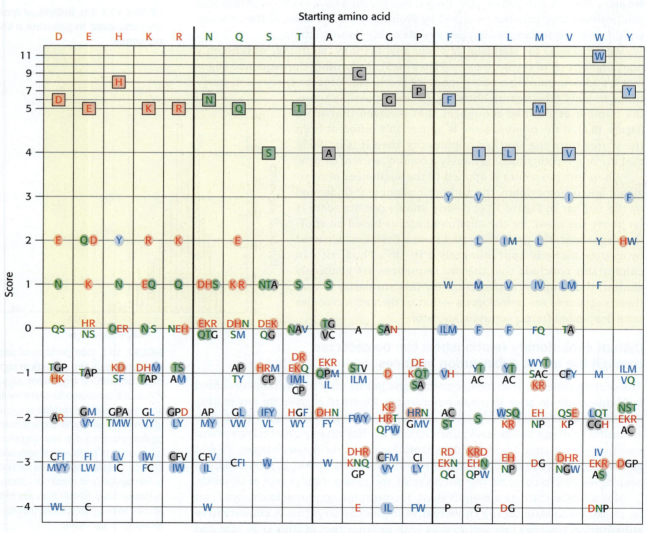

FIGURE 10.7 The Blosum-62 substitution matrix enables scoring based on amino acid similarity. This matrix was derived by examining substitutions within aligned sequence blocks in related proteins. In this graphical view, amino acids are classified into four groups (charged, red; polar, green; large and hydrophobic, blue; other, black). Substitutions that require the change of only a single nucleotide are shaded. Identities are boxed. To find the score for a substitution of, for instance, a Y for an H, you find the Y in the column having H at the top and check the number at the left. In this case, the resulting score is 2.

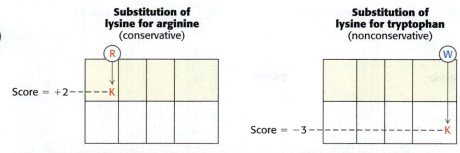

FIGURE 10.8 Conservative and nonconservative substitutions yield very different scores in the Blosum-62 matrix. A conservative substitution (lysine for arginine) receives a positive score, whereas a nonconservative substitution (lysine for tryptophan) is scored negatively. The matrix is depicted as an abbreviated form of Figure 10.7.

corresponding substitution. Notice that scores corresponding to identity (the boxed codes at the top of each column) are not the same for each residue, owing to the fact that less frequently occurring amino acids such as cysteine (C) and tryptophan (W) will align by chance less often than the more common residues. Furthermore, structurally conservative substitutions such as lysine (K) for arginine (R) and isoleucine (I) for valine (V) have relatively high scores, whereas nonconservative substitutions such as lysine for tryptophan result in negative scores (**Figure 10.8**).

When two sequences are compared, each pair of aligned residues is assigned a score based on the matrix. In addition, gap penalties are often assessed. For example, the introduction of a single-residue gap lowers the alignment score by 12 points, and the extension of an existing gap costs 2 points per residue. With the use of the Blosum-62 scoring system, the alignment between human α-hemoglobin and human myoglobin shown in Figure 10.4 receives a score of 115. In many regions, most substitutions are conservative (defined as those substitutions with scores greater than 0), and relatively few are strongly disfavored (**Figure 10.9**).

FIGURE 10.9 Counting conservative substitutions reveals strong similarity between α-hemoglobin and myoglobin. Conservative substitutions are indicated by yellow shading and identities by the orange boxes.

This scoring system detects homology between less obviously related sequences with greater sensitivity than would a comparison of identities only. Consider, for example, the protein leghemoglobin, an oxygen-binding protein found in the roots of leguminous plants. The amino acid sequence of leghemoglobin from the lupine plant can be aligned with that of human myoglobin and scored by using either the simple scoring scheme based on identities only or the Blosum-62 (Figure 10.7). Repeated shuffling and scoring provides a distribution of alignment scores, as shown in **Figure 10.10**.

Scoring based solely on identities (Figure 10.10A) indicates that the probability of the alignment between myoglobin and leghemoglobin occurring by chance alone is 1 in 20. Thus, although the level of similarity suggests a relationship, there is a 5% chance that the similarity is accidental on the basis of this analysis. In contrast, scoring based on the substitution matrix, which accounts for

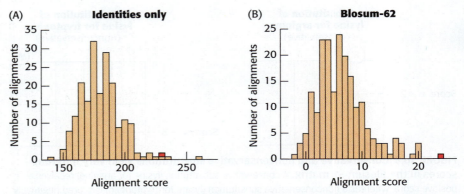

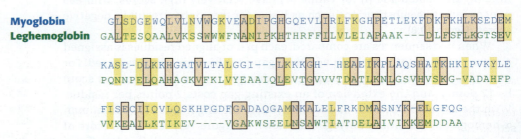

FIGURE 10.10 Alignment using Blosum-62 can uncover more distant evolutionary relationships. Repeated shuffling and scoring reveal the significance of sequence alignment for human myoglobin versus lupine leghemoglobin with the use of either (A) the simple, identity-based scoring system or (B) the Blosum-62. The scores for the alignment of the authentic sequences are shown in red. Accounting for amino acid similarity in addition to identity reveals a greater separation between the authentic alignment and the population of shuffled alignments. (Note that these two scoring systems use different scales on their respective x-axes, so the raw scores cannot be directly compared.)

Myoglobin   GLSDGEWQLVLNVWGKVEADIPGHGQEVLIRLFKGHPETLEKFDKFKHLKSEDEM
Leghemoglobin   GALTESQAALVKSSWWWFNANIPKHTHRFFILVLEIAPAAK---DLFSFLKGTSEV

                     KASE-DLKKHGATVLTALGGI---LKKKGH--HEAEIKPLAQSHATKHKIPVKYLE
                     PQNNPELQAHAGKVFKLVYEAAIQLEVTGVVVTDATLKNLGSVHVSKG-VADAHFP

                     FISECIIQVLQSKHPGDFGADAQGAMNKALELFRKDMASNYK-ELGFQG
                     VVKEAILKTIKEV----VGAKWSEELNSAWTIATDELAIVIKKEMDDAA

FIGURE 10.11 Alignment of human myoglobin and lupine leghemoglobin reveals many conservative substitutions. The use of Blosum-62 yields the alignment shown between human myoglobin and lupine leghemoglobin, illustrating identities (orange boxes) and conservative substitutions (yellow). These sequences are 23% identical.

conservative substitutions, determines that the odds of the alignment occurring by chance are calculated to be approximately 1 in 300 (Figure 10.10B). Thus, an analysis performed with the substitution matrix reaches a much firmer conclusion about the evolutionary relationship between these proteins (**Figure 10.11**).

Experience with sequence analysis has led to the development of simpler rules of thumb. For sequences longer than 100 amino acids, sequence identities greater than 25% are almost certainly not the result of chance alone; such sequences are probably homologous. In contrast, if two sequences are less than 15% identical, their alignment alone is unlikely to indicate statistically significant similarity. For sequences that are between 15% and 25% identical, further analysis is necessary to determine the statistical significance of the alignment.

It must be emphasized that the lack of a statistically significant degree of sequence similarity does not rule out homology. The sequences of many proteins that have descended from common ancestors have diverged to such an extent that the relationship between the proteins can no longer be detected from their sequences alone. As we will see, such homologous proteins can often be identified by examining their three-dimensional structures.

Databases can be searched to identify homologous sequences

When the sequence of a protein is first determined, comparing it with all previously characterized sequences can provide tremendous insight into its evolutionary relatives and, hence, its structure and function. Indeed, an extensive

Name of homologous protein

Homo sapiens ribose 5-phosphate isomerase mRNA complete cds
Sequence ID: AY050633.1 **Length:** 1818 **Number of matches:** 1

Range 1: 252 to 920   *Number of sequences with this similarity expected by chance*

Sequence of queried protein

Sequence of homologous protein

Score	Expect	Method	Identities	Positives	Gaps	Frame
113 bits (283)	2e-25	Compositional matrix adjust.	82/224(37%)	118/224(52%)	15/224(6%)	+3

```
Query    4    DELKKAVGWAALQ-YVQPGTIVGVGTGSTAAHFIDALGTMKGQIE---GAVSSSDASTEK    59
              +E KK  G AA++ +V+    ++G+G+GST   H + +     Q       + +S  + +
Sbjct  252    EEAKKLAGRAAVENHVRNNQVLGIGSGSTIVHAVQRIAERVKQENLNLVCIPTSFQARQL   431

Query   60    LKSLGIHVFDLNEVDSLGIYVDGADEINGHMQMIKGGGAALTREKIIASVAEKFICIADA   119
              +   G+ + DL+    + + +DGADE++  + +IKGGG  LT+EKI+A  A +FI IAD
Sbjct  432    ILQYGLYLSDLDRHPEIDLAIDGADEVDADLNLIKGGGGCLTQEKIVAGYASRFIVIADF   611

Query  120    SKQVDILG---KFPLPVEVIPMARSAVARQLV-KLGGRPEYRQG------VVTDNGNVIL   169
               K   LG      +P+EVIPMA   V+R +  K GG  E R          VVTDNGN IL
Sbjct  612    RKDSKNLGDQWHKGIPIEVIPMAYVPVSRAVSQKFGGVVELRMAVNKAGPVVTDNGNFIL   791

Query  170    DVHGMEILDPIAMENAINAIPGVVTVGLFANRGADVALIGTPDG    213
              D    +    +   AI  IPGVV  GLF N  A+    G  DG
Sbjct  792    DWKFDRVHKWSEVNTAIKMIPGVVDTGLFINM-AERVYFGMQDG    920
```

Plus sign = conservative substitution   *Letter = identity*

FIGURE 10.12 Related protein sequences can be rapidly identified using online BLAST searches. Part of the results from a BLAST search of the nonredundant (nr) protein sequence database using the sequence of ribose 5-phosphate isomerase (also called phosphopentose isomerase, Chapter 20) from *E. coli* as a query. Among the thousands of sequences found is the orthologous sequence from humans, and the alignment between these sequences is shown. The number of sequences with this level of similarity expected to be in the database by chance is 2×10^{-25} as shown by the E value (highlighted in red). Because this value is much less than 1, we can confidently conclude that the observed sequence alignment is highly significant.

sequence comparison is almost always the first analysis performed on a newly elucidated sequence. The sequence-alignment methods just described are used to compare an individual sequence with all members of a database of known sequences.

Database searches for homologous sequences are most often accomplished by using resources available on the Internet at the National Center for Biotechnology Information (www.ncbi.nlm.nih.gov). The procedure used is referred to as a **basic local alignment search tool (BLAST) search**. An amino acid sequence is entered into a browser (blast.ncbi.nlm.nih.gov), and a search is performed, most often against a nonredundant database of all known sequences. In 2021, this database included more than 400 million sequences. A BLAST search yields a list of sequence alignments, each accompanied by an estimate giving the likelihood that the alignment occurred by chance (**Figure 10.12**).

In 1995, investigators reported the first complete sequence of the genome of a free-living organism, the bacterium *Haemophilus influenzae* (Figure 9.25). With the sequences available, they performed a BLAST search with each deduced protein sequence. Of 1743 identified protein-coding regions, also referred to as **open reading frames (ORFs)**, 1007 (58%) could be linked to some protein of known function that had been previously characterized in another organism. An additional 347 ORFs could be linked to sequences in the database for which no function had yet been assigned ("hypothetical proteins"). The remaining 389 sequences did not match any sequence present in the database at that time. Thus, investigators were able to identify likely functions for more than half the proteins within this organism solely by sequence comparisons.

❖ SELF–CHECK QUESTION

Using the National Center for Biotechnology Information website (www.ncbi.nlm.nih.gov), find the sequences of the enzyme triose phosphate isomerase from *E. coli* strain K-12 and from *Homo sapiens* (use isoform #1). Use the Global Align tool on the NCBI BLAST website to align these two sequences. What is the percent identity between these two proteins?

10.2 Examination of Three-Dimensional Structure Enhances Our Understanding of Evolutionary Relationships

Sequence comparison is a powerful tool for extending our knowledge of protein function and kinship. However, biomolecules generally function as intricate three-dimensional structures rather than as linear polymers. Mutations occur at the level of sequence, but the effects of the mutations are at the level of function, and function is directly related to tertiary structure. Consequently, to gain a deeper understanding of evolutionary relationships between proteins, we must examine three-dimensional structures, especially in conjunction with sequence information. The techniques of structural determination were presented in Section 4.5.

Tertiary structure is more conserved than primary structure

Because three-dimensional structure is much more closely associated with function than is sequence, tertiary structure is more evolutionarily conserved than is primary structure. This conservation is apparent in the tertiary structures of the globins (**Figure 10.13**), which are extremely similar even though the similarity between human myoglobin and lupine leghemoglobin is weakly detectable at the sequence level (Figure 10.10) and that between human α-hemoglobin and lupine leghemoglobin is not statistically significant (15% identity). This structural similarity firmly establishes that the framework that binds the heme group and facilitates the reversible binding of oxygen has been conserved over a long evolutionary period.

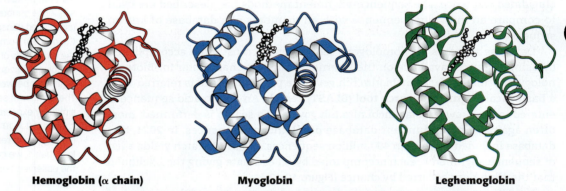

Hemoglobin (α chain)   **Myoglobin**   **Leghemoglobin**

INTERACT with these models in
Achie/e

FIGURE 10.13 Three-dimensional structure is more conserved than primary structure. The tertiary structures of human hemoglobin (α chain), human myoglobin, and lupine leghemoglobin are conserved. Each heme group contains an iron atom to which oxygen binds. [Drawn from 1HBB.pdb, 1MBD.pdb, and 1GDJ.pdb.]

Anyone aware of the similar biochemical functions of hemoglobin, myoglobin, and leghemoglobin could expect the structural similarities. In a growing number of other cases, however, a comparison of three-dimensional structures has revealed striking similarities between proteins that were *not* expected to be related, on the basis of their diverse functions. A case in point is the protein actin, a major component of the cytoskeleton (Section 6.5) and heat shock protein 70 (Hsp70), which assists protein folding inside cells. These two proteins were found to be noticeably similar in structure despite only 16% sequence identity (**Figure 10.14**). On the basis of their three-dimensional structures, actin and Hsp70 are paralogs. The level of structural similarity strongly suggests that, despite their different biological roles in modern organisms, these proteins descended from a common ancestor.

As the three-dimensional structures of more proteins are determined, such unexpected kinships are being discovered with increasing frequency. The search

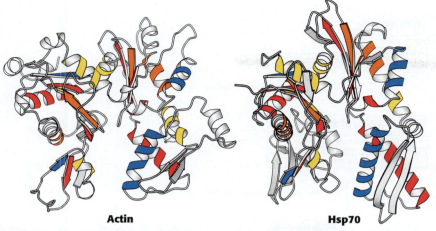

Actin   **Hsp70**

FIGURE 10.14 The structures of actin and a large fragment of heat shock protein 70 (Hsp70) reveal an unexpected relationship. A comparison of the identically colored elements of secondary structure reveals the overall similarity in structure despite the difference in biochemical activities. [Drawn from 1ATN.pdb and 1ATR.pdb.]

INTERACT with these models in
Achieve

for such kinships relies ever more frequently on computer-based searches that are able to compare the three-dimensional structure of any protein with all other known structures.

Knowledge of three-dimensional structures can aid in the evaluation of sequence alignments

The sequence-comparison methods described thus far treat all positions within a sequence equally. However, we have learned from examining families of homologous proteins for which at least one three-dimensional structure is known that regions and residues critical to protein function are more strongly conserved than are other residues. For example, each type of globin contains a bound heme group with an iron atom at its center. A histidine residue that interacts directly with this iron atom (Section 3.2) is conserved in all globins.

After we have identified key residues or highly conserved sequences within a family of proteins, we can generate a **sequence template**—a map of conserved residues that are structurally and functionally important and are characteristic of particular families of proteins. By preferentially scoring residues known to be important for structure or function, we can sometimes identify other family members even when the overall level of sequence similarity is below statistical significance and might be undetectable by other means.

Still other methods are able to identify conserved residues within a family of homologous proteins, even without a known three-dimensional structure. These methods often use substitution matrices that differ at each position within a family of aligned sequences. Such methods can often detect quite distant evolutionary relationships.

Repeated motifs can be detected by aligning sequences with themselves

As discussed in Section 2.5, domains are compact regions of structure that can be found in a wide variety of proteins. In fact, more than 10% of all proteins contain sets of two or more domains that are similar to one another. Sequence search methods can often detect these internally repeated sequences that have been characterized in other proteins.

Often, however, internally repeated sequences do not correspond to previously identified domains. In these cases, their presence can be detected by

(A)

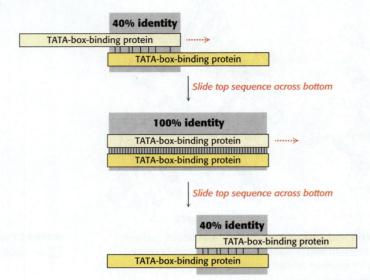

40% identity

TATA-box-binding protein

Slide top sequence across bottom

100% identity

TATA-box-binding protein

TATA-box-binding protein

Slide top sequence across bottom

40% identity

TATA-box-binding protein

TATA-box-binding protein

(B) **Sequence alignment within internal region of similarity**

```
 25  LQNIVSTVNLDCKLDLKAIALQ-ARNAEYNPKRFAAVIMRIREPKTTALIFASGKMVCTGAKSED  88
115  IQNIVGSCDVKFPIRLEGLAYSHAAFSSYEPELFPGLIYRMKVPKIVLLIFVSGKIVITGAKMRD  179
```

(C) **Structure of TATA-box-binding protein**

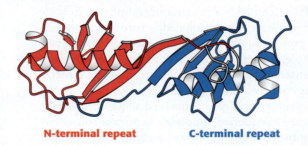

N-terminal repeat   **C-terminal repeat**

FIGURE 10.15 Internal repeats can be identified using a self-alignment.
(A) When the TATA-box-binding protein is aligned against itself, a region of significant identity between the N-terminal and C-terminal halves is revealed. (B) Alignment of these N-terminal and C-terminal sequences shows 40% identity and considerable similarity. The N-terminal region is shown in red and the C-terminal region in blue. (C) Consistent with the sequence alignment data, the three-dimensional structure of the TATA-box-binding protein reveals the presence of two structurally similar repeats. The N-terminal domain is shown in red and the C-terminal domain in blue. [Drawn from 1VOK.pdb.]

INTERACT with this model in
⬚ Achieve

attempting to align a given sequence with itself. The statistical significance of such repeats can be tested by aligning the regions in question as if these regions were sequences from separate proteins. Of course, in a self-alignment, there will be one "hit" that will yield a perfect score. However, additional regions where internal alignment achieves significance may indicate the presence of repeated domains within a given protein.

Consider, for example, the TATA-box-binding protein, a key protein in controlling eukaryotic gene transcription (Section 29.3). Alignment of this protein with itself reveals a significant region of homology between the N-terminal half of the protein with its C-terminal half (**Figure 10.15A**). Comparison of two 65-residue segments from these regions reveals that 40% of the amino acids are identical (**Figure 10.15B**). The estimated probability of such an alignment occurring by chance is less than 1 in 10^{10}. The determination of the three-dimensional structure of the TATA-box-binding protein confirmed the presence of repeated structures; the protein is formed of two nearly identical domains (**Figure 10.15C**). The evidence is convincing that the gene encoding this protein evolved by duplication of a gene encoding a single domain.

Convergent evolution illustrates common solutions to biochemical challenges

Thus far, we have been using sequence analysis methods to identify occurrences of **divergent evolution**, the process by which proteins derived from a common ancestor accumulate differences over time, potentially acquiring new functions. In other examples, proteins have been identified that are structurally similar in important ways but are not descended from a common ancestor. How might two unrelated proteins come to resemble each other structurally? Two proteins evolving independently may have converged on similar structural features to perform a similar biochemical activity. Perhaps that structure was an especially effective solution to a biochemical problem. The process by which very different evolutionary pathways lead to the same solution is called **convergent evolution**.

An example of convergent evolution is found among the serine proteases. Recall that these enzymes cleave peptide bonds by hydrolysis (Section 6.2). The active sites for two such enzymes, mammalian chymotrypsin and bacterial subtilisin, are remarkably similar (**Figure 10.16, top**). In each case, the catalytic triad of a serine residue, a histidine residue, and an aspartate residue is positioned in space in nearly identical arrangements. This conserved spatial arrangement is critical for the activity of these enzymes and affords the same mechanistic solution to the problem of peptide bond hydrolysis.

At first glance, this similarity between chymotrypsin and subtilisin might suggest that these proteins are homologous. However, striking differences in the overall structures of these proteins make an evolutionary relationship extremely unlikely (**Figure 10.16, bottom**). Whereas chymotrypsin consists almost entirely of β sheets, subtilisin contains extensive α-helical structure. Moreover, the key serine, histidine, and aspartic acid residues do not even appear in the same order within the two sequences. It is extremely unlikely that two proteins evolving from a common ancestor could have retained similar active-site structures while other aspects of the structure changed so dramatically.

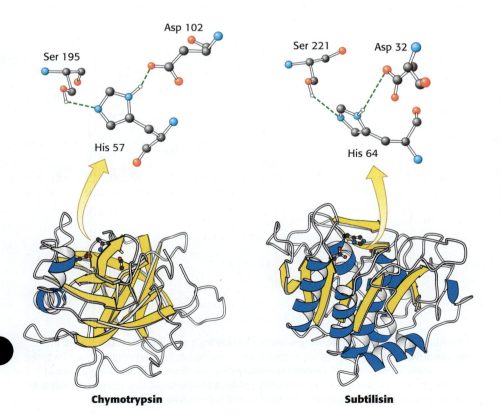

Chymotrypsin

Subtilisin

FIGURE 10.16 Mammalian chymotrypsin and bacterial subtilisin are examples of convergent evolution. The relative positions of the three key catalytic residues are nearly identical in the active sites of the serine proteases chymotrypsin and subtilisin. Yet, the overall protein structures are quite dissimilar, suggesting that although these proteins derived from different ancestors, they arrived at a similar solution to the challenge of peptide bond hydrolysis. [Drawn from 1GCT.pdb and 1SUP.pdb.]

INTERACT with these models in
 Achieve

Comparison of RNA sequences can be a source of insight into RNA secondary structures

The methods and interpretation of sequence alignments described above are not limited to proteins. Homologous RNA sequences can also be similarly studied, yielding important insights into evolutionary relationships. Additionally, these analyses provide clues to the three-dimensional structure of the RNA itself. As noted in Chapter 8, single-stranded nucleic acid molecules fold back on themselves to form elaborate structures held together by Watson–Crick base-pairing and other noncovalent interactions. In a family of sequences that form similar base-paired structures, base sequences may vary, but base-pairing ability is conserved.

Consider, for example, a region from a large RNA molecule present in the ribosomes of all organisms (**Figure 10.17**). In the region shown, the *E. coli* sequence has a guanine (G) residue in position 9 and a cytosine (C) residue in position 22, whereas the human sequence has uracil (U) in position 9 and adenine (A) in position 22. Examination of the six sequences shown in Figure 10.17A reveals that the bases in positions 9 and 22, as well as several of the neighboring positions, retain the ability to form Watson–Crick base pairs even though the identities of the bases in these positions vary. We can deduce that two segments with paired mutations that maintain base-pairing ability are likely to form a double helix (Figure 10.17B), a prediction that was confirmed when the three-dimensional structure of this particular rRNA was determined (Section 30.3). Where sequences are known for several homologous RNA molecules, this type of sequence analysis can often suggest complete secondary structures as well as interactions that do not follow the standard Watson–Crick base-pairing rules (Section 8.2).

❖ SELF–CHECK QUESTION

When RNA alignments are used to determine secondary structure, it is advantageous to have many sequences representing a wide variety of species. Why?

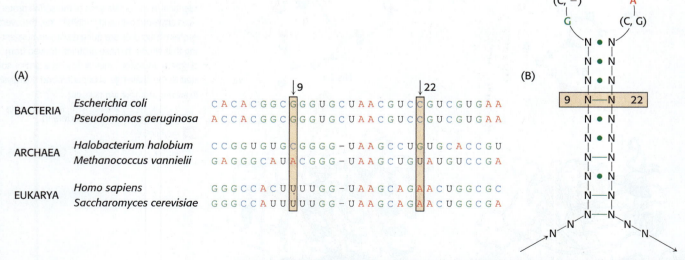

(A)

		9	22
BACTERIA	*Escherichia coli*	C A C A C G G C G G G U G C U A A C G U C C G U C G U G A A	
	Pseudomonas aeruginosa	A C C A C G G C G G G U G C U A A C G U C C G U C G U G A A	
ARCHAEA	*Halobacterium halobium*	C C G G U G U G C G G G G – U A A G C C U G U G C A C C G U	
	Methanococcus vannielii	G A G G G C A U A C G G G – U A A G C U G U A U G U C C G A	
EUKARYA	*Homo sapiens*	G G G C C A C U U U U G G – U A A G C A G A A C U G G C G C	
	Saccharomyces cerevisiae	G G G C C A U U U U U G G – U A A G C A G A A C U G G C G A	

FIGURE 10.17 RNA sequence alignments provide insights into structure based on conservation of base-pairing. (A) A comparison of sequences in a part of ribosomal RNA taken from a variety of species. (B) The implied secondary structure. Green lines indicate positions at which Watson–Crick base-pairing is completely conserved in the sequences shown, whereas dots indicate positions at which Watson–Crick base-pairing is conserved in most cases.

EXAMPLE   Interpreting an RNA Alignment

PROBLEM: Sequences of an RNA fragment from five species have been determined and aligned. Propose a likely secondary structure for these fragments:

(1)   UUGGAGAUUCGGUAGAAUCUCCC
(2)   GCCGGGAAUCGACAGAUUCCCCG
(3)   CCCAAGUCCCGGCAGGGACUUAC
(4)   CUCACCUGCCGAUAGGCAGGUCA
(5)   AAUACCACCCGGUAGGGUGGUUC

GETTING STARTED: For a simple alignment over a short RNA fragment such as the one featured in this problem, the trick is to look for two positions that could form a base pair in each of the aligned sequences. If the possibility of a base pair is conserved through all the sequences, then it is likely that this pairing is a component of the predicted secondary structure.

Of particular interest in this alignment is the base highlighted in red:

(1)   UUGGAGAUUCGGUAGAAUCUCCC
(2)   GCCGGGAAUCGACAGAUUCCCCG
(3)   CCCAAGUCCCGGCAGGGACUUAC
(4)   CUCACCUGCCGAUAGGCAGGUCA
(5)   AAUACCACCCGGUAGGGUGGUUC

In all five sequences, this is an absolutely conserved cytosine. In this alignment, there are only two absolutely conserved guanine bases. One is immediately 3′ to this cytosine, and we know from oligonucleotide structure that a base pair between adjacent positions is highly unlikely. The other is the guanine highlighted in blue, which is separated from the cytosine by four nucleotides. An intriguing possibility!

SOLVE: Now, we have an "anchor" from which we can continue looking for additional pairings. In order for this red–blue pairing to form, the RNA fragment would need to fold on itself. Hence, we should look outward from this pair to identify additional base pairs:

(1)   UUGGAGAUUCGGUAGAAUCUCCC
(2)   GCCGGGAAUCGACAGAUUCCCCG
(3)   CCCAAGUCCCGGCAGGGACUUAC
(4)   CUCACCUGCCGAUAGGCAGGUCA
(5)   AAUACCACCCGGUAGGGUGGUUC

Again, we have a conserved pairing. In sequences (1) and (2), the red and blue positions feature a U–A pair, while in sequences (3), (4), and (5), these positions form a C–G pair. If we continue to "walk" along the sequence in this manner, a clear stretch of base-pairing emerges. Moreover, since several of these positions are absolutely conserved, we can include them in our proposed structure (shown at right, with the original C–G base pair in red/blue).

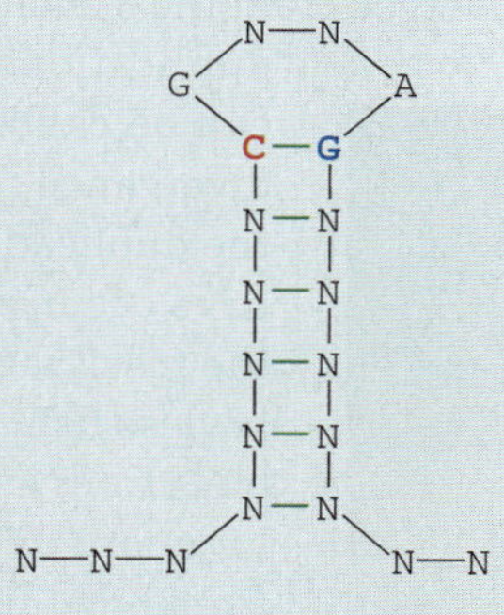

REFLECT: Determining secondary structure predictions from oligonucleotide sequences can appear difficult at first. There are only four possible bases, so patterns may not be easily discernible. The key strategy is to look for potential base pairs and work one's way in either direction from that point. For structure-forming base pairs, while the identities of the bases may not be conserved across the sequence alignments, the possibility of forming a base pair will be absolutely conserved.

FIGURE 10.18 Sequence alignments enable the construction of an evolutionary tree. In the case of the globins, the branching structure was deduced by sequence comparison, whereas the results of fossil studies provided the overall time scale showing when divergence occurred.

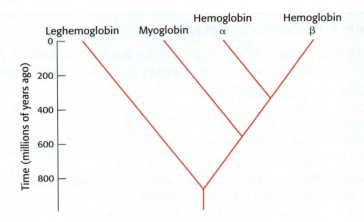

10.3 Evolutionary Trees Can Be Constructed on the Basis of Sequence Information

We have discussed previously how measurement of sequence similarity can provide us with information about the evolutionary relationship between two proteins. This concept suggests that by comparing multiple sequences within a protein family, we can deduce their evolutionary history. As we introduced in Section 1.4, this approach is based on the *molecular clock hypothesis*: the notion that sequences that are more similar to one another have had less evolutionary time to diverge than have sequences that are less similar.

This method can be illustrated by using the three globin sequences in Figures 10.9 and 10.11, as well as the sequence for the human hemoglobin β chain. These sequences can be aligned with the additional constraint that gaps, if present, should be at the same positions in all of the proteins. These aligned sequences can be used to construct an **evolutionary tree** in which the length of the branch connecting each pair of proteins is proportional to the number of amino acid differences between the sequences (**Figure 10.18**).

Evolutionary trees can be calibrated using fossil record data

Such comparisons reveal only the relative divergence times—for example, that myoglobin diverged from hemoglobin twice as long ago as the α chain diverged from the β chain. How can we estimate the approximate dates of gene duplications and other evolutionary events? Evolutionary trees can be calibrated by comparing the deduced branch points with divergence times determined from the fossil record. For example, the duplication leading to the two chains of hemoglobin appears to have occurred 350 million years ago. This estimate is supported by the fossil records on the divergence of jawless fish from bony fish. Jawless fish, such as the lamprey (**Figure 10.19**), contain hemoglobin built from a single subunit. If the fossil record and our globin evolutionary trees are in agreement, jawless fish should have diverged from bony fish at roughly the same time. This is, in fact, the case, as the fossil record suggests this point of divergence was approximately 400 million years ago.

These methods can be applied to both relatively modern and very ancient molecules, such as the ribosomal RNAs that are found in all organisms. Indeed, such an RNA sequence analysis led to the realization that Archaea are a distinct group of organisms that diverged from Bacteria very early in evolutionary history (Figure 1.3).

FIGURE 10.19 The lamprey is a jawless fish whose ancestors diverged from bony fishes approximately 400 million years ago. Lamprey hemoglobin molecules contain only a single type of polypeptide chain.

FIGURE 10.20 **The unicellular red alga** *Galdieria sulphuraria* **is a eukaryote that can survive extreme environments.** This species has been isolated from hot springs such as the Grand Prismatic Spring in Yellowstone National Park, pictured here.

Jennifer Angel

Horizontal gene transfer events may explain unexpected branches of the evolutionary tree

Evolutionary trees that encompass orthologs of a particular protein across a range of species can lead to unexpected findings. Consider the unicellular red alga *Galdieria sulphuraria*, a remarkable eukaryote that can thrive in extreme environments, including at temperatures up to 56°C, at pH values between 0 and 4, and in the presence of high concentrations of toxic metals (**Figure 10.20**). *G. sulphuraria* belongs to the order Cyanidiales, clearly within the eukaryotic branch of the evolutionary tree (**Figure 10.21A**). However, the complete genome sequence of this organism revealed that nearly 5% of the *G. sulphuraria* ORFs encode proteins that are more closely related to bacterial or archaeal, not eukaryotic, orthologs. Furthermore, the proteins that exhibited these unexpected evolutionary relationships possess functions that are likely to confer a survival advantage in extreme environments, such as the removal of metal ions from inside the cell (**Figure 10.21B**).

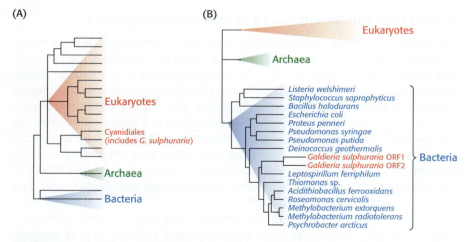

FIGURE 10.21 **Proteins encoded within the *G. sulphuraria* genome with high homology to prokaryotic sequences are suggestive of horizontal gene transfer.** (A) *Galdieria sulphuraria* belongs to the order Cyanidiales, clearly within the eukaryotic branch of the evolutionary tree. (B) Within the completely sequenced *G. sulphuraria* genome, two ORFs encode proteins involved in transport of arsenate ions across membranes. Alignment of these ORFs against orthologs from a variety of species reveals that these pumps are most closely related to their bacterial counterparts, suggesting that a horizontal gene transfer event occurred during the evolution of this species. [(A) Information from Dr. Gerald Schönknecht; (B) Information from G. Schönknecht et al., *Science* 339:1207–1210, 2013, Fig. 3.]

RUSSELL DOOLITTLE A pioneer in bioinformatics and the study of molecular evolution, Russell Doolittle compared the sequences of fibrinopeptides (Section 7.5) across animal species and described how this information could illuminate evolutionary relationships during his postdoctoral fellowship in Sweden. Then, as a faculty member at the University of California, San Diego, Doolittle collected blood samples from a variety of mammals at the San Diego Zoo and constructed a detailed evolutionary tree from their fibrinopeptide sequences. Additionally, well before the existence of online sequence databases, Prof. Doolittle established a compendium of known amino acid sequences, single-handedly compiling sequence data from the scientific literature. With his colleagues, he was able to search this information, perform sequence alignments, and construct phylogenetic trees. Prof. Doolittle made numerous additional seminal contributions throughout his career, including the assessment of sequences for their hydrophobic character (the hydropathy plot, Section 12.4).

One likely explanation for these observations is **horizontal gene transfer**, or the exchange of DNA between species that provides a selective advantage to the recipient. Amongst prokaryotes, horizontal gene transfer is a well-characterized and important evolutionary mechanism. For example, exchange of plasmid DNA between bacterial species likely enabled the transfer of genes encoding restriction endonucleases between bacterial species (Section 6.3). For example, EcoRI (from *E. coli*) and RsrI (from *Rhodobacter sphaeroides*) are 50% identical in sequence over 266 amino acids, clearly indicating a close evolutionary relationship. However, these species of bacteria are not as closely related. These species appear to have obtained the gene for these restriction endonucleases from a common source more recently than the time of their evolutionary divergence.

Recent studies such as those on *G. sulphuraria* described above were made possible by the expansive growth of complete genome sequence information. The results of such studies suggest that horizontal gene transfer from prokaryotes to eukaryotes—that is, between different domains of life—may also have led to evolutionarily significant events.

10.4 Modern Techniques Make the Experimental Exploration of Evolution Possible

Two techniques of biochemistry have made it possible to examine the course of evolution directly and not simply by inference. (1) The polymerase chain reaction (Section 9.1) allows the direct amplification and examination of ancient DNA sequences. If DNA samples can be obtained from preserved remains, we can explore genomes from species that are no longer living. (2) Molecular evolution may also be investigated through the process of synthesizing highly diverse populations of molecules and selecting for a biochemical property. The combination of these two techniques provides a glimpse into the types of molecules that may have existed very early in evolution.

Ancient DNA can sometimes be amplified and sequenced

The tremendous chemical stability of DNA makes the molecule well-suited to its role as the storage site of genetic information. So stable is the molecule that samples of DNA have survived for many thousands of years under appropriate conditions. With the development of PCR and advanced DNA-sequencing methods, such ancient DNA can be amplified and sequenced. This approach was first applied to mitochondrial DNA isolated from a Neanderthal humerus bone estimated at 38,000 years of age. Comparison of the complete Neanderthal mitochondrial sequence with those from human beings (*Homo sapiens*) revealed between 201 and 234 nucleotide substitutions, considerably fewer than the approximately 1500 differences between human beings and chimpanzees over the same region of mitochondrial DNA.

Since the initial sequencing of Neanderthal mitochondrial DNA, the complete genome sequences of a Neanderthal and a closely related hominin known as a Denisovan have been obtained using DNA isolated from nearly 50,000-year-old fossils. Comparison of these sequences suggests that the common ancestor of modern human beings and Neanderthals lived approximately 570,000 years ago, while the common ancestor between Neanderthals and Denisovans lived nearly 380,000 years ago. An evolutionary tree constructed from these data revealed that the Neanderthal was not an intermediate between chimpanzees and human beings but, instead, was an evolutionary "dead end" that became extinct (**Figure 10.22**). Further analysis of these sequences has enabled researchers to determine the extent of interbreeding

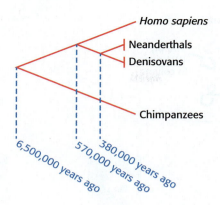

FIGURE 10.22 DNA sequence comparisons enable the placement of Neanderthals and Denisovans on an evolutionary tree. Sequence data suggest that neither Neanderthals nor the Denisovans are on the line of direct descent leading to *Homo sapiens* but, instead, branched off earlier and then became extinct.

between these groups, map the geographic history of these populations, and make inferences about additional ancestors whose DNA has not yet been sequenced.

A few earlier studies claimed to determine the sequences of far more ancient DNA such as that found in insects trapped in amber, but these studies appear to have been flawed. The source of these sequences turned out to be contaminating modern DNA. Successful sequencing of ancient DNA requires sufficient DNA for reliable amplification and the rigorous exclusion of all sources of contamination.

Molecular evolution can be examined experimentally

For a population to evolve, the following conditions must be met: (1) the generation of a diverse population, (2) the selection of members based on some criterion of fitness, and (3) the reproduction of that population to subsequent generations. This last condition determines, in part, the time under which evolution can occur: organisms with long generation times (such as humans) will evolve much more slowly than those that reproduce rapidly (such as bacteria).

With the advancement of the methods of molecular biology, such as those described in Chapter 9, it is possible to meet these evolutionary conditions using nucleic acid molecules in a controlled experimental setting under convenient time-scales. The results of such studies enable us to glimpse how evolutionary processes might have generated catalytic activities and specific binding abilities—important biochemical functions in all living systems.

Biochemists can synthesize a large, diverse population of nucleic acid molecules in the laboratory using **combinatorial chemistry**, a process by which millions of compounds are synthesized in a single process. A population of molecules of a given size can be generated randomly so that many or all possible sequences are present in the mixture. Once an initial population has been prepared, it is subjected to a selection process that isolates specific molecules with desired binding or reactivity properties. Finally, molecules that have survived the selection process are replicated through the use of PCR; primers are directed toward specific sequences included at the ends of each member of the population. Errors that occur naturally in the course of the replication process introduce additional variation into the population in each generation.

Let us consider an application of this approach. Scientists hypothesize that early in evolution of life, before the emergence of proteins, RNA molecules may have played all major roles in biological catalysis. To understand the properties of potential RNA catalysts, researchers have used the methods described above to create an RNA molecule capable of binding adenosine triphosphate and related nucleotides.

FIGURE 10.23 Evolution can be studied in the laboratory setting. A collection of RNA molecules of random sequences is synthesized by combinatorial chemistry. This collection is selected for the ability to bind ATP by passing the RNA through an ATP affinity column (Section 4.1). The ATP-binding RNA molecules are released from the column by washing with excess ATP and then replicated. The process of selection and replication is then repeated several times. The final RNA products with significant ATP-binding ability are isolated and characterized.

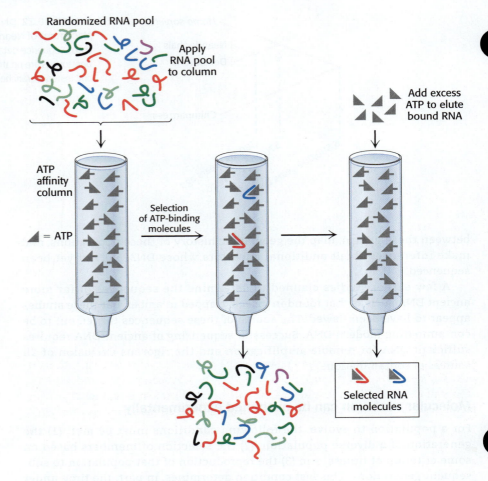

Randomized RNA pool

Apply RNA pool to column

ATP affinity column

▲ = ATP

Selection of ATP-binding molecules

Add excess ATP to elute bound RNA

Selected RNA molecules

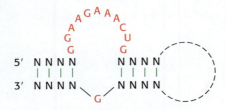

FIGURE 10.24 A conserved secondary structure is common to RNA molecules selected for ATP binding. Bases important for ATP recognition are shown in red. The dashed line indicates a region which varied in length among the various RNA products.

First, an initial population of RNA molecules 169 nucleotides long was created. In this pool of RNA molecules, 120 of the positions differed randomly, with equimolar mixtures of adenine, cytosine, guanine, and uracil. The initial synthetic pool that was used contained approximately 10^{14} unique RNA molecules—a huge number, but still a very small fraction of the total possible pool of random 120-base sequences. From this pool, affinity chromatography (Section 4.1) was used to select those molecules that bound to ATP (**Figure 10.23**).

The molecules that bound well to the ATP affinity column were replicated by reverse transcription into DNA, amplified by PCR, and transcribed back into RNA. Additional diversity was introduced into the pool by the use of a somewhat error-prone reverse transcriptase, which introduces additional mutations into the population during each cycle. The new population was subjected to additional rounds of selection for ATP-binding activity, and after eight generations, members of the selected population were characterized by sequencing. Seventeen different sequences were obtained, 16 of which could form the structure shown in **Figure 10.24**. Each of these molecules bound ATP with dissociation constants less than 50 μM.

The folded structure of the ATP-binding region from one of these RNAs was determined by nuclear magnetic resonance (NMR) methods (Section 4.5). As expected, this 40-nucleotide molecule is composed of two Watson–Crick base-paired helical regions separated by an 11-nucleotide loop (**Figure 10.25A**). This loop folds back on itself in an intricate way (**Figure 10.25B**) to form a deep pocket into which the adenine ring can fit (**Figure 10.25C**). Thus, a structure had evolved in vitro that was capable of a specific interaction.

Synthetic oligonucleotides that can specifically bind ligands, such as the ATP-binding RNA molecules described above, are referred to as **aptamers**.

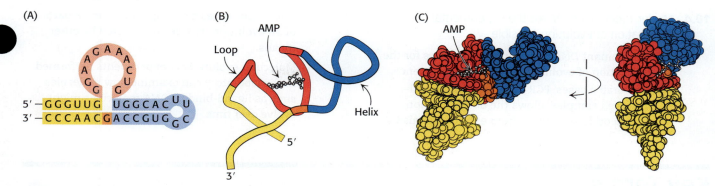

FIGURE 10.25 Experimental molecular evolution yields an ATP-binding RNA molecule.
(A) The Watson–Crick base-pairing pattern of an RNA molecule selected to bind adenosine nucleotides is shown. (B) The NMR structure of this RNA molecule reveals how the nucleotide molecule (here, AMP) binds largely to the loop region (red). (C) In this representation, the tight pocket into which the AMP molecule binds is quite clear. [Drawn from 1RAW.pdb.]

In addition to their role in understanding molecular evolution, aptamers have shown promise as versatile tools for biotechnology and medicine. For example, they have been developed for diagnostic applications, serving as sensors for ligands ranging from small organic molecules, such as cocaine, to larger proteins, such as thrombin. Several aptamers are also being tested in clinical trials as therapies for diseases ranging from leukemia to diabetes. Macugen (pegaptanib sodium), an aptamer which binds to and inhibits the protein vascular endothelial growth factor, has been approved for the treatment of age-related macular degeneration.

Summary

10.1 Homologs Are Descended from a Common Ancestor and Can Be Detected by Sequence Alignments

- Two molecules are homologous if they are derived from a common ancestor.
- Paralogs are homologous molecules that are found in one species and have acquired different functions through evolutionary time.
- Orthologs are homologous molecules that are found in different species and have similar or identical functions.
- Sequences can be aligned to maximize their similarity, and the significance of these alignments can be judged by statistical tests.
- Sequence alignments can be scored based on the number of identities or by the degree of similarity, using a substitution matrix such as Blosum-62.

10.2 Examination of Three-Dimensional Structure Enhances Our Understanding of Evolutionary Relationships

- Comparison of three-dimensional structure may reveal evolutionary relationships not detectable by sequence alignments.

- Alignment of a protein sequence against itself may reveal internal repeats, which suggests a gene duplication event occurred in its evolutionary history.
- Proteins may be related by convergent evolution: they evolved from distinct ancestors but use similar structural features to solve biochemical challenges.
- Nucleic acid sequence alignments enable structural predictions through the identification of base-pairing relationships.

10.3 Evolutionary Trees Can Be Constructed on the Basis of Sequence Information

- Evolutionary trees can be constructed with the assumption that the number of sequence differences corresponds to the time since the two sequences diverged.
- Comparison of evolutionary trees based on sequence alignments with those based on fossil records provides a clearer understanding of the timeline of divergences.
- Horizontal gene transfer events can appear as unexpected branches on the evolutionary tree.

10.4 Modern Techniques Make the Experimental Exploration of Evolution Possible

- Recombinant DNA technology now allows for the exploration of evolution as a laboratory science.
- In favorable cases, PCR amplification of well-preserved samples allows the determination of nucleotide sequences from extinct organisms.

These sequences can help authenticate parts of an evolutionary tree constructed by other means.

- Molecular evolutionary experiments performed in the laboratory can examine how molecules such as ligand-binding RNA molecules might evolve over time.

Key Terms

homologs (p. 302)

orthologs (p. 302)

paralogs (p. 302)

sequence alignment (p. 303)

identity (p. 303)

conservative substitution (p. 305)

nonconservative substitution (p. 305)

substitution matrix (p. 306)

basic local alignment search tool (BLAST) search (p. 309)

open reading frame (ORF) (p. 309)

sequence template (p. 311)

divergent evolution (p. 313)

convergent evolution (p. 313)

evolutionary tree (p. 316)

horizontal gene transfer (p. 318)

combinatorial chemistry (p. 319)

aptamers (p. 320)

Problems

1. Distinguish between paralogs and orthologs. ❖ **1**

2. A evolutionary tree is shown in the image. Similar genes are represented by boxes. The letters X, Y, and Z represent different species. Similarly colored genes designate genes present in the same species (species X, Y, or Z).

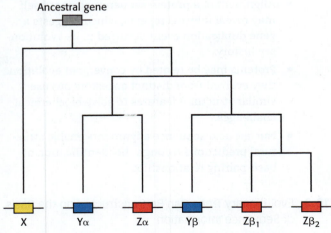

Ancestral gene

Indicate whether the following pairs of genes are orthologs or paralogs: ❖ **1**

(a) $Z\alpha$ and $Z\beta_2$

(b) X and $Z\beta_2$

(c) $Y\alpha$ and $Y\beta$

(d) $Z\beta_1$ and $Z\beta_2$

(e) $Y\alpha$ and $Z\alpha$

(f) X and $Y\alpha$

3. Using the identity-based scoring system applied in Figure 10.4, calculate the score for the following alignment. What is the percent identity between the two sequences? Do you think these two proteins are related? ❖ **2**

(1) WYLGKITRMDAEVLLKKPTVRDGHFLVTQCESSPGEF

(2) WYFGKITRRESERLLLNPENPRGTFLVRESETTKGAY

SISVRFGDSVQ-----HFKVLRDQNGKYYLWAVK-FN

CLSVSDFDNAKGLNVKHYKIRKLDSGGFYITSRTQFS

SLNELVAYHRTASVSRTHTILLSDMNV

SSLQQLVAYYSKHADGLCHRLTNV

4. Explain why protein shuffling is a helpful tool for predicting the significance of homology between two protein sequences. ❖ **2**

5. Using the identity-based scoring system (Section 10.1), calculate the alignment score for the alignment of the following two short sequences:

(1) ASNFLDKAGK

(2) ATDYLEKAGK

Generate a shuffled version of sequence 2 by randomly reordering these 10 amino acids. Align your shuffled sequence with sequence 1 without allowing gaps, and calculate the alignment score between sequence 1 and your shuffled sequence. ❖ **2**

6. Consider the Blosum-62 matrix in Figure 10.7. Replacement of which three amino acids never yields a positive score? What features of these residues might contribute to this observation? ❖ **3**

7. The protein sequence of myoglobin has aspartate (D) in a certain position, whereas the α chain of hemoglobin has glutamate (E) in the same position. How would this substitution be scored in the Blosum-62 matrix? ❖ **3**

8. Consider the following two sequence alignments:

(1) A–SNLFDIRLIG   (2) ASNLFDIRLI–G
    GSNDFYEVKIMD   GSNDFYEVKIMD

Which alignment has a higher score if the identity-based scoring system (Figure 10.4) is used? Which alignment has a higher score if the Blosum-62 substitution matrix (Figure 10.7) is used? For the Blosum scoring, you need not apply a gap penalty. ❖ 3

9. Suppose that the sequences of two proteins each consisting of 200 amino acids are aligned and that the percentage of identical residues has been calculated. How would you interpret each of the following results in regard to the possible divergence of the two proteins from a common ancestor? ❖ 4

(a) 80%   (c) 20%

(b) 50%   (d) 10%

10. Suppose that you wish to synthesize a pool of RNA molecules that contain all four bases at each of 40 positions. How much RNA must you have in grams if the pool is to have at least a single molecule of each sequence? The average molecular weight of a nucleotide is 330 g mol^{-1}. ❖ 5

11. Below is an evolutionary tree for a fictitious protein family, the Solvins. The location of Solvin E is shown.

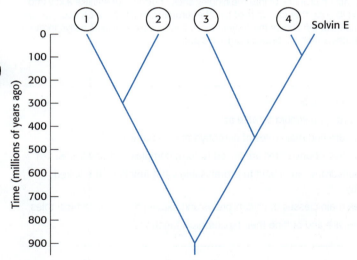

Place Solvins A, B, C, and D on the tree in the numbered positions according to the following statements: ❖ 1

- The Solvin family results from divergent evolution.

- Solvin A diverged from Solvin E three times as long ago as Solvin D evolved from Solvin A.

- Solvin C diverged from Solvin E 500 million years ago.

- Solvin B is not a direct descendent of Solvin A.

Which two Solvins are most similar? When did Solvin B diverge from Solvin E?

12. Classify the following statements as describing convergent evolution, divergent evolution, or both: ❖ 4

(a) Proteins with similar functions that are derived from unrelated ancestors

(b) May result in proteins with similar functions

(c) Change in protein sequence over time

(d) Results in homologous proteins

(e) Proteins with similar functions that are derived from a common ancestor

(f) Chymotrypsin and trypsin are an example.

(g) Chymotrypsin and subtilisin are an example.

13. Classify the following evolutionary processes as examples of either convergent of divergent evolution:

(a) The gene encoding malate dehydrogenase mutates in a way that confers increased specificity for lactate rather than malate.

(b) Coffee and tea plants both have genes to produce caffeine, whereas their most recent common ancestor does not.

(c) Some members of the β-helix fold superfamily of proteins are iron-dependent, whereas others are not.

14. The sequences of three proteins (A, B, and C) are compared with one another, yielding the following levels of identity:

	A	B	C
A	100%	65%	15%
B	65%	100%	55%
C	15%	55%	100%

Assume that the sequence matches are distributed uniformly along each aligned sequence pair. Would you expect protein A and protein C to have similar three-dimensional structures? Explain.

15. You have discovered a mutant form of a thermostable DNA polymerase with significantly reduced fidelity in adding the appropriate nucleotide to the growing DNA strand, compared with wild-type DNA polymerase. How might this mutant be useful in the molecular-evolution experiments described in Section 10.4? ❖ 5

Carbohydrates and Glycoproteins

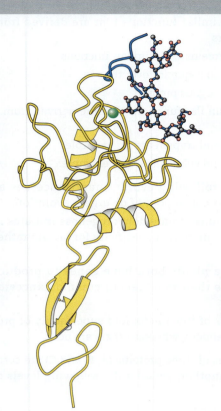

Shutterstock

Carbohydrates are important fuel molecules, but they play many other biochemical roles. Carbohydrate attachments on proteins fit into the binding sites of other proteins like a key into a lock and can act as an ID code that can be read by other proteins, much like a QR code can be read by a phone. The model at left shows the protein P-selectin binding its ligand, a glycoprotein with specific carbohydrates attached. [Drawn from 1G1S.pdb.]

OUTLINE

11.1 Monosaccharides Are the Simplest Carbohydrates

11.2 Monosaccharides Are Linked to Form Complex Carbohydrates

11.3 Carbohydrates Can Be Linked to Proteins to Form Glycoproteins

11.4 Lectins Are Specific Carbohydrate-Binding Proteins

✦ LEARNING GOALS

By the end of this chapter, you should be able to:

1. Describe the structure and main roles of carbohydrates in nature.

2. Describe how simple carbohydrates are linked to form complex carbohydrates.

3. Explain how carbohydrates are linked to proteins and what functions the linked carbohydrates play.

4. Describe the three main classes of glycoproteins and explain their biochemical roles.

5. Define what lectins are and outline their biochemical functions.

For decades, carbohydrates were recognized as important fuels and structural components but were thought to be peripheral to most key activities of the cell. This view has changed dramatically in recent years, as we now know that cells of all organisms are coated in a dense and complex coat of carbohydrates—frequently attached to lipids and proteins—that are essential to cell survival and cell-to-cell interactions and communication. The attachment of a carbohydrate to a protein is the most common post-translational modification of proteins—typically determining where a protein will be localized—and secreted proteins are often extensively decorated with carbohydrates. The extracellular matrix in higher eukaryotes—the environment in which the cells live—is also rich in secreted carbohydrates. Indeed, rather than being mere energy and infrastructure components, carbohydrates supply details and enhancements to the biochemical architecture of the cell, helping to define the functionality and uniqueness of the cell.

11.1 Monosaccharides Are the Simplest Carbohydrates

Carbohydrates are carbon-based molecules that are rich in hydroxyl groups; the empirical formula for many carbohydrates is $(CH_2O)_n$—literally, a carbon hydrate. However, many carbohydrates have additional groups or modifications that deviate from this simple empirical formula. Thus, a better definition for carbohydrates is that they are polyhydroxy aldehydes and ketones, and their derivatives.

There are many monosaccharides but they are structurally similar

Simple carbohydrates are **monosaccharides**, also commonly called *sugars*, which are three to seven carbons in length. These simple sugars serve not only as fuel molecules but also as fundamental constituents of living systems. For instance, DNA has a backbone consisting of alternating phosphoryl groups and deoxyribose, the cyclic form of a five-carbon aldehyde sugar. The smallest monosaccharides, composed of three carbon atoms, are dihydroxyacetone and D- and L-glyceraldehyde.

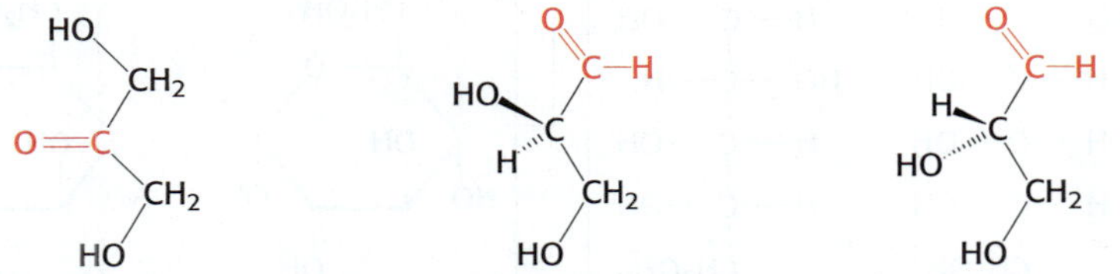

Dihydroxyacetone
(a ketose)

D-Glyceraldehyde
(an aldose)

L-Glyceraldehyde
(an aldose)

Nomenclature. Monosaccharides are typically classified based upon both carbon-chain length and the identity of the most oxidized group. Dihydroxyacetone is called a *ketose* because it contains a keto group (in red, above), whereas glyceraldehyde is called an *aldose* because it contains an aldehyde group (also in red). Both these molecules are referred to as *trioses* (tri- for three, referring to the three carbon atoms that they contain). Similarly, simple monosaccharides with four, five, six, and seven carbon atoms are called *tetroses*, *pentoses*, *hexoses*, and *heptoses*, respectively. Perhaps the monosaccharides of which we are most aware are the hexoses, such as glucose and fructose. Glucose is an essential energy source for virtually all forms of life, and fructose is found commonly in many foods and used as a sweetener; fructose is converted into glucose derivatives inside the cell.

Isomers. Carbohydrates can exist in a dazzling variety of isomeric forms (**Figure 11.1**). Dihydroxyacetone and glyceraldehyde are *constitutional isomers* because they have identical molecular formulas but differ in how the atoms are ordered. These are in contrast to *stereoisomers*, which differ in spatial arrangement but not bonding order. Recall from the discussion of amino acids (Section 2.2) that stereoisomers are designated as having either D or L configuration. Glyceraldehyde has a single asymmetric carbon atom and, thus, there are two stereoisomers of this sugar: D-glyceraldehyde and L-glyceraldehyde. These molecules are a type of stereoisomer called *enantiomers*, which are mirror images of each other. Most monosaccharides from vertebrates have the D configuration. According to convention, the D and L isomers are determined by the configuration of the asymmetric carbon atom farthest from the aldehyde or keto group. Dihydroxyacetone is the only monosaccharide without at least one asymmetric carbon atom.

Monosaccharides made up of more than three carbon atoms have multiple asymmetric carbons, and so they can exist not only as enantiomers but also as *diastereoisomers*, stereoisomers that are not mirror images of each other. The number of possible stereoisomers equals 2^n, where n is the number of asymmetric carbon atoms. Thus, a six-carbon aldose with four asymmetric

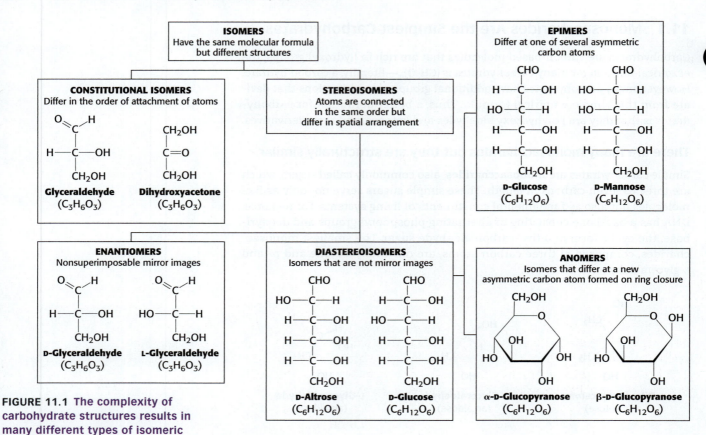

FIGURE 11.1 The complexity of carbohydrate structures results in many different types of isomeric forms.

carbon atoms can exist as eight possible diastereoisomers, each of which can exist in two different enantiomeric forms, for a total of 16 stereoisomers. D-Glucose is one such aldohexose. Note that ketoses have one less asymmetric center than aldoses with the same number of carbon atoms. D-Fructose is the most abundant ketohexose.

Common monosaccharides. The monosaccharides that we will see most frequently in our study of biochemistry are shown in **Figure 11.2**. D-Ribose, the

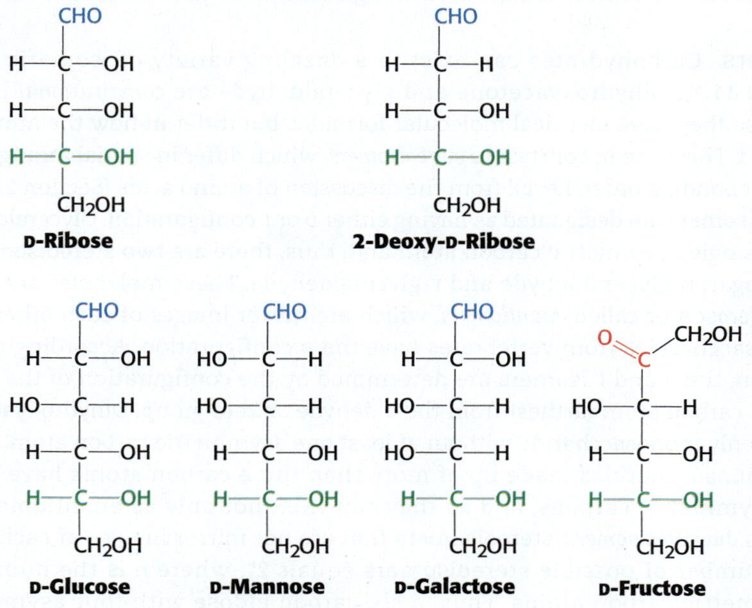

FIGURE 11.2 The most common monosaccharides are pentoses and hexoses. Aldoses contain an aldehyde group (shown in blue), whereas ketoses, such as D-fructose, contain a keto group (shown in red). The asymmetric carbon atom (shown in green) farthest from the aldehyde or keto group designates the structures as being in the D configuration.

carbohydrate component of RNA, is a five-carbon aldose, as is deoxyribose, the monosaccharide component of deoxynucleotides. D-Glucose, D-mannose, and D-galactose are abundant six-carbon aldoses. Note that D-glucose and D-mannose differ in configuration only at C-2, the carbon atom in the second position. Sugars that are diastereoisomers differing in configuration at only a single asymmetric center are **epimers**. Thus, D-glucose and D-mannose are epimeric at C-2; D-glucose and D-galactose are epimeric at C-4.

Most monosaccharides exist as interchanging cyclic forms

The predominant forms of ribose, glucose, fructose, and many other sugars in solution, as is the case inside the cell, are not open chains. Rather, the open-chain forms of these sugars cyclize into rings. The chemical basis for ring formation is that an aldehyde can react with an alcohol to form a *hemiacetal*.

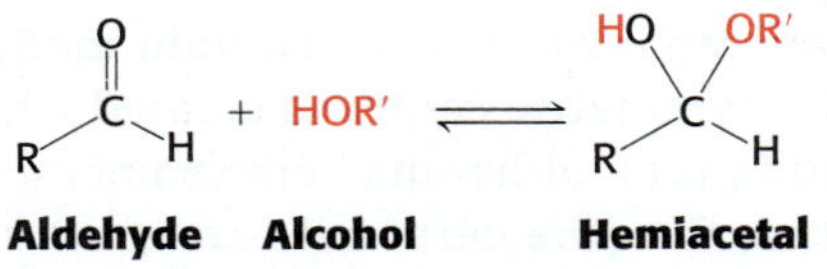

Aldehyde   Alcohol   Hemiacetal

For an aldohexose such as glucose, a single molecule provides both the aldehyde and the alcohol: the C-1 aldehyde in the open-chain form of glucose reacts with the C-5 hydroxyl group to form an intramolecular hemiacetal (**Figure 11.3**). The resulting cyclic hemiacetal, a six-membered ring, is called *pyranose* because of its similarity to the cyclic molecule *pyran*.

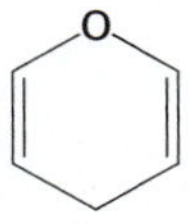

Pyran

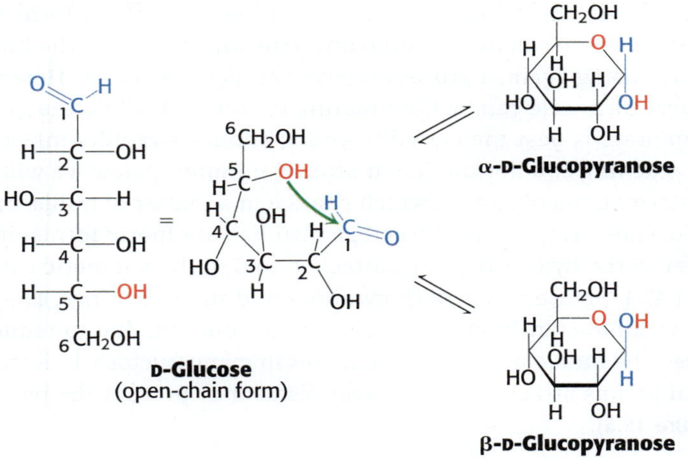

α-D-**Glucopyranose**

D-Glucose
(open-chain form)

β-D-**Glucopyranose**

FIGURE 11.3 D-Glucose and other monosaccharides can temporarily form a pyranose ring. The open-chain form of glucose cyclizes when the C-5 hydroxyl group attacks the oxygen atom of the C-1 aldehyde group to form an intramolecular hemiacetal. Two interchanging anomeric forms of the pyranose ring, designated α and β, can result; both in rapid equilibrium with the open-chain (linear) form.

Similarly, a ketone can react with an alcohol to form a *hemiketal*.

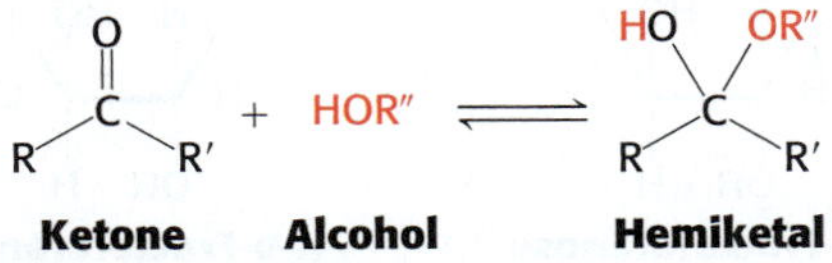

Ketone   Alcohol   Hemiketal

The C-2 keto group in the open-chain form of a ketohexose, such as fructose, can form an intramolecular hemiketal by reacting with either the C-6 hydroxyl group to form a six-membered cyclic hemiketal or the C-5 hydroxyl group to form a five-membered cyclic hemiketal (**Figure 11.4**). The five-membered ring is called a *furanose* because of its similarity to the five-membered heterocycle *furan*.

Furan

The depictions of glucopyranose (glucose) and fructofuranose (fructose) shown in Figures 11.3 and 11.4 are examples of Haworth projections. In such

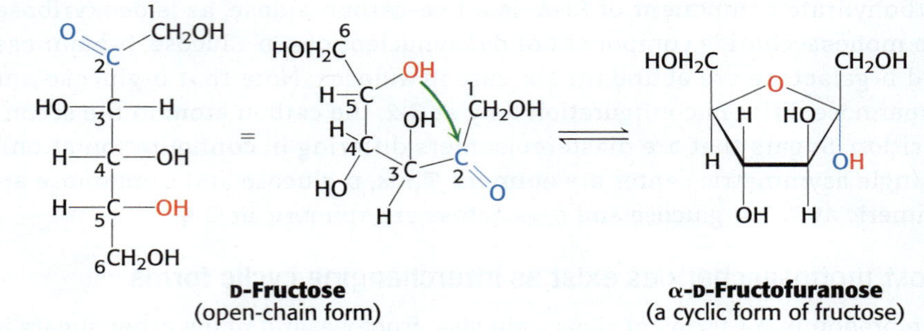

FIGURE 11.4 D-Fructose and other monosaccharides can temporarily form a furanose ring. The open-chain form of fructose cyclizes to a five-membered ring when the C-5 hydroxyl group attacks the C-2 ketone to form an intramolecular hemiketal. Two interchanging anomers are formed, but only the α anomer is shown.

D-Fructose
(open-chain form)

α-D-Fructofuranose
(a cyclic form of fructose)

projections, the carbon atoms in the ring are not written out. The approximate plane of the ring is perpendicular to the plane of the paper, with the heavy line on the ring projecting toward the reader.

We have seen that carbohydrates can contain many asymmetric carbon atoms. An additional asymmetric center is created when a cyclic hemiacetal is formed, generating yet another diastereoisomeric form of sugars called an **anomer**. In glucose, C-1 (the carbonyl carbon atom in the open-chain form) becomes an asymmetric center. Thus, two ring structures are formed: α-D-glucopyranose and β-D-glucopyranose (Figure 11.3). For D sugars drawn as Haworth projections in the standard orientation, as shown in Figure 11.3, the designation α means that the hydroxyl group attached to C-1 is on the opposite side of the ring as C-6; β means that the hydroxyl group is on the same side of the ring as C-6. The C-1 carbon atom is called the *anomeric carbon atom*. An equilibrium mixture of glucose contains approximately one-third α anomer, two-thirds β anomer, and < 1% of the open-chain form. Note that these structures are constantly interconverting by opening to the linear structure and reclosing again, a process referred to as *mutarotation*. Here again, the biochemical timescale can be illuminating to consider: while the interconversion of epimers is best measured in years, anomers rapidly interconvert in seconds, reaching equilibrium in a matter of minutes. Below we will also consider conformations of sugars, which change in a matter of milliseconds.

The furanose-ring form of fructose also has anomeric forms, in which α and β refer to the hydroxyl groups attached to C-2, the anomeric carbon atom (Figure 11.4). Fructose forms both pyranose and furanose rings. The pyranose form predominates in fructose that is free in solution, due to reduced steric hindrances; the furanose form predominates in many fructose derivatives where additional groups attached to the sugar disfavor or prevent the pyranose ring form (**Figure 11.5**).

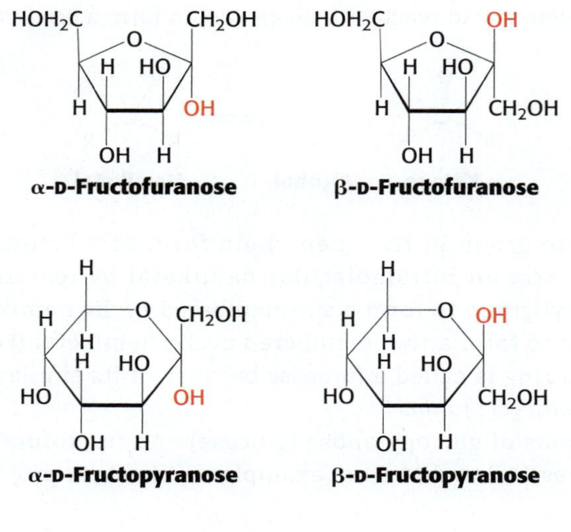

α-D-Fructofuranose

β-D-Fructofuranose

FIGURE 11.5 D-Fructose rapidly interchanges between four distinct ring structures in solution. D-Fructose can form both five-membered furanose (top) and six-membered pyranose (bottom) rings. In each case, both α and β anomers are possible and all are in rapid equilibrium with the open-chain form.

α-D-Fructopyranose

β-D-Fructopyranose

β-D-**Ribofuranose**
(Ribf)

β-2-Deoxy-D-**Ribofuranose**
(dRib)

α-D-**Glucopyranose**
(Glc)

α-D-**Fructofuranose**
(Fru)

α-D-**Galactopyranose**
(Gal)

α-D-**Mannopyranose**
(Man)

FIGURE 11.6 The most common monosaccharides exist primarily in their ring forms. Only one anomer of each is shown. The shorthand carbohydrate code for each sugar is given in parentheses.

β-D-Fructopyranose, found in honey and many fruits, is the sweetest common monosaccharide, while the β-D-fructofuranose form is not nearly as sweet. Because heating converts β-fructopyranose into the β-fructofuranose form, corn syrup with a high concentration of fructose in the β-D-pyranose form is used as a sweetener in cold, but not hot, drinks. **Figure 11.6** shows common sugars in their ring forms.

Pyranose and furanose rings can assume different conformations

The six-membered pyranose ring is not planar because of the tetrahedral geometry of its saturated carbon atoms. Instead, pyranose rings adopt two classes of conformations, termed *chair* and *boat* because of the resemblance to these objects (**Figure 11.7**). In the chair form, the substituents on the ring carbon atoms have two orientations: axial and equatorial. *Axial* bonds are nearly perpendicular to the average plane of the ring, whereas *equatorial* bonds are nearly parallel to this plane. Axial substituents sterically hinder each other if they emerge on the same side of the ring (e.g., 1,3-diaxial groups). In contrast, equatorial substituents are less crowded.

The chair form of β-D-glucopyranose predominates because all axial positions are occupied by hydrogen atoms. The bulkier —OH and —CH_2OH groups emerge at the less-hindered periphery. The boat form of glucose is disfavored because it is quite sterically hindered.

Steric hindrance

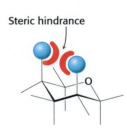

Chair form

Boat form

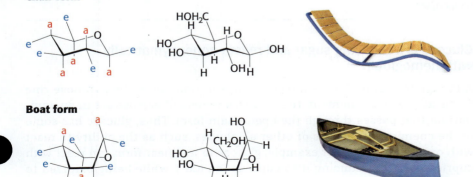

FIGURE 11.7 Pyranose rings like β-D-glucopyranose are not actually planar but adopt chair and boat forms. The chair form is more stable because the less bulky hydrogen atoms occupy all of the axial positions, resulting in less steric hindrance. Abbreviations: a, axial; e, equatorial.
[Photos credit: (Top) Douglas Baldan/Shutterstock; (Bottom) marekuliasz/Shutterstock.]

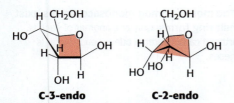

FIGURE 11.8 Furanose rings like β-D-ribofuranose are nonplanar and adopt alternative envelope conformations. The shading indicates the four atoms that lie approximately in a plane.

Furanose rings, like pyranose rings, are not planar. They can be puckered so that four atoms are nearly coplanar and the fifth is about 0.5 Å away from this plane (Figure 11.8). This conformation is called an *envelope form* because the structure resembles an opened envelope with the back flap raised. In the ribose moiety of most biomolecules, either C-2 or C-3 is out of the plane on the same side as C-5. These conformations are called *C-2-endo* and *C-3-endo*, respectively.

D-Glucose is an important fuel for most organisms

D-Glucose is a common and important fuel molecule in nearly all organisms. In mammals, D-glucose is the predominant form of carbohydrate circulated in the blood—commonly called *blood sugar*—and is the only fuel that the brain uses under non-starvation conditions, and the only fuel used by red blood cells. Why is D-glucose, instead of any of the many other monosaccharides, such a prominent biological fuel molecule? We can speculate on the reasons. First, glucose is one of several monosaccharides formed from formaldehyde under prebiotic conditions, and so it may have been available as a fuel source for primitive biochemical systems. Second, glucose is relatively non-reactive (inert) compared to other monosaccharides and is therefore less likely to react nonenzymatically with proteins, potentially rendering them nonfunctional.

A consideration of glucose's specific chemical structure holds the key to understanding why its selection as life's primary sugar was not arbitrary. From the perspective of chemical stability, six-membered rings are preferable to five-membered rings due to decreased ring strain. Likewise, as noted above, the boat conformation is preferable for a six-membered ring, and the most stable boat conformation will have all bulky groups (—OH and —CH₂OH) in the equatorial positions. Of all possible monosaccharides, the most stable six-membered ring structure that can be formed is that of D-glucose in the pyranose form with the anomeric carbon in the β orientation, as this is the only six-membered ring with all five bulky groups in the equatorial positions. This structural observation also explains the roughly 2:1 ratio of β-to-α anomer conformations for D-glucose in an equilibrium solution. Because glucose has such a strong tendency to exist in this stable ring conformation, it consequently has relatively little tendency to modify proteins in a reaction that results from the linear form compared to other sugars, though these reactions do still occur, as described in the next section.

❖ SELF–CHECK QUESTION

Based on ring strain and steric clashes, which is the more stable structure in each pair: D-fructopyranose or D-fructofuranose; boat or chair conformations of D-fructopyranose; bulky groups in axial or equatorial positions about a ring structure; β or α anomer of D-glucopyranose?

Glucose is a reducing sugar and reacts nonenzymatically with hemoglobin

Although 99% of D-glucose molecules in solution are in the pyranose ring forms at any given moment, the α and β anomers of glucose are in an equilibrium that passes through the open-chain form. Thus, glucose has some of the chemical properties of other aldehydes, such as the ability to react with oxidizing agents. For example, glucose in its linear form can react with cupric ion (Cu^{2+}), reducing it to cuprous ion (Cu^+), while being oxidized to gluconic acid.

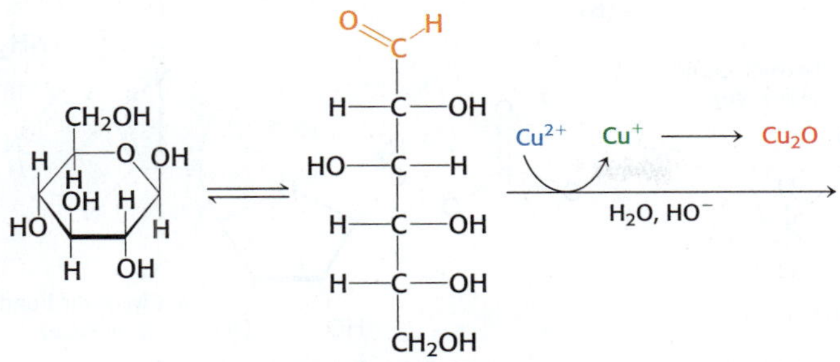

Solutions of cupric ion (known as *Fehling's solution*) provide a simple test for the presence of sugars, such as D-glucose, that at least temporarily adopt an open structure. Sugars that react are called **reducing sugars**. Because ketose sugars can tautomerize to a small degree into temporary structures that have aldehydes, all monosaccharides that can access their open (linear) structures in solution are reducing sugars.

Reducing sugars often nonspecifically react with free amino groups on proteins, usually at a lysine or arginine residue, to form a stable covalent bond. This nonenzymatic (random) addition of a carbohydrate to another molecule is called *glycation*, as opposed to *glycosylation*, which is enzyme catalyzed. As emphasized above, D-glucose has a low tendency to glycate proteins; however, when the concentrations of both the sugar and the protein are very high for long periods of time—as is the case with both D-glucose and hemoglobin in red blood cells—even unlikely reactions can occur at an appreciable rate. As such, the reducing sugar D-glucose reacts with hemoglobin to form glycated hemoglobin (called *hemoglobin A1c* or often just A1C).

Monitoring changes in A1C levels—that is, the amount of *glycated hemoglobin* (often incorrectly called *glycosylated hemoglobin*) in the blood—is an especially useful means of assessing the effectiveness of treatments for diabetes mellitus, a serious condition in which insulin, a key hormone in regulating glucose homeostasis, is either absent (type 1 diabetes) or ineffective (type 2 diabetes), resulting in high levels of blood glucose (Section 24.3). Because the glycated hemoglobin remains in circulation, the amount of the modified hemoglobin corresponds to the long-term regulation—over several months—of glucose levels. This kind of measurable sign of an illness that correlates with a disease regardless of whether it is actually mechanistically involved is called a *biomarker*. In nondiabetic individuals, less than 6% of hemoglobin is glycated, whereas in patients with uncontrolled diabetes, almost 10% of hemoglobin is glycated. Thus, glycated hemoglobin (A1C) is a widely used biomarker for the early onset and treatment of diabetes, among other disorders.

Although the glycation of hemoglobin has no effect on oxygen binding and is thus benign, similar reducing reactions with other proteins such as collagen are often detrimental, because the glycations alter the normal biochemical function of the modified proteins. Following the primary modification, cross-linking may occur between the site of the first modification and elsewhere in the protein, further compromising function. These modifications, known as *advanced glycation end products* (AGEs), have been implicated in aging, arteriosclerosis, and diabetes, as well as other pathological conditions.

Monosaccharides are joined to alcohols and amines through glycosidic linkages by specific enzymes

The biochemical properties of monosaccharides can by modified by reaction with other molecules. These modifications increase the biochemical versatility of carbohydrates, enabling them to serve as signal molecules or

FIGURE 11.9 Carbohydrates are often attached to other molecules through O- and N-glycosidic linkages. (A) An O-glycosidic linkage connects glucose to a methyl group in α-D-methylglucopyranose. (B) An N-glycosidic linkage joins ribose to the base adenine in adenosine monophosphate.

(A)

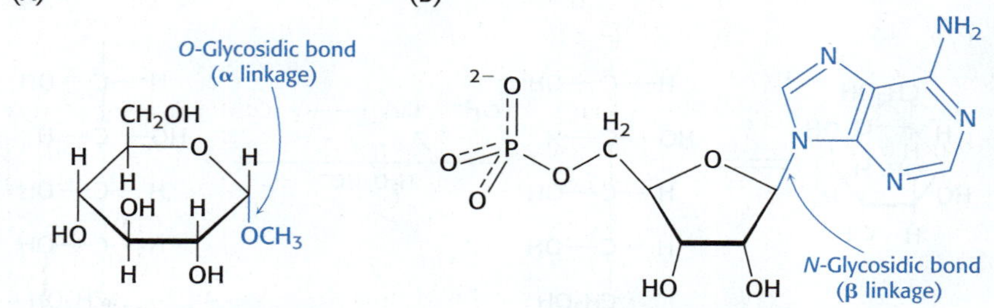

α-D-Methylglucopyranose

(B)

Adenosine monophosphate

facilitating their metabolism. Three common reactants are alcohols, amines, and phosphates.

The covalent linkage formed between the anomeric carbon atom of a carbohydrate and the oxygen atom of an alcohol is called a **glycosidic linkage**—specifically, an O-*glycosidic linkage* (**Figure 11.9A**). O-Glycosidic linkages are prominent when carbohydrates are joined together to form long polymers. In addition, the anomeric carbon atom of a sugar can be linked to the nitrogen atom of an amine to form an N-*glycosidic linkage*, such as when nitrogenous bases are attached to ribose units to form nucleotides (**Figure 11.9B**). Both types of linkages are used to attach carbohydrates to proteins. Carbohydrates can also be modified by the attachment of functional groups to carbons other than the anomeric carbon (**Figure 11.10**). For example, N-acetylgalactosamine is called an *amino sugar*, so named because an amino group replaces a hydroxyl group.

FIGURE 11.10 Monosaccharides can be modified by the addition of substituents other than hydroxyl groups. Additional substituents are shown in red.

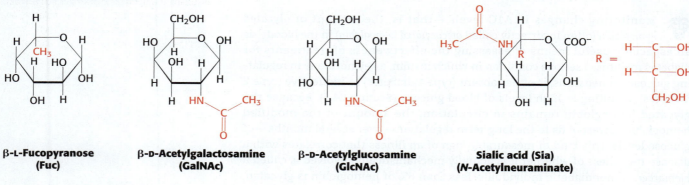

β-L-Fucopyranose (Fuc)   **β-D-Acetylgalactosamine (GalNAc)**   **β-D-Acetylglucosamine (GlcNAc)**   **Sialic acid (Sia) (N-Acetylneuraminate)**

Phosphorylated sugars are key intermediates in metabolism

The addition of phosphoryl groups is a common modification of sugars in metabolic reactions. For instance, the first step in the breakdown of glucose to obtain energy is its conversion into glucose 6-phosphate. Several subsequent intermediates in this metabolic pathway, such as dihydroxyacetone phosphate and glyceraldehyde 3-phosphate, are phosphorylated sugars.

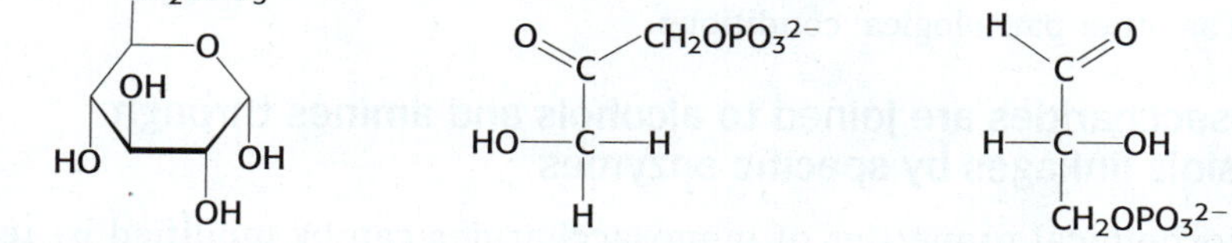

Glucose 6-phosphate (G-6P)   **Dihydroxyacetone phosphate (DHAP)**   **Glyceraldehyde 3-phosphate (GAP)**

By making sugars anionic (negatively charged), phosphorylation not only prevents these sugars from spontaneously leaving the cell by crossing lipid-bilayer membranes but also prevents them from interacting with transporters of the unmodified sugar. Moreover, phosphorylation blocks the formation of alternative ring conformations; for example, phosphorylation of fructose at C-6 prevents the formation of both pyranose anomers, effectively increasing the concentration of the furanose forms. Finally, phosphorylation creates reactive intermediates that will more readily undergo metabolism. For example, a multiply phosphorylated derivative of ribose plays key roles in the biosyntheses of purine and pyrimidine nucleotides (Chapter 25).

❖ SELF–CHECK QUESTION

Which of the following—D-fructose, D-galactose, β-D-ribofuranose—are reducing sugars like D-glucose?

11.2 Monosaccharides Are Linked to Form Complex Carbohydrates

Because sugars contain hydroxyl groups, glycosidic linkages can join one monosaccharide to another. **Oligosaccharides** are built by the linkage of two or more monosaccharides by *O*-glycosidic bonds (**Figure 11.11**). In the disaccharide maltose, for example, two D-glucose residues are joined by a glycosidic linkage between the α-anomeric form of C-1 on one sugar and the hydroxyl oxygen atom on C-4 of the adjacent sugar. This is called an α-1,4-*glycosidic linkage.*

Just as proteins have a directionality defined by the amino and carboxyl ends, oligosaccharides have a directionality defined by their reducing and nonreducing ends. Anomeric carbons in glycosidic linkages are locked in their specific anomeric form and can no longer convert to the open-chain form, or mutarotate. However, the sugar at the reducing end has a free anomeric carbon atom that has reducing activity because it can form the open-chain form. By convention, this end of the oligosaccharide is still called the *reducing end* even when it is bound to another molecule such as a protein and thus no longer has reducing properties.

The fact that monosaccharides have multiple hydroxyl groups means that many different glycosidic linkages are possible. For example, consider three monosaccharides: glucose, mannose, and galactose. In principle, these molecules could be linked together to form more than 12,000 structures differing in the order of the monosaccharides and the hydroxyl groups participating in the glycosidic linkages. In this section, we will look at some of the most common oligosaccharides that actually occur in nature.

Sucrose, lactose, and maltose are common disaccharides

A **disaccharide** consists of two sugars joined by an *O*-glycosidic linkage. **Figure 11.12** illustrates three abundant disaccharides.

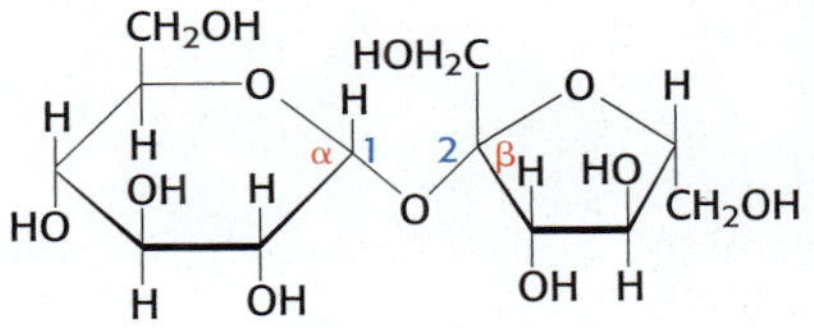

FIGURE 11.12 Sucrose, lactose, and maltose are common dietary disaccharides. In sucrose, neither sugar can access the open chain form since both anomeric carbons are in the glycosidic linkage; hence, sucrose is not a reducing sugar, and there are no anomeric forms. For lactose and maltose, only one anomer is shown. In lactose, the curved lines indicate the connectivity of the atoms but are normal covalent bonds. One sugar is upside down with respect to the other in the structure; however, for viewing convenience both sugars here are shown in the normal orientation for a Haworth projection, necessitating curved lines.

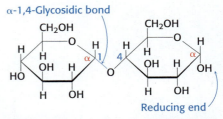

FIGURE 11.11 Maltose is a disaccharide of only D-glucose. Two molecules of glucose are linked by an α-1,4-glycosidic linkage to form the disaccharide maltose. The glucose molecule on the right is capable of assuming the open-chain form, which can act as a reducing agent. The glucose molecule on the left cannot assume the open-chain form, because the C-1 carbon atom is bound to another molecule.

Sucrose
(α-D-Glucopyranosyl-(1→2)-β-D-Fructofuranose)

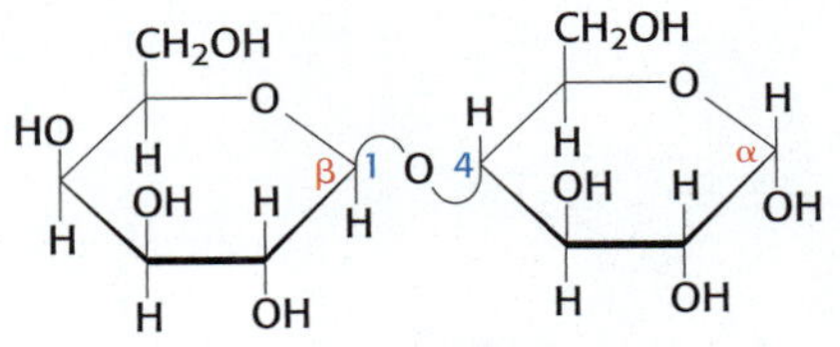

α-Lactose
(β-D-Galactopyranosyl-(1→4)-α-D-Glucopyranose)

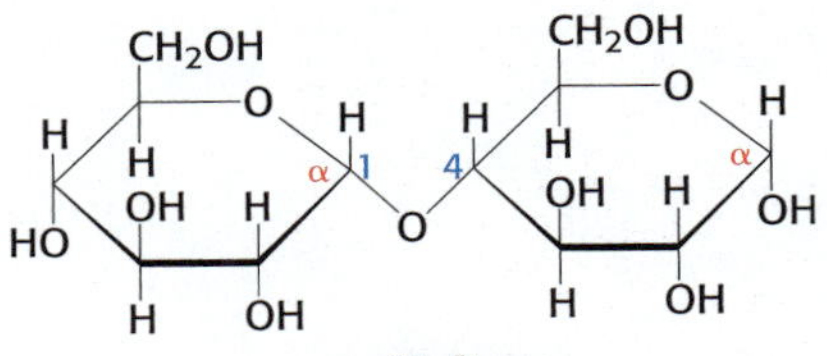

α-Maltose
(α-D-Glucopyranosyl-(1→4)-α-D-Glucopyranose)

- In *sucrose*, obtained commercially from sugar cane or sugar beets and refined into table sugar, both the anomeric carbons of a glucose unit and a fructose unit are joined; the configuration of this glycosidic linkage is α for glucose and β for fructose. Note that because both anomeric carbons are involved in the glycosidic linkage, sucrose is not a reducing sugar. Sucrose can be cleaved into its component monosaccharides by the enzyme *sucrase*, also called *invertase*.

- The disaccharide found in milk, *lactose*, consists of galactose joined to glucose by a β-1,4-glycosidic linkage. Lactose is hydrolyzed to these monosaccharides by *lactase* in human beings and by *β-galactosidase* in bacteria.

- In *maltose*, two glucose units are joined by an α-1,4-glycosidic linkage. Maltose comes from the hydrolysis of large polymeric oligosaccharides such as starch and glycogen and is in turn hydrolyzed to glucose by *maltase* (α-glucosidase). Maltase will also degrade oligosaccharides linked by α-1,4-glycosidic linkages.

The enzymes sucrase, lactase, and maltase are located on the outer surfaces of epithelial cells lining the small intestine. The cleavage products of sucrose, lactose, and maltose can be further processed to provide energy in the form of ATP.

Maltase inhibitors can help to maintain blood glucose homeostasis

Maintaining the proper concentration of glucose in the blood (3.9–5.5 mM)—glucose homeostasis—is vital, since glucose can form advanced glycation products. Moreover, as we noted earlier, elevated blood glucose (hyperglycemia) is characteristic of diabetes. Pharmacological intervention is sometimes required, including the administration of insulin or other drugs that control blood glucose, and maltase is one potential drug target. Following a meal, starch and glycogen are initially degraded by α-amylase secreted by the salivary glands and pancreas. The oligosaccharides generated by α-amylase are further digested by α-glucosidase. Two drugs commonly used to inhibit maltase are acarbose (Precose) and miglitol (Glyset):

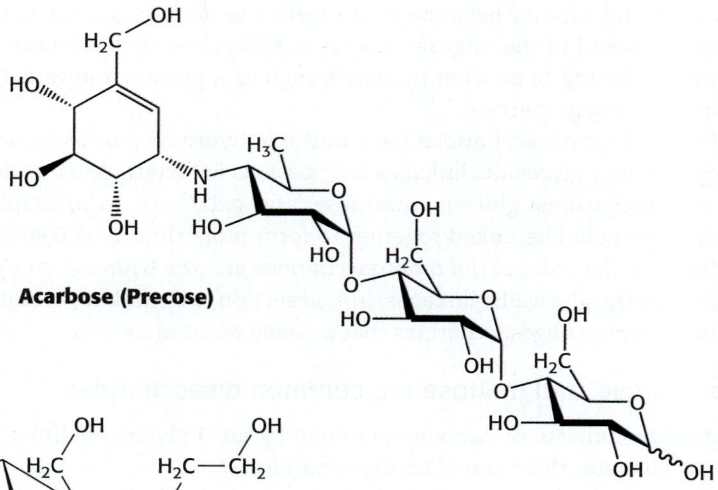

Acarbose (Precose)

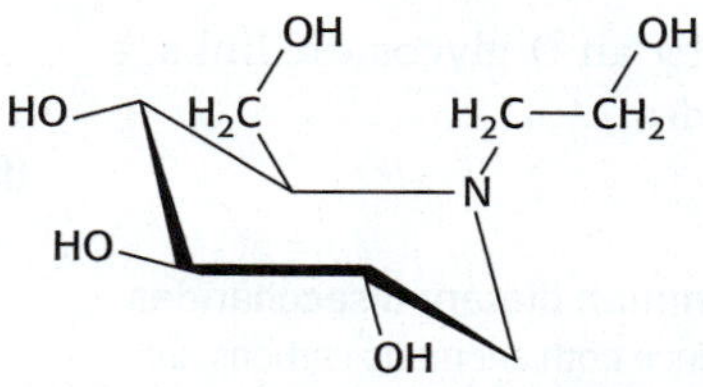

Miglitol (Glyset)

Both acarbose and miglitol are competitive inhibitors (Section 5.6) of maltase and are usually ingested at the start of a meal to slow postprandial (after a meal) glucose absorption in the treatment of diabetes.

🩺 Human milk oligosaccharides protect newborns from infection

In addition to the disaccharide lactose, more than 150 different oligosaccharides have been identified in human milk; the amount and composition of these sugars vary among lactating women. Interestingly, these carbohydrates are not digested by breastfed infants but play a significant role in protecting them against bacterial infection. In particular, they may protect infants from infections caused by certain types of *Streptococcus* bacteria that are sometimes transmitted during vaginal childbirth. How the protection arises is not firmly established, but the oligosaccharides may serve as a fuel source for beneficial bacteria. Alternatively, or in addition, these carbohydrates may prevent the attachment of microbial pathogens to the intestinal wall of the newborn. Biochemists continue to explore the therapeutic potential of these oligosaccharides.

Glycogen and starch are storage polysaccharides of glucose

As we know, glucose is an important energy source in virtually all life forms. However, free glucose molecules cannot be stored; in high concentrations, glucose will disturb the osmotic balance of the cell, potentially resulting in cell death. The solution is to store glucose as units in a large polymer, which is less osmotically active.

Large polymeric oligosaccharides, formed by the linkage of multiple monosaccharides, are called **polysaccharides**, or *glycans*, and play vital roles in energy storage and in maintaining the structural integrity of an organism. If all of the monosaccharide units in a polysaccharide are the same, the polymer is called a *homopolymer*.

The most common homopolymer in animal cells is **glycogen**, the storage form of glucose. Glycogen is present in most of our tissues but is most abundant in muscle and liver. As will be considered in detail in Chapter 21, glycogen is a large, branched polymer of glucose residues. Most of the glucose units in glycogen are linked by α-1,4-glycosidic linkages. The branches are formed by α-1,6-glycosidic linkages, present about once in 12 units (**Figure 11.13**). The great degree of branching in glycogen increases the surface area of a glycogen particle, allowing better access for enzymes to rapidly break it down when glucose is needed in animal cells.

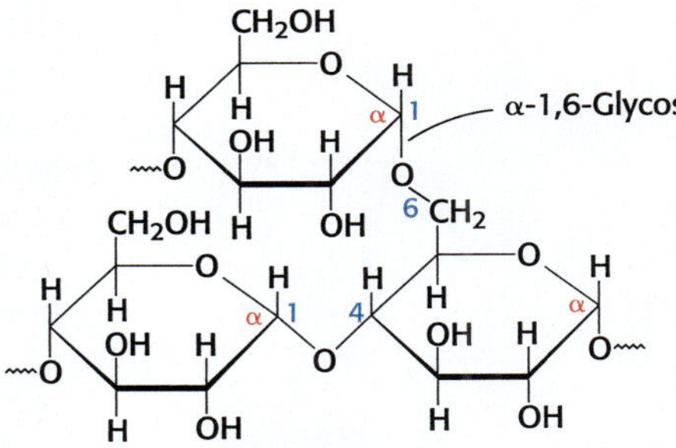

The nutritional reservoir in plants is the homopolymer **starch**, of which there are two forms. *Amylose*, the unbranched type of starch, consists of glucose residues in α-1,4 linkage. *Amylopectin*, the branched form, has about one α-1,6 linkage per 30 α-1,4 linkages. The structure of amylopectin is therefore identical to that of glycogen except for its lower degree of branching, which results in a more compacted structure known as a starch granule. Plants do not need to break down starch as quickly as animals need to degrade glycogen. More than half the carbohydrate ingested by human beings is starch, which is found in wheat, potatoes, rice, and other plant sources. Amylopectin, amylose, and glycogen are all hydrolyzed by *α-amylase*, an enzyme secreted by the salivary glands and the pancreas that can cleave the α-1,4 linkages common to all three glucose-storing homopolymers.

FIGURE 11.13 Glycogen consists of linear polymers with many branch points. Two chains of glucose molecules joined by α-1,4-glycosidic linkages are joined by an α-1,6-glycosidic linkage to create a branch point. Such an α-1,6-glycosidic linkage forms at approximately every 12 glucose units, making glycogen a highly branched molecule.

Cellulose is the main structural polysaccharide of plants

Another major polysaccharide of glucose found in plants, **cellulose**, serves a structural rather than a nutritional role as an important component of the plant cell wall. As one of the most abundant organic compounds in the biosphere, some 10^{15} kg of cellulose is synthesized and degraded on Earth each year, an amount 1000 times as great as the combined weight of the total human population.

Cellulose is an unbranched polymer of glucose residues joined by β-1,4 linkages, in contrast with the α-1,4 linkage seen in starch and glycogen. This simple difference in stereochemistry creates two molecules with very different

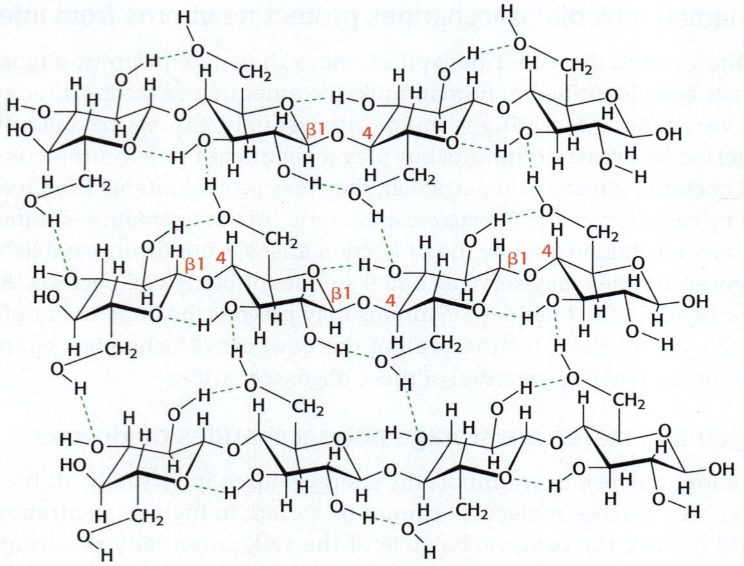

FIGURE 11.14 Glycosidic linkages determine polysaccharide structure. The β-1,4 linkages favor straight chains, which are optimal for structural purposes. The α-1,4 linkages favor bent, helical structures, which are more suitable for storage.

Cellulose
(β-1,4 linkages)

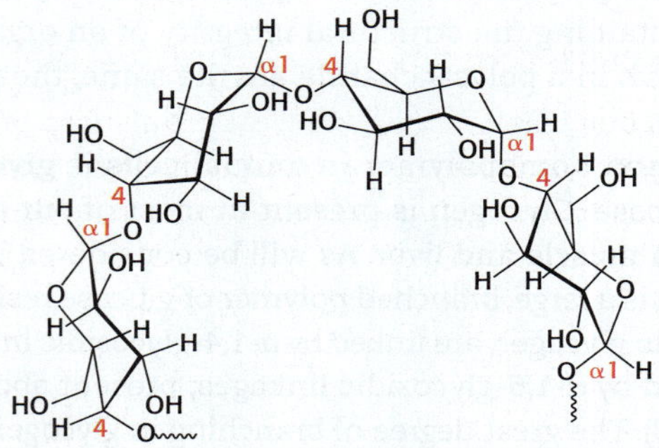

Starch and glycogen
(α-1,4 linkages)

Galacturonic acid

properties and biological functions. The β configuration allows cellulose to form very long, straight chains. Fibrils of cellulose are formed by parallel chains that interact with one another through hydrogen bonds, generating a rigid, support-ive structure (**Figure 11.14, top**). The straight chains formed by β linkages are opti-mal for the construction of fibers having a high tensile strength. In contrast, the α-1,4 linkages in glycogen and starch produce a very different molecular archi-tecture: a hollow helix is formed instead of a straight chain (**Figure 11.14, bottom**). The hollow helix formed by α linkages is well suited to the formation of a more compact, accessible store of sugar.

Although animals lack cellulases and therefore cannot digest wood and veg-etable fibers, cellulose and other plant fibers are still an important constituent of the mammalian diet as a component of dietary fiber. Soluble fiber such as pectin (polygalacturonic acid) slows the movement of food through the gastrointesti-nal tract, allowing improved digestion and the absorption of nutrients. Insolu-ble fibers, such as cellulose, increase the rate at which digestion products pass through the large intestine, softening stools and making them easier to pass.

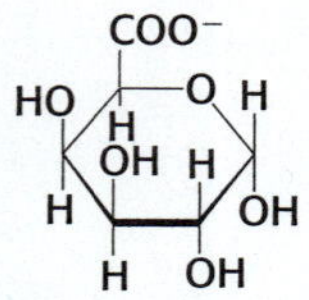

Chitin is the main structural polysaccharide of fungi and arthropods

Likely the second most abundant polysaccharide on the planet, chitin is found in the cell walls of fungi, in the exoskeletons and shells of arthropods such as insects, arachnids, mollusks, and crustaceans, and in numerous other organisms.

**FIGURE 11.15 Chitin, a homopolymer of
N-Acetylglucosamine, is a structural glycan
found in insect wings and the exoskeletons
of arthropods, including lobster shells.**

Charles Curtis/Shutterstock

The structure of chitin is very similar to cellulose with the exception that each
glucose sugar has the C-2 hydroxyl replaced with an amino group which is acetylated such that chitin is a homopolymer of β-1,4 linked N-acetylglucosamine, rather
than glucose (**Figure 11.15**). Chitin fibers are also often crosslinked to one another
and composited with minerals and proteins to further increase its rigidity and
strength. For example, cephalopods such as squid use their razor-sharp beaks,
which are made of extensively cross-linked chitin, to disable and consume prey.

Chitin can be processed to a molecule with a variety of uses

We are all aware, in one way or another, of the many uses of the polysaccharide
cellulose. It is a major constituent of paper, bioadhesives, and the clothes you are
wearing. Less obvious to most of us are uses of chitin. Estimates are that 10,000
tons of chitin could be recovered from the shellfishing industry by processing the
shells into a more versatile molecule called *chitosan*.

Chitosan has a variety of uses. For instance, it is used as a carrier to assist
in drug delivery, as surgical stitches, and as a component of cosmetic and food
products. Recent research shows that chitosan can be used as a component of an
adhesive that strongly adheres to wet tissues, allowing the adhesive to be used
as a surgical dressing.

One way to process chitin, which is water-insoluble, to water-soluble chitosan
is with a series of microbial/enzymatic processes. First, shells are washed and
ground to a fine powder, which is then exposed to lactic acid-generating bacteria. The lactic acid removes minerals from the chitin. Following demineraliza-
tion, the chitin is treated with protease-secreting bacteria to degrade the protein
components. Finally, chitin is treated with chitin deacetylase to remove approximately half of the acetyl groups, generating the more versatile chitosan.

❖ SELF–CHECK QUESTION
Which of these carbohydrates—sucrose, maltose, lactose, cellulose, starch, amylose,
glycogen, chitin—are composed entirely of ᴅ-glucose monomers? Which are structural
(rather than storage) glycans?

11.3 Carbohydrates Can Be Linked to Proteins to Form
Glycoproteins

A carbohydrate group can be covalently attached to a protein to form a **glycoprotein**.
Such modifications are common; 50% of the human proteome consists of
glycoproteins. Many of the proteins secreted from cells are glycosylated, or modified by the attachment of carbohydrates, including most proteins present in the
serum component of blood.

Glycosylation greatly increases the complexity of the proteome. A given protein with several potential glycosylation sites can have many different glycosylated forms, called *glycoforms*. Like protein isoforms (Section 7.2), each is typically generated only in a specific cell type or developmental stage.

In this section, we will examine three classes of glycoproteins:

1. In members of the first class, simply referred to as *glycoproteins*, the protein constituent is the largest component by weight. This versatile class plays a variety of biochemical roles. Many glycoproteins are components of cell membranes, where they take part in processes such as cell adhesion, including the binding of sperm to eggs. Other glycoproteins are formed by linking carbohydrates to soluble proteins.

2. In **proteoglycans**, the protein component of the glycoprotein is conjugated to a particular type of polysaccharide called a **glycosaminoglycan**. Carbohydrates make up a much larger percentage by weight of the proteoglycan compared with simple glycoproteins. Proteoglycans function as structural components and lubricants.

3. **Mucins** (or **mucoproteins**) are, like proteoglycans, predominantly carbohydrate, and they are a key component of mucus. In mucins, the protein component is extensively glycosylated at serine or threonine residues, typically by N-acetylgalactosamine (Figure 11.10). Mucins serve as lubricants.

Carbohydrates can be linked to proteins through asparagine (*N*-linked) or through serine or threonine (*O*-linked) residues

Sugars in glycoproteins are attached either to the amide nitrogen atom in the side chain of asparagine (termed an N-*linkage*) or to the oxygen atom in the side chain of serine or threonine (termed an O-*linkage*), as shown in **Figure 11.16**. An asparagine residue can accept an oligosaccharide only if the residue is part of an Asn-X-Ser or Asn-X-Thr sequence, in which X can be any residue, except proline.

However, not all potential sites are glycosylated. Which sites are glycosylated depends on other aspects of the protein structure and on the cell type in which the protein is expressed. All N-linked oligosaccharides have a pentasaccharide core consisting of three mannose and two N-acetylglucosamine residues. Additional sugars are attached to this core to form the great variety of oligosaccharide patterns found in glycoproteins (**Figure 11.17**).

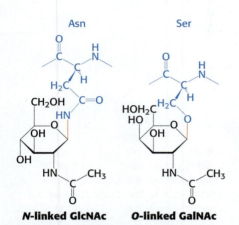

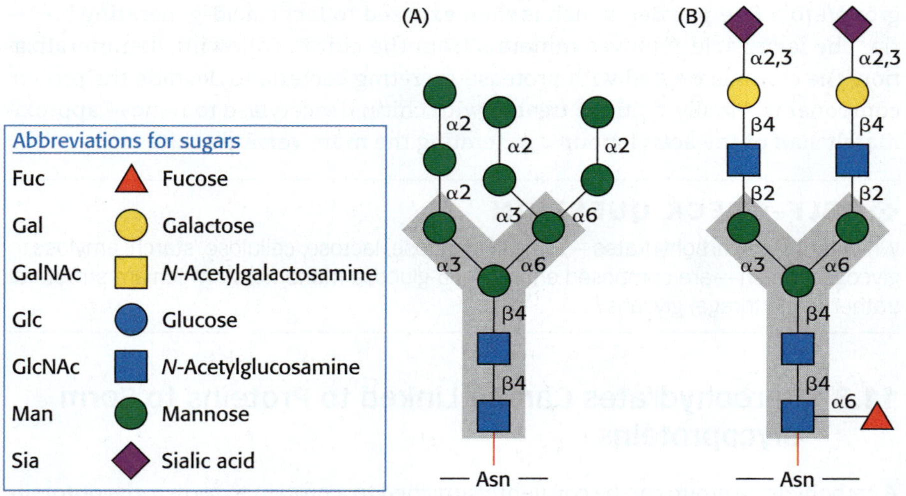

FIGURE 11.16 Glycosidic linkages connect proteins and carbohydrates. A glycosidic linkage connects a carbohydrate to the side chain of asparagine (*N*-linked) or to the side chain of serine or threonine (*O*-linked). The glycosidic linkages are shown in red.

Abbreviations for sugars		
Fuc	▲ (red triangle)	Fucose
Gal	● (yellow circle)	Galactose
GalNAc	■ (yellow square)	N-Acetylgalactosamine
Glc	● (blue circle)	Glucose
GlcNAc	■ (blue square)	N-Acetylglucosamine
Man	● (green circle)	Mannose
Sia	◆ (purple diamond)	Sialic acid

FIGURE 11.17 *N*-linked oligosaccharides have a common core but diverse structures. A pentasaccharide core (shaded gray) is common to all *N*-linked oligosaccharides and serves as the foundation for a wide variety of *N*-linked oligosaccharides, two of which are illustrated: (A) high-mannose type; (B) complex type. The carbohydrate structures are depicted symbolically with a scheme widely used by biochemists.

The glycoprotein erythropoietin is a vital hormone

Let us look at a glycoprotein present in the blood serum called *erythropoietin* (EPO), whose cloned recombinant form has dramatically improved treatment for anemia, particularly that induced by cancer chemotherapy. EPO, which is secreted by the kidneys and stimulates the production of red blood cells, is composed of 165 amino acids and is *N*-glycosylated at three asparagine residues and *O*-glycosylated on a serine residue, resulting in a mature protein that is 40% carbohydrate by weight (**Figure 11.18**). Glycosylation enhances the stability of the protein in the blood; unglycosylated protein has only about 10% of the bioactivity of the glycosylated form because the protein is rapidly removed from the blood by the kidneys.

Although the availability of recombinant human EPO has greatly aided the treatment of anemias, some endurance athletes have abused recombinant human EPO to increase the red-blood-cell count and hence their oxygen-carrying capacity. Drug-testing laboratories are able to distinguish some forms of prohibited human recombinant EPO from natural EPO in athletes by detecting differences in their glycosylation patterns.

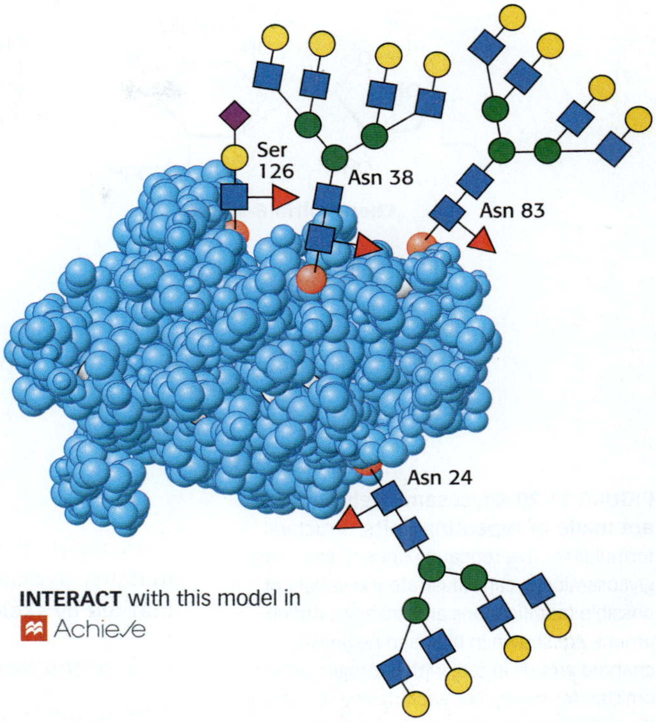

Glycosylation functions in nutrient sensing

An especially important glycosylation reaction is the covalent attachment of a single *N*-acetylglucosamine (GlcNAc) to serine or threonine residues of cytoplasmic, nuclear, and mitochondrial proteins. The reaction, one of the most common post-translational modifications, is catalyzed by *O*-GlcNAc transferase, a highly conserved enzyme found in all multicellular animals. The concentration of GlcNAc reflects the active metabolism of carbohydrates, amino acids, and fats, indicating that nutrients are abundant (**Figure 11.19**).

More than 4000 proteins are modified by GlcNAcylation, including transcription factors and components of signaling pathways. Interestingly, because the GlcNAcylation sites are also potential phosphorylation sites, *O*-GlcNAc transferase and protein kinases may be involved in cross talk to modulate one another's signaling activity. Like phosphorylation, GlcNAcylation is reversible, with another enzyme, called *GlcNAcase*, catalyzing the removal of the carbohydrate. Dysregulation of GlcNAc transferase has been linked to insulin resistance, diabetes, cancer, and neurological pathologies.

Proteoglycans have important structural roles

The glycosaminoglycan component of a proteoglycan makes up as much as 95% of the biomolecule by weight, and so a proteoglycan resembles a polysaccharide more than a protein. Proteoglycans not only function as lubricants and structural components in connective tissue but also mediate the adhesion of cells to the extracellular matrix and bind factors that regulate cell proliferation.

The properties of proteoglycans are determined primarily by the glycosaminoglycan component. Many glycosaminoglycans are made of repeating units of disaccharides containing a derivative of an amino sugar, either glucosamine or galactosamine (**Figure 11.20**). At least one of the two sugars in the repeating unit has a negatively charged carboxylate or sulfate group. The major glycosaminoglycans in animals are chondroitin sulfate, keratan sulfate, heparin, dermatan sulfate, and hyaluronate. Recall that heparin acts as an

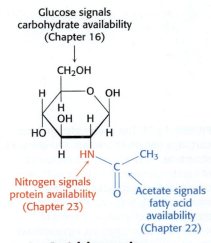

FIGURE 11.18 Erythropoietin has large oligosaccharides linked to three asparagine residues and one serine residue. The structures shown are approximately to scale. See Figure 11.17 for the carbohydrate key. [Drawn from 1BUY.pdb.]

INTERACT with this model in **Achieve**

Glucose signals carbohydrate availability (Chapter 16)

Nitrogen signals protein availability (Chapter 23)

Acetate signals fatty acid availability (Chapter 22)

β-D-Acetylglucosamine (GlcNAc)

FIGURE 11.19 Glycosylation can serve as a nutrient sensor. *N*-Acetylglucosamine is attached to proteins when nutrients are abundant.

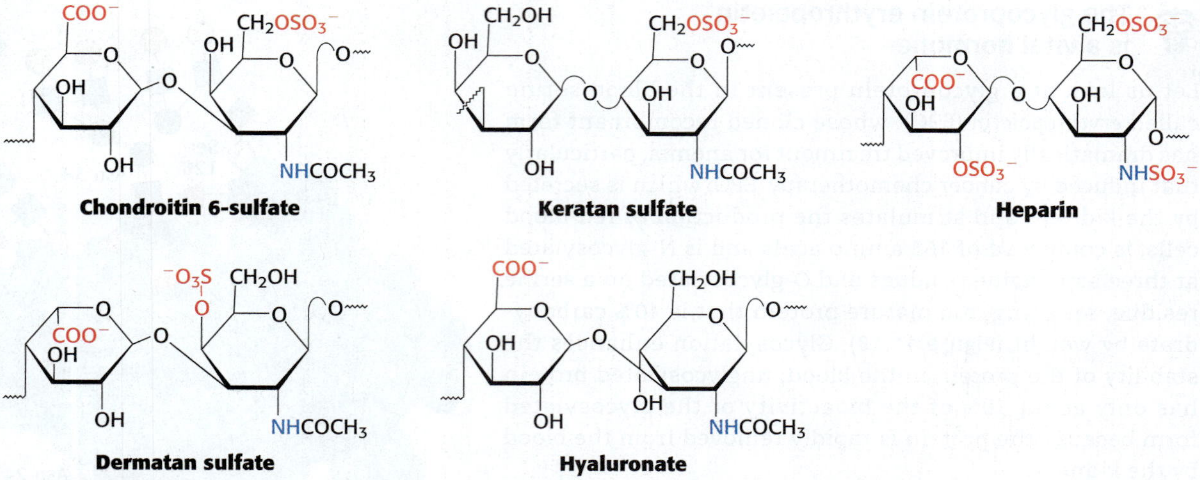

Chondroitin 6-sulfate

Keratan sulfate

Heparin

Dermatan sulfate

Hyaluronate

FIGURE 11.20 Glycosaminoglycans are made of repeating units. Structural formulas for five repeating units of important glycosaminoglycans illustrate the variety of possible modifications and linkages. Amino groups are shown in blue and negatively charged groups in red, with hydrogen atoms omitted for clarity. For each molecule shown, the right-hand hexose is a glucosamine derivative.

anticoagulant to assist the termination of blood clotting (Section 7.5). The inability to degrade glycosaminoglycans is the cause of a collection of diseases marked by skeletal deformities and reduced life expectancies.

Proteoglycans are important components of cartilage

Among the best-characterized proteoglycans are those found in the extracellular matrix of cartilage, where they complement the functions of proteins. For example, the triple helix of collagen protein (Section 2.5) provides structure and tensile strength, whereas the proteoglycan aggrecan serves as a shock absorber. The protein component of aggrecan is a large molecule composed of 2397 amino acids with three globular domains (G1, G2, and G3), and the site of glycosaminoglycan attachment is in the extended region between G2 and G3 (**Figure 11.21**). This linear region contains highly repetitive amino acid sequences, which are sites for the attachment of keratan sulfate and chondroitin sulfate. Many molecules of aggrecan are in turn noncovalently bound through the first globular domain to a very long filament formed by linking together molecules of the sugar component of aggrecan, the glycosaminoglycan hyaluronate (Figure 11.21). In addition, water is bound to the glycosaminoglycans, attracted by the many negative charges.

FIGURE 11.21 The proteoglycan from cartilage has an enormous and complex structure. (A) In this electron micrograph of a proteoglycan from cartilage (with false color added), proteoglycan monomers emerge laterally at regular intervals from opposite sides of a central filament of hyaluronate. (B) A schematic representation. The G1 globular domains of multiple aggrecan proteins bind noncovalently to a central polymer of hyaluronate. Multiple polymers of keratan sulfate and chondroitin sulfate glycosaminoglycans are attached to the extended region between globular domains G2 and G3 of each aggrecan molecule. [Photo credit: Courtesy of Dr. Lawrence Rosenberg and Joseph A. Buckwalter, See: J.A. Buckwalter and L. Rosenberg. "Structural Changes during Development in Bovine Fetal Epiphyseal Cartilage." *Collagen and Related Research*, 3(1983): 489–504 ©1983, with permission from Elsevier.]

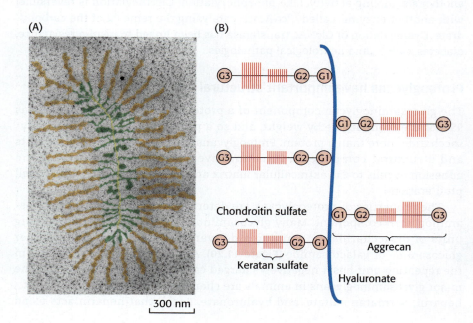

(A)

(B)

Chondroitin sulfate

Keratan sulfate

Aggrecan

Hyaluronate

300 nm

Aggrecan can cushion compressive forces acting on cartilage because the adsorbed water enables it to spring back after having been deformed. When pressure is exerted, as when the foot hits the ground while walking, water is squeezed from the glycosaminoglycan, cushioning the impact. When the pressure is released, the water rebinds. Osteoarthritis, the most common form of arthritis, results when water is lost from proteoglycan with aging. Other forms of arthritis can result from the proteolytic degradation of aggrecan and collagen in the cartilage.

Mucins are glycoprotein components of mucus

Mucins form large polymeric structures that serve as lubricants in mucous and salivary secretions. These glycoproteins are synthesized by specialized cells in the tracheobronchial, gastrointestinal, and genitourinary tracts.

A model of a mucin is shown in **Figure 11.22A**. The defining feature of the mucins is a region of the protein backbone termed the *variable number of tandem repeats* (VNTR) region, which is rich in serine and threonine residues that are O-glycosylated. Indeed, the carbohydrate moiety can account for as much as 80% of the molecule by weight. A number of core carbohydrate structures are conjugated to the protein component of mucin; **Figure 11.22B** shows one example.

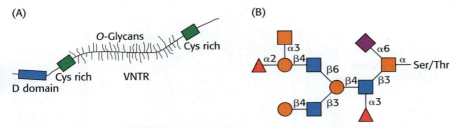

FIGURE 11.22 Mucins have common regions that are heavily *O*-glycosylated. (A) A schematic representation of a mucoprotein. The VNTR region is highly glycosylated, forcing the molecule into an extended conformation. The Cys-rich domains and the D domain facilitate the polymerization of many such molecules. (B) An example of an oligosaccharide that is bound to the VNTR region of the protein. See Figure 11.17 for the carbohydrate key. [Data from A. Varki et al. (Eds.), *Essentials of Glycobiology*, 2nd ed. (Cold Spring Harbor Press, 2009), pp. 117, 118.]

Mucins adhere to epithelial cells and act as a protective barrier; they also hydrate the underlying cells. In addition to protecting cells from environmental insults, such as stomach acid, inhaled chemicals in the lungs, and bacterial infections, mucins have roles in fertilization, the immune response, and cell adhesion. Mucins are overexpressed in bronchitis and cystic fibrosis, and the overexpression of mucins is characteristic of adenocarcinomas—cancers of the glandular cells of epithelial origin.

Protein glycosylation takes place in the lumen of the endoplasmic reticulum and in the Golgi complex

The major pathway for protein glycosylation takes place inside the lumen of the **endoplasmic reticulum (ER)** and in the **Golgi complex**, organelles that play central roles in protein trafficking (**Figure 11.23**). The protein is synthesized by ribosomes attached to the cytoplasmic face of the ER membrane, and the peptide chain is inserted into the lumen of the ER (Section 30.4). The N-linked glycosylation begins in the ER and continues in the Golgi complex, whereas the O-linked glycosylation takes place exclusively in the Golgi complex.

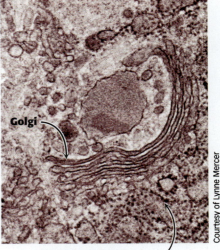

Courtesy of Lynne Mercer

FIGURE 11.23 The Golgi complex and endoplasmic reticulum are the sites of protein glycosylation. The electron micrograph shows the Golgi complex and adjacent endoplasmic reticulum. The black dots on the cytoplasmic surface of the ER membrane are ribosomes.

A large oligosaccharide destined for attachment to the asparagine residue of a protein is assembled on a specialized lipid molecule called *dolichol phosphate*, located in the ER membrane and containing about 20 isoprene (C_5) units.

Isoprene

$n = 15–19$

Dolichol phosphate

The terminal phosphate group of the dolichol phosphate is the site of attachment of the oligosaccharide. This activated (energy-rich) form of the oligosaccharide is subsequently transferred to a specific asparagine residue of the growing polypeptide chain by an enzyme located on the lumenal (interior) side of the ER.

Proteins in the lumen of the ER and in the ER membrane are transported to the Golgi complex, which is a stack of flattened membranous sacs that are the major sorting center of the cell. The O-linked sugar units are fashioned there, and the N-linked sugars, arriving from the ER as a component of a glycoprotein, are modified in many different ways. Proteins proceed from the Golgi complex to lysosomes, secretory granules, or the plasma membrane, according to signals encoded within their amino acid sequences and three-dimensional structures **(Figure 11.24)**.

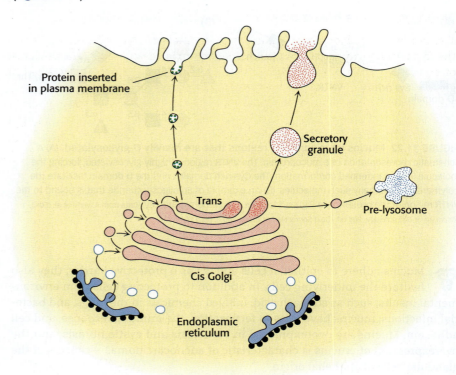

FIGURE 11.24 The Golgi complex is a sorting center. The Golgi complex is the sorting center in the targeting of proteins to lysosomes, secretory vesicles, and the plasma membrane. The cis face of the Golgi complex receives vesicles from the endoplasmic reticulum, and the trans face sends a different set of vesicles to target sites. Vesicles also transfer proteins from one compartment of the Golgi complex to another. [Courtesy of Dr. Marilyn Farquhar.]

Specific enzymes are responsible for oligosaccharide assembly

Complex carbohydrates are synthesized through the action of specific enzymes called **glycosyltransferases**, which catalyze the formation of glycosidic linkages. Given the diversity of known glycosidic linkages, many different enzymes are required. Indeed, glycosyltransferases account for 1% to 2% of gene products in all organisms examined.

While dolichol phosphate-linked oligosaccharides are substrates for some glycosyltransferases, the most common carbohydrate donors for

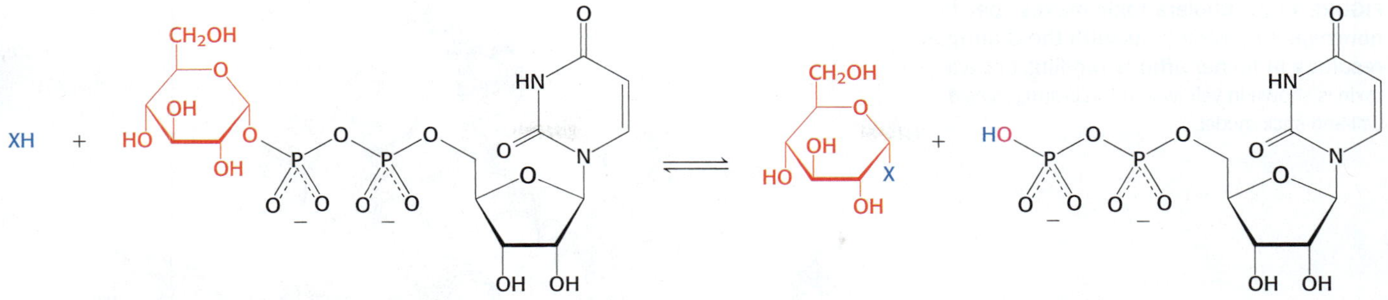

UDP-glucose

UDP

FIGURE 11.25 Glycosyltransferase reactions follow a general form. The sugar to be added comes from a sugar nucleotide—in this case, UDP-glucose. The acceptor, designated X in this illustration, can be one of a variety of biomolecules, including other carbohydrates or proteins.

glycosyltransferases are activated sugar nucleotides, such as UDP-glucose (UDP is the abbreviation for uridine diphosphate; **Figure 11.25**). The attachment of a nucleotide to enhance the energy content of a molecule is a common strategy in biosynthesis that we will see many times in our study of biochemistry. The acceptor substrates for glycosyltransferases are quite varied and include carbohydrates, serine, threonine, and asparagine residues of proteins, lipids, and even nucleic acids.

Blood groups are based on protein glycosylation patterns

The human ABO blood groups illustrate the effects of glycosyltransferases on the formation of glycoproteins. Each blood group is designated by the presence of one of the three different carbohydrates—termed A, B, or O—attached to glycoproteins and glycolipids on the surfaces of red blood cells (**Figure 11.26**). These structures have in common an oligosaccharide foundation called the O (or sometimes H) antigen. The A and B antigens differ from the O antigen by the addition of one extra monosaccharide, either N-acetylgalactosamine (for A) or galactose (for B) through an α-1,3 linkage to a galactose moiety of the O antigen.

Specific glycosyltransferases add the extra monosaccharide to the O antigen. Each person inherits the gene for one glycosyltransferase of this type from each parent. The type A transferase specifically adds N-acetylgalactosamine, whereas the type B transferase adds galactose. These enzymes are identical in all but 4 of 354 positions. In the AB blood group, both enzymes are present, whereas the O phenotype results if both enzymes are absent.

These structures have important implications for blood transfusions and other transplantation procedures. If an antigen not normally present in a person is introduced, the person's immune system recognizes it as foreign. Red-blood-cell lysis occurs rapidly, leading to a severe drop in blood pressure (hypotension), shock, kidney failure, and death from circulatory collapse.

Why are different blood types present in the human population? Suppose that a pathogenic organism such as a parasite expresses on its cell surface a carbohydrate antigen similar to one of the blood-group antigens. Since this antigen is not detected as foreign in a person whose blood type matches the parasite antigen, the parasite may flourish. However, other people with different blood types will be protected. Hence, there will be selective pressure on human beings to vary blood type to prevent parasitic mimicry and a corresponding selective pressure on parasites to enhance mimicry. The constant "arms race" between pathogenic microorganisms and human beings drives the evolution of diversity of surface antigens within the human population.

FIGURE 11.26 The A, B, and O oligosaccharide antigens share a common core structure, the O antigen. Different glycosylation patterns account for the different human blood groups.

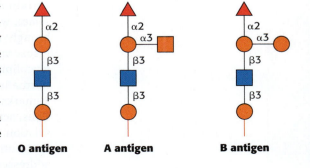

O antigen   **A antigen**   **B antigen**

FIGURE 11.27 Cholera toxin makes specific noncovalent interactions with the O antigen, resulting in higher affinity binding. Cholera toxin is shown in yellow and the O antigen as a ball-and-stick model.

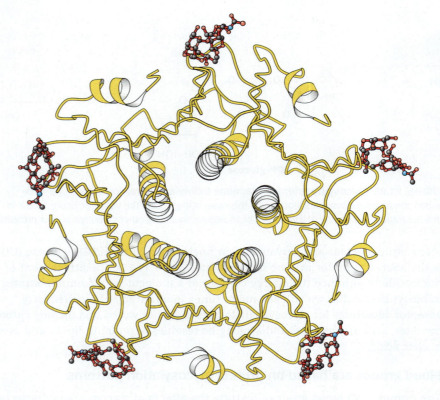

One example of a disease affected by blood type is cholera. Millions of cases of this disease, caused by the organism *Vibrio cholerae*, occur around the world each year, leading to tens of thousands of deaths. Blood type does not appear to influence the likelihood of contracting cholera, but individuals with blood type O are approximately eight times more likely to have severe disease. Biochemical studies show that the O antigen binds more tightly to the toxin responsible for disease symptoms than do the other blood type antigens (**Figure 11.27**). The impact of these interactions on human society is revealed by the observation that the frequency of blood type O is notably low in regions where cholera has been endemic for centuries.

Errors in glycosylation can result in pathological conditions

Protein function is often compromised when proteins are improperly glycosylated, sometimes resulting in disease. For instance, certain types of muscular dystrophy can be traced to improper glycosylation of dystroglycan, a membrane protein that links the extracellular matrix with the cytoskeleton. Indeed, an entire family of severe inherited human diseases called *congenital disorders of glycosylation* has been identified. These pathological conditions reveal the importance of proper modification of proteins by carbohydrates and their derivatives.

An especially clear example of the role of glycosylation is provided by I-cell disease, a lysosomal storage disease. Normally, a carbohydrate marker directs certain digestive enzymes from the Golgi complex to organelles called *lysosomes*, which degrade and recycle damaged cellular components or material brought into the cell by endocytosis. In patients with I-cell disease, affected lysosomes contain large inclusions of undigested glycosaminoglycans and glycolipids—hence the "I" in the name of the disease—because the enzymes normally responsible for the degradation of glycosaminoglycans are missing. Remarkably, the enzymes are present at very high levels in the blood and urine. Thus, active enzymes are synthesized, but, in the absence of appropriate glycosylation, they are exported instead of being sequestered in lysosomes.

In other words, in I-cell disease, a whole series of enzymes are incorrectly addressed and delivered to the wrong location. Normally, these enzymes contain a mannose 6-phosphate residue as a component of an N-oligosaccharide that serves as the marker directing the enzymes from the Golgi complex to lysosomes. In I-cell disease, however, the attached mannose lacks a phosphate. I-cell patients are deficient in the N-acetylglucosamine phosphotransferase catalyzing the first step in the addition of the phosphoryl group; the consequence is the mistargeting of eight essential enzymes (**Figure 11.28**). I-cell disease causes the patient to suffer severe psychomotor impairment and skeletal deformities. Interestingly, mutations in the phosphotransferase have also been linked to stuttering. Why some mutations cause stuttering while others cause I-cell disease is a mystery.

Biochemists use several techniques to analyze the oligosaccharide components of glycoproteins

How is it possible to determine the structure of a glycoprotein—the oligosaccharide structures and their points of attachment? Most approaches make use of enzymes that cleave oligosaccharides at specific types of linkages.

The first step is to detach the oligosaccharide from the protein. For example, N-linked oligosaccharides can be released from proteins by an enzyme such as peptide N-glycosidase F, which cleaves the N-glycosidic linkages connecting the oligosaccharide to the protein. The oligosaccharides can then be isolated and analyzed.

Matrix-assisted laser desorption/ionization/time-of-flight (MALDI-TOF) or other mass spectrometric techniques (Section 4.3) provide the mass of an oligosaccharide fragment; however, many possible oligosaccharide structures are consistent with a given mass. More complete information can be obtained by cleaving the oligosaccharide with enzymes of varying specificities. For example, β-1,4-galactosidase cleaves β-glycosidic linkages exclusively at galactose residues. The products can again be analyzed by mass spectrometry (**Figure 11.29**).

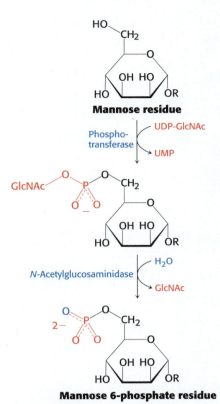

FIGURE 11.28 In a two-step process, a glycoprotein destined for delivery to lysosomes acquires a mannose 6-phosphate marker in the Golgi compartment. First, GlcNAc phosphotransferase adds a phospho-N-acetylglucosamine unit to the 6-OH group of a mannose; second, an N-acetylglucosaminidase removes the added sugar to generate a mannose 6-phosphate residue in the core oligosaccharide.

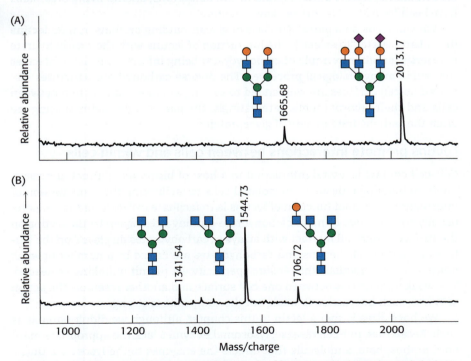

FIGURE 11.29 Oligosaccharides can be characterized by mass spectrometry. Carbohydrate-cleaving enzymes were used to release and specifically cleave the oligosaccharide component of the glycoprotein fetuin. Parts A and B show the masses obtained with MALDI-TOF spectrometry as well as the corresponding structures of the oligosaccharide-digestion products: (A) digestion with peptide N-glycosidase F (to release the oligosaccharide from the protein) and neuraminidase (which cleaves at sialic acid residues); (B) digestion with peptide N-glycosidase F, neuraminidase, and β-1,4-galactosidase. Knowledge of the enzyme specificities and the masses of the products permits the characterization of the oligosaccharide. See Figure 11.17 for the carbohydrate key. [Data from A. Varki, R. D. Cummings, J. D. Esko, H. H. Freeze, G. W. Hart, and J. Marth (Eds.), *Essentials of Glycobiology* (Cold Spring Harbor Laboratory Press, 1999), p. 596.]

The repetition of this process with the use of an array of enzymes of different specificity will eventually reveal the structure of the oligosaccharide.

Biochemists can also locate the points of oligosaccharide attachment by applying proteases to glycoproteins. Cleavage by a specific protease yields a characteristic pattern of peptide fragments that can be analyzed chromatographically. Fragments attached to oligosaccharides can be picked out because their chromatographic properties will change on glycosidase treatment. Mass spectrometric analysis or direct peptide sequencing can reveal the identity of the peptide in question and, with additional effort, the exact site of oligosaccharide attachment.

❖ SELF–CHECK QUESTION

Just as all the proteins made by an organism at any given moment are called the *proteome*, all the carbohydrates and carbohydrate-associated molecules can be called the *glycome*. Compare the amount of information inherent in the genome, the proteome, and the glycome.

11.4 Lectins Are Specific Carbohydrate-Binding Proteins

The diversity and complexity of the carbohydrate units and the variety of ways in which they can be joined in oligosaccharides and polysaccharides attest to their biochemical importance. Evolution rarely results in complex patterns when simpler ones are equally effective, so why all this intricacy and diversity? It is now clear that these carbohydrate structures are the recognition sites for a distinct class of proteins. Such proteins, termed *glycan-binding proteins*, bind specific carbohydrate structures on neighboring cell surfaces. Originally discovered in plants, glycan-binding proteins are ubiquitous, and no living organisms have been found that lack these key proteins.

We will focus on a particular class of glycan-binding proteins termed **lectins** (from Latin *legere*, "to select"). The interaction of lectins with their carbohydrate partners is another example of carbohydrates being information-rich molecules that guide many biological processes. The diverse carbohydrate structures displayed on cell surfaces are well-suited to serving as sites of interaction between cells and their environments. Interestingly, the partners for lectin binding are often the carbohydrate moiety of glycoproteins.

Lectins promote interactions between cells and within cells

Cell–cell contact is a vital interaction in a host of biochemical functions, ranging from building a tissue from isolated cells to facilitating the transmission of information. The chief function of lectins is to facilitate cell–cell contact. A lectin usually contains two or more carbohydrate-binding sites, whereby the lectins on the surface of one cell interact with arrays of carbohydrates displayed on the surface of another cell. Lectins and carbohydrates are linked by a number of weak noncovalent interactions that ensure specificity yet permit unlinking as needed. The weak interactions between one cell surface and another resemble the action of Velcro; each interaction is weak, but the composite is strong.

We have already met a lectin in this chapter, although we didn't name it as such. Recall that, in I-cell disease, lysosomal enzymes lack the appropriate mannose 6-phosphate, a molecule that directs the enzymes to the lysosome. Under normal circumstances, the mannose 6-phosphate receptor, a lectin, binds the enzymes in the Golgi apparatus and directs them to the lysosome. Not surprisingly, I-cell disease can also be caused by a loss of the mannose 6-phosphate receptor.

Lectins are organized into two large classes

Lectins can be divided into two large classes on the basis of their amino acid sequences and biochemical properties.

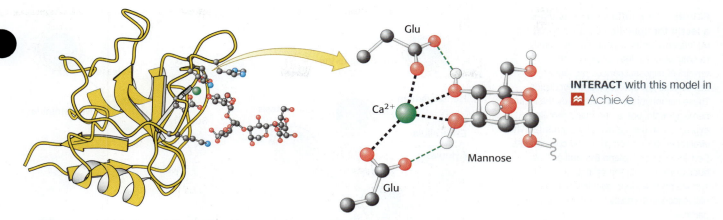

FIGURE 11.30 C-type lectins use calcium ions to bind carbohydrates. Selected interactions of the carbohydrate-binding domain of an animal C-type lectin are shown, with some hydrogen atoms omitted for clarity. A calcium ion tightly bound to two Glu residues of the lectin binds to a mannose residue of the carbohydrate moiety. [Drawn from 2MSB.pdb.]

C-type lectins.

C-type (for <u>c</u>alcium-requiring) lectins are found in animals. These proteins each have a homologous domain of 120 amino acids that is responsible for carbohydrate binding. The structure of one such domain bound to a carbohydrate target is shown in **Figure 11.30**.

A calcium ion on the protein acts as a bridge between the protein and the sugar through direct interactions with sugar OH groups. In addition, two glutamate residues in the protein bind to both the calcium ion and the sugar, and other protein side chains form hydrogen bonds with other OH groups on the carbohydrate. The carbohydrate-binding specificity of a particular lectin is determined by the amino acid residues that bind the carbohydrate. C-type lectins function in a variety of cellular activities, including receptor-mediated endocytosis—a process by which soluble molecules are bound to the cell surface and subsequently internalized (Section 27.3)—and cell–cell recognition.

C-type proteins known as **selectins** bind immune-system cells to sites of injury in the inflammatory response. The L, E, and P forms of selectins bind specifically to carbohydrates on <u>l</u>ymph-node vessels, <u>e</u>ndothelium, or activated blood <u>p</u>latelets, respectively. Selectins play central roles in recruiting leukocytes (white blood cells) to sites of inflammation. By participating in a large set of interactions, each of which is relatively weak, selectins can bring cells together where other molecules of the surfaces of the two cells are positioned to interact if they have appropriate affinities for one another.

L-type lectins.

The other large class of lectins comprises the L-type lectins. These lectins are especially rich in the seeds of leguminous plants such as beans and peas; many of the initial biochemical characterizations of lectins were performed on these particular lectins because they are so readily available. Although the exact role of lectins in plants is unclear, they can serve as potent toxins to herbivorous insects. Other L-type lectins, such as calnexin and calreticulin, are prominent chaperones in the eukaryotic endoplasmic reticulum. Recall that chaperones are proteins that facilitate the folding of other proteins (Section 2.7).

Influenza virus binds to sialic acid residues

Many pathogens gain entry into specific host cells by adhering to cell-surface carbohydrates. For example, influenza virus recognizes sialic acid residues (Figure 11.10) linked to galactose residues that are present on cell-surface glycoproteins. The viral protein that binds to these sugars is a lectin called *hemagglutinin* (**Figure 11.31A**).

CAROLYN BERTOZZI Chemist Carolyn Bertozzi is a leader in developing the field of glycobiology and communicating its exciting possibilities. Initially planning to go to medical school, her interest in chemistry developed in college, and she went on to complete a PhD focused on synthesizing carbohydrate-containing molecules and using them as probes for biological systems. After postdoctoral training, she joined the faculty at the University of California, Berkeley, where her training as a synthetic chemist allowed her to think about how to make new molecules with desired properties. She and her coworkers invented new tools for probing the roles of carbohydrates in biology, including those that allow the synthesis of new molecules in living organisms. Her talent was recognized with a MacArthur "genius" grant early in her career and with a Nobel Prize in Chemistry in 2022. The Prize honored her role in developing a suite of reactions that can be introduced and monitored in living cells, allowing deep insight into cellular processes. She now teaches and does research at Stanford University.

FIGURE 11.31 Influenza virus uses a lectin for specific cell binding. (A) Influenza virus targets cells by binding to sialic acid residues located at the termini of oligosaccharides present on cell-surface glycoproteins and glycolipids. These carbohydrates are bound by the lectin hemagglutinin, one of the major proteins expressed on the surface of the virus. (B) When viral replication is complete, the components (red lines) assemble and the viral particle buds from the cell. The other major viral-surface protein, neuraminidase, cleaves oligosaccharide chains to release the viral particle.

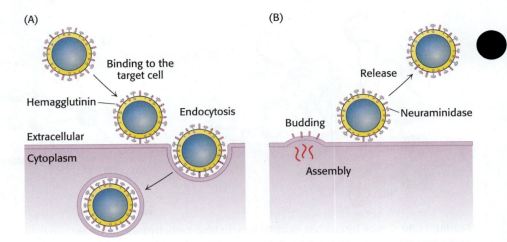

After binding hemagglutinin, the virus is engulfed by the cell and begins to replicate. Viral assembly concludes with the budding of the viral particle from the cell (**Figure 11.31B**). Upon complete assembly, the viral particle is still attached to sialic acid residues of the cell membrane by hemagglutinin on the surface of the new virions. Another viral protein, neuraminidase (sialidase), cleaves the glycosidic linkages between the sialic acid residues and the rest of the cellular glycoprotein, freeing the virus to infect new cells, and thus spreading the infection throughout the respiratory tract. Inhibitors of this enzyme such as oseltamivir (Tamiflu) and zanamivir (Relenza) are important anti-influenza agents.

Viral hemagglutinin's carbohydrate-binding specificity may play an important role in species specificity of infection and ease of transmission. For instance, avian influenza H5N1 (bird flu) is especially lethal and is readily spread between birds. Although human beings can be infected by this virus, infection is rare and human-to-human transmission rarer still. The biochemical basis of these characteristics is that the avian virus hemagglutinin recognizes a different carbohydrate moiety from that recognized by human influenza hemagglutinin. Although human beings have carbohydrate residues that the avian virus might recognize and bind, they are located deep in the lungs. Infection by the avian virus is thus difficult, and, when it does occur, the avian virus is not readily transmitted by sneezing or coughing.

Summary

11.1 Monosaccharides Are the Simplest Carbohydrates

- Carbohydrates are aldoses or ketoses that are rich in hydroxyl groups, and their derivatives. The simplest are monosaccharides—sugars that are three to seven carbons in length.

- Pentoses and hexoses can close into five-membered furanose rings or six-membered pyranose rings. Pyranose rings usually adopt the chair conformation, and furanose rings usually adopt the envelope conformation. These ring structures are in rapid equilibrium with the linear forms in aqueous solutions.

- In sugars undergoing cyclization, an additional asymmetric center is formed, resulting in α and β anomer forms that differ by the position of

the hydroxyl group on the anomeric carbon. In monosaccharides, these anomers rapidly interchange via the linear form.

- D-Glucose is the most common sugar in living organisms and is the human "blood sugar" most likely because it can form a six-membered ring with all bulky groups in the equatorial positions. As a reducing sugar, it can nonenzymatically react with hemoglobin to form glycated hemoglobin, an important biomarker for diabetes.

- Sugars are joined to alcohols and amines by glycosidic linkages from the anomeric carbon atom. For example, N-glycosidic linkages connect sugars to purines and pyrimidines in nucleotides, RNA, and DNA.

11.2 Monosaccharides Are Linked to Form Complex Carbohydrates

- Sugars are linked to one another in disaccharides and polysaccharides by O-glycosidic linkages.
- Sucrose, lactose, and maltose are common disaccharides.
- Sucrose (table sugar) consists of α-glucose and β-fructose joined by a glycosidic linkage between their anomeric carbon atoms.
- Lactose (in milk) consists of galactose joined to glucose by a β-1,4 linkage.
- Maltose (from starch) consists of two glucoses joined by an α-1,4 linkage.
- Starch is a storage homopolymer of glucose in plants; glycogen serves a similar role in animals. Glucose units in starch and glycogen are in α-1,4 linkages with α-1,6 linked branches.
- Cellulose, the major structural polymer of plant cell walls, consists of glucose units joined by β-1,4 linkages, which give rise to long straight chains that form fibrils with high tensile strength.
- Chitin is the major structural polymer of fungal cell walls and the exoskeletons and shells of arthropods. Like cellulose, it is a homopolymer composed of β-1,4 linkages, but the monomeric unit is GlcNAc rather than glucose.

11.3 Carbohydrates Can Be Linked to Proteins to Form Glycoproteins

- Carbohydrates are commonly conjugated to proteins; in glycoproteins, the protein component is predominant.

- Most secreted proteins, such as the signal molecule erythropoietin, are glycoproteins that are prominent on the external surface of the plasma membrane.
- Glycosaminoglycans are polymers of repeating disaccharides with a derivative of glucosamine or galactosamine and a high density of carboxylate or sulfate groups.
- Proteoglycans—proteins bearing covalently linked glycosaminoglycans—are found in the extracellular matrices of animals and are key components of cartilage.
- Mucoproteins, like proteoglycans, serve as lubricants and are predominantly carbohydrate by weight, with the protein component heavily O-glycosylated with N-acetylgalactosamine joining the oligosaccharide to the protein.
- Glycosyltransferases link the oligosaccharide units on proteins in the lumen of the endoplasmic reticulum. Additional sugars are attached in the Golgi complex to form diverse patterns.

11.4 Lectins Are Specific Carbohydrate-Binding Proteins

- Carbohydrates on cell surfaces are recognized by proteins called lectins. In animals, the interplay of lectins and their sugar targets guides cell–cell contact as well as viral entry.
- Lectins also play a role inside the cell, such as directing proteins to their proper cellular locations and assisting in protein folding.

Key Terms

monosaccharide (p. 325)
epimers (p. 327)
anomers (p. 328)
reducing sugar (p. 331)
glycosidic linkage (p. 332)
oligosaccharide (p. 333)
disaccharide (p. 333)

polysaccharide (p. 335)
glycogen (p. 335)
starch (p. 335)
cellulose (p. 335)
glycoprotein (p. 337)
proteoglycan (p. 338)
glycosaminoglycan (p. 338)

mucin (mucoprotein) (p. 338)
endoplasmic reticulum (ER)
  (p. 341)
Golgi complex (p. 341)
glycosyltransferase (p. 342)
lectin (p. 346)
selectin (p. 347)

Problems

1. Which of the following are aldoses? ❖ 1

(a) glyceraldehyde   (c) fucose

(b) glucose   (d) fructose

2. Which of the following are reducing sugars? ❖ 1

(a) sucrose   (c) fructose   (e) lactose

(b) ribose   (d) glucose

3. Which of the following is found in glycogen and starch? ❖ 2

(a) fructose

(b) β-D-glucose

(c) mannose

(d) α-D-glucose

4. How many different oligosaccharides can be made by linking one glucose, one mannose, and one galactose? Assume that each sugar is in its pyranose form. Compare this number with the number of tripeptides that can be made from three different amino acids. ❖ 2

5. Match each term with its description. ❖ 1, ❖ 2

(a) Enantiomers
(b) Cellulose
(c) Lectins
(d) Glycosyltransferases
(e) Epimers
(f) Starch
(g) Carbohydrates
(h) Proteoglycan
(i) Mucoprotein
(j) Glycogen

1. Has the molecular formula $(CH_2O)_n$
2. Monosaccharides that differ at a single asymmetric carbon atom
3. The storage form of glucose in animals
4. The storage form of glucose in plants
5. Glycoprotein containing glycosaminoglycans
6. The most abundant organic molecule in the biosphere
7. N-Acetylgalactosamine is a key component of this glycoprotein
8. Carbohydrate-binding proteins
9. Enzymes that synthesize oligosaccharides
10. Stereoisomers that are mirror images of each other

6. Indicate whether each of the following pairs of sugars consists of anomers, epimers, or an aldose–ketose pair: ❖ 1

(a) D-glyceraldehyde and dihydroxyacetone
(b) D-glucose and D-mannose
(c) D-glucose and D-fructose
(d) α-D-glucose and β-D-glucose
(e) D-ribose and D-ribulose
(f) D-galactose and D-glucose

7. Analyze the pair of compounds. ❖ 1

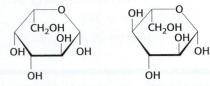

β-L-Galactopyranose   β-L-Glucopyranose

Which of the terms explains the relationship between the two compounds?

(a) Anomers
(b) Epimers
(c) Diastereomers
(d) Enantiomers

8. To which classes of sugars do the monosaccharides shown here belong? ❖ 1

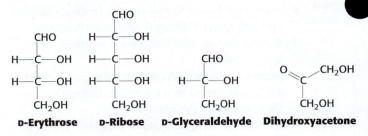

D-Erythrose   D-Ribose   D-Glyceraldehyde   Dihydroxyacetone

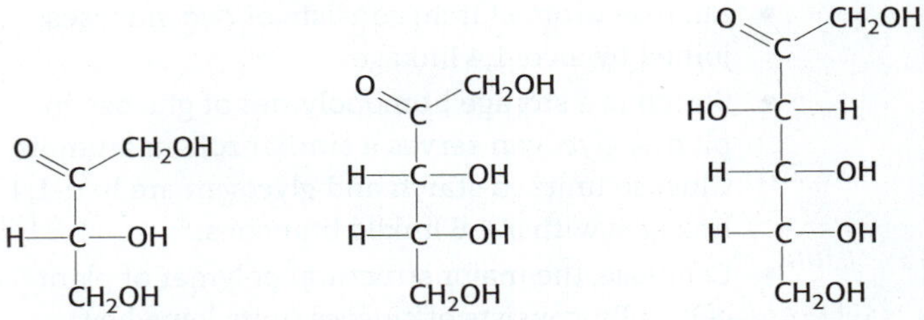

D-Erythrulose   D-Ribulose   D-Fructose

9. The specific rotations of the α and β anomers of D-glucose are +112 degrees and +18.7 degrees, respectively. Specific rotation, $[\alpha]_D$, is defined as the observed rotation of light of wavelength 589 nm (the D line of a sodium lamp) passing through 10 cm of a 1 g ml^{-1} solution of a sample. When a crystalline sample of α-D-glucose is dissolved in water, the specific rotation decreases from 112 degrees to an equilibrium value of 52.7 degrees. On the basis of this result, what are the proportions of the α and β anomers at equilibrium? Assume that the concentration of the open-chain form is negligible. ❖ 1

10. Raffinose is a trisaccharide and a minor constituent in sugar beets. ❖ 2

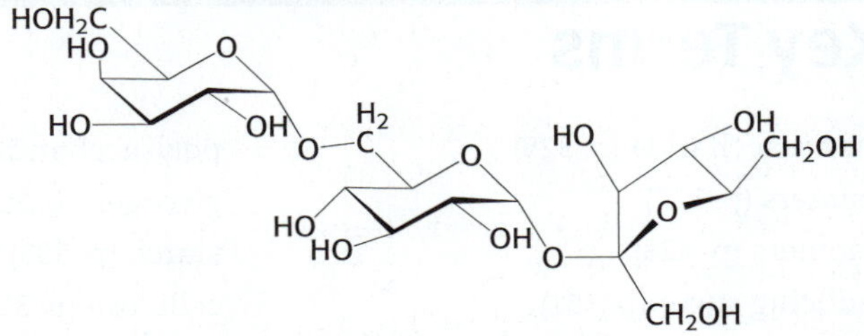

Raffinose

(a) Is raffinose a reducing sugar? Explain.
(b) What are the monosaccharides that compose raffinose?
(c) β-Galactosidase is an enzyme that will remove galactose residues from an oligosaccharide. What are the products of β-galactosidase treatment of raffinose?

11. (a) Compare the number of reducing ends to nonreducing ends in a molecule of glycogen.

(b) As we will see in Chapter 21, glycogen is an important fuel-storage form that is rapidly mobilized. At which end — the reducing or nonreducing — would you expect most metabolism to take place? ❖ 2

12. Glucose and fructose are reducing sugars. Sucrose, or table sugar, is a disaccharide consisting of both fructose and glucose. Is sucrose a reducing sugar? Which statements about reducing sugars are true? ❖ **1**, ❖ **2**

(a) A reducing sugar will not react with the Cu^{2+} in Fehling's reagent.

(b) D-Arabinose (an aldose) is a reducing sugar.

(c) The oxidation of a reducing sugar forms a carboxylic acid sugar.

(d) Reducing sugars contain ketone groups instead of aldehyde groups.

(e) A disaccharide with its anomeric carbons joined by the glycosidic linkage (like sucrose) cannot be a reducing sugar.

13. For each statement below, write which polymers apply from the following list: chitin, glycogen, cellulose, starch. ❖ **2**

(a) Provides structural support for plants

(b) Consists of N-acetylglucosamine monomers

(c) Is a storage form of carbohydrate in animal cells

(d) Is a homopolymer

(e) Is branched

14. List the key classes of glycoproteins, their defining characteristics, and their biological functions. ❖ **1**, ❖ **4**

15. Proteoglycans are part of the extracellular matrix; they provide structure, viscosity and lubrication, and adhesiveness. They are composed of proteins conjugated to carbohydrate components called glycosaminoglycans. The glycosaminoglycan component makes up the majority of the mass of a proteoglycan. Which statements about glycosaminoglycans are true? ❖ **3**, ❖ **4**

(a) β-D-Fructofuranose and amylose are possible components of glycosaminoglycans.

(b) Glycosaminoglycans are heteropolysaccharides composed of repeating disaccharide units.

(c) Chondroitin sulfate and heparin are examples of glycosaminoglycans.

(d) Glycosaminoglycans are homopolysaccharides composed of repeating glucosamine or galactosamine residues.

(e) Because glycosaminoglycans consist of only two residues, they generally have low molecular weights.

16. Which amino acids are used for the attachment of carbohydrates to proteins? ❖ **3**

17. Differentiate between a glycoprotein and a lectin. ❖ **4**, ❖ **5**

18. Suppose that a protein contains six potential N-linked glycosylation sites. How many possible proteins can be generated, depending on which of these sites is actually glycosylated? Do not include the effects of diversity within the carbohydrate added. ❖ **3**

19. How might the technique of affinity chromatography be used to purify lectins? ❖ **5**

20. A contributing factor to the development of arthritis is the inappropriate proteolytic destruction of the aggrecan component of cartilage by the proteolytic enzyme aggrecanase. The immune-system signal molecule interleukin 2 (IL-2) activates aggrecanase; in fact, IL-2 blockers are sometimes used to treat arthritis. Studies were undertaken to determine whether inhibitors of aggrecanase could counteract the effects of IL-2. Pieces of cartilage were incubated in media with various additions, and the amount of aggrecan destruction was measured as a function of time.

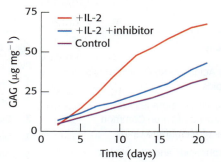

(a) Aggrecan degradation was measured by the release of glycosaminoglycan. What is the rationale for this assay?

(b) Why might glycosaminoglycan release not indicate aggrecan degradation?

(c) What is the purpose of the control (cartilage incubated with no additions)?

(d) What is the effect of adding IL-2 to the system?

(e) What is the response when an aggrecanase inhibitor is added in addition to IL-2?

(f) Why is there some aggrecan destruction in the control with the passage of time?

Lipids and Biological Membranes

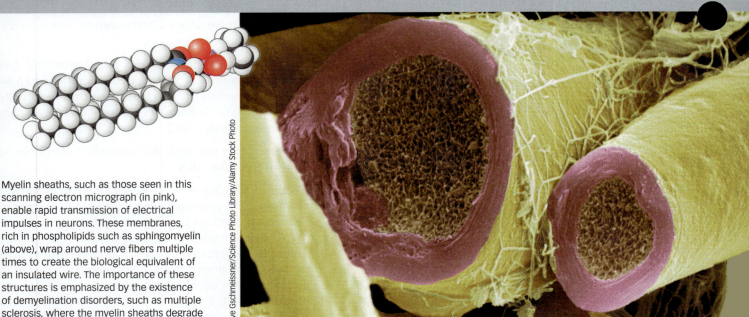

Myelin sheaths, such as those seen in this scanning electron micrograph (in pink), enable rapid transmission of electrical impulses in neurons. These membranes, rich in phospholipids such as sphingomyelin (above), wrap around nerve fibers multiple times to create the biological equivalent of an insulated wire. The importance of these structures is emphasized by the existence of demyelination disorders, such as multiple sclerosis, where the myelin sheaths degrade over time.

Steve Gschmeissner/Science Photo Library/Alamy Stock Photo

OUTLINE

12.1 Fatty Acids Are Key Constituents of Lipids

12.2 Biological Membranes Are Composed of Three Common Types of Membrane Lipids

12.3 Phospholipids and Glycolipids Readily Form Bimolecular Sheets in Aqueous Media

12.4 Proteins Carry Out Most Membrane Processes

12.5 Lipids and Many Membrane Proteins Diffuse Rapidly in the Plane of the Membrane

12.6 Prokaryotes and Eukaryotes Differ in Their Use of Biological Membranes

❖ LEARNING GOALS

By the end of this chapter, you should be able to:

1. Describe the structure of a fatty acid and explain how it is named.
2. Define the three most common types of membrane lipids.
3. Explain how membrane lipids self-assemble into bimolecular sheets.
4. Distinguish between integral and peripheral membrane proteins.
5. Describe how proteins interact with membranes and how membrane-spanning motifs within proteins may be identified.
6. Explain the properties of membrane fluidity and its dependence upon lipid composition.
7. Explain the importance of membranes in eukaryotic cells.

The boundaries of all cells are defined by biological membranes, dynamic structures in which proteins float in a sea of lipids. The lipid component prevents molecules generated inside the cell from leaking out and unwanted molecules from diffusing in, while the protein components act as transport systems that allow the cell to take up specific molecules and remove unwanted ones. In addition to an external cell membrane, eukaryotic cells also contain internal membranes that form the boundaries of organelles such as mitochondria, chloroplasts, peroxisomes, and lysosomes.

Biological membranes serve several additional functions indispensable for life, such as energy storage and information transduction. The proteins associated with the membrane define these functions for any given cell. In this chapter, we will examine the properties of membrane proteins that enable them to exist in the hydrophobic environment of the membrane while connecting two hydrophilic environments. In the next chapter, we will discuss the functions of these proteins.

12.1 Fatty Acids Are Key Constituents of Lipids

By definition, **lipids** are water-insoluble biomolecules that are highly soluble in organic solvents such as chloroform. The hydrophobic properties of lipids are essential to their ability to form membranes. Most lipids owe their hydrophobic properties to one component, their fatty acids.

Fatty acid names are based on their parent hydrocarbons

Let us first consider how fatty acids are named and distinguished from one another. **Fatty acids** are long hydrocarbon chains of various lengths and numbers of double bonds that terminate with carboxylic acid groups. The systematic name for a fatty acid is derived from the name of its parent hydrocarbon by the substitution of *oic* for the final *e*. For example, the C_{18} saturated fatty acid—that is, containing no double bonds—is called *octadecanoic acid* because the parent hydrocarbon is octadecane. A C_{18} fatty acid with one double bond is called *octadecenoic acid*; with two double bonds, *octadecadienoic acid*; and with three double bonds, *octadecatrienoic acid*. The notation 18:0 denotes a C_{18} fatty acid with no double bonds, whereas 18:2 signifies that there are two double bonds. The structures of the ionized forms of two common fatty acids—palmitic acid (16:0) and oleic acid (18:1)—are shown in **Figure 12.1**.

Fatty acid carbon atoms are numbered starting at the carboxyl terminus (**Figure 12.2**). Carbon atoms 2 and 3 are often referred to as α and β, respectively.

Palmitate (16:0)
(ionized form of palmitic acid)

Oleate (18:1)
(ionized form of oleic acid)

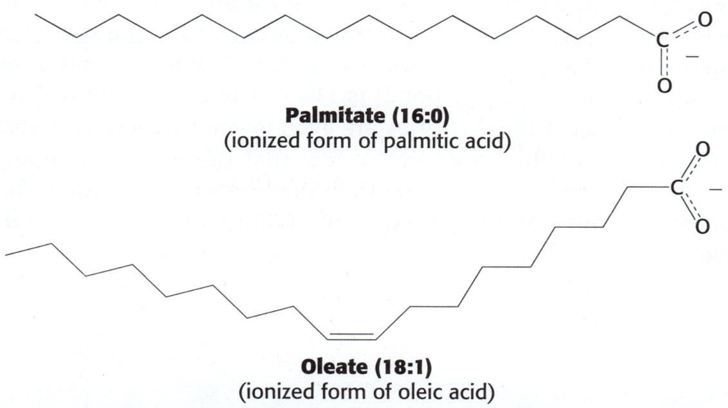

FIGURE 12.1 Fatty acids vary in their chain length and degree of saturation. For example, palmitate is a 16-carbon, saturated fatty acid (16:0), while oleate is an 18-carbon fatty acid with a single cis double bond (18:1).

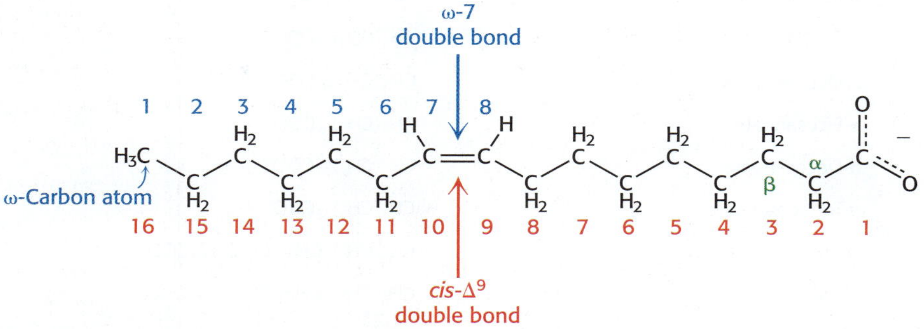

FIGURE 12.2 Fatty acid carbon atoms can be numbered in two ways. In one method (red), the carbon atoms are numbered starting with the carbonyl carbon. Carbon atoms numbered 2 and 3 are often referred to as α and β, respectively (green). The position of a double bond is indicated by the number of the carbon closest to carbon atom 1. Alternatively, carbons can be counted from the distal end, or the ω carbon (blue). For example, a double bond between the ω-7 and ω-8 carbons would be called a ω-7 double bond, as shown.

The methyl carbon atom at the distal end of the chain is called the *ω-carbon atom*. The position of a double bond is represented by the symbol Δ followed by a superscript number. For example, *cis*-Δ^9 means that there is a cis double bond between carbon atoms 9 and 10; *trans*-Δ^2 means that there is a trans double bond between carbon atoms 2 and 3. Alternatively, the position of a double bond can be denoted by counting from the distal end, with the ω-carbon atom (the methyl carbon) as number 1. Fatty acids are ionized at physiological pH, and so it is appropriate to refer to them according to their carboxylate form; for example, palmitate or hexadecanoate.

Chain length and degree of unsaturation affect fatty acid properties

Fatty acids in biological systems usually contain an even number of carbon atoms, typically between 14 and 24 (**Table 12.1**), with the 16- and 18-carbon fatty acids being the most common. The dominance of fatty acid chains containing an even number of carbon atoms reflects the manner in which fatty acids are biosynthesized (Chapter 27). The hydrocarbon chain is almost invariably unbranched in animal fatty acids. The alkyl chain may be saturated, or it may contain one or more double bonds. The configuration of the double bonds in most unsaturated fatty acids is cis. The double bonds in polyunsaturated fatty acids are separated by at least one methylene group.

The properties of fatty acids and of lipids derived from them are markedly dependent on chain length and degree of saturation. Unsaturated fatty acids have lower melting points than do saturated fatty acids of the same length. For example, the melting point of stearic acid is 69.6°C, whereas that of oleic acid (which contains one cis double bond) is 13.4°C. The melting points of polyunsaturated fatty acids of the C_{18} series are even lower. Chain length also affects the melting point, as illustrated by the fact that the melting temperature of palmitic acid (C_{16}) is 6.5 degrees lower than that of stearic acid (C_{18}). Thus, short chain length and unsaturation enhance the fluidity of fatty acids and of their derivatives.

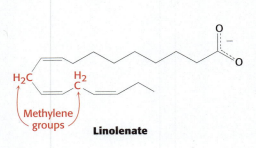

Methylene groups

Linolenate

TABLE 12.1 Some naturally occurring fatty acids in animals

Number of carbons	Number of double bonds	Common name	Systematic name	Formula
12	0	Laurate	*n*-Dodecanoate	$CH_3(CH_2)_{10}COO^-$
14	0	Myristate	*n*-Tetradecanoate	$CH_3(CH_2)_{12}COO^-$
16	0	Palmitate	*n*-Hexadecanoate	$CH_3(CH_2)_{14}COO^-$
18	0	Stearate	*n*-Octadecanoate	$CH_3(CH_2)_{16}COO^-$
20	0	Arachidate	*n*-Eicosanoate	$CH_3(CH_2)_{18}COO^-$
22	0	Behenate	*n*-Docosanoate	$CH_3(CH_2)_{20}COO^-$
24	0	Lignocerate	*n*-Tetracosanoate	$CH_3(CH_2)_{22}COO^-$
16	1	Palmitoleate	*cis*-Δ^9-Hexadecenoate	$CH_3(CH_2)_5CH{=}CH(CH_2)_7COO^-$
18	1	Oleate	*cis*-Δ^9-Octadecenoate	$CH_3(CH_2)_7CH{=}CH(CH_2)_7COO^-$
18	2	Linoleate	*cis, cis*-Δ^9, Δ^{12}-Octadecadienoate	$CH_3(CH_2)_4(CH{=}CHCH_2)_2(CH)_6COO^-$
18	3	Linolenate	all-*cis*-$\Delta^9, \Delta^{12}, \Delta^{15}$-Octadecatrienoate	$CH_3CH_2(CH{=}CHCH_2)_3(CH_2)_6COO^-$
20	4	Arachidonate	all-*cis* $\Delta^5, \Delta^8, \Delta^{11}, \Delta^{14}$-Eicosatetraenoate	$CH_3(CH_2)_4(CH{=}CHCH_2)_4(CH_2)_2COO^-$

12.2 Biological Membranes Are Composed of Three Common Types of Membrane Lipids

Lipids have a variety of biological roles: they serve as fuel molecules, highly concentrated energy stores, signal molecules and messengers in signal-transduction pathways, and components of membranes. The first three roles of lipids will be considered in later chapters. Here, we will explore the role of lipids as the essential component of biological membranes.

In order to appreciate the structural and chemical properties of membrane lipids, let us first consider the membranes themselves. While biological membranes are as diverse in structure as they are in function, they do have in common a number of important attributes:

- Membranes are sheetlike structures, only two molecules thick, that form closed boundaries between different compartments. The thickness of most membranes is between 60 Å (6 nm) and 100 Å (10 nm).

- Membranes consist mainly of lipids and proteins. The mass ratio of lipids to proteins ranges from 1:4 to 4:1. Membranes also contain carbohydrates that are linked to lipids and proteins.

- Membrane lipids are small molecules that have both hydrophilic and hydrophobic moieties. These lipids spontaneously form closed bimolecular sheets in aqueous media. These sheets, called *lipid bilayers*, are barriers to the flow of polar molecules.

- Specific proteins mediate distinctive functions of membranes. Proteins serve as pumps, channels, receptors, energy transducers, and enzymes. Membrane proteins are embedded in lipid bilayers, which create suitable environments for their action.

- Membranes are noncovalent assemblies. The constituent protein and lipid molecules are held together by many noncovalent interactions, which act cooperatively.

- Membranes are asymmetric. The two faces of biological membranes always differ from each other.

- Membranes are fluid structures. Lipid molecules diffuse rapidly in the plane of the membrane, as do proteins, unless they are anchored by specific interactions. In contrast, lipid molecules and proteins do not readily rotate across the membrane. Membranes can be regarded as two-dimensional solutions of oriented proteins and lipids.

- Most cell membranes are electrically polarized, such that the inside is negatively charged. Membrane potential plays a key role in transport, energy conversion, and excitability (Chapter 13).

Here, our focus is on the principal lipids found in the membranes of eukaryotes and bacteria: *phospholipids*, *glycolipids*, and *cholesterol*. While the membrane lipids of archaea are distinct, they have many features related to membrane formation in common with lipids of other organisms.

Phospholipids are the major class of membrane lipids

Phospholipids are abundant in all biological membranes. A **phospholipid** molecule is constructed from four components (**Figure 12.3**):

- One or more fatty acids
- A platform to which the fatty acids are attached
- A phosphate
- An alcohol attached to the phosphate

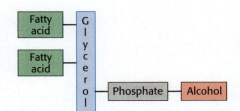

FIGURE 12.3 A phospholipid molecule contains four components. These components are shown here in schematic form: (1) fatty acids (green), (2) a platform for fatty acid attachment (here, glycerol, shown in blue), (3) a phosphate (gray), and (4) an alcohol (red).

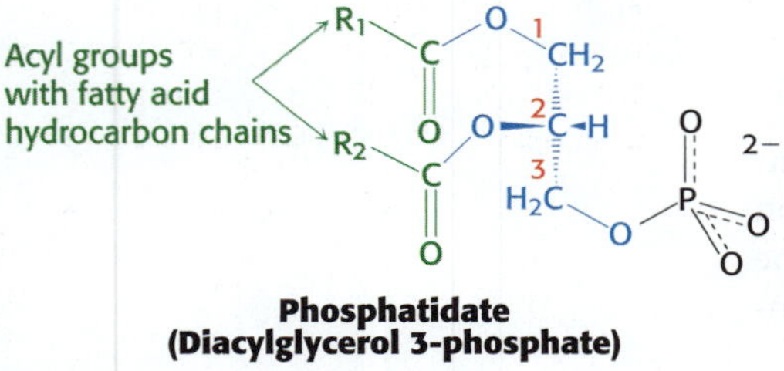

**Phosphatidate
(Diacylglycerol 3-phosphate)**

Acyl groups with fatty acid hydrocarbon chains

FIGURE 12.4 Phosphatidate, the simplest phosphoglyceride, does not contain an alcohol component. The absolute configuration of the center carbon (C-2) is shown.

The fatty acid components provide a hydrophobic barrier, whereas the remainder of the molecule has hydrophilic properties that enable interaction with the aqueous environment.

The platform on which phospholipids are built may be glycerol, a three-carbon alcohol, or sphingosine, a more complex alcohol. Phospholipids derived from glycerol are called **phosphoglycerides**. A phosphoglyceride consists of a glycerol backbone to which are attached two fatty acid chains and a phosphorylated alcohol.

In phosphoglycerides, the hydroxyl groups at C-1 and C-2 of glycerol are esterified to the carboxyl groups of the two fatty acid chains (**Figure 12.4**). The C-3 hydroxyl group of the glycerol backbone is esterified to phosphoric acid. When no further additions are made, the resulting compound is phosphatidate (diacylglycerol 3-phosphate), the simplest phosphoglyceride. Although only small amounts of phosphatidate are present in membranes, it is a key intermediate in the biosynthesis of the other phosphoglycerides (Section 27.1). The absolute configuration of the glycerol 3-phosphate moiety of membrane lipids is shown in Figure 12.4.

The major phosphoglycerides are derived from phosphatidate by the formation of an ester bond between the phosphate group of phosphatidate and the hydroxyl group of one of several alcohols (**Figure 12.5**).

FIGURE 12.5 The common phosphoglycerides found in membranes contain an ester bond between phosphatidate and an alcohol. These structures are colored to match the four phospholipid components depicted in Figure 12.3.

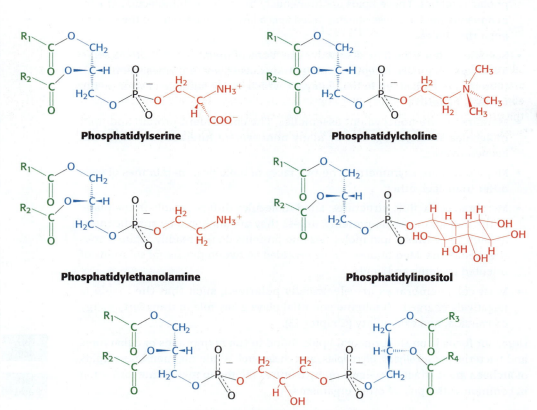

Phosphatidylserine

Phosphatidylcholine

Phosphatidylethanolamine

Phosphatidylinositol

Diphosphatidylglycerol (cardiolipin)

The common alcohol moieties of phosphoglycerides are the amino acid serine, ethanolamine, choline, glycerol, and inositol.

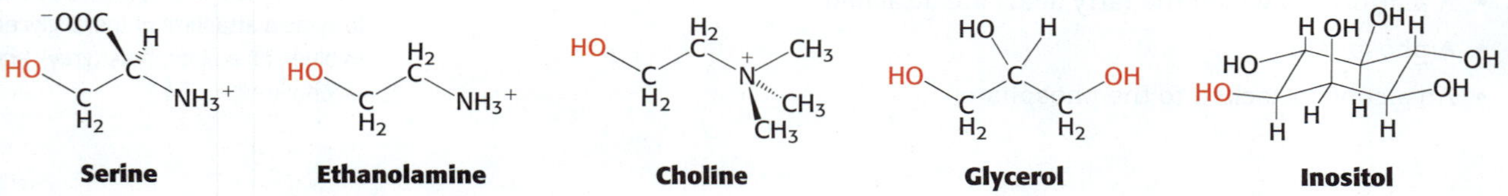

Serine

Ethanolamine

Choline

Glycerol

Inositol

Sphingosine

FIGURE 12.6 Sphingosine forms the backbone of the phospholipid sphingomyelin. The sphingosine moiety of sphingomyelin is highlighted in blue.

Sphingomyelin

Sphingomyelin is a phospholipid found in membranes that is not derived from glycerol. Instead, the backbone in sphingomyelin is **sphingosine**, an amino alcohol that contains a long, unsaturated hydrocarbon chain (**Figure 12.6**). In sphingomyelin, the amino group of the sphingosine backbone is linked to a fatty acid by an amide bond. In addition, the primary hydroxyl group of sphingosine is esterified to phosphorylcholine.

Glycolipids include carbohydrate moieties

The second major class of membrane lipids, **glycolipids**, are sugar-containing lipids. Like sphingomyelin, the glycolipids in animal cells are derived from sphingosine. The amino group of the sphingosine backbone is acylated by a fatty acid, as in sphingomyelin. Glycolipids differ from sphingomyelin in the identity of the unit that is linked to the primary hydroxyl group of the sphingosine backbone. In glycolipids, one or more sugars (rather than phosphorylcholine) are attached to this group. The simplest glycolipid, called a *cerebroside*, contains a single sugar residue, either glucose or galactose.

Sugar unit

Glucose or galactose

Cerebroside
(a glycolipid)

More complex glycolipids contain a branched chain of as many as seven sugar residues. Glycolipids are oriented in a completely asymmetric fashion with the sugar residues always on the extracellular side of the membrane.

Cholesterol is a lipid based on a steroid nucleus

The third major type of membrane lipid, **cholesterol**, has a structure that is quite different from that of phospholipids. It is a steroid, built from four linked hydrocarbon rings.

Cholesterol

MARIE M. DALY In 1947, Marie M. Daly became the first Black woman in the United States to earn a PhD in chemistry. Soon after, she worked with others at Columbia University to examine the role of dietary cholesterol in contributing to hypertension in an animal model system. Later, in an independent faculty position at Albert Einstein College of Medicine, where she remained for more than 25 years, she continued her work on lipids and heart disease and other topics. Dr. Daly demonstrated an interest in promoting the careers of minority women in science. In 1975, she participated in a groundbreaking meeting that led to a report—*The Double Bind: The Price of Being a Minority Woman in Science*—that laid out challenges faced at different career stages. Her energetic pursuit of important scientific questions and her commitment to help others follow in her footsteps illustrate some of the many dimensions of a productive scientific career.

A hydrocarbon tail is linked to the steroid at one end, and a hydroxyl group is attached at the other end. In membranes, the orientation of the molecule is parallel to the fatty acid chains of the phospholipids, and the hydroxyl group interacts with the nearby phospholipid head groups. Cholesterol is absent from prokaryotes but is found to varying degrees in virtually all animal membranes. It constitutes almost 25% of the membrane lipids in certain nerve cells but is essentially absent from some intracellular membranes.

Archaeal membranes are built from ether lipids with branched chains

The membranes of archaea differ in composition from those of eukaryotes or bacteria in three important ways. The first two of these differences clearly relate to the hostile living conditions of many archaea (**Figure 12.7**):

1. The nonpolar chains are joined to a glycerol backbone by ether rather than ester linkages. The ether linkage is more resistant to hydrolysis.

2. The alkyl chains are branched rather than linear. They are built up from repeats of a fully saturated five-carbon fragment. These branched, saturated hydrocarbons are more resistant to oxidation than the unbranched chains of eukaryotic and bacterial membrane lipids. The ability of archaeal lipids to resist hydrolysis and oxidation may help these organisms to withstand the extreme conditions, such as high temperature, low pH, or high salt concentration, under which some of these archaea grow.

3. The stereochemistry of the central glycerol is inverted compared with that shown in Figure 12.4.

FIGURE 12.7 Archaea can thrive in harsh habitats. Here, thermophilic archaea are responsible, in part, for the bright colors of the Morning Glory Pool hot spring in Yellowstone National Park.

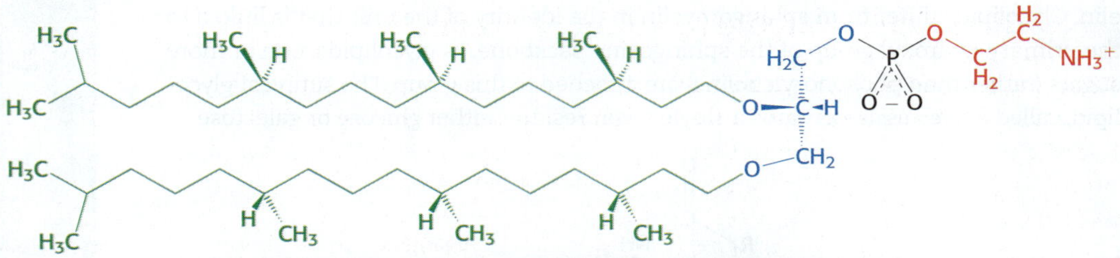

Daniel Viñé Garcia/Getty Images

Membrane lipid from the archaeon _Methanococcus jannaschii_

A membrane lipid is an amphipathic molecule containing a hydrophilic and a hydrophobic moiety

The repertoire of membrane lipids is extensive. However, these lipids possess a critical common structural theme: membrane lipids are **amphipathic molecules (amphiphilic molecules)**, that is, they contain both a hydrophilic and a hydrophobic moiety.

Let us look at a model of a phosphoglyceride, such as phosphatidylcholine. Its overall shape is roughly rectangular (**Figure 12.8A**). The hydrophilic phosphorylcholine moiety lies at one end of the molecule, while the two hydrophobic fatty acid chains project approximately parallel to each other. Sphingomyelin has a similar conformation, as does the archaeal lipid depicted. Therefore, the following shorthand has been adopted to represent these membrane lipids: the hydrophilic unit, also called the _polar head group,_ is represented by a circle, and the hydrocarbon tails are depicted by straight or wavy lines (**Figure 12.8B**).

❖ **SELF–CHECK QUESTION**

What are the three primary types of membrane lipids? What are some molecules that form the polar head groups of phospholipids?

(A)

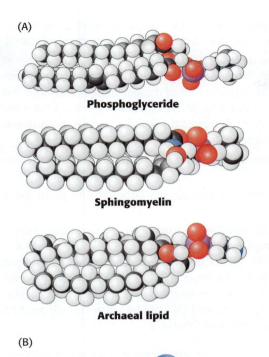

Phosphoglyceride

Sphingomyelin

Archaeal lipid

(B)

FIGURE 12.8 The structures of membrane lipids can be drawn using a shorthand depiction. (A) Space-filling models of a phosphoglyceride, sphingomyelin, and an archaeal lipid show their overall rectangular shapes and distribution of hydrophilic and hydrophobic moieties. (B) A shorthand depiction of a membrane lipid.

12.3 Phospholipids and Glycolipids Readily Form Bimolecular Sheets in Aqueous Media

What properties enable phospholipids to form membranes? Membrane formation is a consequence of the amphipathic nature of the molecules. Their polar head groups favor contact with water, whereas their hydrocarbon tails interact with one another in preference to water. How can molecules with these preferences arrange themselves in aqueous solutions? One way is to form a globular structure called a **micelle**. The polar head groups form the outside surface of the micelle, which is surrounded by water, and the hydrocarbon tails are sequestered inside, interacting with one another (**Figure 12.9**).

Alternatively, the strongly opposed preferences of the hydrophilic and hydrophobic moieties of membrane lipids can be satisfied by forming a **lipid bilayer**, composed of two lipid sheets (**Figure 12.10**). A lipid bilayer is also called a *bimolecular sheet*. The hydrophobic tails of each individual sheet interact with one another, forming a hydrophobic interior that acts as a permeability barrier. The hydrophilic head groups interact with the aqueous medium on each side of the bilayer. The two opposing sheets are called leaflets.

The favored structure for most phospholipids and glycolipids in aqueous media is a bimolecular sheet rather than a micelle. The reason is that the two fatty acid chains of a phospholipid or a glycolipid are too bulky to fit into the interior of a micelle. In contrast, salts of fatty acids (such as sodium palmitate, a constituent of soap) readily form micelles because they contain only one chain. The formation of bilayers instead of micelles by phospholipids is of critical biological importance. A micelle is a limited structure, usually less than 200 Å (20 nm) in diameter. In contrast, a bimolecular sheet can extend to macroscopic dimensions, as much as a millimeter (10^7 Å, or 10^6 nm) or more. Phospholipids and related molecules are important membrane constituents because they readily form extensive bimolecular sheets.

Lipid bilayers form spontaneously by a self-assembly process. In other words, the structure of a bimolecular sheet is inherent in the structure of the constituent lipid molecules. The growth of lipid bilayers from phospholipids

FIGURE 12.9 Micelles are globular structures formed from ionized fatty acids with single tails. Ionized fatty acids readily form such structures, but most phospholipids do not.

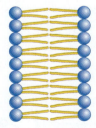

FIGURE 12.10 A lipid bilayer is formed from two opposing lipid sheets.

is rapid and spontaneous in water. Bimolecular sheets are stabilized by the full array of forces that mediate molecular interactions in biological systems:

- Hydrophobic interactions are the major driving force for the formation of lipid bilayers. We have already encountered the dominant role that hydrophobic interactions play in the stabilization of double-helical DNA and the folding of proteins (Sections 1.3 and 2.4). Water molecules are released from the hydrocarbon tails of membrane lipids as these tails become sequestered in the nonpolar interior of the bilayer.

- van der Waals attractive forces between the hydrocarbon tails favor close packing of the tails.

- Electrostatic and hydrogen-bonding attractions between the polar head groups and water molecules.

Because lipid bilayers are held together by many reinforcing, noncovalent interactions (predominantly hydrophobic), they are cooperative structures. These hydrophobic interactions have three significant biological consequences: (1) lipid bilayers have an inherent tendency to be extensive; (2) lipid bilayers will tend to close on themselves so that there are no edges with exposed hydrocarbon chains, and so they form compartments; and (3) lipid bilayers are self-sealing because a hole in a bilayer is energetically unfavorable.

❖ SELF–CHECK QUESTION

Phospholipids form lipid bilayers in water. What structure might form if phospholipids were placed in an organic solvent?

Lipid vesicles can be formed from phospholipids

The propensity of phospholipids to form membranes has been used to create important experimental and clinical tools. **Lipid vesicles**, or **liposomes**, are aqueous compartments enclosed by a lipid bilayer (**Figure 12.11**), which can be used to study membrane permeability or to deliver chemicals to cells.

Liposomes are formed by suspending a suitable lipid, such as phosphatidylcholine, in an aqueous medium, and then sonicating (i.e., agitating by high-frequency sound waves) to give a dispersion of closed vesicles that are quite uniform in size (**Figure 12.12**). Vesicles formed by this method are nearly spherical and have a diameter of about 500 Å (50 nm). Larger vesicles (of the order of 1 μm or 10^4 Å in diameter) can be prepared by slowly evaporating the organic solvent from a suspension of phospholipid in a mixed-solvent system.

Ions or molecules can be trapped in the aqueous compartments of lipid vesicles by forming the vesicles in the presence of these substances, as shown in Figure 12.12. For example, 500-Å-diameter vesicles formed in a 0.1 M glycine solution will trap about 2000 molecules of glycine in each inner aqueous compartment. These glycine-containing vesicles can be separated from the surrounding solution of glycine by dialysis or by gel-filtration chromatography

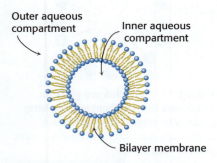

Outer aqueous compartment

Inner aqueous compartment

Bilayer membrane

FIGURE 12.11 A liposome is a small aqueous compartment surrounded by a lipid bilayer.

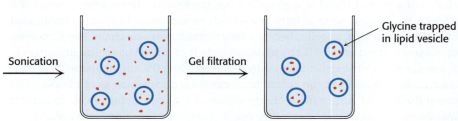

Glycine in H₂O

Phospholipid

Sonication

Gel filtration

Glycine trapped in lipid vesicle

FIGURE 12.12 Liposomes containing trapped ions or molecules can be synthesized. Liposomes containing glycine are formed by the sonication of phospholipids in the presence of glycine. Free glycine is removed by gel filtration.

(Section 4.1). The permeability of the bilayer membrane to glycine can then be determined by measuring the rate of efflux of glycine from the inner compartment of the vesicle to the ambient solution.

Liposomes can be formed with specific membrane proteins embedded in them by solubilizing the proteins in the presence of detergents and then adding them to the phospholipids from which liposomes will be formed. Protein–liposome complexes provide valuable experimental tools for examining a range of membrane-protein functions.

Therapeutic applications for liposomes are currently under active investigation. For example, liposomes containing drugs or DNA can be injected into patients. These liposomes fuse with the plasma membrane of many kinds of cells, introducing into the cells the molecules they contain. Drug delivery with liposomes often lessens its toxicity. Less of the drug is distributed to normal tissues because long-circulating liposomes concentrate in regions of increased blood circulation, such as solid tumors and sites of inflammation. Moreover, the selective fusion of lipid vesicles with particular kinds of cells is a promising means of controlling the delivery of drugs to target cells.

Another well-defined synthetic membrane adopts the structure of a planar bilayer. This membrane can be formed across a 1-mm hole in a partition between two aqueous compartments by dipping a fine paintbrush into a membrane-forming solution, such as phosphatidylcholine in decane, and stroking the tip of the brush across the hole. The lipid film across the hole thins spontaneously into a lipid bilayer. The electrical conduction properties of this macroscopic bilayer membrane are readily studied by inserting electrodes into each aqueous compartment (**Figure 12.13**). For example, the permeability of the membrane to ions is determined by measuring the current across the membrane as a function of the applied voltage.

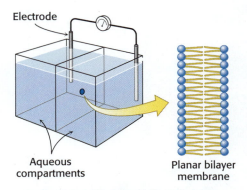

FIGURE 12.13 The properties of planar bilayer membranes can be studied experimentally. A bilayer membrane is formed across a 1-mm hole in a septum that separates two aqueous compartments. This arrangement permits measurements of the permeability and electrical conductance of lipid bilayers.

Lipid bilayers are highly impermeable to ions and most polar molecules

Permeability studies of lipid vesicles and electrical-conductance measurements of planar bilayers have revealed several important properties:

- *Lipid bilayer membranes have a very low permeability for ions and most polar molecules.* Water is a conspicuous exception to this generalization; it traverses such membranes relatively easily because of its low molecular weight, high concentration, and lack of a complete charge.

- *The range of measured permeability coefficients is very wide* (**Figure 12.14**). For example, Na^+ and K^+ traverse lipid bilayer membranes 10^9 times as slowly as does H_2O. Tryptophan, a zwitterion at pH 7, crosses the membrane 10^3 times as slowly as does indole, a structurally related molecule that lacks ionic groups.

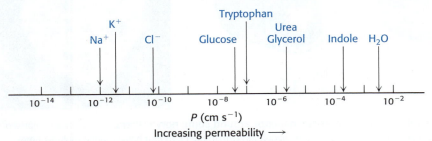

FIGURE 12.14 The permeability coefficients (*P*) of ions and molecules in a lipid bilayer span a wide range of values.

• *The permeability of small molecules is correlated with their solubility in a nonpolar solvent relative to their solubility in water.* This relationship suggests that a small molecule might traverse a lipid bilayer membrane in the following way: first, it sheds any interacting water molecules (Section 1.3); then, it is dissolved in the hydrocarbon core of the membrane; finally, it diffuses through this core to the other side of the membrane, where it becomes resolvated by water. An ion such as Na⁺ traverses membranes very slowly because the replacement of its numerous interactions with polar water molecules by nonpolar interactions with the membrane interior is highly unfavorable energetically.

12.4 Proteins Carry Out Most Membrane Processes

We now turn to membrane proteins, which are responsible for most of the dynamic processes carried out by membranes. Whereas membrane lipids form a permeability barrier and thereby establish compartments, specific proteins mediate nearly all other membrane functions. In particular, proteins transport chemicals and information across a membrane. Membrane lipids create the appropriate environment for the action of such proteins.

Membranes differ in their protein content. For example, relatively pure lipids are well suited for insulation. **Myelin**, a membrane that serves as an electrical insulator around certain nerve fibers, has a low content of protein (18%). As such, myelin plays a critical role in enabling the rapid transmission of nerve impulses, or action potentials (Section 13.4). Its structure is quite unique: the cells that generate myelin wrap their plasma membranes multiple times around the axon, the part of the neuron that conducts the electrical signal (**Figure 12.15**).

Significant myelination of neurons in the brain occurs during infancy but persists throughout adolescence, a reminder that the brain is an actively developing organ throughout childhood. The importance of myelination is further underscored by the existence of demyelinating diseases, such as multiple sclerosis, where myelin assembly is impaired or existing myelin is damaged.

In contrast to myelin, the plasma membranes, or exterior membranes, of most other cells are much more metabolically active. They contain many pumps, channels, receptors, and enzymes. The protein content of these plasma membranes is typically 50%. Energy-transduction membranes, such as the internal membranes of mitochondria and chloroplasts, have the highest content of protein, around 75%.

(A)

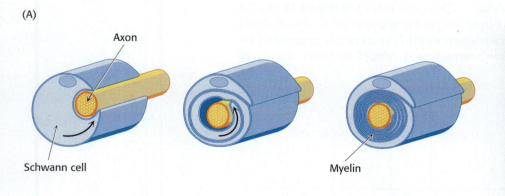

(B)

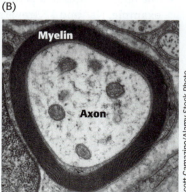

FIGURE 12.15 The lipid-rich membranes of myelin form an electrical insulator around some nerve fibers. (A) Myelin is comprised of the lipid-rich plasma membrane of a Schwann cell in peripheral nerves or the oligodendrocyte in the brain that has wrapped multiple times around the axon, or the portion of the neuron that conducts electrical impulses. [https://biologydictionary.net/myelin-sheath] (B) The myelin sheath around an axon is readily apparent by electron microscopy.

The protein components of a membrane can be readily visualized by SDS–polyacrylamide gel electrophoresis. As we discussed in Section 4.1, the electrophoretic mobility of many proteins in SDS-containing gels depends on the mass rather than on the net charge of the protein. The gel-electrophoresis patterns of three membranes are shown in **Figure 12.16**. It is evident that each of these three membranes contains many proteins but has a distinct protein composition. In general, membranes performing different functions contain different repertoires of proteins.

Proteins associate with the lipid bilayer in a variety of ways

The ease with which a protein can be dissociated from a membrane indicates how intimately it is associated with the membrane. Some membrane proteins can be solubilized by relatively mild means, such as extraction by a solution of high ionic strength (e.g., 1 M NaCl). Other membrane proteins are bound much more tenaciously; they can be solubilized only by using a detergent or an organic solvent. Membrane proteins can be classified as being either peripheral or integral on the basis of their difference in dissociability (**Figure 12.17**):

- **Integral membrane proteins** interact extensively with the hydrocarbon chains of membrane lipids, and they can be released only by agents that compete for these nonpolar interactions. In fact, most integral membrane proteins span the lipid bilayer.

- **Peripheral membrane proteins** are bound to membranes primarily by electrostatic and hydrogen-bond interactions with the head groups of lipids. These polar interactions can be disrupted by adding salts or by changing the pH. Many peripheral membrane proteins are bound to the surfaces of integral proteins, on either the cytoplasmic or the extracellular side of the membrane. Others are anchored to the lipid bilayer by a covalently attached hydrophobic chain, such as a fatty acid.

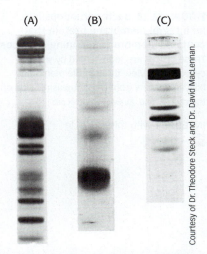

(A) (B) (C)

Courtesy of Dr. Theodore Steck and Dr. David MacLennan.

FIGURE 12.16 The protein composition of plasma membranes varies with cell type. Here, the SDS–acrylamide gel patterns of membrane proteins from three different cell types are shown: (A) The plasma membrane of red blood cells. (B) The photoreceptor membranes of retinal rod cells. (C) The sarcoplasmic reticulum membrane of muscle cells.

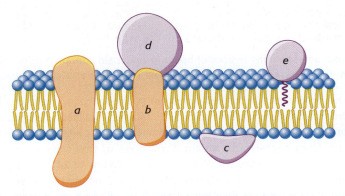

FIGURE 12.17 Membrane proteins interact with lipid bilayers in different ways. Integral membrane proteins (*a* and *b*) interact extensively with the hydrocarbon region of the bilayer. Most known integral membrane proteins traverse the lipid bilayer. Peripheral membrane proteins interact with the polar head groups of the lipids (*c*) or bind to the surfaces of integral proteins (*d*). Other proteins are tightly anchored to the membrane by a covalently attached lipid molecule (*e*).

Proteins interact with membranes in a variety of ways

Membrane proteins are more difficult to purify and crystallize than are water-soluble proteins. Nonetheless, researchers using x-ray crystallography or cryo-electron microscopy have determined the three-dimensional structures of more than 4000 such proteins at sufficiently high resolution to discern the molecular details. As noted in Chapter 2, membrane proteins differ from soluble proteins in the distribution of hydrophobic and hydrophilic groups. We will consider the structures of three membrane proteins in some detail.

FIGURE 12.18 Bacteriorhodopsin is an integral membrane protein that contains membrane-spanning α helices. These helices are represented by yellow cylinders. (A) View through the membrane bilayer. (B) View from the cytoplasmic side of the membrane. [Drawn from 1BRX.pdb.]

INTERACT with this model in
 Achie√e

(A)

(B)

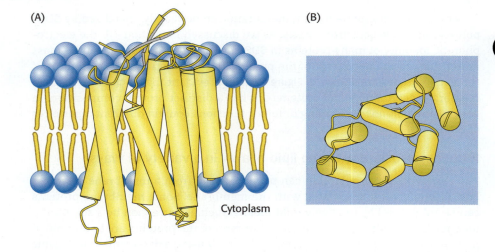

Cytoplasm

- *Membrane-spanning alpha helices are the most common structural motif in membrane proteins.* The first membrane protein that we consider is the archaeal protein bacteriorhodopsin (**Figure 12.18**). We will discuss the function of this important protein in Section 19.1. Bacteriorhodopsin is built almost entirely of α helices; seven closely packed α helices, nearly perpendicular to the plane of the cell membrane, span its 45-Å width. Examination of the primary structure of bacteriorhodopsin reveals that most of the amino acids in these membrane-spanning α helices are nonpolar and only a very few are charged (**Figure 12.19**). This distribution of nonpolar amino acids is sensible because these residues are either in contact with the hydrocarbon core of the membrane or with one another. As will be considered in Section 12.5, membrane-spanning α helices can often be detected by examining amino acid sequence alone.

FIGURE 12.19 The membrane-spanning helices of bacteriorhodopsin contain predominantly nonpolar amino acids. The seven helical regions are highlighted in yellow and the charged residues in red.

- *A channel protein can be formed from beta strands.* Porin, a protein from the outer membranes of bacteria such as *E. coli* and *Rhodobacter capsulatus*, represents a class of membrane proteins with a completely different type of structure from bacteriorhodopsin. Structures of this type are built from β strands and contain essentially no α helices (**Figure 12.20**). The arrangement of β strands is quite simple: each strand is hydrogen bonded to its neighbor in an antiparallel arrangement, forming a single β sheet. The β sheet curls up to form

(A)

(B)

INTERACT with this model in
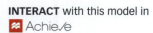 Achie√e

FIGURE 12.20 The bacterial integral membrane protein porin from *Rhodopseudomonas blastica* is built entirely of β strands. (A) Side view. (B) View from the periplasmic space. Only one monomer of the trimeric protein is shown. [Drawn from 1PRN.pdb.]

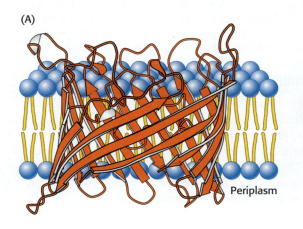

Periplasm

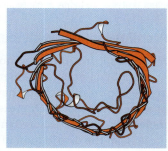

a hollow cylinder that, as its name suggests, forms a pore, or channel, in the membrane. The outside surface of porin is appropriately nonpolar, given that it interacts with the hydrocarbon core of the membrane. In contrast, the inside of the channel is quite hydrophilic and is filled with water. This arrangement of nonpolar and polar surfaces is accomplished by the alternation of hydrophobic and hydrophilic amino acids along each β strand (**Figure 12.21**).

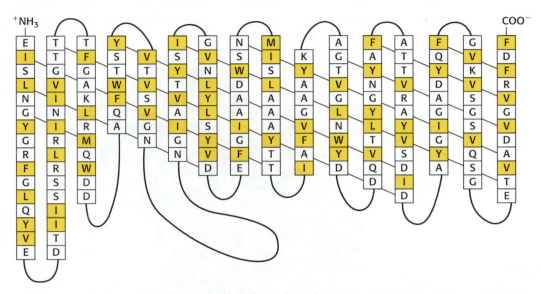

FIGURE 12.21 Porin contains alternating hydrophobic and hydrophilic amino acids along each β strand. The secondary structure of porin from *Rhodopseudomonas blastica* is shown, with the diagonal lines indicating the direction of hydrogen bonding along the β sheet. Hydrophobic residues (F, I, L, M, V, W, and Y) are shown in yellow. These residues tend to lie on the outside of the structure, in contact with the hydrophobic core of the membrane.

- *Embedding part of a protein in a membrane can link the protein to the membrane surface.* The structure of the endoplasmic reticulum membrane-bound enzyme prostaglandin H_2 synthase-1 reveals a rather different role for α helices in protein–membrane associations. This enzyme is a homodimer with a complicated structure consisting primarily of α helices. Unlike bacteriorhodopsin, this protein is not largely embedded in the membrane. Instead, it lies along the outer surface of the membrane, firmly bound by a set of α helices with hydrophobic surfaces that extend from the bottom of the protein into the membrane (**Figure 12.22**). This linkage is sufficiently strong that only the action of detergents can release the protein from the membrane. Thus, this enzyme is classified as an integral membrane protein, although it does not span the membrane.

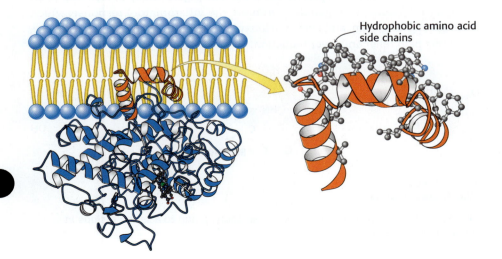

Hydrophobic amino acid side chains

INTERACT with this model in
Achieve

FIGURE 12.22 Prostaglandin H_2 synthase-1 is an integral membrane protein attached to the membrane surface. This enzyme is held in the membrane by a set of α helices (orange) coated with hydrophobic side chains. One monomer of the dimeric enzyme is shown. [Drawn from 1PTH.pdb.]

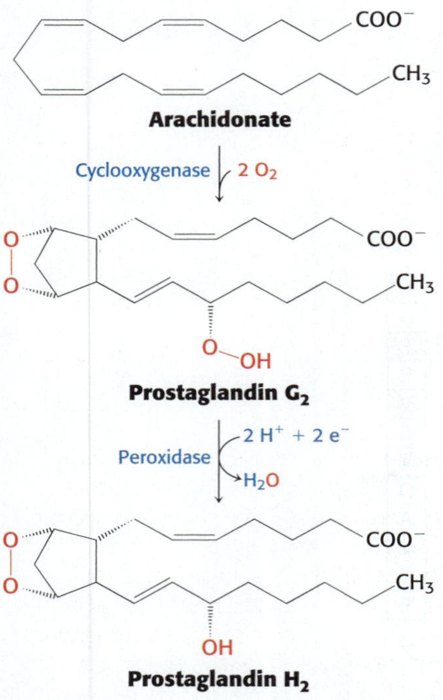

Arachidonate

Cyclooxygenase | $2 O_2$

Prostaglandin G_2

Peroxidase | $2 H^+ + 2 e^-$
→ H_2O

Prostaglandin H_2

FIGURE 12.23 Prostaglandin H_2 synthase-1 catalyzes the formation of prostaglandin H_2 from arachidonic acid in two steps.

The localization of prostaglandin H_2 synthase-1 in the membrane is crucial to its function. This enzyme catalyzes the conversion of arachidonic acid into prostaglandin H_2 in two steps: (1) a cyclooxygenase reaction, and (2) a peroxidase reaction (**Figure 12.23**). The substrate for this enzyme, arachidonic acid, is a hydrophobic molecule generated by the hydrolysis of membrane lipids. Arachidonic acid reaches the active site of the enzyme from the membrane without entering an aqueous environment by traveling through a hydrophobic channel in the protein (**Figure 12.24**). The reaction product, prostaglandin H_2, promotes inflammation and modulates gastric acid secretion.

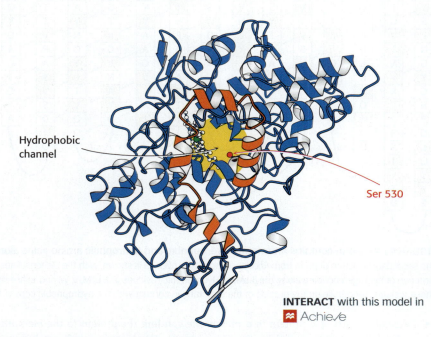

Hydrophobic channel

Ser 530

INTERACT with this model in
Achie√e

FIGURE 12.24 Arachidonic acid reaches the prostaglandin H_2 synthase-1 active site through a hydrophobic channel. A view of prostaglandin H_2 synthase-1 from the membrane-facing side shows this channel more clearly. The membrane-anchoring helices are shown in orange. [Drawn from 1PTH.pdb.]

Indeed, nearly all of us have experienced the importance of this channel: drugs such as aspirin and ibuprofen block the channel and prevent prostaglandin synthesis by inhibiting the cyclooxygenase activity of the synthase. In particular, aspirin acts through the transfer of its acetyl group to a serine residue (Ser 530) that lies along the path to the active site (**Figure 12.25**). In response to an injury or infection, prostaglandins promote the inflammatory response, which includes swelling, pain, and fever. Cyclooxygenase inhibitors blunt this response, providing pain relief and fever reduction.

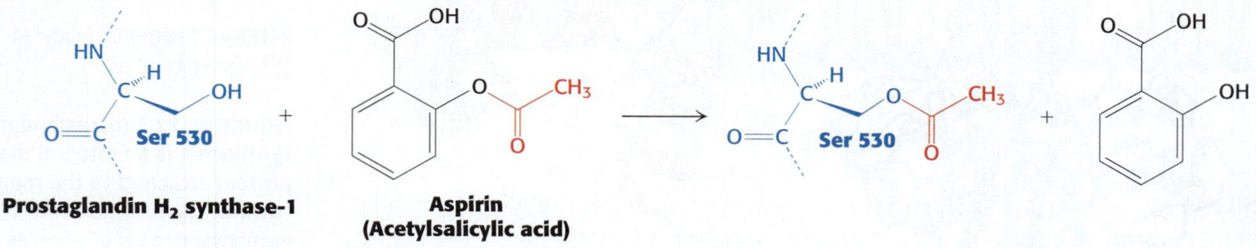

Prostaglandin H_2 synthase-1   +   **Aspirin (Acetylsalicylic acid)**   →   **Ser 530**   +

FIGURE 12.25 Aspirin acts by transferring an acetyl group to a serine residue in prostaglandin H_2 synthase-1.

Two important features emerge from our examination of these three examples of membrane-protein structure. First, the parts of the protein that interact with the hydrophobic parts of the membrane are coated with nonpolar amino acid side chains, whereas those parts that interact with the aqueous environment are much more hydrophilic. Second, the structures positioned within the membrane are quite regular and, in particular, all backbone hydrogen-bond donors and acceptors participate in hydrogen bonds. Breaking a hydrogen bond within a membrane is quite unfavorable, because little or no water is present to compete for the polar groups.

Some proteins associate with membranes through covalently attached hydrophobic groups

The membrane proteins considered thus far associate with the membrane through surfaces generated by hydrophobic amino acid side chains. However, even otherwise soluble proteins can associate with membranes if hydrophobic groups are attached to the proteins. Three such groups are shown in **Figure 12.26**: (1) a palmitoyl group attached to a specific cysteine residue by a thioester bond, (2) a farnesyl group attached to a cysteine residue at the carboxyl terminus, and (3) a glycolipid structure termed a glycosylphosphatidylinositol (GPI) anchor attached to the carboxyl terminus. These modifications are attached by enzyme systems that recognize specific signal sequences near the attachment site.

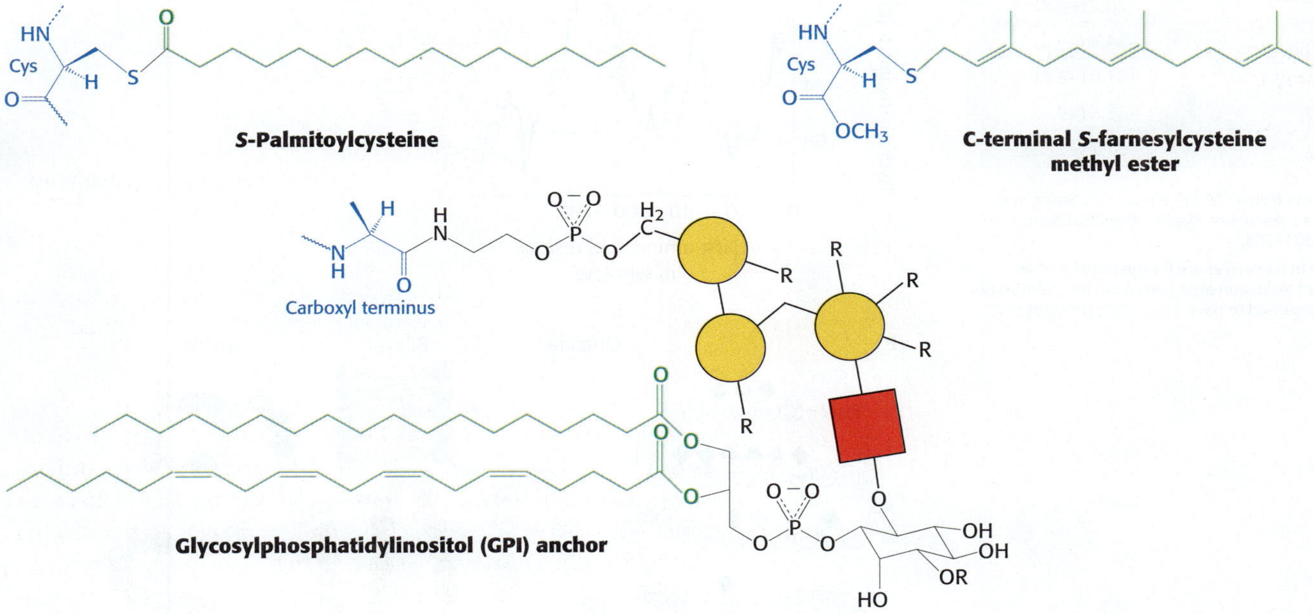

FIGURE 12.26 Membrane anchors are hydrophobic groups that tether proteins to the membrane. The anchors are attached to the proteins (blue) by covalent bonds. The yellow circles and red square correspond to mannose and β-D-acetylglucosamine (GlcNAc), respectively. R groups represent points of additional modification.

S-Palmitoylcysteine

C-terminal S-farnesylcysteine methyl ester

Glycosylphosphatidylinositol (GPI) anchor

Transmembrane helices can be accurately predicted from amino acid sequences

Many membrane proteins, like bacteriorhodopsin, employ α helices to span the hydrophobic part of a membrane. As noted earlier, most of the residues in these α helices are nonpolar and almost none of them are charged. Can we use this information to identify likely membrane-spanning regions from sequence data alone?

One approach to identifying transmembrane helices is to ask whether a postulated helical segment is likely to be more stable in a hydrocarbon environment or in water. Specifically, we want to estimate the free-energy change when a helical segment is transferred from the interior of a membrane to water. Free-energy changes for the transfer of individual amino acid residues from a hydrophobic to

TABLE 12.2 Polarity scale for identifying transmembrane helices

Amino acid residue	Transfer free energy in kJ mol⁻¹ (kcal mol⁻¹)
Phe	15.5 (3.7)
Met	14.3 (3.4)
Ile	13.0 (3.1)
Leu	11.8 (2.8)
Val	10.9 (2.6)
Cys	8.4 (2.0)
Trp	8.0 (1.9)
Ala	6.7 (1.6)
Thr	5.0 (1.2)
Gly	4.2 (1.0)
Ser	2.5 (0.6)
Pro	−0.8 (−0.2)
Tyr	−2.9 (−0.7)
His	−12.6 (−3.0)
Gln	−17.2 (−4.1)
Asn	−20.2 (−4.8)
Glu	−34.4 (−8.2)
Lys	−37.0 (−8.8)
Asp	−38.6 (−9.2)
Arg	−51.7 (−12.3)

Source: Data from D. M. Engelman, T. A. Steitz, and A. Goldman. *Annu. Rev. Biophys. Biophys. Chem.* 15(1986):321–353.

Note: The free energies are for the transfer of an amino acid residue in an α helix from the membrane interior (assumed to have a dielectric constant of 2) to water.

an aqueous environment are given in **Table 12.2**. For example, the transfer of a helix formed entirely of L-arginine residues—a positively charged amino acid—from the interior of a membrane to water would be highly favorable, with a negative transfer free energy of −51.5 kJ mol⁻¹ (−12.3 kcal mol⁻¹) per arginine residue in the helix. In contrast, the transfer of a helix formed entirely of L-phenylalanine—a hydrophobic amino acid—would be unfavorable, with a positive transfer free energy value of +15.5 kJ mol⁻¹ (+3.7 kcal mol⁻¹) per phenylalanine residue in the helix.

The hydrocarbon core of a membrane is typically 30 Å wide, a length that can be traversed by an α helix of about 20 residues. We can take the amino acid sequence of a protein and estimate the free-energy change that takes place when a hypothetical α helix formed of residues 1 through 20 is transferred from the membrane interior to water. The same calculation can be made for residues 2 through 21, 3 through 22, and so forth, until we reach the end of the sequence. The span of 20 residues chosen for this calculation is called a *window*. The free-energy change for each window, termed the *hydropathy index*, is plotted against the first amino acid at the window to create a **hydropathy plot** (**Figure 12.27A**).

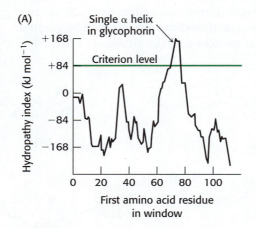

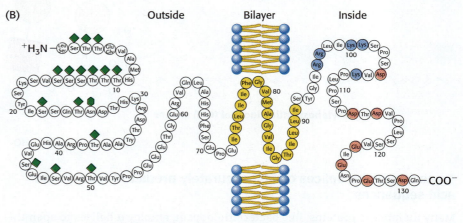

FIGURE 12.27 Hydropathy plots can be used to locate the membrane-spanning helices from protein sequences. (A) The hydropathy plot for glycophorin A is shown. Peaks of greater than +84 kJ mol⁻¹ (+20 kcal mol⁻¹) in hydropathy plots are indicative of potential transmembrane helices. (B) The amino acid sequence and transmembrane disposition of glycophorin A from the red-blood-cell membrane are shown. Sites of glycosylation are indicated by the green shapes. The hydrophobic residues (yellow) buried in the bilayer form a transmembrane α helix. The negatively charged (red) and positively charged (blue) residues on the interior side of the membrane are indicated. [(A) Information from Dr. Vincent Marchesi; (B) Data from D. M. Engelman, T. A. Steitz, and A. Goldman, *Annu. Rev. Biophys. Biophys. Chem.* 15:321–353, 1986. Copyright © 1986 by Annual Reviews, Inc. All rights reserved.]

Empirically, a peak of +84 kJ mol^{-1} (+20 kcal mol^{-1}) or more in a hydropathy plot based on a window of 20 residues indicates that a polypeptide segment could be a membrane-spanning α helix. For example, glycophorin—a protein found in the membranes of red blood cells—is predicted by this criterion to have one membrane-spanning helix, in agreement with experimental findings (**Figure 12.27B**). Note, however, that a peak in the hydropathy plot does not prove that a segment is a transmembrane helix. Even soluble proteins may have highly nonpolar regions. Conversely, some membrane proteins contain membrane-spanning features (such as a set of cylinder-forming β strands) that escape detection by these plots (**Figure 12.28**).

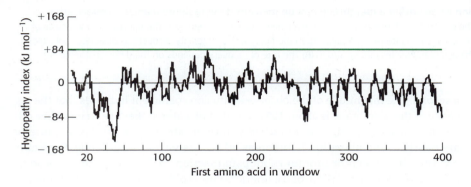

FIGURE 12.28 No strong peaks are observed in the hydropathy plot for porin. Porin is constructed from membrane-spanning β strands rather than α helices (Figure 12.20); hence, it escapes detection using this method.

❖ SELF–CHECK QUESTION

Would a homopolymer of alanine be more likely to form an α helix in water or in a hydrophobic medium? Explain.

12.5 Lipids and Many Membrane Proteins Diffuse Rapidly in the Plane of the Membrane

Biological membranes are not rigid, static structures. On the contrary, lipids and many membrane proteins are constantly in lateral motion, a process called **lateral diffusion**. The rapid lateral movement of membrane proteins has been visualized by means of fluorescence microscopy using the technique of fluorescence recovery after photobleaching (FRAP; **Figure 12.29**):

1. A cell-surface component is specifically labeled with a fluorescent chromophore, and a small region of the cell surface ($\sim 3\ \mu m^2$) is viewed through a fluorescence microscope.

2. The fluorescent molecules in this region are then destroyed (bleached) by a very intense light pulse from a laser, as indicated by the pale spot in Figure 12.29A. The fluorescence of this region is subsequently monitored as a function of time by using a light level sufficiently low to prevent further bleaching.

3. If the labeled component is mobile, bleached molecules leave and unbleached molecules enter the illuminated region, resulting in an increase in the fluorescence intensity, a process called *recovery*. The rate of recovery depends on the lateral mobility of the fluorescence-labeled component, which can be expressed in terms of a diffusion coefficient, D. The average distance S traversed in time t depends on D according to the expression

$$S = (4Dt)^{1/2}$$

The diffusion coefficient of lipids in a variety of membranes is about 1 μm^2 s^{-1}. Thus, a phospholipid molecule diffuses an average distance of 2 μm in 1 s. This rate means that a lipid molecule can travel from one end of a bacterium to the other in

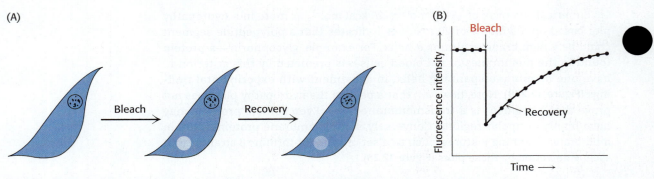

FIGURE 12.29 Lateral diffusion rates within a membrane can be measured using fluorescence recovery after photobleaching (FRAP). (A) The cell surface fluoresces because of a labeled surface component. The fluorescent molecules of a small part of the surface are bleached by an intense light pulse; then the fluorescence intensity recovers as bleached molecules diffuse out of the region and unbleached molecules diffuse into it. (B) The rate of recovery depends on the diffusion coefficient.

a second. The magnitude of the observed diffusion coefficient indicates that the viscosity of the membrane is about 100 times that of water, rather like that of olive oil.

In contrast, proteins vary markedly in their lateral mobility. Some proteins are nearly as mobile as lipids, whereas others are virtually immobile. For example, the photoreceptor protein rhodopsin (Section 14.6), a very mobile protein, has a diffusion coefficient of $0.4 \ \mu m^2 \ s^{-1}$. The rapid movement of rhodopsin is essential for fast signaling. At the other extreme is fibronectin, a peripheral glycoprotein that interacts with the extracellular matrix. For fibronectin, D is less than $10^{-4} \mu m^2 \ s^{-1}$. Fibronectin has a very low mobility because it is anchored to actin filaments on the inside of the plasma membrane through integrin, a transmembrane protein that links the extracellular matrix to the cytoskeleton.

The fluid mosaic model allows lateral movement but not rotation through the membrane

On the basis of the mobility of proteins in membranes, in 1972 S. Jonathan Singer and Garth Nicolson proposed a **fluid mosaic model** to describe the overall organization of biological membranes. The essence of their model is that membranes are two-dimensional solutions of oriented lipids and globular proteins. The lipid bilayer has a dual role: it is both a solvent for integral membrane proteins and a permeability barrier. Membrane proteins are free to diffuse laterally in the lipid matrix unless restricted by special interactions.

Although the lateral diffusion of membrane components can be rapid, the spontaneous rotation of lipids from one face of a membrane to the other is a very slow process. The transition of a molecule from one membrane surface to the other is called *transverse diffusion* or *flip-flop* (**Figure 12.30**). The flip-flop of phospholipid molecules in phosphatidylcholine vesicles has been directly measured by electron spin resonance techniques, which show that a phospholipid molecule flip-flops once in several hours. Thus, a phospholipid molecule takes about 10^9 times as long to flip-flop across a membrane as it takes to diffuse a distance of 50 Å in the lateral direction. The free-energy barriers to flip-flopping are even larger for protein molecules than for lipids because proteins have more-extensive polar regions. In fact, the flip-flop of a protein molecule has not been observed. Hence, membrane asymmetry can be preserved for long periods.

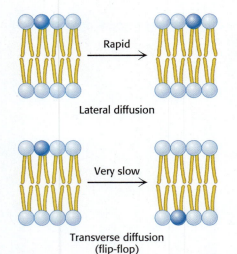

FIGURE 12.30 Lateral diffusion of lipids in membranes is much more rapid than transverse diffusion (flip-flop).

Membrane fluidity is controlled by fatty acid composition and cholesterol content

Many membrane processes, such as transport or signal transduction, depend on the fluidity of the membrane lipids, which in turn depends on the properties of fatty acid chains. Fatty acid chains in membrane bilayers can exist in an ordered, rigid state or in a relatively disordered, fluid state. The transition from the rigid to the fluid state takes place abruptly as the temperature is raised above T_m, the melting temperature (**Figure 12.31**).

The melting temperature of a fluid membrane depends on the length of the fatty acid chains and on their degree of unsaturation (**Table 12.3**). The presence of saturated fatty acid residues favors the rigid state because their straight hydrocarbon chains interact very favorably with one another. On the other hand, a cis double bond produces a bend in the hydrocarbon chain that interferes with a highly ordered packing of fatty acid chains, and so T_m is lowered (**Figure 12.32**). The length of the fatty acid chain also affects the transition temperature. Long hydrocarbon chains interact more strongly than do short ones. Specifically, each additional $-CH_2-$ group makes a favorable contribution of about -2 kJ mol^{-1} (-0.5 kcal mol^{-1}) to the free energy of interaction of two adjacent hydrocarbon chains.

Bacteria regulate the fluidity of their membranes by varying the number of double bonds and the length of their fatty acid chains. For example, the ratio of saturated to unsaturated fatty acid chains in the E. coli membrane decreases from 1.6 to 1.0 as the growth temperature is lowered from 42°C to 27°C. This decrease in the proportion of saturated residues prevents the membrane from becoming too rigid at the lower temperature.

In animals, cholesterol is the key regulator of membrane fluidity. Cholesterol contains a bulky steroid nucleus with a hydroxyl group at one end and a flexible hydrocarbon tail at the other end. Cholesterol inserts into bilayers with its long axis perpendicular to the plane of the membrane. The hydroxyl group of cholesterol forms a hydrogen bond with a carbonyl oxygen atom of a phospholipid head group, whereas the hydrocarbon tail of cholesterol is located in the

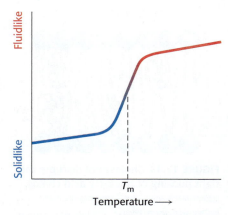

FIGURE 12.31 The melting temperature (T_m) for a phospholipid membrane represents the midpoint along the phase-transition curve. As the temperature is raised, the phospholipid membrane changes from a packed, ordered state to a more random one.

TABLE 12.3 The melting temperature of phosphatidylcholine containing different pairs of identical fatty acid chains

		Fatty acid		
Number of carbons	Number of double bonds	Common name	Systematic name	T_m (°C)
22	0	Behenate	*n*-Docosanoate	75
18	0	Stearate	*n*-Octadecanoate	58
16	0	Palmitate	*n*-Hexadecanoate	41
14	0	Myristate	*n*-Tetradecanoate	24
18	1	Oleate	*cis*-Δ⁹-Octadecenoate	−22

(A)　　　　　　　　　　　　　　(B)

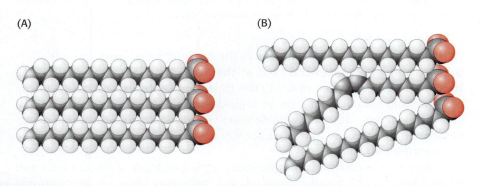

FIGURE 12.32 The highly ordered packing of fatty acid chains in a membrane is disrupted by the presence of cis double bonds. The space-filling models show the packing of (A) three molecules of stearate (C_{18}, saturated) and (B) a molecule of oleate (C_{18}, unsaturated) between two molecules of stearate.

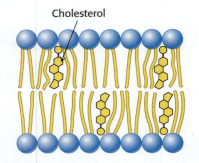

FIGURE 12.33 Cholesterol disrupts the tight packing of the fatty acid chains.
[Information from S. L. Wolfe, *Molecular and Cellular Biology* (Wadsworth, 1993).]

nonpolar core of the bilayer. The different shape of cholesterol compared with that of phospholipids disrupts the regular interactions between fatty acid chains (**Figure 12.33**).

❖ SELF–CHECK QUESTION

Explain why oleic acid (18 carbons, one cis bond) has a lower melting point than stearic acid, which has the same number of carbon atoms but is saturated. How would you expect the melting point of *trans*-oleic acid to compare with that of *cis*-oleic acid? Why might most unsaturated fatty acids in phospholipids be in the cis rather than the trans conformation?

Lipid rafts are highly dynamic complexes formed between cholesterol and specific lipids

In addition to its nonspecific effects on membrane fluidity, cholesterol can form complexes with lipids that contain the sphingosine backbone, including sphingomyelin and certain glycolipids, and with lipid-anchored membrane proteins. These complexes concentrate within small (10–200 nm) and highly dynamic regions within membranes, which have the potential to cluster into larger domains of greater than 300 nm. These structures are referred to as **lipid rafts** (**Figure 12.34A**).

(A)   (B)

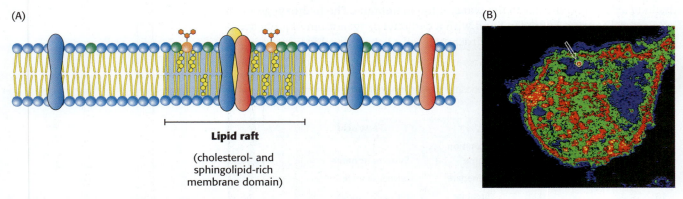

Lipid raft
(cholesterol- and sphingolipid-rich membrane domain)

FIGURE 12.34 Lipid rafts are membrane domains exhibiting reduced fluidity. (A) Cholesterol molecules, sphingomyelin (green head groups), and glycolipids (orange head groups) tend to be concentrated within the lipid rafts. These regions (shaded blue) provide the opportunity for changes in protein structure and oligomerization that may have significant functional consequences. (B) Image of a human monocytic cell, stained with a fluorescent dye sensitive to the level of disorder in its environment. Highly ordered regions have been colored red/yellow. Small regions of increased order, consistent with the presence of lipid raft domains, are apparent. [Photo credit: Katharina Gaus, Enrico Gratton, Eleanor P. W. Kable, et al., "Visualizing lipid structure and raft domains in living cells with two-photon microscopy." PNAS December 23, 2003 100 (26) 15554–15559; https://doi.org/10.1073/pnas.2534386100, Figure 1B. Copyright (2003) National Academy of Sciences, U.S.A.]

One result of these interactions is the moderation of membrane fluidity, making membranes less fluid but at the same time less subject to phase transitions. The presence of lipid rafts thus represents a modification of the original fluid mosaic model for biological membranes. Although their small size and dynamic nature have made them very difficult to study, recent developments in microscopy have significantly improved the ability to study these domains (**Figure 12.34B**).

What are the functions of lipid rafts? By segregation of certain lipid and protein species into smaller domains, lipid rafts may induce conformational changes in membrane proteins, regulating their functional activities. In addition, lipid rafts provide an environment favorable for specific protein–protein interactions, which can be used to promote signal transduction events (Chapter 14).

Furthermore, it appears that pathogens, such as viruses, preferentially bind to lipid rafts to obtain entry into the cell and use lipid rafts to induce membrane budding and subsequent cellular escape. As technology advances, our appreciation of the importance of these membrane domains will likely expand.

All biological membranes are asymmetric

Membranes are structurally and functionally asymmetric: the outer and inner surfaces of all known biological membranes have different components and different enzymatic activities. A clear-cut example is the pump that regulates the concentration of Na$^+$ and K$^+$ ions in cells (**Figure 12.35**). This transport protein is located in the plasma membrane of nearly all cells in higher organisms. The Na$^+$–K$^+$ pump is oriented so that it pumps Na$^+$ out of the cell and K$^+$ into it. Furthermore, ATP must be on the inside of the cell to drive the pump. Ouabain, a specific inhibitor of the pump, is effective only if it is located outside. We shall consider the mechanism of this important and fascinating family of pumps in Chapter 13.

Membrane proteins have a unique orientation because, after synthesis, they are inserted into the membrane in an asymmetric manner. This absolute asymmetry is preserved because membrane proteins do not rotate from one side of the membrane to the other and because membranes are always synthesized by the growth of preexisting membranes.

Lipids, too, are asymmetrically distributed between the two membrane leaflets. For example, in the red-blood-cell membrane, sphingomyelin and phosphatidylcholine are preferentially located in the outer leaflet of the bilayer, whereas phosphatidylethanolamine and phosphatidylserine are located mainly in the inner leaflet. Large amounts of cholesterol are present in both leaflets.

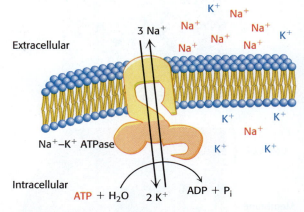

FIGURE 12.35 The orientation of the Na$^+$–K$^+$ transport system in the lipid bilayer exemplifies membrane asymmetry. The Na$^+$–K$^+$ transport system pumps Na$^+$ out of the cell and K$^+$ into the cell by hydrolyzing ATP on the intracellular side of the membrane.

12.6 Prokaryotes and Eukaryotes Differ in Their Use of Biological Membranes

Thus far, we have considered only the plasma membrane of eukaryotic cells. Some bacteria and archaea have only this single membrane, surrounded by a thick cell wall (**Figure 12.36A**). Other bacteria, such as *E. coli*, have two membranes separated by a cell wall (made of proteins, peptides, and carbohydrates) lying between them (**Figure 12.36B**). The inner membrane acts as the permeability barrier, and the outer membrane and the cell wall provide additional protection. The outer membrane is quite permeable to small molecules, owing to the presence of porins. The region between the two membranes containing the cell wall is called the *periplasm*.

The two types of bacterial membranes depicted in Figure 12.36 can be distinguished microscopically using a technique known as **Gram staining**, named for its inventor Hans Christian Gram. The dye called *crystal violet* is added to a fixed sample of bacteria, followed by iodine to trap the dye in the cell. Then, alcohol is used to wash out the dye. In those cells with a thick cell wall, as shown in Figure 12.36A, the trapped dye binds quite tightly and does not wash out. These bacteria stain dark purple, and are referred to as *Gram positive*. In those cells where the cell wall is very thin, the dye washes out quickly; these bacteria appear pink (due to the addition of a second stain at the conclusion of the experiment) and are referred to as *Gram negative* (Figure 12.36B). Gram staining is an important method for the initial classification of a bacterial sample, particularly in fluid samples from an infected patient, where rapid identification of bacterial type is critical for the selection of antibiotic treatment.

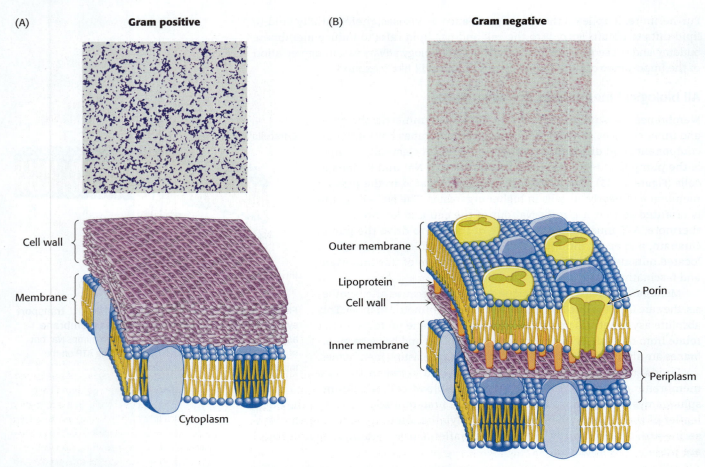

(A) **Gram positive**

Cell wall

Membrane

Cytoplasm

(B) **Gram negative**

Outer membrane

Lipoprotein

Cell wall

Inner membrane

Porin

Periplasm

FIGURE 12.36 Bacteria can be classified based on their plasma membranes. (A) Some bacteria are surrounded by a single lipid bilayer encased in a thick cell wall. These bacteria are termed Gram positive since they retain the crystal violet stain in their cell walls. (B) Other bacteria are surrounded by two separate lipid bilayers. These bacteria are termed Gram negative because they do not retain crystal violet stain and appear pink.
[Photos credit: Richard J. Green/Science Source]

Eukaryotic cells contain compartments bounded by internal membranes

Eukaryotic cells, with the exception of plant cells, do not have cell walls, and their plasma membranes consist of a single lipid bilayer. In plant cells, the cell wall is on the outside of the plasma membrane. Eukaryotic cells are distinguished from prokaryotic cells by the presence of membranes inside the cell that form internal compartments. For example, peroxisomes, organelles that play a major role in the oxidation of fatty acids for energy conversion, are defined by a single membrane. The nucleus is surrounded by a double membrane, the *nuclear envelope*, that consists of a set of closed membranes that come together at structures called *nuclear pores* (**Figure 12.37**). These pores regulate transport into and out of the nucleus. The nuclear envelope is linked to another membrane-defined structure, the *endoplasmic reticulum*, which plays a host of cellular roles, including drug detoxification and the modification of proteins for secretion.

Mitochondria, the organelles in which ATP is synthesized, are also surrounded by two membranes. As in the case for a bacterium, the outer membrane is quite permeable to small molecules, whereas the inner membrane is not. Indeed, considerable evidence now indicates that mitochondria evolved from bacteria by endosymbiosis (Section 18.1).

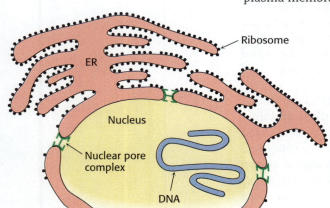

Ribosome

ER

Nucleus

Nuclear pore complex

DNA

FIGURE 12.37 The nuclear envelope is a double membrane connected to another membrane system of eukaryotes, the endoplasmic reticulum.
[Information from E. C. Schirmer and L. Gerace, *Genome Biol.* 3(4):1008.1–1008.4, 2002, reviews, Fig.1.]

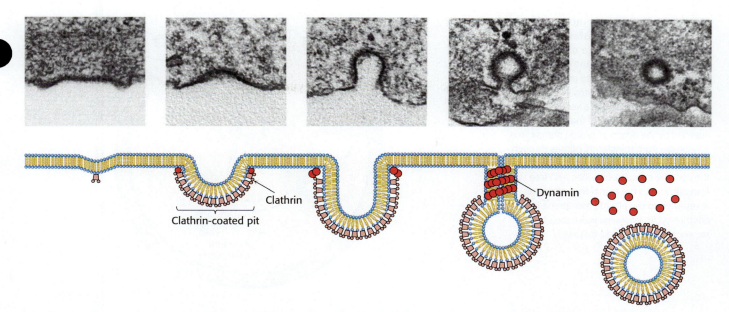

Membrane budding and fusion are highly controlled processes

Membranes must be able to separate (by budding) or join together (by fusion) so that cells and compartments may take up, transport, and release molecules. Let us consider an example of membrane budding using an example at the plasma membrane. Many cells take up molecules through the process of **receptor-mediated endocytosis**. Here, a protein or larger complex initially binds to a receptor on the cell surface. After the receptor is bound, specialized proteins act to cause the membrane in this region to invaginate. One of these specialized proteins is clathrin, which polymerizes into a lattice network around the growing membrane bud, often referred to as a *clathrin-coated pit* (**Figure 12.38**). Through the action of a GTPase enzyme called *dynamin*, the invaginated membrane eventually breaks off and fuses to form a vesicle. Various hormones, transport proteins, and antibodies employ receptor-mediated endocytosis to gain entry into a cell. A less-advantageous consequence is that this pathway is available to viruses and toxins as a means of invading cells. The reverse process—the fusion of a vesicle to a membrane—is a key step in the release of neurotransmitters from a neuron into the synaptic cleft (**Figure 12.39**).

Let's consider one example of receptor-mediated endocytosis. Iron is a critical element for the function and structure of many proteins, including hemoglobin and myoglobin (Chapter 3). However, free iron ions are highly toxic to cells, owing

FIGURE 12.38 Many cells take up molecules by receptor-mediated endocytosis. Receptor binding on the surface of the cell induces the membrane to invaginate, with the assistance of specialized intracellular proteins such as clathrin and dynamin. The process results in the formation of a vesicle within the cell. [Credit: *Top:* Republished with permission of Company of Biologists, from Membrane remodeling in clathrin-mediated endocytosis. Volker Haucke, Michael M. Kozlov, J Cell Sci (2018) 131 (17): jcs216812. Figure 1. Permission conveyed through Copyright Clearance Center, Inc.; *Bottom:* Schmid SL. Reciprocal regulation of signaling and endocytosis: Implications for the evolving cancer cell. J Cell Biol. 2017 Sep 4;216(9):2623-2632. doi: 10.1083/jcb.201705017. Epub 2017 Jul 3. PMID: 28674108; PMCID: PMC5584184. Copyright © 2017 Schmid. Illustration by Marcel Mettlen. Used with permission.]

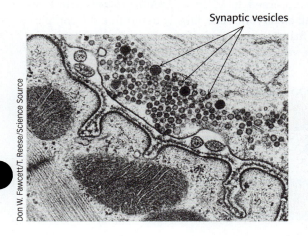

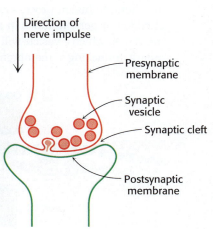

FIGURE 12.39 Vesicle fusion to the plasma membrane is critical for neurotransmitter release. Neurotransmitter-containing synaptic vesicles (indicated by the arrows on the micrograph at left) are arrayed near the plasma membrane of a nerve cell. Synaptic vesicles fuse with the plasma membrane, releasing the neurotransmitter into the synaptic cleft.

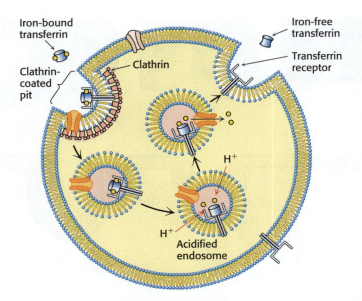

to their ability to catalyze the formation of free radicals. Hence, the transport of iron atoms from the digestive tract to the cells where they are most needed must be tightly controlled. In the bloodstream, iron is bound very tightly by the protein transferrin, which can bind two Fe^{3+} ions with a dissociation constant of 10^{-23} M at neutral pH. Cells requiring iron express the transferrin receptor in their plasma membranes (Section 31.6).

Formation of a complex between the transferrin receptor and iron-bound transferrin initiates receptor-mediated endocytosis, internalizing these complexes within vesicles called **endosomes** (**Figure 12.40**). As the endosomes mature, proton pumps within the vesicle membrane lower the lumenal pH to about 5.5. Under these conditions, the affinity of iron ions for transferrin is reduced; these ions are released and are free to pass through channels in the endosomal membranes into the cytoplasm. The iron-free transferrin complex is recycled to the plasma membrane, where transferrin is released back into the bloodstream and the transferrin receptor can participate in another uptake cycle.

Membrane budding and fusion events throughout the cell must be highly specific in order to prevent incorrect membrane fusion events. The structures of the intermediates in these processes and the detailed mechanisms remain ongoing areas of investigation. Researchers have identified *SNARE proteins*—short for <u>s</u>oluble <u>N</u>-ethylmaleimide-sensitive-factor <u>a</u>ttachment protein <u>re</u>ceptor—as essential to these processes. SNARE proteins guide membrane fusion by drawing appropriate lipid bilayers together through the formation of tightly coiled four-helical bundles (**Figure 12.41**). Once these membranes are brought into close proximity with one another, the fusion process can proceed. In addition, SNARE proteins, which are encoded by gene families in all eukaryotic cells, largely determine the compartment with which a vesicle will fuse. The specificity of membrane fusion ensures the orderly trafficking of membrane vesicles and their cargos through eukaryotic cells.

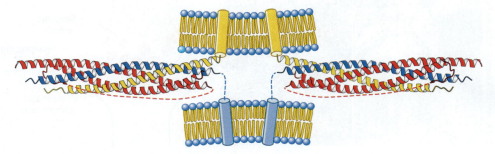

Summary

12.1 Fatty Acids Are Key Constituents of Lipids

- Fatty acids are hydrocarbon chains of various lengths and degrees of unsaturation that terminate with a carboxylic acid group.
- The fatty acid chains in membranes usually contain between 14 and 24 carbon atoms; they may be saturated or unsaturated.
- Short chain length and unsaturation enhance the fluidity of fatty acids and their derivatives by lowering the melting temperature.

12.2 Biological Membranes Are Composed of Three Common Types of Membrane Lipids

- The major types of membrane lipids are phospholipids, glycolipids, and cholesterol.
- Phosphoglycerides, a type of phospholipid, consist of a glycerol backbone, two fatty acid chains, and a phosphorylated alcohol.
- Phosphatidylcholine, phosphatidylserine, and phosphatidylethanolamine are major phosphoglycerides.
- Sphingomyelin, a different type of phospholipid, contains a sphingosine backbone instead of glycerol.
- Glycolipids are sugar-containing lipids derived from sphingosine.
- Cholesterol, which modulates membrane fluidity, is constructed from a steroid nucleus.

12.3 Phospholipids and Glycolipids Readily Form Bimolecular Sheets in Aqueous Media

- Membrane lipids spontaneously form extensive bimolecular sheets in aqueous solutions.
- The driving force for membrane formation is the hydrophobic interactions among the fatty acid tails of membrane lipids. The hydrophilic head groups interact with the aqueous medium.
- Lipid bilayers are cooperative structures, held together by many weak bonds.
- These lipid bilayers are highly impermeable to ions and most polar molecules, yet they are quite fluid, which enables them to act as a solvent for membrane proteins.

12.4 Proteins Carry Out Most Membrane Processes

- Biological membranes vary in their protein content, which reflects the diverse functions of various cell types.

- Integral membrane proteins interact extensively with the hydrocarbon chains of membrane lipids, and they can be released only by agents that compete for these nonpolar interactions.
- Peripheral membrane proteins are bound to membranes primarily by electrostatic and hydrogen-bond interactions with the head groups of lipids and can be released by adding salt or changing the pH.
- Membrane-spanning proteins have regular structures, including β strands, although the α helix is most common.
- Membrane-spanning α helices can often be predicted from amino acid sequence using a hydropathy plot.

12.5 Lipids and Many Membrane Proteins Diffuse Rapidly in the Plane of the Membrane

- Membranes are dynamic structures in which proteins and lipids diffuse rapidly in the plane of the membrane.
- The rotation of lipids from one face of a membrane to the other is usually very slow.
- Proteins do not rotate across bilayers; hence, membrane asymmetry can be preserved.
- The degree of fluidity of a membrane depends on the chain length of its lipids and on the extent to which their constituent fatty acids are unsaturated.
- In animals, cholesterol content also regulates membrane fluidity.

12.6 Prokaryotes and Eukaryotes Differ in Their Use of Biological Membranes

- Most bacteria exhibit one of two types of membranes: a single membrane surrounded by a cell wall, or a double membrane, where the inner membrane serves as the permeability barrier.
- An extensive array of internal membranes in eukaryotes creates compartments within a cell for distinct biochemical functions.
- The budding and fusion of membranes are highly controlled processes.
- Receptor-mediated endocytosis enables the formation of intracellular vesicles when ligands bind to their corresponding receptor proteins in the plasma membrane.
- The reverse process—the fusion of a vesicle to a membrane—is a key step in the release of signaling molecules outside the cell.

Key Terms

lipids (p. 353)

fatty acid (p. 353)

phospholipid (p. 355)

phosphoglyceride (p. 356)

sphingomyelin (p. 357)

sphingosine (p. 357)

glycolipid (p. 357)

cholesterol (p. 357)

amphipathic (amphiphilic) molecule (p. 358)

micelle (p. 359)

lipid bilayer (p. 359)

liposome (lipid vesicle) (p. 360)

myelin (p. 362)

integral membrane protein (p. 363)

peripheral membrane protein (p. 363)

hydropathy plot (p. 368)

lateral diffusion (p. 369)

fluid mosaic model (p. 370)

lipid rafts (p. 372)

Gram staining (p. 373)

receptor-mediated endocytosis (p. 375)

endosomes (p. 376)

Problems

1. Consider the following fatty acid:

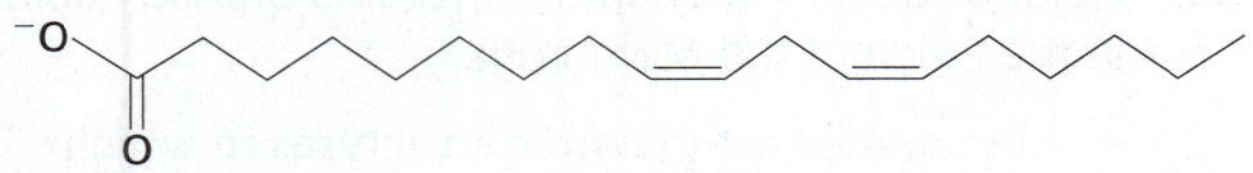

What is an accurate designation of this structure that indicates the length, degree of unsaturation, and location of double bonds? What is the common name for this fatty acid (see Table 12.1)? What are the positions of the double bonds expressed in relation to the distal carbon atom? ❖ **1**

2. Draw the chemical structure of cis-Δ^9-hexadecenoic acid. What is the common name for this fatty acid? ❖ **1**

3. Platelet-activating factor (PAF) is a phospholipid that plays a role in allergic and inflammatory responses, as well as in toxic shock syndrome. The structure of PAF is shown here. How does it differ from the structures of the phospholipids discussed in this chapter? ❖ **2**

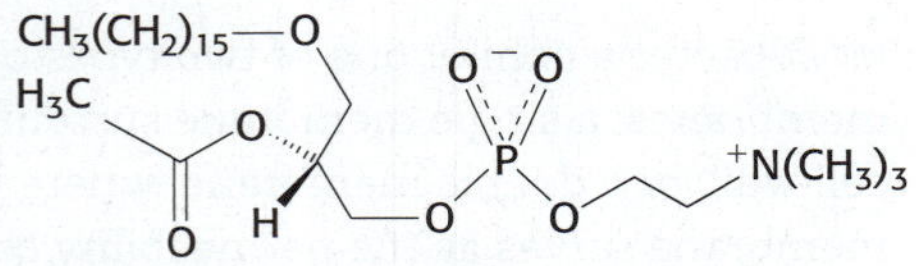

$$CH_3(CH_2)_{15}\!-\!O$$

Platelet-activating factor (PAF)

4. Match the following statements with the type of lipid it best describes: phospholipids, sphingolipids, glycolipids, archaeal lipids, or cholesterol. Some statements may describe more than one lipid type: ❖ **2**

(a) Includes two fatty acids joined to glycerol by ester linkages

(b) Does not contain glycerol

(c) Is a steroid

(d) Contains a sphingosine backbone

(e) Contains one or more sugars

(f) Usually have branched alkyl chains

5. What is the average distance traversed by a membrane lipid in 1 μs, 1 ms, and 1 s? Assume a diffusion coefficient of 10^{-8} cm^2 s^{-1}. ❖ **6**

6. Which of the following molecules could align side-by-side to form a lipid bilayer: phospholipids, fatty acids, triacylglycerol, sphingolipids, cholesterol? ❖ **3**

7. Label the molecules in the diagram below as either integral membrane proteins, peripheral membrane proteins, or lipid-anchored membrane proteins: ❖ **4**

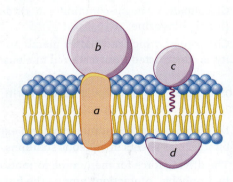

8. Match each statement to the category of membrane-associated protein it describes: membrane-spanning α helix, membrane spanning β strands, lipid-anchored membrane protein, or peripheral membrane protein: ❖ **4**

(a) Contains residue(s) with covalently attached fatty acyl groups

(b) Often tethered to the membrane via a membrane-embedded protein

(c) Composed of about 20 hydrophobic residues

(d) Can usually be released from the membrane with concentrated salt solutions

(e) Features a pattern of alternating nonpolar and polar side chains

9. On the basis of the following hydropathy plots for three proteins (A–C), predict which would be membrane proteins. What are the ambiguities with respect to using such plots to determine if a protein is a membrane protein? ❖ **5**

(A)

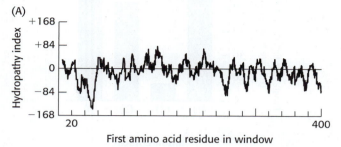

First amino acid residue in window

(B)

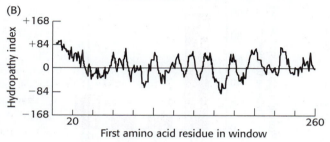

First amino acid residue in window

(C)

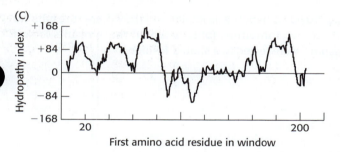

First amino acid residue in window

10. Hydropathy plot analysis of your protein of interest reveals a single, prominent hydrophobic peak. However, you later discover that this protein is soluble and not membrane associated. Explain how the hydropathy plot may have been misleading. ❖ **5**

11. The diffusion coefficient, D, of a rigid spherical molecule is given by

$$D = kT/6\pi\eta r$$

in which η is the viscosity of the solvent, r is the radius of the sphere, k is the Boltzmann constant (1.38×10^{-16} erg degree^{-1}), and T is the absolute temperature. What is the diffusion coefficient at 37°C of a 100-kDa protein in a membrane that has an effective viscosity of 1 poise (1 poise = 1 erg s^{-1} cm^{-3})? What is the average distance traversed by this protein in 1 μs, 1 ms, and 1 s? Assume that this protein is an unhydrated, rigid sphere with a density of 1.35 g cm^{-3}. ❖ **5**

12. The red curve on the following graph shows the fluidity of the fatty acids of a phospholipid bilayer as a function of temperature. The blue curve shows the fluidity in the presence of cholesterol. ❖ **6**

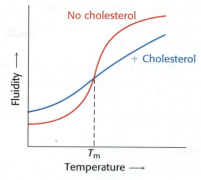

(a) What is the effect of cholesterol?

(b) Why might this effect be biologically important?

13. Calculate the distance olive oil (a lipid) could move in a membrane in 13 seconds assuming the diffusion coefficient is 1 μm^2 s^{-1}. Use the equation $S = (4Dt)^{1/2}$ where S is the distance traveled, t is time, and D is the diffusion coefficient. ❖ **6**

14. Arrange the fatty acids in order of increasing melting point:

(a) Stearic acid (18:0)

(b) Oleic acid (18:1)

(c) Palmitoleic acid (16:1)

(d) α-Linoleic acid (18:3)

(e) Palmitic acid (16:0)

15. Each intracellular fusion of a vesicle with a membrane requires a SNARE protein on the vesicle (called the v-SNARE) and a SNARE protein on the target membrane (called the t-SNARE). Assume that a genome encodes 21 members of the v-SNARE family and 7 members of the t-SNARE family. With the assumption of no specificity, how many potential v-SNARE–t-SNARE interactions could take place? ❖ **7**

16. Some antibiotics act as carriers that bind an ion on one side of a membrane, diffuse through the membrane, and release the ion on the other side. The conductance of a lipid-bilayer membrane containing a carrier antibiotic decreased abruptly when the temperature was lowered from 40°C to 36°C. In contrast, there was little change in conductance of the same bilayer membrane when it contained a channel-forming antibiotic. Why?

17. Of the following pairs of ions and/or molecules, identify which one is more permeable across a lipid bilayer:

(a) K$^+$ vs. H$_2$O

(b) Tryptophan vs. indole

(c) H$_2$O vs. Cl$^-$

In general the permeability coefficient is most directly correlated with which property: size, flexibility, solubility in nonpolar solvents, or interaction with polar head groups?

18. Small mammalian hibernators can withstand body temperatures of 0–5°C without injury. However, the body fats of most mammals have melting temperatures of approximately 25°C. Predict how the composition of the body fat of hibernators might differ from that of their nonhibernating cousins. ❖ 6

19. Classify each of these statements as to whether they describe saturated phospholipids, unsaturated phospholipids, or both:

(a) Contain one or more double bonds within their fatty acid tails

(b) Are built upon a glycerol backbone

(c) Produce a fairly inflexible membrane at low temperatures

(d) Have bent fatty acid tails

20. Ibuprofen and indomethacin are clinically important inhibitors of prostaglandin H_2 synthase-1. Cells expressing this enzyme were incubated under the following conditions, after which the activity of the enzyme was measured by adding radiolabeled arachidonic acid and detecting newly produced prostaglandin H_2:

(1) 40 min *without* inhibitor (control)

(2) 40 min with inhibitor

(3) 40 min with inhibitor, after which the cells were resuspended in medium *without* inhibitor

(4) 40 min with inhibitor, after which the cells were resuspended in medium *without* inhibitor and incubated for an additional 30 minutes

The results are shown below:

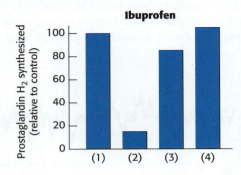

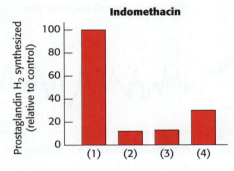

(a) Provide a hypothesis explaining the different results for these two inhibitors. **(b)** How would these results look if aspirin were tested in a similar fashion?

Membrane Channels and Pumps

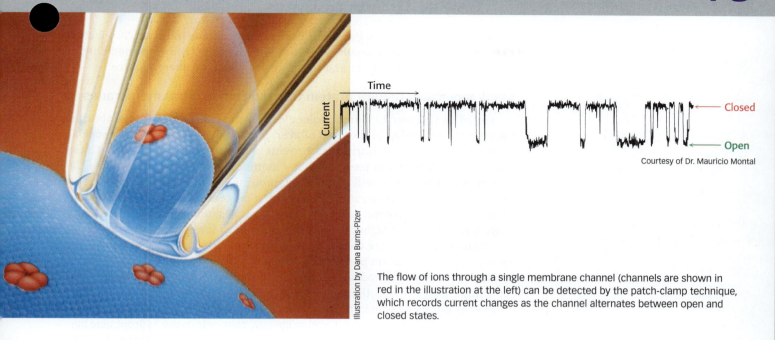

Illustration by Dana Burns-Pizer

Courtesy of Dr. Mauricio Montal

The flow of ions through a single membrane channel (channels are shown in red in the illustration at the left) can be detected by the patch-clamp technique, which records current changes as the channel alternates between open and closed states.

❖ LEARNING GOALS

By the end of this chapter, you should be able to:

1. Distinguish between passive and active transport.
2. Calculate the free energy stored in a concentration or electrochemical gradient.
3. Compare the mechanisms of the P-type ATPase pump and the ABC transporter pump.
4. Contrast primary and secondary active transport.
5. Explain how channels are able to rapidly and selectively move ions across membranes.
6. Describe two mechanisms by which ion channels are gated.
7. Explain the steps that lead to the formation of an action potential.

OUTLINE

13.1 The Transport of Molecules Across a Membrane May Be Active or Passive

13.2 Two Families of Membrane Proteins Use ATP Hydrolysis to Actively Transport Ions and Molecules Across Membranes

13.3 Lactose Permease Is an Archetype of Secondary Transporters That Use One Concentration Gradient to Power the Formation of Another

13.4 Specific Channels Can Rapidly Transport Ions Across Membranes

13.5 Gap Junctions Allow Ions and Small Molecules to Flow Between Communicating Cells

13.6 Specific Channels Increase the Permeability of Some Membranes to Water

The lipid bilayer of biological membranes is intrinsically impermeable to ions and polar molecules, yet these species must be able to cross these membranes for normal cell function. As we will discuss in this chapter, permeability is conferred by three classes of membrane transporter proteins: pumps, carriers, and channels. These proteins are critical for many basic cellular functions, including access to ions and solutes for metabolism, conduction of nerve impulses, contraction of muscles, and communication of signals within cells.

Consider, for example, the rhythmic beating of the heart: the coordinated movement of ions across cardiac cellular membranes can sustain the contraction of heart muscle for, on average, 60 times each minute for over 70 years—a total of over 2.2 billion contractions! The remarkable consistency of cardiac contraction is just one example of the importance of these protein classes.

13.1 The Transport of Molecules Across a Membrane May Be Active or Passive

We will first consider some general principles of membrane transport. Two factors determine whether a molecule will cross a membrane: (1) the permeability of the molecule in a lipid bilayer and (2) the availability of an energy source.

Many molecules require protein transporters to cross membranes

As stated in Chapter 12, some molecules can pass through cell membranes because they dissolve in the lipid bilayer. Such molecules are called *lipophilic molecules*. The steroid hormones provide a physiological example. These cholesterol relatives can pass through a membrane, but what determines the direction in which they will move? Such molecules will pass through a membrane down their concentration gradient in a process called **simple diffusion**. In accord with the Second Law of Thermodynamics, molecules spontaneously move from a region of higher concentration to one of lower concentration.

Matters become more complicated when the molecule is highly polar. For example, sodium ions are present at 143 mM outside a typical cell and at 14 mM inside the cell. However, sodium does not freely enter the cell, because the charged ion cannot pass through the hydrophobic membrane interior. In some circumstances, as during a nerve impulse, sodium ions must enter the cell. How are they able to do so? Sodium ions pass through specific channels in the hydrophobic barrier formed by membrane proteins. This means of crossing the membrane is called **facilitated diffusion** because the diffusion across the membrane is facilitated by the channel. It is also called **passive transport** because the energy driving the ion movement originates from the ion gradient itself, without any contribution by the transport system. Channels, like enzymes, display substrate specificity in that they facilitate the transport of some ions, but not other, even closely related, ions.

How is the sodium gradient established in the first place? In this case, sodium must move, or be pumped, *against* a concentration gradient. Because moving the ion from a low concentration to a higher concentration results in a decrease in entropy, it requires an input of free energy. Protein transporters embedded in the membrane are capable of using an energy source to move the molecule up a concentration gradient. Because an input of energy from another source is required, this means of crossing the membrane is called **active transport**.

Free energy stored in concentration gradients can be quantified

An unequal distribution of molecules is an energy-rich condition because free energy is minimized when all concentrations are equal. Consequently, to attain such an unequal distribution of molecules requires an input of free energy. How can we quantify the amount of energy required to generate a concentration gradient (**Figure 13.1**)? Consider an *uncharged* solute molecule. The free-energy change in transporting this species from side 1, where it is present at a concentration of c_1, to side 2, where it is present at concentration c_2, is

$$\Delta G = RT \ln (c_2/c_1)$$

where R is the gas constant (8.315×10^{-3} kJ mol^{-1} deg^{-1}, or 1.987×10^{-3} kcal mol^{-1} deg^{-1}) and T is the temperature in kelvins. A transport process must be active when ΔG is positive, whereas it can be passive when ΔG is negative. For example, consider the transport of an uncharged molecule from $c_1 = 10^{-3}$ M to $c_2 = 10^{-1}$ M at 25°C (298 K).

$$\Delta G = RT \ln (10^{-1}/10^{-3})$$
$$= (8.315 \times 10^{-3}) \times 298 \times \ln (10^2)$$
$$= +11.4 \text{ kJ } mol^{-1} (+2.7 \text{ kcal } mol^{-1})$$

(A)

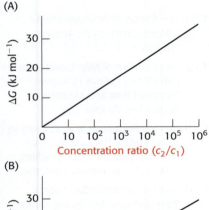

(B)

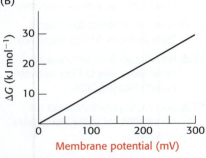

FIGURE 13.1 Transport of solutes and ions across a membrane imposes an energetic cost. The free-energy change in transporting (A) an uncharged solute from a compartment at concentration c_1 to one at c_2 and (B) a singly charged species across a membrane to the side having the same charge as that of the transported ion are shown. The free-energy change imposed by a membrane potential of 59 mV is equivalent to that imposed by a concentration ratio of 10 for a singly charged ion at 25°C.

The ΔG is +11.4 kJ mol^{-1} (+2.7 kcal mol^{-1}), indicating that this transport process requires an input of free energy.

For a *charged* species, the unequal distribution across the membrane generates an electrical potential that also must be considered because the ions will be repelled by like charges. The sum of the concentration and electrical terms is called the **electrochemical potential** or **membrane potential**. The free-energy change is then given by

$$\Delta G = RT \ln (c_2/c_1) + ZF\Delta V$$

in which Z is the electrical charge of the transported species, ΔV is the potential in volts across the membrane (expressed as $V_{\text{side 2}} - V_{\text{side 1}}$), and F is the Faraday constant (96.5 kJ V^{-1} mol^{-1}, or 23.1 kcal V^{-1} mol^{-1}).

EXAMPLE Calculating the Energetic Cost of Ion Transport

PROBLEM: Suppose that the concentrations of Na$^+$ outside and inside the cell are 143 and 14 mM, respectively, and the corresponding values for K$^+$ are 4 and 157 mM. At a membrane potential of –50 mV and a temperature of 37°C, what are the energetic costs of transporting a K$^+$ ion into the cell and a Na$^+$ ion out of the cell?

GETTING STARTED: Since potassium and sodium ions are both charged species, we must use the electrochemical potential equation described in this section to determine the free energy change. As both ions carry a single positive charge, Z will be +1 for each calculation. Applying the equation is rather straightforward, as long as we make sure we're keeping each side of the membrane and the direction of transport consistent. It is important to remember that, by convention, the membrane potential ΔV is expressed as the voltage of the inside of the cell minus the voltage of the outside of the cell ($V_{\text{inside}} - V_{\text{outside}}$).

CALCULATE: For K$^+$ transport, the problem asks us to move the ion into the cell. Hence, the inside of the cell will be side 2 and the outside of the cell will be side 1.

$$\Delta G = RT \ln (c_2/c_1) + ZF\Delta V = RT \ln(c_2/c_1) + ZF(V_2 - V_1)$$

$$\Delta G = (8.315 \times 10^{-3} \text{ kJ mol}^{-1} \text{ deg}^{-1}) \times (310 \text{ deg}) \times \ln (157/4)$$

$$+(1) \times (96.5 \text{ kJ mol}^{-1} \text{ V}^{-1}) \times (-0.05 \text{ V})$$

$$\Delta G = 4.64 \text{ kJ mol}^{-1}$$

For Na$^+$ transport, we are calculating the cost of moving the ion out of the cell, so the *outside* of the cell is side 2 and the *inside* is side 1. Thus, the sign of the ΔV term is opposite that of the first part of the question.

$$\Delta G = RT \ln(c_2/c_1) + ZF\Delta V = RT \ln(c_2/c_1) + ZF(V_2 - V_1)$$

$$\Delta G = (8.315 \times 10^{-3} \text{ kJ mol}^{-1} \text{ deg}^{-1}) \times (310 \text{ deg}) \times \ln (143/14)$$

$$+(1) \times (96.5 \text{ kJ mol}^{-1} \text{ V}^{-1}) \times (0.05 \text{ V})$$

$$\Delta G = 10.82 \text{ kJ mol}^{-1}$$

REFLECT: For both ions, moving the charge against its concentration gradient is energetically unfavorable. However, the cost is significantly greater for Na$^+$ than K$^+$. Why is this the case? Note that while that the chemical term is positive (unfavorable) for both ions, the electrical term is favorable for K$^+$ (ΔV is negative) and unfavorable for Na$^+$ (ΔV is positive) as the membrane has net negative charge on its inner surface. Essentially, transporting Na$^+$ out of the cell requires movement against its chemical gradient and in an electrically unfavorable direction.

13.2 Two Families of Membrane Proteins Use ATP Hydrolysis to Actively Transport Ions and Molecules Across Membranes

The extracellular fluid of animal cells has a salt concentration similar to that of seawater. However, cells must control their intracellular salt concentrations to facilitate specific processes, such as signal transduction and action potential propagation, and prevent unfavorable interactions with high concentrations of ions such as Ca^{2+}. For instance, most animal cells contain a high concentration of K^+ and a low concentration of Na^+ relative to the external medium. To generate these gradients, energy expenditure is required. **Pumps** are membrane proteins that catalyze the active transport of ions against their electrochemical gradients. The sodium and potassium gradients across the plasma membrane are generated by a specific transport system, an enzyme that is called the *Na^+–K^+ pump*, or the *Na^+–K^+ ATPase*. The hydrolysis of ATP by the pump provides the energy needed for the active transport of Na^+ out of the cell and K^+ into the cell, generating the gradients. The pump is called the Na^+–K^+ ATPase because the hydrolysis of ATP takes place only when Na^+ and K^+ are present. This ATPase, like all such enzymes, requires Mg^{2+}.

The Na^+–K^+ ATPase transports ions with a stoichiometry of 3 Na^+ ions and 2 K^+ ions per catalytic cycle. In the example at the end of the last section, we calculated change in free energy accompanying the transport of Na^+ and K^+. Taking into account the transport of 3 Na^+ ions and 2 K^+ ions for every transport cycle, the free energy required to operate the pump is:

$$\Delta G = 3 \times (10.82 \text{ kJ mol}^{-1}) + 2 \times (4.64 \text{ kJ mol}^{-1})$$

$$\Delta G = 41.7 \text{ kJ mol}^{-1}$$

Under typical cellular conditions, the hydrolysis of a single ATP molecule per transport cycle provides sufficient free energy, about -50 kJ mol^{-1} (-12 kcal mol^{-1}) to drive the uphill transport of these ions. The active transport of Na^+ and K^+ is of great physiological significance. Indeed, more than a third of the ATP consumed by a resting animal is used to pump these ions. The Na^+–K^+ gradient in animal cells controls cell volume, renders neurons and muscle cells electrically excitable, and drives the active transport of sugars and amino acids. The purification of other ion pumps has revealed a large family of evolutionarily related ion pumps including proteins from bacteria, archaea, and all eukaryotes. Each of these pumps is specific for a particular ion or set of ions. Two are of particular interest: the *sarcoplasmic reticulum Ca^{2+} ATPase* (or SERCA) transports Ca^{2+} out of the cytoplasm and into the sarcoplasmic reticulum of muscle cells, and the *gastric H^+–K^+ ATPase* is the enzyme responsible for pumping sufficient protons into the stomach to lower the pH to 1.0. These enzymes and the hundreds of known homologs, including the Na^+–K^+ ATPase, are referred to as **P-type ATPases** because they form a key phosphorylated intermediate. In the formation of this intermediate, a phosphoryl group from ATP is linked to the side chain of a specific conserved aspartate residue in the ATPase to form phosphorylaspartate.

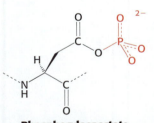

Phosphorylaspartate

P-type ATPases couple phosphorylation and conformational changes to pump calcium ions across membranes

Membrane pumps function by mechanisms that are simple in principle but often complex in detail. Fundamentally, each pump protein can exist in two principal conformational states, one with ion-binding sites open to one side of the membrane and the other with ion-binding sites open to the other side (**Figure 13.2**). To pump ions in a single direction across a membrane, the free energy of ATP hydrolysis must be coupled to the interconversion between these conformational states.

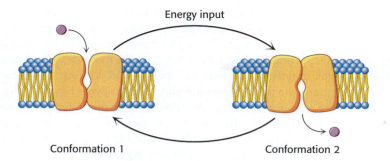

FIGURE 13.2 Pumps use the input of energy to transport ions across a membrane. This simple scheme illustrates the pumping of a molecule across a membrane. The pump interconverts between two conformational states, each with a binding site accessible to a different side of the membrane.

We will consider the structural and mechanistic features of P-type ATPases by examining SERCA. The properties of this P-type ATPase have been established in great detail by relying on crystal structures of the pump in five different states. This enzyme, which constitutes 80% of the total protein in the sarcoplasmic reticulum membrane, plays an important role in relaxation of contracted muscle. Muscle contraction is triggered by an abrupt rise in the cytoplasmic calcium ion level. Subsequent muscle relaxation depends on the rapid removal of Ca^{2+} from the cytoplasm into the sarcoplasmic reticulum, a specialized compartment for Ca^{2+} storage, by SERCA. This pump maintains a Ca^{2+} concentration of approximately 0.1 μM in the cytoplasm, compared to nearly 1.5 mM in the sarcoplasmic reticulum.

The first structure of SERCA to be determined had Ca^{2+} bound, but no nucleotides present (**Figure 13.3**). SERCA is a single 110-kDa polypeptide with a transmembrane domain consisting of 10 α helices. The transmembrane domain includes sites for binding two calcium ions. Each calcium ion is coordinated to seven oxygen atoms coming from a combination of side-chain glutamate, aspartate, threonine, and asparagine residues, backbone carbonyl groups, and water molecules. A large cytoplasmic headpiece constitutes nearly half the molecular weight of the protein and consists of three distinct domains, each with a distinct function. One domain (N) binds the ATP <u>n</u>ucleotide, another (P) accepts the <u>p</u>hosphoryl group on a conserved aspartate residue, and the third (A) serves as an <u>a</u>ctuator, linking changes in the N and P domains to the transmembrane part of the enzyme.

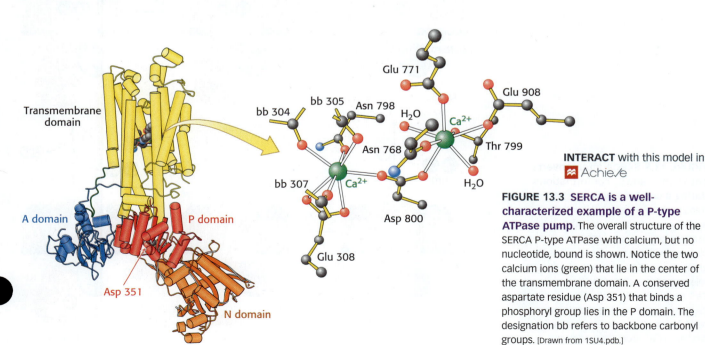

INTERACT with this model in **Achieve**

FIGURE 13.3 SERCA is a well-characterized example of a P-type ATPase pump. The overall structure of the SERCA P-type ATPase with calcium, but no nucleotide, bound is shown. Notice the two calcium ions (green) that lie in the center of the transmembrane domain. A conserved aspartate residue (Asp 351) that binds a phosphoryl group lies in the P domain. The designation bb refers to backbone carbonyl groups. [Drawn from 1SU4.pdb.]

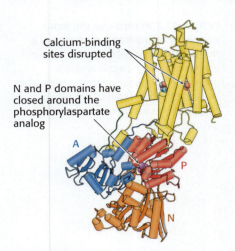

Calcium-binding sites disrupted

N and P domains have closed around the phosphorylaspartate analog

A

P

N

FIGURE 13.4 SERCA undergoes dynamic conformational changes. This structure was determined in the absence of bound calcium but with a phosphorylaspartate analog present in the P domain. Notice how different this structure is from the calcium-bound form shown in Figure 13.3: both the transmembrane part (yellow) and the A, P, and N domains have substantially rearranged. [Drawn from 1WPG.pdb.]

INTERACT with this model in

Achieve

SERCA is a remarkably dynamic protein. For example, the structure of SERCA without bound Ca^{2+}, but with a phosphorylaspartate analog present in the P domain, is shown in **Figure 13.4**. The N and P domains are now closed around the phosphorylaspartate analog, and the A domain has rotated substantially relative to its position in SERCA with Ca^{2+} bound and without the phosphoryl analog. Furthermore, the transmembrane part of the enzyme has rearranged significantly and the well-organized Ca^{2+}-binding sites are disrupted. These sites are now accessible from the side of the membrane opposite the N, P, and A domains.

The structural results can be combined with other studies to construct a detailed mechanism for Ca^{2+} pumping by SERCA (**Figure 13.5**):

1. The catalytic cycle begins with the enzyme in its unphosphorylated state with two calcium ions bound. We will refer to the overall enzyme conformation in this state as E_1; with Ca^{2+} bound, it is E_1-$(Ca^{2+})_2$. In this conformation, SERCA can bind calcium ions only on the cytoplasmic side of the membrane. This conformation is shown in Figure 13.3.

2. In the E_1 conformation, the enzyme can bind ATP. The N, P, and A domains undergo substantial rearrangement as they close around the bound ATP, but there is no substantial conformational change in the transmembrane domain. The calcium ions are now trapped inside the enzyme.

3. The phosphoryl group is then transferred from ATP to Asp 351.

4. Upon ADP release, the enzyme again changes its overall conformation, including the membrane domain this time. This new conformation is referred to as E_2 or E_2-P in its phosphorylated form. The process of interconverting the E_1 and E_2 conformations is sometimes referred to as **eversion**. In the E_2-P conformation, the Ca^{2+}-binding sites become disrupted and the calcium ions are released to the side of the membrane opposite that at which they entered; ion transport has been achieved. This conformation is shown in Figure 13.4.

5. The phosphorylaspartate residue is hydrolyzed to release inorganic phosphate.

FIGURE 13.5 SERCA interconverts between E_1 and E_2 conformations during a catalytic cycle. Ca^{2+} ATPase transports Ca^{2+} through the membrane by a mechanism that includes (1) Ca^{2+} binding from the cytoplasm, (2) ATP binding, (3) ATP cleavage with the transfer of a phosphoryl group to Asp 351 on the enzyme, (4) ADP release and eversion of the enzyme to release Ca^{2+} on the opposite side of the membrane, (5) hydrolysis of the phosphorylaspartate residue, and (6) eversion to prepare for the binding of Ca^{2+} from the cytoplasm.

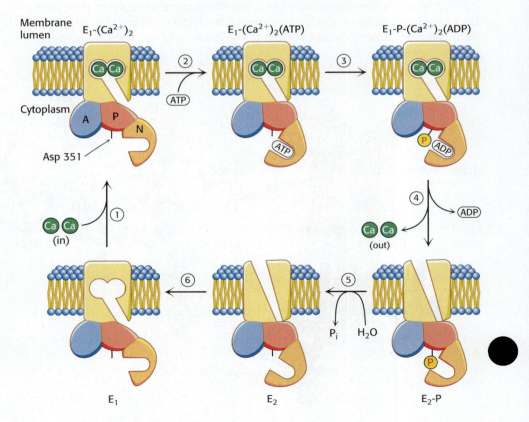

Membrane lumen

E_1-$(Ca^{2+})_2$ E_1-$(Ca^{2+})_2$(ATP) E_1-P-$(Ca^{2+})_2$(ADP)

Cytoplasm   A   P   N

Asp 351

ATP

ADP

Ca Ca (in)   Ca Ca (out)

P_i H_2O

E_1 E_2 E_2-P

6. With the release of phosphate, the interactions stabilizing the E_2 conformation are lost, and the enzyme everts to the E_1 conformation. The binding of two calcium ions from the cytoplasmic side of the membrane completes the cycle.

This mechanism likely applies to other P-type ATPases. For example, Na^+–K^+ ATPase is an $\alpha_2\beta_2$ tetramer. Its α subunit is homologous to SERCA and includes a key aspartate residue analogous to Asp 351. The β subunit does not directly take part in ion transport. A mechanism analogous to that shown in Figure 13.5 applies, with three Na^+ ions binding from the inside of the cell to the E_1 conformation and two K^+ ions binding from outside the cell to the E_2 conformation.

Digoxin specifically inhibits the Na^+–K^+ pump by blocking its dephosphorylation

Certain steroids derived from plants are potent inhibitors ($K_i \approx 10$ nM) of the Na^+–K^+ pump. Digoxin and ouabain are members of this class of inhibitors, which are known as *cardiotonic steroids* because of their strong effects on the heart (**Figure 13.6**). These compounds inhibit the dephosphorylation of the E_2-P form of the ATPase when applied on the extracellular face of the membrane.

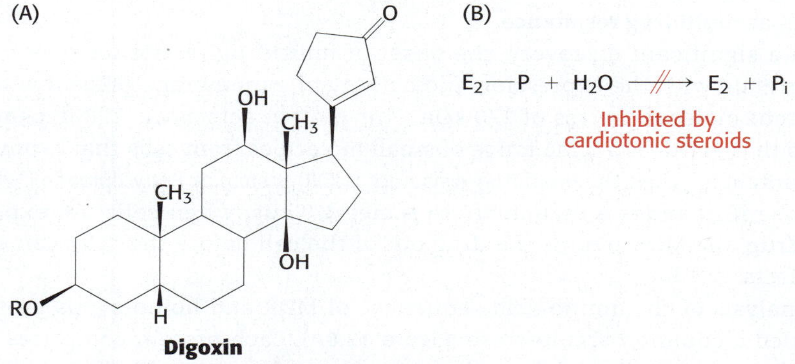

(A)

Digoxin

(B)

$$E_2\!-\!P + H_2O \xrightarrow{\;\;/\!/\;\;} E_2 + P_i$$

Inhibited by cardiotonic steroids

FIGURE 13.6 Digoxin is an example of a cardiotonic steroid. (A) The chemical structure of the cardiotonic steroid digoxin is shown. (B) Cardiotonic steroids inhibit the Na^+–K^+ pump by blocking the dephosphorylation of E_2-P.

Digoxin is one of the cardiotonic steroids derived from the dried leaf of the foxglove plant, *Digitalis purpurea* (**Figure 13.7**). The compound increases the force of contraction of heart muscle and is consequently administered for the treatment of congestive heart failure. Inhibition of the Na^+–K^+ pump by digoxin leads to a higher level of Na^+ inside the cell. As we shall see (Section 13.3), the diminished Na^+ gradient results in slower removal of Ca^{2+} from the cytoplasm. The subsequent increase in the intracellular level of Ca^{2+} enhances the ability of cardiac muscle to contract.

It is interesting to note that cardiotonic steroids were used effectively long before the discovery of the Na^+–K^+ ATPase. In 1785, William Withering, a British physician, heard tales of an elderly woman, known as "the old woman of Shropshire," who cured people of "dropsy" (which today would be recognized as congestive heart failure) with an extract of foxglove. Withering conducted the first scientific study of the effects of foxglove on congestive heart failure and documented its effectiveness.

FIGURE 13.7 Foxglove (*Digitalis purpurea*) is a naturally occurring source of digoxin.

Roger Hall/Shutterstock

P-type ATPases are evolutionarily conserved and play a wide range of roles

Analysis of the complete yeast genome revealed the presence of 16 proteins that clearly belong to the P-type ATPase family. More-detailed sequence analysis suggests that two of these proteins transport H^+ ions, two transport Ca^{2+}, three transport Na^+, and two transport metals such as Cu^{2+}. In addition, five members of this family appear to participate in the transport of phospholipids with amino acid head groups. These five proteins help maintain membrane asymmetry by transporting lipids such as phosphatidylserine from the outer to the inner leaflet of the bilayer membrane. Such enzymes have been termed *flippases*. Remarkably, the human genome encodes 70 P-type ATPases. All members of this protein family employ the same fundamental mechanism: the free energy of ATP hydrolysis drives membrane transport by means of conformational changes, which are induced by the addition and removal of a phosphoryl group at an analogous aspartate site in each protein.

Multidrug resistance highlights a family of membrane pumps with ATP-binding cassette domains

Studies of human disease have revealed another large and important family of active-transport proteins, with structures and mechanisms quite different from those of the P-type ATPase family. These pumps were identified from studies on tumor cells in culture that developed resistance to drugs that had been initially quite toxic to the cells. Remarkably, the development of resistance to one drug had made the cells less sensitive to a range of other compounds, a phenomenon known as **multidrug resistance**.

In a significant discovery, the onset of multidrug resistance was found to correlate with the expression and activity of a membrane protein with an apparent molecular mass of 170 kDa. This protein acts as an ATP-dependent pump that extrudes a wide range of small molecules from cells that express it. The protein is called the *multidrug-resistance (MDR) protein* or *P-glycoprotein* ("glyco" because it includes a carbohydrate moiety). Thus, when cells are exposed to a drug, the MDR pumps the drug out of the cell before the drug can exert its effects.

Analysis of the amino acid sequences of MDR and homologous proteins revealed a common architecture (**Figure 13.8A**). Each protein comprises four domains: two membrane-spanning domains and two ATP-binding domains. The ATP-binding domains of these proteins are called **ATP-binding cassettes (ABCs)** and are homologous to domains in a large family of transport proteins in bacteria and archaea. Transporters that include these domains are called **ABC transporters**. With 79 members, the ABC transporters are the largest single family identified in the *E. coli* genome. The human genome includes more than 150 ABC transporter genes.

The three-dimensional structures of several members of the ABC transporter family have now been determined, including that of the bacterial lipid transporter MsbA. In contrast with the eukaryotic MDR protein, this protein is a dimer of 62-kDa chains: the amino-terminal half of each protein contains the membrane-spanning domain, and the carboxyl-terminal half contains the ATP-binding cassette (**Figure 13.8B**). Prokaryotic ABC proteins are often made up of multiple subunits, such as a dimer of identical chains, or as a heterotetramer of two membrane-spanning domain subunits and two ATP-binding-cassette subunits. The two ATP-binding cassettes are in contact, but they do not interact strongly in the absence of bound ATP (**Figure 13.9A**). The consolidation of the enzymatic activities of several polypeptide chains in prokaryotes to a single chain in eukaryotes is a theme that we will see again.

On the basis of several solved structures of ABC transporters, combined with data from other experiments, a mechanism for active transport by these proteins has been proposed (**Figure 13.10**):

1. The catalytic cycle begins with the transporter free of both ATP and substrate. While the distance between the ATP-binding cassettes in this form may vary with the individual transporter, the substrate binding region of the transporter faces inward.

(A) **Multidrug-resistance protein (MDR)**

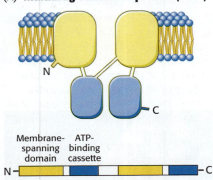

Membrane-spanning domain   ATP-binding cassette

(B) **Bacterial lipid transporter (MsbA)**

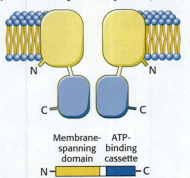

Membrane-spanning domain   ATP-binding cassette

FIGURE 13.8 ABC transporter are a large family of membrane pumps that share a common architecture. These proteins are composed of two transmembrane domains and two ATP-binding domains called ATP-binding cassettes (ABCs). (A) The multidrug-resistance protein is a single polypeptide chain containing all four domains, whereas (B) the bacterial lipid transporter MsbA consists of a dimer of two identical chains, containing one of each domain.

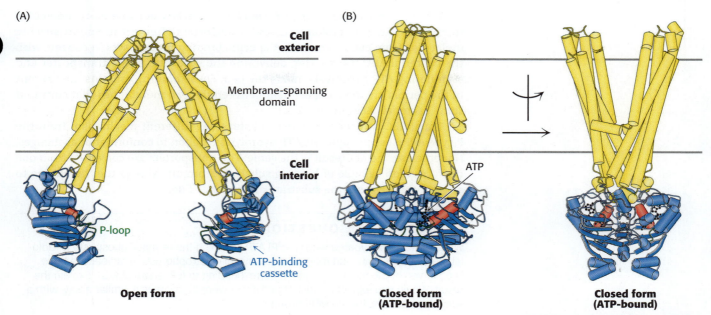

FIGURE 13.9 ABC transporters undergo significant conformational change upon ATP binding. Two structures of the bacterial lipid transporter MsbA, a representative ABC transporter: (A) The nucleotide-free, inward-facing form and (B) the ATP-bound, outward-facing form is shown in two views (rotated by 90 degrees). Upon nucleotide binding, the two ATP-binding cassettes (blue) interact strongly with one another, resulting in reorientation of the substrate binding site. The gray lines indicate the extent of the plasma membrane. [Drawn from 3B5W.pdb and 3B60.pdb.]

INTERACT with this model in
Achieve

2. Substrate enters the central cavity of the transporter from inside the cell. Substrate binding induces conformational changes in the ATP-binding cassettes that increase their affinity for ATP.

3. ATP binds to the ATP-binding cassettes, changing their conformations so that the two domains interact strongly with one another. The close interaction of the ABCs reorients the transmembrane helices such that the substrate binding site is now facing outside the cell (**Figure 13.9B**).

4. The outward facing conformation of the transporter has reduced affinity for the substrate, enabling the release of the substrate on the opposite face of the membrane.

5. The hydrolysis of ATP and the release of ADP and inorganic phosphate reset the transporter for another cycle.

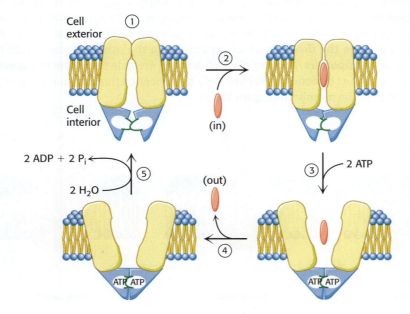

FIGURE 13.10 The open surface of ABC transporters alternates between conformations based on the presence of bound nucleotide. The mechanism includes the following steps: (1) opening of the channel toward the inside of the cell, (2) substrate binding and conformational changes in the ATP-binding cassettes, (3) ATP binding and opening of the channel to the opposite face of the membrane, (4) release of the substrate to the outside of the cell, and (5) ATP hydrolysis to reset the transporter to its initial state.

Whereas eukaryotic ABC transporters generally act to export molecules from inside the cell, prokaryotic ABC transporters often act to import specific molecules from *outside* the cell. A specific binding protein acts in concert with the bacterial ABC transporter, delivering the substrate to the transporter and stimulating ATP hydrolysis inside the cell. These binding proteins are present in the periplasm, the compartment between the two membranes that surround some bacterial cells (Figure 12.36B).

Thus, ABC transporters use a substantially different mechanism from the P-type ATPases to couple the ATP hydrolysis reaction to conformational changes. Nonetheless, the net result is the same: The transporters are converted from one conformation capable of binding substrate from one side of the membrane to another that releases the substrate on the other side.

❖ SELF–CHECK QUESTION

When SERCA is incubated with [γ–^{32}P] ATP (in which the terminal phosphate is radio-labeled) and calcium, and then analyzed by gel electrophoresis, a radioactive band is observed at the molecular weight corresponding to full-length SERCA. Explain the result. Would you expect a similar band if you were performing a similar assay, with a suitable substrate, for the MDR protein?

13.3 Lactose Permease Is an Archetype of Secondary Transporters That Use One Concentration Gradient to Power the Formation of Another

Carriers are proteins that transport ions or molecules across the membrane without hydrolysis of ATP. The mechanism of carriers involves both large conformational changes and the interaction of the protein with only a few molecules per transport cycle, limiting the maximum rate at which transport can occur. Although carriers cannot mediate primary active transport—owing to their inability to hydrolyze ATP—they can couple the thermodynamically unfavorable flow of one species of ion or molecule *up* a concentration gradient to the favorable flow of a different species *down* a concentration gradient, a process referred to as **secondary active transport**. Carriers that move ions or molecules "uphill" by this means are termed **secondary transporters** or **cotransporters**.

Secondary transporters can be classified as either antiporters or symporters. **Antiporters** couple the downhill flow of one species to the uphill flow of another in the *opposite direction* across the membrane; **symporters** use the flow of one species to drive the flow of a different species in the *same direction* across the membrane. **Uniporters**, another class of carriers, are able to transport a specific species in either direction governed only by concentrations of that species on either side of the membrane (**Figure 13.11**).

FIGURE 13.11 The three classes of secondary transporters are antiporters, symporters, and uniporters. Secondary transporters can transport two substrates in opposite directions (antiporters), two substrates in the same direction (symporters), or one substrate in either direction (uniporter).

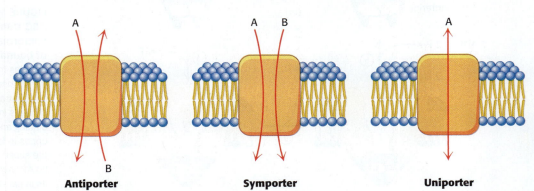

Antiporter   **Symporter**   **Uniporter**

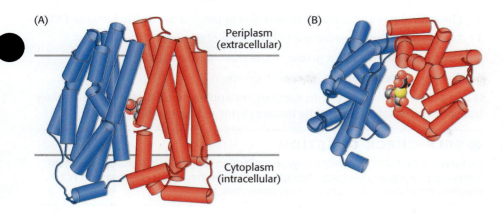

(A)

Periplasm
(extracellular)

Cytoplasm
(intracellular)

(B)

FIGURE 13.12 Lactose permease from *E. coli* is a well-studied example of a secondary transporter. The amino-terminal half of the protein is shown in blue and the carboxyl-terminal half in red. (A) Side view. The gray lines indicate the extent of the plasma membrane. (B) Bottom view (from inside the cell). The protein consists of two halves that surround the sugar and are linked to one another by only a single stretch of polypeptide. [Drawn from 1PV7.pdb.]

INTERACT with this model in
Achieve

Secondary transporters are ancient molecular machines, common today in bacteria and archaea as well as in eukaryotes. For example, approximately 160 (of around 4000) proteins encoded by the E. coli genome are secondary transporters. Sequence comparison and hydropathy analysis suggest that members of the largest family have 12 transmembrane helices that appear to have arisen by duplication and fusion of a membrane protein containing 6 transmembrane helices.

Lactose permease, an E. coli symporter, uses the H^+ gradient across the E. coli membrane (outside has higher H^+ concentration) generated by the oxidation of fuel molecules to drive the uptake of lactose and other sugars against a concentration gradient. This transporter has been extensively studied for many decades and is a useful archetype for this family. As expected from the sequence analysis, the lactose permease consists of two domains, each of which comprises six membrane-spanning α helices (**Figure 13.12**). The two domains are well separated and are joined by a single polypeptide. In this structure, a sugar molecule lies in a pocket in the center of the protein and is accessible from a path that leads from the interior of the cell.

On the basis of this structure and a wide range of other experiments, a mechanism for symporter action has been developed. This mechanism (**Figure 13.13**) has many features similar to those for P-type ATPases and ABC transporters:

1. The cycle begins with the two domains oriented so that the opening to the binding pocket faces outside the cell, in a conformation different from that observed in the structures solved to date. A proton from outside the cell binds to a residue in the permease, quite possibly Glu 325. Although not shown in the figure, the site of protonation likely changes in the course of this cycle.

2. In the protonated form, the permease binds lactose from outside the cell.

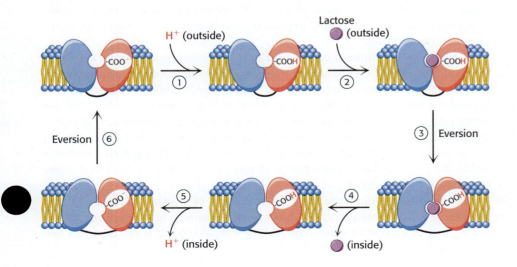

Lactose
(outside)

H^+ (outside)

Eversion ⑥

Eversion ③

H^+ (inside)

(inside)

FIGURE 13.13 Lactose permease catalyzes the symport of a proton and lactose across the membrane. The mechanism begins with the permease open to the outside of the cell (upper left). The permease binds a proton from the outside of the cell (1) and then binds its substrate (2). The permease everts (3) and then releases its substrate (4) and a proton (5) to the inside of the cell. It then everts (6) to complete the cycle.

3. The structure everts to the form observed in the crystal structure (Figure 13.12).

4. The permease releases lactose to the inside of the cell.

5. The permease releases a proton to the inside of the cell.

6. The permease everts to complete the cycle.

It is thought that this eversion mechanism applies to all classes of secondary transporters, which resemble the lactose permease in overall architecture.

❖ SELF–CHECK QUESTION

Carbonyl cyanide 4-(trifluoromethoxy) phenylhydrazone (FCCP) is a proton ionophore: it enables protons to pass freely through membranes. Treatment of *E. coli* with FCCP prevents the accumulation of lactose in these cells. Explain.

13.4 Specific Channels Can Rapidly Transport Ions Across Membranes

Pumps and carriers can move ions and substrates across the membrane at rates approaching several thousand molecules per second. Other membrane proteins, the passive-transport systems called **ion channels**, are capable of ion-transport rates that are more than 1000 times as fast. These rates of transport through ion channels are close to rates expected for ions diffusing freely through aqueous solution. Yet ion channels are not simply tubes that span membranes through which ions can rapidly flow. Instead, they are highly sophisticated molecular machines that respond to chemical and physical changes in their environments and undergo precisely timed conformational changes.

Action potentials are mediated by transient changes in Na⁺ and K⁺ permeability

One of the most important manifestations of ion-channel action is the nerve impulse, which is the fundamental means of communication in the nervous system. A **nerve impulse**, or **action potential**, is an electrical signal produced by the flow of ions across the plasma membrane of a neuron. The interior of a neuron, like that of most other cells, contains a high concentration of K^+ and a low concentration of Na^+. These ionic gradients are generated by the Na^+–K^+ ATPase. The cell membrane has an electrical potential determined by the ratio of the internal to the external concentration of ions. In the resting state, the membrane potential is typically –60 mV. An action potential is generated when the membrane potential is depolarized beyond a critical threshold value (e.g., from –60 to –40 mV). The membrane potential becomes positive within about a millisecond and attains a value of about +30 mV before turning negative again (repolarization). This amplified depolarization is propagated along the nerve terminal (**Figure 13.14**).

Ingenious experiments carried out by Alan Hodgkin and Andrew Huxley revealed that action potentials arise from large, transient changes in the permeability of the axon membrane to Na^+ and K^+ ions. Depolarization of the membrane beyond the threshold level leads to an increase in permeability to Na^+. Sodium ions begin to flow into the cell because of the large electrochemical gradient across the plasma membrane. The entry of Na^+ further depolarizes the membrane, leading to an added increase in Na^+ permeability. This positive feedback yields the very rapid and large change in membrane potential described above and shown in Figure 13.14.

The membrane spontaneously becomes less permeable to Na^+ and more permeable to K^+. Consequently, K^+ flows outward, and so the membrane potential returns to a negative value. The resting level of –60 mV is restored in a few

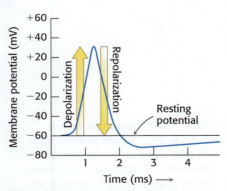

FIGURE 13.14 The action potential is a rapid electrical signal across a plasma membrane. Signals are sent along neurons by the transient depolarization and repolarization of the membrane.

milliseconds as the K^+ conductance decreases to the value characteristic of the unstimulated state. The wave of depolarization followed by repolarization moves rapidly along a nerve cell. The propagation of these waves allows a touch at the tip of your toe to be detected in your brain in a few milliseconds.

The proposal that action potentials arise from changes in Na^+ and K^+ membrane permeability postulated the existence of channels specific for these ions. These channels must open in response to changes in membrane potential and then close after having remained open for a brief period of time. The bold hypotheses of Hodgkin and Huxley predicted the existence of molecules with a well-defined set of properties long before tools existed for their direct detection and characterization.

Patch-clamp conductance measurements reveal the activities of single channels

Direct evidence for the existence of these channels was elucidated using the **patch-clamp technique**. This powerful method enables the measurement of the ion conductance through a small patch of cell membrane (**Figure 13.15**). A clean glass pipette with a tip diameter of about 1 μm is pressed against an intact cell to form a seal. Slight suction leads to the formation of a very tight seal so that the resistance between the inside of the pipette and the bathing solution is many gigaohms (1 gigaohm is equal to 10^9 ohms). Thus, a gigaohm seal (called a *gigaseal*) ensures that an electric current flowing through the pipette is identical with the current flowing through the membrane covered by the pipette. The gigaseal makes possible high-resolution current measurements while a known voltage is applied across the membrane.

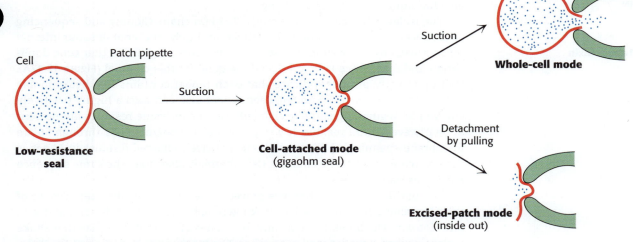

FIGURE 13.15 The patch-clamp technique for monitoring channel activity is highly versatile. A high-resistance seal (gigaseal) is formed between the pipette and a small patch of plasma membrane. This configuration is called the *cell-attached mode*. The breaking of the membrane patch by increased suction produces a low-resistance pathway between the pipette and the interior of the cell. The activity of the channels in the entire plasma membrane can be monitored in this *whole-cell mode*. To prepare a membrane in the *excised-patch mode*, the pipette is pulled away from the cell. A piece of plasma membrane with its cytoplasmic side now facing the medium is monitored by the patch pipette.

Remarkably, the flow of ions through a single channel and transitions between the open and closed states of a channel can be monitored with a time resolution of microseconds (**Figure 13.16**). Furthermore, the activity of a channel in its native membrane environment, even in an intact cell, can be directly observed. Patch-clamp methods provided one of the first views of single

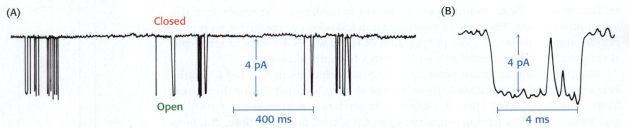

(A)

Closed

4 pA

Open

400 ms

(B)

4 pA

4 ms

FIGURE 13.16 Patch-clamp methods enable the observation of ion movement across single channels. (A) The results of a patch-clamp experiment reveal the small amount of current, measured in picoamperes (pA, 10^{-12} amperes), passing through a single ion channel. The downward spikes indicate transitions between closed and open states. (B) Closer inspection of one of the spikes in (A) reveals the length of time the channel is in the open state.

biomolecules in action. Subsequently, other methods for observing single molecules were invented, opening new vistas on biochemistry at its most fundamental level.

The structure of a potassium ion channel is an archetype for many ion-channel structures

With the existence of ion channels firmly established by patch-clamp methods, scientists sought to identify the molecules that form ion channels. The Na^+ channel was first purified from the electric organ of the electric eel, which is a rich source of the protein forming this channel. The channel was purified on the basis of its ability to bind tetrodotoxin, a neurotoxin from the puffer fish that binds to Na^+ channels very tightly ($K_i \approx 1$ nM). The lethal dose of this poison for an adult human being is about 10 ng.

The isolated Na^+ channel is a single 260-kDa chain. Cloning and sequencing of cDNAs encoding Na^+ channels revealed that the channel contains four internal repeats, each having a similar amino acid sequence, suggesting that gene duplication and divergence have produced the gene for this channel (**Figure 13.17A**). Hydrophobicity profiles indicate that each repeat contains five hydrophobic segments (S1, S2, S3, S5, and S6). Each repeat also contains a highly positively charged S4 segment; positively charged arginine or lysine residues are present at nearly every third residue. It was proposed that segments S1 through S6 are membrane-spanning α helices, while the positively charged residues in S4 act as the voltage sensors of the channel. Ca^{2+} channels also share the same sequence pattern as Na^+ channels.

The purification of K^+ channels proved to be much more difficult because of their low abundance and the lack of known high-affinity ligands comparable to tetrodotoxin. The breakthrough came in studies of mutant fruit flies that shake violently when anesthetized with ether. The mapping and cloning of the gene, termed *Shaker*, responsible for this phenotype revealed the amino acid sequence

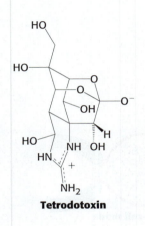

Tetrodotoxin

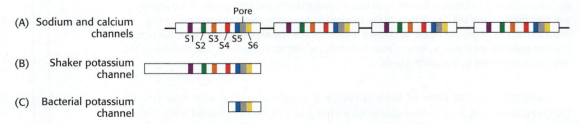

(A)   Sodium and calcium channels

Pore

S1 S3 S5
  S2 S4 S6

(B)   Shaker potassium channel

(C)   Bacterial potassium channel

FIGURE 13.17 Ion channels share structural features. Like colors indicate structurally similar regions of the (A) sodium and calcium channels, (B) the Shaker potassium channel, and (C) the bacterial potassium channel. Each of these channels exhibits approximate fourfold symmetry, either within one chain (sodium, calcium channels) or by forming tetramers (potassium channels).

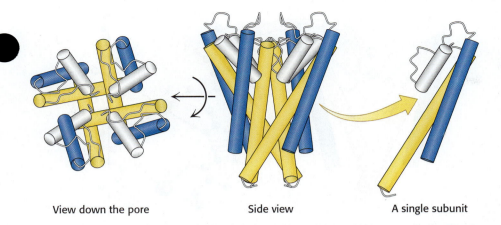

View down the pore   Side view   A single subunit

FIGURE 13.18 The bacterial potassium ion channel forms a pore at the center of a tetrameric complex. The K⁺ channel, composed of four identical subunits, is cone shaped, with the larger opening facing the inside of the cell (center). A view down the pore, looking toward the outside of the cell, shows the relations of the individual subunits (left). One of the four identical subunits of the pore is illustrated at the right, with the pore-forming region shown in gray. [Drawn from 1K4C.pdb.]

INTERACT with this model in
 Achieve

encoded by a K⁺-channel gene. The *Shaker* gene encodes a 70-kDa protein that contains sequences corresponding to segments S1 through S6 in one of the repeated units of the Na⁺ channel (**Figure 13.17B**). Thus, a K⁺-channel subunit is homologous to one of the repeated units of Na⁺ channels. Consistent with this homology, four Shaker polypeptides come together to form a functional channel. Bacterial K⁺ channels have also been discovered that contain only the two membrane-spanning regions corresponding to segments S5 and S6 (**Figure 13.17C**). This and other information suggested that S5 and S6—including the region between them—form the actual pore in the K⁺ channel and that segments S1 through S4 contain the apparatus that opens the pore.

In 1998, Roderick MacKinnon and coworkers determined the structure of a K⁺ channel from the bacterium *Streptomyces lividans* by x-ray crystallography. This channel contains only the pore-forming segments S5 and S6. As expected, the K⁺ channel is a tetramer of identical subunits, each of which includes two membrane-spanning α helices (**Figure 13.18**). The four subunits come together to form a pore in the shape of a cone that runs through the center of the structure.

The structure of the potassium ion channel reveals the basis of ion specificity

The structure of the bacterial potassium channel revealed much about how these proteins efficiently and selectively transport ions across a membrane. Beginning from the inside of the cell, the pore starts with a diameter of approximately 10 Å and then constricts to a smaller cavity with a diameter of 8 Å (**Figure 13.19**). Both the opening to the outside and the central cavity of the pore

BALDOMERO OLIVERA After completing his PhD and postdoctoral training in the United States, Dr. Baldomero "Toto" Olivera returned to his native Philippines to start a faculty position. There, he began studying toxins from readily available cone snails, with which he was familiar from shell-collecting as a child. Cone snails deploy these potent toxins to paralyze their prey. Using chromatography, Dr. Olivera and his coworkers separated the toxins into hundreds of components that they then injected directly into the nervous system of mice. They discovered that each component could induce specific behavioral responses, and subsequent studies revealed that many components can bind and inhibit a single ion channel or pump in a highly specific manner. The different peptide toxins, now called *conotoxins*, are powerful tools, and one has been developed into a drug to treat intractable pain. In the mid-1970s, Dr. Olivera moved to the University of Utah, where he is now Distinguished Professor of Biology.

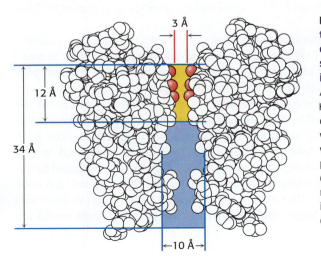

FIGURE 13.19 The path through the potassium channel features only a short region where the ion must be desolvated. A potassium ion entering the K⁺ channel can pass a distance of 22 Å into the membrane while remaining solvated with water (blue). At this point, the pore diameter narrows to 3 Å (yellow), and potassium ions must shed their water and interact with carbonyl groups (red) of the pore amino acids.

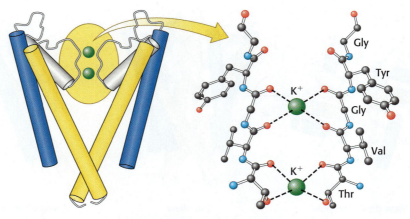

FIGURE 13.20 The selectivity filter determines the preference of the channel for K⁺ over other ions. Potassium ions interact with the carbonyl groups of the TVGYG sequence of the selectivity filter, located at the 3-Å-diameter pore of the K⁺ channel. Only two of the four channel subunits are shown.

TABLE 13.1 Properties of alkali cations

Ion	Ionic radius (Å)	Hydration free energy in kJ mol⁻¹(kcal mol⁻¹)
Li⁺	0.60	−410 (−98)
Na⁺	0.95	−301 (−72)
K⁺	1.33	−230 (−55)
Rb⁺	1.48	−213 (−51)
Cs⁺	1.69	−197 (−47)

are filled with water, and a K⁺ ion can fit in the pore without losing its shell of bound water molecules. Approximately two-thirds of the way through the membrane, the pore becomes more constricted (3-Å diameter, shaded yellow in Figure 13.19), and at that point, any K⁺ ions must give up their water molecules and interact directly with groups from the protein. The channel structure effectively reduces the thickness of the membrane from 34 Å to 12 Å by allowing the solvated ions to penetrate into the membrane before the ions must directly interact with the channel.

The restricted part of the pore is built from residues contributed by the two transmembrane α helices. In particular, a five-amino-acid stretch within this region functions as the **selectivity filter** that determines the preference for K⁺ over other ions (**Figure 13.20**). The stretch has the sequence Thr-Val-Gly-Tyr-Gly (TVGYG), and is nearly completely conserved in all K⁺ channels. The region of the strand containing the conserved sequence lies in an extended conformation and is oriented such that the peptide carbonyl groups are directed into the channel, in good position to interact with the desolvated potassium ions.

Potassium ion channels are 100-fold more permeable to K⁺ than to Na⁺. How is this high degree of selectivity achieved? Ions having a radius larger than 1.5 Å cannot pass into the narrow diameter (3 Å) of the selectivity filter of the K⁺ channel. However, a desolvated Na⁺ is small enough (**Table 13.1**) to pass through the pore. Indeed, the ionic radius of Na⁺ is substantially smaller than that of K⁺. How then is Na⁺ rejected?

The key point is that the free-energy costs of dehydrating ions are considerable. For example, Table 3.1 reveals that the cost of dehydrating Na⁺ is 301 kJ mol⁻¹ (72 kcal mol⁻¹), while that of K⁺ is 230 kJ mol⁻¹ (55 kcal mol⁻¹). The channel pays the cost of desolvating K⁺ by providing optimal compensating interactions with the carbonyl oxygen atoms lining the selectivity filter. Careful studies of the potassium channel, enabled by the determination of its three-dimensional structure, have revealed that the interior of the pore is a highly dynamic, fluid environment. The favorable resolution interactions between the carbonyl oxygen atoms, which carry a partial negative charge, with the cation are balanced by the repulsion of these oxygen atoms from one another. For this channel, the ideal balance is achieved with K⁺, but not with Na⁺ (**Figure 13.21**). Hence, sodium ions are rejected because the higher energetic cost of dehydrating them would not be recovered.

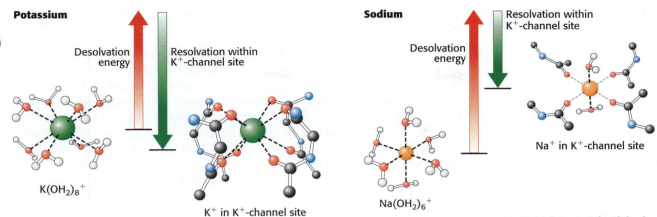

Potassium

Desolvation energy

Resolution within K⁺-channel site

$K(OH_2)_8^+$

K^+ in K^+-channel site

Sodium

Desolvation energy

Resolution within K⁺-channel site

$Na(OH_2)_6^+$

Na^+ in K^+-channel site

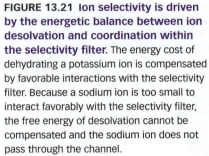

FIGURE 13.21 Ion selectivity is driven by the energetic balance between ion desolvation and coordination within the selectivity filter. The energy cost of dehydrating a potassium ion is compensated by favorable interactions with the selectivity filter. Because a sodium ion is too small to interact favorably with the selectivity filter, the free energy of desolvation cannot be compensated and the sodium ion does not pass through the channel.

The K^+ channel structure enables a clearer understanding of the structure and function of Na^+ and Ca^{2+} channels because of their homology to K^+ channels (Figure 13.17). Sequence comparisons and the results of mutagenesis experiments have implicated the region between segments S5 and S6 in ion selectivity in the Ca^{2+} channel. In Ca^{2+} channels, one glutamate residue of this region in each of the four repeated units plays a major role in determining ion selectivity (**Figure 13.22**). Residues in the positions corresponding to the glutamate residues in Ca^{2+} channels are major components of the selectivity filter of the Na^+ channel. These residues—aspartate, glutamate, lysine, and alanine—are located in each of the internal repeats of the Na^+ channel, forming a region termed the DEKA locus. Thus, the potential fourfold symmetry of the channel is clearly broken in this region, which explains why Na^+ channels consist of a single large polypeptide chain rather than a noncovalent assembly of four identical subunits. The preference of the Na^+ channel for Na^+ over K^+ depends on ionic radius; the diameter of the pore determined by these residues and others is sufficiently restricted that small ions such as Na^+ and Li^+ can pass through the channel, but larger ions such as K^+ are significantly hindered.

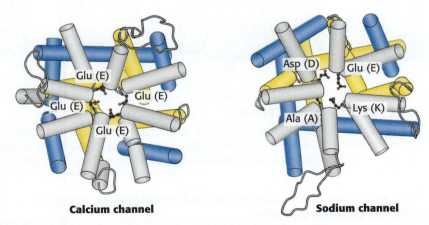

Glu (E)

Glu (E)

Glu (E)

Glu (E)

Calcium channel

Asp (D)

Glu (E)

Ala (A)

Lys (K)

Sodium channel

FIGURE 13.22 The amino acids defining the selectivity filter differ between calcium and sodium channels. The pores of eukaryotic calcium and sodium channels are comprised of single polypeptide chains. In these views, we are looking down the pore; for simplicity, only the S5 (blue), S6 (yellow), and pore-forming (white) regions are drawn. The selectivity filter for the calcium channel contains four glutamate (E) residues, while the filter for the sodium channel contains four different residues (the DEKA locus): aspartate (D), glutamate (E), lysine (K), and alanine (A). [Drawn from 5GJV.pdb and 5X0M.pdb.]

The structure of the potassium ion channel explains its rapid rate of transport

The tight binding sites required for ion selectivity should slow the progress of ions through a channel, yet ion channels achieve rapid rates of ion transport. How is this paradox resolved? A structural analysis of the channel at high resolution provides an appealing explanation. Consider the process of ion conductance starting from inside the cell (**Figure 13.23**). Four K^+-binding sites crucial for rapid ion flow are present in the constricted region of the K^+ channel. A hydrated potassium ion proceeds into the channel and through the relatively unrestricted part of the channel. The ion then gives up its coordinated water molecules and binds to a site within the selectivity-filter region. The ion can move between the four sites within the selectivity filter because they have similar ion affinities. As each subsequent potassium ion moves into the selectivity filter, its positive charge will repel the potassium ion at the nearest site, causing it to shift to a site farther up the channel and in turn push upward any potassium ion already bound to a site farther up. Thus, each ion that binds anew favors the release of an ion from the other side of the channel. This multiple-binding-site mechanism solves the paradox of high ion selectivity and rapid flow.

FIGURE 13.23 The model for K^+-channel ion transport relies on multiple ion binding sites within the pore. The selectivity filter has four binding sites. Hydrated potassium ions can enter these sites, one at a time, losing their hydration shells. When two ions occupy adjacent sites, electrostatic repulsion forces them apart. Thus, as ions enter the channel from one side, other ions are pushed out the other side.

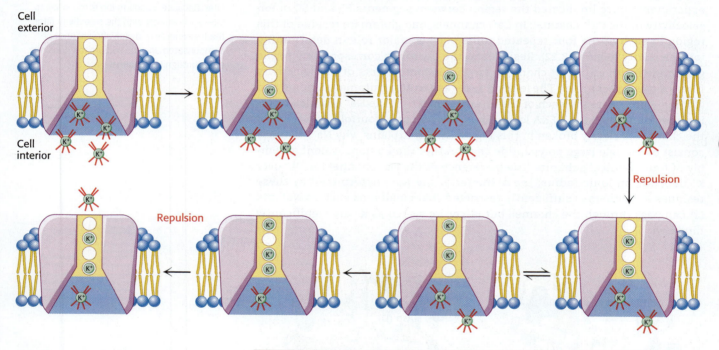

❖ SELF−CHECK QUESTION

The K^+ channel and the Na^+ channel have similar structures and are arranged in the same orientation in the cell membrane. Yet the Na^+ channel allows sodium ions to flow into the cell and the K^+ channel allows potassium ions to flow out of the cell. Explain.

Voltage gating requires substantial conformational changes in specific ion-channel domains

Some Na^+ and K^+ channels are gated by membrane potential; that is, they change conformation to a highly conducting form in response to changes in voltage across the membrane. As already noted, these **voltage-gated channels** include segments S1 through S4 in addition to the pore itself formed by S5 and S6. The structure of a voltage-gated K^+ channel from *Aeropyrum pernix* reveals that the segments S1 through S4 form domains, termed *paddles*, that extend from the core of the channel (**Figure 13.24**). These paddles include the segment S4, the

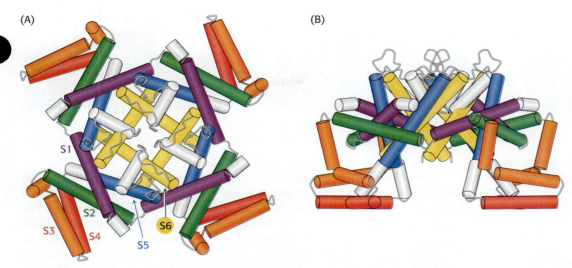

S1

S2

S3   S4

S5

S6

FIGURE 13.24 The S4 region lies on the outer surface of the voltage-gated potassium channel. (A) A view looking down through the pore and (B) a side view of the structure are shown. The positively charged S4 region (red) lies on the outside of the structure at the bottom of the pore. [Drawn from 1ORQ.pdb.]

INTERACT with this model in
⋙ Achie∕e

voltage sensor itself. Segment S4 forms an α helix lined with positively charged residues. In contrast with expectations, segments S1 through S4 are not enclosed within the protein but, instead, are positioned to lie in the membrane itself.

A model for voltage gating has been proposed by Roderick MacKinnon and coworkers on the basis of this structure and a range of other experiments (**Figure 13.25**). In the closed state, the paddles lie in a "down" position. On membrane depolarization, the cytoplasmic side of the membrane becomes more positively charged, repelling the paddles through the membrane into an "up" position. In this position, they pull the four sides of the base on the pore apart, increasing access to the selectivity filter and opening the channel.

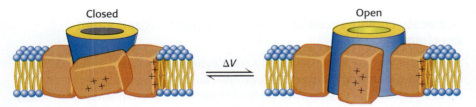

Closed   Open

ΔV

FIGURE 13.25 Movement of the S4 region controls the opening of the voltage-gated potassium channel. The voltage-sensing paddles lie in the "down" position below the closed channel (left). Membrane depolarization pulls these paddles through the membrane. The motion pulls the base of the channel apart, opening the channel (right).

A channel can be inactivated by occlusion of the pore: the ball-and-chain model

The K^+ channel and the Na^+ channel undergo inactivation within milliseconds of opening (**Figure 13.26A**). A first clue to one mechanism of inactivation came from exposing the cytoplasmic side of either channel to the protease trypsin. Cleavage by trypsin produced trimmed channels that stayed persistently open after depolarization, suggesting that a flexible (i.e., accessible to protease) region of the protein was responsible for inactivation. Furthermore, a mutant Shaker channel lacking 42 amino acids near the amino terminus opened in response to depolarization but did not inactivate (**Figure 13.26B**). Remarkably, inactivation was restored by adding a synthetic peptide corresponding to the first 20 residues of the native channel (**Figure 13.26C**).

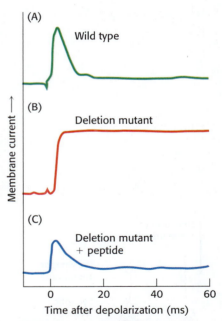

(A)   Wild type

(B)   Deletion mutant

(C)   Deletion mutant + peptide

Membrane current →

0   20   40   60
Time after depolarization (ms)

FIGURE 13.26 The amino-terminal region of the K^+ channel is critical for inactivation. (A) The wild-type Shaker K^+ channel displays rapid inactivation after opening. (B) A mutant channel lacking residues 6 through 46 does not inactivate. (C) Inactivation can be restored by adding a peptide consisting of residues 1 through 20 at a concentration of 100 μM. [Data from W. N. Zagotta, T. Hoshi, and R. W. Aldrich, *Science* 250:568–571, 1990.]

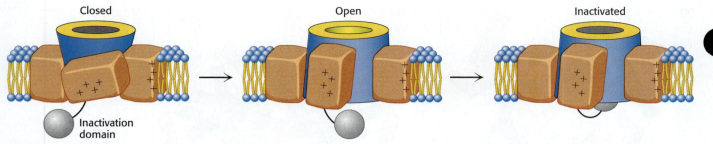

Closed   Open   Inactivated

Inactivation domain

FIGURE 13.27 The ball-and-chain model provides a mechanism for channel inactivation. The inactivation domain, or "ball" (gray), is tethered to the channel by a flexible "chain." In the closed state, the ball is located in the cytoplasm. Depolarization opens the channel and creates a binding site for the positively charged ball in the mouth of the pore. Movement of the ball into this site inactivates the channel by occluding it.

These experiments strongly support the **ball-and-chain model** for channel inactivation that had been proposed years earlier (**Figure 13.27**). According to this model, the first 20 residues of the K^+ channel form a cytoplasmic unit (the *ball*) that is attached to a flexible segment of the polypeptide (the *chain*). When the channel is closed, the ball rotates freely in the aqueous solution. When the channel opens, the ball quickly finds a complementary site in the open pore and occludes it. Hence, the channel opens for only a brief interval before it undergoes inactivation by occlusion. Shortening the chain speeds inactivation because the ball finds its target more quickly. Conversely, lengthening the chain slows inactivation. Thus, the duration of the open state can be controlled by the length and flexibility of the tether. In some senses, the "ball" domains, which include substantial regions of positive charge, can be thought of as large, tethered cations that are pulled into the open channel but get stuck and block further ion conductance.

It is apparent from the example in Figure 13.27 that the closed and inactivated states of the ion channel are not identical. The process of returning the inactivated channel to its initial closed state, termed **recovery**, requires disengagement of the ball domain from the pore opening as well as realignment of the voltage-sensing paddles. Depending upon the mechanism of inactivation, recovery can take milliseconds to seconds. Importantly, while inactivated, the channel cannot be opened by further membrane depolarization. As we shall see, the time needed for channel recovery has implications for repeated action potential generation.

The acetylcholine receptor is an archetype for ligand-gated ion channels

Nerve impulses are communicated across synapses by small, diffusible molecules called **neurotransmitters**. For example, one function of the neurotransmitter acetylcholine is to stimulate skeletal muscle contraction when released by contacting neurons. The presynaptic membrane of a synapse is separated from the postsynaptic membrane by a gap of about 50 nm called the **synaptic cleft**. The arrival of a nerve impulse at the end of an axon leads to the synchronous export of the contents of some 300 membrane-bound compartments, or vesicles, of acetylcholine into the cleft (**Figure 13.28**). The binding of acetylcholine to the postsynaptic membrane markedly changes its ionic permeability, triggering an action potential. Acetylcholine opens a single kind of cation channel, called the *acetylcholine receptor*, which is almost equally permeable to Na^+ and K^+.

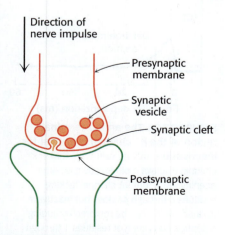

Direction of nerve impulse

Presynaptic membrane

Synaptic vesicle

Synaptic cleft

Postsynaptic membrane

FIGURE 13.28 The synapse is the gap between cells where nerve impulses are transmitted.

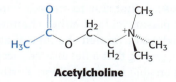

Acetylcholine

The acetylcholine receptor is the best-understood **ligand-gated channel**, a channel that is gated not by voltage but by the presence of specific ligands. The binding of acetylcholine to the channel is followed by its transient opening.

The electric organ of *Torpedo marmorata*, an electric ray (**Figure 13.29**), is a choice source of acetylcholine receptors for study because its electroplaxes (voltage-generating cells) are very rich in postsynaptic membranes densely packed with acetylcholine receptors (~20,000 μm^{-2}) and therefore highly responsive to this neurotransmitter.

In order to study the acetylcholine receptor in molecular detail, biochemists solubilized the protein by adding a nonionic detergent to a postsynaptic membrane preparation from the electric organ and purified it by affinity chromatography on a column bearing covalently attached cobratoxin, a small protein toxin from snakes that has a high affinity for acetylcholine receptors. With the use of techniques presented in Chapter 4, the 268-kDa receptor was identified as a pentamer of four kinds of membrane-spanning subunits—α_2, β, γ, and δ. The cloning and sequencing of the cDNAs for the four kinds of subunits (50–58 kDa) showed that they have clearly similar sequences; the genes for the α, β, γ, and δ subunits arose by duplication and divergence of a common ancestral gene. Each subunit has a large extracellular domain, followed at the carboxyl end by four predominantly hydrophobic segments that span the bilayer membrane.

Using cryo-EM and x-ray crystallographic methods (Section 4.5), the structure of purified acetylcholine receptors has been determined to have approximate fivefold symmetry, in harmony with the similarity of its five constituent subunits (**Figure 13.30A**). The subunits are arranged in the form of a ring that creates a pore through the membrane. Acetylcholine binds at the $\alpha-\gamma$ and $\alpha-\delta$ interfaces (**Figure 13.30B**).

What is the basis of channel opening in the acetylcholine receptor? A high-resolution answer to this question has remained elusive, but the details of this mechanism are becoming apparent. Cryo-EM images of the receptor in the open and closed states have been determined, albeit at low resolution. These structures indicate that the binding of acetylcholine to the extracellular domain ultimately leads to the straightening of the α-helices from the α and δ subunits

FIGURE 13.29 The marbled electric ray (*Torpedo marmorata*) has an electric organ rich in acetylcholine receptors. These organs, called electroplaxes, can deliver a shock of as much as 200 V for approximately 1 second. [A. Martin UW Photography/Moment/ Getty Images]

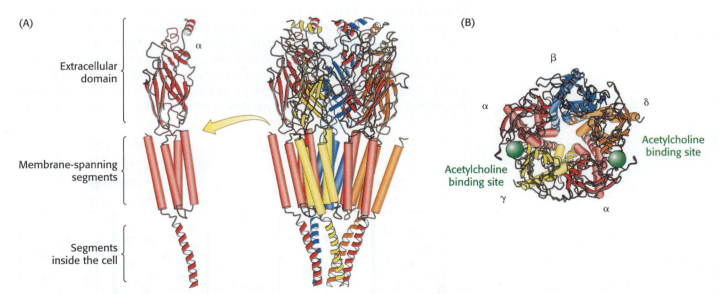

FIGURE 13.30 The acetylcholine receptor is a pentamer formed from four different subunit proteins. A model for the structure of the acetylcholine receptor deduced from high-resolution electron microscopic studies reveals that each subunit consists of a large extracellular domain consisting primarily of β strands, four membrane-spanning α helices, and a final α helix inside the cell. (A) A side view shows the pentameric receptor with each subunit type in a different color. One copy of the α subunit is shown in isolation. (B) A view down the channel from outside the cell shows the pore opening as well as the binding sites for the acetylcholine ligand, indicated in green. [Drawn from 2BG9.pdb.]

INTERACT with this model in
🏠 Achie√e

FIGURE 13.31 Ligand binding induces the opening of the acetylcholine receptor. Binding of acetylcholine to the extracellular region of the receptor leads to a series of conformational changes that are transmitted to the pore-lining helices. Notice that upon ligand binding, the pore-lining helix of the α subunit straightens (green bar). This subtle structural adjustment changes the gating properties of the pore, enabling cations to pass through. [Drawn from 4AQ5.pdb and 4AQ9.pdb.]

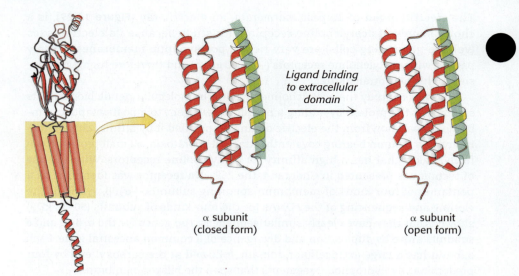

Ligand binding to extracellular domain

α subunit (closed form)

α subunit (open form)

that line the pore (**Figure 13.31**). In the closed state, the region of narrowest diameter in the pore (the "gate") is located about midway through the membrane. This region is lined with nonpolar residues, which cannot form favorable interactions with K⁺ and Na⁺ ions. However, once the helices are straightened, the region of narrowest diameter shifts closer to the inner membrane leaflet. This region is lined with polar residues and can conduct K⁺ and Na⁺ ions freely.

Action potentials integrate the activities of several ion channels working in concert

To see how ligand-gated and voltage-gated channels work together to generate a sophisticated physiological response, we now revisit the action potential introduced at the beginning of this section. First, we need to introduce the concept of **equilibrium potential**. Suppose that a membrane separates two solutions that contain different concentrations of some cation X^+, as well as an equivalent amount of anions to balance the charge in each solution (**Figure 13.32**). Let $[X^+]_{in}$ be the concentration of X^+ on one side of the membrane (corresponding to the inside of a cell) and $[X^+]_{out}$ be the concentration of X^+ on the other side (corresponding to the outside of a cell). Now let us assume that an ion channel opens that allows X^+ to move across the membrane. What will happen? It seems clear

FIGURE 13.32 The equilibrium potential represents the balance between chemical and electrical gradients. The membrane potential reaches equilibrium when the driving force due to the concentration gradient is exactly balanced by the opposing force due to the repulsion of like charges.

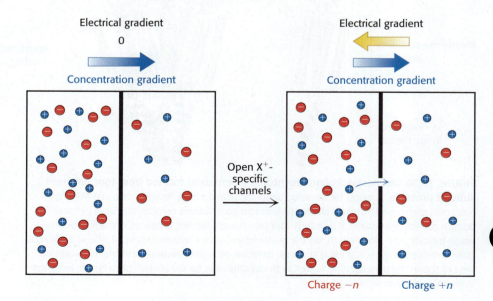

Electrical gradient
0
Concentration gradient

Open X^+-specific channels

Electrical gradient

Concentration gradient

Charge $-n$ Charge $+n$

that X^+ will move through the channel from the side with the higher concentration to the side with the lower concentration. However, positive charges will start to accumulate on the side with the lower concentration, making it more difficult to move each additional positively charged ion. An equilibrium will be achieved when the driving force due to the concentration gradient is balanced by the electrostatic force resisting the motion of an additional charge. In these circumstances, the membrane potential is given by the Nernst equation:

$$V_{eq} = -(RT/zF) \ln([X]_{in}/[X]_{out})$$

where R is the gas constant and F is the Faraday constant (96.5 kJ V^{-1} mol^{-1}, or 23.1 kcal V^{-1} mol^{-1}) and z is the charge on the ion X (e.g., +1 for X^+).

The membrane potential at equilibrium is called the equilibrium potential for a given ion at a given concentration ratio across a membrane. For sodium with $[Na^+]_{in}$ = 14 mM and $[Na^+]_{out}$ = 143 mM, the equilibrium potential is +62 mV at 37°C. Similarly, for potassium with $[K^+]_{in}$ = 157 mM and $[K^+]_{out}$ = 4 mM, the equilibrium potential is –98 mV. In the absence of stimulation, the resting potential for a typical neuron is –60 mV. This value is close to the equilibrium potential for K^+ owing to the fact that a small number of K^+ channels are open.

EXAMPLE Calculating Equilibrium Potentials

PROBLEM: For a typical mammalian cell, the intracellular and extracellular concentrations are of the calcium ion (Ca^{2+}) are 0.2 μM and 1.8 mM, respectively. Calculate the equilibrium potential at 37°C for Ca^{2+}.

GETTING STARTED: This problem represents a fairly straightforward application of the Nernst equation:

$$V_{eq} = -(RT/zF)\ln([X]_{in}/[X]_{out})$$

The most common pitfalls made in the application of this equation are mixing up the inside and outside concentrations and applying the incorrect ion charge (z). The calcium ion is a divalent cation, so $z = +2$.

CALCULATE: Using the concentrations given (note that the two concentrations are given in different units!), substitute into the Nernst equation:

$$V_{eq} = -(RT/zF)\ln([X]_{in}/[X]_{out})$$

$$V_{eq} = -\left(\frac{(8.315 \times 10^{-3} \text{ kJ mol}^{-1} \text{ deg}^{-1}) \times (310 \text{ deg})}{(+2) \times (96.5 \text{ kJ V}^{-1} \text{ mol}^{-1})}\right)\ln\left(\frac{0.0002 \text{ mM}}{1.8 \text{ mM}}\right)$$

$$V_{eq} = +122 \text{ mV}$$

REFLECT: Calcium ions are maintained at a very low intracellular concentration, owing to the activity of Ca^{2+} pumps such as SERCA. Hence, the equilibrium potential is quite high, over 120 millivolts. Cells are highly sensitive to changes in intracellular Ca^{2+}, which can be exploited by signal transduction pathways (Section 14.2).

We are now prepared to consider what happens in the generation of an action potential (**Figure 13.33**):

- Initially, a neurotransmitter such as acetylcholine is released into the synaptic cleft from a presynaptic membrane (Figure 13.28). The released acetylcholine binds to the acetylcholine receptor on the postsynaptic membrane, causing it to open within less than a millisecond. The acetylcholine receptor is a nonspecific cation channel. Sodium ions flow into the cell and potassium

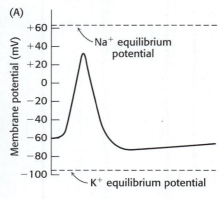

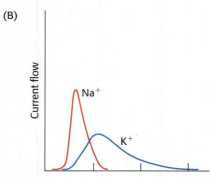

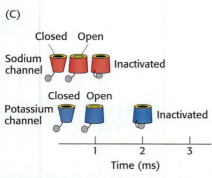

FIGURE 13.33 The action potential is generated by the coordinated opening, closing, and inactivation of Na⁺ and K⁺ channels. (A) On the initiation of an action potential, the membrane potential moves from the resting potential upward toward the Na⁺ equilibrium potential and then downward toward the K⁺ equilibrium potential. (B) The currents through the Na⁺ and K⁺ channels underlying the action potential. (C) The states of the Na⁺ and K⁺ channels during the action potential.

ions flow out of the cell. Without any further events, the membrane potential would move to a value corresponding to the average of the equilibrium potentials for Na⁺ and K⁺, approximately –20 mV.

- As the membrane potential approaches –40 mV, the voltage-sensing paddles of Na⁺ channels are pulled into the membrane, opening the Na⁺ channels. With these channels open, sodium ions flow rapidly into the cell and the membrane potential rises rapidly toward the Na⁺ equilibrium potential (Figure 13.33B, red curve).

- The voltage-sensing paddles of K⁺ channels also are pulled into the membrane by the changed membrane potential, but more slowly than Na⁺ channel paddles. Nonetheless, after approximately 1 ms, many K⁺ channels start to open.

- As the K⁺ channels open, inactivation "ball" domains plug the open Na⁺ channels, decreasing the Na⁺ current. The acetylcholine receptors that initiated these events are also inactivated on this time scale.

- With the Na⁺ channels inactivated and only the K⁺ channels open, the membrane potential drops rapidly toward the K⁺ equilibrium potential (Figure 13.33B, blue curve). The open K⁺ channels are susceptible to inactivation by their "ball" domains, and these K⁺ currents, too, are blocked.

- With the membrane potential returned to close to its initial value, the inactivation domains are released and the channels undergo recovery to return to their original closed states. During the time preceding this recovery, the Na⁺ channels cannot be reopened by membrane depolarization. Hence, a new action potential cannot be induced for a short interval, termed the **refractory period**, which usually lasts around one millisecond in neurons.

These events propagate along the neuron as the depolarization of the membrane opens channels in nearby patches of membrane. Note that the refractory period ensures that an action potential does not recur in parts of the membrane that were just depolarized. The net result is that the action potential travels along the membrane in one direction.

How much current actually flows across the membrane over the course of an action potential? Consider that a typical nerve cell contains 100 Na⁺ channels per square micrometer. At a membrane potential of +20 mV, each channel conducts 10^7 ions per second. Thus, in a period of 1 millisecond, approximately 10^5 ions flow through each square micrometer of membrane surface. Assuming a cell volume of 10^4 μm³ and a surface area of 10^4 μm², this rate of ion flow corresponds to an increase in the Na⁺ concentration of less than 1%. How can this be? A robust action potential is generated because the membrane potential is very sensitive to even a slight change in the distribution of charge. This sensitivity makes the action potential a very efficient means of signaling over long distances and with rapid repetition rates.

Disruption of ion channels by mutations or chemicals can be potentially life-threatening

The generation of an action potential requires the precise coordination of gating events of a collection of ion channels. Perturbation of this timing can have devastating effects. For example, the rhythmic generation of action potentials by the heart is absolutely essential to maintain delivery of oxygenated blood to the peripheral tissues. **Long QT syndrome** (**LQTS**) is a genetic disorder in which the recovery of the action potential from its peak potential to the resting equilibrium potential is delayed. The term QT refers to a specific feature of the cardiac electrical activity pattern as measured by electrocardiography. LQTS can lead to brief losses of consciousness (syncope), disruption of normal cardiac rhythm (arrhythmia), and sudden death. The most common mutations identified in LQTS patients inactivate K⁺ channels or prevent the proper trafficking of these channels to the plasma membrane. The resulting loss in potassium permeability slows the repolarization of the membrane and delays the induction of the subsequent cardiac contraction, rendering the cardiac tissue susceptible to arrhythmias.

Prolongation of the cardiac action potential in this manner can also be a potentially dangerous side effect of certain therapeutic drugs. In particular, the K⁺ channel hERG (for human ether-a-go-go-related gene, named for its ortholog in *Drosophila melanogaster*) is highly susceptible to interactions with certain drugs. The hydrophobic regions of these drugs can block hERG by binding to two nonconserved aromatic residues on the internal surface of the channel cavity. In addition, this cavity is predicted to be wider than other K⁺ channels because of the absence of a conserved Pro-X-Pro motif within the S6 hydrophobic segment. Inhibition of hERG by these drugs can lead to an increased risk of cardiac arrhythmias and sudden death. Accordingly, a number of these agents, such as the antihistamine terfenadine, have been withdrawn from the market. Screening for the inhibition of hERG is now a critical safety hurdle for the pharmaceutical advancement of a molecule to an approved drug.

Hyperpolarization-activated ion channels enable pacemaker activity in the heart

The heart utilizes coordinated changes in the membrane potential to facilitate the efficient muscular contractions necessary to pump blood throughout the body effectively. However, the heart can beat spontaneously, without the use of a neurotransmitter such as acetylcholine. How does the heart initiate an action potential without such an input?

The answer lies within a collection of cardiac cells known as the pacemaker. While there are several pacemakers in the human heart, the most important of these is the sinoatrial (SA) node, located on the posterior wall of the right atrium. The cells of the SA node spontaneously generate action potentials at rates from 60 to 100 per second. Researchers have known for many years that these cells possess an unusual capacity for mixed Na⁺/K⁺ conductance when their membranes are hyperpolarized. This membrane permeability has been referred to as the *funny current* (I_f). At rest, this current is active, leading to a gradual depolarization of the membrane to a level that triggers the initiation of an action potential. During the action potential, this current shuts off. However, once the action potential is complete and the membrane returns to its resting potential, the funny current turns back on, repeating the cycle. The action potential propagates from the SA node to the rest of the cardiac tissue in a highly organized manner, stimulating muscle contraction and the movement of blood.

The channels responsible for the funny current have been identified. The hyperpolarization-activated cyclic nucleotide-gated (HCN) channels comprise four isoforms. Functional channels are composed of tetramers of these isoforms, either all of the same type (homotetramers) or of a combination of types (heterotetramers). HCN4 is the most prominently expressed in the SA node. These channels are unusual in that they are opened by hyperpolarization, and close when the membrane is depolarized. Moreover, these channels are responsive to changing levels of intracellular cyclic AMP (cAMP), as expected: while the heart can beat spontaneously, neuronal and hormonal input can affect this rate to accommodate the body's changing energetic needs.

Mutations in the *HCN4* gene have been identified in patients with sick sinus syndrome, a condition where patients exhibit an abnormally low heart rate (bradycardia) that may manifest with symptoms of dizziness, fainting, headaches, and fatigue. In one family, the identified mutation resulted in a greater hyperpolarization required to achieve HCN4 opening (**Figure 13.34**) when these

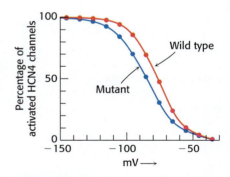

FIGURE 13.34 A disease-causing HCN4 mutant requires greater membrane hyperpolarization to open the channel. This graph shows the percentage of open channels at a given membrane potential. HCN4 channels are open at hyperpolarized membrane potentials, and close as the membrane depolarizes. For the mutant HCN4 channels (blue circles), the channel requires a more hyperpolarized state to be open to the same extent as the wild-type channels (red circles). [R. Milanese, M. Baruscotti, T. Gnecchi-Ruscone, and D. DiFrancesco, *New Eng. J. Med.* 354:151–157, 2006, Fig. 3A.]

channels were expressed in cultured cells. In a patient carrying this mutation, fewer HCN4 channels are open at a given resting potential, leading to a slower depolarization and a reduced heart rate.

13.5 Gap Junctions Allow Ions and Small Molecules to Flow Between Communicating Cells

The ion channels we have considered thus far have narrow pores and are moderately to highly selective in the ions they allow to pass through them. They are closed in the resting state and have short lifetimes in the open state, typically a millisecond, that enable them to transmit frequent neural signals. Let us turn now to a channel with a very different role. **Gap junctions**, also known as **cell-to-cell channels**, serve as passageways between the interiors of contiguous cells. Gap junctions are clustered in discrete regions of the plasma membranes of apposed cells. Electron micrographs of sheets of gap junctions show them tightly packed in a regular hexagonal array (**Figure 13.35**). An approximately 20-Å central hole, the lumen of the channel, is prominent in each gap junction. These channels span the intervening space, or gap, between apposed cells (hence, the name *gap junction*).

Small hydrophilic molecules as well as ions can pass through gap junctions. The pore size of the junctions was determined by microinjecting a series of fluorescent molecules into cells and observing their passage into adjoining cells. All polar molecules with a mass of less than about 1 kDa can readily pass through these cell-to-cell channels. Thus, inorganic ions and most metabolites (e.g., sugars, amino acids, and nucleotides) can flow between the interiors of cells joined by gap junctions. In contrast, proteins, nucleic acids, and polysaccharides are too large to traverse these channels.

Gap junctions are important for intercellular communication. Cells in some excitable tissues, such as heart muscle, are coupled by the rapid flow of ions through these junctions, which ensures a speedy and synchronous response to stimuli. Gap junctions are also essential for the nourishment of cells that are distant from blood vessels, as in bone and the lens of the eye. Moreover, communicating channels are important in development and differentiation. For example, the quiescent (calm) uterus transforms to a forcefully contracting organ at the onset of labor; the formation of functional gap junctions at that time creates a syncytium of muscle cells that contract in synchrony.

A gap junction is made of 12 molecules of connexin, one of a family of transmembrane proteins with molecular masses ranging from 30 to 42 kDa. Each connexin molecule contains four membrane-spanning helices (**Figure 13.36A**). Six connexin molecules are hexagonally arrayed to form a half-channel, called a **connexon** or **hemichannel**. Two connexons join end to end in the intercellular space to form a functional channel between the communicating cells (**Figure 13.36B**). Each connexon adopts a funnel shape: At the cytoplasmic face, the inner diameter of the channel is 35 Å, while at its innermost point, the pore narrows to a diameter of 14 Å (**Figure 13.36C**). Cell-to-cell channels differ from other membrane channels in three respects:

1. They traverse *two* membranes rather than one.

2. They connect cytoplasm to cytoplasm, rather than to the extracellular space or the lumen of an organelle.

3. The connexons forming a channel are synthesized by different cells.

Gap junctions form readily when cells are brought together. A cell-to-cell channel, once formed, tends to stay open for seconds to minutes. They are closed by high concentrations of calcium ion and by low pH. The closing of gap junctions

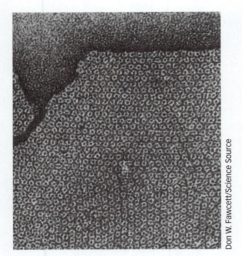

FIGURE 13.35 Gap junctions form tightly packed arrays. This electron micrograph shows a sheet of isolated gap junctions. The cylindrical connexons form a hexagonal lattice having a unit-cell length of 85 Å. The densely stained central hole has a diameter of about 20 Å.

Don W. Fawcett/Science Source

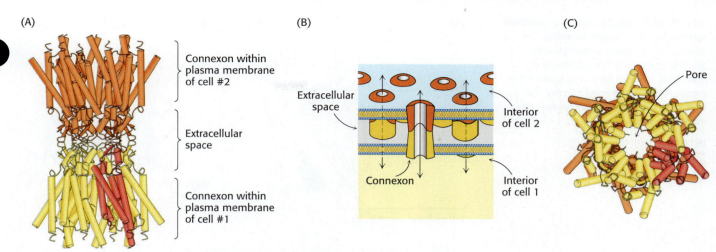

FIGURE 13.36 Gap junctions are formed from connexons of neighboring cells. (A) Six connexins join to form a connexon, or hemichannel, within the plasma membrane (yellow). A single connexin monomer is highlighted in red. The extracellular region of one connexon binds to the same region of a connexon from another cell (orange), forming a complete gap junction. (B) A schematic view of the gap junction, oriented in the same direction as in (A). (C) A bottom-up view looking through the pore of a gap junction. This perspective is visualized in Figure 13.35. [(A) and (C) Drawn from 2ZW3.pdb; (B) Information from Dr. Werner Loewenstein.]

by Ca^{2+} and H^+ serves to seal normal cells from injured or dying neighbors. Gap junctions are also controlled by membrane potential and by hormone-induced phosphorylation.

❖ SELF–CHECK QUESTION

The human genome contains more than 20 connexin-encoding genes. Several of these genes are expressed in high levels in the heart. Why are connexins so highly expressed in cardiac tissue?

13.6 Specific Channels Increase the Permeability of Some Membranes to Water

One more important class of channels does not take part in ion transport at all. Instead, these channels increase the rate at which water flows through membranes. As noted in Section 12.3, membranes are reasonably permeable to water. Why, then, are water-specific channels required? In certain tissues, in some circumstances, rapid water transport through membranes is necessary. In the kidney, for example, water must be rapidly reabsorbed into the bloodstream after filtration. Similarly, in the secretion of saliva and tears, water must flow quickly through membranes. These observations suggested the existence of specific water channels, but initially the channels could not be identified.

The channels, now called **aquaporins**, were discovered serendipitously. Peter Agre (pictured at the beginning of Chapter 1) noticed a protein present at high levels in red-blood-cell membranes that had not been previously identified because the protein does not stain well with Coomassie blue, the stain commonly used in protein SDS-PAGE (Section 4.1). In addition to red blood cells, this protein was found in large quantities in tissues such as the kidney and the cornea, precisely the tissues hypothesized to contain water channels. On the basis of this observation, further studies were designed, revealing that this 24-kDa membrane protein is, indeed, a water channel.

FIGURE 13.37 Aquaporins are channels specifically permeable to water. The structure of aquaporin viewed (A) from the side and (B) from the top. Notice the hydrophilic residues (shown as space-filling models) that line the water channel. The gray lines in (A) indicate the extent of the plasma membrane. [Drawn from 1J4N.pdb.]

INTERACT with this model in

 Achie√e

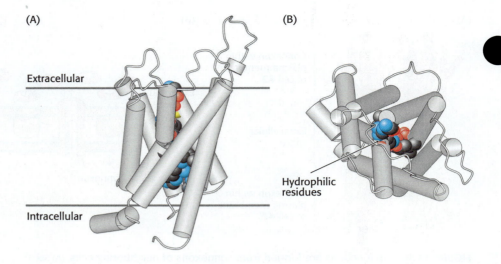

(A)

Extracellular

Intracellular

(B)

Hydrophilic residues

The structure of aquaporin consists of six membrane-spanning α helices (**Figure 13.37**). Two loops containing hydrophilic residues line the actual channel. Water molecules pass through in single file at a rate of 10^6 molecules per second. Importantly, specific positively charged residues toward the center of the channel prevent the transport of protons through aquaporin. Thus, aquaporin channels will not disrupt proton gradients, which play fundamental roles in energy transduction, as we will see in Chapter 18. Remarkably, the aquaporins are channels that have evolved specifically to conduct uncharged substrates.

Summary

13.1 The Transport of Molecules Across a Membrane May Be Active or Passive

- Lipophilic molecules can pass through a membrane's hydrophobic interior by simple diffusion.

- Passive transport or facilitated diffusion takes place when an ion or polar molecule moves down its concentration gradient with the assistance of a channel.

- Active transport uses an input of energy to move molecules against a concentration gradient.

- The electrochemical potential measures the combined ability of a concentration gradient and an uneven distribution of charge to drive species across a membrane.

13.2 Two Families of Membrane Proteins Use ATP Hydrolysis to Actively Transport Ions and Molecules Across Membranes

- Pumps are membrane proteins that catalyze the active transport of ions against their electrochemical gradients.

- P-type ATPases are pumps that are transiently phosphorylated on an aspartic acid residue during the transport cycle.

- ABC transporters are pumps that use ATP-binding cassettes to control protein conformation.

13.3 Lactose Permease Is an Archetype of Secondary Transporters That Use One Concentration Gradient to Power the Formation of Another

- Carriers are proteins that transport ions or molecules across the membrane without hydrolysis of ATP.

- Uniporters transport a substrate in either direction, determined by the concentration gradient.

- Antiporters couple the downhill flow of one substrate in one direction to the uphill flow of another in the opposite direction.

- Symporters couple the downhill flow of one substrate in one direction to the uphill flow of another in the same direction.

13.4 Specific Channels Can Rapidly Transport Ions Across Membranes

- Ion channels allow the rapid movement of ions across the hydrophobic barrier of the membrane.

- Ions must transiently lose their coordinated water molecules as they move to the narrowest part of an ion channel—the selectivity filter—which determines the specificity of the ion channel.

- Voltage-gated ion channels are opened by changes in membrane potential.
- Many channels spontaneously inactivate after having been open for a short period of time.
- Ligand-gated channels are opened or closed by the binding of ligands.
- Nerve impulses, or action potentials, are electrical signals produced by the coordinated flow of K^+ and Na^+ ions through their respective ion channels.

13.5 Gap Junctions Allow Ions and Small Molecules to Flow Between Communicating Cells

- Gap junctions, or cell-to-cell channels, serve to connect the interiors of contiguous groups of cells.

- Gap junctions are important for intercellular communication, enabling the coordinated contraction of the heart during a beat or the uterus during labor.

13.6 Specific Channels Increase the Permeability of Some Membranes to Water

- Aquaporin, a water-channel-forming protein, consists of six membrane-spanning α helices and a central channel lined with hydrophilic residues that allow water molecules to pass in single file.
- Aquaporins do not transport protons, and thus do not disrupt protein gradients across membranes.

Key Terms

simple diffusion (p. 382)

facilitated diffusion (passive transport) (p. 382)

active transport (p. 382)

electrochemical potential (membrane potential) (p. 383)

pumps (p. 384)

P-type ATPase (p. 384)

eversion (p. 386)

multidrug resistance (p. 388)

ATP-binding cassette (ABC) (p. 388)

ABC transporter (p. 388)

carrier (p. 390)

secondary active transport (p. 390)

secondary transporter (cotransporter) (p. 390)

antiporter (p. 390)

symporter (p. 390)

uniporter (p. 390)

ion channel (p. 392)

nerve impulse (action potential) (p. 392)

patch-clamp technique (p. 393)

selectivity filter (p. 396)

voltage-gated channel (p. 398)

ball-and-chain model (p. 400)

recovery (p. 400)

neurotransmitters (p. 400)

synaptic cleft (p. 400)

ligand-gated channel (p. 400)

equilibrium potential (p. 402)

refractory period (p. 404)

long QT syndrome (LQTS) (p. 404)

gap junction (cell-to-cell channels) (p. 406)

connexon (hemichannel) (p. 406)

aquaporin (p. 407)

Problems

1. Differentiate between simple diffusion and facilitated diffusion. ❖ 1

2. Distinguish the mechanisms by which uniporters and channels transport ions or molecules across the membrane. ❖ 1, ❖ 4

3. Which of the following statements applies to passive transport or active transport? ❖ 1

(a) Moves substances from an area of low concentration to an area of higher concentration

(b) Does not require energy

(c) Requires energy

(d) Moves substances from an area of high concentration to an area of lower concentration

4. For a typical mammalian cell, the intracellular and extracellular concentrations of the chloride ion (Cl^-) are 4 µM and 150 mM, respectively. Calculate the equilibrium potential at 37°C for the chloride ion. ❖ 2

5. Intestinal epithelial cells pump glucose into the cell against its concentration gradient using the Na^+–glucose symporter. The symporter couples the "downhill" transport of two Na^+ ions into the cell to the "uphill" transport of glucose into the cell.

(a) If the Na^+ concentration outside the cell is 155 mM and that inside the cell is 17 mM, and the membrane potential is −55 mV, what is the maximum energy available for pumping a mole of glucose into the cell?

(b) What is the maximum ratio of [glucose]$_{in}$/[glucose]$_{out}$ that could theoretically be produced if the energy coupling were 100% efficient? ❖ 2

6. If the pH of the interior of the stomach is pH 2, what is the amount of energy required for the transport of protons from the interior of the cell (pH 7.4) into the stomach lumen? Assume the membrane potential across this membrane is –70 mV and the temperature is 37°C. ❖ 2

7. Classify each phrase as a descriptor of ABC transporters, P-type ATPases, or both: ❖ 3

(a) Substrate binds before ATP

(b) The multidrug resistance protein is an example.

(c) An aspartate residue in the membrane pump is phosphorylated.

(d) The gastric H$^+$–K$^+$ ATPase is an example.

(e) ATP-dependent

(f) Two ATP-binding domains

8. Which of the following binding states are present in the SERCA catalytic cycle? For each state, assume that only the species mentioned are bound. ❖ 3

(a) AMP bound

(b) 2 Ca^{2+}, ADP, and P$_i$ bound

(c) 2 Ca^{2+} bound

(d) 2 Ca^{2+} and ADP bound

(e) ADP bound

(f) 2 Ca^{2+} and ATP bound

(g) P$_i$ bound

9. To study the mechanism of SERCA, you prepare membrane vesicles containing this protein oriented such that its ATP binding site is on the outer surface of the vesicle. To measure pump activity, you use an assay that detects the formation of inorganic phosphate in the medium. When you add calcium and ATP to the medium, you observe phosphate production for only a short period of time. Only after the addition of calcimycin, a molecule that makes membranes selectively permeable to calcium, do you observe sustained phosphate production. Explain. ❖ 3

10. Name the three types of carrier proteins. Which of these can mediate secondary active transport? ❖ 4

11. Classify each description as characterizing facilitated diffusion, primary active transport, secondary active transport, or both primary and secondary active transport:

(a) Requires ATP

(b) Includes the Na$^+$–K$^+$ ATPase pump

(c) Uses energy stored in electrochemical gradients generated by pumps

(d) Always moves more than one substance at a time

(e) Does not require energy input

(f) Moves substances against a chemical gradient

(g) Directly uses ATP hydrolysis to pump substances across the membrane

(h) Includes lactose permease

(i) Includes uniporters

12. In energetic terms, why do K$^+$ channels not enable Na$^+$ ions to cross the membrane? ❖ 5

13. How might a mutation in a cardiac voltage-dependent sodium channel cause long QT syndrome? ❖ 7

14. Batrachotoxin (BTX) is a steroidal alkaloid from the skin of *Phyllobates terribilis*, a poisonous Colombian frog (the source of the poison used on blowgun darts). In the presence of BTX, Na$^+$ channels in an excised patch stay persistently open when the membrane is depolarized. They close when the membrane is repolarized. Which transition is blocked by BTX?

15. Acid sensing is associated with pain, tasting, and other biological activities. Acid sensing is carried out by a ligand-gated channel that permits Na$^+$ influx in response to H$^+$. This family of acid-sensitive ion channels (ASICs) includes a number of members. Psalmotoxin 1 (PcTX1), a venom from the tarantula, inhibits some members of this family. The following electrophysiological recordings of cells containing several members of the ASIC family were made in the presence of the toxin at a concentration of 10 nM. The channels were opened by changing the pH from 7.4 to the indicated values. The PcTX1 was present for a short time (indicated by the black bar above the recordings below), after which time it was rapidly washed from the system.

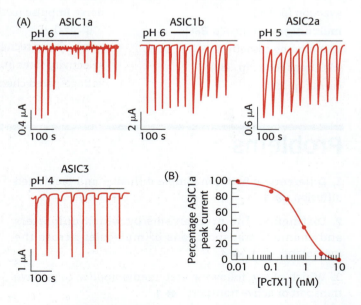

(a) Which member of the ASIC family—ASIC1a, ASIC1b, ASIC2a, or ASIC3—is most sensitive to the toxin?

(b) Is the effect of the toxin reversible? Explain.

(c) What concentration of PcTX1 yields 50% inhibition of the sensitive channel?

16. Cone snails are carnivores that inject a powerful set of toxins into their prey, leading to rapid paralysis. Many of these toxins are found to bind to specific ion-channel proteins. Why are such molecules so toxic? How might such toxins be useful for biochemical studies? ❖ **7**

17. Determine whether each phrase describes ligand-gated ion channels, voltage-gated ion channels, or both: ❖ **6**

(a) Change conformation in response to changing membrane potential

(b) May participate in an action potential

(c) An example is the acetylcholine receptor

(d) Is a form of passive transport

(e) Change conformation in response to a signal-molecule binding

18. Immediately after the repolarization phase of an action potential, the neuronal membrane is temporarily unable to respond to the stimulation of a second action potential, a phenomenon referred to as the *refractory period*. What is the mechanistic basis for the refractory period? ❖ **7**

19. Place the events of an action potential in order, starting and ending with a cell at its resting membrane potential. ❖ **7**

(a) K^+ rushes out of the cell, causing repolarization.

(b) Ligand activation of the acetylcholine receptor depolarizes the membrane.

(c) K^+ channels fully open, and Na^+ channels are inactivated.

(d) Fast Na^+ and slow K^+ channels are activated.

(e) Na^+ rushes into the cell, causing membrane depolarization.

(f) K^+ channels close slowly, resulting in hyperpolarization. Na^+ channel gates reset.

20. Drugs like tetrodotoxin (TTX) and tetraethylammonium (TEA) alter action potentials in neurons. TTX is able to block voltage-gated sodium (Na^+) channels, whereas TEA is able to block voltage-gated potassium (K^+) channels. The first graph shows an action potential in the absence of any of these drugs. The next two graphs show the effect of either TTX or TEA on the action potential. Match these graphs with the likely drug that was used. ❖ **7**

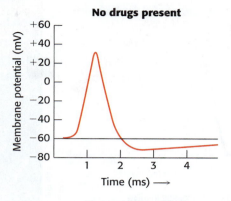

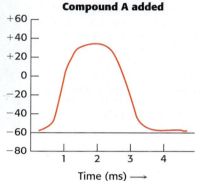

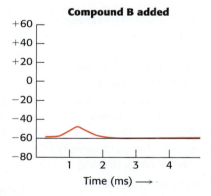

Signal-Transduction Pathways

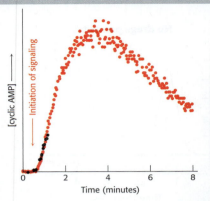

Many biological processes involve biochemical signaling. The "fight-or-flight" response triggered by fear or shock is mediated by the release of the hormone adrenaline (epinephrine). Information about the arrival of this molecule at specific receptors on cell surfaces is converted rapidly into increases in the concentration of cyclic adenosine monophosphate (cAMP) inside cells. This, in turn, drives other biochemical responses.

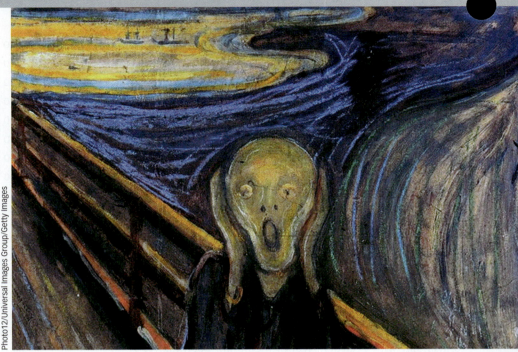

Photo12/Universal Images Group/Getty Images

OUTLINE

14.1 Many Signal-Transduction Pathways Share Common Themes

14.2 Epinephrine Signaling: Heterotrimeric G Proteins Transmit Signals and Reset Themselves

14.3 Insulin Signaling: Phosphorylation Cascades Are Central to Many Signal-Transduction Processes

14.4 Epidermal Growth Factor: Receptor Dimerization Can Drive Signaling

14.5 Defects in Signal-Transduction Pathways Can Lead to Cancer and Other Diseases

14.6 Sensory Systems Are Based on Specialized Signal-Transduction Pathways

❖ LEARNING GOALS

By the end of this chapter, you should be able to:

1. Describe the essential features of a signal-transduction circuit.

2. Define what is meant by a second messenger and give several examples.

3. Compare and contrast how the adrenergic receptors, the insulin receptor, and the epidermal growth factor receptors transmit a ligand-binding event across the plasma membrane to initiate an intracellular signaling cascade.

4. Give examples of how mutations in signaling proteins can result in cancer.

5. Discuss the signal-transduction pathways that are central to the senses of smell, vision, and hearing.

Imagine you and a group of friends are sitting around a firepit, waiting to finish cooking your dinner. Unexpectedly, you hear a rustling sound and fear that you may have an uninvited animal visitor. Your heart pounds, your eyes grow wide, and you turn pale, as you and your friends quickly run away from the sound. After 15 minutes, when you have calmed down and return to the fire; everything seems untouched. The same cannot be said for your leg, however, which sustained a deep scratch as you made your escape. After your friends help you clean and dress your wound, your thoughts return to dinner. The food smells, looks, and tastes delicious. After enjoying a nice meal, you feel satisfied.

Many of the steps in this encounter involve biochemical signaling processes within your cells. These chains of events, termed *signal-transduction pathways*, convert molecular messages into a range of physiological responses. In this chapter, we will focus on sets of signal-transduction pathways that affect circulation, metabolism, cell growth and division, and sensory perception.

14.1 Many Signal-Transduction Pathways Share Common Themes

The conversion of information of the presence or concentration of a signal molecule into other forms is called **transduction**. Signal-transduction pathways often include many components and branches. Moreover, they show variations that are used in different tissues and cell types. They can be immensely complicated and confusing. Nonetheless, the logic of signal transduction can be simplified by examining the common strategies and classes of molecules that recur in these pathways. These principles of signal-transduction pathways introduced here are at play in essentially all the metabolic pathways we will be exploring throughout the rest of the book.

Signal transduction depends on molecular circuits

Signal-transduction pathways follow a broadly similar course that can be viewed as a molecular circuit (**Figure 14.1**). All such circuits contain certain key steps:

1. *Release of the primary messenger ("signal").* A stimulus such as stress, a wound, or a digested meal triggers the release of the signal molecule, also called the **primary messenger**.

2. *Reception of the primary messenger.* Most signal molecules do not enter cells. Instead, proteins in the cell membrane act as **receptors** that convert signal molecule binding on the cell surface to a structural change within the cell's interior. Receptors span the cell membrane and thus have both extracellular and intracellular components. A binding site on the extracellular side specifically recognizes the signal molecule. Such binding sites are analogous to enzyme active sites except that no catalysis takes place within them. The interaction of the ligand and the receptor alters the tertiary or quaternary structure of the receptor to induce a structural change on the intracellular side.

3. *Delivery of the message inside the cell by the second messenger.* Other small molecules, called **second messengers**, are used to relay information from receptor–ligand complexes. Second messengers are intracellular molecules that change in concentration in response to environmental signals and mediate the next step in the molecular information circuit. Some particularly important second messengers are cyclic AMP (cAMP) and cyclic GMP (cGMP), calcium ion, inositol 1,4,5-trisphosphate (IP_3), and diacylglycerol (DAG), shown in **Figure 14.2**.

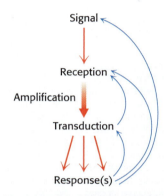

FIGURE 14.1 Signals such as changes in hormone levels lead to responses through signal-transduction pathways. An environmental signal is first received by interaction with a cellular component, most often a cell-surface receptor. Typically, the transduction process comprises many steps. The signal is often amplified before evoking one or more responses. Feedback pathways regulate the entire signaling process.

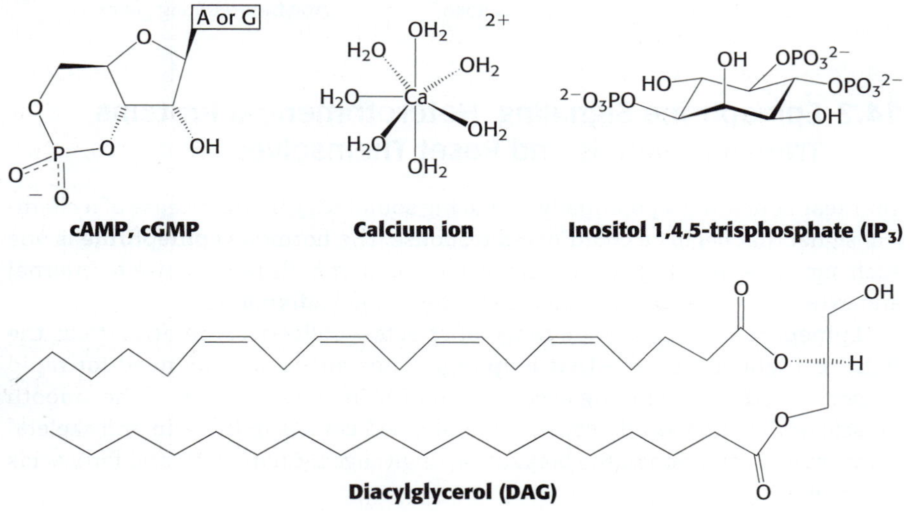

FIGURE 14.2 Several molecules function as second messengers common to many signal-transduction pathways. As intracellular molecules that change in concentration in response to environmental signals, second messengers convey information inside the cell.

cAMP, cGMP

Calcium ion

Inositol 1,4,5-trisphosphate (IP_3)

Diacylglycerol (DAG)

The use of second messengers has several consequences. First, the signal may be amplified significantly: Only a small number of receptor molecules may be activated by the direct binding of signal molecules, but each activated receptor molecule can lead to the generation of many second messengers. Thus, a low concentration of signal in the environment, even as little as a single molecule, can yield a large intracellular signal and response. Second, these messengers are often free to diffuse so that they can influence processes throughout the cell. Third, the use of common second messengers in multiple signaling pathways creates both opportunities and potential challenges. Input from several signaling pathways, often called cross talk, may alter the concentration of a common second messenger. Cross talk permits more finely tuned regulation of cell activity than would the action of individual independent pathways. However, inappropriate cross talk has the potential for misinterpretation of changes in second-messenger concentration.

4. *Activation of other molecules that directly alter the physiological response.* The ultimate effect of the signal pathway is to activate (or inhibit) the enzymes, ion channels, pumps, and transcription factors that directly control metabolic pathways, the permeability of membranes to specific ions, and gene expression.

5. *Termination of the signal.* After a cell has completed its response to a signal, the signaling process must be terminated or the cell would lose its responsiveness to incoming signals. Moreover, signaling processes that fail to terminate properly can have highly undesirable consequences. Many cancers are associated with signal-transduction processes that are overly active, especially processes that control cell growth.

We will examine components of the three signal-transduction pathways shown in **Figure 14.3**. In doing so, we will encounter many classes of proteins and other molecules that are present in many other signal-transduction pathways as well as some common strategies. It is important to keep in mind that each specific example reveals components present in many other pathways.

FIGURE 14.3 Three pathways illustrate many of the principles of signal transduction. The binding of signaling molecules to their receptors initiates pathways that lead to important physiological responses.

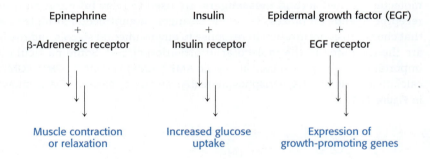

14.2 Epinephrine Signaling: Heterotrimeric G Proteins Transmit Signals and Reset Themselves

Your fear at hearing a potentially menacing sound triggers the release of a chemical signal that begins a coordinated response. The hormone **epinephrine** is one such signal, secreted by the adrenal glands of mammals in response to internal and external stressors. The molecule is often called **adrenaline**.

Epinephrine exerts a wide range of effects—collectively referred to as the fight-or-flight response—that help organisms anticipate the need for rapid muscular activity, including acceleration of heart rate, dilation of the smooth muscle of the airways and constriction of blood vessels in the skin and skeletal muscles, and initiation of the breakdown of glycogen (Chapter 21) and fatty acids (Chapter 22).

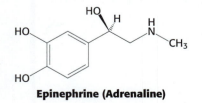

Epinephrine (Adrenaline)

Epinephrine signaling begins with ligand binding to cell-surface proteins called **adrenergic receptors**. Nine adrenergic receptors are encoded within the human genome, and they fall into two classes: α-adrenergic receptors and β-adrenergic receptors. These initiate analogous and, in many ways, complementary signaling pathways, as we shall see shortly. These receptors are members of the largest class of cell-surface receptors, called the **seven-transmembrane-helix (7TM) receptors**. Members of this family are responsible for transmitting information initiated by signals as diverse as hormones, neurotransmitters, odorants, and even light. Tens of thousands of such receptors are now known, including nearly 1000 encoded in the human genome. Furthermore, about one-third of the marketed therapeutic drugs target receptors of this class. As the name indicates, these receptors contain seven helices that span the membrane bilayer, as originally deduced from the patterns of hydrophobic residues in the amino acid sequences of these receptors (**Figure 14.4**).

We begin our more-detailed discussion of 7TM receptors with the β-adrenergic receptors that are typically present in tissues such as the heart and the airways. The three-dimensional structure of human β-adrenergic receptors (β-AR) have been determined in a variety of forms by x-ray crystallography. As anticipated, the receptor structure consists of a bundle of seven membrane-spanning alpha helices, with helices adjacent in the sequence next to one another, circling in a counterclockwise direction when viewed from the extracellular side (**Figure 14.5**).

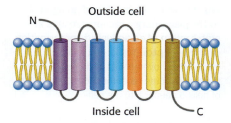

FIGURE 14.4 A large and important family of receptor share a common structure with seven transmembrane (7TM) helices. This schematic representation of a 7TM helix receptor shows its passage through the membrane seven times.

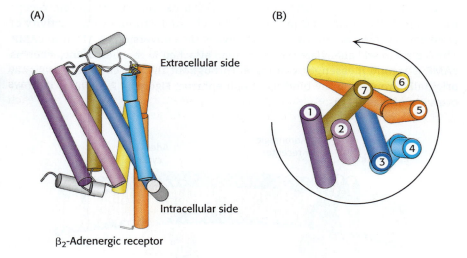

(A)
Extracellular side
Intracellular side
β₂-Adrenergic receptor

(B)

FIGURE 14.5 The three-dimensional structure of the β-adrenergic receptor reveals the arrangement of helices in 7TM receptors. (A) A side view shows the 7TM helices spanning the membrane. (B) A top view from the extracellular side shows the 7TM helices arranged in a counterclockwise manner.

The binding site for ligands such as epinephrine that activate the receptor as well as other compounds that block this site and prevent activation is present toward the center of the bundle of helices, about 20% of the way into the receptor from the extracellular side. Ligands that activate receptors are referred to as **agonists**. The presence of a bound agonist stabilizes a modified conformation of the receptor (**Figure 14.6**). Most notably, helix 5 extends deeper into the cell, and helix 6 moves outward from the center of the receptor. This change in conformational preference is crucial since it represents the transmission of information from a ligand increasing in concentration outside the cell to structural changes inside the cell.

Ligand binding to 7TM receptors leads to the activation of heterotrimeric G proteins

What are the next steps in signaling after the binding of epinephrine? Biochemical studies had revealed that this signaling process required a specific nucleotide, GTP. The conformational change on the cytoplasmic side of the receptor activates

FIGURE 14.6 The binding of an agonist to the β-adrenergic receptor results in stabilization of an activated conformation of the receptor. The largest changes involve the movement of helix 6 away from the center of the receptor and the extension of helix 5 deeper into the cell.

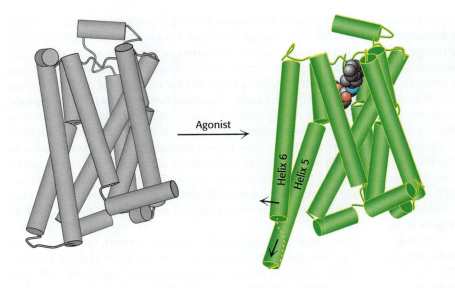

Agonist

Helix 6

Helix 5

a protein called a **G protein**, named for its requirement for guanyl nucleotides. We will examine this process in three types of muscle cells: cardiac muscle cells, cells of the constrictor muscle that controls dilation of the pupil of the eye, and smooth muscle cells in the small blood vessels.

Let us begin with cardiac muscles, which express β-adrenergic receptors. Here, the associated G protein, when activated, stimulates the activity of adenylate cyclase, an enzyme that catalyzes the conversion of ATP into cAMP. The G protein and adenylate cyclase remain attached to the membrane, whereas cAMP, a second messenger, can travel throughout the cell carrying the signal originally brought by the binding of epinephrine. **Figure 14.7** provides a broad overview of these steps.

FIGURE 14.7 The signal-transduction pathway initiated by epinephrine in cardiac muscle cells results in an increase in cAMP concentration, increasing contraction. Epinephrine binding to the 7TM receptor initiates a signal-transduction pathway that acts through a G protein and cAMP to activate protein kinase A. Phosphorylation of certain muscle proteins promotes muscle contraction.

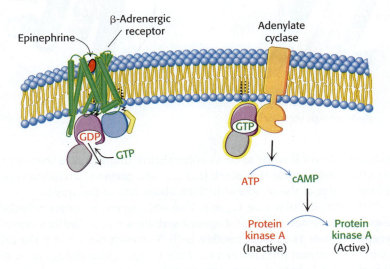

Epinephrine

β-Adrenergic receptor

Adenylate cyclase

GDP

GTP

GTP

ATP → cAMP

Protein kinase A (Inactive) → Protein kinase A (Active)

- *The role of the hormone-bound receptor is to catalyze the exchange of GTP for bound GDP.* Let us consider the role of the G protein in this signaling pathway in more detail. In its inactive state, the G protein is bound to GDP. In this form, the G protein exists as a heterotrimer consisting of α, β, and γ subunits; the α subunit (referred to as G_α) binds the nucleotide at the interface between two domains (**Figure 14.8**). The α and γ subunits are usually anchored to the membrane by covalently attached fatty acids.

(A)

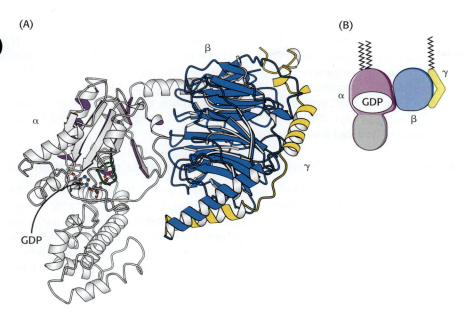

(B)

FIGURE 14.8 An inactive G protein exists as a heterotrimer consisting of α, β, and γ subunits. (A) A ribbon diagram shows the relation between the three subunits. In this complex, GDP is bound to the α subunit (gray and purple) in a pocket close to the surface through which the α subunit interacts with the βγ dimer. (B) A schematic representation of the heterotrimeric G protein is shown. [Drawn from 1GOT.pdb.]

INTERACT with this model in
Achieve

• *Analysis of the hormone-G-protein complex structure reveals its mechanism.* The structure of the complex between the hormone–β_2–AR complex and the heterotrimeric G protein shows that the G_α subunit interacts with the surface of the receptor, primarily with the two helices (helix 5 and helix 6) that changes most dramatically upon agonist binding (**Figure 14.9**). When bound to the receptor, the two domains of the G protein are relatively free to move with respect to one another, allowing G_α to open substantially, enabling the replacement of GDP with GTP from solution.

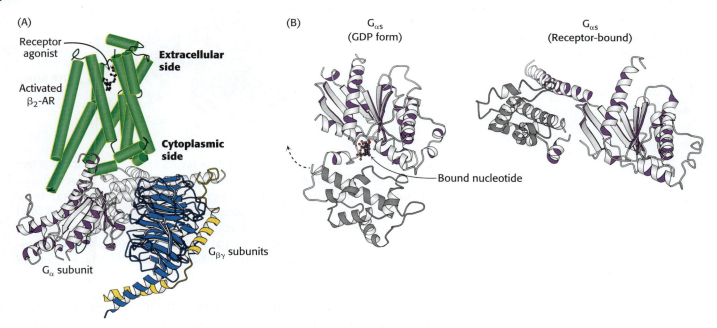

FIGURE 14.9 The activated β_2-AR activates a heterotrimeric G protein by allowing the α subunit to open and release GDP and bind GTP. (A) When the β_2-AR (green) binds a receptor agonist, the cytoplasmic face of the receptor forms an interaction surface with the G_α subunit of a heterotrimeric G protein. (B) The interaction with the activated receptor leads to a substantial conformational change in the G_α protein, in which the guanine nucleotide binding site is opened, enabling nucleotide exchange. [Drawn from 3SN6.pdb and 1AZT.pdb.]

- A *single hormone–receptor complex can stimulate nucleotide exchange in many G-protein heterotrimers.* On GTP binding, the α subunit dissociates from the βγ dimer ($G_{\beta\gamma}$) and from the receptor. The freed receptor is available to bind other G-protein heterotrimers and to catalyze the exchange of GTP for GDP. Thus, hundreds of G_α molecules can be converted from their GDP form into their GTP form for each bound molecule of hormone, amplifying the response. Because they signal through G proteins, 7TM receptors are often called **G-protein-coupled receptors (GPCRs).**

Activated G proteins transmit signals by binding to other proteins

In the GTP form, the surface of G_α that had been bound to $G_{\beta\gamma}$ has changed its conformation from the GDP form so that it no longer has a high affinity for $G_{\beta\gamma}$. This surface is now exposed for binding to other proteins. In the β-AR pathway, the new binding partner is **adenylate cyclase**, the enzyme that converts ATP into cAMP. This enzyme includes 12 membrane-spanning helices with two cytoplasmic domains that come together to form the catalytic part of the enzyme (**Figure 14.10**).

(A)

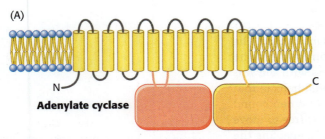

(B)

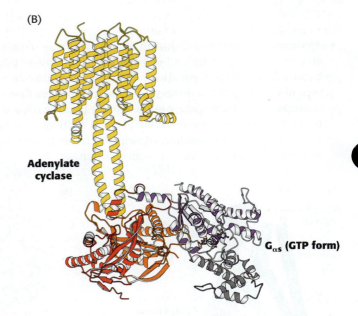

Adenylate cyclase

$G_{\alpha s}$ (GTP form)

FIGURE 14.10 Adenylate cyclase is activated by binding of $G_{\alpha s}$ (GTP). (A) Adenylate cyclase is a membrane protein with two intracellular domains that come together to form the catalytic apparatus. (B) The structure of a complex between G_α in its GTP form bound to a adenylate cyclase is shown. The surface of $G_{\alpha s}$ that had been bound to the βγ dimer now binds adenylate cyclase. [Drawn from 1AZS.pdb.]

INTERACT with this model in
 Achieve

The interaction of G_α with adenylate cyclase favors a more catalytically active conformation of the enzyme, thus stimulating cAMP production. Indeed, the G_α subunit that participates in the β-AR pathway is called $G_{\alpha s}$, where "s" stands for *stimulatory*. The generation of cAMP by adenylate cyclase provides a second level of amplification because each activated adenylate cyclase molecule can convert ATP into cAMP at a rate of approximately $100\ s^{-1}$. The net result is that the binding of epinephrine to the receptor on the cardiac muscle cell surface increases the level of cAMP inside the cell rapidly.

Cyclic AMP stimulates the phosphorylation of many target proteins by activating protein kinase A

In all eukaryotic cells, most effects of cAMP are mediated by the activation of a single protein kinase, protein kinase A (PKA). As described in Chapter 7, PKA consists of two regulatory (R) chains and two catalytic (C) chains (R_2C_2). In the absence of cAMP, the R_2C_2 complex is catalytically inactive. The binding of cAMP to the regulatory chains decreases their affinity for the catalytic chains, which are catalytically active when freed. Activated PKA then phosphorylates specific serine and threonine residues in many targets to alter their activity.

In cardiac muscle cells, increases in cAMP levels leads to increased contraction. In these cells, one of these PKA substrates is troponin I. This protein is part of the troponin complex found in cardiac and skeletal muscle but not smooth muscle. The troponin complex blocks the ability of myosin to bind to actin, preventing muscle contraction (see Chapter 6). Upon phosphorylation of troponin I, the binding of troponin to actin is weakened and myosin is more likely to bind to actin, driving muscle contraction. PKA also phosphorylates certain calcium-specific ion channels and activates them. The increased level of calcium ions binds to another troponin subunit and stimulates muscle contraction. The signal-transduction pathway initiation by epinephrine with the β-adrenergic receptor is summarized in **Figure 14.11**.

G proteins spontaneously reset themselves through GTP hydrolysis

How is the signal initiated by epinephrine switched off? G_α subunits have intrinsic GTPase activity, that is, by themselves they can promote the hydrolysis of bound GTP to GDP and P_i. This hydrolysis reaction is slow, however, requiring from seconds to minutes. Thus, the GTP form of G_α is able to activate downstream components of the signal-transduction pathway before it is deactivated by GTP hydrolysis. In essence, the bound GTP acts as a built-in clock that spontaneously resets the G_α subunit after a short time. After GTP hydrolysis and the release of P_i, the GDP-bound form of G_α then reassociates with $G_{\beta\gamma}$ to re-form the inactive heterotrimeric protein (**Figure 14.12**). The GTP hydrolysis step also provides an additional opportunity for regulation; Some G-protein signaling pathways include additional components called GTPase-activating proteins that increase the rate of GTP hydrolysis by G-protein α subunits.

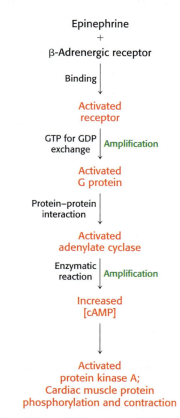

Epinephrine
+
β-Adrenergic receptor

Binding

Activated receptor

GTP for GDP exchange | Amplification

Activated G protein

Protein–protein interaction

Activated adenylate cyclase

Enzymatic reaction | Amplification

Increased [cAMP]

Activated protein kinase A; Cardiac muscle protein phosphorylation and contraction

FIGURE 14.11 Epinephrine binding promotes cardiac muscle contraction. The binding of epinephrine to the β-adrenergic receptor initiates the signal-transduction pathway. The process in each step is indicated in black at the left of each arrow. Steps that have the potential for signal amplification are indicated at the right in green. The final step is phosphorylation of key cardiac muscle proteins, promoting muscle contraction.

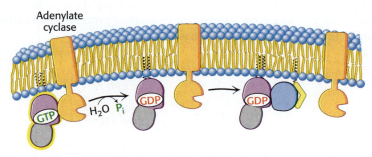

FIGURE 14.12 G proteins reset themselves by GTP hydrolysis. On hydrolysis of the bound GTP promoted by the GTPase activity of G_α, G_α dissociates from adenylate cyclase, reversing its stimulation. G_α in its GDP form then reassociates with the βγ dimer to reform the heterotrimeric G protein.

The hormone-bound activated receptor must be reset as well to prevent the continuous activation of G proteins. This resetting is accomplished by two processes (**Figure 14.13**). First, the hormone can dissociate, returning the

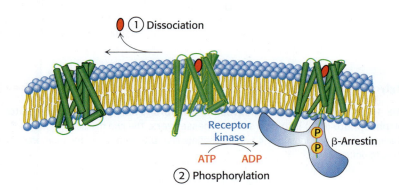

FIGURE 14.13 β-Adrenergic receptor signaling is terminated by two processes. Signal transduction by the 7TM receptor is halted (1) by dissociation of the signal molecule from the receptor and (2) by phosphorylation of the cytoplasmic C-terminal tail of the receptor and the subsequent binding of β-arrestin.

receptor to its initial inactive state. The likelihood that the receptor remains in its unbound state depends on the extracellular concentration of hormone. Second, members of a specialized family of enzymes, the G-protein receptor kinases, are bound by activated receptors and phosphorylates serine and threonine residues in their carboxyl-terminal tails. Finally, proteins called β-arrestins bind to the phosphorylated receptors and block their ability to activate G proteins.

Some 7TM receptors activate the phosphoinositide cascade

We now turn to the muscle cells that control dilation of the pupils of our eyes and those that line small blood vessels in our skin. These cells express α1-adrenergic receptors rather than β-adrenergic receptors. α-Adrenergic receptors are homologous to β-adrenergic receptors, also bind epinephrine, and have very similar three-dimensional structures. They differ, however, in the G proteins to which they are coupled in their signal-transduction pathway by virtue of differences in parts of the intracellular face of the receptors. α1-Adrenergic receptors activate a G protein called $G_{\alpha q}$. In its GTP form, $G_{\alpha q}$ binds to and activates a form of the enzyme phospholipase C.

Phospholipase C catalyzes the cleavage of a phospholipid present in cell membranes, phosphatidylinositol 4,5-bisphosphate (PIP_2), into the two second messengers—inositol 1,4,5-trisphosphate (IP_3) and diacylglycerol (DAG; **Figure 14.14**). IP_3 is soluble and diffuses away from the membrane. This second messenger causes the rapid release of Ca^{2+} from the intracellular stores in the endoplasmic reticulum (ER), which accumulates a reservoir of Ca^{2+} through the action of transporters such as the Ca^{2+} ATPase (Section 13.2). On binding IP_3, specific IP_3-gated Ca^{2+}–channel proteins in the ER membrane open to allow calcium ions to flow from the ER into the cytoplasm. Calcium ion is itself a signaling molecule: it can bind proteins, including a ubiquitous signaling protein called **calmodulin (CaM)**, and enzymes such as protein kinase C.

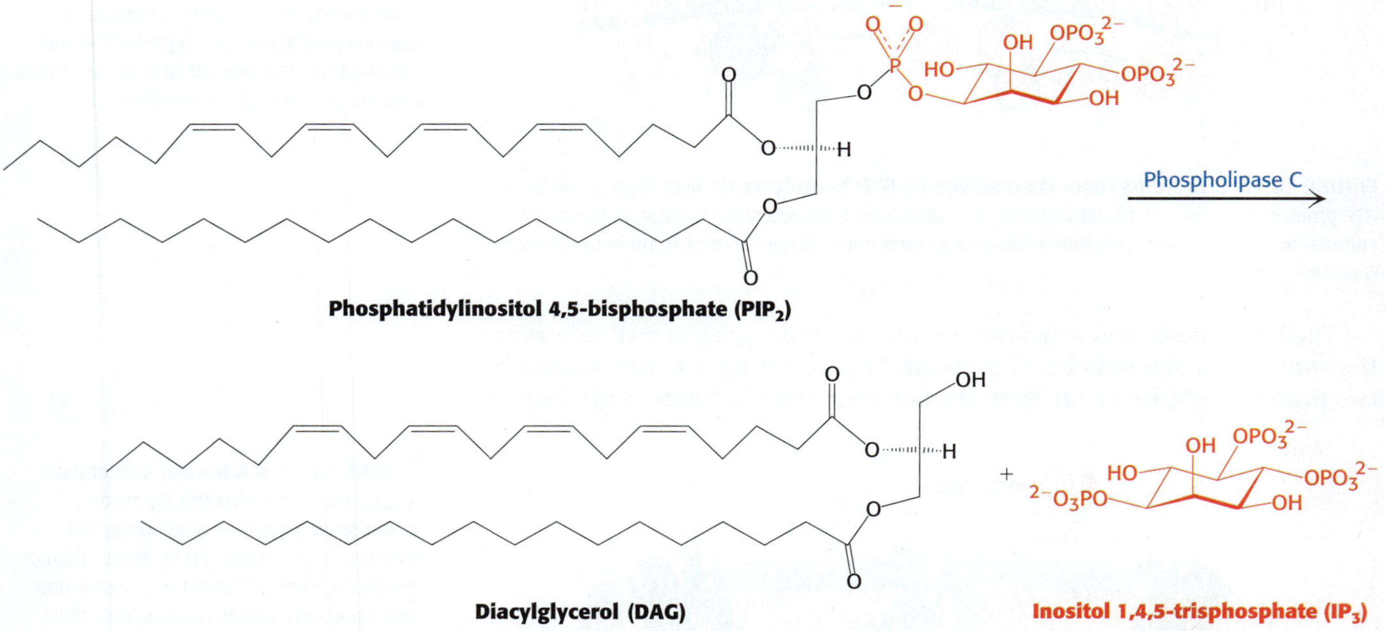

Phosphatidylinositol 4,5-bisphosphate (PIP₂)

Diacylglycerol (DAG)

Inositol 1,4,5-trisphosphate (IP₃)

FIGURE 14.14 Phospholipase C cleaves the membrane lipid phosphatidylinositol 4,5-bisphosphate (PIP₂) into two second messengers. The two messengers are diacylglycerol (DAG) and inositol 1,4,5-trisphosphate (IP₃). DAG remains in the membrane while IP₃ is free to diffuse away.

DAG remains in the plasma membrane where it binds and activates protein kinase C (PKC), a protein kinase that phosphorylates serine and threonine residues in many target proteins. Specialized domains of this kinase bind DAG and require bound calcium to do so. Thus, DAG and IP$_3$ work in tandem: IP$_3$ increases the Ca^{2+} concentration, and Ca^{2+} facilitates the DAG-mediated activation of protein kinase C. Like PKA (Chapter 7), activation involves removal of a pseudo-substrate in a regulatory domain from the active site. The overall signal-transduction pathway, referred to as the phosphoinositide cascade, is summarized in **Figure 14.15**. Both IP$_3$ and DAG act transiently because they are converted into other species by phosphorylation or other processes.

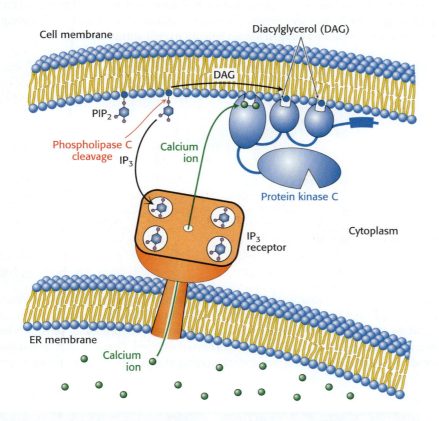

FIGURE 14.15 The phosphoinositide cascade results in protein kinase C activation. Phospholipase C cleaves PIP$_2$ into IP$_3$ and DAG. IP$_3$ release results in the influx of calcium ions (owing to the opening of the IP$_3$ receptor ion channels), and DAG accumulation results in the activation of protein kinase C when protein kinase C binds to DAG in the membrane. Calcium ions bind to domains in protein kinase C and help facilitate its binding to DAG.

Calcium ion is a widely used second messenger

Calcium ion participates in many signaling processes in addition to the phosphoinositide cascade. Two properties of this ion account for its widespread use as an intracellular messenger:

- *Fleeting changes in Ca^{2+} concentration are readily detected.* At steady state, intracellular levels of Ca^{2+} must be kept low to prevent the precipitation of carboxylated and phosphorylated compounds, which form poorly soluble salts with Ca^{2+}. Transport systems extrude Ca^{2+} from the cytoplasm, maintaining the cytoplasmic concentration of Ca^{2+} at approximately 100 nM—several orders of magnitude lower than that of the extracellular medium. Given this low steady-state level, transient increases in Ca^{2+} concentration produced by signaling events can be readily sensed.

- *Ca^{2+} can bind tightly to proteins and induce substantial structural rearrangements.* A second property of Ca^{2+} that makes it a highly suitable intracellular messenger is its ability to interact with protein through many points of contact. Calcium ions bind well to negatively charged oxygen atoms (from the side

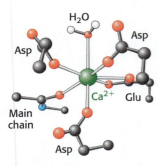

FIGURE 14.16 Calcium-binding sites typically include seven oxygen atoms. In one common mode of binding, calcium is coordinated to six oxygen atoms from a protein and one from water. The multiple protein-based ligands allow calcium ions to stabilize particular protein structures.

chains of glutamate and aspartate) and uncharged oxygen atoms (main-chain carbonyl groups and side-chain oxygen atoms from glutamine and asparagine; **Figure 14.16**). The capacity of Ca^{2+} to be coordinated to multiple ligands—from six to eight oxygen atoms—enables it to cross-link different segments of a protein and induce significant conformational changes.

Our understanding of the role of Ca^{2+} in cellular processes has been greatly enhanced by our ability to detect changes in Ca^{2+} concentrations inside cells and even monitor these changes in real time. In Section 4.2, we discussed how intact proteins can be detected in live cells using fluorescence microscopy. A similar approach can be used for the visualization of changes in Ca^{2+} concentrations. This ability depends on the use of specially designed dyes such as Fura-2 that bind Ca^{2+} and change their fluorescent properties on Ca^{2+} binding. Fura-2 binds Ca^{2+} through appropriately positioned oxygen atoms (shown in red) within its structure:

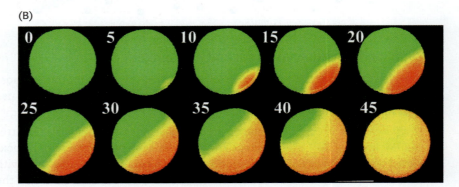

Fura-2

When such a dye is introduced into cells, changes in available Ca^{2+} concentration can be monitored with microscopes capable of detecting changes in fluorescence (**Figure 14.17**). Following pioneering work on calcium sensors, probes for sensing other second messengers such as cAMP also have been developed. These molecular-imaging agents are greatly enhancing our understanding of signal-transduction processes.

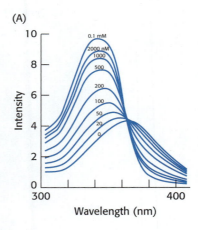

(A)

(B)

FIGURE 14.17 Free calcium ion concentrations inside cells can be imaged using fluorescence microscopy with appropriate dyes. (A) The fluorescence spectra of the calcium-binding dye Fura-2 can be used to measure available calcium ion concentrations in solution and in cells. (B) A series of images show Ca^{2+} spreading across an egg cell following fertilization by sperm, with the numbers indicating time elapsed in seconds. These images were obtained through the use of Fura-2. The images are false colored: orange represents high Ca^{2+} concentrations, and green represents low Ca^{2+} concentrations. [(A) Information from S. J. Lippard and J. M. Berg, *Principles of Bioinorganic Chemistry* (University Science Books, 1994), p. 193; (B) Credit: Republished with permission of Company of Biologists, from Exploring the mechanism of action of the sperm-triggered calcium-wave pacemaker in ascidian zygotes, Carroll, M. et al., December 15, 2003, *J Cell Sci* 116: 4997–5004; permission conveyed through Copyright Clearance Center, Inc. Courtesy Alex McDougall.]

Calcium ion often activates the regulatory protein calmodulin

Calmodulin (CaM), a 17-kDa protein with four Ca^{2+}-binding sites, serves as a calcium sensor in nearly all eukaryotic cells. At cytoplasmic concentrations above about 500 nM, Ca^{2+} binds to and activates calmodulin. Calmodulin is a member of the EF-hand protein family. The **EF hand** is a Ca^{2+}-binding motif that consists of a helix, a loop, and a second helix. This motif was named the *EF hand* because the two key helices designated *E* and *F* in parvalbumin, the first protein where the motif was observed, are positioned like the forefinger and thumb of the right hand (**Figure 14.18**). The two helices and the intervening loop of the EF hand form the Ca^{2+}-binding motif. Seven oxygen atoms are coordinated to each Ca^{2+}, six from the protein and one from a bound water molecule. Calmodulin is made up of four EF-hand motifs, each of which can bind a single Ca^{2+} ion.

The binding of Ca^{2+} to calmodulin induces substantial conformational changes in its EF hands, exposing hydrophobic surfaces that can be used to bind other proteins. Using its two sets of two EF hands, calmodulin clamps down around specific regions of target proteins—usually exposed α helices with appropriately positioned hydrophobic and charged groups (**Figure 14.19**). Some protein kinases, for example, are activated when calmodulin binds to a target α helix and removes it from a position when it inhibits kinase function.

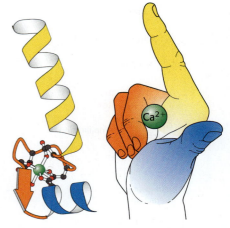

FIGURE 14.18 An EF hand is a binding site for Ca^{2+} in many calcium-sensing proteins. The structure, resembling a hand, is formed by a helix-loop-helix unit with the E helix is yellow, the loop in red, the F helix is blue, and calcium is represented by the green sphere. [Drawn from 1CLL.pdb.]

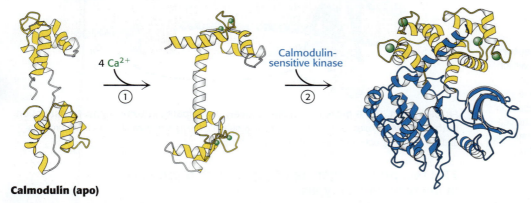

Calmodulin (apo)

FIGURE 14.19 Calmodulin binds calcium and activates some protein kinases. On Ca^{2+} binding to the apo, or calcium-free, form of calmodulin, the two halves of calmodulin change conformation. Calmodulin can then clamp down around an α helix through surfaces that are exposed. Some kinases have an α helix that inhibits enzyme activity and calmodulin binding can relieve this inhibition. [Drawn from 1CFD.pdb, 1CLL.pdb, and 6PAW.pdb.]

INTERACT with the models shown in Figure 14.18 and 14.19 in
Achie√e

How, then, does initiation of the phosphoinositide cascade lead to smooth muscle contraction? Smooth muscle does not contain the troponin complex. However, myosin binding to actin and subsequent contraction is stimulated by phosphorylation of a component of myosin, the regulatory light chain. The increased level of calcium activates calmodulin, which, in turn, binds to and activates myosin light chain kinase. The activated kinase phosphorylates the myosin light chain and facilitates contraction. This signal-transduction-pathway-induced contraction leads to pupil dilation and pale skin as blood flow is restricted.

Some receptors signal through G proteins that inhibit rather than stimulate adenylate cyclase

Some blood vessel smooth muscle cells express a different class of 7TM receptors termed α_2-adrenergic receptors. These receptors are quite similar to the α_1- and β-adrenergic receptors, but signal through a third class of G protein, $G_{\alpha i}$. Like $G_{\alpha s}$, this G-protein α subunit binds adenylate cyclase, but $G_{\alpha i}$-binding inhibits

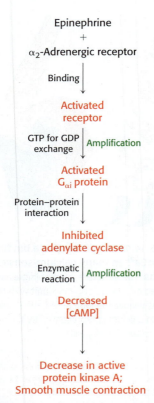

Epinephrine
+
α_2-Adrenergic receptor

Binding ↓

**Activated
receptor**

GTP for GDP
exchange | Amplification

**Activated
$G_{\alpha i}$ protein**

Protein–protein
interaction

**Inhibited
adenylate cyclase**

Enzymatic
reaction | Amplification

**Decreased
[cAMP]**

↓

**Decrease in active
protein kinase A;
Smooth muscle contraction**

FIGURE 14.20 α2-Adrenergic receptors signal through the G protein $G_{\alpha i}$, which inhibits adenylate cyclase. The signal-transduction pathway from epinephrine binding to α2-adrenergic receptors results in smooth muscle contraction. This is achieved through a decrease in cAMP concentration and the associated reduction in protein kinase A activity.

Adenosine

Caffeine

cyclase activity rather than simulating it. Thus, activation of this signaling pathway results in a decrease in cAMP levels. Such lower cAMP levels influence a protein kinase cascade (**Figure 14.20**). The net result is increased smooth muscle contraction.

❖ SELF–CHECK QUESTION

Why is increased cardiac muscle contraction favored by increased cAMP levels but increased smooth muscle contraction favored by reduced cAMP levels?

G-protein $\beta\gamma$-dimers can also directly participate in signaling

When a heterotrimeric G protein is activated, it dissociates into a G_{α} subunit and a $G_{\beta\gamma}$ dimer. It was widely believed that the only role of the $G_{\beta\gamma}$ dimer was to sequester G_{α} subunits. However, through studies using purified G-protein subunits led to the discovery that the $G_{\beta\gamma}$ dimer was the active component in activating a key ion channel in the heart (**Figure 14.21**). Subsequent studies have revealed that these subunits play active roles in a wide range of signaling pathways, revealing an additional level of complexity and control.

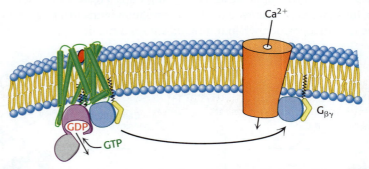

FIGURE 14.21 The G protein $\beta\gamma$ dimer actively participates in some signal-transduction pathways. When released from the G protein α subunit, the $\beta\gamma$ dimer can bind to other proteins and, for example, open certain ion channels.

7TM receptors trigger signaling through G proteins in many other cell types

We noted earlier that the human genome encodes nearly 1000 7TM receptors. Approximately 1/3 of these participate in our sense of smell, and a few others are central to vision. We will explore these specialized receptors at the end of this chapter. The others participate in one or more processes in almost all aspects of physiology.

This array of signaling molecules, receptors, and pathways can be dizzyingly complex. However, once we know that a pathway involves a 7TM receptor and we identify the participating G protein, we can deduce the participating second messengers and key steps in the pathway (**Figure 14.22**). For example, the peptide hormone glucagon, which plays a central role in control of metabolism during fasting, has a receptor that signals through $G_{\alpha s}$. We can deduce that cAMP levels will rise in cells that express the receptor, and this will activate PKA. PKA phosphorylates two enzymes that lead to the breakdown of glycogen, the polymeric store of glucose, and the inhibition of further glycogen synthesis (Section 21.4). Our genome encodes four receptors for the nucleoside adenosine that are expressed throughout the body, including the heart and the brain. One of these, the adenosine A_1 receptor, signals through $G_{\alpha i}$. The A_1 pathway slows the heartbeat as we would anticipate due to a decreased level of cAMP. In the brain, it decreases the release of certain neurotransmitters. Compounds that block receptors without activating them are referred to as **antagonists**. You are likely familiar with the actions of a widely occurring adenosine receptor antagonist, caffeine.

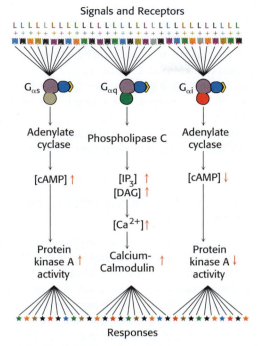

FIGURE 14.22 Common pathways recur in many different signaling processes involving G proteins. A very large number of 7TM receptors and their associated ligands signal through the three classes of G protein α subunits: $G_{\alpha s}$, $G_{\alpha q}$, and $G_{\alpha i}$. The occurrence of these α subunits in the pathways dictate several steps in the pathways. After these steps, a wide variety of responses can occur.

14.3 Insulin Signaling: Phosphorylation Cascades Are Central to Many Signal-Transduction Processes

We now turn to a signaling pathway that comes into play after you return to complete your meal after your quick retreat. The maintenance of blood glucose levels with a relatively narrow range in the face of substantial changes in food intake and activity is a major challenge that we will explore extensively later in the book. For now, we will explore the signal-transduction pathway initiated by **insulin**, the hormone released in response to increased blood-glucose levels after a meal (**Figure 14.23**). In all of its detail, this multifaceted pathway is quite complex.

Barbara Steiner

EVA NEER Born in Poland in 1938, Eva Neer escaped from the Nazis at the age of one, hiding under her mother's skirt. Her family emigrated first to Brazil, then to the United States. After completing her MD degree, she began research purifying the membrane protein adenylate cyclase; this was an enzyme of much interest, given research by Earl Sutherland, Martin Rodbell, and Alfred Gilman implicating cAMP as a key intracellular messenger. Dr. Neer continued her career as a faculty member, working on many aspects of G-protein signaling and sharing some of her partially purified G_{as} with David Clapham, a colleague interested in ion-channel activation. Her preparation potently activated the ion channels, but subsequent studies revealed that this was due not to $G_{\alpha s}$ but to contaminating $G_{\beta\gamma}$ that was present. This serendipitous discovery revealed the active role $G_{\beta\gamma}$ plays in some signaling systems and laid the foundation for many future studies. Dr. Neer was known for her great integrity; she trained many graduate students and postdoctoral fellows before she died prematurely of breast cancer at the age of 62.

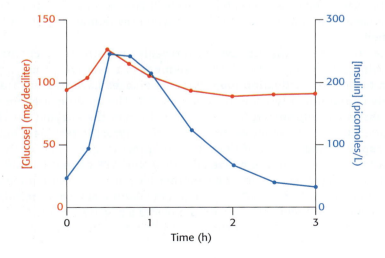

FIGURE 14.23 Glucose and insulin levels increase and then decrease following a meal. Shortly after eating, glucose levels in the bloodstream increase. These increased levels are sensed and insulin is released into the bloodstream from the pancreas. Insulin then stimulates absorption of glucose by muscle and fat cells, decreasing the glucose levels.

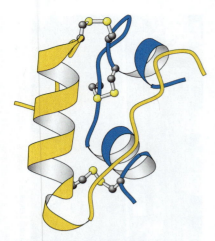

FIGURE 14.24 Insulin comprises two chains linked by two interchain disulfide bonds. The α chain (blue) also has an intrachain disulfide bond. The β chain is shown in yellow. [Drawn from 1B2F.pdb.]

Hence, we will focus solely on the major branch, which leads to the movement of glucose transporters to the surfaces of fat and muscle cells. These transporters allow these cells to take up the glucose that is plentiful in the bloodstream after a meal. We will discuss the mechanisms by which pancreatic β-cells sense glucose levels and release insulin in Chapter 15.

The signaling pathways we have examined so far have activated protein kinases as downstream components. The receptor for insulin is a member of a large family of receptors that include protein kinases as a key part of their covalent structures that are inactive under resting conditions. The activation of these protein kinases sets in motion other processes that ultimately modify the effectors of these pathways.

The insulin receptor is a protein kinase that is autoinhibited prior to insulin binding

Insulin is a peptide hormone, consisting of two chains that are linked by three disulfide bonds (**Figure 14.24**). Its receptor has a quite different structure from that of the β-AR. The insulin receptor is a homodimer of heterodimers. Each component of the homodimer consists of one α chain and one β chain (**Figure 14.25**). The α chain has an elaborate structure with six domains while the β chain has two domains in addition to a protein kinase domain. Each α chain lies completely outside the cell, whereas each β chain lies primarily inside the cell, spanning the membrane with a single transmembrane segment. The kinase domain in the β chain is homologous to that from protein kinase A but differs in two important ways. First, the insulin-receptor kinase is a tyrosine kinase; that is, it catalyzes the transfer of a phosphoryl group from ATP to the hydroxyl group of tyrosine, rather than serine or threonine.

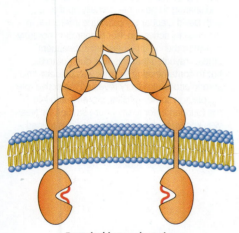

Protein kinase domains

FIGURE 14.25 The insulin receptor consists of two subunits, each of which consists of an α chain and a β chain with protein kinase domains inside the cell. The two α chains are depicted in a lighter shade of orange than are the β chains. Each α chain is entirely outside the cell, while each β chain lies primarily inside the cell and includes a protein kinase domain.

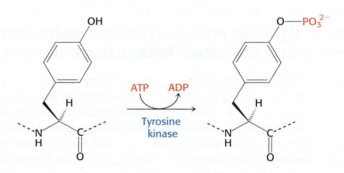

Because this tyrosine kinase is a component of the receptor itself, the insulin receptor is referred to as a receptor tyrosine kinase. Second, the insulin-receptor kinase is in an inactive conformation when the domain is not covalently modified.

The structure of this inactive form reveals an intricate mechanism for maintaining this inactivity. The key element is a relatively unstructured region termed the **activation loop**. Tyrosine kinases have two substrates, a peptide containing a tyrosine residue and ATP. The binding sites for these two substrates, determined from the active form of the enzyme, are shown in **Figure 14.26**. In the inactive form, one of the tyrosine residues that is to be phosphorylated as part of the activation process and is part of the activation loop is bound in the active site. However, ATP cannot bind because the ATP binding site is occupied by another part of the activation loop with the sequence Asp-Phe-Gly (DFG). The phenylalanine sidechain mimics the adenine ring of ATP while the aspartate provides a negative charge. Thus, even though a potential substrate is the active site, no reaction can take place as

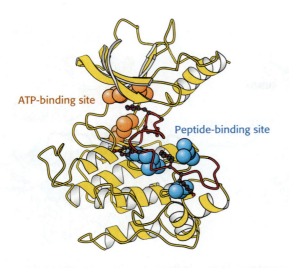

ATP-binding site

Peptide-binding site

FIGURE 14.26 The insulin receptor kinase is inactive due to part of the activation loop blocking the ATP-binding site. The structure of the unactivated form of the insulin receptor kinase is shown with residues from the ATP-binding site in orange and residues from the peptide substrate binding site in blue. The activation loop, shown in red, has a tyrosine residue in the active site, but it cannot be phosphorylated because the ATP-binding site is blocked. [Drawn from 1IRK.pdb.]

the DFG sequence is tethered closely enough to the protein structure that it cannot move out of the way. This is analogous to the impossibility of scratching your right elbow with your right hand.

Insulin binding results in the cross-phosphorylation and activation of the insulin receptor

In the absence of bound insulin, the insulin receptor dimer has an inverted "V"-like structure. Most importantly, this structure holds the kinase domains apart so that they cannot interact with one another. Insulin binds first to a site that includes elements from both subunits with a substantial conformation change induced upon binding. Overall, each receptor can bind up to four insulin molecules, and the final dimer has a "T"-like structure (**Figure 14.27**).

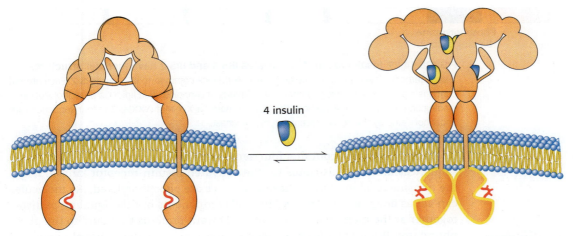

4 insulin

In this structure, the two protein kinase domains inside of the cell are pushed together. The flexible activation loop of one kinase subunit can fit into the active site of the other kinase subunit within the dimer in place of the domain's own activation loop, opening up an ATP-binding site, and allowing the phosphorylation reaction to occur. When these tyrosine residues are phosphorylated, they can no longer fit into the active site and the activation loops swings out into a different conformation with both the ATP- and peptide-binding sites available to phosphorylated additional substrates. Three tyrosine residues in each activation loop are phosphorylated (**Figure 14.28**). The net result of this overall process is that insulin binding on the outside of the cell results in the activation of a membrane-associated kinase within the cell.

FIGURE 14.27 The insulin receptor dramatically changes conformation when it binds insulin. Upon binding insulin, the extracellular domains of the insulin receptor change conformation, causing the overall receptor to convert from an inverted "V" shape to a "T" shape. This allows the protein kinase domains to come together. The receptor dimer can bind up to four insulin molecules.

FIGURE 14.28 Activation of the insulin receptor involves phosphorylation of tyrosine residues in the kinase domain. One kinase domain phosphorylates tyrosine residues in the other. When tyrosine residues in the activation loop are phosphorylated, they can no longer fit into the active site; the activation loop is released, opening up the ATP-binding site and activating the kinase. [Drawn from 1IRK.pdb and 1IR3.pdb.]

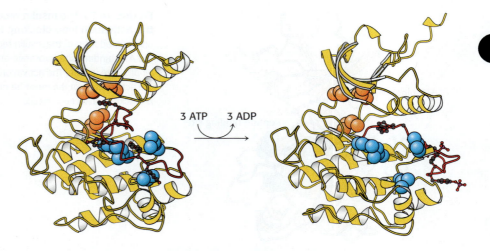

The activated insulin-receptor kinase initiates a kinase cascade

With the kinase activated, additional sites within the receptor are also as the two subunits of the receptor are in close proximity to one another. These phosphorylated sites act as docking sites for other substrates, including a class of molecules referred to as insulin-receptor substrates (IRS). IRS-1 and IRS-2 are two homologous proteins with a common modular structure (**Figure 14.29**). The amino-terminal part includes a **pleckstrin homology domain**, which binds phosphoinositide, and a phosphotyrosine-binding domain. These domains act together to anchor the IRS protein to the insulin receptor and the associated membrane.

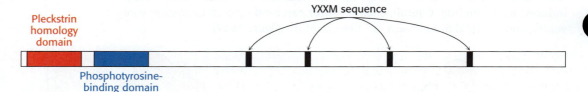

FIGURE 14.29 Insulin-receptor substrates IRS-1 and IRS-2 have modular structures. This schematic view represents the amino acid sequence common to IRS-1 and IRS-2. Each protein contains a pleckstrin homology domain (which binds phosphoinositide lipids), a phosphotyrosine-binding domain, and four sequences that approximate Tyr-X-X-Met (YXXM). The four sequences are phosphorylated by the insulin-receptor tyrosine kinase.

Each IRS protein contains four sequences of the form Tyr-X-X-Met. These sequences are also substrates for the activated insulin-receptor kinase. When the tyrosine residues within these sequences are phosphorylated, IRS molecules can act as adaptors, bringing additional components of this signaling pathway together at the membrane. The most important proteins that bind to the phosphorylated IRS molecules are specific lipid kinases, called phosphoinositide 3-kinases (PI3Ks), that add a phosphoryl group to the 3-position of inositol in phosphatidylinositol 4,5-bisphosphate (PIP_2) to form phosphatidylinositol 3,4,5-trisphosphate (PIP_3) (**Figure 14.30**). These enzymes are dimers of 110-kDa

FIGURE 14.30 Phosphoinositide 3-kinase adds a phosphoryl group to PIP_2, forming PIP_3. This generates a modified lipid headgroup that can be recognized by other enzymes.

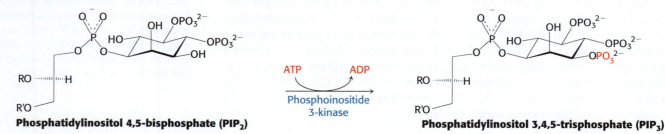

Phosphatidylinositol 4,5-bisphosphate (PIP_2)

Phosphoinositide 3-kinase

Phosphatidylinositol 3,4,5-trisphosphate (PIP_3)

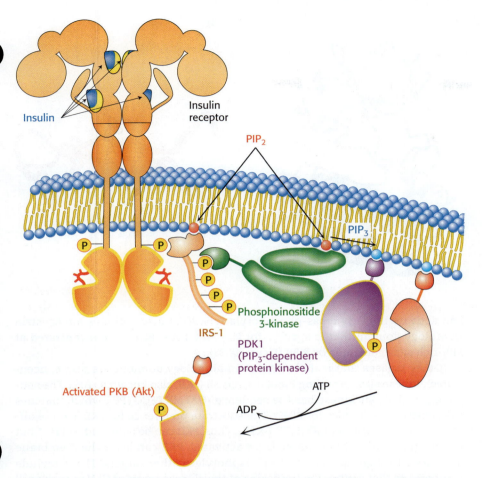

FIGURE 14.31 The insulin signaling pathway results in protein kinase activation.
The binding of insulin results in the cross-phosphorylation and activation of the insulin
receptor. Phosphorylated sites on the receptor act as binding sites for insulin-receptor
substrates such as IRS-1. The lipid kinase phosphoinositide 3-kinase binds to phosphorylated
sites on IRS-1 through its regulatory domain and converts PIP_2 into PIP_3. PIP_3 then serves as
a binding site for PIP_3-dependent protein kinase (PDK1), which phosphorylates and activates
kinases such as PKB. Activated PKB can then diffuse throughout the cell to continue the
signal-transduction pathway.

catalytic subunits and 85-kDa regulatory subunits. These enzymes bind to the
phosphorylated IRS proteins and are drawn to the membrane where they can
phosphorylate PIP_2 to form PIP_3. The PIP_3 molecules serve as gathering points for
other proteins, continuing a kinase cascade (**Figure 14.31**).

The first protein recruited to the PIP_3 molecules is phosphoinositide-dependent
kinase-1 (PDK1). PDK1 is a member of a large class of serine/threonine-specific
kinases that require phosphorylation on their activation loops in a manner similar to the insulin-receptor kinase. However, the activation mechanism is different.
For the serine-threonine kinases, the enzymes adapt an inactivate conformation
even through the ATP- and peptide-binding sites are accessible. Phosphorylation
results in a change to an active conformation (**Figure 14.32**).

The most important characteristics of a protein kinase are its substrate
sequence preferences and the sequence around its activation site. For PDK1, its
substrates approximate the sequence -XXTX(**S,T**)FCGT-, where X represents relatively variable positions. Its activation site has the sequence -QARAN**S**FGV**T**A-,
where the underlined residues match the preferred substrate sequence. The
match between this sequence and the substrate preference indicates that PDK1

FIGURE 14.32 Phosphorylation activates phosphoinositide-dependent protein kinase. (A) The structure of phosphoinositide-dependent protein kinase (PDK1) is shown in an inactive conformation (with an alanine replacing a key serine to prevent self-activation). (B) A comparison of the structures of the activation loops of PDK1, in inactive and active forms, shows the conformation change associated with activation. [Drawn from 2BIY.pdb and 3HRF.pdb.]

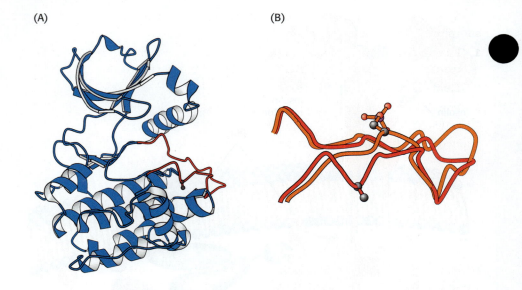

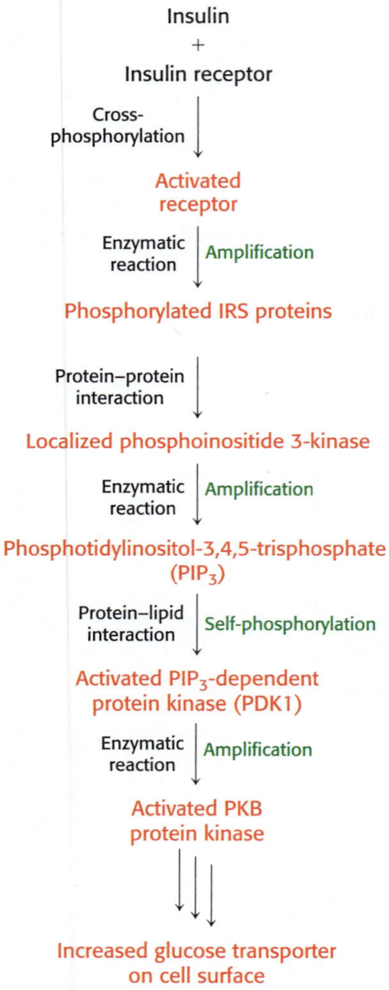

(A) (B)

Insulin
+
Insulin receptor

Cross-phosphorylation

Activated receptor

Enzymatic reaction | Amplification

Phosphorylated IRS proteins

Protein–protein interaction

Localized phosphoinositide 3-kinase

Enzymatic reaction | Amplification

Phosphotidylinositol-3,4,5-trisphosphate (PIP₃)

Protein–lipid interaction | Self-phosphorylation

Activated PIP₃-dependent protein kinase (PDK1)

Enzymatic reaction | Amplification

Activated PKB protein kinase

Increased glucose transporter on cell surface

FIGURE 14.33 Insulin binding initiates a signaling pathway leading to a protein kinase cascade. Key steps in the signal-transduction pathway are shown. The result is activation of protein kinases that phosphorylate targets, leading to an increase in glucose transporters on cell surfaces, among other effects.

can activate itself by cross-phosphorylation. PDK1 also includes a pleckstrin homology domain that is specific for PIP_3 and, thus, PDK1 is concentrated at PIP_3-rich sites generated by PI3K (Figure 14.31).

Other kinases that include pleckstrin homology domains are also concentrated at these sites including Protein Kinase B (PKB, also known as Akt). The activation loop of this kinase has the sequence -GATMKT_F_C_GT_PE-, which matches the substrate preference for PDK1. The substrate preference for PKB is quite different with the form -RRRXX(**S,T**)XSXSX-. Thus, PKB can be activated by PDK1 but is not capable of self-activation. Once activated, PKB can leave the membrane and move throughout the cell to phosphorylate other targets. These include components that control the trafficking of the glucose receptor GLUT4 to the cell surface. This completes the part of the insulin-signaling pathway we have been considering after our meal, namely, the process from an increase in insulin concentration in the bloodstream to enhanced glucose uptake.

The cascade initiated by the binding of insulin to the insulin receptor is summarized in **Figure 14.33**. The signal is amplified at several stages along this pathway. Because the activated insulin receptor itself is a protein kinase, each activated receptor can phosphorylate multiple IRS molecules. Activated enzymes further amplify the signal in at least two of the subsequent steps. Thus, a small increase in the concentration of circulating insulin can produce a robust intracellular response.

❖ SELF-CHECK QUESTION

Can a kinase with substrate preference XXXPX(**S,T**)PXXXX and an activation loop sequence of -HTGFL**T**EYVAT- activate itself?

Insulin signaling is terminated by the action of phosphatases

We have seen that the activated G protein promotes its own inactivation by the catalytic release of a phosphoryl group from GTP. In contrast, proteins phosphorylated on serine, threonine, or tyrosine residues are extremely stable kinetically. Specific enzymes, called protein phosphatases, are required to hydrolyze these phosphorylated proteins and return them to their initial states. Similarly, enzymes called lipid phosphatases are required to remove phosphoryl groups from inositol lipids that had been activated by lipid kinases.

In insulin signaling, three classes of enzymes are of particular importance in shutting off the signaling pathway:

1. Protein tyrosine phosphatases that remove phosphoryl groups from tyrosine residues on the insulin receptor and the IRS adaptor proteins

2. Lipid phosphatases that hydrolyze PIP_3 to PIP_2

3. Protein serine phosphatases that remove phosphoryl groups from activated protein kinases such as PKB

Many of these phosphatases are activated or recruited as part of the response to insulin. Thus, the binding of the initial signal sets the stage for the eventual termination of the response.

14.4 Epidermal Growth Factor: Receptor Dimerization Can Drive Signaling

Next, we will look at the signaling that takes place shortly after you are wounded. Within a few hours, signaling molecules such as epidermal growth factor (EGF) are released into the affected region. EGF is a 6-kDa polypeptide that stimulates the growth of epidermal and epithelial cells (**Figure 14.34**). EGF is bound by a family of receptor tyrosine kinases, similar in some ways to the insulin receptor, but quite different in others. Unlike those of the insulin receptor, however, these receptors exist as monomers. Each EGF receptor monomer binds a single molecule of EGF in its extracellular domain. EGF binding results in a substantial rearrangement of the extracellular domain (**Figure 14.35**). After this rearrangement, the receptors dimerize, stabilized by a structure termed the *dimerization arm* from each monomer. This arm is buried within the monomer structure prior to EGF binding but inserts into a binding pocket on the other monomer after EGF binding.

Epidermal growth factor (EGF)

FIGURE 14.34 Epidermal growth factor consists of a single polypeptide chain stabilized by three disulfide bonds.
[Drawn from 1EGF.pdb.]

INTERACT with this model in
🌊 Achie√e

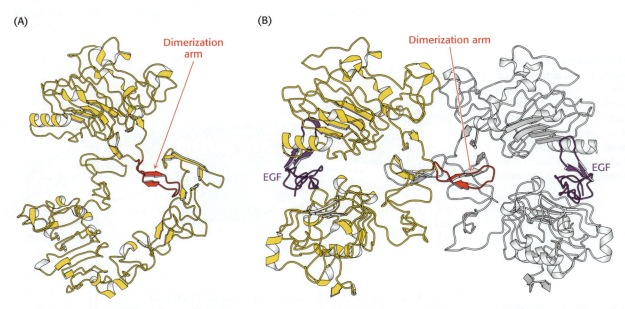

(A)

Dimerization arm

(B)

Dimerization arm

EGF

EGF

FIGURE 14.35 EGF binding to the EGF receptor results in a substantial rearrangement and facilitates receptor dimerization. (A) The structure of the extracellular domain of the EGF receptor is folded so that a region termed the dimerization loop is facing toward the interior of the structure. (B) Upon EGF binding, the structure rearranges so that the dimerization arm points outward and participates with dimerization with another receptor molecule. [Drawn from 1NQL.pdb and 1IVO.pdb.]

INTERACT with this model in
🌊 Achie√e

The EGF receptor undergoes phosphorylation of its carboxyl-terminal tail

Like the insulin receptor, the kinase domains of the EGF receptor are inactive prior to dimerization. When an EGF receptor dimerizes, the kinase domains come together to form an asymmetric dimer with one kinase domain functioning as an allosteric activator of the other, stabilizing an active conformation. This active kinase phosphorylates tyrosine residues on the other subunit, but in a region that lies on the C-terminal side of the kinase domain, rather in an activation loop. As many as five tyrosine residues in this region are phosphorylated.

EGF signaling leads to the activation of Ras, a small G protein

The phosphotyrosines on the EGF receptors act as docking sites for phosphotyrosine-binding domains on other proteins. The intracellular signaling cascade begins with the binding of Grb2, a key adaptor protein that contains one phosphotyrosine-binding domain and two domains that recognize extended, proline-rich sequences. On phosphorylation of the receptor, Grb2 binds to the phosphotyrosine residues of the EGF receptor. Through its other domains, Grb2 then binds polyproline-rich polypeptides within a protein called Sos. Sos, in turn, binds to a protein called Ras and activates it.

A very prominent signal-transduction component, Ras is a member of a class of proteins called the small G proteins. These proteins have many key mechanistic and structural motifs in common with the G_α subunit of the heterotrimeric G proteins and are related by divergent evolution. In keeping with this relationship, Ras contains bound GDP in its unactivated form. Sos opens up the nucleotide-binding pocket of Ras, allowing GDP to escape and GTP to enter in its place. Because of its effect on Ras, Sos is referred to as a guanine-nucleotide-exchange factor (GEF). Thus, the binding of EGF to its receptor leads to the conversion of Ras into its GTP form through the intermediacy of Grb2 and Sos (**Figure 14.36**). This occurs on the membrane surface as Ras is anchored via covalently attached lipids.

FIGURE 14.36 The binding of EGF to the EGF receptor results in the activation of Ras. The dimerization of the EGF receptor due to EGF binding leads to the phosphorylation of the C-terminal tails of the receptor. This, in turn, results in the recruitment of Grb2 and Sos, and the exchange of GTP for GDP in Ras. This signal-transduction pathway results in the conversion of Ras into its activated GTP-bound form.

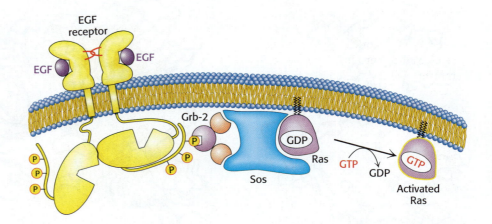

Activated Ras initiates a protein kinase cascade

Ras changes conformation when it is transformed from its GDP into its GTP form. In the GTP form, Ras binds other proteins, including a membrane-anchored protein kinase termed Raf. When bound to Ras, Raf undergoes a conformational change that activates the Raf protein kinase. Activated Raf then phosphorylates other proteins, including protein kinases termed MEKs. In turn, MEKs activate kinases called extracellular signal-regulated kinases (ERKs). ERKs then phosphorylate numerous substrates, including transcription factors in the nucleus as well as other protein kinases. The flow of information from the arrival of EGF at the cell surface to changes in gene expression is summarized in **Figure 14.37**.

EGF signaling is terminated by protein phosphatases and the intrinsic GTPase activity of Ras

Because so many components of the EGF signal-transduction pathway are activated by phosphorylation, we can expect protein phosphatases to play key roles in the termination of EGF signaling. Indeed, crucial phosphatases remove phosphoryl groups from tyrosine residues on the EGF receptor and from serine, threonine, and tyrosine residues in the protein kinases that participate in the signaling cascade. The signaling process itself sets in motion the events that activate many of these phosphatases. Consequently, signal activation also initiates signal termination.

Like the G proteins activated by 7TM receptors, Ras possesses intrinsic GTPase activity. Thus, the activated GTP form of Ras spontaneously converts into the inactive GDP form. The rate of conversion can be accelerated in the presence of GTPase-activating proteins (GAPs), which interact with small G proteins in the GTP form and facilitate GTP hydrolysis. Thus, the lifetime of activated Ras is regulated by accessory proteins in the cell.

 ❖ SELF–CHECK QUESTION

List one important similarity and one important difference between Ras and a heterotrimeric G protein.

14.5 Defects in Signal-Transduction Pathways Can Lead to Cancer and Other Diseases

In light of their complexity, it comes as no surprise that signal-transduction pathways occasionally fail, leading to disease states. Cancer, a set of diseases characterized by uncontrolled or inappropriate cell growth, is strongly associated with defects in signal-transduction proteins, particularly those that play key roles in growth-control pathways. Indeed, the study of cancer—particularly cancers caused by certain viruses—has contributed greatly to our understanding of signal-transduction proteins and pathways.

Mammalian cells may contain three distinct Ras proteins (H-, K-, and N-Ras), each of which cycles between inactive GDP and active GTP forms. The genes encoding members of the Ras family are among the most commonly mutated in human tumors. The most common mutations in tumors lead to a loss of the ability to hydrolyze GTP. Thus, the Ras protein is trapped in the "on" position and continues to stimulate cell growth, even in the absence of a continuing signal. Genes such as these mutated Ras genes are referred to **oncogenes** because they can induce cancer under appropriate circumstances. Unmutated Ras genes are referred to as **proto-oncogenes** since they can be converted into oncogenes through mutation.

Other genes can contribute to cancer development only when both copies of the gene normally present in a cell are deleted or otherwise damaged. Such genes are called **tumor-suppressor genes**. For example, genes for some of the phosphatases that participate in the termination of EGF signaling are tumor suppressors. Without any functional phosphatase present, EGF signaling persists once initiated, stimulating inappropriate cell growth.

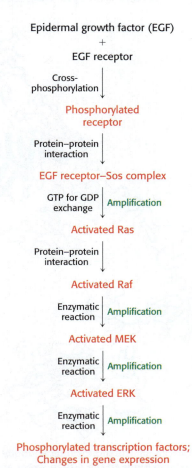

FIGURE 14.37 The EGF signaling pathway results in a protein kinase cascade. The key steps in the pathway are shown. A kinase cascade leads to the phosphorylation of transcription factors and concomitant changes in gene expression.

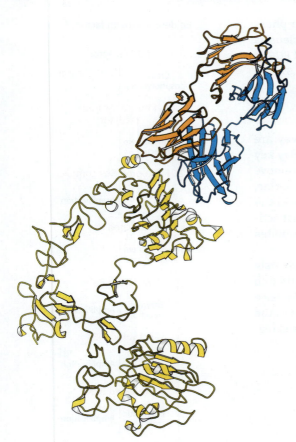

FIGURE 14.38 The monoclonal antibody drug cetuximab binds to the EGF receptor and prevents receptor signaling. The structure of the complex between the extracellular domain of the EGF receptor and an F$_{ab}$ fragment from cetuximab reveals that the antibody binds in a manner that blocks the conformational changes necessary for productive dimerization and signaling. [Drawn from 1YY9.pdb.]

Monoclonal antibodies can be used to inhibit signal-transduction pathways activated in tumors

Mutated or overexpressed receptor tyrosine kinases are frequently observed in tumors. For instance, the epidermal-growth-factor receptor (EGFR) is overexpressed in some human epithelial cancers, including breast, ovarian, and colorectal cancer. Because some small amount of the receptor can dimerize and activate the signaling pathway even without binding to EGF, overexpression of the receptor increases the likelihood that a "grow and divide" signal will be inappropriately sent to the cell.

This understanding of cancer-related signal-transduction pathways has led to a therapeutic approach that targets the EGFR. The strategy is to produce monoclonal antibodies to the extracellular domains of the offending receptors. One such antibody, cetuximab (Erbitux), has effectively targeted the EGFR in colorectal cancers. Cetuximab inhibits the EGFR by competing with EGF for the binding site on the receptor (Figure 14.38). Because the antibody sterically blocks the change in conformation that exposes the dimerization arm, the antibody itself cannot induce dimerization. The result is that the EGFR-controlled pathway is not initiated.

Cetuximab is not the only monoclonal antibody that has been developed to target a receptor tyrosine kinase. Trastuzumab (Herceptin) inhibits another EGFR family member, HER2, that is overexpressed in approximately 30% of breast cancers (Figure 14.39). HER2 normally forms heterodimers with other members of the EGFR family, but it can signal even in the absence of ligand when it homodimerizes. Thus, overexpression of HER2 can stimulate cell proliferation. Breast tumors are now routinely screened for HER2 overexpression, and patients showing such overexpression may be treated with Herceptin if appropriate. Thus, this cancer treatment is tailored to the genetic characteristics of the tumor.

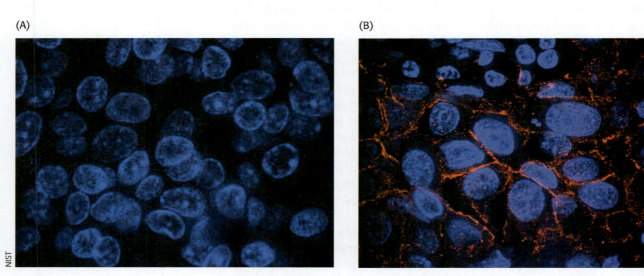

(A)

(B)

FIGURE 14.39 The EGF receptor family member HER2 is overexpressed in some breast cancer cells. (A) A panel of breast cancer cells expressing HER2 at normal levels. Cell nuclei are blue, and HER2 would be red if detectable. (B) A panel of breast cancer cells overexpressing HER2. The overexpression is due to amplification of the *HER2* gene such that many more than the normal two copies are present.

Protein kinase inhibitors can be effective anticancer drugs

The widespread occurrence of overactive protein kinases in cancer cells suggests that molecules that inhibit these enzymes might act as antitumor agents. For example, more than 90% of patients with chronic myelogenous leukemia (CML) show a specific chromosomal defect in cancer cells (**Figure 14.40**). The translocation of genetic material between chromosomes 9 and 22 causes the c-*abl* gene, which encodes a tyrosine kinase of the Src family, to be inserted into the *bcr* gene on chromosome 22. The result is the production of a fusion protein called Bcr-Abl that consists primarily of sequences for the c-Abl kinase. However, the *bcr-abl* gene is not regulated appropriately; it is expressed at higher levels than that of the gene encoding the normal c-Abl kinase, stimulating a growth-promoting pathway.

Because of this overexpression, leukemia cells express a unique target for chemotherapy. A specific inhibitor of the Bcr-Abl kinase, Gleevec (STI-571, imatinib mesylate), has proved to be a highly effective treatment for patients with CML (**Figure 14.41**). This approach to cancer chemotherapy is fundamentally distinct from most approaches, which target all rapidly growing cells, including normal ones. Because Gleevec targets tumor cells specifically, side effects caused by the impairment of normal dividing cells can be minimized. Subsequently, drugs that target other protein kinases have been developed for treatment of other cancer types. Thus, our understanding of signal-transduction pathways has enabled conceptually new disease treatment strategies.

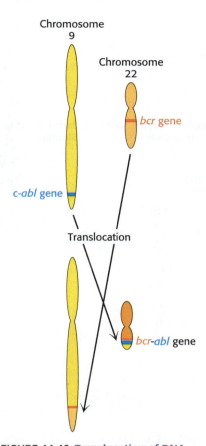

FIGURE 14.40 Translocation of DNA from one chromosome to another results in the formation of the *bcr-abl* gene. In chronic myelogenous leukemia, parts of chromosomes 9 and 22 are reciprocally exchanged, causing the *bcr* and *abl* genes to fuse. The protein kinase encoded by the *bcr-abl* gene is expressed at higher levels in tumor cells than is the c-*abl* gene in normal cells leading to inappropriate cell proliferation.

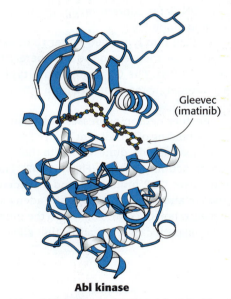

Abl kinase

FIGURE 14.41 The drug imatinib inhibits Abl kinase by blocking the ATP-binding site. The structure of the kinase domain of Abl with imatinib (Gleevec) bound is shown. The inhibitor occupies the ATP-binding site in the kinase. [Drawn from 1IEP.pdb.]

14.6 Sensory Systems Are Based on Specialized Signal-Transduction Pathways

Recall once again the scene you imagined at the start of this chapter. One enticement to eating dinner after your return to the campfire was the delicious smell of the cooked food. Many organic compounds can evaporate from the surface of food, and both the number and amount of these increases as food is heated. How are these compounds sensed and distinguished from one another? In this section, we will explore the signal-transduction pathways behind smell and other senses as well, including vision and hearing.

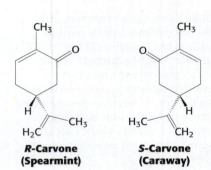

Benzaldehyde
(Almond)

3-Methylbutane-1-thiol
(Skunk)

Geraniol
(Rose)

Zingiberene
(Ginger)

R-Carvone
(Spearmint)

S-Carvone
(Caraway)

A huge family of 7TM receptors detect a wide variety of organic compounds

Human beings can detect and distinguish thousands of different compounds by smell, often with considerable sensitivity and specificity. Most odorants are small organic compounds with sufficient volatility that they can be carried as vapors into the nose. For example, a major component responsible for the odor of almonds is the simple aromatic compound benzaldehyde, whereas the sulfhydryl compound 3-methylbutane-1-thiol is a major component of the odor of skunks.

What properties of these molecules are responsible for their odors? First, the shape of the molecule rather than its other physical properties is crucial. We can most clearly see the importance of shape by comparing molecules such as those responsible for the odors of spearmint and caraway. These compounds are identical in essentially all physical properties such as hydrophobicity because they are exact mirror images of one another. Thus, the odor produced by an odorant depends not on a physical property but on the compound's interaction with specific receptors.

Investigations revealed that olfaction depends on cAMP and GTP, and a G-protein α subunit, termed $G_{(olf)}$, was discovered that is uniquely expressed in the olfactory system. From our knowledge, these factors should strongly suggest the involvement of 7TM receptors. Complementary DNAs were sought that (1) were expressed primarily in the sensory neurons lining the nasal epithelium, (2) encoded members of the 7TM-receptor family, and (3) were present as a large and diverse family to account for the range of odorants. Based on these criteria, cDNAs for odorant receptors from rats were identified in 1991 by Richard Axel and Linda Buck.

The odorant receptor (OR) family is even larger than expected: more than 1000 OR genes are present in the mouse and the rat, whereas the human genome encodes approximately 400 ORs. The OR proteins are typically 20% identical in sequence with the β-adrenergic receptor and from 30% to 60% identical with one another.

Interestingly, each olfactory neuron expresses only a single OR gene, among the hundreds available. The binding of an odorant to an OR on the neuronal surface initiates a signal-transduction cascade, quite similar to that from the β-adrenergic receptor, that results in an action potential (**Figure 14.42**). The ligand-bound OR activates $G_{(olf)}$, which releases GDP, binds GTP, and dissociates its βγ subunits. The α subunit then binds to and stimulates a specific adenylate cyclase, increasing the intracellular concentration of cAMP. The rise in the intracellular concentration of cAMP activates a nonspecific cation channel that allows calcium and other

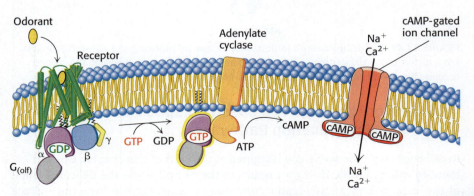

FIGURE 14.42 The olfactory signal-transduction cascade is similar to that for the β-adrenergic receptor. The binding of odorant to an olfactory receptor results in stimulating the activity of adenylate cyclase. The final result is the opening of cAMP-gated ion channels and the initiation of an action potential.

cations into the cell. The flow of cations through the channel depolarizes the neuronal membrane and initiates an action potential. This action potential, combined with those from other olfactory neurons, leads to the perception of a specific odor.

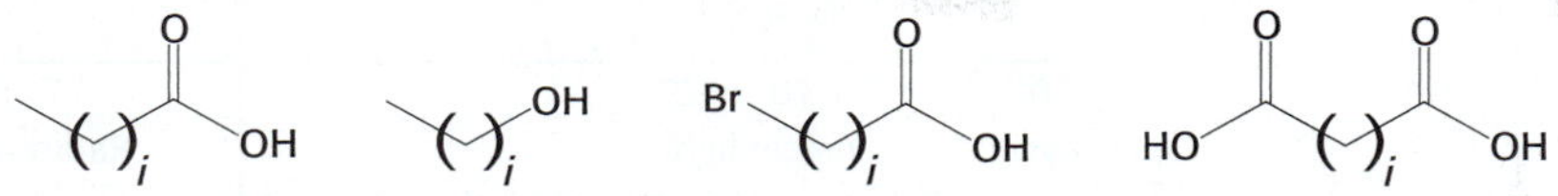

Carboxylic acids (i = 2–7)   **Alcohols** (i = 4–8)   **Bromocarboxylic acids** (i = 3–7)   **Dicarboxylic acids** (i = 4–7)

FIGURE 14.43 Olfactory-receptor activation has been tested with sets of closely related compounds. The structures of four series of compounds are shown.

An obvious challenge presented to an investigator by the large size of the OR family is to match each OR with the one or more odorant molecules to which it binds. Dramatic progress was made by taking advantage of our knowledge of the OR signal-transduction pathway. A section of nasal epithelium from a mouse was loaded with the calcium-sensitive dye Fura-2. The tissue was then treated with different odorants, one at a time, at a specific concentration (**Figure 14.43**). If the odorant had bound to an OR and activated it, that neuron could be detected under a microscope by the change in fluorescence caused by the influx of calcium that takes place as part of the signal-transduction process.

To determine which OR was responsible for the response, cDNA was generated from mRNA that had been isolated from single identified neurons, and cDNAs were produced that encoded ORs. These studies revealed there is not a simple 1:1 correspondence between odorants and receptors. Almost every odorant activates a number of receptors (usually to different extents) and almost every receptor is activated by more than one odorant (**Figure 14.44**). Note, however, that each odorant activates a unique combination of receptors. This combinatorial mechanism allows even a small array of receptors to distinguish a vast number of odorants.

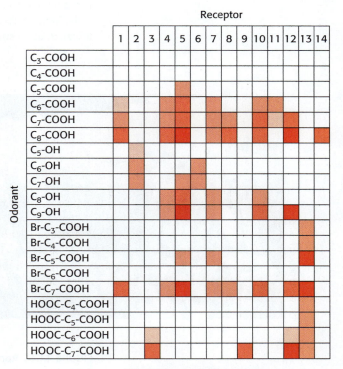

FIGURE 14.44 Different compounds result in distinct patterns of olfactory-receptor activation. Fourteen different receptors were tested for responsiveness to the compounds shown in Figure 14.43. A colored box indicates that the receptor at the top responded to the compound at the left. Darker colors indicate that the receptor was activated at a lower concentration of odorant.

One of the manifestations of COVID-19 in many people is a loss of the sense of smell. While the cause of this is still under investigation, a clue has come from genomic studies of populations of individuals who either have or have not lost their sense of smell. These studies have suggested that enzymes that add sugars to some odorants to help terminate olfactory signaling may differ in these two groups of people. This example highlights another mechanism for signal termination, but also another role of carbohydrates that we will see again when we discuss drug metabolism.

❖ SELF–CHECK QUESTION

What symptoms would you expect if an individual had a mutation that disrupted an olfactory receptor gene?

Vision relies on a specialized 7TM receptor to signal in response to absorbed light

We now turn from olfaction to vision. Vision is based on the absorption of light by photoreceptor cells in the eye. These cells are sensitive to light in a narrow region of the electromagnetic spectrum—the region with wavelengths between

FIGURE 14.45 Visible light represents a small region of the electromagnetic spectrum. Visible light has wavelengths between 390 and 750 nm.

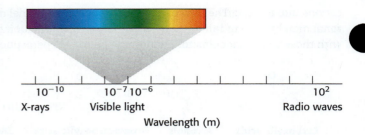

390 and 750 nm (**Figure 14.45**). Vertebrates have two kinds of photoreceptor cells, called rods and cones because of their distinctive shapes. Cones function in bright light and are responsible for color vision, whereas rods function in dim light but do not perceive color. A human retina contains about 3 million cones and 100 million rods. Remarkably, a rod cell can respond to a single photon, and the brain requires fewer than 10 such responses to register the sensation of a flash of light.

Courtesy of Dr. Deric Bownds

Discs

Outer segment

FIGURE 14.46 Rod cells have elongated structures including stacked discs where light absorption occurs. (Left) A scanning electron micrograph of retinal rod cells. (Right) A schematic representation of a rod cell.

Rods are slender, elongated structures; the outer segment is specialized for photoreception (**Figure 14.46**). It contains a stack of about 1000 discs, which are membrane-enclosed sacs densely packed with photoreceptor molecules. The photosensitive molecule is often called a visual pigment because it is highly colored owing to its ability to absorb light. The photoreceptor molecule in rods is rhodopsin, which consists of the protein opsin linked to 11-cis-retinal, a prosthetic group, via a lysine residue.

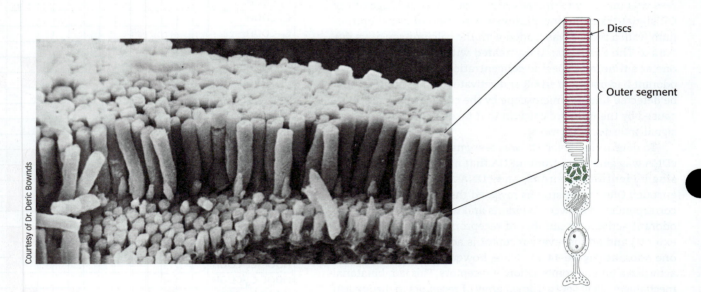

Schiff base

11-*cis*-Retinal

(11-*cis*-Retinal)   Lysine

Rhodopsin absorbs light very efficiently in the middle of the visible spectrum, its absorption being centered on 500 nm, which nicely matches the solar output (**Figure 14.47**). A rhodopsin molecule will absorb a high percentage of the photons of the correct wavelength that strike it. Opsin is a member of the 7TM-receptor family. Indeed, rhodopsin was the first member of this family to be purified, its gene was the first to be cloned and sequenced, and its three-dimensional structure was the first to be determined. The color of rhodopsin and its responsiveness to light depend on the presence of the light-absorbing group 11-cis-retinal.

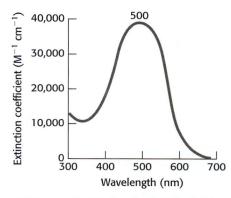

FIGURE 14.47 Rhodopsin absorbs light quite effectively toward the blue end of the visible spectrum. The absorption spectrum of rhodopsin is shown. The large ($>10,000$ M^{-1} cm^{-1}) extinction coefficients reveal that most of the light striking a rhodopsin molecule is absorbed.

Light absorption induces a specific isomerization of bound 11-*cis*-retinal

How does the absorption of light by retinal generate a signal? George Wald and his coworkers discovered that light absorption results in the isomerization of the 11-cis-retinal group of rhodopsin to its all-trans form (**Figure 14.48**). This isomerization causes the nitrogen atom from the linked lysine residue to move approximately 5Å, assuming that the cyclohexane ring of the retinal group remains fixed. In essence, the light energy of a photon is converted into atomic motion. The change in atomic positions, like the binding of a ligand to other 7TM receptors, sets in motion a series of events that lead to the closing of ion channels and the generation of a nerve impulse.

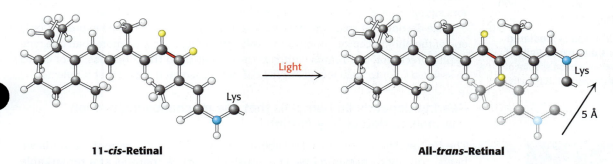

11-*cis*-Retinal   **All-*trans*-Retinal**

FIGURE 14.48 Isomerization of retinal results in a substantial motion. The nitrogen atom of the lysine residue linked to retinal moves 5Å as a consequence of the light-induced isomerization of 11-*cis*-retinal to all-*trans*-retinal by rotation about the bond shown in red.

Like these receptors, this form of rhodopsin activates a heterotrimeric G protein that propagates the signal. The G protein associated with rhodopsin is called transducin. Light absorption and retinal isomerization triggers the exchange of GDP for GTP by the α subunit of transducin (**Figure 14.49**). On the binding of GTP, the βγ subunits of transducin are released and the GTP form of the α subunit is free to bind

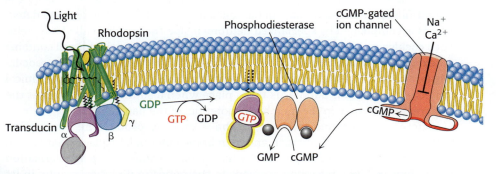

FIGURE 14.49 Visual signal transduction involves G-protein activation of cGMP phosphodiesterase. The light-induced activation of rhodopsin activates a novel G protein, transducin, which leads to the hydrolysis of cGMP. This, in turn, leads to ion-channel closing and the initiation of an action potential.

to its target. The target for transducing is not adenylate cyclase, but rather a specific cGMP phosphodiesterase. Transducin binds to an inhibitory subunit of this enzyme and removes it. The activated phosphodiesterase is a potent enzyme that rapidly hydrolyzes cGMP to GMP. The reduction in cGMP concentration causes cGMP-gated ion channels to close, leading to the hyperpolarization of the membrane and neuronal signaling. At each step in this process, the initial signal—the absorption of a single photon—is amplified so that it leads to sufficient membrane hyperpolarization to result in signaling. For a multistep, enzyme-based process, the visual signal-transduction system is strikingly fast, responding in approximately 1 ms.

Color vision is mediated by three cone receptors that are homologs of rhodopsin

Cone cells, like rod cells, contain visual pigments. Like rhodopsin, these photoreceptor proteins are members of the 7TM-receptor family and use 11-cis-retinal as their chromophore. In human cone cells, there are three distinct photoreceptor proteins with absorption maxima at 426, 530, and ~560 nm (**Figure 14.50**). These absorbances correspond to the blue, green, and yellow-green regions of the spectrum. Recall that the absorption maximum for rhodopsin is 500 nm. The photoreceptor protein with its absorption maximum near 560 nm absorbs red light (with wavelength > 620 nm), whereas the other two photoreceptors do not. These three proteins are often referred to as the blue, green, and red photoreceptors. The cone photoreceptors signal through pathways that are identical to that used by rhodopsin.

The decoding of color vision is analogous to olfaction. Each cone expresses only one of the photoreceptors, but most wavelengths of light activate more than one type of photoreceptor. Our brains compare the signals from the three classes of cones to assign a color to an object in our field of view based on the light that it reflects.

Hearing depends on hair cells that use mechanosensitive ion channels to detect tiny motions

The first indications that we had about our animal intruder were the sounds we heard coming from behind us. Our sense of hearing operates at a remarkable speed. We hear frequencies ranging from 200 to 20,000 Hz (cycles per second), corresponding to times of 5 to 0.05 ms. Furthermore, our ability to locate sound sources—one of the most important functions of hearing—depends on the ability to detect the time delay between the arrival of a sound at one ear and its arrival at the other. Given the separation of our ears and the speed of sound, we must be able to accurately sense time differences of 0.7 ms. In fact, human beings can locate sound sources associated with temporal delays as short as 0.02 ms. This high time resolution implies that hearing depends on direct transduction mechanisms that do not use enzymes or second messengers.

Sound waves are detected inside the cochlea of the inner ear. The cochlea is a fluid-filled, membranous sac that is coiled like a snail shell. The primary detection is accomplished by specialized neurons inside the cochlea called hair cells. Each cochlea contains approximately 16,000 hair cells, and each hair cell contains a hexagonally shaped bundle of 20 to 300 hairlike projections called stereocilia (**Figure 14.51**). These stereocilia are graded in length across the bundle. Mechanical deflection of the hair bundle, as takes place when a sound wave arrives at the ear, creates a change in the membrane potential of the hair cell.

Micromanipulation experiments have directly probed the connection between mechanical stimulation and membrane potential. Displacement toward the direction of the tallest part of the hair bundle results in the depolarization of the hair cell, whereas displacement in the opposite direction results in its hyperpolarization. Remarkably, displacement of the hair bundle by as little as 3Å (0.3 nm) results in a measurable (and functionally important) change in membrane potential. This motion of 0.001 degree corresponds to a 0.3-inch movement of the top of a 100-story building.

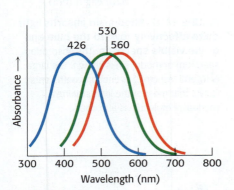

FIGURE 14.50 Color vision is mediated by visual pigments that absorb light of different colors. The absorption spectra of the three types of light receptors from human cone cells are shown.

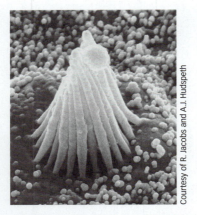

Courtesy of R. Jacobs and A.J. Hudspeth

FIGURE 14.51 Hearing is mediated by bundles of stereocilia on hair cells. An electron micrograph of a stereocilia bundle is shown.

The rapid response, within microseconds, suggests that the movement of the hair bundle acts on ion channels directly. An important observation is that adjacent stereocilia are linked by individual filaments called tip links (**Figure 14.52**). The presence of these tip links suggests a simple mechanical model for transduction by hair cells (**Figure 14.53**). The tip links are coupled to ion channels in the membranes of the stereocilia that are gated by mechanical stress. In the absence of a stimulus, approximately 15% of these channels are open. When the hair bundle is displaced toward its tallest part, the stereocilia slide across one another and the tension on the tip links increases, causing additional channels to open. The flow of ions through the newly opened channels depolarizes the membrane. Conversely, if the displacement is in the opposite direction, the tension on the tip links decreases, the open channels close, and the membrane hyperpolarizes. Thus, the mechanical motion of the hair bundle is directly converted into current flow across the hair-cell membrane. Many of the molecular components of this elegant micromechanical system have now been identified, including those from the mechanosensitive ion channel and the tip link filaments, as well as specialized myosin motors that maintain appropriation tension in the system.

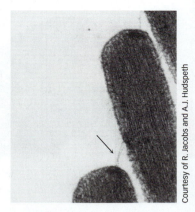

FIGURE 14.52 Tip links that connect adjacent stereocilia are crucial for signal detection in hearing. An electron micrograph of tip links is shown. The tip link between two stereocilia is marked by an arrow.

❖ SELF–CHECK QUESTION

What are the primary signals in olfaction, vision, and hearing? Compare the mechanisms by which these primary signals are first detected in the three sensory systems.

Comparison of different organisms yields insights into sensory system evolution

In addition to representing instructive examples of applying the knowledge we have gained about signal-transduction pathways, sensory systems also provide compelling glimpses into evolution at the molecular level. As we noted, rodents have more than 1000 ORs, whereas our genomes only include only approximately 400 functional genes. Further examination reveals that our genomes contain hundreds of OR pseudogenes, DNA segments that encode ORs but have one or more defects so that they will not produce functional receptors. Examination of other organisms reveals a steady conversion from functional OR genes into pseudogenes over the course of the primate evolution (**Figure 14.54**).

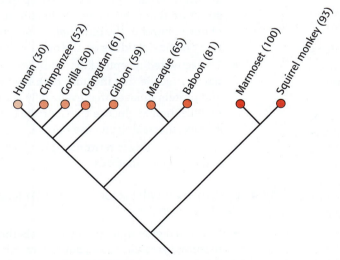

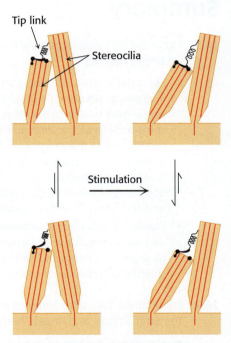

FIGURE 14.53 Hair cell signal detection involves ion-channel opening in response to small displacements of stereocilia bundles. When the hair bundle is tipped toward the tallest part, the tip link pulls on an ion channel and opens it. Movement in the opposite direction relaxes the tension in the tip link, increasing the probability that any open channels will close. [Information from A. J. Hudspeth, *Nature* 341:397–404, 1989 and M. Beurg, R. Fettiplace, J. H. Nam, and A. J. Ricci, *Nature Neuroscience* 12:553–558, 2009.]

FIGURE 14.54 Functional odorant receptor genes have been lost over the course of primate evolution. Odorant receptors appear to have lost function through conversion into pseudogenes during primate evolution. The percentage of OR genes that appear to be functional for each species is given in parentheses.

Viktar Malyshchyts/Shutterstock

FIGURE 14.55 More colors can be distinguished with three versus two cone pigments. The left side of the image shows its appearance with three cone pigments, while the right side show its appearance with two. Most human beings have three functional pigments, whereas mammals such as rodents and dogs have two.

The opposite has occurred with regard to color vision. Human beings and many other primates have three cone receptors. Most other placental mammals have only two. Thus, rodents as well as organisms such as dogs do sense color but are not capable of distinguishing color such as red and green as well as we are. Indeed, some human beings have variations in their visual pigments (so-called "color blindness") so that they see colors differently (**Figure 14.55**). Earlier ancestors of all animals had four cone pigments as do some modern organisms such as many birds. In addition to better color discrimination, the absorption spectra extend further into the ultraviolet region so that they can sense colors that we cannot. These differences provide vivid examples of how signal-transduction pathways have a profound influence on how we experience the world.

Summary

14.1 Many Signal-Transduction Pathways Share Common Themes

- Most signal-transduction pathways involve a common set of steps: Binding of a primary messenger to a receptor on the cell surface; transduction of the information across the cell membrane to the interior of the cell; a change in the concentration of one or more second messenger molecules inside the cell; initiation of other processes by the second messenger; and termination of the signaling.

- Signal-transduction processes can be very intricate as they occur with variation in different biological contexts and cell types, but common themes and classes of molecules do occur.

14.2 Epinephrine Signaling: Heterotrimeric G Proteins Transmit Signals and Reset Themselves

- Hormones such as epinephrine bind to members of a large family of cell-surface receptors that have a common structure of seven transmembrane (7TM) helices.

- When activated by ligand binding, these 7TM helix receptors bind to specific heterotrimeric G proteins and stimulate the release of bound GDP and the binding of GTP in the α subunit of the G protein.

- When activated by GTP binding, the α subunit of the G protein releases the heterodimer consisting of the β and γ subunits and then binds to other targets such as the enzyme adenylate cyclase.

- Stimulation of adenylate cyclase increases the concentration of the second messenger cyclic AMP, which, in turn, activates protein kinase A.

- Some 7TM receptors signal through different heterotrimeric G proteins that inhibit the activity of adenylate cyclase or activate the phosphoinositide pathway, starting with the activation of phospholipase C, which cleaves a membrane lipid to produce two secondary messengers, diacylglycerol and inositol 1,4,5-trisphosphate.

- G-protein signaling is terminated by the hydrolysis of the bound GTP to GDP.

14.3 Insulin Signaling: Phosphorylation Cascades Are Central to Many Signal-Transduction Processes

- Protein kinases are key components in many signal-transduction pathways, including some for which the protein kinase is an integral component of the initial receptor.

- Insulin binds to a dimeric receptor that includes an intracellular tyrosine kinase as part of its structure that is activated by the kinase domain of one subunit of the dimer phosphorylating certain tyrosine residues on the other subunit.

- The activated receptor kinase initiates a signaling cascade that includes both lipid kinases and protein kinases that eventually leads to the mobilization of glucose transporters to the cell surface, increasing glucose uptake.

- Insulin signaling is terminated through the action of phosphatases.

14.4 Epidermal Growth Factor: Receptor Dimerization Can Drive Signaling

- Epidermal growth factor also signals through a receptor tyrosine kinase, but this receptor is a monomer prior to EGF binding.

- EGF binding induces a conformational change that allows receptor dimerization and cross-phosphorylation.

- The phosphorylated receptor binds adaptor proteins that mediate the activation of Ras, a small G protein that initiates a protein kinase cascade that eventually leads to the phosphorylation of transcription factors and changes in gene expression.
- EGF signaling is terminated by the action of phosphatases and the hydrolysis of GTP by Ras.

14.5 Defects in Signal-Transduction Pathways Can Lead to Cancer and Other Diseases

- Genes encoding components of signal-transduction pathways that control cell growth are often mutated in cancer.
- Some genes can be mutated to forms called oncogenes that are active regardless of appropriate signals.
- Monoclonal antibodies directed against cell-surface receptors that participate in signaling have been developed for use in cancer treatment.
- Our understanding of the molecular basis of cancer has enabled the development of anticancer drugs directed against specific targets, such as specific protein kinase inhibitors.

14.6 Sensory Systems Are Based on Specialized Signal-Transduction Pathways

- Olfaction involves a large family of 7TM receptors that signal through a pathway involving adenylate cyclase, leading to ion-channel opening.
- Vision uses a specialized 7TM receptor called rhodopsin that includes a covalently bound ligand, 11-cis-retinal, that isomerizes upon absorbing light, activating the receptor.
- Color vision is mediated by a set of three receptors similar to rhodopsin that tune the absorption of light to different wavelengths.
- Hearing involves specialized cells called hair cells that have bundles of stereocilia that move subtly in response to sound waves in the air.
- Signal transduction is very fast, depending on the direct coupling of stereocilia motion to ion-channel opening, without any chemical steps.

Key Terms

transduction (p. 413)
primary messenger (p. 413)
receptor (p. 413)
second messenger (p. 413)
epinephrine (adrenaline) (p. 414)
adrenergic receptor (p. 415)
seven-transmembrane-helix (7TM) receptor (p. 415)

agonist (p. 415)
G protein (p. 416)
G-protein-coupled receptor (GPCR) (p. 418)
adenylate cyclase (p. 418)
calmodulin (CaM) (p. 420)
EF hand (p. 423)
antagonist (p. 424)

insulin (p. 425)
activation loop (p. 426)
pleckstrin homology domain (p. 428)
oncogene (p. 433)
proto-oncogene (p. 433)
tumor-suppressor gene (p. 433)

Problems

1. Place the steps of the insulin-signaling pathway in the correct order: ❖ 1
(a) Phosphorylation of IRS proteins
(b) Alteration of glucose metabolism in response to insulin
(c) Binding of insulin to the α subunit of the insulin receptor
(d) Activation of insulin receptor tyrosine kinase
(e) Activation of PKB by PDK1
(f) Recruitment of PI3K to the cell membrane
(g) Conversion of PIP_2 to PIP_3

2. Glucagon is a hormone that is released when the level of circulating glucose in the blood is low. Glucagon signals the cell via a $G_{\alpha s}$ signaling pathway. Put the steps of the glucagon signal-transduction pathway in order. ❖ 1
(a) Adenylate cyclase converts ATP to cAMP.
(b) $G_{\alpha s}$ releases GDP and binds GTP.
(c) Glucagon binds extracellularly to the G-protein-coupled receptor.
(d) Activated $G_{\alpha s}$ activates adenylate cyclase.
(e) cAMP activates protein kinase A (PKA).
(f) PKA activates enzymes needed for glucose release.

3. Suppose that each β-adrenergic receptor bound to epinephrine converts 100 molecules of $G_{\alpha s}$ into their GTP forms and that each molecule of activated adenylate cyclase produces 1000 molecules of cAMP per second. With the assumption of a full response, how many molecules of cAMP will be produced in 10 s after the formation of a single complex between epinephrine and the β-adrenergic receptor? Assume the pathway is fully responsive. ❖ 1

4. A mutated form of the α subunit of the heterotrimeric G protein has been identified; this form readily exchanges nucleotides even in the absence of an activated receptor. What would be the effect on a signaling pathway containing the mutated α subunit? ❖ 5

5. Glucose is mobilized for ATP generation in muscle in response to epinephrine, which activates $G_{\alpha s}$. Cyclic AMP phosphodiesterase is an enzyme that converts cAMP into AMP. How would inhibitors of cAMP phosphodiesterase affect glucose mobilization in muscle? ❖ 2

6. Some G-protein-coupled receptors are sensitive to hormones such as angiotensin II and oxytocin and act through molecules such as phospholipase C (PLC) and IP_3. ❖ 1

(a) Complete the flowchart showing the cleavage of a membrane lipid to form compounds that ultimately cause the phosphorylation of specific proteins. Use the following terms to complete the steps: IP_3, Ser/Thr residues, DAG, PIP_3, PKC, Ca^{2+}.

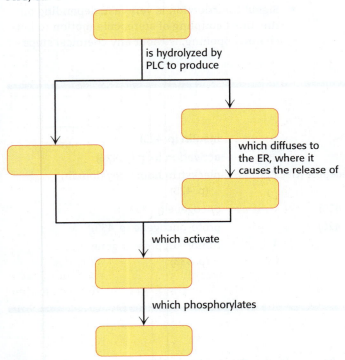

(b) Of the terms you used to complete the chart, which can be considered second messengers?

7. Consider the structure of calmodulin (Figure 14.19). Which of the following statements are correct?

(a) Calmodulin contains two Ca^{2+} binding sites.

(b) Binding of Ca^{2+} activates calmodulin.

(c) Each calcium-binding domain in calmodulin is formed from a helix-loop-helix motif.

(d) Calmodulin is a structural subunit of proteins such as CaM kinase.

(e) Calmodulin contains a central, flexible β sheet that enables it to wrap around another protein, inhibiting the protein's activity.

(f) A calcium ion is bound in the loop of an EF hand.

8. Classify each statement as describing G-protein-coupled receptors, receptor tyrosine kinases, both, or neither: ❖ 3

(a) An example is the insulin receptor.

(b) They transport some ligands through the membrane.

(c) Their structure contains seven transmembrane helices.

(d) An example is the adrenergic receptor.

(e) Receptor activation causes phosphorylation of their cytosolic subunits.

(f) Ligand binding induces conformational change in these receptors.

(g) They activate heterotrimeric G proteins directly.

(h) Autophosphorylation of these receptors can initiate signal.

(i) Phosphorylation of these receptors can terminate signal.

9. Drugs like imatinib (Gleevec) and cetuximab (Erbitux) are approved for the treatment of various types of cancer. Contrast these two drugs based on their mechanism of action. ❖ 4

10. Which of the following events contribute to the termination of a response in a GPCR signaling pathway? ❖ 1

(a) The ligand dissociates from the receptor, which resumes its inactive conformation.

(b) G_α dissociates from the $G_{\beta\gamma}$ subunits.

(c) G_α releases GDP and binds GTP.

(d) G_α hydrolyzes GTP to GDP and P_i.

(e) The receptor is inactivated by phosphorylation of Ser and other residues on its intracellular domain.

11. The structure of adenylate cyclase is similar to the structures of some types of DNA polymerases, suggesting that these enzymes derived from a common ancestor. Compare the reactions catalyzed by these two enzymes. In what ways are they similar?

12. Suppose you were investigating a newly discovered growth-factor signal-transduction pathway, involving a G-protein-coupled receptor system. When you add $GTP_{\gamma S}$, a nonhydrolyzable analog of GTP, the duration of the hormonal response increased. Why does the response last longer in the presence of $GTP_{\gamma S}$? ❖ 2

13. Consider the structure of ATP, and the identification of the individual phosphate groups.

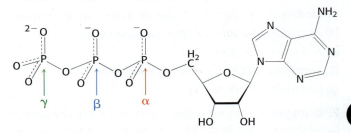

(a) If you wanted to identify the substrates of a particular kinase in a signaling pathway, you should use a version of ATP with ^{33}P-labeling at which phosphate?

(b) If you wanted to generate a labeled form of cAMP by the enzymatic action of adenylate cyclase, you should use a version of ATP with ^{33}P labeling at which phosphate?

14. You wish to determine the hormone-binding specificity of a newly identified membrane receptor. Three different hormones, X, Y, and Z, were mixed with the receptor in separate experiments, and the percentage of binding capacity of the receptor was determined as a function of hormone concentration, as shown in graph A.

(A)

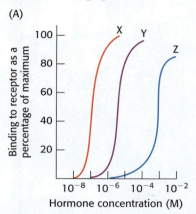

(a) What concentrations of each hormone yield 50% maximal binding?

(b) Which hormone shows the highest binding affinity for the receptor?

You next wish to determine whether the hormone–receptor complex stimulates the adenylate cyclase cascade. To do so, you measure adenylate cyclase activity as a function of hormone concentration, as shown in graph B.

(B)

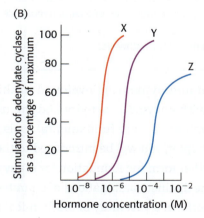

(c) What is the relationship between the binding affinity of the hormone–receptor complex and the ability of the hormone to enhance adenylate cyclase activity? What can you conclude about the mechanism of action of the hormone–receptor complex?

(d) Suggest experiments that would determine whether a $G_{\alpha s}$ protein is a component of the signal-transduction pathway.

15. Unlike the olfactory neurons in the mammalian systems discussed herein, olfactory neurons in the nematode *C. elegans* express multiple olfactory receptors. In particular, one neuron (called AWA) expresses receptors for compounds to which the nematode is attracted, whereas a different neuron (called AWB) expresses receptors for compounds that the nematode avoids. Suppose that a transgenic nematode is generated such that one of the receptors for an attractant is expressed in AWB rather than AWA. What behavior would you expect in the presence of the corresponding attractant? ❖ 5

16. A mixture of two compounds is applied to a section of olfactory epithelium. Only receptors 3, 5, 9, 12, and 13 are activated. Using the figure, identify the likely compounds in the mixture. ❖ 5

	Receptor														
	1	2	3	4	5	6	7	8	9	10	11	12	13	14
C_3-COOH														
C_4-COOH														
C_5-COOH					▨									
C_6-COOH	▨			▨	▨		▨			▨	▨			
C_7-COOH	▨		▨	▨	▨	▨	▨	▨		▨	▨	▨	▨	
C_8-COOH	▨		▨		▨	▨	▨	▨		▨		▨	▨	▨
C_5-OH		▨												
C_6-OH		▨			▨									
C_7-OH		▨		▨	▨									
C_8-OH				▨	▨		▨			▨				
C_9-OH				▨	▨		▨					▨		
Br-C_3-COOH													▨	
Br-C_4-COOH														
Br-C_5-COOH					▨		▨						▨	
Br-C_6-COOH														
Br-C_7-COOH	▨		▨		▨	▨	▨		▨	▨	▨	▨	▨	
HOOC-C_4-COOH														
HOOC-C_5-COOH														
HOOC-C_6-COOH			▨									▨	▨	
HOOC-C_7-COOH			▨			▨			▨		▨		▨	▨

17. Our ability to determine the direction from which a sound is coming is partly based on the difference in time at which our two ears detect the sound. Given the speed of sound (350 ms^{-1}) and the separation between our ears (0.15 m), what difference is expected in the times at which a sound arrives at our two ears? How does this difference compare with the time resolution of the human hearing system? Would a sensory system that utilized 7TM receptors and G proteins be capable of adequate time resolution? ❖ 5

18. Suppose humans have 380 different odorant receptors.

(a) How many different odorants could a person distinguish if each odorant is bound by a single type of receptor and each receptor binds only a single odorant? ❖ 5

(b) How many if each odorant binds to two different odorant receptors?

(c) How many if each odorant binds to three different odorant receptors?

19. Describe the effect of light absorption on 11-*cis*-retinal bound within rhodopsin. ❖ 5

Metabolism: Basic Concepts and Themes

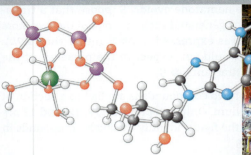

Dani Salvà/VWPics/Alamy

Currencies are used as a "go-between" in the exchange of labor and resources: people first earn money and later exchange it for goods and services. In an analogous way, organisms extract and store energy and later use it to power almost all biological processes requiring the input of energy, including homeostasis, biosynthesis, and motion. Within all organisms, this exchange is accomplished during cellular metabolism using a single energy currency: ATP.

OUTLINE

15.1 Metabolism Is Composed of Many Interconnected Reactions

15.2 ATP Is the Universal Currency of Free Energy in Biological Systems

15.3 The Oxidation of Carbon Fuels Is an Important Source of Cellular Energy

15.4 Metabolic Pathways Contain Many Recurring Motifs

❖ LEARNING GOALS

By the end of this chapter, you should be able to:

1. Explain what is meant by metabolism in terms of both catabolic and anabolic processes.

2. Identify the factors that make ATP a useful molecule for capturing and transferring chemical energy.

3. Explain how ATP can power reactions that would otherwise not take place.

4. Describe the relationship between the oxidation state of a carbon molecule and its usefulness as a fuel.

5. Summarize the recurring motifs in metabolic pathways.

The concepts of conformation and dynamics developed in this book so far—especially those dealing with enzymatic activity and the transport of molecules and ions across membranes—are fundamental tools for understanding biochemistry. In this chapter, you will be introduced to the recurring themes of metabolism that together form the foundation for many of the concepts in the rest of this book. The distinct combinations of a limited selection of enzymes and cofactors can perform a vast array of biochemical feats using the universal energy currency of life, ATP. Combining knowledge of this currency and the themes of metabolism with the concepts of the previous chapters will allow us to tackle two of the fundamental questions of biochemistry:

1. How does a cell extract energy and reducing power from its environment?

2. How does a cell synthesize the building blocks of its macromolecules and then the macromolecules themselves?

While this chapter gives many specific examples of reaction classes and patterns, it's important to first focus on familiarizing yourself with the broad patterns; the examples will be covered in greater depth in subsequent chapters.

15.1 Metabolism Is Composed of Many Interconnected Reactions

Living organisms require a continual input of free energy for three major purposes: (1) mechanical work in muscle contraction and cellular movements, (2) the active transport of molecules and ions, and (3) the synthesis of macromolecules and other biomolecules from simple precursors. The free energy used in these processes, which maintain an organism in a steady state that is far from equilibrium, is derived from the environment. **Phototrophs**, which include all photosynthetic organisms, capture energy from sunlight, whereas **chemotrophs**, which include all animals, capture energy through the oxidation of chemicals, often generated by phototrophs. Despite this difference, all living organisms share a tremendous number of metabolic reactions to utilize and produce lipids, proteins, nucleic acids, and common carbohydrates like glucose.

The fundamental cellular processes of extracting energy and synthesizing new material are carried out by a highly integrated network of chemical reactions collectively known as **metabolism (or intermediary metabolism)**. Metabolism is essentially a sequence of chemical reactions that begins with a particular molecule and results in the formation of some other molecule or molecules in a carefully defined fashion (**Figure 15.1**). There are many such defined pathways in the cell (**Figure 15.2**), and we will examine a few of them in some detail later. Advances in analytical techniques such as mass spectrometry (Section 4.3) have greatly facilitated the study of metabolism and generated yet another "omics"—metabolomics.

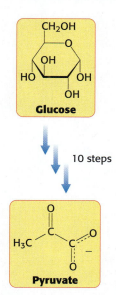

FIGURE 15.1 Glucose metabolism in human beings starts with glycolysis. Glycolysis is a metabolic pathway consisting of 10 linked reactions (steps) transforming the molecules into pyruvate.

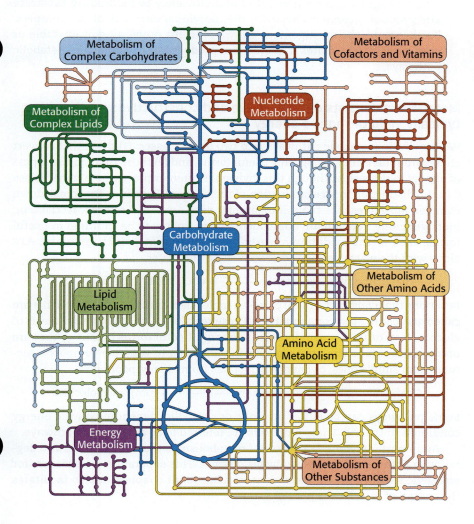

FIGURE 15.2 Metabolic pathways are interconnected series of enzyme-catalyzed reactions. Each node in the diagram represents a specific metabolite, whereas lines connecting nodes represent enzyme-catalyzed reactions. These reactions and the pathways they make up form a deeply integrated network in which reactions communicate with one another through multiple modes of regulation. The details of this figure will be explored over the next 12 chapters. [From the Kyoto Encyclopedia of Genes and Genomes (www.genome.ad.jp/kegg).]

EXPLORE this pathway further in the Metabolic Map in
 Achieve

This chapter highlights unifying themes common to all metabolic reactions:

- *Metabolism is a coherent network containing many common motifs.* More than a thousand chemical reactions take place in even as simple an organism as *Escherichia coli.* The array of reactions may seem overwhelming at first glance, but closer scrutiny reveals common motifs. One motif across all life forms is the use of a common energy currency, **adenosine triphosphate (ATP)**, which links energy-releasing (exergonic) pathways with energy-requiring (endergonic) pathways. In all organisms, either sunlight or the oxidation of chemical fuels powers the formation of ATP. Another motif is the repeated appearance of a limited number of activated intermediates; about 100 molecules play central roles in all forms of life.

- *While metabolism involves many different reactions, it uses only a few kinds of mechanisms.* Furthermore, these mechanisms are usually quite simple.

- *Metabolic reactions are highly regulated.* Fuels are degraded and large molecules are constructed step by step in a series of linked reactions called *metabolic pathways.* The total number of intricately connected metabolic pathways taking place in your cells is mind-boggling (Figure 15.2); for obvious reasons, a high degree of coordination is required to keep all the reactions running smoothly. Metabolic pathways are interdependent, and their activity is carefully coordinated by exquisitely sensitive means of communication in which allosteric enzymes are regulated by small molecules (Chapter 7) as well as being subject to external hormonal control (Chapter 14).

- *Many of the enzymes involved in metabolism are organized into large complexes.* This arrangement increases speed and efficiency by facilitating substrate and product movement between the individual enzymes of the complex. Furthermore, the arrangement allows efficient processing of unstable or even toxic intermediates that are often produced or consumed in metabolic pathways.

Metabolism consists of destructive and constructive reactions that typically yield or require energy

We can divide metabolic pathways into two broad classes: (1) those that convert energy from fuels into biologically useful forms, and (2) those that require inputs of energy to proceed. Although this division is often imprecise, it is nonetheless a useful distinction in an examination of metabolism. Reactions that break down complex molecules, like fuels, into simpler ones are called *catabolic reactions* or, more generally, **catabolism**. Catabolism allows energy to be captured in useful forms, such as the oxidation of carbohydrates and fats which generates ATP along with the waste products carbon dioxide and water.

$$\text{Fuel (carbohydrates, fats)} \xrightarrow{\text{Catabolism}} CO_2 + H_2O + \text{useful energy}$$

Reactions that construct a more complex molecule from simpler components are called *anabolic reactions* or **anabolism**. These reactions—such as the synthesis of glucose, fats, or DNA—usually require energy. Useful forms of energy that are produced in catabolism are employed in anabolism to generate complex structures from simple ones, or energy-rich states from energy-poor ones.

$$\text{Useful energy} + \text{simple precursors} \xrightarrow{\text{Anabolism}} \text{complex molecules}$$

Some pathways can be either anabolic or catabolic, depending on the energy conditions in the cell. These pathways are referred to as **amphibolic pathways**.

An important general principle of metabolism is that biosynthetic and degradative pathways are almost always distinct. This separation is necessary for energetic reasons, as will be evident in subsequent chapters. It also facilitates the control of metabolism.

A thermodynamically unfavorable reaction can be driven by a favorable reaction

How are specific pathways constructed from individual reactions? A pathway must satisfy minimally two criteria: (1) the individual reactions must be specific, and (2) each of the reactions that constitutes the pathway must be thermodynamically favored under real, rather than standard, conditions. A reaction that is specific will yield only one particular product or set of products from its reactants. As discussed in Section 5.1, enzymes provide this specificity.

The thermodynamics of metabolism are most readily approached in relation to free energy, which was discussed in Sections 1.3 and 5.2. A reaction can occur spontaneously only if ΔG, the change in free energy, is negative. Recall that ΔG for the formation of products C and D from substrates A and B is given by

$$\Delta G = \Delta G^{\circ\prime} + RT \ln \frac{[C][D]}{[A][B]}$$

Thus, the ΔG of a reaction depends on the nature of the reactants and products (expressed by the $\Delta G^{\circ\prime}$ term, the standard free-energy change) and on their concentrations (expressed by the second term). In a metabolic pathway, reactions with a positive $\Delta G^{\circ\prime}$ can proceed under physiological conditions because the concentrations of reactants and products are far from standard conditions, such that the ΔG of the reaction is negative. Another important thermodynamic fact is that the overall free-energy change for a chemically coupled series of reactions is equal to the sum of the free-energy changes of the individual steps. Consider the following reactions:

$$
\begin{array}{lll}
A \rightleftharpoons B + C & \Delta G^{\circ\prime} = +21 \text{ kJ mol}^{-1} & (+5 \text{ kcal mol}^{-1}) \\
B \rightleftharpoons D & \Delta G^{\circ\prime} = -34 \text{ kJ mol}^{-1} & (-8 \text{ kcal mol}^{-1}) \\
\hline
A \rightleftharpoons C + D & \Delta G^{\circ\prime} = -13 \text{ kJ mol}^{-1} & (-3 \text{ kcal mol}^{-1})
\end{array}
$$

Under standard conditions, A cannot be spontaneously converted into B and C, because $\Delta G^{\circ\prime}$ is positive. However, the conversion of B into D under standard conditions is thermodynamically feasible; if the two reactions are coupled such that one cannot occur without the other, then the free-energy changes are additive. Thus, the conversion of A into C and D has a $\Delta G^{\circ\prime}$ of -13 kJ mol^{-1} (-3 kcal mol^{-1}), which means that it can occur spontaneously under standard conditions.

Thus, a thermodynamically unfavorable reaction can be driven by a thermodynamically favorable reaction to which it is coupled. This kind of coupling occurs in the active sites of enzymes; in this example, the reactions are coupled by the shared chemical intermediate B, which is both formed and destroyed in the course of the reaction. Metabolic pathways often employ the coupling of thermodynamically favorable reactions to unfavorable reactions to create new enzymatically catalyzed reactions that have an overall negative change in free energy.

❖ SELF–CHECK QUESTION

Imagine that a cell needs to produce a compound B, but the reaction converting A into B has a positive $\Delta G^{\circ\prime}$ value. There are two ways that A can still be converted to B; one involves changing the reaction itself, and the other does not. What are they?

15.2 ATP Is the Universal Currency of Free Energy in Biological Systems

Just as commerce is facilitated by the use of a common currency, the commerce of the cell—metabolism—is facilitated by the use of a common energy currency, adenosine triphosphate (ATP). Part of the free energy derived from light or the oxidation of food is transformed into this highly accessible molecule, which acts as the free-energy donor in most energy-requiring processes such as motion, active transport, and biosynthesis. Indeed, most of catabolism consists of reactions that extract energy from fuels such as carbohydrates and fats and capture it through the production of ATP.

ATP hydrolysis is exergonic

ATP is a nucleotide consisting of adenine, a ribose, and a triphosphate unit (Figure 15.3). The active form of ATP is usually a complex of ATP with Mg^{2+} or Mn^{2+}. In considering the role of ATP as an energy carrier, we can focus on its triphosphate moiety.

- *ATP is an energy-rich molecule because its triphosphate unit contains two phosphoanhydride linkages.* A large amount of free energy is released when ATP is hydrolyzed to adenosine diphosphate (ADP) and orthophosphate (P_i), or when ATP is hydrolyzed to adenosine monophosphate (AMP) and pyrophosphate (PP_i).

$$ATP + H_2O \rightleftharpoons ADP + P_i$$
$$\Delta G^{\circ\prime} = -30.5 \text{ kJ mol}^{-1} \ (-7.3 \text{ kcal mol}^{-1})$$

$$ATP + H_2O \rightleftharpoons AMP + PP_i$$
$$\Delta G^{\circ\prime} = -45.6 \text{ kJ mol}^{-1}(-10.9 \text{ kcal mol}^{-1})$$

However, this hydrolysis reaction is frequently misunderstood. It is imperative to note that this release of free energy does *not* derive from the cleavage of the any of the covalent bonds in ATP, as it requires energy to break a covalent bond. Instead, the energy is released by the formation of new covalent bonds and noncovalent interactions with water, and from the increase in entropy of

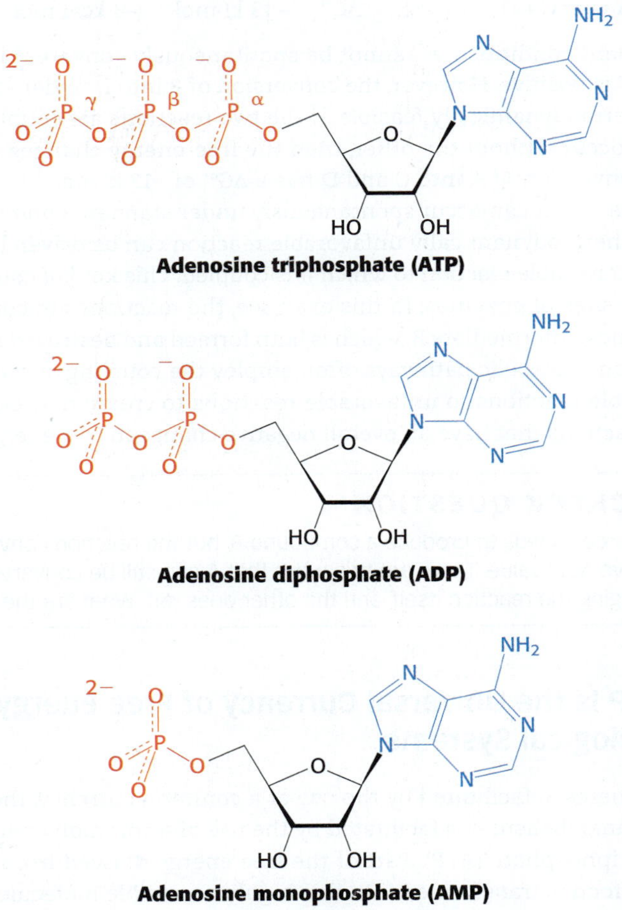

FIGURE 15.3 Structures of ATP, ADP, and AMP differ only by the number of phosphates. These adenylates consist of adenine (blue), a ribose (black), and a tri-, di-, or monophosphate unit (red). The innermost phosphorus atom of ATP is designated P_α, the middle one P_β, and the outermost one P_γ.

Adenosine triphosphate (ATP)

Adenosine diphosphate (ADP)

Adenosine monophosphate (AMP)

the products relative to the reactants. Furthermore, the precise ΔG for these reactions depends on the ionic strength of the medium and on the concentrations of Mg^{2+} and other metal ions. Under typical cellular concentrations, the ΔG for these hydrolyses is approximately -50 kJ mol^{-1}(-12 kcal mol^{-1}).

- *The ATP–ADP cycle is the fundamental mode of energy exchange in biological systems.* The free energy released in the hydrolysis of ATP is harnessed to drive reactions that require an input of free energy, such as muscle contraction. In turn, ATP is formed from ADP and P_i when fuel molecules are oxidized in chemotrophs or when light is trapped by phototrophs.

- *Enzymes catalyze the exchange of phosphoryl groups from one nucleotide to another.* Some biosynthetic reactions are driven not by ATP, but by guanosine triphosphate (GTP), uridine triphosphate (UTP), and cytidine triphosphate (CTP). The diphosphate forms of these nucleotides are denoted by GDP, UDP, and CDP, and the monophosphate forms by GMP, UMP, and CMP. The phosphorylation of nucleoside monophosphates is catalyzed by a family of nucleoside monophosphate kinases. The phosphorylation of nucleoside diphosphates is catalyzed by nucleoside diphosphate kinase, an enzyme with broad specificity.

$$\underset{\substack{\text{Nucleoside} \\ \text{monophosphate}}}{\text{NMP}} + \text{ATP} \xrightleftharpoons{\substack{\text{nucleoside monophosphate} \\ \text{kinase}}} \text{NDP} + \text{ADP}$$

$$\underset{\substack{\text{Nucleoside} \\ \text{diphosphate}}}{\text{NDP}} + \text{ATP} \xrightleftharpoons{\substack{\text{Nucleoside diphosphate} \\ \text{kinase}}} \text{NTP} + \text{ADP}$$

- *Although all of the nucleoside triphosphates are energetically equivalent, ATP is nonetheless the primary cellular energy carrier and therefore the energy currency.* In addition, two important electron carriers—NAD$^+$ and FAD—as well the acyl group carrier, coenzyme A, are derivatives of ATP. It is intriguing to consider why evolution selected adenosine derivatives, but it may be due to differences in the stability of adenine compared to the other bases in prebiotic conditions. Regardless of the reason, the role of adenosine phosphates—particularly ATP—in energy metabolism is paramount.

ATP hydrolysis drives metabolism by shifting the equilibrium of coupled reactions

An otherwise unfavorable reaction can be made possible by coupling to ATP hydrolysis. Consider a reaction that is thermodynamically unfavorable without an input of free energy, a situation common to most biosynthetic reactions. Suppose that the standard free energy of the conversion of compound A into compound B is $+16.7$ kJ mol^{-1} or $+4.0$ kcal mol^{-1}

$$\text{A} \rightleftharpoons \text{B} \quad \Delta G^{\circ\prime} = +16.7 \text{ kJ mol}^{-1}\,(+4.0 \text{ kcal mol}^{-1})$$

The equilibrium constant K'_{eq} of this reaction at 25°C is related to $\Delta G^{\circ\prime}$ (in units of kilojoules per mole) by

$$K'_{eq} = [\text{B}]_{eq}/[\text{A}]_{eq} = e^{-\Delta G^{\circ\prime}/2.47} = 1.15 \times 10^{-3}$$

Thus, net conversion of A into B cannot take place when the molar ratio of B to A is equal to or greater than 1.15×10^{-3}.

However, A can be converted into B under these conditions if the reaction is coupled to the hydrolysis of ATP. Under standard conditions, the $\Delta G^{\circ\prime}$ of hydrolysis is approximately -30.5 kJ mol^{-1} or -7.3 kcal mol^{-1}. The new overall reaction is

$$\text{A} + \text{ATP} + \text{H}_2\text{O} \rightleftharpoons \text{B} + \text{ADP} + \text{P}_i$$

$$\Delta G^{\circ\prime} = -13.8 \text{ kJ mol}^{-1}\,(-3.3 \text{ kcal mol}^{-1})$$

Its free-energy change of -13.8 kJ mol^{-1} or -3.3 kcal mol^{-1} is the sum of the value of $\Delta G^{\circ\prime}$ for the conversion of A into B ($+16.7$ kJ mol^{-1} or $+4.0$ kcal mol^{-1}) and the value of $\Delta G^{\circ\prime}$ for the hydrolysis of ATP (-30.5 kJ mol^{-1} or -7.3 kcal mol^{-1}). At pH 7, the equilibrium constant of this coupled reaction is

$$K'_{eq} = \frac{[B]_{eq}}{[A]_{eq}} \times \frac{[ADP]_{eq}[P_i]_{eq}}{[ATP]_{eq}} = e^{13.8/2.47} = 2.67 \times 10^2$$

At equilibrium, the ratio of [B] to [A] is given by

$$\frac{[B]_{eq}}{[A]_{eq}} = K'_{eq} \frac{[ATP]_{eq}}{[ADP]_{eq}[P_i]_{eq}}$$

which means that the hydrolysis of ATP enables A to be converted into B until the [B]/[A] ratio reaches a value of 2.67×10^2.

This equilibrium ratio is strikingly different from the value of 1.15×10^{-3} for the reaction A $\longrightarrow$ B in the absence of ATP hydrolysis. In other words, coupling the hydrolysis of ATP with the conversion of A into B under standard conditions has changed the equilibrium ratio of B to A by a factor of about 10^5. If we were to use the ΔG of hydrolysis of ATP under cellular conditions (-50.2 kJ mol^{-1} or -12 kcal mol^{-1}) in our calculations instead of $\Delta G^{\circ\prime}$, the change in the equilibrium ratio would be even more dramatic, on the order of 10^8.

We see here the thermodynamic essence of ATP's action as an energy-coupling agent. Cells maintain ATP levels by using oxidizable substrates or light as sources of free energy for synthesizing the molecule. In the cell, the hydrolysis of an ATP molecule in a coupled reaction then changes the equilibrium ratio of products to reactants by a very large factor, of the order of 10^8. More generally, the hydrolysis of n ATP molecules changes the equilibrium ratio of a coupled reaction (or sequence of reactions) by a factor of 10^{8n}. For example, the theoretical hydrolysis of three ATP molecules in a coupled reaction changes the equilibrium ratio by a factor of 10^{24}. Thus, a thermodynamically unfavorable process can be converted into a favorable one by coupling it to the hydrolysis of a sufficient number of ATP molecules in a new reaction.

It should also be emphasized that A and B in the preceding coupled reaction may be interpreted very generally, not only as different chemical species. For example, A and B may represent activated and unactivated conformations of a protein that is activated by phosphorylation with ATP. Through such changes in protein conformation, molecular motors such as myosin, kinesin, and dynein convert the chemical energy of ATP into mechanical energy (Section 6.5). Indeed, this conversion is the basis of muscle contraction.

Alternatively, A and B may refer to the concentrations of an ion or molecule on the outside and inside of a cell, as in the active transport of a nutrient. The active transport of Na$^+$ and K$^+$ across membranes is driven by the phosphorylation of the sodium–potassium pump by ATP and its subsequent dephosphorylation (Section 13.2).

The high phosphoryl potential of ATP results from structural differences between ATP and its hydrolysis products

What makes ATP an efficient phosphoryl-group donor? Let us compare the standard free energy of hydrolysis of ATP with that of a phosphate ester, such as glycerol 3-phosphate:

$$ATP + H_2O \rightleftharpoons ADP + P_i$$
$$\Delta G^{\circ\prime} = -30.5 \text{ kJ mol}^{-1}(-7.3 \text{ kcal mol}^{-1})$$

$$\text{Glycerol 3-phosphate} + H_2O \rightleftharpoons \text{glycerol} + P_i$$
$$\Delta G^{\circ\prime} = -9.2 \text{ kJ mol}^{-1} (-2.2 \text{ kcal mol}^{-1})$$

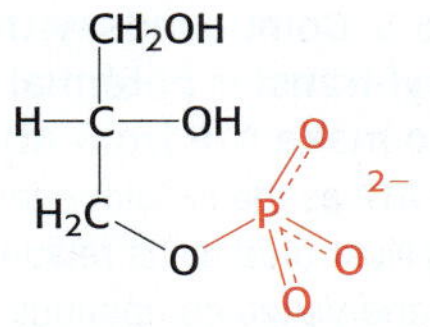

The magnitude of $\Delta G^{\circ\prime}$ for the hydrolysis of glycerol 3-phosphate is much smaller than that of ATP, which means that ATP has a stronger tendency to transfer its terminal phosphoryl group to water than does glycerol 3-phosphate. In other words, ATP has a higher **phosphoryl-transfer potential** than glycerol 3-phosphate.

The high phosphoryl-transfer potential of ATP can be explained by features of the ATP structure. Because $\Delta G^{\circ\prime}$ depends on the difference in free energies of the products and reactants, we need to examine the structures of both ATP and its hydrolysis products, ADP and P_i, to answer this question. Four factors are important:

1. *Resonance stabilization.* Orthophosphate (P_i), one of the products of ATP hydrolysis, has greater resonance stabilization than do any of the ATP phosphoryl groups. Orthophosphate has several resonance forms of similar energy (**Figure 15.4**), whereas the γ phosphoryl group of ATP has a smaller number.

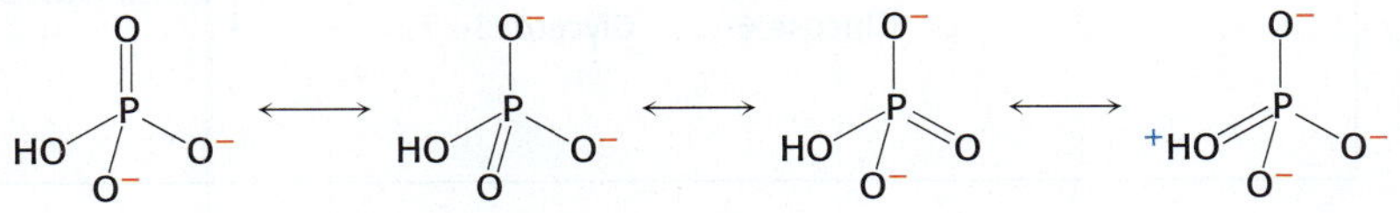

Glycerol 3-phosphate

FIGURE 15.4 **Orthophosphate has four favorable resonance structures.**

2. *Electrostatic repulsion.* At pH 7, the triphosphate unit of ATP carries about four negative charges. These charges repel one another because they are in close proximity. The repulsion between them is reduced when ATP is hydrolyzed.

3. *Increase in entropy.* The entropy of the products of ATP hydrolysis is greater, in that there are now two molecules instead of a single ATP molecule. We disregard the molecule of water used to hydrolyze the ATP; given the high concentration (55.5 M), there is effectively no change in the concentration of water during the reaction.

4. *Stabilization due to hydration.* Water binds to ADP and P_i, stabilizing these molecules, and rendering the reverse reaction — the synthesis of ATP — more unfavorable.

ATP is often called a high-energy phosphate compound, and its phosphoanhydride linkages are sometimes referred to as high-energy bonds. Indeed, a "squiggle" ($\sim P$) is often used to indicate the bonds that can be broken in ATP hydrolysis. This is a misnomer, however, as there is nothing special about the bonds themselves. The phosphoanhydride linkages are high-energy in the sense that much free energy is released when they are hydrolyzed — a reaction with water — for the reasons listed above, but the energy release comes from entropy and new bond formation, rather than bond cleavage.

Phosphoryl-transfer potential is an important form of cellular energy transformation

The standard free energies of hydrolysis provide a convenient means of comparing the phosphoryl-transfer potential of phosphorylated compounds. Such comparisons reveal that ATP is not the only compound with a high phosphoryl-transfer potential. In fact, some compounds in biological systems have a higher phosphoryl-transfer potential than that of ATP. These compounds include phosphoenolpyruvate (PEP), 1,3-bisphosphoglycerate (1,3-BPG), and creatine phosphate (**Figure 15.5**).

PEP can transfer its phosphoryl group to ADP to form ATP. Indeed, this transfer is one of the ways in which ATP is generated in the breakdown of sugars

FIGURE 15.5 Compounds with high phosphoryl-transfer potential can be used to make ATP from ADP.
The role of ATP as the cellular energy currency is illustrated by its relation to other phosphorylated compounds. ATP has a phosphoryl-transfer potential that is intermediate among the biologically important phosphorylated molecules. High-phosphoryl-transfer-potential compounds (1,3-BPG, PEP, and creatine phosphate) derived from the metabolism of fuel molecules are used to power ATP synthesis. In turn, ATP donates a phosphoryl group to other biomolecules to facilitate their metabolism. [Data from D. L. Nelson and M. M. Cox, *Lehninger Principles of Biochemistry*, 5th ed. (W. H. Freeman and Company, 2009), Fig. 13-19.]

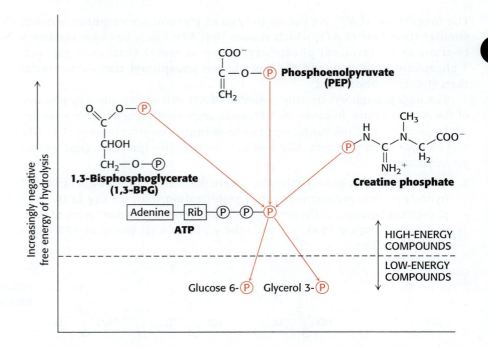

TABLE 15.1 Standard free energies of hydrolysis of some phosphorylated compounds

Compound	kJ mol^{-1}	kcal mol^{-1}
Phosphoenolpyruvate	−61.9	−14.8
1,3-Bisphosphoglycerate	−49.4	−11.8
Creatine phosphate	−43.1	−10.3
ATP (to ADP)	−30.5	−7.3
Glucose 1-phosphate	−20.9	−5.0
Pyrophosphate	−19.3	−4.6
Glucose 6-phosphate	−13.8	−3.3
Glycerol 3-phosphate	−9.2	−2.2

(Section 16.1). It is significant that ATP has a phosphoryl-transfer potential that is intermediate among the biologically important phosphorylated molecules (**Table 15.1**). This intermediate position enables ATP to function efficiently as a carrier of phosphoryl groups.

The amount of ATP in muscle suffices to sustain contractile activity for less than a second. Creatine phosphate in vertebrate muscle serves as a reservoir of high-potential phosphoryl groups that can be readily transferred to ADP. Indeed, we use creatine phosphate to regenerate ATP from ADP every time we exercise strenuously. This reaction is catalyzed by creatine kinase.

$$\text{Creatine phosphate} + \text{ADP} \xrightleftharpoons[\quad]{\substack{\text{creatine}\\\text{kinase}}} \text{ATP} + \text{creatine}$$

At pH 7, the standard free energy of hydrolysis of creatine phosphate is −43.1 kJ mol^{-1} (−10.3 kcal mol^{-1}), compared with −30.5 kJ mol^{-1} (−7.3 kcal mol^{-1}) for ATP.

As illustrated in the next boxed example, there is a large amount of creatine phosphate present in resting muscle. Because of its abundance and high phosphoryl-transfer potential relative to that of ATP, creatine phosphate is a highly effective phosphoryl buffer. Indeed, creatine phosphate is the major source of phosphoryl groups for ATP regeneration for a runner during the first 4 seconds of a 100-meter sprint. The fact that creatine phosphate can replenish

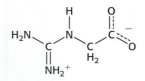

Creatine

Creatine phosphate

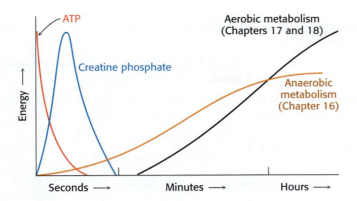

FIGURE 15.6 The sources of ATP change as exercise duration increases, even within the first few seconds. In the initial seconds of exertion, power is generated by existing high-phosphoryl-transfer compounds (ATP and creatine phosphate). Subsequently, the ATP must be regenerated by metabolic pathways.

ATP pools is the basis of the use of creatine as a dietary supplement by athletes in sports requiring short bursts of intense activity. After the creatine phosphate pool is depleted, ATP must be generated through anaerobic or aerobic metabolism (**Figure 15.6**).

EXAMPLE **Calculating ΔG for a Coupled Reaction Under Real Conditions**

PROBLEM: Creatine phosphate is used as a phosphoryl donor for ATP synthesis in muscle in the following reaction:

$$\text{Creatine phosphate} + \text{ADP} \underset{}{\overset{\text{creatine kinase}}{\rightleftharpoons}} \text{ATP} + \text{creatine}$$

When muscles are at rest, the reaction proceeds in the reverse direction, synthesizing creatine phosphate. As noted above, for a skeletal muscle at rest, the following metabolites are present at the indicated concentrations.

$$[\text{ATP}] = 4 \text{ mM}; \text{ADP} = 0.013 \text{ mM}$$

$$[\text{creatine phosphate}] = 25 \text{ mM}; [\text{creatine}] = 13 \text{ mM}$$

Calculate the ΔG for the creatine kinase reaction as written above for a muscle at rest assuming body temperature is 37°C.

GETTING STARTED: What formula do we need to calculate ΔG, and what information are we missing to complete this calculation?

$$\Delta G = \Delta G^{\circ\prime} + RT \ln \frac{[\text{C}][\text{D}]}{[\text{A}][\text{B}]}$$

ANALYZE: The values necessary for the right-hand term are provided, but we need to know $\Delta G^{\circ\prime}$ for the kinase reaction. How do we determine $\Delta G^{\circ\prime}$? Recall from this chapter that two component parts of a reaction have additive $\Delta G^{\circ\prime}$ values. We need to dissect the creatine kinase reaction into its two component reactions and look up the $\Delta G^{\circ\prime}$ values for each of the component reactions. These are:

$$\text{ADP} + \text{P}_i \longrightarrow \text{ATP} + \text{H}_2\text{O}$$

$$\text{Creatine phosphate} + \text{H}_2\text{O} \longrightarrow \text{creatine} + \text{P}_i$$

Using Table 15.1, we can determine the $\Delta G^{\circ\prime}$ values for both reactions.

For ATP synthesis, $\Delta G^{\circ\prime} = +30.5$ kJ mol^{-1}. For creatine phosphate hydrolysis, $\Delta G^{\circ\prime} = -43.1$ kJ mol^{-1}.

CALCULATE: What is the $\Delta G^{\circ\prime}$ of the overall creatine kinase reaction in the direction written?

$\Delta G^{\circ\prime}$ values are additive, so the $\Delta G^{\circ\prime}$ for the kinase reaction is -43.1 kJ mol^{-1} + 30.5 kJ mol^{-1} = -12.6 kJ mol^{-1}.

Now, having determined $\Delta G^{\circ\prime}$, we can use the equation above and the values provided to determine ΔG.

$$\Delta G = -12.6 \text{ kJ mol}^{-1} + RT \ln\left(\frac{[\text{ATP}][\text{creatine}]}{[\text{ADP}][\text{creatine phosphate}]}\right)$$

$$= -12.6 \text{ kJ mol}^{-1} + (0.0083 \text{ kJ K}^{-1} \text{ mol}^{-1})(310 \text{ K}) \ln\left(\frac{[4 \text{ mM}][13 \text{ mM}]}{[0.013 \text{ mM}][25 \text{ mM}]}\right)$$

$$= -12.6 \text{ kJ mol}^{-1} + (2.573 \text{ kJ mol}^{-1}) \ln(160)$$

$$= 0.5 \text{ kJ mol}^{-1}$$

REFLECT: Note that this value is close to zero, indicating that the reaction is very near equilibrium. Does this make sense to you, for resting muscle? Consider that upon sudden muscle contraction during exercise, the depletion of ATP shifts the reaction toward creatine and ATP. Conversely, during the recovery from exercise, the replenishment of ATP from oxidative phosphorylation shifts the reaction back toward the formation of creatine phosphate. Once replenished, and when the muscle is at rest, we should expect that the reaction will eventually settle at equilibrium concentrations and, therefore, ΔG should be near zero.

15.3 The Oxidation of Carbon Fuels Is an Important Source of Cellular Energy

ATP serves as the principal immediate donor of free energy in biological systems rather than as a long-term storage form of free energy. In a typical cell, an ATP molecule is consumed within a minute of its formation.

- *Having mechanisms for regenerating ATP is vital.* Although the total quantity of ATP in the body is limited to approximately 100 g, the turnover of this small quantity of ATP is very high. For example, a resting human being consumes about 40 kg (88 pounds) of ATP in 24 hours. During strenuous exertion, the rate of utilization of ATP may be as high as 0.5 kg/minute. For a 2-hour run, 60 kg (132 pounds) of ATP is utilized. Of course, this is essentially the same 100 g of ATP being used and regenerated again and again.

- *The regeneration of ATP is one of the primary roles of catabolism.* Motion, active transport, signal amplification, and biosynthesis can take place only if ATP is continually regenerated from ADP (**Figure 15.7**). In most aerobic organisms, the carbon in fuel molecules—such as glucose and fats—is oxidized to CO_2. The resulting electrons are captured and their energy is used to regenerate ATP from ADP and P_i.

- *Oxidation of fuels takes place one carbon at a time.* The ultimate electron acceptor for the aerobic oxidation of carbon is O_2, which is reduced to water, and the oxidation product is CO_2. Consequently, the more reduced a carbon is to begin with, the more free energy is released by its oxidation. **Figure 15.8** shows the $\Delta G^{\circ\prime}$ of oxidation for one-carbon compounds. While fuel molecules are more complex (**Figure 15.9**) than these single-carbon compounds, their oxidation nevertheless takes place one carbon at a time. The carbon-oxidation energy is used in some cases to create a compound with high phosphoryl-transfer potential and in other cases to create an ion gradient. In either case, the end point is the formation of ATP.

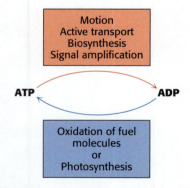

FIGURE 15.7 The ATP–ADP cycle is the fundamental mode of energy exchange in biological systems.

Most energy ────────────────────────────────→ Least energy

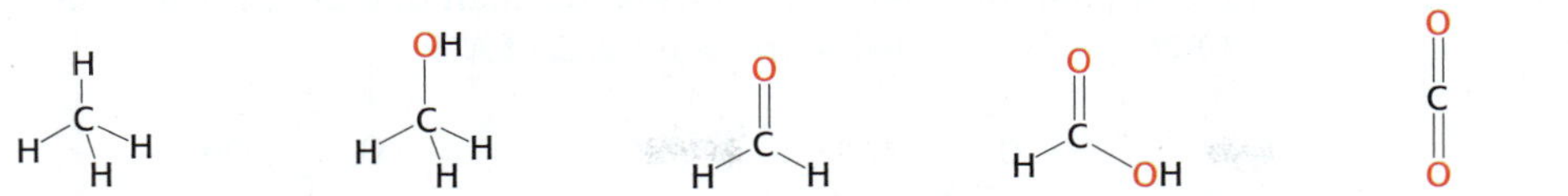

Methane	**Methanol**	**Formaldehyde**	**Formic acid**	**Carbon dioxide**
$G'_{oxidation}$ (kJ mol^{-1}) −820	−703	−523	−285	0
$G'_{oxidation}$ (kcal mol^{-1}) −196	−168	−125	−68	0

FIGURE 15.8 The free energy of oxidation of single-carbon compounds increases with the degree of hydrogen saturation.

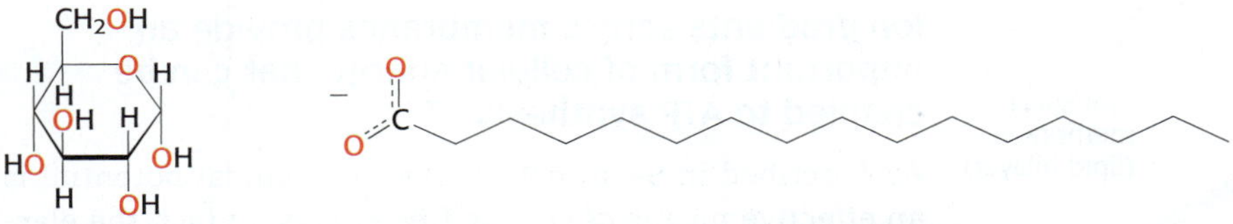

Glucose   **Saturated fatty acid**

FIGURE 15.9 Fats are a more efficient fuel source than carbohydrates such as glucose because the carbon in fats is more reduced.

Compounds with high phosphoryl-transfer potential can couple carbon oxidation to ATP synthesis

How is the energy released from the oxidation of a carbon compound captured in ATP? As an example, consider glyceraldehyde 3-phosphate (shown at left below), which is a metabolite of glucose formed during the oxidation of that sugar. The C-1 carbon (shown in red) is at the aldehyde-oxidation level and is not in its most oxidized state. Oxidation of the aldehyde to an acid will release energy.

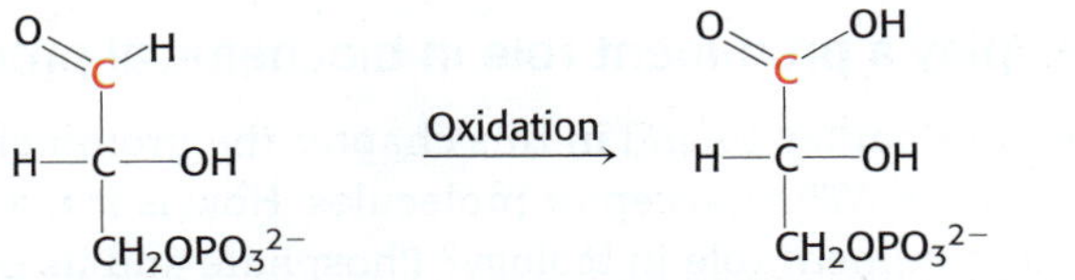

$$\xrightarrow{\text{Oxidation}}$$

Glyceraldehyde 3-phosphate   **3-Phosphoglyceric acid**

However, the oxidation does not take place directly. Instead, the carbon oxidation generates an acyl phosphate, 1,3-bisphosphoglycerate (1,3-BPG). The electrons released are captured by NAD^+, forming NADH, in a process that we will consider shortly.

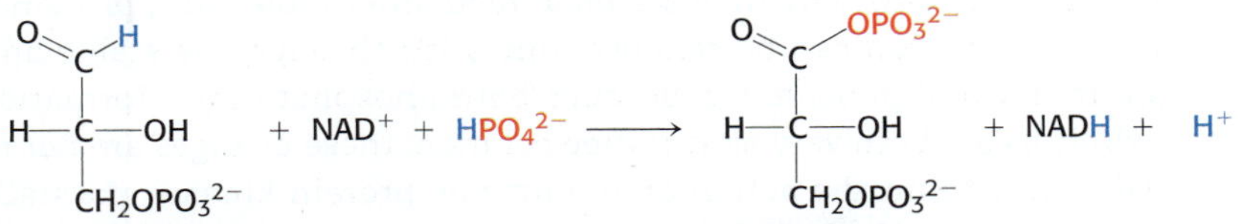

$$+ NAD^+ + HPO_4^{2-} \longrightarrow + NADH + H^+$$

Glyceraldehyde 3-phosphate (GAP)   **1,3-Bisphosphoglycerate (1,3-BPG)**

For reasons similar to those discussed for ATP, 1,3-BPG has a high phosphoryl-transfer potential that is, in fact, greater than that of ATP. Thus, the hydrolysis of 1,3-BPG can be coupled to the synthesis of ATP.

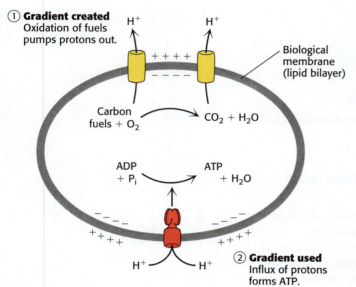

The energy of oxidation is initially trapped as a high-phosphoryl-transfer-potential compound and then used to form ATP. The oxidation energy of a carbon atom is transformed into phosphoryl-transfer potential, first as 1,3-BPG and ultimately as ATP. We will consider these reactions in mechanistic detail in Chapter 16.

Ion gradients across membranes provide an important form of cellular energy that can be coupled to ATP synthesis

As described in Section 13.1, electrochemical potential is an effective means of storing free energy. In fact, the electrochemical potential of ion gradients across membranes—produced by the oxidation of fuel molecules or by phototrophy—ultimately powers the synthesis of most of the ATP in cells. In general, ion gradients are versatile means of coupling thermodynamically unfavorable reactions to favorable ones.

In animals, proton gradients generated by the oxidation of carbon fuels account for more than 90% of ATP generation (**Figure 15.10**). This process, called *oxidative phosphorylation*, will be explored in depth in Chapter 18. ATP hydrolysis can then be used to form ion gradients of different types and functions. The electrochemical potential of a Na^+ gradient, for example, can be tapped to pump Ca^{2+} out of cells or to transport nutrients such as sugars and amino acids into cells.

Phosphates play a prominent role in biochemical processes

We have seen in Chapter 14 and in this chapter the prominence of phosphoryl group transfer from ATP to acceptor molecules. How is it that phosphate came to play such a prominent role in biology? Phosphate and its esters have several characteristics that render them useful for biochemical systems:

- *The energy release from phosphate esters can be manipulated by enzymes, making them ideal regulatory molecules.* Phosphate esters have an important chemical characteristic by being thermodynamically unstable yet kinetically stable in water. Phosphate esters are thus molecules whose energy release can be manipulated by enzymes. The kinetic stability of phosphate esters is due to the negative charges that make them resistant to hydrolysis in the absence of enzymes. This accounts for the presence of phosphate in the backbone of DNA and explains why they are ideal regulatory molecules; protein structure and function can be reliably controlled through the stable addition and removal of phosphates. Because both phosphate ester formation and hydrolysis are both very slow in free solution, these changes are not readily reversed without the action of an enzyme: protein kinases normally add phosphates to proteins while protein phosphatases typically remove them, as we explored in Section 7.3.

FIGURE 15.10 Proton gradients formed using the energy from either sunlight or chemical oxidation can power ATP synthesis. Either the oxidation of fuels or the capture of the energy from sunlight can power the formation of proton gradients by the action of specific proton pumps (yellow cylinders). These proton gradients can in turn drive the synthesis of ATP when the protons flow through an ATP-synthesizing enzyme (red complex).

- *The addition of a phosphate group changes molecule conformation and behavior.* Phosphates are also frequently added to metabolites that might otherwise diffuse through the cell membrane. Even when transporters exist for unphosphorylated forms of a metabolite, the addition of a phosphate changes the geometry and polarity of the molecules so that they no longer fit in the binding sites of the transporters.

- *All other ions fall short.* No other ions have the chemical characteristics of phosphate. Citrate is not sufficiently charged to prevent hydrolysis. Arsenate forms esters that are unstable and susceptible to spontaneous hydrolysis. Indeed, arsenate is poisonous to cells because it can replace phosphate in reactions required for ATP synthesis, generating unstable compounds and preventing ATP synthesis. Silicate is more abundant than phosphate, but silicate salts are virtually insoluble and are used for biomineralization, as in diatoms and certain sponges. Only phosphate has the chemical properties to meet the needs of living systems.

Energy from food is extracted in three stages

Let us take an overall view of the processes of energy conversion in animals before considering them in detail in subsequent chapters. Hans Krebs described three stages in the generation of energy from the oxidation of food (**Figure 15.11**).

1. *Large molecules in food are broken down into smaller units in the process of digestion.* Proteins are hydrolyzed to their 20 different amino acids, polysaccharides are hydrolyzed to simple sugars such as glucose, and lipids are hydrolyzed to glycerol and fatty acids. The degradation products are then absorbed by the cells of the intestine and distributed throughout the body. This stage is strictly a preparation stage; no useful energy is captured in this phase.

2. *Numerous small molecules resulting from digestion are degraded to a few simple units that play a central role in metabolism.* In fact, most of them—sugars, fatty acids, glycerol, and several amino acids—are converted into the acetyl unit of acetyl CoA. Some ATP is generated in this stage, but the amount is small compared with that obtained in the third stage.

3. *ATP is produced from the complete oxidation of the acetyl unit of acetyl CoA.* The third stage consists of the citric acid cycle and oxidative phosphorylation, which are the final common pathways in the oxidation of fuel molecules. Acetyl CoA brings acetyl units into the citric acid cycle—also called the CAC, the tricarboxylic acid (TCA) cycle, or the Krebs cycle—where they are completely oxidized to CO_2. Four pairs of electrons are transferred (three to NAD^+ and one to FAD) for each acetyl group that is oxidized (Chapter 17). Then, a proton gradient is generated as electrons flow from the reduced forms of these carriers to O_2. This gradient is used to synthesize ATP (Chapter 18).

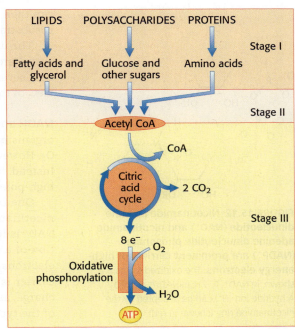

FIGURE 15.11 The extraction of energy from food molecules by aerobic organisms can be described as taking place in three stages.

❖ **SELF-CHECK QUESTION**

Which of the following statements is/are true? (1) Energy is released when the covalent bond linking two phosphates in ATP is broken. (2) ATP is the cellular energy currency because it has the highest phosphoryl transfer potential of any biological molecule. (3) ATP has two phosphoanhydride linkages, each of which can be hydrolyzed to provide useable energy for an organism.

15.4 Metabolic Pathways Contain Many Recurring Motifs

The sheer number of reactants and reactions of metabolism makes it seem very complex. Fortunately, we can look to several unifying themes to help make sense of it all. These themes include common metabolites, reactions, and regulatory schemes that stem from a common evolutionary heritage.

Activated carriers exemplify the modular structure and economy of metabolism

We have seen that phosphoryl transfer can be used to drive otherwise endergonic reactions, alter the conformation energy of a protein, or serve as a signal to alter the activity of a protein. The phosphoryl-group donor in all these reactions is ATP. In summary, ATP is an **activated carrier** of phosphoryl groups because phosphoryl transfer from ATP is an exergonic process. Activated carriers are small molecules to which a chemical group or electrons have been added, which can then be donated to another molecule; they frequently act as coenzymes or cosubstrates in enzyme-catalyzed reactions.

The use of ATP and other activated carriers is a recurring motif in biochemistry. Here we introduce the activated carriers that will appear repeatedly in the next several chapters, often functioning as coenzymes. Next, we provide an overview of their key roles.

NADH and FADH$_2$: Activated carriers of electrons for fuel oxidation. In aerobic organisms, the ultimate electron acceptor in the oxidation of fuel molecules is O$_2$. However, electrons are not transferred directly from the fuel molecule to O$_2$. Instead, fuel molecules transfer electrons to carriers, which then transfer their high-potential electrons to O$_2$.

One of the major electron carriers in the oxidation of fuel molecules is nicotinamide adenine dinucleotide (NAD$^+$; **Figure 15.12**). The reactive part of NAD$^+$ is its nicotinamide ring, derived from vitamin B$_3$ (niacin). In the oxidation of a substrate, the nicotinamide ring of NAD$^+$ accepts a proton and two electrons, which are equivalent to a hydride ion (H:$^-$). The reduced form of this carrier is NADH. In the oxidized form, the nitrogen atom carries a positive charge, as indicated by NAD$^+$. NAD$^+$ is the electron acceptor in many reactions of the type

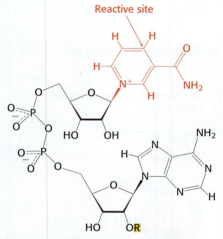

FIGURE 15.12 Nicotinamide adenine dinucleotide (NAD$^+$) and nicotinamide adenine dinucleotide phosphate (NADP$^+$) are prominent carriers of high-energy electrons. The oxidized forms are shown; in NAD$^+$, R = H; in NADP$^+$, R = PO$_3^{2-}$. A hydride ion (H$^-$) can be accepted by the nicotinamide ring (shown in red).

In this dehydrogenation, one hydrogen atom of the substrate is directly transferred to NAD$^+$, whereas the other appears in the solvent as a proton. Both electrons lost by the substrate are transferred to the nicotinamide ring.

The other major electron carrier in the oxidation of fuel molecules is the coenzyme flavin adenine dinucleotide (FAD; **Figure 15.13**). The abbreviations for the oxidized and reduced forms of this carrier are FAD and FADH$_2$, respectively. FAD is the electron acceptor in reactions of this type:

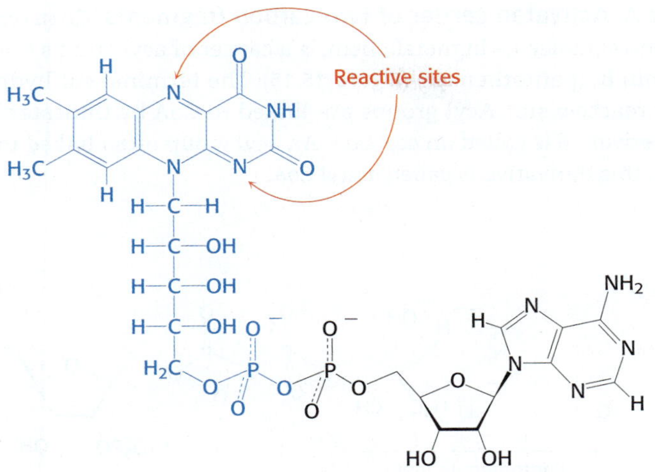

FIGURE 15.13 The oxidized form of flavin adenine dinucleotide (FAD) consists of two units. FAD has a flavin mononucleotide (FMN) unit (shown in blue) and an AMP unit (shown in black).

The reactive part of FAD is its isoalloxazine ring, a derivative of the vitamin riboflavin (**Figure 15.14**). FAD, like NAD⁺, can accept two electrons. In doing so, FAD, unlike NAD⁺, takes up two protons; its reduced form is FADH₂. These carriers of high-potential electrons as well as flavin mononucleotide (FMN)—an electron carrier similar to FAD but lacking the adenine nucleotide—will be considered further in Chapter 18.

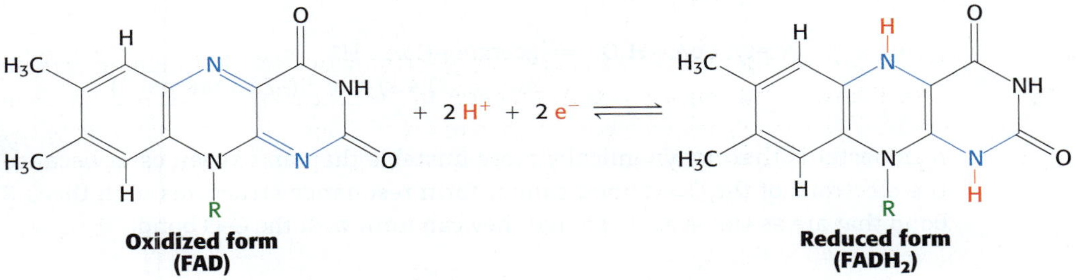

FIGURE 15.14 Structures of the reactive components of FAD and FADH₂ reveal that electrons and protons are carried by the reactive isoalloxazine ring component.

NADPH: Activated carrier of electrons for reductive biosynthesis. High-potential electrons are required in most biosyntheses because the precursors are more oxidized than the products. Hence, reducing power is needed in addition to ATP. For example, in the biosynthesis of fatty acids, a keto group is reduced to a methylene group in several steps. This sequence of reactions requires an input of four electrons.

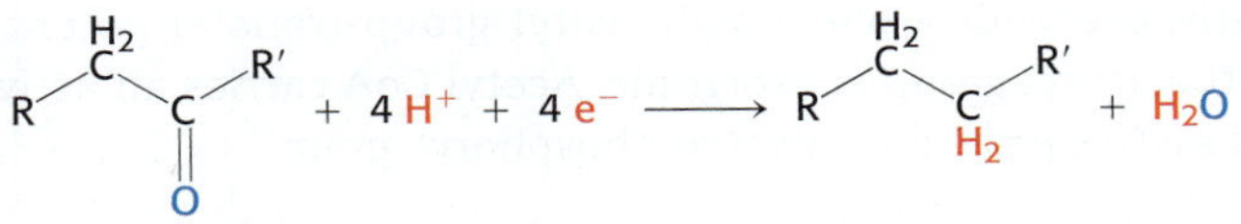

The electron donor in most reductive biosyntheses is NADPH, the reduced form of nicotinamide adenine dinucleotide phosphate (Figure 15.12). Its oxidized form is NADP⁺. NADPH differs from NADH in that the 2′-hydroxyl group of its adenosine moiety is esterified with phosphate. NADPH carries electrons in the same way as NADH. However, NADPH is used almost exclusively for reductive biosyntheses, whereas NADH is used primarily for the generation of ATP. The extra phosphoryl group on NADPH is a tag that enables enzymes to distinguish between high-potential electrons to be used in anabolism and those to be used in catabolism.

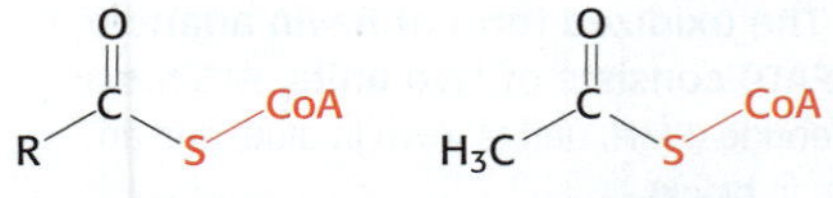

Acyl CoA   **Acetyl CoA**

Coenzyme A: Activated carrier of two-carbon fragments. Coenzyme A (CoA), another central molecule in metabolism, is a carrier of acyl groups that is derived from vitamin B$_5$ (pantothenate) (**Figure 15.15**). The terminal sulfhydryl group in CoA is the reactive site. Acyl groups are linked to CoA by thioester bonds. The resulting derivative is called an *acyl CoA*. An acyl group often linked to CoA is the acetyl unit; this derivative is called *acetyl CoA*.

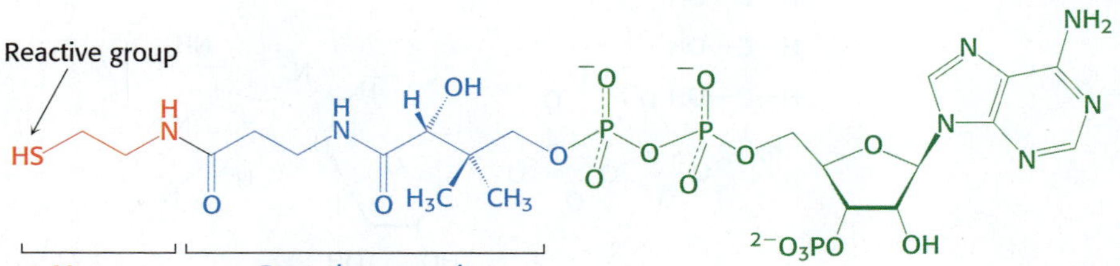

FIGURE 15.15 Structure of coenzyme A (CoA) reveals its derivation from both ADP and the vitamin pantothenate.

Acyl groups are important constituents both in catabolism, as in the oxidation of fatty acids, and in anabolism, as in the synthesis of membrane lipids.

The $\Delta G^{\circ\prime}$ for the hydrolysis of acetyl CoA has a large negative value:

$$\text{Acetyl CoA} + H_2O \rightleftharpoons \text{acetate} + \text{CoA} + H^+$$
$$\Delta G^{\circ\prime} = -31.4 \text{ kJ mol}^{-1} \ (-7.5 \text{ kcal mol}^{-1})$$

A thioester is thermodynamically more unstable than an oxygen ester because the electrons of the C=O bond cannot form resonance structures with the C–S bond that are as stable as those that they can form with the C–O bond.

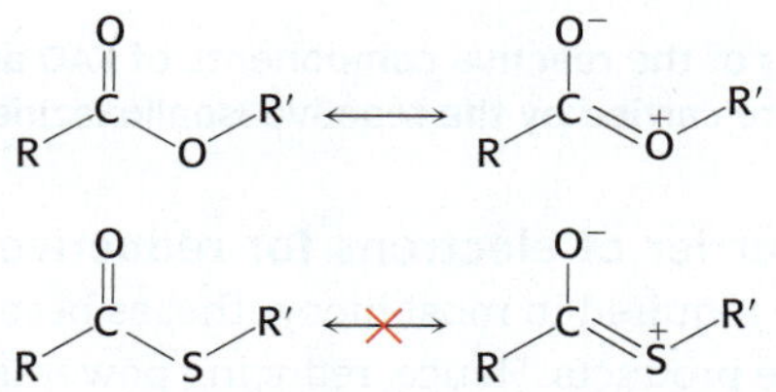

Oxygen esters are stabilized by resonance structures not available to thioesters.

Consequently, acetyl CoA has a high acetyl-group-transfer potential because transfer of the acetyl group is exergonic. Acetyl CoA carries an activated acetyl group, just as ATP carries an activated phosphoryl group.

The use of activated carriers illustrates two key aspects of metabolism:

1. *Kinetic stability allows enzymatic control over the flow of energy.* First, NADH, NADPH, and FADH$_2$ react slowly with O$_2$ in the absence of a catalyst. Likewise, ATP and acetyl CoA are hydrolyzed slowly (over many hours or even days) in the absence of a catalyst. These molecules are kinetically quite stable in the face of a large thermodynamic driving force for reaction with O$_2$ (in regard to the electron carriers) and H$_2$O (for ATP and acetyl CoA). The kinetic stability of these molecules in the absence of specific catalysts is essential for their biological function because it enables enzymes to control the flow of free energy and reducing power.

TABLE 15.2 Some activated carriers in metabolism

Carrier molecule in activated form	Group carried	Vitamin precursor
ATP	Phosphoryl	
NADH and NADPH	Electrons	Nicotinate (niacin) (vitamin B_3)
$FADH_2$ and $FMNH_2$	Electrons	Riboflavin (vitamin B_2)
Coenzyme A	Acyl	Pantothenate (vitamin B_5)
Lipoamide	Acyl	
Thiamine pyrophosphate	Aldehyde	Thiamine (vitamin B_1)
Biotin	CO_2	Biotin (vitamin B_7)
Tetrahydrofolate	One-carbon units	Folate (vitamin B_9)
S-Adenosylmethionine	Methyl	
Uridine diphosphate glucose	Glucose	
Cytidine diphosphate diacylglycerol	Phosphatidate	
Nucleoside triphosphates	Nucleotides	

Note: Many of the activated carriers are coenzymes that are derived from water-soluble vitamins.

2. *A small set of carriers accomplishes the majority of the exchanges of activated groups in metabolic pathways* (**Table 15.2**). The existence of a recurring "cast of characters"—a small set of activated carriers in all organisms—illustrates the modular design of metabolism. A small set of molecules carries out a very wide range of tasks.

Many activated carriers are derived from vitamins

As noted in Table 15.2, most activated carriers that act as coenzymes are derived from **vitamins**, organic molecules that are needed in small amounts in the diets of some higher animals. In all cases, a vitamin must be modified before it can serve its function. **Table 15.3** lists the vitamins that act as coenzymes, and **Figure 15.16** shows the structures of some of B group vitamins. Not all vitamins function as coenzymes; vitamins designated by the letters A, C, D, E, and K (**Table 15.4**) have a diverse array of functions that we will encounter, along with the B vitamins, in our study of biochemistry.

TABLE 15.3 The B vitamins

Vitamin	Coenzyme	Typical reaction Type	Consequences of Deficiency
Thiamine (B_1)	Thiamine pyrophosphate (TPP)	Aldehyde transfer	Beriberi (weight loss, heart problems, neurological dysfunction)
Riboflavin (B_2)	Flavin adenine dinucleotide (FAD)	Oxidation–reduction	Lesions of the mouth, dermatitis
Pyridoxine (B_6)	Pyridoxal phosphate (PLP)	Group transfer to or from amino acids	Depression, confusion, convulsions
Nicotinic acid (niacin) (B_3)	Nicotinamide adenine dinucleotide (NAD^+)	Oxidation–reduction	Pellagra (dermatitis, depression, diarrhea)
Pantothenic acid (B_5)	Coenzyme A (CoA)	Acyl-group transfer	Hypertension
Biotin (B_7)	Biotin–lysine adducts (biocytin)	ATP-dependent carboxylation and carboxyl-group transfer	Rash about the eyebrows, muscle pain, fatigue (rare)
Folic acid (B_9)	Tetrahydrofolate (THF)	Transfer of one-carbon components; thymine synthesis	Anemia, neural-tube defects in development
Cobalamin (B_{12})	5′-Deoxyadenosyl- cobalamin	Transfer of methyl groups; intramolecular rearrangements	Anemia, pernicious anemia, methylmalonic acidosis

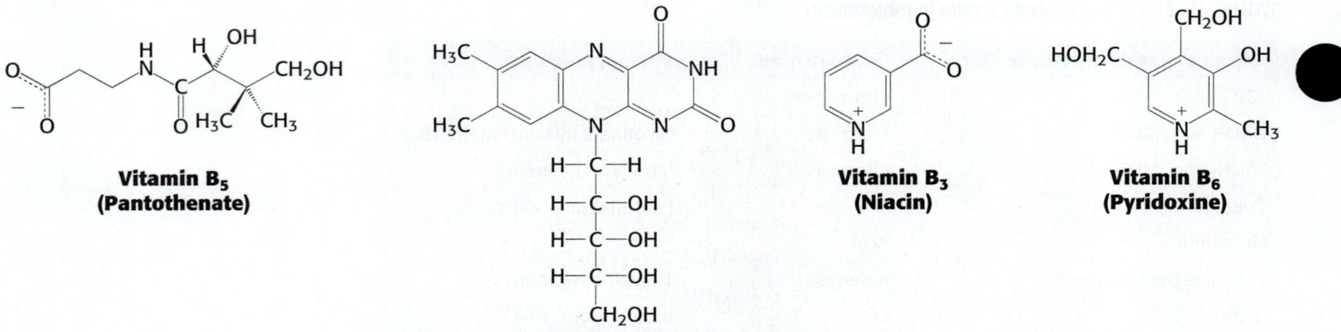

Vitamin B₅ (Pantothenate)

Vitamin B₂ (Riboflavin)

Vitamin B₃ (Niacin)

Vitamin B₆ (Pyridoxine)

FIGURE 15.16 The B vitamins are a diverse group of small, water-soluble molecules. These vitamins, along with vitamin C, are often referred to as water-soluble vitamins because of the ease with which they dissolve in water.

TABLE 15.4 Noncoenzyme vitamins

Vitamin	Function	Deficiency
A	Roles in vision, growth, reproduction	Night blindness, cornea damage, damage to respiratory tract and gastrointestinal tract
C	Antioxidant, essential for collagen production	Scurvy (swollen and bleeding gums, subdermal hemorrhaging)
D	Regulation of calcium and phosphate metabolism	Rickets (children): skeletal deformities, impaired growth
		Osteomalacia (adults): soft, bending bones
E	Antioxidant	Lesions in muscles and nerves (rare)
K	Blood coagulation	Subdermal and cerebral hemorrhaging

Vitamins serve the same roles in nearly all forms of life, but higher animals lost the capacity to synthesize them in the course of evolution. For instance, whereas E. coli can thrive on glucose and organic salts, human beings require at least 12 vitamins in their diet. The biosynthetic pathways for vitamins can be complex; thus, it is biologically more efficient to ingest vitamins than to synthesize the enzymes required to construct them from simple molecules. This efficiency comes at the cost of dependence on other organisms for chemicals essential for life. Indeed, vitamin deficiency can generate diseases in all organisms requiring these molecules (Tables 15.3 and 15.4).

Key reactions are reiterated throughout metabolism

Just as there is an economical use of the same small set of activated carriers, there is also an economical use of a small set of biochemical reactions. In fact, the thousands of metabolic reactions can be subdivided into just six types (Table 15.5). These six types of reactions correspond to the six major classes of enzymes discussed in Section 5.1. Specific reactions of each type appear repeatedly.

TABLE 15.5 Types of chemical reactions in metabolism

Type of reaction	Description
Oxidation–reduction	Electron transfer
Group transfer	Transfer of a functional group from one molecule to another
Hydrolytic	Cleavage of bonds by the addition of water
Carbon bond cleavage by means other than hydrolysis or oxidation	Two substrates yielding one product or vice versa. When H_2O or CO_2 are a product, a double bond is formed.
Isomerization	Rearrangement of atoms to form isomers
Ligation requiring ATP cleavage	Formation of covalent bonds (i.e., carbon–carbon bonds)

1. *Oxidation–reduction reactions are essential components of many pathways.* Useful energy is often derived from the oxidation of carbon compounds. Consider the following two reactions:

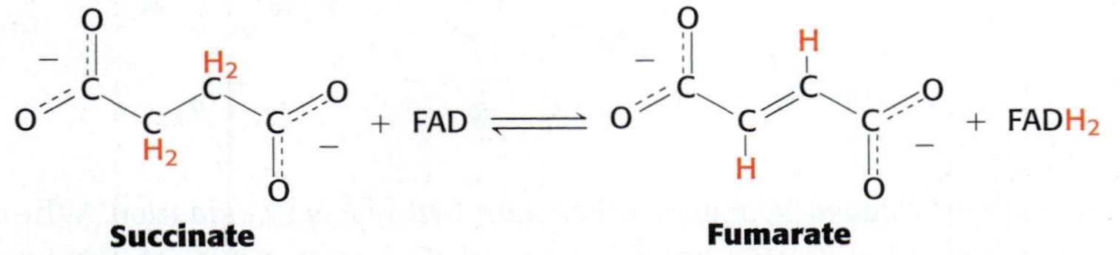

$$\text{Succinate} + \text{FAD} \rightleftharpoons \text{Fumarate} + \text{FADH}_2 \quad (1)$$

$$\text{Malate} + \text{NAD}^+ \rightleftharpoons \text{Oxaloacetate} + \text{NADH} + \text{H}^+ \quad (2)$$

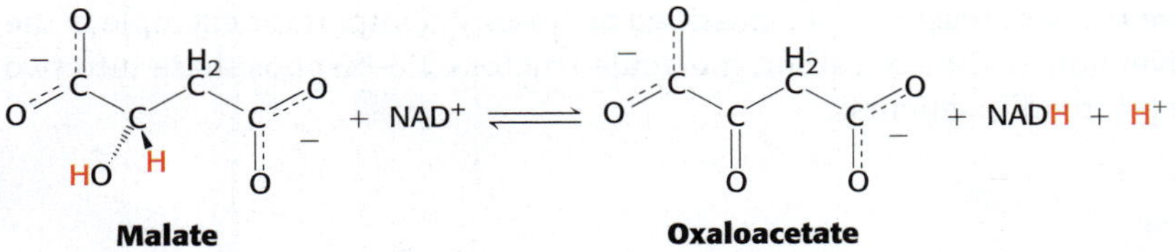

These two oxidation–reduction reactions are components of the citric acid cycle (Chapter 17), which completely oxidizes the activated two-carbon fragment of acetyl CoA to two molecules of CO_2. In reaction 1, $FADH_2$ carries the electrons, whereas in reaction 2, electrons are carried by NADH.

2. *Group-transfer reactions play a variety of roles.* Phosphoryl group transfer is representative of such a reaction:

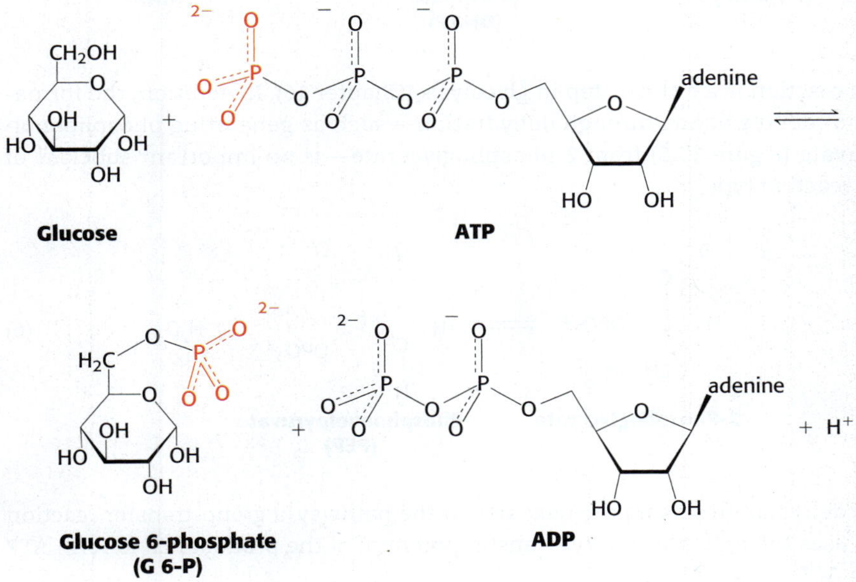

$$\text{Glucose} + \text{ATP} \rightleftharpoons \text{Glucose 6-phosphate (G 6-P)} + \text{ADP} + \text{H}^+ \quad (3)$$

A phosphoryl group is transferred from the activated phosphoryl-group carrier, ATP, to glucose in the initial step in glycolysis, a key pathway for extracting energy from glucose (Chapter 16). This reaction traps glucose in the cell so that further catabolism can take place. As stated earlier, group-transfer reactions are used to synthesize ATP. We also saw examples of their use in signaling pathways (Chapter 14).

3. *Hydrolytic reactions cleave bonds by the addition of water.* Hydrolysis is a common means of degrading large molecules, either to facilitate further metabolism or to reuse some of the components for biosynthetic purposes. For example, recall from Section 6.1 that proteins are digested by hydrolytic cleavage:

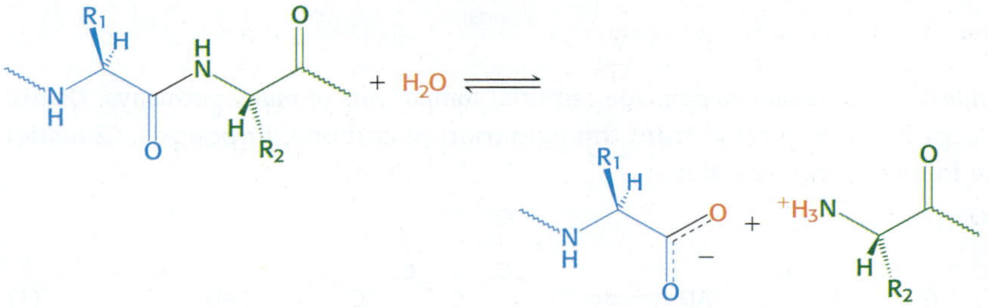

$$(4)$$

4. *Carbon bonds can be cleaved by means other than hydrolysis or oxidation.* When CO_2 or H_2O is released, a double bond is formed. The enzymes that catalyze these types of reactions are classified as *lyases.* An important example is the conversion of the six-carbon molecule fructose 1,6-bisphosphate into two three-carbon fragments:

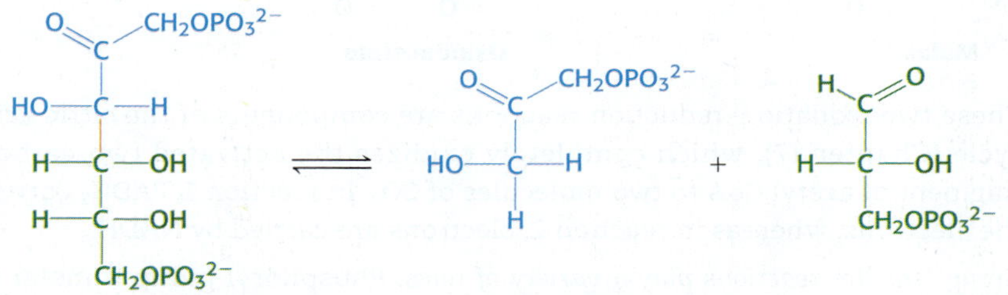

$$(5)$$

Fructose 1,6-bisphosphate (F 1,6-BP)   **Dihydroxyacetone phosphate (DHAP)**   **Glyceraldehyde 3-phosphate (GAP)**

This reaction is a critical step in glycolysis (Chapter 16). In addition, the formation of double bonds through dehydration—such as generating phosphoenolpyruvate (Figure 15.5) from 2-phosphoglycerate—is an important subclass of this reaction type:

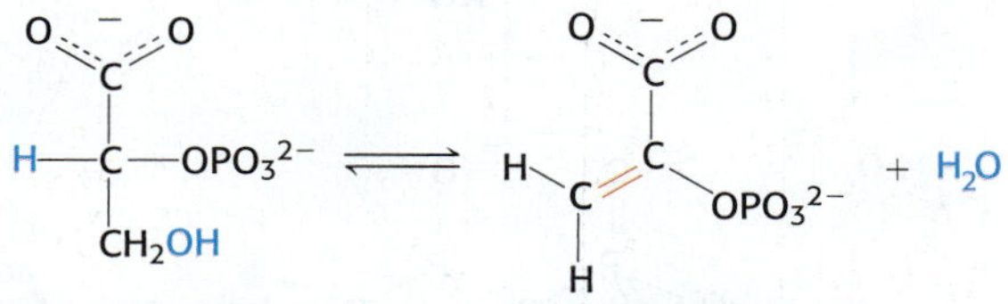

$$(6)$$

2-Phosphoglycerate   **Phosphoenolpyruvate (PEP)**

This dehydration sets up the next step in the pathway, a group-transfer reaction that uses the high phosphoryl-transfer potential of the product PEP to form ATP from ADP.

5. *Isomerization reactions rearrange particular atoms within a molecule.* The role of these reactions is often to prepare the molecule for subsequent reactions such as the oxidation–reduction reactions described above.

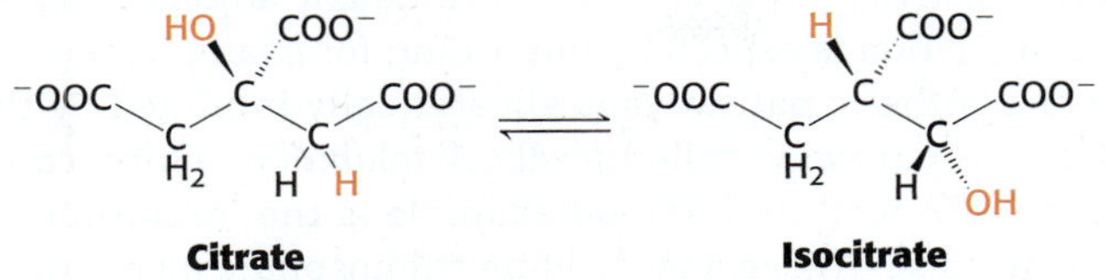

Citrate   **Isocitrate**   (7)

The reaction above is an essential component of the citric acid cycle, analyzed in detail in Chapter 17.

6. *Ligation reactions form bonds by using free energy from ATP hydrolysis.* In many cases, the free energy from ATP hydrolysis is necessary to combine smaller molecules to form larger ones. For example, oxaloacetate is formed from pyruvate and CO_2, using energy from ATP to form the carbon–carbon bond.

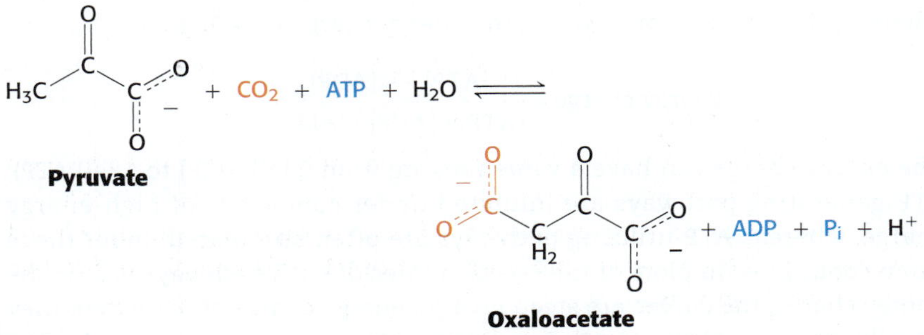

Pyruvate

Oxaloacetate $+ \; ADP \; + \; P_i \; + \; H^+$ (8)

The oxaloacetate can be used in the citric acid cycle, or converted into glucose or amino acids such as aspartic acid.

These six fundamental reaction types are the basis of metabolism; you will encounter them repeatedly in the next few chapters. Remember, all six types can proceed in either direction, depending on the standard free energy for the specific reaction and the concentrations of the reactants and products inside the cell.

Metabolic processes are regulated in three principal ways

Because the network of metabolic reactions is so complex, it must be rigorously regulated. Cells have feedback mechanisms that monitor and adjust levels of available nutrients and the activity of metabolic pathways to create *homeostasis*, a stable biochemical environment. At the same time, metabolic control must be flexible, able to adjust to the constantly changing external environments of cells. Metabolism is regulated through three principal mechanisms:

1. *The amounts of enzymes can be altered.* The amount of a particular enzyme depends on both its rate of synthesis and its rate of degradation. For many enzymes, their abundance is principally adjusted by a change in the rate of transcription of the genes encoding them (Section 29.3). In E. coli, for example, within minutes the presence of lactose induces a more-than-50-fold increase in the rate of synthesis of β-galactosidase, the enzyme required for the breakdown of this disaccharide.

2. *The accessibility of substrates by enzymes can be restricted.* Limiting the availability of substrates is another means of regulating metabolism in all organisms. For instance, glucose breakdown can take place in many cells only if insulin is present to promote the entry of glucose into the cell. In eukaryotes, access to substrates is regulated in part by compartmentalization. For example, fatty acid oxidation takes place in mitochondria, whereas fatty

acid synthesis takes place in the cytoplasm. Compartmentalization often segregates opposed reactions.

3. *The catalytic activity of enzymes can be internally regulated or externally controlled.* The catalytic activity of enzymes is continuously adjusted in several ways. Allosteric regulation is especially important; for example, the first reaction in many biosynthetic pathways is allosterically inhibited by the ultimate product of the pathway, called *feedback inhibition*, which can be almost instantaneous. A well-understood example is the inhibition of aspartate transcarbamoylase (ATCase) by cytidine triphosphate (Section 7.1). Another recurring mechanism is reversible covalent modification (Section 7.3). Hormones like adrenaline externally control the catalytic rates of enzymes often by inducing reversible covalent modification of key enzymes or by changing the concentrations of allosteric activators and inhibitors, as we explored in Chapter 14. Finally, many reactions are regulated by the energy status of the cell. One useful concept that can serve as an index of the energy status is the **energy charge**, which is proportional to the mole fraction of ATP plus half the mole fraction of ADP, given that ATP contains two anhydride bonds, whereas ADP contains one. Hence, the energy charge is defined as

$$\text{Energy charge} = \frac{[\text{ATP}] + \frac{1}{2}[\text{ADP}]}{[\text{ATP}] + [\text{ADP}] + [\text{AMP}]}$$

The energy charge can have a value ranging from 0 (all AMP) to 1 (all ATP). ATP-generating pathways are inhibited under conditions of high-energy charge, whereas ATP-utilizing pathways are often stimulated under those same conditions. In plots of the reaction rates of such pathways versus the energy charge, the curves are steep near an energy charge of 0.9, where they usually intersect (**Figure 15.17**). Regulation of these pathways has evolved to maintain the energy charge within rather narrow limits. In other words, the energy charge, like the pH of a cell, is buffered. The energy charge of most cells ranges from 0.90 to 0.95, but can fall to less than 0.7 in muscle during high-intensity exercise. However, energy charge itself does not directly regulate pathways; rather, enzymes regulating these pathways are allosterically inhibited or activated by binding to ATP or AMP, specifically.

Looking at the commonalities in diverse metabolic pathways exposes a chemical logic that renders the complexity of the chemistry of living systems more manageable and reveals its elegance. As you delve into any of the next 12 chapters, you might find it useful to return to this chapter to review the common themes.

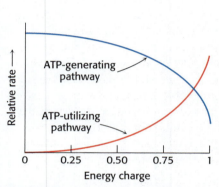

FIGURE 15.17 Energy charge is a useful concept for understanding how metabolism is regulated. When the energy charge is high, ATP commonly inhibits the relative rates of a typical ATP-generating pathway and stimulates the typical ATP-utilizing pathway, while AMP has the opposite effects.

EXPLORE this pathway further in the Metabolic Map in
 Achieve

❖ SELF–CHECK QUESTION

Name the common cellular carrier of activated two-carbon fragments, and name the common carrier of electrons for biosynthesis. Both these molecules have parts derived from which group of small organic molecules that are precursors to coenzymes?

Summary

15.1 Metabolism Is Composed of Many Interconnected Reactions

- All cells transform energy. They extract energy from their environment and use this energy to convert simple molecules into cellular components.

- The process of energy transduction takes place through metabolism, a highly integrated network of chemical reactions.

- Metabolism can be subdivided into catabolism (processes that break complex things into simpler ones and typically extract energy from fuels) and anabolism (biosynthetic processes that typically require energy).

- The most valuable thermodynamic concept for understanding bioenergetics is free energy. A reaction can occur spontaneously only if the change in free energy (ΔG) is negative.

- A thermodynamically unfavorable reaction can be driven by a thermodynamically favorable one, which is the hydrolysis of ATP in many cases.

15.2 ATP Is the Universal Currency of Free Energy in Biological Systems

- The energy derived from catabolism is captured in the form of adenosine triphosphate (ATP).
- ATP hydrolysis is exergonic and the energy released can be used to power cellular processes, including motion, active transport, and biosynthesis.
- ATP, the universal currency of energy in biological systems, is an energy-rich molecule because it contains two kinetically stable, but thermodynamically unstable, phosphoanhydride linkages.

15.3 The Oxidation of Carbon Fuels Is an Important Source of Cellular Energy

- ATP formation is coupled to the oxidation of chemicals, either directly or through the formation of ion gradients in chemotrophs. Phototrophic organisms, including all photosynthetic organisms, can use light to generate such gradients.
- The extraction of energy from food by aerobic organisms comprises three stages: (1) large molecules are catabolized into smaller ones, such as amino acids, sugars, and fatty acids;

(2) these small molecules are degraded to a few simple units, such as acetyl CoA; (3) the citric acid cycle and oxidative phosphorylation generate ATP as electrons flow to O_2, the ultimate electron acceptor, and fuels are completely oxidized to CO_2.

15.4 Metabolic Pathways Contain Many Recurring Motifs

- Metabolism is characterized by common motifs. A small number of recurring activated carriers, such as ATP, NADH, and acetyl CoA, transfer activated groups in many metabolic pathways. Many activated carriers are derived from vitamins—small organic molecules required in the diets of many higher organisms.
- NADPH, which carries two electrons at a high potential, provides reducing power for reductive biosynthesis of cell components.
- Key reaction types are used repeatedly in metabolic pathways. Metabolism is regulated in a variety of ways: the amounts of enzymes are altered, the catalytic activities of many enzymes are regulated by allosteric interactions and controlled by covalent modification, and the availability of many substrates is also limited.
- The energy charge, which is a measure of the relative amounts of ATP, ADP, and AMP, is a useful concept for approximating the energy status of a cell. ATP generally inhibits ATP-generating pathways and stimulates ATP-utilizing pathways, while AMP often has the opposite effects.

Key Terms

phototroph (p. 447)

chemotroph (p. 447)

metabolism (intermediary metabolism) (p. 447)

adenosine triphosphate (ATP) (p. 448)

catabolism (p. 448)

anabolism (p. 448)

amphibolic pathway (p. 448)

phosphoryl-transfer potential (p. 453)

activated carrier (p. 460)

vitamin (p. 463)

energy charge (p. 468)

Problems

1. Differentiate between anabolism and catabolism. ❖ 1

2. Distinguish between a phototroph and a chemotroph. ❖ 1

3. What are the three major stages in the energy extraction from food? Assess the contribution of each stage to ATP synthesis. ❖ 1

4. What are the three primary uses for cellular energy?

5. What factors account for the high phosphoryl-transfer potential of nucleoside triphosphates? ❖ 2

6. The standard free energy of hydrolysis for ATP is −30.5 kJ mol⁻¹ (−7.3 kcal mol⁻¹).

$$ATP + H_2O \rightleftharpoons ADP + P_i$$

How might physiological conditions, such as rest, intense

exercise, or an alteration of intracellular ionic environment, alter the free energy of hydrolysis? ❖ **3**

7. What is the direction of each of the following reactions when the reactants are initially present in equimolar amounts? Use the data given in Table 15.1. ❖ **3**

(a) ATP + creatine ⇌ creatine phosphate + ADP
(b) ATP + glycerol ⇌ glycerol 3-phosphate + ADP
(c) ATP + pyruvate ⇌ phosphoenolpyruvate + ADP
(d) ATP + glucose ⇌ glucose 6-phosphate + ADP

8. Consider the following reaction:

ATP + pyruvate ⇌ phosphoenolpyruvate + ADP

(a) Calculate $\Delta G^{\circ\prime}$ and K'_{eq} at 25°C for this reaction by using the data given in Table 15.1.

(b) What is the equilibrium ratio of pyruvate to phosphoenolpyruvate if the ratio of ATP to ADP is 10? ❖ **3**

9. Using the information in Table 15.1, calculate $\Delta G^{\circ\prime}$ for the isomerization of glucose 6-phosphate to glucose 1-phosphate. What is the equilibrium ratio of glucose 6-phosphate to glucose 1-phosphate at 25°C?

10. The formation of acetyl CoA from acetate is an ATP-driven reaction:

Acetate + ATP + CoA ⇌ acetyl CoA + AMP + PP_i ❖ **3**

(a) Calculate $\Delta G^{\circ\prime}$ for this reaction by using data given in this chapter.

(b) The PP_i formed in the preceding reaction is rapidly hydrolyzed in vivo because of the ubiquity of inorganic pyrophosphatase. The $\Delta G^{\circ\prime}$ for the hydrolysis of PP_i is −19.2 kJ mol^{-1}(−4.6 kcal mol^{-1}). Calculate the $\Delta G^{\circ\prime}$ for the overall reaction, including pyrophosphate hydrolysis. What effect does the hydrolysis of PP_i have on the formation of acetyl CoA? ❖ **3**, ❖ **5**

11. The muscles of some invertebrates are rich in arginine phosphate (phosphoarginine). Propose a function for this amino acid derivative. ❖ **2**

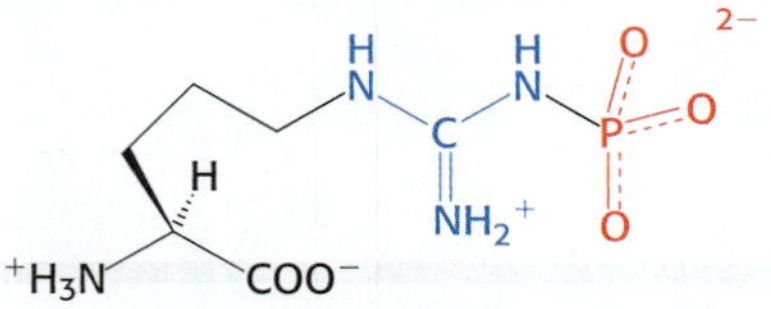

Arginine phosphate

12. What is the structural feature common to ATP, FAD, NAD$^+$, and CoA? ❖ **5**

13. Creatine is a popular dietary supplement.
(a) What is the biochemical rationale for the use of creatine?
(b) What type of exercise would benefit most from creatine supplementation? ❖ **3**

14. The enzyme aldolase catalyzes the following reaction in the glycolytic pathway:

Fructose 1, 6-bisphosphate $\xrightleftharpoons{\text{Aldolase}}$

dihydroxyacetone phosphate +

glyceraldehyde 3-phosphate

The $\Delta G^{\circ\prime}$ for the reaction is +23.8 kJ mol^{-1} (+5.7 kcal mol^{-1}), whereas the ΔG in the cell is −1.3 kJ mol^{-1} (−0.3 kcal mol^{-1}). Calculate the ratio of reactants to products under equilibrium and intracellular conditions. Using your results, explain how the reaction can be endergonic under standard conditions and exergonic under intracellular conditions.

15. The concentrations of ATP, ADP, and P_i differ with cell type. Consequently, the release of free energy with the hydrolysis of ATP will vary with cell type. Using the following table, calculate the ΔG for the hydrolysis of ATP in liver, muscle, and brain cells. In which cell type is the free energy of ATP hydrolysis most negative? ❖ **2**

	ATP (mM)	ADP (mM)	P_i (mM)
Liver	3.5	1.8	5.0
Muscle	8.0	0.9	8.0
Brain	2.6	0.7	2.7

16. Examine the pairs of molecules and identify the more-reduced molecule in each pair. ❖ **4**

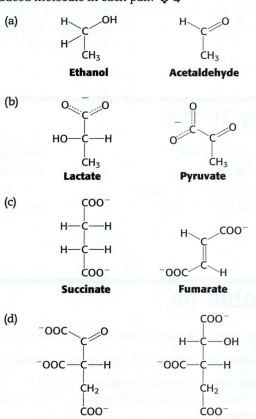

(a) Ethanol | Acetaldehyde

(b) Lactate | Pyruvate

(c) Succinate | Fumarate

(d) Oxalosuccinate | Isocitrate

(e)

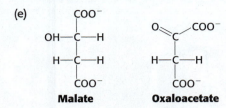

Malate   **Oxaloacetate**

17. What are the activated electron carriers for catabolism? For anabolism? ❖ **5**

18. What are the six common types of reactions seen in biochemistry? ❖ **5**

19. What are the three principal means of controlling metabolic reactions described in this chapter? ❖ **5**

20. The following graph shows how the ΔG for the hydrolysis of ATP varies as a function of the Mg^{2+} concentration, $pMg = -\log[Mg^{2+}]$. ❖ **2**

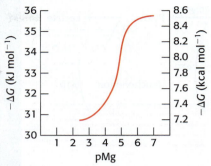

(a) How does decreasing $[Mg^{2+}]$ affect the ΔG of hydrolysis for ATP? **(b)** Explain this effect.

16 Glycolysis and Gluconeogenesis

Glucose metabolism can generate the ATP to power muscle contraction. During a sprint, when the ATP needs outpace oxygen delivery, glucose is metabolized to lactate (pathway A). When oxygen delivery is adequate, glucose is metabolized more efficiently to carbon dioxide and water (pathway B).

Mauro Ujetto/NurPhoto/Getty Images

OUTLINE

16.1 Glycolysis Is an Energy-Conversion Pathway in Most Organisms

16.2 Glycolysis Can Be Divided into Two Parts

16.3 The Glycolytic Pathway Is Tightly Controlled

16.4 Glucose Can Be Synthesized from Noncarbohydrate Precursors

16.5 Gluconeogenesis and Glycolysis Are Reciprocally Regulated

❖ LEARNING GOALS

By the end of this chapter, you should be able to:

1. Describe how ATP is generated in glycolysis.

2. Explain why the regeneration of NAD^+ is crucial to fermentations.

3. Describe how gluconeogenesis is powered in the cell.

4. Describe the coordinated regulation of glycolysis and gluconeogenesis.

Our understanding of carbohydrate metabolism has a rich history. Indeed, the development of biochemistry and the delineation of the metabolic pathways that begin with glucose went hand in hand. A key discovery was made in 1897 by a pair of biochemist brothers, Hans and Eduard Buchner, quite by accident. The Buchners were interested in manufacturing cell-free extracts of baker's yeast for possible therapeutic use. These extracts had to be preserved without the use of antiseptics such as phenol, and so they decided to try sucrose, a commonly used preservative in kitchen chemistry. They obtained a startling result: sucrose was rapidly fermented into ethanol by the yeast juice. The Buchners demonstrated for the first time that fermentation could take place outside living cells.

The significance of this finding was immense. The accepted view of their day, asserted by Louis Pasteur in 1860, was that fermentation is inextricably tied to living cells. The chance discovery by the Buchners refuted this dogma. The Buchners' discovery inspired the search for the biochemicals that catalyze the conversion of sucrose into alcohol and opened the door to modern biochemistry.

16.1 Glycolysis Is an Energy-Conversion Pathway in Most Organisms

The first metabolic pathway that we encounter is glycolysis, an ancient energy-conversion pathway employed by a host of organisms. **Glycolysis** is the sequence of reactions that metabolizes one molecule of glucose to two molecules of pyruvate with the concomitant net production of two molecules of ATP. This process is anaerobic, meaning that it does not require O_2, because it evolved before substantial amounts of oxygen accumulated in the atmosphere. Glycolysis is common to virtually all cells, both prokaryotic and eukaryotic. The complete glycolytic pathway is also known as the *Embden-Meyerhof-Parnas pathway*, named after three pioneers of research on glycolysis. As observed accidentally by the Buchners, pyruvate can be further processed anaerobically by fermentation to lactic acid, ethanol, or a variety of other final products depending on the organism. Under aerobic conditions, pyruvate can be completely oxidized to CO_2, generating much more ATP, as will be described in Chapters 17 and 18. **Figure 16.1** shows a few possible fates of pyruvate produced by glycolysis.

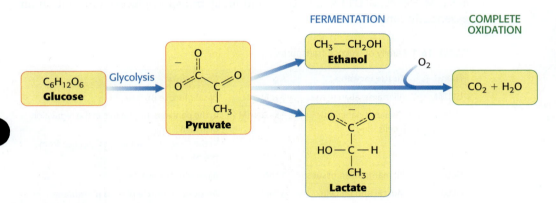

FIGURE 16.1 Glucose can be metabolized to different products depending on the organism and the presence or absence of oxygen.

Recall from Chapter 11 that glucose is a common but yet precious fuel for almost all organisms. In mammals, it is the only fuel that the brain uses under nonstarvation conditions and the only fuel that red blood cells can use at all. Because glucose is such a precious fuel, metabolic products, such as pyruvate and lactate, are salvaged to synthesize glucose in the process of **gluconeogenesis**, but the majority of glucose circulating in our blood each day originates directly from our diet.

Glucose is generated from dietary carbohydrates

In our diets, we typically consume a generous amount of glucose in the form of starch and a smaller amount as glycogen. These complex carbohydrates must be converted into simpler carbohydrates for absorption by the intestine and transport in the blood. Starch and glycogen are digested primarily by the pancreatic and salivary enzyme α-*amylase*. Amylase cleaves the α-1,4 bonds of starch and glycogen, but not the α-1,6 bonds. The products are the di- and trisaccharides maltose and maltotriose. The material not digestible because of the α-1,6 bonds is called the *limit dextrin*.

α-Glucosidase (maltase) cleaves maltose and maltotriose, as well as any other α-1,4-linked oligosaccharides that may have escaped digestion by the amylase, into glucose molecules. α-Dextrinase further digests the limit dextrin.

α-Glucosidase is located on the surface of the intestinal cells, as are sucrase and lactase. Sucrase degrades ingested sucrose to fructose and glucose, while lactase is responsible for degrading the milk sugar lactose into glucose and galactose. The monosaccharides are transported into the endothelial cells lining the intestine by the action of active transporters. Glucose then moves passively down its concentration gradient out of endothelial cells and into the bloodstream, and again passively into cells that will finally utilize it. How is this directional movement from one tissue to another accomplished?

A family of transporters enables glucose to enter and leave animal cells

Several glucose transporters (named GLUT1 to GLUT5) mediate the thermodynamically downhill movement of glucose across the plasma membranes of animal cells. Although all have a 12-transmembrane-helix structure similar to that of lactose permease (Section 13.3), the members of this family have distinctive roles, summarized in **Table 16.1**. GLUT1 and GLUT3, present in nearly all mammalian cells, are responsible for basal glucose uptake. Their K_M value for glucose is about 1 mM, significantly less than the normal blood serum-glucose range of 4–8 mM. Hence, GLUT1 and GLUT3 continually transport glucose into cells at an essentially constant rate.

TABLE 16.1 Family of glucose transporters

Name	Tissue location	K_M	Comments
GLUT1	All mammalian tissues	1 mM	Basal glucose uptake
GLUT2	Liver and pancreatic β cells	15–20 mM	In the pancreas, plays a role in the regulation of insulin In the liver, removes excess glucose from the blood
GLUT3	All mammalian tissues	1 mM	Basal glucose uptake
GLUT4	Muscle and fat cells	5 mM	Amount in muscle plasma membrane increases with endurance training and in response to insulin
GLUT5	Small intestine	—	Primarily a fructose transporter

As described in Section 14.3, the number of GLUT4 transporters in the plasma membrane of muscle and fat cells increases rapidly in the presence of insulin, which signals the fed state. Thus, insulin promotes the uptake of glucose by muscle and fat. Endurance exercise training also increases the amount of this transporter present in muscle membranes. GLUT2 is present in liver and pancreatic β cells but is distinctive in having a very high K_M value for glucose (15–20 mM). Because of this, glucose enters these tissues at a biologically significant rate only when there is much glucose in the blood. The pancreas can sense the glucose level and accordingly adjust the rate of insulin secretion. Lastly, while GLUT5 can also transport glucose, it functions primarily as a fructose transporter in the small intestine. Once finally in the cytoplasm of a glycolytically active cell, glucose and other monosaccharides are ready to enter the glycolytic pathway.

16.2 Glycolysis Can Be Divided into Two Parts

The 10 enzyme-catalyzed reactions that make up the glycolytic pathway can be thought of as comprising two stages (**Figure 16.2**). Stage 1 is the trapping and preparation phase. No ATP is generated in this stage. In stage 1, glucose is

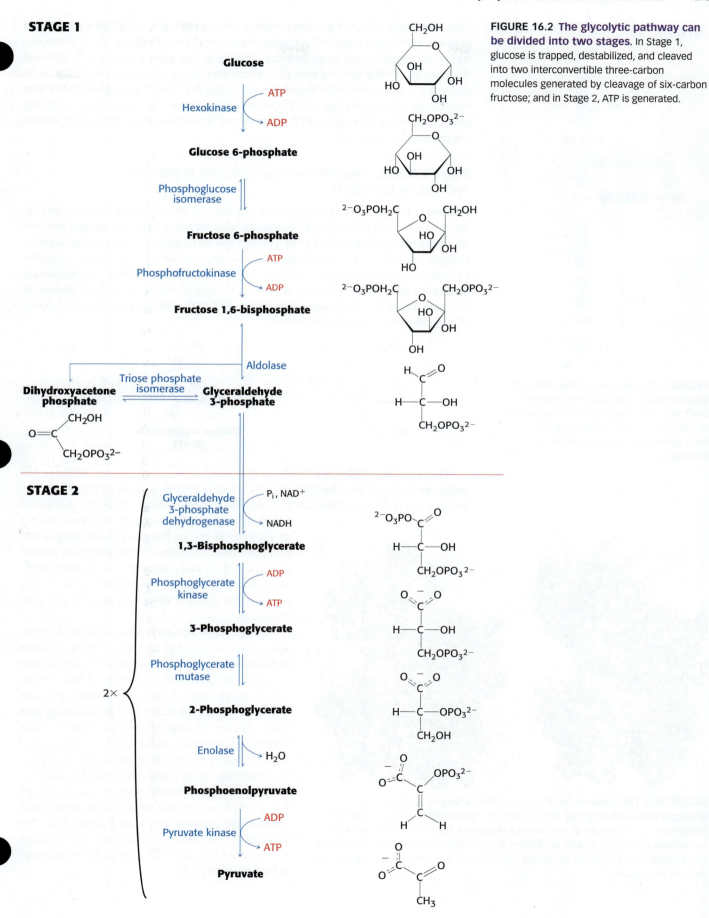

STAGE 1

STAGE 2

FIGURE 16.2 The glycolytic pathway can be divided into two stages. In Stage 1, glucose is trapped, destabilized, and cleaved into two interconvertible three-carbon molecules generated by cleavage of six-carbon fructose; and in Stage 2, ATP is generated.

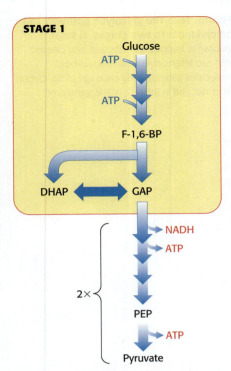

STAGE 1

Glucose

ATP

ATP

F-1,6-BP

DHAP   GAP

2×

NADH

ATP

PEP

ATP

Pyruvate

FIGURE 16.3 The first or "investment" stage of glycolysis uses the energy from two molecules of ATP. The products of this stage are two molecules of GAP for every molecule of glucose entering the pathway.

converted into fructose 1,6-bisphosphate in three steps: a phosphorylation, an isomerization, and a second phosphorylation reaction. The strategy of these initial steps in glycolysis is to trap the glucose in the cell and form a compound that can be readily cleaved into phosphorylated three-carbon units (**Figure 16.3**). Stage 1 is completed with the cleavage of the fructose 1,6-bisphosphate into two three-carbon fragments. These resulting three-carbon units are readily interconvertible. In stage 2, ATP is harvested when the three-carbon fragments are oxidized to pyruvate.

Stage 1 begins: Hexokinase traps glucose in the cell and begins glycolysis

Glucose entering a glycolytically active cell has one principal fate: it is phosphorylated by ATP to form glucose 6-phosphate. This step is notable for several reasons. Glucose 6-phosphate cannot pass through the membrane because of the negative charges on the phosphoryl groups, and it is not a substrate for glucose transporters. Also, the addition of the phosphoryl group facilitates the eventual metabolism of glucose into three-carbon molecules with high phosphoryl-transfer potential. The transfer of the phosphoryl group from ATP to the hydroxyl group on carbon 6 of glucose is catalyzed by hexokinase.

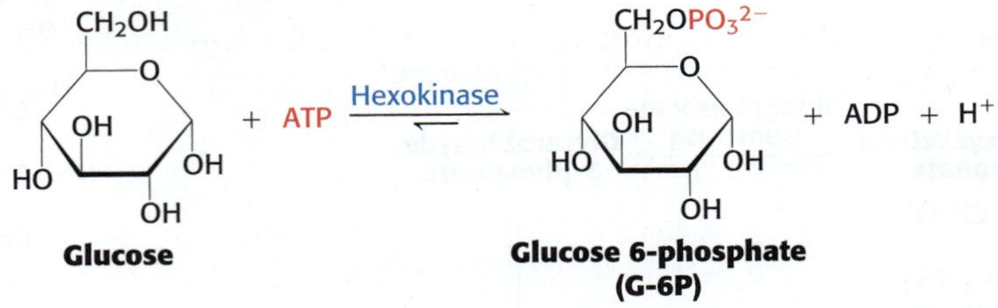

CH_2OH ... $+$ ATP $\xrightarrow{\text{Hexokinase}}$ $CH_2OPO_3^{2-}$ ... $+$ ADP $+$ H$^+$

Glucose   **Glucose 6-phosphate (G-6P)**

Phosphoryl transfer is a fundamental reaction in biochemistry. **Kinases** are enzymes that catalyze the transfer of a phosphoryl group from ATP to an acceptor, or to ADP from a phosphoryl donor. Hexokinase, then, catalyzes the transfer of a phosphoryl group from ATP to a variety of six-carbon sugars (hexoses), such as glucose and mannose. Hexokinase, like adenylate kinase (Section 6.1) and all other kinases, requires Mg^{2+} (or another divalent metal ion such as Mn^{2+}) for activity. The divalent metal ion forms a complex with ATP.

X-ray crystallographic studies of yeast hexokinase revealed that the binding of glucose induces a large conformational change in the enzyme. Hexokinase consists of two lobes, which move toward each other when glucose is bound (**Figure 16.4**). On glucose binding, one lobe rotates 12 degrees with respect to the other, resulting in movements of the polypeptide backbone of as much as 8 Å. The cleft between the lobes closes, and the bound glucose becomes surrounded by protein, except for the hydroxyl group of carbon 6, which will accept the phosphoryl group from ATP. The closing of the cleft in hexokinase is a striking example of the role of induced fit in enzyme action (Section 5.3).

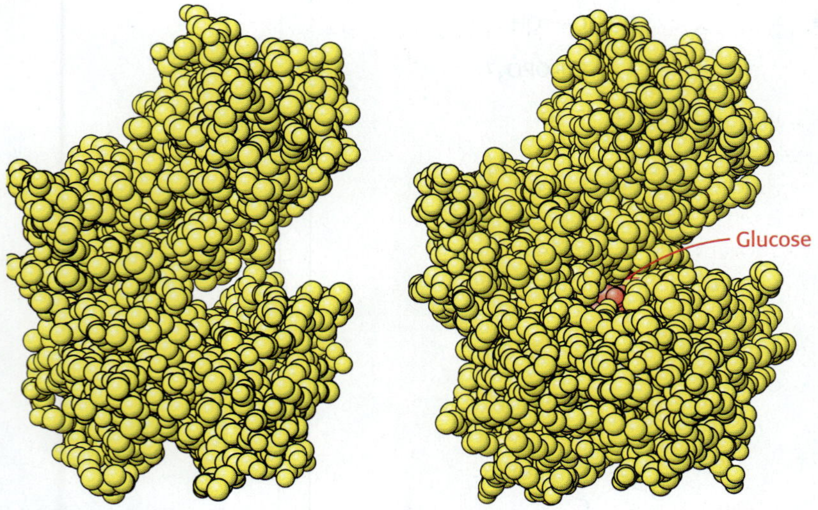

Glucose

FIGURE 16.4 The induced fit of hexokinase is due to large structural changes upon binding the first of its two substrates. The two lobes of hexokinase are separated in the absence of glucose (left). The conformation of hexokinase changes markedly on binding glucose as the two lobes of the enzyme come together, creating the necessary environment for catalysis (right). [Drawn from 2YHX.pdb and 1HKG.pdb.]

The glucose-induced structural changes are significant in two respects. First, the environment around the glucose becomes more nonpolar, which favors reaction between the hydrophilic hydroxyl group of glucose and the terminal phosphoryl group of ATP. Second, the conformational changes enable the kinase to discriminate against H_2O as a substrate. The closing of the cleft keeps water molecules away from the active site. If hexokinase were rigid, a molecule of H_2O occupying the binding site for the —CH_2OH of glucose could attack the γ phosphoryl group of ATP, forming ADP and P_i. In other words, a rigid kinase would likely also be an ATPase.

Fructose 1,6-bisphosphate is generated from glucose 6-phosphate

A crucial step toward completion of the first phase of glycolysis—the formation of fructose 1,6-bisphosphate—is the isomerization of glucose 6-phosphate to fructose 6-phosphate. Recall that the open-chain form of glucose has an aldehyde group at carbon 1, whereas the open-chain form of fructose has a keto group at carbon 2. Thus, the isomerization of glucose 6-phosphate to fructose 6-phosphate is a conversion of an aldose into a ketose. The reaction catalyzed by phosphoglucose isomerase takes several steps because both glucose 6-phosphate and fructose 6-phosphate are present primarily in the cyclic forms. The enzyme must first open the six-membered ring of glucose 6-phosphate, catalyze the isomerization, and then promote the formation of the five-membered ring of fructose 6-phosphate.

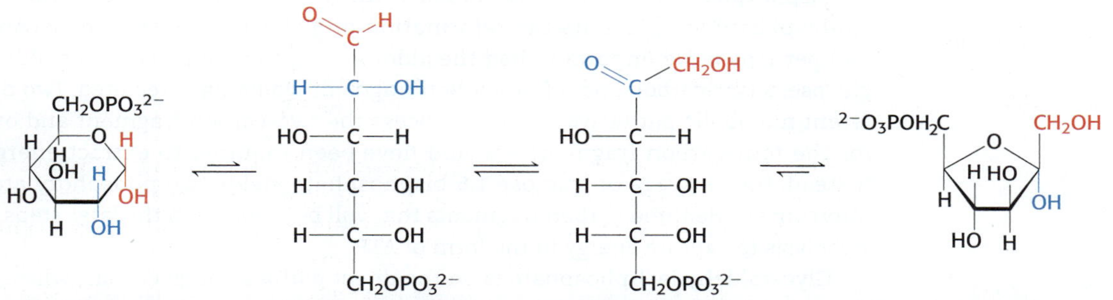

Glucose 6-phosphate (G-6P)   **Glucose 6-phosphate (open-chain form)**   **Fructose 6-phosphate (open-chain form)**   **Fructose 6-phosphate (F-6P)**

A second phosphorylation reaction follows the isomerization step. Fructose 6-phosphate is phosphorylated at the expense of ATP to fructose 1,6-bisphosphate (F-1,6-BP). The prefix *bis*- in bisphosphate means that two separate monophosphoryl groups are present, whereas the prefix *di*- in diphosphate (as in adenosine diphosphate) means that two phosphoryl groups are present and are connected by an anhydride linkage.

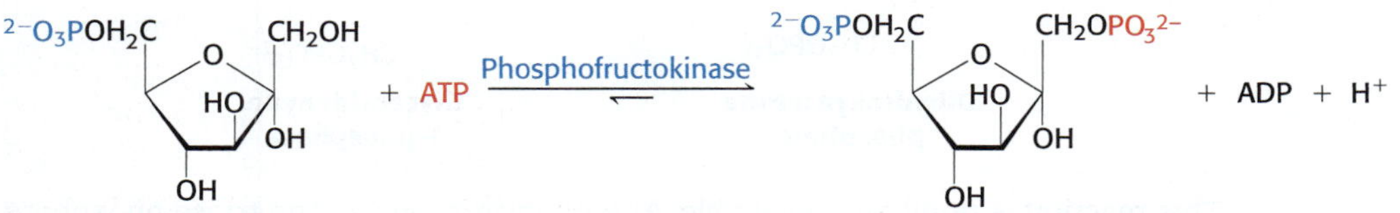

Fructose 6-phosphate (F-6P)   **Fructose 1,6-bisphosphate (F-1,6-BP)**

This reaction is catalyzed by phosphofructokinase (PFK), an allosteric enzyme that sets the pace of glycolysis. As we will learn, this enzyme plays a central role in the metabolism of many molecules in all parts of the body.

The six-carbon sugar is cleaved into two three-carbon fragments

The newly formed fructose 1,6-bisphosphate is cleaved into glyceraldehyde 3-phosphate (GAP) and dihydroxyacetone phosphate (DHAP), completing stage 1 of glycolysis. The products of the remaining steps in glycolysis consist of three-carbon units rather than six-carbon units. This reaction, which is readily reversible, is catalyzed by aldolase. This enzyme derives its name from the nature of the reverse reaction, an aldol addition.

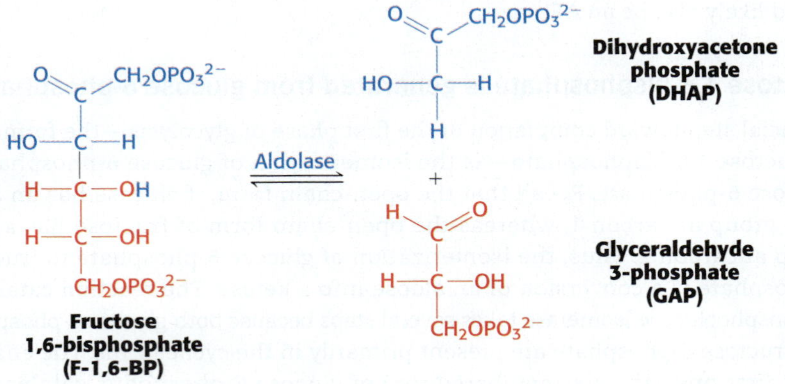

What is the biochemical rationale for the isomerization of glucose 6-phosphate to fructose 6-phosphate and its subsequent phosphorylation to form fructose 1,6-bisphosphate? First, phosphorylation of the fructose 6-phosphate to fructose 1,6-bisphosphate prevents the reformation of glucose 6-phosphate. Second, and perhaps more important, had the aldol cleavage taken place in the aldose glucose, a two-carbon and a four-carbon fragment would have resulted. Two different metabolic pathways—one to process the two-carbon fragment and one for the four-carbon fragment—would have been required to extract energy. Instead, the cleavage of fructose 1,6-bisphosphate yields two phosphorylated interconvertible three-carbon fragments that will be oxidized in the later steps of glycolysis to capture energy in the form of ATP.

Glyceraldehyde 3-phosphate is on the direct pathway of glycolysis, whereas dihydroxyacetone phosphate is not. Unless a means exists to convert dihydroxyacetone phosphate into glyceraldehyde 3-phosphate, a three-carbon fragment useful for generating ATP will be lost. These compounds are isomers that can be readily interconverted: dihydroxyacetone phosphate is a ketose, whereas glyceraldehyde 3-phosphate is an aldose. The isomerization of these three-carbon phosphorylated sugars is catalyzed by triose phosphate isomerase (TPI, sometimes abbreviated TIM).

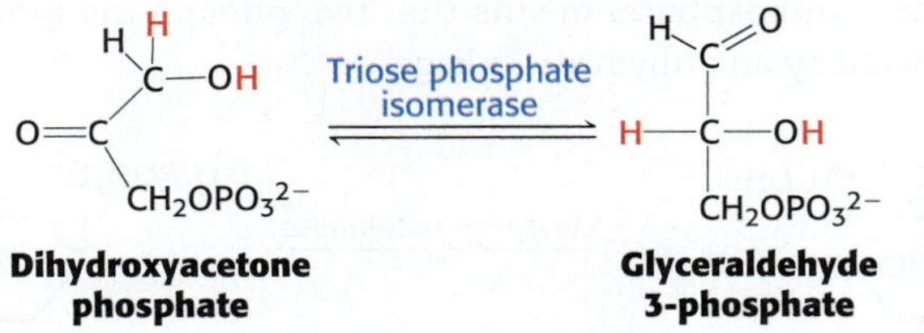

This reaction is rapid and reversible. At equilibrium, 96% of the triose phosphate is dihydroxyacetone phosphate. However, the reaction proceeds readily from dihydroxyacetone phosphate to glyceraldehyde 3-phosphate because the subsequent reactions of glycolysis remove this product.

Mechanism: Triose phosphate isomerase salvages a three-carbon fragment

Much is known about the catalytic mechanism of triose phosphate isomerase (**Figure 16.5**). TPI catalyzes the transfer of a hydrogen atom from carbon 1 to

FIGURE 16.5 Triose phosphate isomerase is an example of a common "barrel" motif, with the active site buried in the center. This enzyme consists of a central core of eight parallel β strands (orange) surrounded by eight α helices (blue). This structural motif, called an αβ barrel or a TIM barrel because of this enzyme, is also found in the glycolytic enzymes aldolase, enolase, and pyruvate kinase. Notice that histidine 95 and glutamate 165, essential components of the active site of triose phosphate isomerase, are located in the barrel. A loop (red) closes off the active site on substrate binding. [Drawn from 2YPI.pdb.]

INTERACT with this model in
⚡ Achie⌁e

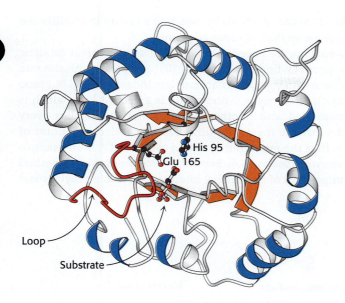

carbon 2, an intramolecular oxidation–reduction. This isomerization of a ketose into an aldose proceeds through an enediol intermediate (**Figure 16.6**).

X-ray crystallographic and other studies showed that glutamate 165 plays the role of a general acid–base catalyst: it abstracts a proton (H⁺) from carbon 1 and then donates it to carbon 2. However, the carboxylate group of glutamate 165 by itself is not basic enough to pull a proton away from a carbon atom adjacent to a

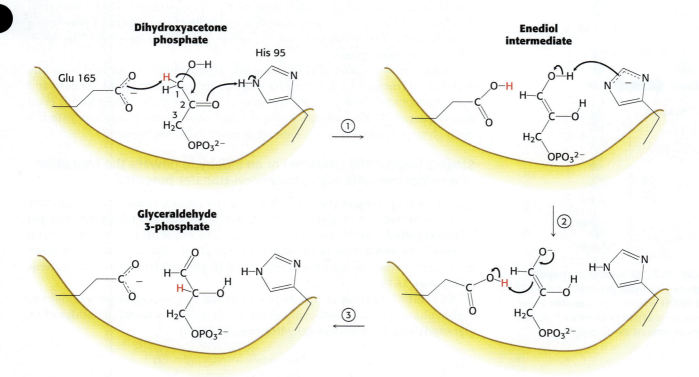

FIGURE 16.6 The catalytic mechanism of triose phosphate isomerase is an acid/base mechanism proceeding through an enediol intermediate. (1) Glutamate 165 acts as a general base by abstracting a proton (H⁺) from carbon 1. Histidine 95, acting as a general acid, donates a proton to the oxygen atom bonded to carbon 2, forming the enediol intermediate. (2) Glutamic acid, now acting as a general acid, donates a proton to C-2, while histidine removes a proton from the OH group of C-1. (3) The product is formed, and glutamate and histidine are returned to their ionized and neutral forms, respectively.

carbonyl group. Histidine 95 assists catalysis by donating a proton to stabilize the negative charge that develops on the C-2 carbonyl group.

Two features of this enzyme are noteworthy. First, TPI displays great catalytic prowess. It accelerates isomerization by a factor of 10^{10} compared with the rate obtained with a simple base catalyst such as acetate ion. Indeed, the k_{cat}/K_M ratio for the isomerization of glyceraldehyde 3-phosphate is 2×10^8 M^{-1} s^{-1}, which is close to the diffusion-controlled limit. In other words, catalysis takes place every time that enzyme and substrate meet. The diffusion-controlled encounter of substrate and enzyme is thus the rate-limiting step in catalysis. TPI is an example of a kinetically perfect enzyme (Section 5.4). Second, TPI suppresses an undesired side reaction, the decomposition of the enediol intermediate into methyl glyoxal and orthophosphate.

Enediol intermediate   **Methyl glyoxal**

In solution, this physiologically useless reaction is 100 times as fast as isomerization. Moreover, methyl glyoxal is a highly reactive compound that can modify the structure and function of a variety of biomolecules, including proteins and DNA. The reaction of methyl glyoxal with a biomolecule is an example of deleterious reactions called advanced glycation end products, discussed earlier (AGEs, Section 11.1). Hence, TPI must prevent the enediol from leaving the enzyme. This labile intermediate is trapped in the active site by the movement of a loop of 10 residues (Figure 16.5). This loop serves as a lid on the active site, shutting it when the enediol is present and reopening it when isomerization is completed. We see here a striking example of one means of preventing an undesirable alternative reaction: the active site is kept closed until the desirable reaction takes place.

Thus, two molecules of glyceraldehyde 3-phosphate are formed from one molecule of fructose 1,6-bisphosphate by the sequential action of aldolase and triose phosphate isomerase. The economy of metabolism is evident in this reaction sequence. The isomerase funnels dihydroxyacetone phosphate into the main glycolytic pathway; a separate set of reactions is not needed.

Stage 2 begins: The oxidation of an aldehyde powers the formation of a compound with high phosphoryl-transfer potential

The preceding steps in glycolysis have transformed one molecule of glucose into two molecules of glyceraldehyde 3-phosphate, but no energy has yet been extracted. On the contrary, thus far, two molecules of ATP have been invested. We come now to the second stage of glycolysis—a series of steps that harvest some of the energy contained in glyceraldehyde 3-phosphate as ATP (Figure 16.7).

The initial reaction in this sequence is the conversion of glyceraldehyde 3-phosphate into 1,3-bisphosphoglycerate (1,3-BPG), a reaction catalyzed by glyceraldehyde 3-phosphate dehydrogenase (GAPDH).

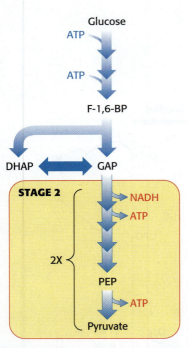

FIGURE 16.7 The second or "payoff" stage of glycolysis generates both ATP and NADH. Note that each reaction in the second stage occurs twice for each glucose molecule entering the pathway.

Glyceraldehyde 3-phosphate (GAP)   **1,3-Bisphosphoglycerate (1,3-BPG)**

1,3-Bisphosphoglycerate is an acyl phosphate, which is a mixed anhydride of phosphoric acid and a carboxylic acid. Such compounds have a high phosphoryl-transfer potential (Section 15.2); one of its phosphoryl groups is transferred to ADP in the next step in glycolysis.

The reaction catalyzed by GAPDH can be viewed as the sum of two processes: the oxidation of the aldehyde to a carboxylic acid by NAD^+, and the joining of the carboxylic acid and orthophosphate to form the acyl-phosphate product.

The first reaction is thermodynamically quite favorable, with a standard free-energy change, $\Delta G^{\circ\prime}$, of approximately -50 kJ mol^{-1} (-12 kcal mol^{-1}), whereas the second reaction is quite unfavorable, with a standard free-energy change of the same magnitude but the opposite sign. If these two reactions simply took place in succession, the second reaction would have a very large activation energy and thus not take place at a biologically significant rate. These two processes must be coupled so the favorable aldehyde oxidation can be used to drive the formation of the acyl phosphate.

How are these reactions coupled? The key is an intermediate, formed as a result of the aldehyde oxidation, that is connected to the enzyme by a thioester linkage (Section 15.4). The thioester intermediate, which is higher in free energy than the free carboxylic acid is, couples the favorable oxidation and the unfavorable phosphorylation reactions. This coupling preserves much of the free energy released in the oxidation reaction. The thioester intermediate reacts with orthophosphate to form the high-energy compound 1,3-bisphosphoglycerate.

We see here the use of a covalent enzyme-bound intermediate as a mechanism of energy coupling. A free-energy profile of the GAPDH reaction, compared with a hypothetical process in which the reaction proceeds without this intermediate, reveals how this intermediate allows a favorable process to drive an unfavorable one (**Figure 16.8**).

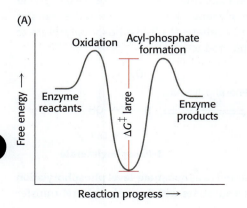

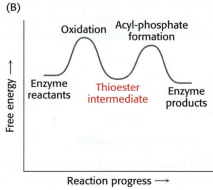

FIGURE 16.8 Free-energy profiles for glyceraldehyde oxidation followed by acyl-phosphate formation demonstrate the importance of the reaction coupling. (A) A hypothetical case with no coupling between the two processes. The second step would have a very large activation barrier, making the reaction very slow. (B) The actual case with the two reactions coupled through a thioester intermediate vastly accelerates the reaction and allows a favorable reaction to drive an unfavorable one.

FIGURE 16.9 The active site of GAPDH includes cysteine and histidine residues adjacent to a bound NAD⁺ molecule. The sulfur atom of cysteine will react with the substrate to form a transitory thioester intermediate. [Drawn from 1GAD.pdb.]

INTERACT with this model in
 Achieve

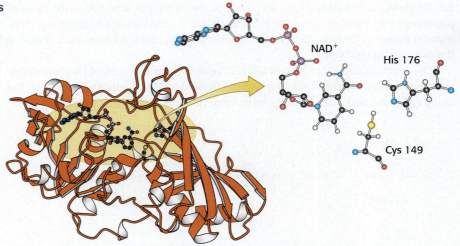

Mechanism: Phosphorylation is coupled to the oxidation of glyceraldehyde 3-phosphate by a thioester intermediate

The active site of GAPDH includes a reactive cysteine residue, as well as NAD⁺ and a crucial histidine (**Figure 16.9**). Let's consider in detail how these components cooperate in the reaction mechanism (**Figure 16.10**) in the following four steps:

1. The aldehyde substrate reacts with the sulfhydryl group of cysteine 149 on the enzyme to form a hemithioacetal.

2. The hydride ion is transferred to a molecule of NAD⁺ that is bound to the enzyme and is adjacent to the cysteine residue. This reaction is favored by the deprotonation of the hemithioacetal by histidine 176. The products are the reduced coenzyme NADH and a thioester intermediate. This thioester intermediate has a free energy close to that of the reactants (Figure 16.8).

3. The NADH formed from the aldehyde oxidation leaves the enzyme and is replaced by a second molecule of NAD⁺. This step is important because the positive charge on NAD⁺ polarizes the thioester intermediate to facilitate the attack by orthophosphate.

4. The orthophosphate attacks the thioester to form 1,3-BPG and free the cysteine residue.

This example illustrates the essence of energy transformations and of metabolism itself: energy released by carbon oxidation is converted into high phosphoryl-transfer potential.

ATP is formed by phosphoryl transfer from 1,3-bisphosphoglycerate

1,3-Bisphosphoglycerate is an energy-rich molecule with a greater phosphoryl-transfer potential than that of ATP (Section 15.2). Thus, 1,3-BPG can be used to power the synthesis of ATP from ADP. Phosphoglycerate kinase catalyzes the transfer of the phosphoryl group from the acyl phosphate of 1,3-bisphosphoglycerate to ADP; ATP and 3-phosphoglycerate are the products.

The formation of ATP in this manner is referred to as **substrate-level phosphorylation** because the phosphate donor, 1,3-BPG, is a substrate with high phosphoryl-transfer

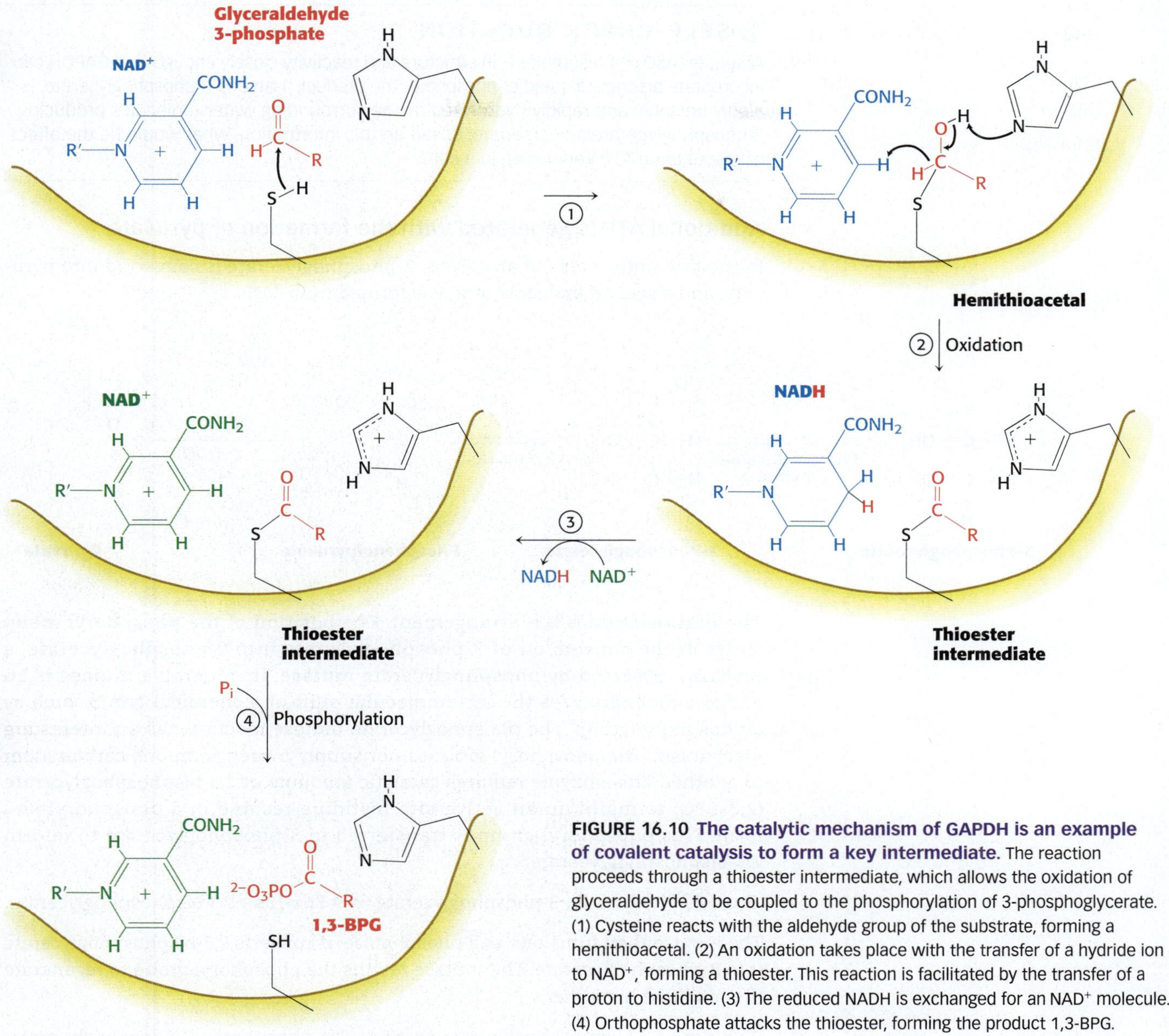

FIGURE 16.10 The catalytic mechanism of GAPDH is an example of covalent catalysis to form a key intermediate. The reaction proceeds through a thioester intermediate, which allows the oxidation of glyceraldehyde to be coupled to the phosphorylation of 3-phosphoglycerate. (1) Cysteine reacts with the aldehyde group of the substrate, forming a hemithioacetal. (2) An oxidation takes place with the transfer of a hydride ion to NAD^+, forming a thioester. This reaction is facilitated by the transfer of a proton to histidine. (3) The reduced NADH is exchanged for an NAD^+ molecule. (4) Orthophosphate attacks the thioester, forming the product 1,3-BPG.

potential. We will contrast this manner of ATP formation with the formation of ATP from ionic gradients in Chapters 18 and 19.

Thus, the outcomes of the reactions catalyzed by glyceraldehyde 3-phosphate dehydrogenase and phosphoglycerate kinase are:

1. Glyceraldehyde 3-phosphate, an aldehyde, is oxidized to 3-phosphoglycerate, a carboxylic acid.

2. NAD^+ is concomitantly reduced to NADH.

3. ATP is formed from P_i and ADP at the expense of carbon-oxidation energy.

In essence, the energy released during the oxidation of glyceraldehyde 3-phosphate to 3-phosphoglycerate is temporarily trapped as 1,3-bisphosphoglycerate. This energy powers the transfer of a phosphoryl group from 1,3-bisphosphoglycerate to ADP to yield ATP. Keep in mind that, because of the actions of aldolase and triose phosphate isomerase, two molecules of glyceraldehyde 3-phosphate were formed, and hence two molecules of ATP were generated. These ATP molecules make up for the two molecules of ATP consumed in the first stage of glycolysis.

❖ **SELF–CHECK QUESTION**

Arsenate (AsO_4^{3-}) resembles P_i in structure and reactivity closely enough that GAPDH can incorporate arsenate instead of phosphate. The product, 1-arseno-3-phosphoglycerate, is highly unstable and rapidly hydrolyzed by the surrounding water molecules producing 3-phosphoglycerate and arsenate. Based on this information, what would be the effect of arsenate on ATP generation in a cell?

Additional ATP is generated with the formation of pyruvate

In the remaining steps of glycolysis, 3-phosphoglycerate is converted into pyruvate, and a second molecule of ATP is formed from ADP.

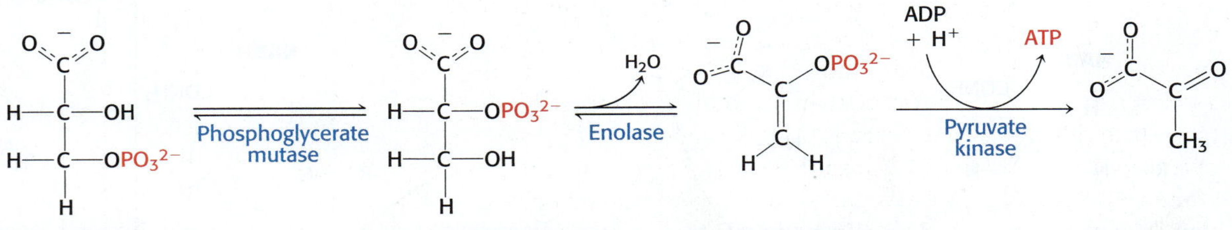

3-Phosphoglycerate   **2-Phosphoglycerate**   **Phosphenolpyruvate**   **Pyruvate**

The first reaction is a rearrangement. The position of the phosphoryl group shifts in the conversion of 3-phosphoglycerate into 2-phosphoglycerate, a reaction catalyzed by phosphoglycerate mutase. In general, a mutase is an enzyme that catalyzes the intramolecular shift of a chemical group, such as a phosphoryl group. The phosphoglycerate mutase reaction has an interesting mechanism: the phosphoryl group is not simply moved from one carbon atom to another. This enzyme requires catalytic amounts of 2,3-bisphosphoglycerate (2,3-BPG) to maintain an active-site histidine residue in a phosphorylated form. This phosphoryl group is transferred to 3-phosphoglycerate to reform 2,3-bisphosphoglycerate.

Enz-His-phosphate + 3-phosphoglycerate ⇌ Enz-His + 2,3-bisphosphoglycerate

The mutase then functions as a phosphatase: it converts 2,3-bisphosphoglycerate into 2-phosphoglycerate. The mutase retains the phosphoryl group to regenerate the modified histidine.

Enz-His + 2,3-bisphosphoglycerate ⇌ Enz-His-phosphate + 2-phosphoglycerate

The sum of these reactions yields the mutase reaction:

3-Phosphoglycerate ⇌ 2-phosphoglycerate

In the next reaction, the dehydration of 2-phosphoglycerate introduces a double bond, creating an enol. Enolase catalyzes this formation of the enol phosphate phosphoenolpyruvate (PEP). This dehydration markedly elevates the transfer potential of the phosphoryl group. An enol phosphate has a high phosphoryl-transfer potential, whereas the phosphate ester of an ordinary alcohol, such as 2-phosphoglycerate, has a low one. The $\Delta G^{\circ\prime}$ of the hydrolysis of a phosphate ester of an ordinary alcohol is -13 kJ mol^{-1} (-3 kcal mol^{-1}), while that of phosphoenolpyruvate is -62 kJ mol^{-1} (-15 kcal mol^{-1}).

Why does phosphoenolpyruvate have such a high phosphoryl-transfer potential? The phosphoryl group traps the molecule in its unstable enol form. When the phosphoryl group has been donated to ATP, the enol undergoes a conversion into the more stable ketone—namely, pyruvate.

Phosphenolpyruvate   **Pyruvate** (enol form)   **Pyruvate**

Thus, the high phosphoryl-transfer potential of phosphoenolpyruvate arises primarily from the large driving force of the subsequent enol–ketone conversion. Hence, pyruvate is formed, and ATP is generated concomitantly. The virtually irreversible transfer of a phosphoryl group from phosphoenolpyruvate to ADP is catalyzed by pyruvate kinase.

What is the energy source for the formation of phosphoenolpyruvate? The answer to this question becomes clear when we compare the structures of 2-phosphoglycerate and pyruvate. The formation of pyruvate from 2-phosphoglycerate is, in essence, an internal oxidation–reduction reaction; carbon 3 takes electrons from carbon 2 in the conversion of 2-phosphoglycerate into pyruvate. Compared with 2-phosphoglycerate, C-3 is more reduced in pyruvate, whereas C-2 is more oxidized. Once again, carbon oxidation powers the synthesis of a compound with high phosphoryl-transfer potential, phosphoenolpyruvate here and 1,3-bisphosphoglycerate earlier, which allows the synthesis of ATP.

Because the molecules of ATP used in forming fructose 1,6-bisphosphate have already been regenerated, the two molecules of ATP generated from phosphoenolpyruvate are "profit."

Two ATP molecules are formed in the conversion of glucose into pyruvate

The net reaction in the transformation of glucose into pyruvate is

$$\text{Glucose} + 2\ P_i + 2\ ADP + 2\ NAD^+ \longrightarrow 2\ \text{pyruvate} + 2\ ATP + 2\ NADH + 2\ H^+ + 2\ H_2O$$

Thus, two molecules of ATP are generated in the conversion of glucose into two molecules of pyruvate. The reactions of glycolysis are summarized in **Table 16.2**.

TABLE 16.2 Reactions of glycolysis

Step	Reaction	Enzyme	Reaction type	$\Delta G^{\circ\prime}$ in kJ mol^{-1} (kcal mol^{-1})	ΔG in kJ mol^{-1} (kcal mol^{-1})
1	Glucose + ATP $\longrightarrow$ glucose 6-phosphate + ADP + H$^+$	Hexokinase	Phosphoryl transfer	−16.7 (−4.0)	−33.5 (−8.0)
2	Glucose 6-phosphate $\rightleftharpoons$ fructose 6-phosphate	Phosphoglucose isomerase	Isomerization	+1.7 (+0.4)	−2.5 (−0.6)
3	Fructose 6-phosphate + ATP $\longrightarrow$ fructose 1,6-bisphosphate + ADP + H$^+$	Phosphofructokinase	Phosphoryl transfer	−14.2 (−3.4)	−22.2 (−5.3)
4	Fructose 1,6-bisphosphate $\rightleftharpoons$ dihydroxyacetone phosphate + glyceraldehyde 3-phosphate	Aldolase	Aldol cleavage	+23.8 (+5.7)	−1.3 (−0.3)
5	Dihydroxyacetone phosphate $\rightleftharpoons$ glyceraldehyde 3-phosphate	Triose phosphate isomerase	Isomerization	+7.5 (+1.8)	+2.5 (+0.6)
6	Glyceraldehyde 3-phosphate + P$_i$ + NAD$^+$ $\rightleftharpoons$ 1,3-bisphosphoglycerate + NADH + H$^+$	Glyceraldehyde 3-phosphate dehydrogenase	Phosphorylation coupled to oxidation	+6.3 (+1.5)	−1.7 (−0.4)
7	1,3-Bisphosphoglycerate + ADP $\rightleftharpoons$ 3-phosphoglycerate + ATP	Phosphoglycerate kinase	Phosphoryl transfer	−18.8 (−4.5)	+1.3 (+0.3)
8	3-Phosphoglycerate $\rightleftharpoons$ 2-phosphoglycerate	Phosphoglycerate mutase	Phosphoryl shift	+4.6 (+1.1)	+0.8 (+0.2)
9	2-Phosphoglycerate $\rightleftharpoons$ phosphoenolpyruvate + H$_2$O	Enolase	Dehydration	+1.7 (+0.4)	−3.3 (−0.8)
10	Phosphoenolpyruvate + ADP + H$^+$ $\longrightarrow$ pyruvate + ATP	Pyruvate kinase	Phosphoryl transfer	−31.4 (−7.5)	−16.7 (−4.0)

Note: ΔG, the actual free-energy change, has been calculated from $\Delta G^{\circ\prime}$ and known concentrations of reactants under typical physiological conditions. Glycolysis can proceed only if the ΔG values of all reactions are negative. The small positive ΔG values of three of the above reactions indicate that the concentrations of metabolites in vivo in cells undergoing glycolysis are not precisely known.

The energy released in the anaerobic conversion of glucose into two molecules of pyruvate is about -90 kJ mol^{-1} (-22 kcal mol^{-1}). We shall see in Chapters 17 and 18 that much more energy can be released from glucose in the presence of oxygen.

❖ **SELF–CHECK QUESTION**

The conversion of one molecule of fructose 1,6-bisphosphate into two molecules of pyruvate results in the net synthesis of how many molecules each of NADH and ATP?

NAD⁺ is regenerated from the metabolism of pyruvate

The conversion of glucose into two molecules of pyruvate has resulted in the net synthesis of ATP. However, an energy-converting pathway that stops at pyruvate will not proceed for long, because redox balance has not been maintained. As we have seen, the activity of glyceraldehyde 3-phosphate dehydrogenase, in addition to generating a compound with high phosphoryl-transfer potential, reduces NAD⁺ to NADH. In the cell, there are limited amounts of NAD⁺, which is derived from the vitamin niacin (B$_3$), a dietary requirement for human beings. Consequently, NAD⁺ must be regenerated for glycolysis to proceed. Thus, the final process in the pathway is the regeneration of NAD⁺ through the metabolism of pyruvate.

The sequence of reactions from glucose to pyruvate is similar in most organisms and most types of cells. In contrast, the fate of pyruvate is variable. Three reactions of pyruvate are of primary importance: conversion into ethanol, lactate, or carbon dioxide (**Figure 16.11**). The first two reactions are fermentations that take place in the absence of oxygen. A **fermentation** is an ATP-generating process in which organic compounds act both as the donor and as the acceptor of electrons; thus, they are redox neutral pathways for the production of ATP from an organic fuel molecule. In contrast, in the presence of oxygen—the most common situation in multicellular organisms and in many unicellular ones—pyruvate is oxidized to carbon dioxide and water through the citric acid cycle and the electron-transport chain, with oxygen serving as the final electron acceptor. We now take a closer look at these three examples of possible fates of pyruvate.

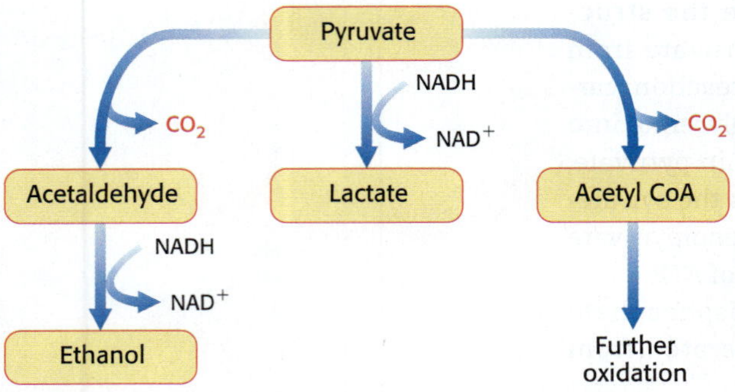

Ethanol fermentation. Ethanol is formed from pyruvate in yeast and several other microorganisms. The conversion of glucose into ethanol in anerobic conditions is called **ethanol fermentation.** The first step is the decarboxylation of pyruvate. This reaction is catalyzed by pyruvate decarboxylase, which requires the coenzyme thiamine pyrophosphate that is derived from the vitamin thiamine (B$_1$). The second step is the reduction of acetaldehyde to ethanol by NADH, in a reaction catalyzed by alcohol dehydrogenase. This reaction regenerates NAD⁺.

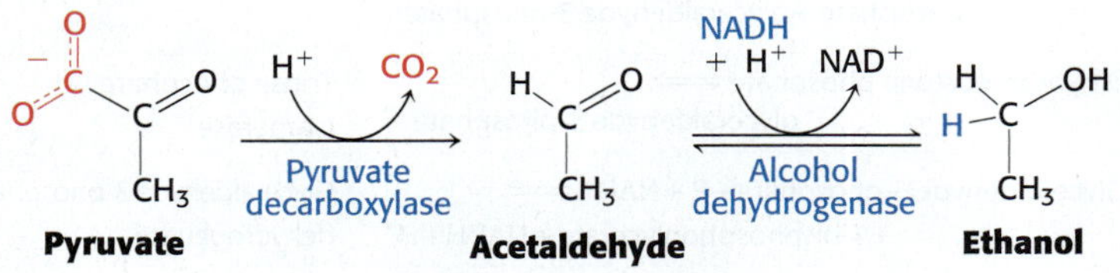

FIGURE 16.11 Pyruvate has many possible fates. Shown are just three possible fates, but many others exist. Ethanol and lactate can be formed by reactions that include NADH. Alternatively, a two-carbon unit from pyruvate can be coupled to coenzyme A (Chapter 17) to form acetyl CoA.

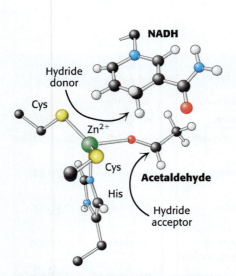

FIGURE 16.12 The active site of alcohol dehydrogenase contains a critical zinc ion bound to two cysteine residues and one histidine residue. The zinc ion binds the acetaldehyde substrate through its oxygen atom, polarizing the substrate so that it more easily accepts a hydride from NADH. Only the nicotinamide ring of NADH is shown.

The active site of alcohol dehydrogenase contains a zinc ion that is coordinated to the sulfur atoms of two cysteine residues and a nitrogen atom of histidine (**Figure 16.12**). This zinc ion polarizes the carbonyl group of the substrate to favor the transfer of a hydride from NADH.

The net result of this anaerobic process is

$$\text{Glucose} + 2\,\text{P}_i + 2\,\text{ADP} + 2\,\text{H}^+ \longrightarrow 2\,\text{ethanol} + 2\,\text{CO}_2 + 2\,\text{ATP} + 2\,\text{H}_2\text{O}$$

Note that NAD⁺ and NADH do not appear in this equation, even though they are crucial for the overall process. NADH generated by the oxidation of glyceraldehyde 3-phosphate is consumed in the reduction of acetaldehyde to ethanol. Thus, there is no net oxidation–reduction in the conversion of glucose into ethanol (**Figure 16.13**). The ethanol formed during fermentation provides a key ingredient for brewing and winemaking.

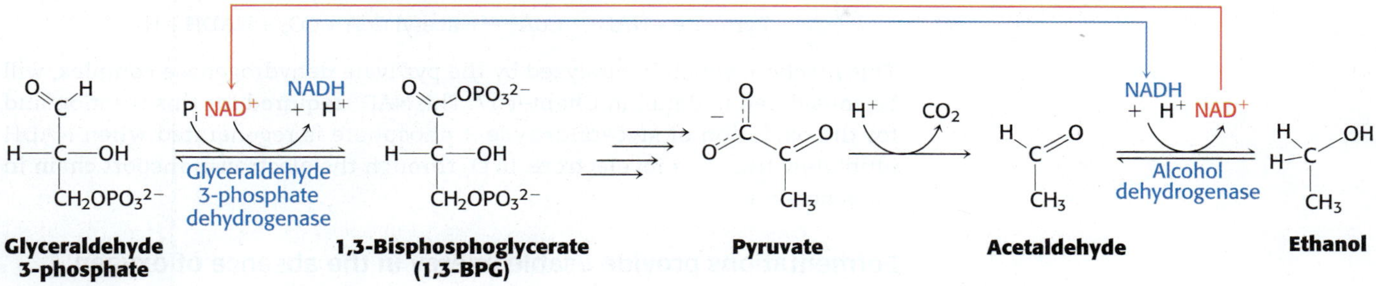

Glyceraldehyde 3-phosphate → P_i, NAD⁺ / NADH + H⁺ (Glyceraldehyde 3-phosphate dehydrogenase) → **1,3-Bisphosphoglycerate (1,3-BPG)** → → → **Pyruvate** → H⁺, CO_2 → **Acetaldehyde** → NADH + H⁺, NAD⁺ (Alcohol dehydrogenase) → **Ethanol**

FIGURE 16.13 The NADH produced in glycolysis must be reoxidized to NAD⁺ for the glycolytic pathway to continue. NADH is constantly generated by the glyceraldehyde 3-phosphate dehydrogenase reaction. In ethanol fermentation, alcohol dehydrogenase oxidizes NADH and generates ethanol.

Lactic acid fermentation. Lactate is formed from pyruvate under anaerobic conditions in a variety of microorganisms and in a variety of animal tissues. The conversion of glucose to lactate in anaerobic conditions is called **lactic acid fermentation**. After the Buchners' original discovery of ethanol fermentation by yeast, studies of muscle extracts later showed that many of the reactions of lactic acid fermentation were the same as those of ethanol fermentation. This second exciting discovery revealed an underlying unity in biochemistry. Lactic acid fermentation occurs regularly in the retina of the eye and in erythrocytes (red blood cells), even in the presence of oxygen, as well as in tissues like the outer layers of the skin that are perpetually limited in oxygen availability. Additionally, certain types of skeletal muscles in most animals can also function anaerobically for short periods. For example, lactic acid fermentation occurs when a specific type of muscle fiber, called fast-twitch or type IIb fibers, performs short bursts of intense exercise because the ATP needs rise faster than the ability of the body to provide oxygen to the muscle. The muscle functions anaerobically until fatigue sets in, which is caused, in part, by lactate buildup. Indeed, the pH of resting type IIb muscle fibers, which is about 7.0, may fall to as low as 6.3 during the bout of exercise. The drop in pH inhibits phosphofructokinase, as we will discuss shortly. A lactate/H⁺ symporter allows the exit of lactate from the muscle cell. The reduction of pyruvate by NADH to form lactate is catalyzed by lactate dehydrogenase.

Pyruvate → NADH + H⁺, NAD⁺ (Lactate dehydrogenase) → **Lactate**

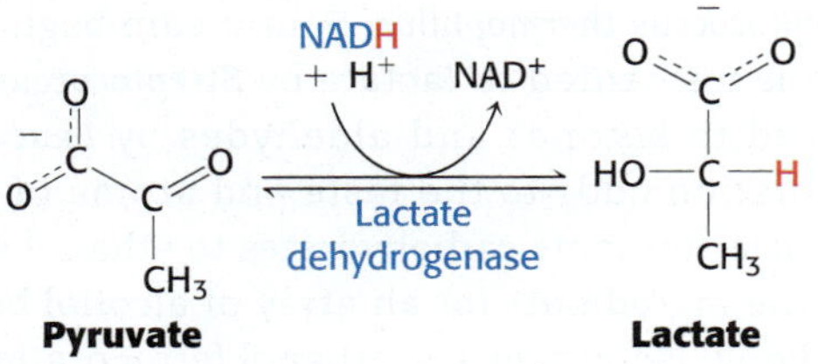

The overall reaction in the conversion of glucose into lactate is

$$\text{Glucose} + 2\ P_i + 2\ \text{ADP} \longrightarrow 2\ \text{lactate} + 2\ \text{ATP} + 2\ H_2O$$

As in ethanol fermentation, there is no net oxidation–reduction. The NADH formed in the oxidation of glyceraldehyde 3-phosphate is consumed in the reduction of pyruvate. The regeneration of NAD⁺ in the reduction of pyruvate to lactate or ethanol sustains the continued process of glycolysis under anaerobic conditions (**Figure 16.14**).

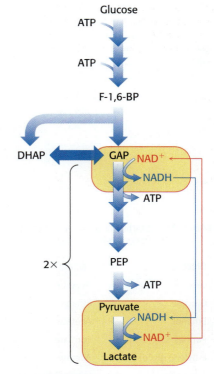

Regeneration of NAD⁺

FIGURE 16.14 NAD⁺ is regenerated by lactate dehydrogenase during lactic acid fermentation. Lactate dehydrogenase oxidizes NADH while generating lactic acid.

Oxidation by the citric acid cycle and the electron transport chain. Carbon dioxide and water are formed, and a great deal of energy extracted, when pyruvate is metabolized under aerobic conditions by means of the citric acid cycle and the electron-transport chain. In contrast, only a fraction of the energy of glucose is released in its anaerobic conversion into ethanol or lactate. The entry point to this oxidative pathway is acetyl coenzyme A (acetyl CoA), which is formed inside mitochondria by the oxidative decarboxylation of pyruvate.

$$\text{Pyruvate} + \text{NAD}^+ + \text{CoA} \longrightarrow \text{acetyl CoA} + \text{CO}_2 + \text{NADH} + \text{H}^+$$

This reaction, which is catalyzed by the pyruvate dehydrogenase complex, will be considered in detail in Chapter 17. The NAD^+ required for this reaction and for the oxidation of glyceraldehyde 3-phosphate is regenerated when NADH ultimately transfers its electrons to O_2 through the electron-transport chain in mitochondria.

Fermentations provide usable energy in the absence of oxygen

Fermentations yield only a fraction of the energy available from the complete combustion of glucose. Why is a relatively inefficient metabolic pathway so extensively used? The fundamental reason is that oxygen is not required. The ability to survive without oxygen affords a host of living accommodations such as soils, deep water, and skin pores. Some organisms, called **obligate anaerobes**, cannot survive in the presence of O_2, a highly reactive compound. The bacterium *Clostridium perfringens*, the cause of gangrene, is an example of an obligate anaerobe. Other pathogenic obligate anaerobes are listed in **Table 16.3**. Some organisms, such as yeast, are **facultative anaerobes** that metabolize glucose aerobically when oxygen is present and perform fermentation when oxygen is absent.

TABLE 16.3 Examples of pathogenic obligate anaerobes

Bacterium	Result of infection
Clostridium tetani	Tetanus (lockjaw)
Clostridium botulinum	Botulism (an especially severe type of food poisoning)
Clostridium perfringens	Gas gangrene (gas is produced as an end point of the fermentation, distorting and destroying the tissue)
Bartonella hensela	Cat-scratch fever (flu-like symptoms)
Bacteroides fragilis	Abdominal, pelvic, pulmonary, and blood infections

Many food products, including sour cream, yogurt, various cheeses, beer, wine, and sauerkraut, result from fermentation. Yogurt is produced by the fermentation of lactose in milk to lactate by a mixed culture of *Lactobacillus acidophilus* and *Streptococcus thermophilus*. Sour cream begins with pasteurized light cream, which is fermented to lactate by *Streptococcus lactis*. The lactate is further fermented to ketones and aldehydes by *Leuconostoc citrovorum*. The second fermentation adds to the taste and aroma of sour cream. Yeast, *Saccharomyces cerevisiae*, ferments carbohydrates to ethanol and carbon dioxide, providing some of the ingredients for an array of alcohol beverages. Although we have considered only lactic acid and ethanol fermentation, microorganisms are capable of generating a wide array of molecules as end points of fermentation (**Table 16.4**).

TABLE 16.4 Starting and ending points of various fermentations

Glucose	$\rightarrow$	Lactate
Lactate	$\rightarrow$	Acetate
Glucose	$\rightarrow$	Ethanol
Ethanol	$\rightarrow$	Acetate
Purines	$\rightarrow$	Formate
Ethylene glycol	$\rightarrow$	Acetate
Threonine	$\rightarrow$	Propionate
Leucine	$\rightarrow$	2-Alkylacetate
Phenylalanine	$\rightarrow$	Propionate

Note: The products of some fermentations are the substrates for others.

❖ SELF–CHECK QUESTION

The conversion of one molecule of glucose into two molecules of lactate results in the net synthesis of how many ATP and how many NADH molecules?

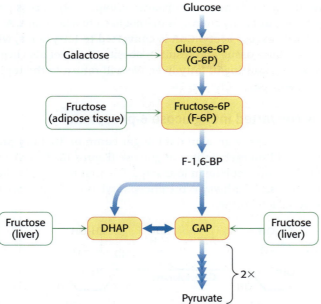

FIGURE 16.15 Additional sugars such as fructose and galactose can enter glycolysis at a variety of points.

Fructose is converted into glycolytic intermediates by fructokinase

Although glucose is the most widely used monosaccharide, others also are important fuels. Let us consider how fructose is funneled into the glycolytic pathway (**Figure 16.15**). There is no separate catabolic pathway for metabolizing fructose, and so the strategy is to convert this sugar into a metabolite of glucose.

The main site of fructose metabolism is the liver, using the fructose 1-phosphate pathway (**Figure 16.16**). The first step is the phosphorylation of fructose to fructose 1-phosphate by fructokinase. Fructose 1-phosphate is then split into glyceraldehyde and dihydroxyacetone phosphate, an intermediate in glycolysis. This aldol cleavage is catalyzed by a specific fructose 1-phosphate aldolase. Glyceraldehyde is then phosphorylated to glyceraldehyde 3-phosphate, a glycolytic intermediate, by triose kinase. In other tissues, such as adipose tissue, fructose can be phosphorylated to fructose 6-phosphate by hexokinase.

Fructose consumption, previously associated with pathological conditions, is likely inconsequential relative to caloric intake. Fructose, a commonly used sweetener, is a component of sucrose and high fructose corn syrup (which contains approximately 55% fructose and 45% glucose). Epidemiological as well as clinical studies previously linked excessive fructose consumption to fatty liver, insulin insensitivity, and obesity, while other studies found negligible or even beneficial effects of fructose consumption, relative to calorically equivalent quantities of other sugars like glucose and sucrose. Recent meta-analyses, large studies which methodically evaluate the results of many other experiments collectively, now indicate that the widely held belief in the negative effects of fructose is no longer supported by the current literature. Interestingly, the hypothesis that overconsumption of fructose, specifically, contributed to poor health outcomes has a biochemical underpinning.

Note that, as shown in Figure 16.16, the actions of fructokinase and triose kinase bypass the most important regulatory step in glycolysis, the phosphofructokinase-catalyzed reaction. Thus, the fructose-derived glyceraldehyde 3-phosphate and dihydroxyacetone phosphate are produced in an unregulated fashion. However, it is likely that additional regulatory enzymes prevent any further unregulated metabolism of the downstream product pyruvate. Although the type of sugar may be unimportant, the overconsumption of sugars and fats is still strongly associated with fatty liver, insulin insensitivity, and obesity, conditions

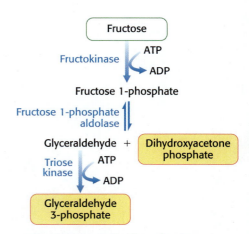

FIGURE 16.16 In the liver, fructose enters the glycolytic pathway through the fructose 1-phosphate pathway.

that may eventually result in type 2 diabetes (Chapter 24). Excess pyruvate produced in glycolysis, from any source, is metabolized to acetyl CoA. As we will see in Chapter 22, this excess acetyl CoA is converted to fatty acids, which can be transported to adipose tissue and result in obesity. The liver also begins to accumulate fatty acids, resulting in fatty liver. We will return to the topic of obesity and caloric homeostasis in Chapter 24.

Galactose is converted into glucose 6-phosphate

Like fructose, galactose is an abundant sugar common in dairy products that must be converted into metabolites of glucose (Figure 16.15). Galactose is converted into glucose 6-phosphate in four steps. The first reaction in the galactose–glucose interconversion pathway is the phosphorylation of galactose to galactose 1-phosphate by galactokinase.

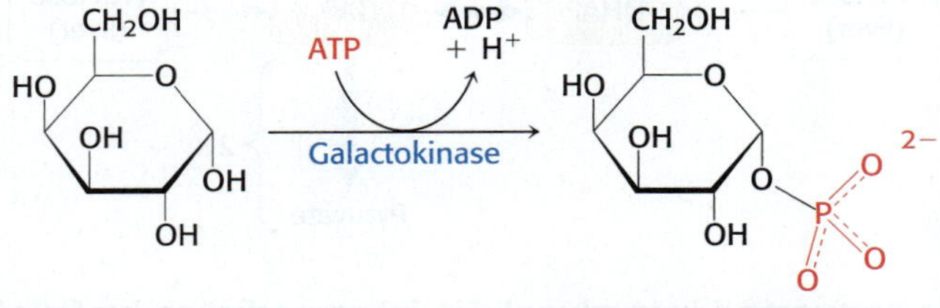

Galactose → **Galactose 1-phosphate**

Galactose 1-phosphate then acquires a uridyl group from uridine diphosphate glucose (UDP-glucose), an activated intermediate in the synthesis of carbohydrates (Section 21.4).

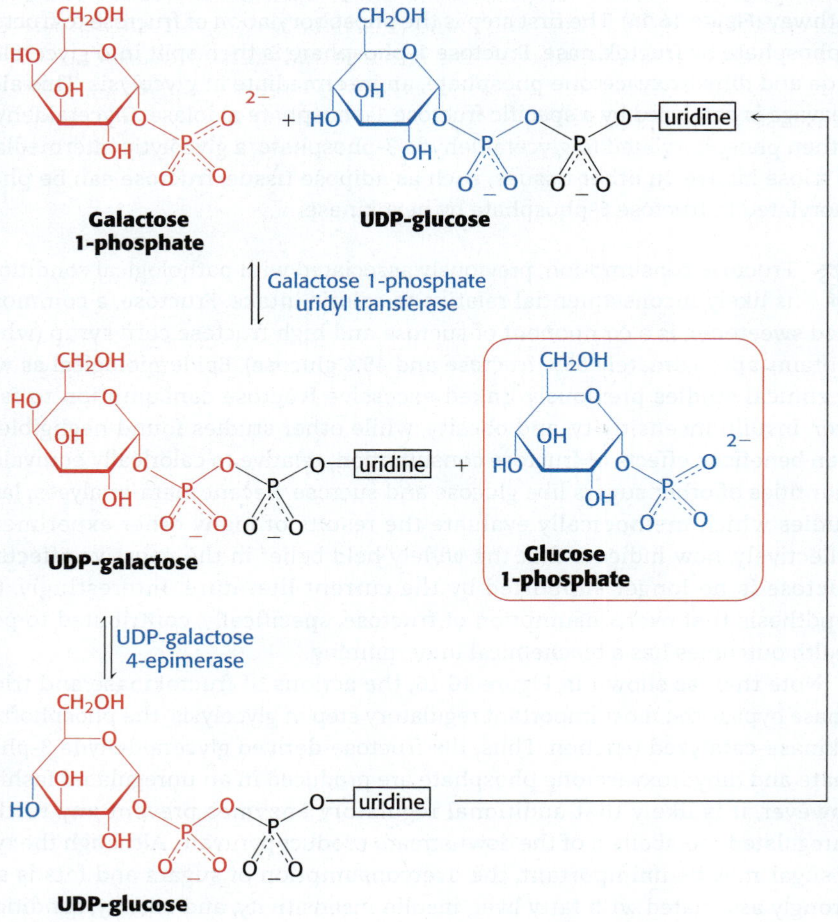

Galactose 1-phosphate + **UDP-glucose**

Galactose 1-phosphate uridyl transferase

UDP-galactose + **Glucose 1-phosphate**

UDP-galactose 4-epimerase

UDP-glucose

The products of this reaction, which is catalyzed by galactose 1-phosphate uridyl transferase, are UDP-galactose and glucose 1-phosphate. The galactose moiety of UDP-galactose is then epimerized to glucose. The configuration of the hydroxyl group at carbon 4 is inverted by UDP-galactose 4-epimerase.

The sum of the reactions catalyzed by galactokinase, the transferase, and the epimerase is

$$\text{Galactose} + \text{ATP} \longrightarrow \text{glucose 1-phosphate} + \text{ADP} + \text{H}^+$$

Note that UDP-glucose is not consumed in the conversion of galactose into glucose, because it is regenerated from UDP-galactose by the epimerase. This reaction is reversible, and the product of the reverse direction also is important. The conversion of UDP-glucose into UDP-galactose is essential for the synthesis of galactosyl residues in complex polysaccharides and glycoproteins if the amount of galactose in the diet is inadequate to meet these needs.

Finally, glucose 1-phosphate, formed from galactose, is isomerized to glucose 6-phosphate by phosphoglucomutase.

$$\text{Glucose 1-phosphate} \underset{}{\overset{\text{Phosphoglucomutase}}{\rightleftharpoons}} \text{glucose 6-phosphate}$$

We shall return to this reaction when we consider the synthesis and degradation of glycogen, which proceeds through glucose 1-phosphate, in Chapter 21.

Galactose can be highly toxic with a defective metabolic pathway

Rare disorders that interfere with the metabolism of galactose are collectively referred to as *galactosemias*. The most common form, called classic galactosemia, is an inherited deficiency in galactose 1-phosphate uridyl transferase activity. Afflicted infants fail to thrive; they vomit or have diarrhea after consuming milk, and enlargement of the liver and jaundice are common, sometimes progressing to cirrhosis. Cataracts will form, and lethargy and delayed neurological development also are common. The blood-galactose level is markedly elevated, and galactose is found in the urine.

The most common treatment is to remove galactose (and lactose) from the diet. An enigma of galactosemia is that, although elimination of galactose from the diet prevents liver disease and cataract development, the majority of patients still suffer from central nervous system malfunction, most commonly a delayed acquisition of language skills. Female patients also display ovarian failure.

Cataract formation is better understood. A cataract is the clouding of the normally clear lens of the eye due to pathological protein aggregation (**Figure 16.17**). If the transferase is not active in the lens of the eye, the presence of aldose reductase causes the accumulating galactose to be reduced to galactitol.

Galactose → (NADPH + H⁺ → NADP⁺, Aldose reductase) → **Galactitol**

Galactitol is poorly metabolized and accumulates in the lens. Water will diffuse into the lens to maintain osmotic balance, triggering the formation of cataracts. In fact, there is a high incidence of cataract formation with age in populations that consume substantial amounts of milk into adulthood.

(A)

(B)

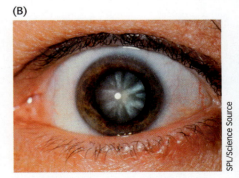

Tim Mainiero/Shutterstock

SPL/Science Source

FIGURE 16.17 Cataracts are evident as the clouding of the lens. (A) A healthy eye. (B) An eye with a cataract.

✚ Many adults worldwide are intolerant of milk because they are deficient in lactase

Many adults are unable to metabolize the milk sugar lactose and experience gastrointestinal disturbances if they drink milk. Lactose intolerance is most commonly caused by a deficiency of the enzyme lactase, which cleaves lactose into glucose and galactose.

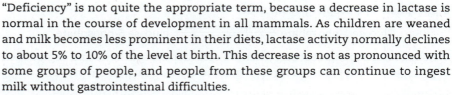

Lactose   Galactose   Glucose

"Deficiency" is not quite the appropriate term, because a decrease in lactase is normal in the course of development in all mammals. As children are weaned and milk becomes less prominent in their diets, lactase activity normally declines to about 5% to 10% of the level at birth. This decrease is not as pronounced with some groups of people, and people from these groups can continue to ingest milk without gastrointestinal difficulties.

How did this tolerance evolve? With the development of dairy farming, an adult with active lactase would have a selective advantage in being able to consume calories from the readily available milk. Indeed, estimates suggest that people with the mutation would have produced almost 20% more fertile offspring. Some form of lactose tolerance evolved independently at least four different times in different human populations in the last 10,000 years, indicating that the evolutionary selective pressure on lactase persistence must have been substantial, attesting to the biochemical value of being able to use milk as an energy source into adulthood.

What happens to the lactose in the intestine of a lactase-deficient person? The lactose is a good energy source for microorganisms in the colon (**Figure 16.18**), which ferment it to lactic acid while generating methane (CH_4) and hydrogen gas (H_2). The gas produced creates the uncomfortable feeling of gut distension and the annoying problem of flatulence. The lactate produced by these microorganisms is osmotically active and draws water into the intestine, as does any undigested lactose, resulting in diarrhea. If severe enough, the gas and diarrhea hinder the absorption of other nutrients such as fats and proteins. The simplest treatment is to avoid the consumption of products containing much lactose. Alternatively, the enzyme lactase can be ingested with milk products.

FIGURE 16.18 *Lactobacillus* **is one example of an industrially useful anaerobic bacterium.** A scanning electron micrograph of an anaerobic bacterial species from the genus *Lactobacillus* is shown. As suggested by its name, this genus ferments glucose into lactic acid. *Lactobacillus* is widely used in the food industry and is an important component of the normal human bacterial flora of the urogenital tract, where it prevents the growth of harmful organisms by creating an acid environment.

SPL/Science Source

❖ SELF–CHECK QUESTION

Which condition is more detrimental to human health: lactose intolerance or classic galactosemia? Also, if a person were galactosemic, would it be better to be lactose tolerant or intolerant?

16.3   The Glycolytic Pathway Is Tightly Controlled

The glycolytic pathway has a dual role: it degrades glucose to generate ATP, and it provides building blocks for biosynthetic reactions. The rate of conversion of glucose into pyruvate is regulated to meet these two major cellular needs. In metabolic pathways, enzymes catalyzing essentially irreversible reactions are potential sites of control. In glycolysis, the reactions catalyzed by hexokinase, phosphofructokinase, and pyruvate kinase are virtually irreversible, and each of them serves as a control site. These enzymes become more active or less so in response to the reversible binding of allosteric effectors or to covalent modification. In addition, the amounts of these important enzymes are varied by the regulation of transcription to meet changing metabolic needs. The time required for allosteric regulation, control by phosphorylation, and transcriptional change is measured typically in milliseconds, seconds, and hours, respectively. We will consider the control of glycolysis in two different tissues—skeletal muscle and liver.

Glycolysis in muscle is regulated to meet the need for ATP

Glycolysis in skeletal muscle provides ATP primarily to power contraction. Consequently, the primary regulation of muscle glycolysis is the ratio of ATP to AMP. Let's examine how each of the key regulatory enzymes responds to changes in the amounts of ATP and AMP present in the cell.

Phosphofructokinase. Phosphofructokinase is the most important control site in the mammalian glycolytic pathway (**Figure 16.19**). High levels of ATP allosterically inhibit the enzyme (a 340-kDa tetramer). ATP binds to a specific regulatory site that is distinct from the catalytic site. The binding of ATP lowers the enzyme's affinity for fructose 6-phosphate. Thus, a high concentration of ATP converts the hyperbolic binding curve of fructose 6-phosphate into a sigmoidal one (**Figure 16.20**). AMP reverses the inhibitory action of ATP, and so the activity of the enzyme increases when the ATP/AMP ratio is lowered. A decrease in pH also inhibits phosphofructokinase activity by augmenting the inhibitory effect of ATP. The pH might fall when fast-twitch muscle is functioning anaerobically, producing excessive quantities of lactic acid. The inhibitory effect protects the muscle from damage that would result from the accumulation of too much acid.

INTERACT with this model in
Achieve

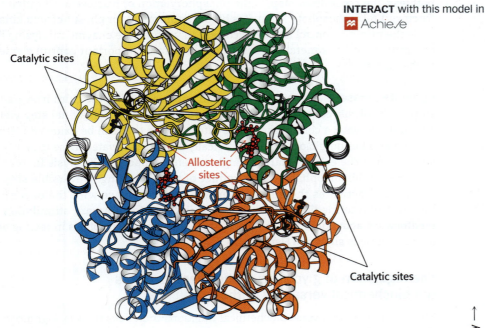

FIGURE 16.19 Phosphofructokinase from *E. coli* is comprised of four identical subunits with separate catalytic and allosteric sites. Each subunit of the human liver enzyme consists of two domains that are similar to the *E. coli* enzyme. [Drawn from 1PFK.pdb.]

Why is AMP and not ADP the positive regulator of phosphofructokinase? When ATP is being utilized rapidly, the enzyme adenylate kinase can form ATP from ADP by the following reaction:

$$ADP + ADP \rightleftharpoons ATP + AMP$$

Thus, some ATP is salvaged from ADP, and AMP becomes the signal for the low-energy state.

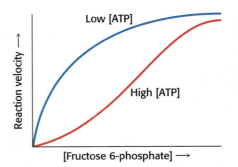

FIGURE 16.20 A high level of ATP allosterically inhibits phosphofructokinase by decreasing its affinity for fructose 6-phosphate.

Moreover, the use of AMP as an allosteric effector provides an especially sensitive regulation. We can understand why by considering the following. First, the total adenylate pool ([ATP], [ADP], [AMP]) in a cell is constant over the short term. Second, the concentration of ATP is greater than that of ADP, and the concentration of ADP is, in turn, greater than that of AMP. Consequently, small percentage changes in [ATP] result in larger percentage changes in the concentrations of the other adenylate nucleotides. This magnification of small changes in [ATP] to larger changes in [AMP] leads to tighter regulation by increasing the range of sensitivity of phosphofructokinase.

Hexokinase. Phosphofructokinase is the most prominent regulatory enzyme in glycolysis, but it is not the only one. Hexokinase, the enzyme catalyzing the first step of glycolysis, is inhibited by its product, glucose 6-phosphate. High concentrations of this molecule signal that the cell no longer requires glucose for energy or for the synthesis of glycogen, a storage form of glucose (Chapter 21), and the glucose will be left in the blood. A rise in glucose 6-phosphate concentration is a means by which phosphofructokinase communicates with hexokinase. When phosphofructokinase is inactive, the concentration of fructose 6-phosphate rises. In turn, the level of glucose 6-phosphate rises because it is in equilibrium with fructose 6-phosphate. Hence, the inhibition of phosphofructokinase leads to the inhibition of hexokinase.

Why is phosphofructokinase rather than hexokinase the pacemaker of glycolysis? The reason becomes evident on noting that glucose 6-phosphate is not solely a glycolytic intermediate. In muscle, glucose 6-phosphate can also be converted into glycogen. The first irreversible reaction unique to the glycolytic pathway, the **committed step**, is the phosphorylation of fructose 6-phosphate to fructose 1,6-bisphosphate. Thus, it is highly appropriate for phosphofructokinase to be the primary control site in glycolysis. In general, the enzyme catalyzing the committed step in a metabolic sequence is the most important control element in the pathway.

Pyruvate kinase. Pyruvate kinase, the enzyme catalyzing the third irreversible step in glycolysis, controls the outflow from this pathway. This final step yields ATP and pyruvate, a central metabolic intermediate that can be oxidized further or used as a building block. ATP allosterically inhibits pyruvate kinase to slow glycolysis when the energy charge is high. When the pace of glycolysis increases, fructose 1,6-bisphosphate—the product of the preceding irreversible step in glycolysis—activates the kinase to enable it to keep pace with the oncoming high flux of intermediates. Such a process is called **feedforward stimulation** or **feedforward activation**. A summary of the regulation of glycolysis in resting and active muscle is shown in **Figure 16.21**.

The regulation of glycolysis in the liver illustrates the biochemical versatility of the liver

The liver has more diverse biochemical functions than does muscle. Significantly, the liver maintains blood-glucose concentration: it stores glucose as glycogen when glucose is plentiful, and it releases glucose when supplies are low. It also uses glucose to generate reducing power for biosynthesis (Section 20.3) as well as to synthesize a host of biochemicals. So, although the liver has many of the regulatory features of muscle glycolysis, the regulation of glycolysis in the liver is more complex.

Phosphofructokinase. Liver phosphofructokinase can be regulated by ATP as in muscle, but such regulation is not as important since the liver does not experience the sudden ATP needs that a contracting muscle does. Likewise, low pH is not an important metabolic signal for the liver enzyme, because lactate is not

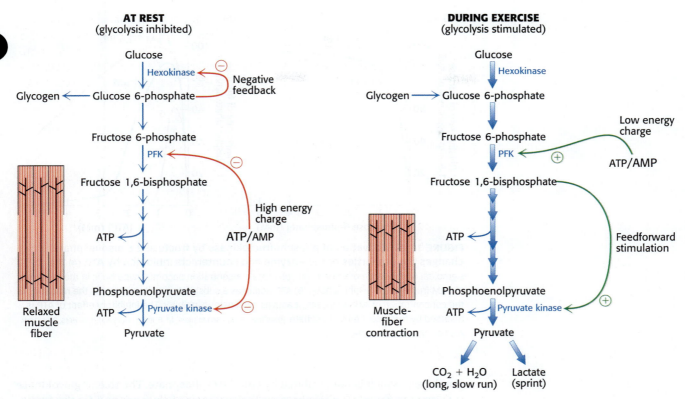

FIGURE 16.21 Regulation of glycolysis in muscle. At rest (left), glycolysis is not very active (thin arrows). The high concentration of ATP inhibits phosphofructokinase (PFK), pyruvate kinase, and hexokinase. Glucose 6-phosphate is converted into glycogen (Chapter 21). During exercise (right), the decrease in the ATP/AMP ratio resulting from muscle contraction activates phosphofructokinase and hence glycolysis. The flux down the pathway is increased, as represented by the thick arrows.

normally produced in the liver. Indeed, as we will see, lactate is converted into glucose in the liver.

Glycolysis in the liver furnishes carbon skeletons for biosyntheses, and so a signal indicating whether building blocks are abundant or scarce should also regulate phosphofructokinase. In the liver, phosphofructokinase is inhibited by citrate, an early intermediate in the citric acid cycle (Chapter 17). A high level of citrate in the cytoplasm means that biosynthetic precursors are abundant, and so there is no need to degrade additional glucose for this purpose. Citrate inhibits phosphofructokinase by enhancing the inhibitory effect of ATP.

The key means by which glycolysis in the liver responds to changes in blood glucose is through the signal molecule fructose 2,6-bisphosphate (F-2,6-BP), a potent activator of phosphofructokinase. In the liver, the concentration of fructose 6-phosphate rises when blood-glucose concentration is high, and the abundance of fructose 6-phosphate accelerates the synthesis of F-2,6-BP (**Figure 16.22**). Hence, an abundance of fructose 6-phosphate leads to a higher concentration of F-2,6-BP. The binding of fructose 2,6-bisphosphate increases the affinity of phosphofructokinase for fructose 6-phosphate and diminishes the inhibitory effect of ATP (**Figure 16.23**). Glycolysis is thus accelerated when glucose is abundant. We will further explore the additional hormonal control over the synthesis and degradation of this important regulatory molecule after we have considered gluconeogenesis.

Hexokinase and glucokinase. The hexokinase reaction in the liver is regulated as in the muscle. However, the liver, in keeping with its role as monitor of blood-glucose levels, possesses another specialized isozyme of hexokinase, called

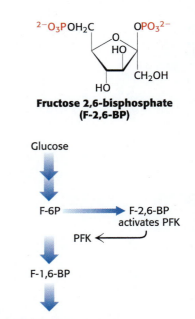

Fructose 2,6-bisphosphate (F-2,6-BP)

FIGURE 16.22 Fructose 2,6-bisphosphate allosterically activates phosphofructokinase. In high concentrations, fructose 6-phosphate (F-6P) activates the enzyme phosphofructokinase (PFK) through an intermediary, fructose 2,6-bisphosphate (F-2,6-BP).

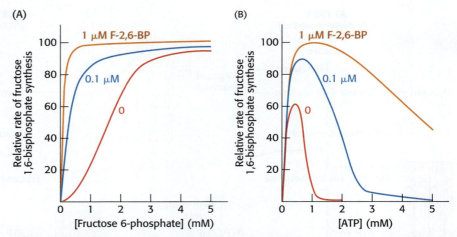

FIGURE 16.23 Activation of phosphofructokinase by fructose 2,6-bisphosphate changes the properties of the enzyme and counteracts inhibition by ATP. (A) The sigmoidal dependence of velocity on substrate concentration becomes hyperbolic in the presence of 1 μM fructose 2,6-bisphosphate. (B) ATP, acting as a substrate, initially stimulates the reaction. As the concentration of ATP increases, it acts as an allosteric inhibitor. The inhibitory effect of ATP is reversed by fructose 2,6-bisphosphate. [Data from E. Van Schaftingen, M. F. Jett, L. Hue, and H. G. Hers, *Proc. Natl. Acad. Sci. U.S.A.* 78:3483–3486, 1981.]

glucokinase, which is not inhibited by glucose 6-phosphate. The role of glucokinase is to provide glucose 6-phosphate for the synthesis of glycogen and for the formation of fatty acids (Chapter 22).

Remarkably, glucokinase displays the sigmoidal kinetics characteristic of an allosteric enzyme even though it functions as a monomer. Glucokinase phosphorylates glucose only when glucose is abundant because the affinity of glucokinase for glucose is about 50-fold lower than that of hexokinase. Moreover, when glucose concentration is low, glucokinase is inhibited by the liver-specific glucokinase regulatory protein (GKRP), which sequesters the kinase in the nucleus until the glucose concentration increases. The low affinity of glucokinase for glucose gives the brain and muscles first call on glucose when its supply is limited, and it ensures that glucose will not be wasted when it is abundant.

Glucokinase is also present in the β cells of the pancreas, which secrete the hormone insulin in response to the increased formation of glucose 6-phosphate by glucokinase when blood-glucose levels are elevated. Insulin signals the need to remove glucose from the blood for storage as glycogen or conversion into fat. Drugs that activate liver glucokinase or disrupt its interaction with GKRP are being evaluated as a treatment for type 2 diabetes, in which the sensitivity to natural levels of insulin has decreased.

Pyruvate kinase. Several isozymic forms of pyruvate kinase (a tetramer of 57-kDa subunits) encoded by different genes are present in mammals: The L type predominates in the liver, and the M type in muscle and the brain. The L and M forms of pyruvate kinase have many properties in common. Indeed, the liver enzyme behaves much like the muscle enzyme with regard to allosteric regulation, except that the liver enzyme is also inhibited by alanine (synthesized in one step from pyruvate), a signal that building blocks are available. Moreover, the isozymic forms differ in their susceptibility to covalent modification. The catalytic properties of the L form—but not of the M form—are also controlled by reversible phosphorylation (**Figure 16.24**). When the blood-glucose level is low, the glucagon-triggered cyclic AMP cascade (Section 14.2) leads to the phosphorylation of pyruvate kinase, which diminishes its activity. This hormone-triggered phosphorylation prevents the liver

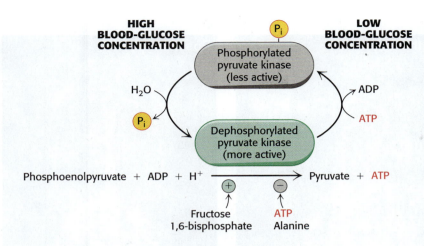

Pyruvate kinase is automatically regulated by allosteric effectors but controlled by extracellular signals through reversible covalent modification. Fructose 1,6-bisphosphate allosterically stimulates the enzyme, while ATP and alanine are allosteric inhibitors. Glucagon, secreted in response to low blood glucose, promotes phosphorylation and inhibition of the enzyme. When blood glucose concentration is adequate, the enzyme is dephosphorylated and activated.

from consuming glucose when it is more urgently needed by the brain and muscle. We see here a clear-cut example of how isoenzymes contribute to the metabolic diversity of different organs.

The enzymes of glycolysis are physically associated with one another

Evidence has accumulated that the enzymes of glycolysis in eukaryotes are organized into complexes. For example, in yeast, the glycolytic enzymes are associated with the mitochondria, while in mammalian erythrocytes, the enzymes are found bound to the inner surface of the cell membrane. Indeed, some level of organization appears to occur in all cell types. This arrangement increases enzyme efficiency by facilitating movement of substrates and products between enzymes—a process called **substrate channeling**—and prevents the release of any toxic intermediates. We will see that the organization of metabolic pathways into large complexes is a common occurrence inside the cell.

Aerobic glycolysis is a property of tumor cells and other rapidly growing cells

Tumors have been known for decades to display enhanced rates of glucose uptake and glycolysis. Indeed, rapidly growing tumor cells will metabolize glucose to lactate even in the presence of oxygen, a process called **aerobic glycolysis** or the *Warburg effect*, after Otto Warburg, the biochemist who first noted this characteristic of cancer cells in the 1920s. In fact, tumors with a high glucose uptake are particularly aggressive, and the cancer is likely to have a poor prognosis. A nonmetabolizable glucose analog, $2\text{-}^{18}\text{F-2-D-deoxyglucose}$, detectable by a combination of positron emission tomography (PET) and computer-aided tomography (CAT), easily visualizes tumors and allows monitoring of treatment effectiveness (**Figure 16.25**).

What selective advantage does aerobic glycolysis offer the tumor over the energetically more efficient oxidative phosphorylation? Researchers are actively pursuing the answer to this question, but we can speculate on the benefits. First, aerobic glycolysis generates lactic acid that is then secreted. Acidification of the tumor environment has been shown to facilitate tumor invasion. Moreover, lactate impairs the activation of $CD8^+$ T and NK immune system cells that normally attack the tumor. However, even leukemias perform aerobic glycolysis, and leukemia is not an invasive cancer. Second, and perhaps more importantly, the increased uptake of glucose and formation of glucose 6-phosphate provides substrates for another metabolic pathway—the pentose phosphate pathway (Section 20.3)—that generates biosynthetic reducing power, NADPH. Finally, cancer cells grow more rapidly than the blood vessels that nourish them; thus, as solid tumors grow, the oxygen concentration in their environment falls. In other

FIGURE 16.25 Tumors can be visualized with 2-¹⁸F-2-ᴅ-deoxyglucose (FDG) and positron emission tomography. (A) A nonmetabolizable glucose analog infused into a patient and detected by a combination of positron emission and computer-aided tomography reveals the presence of a malignant tumor (T). (B) After 4 weeks of treatment with a tyrosine kinase inhibitor (Section 14.5), the tumor shows no uptake of FDG, indicating decreased metabolism. Excess FDG, which is excreted in the urine, also visualizes the kidneys (K) and bladder (B). [Credit: Courtesy of A. D. Van den Abbeele, MD, Dana-Farber Cancer Institute, Boston.]

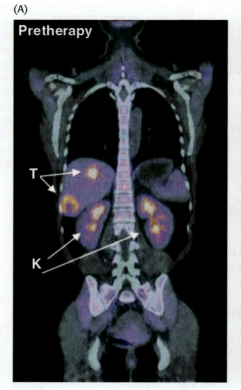
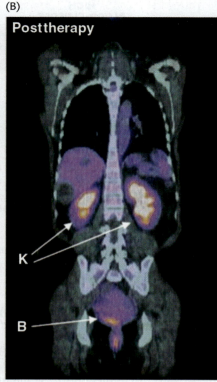

(A) Pretherapy

(B) Posttherapy

TABLE 16.5 Proteins in glucose metabolism encoded by genes regulated by hypoxia-inducible factor

GLUT1
GLUT3
Hexokinase
Phosphofructokinase
Aldolase
Glyceraldehyde 3-phosphate dehydrogenase
Phosphoglycerate kinase
Enolase
Pyruvate kinase

words, they begin to experience a deficiency of oxygen, called *hypoxia*. The use of aerobic glycolysis reduces the dependence of cell growth on oxygen.

What biochemical alterations facilitate the switch to aerobic glycolysis? Again, the answers are not complete, but changes in gene expression of isozymic forms of two glycolytic enzymes may be crucial. Tumor cells express an isozyme of hexokinase that binds to mitochondria. There, the enzyme has ready access to any ATP generated by oxidative phosphorylation and is not susceptible to feedback inhibition by its product, glucose 6-phosphate. More importantly, an embryonic isozyme of pyruvate kinase, pyruvate kinase M, is also expressed. Remarkably, this isozyme has a lower catalytic rate than normal pyruvate kinase and creates a bottleneck, allowing the use of glycolytic intermediates for biosynthetic processes required for cell proliferation. The need for biosynthetic precursors is greater than the need for ATP, suggesting that even glycolysis at a reduced rate produces sufficient ATP to allow cell proliferation. Although originally observed in cancer cells, the Warburg effect is also seen in noncancerous, rapidly dividing cells.

Cancer and endurance training affect glycolysis in a similar fashion

The hypoxia that some tumors experience with rapid growth activates a transcription factor, hypoxia-inducible transcription factor (HIF-1). HIF-1 increases the expression of most glycolytic enzymes and the glucose transporters GLUT1 and GLUT3 (Table 16.5). These adaptations by the cancer cells enable a tumor to survive until blood vessels can grow. HIF-1 also increases the expression of signal molecules, such as vascular endothelial growth factor (VEGF), that facilitate the growth of blood vessels that will provide nutrients to the cells (Figure 16.26). Without new blood vessels, a tumor would cease to grow and either die or remain harmlessly small.

Efforts are underway to develop drugs that inhibit the growth of blood vessels in tumors. Indeed, bevacizumab, a monoclonal antibody that binds to VEGF and prevents activation of angiogenesis, has been approved for treatment of glioblastomas, which are fast-growing cancers of the central nervous system derived from glial cells.

Interestingly, anaerobic exercise training—forcing muscles to rely on lactic acid fermentation for ATP production—also activates HIF-1, producing the same effects as those seen in the tumor—enhanced ability to generate ATP anaerobically and a stimulation of blood-vessel growth. These biochemical effects account for the improved athletic performance that results from training and demonstrate how behavior can affect biochemistry. Other signals from sustained muscle contraction trigger muscle mitochondrial biogenesis, allowing for more efficient aerobic energy generation and forestalling the need to resort to lactic acid fermentation for ATP synthesis.

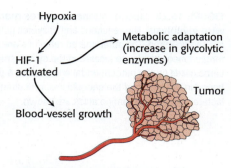

❖ SELF-CHECK QUESTION

Muscle phosphofructokinase activity increases as a function of ATP concentration, but only up to a point, and then it falls rapidly. Explain these results and how they relate to the role of phosphofructokinase in glycolysis.

FIGURE 16.26 Hypoxia alters gene expression in tumors. The hypoxic conditions inside a tumor mass lead to the activation of the hypoxia-inducible transcription factor (HIF-1), which induces metabolic adaptation (an increase in glycolytic enzymes) and activates angiogenic factors that stimulate the growth of new blood vessels. [Information from C. V. Dang and G. L. Semenza, *Trends Biochem. Sci.* 24:68–72, 1999.]

16.4   Glucose Can Be Synthesized from Noncarbohydrate Precursors

We now turn to the synthesis of glucose from noncarbohydrate precursors, a process called *gluconeogenesis*. Maintaining levels of glucose is important because the brain depends on glucose as its primary fuel and red blood cells use glucose as their only fuel. The daily glucose requirement of the brain in a typical adult human being is about 120 g, which accounts for most of the 160 g of glucose needed daily by the whole body. The amount of glucose present in body fluids is about 20 g, and that readily available from glycogen is approximately 190 g. Thus, the direct glucose reserves are sufficient to meet glucose needs for about a day. Gluconeogenesis is especially important during a longer period of fasting or starvation (Section 24.2) and to constantly remove and recycle lactate, which is always being produced by some tissues.

The gluconeogenic pathway converts pyruvate into glucose. The major noncarbohydrate precursors of glucose—lactate, amino acids, and glycerol—are first converted into pyruvate or enter the pathway at later intermediates such as oxaloacetate and dihydroxyacetone phosphate (**Figure 16.27**). Lactate is readily converted into pyruvate by the action of lactate dehydrogenase (Section 16.2). Amino acids are derived from proteins in the diet and, during starvation, from the breakdown of proteins in skeletal muscle (Section 23.1).

The hydrolysis of triacylglycerols (Section 22.2) in fat cells yields glycerol and fatty acids. Glycerol is a precursor of glucose, but animals cannot convert fatty acids into glucose, for reasons that will be discussed in Chapter 17. Glycerol may enter either the gluconeogenic or the glycolytic pathway as dihydroxyacetone phosphate.

$$\underset{\textbf{Glycerol}}{\begin{array}{c}CH_2OH \\ | \\ HO-C-H \\ | \\ CH_2OH \end{array}} \xrightarrow[\substack{\text{Glycerol} \\ \text{kinase}}]{\substack{\text{ATP} \quad ADP \\ +\,H^+}} \underset{\substack{\textbf{Glycerol} \\ \textbf{phosphate}}}{\begin{array}{c}CH_2OH \\ | \\ HO-C-H \\ | \\ CH_2OPO_3^{2-} \end{array}} \xrightarrow[\substack{\text{Glycerol} \\ \text{phosphate} \\ \text{dehydrogenase}}]{\substack{NAD^+ \quad NADH \\ +\,H^+}} \underset{\substack{\textbf{Dihydroxyacetone} \\ \textbf{phosphate}}}{\begin{array}{c}CH_2OH \\ | \\ O=C \\ | \\ CH_2OPO_3^{2-} \end{array}}$$

The major site of gluconeogenesis is the liver, with a small amount also taking place in the kidney and other tissues. Gluconeogenesis in the liver and kidney helps to maintain the glucose concentration in the blood so that the brain and muscle can extract sufficient glucose from it to meet their metabolic demands.

Gluconeogenesis is not a reversal of glycolysis

In glycolysis, glucose is converted into pyruvate; in gluconeogenesis, pyruvate is converted into glucose. However, gluconeogenesis is not a reversal of glycolysis.

FIGURE 16.27 Gluconeogenesis shares many enzymatic steps with glycolysis. The reactions and enzymes unique to gluconeogenesis are shown in red. The other shared reactions are in black. The enzymes for gluconeogenesis are located in the cytoplasm, except for pyruvate carboxylase (in the mitochondria) and glucose 6-phosphatase (membrane bound in the lumen of the endoplasmic reticulum). The entry points for lactate, glycerol, and amino acids are shown.

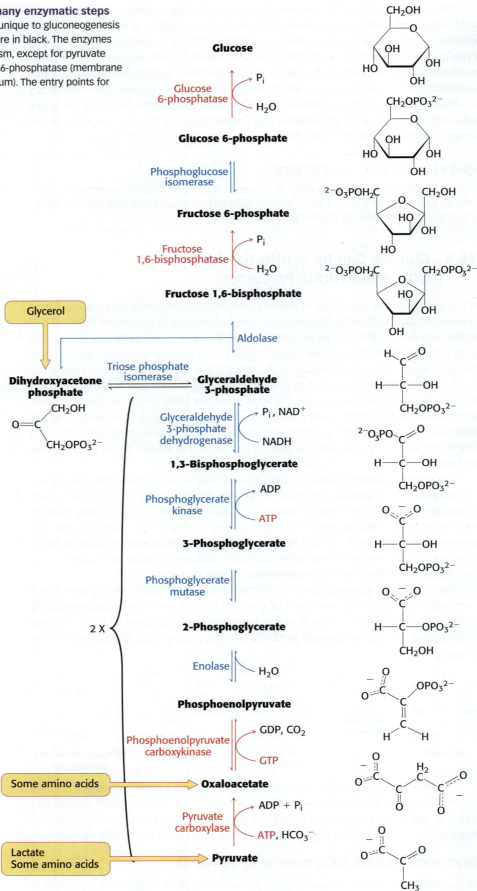

Several reactions must differ because the equilibrium of glycolysis lies far on the side of pyruvate formation. The actual free-energy change for the formation of pyruvate from glucose is about -90 kJ mol^{-1} (-22 kcal mol^{-1}) under typical cellular conditions. Most of the decrease in free energy in glycolysis takes place in the three essentially irreversible steps catalyzed by hexokinase, phosphofructokinase, and pyruvate kinase.

$$\text{Glucose} + \text{ATP} \xrightarrow{\text{Hexokinase}} \text{glucose 6-phosphate} + \text{ADP}$$
$$\Delta G = -33 \text{ kJ mol}^{-1} \ (-8.0 \text{ kcal mol}^{-1})$$

$$\text{Fructose 6-phosphate} + \text{ATP} \xrightarrow{\text{Phosphofructokinase}} \text{fructose 1,6-bisphosphate} + \text{ADP}$$
$$\Delta G = -22 \text{ kJ mol}^{-1} \ (-5.3 \text{ kcal mol}^{-1})$$

$$\text{Phosphoenolpyruvate} + \text{ADP} \xrightarrow{\text{Pyruvate kinase}} \text{pyruvate} + \text{ATP}$$
$$\Delta G = -17 \text{ kJ mol}^{-1} \ (-4.0 \text{ kcal mol}^{-1})$$

In gluconeogenesis, these virtually irreversible reactions of glycolysis must be bypassed.

The conversion of pyruvate into phosphoenolpyruvate begins with the formation of oxaloacetate

The first step in gluconeogenesis is the carboxylation of pyruvate to form oxaloacetate at the expense of a molecule of ATP, a reaction catalyzed by pyruvate carboxylase. This reaction occurs in the mitochondria.

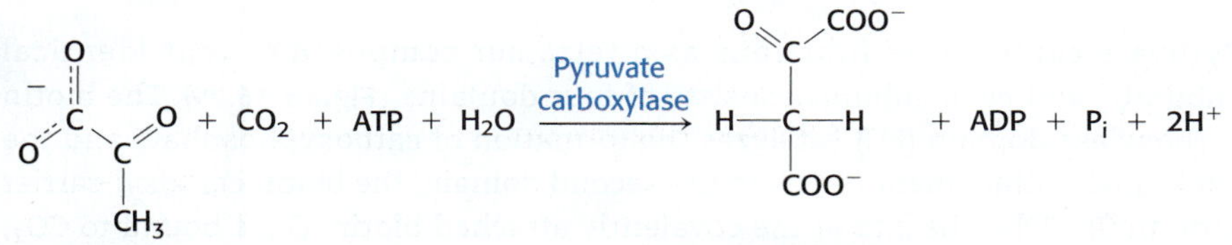

Pyruvate   **Oxaloacetate**

Pyruvate carboxylase requires biotin, a covalently attached prosthetic group, which serves as the carrier of activated CO_2. The carboxylate group of biotin is linked to the ε-amino group of a specific lysine residue by an amide bond (**Figure 16.28**). Recall that, in aqueous solutions, CO_2 exists primarily as HCO_3^- with the aid of carbonic anhydrase (Section 6.2).

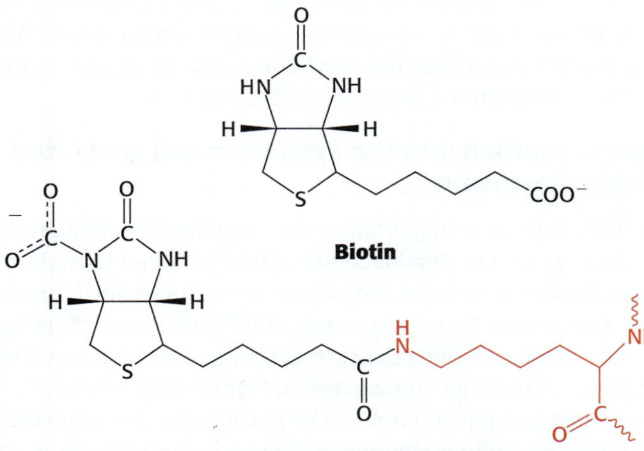

Biotin

Carboxybiotin covalently bound to ε-amino group of a lysine

FIGURE 16.28 Carboxybiotin linked to lysine can act as a movable "arm" carrying activated CO$_2$ groups. Notice the great distance between the bound CO_2 group and where the lysine connects to the rest of the protein. This allows the lysyl-carboxybiotin moiety to act like a long lever arm to move CO_2 from one location in a protein to another. This movement is usually from one distinct active site to a second active site within the same or, in the case of pyruvate carboxylase, in a different subunit.

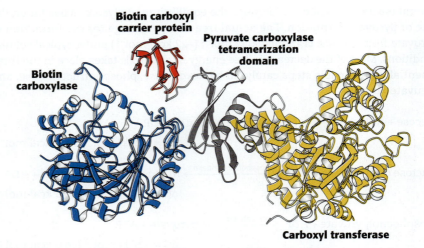

FIGURE 16.29 A subunit of pyruvate carboxylase shows the complexity of the enzyme. The functional enzyme is a tetramer composed of these complex subunits which each have four distinct domains as shown. Biotin, covalently attached to the biotin carboxyl carrier protein, transports CO_2 from the biotin carboxylase active site to the carboxyl transferase active site. [Information from G. Lasso, L. P. C. Yu, D. Gil, S. Xiang, L. Tong, and M. Valle, *Structure* 18:1300–1310, 2010.]

The carboxylation of pyruvate takes place in three stages:

$$HCO_3^- + ATP \rightleftharpoons HOCO_2 - PO_3^{2-} + ADP$$

$$\text{Biotin-enzyme} + HOCO_2 - PO_3^{2-} \longrightarrow CO_2 - \text{biotin-enzyme} + P_i$$

$$CO_2 - \text{biotin-enzyme} + \text{pyruvate} \rightleftharpoons \text{biotin-enzyme} + \text{oxaloacetate}$$

Pyruvate carboxylase functions as a tetramer composed of four identical subunits, and each subunit consists of four domains (**Figure 16.29**). The biotin carboxylase domain (BC) catalyzes the formation of carboxyphosphate and the subsequent attachment of CO_2 to the second domain, the biotin carboxyl carrier protein (BCCP)—the site of the covalently attached biotin. Once bound to CO_2, BCCP leaves the biotin carboxylase active site and swings almost the entire length of the subunit (≈ 75 Å) to the active site of the carboxyl transferase domain (CT), which transfers the CO_2 to pyruvate to form oxaloacetate. BCCP in one subunit interacts with the active sites on an adjacent subunit. The fourth domain (PT) facilitates the formation of the tetramer and is the binding site for acetyl CoA, a required allosteric activator.

How acetyl CoA facilitates the carboxylase reaction is under investigation. Recent research suggests that by binding to the enzyme, acetyl CoA enhances conformational communication between the BC and CT domains that enables the movement of the BCCP domain between two catalytic sites. The allosteric activation of pyruvate carboxylase by acetyl CoA is also an important physiological control mechanism that will be discussed in Section 17.5.

Oxaloacetate is shuttled into the cytoplasm and converted into phosphoenolpyruvate

Oxaloacetate must thus be transported to the cytoplasm to complete the synthesis of phosphoenolpyruvate. Oxaloacetate is first reduced to malate by malate dehydrogenase. Malate is transported across the mitochondrial membrane and reoxidized to oxaloacetate by a cytoplasmic NAD^+-linked malate dehydrogenase (**Figure 16.30**). The formation of oxaloacetate from malate also provides NADH for use in subsequent steps in gluconeogenesis. Finally, oxaloacetate is simultaneously decarboxylated and phosphorylated by phosphoenolpyruvate carboxykinase (PEPCK) to generate phosphoenolpyruvate. The phosphoryl donor is GTP. The CO_2 that was added to pyruvate by pyruvate carboxylase comes off in this step.

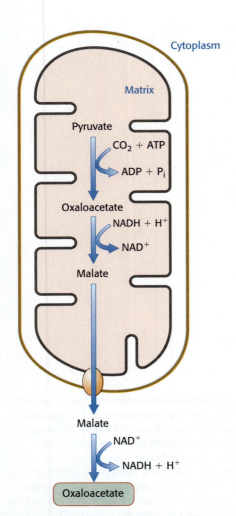

FIGURE 16.30 Oxaloacetate used in the cytoplasm for gluconeogenesis is formed in the mitochondrial matrix. Pyruvate is carboxylated to form oxaloacetate that then leaves the mitochondrion by a specific transport system (oval structure in the mitochondrial membranes) in the form of malate. Malate is reoxidized to oxaloacetate in the cytoplasm.

The chemical structures show the reaction:

Oxaloacetate + GTP $\rightleftharpoons$ (Phosphoenolpyruvate carboxykinase) Phosphoenolpyruvate + GDP + CO_2

Oxaloacetate ... **Phosphoenolpyruvate**

The sum of the reactions catalyzed by pyruvate carboxylase and phosphoenolpyruvate carboxykinase is

$$\text{Pyruvate} + \text{ATP} + \text{GTP} + \text{H}_2\text{O} \longrightarrow \text{phosphoenolpyruvate} + \text{ADP} + \text{GDP} + \text{P}_i + 2\text{H}^+$$

This pair of reactions bypasses the irreversible reaction catalyzed by pyruvate kinase in glycolysis.

Why is a carboxylation and a decarboxylation required to form phosphoenolpyruvate from pyruvate? Recall that, in glycolysis, the presence of a phosphoryl group traps the unstable enol isomer of pyruvate as phosphoenolpyruvate (Section 16.2). However, the addition of a phosphoryl group to pyruvate is a highly unfavorable reaction: the $\Delta G^{\circ\prime}$ of the reverse of the glycolytic reaction catalyzed by pyruvate kinase is +31 kJ mol^{-1} (+7.5 kcal mol^{-1}). In gluconeogenesis, the use of the carboxylation and decarboxylation steps results in a much more favorable $\Delta G^{\circ\prime}$. The formation of phosphoenolpyruvate from pyruvate in the gluconeogenic pathway has a $\Delta G^{\circ\prime}$ of + 0.8 kJ mol^{-1} (+0.2 kcal mol^{-1}). A molecule of ATP is used to power the addition of a molecule of CO_2 to pyruvate in the carboxylation step. That CO_2 is then removed to power the formation of phosphoenolpyruvate in the decarboxylation step.

Decarboxylations often drive reactions that are otherwise highly endergonic. This metabolic motif is used in the citric acid cycle (Chapter 17), the pentose phosphate pathway (Chapter 20), and fatty acid synthesis (Section 22.4).

The conversion of fructose 1,6-bisphosphate into fructose 6-phosphate and orthophosphate is an irreversible step

On formation, phosphoenolpyruvate is metabolized by the enzymes of glycolysis but in the reverse direction. These reactions are near equilibrium under intracellular conditions; so, when conditions favor gluconeogenesis, the reverse reactions will take place until the next irreversible step is reached. This step is the hydrolysis of fructose 1,6-bisphosphate to fructose 6-phosphate and P_i.

$$\text{Fructose 1,6-bisphosphate} + \text{H}_2\text{O} \xrightarrow{\text{Fructose 1,6-bisphosphatase}} \text{fructose 6-phosphate} + \text{P}_i$$

The enzyme responsible for this step is fructose 1,6-bisphosphatase (FBPase). Like its glycolytic counterpart, it is an allosteric enzyme that participates in the regulation of gluconeogenesis. It is also an example of a **phosphatase**, an enzyme that catalyzes the hydrolysis of a phosphate to form inorganic phosphate and the reverse reaction. Note that the difference between kinases and phosphatases is not the physiologically relevant direction of the reaction but whether the phosphate is transferred to water or ADP.

The generation of free glucose occurs only in some tissues and is an important control point

The fructose 6-phosphate generated by fructose 1,6-bisphosphatase is readily converted into glucose 6-phosphate. The final step in the generation of free glucose takes place primarily in the liver, a tissue whose metabolic duty is to maintain adequate glucose concentration in the blood for use by other tissues. Free glucose is not formed in the cytoplasm. Rather, glucose 6-phosphate is transported into the lumen of the endoplasmic reticulum, where it is hydrolyzed to

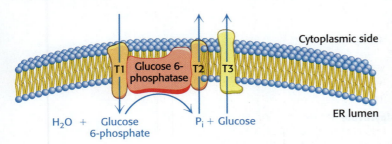

Cytoplasmic side

ER lumen

H_2O + Glucose 6-phosphate P_i + Glucose

FIGURE 16.31 In the liver, free glucose is generated from glucose 6-phosphate by glucose 6-phosphatase. Several endoplasmic reticulum (ER) proteins play a role in the generation of glucose from glucose 6-phosphate in the liver. T1 transports glucose 6-phosphate into the lumen of the ER, whereas T2 and T3 transport P_i and glucose, respectively, back into the cytoplasm. [Information from A. Buchell and I. D. Waddel, *Biochem. Biophys. Acta* 1092:129–137, 1991.]

glucose by glucose 6-phosphatase, which is bound to the ER membrane (**Figure 16.31**). Glucose and P_i are then shuttled back to the cytoplasm by a pair of transporters.

In most tissues, gluconeogenesis ends with the formation of glucose 6-phosphate. Free glucose is not generated because most tissues lack glucose 6-phosphatase. Rather, the glucose 6-phosphate is commonly converted into glycogen, the storage form of glucose (Chapter 21).

Six high-transfer-potential phosphoryl groups are spent in synthesizing glucose from pyruvate

The formation of glucose from pyruvate is energetically unfavorable unless it is coupled to reactions that are favorable. Compare the stoichiometry of gluconeogenesis with that of the reverse of glycolysis.

The stoichiometry of gluconeogenesis is

$$2 \text{ Pyruvate} + 4 \text{ ATP} + 2 \text{ GTP} + 2 \text{ NADH} + 6 H_2O \longrightarrow$$
$$\text{glucose} + 4 \text{ ADP} + 2 \text{ GDP} + 6 P_i + 2 \text{ NAD}^+ + 2 H^+$$
$$\Delta G^{\circ\prime} = -48 \text{ kJ mol}^{-1} (-11 \text{ kcal mol}^{-1})$$

In contrast, the stoichiometry for the reversal of glycolysis is

$$2 \text{ Pyruvate} + 2 \text{ ATP} + \text{NADH} + 2 H_2O \longrightarrow \text{glucose} + 2 \text{ ADP} + 2 P_i + 2 \text{ NAD}^+ + 2 H^+$$
$$\Delta G^{\circ\prime} = +90 \text{ kJ mol}^{-1} (+22 \text{ kcal mol}^{-1})$$

Note that six nucleoside triphosphate molecules are hydrolyzed to synthesize glucose from pyruvate in gluconeogenesis, whereas only *two* molecules of ATP are generated in glycolysis in the conversion of glucose into pyruvate. Thus, the extra cost of gluconeogenesis is four high-phosphoryl-transfer-potential molecules for each molecule of glucose synthesized from pyruvate. The four additional molecules having high phosphoryl-transfer potential are needed to turn an energetically unfavorable process (the reversal of glycolysis) into a favorable one (gluconeogenesis). Here we have a clear example of the coupling of reactions: NTP hydrolysis is used to power an energetically unfavorable reaction. The reactions of gluconeogenesis are summarized in **Table 16.6**.

TABLE 16.6 Reactions of gluconeogenesis

Step	Reaction	Enzyme	Reaction type	$\Delta G^{\circ\prime}$ in kJ mol^{-1}
1	Pyruvate + CO_2 + ATP + $H_2O \longrightarrow$ oxaloacetate + ADP + P_i + 2H$^+$	Pyruvate carboxylase	Phosphoryl transfer	−2.1
2	Oxaloacetate + GTP $\rightleftharpoons$ phosphoenolpyruvate + GDP + CO_2	Phosphoenolpyruvate carboxykinase	Phosphoryl-coupled oxidation	+2.9
3	Phosphoenolpyruvate + H_2O $\rightleftharpoons$ 2-phosphoglycerate	Enolase	Hydration	−1.7
4	2-Phosphoglycerate $\rightleftharpoons$ 3-phosphoglycerate	Phosphoglycerate mutase	Phosphoryl shift	−4.6
5	3-Phosphoglycerate + ATP $\rightleftharpoons$ 1,3-bisphosphoglycerate + ADP	Phosphoglycerate kinase	Phosphoryl transfer	+18.8
6	1,3-Bisphosphoglycerate + NADH + H$^+$ $\rightleftharpoons$ glyceraldehyde 3-phosphate + NAD$^+$ + P_i	Glyceraldehyde 3-phosphate dehydrogenase	Phosphoryl-coupled reduction	−6.3
7	Glyceraldehyde 3-phosphate $\rightleftharpoons$ dihydroxyacetonephosphate	Triose phosphate isomerase	Isomerization	−7.5
8	Glyceraldehyde 3-phosphate + dihydroxyacetone phosphate $\rightleftharpoons$ fructose 1,6-bisphosphate	Aldolase	Aldol addition	−23.8
9	Fructose 1,6-bisphosphate + $H_2O \longrightarrow$ fructose 6-phosphate + P_i	Fructose 1,6-bisphosphatase	Phosphoryl transfer	−16.3
10	Fructose 6-phosphate $\rightleftharpoons$ glucose 6-phosphate	Phosphoglucose isomerase	Isomerization	−1.7
11	Glucose 6-phosphate + $H_2O \longrightarrow$ glucose + P_i	Glucose 6-phosphatase	Phosphoryl transfer	−12.1

16.5 Gluconeogenesis and Glycolysis Are Reciprocally Regulated

Gluconeogenesis and glycolysis are coordinated so that, within a cell, one pathway is relatively inactive while the other is highly active. If both sets of reactions were highly active at the same time, the net result would be the hydrolysis of four nucleoside triphosphates (two ATP molecules plus two GTP molecules) per reaction cycle. Both glycolysis and gluconeogenesis are highly exergonic under cellular conditions, and so there is no thermodynamic barrier to such simultaneous activity. However, the amounts and activities of the distinctive enzymes of each pathway are controlled so that both pathways are not highly active at the same time. The rate of glycolysis is also determined by the concentration of glucose, and the rate of gluconeogenesis by the concentrations of lactate and other precursors of glucose. The basic premise of the reciprocal regulation is that, when energy is needed or glycolytic intermediates are needed for biosynthesis, glycolysis will predominate. When there is a surplus of energy and glucose precursors, gluconeogenesis will take over.

Glycolysis and gluconeogenesis are regulated by adenosine nucleotides and other metabolic intermediates

The first important regulation site is the interconversion of fructose 6-phosphate and fructose 1,6-bisphosphate (**Figure 16.32**). Consider first a situation in which energy is needed, as in active muscle. In this case, the concentration of AMP is high. Under this condition, AMP stimulates phosphofructokinase but inhibits fructose 1,6-bisphosphatase. Thus, glycolysis is turned on and gluconeogenesis is inhibited. Conversely, high levels of ATP and citrate, and therefore low levels of AMP, indicate that the energy charge is high and that biosynthetic intermediates are abundant. ATP and citrate both inhibit phosphofructokinase, whereas the

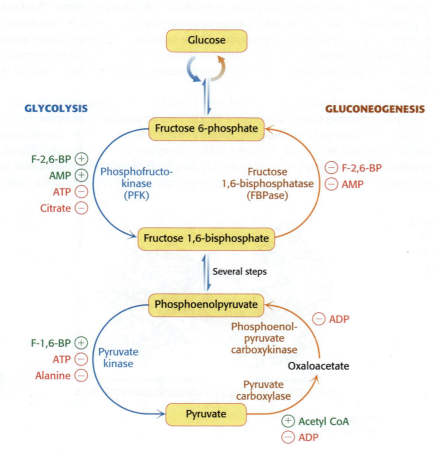

FIGURE 16.32 Gluconeogenesis and glycolysis in the liver are under reciprocal regulation. The concentration of fructose 2,6-bisphosphate is high in the fed state and low in starvation. Another important control is the inhibition of pyruvate kinase by phosphorylation during starvation.

decrease in AMP relieves inhibition of fructose 1,6-bisphosphatase. Under these conditions, glycolysis is nearly switched off and gluconeogenesis is promoted. Why does citrate take part in this regulatory scheme? As we will see in Chapter 17, citrate reports on the status of the citric acid cycle, the primary pathway for oxidizing fuels in the presence of oxygen. High levels of citrate indicate an energy-rich situation and the presence of precursors for biosynthesis.

Glycolysis and gluconeogenesis are also reciprocally regulated at the inter-conversion of phosphoenolpyruvate and pyruvate in the liver. The glycolytic enzyme pyruvate kinase is inhibited by allosteric effectors ATP and alanine, which signal that the energy charge is high and that building blocks are abundant. Conversely, pyruvate carboxylase and phosphoenolpyruvate carboxykinase, which catalyze the first two steps of gluconeogenesis from pyruvate, are inhibited by ADP. Pyruvate carboxylase is also activated by acetyl CoA, which, like citrate, indicates that the citric acid cycle is producing energy and biosynthetic intermediates (Chapter 17). Hence, gluconeogenesis is favored when the cell is rich in biosynthetic precursors and ATP.

In mammals, glycolysis and gluconeogenesis in the liver are controlled by hormones sensitive to blood-glucose concentration

In the liver, rates of glycolysis and gluconeogenesis are adjusted to maintain the blood-glucose concentration. The signal molecule fructose 2,6-bisphosphate strongly stimulates phosphofructokinase (PFK) and inhibits fructose 1,6-bisphosphatase (FBPase). When blood glucose is low, fructose 2,6-bisphosphate (F-2,6-BP) loses a phosphoryl group to form fructose 6-phosphate, which is not an allosteric effector of either PFK or FBPase.

How is the concentration of fructose 2,6-bisphosphate controlled to rise and fall with blood-glucose levels?

- *A bifunctional enzyme determines fructose 2,6-bisphosphate levels.* Two enzyme activities regulate the concentration of F-2,6-BP: one phosphorylates fructose 6-phosphate to form F-2,6-BP, and the other dephosphorylates F-2,6-BP, converting it back to fructose 6-phosphate. The formation reaction is catalyzed by phosphofructokinase 2 (PFK2), a different enzyme from phosphofructokinase. F-2,6-BP is degraded by a specific phosphatase, fructose bisphosphatase 2 (FBPase2).

The striking finding is that both PFK2 and FBPase2 are present in a single 55-kDa polypeptide chain (**Figure 16.33**). This **bifunctional enzyme**, PFK2/FBPase2, contains an N-terminal regulatory domain, followed by a kinase domain and a phosphatase domain. PFK2 resembles adenylate kinase, whereas FBPase2 resembles phosphoglycerate mutase (Section 16.2). Recall that the mutase is essentially a phosphatase. In the bifunctional enzyme, the phosphatase activity

FIGURE 16.33 The bifunctional enzyme PFK2/FBPase2 has two distinct domains. The kinase domain (yellow) is fused to the phosphatase domain (red). The bar represents the amino acid sequence of the enzyme. [Drawn from 1BIF.pdb.]

INTERACT with this model in
 Achieve

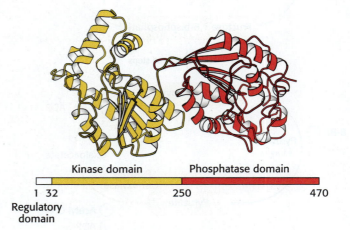

evolved to become specific for F-2,6-BP. The bifunctional enzyme itself probably arose by the fusion of genes encoding the kinase and phosphatase domains.

- *The bifunctional enzyme is controlled by reversible phosphorylation.* What controls whether PFK2 or FBPase2 dominates the bifunctional enzyme's activities in the liver? The activities of PFK2 and FBPase2 are reciprocally controlled by phosphorylation of a single serine residue. When glucose is scarce, such as during a night's fast, a rise in the blood concentration of the hormone glucagon triggers a cyclic AMP signal cascade (Section 14.2), leading to the phosphorylation of this bifunctional enzyme by protein kinase A (**Figure 16.34**). This covalent modification activates FBPase2 and inhibits PFK2, lowering the level of F-2,6-BP. Gluconeogenesis predominates. Glucose formed by the liver under these conditions is essential for the viability of the brain. Glucagon stimulation of protein kinase A also inactivates pyruvate kinase in the liver.

   Conversely, when the blood-glucose concentration is high, such as after a meal, gluconeogenesis is not needed. Insulin is secreted and initiates a signal pathway that activates a protein phosphatase, which removes the phosphoryl group from the bifunctional enzyme. This covalent modification activates PFK2 and inhibits FBPase2. The resulting rise in the level of F-2,6-BP accelerates glycolysis. In liver, a key role of glycolysis is to generate metabolites for biosynthesis. The coordinated control of glycolysis and gluconeogenesis is facilitated by the location of the kinase and phosphatase domains on the same polypeptide chain as the regulatory domain.

- *The bifunctional enzyme is also under transcriptional control by glucagon and insulin.* The hormones glucagon and insulin also regulate the amounts of essential enzymes by altering gene expression, primarily by changing the rate of transcription. Transcriptional control in eukaryotes is much slower than allosteric control, taking hours or days instead of seconds to minutes.

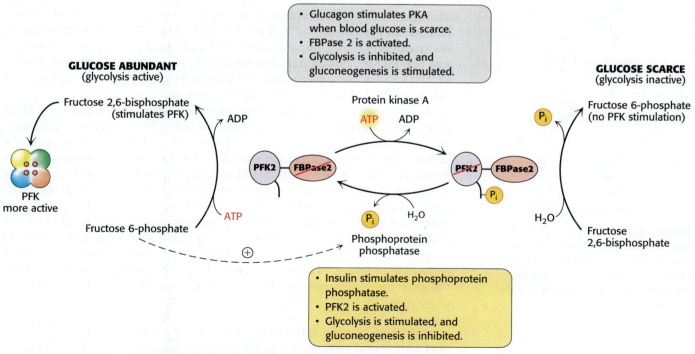

FIGURE 16.34 The synthesis and degradation of fructose 2,6-bisphosphate is hormonally controlled. A low blood-glucose concentration as signaled by glucagon leads to the phosphorylation of the bifunctional enzyme and hence to a lower concentration of fructose 2,6-bisphosphate, slowing glycolysis. Insulin accelerates the formation of fructose 2,6-bisphosphate by facilitating the dephosphorylation of the bifunctional enzyme.

Glucagon rises during fasting, when gluconeogenesis is needed to replace scarce glucose. To encourage gluconeogenesis, glucagon inhibits the expression of the three regulated glycolytic enzymes and stimulates instead the production of two key gluconeogenic enzymes, phosphoenolpyruvate carboxykinase and fructose 1,6-bisphosphatase.

In contrast, insulin levels rise subsequent to eating, when there is plenty of glucose for glycolysis. To encourage glycolysis, insulin stimulates the expression of phosphofructokinase, pyruvate kinase, and the bifunctional enzyme that makes and degrades F-2,6-BP.

❖ **SELF–CHECK QUESTION**

Why is the regulation of phosphofructokinase by ATP and AMP not as important in the liver as it is in muscle?

Substrate cycles amplify metabolic signals and produce heat

A pair of reactions such as the phosphorylation of fructose 6-phosphate to fructose 1,6-bisphosphate and its hydrolysis back to fructose 6-phosphate is called a **substrate cycle**. As already mentioned, both reactions are not fully active at the same time because of reciprocal regulation. However, isotope-labeling studies have shown that some fructose 6-phosphate is phosphorylated to fructose 1,6-bisphosphate even during gluconeogenesis. There also is a limited degree of cycling in other pairs of opposed irreversible reactions. This cycling was regarded as an imperfection in metabolic control, and so substrate cycles have sometimes been called *futile cycles*.

Pathological conditions may result from futile cycles. A clear example is malignant hyperthermia, which occurs predominantly in muscle tissue. Susceptible individuals are sensitive to certain anesthetics which cause uncontrolled calcium release from the sarcoplasmic reticulum through the calcium channel. The rapid rise in cytoplasmic calcium activates the sarcoplasmic reticulum calcium pump, the Ca^{2+} ATPase, a P-type ATPase, in an attempt to remove the calcium from the cytoplasm. The constant release and reuptake of calcium, and the consequent rapid hydrolysis of ATP, raises the body temperature to 44°C (111°F) and depletes muscle ATP. Muscle damage and death may occur if untreated with an inhibitor of the calcium channel.

Despite such extraordinary circumstances, substrate cycles now seem likely to be biologically important. One possibility is that substrate cycles amplify metabolic signals. Suppose that the rate of conversion of A into B is 100 and of B into A is 90, giving an initial net flux of 10. Assume that an allosteric effector increases the A → B rate by 20% to 120 and reciprocally decreases the B → A rate by 20% to 72. The new net flux is 48, and so a 20% change in the rates of the opposing reactions has led to a 380% increase in the net flux. In the example shown in **Figure 16.35**, this amplification is made possible by the rapid hydrolysis of ATP. The flux down the glycolytic pathway has been suggested to increase as much as 1000-fold at the initiation of intense exercise. Because the allosteric activation of enzymes alone seems unlikely to explain this increased flux, the existence of substrate cycles may partly account for the rapid rise in the rate of glycolysis.

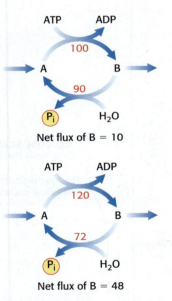

FIGURE 16.35 Substrate cycles can have advantages. This ATP-driven cycle operates at two different rates. A small change in the rates of the two opposing reactions results in a large change in the *net* flux of product B.

Lactate and alanine formed by contracting muscle and peripheral tissues are used by other organs

Lactate produced by active skeletal muscle, erythrocytes, skin, and other tissues is a source of energy for other organs. Erythrocytes lack mitochondria and can never oxidize glucose completely, whereas other tissues like the skin remain largely anaerobic at all times due to the distance from the arterial blood supply.

In contracting fast-twitch skeletal muscle fibers during vigorous exercise, the rate at which glycolysis produces pyruvate exceeds the rate at which the citric acid cycle oxidizes it. In these cells, lactate dehydrogenase reduces excess pyruvate to lactate to restore redox balance (Section 16.2). However, lactate is a dead-end in metabolism. It must be converted back into pyruvate before it can be metabolized. Both pyruvate and lactate diffuse out of these cells through carriers into the blood. In contracting skeletal muscle, the formation and release of lactate lets the muscle generate ATP in the absence of oxygen and shifts the burden of metabolizing lactate from muscle to other organs.

The pyruvate and lactate in the bloodstream have two fates. In one fate, the plasma membranes of some cells—particularly cells in cardiac muscle and slow-twitch (type 1) skeletal muscle—contain carriers that make the cells highly permeable to lactate and pyruvate. These molecules diffuse from the blood into such permeable cells. Once inside these well-oxygenated cells, lactate can be reverted back to pyruvate and metabolized through the citric acid cycle and oxidative phosphorylation to generate ATP. The use of lactate in place of glucose by these cells makes more circulating glucose available to the active muscle cells. In the other fate, excess lactate enters the liver and is converted first into pyruvate and then into glucose by the gluconeogenic pathway. Contracting skeletal muscle supplies lactate to the liver, which uses it to synthesize and release glucose. Thus, the liver restores the level of glucose necessary for active muscle cells, which derive ATP from the glycolytic conversion of glucose into lactate. Because this multi-tissue cooperation was elucidated by Gerty and Carl Cori, it is known as the **Cori cycle** (**Figure 16.36**).

Studies have shown that alanine, like lactate, is a major precursor of glucose in the liver. The alanine is generated in muscle when the carbon skeletons of some amino acids are used as fuels. The nitrogens from these amino acids are transferred to pyruvate to form alanine; the reverse reaction takes place in the liver. This process also helps maintain nitrogen balance. The interplay between glycolysis and gluconeogenesis is summarized in **Figure 16.37**, which shows how these pathways help meet the energy needs of different cell types.

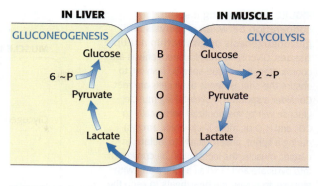

FIGURE 16.36 The Cori cycle allows the liver to better support other tissues, such as skeletal muscle. Lactate formed by active muscle is converted into glucose by the liver. This cycle shifts part of the metabolic burden of active muscle to the liver. The symbol, ~P represents nucleoside triphosphates.

EXPLORE this pathway further in the Metabolic Map in
Achieve

❖ SELF–CHECK QUESTION

If lactate is a dead-end product in that its sole fate is to be converted back into glucose, then why do cells bother to form lactate at all?

⚕ Deficiencies in glycolytic or gluconeogenic enzymes are rare genetic disorders

Genetically inherited deficiencies in the activity or regulation of specific enzymes are sometimes called *inborn errors of metabolism*. Because the enzymes of glycolysis and gluconeogenesis are so central to human metabolism, nonlethal deficiencies in these enzymes are rare but do still occur. Here we will briefly discuss two:

Triose phosphate isomerase deficiency. TPID is a multisystem disorder presenting in early childhood. Symptoms include congenital hemolytic anemia (deficiency of red cells due to premature destruction of the cell that is present at birth) and progressive neuromuscular disorder, including cardiomyopathy (inflammation and damage to the heart muscle). In severe cases, TPID can lead to death in early childhood. Dihydroxyacetone phosphate accumulates in cells, especially red blood cells. TPID is rare, so the causal relationship between the biochemical defect and the clinical features remains to be established. Let's consider a hypothesis regarding the pathologies of TPID.

Alfred Eisenstaedt/The LIFE Picture Collection/Shutterstock

GERTY CORI After growing up in a Jewish family in Austria-Hungary, Gerty Cori (*née* Radnitz) attended medical school at the German University of Prague. During medical school she met Carl Cori, and the two were married and became lifelong research partners, working together as equals. Despite this, Carl often received more recognition than Gerty due to gender discrimination, something he ardently opposed his entire career. They relocated to the United States in 1922 due to rising anti-Semitism in Europe. Dr. Cori published numerous papers with Carl, and independently, on tumor and carbohydrate metabolism. She became the first woman to win the Nobel Prize in Physiology or Medicine, which she shared with Carl for their work on glycogen metabolism and the elaboration of the Cori Cycle. Notably, six other scientists who were mentored by the Coris also went on to win Nobel Prizes.

FIGURE 16.37 During a sprint, cooperation between glycolysis and gluconeogenesis occurs within multiple tissues. Glycolysis and gluconeogenesis are coordinated, in a tissue-specific fashion, to ensure the energy needs of all cells are met. Consider a sprinter. In skeletal leg muscle, glucose will be metabolized aerobically to CO_2 and H_2O or, more likely (thick arrows) during a sprint, anaerobically to lactate. In cardiac muscle, the lactate can be converted into pyruvate and used as a fuel, along with glucose, to power the heartbeats to keep the sprinter's blood flowing. Gluconeogenesis, a primary function of the liver, will be taking place rapidly (thick arrows) to ensure that enough glucose is present in the blood for skeletal and cardiac muscle, as well as for other tissues. Glycogen, glycerol, and amino acids are other sources of energy that we will learn about in later chapters.

EXPLORE this pathway further in the Metabolic Map in

 Achieve

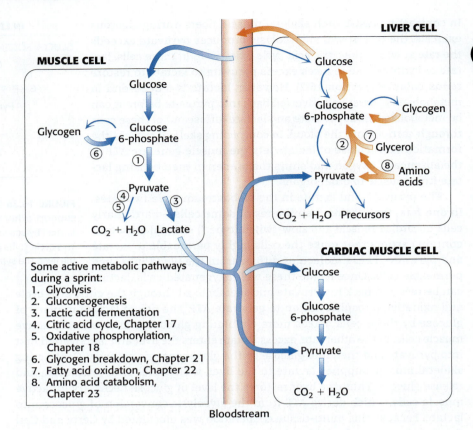

Methylglyoxal

First of all, recall that central nervous system and red blood cells rely on glucose metabolism exclusively for energy. Thus, any disruption in glycolysis would impact these tissues. Also, though resting skeletal muscles utilize fats as fuels, rapidly contracting muscle quickly becomes reliant on glucose for lactic acid fermentation.

Consider the consequences of TPID by reviewing its role in glycolysis in Figure 16.2. If the isomerase activity is missing, then half of the carbons of glucose cannot be metabolized to yield ATP. These observations account for the nature of the symptoms: without adequate ATP, neuromuscular function would be compromised. However, research suggests that disruption of energy metabolism is not the cause of the more serious consequences of TPID. Even if the cells try to compensate for lack of isomerase activity by processing more glucose, there will be an unavoidable buildup of dihydroxyacetone phosphate. Strong evidence now exists that this buildup of dihydroxyacetone phosphate is to blame, as it can be converted into methylglyoxal.

Methylglyoxal, a highly reactive molecule, covalently binds to available amino groups on proteins, yielding advanced glycation end products (AGEs) (Section 11.1). These modifications inhibit protein function. Extensive loss of protein function would then contribute to the pathologies observed in TPID, including early death. AGEs have also been implicated in aging, arteriosclerosis (thickening and hardening of artery walls), and diabetes. Clearly establishing the causes of all TPID pathologies awaits further research.

Pyruvate carboxylase deficiency. PCD is another rare disorder of carbohydrate metabolism. Several types vary in the seriousness of clinical outcomes, but all are characterized to some extent by hypoglycemia (low concentration of blood glucose) and lactic acidosis (excessive lactic acid in the blood). Symptoms include lethargy, seizures, and, in severe cases, death in the first few months of life. In this case the two primary biochemical characteristics, hypoglycemia and lactic acidosis, are reasonably easy to explain. Again, as with TPID, lack of glucose would impair neuromuscular functions.

Because pyruvate carboxylase is a key regulatory enzyme in gluconeogenesis, which occurs primarily in the liver, we can surmise that the lack of the pyruvate carboxylase would account for the hypoglycemia. Lacking this enzyme, the liver cannot serve its role in maintaining adequate blood glucose concentration for tissues that depend on glucose. Likewise, an important role of liver is to remove lactic acid from the blood and use it as a gluconeogenic precursor. However, lactic acid can't be used to form glucose due to the lack of pyruvate carboxylase, causing it to remain in the blood, leading to a fall in blood pH (acidosis). In later chapters we will see that the oxaloacetate formed by pyruvate carboxylase is an important and versatile molecule, so the lack of carboxylase activity has greater effects than discussed here.

Glycolysis and gluconeogenesis are evolutionarily intertwined

The metabolism of glucose has ancient origins. Organisms living in the early biosphere depended on the anaerobic generation of energy until significant amounts of oxygen began to accumulate 2 billion years ago. Glycolytic enzymes were most likely derived independently rather than by gene duplication, because glycolytic enzymes with similar properties do not have similar amino acid sequences. Although there are four kinases and two isomerases in the pathway, sequence and structural comparisons suggest that these sets of enzymes are not related to one another by divergent evolution.

We can speculate on the relationship between glycolysis and gluconeogenesis if we think of glycolysis as consisting of two segments: the metabolism of hexoses (the upper segment of Figure 16.2) and the metabolism of trioses (the lower segment of Figure 16.2). The enzymes of the upper segment are different in some species and are missing entirely in some archaea, whereas enzymes of the lower segment are quite conserved. In fact, four enzymes of the lower segment are present in all species. The lower part of the glycolytic pathway is used by gluconeogenesis, too. This common part of the two pathways may be the oldest part, constituting the core to which the other steps were added. The upper part would have varied according to the sugars that were available to evolving organisms in particular niches. Interestingly, this core part of carbohydrate metabolism can generate triose precursors for ribose sugars, a component of RNA and a critical requirement for the RNA world. Thus, we are left with the unanswered question: Was the original core pathway used for energy conversion or biosynthesis?

❖ SELF–CHECK QUESTION

What are the likely consequences of a genetic disorder rendering fructose 1,6-bisphosphatase in the liver less sensitive to regulation by fructose 2,6-bisphosphate?

Summary

16.1 Glycolysis Is an Energy-Conversion Pathway in Most Organisms

- Glycolysis is a highly common pathway among most organisms; it consists of a set of 10 cytoplasmic reactions that convert glucose into pyruvate.
- Glucose and other monosaccharides are generated from the breakdown of dietary starch, glycogen, sucrose, and other nutrients.
- A family of glucose transporters facilitates the movement of glucose throughout the body.

16.2 Glycolysis Can Be Divided into Two Parts

- The first stage consumes two molecules of ATP as glucose is converted into dihydroxyacetone phosphate and glyceraldehyde 3-phosphate, which are readily interconvertible.
- In the second stage, glyceraldehyde 3-phosphate is oxidized and ultimately converted into pyruvate, generating two molecules of ATP.
- There is a net gain of two molecules of ATP in the formation of two molecules of

pyruvate from one molecule of glucose because each of the ATP-generating second stage steps occur twice for each molecule of glucose metabolized.

- The electron acceptor in the oxidation of glyceraldehyde 3-phosphate is NAD^+, which must be regenerated for glycolysis to continue.

- In aerobic organisms, the NADH formed in glycolysis transfers its electrons to O_2 through the electron-transport chain, which thereby regenerates NAD^+.

- Under anaerobic conditions and in some microorganisms, NAD^+ is regenerated by the reduction of pyruvate to lactate. In other microorganisms, NAD^+ is regenerated by the reduction of pyruvate to ethanol. These two processes are examples of fermentations.

16.3 The Glycolytic Pathway Is Tightly Controlled

- The rate of conversion of glucose into pyruvate is tightly regulated at specific steps.

- Phosphofructokinase—the most important control element in glycolysis—is inhibited by high levels of ATP and citrate and activated by AMP and fructose 2,6-bisphosphate.

- Hexokinase is inhibited by glucose 6-phosphate, which accumulates when phosphofructokinase is inactive.

- ATP and alanine allosterically inhibit pyruvate kinase, whereas fructose 1,6-bisphosphate activates it.

16.4 Glucose Can Be Synthesized from Noncarbohydrate Precursors

- Gluconeogenesis, which occurs primarily in the liver, is the synthesis of glucose from noncarbohydrate sources, such as lactate, amino acids,

glycerol, and alanine produced from pyruvate by active skeletal muscle.

- Gluconeogenesis requires four new reactions to bypass the essential irreversibility of three reactions in glycolysis.

- In two of the new reactions, pyruvate is carboxylated in mitochondria to oxaloacetate, which in turn is decarboxylated and phosphorylated in the cytoplasm to phosphoenolpyruvate, at the expense of one ATP and one GTP.

- The other distinctive reactions of gluconeogenesis are the hydrolyses of fructose 1,6-bisphosphate and glucose 6-phosphate, which are catalyzed by specific phosphatases.

16.5 Gluconeogenesis and Glycolysis Are Reciprocally Regulated

- Gluconeogenesis and glycolysis are reciprocally regulated so that one pathway is relatively inactive while the other is highly active.

- Phosphofructokinase and fructose 1,6-bisphosphatase are key control elements of gluconeogenesis and glycolysis.

- Fructose 2,6-bisphosphate, an intracellular signal molecule present at higher levels when glucose is abundant, activates glycolysis and inhibits gluconeogenesis by regulating these enzymes.

- Pyruvate kinase and pyruvate carboxylase are regulated by other effectors so that both are not maximally active at the same time.

- Allosteric regulation and reversible phosphorylation, which are rapid, are complemented by transcriptional control, which takes place in hours or days.

Key Terms

glycolysis (p. 473)

gluconeogenesis (p. 473)

kinase (p. 476)

substrate-level phosphorylation (p. 482)

fermentation (p. 486)

ethanol fermentation (p. 486)

lactic acid fermentation (p. 487)

obligate anaerobe (p. 488)

facultative anaerobe (p. 488)

committed step (p. 494)

feedforward stimulation (feedforward activation) (p. 494)

substrate channeling (p. 497)

aerobic glycolysis (p. 497)

phosphatase (p. 503)

bifunctional enzyme (p. 506)

substrate cycle (p. 508)

Cori cycle (p. 509)

Problems

1. Select all the steps of glycolysis in which ATP is produced.
(a) 1,3-bisphosphoglycerate → 3-phosphoglycerate
(b) Glyceraldehyde 3-phosphate → 1,3-bisphosphoglycerate
(c) Fructose 6-phosphate → fructose 1,6-bisphosphate
(d) Glucose → glucose 6-phosphate
(e) Phosphoenolpyruvate → pyruvate

2. Which of the following statements is (are) true for a muscle performing lactic acid fermentation?
(a) The process is inhibited by AMP.
(b) The process is inhibited by ATP.
(c) There is a net synthesis of NADH.
(d) The process is exergonic.

3. Match each term with its description. ❖ 1

(a) Hexokinase
(b) Phosphoglucose isomerase
(c) Phosphofructokinase
(d) Aldolase
(e) Triose phosphate isomerase
(f) Glyceraldehyde 3-phosphate dehydrogenase
(g) Phosphoglycerate kinase
(h) Phosphoglycerate mutase
(i) Enolase
(j) Pyruvate kinase

1. Forms fructose 1,6-bisphosphate
2. Generates the first high-phosphoryl-transfer-potential compound that is not ATP
3. Converts glucose 6-phosphate into fructose 6-phosphate
4. Phosphorylates glucose
5. Generates the second molecule of ATP
6. Cleaves fructose 1,6-bisphosphate
7. Generates the second high-phosphoryl-transfer-potential compound that is not ATP
8. Catalyzes the interconversion of three-carbon isomers
9. Converts 3-phosphoglycerate into 2-phosphoglycerate
10. Generates the first molecule of ATP

4. Lactic acid fermentation and ethanol fermentation are oxidation–reduction reactions. Identify the ultimate electron donor and electron acceptor. ❖ 2

5. Why is it advantageous for the liver to have both hexokinase and glucokinase to phosphorylate glucose?

6. The interconversion of DHAP and GAP greatly favors the formation of DHAP at equilibrium. Yet the conversion of DHAP by triose phosphate isomerase proceeds readily. Why?

7. Although both hexokinase and phosphofructokinase catalyze irreversible steps in glycolysis and the hexokinase-catalyzed step is first, phosphofructokinase is nonetheless the pacemaker of glycolysis. What does this information tell you about the fate of the glucose 6-phosphate formed by hexokinase?

8. What reactions of glycolysis are not readily reversible under intracellular conditions? ❖ 4

9. Why is it physiologically advantageous for the pancreas to use GLUT2, with a high K_M, as the transporter that allows glucose entry into β cells?

10. Suppose that a microorganism that was an obligate anaerobe suffered a mutation that resulted in the loss of triose phosphate isomerase activity. How would this loss affect the ATP yield of fermentation? Could such an organism survive? ❖ 1

11. What are the equilibrium concentrations of fructose 1,6-bisphosphate, dihydroxyacetone phosphate, and glyceraldehyde 3-phosphate when 1 mM fructose 1,6-bisphosphate is incubated with aldolase under standard conditions?

12. Indicate which of the conditions listed in the right-hand column increase the activity of the glycolytic and gluconeogenic pathway in the liver. ❖ 4
(a) Glycolysis
(b) Gluconeogenesis

1. Increase in ATP
2. Increase in AMP
3. Increase in fructose 2,6-bisphosphate
4. Increase in citrate
5. Increase in acetyl CoA
6. Increase in insulin
7. Increase in glucagon
8. Fasting
9. Fed

13. Why does the lack of glucose 6-phosphatase activity in the brain and muscle make good physiological sense?

14. How many NTP molecules are required for the synthesis of one molecule of glucose from two molecules of pyruvate? How many NADH molecules? ❖ 3

15. Suppose 25 glucose molecules are formed during gluconeogenesis. Calculate the amount of pyruvate, ATP, and NADH molecules required.

16. Predict the effect of each of the following mutations on the pace of glycolysis in liver cells: ❖ 4
(a) Loss of the allosteric site for ATP in phosphofructokinase
(b) Loss of the binding site for citrate in phosphofructokinase

(c) Loss of the phosphatase domain of the bifunctional enzyme that controls the level of fructose 2,6-bisphosphate

(d) Loss of the binding site for fructose 1,6-bisphosphate in pyruvate kinase

17. Avidin, a 70-kDa protein in egg white, has very high affinity for biotin. In fact, it is a highly specific inhibitor of biotin enzymes. Which of the following conversions would be blocked by the addition of avidin to a cell homogenate?

(a) Glucose $\longrightarrow$ pyruvate

(b) Pyruvate $\longrightarrow$ glucose

(c) Oxaloacetate $\longrightarrow$ glucose

(d) Malate $\longrightarrow$ oxaloacetate

(e) Pyruvate $\longrightarrow$ oxaloacetate

(f) Glyceraldehyde 3-phosphate $\longrightarrow$

$$\text{fructose 1,6-bisphosphate}$$

18. In the conversion of glucose into two molecules of lactate, the NADH generated earlier in the pathway is oxidized to NAD^+. Why is it not to the cell's advantage to simply make more NAD^+ so that the regeneration would not be necessary? After all, the cell would save much energy because it would no longer need to synthesize lactic acid dehydrogenase. ❖ 2

19. Hexokinase in red blood cells has a K_M of approximately 50 µM. Because life is hard enough as it is, let's assume that the hexokinase displays Michaelis–Menten kinetics. What concentration of blood glucose would yield v_o equal to 90% V_{max}? What does this result tell you if normal blood glucose levels range between approximately 3.6 and 6.1 mM?

20. In principle, a futile cycle that includes phosphofructokinase and fructose 1,6-bisphosphatase could be used to generate heat. The heat could be used to warm tissues.

For instance, certain bumblebees have been reported to use such a futile cycle to warm their flight muscles on cool mornings. Scientists undertook a series of experiments to determine if a number of species of bumblebee use this futile cycle. Their approach was to measure the activity of PFK and F-1,6-BPase in flight muscle.

(a) What was the rationale for comparing the activities of these two enzymes?

(b) The data below show the activities of both enzymes for a variety of bumblebee species (genera *Bombus* and *Psithyrus*). Do these results support the notion that bumblebees use futile cycles to generate heat? Explain.

(c) In which species might futile cycling take place? Explain your reasoning.

(d) Do these results prove that futile cycling does not participate in heat generation?

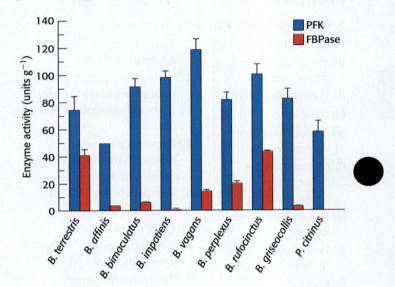

Pyruvate Dehydrogenase and the Citric Acid Cycle

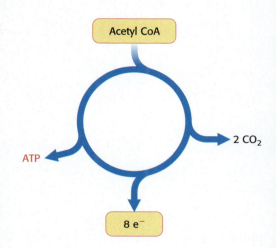

Roundabouts, or traffic circles, function as hubs to facilitate traffic flow. Likewise, the citric acid cycle is the biochemical hub of the cell, oxidizing carbon fuels, usually in the form of acetyl CoA, as well as serving as a source of precursors for biosynthesis.

Chalermkiat Seedokmal/Getty Images

❖ LEARNING GOALS

By the end of this chapter, you should be able to:

1. Explain why the reaction catalyzed by the pyruvate dehydrogenase complex is a crucial juncture in metabolism.
2. Identify the means by which the pyruvate dehydrogenase complex is regulated.
3. Identify the primary catabolic purpose of the citric acid cycle.
4. Explain the efficiency of using the citric acid cycle to oxidize acetyl CoA.
5. Describe how the citric acid cycle is regulated.
6. Describe the role of the citric acid cycle in biosynthesis.
7. Identify the biochemical advantages that the glyoxylate cycle provides.

OUTLINE

17.1 The Citric Acid Cycle Harvests High-Energy Electrons

17.2 The Pyruvate Dehydrogenase Complex Links Glycolysis to the Citric Acid Cycle

17.3 The Citric Acid Cycle Oxidizes Two-Carbon Units

17.4 Entry to the Citric Acid Cycle and Metabolism Through It Are Controlled

17.5 The Citric Acid Cycle Is a Source of Biosynthetic Precursors

17.6 The Glyoxylate Cycle Enables Plants and Bacteria to Grow on Acetate

The metabolism of glucose to pyruvate in glycolysis, an anaerobic process, harvests but a fraction of the ATP available from glucose. Most of the ATP generated in metabolism is provided by the *aerobic* processing of glucose. This process starts with the complete oxidation of glucose derivatives to carbon dioxide. This oxidation begins with pyruvate and takes place in a series of reactions catalyzed by an enormous enzyme cluster called the *pyruvate dehydrogenase complex*. The oxidation ends with the citric acid cycle, a set of eight reactions that forms essentially the center of the complex web of catabolic and biosynthetic transformations known as *metabolism*. The electrons removed in the conversion of pyruvate to carbon dioxide ultimately provide the energy for the synthesis of most of the ATP within a cell.

515

17.1 The Citric Acid Cycle Harvests High-Energy Electrons

The **citric acid cycle (CAC)**, also known as the **tricarboxylic acid (TCA) cycle** or the **Krebs cycle**, is the final pathway for the oxidation of fuel molecules—carbohydrates, fatty acids, and amino acids. Most fuel molecules enter the cycle as **acetyl CoA** (acetyl coenzyme A).

Acetyl coenzyme A (Acetyl CoA)

Under aerobic conditions, the pyruvate generated from glucose is first oxidatively decarboxylated to form acetyl CoA by a large enzyme complex called the **pyruvate dehydrogenase complex**. The acetyl CoA then enters the citric acid cycle, where all remaining carbons originally derived from glucose are completely oxidized to CO_2. In eukaryotes, the reactions of the pyruvate dehydrogenase complex and the citric acid cycle take place in the matrix of the mitochondria (**Figure 17.1**), in contrast with those of glycolysis, which take place in the cytoplasm.

FIGURE 17.1 Mitochondria have distinct compartments defined by two membranes. The double membrane of a single mitochondrion is evident in this electron micrograph. The numerous invaginations of the inner mitochondrial membrane are called *cristae*. The oxidative decarboxylation of pyruvate and the sequence of reactions in the citric acid cycle take place within the matrix. [Omikron/Science Source.]

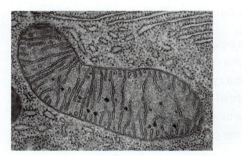

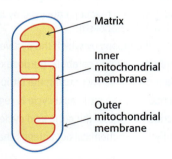

What is the function of the citric acid cycle in transforming fuel molecules into ATP? Recall that fuel molecules are carbon compounds that are capable of being oxidized—that is, of losing electrons (Section 15.3). The citric acid cycle includes a series of oxidation–reduction reactions that result in the oxidation of an acetyl group to two molecules of carbon dioxide. This oxidation generates high-energy electrons that will be used to power the synthesis of ATP. The catabolic function of the citric acid cycle is the harvesting of high-energy electrons from carbon fuels.

The citric acid cycle is also the central metabolic hub of the cell, providing a link that connects both catabolic and anabolic pathways. It is the gateway to the aerobic metabolism of any molecule that can be transformed into an acetyl group or a component of the citric acid cycle. The cycle is also an important source of precursors for the building blocks of many other molecules such as amino acids, nucleotide bases, and porphyrin (the organic component of heme). The citric acid cycle component, oxaloacetate, is also an important precursor to glucose (Section 16.4).

The overall pattern of the citric acid cycle is shown in **Figure 17.2**. A four-carbon compound (oxaloacetate) condenses with a two-carbon acetyl unit to yield a six-carbon tricarboxylic acid. The six-carbon compound releases CO_2 twice in two successive oxidative decarboxylations that yield high-energy electrons. What remains is a four-carbon compound that is further processed to regenerate oxaloacetate, which can initiate another round of the cycle. Two carbon atoms enter the cycle as an acetyl unit, and two carbon atoms leave the cycle in the form of two molecules of CO_2.

Note that the citric acid cycle itself neither generates much ATP nor includes oxygen as a reactant (**Figure 17.3**). Instead, the citric acid cycle removes electrons from acetyl CoA and uses these electrons to reduce NAD^+ and FAD to form NADH and $FADH_2$. Three hydride ions (hence, six electrons) are transferred to three molecules of nicotinamide adenine dinucleotide (NAD^+), and one pair of hydrogen atoms (hence, two electrons) is transferred to one molecule of flavin adenine dinucleotide (FAD) each time an acetyl CoA is processed by the cycle. Electrons released in the reoxidation of NADH and $FADH_2$ flow through a series of membrane proteins (referred to as the *electron-transport chain*) to generate a proton gradient across the inner mitochondrial membrane. These protons then flow through ATP synthase to generate ATP from ADP and inorganic phosphate. These electron carriers yield nine molecules of ATP when they are oxidized by O_2 through the process of oxidative phosphorylation (Chapter 18).

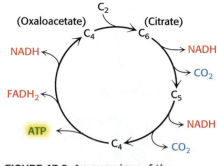

FIGURE 17.2 An overview of the citric acid cycle reveals the flow of carbon atoms and the major products of the cycle. A two-carbon group (C_2) combines with a four-carbon molecule (C_4, oxaloacetate) to form a six-carbon molecule (C_6, citrate). Two carbons are lost as CO_2 in the cycle, ending with the regeneration of oxaloacetate for another round. As the result, each round of the citric acid cycle oxidizes two-carbon units, producing two molecules of CO_2, one molecule of ATP, and high-energy electrons in the form of NADH and $FADH_2$.

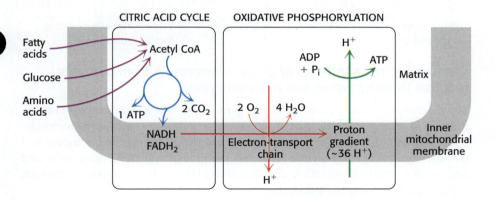

FIGURE 17.3 Cellular respiration removes high-energy electrons from carbon fuel molecules to generate ATP. The citric acid cycle constitutes the first stage in cellular respiration, the removal of high-energy electrons from carbon fuels in the form of NADH and $FADH_2$ (blue pathway). These electrons reduce O_2 to generate a proton gradient (red pathway), which is used to synthesize ATP (green pathway). The reduction of O_2 and the synthesis of ATP constitute oxidative phosphorylation.

The citric acid cycle, in conjunction with oxidative phosphorylation, provides the bulk of energy used by aerobic cells—in human beings, greater than 90%. It is highly efficient because the oxidation of a limited number of citric acid cycle molecules can generate large amounts of NADH and $FADH_2$. Note in Figure 17.2 that the four-carbon molecule, oxaloacetate, that initiates the first step in the citric acid cycle is regenerated at the end of one passage through the cycle. Thus, one molecule of oxaloacetate is capable of participating in the oxidation of many acetyl groups.

17.2 The Pyruvate Dehydrogenase Complex Links Glycolysis to the Citric Acid Cycle

Carbohydrates, most notably glucose, are processed by glycolysis into pyruvate (Chapter 16). Under anaerobic conditions, the pyruvate is converted into lactate or ethanol, depending on the organism. Under aerobic conditions, the pyruvate is transported into mitochondria by a specific carrier protein embedded in the mitochondrial membrane. In the mitochondrial matrix, pyruvate is oxidatively

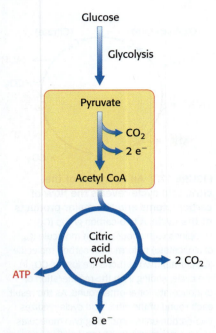

FIGURE 17.4 The pyruvate dehydrogenase complex connects glycolysis and the citric acid cycle. Pyruvate produced by glycolysis is converted into acetyl CoA, the fuel of the citric acid cycle.

decarboxylated by the pyruvate dehydrogenase complex—a highly integrated unit of three distinct enzymes—to form acetyl CoA.

$$\text{Pyruvate} + \text{CoA} + \text{NAD}^+ \longrightarrow \text{acetyl CoA} + \text{CO}_2 + \text{NADH} + \text{H}^+$$

This irreversible reaction is the link between glycolysis and the citric acid cycle (**Figure 17.4**). Note that the pyruvate dehydrogenase complex produces CO_2 and captures high-transfer-potential electrons in the form of NADH. Thus, the pyruvate dehydrogenase reaction has many of the key features of the reactions of the citric acid cycle itself.

The pyruvate dehydrogenase complex, the components of which are detailed in **Table 17.1**, is another example of the organization of enzymes into supramolecular structures (Section 16.3). Pyruvate dehydrogenase complex is a member of a family of homologous complexes that include the citric acid cycle enzyme α-ketoglutarate dehydrogenase complex (Section 17.3). These complexes are giant, larger than ribosomes, with molecular masses ranging from 4 million to 10 million Da. As we will see, their elaborate structures allow groups to travel from one active site to another, connected by tethers to the core of the structure.

TABLE 17.1 Pyruvate dehydrogenase complex of *E. coli*

Enzyme	Abbreviation	Prosthetic group	Reaction catalyzed
Pyruvate dehydrogenase component	E_1	TPP	Oxidative decarboxylation of pyruvate
Dihydrolipoyl transacetylase	E_2	Lipoamide	Transfer of acetyl group to CoA
Dihydrolipoyl dehydrogenase	E_3	FAD	Regeneration of the oxidized form of lipoamide

Mechanism: The synthesis of acetyl coenzyme A from pyruvate requires three enzymes and five coenzymes

The mechanism of the pyruvate dehydrogenase reaction is wonderfully complex, more so than is suggested by its simple stoichiometry. The reaction requires the participation of the three enzymes of the pyruvate dehydrogenase complex and five coenzymes. The coenzymes thiamine pyrophosphate (TPP), lipoic acid, and FAD serve as catalytic cofactors, and CoA and NAD^+ are stoichiometric cofactors, cofactors that function as substrates.

Thiamine pyrophosphate (TPP)   **Lipoic acid**

The conversion of pyruvate into acetyl CoA consists of decarboxylation, oxidation, and transfer of the resultant acetyl group to CoA. A fourth step is required to regenerate the active enzyme.

Pyruvate — (Decarboxylation, CO_2) → (Oxidation, $2 e^-$) → (Transfer to CoA, CoA) → **Acetyl CoA**

These steps must be coupled to preserve the free energy derived from the decarboxylation step to drive the formation of NADH and acetyl CoA. Let us look into each step in more detail:

1. *Decarboxylation.* Pyruvate combines with TPP and is then decarboxylated to yield hydroxyethyl-TPP.

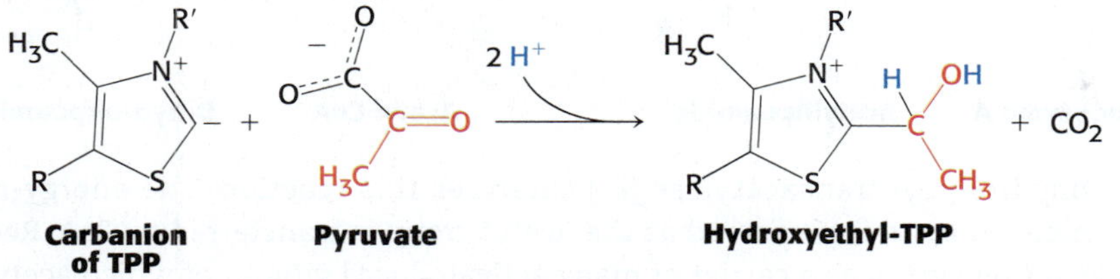

This reaction, the rate-limiting step in acetyl CoA synthesis, is catalyzed by the pyruvate dehydrogenase component (E_1) of the multienzyme complex. TPP is the prosthetic group of the pyruvate dehydrogenase component (**Figure 17.5**).

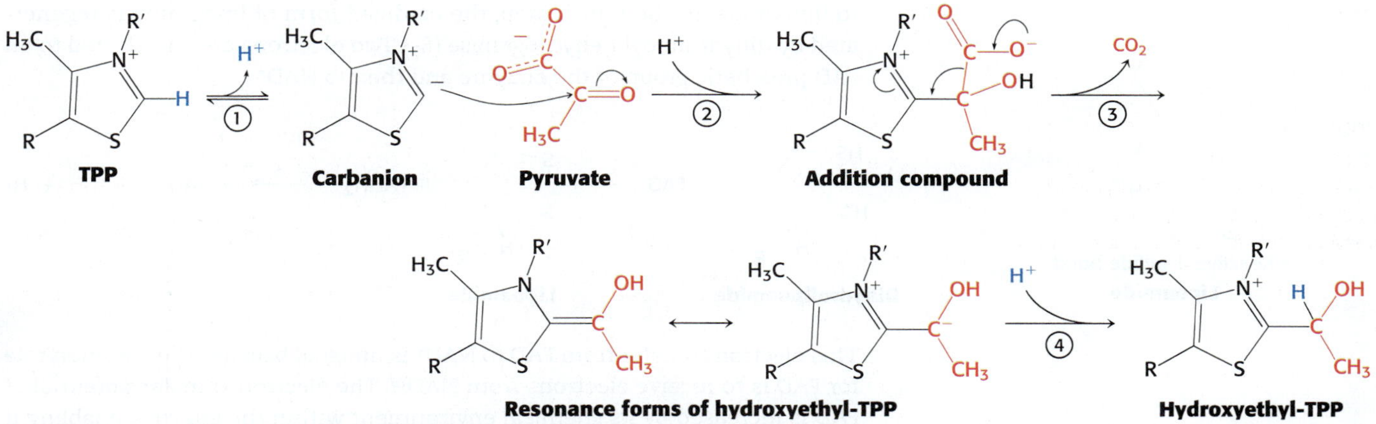

FIGURE 17.5 The mechanism of the E_1 decarboxylation reaction uses a critical thiamine-derived prosthetic group. E_1 is the pyruvate dehydrogenase component of the pyruvate dehydrogenase complex. A key feature of the prosthetic group, TPP, is that the carbon atom between the nitrogen and sulfur atoms in the thiazole ring is much more acidic than most =C— groups, with a pK_a value near 10. (1) This carbon center ionizes to form a *carbanion*. (2) The carbanion readily adds to the carbonyl group of pyruvate. (3) This addition is followed by the decarboxylation of pyruvate. The positively charged ring of TPP acts as an electron sink that stabilizes the negative charge that is transferred to the ring as part of the decarboxylation. (4) Protonation yields hydroxyethyl-TPP.

2. *Oxidation.* The hydroxyethyl group attached to TPP is oxidized to form an acetyl group while being simultaneously transferred to lipoamide, a derivative of lipoic acid that is linked to the side chain of a lysine residue by an amide linkage. Note that this transfer results in the formation of an energy-rich thioester bond.

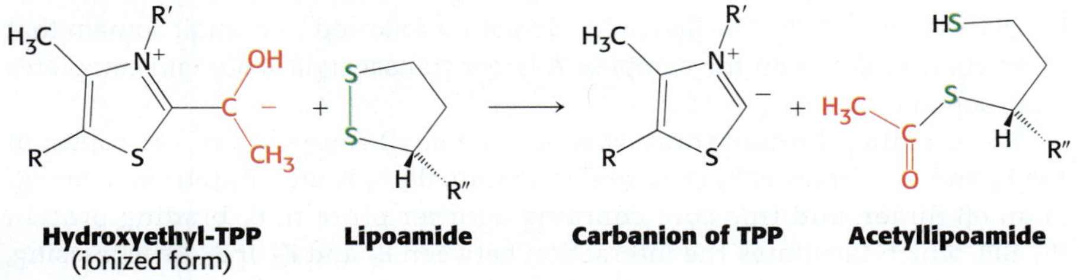

The oxidant in this reaction is the disulfide group of lipoamide, which is reduced to its disulfhydryl form. This reaction, also catalyzed by the pyruvate dehydrogenase component E_1, yields acetyllipoamide.

3. *Formation of acetyl CoA.* The acetyl group is transferred from acetyllipoamide to CoA to form acetyl CoA.

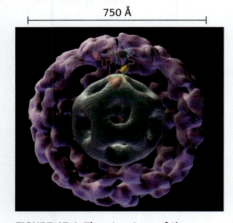

Coenzyme A   **Acetyllipoamide**   **Acetyl CoA**   **Dihydrolipoamide**

Dihydrolipoyl transacetylase (E_2) catalyzes this reaction. The energy-rich thioester bond is preserved as the acetyl group is transferred to CoA. Recall that CoA serves as a carrier of many activated acyl groups, of which acetyl is the simplest (Section 15.4). Acetyl CoA, the fuel for the citric acid cycle, has now been generated from pyruvate.

4. *Regeneration of oxidized lipoamide.* The pyruvate dehydrogenase complex cannot complete another catalytic cycle until the dihydrolipoamide is oxidized to lipoamide. In the fourth step, the oxidized form of lipoamide is regenerated by dihydrolipoyl dehydrogenase (E_3). Two electrons are transferred to an FAD prosthetic group of the enzyme and then to NAD^+.

Dihydrolipoamide   **Lipoamide**

This electron transfer from FAD to NAD^+ is unusual because the common role for FAD is to receive electrons from NADH. The electron-transfer potential of FAD is increased by its chemical environment within the enzyme, enabling it to transfer electrons to NAD^+. Proteins tightly associated with FAD or flavin mononucleotide (FMN) are called **flavoproteins**.

Flexible linkages allow lipoamide to move between different active sites

The structures and precise composition of the component enzymes of the pyruvate dehydrogenase complex vary among species. However, there are common features.

The core of the complex is formed by 60 molecules of the transacetylase component E_2 (**Figure 17.6**). Transacetylase consists of 20 catalytic trimers assembled to form a hollow cube. Each of the three subunits forming a trimer has three major domains (**Figure 17.7**). At the amino terminus is a small domain that contains a bound flexible lipoamide cofactor attached to a lysine residue. This domain is homologous to biotin-binding domains such as that of pyruvate carboxylase (Figure 16.29). The lipoamide domain is followed by a small domain that interacts with E_3 within the complex. A larger transacetylase domain completes an E_2 subunit.

Surrounding the core transacetylase is a shell composed of ~45 copies of the E_1 and ~10 copies of E_3 enzymes. In mammals, E_1 is an $\alpha_2\beta_2$ tetramer, and E_3 is an $\alpha\beta$ dimer, and this core contains another protein, E_3-binding protein (E_3-BP), which facilitates the interaction between E_2 and E_3. If E_3-BP is missing, the complex has greatly reduced activity. The gap between the outer shell and the transacetylase core allows the lipoamide arms to visit the various active sites (Figure 17.6).

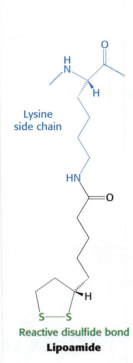

Lysine side chain

Reactive disulfide bond
Lipoamide

750 Å

FIGURE 17.6 The structure of the pyruvate dehydrogenase complex from bacteria reveals a massive protein complex. The image of the complex from *B. stearothermophilus*, which was derived from cryo-electron microscopic data (Section 4.5), shows an inner core consisting of the E_2 enzyme. The shell surrounding the core consists of E_1 and E_3 enzymes, although only the E_1 enzymes are shown in this structure. Two of the 60 lipoamide arms are shown (red and yellow).

[Donald Bliss, National Library of Medicine.]

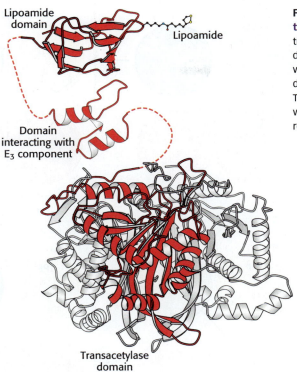

Lipoamide domain

Lipoamide

Domain interacting with E_3 component

Transacetylase domain

FIGURE 17.7 The transacetylase (E_2) core is made up of three distinct domains. The figure shows one subunit of the transacetylase trimer. *Notice* that each subunit consists of three domains: a lipoamide-binding domain, a small domain that interacts with E_3, and a large transacetylase catalytic domain. The catalytic domains interact with one another to form the catalytic trimer. Transacetylase domains of three identical subunits are shown, with one depicted in red and the others in white in the ribbon representation.

How do the three distinct active sites work in concert? The key is the long, flexible lipoamide arm of the E_2 subunit, which carries substrate from active site to active site (**Figure 17.8**):

1. Pyruvate is decarboxylated at the active site of E_1, forming the hydroxyethyl-TPP intermediate, and CO_2 leaves as the first product. This active site lies deep within the E_1 complex, connected to the enzyme surface by a 20-Å-long hydrophobic channel.

2. E_2 inserts the lipoamide arm of the lipoamide domain into the deep channel in E_1 leading to the active site.

3. E_1 catalyzes the transfer of the acetyl group to the lipoamide. The acetylated arm then leaves E_1 and enters the E_2 cube to visit the active site of E_2, located deep in the cube at the subunit interface.

4. The acetyl moiety is then transferred to CoA, and the second product, acetyl CoA, leaves the cube. The reduced lipoamide arm then swings to the active site of the E_3 flavoprotein.

5. At the E_3 active site, the lipoamide is oxidized by coenzyme FAD. The reactivated lipoamide is ready to begin another reaction cycle.

6. The final product, NADH, is produced with the reoxidation of $FADH_2$ to FAD.

The structural integration of three kinds of enzymes and the long, flexible lipoamide arm make the coordinated catalysis of a complex reaction possible. The proximity of one enzyme to another increases the overall reaction rate and minimizes side reactions. All the intermediates in the oxidative decarboxylation of pyruvate remain bound to the complex throughout the reaction sequence and are readily transferred as the flexible arm of E_2 calls on each active site in turn.

❖ SELF-CHECK QUESTION

A key product of pyruvate dehydrogenase complex reaction, acetyl CoA, is released after the fourth step. What is the purpose of the remaining steps?

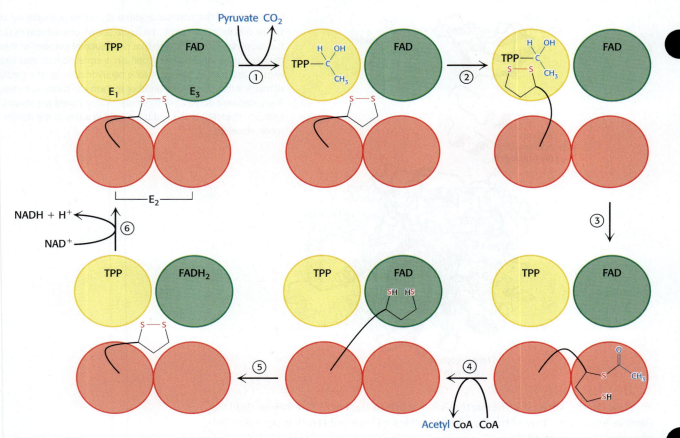

FIGURE 17.8 Three enzymes cooperate in the full reactions of the pyruvate dehydrogenase complex. At the top left, the enzyme (represented by a yellow, a green, and two red spheres) is unmodified and ready for a catalytic cycle. (1) Pyruvate is decarboxylated to form hydroxyethyl-TPP. (2) The lipoamide arm of E_2 moves into the active site of E_1. (3) E_1 catalyzes the transfer of the two-carbon group to the lipoamide group to form the acetyl–lipoamide complex. (4) E_2 catalyzes the transfer of the acetyl moiety to CoA to form the product acetyl CoA. The dihydrolipoamide arm then swings to the active site of E_3. E_3 catalyzes (5) the oxidation of the dihydrolipoamide and (6) the transfer of the protons and electrons to NAD^+ to complete the reaction cycle.

17.3 The Citric Acid Cycle Oxidizes Two-Carbon Units

The conversion of pyruvate into acetyl CoA by the pyruvate dehydrogenase complex is the link between glycolysis and cellular respiration because acetyl CoA is the fuel for the citric acid cycle. Indeed, under aerobic conditions, all fuels are ultimately metabolized to acetyl CoA or components of the citric acid cycle.

Citrate synthase forms citrate from oxaloacetate and the acetyl group from acetyl coenzyme A

The citric acid cycle begins with the addition of a four-carbon unit, oxaloacetate, and a two-carbon unit, the acetyl group of acetyl CoA. Oxaloacetate reacts with acetyl CoA and H_2O to yield citrate and CoA.

This reaction, which is an aldol addition followed by a hydrolysis, is catalyzed by citrate synthase. Oxaloacetate first combines with acetyl CoA to form citryl CoA, a molecule that is energy rich because it contains the thioester bond that originated in acetyl CoA. The hydrolysis of citryl CoA thioester to citrate and CoA drives the overall reaction far in the direction of the synthesis of citrate. In essence, the hydrolysis of the thioester powers the synthesis of a new molecule from two precursors.

Mechanism: The mechanism of citrate synthase prevents undesirable reactions

Because the condensation of acetyl CoA and oxaloacetate initiates the citric acid cycle, it is very important that side reactions, notably the hydrolysis of acetyl CoA to acetate and CoA, be minimized. Let us briefly consider how citrate synthase prevents the wasteful hydrolysis of acetyl CoA.

Mammalian citrate synthase is a dimer of identical 49-kDa subunits. Each active site is located in a cleft between the large and the small domains of a subunit, adjacent to the subunit interface. X-ray crystallographic studies of citrate synthase and its complexes with several substrates and inhibitors have revealed that the enzyme undergoes large conformational changes in the course of catalysis. Citrate synthase exhibits sequential, ordered kinetics: oxaloacetate binds first, followed by acetyl CoA.

The reason for the ordered binding is that oxaloacetate induces a major structural rearrangement leading to the creation of a binding site for acetyl CoA. The binding of oxaloacetate converts the open form of the enzyme into a more closed form (**Figure 17.9**). In each subunit, the small domain rotates 19 degrees relative to the large domain. Movements as large as 15 Å are produced by the rotation of α helices elicited by quite small shifts of side chains around bound oxaloacetate. These structural changes create a binding site for acetyl CoA.

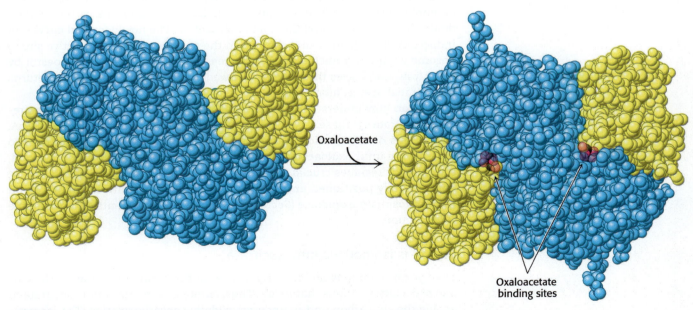

Oxaloacetate

Oxaloacetate binding sites

FIGURE 17.9 The ordered binding of substrates by citrate synthase is explained by conformational changes upon binding oxaloacetate. The small domain of each subunit of the homodimer is shown in yellow; the large domains are shown in blue. (Left) The open form of enzyme alone is shown. (Right) The closed form of the oxaloacetate-enzyme complex reveals significant structural changes upon oxaloacetate binding that create a binding site with high affinity for acetyl CoA, the second substrate to bind. [Drawn from 5CSC.pdb and 4CTS.pdb.]

INTERACT with this model in
≋ Achie√e

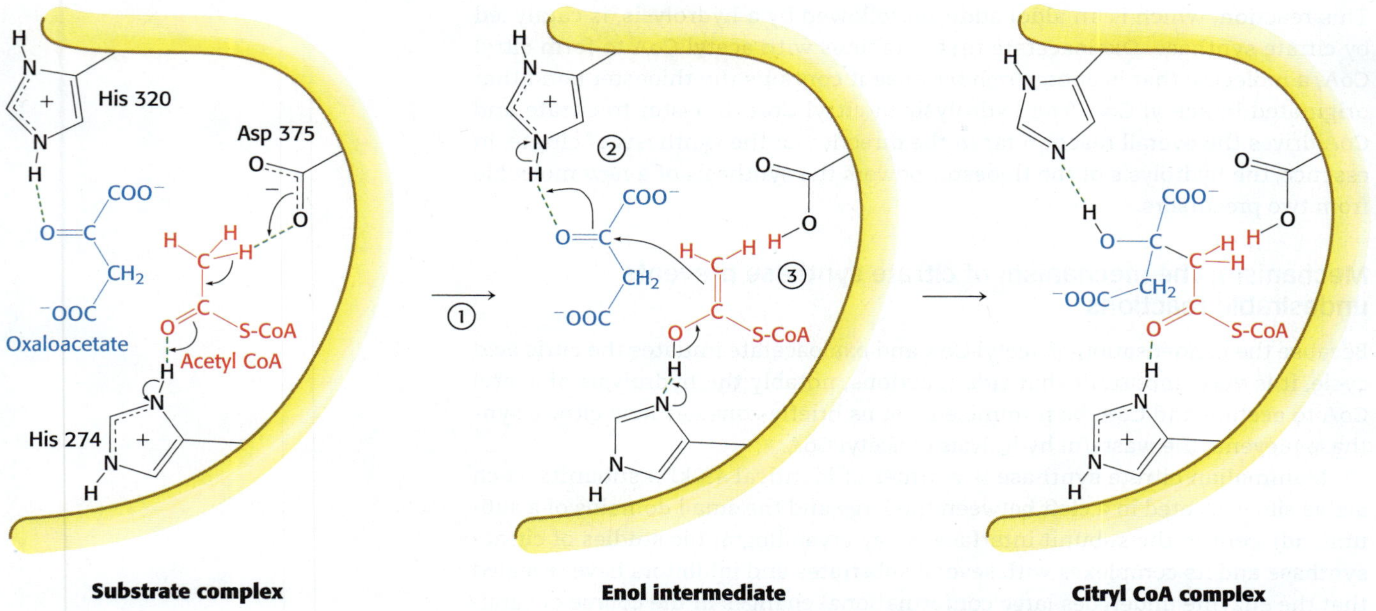

FIGURE 17.10 The first part of the citrate synthase mechanism forms citryl CoA.
(1) In the substrate complex (left), His 274 donates a proton to the carbonyl oxygen of acetyl CoA to promote the removal of a methyl proton by Asp 375 to form the enol intermediate (center).
(2) Oxaloacetate is activated by the transfer of a proton from His 320 to its carbonyl carbon atom.
(3) Simultaneously, the enol of acetyl CoA attacks the carbonyl carbon of oxaloacetate to form a carbon–carbon bond linking acetyl CoA and oxaloacetate. His 274 is reprotonated, and citryl CoA is formed. His 274 participates again as a proton donor to hydrolyze the thioester (not shown), yielding citrate and CoA.

Citrate synthase catalyzes the condensation reaction by bringing the substrates into close proximity, orienting them, and polarizing certain bonds (**Figure 17.10**). The donation and removal of protons transforms acetyl CoA into an enol intermediate. The enol attacks oxaloacetate to form a carbon–carbon double bond linking acetyl CoA and oxaloacetate. The newly formed citryl CoA induces additional structural changes in the enzyme, causing the active site to become completely enclosed. The enzyme cleaves the citryl CoA thioester by hydrolysis. CoA leaves the enzyme, followed by citrate, and the enzyme returns to the initial open conformation.

We can now understand how the wasteful hydrolysis of acetyl CoA is prevented. Citrate synthase is well suited to hydrolyze *citryl* CoA but not *acetyl* CoA. How is this discrimination accomplished? First, acetyl CoA does not bind to the enzyme until oxaloacetate is bound and ready for condensation. Second, the catalytic residues crucial for the hydrolysis of the thioester linkage are not appropriately positioned until citryl CoA is formed. As with hexokinase and triose phosphate isomerase (Section 16.2), induced fit prevents an undesirable side reaction.

Citrate is isomerized into isocitrate

The hydroxyl group is not properly located in the citrate molecule for the oxidative decarboxylations that follow. Thus, citrate is isomerized into isocitrate to enable the six-carbon unit to undergo oxidative decarboxylation. The isomerization of citrate is accomplished by a dehydration step followed by a hydration step. The result is an interchange of an H and an OH. The enzyme catalyzing both steps is called *aconitase* because *cis*-aconitate is an intermediate.

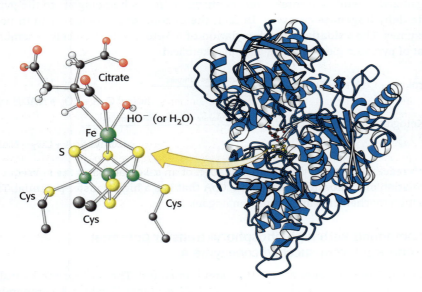

Aconitase is an **iron–sulfur protein**, or **nonheme iron protein**, in that it contains iron that is not bonded to heme. Rather, its four iron atoms are complexed to four inorganic sulfides and three cysteine sulfur atoms, leaving one iron atom available to bind citrate through one of its COO^- groups and an OH group (**Figure 17.11**). This Fe–S cluster participates in dehydrating and rehydrating the bound substrate.

FIGURE 17.11 Citrate binds directly to the iron–sulfur complex of aconitase. A 4Fe–4S iron–sulfur cluster is a component of the active site of aconitase. Notice that one of the iron atoms of the cluster binds to a COO^- group and an OH group of citrate. [Drawn from 1C96.pdb.]

INTERACT with this model in
✄ Achieve

Isocitrate is oxidized and decarboxylated to alpha-ketoglutarate

We come now to the first of four oxidation–reduction reactions in the citric acid cycle. The oxidative decarboxylation of isocitrate is catalyzed by isocitrate dehydrogenase.

$$\text{Isocitrate} + NAD^+ \longrightarrow \alpha\text{-ketoglutarate} + CO_2 + NADH$$

The intermediate in this reaction is oxalosuccinate, an unstable β-ketoacid. While bound to the enzyme, it loses CO_2 to form α-ketoglutarate.

This oxidation generates the first high-transfer-potential electron carrier, NADH, in the cycle.

Succinyl coenzyme A is formed by the oxidative decarboxylation of alpha-ketoglutarate

The conversion of isocitrate into α-ketoglutarate is followed by a second oxidative decarboxylation reaction, the formation of succinyl CoA from α-ketoglutarate.

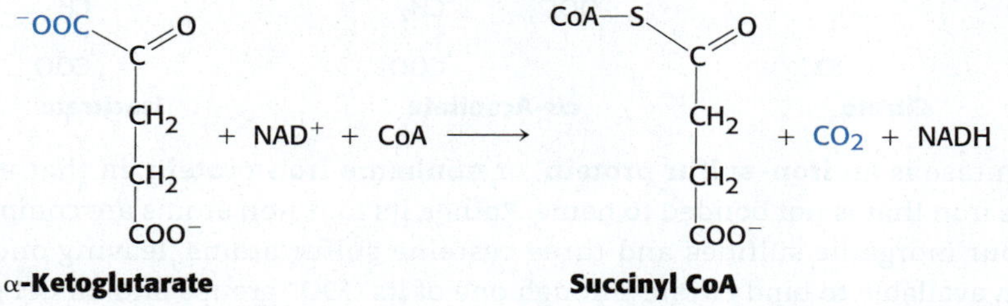

α-**Ketoglutarate** $+ NAD^+ + CoA \longrightarrow$ **Succinyl CoA** $+ CO_2 + NADH$

This reaction is catalyzed by the α-**ketoglutarate dehydrogenase complex**, an organized assembly of three kinds of enzymes that is homologous to the pyruvate dehydrogenase complex. In fact, the E_3 component is identical in both enzymes. The oxidative decarboxylation of α-ketoglutarate closely resembles that of pyruvate (Section 17.2), also an α-ketoacid.

$$\text{Pyruvate} + \text{CoA} + \text{NAD}^+ \xrightarrow{\text{Pyruvate dehydrogenase complex}}$$
$$\text{acetyl CoA} + CO_2 + \text{NADH} + H^+$$

$$\text{α-Ketoglutarate} + \text{CoA} + \text{NAD}^+ \xrightarrow{\text{α-Ketoglutarate dehydrogenase complex}}$$
$$\text{succinyl CoA} + CO_2 + \text{NADH}$$

Both reactions include the decarboxylation of an α-ketoacid and the subsequent formation of a thioester linkage with CoA that has a high transfer potential. The reaction mechanisms are entirely analogous.

A compound with high phosphoryl-transfer potential is generated from succinyl coenzyme A

Succinyl CoA is an energy-rich thioester compound. The $\Delta G^{\circ\prime}$ for the hydrolysis of succinyl CoA is about -33.5 kJ mol^{-1} (-8.0 kcal mol^{-1}), which is comparable to that of ATP (-30.5 kJ mol^{-1}, or -7.3 kcal mol^{-1}). In the citrate synthase reaction, the cleavage of the thioester bond powers the synthesis of the six-carbon citrate from the four-carbon oxaloacetate and the two-carbon fragment. The cleavage of the thioester bond of succinyl CoA is coupled to the phosphorylation of a purine nucleoside diphosphate, usually ADP. This reaction, which is readily reversible, is catalyzed by succinyl CoA synthetase.

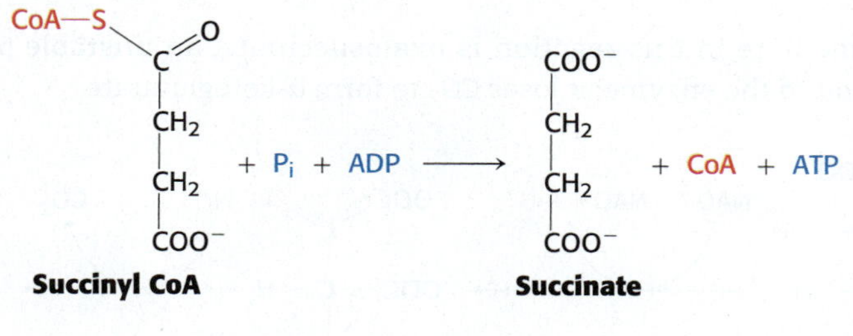

Succinyl CoA $+ P_i + ADP \longrightarrow$ **Succinate** $+ CoA + ATP$

This reaction is the only step in the citric acid cycle that directly yields a compound with high phosphoryl-transfer potential. In mammals, there are two isozymic forms of the enzyme, one specific for ADP and one for GDP. In tissues that perform large amounts of cellular respiration, such as skeletal and heart muscle, the ADP-requiring isozyme predominates. In tissues that perform many anabolic reactions, such as the liver, the GDP-requiring enzyme is common.

The GDP-requiring enzyme is believed to work in reverse of the direction observed in the TCA cycle; that is, GTP is used to power the synthesis of succinyl CoA, which is a precursor for heme synthesis (Section 25.4). The specificity and nucleoside used varies also by species. For example, the enzyme in E. coli can use either GDP or ADP as the phosphoryl-group acceptor.

Note that the enzyme **nucleoside diphosphokinase**, which catalyzes the following reaction,

$$XTP + YDP \rightleftharpoons XDP + YTP$$

allows the γ phosphoryl group to be readily transferred from any nucleotide triphosphate (XTP) to any other nucleotide diphosphate (YDP); for example, from GTP to form ATP from ADP. This reaction thereby allows the adjustment of the concentration of GTP or ATP to meet the cell's need and keeps the concentrations of all the various nucleoside triphosphates in the cell near equilibrium with one another.

Mechanism: Succinyl coenzyme A synthetase transforms types of biochemical energy

The mechanism of this reaction is a clear example of an energy transformation: energy inherent in the thioester molecule is transformed into phosphoryl-group-transfer potential (**Figure 17.12**). First, coenzyme A is displaced by orthophosphate (step 1), which generates another energy-rich compound, succinyl phosphate. A histidine residue plays a key role as a moving arm that detaches the phosphoryl group (step 2), then swings over to a bound ADP (step 3) and transfers the group to form ATP (step 4). The participation of high-energy compounds in all the steps is attested to by the fact that the reaction is readily reversible: $\Delta G^{\circ\prime} = -3.4$ kJ mol^{-1} (-0.8 kcal mol^{-1}). The formation of ATP at the expense of succinyl CoA is an example of substrate-level phosphorylation (Section 16.2).

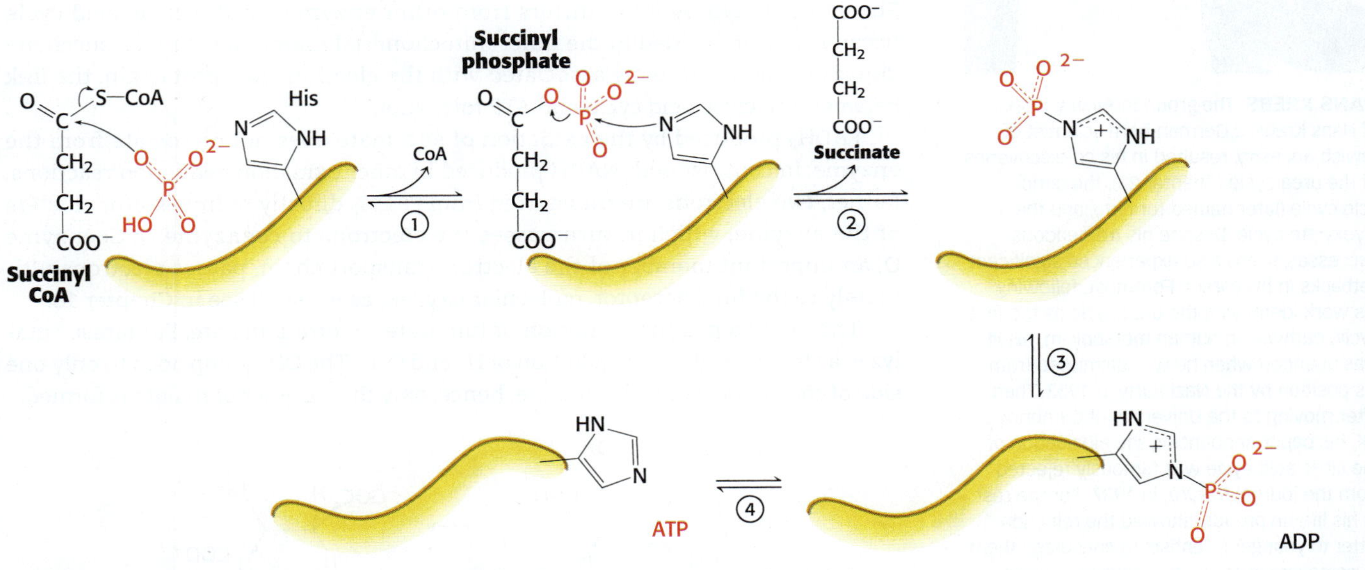

FIGURE 17.12 The mechanism of succinyl CoA synthetase allows the formation of a phosphoanhydride through a phosphorylated enzyme intermediate. (1) Orthophosphate displaces coenzyme A, which generates another energy-rich compound, succinyl phosphate. (2) A histidine residue removes the phosphoryl group with the concomitant generation of succinate and phosphohistidine. (3) The phosphohistidine residue then swings over to a bound ADP, and (4) the phosphoryl group is transferred to form ATP.

Oxaloacetate is regenerated by the oxidation of succinate

Reactions of four-carbon compounds constitute the final stage of the citric acid cycle: the regeneration of oxaloacetate.

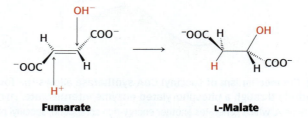

The reactions constitute a metabolic motif that we will see again in fatty acid synthesis and degradation as well as in the degradation of some amino acids. A methylene group (CH_2) is converted into a carbonyl group C=O in three steps: an oxidation, a hydration, and a second oxidation reaction. Oxaloacetate is thereby regenerated for another round of the cycle, and more energy is extracted in the form of $FADH_2$ and NADH.

Succinate is oxidized to fumarate by succinate dehydrogenase. The hydrogen acceptor is FAD rather than NAD^+, which is used in the other three oxidation reactions in the cycle. FAD is the electron acceptor in this reaction because the free-energy change is insufficient to reduce NAD^+. FAD is nearly always the electron acceptor in oxidations that remove two hydrogen atoms from a substrate. In succinate dehydrogenase, the isoalloxazine ring of FAD is covalently attached to a histidine side chain of the enzyme (denoted E-FAD).

$$E\text{-}FAD + succinate \rightleftharpoons E\text{-}FADH_2 + fumarate$$

Succinate dehydrogenase, like aconitase, is an iron–sulfur protein. Indeed, succinate dehydrogenase contains three different kinds of iron–sulfur clusters: 2Fe–2S (two iron atoms bonded to two inorganic sulfides), 3Fe–4S, and 4Fe–4S. Succinate dehydrogenase differs from other enzymes in the citric acid cycle because it is embedded in the inner mitochondrial membrane. In fact, succinate dehydrogenase is directly associated with the electron-transport chain, the link between the citric acid cycle and ATP formation.

$FADH_2$ produced by the oxidation of succinate does not dissociate from the enzyme, in contrast with NADH produced in other oxidation–reduction reactions. Rather, two electrons are transferred from $FADH_2$ directly to iron–sulfur clusters of the enzyme, which in turn passes the electrons to coenzyme Q. Coenzyme Q, an important member of the electron-transport chain, passes electrons ultimately to the final acceptor, molecular oxygen, as we shall see in Chapter 18.

The next step is the hydration of fumarate to form L-malate. Fumarase catalyzes a stereospecific trans addition of H^+ and OH^-. The OH^- group adds to only one side of the double bond of fumarate; hence, only the L isomer of malate is formed.

HANS KREBS The groundbreaking work of Hans Krebs, a German-born scientist of Jewish ancestry, resulted in his co-discoveries of the urea cycle (Chapter 23), the citric acid cycle (later named for him), and the glyoxylate cycle. Despite his tremendous successes, Krebs also experienced significant setbacks in his career. Foremost, following his work identifying the urea cycle as the first cyclic pathway in human metabolism, his life was upended when he was dismissed from his position by the Nazi Party in 1933. Then, after moving to the University of Cambridge, UK, his paper announcing the elucidation of the citric acid cycle was famously rejected from the journal *Nature*, in 1937. For the rest of his life he proudly showed the rejection letter to younger scientists to encourage them to persevere in the face of adversity. Finally, experiments in 1941 appeared to discredit his work on the citric acid cycle (see problem 18 at the end of the chapter). This issue plagued him until it was finally resolved by another investigator in 1948. In 1953 Krebs received the Nobel Prize in Physiology or Medicine for his work in determining the pathways of central metabolism.

Finally, malate is oxidized to form oxaloacetate. This reaction is catalyzed by malate dehydrogenase, and NAD^+ is again the hydrogen acceptor.

$$Malate + NAD^+ \rightleftharpoons oxaloacetate + NADH + H^+$$

The standard free energy for this reaction, unlike that for the other steps in the citric acid cycle, is significantly positive ($\Delta G^{\circ\prime}$ = +29.7 kJ mol^{-1}, or + 7.1 kcal mol^{-1}). The oxidation of malate is driven by the use of the products—oxaloacetate by citrate synthase and NADH by the electron-transport chain.

❖ SELF-CHECK QUESTION

Fluoroacetate is a toxic molecule that enters the citric acid cycle (CAC) due to its similarity to an acetyl group, but ultimately it inhibits the CAC. When fluoroacetate is added to mitochondria, fluorocitrate builds up. What step of the CAC is inhibited by fluoroacetate?

The citric acid cycle produces high-transfer-potential electrons, ATP, and CO$_2$

The net reaction of the citric acid cycle is

Acetyl CoA + 3 NAD$^+$ + FAD + ADP + P$_i$ + 2 H$_2$O $\longrightarrow$

$\qquad\qquad\qquad$ 2 CO$_2$ + 3 NADH + FADH$_2$ + ATP + 2 H$^+$ + CoA

Let us recapitulate the reactions that give this stoichiometry (**Figure 17.13** and **Table 17.2**) and summarize.

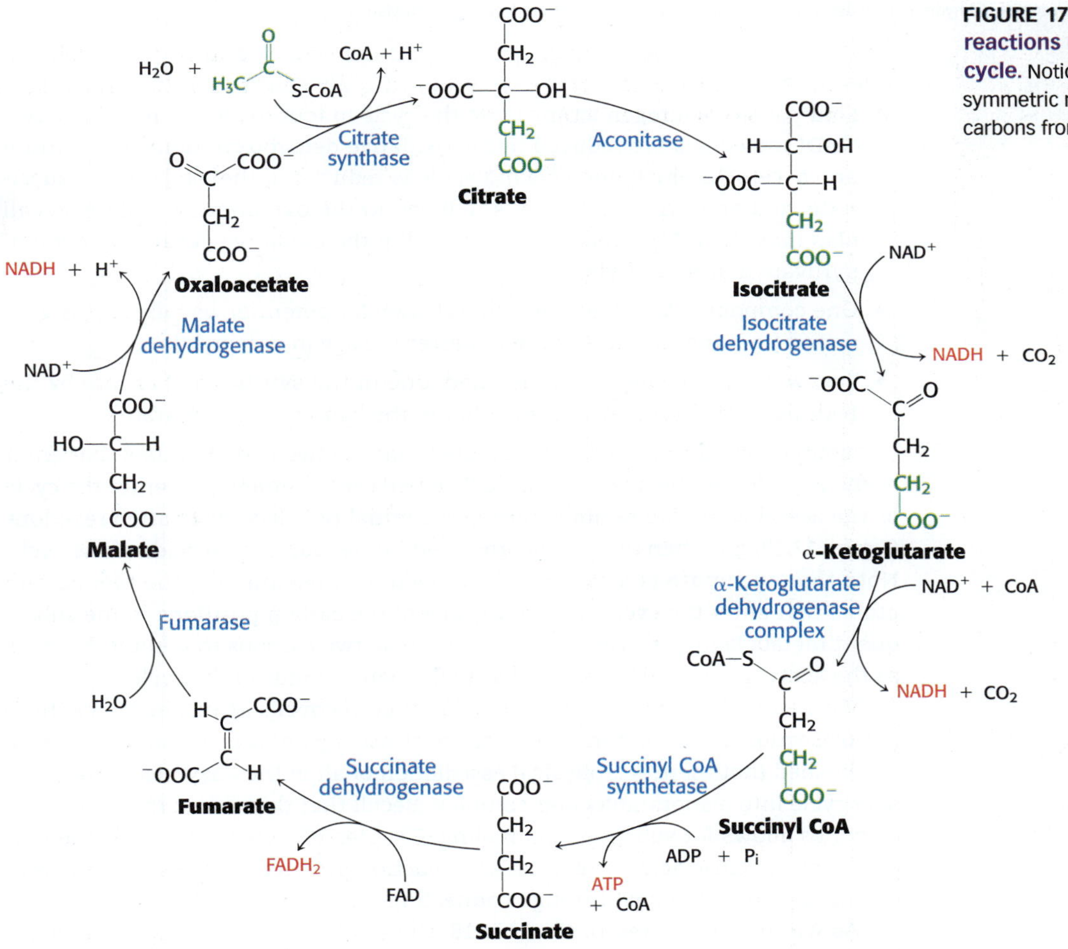

FIGURE 17.13 Eight enzyme-catalyzed reactions make up the full citric acid cycle. Notice that since succinate is a symmetric molecule, the identity of the carbons from the acetyl unit is lost.

- Two carbon atoms enter the cycle in the condensation of an acetyl unit (from acetyl CoA) with oxaloacetate. Two carbon atoms leave the cycle in the form of CO$_2$ in the successive decarboxylations catalyzed by isocitrate dehydrogenase and α-ketoglutarate dehydrogenase.

TABLE 17.2 Citric acid cycle

Step	Reaction	Enzyme	Prosthetic group	Type*	$\Delta G^{\circ\prime}$ kJ mol^{-1}	$\Delta G^{\circ\prime}$ kcal mol^{-1}
1	Acetyl CoA + oxaloacetate + H_2O $\longrightarrow$ citrate + CoA + H^+	Citrate synthase		a	−31.4	−7.5
2a	Citrate $\rightleftharpoons$ cis-aconitate + H_2O	Aconitase	Fe–S	b	+8.4	+2.0
2b	cis-Aconitate + H_2O $\rightleftharpoons$ isocitrate	Aconitase	Fe–S	c	−2.1	−0.5
3	Isocitrate + NAD$^-$ $\rightleftharpoons$ α-ketoglutarate + CO_2 + NADH	Isocitrate dehydrogenase		d + e	−8.4	−2.0
4	α-Ketoglutarate + NAD$^+$ + CoA $\rightleftharpoons$ succinyl CoA + CO_2 + NADH	α-Ketoglutarate dehydrogenase complex	Lipoic acid, FAD, TPP	d + e	−30.1	−7.2
5	Succinyl CoA + P$_i$ + ADP $\rightleftharpoons$ succinate + ATP + CoA	Succinyl CoA synthetase		f	−3.3	−0.8
6	Succinate + FAD (enzyme-bound) $\rightleftharpoons$ fumarate + FADH$_2$ (enzyme-bound)	Succinate dehydrogenase	FAD, Fe–S	e	0	0
7	Fumarate + H_2O $\rightleftharpoons$ L-malate	Fumarase		c	−3.8	−0.9
8	L-Malate + NAD$^+$ $\rightleftharpoons$ oxaloacetate + NADH + H^+	Malate dehydrogenase		e	+29.7	+7.1

*Reaction type: (a) condensation; (b) dehydration; (c) hydration; (d) decarboxylation; (e) oxidation; (f) substrate-level phosphorylation.

- Four pairs of hydrogen atoms leave the cycle in four oxidation reactions. Two NAD$^+$ molecules are reduced in the oxidative decarboxylations of isocitrate and α-ketoglutarate, one FAD molecule is reduced in the oxidation of succinate, and one NAD$^+$ molecule is reduced in the oxidation of malate. Recall also that one NAD$^+$ molecule is reduced in the oxidative decarboxylation of pyruvate to form acetyl CoA.

- One compound with high phosphoryl-transfer potential, usually ATP, is generated from the cleavage of the thioester linkage in succinyl CoA.

- Two water molecules are consumed: one in the synthesis of citrate by the hydrolysis of citryl CoA and the other in the hydration of fumarate.

Isotope-labeling studies have revealed that the two carbon atoms that enter each cycle are not the ones that leave. The two carbon atoms that enter the cycle as the acetyl group are retained during the initial two decarboxylation reactions (Figure 17.13) and then remain incorporated in the four-carbon acids of the cycle. Note that succinate is a symmetric molecule. Consequently, the two carbon atoms that enter the cycle can occupy any of the carbon positions in the subsequent metabolism of the four-carbon acids. The two carbons that enter the cycle as the acetyl group will be released as CO_2 in later rounds of the cycle.

Various techniques, such as fluorescence recovery after photobleaching (FRAP, Section 12.5) and tandem mass spectroscopy analysis (Section 4.3), have established that there is a physical association of all of the enzymes of the citric acid cycle into a supramolecular complex. Recall that this close arrangement of enzymes allows for substrate channeling (Chapter 16), which enhances the efficiency of the citric acid cycle because a reaction product can pass directly from one active site to the next through connecting channels.

As will be considered in Chapter 18, the electron-transport chain oxidizes the NADH and FADH$_2$ formed in the citric acid cycle. The transfer of electrons from these carriers to O_2, the final electron acceptor, leads to the generation of a proton gradient across the inner mitochondrial membrane. This proton-motive force then powers the generation of ATP; the net stoichiometry is about 2.5 ATP per NADH, and 1.5 ATP per FADH$_2$. Consequently, nine high-transfer-potential phosphoryl groups are generated when the electron-transport chain oxidizes 3 NADH molecules and 1 $FADH_2$ molecule, and one ATP is directly formed in one round of the citric acid cycle. Thus, one acetyl unit generates approximately 10 molecules of ATP. In dramatic contrast, the anaerobic glycolysis of an entire glucose molecule generates only 2 molecules of ATP (and 2 molecules of lactate).

Recall that molecular oxygen does not participate directly in the citric acid cycle. However, the cycle operates only under aerobic conditions because NAD^+ and FAD can be regenerated in the mitochondrion only by the transfer of electrons to molecular oxygen. In contrast, glycolysis can proceed under anaerobic conditions because NAD^+ is regenerated in the conversion of pyruvate into lactate. Glycolysis has both an aerobic and an anaerobic mode, whereas the citric acid cycle is strictly aerobic.

❖ SELF–CHECK QUESTION

The nucleoside trisphosphate produced directly in the citric acid cycle is sometimes GTP. Explain why a GTP molecule, or another nucleoside triphosphate, is energetically equivalent to an ATP molecule in metabolism.

17.4 Entry to the Citric Acid Cycle and Metabolism Through It Are Controlled

The citric acid cycle is the final common pathway for the aerobic oxidation of fuel molecules. Moreover, as we will see shortly (Section 17.5) and repeatedly elsewhere in our study of biochemistry, the cycle is an important source of building blocks for a host of important biomolecules. As befits its role as the metabolic hub of the cell, entry into the cycle and the rate of the cycle itself are controlled at several stages.

The pyruvate dehydrogenase complex is regulated allosterically and by reversible phosphorylation

As stated earlier, glucose can be formed from pyruvate. However, the formation of acetyl CoA from pyruvate is an irreversible step in animals and thus they are unable to convert acetyl CoA back into glucose. The oxidative decarboxylation of pyruvate to acetyl CoA commits the carbon atoms of glucose to one of two principal fates: oxidation to CO_2 by the citric acid cycle, with the concomitant generation of energy, or incorporation into lipids (**Figure 17.14**).

As expected of an enzyme at a critical branch point in metabolism, the activity of the pyruvate dehydrogenase complex is tightly regulated. High concentrations of reaction products inhibit the reaction: acetyl CoA inhibits the transacetylase component (E_2) by binding directly, whereas NADH inhibits the dihydrolipoyl dehydrogenase (E_3). High concentrations of NADH and acetyl CoA inform the enzyme that the energy needs of the cell have been met or that fatty acids are being degraded to produce acetyl CoA and NADH. In either case, there is no need to metabolize pyruvate to acetyl CoA. This inhibition has the effect of conserving glucose, because most pyruvate is derived from glucose by glycolysis (Section 16.2).

The key means of regulation of the complex in eukaryotes is reversible covalent modification (**Figure 17.15**). Phosphorylation of the pyruvate dehydrogenase component (E_1) by one of four tissue-specific isozymes of pyruvate dehydrogenase kinase (PDK) switches off the activity of the complex. Deactivation is then reversed by one of two isozymes of pyruvate dehydrogenase phosphatase (PDP). In mammals, both the kinase and phosphatase are associated with the E_2-E_3-BP core complex, and the activities of both are regulated and subject to tissue-specific hormonal control.

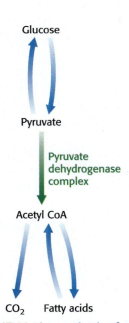

FIGURE 17.14 The synthesis of acetyl CoA by the pyruvate dehydrogenase complex is a key irreversible step in glucose metabolism.

EXPLORE this pathway further in the Metabolic Map in
🗺 Achie∕e

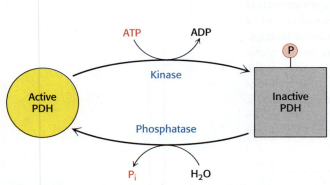

FIGURE 17.15 The activity of the pyruvate dehydrogenase complex is regulated by reversible phosphorylation. A specific kinase phosphorylates and inactivates pyruvate dehydrogenase (PDH), and a phosphatase activates the dehydrogenase by removing the phosphoryl group. The kinase and the phosphatase also are highly regulated enzymes.

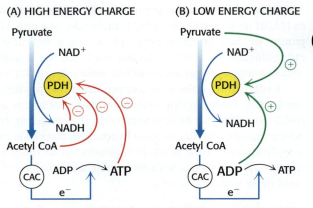

FIGURE 17.16 The pyruvate dehydrogenase complex responds to changes in the energy charge of the cell. (A) The complex (PDH) is inhibited by its immediate products, NADH and acetyl CoA, as well as by the ultimate product of the citric acid cycle (CAC) and oxidative phosphorylation, ATP. (B) PDH is activated by pyruvate and ADP, which inhibit the kinase that phosphorylates PDH.

To see how this regulation works in biological conditions, consider muscle that is becoming active after a period of rest (**Figure 17.16**). At rest, the muscle will not have significant energy demands. Consequently, the NADH/NAD+, acetyl CoA/CoA, and ATP/ADP ratios will be high. These high ratios promote phosphorylation and inactivation of the complex by activating PDK. In other words, high concentrations of immediate products of the reaction (acetyl CoA and NADH) and a final product of the pathway (ATP) inhibit the activity. Thus, pyruvate dehydrogenase is switched off when the energy charge is high.

As exercise begins, the concentrations of ADP and pyruvate will increase as muscle contraction consumes ATP and glucose is converted into pyruvate to meet the energy demands. Both ADP and pyruvate activate the dehydrogenase by inhibiting the kinase. Moreover, the phosphatase is stimulated by Ca^{2+}, the same signal that initiates muscle contraction. A rise in the cytoplasmic Ca^{2+} concentration increases the mitochondrial Ca^{2+} concentration. The rise in mitochondrial Ca^{2+} activates the phosphatase, enhancing pyruvate dehydrogenase activity.

In some tissues, the phosphatase is also controlled by hormones. For example, in liver tissue, epinephrine binds to the α-adrenergic receptor to initiate the phosphatidylinositol pathway (Section 14.2), causing an increase in Ca^{2+} concentration that activates the phosphatase. In tissues capable of fatty acid synthesis, such as liver and adipose tissue, insulin—the hormone that signifies the fed state—stimulates the phosphatase, increasing the conversion of pyruvate into acetyl CoA. Acetyl CoA is the precursor for fatty acid synthesis (Section 22.5). In these tissues, the pyruvate dehydrogenase complex is activated to funnel glucose to pyruvate and then to acetyl CoA and ultimately to fatty acids. In people with a pyruvate dehydrogenase phosphatase deficiency, pyruvate dehydrogenase is always phosphorylated and thus inactive. Consequently, glucose is processed to lactate rather than acetyl CoA, resulting in lactic acidosis. Lactic acidosis causes many tissues to malfunction, including the central nervous system.

Diabetic neuropathy may be due to inhibition of the pyruvate dehydrogenase complex

Diabetic neuropathy—a numbness, tingling, or pain in the hands, arm, fingers, toes, feet, and legs—is a common complication of both type 1 and type 2 diabetes, affecting approximately 50% of patients. There is no cure for the condition, and treatment relies on painkillers. Recent research in mice and tissue culture suggests that overproduction of lactic acid by cells in the dorsal root ganglion, a part of the nervous system responsible for pain perception, may be a significant contributor.

Lactate, which is produced by some cells of the nervous system, is a common fuel for neurons, which import and convert it to pyruvate for use in cellular respiration. However, it appears that hyperglycemia (high glucose concentration), the defining feature of diabetes, increases pyruvate dehydrogenase kinase activity in the cells of the dorsal root ganglion, leading to inhibition of the pyruvate dehydrogenase complex. Glycolytically produced pyruvate is then processed to lactate. The overabundance of lactate leads to an increase in acid-sensing nociceptors (pain receptors), a type of G-protein-coupled receptor (Section 14.2), resulting in diabetic neuropathy. In experimental systems, three approaches—pharmacological inhibition of either pyruvate dehydrogenase kinase or lactic acid dehydrogenase, or genetic elimination of the kinase gene—greatly reduce, but do not eliminate, diabetic neuropathy. Thus, the search for good therapeutics shows promise but must continue.

The citric acid cycle is regulated at several points

The rate of the citric acid cycle is precisely adjusted to meet an animal cell's needs for ATP (**Figure 17.17**). Two allosteric enzymes primarily regulate the rate of cycling: isocitrate dehydrogenase and α-ketoglutarate dehydrogenase, the first two enzymes in the cycle to generate high-energy electrons.

1. *Isocitrate dehydrogenase is allosterically stimulated by ADP, which enhances the enzyme's affinity for substrates.* The binding of isocitrate, NAD^+, Mg^{2+}, and ADP are all mutually synergistic, whereas ATP is inhibitory. The reaction product NADH also inhibits isocitrate dehydrogenase by directly displacing NAD^+. It is important to note that several steps in the cycle require NAD^+ or FAD, which are abundant only when the energy charge is low.

2. *The α-ketoglutarate dehydrogenase complex catalyzes the rate-limiting step.* Some aspects of this enzyme complex's regulation are like those of the pyruvate dehydrogenase complex, as might be expected from the homology of the two enzymes. α-Ketoglutarate dehydrogenase is inhibited by succinyl CoA and NADH, the products of the reaction that it catalyzes. In addition, α-ketoglutarate dehydrogenase is inhibited by ATP, so that the rate of the citric acid cycle is reduced when cellular ATP is abundant. α-Ketoglutarate dehydrogenase deficiency is observed in a number of neurological disorders, including Alzheimer's disease.

In many bacteria, the synthesis of citrate from oxaloacetate and acetyl CoA carbon units is another important control point. In these organisms, ATP is an allosteric inhibitor of citrate synthase, causing an increase in the apparent value of K_M for acetyl CoA. The result is that as the level of ATP increases, less citrate synthase is bound to acetyl CoA and so less citrate is formed.

The use of isocitrate dehydrogenase and α-ketoglutarate dehydrogenase as points of regulation in human metabolism integrates the citric acid cycle with other pathways and highlights the central role of the citric acid cycle in metabolism. For instance, the inhibition of isocitrate dehydrogenase leads to a buildup of citrate, because the interconversion of isocitrate and citrate is readily reversible under intracellular conditions. Citrate can be transported to the cytoplasm, where it signals phosphofructokinase to halt glycolysis (Section 16.3) and where it can serve as a source of acetyl CoA for fatty acid synthesis (Section 22.5). The α-ketoglutarate that accumulates when α-ketoglutarate dehydrogenase is inhibited can be used as a precursor for several amino acids and the purine bases (Chapters 25 and 26). Finally, as we will explore in Chapter 24, defects in the citric acid cycle and related enzymes can lead to the formation of cancer.

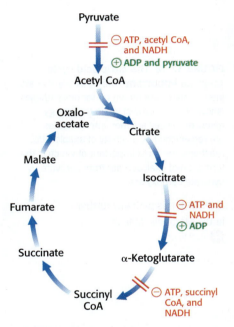

FIGURE 17.17 The citric acid cycle is regulated primarily by the concentrations of ATP and NADH. The key control points are the enzymes isocitrate dehydrogenase and α-ketoglutarate dehydrogenase.

❖ **SELF–CHECK QUESTION**

The citric acid cycle is part of aerobic respiration, but no O_2 is required for the cycle. Explain this apparent paradox.

17.5 The Citric Acid Cycle Is a Source of Biosynthetic Precursors

Thus far, discussion has focused on the citric acid cycle as the major degradative pathway for the generation of ATP. As a major metabolic hub of the cell, the citric acid cycle also integrates many of the cell's other metabolic pathways, including those of carbohydrates, fats, amino acids, and porphyrins. This integration is aided by the fact that the cytoplasmic and mitochondrial pools of citric acid cycle components are shared, allowing the use of citric acid intermediates for biosyntheses taking place throughout the cell (**Figure 17.18**). For example, most of the carbon atoms in porphyrins come from succinyl CoA in a pathway that occurs both in the cytoplasm and mitochondria; fats are synthesized in the cytoplasm from mitochondrial citrate; and many of the amino acids, used throughout the cell, are derived from α-ketoglutarate and oxaloacetate. The details of these biosynthetic processes will be considered in subsequent chapters.

FIGURE 17.18 The citric acid cycle plays an important role in biosynthesis. Intermediates are drawn off for biosyntheses (shown by red arrows) when the energy needs of the cell are met. Intermediates are replenished by a variety of anaplerotic reactions, the most important of which is the formation of oxaloacetate from pyruvate by pyruvate carboxylase.

EXPLORE this pathway further in the Metabolic Map in Achieve

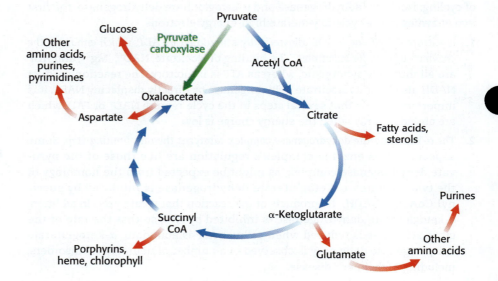

The citric acid cycle must be capable of being rapidly replenished

An important consideration is that citric acid cycle intermediates must be replenished if any are drawn off for biosyntheses. Suppose that much oxaloacetate is converted into amino acids for protein synthesis and, subsequently, the energy needs of the cell rise. The citric acid cycle will operate to a reduced extent unless new oxaloacetate is formed, because acetyl CoA cannot enter the cycle unless it condenses with oxaloacetate. Even though oxaloacetate is recycled, a minimal level must be maintained to allow the cycle to function.

How is oxaloacetate replenished? Mammals lack the enzymes for the net conversion of acetyl CoA into oxaloacetate or any other citric acid cycle intermediate, so no matter how much acetyl CoA enters the cycle, there is never any *net* increase in the number of oxaloacetate molecules. Rather, oxaloacetate is formed by the carboxylation of pyruvate, in a reaction catalyzed by the biotin-dependent enzyme pyruvate carboxylase (Figure 17.18).

$$\text{Pyruvate} + CO_2 + \text{ATP} + H_2O \longrightarrow \text{oxaloacetate} + \text{ADP} + P_i + 2H^+$$

Recall that this enzyme plays a crucial role in gluconeogenesis (Section 16.4). It is active only in the presence of acetyl CoA, which signifies the need for more oxaloacetate. If the energy charge is high, oxaloacetate is converted into glucose. If the energy charge is low, oxaloacetate replenishes the citric acid cycle. The synthesis of oxaloacetate by the carboxylation of pyruvate is an example of an **anaplerotic reaction** (*anaplerotic* is of Greek origin, meaning to "fill up"), a reaction that leads to the net synthesis, or replenishment, of pathway components. Note that because the citric acid cycle is a cycle, it can be replenished by the generation of any of the intermediates. For example, aspartate can be deaminated to also generate oxaloacetate. Glutamine is an especially important source of citric acid cycle intermediates in rapidly growing cells, including cancer cells. Glutamine is converted into glutamate and then into α-ketoglutarate (Section 23.5).

The disruption of pyruvate metabolism is the cause of beriberi and poisoning by mercury and arsenic

A neurologic and cardiovascular disorder known as **beriberi** is caused by a dietary deficiency of thiamine (also called *vitamin B$_1$*). The disease is prevalent in areas of the world where rice is the major food staple, because white rice has a rather low content of thiamine. This deficiency is partly overcome if the whole rice grain is soaked in water before milling, because some of the thiamine in the husk then leaches into the rice kernel. The problem is worsened if the rice is polished to remove the outer layer (that is, converted from brown to white rice), because only the outer layer contains significant amounts of thiamine. A common form of beriberi called *wet beriberi* is characterized by swelling in the abdomen and legs due to cardiac abnormalities.

Another form of beriberi, called *Wernicke's encephalopathy*, is also occasionally seen in alcoholics who are severely malnourished and thus thiamine deficient. The disease is characterized by neurologic and cardiac symptoms. Damage to the peripheral nervous system is expressed as pain in the limbs, weakness of the musculature, and distorted skin sensation. The heart may be enlarged and the cardiac output inadequate.

Which biochemical processes might be affected by a deficiency of thiamine? Thiamine is the precursor of the cofactor thiamine pyrophosphate (TPP). This cofactor is the prosthetic group of three important enzymes: pyruvate dehydrogenase, α-ketoglutarate dehydrogenase, and transketolase. (Transketolase functions in the pentose phosphate pathway, which will be considered in Chapter 20.) The common feature of enzymatic reactions using TPP is the transfer of an activated aldehyde unit.

In patients with beriberi, the levels of pyruvate and α-ketoglutarate in the blood are higher than normal, and the increase in the level of pyruvate in the blood is especially pronounced after the ingestion of glucose. A related finding is that the activities of the pyruvate and α-ketoglutarate dehydrogenase complexes are abnormally low. The low transketolase activity of red blood cells in beriberi is an easily measured and reliable diagnostic indicator of the disease.

Why does TPP deficiency lead primarily to cardiac and neurological disorders? Heart tissue is highly sensitive to metabolic disruptions, as its need for ATP is never-ending and it has no appreciable stores of energy. The nervous system relies essentially on glucose as its only fuel. The product of glycolysis, pyruvate, can enter the citric acid cycle only through the pyruvate dehydrogenase complex. With that enzyme deactivated, the nervous system has no source of fuel. In contrast, most other tissues can use fats as a source of fuel for the citric acid cycle.

Symptoms similar to those of beriberi appear in organisms exposed to mercury or arsenite (AsO_3^{3-}). Both materials have a high affinity for neighboring (vicinal) sulfhydryls, such as those in the reduced dihydrolipoyl groups of the E$_3$ component of

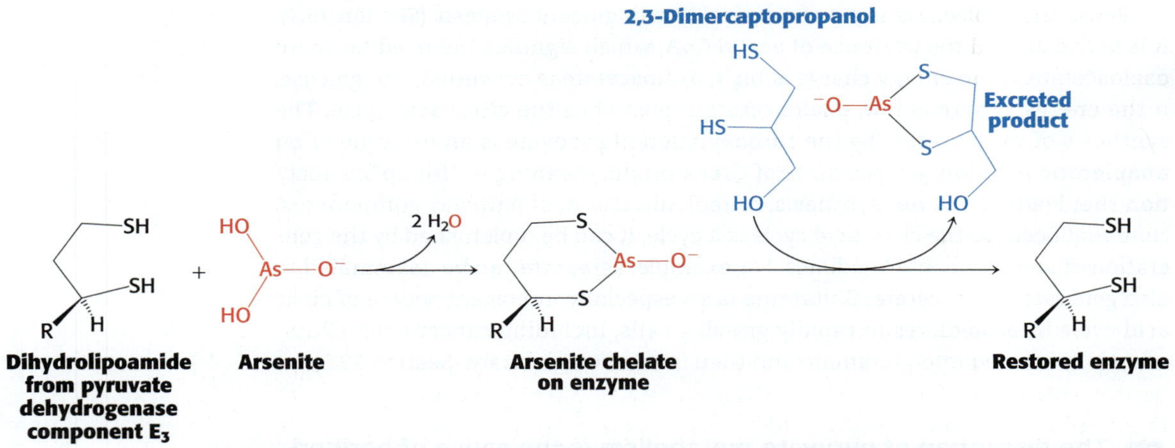

FIGURE 17.19 Arsenite inhibits the pyruvate dehydrogenase complex by inactivating the dihydrolipoamide component of the transacetylase. Some sulfhydryl reagents, such as 2,3-dimercaptoethanol, relieve the inhibition by forming a complex with the arsenite that can be excreted.

the pyruvate dehydrogenase complex (Figure 17.19). The binding of mercury or arsenite to the dihydrolipoyl groups inhibits the complex and leads to central nervous system pathologies. The proverbial phrase "mad as a hatter" refers to the strange behavior of poisoned hatmakers who used mercury nitrate, which is absorbed through the skin, to soften and shape animal furs. Similar symptoms afflicted the early photographers, who used vaporized mercury to create daguerreotypes.

Treatment for these poisons is the administration of sulfhydryl reagents with adjacent sulfhydryl groups to compete with the dihydrolipoyl residues for binding with the metal ion. The reagent–metal complex is then excreted in the urine. Indeed, 2,3-dimercaptopropanol (Figure 17.19) was developed after World War I as an antidote to an arsenic-based chemical weapon.

The citric acid cycle likely evolved from preexisting pathways

How did the citric acid cycle come into being? Although definitive answers are elusive, informed speculation is possible. We can hypothesize how evolution might work at the level of biochemical pathways.

The citric acid cycle was most likely assembled from preexisting reaction pathways. As noted earlier, many of the intermediates formed in the citric acid cycle are used in metabolic pathways for amino acids and porphyrins. Thus, compounds such as pyruvate, α-ketoglutarate, and oxaloacetate were likely present early in evolution for biosynthetic purposes. The oxidative decarboxylation of these α-ketoacids is quite favorable thermodynamically and can be used to drive the synthesis of both acyl CoA derivatives and NADH. These reactions almost certainly formed the core of processes that preceded the citric acid cycle evolutionarily. Interestingly, α-ketoglutarate and oxaloacetate can be interconverted by transamination of the respective amino acids by aspartate aminotransferase, another key biosynthetic enzyme. Thus, cycles comprising smaller numbers of intermediates used for a variety of biochemical purposes could have existed before the present form evolved.

17.6 The Glyoxylate Cycle Enables Plants and Bacteria to Grow on Acetate

Acetyl CoA that enters the citric acid cycle has but one fate: oxidation to CO_2 and H_2O. The citric acid cycle cannot convert acetyl CoA into glucose. Although oxaloacetate, a key precursor to glucose, is formed in the citric acid cycle, the two decarboxylations that take place before the regeneration of oxaloacetate preclude the *net* conversion of acetyl CoA into glucose.

However, in plants and in some microorganisms, there is a metabolic pathway that allows the conversion of acetyl CoA generated from fat stores into glucose. This reaction sequence, called the **glyoxylate cycle**, is similar to the citric acid cycle in some ways but different in that it bypasses the two decarboxylation steps of the cycle. Another important difference is that two molecules of acetyl CoA enter per turn of the glyoxylate cycle, compared with one in the citric acid cycle.

The glyoxylate cycle (**Figure 17.20**), like the citric acid cycle, begins with the condensation of acetyl CoA and oxaloacetate to form citrate, which is then isomerized to isocitrate. Instead of being decarboxylated, as in the citric acid cycle, isocitrate is cleaved by isocitrate lyase into succinate and glyoxylate. The ensuing steps regenerate oxaloacetate from glyoxylate. First, acetyl CoA condenses with glyoxylate to form malate in a reaction catalyzed by malate synthase, and then malate is oxidized to oxaloacetate, as in the citric acid cycle. The sum of these reactions is

$$2 \text{ Acetyl CoA} + \text{NAD}^+ + 2 \text{ H}_2\text{O} \longrightarrow \text{succinate} + 2 \text{ CoA} + \text{NADH} + 2 \text{ H}^+$$

The glyoxylate cycle is especially prominent in oil-rich seeds, such as those from sunflowers, cucumbers, and castor beans, where it takes place in organelles called **glyoxysomes**. Succinate, released midcycle, can be converted into carbohydrates by a combination of the citric acid cycle and gluconeogenesis. The carbohydrates power seedling growth until the cell can begin photosynthesis. Thus, organisms with the glyoxylate cycle gain metabolic versatility because they can use acetyl CoA as a precursor of glucose and other biomolecules.

FIGURE 17.20 The glyoxylate cycle allows plants to use acetyl groups for biosynthesis. Plant seeds germinate using acetyl groups derived from stored triacylglycerides. The cycle bypasses the decarboxylation steps of the citric acid cycle and incorporates two acetyl groups per cycle. The reactions of this cycle are the same as those of the citric acid cycle except for the ones catalyzed by isocitrate lyase and malate synthase, which are boxed in blue.

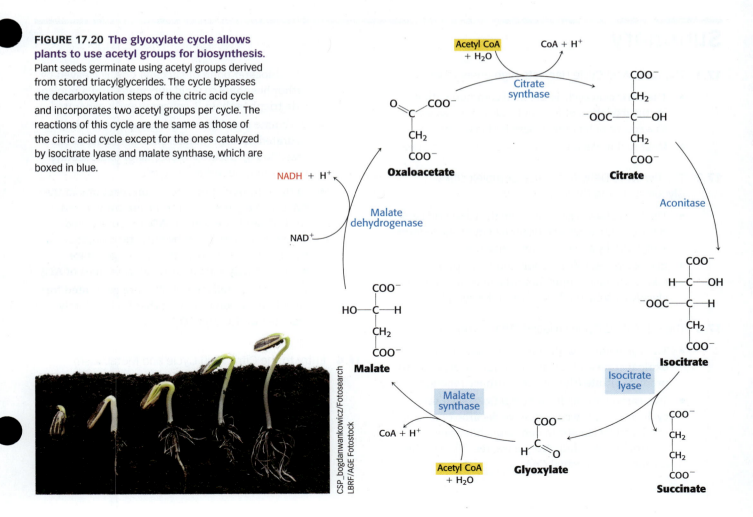

CSP_bogdanwankowicz/Fotosearch
LBRF/AGE Fotostock

☤ Blocking the glyoxylate cycle may lead to new treatments for tuberculosis

Tuberculosis (TB) is one of the leading causes of death throughout the world. In 2020, according to the World Health Organization, 10 million people contracted TB, including 1.1 million children, and 1.5 million people died from the disease. The bacterium responsible, *Mycobacterium tuberculosis*, is transmitted by people with active lung infections by coughing and sneezing. A common treatment for tuberculosis is the antibiotic rifampicin, an inhibitor of bacterial RNA synthesis (Chapter 31). However, strains of *M. tuberculosis* are developing resistance to rifampicin, so new treatments are urgently needed.

Some exciting new research suggests a possible treatment that relies on the bacterium's dependence on the glyoxylate cycle to convert fats to glucose, especially when in a latent (not actively dividing) state in the lungs. A suicide inhibitor or mechanism-based inhibitor (Section 5.6) for isocitrate lyase, 2-vinyl-isocitrate, has been synthesized that may halt the bacterium's metabolism.

During catalysis, a reactive thiolate is formed on Cys_{191} in the active site. When the lyase reacts with the inhibitor, succinate is released as in the normal reaction, but a thioether-linked homopyruvoyl moiety is covalently linked to Cys_{191}, inhibiting the enzyme.

This development may be an important opening into TB therapy because the active site Cys_{191} is conserved in all strains of *M. tuberculosis*, which reduces the likelihood of the development of resistance to the treatment. Although this research is quite new, it will be interesting to see if this ultimately leads to an effective medication for TB.

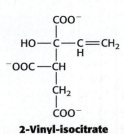

2-Vinyl-isocitrate

Homopyruvoyl-modified isocitrate lyase

Summary

17.1 The Citric Acid Cycle Harvests High-Energy Electrons

- The citric acid cycle is the final common pathway for the oxidation of fuel molecules and also serves as a source of building blocks for biosyntheses.
- Most fuel molecules enter the cycle as acetyl CoA.

17.2 The Pyruvate Dehydrogenase Complex Links Glycolysis to the Citric Acid Cycle

- The link between glycolysis and the citric acid cycle is the oxidative decarboxylation of pyruvate to form acetyl CoA by a massive three-enzyme complex.
- In eukaryotes, this reaction and those of the cycle take place inside mitochondria, in contrast with glycolysis, which takes place in the cytoplasm.

17.3 The Citric Acid Cycle Oxidizes Two-Carbon Units

- The cycle starts with the condensation of oxaloacetate and the acetyl unit of acetyl CoA to produce citrate (the citrate synthase reaction).
- Two carbon atoms from acetyl CoA leave the cycle as CO_2 in two successive decarboxylations catalyzed by isocitrate dehydrogenase and α-ketoglutarate dehydrogenase, resulting in the formation of succinyl CoA.

- The thioester bond of succinyl CoA is cleaved by orthophosphate to yield succinate, and a nucleoside trisphosphate is concomitantly generated.
- Succinate is oxidized to fumarate, which is then hydrated to form malate. Finally, malate is oxidized to regenerate a molecule of oxaloacetate for the next round of the cycle.
- In the four oxidation–reduction reactions in the cycle, three pairs of electrons are transferred to NAD^+ and one pair to FAD. These reduced electron carriers are subsequently oxidized by the electron-transport chain to generate approximately nine additional molecules of ATP.
- A total of 10 molecules of ATP are generated for each two-carbon fragment that is completely oxidized to H_2O and CO_2.

17.4 Entry to the Citric Acid Cycle and Metabolism Through It Are Controlled

- The citric acid cycle operates only under aerobic conditions because it requires a supply of NAD^+ and FAD, which are regenerated in oxidative phosphorylation. Consequently, the rate of the citric acid cycle depends on the need for ATP.

- The irreversible formation of acetyl CoA from pyruvate is an important regulatory point: the activity of the pyruvate dehydrogenase complex is stringently regulated by reversible phosphorylation.

- In eukaryotes, two enzymes in the citric acid cycle are also important for regulation: isocitrate dehydrogenase and α-ketoglutarate dehydrogenase. Energy-rich molecules decrease the activities of these regulatory enzymes.

- These mechanisms complement each other in reducing the rate of formation of acetyl CoA when the energy charge of the cell is high and when biosynthetic intermediates are abundant.

17.5 The Citric Acid Cycle Is a Source of Biosynthetic Precursors

- When the cell has adequate energy available, the citric acid cycle can provide a source of

building blocks for a host of important biomolecules, such as nucleotide bases, proteins, and heme groups. This use depletes the cycle of intermediates.

- When the cycle again needs to metabolize fuel, anaplerotic reactions replenish the cycle intermediates.

17.6 The Glyoxylate Cycle Enables Plants and Bacteria to Grow on Acetate

- The glyoxylate cycle enables plants and bacteria to subsist on acetate because it bypasses the two decarboxylation steps of the citric acid cycle.

- The glyoxylate cycle is especially prominent in oil-rich seeds where it takes place in glyoxysomes. The succinate produced can be converted into carbohydrates by a combination of the citric acid cycle and gluconeogenesis.

Key Terms

citric acid (tricarboxylic acid, TCA; Krebs) cycle (CAC) (p. 516)

acetyl CoA (p. 516)

pyruvate dehydrogenase complex (p. 516)

flavoprotein (p. 520)

iron–sulfur protein (nonheme iron protein) (p. 525)

α-ketoglutarate dehydrogenase complex (p. 526)

nucleoside diphosphokinase (p. 527)

anaplerotic reaction (p. 535)

beriberi (p. 535)

glyoxylate cycle (p. 537)

glyoxysome (p. 537)

Problems

1. Write the reaction that links glycolysis and the citric acid cycle. What is the enzyme that catalyzes the reaction? ❖ 1

2. Which of the following catalytic cofactors are required by the pyruvate dehydrogenase complex?

(a) NAD⁺, lipoic acid, thiamine pyrophosphate

(b) NAD⁺, biotin, thiamine pyrophosphate

(c) NAD⁺, FAD, lipoic acid

(d) FAD, lipoic acid, thiamine pyrophosphate

3. The pyruvate dehydrogenase complex is mechanistically and structurally similar to which of the following?

(a) Isocitrate dehydrogenase

(b) α-Ketoglutarate dehydrogenase

(c) Glyceraldehyde 3-phosphate dehydrogenase

(d) Fumarase

4. What coenzymes are required by the pyruvate dehydrogenase complex? What are their roles? ❖ 1

5. Distinguish between catalytic coenzymes and stoichiometric coenzymes in the pyruvate dehydrogenase complex.

6. What is the $\Delta G°'$ for the complete oxidation of the acetyl unit of acetyl CoA by the citric acid cycle?

7. Patients with pyruvate dehydrogenase deficiency show high levels of lactic acid in the blood. However, in some cases, treatment with dichloroacetate (DCA), which inhibits the kinase associated with the pyruvate dehydrogenase complex, lowers lactic acid levels. ❖ 2

(a) How does DCA act to stimulate pyruvate dehydrogenase activity?

(b) What does this suggest about pyruvate dehydrogenase activity in patients who respond to DCA?

8. (a) Predict the effect of a mutation that enhances the activity of the kinase associated with the pyruvate dehydrogenase complex. ❖ 2

(b) Predict the effect of a mutation that reduces the activity of the phosphatase associated with the pyruvate dehydrogenase complex.

9. Match each term with its description.

(a) Acetyl CoA
(b) Citric acid cycle
(c) Pyruvate dehydrogenase complex
(d) Thiamine pyrophosphate
(e) Lipoic acid
(f) Pyruvate dehydrogenase
(g) Acetyllipoamide
(h) Dihydrolipoyl transacetylase
(i) Dihydrolipoyl dehydrogenase
(j) Beriberi

1. Catalyzes the link between glycolysis and the citric acid cycle
2. Coenzyme required by transacetylase
3. Final product of pyruvate dehydrogenase
4. Catalyzes the formation of acetyl CoA
5. Regenerates active transacetylase
6. Fuel for the citric acid cycle
7. Coenzyme required by pyruvate dehydrogenase
8. Catalyzes the oxidative decarboxylation of pyruvate
9. Due to a deficiency of thiamine
10. Central metabolic hub

10. Match each enzyme with its description.

(a) Pyruvate dehydrogenase complex
(b) Citrate synthase
(c) Aconitase
(d) Isocitrate dehydrogenase
(e) α-Ketoglutarate dehydrogenase
(f) Succinyl CoA synthetase
(g) Succinate dehydrogenase
(h) Fumarase
(i) Malate dehydrogenase
(j) Pyruvate carboxylase

1. Catalyzes the formation of isocitrate
2. Synthesizes succinyl CoA
3. Generates malate
4. Generates ATP
5. Converts pyruvate into acetyl CoA
6. Converts pyruvate into oxaloacetate
7. Condenses oxaloacetate and acetyl CoA
8. Catalyzes the formation of oxaloacetate
9. Synthesizes fumarate
10. Catalyzes the formation of α-ketoglutarate

11. How is succinate dehydrogenase unique compared with the other enzymes in the citric acid cycle?

12. The oxidation of malate by NAD^+ to form oxaloacetate is a highly endergonic reaction under standard conditions $[\Delta G^{\circ\prime} = 29$ kJ mol^{-1}(7 kcal mol^{-1})]. The reaction proceeds readily under physiological conditions. ❖ **4**

(a) Explain why the reaction proceeds as written under physiological conditions.

(b) Assuming an $[NAD^+]/[NADH]$ ratio of 8 and a pH of 7, what is the lowest [malate]/[oxaloacetate] ratio at which oxaloacetate can be formed from malate?

13. It is possible, with the use of the reactions and enzymes considered in this chapter, to convert pyruvate into α-ketoglutarate without depleting any of the citric acid cycle components. Write a balanced reaction scheme for this conversion, showing cofactors and identifying the required enzymes. ❖ **6**

14. Malonate is a competitive inhibitor of succinate dehydrogenase. How will the concentrations of citric acid cycle intermediates change immediately after the addition of malonate? Why is malonate not a substrate for succinate dehydrogenase?

$$COO^-$$
$$|$$
$$CH_2$$
$$|$$
$$COO^-$$
Malonate

15. What is the chief benefit of being able to perform the glyoxylate cycle? ❖ **7**

16. Fats are usually metabolized into acetyl CoA and then further processed through the citric acid cycle. In Chapter 16, we saw that glucose can be synthesized from oxaloacetate, a citric acid cycle intermediate. Why, then, after a long bout of exercise depletes our carbohydrate stores, do we need to replenish those stores by eating carbohydrates? Why do we not simply replace them by converting fats into carbohydrates?

17. As we will see (Chapter 22), fatty acid breakdown generates a large amount of acetyl CoA. What will be the effect of fatty acid breakdown on pyruvate dehydrogenase complex activity? On glycolysis? ❖ **2**

18. In experiments carried out in 1941 to investigate the citric acid cycle, oxaloacetate labeled with ^{14}C in the carboxyl carbon atom farthest from the keto group was introduced to an active preparation of mitochondria.

Oxaloacetate

Analysis of the α-ketoglutarate formed showed that none of the radioactive label had been lost. Decarboxylation of α-ketoglutarate then yielded succinate devoid of radioactivity. All the label was in the released CO_2.

(a) Why were the early investigators of the citric acid cycle surprised that *all* the label emerged in the CO_2? (*Note: you should answer this question first before reading any further.*)

(b) The interpretation of these experiments was that citrate (or any other symmetric compound) cannot be an intermediate in the formation of α-ketoglutarate, because of the asymmetric fate of the label, thus appearing to disprove Hans Krebs's assertions about the cycle. This view seemed compelling until Alexander Ogston incisively pointed out in 1948 that "it is possible that an asymmetric enzyme which attacks a symmetrical compound can distinguish between its identical groups." For simplicity, consider a molecule in which two hydrogen atoms, a group X and a different group Y, are bonded to a tetrahedral carbon atom as a model for citrate. Explain how a symmetric molecule can react with an enzyme in an asymmetric way.

19. As will become clearer in Chapter 18, the activity of the citric acid cycle can be monitored by measuring the amount of O_2 consumed. The greater the rate of O_2 consumption, the faster the rate of the cycle. Hans Krebs used this assay to investigate the cycle in 1937. He used as his experimental system minced pigeon-breast muscle, which is rich in mitochondria. In one set of experiments, Krebs measured the O_2 consumption in the presence of carbohydrate only and in the presence of carbohydrate and citrate. The results are shown in the following table. ❖ 5

Effect of citrate on oxygen consumption by minced pigeon-breast muscle

Time (min)	Micromoles of oxygen consumed	
	Carbohydrate only	Carbohydrate plus 3 μmol of citrate
10	26	28
60	43	62
90	46	77
150	49	85

(a) How much O_2 would be absorbed if the added citrate were completely oxidized to H_2O and CO_2?

(b) On the basis of your answer to part (a), what do the results given in the table suggest?

20. The bacterium *Mycobacterium tuberculosis*, the cause of tuberculosis, can invade the lungs and persist in a latent state for years. During this time, the bacteria reside in granulomas—nodular scars containing bacteria and host-cell debris in the center and surrounded by immune cells. The granulomas are lipid-rich, oxygen-poor environments. How these bacteria manage to persist is something of a mystery. The results of research suggest that the glyoxylate cycle is required for the persistence. The following data show the amount of bacteria, presented as colony-forming units (cfu), in mice lungs in the weeks after an infection. ❖ 7

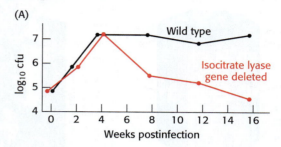

(A)

In graph A, the black circles represent the results for wild-type bacteria and the red circles represent the results for bacteria from which the gene for isocitrate lyase was deleted.

(a) What is the effect of the absence of isocitrate lyase?

The techniques described in Chapter 9 were used to reinsert the gene encoding isocitrate lyase into bacteria from which it had previously been deleted. In graph B, black circles represent bacteria into which the gene was reinserted and red circles represent bacteria in which the gene was still missing.

(b) Do these results support those obtained in part (a)?

(c) What is the purpose of the experiment in part (b)?

(d) Why do these bacteria perish in the absence of the glyoxylate cycle?

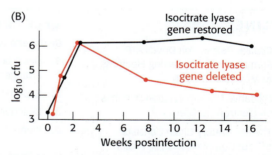

(B)

[Data from McKinney et al., *Nature* 406:735–738, 2000.]

18 Oxidative Phosphorylation

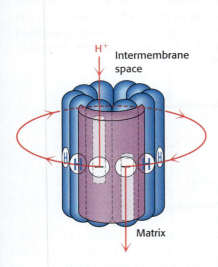

H⁺

Intermembrane space

Matrix

Melvyn Longhurst/Corbis Documentary/Getty Images

Waterwheels capture and convert the kinetic energy of water by rotating a wheel which spins a central axis, or rotor. While humans have used waterwheels for millennia to mill grains and accomplish many other tasks, nature beat humans in the race to develop a rotational motor by at least 1.5 billion years, prior to the divergence of eukaryotes and prokaryotes. Remarkably, ATP synthase, the final complex of the mitochondrial respiratory machinery, uses the flow of protons across a membrane to spin a protein wheel and rotor to generate ATP. This enzyme is the world's smallest rotary motor.

OUTLINE

18.1 Cellular Respiration Drives ATP Formation by Transferring Electrons to Molecular Oxygen

18.2 Oxidative Phosphorylation Depends on Electron Transfer

18.3 The Respiratory Chain Consists of Four Complexes: Three Proton Pumps and a Physical Link to the Citric Acid Cycle

18.4 A Proton Gradient Powers the Synthesis of ATP

18.5 Many Shuttles Allow Movement Across Mitochondrial Membranes

18.6 The Regulation of Cellular Respiration Is Governed Primarily by the Need for ATP

18.7 Proton Gradients Generated by Respiratory Chains Drive Many Biochemical Processes

🔷 LEARNING GOALS

By the end of this chapter, you should be able to:

1. Describe the key components of the electron-transport chain and how they are arranged.

2. Explain the benefits of having the electron-transport chain located in a membrane.

3. Describe how the proton-motive force is converted into ATP.

4. Identify the ultimate determinant of the rate of cellular respiration.

The amount of ATP that human beings require to go about their lives is staggering: a sedentary person of 70 kg (154 lbs) requires about 8400 kJ (2000 kcal) for a day's worth of activity. To provide this much energy requires 83 kg of ATP. However, human beings possess only about 250 g of ATP at any given moment.

The disparity between the amount of ATP that we have and the amount that we require is compensated by recycling ADP back to ATP. Each ATP molecule is recycled approximately 300 times per day. This recycling takes place primarily through oxidative phosphorylation. The process of oxidative phosphorylation finally completes the capture of chemical energy from glucose and its derivatives that originated with glycolysis.

18.1 Cellular Respiration Drives ATP Formation by Transferring Electrons to Molecular Oxygen

We begin our study of **oxidative phosphorylation** by examining the oxidation–reduction reactions that allow the flow of electrons from NADH and $FADH_2$ to oxygen. The electron flow takes place in four large protein complexes that are embedded in the inner mitochondrial membrane, together called the **respiratory chain** or the **electron-transport chain**.

$$NADH + \tfrac{1}{2}O_2 + H^+ \longrightarrow H_2O + NAD^+$$
$$\Delta G^{\circ\prime} = -220.1 \text{ kJ mol}^{-1} (-52.6 \text{ kcal mol}^{-1})$$

The overall reaction is exergonic. Importantly, three of the complexes of the electron-transport chain use the energy released by the electron flow to pump protons out of the mitochondrial matrix. In essence, energy is transformed. The resulting unequal distribution of protons generates a pH gradient and a transmembrane electrical potential that creates a **proton-motive force**. ATP is synthesized when protons flow back to the mitochondrial matrix through an enzyme complex.

$$ADP + P_i + H^+ \longrightarrow ATP + H_2O$$
$$\Delta G^{\circ\prime} = +30.5 \text{ kJ mol}^{-1} (+7.3 \text{ kcal mol}^{-1})$$

Thus, the oxidation of fuels and the phosphorylation of ADP are coupled by a proton gradient across the inner mitochondrial membrane (**Figure 18.1**). Collectively, the generation of high-transfer-potential electrons by the citric acid cycle, their flow through the respiratory chain, and the accompanying synthesis of ATP is called **respiration** or **cellular respiration**.

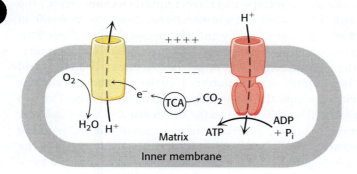

FIGURE 18.1 Oxidative phosphorylation is fundamentally the combination of two processes. Oxidation and ATP synthesis are coupled by transmembrane proton fluxes. Electrons from NADH and $FADH_2$, formed in the TCA cycle, flow through the electron-transport chain to reduce oxygen to water (yellow cylinder). Some components of the chain pump protons from the mitochondrial matrix to the intermembrane space. The protons return to the matrix by flowing through another protein complex, ATP synthase (red structure), powering the synthesis of ATP.

Eukaryotic oxidative phosphorylation takes place in mitochondria

Recall that a biochemical role of the citric acid cycle, which takes place in the mitochondrial matrix, is the generation of high-energy electrons. It is fitting, therefore, that oxidative phosphorylation, which will utilize the energy of these electrons to create ATP, also takes place in mitochondria.

Electron microscopic studies revealed that mitochondria have two membranes: an *outer membrane* and an extensive, highly folded *inner membrane*. Hence, there are two compartments in mitochondria: (1) the **intermembrane space (IM space)** between the outer and the inner membranes and (2) the **matrix**, which is bounded by the inner membrane (**Figure 18.2**). The mitochondrial matrix is the site of most of the reactions of the citric acid cycle and fatty acid oxidation, whereas oxidative phosphorylation takes place in the inner mitochondrial membrane.

The inner membrane is folded into a series of internal ridges and tube-like structures collectively called *cristae*. The increase in surface area of the

FIGURE 18.2 Mitochondria are bound by a double membrane, which creates two distinct internal compartments. (A) An electron micrograph and (B) a diagram of a mitochondrion. Mitochondria are oval-shaped organelles, typically ~2 μm in length and 0.5 μm in diameter, or approximately the size of a bacterium.

[(A) Keith R. Porter/Science Source.]

(A)

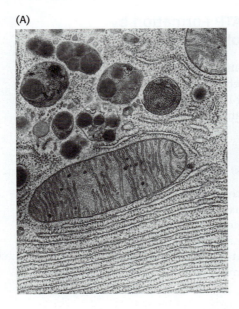

(B)

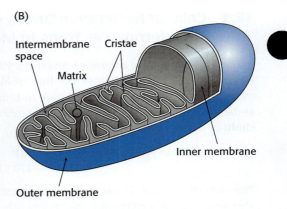

inner mitochondrial membrane provided by the cristae creates more sites for oxidative phosphorylation than would be the case with a simple, unfolded membrane. Humans contain an estimated 14,000 m² of inner mitochondrial membrane, which is the approximate equivalent of three football fields in the United States.

The outer membrane is quite permeable to most small molecules and ions because it contains mitochondrial porin, a 30- to 35-kDa pore-forming protein also known as VDAC, for voltage-dependent anion channel. VDAC, the most prevalent protein in the outer mitochondrial membrane, plays a role in the regulated flux of metabolites—usually anionic species such as phosphate, chloride, organic anions, and the adenine nucleotides—across the outer membrane. In contrast, the inner membrane is impermeable to nearly all ions and polar molecules. A large family of transporters shuttles metabolites such as ATP, pyruvate, and citrate across the inner mitochondrial membrane. The two faces of this membrane will be referred to as the *matrix side* and the *cytoplasmic side* (the latter because it is freely accessible to most small molecules in the cytoplasm). They are also called the *N* and *P* sides, respectively, because the membrane potential is negative on the matrix side and positive on the cytoplasmic side (Figure 18.1).

In bacteria, the electron-driven proton pumps and ATP-synthesizing complexes are located in the cytoplasmic membrane, the inner of two membranes. In many bacteria, the outer membrane, like that of mitochondria, is permeable to most small metabolites because of the presence of porins.

Mitochondria are the result of an endosymbiotic event

Mitochondria are semiautonomous organelles that live in an endosymbiotic relationship with the host cell. These organelles contain their own DNA, which encodes a variety of different proteins and RNAs. The genomes of mitochondria range broadly in size across species. Human mitochondrial DNA comprises 16,569 bp and encodes 13 respiratory-chain proteins as well as the small and large ribosomal RNAs and enough tRNAs to translate all codons. However, mitochondria also contain many proteins encoded by nuclear DNA. Cells that contain mitochondria depend on these organelles for oxidative phosphorylation, and the mitochondria in turn depend on the cell for their very existence. How did this intimate symbiotic relationship come to exist?

It is now well established that an endosymbiotic event occurred in the evolution of eukaryotic organisms in which a free-living organism capable of oxidative phosphorylation was engulfed by another cell. Formerly a hypothesis famously championed by biologist Lynn Margulis, this explanation of the origins of both mitochondria and chloroplasts is now widely accepted as the *endosymbiotic theory*. The shape, size, double-membrane, and circular DNA (with exceptions) all resemble characteristics of some gram-negative bacteria; however, it is the evidence from DNA sequences that is conclusive. Thanks to the rapid accumulation of sequence data for mitochondrial and bacterial genomes, speculation on the origin of the first mitochondrion with some authority is now possible. The most mitochondrial-like bacterial genome is that of *Rickettsia prowazekii*, the cause of louse-borne typhus. The genome for this organism is more than 1 million base pairs in size and contains 834 protein-encoding genes. Sequence data suggest that all extant mitochondria are derived from an ancestor of *R. prowazekii* as the result of a single endosymbiotic event.

The evidence that modern mitochondria result from a single event comes from examination of the most bacteria-like mitochondrial genome, that of the protozoan *Reclinomonas americana*. Its genome contains 97 genes, of which 62 specify proteins. The genes encoding these proteins include all of the protein-coding genes found in all of the sequenced mitochondrial genomes. Yet, this genome encodes less than 2% of the protein-coding genes in the bacterium *E. coli*. In other words, a small fraction of bacterial genes—2%—is found in all examined mitochondria.

How is it possible that all mitochondria have the same 2% of the bacterial genome? It seems unlikely that mitochondrial genomes resulting from several endosymbiotic events could have been independently reduced to the same set of genes found in *R. americana*. What was the fate of the mitochondrial genes no longer harbored by the mitochondria? The reduced genomes of the mitochondria are the result of endosymbiotic gene transfer, in which the mitochondrial genes became part of the nuclear genome. Thus, the original bacterial cell lost DNA, making it incapable of independent living, and the host cell became dependent on the ATP generated by its tenant.

LYNN MARGULIS The concept of an endosymbiotic origin for mitochondria and chloroplasts, first proposed over a century ago, was quite controversial for decades. In 1967, Professor Lynn Margulis (then Lynn Sagan) published a 50-page paper, "On the Origin of Mitosing Cells," with an extensive analysis of the characteristics of different groups of unicellular organisms and possible endosymbiotic events. She and others continued to collect evidence to support the claims in the paper, but the scientific community remained skeptical of the endosymbiosis hypothesis for many years. Eventually, many lines of evidence came together to support the conclusion that chloroplasts and mitochondria each arose from independent endosymbiotic events more than 1 billion years ago. Margulis received the National Medal of Science in 1999 and has been described as a "vindicated heretic."

Nancy R. Schiff/Getty Images

❖ SELF-CHECK QUESTION

Name the cellular compartments and membranes, in order, that a molecule passes through when traveling from the cytosol into the center of a mitochondrion. Additionally, what are the ridges and tube-like structures in a mitochondrion called, and what membrane are they made from?

18.2 Oxidative Phosphorylation Depends on Electron Transfer

In Chapter 17, the generation of NADH and FADH$_2$ by the oxidation of acetyl CoA was identified as an important function of the citric acid cycle. In oxidative phosphorylation, electrons from NADH and FADH$_2$ are used to reduce molecular oxygen to water. The highly exergonic reduction is accomplished by a number of electron-transfer reactions, which take place in a set of membrane proteins known as the electron-transport chain.

The electron-transfer potential of an electron is measured as redox potential

In oxidative phosphorylation, the *electron*-transfer potential of NADH or FADH$_2$ is converted into the *phosphoryl*-transfer potential of ATP. To better understand

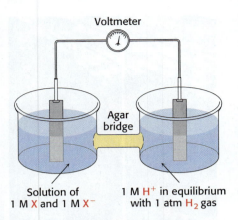

Voltmeter

Agar bridge

Solution of 1 M X and 1 M X⁻   1 M H⁺ in equilibrium with 1 atm H₂ gas

FIGURE 18.3 Redox potential can be precisely measured. In the apparatus for the measurement of the standard oxidation–reduction potential of a redox couple, electrons flow through the wire connecting the cells, whereas ions flow through the agar bridge.

this conversion, we need quantitative expressions for these forms of free energy. The measure of phosphoryl-transfer potential is already familiar to us: it is given by $\Delta G^{\circ\prime}$ for the hydrolysis of the activated phosphoryl compound. The corresponding expression for the electron-transfer potential is E'_0, the **reduction potential** (also called the **redox potential** or **oxidation–reduction potential**).

Consider a substance that can exist in an oxidized form X and a reduced form X⁻. Such a pair is called a *redox couple* and is designated X : X⁻. The reduction potential of this couple can be determined by measuring the electromotive force generated by an apparatus called a *sample half-cell* connected to a *standard reference half-cell* (**Figure 18.3**). The sample half-cell consists of an electrode immersed in a solution of 1 M oxidant (X) and 1 M reductant (X⁻). The standard reference half-cell consists of an electrode immersed in a 1 M H⁺ solution that is in equilibrium with H_2 gas at 1 atm of pressure. The electrodes are connected to a voltmeter, and an agar bridge allows ions to move from one half-cell to the other, establishing electrical continuity between the half-cells. Electrons then flow from one half-cell to the other through the wire connecting the two half-cells to the voltmeter. If the reaction proceeds in the direction

$$X^- + H^+ \rightarrow X + \tfrac{1}{2}H_2$$

the reactions in the half-cells (referred to as *half-reactions* or *couples*) must be

$$X^- \rightarrow X + e^- \qquad H^+ + e^- \rightarrow \tfrac{1}{2}H_2$$

Thus, electrons flow from the sample half-cell to the standard reference half-cell, and the sample-cell electrode is taken to be negative with respect to the standard-cell electrode. The reduction potential of the X : X⁻ couple is the observed voltage at the start of the experiment (when X, X⁻, and H⁺ are 1 M with 1 atm of H_2). The reduction potential of the H⁺ : H_2 couple is defined to be 0 volts. In oxidation–reduction reactions, the donor of electrons, in this case X⁻, is called the *reductant* or *reducing agent*, whereas the acceptor of electrons, H⁺ here, is called the *oxidant* or *oxidizing agent*.

The meaning of the reduction potential is now evident. A negative reduction potential means that the oxidized form of a substance has lower affinity for electrons than does H_2, as in the preceding example. A positive reduction potential means that the oxidized form of a substance has higher affinity for electrons than does H_2. These comparisons refer to standard conditions—namely, 1 M oxidant, 1 M reductant, 1 M H⁺, and 1 atm H_2. Thus, a strong reducing agent (such as NADH) is poised to donate electrons and has a negative reduction potential, whereas a strong oxidizing agent (such as O_2) is ready to accept electrons and has a positive reduction potential.

The reduction potentials of many biologically important redox couples are known (**Table 18.1**). Table 18.1 is like those presented in chemistry textbooks, except that a hydrogen ion concentration of 10^{-7} M (pH 7) instead of 1 M (pH 0) is the standard state adopted by biochemists. This difference is denoted by the prime in E'_0. Recall that the prime in $\Delta G^{\circ\prime}$ denotes a standard free-energy change at pH 7.

The standard free-energy change $\Delta G^{\circ\prime}$ is related to the change in reduction potential $\Delta E'_0$ by

$$\Delta G^{\circ\prime} = -nF\Delta E'_0 \qquad (1)$$

in which n is the number of electrons transferred, F is a proportionality constant called the *Faraday constant* (96.48 kJ mol⁻¹V⁻¹ or 23.06 kcal mol⁻¹V⁻¹), $\Delta E'_0$ is in volts, and $\Delta G^{\circ\prime}$ is in kilojoules or kilocalories per mole.

The free-energy change of an oxidation–reduction reaction can be readily calculated from the reduction potentials of the reactants.

TABLE 18.1 Standard reduction potentials of some reactions

Oxidant	Reductant	n	E_0' (V)
Succinate $+CO_2$	α-Ketoglutarate	2	−0.67
Acetate	Acetaldehyde	2	−0.60
Ferredoxin (oxidized)	Ferredoxin (reduced)	1	−0.43
$2 H^+$	H_2	2	−0.42
NAD^+	$NADH + H^+$	2	−0.32
$NADP^+$	$NADPH + H^+$	2	−0.32
Lipoate (oxidized)	Lipoate (reduced)	2	−0.29
Glutathione (oxidized)	Giutathione (reduced)	2	−0.23
FAD	$FADH_2$	2	−0.22
Acetaldehyde	Ethanol	2	−0.20
Pyruvate	Lactate	2	−0.19
$2 H^+$	H_2	2	0.00^1
Fumarate	Succinate	2	−0.03
Cytochrome b (+3)	Cytochrome b (+2)	1	+0.07
Dehydroascorbate	Ascorbate	2	+0.08
Ubiquinone (oxidized)	Ubiquinone (reduced)	2	+0.10
Cytochrome c (+3)	Cytochrome c (+2)	1	+0.22
Fe (+3)	Fe (+2)	1	+0.77
$\frac{1}{2} O_2 + 2 H^+$	H_2O	2	+0.82

Note: E_0' is the standard oxidation–reduction potential (pH 7, 25°C), and n is the number of electrons transferred. E_0' refers to the partial reaction written as Oxidant $+ e^- \rightarrow$ reductant.

[1] Standard oxidation – reduction potential at pH = 0. Compare with $E_0' = -0.42$ at pH = 7.

EXAMPLE **Calculating the Standard Free Energy of a Reaction from Reduction Potentials**

PROBLEM: Calculate $\Delta G^{\circ\prime}$ for the reduction of pyruvate by NADH, catalyzed by lactate dehydrogenase. Recall that this reaction maintains redox balance in lactic acid fermentation (see Figure 16.11).

$$\text{Pyruvate} + \text{NADH} + H^+ \rightleftharpoons \text{lactate} + NAD^+ \tag{A}$$

GETTING STARTED: From Table 18.1 we see that the reduction potential of the NAD^+ : NADH couple, or half-reaction, is –0.32 V, whereas that of the pyruvate:lactate couple is –0.19 V. By convention, reduction potentials (as in Table 18.1) refer to partial reactions written as reductions: oxidant $+ e^- \rightarrow$ reductant. Hence,

$$\text{Pyruvate} + 2 H^+ + 2 e^- \rightarrow \text{lactate} \qquad E_0' = -0.19 \text{ V} \tag{B}$$
$$NAD^+ + H^+ + 2 e^- \rightarrow \text{NADH} \qquad E_0' = -0.32 \text{ V} \tag{C}$$

To obtain reaction A from reactions B and C, we need to reverse the direction of reaction C so that NADH appears on the left side of the arrow. In doing so, the sign of E_0' must be changed.

$$\text{Pyruvate} + 2 H^+ + 2 e^- \rightarrow \text{lactate} \qquad E_0' = -0.19 \text{ V} \tag{B}$$
$$\text{NADH} \rightarrow NAD^+ + H^+ + 2 e^- \qquad E_0' = +0.32 \text{ V} \tag{D}$$

Finally, recall that the Faraday constant is 96.48 kJ mol^{-1} K^{-1}.

CALCULATE: Let's begin by calculating $\Delta G^{\circ\prime}$ for reaction B using equation 1 and substituting in the value of the Faraday constant and $n = 2$ since two electrons are exchanged.

$$\Delta G^{\circ\prime} = -2 \times 96.48 \text{ kJ mol}^{-1} \text{ V}^{-1} \times -0.19 \text{ V}$$
$$= +36.7 \text{ kJ mol}^{-1}$$

Likewise, for reaction D,

$$\Delta G^{\circ\prime} = -2 \times 96.48 \text{ kJ mol}^{-1} \text{ V}^{-1} \times +0.32 \text{ V}$$
$$= -61.7 \text{ kJ mol}^{-1}$$

Thus, the free energy for reaction A is given by

$$\Delta G^{\circ\prime} = \Delta G^{\circ\prime}(\text{for reaction B}) + \Delta G^{\circ\prime}(\text{for reaction D})$$
$$= +36.7 \text{ kJ mol}^{-1} - 61.7 \text{ kJ mol}^{-1}$$
$$= -25.0 \text{ kJ mol}^{-1} \text{ or } -6.0 \text{ kcal mol}^{-1}$$

REFLECT: What does $\Delta G^{\circ\prime}$ tell us? We can see from this result that under standard conditions the reaction is exergonic, releasing 25.0 kJ per mole. Thus, the reduction of pyruvate to lactate by NADH is spontaneous because electrons are more attracted to pyruvate than NAD$^+$. Note that all of this information was deduced entirely from the known reduction potentials.

Electron flow from NADH to molecular oxygen powers the formation of a proton gradient

Let's now consider the driving force of oxidative phosphorylation, which is the electron-transfer potential of NADH or FADH$_2$ relative to that of O$_2$. How much energy is released by the reduction of O$_2$ with NADH? Let us calculate $\Delta G^{\circ\prime}$ for this reaction. The pertinent half-reactions are

$$\tfrac{1}{2}O_2 + 2 H^+ + 2e^- \rightarrow H_2O \qquad E'_0 = +0.82 \text{ V}$$
$$NAD^+ + H^+ + 2e^- \rightarrow NADH \qquad E'_0 = -0.32 \text{ V}$$

The combination of the two half-reactions, as it proceeds in the electron-transport chain, yields

$$\tfrac{1}{2}O_2 + NADH + H^+ \rightarrow H_2O + NAD^+$$

The standard free energy for this reaction is then given by

$$\Delta G^{\circ\prime} = (-2 \times 96.48 \text{ kJ mol}^{-1}\text{V}^{-1} \times +0.82 \text{ V})$$
$$+ (-2 \times 96.48 \text{ kJ mol}^{-1}\text{V}^{-1} \times +0.32 \text{ V})$$
$$= -158.2 \text{ kJ mol}^{-1} + (-61.9) \text{ kJ mol}^{-1}$$
$$= -220.1 \text{ kJ mol}^{-1} \text{ or } -52.6 \text{ kcal mol}^{-1}$$

This release of free energy is substantial. Recall that $\Delta G^{\circ\prime}$ for the synthesis of ATP is 30.5 kJ mol^{-1} (7.3 kcal mol^{-1}). The released energy is initially used to generate a proton gradient that is then used for the synthesis of ATP and the transport of metabolites across the mitochondrial membrane.

How can the energy associated with a proton gradient be quantified? Recall from Section 13.1 that the free-energy change for a species moving from one side of a membrane where it is at concentration c_1 to the other side where it is at a concentration c_2 is given by

$$\Delta G = RT \ln(c_2/c_1) + ZF\Delta V \qquad (2)$$

in which Z is the electrical charge of the transported species, and ΔV is the potential in volts across the membrane. Under typical conditions for the inner mitochondrial membrane, the pH outside is 1.4 units lower than inside [corresponding to $\ln(c_2/c_1)$ of 3.2], and the membrane potential is 0.14 V, the outside being positive. Finally, recall that the gas constant, R, is -8.315×10^{-3} kJ mol^{-1} K^{-1}. Because $Z = +1$ for protons, from equation 2 we see that the free-energy change is $(8.32 \times 10^{-3}$ kJ mol^{-1} K$^{-1} \times 310$ K $\times 3.2) + (+1 \times 96.48$ kJ mol^{-1} V$^{-1} \times 0.14$ V$) = 21.8$ kJ mol^{-1} or 5.2 kcal mol^{-1}. Thus, each proton that is transported out of the matrix to the cytoplasmic side corresponds to 21.8 kJ mol^{-1} of free energy.

❖ SELF–CHECK QUESTION

Identify the oxidant and the reductant in the following reaction:
Pyruvate + NADH + H$^+$ ⇌ lactate + NAD$^+$

18.3 The Respiratory Chain Consists of Four Complexes: Three Proton Pumps and a Physical Link to the Citric Acid Cycle

Electrons are transferred from NADH to O_2 through a chain of three large protein complexes called *NADH-Q oxidoreductase*, *Q-cytochrome c oxidoreductase*, and *cytochrome c oxidase* (**Figure 18.4** and **Table 18.2**). Electron flow within these transmembrane complexes is highly exergonic and powers the transport of protons across the inner mitochondrial membrane. A fourth large protein complex, called *succinate-Q reductase*, contains the succinate dehydrogenase that generates FADH$_2$ in the citric acid cycle. Electrons from this FADH$_2$ enter the electron-transport chain at Q-cytochrome c oxidoreductase. Succinate-Q reductase, in contrast with the other complexes, does not pump protons. NADH-Q oxidoreductase, succinate-Q reductase, Q-cytochrome c oxidoreductase, and cytochrome c oxidase are also called *Complex I, II, III,* and *IV*, respectively. Complexes I, III, and IV appear to be associated in a supramolecular complex which facilitates the rapid transfer of substrate and prevents the release of reaction intermediates, as will be discussed further at the end of this section.

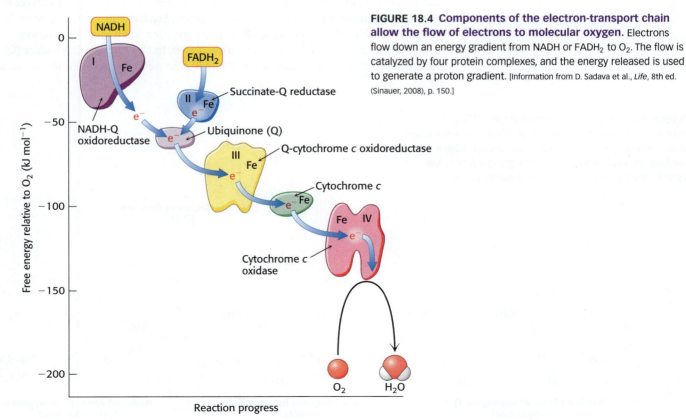

FIGURE 18.4 Components of the electron-transport chain allow the flow of electrons to molecular oxygen. Electrons flow down an energy gradient from NADH or FADH$_2$ to O$_2$. The flow is catalyzed by four protein complexes, and the energy released is used to generate a proton gradient. [Information from D. Sadava et al., *Life*, 8th ed. (Sinauer, 2008), p. 150.]

TABLE 18.2 Components of the mitochondrial electron-transport chain

Enzyme complex	Mass (kDa)	Prosthetic group	Oxidant or reductant		
			Matrix side	**Membrane core**	**Intermembrane side**
NADH-Q oxidoreductase	> 900	FMN Fe–S	NADH	Q	
Succinate-Q reductase	140	FAD Fe–S	Succinate	Q	
Q-cytochrome c oxidoreductase	250	Heme b_H Heme b_L Heme c_1 Fe–S		Q	Cytochrome c
Cytochrome c oxidase	160	Heme a Heme a_3 Cu_A and Cu_B			Cytochrome c

Information from: J. W. DePierre and L. Ernster, *Annu. Rev. Biochem.* 46:215, 1977; Y. Hatefi, *Annu. Rev. Biochem.* 54:1015, 1985; and J. E. Walker, *Q. Rev. Biophys.* 25:253, 1992.

Two electron carriers ferry the electrons from one complex to the next. The first is **coenzyme Q** (or **Q,** for short), which is also known as *ubiquinone* because it is a ubiquitous quinone in biological systems. Coenzyme Q is a hydrophobic quinone that diffuses rapidly within the inner mitochondrial membrane. Electrons are carried from NADH-Q oxidoreductase to Q-cytochrome c oxidoreductase, the third complex of the chain, by the reduced form of Q. Electrons from the $FADH_2$ generated by the citric acid cycle are transferred first to ubiquinone and then to the Q-cytochrome c oxidoreductase complex.

Coenzyme Q is a quinone derivative with a long tail consisting of five-carbon isoprene units that account for its hydrophobic nature. The number of isoprene units in the tail depends on the species. The most common mammalian form contains 10 isoprene units (coenzyme Q_{10}). For simplicity, the subscript will be omitted from this abbreviation because all varieties function in an identical manner.

Quinones can exist in several oxidation states. In the fully oxidized state (Q), coenzyme Q has two keto groups (**Figure 18.5**). The addition of one electron and

FIGURE 18.5 Quinones can have three different oxidation states, two of which are stable. The reduction of ubiquinone (Q) to ubiquinol (QH_2) proceeds through a semiquinone intermediate (QH·). The R group indicates the same isoprenoid units shown in the image on the left.

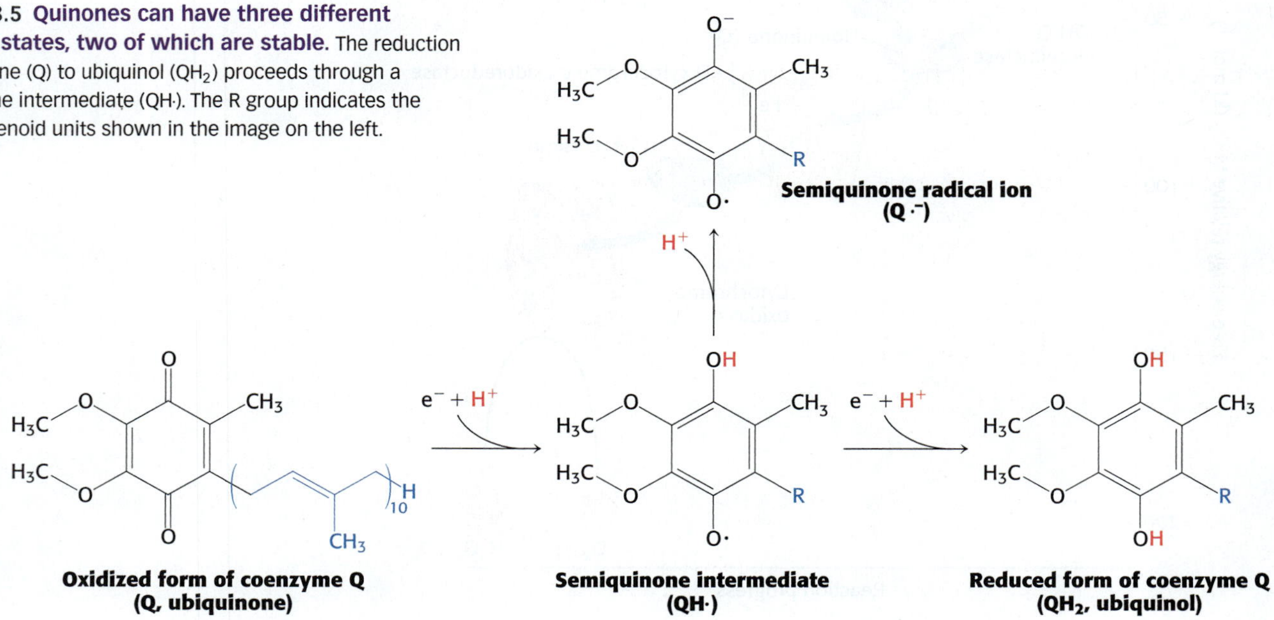

Oxidized form of coenzyme Q
(Q, ubiquinone)

Semiquinone radical ion
(Q·⁻)

Semiquinone intermediate
(QH·)

Reduced form of coenzyme Q
(QH_2, ubiquinol)

one proton results in the semiquinone form (QH·). The semiquinone can lose a proton to form a semiquinone radical anion (Q ·⁻). The addition of a second electron and proton to the semiquinone generates ubiquinol (QH_2), the fully reduced form of coenzyme Q, which holds its protons more tightly. Thus, for quinones, electron-transfer reactions are coupled to proton binding and release, a property that is key to transmembrane proton transport. Because ubiquinone is soluble in the membrane, a pool of Q and QH_2—called the Q *pool*—is thought to exist in the inner mitochondrial membrane, although recent research suggests that the Q pool is confined to the protein complexes of the electron-transport chain.

In contrast with Q, the second special electron carrier is a protein called a *cytochrome*. *Cytochromes* are electron-transferring proteins that contain a heme prosthetic group. **Cytochrome c**, a small soluble protein that is loosely associated with the inner mitochondrial membrane, shuttles electrons from Q-cytochrome c oxidoreductase to cytochrome c oxidase, the final component in the chain and the one that catalyzes the reduction of O_2. Cytochrome c is present in all organisms having mitochondrial respiratory chains, evolving more than 1.5 billion years ago, before the divergence of plants and animals. Its structure and function have been conserved throughout this period, as evidenced by the fact that the cytochrome c of any eukaryotic species reacts in vitro with the cytochrome c oxidase of any other species. For example, wheat-germ cytochrome c reacts with human cytochrome c oxidase. Even bacterial cytochromes closely resemble cytochrome c from tuna-heart mitochondria (**Figure 18.6**). This evidence attests to an efficient evolutionary solution to electron transfer bestowed by the structural and functional characteristics of cytochrome c.

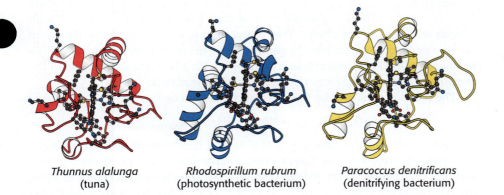

Thunnus alalunga
(tuna)

Rhodospirillum rubrum
(photosynthetic bacterium)

Paracoccus denitrificans
(denitrifying bacterium)

FIGURE 18.6 The three-dimensional structure of cytochrome c is remarkably conserved among distantly related species. The side chains are shown for 21 conserved amino acids as well as the centrally located planar heme. [Drawn from 3CYT.pdb, 3C2C.pdb, and 155C.pdb.]

INTERACT with these models in
☆ Achieve

Iron–sulfur clusters are common components of the electron-transport chain

Iron–sulfur clusters in **iron–sulfur proteins** (also called **nonheme iron proteins**) play a critical role in a wide range of reduction reactions in biological systems. Several types of Fe–S clusters are known (**Figure 18.7**). In the simplest kind, a single iron ion

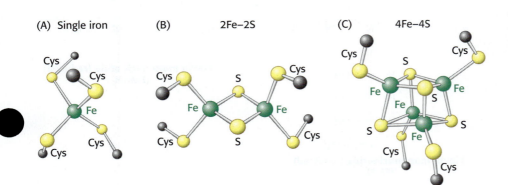

(A) Single iron   (B)   2Fe–2S   (C)   4Fe–4S

FIGURE 18.7 Three different types of iron–sulfur clusters are commonly found in biological systems. Each of these clusters can undergo oxidation–reduction reactions. (A) A single iron ion bound by four cysteine residues. (B) 2Fe–2S cluster with iron ions bridged by sulfide ions. (C) 4Fe–4S cluster.

is tetrahedrally coordinated to the sulfhydryl groups of four cysteine residues of the protein (Figure 18.7A). A second kind, denoted by 2Fe–2S, contains two iron ions, two inorganic sulfides, and usually four cysteine residues (Figure 18.7B). A third type, designated 4Fe–4S, contains four iron ions, four inorganic sulfides, and four cysteine residues (Figure 18.7C). NADH-Q oxidoreductase contains both 2Fe–2S and 4Fe–4S clusters. Iron ions in these Fe–S complexes cycle between Fe^{2+} (reduced) and Fe^{3+} (oxidized) states. Unlike quinones and flavins, iron–sulfur clusters generally undergo oxidation–reduction reactions without releasing or binding protons.

The importance of Fe–S clusters is illustrated by the loss of function of the protein frataxin. Frataxin is a small (14.2-kDa) mitochondrial protein that is crucial for the synthesis of Fe–S clusters. Mutations in frataxin causes Friedreich's ataxia, a disease of varying severity that affects the central and peripheral nervous system as well as the heart and skeletal system. Severe cases lead to death in young adult life. The most common mutation is trinucleotide expansion (Section 28.4) in the gene for frataxin.

❖ SELF–CHECK QUESTION

Explain why the structure of coenzyme Q makes it an effective mobile electron carrier in the electron-transport chain.

The high-potential energy electrons of NADH enter the respiratory chain at NADH-Q oxidoreductase

The electrons of NADH enter the chain at **NADH-Q oxidoreductase** (also called **Complex I**), an enormous enzyme (>900 kDa) consisting of approximately 45 polypeptide chains organized into 14 core subunits that are found in all species with respiratory chains. This proton pump, like that of the other two in the respiratory chain, is encoded by genes residing in both the mitochondria and the nucleus. NADH-Q oxidoreductase is L-shaped, with a horizontal arm lying in the membrane and a vertical arm that projects into the matrix.

The reaction catalyzed by this enzyme appears to be

$$NADH + Q + 5\,H^+_{matrix} \rightarrow NAD^+ + QH_2 + 4\,H^+_{intermembrane\ space}$$

The initial step is the binding of NADH and the transfer of its two high-potential electrons to the flavin mononucleotide (FMN) prosthetic group, yielding the reduced form, $FMNH_2$ (**Figure 18.8**). The electron acceptor of FMN, the isoalloxazine ring, is identical with that of FAD. Electrons are then transferred from $FMNH_2$ to a series of iron–sulfur clusters, the second type of prosthetic group in NADH-Q oxidoreductase.

FIGURE 18.8 Flavins have two oxidation states. The states are exchanged by the binding or loss of two electrons and two protons.

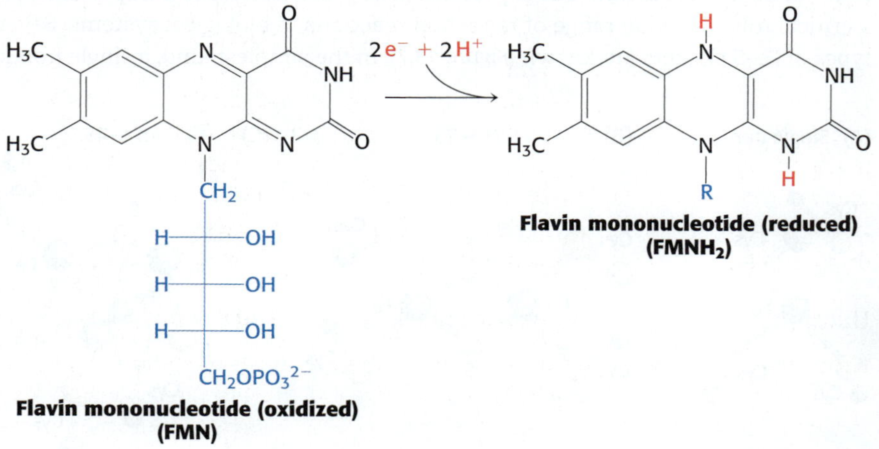

Flavin mononucleotide (oxidized) (FMN)

$2\,e^- + 2\,H^+$

Flavin mononucleotide (reduced) (FMNH$_2$)

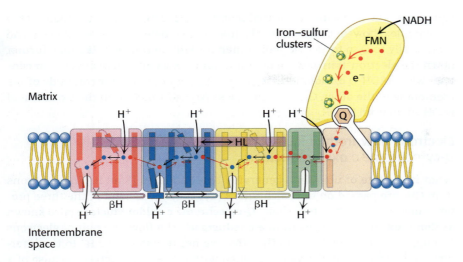

FIGURE 18.9 Electron transfer through NADH-Q oxidoreductase is coupled to proton transfer reactions. Electrons flow in Complex I from NADH through FMN and a series of iron–sulfur clusters to ubiquinone (Q), forming Q^{2-}. The charges on Q^{2-} are electrostatically transmitted to hydrophilic amino acid residues—shown as red (glutamate) and blue (lysine or histidine) balls—that power the movement of HL and βH components. This movement changes the conformation of the transmembrane helices and results in the transport of four protons out of the mitochondrial matrix. [Information from R. Baradaran et al., *Nature* 494:443–448, 2013.]

Recent structural studies have suggested how Complex I acts as a proton pump. What are the structural elements required for proton-pumping? The membrane-embedded part of the complex has four proton half-channels consisting, in part, of vertical helices. One set of half-channels is exposed to the matrix and the other to the intermembrane space (**Figure 18.9**). The vertical helices are linked on the matrix side by a long horizontal helix (HL) that connects the matrix half-channels, while the intermembrane space half-channels are joined by a series of β-hairpin-helix connecting elements (βH). An enclosed Q chamber, the site where Q accepts electrons from NADH, exists near the junction of the hydrophilic portion and the membrane-embedded portion. Finally, a hydrophilic funnel connects the Q chamber to a water-lined channel (into which the half channels open) that extends the entire length of the membrane-embedded portion.

How do these structural elements cooperate to pump protons out of the matrix? When Q accepts two electrons from NADH, generating Q^{2-}, the negative charges on Q^{2-} interact electrostatically with negatively charged amino acid residues in the membrane-embedded arm, causing conformational changes in the long horizontal helix and the βH elements. These changes in turn alter the structures of the connected vertical helices that change the pK_a of amino acids, allowing protons from the matrix to first bind to the amino acids, then dissociate into the water-lined channel, and finally enter the intermembrane space. Thus, the flow of two electrons from NADH to coenzyme Q through NADH-Q oxidoreductase leads to the pumping of four hydrogen ions out of the matrix of the mitochondrion. Q^{2-} subsequently takes up two protons from the matrix as it is reduced to QH_2. The removal of these protons from the matrix contributes to the formation of the proton-motive force. The QH_2 subsequently leaves the enzyme for the Q pool, allowing another reaction cycle to occur.

It is important to note that the citric acid cycle and the pyruvate dehydrogenase complex are not the only sources of mitochondrial NADH. As we will see in Chapter 22, fatty acid degradation, which also takes place in mitochondria, is another crucial source of NADH for the electron-transport chain. Moreover, electrons from NADH generated in the cytoplasm can be transported into mitochondria for use by the electron-transport chain, as we will see in Section 18.5.

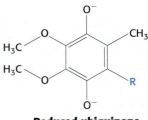

Reduced ubiquinone
(Q^{2-})

Ubiquinol is the entry point for electrons from $FADH_2$ of flavoproteins

Recall that $FADH_2$ is formed in the citric acid cycle, in the oxidation of succinate to fumarate by succinate dehydrogenase (Section 17.3). Succinate dehydrogenase is part of **succinate-Q reductase (Complex II)**, an integral membrane protein complex of the inner mitochondrial membrane. $FADH_2$ does not actually leave the complex, however. Rather, its electrons are transferred to Fe–S centers and then finally to Q to form QH_2, which then is ready to transfer electrons further down the electron-transport chain. Importantly, succinate-Q reductase, in contrast with NADH-Q oxidoreductase, does not pump protons from one side of the membrane to the other. Consequently, less ATP is formed from the oxidation of $FADH_2$ than from NADH.

Electrons flow from ubiquinol to cytochrome *c* through Q-cytochrome *c* oxidoreductase

What is the fate of ubiquinol generated by Complexes I and II? The electrons from QH_2 are passed on to cytochrome *c* (Cyt *c*) by the second of the three proton pumps in the respiratory chain, **Q-cytochrome *c* oxidoreductase** (also known as **Complex III** and as **cytochrome *c* reductase**). The flow of a pair of electrons through this complex leads to the effective net transport of 2 H^+ to the intermembrane space, half the yield obtained with NADH-Q reductase because of a smaller thermodynamic driving force.

$$QH_2 + 2\ Cyt\ c_{ox} + 2\ H^+_{matrix} \rightarrow Q + 2\ Cyt\ c_{red} + 4\ H^+_{intermembrane\ space}$$

Q-cytochrome *c* oxidoreductase itself contains two types of cytochromes, named *b* and c_1 (**Figure 18.10**). The heme prosthetic group in cytochromes *b*, c_1, and *c* is iron-protoporphyrin IX, the same heme present in myoglobin and hemoglobin (Section 3.2). The iron ion of a cytochrome, in contrast to hemoglobin and myoglobin, alternates between a reduced ferrous (+2) state and an oxidized ferric (+3) state during electron transport. The two cytochrome subunits

FIGURE 18.10 Q-cytochrome c oxidoreductase is a homodimer, with each monomer consisting of 11 distinct polypeptide chains. Although both monomers are identical, key parts are shown in color in only one monomer for ease of viewing. *Notice* that the major prosthetic groups—three hemes and a 2Fe–2S cluster—are located either near the edge of the complex bordering the intermembrane space (top) or in the region embedded in the membrane (α helices represented by tubes). They are well-positioned to mediate the electron-transfer reactions between quinones in the membrane and cytochrome *c* in the intermembrane space. [Drawn from 1BCC.pdb.]

INTERACT with this model in
Achieve

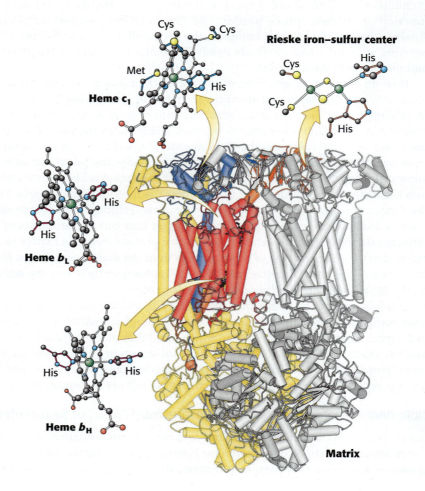

of Q-cytochrome c oxidoreductase contain a total of three hemes: two hemes within cytochrome b, termed heme b_L (L for low affinity) and heme b_H (H for high affinity), and one heme within cytochrome c_1. These identical hemes have different electron affinities because they are in different polypeptide environments. For example, heme b_L, which is located in a cluster of helices near the intermembrane face of the membrane, has lower affinity for an electron than does heme b_H, which is near the matrix side.

In addition to the hemes, the enzyme contains an iron–sulfur protein with a 2Fe–2S center. This center, termed the *Rieske center*, is unusual in that one of the iron ions is coordinated by two histidine residues rather than two cysteine residues. This coordination stabilizes the center in its reduced form, raising its reduction potential so that it can readily accept electrons from QH_2.

The Q cycle funnels electrons from a two-electron carrier to a one-electron carrier while pumping protons

QH_2 passes two electrons to Q-cytochrome c oxidoreductase, but the acceptor of electrons in this complex, cytochrome c, can accept only one electron. How does the switch from the two-electron carrier ubiquinol to the one-electron carrier cytochrome c take place without the release of dangerous radicals? The mechanism for the coupling of electron transfer from QH_2 to cytochrome c to transmembrane proton transport is known as the **Q cycle (Figure 18.11)**. Two QH_2 molecules bind to the complex consecutively, each giving up two electrons and two H^+. These protons are released to the intermembrane space. Additionally, one molecule of oxidized Q binds and it is this molecule, along with two molecules of cytochrome c, that accept the four electrons. The Q cycle occurs in two halves:

1. In the first half of the cycle, two molecules of Q from the Q pool—one reduced (QH_2) and one oxidized (Q)—first bind to two distinct sites called Q_o and Q_i, respectively. The QH_2 bound to Q_o passes its two electrons through the complex to different destinations. One electron flows first to the Rieske 2Fe–2S cluster, then to cytochrome c_1, and finally to the first molecule of oxidized cytochrome c, converting it into its reduced form. The reduced cytochrome c molecule is free to diffuse away from the enzyme to continue down the respiratory chain. The second electron passes through two heme groups of

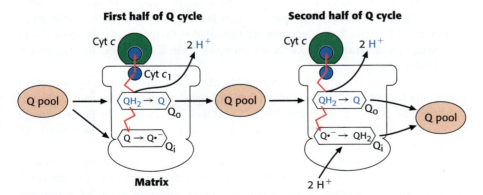

FIGURE 18.11 The Q cycle allows the safe transfer of electrons from QH_2 to cytochrome c. The Q cycle takes place in Complex III, which is represented in outline form. In the first half of the cycle, two molecules of Q bind, one oxidized (Q) and the other reduced (QH_2). Two electrons of the bound QH_2 are transferred, one to cytochrome c and the other to the bound Q in the second binding site to form the semiquinone radical anion $Q^{\cdot-}$. The newly formed Q dissociates and enters the Q pool. In the second half of the cycle, a second QH_2 also gives up its electrons to Complex III, one to a second molecule of cytochrome c and the other to reduce $Q^{\cdot-}$ to QH_2. This second electron transfer results in the uptake of two protons from the matrix. The path of electron transfer is shown in red.

cytochrome *b* to the oxidized Q molecule bound at the Q_i site, reducing it to a dangerous semiquinone radical anion ($Q \cdot^-$) which is held tightly within the complex to prevent its release. The now fully oxidized Q leaves the Q_o site and reenters the Q pool.

2. In the second half of the cycle, a second molecule of QH_2 binds to the Q_o site and reacts in the same way as the first. One of the electrons is transferred to cytochrome *c*, which becomes reduced and leaves, while the second electron is transferred to the semiquinone radical anion still bound in the Q_i binding site from the first half of the cycle. On the addition of the second electron, the radical accepts two protons from the matrix and forms QH_2, which then rejoins the Q pool along with the second oxidized Q molecule from the Q_o site. The removal of these two protons from the matrix contributes to the formation of the proton gradient.

Thus, with each full cycle, two reduced QH_2 and one oxidized Q enter the complex and two oxidized Q and one QH_2 leave. The two electrons that are extracted from the Q pool each cycle pass to two molecules of cytochrome *c*, without releasing a radical to solution. Finally, four protons are released into the intermembrane space, and two protons are removed from the mitochondrial matrix. The overall reaction is:

$$2 \, QH_2 + Q + 2 \, \text{Cyt } c_{ox} + 2 \, H^+_{matrix} \rightarrow 2 \, Q + QH_2 + 2 \, \text{Cyt } c_{red} + 4 \, H^+_{intermembrane \, space}$$

The problem of how to funnel electrons safely and efficiently from a two-electron carrier (QH_2) to a one-electron carrier (cytochrome *c*) is solved by the Q cycle. The cytochrome *b* component of the reductase is in essence a recycling device that enables both electrons of QH_2 to be used effectively.

Cytochrome *c* oxidase catalyzes the reduction of molecular oxygen to water

The last of the three proton-pumping assemblies of the respiratory chain is **cytochrome *c* oxidase (Complex IV)**. Cytochrome *c* oxidase catalyzes the transfer of electrons from the reduced form of cytochrome *c* to molecular oxygen, the final acceptor.

$$4 \, \text{Cyt } c_{red} + 8 \, H^+_{matrix} + O_2 \rightarrow 4 \, \text{Cyt } c_{ox} + 2 \, H_2O + 4 \, H^+_{intermembrane \, space}$$

The requirement of oxygen for this reaction is what makes "aerobic" organisms aerobic. To obtain oxygen for this reaction is the primary reason that human beings must breathe. Four electrons are funneled to O_2 to completely reduce it to H_2O, and, concomitantly, protons are pumped from the matrix into the intermembrane space.

This reaction is quite thermodynamically favorable. From the reduction potentials in Table 18.1, the standard free-energy change for this reaction is calculated to be $\Delta G^{\circ\prime} = -231.8$ kJ mol^{-1} (-55.4 kcal mol^{-1}). As much of this free energy as possible must be captured in the form of a proton gradient for subsequent use in ATP synthesis.

Our understanding of mammalian cytochrome *c* oxidase is reasonably well understood at the structural level because of extensive studies of bovine cytochrome *c* oxidase (**Figure 18.12**). It consists of 13 subunits, three of which are encoded by the mitochondrion's own genome. Cytochrome *c* oxidase contains two heme A groups and three copper ions, arranged as two copper centers, designated A and B. One center, Cu_A/Cu_A, contains two copper ions linked by two bridging cysteine residues. This center initially accepts electrons from reduced cytochrome *c*. The remaining copper ion, Cu_B, is bonded to three histidine residues, one of which is modified by covalent linkage to a tyrosine residue. The copper centers alternate between the reduced Cu^+ (cuprous) form and the oxidized Cu^{2+} (cupric) form as they accept and donate electrons.

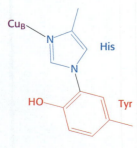

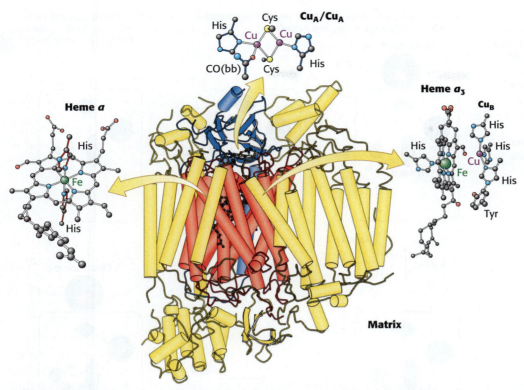

FIGURE 18.12 The structure of bovine cytochrome *c* oxidase reveals 13 polypeptide chains and numerous electron carriers. Notice that most of the complex, as well as two major prosthetic groups (heme *a* and heme a_3 – Cu_B) are embedded in the membrane (α helices are represented by vertical cylinders). Heme a_3 – Cu_B is the site of the reduction of oxygen to water. The Cu_A/Cu_A prosthetic group is positioned near the intermembrane space to better accept electrons from cytochrome *c*. CO(bb) is a carbonyl group of the peptide backbone. [Drawn from 2OCC.pdb.]

INTERACT with this model in
Achieve

In cytochrome *c* oxidase, there are two heme A molecules, called *heme* a and *heme* a_3, which differ from the heme in cytochrome *c* and c_1 in three ways: (1) a formyl group replaces a methyl group, (2) a C_{17} hydrocarbon chain replaces one of the vinyl groups, and (3) the heme is not covalently attached to the protein.

Heme A

Heme *a* and heme a_3 have distinct redox potentials because they are located in different environments within cytochrome *c* oxidase. An electron flows from cytochrome *c* to Cu_A/Cu_A, to heme *a*, to heme a_3, to Cu_B, and finally to O_2. Heme a_3 and Cu_B are directly adjacent. Together, heme a_3 and Cu_B form the active center at which O_2 is reduced to H_2O.

FIGURE 18.13 The mechanism of cytochrome _c_ oxidase prevents early oxygen release. The cycle begins and ends with all prosthetic groups in their oxidized forms (shown in blue). Reduced forms are in red. Four cytochrome _c_ molecules donate four electrons, allowing the binding and cleavage of an O_2 molecule and the import of four H^+ from the matrix to form two molecules of H_2O, which are released from the enzyme to regenerate the initial state.

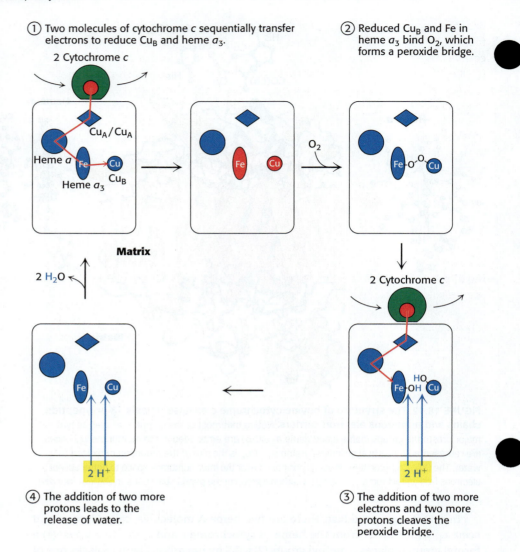

① Two molecules of cytochrome _c_ sequentially transfer electrons to reduce Cu_B and heme a_3.

② Reduced Cu_B and Fe in heme a_3 bind O_2, which forms a peroxide bridge.

④ The addition of two more protons leads to the release of water.

③ The addition of two more electrons and two more protons cleaves the peroxide bridge.

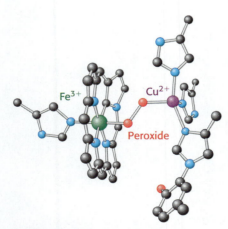

FIGURE 18.14 Two metals form a peroxide bridge in cytochrome _c_ oxidase. The oxygen bound to heme a_3 is reduced to peroxide by the presence of Cu_B.

Four molecules of cytochrome _c_ bind consecutively to the enzyme and transfer an electron to reduce one molecule of O_2 to H_2O (**Figure 18.13**):

1. Electrons from two molecules of reduced cytochrome _c_ flow down an electron-transfer pathway within cytochrome _c_ oxidase, one stopping at Cu_B and the other at heme a_3. With both centers in the reduced state, they together can now bind an oxygen molecule.

2. As molecular oxygen binds, it abstracts an electron from each of the nearby ions in the active center to form a peroxide (O_2^{2-}) bridge between them (**Figure 18.14**).

3. Two more molecules of cytochrome _c_ bind and release electrons that travel to the active center. The addition of an electron as well as H^+ to each oxygen atom reduces the two ion–oxygen groups to Cu_B^{2+}—OH and Fe^{3+}—OH.

4. Reaction with two more H^+ ions allows the release of two molecules of H_2O and resets the enzyme to its initial, fully oxidized form.

$$4 \text{ Cyt } c_{red} + 4 \text{ H}^+_{matrix} + O_2 \rightarrow 4 \text{ Cyt } c_{ox} + 2 \text{ H}_2O$$

The four protons in the reaction shown in Step 4 come exclusively from the matrix. Thus, the consumption of these four protons contributes directly to the proton gradient. Recall that each proton contributes 21.8 kJ mol^{-1} (5.2 kcal mol^{-1}) to the free energy associated with the proton gradient; so these four protons contribute 87.2 kJ mol^{-1} (20.8 kcal mol^{-1}), an amount substantially less than the free energy available from the reduction of oxygen to water. What is the fate of this missing energy? Remarkably, cytochrome c oxidase uses this energy to pump four additional protons from the matrix into the intermembrane space in the course of each reaction cycle for a total of eight protons removed from the matrix (**Figure 18.15**). The details of how these protons are transported through the protein are still under study. However, two effects contribute to the mechanism. First, charge neutrality tends to be maintained in the interior of proteins. Thus, the addition of an electron to a site inside a protein tends to favor the binding of H^+ to a nearby site. Second, conformational changes take place, particularly around the heme a_3–Cu_B center, in the course of the reaction cycle. Presumably, in one conformation, protons may enter the protein exclusively from the matrix side, whereas, in another, they may exit exclusively to the intermembrane space. Thus, the overall process catalyzed by cytochrome c oxidase is

$$4 \text{ Cyt } c_{red} + 8 \text{ H}^+_{matrix} + O_2 \rightarrow 4 \text{ Cyt } c_{ox} + 2 \text{ H}_2O + 4 \text{ H}^+_{intermembrane space}$$

Figure 18.16 summarizes the flow of electrons from NADH and $FADH_2$ through the respiratory chain. This series of exergonic reactions is coupled to the pumping of protons from the matrix. As we will see shortly, the energy inherent in the proton gradient will be used to synthesize ATP.

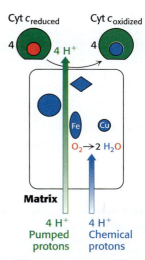

FIGURE 18.15 Proton transport by cytochrome c oxidase has two components. Four protons are taken up from the matrix side to reduce one molecule of O_2 to two molecules of H_2O. These protons are called *chemical protons* because they participate in a clearly defined reaction with O_2. Four additional *pumped protons* are transported out of the matrix and released into the intermembrane space in the course of the reaction.

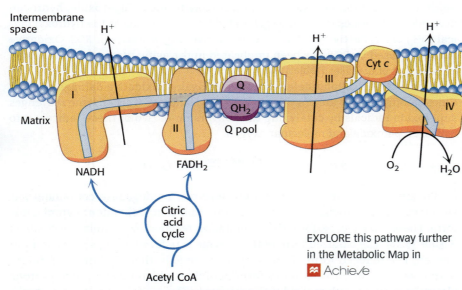

FIGURE 18.16 Electrons flow via two pathways through the electron-transport chain. High-energy electrons in the form of either NADH or $FADH_2$ are generated by the citric acid cycle. These electrons flow through the respiratory chain, which powers proton pumping and results in the reduction of O_2.

EXPLORE this pathway further in the Metabolic Map in 📚 Achieve

Most of the electron-transport chain is organized into a larger complex called the respirasome

The portrayal of the electron-transport chain in Figure 18.16 shows the individual components as isolated units. Although fine for illustrative purposes, this depiction is too simplistic. Recent research has established that three of the components of the electron-transport chain are arranged into a massive complex called the **respirasome**. The human respirasome consists of two copies of Complex I, Complex III, and Complex IV (**Figure 18.17**). The respirasome forms a circular structure with two copies each of Complex I and Complex IV surrounding two copies of Complex III. Although not experimentally established, the structure allows for Complex II to associate in a gap between Complexes I and IV. The structure of the respirasome was solved using the technique of cryo-electron microscopy (cryo-EM) (Section 4.5). The respirasome provides yet another example of multienzyme pathways that are organized into large complexes to enhance efficiency.

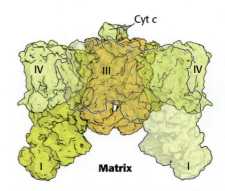

FIGURE 18.17 Three complexes of the ETC associate to form the respirasome. The structure of the electron-transport-chain supercomplex (or respirasome) has been determined by cryo-electron microscopy.

Toxic derivatives of molecular oxygen such as superoxide radicals are scavenged by protective enzymes

As discussed earlier, molecular oxygen is an ideal terminal electron acceptor, because its high affinity for electrons provides a large thermodynamic driving force. However, danger lurks in the reduction of O_2. The transfer of four electrons leads to safe products (two molecules of H_2O), but partial reduction generates hazardous compounds. In particular, the transfer of a single electron to O_2 forms a superoxide ion, whereas the transfer of two electrons yields peroxide.

$$O_2 \xrightarrow{e^-} \underset{\substack{\text{Superoxide} \\ \text{ion}}}{O^{\cdot-}_2} \xrightarrow{e^-} \underset{\text{Peroxide}}{O_2^{2-}}$$

TABLE 18.3 Pathological conditions that may entail free-radical injury

Atherogenesis
Emphysema; bronchitis
Parkinson's disease
Duchenne muscular dystrophy
Cervical cancer
Alcoholic liver disease
Diabetes
Acute renal failure
Down syndrome
Retrolental fibroplasia
Cerebrovascular disorders
Ischemia; reperfusion injury

Information from M. Lieberman and A. D. Marks, *Basic Medical Biochemistry: A Clinical Approach,* 4th ed. (Lippincott, Williams & Wilkins, 2012), p. 437.

Both compounds are potentially destructive. The strategy for the safe reduction of O_2 is clear: the catalyst avoids releasing partly reduced intermediates. Cytochrome c oxidase meets this crucial criterion by holding O_2 tightly between Fe and Cu ions.

Although cytochrome c oxidase and other proteins that reduce O_2 are remarkably successful in not releasing intermediates, small amounts of superoxide anion and hydrogen peroxide are unavoidably formed. Superoxide, hydrogen peroxide, and species that can be generated from them such as the hydroxyl radical (OH·) are collectively referred to as *reactive oxygen species* or *ROS*. Oxidative damage caused by ROS has been implicated in the aging process as well as in a growing list of diseases (**Table 18.3**).

What are the cellular defense strategies against oxidative damage by ROS? Chief among them is the enzyme superoxide dismutase. This enzyme scavenges superoxide radicals by catalyzing the conversion of two of these radicals into hydrogen peroxide and molecular oxygen.

$$2\,O^{\cdot-}_2 + 2\,H^+ \xrightleftharpoons{\text{Superoxide dismutase}} O_2 + H_2O_2$$

Eukaryotes contain two forms of this enzyme, a manganese-containing version located in mitochondria and a copper-and-zinc-dependent cytoplasmic form. These enzymes perform the dismutation reaction by a similar mechanism (**Figure 18.18**). The oxidized form of the enzyme is reduced by superoxide to form oxygen. The reduced form of the enzyme, formed in this reaction, then reacts with a second superoxide ion to form peroxide, which takes up two protons along the reaction path to yield hydrogen peroxide.

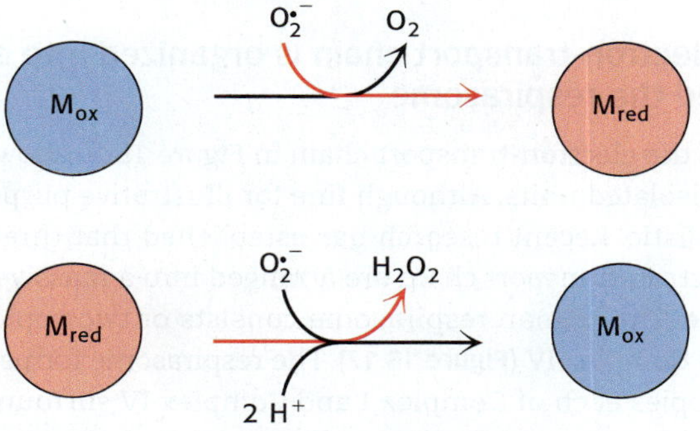

FIGURE 18.18 The superoxide dismutase mechanism has two phases. The oxidized form of superoxide dismutase (M_{ox}) reacts with one superoxide ion to form O_2 and generate the reduced form of the enzyme (M_{red}). The reduced form then reacts with a second superoxide and two protons to form hydrogen peroxide and regenerate the oxidized form of the enzyme.

The hydrogen peroxide formed by superoxide dismutase and by other processes is scavenged by *catalase*, a ubiquitous heme protein that catalyzes the dismutation of hydrogen peroxide into water and molecular oxygen.

$$2H_2O_2 \xrightleftharpoons{\text{Catalase}} O_2 + 2\,H_2O$$

Superoxide dismutase and catalase are remarkably efficient, performing their reactions at or near the diffusion-limited rate (Section 5.4). Glutathione peroxidase also plays a role in scavenging H_2O_2 (Section 20.5). Other cellular defenses against oxidative damage include the antioxidant vitamins, vitamins E and C. Because it is lipophilic, vitamin E is especially useful in protecting membranes from lipid peroxidation.

A long-term benefit of exercise may be to increase the amount of superoxide dismutase in the cell. The elevated aerobic metabolism during exercise causes more ROS to be generated. In response, the cell synthesizes more protective enzymes. The net effect is one of protection, because the increase in superoxide dismutase more effectively protects the cell during periods of rest.

Despite the fact that reactive oxygen species are known hazards, recent evidence suggests that, under certain circumstances, the controlled generation of these molecules may be an important component of signal-transduction pathways. For instance, growth factors have been shown to increase ROS levels as part of their signaling pathway, which in turn regulate channels and transcription factors. ROS have been implicated in the control of cell differentiation, the immune response, and autophagy, as well as other metabolic activities. The dual role of ROS is an excellent example of the wondrous complexity of biochemistry of living systems: even potentially harmful substances can be harnessed to play useful roles.

❖ SELF–CHECK QUESTION

Which electron-transport chain complex is not a member of the respirasome? Which one contains an enzyme of the citric acid cycle? Finally, which one uses FAD as a coenzyme?

Electrons can be transferred between groups that are not in contact

How are electrons transferred between electron-carrying groups of the respiratory chain? This question is intriguing because these groups are frequently buried in the interior of a protein in fixed positions and are therefore not directly in contact with one another. Electrons can move through space, even through a vacuum. However, the rate of electron transfer through space falls off rapidly as the electron donor and electron acceptor move apart from each other, decreasing by a factor of 10 for each increase in separation of 0.8 Å.

The protein environment provides more-efficient pathways for electron conduction: typically, the rate of electron transfer decreases by a factor of 10 every 1.7 Å (**Figure 18.19**). For groups in contact, electron-transfer reactions can be quite fast, with rates of approximately $10^{13}\ s^{-1}$. Within proteins in the electron-transport chain, electron-carrying groups are typically separated by 15 Å beyond their van der Waals contact distance. For such separations, we expect electron-transfer rates of approximately $10^4\ s^{-1}$ (i.e., electron transfer in less than 1 ms), assuming that all other factors are optimal. Without the mediation of the protein, an electron transfer over this distance would take approximately 1 day.

The case is more complicated when electrons must be transferred between two distinct proteins, such as when cytochrome *c* accepts electrons from Complex III or passes them on to Complex IV. A series of hydrophobic interactions bring the heme groups of cytochrome *c* and c_1 to within 4.5 Å of each other, with the iron atoms separated by 17.4 Å. This distance could allow cytochrome *c* reduction at a rate of $8.3 \times 10^6\ s^{-1}$.

FIGURE 18.19 Proteins dramatically improve the rate of electron transfer between atoms at a distance. The rate of electron transfer decreases as the electron donor and the electron acceptor move apart. In a vacuum, the rate decreases by a factor of 10 for every increase of 0.8 Å. In proteins, the rate decreases more gradually, by a factor of 10 for every increase of 1.7 Å; however, this rate is only approximate because variations in the structure of the intervening protein medium can affect it.

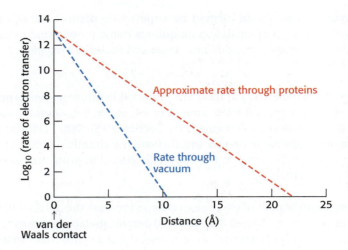

♦ **SELF-CHECK QUESTION**

Amytal is a barbiturate sedative that inhibits electron flow through Complex I. How would the addition of amytal to actively respiring mitochondria affect the relative oxidation–reduction states of the components of the electron-transport chain and the citric acid cycle?

18.4   A Proton Gradient Powers the Synthesis of ATP

Thus far, we have considered the flow of electrons from NADH to O_2, an *exergonic* process.

$$NADH + \tfrac{1}{2} O_2 + H^+ \rightleftharpoons H_2O + NAD^+$$
$$\Delta G^{\circ\prime} = -220.1 \text{ kJ mol}^{-1} \ (-52.6 \text{ kcal mol}^{-1})$$

Next, we consider how this process is coupled to the synthesis of ATP, an *endergonic* process.

$$ADP + P_i + H^+ \rightleftharpoons ATP + H_2O$$
$$\Delta G^{\circ\prime} = +30.5 \text{ kJ mol}^{-1} \ (+7.3 \text{ kcal mol}^{-1})$$

A molecular assembly in the inner mitochondrial membrane carries out the synthesis of ATP. This enzyme complex was originally called the *mitochondrial ATPase* or F_1F_0 *ATPase* because it was discovered through its catalysis of the reverse reaction, the hydrolysis of ATP. **ATP synthase**, its preferred name, emphasizes its actual role in the mitochondrion. It is also called **Complex V**.

How is the oxidation of NADH coupled to the phosphorylation of ADP? Electron transfer was first suggested to lead to the formation of a covalent high-energy intermediate that serves as a compound having a high phosphoryl-transfer potential, analogous to the generation of ATP by the formation of 1,3-bisphosphoglycerate in glycolysis (Section 16.2). An alternative proposal was that electron transfer aids the formation of an activated protein conformation, which then drives ATP synthesis. However, the search for such intermediates for several decades proved fruitless.

The chemiosmotic hypothesis suggested that ATP formation is powered by a proton gradient

In 1961, Peter Mitchell suggested a radically different mechanism called the *chemiosmotic hypothesis*. He proposed that electron transport and ATP synthesis are coupled by a proton gradient across the inner mitochondrial membrane.

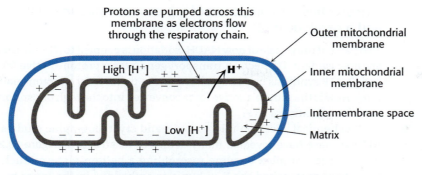

FIGURE 18.20 **The chemiosmotic hypothesis states that electron-driven proton transport and ATP synthesis are coupled.** Electron transfer through the respiratory chain leads to the pumping of protons from the matrix to the intermembrane space. The pH gradient and membrane potential constitute a proton-motive force that is used to drive ATP synthesis.

In his model, the transfer of electrons through the respiratory chain leads to the pumping of protons from the matrix to the intermembrane space. The H^+ concentration becomes lower in the matrix, and an electric field with the matrix side negative is generated (**Figure 18.20**). Protons then flow back into the matrix to equalize the distribution. Mitchell's idea was that this flow of protons drives the synthesis of ATP by ATP synthase.

As noted earlier, the energy-rich unequal distribution of protons across a membrane is called the *proton-motive force*. The proton-motive force is composed of two components: a chemical gradient and a charge gradient. The chemical gradient for protons can be represented as a pH gradient. The charge gradient is created by the positive charge on the unequally distributed protons forming the chemical gradient. Mitchell proposed that both components power the synthesis of ATP.

$$\text{Proton-motive force } (\Delta p) = \text{chemical gradient } (\Delta pH) + \text{charge gradient}(\Delta\psi)$$

Mitchell's highly innovative hypothesis is now supported by a wealth of evidence. Indeed, electron transport does generate a proton gradient across the inner mitochondrial membrane. The pH outside is 1.4 units lower than inside, and the membrane potential is 0.14 V, the outside being positive. As we calculated in Section 18.2, this membrane potential corresponds to a free energy of 21.8 kJ (5.2 kcal) per mole of protons.

An artificial system was created to elegantly demonstrate the basic principle of the chemiosmotic hypothesis. The role of the respiratory chain was played by bacteriorhodopsin, a membrane protein from halobacteria that pumps protons when illuminated (Chapter 19). Synthetic vesicles containing bacteriorhodopsin and mitochondrial ATP synthase purified from beef heart were created (**Figure 18.21**). When the vesicles were exposed to light, ATP was formed. This key experiment clearly showed that the respiratory chain and ATP synthase are biochemically separate systems, linked only by a proton-motive force.

Some have argued that, along with the elucidation of the structure of DNA, the discovery that ATP synthesis is powered by a proton gradient is one of the two major advances in biology in the twentieth century. Mitchell's initial postulation of the chemiosmotic theory was not warmly received by all: some scientists thought of Mitchell as a "court jester," whose work was of no consequence. In Chapter 19 we will discuss how collaboration with another scientist changed this perception. Eventually, Peter Mitchell was awarded the Nobel Prize in Chemistry for his contributions to understanding oxidative phosphorylation.

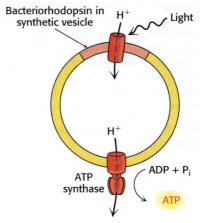

FIGURE 18.21 **The chemiosmotic hypothesis is supported by experiments using bacterial proton pumps.** ATP is synthesized when reconstituted membrane vesicles containing bacteriorhodopsin (a light-driven proton pump) and ATP synthase are exposed to light.

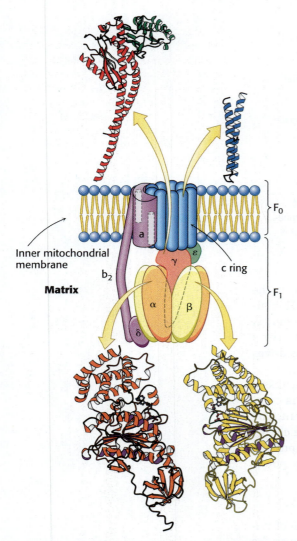

FIGURE 18.22 The structure of ATP synthase reveals a complex molecular rotational motor. A schematic structure is shown along with representations of the components for which structures have been determined to high resolution. The P-loop NTPase domains of the α and β subunits are indicated by purple shading. Notice that part of the enzyme complex is embedded in the inner mitochondrial membrane, whereas the remainder resides in the matrix. [Drawn from 1E79.pdb and 1C0V.pdb.]

INTERACT with the above model in
≈ Achie√e

ATP synthase is composed of a proton-conducting unit and a catalytic unit

Two parts of the puzzle of how NADH oxidation is coupled to ATP synthesis are now evident: (1) electron transport generates a proton-motive force, and (2) ATP synthesis by ATP synthase can be powered by a proton-motive force. How is the proton-motive force converted into the high phosphoryl-transfer potential of ATP?

Biochemical, electron microscopic, and crystallographic studies of ATP synthase have revealed many details of its structure (**Figure 18.22**). It is a large, complex enzyme resembling a ball on a stick. Much of the "stick" part, called the F_0 subunit, is embedded in the inner mito-chondrial membrane. The 85-Å-diameter ball, called the F_1 subunit, protrudes into the mitochondrial matrix. The F_1 subunit contains the catalytic activity of the synthase. In fact, isolated F_1 subunits display ATPase activity.

The F_1 subunit consists of five types of polypeptide chains (α_3, β_3, γ, δ, and ϵ) with the indicated stoichiometry. The α and β subunits, which make up the bulk of the F_1, are arranged alternately in a hexa-meric ring; they are homologous to one another and are members of the P-loop NTPase family (Section 9.4). Both bind nucleotides but only the β subunits are catalytically active. Just below the α and β subunits is a central stalk consisting of the γ and ε proteins. The γ subunit includes a long helical coiled coil that extends into the center of the $\alpha_3\beta_3$ hexamer. The γ subunit breaks the symmetry of the $\alpha_3\beta_3$ hexamer: each of the β subunits is distinct by virtue of its interaction with a different, asymmetrical, face of γ. Distinguishing the three β subunits is crucial for understanding the mechanism of ATP synthesis.

The F_0 subunit is a hydrophobic segment that spans the inner mito-chondrial membrane. F_0 contains the proton channel of the complex. This channel consists of a ring comprising from 8 to 14 **c** subunits that are embedded in the membrane. A single **a** subunit binds to the outside of the ring. The F_0 and F_1 subunits are connected in two ways: by the central γε stalk and by an exterior column. The exterior column consists of one **a** subunit, two **b** subunits, and the δ subunit.

ATP synthases interact with one another to form dimers, which then associate to form large oligomers of dimers (**Figure 18.23**). This association stabilizes the individual enzymes to the rotational forces required for catal-ysis, which we will examine shortly, and facilitates the curvature of the inner mitochondrial membrane. The formation of the cristae allows the proton pumps of the electron-transport chain to localize the proton gra-dient in the vicinity of the synthases, which are located at the tips of the cristae, thereby enhancing the efficiency of ATP synthesis (**Figure 18.24**).

FIGURE 18.23 Mitochondrial ATP synthase forms homodimers. A schematic representation of a dimer of mitochondria ATP synthase is shown embedded in the inner mitochondrial membrane. The dimer, joined by an assortment of proteins, assists in bending the inner mitochondria membrane. The structure of the dimer was determined by cryo-electron microscopy.

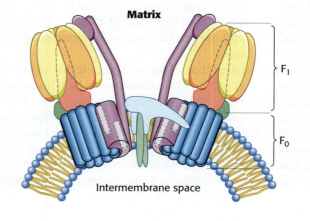

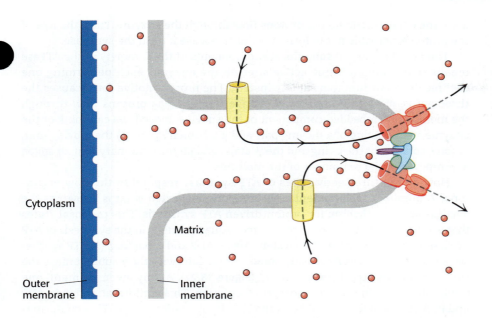

FIGURE 18.24 ATP synthase assists in the formation of cristae. Oligomers of ATP synthase dimers facilitate cristae formation, creating an area where the protons (shown as red balls) are concentrated and have ready access to the F_0 portion of the ATP synthase. The electron-transport chain is represented by the yellow cylinders embedded in the inner mitochondrial membrane. [Information from K. M. Davies et al., *Proc. Natl. Acad. Sci. U.S.A.* 108:14121–14126, 2011.]

Cytoplasm

Matrix

Outer membrane

Inner membrane

Proton flow through ATP synthase leads to the release of tightly bound ATP via the binding-change mechanism

ATP synthase catalyzes the formation of ATP from ADP and orthophosphate.

$$ADP^{3-} + HPO_4^{2-} + H^+ \rightleftharpoons ATP^{4-} + H_2O$$

The actual substrates are ADP and ATP complexed with Mg^{2+}, as in all known phosphoryl-transfer reactions with these nucleotides. A terminal oxygen atom of ADP attacks the phosphorus atom of P_i to form a pentacovalent intermediate, which then dissociates into ATP and H_2O (**Figure 18.25**).

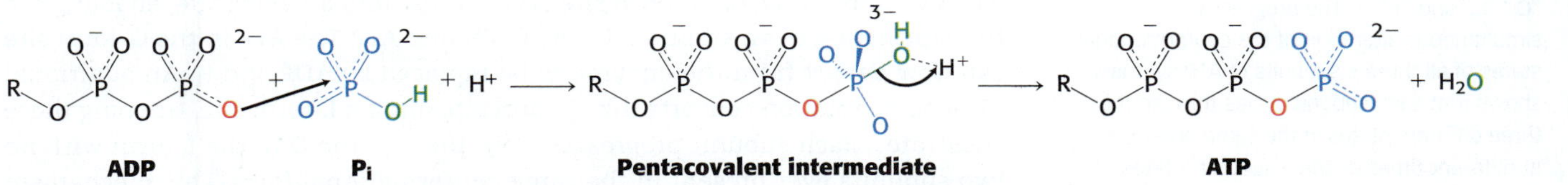

ADP   **P_i**   **Pentacovalent intermediate**   **ATP**

FIGURE 18.25 ATP-synthesis is catalyzed by ATP synthase. One of the oxygen atoms of ADP attacks the phosphorus atom of P_i to form a pentacovalent intermediate, which then forms ATP and releases a molecule of H_2O.

How does the flow of protons drive the synthesis of ATP? Isotopic-exchange experiments unexpectedly revealed that enzyme-bound ATP forms readily in the absence of a proton-motive force. When ADP and P_i were added to ATP synthase in $H_2^{18}O$, ^{18}O became incorporated into P_i through the synthesis of ATP and its subsequent hydrolysis (**Figure 18.26**). The rate of incorporation of ^{18}O into P_i showed that about equal amounts of bound ATP and ADP are in equilibrium at the catalytic site, even in the absence of a proton gradient. However, ATP does not

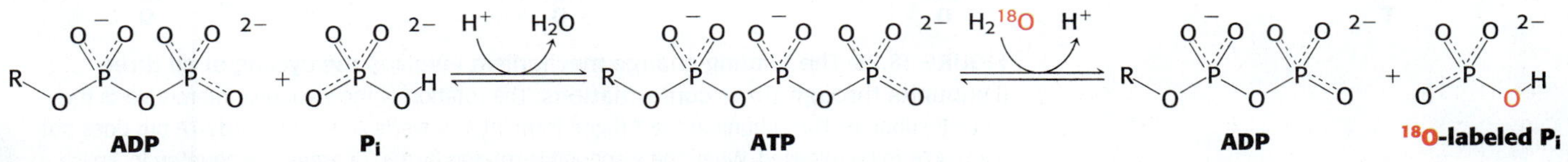

ADP   **P_i**   **ATP**   **ADP**   **^{18}O-labeled P_i**

FIGURE 18.26 ATP forms without a proton-motive force but is not released. The results of isotopic-exchange experiments indicate that enzyme-bound ATP is formed from ADP and P_i in the absence of a proton-motive force.

(A)

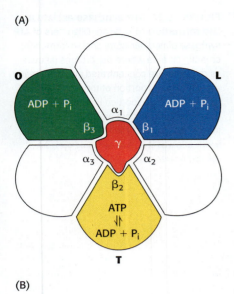

(B)

β subunit 1   L → T → O → L → T → O...

β subunit 2   O → L → T → O → L → T...

β subunit 3   T → O → L → T → O → L...

FIGURE 18.27 ATP synthase nucleotide-binding sites cycle through distinct conformational states. (A) The γ subunit passes through the center of the α₃β₃ hexamer and makes the nucleotide-binding sites in the β subunits distinct from one another. The β subunits are colored to distinguish them from one another, whereas the conformational states are denoted "O," "L," and "T." (B) The progressive and simultaneous alteration of the conformational states of all three β subunits of ATP synthase shows that each subunit cycles through the three different states in the same order but at different times as the γ subunit rotates.

leave the catalytic site unless protons flow through the enzyme. Thus, the role of the proton gradient is not to form ATP but to release it from the synthase.

The fact that three β subunits are components of the F_1 moiety of the ATPase means that there are three active sites on the enzyme, each performing one of three different functions at any instant. The proton-motive force causes the three active sites to sequentially change functions as protons flow through the membrane-embedded component of the enzyme. Indeed, we can think of the enzyme as consisting of a moving part and a stationary part: (1) the moving unit, or *rotor*, consists of the **c** ring and the γε stalk; and (2) the stationary unit, or *stator*, is composed of the remainder of the molecule.

How do the three active sites of ATP synthase respond to the flow of protons? A number of experimental observations suggested what is known as the *binding-change mechanism* for proton-driven ATP synthesis. This proposal states that a β subunit can perform each of three sequential steps in the synthesis of ATP by changing conformation. These steps are (1) ADP and P_i binding, (2) ATP synthesis, and (3) ATP release. As already noted, interactions with the γ subunit make the three β subunits structurally distinct (**Figure 18.27**). At any given moment, one β subunit will be in the L, or loose, conformation. This conformation binds ADP and P_i. A second subunit will be in the T, or tight, conformation. This conformation binds ATP with great affinity, so much so that it will convert bound ADP and P_i into ATP. Both the T and L conformations are sufficiently constrained that they cannot release bound nucleotides. The final subunit will be in the O, or open, form. This form has a more open conformation and can bind or release adenine nucleotides.

The rotation of the γ subunit drives the interconversion of these three forms (**Figure 18.28**). Consider a single subunit, like the one shown in yellow in Figure 18.28, which is initially in the T form in which bound ADP and P_i are transiently combining to form ATP. Suppose that the γ subunit is rotated by 120 degrees in a counterclockwise direction (as viewed from the top). This rotation converts the T-form site into an O-form site with the nucleotide bound as ATP. Concomitantly, the L-form site is converted into a T-form site, enabling the transformation of an additional ADP and P_i into ATP. The ATP in the O-form site can now depart from the enzyme to be replaced by ADP and P_i. An additional 120-degree rotation converts this O-form site into an L-form site, trapping these substrates. Each subunit progresses from the T to the O to the L form with no two subunits ever present in the same conformational form. This mechanism suggests that ATP can be synthesized and released by driving the rotation of the γ subunit in the appropriate direction.

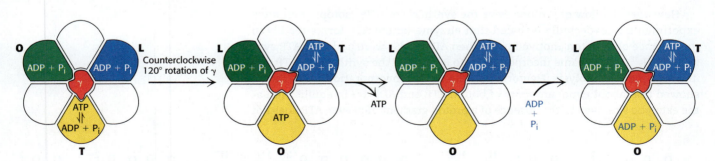

FIGURE 18.28 The binding-change mechanism involves the cycling of all three β subunits through three conformations. The rotation of the γ subunit interconverts the three β subunits. The subunit in the T (tight) form interconverts ADP and P_i and ATP but does not allow ATP to be released. When the γ subunit is rotated by 120 degrees in a counterclockwise direction, the T-form subunit is converted into the O form, allowing ATP release. ADP and P_i can then bind to the O-form subunit. An additional 120-degree rotation (not shown) traps these substrates in an L-form subunit.

Rotational catalysis is the world's smallest molecular motor

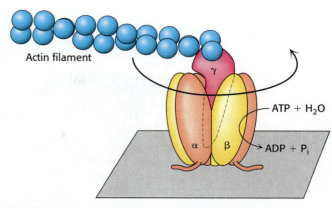

Actin filament

Is it possible to observe the proposed rotation directly? Elegant experiments, using single-molecule techniques (Section 5.5), have demonstrated the rotation through the use of a simple experimental system consisting solely of cloned $\alpha_3\beta_3\gamma$ subunits (**Figure 18.29**). The β subunits were engineered to contain amino-terminal polyhistidine tags, which have a high affinity for nickel ions. This property of the tags allowed the $\alpha_3\beta_3$ assembly to be immobilized on a glass surface that had been coated with nickel ions. The γ subunit was linked to a fluorescently labeled actin filament to provide a long segment that could be observed under a fluorescence microscope. Remarkably, the addition of ATP caused the actin filament to rotate unidirectionally in a counterclockwise direction, confirming visually that the γ subunit was rotating, driven by the hydrolysis of ATP. Thus, the catalytic activity of an individual molecule could be observed. The counterclockwise rotation is consistent with the predicted mechanism for hydrolysis because the molecule was viewed from below relative to the view shown in Figure 18.29.

FIGURE 18.29 The ATP-driven rotation in ATP synthase has been directly observed. The $\alpha_3\beta_3$ hexamer of ATP synthase is fixed to a surface, with the γ subunit projecting upward and linked to a fluorescently labeled actin filament. The addition and subsequent hydrolysis of ATP result in the counterclockwise rotation of the γ subunit, which can be directly seen under a fluorescence microscope.

More-detailed analysis in the presence of lower concentrations of ATP revealed that the γ subunit rotates in 120-degree increments. Each increment corresponds to the hydrolysis of a single ATP molecule. In addition, from the results obtained by varying the length of the actin filament and measuring the rate of rotation, the enzyme appears to operate near 100% efficiency; that is, essentially all of the energy released by ATP hydrolysis is converted into rotational motion.

❖ SELF–CHECK QUESTION

Recall our study of enzymes in previous chapters: Why do isolated F_1 subunits of ATP synthase catalyze ATP hydrolysis?

Proton flow around the c ring powers ATP synthesis

The direct observation of rotary motion of the γ subunit is strong evidence for the rotational mechanism for ATP synthesis. The last remaining question is: How does proton flow through F_0 drive the rotation of the γ subunit? The mechanism depends on the structures of the **a** and **c** subunits of F_0 (**Figure 18.30**). The stationary **a** subunit directly abuts the membrane-spanning ring formed by 8 to 14 **c** subunits. The **a** subunit includes two hydrophilic half-channels that do not span the membrane (Figure 18.30). Thus, protons can pass into either of these channels, but they cannot move completely across the membrane. The **a** subunit is positioned such that each half-channel directly interacts with one **c** subunit.

The structure of the **c** subunit was determined both by NMR methods and by x-ray crystallography. Each polypeptide chain forms a pair of α helices that span the membrane. A glutamic acid (or aspartic acid) residue is found in the middle of one of the helices. If the glutamate is charged (unprotonated), the **c** subunit will not move into the membrane. The key to proton movement across

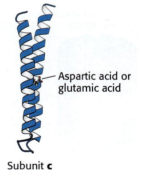

Aspartic acid or glutamic acid

Subunit c

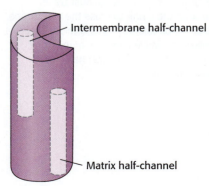

Intermembrane half-channel

Matrix half-channel

Subunit a

FIGURE 18.30 Components of the proton-conducting unit of ATP synthase include two half channels and a critical charged residue. The **c** subunit consists of two α helices that span the membrane. In *E. coli,* an aspartic or glutamic acid residue in one of the helices lies on the center of the membrane. The structure of the **a** subunit has not yet been directly observed, but it appears to include two half-channels that allow protons to enter and pass partway but not completely through the membrane.

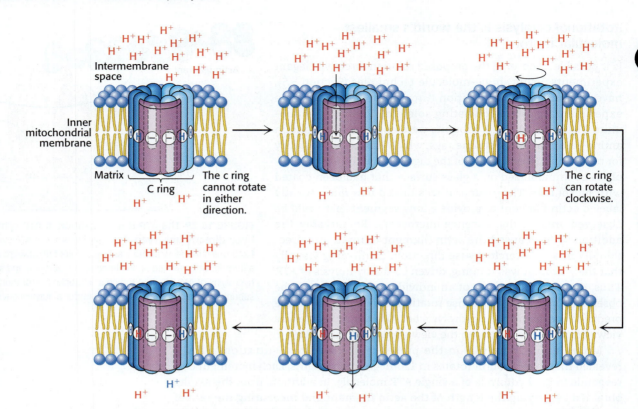

FIGURE 18.31 Proton motion across the membrane drives rotation of the c ring.
A proton enters from the intermembrane space into the cytoplasmic half-channel to neutralize the charge on an aspartate or glutamate residue in a **c** subunit. With this charge neutralized, the **c** ring can rotate clockwise by one **c** subunit, moving an aspartic or glutamic acid residue out of the membrane into the matrix half-channel. This proton can move into the matrix, resetting the system to its initial state.

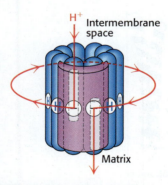

FIGURE 18.32 Protons must travel around the c ring to cross the membrane.
Each proton enters the cytoplasmic half-channel, follows a complete rotation of the **c** ring, and exits through the other half-channel into the matrix.

the membrane is that, in a proton-rich environment, such as the intermembrane space, a proton will enter a channel and bind the glutamate residue, while the glutamic acid in a proton-poor environment of the other half-channel will release a proton (**Figure 18.31**). The **c** subunit with the bound proton then moves into the membrane as the ring rotates by one **c** subunit. This rotation brings the newly deprotonated **c** subunit from the matrix half-channel to the proton-rich intermembrane space half-channel, where it can bind a proton. The movement of protons through the half-channels from the high proton concentration of the intermembrane space to the low proton concentration of the matrix powers the rotation of the **c** ring. The **a** unit remains stationary as the **c** ring rotates. Each proton that enters the intermembrane space half-channel of the **a** unit moves through the membrane by riding around on the rotating **c** ring to exit through the matrix half-channel into the proton-poor environment of the matrix (**Figure 18.32**).

How does the rotation of the **c** ring lead to the synthesis of ATP? The **c** ring is tightly linked to the γ and ε subunits. Thus, as the **c** ring turns, the γ and ε subunits are turned inside the $\alpha_3\beta_3$ hexamer unit of F_1. The rotation of the γ subunit in turn promotes the synthesis of ATP through the binding-change mechanism. The exterior column formed by the two **b** chains and the δ subunit prevents the $\alpha_3\beta_3$ hexamer from rotating along with the **c** ring. Recall that the dimerization and subsequent oligomerization of the synthase in the cristae also stabilizes the enzyme to rotational forces.

As stated earlier, the number of **c** subunits in the **c** ring appears to range between 8 and 14. This number is significant because it determines the number of protons that must be transported to generate a molecule of ATP. Each 360-degree rotation of the γ subunit leads to the synthesis and release of three molecules of ATP. Thus, if there are 10 **c** subunits in the ring (as was observed in a crystal structure of yeast mitochondrial ATP synthase), each ATP generated requires the transport of $10/3 = 3.33$ protons. Recent evidence shows that the **c** rings of all vertebrates are composed of 8 subunits, making vertebrate ATP synthase the most efficient ATP synthase known, with the transport of only 2.7 protons required for ATP synthesis. For simplicity in calculations, we will assume a rounded average of these two numbers—3 protons—must flow into the matrix for each ATP formed, but we must keep in mind that the true value may differ. As we will see, the electrons from NADH pump enough protons to generate 2.5 molecules of ATP, whereas those from $FADH_2$ yield 1.5 molecules of ATP.

Let us return for a moment to the example with which we began this chapter. If a resting human being requires 85 kg of ATP per day for bodily functions, then 3.3×10^{25} protons must flow through the ATP synthase per day, or 3.9×10^{20} protons per second. **Figure 18.33** summarizes the process of oxidative phosphorylation.

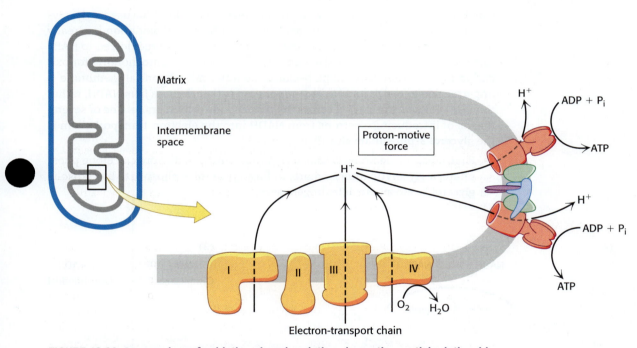

FIGURE 18.33 **An overview of oxidative phosphorylation shows the spatial relationship of the components.** The electron-transport chain generates a proton gradient by pumping protons into the cristae, which is used to synthesize ATP by ATP synthase concentrated on the ends of the cristae.

ATP synthase and G proteins have several common features

The α and β subunits of ATP synthase are members of the P-loop NTPase family of proteins. In Section 14.2, we learned that the signaling properties of other members of this family, the G proteins, depend on their ability to bind nucleoside triphosphates and diphosphates with great tenacity. They do not exchange nucleotides unless they are stimulated to do so by interaction with other proteins. The binding-change mechanism of ATP synthase is a variation on this theme. The P-loop regions of the β subunits will bind either ADP or ATP (or release ATP), depending on which of three different faces of the γ subunit they interact with. The conformational changes take place in an orderly way, driven by the rotation of the γ subunit.

18.5 Many Shuttles Allow Movement Across Mitochondrial Membranes

The inner mitochondrial membrane must be impermeable to most molecules, yet much exchange has to take place between the cytoplasm and the mitochondria. This exchange is mediated by an array of membrane-spanning transporter proteins.

Electrons from cytoplasmic NADH enter mitochondria by shuttles

One function of the respiratory chain is to regenerate NAD$^+$ for use in glycolysis. NADH generated by the citric acid cycle and fatty acid oxidation is already in the mitochondrial matrix, but how is cytoplasmic NADH reoxidized to NAD$^+$ under aerobic conditions? NADH cannot simply pass into mitochondria for oxidation by the respiratory chain, because the inner mitochondrial membrane is impermeable to NADH and NAD$^+$. The solution is that *electrons from NADH,* rather than NADH itself, are carried across the mitochondrial membrane. One of several means of introducing electrons from NADH into the electron-transport chain is the **glycerol 3-phosphate shuttle** (Figure 18.34):

- First, a pair of electrons is transferred from NADH to dihydroxyacetone phosphate, a glycolytic intermediate, to form glycerol 3-phosphate by cytosolic glycerol 3-phosphate dehydrogenase.

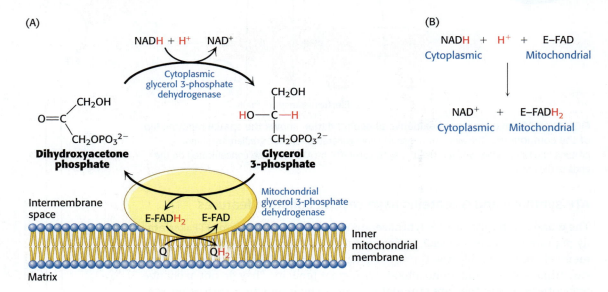

FIGURE 18.34 The glycerol 3-phosphate shuttle allows for rapid rates of oxidative phosphorylation at the cost of energetic efficiency. Electrons from NADH can enter the mitochondrial electron-transport chain by being used to reduce dihydroxyacetone phosphate to glycerol 3-phosphate. Glycerol 3-phosphate is reoxidized by electron transfer to an FAD prosthetic group in a membrane-bound glycerol 3-phosphate dehydrogenase. Subsequent electron transfer to Q to form QH$_2$ allows these electrons to enter the electron-transport chain.

- Next, glycerol 3-phosphate moves into the mitochondrion where it is reoxidized to dihydroxyacetone phosphate on the outer surface of the inner mitochondrial membrane by a membrane-bound isozyme of glycerol 3-phosphate dehydrogenase. The reaction passes the electron pair to an FAD prosthetic group in this enzyme to form $FADH_2$, and regenerates dihydroxyacetone phosphate.

- Finally, the reduced flavin transfers its electrons to a molecule of Q, which then enters the respiratory chain as QH_2, and thus ultimately delivers the electrons to Complex III.

When cytoplasmic NADH transported by the glycerol 3-phosphate shuttle is oxidized by the respiratory chain, 1.5 rather than 2.5 molecules of ATP are formed. The yield is lower because FAD rather than NAD^+ is the electron acceptor in mitochondrial glycerol 3-phosphate dehydrogenase. The use of FAD enables electrons from cytoplasmic NADH to be transported into mitochondria against an NADH concentration gradient. The price of this transport is one molecule of ATP per two electrons. This glycerol 3-phosphate shuttle is especially prominent in muscle, enabling it to sustain a very high rate of oxidative phosphorylation.

In the heart and liver, electrons from cytoplasmic NADH are brought into mitochondria by the **malate–aspartate shuttle**, which is mediated by two membrane carriers and four enzymes (**Figure 18.35**). Electrons are transferred from NADH in the cytoplasm to oxaloacetate, forming malate, which traverses the inner mitochondrial membrane in exchange for α-ketoglutarate and is then reoxidized by NAD^+ in the matrix to form NADH in a reaction catalyzed by the citric acid cycle enzyme malate dehydrogenase. The resulting oxaloacetate does not readily cross the inner mitochondrial membrane, and so a transamination reaction (Section 23.3) is needed to form aspartate, which can be transported to the cytoplasmic side in exchange for glutamate. Glutamate donates an amino group to oxaloacetate, forming aspartate and α-ketoglutarate. In the cytoplasm, aspartate is then deaminated to form oxaloacetate, and the cycle is restarted.

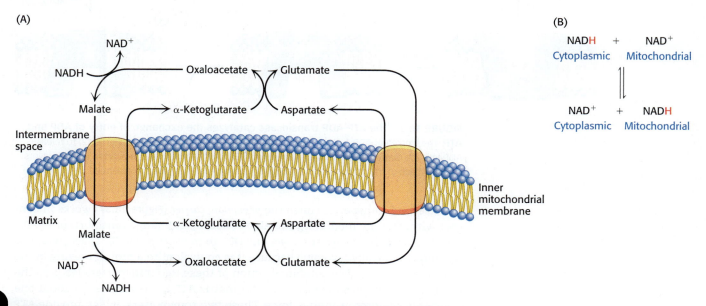

FIGURE 18.35 The malate–aspartate shuttle preserves all the chemical potential energy of cytoplasmically generated NADH. NADH produced in the matrix as a result of the shuttle is free to enter the ETC at Complex I.

The entry of ADP into mitochondria is coupled to the exit of ATP by ATP-ADP translocase

The major function of oxidative phosphorylation is to generate ATP from ADP. ATP and ADP do not diffuse freely across the inner mitochondrial membrane. How are these highly charged molecules moved across the inner membrane into the cytoplasm? A specific transport protein, **ATP-ADP translocase** (also called **adenine nucleotide translocase** or **ANT**), enables these molecules to transverse this permeability barrier. Most important, the flows of ATP and ADP are coupled. ADP enters the mitochondrial matrix only if ATP exits, and vice versa. This process is carried out by the translocase, an antiporter:

$$ADP^{3-}_{\text{cytoplasm}} + ATP^{4-}_{\text{matrix}} \rightarrow ADP^{3-}_{\text{matrix}} + ATP^{4-}_{\text{cytoplasm}}$$

ANT is highly abundant, constituting about 15% of the protein in the inner mitochondrial membrane. The abundance is a manifestation of the fact that human beings exchange the equivalent of their weight in ATP each day.

The 30-kDa translocase contains a single nucleotide-binding site that alternately faces the matrix and the cytoplasmic sides of the membrane (**Figure 18.36**). ATP and ADP bind to ANT without Mg^{2+}, and ATP has one more negative charge than that of ADP. Thus, in an actively respiring mitochondrion with a positive membrane potential, ATP transport out of the mitochondrial matrix is favored, but the additional negative charge moving out dissipates some of the proton-motive force.

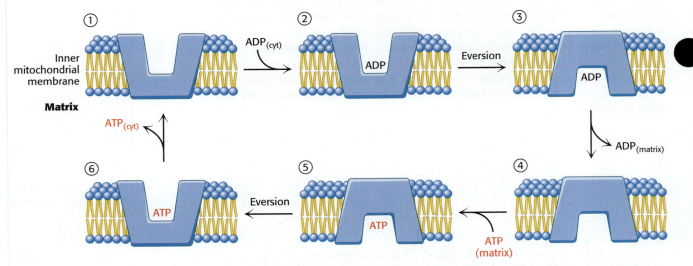

FIGURE 18.36 The ATP-ADP translocase catalyzes the exchange of entry of ADP and ATP. The binding of ADP from the cytoplasm (1) favors eversion of the transporter (2) to release ADP into the matrix (3). Subsequent binding of ATP from the matrix to the everted form (4) favors eversion back to the original conformation (5), releasing ATP into the cytoplasm (6).

A second transporter called the *phosphate carrier* (**Figure 18.37**) works in concert with ANT. By mediating the electroneutral exchange (antiport) of $H_2PO_4^-$ for OH^-, phosphate carrier allows one OH^- to move into the IM space, which is thermodynamically favorable because it is equivalent to a proton moving in the opposite direction. The combined action of these two transporters leads to the exchange of cytoplasmic ADP and P_i for matrix ATP at the expense of about one quarter of the proton-motive force. These two transporters, which provide ATP synthase with its substrates, are associated with the synthase to form a large complex called the *ATP synthasome*. The inhibition of ANT leads to the subsequent inhibition of cellular respiration as well, as we will see in the next section.

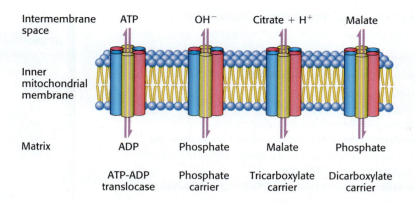

FIGURE 18.37 Multiple structurally homologous mitochondrial transporters carry specific metabolites across the inner mitochondrial membrane.

ANT and phosphate carrier are just two of many structurally homologous mitochondrial transporters for ions and charged metabolites present in the inner mitochondrial membrane (Figure 18.37). The dicarboxylate carrier enables malate, succinate, and fumarate to be exported from the mitochondrial matrix in exchange for P_i. The tricarboxylate carrier exchanges citrate and H^+ for malate. In all, more than 40 such carriers are encoded in the human genome.

❖ SELF–CHECK QUESTION

Recall what thermodynamically drives the action of the ATP-ADP translocase (ANT). Why must this enzyme use Mg^{2+}-free forms of ATP and ADP?

18.6 The Regulation of Cellular Respiration Is Governed Primarily by the Need for ATP

Because ATP is the end product of cellular respiration, the ATP needs of the cell are the ultimate determinant of the rate of respiratory pathways and their components.

The complete oxidation of glucose yields about 30 molecules of ATP

We can now estimate how many molecules of ATP are formed when glucose is completely oxidized to CO_2. The number of ATP molecules formed in glycolysis and the citric acid cycle is unequivocally known because it is determined by the stoichiometries of chemical reactions. In contrast, the ATP yield of oxidative phosphorylation is less certain because the stoichiometries of proton pumping, ATP synthesis, and metabolite-transport processes need not be an integer or even have fixed values.

Recall from Section 18.3 that the best current estimates for the number of protons pumped out of the matrix per electron pair by Complexes I, III, and IV are four, two, and four, respectively. The synthesis of a molecule of ATP is driven by the flow of about three protons through ATP synthase. An additional proton is consumed in transporting ATP from the matrix to the cytoplasm. Hence, about 2.5 molecules of cytoplasmic ATP are generated as a result of the flow of a pair of electrons from NADH to O_2. For electrons that enter at Complex III, such as those from the oxidation of succinate or cytoplasmic NADH transferred by the glycerol-phosphate shuttle, the yield is about 1.5 molecules of ATP per electron pair.

Hence, as tallied in **Table 18.4**, about 30 molecules of ATP are formed when glucose is completely oxidized to CO_2. Most of the ATP, 26 of 30 molecules formed, is generated by oxidative phosphorylation, while two of the remaining four are made in the citric acid cycle, which is oxygen-dependent (aerobic). Recall that the anaerobic metabolism of glucose yields only two molecules of ATP. One of the effects of endurance exercise, a practice that calls for much ATP for an

TABLE 18.4 ATP yield from the complete oxidation of glucose

Reaction sequence	ATP yield per glucose molecule
Glycolysis: Conversion of glucose into pyruvate (in the cytoplasm)	
Phosphorylation of glucose	−1
Phosphorylation of fructose 6-phosphate	−1
Dephosphorylation of 2 molecules of 1,3-BPG	+2
Dephosphorylation of 2 molecules of phosphoenolpyruvate	+2
2 molecules of NADH are formed in the oxidation of 2 molecules of glyceraldehyde 3-phosphate	
Conversion of pyruvate into acetyl CoA (inside mitochondria)	
2 molecules of NADH are formed	
Citric acid cycle (inside mitochondria)	
2 molecules of adenosine triphosphate are formed from 2 molecules of succinyl CoA	+2
6 molecules of NADH are formed in the oxidation of 2 molecules each of isocitrate, α-ketoglutarate, and malate	
2 molecules of $FADH_2$ are formed in the oxidation of 2 molecules of succinate	
Oxidative phosphorylation (inside mitochondria)	
2 molecules of NADH formed in glycolysis; each yields 1.5 molecules of ATP (assuming transport of NADH by the glycerol 3-phosphate shuttle)	+3
2 molecules of NADH formed in the oxidative decarboxylation of pyruvate; each yields 2.5 molecules of ATP	+5
2 molecules of $FADH_2$ formed in the citric acid cycle; each yields 1.5 molecules of ATP	+3
6 molecules of NADH formed in the citric acid cycle; each yields 2.5 molecules of ATP	+15
Net Yield per Molecule of Glucose	+30

Information on the ATP yield of oxidative phosphorylation is from values given in P. C. Hinkle, M. A. Kumar, A. Resetar, and D. L. Harris, *Biochemistry* 30:3576, 1991.

Note: The current value of 30 molecules of ATP per molecule of glucose supersedes the earlier value of 36 molecules of ATP. The stoichiometries of proton pumping, ATP synthesis, and metabolite transport should be regarded as estimates. About 2 more molecules of ATP are formed per molecule of glucose oxidized when the malate–aspartate shuttle rather than the glycerol 3-phosphate shuttle is used.

extended period of time, is to increase the number of mitochondria and blood vessels in muscle and thus increase the extent of ATP generation by oxidative phosphorylation.

The rate of oxidative phosphorylation is determined by the need for ATP

How is the rate of the electron transport through the chain controlled? Under most physiological conditions, electron transport is tightly coupled to phosphorylation. Electrons do not usually flow through the electron-transport chain to O_2 unless ADP is simultaneously phosphorylated to ATP. When ADP concentration rises, as would be the case in active muscle, the rate of oxidative phosphorylation increases to meet the ATP needs of the muscle. The regulation of the rate of oxidative phosphorylation by the ADP level is called *respiratory control* or *acceptor control*.

The need for ATP likewise affects the rate of the citric acid cycle. At low concentrations of ADP, as in a resting muscle, NADH and $FADH_2$ are not consumed by the electron-transport chain. The citric acid cycle slows because there is less NAD^+ and FAD to feed the cycle. As the ADP level rises and oxidative phosphorylation speeds up, NADH and $FADH_2$ are oxidized, and the citric acid cycle becomes more active. The coordination of the components of cellular respiration, as shown in **Figure 18.38**, makes such regulation possible.

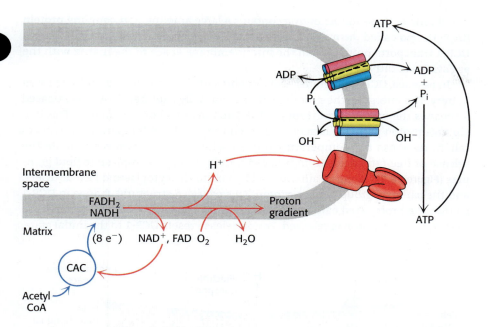

FIGURE 18.38 ATP synthase depends upon the rest of the ATP synthesome, as well as the ETC and CAC. The citric acid cycle (CAC) generates high-energy electrons by oxidizing acetyl CoA. The electrons are used to reduce oxygen to water and in the process develop a proton gradient. The proton gradient powers the synthesis of ATP, which requires the coordination of ATP synthase, the ATP-ADP translocase, and the phosphate carrier.

ATP synthase can be regulated

Mitochondria contain an evolutionarily conserved protein, *inhibitory factor 1* (IF1), that specifically inhibits the potential hydrolytic activity of the F_0F_1 ATP synthase. What is the function of IF1? Consider a circumstance where tissues may be deprived of oxygen (ischemia), as might happen during a stroke or heart attack. Without oxygen as the electron acceptor, the electron-transport chain will be unable to generate the proton-motive force. The ATP in the mitochondria would be hydrolyzed by the synthase, working in reverse. The role of IF1 is to prevent the wasteful hydrolysis of ATP by inhibiting the hydrolytic activity of the synthase.

How is this inhibition regulated? IF1 dimerizes when the pH in the matrix decreases due to the dissipation of the proton-motive force. IF1 dimers then bind tightly to ATP synthase dimers, preventing the β subunits from changing conformation and thus halting any action of ATP synthase until the pH of the matrix is returned to normal. IF1 is also overexpressed in many types of cancer. This overexpression plays a role in the induction of the Warburg effect, the switch from oxidative phosphorylation to aerobic glycolysis as the principal means for ATP synthesis (Section 16.3).

Regulated uncoupling leads to the generation of heat

Some organisms possess the ability to uncouple oxidative phosphorylation from ATP synthesis to generate heat. Such uncoupling is a means to maintain body temperature in hibernating animals, in some newborn animals (including human beings), and in many adult mammals, especially those adapted to cold. The skunk cabbage uses an analogous mechanism to heat its floral spikes in early spring, increasing the evaporation of odoriferous molecules that attract insects to fertilize its flowers. In animals, the uncoupling is in brown adipose tissue, which is specialized tissue for the process called *nonshivering thermogenesis,* or the ability to generate heat without using the rapid, unconscious contraction of skeletal muscles (shivering). In contrast, white adipose tissue, which constitutes the bulk of adipose tissue, plays no role in thermogenesis directly but serves as an insulating layer, an energy source, and an endocrine gland (Chapters 24 and 27).

Brown adipose tissue is very rich in mitochondria. The tissue appears brown from the combination of the greenish-colored cytochromes in the numerous mitochondria and the red hemoglobin present in the extensive blood supply, which helps to carry the heat through the body. The inner mitochondrial membrane of these mitochondria contains a large amount of **uncoupling protein (UCP-1)**, also called *thermogenin*, a dimer of 33-kDa subunits that resembles ANT. UCP-1 transports protons from the intermembrane space to the matrix with the assistance of fatty acids.

In essence, UCP-1 generates heat by short-circuiting the mitochondrial proton battery. The energy of the proton gradient, normally captured as ATP, is released as heat as the protons flow through UCP-1 to the mitochondrial matrix. This dissipative proton pathway is activated when the core body temperature begins to fall. In response to a temperature drop, α-adrenergic hormones stimulate the release of free fatty acids from triacylglycerols stored in cytoplasmic lipid granules (**Figure 18.39**). Long-chain fatty acids bind to the cytoplasmic face of UCP-1, and the carboxyl group binds a proton. This causes a structural change in UCP-1 so that the protonated carboxyl now faces the proton-poor environment of the matrix, and the proton is released. Proton release resets UCP-1 to the initial state.

FIGURE 18.39 UCP-1 generates heat by transporting protons without the synthesis of ATP.

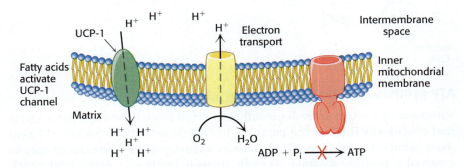

In addition to UCP-1, two other uncoupling proteins have been identified: UCP-2, which is found in a wide variety of tissues, and UCP-3, which is localized to skeletal muscle and brown fat. Both proteins share greater than 50% identity with UCP-1. This family of uncoupling proteins may play a role in energy homeostasis, and the genes for UCP-2 and UCP-3 map to regions of the human and mouse chromosomes that have been linked to obesity, supporting the notion that they function as a means of regulating body weight. Until recently, adult humans were believed to lack brown fat tissue. However, new studies have established that adults have brown adipose tissue in the neck and upper chest regions that is activated by cold (**Figure 18.40**). Obesity, which is predominantly caused by the swelling of intracellular lipid droplets in white adipose tissue, also leads to a decrease in brown adipose tissue.

FIGURE 18.40 Brown adipose tissue is revealed on exposure to cold. The results of PET–CT scanning show the uptake and distribution of ^{18}F-fluorodeoxyglucose (^{18}F-FDG) in adipose tissue. The patterns of ^{18}F-FDG uptake in the same subject are dramatically different under thermoneutral conditions (A) and after exposure to cold (B).

[Republished with permission of American Society for Clinical Investigation, from *J Clin Invest*. 2013; 123(8):3395–3403. doi:10.1172/JCI68993. Permission conveyed through Copyright Clearance Center, Inc.]

(A)   (B)

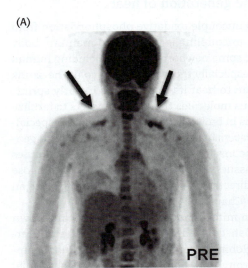

PRE   POST

Reintroduction of UCP-1 into pigs may be economically valuable

We can witness the effects of a lack of nonshivering thermogenesis by examining pig behavior. The ancestors of pigs are believed to have lost the gene for UCP-1 ~20 million years ago when they inhabited tropical and subtropical environments where they could survive without nonshivering thermogenesis. As the range of pigs expanded, however, the lack of UCP-1 (and hence, the absence of functional brown fat) became a liability to which pigs have adapted with distinctive behavioral traits. Among mammals, pigs have unusually large litters, and they are the only ungulates (hoofed mammals) that build nests for birth; piglets generate heat by huddling with their littermates in an insulated nest and shivering.

The deficiency of UCP-1 has important implications for pig farming. Even with the use of heat lamps, which account for 35% of the energy cost of raising pigs, piglet mortality may be as high as 20% because of hypothermia. Furthermore, because of their lack of UCP-1, pigs accumulate fat in white adipose tissue as a thermal insulator, meaning there is less lean meat available.

Recently, researchers have inserted the gene for UCP-1 into pig embryos. The resulting pigs are cheaper to raise due to the decrease in energy costs and provide more high-quality lean meat. The gene for UCP-1 was inserted into the pig embryos using the CRISPR technique described in Chapter 9. We see here another example of the expanding reach of biochemistry into areas of the economy beyond just the biomedical arena.

Oxidative phosphorylation can be inhibited at many stages

Many potent and lethal poisons exert their effect by inhibiting oxidative phosphorylation at one of a number of different locations (**Figure 18.41**).

Inhibition of the electron-transport chain. Rotenone, which is used as a fish and insect poison, and amytal, a barbiturate sedative, block electron transfer in NADH-Q oxidoreductase and thereby prevent the use of NADH as a substrate. In the presence of rotenone and amytal, electron flow resulting from the oxidation of succinate is unimpaired, because these electrons enter through QH_2, beyond the block. As an electron-transport-chain inhibitor, rotenone may play a role, along with genetic susceptibility, in the development of Parkinson's disease.

Other drugs block the chain at different points. For example, Antimycin A interferes with electron flow from cytochrome b_H in Q-cytochrome c oxidoreductase. Furthermore, electron flow in cytochrome c oxidase can be blocked by cyanide (CN^-), azide (N_3^-), and carbon monoxide (CO). Cyanide and azide react with the ferric form of heme a_3, whereas carbon monoxide inhibits the ferrous form. Inhibition of the electron-transport chain also inhibits ATP synthesis because the proton-motive force can no longer be generated.

Inhibition of ATP synthase. Oligomycin, an antibiotic used as an antifungal agent, and dicyclohexylcarbodiimide (DCC) prevent the influx of protons through ATP synthase by binding to the carboxylate group of the **c** subunits required for proton binding. Modification of only one **c** subunit by DCC is sufficient to inhibit the rotation of the entire **c** ring and hence ATP synthesis. If actively respiring mitochondria are exposed to an inhibitor of ATP synthase, the electron-transport chain ceases to operate. This observation clearly illustrates that electron transport and ATP synthesis are normally tightly coupled.

Uncoupling electron transport from ATP synthesis. The tight coupling of electron transport and phosphorylation in mitochondria can be uncoupled *pathologically*, in an unregulated way, by small molecules instead of *beneficially*, in a regulated way, by uncoupling proteins. Certain acidic aromatic compounds, including 2,4-dinitrophenol (DNP), carry protons across the inner mitochondrial membrane,

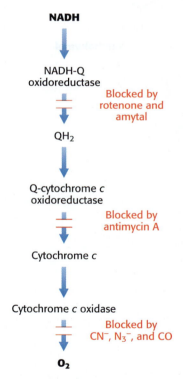

FIGURE 18.41 Inhibitors of electron transport block the process at a variety of sites.

down their concentration gradient. In the presence of these uncouplers, electron transport from NADH to O_2 proceeds in a normal fashion, but ATP is not formed by mitochondrial ATP synthase, because the proton-motive force across the inner mitochondrial membrane is continuously dissipated. This loss of respiratory control leads to increased oxygen consumption and oxidation of NADH. Indeed, in the accidental or intentional ingestion of uncouplers, large amounts of metabolic fuels are consumed, but no energy is captured as ATP. Rather, energy is released as heat.

DNP is the active ingredient in some herbicides and fungicides. There are reports that Soviet soldiers were given DNP to keep them warm during the long Russian winters. Remarkably, some people consume DNP today as a weight-loss drug, despite the fact that the FDA banned its use in 1938, calling it "extremely dangerous and not fit for human consumption." Its use can lead to death by hyperthermia, with body temperatures reaching in excess of 43°C prior to death.

Drugs are being sought that would function as mild uncouplers, uncouplers not as potentially lethal as DNP, for use in treatment of obesity and related pathologies. Xanthohumol, a prenylated chalcone found in hops and beer, shows promise in this regard. Xanthohumol also scavenges free radicals and is used for treatment of certain types of cancers.

Inhibition of ATP export. ANT is specifically inhibited by very low concentrations of atractyloside (a plant glycoside) or bongkrekic acid (an antibiotic from a mold). Atractyloside binds to the translocase when its nucleotide site faces the intermembrane space, whereas bongkrekic acid binds when this site faces the mitochondrial matrix. Oxidative phosphorylation stops soon after either inhibitor is added, demonstrating the essentiality of ANT for maintaining adequate amounts of ADP to accept the energy associated with the proton-motive force.

The effects of various inhibitors can now be reliably measured using Seahorse XF technology (Agilent Technologies), which now allows the measurement of the rate of aerobic respiration and lactic acid fermentation simultaneously in real time in cultured cells. The extent of aerobic respiration is determined by measuring the oxygen consumption rate, while the rate of glycolysis correlates with the extracellular acidification rate. This method is now widely used in both industry and academia in drug discovery, cancer research, and many other areas.

New mitochondrial diseases are constantly being discovered

The number of diseases that can be attributed to mitochondrial mutations is steadily growing in step with our growing understanding of the biochemistry and genetics of mitochondria. Mitochondrial diseases are estimated to affect from 10 to 15 per 100,000 people, roughly equivalent to the prevalence of the muscular dystrophies. Some of these mutations impair the use of NADH, whereas others block electron transfer to Q. Mutations in Complex I are the most frequent cause of mitochondrial diseases; indeed, the first mitochondrial disease to be understood was Leber hereditary optic neuropathy (LHON), a form of blindness that strikes in midlife as a result of mutations in Complex I. The accumulation of mutations in mitochondrial genes in a span of several decades may contribute to aging, degenerative disorders, and cancer.

Mitochondrial disease is primarily inherited maternally, because a human egg harbors several hundred thousand molecules of mitochondrial DNA, whereas a sperm contributes only a few hundred. Thus, the paternal contribution has little effect on the mitochondrial genotype. Because the maternally inherited mitochondria are present in large numbers but not all of the mitochondria may be affected, the pathologies of mitochondrial mutants can be quite complex. Even within a single family carrying an identical mutation, chance variations in the percentage of mitochondria with the mutation lead to large variations in the nature and severity of the symptoms of the pathological condition as well as the time of onset. As the percentage of defective mitochondria increases, energy-generating capacity diminishes until, at some threshold, the cell can no longer function properly. Defects in cellular respiration are doubly dangerous: not only does energy transduction decrease, but also the likelihood that reactive oxygen species will be generated increases. Organs that are highly dependent on oxidative phosphorylation, such as the nervous system, the retina, and the heart, are most vulnerable to mutations in mitochondrial DNA.

❖ SELF–CHECK QUESTION

4,6-Dinitro-o-cresol (DNOC) was used as a pesticide until it was banned because it causes a dramatic rise in body temperature, profuse sweating, rapid breathing, and a decrease in body fat. When tested, it affected isolated mitochondria. What do you expect is its effect on oxygen consumption by isolated mitochondria, and why?

Mitochondria play a key role in apoptosis

In the course of development or in cases of significant cell damage, individual cells within multicellular organisms undergo a form of programmed cell death called **apoptosis**, with mitochondria acting as control centers regulating this process. Although the details continue to be established, the outer membrane of damaged mitochondria becomes highly permeable, a process referred to as *mitochondrial outer membrane permeabilization* (MOMP). This permeabilization is instigated by a family of proteins (Bcl family) that were initially discovered because of their role in cancer.

One of the most potent activators of apoptosis, cytochrome *c*, exits the mitochondria and interacts with apoptotic peptidase-activating factor 1 (APAF-1), which leads to the formation of a protein complex called the *apoptosome*. The apoptosome recruits and activates a proteolytic enzyme called *caspase 9*, a member of the cysteine protease family (Section 6.2) that in turn activates a cascade of other caspases. Each caspase type destroys a particular target, such as the proteins that maintain cell structure. Another target is a protein that inhibits an enzyme that destroys DNA (an enzyme called caspase-activated DNAse or CAD), freeing CAD to cleave the genetic material. This cascade of proteolytic enzymes has been called "death by a thousand tiny cuts."

❖ SELF–CHECK QUESTION

The most common metabolic sign of mitochondrial disorders is lactic acidosis. Why?

18.7 Proton Gradients Generated by Respiratory Chains Drive Many Biochemical Processes

The main concept presented in this chapter is that mitochondrial electron transfer and ATP synthesis are linked by a transmembrane proton gradient. ATP synthesis in bacteria and chloroplasts also is driven by proton gradients. Here we will briefly consider how these gradients can be used in additional ways, sometimes with startling effects. As a case in point, in many bacteria, the proton gradient is responsible for providing the power for a simple but important aspect of these organisms: how they move.

Proton flow through a rotary motor allows bacteria to swim

In one second, a motile bacterium can move 25 µm, or about 10 body lengths. A human being sprinting at a proportional rate would complete the 100-meter dash in slightly more than 5 s. Bacteria such as *Escherichia coli* and *Salmonella typhimurium* swim at such speed by rotating flagella that lie on their surfaces (**Figure 18.42**).

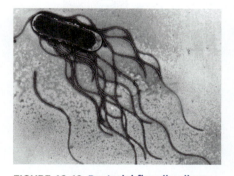

FIGURE 18.42 Bacterial flagella allow bacteria to swim rapidly. An electron micrograph of *Salmonella typhimurium* shows flagella directed backwards, where they can form a bundle under appropriate circumstances. Bacterial flagella are polymers approximately 15 nm in diameter and as much as 15 µm in length, composed of subunits of a protein called flagellin. These subunits associate into a left-handed helical structure with a hollow core. At its base, each flagellum has a rotary motor. [USDA/Science Source.]

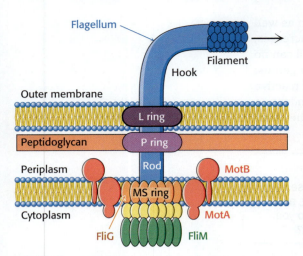

FIGURE 18.43 The flagellar motor is a complex structure containing as many as 40 distinct types of proteins. A schematic view of the flagellar motor is shown with the approximate positions of the proteins MotA and MotB (red), FliG (orange), FliN (yellow), and FliM (green).

When the flagella rotate in a counterclockwise direction (viewed from outside a bacterium), the separate flagella form a bundle that very efficiently propels the bacterium through solution. The motors that power this impressive motion are strikingly different from the eukaryotic motors that we explored in Section 6.5. In the bacterial motor, an element spins around a central axis rather than moving along a polymeric track. Early experiments by Julius Adler demonstrated that ATP is *not* required for flagellar motion. Rather, the necessary free energy is derived from the proton gradient that exists across the plasma membrane, established by the electron-transport chains of these organisms that pump protons into the periplasm.

Although the flagellar motor is quite complex, five components particularly crucial to motor function have been identified through genetic studies (**Figure 18.43**). MotA and MotB are membrane proteins, each with cytoplasmic and periplasmic domains, respectively. Approximately 11 MotA–MotB pairs form a ring around the base of the flagellum. The proteins FliG, FliM, and FliN are part of a disc-like structure called the MS (membrane and supramembrane) ring, with approximately 30 FliG subunits coming together to form the ring. The MotA–MotB pair and FliG combine to create a proton channel that drives the rotation of the flagellum similar to the proton-driven rotary motion of ATP synthase. In fact, such a mechanism was first proposed by Howard Berg to explain flagellar rotation before the rotary mechanism of ATP synthase was elucidated. Each MotA–MotB pair is conjectured to form a structure that has two half-channels (**Figure 18.44A**); FliG serves as the rotating proton carrier. In this scenario, a proton from the periplasmic space passes into the outer half-channel and is transferred to an FliG subunit. The MS ring rotates, rotating the flagellum with it and allowing the proton to pass into the inner half-channel and into the cell (**Figure 18.44B**). Ongoing structural and mutagenesis studies are testing and refining this hypothesis.

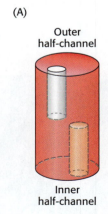

(A)

Outer half-channel

Inner half-channel

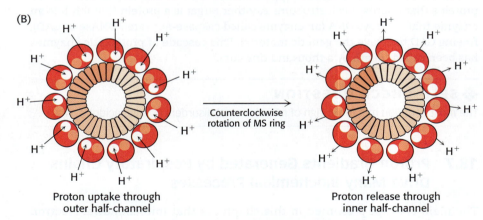

(B)

Counterclockwise rotation of MS ring

Proton uptake through outer half-channel

Proton release through inner half-channel

FIGURE 18.44 Rotation of the flagellum involves proton half-channels like ATP synthase. (A) MotA–MotB may form a structure having two half-channels. (B) One model for the mechanism of coupling rotation to a proton gradient requires protons to be taken up into the outer half-channel and transferred to the MS ring. The MS ring rotates in a counterclockwise direction, and the protons are released into the inner half-channel. The flagellum is linked to the MS ring and so the flagellum rotates as well.

Power transmission by proton gradients is a central motif of bioenergetics

Proton gradients power a variety of energy-requiring processes, some of which we have explored already, such as the rotation of bacterial flagella, the generation of heat by nonshivering thermogenesis, the active transport of calcium

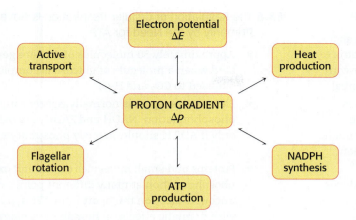

FIGURE 18.45 The proton gradient is an interconvertible form of free energy.

ions by mitochondria, the entry of some amino acids and sugars into bacteria, and the transfer of electrons from $NADP^+$ to NADPH. It's clear that proton gradients are a central, interconvertible currency of free energy in biological systems (**Figure 18.45**). Peter Mitchell, who established the chemiosmotic hypothesis (Section 18.4), noted that the proton-motive force is a marvelously simple and effective store of free energy because it requires only a thin, closed lipid membrane between two aqueous phases.

❖ SELF–CHECK QUESTION

When bacteria such as *E. coli* are starved to a sufficient extent, they become nonmotile. However, when such bacteria are placed in an acidic solution, they resume swimming. Explain.

Summary

18.1 Cellular Respiration Drives ATP Formation by Transferring Electrons to Molecular Oxygen

- Mitochondria generate most of the ATP required by aerobic cells through a combination of the reactions of the citric acid cycle and oxidative phosphorylation.
- Mitochondria are descendants of a free-living bacterium that established a symbiotic relationship with another cell.

18.2 Oxidative Phosphorylation Depends on Electron Transfer

- In oxidative phosphorylation, the synthesis of ATP is coupled to the flow of electrons from NADH or $FADH_2$ to O_2 by a proton gradient across the inner mitochondrial membrane.
- Electron flow through three asymmetrically oriented transmembrane complexes results in the pumping of protons out of the mitochondrial matrix and the generation of a membrane potential. ATP is synthesized when protons flow back to the matrix through ATP synthase.

18.3 The Respiratory Chain Consists of Four Complexes: Three Proton Pumps and a Physical Link to the Citric Acid Cycle

- The respiratory assembly of the inner mitochondrial membrane has numerous different electron carriers. Electrons from NADH are transferred to NADH-Q oxidoreductase (Complex I), the first of four complexes with the electrons emerging on QH_2, the reduced form of ubiquinone (Q).
- The citric acid cycle enzyme succinate dehydrogenase is a component of the succinate-Q reductase complex (Complex II), which donates electrons from $FADH_2$ to Q to form QH_2. Regardless of the source, QH_2 transfers its electrons to Q-cytochrome c oxidoreductase (Complex III) which reduces cytochrome c, a water-soluble peripheral membrane protein.
- Cytochrome c transfers electrons to cytochrome c oxidase (Complex IV). A heme iron ion and a copper ion in this oxidase transfer electrons to O_2, the ultimate acceptor, to form H_2O. Complexes I, III, and IV are organized into a large molecular structure called the respirasome.

18.4 A Proton Gradient Powers the Synthesis of ATP

- The flow of electrons through Complexes I, III, and IV leads to the transfer of protons from the matrix to the IM space, generating a proton-motive force consisting of a pH gradient and a membrane potential.
- ATP synthesis is driven by the flow of protons back to the matrix through ATP synthase, an enzyme complex that is a molecular motor made of two operational units: a rotating component and a stationary component.
- The rotation of ATP synthase's γ subunit induces structural changes in the β subunit that result in the synthesis and release of ATP from the enzyme.

18.5 Many Shuttles Allow Movement Across Mitochondrial Membranes

- The electrons of cytoplasmic NADH are transferred into mitochondria by the glycerol phosphate shuttle to form $FADH_2$ from FAD, or by the malate–aspartate shuttle to form mitochondrial NADH.
- The exchange of ADP and ATP between the matrix and IM space is mediated by two transporters, ANT and phosphate carrier, driven by membrane potential.

18.6 The Regulation of Cellular Respiration Is Governed Primarily by the Need for ATP

- Approximately 30 molecules of ATP are generated when a molecule of glucose is completely oxidized to CO_2 and H_2O.
- Electron transport is normally tightly coupled to phosphorylation. NADH and $FADH_2$ are oxidized only if ADP is simultaneously phosphorylated to ATP.
- Proteins and small molecules can inhibit oxidative phosphorylation at many different points, or can uncouple electron transport from ATP synthesis, with dramatic effects on human physiology.

18.7 Proton Gradients Generated by Respiratory Chains Drive Many Biochemical Processes

- A proton gradient across the plasma membrane, rather than ATP hydrolysis, powers the rotational flagellar motor of motile bacteria in a manner analogous to the way ATP synthase functions.
- Proton gradients drive diverse physiological processes because the mechanism is fundamentally simple and can be used in many ways by proteins.

Key Terms

oxidative phosphorylation (p. 543)

electron-transport chain (respiratory chain) (p. 543)

proton-motive force (p. 543)

cellular respiration (respiration) (p. 543)

intermembrane space (IM space) (p. 543)

matrix (p. 543)

reduction potential (redox or oxidation–reduction potential) (p. 546)

coenzyme Q (Q, ubiquinone) (p. 550)

cytochrome c (p. 551)

iron–sulfur proteins (nonheme iron proteins) (p. 551)

NADH-Q oxidoreductase (Complex I) (p. 552)

succinate-Q reductase (Complex II) (p. 553)

Q-cytochrome c oxidoreductase (Complex III) (p. 554)

Q cycle (p. 555)

cytochrome c oxidase (Complex IV) (p. 556)

respirasome (p. 559)

ATP synthase (Complex V) (p. 562)

glycerol 3-phosphate shuttle (p. 570)

malate–aspartate shuttle (p. 571)

ATP-ADP translocase (adenine nucleotide translocase, ANT) (p. 572)

uncoupling protein (UCP-1) (p. 576)

apoptosis (p. 579)

Problems

1. Which of the following electron-transport chain complexes is also a member of the citric acid cycle and not a component of the respirasome? ❖ 1

(a) Complex I

(b) Complex II

(c) Complex III

(d) Complex IV

2. The final electron acceptor for the electron-transport chain is which of the following? ❖ 1

(a) O_2

(b) Coenzyme Q

(c) CO_2

(d) NAD^+

3. Electrons are passed on to cytochrome c using which of the following? ❖ 1

(a) The Cori cycle

(b) The Q cycle

(c) The glucose-alanine cycle

(d) The Krebs cycle

4. Which of the following statements about the proton-motive force is not true? ❖ 2, ❖ 3

(a) It requires an intact membrane.

(b) It consists of an unequal distribution of protons.

(c) It is used to synthesize ATP.

(d) It is powered by the hydrolysis of ATP.

5. The standard oxidation–reduction potential (E'_0) for the reduction of O_2 to H_2O is given as 0.82 V in Table 18.1. However, the value given in textbooks of chemistry is 1.23 V. Account for this difference.

6. Why are electrons carried by $FADH_2$ not as energy rich as those carried by NADH? What is the consequence of this difference? Calculate the energy released by the reduction of O_2 with $FADH_2$.

7. Compare the $\Delta G^{\circ\prime}$ values for the oxidation of succinate by NAD^+ and by FAD. Use the data given in Table 18.1 to find the E'_0 of the NAD^+ – NADH and fumarate-succinate couples, and assume that E'_0 for the FAD – $FADH_2$ redox couple is nearly 0.05 V. Why is FAD rather than NAD^+ the electron acceptor in the reaction catalyzed by succinate dehydrogenase?

8. Place the following components of the electron-transport chain in their proper order: ❖ 1

(a) Cytochrome c

(b) Q-cytochrome c oxidoreductase

(c) NADH-Q reductase

(d) Cytochrome c oxidase

(e) Ubiquinone

9. Match each term below with its alternative name. ❖ 1

(a) Complex I   1. Q-cytochrome c oxidoreductase

(b) Complex II   2. Coenzyme Q

(c) Complex III   3. Succinate-Q reductase

(d) Complex IV   4. NADH-Q oxidoreductase

(e) Ubiquinone   5. Cytochrome c oxidase

10. Rotenone inhibits electron flow through NADH-Q oxidoreductase. Antimycin A blocks electron flow between cytochromes b and c_1. Cyanide blocks electron flow through cytochrome c oxidase to O_2. Predict the relative oxidation–reduction state of each of the following respiratory-chain components in mitochondria that are treated with each of the inhibitors: ❖ 1

(a) NAD^+

(b) NADH-Q oxidoreductase

(c) Coenzyme Q

(d) Cytochrome c_1

(e) Cytochrome c

(f) Cytochrome a

11. What is the yield of ATP when each of the following substrates is completely oxidized to CO_2 by a mammalian cell homogenate? Assume that glycolysis, the citric acid cycle, and oxidative phosphorylation are fully active.

(a) Pyruvate

(b) Lactate

(c) Fructose 1,6-bisphosphate

(d) Phosphoenolpyruvate

(e) Galactose

(f) Dihydroxyacetone phosphate

12. What is the effect of each of the following inhibitors on electron-transport and ATP formation by the respiratory chain? ❖ 1, ❖ 3

(a) Azide

(b) Atractyloside

(c) Rotenone

(d) DNP

(e) Carbon monoxide

(f) Antimycin A

13. Arsenate (AsO_4^{3-}) closely resembles phosphate in structure and reactivity. However, arsenate esters are unstable and are spontaneously hydrolyzed. If arsenate is added to actively respiring mitochondria, what would be the effect on ATP synthesis? On the rate of the electron-transport chain? Briefly explain.

14. Oxidative phosphorylation in mitochondria is often monitored by measuring oxygen consumption. When oxidative phosphorylation is proceeding rapidly, the mitochondria will rapidly consume oxygen. If there is little oxidative phosphorylation, only small amounts of oxygen will be used. You are given a suspension of isolated mitochondria and directed to add the following compounds in the order from a to h. With the addition of each compound, all of the previously added compounds remain present. Predict the effect of each addition on oxygen consumption by the isolated mitochondria. ❖ 1, ❖ 2, ❖ 3, ❖ 4

(a) Glucose

(b) ADP + P_i

(c) Citrate

(d) Oligomycin

(e) Succinate

(f) Dinitrophenol

(g) Rotenone

(h) Cyanide

15. The number of molecules of inorganic phosphate incorporated into organic form per atom of oxygen consumed, termed the *P:O ratio*, was frequently used as an index of oxidative phosphorylation.

(a) What are the relationships among the P:O ratio to the ratio of the number of protons translocated per electron pair ($H^+/2e^-$) and the ratio of the number of protons needed to synthesize ATP and transport it to the cytoplasm (P/H^+)?

(b) What are the P:O ratios for electrons donated by matrix NADH and by succinate?

16. The precise site of action of a respiratory-chain inhibitor can be revealed by the *crossover technique*. Britton Chance devised elegant spectroscopic methods for determining the proportions of the oxidized and reduced forms of each carrier. This determination is feasible because the forms have distinctive absorption spectra, as illustrated in the graph for cytochrome *c*.

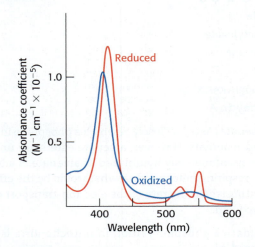

Suppose you are given a new inhibitor and find that its addition to respiring mitochondria causes the carriers between NADH and QH_2 to become more reduced and those between cytochrome *c* and O_2 to become more oxidized. Where does your inhibitor act? ❖ **1,** ❖ **2**

17. Years ago, uncouplers were suggested to make wonderful diet drugs. Explain why this idea was proposed and why it was rejected. Why might the producers of antiperspirants be supportive of the idea? ❖ **3,** ❖ **4**

18. Give an example of the use of the proton-motive force in ways other than for the synthesis of ATP.

19. XF technology (Seahorse Bioscience) now allows the measurement of the rate of aerobic respiration and lactic acid fermentation simultaneously in real time in cultured cells. The extent of aerobic respiration is determined by measuring the oxygen consumption rate (OCR, measured in picomoles of oxygen consumed per minute), while the rate of glycolysis correlates with the extracellular acidification rate (ECAR-milli pH per minute, or the changes in pH that occur over time). The next graph shows the results of an experiment using the new technology.

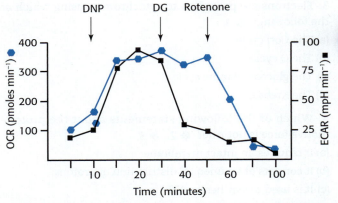

Dinitrophenol (DNP), the glycolysis inhibitor 2-deoxyglucose (DG), and rotenone were added sequentially to cell cultures. ❖ **1,** ❖ **2**

(a) What is the effect on OCR and ECAR of adding DNP to the cell culture? Explain these results.

(b) Explain the effect of the addition of 2-deoxyglucose.

(c) Explain how 2-deoxyglucose acts as an inhibitor of glycolysis?

(d) Explain the effect of the addition of rotenone.

20. A mutation in a mitochondrial gene encoding a component of ATP synthase has been identified. People who have this mutation suffer from muscle weakness, ataxia (a loss of coordination), and retinitis pigmentosa (retinal degeneration). A tissue biopsy was performed on each of three patients having this mutation, and submitochondrial particles were isolated that were capable of succinate-sustained ATP synthesis. First, the activity of the ATP synthase was measured on the addition of succinate and the following results were obtained. ❖ **3,** ❖ **4**

	ATP synthase activity (nmol of ATP formed min^{-1} mg^{-1})
Controls	3.0
Patient 1	0.25
Patient 2	0.11
Patient 3	0.17

(a) What was the purpose of the addition of succinate?

(b) What is the effect of the mutation on succinate-coupled ATP synthesis?

Next, the ATPase activity of the enzyme was measured by incubating the submitochondrial particles with ATP in the absence of succinate.

	ATP hydrolysis (nmol of ATP hydrolyzed min^{-1} mg^{-1})
Controls	33
Patient 1	30
Patient 2	25
Patient 3	31

(c) Why was succinate omitted from the reaction?

(d) What is the effect of the mutation on ATP hydrolysis?

(e) What do these results, in conjunction with those obtained in the first experiment, tell you about the nature of the mutation?

Phototrophy and the Light Reactions of Photosynthesis

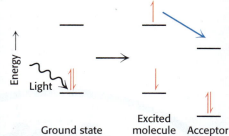

Ground state · Excited molecule · Acceptor

Phototrophic halobacteria are visible from the air in salt ponds near the San Francisco Bay. Although nonphotosynthetic, these reddish-purple archaea have the ability to produce additional ATP with energy from sunlight using an evolutionary relative of rhodopsin, a light-activated protein found in your eyes. Photosynthetic green algae are also visible in nearby ponds, which, in contrast, use sunlight exclusively for metabolic energy needs and can produce sugar from carbon dioxide. Both organisms accomplish phototrophy, the key to which is the ability of electrons to absorb visible light and then transfer from one molecule to another using the process of photoinduced charge separation diagrammed above.

flowertiare/Deposit Photos

❖ LEARNING GOALS

By the end of this chapter, you should be able to:

1. Describe phototrophy and the light reactions.
2. Identify the key products of the light reactions.
3. Explain how redox balance is maintained during the light reactions.
4. Explain how ATP is synthesized in chloroplasts.
5. Describe the function of the light-harvesting complex.

Every year, Earth is bathed in photons with a total energy content of approximately 10^{24} kJ. By comparison, a major hurricane has one billionth that amount of energy. The source of the energy is, of course, the electromagnetic radiation from the sun. On our planet, there are organisms capable of collecting a fraction, approximately 2%, of this solar energy, and converting it into chemical energy. Green plants are the most obvious of these organisms, although 60% of this conversion is carried out by algae, bacteria, and archaea. Although only a small amount of the incident energy is captured, it powers almost all life on Earth. This transformation is perhaps the most important of all of the energy transformations that we will see in our study of biochemistry; without it, the vast majority of life as we know it on our planet simply could not exist.

OUTLINE

19.1 Phototrophy Converts Light Energy into Chemical Energy

19.2 In Eukaryotes, Photosynthesis Takes Place in Chloroplasts

19.3 Light Absorption by Chlorophyll Molecules Induces Electron Transfer

19.4 Two Photosystems Generate a Proton Gradient and Reducing Power in Cyanobacteria and Photosynthetic Eukaryotes

19.5 A Proton Gradient across the Thylakoid Membrane Drives ATP Synthesis

19.6 Accessory Pigments Funnel Energy into Reaction Centers

19.7 The Ability to Convert Light into Chemical Energy Is Ancient

19.1 Phototrophy Converts Light Energy into Chemical Energy

The metabolisms of living organisms can be subdivided into two pairs of fundamental groups based upon how they capture energy and obtain carbon. The process of converting electromagnetic radiation into chemical energy is called *phototrophy*, and, as noted in Chapter 15, organisms which utilize this mechanism to capture energy are called **phototrophs** (literally, "light-feeders").

An example of a phototroph is the archaeon *Halobacterium salinarum* that grows in extremely salty water such as that found in the Dead Sea, the Great Salt Lake, or in evaporation ponds used to isolate salt, as shown at the start of this chapter. When grown under low oxygen conditions, this organism generates additional ATP by using just two membrane proteins (**Figure 19.1**), relatives of which we have already encountered in other contexts. The first is ATP synthase, which utilizes a proton gradient to produce ATP (Section 18.4). The second is **bacteriorhodopsin**, which we discussed briefly in Section 12.4.

Bacteriorhodopsin is similar to the 7-transmembrane helix protein rhodopsin involved in vision (Section 14.6). Like rhodopsin, bacteriorhodopsin includes bound 11-cis retinal, absorbing sunlight efficiently and producing all-trans retinal. Unlike rhodopsin, bacteriorhodopsin does not signal through other proteins but acts as a proton pump, transporting one proton—for each photon absorbed—from the cytoplasm to the periplasm, the space between the cell membrane and the cell wall.

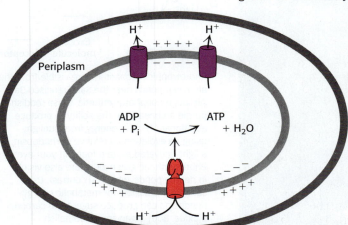

FIGURE 19.1 Phototrophy can be accomplished by the cooperation of just two proteins. The archaeon *Halobacterium salinarum* accomplishes phototrophy by expressing the light-absorbing protein bacteriorhodopsin in its inner cell membrane. Bacteriorhodopsin (purple structures) uses the energy from sunlight to transport protons across the membrane and into the periplasm. These protons return through ATP synthase (red structure) to make ATP.

Many phototrophic organisms use captured light energy, along with electrons harvested from a donor molecule, to convert carbon dioxide into carbohydrate in a complex process known as **photosynthesis**. The most common form of photosynthesis on Earth today, used by all plants, algae, and many bacteria, is called *oxygenic photosynthesis* and is the primary subject of this chapter. Oxygenic photosynthesis is so named because water is the electron donor and oxygen is produced as a waste product.

$$CO_2 + H_2O \xrightarrow{\text{Light}} (CH_2O) + O_2$$

In accordance with how the term is most commonly used, we will simply use the term "photosynthesis" to refer to oxygenic photosynthesis and will explicitly refer to alternative types as *anoxygenic photosynthesis* when they arise. In the above reaction, CH_2O represents carbohydrate, primarily sucrose and starch. Photosynthetic carbohydrate synthesis constantly regenerates organic molecules from carbon dioxide. These can then be used as energy sources for **chemotrophs**, like humans, which derive energy from chemicals as opposed to sunlight.

Photosynthetic organisms are also **autotrophs** ("self-feeders") because they obtain carbon primarily from carbon dioxide, rather than from organic molecules. Autotrophs can synthesize chemical fuels such as glucose from carbon dioxide and water. Photosynthetic organisms capture and store energy from sunlight and then later utilize this energy from the synthesized glucose through the glycolytic pathway and aerobic metabolism. In contrast to autotrophs, organisms that obtain carbon primarily from organic chemicals are called **heterotrophs**, which ultimately depend on autotrophs for these molecules. Thus, the carbohydrates produced by photosynthetic organisms provide not only energy to run much of the biological world, but also the carbon molecules used by many organisms to make a wide array of biomolecules.

Photosynthesis comprises light reactions and dark reactions

Photosynthesis is composed of two parts: the light reactions and the dark reactions. In the **light reactions**, light energy is captured as chemical energy through the formation of ATP and electrons are extracted to provide reducing power for biosynthetic reactions. The products of the light reactions are then used in the so-called *dark reactions* to drive the reduction of CO_2 and its conversion into glucose and other sugars. The dark reactions are also called the *Calvin–Benson cycle* or *light-independent reactions* and will be discussed in Chapter 20.

The same biochemical principles govern both respiration and photosynthesis

As we will see soon, the light reactions of photosynthesis closely resemble the events of oxidative phosphorylation, whereas the dark reactions accomplish the reversal of cellular respiration. In Chapters 17 and 18, we learned that cellular respiration is the oxidation of glucose to CO_2 with the reduction of O_2 to water, a process that generates ATP.

$$C_6H_{12}O_6 + 6\,O_2 \xrightarrow{\text{Cellular respiration}} 6\,CO_2 + 6\,H_2O + \text{energy}$$

In photosynthesis, this process must be reversed—reducing CO_2 and oxidizing H_2O to synthesize glucose.

$$\text{Energy} + 6\,H_2O + 6\,CO_2 \xrightarrow{\text{Photosynthesis}} C_6H_{12}O_6 + 6\,O_2$$

Although the processes of respiration and photosynthesis are chemically opposite each other, the biochemical principles governing the two processes are nearly identical. The key to both processes is the generation of high-energy electrons.

- *During respiration, the citric acid cycle oxidizes carbon fuels to CO_2 to generate high-energy electrons.* The flow of these high-energy electrons down an electron-transport chain generates a proton-motive force. This proton-motive force is then transduced by ATP synthase to form ATP.

- *During photosynthesis, energy from light boosts electrons from a low-energy state to a high-energy state.* In the high-energy, unstable state, these electrons can be passed to nearby molecules. These electrons are then used to (1) produce reducing power in the form of NADPH and (2) generate a proton-motive force across a membrane, which subsequently drives the synthesis of ATP (**Figure 19.2**).

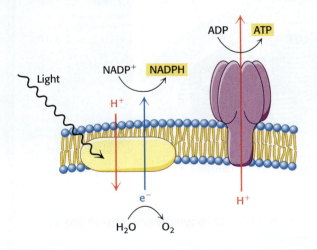

FIGURE 19.2 The light reactions of oxygenic photosynthesis generate both ATP and reducing power. Light is absorbed and the energy is used to drive electrons from water to generate NADPH, which can be used later for reduction reactions, and to drive protons across a membrane. In the same fashion as shown for the halobacterium in Figure 19.1, these protons return through ATP synthase (purple structure) to make ATP.

❖ SELF–CHECK QUESTION

Photosynthesis is a complex metabolic process that is the sum of which two of the following four metabolic schemes for carbon and energy capture: chemotrophy, heterotrophy, phototrophy, or autotrophy?

Two kinds of light reactions take place in the green plants

Photosynthesis in green plants is mediated by two kinds of light reactions, which take place in two distinct transmembrane protein complexes, called *photosystems*. **Photosystem I** generates reducing power in the form of NADPH but, in the process, becomes electron deficient. **Photosystem II** oxidizes water and transfers the electrons to replenish the electrons lost by photosystem I. A side product of these reactions is O_2. Electron flow from photosystem II to photosystem I generates the transmembrane proton gradient, augmented by the protons released by the oxidation of water, that drives the synthesis of ATP. In keeping with the similarity of their principles of operation, in eukaryotes both respiration and photosynthesis take place in double-membrane organelles: mitochondria for cellular respiration and chloroplasts for photosynthesis.

19.2 In Eukaryotes, Photosynthesis Takes Place in Chloroplasts

In eukaryotes, photosynthesis takes place in specialized organelles called **chloroplasts**, typically 5 μm long. Like a mitochondrion, a chloroplast has an outer membrane and an inner membrane, with an intervening intermembrane space (**Figure 19.3**). The inner membrane surrounds a space called the **stroma**, which is the site of the dark reactions of photosynthesis (Chapter 20). In the stroma are membranous structures called **thylakoids**, which are flattened sacs, or discs. The thylakoid sacs are stacked to form a **granum**. Different grana are linked by regions of thylakoid membrane called *stroma lamellae*. The thylakoid membranes separate the thylakoid space from the stroma space. Thus, chloroplasts have three different membranes (*outer, inner,* and *thylakoid membranes*) and three separate spaces (*intermembrane, stroma,* and *thylakoid lumen*). In a developing chloroplast, thylakoids arise from budding of the inner membrane, and so they are analogous to the mitochondrial cristae. Like the mitochondrial cristae, they are the site of coupled oxidation–reduction reactions of the light reactions that generate the proton-motive force.

FIGURE 19.3 The anatomy of a chloroplast reveals complex membrane-enclosed structures. Plant-leaf cells contain between 1 and 100 chloroplasts, depending on the species, cell type, and growth conditions. At far right, an electron micrograph of a chloroplast from a spinach leaf shows thylakoid membranes packed together to form grana.

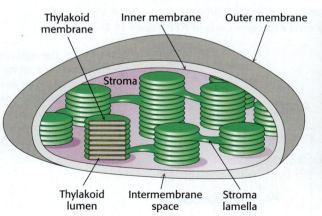
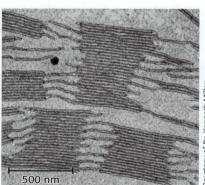

Thylakoid membrane · Inner membrane · Outer membrane · Stroma · Thylakoid lumen · Intermembrane space · Stroma lamella

500 nm

Courtesy of Dr. Kenneth Miller

The primary events of photosynthesis take place in thylakoid membranes

The thylakoid membranes contain the energy-transforming machinery: light-harvesting proteins, reaction centers, electron-transport chains, and ATP synthase. These membranes contain nearly equal amounts of lipids and proteins. The lipid composition is highly distinctive: about 75% of the total lipids are galactolipids and 10% are sulfolipids, whereas only 10% are phospholipids. The thylakoid membrane and the inner membrane, like the inner mitochondrial membrane, are impermeable to most molecules and ions. The outer membrane of a chloroplast, like that of a mitochondrion, is highly permeable to small molecules and ions. The stroma contains the soluble enzymes that utilize the NADPH and ATP synthesized by the thylakoids to convert CO_2 into sugar.

Chloroplasts arose from an endosymbiotic event

Chloroplasts contain their own DNA and the machinery for replicating and expressing it. However, chloroplasts are not independent from the rest of the cell; they also contain many proteins encoded by nuclear DNA. How did the intriguing relationship between the cell and its chloroplasts develop? It is now widely accepted that, in a manner analogous to the evolution of mitochondria (Section 18.1), chloroplasts are also the result of endosymbiotic events in which a photosynthetic microorganism was engulfed by a eukaryotic host. Evidence suggests that chloroplasts in both higher plants and green algae are derived from a single endosymbiotic event, when an ancestor of a modern-day cyanobacterium was engulfed (**Figure 19.4**), whereas those in red and brown algae are derived from at least one additional event.

The chloroplast genome is smaller than that of a cyanobacterium, but the two genomes have key features in common. Both are circular and have a single start site for DNA replication, and the genes of both are arranged in operons—sequences of functionally related genes under common control. In the course of evolution, many of the genes of the chloroplast ancestor were transferred to the plant cell's nucleus or, in some cases, lost entirely, thus establishing a fully dependent relationship.

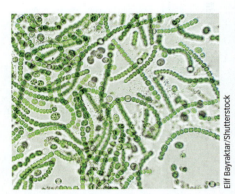

FIGURE 19.4 Cyanobacteria are the prokaryotic photosynthesizers that are the most similar to plant chloroplasts. A colony of the photosynthetic filamentous cyanobacterium *Anabaena* is shown at 500× magnification. Ancestors of these bacteria are thought to have evolved into present-day chloroplasts of plants.

Elif Bayraktar/Shutterstock

❖ SELF–CHECK QUESTION
Explain the relationship between the following terms: thylakoid, stroma, chloroplast, and endosymbiosis.

19.3 Light Absorption by Chlorophyll Molecules Induces Electron Transfer

The trapping of light energy is the key to phototrophy, and therefore also photosynthesis. The first event is the absorption of light by a photoreceptor molecule. The principal photoreceptors in the chloroplasts of most green plants are collectively chlorophylls, and of particular importance among these is the pigment molecule **chlorophyll *a***, a substituted tetrapyrrole (**Figure 19.5**). The four nitrogen atoms of the pyrroles are coordinated to a magnesium ion. Unlike a porphyrin such as heme, chlorophyll *a* has a reduced pyrrole ring and an additional 5-carbon ring fused to one of the pyrrole rings. Another distinctive feature of chlorophyll *a* is the presence of *phytol*, a highly hydrophobic 20-carbon alcohol, esterified to an acid side chain.

Chlorophylls are very effective photoreceptors because they contain networks of conjugated double bonds—alternating single and double bonds. Such compounds are called conjugated polyenes. In polyenes, the electrons are not localized to a particular atomic nucleus. Upon the absorption of light energy, the electron transitions from a low-energy molecular orbital to a higher-energy orbital.

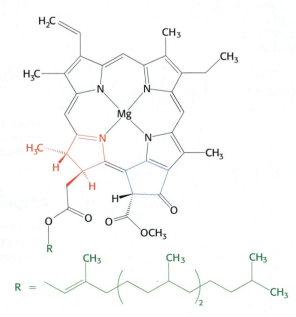

FIGURE 19.5 Chlorophyll *a* is structurally similar to heme. Like heme, chlorophyll *a* is a cyclic tetrapyrrole. One of the pyrrole rings (shown in red) is reduced, and an additional five-carbon ring (shown in blue) is fused to another pyrrole ring. A phytol chain (shown in green) is connected by an ester linkage. Magnesium ion binds at the center of the structure.

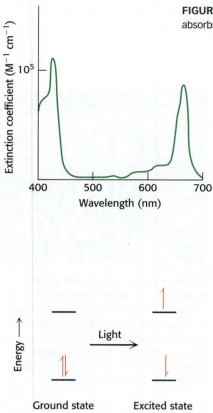

Chlorophylls have very strong absorption bands in the visible region of the spectrum, where the solar output reaching Earth is maximal (**Figure 19.6**). Chlorophyll *a*'s peak molar extinction coefficient (ε), a measure of a compound's ability to absorb light, is greater than 10^5 M^{-1} cm^{-1}, among the highest observed for organic compounds.

Transferring electrons allows energy to be captured instead of lost as heat

What happens when light is absorbed by a pigment molecule? The energy from the light excites an electron from its ground energy level to an excited energy level (**Figure 19.7**). This high-energy electron can have one of several fates. For the vast majority of compounds that absorb light, the electron simply returns to the ground state and the absorbed energy is converted into heat; however, the energy can also be emitted as light (fluorescence or phosphorescence) or transferred to another nearby electron (resonance energy transfer). However, one remaining possibility is the key to phototrophy: if a suitable electron acceptor is nearby, as is the case for chlorophylls in photosynthetic systems, the excited electron can move from the initial molecule to the acceptor (**Figure 19.8**). A positive charge forms on the initial molecule, owing to the loss of an electron, and a negative charge forms on the acceptor, owing to the gain of an electron. Hence, this process is referred to as **photoinduced charge separation**.

In chloroplasts, the arrangement of the photosynthetic apparatus maximizes photoinduced charge separation and minimizes an unproductive return of the electron to its ground state. The electron, extracted from its initial site by the absorption of light, now has reducing power: it can reduce other molecules to store the energy originally obtained from light in chemical forms.

FIGURE 19.7 The absorption of light leads to the excitation of an electron from its ground state to a higher energy level.

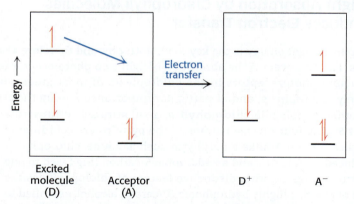

FIGURE 19.8 The key to photosynthesis is photoinduced charge separation. If a suitable electron acceptor is nearby, an electron that has been moved to a high energy level by light absorption can move from the excited molecule to the acceptor.

A "special pair" of chlorophylls initiate charge separation

The specific site at which photoinduced charge separation occurs is called the **reaction center**. The structure of the reaction center from anoxygenic photosynthetic bacteria—such as the nonsulfur purple bacterium *Rhodopseudomonas viridis*—was the first one determined. This reaction center consists of four

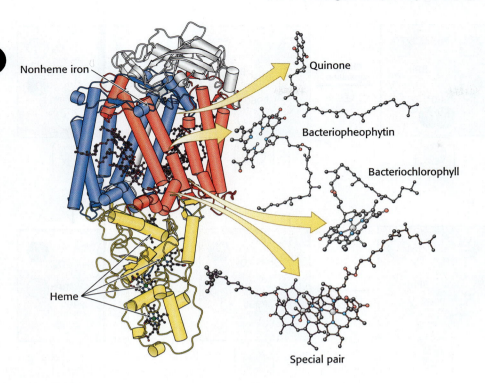

FIGURE 19.9 The photosynthetic reaction center in a nonsulfur purple bacterium is a large complex with many electron-carrying groups and one special pair. The core of the reaction center from *Rhodopseudomonas viridis* consists of two similar chains: L (shown in red) and M (blue). An H chain (white) and a cytochrome subunit (yellow) complete the structure. Notice that the L and M subunits are composed largely of α helices that span the membrane. Also notice that a chain of electron-carrying prosthetic groups, beginning with a special pair of bacteriochlorophylls and ending at a bound quinone, runs through the structure from bottom to top in this view. [Drawn from 1PRC.pdb.]

INTERACT with this model in
≈ Achieve

**Bacteriochlorophyll *b*
(BChl-*b*)**

**Bacteriopheophytin
(BPh)**

polypeptides, abbreviated L, M, H, and C (**Figure 19.9**). Sequence comparisons and subsequent structural studies have confirmed that this bacterial reaction center is homologous to the more complex cyanobacterial and plant systems. Thus, many of our observations of the purple bacterial system also apply these other systems.

The L and M subunits form the structural and functional core of the purple bacterial photosynthetic reaction center. Each of these homologous subunits contains five transmembrane helices, in contrast with the H subunit, which has just one. The H subunit lies on the cytoplasmic side of the cell membrane, and the cytochrome subunit lies on the exterior face of the cell membrane, called the *periplasmic side* because it faces the periplasm—the space between the cell membrane and the cell wall. Four bacteriochlorophyll *b* (BChl-*b*) molecules, two bacteriopheophytin *b* (BPh) molecules, two quinones (Q_A and Q_B), and a ferrous ion are associated with the L and M subunits.

Bacteriochlorophylls are photoreceptors similar to chlorophylls, except for the reduction of an additional pyrrole ring and other minor differences that shift their absorption maxima to the near infrared, to wavelengths as long as 1000 nm. *Bacteriopheophytin* is the term for a bacteriochlorophyll that has two protons instead of a magnesium ion at its center.

The reaction begins with light absorption by a pair of BChl-*b* molecules that lie near the periplasmic side of the membrane in the L–M dimer. The pair of BChl-*b* molecules is called the **special pair** because of its fundamental role in photosynthesis. The special pair absorbs light maximally at 960 nm and, for this reason, is often called P960 (P stands for pigment). After absorbing light, the excited special pair ejects an electron, which is transferred through another BChl-*b* to the bacteriopheophytin in the L chain (**Figure 19.10**, steps 1 and 2). This initial charge separation yields a positive charge on the special pair (P960$^+$) and a negative charge on BPh (BPh$^-$). The electron ejection and transfer take place in less than 10 picoseconds (10^{-12} s), again illustrating the incredibly short time-frames of molecular interactions at the biochemical timescale.

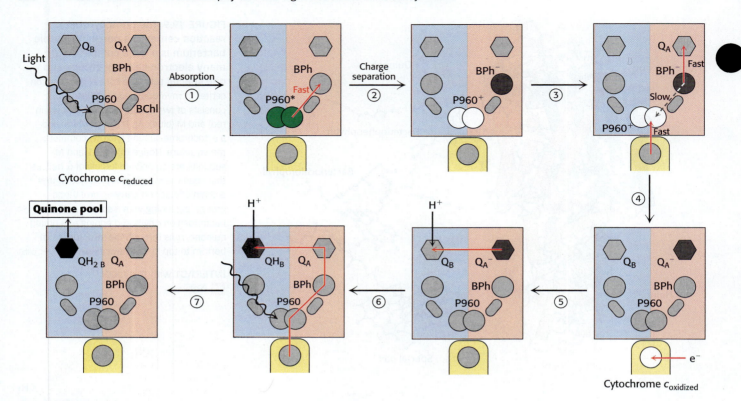

FIGURE 19.10 In the reaction center of nonsulfur purple bacteria, electrons flow from the special pair to a quinone. The absorption of light by the special pair (P960) results in the rapid transfer of an electron from this site to a bacteriopheophytin (BPh), creating a photoinduced charge separation (steps 1 and 2). (The asterisk on P960 stands for excited state.) The possible return of the electron from the pheophytin to the oxidized special pair is suppressed by the "hole" in the special pair being refilled with an electron from the cytochrome subunit and the electron from the pheophytin being transferred to a quinone (Q_A) that is farther away from the special pair (steps 3 and 4). Q_A passes the electron to Q_B. The reduction of a quinone (Q_B) on the cytoplasmic side of the membrane results in the uptake of two protons from the cytoplasm (steps 5 and 6). The reduced quinone can move into the quinone pool in the membrane (step 7).

A proton gradient across the membrane is established

A nearby electron acceptor, a tightly bound quinone (Q_A), quickly grabs the electron away from BPh^- before the electron has a chance to fall back to the P960 special pair. From Q_A, the electron moves to a more loosely associated quinone, Q_B. The absorption of a second photon and the movement of a second electron from the special pair through the bacteriopheophytin to the quinones completes the two-electron reduction of Q_B from Q to QH_2. Because the Q_B-binding site lies near the cytoplasmic side of the membrane, two protons are taken up from the cytoplasm, contributing to the development of a proton gradient across the cell membrane (Figure 19.10, steps 5, 6, and 7).

In their high-energy states, $P960^+$ and BPh^- could undergo charge recombination; that is, the electron on BPh^- could move back to neutralize the positive charge on the special pair. Its return to the special pair would waste a valuable high-energy electron and simply convert the absorbed light energy into heat. How is charge recombination prevented? Two factors in the structure of the reaction center work together to suppress charge recombination nearly completely (Figure 19.10, steps 3 and 4). First, the next electron acceptor (Q_A) is less than 10 Å away from BPh^-, and so the electron is rapidly transferred farther away from the special pair. Second, one of the hemes of the cytochrome subunit is less than 10 Å away from the special pair, and so the positive charge on P960 is neutralized by the transfer of an electron from the reduced cytochrome.

Cyclic electron flow reduces the cytochrome of the reaction center

The cytochrome subunit of the reaction center must regain an electron to complete the cycle. It does so by taking back two electrons from reduced quinone (QH_2). QH_2 first enters the Q pool in the membrane where it is reoxidized to Q by complex bc_1, which is homologous to Complex III of the respiratory electron-transport chain. Complex bc_1 transfers the electrons from QH_2 to cytochrome c_2, a water-soluble protein in the periplasm, and in the process pumps protons into the periplasmic space. The electrons now on cytochrome c_2 flow to the cytochrome subunit of the reaction center. The flow of electrons is thus cyclic in these bacteria, and oxygen is not evolved (**Figure 19.11**). The proton gradient generated in the course of this cycle drives the synthesis of ATP through the action of ATP synthase.

❖ **SELF-CHECK QUESTION**

What is meant by a "special pair" in the context of photosynthesis, and what important role does it play?

19.4 Two Photosystems Generate a Proton Gradient and Reducing Power in Cyanobacteria and Photosynthetic Eukaryotes

Photosynthesis is more complicated in eukaryotes and cyanobacteria than in the purple photosynthetic bacteria discussed in the previous section. In these other organisms including all green plants, photosynthesis depends on the interplay of two kinds of membrane-bound, light-sensitive complexes—photosystem I (PS I) and photosystem II (PS II), as shown in **Figure 19.12**. Note that photosystem I is so named because it was discovered first, but it is actually photosystem II that acts first in the sequence of events that take place during photosynthesis.

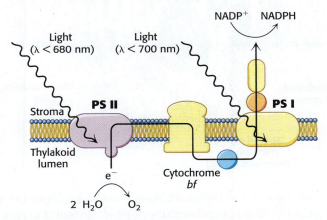

FIGURE 19.12 In green plants, electrons can flow through two photosystems. The absorption of photons by two distinct photosystems (PS II and PS I) is required for complete electron flow from water to $NADP^+$.

There are similarities in photosynthesis between green plants and purple photosynthetic bacteria. Both require light to energize reaction centers consisting of special pairs, and both transfer electrons by using electron-transport chains. However, in plants, electron flow can be cyclic, but it progresses linearly from photosystem II to photosystem I under most circumstances.

Photosystem II transfers electrons from water to plastoquinone and generates a proton gradient

Photosystem II, an enormous transmembrane assembly of more than 20 subunits, catalyzes the light-driven transfer of electrons from water to plastoquinone. This electron acceptor closely resembles ubiquinone, a component of the mitochondrial electron-transport chain. Plastoquinone cycles between an oxidized form (Q) and a reduced form (QH_2, plastoquinol). The overall reaction catalyzed by photosystem II is

$$2Q + 2H_2O \xrightarrow{\text{Light}} O_2 + 2QH_2$$

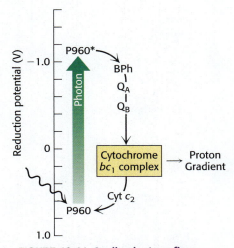

FIGURE 19.11 Cyclic electron flow in the bacterial reaction center allows the generation of ATP but not reducing power. Excited electrons from the P960 reaction center flow through bacteriopheophytin (BPh), a pair of quinone molecules (Q_A and Q_B), cytochrome bc_1 complex, and finally through cytochrome c_2 to the reaction center. The cytochrome bc_1 complex pumps protons as a result of electron flow, which powers the formation of ATP.

The electrons in QH_2 are at a higher redox potential than those in H_2O. Recall that, in oxidative phosphorylation, electrons flow downhill from ubiquinol to an acceptor, O_2, which is at a *lower* potential. In contrast, photosystem II drives the reaction in a thermodynamically uphill direction by using the energy of light.

Photosystem II is comparable to the purple bacterial reaction center

This light-driven transfer of elections is similar to the transfer reactions catalyzed by the purple bacterial system in that a quinone is converted from its oxidized into its reduced form. The core of photosystem II is reasonably similar to the purple bacterial reaction center (**Figure 19.13**). The core of the photosystem is formed by D1 and D2, a pair of similar 32-kDa subunits spanning the thylakoid membrane that are homologous to the L and M chains of the bacterial reaction center. Unlike the purple bacterial system, photosystem II contains a large number of additional subunits that bind more than 30 chlorophyll molecules altogether and increase the efficiency with which light energy is absorbed and transferred to the reaction center.

FIGURE 19.13 Photosystem II is a large membrane protein complex with many chlorophyll molecules, one special pair, and the water-oxidizing complex. The D_1 and D_2 subunits and the numerous bound chlorophyll molecules (green) are shown. Notice that the special pair and the water-oxidizing complex lie toward the thylakoid-lumen side of the membrane. [Drawn from 1S5L.pdb.]

INTERACT with this model in
 Achieve

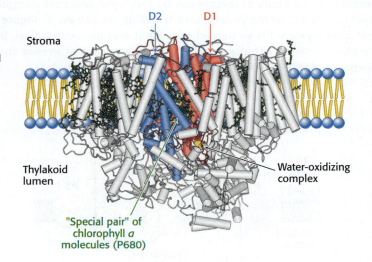

The electron flow in the thylakoid membrane is analogous to that in the bacterial membrane. The photochemistry of photosystem II begins with excitation of a special pair of chlorophyll molecules that are bound by the D1 and D2 subunits (**Figure 19.14**). Because the chlorophyll *a* molecules of the special pair absorb light at 680 nm, the special pair is often called *P680*. On excitation, P680 rapidly transfers an electron to a nearby pheophytin. From there, the electron is transferred first to a tightly bound plastoquinone at site Q_A and then to a mobile plastoquinone at site Q_B. With the arrival of a second electron and the uptake of two protons, the mobile plastoquinone is reduced to QH_2. At this point, the energy of two photons has been safely and efficiently stored in the reducing potential of QH_2.

Plastoquinone
(oxidized form, Q)

Plastoquinol
(reduced form, QH_2)

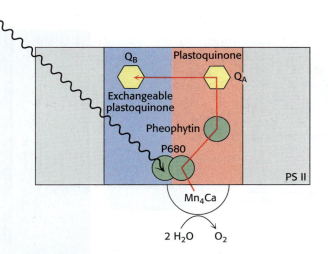

FIGURE 19.14 In photosystem II, electrons flow from manganese atoms to plastoquinone. Light absorption induces electron transfer from P680 down an electron-transfer pathway to an exchangeable plastoquinone. The positive charge on P680 is neutralized by electron flow from water molecules bound at the manganese center. Gray represents the rest of the photosystem beyond the reaction center core.

However, the source of electrons is different from that of the bacterial reaction center. The major difference between the bacterial system and photosystem II is the source of the electrons that are used to neutralize the positive charge formed on the special pair. P680+, a very strong oxidant, extracts electrons from water molecules bound at the **water-oxidizing complex (WOC)**, also called the **manganese center**. The core of this complex includes four manganese ions, one calcium ion, five oxide (O^{2-}) ions, and four water molecules (**Figure 19.15**). Manganese is well suited for a role in this process because of its ability to exist in multiple oxidation states and to form strong bonds with oxygen-containing species.

Each time the absorption of a photon powers the removal of an electron to form P680+, the positively charged special pair extracts an electron from a tyrosine residue that lies between P680 and the WOC to form a tyrosine radical. The tyrosine radical then removes an electron from the manganese center. Repeats of this process produce a center that is capable of carrying out the remarkable water oxidation reaction that generates oxygen.

The ability to control light pulses in the laboratory provides a powerful probe of the oxygen-generation process (**Figure 19.16**). Starting with chloroplasts that have remained in the dark for some time, an initial short flash of light produces very little oxygen. Application of a second flash produces the same result. However, with a third flash, a substantial amount of oxygen is produced. Additional flashes reveal that large amounts of oxygen are produced with every fourth flash thereafter.

Scientists have combined these observations with the results from other experiments to propose a mechanism for the oxygen-generating process (**Figure 19.17**). The formation of the oxygen–oxygen bond is proposed to involve an attack of a hydroxide ion bound to the calcium ion on a highly electrophilic oxo group bound to a manganese ion. The formation of the oxygen–oxygen bond requires the release of four electrons. An essential feature of the manganese center is its ability to develop and maintain a sufficiently oxidized state to take up these four electrons. The rates for the steps in this cycle have also been measured, revealing that the process that generates the oxygen powering so much of the life on our planet takes place in less than two milliseconds.

In summary, all oxygenic phototrophs—the most common type of photosynthetic organism—use the same inorganic core and protein components for capturing the energy of sunlight. A single solution to the biochemical problem of extracting electrons from water evolved billions of years ago, and has been conserved for use under a wide variety of phylogenetic and ecological circumstances.

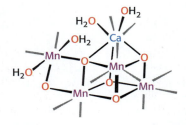

FIGURE 19.15 The core of the water-oxidizing complex includes four manganese ions and one calcium ion. The center is oxidized, one electron at a time, until two H_2O molecules are oxidized to form a molecule of O_2, which is then released from the complex.

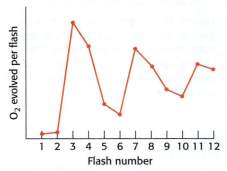

FIGURE 19.16 Four photons are required to generate one oxygen molecule after the first cycle. When dark-adapted chloroplasts are exposed to a brief flash of light, one electron passes through photosystem II. Monitoring the O_2 released after each flash reveals that O_2 is released after the 3rd, 7th, and 11th flashes.

(A)

FIGURE 19.17 The model for the water-oxidizing complex explains the light-flash experiments. (A) The absorption of each photon by the reaction center generates a tyrosine radical that extracts one electron at a time from the manganese ions. The structures are designated S_0 to S_4, indicating the number of electrons removed. The model additionally explains why the first O_2 release occurs after three flashes: the dark-adapted chloroplasts start in the S_1 state—that is, the one-electron reduced state. (B) Evolution of oxygen is evident by the generation of bubbles in the aquatic plant *Elodea*.

Similar to those of the purple bacterial reaction center, the reactions of photosystem II establish a proton gradient. Photosystem II spans the thylakoid membrane such that the site of quinone reduction is on the side of the stroma, whereas WOC lies in the thylakoid lumen. Thus, the two protons that are taken up with the reduction of Q to QH_2 come from the stroma, and the four protons that are liberated in the course of water oxidation are released into the lumen. This distribution of protons generates a proton gradient across the thylakoid membrane characterized by an excess of protons in the thylakoid lumen compared with the stroma (**Figure 19.18**).

Cytochrome *bf* links photosystem II to photosystem I

Electrons flow from photosystem II to photosystem I through the **cytochrome *bf*** complex. This complex catalyzes the transfer of electrons from plastoquinol (QH_2) to plastocyanin (Pc), a small, soluble copper protein in the thylakoid lumen.

$$QH_2 + 2\,Pc(Cu^{2+}) \rightarrow Q + 2\,Pc(Cu^{+}) + 2\,H^{+}_{\text{thylakoid lumen}}$$

The two protons from plastoquinol are released into the thylakoid lumen. This reaction is reminiscent of that catalyzed by Complex III in oxidative phosphorylation, and most components of the cytochrome *bf* complex are homologous to those of Complex III.

This complex catalyzes the reaction by proceeding through the Q cycle (Figure 18.11). In the first half of the Q cycle, plastoquinol (QH_2) is oxidized to plastoquinone (Q), one electron at a time. The electrons from plastoquinol flow through the cytochrome *bf* complex to convert oxidized plastocyanin (Pc) into its reduced form (**Figure 19.19**).

In the second half of the Q cycle, cytochrome *bf* reduces a molecule of plastoquinone from the Q pool to plastoquinol, taking up two protons from one side of the membrane, and then reoxidizes plastoquinol to release these protons on the other side. The enzyme is oriented so that protons are released into the thylakoid lumen and taken up from the stroma, contributing further to the proton gradient across the thylakoid membrane (Figure 19.19).

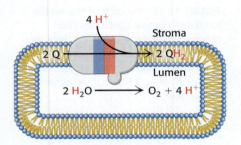

FIGURE 19.18 Photosystem II releases protons into the thylakoid lumen and takes them up from the stroma. The result is a pH gradient across the thylakoid membrane with an excess of protons (low pH) inside.

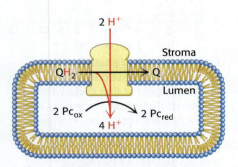

FIGURE 19.19 Electron flow through cytochrome *bf* contributes to the proton gradient. The cytochrome *bf* complex oxidizes QH_2 to Q through the Q cycle. Four protons are released into the thylakoid lumen in each cycle.

Photosystem I uses light energy to generate reduced ferredoxin, a powerful reductant

The final stage of the light reactions is catalyzed by photosystem I, a transmembrane complex consisting of about 15 polypeptide chains and multiple associated proteins and cofactors (**Figure 19.20**). The core of this system is a pair of similar subunits—psaA and psaB—which bind 80 chlorophyll molecules as well as other redox factors. These subunits are quite a bit larger than the core subunits of photosystem II and the purple bacterial reaction center. Nonetheless, they appear to be homologous; the terminal 40% of each subunit is similar to a corresponding subunit of photosystem II.

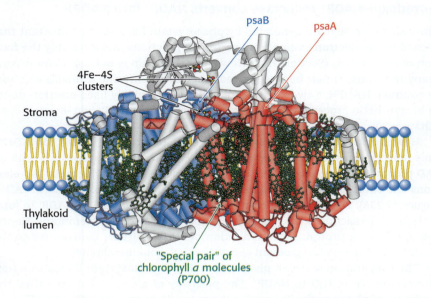

FIGURE 19.20 Photosystem I is a protein complex with numerous chlorophyll molecules and one special pair. Notice the numerous bound chlorophyll molecules, shown in green, including the special pair, as well as the iron–sulfur clusters that facilitate electron transfer from the stroma. [Drawn from 1JB0.pdb.]

INTERACT with this model in

Achieve

A special pair of chlorophyll *a* molecules lies at the center of photosystem I, where they absorb light maximally at 700 nm. This center, called *P700*, initiates photoinduced charge separation (**Figure 19.21**). The electron travels from P700 down a pathway through chlorophyll at site A_0 and quinone at site A_1 to a set of 4Fe–4S clusters. The next step is the transfer of the electron to ferredoxin (Fd), a soluble protein containing a 2Fe–2S cluster coordinated to four cysteine residues (**Figure 19.22**). Ferredoxin transfers electrons to $NADP^+$. Meanwhile, $P700^+$ captures

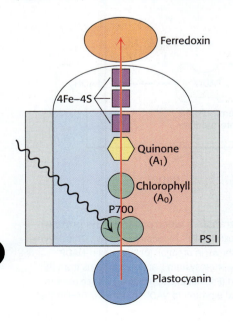

FIGURE 19.21 In photosystem I, electrons flow to ferredoxin. Light absorption induces electron transfer from P700 down an electron-transfer pathway that includes a chlorophyll molecule, a quinone molecule, and three 4Fe–4S clusters to reach ferredoxin. The positive charge left on P700 is neutralized by electron transfer from reduced plastocyanin. Gray represents the rest of the photosystem beyond the reaction center core.

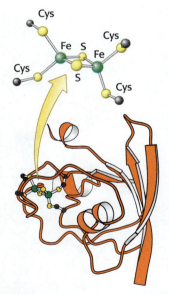

FIGURE 19.22 Ferredoxin is an electron-carrying, iron-sulfur protein. In plants, ferredoxin contains a 2Fe–2S cluster. This protein accepts electrons from photosystem I and carries them to ferredoxin–NADP⁺ reductase. [Drawn from 1FXA.pdb.]

INTERACT with this model in

 Achieve

an electron from reduced plastocyanin provided by photosystem II to return to P700 so that P700 can be excited again. Thus, the overall reaction catalyzed by photosystem I is a simple one-electron oxidation–reduction reaction.

$$Pc(Cu^+) + Fd_{ox} \xrightarrow{\text{Light}} Pc(Cu^{2+}) + Fd_{red}$$

Given that the reduction potentials for plastocyanin and ferredoxin are +0.37 V and –0.45 V, respectively, the standard free energy for this reaction is +79.1 kJ mol^{-1} (+18.9 kcal mol^{-1}). This uphill reaction is driven by the absorption of a 700-nm photon, which has an energy of 171 kJ mol^{-1} (40.9 kcal mol^{-1}).

Ferredoxin–NADP$^+$ reductase converts NADP$^+$ into NADPH

The reduced ferredoxin generated by photosystem I is a strong reductant that is used as an electron source for a variety of reactions, most notably the fixation of N$_2$ into NH$_3$ (Section 25.1). However, ferredoxin is not useful for driving many reactions, in part because ferredoxin carries only one available electron. In contrast, NADPH, a two-electron reductant, is a widely used electron donor in biosynthetic processes, including the reactions of the Calvin–Benson cycle (Chapter 20).

How is reduced ferredoxin used to drive the reduction of NADP$^+$ to NADPH? This reaction is catalyzed by *ferredoxin–NADP$^+$ reductase,* a flavoprotein with an FAD prosthetic group (**Figure 19.23A**). The bound FAD moiety accepts two electrons and two protons from two molecules of reduced ferredoxin to form FADH$_2$ (**Figure 19.23B**). The enzyme then transfers a hydride ion (H$^-$) to NADP$^+$ to form NADPH. This reaction takes place on the stromal side of the membrane. Hence, the uptake of a proton in the reduction of NADP$^+$ further contributes to the generation of the proton gradient across the thylakoid membrane.

The cooperation between photosystem I and photosystem II creates a flow of electrons from H$_2$O to NADP$^+$. The pathway of electron flow is called the **Z scheme of photosynthesis** because the redox diagram from P680 to P700* looks like a letter Z on its side (**Figure 19.24**).

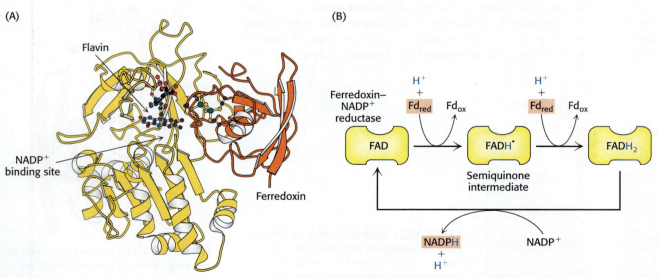

(A)

INTERACT with this model in
✿ Achie√e

(B)

FIGURE 19.23 Ferredoxin–NADP$^+$ reductase transfers electrons from the protein ferredoxin to NADP$^+$. (A) The structure of ferredoxin–NADP$^+$ reductase is shown. This enzyme accepts electrons, one at a time, from ferredoxin (shown in orange). (B) Ferredoxin–NADP$^+$ reductase first accepts two electrons, generated by photosystem I, and two protons from the lumen to form two molecules of reduced ferredoxin (Fd). Reduced ferredoxin is then used to form FADH$_2$, which then transfers two electrons and a proton to NADP$^+$ to form NADPH in the stroma. [Drawn from 1EWY.pdb.]

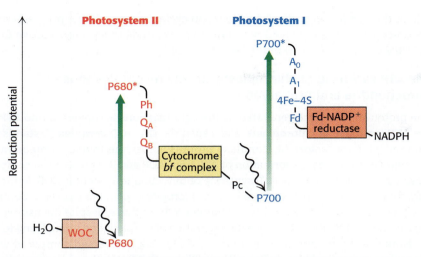

FIGURE 19.24 The "Z" scheme of photosynthesis is the linear path of electrons from H_2O to $NADP^+$. This endergonic reaction is made possible by the absorption of light by photosystem II (P680) and photosystem I (P700). Abbreviations: Ph, pheophytin; Q_A and Q_B, plastoquinone-binding proteins; Pc, plastocyanin; A_0 and A_1, acceptors of electrons from P700*; Fd, ferredoxin; WOC, water-oxidizing complex.

❖ SELF–CHECK QUESTION

PS I creates a powerful reductant, while PS II creates a powerful oxidant. Identify each and describe their roles.

19.5 A Proton Gradient across the Thylakoid Membrane Drives ATP Synthesis

In 1966, André Jagendorf showed that chloroplasts synthesize ATP in the dark when an artificial pH gradient is imposed across the thylakoid membrane. He soaked chloroplasts in a pH 4.0 buffer and then rapidly mixed them with a pH 8.0 buffer containing ADP and P_i. While the pH of the thylakoid lumen remained at 4.0, the pH of the stroma suddenly increased to 8.0, resulting in a burst of ATP synthesis that accompanied the disappearance of the pH gradient. This incisive experiment was one of the first to unequivocally support the hypothesis put forth by Peter Mitchell that ATP synthesis is driven by proton-motive force (Section 18.4).

The principles of ATP synthesis in chloroplasts are nearly identical to those in mitochondria. ATP formation is driven by a proton-motive force in both photophosphorylation and oxidative phosphorylation. We have seen how light induces electron transfer through photosystems II and I and the cytochrome *bf* complex. At various stages in this process, protons are released into the thylakoid lumen or taken up from the stroma, generating a proton gradient.

As discussed in Section 18.4, energy inherent in the proton gradient, called the *proton-motive force* (Δp), is described as the sum of two components: a charge gradient and a chemical gradient. The gradient is maintained because the thylakoid membrane is essentially impermeable to protons. The thylakoid lumen becomes markedly acidic, with the pH approaching 4.0, whereas the stroma reaches a pH around 8.0. The light-induced transmembrane proton gradient is about 3.5 pH units. In chloroplasts, nearly all of Δp arises from the pH gradient, whereas in mitochondria, the contribution from the membrane potential is larger. The reason for this difference is that the thylakoid membrane is quite permeable to Cl^- and Mg^{2+}. The light-induced transfer of H^+ into the thylakoid space is accompanied by the transfer of either Cl^- in the same direction or Mg^{2+} (1 Mg^{2+} per 2 H^+) in the opposite direction. Consequently, electrical neutrality is maintained and no membrane potential is generated. The influx of Mg^{2+} into the stroma plays a

PETER MITCHELL AND ANDRÉ JAGENDORF
The chemiosmotic hypothesis proposed by Peter Mitchell (shown at top) was initially poorly received by other scientists. But this changed dramatically when André Jagendorf (shown below Mitchell) demonstrated that chloroplasts can synthesize ATP in the dark if a pH gradient is created. Jagendorf credits a third scientist, Geoffrey Hind, for helping him make the connection. Jagendorf said, "I had heard Peter Mitchell talk about chemiosmosis. . . . His words went into one of my ears and out the other, leaving me feeling annoyed [that the conference organizers] had allowed such a ridiculous and incomprehensible speaker in. But Geoffrey [Hind] read *Nature*. He read Mitchell's paper and came to me. At this point I began to communicate with Peter [and] later that summer I did the experiment . . ." Jagendorf's results only made sense through the lens of Mitchell's chemiosmotic hypothesis, which eventually won Mitchell the Nobel Prize in Chemistry. Their mutually beneficial relationship illustrates the importance of open communication and collaboration in science.

role in the regulation of the Calvin–Benson cycle (Section 20.2). A pH gradient of 3.5 units across the thylakoid membrane corresponds to a proton-motive force of 0.20 V or a ΔG of -20.0 kJ mol^{-1} (-4.8 kcal mol^{-1}).

The ATP synthase of chloroplasts closely resembles those of mitochondria and prokaryotes

The proton-motive force generated by the light reactions is converted into ATP by the **ATP synthase of chloroplasts**, also called the **CF$_1$ – CF$_0$ complex** (C stands for chloroplast; F for factor). CF$_0$ conducts protons across the thylakoid membrane, whereas CF$_1$ catalyzes the formation of ATP from ADP and P$_i$. The CF$_1$ – CF$_0$ complex closely resembles the F$_1$ – F$_0$ ATP synthase of mitochondria (Section 18.4). Remarkably, the β subunits of ATP synthase in corn chloroplasts are more than 60% identical in amino acid sequence with those of human ATP synthase, despite the passage of approximately 1 billion years since the separation of the plant and animal kingdoms.

Note that the membrane orientation of CF$_1$ – CF$_0$ is reversed compared with that of the mitochondrial ATP synthase (**Figure 19.25**). However, the functional orientation of the two synthases is identical: protons flow from the lumen through the enzyme to either the thylakoid stroma or the mitochondrial matrix, where ATP is synthesized. Because CF$_1$ is on the stromal surface of the thylakoid membrane, the newly synthesized ATP is released directly into the stromal space. Likewise, NADPH formed by photosystem I is released into the stromal space. Thus, ATP and NADPH, the products of the light reactions of photosynthesis, are appropriately positioned for the subsequent dark reactions in which CO$_2$ is converted into carbohydrate (Section 20.1).

FIGURE 19.25 Photosynthesis and oxidative phosphorylation have a similar layout but differ in a number of ways. (A) The light-induced electron transfer in photosynthesis drives protons into the thylakoid lumen. The excess protons flow out of the lumen through ATP synthase to generate ATP in the stroma. (B) In oxidative phosphorylation, electron flow down the electron-transport chain pumps protons out of the mitochondrial matrix. Excess protons from the intermembrane space flow into the matrix through ATP synthase to generate ATP in the matrix.

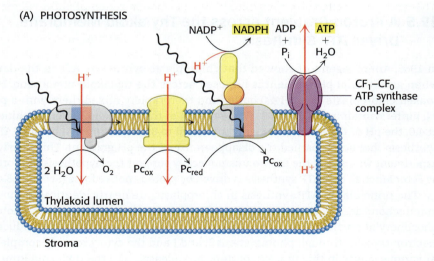

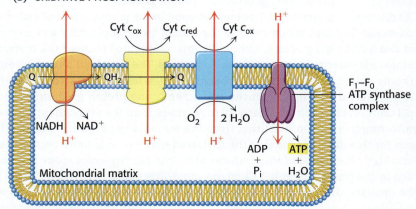

The activity of chloroplast ATP synthase is regulated

The activity of the ATP synthase is sensitive to the redox conditions in the chloroplast. For maximal activity, a specific disulfide bond in the γ subunit must be reduced to two cysteines. The reductant is reduced thioredoxin, which is formed from ferredoxin generated in photosystem I, by ferredoxin–thioredoxin reductase, an iron–sulfur containing enzyme.

$$2 \text{ Reduced ferredoxin} + \text{thioredoxin disulfide} \xrightleftharpoons{\text{Ferredoxin-thioredoxin reductase}}$$

$$2 \text{ oxidized ferredoxin} + \text{reduced thioredoxin} + 2 \text{ H}^+$$

Conformational changes in the ε subunit also contribute to synthase regulation. The ε subunit appears to exist in two conformations. One conformation inhibits ATP hydrolysis by the synthase, while the other, which is generated by an increase in the proton-motive force, allows ATP synthesis and facilitates the reduction of the disulfide bond in the γ subunit. Thus, synthase activity is maximal when biosynthetic reducing power and a proton gradient are available. We will see in Chapter 20 that redox regulation is also important in photosynthetic carbon metabolism.

Cyclic electron flow through photosystem I leads to the production of ATP instead of NADPH

On occasion, when the ratio of NADPH to NADP$^+$ is very high, as might be the case if there was another source of electrons to form NADPH (Section 20.3), NADP$^+$ may be unavailable to accept electrons from reduced ferredoxin. In these circumstances, specific large protein complexes allow cyclic electron flow that powers ATP synthesis. Excited electrons from P700, the reaction center of photosystem I, generate reduced ferredoxin. The electron in reduced ferredoxin is transferred to the cytochrome *bf* complex rather than to NADP$^+$. This electron then flows back through the cytochrome *bf* complex to reduce plastocyanin, which can then be reoxidized by P700$^+$ to complete a cycle. The net outcome of this cyclic flow of electrons is the pumping of protons by the cytochrome *bf* complex. The resulting proton gradient then drives the synthesis of ATP. In this process, called **cyclic photophosphorylation**, ATP is generated without the concomitant formation of NADPH (**Figure 19.26**), reminiscent of the cyclic phototrophic scheme of the nonsulfur purple bacteria (Section 19.3). Note that because photosystem II does not participate in cyclic photophosphorylation, O$_2$ is not formed from H$_2$O.

FIGURE 19.26 Cyclic photophosphorylation allows ATP production without a need for an electron source. (A) In this pathway, electrons from reduced ferredoxin are transferred to cytochrome *bf* rather than to ferredoxin–NADP$^+$ reductase, and thus can be reused indefinitely without the need for an external electron source. The flow of electrons through cytochrome *bf* pumps protons into the thylakoid lumen. These protons flow through ATP synthase to generate ATP. Neither NADPH nor O$_2$ is generated by this pathway. Abbreviations: Fd, ferredoxin; Pc, plastocyanin. (B) A scheme showing the energetic basis for cyclic photophosphorylation.

(A)

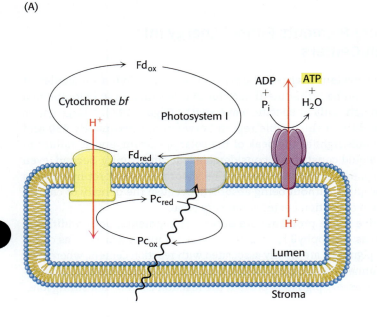

(B)

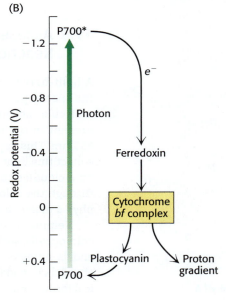

The absorption of eight photons yields one O_2, two NADPH, and three ATP molecules

We can now estimate the overall stoichiometry for the light reactions. The absorption of four photons by photosystem II generates one molecule of O_2 and releases four protons into the thylakoid lumen. The two molecules of plastoquinol are oxidized by the Q cycle of the cytochrome *bf* complex to release eight protons into the lumen. Finally, the electrons from four molecules of reduced plastocyanin are driven to ferredoxin by the absorption of four additional photons. The four molecules of reduced ferredoxin generate two molecules of NADPH. Thus, the overall reaction is

$$2\ H_2O + 2\ NADP^+ + 10\ H^+_{stroma} \rightarrow O_2 + 2\ NADPH + 12\ H^+_{lumen}$$

The 12 protons released in the lumen can then flow through ATP synthase. Let's assume that 12 protons must pass through CF_0 to complete one full rotation of CF_1. A single rotation generates three molecules of ATP. Given the ratio of 3 ATP for 12 protons, the overall reaction is

$$2\ H_2O + 2\ NADP^+ + 10\ H^+_{stroma} \longrightarrow O_2 + 2\ NADPH + 12\ H^+_{lumen}$$
$$\underline{3\ ADP^{3-} + 3\ P_i^{2+} + 3\ H^+ + 12\ H^+_{lumen} \longrightarrow 3\ ATP^{4-} + 3\ H_2O + 12\ H^+_{stroma}}$$
$$2\ NADP^+ + 3\ ADP^{3-} + 3\ P_i^{2-} + H^+ \longrightarrow O_2 + 2\ NADPH + 3\ ATP^{4-} + H_2O$$

Thus, eight photons are required to yield three molecules of ATP (2.7 photons/ATP).

Cyclic photophosphorylation is a somewhat more productive way to synthesize ATP than noncyclic photophosphorylation (the Z scheme shown in Figure 19.24). The absorption of four photons by photosystem I leads to the release of eight protons into the lumen by the cytochrome *bf* system. These protons flow through ATP synthase to yield two molecules of ATP. Thus, each two absorbed photons yield one molecule of ATP. No NADPH is produced.

❖ SELF–CHECK QUESTION

The relative proportions of the components of the proton-motive force that drive ATP synthesis in chloroplasts are different from those in mitochondria. What are these two components, and which dominates in chloroplasts? Given the similarities between the synthases in both organelles, explain the basis of this difference.

19.6 Accessory Pigments Funnel Energy into Reaction Centers

A light-harvesting system that relied only on the chlorophyll *a* molecules of the special pair would be rather inefficient for two reasons. First, chlorophyll *a* molecules absorb light only at specific wavelengths (Figure 19.6). A large gap is present in the middle of the visible region between approximately 450 and 650 nm. This gap falls right at the peak of the solar spectrum, and so failure to collect this light would constitute a considerable lost opportunity. Second, even on a cloudless day, many photons that can be absorbed by chlorophyll *a* pass through the chloroplast without being absorbed, because the density of chlorophyll *a* molecules in a reaction center is not very great.

Helping to solve these problems are accessory pigments, both additional chlorophylls, such as chlorophyll *b*, and other classes of molecules, such as carotenoids. Accessory pigments are closely associated with reaction centers, where they absorb light and funnel the energy to the reaction center for conversion into chemical forms. Thus, accessory pigments prevent the reaction center from sitting idle.

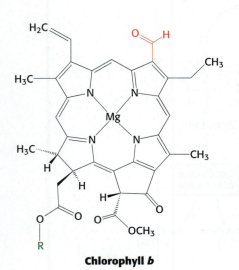

Chlorophyll *b*

Chlorophyll *b* and **carotenoids** are important accessory pigments that funnel energy to the reaction center. Chlorophyll *b* differs from chlorophyll *a* in having a formyl group in place of a methyl group. This small difference shifts its two major absorption peaks toward the center of the visible region. In particular, chlorophyll *b* efficiently absorbs light with wavelengths between 450 and 500 nm (**Figure 19.27**).

Carotenoids are extended polyenes that absorb light between 400 and 500 nm. The carotenoids are responsible for most of the yellow and red colors of fruits and flowers, and they provide the brilliance of fall, when the chlorophyll molecules degrade first, revealing the carotenoids. Tomatoes are especially rich in the carotenoid lycopene, which accounts for their red color, whereas carrots and pumpkins have abundant amounts of the carotenoid β-carotene, accounting for their orange color.

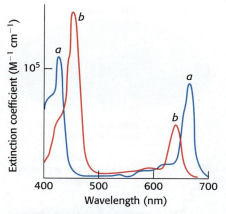

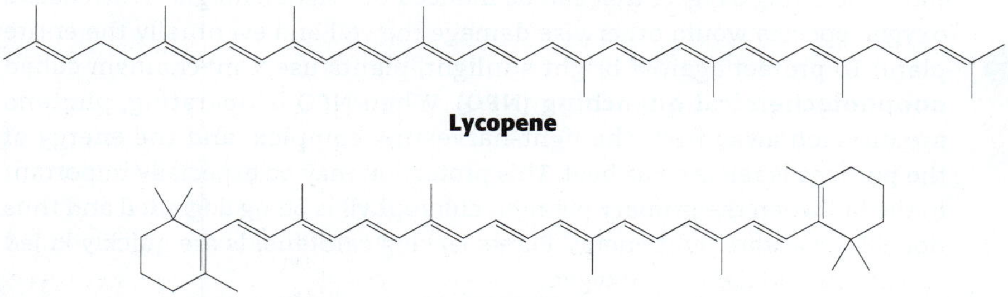

FIGURE 19.27 Chlorophylls *a* and *b* absorb light maximally at different wavelengths. Shown are the absorption spectra of chlorophylls *a* and *b*.

Lycopene

β-Carotene

Resonance energy transfer allows energy to move from the site of initial absorbance to the reaction center

How is energy funneled from accessory pigments to a reaction center? As discussed at the beginning of the chapter, the absorption of a photon does not always lead to electron excitation and transfer. More commonly, excitation energy is transferred from one molecule to a nearby molecule through electromagnetic interactions through space (**Figure 19.28**). The rate of this process, called *resonance energy transfer*, depends strongly on the distance between the energy-donor and the energy-acceptor molecules; an increase in the distance between the donor and the acceptor by a factor of 2 typically results in a decrease in the energy-transfer rate by a factor of $2^6 = 64$. For reasons of energy conservation, energy transfer must be from a donor in the excited state to an acceptor of equal or lower energy. The excited state of the special pair of chlorophyll molecules is lower in energy than that of single chlorophyll molecules, allowing reaction centers to trap the energy transferred from other molecules.

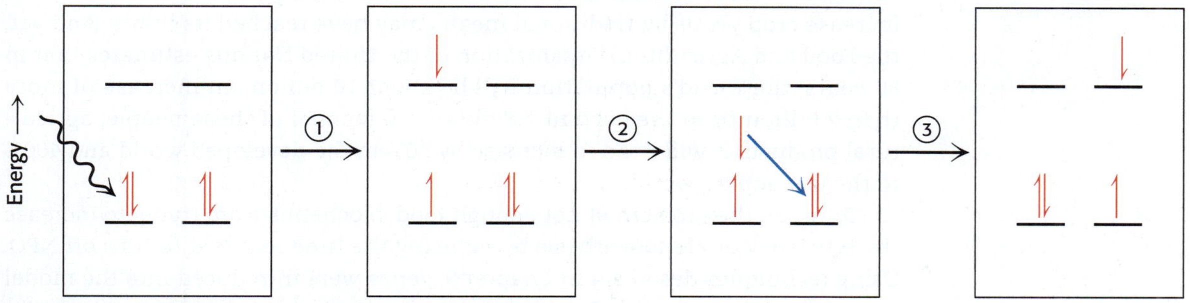

FIGURE 19.28 Resonance energy transfer allows electrons to transfer energy over a distance. (1) An electron can accept energy from electromagnetic radiation of appropriate wavelength and jump to a higher energy state. (2) When the excited electron falls back to its lower energy state, the absorbed energy is released. (3) The released energy can be absorbed by an electron in a nearby molecule, and this electron jumps to a high energy state.

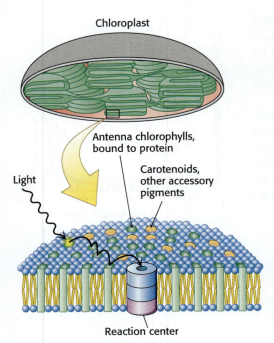

Chloroplast

Antenna chlorophylls, bound to protein

Carotenoids, other accessory pigments

Light

Reaction center

FIGURE 19.29 Energy is transferred from accessory pigments to reaction centers. Light energy absorbed by accessory chlorophyll molecules or other pigments can be transferred to reaction centers by resonance energy transfer (black arrows), where it drives photoinduced charge separation. [Source: D. L. Nelson and M. M. Cox, *Lehninger Principles of Biochemistry*, 5th ed. (W. H. Freeman and Company, 2008), Fig. 19.52.]

What is the structural relationship between the accessory pigment and the reaction center? The accessory pigments are arranged in numerous **light-harvesting complexes** that completely surround the reaction center (Figure 19.29). The 26-kDa subunit of light-harvesting complex II (LHC-II) is the most abundant membrane protein in chloroplasts. This subunit binds seven chlorophyll *a* molecules, six chlorophyll *b* molecules, and two carotenoid molecules. Similar light-harvesting assemblies exist in many photosynthetic bacteria.

Accessory pigments also protect plants from reactive oxygen

In addition to their role in transferring energy to reaction centers, the carotenoids and other accessory pigments serve a safeguarding function. Carotenoids suppress harmful photochemical reactions, particularly those including oxygen, that can be induced by bright sunlight. The reactive oxygen species would otherwise damage the cell and eventually the entire plant. To protect against bright sunlight, plants use a mechanism called **nonphotochemical quenching (NPQ)**. When NPQ is operating, photons are directed away from the light-harvesting complex, and the energy of the photons is released as heat. This protection may be especially important in the fall when the primary pigment chlorophyll is being degraded and thus not able to absorb light energy. Plants lacking carotenoids are quickly killed on exposure to light and oxygen.

Increasing the efficiency of photosynthesis will increase crop yields

NPQ is a means by which plants protect themselves from damaging reactive oxygen species that are generated by intense sunlight—essentially a botanical sunscreen. Understanding NPQ, a complex process consisting of several components, is an active area of research. It is established that NPQ requires a proton gradient across the thylakoid membrane and, in higher photosynthetic organisms such as plants, a particular protein, photosystem II subunit S (PsbS). PsbS, in the presence of a pH gradient, acts in seconds to help to initiate NPQ by interacting with protein components of the light-harvesting complex. This interaction redirects the energy absorbed by the light-harvesting complex from the reaction centers so that the energy is lost as heat. Although quick to be activated, NPQ is slow to turn off, depressing photosynthetic yields to be as much as 30% in some crops. This slow turn off is a target for biochemists hoping to increase the efficiency of photosynthesis and, consequently, crop yield.

Why is it necessary to boost photosynthesis? Traditional plant breeding has greatly increased crop yield over the past century. Additionally, modern farming practices and tools such as fertilizers, pesticides, and irrigation have contributed to increased yield. However, studies suggest that the ability to increase crop yields by traditional means may have reached its limits. And yet, the Food and Agricultural Organization of the United Nations estimates that in 30 years, the world's population will be about 10 billion, an increase of more than 2 billion from the current 7.8 billion. To feed all of these people, agricultural production will need to increase by 50% in the developed world and 400% in the developing world.

To tackle the problem of not enough food, biochemists are trying to increase the efficiency of photosynthesis by reducing the time required to turn off NPQ. Using techniques described in Chapter 9, genes were introduced into the model organism—the tobacco plant—that increased the speed of NPQ shutdown. The result was a 15% increase in plant size. Similar results have now been observed in rice, and the next steps are to modify other crops such as wheat and cowpea—important sources of protein in sub-Saharan Africa.

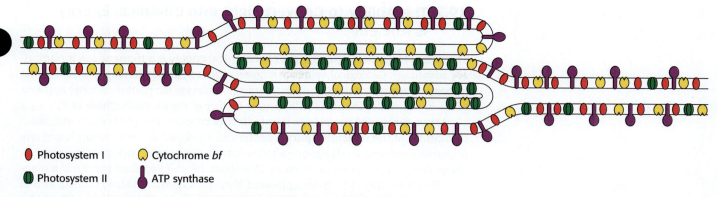

● Photosystem I   ○ Cytochrome *bf*

◉ Photosystem II   ♦ ATP synthase

FIGURE 19.30 Photosynthetic assemblies are differentially distributed in the stacked and unstacked regions of thylakoid membranes. [Information from Dr. Jan M. Anderson and Dr. Bertil Andersson.]

The components of photosynthesis are highly organized

The intricacy of photosynthesis, seen already in the elaborate interplay of complex components, extends even to the placement of the components in the thylakoid membranes. Thylakoid membranes of most plants are differentiated into stacked and unstacked regions (**Figure 19.30**). Stacking increases the amount of thylakoid membrane in a given chloroplast volume. Both regions surround a common internal thylakoid space, but only unstacked regions make direct contact with the chloroplast stroma.

Stacked and unstacked regions differ in the nature of their photosynthetic assemblies (Figure 19.30). Photosystem I and ATP synthase are located almost exclusively in unstacked regions, whereas photosystem II is present mostly in stacked regions. The cytochrome *bf* complex is found in both regions. A common internal thylakoid space enables protons liberated by photosystem II in stacked membranes to be used by ATP synthase molecules that are located far away in unstacked membranes. Plastoquinone and plastocyanin are the mobile carriers of electrons between assemblies located in different regions of the thylakoid membrane.

What is the functional significance of the structural variety in the thylakoid membrane system? The positioning of photosystem I in the unstacked membranes gives it direct access to the stroma for the reduction of $NADP^+$. ATP synthase is also located in the unstacked region, which provides the space needed for its large CF_1 globule and gives it access to ADP. In contrast, the tight quarters of the stacked region pose no problem for photosystem II, which interacts with a small polar electron donor (H_2O) and a highly lipid-soluble electron carrier (plastoquinone).

Many herbicides inhibit the light reactions of photosynthesis

Many commercial herbicides kill weeds by interfering with the action of photosystem II or photosystem I. Inhibitors of PS II block electron flow, whereas inhibitors of PS I divert electrons from its terminal part. PS II inhibitors include urea derivatives such as *diuron* and triazine derivatives such as *atrazine*. These chemicals bind to the Q_B site of the D1 subunit of photosystem II and block the formation of plastoquinol (QH_2). PS I inhibitors include Paraquat (1,1′-dimethyl-4-4′-bipyridinium), a dication, which can accept electrons from photosystem I to become a radical. This radical reacts with O_2 to produce reactive oxygen species such as superoxide (O_2^-) and hydroxyl radical (OH•). Such reactive oxygen species react with many biomolecules, including double bonds in membrane lipids, damaging the membrane.

Diuron

Atrazine

19.7 The Ability to Convert Light into Chemical Energy Is Ancient

The ability to convert light energy into chemical energy is a tremendous evolutionary advantage. Geological evidence suggests that oxygenic photosynthesis became important approximately 2 billion years ago. Anoxygenic photosynthetic systems arose much earlier in the 3.5-billion-year history of life on Earth (**Table 19.1**).

As we discussed at various points in this chapter, the photosynthetic reaction center of the nonsulfur purple bacterium *Rhodopseudomonas viridis* has many features common to oxygenic photosynthetic systems and clearly predates them. Green sulfur bacteria such as *Chlorobium thiosulfatophilum* carry out a reaction that also seems to have appeared before oxygenic photosynthesis and is even more similar to oxygenic photosynthesis than the photosystem of *R. viridis*. Reduced sulfur species such as H_2S are electron donors in the overall photosynthetic reaction

$$CO_2 + 2\,H_2S \xrightarrow{\text{Light}} (CH_2O) + 2\,S + H_2O$$

Nonetheless, photosynthesis did not evolve immediately at the origin of life. No photosynthetic organisms have been discovered in the domain Archaea, implying that photosynthesis evolved in the domain Bacteria after archaea and bacteria diverged from a common ancestor. All domains of life do have electron-transport chains in common, however. As we have seen, components such as the ubiquinone–cytochrome c oxidoreductase (Section 18.3) and cytochrome *bf* family (Section 19.4) are present in both respiratory and photosynthetic electron-transport chains. These components were the foundations on which light-energy-capturing systems evolved.

Artificial photosynthetic systems may provide clean, renewable energy

As we have learned, oxygenic photosynthetic organisms use sunlight to oxidize H_2O, producing O_2 and protons used to power ATP synthesis and generate NADPH. Research is currently underway to try to mimic this process in order to provide clean energy. The idea is that photovoltaic cells could use light energy to oxidize water, producing O_2 as well as H_2. Hydrogen gas is a fuel that, upon reaction with oxygen, releases energy and only water as a waste product. Recent work suggests that semiconductors composed of organic–inorganic material show great promise in this application.

Photovoltaic cells are now being paired with microorganisms to generate a variety of biomolecules, including fuels, in a clean, renewable fashion. One example of such a hybrid-biological-inorganic (HBI) system is shown in **Figure 19.31**). Sunlight is converted into an electrical current by the photovoltaic cell. The current flows to the anode, where water is oxidized and the resulting electrons are taken up by the cathode. Bacteria, such as *Cupriavidus necator* (also used in bioremediation because of its ability to degrade chlorinated hydrocarbons) live in direct contact with the cathode. The bacteria use the electrons generated by

Table 19.1 Major groups of photosynthetic prokaryotes

Bacteria	Photosynthetic electron donor	O_2 use
Green sulfur	H_2, H_2S, S	Anoxygenic
Green nonsulfur	Variety of amino acids and organic acids	Anoxygenic
Purple sulfur	H_2, H_2S, S	Anoxygenic
Purple nonsulfur	Usually organic molecules	Anoxygenic
Cyanobacteria	H_2O	Oxygenic

the oxidation of water to reduce CO_2 to industrially important biomolecules. The efficiency of such HBI systems for converting sunlight into chemical energy is nearly 10%, with the potential for improvement, surpassing the efficiency of natural photosynthesis, which has an efficiency of 2%.

Photosensitive proteins are transforming other fields

Phototrophic organisms use a range of light-sensitive proteins in addition to those discussed thus far. Scientists have taken advantage of these proteins to develop new research tools and even therapies. A key example is channelrhodopsin first characterized from green algae, where it plays a role in light sensing. Channelrhodopsin is homologous to bacteriorhodopsin but functions as an ion channel rather than a proton pump. Thus, exposure of cells with channelrhodopsin present in their cell membranes to light results in ion channel opening and the flow of ions down any gradients.

Scientists have expressed channelrhodopsin in neurons, making these neurons sensitive to light. The firing of these neurons can then be controlled by applying short flashes of light to initiate action potentials (**Figure 19.32**) for cells in culture but also in engineered experimental animals. These methods, termed *optogenetics*, facilitate exploration of the mechanisms of nervous system function with unprecedented precision, again demonstrating the power of biochemical discoveries to transform other fields.

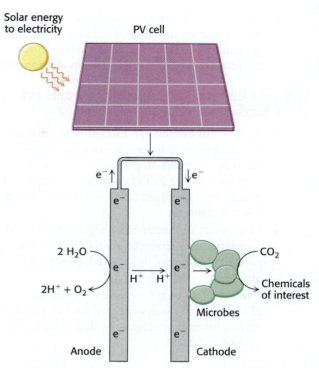

FIGURE 19.31 Photovoltaic cells can drive artificial photosynthesis. After a PV cell absorbs light, the resulting current flows to the anode to oxidize water, with the electrons then taken up by the cathode. Bacteria in contact with the cathode use the electrons to reduce CO_2 to biomolecules.

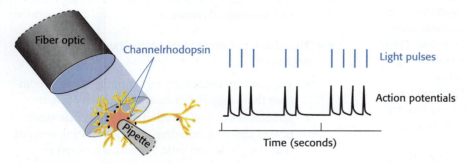

FIGURE 19.32 Optogenetics allows neurons to be controlled using light. Neurons engineered to express the light-sensitive ion channel channelrhodopsin in their cell membrane respond to pulses of light. Light causes channelrhodopsin to open and allow the flow of ions across the membrane. This flow of ions triggers neuron action potentials, which can be detected with a small pipette in a setup known as a patch clamp.

Summary

19.1 Phototrophy Converts Light Energy into Chemical Energy

- The process of living organisms capturing light energy as chemical energy is called phototrophy.
- Photosynthesis is the combination of phototrophy with the reduction of carbon dioxide to form carbohydrate.

19.2 In Eukaryotes, Photosynthesis Takes Place in Chloroplasts

- The proteins that participate in the light reactions of photosynthesis are located in the thylakoid membranes of chloroplasts.

- The light reactions result in (1) the creation of reducing power for the production of NADPH, (2) the generation of a transmembrane proton gradient for the formation of ATP, and (3) the production of O_2.

19.3 Light Absorption by Chlorophyll Molecules Induces Electron Transfer

- An electron excited to a high-energy state by the absorption of a photon can move to nearby electron acceptors. In photosynthesis, an excited electron leaves a pair of associated chlorophyll molecules—the special pair—that are located in reaction centers.
- In the reaction center of a purple bacterium, the electron moves from the special pair containing bacteriochlorophyll to a bacteriopheophytin to quinones.
- The reduction of quinones leads to the generation of a proton gradient, which drives ATP synthesis in a manner analogous to that of oxidative phosphorylation.

19.4 Two Photosystems Generate a Proton Gradient and Reducing Power in Cyanobacteria and Photosynthetic Eukaryotes

- Photosynthesis in green plants, algae, and cyanobacteria is mediated by two linked photosystems. In photosystem II (PS II), the excitation of a special pair of chlorophyll molecules called P680 leads to electron transfer to plastoquinone. The electrons are replenished by the extraction of electrons from a water molecule at the manganese-containing water-oxidizing complex, generating O_2.
- The plastoquinol produced by PS II is reoxidized by the cytochrome *bf* complex. The electrons are transferred to plastocyanin, and then photosystem I (PS I).
- In PS I, the excitation of special pair P700 releases electrons that flow to ferredoxin, a powerful

reductant. Ferredoxin–$NADP^+$ reductase, a flavoprotein, then catalyzes the formation of NADPH.
- A proton gradient is generated as electrons pass through PS II, through the cytochrome *bf* complex, and through ferredoxin–$NADP^+$ reductase.

19.5 A Proton Gradient across the Thylakoid Membrane Drives ATP Synthesis

- The proton gradient across the thylakoid membrane creates a proton-motive force that is used by the CF_0–CF_1 ATP synthase of chloroplasts synthase to form ATP.
- If the NADPH:$NADP^+$ ratio is high, electrons transferred to ferredoxin by PS I can reenter the cytochrome *bf* complex during cyclic photophosphorylation, which leads to the generation of a proton gradient by the cytochrome *bf* complex without the formation of NADPH or O_2.

19.6 Accessory Pigments Funnel Energy into Reaction Centers

- Light-harvesting complexes that surround the reaction centers contain additional chlorophyll and carotenoid molecules that absorb light in the center of the visible spectrum.
- These accessory pigments increase the efficiency of light capture by absorbing light and transferring the energy to reaction centers through resonance energy transfer.

19.7 The Ability to Convert Light into Chemical Energy Is Ancient

- The photosystems have structural features in common that suggest a mutual evolutionary origin.
- Similarities in organization and molecular structure to those of oxidative phosphorylation suggest that the photosynthetic apparatus evolved from an early energy-transduction system.

Key Terms

phototroph (p. 586)
bacteriorhodopsin (p. 586)
photosynthesis (p. 586)
chemotroph (p. 586)
autotroph (p. 586)
heterotroph (p. 586)
light reactions (p. 587)
photosystem I (PS I) (p. 588)
photosystem II (PS II) (p. 588)
chloroplast (p. 588)

stroma (p. 588)
thylakoid (p. 588)
granum (p. 588)
chlorophyll *a* (p. 589)
photoinduced charge separation (p. 590)
reaction center (p. 590)
special pair (p. 591)
water-oxidizing complex (WOC) (manganese center) (p. 595)

cytochrome *bf* (p. 596)
Z scheme of photosynthesis (p. 598)
ATP synthase of chloroplasts (CF_1–CF_0 complex) (p. 600)
cyclic photophosphorylation (p. 601)
carotenoid (p. 603)
light-harvesting complex (p. 604)
nonphotochemical quenching (NPQ) (p. 604)

Problems

1. Under normal circumstances, photosystem I donates electrons to:

(a) Photosystem II

(b) ATP synthase

(c) P700

(d) NADP$^+$

2. The water-oxidizing complex provides electrons to:

(a) P680

(b) P450

(c) P700

(d) Cytochrome c

3. Cyclic photophosphorylation allows the generation of _____ without the synthesis of _____.

(a) ATP, NADPH

(b) NADPH, ATP

(c) ATP, NADH

(d) ATP, O_2

4. Which description of resonance energy transfer is true?

(a) Results in separation of charge

(b) Directly powers ATP synthesis

(c) Allows energy flux in the light-harvesting complex

(d) Does not exist

5. Human beings do not capture energy by photosynthesis, yet this process is critical to our survival. Explain.

6. What is the overall reaction for the light reactions of photosynthesis? ❖ 1

7. Indicate whether each of the following applies to photosystem I or photosystem II: ❖ 1

(a) Generates NADPH

(b) Establishes a proton gradient

(c) Oxidizes water

(d) Produces O_2

(e) Uses reduced ferredoxin to generate a biosynthetic reductant

8. Photosynthesis can be measured by measuring the rate of oxygen production. When plants are exposed to light of wavelength 680 nm, more oxygen is evolved than if the plants are exposed to light of 700 nm. If plants are illuminated by a combination of light of 680 nm and 700 nm, the oxygen production exceeds that of either wavelength alone. Explain both results. ❖ 1

9. What is the advantage of having an extensive set of thylakoid membranes in the chloroplasts? ❖ 1, ❖ 2

10. Explain how light-harvesting complexes enhance the efficiency of photosynthesis. ❖ 5

11. What is the ultimate electron acceptor in oxygenic photosynthesis? The ultimate electron donor? What powers the electron flow between the donor and the acceptor? ❖ 3

12. What are the various sources of protons that contribute to the generation of a proton gradient in chloroplasts? ❖ 4

13. Calculate the $\Delta E'_0$ and $\Delta G^{\circ\prime}$ for the reduction of NADP$^+$ by ferredoxin. Use data given in Table 18.1.

14. It can be argued that, if life were to exist elsewhere in the universe, it would require some process like photosynthesis. If a starship were to land on a distant planet and find no measurable oxygen in the atmosphere, could the crew conclude that photosynthesis is not taking place? ❖ 1

15. Dichlorophenyldimethylurea (DCMU), a herbicide, interferes with photophosphorylation and O_2 evolution. However, it does not block O_2 evolution in the presence of an artificial electron acceptor such as ferricyanide. Propose a site for the inhibitory action of DCMU.

16. Predict the effect of the herbicide dichlorophenyldimethylurea (DCMU) on a plant's ability to perform cyclic photophosphorylation.

17. Consider the relation between the energy of a photon and its wavelength.

(a) Some bacteria are able to harvest 1000-nm light. What is the energy (in kilojoules or kilocalories) of a mole (also called an *einstein*) of 1000-nm photons?

(b) What is the maximum increase in redox potential that can be induced by a 1000-nm photon?

(c) What is the minimum number of 1000-nm photons needed to form ATP from ADP and P_i? Assume a ΔG of 50 kJ mol^{-1} (12 kcal mol^{-1}) for the phosphorylation reaction.

18. What structural features of mitochondria corresponds to the thylakoid membranes and the stroma? ❖ 1, ❖ 2, ❖ 3, ❖ 4

19. Compare and contrast oxidative phosphorylation and photosynthesis. ❖ 1, ❖ 2, ❖ 3, ❖ 4

The Calvin–Benson Cycle and the Pentose Phosphate Pathway

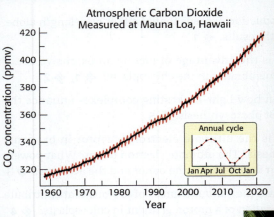

Atmospheric Carbon Dioxide Measured at Mauna Loa, Hawaii

Atmospheric carbon dioxide measurements at Mauna Loa, Hawaii, show the increase in atmospheric carbon dioxide concentration since 1960 in units of parts per million by volume (ppmv). The sawtooth appearance of the curve is the result of annual cycles resulting from seasonal variation in CO_2 fixation by the Calvin–Benson cycle in terrestrial plants (graph inset). Much of this fixation takes place in rain forests, which account for approximately 50% of terrestrial fixation. [Data from http://www.esrl.noaa.gov/gmd/ccgg/trends]

Travelpix Ltd/Photographer's Choice RF/Getty Images

OUTLINE

20.1 The Calvin–Benson Cycle Synthesizes Hexoses from Carbon Dioxide and Water

20.2 The Activity of the Calvin–Benson Cycle Depends on Environmental Conditions

20.3 The Pentose Phosphate Pathway Generates NADPH and Synthesizes Pentoses

20.4 The Metabolism of Glucose 6-Phosphate by the Pentose Phosphate Pathway Is Coordinated with Glycolysis

20.5 Glucose 6-Phosphate Dehydrogenase Plays a Key Role in Protection Against Reactive Oxygen Species

❖ LEARNING GOALS

By the end of this chapter, you should be able to:

1. Explain the function of the Calvin–Benson cycle.

2. Describe how the light reactions and the Calvin–Benson cycle are coordinated.

3. Identify the two stages of the pentose phosphate pathway and explain how the pathway is coordinated with glycolysis and gluconeogenesis.

4. Identify the regulation of the oxidative phase of the pentose phosphate pathway, including the key enzyme and its importance in health and disease.

Photosynthesis proceeds in two parts. The light reactions, discussed in Chapter 19, transform light energy into ATP and biosynthetic reducing power, NADPH. The dark reactions use the ATP and NADPH produced by the light reactions to reduce carbon atoms from their fully oxidized state as carbon dioxide to a more reduced state as a hexose. The dark reactions are called the **Calvin–Benson cycle**, after Melvin Calvin and Andrew Benson, two of the biochemists who helped elucidated the pathway, or sometimes simply the Calvin cycle.

This chapter also examines the pentose phosphate pathway that allows glucose to be oxidized to generate NADPH. The pentose phosphate pathway and Calvin–Benson cycle have several enzymes and intermediates in common. The Calvin–Benson cycle uses NADPH to reduce carbon dioxide to generate hexoses, whereas the pentose phosphate pathway oxidizes glucose to carbon dioxide to generate NADPH.

20.1 The Calvin–Benson Cycle Synthesizes Hexoses from Carbon Dioxide and Water

As we learned in Chapter 16, glucose can be formed from noncarbohydrate precursors, such as lactate and amino acids, by gluconeogenesis. The energy powering gluconeogenesis ultimately comes from previous catabolism of carbon fuels. In contrast, photosynthetic organisms can use the Calvin–Benson cycle to synthesize glucose from carbon dioxide gas and water, by using sunlight as an energy source. The Calvin–Benson cycle introduces into life all the carbon atoms that will be used as fuel and as the carbon backbones of biomolecules. Photosynthetic organisms are autotrophs as they can convert sunlight into chemical energy, which they subsequently use to power their biosynthetic processes.

The Calvin–Benson cycle was elucidated through studies that depended on the availability of radioactive ^{14}C through cyclotrons at the University of California, Berkeley, starting just after World War II. Unicellular algae were exposed to $^{14}CO_2$ for short periods of time, and the radioactive compounds were separated, identified, and quantified.

These studies revealed that the Calvin–Benson cycle comprises three stages (**Figure 20.1**):

1. The fixation of CO_2 to ribulose 1,5-bisphosphate to form two molecules of 3-phosphoglycerate;

2. The reduction of 3-phosphoglycerate to form hexose sugars; and

3. The regeneration of ribulose 1,5-bisphosphate so that more CO_2 can be fixed.

This set of reactions takes place in the stroma of chloroplasts—the photosynthetic organelles (Figure 19.3).

Arthur Nonomura, Ph. D.

ANDREW BENSON After completing his PhD in carbohydrate chemistry from Caltech, Andrew Benson moved to the University of California, Berkeley, where he was exposed to research in photosynthesis. Using radioisotopes of carbon including ^{14}C, he worked as part of a team with Melvin Calvin on the pathway of CO_2 fixation. He was the first to identify ribulose 1,5-bisphosphate as the key component that captures CO_2, among other discoveries.

The elucidation of this pathway represents a good example of how a team of scientists from different backgrounds can solve important problems. Unfortunately, it also revealed the challenges of such collaborations. Calvin dismissed Benson in 1954, shortly after the pathway was elucidated. Calvin was awarded the Nobel Prize by himself in 1961 and barely mentioned Benson in his Nobel lecture. After leaving Berkeley, Dr. Benson went on to have an impressive career working on plant lipids and a range of topics in marine biology, primarily at the Scripps Institution for Oceanography, where he also served as an important scientific leader.

Stage 1: Carbon dioxide reacts with ribulose 1,5-bisphosphate to form two molecules of 3-phosphoglycerate

The first step in the Calvin–Benson cycle is the fixation of CO_2. This begins with the conversion of ribulose 1,5-bisphosphate into a highly reactive enediolate intermediate. The CO_2 molecule condenses with the enediolate intermediate to form an unstable six-carbon compound, which is rapidly hydrolyzed to two molecules of 3-phosphoglycerate.

$$CH_2OPO_3^{2-} \quad C=O \quad H-C-OH \quad H-C-OH \quad CH_2OPO_3^{2-}$$

Ribulose 1,5-bisphosphate

H^+

$$CH_2OPO_3^{2-} \quad C-O^- \quad C-OH \quad H-C-OH \quad CH_2OPO_3^{2-}$$

Enediolate intermediate

CO_2

$$CH_2OPO_3^{2-} \quad HO-C-COO^- \quad C=O \quad H-C-OH \quad CH_2OPO_3^{2-}$$

Unstable intermediate

H_2O

$$2 \ HO-C-H \quad CO_2^-$$
$$CH_2OPO_3^{2-}$$

3-Phosphoglycerate

This highly exergonic reaction [$\Delta G^{\circ\prime} = -51.9$ kJ mol^{-1} (-12.4 kcal mol^{-1})] is catalyzed by **ribulose 1,5-bisphosphate carboxylase/oxygenase** (often called *Rubisco* or *RuBisCo*), an enzyme located on the stromal surface of the thylakoid membranes of chloroplasts. This important reaction is the rate-limiting step in hexose synthesis.

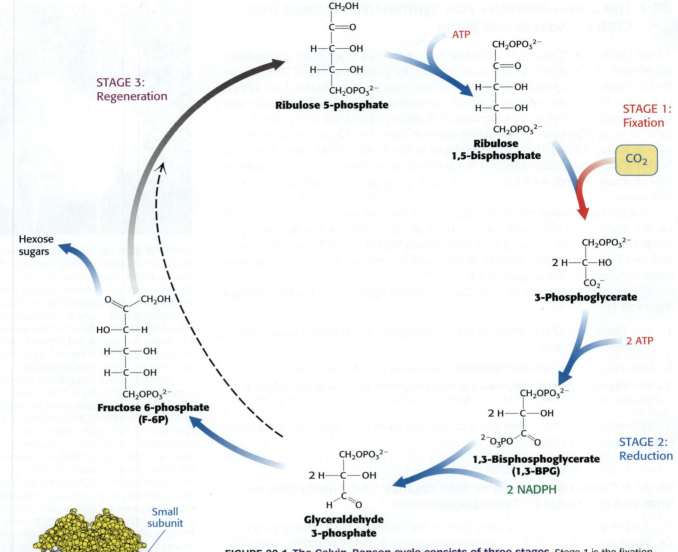

FIGURE 20.1 The Calvin–Benson cycle consists of three stages. Stage 1 is the fixation of carbon by the carboxylation of ribulose 1,5-bisphosphate. Stage 2 is the reduction of the fixed carbon to begin the synthesis of hexose. Stage 3 is the regeneration of the starting compound ribulose 1,5-bisphosphate.

FIGURE 20.2 The enzyme ribulose 1,5-bisphosphate carboxylase/oxygenase (Rubisco) comprises 16 subunits. There are eight small subunits, one shown in blue and the others in white, and eight large subunits (partially obscured by the small units), one shown in red and the others in yellow. The active sites lie in the large subunits. [Drawn from 1RXO.pdb.]

INTERACT with this model in
 Achieve

In plants and green algae, Rubisco consists of eight large (L, 55-kDa) subunits and eight small (S, 15-kDa) ones (**Figure 20.2**). Each L chain contains a catalytic site and a regulatory site. The S chains enhance the catalytic activity of the L chains. This enzyme is abundant in chloroplasts, accounting for approximately 30% of the total leaf protein in some plants. In fact, Rubisco is the most abundant enzyme and perhaps the most abundant protein in the entire biosphere. Large amounts are present because Rubisco is a relatively slow enzyme; its maximal catalytic rate is only 3 s^{-1}.

Rubisco activity depends on magnesium and carbamate

Rubisco requires a bound divalent metal ion for activity, usually magnesium ion. Like the zinc ion in the active site of carbonic anhydrase (Section 6.3), this metal ion serves to activate a bound substrate molecule by stabilizing a negative charge. Interestingly, a CO_2 molecule other than the substrate is required to complete the assembly of the Mg^{2+}-binding site in Rubisco. This CO_2 molecule adds to

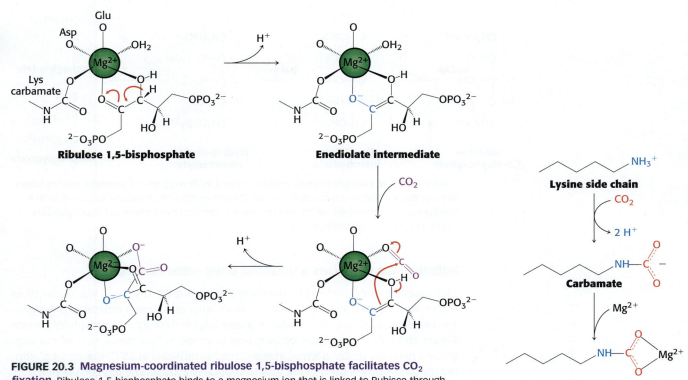

FIGURE 20.3 Magnesium-coordinated ribulose 1,5-bisphosphate facilitates CO₂ fixation. Ribulose 1,5-bisphosphate binds to a magnesium ion that is linked to Rubisco through a glutamate residue, an aspartate residue, and the lysine carbamate. The coordinated ribulose 1,5-bisphosphate gives up a proton to form a reactive enediolate species that reacts with CO_2 to form a new carbon–carbon bond.

the uncharged ε-amino group of lysine 201 to form a carbamate. This negatively charged carbamate then binds the Mg^{2+} ion.

The metal center plays a key role in binding ribulose 1,5-bisphosphate and activating it so that it reacts with CO_2 (**Figure 20.3**). Ribulose 1,5-bisphosphate binds to Mg^{2+} through its keto group and an adjacent hydroxyl group. This complex is readily deprotonated to form an enediolate intermediate. This reactive species, analogous to the zinc-hydroxide species in carbonic anhydrase, reacts with CO_2, forming the new carbon–carbon bond. The resulting 2-carboxy-3-keto-D-arabinitol 1,5-bisphosphate is coordinated to the Mg^{2+} ion through three groups, including the newly formed carboxylate. A molecule of H_2O is then added to this β-ketoacid to form an intermediate that cleaves to form two molecules of 3-phosphoglycerate (**Figure 20.4**).

Ribulose 1,5-bisphosphate	Enediolate intermediate	2-Carboxy-3-keto-D-arabinitol 1,5-bisphosphate	Hydrated intermediate	3-Phosphoglycerate

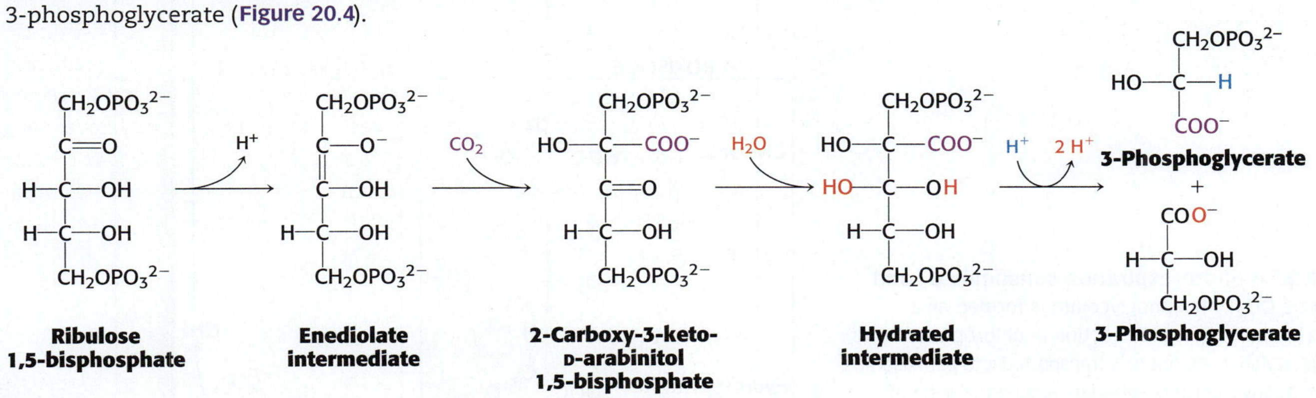

FIGURE 20.4 In stage 1 of the Calvin–Benson cycle, 3-phosphoglycerate is formed. The pathway for the conversion of ribulose 1,5-bisphosphate and CO_2 into two molecules of 3-phosphoglycerate is shown. Although the free species are shown, these steps take place on the magnesium ion.

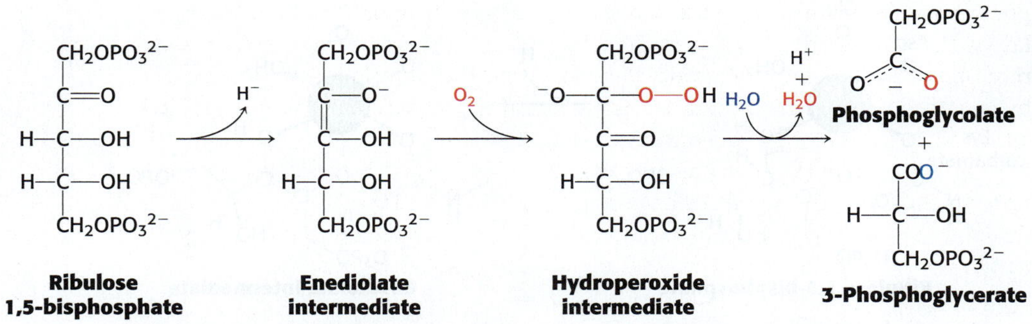

FIGURE 20.5 The enediolate on Rubisco can react with oxygen in a wasteful side reaction. The reactive enediolate intermediate on Rubisco also reacts with molecular oxygen to form a hydroperoxide intermediate, which then proceeds to form one molecule of 3-phosphoglycerate and one molecule of phosphoglycolate.

Rubisco also catalyzes a wasteful oxygenase reaction

The reactive enediolate intermediate generated on the Mg^{2+} ion sometimes reacts with O_2 instead of CO_2; thus, Rubisco also catalyzes a deleterious oxygenase reaction, yielding the products phosphoglycolate and 3-phosphoglycerate (**Figure 20.5**). The rate of the carboxylase reaction is four times that of the oxygenase reaction under normal atmospheric conditions at 25°C; the stromal concentration of CO_2 is then 10 μM and that of O_2 is 250 μM. The oxygenase reaction, like the carboxylase reaction, requires that lysine 201 be in the carbamate form. Because this carbamate forms only in the presence of CO_2, Rubisco is prevented from catalyzing the oxygenase reaction exclusively when CO_2 is absent.

Phosphoglycolate is not a versatile metabolite. A salvage pathway recovers part of its carbon skeleton (**Figure 20.6**). A specific phosphatase converts

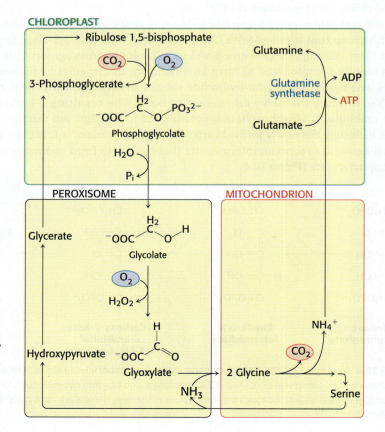

FIGURE 20.6 Photorespiration consumes O_2 and produces CO_2. Phosphoglycolate is formed as a product of the oxygenase reaction in chloroplasts. After dephosphorylation, glycolate is transported into peroxisomes, where it is converted into glyoxylate and then glycine. In mitochondria, two glycines are converted into serine, after losing a carbon as CO_2 and ammonium ion. The serine is converted back into 3-phosphoglycerate, and the ammonium ion is salvaged in chloroplasts.

phosphoglycolate into glycolate, which enters **peroxisomes** (**Figure 20.7**). Glycolate is then oxidized to glyoxylate by glycolate oxidase, an enzyme with a flavin mononucleotide prosthetic group. The H_2O_2 produced in this reaction is cleaved by catalase to H_2O and O_2. Transamination of glyoxylate then yields glycine, which then enters the mitochondria.

Two glycine molecules can unite to form serine with the release of CO_2 and ammonium ion (NH_4^+). The ammonium ion, used in the synthesis of nitrogen-containing compounds, is salvaged by a glutamine synthetase reaction. Serine reenters the peroxisome, where it donates its ammonia to glyoxylate and is subsequently converted back into 3-phosphoglycerate.

This salvage pathway serves to recycle three of the four carbon atoms of two molecules of glycolate. However, one carbon atom is lost as CO_2. This process is called **photorespiration** because O_2 is consumed and CO_2 is released. Photorespiration is wasteful because organic carbon is converted into CO_2 without the production of ATP, NADPH, or another energy-rich metabolite. Although evolutionary processes have presumably enhanced the preference of Rubisco for carboxylation—for instance, the Rubisco of higher plants is eight times more specific for carboxylation than that of photosynthetic bacteria—photorespiration still accounts for the loss of up to 25% of the carbon fixed.

❖ SELF–CHECK QUESTION

Suppose ribulose 1,5-bisphosphate reacts with pure $^{14}CO_2$ in the absence of oxygen. One half of the 3-phosphoglycerate is labeled with ^{14}C. Why?

Stage 2: Hexose phosphates are made from phosphoglycerate

The 3-phosphoglycerate product of Rubisco is used to synthesize 6-carbon sugars. First, 3-phosphoglycerate is converted into fructose 6-phosphate, which is readily converted to glucose 1-phosphate and then to glucose 6-phosphate. The mixture of the three phosphorylated hexoses is called the hexose monophosphate pool. The steps in this conversion (**Figure 20.8**) are like those of the gluconeogenic pathway, except that glyceraldehyde 3-phosphate dehydrogenase in chloroplasts—which generates glyceraldehyde 3-phosphate (GAP)—is specific for NADPH rather than NADH. These reactions and that catalyzed by Rubisco bring CO_2 to the level of a hexose, converting CO_2 into a chemical fuel at the expense of NADPH and ATP generated from the light reactions.

Stage 3: Ribulose 1,5-bisphosphate is regenerated

The third phase of the Calvin–Benson cycle is the regeneration of ribulose 1,5-bisphosphate, the acceptor of CO_2 in the first step. The problem is to construct a five-carbon sugar from six-carbon and three-carbon sugars. A transketolase and an aldolase play the major role in the rearrangement of the carbon atoms. The transketolase, which we will see again in the pentose phosphate pathway, requires the coenzyme thiamine pyrophosphate (TPP) to transfer a two-carbon unit (CO—CH_2OH) from a ketose to an aldose.

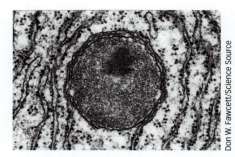

FIGURE 20.7 Peroxisomes are organelles defined by a single membrane. An electron micrograph of a peroxisome is shown. The peroxisomes are surrounded by rough endoplasmic reticulum.

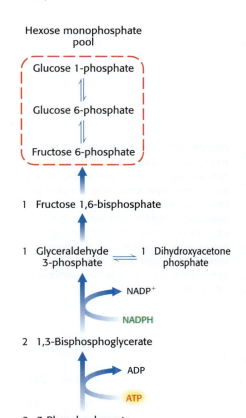

FIGURE 20.8 In stage 2 of the Calvin–Benson cycle, hexose phosphate is formed. 3-Phosphoglycerate is converted into fructose 6-phosphate in a pathway parallel to that of gluconeogenesis.

The **transketolase** converts fructose 6-phosphate and glyceraldehyde 3-phosphate into erythrose 4-phosphate and xylulose 5-phosphate.

Aldolase, which we have already encountered in glycolysis, catalyzes an aldol condensation between dihydroxyacetone phosphate (DHAP) and an aldehyde. This enzyme is highly specific for dihydroxyacetone phosphate, but it accepts a wide variety of aldehydes.

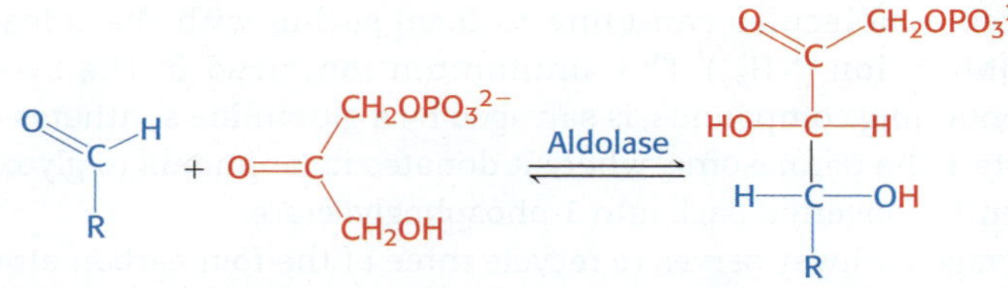

Aldose (n carbons) **Dihydroxyacetone phosphate** **Ketose** ($n + 3$ carbons)

Aldolase generates sedoheptulose 1,7-bisphosphate from erythrose 4-phosphate and dihydroxyacetone phosphate. A phosphatase removes a phosphate from sedoheptulose 1,7-bisphosphosphate to form sedoheptulose 7-phosphate. Transketolase again enters the action to combine sedoheptulose 7-phosphate with another molecule of glyceraldehyde 3-phosphate to form the five-carbon sugar ribose 5-phosphate as well as a second molecule of xylulose 5-phosphate (**Figure 20.9**).

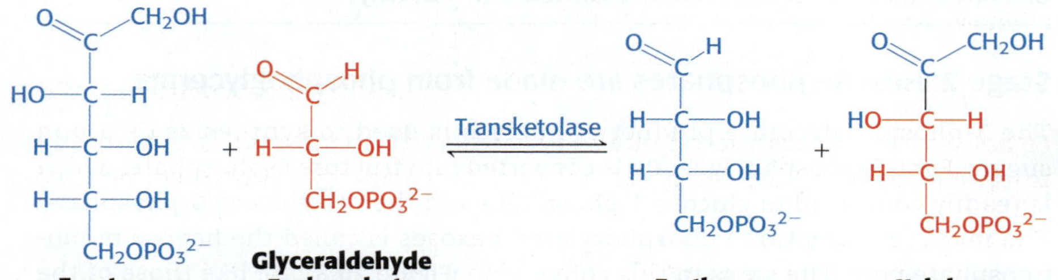

Fructose 6-phosphate **Glyceraldehyde 3-phosphate (GAP)** **Erythrose 4-phosphate** **Xylulose 5-phosphate**

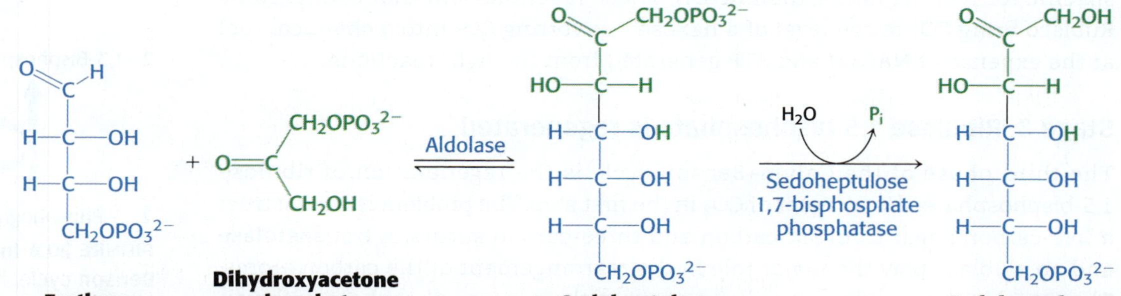

Erythrose 4-phosphate **Dihydroxyacetone phosphate (DHAP)** **Sedoheptulose 1,7-bisphosphate** **Sedoheptulose 7-phosphate**

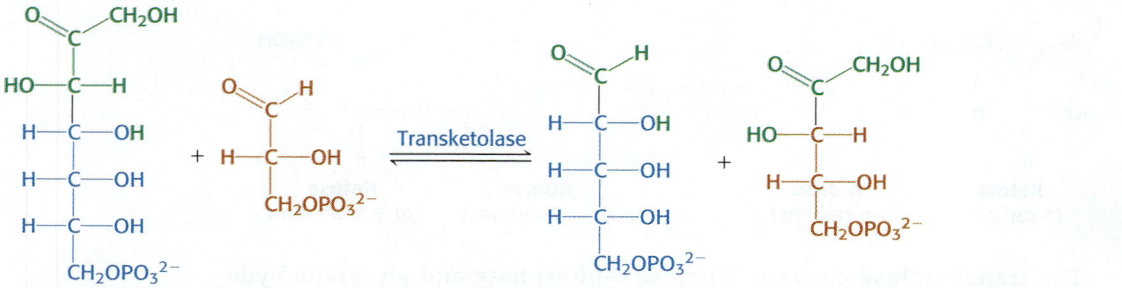

Sedoheptulose 7-phosphate **Glyceraldehyde 3-phosphate** **Ribose 5-phosphate** **Xylulose 5-phosphate**

FIGURE 20.9 Five-carbon sugars are formed from six-carbon and three-carbon sugars. First, transketolase converts a six-carbon sugar and a three-carbon sugar into a four-carbon sugar and a five-carbon sugar. Then, aldolase combines the four-carbon product and a three-carbon sugar to form a seven-carbon sugar. Finally, this seven-carbon sugar reacts with another three-carbon sugar to form two additional five-carbon sugars.

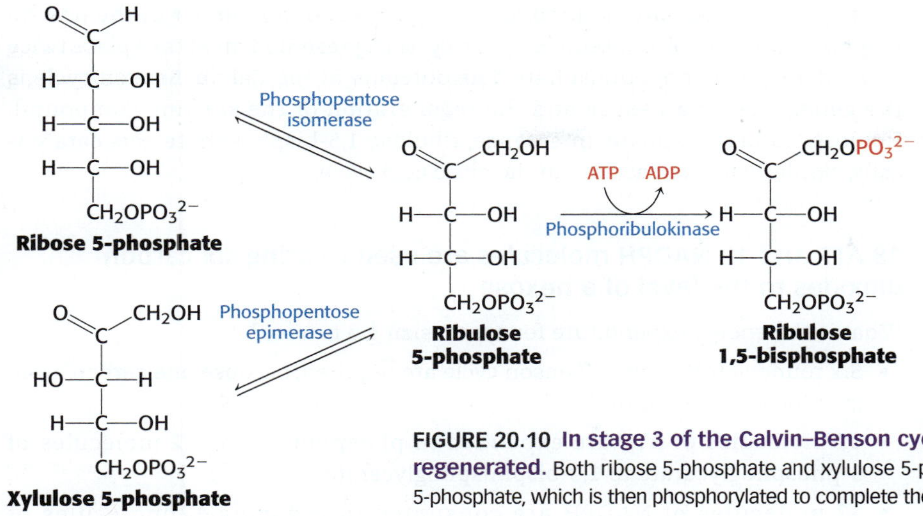

O=C—H
H—C—OH
H—C—OH
H—C—OH
CH$_2$OPO$_3^{2-}$
Ribose 5-phosphate

Phosphopentose isomerase

O=C—CH$_2$OH
H—C—OH
H—C—OH
CH$_2$OPO$_3^{2-}$
Ribulose 5-phosphate

ATP   ADP
Phosphoribulokinase

O=C—CH$_2$OPO$_3^{2-}$
H—C—OH
H—C—OH
CH$_2$OPO$_3^{2-}$
Ribulose 1,5-bisphosphate

O=C—CH$_2$OH
HO—C—H
H—C—OH
CH$_2$OPO$_3^{2-}$
Xylulose 5-phosphate

Phosphopentose epimerase

FIGURE 20.10 In stage 3 of the Calvin–Benson cycle, ribulose 1,5-bisphosphate is regenerated. Both ribose 5-phosphate and xylulose 5-phosphate are converted into ribulose 5-phosphate, which is then phosphorylated to complete the regeneration of ribulose 1,5-bisphosphate.

Finally, ribose 5-phosphate is converted into ribulose 5-phosphate by phospho-pentose isomerase, whereas the two molecules of xylulose 5-phosphate are converted into ribulose 5-phosphate by phosphopentose epimerase. Ribulose 5-phosphate is converted into ribulose 1,5-bisphosphate through the action of phosphoribulokinase (**Figure 20.10**). The sum of these reactions shown in Figures 20.9 and 20.10 is:

Fructose 6-phosphate + 2 glyceraldehyde 3-phosphate
    + dihydroxyacetone phosphate + 3 ATP ⟶
                3 ribulose 1,5-bisphosphate + 3 ADP

Figure 20.11 presents the required reactions with the proper stoichiometry to convert three molecules of CO_2 into one molecule of DHAP. However, two

FIGURE 20.11 The full Calvin–Benson cycle process for converting three molecules of CO_2 to dihydroxyacetone phosphate is complex. The cycle is not as simple as presented in Figure 20.1; rather, it entails many reactions that lead ultimately to the synthesis of glucose and the regeneration of ribulose 1,5-bisphosphate. [Information from J. R. Bowyer and R. C. Leegood. "Photosynthesis," in *Plant Biochemistry*, P. M. Dey and J. B. Harborne, Eds. (Academic Press, 1997), p. 85.]

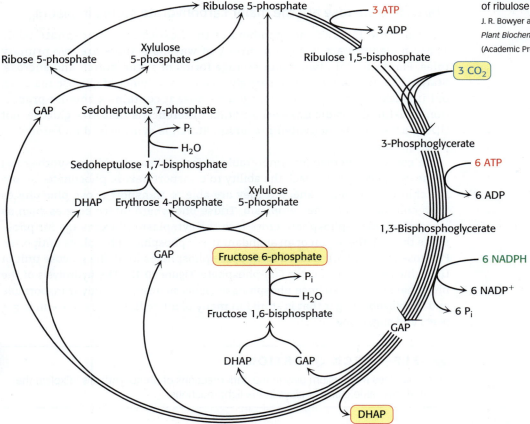

molecules of DHAP are required for the synthesis of a member of the hexose monophosphate pool. Consequently, the cycle as presented must take place twice to yield a hexose monophosphate. The outcome of the Calvin–Benson cycle is the generation of a hexose and the regeneration of the starting compound, ribulose 1,5-bisphosphate. In essence, ribulose 1,5-bisphosphate acts catalytically, similarly to oxaloacetate in the citric acid cycle.

18 ATP and 12 NADPH molecules are used to bring six carbon dioxides to the level of a hexose

What is the energy expenditure for synthesizing a hexose?

- Six rounds of the Calvin–Benson cycle are required, because one carbon atom is reduced in each round.
- 12 molecules of ATP are expended in phosphorylating 12 molecules of 3-phosphoglycerate to 1,3-bisphosphoglycerate.
- 12 molecules of NADPH are consumed in reducing 12 molecules of 1,3-bisphosphoglycerate to glyceraldehyde 3-phosphate.
- An additional six molecules of ATP are spent in regenerating ribulose 1,5-bisphosphate.

We can now write a balanced equation for the net reaction of the Calvin–Benson cycle:

$$6\ CO_2 + 18\ ATP + 12\ NADPH + 12\ H_2O \longrightarrow$$
$$C_6H_{12}O_6 + 18\ ADP + 18\ P_i + 12\ NADP^+ + 6\ H^+$$

Thus, three molecules of ATP and two molecules of NADPH are consumed in incorporating a single CO_2 molecule into a hexose such as glucose or fructose.

Starch and sucrose are the major carbohydrate stores in plants

What are the fates of the members of the hexose monophosphate pool? These molecules are used in a variety of ways, but there are two primary roles. Plants contain two major storage forms of sugar: starch and sucrose. **Starch**, like its animal counterpart glycogen, is a polymer of glucose residues, but it is less branched than glycogen because it contains a smaller proportion of α-1,6-glycosidic linkages. Another difference is that ADP-glucose, not UDP-glucose, is the activated precursor. Starch is synthesized and stored in chloroplasts.

In contrast, **sucrose** (common table sugar), a disaccharide, is synthesized in the cytoplasm. Plants lack the ability to transport hexose phosphates across the chloroplast membrane, but they are able to transport triose phosphates from chloroplasts to the cytoplasm. Triose phosphate intermediates such as glyceraldehyde 3-phosphate cross into the cytoplasm in exchange for phosphate through the action of an abundant triose phosphate-phosphate antiporter. Fructose 6-phosphate formed from triose phosphates joins the glucose unit of UDP-glucose to form sucrose 6(F)-phosphate (**Figure 20.12**). The hydrolysis of the phosphate ester by sucrose phosphatase yields sucrose, a readily transportable and mobilizable sugar that is stored in many plant cells, such as those in sugar beets and sugarcane.

❖ SELF–CHECK QUESTION

What role does magnesium play in the dark reactions of photosynthesis? Explain the role that this same element plays in the light reactions.

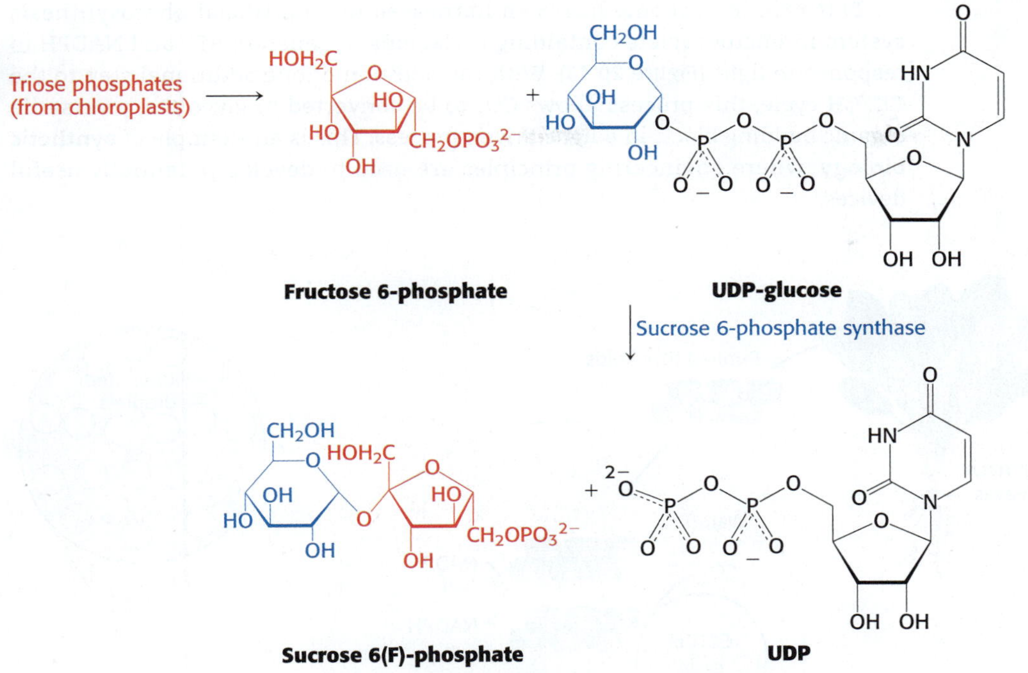

Fructose 6-phosphate   **UDP-glucose**

Sucrose 6-phosphate synthase

Sucrose 6(F)-phosphate   **UDP**

FIGURE 20.12 Carbohydrates are stored as sucrose. Sucrose 6-phosphate is formed by the reaction between fructose 6-phosphate and the activated intermediate uridine diphosphate glucose (UDP-glucose). Sucrose phosphatase subsequently generates free sucrose (not shown).

Inspired by the Calvin–Benson cycle, scientists are developing new methods for fixing carbon dioxide

Carbon dioxide is a potential plentiful carbon source for the production of organic compounds. Applying the knowledge of biochemical pathways from a wide range of organisms, scientists have conceptually developed and, in some cases, implemented cycles that fix carbon dioxide into organic compounds. One such cycle referred to as the CETCH cycle—for crotonyl–coenzyme A (CoA)/ethylmalonyl-CoA/hydroxybutyryl-CoA—catalyzes the conversion of 2 molecules of CO_2 to 1 molecule of glyoxylate, $^-O_4$ C-CHO. This cycle involves 12 steps and 15 enzymes. The balanced chemical reaction is

$$2\ CO_2 + 4\ NADPH + 2\ FADH2 + 2\ ATP + 2\ O_2 + 3\ H^+ \longrightarrow$$
$$^-O_2C\text{-CHO} + 4\ NADP^+ + 2\ FAD + 4\ ADP + 2\ P_i + 5\ H_2O$$

Carbon fixation occurs via the reaction catalyzed by enoyl-CoA carboxylases/reductases.

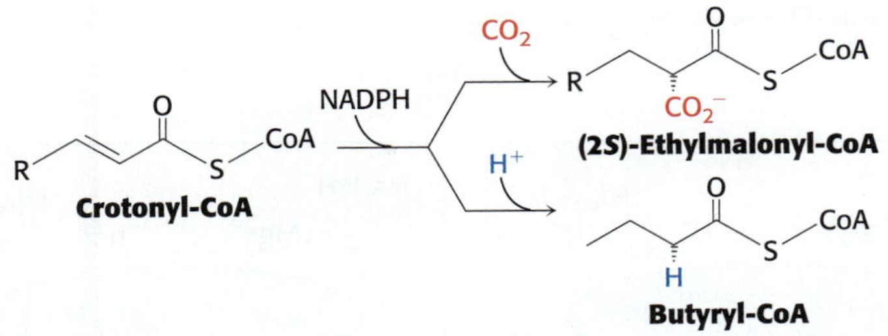

These enzymes are several times more efficient at fixing CO_2 than is Rubisco and do not react with oxygen.

This reaction scheme has been harnessed in an artificial photosynthesis system in microdroplets containing thylakoids to generate ATP and NADPH in response to light (**Figure 20.13**). With the addition of one additional step to the CETCH cycle, this process allows CO_2 to be converted to glycolate, a versatile organic building block, in a light-driven process. This is an example of **synthetic biology**, where engineering principles are used to develop potentially useful devices.

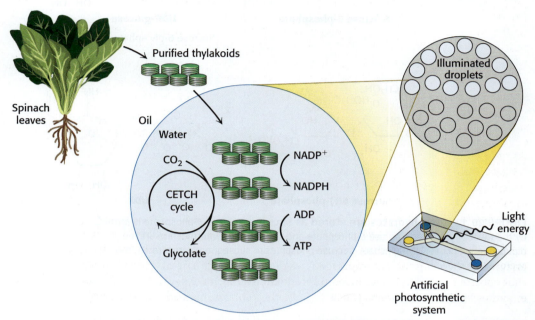

FIGURE 20.13 Scientists have engineered artificial photosynthesis using alternative CO_2 fixation pathways. A synthetic photosynthetic system is shown using purified thylakoids to drive the light reactions and a designed CO_2 fixation system for the dark reactions. [From N. J. Gaut and K. P. Adamala, "Toward Artificial Photosynthesis," *Science* 368:587–588, 2020.]

20.2 The Activity of the Calvin–Benson Cycle Depends on Environmental Conditions

Light levels, temperature, and availability of CO_2 in different environments are crucial factors for effective carbon fixation and avoidance of processes damaging to plants. How are the light reactions linked to the dark reactions to regulate fixing CO_2 into biomolecules? The principal means of regulation is alteration of the stromal environment by the light reactions. The light reactions lead to an increase in stromal pH (a decrease in the H^+ concentration), as well as an increase in the stromal concentrations of Mg^{2+}, NADPH, and reduced ferredoxin—changes that contribute to the activation of certain Calvin–Benson cycle enzymes, located in the stroma (**Figure 20.14**).

FIGURE 20.14 Light regulates the Calvin–Benson cycle. The light reactions of photosynthesis transfer electrons out of the thylakoid lumen into the stroma and transfer protons from the stroma into the thylakoid lumen. As a consequence of these processes, the concentrations of NADPH, reduced ferredoxin (Fd), and Mg^{2+} in the stroma are higher in the light than in the dark. The stromal pH is also increased (the concentration of H^+ is lowered) in the light as a result of proton pumping from the stroma to the thylakoid lumen. Each of these concentration changes helps couple the Calvin–Benson cycle reactions to the light reactions.

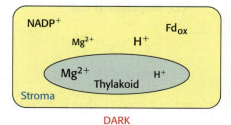

DARK

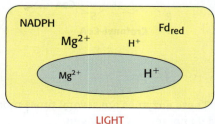

LIGHT

Rubisco is activated by light-driven changes in proton and magnesium ion concentrations

As stated earlier, the rate-limiting step in the Calvin–Benson cycle is the carboxylation of ribulose 1,5-bisphosphate to form two molecules of 3-phosphoglycerate. The activity of Rubisco increases markedly on illumination because light facilitates the carbamate formation necessary for enzyme activity (Figure 20.3). In the stroma, the pH increases from 7 to 8, and the level of Mg^{2+} rises. Both effects are consequences of the light-driven pumping of protons into the thylakoid space. Mg^{2+} ions from the thylakoid space are released into the stroma to compensate for the influx of protons. Carbamate formation is favored at alkaline pH. CO_2 adds to a deprotonated form of lysine 201 of Rubisco, and Mg^{2+} ion binds to the carbamate to generate the active form of the enzyme. Thus, light leads not only to the generation of ATP and NADPH, but also to regulatory signals that stimulate CO_2 fixation.

Thioredoxin plays a key role in regulating the Calvin–Benson cycle

As we saw in Chapter 19, light-driven reactions lead to electron transfer from water to ferredoxin and, eventually, to NADPH in photosystem I. The presence of reduced ferredoxin and NADPH are good signals that conditions are right for biosynthesis. One way in which this information is conveyed to biosynthetic enzymes is by thioredoxin, a 12-kDa protein containing neighboring cysteine residues that cycle between a reduced sulfhydryl and an oxidized disulfide form (**Figure 20.15**). The reduced form of thioredoxin activates many biosynthetic enzymes, including the chloroplast ATP synthase, by reducing disulfide bridges that control their activity. Reduced thioredoxin also inhibits several degradative enzymes by the same means (**Table 20.1**). In chloroplasts, oxidized thioredoxin is

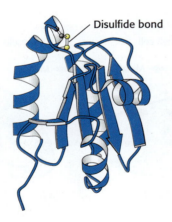

Disulfide bond

FIGURE 20.15 Thioredoxin has a redox-active disulfide bond. The oxidized form of thioredoxin contains a disulfide bond. When thioredoxin is reduced by reduced ferredoxin, the disulfide bond is converted into two free sulfhydryl groups. Reduced thioredoxin can cleave disulfide bonds in enzymes, activating certain Calvin–Benson cycle enzymes and inactivating some degradative enzymes. [Drawn from 1F9M.pdb.]

INTERACT with this model in
🌊 Achieve

TABLE 20.1 Enzymes regulated by thioredoxin

Enzyme	Pathway
Rubisco	Carbon fixation in the Calvin–Benson cycle
Fructose 1,6-bisphosphatase	Gluconeogenesis
Glyceraldehyde 3-phosphate dehydrogenase	Calvin–Benson cycle, gluconeogenesis, glycolysis
Sedoheptulose 1,7-bisphosphatase	Calvin–Benson cycle
Glucose 6-phosphate dehydrogenase	Pentose phosphate pathway
Phenylalanine ammonia lyase	Lignin synthesis
Phosphoribulokinase	Calvin–Benson cycle
$NADP^+$-malate dehydrogenase	C_4 pathway
CF_1–CF_0 ATP synthase	The light reactions

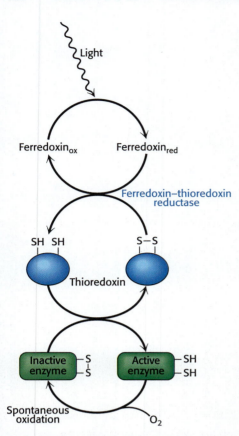

FIGURE 20.16 Thioredoxin can activate enzymes by reducing them. Reduced thioredoxin activates certain Calvin–Benson cycle enzymes by cleaving regulatory disulfide bonds.

reduced by ferredoxin in a reaction catalyzed by ferredoxin–thioredoxin reductase. Thus, the activities of the light and dark reactions of photosynthesis are coordinated through electron transfer from reduced ferredoxin to thioredoxin and then to component enzymes containing regulatory disulfide bonds (**Figure 20.16**).

NADPH is a signal molecule that activates two biosynthetic enzymes, phosphoribulokinase and glyceraldehyde 3-phosphate dehydrogenase. In the dark, these enzymes are inhibited by association with an 8.5-kDa protein called CP12, an intrinsically disordered protein. NADPH disrupts this association by promoting the formation of two disulfide bonds in CP12, leading to the release of the active enzymes.

The C$_4$ pathway of tropical plants and grasses accelerates photosynthesis by concentrating carbon dioxide

The oxygenase activity of Rubisco presents a biochemical challenge to tropical plants because the oxygenase activity increases more rapidly with temperature than does the carboxylase activity. How, then, do plants that grow in hot climates, such as sugarcane, prevent very high rates of wasteful photorespiration? Their solution to this problem is to achieve a high local concentration of CO_2 at the site of the Calvin–Benson cycle in their photosynthetic cells. The essence of this process is that four-carbon (C$_4$) compounds such as oxaloacetate and malate carry CO_2 from mesophyll cells, which are in contact with air, to bundle-sheath cells, which are the major sites of photosynthesis (**Figure 20.17**). The decarboxylation of the four-carbon compound in a bundle-sheath cell maintains a high concentration of CO_2 at the site of the Calvin–Benson cycle. The three-carbon product returns to the mesophyll cell for another round of carboxylation. This metabolic pathway is called the **C$_4$ pathway** or the **Hatch–Slack pathway**.

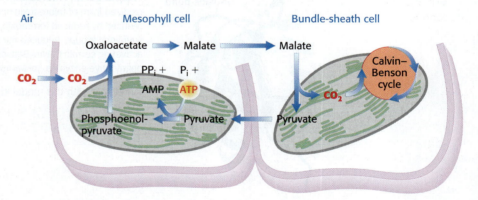

FIGURE 20.17 The C$_4$ pathway can concentrate CO_2. Carbon dioxide is concentrated in bundle-sheath cells by the expenditure of ATP in mesophyll cells.

In addition to tropical plants, grasses and sedges in temperate climates employ the C$_4$ pathway. In fact, they are the most common users of the pathway. Corn or maize, domesticated by selective breeding from the Mexican grass teosinte about 9,000 years ago, is a C$_4$ plant of tremendous economic and agricultural importance, accounting for approximately 20% of human nutrition worldwide.

The C$_4$ pathway for the transport of CO_2 starts in a mesophyll cell, with the condensation of CO_2 in the form of bicarbonate and phosphoenolpyruvate to form *oxaloacetate* in a reaction catalyzed by phosphoenolpyruvate carboxylase. Oxaloacetate is converted into malate by an NADP$^+$-linked malate dehydrogenase.

Malate enters the bundle-sheath cell and is oxidatively decarboxylated within the chloroplasts by an $NADP^+$-linked malate dehydrogenase. The released CO_2 enters the Calvin–Benson cycle in the usual way by condensing with ribulose 1,5-bisphosphate. Pyruvate formed in this decarboxylation reaction returns to the mesophyll cell. Finally, phosphoenolpyruvate is formed from pyruvate by pyruvate-P_i dikinase.

The net reaction of this C_4 pathway is

$$CO_2 \text{ (in mesophyll cell)} + ATP + 2\ H_2O \longrightarrow$$
$$CO_2 \text{ (in bundle-sheath cell)} + AMP + 2\ P_i + 2\ H^+$$

Since the products are AMP and two equivalents of inorganic phosphate, the energetic equivalent of two ATP molecules is consumed in transporting CO_2 to the chloroplasts of the bundle-sheath cells. Note that six molecules of CO_2 are required to synthesize glucose so the C_4 pathway requires an extra 12 ATP when compared to the C_3 pathway. In essence, this process is active transport: the pumping of CO_2 into the bundle-sheath cell is driven by the hydrolysis of one molecule of ATP to one molecule of AMP and two molecules of orthophosphate. The CO_2 concentration can be 20 times as great in the bundle-sheath cells as in the mesophyll cells. Interestingly, phosphoenolpyruvate carboxylase has a higher affinity for CO_2 (as bicarbonate) than does Rubisco. C_4 plants use this affinity difference to their advantage. The high CO_2 affinity of phosphoenolpyruvate carboxylase means Rubisco can be adequately supplied with CO_2, while the stomata — the openings in the leaves that allow gas exchange (Figure 20.18) — do not have to open as completely in the heat of the tropical day, thereby preventing water loss.

When the C_4 pathway and the Calvin–Benson cycle operate together, the net reaction is

$$6\ CO_2 + 30\ ATP + 12\ NADPH + 24\ H_2O \longrightarrow$$
$$C_6H_{12}O_6 + 30\ ADP + 30\ P_i + 12\ NADP^+ + 18\ H^+$$

Note that 30 molecules of ATP are consumed per hexose molecule formed when the C_4 pathway delivers CO_2 to the Calvin–Benson cycle, in contrast with 18 molecules of ATP per hexose molecule in the absence of the C_4 pathway. The high concentration of CO_2 in the bundle-sheath cells of C_4 plants — which is the result of the expenditure of the additional 12 molecules of ATP — is critical for their rapid photosynthetic rate, because CO_2 is limiting when light is abundant. A high CO_2 concentration also minimizes the energy loss caused by photorespiration.

Tropical plants with a C_4 pathway do little photorespiration because the high concentration of CO_2 in their bundle-sheath cells accelerates the carboxylase reaction relative to the oxygenase reaction. This effect is especially important at higher temperatures. The geographic distribution of plants having this pathway (C_4 plants) and those lacking it (C_3 plants) can now be understood in molecular terms. C_4 plants have the advantage in a hot environment and under high illumination, which accounts for their prevalence in the tropics. C_3 plants, which consume only 18 molecules of ATP per hexose molecule formed in the absence of photorespiration (compared with 30 molecules of ATP for C_4 plants), are more efficient at temperatures lower than about 28°C, and so they predominate in temperate environments (Figure 20.19). Although C_4 plants comprise only 3% of all flowering plants, they account for 25% of all terrestrial carbon fixation because of their minimization of photorespiration.

FIGURE 20.18 Stomata can open to facilitate gas exchange. A scanning electron micrograph shows a closed stoma and an open stoma.

Power and Syred/Science Source

(A)

(B)

FIGURE 20.19 **Both C$_3$ and C$_4$ plants are important.** (A) C$_3$ plants, such as trees, account for approximately 95% of plant species. (B) Corn is a C$_4$ plant of tremendous agricultural importance.

Crassulacean acid metabolism permits growth in arid ecosystems

Many plants, including some that grow in hot, dry climates, keep the stomata of their leaves closed in the heat of the day to prevent water loss (see Figure 20.18). As a consequence, CO$_2$ cannot be absorbed during the daylight hours, when it is needed for glucose synthesis. Rather, CO$_2$ enters the leaf when the stomata open at the cooler temperatures of night.

To store the CO$_2$ until it can be used during the day, such plants make use of an adaptation called **crassulacean acid metabolism (CAM)**, named after the genus *Crassulacea* (the succulents). Carbon dioxide is fixed by the C$_4$ pathway into malate, which is stored in vacuoles. During the day, malate is decarboxylated and the CO$_2$ becomes available to the Calvin–Benson cycle. In contrast with C$_4$ plants, CAM plants separate CO$_2$ accumulation from CO$_2$ utilization temporally rather than spatially.

Although CAM plants do prevent water loss, the use of malate as the sole source of CO$_2$ comes at a metabolic cost. Because malate storage is limited, CAM plants cannot generate CO$_2$ as rapidly as it can be imported by C$_3$ and C$_4$ plants. Consequently, the growth rate of CAM plants is slower than that of C$_3$ and C$_4$ plants. The saguaro cactus, a CAM plant that can live up to 200 years and reach a height of 60 feet, can take 15 years to grow to only 1 foot in height (**Figure 20.20**).

FIGURE 20.20 **Crassulacean acid metabolism facilitates growth in arid climates.** Because of crassulacean acid metabolism, plants such as cacti are well suited to life in the desert.

20.3 The Pentose Phosphate Pathway Generates NADPH and Synthesizes Pentoses

While photosynthetic organisms can use the light reactions for generation of some NADPH, the **pentose phosphate pathway** meets the NADPH needs of non-photosynthetic organisms and of the non-photosynthetic tissues in plants. The pentose phosphate pathway occurs in the cytoplasm of all organisms and is a crucial source of NADPH for use in reductive biosynthesis (**Table 20.2**) as well as for protection against oxidative stress. It consists of two phases (**Figure 20.21**):

1. *The oxidative generation of NADPH.* In the oxidative phase, NADPH is generated when glucose 6-phosphate is oxidized to ribulose 5-phosphate.

$$\text{Glucose 6-phosphate} + 2\ \text{NADP}^+ + \text{H}_2\text{O} \longrightarrow$$
$$\text{ribulose 5-phosphate} + 2\ \text{NADPH} + 2\ \text{H}^+ + \text{CO}_2$$

2. *The nonoxidative interconversion of sugars.* In the nonoxidative phase, the pathway catalyzes the interconversion of three-, four-, five-, six-, and seven-carbon sugars in a series of nonoxidative reactions. Excess five-carbon sugars may be converted into intermediates of the glycolytic pathway. These interconversions rely on the same reactions that lead to the regeneration of ribulose 1,5-bisphosphate in the Calvin–Benson cycle.

TABLE 20.2 Pathways requiring NADPH

Reductive biosynthesis
Fatty acid biosynthesis
Cholesterol biosynthesis
Neurotransmitter biosynthesis
Nucleotide biosynthesis
Protection from oxidative stress
Reduction of oxidized glutathione
Cytochrome P450 monooxygenases

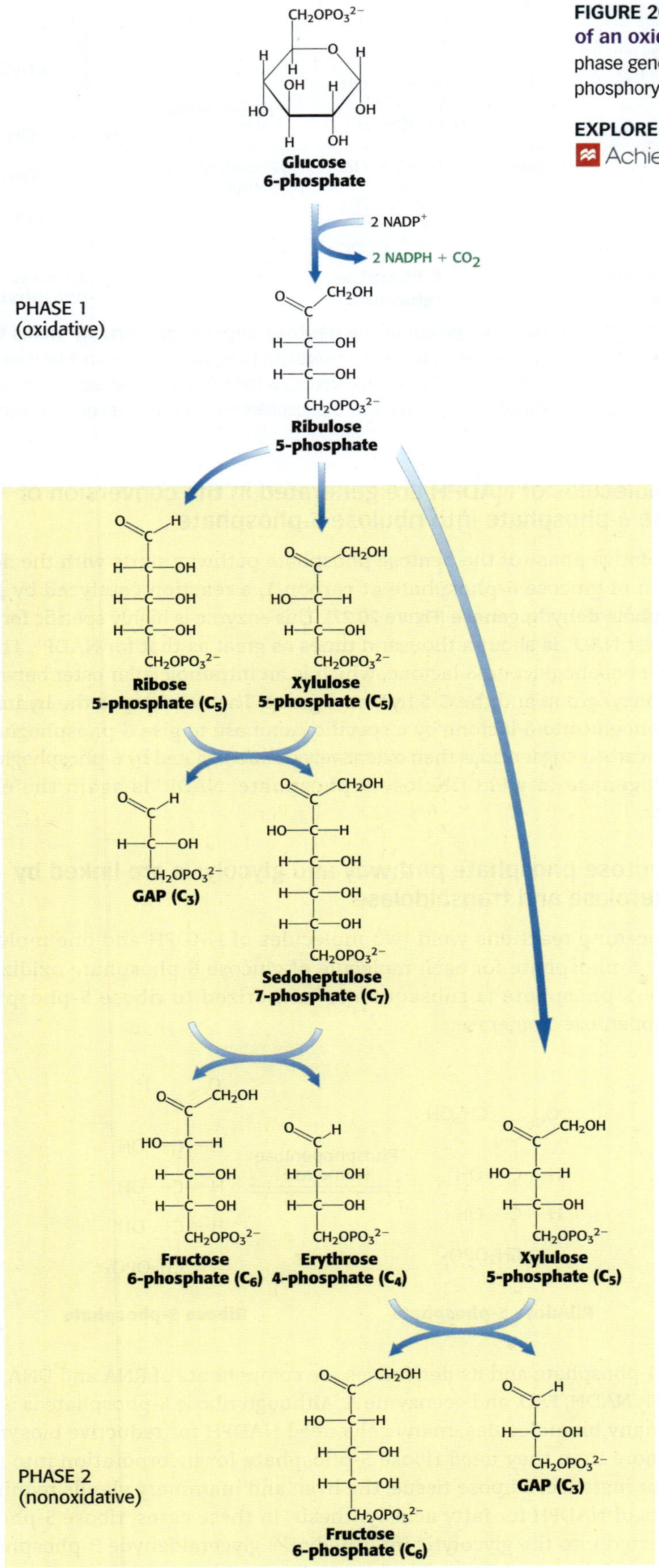

FIGURE 20.21 The pentose phosphate pathway consists of an oxidative and a nonoxidative phase. The oxidative phase generates NADPH, and the nonoxidative phase interconverts phosphorylated sugars.

EXPLORE this pathway further in the Metabolic Map in
≋ Achieve

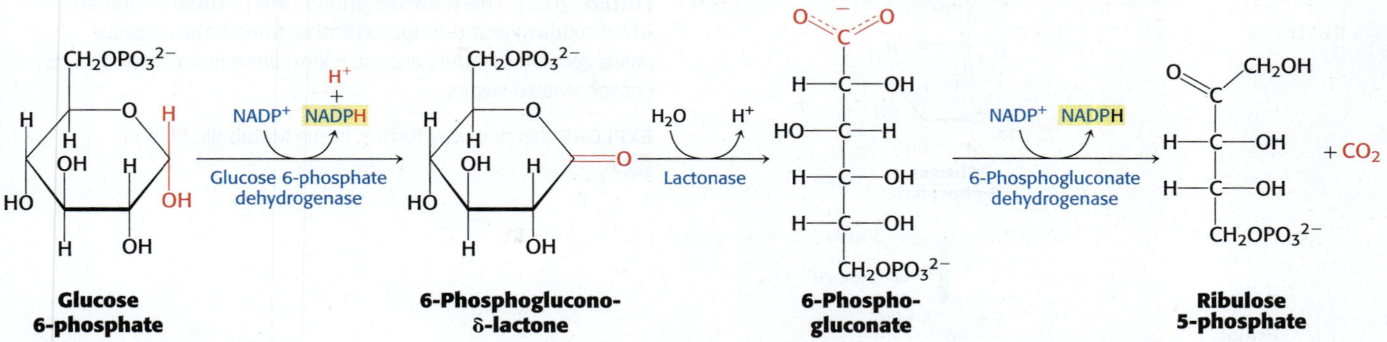

Glucose 6-phosphate → (Glucose 6-phosphate dehydrogenase; NADP⁺ → NADPH + H⁺) → **6-Phosphoglucono-δ-lactone** → (Lactonase; H₂O → H⁺) → **6-Phospho-gluconate** → (6-Phosphogluconate dehydrogenase; NADP⁺ → NADPH) → **Ribulose 5-phosphate** + CO₂

EXPLORE this pathway further in the Metabolic Map in
Achieve

FIGURE 20.22 The oxidative phase of the pentose phosphate pathway yields two molecules of NADPH. Glucose 6-phosphate is oxidized to 6-phosphoglucono-δ-lactone to generate one molecule of NADPH. The lactone product is hydrolyzed to 6-phosphogluconate, which is oxidatively decarboxylated to ribulose 5-phosphate with the generation of a second molecule of NADPH.

Two molecules of NADPH are generated in the conversion of glucose 6-phosphate into ribulose 5-phosphate

The oxidative phase of the pentose phosphate pathway starts with the dehydrogenation of glucose 6-phosphate at carbon 1, a reaction catalyzed by **glucose 6-phosphate dehydrogenase** (Figure 20.22). This enzyme is highly specific for NADP⁺; the K_M for NAD⁺ is about a thousand times as great as that for NADP⁺. The product is 6-phosphoglucono-δ-lactone, which is an intramolecular ester between the C-1 carboxyl group and the C-5 hydroxyl group. The next step is the hydrolysis of 6-phosphoglucono-δ-lactone by a specific lactonase to give 6-phosphogluconate. This six-carbon sugar acid is then oxidatively decarboxylated by 6-phosphogluconate dehydrogenase to yield ribulose 5-phosphate. NADP⁺ is again the electron acceptor.

The pentose phosphate pathway and glycolysis are linked by transketolase and transaldolase

The preceding reactions yield two molecules of NADPH and one molecule of ribulose 5-phosphate for each molecule of glucose 6-phosphate oxidized. The ribulose 5-phosphate is subsequently isomerized to ribose 5-phosphate by phosphopentose isomerase.

Ribulose 5-phosphate → (Phosphopentose isomerase) → **Ribose 5-phosphate**

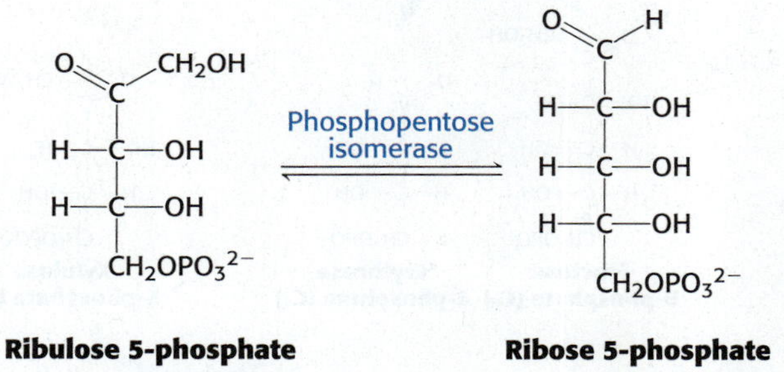

Ribose 5-phosphate and its derivatives are components of RNA and DNA, as well as of ATP, NADH, FAD, and coenzyme A. Although ribose 5-phosphate is a precursor to many biomolecules, many cells need NADPH for reductive biosyntheses much more than they need ribose 5-phosphate for incorporation into nucleotides. For instance, adipose tissue, the liver, and mammary glands require large amounts of NADPH for fatty acid synthesis. In these cases, ribose 5-phosphate is converted into the glycolytic intermediates glyceraldehyde 3-phosphate and fructose 6-phosphate by transketolase and transaldolase. These enzymes create a reversible link between the pentose phosphate pathway and glycolysis by catalyzing three successive reactions:

$$C_5 + C_5 \underset{\text{Transketolase}}{\rightleftharpoons} C_3 + C_7$$

$$C_3 + C_7 \underset{\text{Transketolase}}{\rightleftharpoons} C_6 + C_4$$

$$C_4 + C_5 \underset{\text{Transketolase}}{\rightleftharpoons} C_6 + C_3$$

The net result of these reactions is the formation of two hexoses and one triose from three pentoses:

$$3\,C_5 \rightleftharpoons 2\,C_6 + C_3$$

The first of the three reactions linking the pentose phosphate pathway and glycolysis is the formation of glyceraldehyde 3-phosphate and sedoheptulose 7-phosphate from two pentoses.

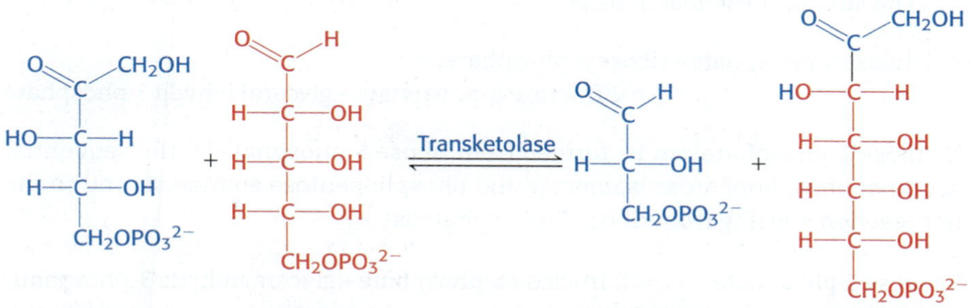

Xylulose 5-phosphate   **Ribose 5-phosphate**   **Glyceraldehyde 3-phosphate**   **Sedoheptulose 7-phosphate**

The donor of the two-carbon unit in this reaction is xylulose 5-phosphate, an epimer of ribulose 5-phosphate. A ketose is a substrate of transketolase only if its hydroxyl group at C-3 has the configuration of xylulose rather than ribulose. Ribulose 5-phosphate is converted into the appropriate epimer for the transketolase reaction by phosphopentose epimerase in the reverse reaction of that which takes place in the Calvin–Benson cycle.

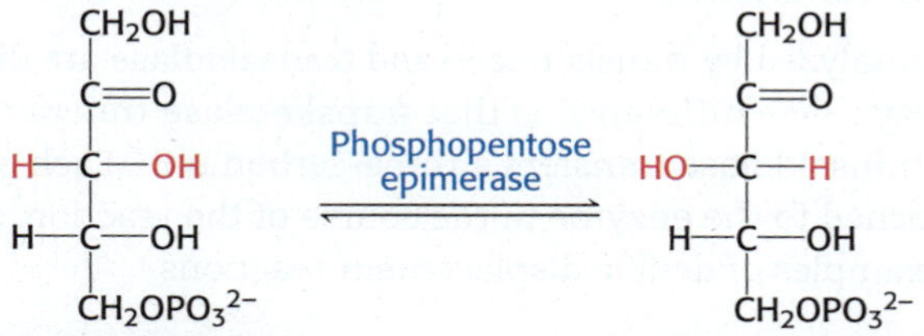

Ribulose 5-phosphate   **Xylulose 5-phosphate**

Glyceraldehyde 3-phosphate and sedoheptulose 7-phosphate generated by the transketolase then react to form fructose 6-phosphate and erythrose 4-phosphate. This synthesis of a four-carbon sugar and a six-carbon sugar is catalyzed by transaldolase.

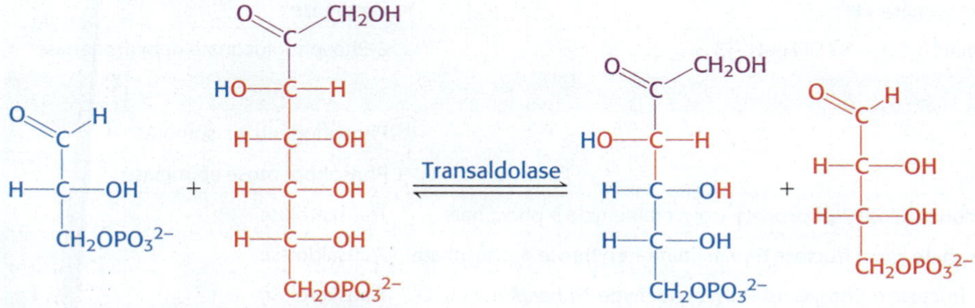

Glyceraldehyde 3-phosphate   **Sedoheptulose 7-phosphate**   **Fructose 6-phosphate**   **Erythrose 4-phosphate**

In the third reaction, transketolase catalyzes the synthesis of fructose 6-phosphate and glyceraldehyde 3-phosphate from erythrose 4-phosphate and xylulose 5-phosphate.

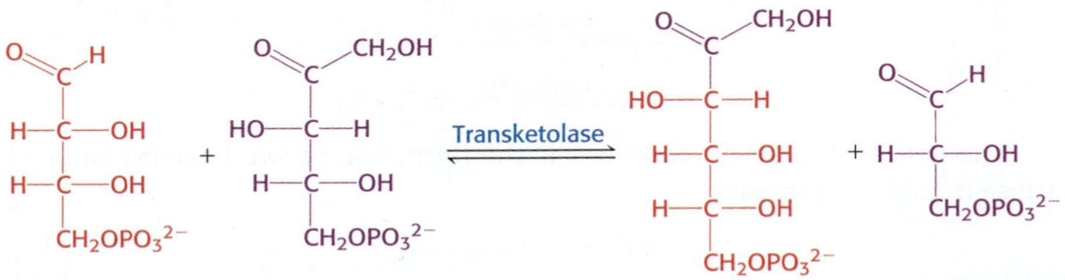

Erythrose   **Xylulose**   **Fructose**   **Glyceraldehyde**
4-phosphate   **5-phosphate**   **6-phosphate**   **3-phosphate**

The sum of these reactions is

2 Xylulose 5-phosphate + ribose 5-phosphate $\rightleftharpoons$
                    2 fructose 6-phosphate + glyceraldehyde 3-phosphate

Xylulose 5-phosphate can be formed from ribose 5-phosphate by the sequential action of phosphopentose isomerase and phosphopentose epimerase, and so the net reaction starting from ribose 5-phosphate is

3 Ribose 5-phosphate $\rightleftharpoons$ 2 fructose 6-phosphate + glyceraldehyde 3-phosphate

Thus, excess ribose 5-phosphate formed by the pentose phosphate pathway can be completely converted into glycolytic intermediates. Moreover, any ribose ingested in the diet can be processed into glycolytic intermediates by this pathway. The carbon skeletons of sugars can be extensively rearranged to meet physiological needs (Table 20.3).

Transketolase and transaldolase stabilize carbanionic intermediates by different mechanisms

The reactions catalyzed by transketolase and transaldolase are distinct, yet similar in many ways. One difference is that transketolase transfers a two-carbon unit, whereas transaldolase transfers a three-carbon unit. Each of these units is transiently attached to the enzyme in the course of the reaction, and thus, these enzymes are examples of double displacement reactions.

TABLE 20.3 Pentose phosphate pathway

Reaction	Enzyme
Oxidative phase	
Glucose 6-phosphate + NADP$^+$ $\longrightarrow$ 6-phosphoglucono-δ-lactone + NADPH + H$^+$	Glucose 6-phosphate dehydrogenase
6-Phosphoglucono-δ-lactone + H$_2$O $\longrightarrow$ 6-phosphogluconate + H$^+$	Lactonase
6-Phosphogluconate + NADP$^+$ $\longrightarrow$ ribulose 5-phosphate + CO$_2$ + NADPH + H$^+$	6-Phosphogluconate dehydrogenase
Nonoxidative phase	
Ribulose 5-phosphate $\rightleftharpoons$ ribose 5-phosphate	Phosphopentose isomerase
Ribulose 5-phosphate $\rightleftharpoons$ xylulose 5-phosphate	Phosphopentose epimerase
Xylulose 5-phosphate + ribose 5-phosphate $\rightleftharpoons$ sedoheptulose 7-phosphate + glyceraldehyde 3-phosphate	Transketolase
Sedoheptulose 7-phosphate + glyceraldehyde 3-phosphate $\rightleftharpoons$ fructose 6-phosphate + erythrose 4-phosphate	Transaldolase
Xylulose 5-phosphate + erythrose 4-phosphate $\rightleftharpoons$ fructose 6-phosphate + glyceraldehyde 3-phosphate	Transketolase

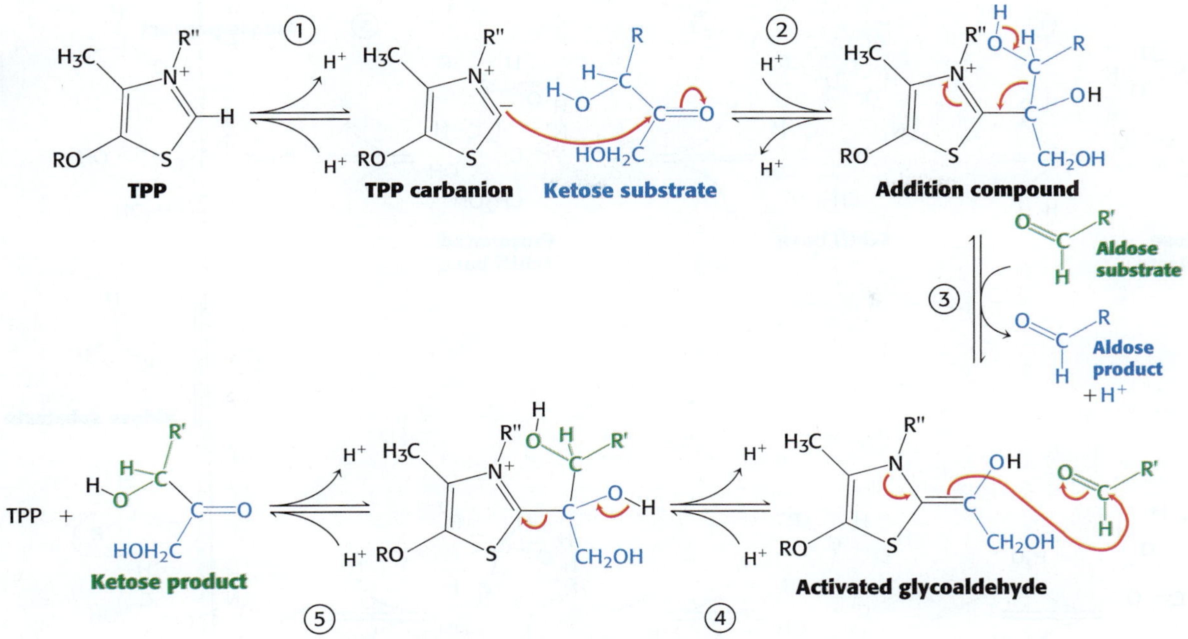

FIGURE 20.23 The transketolase reaction takes place on thiamine pyrophosphate.
(1) Thiamine pyrophosphate (TPP) ionizes to form a carbanion. (2) The carbanion of TPP attacks the ketose substrate. (3) Cleavage of a carbon–carbon bond frees the aldose product and leaves a two-carbon fragment joined to TPP. (4) This activated glycoaldehyde intermediate attacks the aldose substrate to form a new carbon–carbon bond. (5) The ketose product is released, freeing the TPP for the next reaction cycle.

Transketolase reaction. Transketolase contains a tightly bound thiamine pyrophosphate as its prosthetic group. The enzyme transfers a two-carbon glycoaldehyde from a ketose donor to an aldose acceptor. The site of the addition of the two-carbon unit is the thiazole ring of thiamine pyrophosphate (TPP).

Transketolase is homologous to the E_1 subunit of the pyruvate dehydrogenase complex, and the reaction mechanism is similar (**Figure 20.23**), proceeding as follows:

1. The C-2 carbon atom of bound TPP readily ionizes to give a carbanion.

2. The negatively charged carbon atom of this reactive intermediate attacks the carbonyl group of the ketose substrate.

3. The resulting addition compound releases the aldose product to yield an activated glycoaldehyde unit.

4. The positively charged nitrogen atom in the thiazole ring acts as an electron sink in the development of this activated intermediate. The carbonyl group of a suitable aldose acceptor then condenses with the activated glycoaldehyde unit to form a new ketose.

5. The ketose is then released from the enzyme.

Transaldolase reaction. Transaldolase transfers a three-carbon dihydroxyacetone unit from a ketose donor to an aldose acceptor. Transaldolase, in contrast to transketolase, does not contain a prosthetic group. Rather, a Schiff base is formed between the carbonyl group of the ketose substrate and the ε-amino group of a lysine residue at the active site of the enzyme (**Figure 20.24**). This kind of covalent enzyme–substrate intermediate is like that formed in fructose 1,6-bisphosphate

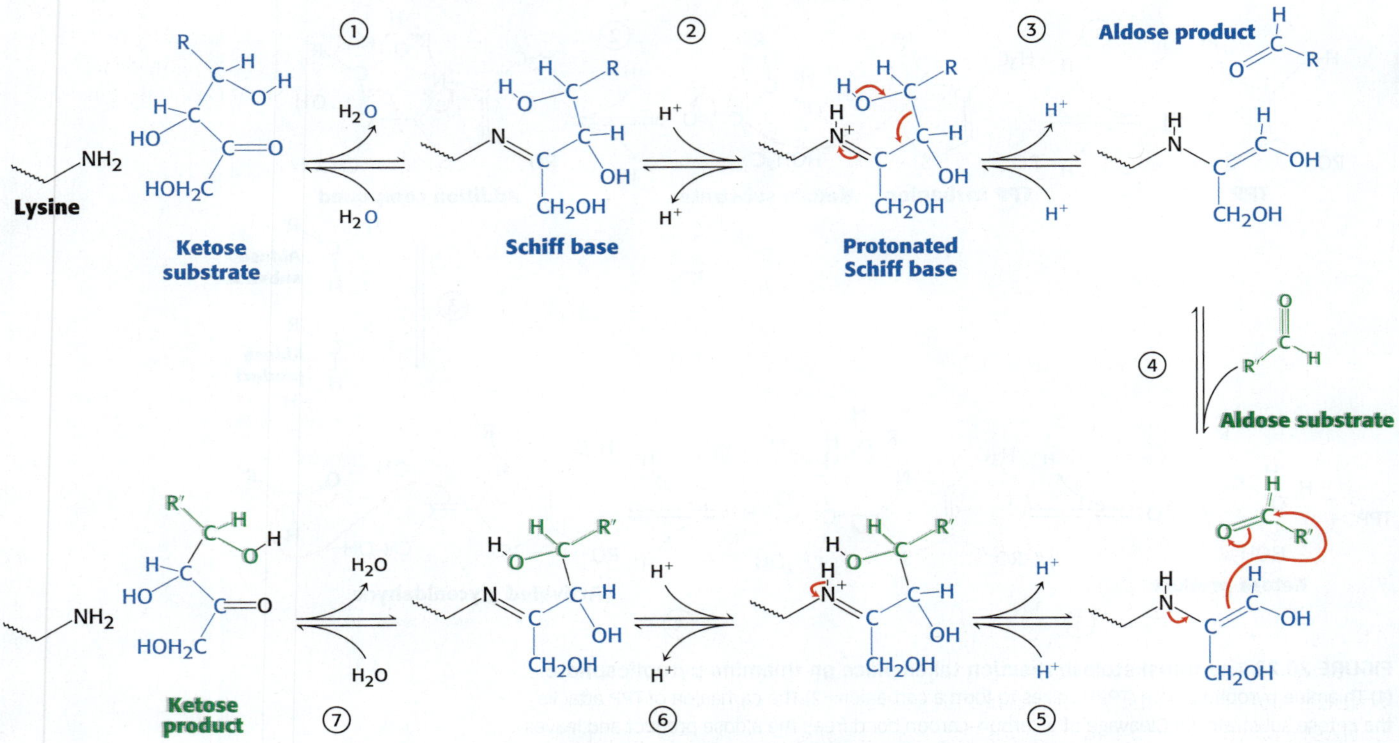

FIGURE 20.24 The transaldolase reaction involves a Schiff base intermediate.
(1) The reaction begins with the formation of a Schiff base between a lysine residue in transaldolase and the ketose substrate. (2) Protonation of the Schiff base occurs. (3) Deprotonation leads to the release of the aldose product, leaving a three-carbon fragment attached to the lysine residue. (4) This intermediate adds to the aldose substrate. (5) Protonation occurs, forming a new carbon–carbon bond. (6) Subsequent deprotonation and (7) hydrolysis of the Schiff base releases the ketose product from the lysine side chain, completing the reaction cycle.

aldolase in the glycolytic pathway and, indeed, the enzymes are homologous. The reaction proceeds as follows:

1. The first step is the formation of the Schiff base.

2. The Schiff base becomes protonated and the bond between C-3 and C-4 is split.

3. Upon reprotonation, the aldose product is released, leaving a three-carbon fragment bound to the enzyme. The negative charge on the Schiff-base carbanion moiety is stabilized by resonance (Figure 20.24). The positively charged nitrogen atom of the protonated Schiff base acts as an electron sink.

4. The Schiff-base adduct is stable until a suitable aldose becomes bound.

5. The dihydroxyacetone moiety then reacts with the carbonyl group of the aldose. Protonation allows the formation of a new carbon–carbon bond.

6. Subsequent deprotonation occurs.

7. Following deprotonation, hydrolysis of the Schiff base releases the ketose product.

The nitrogen atom of the protonated Schiff base plays the same role in transaldolase as the thiazole-ring nitrogen atom does in transketolase. In each enzyme, a group within an intermediate reacts like a carbanion in attacking a carbonyl group to form a new carbon–carbon bond. In each case, the charge on the carbanion is stabilized by resonance (**Figure 20.25**).

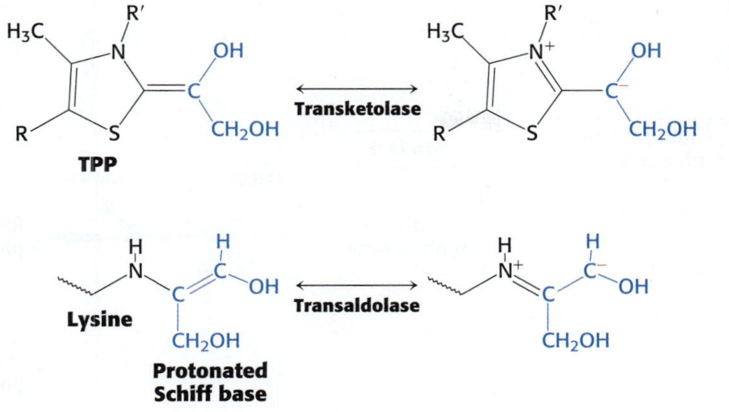

FIGURE 20.25 For transketolase and transaldolase, a carbanion intermediate is stabilized by resonance. In transketolase, TPP stabilizes this intermediate; in transaldolase, a protonated Schiff base plays this role.

20.4 The Metabolism of Glucose 6-Phosphate by the Pentose Phosphate Pathway Is Coordinated with Glycolysis

Glucose 6-phosphate is metabolized by both the glycolytic pathway and the pentose phosphate pathway. How is the processing of this important metabolite partitioned between these two metabolic routes? The cytoplasmic concentration of NADP$^+$ plays a key role in determining the fate of glucose 6-phosphate.

The rate of the oxidative phase of the pentose phosphate pathway is controlled by the level of NADP$^+$

The first reaction in the oxidative branch of the pentose phosphate pathway—the dehydrogenation of glucose 6-phosphate—is essentially irreversible. In fact, this reaction is rate limiting under physiological conditions and serves as the control site for the oxidative branch of the pathway. The most important regulatory factor is the level of NADP$^+$. Low levels of NADP$^+$ limit the dehydrogenation of glucose 6-phosphate because it is needed as the electron acceptor. The effect of low levels of NADP$^+$ is intensified by the fact that NADPH competes with NADP$^+$ in binding to the enzyme. The ratio of NADP$^+$ to NADPH in the cytoplasm of a liver cell from a well-fed rat is about 0.014.

The marked effect of the NADP$^+$ level on the rate of the oxidative phase ensures that NADPH is not generated unless the supply needed for reductive biosyntheses or protection against oxidative stress is low. The nonoxidative phase of the pentose phosphate pathway is controlled primarily by the availability of substrates.

The flow of glucose 6-phosphate depends on the need for NADPH, ribose 5-phosphate, and ATP

We can grasp the intricate interplay between glycolysis and the pentose phosphate pathway by examining the metabolism of glucose 6-phosphate in four different metabolic situations (**Figure 20.26**).

Mode 1. *Much more ribose 5-phosphate than NADPH is required.* For example, rapidly dividing cells need ribose 5-phosphate for the synthesis of nucleotide precursors of DNA. Most of the glucose 6-phosphate is converted into fructose 6-phosphate and glyceraldehyde 3-phosphate by the glycolytic pathway. Transaldolase and transketolase then convert two molecules of fructose 6-phosphate and one molecule of glyceraldehyde 3-phosphate into three molecules of ribose 5-phosphate by a reversal of the reactions described earlier. The stoichiometry of mode 1 is

5 Glucose 6-phosphate + ATP $\longrightarrow$ 6 ribose 5-phosphate + ADP + H$^+$

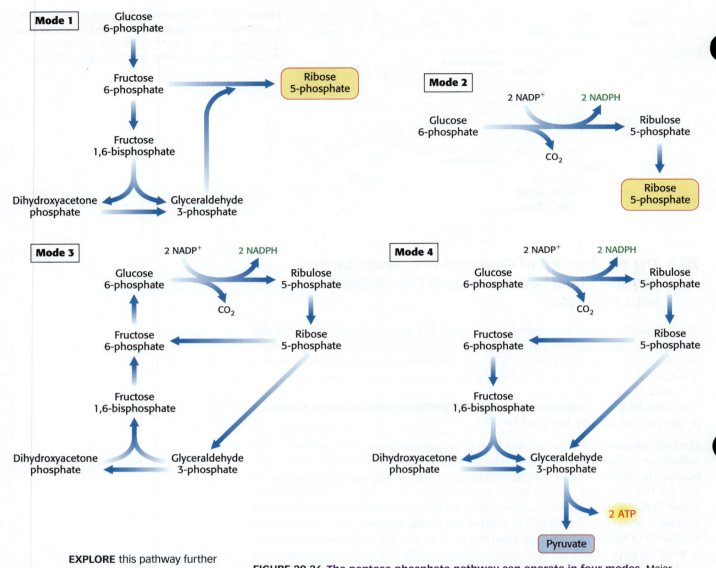

EXPLORE this pathway further in the Metabolic Map in
Achieve

FIGURE 20.26 The pentose phosphate pathway can operate in four modes. Major products are shown in color.

TABLE 20.4 Tissues with active pentose phosphate pathways

Tissue	Function
Adrenal gland	Steroid synthesis
Liver	Fatty acid and cholesterol synthesis
Testes	Steroid synthesis
Adipose tissue	Fatty acid synthesis
Ovary	Steroid synthesis
Mammary gland	Fatty acid synthesis
Red blood cells	Maintenance of reduced glutathione

Mode 2. *The needs for NADPH and for ribose 5-phosphate are balanced.* Under these conditions, glucose 6-phosphate is processed to one molecule of ribulose 5-phosphate while generating two molecules of NADPH. Ribulose 5-phosphate is then converted into ribose 5-phosphate. The stoichiometry of mode 2 is

Glucose 6-phosphate + 2 NADP$^+$ + H$_2$O $\longrightarrow$

$$\text{ribose 5-phosphate} + 2\,\text{NADPH} + 2\,\text{H}^+ + \text{CO}_2$$

Mode 3. *Much more NADPH than ribose 5-phosphate is required.* For example, adipose tissue requires a high level of NADPH for the synthesis of fatty acids (Table 20.4). In this case, glucose 6-phosphate is completely oxidized to CO$_2$.

Three groups of reactions are active in this situation. First, the oxidative phase of the pentose phosphate pathway forms two molecules of NADPH and one molecule of ribulose 5-phosphate. Then, ribulose 5-phosphate is converted into fructose 6-phosphate and glyceraldehyde 3-phosphate by transketolase and transaldolase. Finally, glucose 6-phosphate is resynthesized from fructose

6-phosphate and glyceraldehyde 3-phosphate by the gluconeogenic pathway. The stoichiometries of these three sets of reactions are

$$6 \text{ Glucose 6-phosphate} + 12 \text{ NADP}^+ + 6 \text{ H}_2\text{O} \longrightarrow$$
$$6 \text{ Ribose 5-phosphate} + 12 \text{ NADPH} + 12 \text{ H}^+ + 6 \text{ CO}_2$$

$$6 \text{ Ribose 5-phosphate} \longrightarrow$$
$$4 \text{ Fructose 6-phosphate} + 2 \text{ glyceraldehyde 3-phosphate}$$

$$4 \text{ Fructose 6-phosphate} + 2 \text{ glyceraldehyde 3-phosphate} + \text{H}_2\text{O} \longrightarrow$$
$$5\text{-Glucose 6-phosphate} + \text{P}_i$$

The sum of the mode 3 reactions is

$$\text{Glucose 6-phosphate} + 12 \text{ NADP}^+ + 7 \text{ H}_2\text{O} \longrightarrow 6 \text{ CO}_2 + 12 \text{ NADPH} + 12 \text{ H}^+ + \text{P}_i$$

Thus, the equivalent of glucose 6-phosphate can be completely oxidized to CO_2 with the concomitant generation of NADPH. In essence, ribose 5-phosphate produced by the pentose phosphate pathway is recycled into glucose 6-phosphate by transketolase, transaldolase, and some of the enzymes of the gluconeogenic pathway.

Mode 4. *Both NADPH and ATP are required.* Alternatively, ribulose 5-phosphate formed from glucose 6-phosphate can be converted into pyruvate. Fructose 6-phosphate and glyceraldehyde 3-phosphate derived from ribose 5-phosphate enter the glycolytic pathway rather than reverting to glucose 6-phosphate. In this mode, ATP and NADPH are concomitantly generated, and five of the six carbons of glucose 6-phosphate emerge in pyruvate.

$$3 \text{ Glucose 6-phosphate} + 6 \text{ NADP}^+ + 5 \text{ NAD}^+ + 5 \text{ P}_i + 8 \text{ ADP} \longrightarrow$$
$$5 \text{ pyruvate} + 3 \text{ CO}_2 + 6 \text{ NADPH} + 5 \text{ NADH} + 8 \text{ ATP} + 2 \text{ H}_2\text{O} + 8 \text{ H}^+$$

Pyruvate formed by these reactions can be oxidized to generate more ATP, or it can be used as a building block in a variety of biosyntheses.

The pentose phosphate pathway is required for rapid cell growth

Rapidly dividing cells, such as cancer cells, require ribose 5-phosphate for nucleic acid synthesis and NADPH for fatty acid synthesis, which in turn is required to form membrane lipids. Recall that rapidly dividing cells switch to aerobic glycolysis to meet their ATP needs. Glucose 6-phosphate and glycolytic intermediates are then used to generate NADPH and ribose 5-phosphate using the nonoxidative phase of the pentose phosphate pathway.

The diversion of glycolytic intermediates into the nonoxidative phase is facilitated by the expression of the gene for a pyruvate kinase isozyme, PKM. PKM has a low catalytic activity, creating a bottleneck in the glycolytic pathway. Glycolytic intermediates accumulate and are then used by the pentose phosphate pathway to synthesize NADPH and ribose 5-phosphate. The shunting of phosphorylated intermediates into the nonoxidative phase of the pentose phosphate pathway is further enabled by the inhibition of triose phosphate isomerase by phosphoenolpyruvate, the substrate of PKM.

The Calvin–Benson cycle and the pentose phosphate pathway are essentially mirror images of one another

The complexities of the Calvin–Benson cycle and the pentose phosphate pathway are easier to comprehend if we consider them as functional mirror images of each other. The Calvin–Benson cycle begins with the fixation of CO_2 and proceeds to use NADPH in the synthesis of glucose. The pentose phosphate pathway begins with the oxidation of a glucose-derived carbon atom to CO_2 and concomitantly generates NADPH. The regeneration phase of the Calvin–Benson cycle converts C_6 and C_3 molecules back into the starting material—the C_5 molecule ribulose 1,5-bisphosphate. The pentose phosphate pathway converts a C_5 molecule, ribulose 5-phosphate, into C_6 and C_3 intermediates of the glycolytic pathway. Not surprisingly, in photosynthetic organisms, many enzymes are common to the two pathways. We see here the economy of evolution: the use of identical enzymes for similar reactions with different ends.

❖ SELF–CHECK QUESTION

Name a reaction in the pentose phosphate pathway that involves erythrose 4-phosphate, and name the associated enzyme. Which reaction in the Calvin–Benson cycle involves erythrose 4-phosphate and the same enzyme?

20.5 Glucose 6-Phosphate Dehydrogenase Plays a Key Role in Protection Against Reactive Oxygen Species

The NADPH generated by the pentose phosphate pathway plays a vital role in protecting the cells from reactive oxygen species (ROS). Reactive oxygen species generated in oxidative metabolism inflict damage on all classes of macromolecules and can ultimately lead to cell death. Indeed, ROS are implicated in a number of human diseases, such as diabetes. Reduced **glutathione** (GSH), a tripeptide with a free sulfhydryl group, combats oxidative stress by reducing ROS to harmless forms. Its task accomplished, the glutathione is now in the oxidized form (GSSG) and must be reduced to regenerate GSH. The reducing power is supplied by the NADPH generated by glucose 6-phosphate dehydrogenase in the pentose phosphate pathway. Indeed, cells with reduced levels of glucose 6-phosphate dehydrogenase are especially sensitive to oxidative stress.

⚕ Glucose 6-phosphate dehydrogenase deficiency causes a drug-induced hemolytic anemia

The importance of the pentose phosphate pathway is highlighted by some people's atypical responses to certain drugs. For instance, pamaquine, the first synthetic antimalarial drug introduced in 1926, was associated with the appearance of severe and mysterious ailments. Most patients tolerated the drug well, but a few developed severe symptoms within a few days after therapy was started. Their urine turned black, jaundice developed, and the hemoglobin content of the blood dropped sharply. In some cases, massive destruction of red blood cells caused death.

This drug-induced hemolytic anemia was shown 30 years later to be caused by a deficiency of glucose 6-phosphate dehydrogenase, the enzyme catalyzing the first step in the oxidative branch of the pentose phosphate pathway. The result is a dearth of NADPH in all cells, but this deficiency is most acute in red blood cells because they lack mitochondria and have no alternative means of generating reducing power. This defect, which is inherited on the X chromosome, is the most common condition that results from an enzyme malfunction, affecting hundreds of millions of people.

Pamaquine sensitivity is not simply a historical oddity about malaria treatment many decades ago. Primaquine, an antimalarial closely related to pamaquine, is widely used in malaria-prone regions of the world. Vicine, a pyrimidine glycoside of fava beans (**Figure 20.27**), can also induce hemolysis. People deficient in glucose 6-phosphate dehydrogenase suffer hemolysis from eating fava beans or inhaling the pollen of the fava flowers, a response called favism.

How can we explain hemolysis caused by pamaquine, primaquine, and vicine biochemically? These chemicals are oxidative agents that generate peroxides, reactive oxygen species that can damage membranes as well as other

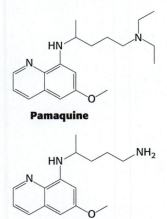

γ-Glutamate

Cysteine

Glycine

Glutathione (reduced)
(γ-Glutamylcysteinylglycine)

Pamaquine

Primaquine

Vicine

FIGURE 20.27 Fava beans produce a pyrimidine glycoside. The Mediterranean plant *Vicia fabia* is the source of fava beans that contain the pyrimidine glycoside vicine.

biomolecules. Peroxides are usually eliminated by the enzyme glutathione peroxidase, which uses reduced glutathione as a reducing agent.

$$2\ GSH + ROOH \xrightarrow{\text{Glutathione peroxidase}} GSSG + H_2O + ROH$$

The major role of NADPH in red blood cells is to reduce the disulfide form of glutathione to the sulfhydryl form. The enzyme that catalyzes the regeneration of reduced glutathione is glutathione reductase.

γ-Glu—Cys—Gly
|
S
|
S   + NADPH + H⁺ ⇌ Glutathione reductase 2 γ-Glu—Cys—Gly   + NADP⁺
| |
γ-Glu—Cys—Gly   SH

Oxidized glutathione (GSSG)   **Reduced glutathione (GSH)**

Red blood cells with a lowered level of reduced glutathione are more susceptible to hemolysis. In the absence of glucose 6-phosphate dehydrogenase, peroxides continue to damage membranes because no NADPH is being produced to restore reduced glutathione. Thus, the answer to our question is that glucose 6-phosphate dehydrogenase is required to maintain reduced glutathione levels to protect against oxidative stress. In the absence of oxidative stress, however, the deficiency is quite benign. The sensitivity to oxidative agents of people having this dehydrogenase deficiency also clearly demonstrates that atypical reactions to drugs may have a genetic basis.

Reduced glutathione is also essential for maintaining the normal structure of red blood cells by maintaining the structure of hemoglobin. The reduced form of glutathione serves as a sulfhydryl buffer that keeps the residues of hemoglobin in the reduced sulfhydryl form. Without adequate levels of reduced glutathione, the hemoglobin sulfhydryl groups can no longer be maintained in the reduced form. Hemoglobin molecules then cross-link with one another to form aggregates called *Heinz bodies* on cell membranes (**Figure 20.28**). Membranes damaged by Heinz bodies and reactive oxygen species become deformed, and the cell is likely to undergo lysis.

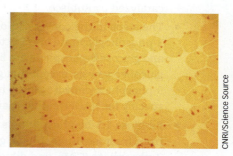

FIGURE 20.28 Heinz bodies are made of denatured hemoglobin. The light micrograph shows red blood cells obtained from a person deficient in glucose 6-phosphate dehydrogenase. The dark particles, called Heinz bodies, inside the cells are clumps of denatured hemoglobin that adhere to the plasma membrane and stain with basic dyes. Red blood cells in such people are highly susceptible to oxidative damage.

A deficiency of glucose 6-phosphate dehydrogenase can be protective against malaria

The incidence of the most common form of glucose 6-phosphate dehydrogenase deficiency, characterized by a 10-fold reduction in enzymatic activity in red blood cells, is 11% among Americans of African heritage compared to less than 1% in Americans of European descent. This high frequency suggests that the deficiency may be advantageous under certain environmental conditions found in Africa but not Europe. Indeed, glucose 6-phosphate dehydrogenase deficiency protects against the deadliest form of malaria, which is prevalent in Africa's subtropical regions. The parasites causing this disease require NADPH for growth. Moreover, infection by the parasites induces oxidative stress in infected human cells. Because the pentose phosphate pathway is compromised, the cells and parasite die from oxidative damage. Thus, glucose 6-phosphate dehydrogenase deficiency is a mechanism of protection against malaria, which accounts for its high frequency in malaria-prone regions of the world. We see here once again the interplay of heredity and environment in the effect of drugs and diet on physiology.

The ability of glucose 6-phosphate dehydrogenase deficiency to protect against malaria does, however, create a public health conundrum. Primaquine is a commonly used and effective antimalarial drug. However, indiscriminate use of primaquine causes hemolysis in individuals deficient in glucose 6-phosphate dehydrogenase. Scientists are actively pursuing alternative ways of addressing malaria, and the first antimalarial vaccine was approved in late 2021.

Summary

20.1 The Calvin–Benson Cycle Synthesizes Hexoses from Carbon Dioxide and Water

- The dark phase of photosynthesis, called the Calvin–Benson cycle, starts with the reaction of CO_2 and ribulose 1,5-bisphosphate to form two molecules of 3-phosphoglycerate, in a reaction catalyzed by Rubisco.

- The steps in the conversion of 3-phosphoglycerate into fructose 6-phosphate and glucose 6-phosphate are like those of gluconeogenesis, except that glyceraldehyde 3-phosphate dehydrogenase in chloroplasts is specific for NADPH rather than NADH.

- Ribulose 1,5-bisphosphate is regenerated from fructose 6-phosphate, glyceraldehyde 3-phosphate (GAP), and dihydroxyacetone phosphate by a complex series of reactions.

- Three molecules of ATP and two molecules of NADPH are consumed for each molecule of CO_2 converted into a hexose.

- Rubisco also catalyzes a competing oxygenase reaction, which produces phosphoglycolate and 3-phosphoglycerate. The process of photorespiration recycles phosphoglycolate, which leads to the release of CO_2 and further consumption of O_2.

20.2 The Activity of the Calvin–Benson Cycle Depends on Environmental Conditions

- Reduced thioredoxin formed by the light-driven transfer of electrons from ferredoxin activates

enzymes of the Calvin–Benson cycle by reducing disulfide bridges.

- The light-induced increase in pH and Mg^{2+} levels of the stroma are important in stimulating the carboxylation of ribulose 1,5-bisphosphate by Rubisco.

- Photorespiration is minimized in tropical plants and grasses, which have an accessory pathway — the C_4 pathway — for concentrating CO_2 at the site of the Calvin–Benson cycle.

- Some plants use crassulacean acid metabolism (CAM) to prevent dehydration. In CAM plants, the C_4 pathway is active during the night; during the day, gas exchange is eliminated and CO_2 is generated from malate stored in vacuoles.

20.3 The Pentose Phosphate Pathway Generates NADPH and Synthesizes Pentoses

- The pentose phosphate pathway consists of two phases: an oxidative phase that generates NADPH and ribulose 5-phosphate in the cytoplasm, and a nonoxidative phase.

- In the nonoxidative phase, ribose 5-phosphate is converted into GAP and fructose 6-phosphate by transketolase and transaldolase. These two enzymes create a reversible link between the pentose phosphate pathway and glycolysis.

- When NADPH needs are paramount, fructose 6-phosphate and GAP from the nonoxidative phase are used to synthesize glucose 6-phosphate for metabolism in the oxidative phase, generating 12 molecules of NADPH for each molecule of glucose 6-phosphate that is completely oxidized to CO_2.

20.4 The Metabolism of Glucose 6-Phosphate by the Pentose Phosphate Pathway Is Coordinated with Glycolysis

- Only the nonoxidative branch of the pathway is significantly active when much more ribose 5-phosphate than NADPH must be synthesized.
- Under these conditions, fructose 6-phosphate and GAP are converted into ribose 5-phosphate without the formation of NADPH. Alternatively, ribose 5-phosphate formed by the oxidative branch can be converted into pyruvate through fructose 6-phosphate and GAP. In this mode, ATP and NADPH are generated.

- The interplay of the glycolytic and pentose phosphate pathways enables the levels of NADPH, ATP, and building blocks such as ribose 5-phosphate and pyruvate to be continuously adjusted to meet cellular needs.

20.5 Glucose 6-Phosphate Dehydrogenase Plays a Key Role in Protection Against Reactive Oxygen Species

- NADPH generated by glucose 6-phosphate dehydrogenase maintains the appropriate levels of reduced glutathione required to combat oxidative stress, and it ensures the proper reducing environment in the cell.
- Cells with diminished glucose 6-phosphate dehydrogenase activity are especially sensitive to oxidative stress.

Key Terms

Calvin–Benson cycle (p. 610)

ribulose 1,5-bisphosphate carboxylase/oxygenase (Rubisco) (p. 611)

peroxisome (p. 615)

photorespiration (p. 615)

transketolase (p. 615)

aldolase (p. 616)

starch (p. 618)

sucrose (p. 618)

synthetic biology (p. 620)

C_4 pathway (Hatch–Slack pathway) (p. 622)

crassulacean acid metabolism (CAM) (p. 624)

pentose phosphate pathway (p. 624)

glucose 6-phosphate dehydrogenase (p. 626)

glutathione (p. 634)

Problems

1. In chloroplasts, glyceraldehyde 3-phosphate dehydrogenase uses NADPH as a cofactor in the synthesis of glucose. In cytoplasmic gluconeogenesis, however, the isozyme of this dehydrogenase uses NADH. Why is it advantageous for the enzyme in the chloroplast to use NADPH? ❖ 1

(a) NADPH is harmful to chloroplasts when levels become too high.

(b) NADPH is abundant in chloroplasts, because it is generated by the light reactions.

(c) The presence of NADH increases the rate of wasteful photorespiration.

2. Arrange the steps to outline the Calvin–Benson cycle (first stage to last stage) ❖ 1

(a) Conversion of 3-phosphoglycerate into hexose

(b) Regeneration of ribulose 1,5-bisphosphate

(c) Fixation of CO_2 with ribulose 1,5-bisphosphate

(d) Formation of 3-phosphoglycerate

3. Match each characteristic to either the Calvin–Benson cycle or the Krebs cycle. ❖ 1

(a) Matrix

(b) Requires high-energy electrons from NADPH

(c) Regenerates ribose 1,5-bisphosphate

(d) Regenerates oxaloacetate

(e) Releases CO_2

(f) Stroma

(g) Carbon chemistry for photosynthesis

(h) Fixes CO_2

(i) Harvests high-energy electrons to form NADH

(j) Carbon chemistry for cellular respiration

4. What statement best describes photorespiration? ❖ 1

(a) Plants transform light energy into ATP.

(b) Plants consume NADPH to reduce carbon atoms.

(c) Plants consume O_2 and release CO_2.

(d) Plants oxidize glucose to generate NADPH.

5. Rubisco catalyzes both a carboxylase reaction and a wasteful oxygenase reaction. Use the kinetic parameters for the two reactions in the table to determine the specificity constant, k_{cat}/K_M, for each substrate in units of $M^{-1}s^{-1}$. ❖ 1

Substrate	$k_{cat}(s^{-1})$	$K_M(mM)$
CO_2	3	10
O_2	2	500

6. Which of the following stromal changes occur in response to light that regulate the Calvin–Benson cycle? ❖ 2

(a) Increased pH

(b) Increased levels of Mg^{2+}

(c) Decreased levels of carbamate

(d) Decreased concentration of NADPH

(e) Increased amounts of the reduced form of ferredoxin

7. Rubisco requires a molecule of CO_2 covalently bound to lysine 201 for catalytic activity. The carboxylation of Rubisco is favored by high pH and high Mg^{2+} concentration in the stroma. Why does it make physiological sense for these conditions to favor Rubisco carboxylation? ❖ 2

(a) They are present when the light reactions occur.

(b) They are the normal environment in the stroma.

(c) They are present in the dark, when Calvin–Benson cycle reactions occur.

(d) They concentrate CO_2 and prevent photorespiration.

8. The enzyme ribose 5-phosphate isomerase catalyzes the conversion between ribose 5-phosphate (R5P) and ribulose 5-phosphate (Ru5P) through an enediolate intermediate. In the Calvin–Benson cycle, Ru5P is used to replenish ribulose 1,5-bisphosphate, a substrate for Rubisco.

For the conversion of R5P to Ru5P, if $G^{\circ\prime} = 0.460$ kJ/mol and $\Delta G = 3.80$ kJ/mol, what is the ratio of Ru5P to R5P at 298 K? Which of the statements is true?

(a) This reaction is favorable, and it is likely regulated.

(b) This reaction is not favorable, and it is likely regulated.

(c) This reaction is favorable, and it is not likely regulated.

(d) This reaction is not favorable, and it is not likely regulated.

9. Which phrases describe the light reactions of photosynthesis, which describe the Calvin–Benson cycle, and which describe both? ❖ 2

(a) Capture light energy using chlorophyll

(b) Use CO_2

(c) Require water

(d) Take place in the chloroplasts of plants

(e) Generate oxygen gas

(f) Synthesize ATP and NADPH

(g) Produce sugars

(h) Need ribulose 1,5-bisphosphate

10. The Calvin–Benson cycle must repeat six times in order for a cell to have enough material to construct a complete glucose molecule. A variety of molecules are used and produced during this process. Sort the molecules into three categories: molecules brought in and used in six turns of the Calvin–Benson cycle, molecules produced during six turns of the Calvin–Benson cycle that leave the cycle, and molecules used and regenerated within the Calvin–Benson cycle during six turns. ❖ 1

(a) 18 ATP

(b) 18 ADP

(c) 6 Ribulose 1,5-bisphosphate

(d) 12 NADPH

(e) 12 $NADP^+ + 12\ H^+$

(f) 6 CO_2

(g) 2 glyceraldehyde 3-phosphate

(h) 10 glyceraldehyde 3-phosphate

11. Fill in the blanks: Insulin _____ glucose uptake and oxidation. Entry of glucose into cells is accompanied by its phosphorylation to form glucose 6-phosphate (G6P). Not only is G6P able to be processed glycolytically to generate ATP, but it can also be shunted into the pentose phosphate pathway (PPP) to generate _____ and _____. An abundance of G6P, as would be present upon insulin signaling, would likely _____ the flux of G6P into the PPP. ❖ 3

12. The net equation for the oxidative reactions of the pentose phosphate pathway is

$$\text{Glucose 6-phosphate} + 2\ NADP^+ + H_2O \longrightarrow$$
$$\text{ribulose 5-phosphate} + 2\ NADPH + CO_2 + 2\ H^+$$

Select true statements about the pentose phosphate pathway. ❖ 3

(a) Glucose is a precursor of the pentose phosphate pathway.

(b) The pentose phosphate pathway contributes to nucleotide synthesis.

(c) Carbon atoms from the pentose sugar products may enter the glycolytic pathway.

(d) Glucose 6-phosphate is reduced to ribulose 5-phosphate in this series of reactions.

13. Consider a cell that requires much more ribose 5-phosphate than NADPH. The cell needs ribose 5-phosphate but has a relatively high concentration of NADPH and a low concentration of $NADP^+$. These conditions may occur in rapidly dividing cells. Choose the statements that describe the fate of glucose 6-phosphate, glycolytic intermediates, and pentose phosphate pathway intermediates in this cell. ❖ 3

(a) Conversion of glycolytic intermediates to ribose 5-phosphate requires transketolase and transaldolase.

(b) Most of the glucose 6-phosphate enters the glycolytic pathway and is converted to fructose 6-phosphate and glyceraldehyde 3-phosphate.

(c) The oxidative pentose phosphate pathway reaction catalyzed by glucose 6-phosphate dehydrogenase is slowed down.

(d) Three molecules of glyceraldehyde 3-phosphate and two molecules of fructose 6-phosphate are used to generate five molecules of ribose 5-phosphate.

(e) Under the given conditions, all triose phosphates are converted to pyruvate by the glycolytic pathway.

(f) Most of the glucose 6-phosphate enters the pentose phosphate pathway.

14. Radioactive-labeling experiments can yield estimates of the amount of glucose 6-phosphate metabolized by the pentose phosphate pathway (PPP) and the amount metabolized aerobically by the combined action of glycolysis, the pyruvate dehydrogenase complex, and the citric acid cycle. A researcher has a tissue sample and two radioactively labeled glucose samples, one labeled with ^{14}C at C-1 and the other labeled with ^{14}C at C-6. To determine the relative activity of aerobic glucose metabolism compared with glucose metabolism by the PPP, the research should compare the radioactivity of which product? What is the rationale for the experiment? ❖ 3

(a) Aerobic glucose metabolism decarboxylates C-1 and C-6, whereas the PPP only decarboxylates C-1.

(b) Glucose 6-phosphate is only metabolized by the PPP.

(c) Aerobic glucose metabolism in oxidative, whereas the PPP is completely nonoxidative.

(d) Aerobic glucose metabolism reduces NAD^+, whereas the PPP reduces $NADP^+$.

(e) In the PPP, C-1 of glucose 6-phosphate becomes C-1 of ribose 5-phosphate.

15. Which molecule controls the rate of the pentose phosphate pathway? ❖ 3

(a) $NADP^+$/NADPH   (d) Glucose

(b) Ribose 5-phosphate   (e) NAD^+/NADH

(c) ADP/ATP

16. Think about which reaction is influenced by the molecule in Question 15. Choose the enzyme that catalyzes the reaction.

(a) Glucose 6-phosphate dehydrogenase

(b) Transketolase

(c) Transaldolase

(d) Lactonase

(e) Phosphopentose isomerase

(f) Phosphopentose epimerase

17. Provide an example of pathogen resistance provided by the loss of the activity of an enzyme (name the pathogen and the enzyme). Provide an example of another genetic modification associated with resistance to this pathogen. ❖ 4

18. Why might a diet that includes fava beans pose a risk for some individuals? ❖ 4

19. What is the stoichiometry of the synthesis of (a) ribose 5-phosphate from glucose 6-phosphate without the concomitant generation of NADPH, and (b) NADPH from glucose 6-phosphate without concomitant formation of pentose sugars? ❖ 3

20. Why is the C_4 pathway valuable for tropical plants? ❖ 1

21. Which of the following is not a component of the pentose phosphate pathway? ❖ 3

(a) NADPH

(b) CO_2

(c) Ribose 5-phosphate

(d) Phosphoenolpyruvate

(e) Erythrose 4-phosphate

Glycogen Metabolism

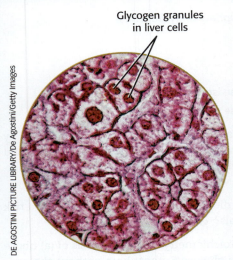

Glycogen granules in liver cells

DE AGOSTINI PICTURE LIBRARY/De Agostini/Getty Images

WESTOCK PRODUCTIONS/Shutterstock

The brain requires glucose at all times, while muscles require extra glucose during bouts of exercise. Glycogen, a molecule used to store glucose, is used by the liver to keep the brain fed between meals, including while you sleep. The same molecule is used by muscles to power intense exercise.

OUTLINE

21.1 Glycogen Metabolism Is the Regulated Release and Storage of Glucose in Multiple Tissues

21.2 Glycogen Breakdown Requires the Interplay of Several Enzymes

21.3 Phosphorylase Is Regulated by Allosteric Interactions and Controlled by Reversible Phosphorylation

21.4 Glucagon and Epinephrine Signal the Need for Glycogen Breakdown

21.5 Glycogen Synthesis Requires Several Enzymes and Uridine Diphosphate Glucose

21.6 Glycogen Breakdown and Synthesis Are Reciprocally Controlled by Hormones

❖ LEARNING GOALS

By the end of this chapter, you should be able to:

1. List and describe the steps of glycogen breakdown and identify the enzymes required.
2. Explain the regulation of glycogen breakdown.
3. Describe the steps of glycogen synthesis and identify the enzymes required.
4. Describe how glycogen degradation and synthesis are coordinated and regulated.
5. Compare the different roles of glycogen metabolism in liver and muscle.

Glucose is an important fuel and a key precursor for the biosynthesis of many molecules. However, glucose itself cannot be stored, because high concentrations of glucose disrupt the osmotic balance of the cell, which would cause cell damage or death. Instead, glucose is stored as glycogen, a significantly less osmotically active and highly branched polymer that can be rapidly broken down to yield glucose molecules when energy is needed. Glycogen is present in bacteria, archaea, protists, and animals, while plants store glucose primarily as starch, a similar, but less-branched polymer.

Glycogen is not as reduced as fatty acids are and consequently is not as energy-rich, so why isn't all excess fuel stored as fatty acids rather than as glycogen? In animals, the controlled release of glucose from glycogen maintains blood-glucose concentration between meals, including during sleep. The circulating blood keeps the brain supplied with glucose, which is virtually the only fuel it uses, except during prolonged starvation (Chapter 24). Moreover, the readily mobilized glucose from glycogen is a good source of energy for sudden, strenuous activity as, unlike fatty acids, it can be metabolized in the absence of oxygen and can thus supply energy for anaerobic activity (Section 16.2).

21.1 Glycogen Metabolism Is the Regulated Release and Storage of Glucose in Multiple Tissues

In humans, most tissues have some glycogen, although the two major sites of glycogen storage are the liver and skeletal muscle. The concentration of glycogen is higher in the liver than in muscle (10% versus 2% by weight), but more glycogen is stored in skeletal muscle overall because of muscle's much greater mass. Glycogen is present in the cytoplasm, with the molecule appearing as granules consisting of multiple glycogen molecules (**Figure 21.1**). An individual glycogen molecule has approximately 12 layers of glucose molecules and can be as large as 40 nm, containing approximately 55,000 glucose residues and a single protein, called *glycogenin*, at the core (**Figure 21.2**).

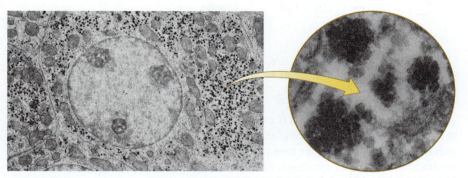

FIGURE 21.1 Liver and muscle tissue are the primary storage sites for glycogen. An electron micrograph of a liver cell reveals a high density of glycogen granules (the dark spots) in the cytoplasm. [Credits: (Left) Courtesy of Dr. George Palade/Yale University, Harvey Cushing/John Hay Whitney Medical Library; (Right) H. Jastrow from Dr. Jastrow's microscopic atlas (www.drjastrow.de).]

FIGURE 21.2 Glycogen is a complex homopolymer of glucose with a protein core. At the core of the glycogen molecule is the protein glycogenin (yellow). Each line represents glucose molecules joined by α-1,4-glycosidic linkages. The nonreducing ends of the glycogen molecule form the surface of the glycogen granule, where degradation takes place. [Information from R. Melendez et al., *Biophys. J.* 77:1327–1332, 1999.]

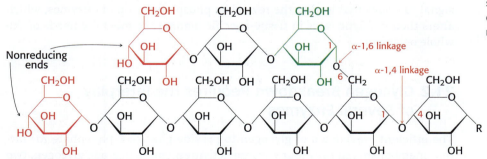

FIGURE 21.3 The fundamental structure of glycogen is a linear polymer with occasional branches. In this structure of two outer branches of a glycogen molecule, the residues at the nonreducing ends are shown in red and the residue that starts a branch is shown in green. The rest of the glycogen molecule is represented by R.

Recall that most of the glucose residues in glycogen are linked by α-1,4-glycosidic bonds (**Figure 21.3**); but branches at about every 12 residues are created by α-1,6-glycosidic bonds. α-Glycosidic linkages form open helical polymers, whereas β linkages produce nearly straight strands that form structural fibrils, as in cellulose (Figure 11.14).

Glycogen degradation and synthesis are relatively simple biochemical processes. Glycogen degradation consists of three steps, ultimately producing glucose 6-phosphate for further metabolism. Glucose 6-phosphate has three

FIGURE 21.4 Glucose 6-phosphate can have many metabolic fates. Glucose 6-phosphate derived from glycogen can be (1) used as a fuel for anaerobic or aerobic metabolism as in, for instance, muscle; (2) converted into free glucose in the liver and subsequently released into the blood; or (3) processed by the pentose phosphate pathway to generate NADPH and ribose in a variety of tissues.

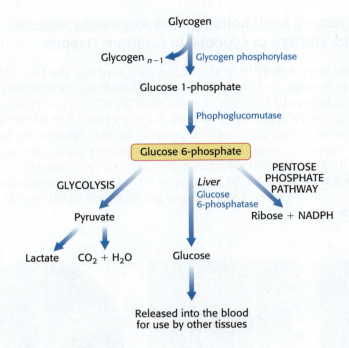

possible fates (**Figure 21.4**): (1) it can be metabolized by glycolysis, (2) in the liver it can be converted into free glucose for release into the bloodstream, or (3) it can be processed by the pentose phosphate pathway to yield NADPH and ribose derivatives. Glycogen synthesis also requires several steps; it takes place when glucose is abundant and glycogen is depleted.

The regulation of glycogen degradation and synthesis is complex, in part because all of the enzymes involved in glycogen metabolism and its regulation are associated with the glycogen particle. Several enzymes taking part in glycogen metabolism allosterically respond to metabolites that signal the energy needs of the cell. Through these allosteric responses, enzyme activity is adjusted to meet the needs of the individual cell. In contrast, hormones may initiate signal cascades that lead to the reversible phosphorylation of enzymes, which alters their catalytic rates in tissue-specific manners to meet the needs of the whole organism.

21.2 Glycogen Breakdown Requires the Interplay of Several Enzymes

The efficient breakdown of glycogen to provide glucose 6-phosphate for further metabolism requires four enzyme activities: one to degrade glycogen, two to remodel glycogen so that it can remain a substrate for degradation, and one to convert the product of glycogen breakdown into a form suitable for further metabolism. We will examine each of these activities in turn.

Phosphorylase catalyzes the phosphorolytic cleavage of glycogen to release glucose 1-phosphate

The key enzyme in glycogen breakdown, **glycogen phosphorylase** (often just called **phosphorylase**), cleaves its substrate by the addition of orthophosphate (P_i) to yield glucose 1-phosphate. The cleavage of a bond by the addition of orthophosphate is referred to as **phosphorolysis**.

$$\text{Glycogen} + P_i \rightleftharpoons \text{glucose 1-phosphate} + \text{glycogen}$$
$$(n \text{ residues}) \qquad\qquad\qquad\qquad (n-1 \text{ residues})$$

Phosphorylase catalyzes the sequential removal of glucosyl residues from the nonreducing ends of the glycogen molecule (the ends with a free OH group on carbon 4). Orthophosphate splits the glycosidic linkage between C-1 of the terminal residue and C-4 of the adjacent one. Specifically, it cleaves the bond between the C-1 carbon atom and the glycosidic oxygen atom, and the α configuration at C-1 is retained.

Glycogen
(*n* residues)   **Glucose 1-phosphate**   **Glycogen**
(*n* − 1 residues)

Glucose 1-phosphate released from glycogen can be readily converted into glucose 6-phosphate, an important metabolic intermediate, by the enzyme phosphoglucomutase.

The reaction catalyzed by phosphorylase is readily reversible in vitro. At pH 6.8, the equilibrium ratio of orthophosphate to glucose 1-phosphate is 3.6. The value of $\Delta G^{\circ\prime}$ for this reaction is small because a glycosidic bond is replaced by a phosphoryl ester bond that has a nearly equal transfer potential. However, phosphorolysis proceeds far in the direction of glycogen breakdown in vivo because the $[P_i]$/[glucose 1-phosphate] ratio is usually greater than 100, substantially favoring phosphorolysis. We see here an example of how the cell can alter the free-energy change to favor a reaction's occurrence by altering the ratio of substrate and product.

The phosphorolytic cleavage of glycogen is energetically advantageous because the released sugar is already phosphorylated. In contrast, a hydrolytic cleavage would yield glucose, which would then have to be phosphorylated at the expense of a molecule of ATP to enter the glycolytic pathway. An additional advantage of phosphorolytic cleavage for muscle cells is that no transporters exist for glucose 1-phosphate, which is negatively charged under physiological conditions, and so it cannot be transported or diffuse out of the cell.

Mechanism: Pyridoxal phosphate participates in the phosphorolytic cleavage of glycogen

The special challenge faced by phosphorylase is to cleave glycogen phosphorolytically rather than hydrolytically to save the ATP required to phosphorylate free glucose. Thus, water must be excluded from the active site. Phosphorylase is a dimer of two identical 97-kDa subunits, each compactly folded into an amino-terminal domain containing a glycogen-binding site and a carboxyl-terminal domain (**Figure 21.5**). The catalytic site in each subunit is located in a deep crevice formed by residues from both domains. Substrates bind synergistically, causing the crevice to narrow, thereby excluding water.

What is the mechanistic basis of the phosphorolytic cleavage of glycogen? Several clues suggest a mechanism. First, both the glycogen substrate and the glucose 1-phosphate product have an α configuration at C-1. A direct attack by phosphate on C-1 of a sugar would invert the configuration at this carbon atom because the reaction would proceed through a pentacovalent transition state. The observation that the glucose 1-phosphate formed has an α rather than a β configuration suggests that an even number of steps (most simply, two) is required. A likely explanation for these results is that a carbocation intermediate is formed from the glucose residue.

A second clue to the catalytic mechanism of phosphorylase is its requirement for the coenzyme **pyridoxal phosphate (PLP)**, a derivative of pyridoxine (vitamin B_6, Section 15.4). The aldehyde group of this coenzyme forms a Schiff-base linkage

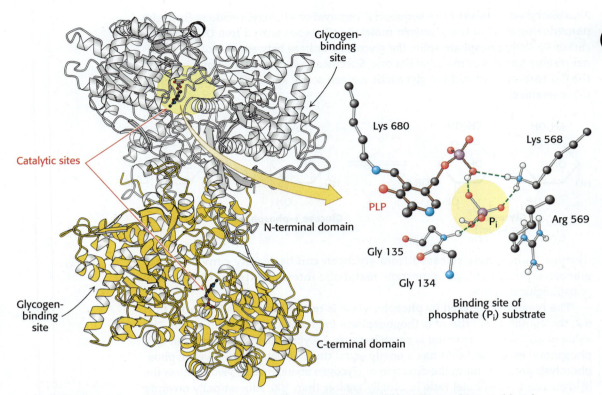

Catalytic sites

Glycogen-binding site

PLP

N-terminal domain

Glycogen-binding site

C-terminal domain

Lys 680

Lys 568

Arg 569

Gly 135

P_i

Gly 134

Binding site of phosphate (P_i) substrate

FIGURE 21.5 Glycogen phosphorylase forms a homodimer with a pyridoxal phosphate (PLP) group in each active site. One subunit is shown in white and the other in yellow. Each catalytic site includes a PLP group linked to lysine 680 of the enzyme. The binding site for the phosphate (P_i) substrate is shown. Notice that the catalytic site lies between the C-terminal domain and the glycogen-binding site. A narrow crevice, which binds four or five glucose units of glycogen, connects the two sites. The separation of the sites allows the catalytic site to phosphorolyze several glucose units before the enzyme must rebind the glycogen substrate. [Drawn from 1NOI.pdb.]

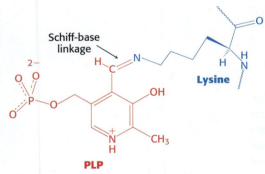

Schiff-base linkage

Lysine

PLP

FIGURE 21.6 PLP forms a Schiff-base linkage with lysine. A pyridoxal phosphate (PLP) group (red) forms a Schiff base with a lysine residue (blue) at the active site of phosphorylase, where it functions as a general acid–base catalyst.

with a specific lysine side chain of the enzyme (**Figure 21.6**). Structural studies indicate that the reacting orthophosphate group takes a position between the 5′-phosphate group of PLP and the glycogen substrate (**Figure 21.7**). The 5′-phosphate group of PLP acts in tandem with orthophosphate by serving as a proton donor and then as a proton acceptor (i.e., as a general acid–base catalyst). Orthophosphate (in the HPO_4^{2-} form) donates a proton to the oxygen atom attached to carbon 4 of the departing glycogen chain and simultaneously acquires a proton from PLP. The carbocation intermediate formed in this step is then attacked by orthophosphate to form α-glucose 1-phosphate, with the concomitant return of a proton to pyridoxal phosphate.

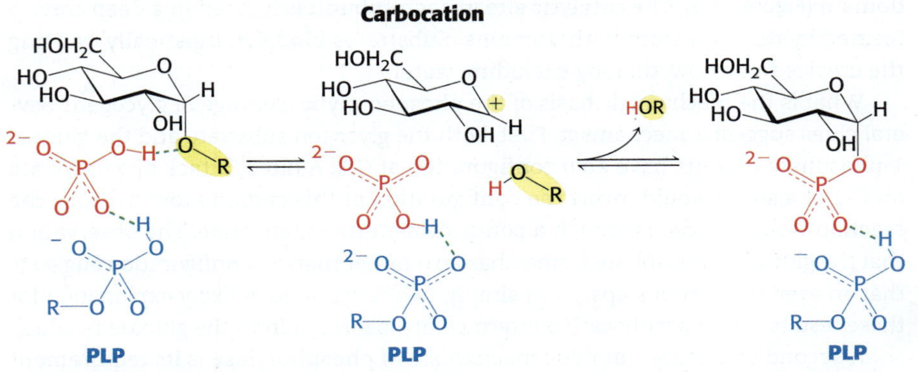

Carbocation

PLP

PLP

PLP

FIGURE 21.7 The mechanism of phosphorylase involves PLP and the formation of a carbocation. A bound HPO_4^{2-} group (red) favors the cleavage of the glycosidic bond by donating a proton to the C-4 oxygen of the departing glycosyl group (black).

The glycogen-binding site is 30 Å away from the catalytic site (Figure 21.5), but it is connected to the catalytic site by a narrow crevice able to accommodate four or five glucose units. The large separation between the binding site and the catalytic site enables the enzyme to phosphorolyze many residues without having to dissociate and reassociate after each catalytic cycle. An enzyme that can catalyze many reactions without having to dissociate and reassociate after each catalytic step is said to be *processive*—a property of enzymes that synthesize and degrade large polymers. We will see such enzymes again when we consider DNA and RNA synthesis.

Debranching enzyme also is needed for the breakdown of glycogen

Glycogen phosphorylase acting alone degrades glycogen to a limited extent. The enzyme can break α-1,4-glycosidic bonds on glycogen branches but soon encounters an obstacle. The α-1,6-glycosidic bonds at the branch points are not susceptible to cleavage by phosphorylase. Indeed, phosphorylase stops cleaving α-1,4 linkages when it reaches a terminal residue four residues away from a branch point. Because about 1 in 12 residues is branched, cleavage by the phosphorylase alone would come to a halt after the release of eight glucose molecules per branch.

How can the remainder of the glycogen molecule be mobilized for use as a fuel? Two additional enzyme activities are necessary to remodel glycogen for continued degradation by the phosphorylase (**Figure 21.8**).

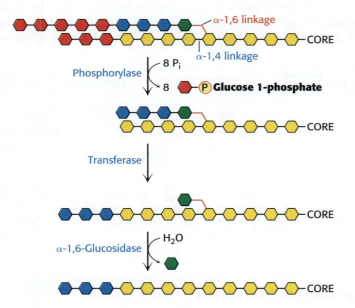

FIGURE 21.8 Glycogen remodeling requires three distinct catalytic activities. First, α-1,4-glycosidic bonds on each branch are cleaved by phosphorylase, leaving four residues along each branch. The transferase shifts a block of three glucosyl residues from one outer branch to the other. In this reaction, the α-1,4-glycosidic link between the blue and the green residues is broken and a new α-1,4 link between the blue and the yellow residues is formed. The green residue is then removed by α-1,6-glucosidase, leaving a linear chain with all α-1,4 linkages, suitable for further cleavage by phosphorylase.

A transferase shifts a block of three glucosyl residues from one outer branch to another, thus exposing a single glucose residue joined by an α-1,6-glycosidic linkage. The α-1,6-glucosidase activity then hydrolyzes the α-1,6-glycosidic bond.

Glycogen
(*n* residues)

Glucose

Glycogen
(*n* − 1 residues)

In eukaryotes, the transferase and the α-1,6-glucosidase activities are present in a single 160-kDa protein, called *debranching enzyme*, providing yet another example of a bifunctional enzyme. A free glucose molecule is released and then phosphorylated by the glycolytic enzyme hexokinase, if the glucose is to be processed by glycolysis or the pentose phosphate pathway. Thus, the transferase and α-1,6-glucosidase activities of debranching enzyme convert the branched structure into a linear one, which paves the way for further cleavage by phosphorylase.

Phosphoglucomutase converts glucose 1-phosphate into glucose 6-phosphate

Glucose 1-phosphate formed in the phosphorolytic cleavage of glycogen must be converted into glucose 6-phosphate to enter the metabolic mainstream. This shift of a phosphoryl group is catalyzed by phosphoglucomutase. Recall that this enzyme is also used in galactose metabolism (Section 16.2). To cause this shift, the enzyme exchanges a phosphoryl group with the substrate (Figure 21.9). The key to this mechanism is that the catalytic site of an active mutase molecule already contains a phosphorylated serine residue. This phosphoryl group is transferred from the serine residue to the C-6 hydroxyl group of glucose 1-phosphate to form glucose 1,6-bisphosphate. To regenerate the phosphoserine and complete the mechanism, the C-1 phosphoryl group of the bisphosphate intermediate is then shuttled to the same serine residue, resulting in the formation of glucose 6-phosphate and the regeneration of the phosphoenzyme.

FIGURE 21.9 Phosphoglucomutase transfers a phosphoryl group to the substrate. A different phosphoryl group is transferred back to restore the enzyme to its initial state.

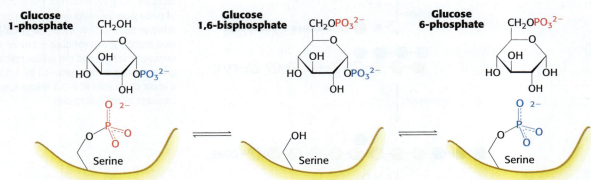

❖ **SELF–CHECK QUESTION**

The complete oxidation of glucose 6-phosphate derived from free glucose yields 30 molecules ATP, whereas glucose 6-phosphate derived from glycogen yields 31 molecules of ATP. How can you account for this difference?

The liver contains glucose 6-phosphatase, a hydrolytic enzyme absent from muscle

A major function of the liver is to maintain a nearly constant concentration of glucose in the blood to keep the brain, red blood cells, and other tissues adequately supplied. The liver achieves this by releasing glucose into the blood between meals and during muscular activity through the action of glucose 6-phosphatase, the last enzyme in gluconeogenesis (Section 16.4). This enzyme hydrolytically cleaves the phosphoester linkage to form orthophosphate and free glucose, which can then exit the liver by the action of GLUT transporters (Section 16.1). Recall that glucose 6-phosphatase is located on the lumenal side of the smooth endoplasmic reticulum membrane, and so glucose 6-phosphate is transported into the endoplasmic reticulum while glucose and orthophosphate are then shuttled back into the cytoplasm (Section 16.4).

$$\text{Glucose 6-phosphate} + H_2O \longrightarrow \text{glucose} + P_i$$

Glucose 6-phosphatase is absent from most other tissues. Muscle tissues retain glucose 6-phosphate for the generation of ATP. In contrast, glucose is not a major fuel for the liver.

❖ SELF–CHECK QUESTION

What four enzymes are required for the liver to release glucose into the blood when an animal is asleep and fasting?

21.3 Phosphorylase Is Regulated by Allosteric Interactions and Controlled by Reversible Phosphorylation

Glycogen degradation is precisely controlled by multiple interlocking mechanisms, with the enzyme glycogen phosphorylase playing a central role. Phosphorylase is regulated by several allosteric effectors that signal the energy state of the cell, as well as controlled by reversible phosphorylation, which is responsive to hormones such as insulin, epinephrine, and glucagon. We will examine the differences in the control of two isozymic forms of glycogen phosphorylase: one specific to liver and one specific to skeletal muscle. These differences are because the liver maintains glucose homeostasis of the organism as a whole, whereas the muscle uses glucose to produce energy for itself.

Liver phosphorylase produces glucose for use by other tissues

The dimeric phosphorylase exists in two interconvertible forms: a usually *active* phosphorylated *a* form and a usually *inactive* unphosphorylated *b* form (**Figure 21.10**). Each of these two forms exists in equilibrium between an active relaxed (R) state

FIGURE 21.10 Phosphorylase exists in two distinct quaternary states. Phosphorylase *a* is phosphorylated on serine 14 of each subunit. This modification favors the structure of the more-active R state. One subunit is shown in white, with helices and loops important for regulation shown in blue and red. The other subunit is shown in yellow, with the regulatory structures shown in orange and green. Phosphorylase *b* is not phosphorylated and exists predominantly in the T state. Notice that the catalytic sites are partly occluded in the T state. [Drawn from 1GPA.pdb and 1NOJ.pdb.]

INTERACT with these models in
 Achieve

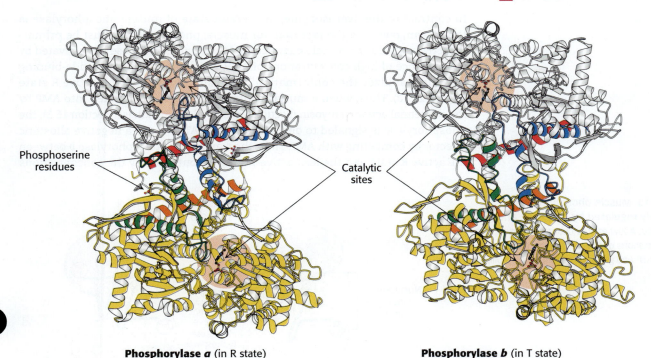

Phosphoserine residues

Catalytic sites

Phosphorylase *a* (in R state)   **Phosphorylase *b*** (in T state)

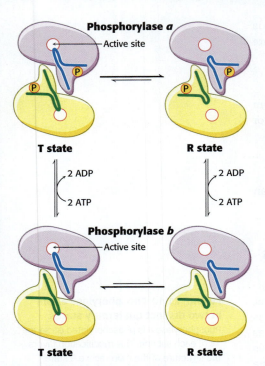

T state   **R state**

FIGURE 21.11 Phosphorylase is regulated by phosphorylation. Both phosphorylase *b* and phosphorylase *a* exist as equilibria between an active R state and a less-active T state. Phosphorylase *b* is usually inactive because the equilibrium favors the T state. Phosphorylase *a* is usually active because the equilibrium favors the R state. Regulatory structures are shown in blue and green.

and a much less active tense (T) state, but the equilibrium for phosphorylase *a* favors the active R state, whereas the equilibrium for phosphorylase *b* favors the less-active T state (**Figure 21.11**).

The role of glycogen degradation in the liver is to form glucose for export to other tissues when the blood-glucose concentration is low. Consequently, we can think of the default state of liver phosphorylase as being the *a* form: glucose is to be generated unless the enzyme is signaled otherwise. The liver phosphorylase *a* form thus exhibits the most responsive R ↔ T transition (**Figure 21.12**). The binding of glucose to the active site shifts the *a* form from the active R state to the less-active T state. In essence, the enzyme reverts to the low-activity T state only when it detects the presence of sufficient glucose. If glucose is present in the diet, there is no need to degrade glycogen.

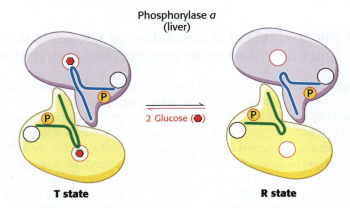

T state   **R state**

FIGURE 21.12 Liver phosphorylase is allosterically inhibited by glucose. The binding of glucose to phosphorylase *a* shifts the equilibrium to the T state and inactivates the enzyme. Thus, glycogen is not mobilized when glucose is already abundant.

Muscle phosphorylase is regulated by changes in AMP and ATP concentrations

In contrast to the liver isozyme, the default state of muscle phosphorylase is the *b* form, owing to the fact that, for muscle, phosphorylase must be primarily active only during muscle contraction. Muscle phosphorylase *b* is activated by the presence of high concentrations of AMP, which binds to a nucleotide-binding site and stabilizes the conformation of phosphorylase *b* in the active R state (**Figure 21.13**). Thus, when a muscle contracts and ATP is converted into AMP by the sequential action of myosin (Section 6.4) and adenylate kinase (Section 16.3), the phosphorylase is signaled to degrade glycogen. ATP acts as a negative allosteric effector by competing with AMP. Thus, the transition of phosphorylase *b* between the active R state and the less-active T state is controlled by the concentrations

FIGURE 21.13 Muscle phosphorylase is allosterically regulated by ATP, AMP, and glucose 6-phosphate. A low energy charge, represented by high concentrations of AMP, favors the transition to the R state. ATP and glucose 6-phosphate stabilize the T state.

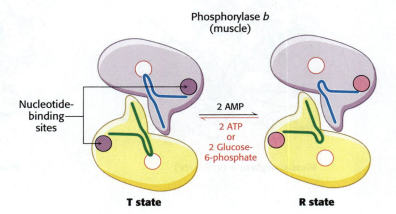

T state   **R state**

of these metabolites in muscle cells. If ATP is unavailable, glucose 6-phosphate may bind at the ATP-binding site and stabilize the less-active state of phosphorylase *b*, an example of feedback inhibition. In resting muscle, phosphorylase *b* is inactive because of the inhibitory effects of ATP and glucose 6-phosphate. In contrast, phosphorylase *a* is fully active, regardless of the concentration of AMP, ATP, and glucose 6-phosphate.

Unlike the enzyme in muscle, the liver phosphorylase is insensitive to regulation by AMP because the liver does not undergo the dramatic changes in the concentrations of AMP and ATP seen in a contracting muscle. We see here a clear example of the use of isozymes to establish the tissue-specific biochemical properties of the liver and muscle. In human beings, liver phosphorylase and muscle phosphorylase are approximately 90% identical in amino acid sequence, yet the 10% difference results in subtle but important shifts in the regulation of the two enzymes.

❖ SELF-CHECK QUESTION

Crystals of phosphorylase *a* grown in the presence of glucose shatter when a substrate such as glucose 1-phosphate is added. Why?

Biochemical characteristics of muscle fiber types differ

Not only do the biochemical needs of liver and muscle differ with respect to glycogen metabolism, but the biochemical needs of different muscle fiber types also vary. Skeletal muscle consists of three different fiber types: type I, or slow-twitch muscle; type IIb (also called type IIx), or fast-twitch fibers; and type IIa fibers that have properties intermediate between the other two fiber types.

- *Type I (slow-twitch) fibers rely predominantly on cellular respiration to derive energy.* These fibers are powered by fatty acid degradation and are rich in mitochondria, the site of fatty acid degradation and the citric acid cycle. As we will see in Chapter 22, fatty acids are an excellent energy storage form, but generating ATP from fatty acids is slower than from glycogen. Glycogen is not an important fuel for type I fibers, and consequently the amount of glycogen phosphorylase is low. Type I fibers power endurance activities.

- *Type IIb (fast-twitch) fibers use glycogen as their main fuel.* Consequently, the glycogen and glycogen phosphorylase are abundant. These fibers are also rich in glycolytic enzymes needed to process glucose quickly in the absence of oxygen and poor in mitochondria. Type IIb fibers power burst activities such as sprinting and weight lifting.

No amount of training can interconvert type I fibers and type IIb fibers. However, there is some evidence that type IIa fibers are "trainable"; that is, endurance training enhances the oxidative capacity of type IIa fibers, while burst activity training enhances the glycolytic capacity. Table 21.1 shows the biochemical profile of the fiber types.

TABLE 21.1 Biochemical characteristics of muscle fiber types

Characteristic	Type I	Type IIa	Type IIb
Fatigue resistance	High	Intermediate	Low
Mitochondrial density	High	Intermediate	Low
Metabolic type	Oxidative	Oxidative/glycolytic	Glycolytic
Myoglobin content	High	Intermediate	Low
Glycogen content	Low	Intermediate	High
Triacylglycerol content	High	Intermediate	Low
Glycogen phosphorylase activity	Low	Intermediate	High
Phosphofructokinase activity	Low	Intermediate	High
Citrate synthase activity	High	Intermediate	Low

^+H_3N–His–Ser–Gln–Gly–Thr–Phe–Thr–Ser–Asp–Tyr–
5   10

–Ser–Lys–Tyr–Leu–Asp–Ser–Arg–Arg–Ala–Gln–
15   20

–Asp–Phe–Val–Gln–Trp–Leu–Met–Asn–Thr–COO$^-$
25   29

Glucagon

Epinephrine

Phosphorylation promotes the conversion of phosphorylase *b* to phosphorylase *a*

In both liver and muscle, hormones initiate the conversion of phosphorylase *b* into phosphorylase *a* by the phosphorylation of a single serine residue (serine 14) in each subunit. Low blood concentration of glucose leads to the secretion of the peptide hormone **glucagon** from the α cells of the pancreas, which results in phosphorylation of the enzyme, converting it to the phosphorylase *a* form in liver. The hormone **epinephrine** (**adrenaline**), a catecholamine derived from tyrosine, also induces this conversion when it binds to receptors in muscle and liver.

Emotions like the excitement of exercise or fear will cause release of epinephrine from the adrenal medulla into the blood. Epinephrine binds to receptors in muscle (and in liver), again inducing the phosphorylation of phosphorylase *b* to phosphorylase *a*. The regulatory enzyme **phosphorylase kinase** catalyzes this covalent modification in response to either hormone.

Comparison of the structures of phosphorylase *a* in the R state and phosphorylase *b* in the T state reveals that subtle structural changes at the subunit interfaces are transmitted to the active sites (Figure 21.10). The transition from the T state (the prevalent state of phosphorylase *b*) to the R state (the prevalent state of phosphorylase *a*) entails a 10-degree rotation around the twofold axis of the dimer. Most important, this transition is associated with structural changes in α helices that move a loop out of the active site of each subunit. Thus, the T state is less active because the catalytic site is partly blocked. In contrast, in the R state, the catalytic site is more accessible, and a binding site for orthophosphate is well organized.

Phosphorylase kinase is activated by phosphorylation and calcium ions

Phosphorylase kinase converts phosphorylase *b* into the *a* form by attaching a phosphoryl group. The subunit composition of phosphorylase kinase in skeletal muscle is $(\alpha\beta\gamma\delta)_4$, and the mass of this very large protein is 1300 kDa. The enzyme consists of two $(\alpha\beta\gamma\delta)_2$ lobes that are joined by a β_4 bridge that is the core of the enzyme and serves as a scaffold for the other subunits (**Figure 21.14**). The γ subunit

FIGURE 21.14 Phosphorylase kinase is activated by calcium ions and hormonally controlled phosphorylation. Phosphorylase kinase, an $(\alpha\beta\gamma\delta)_4$ assembly, is partly activated by Ca^{2+} binding to the δ subunits. Activation is maximal when the β and α subunits are phosphorylated in response to hormonal signals. When active, the enzyme converts phosphorylase *b* into phosphorylase *a*.

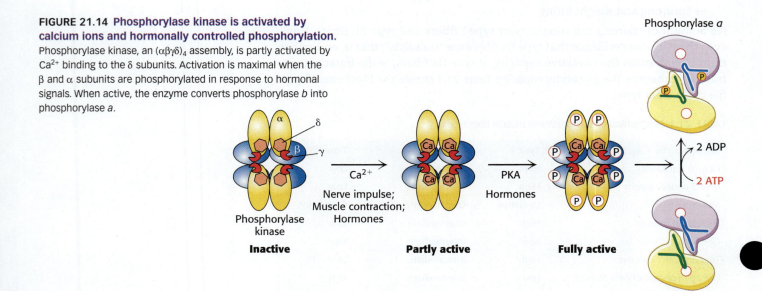

contains the active site, while all of the remaining subunits (≈90% by mass) play regulatory roles. The δ subunit is the Ca^{2+} binding protein calmodulin, a calcium sensor that stimulates many enzymes in eukaryotes (Section 14.2).

Activation of phosphorylase kinase is initiated when Ca^{2+} binds to the δ subunit. This mode of activation of the kinase is especially noteworthy in muscle, where muscle contraction is triggered by the release of Ca^{2+} from the sarcoplasmic reticulum (Figure 21.14). Maximal activation is achieved with the phosphorylation of the β and α subunits of the Ca^{2+}-bound kinase. The α and β subunits of the calcium-bound enzyme are targets of protein kinase A. The β subunit is phosphorylated first, followed by the phosphorylation of the α subunit. The stimulation of phosphorylase kinase is one step in a signal-transduction cascade initiated by signal molecules such as glucagon and epinephrine.

21.4 Glucagon and Epinephrine Signal the Need for Glycogen Breakdown

Protein kinase A (PKA) activates phosphorylase kinase, which in turn activates glycogen phosphorylase. But what activates PKA? What is the signal that ultimately triggers an increase in glycogen breakdown?

G proteins transmit the signal for the initiation of glycogen breakdown

Let's now explore in greater depth the signaling mechanisms of the two hormones that trigger glycogen breakdown: glucagon and epinephrine. Their activities are tissue-dependent. The liver is highly responsive to glucagon, which is released in response to low glucose concentrations in the blood. Physiologically, glucagon signifies the fasted state (**Figure 21.15**) and stimulates the liver to release glucose into the blood. In contrast, because muscle plays no regulatory role in maintaining glucose concentrations and cannot release glucose into the blood, muscle

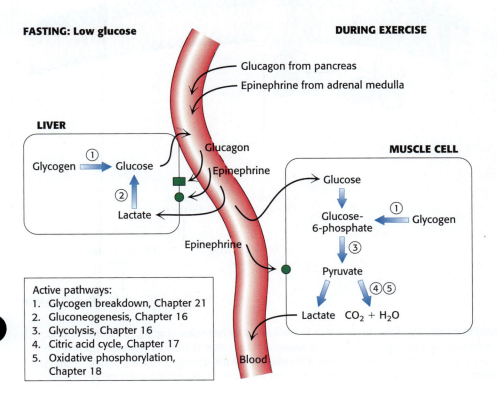

FASTING: Low glucose

DURING EXERCISE

Glucagon from pancreas
Epinephrine from adrenal medulla

LIVER

Glycogen → Glucose ①

② Lactate

Glucagon
Epinephrine

Epinephrine

Active pathways:
1. Glycogen breakdown, Chapter 21
2. Gluconeogenesis, Chapter 16
3. Glycolysis, Chapter 16
4. Citric acid cycle, Chapter 17
5. Oxidative phosphorylation, Chapter 18

Blood

MUSCLE CELL

Glucose → Glucose-6-phosphate ← Glycogen ①
Glucose-6-phosphate ③ → Pyruvate
Pyruvate → Lactate / CO_2 + H_2O ④⑤

FIGURE 21.15 Glycogen breakdown is controlled by the combined effects of multiple hormones in a tissue-specific manner. Glucagon stimulates liver-glycogen breakdown when blood glucose is low. Epinephrine enhances glycogen breakdown in muscle and the liver to provide fuel for muscle contraction.

EXPLORE this pathway further in the Metabolic Map in
≋ Achieve

is insensitive to glucagon. Instead, muscular activity or its anticipation leads to the release of epinephrine, which markedly stimulates glycogen breakdown in muscle and, to a lesser extent, in the liver. The breakdown of liver glycogen in response to epinephrine is meant to provide adequate glucose to the muscles for the expected increase in muscle contraction following the initiation of the "fight-or-flight" response.

How do glucagon and epinephrine trigger the breakdown of glycogen? They initiate a cyclic AMP signal-transduction cascade (**Figure 21.16**):

1. The signal molecules glucagon and epinephrine bind to specific seven-transmembrane (7TM) receptors in the plasma membranes of target cells (Section 14.2). Epinephrine binds to the β-adrenergic receptor in muscle, whereas glucagon binds to the glucagon receptor in the liver. These binding events activate the G_s protein: a specific external signal has been transmitted into the cell through structural changes, first in the receptor and then in the G protein.

2. The GTP-bound subunit of G_s activates the transmembrane protein adenylate cyclase. This enzyme catalyzes the formation of the second messenger cyclic AMP (cAMP) from ATP.

3. The elevated cytoplasmic concentration of cAMP activates PKA (Section 7.3). The binding of cAMP to inhibitory regulatory subunits of PKA triggers their dissociation from the catalytic subunits. The free catalytic subunits are now active.

4. PKA phosphorylates phosphorylase kinase, first on the β subunit and then on the α subunit, which activates glycogen phosphorylase.

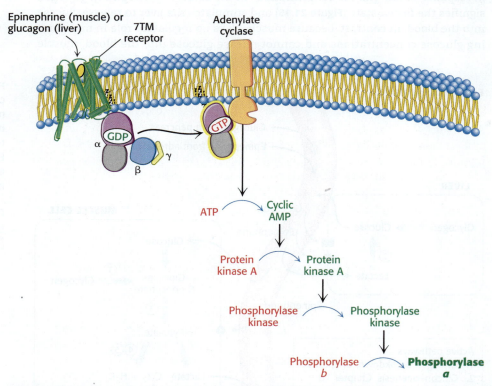

FIGURE 21.16 Glycogen degradation is stimulated by hormone binding to 7TM receptors. Hormone binding to 7TM receptors in the appropriate cells (liver or muscle) initiates a G-protein-dependent signal-transduction pathway that results in the phosphorylation and activation of glycogen phosphorylase. The inactive forms of the enzymes are shown in red and the active forms in green.

The signal-transduction processes in the liver are more complex than those in muscle. Epinephrine can also elicit glycogen degradation in the liver. However, in addition to binding to the β-adrenergic receptor, it binds to the 7TMα-adrenergic receptor, which then initiates the *phosphoinositide cascade* (Section 14.2) that induces the release of Ca^{2+} from endoplasmic reticulum stores. Recall that the δ subunit of phosphorylase kinase is the Ca^{2+} sensor calmodulin. The binding of Ca^{2+} to calmodulin leads to a partial activation of phosphorylase kinase. Stimulation by both glucagon and epinephrine leads to maximal mobilization of liver glycogen.

The cyclic AMP cascade highly amplifies the effects of hormones, resulting in a high-gain system. Gain is the ratio of the output signal to the input signal. In the case of glycogen metabolism, the binding of a small number of hormone molecules to cell-surface receptors leads to the release of a very large number of glucose units. Indeed, much of the stored glycogen would be mobilized within seconds were it not for a counterregulatory system.

Glycogen breakdown must be rapidly turned off when necessary

It is crucial that the high-gain system of glycogen breakdown be terminated quickly to prevent the wasteful depletion of glycogen after energy needs have been met and to prevent dangerously high blood-glucose concentrations. When glucose needs have been satisfied, phosphorylase kinase and glycogen phosphorylase are dephosphorylated and inactivated. Simultaneously, glycogen synthesis is activated.

The signal-transduction pathway leading to the activation of glycogen phosphorylase is shut down automatically when secretion of the initiating hormone ceases. The inherent GTPase activity of the G protein converts the bound GTP into inactive GDP, and phosphodiesterases that are always present in the cell convert cyclic AMP into AMP. **Protein phosphatase 1 (PP1)** removes the phosphoryl groups from phosphorylase kinase, thereby inactivating the enzyme. Finally, PP1 also removes the phosphoryl group from glycogen phosphorylase, converting the enzyme into the usually inactive *b* form.

21.5 Glycogen Synthesis Requires Several Enzymes and Uridine Diphosphate Glucose

As we saw with glycolysis and gluconeogenesis (Chapter 16), related degradative and biosynthetic pathways rarely operate by precisely the same reactions in the forward and reverse directions. Glycogen metabolism provided the first known example of this important principle: separate pathways afford much greater flexibility, both in energetics and in control.

Glycogen is synthesized by a pathway that utilizes **uridine diphosphate glucose (UDP-glucose)** as the activated glucose donor.

Synthesis: $\text{Glycogen}_n + \text{UDP-glucose} \longrightarrow \text{glycogen}_{n+1} + \text{UDP}$

Degradation: $\text{Glycogen}_{n+1} + P_i \longrightarrow \text{glycogen}_n + \text{glucose 1-phosphate}$

UDP-glucose is an activated form of glucose

UDP-glucose, the glucose donor in the biosynthesis of glycogen, is called an *activated* form of glucose, just as ATP and acetyl CoA are activated forms of orthophosphate and acetate, respectively. The C-1 carbon atom of the glucosyl unit of UDP-glucose is activated because its hydroxyl group is esterified to the diphosphate moiety of UDP.

UDP-glucose is synthesized from glucose 1-phosphate and uridine triphosphate (UTP) in a reaction catalyzed by UDP-glucose pyrophosphorylase. This reaction liberates the outer two phosphoryl residues of UTP as pyrophosphate.

**Uridine diphosphate glucose
(UDP-glucose)**

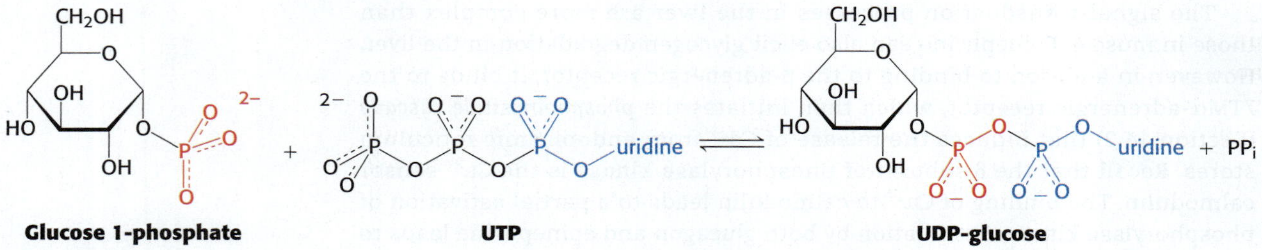

Glucose 1-phosphate + **UTP** ⇌ **UDP-glucose** + PP$_i$

This reaction is readily reversible. However, pyrophosphate is rapidly hydrolyzed in vivo to orthophosphate by an inorganic pyrophosphatase, keeping the concentration of pyrophosphate consistently low. Thus, the essentially irreversible hydrolysis of pyrophosphate drives the synthesis of UDP-glucose because the concentration of one of the products is kept near zero, making ΔG large and negative.

$$\text{Glucose 1-phosphate} + \text{UTP} \rightleftharpoons \text{UDP-glucose} + \text{PP}_i$$

$$\underline{\text{PP}_i + \text{H}_2\text{O} \longrightarrow 2\,\text{P}_i}$$

$$\text{Glucose 1-phosphate} + \text{UTP} + \text{H}_2\text{O} \longrightarrow \text{UDP-glucose} + 2\,\text{P}_i$$

The synthesis of UDP-glucose exemplifies another recurring theme in biochemistry: many biosynthetic reactions are driven by the hydrolysis of pyrophosphate.

Glycogen synthase catalyzes the transfer of glucose from UDP-glucose to growing chains

The key regulatory enzyme in glycogen synthesis is **glycogen synthase**, which adds new glucosyl units to the nonreducing terminal residues of glycogen. Humans have two isozymes of glycogen synthase: one is specific to the liver, while the other is expressed in muscle and other tissues. The activated glucosyl unit of UDP-glucose is transferred to the hydroxyl group at C-4 of a terminal residue to form an α-1,4-glycosidic linkage. UDP is displaced by the terminal hydroxyl group of the growing glycogen molecule.

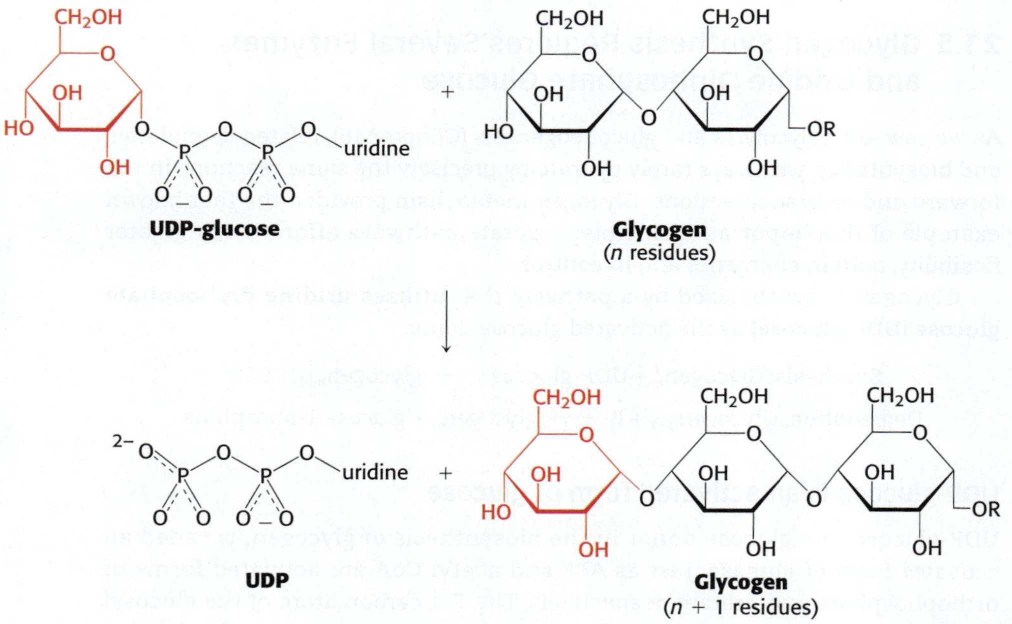

UDP-glucose + **Glycogen** (*n* residues)

UDP + **Glycogen** (*n* + 1 residues)

Glycogen synthase can add glucosyl residues only to a polysaccharide chain already containing at least four residues. Thus, glycogen synthesis requires a primer. This priming function is carried out by **glycogenin**, a Mn^{2+}-requiring glycosyltransferase composed of two identical 37-kDa subunits. Each subunit of glycogenin catalyzes the formation of α-1,4-glucose polymers on the other subunit (*intermolecular* autoglycosylation), or within the same subunit (*intramolecular* autoglycosylation). UDP-glucose is used to processively add glucosyl units to the phenolic hydroxyl group of a specific tyrosine residue in each glycogenin subunit. This proceeds until the chains reach 10 to 20 glucosyl units in length, at which point glycogen synthase takes over to extend the glycogen molecules attached to each subunit. Thus, every glycogen molecule has a glycogenin monomer covalently attached at its core (Figure 21.2).

A branching enzyme forms α-1,6 linkages

Glycogen synthase catalyzes only the synthesis of α-1,4 linkages; another enzyme is required to form the α-1,6 linkages that make glycogen a branched polymer. Branching takes place after a number of glucosyl residues are joined in α-1,4 linkages by glycogen synthase (**Figure 21.17**).

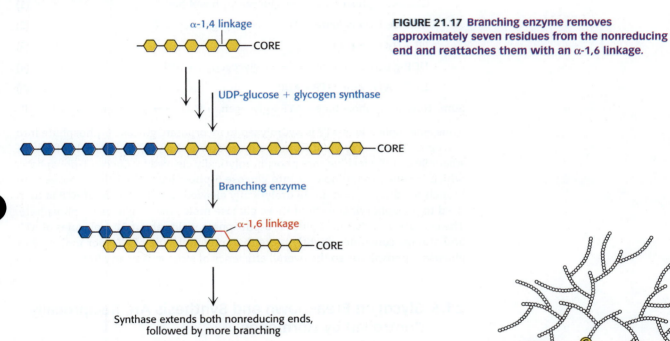

FIGURE 21.17 Branching enzyme removes approximately seven residues from the nonreducing end and reattaches them with an α-1,6 linkage.

A branch is created by the breaking of an α-1,4 link and the formation of an α-1,6 link, transferring a block of residues, typically seven, to a more interior site. The branching enzyme catalyzing this reaction requires that the block of residues must include the nonreducing terminus, and must come from a chain at least 11 residues long. In addition, the new branch point must be at least four residues away from a preexisting one.

Branching is important because it increases the solubility of glycogen. Furthermore, branching creates a large number of terminal residues, the sites of action of glycogen phosphorylase and synthase (**Figure 21.18**). Thus, branching increases the rate of glycogen synthesis and degradation.

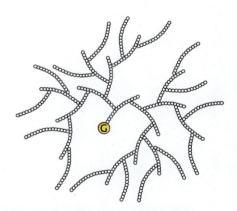

FIGURE 21.18 A cross section of a glycogen molecule reveals the basic branching structure. The component labeled G is glycogenin. Note that this is a cross section of the three-dimensional structure illustrated in Figure 21.2.

Glycogen synthase is the key regulatory enzyme in glycogen synthesis

Glycogen synthase, like phosphorylase, exists in two forms: an active nonphosphorylated *a* form and a usually inactive phosphorylated *b* form. Again, like the phosphorylase, the interconversion of the two forms is controlled extracellularly by the action of hormones; however, note that phosphorylation has opposite effects on the enzymatic activities of glycogen synthase and glycogen phosphorylase.

Glycogen synthase is phosphorylated at multiple sites by several protein kinases—notably, **glycogen synthase kinase**, which is under the control of insulin, and PKA. The function of the multiple phosphorylation sites is still under investigation. In contrast, the key means of automatically regulating glycogen synthase activity is by allosteric regulation of the phosphorylated form of the enzyme, glycogen synthase b. Glucose 6-phosphate is a powerful activator of the enzyme, stabilizing the R state of the enzyme relative to the T state and providing automatic activation of the enzyme when the concentrations of glucose-phosphates are high within a given cell.

Glycogen is an efficient storage form of glucose

What is the cost of converting glucose 6-phosphate into glycogen and back into glucose 6-phosphate? The pertinent reactions have already been described except for reaction 5 below, which is the regeneration of UTP. ATP phosphorylates UDP in a reaction catalyzed by nucleoside diphosphokinase.

$$\text{Glucose 6-phosphate} \longrightarrow \text{glucose 1-phosphate} \tag{1}$$

$$\text{Glucose 1-phosphate} + \text{UTP} \longrightarrow \text{UDP-glucose} + \text{PP}_i \tag{2}$$

$$\text{PP}_i + \text{H}_2\text{O} \longrightarrow 2\,\text{P}_i \tag{3}$$

$$\text{UDP-glucose} + \text{glycogen}_n \longrightarrow \text{glycogen}_{n+1} + \text{UDP} \tag{4}$$

$$\text{UDP} + \text{ATP} \longrightarrow \text{UTP} + \text{ADP} \tag{5}$$

Sum: $\text{Glucose 6-phosphate} + \text{ATP} + \text{glycogen}_n + \text{H}_2\text{O} \longrightarrow \text{glycogen}_{n+1} + \text{ADP} + 2\,\text{P}_i$

Thus, one molecule of ATP is hydrolyzed to incorporate glucose 6-phosphate into glycogen. The energy yield from the breakdown of glycogen is highly efficient. About 90% of the residues are phosphorolytically cleaved to glucose 1-phosphate, which is converted at no cost into glucose 6-phosphate. The other residues are branch residues, which are hydrolytically cleaved. One molecule of ATP is then used to phosphorylate each of these glucose molecules to glucose 6-phosphate. The complete oxidation of glucose 6-phosphate yields about 31 molecules of ATP, and storage consumes slightly more than one molecule of ATP per molecule of glucose 6-phosphate; so the overall efficiency of storage is nearly 97%.

21.6 Glycogen Breakdown and Synthesis Are Reciprocally Controlled by Hormones

An important control mechanism prevents glycogen from being synthesized at the same time as it is being broken down. The same glucagon- and epinephrine-triggered cAMP cascades that initiate glycogen breakdown in the liver and muscle, respectively, also shut off glycogen synthesis. Glucagon and epinephrine control both glycogen breakdown and glycogen synthesis through PKA (**Figure 21.19**). Recall that PKA adds a phosphoryl group to phosphorylase kinase, activating that enzyme and initiating glycogen breakdown. Both glycogen synthase kinase and PKA add phosphoryl groups to glycogen synthase, but this phosphorylation leads to a *decrease* in enzymatic activity. In this way, glycogen breakdown and synthesis are reciprocally regulated. How is the enzymatic activity reversed so that glycogen breakdown halts and glycogen synthesis begins?

Protein phosphatase 1 reverses the effects of kinases on glycogen metabolism

After a bout of exercise, muscle must shift from a glycogen-degrading mode to one of glycogen replenishment. A first step in this metabolic task is to shut down the phosphorylated proteins that stimulate glycogen breakdown. This task is accomplished by protein phosphatases that catalyze the hydrolysis of phosphorylated serine and threonine residues in proteins.

DURING EXERCISE OR FASTING

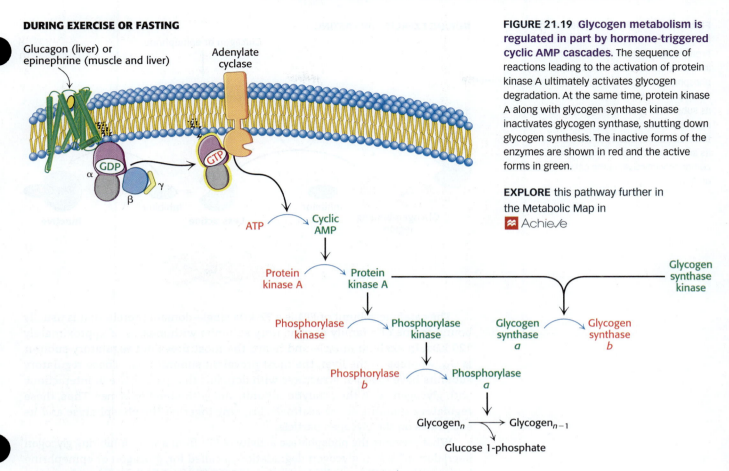

FIGURE 21.19 Glycogen metabolism is regulated in part by hormone-triggered cyclic AMP cascades. The sequence of reactions leading to the activation of protein kinase A ultimately activates glycogen degradation. At the same time, protein kinase A along with glycogen synthase kinase inactivates glycogen synthase, shutting down glycogen synthesis. The inactive forms of the enzymes are shown in red and the active forms in green.

EXPLORE this pathway further in the Metabolic Map in Achieve

Protein phosphatase 1 (PP1) plays key roles in regulating glycogen metabolism (**Figure 21.20**). PP1 decreases the rate of glycogen breakdown by reversing the effects of the phosphorylation cascade: it inactivates phosphorylase a and phosphorylase kinase by dephosphorylating them. PP1 also removes phosphoryl groups from glycogen synthase b to convert it into the more active glycogen synthase a form. Thus, PP1 accelerates glycogen synthesis; it is yet another molecular device for coordinating carbohydrate storage.

AFTER A MEAL OR WHEN AT REST

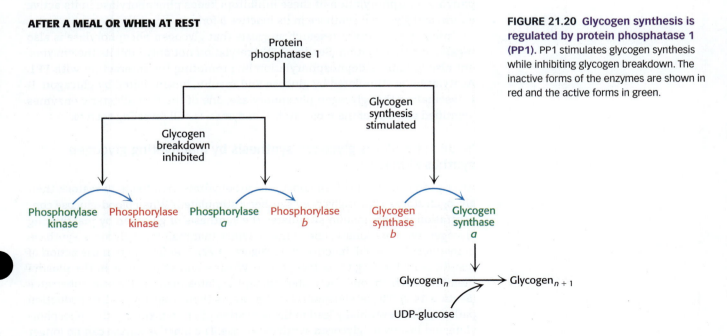

FIGURE 21.20 Glycogen synthesis is regulated by protein phosphatase 1 (PP1). PP1 stimulates glycogen synthesis while inhibiting glycogen breakdown. The inactive forms of the enzymes are shown in red and the active forms in green.

FIGURE 21.21 PP1 is regulated by hormonally controlled cascades. Protein phosphatase 1 (PP1) regulation in muscle takes place in two steps. First, phosphorylation of G_M by protein kinase A dissociates the catalytic subunit from its substrates in the glycogen particle and reduces its activity. Then, phosphorylation of the inhibitor subunit by protein kinase A and its subsequent binding by the phosphatase completely inactivates the catalytic subunit of PP1.

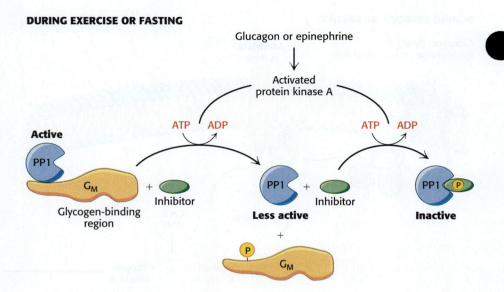

DURING EXERCISE OR FASTING

The catalytic subunit of PP1 is a 37-kDa single-domain protein that is usually bound to one of a family of regulatory subunits with masses of approximately 120 kDa. In skeletal muscle and heart, the most prevalent regulatory subunit is G_M, whereas in the liver, the most prevalent subunit is G_L. These regulatory subunits have modular structures with domains that participate in interactions with glycogen, with the catalytic subunit, and with target enzymes. Thus, these regulatory subunits act as scaffolds, bringing together the phosphatase and its substrates on the glycogen particle.

What prevents the phosphatase activity of PP1 from always inhibiting glycogen degradation? When glycogen degradation is called for, glucagon or epinephrine initiates the cAMP cascade that activates PKA (**Figure 21.21**), which reduces the activity of PP1 by two mechanisms. First, in muscle, G_M is phosphorylated in the domain responsible for binding the catalytic subunit. The catalytic subunit is released from glycogen and from its substrates, and its phosphatase activity is greatly reduced. Second, almost all tissues contain small proteins (Inhibitor in Figure 21.21) that, when phosphorylated, bind to the catalytic subunit of PP1 and inhibit it. Thus, when glycogen degradation is switched on by cAMP, the accompanying phosphorylation of these inhibitors keeps phosphorylase in its active *a* form and glycogen synthase in its inactive *b* form.

Interestingly, recent research suggests that glycogen phosphorylase is also regulated by acetylation (Section 7.3). Acetylation not only inhibits the enzyme, but also enhances dephosphorylation by promoting the interaction with PP1. Acetylation is stimulated by glucose and insulin, but inhibited by glucagon. It is fascinating that glycogen phosphorylase, one of the first allosteric enzymes identified and one of the most studied enzymes, is still revealing secrets.

Insulin stimulates glycogen synthesis by inactivating glycogen synthase kinase

After exercise, people often consume carbohydrate-rich foods to restock their glycogen stores. How is glycogen synthesis stimulated? When blood-glucose concentration is high, insulin stimulates the synthesis of glycogen by inactivating glycogen synthase kinase, one of the enzymes that maintains glycogen synthase in its phosphorylated, inactive state (**Figure 21.22**). The first step in the action of insulin is its binding to its receptor, a tyrosine kinase receptor in the plasma membrane which, once activated, phosphorylates insulin-receptor substrates (Section 14.3). These phosphorylated proteins then trigger signal-transduction pathways that eventually lead to the activation of protein kinases that phosphorylate and inactivate glycogen synthase kinase. The inactive kinase can no longer

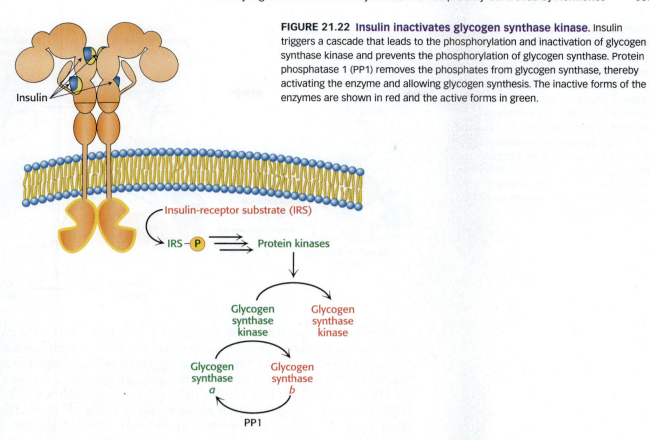

FIGURE 21.22 Insulin inactivates glycogen synthase kinase. Insulin triggers a cascade that leads to the phosphorylation and inactivation of glycogen synthase kinase and prevents the phosphorylation of glycogen synthase. Protein phosphatase 1 (PP1) removes the phosphates from glycogen synthase, thereby activating the enzyme and allowing glycogen synthesis. The inactive forms of the enzymes are shown in red and the active forms in green.

maintain glycogen synthase in its phosphorylated, inactive state. Additionally, PP1 dephosphorylates glycogen synthase, activating it, and restoring glycogen reserves. Finally, recall that insulin also increases the amount of glucose in the cell by increasing the number of glucose transporters in the membrane (Section 16.1). The net effect of insulin is thus the replenishment of glycogen stores.

Glycogen metabolism in the liver regulates the blood-glucose concentration

After a meal rich in carbohydrates, blood-glucose concentration rises, and glycogen synthesis is stepped up in the liver. Although insulin is the primary signal for glycogen synthesis, another is the concentration of glucose in the blood, which normally ranges from about 4.4–6.7 mM. The liver senses the concentration of glucose in the blood and takes up or releases glucose accordingly. Experiments performed in rodents established that the amount of liver phosphorylase *a* decreases rapidly when glucose is infused (**Figure 21.23**). After a lag period, the amount of glycogen synthase *a* increases, which results in glycogen synthesis. How does phosphorylase perform its function as the glucose sensor in liver cells, facilitating the switch from degradation to synthesis?

Phosphorylase *a* and PP1 are localized to the glycogen particle by interactions with the G_L subunit of PP1. The binding of glucose to phosphorylase *a* shifts its allosteric equilibrium from the active R form to the inactive T form (Figure 21.12). This conformational change renders the phosphoryl group on serine 14 a substrate for PP1. PP1 binds tightly to phosphorylase *a* only when the phosphorylase is in the R state, but PP1 is inactive when bound. When glucose induces the transition to the T form, PP1 and the phosphorylase dissociate from each other and the glycogen particle, and PP1 becomes active, converting phosphorylase *a* to the *b* form (Figure 21.11). Recall that the R ↔ T transition of muscle phosphorylase *a* is unaffected by glucose and is thus unaffected by the rise in blood-glucose concentration (Section 21.3).

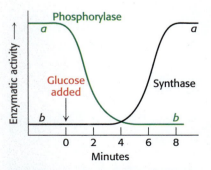

FIGURE 21.23 Blood glucose regulates liver-glycogen metabolism. The infusion of glucose into the bloodstream leads to the inactivation of phosphorylase, followed by the activation of glycogen synthase, in the liver.

[Data from W. Stalmans, H. De Wulf, L. Hue, and H.-G. Hers, *Eur. J. Biochem.* 41:117–134, 1974.]

FIGURE 21.24 Glucose regulates liver-glycogen metabolism. After a carbohydrate-rich meal, glucose binds to and inhibits glycogen phosphorylase *a* in the liver, facilitating the formation of the T state of phosphorylase *a*. The T state of phosphorylase *a* does not bind protein phosphate 1 (PP1), leading to the dissociation of PP1 from glycogen phosphorylase *a*. The free PP1 is no longer inhibited and dephosphorylates glycogen phosphorylase *a* and glycogen synthase *b*, leading to the inactivation of glycogen breakdown and the activation of glycogen synthesis.

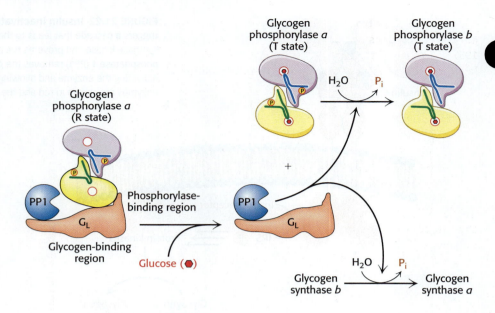

How does glucose binding to glycogen phosphorylase stimulate glycogen synthesis? As mentioned above, the conversion of *a* into *b* is accompanied by the release of PP1, which is then free to activate glycogen synthase (**Figure 21.24**). The removal of the phosphoryl group of inactive glycogen synthase *b* converts it into the active *a* form. What accounts for the lag between termination of glycogen degradation and the beginnings of glycogen synthesis (Figure 21.23)? There are about 10 phosphorylase *a* molecules per molecule of PP1. Consequently, the activity of glycogen synthase begins to increase only after most of phosphorylase *a* is converted into *b*. The lag between the decrease in glycogen degradation and the increase in glycogen synthesis prevents the two pathways from operating simultaneously. This remarkable glucose-sensing system depends on three key elements: (1) communication within phosphorylase between the allosteric site for glucose and the serine phosphate, (2) the use of PP1 to inactivate phosphorylase and activate glycogen synthase, and (3) the binding of PP1 to phosphorylase *a* to prevent the premature activation of glycogen synthase.

Efforts are underway to develop drugs that disrupt the interaction of liver phosphorylase with the G_L subunit as a treatment for type 2 diabetes, a disease characterized by excess blood glucose (Section 24.3). One experimental drug binds at a unique site on the enzyme that, synergistically with glucose, stabilizes the T state of phosphorylase and thereby enhances the transition to glycogen synthesis, as described earlier. Hence, disrupting the association of phosphorylase with the G_L would render it a substrate for PP1. Glycogen degradation would decrease and glucose release into the blood would be inhibited.

❖ **SELF–CHECK QUESTION**

What pathway, in addition to the cAMP-induced signal-transduction pathway, is used in the liver to maximize glycogen breakdown?

Biochemists have uncovered the biochemical basis of multiple glycogen-storage diseases

In 1929, Edgar von Gierke described the first glycogen-storage disease, which subsequently bore his name. Patients with von Gierke disease have a distended abdomen, caused by a massive enlargement of the liver, and pronounced hypoglycemia (low blood glucose) between meals. Furthermore, the blood-glucose concentration does not rise on administration of epinephrine and glucagon. In 1952, Gerty and Carl Cori discovered the mechanism for this disease: the liver lacks glucose 6-phosphatase (Figure 21.4). This finding was the first demonstration of an inherited deficiency of a liver enzyme.

The absence of glucose 6-phosphatase in the liver causes hypoglycemia because glucose cannot be formed from glucose 6-phosphate, which cannot cross the plasma membrane to leave the liver. Thus, the glycogen in the liver of a patient with von Gierke disease is normal in structure but is present in abnormally large amounts. The presence of excess glucose 6-phosphate also triggers an increase in glycolysis in the liver, leading to high concentrations of lactate and pyruvate in the blood. This disease can also be produced by a mutation in the gene that encodes the glucose 6-phosphate transporter. Recall that glucose 6-phosphate must be transported into the lumen of the endoplasmic reticulum to be hydrolyzed by phosphatase (Section 16.4). Mutations in the other three essential proteins of this system can likewise lead to von Gierke disease.

Seven other glycogen-storage diseases have been characterized (Table 21.2). In Pompe disease (type II), lysosomes become engorged with glycogen because they lack α-1,4-glucosidase, a hydrolytic enzyme confined to these organelles (Figure 21.25). There are several subtypes of glycogen-storage disease III, affecting either muscle or liver. Regardless of subtype, the outer branches of the glycogen are very short. Patients with type III lack the debranching enzyme (α-1,6-glucosidase), and so only the outermost branches of glycogen can be effectively utilized. Thus, only a small fraction of this abnormal glycogen is functionally active as an accessible store of glucose.

A defect in glycogen metabolism confined to muscle is found in McArdle disease (type V) when muscle phosphorylase activity is absent, limiting the capacity to perform strenuous exercise. Despite this limitation, patients with McArdle disease are otherwise normal and well developed, illustrating that effective utilization of muscle glycogen is not essential for life. We will examine McArdle disease further in Chapter 24.

FIGURE 21.25 An electron micrograph of a muscle cell from a patient with Pompe disease reveals enlarged lysosomes. Glycogen-engorged lysosomes are seen throughout the cell, including in the myofibrils (white arrow) and at the cell periphery (dashed black arrow). As the disease progresses, lysosomes may rupture, releasing large amounts of glycogen into the cytoplasm. Such accumulations of cytoplasmic glycogen are called *glycogen lakes* (solid black arrow). [Credit: Reproduced with permission of the author, from Thurberg, B.L., et al., Characterization of pre- and post-treatment pathology after enzyme replacement for pompe disease. *Lab. Invest*, 2006 Dec: 86(12):1208–20.]

TABLE 21.2 Glycogen-storage diseases

Type	Defective enzyme	Organ affected	Glycogen in the affected organ	Clinical features
I von Gierke	Glucose 6-phosphatase or transport system	Liver and kidney	Increased amount; normal structure	Massive enlargement of the liver; failure to thrive; severe hypoglycemia, ketosis, hyperuricemia, hyperlipemia
II Pompe	α-1,4-glucosidase (lysosomal)	All organs	Massive increase in amount; normal structure	Cardiorespiratory failure causes death, usually before age 2
III Cori	α-1,6-glucosidase (debranching enzyme)	Muscle and liver	Increased amount; short outer branches	Like type I, but milder course
IV Andersen	Branching enzyme (α-1,4 → α-1,6)	Liver and spleen	Normal amount; very long outer branches	Progressive cirrhosis of the liver; liver failure causes death, usually before age 2
V McArdle	Phosphorylase	Muscle	Moderately increased amount; normal structure	Limited ability to perform strenuous exercise because of painful muscle cramps; otherwise patient is normal and well developed
VI Hers	Phosphorylase	Liver	Increased amount	Like type I, but milder course
VII	Phosphofructokinase	Muscle	Increased amount; normal structure	Like type V
VIII	Phosphorylase kinase	Liver	Increased amount; normal structure	Mild liver enlargement; mild hypoglycemia

Note: Types I through VII are inherited as autosomal recessives. Type VIII is sex-linked.

Summary

21.1 Glycogen Metabolism Is the Regulated Release and Storage of Glucose in Multiple Tissues

- Glycogen, a readily mobilized fuel store, is a branched polymer of glucose residues with α-1,4-glycosidic linkages and branches approximately every 12th residue created by an α-1,6-glycosidic bond.

- Glycogen is present to some extent in nearly all tissues but found in large amounts in muscle cells and in liver cells, where it is stored in the cytoplasm in the form of hydrated granules.

21.2 Glycogen Breakdown Requires the Interplay of Several Enzymes

- Most of the glycogen molecule is degraded to glucose 1-phosphate by the action of glycogen phosphorylase, which uses orthophosphate to split the α-1,4-glycosidic linkages; glucose 1-phosphate can then be reversibly converted into glucose 6-phosphate.

- Branch points are degraded by the combined actions of an oligosaccharide transferase and an α-1,6-glucosidase, both part of the bifunctional enzyme called *debranching enzyme*.

21.3 Phosphorylase Is Regulated by Allosteric Interactions and Controlled by Reversible Phosphorylation

- Phosphorylase *b*, which is usually inactive, is converted into active phosphorylase *a* by the phosphorylation of a single serine residue by phosphorylase kinase.

- In the liver, phosphorylase *a* liberates glucose for export to other organs, such as skeletal muscle and the brain; phosphorylase *a* is inhibited by the presence of glucose.

- In muscle, phosphorylase *b* can be activated by the binding of AMP to generate glucose for use inside the cells as a fuel for contractile activity. The effect of AMP is counteracted by ATP and glucose 6-phosphate.

21.4 Glucagon and Epinephrine Signal the Need for Glycogen Breakdown

- Epinephrine and glucagon stimulate glycogen breakdown through specific 7TM receptors.

- Muscle is responsive to epinephrine, whereas the liver is responsive to both epinephrine and glucagon.

- Both signal molecules initiate a kinase cascade that leads to the activation of phosphorylase kinase, which in turn converts glycogen phosphorylase *b* to the phosphorylated *a* form.

21.5 Glycogen Synthesis Requires Several Enzymes and Uridine Diphosphate Glucose

- The pathway for glycogen synthesis differs from that for glycogen breakdown: Glycogen synthase catalyzes the transfer of glucose from UDP-glucose to the C-4 hydroxyl group of a terminal residue in the growing glycogen molecule.

- Synthesis is primed by glycogenin, an autoglycosylating protein that contains a covalently attached oligosaccharide unit on a specific tyrosine residue.

- Branching enzyme converts some of the α-1,4 linkages into α-1,6 linkages, thus increasing the number of ends and allowing glycogen to be made and degraded more rapidly.

21.6 Glycogen Breakdown and Synthesis Are Reciprocally Controlled by Hormones

- Glycogen synthesis and degradation are coordinated by several amplifying reaction cascades: epinephrine and glucagon stimulate glycogen breakdown and inhibit its synthesis by increasing the cytoplasmic concentration of cyclic AMP, which activates PKA.

- PKA activates glycogen breakdown by attaching a phosphate to phosphorylase kinase, and it inhibits glycogen synthesis by phosphorylating glycogen synthase. Glycogen synthase kinase also inhibits synthesis by phosphorylating the synthase.

- The glycogen-mobilizing actions of PKA are reversed by PP1, which is regulated by several hormones: epinephrine inhibits PP1, while insulin triggers a cascade that phosphorylates and inactivates glycogen synthase kinase. Hence, glycogen synthesis is decreased by epinephrine and increased by insulin.

Key Terms

glycogen phosphorylase (p. 642)

phosphorolysis (p. 642)

pyridoxal phosphate (PLP)
(p. 643)

glucagon (p. 650)

epinephrine (adrenaline) (p. 650)

phosphorylase kinase (p. 650)

protein phosphatase 1 (PP1) (p. 653)

uridine diphosphate glucose
(UDP-glucose) (p. 653)

glycogen synthase (p. 654)

glycogenin (p. 654)

glycogen synthase kinase
(p. 655)

Problems

1. What are the three steps in glycogen degradation, and what enzymes are required? ❖ **1**

2. Which molecule is the immediate product of glycogen phosphorylase? ❖ **1**

(a) Glucose 6-phosphate

(b) Glucose 1-phosphate

(c) Fructose 1-phosphate

(d) Glucose 1,6-bisphosphate

3. Which of the following is a substrate for glycogen synthase? ❖ **3**

(a) UTP-glucose

(b) Glucose 1-phosphate

(c) CDP-glucose

(d) UDP-glucose

4. α-Amylose is an unbranched glucose polymer. Why would this polymer not be as effective a storage form of glucose as glycogen?

5. Compare the allosteric regulation of phosphorylase in the liver and in muscle, and explain the significance of the difference. ❖ **2,** ❖ **5**

6. Why is water excluded from the active site of phosphorylase? Predict the effect of a mutation that allows water molecules to enter. ❖ **1**

7. Outline the signal-transduction cascade for glycogen degradation in muscle. ❖ **2**

8. There must be a way to shut down glycogen breakdown quickly to prevent the wasteful depletion of glycogen after energy needs have been met. What mechanisms are employed to turn off glycogen breakdown?

9. Phosphorylation has opposite effects on glycogen synthesis and breakdown. What is the advantage of its having opposing effects? ❖ **4**

10. What enzymes are required for the synthesis of a glycogen particle starting from glucose 6-phosphate? ❖ **3**

11. The following reaction accounts for the synthesis of UDP-glucose. This reaction is readily reversible. How is it made irreversible in vivo?

$$\text{Glucose 1-phosphate} + \text{UTP} \rightleftharpoons \text{UDP-glucose} + \text{PP}_i$$

12. Why does activation of the phosphorylated b form of glycogen synthase by high concentrations of glucose 6-phosphate make good biochemical sense? ❖ **4**

13. Write a balanced equation showing the effect of simultaneous activation of glycogen phosphorylase and glycogen synthase. Include the reactions catalyzed by phosphoglucomutase and UDP-glucose pyrophosphory-lase. ❖ **4**

14. How does insulin stimulate glycogen synthesis? ❖ **4**

15. Write a balanced equation for the formation of glycogen from galactose. ❖ **3**

16. In exercising muscle, glycogen degradation supplies the muscle with glucose 6-phosphate. In order to stimulate muscle glycogen degradation, protein phosphatase 1 (PP1) must be inhibited. Four of the five following events are involved in the inactivation of PP1 in exercising muscle. Which are they? ❖ **2,** ❖ **4**

(a) PKA phosphorylates an inhibitor of PP1.

(b) Phosphorylated PP1 inhibitor binds to PP1, facilitating glycogen degradation by phosphorylase a.

(c) PKA phosphorylates G_M in the G_M-PP1 complex, resulting in its dissociation.

(d) Epinephrine initiates a cAMP signal-transduction cascade that utilizes PKA.

(e) Insulin initiates a protein kinase cascade that utilizes glycogen synthase kinase.

17. Suggest another mutation in glucose metabolism that causes symptoms similar to those of von Gierke disease.

18. Experiments were performed in which serine (S)14 of glycogen phosphorylase was replaced by glutamate (E).

The V_{max} of the mutant enzyme was then compared with the wild-type phosphorylase in both the *a* and the *b* form.

V_{max} μmoles of glucose 1-PO$_4$ released min^{-1} mg^{-1}	
Wild-type phosphorylase *b*	25 ± 0.4
Wild-type phosphorylase *a*	100 ± 5
S to E mutant	60 ± 3

(a) Explain the results obtained with the mutant.

(b) Predict the effect of substituting aspartic acid for the serine.

19. The liver is a major storage site for glycogen. Purified from two samples of human liver, glycogen was either treated or not treated with α-amylase and subsequently analyzed by SDS-PAGE and western blotting with the use of antibodies to glycogenin (see Chapter 4 to review these techniques). The results are presented below.

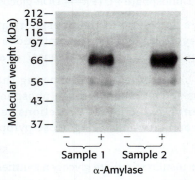

Credit: Courtesy of Dr. Peter J. Roach, Indiana University School of Medicine

(a) Why are no proteins visible in the lanes without amylase treatment?

(b) What is the effect of treating the samples with α-amylase? Explain the results.

(c) List other proteins that you might expect to be associated with glycogen. Why are other proteins not visible?

20. The gene for glycogenin was transfected into a cell line that normally stores only small amounts of glycogen. The cells were then manipulated according to the following protocol, and glycogen was isolated and analyzed by SDS-PAGE and western blotting by using an antibody to glycogenin with and without α-amylase treatment. The results are presented below.

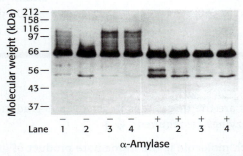

Credit: Courtesy of Dr. Peter J. Roach, Indiana University School of Medicine

The protocol: Cells cultured in growth medium and 25 mM glucose (lane 1) were switched to medium containing no glucose for 24 hours (lane 2). Glucose-starved cells were refed with medium containing 25 mM glucose for 1 hour (lane 3) or 3 hours (lane 4). Samples (12 μg of protein) were either treated or not treated with α-amylase, as indicated, before being loaded on the gel. ❖ 3

(a) Why did the western analysis produce a "smear"—that is, the high-molecular-weight staining in lane 1(−)?

(b) What is the significance of the decrease in high-molecular-weight staining in lane 2(−)?

(c) What is the significance of the difference between lanes 2(−) and 3(−)?

(d) Suggest a plausible reason why there is essentially no difference between lanes 3(−) and 4(−).

(e) Why are the bands at 66 kDa the same in the lanes treated with amylase, despite the fact that the cells were treated differently?

Fatty Acid and Triacylglycerol Metabolism

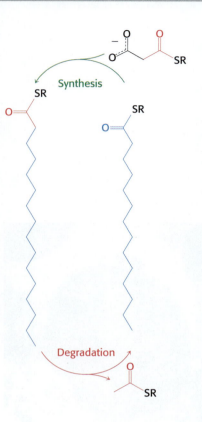

Like mammals, birds such as the ruby-throated hummingbird store energy for later use in the most weight-efficient way using fats. This bird beats its wings approximately 50 times per second nonstop to fly across the Gulf of Mexico using the energy stored in its tiny body as fat. Shown at right, the processes of fatty acid synthesis (preparation for energy storage) and fatty acid degradation (preparation for energy use) are, in many ways, the reverse of each other.

❖ LEARNING GOALS

By the end of this chapter, you should be able to:

1. Identify the repeated steps of fatty acid degradation.

2. Describe ketone bodies and their role in metabolism.

3. Explain how fatty acids are synthesized.

4. Explain how fatty acid metabolism is regulated.

OUTLINE

22.1 Triacylglycerols Are Highly Concentrated Energy Stores

22.2 The Use of Fatty Acids as Fuel Requires Three Stages of Processing

22.3 Unsaturated and Odd-Chain Fatty Acids Require Additional Steps for Degradation

22.4 Ketone Bodies Are a Fuel Source Derived from Fats

22.5 Fatty Acids Are Synthesized by Fatty Acid Synthase

22.6 The Elongation and Unsaturation of Fatty Acids Are Accomplished by Accessory Enzyme Systems

22.7 Acetyl CoA Carboxylase Plays a Key Role in Controlling Fatty Acid Metabolism

We turn now from the metabolism of carbohydrates to that of fatty acids, which have four major physiological roles:

1. *Fatty acids are fuel molecules.* Fatty acids are mobilized from storage in fat tissue and oxidized to meet the energy needs of a cell or organism. During rest or moderate exercise, such as walking, fatty acids are our primary source of energy.

2. *Fatty acids are building blocks of phospholipids and glycolipids.* These amphipathic molecules are important components of biological membranes, as discussed in Chapter 12.

3. *Many proteins are modified by the covalent attachment of fatty acids.* The attached fatty acids target proteins to membrane locations (Section 12.4).

4. *Fatty acid derivatives serve as hormones and intracellular messengers.*

In this chapter, we focus on the degradation and synthesis of fatty acids.

22.1 Triacylglycerols Are Highly Concentrated Energy Stores

Fatty acids are stored as **triacylglycerols** (also called **neutral fats** or **triglycerides**), which are uncharged esters of fatty acids with glycerol.

A triacylglycerol

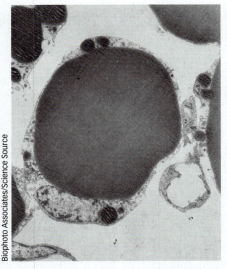

FIGURE 22.1 An electron micrograph of an adipocyte reveals an enormous intracellular lipid droplet. A small band of cytoplasm surrounds the large deposit of triacylglycerols.

Triacylglycerols are stored mainly in adipose tissue, composed of cells called *adipocytes* (**Figure 22.1**). Triacylglycerols are highly concentrated stores of metabolic energy because they are reduced and anhydrous (containing no water). The yield from the complete oxidation of fatty acids is about 38 kJ g^{-1} (9 kcal g^{-1}), in contrast with about 17 kJ g^{-1} (4 kcal g^{-1}) for carbohydrates and proteins. The basis of this large difference in caloric yield is that fatty acids are much more reduced than carbohydrates or proteins. Furthermore, triacylglycerols are nonpolar, and so they are stored in a nearly anhydrous form, whereas much more polar carbohydrates are more highly hydrated. In fact, 1 g of dry glycogen binds about 2 g of water. Consequently, a gram of nearly anhydrous fat stores 6.75 times as much energy as a gram of hydrated glycogen, which is likely the reason that triacylglycerols rather than glycogen were selected in evolution as the major energy reservoir.

Consider a typical 70-kg man, who has fuel reserves of 420,000 kJ (100,000 kcal) in triacylglycerols, 100,000 kJ (24,000 kcal) in protein (mostly in muscle), 2500 kJ (600 kcal) in glycogen, and 170 kJ (40 kcal) in glucose. Triacylglycerols constitute about 11 kg of his total body weight. If this amount of energy were stored in glycogen, his total body weight would be 64 kg greater. The glycogen and glucose stores provide enough energy to sustain physiological function for about 24 hours, whereas the triacylglycerol stores allow survival for several weeks.

In mammals, the major site of triacylglycerol accumulation is the cytoplasm of adipocytes (fat cells). This fuel-rich, white adipose tissue is located throughout the body, notably under the skin (subcutaneous fat) and surrounding the internal organs (visceral fat). Triacylglycerols coalesce to form a large globule, called a *lipid droplet*, which may occupy most of the cell volume (Figure 22.1). The lipid droplet is surrounded by a monolayer of phospholipids and numerous proteins required for triacylglycerol metabolism. Originally believed to be inert-lipid deposits, lipid droplets are now understood to be dynamic organelles essential for the regulation of lipid metabolism. Adipose cells are specialized for the synthesis and storage of triacylglycerols and for their mobilization into fuel molecules, which are transported in the blood to other tissues. Muscle also stores triacylglycerols for its own energy needs. Indeed, triacylglycerols are evident as the "marbling" of expensive cuts of beef.

It's generally well known that many mammals like mice and bears hibernate over the long winter months. Although their metabolism slows during hibernation, the energy needs of the animal must still be met, primarily by fat metabolism. However, the utility of triacylglycerols as an energy source is perhaps most dramatically illustrated by the abilities of migratory birds like the ruby-throated hummingbird featured on the first page of the chapter, which can fly great distances without eating. Another example is the American golden plover, which flies from Alaska to the southern tip of South America; a large segment of the flight (3800 km, or 2400 miles) is over open ocean, where the birds cannot feed. Fatty acids provide the energy source for these prodigious feats.

Dietary lipids are digested by pancreatic lipases

Most lipids are ingested in the form of triacylglycerols and must be degraded to fatty acids for absorption across the intestinal epithelium. Intestinal enzymes called **lipases**, secreted by the pancreas, degrade triacylglycerols to free fatty acids and monoacylglycerol (**Figure 22.2**). Lipids present a special problem because, unlike carbohydrates and proteins, these molecules are not soluble in water. They exit the stomach as an emulsion—particles with a triacylglycerol core surrounded by cholesterol and cholesterol esters.

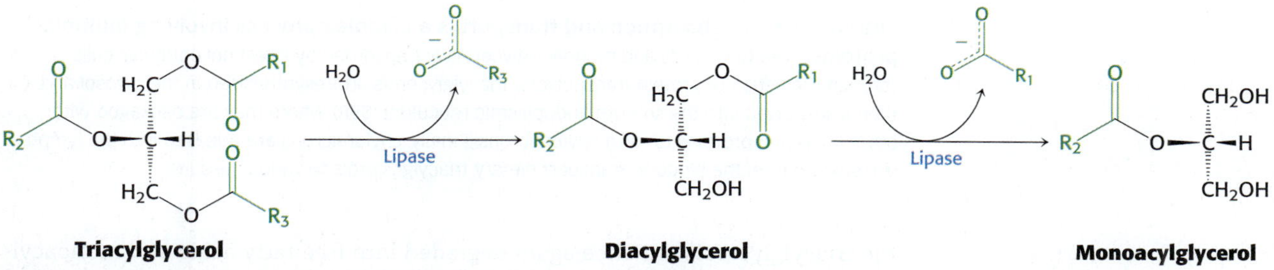

Triacylglycerol   **Diacylglycerol**   **Monoacylglycerol**

FIGURE 22.2 Pancreatic lipases convert triacylglycerols into fatty acids and monoacylglycerol for absorption into the intestine.

How are the lipids made accessible to the lipases, which are in aqueous solution? First, the particles are coated with bile acids (**Figure 22.3**), amphipathic molecules synthesized from cholesterol in the liver and secreted from the gallbladder. The ester bonds of each lipid are oriented toward the surface of the bile salt-coated particle, rendering the bond more accessible to digestion by lipases in aqueous solution. However, in this form, the particles are still not substrates for digestion. The protein colipase (another pancreatic secretory product) must bind the lipase to the particle to permit lipid degradation. If the production of bile salts is inadequate due to liver disease, large amounts of fats (as much as 30 g day^{-1}) are excreted in the feces. This condition is referred to as *steatorrhea*, after stearic acid, a common fatty acid.

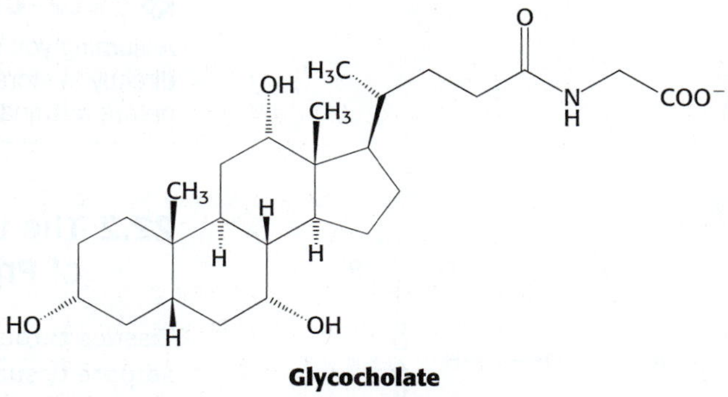

Glycocholate

FIGURE 22.3 Bile acids, such as glycocholate, facilitate lipid digestion in the intestine.

Dietary lipids are transported in chylomicrons

The final digestion products of intestinal lipases, free fatty acids and monoacylglycerol, are carried in micelles to the plasma membrane of intestinal epithelial cells where they are transported inside by membrane proteins such as fatty-acid transport proteins (FATPs) (**Figure 22.4**). Once inside the cell, fatty-acid binding proteins (FABPs) ferry them to the cytosolic face of the smooth endoplasmic reticulum (SER), where triacylglycerols are resynthesized. After transport into the lumen of the SER, newly synthesized triacylglycerols associate with specific proteins, phospholipids, and cholesterol to form lipoprotein transport particles called **chylomicrons**. Chylomicrons are stable particles approximately 2000 Å (200 nm) in diameter that are composed of 98% triacylglycerols, with the proteins, phospholipids, and cholesterol on the surface (Figure 22.4).

Chylomicrons, which also transport fat-soluble vitamins and cholesterol, are next released into the lymph system and then into the blood. These particles bind to membrane-bound lipases, primarily at adipose tissue and muscle, where

LUMEN   INTESTINAL CELL

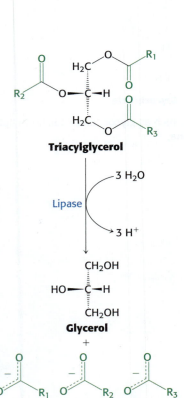

FIGURE 22.4 Fat absorption and transport is a complex process involving multiple proteins. Free fatty acids and monoacylglycerols are absorbed by intestinal epithelial cells through the action of protein transporters. Triacylglycerols are resynthesized in the cytosol and then transported into the smooth endoplasmic reticulum (SER) where they are packaged with other lipids and proteins to form chylomicrons. Finally, chylomicrons are released into the lymph system and then the blood to transport dietary triacylglycerols to various tissues.

the triacylglycerols are once again degraded into free fatty acids and monoacylglycerols for transport into the tissue. The triacylglycerols are then resynthesized inside the cell and stored. In the muscle, they can be oxidized to provide energy.

❖ SELF-CHECK QUESTION

Assuming you start in the intestinal lumen with a dietary triacylglyceride that will go directly to storage, how many times will this molecule be hydrolyzed and resynthesized before it is finally stored in an adipocyte?

22.2 The Use of Fatty Acids as Fuel Requires Three Stages of Processing

Tissues throughout the body gain access to the lipid energy reserves stored in adipose tissue through three stages of processing.

1. *Mobilization.* Lipids must first be mobilized. In this process, triacylglycerols are degraded to fatty acids and glycerol, which are released from the adipose tissue and transported to the energy-requiring tissues.

2. *Activation and transport.* In the energy-requiring tissues, the fatty acids must be activated and transported into mitochondria for degradation.

3. *Breakdown into acetyl CoA.* Third, the fatty acids are broken down in a step-by-step fashion into acetyl CoA, which is then processed in the citric acid cycle.

Mobilization: Triacylglycerols are hydrolyzed by hormone-stimulated lipases

Consider someone who has just awakened from a night's sleep and begins a bout of exercise. Glycogen stores will be low, but lipids are readily available. How are these lipid stores mobilized?

Before fats can be used as fuels, the triacylglycerol storage form must be hydrolyzed to yield isolated fatty acids. This reaction is catalyzed by hormonally controlled lipases. Under the physiological conditions facing an early-morning runner, glucagon and epinephrine will be present. In adipose tissue, these hormones trigger 7TM receptors that activate adenylate cyclase (Section 14.2).

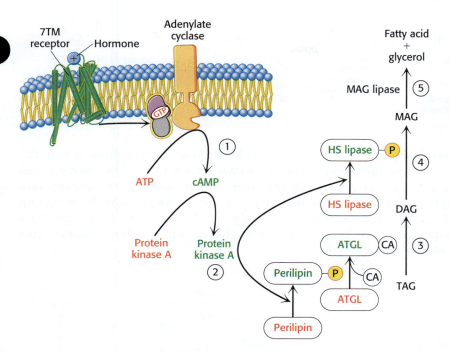

FIGURE 22.5 Triacylglycerols in adipose tissue are converted into free fatty acids in response to hormonal signals. (1) The hormones activate protein kinase A through the cAMP cascade. (2) Protein kinase A phosphorylates perilipin, resulting in the restructuring of the lipid droplet and release of the coactivator of ATGL, thereby activating ATGL. (3) ATGL converts triacylglycerol into diacylglycerol. (4) Hormone-sensitive lipase releases a fatty acid from diacylglycerol, generating monoacylglycerol. (5) Monoacylglycerol lipase completes the mobilization process. Abbreviations: 7TM, seven transmembrane receptor; ATGL, adipose triglyceride lipase; CA, coactivator; HS lipase, hormone-sensitive lipase; MAG lipase, monoacylglycerol lipase; DAG, diacylglycerol; TAG, triacylglycerol.

The increased level of cyclic AMP then stimulates protein kinase A, which phosphorylates two key proteins: perilipin, a fat-droplet-associated protein, and hormone-sensitive lipase (**Figure 22.5**). The phosphorylation of perilipin has two crucial effects. First, it restructures the fat droplet so the triacylglycerols are more accessible to the mobilization. Second, the phosphorylation of perilipin triggers the release of a coactivator for adipose triglyceride lipase (ATGL). Once bound to the coactivator, ATGL initiates the mobilization of triacylglycerols by releasing a fatty acid from triacylglycerol, forming diacylglycerol. Diacylglycerol is converted into a free fatty acid and monoacylglycerol by the hormone-sensitive lipase. Finally, a monoacylglycerol lipase completes the mobilization of fatty acids with the production of a free fatty acid and glycerol. Thus, epinephrine and glucagon induce lipolysis. Although their role in muscle is not as firmly established, these hormones probably also regulate the use of triacylglycerol stores in that tissue.

Although adipocytes are the primary site of triacylglycerol metabolism, the liver plays a crucial role in lipid metabolism, as we will examine in more detail in Chapters 24 and 27. Hepatocytes are the most common type of liver cells, and these cells function in all aspects of lipid metabolism, including import, synthesis, storage, and secretion of lipids. These processes are responsive to diet and energy needs of the liver and other tissues. Recent research suggests that the epinephrine/cAMP pathway also regulates lipolysis from lipid droplets in cultured hepatocytes. Interestingly, this signaling pathway is disrupted by ethanol, and this disruption may play a role in the development of fatty liver, a risk factor for obesity and diabetes (Section 24.3).

Mobilization continues: Free fatty acids and glycerol are released into the blood

Fatty acids are not soluble in aqueous solutions. In order to reach tissues that require fatty acids, the released fatty acids bind to the blood protein albumin, which has seven binding sites for fatty acids of varying affinity, and delivers them to tissues in need of fuel.

Glycerol formed by lipolysis is absorbed by the liver and phosphorylated. It is then oxidized to dihydroxyacetone phosphate, which is isomerized to

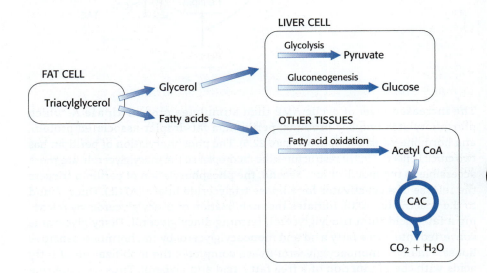

glyceraldehyde 3-phosphate. This molecule is an intermediate in both the glycolytic and the gluconeogenic pathways. Hence, glycerol can be converted into pyruvate or glucose in the liver, which contains the appropriate enzymes (**Figure 22.6**). The reverse process can take place by the reduction of dihydroxyacetone phosphate to glycerol 3-phosphate. Hydrolysis by a phosphatase then gives glycerol. Thus, glycerol and glycolytic intermediates are readily interconvertible.

FIGURE 22.6 Lipolysis generates fatty acids and glycerol. The fatty acids are used as fuel by many tissues. The liver processes glycerol by either the glycolytic or the gluconeogenic pathway, depending on its metabolic circumstances. Abbreviation: CAC, citric acid cycle.

Activation: Fatty acids are linked to coenzyme A before they are oxidized

Fatty acids separate from the albumin in the blood stream and diffuse across the cell membrane with the assistance of fatty-acid transport proteins, the same proteins that facilitate their uptake into intestinal cells (Figure 22.4). Within the cell, fatty acids are transported while bound to fatty-acid binding proteins.

Fatty acid oxidation occurs in the mitochondrial matrix, but in order to enter the mitochondria, the fatty acids are first activated through the formation of a thioester linkage to coenzyme A. ATP drives the formation of the thioester linkage between the carboxyl group of a fatty acid and the sulfhydryl group of coenzyme A. This activation reaction takes place on the outer mitochondrial membrane, where it is catalyzed by acyl CoA synthetase.

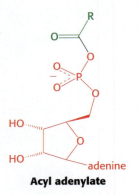

Acyl adenylate

Acyl CoA synthetase accomplishes the activation of a fatty acid in two steps. First, the fatty acid reacts with ATP to form an **acyl adenylate**. In this mixed anhydride, the carboxyl group of a fatty acid is bonded to the phosphoryl group of AMP. The other two phosphoryl groups of the ATP substrate are released as pyrophosphate. In the second step, the sulfhydryl group of coenzyme A attacks the acyl adenylate, which is tightly bound to the enzyme, to form acyl CoA and AMP.

$$\text{Fatty acid} + \text{ATP} \rightleftharpoons \text{Acyl adenylate} + \text{PP}_i$$

$$\text{Acyl adenylate} + \text{HS—CoA} \rightleftharpoons \text{Acyl CoA} + \text{AMP} + \text{H}^+$$

These partial reactions are freely reversible. In fact, the equilibrium constant for the sum of these reactions is close to 1. One high-transfer-potential compound is cleaved (between PP$_i$ and AMP), and one high-transfer-potential compound is formed (the thioester acyl CoA). How is the overall reaction driven forward? The answer is that pyrophosphate is rapidly hydrolyzed by a pyrophosphatase. The complete reaction is

$$\text{RCOO}^- + \text{CoA} + \text{ATP} + \text{H}_2\text{O} \longrightarrow \text{RCO-CoA} + \text{AMP} + 2\,\text{P}_i + 2\,\text{H}^+$$

This reaction is quite favorable because the equivalent of two molecules of ATP is hydrolyzed, whereas only one high-transfer-potential compound is formed. We see here another example of a recurring theme in biochemistry: many biosynthetic reactions are made irreversible by the hydrolysis of inorganic pyrophosphate.

Another motif recurs in this activation reaction. The enzyme-bound acyl adenylate intermediate is not unique to the synthesis of acyl CoA. Acyl adenylates are frequently formed when carboxyl groups are activated in biochemical reactions. Amino acids are activated for protein synthesis by a similar mechanism (Section 30.2), although the enzymes that catalyze this process are not homologous to acyl CoA synthetase. Thus, activation by adenylation recurs in part because of convergent evolution.

❖ SELF–CHECK QUESTION

Suppose that a promoter mutation leads to the overproduction of protein kinase A in adipose cells. How might fatty acid metabolism be altered by this mutation?

Transport: Carnitine carries long-chain activated fatty acids into the mitochondrial matrix

Fatty acids are activated on the outer mitochondrial membrane, whereas they are oxidized in the mitochondrial matrix. A special transport mechanism is needed to carry activated long-chain fatty acids across the inner mitochondrial membrane. These fatty acids must be conjugated to **carnitine**, an alcohol with both a positive and a negative charge (a zwitterion). The acyl group is transferred from the sulfur atom of coenzyme A to the hydroxyl group of carnitine to form acyl carnitine. This reaction is catalyzed by carnitine acyltransferase I (CATI), which is bound to the outer mitochondrial membrane.

$$\text{Acyl CoA} + \text{Carnitine} \rightleftharpoons \text{Acyl carnitine} + \text{HS—CoA}$$

Acyl carnitine is then shuttled across the inner mitochondrial membrane by a translocase (**Figure 22.7**). The acyl group is transferred back to coenzyme A on the

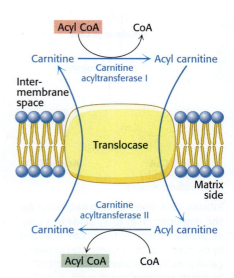

FIGURE 22.7 Acyl carnitine translocase mediates the entry of acyl carnitine into the mitochondrial matrix. Carnitine returns to the cytoplasmic side of the inner mitochondrial membrane in exchange for acyl carnitine.

matrix side of the membrane. This reaction, which is catalyzed by carnitine acyltransferase II (CATII), is simply the reverse of the reaction that takes place in the cytoplasm. The reaction is thermodynamically feasible because of the zwitterionic nature of carnitine. The O-acyl link in carnitine has a high group-transfer potential, apparently because, being zwitterions, carnitine and its esters are solvated differently from most other alcohols and their esters. Finally, the translocase returns carnitine to the cytoplasmic side in exchange for an incoming acyl carnitine.

A number of diseases have been traced to a deficiency of carnitine, the transferase, or the translocase. Inability to synthesize carnitine may be a contributing factor to the development of autism in some people. The symptoms of carnitine deficiency range from mild muscle cramping to severe weakness and even death. In general, muscle, kidney, and heart are the tissues primarily impaired. Muscle weakness during prolonged exercise is a symptom of a deficiency of carnitine acyltransferases because muscle relies on fatty acids as a long-term source of energy. Medium-chain ($C_8 - C_{10}$) fatty acids are oxidized normally in these patients because these fatty acids can enter the mitochondria, to some degree, in the absence of carnitine. These diseases illustrate that the impaired flow of a metabolite from one compartment of a cell to another can lead to a pathological condition.

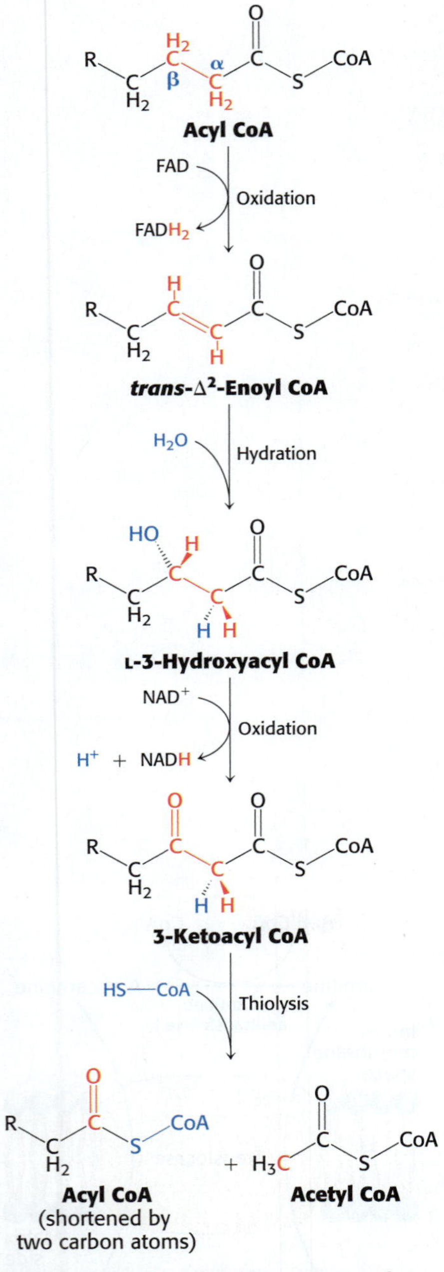
Acyl CoA
FAD
Oxidation
FADH2
trans-Δ2-Enoyl CoA
H2O
Hydration
L-3-Hydroxyacyl CoA
NAD+
Oxidation
H+ + NADH
3-Ketoacyl CoA
HS—CoA
Thiolysis
Acyl CoA
(shortened by
two carbon atoms)
Acetyl CoA

FIGURE 22.8 The reaction sequence for the degradation of fatty acids has four sequential steps. Fatty acids are degraded by the repetition of a four-reaction sequence called *β-oxidation*, consisting of oxidation, hydration, oxidation, and thiolysis.

EXPLORE this pathway further in the Metabolic Map in Achieve

Breakdown: Acetyl CoA, NADH, and $FADH_2$ are generated in each round of fatty acid oxidation

A saturated acyl CoA is degraded by a recurring sequence of four reactions: oxidation by flavin adenine dinucleotide (FAD), hydration, oxidation by NAD^+, and thiolysis by coenzyme A (**Figure 22.8**). The fatty acid chain is shortened by two carbon atoms as a result of these reactions, and $FADH_2$, NADH, and acetyl CoA are generated. Because oxidation takes place at the β carbon atom, this series of reactions is called the **β-oxidation pathway**. The four reactions of each round of β-oxidation are described below and summarized in **Table 22.1**.

1. *Oxidation of acyl CoA, generating $FADH_2$.* This reaction is catalyzed by acyl CoA dehydrogenase to give an enoyl CoA with a trans double bond between C-2 and C-3.

$$\text{Acyl CoA} + \text{E-FAD} \longrightarrow trans\text{-}\Delta^2\text{-enoyl CoA} + \text{E-FADH}_2$$

As in the dehydrogenation of succinate in the citric acid cycle, FAD rather than NAD^+ is the electron acceptor because the ΔG for this reaction is insufficient to drive the reduction of NAD^+. Electrons from the $FADH_2$ prosthetic group of the reduced acyl CoA dehydrogenase are transferred to a second flavoprotein called *electron-transferring flavoprotein* (ETF). In turn, ETF donates electrons to ETF: ubiquinone reductase, an iron–sulfur protein. Ubiquinone is thereby reduced to ubiquinol, which delivers its high-potential electrons to the second proton-pumping site of the respiratory chain (Section 18.3). Consequently, 1.5 molecules of ATP are generated per molecule of $FADH_2$ formed in this dehydrogenation step, as in the oxidation of succinate to fumarate.

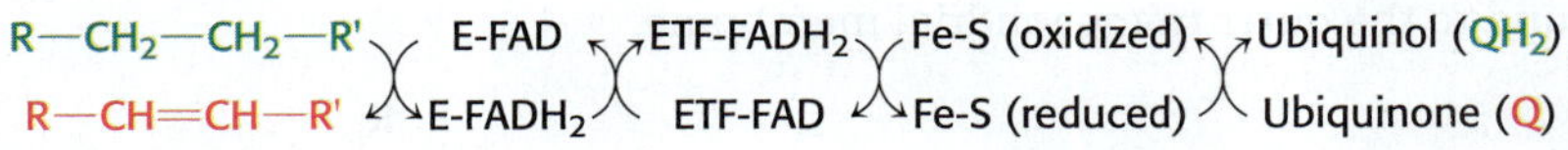
R—CH2—CH2—R'
R—CH=CH—R'
E-FAD
E-FADH2
ETF-FADH2
ETF-FAD
Fe-S (oxidized)
Fe-S (reduced)
Ubiquinol (QH2)
Ubiquinone (Q)

2. *Hydration of enyol CoA.* The next step is the hydration of the double bond between C-2 and C-3 by enoyl CoA hydratase.

$$trans\text{-}\Delta^2\text{-Enoyl CoA} + H_2O \rightleftharpoons \text{L-3-hydroxyacyl CoA}$$

The hydration of enoyl CoA is stereospecific. Only the L isomer of 3-hydroxyacyl CoA is formed when the *trans*-Δ^2 double bond is hydrated.

TABLE 22.1 Principal reactions in fatty acid oxidation

Step	Reaction	Enzyme
Activation	Fatty acid + CoA + ATP $\rightleftharpoons$ acyl CoA + AMP + PP$_i$	Acyl CoA synthetase (also called fatty acid thiokinase and fatty acid:CoA ligase)*
Transport	Carnitine + acyl CoA $\rightleftharpoons$ acyl carnitine + CoA	Carnitine acyltransferase (also called carnitine palmitoyl transferase)
1st Oxidation	Acyl CoA + E-FAD $\longrightarrow$ trans-Δ^2-enoyl CoA + E-FADH$_2$	Acyl CoA dehydrogenases (several isozymes having different chain-length specificity)
Hydration	trans-Δ^2-Enoyl CoA + H$_2$O $\rightleftharpoons$ L-3-Hydroxyacyl CoA	Enoyl CoA hydratase (also called crotonase or 3-hydroxyacyl CoA hydrolyase)
2nd Oxidation	L-3-Hydroxyacyl CoA + NAD$^+$ $\rightleftharpoons$ 3-ketoacyl CoA + NADH + H$^+$	L-3-Hydroxyacyl CoA dehydrogenase
Cleavage	3-Ketoacyl CoA + CoA $\rightleftharpoons$ acetyl CoA + acyl CoA (shortened by C$_2$)	β-Ketothiolase (also called thiolase)

*An AMP-forming ligase.

The enzyme also hydrates a cis-Δ^2 double bond, but the product then is the D isomer. We shall return to this point shortly in considering how unsaturated fatty acids are oxidized. The hydration of enoyl CoA is a prelude to the next reaction.

3. *Second oxidation, generating NADH.* This reaction converts the hydroxyl group at C-3 into a keto group and generates NADH. This oxidation is catalyzed by L-3-hydroxyacyl CoA dehydrogenase, which is specific for the L isomer of the hydroxyacyl substrate.

L-3-Hydroxyacyl CoA + NAD$^+$ $\rightleftharpoons$ 3-ketoacyl CoA + NADH + H$^+$

The preceding reactions have oxidized the methylene group at C-3 to a keto group, setting up the final step.

4. *Cleavage, generating a shortened acyl CoA and acetyl CoA.* The final step is the cleavage of 3-ketoacyl CoA by the thiol group of a second molecule of coenzyme A, which yields acetyl CoA and an acyl CoA shortened by two carbon atoms. This thiolytic cleavage is catalyzed by β-ketothiolase.

3-Ketoacyl CoA + CoA $\rightleftharpoons$ acetyl CoA + acyl CoA

(n carbons)   ($n-2$ carbons)

The shortened acyl CoA then undergoes another cycle of oxidation, starting with the reaction catalyzed by acyl CoA dehydrogenase (**Figure 22.9**). Fatty acid chains containing from 12 to 18 carbon atoms are oxidized by the long-chain acyl CoA dehydrogenase. The medium-chain acyl CoA dehydrogenase oxidizes fatty acid chains having from 4 to 14 carbons, whereas the short-chain acyl CoA dehydrogenase acts only on 4- and 6-carbon fatty acid chains. In contrast, β-ketothiolase, hydroxyacyl dehydrogenase, and enoyl CoA hydratase act on fatty acid molecules of almost any length.

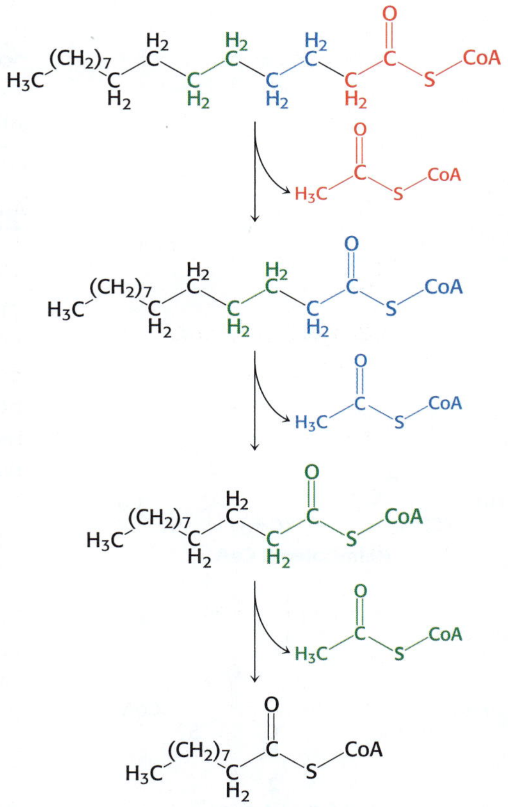

FIGURE 22.9 Each round of fatty acid degradation removes two carbons. The first three rounds in the degradation of palmitate are shown. Notice that two-carbon units are sequentially removed from the carboxyl end of the fatty acid.

The complete oxidation of palmitate yields 106 molecules of ATP

We can now calculate the energy yield derived from the oxidation of a fatty acid. In each reaction cycle, an acyl CoA is shortened by two carbon atoms, and one molecule each of FADH$_2$, NADH, and acetyl CoA are formed.

$$C_n\text{-acyl CoA} + FAD + NAD^+ + H_2O + CoA \longrightarrow$$

$$C_{n-2}\text{-acyl CoA} + FADH_2 + NADH + \text{acetyl CoA} + H^+$$

The degradation of palmitoyl CoA (C_{16}-acyl CoA) requires seven reaction cycles. In the seventh cycle, the C_4-ketoacyl CoA is thiolyzed to two molecules of acetyl CoA. Hence, the stoichiometry of the oxidation of palmitoyl CoA is

Palmitoyl CoA + 7 FAD + 7 NAD$^+$ + 7 CoA + 7 H$_2$O $\longrightarrow$

8 acetyl CoA + 7 FADH$_2$ + 7 NADH + 7 H$^+$

Approximately 2.5 molecules of ATP are generated when the respiratory chain oxidizes each of these NADH molecules, whereas 1.5 molecules of ATP are formed for each FADH$_2$ because their electrons enter the chain at the level of ubiquinol. Recall that the oxidation of acetyl CoA by the citric acid cycle yields 10 molecules of ATP. Hence, the number of ATP molecules formed in the oxidation of palmitoyl CoA is 10.5 from the seven FADH$_2$, 17.5 from the seven NADH, and 80 from the eight acetyl CoA molecules, which gives a total of 108. The equivalent of 2 molecules of ATP is consumed in the activation of palmitate, in which ATP is split into AMP and 2 molecules of orthophosphate. Thus, the complete oxidation of a molecule of palmitate yields 106 molecules of ATP.

❖ SELF–CHECK QUESTION

Explain why people with a hereditary deficiency of carnitine acyltransferase II have muscle weakness. Why are the symptoms more severe during fasting?

22.3 Unsaturated and Odd-Chain Fatty Acids Require Additional Steps for Degradation

The β-oxidation pathway accomplishes the complete degradation of saturated fatty acids having an even number of carbon atoms. Most fatty acids have such structures because of their mode of synthesis (to be addressed later in this chapter). However, not all fatty acids are so simple. The oxidation of fatty acids containing double bonds requires additional steps, as does the oxidation of fatty acids containing an odd number of carbon atoms.

An isomerase and a reductase are required for the oxidation of unsaturated fatty acids

The oxidation of unsaturated fatty acids presents some difficulties, yet many such fatty acids are available in the diet. Most of the reactions are the same as those for saturated fatty acids. In fact, only two additional enzymes—an isomerase and a reductase—are needed to degrade a wide range of unsaturated fatty acids.

Consider the oxidation of palmitoleate (**Figure 22.10**). This C_{16} unsaturated fatty acid, which has one double bond between C-9 and C-10, is activated and transported across the inner mitochondrial membrane in the same way as saturated fatty acids are. Palmitoleoyl CoA then undergoes three cycles of degradation, which are carried out by the same enzymes as those in the oxidation of saturated fatty acids. However, the cis-Δ^3-enoyl CoA formed in the third round is not a substrate for acyl CoA dehydrogenase. The presence of a double bond between C-3 and C-4 prevents the formation of another double bond between C-2 and C-3. This impasse is resolved by a new reaction that shifts the position and configuration of the cis-Δ^3 double bond. cis-Δ^3-Enoyl CoA isomerase converts this double bond into a trans-Δ^2 double bond (Figure 22.10). The double bond is now between C-2 and C-3. The subsequent reactions are those of the saturated fatty acid oxidation pathway, in which the trans-Δ^2-enoyl CoA is a regular substrate.

Human beings require polyunsaturated fatty acids (Section 22.6), which have multiple double bonds, as important precursors for signal molecules. Excess polyunsaturated fatty acids are degraded by β-oxidation. However, another problem

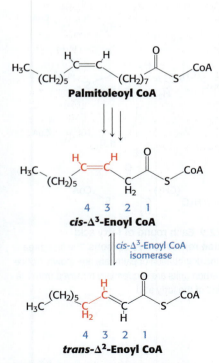

FIGURE 22.10 The degradation of a monounsaturated fatty acid involves an additional enzymatic step. Cis-Δ^3-Enoyl CoA isomerase allows continued β-oxidation of fatty acids with a single double bond.

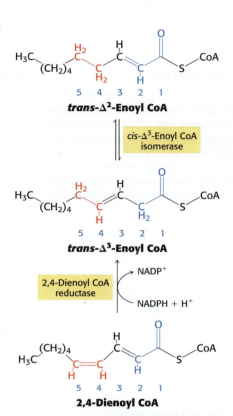

Linoleoyl CoA

cis-Δ^3-Enoyl CoA isomerase

FAD   FADH$_2$

Acyl CoA dehydrogenase

trans-Δ^2-Enoyl CoA

cis-Δ^3-Enoyl CoA isomerase

trans-Δ^3-Enoyl CoA

2,4-Dienoyl CoA reductase

NADP$^+$

NADPH + H$^+$

2,4-Dienoyl CoA

FIGURE 22.11 Multiple additional enzymes are necessary to oxidize polyunsaturated fatty acids. The oxidation of linoleoyl CoA is shown. The complete oxidation of the diunsaturated fatty acid linoleate is facilitated by the activity of enoyl CoA isomerase and 2,4-dienoyl CoA reductase. Eventually, the action of *cis*-Δ^3-enoyl CoA isomerase shifts the double bond to the Δ^2 position, making the molecule identical to an intermediate of saturated fatty acid degradation.

arises with the oxidation of polyunsaturated fatty acids. Consider linoleate, a C$_{18}$ polyunsaturated fatty acid with *cis*-Δ^9 and *cis*-Δ^{12} double bonds (**Figure 22.11**). The *cis*-Δ^3 double bond (between carbons 3 and 4) formed after three rounds of β-oxidation is converted into a *trans*-Δ^2 double bond (between carbons 2 and 3) by the aforementioned isomerase. The acyl CoA produced by another round of β-oxidation contains a *cis*-Δ^4 double bond. Dehydrogenation of this species by acyl CoA dehydrogenase yields a 2,4-dienoyl intermediate (double bond between carbons 2 and 3 and carbons 4 and 5), which is not a substrate for the next enzyme in the β-oxidation pathway. This impasse is circumvented by 2,4-dienoyl CoA reductase, an enzyme that uses NADPH to reduce the 2,4-dienoyl intermediate to *trans*-Δ^3-enoyl CoA. *cis*-Δ^3-Enoyl CoA isomerase then converts *trans*-Δ^3-enoyl CoA into the *trans*-Δ^2 form, a customary intermediate in the β-oxidation pathway (Figure 22.11). These catalytic strategies are elegant and economical. Only two extra enzymes are needed for the oxidation of *any* polyunsaturated fatty acid. Odd-numbered double bonds are handled by the isomerase, and even-numbered ones by the reductase and the isomerase.

Odd-chain fatty acids yield propionyl CoA in the final thiolysis step

Fatty acids having an odd number of carbon atoms are minor species. They are oxidized in the same way as fatty acids having an even number, except that propionyl CoA and acetyl CoA, rather than two molecules of acetyl CoA, are produced in the final round of degradation. The activated three-carbon unit in propionyl CoA enters the citric acid cycle after it has been converted into succinyl CoA (**Figure 22.12**). Interestingly, this is the only product of fatty acid catabolism that can serve as a gluconeogenic substrate. Succinyl CoA can be converted into oxaloacetate to provide carbon for glucose biosynthesis, a unique benefit of odd-numbered fatty acids.

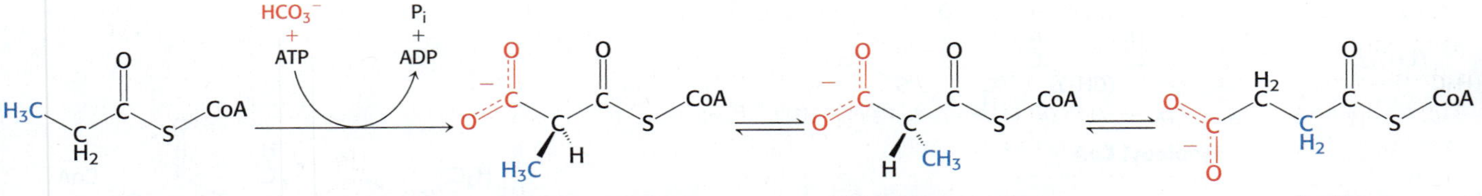

Propionyl CoA   **D-Methylmalonyl CoA**   **L-Methylmalonyl CoA**   **Succinyl CoA**

FIGURE 22.12 The final product of odd-numbered fatty acid degradation is converted into a CAC intermediate. Propionyl CoA, generated from fatty acids with an odd number of carbons as well as some amino acids, is converted into the citric acid cycle intermediate succinyl CoA, which can be a substrate for gluconeogenesis.

The pathway from propionyl CoA to succinyl CoA is especially interesting because it entails a rearrangement that requires **vitamin B$_{12}$** (also known as **cobalamin**). Propionyl CoA is carboxylated at the expense of the hydrolysis of a molecule of ATP to yield the D isomer of methylmalonyl CoA (Figure 22.12). This carboxylation reaction is catalyzed by propionyl CoA carboxylase, a biotin enzyme that has a catalytic mechanism like that of the homologous enzyme pyruvate carboxylase (Section 16.4). The D isomer of methylmalonyl CoA is racemized to the L isomer, the substrate for a mutase that converts it into succinyl CoA by an intramolecular rearrangement. The —CO—S—CoA group migrates from C-2 to a methyl group in exchange for a hydrogen atom. This very unusual isomerization is catalyzed by methylmalonyl CoA mutase, which contains a derivative of cobalamin as its coenzyme.

Vitamin B$_{12}$ contains a corrin ring and a cobalt atom

Cobalamin enzymes, which are present in most organisms, catalyze three types of reactions: (1) intramolecular rearrangements; (2) methylations, as in the synthesis of methionine (Section 25.2); and (3) the reduction of ribonucleotides to deoxyribonucleotides (Section 26.3). In mammals, only two reactions are known to require coenzyme B$_{12}$. The conversion of L-methylmalonyl CoA into succinyl CoA is one, and the formation of methionine by the methylation of homocysteine is the other (Section 25.2). The latter reaction is especially important because methionine is required for the generation of coenzymes that participate in the synthesis of purines and thymine, which are needed for nucleic acid synthesis. Interestingly, vitamin B$_{12}$ is a molecule that is made only by prokaryotic organisms; all animals obtain coenzyme B$_{12}$ either from diet or absorption from gut microorganisms, and plants and algae obtain it via their symbiosis with prokaryotes.

The core of cobalamin consists of a corrin ring with a central cobalt atom (**Figure 22.13**). The corrin ring, like a porphyrin, has four pyrrole units. Two of them

FIGURE 22.13 Coenzyme B$_{12}$ (cobalamin) is a class of molecules with a complex structure. Forms of coenzyme B$_{12}$ vary depending on the component designated X in the left-hand structure. 5′-Deoxyadenosylcobalamin is the form of the coenzyme in methylmalonyl mutase. Substitution of methyl and cyano groups for X creates methylcobalamin and cyanocobalamin, respectively.

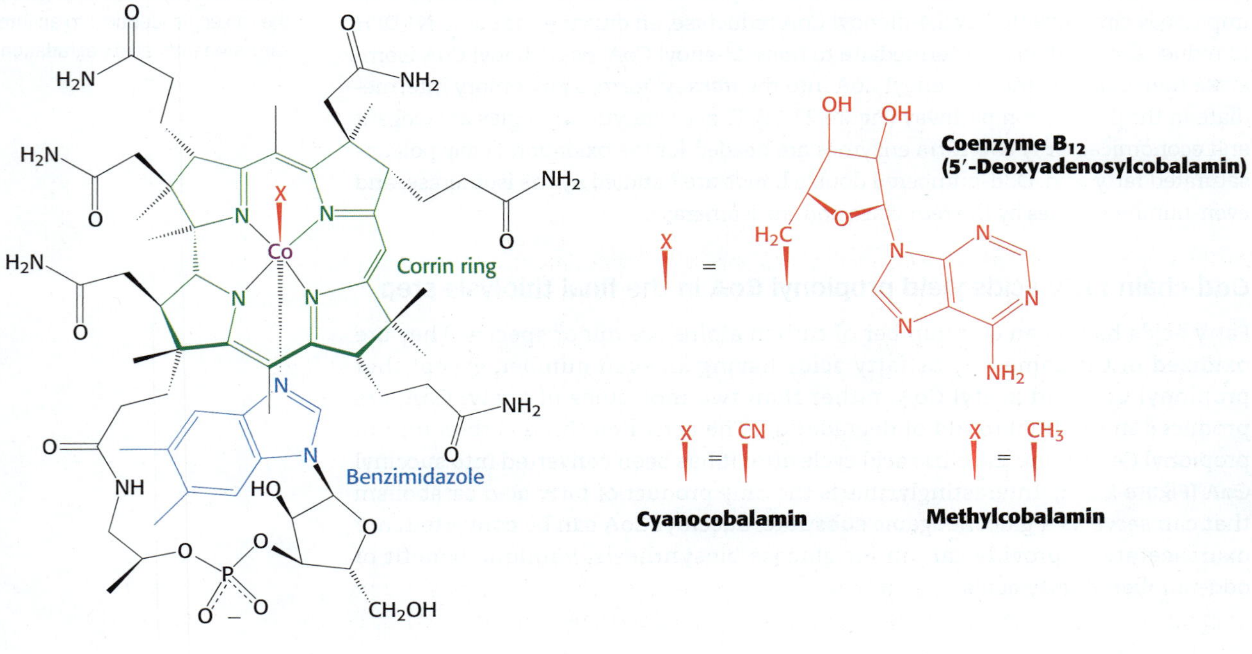

Coenzyme B$_{12}$ (5′-Deoxyadenosylcobalamin)

Corrin ring

Benzimidazole

Cyanocobalamin   **Methylcobalamin**

are directly bonded to each other, whereas the others are joined by methine bridges, as in porphyrins. The corrin ring is more reduced than that of porphyrins, and the substituents are different. A cobalt atom is bonded to the four pyrrole nitrogens. The fifth substituent linked to the cobalt atom is a derivative of dimethylbenzimidazole that contains ribose 3-phosphate and aminoisopropanol. One of the nitrogen atoms of dimethylbenzimidazole is linked to the cobalt atom. In coenzyme B_{12}, the sixth substituent linked to the cobalt atom is a 5′-deoxyadenosyl unit or a methyl group. This position can also be occupied by a cyano group. Cyanocobalamin is the form of the coenzyme administered to treat B_{12} deficiency. In all of these compounds, the cobalt is in the +3 oxidation state. Our knowledge of vitamin B_{12}, and countless other molecules including pepsin, penicillin, and insulin, was made possible by the pioneering work of Prof. Dorothy Hodgkin, sometimes called the "Queen of Crystallography" for the numerous advances she made over a six-decade career.

Mechanism: Methylmalonyl CoA mutase catalyzes a rearrangement to form succinyl CoA

The rearrangement reactions catalyzed by coenzyme B_{12} are exchanges of two groups attached to adjacent carbon atoms of the substrate (**Figure 22.14**).

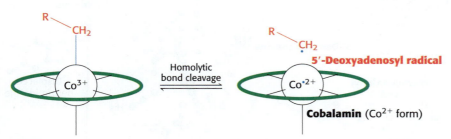

FIGURE 22.14 Cobalamin enzymes catalyze specific rearrangement reactions. A group on one carbon is exchanged with a proton on an adjacent carbon. The R group can be an amino group, a hydroxyl group, or a substituted carbon.

A hydrogen atom migrates from one carbon atom to the next, and an R group (such as the —CO—S—CoA group of methylmalonyl CoA) concomitantly moves in the reverse direction. The first step in these intramolecular rearrangements is the cleavage of the carbon–cobalt bond of 5′-deoxyadenosylcobalamin to generate the Co^{2+} form of the coenzyme and a 5′-deoxyadenosyl radical, —CH_2· (**Figure 22.15**). In this homolytic cleavage reaction, one electron of the Co–C bond stays with Co (reducing it from the +3 to the +2 oxidation state), whereas the other electron stays with the carbon atom, generating a free radical. In contrast, nearly all other cleavage reactions in biological systems are heterolytic: an electron *pair* is transferred to one of the two atoms that were bonded together.

Harold Clements/Hulton Archive/Getty Images

DOROTHY HODGKIN As a teenager in England in the mid-1920s, Dorothy Hodgkin petitioned her grammar school to be educated in chemistry along with her male peers. During her PhD training she published the first data from hydrated protein crystals in 1934, and then captured the first images of insulin in her own lab in 1935. She went on to determine the structure of penicillin, which paved the way for its mass production. In 1956, she determined the structure of vitamin B_{12}, by far the most complex structure ever solved, an accomplishment that Nobel laureate Sir Lawrence Bragg likened to "breaking the sound barrier." Dr. Hodgkin was nominated for a Nobel Prize 32 times in either Physics or Chemistry before finally winning in Chemistry in 1964. Five years later, she solved the structure of insulin, after 34 years of work. She accomplished this, and much more, all while suffering from rheumatoid arthritis since the age of 28. Despite her arthritis dramatically affecting her hands later in life, she worked until age 84 and died in 1994. One of her former students, Margaret Thatcher, famously hung a picture of Prof. Hodgkin in her office while serving as Prime Minister of the UK.

FIGURE 22.15 Cobalamin enzymes accomplish homolytic bond cleavage to form a radical. The methylmalonyl CoA mutase reaction begins with the homolytic cleavage of the bond joining Co^{3+} of coenzyme B_{12} to a carbon atom of the ribose of the adenosine moiety of the enzyme. The cleavage generates a 5′-deoxyadenosyl radical and leads to the reduction of Co^{3+} to Co^{2+}. The letter R represents the 5′-deoxyadenosyl component of the coenzyme, and the green oval represents the remainder of the coenzyme.

What is the role of this very unusual —CH_2· radical? This highly reactive species abstracts a hydrogen atom from the substrate to form 5′-deoxyadenosine and a substrate radical (**Figure 22.16**). This substrate radical spontaneously rearranges: the carbonyl CoA group migrates to the position formerly occupied by H on the neighboring carbon atom to produce a different radical. This product radical abstracts a hydrogen atom from the methyl group of 5′-deoxyadenosine

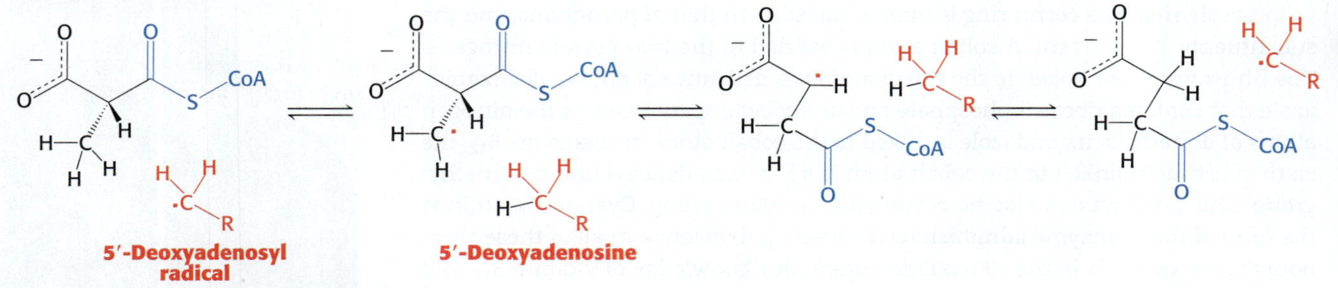

L-Methylmalonyl CoA

Succinyl CoA

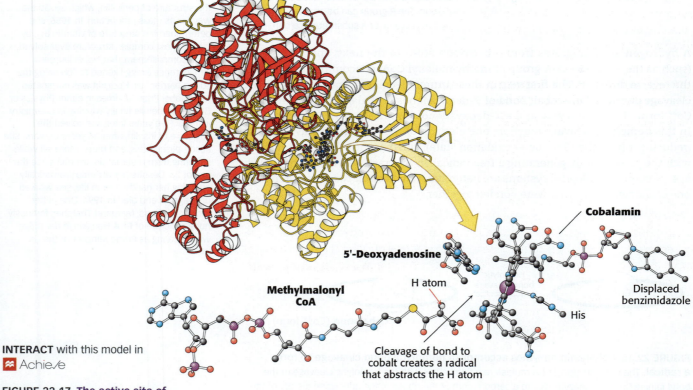

FIGURE 22.16 Succinyl CoA is formed from methylmalonyl CoA by a rearrangement reaction. A free radical abstracts a hydrogen atom in the rearrangement of methylmalonyl CoA to succinyl CoA.

to complete the rearrangement and return the deoxyadenosyl unit to the radical form. The role of coenzyme B_{12} in such intramolecular migrations is to serve as a source of free radicals for the abstraction of hydrogen atoms.

An essential property of coenzyme B_{12} is the weakness of its cobalt–carbon bond, which is readily cleaved to generate a radical. To facilitate the cleavage of this bond, enzymes such as methylmalonyl CoA mutase displace the benzimidazole group from the cobalamin and bind to the cobalt atom through a histidine residue (**Figure 22.17**). The steric crowding around the cobalt–carbon bond within the corrin ring system contributes to the bond weakness.

Cobalamin

5'-Deoxyadenosine

Methylmalonyl CoA

H atom

Displaced benzimidazole

His

Cleavage of bond to cobalt creates a radical that abstracts the H atom

INTERACT with this model in
 Achieve

FIGURE 22.17 The active site of methylmalonyl CoA mutase reveals how radical formation is accomplished. Notice that a histidine residue from the enzyme binds to cobalt in place of benzimidazole. This arrangement of substrate and coenzyme in the active site facilitates the cleavage of the cobalt–carbon bond and the subsequent abstraction of a hydrogen atom from the substrate. [Drawn from 4REQ.pdb.]

Fatty acids are also oxidized in peroxisomes

Although most fatty acid oxidation takes place in mitochondria, oxidation of long-chain and branched fatty acids takes place in cellular organelles called **peroxisomes**. These organelles are small membrane-bounded compartments that are present in the cells of most eukaryotes. Fatty acid oxidation in these organelles, which halts at octanoyl CoA, serves to shorten very long chains (C_{26}) to make them better substrates of β-oxidation in mitochondria. Peroxisomal oxidation differs

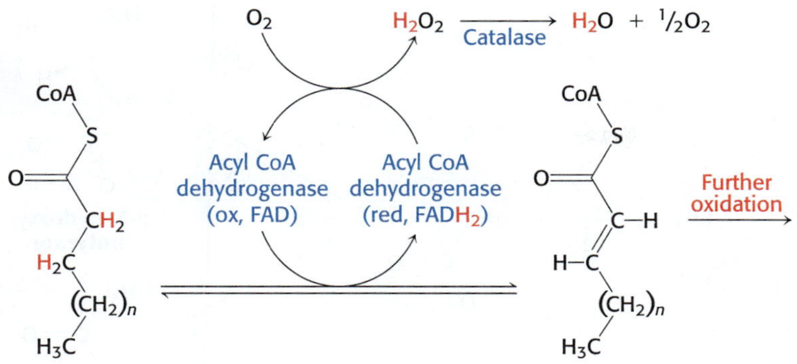

FIGURE 22.18   Initiation of peroxisomal fatty acid degradation requires a flavoprotein. The first dehydrogenation in the degradation of fatty acids in peroxisomes requires a flavoprotein dehydrogenase that transfers electrons from its $FADH_2$ moiety to O_2 to yield H_2O_2.

from β-oxidation in the initial dehydrogenation reaction (**Figure 22.18**). In peroxisomes, acyl CoA dehydrogenase—a flavoprotein—transfers electrons from the substrate to $FADH_2$ and then to O_2 to yield H_2O_2. In mitochondrial β-oxidation, the high-energy electrons would be captured as $FADH_2$ for use in the electron-transport chain. Because H_2O_2 is produced instead of $FADH_2$, peroxisomes contain high concentrations of the enzyme catalase to degrade H_2O_2 into water and O_2. Subsequent steps are identical with those of their mitochondrial counterparts, although they are carried out by different isoforms of the enzymes.

Peroxisomes do not function in patients with Zellweger syndrome. Liver, kidney, and muscle abnormalities usually lead to death by age 6 years. The syndrome is caused by a defect in the import of enzymes into the peroxisomes. Here we see a pathological condition resulting from an inappropriate cellular distribution of enzymes.

Some fatty acids contribute to the development of pathological conditions

As we will see shortly (Section 22.5), certain polyunsaturated fatty acids are essential for life, serving as precursors to various signal molecules. Vegetable oils, used commonly in food preparation, are rich in polyunsaturated fatty acids. However, polyunsaturated fatty acids are unstable and are readily oxidized. This tendency to become rancid reduces their shelf life and renders them undesirable for cooking. To circumvent this problem, polyunsaturated fatty acids are sometimes hydrogenated, converting them to saturated and monounsaturated fats. However, the conditions of hydrogenation also cause an undesirable isomerization of some cis-double bonds, resulting in the unwanted production of trans unsaturated fatty acids (popularly known as "trans fats"), a variety of fat that is rare in nature.

Epidemiological evidence suggests that consumption of large amounts of saturated fatty acids and trans fat promotes obesity, type 2 diabetes, and atherosclerosis. The mechanism by which these fats exert these effects is under active investigation. Some evidence suggests that they slow β-oxidation directly, while other evidence suggests that they promote an inflammatory response and may mute the action of insulin and other hormones.

22.4 Ketone Bodies Are a Fuel Source Derived from Fats

The acetyl CoA formed in fatty acid oxidation enters the citric acid cycle only if fat and carbohydrate degradation are appropriately balanced. Acetyl CoA must combine with oxaloacetate to gain entry to the citric acid cycle. The availability of oxaloacetate, however, depends on an adequate supply of carbohydrate. Recall that oxaloacetate is normally formed from pyruvate, the product of glucose degradation in glycolysis, by pyruvate carboxylase (Section 16.4). If carbohydrate is

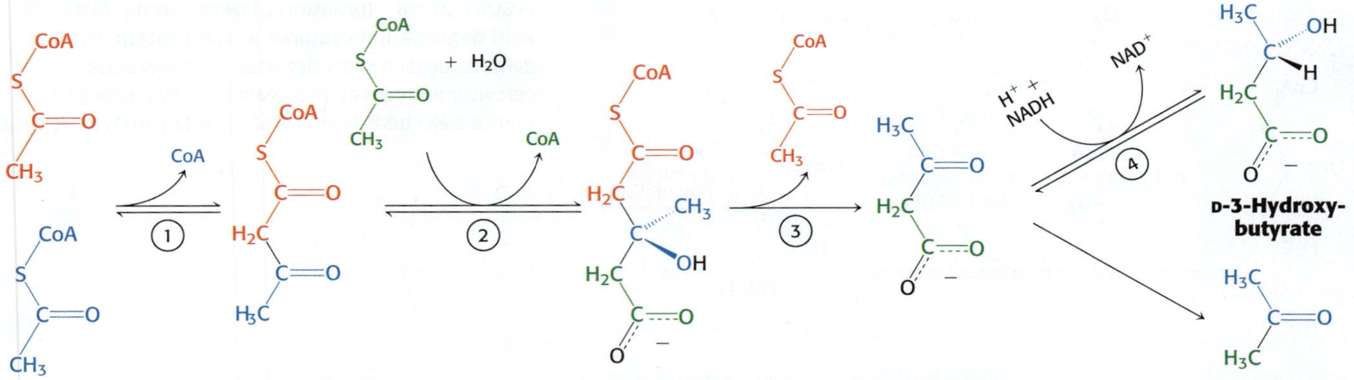

FIGURE 22.19 Ketone bodies are formed from acetyl CoA primarily in the liver.
Three molecules are known as ketone bodies: acetoacetate, D-3-hydroxybutyrate, and
acetone. Liver enzymes catalyzing the ketone body formation reactions are (1) 3-ketothiolase,
(2) hydroxymethylglutaryl CoA synthase, (3) hydroxymethylglutaryl CoA cleavage enzyme, and
(4) D-3-hydroxybutyrate dehydrogenase. Acetoacetate spontaneously decarboxylates to form acetone.

unavailable or improperly utilized, the concentration of oxaloacetate is lowered
and acetyl CoA cannot enter the citric acid cycle.

In fasting or diabetes, oxaloacetate is consumed to form glucose by the
gluconeogenic pathway (Section 16.4) and hence is unavailable for condensation
with acetyl CoA. Under these conditions, acetyl CoA is diverted to the formation
of acetoacetate and D-3-hydroxybutyrate. Acetoacetate, D-3-hydroxybutyrate,
and acetone are often referred to as **ketone bodies**. Abnormally high levels of
ketone bodies are present in the blood of untreated diabetics.

Acetoacetate is formed from acetyl CoA in three steps (**Figure 22.19**). Two
molecules of acetyl CoA condense to form acetoacetyl CoA. This reaction, which
is catalyzed by thiolase, is the reverse of the thiolysis step in the oxidation
of fatty acids. Acetoacetyl CoA then reacts with acetyl CoA and water to give
3-hydroxy-3-methylglutaryl CoA (HMG-CoA) and CoA. This condensation resembles the one catalyzed by citrate synthase (Section 17.3). This reaction, which
has a favorable equilibrium owing to the hydrolysis of a thioester linkage, com-
pensates for the unfavorable equilibrium in the formation of acetoacetyl CoA.
3-Hydroxy-3-methylglutaryl CoA is then cleaved to acetyl CoA and acetoacetate.
The sum of these reactions is

$$2 \text{ Acetyl CoA} + H_2O \longrightarrow \text{acetoacetate} + 2 \text{ CoA} + H^+$$

D-3-Hydroxybutyrate is formed by the reduction of acetoacetate in the mitochondrial matrix by D-3-hydroxybutyrate dehydrogenase. The ratio of hydroxy-
butyrate to acetoacetate depends on the $NADH/NAD^+$ ratio inside mitochondria.

Because it is a β-ketoacid, acetoacetate also undergoes a slow, spontaneous
decarboxylation to acetone. The odor of acetone may be detected in the breath of a
person who has a high level of acetoacetate in the blood. Under starvation conditions, some of the acetone may be captured to synthesize small amounts of glucose.

Ketone bodies are a major fuel in some tissues

The liver is the major site of the production of acetoacetate and 3-hydroxy-
butyrate. These molecules are carried from the liver mitochondria into the blood
by transport proteins and are conveyed to other tissues such as heart and kidney
(**Figure 22.20**). Acetoacetate and 3-hydroxybutyrate are normal fuels of respiration
and are quantitatively important as sources of energy. Indeed, heart muscle and
the renal cortex use acetoacetate in preference to glucose. In contrast, glucose

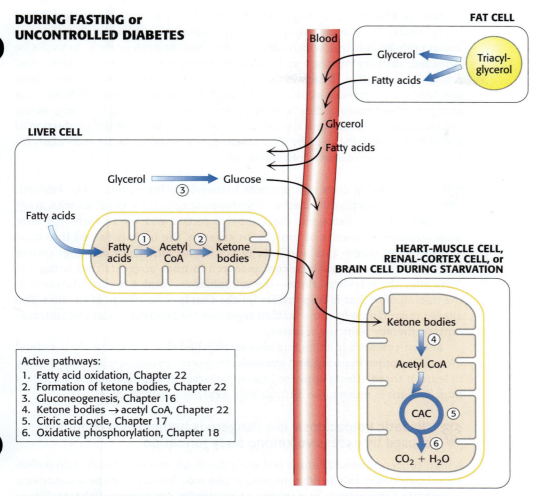

FIGURE 22.20 Pathway Integration: The liver supplies ketone bodies to the peripheral tissues. During fasting or in untreated diabetics, the liver converts fatty acids into ketone bodies, which are a fuel source for a number of tissues. Ketone-body production is especially important during starvation, when ketone bodies are the predominant fuel.

is the major fuel for the brain and red blood cells in well-nourished people on a balanced diet. However, the brain adapts to the utilization of acetoacetate during starvation and diabetes. In prolonged starvation, 75% of the fuel needs of the brain are met by ketone bodies (Section 24.5).

Acetoacetate is converted into acetyl CoA in two steps. First, acetoacetate is activated by the transfer of CoA from succinyl CoA in a reaction catalyzed by a specific CoA transferase. Second, acetoacetyl CoA is cleaved by thiolase to yield two molecules of acetyl CoA, which can then enter the citric acid cycle (**Figure 22.21**). The liver has acetoacetate available to supply to other organs because it lacks this particular CoA transferase. 3-Hydroxybutyrate requires an additional step to yield acetyl CoA. It is first oxidized to produce acetoacetate, which is processed as described up to this point, and NADH for use in oxidative phosphorylation.

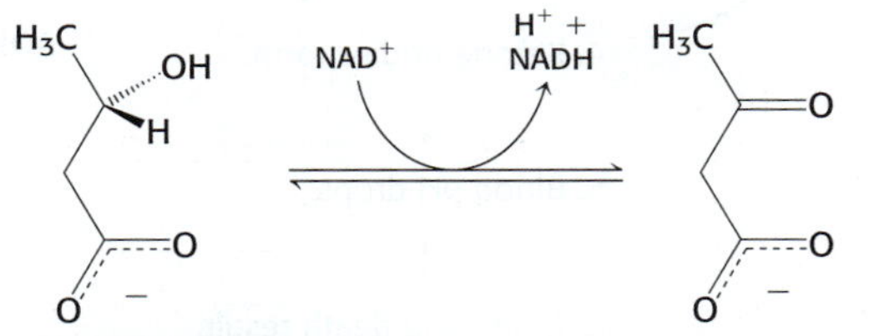

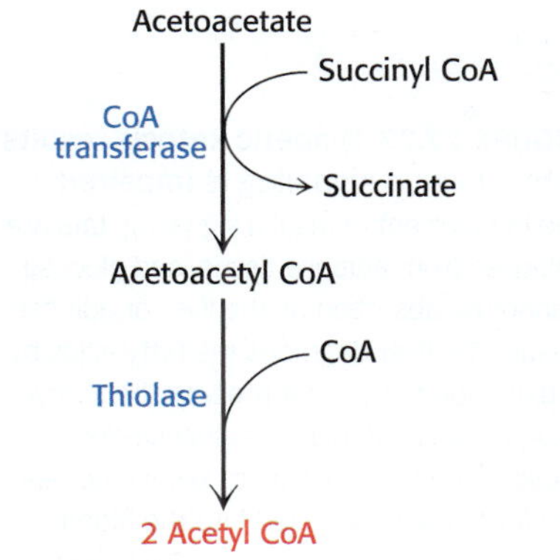

FIGURE 22.21 Acetoacetate can be used as a fuel to drive ATP formation. Acetoacetate can be converted into two molecules of acetyl CoA, which then enter the citric acid cycle.

D-3-Hydroxybutyrate   Acetoacetate

Ketone bodies can be regarded as a water-soluble, transportable form of acetyl units. Fatty acids are released by adipose tissue and converted into acetyl units by the liver, which then exports them as acetoacetate. As might be expected, both acetoacetate and 3-hydroxybutyrate also have regulatory roles. A high concentration of each in the blood signifies an abundance of acetyl units and leads to a decrease in the rate of lipolysis in adipose tissue. Although the liver produces ketone bodies for other tissues to use, it also gains energy in the process of synthesizing and releasing ketone bodies, because the production of acetyl CoA from fatty acids by β-oxidation produces both NADH and FADH$_2$ for ATP production by oxidative phosphorylation.

Interestingly, diets that promote ketone-body formation, called *ketogenic diets*, are frequently used as a therapeutic option for children with drug-resistant epilepsy. Ketogenic diets are rich in fats and low in carbohydrates, with adequate amounts of protein. In essence, the body is forced into starvation mode, where fats and ketone bodies become the main fuel source (Section 24.5). Remarkably, recent research in mice suggests that ketogenic diets alter the intestinal flora—the microbiome—and it is this alteration in the microbiome that is responsible for the therapeutic effects of the diet. The altered microbiome in some fashion regulates the levels of certain neurotransmitters, thereby reducing seizures.

Recent research in mice has also established that ketogenic diets extend lifespan, improve memory, and maintain long-term health. It will be interesting to learn if these effects apply to human beings, although gathering such data from humans is much more difficult than experiments with mice.

Diabetic ketoacidosis is a dangerous pathological condition caused by excessive ketone body formation

A high concentration of ketone bodies in the blood, the result of certain pathological conditions, can be life threatening. The most common of these conditions is diabetic ketoacidosis in patients with insulin-dependent diabetes. These patients are unable to produce insulin, which would normally be released after meals, signaling tissues to take up glucose and curtailing fatty acid mobilization by adipose tissue. The absence of insulin has two major biochemical consequences (**Figure 22.22**). First, the liver cannot absorb glucose and consequently cannot provide oxaloacetate to process fatty acid-derived acetyl CoA.

FIGURE 22.22 Diabetic ketosis results when insulin signaling is impaired. In the impairment of insulin signaling, fats are released from adipose tissue, and glucose cannot be absorbed by the liver or adipose tissue. The liver degrades the fatty acids by β-oxidation but cannot process the acetyl CoA, because of a lack of glucose-derived oxaloacetate (OAA). Excess ketone bodies are formed and released into the blood.

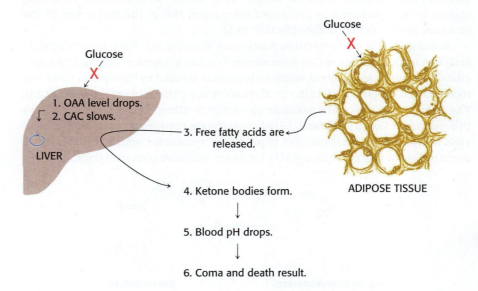

Second, adipose cells continue to release fatty acids into the bloodstream; these fatty acids are then taken up by the liver and converted into ketone bodies. The liver thus produces large amounts of ketone bodies, which are acids, resulting in severe acidosis. The decrease in pH impairs tissue function, most importantly in the central nervous system.

❖ SELF–CHECK QUESTION

Calculate the number of molecules of ATP generated by the liver in the conversion of palmitate, a C_{16} fatty acid, into acetoacetate.

Animals cannot convert fatty acids into glucose

A typical human being has far greater fat stores than glycogen stores. However, glycogen is necessary to fuel very active muscle, as well as the brain, which normally uses only glucose as a fuel. When glycogen stores are low, why can't the body make use of fat stores and convert fatty acids into glucose? Because animals are unable to accomplish the net synthesis of glucose from fatty acids. Specifically, acetyl CoA cannot be converted into gluconeogenic substrates—pyruvate or oxaloacetate—in animals.

Recall that the reaction that generates acetyl CoA from pyruvate, catalyzed by the pyruvate dehydrogenase complex, is irreversible (Section 17.2). Additionally, when the two carbon atoms of the acetyl group of acetyl CoA enter the citric acid cycle, two carbon atoms also leave the cycle in the two decarboxylation reactions catalyzed by isocitrate dehydrogenase and α-ketoglutarate dehydrogenase (Section 17.3). Consequently, oxaloacetate is regenerated, but it is not formed de novo when the acetyl unit of acetyl CoA is oxidized by the citric acid cycle. In essence, two carbon atoms enter the cycle as an acetyl group, but two carbons leave the cycle as CO_2 before oxaloacetate is generated. As a result, no net synthesis of oxaloacetate is possible, and thus humans have no way of using the carbon in an acetyl group to synthesize glucose.

Also recall that, in contrast, plants and many microorganisms have the glyoxylate cycle; they have two additional enzymes that enable them to convert the carbon atoms of acetyl CoA into oxaloacetate (Section 17.6). In conclusion, although a tremendous amount of human metabolism consists of reactions that are interconnected such that the carbon from one molecule can be used to make most anything else, this inability to convert acetyl groups into glucose can be viewed as the central disconnect in animal metabolism. We will explore the consequences of this inability more in Chapter 24.

22.5 Fatty Acids Are Synthesized by Fatty Acid Synthase

Fatty acids are synthesized by a complex of enzymes that together are called **fatty acid synthase**. Because eating most typical diets meets our physiological needs for fats and lipids, adult human beings have little need for de novo fatty acid synthesis. However, many tissues, such as liver and adipose tissue, are capable of synthesizing fatty acids, and this synthesis is required under certain physiological conditions. For instance, fatty acid synthesis is necessary during embryonic development and during lactation in mammary glands. Inappropriate fatty acid synthesis in the liver from excess caloric consumption or alcohol contributes to liver failure.

Acetyl CoA, the end product of fatty acid degradation, is the precursor for virtually all fatty acids. The biochemical challenge is to link the two carbon units together and reduce the carbons to produce palmitate, a C_{16} fatty acid. Palmitate then serves as a precursor for the variety of other fatty acids.

Fatty acid degradation and synthesis mirror each other in their chemical reactions

Fatty acid degradation and synthesis each consist of four steps that are the reverse of each other in their basic chemistry (**Figure 22.23**). As we have already seen, degradation is an oxidative process that converts a fatty acid into a set of activated acetyl units. Fatty acid synthesis is essentially the reverse of this process. The process starts with the individual units to be assembled—in this case, with an activated acyl group (most simply, an acetyl unit) and a malonyl unit (Figure 22.23). The malonyl unit condenses with the acetyl unit to form a four-carbon fragment. To produce the required hydrocarbon chain, the carbonyl group is reduced to a methylene group in three steps: a reduction, a dehydration, and another reduction, exactly the opposite of degradation. The product of the reduction is butyryl CoA. Another activated malonyl group condenses with the butyryl unit, and the process is repeated until a C_{16} or shorter fatty acid is synthesized.

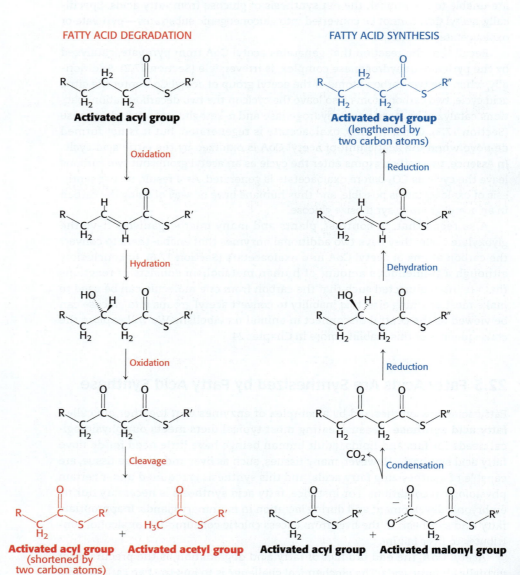

FATTY ACID DEGRADATION

FATTY ACID SYNTHESIS

FIGURE 22.23 **The steps in fatty acid degradation and synthesis are chemically similar.** The two processes are in many ways mirror images of each other.

EXPLORE this pathway further in the Metabolic Map in

Achieve

Fatty acids are synthesized and degraded by different pathways

Although fatty acid synthesis is the reversal of the degradative pathway in regard to basic chemical reactions, the synthetic and degradative pathways are different mechanistically, again exemplifying the principle that synthetic and degradative pathways are almost always distinct. Some important differences between the pathways are as follows:

1. Synthesis takes place in the cytoplasm, in contrast with degradation, which takes place primarily in the mitochondrial matrix.

2. Intermediates in fatty acid synthesis are covalently linked to the sulfhydryl groups of an **acyl carrier protein (ACP)**, whereas intermediates in fatty acid breakdown are covalently attached to the sulfhydryl group of coenzyme A.

3. The enzymes of fatty acid synthesis in higher organisms are joined in a single polypeptide chain called fatty acid synthase. In contrast, the degradative enzymes are not linked covalently.

4. The growing fatty acid chain is elongated by the sequential addition of two-carbon units derived from acetyl CoA. The activated donor of two-carbon units in the elongation step is malonyl ACP. The elongation reaction is driven by the release of CO_2.

5. The reductant in fatty acid synthesis is NADPH, whereas the oxidants in fatty acid degradation are NAD^+ and FAD.

6. The isomeric form of the hydroxyacyl intermediate in degradation is L, while the D form is used in synthesis.

The formation of malonyl CoA is the committed step in fatty acid synthesis

Fatty acid synthesis starts with the carboxylation of acetyl CoA to **malonyl CoA**. This irreversible reaction is the committed step in fatty acid synthesis.

The synthesis of malonyl CoA is catalyzed by the cytosolic isozyme of a critical regulatory enzyme, **acetyl CoA carboxylase (ACC)**. ACC contains a biotin prosthetic group, with the carboxyl group of biotin covalently attached to the ε amino group of a lysine residue, as in pyruvate carboxylase (Figure 16.28) and propionyl CoA carboxylase (Section 22.3). As with these other enzymes, a carboxybiotin intermediate is formed at the expense of the hydrolysis of a molecule of ATP. The activated CO_2 group in this intermediate is then transferred to acetyl CoA to form malonyl CoA.

$$\text{Biotin-enzyme} + \text{ATP} + \text{HCO}_3^- \rightleftharpoons \text{CO}_2\text{-biotin-enzyme} + \text{ADP} + \text{P}_i$$

$$\text{CO}_2\text{-biotin-enzyme} + \text{acetyl CoA} \longrightarrow \text{malonyl CoA} + \text{biotin-enzyme}$$

ACC has two isozymes—ACC1, the cytosolic form, and ACC2, which is located in the mitochondria. We will discuss the regulatory roles of these proteins in Section 22.7.

Intermediates in fatty acid synthesis are attached to an acyl carrier protein

The intermediates in fatty acid synthesis are linked to an acyl carrier protein. Specifically, they are linked to the sulfhydryl terminus of a phosphopantetheine group. In the degradation of fatty acids, this unit is present as part of coenzyme A, whereas, in their synthesis, it is attached to a serine residue of the acyl carrier protein (**Figure 22.24**). Thus, ACP—a single polypeptide chain of 77 residues—can be regarded as a giant prosthetic group, a "macro CoA."

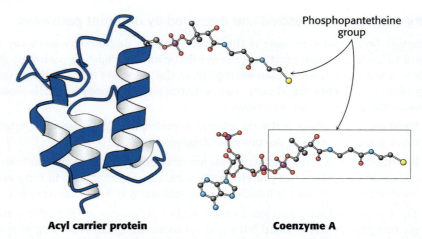

Phosphopantetheine
group

Acyl carrier protein   **Coenzyme A**

FIGURE 22.24 **Both acyl carrier protein and coenzyme A include phosphopantetheine as their reactive units.** [Drawn from 1ACP.pdb.]

Fatty acid synthesis consists of a series of condensation, reduction, dehydration, and reduction reactions

The enzyme system that catalyzes the synthesis of saturated long-chain fatty acids from acetyl CoA, malonyl CoA, and NADPH is called *fatty acid synthase*. The synthase is actually a complex of distinct enzymes. The fatty acid synthase complex in bacteria is readily dissociated into individual enzymes when the cells are broken apart. The availability of these isolated enzymes has helped biochemists elucidate the steps in fatty acid synthesis (**Table 22.2**). In fact, the reactions leading to fatty acid synthesis in higher organisms are very much like those of bacteria.

TABLE 22.2 Principal reactions in fatty acid synthesis in bacteria

Step	Reaction	Enzyme
1	Acetyl CoA + HCO$_3^-$ + ATP $\longrightarrow$ malonyl CoA + ADP + P$_i$ + H$^+$	Acetyl CoA carboxylase
2	Acetyl CoA + ACP $\rightleftharpoons$ acetyl ACP + Co A	Acetyl transacylase
3	Malonyl CoA + ACP $\rightleftharpoons$ malonyl ACP + CoA	Malonyl transacylase
4	Acetyl ACP + malonyl ACP $\longrightarrow$ acetoacetyl ACP + ACP + CO$_2$	β-Ketoacyl synthase
5	Acetoacetyl ACP + NADPH + H$^+$ $\rightleftharpoons$ D-3-hydroxybutyryl ACP + NADP$^+$	β-Ketoacyl reductase
6	D-3-Hydroxybutyryl ACP $\rightleftharpoons$ crotonyl ACP + H$_2$O	3-Hydroxyacyl dehydratase
7	Crotonyl ACP + NADPH + H$^+$ $\longrightarrow$ butyryl ACP + NADP$^+$	Enoyl reductase

The elongation phase of fatty acid synthesis starts with the formation of acetyl ACP and malonyl ACP. Acetyl transacylase and malonyl transacylase catalyze these reactions.

$$\text{Acetyl CoA + ACP} \rightleftharpoons \text{acetyl ACP + CoA}$$

$$\text{Malonyl CoA + ACP} \rightleftharpoons \text{malonyl ACP + CoA}$$

Malonyl transacylase is highly specific, whereas acetyl transacylase can transfer acyl groups other than the acetyl unit, though at a much slower rate. The synthesis of fatty acids with an odd number of carbon atoms starts with propionyl ACP, which is formed from propionyl CoA by acetyl transacylase.

Acetyl ACP and malonyl ACP react to form acetoacetyl ACP (**Figure 22.25**). The β-ketoacyl synthase, also called the condensing enzyme, catalyzes this condensation reaction.

$$\text{Acetyl ACP} + \text{malonyl ACP} \longrightarrow \text{acetoacetyl ACP} + \text{ACP} + CO_2$$

In the condensation reaction, a four-carbon unit is formed from a two-carbon unit and a three-carbon unit, and CO_2 is released. Why is the four-carbon unit not formed from two 2-carbon units—say, two molecules of acetyl ACP? The answer is that the equilibrium for the synthesis of acetoacetyl ACP from two molecules of acetyl ACP is highly unfavorable. In contrast, the equilibrium is favorable if malonyl ACP is a reactant because its decarboxylation contributes a substantial release of free energy. In effect, ATP drives the condensation reaction, though ATP does not directly participate in the condensation reaction. Instead, ATP is used to carboxylate acetyl CoA to malonyl CoA. The free energy thus stored in malonyl CoA is released in the decarboxylation accompanying the formation of acetoacetyl ACP. Although HCO_3^- is required for fatty acid synthesis, its carbon atom does not appear in the product because it is released in the subsequent decarboxylation. Rather, all the carbon atoms of fatty acids containing an even number of carbon atoms are derived only from acetyl CoA.

The next three steps in fatty acid synthesis reduce the keto group at C-3 to a methylene group (Figure 22.25). First, acetoacetyl ACP is reduced to D-3-hydroxybutyryl ACP by β-ketoacyl reductase. This reaction differs from the corresponding one in fatty acid degradation in two respects: (1) the D rather than the L isomer is formed; and (2) NADPH is the reducing agent, whereas NAD^+ is the oxidizing agent in β-oxidation. This difference exemplifies the general principle that NADPH is typically consumed in biosynthetic reactions, whereas NADH is typically generated in energy-yielding reactions. Then, D-3-hydroxybutyryl ACP is dehydrated to form crotonyl ACP, which is a *trans*-Δ^2-enoyl ACP by 3-hydroxyacyl dehydratase. The final step in the cycle reduces crotonyl ACP to butyryl ACP. NADPH is again the reductant, whereas FAD is the oxidant in the corresponding reaction in β-oxidation. The bacterial enzyme that catalyzes this step, enoyl reductase, can be inhibited by triclosan, a broad-spectrum antibacterial agent that is added to a variety of products such as toothpaste, soaps, and skin creams. These last three reactions—a reduction, a dehydration, and a second reduction—convert acetoacetyl ACP into butyryl ACP, which completes the first elongation cycle.

In the second round of fatty acid synthesis, butyryl ACP condenses with malonyl ACP to form a C_6-β-ketoacyl ACP. This reaction is like the one in the first round, in which acetyl ACP condenses with malonyl ACP to form a C_4-β-ketoacyl ACP. Reduction, dehydration, and a second reduction convert the C_6-β-ketoacyl ACP into a C_6-acyl ACP, which is ready for a third round of elongation. The elongation cycles continue until C_{16}-acyl ACP is formed. This intermediate is a good substrate for a thioesterase that hydrolyzes C_{16}-acyl ACP to yield palmitate and ACP. The thioesterase acts as a ruler to determine fatty acid chain length. The synthesis of longer-chain fatty acids is discussed in Section 22.6.

❖ SELF–CHECK QUESTION

Although HCO_3^- is required for fatty acid synthesis, its carbon atom does not appear in the product. Explain how this omission is possible.

Fatty acids are synthesized by a multifunctional enzyme complex in animals

Although the basic biochemical reactions in fatty acid synthesis are very similar in E. coli and eukaryotes, the structure of the synthase varies considerably. The component enzymes of animal fatty acid synthases, in contrast with those of E. coli and plants, are linked in a large polypeptide chain.

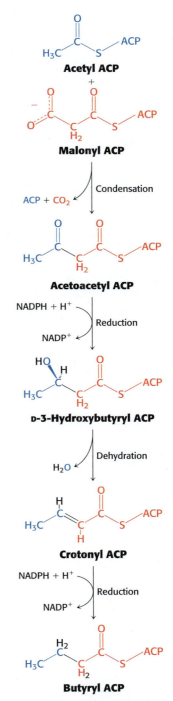

FIGURE 22.25 Like β-oxidation, saturated fatty acid synthesis consists of four steps. Saturated fatty acid synthesis begins with the condensation of malonyl ACP and acetyl ACP to form acetoacetyl ACP. Acetoacetyl ACP is then reduced, dehydrated, and reduced again to form butyryl ACP. Another cycle begins with the condensation of butyryl ACP and malonyl ACP. The sequence of reactions is repeated until the final product palmitate is formed.

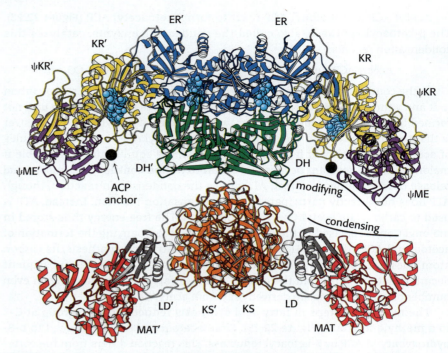

FIGURE 22.26 Mammalian fatty acid synthase is a large, dimeric, multidomain protein complex. The complex is a dimer of two multidomain monomers and can be described as consisting of two main parts: the body of the lower part is the β-ketosynthase (KS), while the malonylacetyl transferase (MAT) domains comprise the legs. These structures are joined by a linker domain (LD). The lower part is connected at the waist to the upper part, which consists of dehydratase (DH) and enoyl reductase (ER), which together comprise the upper body. The ketoreductase (KR) domains form the arms. The noncatalytic domains are ψKR and ψME. Bound NADP+ cofactors are shown as blue spheres and the attachment site for the acyl carrier protein, and thioesterase is shown as a black sphere. The domains of the second chain are indicated by the abbreviation followed by a prime. [Republished with permission of American Assn for the Advancement of Science, from Maier et al., The crystal structure of a mammalian fatty acid synthase, *Science* 321:1315–1322, September 5, 2008, Figure 1a. Permission conveyed through Copyright Clearance Center, Inc. Image courtesy Maier and Ban, ETH Zurich.]

The structure of a large part of the mammalian fatty acid synthase has been determined, with the acyl carrier protein and thioesterase remaining to be resolved. The enzyme is a dimer of identical 270-kd subunits. Each chain contains all of the active sites required for activity, as well as an acyl carrier protein tethered to the complex (**Figure 22.26**). Despite the fact that each chain possesses all of the enzymes required for fatty acid synthesis, the monomers are not active. A dimer is required.

The two component chains interact such that the enzyme activities are partitioned into two distinct compartments, a lower body and an upper body, joined by a waist. The lower body contains the β-ketosynthase, while the legs comprise the malonylacetyl transferase domains which are joined to the body by a linker domain. The body and the legs catalyze the condensing reactions. The upper body contains dehydratase and enoyl reductase, and the arms contain the ketoreductase domains. The upper body and the arms catalyze the modification reactions, the reduction and dehydration activities that result in the saturated fatty acid product. Interestingly, the mammalian fatty acid synthase has one active site—malonyl/acetyl transacylase—that adds both acetyl CoA and malonyl CoA. In contrast, most other fatty acid synthases have two separate enzyme activities—one for acetyl CoA and one for malonyl CoA. The enzyme also contains noncatalytic domains, designated by symbol ψ. The ψKR resembles the ketoreductase domain, and the ψME domain is homologous

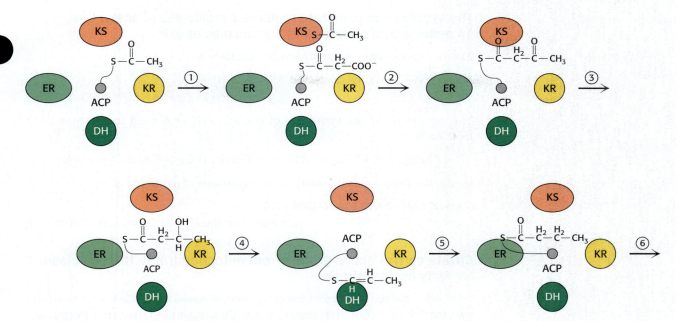

FIGURE 22.27 One catalytic cycle of mammalian fatty acid synthesis consists of seven steps. The cycle begins after MAT (not shown) attaches an acetyl unit to ACP. (1) ACP delivers the acetyl unit to KS, and MAT then attaches a malonyl unit to ACP. (2) ACP visits KS again, which condenses the acetyl and malonyl units to form the β-ketoacyl product, attached to the ACP. (3) ACP delivers the β-ketoacyl product to the KR enzyme, which reduces the keto group to an alcohol. (4) The β-hydroxyl product then visits DH, which introduces a double bond with the loss of water. (5) The enoyl product is delivered to the ER enzyme, where the double bond is reduced. (6) ACP hands the reduced product to KS and is recharged with malonyl CoA by MAT. (7) KS condenses the two molecules on ACP, which is now ready to begin another cycle. See Figure 22.26 for abbreviations.

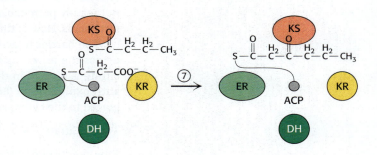

to methyltransferase enzymes. Both the ψKR and ψME domains play a role in maintaining enzyme structure.

Let us consider one catalytic cycle of the fatty acid synthase complex (**Figure 22.27**). An elongation cycle begins after malonyl/acetyl transacylase (MAT) moves an acetyl unit from coenzyme A to the acyl carrier protein (ACP). β-Ketosynthase (β-KS) accepts the acetyl unit, which forms a thioester with a cysteine residue at the β-KS active site. The vacant ACP is reloaded by MAT, this time with a malonyl moiety. Malonyl ACP visits the active site of β-KS where the condensation of the two 2-carbon fragments takes place on the ACP with the concomitant release of CO_2. The selecting and condensing process concludes with the β-ketoacyl product attached to the ACP.

The loaded ACP then sequentially visits the active sites of the modification compartment of the enzyme. Here, the β-keto group of the substrate is reduced to —OH, dehydrated, and finally reduced to yield the saturated acyl product, still attached to the ACP. With the completion of the modification process, the reduced product is transferred to the β-KS while the ACP accepts another malonyl unit. Condensation takes place and is followed by another modification cycle. The process is repeated until the thioesterase releases the final C_{16} palmitic acid product.

Many eukaryotic multienzyme complexes are multifunctional proteins in which different enzymes are linked covalently. Multifunctional enzymes such as fatty acid synthase seem likely to have arisen in eukaryotic evolution by fusion of the individual genes of evolutionary ancestors.

The synthesis of palmitate requires 8 molecules of acetyl CoA, 14 molecules of NADPH, and 7 molecules of ATP

The stoichiometry of the synthesis of palmitate is

$$\text{Acetyl CoA} + 7 \text{ malonyl CoA} + 14 \text{ NADPH} + 20 \text{ H}^+ \longrightarrow$$
$$\text{palmitate} + 7 \text{ CO}_2 + 14 \text{ NADP}^+ + 8 \text{ CoA} + 6 \text{ H}_2\text{O}$$

The equation for the synthesis of the malonyl CoA used in the preceding reaction is

$$7 \text{ Acetyl CoA} + 7 \text{ CO}_2 + 7 \text{ ATP} \longrightarrow 7 \text{ malonyl CoA} + 7 \text{ ADP} + 7 \text{ P}_i + 14 \text{ H}^+$$

Hence, the overall stoichiometry for the synthesis of palmitate is

$$8 \text{ Acetyl CoA} + 7 \text{ ATP} + 14 \text{ NADPH} + 6 \text{ H}^+ \longrightarrow$$
$$\text{palmitate} + 14 \text{ NADP}^+ + 8 \text{ CoA} + 6 \text{ H}_2\text{O} + 7 \text{ ADP} + 7 \text{ P}_i$$

Citrate carries acetyl groups from mitochondria to the cytoplasm for fatty acid synthesis

Fatty acids are synthesized in the cytoplasm, whereas acetyl CoA is formed from pyruvate in mitochondria. Hence, acetyl CoA must be transferred from mitochondria to the cytoplasm for fatty acid synthesis. Mitochondria, however, are not readily permeable to acetyl CoA. Recall that carnitine carries only long-chain fatty acids. How is this obstacle overcome?

The barrier to acetyl CoA is bypassed by citrate, which carries acetyl groups across the inner mitochondrial membrane. As you will recall from Chapter 17, citrate is formed in the mitochondrial matrix by the condensation of acetyl CoA with oxaloacetate (**Figure 22.28**). When present at high levels, citrate is transported to the cytoplasm, where it is cleaved by ATP-citrate lyase.

$$\text{Citrate} + \text{ATP} + \text{CoA} + \text{H}_2\text{O} \longrightarrow \text{acetyl CoA} + \text{ADP} + \text{P}_i + \text{oxaloacetate}$$

This reaction occurs in three steps: (1) The formation of a phospho-enzyme with the donation of a phosphoryl group from ATP; (2) binding of citrate and CoA followed by the formation of citroyl CoA and release of the phosphate; and (3) cleavage of citroyl CoA to yield acetyl CoA and oxaloacetate. As we will see shortly (Section 22.6), citrate stimulates acetyl CoA carboxylase, the enzyme that regulates fatty acid metabolism. Recall also, the presence of citrate in the cytoplasm inhibits phosphofructokinase, the enzyme that controls the glycolytic pathway.

ATP-citrate lyase is stimulated by insulin, which initiates a signal transduction pathway that ultimately results in the phosphorylation and activation of the lyase by protein kinase B (also called Akt).

FIGURE 22.28 Acetyl CoA can be transferred from the mitochondrial matrix to the cytoplasm. Acetyl CoA is transferred from mitochondria to the cytoplasm, and the reducing potential of NADH is concomitantly converted into that of NADPH by this series of reactions.

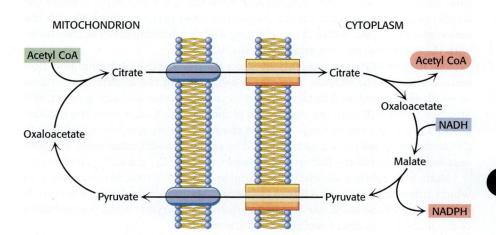

Several sources supply NADPH for fatty acid synthesis

Oxaloacetate formed in the transfer of acetyl groups to the cytoplasm must now be returned to the mitochondria. The inner mitochondrial membrane is impermeable to oxaloacetate. Hence, a series of bypass reactions are needed. These reactions also generate much of the NADPH needed for fatty acid synthesis. First, oxaloacetate is reduced to malate by NADH. This reaction is catalyzed by a malate dehydrogenase in the cytoplasm.

$$\text{Oxaloacetate} + \text{NADH} + \text{H}^+ \rightleftharpoons \text{malate} + \text{NAD}^+$$

Second, malate is oxidatively decarboxylated by NADP$^+$-linked malate enzyme (also called *malic enzyme*).

$$\text{Malate} + \text{NADP}^+ \longrightarrow \text{pyruvate} + \text{CO}_2 + \text{NADPH}$$

The pyruvate formed in this reaction readily enters mitochondria, where it is carboxylated to oxaloacetate by pyruvate carboxylase.

$$\text{Pyruvate} + \text{CO}_2 + \text{ATP} + \text{H}_2\text{O} \longrightarrow \text{oxaloacetate} + \text{ADP} + \text{P}_i + 2\text{H}^+$$

The sum of these three reactions is

$$\text{NADP}^+ + \text{NADH} + \text{ATP} + \text{H}_2\text{O} \longrightarrow \text{NADPH} + \text{NAD}^+ + \text{ADP} + \text{P}_i + \text{H}^+$$

Thus, one molecule of NADPH is generated for each molecule of acetyl CoA that is transferred from mitochondria to the cytoplasm. Hence, eight molecules of NADPH are formed when eight molecules of acetyl CoA are transferred to the cytoplasm for the synthesis of palmitate. The additional six molecules of NADPH required for this process come from the pentose phosphate pathway (Section 20.3).

The accumulation of the precursors for fatty acid synthesis is a wonderful example of the coordinated use of multiple pathways. The citric acid cycle, transport of oxaloacetate from the mitochondria, and pentose phosphate pathway provide the carbon atoms and reducing power, whereas glycolysis and oxidative phosphorylation provide the ATP to meet the needs for fatty acid synthesis (**Figure 22.29**).

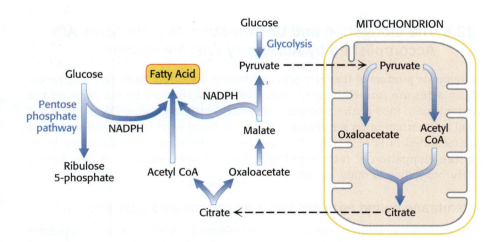

FIGURE 22.29 Pathway Integration: Fatty acid synthesis requires the cooperation of various pathways in different cellular compartments. The combination of glycolysis, the citric acid cycle, and the pentose phosphate pathway can be used to efficiently convert glucose into fatty acids.

EXPLORE this pathway further in the Metabolic Map in
Achie√e

Fatty acid metabolism is altered in tumor cells

We have previously seen that cancer cells alter glucose metabolism to meet the needs of rapid cell growth. Cancer cells must also increase fatty acid synthesis for use as signal molecules, as well as for incorporation into membrane phospholipids for the construction of new cell membranes in rapidly dividing cells. Many of the enzymes of fatty acid synthesis are overexpressed in most human cancers, and this expression is correlated with tumor malignancy. Recall that normal cells do little de novo fatty acid synthesis, relying instead on dietary intake to meet their fatty needs.

The dependence of de novo fatty acid synthesis provides possible therapeutic targets to inhibit cancer cell growth. Inhibition of β-ketoacyl ACP synthase—the enzyme that catalyzes the condensation step of fatty acid synthesis—does indeed inhibit phospholipid synthesis and subsequent cell growth in some cancers, apparently by inducing apoptosis. However, another startling observation was made: mice treated with inhibitors of the β-ketoacyl ACP synthase showed remarkable weight loss because they ate less. Thus, fatty acid synthase inhibitors are exciting candidates both as antitumor and as anti-obesity drugs.

ACC is also being investigated as a possible target for inhibiting cancer cell growth. Inhibition of the carboxylase in prostate and breast cancer cell lines induces apoptosis in the cancer cells, and yet is without effect in normal cells. Understanding the alteration of fatty acid metabolism in cancer cells is a developing area of research that holds promise of generating new cancer therapies.

Triacylglycerols may become an important renewable energy source

Work is underway to develop efficient means of generating triacylglycerols for use as renewable biodiesel fuel. Briefly, CO_2, CO, and H_2 (collectively called *syngas* because they can be used to make synthetic natural gas) are captured from municipal waste or generated from other waste sources and fed to anaerobic acetogenic bacteria. Acetogenic bacteria contain a complex pathway, called the *Wood-Ljungdahl pathway*, that can synthesize acetate from syngas. In this pathway, CO_2 and CO are reduced to a methyl group which then reacts with CO and coenzyme A to form acetyl CoA. The acetate can be harvested from the bacteria, then used as a carbon source for oleaginous yeast, which can synthesize and accumulate triacylglycerols to 20% to 70% of their cell mass, depending on culture conditions. Using large fermenters and complex algorithms to control bacterial and yeast growth, triacylglycerols can be synthesized in sufficient quantities for biodiesel and other renewable products.

22.6 The Elongation and Unsaturation of Fatty Acids Are Accomplished by Accessory Enzyme Systems

The major product of the fatty acid synthase is palmitate. In eukaryotes, longer fatty acids are formed by elongation reactions catalyzed by enzymes on the cytoplasmic face of the endoplasmic reticulum membrane. These reactions add two-carbon units sequentially to the carboxyl ends of both saturated and unsaturated fatty acyl CoA substrates. Malonyl CoA is the two-carbon donor in the elongation of fatty acyl CoAs. Again, condensation is driven by the decarboxylation of malonyl CoA.

Membrane-bound enzymes generate unsaturated fatty acids

Endoplasmic reticulum enzymes also introduce double bonds into long-chain acyl CoAs. For example, in the conversion of stearoyl CoA into oleoyl CoA, a cis-Δ^9 double bond is inserted by an oxidase that employs molecular oxygen and NADH (or NADPH).

$$\text{Stearoyl CoA} + \text{NADH} + \text{H}^+ + \text{O}_2 \longrightarrow \text{oleoyl CoA} + \text{NAD}^+ + 2\,\text{H}_2\text{O}$$

This reaction is catalyzed by a complex of three membrane-bound proteins: NADH-cytochrome b_5 reductase, cytochrome b_5, and stearoyl CoA desaturase (**Figure 22.30**). First, electrons are transferred from NADH to the FAD moiety of NADH-cytochrome b_5 reductase. The heme iron atom of cytochrome b_5 is then

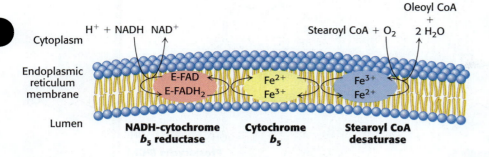

FIGURE 22.30 Fatty acids are desaturated by an electron-transport chain in the ER membrane.

reduced to the Fe^{2+} state. The nonheme iron atom of the desaturase is subsequently converted into the Fe^{2+} state, which enables it to interact with O_2 and the saturated fatty acyl CoA substrate. A double bond is formed and two molecules of H_2O are released. Two electrons come from NADH and two from the single bond of the fatty acyl substrate.

A variety of unsaturated fatty acids can be formed from oleate by a combination of elongation and desaturation reactions. For example, oleate can be elongated to a 20:1 cis-Δ^{11} fatty acid. Alternatively, a second double bond can be inserted to yield an 18:2 cis-Δ^6, Δ^9 fatty acid. Similarly, palmitate (16:0) can be oxidized to palmitoleate (16:1 cis-Δ^9), which can then be elongated to cis-vaccenate (18:1 cis-Δ^{11}).

Unsaturated fatty acids in mammals are derived from either palmitoleate (16:1 cis-Δ^9), oleate (18:1 cis-Δ^9), linoleate (18:2 cis-Δ^9, Δ^{12}), or linolenate (18:3 cis-Δ^9, Δ^{12}, Δ^{15}). However, mammals lack the enzymes to introduce double bonds at carbon atoms beyond C-9 in the fatty acid chain. Because mammals cannot synthesize linoleate and linolenate, these are essential fatty acids in human metabolism. The term *essential* means that they must be supplied in the diet because they are required by an organism and cannot be synthesized by the organism itself. It should be noted that this terminology has no relationship to the term *essential oil*, which has no specific chemical definition. Linoleate and linolenate furnished by the diet are the starting points for the synthesis of a variety of other unsaturated fatty acids.

Eicosanoid hormones are derived from polyunsaturated fatty acids

A 20:4 fatty acid derived from linoleate called *arachidonate* (20:4 cis-Δ^5, Δ^8, Δ^{11}, Δ^{14}) is the major precursor of several classes of signal molecules: prostaglandins, prostacyclins, thromboxanes, and leukotrienes (**Figure 22.31**).

A **prostaglandin** is a 20-carbon fatty acid containing a 5-carbon ring (**Figure 22.32**). This basic compound is modified by reductases and isomerases to yield nine major classes of prostaglandins, designated PGA through PGI; a subscript denotes the number of carbon–carbon double bonds outside the ring. Prostaglandins with two double bonds, such as PGE_2, are derived from arachidonate; the other two double bonds of this precursor are lost in forming a 5-membered ring. Prostacyclin and thromboxanes are related compounds that arise from a nascent prostaglandin. They are generated by prostacyclin synthase and thromboxane synthase, respectively. Alternatively, arachidonate can be converted into leukotrienes by the action of lipoxygenase. Leukotrienes, first found in leukocytes, contain three conjugated double bonds—hence the name. Prostaglandins, prostacyclin, thromboxanes, and leukotrienes are called **eicosanoids** (from the Greek *eikosi*, "twenty") because they contain 20 carbon atoms.

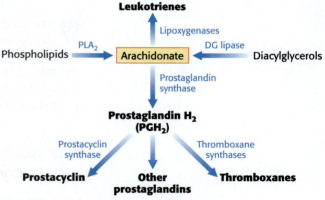

FIGURE 22.31 Several eicosanoid hormones are produced from arachidonate by distinct enzymatic pathways. Prostaglandin synthase catalyzes the first step in a pathway leading eventually to all prostaglandins, prostacyclins, and thromboxanes. Lipoxygenase catalyzes the initial step in a separate pathway leading to leukotrienes.

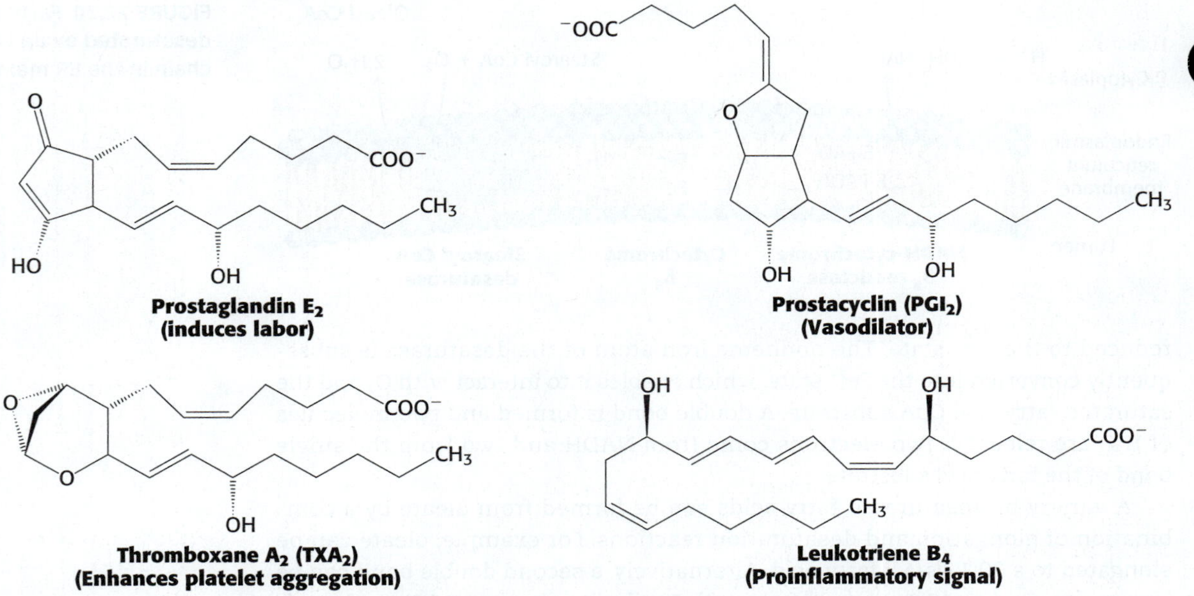

FIGURE 22.32 Eicosanoids have diverse functions but similar 20-carbon structures.

Prostaglandins and other eicosanoids are local hormones because they are short-lived. They alter the activities both of the cells in which they are synthesized and of adjoining cells by binding to 7TM receptors. Their effects may vary from one cell type to another, in contrast with the more-uniform actions of global hormones such as insulin and glucagon. Prostaglandins stimulate inflammation, regulate blood flow to particular organs, control ion transport across membranes, modulate synaptic transmission, and induce sleep.

Recall that aspirin (acetylsalicylate) blocks access to the active site of the enzyme that converts arachidonate into prostaglandin H_2 (Section 12.4). Because arachidonate is the precursor of other prostaglandins, prostacyclin, and thromboxanes, blocking this step interferes with many signaling pathways. Aspirin's ability to obstruct these pathways accounts for its wide-ranging effects on inflammation, fever, pain, and blood clotting.

22.7 Acetyl CoA Carboxylase Plays a Key Role in Controlling Fatty Acid Metabolism

Fatty acid metabolism is stringently controlled so that synthesis and degradation are highly responsive to physiological needs. Fatty acid synthesis is maximal when carbohydrates and energy are plentiful and when fatty acids are scarce. Both ACC1 and ACC2 play essential roles in regulating fatty acid synthesis and degradation. Recall that this enzyme catalyzes the committed step in fatty acid synthesis: the production of malonyl CoA (the activated two-carbon donor). This important enzyme is subject to both local and hormonal regulation. We will examine each of these levels of regulation in turn.

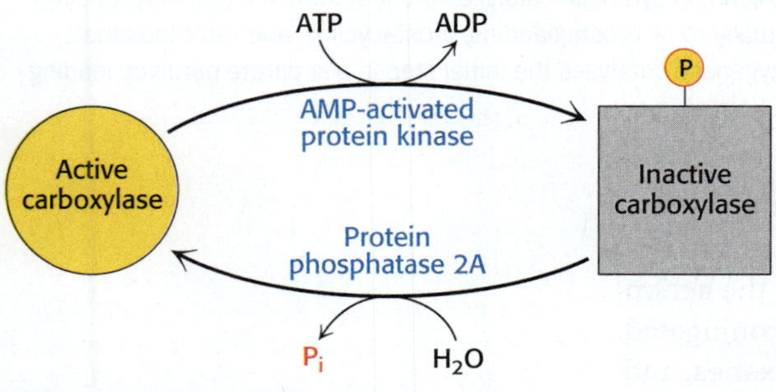

Acetyl CoA carboxylase is regulated by conditions in the cell

FIGURE 22.33 Acetyl CoA carboxylase is inhibited by phosphorylation by AMPK.

ACC1 responds to changes in the cytosol; it is switched off by phosphorylation and activated by dephosphorylation (**Figure 22.33**).

AMP-activated protein kinase (AMPK) converts the carboxylase into an inactive form by modifying three serine residues. AMPK is essentially a cellular fuel gauge; it is activated by AMP and inhibited by ATP.

The carboxylase is also allosterically stimulated by citrate. The level of citrate is high when both acetyl CoA and ATP are abundant, signifying that raw materials and energy are available for fatty acid synthesis. Citrate acts in an unusual way on inactive ACC1, which exists as isolated inactive dimers. Citrate facilitates the polymerization of the inactive dimers into active filaments (**Figure 22.34**). However, polymerization by citrate alone requires supraphysiological concentrations. In the cell, citrate-induced polymerization is facilitated by the protein MIG12, which greatly reduces the amount of citrate required. Polymerization can partly reverse the inhibition produced by phosphorylation (**Figure 22.35**). The stimulatory effect of citrate on the carboxylase is counteracted by palmitoyl CoA, which is abundant when there is an excess of fatty acids. Palmitoyl CoA causes the filaments to disassemble into the inactive subunits. Palmitoyl CoA also inhibits the translocase that transports citrate from mitochondria to the cytoplasm, as well as glucose 6-phosphate dehydrogenase, which generates NADPH in the pentose phosphate pathway.

The isozyme ACC2, located in the mitochondria, plays a role in the regulation of fatty acid degradation. Malonyl CoA, the product of the carboxylase reaction, is present at a high level when fuel molecules are abundant. Malonyl CoA inhibits CATI, preventing the entry of fatty acyl CoAs into the mitochondrial matrix in times of plenty. Malonyl CoA is an especially effective inhibitor of CATI in heart and muscle, tissues that have little fatty acid synthesis capacity of their own. In these tissues, ACC2 may be a purely regulatory enzyme. ACC2 is also phosphorylated and inhibited by AMPK. The reduction in the amount of mitochondrial malonyl CoA following the inhibition of AMPK allows fatty acid transport into the mitochondria for β-oxidation. Thus, AMPK inhibits fatty acid synthesis while stimulating fatty acid oxidation. Activators of AMPK are being investigated to treat nonalcoholic fatty liver disease (NAFLD), a condition due to overnutrition where fats accumulate in the liver. Untreated NAFLD is often a precursor to type 2 diabetes (Section 24.3).

Acetyl CoA carboxylase is controlled by a variety of hormones

ACC is controlled by the hormones glucagon, epinephrine, and insulin, which denote the overall energy status of the organism. Insulin stimulates fatty acid synthesis by activating the carboxylase, whereas glucagon and epinephrine have the reverse effect.

Regulation by glucagon and epinephrine. Consider, again, a person who has just awakened from a night's sleep and begins a bout of exercise. As mentioned, glycogen stores will be low, but lipids are readily available for mobilization.

As stated earlier, the hormones glucagon and epinephrine, present under conditions of fasting and exercise, will stimulate the mobilization of fatty acids from triacylglycerols in fat cells, which will be released into the blood, and probably from muscle cells, where they will be used immediately as fuel. These same hormones will inhibit fatty acid synthesis by inhibiting ACC. Although the exact mechanism by which these hormones exert their effects is not known, the net result is to augment the inhibition by the AMPK. This result makes sound physiological sense: when the energy level of the cell is low, as signified by a high concentration of AMP, and the energy level of the organism is low, as signaled by glucagon, fats should not be synthesized. Epinephrine, which signals the need for immediate energy, enhances this effect. Hence, these catabolic hormones switch off fatty acid synthesis by keeping the carboxylase in the inactive phosphorylated state.

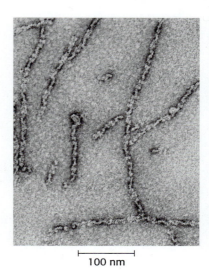

100 nm

FIGURE 22.34 Filaments of acetyl CoA carboxylase are enzymatically active. The electron micrograph shows the enzymatically active filamentous form of acetyl CoA carboxylase from chicken liver. The inactive form is a dimer of 265-kDa subunits. [Courtesy of Dr. M. Daniel Lane.]

(A)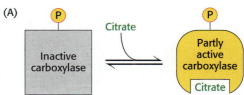

(B)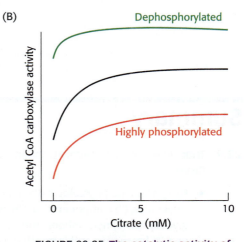

FIGURE 22.35 The catalytic activity of acetyl CoA carboxylase depends upon the concentration of citrate. (A) Citrate can partly activate the phosphorylated carboxylase. (B) The dephosphorylated form of the carboxylase is highly active even when citrate is absent. Citrate partly overcomes the inhibition produced by phosphorylation. [Information from G. M. Mabrouk, I. M. Helmy, K. G. Thampy, and S. J. Wakil. *J. Biol. Chem.* 265:6330–6338, 1990.]

Regulation by insulin. Now consider the situation after the exercise has ended and the runner has had a meal. In this case, the hormone insulin inhibits the mobilization of fatty acids and stimulates their accumulation as triacylglycerols by muscle and adipose tissue. Insulin also stimulates fatty acid synthesis by activating ACC. Insulin activates the carboxylase by enhancing the phosphorylation and inactivation of AMPK by protein kinase B. Insulin also promotes the activity of a protein phosphatase that dephosphorylates and activates ACC. Thus, the signal molecules glucagon, epinephrine, and insulin act in concert on triacylglycerol metabolism and acetyl CoA carboxylase to carefully regulate the utilization and storage of fatty acids.

Response to diet. Long-term control is mediated by changes in the rates of synthesis and degradation of the enzymes participating in fatty acid synthesis. Animals that have fasted and are then fed high-carbohydrate, low-fat diets show marked increases in their amounts of ACC and fatty acid synthase within a few days. This type of regulation is known as *adaptive control*. This regulation, which is mediated both by insulin and by glucose, is at the level of gene transcription.

❖ SELF–CHECK QUESTION

Consider citrate's role as a regulator of fatty acid metabolism and recall another regulatory connection we saw in Chapter 16 but could not fully explain at that point: citrate is an inhibitor of phosphofructokinase. Explain now why this is appropriate.

AMP-activated protein kinase is a key regulator of metabolism

As we have already seen, AMPK inhibits fatty acid synthesis while simultaneously stimulating fatty acid oxidation. However, this ubiquitous trimeric enzyme ($\alpha\beta\gamma$), which exists in several isozymic forms, regulates a number of other metabolic processes. As in fatty acid metabolism, AMPK in general activates ATP-generating pathways and inhibits pathways that require ATP. Thus, AMPK stimulates glucose uptake and mitochondria biogenesis, while inhibiting cholesterol synthesis and protein synthesis. In some immune cells AMPK moderates the inflammatory response. AMPK also helps initiate nonshivering thermogenesis in brown adipose tissue (Section 18.6). Finally, its importance to growth is illustrated by the observation that its absence is lethal in mouse embryogenesis, suggesting that AMPK is crucial for early development. Interestingly, the isozymic forms of this key enzyme are themselves regulated by phosphorylation by a number of kinases. Understanding the intricacies of this enzyme will surely keep biochemists engaged for years to come.

Summary

22.1 Triacylglycerols Are Highly Concentrated Energy Stores

- Fatty acids are physiologically important as (1) fuel molecules, (2) components of phospholipids and glycolipids, (3) hydrophobic modifiers of proteins, and (4) hormones and intracellular messengers.

- Fatty acids are stored as triacylglycerols in lipid droplets within adipocytes of adipose tissue.

- Dietary triacylglycerols are coated with bile acids, rendering them more easily digested by pancreatic lipases. The digestion products are transported inside intestinal cells where triacylglycerols are resynthesized. New triacylglycerols are packaged into chylomicrons which are released into the lymph system and then the blood for transport.

22.2 The Use of Fatty Acids as Fuel Requires Three Stages of Processing

- Triacylglycerols can be mobilized by the action of hormonally controlled lipases. Glucagon and epinephrine stimulate triacylglycerol breakdown, while insulin inhibits it.

- Fatty acids are activated to acyl CoAs, transported across the inner mitochondrial membrane while bound to carnitine, and degraded in the mitochondrial matrix by a recurring sequence of four reactions: oxidation by FAD, hydration, oxidation by NAD$^+$, and thiolysis by coenzyme A.

- The FADH$_2$ and NADH formed in the oxidation steps transfer their electrons to O$_2$ by means of the ETC, whereas the acetyl CoA formed in the thiolysis step normally enters the citric acid cycle.

22.3 Unsaturated and Odd-Chain Fatty Acids Require Additional Steps for Degradation

- Fatty acids that contain double bonds or odd numbers of carbon atoms require ancillary steps to be degraded.

- An isomerase and a reductase are required for the oxidation of unsaturated fatty acids, whereas propionyl CoA derived from chains with odd numbers of carbon atoms requires a vitamin B_{12}-dependent enzyme to be converted into succinyl CoA.

22.4 Ketone Bodies Are a Fuel Source Derived from Fats

- Ketone bodies are an important fuel source for some tissues. The primary ketone bodies—acetoacetate and β-hydroxybutyrate—are formed in the liver by condensation of acetyl CoA units.

- Mammals are unable to convert fatty acids into glucose because they lack a pathway for the net production of oxaloacetate, pyruvate, and other gluconeogenic intermediates from acetyl CoA.

22.5 Fatty Acids Are Synthesized by Fatty Acid Synthase

- Fatty acids are synthesized in the cytoplasm by fatty acid synthase. Synthesis starts with the carboxylation of acetyl CoA to malonyl CoA, the committed step. This ATP-driven reaction is catalyzed by acetyl CoA carboxylase (ACC), a biotin enzyme.

- The intermediates in fatty acid synthesis are linked to an acyl carrier protein (ACP). Acetyl ACP and malonyl ACP condense to form acetoacetyl ACP, with the release of CO_2. NADPH is used as a reductant in subsequent reactions which mirror the reverse of β-oxidation. Additional two-carbon units from malonyl ACP are added, typically ending at palmitoyl ACP, which is hydrolyzed to palmitate.

- A reaction cycle based on the formation and cleavage of citrate carries acetyl groups from mitochondria to the cytoplasm, while the NADPH needed for synthesis is generated in the transfer of reducing equivalents from mitochondria.

22.6 The Elongation and Unsaturation of Fatty Acids Are Accomplished by Accessory Enzyme Systems

- Fatty acids are elongated and desaturated by enzyme systems in the ER membrane. Desaturation requires NADH and O_2 and is carried out by a complex consisting of a flavoprotein, a cytochrome, and a nonheme iron protein.

- Mammals lack the enzymes to introduce double bonds distal to C-9, and so they require linoleate and linolenate in their diets.

- Arachidonate is the precursor of several classes of signal molecules—prostaglandins, prostacyclins, thromboxanes, and leukotrienes—that act as messengers and local hormones because of their transience and are called *eicosanoids* because they contain 20 carbon atoms.

- Aspirin (acetylsalicylate), an anti-inflammatory and antithrombotic drug, irreversibly blocks the synthesis of some of these eicosanoids.

22.7 Acetyl CoA Carboxylase Plays a Key Role in Controlling Fatty Acid Metabolism

- Fatty acid synthesis and degradation are reciprocally regulated so that both are not simultaneously active. Acetyl CoA carboxylase (ACC), the essential control site, is phosphorylated and inactivated by AMPK. The phosphorylation is reversed by a protein phosphatase.

- Citrate, which signals an abundance of building blocks and energy, partly reverses the inhibition by phosphorylation. Carboxylase activity is stimulated by insulin and inhibited by glucagon and epinephrine.

- In times of plenty, fatty acyl CoAs do not enter the mitochondrial matrix, because malonyl CoA inhibits carnitine acyltransferase I.

Key Terms

triacylglycerol (neutral fat, triglyceride) (p. 666)
lipase (p. 667)
chylomicron (p. 667)
acyl adenylate (p. 670)
carnitine (p. 671)

β-oxidation pathway (p. 672)
vitamin B_{12} (cobalamin) (p. 676)
peroxisomes (p. 678)
ketone bodies (p. 680)
fatty acid synthase (p. 683)
acyl carrier protein (ACP) (p. 685)

malonyl CoA (p. 685)
acetyl CoA carboxylase (ACC) (p. 685)
prostaglandin (p. 693)
eicosanoid (p. 693)
AMP-activated protein kinase (AMPK) (p. 695)

Problems

1. Write a balanced equation for the conversion of glycerol into pyruvate. Which enzymes are required in addition to those of the glycolytic pathway?

2. Place the following list of reactions or relevant locations in the β-oxidation of fatty acids in the proper order. ❖ 1

(a) Reaction with carnitine

(b) Fatty acid in the cytoplasm

(c) Activation of fatty acid by joining to CoA

(d) Hydration

(e) NAD⁺-linked oxidation

(f) Thiolysis

(g) Acyl CoA in mitochondrion

(h) FAD-linked oxidation

3. We have encountered reactions similar to the oxidation, hydration, and oxidation reactions of fatty acid degradation earlier in our study of biochemistry. What other pathway employs this set of reactions? ❖ 1

4. Compare the ATP yields from palmitic acid and palmitoleic acid. ❖ 1

5. Stearic acid is a C_{18} fatty acid component of chocolate. Suppose you had a depressing day and decided to settle matters by gorging on chocolate. How much ATP would you derive from the complete oxidation of stearic acid to CO_2? ❖ 1

6. Write a balanced equation for the conversion of stearate into acetoacetate. ❖ 2

7. The β-oxidation pathway degrades activated fatty acids (acyl CoA) to acetyl CoA, which then enters the citric acid cycle. Additional enzymes are required to oxidize unsaturated, odd–chain, long–chain, and branched fatty acids. Which of the following statements is/are true?

(a) Even–chain saturated fatty acids are oxidized to acetyl CoA in the β–oxidation pathway.

(b) Complete catabolism of the three–carbon remnant of a 15–carbon fatty acid requires some citric acid cycle enzymes.

(c) Enoyl–CoA isomerase, an enzyme that converts cis double bonds to trans double bonds in fatty acid metabolism, bypasses a step that reduces Q, resulting in a higher ATP yield.

(d) The final round of β–oxidation for a 13–carbon saturated fatty acid yields acetyl CoA and propionyl–CoA, a three–carbon fragment.

(e) A 14–carbon monounsaturated fatty acid with cis configuration yields more ATP than a 14–carbon saturated fatty acid.

(f) Trans double bonds in unsaturated fatty acids are not well recognized by β–oxidation enzymes.

8. Which of the following statements regarding the body's inability to convert fat to carbohydrates is/are true?

(a) The acetyl CoA that is produced from the oxidation of fatty acids enters the citric acid cycle but does not lead to a net gain of oxaloacetate.

(b) The decarboxylation steps of the citric acid cycle prevent the net synthesis of glucose precursors from acetyl CoA in animal tissues.

(c) Metabolism of odd-chain fatty acids yields propionyl CoA, which is converted to succinyl CoA and can enter the citric acid cycle.

(d) Acetyl CoA that is produced from the oxidation of fatty acids and enters the citric acid cycle is used in the *de novo* synthesis of pyruvate.

(e) Acetoacetate is a normal, major energy source for the brain, heart muscle, and renal cortex.

9. What is the committed step in fatty acid synthesis, and which enzyme catalyzes the step? ❖ 3

10. Compare and contrast fatty acid oxidation and synthesis with respect to

(a) site of the process

(b) acyl carrier

(c) reductants and oxidants

(d) stereochemistry of the intermediates

(e) direction of synthesis or degradation

(f) organization of the enzyme system

11. Lauric acid is a 12-carbon fatty acid with no double bonds. The sodium salt of lauric acid (sodium laurate) is a common detergent used in a variety of products, including laundry detergent, shampoo, and toothpaste. How many molecules of ATP and NADPH are required to synthesize lauric acid? ❖ 3

12. Arrange the following steps in fatty acid synthesis in their proper order. ❖ 3

(a) Dehydration

(b) Condensation

(c) Release of a C_{16} fatty acid

(d) Reduction of a carbonyl group

(e) Formation of malonyl ACP

13. True or False? If false, explain. ❖ 3

(a) Biotin is required for fatty acid synthase activity.

(b) The condensation reaction in fatty acid synthesis is powered by the decarboxylation of malonyl CoA.

(c) Fatty acid synthesis does not depend on ATP.

(d) Palmitate is the end product of fatty acid synthase.

(e) All of the enzyme activities required for fatty acid synthesis in mammals are contained in a single polypeptide chain.

(f) Fatty acid synthase in mammals is active as a monomer.

(g) The fatty acid arachidonate is a precursor for signal molecules.

(h) Acetyl CoA carboxylase is inhibited by citrate.

14. Suppose that you had an in vitro fatty acid–synthesizing system with all of the enzymes and cofactors required for fatty acid synthesis except for acetyl CoA. To this system, you added acetyl CoA that contained radioactive hydrogen (^{3}H, tritium) and carbon 14(^{14}C), as shown here.

$$^{3}\text{H} - \overset{\overset{\displaystyle ^{3}\text{H}}{|}}{\underset{\underset{\displaystyle ^{3}\text{H}}{|}}{\overset{14}{\text{C}}}} - \overset{\overset{\displaystyle \text{O}}{\|}}{\text{C}} - \text{SCoA}$$

The ratio of ^{3}H/^{14}C is 3. What would the ^{3}H/^{14}C ratio be after the synthesis of palmitic acid (C_{16}) with the use of the radioactive acetyl CoA? ❖ **3**

15. The serine residues in acetyl CoA carboxylase that are the target of the AMP-activated protein kinase are mutated to alanine. What is a likely consequence of this mutation? ❖ **4**

16. For each of the following unsaturated fatty acids, indicate whether the biosynthetic precursor in animals is palmitoleate, oleate, linoleate, or linolenate.

(a) 18:1 cis-Δ^{11}

(b) 18:3 cis-Δ^6, Δ^9, Δ^{12}

(c) 20:2 cis-Δ^{11}, Δ^{14}

(d) 20:3 cis-Δ^5, Δ^8, Δ^{11}

(e) 22:1 cis-Δ^{13}

(f) 22:6 cis-Δ^4, Δ^7, Δ^{10}, Δ^{13}, Δ^{16}, Δ^{19}

17. Figure 22.35 shows the response of acetyl CoA carboxylase to varying amounts of citrate. Explain this effect in light of the allosteric effects that citrate has on the enzyme. Predict the effects of increasing concentrations of palmitoyl CoA. ❖ **4**

18. Classify each event involved in the regulation of fatty acid metabolism as an outcome of hormonal regulation by insulin or by glucagon and epinephrine:

(a) Stimulates fatty acid synthesis

(b) Activates acetyl CoA carboxylase

(c) Inhibits fatty acid synthesis

(d) Inhibits acetyl CoA carboxylase

(e) Stimulates protein phosphatase activity

(f) Increases activity of AMPK

19. What are the three reactions that allow the conversion of cytoplasmic NADH into NADPH? What enzymes are required? Show the sum of the three reactions. ❖ **3**

20. Carnitine acyltransferase I (CATI) catalyzes the conversion of long-chain acyl CoA into acyl carnitine, a prerequisite for transport into mitochondria and subsequent degradation. A mutant enzyme was constructed with a single amino acid change at position 3 of glutamic acid for alanine. Graphs A through C show data from studies that were performed to identify the effect of the mutation. ❖ **1**

(a) What is the effect of the mutation on enzyme activity when the concentration of carnitine is varied? (Results are shown below.) What are the K_M and V_{max} values for the wild-type and mutant enzymes?

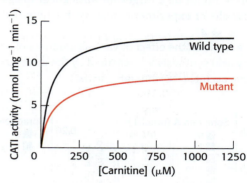

(b) What is the effect when the experiment is repeated with varying concentrations of palmitoyl CoA? What are the K_M and V_{max} values for the wild-type and mutant enzymes?

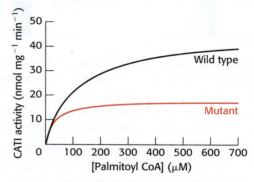

(c) The next graph shows the inhibitory effect of malonyl CoA on the wild-type and mutant enzymes. Which enzyme is more sensitive to malonyl CoA inhibition?

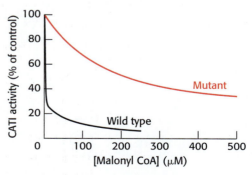

(d) Suppose that the concentration of palmitoyl CoA = 100 μM, that of carnitine = 100 μM, and that of malonyl CoA = 10 μM.

Under these conditions, what is the most prominent effect of the mutation on the properties of the enzyme?

(e) What can you conclude about the role of glutamate 3 in CATI function?

21. Soraphen A is a natural antifungal agent isolated from a myxobacterium. Soraphen A is also a potent inhibitor of acetyl CoA carboxylase, the regulatory enzyme that initiates fatty acid synthesis. As discussed in Section 22.5, cancer cells must produce a large amount of fatty acids to generate phospholipids for membrane synthesis. Thus, acetyl CoA carboxylase might be a target for anticancer drugs. Below are the results of experiments testing this hypothesis on a cancer line. ❖ **4**

Figure A shows the effect of various amounts of soraphen A on fatty acid synthesis as measured by the incorporation of radioactive acetate (^{14}C) into fatty acids.

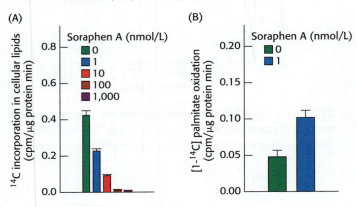

(a) How did soraphen A alter fatty acid synthesis? How was synthesis affected by increasing concentrations of soraphen A?

Figure B shows the results obtained when fatty acid oxidation was measured, as the release of radioactive CO_2 from added radioactive (^{14}C) palmitate, in the presence of soraphen A.

(b) How did soraphen A alter fatty acid oxidation?

(c) Explain the results obtained in B in light of the fact that soraphen A inhibits acetyl CoA carboxylase.

More experiments were undertaken to assess whether carboxylase inhibition did indeed prevent phospholipid

synthesis. Again, cells were grown in the presence of radioactive acetate, phospholipids were subsequently isolated, and the amount of ^{14}C incorporated into the phospholipid was determined. The effect of soraphen A on phospholipid synthesis is shown in Figure C.

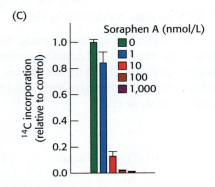

(d) Did the inhibition of carboxylase by the drug alter phospholipid synthesis?

(e) How might phospholipid synthesis affect cell viability?

Finally, as shown in Figure D, experiments were performed to determine if the drug inhibits cancer growth in vitro by determining the number of cancer cells surviving upon exposure to soraphen A.

(f) How did soraphen A affect cancer cell viability?

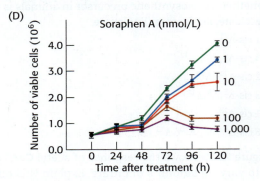

[Data from: Beckers, A., Organe, S., Timmermans, L., Scheys, K., Peeters, A., Brusselmans, K., Verhoeven, G., and Swinnen, J. V., Chemical inhibition of acetyl-CoA carboxylase induces growth arrest and cytotoxicity selectively in cancer cells, *Cancer Res.* 67:8180–8187, 2007.]

Protein Turnover and Amino Acid Catabolism

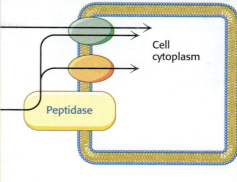

Amino acids

Proteolytic enzymes

Proteins

Oligopeptides

Cell cytoplasm

Peptidase

Genevieve Vallee/Alamy

Nitrogen is a critical component of countless biomolecules. Some plants have adapted to live in nitrogen-depleted soil by capturing and digesting insects. The biochemical mechanisms used by such plants to degrade proteins into amino acids mirror those used by animals and highlight the important role of nitrogen in biochemical biosynthesis.

❖ LEARNING GOALS

By the end of this chapter, you should be able to:

1. Explain the importance of the regulation of protein turnover.
2. Identify the primary role of ubiquitin and describe the enzymes required for ubiquitination.
3. Explain the function of the proteasome.
4. Describe the fate of nitrogen that is removed when amino acids are used as fuels.
5. Explain how the carbon skeletons of the amino acids are metabolized after nitrogen removal.
6. Identify metabolic errors in amino acid degradation.

OUTLINE

23.1 Proteins Are Degraded to Amino Acids

23.2 Protein Turnover Is Tightly Regulated

23.3 The First Step in Amino Acid Degradation Is the Removal of Nitrogen

23.4 Ammonium Ions Are Converted into Urea in Most Terrestrial Vertebrates

23.5 Carbon Atoms of Degraded Amino Acids Emerge as Major Metabolic Intermediates

23.6 Inborn Errors of Metabolism Can Disrupt Amino Acid Degradation

Our study of metabolism has, up to this point, focused on carbohydrates and lipids that contain only carbon, oxygen, and hydrogen. We now turn to the subject of nitrogen metabolism, focusing first on catabolic pathways involving proteins and amino acids. In animals, the digestion of dietary proteins in the intestine and the degradation of proteins within the cell provide a steady supply of amino acids for the biosynthesis of new proteins and nucleotide bases. Cellular proteins are also constantly degraded in response to changing metabolic demands or due to misfolding or damage. Excess amino acids can neither be stored nor excreted; instead, they must be used as metabolic fuel. In this chapter we will also consider several genetic errors of amino acid degradation. The study of amino acid metabolism is especially rewarding because it is rich in connections between basic biochemistry and clinical medicine.

TABLE 23.1 Essential amino acids in human beings

Histidine	Phenylalanine
Isoleucine	Threonine
Leucine	Tryptophan
Lysine	Valine
Methionine	

23.1 Proteins Are Degraded to Amino Acids

Dietary protein is a vital source of amino acids. Especially important dietary proteins are those containing the essential amino acids—amino acids that cannot be synthesized and must be acquired in the diet (Table 23.1). Proteins ingested in the diet are digested into amino acids or small peptides that can be absorbed by the intestine and transported in the blood. Another crucial source of amino acids is the degradation of cellular proteins.

The digestion of dietary proteins begins in the stomach and is completed in the intestine

Protein digestion begins in the stomach, where the acidic environment denatures proteins into random coils. Denatured proteins are more accessible as substrates for proteolysis than are native proteins. The primary proteolytic enzyme of the stomach is pepsin, a nonspecific protease that, remarkably, is maximally active at pH 2. Thus, pepsin can function in the highly acidic environment of the stomach that disables other proteins.

The partly digested proteins then move from the acidic environment of the stomach to the beginning of the small intestine, called the *duodenum*. The low pH of the food as well as the polypeptide products of pepsin digestion stimulate the release of hormones that promote the secretion from the pancreas of sodium bicarbonate ($NaHCO_3$), which neutralizes the pH of the food, and a variety of pancreatic proteolytic enzymes. Recall that these enzymes are secreted as inactive zymogens that are then converted into active enzymes (Sections 6.1 and 7.4). The battery of enzymes displays a wide array of specificity, and so the substrates are degraded into free amino acids as well as di- and tripeptides (Figure 23.1). Digestion is further enhanced by proteolytic enzymes, such as aminopeptidase N, that are located in the plasma membrane of the intestinal cells. Aminopeptidases digest proteins from the amino-terminal end, and the resulting free amino acids—as well as di- and tripeptides—are transported into the intestinal cells.

At least seven different transporters exist, each specific to a different group of amino acids. A number of inherited disorders result from mutations in these transporters. For example, Hartnup disease, a rare disorder characterized by rashes, ataxia (lack of muscle control), delayed mental development, and diarrhea, results from a defect in the transporter for tryptophan and other nonpolar amino acids. The absorbed amino acids are subsequently released into the blood by a number of Na^+–amino acid antiporters for use by other tissues (Figure 23.1).

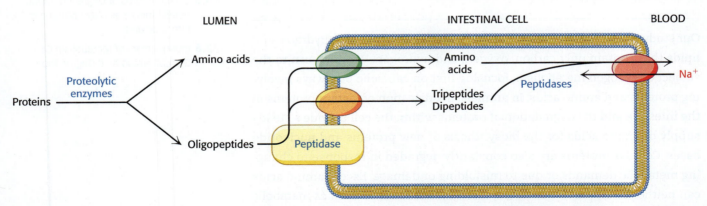

FIGURE 23.1 The products of protein digestion are absorbed by the small intestine. Protein digestion is primarily a result of the activity of enzymes secreted by the pancreas. Aminopeptidases associated with the intestinal epithelium further digest proteins. The amino acids and di- and tripeptides are absorbed into the intestinal cells by specific transporters (green and orange ovals). Free amino acids are then released into the blood by transporters (red oval) for use by other tissues.

Cellular proteins are degraded at different rates

Protein turnover—the degradation and resynthesis of proteins—takes place constantly in cells. Although some proteins are very stable, many proteins are short lived, particularly those that participate in metabolic regulation. These proteins can be quickly degraded to activate or shut down a signaling pathway. In addition, cells must eliminate damaged proteins. A significant proportion of newly synthesized protein molecules are defective because of errors in translation or misfolding. Even proteins that are normal when first synthesized may undergo oxidative damage or be altered in other ways with the passage of time. These proteins must be removed before they accumulate and aggregate. Indeed, a number of pathological conditions, such as certain forms of Parkinson's disease and Huntington's disease, are associated with protein aggregation.

The half-lives of proteins range over several orders of magnitude. Ornithine decarboxylase, at approximately 11 minutes, has one of the shortest half-lives of any mammalian protein. This enzyme participates in the synthesis of polyamines, which are cellular cations essential for growth and differentiation. The life of hemoglobin, however, is limited only by the life of the red blood cell, and the lens protein crystallin is limited by the life of the organism.

❖ SELF–CHECK QUESTION

Where are dietary amino acids absorbed, and what is the primary fate of both dietary and cellular derived amino acids?

23.2 Protein Turnover Is Tightly Regulated

How can a cell distinguish proteins that should be degraded? **Ubiquitin**, a small, 76-amino-acid protein, is so named because it is ubiquitously present in all eukaryotic cells. Ubiquitin serves many functional roles, but its primary role is to serve as a tag that marks proteins for destruction (**Figure 23.2**). Ubiquitin, in this role then, is the cellular equivalent of the "black spot" of Robert Louis Stevenson's *Treasure Island*—the signal for death.

Ubiquitin tags proteins for destruction

Ubiquitin is highly conserved in eukaryotes: yeast and human ubiquitin differ at only 3 of 76 residues. A protein destined to be degraded will be covalently modified by the attachment of ubiquitin—by its carboxyl-terminal glycine residue—to the ε-amino groups of one or more lysine residues on the target protein. The energy for the formation of these isopeptide bonds (iso because ε- rather than α-amino groups are targeted) comes from ATP hydrolysis.

Three enzymes participate in the attachment of ubiquitin to a protein (**Figure 23.3**):

- *Ubiquitin-activating enzyme* (E1). First, the C-terminal carboxylate group of ubiquitin becomes linked to a sulfhydryl group of E1 by a thioester bond. This ATP-driven reaction is reminiscent of fatty acid activation (Section 22.2). In this reaction, an acyl adenylate is formed at the C-terminal carboxylate of ubiquitin with the release of pyrophosphate, and the ubiquitin is subsequently transferred to a sulfhydryl group of a key cysteine residue in E1.

- *Ubiquitin-conjugating enzyme* (E2). The activated ubiquitin is then shuttled to a sulfhydryl group of E2, a reaction catalyzed by E2 itself.

- *Ubiquitin–protein ligase* (E3). Finally, E3 catalyzes the transfer of ubiquitin from E2 to an ε-amino group on the target protein. E3 does this by bringing E2 and the target protein together. In some cases, the ubiquitin is passed to a cysteine residue of E3 first, whereas in other cases it is transferred directly from the E2 to the target protein.

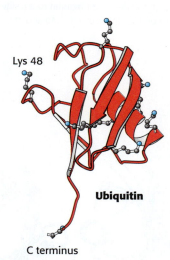

FIGURE 23.2 Ubiquitin is a small, compact protein with seven lysines. Notice that ubiquitin has an extended carboxyl terminus, which is activated and linked to proteins targeted for destruction. Lysine residues, including lysine 48, the major site for linking additional ubiquitin molecules, are shown as ball-and-stick models. [Drawn from 1UBI.pdb.]

INTERACT with this model in
📱 Achieve

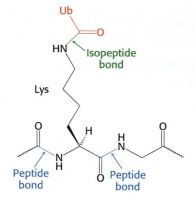

FIGURE 23.3 Ubiquitin conjugation requires three enzymes. The ubiquitin-activating enzyme E1 adenylates ubiquitin (Ub) (1) and transfers the ubiquitin to one of its own cysteine residues (2). Ubiquitin is then transferred to a cysteine residue in the ubiquitin-conjugating enzyme E2 by the E2 enzyme (3). Finally, the ubiquitin–protein ligase E3 transfers the ubiquitin to a lysine residue on the target protein (4a and 4b).

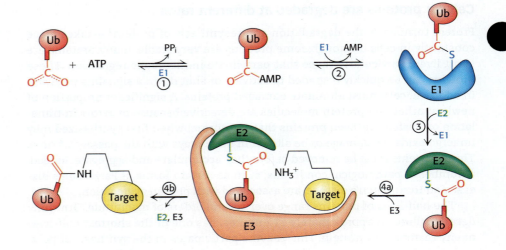

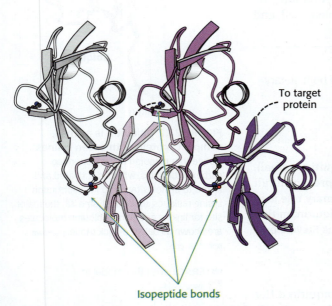

Isopeptide bonds

FIGURE 23.4 In a tetraubiquitin chain, four ubiquitin molecules are linked by isopeptide bonds. Notice that each isopeptide bond is formed by the linkage of the carboxylate group at the end of the extended C terminus with the ε-amino group of a lysine residue, in this case lysine 48. Dashed lines indicate the positions of the extended C-termini that were not observed in the crystal structure. This unit is the primary signal for degradation when linked to a target protein. [Drawn from 1TBE.pdb.]

INTERACT with this model in
 Achieve

The ubiquitination reaction can be processive: E3 can remain bound to the target protein and generate a chain of ubiquitin molecules by linking the ε-amino group of a lysine residue of one ubiquitin molecule to the terminal carboxylate of another. Alternatively, in some cases E3 will dissociate after the first ubiquitin addition, and a chain can be extended by another E2/E3 pair. Importantly, ubiquitin can be added onto any of the seven lysines or the N-terminus of the previous ubiquitin, forming chains with different linkages, lengths, and degrees of branching. Some of the linkages signal for proteasomal degradation, while others do not; although the outcome may be context-dependent in different cells. For example, a chain of four or more ubiquitin molecules linked via lysine 48 of ubiquitin has long been known to be an especially effective signal for protein degradation (**Figure 23.4**).

What determines whether a protein becomes ubiquitinated? A specific sequence of amino acids, termed a **degron**, indicates that a protein should be degraded. One such signal appears to be quite simple: the half-life of a cytoplasmic protein is determined to a large extent by its amino-terminal residue (**Table 23.2**). This dependency is referred to as the *N-end rule* or the *N-degron*. A yeast protein with methionine at its N terminus typically has a half-life of more than 20 hours, whereas one with arginine at this position has a half-life of about 2 minutes. A highly destabilizing N-terminal residue such as arginine or leucine favors rapid ubiquitination, whereas a stabilizing residue such as methionine or proline does not.

Why is the N-end rule only *apparently* simple? Sometimes the N-degron is exposed only after the protein is proteolytically cleaved. Such degrons are called *pre-N-degrons* or *pro-N-degrons*, in analogy to proenzymes (Section 7.4), because the protein must be cleaved to expose the signal. In other cases, the destabilizing amino acid is added to the protein after the protein is synthesized. Other modifications, notably N-terminal acetylation, can also activate a degron. Additional degrons thought to identify proteins for degradation include cyclin destruction boxes, which are amino acid sequences that mark cell-cycle proteins for destruction, and PEST sequences, which contain the amino acid sequence proline (P, single-letter abbreviation), glutamic acid (E), serine (S), and threonine (T).

E3 enzymes are the readers of N-terminal residues. Although most eukaryotes have only one or a small number of distinct E1 enzymes, all eukaryotes have many distinct E2 enzymes and even more E3 enzymes. For example, in baker's yeast there is only one E3 but ten E2 enzymes and approximately 80 E3 enzymes. Indeed, the E3 family is one of the largest gene families in human beings, with over

600 predicted E3 ligases in the human genome. The diversity of target proteins that must be tagged for destruction requires a large number of E3 proteins as readers.

Three examples demonstrate the importance of E3 proteins to normal cell function. First, proteins that are not degraded because of a defective E3 may accumulate, causing a disease of protein aggregation such as juvenile or early-onset Parkinson's disease. Second, a defect in another member of the E3 family causes Angelman syndrome, a severe neurological disorder characterized by an unusually happy disposition, cognitive disability, absence of speech, uncoordinated movement, and hyperactivity. Fascinatingly, if the same ligase is overexpressed, the result is autism. Third, inappropriate protein turnover also can lead to cancer. For example, human papillomavirus (HPV) encodes a protein that activates a specific E3 enzyme. The enzyme ubiquitinates the tumor suppressor p53 and other proteins that control DNA repair, which are then destroyed. The activation of this E3 enzyme is observed in more than 90% of cervical carcinomas. Thus, the inappropriate marking of key regulatory proteins for destruction can trigger further events, leading to tumor formation.

As noted above, the role of ubiquitin is much broader than merely marking proteins for destruction. Ubiquitination also regulates proteins involved in DNA repair, chromatin remodeling, innate immunity, membrane trafficking, and autophagy, among other biochemical processes.

The proteasome digests the ubiquitin-tagged proteins

If ubiquitin is the mark of death, what is the executioner? A large protease complex called the **proteasome** (more precisely, the *26S proteasome*) digests the ubiquitinated proteins. This ATP-driven multisubunit protease spares ubiquitin, which is then recycled. The 26S proteasome is a complex of two components: a 20S catalytic unit and a 19S regulatory unit.

The 20S unit is composed of 28 subunits, encoded by 14 genes, arranged into 4 heteroheptameric rings stacked to form a structure resembling a barrel (**Figure 23.5**). The outer two rings of the barrel are made up of α-type subunits and the inner two rings of β-type subunits. The 20S catalytic core is a sealed barrel. Access to its interior is controlled by a 19S regulatory unit, itself a 700-kDa complex made up of 19 subunits. Two such 19S complexes bind to the 20S proteasome core, one at each end, to form the complete 26S proteasome (**Figure 23.6**).

The 19S regulatory unit has three functions. First, components of the 19S unit are ubiquitin receptors that bind specifically to polyubiquitin chains, thereby ensuring that only ubiquitinated proteins are degraded. Second, the doomed protein is unfolded and directed into the catalytic core. Finally, an isopeptidase in the 19S unit cleaves off intact ubiquitin molecules from the proteins so that they can be reused. Key components of the 19S complex are six ATPases of a type

TABLE 23.2 Dependence of the half-lives of cytoplasmic yeast proteins on the identity of their amino-terminal residues

Highly stabilizing residues ($t_{1/2} > 20$ hours)			
Ala	Cys	Gly	Met
Pro	Ser	Thr	Val
Intrinsically destabilizing residues ($t_{1/2} = 2$ to 30 minutes)			
Arg	His	Ile	Leu
Lys	Phe	Trp	Tyr
Destabilizing residues after chemical modification ($t_{1/2} = 3$ to 30 minutes)			
Asn	Asp	Gln	Glu

Data from J. W. Tobias, T. E. Schrader, G. Rocap, and A. Varshavsky, *Science* 254(5036):1374–1377, 1991.

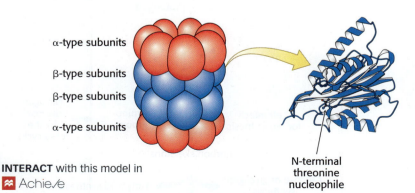

α-type subunits
β-type subunits
β-type subunits
α-type subunits

N-terminal threonine nucleophile

INTERACT with this model in
Achieve

FIGURE 23.5 The 20S proteasome is barrel-shaped and made up of 28 homologous subunits. The subunits (α-type, red; β-type, blue) are arranged in four rings of 7 subunits each. Some of the β-type subunits (right) include protease active sites at their amino termini. [Subunit drawn from 1RYP.pdb.]

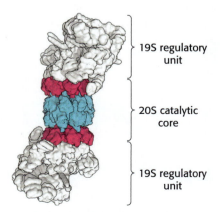

19S regulatory unit

20S catalytic core

19S regulatory unit

FIGURE 23.6 The 26S proteasome consists of one 20S and two 19S subunits. The 19S are regulatory and each attached at either end of the 20S catalytic unit.

called the *AAA + class* (<u>A</u>TPases <u>a</u>ssociated with various cellular <u>a</u>ctivities). ATP hydrolysis assists the 19S complex to unfold the substrate and induce conformational changes in the 20S catalytic core (called *gate opening*) so that the substrate can be passed into the center of the complex. Alternative ways to enter the 20S barrel have also been discovered. In some cases, partially unfolded proteins can themselves induce gate opening without the 19S, and there are alternative proteasome caps that also allow the degradation of unstructured proteins.

The proteolytic active sites are sequestered in the interior of the 20S barrel to protect potential substrates until they are directed into the barrel. There are three types of active sites in the β subunits, each with a different specificity, but all employ an N-terminal threonine. The hydroxyl group of the threonine residue is converted into a nucleophile that attacks the carbonyl groups of peptide bonds to form acyl-enzyme intermediates. Substrates are degraded in a processive manner without the release of degradation intermediates, until the substrate is digested to peptides ranging in length from seven to nine residues. These peptide products are released from the proteasome and further degraded by other cellular proteases to yield individual amino acids. Thus, the ubiquitination pathway and the proteasome cooperate to degrade unwanted proteins. **Figure 23.7** presents an overview of the fates of amino acids following proteasomal digestion. A separate process

FIGURE 23.7 The proteasome and other proteases generate free amino acids. Ubiquitinated proteins are processed to peptide fragments. Ubiquitin is removed and recycled prior to protein degradation. The peptide fragments are further digested to yield free amino acids, which can be used for biosynthetic reactions, most notably protein synthesis. Alternatively, the amino group can be removed and processed to urea (Figure 23.15) and the carbon skeleton can be used to synthesize carbohydrate or fats or used directly as a fuel for cellular respiration (Section 23.5).

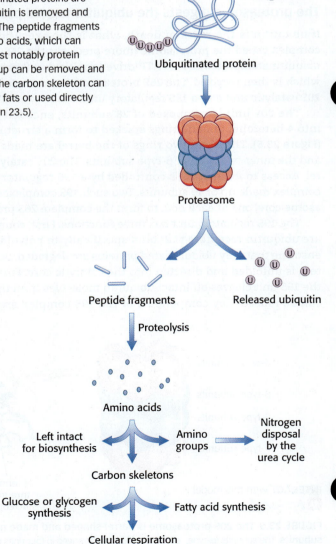

called *autophagy* serves as a second mechanism for protein turnover, although the details of that system are beyond the scope of this chapter.

The ubiquitin pathway and the proteasome have prokaryotic counterparts

Both the ubiquitin pathway and the proteasome appear to be present in all eukaryotes. Homologs of the proteasome are also found in some prokaryotes. The proteasomes of some archaea are quite similar in overall structure to their eukaryotic counterparts and similarly have 28 subunits (**Figure 23.8**). In the archaeal proteasome, however, all α outer-ring subunits and all β inner-ring subunits are identical; in eukaryotes, each α or β subunit is one of seven different isoforms. This specialization provides distinct substrate specificity.

Protein degradation can be used to regulate biological function

A number of physiological processes are controlled at least in part by protein degradation through the ubiquitin–proteasome pathway (**Table 23.3**). In each case, the proteins being degraded are regulatory proteins. Consider, for example, control of the inflammatory response. A transcription factor, NF-κB (NF for nuclear factor), initiates the expression of a number of the genes that take part in this response. This factor is itself activated by the degradation of an attached inhibitory protein, I-κB (I for inhibitor). In response to inflammatory signals that bind to membrane-bound receptors, I-κB is phosphorylated at two serine residues, creating an E3 binding site. The binding of E3 leads to the ubiquitination and degradation of I-κB, unleashing NF-κB. The liberated transcription factor migrates from the cytoplasm to the nucleus to stimulate the transcription of the target genes. The NF-κB–I-κB system illustrates the interplay of several key regulatory motifs: receptor-mediated signal transduction, phosphorylation, compartmentalization, controlled and specific degradation, and selective gene expression. The importance of the ubiquitin–proteasome system for the regulation of gene expression is highlighted by the use of bortezomib (Velcade), a potent inhibitor of the proteasome, as a therapy for multiple myeloma. Bortezomib is a dipeptidyl boronic acid inhibitor of the proteasome.

Degrons are also used as regulatory mechanisms for protein expression. For example, degradation signals are commonly located in protein regions that also facilitate protein–protein interactions. Why might the coexistence of these two functions in the same region be useful? Exposure of such a region to solvent, and therefore to the binding of E3 enzyme, suggests that a component of a multiprotein complex has failed to form properly or that one component has been synthesized in excess. This exposure leads to rapid degradation and the restoration of appropriate stoichiometries.

The evolutionary studies of proteasomes described previously have also yielded potential clinical benefits. The bacterial pathogen *Mycobacterium tuberculosis*, the cause of tuberculosis, harbors a proteasome that is very similar to the human counterpart. Nevertheless, recent work has shown that it is possible to exploit the differences between the human and the bacterial proteasomes to develop specific inhibitors of the *M. tuberculosis* complex. Oxathiazol-2-one compounds such as HT1171 are suicide inhibitors (Section 5.6) of the proteolytic activity of the *M. tuberculosis* proteasome, but have no effect on the proteasomes of the human host. This is especially exciting because these drugs kill the nonreplicating form of *M. tuberculosis*, and thus may not require the prolonged treatment required with conventional drugs, thereby reducing the likelihood of drug resistance due to interruption of the treatment regime.

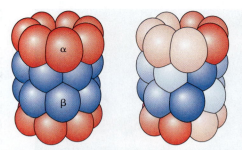

Archaeal proteasome   Eukaryotic proteasome

FIGURE 23.8 Proteasomes have become more complex in eukaryotes. The archaeal proteasome consists of 14 identical α subunits and 14 identical β subunits. In the eukaryotic proteasome, gene duplication and specialization has led to 7 distinct subunits of each type. The overall architecture of the proteasome is conserved.

TABLE 23.3 Processes regulated by protein degradation

Gene transcription
Cell-cycle progression
Organ formation
Circadian rhythms
Inflammatory response
Tumor suppression
Cholesterol metabolism
Antigen processing

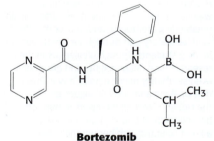

Bortezomib
(a dipeptidyl boronic acid)

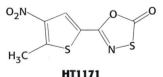

HT1171
5-(2-methyl-3-nitrothiophen-2-yl)-
1,3,4-oxathiazol-2-one

❖ SELF–CHECK QUESTION

Degrons are often found in parts of proteins that facilitate protein complex formation. Explain why this might be useful.

CECILE PICKART The observation that some protein degradation processes required ATP hydrolysis suggested that a novel enzymatic system was at work. Several research teams set to work to characterize this system, one of which included Dr. Cecile Pickart. Dr. Pickart had completed her PhD at Brandeis University, working with a leading enzymologist. As a postdoctoral fellow, she worked with Dr. Irwin Rose at Fox Chase Cancer, where she helped characterize the ubiquitin-based protein degradation system. She was so sure-footed that several of her colleagues agreed with the statement by another that she "described the enzymes to the world as though she had obtained the answer sheet." Later, as an independent researcher at SUNY-Buffalo and then Johns Hopkins, she focused completely on various aspects of ubiquitin biochemistry. She was an outstanding mentor to students, both in her own laboratory and throughout the university, helping students through challenges in their research, with deep insights and attention to even seemingly minor details. It was fitting that Dr. Rose and two colleagues shared the Nobel Prize in 2004 based, in part, on her research. The field of biochemistry lost a great scientist with her untimely death in 2006 at the age of 51.

23.3 The First Step in Amino Acid Degradation Is the Removal of Nitrogen

What is the fate of amino acids released on protein digestion or turnover? The first call is for their use as building blocks for biosynthetic reactions. However, any amino acids not needed as building blocks are degraded to compounds able to enter the metabolic mainstream. The amino group is first removed, and then the remaining carbon skeleton is metabolized to a glycolytic intermediate, to one of several citric acid cycle intermediates, or to acetyl CoA. The major site of amino acid degradation in mammals is the liver, although muscles also readily degrade the branched-chain amino acids (Leu, Ile, and Val). The fate of the α-amino group will be considered first, followed by that of the carbon skeleton (Section 23.5).

Alpha-amino groups are converted into ammonium ions by the oxidative deamination of glutamate in the liver

The α-amino group of many amino acids is transferred to α-ketoglutarate to form glutamate, which is then oxidatively deaminated in the liver to yield ammonium ion (NH_4^+).

Aminotransferases (also called **transaminases**) catalyze the transfer of an α-amino group from an α-amino acid to an α-ketoacid.

Aminotransferases generally funnel α-amino groups from a variety of amino acids to α-ketoglutarate for conversion into NH_4^+. Aspartate aminotransferase, one of the most important of these enzymes, catalyzes the transfer of the amino group of aspartate to α-ketoglutarate.

$$\text{Aspartate} + \text{α-ketoglutarate} \rightleftharpoons \text{oxaloacetate} + \text{glutamate}$$

Alanine aminotransferase catalyzes the transfer of the amino group of alanine to α-ketoglutarate.

$$\text{Alanine} + \text{α-ketoglutarate} \rightleftharpoons \text{pyruvate} + \text{glutamate}$$

Transamination reactions are reversible and can thus be used to synthesize amino acids from α-ketoacids, as we shall see in Section 25.2

The nitrogen atom in glutamate is converted into free ammonium ion by oxidative deamination, a reaction catalyzed by **glutamate dehydrogenase**. This enzyme is unusual in being able to utilize either NAD^+ or $NADP^+$, at least in some species. The reaction proceeds by dehydrogenation of the C—N bond, followed by hydrolysis of the resulting ketimine.

This reaction equilibrium constant is close to 1 in the liver, so the direction of the reaction is determined by the concentrations of reactants and products. Normally, the reaction is driven forward by the rapid removal of ammonium ion. Ammonium ions are toxic, and particularly detrimental to the brain, so ammonium is formed only in large amounts in the liver, where it is immediately eliminated by its conversion to urea. Thus, glutamate dehydrogenase is essentially a liver-specific enzyme. To further prevent the release of free ammonium, glutamate dehydrogenase is located in mitochondria along with some of the other enzymes required for the production of urea.

In mammals, but not in other organisms, glutamate dehydrogenase is allosterically inhibited by GTP and stimulated by ADP. These nucleotides exert their regulatory effects in a unique manner: an abortive complex, a set of molecules bound to the enzyme that cannot move forward toward the final products, is formed when a product is replaced by substrate in the active site before the reaction is complete. For instance, the enzyme bound to glutamate and NAD(P)H is an abortive complex. GTP facilitates the formation of such complexes while ADP destabilizes them.

The sum of the reactions catalyzed by aminotransferases and glutamate dehydrogenase is

$$\alpha\text{-Amino acid} + NAD^+ + H_2O \rightleftharpoons \alpha\text{-ketoacid} + NH_4^+ + NADH + H^+$$
$$\text{(or } NADP^+) \hspace{4cm} \text{(or NADPH)}$$

In most terrestrial vertebrates, NH_4^+ is converted into urea, which is excreted.

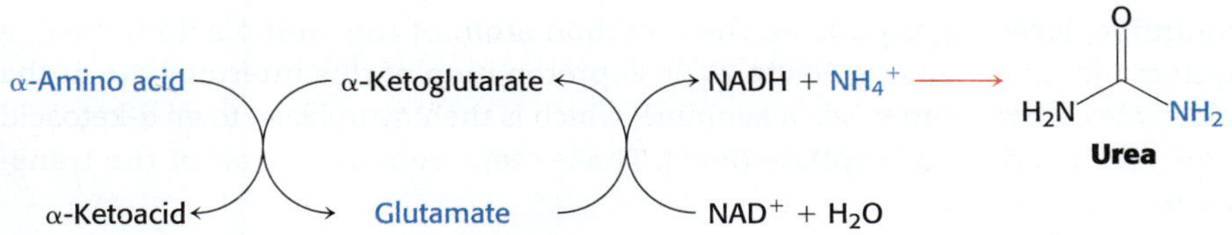

Urea

❖ **SELF–CHECK QUESTION**

How do aminotransferases and glutamate dehydrogenase cooperate in the metabolism of the amino group of amino acids?

Mechanism: Pyridoxal phosphate forms Schiff-base intermediates in aminotransferases

All aminotransferases contain the prosthetic group **pyridoxal phosphate (PLP)**, which is derived from pyridoxine (vitamin B$_6$). Pyridoxal phosphate includes a slightly basic pyridine ring to which a slightly acidic OH group is attached. Thus, pyridoxal phosphate derivatives can form a stable tautomeric form in which the pyridine nitrogen atom is protonated (hence, positively charged), while the OH group loses a proton and hence is negatively charged, forming a phenolate.

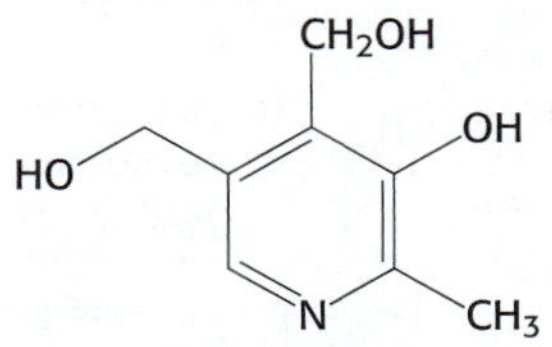

Pyridoxine (vitamin B$_6$)

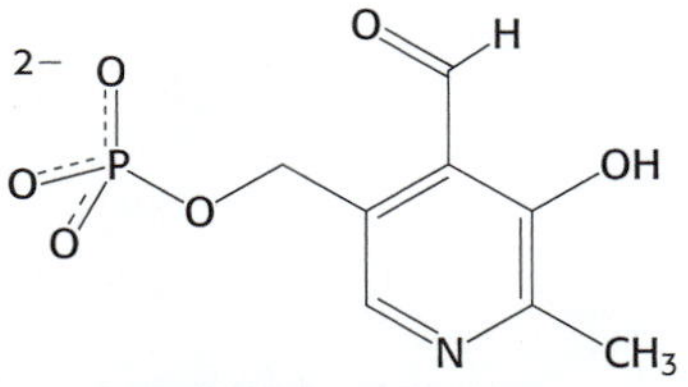

Pyridoxal phosphate (PLP)

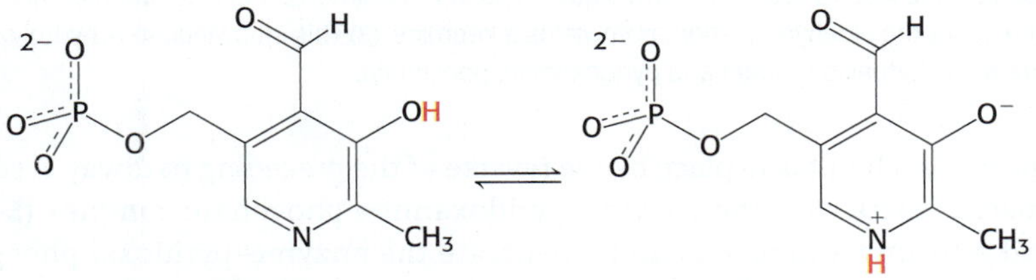

PLP (phenol)   **PLP (phenolate)**

The most important functional group on PLP is the aldehyde. This group forms covalent Schiff-base intermediates with amino acid substrates. Indeed, even in the absence of substrate, the aldehyde group of PLP usually forms a Schiff-base linkage with the ε-amino group of a specific lysine residue at the enzyme's active site. A new Schiff-base linkage is formed on addition of an amino acid substrate.

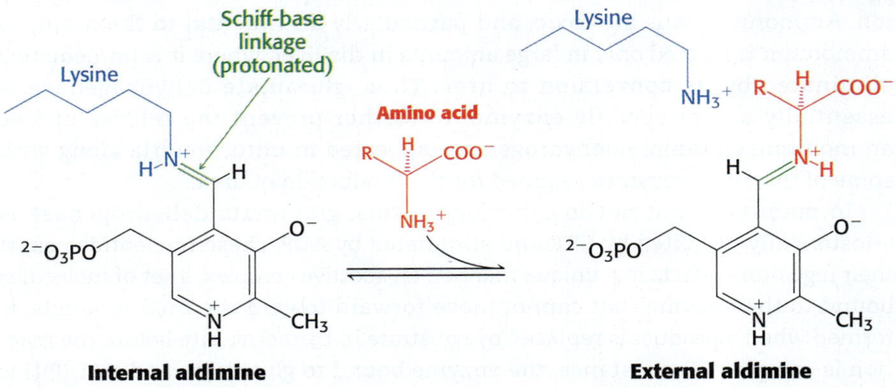

Internal aldimine   **External aldimine**

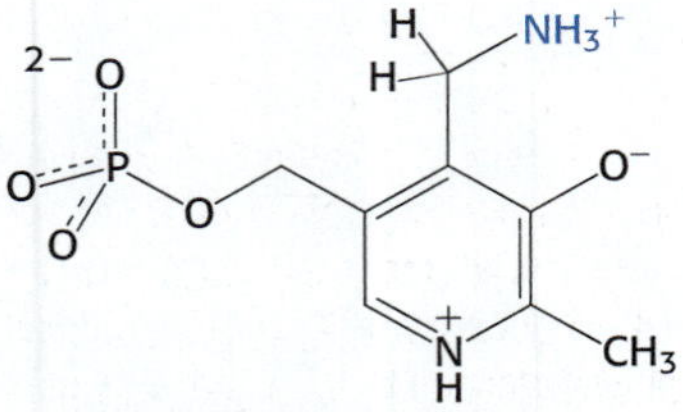

Pyridoxamine phosphate (PMP)

The α-amino group of the amino acid substrate displaces the ε-amino group of the active-site lysine residue. In other words, an *internal* aldimine becomes an *external* aldimine. The amino acid–PLP Schiff base that is formed remains tightly bound to the enzyme by multiple noncovalent interactions. The Schiff-base linkage often accepts a proton at the nitrogen of the pyridine ring, with the positive charge stabilized by interaction with the negatively charged phenolate group of PLP.

The Schiff base between the amino acid substrate and PLP, the external aldimine, loses a proton from the α-carbon atom of the amino acid to form a quinonoid intermediate (**Figure 23.9**). Reprotonation of this intermediate at the aldehyde carbon atom yields a ketimine, which is then hydrolyzed to an α-ketoacid and pyridoxamine phosphate (PMP). These steps constitute half of the transamination reaction.

$$\text{Amino acid}_1 + \text{E-PLP} \rightleftharpoons \alpha\text{-ketoacid}_1 + \text{E-PMP}$$

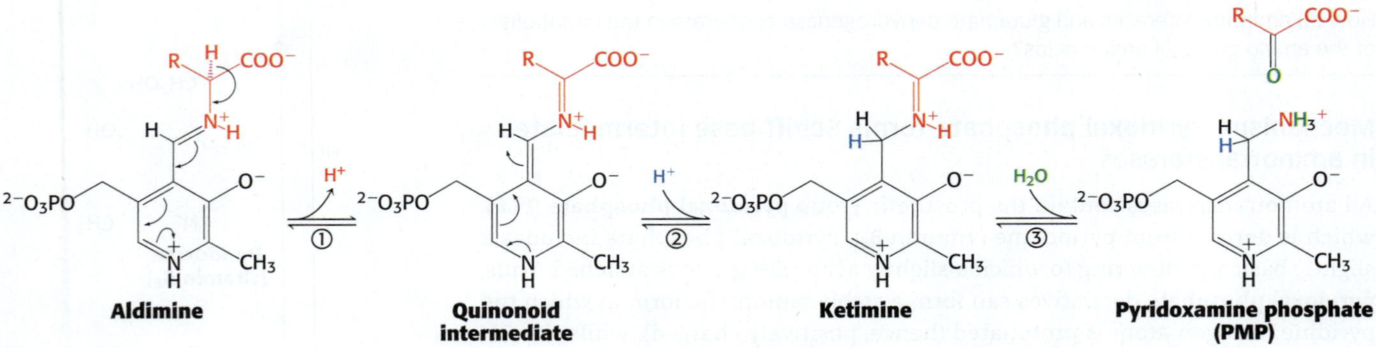

Aldimine   **Quinonoid intermediate**   **Ketimine**   **Pyridoxamine phosphate (PMP)**

FIGURE 23.9 Transaminations proceed through three steps to form PMP. (1) The external aldimine loses a proton to form a quinonoid intermediate. (2) Reprotonation of this intermediate at the aldehyde carbon atom yields a ketimine. (3) This intermediate is hydrolyzed to generate the α-ketoacid product and pyridoxamine phosphate.

The second half takes place by the reverse of the preceding pathway. A second α-ketoacid reacts with the enzyme–pyridoxamine phosphate complex (E-PMP) to yield a second amino acid and regenerate the enzyme–pyridoxal phosphate complex (E-PLP).

$$\alpha\text{-Ketoacid}_2 + \text{E-PMP} \rightleftharpoons \text{amino acid}_2 + \text{E-PLP}$$

The sum of these partial reactions is

$$\text{Amino acid}_1 + \alpha\text{-ketoacid}_2 \rightleftharpoons \text{amino acid}_2 + \alpha\text{-ketoacid}_1$$

Aspartate aminotransferase is an archetypal pyridoxal-dependent transaminase

The mitochondrial enzyme aspartate aminotransferase provides an especially well-studied example of PLP as a coenzyme for transamination reactions. X-ray crystallographic studies have provided detailed views of how PLP and substrates are bound, confirming much of the proposed catalytic mechanism. Each of the identical 45-kDa subunits of this dimer consists of a large domain and a small one (**Figure 23.10**). PLP is bound to the large domain, in a pocket near the domain interface. In the absence of substrate, the aldehyde group of PLP is in a Schiff-base linkage with lysine 258, as expected. Adjacent to the coenzyme's binding site is a conserved arginine residue that interacts with the α-carboxylate group of the amino acid substrate, helping to orient the substrate appropriately in the active site. A base is necessary to remove a proton from the α-carbon group of the amino acid and to transfer it to the aldehyde carbon atom of PLP (Figure 23.9, steps 1 and 2). The lysine amino group that was initially in Schiff-base linkage with PLP appears to serve as the proton donor and acceptor.

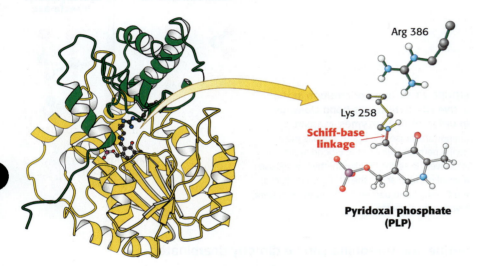

FIGURE 23.10 The active site of aspartate aminotransferase reveals PLP bound to a lysine through a Schiff-base linkage. Aspartate aminotransferase, a prototypical PLP-dependent enzyme, includes pyridoxal phosphate attached to the enzyme by a Schiff-base linkage with lysine 258. An arginine residue in the active site helps orient substrates by binding to their α-carboxylate groups. Only one of the enzyme's two subunits is shown. [Drawn from 1AAW.pdb.]

INTERACT with this model in
Achieve

Blood levels of aminotransferases serve a diagnostic function

The presence of alanine and aspartate aminotransferase in the blood is an indication of liver damage. Liver damage can occur for a number of reasons, including viral hepatitis, long-term excessive alcohol consumption, and reaction to drugs such as acetaminophen (Section 32.3). Under these conditions, liver cell membranes are damaged, and some cellular proteins—including the aminotransferases—leak into the blood. Normal blood values for alanine and aspartate aminotransferase activity are 5–30 units/l and 40–125 units/l, respectively. Depending on the extent of liver damage, the values will reach 200–300 units/l.

Pyridoxal phosphate enzymes catalyze a wide array of reactions

Transamination is just one of a wide range of amino acid transformations that are catalyzed by PLP enzymes. The other reactions catalyzed by PLP enzymes at the α-carbon atom of amino acids are decarboxylations, deaminations, racemizations, and aldol cleavages (**Figure 23.11**). In addition, PLP enzymes catalyze elimination and replacement reactions at the β-carbon atom (e.g., tryptophan synthetase in the synthesis of tryptophan) and the γ-carbon atom (e.g., cystathionine β-synthase in the synthesis of cysteine) of amino acid substrates. Three common features of PLP catalysis underlie these diverse reactions.

1. A Schiff base is formed by the amino acid substrate (the amine component) and PLP (the carbonyl component).

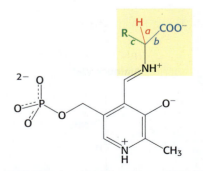

FIGURE 23.11 PLP enzymes labilize one of three bonds at the α-carbon atom of an amino acid substrate. For example, bond *a* is labilized by aminotransferases, bond *b* by decarboxylases, and bond *c* by aldolases (such as threonine aldolases). PLP enzymes also catalyze reactions at the β- and γ-carbon atoms of amino acids.

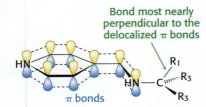

FIGURE 23.12 Stereoelectronic effects determine specific bond cleavage. The orientation about the N—C$_\alpha$ bond determines the most favored reaction catalyzed by a pyridoxal phosphate enzyme. The bond that is most nearly perpendicular to the plane of delocalized π bonds (represented by dashed lines) of the pyridoxal phosphate electron sink is most easily cleaved.

2. The protonated form of PLP acts as an electron sink to stabilize catalytic intermediates that are negatively charged. Electrons from these intermediates are attracted to the positive charge on the ring nitrogen atom.

3. The product Schiff base is cleaved at the completion of the reaction.

How does an enzyme selectively break a particular one of three bonds at the α-carbon atom of an amino acid substrate? An important principle is that the bond being broken must be perpendicular to the π bonds of the electron sink (**Figure 23.12**). An aminotransferase, for example, binds the amino acid substrate so that the C$_\alpha$—H bond is perpendicular to the PLP ring (**Figure 23.13**). In serine hydroxymethyltransferase, the enzyme that converts serine into glycine, the N—C$_\alpha$ bond is rotated so that the C$_\alpha$—C$_\beta$ bond is most nearly perpendicular to the plane of the PLP ring, favoring its cleavage. This means of choosing one of several possible catalytic outcomes is called *stereoelectronic control.*

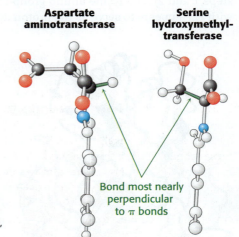

FIGURE 23.13 Bond orientation in the active site determines bond cleavage in different PLP enzymes. In aspartate aminotransferase, the C$_\alpha$—H bond is most nearly perpendicular to the π-bond system and is cleaved. In serine hydroxymethyltransferase, a small rotation about the N—C$_\alpha$ bond places the C$_\alpha$—C$_\beta$ bond perpendicular to the π system, favoring its cleavage.

Serine and threonine can be directly deaminated

The α-amino groups of serine and threonine can be directly converted into NH$_4^+$ without first being transferred to α-ketoglutarate. These direct deaminations are catalyzed by serine dehydratase and threonine dehydratase, in which PLP is the prosthetic group.

$$\text{Serine} \longrightarrow \text{pyruvate} + \text{NH}_4^+$$

$$\text{Threonine} \longrightarrow \alpha\text{-ketobutyrate} + \text{NH}_4^+$$

These enzymes are called *dehydratases* because dehydration precedes deamination. Serine loses a hydrogen ion from its α-carbon atom and a hydroxide ion group from its β-carbon atom to yield aminoacrylate. This unstable compound reacts with H$_2$O to give pyruvate and NH$_4^+$. Thus, the presence of a hydroxyl group attached to the β-carbon atom in each of these amino acids permits the direct deamination.

Peripheral tissues transport nitrogen to the liver

Although most amino acid degradation takes place in the liver, other tissues can degrade amino acids. For instance, muscle uses branched-chain amino acids as a source of fuel during prolonged exercise and fasting. How is the nitrogen processed in these other tissues? As in the liver, the first step is the removal of the nitrogen from the amino acid. However, muscle lacks the enzymes of the urea cycle, and so the nitrogen must be released in a nontoxic form that can be absorbed by the liver and converted into urea.

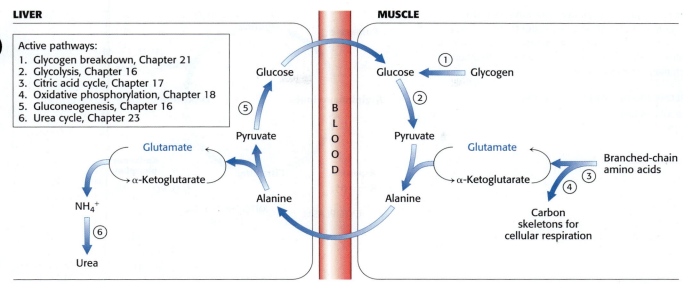

FIGURE 23.14 PATHWAY INTEGRATION: The glucose–alanine cycle allows muscle cells to use amino acids as fuel. During prolonged exercise and fasting, muscle uses branched-chain amino acids as fuel. The nitrogen removed is transferred (through glutamate) to alanine, which is released into the blood stream. In the liver, alanine is taken up and converted into pyruvate for the subsequent synthesis of glucose.

EXPLORE this pathway further in the Metabolic Map in
🔖 Achie√e

Nitrogen is transported from muscle to the liver in two principal transport forms: alanine and glutamine. To form alanine, first, glutamate is formed by transamination reactions, but the nitrogen is then transferred to pyruvate. Alanine is then released into the blood (**Figure 23.14**). The liver takes up the alanine and, by transamination, converts it back into pyruvate that can be used for gluconeogenesis; the amino group eventually appears as urea. This transport is referred to as the **glucose–alanine cycle**. It is reminiscent of the Cori cycle discussed earlier (Figure 16.36). However, in contrast with the Cori cycle, pyruvate is not reduced to lactate by NADH, and thus more high-energy electrons are conserved for oxidative phosphorylation in the muscle by using alanine.

Glutamine is also a key transport form of nitrogen. Glutamine synthetase catalyzes the synthesis of glutamine from glutamate and NH_4^+ in an ATP-dependent reaction:

$$NH_4^+ + \text{glutamate} + \text{ATP} \xrightarrow{\text{Glutamine synthetase}} \text{glutamine} + \text{ADP} + P_i$$

The nitrogens of glutamine can be eliminated by incorporation into urea in the liver.

❖ SELF–CHECK QUESTION

What amino acids can be deaminated directly to produce ammonium without the use of glutamate dehydrogenase, and what structural feature makes this possible?

23.4 Ammonium Ions Are Converted into Urea in Most Terrestrial Vertebrates

Some of the NH_4^+ formed in the breakdown of amino acids is consumed in the biosynthesis of nitrogen compounds. What happens to the rest of it? In most terrestrial vertebrates, the excess NH_4^+ is converted into urea and then excreted. Such organisms are referred to as *ureotelic*.

In terrestrial vertebrates, urea is synthesized by the **urea cycle** (**Figure 23.15**). One of the nitrogen atoms of urea is transferred from an amino acid, aspartate. The other nitrogen atom is derived directly from free NH_4^+, and the carbon atom comes from HCO_3^- (derived by the hydration of CO_2 through the action of carbonic anhydrase).

FIGURE 23.15 The urea cycle eliminates both nitrogen and carbon waste products. Two nitrogen atoms enter the cycle and leave as urea. Carbon dioxide is simultaneously eliminated as it is hydrated to bicarbonate, which then enters the cycle.

EXPLORE this pathway further in the Metabolic Map in

Achieve

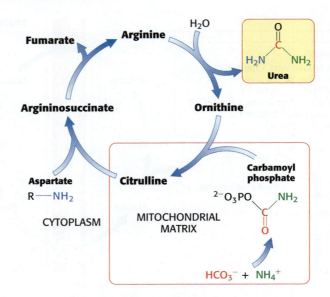

The urea cycle begins with the formation of carbamoyl phosphate

The urea cycle begins with the coupling of free NH_3 with HCO_3^- to form carbamoyl phosphate; this is the committed reaction of the urea cycle, which is catalyzed by **carbamoyl phosphate synthetase I**. In mammals, two distinct isozymes of carbamoyl phosphate synthetase are present. Carbamoyl phosphate synthetase I generates carbamoyl phosphate for both the urea cycle and the first step in pyrimidine biosynthesis (Section 26.1). Carbamoyl phosphate synthetase II is active in other steps of pyrimidine biosynthesis.

Carbamoyl phosphate is a simple molecule, but its synthesis is complex, requiring three steps.

Bicarbonate   **Carboxyphosphate**   **Carbamic acid**   **Carbamoyl phosphate**

Note that NH_3, because it is a base, normally exists as NH_4^+ in aqueous solution. However, carbamoyl phosphate synthetase I uses only NH_3 as a substrate. The reaction begins with the phosphorylation of HCO_3^- to form carboxyphosphate, which then reacts with NH_3 to form carbamic acid. Finally, a second molecule of ATP phosphorylates carbamic acid to form carbamoyl phosphate. The structure and mechanism of the enzyme that catalyzes these reactions will be presented in Chapter 26. The consumption of two molecules of ATP makes this synthesis of carbamoyl phosphate essentially irreversible.

Carbamoyl phosphate synthetase I is the key regulatory enzyme for urea synthesis

Carbamoyl phosphate synthetase I is regulated both allosterically and by covalent modification so that it is maximally active when amino acids are being metabolized for fuel use. The allosteric regulator N-acetylglutamate is required for synthetase activity. This molecule is synthesized by N-acetylglutamate synthase.

Acetyl CoA   +   **Glutamate**   *N*-Acetylglutamate synthase   ***N*-Acetylglutamate**

N-acetylglutamate synthase is itself activated by arginine, an intermediate in the urea cycle. Thus, N-acetylglutamate is synthesized when amino acids, as represented by arginine and glutamate, are readily available, and carbamoyl phosphate synthetase I is then activated to process the generated ammonia. When ammonia is not being generated, the synthetase is inhibited by acetylation. A rise in the concentration of mitochondrial NAD$^+$, indicative of an energy-poor state, stimulates a deacetylase that removes the acetyl group, activating the synthetase and readying the enzyme for processing ammonia from protein degradation. Biochemists have not yet determined how synthetase acetylation is controlled, serving again as a reminder that many discoveries are yet to be made in biochemistry.

Carbamoyl phosphate reacts with ornithine to begin the urea cycle

The carbamoyl group of carbamoyl phosphate has a high transfer potential because of its anhydride linkage. The carbamoyl group is transferred to ornithine to form citrulline, in a reaction catalyzed by ornithine transcarbamoylase.

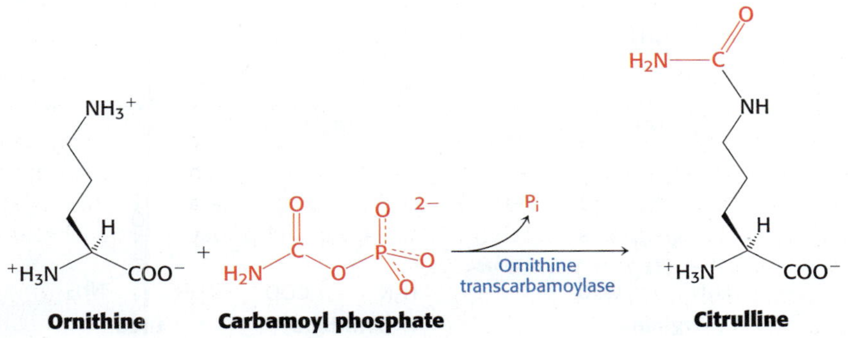

Ornithine   **Carbamoyl phosphate**   **Citrulline**

Ornithine and citrulline are amino acids, but they are not proteogenic, meaning they are not used as building blocks of proteins. The formation of NH$_4^+$ by glutamate dehydrogenase, its incorporation into carbamoyl phosphate as NH$_3$, and the subsequent synthesis of citrulline take place in the mitochondrial matrix. In contrast, the next three reactions of the urea cycle, which lead to the formation of urea, take place in the cytoplasm.

Citrulline is transported to the cytoplasm, where it condenses with aspartate, the donor of the second amino group of urea. This synthesis of argininosuccinate, catalyzed by argininosuccinate synthetase, is driven by the cleavage of ATP into AMP and pyrophosphate and by the subsequent hydrolysis of pyrophosphate.

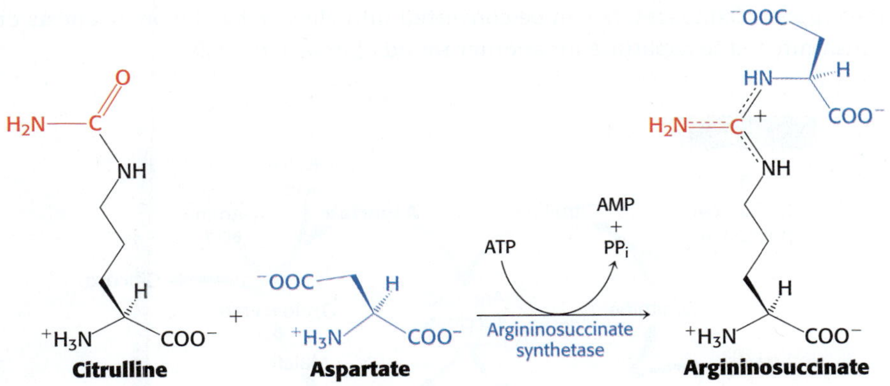

Citrulline   **Aspartate**   **Argininosuccinate**

Argininosuccinase (also called *argininosuccinate lyase*) cleaves argininosuccinate into arginine and fumarate. Thus, the carbon skeleton of aspartate is preserved in the form of fumarate.

Argininosuccinate → (Argininosuccinase) → **Arginine** + **Fumarate**

Finally, arginine is hydrolyzed to generate urea and ornithine in a reaction catalyzed by arginase. Ornithine is then transported back into the mitochondrion to begin another cycle, and the urea is excreted. Indeed, human beings excrete about 10 kg (22 pounds) of urea per year.

Arginine + H_2O → (Arginase) → **Ornithine** + **Urea**

The urea cycle is linked to gluconeogenesis

The stoichiometry of urea synthesis is

$$CO_2 + NH_4^+ + 3ATP + \text{aspartate} + 2H_2O \longrightarrow$$
$$\text{urea} + 2ADP + 2P_i + AMP + PP_i + \text{fumarate}$$

Pyrophosphate is rapidly hydrolyzed, and so the equivalent of four molecules of ATP are consumed in these reactions to synthesize one molecule of urea. The synthesis of fumarate by the urea cycle is important because it is a precursor for glucose synthesis (**Figure 23.16**). Fumarate is hydrated to malate, which is in turn oxidized to oxaloacetate, producing one NADH that can be used to generate 2.5 ATP molecules; thus, the actual energetic cost of urea synthesis is not as great as it may appear. Oxaloacetate can be converted into glucose by gluconeogenesis or transaminated to aspartate for another round of urea synthesis.

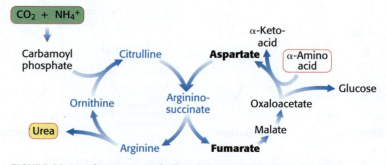

FIGURE 23.16 Nitrogen metabolism is integrated with other pathways. The urea cycle, gluconeogenesis, and the transamination of oxaloacetate are linked by fumarate and aspartate.

Inherited defects of the urea cycle cause hyperammonemia and can lead to brain damage

Genetic disorders of the urea cycle that block carbamoyl phosphate synthesis or of any of the four steps of the urea cycle have devastating consequences. The synthesis of urea in the liver is the major route for the removal of NH_4^+; thus, any defect in the urea cycle leads to an elevated level of NH_4^+ in the blood (a condition called *hyperammonemia*). Urea cycle disorders occur with a prevalence of about 1 in 15,000, with some cases becoming evident a day or two after birth, when the afflicted infant becomes lethargic and vomits periodically. Coma and irreversible brain damage may soon follow.

Why are high levels of NH_4^+ toxic? Biochemists do not yet know for certain, but recent work suggests that NH_4^+ may inappropriately activate a sodium-potassium-chloride cotransporter. This activation disrupts the osmotic balance of the nerve cell, causing swelling that damages the cell and results in neurological disorders. Other studies have found that ammonium disrupts neurotransmitter systems, which is reasonable since many neurotransmitters either are amino acids or are derived from amino acids. Additionally, ammonium has been found to impact energy metabolism, levels of oxidative stress, nitric oxide synthesis, and signal-transduction pathways. Due to these many impacts, the effects of ammonium on the brain are probably complex.

Ingenious strategies for coping with deficiencies in urea synthesis have been devised on the basis of a thorough understanding of the underlying biochemistry. Consider, for example, argininosuccinase deficiency. This disorder can be partly bypassed by providing a surplus of arginine in the diet and restricting the total protein intake. In the liver, arginine is split into urea and ornithine, which then reacts with carbamoyl phosphate to form citrulline (Figure 23.17). This urea-cycle intermediate condenses with aspartate to yield argininosuccinate, which is then excreted. Note that two nitrogen atoms—one from carbamoyl phosphate and the other from aspartate—are eliminated from the body per molecule of arginine provided in the diet. In essence, argininosuccinate substitutes for urea in carrying nitrogen out of the body.

The treatment of carbamoyl phosphate synthetase deficiency or ornithine transcarbamoylase deficiency illustrates a different strategy for circumventing a metabolic block. Citrulline and argininosuccinate cannot be used to dispose of nitrogen atoms because their formation is impaired. Under these conditions, excess nitrogen accumulates in glycine and glutamine. The challenge then is to rid the body of the nitrogen accumulating in these two amino acids. That goal is accomplished by supplementing a protein-restricted diet with large amounts of benzoate and phenylacetate. Benzoate is activated to benzoyl CoA, which reacts with glycine to form hippurate, which is excreted (Figure 23.18). Likewise, phenylacetate is activated to

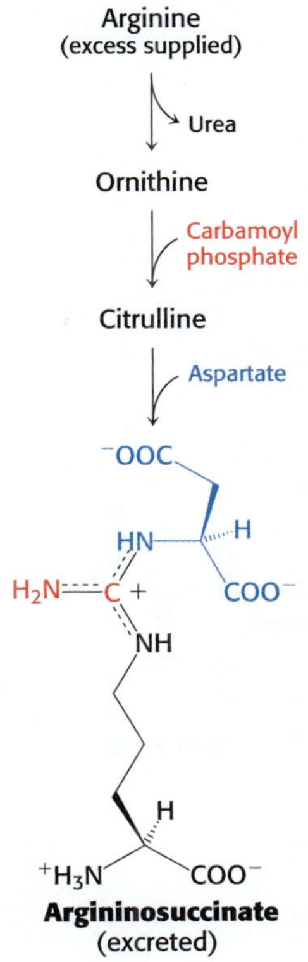

FIGURE 23.17 Argininosuccinase deficiency can be managed by supplementing the diet with arginine. Nitrogen is excreted in the form of argininosuccinate in the absence of functional argininosuccinase.

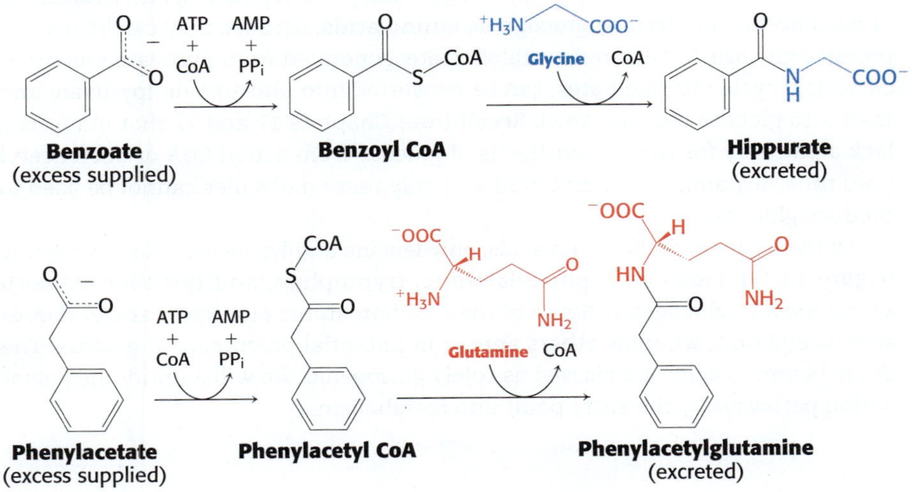

FIGURE 23.18 Both carbamoyl phosphate synthetase and ornithine transcarbamoylase deficiencies can be treated the same way. By supplementing the diet with benzoate and phenylacetate, excess nitrogen can be excreted in the form of hippurate and phenylacetylglutamine.

phenylacetyl CoA, which reacts with glutamine to form phenylacetylglutamine, which is also excreted. These conjugates substitute for urea in the disposal of nitrogen. Thus, latent biochemical pathways can be activated to partly bypass a genetic defect.

Urea is not the only means of disposing of excess nitrogen

As stated earlier, most terrestrial vertebrates are *ureotelic*; they excrete excess nitrogen as urea. However, urea is not the only excretable form of nitrogen. *Ammonotelic* organisms, such as aquatic vertebrates and invertebrates, release nitrogen as NH_4^+ and rely on the aqueous environment to dilute this toxic substance. Interestingly, lungfish, which are normally ammonotelic, become ureotelic in time of drought, when they live out of the water.

While both ureotelic and ammonotelic organisms require water (to varying degrees) for nitrogen excretion, *uricotelic* organisms, such as birds and reptiles, secrete nitrogen in an almost-solid slurry, as the purine uric acid. Besides requiring very little water, the secretion of uric acid also has the advantage of removing four atoms of nitrogen per molecule. The pathway for nitrogen excretion observed in different organisms developed over the course of evolution, depending on the habitat of each organism.

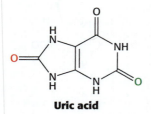

Uric acid

❖ SELF-CHECK QUESTION

What are the immediate biochemical sources for the two nitrogen atoms in urea?

23.5 Carbon Atoms of Degraded Amino Acids Emerge as Major Metabolic Intermediates

We now turn to the fates of the carbon skeletons of amino acids after the removal of the α-amino group. The strategy of amino acid degradation is to transform the carbon skeletons into major metabolic intermediates that can be converted into glucose or oxidized by the citric acid cycle. The conversion pathways range from extremely simple to quite complex. In fact, some amino acids can be degraded by alternate pathways. For instance, threonine can be metabolized to succinyl CoA or pyruvate. In all cases, however, the carbon skeletons of the diverse set of 20 fundamental amino acids are funneled into only seven molecules: pyruvate, acetyl CoA, acetoacetyl CoA, α-ketoglutarate, succinyl CoA, fumarate, and oxaloacetate. We see here an example of the remarkable economy of metabolic conversions.

Amino acids that are degraded to acetyl CoA or acetoacetyl CoA are termed **ketogenic amino acids** because they can give rise to ketone bodies or fatty acids. Amino acids that are degraded to pyruvate, α-ketoglutarate, succinyl CoA, fumarate, or oxaloacetate are termed **glucogenic amino acids**, because they can be used to produce glucose. For example, oxaloacetate, generated from pyruvate and other citric acid cycle intermediates, can be converted into phosphoenolpyruvate and then into glucose (Section 16.4). Recall from Chapters 17 and 22 that mammals lack a pathway for the net synthesis of glucose from acetyl CoA or acetoacetyl CoA; thus, any amino acid that produces only these molecules cannot be used to produce glucose.

Of the basic set of 20 amino acids, only leucine and lysine are solely ketogenic (**Figure 23.19**). Isoleucine, phenylalanine, tryptophan, and tyrosine are both ketogenic and glucogenic. Some of their carbon atoms emerge in acetyl CoA or acetoacetyl CoA, whereas others appear in potential precursors of glucose. The other 14 amino acids are classed as solely glucogenic. We will identify the degradation pathways by the entry point into metabolism.

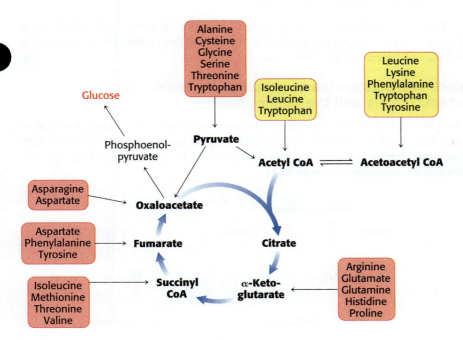

FIGURE 23.19 The carbon skeletons of all 20 proteogenic amino acids enter central metabolism as just seven different molecules. Glucogenic amino acids are shaded red, and ketogenic amino acids are shaded yellow. Several amino acids are both glucogenic and ketogenic.

Pyruvate is an entry point into metabolism for a number of amino acids

Pyruvate is the entry point of the three-carbon amino acids—alanine, serine, and cysteine—into the metabolic mainstream (**Figure 23.20**). The transamination of alanine directly yields pyruvate.

$$\text{Alanine} + \alpha\text{-ketoglutarate} \rightleftharpoons \text{pyruvate} + \text{glutamate}$$

As mentioned earlier in the chapter, glutamate is then oxidatively deaminated in the liver, yielding NH_4^+ and regenerating α-ketoglutarate. The sum of these reactions is

$$\text{Alanine} + \text{NAD(P)}^+ + H_2O \rightleftharpoons \text{pyruvate} + NH_4^+ + \text{NAD(P)H} + H^+$$

Another simple reaction in the degradation of amino acids is the deamination of serine to pyruvate by serine dehydratase (Section 23.3).

$$\text{Serine} \longrightarrow \text{pyruvate} + NH_4^+$$

Cysteine can be converted into pyruvate by several pathways, with its sulfur atom emerging in H_2S, SCN^-, or SO_3^{2-}.

The carbon atoms of three other amino acids can be converted into pyruvate. Glycine can be converted into serine by the enzymatic addition of a hydroxymethyl group, or it can be cleaved to give CO_2, NH_4^+, and an activated one-carbon unit. Threonine can give rise to pyruvate through the intermediate 2-amino-3-ketobutyrate. Three carbon atoms of tryptophan can emerge in alanine, which can be converted into pyruvate.

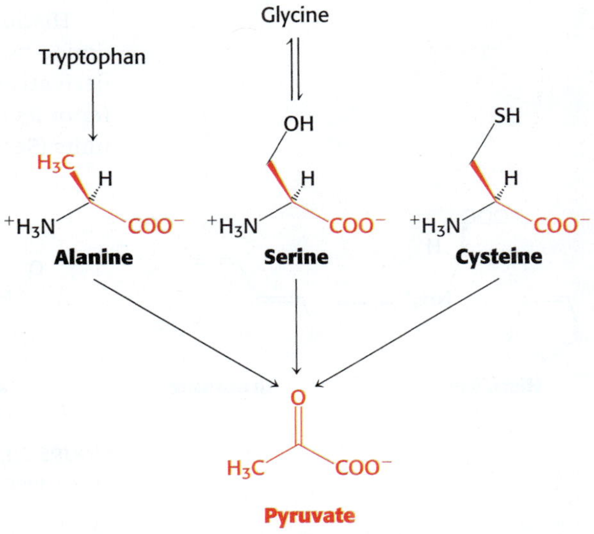

FIGURE 23.20 Pyruvate is the metabolic entry point for the carbon skeletons from three-carbon amino acids. Pyruvate is the point of entry for alanine, serine, cysteine, and glycine. Note that alanine is eventually produced in the degradation of tryptophan, and that carbon skeleton can then be used to form pyruvate.

Oxaloacetate is an entry point into metabolism for aspartate and asparagine

Aspartate and asparagine are converted into oxaloacetate, a citric acid cycle intermediate. Aspartate, a four-carbon amino acid, is directly transaminated to oxaloacetate.

$$\text{Aspartate} + \alpha\text{-ketoglutarate} \rightleftharpoons \text{oxaloacetate} + \text{glutamate}$$

Asparagine is hydrolyzed by asparaginase to NH_4^+ and aspartate, which is then transaminated.

Recall that aspartate can also be converted into fumarate by the urea cycle (Figure 23.16). Fumarate is a point of entry for half the carbon atoms of tyrosine and phenylalanine, as will be discussed shortly.

Alpha-ketoglutarate is an entry point into metabolism for amino acids with five-carbon chains

The carbon skeletons of several amino acids with five-carbon chains enter the citric acid cycle as α-ketoglutarate. These amino acids are first converted into glutamate, which is then oxidatively deaminated by glutamate dehydrogenase to yield α-ketoglutarate (**Figure 23.21**).

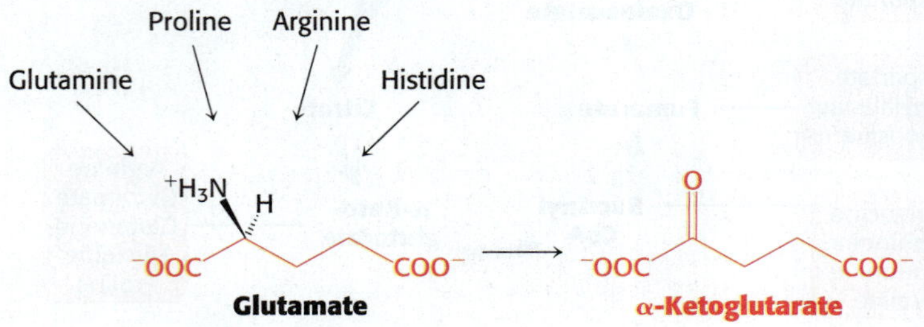

FIGURE 23.21 Four amino acids produce carbon skeletons that enter central metabolism as α-ketoglutarate. These amino acids with five-carbon chains are first converted into glutamate.

Histidine is converted into 4-imidazolone 5-propionate (**Figure 23.22**). The amide bond in the ring of this intermediate is hydrolyzed to the N-formimino derivative of glutamate, which is then converted into glutamate by the transfer of its formimino group to tetrahydrofolate, a carrier of activated one-carbon units (Section 25.2).

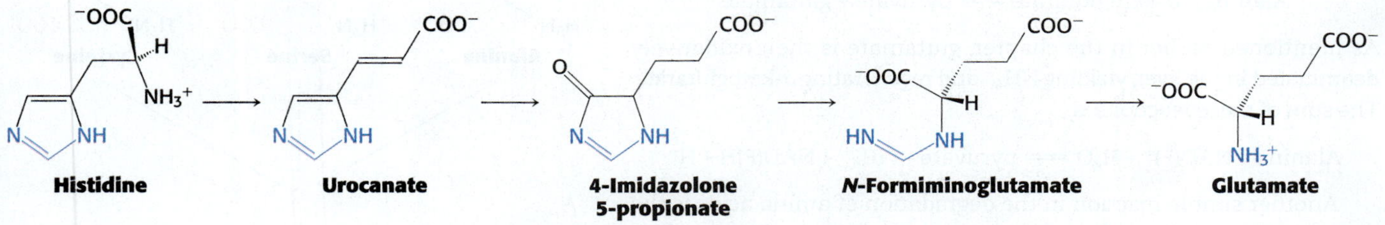

FIGURE 23.22 Histidine is converted to glutamate in four steps. The glutamate produced here is then deaminated to form α-ketoglutarate.

Glutamine is hydrolyzed to glutamate and NH_4^+ by glutaminase. Proline and arginine are each converted into glutamate γ-semialdehyde, which is then oxidized to glutamate (**Figure 23.23**).

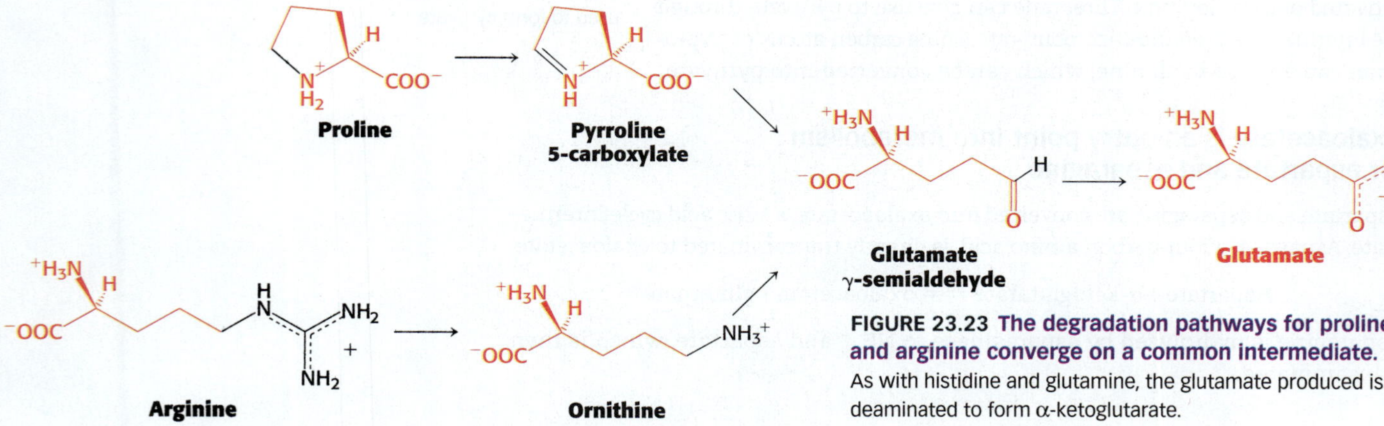

FIGURE 23.23 The degradation pathways for proline and arginine converge on a common intermediate. As with histidine and glutamine, the glutamate produced is deaminated to form α-ketoglutarate.

Succinyl coenzyme A is a point of entry for several amino acids

Succinyl CoA is a point of entry for some of the carbon atoms of methionine, iso-leucine, threonine, and valine. Propionyl CoA and then methylmalonyl CoA are intermediates in the breakdown of these four amino acids (**Figure 23.24**). This path-way from propionyl CoA to succinyl CoA is also used in the oxidation of fatty acids that have an odd number of carbon atoms. The mechanism for the interconversion of propionyl CoA and methylmalonyl CoA was presented in Section 22.3.

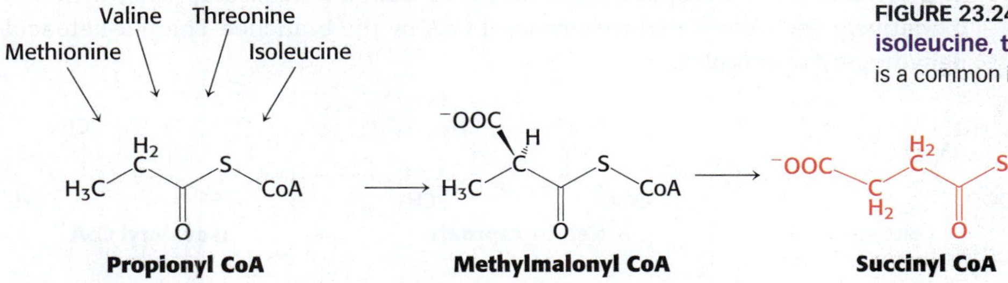

FIGURE 23.24 Succinyl CoA is formed from methionine, isoleucine, threonine, and valine. Propionyl CoA is a common intermediate in all four pathways.

Methionine degradation requires the formation of a key methyl donor, S-adenosylmethionine

Methionine is converted into succinyl CoA in nine steps (**Figure 23.25**). The first step is the adenylation of methionine to form S-adenosylmethionine (SAM), a common methyl donor in the cell (Section 25.2). Loss of the methyl and adenosyl groups yields homocysteine, which is eventually processed to α-ketobutyrate. This α-ketoacid is oxidatively decarboxylated by the α-ketoacid dehydrogenase complex to propionyl CoA, which is processed to succinyl CoA, as described in Section 22.3.

FIGURE 23.25 Methionine catabolism proceeds through the formation of SAM. The pathway for the conversion of methionine into succinyl CoA forms S-adenosylmethionine (SAM), an important methyl donor.

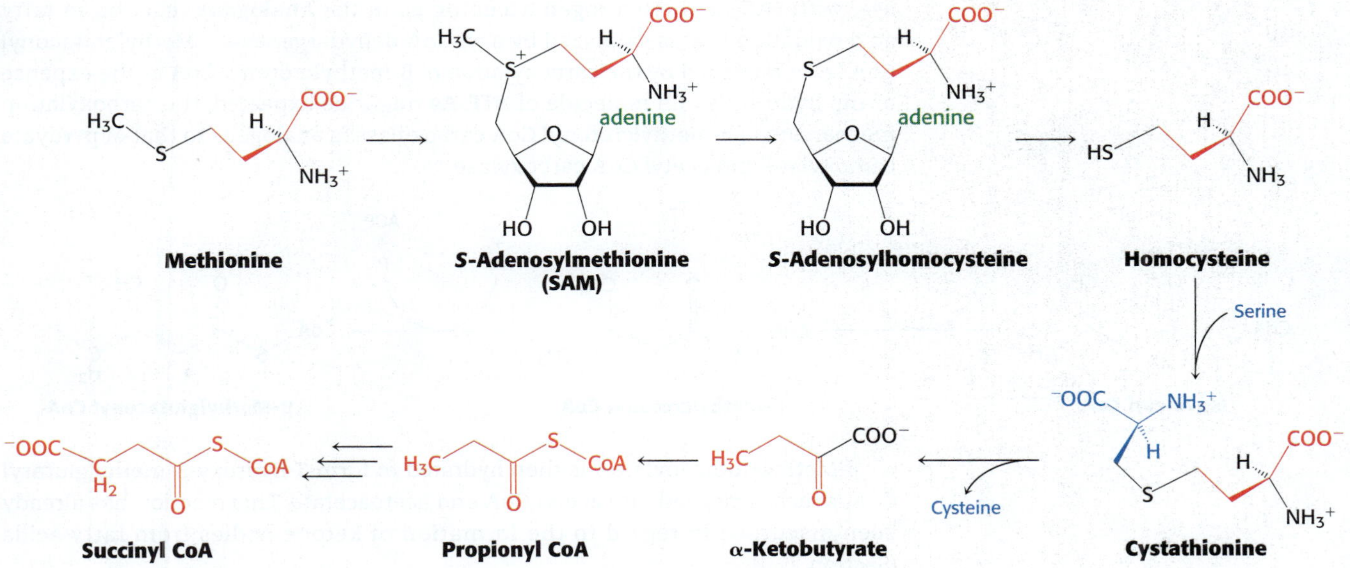

Threonine deaminase initiates the degradation of threonine

Threonine can be degraded by a number of pathways. One requires the enzyme threonine deaminase, also known as threonine dehydratase. This pyridoxal-dependent enzyme converts threonine into α-ketobutyrate.

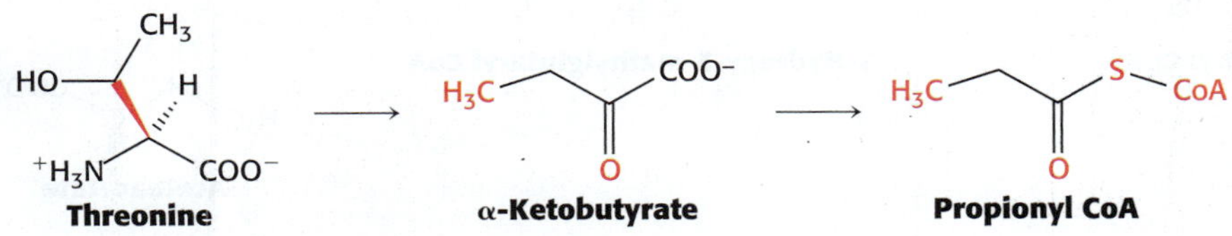

α-Ketobutyrate then forms propionyl CoA, which is subsequently metabolized to succinyl CoA, as shown in Figure 23.24.

The branched-chain amino acids yield acetyl CoA, acetoacetate, or propionyl CoA

The branched-chain amino acids are degraded by reactions that we have already encountered in the citric acid cycle and fatty acid oxidation. Leucine is transaminated to the corresponding α-ketoacid, called *α-ketoisocaproate*, which is oxidatively decarboxylated to isovaleryl CoA by the branched-chain α-ketoacid dehydrogenase complex.

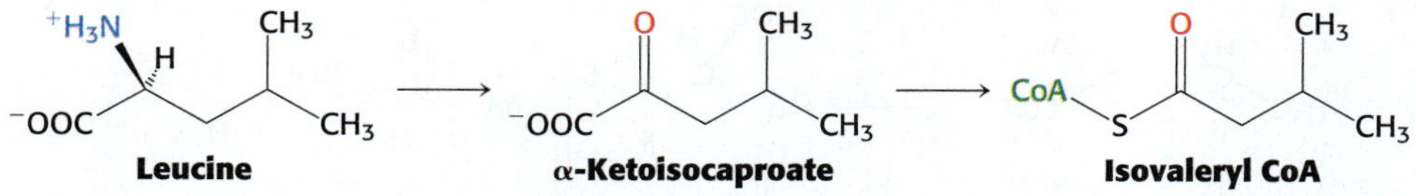

The α-ketoacids of valine and isoleucine, the other two branched-chain aliphatic amino acids, also are substrates (as is α-ketobutyrate derived from methionine). The oxidative decarboxylation of these α-ketoacids is analogous to that of pyruvate to acetyl CoA and of α-ketoglutarate to succinyl CoA. The branched-chain α-ketoacid dehydrogenase, a multienzyme complex, is a homolog of pyruvate dehydrogenase complex (Section 17.2) and α-ketoglutarate dehydrogenase complex (Section 17.3). Indeed, the E3 components of these enzymes, which regenerate the oxidized form of lipoamide, are identical.

The isovaleryl CoA derived from leucine is dehydrogenated to yield β-methylcrotonyl CoA. This oxidation is catalyzed by isovaleryl CoA dehydrogenase, with FAD as the hydrogen acceptor, as in the analogous reaction in fatty acid oxidation that is catalyzed by acyl CoA dehydrogenase. β-Methylglutaconyl CoA is then formed by the carboxylation of β-methylcrotonyl CoA at the expense of the hydrolysis of a molecule of ATP. As might be expected, the carboxylation mechanism of β-methylcrotonyl CoA carboxylase is analogous to that of pyruvate carboxylase and acetyl CoA carboxylase.

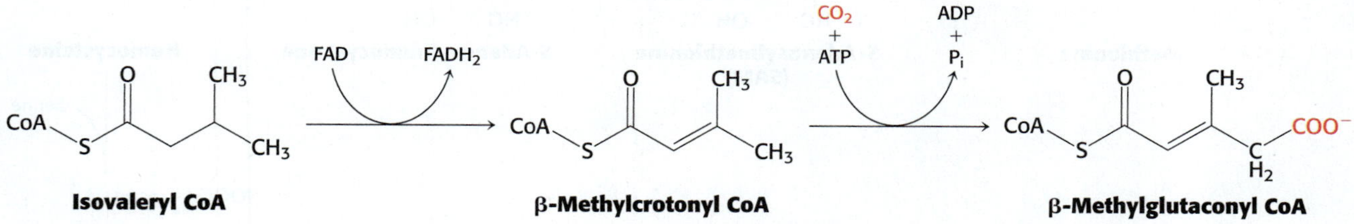

β-Methylglutaconyl CoA is then hydrated to form 3-hydroxy-3-methylglutaryl CoA, which is cleaved into acetyl CoA and acetoacetate. This reaction has already been discussed in regard to the formation of ketone bodies from fatty acids (Section 22.4).

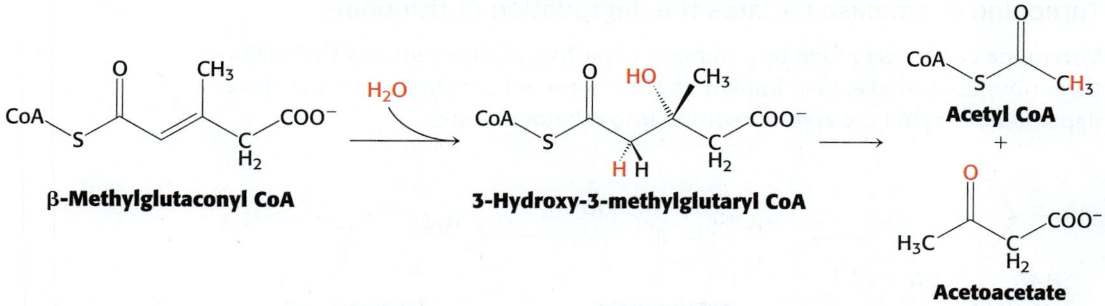

The degradative pathways of valine and isoleucine resemble that of leucine. After transamination and oxidative decarboxylation to yield a CoA derivative, the subsequent reactions are like those of fatty acid oxidation. Isoleucine yields acetyl CoA and propionyl CoA, whereas valine yields CO_2 and propionyl CoA. The degradation of leucine, valine, and isoleucine validate a point made in Chapter 15: the number of reactions in metabolism is large, but the number of *kinds* of reactions is relatively small. The degradation of leucine, valine, and isoleucine provides a striking illustration of the underlying simplicity and elegance of metabolism.

Oxygenases are required for the degradation of aromatic amino acids

The degradation pathways of the aromatic amino acids, which yield the common intermediates acetoacetate, fumarate, and pyruvate, are not as straightforward as that of the amino acids previously discussed. For the aromatic amino acids, molecular oxygen is used to break the aromatic ring.

The degradation of phenylalanine begins with its hydroxylation to tyrosine, a reaction catalyzed by phenylalanine hydroxylase. This enzyme is called a *monooxygenase* (or *mixed-function oxygenase*) because *one* atom of O_2 appears in the product and the other atom in H_2O.

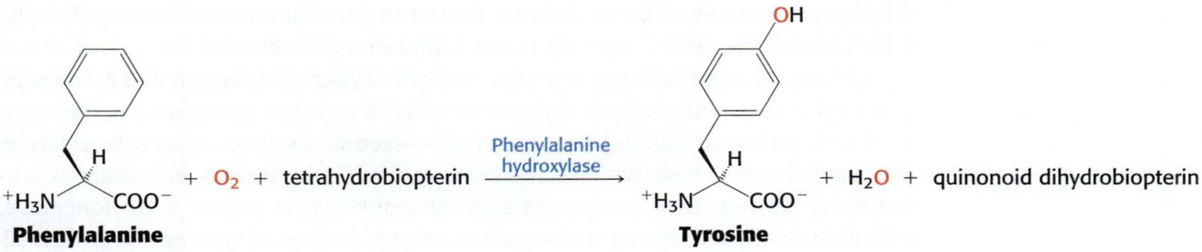

Phenylalanine $+ O_2 +$ tetrahydrobiopterin $\xrightarrow{\text{Phenylalanine hydroxylase}}$ **Tyrosine** $+ H_2O +$ quinonoid dihydrobiopterin

The reductant here is tetrahydrobiopterin, an electron carrier that has not been previously discussed and is derived from the cofactor biopterin, which human beings synthesize de novo. The quinonoid form of dihydrobiopterin is produced in the hydroxylation of phenylalanine. It is reduced back to tetrahydrobiopterin by NADPH in a reaction catalyzed by dihydropteridine reductase.

Quinonoid dihydrobiopterin $\xrightarrow[\text{Dihydropteridine reductase}]{\text{NADPH} + \text{H}^+ \quad \text{NADP}^+}$ **Tetrahydrobiopterin**

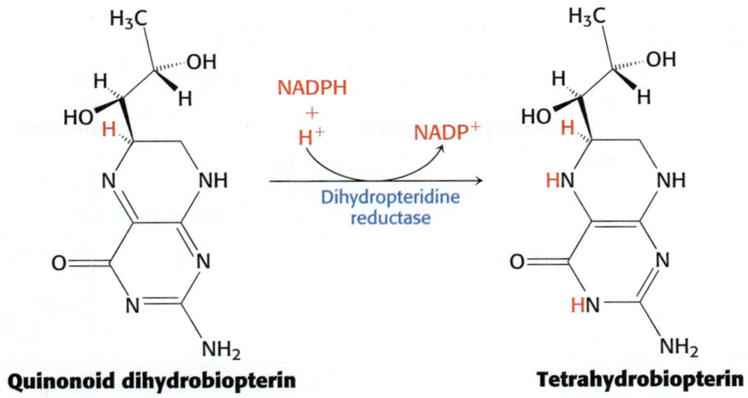

The sum of the reactions catalyzed by phenylalanine hydroxylase and dihydropteridine reductase is

$$\text{Phenylalanine} + O_2 + \text{NADPH} + H^+ \longrightarrow \text{tyrosine} + \text{NADP}^+ + H_2O$$

Note that these reactions can also be used to synthesize tyrosine from phenylalanine.

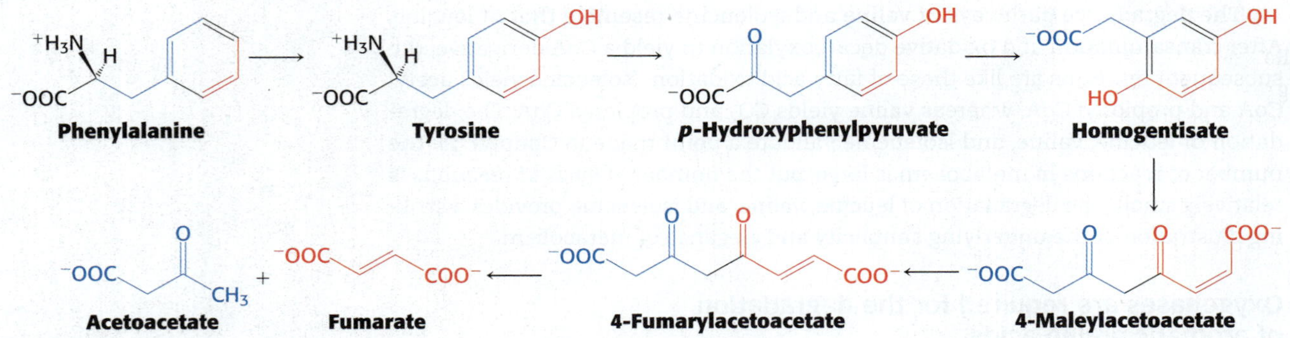

FIGURE 23.26 Phenylalanine and tyrosine degradation produce both acetoacetate and fumarate. This process requires both dioxygenases and monooxygenases.

The next step in the degradation of phenylalanine and tyrosine is the transamination of tyrosine to *p*-hydroxyphenylpyruvate (**Figure 23.26**). This α-ketoacid then reacts with O_2 to form homogentisate. The enzyme catalyzing this complex reaction, *p*-hydroxyphenylpyruvate hydroxylase, is called a *dioxygenase* because both atoms of O_2 become incorporated into the product, one on the ring and one in the carboxyl group. The aromatic ring of homogentisate is then cleaved by O_2, which yields 4-maleylacetoacetate. This reaction is catalyzed by homogentisate oxidase, another dioxygenase. 4-Maleylacetoacetate is then isomerized to 4-fumarylacetoacetate by an enzyme that uses glutathione as a cofactor. Finally, 4-fumarylacetoacetate is hydrolyzed to fumarate and acetoacetate.

Tryptophan degradation requires several oxygenases (**Figure 23.27**). Tryptophan 2,3-dioxygenase cleaves the pyrrole ring, and kynurenine 3-monooxygenase hydroxylates the remaining benzene ring, a reaction similar to the hydroxylation of phenylalanine to form tyrosine. Alanine is removed and the 3-hydroxyanthranilate is cleaved by another dioxygenase and subsequently processed to acetoacetate. Note that nearly all cleavages of aromatic rings in biological systems are catalyzed by dioxygenases. The active sites of these enzymes contain iron that is not part of heme or an iron–sulfur cluster.

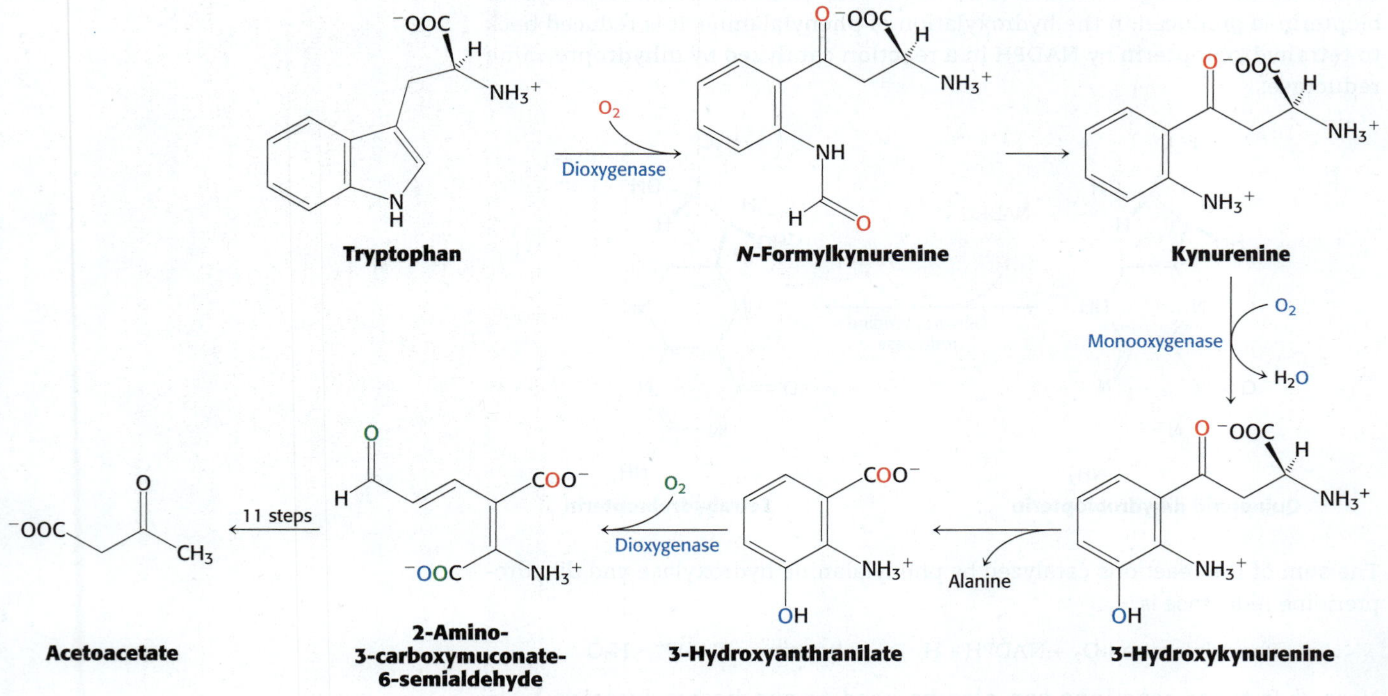

FIGURE 23.27 Tryptophan degradation requires multiple oxygenases and numerous steps. Note that both alanine and acetoacetate are ultimately formed.

Protein metabolism helps to power the flight of migratory birds

Earlier we learned that fats are the primary fuel for migratory birds that fly great distances without feeding (Section 22.1). Proteins are also degraded to contribute fuel for these impressive feats of endurance. Why would proteins be degraded? There are several likely reasons. First, since the birds are not feeding, proteins involved with digestion and transport of nutrients from the gut are not needed and thus can be degraded. Second, gluconeogenesis is required to provide glucose for the central nervous system, and the glucogenic amino acids will meet this need. Moreover, some amino acid carbon skeletons can replenish citric acid cycle intermediates to facilitate fat metabolism. Third, the reduction in body mass resulting from protein degradation will reduce the energy cost of flight. There is one final benefit. In addition to not feeding during flight, the migratory birds cannot hydrate, yet they still require water. While the catabolism of fats yields little water—0.029 g H_2O kJ^{-1}—protein provides fivefold more water than fats—0.155 g H_2O kJ^{-1}—and may be a vital source of water for migratory flights.

EXAMPLE **Determining Metabolic Products of Amino Acid Degradation**

PROBLEM: Isoleucine is said to be both a ketogenic and glucogenic amino acid. Starting from 2-methylbutyryl CoA, a degradation intermediate of isoleucine, show that it is both ketogenic and glucogenic.

GETTING STARTED: Let's begin by thinking about the structure of 2-methylbutyryl CoA. We have seen similar molecules before. What general type of biochemical does this look like?

It looks like a fatty acyl CoA, except that there is a methyl group where a hydrogen would be in a fatty acyl CoA. To see this more clearly, let's look at the molecule with the substitution of a hydrogen for the methyl group, butyryl CoA.

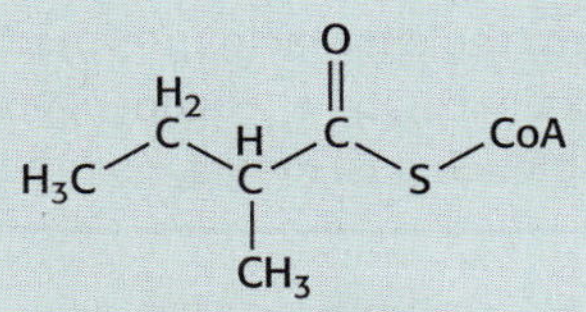

2-Methylbutyryl CoA

ANALYZE: Thinking back to Chapter 22, how would you degrade butyryl CoA? β-Oxidation would be the way to go. So, let's apply β-oxidation to 2-methylbutyryl CoA. Recall that the first step in the β-oxidation is the introduction of a double bond at the β-carbon, accompanied with reduction of FAD. Recall that we saw the reaction in the conversion of succinate into fumarate in the citric acid cycle as well as in β-oxidation.

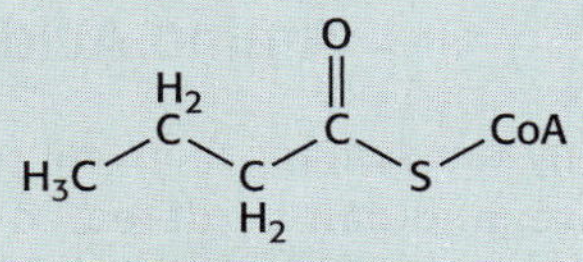

Butyryl CoA

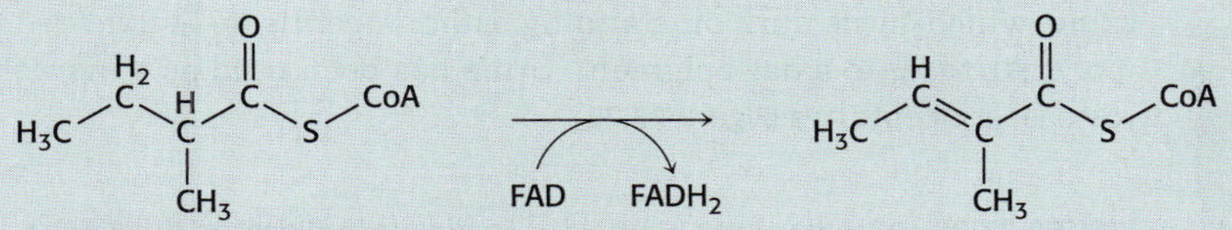

2-Methylbutyryl CoA   FAD → FADH₂

Next, recall that the next two reactions of β-oxidation are a hydration followed by an oxidation, with the concomitant reduction of NAD⁺.

Finally, the last reaction is thiolysis by addition of CoA.

Acetyl CoA

Propionyl CoA

Thiolysis yields acetyl CoA and propionyl CoA.

REFLECT: Acetyl CoA can be used for ketone body production, accounting for the fact that isoleucine is a ketogenic amino acid. What is the metabolic fate of propionyl CoA? As we saw in Section 22.3, propionyl CoA is converted into the citric acid intermediate succinyl CoA, which can be metabolized to oxaloacetate, a gluconeogenic substrate (Section 16.4). So we have shown that isoleucine is, in fact, both a ketogenic and a glucogenic amino acid.

Homogentisate

Air

Highly colored polymer

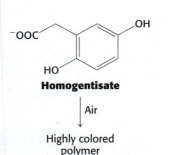

23.6 Inborn Errors of Metabolism Can Disrupt Amino Acid Degradation

Many unusual biochemical products are produced or overproduced in a disease state, and can be detected in the urine. Additionally, errors in amino acid metabolism provided some of the first examples of biochemical defects linked to pathological conditions. For instance, *alkaptonuria*, an inherited metabolic disorder caused by the absence of homogentisate oxidase, was described in 1902. Homogentisate, a normal intermediate in the degradation of phenylalanine and tyrosine (Figure 23.26), accumulates in alkaptonuria because its degradation is blocked. Homogentisate is excreted in the urine, which turns dark on standing as homogentisate is oxidized and polymerizes into a dark pigment. Urine has been used as a marker for disease for centuries (Figure 23.28).

FIGURE 23.28 Urine has long been used to diagnose diseases due to the excretion of unusual biochemical products. This fourteenth-century hand-colored woodcut from Germany depicts a wheel that classifies urine samples according to their color and consistency. In the middle of the wheel, a doctor inspects a patient's urine by sight, smell, and taste. The vials on the wheel aided physicians in diagnosing diseases. [Ulrich Pinder. Epiphanie Medicorum. Speculum videndi urinas hominum. Clavis aperiendi portas pulsuum. Berillus discernendi causas & differentias febrium. Nuremberg: 1506. Rosenwald Collection. Rare Book and Special Collections Division, Library of Congress (128.2).]

Branched-chain ketoaciduria is a serious disorder of branched-chain amino acid degradation

Although alkaptonuria is a relatively harmless condition, such is not the case with other errors in amino acid metabolism. In branched-chain ketoaciduria, most commonly called *maple syrup urine disease*, the oxidative decarboxylation of α-ketoacids derived from valine, isoleucine, and leucine is blocked because the branched-chain dehydrogenase is missing or defective. Hence, the levels of these α-ketoacids and the branched-chain amino acids that give rise to them are markedly elevated in both blood and urine. The urine of patients has the odor of maple syrup, hence the common name of the disease. The disease usually leads to cognitive and physical disabilities unless the patient is placed on a diet low in valine, isoleucine, and leucine early in life. Thankfully, the disease can be readily detected in newborns by screening urine samples with 2,4-dinitrophenylhydrazine, which reacts with α-ketoacids to form 2,4-dinitrophenylhydrazone derivatives. A definitive diagnosis can be made by mass spectrometry. **Table 23.4** lists some other diseases of amino acid metabolism.

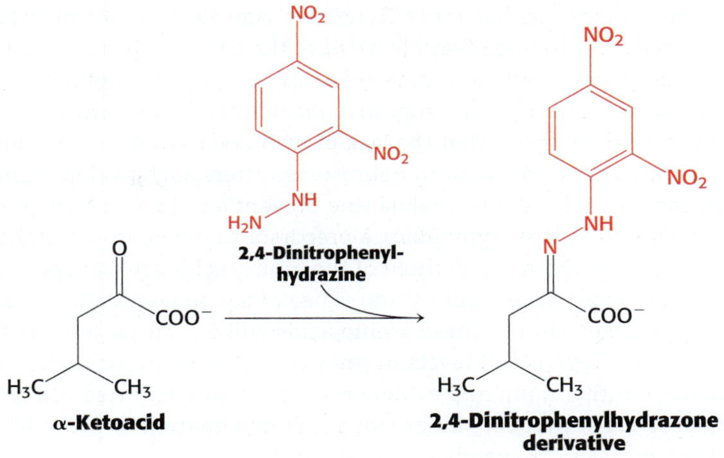

TABLE 23.4 Inborn errors of amino acid metabolism

Disease	Enzyme deficiency	Symptoms
Citrullinemia	Arginosuccinate lyase	Lethargy, seizures, reduced muscle tension
Tyrosinemia	Various enzymes of tyrosine degradation	Weakness, liver damage, cognitive disability
Albinism	Tyrosinase	Absence of pigmentation
Homocystinuria	Cystathionine β-synthase	Scoliosis, muscle weakness, cognitive disability, thin blond hair
Hyperlysinemia	α-Aminoadipic semialdehyde dehydrogenase	Seizures, cognitive disability, lack of muscle tone, ataxia

Phenylketonuria is one of the most common metabolic disorders

Perhaps the best known of the diseases of amino acid metabolism is **phenylketonuria**, which occurs with a prevalence of 1 in 10,000 births. Phenylketonuria is caused by an absence or deficiency of phenylalanine hydroxylase or, more rarely, of its tetrahydrobiopterin cofactor. Phenylalanine accumulates in all body fluids because it cannot be converted into tyrosine. Normally, three-quarters of phenylalanine molecules are converted into tyrosine, and the other quarter become incorporated into proteins. Because the major outflow pathway is blocked in phenylketonuria, the blood level of phenylalanine is typically at least

Phenylalanine

Phenylpyruvate

α-Ketoacid

α-Amino acid

20-fold as high as in unaffected people. Minor fates of phenylalanine in people who process phenylalanine normally, such as the formation of phenylpyruvate which is a phenyl ketone, become major products which are then excreted in the urine. Indeed, the initial description of phenylketonuria in 1934 was made by observing the reaction of phenylpyruvate in the urine of phenylketonurics with $FeCl_3$, which turns the urine olive green.

Untreated phenylketonuria results almost inexorably in severe cognitive disabilities. Babies with phenylketonuria do not show symptoms at birth but are severely affected by the disease by age 1 if untreated, and if they remain untreated, life expectancy is between 20 and 30 years. The therapy for phenyl-ketonuria is a low-phenylalanine diet supplemented with tyrosine, because tyrosine is normally synthesized from phenylalanine. The aim is to provide just enough phenylalanine to meet the needs for growth and replacement. Proteins that have a low content of phenylalanine, such as casein from milk, are hydrolyzed and phenylalanine is removed by adsorption.

Early diagnosis of phenylketonuria is essential and has been accomplished by mass screening programs of all babies born in the United States and Canada. The phenylalanine level in the blood is the preferred diagnostic criterion because it is more sensitive and reliable than the $FeCl_3$ test. Prenatal diagnosis of phenylketonuria with DNA probes has become feasible because the gene has been cloned and the exact locations of many mutations have been discovered in the protein.

The biochemical basis of the cognitive disabilities is not firmly established, but one hypothesis suggests that the lack of hydroxylase reduces the amount of tyrosine, an important precursor to neurotransmitters such as dopamine. Moreover, high concentrations of phenylalanine prevent amino acid transport of any tyrosine present as well as tryptophan, a precursor to the neurotransmitter serotonin, into the brain. Because all three of the amino acids are transported by the same carrier, phenylalanine will saturate the carrier, preventing access to tyrosine and tryptophan. Lack of these amino acids will impair protein synthesis in the brain. Finally, high blood levels of phenylalanine result in higher levels of phenylalanine in the brain, and evidence suggests this elevated concentration inhibits glycolysis at pyruvate kinase, disrupts myelination of nerve fibers, and reduces the synthesis of several neurotransmitters.

❖ SELF–CHECK QUESTION

The end products of tryptophan degradation are acetyl CoA and acetoacetyl CoA, yet tryptophan is a gluconeogenic amino acid in animals. Explain.

Summary

23.1 Proteins Are Degraded to Amino Acids

- Dietary protein is digested in the intestine, producing amino acids that are transported throughout the body.
- Cellular proteins are degraded at widely variable rates, ranging from minutes to the life of the organism. The turnover of cellular proteins is a regulated process requiring complex enzyme systems.

23.2 Protein Turnover Is Tightly Regulated

- Proteins to be degraded by the proteasome are conjugated with ubiquitin, a small conserved protein. The ubiquitin-conjugating system is composed of three distinct enzymes.
- The proteasome is a large, barrel-shaped complex that digests ubiquitinated proteins at the expense of ATP. The resulting amino acids provide a source of precursors for protein, nucleotide bases, and other nitrogenous compounds.

23.3 The First Step in Amino Acid Degradation Is the Removal of Nitrogen

- Surplus amino acids are used as building blocks and as metabolic fuel. The first step in their degradation is the removal of their α-amino groups by transamination to yield α-ketoacids. Transamination is catalyzed by aminotransferases that all use pyridoxal phosphate as a coenzyme.

- The α-amino group funnels into α-ketoglutarate to form glutamate, which is then oxidatively deaminated in the liver by glutamate dehydrogenase to yield NH_4^+ and α-ketoglutarate.

23.4 Ammonium Ions Are Converted into Urea in Most Terrestrial Vertebrates

- The first step in the synthesis of urea is the formation of carbamoyl phosphate, which is synthesized from HCO_3^-, NH_3, and two molecules of ATP by carbamoyl phosphate synthetase I.

- A second reaction in the mitochondria produces citrulline, which leaves the mitochondrion and condenses with aspartate, which provides the second nitrogen of what will become urea. Two subsequent reactions in the cytosol produce urea and regenerate the intermediates of the cycle.

23.5 Carbon Atoms of Degraded Amino Acids Emerge as Major Metabolic Intermediates

- The carbon atoms of degraded amino acids are converted into pyruvate, acetyl CoA, acetoacetate, or an intermediate of the citric acid cycle.

- Most amino acids are solely glucogenic, two are solely ketogenic, and a few are both ketogenic and glucogenic.

- Carbon skeletons of glucogenic amino acids enter major metabolic pathways as pyruvate, oxaloacetate, α-ketoglutarate, or succinyl CoA. Ketogenic amino acids produce acetoacetate or acetyl CoA.

- The rings of aromatic amino acids are degraded by oxygenases. Phenylalanine hydroxylase, a monooxygenase, uses tetrahydrobiopterin as the reductant. One of the oxygen atoms of O_2 emerges in tyrosine and the other in water.

- Subsequent steps in the degradation of these aromatic amino acids are catalyzed by dioxygenases, which catalyze the insertion of both atoms of O_2 into organic products.

23.6 Inborn Errors of Metabolism Can Disrupt Amino Acid Degradation

- Errors in amino acid metabolism were sources of some of the first insights into the correlation between pathology and biochemistry.

- Phenylketonuria, the best known of the many hereditary errors of amino acid metabolism, results from the accumulation of high levels of phenylalanine in the body fluids. This accumulation leads to cognitive disabilities unless the patient is placed on low-phenylalanine diets immediately after birth.

Key Terms

ubiquitin (p. 703)

degron (p. 704)

proteasome (p. 705)

aminotransferases (transaminases) (p. 708)

glutamate dehydrogenase (p. 708)

pyridoxal phosphate (PLP) (p. 709)

glucose–alanine cycle (p. 713)

urea cycle (p. 713)

carbamoyl phosphate synthetase I (p. 714)

ketogenic amino acid (p. 718)

glucogenic amino acid (p. 718)

phenylketonuria (p. 727)

Problems

1. What are the steps required to attach ubiquitin to a target protein? ❖ 1, ❖ 2

2. Protein hydrolysis is an exergonic process, yet the 26S proteasome is dependent on ATP hydrolysis for activity. ❖ 1, ❖ 3

(a) Explain why ATP hydrolysis is required by the 26S proteasome.

(b) Small peptides can be hydrolyzed without the expenditure of ATP. How does this information concur with your answer to part (a)?

3. Aminotransferases require which of the following cofactors?

(a) NAD⁺/NADP⁺

(a) $NAD^+/NADP^+$

(b) Pyridoxal phosphate

(c) Thiamine pyrophosphate

(d) Biopterin

4. Which of the following compounds readily accepts amino groups from amino acids?

(a) Glutamine

(b) Isocitrate

(c) Malate

(d) α-Ketoglutarate

5. The immediate donors of the nitrogen atoms of urea are:

(a) Aspartate and glutamate

(b) Glutamate and carbamoyl phosphate

(c) Aspartate and carbamoyl phosphate

(d) Glutamine and aspartate

6. Which compound of the urea cycle is synthesized in the mitochondria and transported into the cytoplasm?

(a) Ornithine

(b) Citrulline

(c) Argininosuccinate

(d) Arginine

7. Name the α-ketoacid that is formed by the transamination of each of the following amino acids: ❖ 4, ❖ 5

(a) Alanine

(b) Aspartate

(c) Glutamate

(d) Leucine

(e) Phenylalanine

(f) Tyrosine

8. (a) Write a balanced equation for the conversion of aspartate into glucose through the intermediate oxaloacetate. Which coenzymes participate in this transformation?

(b) Write a balanced equation for the conversion of aspartate into oxaloacetate through the intermediate fumarate.

9. What amino acids yield citric acid cycle components and glycolysis intermediates when deaminated? ❖ 4, ❖ 5

10. Match the property on the left with the biochemical on the right. ❖ 4

(a) Formed from NH_4^+

(b) Hydrolyzed to yield urea

(c) A second source of nitrogen

(d) Reacts with aspartate

(e) Cleavage yields fumarate

(f) Accepts the first nitrogen

(g) Final product

1. Aspartate

2. Urea

3. Ornithine

4. Carbamoyl phosphate

5. Arginine

6. Citrulline

7. Argininosuccinate

11. Identify structures A–D, and place them in the order that they appear in the urea cycle. ❖ 4

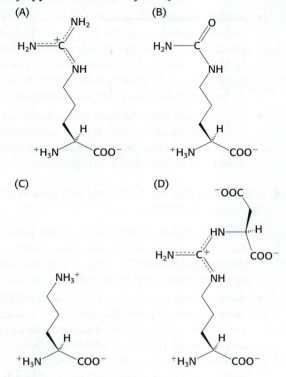

12. Four high-transfer-potential phosphoryl groups are consumed in the synthesis of urea according to the stoichiometry given in Section 23.4. In this reaction, aspartate is converted into fumarate. Suppose that fumarate is converted into oxaloacetate. What is the resulting stoichiometry of urea synthesis? How many high-transfer-potential phosphoryl groups are spent?

13. A friend bets you a million dollars that you can't prove that the urea cycle is linked to the citric acid cycle and other metabolic pathways. Can you prove it?

14. Why should phenylketonurics avoid using aspartame, an artificial sweetener? (Hint: Aspartame is L-aspartyl-L-phenylalanine methyl ester.) ❖ 6

15. What is meant by the terms *ketogenic amino acids* and *glucogenic amino acids*? ❖ 5

16. Pyruvate dehydrogenase complex and α-ketoglutarate dehydrogenase complex are huge enzymes consisting of three discrete enzymatic activities. Which amino acids require a related enzyme complex, and what is the name of the enzyme?

17. The carbon skeletons of the 20 common amino acids can be degraded into a limited number of end products. What are the end products, and in what metabolic pathway are they commonly found? ❖ 5

18. In Chapter 5, we learned that there are two types of bisubstrate reactions, sequential and double-displacement. Which type characterizes the action of aminotransferases? Explain your answer.

19. Within a few days after a fast begins, nitrogen excretion accelerates to a higher-than-normal level. After a few weeks, the rate of nitrogen excretion falls to a lower level and continues at this low rate. However, after the fat stores have been depleted, nitrogen excretion rises to a high level. ❖ **4,** ❖ **5** **(a)** What events trigger the initial surge of nitrogen excretion? **(b)** Why does nitrogen excretion fall after several weeks of fasting? **(c)** Explain the increase in nitrogen excretion when the lipid stores have been depleted.

20. In eukaryotes, the 20S proteasome component in conjunction with the 19S component degrades ubiquitinated proteins with the hydrolysis of a molecule of ATP. Archaea lack ubiquitin and the 26S proteasome but do contain a 20S proteasome. Some archaea also contain an ATPase that is homologous to the ATPases of the eukaryotic 19S component. This archaeal ATPase activity was isolated as a 650-kDa complex (called PAN) from the archaeon *Thermoplasma*, and experiments were performed to determine if PAN could enhance the activity of the 20S proteasome from *Thermoplasma* as well as other 20S proteasomes.

Protein degradation was measured as a function of time and in the presence of various combinations of components. Graph A shows the results. ❖ **1,** ❖ **2,** ❖ **3**

(A)

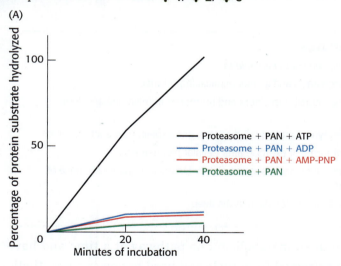

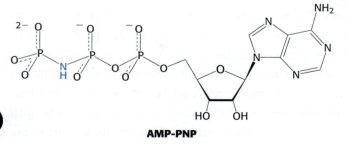

AMP-PNP

(a) What is the effect of PAN on archaeal proteasome activity in the absence of nucleotides?

(b) Which nucleotide stimulates protein digestion?

(c) What evidence suggests that ATP hydrolysis, and not just the presence of ATP, is required for digestion?

A similar experiment was performed with a small peptide as a substrate for the proteasome instead of a protein. The results obtained are shown in graph B.

(B)

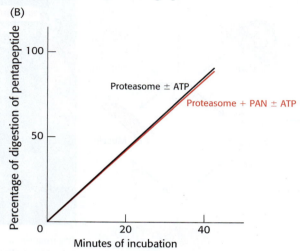

(d) How do the requirements for peptide digestion differ from those of protein digestion?

(e) Suggest some reasons for the difference.

The ability of PAN from the archaeon *Thermoplasma* to support protein degradation by the 20S proteasomes from the archaeon *Methanosarcina* and rabbit muscle was then examined.

Percentage of digestion of protein substrate (Source of the 20S proteasome)

Additions	Thermoplasma	Methanosarcina	Rabbit muscle
None	11	10	10
PAN	8	8	8
PAN+ATP	100	40	30
PAN+ADP	12	9	10

[Data from P. Zwickl, D. Ng, K. M. Woo, H.-P. Klenk, and A. L. Goldberg, An archaebacterial ATPase, homologous to ATPase in the eukaryotic 26S proteasome, activates protein breakdown by 20S proteasomes, *J. Biol. Chem.* 274(37):26008–26014, 1999.]

(f) Can the *Thermoplasma* PAN augment protein digestion by the proteasomes from other organisms?

(g) What is the significance of the stimulation of rabbit muscle proteasome by *Thermoplasma* PAN?

Integration of Energy Metabolism

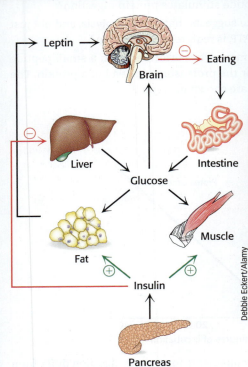

Debbie Eckert/Alamy

The maintenance of blood-glucose levels requires the integration of many elaborate metabolic pathways. Disorders affecting the principal organs, hormones, and pathways involved can lead to disease. Type 1 diabetes, for example, results from the dysfunction of specific pancreatic cells that normally release the hormone insulin after meals. Fundraising events support research for better treatments for the approximately 9 million people worldwide who are affected by type 1 diabetes.

OUTLINE

24.1 Caloric Homeostasis Is a Means of Regulating Body Weight

24.2 The Fasted–Fed Cycle Is a Response to Eating and Sleeping Behaviors

24.3 Diabetes Is a Common Metabolic Disease Often Resulting from Obesity

24.4 Exercise Beneficially Alters the Biochemistry of Cells

24.5 Starvation Induces Protein Wasting and Ketone Body Formation

24.6 Ethanol Alters Energy Metabolism in the Liver

❖ LEARNING GOALS

By the end of this chapter, you should be able to:

1. Describe caloric homeostasis and its biochemical importance.

2. Explain the role of the neurotransmitters and hormones in maintaining caloric homeostasis.

3. Differentiate between type 1 and type 2 diabetes, and explain how each develops.

4. Explain how different forms of exercise alter cellular biochemistry.

5. Contrast the biochemical responses to the nightly fasted–fed cycle with those of prolonged starvation.

6. Explain how ethanol alters metabolism in the liver.

So far we have examined metabolism one pathway at a time and have seen how energy is extracted from fuels and used to power biosynthetic reactions. In Chapters 25 through 31, we will extend our study of biosynthetic reactions to the synthesis of nucleic acids, diverse lipids, and proteins. First, however, in this chapter we will take a step back to examine large-scale biochemical interactions that constitute the physiology of the organism and the regulation of energy at the organismal level. We will examine how various biochemical pathways and tissues interact with one another in both resting metabolism and in a variety of states of metabolic perturbation, namely: obesity, fasting, diabetes, intense exercise, starvation, and excessive alcohol consumption.

24.1 Caloric Homeostasis Is a Means of Regulating Body Weight

In this section we will address an apparently simple but actually quite complex question: at the biochemical level, how does an organism know when to eat and when to refrain from eating? The ability to maintain adequate, but not excessive, energy stores is called **caloric homeostasis** or **energy homeostasis**. By now in our study of biochemistry, we are well aware of the fact that carbohydrates, lipids, and the carbon skeletons of amino acids are sources of energy (**Figure 24.1**). We consume these energy sources as foods and capture free energy by re-forming ATP,

FIGURE 24.1 Maintaining homeostasis requires complex metabolic regulation that coordinates the use of nutrient pools. The numbers indicate the chapters in which the topics are discussed. [Information from D. U. Silverthorn, *Human Physiology: An Integrated Approach,* 3rd ed. (Pearson, 2004), Fig. 22-2.]

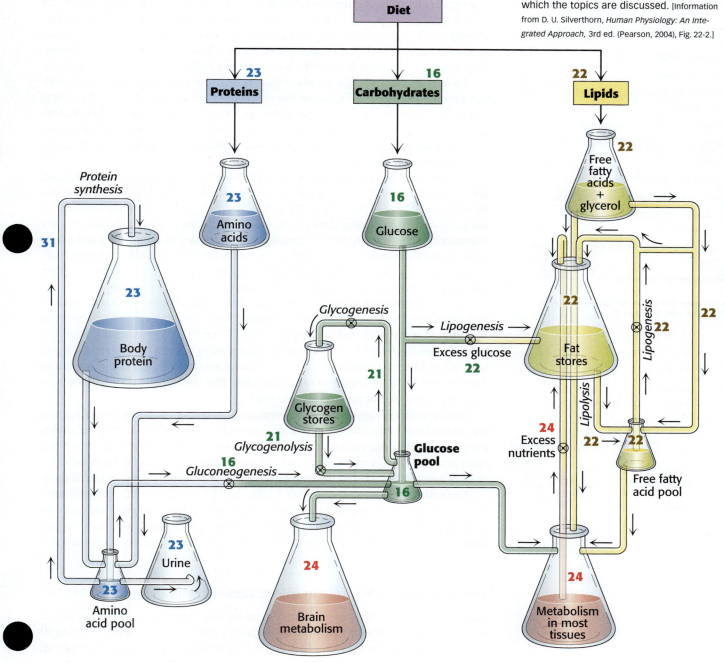

☐ Fats   ☐ Carbohydrates   ☐ Amino acids

which we use to power our lives. As illustrated in Figure 24.1, to do this requires an incredible amount of coordination among numerous metabolic pathways that we have explored in Chapters 16 through 23. However, despite this mechanistic complexity, like all energy transformations our energy consumption and expenditure are governed by the laws of thermodynamics. Recall that the First Law of Thermodynamics states that energy can neither be created nor destroyed. Translated into the practical terms of our diets:

$$\text{Energy consumed} = \text{energy expended} + \text{energy stored}$$

This simple equation has severe physiological and health implications: according to the First Law of Thermodynamics, if we consume more energy than we expend, we will eventually become overweight or obese. Regardless of the original source, any excess energy molecules are ultimately converted to triacylglycerols which are stored in adipocytes. The number of adipocytes becomes fixed in adults, and so obesity results in engorged adipocytes. Indeed, the cells may increase as much as 1000-fold in size. Obesity is generally defined as a body mass index (BMI) of more than 30 kg m^{-2}, whereas overweight is defined as a BMI of more than 25 kg m^{-2}.

Obesity is a worldwide problem that is now considered a pandemic, with the number of people considered obese having tripled since 1975 and totaling 650 million people. Obesity is identified as a risk factor in a host of pathological conditions, including diabetes mellitus, hypertension, and cardiovascular disease (Table 24.1). We will consider the biochemical basis of pathologies caused by obesity later in the chapter.

Before we undertake a biochemical analysis of the results of overconsumption, let us consider why the obesity pandemic is occurring in the first place. There are several overlapping explanations. The first is a commonly held view that our bodies are programmed to rapidly store excess calories in times of plenty, an evolutionary adaptation to the scarcity of food. Consequently, we store calories as if a fast might begin tomorrow, regardless of whether it actually does. We are also no longer subject to predation, which once would have applied selective pressure to maintain low body fats so that our ancestors could more easily flee from an attack.

Another factor is that calorie-dense, highly palatable foods—rich in sugar and fats—are now more readily accessible, stimulating the same reward pathways in the brain that are triggered by drugs such as cocaine. Such reward pathways can be strong enough to override appetite-suppressing signals. These foods, along with other environmental factors such as pollution, can also affect our intestinal microbiome—the bacteria and other microbes that inhabit our guts—that play a significant role in how we process our food.

Finally, although genetics cannot explain the obesity pandemic affecting vast amounts of the entire human *population*, it is evident that *individuals* respond differently to obesity-inducing environmental conditions, and that this difference has a large genetic component. Various studies have indicated the heritability of fat mass to be between 30% and 70%, making it extra difficult for some people to lose weight.

Regardless of why we may have a propensity to gain weight, this predisposition can be counteracted behaviorally—by eating less and exercising more. It also appears that it is easier to prevent weight gain than it is to lose weight. Clearly, caloric homeostasis is a complicated biological phenomenon that will engage biomedical research scientists for some time to come.

The brain plays a key role in caloric homeostasis

As disturbing as the obesity pandemic is, an equally intriguing observation is that many people are able to maintain an approximately constant weight throughout adult life, despite consuming literally tons of food over a lifetime. Although willpower, exercise, and a bathroom scale often play a role in this homeostasis, some favorable biochemical signaling must also be behind this remarkable

TABLE 24.1 Health consequences of obesity

Coronary heart disease
Type 2 diabetes
Cancers (endometrial, breast, colon, and others)
Hypertension (high blood pressure)
Dyslipidemia (disruption of lipid metabolism)
Stroke
Liver and gallbladder disease
Sleep apnea and respiratory problems
Osteoarthritis (degeneration of joint cartilage/bone)
Gynecological problems (abnormal menses, infertility)
Male infertility problems

Information from Centers for Disease Control and Prevention website (www.cdc.gov).

physiological feat. What makes this balance of energy input and output possible? As one might imagine, the answer is complicated, entailing many biochemical signals as well as a host of behavioral factors. We will focus on a few key biochemical signals and divide our discussion into two parts: short-term signals that are active during a meal, and long-term signals that report on the overall energy status of the body. These signals originate in the gastrointestinal tract, the β cells of the pancreas, and fat cells. The primary target of these signals is the brain, in particular a group of neurons in a region of the hypothalamus called the *arcuate nucleus*.

Short-term signals from the gastrointestinal tract induce feelings of satiety

Short-term signals relay feelings of satiety (fullness) from the gut to various regions of the brain and thus reduce the urge to eat during what is known as the *postprandial period*, or the time right after a meal (**Figure 24.2**).

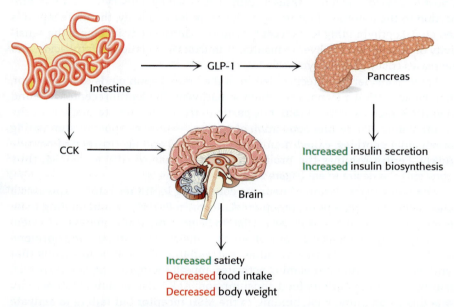

FIGURE 24.2 Signal molecules originating from the gut induce feelings of satiety in the brain. Cholecystokinin (CCK) is secreted by specialized cells of the small intestine in response to a meal and activates satiation pathways in the brain. Glucagon-like peptide 1 (GLP-1), secreted by L cells in the intestine, also activates satiation pathways in the brain and potentiates insulin action in the pancreas. [Information from S. C. Wood, *Cell Metab*. 9:489–498, 2009, Fig. 1.]

The best-studied short-term signals are cholecystokinin (CCK) and glucagon-like peptide 1 (GLP-1). Both are small peptide hormones secreted into the blood by cells in the upper part of the small intestine as postprandial satiation signals. Both bind to their respective G-protein-coupled receptors (Section 14.2) in peripheral neurons, relaying signals to the brain that generate feelings of satiety. CCK and GLP-1 also play important roles in digestion; CCK stimulates the secretion of pancreatic enzymes and bile salts from the gallbladder, while GLP-1 enhances glucose-induced insulin secretion and inhibits glucagon secretion. Recall that **insulin** causes glucose to be removed from the blood and stimulates the synthesis of glycogen and lipids, while **glucagon** has effects opposite those of insulin.

Although we have examined only two short-term signals, many others are believed to exist (**Table 24.2**). Most of the short-term signals thus far identified are appetite suppressants. However, another small peptide hormone called

TABLE 24.2 Gastrointestinal peptides that regulate food intake

Appetite-suppressing signals
Cholecystokinin
Glucagon-like peptide 1
Glucagon-like peptide 2
Amylin
Somatostatin
Bombesin
Enterostatin
Apolipoprotein A-IV
Gastric inhibitory peptide
Appetite-enhancing peptide
Ghrelin

Information from M. H. Stipanuk, ed., *Biochemical, Physiological, Molecular Aspects of Human Nutrition*, 2nd ed. (Saunders/Elsevier, 2006), p. 627, Box 22-1.

ghrelin—secreted by the stomach—acts on regions of the hypothalamus to stimulate appetite through its G-protein-coupled receptor. Ghrelin secretion increases before a meal and decreases afterward.

Leptin and insulin regulate long-term control over caloric homeostasis

Two key signal molecules regulate energy homeostasis over the time scale of hours or days: **leptin**, which is secreted by the adipocytes, and insulin, which is secreted by the pancreatic β cells. Leptin communicates the status of the triacylglycerol stores, whereas insulin communicates the status of glucose in the blood—in other words, carbohydrate availability. We will consider leptin first.

Adipose tissue was formerly considered an inert storage form of triacylglycerols. However, later work showed that adipose tissue is an active endocrine tissue, secreting signal molecules called *adipokines*, such as leptin, that regulate a host of physiological processes. Leptin is secreted by adipocytes in direct proportion to the amount of fat present: the more fat in a body, the more leptin is secreted. Leptin binding to its receptor throughout the body increases the sensitivity of muscle and the liver to insulin, stimulates β oxidation of fatty acids, and decreases triacylglycerol synthesis.

Let us consider the effects of leptin in the brain. Leptin binds to its receptor, thereby activating a signal-transduction pathway. The leptin receptor is found in various regions of the brain, but particularly in the arcuate nucleus of the hypothalamus. There, one population of neurons expresses appetite-stimulating (orexigenic) peptides, called neuropeptide Y (NPY) and agouti-related peptide (AgRP). Leptin inhibits the production and release of NPY and AgRP, thus repressing the desire to eat (**Figure 24.3**).

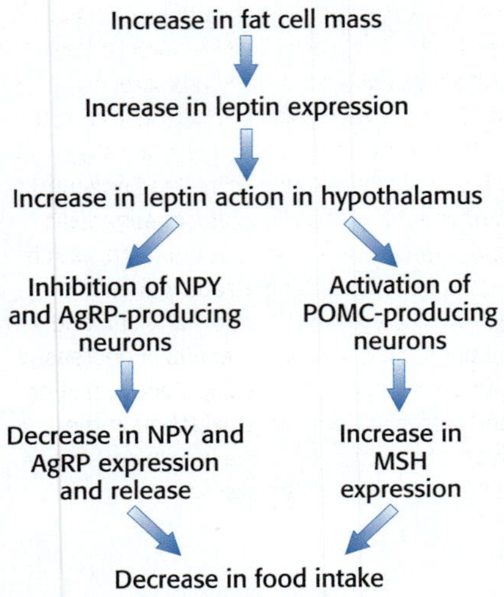

The second population of neurons containing leptin receptors expresses a large precursor polypeptide, proopiomelanocortin (POMC). Leptin binding to its receptor on POMC neurons causes POMC to be proteolytically processed to yield a variety of signal molecules, one of which is melanocyte-stimulating hormone (MSH). MSH, originally discovered as a stimulator of melanocytes (cells that synthesize the pigment melanin), activates appetite-suppressing (anorexigenic) neurons and thus inhibits food consumption. AgRP also inhibits MSH activity by acting as an antagonist, binding to the MSH receptor but failing to activate the receptor (Figure 24.3). Thus, the net effect of leptin binding to its receptor is the initiation of a complex signal-transduction pathway that ultimately curtails food intake. Insulin receptors are also present in the hypothalamus, although the mechanism of insulin action in the brain is less clear than that of leptin. Insulin appears to inhibit NPY/AgRP-producing neurons, thus inhibiting food consumption.

FIGURE 24.3 Leptin suppresses appetite by affecting the brain. Leptin is an adipokine secreted by adipose tissue in direct relation to fat mass. When fat mass increases, leptin inhibits NPY and AgRP secretion while stimulating the release of appetite-suppressing hormone MSH. [Information from M. H. Stipanuk, *Biochemical, Physiological, & Molecular Aspects of Human Nutrition*, 2nd ed. (Saunders/Elsevier, 2006), Fig. 22-2.]

❖ SELF–CHECK QUESTION

Name two short-term signals that are appetite-suppressing (anorexigenic) and one that is appetite-stimulating (orexigenic).

Leptin is one of several hormones secreted by adipose tissue

Leptin was the first adipokine discovered because of the dramatic effects of its absence. Researchers discovered a strain of mice called ob/ob mice, which completely lack leptin and, as a result, are extremely obese. These mice display hyperphagia (overeating), hyperlipidemia (accumulation of triacylglycerols in muscle and liver), and an insensitivity to insulin. Since the discovery of leptin, other adipokines have been detected. For instance, adiponectin is another signal molecule produced by the adipocytes; but, unlike leptin, secretion of adiponectin increases in proportion to decreases in fat mass.

A key function of adiponectin appears to be increasing the sensitivity of the organism to insulin. Both leptin and adiponectin exert their effects through the key regulatory enzyme, AMP-activated protein kinase (AMPK) (**Figure 24.4**). Recall that this enzyme is active when AMP levels are elevated and ATP levels are diminished, and this activation leads to a decrease in anabolism and an increase in catabolism, most notably an increase in fatty acid oxidation (Section 22.7). In insulin-resistant obese animals, leptin levels increase while those of adiponectin decrease.

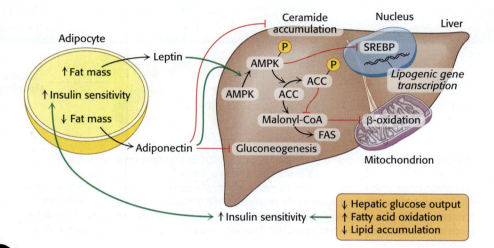

FIGURE 24.4 Adipokines help to maintain systemic lipid and glucose homeostasis. Both leptin and adiponectin act to prevent lipid accumulation in the liver and stimulate fatty acid oxidation by activating AMPK, which decreases the expression of lipogenic transcription factors such as SREBP and inactivates acetyl CoA carboxylase 1 (ACC). Inactivating ACC reduces the concentration of malonyl CoA, which normally inhibits oxidation and stimulates lipogenesis. Adiponectin also inhibits gluconeogenesis and reduces ceramide accumulation, which would otherwise further reduce insulin action. The combined action of leptin and adiponectin on the liver results in decreased glucose output, decreased lipid accumulation, and increased fatty acid oxidation in the liver, all of which enhance insulin sensitivity for the liver and other tissues.

Adipocytes produce two additional hormones, RBP4 (originally discovered as a retinol binding protein) and resistin, that promote insulin resistance. Although it is unclear why adipocytes secrete hormones that facilitate insulin resistance, a pathological condition, we can speculate on the reason. These signal molecules may help to fine-tune the actions of leptin and adiponectin or perhaps to act as "brakes" on the action of leptin and adiponectin to prevent hypoglycemia in the fasted state. Some evidence indicates that enlarged adipocytes that result from obesity may secrete higher levels of insulin-antagonizing hormones and thus contribute to insulin resistance. Resistin has recently been implicated as a causal factor in the increased incidence of cardiovascular disease associated with obesity.

Leptin resistance may be a contributing factor to obesity

If leptin is produced in proportion to body-fat mass and leptin inhibits eating, why do people become obese? Obese people, in most cases, have both functioning leptin receptors and a high blood concentration of leptin. The failure to respond to the anorexigenic effects of leptin is called *leptin resistance*. What is the basis of leptin resistance?

As for most questions in the exciting area of energy homeostasis, the answer is not well worked out, but recent evidence suggests that a group of proteins called *suppressors of cytokine signaling* (SOCS) may take part. These proteins fine-tune some hormonal systems by inhibiting receptor action. SOCS proteins inhibit receptor signaling by a number of means. Consider, for example, the effect of SOCS proteins on the insulin receptor. Recall that insulin stimulates the autophosphorylation of tyrosine residues on the insulin receptor, which in turn phosphorylates IRS-1, initiating the insulin-signaling

(A)

(B)

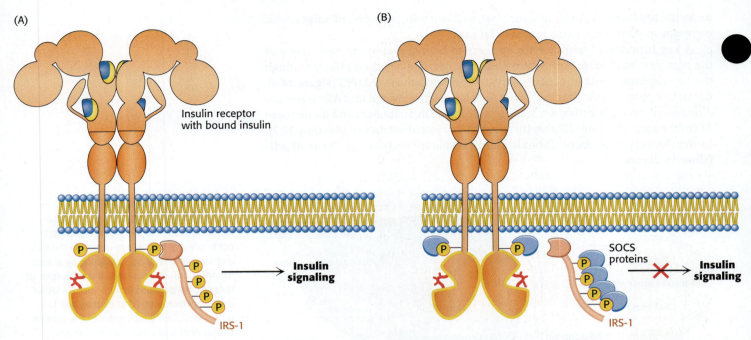

Insulin receptor with bound insulin

Insulin signaling

SOCS proteins

Insulin signaling

IRS-1

IRS-1

FIGURE 24.5 Suppressors of cytokine signaling (SOCS) regulate insulin receptor function. (A) Insulin binding results in phosphorylation of the receptor and subsequent phosphorylation of IRS-1. These processes initiate the insulin-signaling pathway. (B) SOCS proteins disrupt interactions of components of the insulin-signaling pathway by binding phosphorylated proteins and thereby inhibiting the pathway. The binding of a signal component by SOCS results in proteasomal degradation in some cases. (IRS-1, insulin-receptor substrate 1; SOCS, suppressor of cytokine signaling.)

pathway (**Figure 24.5A**). SOCS proteins bind to phosphorylated tyrosine residues on receptors or other members of the signal-transduction pathway, thereby disrupting signal flow and thus altering the cell's biochemical activity (**Figure 24.5B**). In other cases, the binding of SOCS proteins to components of the signal-transduction pathway may also enhance proteolytic degradation of these components by the proteasome (Section 23.2).

Evidence in support of a role for SOCS in leptin resistance comes from mice that have had SOCS selectively deleted from POMC-expressing neurons. These mice display an enhanced sensitivity to leptin and are resistant to weight gain even when fed a high-fat diet. The reason why the activity of SOCS proteins increases, leading to leptin resistance, remains to be determined. Alternatively, leptin resistance can arise by means unrelated to SOCS. For example, to have systemic effects, leptin must enter the brain. Mutations in proteins that transport leptin into the brain or in the brain leptin receptors could also result in leptin resistance.

24.2 The Fasted–Fed Cycle Is a Response to Eating and Sleeping Behaviors

Next, let's examine the natural cycles of human metabolism in a resting and non-pathological state. We begin with a physiological condition called the **fasted–fed cycle**, which we all experience in the hours after an evening meal and through the night's fast. This nightly fasted–fed cycle has three stages: the well-fed (post-prandial) state after a meal, the early fasting (postabsorptive) state during the night, and the refed state after breakfast. A major goal of the many biochemical alterations in this period is to maintain **glucose homeostasis** — that is, a constant blood-glucose concentration. Maintaining glucose homeostasis is crucial because glucose is normally the only fuel source for the brain. The two primary signals regulating the fasted–fed cycle are insulin and glucagon.

The postprandial state follows a meal

After we consume and digest a meal, glucose and amino acids are transported from the intestine to the blood. The dietary lipids are packaged into chylomicrons and transported to the blood by the lymphatic system. This fed condition leads to the secretion of insulin, which signals the postprandial state. Insulin stimulates glycogen synthesis in both muscle and the liver, suppresses gluconeogenesis by the liver, and stimulates protein synthesis. Insulin also accelerates glycolysis in the liver, which in turn increases the synthesis of fatty acids.

The liver helps to limit the amount of glucose in the blood during times of plenty by storing it as glycogen, so as to be able to release glucose in times of scarcity. How is the excess blood glucose present after a meal removed? The liver is able to trap large quantities of glucose because it possesses an isozyme of hexokinase called *glucokinase* that converts glucose into glucose 6-phosphate, which cannot be transported out of the cell. Recall that glucokinase has a high K_M value and is thus highly active only when blood-glucose concentration is high (Section 16.3). Furthermore, glucokinase is not inhibited by glucose 6-phosphate as hexokinase is. Consequently, the liver forms glucose 6-phosphate more rapidly as the blood-glucose concentration rises. The increase in glucose 6-phosphate, which activates glycogen synthase, coupled with the effects of insulin leads to a buildup of glycogen stores.

The hormonal effects on glycogen synthesis and storage are reinforced by a direct action of glucose itself. Phosphorylase *a* is a glucose sensor in addition to being the enzyme that cleaves glycogen. When the glucose level is high, the binding of glucose to phosphorylase *a* renders the enzyme susceptible to the action of a phosphatase that converts it into phosphorylase *b*, which does not readily degrade glycogen (Figure 21.24). Thus, glucose allosterically shifts the glycogen system from a degradative to a synthetic mode.

The high insulin level in the fed state also promotes the entry of glucose into muscle and adipose tissue. Insulin stimulates the synthesis of glycogen by muscle as well as by the liver. The entry of glucose into adipose tissue provides glycerol 3-phosphate for the synthesis of triacylglycerols. The action of insulin also extends to amino acid and protein metabolism. Insulin promotes the uptake of branched-chain amino acids (valine, leucine, and isoleucine) by muscle. Indeed, insulin has a general stimulating effect on protein synthesis, which favors a building up of muscle protein. In addition, it inhibits the intracellular degradation of proteins.

The postabsorptive state occurs at the beginning of a fast

The blood-glucose concentration begins to drop several hours after a meal, leading to a decrease in insulin secretion and a rise in glucagon secretion by the α cells of the pancreas. This period is called the *fasted state* or *postabsorptive state* because it immediately follows the absorption of glucose from the blood that resulted from the previous meal. The regulation of glucagon secretion is poorly understood, but when glucose is abundant, β cells inhibit glucagon secretion. When glucose concentration falls, the inhibition is relieved, and glucagon is secreted by the α cells of the pancreas. Just as insulin signals the fed state, glucagon signals the fasted state, serving to mobilize glycogen stores when there is no dietary intake of glucose.

The main target organ of glucagon is the liver. Glucagon stimulates glycogen breakdown and inhibits glycogen synthesis by triggering the cyclic AMP cascade;

this leads to the phosphorylation and activation of phosphorylase and the inhibition of glycogen synthase (Section 21.6). Glucagon also inhibits fatty acid synthesis by diminishing the production of pyruvate and by maintaining the inactive phosphorylated state of acetyl CoA carboxylase. In addition, glucagon stimulates the gluconeogenic state in the liver and blocks glycolysis by decreasing in the amount of fructose 2,6-bisphosphate (F-2,6-BP), which normally stimulates glycolysis and inhibits gluconeogenesis (Section 16.5). Hence, the decrease in F-2,6-BP results in a decrease in glycolytic activity and an increase in gluconeogenesis because of the opposing effects of F-2,6-BP on phosphofructokinase and fructose-1,6-bisphosphatase (Section 16.5; **Figure 24.6**).

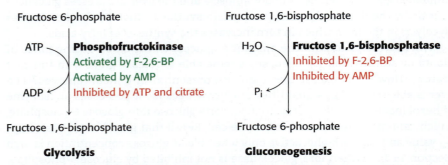

FIGURE 24.6 Glycolysis and gluconeogenesis are reciprocally regulated. Phosphofructokinase is the key enzyme in the regulation of glycolysis, whereas fructose 1,6-bisphosphatase is the principal enzyme controlling the rate of gluconeogenesis. Note the reciprocal relationship between the pathways and the signal molecules.

EXPLORE these pathways further in the Metabolic Map in
≈ Achieve

All known actions of glucagon are mediated by protein kinases that are activated by cyclic AMP. The activation of the cyclic AMP cascade results in a higher level of phosphorylase *a* activity and a lower level of glycogen synthase *a* activity. Glucagon's effect on this cascade is reinforced by the low concentration of glucose in the blood. The diminished binding of glucose to phosphorylase *a* makes the enzyme less susceptible to the hydrolytic action of the phosphatase. Instead, the phosphatase remains bound to phosphorylase *a*, and so the synthase stays in the inactive phosphorylated form. Consequently, there is a rapid mobilization of glycogen.

The large amount of glucose formed by the hydrolysis of glucose 6-phosphate derived from glycogen is then released from the liver into the blood. The entry of glucose into muscle and adipose tissue decreases in response to a low insulin level. The diminished utilization of glucose by muscle and adipose tissue also contributes to the maintenance of the blood-glucose concentration. The net result of these actions of glucagon is to markedly increase the release of glucose by the liver. Both muscle and the liver use fatty acids as fuel when the blood-glucose concentration drops, saving the glucose for use by the brain and red blood cells. Thus, the blood-glucose concentration is kept at or above 4.4 mM by three major factors: (1) the mobilization of glycogen and the release of glucose by the liver, (2) the release of fatty acids by adipose tissue, and (3) the shift in the fuel used by muscle from glucose to fatty acids.

What is the result of the depletion of the liver's glycogen stores? Gluconeogenesis from lactate and alanine continues, but this process merely replaces glucose that had already been converted into lactate and alanine by tissues such as muscle and red blood cells. Moreover, the brain oxidizes glucose completely to CO_2 and H_2O. Thus, for the net synthesis of glucose to take place, another source of carbon is required. Glycerol released from adipose tissue on lipolysis provides some of the carbon atoms, with the remaining carbon atoms coming from the hydrolysis of muscle proteins.

The refed state occurs at the end of a long fast

What are the biochemical responses to a large breakfast? Fat is processed exactly as it is processed in the normal fed state. However, that is not the case for glucose. The difference between the refed state and other postprandial states is the length of the fast. In the case of the first morning meal or any other meal after a long fast, the liver does not initially absorb glucose from the blood, but instead leaves it for the other tissues. Moreover, the liver remains in a gluconeogenic mode. Now, however, the newly synthesized glucose is used to replenish the liver's glycogen stores. As the blood-glucose concentration continues to rise, the liver completes the replenishment of its glycogen stores and begins to process the remaining excess glucose for fatty acid synthesis.

❖ SELF–CHECK QUESTION

What are the three primary sources of glucose 6-phosphate in liver cells?

24.3 Diabetes Is a Common Metabolic Disease Often Resulting from Obesity

Having considered the regulation of body weight and glucose homeostasis, we now examine the biochemical results when regulation fails because of behavior, genetics, or a combination of the two. The most common result of such a failure is obesity. Humans maintain about a day's worth of glycogen and, after these stores have been replenished, excess carbohydrates are converted into fatty acids and then into triacylglycerols. Amino acids are not stored at all, and so excess amino acids are also ultimately converted into triacylglycerols. Thus, regardless of the type of food consumed, excess consumption results in increased fat stores.

We begin our consideration of the effects of disruptions in caloric homeostasis by looking at diabetes mellitus, a complex disease characterized by grossly abnormal fuel usage and named for the excessive urination that is a classic symptom (*diabetes*, from Greek, means "to pass through"). Glucose is overproduced by the liver and underutilized by other organs. As a result, glucose accumulates in the blood and is excreted in the urine when its concentration exceeds the reabsorptive capacity of the renal tubules (*mellitus*, from Latin, means "sweetened with honey"). Water accompanies the excreted glucose, and so an untreated diabetic in the acute phase of the disease is hungry and thirsty. *Mellitus* distinguishes this disease from diabetes *insipidus*, which is caused by impaired renal reabsorption of water. While diabetes insipidus is relatively rare, the incidence of diabetes mellitus is about 6% of the world population, hence it is usually referred to simply as *diabetes*.

Type 1 diabetes is caused by the destruction of the insulin-secreting β cells in the pancreas, usually due to an autoimmune disorder and beginning before age 20. Type 1 diabetes is also referred to as *insulin-dependent diabetes*, meaning that the affected person requires the administration of insulin to live. Most diabetics, in contrast, have a normal or even higher level of insulin in their blood, but they are poorly responsive to the hormone, a characteristic called **insulin resistance**. This form of the disease, known as **type 2 diabetes**, typically arises later in life than does the insulin-dependent form. Although there is a significant genetic component, obesity is a significant predisposing factor for the development of insulin resistance and type 2 diabetes.

In our discussion of diabetes in this chapter, we will consider type 2 first, which accounts for approximately 90% of the diabetes cases throughout the world and is the most common metabolic disease in the world. In the United States, where 10% of the population is affected, it is the leading cause of blindness, kidney failure, and amputation.

❖ **SELF–CHECK QUESTION**

People who consume little fat but excess carbohydrates can still become obese. How is this result possible?

Insulin initiates a complex signal-transduction pathway in muscle

What is the biochemical basis of insulin resistance? How does insulin resistance lead to failure of the pancreatic β cells that results in type 2 diabetes? How does obesity contribute to this progression? To answer these questions and begin to unravel the mysteries of metabolic disorders, let us reexamine the mechanism of action of insulin in muscle, the largest tissue target of insulin. Indeed, muscle uses about 85% of the glucose ingested during a meal.

In a normal cell, insulin binds to its receptor, which then autophosphorylates on tyrosine residues, with each subunit of the receptor phosphorylating its partner. Phosphorylation of the receptor generates binding sites for insulin-receptor substrates (IRSs), such as IRS-1 (**Figure 24.7**). Subsequent phosphorylation of IRS-1 by the tyrosine kinase activity of the insulin receptor engages the insulin-signaling pathway (1). Phosphorylated IRS-1 binds to phosphoinositide 3-kinase (PI3K) and activates it. PI3K catalyzes the conversion of phosphatidylinositol 4,5-bisphosphate (PIP_2) into phosphatidylinositol 3,4,5-trisphosphate (PIP_3), a second messenger (2). PIP_3 activates the phosphatidylinositol-dependent protein kinase (PDK) (3), which in turn activates several other kinases, most notably Akt (4), also known as protein kinase B (PKB).

Akt facilitates the translocation of the glucose transporter (GLUT4)-containing vesicles to the cell membrane, which leads to a more robust absorption of glucose from the blood. Moreover, Akt phosphorylates and inhibits glycogen synthase kinase (GSK3). Recall that GSK3 inhibits glycogen synthase (Section 21.6). Thus, insulin leads to the absorption of glucose from the blood, activation of glycogen synthase, and enhanced glycogen synthesis.

FIGURE 24.7 Insulin signaling activates Akt using the second messenger PIP₃. The binding of insulin results in the cross-phosphorylation and activation of the insulin receptor. Phosphorylated sites on the receptor act as binding sites for insulin-receptor substrates such as IRS-1 (1). The lipid kinase phosphoinositide 3-kinase binds to phosphorylated sites on IRS-1 through its regulatory domain and then converts PIP_2 into PIP_3 (2). Binding to PIP_3 activates PIP_3-dependent protein kinase (3), which phosphorylates and activates kinases such as Akt (4). Activated Akt can then diffuse throughout the cell and continue the signal-transduction pathway.

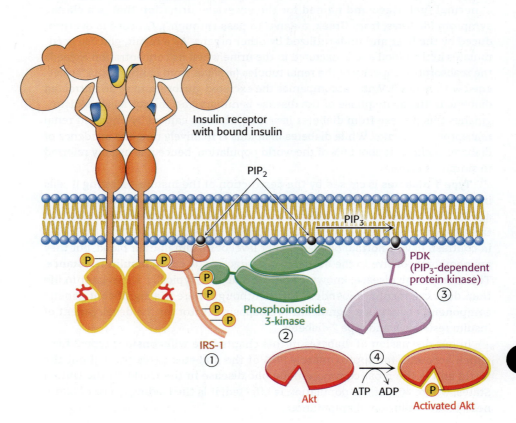

Like all signal pathways, the insulin-signaling cascade must be capable of being turned off. In addition to the breakdown of insulin itself, three different processes notably contribute to the down-regulation of insulin signaling. First, phosphatases deactivate the insulin receptor and destroy a key second messenger. PTP1B (protein tyrosine phosphatase 1B) removes phosphoryl groups from the receptor, thus inactivating it. The second messenger PIP$_3$ is inactivated by the phosphatase PTEN (phosphatase and tensin homolog), which dephosphorylates it, forming PIP$_2$, which is not a second messenger.

Second, the IRS protein can be inactivated by phosphorylation on serine residues by specific Ser/Thr kinases. These kinases are activated by stress signals and may play a role in the development of insulin resistance. Finally, SOCS proteins, the regulatory proteins discussed earlier, interact with the insulin receptor and IRS-1 and apparently facilitate their proteolytic degradation by the proteasome complex.

Metabolic syndrome often precedes type 2 diabetes

With our knowledge of the key components of energy homeostasis, let us begin our investigation of the biochemical basis of insulin resistance and type 2 diabetes. A cluster of pathologies—including insulin resistance, hyperglycemia, and dyslipidemia (characterized by high blood levels of triacylglycerols, cholesterol, and low-density lipoproteins)—often develop together. This clustering, called **metabolic syndrome**, is thought to be a predecessor of type 2 diabetes.

A consequence of obesity is that the amount of triacylglycerols consumed exceeds the adipose tissue's storage capacity. As a result, other tissues begin to accumulate fat, most notably muscle (myosteatosis), and the liver, a condition called *hepatic steatosis* (**Figure 24.8**). This accumulation results in insulin resistance and ultimately in pancreatic failure. We will focus on muscle and the pancreatic β cells.

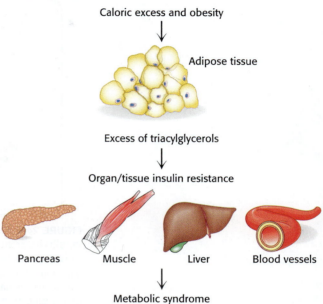

FIGURE 24.8 The storage capacity of fat tissue can be exceeded in obesity. In caloric excess, the storage capacity of adipocytes can be exceeded with deleterious results. The excess fat accumulates in other tissues, resulting in biochemical malfunction of the tissues. When the pancreas, muscle, liver, and cells lining the blood vessels are affected, metabolic syndrome, a condition that often precedes type 2 diabetes, may result. [Information from S. Fröjdö, H. Vidal, and L. Pirola, *Biochim. Biophys. Acta* 1792:83–92, 2009, Fig. 1.]

❖ SELF–CHECK QUESTION

What would be the effect of a mutation in the gene for PTP1B (protein tyrosine phosphatase 1B) that inactivated the enzyme in a person who has type 2 diabetes?

Excess fatty acids in muscle modify metabolism

Myosteatosis and insulin resistance do not always develop with obesity; however, the pathology begins when more fats are present than can be processed by muscle. Although the rate of β oxidation increases in response to the high concentration of fats, the mitochondria are not capable of processing all of the fatty acids by β oxidation; fatty acids accumulate in the mitochondria and eventually spill over into the cytoplasm. The unprocessed fatty acids are reincorporated into triacylglycerols, resulting in the accumulation of fat in the cytoplasm.

In the cytoplasm, levels of diacylglycerol and ceramide (a component of sphingolipids) also increase. Diacylglycerol is a second messenger that activates protein kinase C (PKC) (Section 14.2). When active, PKC and other Ser/Thr protein kinases are capable of phosphorylating IRS and reducing the ability of IRS to propagate the insulin signal. Saturated and trans unsaturated fatty acids may also activate kinases that block the insulin signal. Ceramide or its metabolites

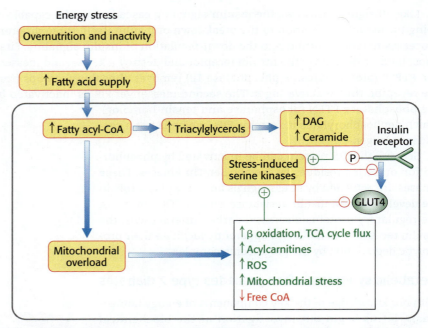

FIGURE 24.9 Excess fat can disrupt signal-transduction pathways and lead to insulin insensitivity. Excess fat accumulation in peripheral tissues, most notably muscle, can disrupt some signal-transduction pathways and inappropriately activate others. In particular, diacylglycerols and ceramide activate stress-induced pathways that interfere with insulin signaling, resulting in insulin resistance. (Abbreviations: DAG, diacylglycerol; GLUT4, glucose transporter 4; TCA, tricarboxylic acid cycle; ROS, reactive oxygen species.)

inhibit glucose uptake and glycogen synthesis, apparently by inhibiting PDK and Akt (Section 14.3). The result is a diet-induced insulin resistance that can, but does not always, develop over time in obese individuals (**Figure 24.9**).

Insulin resistance in muscle facilitates pancreatic failure

What is the effect of overnutrition on the pancreas? This question is important because a primary function of the pancreas is to respond to the presence of glucose in the blood by secreting insulin. Indeed, the β cell is an insulin factory. The mRNA that encodes proinsulin, the precursor to insulin, constitutes 20% of the total mRNA in the pancreas, and proinsulin itself constitutes 50% of the total protein synthesized in the pancreas.

Glucose enters the pancreatic β cells through the glucose transporter GLUT2. Recall that GLUT2 will allow rapid glucose transport only when blood glucose is plentiful, ensuring that insulin is secreted only when glucose is abundant, such as after a meal (Section 16.1). Likewise, pancreatic glucokinase, with its relatively high K_M value, traps large amounts of glucose within the cell only when blood-glucose concentrations are high. These two proteins, in concert, act as a pancreatic glucose sensor.

The β cell metabolizes glucose to CO_2 and H_2O through cellular respiration, generating ATP. The resulting increase in the ATP/ADP ratio closes an ATP-sensitive K^+ channel that, when open, allows potassium to flow out of the cell (**Figure 24.10**). The resulting alteration in the cellular ionic environment opens a voltage-sensitive Ca^{2+} channel, and the resulting influx of Ca^{2+} causes insulin-containing secretory vesicles to fuse with the cell membrane and release insulin into the blood. Thus, the increase in ATP resulting from the metabolism of glucose has been translated by the membrane proteins into a physiological response—the secretion of insulin and the subsequent removal of glucose from the blood.

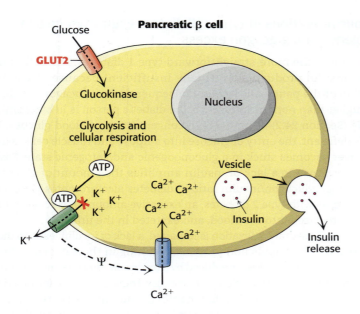

Pancreatic β cell

Glucose

GLUT2

Glucokinase

Nucleus

Glycolysis and cellular respiration

ATP

ATP

K⁺ K⁺

K⁺

K⁺

Ca²⁺ Ca²⁺

Ca²⁺ Ca²⁺

Ca²⁺

Vesicle

Insulin

Insulin release

Ψ

Ca²⁺

FIGURE 24.10 Insulin release by the pancreas is regulated by blood-glucose concentration, using ATP as a second messenger. The metabolism of glucose by glycolysis and cellular respiration increases the concentration of ATP, which causes an ATP-sensitive potassium channel to close. The closure of this channel alters the charge across the membrane (ψ) and causes a calcium channel to open. The influx of calcium causes insulin-containing granules to fuse with the plasma membrane, releasing insulin into the blood.

What aspect of β cell function ultimately fails, causing the transition from insulin resistance to full-fledged type 2 diabetes? Under normal circumstances, proinsulin synthesized by β cells folds into its three-dimensional structure in the endoplasmic reticulum, is processed to insulin, and is subsequently packaged into vesicles for secretion. As insulin resistance develops in the muscle, the β cells respond by synthesizing yet more insulin in a futile attempt to drive insulin action. The ability of the endoplasmic reticulum to process all of the proinsulin and insulin becomes compromised, a condition known as **endoplasmic reticulum (ER) stress**, and unfolded or misfolded proteins accumulate.

ER stress initiates a signal pathway called the **unfolded protein response (UPR)**, a pathway intended to save the cell. UPR consists of several steps. First, general protein synthesis is inhibited so as to prevent more proteins from entering the ER. Second, chaperone synthesis is stimulated. Recall that chaperones are proteins that assist the folding of other proteins (Section 2.7). Third, misfolded proteins are removed from the ER and are subsequently delivered to the proteasome for destruction. Finally, if the described response fails to alleviate the ER stress, programmed cell death is triggered, which ultimately leads to β cell death and full-fledged type 2 diabetes.

What are the treatments for type 2 diabetes? Most are behavioral in nature. Type 2 diabetics are advised to count calories, making sure that energy intake does not exceed energy output; to consume a diet rich in vegetables, fruits, and grains; and to get plenty of aerobic exercise. Note that these guidelines are the same as those for healthy living, even for those not suffering from type 2 diabetes. Monitoring of blood-glucose concentration is important (normal is 3.6 to 6.1 mM) and for patients who are not able to maintain proper glucose levels with behavioral modifications, drug treatments are required. The administration of insulin may be necessary upon β cell failure, and/or treatment with metformin (Glucophage), which activates AMPK, may be effective. Recall that AMPK promotes the oxidation of fats while inhibiting fat synthesis and storage. It also stimulates glucose uptake and storage by muscle while inhibiting gluconeogenesis in the liver.

❖ **SELF–CHECK QUESTION**

What is the relationship between fatty acid oxidation and insulin resistance in the muscle?

Metabolic alterations in type 1 diabetes result from insulin insufficiency and glucagon excess

We now turn to the more-straightforward type 1 diabetes. Recall that in type 1 diabetes, insulin production is usually insufficient because of autoimmune destruction of the pancreatic β cells. Consequently, the glucagon/insulin ratio is always higher than normal. In essence, the diabetic person is in biochemical fasting mode (Section 24.2) despite a high concentration of blood glucose. Because insulin is deficient, the entry of glucose into adipose and muscle cells is impaired, and the liver becomes stuck in a gluconeogenic and ketogenic state. Essentially, the cells' response to a lack of insulin amplifies the amount of glucose in the blood. The high glucagon/insulin ratio in diabetes also promotes glycogen breakdown. Hence, as with type 2 diabetes, an excessive amount of glucose is produced by the liver, released into the blood, and excreted in the urine.

Because carbohydrate utilization is impaired, a lack of insulin leads to the uncontrolled breakdown of lipids and proteins, resulting in the ketogenic state. Large amounts of acetyl CoA are then produced by β oxidation. However, much of the acetyl CoA cannot enter the citric acid cycle because there is insufficient oxaloacetate for the condensation step. Recall that mammals can synthesize oxaloacetate from pyruvate, a product of glycolysis, but not from acetyl CoA; instead, they generate ketone bodies (Section 22.4). A striking feature of diabetes is the shift in fuel usage from carbohydrates to fats; glucose, more abundant than ever, is unable to be used. In high concentrations, ketone bodies overwhelm the kidney's capacity to maintain acid–base balance. The untreated diabetic can go into a coma because of a lowered blood-pH level and dehydration. Interestingly, diabetic ketosis is rarely a problem in type 2 diabetes because, in most cases, insulin signaling is still active enough to prevent excessive lipolysis in liver and adipose tissue.

What is the treatment for type 1 diabetes? As noted earlier, the regular administration of insulin is required for survival. Likewise, blood-glucose concentration must be monitored. Although the behaviors recommended for type 2 diabetes—watching calories, exercising, and eating a healthy diet—do not treat type 1 diabetes, they are beneficial to managing it and to maintaining a healthy lifestyle.

❖ SELF–CHECK QUESTION

Insulin-dependent diabetes is often accompanied by hypertriglyceridemia, which is an excess blood level of triacylglycerols. Suggest a biochemical explanation.

24.4 Exercise Beneficially Alters the Biochemistry of Cells

Thus far, we have been considering metabolism in the context of excess calorie consumption, as in obesity. We now look at the physiological response when there is an extreme caloric need, as occurs during exercise.

Fuel choice during exercise is determined by the intensity and duration of activity

Skeletal muscle accounts for ≈40% of total body mass and ≈35% of resting metabolic activity. Moreover, skeletal muscle is the largest target tissue for insulin. Given its biochemical prominence, increased muscle activity—exercise—coupled with a healthy diet is one of the most effective treatments for diabetes as well as a host of other pathological conditions. These include coronary disease, hypertension, depression, age-related muscle wasting (sarcopenia), and a variety of cancers. With regard to type 2 diabetes, exercise increases insulin sensitivity and is in fact more effective than pharmacological interventions. What is the basis of this beneficial effect? To answer this question, we must first consider what energy sources muscles use during exercise.

The fuels used in anaerobic exercises—sprinting, for example—differ from those used in aerobic exercises—such as distance running. The selection of fuels during these different forms of exercise illustrates many important facets of energy transduction and metabolic integration. ATP directly powers myosin, the protein immediately responsible for converting chemical energy into movement (Section 6.4). However, the amount of ATP in muscle is small. Hence, the power output and, in turn, the velocity of running depend on the rate of ATP production from other fuels.

As shown in **Table 24.3**, creatine phosphate (phosphocreatine) can swiftly transfer its high-potential phosphoryl group to ADP to generate ATP. However, the amount of creatine phosphate, like that of ATP itself, is limited. Creatine phosphate and ATP can power intense muscle contraction for 5 to 6 s. Maximum speed in a sprint can thus be maintained for only 5 to 6 s (Figure 15.6). Thus, the winner in a 100-meter sprint is the runner who both achieves the highest initial velocity and then slows down the least.

Creatine phosphate + ADP

Creatine kinase

Creatine + ATP

TABLE 24.3 Fuel sources for muscle contraction

Fuel source	Maximal rate of ATP production (mmol s^{-1})	Total ~P available (mmol)
Muscle ATP		223
Creatine phosphate	73.3	446
Conversion of muscle glycogen into lactate	39.1	6700
Conversion of muscle glycogen into CO_2	16.7	84,000
Conversion of liver glycogen into CO_2	6.2	19,000
Conversion of adipose-tissue fatty acids into CO_2	6.7	4,000,000

Note: Fuels stored are estimated for a 70-kg person having a muscle mass of 28 kg.

Data from E. Hultman and R. C. Harris, in *Principles of Exercise Biochemistry*, edited by J. R. Poortmans (Karger, 2004), pp. 78–119.

Sprinting. A 100-meter sprint is powered by stored ATP, creatine phosphate, and the anaerobic glycolysis of muscle glycogen. During a ~10-second sprint, the concentration of ATP in muscle drops from 5.2 to 3.7 mM, and that of creatine phosphate decreases from 9.1 to 2.6 mM. Anaerobic glycolysis provides fuel to make up for the loss of ATP and creatine phosphate. The conversion of muscle glycogen into lactate can generate a good deal more ATP, but the rate is slower than that of phosphoryl-group transfer from creatine phosphate. Because of anaerobic glycolysis, the blood-lactate concentration is increased from 1.6 to 8.3 mM. The release of H$^+$ from the intensely active muscle concomitantly lowers the blood pH from 7.42 to 7.24. Sprinting is powered by fast-twitch muscle fiber (type IIb) specialized for anaerobic glycolysis. One of the effects of high-intensity training is to increase the amount of lactate transporters in the membranes of slow-twitch fibers (type I), which remove lactate from the blood and slow the fall in blood pH. Nonetheless, the pace of a 100-meter sprint cannot be sustained in a 1000-meter run (~132 s) for two reasons. First, creatine phosphate is consumed within a few seconds. Second, the lactate produced contributes to acidosis. Thus, alternative fuel sources are needed.

Middle-distance running. Let's examine the fuel use of a middle-distance runner. The complete oxidation of muscle glycogen to CO_2 by aerobic respiration substantially increases the energy available to power the 1000-meter run, but this aerobic process is a good deal slower than anaerobic glycolysis. Because ATP is produced more slowly by oxidative phosphorylation than by glycolysis

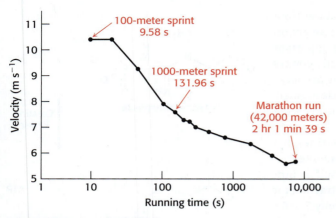

FIGURE 24.11 Running velocity decreases dramatically as a function of the duration of the race. The values shown are world track records. [Data from trackandfieldnews.com.]

FIGURE 24.12 An idealized representation of fuel use and RQ illustrates major changes as a function of aerobic exercise intensity. (A) With increased exercise intensity, the use of fats as fuels falls as the utilization of carbohydrates increases. (B) The respiratory quotient (RQ) measures the alteration in fuel use.

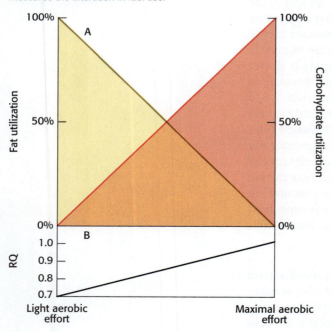

(Table 24.3), the runner's pace is necessarily slower than in a 100-meter sprint. The championship velocity for the 1000-meter run is about 7.6 m s^{-1}, compared with approximately 10.4 m s^{-1} for the 100-meter event (**Figure 24.11**).

Long-distance running. Running a marathon (26.2 miles or 42,000 meters) requires a different selection of fuels and is characterized by cooperation (depending on the capabilities of the runner) between muscle, liver, and adipose tissue. Liver glycogen complements muscle glycogen as an energy store that can be tapped. Elite marathoners can use the aerobic combustion of glucose to power the entire race, provided they consume fuels during the race. However, for most of us, the total body glycogen stores (103 mol of ATP at best) are insufficient to provide the 150 mol of ATP needed for this grueling event. Much larger quantities of ATP can be obtained by the oxidation of fatty acids derived from the breakdown of fat in adipose tissue, but the maximal rate of ATP generation is slower yet than that of glycogen oxidation and is more than 10-fold slower than that with creatine phosphate. Thus, ATP is generated much more slowly from high-capacity stores than from limited ones, accounting for the different velocities of anaerobic and aerobic events. Fats are rapidly consumed in activities such as distance running, explaining why extended aerobic exercise is beneficial for people who are insulin resistant.

Is it possible to determine the contribution of each fuel as a function of exercise intensity? The percentage contribution of each fuel can be measured with the use of a respirometer, which measures the **respiratory quotient (RQ)**, the ratio of CO_2 produced to O_2 consumed. Consider the complete combustion of glucose:

$$\underset{\text{Glucose}}{C_6H_{12}O_6} + 6\,O_2 \longrightarrow 6\,CO_2 + 6\,H_2O$$

The RQ for glucose is 1. Now consider the oxidation of a typical fatty acid, palmitate:

$$\underset{\text{Palmitate}}{C_{16}H_{32}O_2} + 23\,O_2 \longrightarrow 16\,CO_2 + 16\,H_2O$$

The RQ for palmitate oxidation is 0.7. Thus, as aerobic exercise intensity increases, the RQ will rise from 0.7 (only fats are used as fuel) to 1.0 (only glucose is used as fuel). Between these values, a mixture of fuels is used (**Figure 24.12**).

What is the optimal mix of fuels for use during a marathon? As suggested above, this is a complex question that varies with the athlete and level of training. Studies have shown that, when muscle glycogen has been depleted, the power output of the muscle falls to approximately 50% of maximum. Power output decreases despite the fact that ample supplies of fat are available, suggesting that fats can supply only about 50% of maximal aerobic effort. Indeed, depletion of glycogen stores during a race is referred to as "bonking" or "hitting the wall," with the result that the athlete must greatly reduce the pace.

How is an optimal mix of these fuels achieved? A low blood-sugar level leads to a high glucagon/insulin ratio, which in turn mobilizes fatty acids from adipose tissue. Fatty acids readily enter muscle, where they are degraded by β oxidation to acetyl CoA and then to CO_2. The elevated acetyl CoA level decreases the activity of the pyruvate dehydrogenase complex to block the conversion of pyruvate into acetyl CoA. Hence, fatty acid oxidation decreases the funneling of glucose into the citric acid cycle and oxidative phosphorylation. Glucose is spared so that just enough remains available at the end of the marathon to increase the pace as the finish line draws near. The simultaneous use of both fuels gives a higher mean velocity than would be attained if glycogen were totally consumed before the start of fatty acid oxidation. It is important to bear in mind that fuel use is only one of many factors that determine running ability.

The perplexing symptoms of McArdle disease result from the distinct ways skeletal muscle produces ATP

In Section 21.6 we discussed several glycogen storage diseases. We can now explain the complex symptoms of McArdle disease (Type V glycogen storage disease), a rare condition affecting approximately 1 in 100,000 people. Affected individuals experience painful muscle fatigue and have burgundy-colored urine following an attempt at vigorous exercise. The coloration is due to rhabdomyolysis (rapid breakdown of skeletal muscle), which results in myoglobinuria (myoglobin in the blood and urine) as damaged tissue leaks myoglobin into the blood, which is passed to the urine. Two additional observations are interesting: resting and moderate exercise do not induce the symptoms, and vigorous exercise causes an increase in blood pH, rather than the usually observed decrease in pH. What is the biochemical basis for these symptoms?

As discussed in the previous section, during vigorous exercise, skeletal muscle contractions are powered by glycogen mobilization by glycogen phosphorylase. Patients with McArdle disease lack functional muscle glycogen phosphorylase, leading to the inability to rapidly mobilize glucose from glycogen. In an unaffected individual, intracellular and blood pH fall as lactic acid is produced during anaerobic glycolysis (Section 16.2). Because McArdle patients cannot mobilize glucose, lactic acid will not be produced and there will be no fall in pH; rather, it rises. To explain this, recall again that an immediate source of ATP in skeletal muscle is the phosphorylation of ATP at the expense of creatine phosphate (Section 15.2). In a futile attempt to power exercise, skeletal muscles of an affected individual rapidly synthesize ATP at the expense of creatine phosphate. However, the guanidinium group of creatine is a base, leading to an increase in pH.

Finally, why don't affected individuals experience symptoms at rest or when performing leisurely exercise? As was discussed previously in this chapter, under these conditions, skeletal muscle derives most of its energy from fatty acid oxidation rather than from glycogen mobilization.

Mitochondrial biogenesis is stimulated by muscular activity

Hopefully it is now apparent that muscles must utilize a variety of fuel sources for optimal activity. Now let's consider what biochemical changes exercise induces to optimize athletic performance.

When muscle is stimulated to contract during exercise by receiving nerve impulses from motor neurons, calcium is released from the sarcoplasmic reticulum (**Figure 24.13**). Calcium induces muscle contraction, but recall that calcium is also a potent second messenger and frequently works in association with the calcium-binding protein calmodulin (Section 14.1). In its capacity as a second messenger, calcium stimulates various calcium-dependent enzymes, such as calmodulin-dependent protein kinase. The calcium-dependent enzymes, as well as AMPK, subsequently activate particular transcription-factor complexes.

FIGURE 24.13 Exercise results in mitochondrial biogenesis and enhanced fat metabolism.
An action potential causes Ca²⁺ release from the sarcoplasmic reticulum (SR), the muscle-cell equivalent of the endoplasmic reticulum. The Ca²⁺, in addition to instigating muscle contraction, activates nuclear transcription factors that stimulate the expression of specific genes. The products of these genes, in conjunction with the products of mitochondrial genes, are responsible for mitochondrial biogenesis. Fatty acids activate a different set of genes that increase the fatty acid oxidation capability of mitochondria. [Information from D. A. Hood, *J Appl. Physiol.* 90:1137–1157, 2001, Fig. 2.]

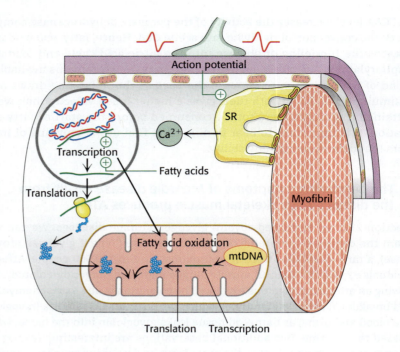

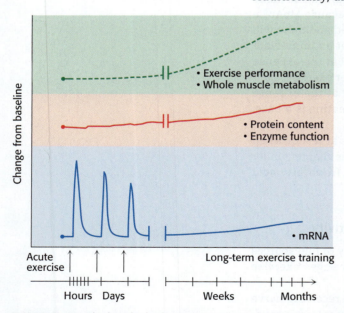

FIGURE 24.14 Biochemical adaptations to exercise vary over a range of time scales. Changes as a function of duration of exercise are shown for mRNA synthesis (bottom), protein synthesis (middle), and exercise performance (top). [From B. Egan and J. R. Zierath, Exercise metabolism and the molecular regulation of skeletal muscle adaptation, *Cell Metab.* 17:162–184, 2013, Fig. 1.]

Two patterns of gene expression, in particular, change in response to regular exercise (Figure 24.13). Regular exercise enhances the production of proteins required for fatty acid metabolism, such as the enzymes of β oxidation. Interestingly, fatty acids themselves function as signal molecules to activate the transcription of enzymes involved in fatty acid metabolism. Additionally, another set of transcription factors activated by the calcium signal cascade initiates a signal pathway that increases mitochondrial biogenesis. In concert, the increase in fatty acid oxidizing capability and additional mitochondria allow for the efficient metabolism of fatty acids. Because an excess of fatty acids results in insulin resistance (Section 24.3), efficient metabolism of fatty acids results in an increase in insulin sensitivity. Indeed, muscles of well-trained athletes may contain high concentrations of triacylglycerols yet still maintain exquisite sensitivity to insulin. These alterations of muscle biochemistry are some of the molecular manifestations of the training effects of exercise.

Exercise alters muscle and whole-body metabolism

In addition to the changes to mitochondrial biogenesis and fatty acid metabolism in muscle tissue, exercise has additional biochemical effects on muscle and other tissues. These effects also vary depending on the type and duration/frequency of the exercise. Here we will give a brief overview of the molecular adaptations to exercise, and then we will compare the effects of endurance and resistance training on muscle and whole-body biochemistry.

Regardless of the type of exercise, the timing of the metabolic response, in general, is the same (**Figure 24.14**). The first response to exercise is an alteration in the patterns of gene expression: a variety of transcription factors are activated, and all of these transcription factors stimulate genes encoding proteins that facilitate exercise similar to the changes described above. In addition to those changes, recall that hypoxia (low oxygen) activates the hypoxia inducible factor (HIF-1), which stimulates the genes encoding glycolytic enzymes, glucose transporters, and angiogenesis-stimulating factors (Section 16.3). Remarkably, as shown in Figure 24.14, even a single bout of exercise can stimulate gene expression several-fold, with levels returning to normal in 24 hours. However, it takes repeated bouts of exercise to accumulate enough mRNA to enhance the synthesis of proteins responding to exercise. In addition to the proteins mentioned above, other examples of such proteins include muscle contractile and structural proteins, and the oxygen-binding proteins hemoglobin and myoglobin (Chapter 3). When these and other proteins accumulate to a sufficient extent, the physiological effects of exercise become apparent. You can run longer and faster, or you can do more repetitions with heavier weights.

Table 24.4 shows some of the molecular adaptations and health benefits accruing from endurance and resistance training. For most of us, some combination of the two will result in maximal health benefits. The American Medical Association and the American College of Sports Medicine launched a global initiative several years ago called *Exercise is Medicine*. As Table 24.4 illustrates, and as we have learned throughout our study of biochemistry, this is indeed the case: exercise is medicine.

TABLE 24.4 Biochemical adaptations and health benefits of aerobic and resistance training

	Aerobic (endurance) training	Resistance (strength) training
Biochemical response in skeletal muscle tissue		
Muscle growth (hypertrophy)	↔	↑↑↑
Anaerobic capacity	↑	↑↑
Muscle protein synthesis	↔↑	↑↑↑
Mitochondria protein synthesis	↑↑	↔↑
Lactate tolerance	↑↑	↔↑
Mitochondria density and oxidative capacity	↑↑↑	↔↑
Endurance capacity	↑↑↑	↔↑
Whole-body and metabolic health		
Bone mineral density	↑↑	↑↑
Percent body fat	↓↓	↓
Lean body mass	↔	↑↑
Insulin sensitivity	↑↑	↑↑
Inflammatory markers	↓↓	↓
Resting heart rate	↓↓	↔
Blood pressure at rest	↔↓	↔
Cardiovascular risk	↓↓↓	↓
Basal metabolic rate	↑	↑↑

↑ Values increase; ↓ values decrease; ↔ no change; ↔↑ or ↔↓ little or no change.

Information from B. Egan and J. R. Zierath, Exercise metabolism and the molecular regulation of skeletal muscle adaptation, *Cell Metab.* 17:162–184, 2013, Table 1.

EXAMPLE Measuring the Impact of a Single Athletic Activity on Caloric Homeostasis

PROBLEM: Estimate how long a person has to jog to offset the calories obtained from eating just 10 macadamia nuts, each with a mass of about 2 g.

GETTING STARTED: Let's make some assumptions about both the energy captured and energy used. First, because a macadamia nut consists mainly of fats (~37 kJ g^{-1}, ~9 kcal g^{-1}), we can assume that each nut provides approximately 75 kJ, or 18 kcal. Second, let's assume that jogging has an incremental power consumption of 400 W.

SOLVE: The ingestion of 10 nuts results in an intake of about 753 kJ (180 kcal). A watt (W) is equal to 1 joule (J) per second (0.239 calorie per second), so running with a typical power consumption rate of 400 W requires 0.4 kJ per second, or 0.4 kJ s^{-1} (0.0956 kcal s^{-1}). Dividing the total energy captured, 753 kJ, by the rate of energy use, 0.4 kJ s^{-1}, gives us 1883 s, the time one would need to run.

REFLECT: So, a person would have to run just over 31 minutes to expend the calories provided by just 10 small nuts! This calculation illustrates why it can be very difficult to offset increased eating with exercise alone and why caloric homeostasis is primarily maintained by hormonal controls on appetite.

24.5 Starvation Induces Protein Wasting and Ketone Body Formation

What happens when the human body becomes deprived of energy sources? This condition, known as starvation, is a circumstance affecting nearly a billion people worldwide. Typically, a well-nourished 70-kg person has fuel reserves totaling about 670,000 kJ (161,000 kcal; Table 24.5). The energy need for a 24-hour period ranges from about 6700 kJ (1600 kcal) to 25,000 kJ (6000 kcal), of course depending on the extent of activity, as we have just examined. Thus, stored fuels suffice to meet caloric needs during starvation for 1 to 3 months. However, the carbohydrate reserves are exhausted in only a day.

TABLE 24.5 Typical fuel reserves in a 70-kg person

Organ	Available energy in kilojoules (kcal)		
	Glucose or glycogen	**Triacylglycerols**	**Mobilizable proteins**
Blood	250 (60)	20 (45)	0 (0)
Liver	1700 (400)	2000 (450)	1700 (400)
Brain	30 (8)	0 (0)	0 (0)
Muscle	5000 (1200)	2000 (450)	100,000 (24,000)
Adipose tissue	330 (80)	560,000 (135,000)	170 (40)

Data from G. F. Cahill, Jr., *Clin. Endocrinol. Metab.* 5:398, 1976.

The first priority during starvation is the maintenance of blood-glucose concentration

Even under starvation conditions, the blood-glucose concentration must be maintained above 2.2 mM. The first priority of metabolism in starvation is to provide sufficient glucose to the brain and other tissues (such as red blood cells) that are absolutely dependent on this fuel. However, precursors of glucose are not abundant. Most energy is stored in the fatty acyl moieties of triacylglycerols. Moreover, recall that fatty acids cannot be converted into glucose, because acetyl CoA resulting from fatty acid breakdown cannot be transformed into pyruvate (Section 22.4). The glycerol moiety of triacylglycerol can be converted into glucose, but only a limited amount is available. The only other potential source of glucose is the carbon skeletons of amino acids derived from the breakdown of proteins. However, proteins are not stored, and so any breakdown will necessitate a loss of function. Thus, the second priority of metabolism in starvation is to preserve protein, which is accomplished by shifting the fuel being used from glucose to fatty acids and ketone bodies (**Figure 24.15**).

The metabolic changes on the first day of starvation are like those after an overnight fast. The low blood-sugar level leads to decreased insulin secretion and increased glucagon secretion. The dominant metabolic processes are the mobilization of triacylglycerols in adipose tissue and gluconeogenesis by the liver. The liver obtains energy for its own needs by oxidizing fatty acids released from adipose tissue. The concentrations of acetyl CoA and citrate consequently increase, which switches off glycolysis. The uptake of glucose by muscle is markedly diminished because of the low insulin level, whereas fatty acids enter freely. Consequently, muscle uses almost no glucose and relies nearly exclusively on fatty acids for fuel. The β oxidation of fatty acids by muscle halts the conversion of pyruvate into acetyl CoA, because acetyl CoA stimulates the phosphorylation of the pyruvate dehydrogenase complex, which renders it inactive (Section 17.4). Hence, any available pyruvate, lactate, and alanine are exported to the liver for conversion into glucose. Glycerol derived from the cleavage of triacylglycerols is another raw material for the synthesis of glucose by the liver.

Proteolysis provides a major source of carbon skeletons for gluconeogenesis. During starvation, degraded proteins are not replenished and instead serve as carbon sources for glucose synthesis. Initial sources of protein are those that turn over rapidly, such as proteins of the intestinal epithelium and the secretions of the pancreas. Proteolysis of muscle protein can provide large amounts of amino acids for gluconeogenesis. However, survival for most animals depends on being able to move rapidly, which requires a large muscle mass, and so muscle protein loss must be minimized.

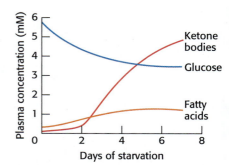

FIGURE 24.15 The blood concentrations of fatty acids and ketone bodies increase in starvation, while glucose decreases.

❖ SELF–CHECK QUESTION

The hormone glucagon signifies the fasted state, yet it inhibits glycolysis in the liver. How do we benefit from this inhibition of an energy-production pathway?

Metabolic adaptations in prolonged starvation minimize protein degradation

How is the loss of muscle and other proteins curtailed? After about 3 days of starvation, the liver forms large amounts of the ketone bodies, acetoacetate, and D-3-hydroxybutyrate (Section 22.4). Their synthesis from acetyl CoA increases markedly because the citric acid cycle is unable to oxidize all the acetyl units generated by fatty acid degradation. Gluconeogenesis depletes the supply of oxaloacetate, which is essential for the entry of acetyl CoA into the citric acid cycle. Consequently, the liver produces large quantities of ketone bodies, which are released into the blood. At this time, the brain begins to consume significant amounts of acetoacetate in place of glucose. After 3 days of starvation, about a quarter of the brain's energy needs are met by ketone bodies (**Table 24.6**). The heart also uses ketone bodies as fuel.

TABLE 24.6 Fuel metabolism during starvation

Fuel exchanges and consumption	Amount formed or consumed in 24 hours (grams)	
	3rd day	**40th day**
Fuel use by the brain		
Glucose	100	40
Ketone bodies	50	100
All other use of glucose	50	40
Fuel mobilization		
Adipose-tissue lipolysis	180	180
Muscle-protein degradation	75	20
Fuel output of the liver		
Glucose	150	80
Ketone bodies	150	150

After several weeks of starvation, ketone bodies become the major fuel of the brain. Acetoacetate is activated by the transfer of CoA from succinyl CoA to give acetoacetyl CoA (Section 22.4). Cleavage by thiolase then yields two molecules of acetyl CoA, which enter the citric acid cycle. In essence, ketone bodies are equivalents of fatty acids that are an accessible fuel source for the brain. Only 40 g of glucose is then needed per day for the brain, compared with about 120 g in the first day of starvation (**Figure 24.16A**). The effective conversion of fatty acids into ketone bodies by the liver and their use by the brain markedly diminishes the need for glucose. Hence, less muscle is degraded than in the first days of starvation. The breakdown of 20 g of muscle daily compared with 75 g early in starvation is most important for survival. How are the amino acids released by muscle processed? The amino acids are transported to the liver where nitrogen is removed as urea, gluconeogenic amino acids are metabolized into glucose, and ketogenic amino acids are used as fuel for the liver or processed to ketone bodies.

What happens after depletion of the triacylglycerol stores? The ketone body contribution disappears, and the only source of fuel that remains is protein (**Figure 24.16B**). Protein degradation accelerates, and a loss of heart, liver, or kidney function inevitably causes death. A person's survival time is mainly determined by the size of the triacylglycerol depot. Examinations of people who have starved themselves to death as a political protest show that lean individuals succumb after about 70 days, whereas obese individuals can survive for 6 or 7 months.

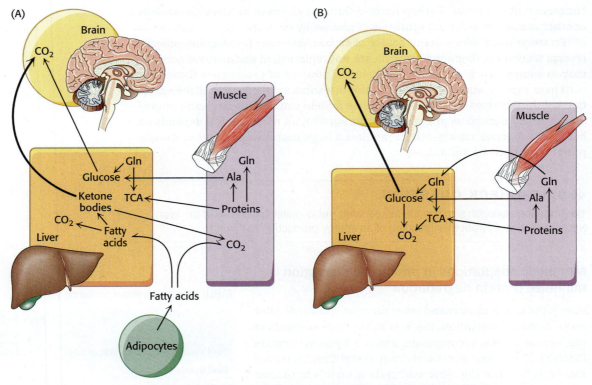

FIGURE 24.16 Fuel use during prolonged starvation changes after fat mass is depleted. A simplified depiction of the relation of adipose tissue, liver, muscle, and brain is shown. (A) Fat mass is adequate. (B) Fat mass is depleted. Only the amino acids alanine and glutamine are shown since they are also important for nitrogen transport.

◆ SELF–CHECK QUESTION

What tissue provides the largest energy supply in a well-nourished human being? What other tissues also store large amounts of energy and in what form?

24.6 Ethanol Alters Energy Metabolism in the Liver

Ethanol has been a part of the human diet for centuries, partly because of its intoxicating effects and partly because alcoholic beverages provided a safe means of hydration when uncontaminated water was scarce. However, its consumption in excess can result in a number of health problems, most notably liver damage. What is the biochemical basis of these health problems?

Ethanol metabolism leads to an excess of NADH

Ethanol cannot be excreted and must be metabolized, primarily by the liver. This metabolism is accomplished by two pathways. The first pathway comprises two steps. The first step, catalyzed by the enzyme *alcohol dehydrogenase*, takes place in the cytoplasm:

$$\underset{\text{Ethanol}}{CH_3CH_2OH} + NAD^+ \xrightarrow{\overset{\text{Alcohol}}{\text{dehydrogenase}}} \underset{\text{Acetaldehyde}}{CH_3CHO} + NADH + H^+$$

The second step, catalyzed by *aldehyde dehydrogenase*, takes place in mitochondria:

$$\underset{\text{Acetaldehyde}}{CH_3CHO} + NAD^+ + H_2O \xrightarrow{\overset{\text{Aldehyde}}{\text{dehydrogenase}}} \underset{\text{Acetate}}{CH_3COO^-} + NADH + 2H^+$$

Note that ethanol consumption leads to an accumulation of NADH. This high concentration of NADH inhibits gluconeogenesis by preventing the oxidation of lactate to pyruvate. In fact, the high concentration of NADH will cause the reverse reaction to predominate, and lactate will accumulate. The consequences may be hypoglycemia and lactic acidosis.

The overabundance of NADH also inhibits fatty acid oxidation. An important metabolic purpose of fatty acid oxidation is to generate NADH for ATP generation by oxidative phosphorylation, but an alcohol consumer's NADH needs are met by ethanol metabolism. In fact, the excess NADH signals that conditions are right for fatty acid synthesis. Hence, triacylglycerols accumulate in the liver, leading to a condition known as "fatty liver" that is exacerbated in obese persons.

The second pathway for ethanol metabolism uses cytochrome P450 enzymes. This means of ethanol metabolism is also called the *microsomal ethanol-oxidizing system* (MEOS). This cytochrome P450-dependent pathway (Section 27.4) generates acetaldehyde and subsequently acetate while oxidizing biosynthetic reducing power, NADPH, to NADP$^+$. Because it uses oxygen, this pathway generates free radicals that damage tissues. Moreover, because the system consumes NADPH, the antioxidant glutathione cannot be regenerated (Section 20.5), exacerbating the oxidative stress.

Ethanol metabolites cause liver damage

What are the effects of the other metabolites of ethanol? Liver mitochondria can convert acetate into acetyl CoA in a reaction requiring ATP. The enzyme is the acyl CoA synthetase that normally activates short-chain fatty acids.

$$\text{Acetate} + \text{coenzyme A} + \text{ATP} \longrightarrow \text{acetyl CoA} + \text{AMP} + \text{PP}_i$$

$$\text{PP}_i \longrightarrow 2\,\text{P}_i$$

However, further processing of the acetyl CoA by the citric acid cycle is blocked, because NADH inhibits two important citric acid cycle regulatory enzymes—isocitrate dehydrogenase and α-ketoglutarate dehydrogenase.

The accumulation of acetyl CoA has several consequences. First, ketone bodies will form and be released into the blood, aggravating the acidic condition already resulting from the high lactate concentration. The processing of the acetate in the liver becomes inefficient, leading to a buildup of acetaldehyde. This very reactive compound forms covalent bonds with many important functional groups in proteins, impairing protein function. If ethanol is consistently consumed at high levels, the acetaldehyde can significantly damage the liver, eventually leading to cell death.

Ethanol also alters gene transcription, further exacerbating liver damage. Ethanol in some fashion stimulates the synthesis of sterol regulatory-element binding proteins (SREBPs) (Section 27.3), a family of transcription factors that upregulates the genes that promote fatty acid synthesis. At the same time, acetaldehyde derived from ethanol inactivates another transcription factor, PPARα (peroxisome proliferator-activated receptor), which normally stimulates the genes involved in fatty acid oxidation. The biochemical effects of ethanol consumption can be quite rapid. For instance, fat accumulates in the liver within a few days of moderate alcohol consumption.

Liver damage from excessive ethanol consumption occurs in three stages. The first stage is the aforementioned development of fatty liver. In the second stage—alcoholic hepatitis—groups of cells die and inflammation results. This stage can itself be fatal. In stage three—cirrhosis—fibrous structure and scar tissue are produced around the dead cells. Cirrhosis impairs many of the liver's biochemical functions. The cirrhotic liver is unable to convert ammonia into urea, and blood levels of ammonia rise. Ammonia is toxic to the nervous system and can cause coma and death. Cirrhosis of the liver arises in about 25% of alcoholics, and about 75% of all cases of liver cirrhosis are the result of alcoholism. Viral hepatitis is a nonalcoholic cause of liver cirrhosis.

Excess ethanol consumption disrupts vitamin metabolism

The adverse effects of ethanol are not limited to the metabolism of ethanol itself. Vitamin A (retinol) is converted into retinoic acid, an important signal molecule for growth and development in vertebrates, by the same dehydrogenases that metabolize ethanol. Consequently, this activation does not take place in the presence of ethanol, which acts as a competitive inhibitor. Moreover, the P450 enzymes induced by ethanol inactivate retinoic acid. These disruptions in the retinoic acid signaling pathway are believed to be responsible, at least in part, for fetal alcohol syndrome as well as the development of a variety of cancers.

The disruption of vitamin A metabolism is a direct result of the biochemical changes induced by excess ethanol consumption. Other disruptions in metabolism result from another common characteristic of alcoholics—malnutrition. Alcoholics will frequently drink instead of eating. A dramatic neurological disorder, referred to as *Wernicke–Korsakoff syndrome*, results from insufficient intake of the vitamin thiamine. Symptoms include mental confusion, unsteady gait, and lack of fine motor skills.

The symptoms of Wernicke–Korsakoff syndrome are similar to those of beriberi (Section 17.5) because both conditions result from a lack of thiamine. Thiamine is converted into the coenzyme thiamine pyrophosphate, a key constituent of the pyruvate dehydrogenase complex. Recall that this complex links glycolysis with the citric acid cycle. Disruptions in the pyruvate dehydrogenase complex are most evident as neuromuscular disorders because the nervous system is normally dependent on glucose for energy generation.

Alcoholic scurvy is occasionally observed because of an insufficient ingestion of ascorbate (vitamin C). Ascorbate is required for the formation of stable collagen fibers. The symptoms of scurvy include skin lesions and blood-vessel fragility. Most notable are bleeding gums, the loss of teeth, and periodontal infections. Gums are especially sensitive to a lack of ascorbate because the collagen in gums turns over rapidly.

What is the biochemical basis for scurvy? Ascorbate is required for the synthesis of 4-hydroxyproline, an amino acid necessary for collagen stability. To form this unusual amino acid, proline residues on the amino side of glycine residues in nascent collagen chains become hydroxylated. One oxygen atom from O_2 becomes attached to C-4 of proline, while the other oxygen atom is taken up by α-ketoglutarate, which is converted into succinate (**Figure 24.17**).

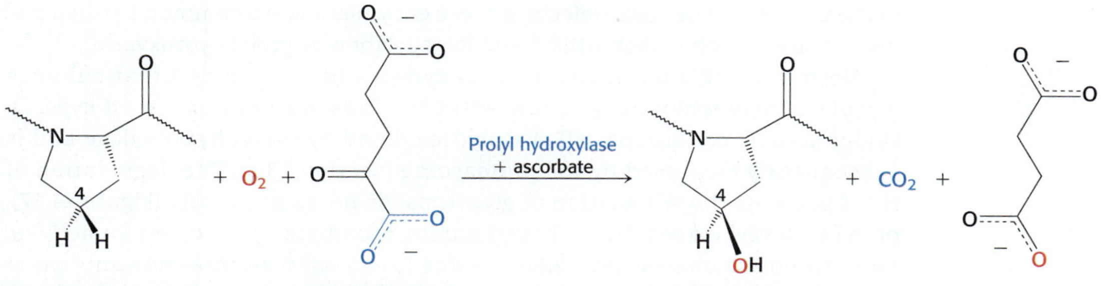

| Prolyl residue | α-Ketoglutarate | 4-Hydroxyprolyl residue | Succinate |

FIGURE 24.17 Formation of 4-hydroxyproline requires ascorbate. Proline is hydroxylated at C-4 by the action of prolyl hydroxylase, an enzyme that activates molecular oxygen.

This reaction is catalyzed by prolyl hydroxylase, a dioxygenase (Section 23.5), which requires an Fe^{2+} ion to activate O_2. The enzyme also converts α-ketoglutarate into succinate without hydroxylating proline. In this partial reaction, an oxidized iron complex is formed, which inactivates the enzyme. How is the active enzyme regenerated? Ascorbate comes to the rescue by reducing the ferric ion of the inactivated enzyme. In the recovery process, ascorbate is oxidized to dehydroascorbic acid (**Figure 24.18**). Thus, ascorbate serves here as a specific antioxidant.

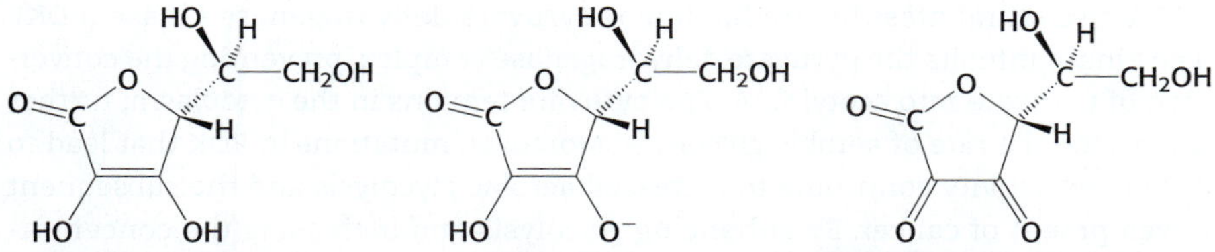

| Ascorbic acid | Ascorbate | Dehydroascorbic acid |

FIGURE 24.18 Ascorbate is the ionized form of vitamin C, and dehydroascorbic acid is the oxidized form of ascorbate.

Why does impaired hydroxylation have such devastating consequences? Collagen synthesized in the absence of ascorbate is less stable than the normal protein. Hydroxyproline stabilizes the collagen triple helix by forming interstrand hydrogen bonds. The abnormal fibers formed by insufficiently hydroxylated collagen account for the symptoms of scurvy. In the next section, we will see that prolyl hydroxylase also plays a critical role in the development of some cancers.

❖ SELF–CHECK QUESTION

What are the two primary means of processing ethanol?

Ethanol and defects in central energy metabolism contribute to the development of cancer

In addition to the numerous detrimental effects outlined above, ethanol is also a well-known carcinogen, although the mechanisms and tissues affected are too numerous to list, and many remain poorly established. There are many other connections between biochemical defects in the enzymes of central metabolism and cancer, some of which are exacerbated by ethanol consumption. For example, mutations that alter the activity of succinate dehydrogenase, fumarase, or pyruvate dehydrogenase kinase all enhance aerobic glycolysis (Section 16.3). Recall that in aerobic glycolysis, cancer cells preferentially metabolize glucose to lactate even in the presence of oxygen. Defects in these enzymes share a common biochemical link: the transcription factor HIF-1 and its regulation by prolyl hydroxylase.

Normally, HIF-1 upregulates the enzymes and transporters that enhance glycolysis only when oxygen concentration falls, a condition called *hypoxia*. Under aerobic conditions, HIF-1 is hydroxylated by prolyl hydroxylase and is subsequently destroyed by the proteasome (Section 23.2). The degradation of HIF-1 prevents the stimulation of glycolysis. As noted previously (Figure 24.17), prolyl hydroxylase requires α-ketoglutarate, ascorbate, and oxygen for activity. Thus, when ascorbate is unavailable—due to excessive alcohol consumption or when oxygen concentration falls—prolyl hydroxylase is inactive, HIF-1 is not hydroxylated and not degraded, and the synthesis of proteins required for glycolysis is stimulated. As a result, the rate of glycolysis is increased.

Recent research suggests that defects in the enzymes of the citric acid cycle may significantly affect the regulation of prolyl hydroxylase. When either succinate dehydrogenase or fumarase is defective, succinate and fumarate accumulate in the mitochondria and spill over into the cytoplasm. Both succinate and fumarate are competitive inhibitors of prolyl hydroxylase. The inhibition of prolyl hydroxylase results in the stabilization of HIF-1, since HIF-1 is no longer hydroxylated. Lactate, the end product of glycolysis, also appears to inhibit prolyl hydroxylase by interfering with the action of ascorbate. These multiple factors reducing prolyl hydroxylase activity are likely exacerbated by the effects of ethanol consumption described previously.

In addition to increasing the amount of the proteins required for glycolysis, HIF-1 also stimulates the production of pyruvate dehydrogenase kinase (PDK). The kinase inhibits the pyruvate dehydrogenase complex, preventing the conversion of pyruvate into acetyl CoA. The pyruvate remains in the cytoplasm, further increasing the rate of aerobic glycolysis. Moreover, mutations in PDK that lead to enhanced activity contribute to increased aerobic glycolysis and the subsequent development of cancer. By enhancing glycolysis and increasing the concentration of lactate, the mutations in PDK result in the inhibition of prolyl hydroxylase and the stabilization of HIF-1.

One additional and particularly fascinating example of the manipulation of cellular control by cancer cells comes from recent findings on the mitochondrial enzyme acetyl CoA acetyltransferase. Under normal circumstances, this enzyme synthesizes ketone bodies; however, in certain cancers, acetyl CoA acetyltransferase is phosphorylated, causing it to form active tetramers. Interestingly, the enzyme then acts as a protein acetyltransferase, adding acetyl groups to pyruvate dehydrogenase and pyruvate dehydrogenase phosphatase. The acetylation inhibits the two enzymes and facilitates the metabolic switch from oxidative phosphorylation to aerobic glycolysis, thereby enhancing the Warburg effect (Section 16.3). In essence, the enzyme is hijacked from its normal role in ketone body formation to further cancer growth.

These observations linking central metabolic enzymes to cancer suggest that cancer is also a metabolic disease, not simply a disease of mutant growth factors and cell cycle control proteins. This realization opens the door to new thinking about cancer treatments. Indeed, preliminary experiments suggest that if cancer cells undergoing aerobic glycolysis are forced by pharmacological manipulation to use oxidative phosphorylation, the cancer cells lose their malignant properties. It seems that even in metabolic pathways as well known as the citric acid cycle, which has been studied for the better part of a century, there are still secrets to be revealed by future biochemists.

Summary

24.1 Caloric Homeostasis Is a Means of Regulating Body Weight

- Maintaining near-constant body weight throughout adult life is accomplished by caloric homeostasis. When energy intake is greater than expenditure, weight gain results, which may lead to obesity. Obesity is at pandemic levels and is implicated as a contributing factor in a host of pathological conditions.

- Various signal molecules act on the brain to control appetite, including short-term signals like CCK and GLP-1, which relay satiety signals to the brain, and signals like leptin and insulin, which regulate appetite over longer periods.

- Leptin is secreted by adipose tissue in proportion to mass and reduces hunger by binding to receptors in the brain. Obesity can develop even with normal amounts of leptin and leptin receptor, suggesting that such individuals are leptin-resistant. Suppressors of cytokine signaling may inhibit leptin signaling, leading to leptin resistance and obesity.

24.2 The Fasted–Fed Cycle Is a Response to Eating and Sleeping Behaviors

- Insulin signals the fed state; it stimulates the formation of glycogen and triacylglycerols. In contrast, glucagon signals a low blood-glucose concentration, i.e., the fasted state; it stimulates glycogen breakdown and gluconeogenesis by the liver, and triacylglycerol hydrolysis by adipose tissue.

- After a meal, the rise in the blood-glucose concentration leads to increased insulin and decreased glucagon secretions, leading to glycogen synthesis in muscle and the liver.

- When the blood-glucose concentration falls, glucose is formed by glycogen degradation and gluconeogenesis, and fatty acids are released by the hydrolysis of triacylglycerols.

- As a fast continues, the liver and muscle increasingly use fatty acids to meet their own energy needs so that glucose is conserved for use by the brain and the red blood cells.

24.3 Diabetes Is a Common Metabolic Disease Often Resulting from Obesity

- Diabetes is the most common metabolic disease in the world, with type 2 being far more common than type 1.

- Type 1 diabetes results when insulin is absent due to destruction of the pancreatic β cells. The body is still responsive to insulin, which is used to treat the disease.

- Type 2 diabetes is characterized by insulin resistance, in which tissues do not respond normally to insulin. Obesity is a significant predisposing factor for type 2 diabetes.

- Excess fats accumulate in the muscle tissue of an obese individual. These fats are processed to second messengers that activate signal-transduction pathways, which inhibit insulin signaling, leading to insulin resistance.

- Insulin resistance ultimately leads to pancreatic β-cell failure. Increased insulin production to compensate for resistance results in endoplasmic reticulum stress and subsequent activation of apoptotic pathways that lead to β cell death.

- Lack of insulin signaling leads to elevated blood-glucose levels, the mobilization of triacylglycerols, and excessive ketone body formation. Accelerated ketone body formation can lead to acidosis, coma, and death in untreated diabetics.

24.4 Exercise Beneficially Alters the Biochemistry of Cells

- Exercise is a useful treatment for insulin resistance and type 2 diabetes. Muscle activity stimulates mitochondrial biogenesis in a calcium-dependent manner. The increase in the number of mitochondria facilitates fatty acid oxidation in the muscle, resulting in increased insulin sensitivity.

- Fuel choice in exercise is determined by the intensity and duration of the bout of exercise. Sprinting is powered by stored ATP, creatine phosphate, and anaerobic glycolysis, while running a marathon requires the oxidation of both muscle glycogen and fatty acids derived from adipose tissue.

24.5 Starvation Induces Protein Wasting and Ketone Body Formation

- The metabolic adaptations during starvation serve to prioritize blood glucose yet minimize protein degradation.
- Ketone bodies are formed by the liver from fatty acids within a few days after the onset of starvation and, after several weeks, they become the major fuel of the brain. The diminished need for glucose decreases the rate of muscle breakdown, prolonging survival.

24.6 Ethanol Alters Energy Metabolism in the Liver

- The oxidation of ethanol results in an unregulated overproduction of NADH, which has several consequences. A rise in the blood concentration of lactic acid and ketone bodies causes acidosis. The liver is damaged because the excess NADH causes excessive fat formation as well as the generation of acetaldehyde, a reactive molecule.
- Alterations in the activity of several transcription factors due to ethanol consumption also contribute to liver pathology. Severe liver damage can result with prolonged ethanol use.
- Ethanol disrupts vitamin absorption and is carcinogenic. Numerous defects in central metabolism lead to cancer by a variety of unexpected mechanisms, indicating that in many cases cancer can be viewed as a metabolic disease.

Key Terms

caloric homeostasis (energy homeostasis) (p. 733)

insulin (p. 735)

glucagon (p. 735)

leptin (p. 736)

fasted–fed cycle (p. 738)

glucose homeostasis (p. 739)

type 1 diabetes (p. 741)

insulin resistance (p. 741)

type 2 diabetes (p. 741)

metabolic syndrome (p. 743)

endoplasmic reticulum (ER) stress (p. 745)

unfolded protein response (UPR) (p. 745)

respiratory quotient (RQ) (p. 748)

Problems

1. Pick one answer for each question: Under normal circumstances, what is the most commonly used fuel by the brain? Which becomes a substantial source of energy for the brain after a week of fasting?

(a) Glycerol

(b) Glucose

(c) Fatty acids

(d) Ketone bodies

2. Which of the following pathways is not stimulated by epinephrine?

(a) Glycogen degradation

(b) Mobilization of fatty acids from adipose tissue

(c) Glycogen synthesis

(d) Glycolysis in muscle

3. As disturbing as the obesity pandemic is, an equally intriguing, almost amazing observation is that many people are able to maintain an approximately constant weight throughout adult life. A few simple calculations of a simplified situation illustrate how remarkable this feat is. Consider a 120-pound woman whose weight does not change significantly between the ages of 25 and 65. Let's say that the woman requires 8400 kJ (2000 kcal) per day^{-1}. For simplicity's sake, let us assume the woman's diet consists predominantly of fatty acids derived from lipids. The energy density of fatty acids is 38 kJ (9 kcal) g^{-1}. How much food has she consumed over the 40 years in question? ❖ 1

4. Suppose that our test subject from problem 3 gained 55 pounds between the ages of 25 and 65 (a common occurrence), and that her weight at 65 years of age is 175 pounds. Calculate how many excess calories she consumed per day to gain the 55 pounds over 40 years. Assume that our test subject is 5 feet, 6 inches tall. What is her BMI? Would she be considered obese at 175 pounds? ❖ 1

5. Adipose tissue was once only considered a storage site for fat. Why is this view no longer considered correct?

6. What are the two key hormones responsible for long-term maintenance of caloric homeostasis? ❖ **2**

7. What two biochemical roles does CCK play? GLP-1? ❖ **2**

8. Match the characteristic (a–i) with the appropriate hormone (1–6). ❖ **2**

(a) Secreted by adipose tissue

(b) Stimulates liver gluconeogenesis

(c) GPCR pathway

(d) Satiety signal

(e) Enhances insulin secretion

(f) Secreted by the pancreas during a fast

(g) Secreted after a meal by the pancreas

(h) Stimulates glycogen synthesis

(i) Missing in type 1 diabetes

1. leptin
2. adiponectin
3. GLP-1
4. CCK
5. insulin
6. glucagon

9. Differentiate between type 1 and type 2 diabetes. ❖ **3**

10. The typical human adult uses about 160 g of glucose per day, 120 g of which is used by the brain. The available reserve of glucose (~20 g of circulating glucose and ~190 g of glycogen) is adequate for about one day. After the reserve has been depleted during starvation, what other sources can be used to produce glucose?

(a) Glycerol from triacylglycerols

(b) Glucogenic amino acids from muscle protein

(c) Pyruvate by carboxylation to form oxaloacetate

(d) Fatty acids from triacylglycerols

(e) Ketogenic amino acids from muscle protein

11. The rate of energy expenditure of a typical 70-kg person at rest is about 70 watts (W), like that of a light bulb.

(a) Express this rate in kilojoules per second and in kilocalories per second.

(b) How many electrons flow through the mitochondrial electron-transport chain per second under these conditions?

(c) Estimate the corresponding rate of ATP production.

(d) The total ATP content of the body is about 50 g. Estimate how often an ATP molecule turns over in a person at rest.

12. Recall that the respiratory quotient (RQ) is defined as the volume of CO_2 released divided by the volume of O_2 consumed.

(a) Calculate the RQ values for the complete oxidation of glucose and of tripalmitoylglycerol. (b) What do RQ measurements reveal about the contributions of different energy sources during intense exercise? (Assume that protein degradation is negligible.) ❖ **4**

13. Ingesting large amounts of glucose immediately before a marathon might seem to be a good way of increasing the fuel stores. However, experienced runners do not ingest glucose before a race. What is the biochemical reason for their avoidance of this potential fuel? (Hint: consider the effect of glucose ingestion on the level of insulin.) ❖ **2**, ❖ **4**

14. The most common form of malnutrition in children in the world, kwashiorkor, is caused by a diet having ample calories but little protein. The high levels of carbohydrate result in high levels of insulin. ❖ **5**

(a) What is the effect of high levels of insulin on lipid utilization? On protein metabolism?

(b) Children suffering from kwashiorkor often have large distended bellies caused by water from the blood leaking into extracellular spaces. Suggest a biochemical basis for this condition.

15. How is the metabolism of the liver coordinated with that of skeletal muscle during strenuous exercise? What is the advantage of converting pyruvate into lactate in skeletal muscle? ❖ **4**

16. What is the major fuel for resting muscle? What is the major fuel for muscle under strenuous work conditions? ❖ **4**

17. (a) After light exercise, the oxygen consumed in recovery is approximately equal to the oxygen deficit, which is the amount of additional oxygen that would have been consumed had oxygen consumption reached steady state immediately. How is the oxygen consumed in recovery used? (b) The oxygen consumed after strenuous exercise stops is significantly greater than the oxygen deficit and is termed *excess post-exercise oxygen consumption* (EPOC). Why is so much more oxygen required after intense exercise? ❖ **4**

18. Alcohol consumption on an empty stomach results in some interesting biochemical as well as embarrassing behavioral alterations. We will ignore the latter. Gluconeogenesis falls; there are increases in intracellular ratios of lactate to pyruvate, of glycerol 3-phosphate to dihydroxy-acetone phosphate, of glutamate to α-ketoglutarate, and of D-3-hydroxybutyrate to acetoacetate. Hypoglycemia develops rapidly. Blood pH also falls. Alcohol consumption by a well-fed individual does not lead to hypoglycemia or alteration in blood pH. ❖ **6**

(a) Why does ethanol consumption result in the altered ratios?

(b) Why does hypoglycemia and blood acidosis result in the hungry individual?

(c) Why does a well-fed person not experience hypoglycemia?

19. Recall that pyruvate carboxylase is the enzyme that catalyzes the most important anaplerotic reaction, which restores CAC intermediates by converting pyruvate into oxaloacetate (Section 17.5). Pyruvate carboxylase deficiency is a fatal disorder. Patients with pyruvate carboxylase deficiency sometimes display some or all of the following symptoms: lactic acidosis, hyperammonemia (excess NH_4^+ in the blood), hypoglycemia, and demyelination of the regions of the brain due to insufficient lipid synthesis. Provide a possible biochemical rationale for each of these symptoms.

20. The graph shows the relation between blood-lactate levels, oxygen consumption, and heart rate during exercise of increasing intensity. The values for oxygen consumption and heart rate are indicators of the degree of exertion. ❖ 4

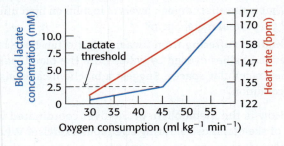

(a) Why is some lactate produced even when exercise is moderate?

(b) Biochemically, what is taking place when the lactate concentration begins to rise rapidly, a point called the lactate threshold?

(c) Endurance athletes will sometimes measure blood-lactate levels during training so that they know their lactate threshold. Then, during events, they will race just at or below their lactate threshold until the late stages of the race. Biochemically, why is this practice wise?

(d) Training can increase the lactate threshold. Explain.

Biosynthesis of Amino Acids

Glutamate

Hugh Spencer/Science Source

While nitrogen, as N_2 gas, makes up the majority of Earth's atmosphere, most organisms cannot directly access it for use in amino acids. They rely on nitrogen-fixing microorganisms, such as bacteria that live in nodules on the roots of yellow clover, to convert N_2 to ammonia (NH_3). Ammonia can then be used to synthesize, first, glutamate, and then other amino acids.

❖ LEARNING GOALS

By the end of this chapter, you should be able to:

1. Explain the centrality of nitrogen fixation to life and describe how atmospheric nitrogen is converted into biologically useful forms of nitrogen.
2. Identify the sources of carbon atoms for amino acid synthesis.
3. Describe the role of feedback inhibition in controlling amino acid synthesis.
4. Identify important biomolecules that are derived from amino acids.

OUTLINE

25.1 Nitrogen Fixation: Microorganisms Use ATP and a Powerful Reductant to Reduce Atmospheric Nitrogen to Ammonia

25.2 Amino Acids Are Made from Intermediates of the Citric Acid Cycle and Other Major Pathways

25.3 Feedback Inhibition Regulates Amino Acid Biosynthesis

25.4 Amino Acids Are Precursors of Many Biomolecules

The assembly of biological macromolecules, including proteins and nucleic acids, requires the generation of appropriate starting materials. In this chapter, we will consider the biosynthesis of amino acids — the building blocks of proteins and the nitrogen source for many other important molecules, including nucleotides, neurotransmitters, and prosthetic groups such as porphyrins.

Amino acid biosynthesis is intimately connected with nutrition because many higher organisms, including human beings, have lost the ability to synthesize some amino acids and must therefore obtain adequate quantities of these essential amino acids in their diets. Furthermore, because some amino acid biosynthetic enzymes are absent in mammals but present in plants and microorganisms, they are useful targets for herbicides and antibiotics.

25.1 Nitrogen Fixation: Microorganisms Use ATP and a Powerful Reductant to Reduce Atmospheric Nitrogen to Ammonia

One of the most challenging biochemical problems organisms face is obtaining nitrogen in a usable form. The nitrogen in amino acids, purines, pyrimidines, and other biomolecules ultimately comes from atmospheric nitrogen, N_2. The extremely strong $N \equiv N$ bond, which has a bond energy of 940 kJ mol^{-1} (225 kcal mol^{-1}), is highly resistant to chemical attack. Indeed, Antoine Lavoisier named nitrogen gas "azote," from Greek words meaning "without life," because it is so unreactive. Nevertheless, the conversion of nitrogen and hydrogen to form ammonia is thermodynamically favorable; the reaction is difficult kinetically because of the activation energy required to form intermediates along the reaction pathway.

Although higher organisms are unable to fix nitrogen, this conversion is carried out by some bacteria and archaea. The biosynthetic process starts with the reduction of N_2 to NH_3 (ammonia), a process called **nitrogen fixation**. For example, symbiotic *Rhizobium* bacteria invade the roots of leguminous plants and form root nodules in which they fix nitrogen, supplying both the bacteria and the plants. The importance of nitrogen fixation by diazotrophic (nitrogen-fixing) microorganisms to the metabolism of all higher eukaryotes cannot be overstated: the amount of N_2 fixed by these species has been estimated to be 10^{11} kilograms per year, about 60% of Earth's newly fixed nitrogen. Lightning and ultraviolet radiation fix another 15%; the other 25% is fixed by industrial processes. The industrial process for nitrogen fixation devised by Fritz Haber in 1910 is still being used in fertilizer factories.

$$N_2 + 3H_2 \rightleftharpoons 2NH_3$$

The fixation of N_2 is typically carried out by mixing it with H_2 gas over an iron catalyst at about 500°C and a pressure of 300 atmospheres.

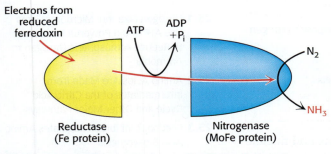

FIGURE 25.1 **The nitrogenase complex fixes nitrogen to ammonia.** Electrons flow from ferredoxin to the reductase (iron protein, or Fe protein) to nitrogenase (molybdenum–iron protein, or MoFe protein) to reduce nitrogen to ammonia. ATP hydrolysis within the reductase drives conformational changes necessary for the efficient transfer of electrons.

Biological nitrogen fixation is catalyzed by the nitrogenase complex

To meet the kinetic challenge, the biological process of nitrogen fixation requires a complex enzyme with multiple redox centers. The **nitrogenase complex**, which carries out this fundamental transformation, consists of two proteins: a *reductase* (also called the iron protein or Fe protein), which provides electrons with high reducing power, and *nitrogenase* (also called the molybdenum–iron protein or MoFe protein), which uses these electrons to reduce N_2 to NH_3. The transfer of electrons from the reductase to the nitrogenase component is coupled to the hydrolysis of ATP by the reductase (**Figure 25.1**).

In principle, the reduction of N_2 to NH_3 is a six-electron process.

$$N_2 + 6\ e^- + 6\ H^+ \longrightarrow 2\ NH_3$$

However, the biological reaction always generates at least 1 mol of H_2 in addition to 2 mol of NH_3 for each mol of $N \equiv N$. Hence, an input of two additional electrons is required.

$$N_2 + 8\ e^- + 8\ H^+ \longrightarrow 2\ NH_3 + H_2$$

In most nitrogen-fixing microorganisms, the eight high-potential electrons come from reduced ferredoxin, generated by oxidative processes. Two molecules of ATP are hydrolyzed for each electron transferred. Thus, at least 16 molecules of ATP are hydrolyzed for each molecule of N_2 reduced.

$$N_2 + 8\ e^- + 8\ H^+ + 16\ ATP + 16\ H_2O \longrightarrow 2\ NH_3 + H_2 + 16\ ADP + 16\ P_i$$

Recall that O_2 is required for oxidative phosphorylation to generate the ATP necessary for nitrogen fixation. However, the nitrogenase complex is exquisitely sensitive to inactivation by O_2. To allow ATP synthesis and nitrogenase to function simultaneously, leguminous plants maintain a very low concentration of free O_2 in their root nodules, the location of the nitrogenase. This is accomplished by binding O_2 to leghemoglobin, a homolog of hemoglobin (Figure 10.13).

The iron–molybdenum cofactor of nitrogenase binds and reduces atmospheric nitrogen

Both the reductase and the nitrogenase components of the complex are iron–sulfur proteins, in which iron is bonded to the sulfur atom of a cysteine residue and to inorganic sulfide. Recall that iron–sulfur clusters act as electron carriers (Section 18.3). The reductase is a dimer of identical 30-kDa subunits bridged by a 4Fe–4S cluster (**Figure 25.2**).

The role of the reductase is to transfer electrons from a suitable donor, such as reduced ferredoxin, to the nitrogenase component. The 4Fe–4S cluster carries the electrons, one at a time, to nitrogenase. The binding and hydrolysis of ATP triggers a conformational change that moves the reductase closer to the nitrogenase component, whence it is able to transfer its electron to the center of nitrogen reduction.

The nitrogenase component is an $\alpha_2\beta_2$ tetramer (240-kDa), in which the α and β subunits are homologous to each other and structurally quite similar (**Figure 25.3**). The nitrogenase requires the FeMo cofactor, which consists of $[Fe_4–S_3]$ and $[Mo–Fe_3–S_3]$ subclusters joined by three disulfide bonds. A carbon atom (the interstitial carbon), donated by S-adenosylmethionine (Section 23.5), sits between the iron atoms of the FeMo cofactor. The FeMo cofactor is also coordinated to a homocitrate moiety and to the α subunit through one histidine residue and one cysteinate residue.

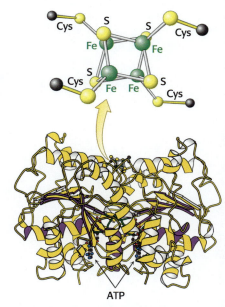

FIGURE 25.2 The Fe protein, or reductase, transfers elections of high reducing power. This protein is a dimer composed of two polypeptide chains linked by a 4Fe–4S cluster. Each monomer is a member of the P-loop NTPase family and contains an ATP-binding site. [Drawn from 1N2C.pdb.]

INTERACT with this model in
 Achieve

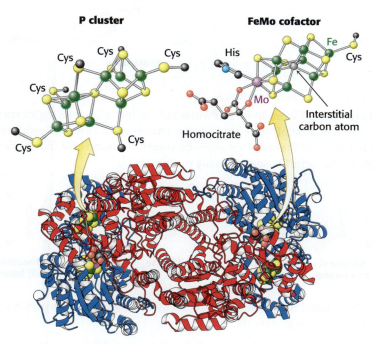

FIGURE 25.3 The MoFe protein, or nitrogenase, catalyzes the reduction of N_2 to NH_3. This protein is a heterotetramer composed of two α subunits (red) and two β subunits (blue). The complex contains two copies each of two types of clusters: P clusters and FeMo cofactors. Each P cluster contains eight iron atoms (green) and seven sulfides linked to the protein by six cysteine residues. Each FeMo cofactor contains one molybdenum atom, seven iron atoms, nine sulfides, the interstitial carbon atom, and a homocitrate, and is linked to the protein by one cysteinate residue and one histidine residue. [Drawn from 1M1N.pdb.]

Electrons from the reductase enter at the *P clusters*, which are located at the α–β interface. The role of the P clusters is to store electrons until they can be used productively to reduce nitrogen at the FeMo cofactor. The FeMo cofactor is the site of nitrogen fixation. One face of the FeMo cofactor is likely to be the site of nitrogen reduction. The electron-transfer reactions from the P cluster take place in concert with the binding of hydrogen ions to nitrogen as it is reduced. The mechanism of this remarkable reaction remains an active focus of biochemical research.

Ammonium ion is assimilated into an amino acid through glutamate and glutamine

The NH_3 generated by the nitrogenase complex is a base, and becomes NH_4^+ in aqueous solutions. The next step in the assimilation of nitrogen into biomolecules is the entry of NH_4^+ into amino acids. The amino acids glutamate and glutamine play pivotal roles in this regard, acting as nitrogen donors for most amino acids. The α-amino group of most amino acids comes from the α-amino group of glutamate by transamination (Section 23.3). Glutamine, the other major nitrogen donor, contributes its side-chain nitrogen atom in the biosynthesis of a wide range of important compounds, including the amino acids tryptophan and histidine.

Glutamate is synthesized from NH_4^+ and α-ketoglutarate, a citric acid cycle intermediate, by the action of glutamate dehydrogenase. We have already encountered this enzyme in the degradation of amino acids (Section 23.3). Recall that NAD^+ is the oxidant in catabolism, whereas NADPH is the reductant in biosyntheses. Glutamate dehydrogenase is unusual in that it does not discriminate between NADH and NADPH, at least in some species.

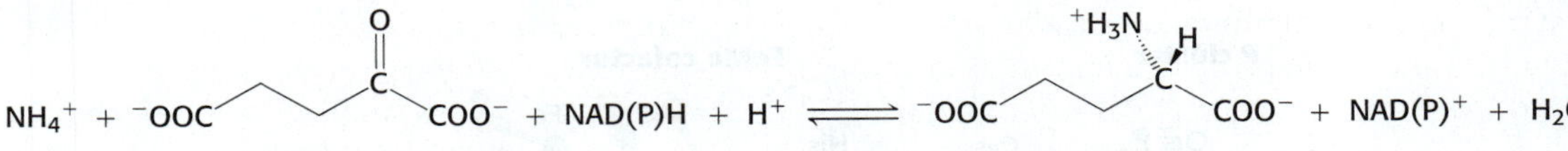

$$NH_4^+ + {}^-OOC\!-\!\!\!-\!\!\!C\!(\!=\!\!O)\!-\!COO^- + NAD(P)H + H^+ \rightleftharpoons {}^-OOC\!-\!\!\!-\!\!\!CH(^+H_3N)(H)\!-\!COO^- + NAD(P)^+ + H_2O$$

α-**Ketoglutarate**   **Glutamate**

The reaction proceeds in two steps:

1. A Schiff base forms between ammonium ion and α-ketoglutarate. The formation of a Schiff base between an amine and a carbonyl compound is a key reaction that takes place at many stages of amino acid biosynthesis and degradation. Schiff bases are easily protonated.

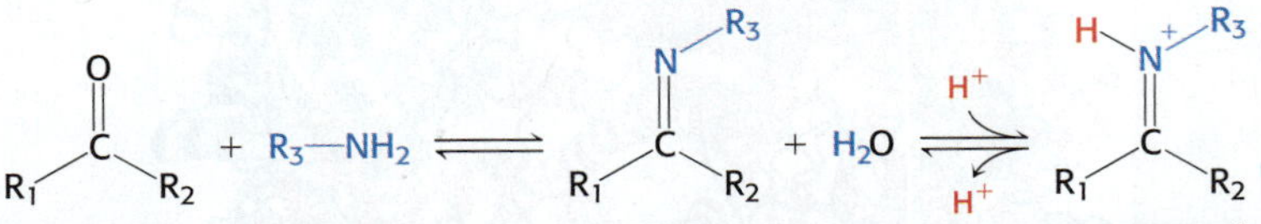

Carbonyl compound	**Amino donor**	**Schiff base**	**Protonated Schiff base**

$$R_1\!-\!C(\!=\!\!O)\!-\!R_2 + R_3\!-\!NH_2 \rightleftharpoons R_1\!-\!C(\!=\!\!N\!-\!R_3)\!-\!R_2 + H_2O \underset{H^+}{\overset{H^+}{\rightleftharpoons}} R_1\!-\!C(\!=\!\!N^+(H)\!-\!R_3)\!-\!R_2$$

2. The protonated Schiff base is reduced by the transfer of a hydride ion from NAD(P)H to form glutamate.

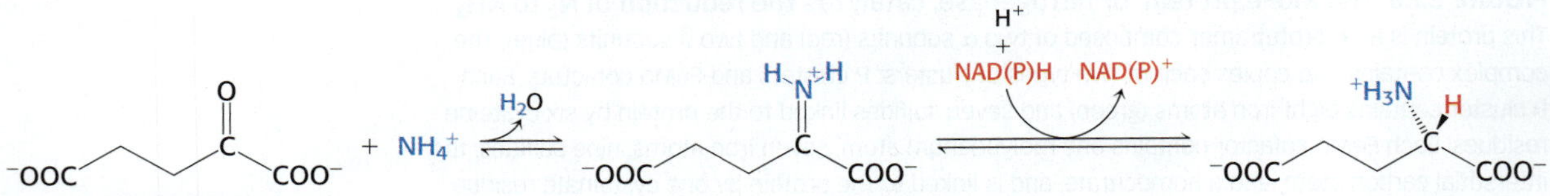

$${}^-OOC\!-\!\!\!C(\!=\!\!O)\!-\!COO^- + NH_4^+ \underset{H_2O}{\rightleftharpoons} {}^-OOC\!-\!\!\!C(\!=\!\!N^+(H)H)\!-\!COO^- \xrightarrow[\;]{H^+ + NAD(P)H \;\; NAD(P)^+} {}^-OOC\!-\!\!\!CH(^+H_3N)(H)\!-\!COO^-$$

α-**Ketoglutarate**   **Glutamate**

The reaction catalyzed by glutamate dehydrogenase is crucial because it establishes the stereochemistry of the α-carbon atom (S absolute configuration) in glutamate. The enzyme binds the α-ketoglutarate substrate in such a way that hydride transferred from NAD(P)H is added to form the L isomer of glutamate (**Figure 25.4**). As we shall see, this stereochemistry is established for other amino acids by transamination reactions that rely on pyridoxal phosphate. Establishing the correct stereochemistry is one of the fundamental challenges of nitrogen incorporation into biomolecules.

A second ammonium ion is incorporated into glutamate to form glutamine by the action of glutamine synthetase. This amidation is driven by the hydrolysis of ATP. ATP participates directly in the reaction by phosphorylating the side chain of glutamate to form an acyl-phosphate intermediate, which then reacts with ammonia to form glutamine.

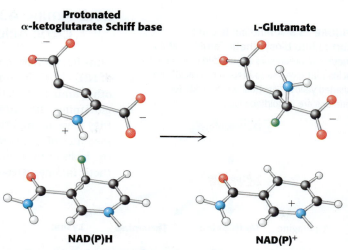

Protonated α-ketoglutarate Schiff base **L-Glutamate**

NAD(P)H **NAD(P)+**

FIGURE 25.4 Glutamate chirality is established by hydride transfer to the Schiff base. In the active site of glutamate dehydrogenase, hydride transfer (green) from NAD(P)H to a specific face of the achiral protonated Schiff base of α-ketoglutarate establishes the L configuration of glutamate.

Glutamate   **Acyl-phosphate intermediate**   **Glutamine**

NH_4^+ initially binds to the enzyme, but upon formation of the acyl-phosphate intermediate, ammonium ion is deprotonated as a high-affinity ammonia-binding site is formed. A specific site for ammonia binding is required to prevent attack by water from hydrolyzing the intermediate and wasting a molecule of ATP. As we shall soon see, the regulation of glutamine synthetase plays a critical role in controlling nitrogen metabolism (Section 25.3).

Glutamate dehydrogenase and glutamine synthetase are present in all organisms. Many organisms also contain an evolutionarily unrelated enzyme, glutamate synthase, which catalyzes the reductive amination of α-ketoglutarate to glutamate. Glutamine is the nitrogen donor.

α-Ketoglutarate + glutamine + NADPH + H$^+$ ⇌ 2 glutamate + NADP$^+$

The side-chain amide of glutamine is hydrolyzed to generate ammonia within the enzyme, a recurring theme throughout nitrogen metabolism. When NH_4^+ is limiting, most of the glutamate is made by the sequential action of glutamine synthetase and glutamate synthase. The sum of these reactions is

$$NH_4^+ + \text{α-ketoglutarate} + NADPH + ATP \longrightarrow \text{glutamate} + NADP^+ + ADP + P_i \quad (1)$$

Note that this stoichiometry differs from that of the glutamate dehydrogenase reaction in that ATP is hydrolyzed. Why do organisms sometimes use this more expensive pathway? The answer is that the value of K_M of glutamate dehydrogenase for NH_4^+ is high (~1 mM), and so this enzyme is not saturated when NH_4^+ is limiting. In contrast, glutamine synthetase has very low K_M for NH_4^+. Thus, ATP hydrolysis is required to capture ammonia when it is scarce.

❖ SELF–CHECK QUESTION

Write out the balanced reactions for glutamine synthetase and glutamate synthase. Show that the sum of these two reactions matches equation 1.

25.2 Amino Acids Are Made from Intermediates of the Citric Acid Cycle and Other Major Pathways

Thus far, we have considered the conversion of N_2 into NH_4^+ and the assimilation of NH_4^+ into glutamate and glutamine. We turn now to the biosynthesis of the other amino acids, the majority of which obtain their nitrogen from glutamate or glutamine. The pathways for the biosynthesis of amino acids are diverse. However, they have an important common feature: their carbon skeletons come from intermediates of glycolysis, the pentose phosphate pathway, or the citric acid cycle. On the basis of these starting materials, amino acids can be grouped into six biosynthetic families (**Figure 25.5**).

FIGURE 25.5 Amino acids can be sorted into biosynthetic families. Major metabolic precursors are shaded blue. Amino acids that give rise to other amino acids are shaded yellow. Essential amino acids for humans are in boldface type.

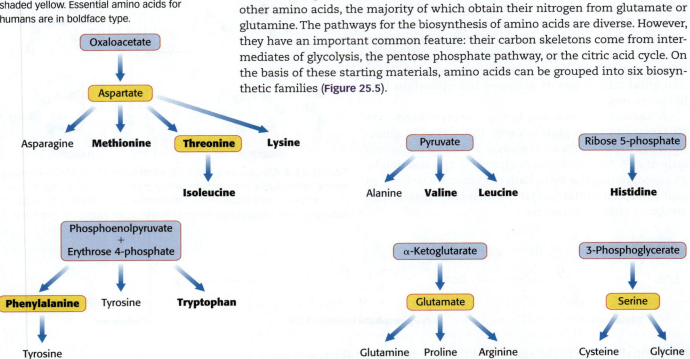

TABLE 25.1 Basic set of 20 amino acids

Nonessential	Essential
Alanine	Histidine
Arginine	Isoleucine
Asparagine	Leucine
Aspartate	Lysine
Cysteine	Methionine
Glutamate	Phenylalanine
Glutamine	Threonine
Glycine	Tryptophan
Proline	Valine
Serine	
Tyrosine	

Human beings can synthesize some amino acids but must obtain others from their diet

Most microorganisms, such as *E. coli*, can synthesize the entire set of 20 amino acids, whereas human beings cannot make 9 of them. The amino acids that must be supplied in the diet are called **essential amino acids**, whereas the others, which can be synthesized if dietary content is insufficient, are termed **nonessential amino acids** (**Table 25.1**). These designations refer to the needs of an organism under a particular set of conditions. For example, enough arginine is synthesized by the urea cycle to meet the needs of an adult, but perhaps not those of a growing child. Likewise, tyrosine is sometimes designated as an essential amino acid, but is not if phenylalanine is present in adequate amounts. In mammals, tyrosine can be synthesized from phenylalanine in one step.

A deficiency of even one amino acid results in a negative nitrogen balance. In this state, more protein is degraded than is synthesized, and so more nitrogen is excreted than is ingested.

The nonessential amino acids are synthesized by quite simple reactions, whereas the pathways for the formation of the essential amino acids are quite complex. For example, the nonessential amino acids alanine and aspartate are synthesized in a single step from pyruvate and oxaloacetate, respectively.

In contrast, the pathways for the essential amino acids require from 5 to 16 steps (**Figure 25.6**). The sole exception to this pattern is arginine, inasmuch as the synthesis of this nonessential amino acid de novo requires 10 steps. Typically, however, it is made in only 3 steps from ornithine as part of the urea cycle. Tyrosine, classified as a nonessential amino acid because it can be synthesized in 1 step from phenylalanine, requires 10 steps to be synthesized from scratch and is essential if phenylalanine is not abundant.

Aspartate, alanine, and glutamate are formed by the addition of an amino group to an alpha-ketoacid

Let us begin with the biosynthesis of nonessential amino acids. Three α-ketoacids—α-ketoglutarate, oxaloacetate, and pyruvate—can be converted into amino acids in one step through the addition of an amino group. We have seen that α-ketoglutarate can be converted into glutamate by reductive amination (Section 23.3). The amino group from glutamate can be transferred to other α-ketoacids by transamination reactions. Thus, aspartate and alanine can be made from the addition of an amino group to oxaloacetate and pyruvate, respectively.

$$\text{Oxaloacetate} + \text{glutamate} \rightleftharpoons \text{aspartate} + \alpha\text{-ketoglutarate}$$

$$\text{Pyruvate} + \text{glutamate} \rightleftharpoons \text{alanine} + \alpha\text{-ketoglutarate}$$

These reactions are carried out by pyridoxal phosphate-dependent aminotransferases. Transamination reactions are required for the synthesis of most amino acids.

In Section 23.3, we considered the mechanism of aminotransferases as applied to the metabolism of amino acids. Let us review the aminotransferase mechanism as it operates in the biosynthesis of amino acids. The reaction pathway begins with **pyridoxal phosphate (PLP)** in a Schiff-base linkage with lysine at the aminotransferase active site, forming an internal aldimine (**Figure 25.7**).

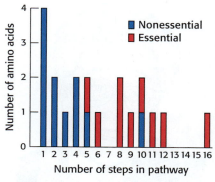

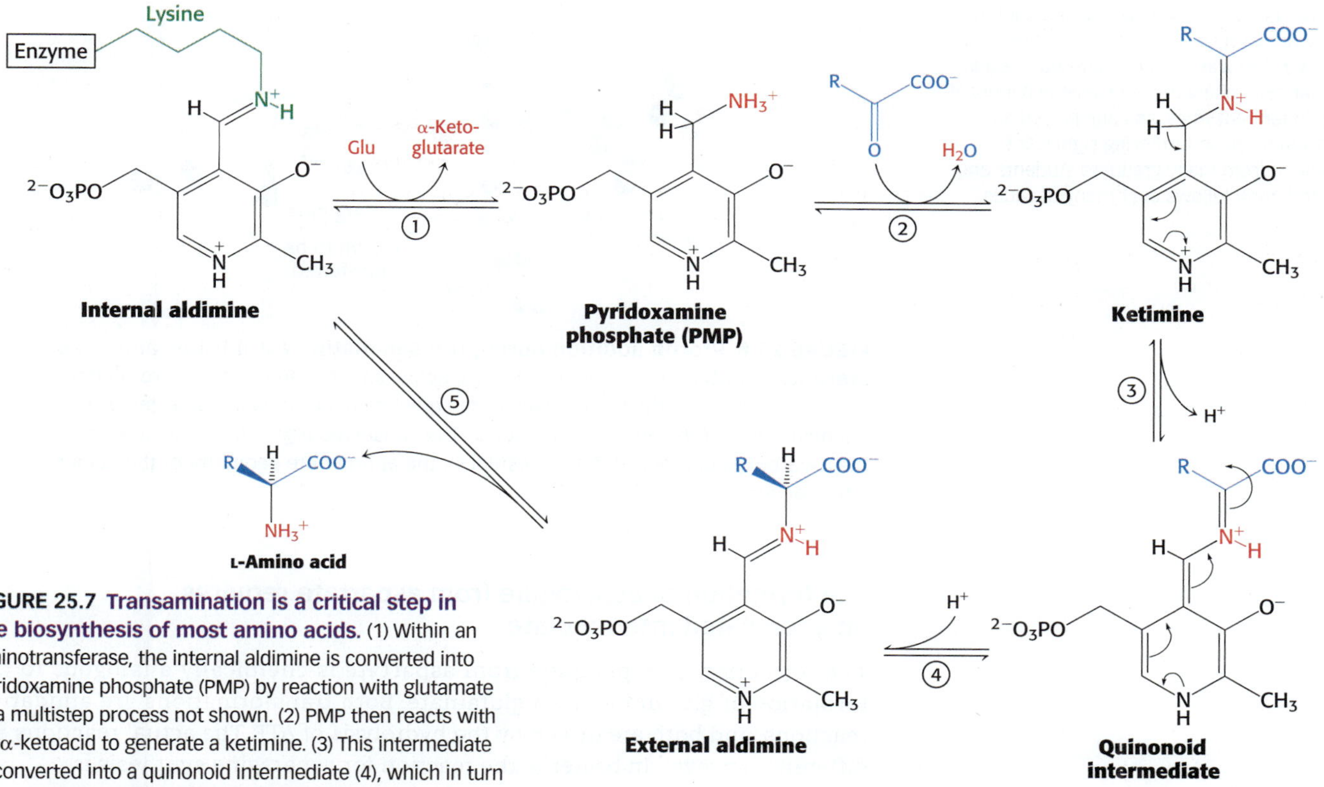

FIGURE 25.6 Essential and nonessential amino acids can be distinguished by the required number of biosynthetic steps. Some amino acids are nonessential to human beings because they can be biosynthesized in a small number of steps. Those amino acids requiring a large number of steps for their synthesis are essential in the diet because some of the enzymes for these steps have been lost in the course of evolution.

FIGURE 25.7 Transamination is a critical step in the biosynthesis of most amino acids. (1) Within an aminotransferase, the internal aldimine is converted into pyridoxamine phosphate (PMP) by reaction with glutamate in a multistep process not shown. (2) PMP then reacts with an α-ketoacid to generate a ketimine. (3) This intermediate is converted into a quinonoid intermediate (4), which in turn yields an external aldimine. (5) The aldimine is cleaved to release the newly formed amino acid to complete the cycle.

1. An amino group is transferred from glutamate to form pyridoxamine phosphate (PMP), the actual amino donor, in a multistep process.

2. PMP then reacts with an incoming α-ketoacid to form a ketimine.

3. Proton loss forms a quinonoid intermediate.

4. Proton addition at a different site forms an external aldimine.

5. The newly formed amino acid is released with the concomitant re-formation of the internal aldimine.

A common step determines the chirality of all amino acids

Aspartate aminotransferase is the prototype of a large family of PLP-dependent enzymes. Comparisons of amino acid sequences as well as several three-dimensional structures reveal that almost all aminotransferases having roles in amino acid biosynthesis are related to aspartate aminotransferase by divergent evolution. An examination of the aligned amino acid sequences reveals that two residues are completely conserved. These residues are the lysine residue that forms the Schiff base with the PLP cofactor (lysine 258 in aspartate aminotransferase) and an arginine residue that interacts with the α-carboxylate group of the ketoacid (Figure 23.10).

An essential step in the transamination reaction is the protonation of the quinonoid intermediate to form the external aldimine. The chirality of the amino acid formed is determined by the direction from which this proton is added to the quinonoid intermediate (**Figure 25.8**). The interaction between the conserved arginine residue and the α-carboxylate group helps orient the substrate so that the lysine residue transfers a proton to the bottom face of the quinonoid intermediate, generating an aldimine with an L configuration at the C_α center.

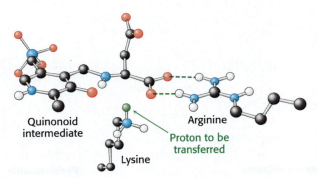

FIGURE 25.8 Proton addition during transamination establishes amino acid stereochemistry. In an aminotransferase active site, the addition of a proton from the lysine residue to the bottom face of the quinonoid intermediate determines the L configuration of the amino acid product. The conserved arginine residue interacts with the α-carboxylate group and helps establish the appropriate geometry of the quinonoid intermediate.

The formation of asparagine from aspartate requires an adenylated intermediate

The formation of asparagine from aspartate is chemically analogous to the formation of glutamine from glutamate. Both transformations are amidation reactions, and both are driven by the hydrolysis of ATP. The actual reactions are different, however. In bacteria, the reaction for asparagine synthesis is

$$\text{Aspartate} + \text{NH}_4^+ + \text{ATP} \longrightarrow \text{asparagine} + \text{AMP} + \text{PP}_i + \text{H}^+$$

Thus, the products of ATP hydrolysis are AMP and PP_i rather than ADP and P_i. Aspartate is activated by adenylation rather than by phosphorylation.

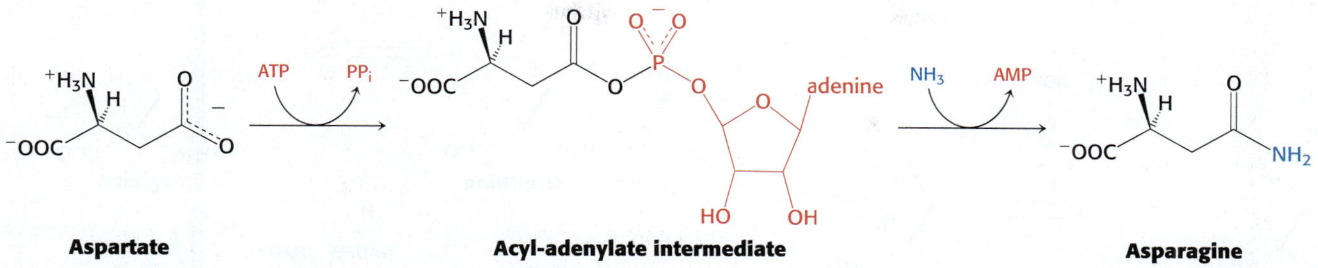

Aspartate **Acyl-adenylate intermediate** **Asparagine**

We have encountered this mode of activation in fatty acid degradation and will see it again in lipid and protein synthesis.

In mammals, the nitrogen donor for asparagine is glutamine rather than ammonia as in bacteria. Ammonia is generated by hydrolysis of the side chain of glutamine and directly transferred to activated aspartate, bound in the active site. An advantage is that the cell is not directly exposed to NH_4^+, which is toxic at high levels to human beings and other mammals. The use of glutamine hydrolysis as a mechanism for generating ammonia for use within the same enzyme is a motif common throughout biosynthetic pathways.

❖ SELF–CHECK QUESTION

Consider the conversion of glutamate to glutamine and aspartate to asparagine in humans. Name two ways in which these reactions are similar, and two ways in which they differ.

Glutamate is the precursor of glutamine, proline, and arginine

The synthesis of glutamate by the reductive amination of α-ketoglutarate has already been discussed, as has the conversion of glutamate into glutamine (Section 25.1). Glutamate is the precursor of two other nonessential amino acids: proline and arginine. First, the γ-carboxyl group of glutamate reacts with ATP to form an acyl phosphate. This mixed anhydride is then reduced by NADPH to an aldehyde.

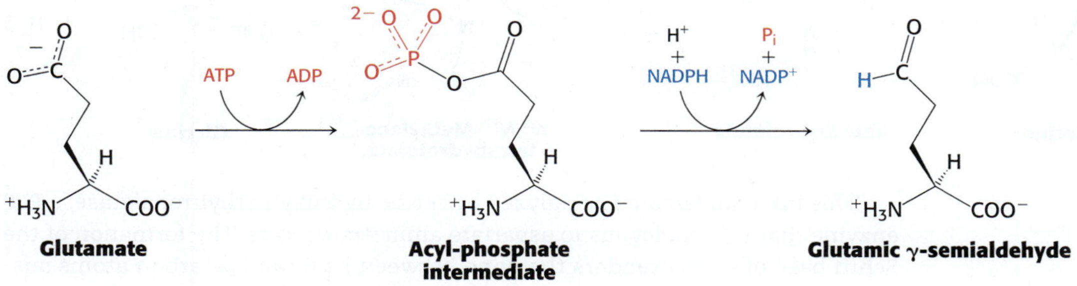

Glutamate **Acyl-phosphate intermediate** **Glutamic γ-semialdehyde**

Glutamic γ-semialdehyde cyclizes with a loss of H_2O in a nonenzymatic process to give Δ^1-pyrroline 5-carboxylate, which is reduced by NADPH to proline. Alternatively, the semialdehyde can be transaminated to ornithine, which is converted in several steps into arginine in the urea cycle (Figure 23.16).

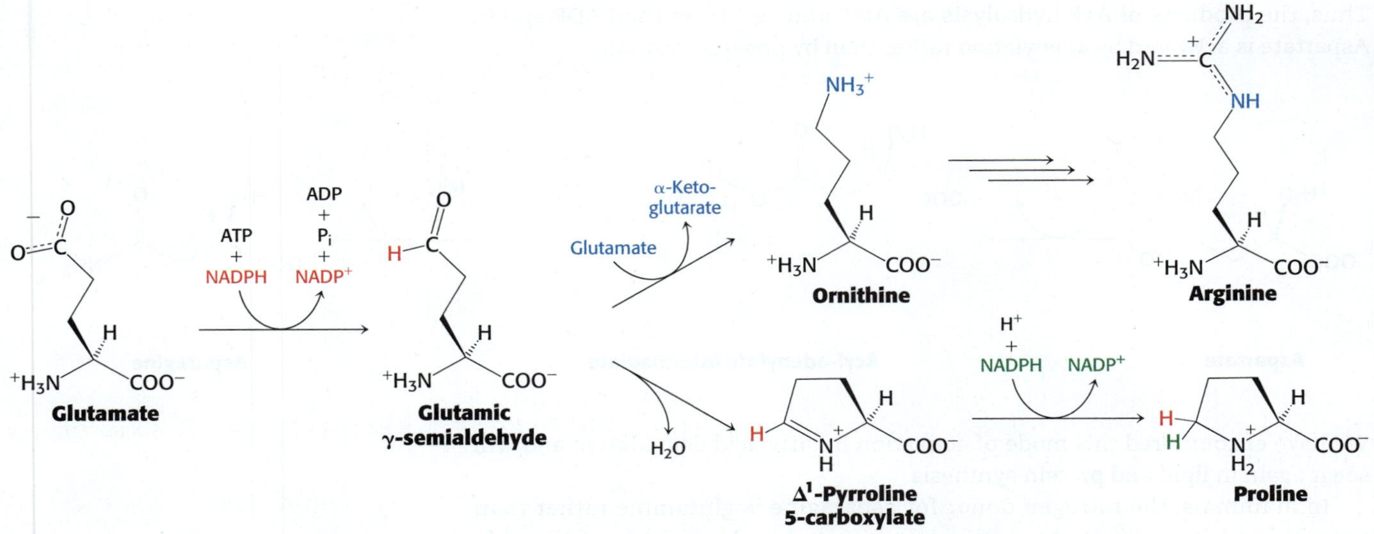

3-Phosphoglycerate is the precursor of serine, cysteine, and glycine

Serine is synthesized from 3-phosphoglycerate, an intermediate in glycolysis. The first step is an oxidation to 3-phosphohydroxypyruvate. This α-ketoacid is transaminated to 3-phosphoserine, which is then hydrolyzed to serine.

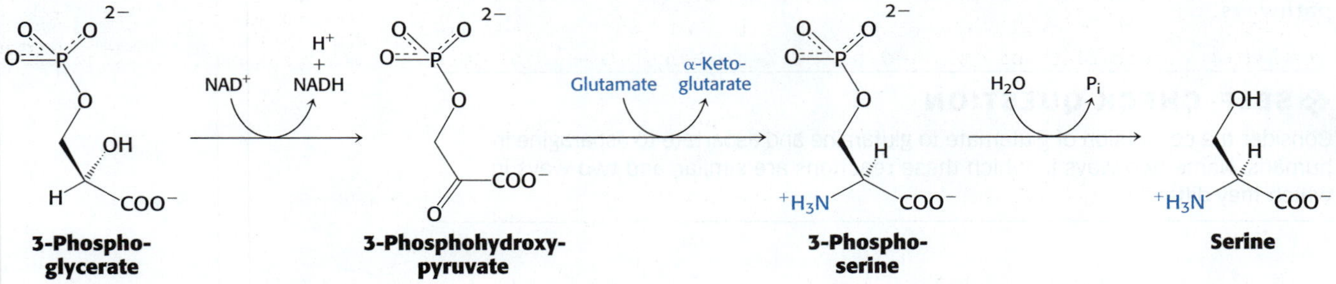

Serine is the precursor of *cysteine* and *glycine*. As we shall see, the conversion of serine into cysteine requires the substitution of a sulfur atom derived from methionine for the side-chain oxygen atom. In the formation of glycine, the side-chain methylene group of serine is transferred to tetrahydrofolate, a carrier of one-carbon units that will be discussed shortly.

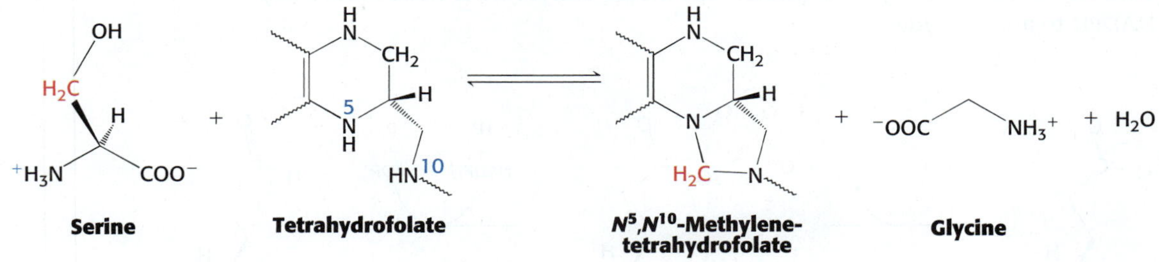

This interconversion is catalyzed by serine hydroxymethyltransferase, a PLP enzyme that is homologous to aspartate aminotransferase. The formation of the Schiff base of serine renders the bond between its α- and β-carbon atoms susceptible to cleavage, enabling the transfer of the β-carbon to tetrahydrofolate and producing the Schiff base of glycine.

Tetrahydrofolate carries activated one-carbon units at several oxidation levels

Tetrahydrofolate is a highly versatile carrier of activated one-carbon units. This cofactor consists of three groups: a substituted pteridine, *p*-aminobenzoate, and

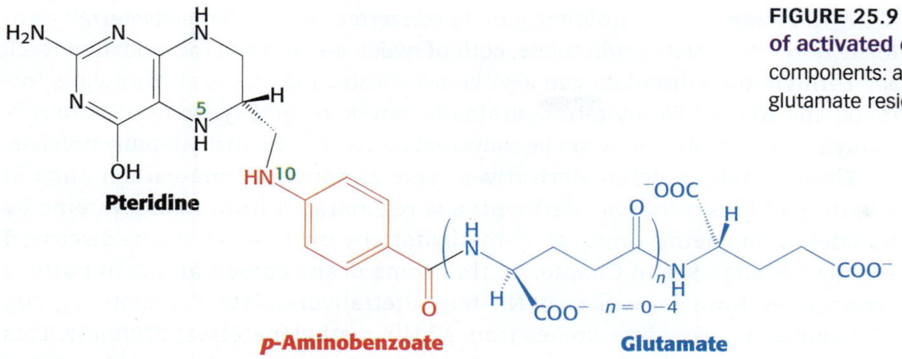

FIGURE 25.9 Tetrahydrofolate is an important carrier of activated one-carbon units. This cofactor includes three components: a pteridine ring, *p*-aminobenzoate, and one or more glutamate residues.

a chain of one or more glutamate residues (**Figure 25.9**). Mammals can synthesize the pteridine ring, but they are unable to conjugate it to the other two units. They obtain tetrahydrofolate from their diets or from microorganisms in their intestinal tracts.

The one-carbon group carried by tetrahydrofolate is bonded to its N-5 or N-10 nitrogen atom (denoted as N^5 and N^{10}) or to both. This unit can exist in three oxidation states (**Table 25.2**). The most-reduced form carries a *methyl* group, whereas the intermediate form carries a *methylene* group. More-oxidized forms carry a *formyl*, *formimino*, or *methenyl* group. The fully oxidized one-carbon unit, CO_2, is carried by biotin rather than by tetrahydrofolate.

The one-carbon units carried by tetrahydrofolate are interconvertible (**Figure 25.10**). N^5,N^{10}-Methylenetetrahydrofolate can be reduced to N^5-methyltetrahydrofolate or oxidized to N^5,N^{10}-methenyltetrahydrofolate.

TABLE 25.2 One-carbon groups carried by tetrahydrofolate

Oxidation state	Group	
	Formula	Name
Most reduced (= methanol)	$-CH_3$	Methyl
Intermediate (= formaldehyde)	$-CH_2-$	Methylene
Most oxidized (= formic acid)	$-CHO$	Formyl
	$-CHNH$	Formimino
	$-CH=$	Methenyl

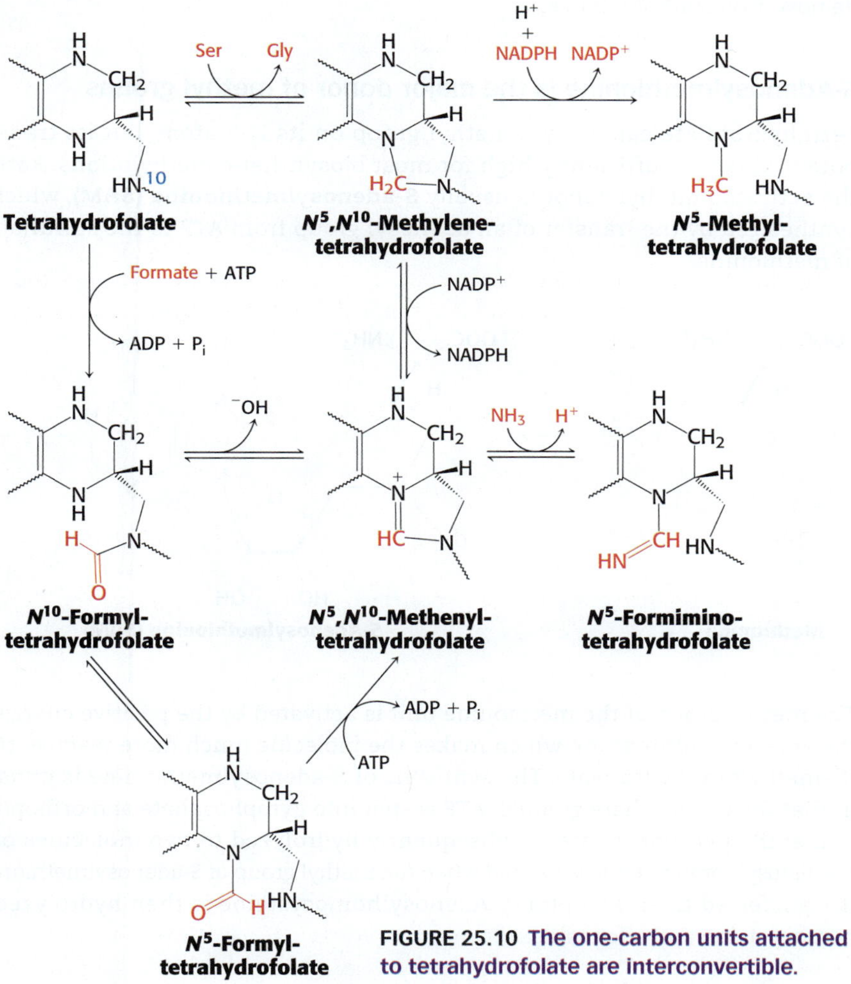

FIGURE 25.10 The one-carbon units attached to tetrahydrofolate are interconvertible.

N^5,N^{10}-Methenyltetrahydrofolate can be converted into N^5-*formiminotetrahydro*-folate or N^{10}-*formyl*tetrahydrofolate, both of which are at the same oxidation level. N^{10}-Formyltetrahydrofolate can also be synthesized from tetrahydrofolate, formate, and ATP. N^5-Formyltetrahydrofolate can be reversibly isomerized to N^{10}-formyltetrahydrofolate, or it can be converted into N^5,N^{10}-methenyltetrahydrofolate.

These tetrahydrofolate derivatives serve as donors of one-carbon units in a variety of biosyntheses. Methionine is regenerated from homocysteine by transfer of the methyl group of N^5-methyltetrahydrofolate, as will be discussed shortly. We shall see in Chapter 26 that some of the carbon atoms of purines are acquired from derivatives of N^{10}-formyltetrahydrofolate. The methyl group of thymine, a pyrimidine, comes from N^5,N^{10}-methylenetetrahydrofolate. This tetrahydrofolate derivative can also donate a one-carbon unit in an alternative synthesis of glycine that starts with CO_2 and NH_4^+, a reaction catalyzed by glycine synthase (called the *glycine cleavage enzyme* when it operates in the reverse direction).

$$CO_2 + NH_4^+ + N^5,N^{10}\text{-methylenetetrahydrofolate} + NADH \rightleftharpoons$$
$$\text{glycine} + \text{tetrahydrofolate} + NAD^+$$

Thus, one-carbon units at each of the three oxidation levels are utilized in biosyntheses.

Furthermore, tetrahydrofolate serves as an acceptor of one-carbon units in degradative reactions. The major source of one-carbon units is the facile conversion of serine into glycine by serine hydroxymethyltransferase, which yields N^5, N^{10}-methylenetetrahydrofolate. Serine can be derived from 3-phosphoglycerate, and so this pathway enables one-carbon units to be formed de novo from carbohydrates.

S-Adenosylmethionine is the major donor of methyl groups

Tetrahydrofolate can carry a methyl group on its N-5 atom, but its transfer potential is not sufficiently high for most biosynthetic methylations. Rather, the activated methyl donor is usually **S-adenosylmethionine (SAM)**, which is synthesized by the transfer of an adenosyl group from ATP to the sulfur atom of methionine.

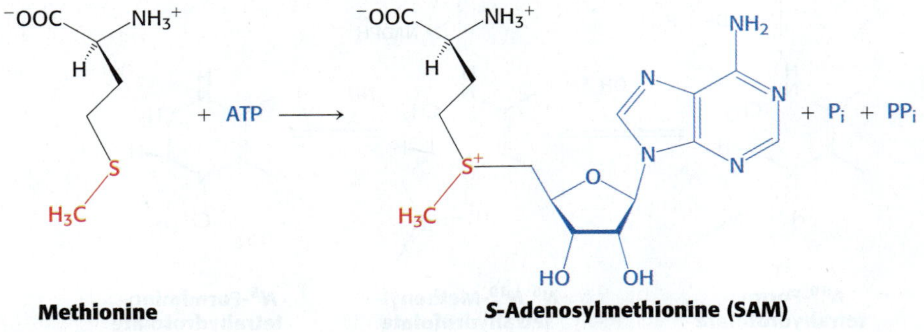

Methionine + ATP $\longrightarrow$ **S-Adenosylmethionine (SAM)** + P_i + PP_i

The methyl group of the methionine unit is activated by the positive charge on the adjacent sulfur atom, which makes the molecule much more reactive than N^5-methyltetrahydrofolate. The synthesis of S-adenosylmethionine is unusual in that the triphosphate group of ATP is split into pyrophosphate and orthophosphate; the pyrophosphate is subsequently hydrolyzed to two molecules of P_i. S-Adenosylhomocysteine is formed when the methyl group of S-adenosylmethionine is transferred to an acceptor. S-Adenosylhomocysteine is then hydrolyzed to homocysteine and adenosine.

S-Adenosylmethionine (SAM)   **S-Adenosylhomocysteine**   **Homocysteine**

Methionine can be regenerated by the transfer of a methyl group to homocysteine from N^5-methyltetrahydrofolate, a reaction catalyzed by methionine synthase.

Homocysteine   **N^5-Methyl-**   **Methionine**   **Tetrahydrofolate**
tetrahydrofolate

The coenzyme that mediates this transfer of a methyl group is methylcobalamin, derived from vitamin B_{12}. In fact, this reaction and the rearrangement of L-methylmalonyl CoA to succinyl CoA (Section 22.3), catalyzed by a homologous enzyme, are the only two B_{12}-dependent reactions known to take place in mammals. Another enzyme that converts homocysteine into methionine without vitamin B_{12} is also present in many organisms.

These reactions constitute the **activated methyl cycle** (**Figure 25.11**). Methyl groups enter the cycle in the conversion of homocysteine into methionine and are then made highly reactive by the addition of an adenosyl group, which makes the sulfur atoms positively charged and the methyl groups much more electrophilic. The high transfer potential of the S-methyl group enables it to be transferred to a wide variety of acceptors. Among the acceptors modified by S-adenosylmethionine are specific bases in bacterial DNA. For instance, the methylation of DNA protects bacterial DNA from cleavage by restriction enzymes (Section 6.4). Methylation is also important for the synthesis of phospholipids (Section 27.1).

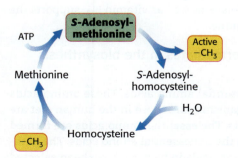

FIGURE 25.11 The activated methyl cycle enables regeneration of methionine. The methyl group of methionine is activated by the formation of S-adenosylmethionine.

❖ SELF–CHECK QUESTION

Which derivative of folate is a reactant in the conversion of **(a)** glycine into serine? **(b)** homocysteine into methionine?

Cysteine is synthesized from serine and homocysteine

In addition to being a precursor of methionine in the activated methyl cycle, homocysteine is an intermediate in the synthesis of cysteine. Serine and homocysteine condense to form cystathionine. This reaction is catalyzed by cystathionine β-synthase. Cystathionine is then deaminated and cleaved to cysteine and α-ketobutyrate by cystathionine γ-lyase or cystathionase.

Both of these enzymes utilize PLP and are homologous to aspartate aminotransferase. The net reaction is

$$\text{Homocysteine} + \text{serine} \rightleftharpoons \text{cysteine} + \alpha\text{-ketobutyrate} + NH_4^+$$

Note that the sulfur atom of cysteine is derived from homocysteine, whereas the carbon skeleton comes from serine.

High homocysteine levels correlate with vascular disease

People with elevated serum levels of homocysteine (homocysteinemia) or the disulfide-linked dimer homocystine (homocystinuria) have an unusually high risk for coronary heart disease and arteriosclerosis. The most common genetic cause of high homocysteine levels is a mutation within the gene encoding cystathionine β-synthase. High levels of homocysteine appear to damage cells lining blood vessels and to increase the growth of vascular smooth muscle. The amino acid raises oxidative stress as well and has also been implicated in the development of type 2 diabetes.

The molecular basis of homocysteine's action has not been clearly identified, but it may result from stimulation of the inflammatory response. Vitamin treatments are sometimes effective in reducing homocysteine levels in some people. Treatment with vitamins maximizes the activity of the two major metabolic pathways processing homocysteine. Pyridoxal phosphate, a vitamin B_6 derivative, is necessary for the activity of cystathionine β-synthase, which converts homocysteine into cystathionine; tetrahydrofolate, as well as vitamin B_{12}, supports the methylation of homocysteine to methionine.

Shikimate and chorismate are intermediates in the biosynthesis of aromatic amino acids

We turn now to the biosynthesis of essential amino acids. These amino acids are synthesized by plants and microorganisms, and those in the human diet are ultimately derived primarily from plants. The essential amino acids are formed by much more complex routes than are the nonessential amino acids. The pathways for the synthesis of aromatic amino acids in bacteria have been selected for discussion here because they are well understood and exemplify recurring mechanistic motifs.

Phenylalanine, tyrosine, and tryptophan are synthesized by a common pathway in *E. coli* (**Figure 25.12**). The initial step is the condensation of phosphoenolpyruvate (a glycolytic intermediate) with erythrose 4-phosphate (a pentose phosphate pathway intermediate). The resulting seven-carbon open-chain

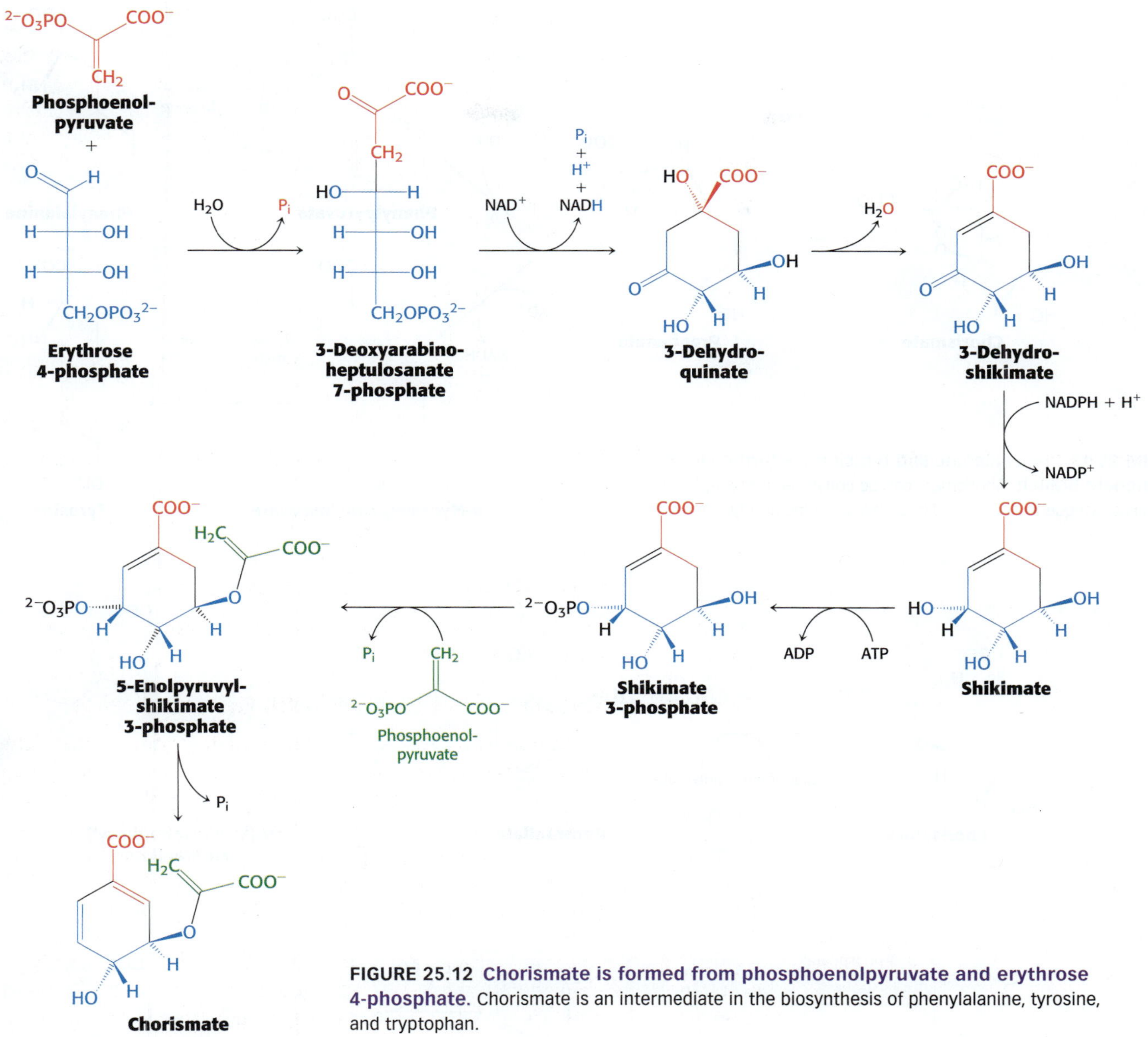

FIGURE 25.12 Chorismate is formed from phosphoenolpyruvate and erythrose 4-phosphate. Chorismate is an intermediate in the biosynthesis of phenylalanine, tyrosine, and tryptophan.

sugar is oxidized, loses its phosphoryl group, and cyclizes to 3-dehydroquinate. Dehydration then yields 3-dehydroshikimate, which is reduced by NADPH to shikimate. The phosphorylation of shikimate by ATP gives shikimate 3-phosphate, which condenses with a second molecule of phosphoenolpyruvate. The resulting 5-enolpyruvyl intermediate loses its phosphoryl group, yielding chorismate, the common precursor of all three aromatic amino acids.

The importance of this pathway is revealed by the effectiveness of glyphosate, a broad-spectrum herbicide. This compound is an uncompetitive inhibitor of the enzyme that produces 5-enolpyruvylshikimate 3-phosphate. It blocks aromatic amino acid biosynthesis in plants but is fairly nontoxic in animals because they lack the enzyme. Recently, there have been claims that glyphosate may cause cancer, although most scientific evidence does not support the claims.

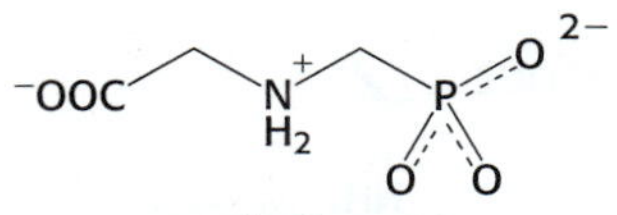

The pathway bifurcates at chorismate. Let us first follow the *prephenate branch* (**Figure 25.13**). The enzyme chorismate mutase converts chorismate into prephenate, the immediate precursor of the aromatic ring of phenylalanine and tyrosine. This fascinating conversion is a rare example of an electrocyclic reaction in biochemistry, mechanistically similar to the well-known Diels–Alger reaction in organic chemistry. Dehydration and decarboxylation

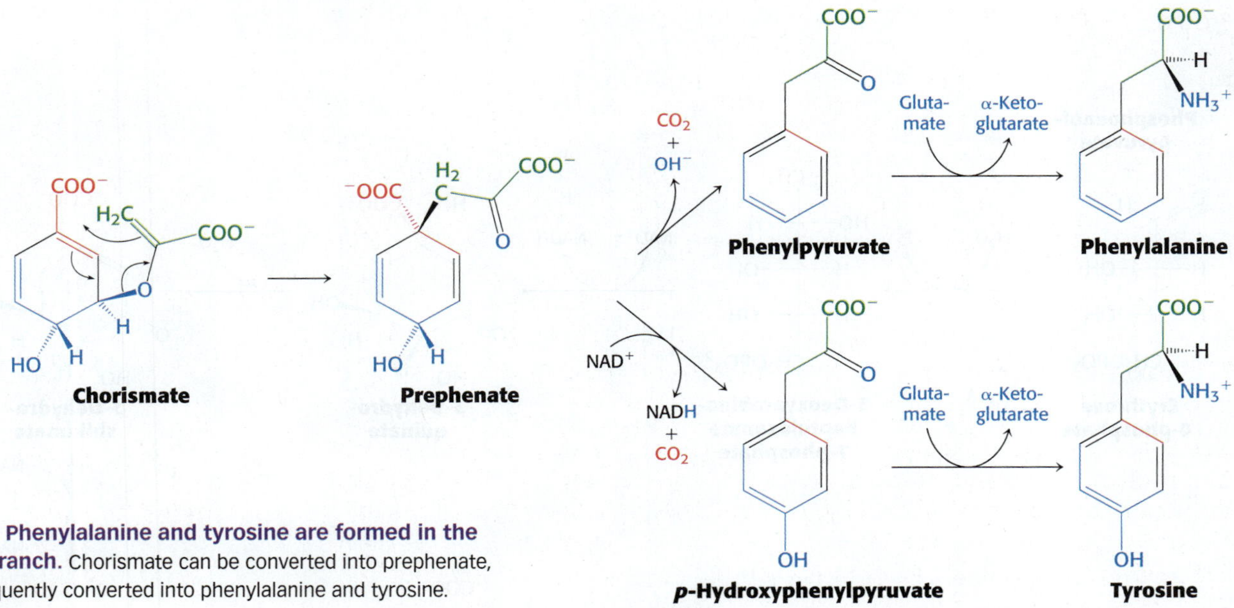

FIGURE 25.13 Phenylalanine and tyrosine are formed in the prephenate branch. Chorismate can be converted into prephenate, which is subsequently converted into phenylalanine and tyrosine.

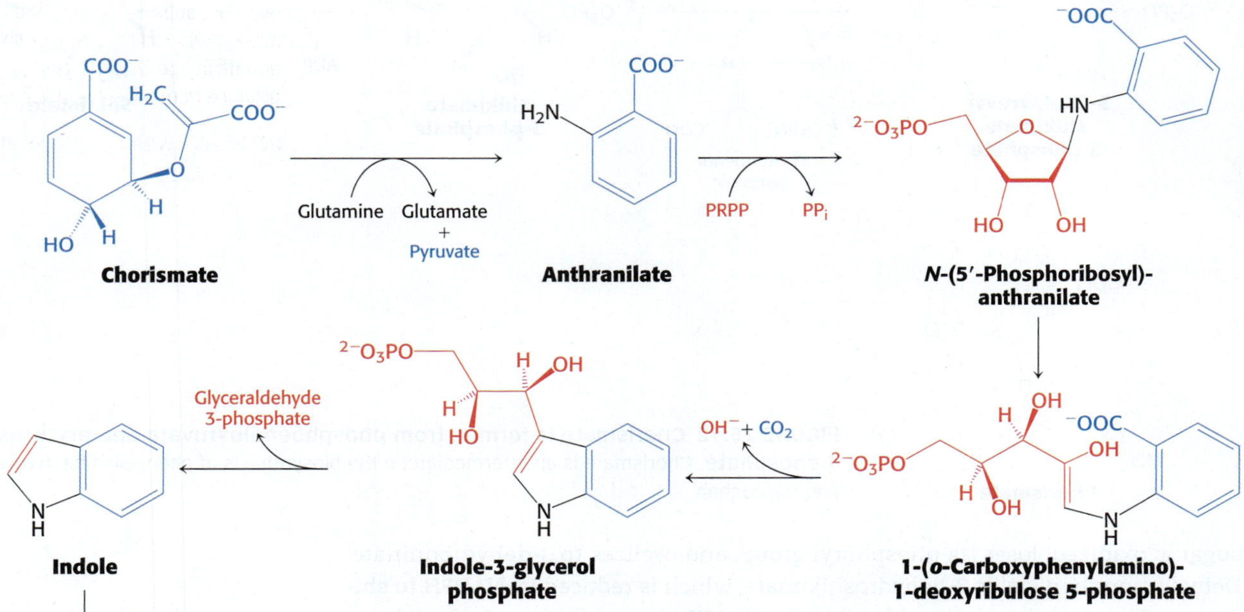

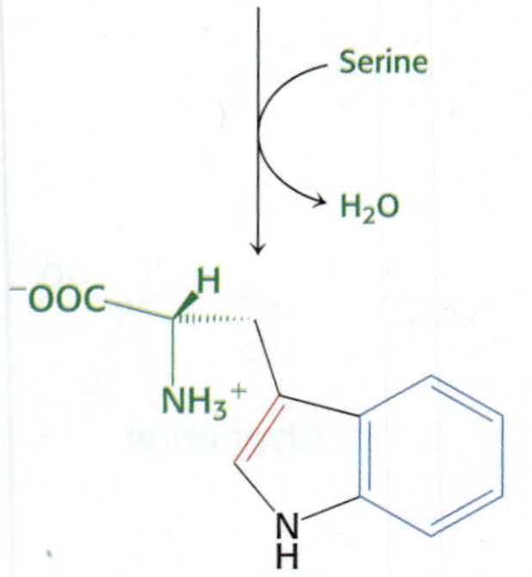

FIGURE 25.14 Tryptophan is formed in the anthranilate branch. Chorismate can be converted into anthranilate, which is subsequently converted into tryptophan.

yield phenylpyruvate. Alternatively, prephenate can be oxidatively decarboxylated to *p*-hydroxyphenylpyruvate. These α-ketoacids are then transaminated to form phenylalanine and tyrosine.

The *anthranilate branch* leads to the synthesis of tryptophan (**Figure 25.14**). Chorismate acquires an amino group derived from the hydrolysis of the side chain of glutamine and releases pyruvate to form anthranilate. Anthranilate then condenses with 5-phosphoribosyl-1-pyrophosphate (PRPP), an activated form of ribose phosphate.

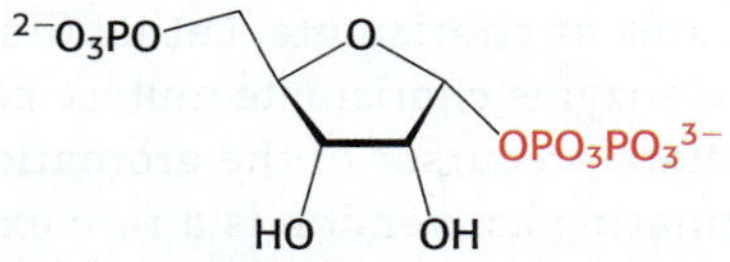

5-Phosphoribosyl-1-pyrophosphate (PRPP)

PRPP is also an important intermediate in the synthesis of histidine, pyrimidine nucleotides, and purine nucleotides (Sections 26.1 and 26.2). The C-1 atom of ribose 5-phosphate becomes bonded to the nitrogen atom of anthranilate in a reaction that is driven by the release and hydrolysis of pyrophosphate. The ribose moiety of phosphoribosylanthranilate undergoes rearrangement to yield 1-(o-carboxyphenylamino)-1-deoxyribulose 5-phosphate. This intermediate is dehydrated and then decarboxylated to indole-3-glycerol phosphate. Tryptophan synthase completes the synthesis of tryptophan with the removal of the side chain of indole-3-glycerol phosphate, yielding glyceraldehyde 3-phosphate, and its replacement with the carbon skeleton of serine.

Tryptophan synthase illustrates substrate channeling in enzymatic catalysis

The *E. coli* enzyme tryptophan synthase, an $\alpha_2\beta_2$ tetramer, can be dissociated into two α subunits and a β_2 dimer (**Figure 25.15**). The α subunit catalyzes the formation of indole from indole-3-glycerol phosphate, whereas each β subunit has a PLP-containing active site that catalyzes the condensation of indole and serine to form tryptophan. Serine forms a Schiff base with this PLP, which is then dehydrated to give the Schiff base of aminoacrylate. This reactive intermediate is attacked by indole to give tryptophan.

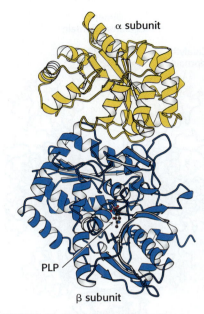

α subunit

PLP

β subunit

FIGURE 25.15 Tryptophan synthase contains two remotely positioned catalytic domains. The complex is formed by one α subunit (yellow) and one β subunit (blue). Pyridoxal phosphate (PLP) is bound deeply inside the β subunit, a considerable distance from the α subunit. [Drawn from 1BKS.pdb.]

INTERACT with this model in
🎵 Achieve

Indole

H_2C COO^-

H N^+ H

$^{2-}O_3PO$ O^-

N^+ CH_3
H

Schiff base of aminoacrylate
(derived from serine)

The synthesis of tryptophan poses a challenge. Indole, a hydrophobic molecule, readily traverses membranes and would be lost from the cell if it were allowed to diffuse away from the enzyme. This problem is solved in an ingenious way. A 25-Å-long channel connects the active site of the α subunit with that of the adjacent β subunit in the $\alpha_2\beta_2$ tetramer (**Figure 25.16**). Thus, indole can diffuse from one active site to the other without being released into bulk solvent. Isotope-labeling experiments showed that indole formed by the α subunit does not leave the enzyme when serine is present. Furthermore, the two partial reactions are coordinated. Indole is not formed by the α subunit until the highly reactive aminoacrylate is ready and waiting in the β subunit.

We see here a clear-cut example of **substrate channeling** in catalysis by a multienzyme complex. Channeling substantially increases the catalytic rate. Furthermore, a deleterious side reaction—in this case, the potential loss of an intermediate—is prevented. We shall encounter other examples of substrate channeling in Chapter 26.

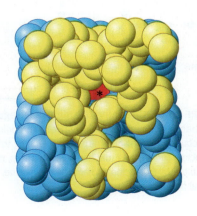

FIGURE 25.16 Tryptophan synthase uses substrate channeling to prevent loss of the indole intermediate. A 25-Å tunnel runs from the active site of the α subunit of tryptophan synthase (yellow) to the PLP cofactor (red) in the active site of the β subunit (blue). The asterisk indicates the center of the tunnel.

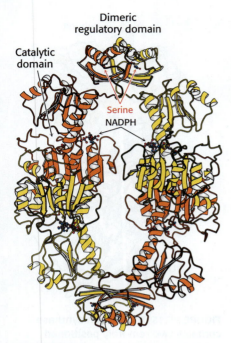

FIGURE 25.17 3-Phosphoglycerate dehydrogenase is a homotetramer. This enzyme, which catalyzes the committed step in the serine biosynthetic pathway, is inhibited by serine. There are two serine-binding dimeric regulatory domains—one at the top and the other at the bottom of the structure. [Drawn from 1PSD.pdb.]

INTERACT with this model in ≈ Achieve

25.3 Feedback Inhibition Regulates Amino Acid Biosynthesis

The rate of synthesis of amino acids depends mainly on the *amounts* of the biosynthetic enzymes and on their *activities*. We now consider the control of enzymatic activity. The regulation of enzyme synthesis in eukaryotes will be discussed in Chapter 31.

In a biosynthetic pathway, the first irreversible reaction is usually an important regulatory site. As we have seen in Section 7.1, this reaction is referred to as the *committed step*. The final product of the pathway (Z) often inhibits the enzyme that catalyzes the committed step ($A \rightarrow B$).

This kind of control is essential for the conservation of building blocks and metabolic energy. Consider the biosynthesis of serine (Section 25.2). The committed step in this pathway is the oxidation of 3-phosphoglycerate, catalyzed by the enzyme 3-phosphoglycerate dehydrogenase. The *E. coli* enzyme is a tetramer of four identical subunits, each comprising a catalytic domain and a serine-binding regulatory domain (**Figure 25.17**). The binding of serine to a regulatory site reduces the value of V_{max} for the enzyme; an enzyme bound to four molecules of serine is essentially inactive. Thus, if serine is abundant in the cell, the enzyme activity is inhibited, and so 3-phosphoglycerate, a key building block that can be used for other processes, is not wasted.

Branched pathways require sophisticated regulation

The regulation of branched pathways is more complicated because the concentration of two products must be taken into account. In fact, several intricate feedback mechanisms have been found in branched biosynthetic pathways.

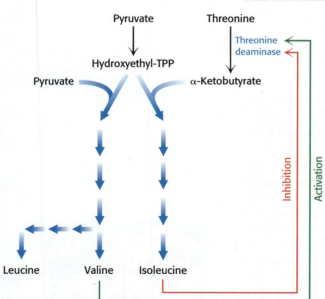

FIGURE 25.18 Threonine deaminase is regulated by feedback inhibition and activation. Threonine is converted into α-ketobutyrate in the committed step, leading to the synthesis of isoleucine. The enzyme that catalyzes this step, threonine deaminase, is inhibited by isoleucine and activated by valine, the product of a parallel pathway.

Feedback inhibition and activation. Two pathways with a common initial step may each be inhibited by its own product and activated by the product of the other pathway. Consider, for example, the biosynthesis of the branched chain amino acids valine, leucine, and isoleucine. A common intermediate, hydroxyethyl thiamine pyrophosphate (hydroxyethyl-TPP; Section 17.2), initiates the pathways leading to all three of these amino acids. Hydroxyethyl-TPP reacts with α-ketobutyrate in the initial step for the synthesis of isoleucine. Alternatively, hydroxyethyl-TPP reacts with pyruvate in the committed step for the pathways leading to valine and leucine. Thus, the relative concentrations of α-ketobutyrate and pyruvate determine how much isoleucine is produced compared with valine and leucine.

Threonine deaminase, the PLP enzyme that catalyzes the formation of α-ketobutyrate, is allosterically inhibited by isoleucine (**Figure 25.18**). This enzyme is also allosterically activated by valine. Thus, this enzyme is inhibited by the end product of the pathway that it initiates and is activated by the end product of a competitive pathway. This mechanism balances the amounts of different amino acids that are synthesized.

The regulatory domain in threonine deaminase is very similar in structure to the regulatory domain in 3-phosphoglycerate dehydrogenase (**Figure 25.19**). In the latter enzyme, regulatory domains of two subunits interact to form a dimeric

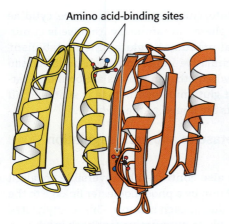

Amino acid-binding sites

**Dimeric regulatory domain of
3-phosphoglycerate dehydrogenase**

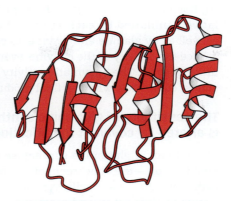

**Single-chain regulatory domain of
threonine deaminase**

FIGURE 25.19 The regulatory domains of 3-phosphoglycerate dehydrogenase and threonine deaminase are related structurally. The regulatory domain of 3-phosphoglycerate dehydrogenase is comprised of two subunits, while the corresponding domain of threonine deaminase is comprised of a single chain. Both overall structures have four α helices and eight β strands in similar locations. Sequence analyses have revealed this amino acid-binding regulatory domain to be present in other enzymes as well. [Drawn from 1PSD.pdb and 1TDJ.pdb.]

INTERACT with these models in
🌊 Achieve

serine-binding regulatory unit, and so the tetrameric enzyme contains two such regulatory units. Each unit is capable of binding two serine molecules. In threonine deaminase, the two regulatory domains are fused into a single unit with two differentiated amino acid-binding sites, one for isoleucine and the other for valine. Sequence analysis shows that similar regulatory domains are present in other amino acid biosynthetic enzymes. The similarities suggest that feedback-inhibition processes may have evolved by the linkage of specific regulatory domains to the catalytic domains of biosynthetic enzymes.

Enzyme multiplicity. In a process known as **enzyme multiplicity**, the committed step can be catalyzed by two or more isozymes, enzymes with essentially identical catalytic mechanisms but different regulatory properties (Section 7.2). For example, the phosphorylation of aspartate is the committed step in the biosynthesis of threonine, methionine, and lysine. Three distinct aspartokinases, which evolved by gene duplication, catalyze this reaction in E. coli (**Figure 25.20**). The catalytic domains of these enzymes show approximately 30% sequence identity. Although the mechanisms of catalysis are the same, their activities are regulated differently: one enzyme is not subject to feedback inhibition, another is inhibited by threonine, and the third is inhibited by lysine.

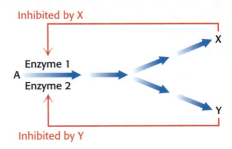

Inhibited by X

A   Enzyme 1
    Enzyme 2

X

Y

Inhibited by Y

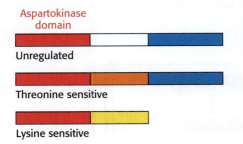

Aspartokinase domain

Unregulated

Threonine sensitive

Lysine sensitive

FIGURE 25.20 The three aspartokinases from E. coli have different regulatory domains. Each catalyzes the committed step in the biosynthesis of a different amino acid: (top) methionine, (middle) threonine, and (bottom) lysine. They have a catalytic domain in common (red) but differ in their regulatory domains (yellow and orange). The blue domain represents another enzyme (homoserine dehydrogenase) involved in aspartate metabolism. Thus, the top two aspartokinases are bifunctional enzymes.

Cumulative feedback inhibition. A common step is partly inhibited by multiple final products, each acting independently. The regulation of glutamine synthetase in E. coli is a striking example of **cumulative feedback inhibition**. Recall that glutamine is synthesized from glutamate, NH_4^+, and ATP. Glutamine synthetase consists of 12 identical 50-kDa subunits arranged in two hexagonal rings that face each other. This enzyme regulates the flow of nitrogen and hence plays a key role in controlling bacterial metabolism. The amide group of glutamine is a source of nitrogen in the biosyntheses of a variety of compounds, such as tryptophan, histidine, carbamoyl phosphate, glucosamine 6-phosphate, cytidine triphosphate, and adenosine monophosphate. Glutamine synthetase is cumulatively inhibited by each of these final products of glutamine metabolism, as well as by alanine and glycine. In cumulative inhibition, each inhibitor can reduce the activity of the enzyme, even when other inhibitors are bound at saturating levels. The enzymatic activity of glutamine synthetase is switched off almost completely when all final products are bound to the enzyme.

The sensitivity of glutamine synthetase to allosteric regulation is altered by covalent modification

The activity of glutamine synthetase is also controlled by reversible covalent modification—the attachment of an AMP unit by a phosphodiester linkage to the hydroxyl group of a specific tyrosine residue in each subunit. This adenylylated enzyme is less active and more susceptible to cumulative feedback inhibition than is the deadenylylated form. The covalently attached AMP unit is removed from the adenylylated enzyme by phosphorolysis.

The adenylylation and phosphorolysis reactions are catalyzed by the same enzyme, adenylyl transferase. This enzyme is composed of two homologous halves, one half catalyzing the adenylylation reaction and the other half the phosphorolytic deadenylylation reaction. What determines whether an AMP unit is added or removed? The specificity of adenylyl transferase is controlled by a regulatory protein P_{II}, a trimeric protein that can exist in two forms, unmodified (P_{II}) or covalently bound to UMP (P_{II}-UMP). The complex of P_{II} and adenylyl transferase catalyzes the attachment of an AMP unit to glutamine synthetase, which reduces its activity. Conversely, the complex of P_{II}-UMP and adenylyl transferase removes AMP from the adenylylated enzyme (**Figure 25.21**).

This scheme of regulation immediately raises the question, how is the modification of P_{II} controlled? P_{II} is converted into P_{II}-UMP by the attachment of uridine monophosphate to a specific tyrosine residue (Figure 25.21). This reaction, which is catalyzed by uridylyl transferase, is stimulated by ATP and α-ketoglutarate, whereas it is inhibited by glutamine. In turn, the UMP units on P_{II} are removed by hydrolysis, a reaction promoted by glutamine and inhibited by α-ketoglutarate. These opposing catalytic activities are present on a single polypeptide chain and are controlled so that the enzyme does not simultaneously catalyze uridylylation and hydrolysis. In essence, if glutamine is present, the covalent modification system favors adenylylation and inactivation

Tyrosine residue

AMP

HO   OH

Tyrosine residue modified by adenylylation

FIGURE 25.21 Glutamine synthetase is regulated by covalent modification. Adenylyl transferase (AT), in association with the regulatory protein P_{II}, adenylylates and inactivates the synthetase. When associated with P_{II} bound to UMP, AT deadenylylates the synthetase, thereby activating the enzyme. Uridylyl transferase (UT), the enzyme that modifies P_{II}, is allosterically regulated by α-ketoglutarate, ATP, and glutamine. [Information from D. L. Nelson and M. M. Cox, Lehninger Principles of Biochemistry, 7th ed. (W. H. Freeman and Company, 2013), Fig. 22.9.]

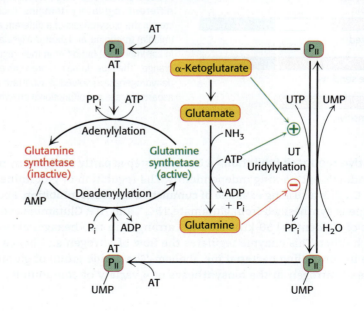

of glutamine synthetase. If glutamine is absent, as indicated by the presence of its precursors α-ketoglutarate and ATP, the control system results in the deadenylylation and activation of the synthetase.

The integration of nitrogen metabolism in a cell requires that a large number of input signals be detected and processed. In addition, the regulatory protein P_{II} also participates in regulating the transcription of genes for glutamine synthetase and other enzymes taking part in nitrogen metabolism. The evolution of covalent regulation superimposed on feedback inhibition provided many more regulatory sites and made possible a finer tuning of the flow of nitrogen in the cell. We have previously seen such a dual regulatory format in the regulation of glycogen metabolism (Section 21.6).

25.4 Amino Acids Are Precursors of Many Biomolecules

In addition to being the building blocks of proteins and peptides, amino acids serve as precursors of many kinds of small molecules that have important and diverse biological roles. Let us briefly survey some of the biomolecules that are derived from amino acids (**Figure 25.22**).

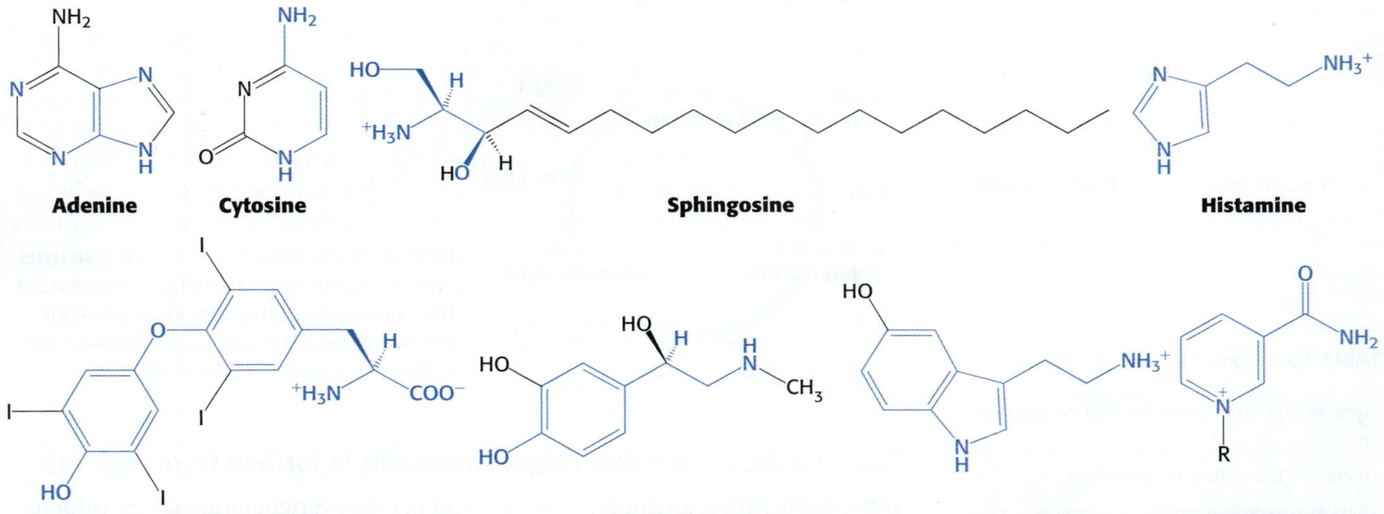

FIGURE 25.22 A diverse array of biomolecules are derived from amino acids. The atoms contributed by amino acids are shown in blue.

Purines and pyrimidines are derived largely from amino acids. The biosynthesis of these precursors of DNA, RNA, and numerous coenzymes will be discussed in detail in Chapter 26. The reactive terminus of sphingosine, an intermediate in the synthesis of sphingolipids (Section 12.2), comes from serine. Histamine, a potent vasodilator, is derived from histidine by decarboxylation. Tyrosine is a precursor of thyroxine (tetraiodothyronine, a hormone that modulates metabolism), epinephrine (Section 14.2), and melanin (a complex polymeric molecule responsible for skin pigmentation). The neurotransmitter serotonin (5-hydroxytryptamine) and the nicotinamide ring of NAD$^+$ are synthesized from tryptophan. Let us now consider in more detail three particularly important biochemicals derived from amino acids.

Glutathione, a gamma-glutamyl peptide, serves as a sulfhydryl buffer and an antioxidant

Glutathione, a tripeptide containing a sulfhydryl group, is a highly distinctive amino acid derivative with several important roles (**Figure 25.23**).

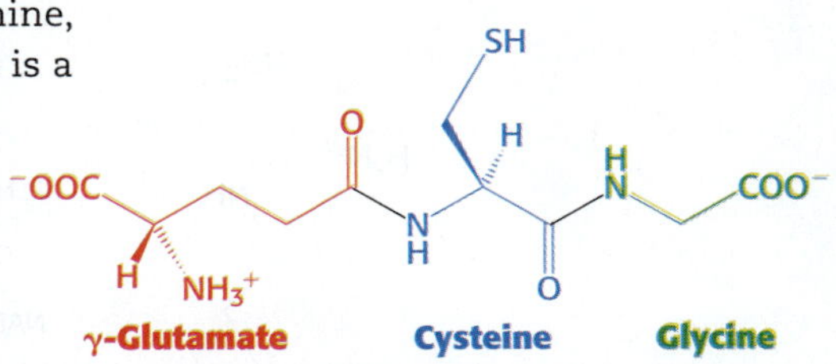

FIGURE 25.23 Glutathione is an important sulfhydryl buffer and antioxidant. This tripeptide consists of a cysteine residue flanked by a glycine residue and a glutamate residue that is linked to cysteine by an isopeptide bond between glutamate's side-chain carboxylate group and cysteine's amino group.

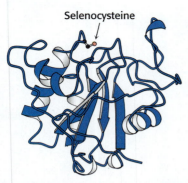

Selenocysteine

FIGURE 25.24 Glutathione peroxidase contains a selenocysteine residue in its active site. [Drawn from 1GP1.pdb.]

INTERACT with this model in
≋ Achie√e

For example, glutathione, present at high levels (~5 mM) in animal cells, protects red blood cells from oxidative damage by serving as a sulfhydryl buffer (Section 20.5). It cycles between a reduced thiol form (GSH) and an oxidized form (GSSG) in which two tripeptides are linked by a disulfide bond.

$$2\,GSH + RO{-}OH \rightleftharpoons GSSG + H_2O + ROH$$

GSSG is reduced to GSH by glutathione reductase, a flavoprotein that uses NADPH as the electron source. The ratio of GSH to GSSG in most cells is greater than 500. Glutathione plays a key role in detoxification by reacting with hydrogen peroxide and organic peroxides, the harmful by-products of aerobic life (Section 20.5).

Glutathione peroxidase, the enzyme catalyzing the reaction of GSH with peroxides, is remarkable in having a modified amino acid containing a selenium (Se) atom (**Figure 25.24**). Specifically, its active site contains the selenium analog of cysteine, in which selenium has replaced sulfur. The selenolate (E-Se⁻) form of this residue reduces the peroxide substrate to an alcohol and is in turn oxidized to selenenic acid (E-SeOH). Glutathione then comes into action by forming a selenosulfide adduct (E-Se-S-G). A second molecule of glutathione then regenerates the active form of the enzyme by attacking the selenosulfide to form oxidized glutathione (**Figure 25.25**).

GSSG
+ H⁺   E-Se⁻   ROOH
            Selenolate + H⁺

GSH   ROH

E-Se-S-G   E-SeOH
Selenosulfide   **Selenenic acid**

              H₂O   GSH

FIGURE 25.25 Glutathione peroxidase uses two molecules of glutathione to regenerate the selenolate active site after peroxide cleavage. [Information from O. Epp, R. Ladenstein, and A. Wendel. *Eur. J. Biochem.* 133:51–69, 1983.]

TABLE 25.3 Some functions of nitric oxide

Reduces blood pressure by relaxing vascular muscle
Increases blood flow to the kidney
Acts as a neurotransmitter to increase cerebral blood flow
Dilates pulmonary vessels
Mediates erectile function
Controls peristalsis of the gastrointestinal tract
Regulates inflammation

Nitric oxide, a short-lived signal molecule, is formed from arginine

Nitric oxide (NO) is an important messenger in many vertebrate signal-transduction processes, identified first as a relaxing factor in the cardiovascular system. It is now known to have a variety of roles not only in the cardiovascular system, but also in the immune and nervous systems (**Table 25.3**). NO has also been shown to stimulate mitochondrial biogenesis. This free-radical gas is produced endogenously from arginine in a complex reaction that is catalyzed by nitric oxide synthase. NADPH and O₂ are required for the synthesis of nitric oxide (**Figure 25.26**).

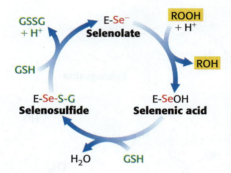

Arginine   **N-ω-Hydroxy-arginine**   **Citrulline**   **Nitric oxide**

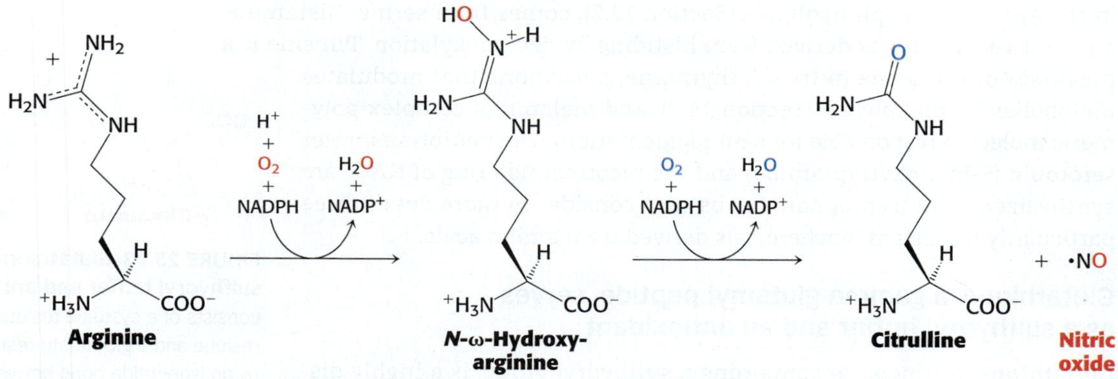

FIGURE 25.26 Nitric oxide is generated by the oxidation of arginine.

Nitric oxide acts by binding to and activating soluble guanylate cyclase, an important enzyme in signal transduction. Soluble guanylate cyclase is homologous to adenylate cyclase (Section 14.2) but includes a heme-containing domain that binds NO.

Amino acids are precursors for a number of neurotransmitters

Catecholamines, signal molecules with a variety of biological roles, are derived from tyrosine (**Figure 25.27**). Norepinephrine and dopamine are neurotransmitters, and epinephrine is familiar to us for the regulation of fuel use, stimulating glycogen breakdown and lipid mobilization (Section 14.2). In general, the catecholamines are associated with stress and play a role in the "fight or flight" response.

Tryptophan is a precursor for two neurotransmitters (**Figure 25.28A**). Serotonin functions in several ways. Most notably, serotonin regulates moods and may be involved in alleviating depression. Serotonin is also found in the gastrointestinal tract, where it regulates intestinal contraction, and platelets, where it facilitates vasoconstriction. Serotonin can be metabolized to another hormone, melatonin, which functions to maintain our natural sleep-wake cycle (**Figure 25.28B**).

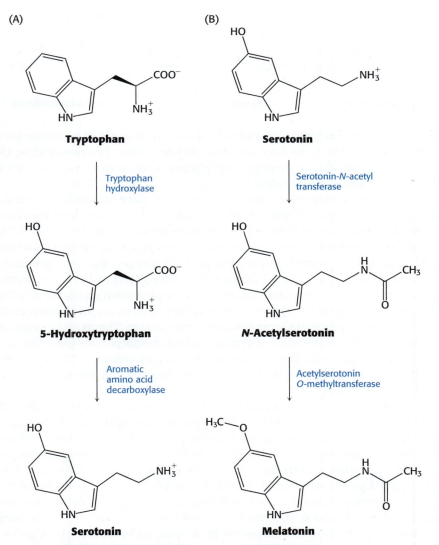

(A)

Tryptophan

↓ Tryptophan hydroxylase

5-Hydroxytryptophan

↓ Aromatic amino acid decarboxylase

Serotonin

(B)

Serotonin

↓ Serotonin-*N*-acetyl transferase

***N*-Acetylserotonin**

↓ Acetylserotonin *O*-methyltransferase

Melatonin

FIGURE 25.28 The neurotransmitters serotonin and melatonin are synthesized from tryptophan. (A) Serotonin is synthesized from tryptophan in two enzymatic steps. (B) Melatonin, in turn, is generated from serotonin in two additional steps.

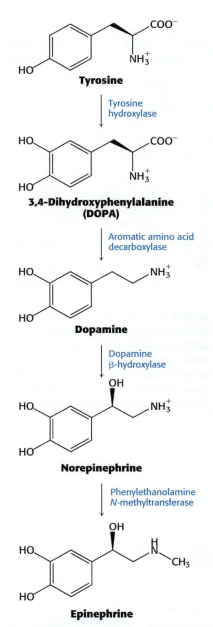

Tyrosine

↓ Tyrosine hydroxylase

3,4-Dihydroxyphenylalanine (DOPA)

↓ Aromatic amino acid decarboxylase

Dopamine

↓ Dopamine β-hydroxylase

Norepinephrine

↓ Phenylethanolamine *N*-methyltransferase

Epinephrine

FIGURE 25.27 Tyrosine is a precursor for the catecholamines. The catecholamines—dopamine, norepinephrine, and epinephrine—function as stress-signal molecules.

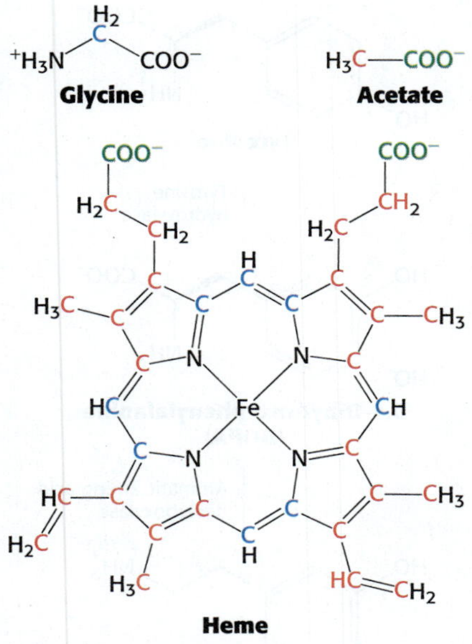

FIGURE 25.29 The origins of atoms in heme were revealed by isotope-labeling studies. The final structure of heme contains 8 carbons from the α-carbon of glycine (blue), 24 carbons from the methyl group of acetate (red), and two carbons from the carboxyl group of acetate (green).

Porphyrins are synthesized from glycine and succinyl coenzyme A

The participation of an amino acid in the biosynthesis of the porphyrin rings of hemes and chlorophylls was first revealed by isotope-labeling experiments carried out by David Shemin and his colleagues. In 1945, they showed that the nitrogen atoms of heme were labeled after the feeding of [^{15}N] glycine to human subjects, whereas the ingestion of [^{15}N] glutamate resulted in very little labeling.

Experiments using ^{14}C, which had just become available, revealed that 8 of the carbon atoms of heme in nucleated duck erythrocytes are derived from the α-carbon atom of glycine and none from the carboxyl carbon atom. Subsequent studies demonstrated that the other 26 carbon atoms of heme can arise from acetate. Moreover, the ^{14}C in methyl-labeled acetate emerged in 24 of these carbon atoms, whereas the ^{14}C in carboxyl-labeled acetate appeared only in the other 2 (Figure 25.29).

This highly distinctive labeling pattern suggested that acetate is converted to succinyl-CoA through enzymes from the citric acid cycle (Section 17.3) and that a heme precursor is formed by the condensation of glycine with succinyl-CoA. Indeed, the first step in the biosynthesis of porphyrins in mammals is the condensation of glycine and succinyl CoA to form δ-aminolevulinate.

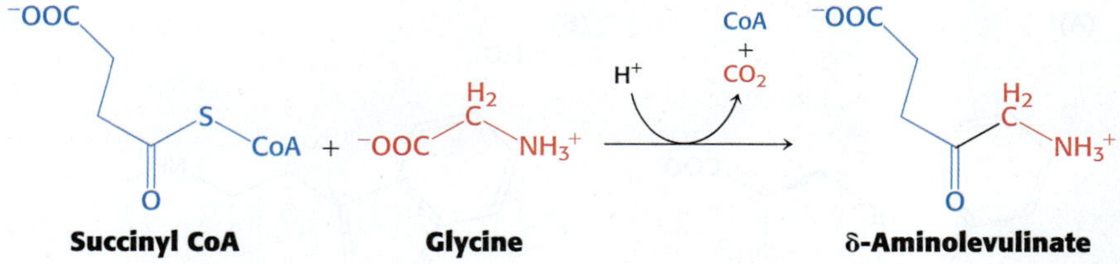

This reaction is catalyzed by δ-aminolevulinate synthase, a PLP enzyme present in mitochondria. Consistent with the labeling studies described earlier, the carbon atom from the carboxyl group of glycine is lost as carbon dioxide, while the α-carbon remains in δ-aminolevulinate.

The reactions required for heme synthesis take place in both the mitochondria and the cytoplasm (Figure 25.30), revealing another example of intercompartmental cooperation (Figure 16.30). δ-Aminolevulinate is generated in the mitochondria and is then transported into the cytoplasm, where two molecules of δ-aminolevulinate condense to form porphobilinogen, the next intermediate. Four molecules of porphobilinogen then condense head to tail to form a linear tetrapyrrole in a reaction catalyzed by porphobilinogen deaminase. The enzyme-bound linear tetrapyrrole then cyclizes to form uroporphyrinogen III, which has an asymmetric arrangement of side chains. This reaction requires a cosynthase. In the presence of synthase alone, uroporphyrinogen I, the nonphysiological symmetric isomer, is produced. Uroporphyrinogen III is also a key intermediate in the synthesis of vitamin B$_{12}$ by bacteria and that of chlorophyll by bacteria and plants.

The porphyrin skeleton is now formed. Subsequent reactions alter the side chains and the degree of saturation of the porphyrin ring (Figure 25.30). Coproporphyrinogen III is formed by the decarboxylation of the acetate side chains and then transported into the mitochondria. There, the desaturation of the porphyrin ring and the conversion of two of the propionate side chains into vinyl groups yield protoporphyrin IX. The chelation of iron finally gives heme, the prosthetic group of proteins such as myoglobin, hemoglobin, catalase, peroxidase, and cytochrome c. The insertion of the ferrous form of iron is catalyzed by ferrochelatase. Iron is transported in the plasma by transferrin, a protein that binds two ferric ions, and is stored in tissues inside molecules of ferritin (Section 31.6).

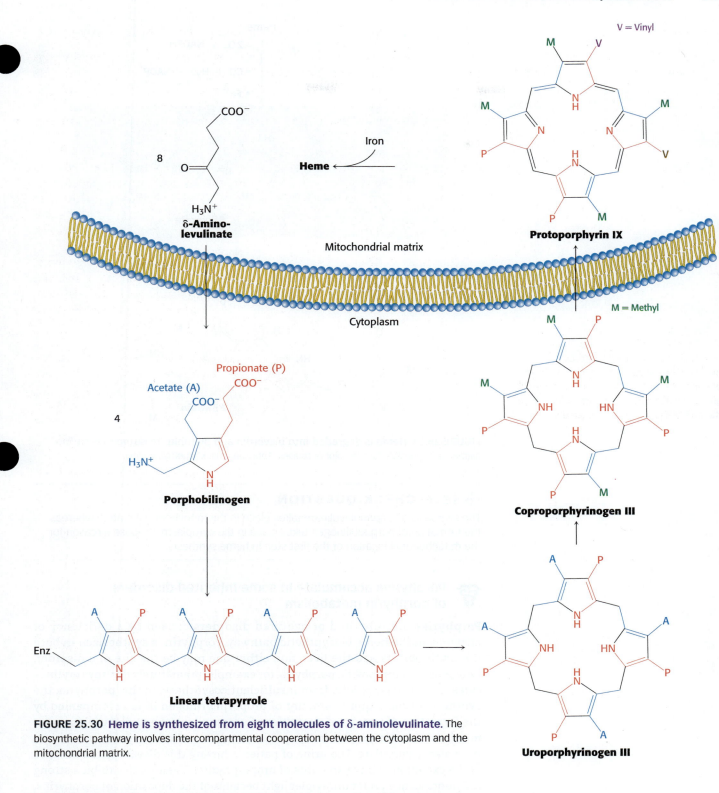

FIGURE 25.30 Heme is synthesized from eight molecules of δ-aminolevulinate. The biosynthetic pathway involves intercompartmental cooperation between the cytoplasm and the mitochondrial matrix.

Studies with ^{15}N-labeled glycine revealed that the normal human erythrocyte has a life span of about 120 days. The first step in the degradation of the heme group is the cleavage of its α-methine bridge to form the green pigment biliverdin, a linear tetrapyrrole (**Figure 25.31**). This reaction is catalyzed by heme oxygenase. The central methine bridge of biliverdin is then reduced by biliverdin reductase to form bilirubin, an orange-yellow pigment. The changing color of a bruise is a highly graphic indicator of these degradative reactions.

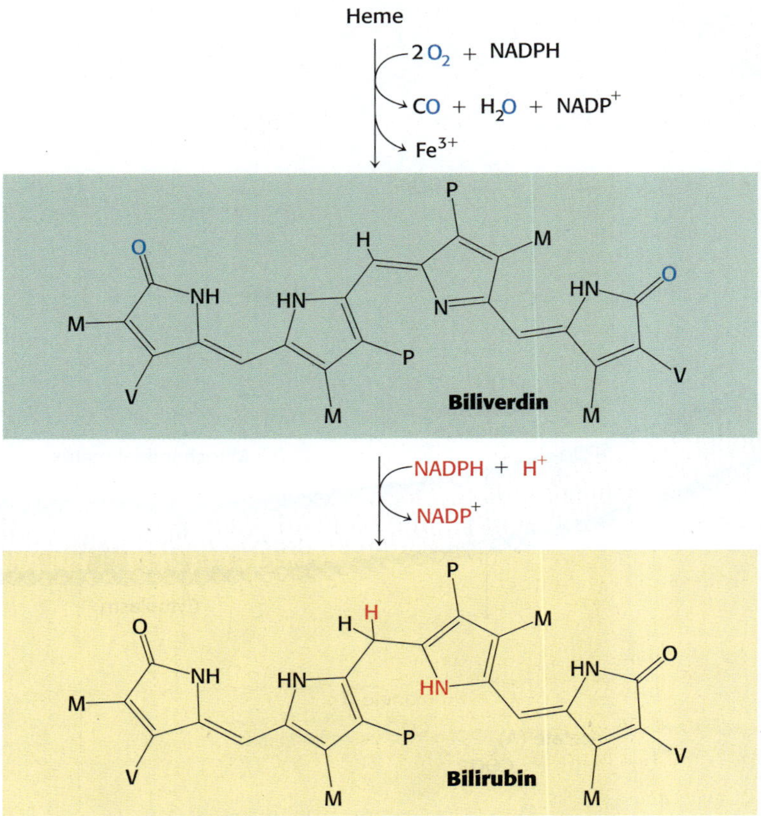

FIGURE 25.31 Heme is degraded into biliverdin and bilirubin. These products are the pigments responsible for the color of bruises. Abbreviations: M, methyl; V, vinyl.

❖ SELF-CHECK QUESTION

The synthesis of δ-aminolevulinate takes place in the mitochondrial matrix, whereas the formation of porphobilinogen takes place in the cytoplasm. Propose a reason for the mitochondrial location of the first step in heme synthesis.

⚕ Porphyrins accumulate in some inherited disorders of porphyrin metabolism

Porphyrias are inherited or acquired disorders caused by a deficiency of enzymes in the heme biosynthetic pathway. Porphyrin is synthesized in both the erythroblasts and the liver, and either one may be the site of a disorder. Congenital erythropoietic porphyria, for example, prematurely destroys erythrocytes. This disease results from insufficient cosynthase. In this porphyria, the synthesis of the required amount of uroporphyrinogen III is accompanied by the formation of very large quantities of uroporphyrinogen I, the useless symmetric isomer. Uroporphyrin I, coproporphyrin I, and other symmetric derivatives also accumulate. The urine of patients having this disease is red because of the excretion of large amounts of uroporphyrin I. Their teeth exhibit a strong red fluorescence under ultraviolet light because of the deposition of porphyrins. Furthermore, their skin is usually very sensitive to light because photoexcited porphyrins are quite reactive.

Acute intermittent porphyria is the most prevalent of the porphyrias affecting the liver. This porphyria is characterized by the overproduction of porphobilinogen and δ-aminolevulinate, which results in severe abdominal pain and neurological dysfunction. The "madness" of George III, king of England during the American Revolution, may have been due to this porphyria.

Summary

25.1 Nitrogen Fixation: Microorganisms Use ATP and a Powerful Reductant to Reduce Atmospheric Nitrogen to Ammonia

- Nitrogen fixation is the process by which N_2 is reduced to NH_3.
- Using ATP and reduced ferredoxin, microorganisms use the nitrogenase complex to catalyze nitrogen fixation.
- Higher organisms use the fixed nitrogen in the form of NH_4^+ to synthesize amino acids, nucleotides, and other nitrogen-containing biomolecules.
- The major points of entry of NH_4^+ into metabolism are glutamine or glutamate.

25.2 Amino Acids Are Made from Intermediates of the Citric Acid Cycle and Other Major Pathways

- Human beings can synthesize 11 of the basic set of 20 amino acids. These amino acids are called *nonessential*.
- The *essential* amino acids must be supplied in the diet.
- The pathways for the synthesis of nonessential amino acids are quite simple, while those of the essential amino acids tend to contain greater than 5 catalytic steps.
- A transamination reaction takes place in the synthesis of most amino acids. At this step, the chirality of the amino acid is established.
- Alanine, aspartate, and glutamate are synthesized in one step by the addition of an amino group to pyruvate, oxaloacetate, and α-ketoglutarate, respectively.
- Glutamine is synthesized from NH_4^+ and glutamate, and asparagine is synthesized similarly

in bacteria. In mammals, the nitrogen donor for asparagine is glutamine.

- Proline and arginine are derived from glutamate.
- Serine, formed from 3-phosphoglycerate, is the precursor of glycine and cysteine.
- Tyrosine is synthesized by the hydroxylation of phenylalanine, an essential amino acid.
- In many of these pathways, tetrahydrofolate serves as an important carrier of one-carbon units, while S-adenosylmethionine acts as a donor of methyl groups.

25.3 Feedback Inhibition Regulates Amino Acid Biosynthesis

- Most of the pathways of amino acid biosynthesis are regulated by feedback inhibition in which the committed step is allosterically inhibited by the final product.
- The regulation of branched pathways requires extensive interaction among the branches that includes both negative and positive regulation.
- The regulation of glutamine synthetase in *E. coli* is a striking demonstration of cumulative feedback inhibition and of control by a cascade of reversible covalent modifications.

25.4 Amino Acids Are Precursors of Many Biomolecules

- Amino acids are precursors of a variety of biomolecules.
- Glutathione (γ-Glu-Cys-Gly) serves as a sulfhydryl buffer and detoxifying agent.
- Nitric oxide, a short-lived messenger, is formed from arginine.
- Porphyrins, the iron-binding groups in heme, are synthesized from glycine and succinyl CoA.

Key Terms

nitrogen fixation (p. 764)

nitrogenase complex (p. 764)

essential amino acids (p. 768)

nonessential amino acids (p. 768)

pyridoxal phosphate (PLP) (p. 769)

tetrahydrofolate (p. 772)

S-adenosylmethionine (SAM) (p. 774)

activated methyl cycle (p. 775)

substrate channeling (p. 779)

enzyme multiplicity (p. 781)

cumulative feedback inhibition (p. 781)

glutathione (p. 783)

nitric oxide (NO) (p. 784)

porphyrias (p. 788)

Problems

1. Identify the two components of the nitrogenase complex and describe their specific tasks. ❖ 1

2. Which of the following statements about nitrogen fixation are correct? ❖ 1

(a) Atmospheric nitrogen is reduced to the biologically useful form NH_3 (or NH_4^+).

(b) Nitrogen fixation in nature and in the lab requires a metal cofactor or catalyst.

(c) Nitrogen fixation is energetically neutral, using a negligible amount of ATP.

(d) The enzyme nitrogenase, which takes part in nitrogen fixation, is inactivated by oxygen.

(e) Ammonium from the atmosphere is fixed to more usable forms of nitrogen, such as nitrate (NO_3^-).

(f) The availability of fixed nitrogen limits biological productivity.

3. The first step in the incorporation of NH_4^+ into amino acids is catalyzed by glutamate dehydrogenase, in which an α-keto acid is converted to glutamate by transamination. ❖ 1

(a) Draw the structure of the α-keto acid precursor to glutamate.

(b) Besides glutamate, what are the other products of this reaction?

4. The entry of nitrogen into the biosphere is not an energetically cheap process. How many molecules of ATP must be hydrolyzed per N_2 molecule fixed? ❖ 1

5. The nitrogenase complex consists of two protein components: the reductase and the nitrogenase. What is the function of each of these proteins? ❖ 1

6. What are the seven precursors of the 20 amino acids? ❖ 2

7. Write a balanced equation for the synthesis of alanine from glucose. ❖ 2

8. Which of the 20 amino acids can be synthesized directly from a common metabolic intermediate by a transamination reaction? ❖ 2

9. Which amino acids are formed from the glycolysis intermediate 3-phosphoglycerate? ❖ 2

10. Which of the following molecules is the most direct source for the sulfur atom in the sulfur-containing amino acids in human beings? ❖ 2

(a) Biotin   (d) Tetrahydrofolate

(b) Homocysteine   (e) Coenzyme A

(c) Glutathione

11. The atoms from tryptophan shaded below are derived from two other amino acids. Name them. ❖ 2

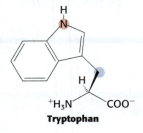

Tryptophan

12. Certain species of bacteria possess an enzyme, ornithine cyclodeaminase, that can catalyze the conversion of L-ornithine into L-proline in a single catalytic cycle.

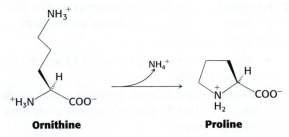

Ornithine   **Proline**

The enzyme lysine cyclodeaminase has also been identified. Predict the product of the reaction catalyzed by lysine cyclodeaminase.

13. For the following example of a branched pathway, propose a feedback inhibition scheme that would result in the production of equal amounts of Y and Z. ❖ 3

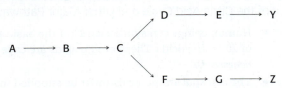

14. Consider the branched pathway in problem 13. The first common step (A ⟶ B) is partly inhibited by both of the final products, each acting independently of the other. Suppose that a high level of Y alone decreased the rate of the A ⟶ B step from 100 to $60\ s^{-1}$ and that a high level of Z alone decreased the rate from 100 to $40\ s^{-1}$. What would the rate be in the presence of high levels of both Y and Z? ❖ 3

15. What are the intermediates in the flow of nitrogen from N_2 to heme? ❖ 4

16. David Shemin and coworkers used acetate-labeling experiments to conclude that succinyl-CoA is a key intermediate in the biosynthesis of heme. Identify the intermediates in the conversion of acetate into succinyl-CoA. ❖ 4

17. Which of the following statements about heme are correct? ❖ 4

(a) Phenylalanine is a precursor to heme.

(b) Heme is found in trypsin.

(c) Heme is found in cytochrome c.

(d) Heme contains a magnesium ion.

(e) Succinyl-CoA is a precursor to heme.

(f) Heme contains an iron ion.

18. In this chapter, we considered three different cofactors/cosubstrates that act as carriers of one-carbon units. Name them.

19. Individuals with an inactivating mutation in phenylalanine hydroxylase require an additional amino acid in their diet which is not an essential amino acid in the ordinary human diet. Which one?

Nucleotide Biosynthesis

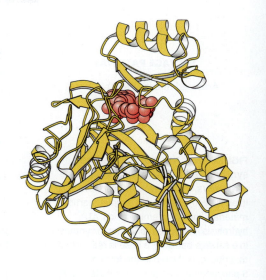

An Arduino board is a versatile electronics component that can be modified to perform a variety of functions. The ability to be adjusted allows this component to participate in many related processes. Similarly, the ATP grasp domain depicted at right is found in many enzymes that participate in nucleotide biosynthesis; evolutionary modifications have allowed enzymes containing such domains to catalyze a range of related reactions.

Sergey Privalov/ Shutterstock

❖ LEARNING GOALS

By the end of this chapter, you should be able to:

1. Describe how pyrimidine nucleotides are synthesized.
2. Describe the pathway for purine nucleotide synthesis.
3. Explain how deoxyribonucleotides are formed.
4. Name the regulatory steps in nucleotide synthesis.
5. Identify the pathological conditions resulting from impaired nucleotide synthesis.

OUTLINE

26.1 Nucleotides Can Be Synthesized by de Novo or Salvage Pathways

26.2 The Pyrimidine Ring Is Assembled from CO_2, Ammonia, and Aspartate

26.3 Purine Bases Can Be Synthesized from Glycine, Aspartate, and Other Components

26.4 Deoxyribonucleotides Are Synthesized by the Reduction of Ribonucleotides

26.5 Key Steps in Nucleotide Biosynthesis Are Regulated by Feedback Inhibition

26.6 Disruptions in Nucleotide Metabolism Can Cause Pathological Conditions

In this chapter, we describe the biosynthesis of nucleotides, continuing along the path begun in Chapter 25, which described the incorporation of nitrogen into amino acids from inorganic sources such as nitrogen gas. The amino acids glycine and aspartate are the scaffolds on which the ring systems present in nucleotides are assembled. Furthermore, aspartate and the side chain of glutamine serve as sources of NH_2 groups in the formation of nucleotides.

Nucleotide biosynthetic pathways are tremendously important as intervention points for therapeutic agents. Many of the most widely used drugs in the treatment of cancer block steps in nucleotide biosynthesis, particularly steps in the synthesis of DNA precursors.

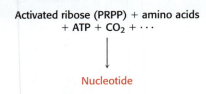

DE NOVO PATHWAY

Activated ribose (PRPP) + amino acids
+ ATP + CO_2 + $\cdots$

↓

Nucleotide

SALVAGE PATHWAY

Activated ribose (PRPP) + base

↓

Nucleotide

FIGURE 26.1 Nucleotides can be synthesized by de novo and salvage pathways. In de novo synthesis, the base itself is synthesized from simpler starting materials, including amino acids. ATP hydrolysis is required for de novo synthesis. In a salvage pathway, a base is reattached to a ribose, activated in the form of 5-phosphoribosyl-1-pyrophosphate (PRPP).

26.1 Nucleotides Can Be Synthesized by de Novo or Salvage Pathways

Nucleotides are key biomolecules required for a variety of life processes.

- Nucleotides are the activated precursors of nucleic acids, necessary for the replication of the genome and the transcription of the genetic information into RNA.
- An adenine nucleotide, ATP, is the universal currency of energy.
- A guanine nucleotide, GTP, also serves as an energy source for a more select group of biological processes.
- Nucleotide derivatives such as UDP-glucose participate in biosynthetic processes.
- Nucleotides such as cyclic AMP and cyclic GMP are essential components of signal-transduction pathways.
- ATP acts as the donor of phosphoryl groups transferred by protein kinases in a variety of signaling pathways.

The pathways for the biosynthesis of nucleotides fall into two classes: *de novo* pathways and *salvage* pathways (**Figure 26.1**). In **de novo pathways**, the nucleobases are assembled from scratch; meaning, from simpler compounds. The framework for a pyrimidine base is assembled first and then attached to ribose. In contrast, the framework for a purine base is synthesized piece by piece directly onto a ribose-based structure. These de novo pathways each comprise a small number of elementary reactions that are repeated with variations to generate different nucleotides, as might be expected for pathways that appeared very early in evolution. In **salvage pathways**, preformed bases are recovered and reconnected to a ribose unit.

De novo pathways lead to the synthesis of *ribonucleotides*. However, DNA is built from *deoxyribonucleotides*. Consistent with the notion that RNA preceded DNA in the course of evolution, all deoxyribonucleotides are synthesized from the corresponding ribonucleotides. The deoxyribose sugar is generated by the reduction of ribose within a fully formed nucleotide. Furthermore, the methyl group that distinguishes the thymine of DNA from the uracil of RNA is added at the last step in the pathway.

Recall from Chapter 8 that a *nucleoside* is a purine or pyrimidine base linked to a sugar and that a *nucleotide* is a phosphate ester of a nucleoside. The names of the major bases of RNA and DNA, and of their nucleoside and nucleotide derivatives, are given in **Table 26.1**.

TABLE 26.1 Nomenclature of bases, nucleosides, and nucleotides

RNA		
Base	**Ribonucleoside**	**Ribonucleotide (5′-monophosphate)**
Adenine (A)	Adenosine	Adenylate (AMP)
Guanine (G)	Guanosine	Guanylate (GMP)
Uracil (U)	Uridine	Uridylate (UMP)
Cytosine (C)	Cytidine	Cytidylate (CMP)
DNA		
Base	**Deoxyribonucleoside**	**Deoxyribonucleotide (5′-monophosphate)**
Adenine (A)	Deoxyadenosine	Deoxyadenylate (dAMP)
Guanine (G)	Deoxyguanosine	Deoxyguanylate (dGMP)
Thymine (T)	Thymidine	Thymidylate (TMP)
Cytosine (C)	Deoxycytidine	Deoxycytidylate (dCMP)

26.2 The Pyrimidine Ring Is Assembled from CO$_2$, Ammonia, and Aspartate

In de novo synthesis of pyrimidines, the ring is synthesized first and then attached to a ribose phosphate to form a pyrimidine nucleotide (**Figure 26.2**). Pyrimidine rings are assembled from bicarbonate, aspartate, and ammonia. Although an ammonia molecule already present in solution can be used, the ammonia is usually produced from the hydrolysis of the side chain of glutamine.

Bicarbonate and other oxygenated carbon compounds are activated by phosphorylation

The first step in de novo pyrimidine biosynthesis is the synthesis of carbamoyl phosphate from bicarbonate and ammonia in a multistep process, requiring the cleavage of two molecules of ATP. This reaction is catalyzed by carbamoyl phosphate synthetase II (CPS II). Recall that carbamoyl phosphate synthetase I facilitates ammonia incorporation into urea (Section 23.4). Carbamoyl phosphate synthetase II is a dimer composed of a small subunit that hydrolyzes glutamine to form NH$_3$ and a large subunit that completes the synthesis of carbamoyl phosphate. Analysis of the structure of the large subunit reveals two homologous domains, each of which catalyzes an ATP-dependent step (**Figure 26.3**).

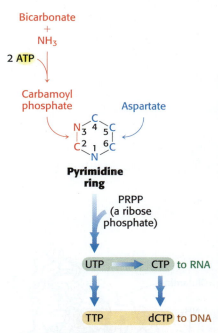

FIGURE 26.2 Pyrimidine nucleotides can be synthesized de novo. The C-2 and N-3 atoms in the pyrimidine ring come from carbamoyl phosphate, whereas the other atoms of the ring come from aspartate.

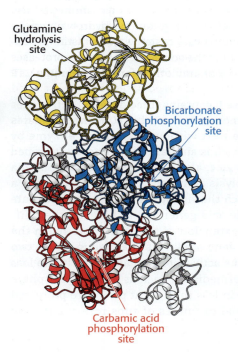

Glutamine hydrolysis site

Bicarbonate phosphorylation site

Carbamic acid phosphorylation site

INTERACT with this model in Achieve

FIGURE 26.3 Carbamoyl phosphate synthetase II includes sites for three reactions. This enzyme consists of two chains. The smaller chain (yellow) contains a site for glutamine hydrolysis to generate ammonia. The larger chain includes two ATP-grasp domains (blue and red). In one ATP-grasp domain (blue), bicarbonate is phosphorylated to carboxyphosphate, which then reacts with ammonia to generate carbamic acid. In the other ATP-grasp domain, the carbamic acid is phosphorylated to produce carbamoyl phosphate. [Drawn from 1JDB.pdb.]

In the first step, bicarbonate is phosphorylated by ATP to form carboxyphosphate and ADP. Ammonia then reacts with carboxyphosphate to form carbamic acid and inorganic phosphate.

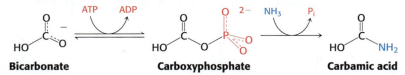

Bicarbonate   **Carboxyphosphate**   **Carbamic acid**

The active site for this reaction lies in a domain formed by the amino-terminal third of CPS. This domain forms a structure called an **ATP-grasp fold**, which surrounds ATP and holds it in an orientation suitable for nucleophilic attack at the γ phosphoryl group. Proteins containing ATP-grasp folds catalyze the formation of carbon–nitrogen bonds through acyl-phosphate intermediates. Such ATP-grasp folds are widely used in nucleotide biosynthesis.

In the second step catalyzed by carbamoyl phosphate synthetase II, carbamic acid is phosphorylated by another molecule of ATP to form carbamoyl phosphate.

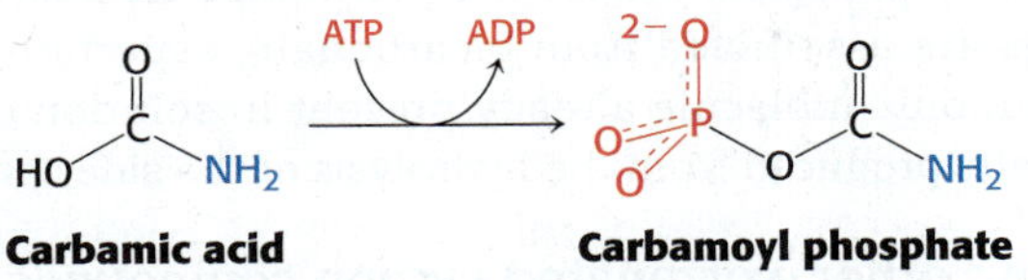

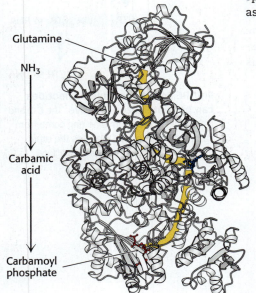

Carbamic acid → **Carbamoyl phosphate** (ATP → ADP)

This reaction takes place in a second ATP-grasp domain within the enzyme. The active sites leading to carbamic acid formation and carbamoyl phosphate formation are very similar, revealing that this enzyme evolved by a gene duplication event. Indeed, duplication of a gene encoding an ATP-grasp domain followed by specialization was central to the evolution of nucleotide biosynthetic processes, as we shall see.

The side chain of glutamine can be hydrolyzed to generate ammonia

Glutamine is the primary source of ammonia for carbamoyl phosphate synthetase II. The small subunit of the enzyme hydrolyzes glutamine to form ammonia and glutamate. The active site of the glutamine-hydrolyzing component contains a catalytic dyad comprising a cysteine and a histidine residue. Such a catalytic dyad, reminiscent of the active site of cysteine proteases (Figure 6.16A), is conserved in a family of amidotransferases, including CTP synthetase and GMP synthetase.

Carbamoyl phosphate synthetase II contains three different active sites (Figure 26.3), separated from one another by a total of 80 Å. Intermediates generated at one site move to the next without leaving the enzyme by means of substrate channeling, which is similar to the process described for tryptophan synthetase (**Figure 26.4**; also Figure 25.16). The ammonia generated in the glutamine-hydrolysis active site travels 45 Å through a channel within the enzyme to reach the site at which carboxyphosphate has been generated. The carbamic acid generated at this site then diffuses an additional 35 Å through an extension of the channel to reach the site at which carbamoyl phosphate is generated. This channeling serves two roles: (1) intermediates generated at one active site are captured with no loss caused by diffusion; and (2) labile intermediates, such as carboxyphosphate and carbamic acid (which decompose in less than 1 s at pH 7), are protected from hydrolysis. We will see additional examples of substrate channeling later in this chapter.

The caption (left column)

Glutamine

NH_3

Carbamic acid

Carbamoyl phosphate

FIGURE 26.4 The three active sites of carbamoyl phosphate synthetase II are linked by a channel through which intermediates pass. The channel is shown in yellow. Glutamine enters one active site, and carbamoyl phosphate, which includes the nitrogen atom from the glutamine side chain, leaves another 80 Å away. [Drawn from 1JDB.pdb.]

INTERACT with this model in
 Achie√e

The pyrimidine ring is completed and coupled to ribose

Carbamoyl phosphate reacts with aspartate to form carbamoylaspartate in a reaction catalyzed by aspartate transcarbamoylase. Carbamoylaspartate then cyclizes to form dihydroorotate, which is then oxidized to form orotate.

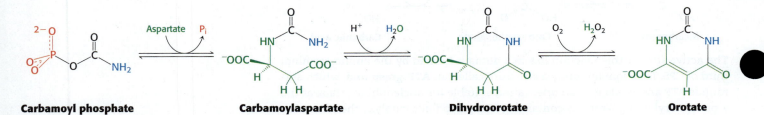

Carbamoyl phosphate — (Aspartate, P_i) → **Carbamoylaspartate** — (H^+, H_2O) → **Dihydroorotate** — (O_2, H_2O_2) → **Orotate**

In mammals, the enzymes that form orotate are part of a single large polypeptide chain called CAD, for <u>c</u>arbamoyl phosphate synthase, <u>a</u>spartate transcarbamoylase and <u>d</u>ihydroorotase.

At this stage, orotate couples to ribose, in the form of **5-phosphoribosyl-1-pyrophosphate (PRPP)**, a form of ribose activated to accept nucleobases. 5-Phosphoribosyl-1-pyrophosphate synthetase synthesizes PRPP by adding a pyrophosphate from ATP to ribose-5-phosphate, which is formed by the pentose phosphate pathway.

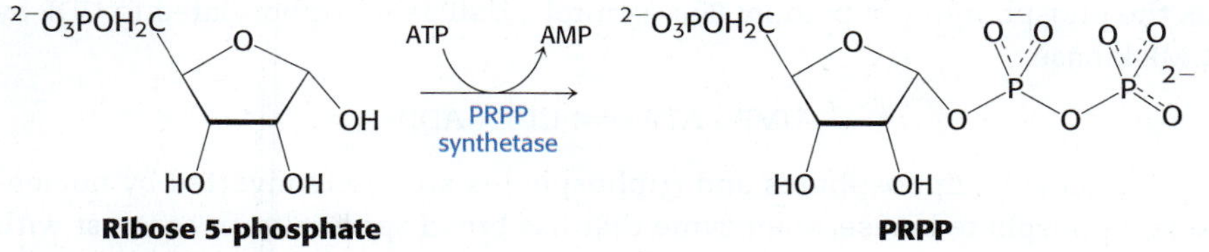

Ribose 5-phosphate   **PRPP**

Orotate reacts with PRPP to form **orotidylate**, a pyrimidine nucleotide. This reaction is driven by the hydrolysis of pyrophosphate. The enzyme that catalyzes this addition, orotate phosphoribosyltransferase, is homologous to a number of other phosphoribosyltransferases that add different groups to PRPP to form the other nucleotides.

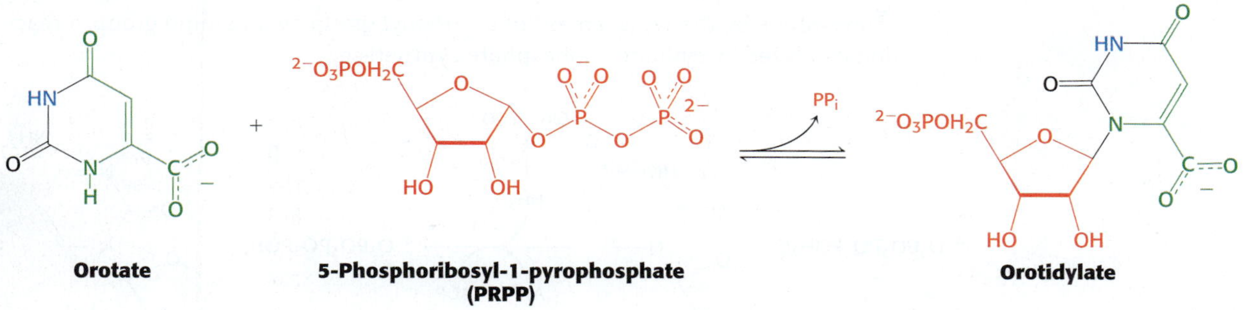

Orotate   **5-Phosphoribosyl-1-pyrophosphate (PRPP)**   **Orotidylate**

Orotidylate is then decarboxylated to form uridylate (UMP), a major pyrimidine nucleotide that is a precursor to RNA. This reaction is catalyzed by orotidylate decarboxylase, also called orotidine 5'-phosphate decarboxylase.

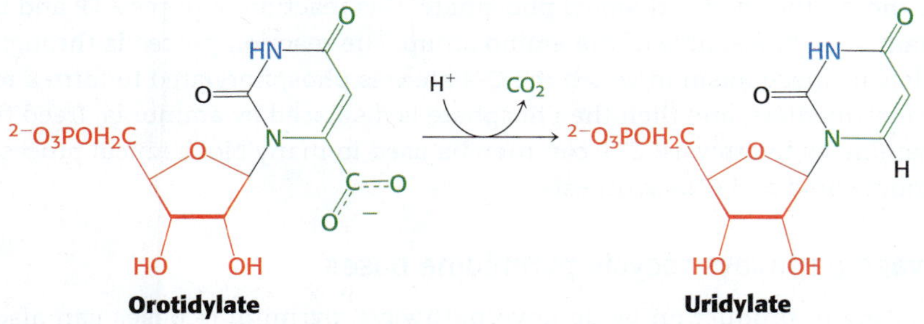

Orotidylate   **Uridylate**

Orotidylate decarboxylase is one of the most proficient enzymes known. In its absence, decarboxylation is extremely slow and is estimated to take place once every 78 million years; with the enzyme present, it takes place approximately once per second, a rate enhancement of 10^{17}-fold. The phosphoribosyltransferase and decarboxylase activities are located on the same polypeptide chain, providing another example of a bifunctional enzyme. The bifunctional enzyme is called *uridine monophosphate synthetase*.

❖ **SELF–CHECK QUESTION**

Identify the sources of the atoms that form the pyrimidine ring.

Nucleotide mono-, di-, and triphosphates are interconvertible

How is the other major pyrimidine ribonucleotide, cytidine, formed? It is synthesized from the uracil base of UMP, but the synthesis can take place only after UMP has been converted into UTP. Recall that the diphosphates and triphosphates are the active forms of nucleotides in biosynthesis and energy conversions. Nucleoside monophosphates are converted into nucleoside triphosphates in stages. First, nucleoside monophosphates are converted into diphosphates by specific nucleoside monophosphate kinases that utilize ATP as the phosphoryl-group donor. For example, UMP is phosphorylated to UDP by UMP kinase.

$$\text{UMP} + \text{ATP} \rightleftharpoons \text{UDP} + \text{ADP}$$

Nucleoside diphosphates and triphosphates are interconverted by nucleoside diphosphate kinase, an enzyme that has broad specificity, in contrast with the monophosphate kinases. X and Y represent any of several ribonucleosides or even deoxyribonucleosides:

$$\text{XDP} + \text{YTP} \rightleftharpoons \text{XTP} + \text{YDP}$$

CTP is formed by amination of UTP

After uridine triphosphate has been formed, it can be transformed into cytidine triphosphate by the replacement of a carbonyl group by an amino group, a reaction catalyzed by cytidine triphosphate synthetase.

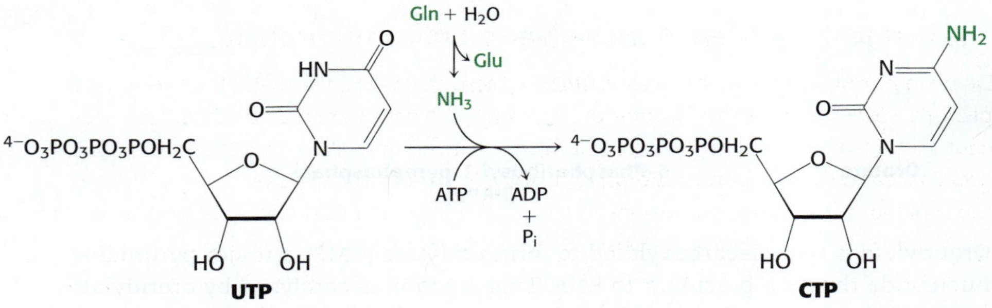

Like the synthesis of carbamoyl phosphate, this reaction requires ATP and uses glutamine as the source of the amino group. The reaction proceeds through an analogous mechanism in which the O-4 atom is phosphorylated to form a reactive intermediate, and then the phosphate is displaced by ammonia, freed from glutamine by hydrolysis. CTP can then be used in many biochemical processes, including lipid and RNA synthesis.

Salvage pathways recycle pyrimidine bases

In adding to production by de novo pathways, pyrimidine bases can also be recovered from the breakdown products of DNA and RNA through salvage pathways, as noted earlier. In these pathways, a preformed base is reincorporated into a nucleotide. We will consider the salvage pathway for the pyrimidine base thymine. Thymine is found in DNA and base-pairs with adenine in the DNA double helix. Thymine released from degraded DNA is salvaged in two steps. First, thymine is converted into the nucleoside thymidine by thymidine phosphorylase.

$$\text{Thymine} + \text{deoxyribose-1-phosphate} \rightleftharpoons \text{thymidine} + \text{P}_i$$

Thymidine is then converted into a nucleotide by thymidine kinase.

$$\text{Thymidine} + \text{ATP} \rightleftharpoons \text{TMP} + \text{ADP}$$

Viral thymidine kinase differs from the mammalian enzyme and thus provides a therapeutic target. For instance, herpes simplex infections are treated with acyclovir (acycloguanosine), which viral thymidine kinase converts into acyclovir monophosphate by adding a phosphate to the hydroxyl group of acyclovir. Viral thymidine kinase binds acyclovir more than 200 times as tightly as cellular thymidine kinase, accounting for its effect on infected cells only. Acyclovir monophosphate is phosphorylated by cellular enzymes to yield acyclovir triphosphate, which competes with dGTP for DNA polymerase. Once incorporated into viral DNA, acyclovir triphosphate acts as a chain terminator since it lacks a 3′-hydroxyl required for chain extension. As we will see shortly, thymidine kinase also plays a role in the de novo synthesis of thymidylate.

26.3 Purine Bases Can Be Synthesized from Glycine, Aspartate, and Other Components

Like pyrimidine nucleotides, purine nucleotides can be synthesized de novo or by salvage pathways. When synthesized de novo, purine synthesis begins with simple starting materials such as amino acids and bicarbonate (**Figure 26.5**). Unlike the bases of pyrimidines, the purine bases are assembled already attached to the ribose ring. Alternatively, purine bases released by the hydrolytic degradation of nucleic acids and nucleotides can be salvaged and recycled. Purine salvage pathways are especially noted for the energy that they save and the remarkable effects of their absence.

The purine ring system is assembled on ribose phosphate

De novo purine biosynthesis, like pyrimidine biosynthesis, requires PRPP. However, for purines, PRPP provides the foundation on which the bases are constructed. This process is started by the introduction of a nitrogen atom onto the ribose structure. This process, catalyzed by glutamine phosphoribosyl amidotransferase, is the displacement of pyrophosphate by ammonia. This reaction, the committed step in purine biosynthesis, produces 5-phosphoribosyl-1-amine, with the amine in the β configuration.

Glutamine phosphoribosyl amidotransferase comprises two domains: the first is homologous to the phosphoribosyltransferases in purine salvage pathways, whereas the second produces ammonia from glutamine by hydrolysis. To prevent wasteful hydrolysis of either substrate, the amidotransferase assumes the active configuration only on binding of both PRPP and glutamine. As is the case with carbamoyl phosphate synthetase II, the ammonia generated at the glutamine-hydrolysis active site passes through a channel to reach PRPP without being released into solution.

The purine ring is assembled by successive steps of activation by phosphorylation followed by displacement

Nine additional steps are required to assemble the purine ring. Remarkably, the first six steps are analogous reactions. Most of these steps are catalyzed by enzymes with ATP-grasp domains that are homologous to those in carbamoyl phosphate synthetase. Each step consists of the activation of a carbon-bound oxygen atom (typically a carbonyl oxygen atom) by phosphorylation, followed by the displacement of the phosphoryl group by ammonia or an amine group acting as a nucleophile (Nu). The hydrolysis of ATP provides the thermodynamic driving force to favor synthesis over degradation.

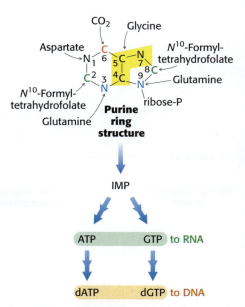

Acyclovir

FIGURE 26.5 Purine nucleotides can be synthesized de novo. The origins of the atoms in the purine ring are indicated.

PRPP

5-Phosphoribosyl-1-amine

De novo purine biosynthesis proceeds as shown in **Figure 26.6**. **Table 26.2** lists the enzymes that catalyze each step of the reaction.

1. The carboxylate group of a glycine residue is activated by phosphorylation and then coupled to the amino group of phosphoribosylamine. A new amide bond is formed, and the amino group of glycine is free to act as a nucleophile in the next step.

2. N^{10}-formyltetrahydrofolate donates a formyl moiety to this amino group to form formylglycinamide ribonucleotide.

3. The inner carbonyl group is activated by phosphorylation and then converted into an amidine by the addition of ammonia derived from glutamine.

4. The product of this reaction, formylglycinamidine ribonucleotide, cyclizes to form the five-membered imidazole ring found in purines. Although this

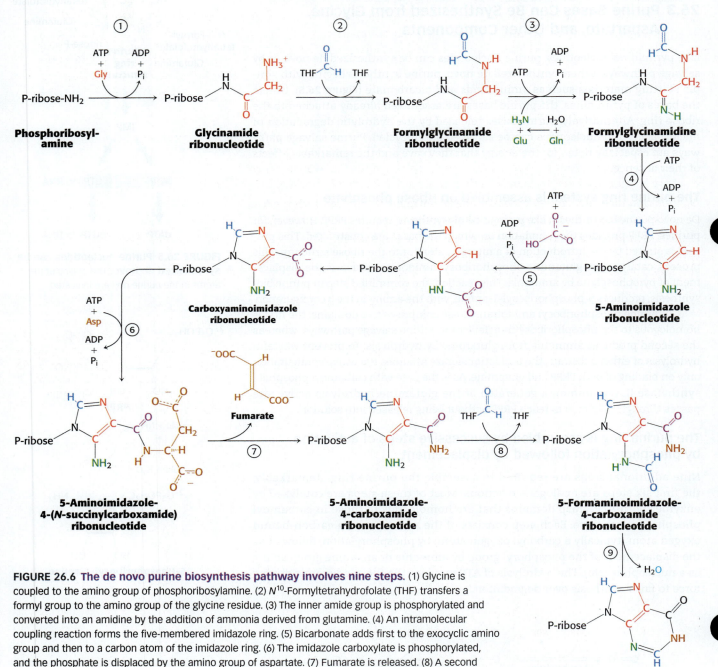

FIGURE 26.6 The de novo purine biosynthesis pathway involves nine steps. (1) Glycine is coupled to the amino group of phosphoribosylamine. (2) N^{10}-Formyltetrahydrofolate (THF) transfers a formyl group to the amino group of the glycine residue. (3) The inner amide group is phosphorylated and converted into an amidine by the addition of ammonia derived from glutamine. (4) An intramolecular coupling reaction forms the five-membered imidazole ring. (5) Bicarbonate adds first to the exocyclic amino group and then to a carbon atom of the imidazole ring. (6) The imidazole carboxylate is phosphorylated, and the phosphate is displaced by the amino group of aspartate. (7) Fumarate is released. (8) A second formyl group is donated from N^{10}-formyltetrahydrofolate (THF). (9) Cyclization completes the synthesis of inosinate, a purine nucleotide.

cyclization is likely to be favorable thermodynamically, a molecule of ATP is consumed to ensure irreversibility. The familiar pattern is repeated: a phosphoryl group from the ATP molecule activates the carbonyl group and is displaced by the nitrogen atom attached to the ribose molecule. Cyclization is thus an intramolecular reaction in which the nucleophile and phosphate-activated carbon atom are present within the same molecule. In higher eukaryotes, the enzymes catalyzing steps 1, 2, and 4 (Table 26.2) are components of a single polypeptide chain.

5. Bicarbonate is activated by phosphorylation and then attacked by the exocyclic amino group. The product of the reaction in step 5 rearranges to transfer the carboxylate group to the imidazole ring. Interestingly, mammals do not require ATP for this step; bicarbonate apparently attaches directly to the exocyclic amino group and is transferred to the imidazole ring.

6. The imidazole carboxylate group is phosphorylated again and the phosphate group is displaced by the amino group of aspartate. Once again, in higher eukaryotes, the enzymes catalyzing steps 5 and 6 (Table 26.2) share a single polypeptide chain.

7. Fumarate, an intermediate in the citric acid cycle, is eliminated, leaving the nitrogen atom from aspartate joined to the imidazole ring. The use of aspartate as an amino-group donor and the concomitant release of fumarate are reminiscent of the conversion of citrulline into arginine in the urea cycle, and these steps are catalyzed by homologous enzymes in the two pathways (Section 23.4).

8. A formyl group from N^{10}-formyltetrahydrofolate is added to this nitrogen atom to form a final 5-formaminoimidazole-4-carboxamide ribonucleotide.

9. 5-Formaminoimidazole-4-carboxamide ribonucleotide cyclizes with the loss of water to form inosine monophosphate (inosinate).

This pathway completes the formation of the purine ring system.

❖ SELF–CHECK QUESTION

Identify the sources of the atoms that form the purine ring.

AMP and GMP are formed from IMP

A few steps convert inosinate into either AMP or GMP (**Figure 26.7**). Adenylate is synthesized from inosinate by the substitution of an amino group for the

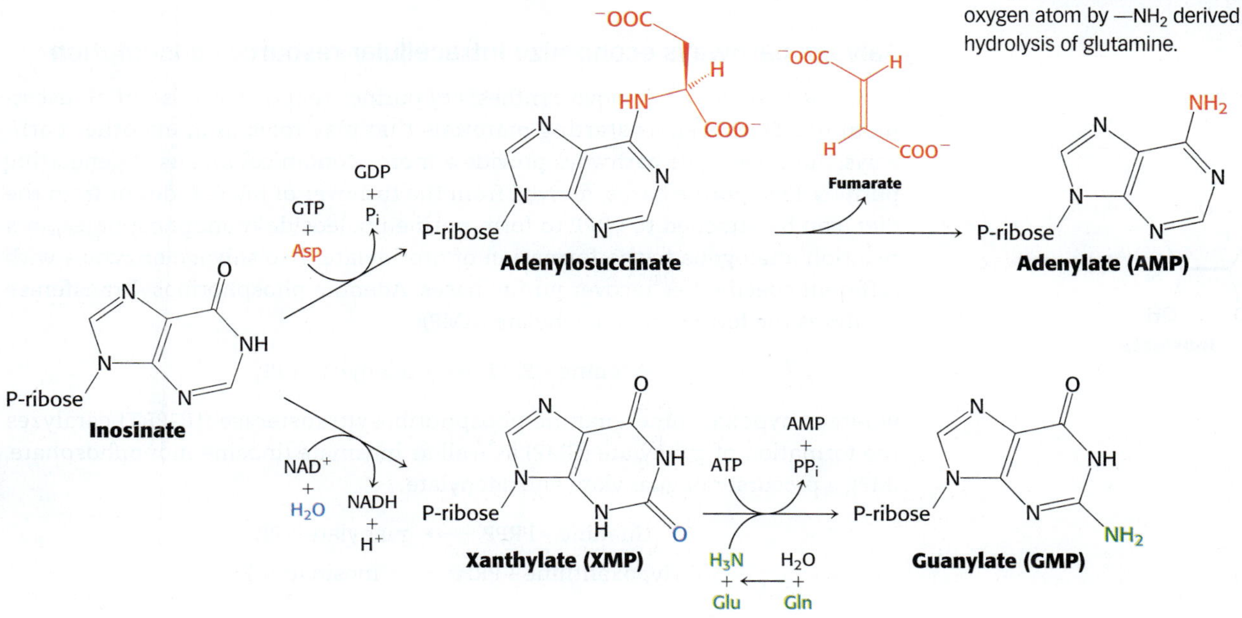

TABLE 26.2 The enzymes of de novo purine synthesis

Step	Enzyme
1	Glycinamide ribonucleotide (GAR) synthetase
2	GAR transformylase
3	Formylglycinamidine synthase
4	Aminoimidazole ribonucleotide synthetase
5	Carboxyaminoimidazole ribonucleotide synthetase
6	Succinylaminoimidazole carboxamide ribonucleotide synthetase
7	Adenylosuccinate lyase
8	Aminoimidazole carboxamide ribonucleotide transformylase
9	Inosine monophosphate cyclohydrolase

FIGURE 26.7 Inosinate can be converted to AMP and GMP. AMP is formed by the addition of aspartate followed by the release of fumarate. GMP is generated by the addition of water, dehydrogenation by NAD⁺, and the replacement of the carbonyl oxygen atom by —NH₂ derived by the hydrolysis of glutamine.

(A)

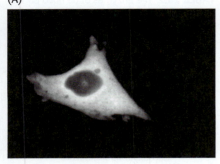

(B)

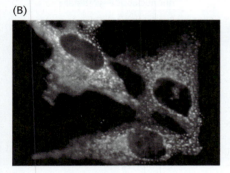

FIGURE 26.8 Purinosomes form when purine synthesis is required. A gene construct encoding a fusion protein consisting of formylglycinamidine synthase and GFP was transfected into and expressed in Hela cells, a human cell line. (A) In the presence of purines (the absence of purine synthesis), the GFP was seen as a diffuse stain throughout the cytoplasm. (B) When the cells were shifted to a purine-free medium, purinosomes formed, seen as cytoplasmic granules, and purine synthesis occurred. [Republished with permission of AAAS, from An, S., Kumar, R., Sheets, E. D., and Benkovic, S. J. 2. Reversible compartmentalization of de novo purine biosynthetic complexes in living cells. *Science* 320: 103–106, 2008; permission conveyed through Copyright Clearance Center, Ltd.]

carbonyl oxygen atom at C-6. Again, the addition of aspartate followed by the elimination of fumarate contributes the amino group. GTP, rather than ATP, is the phosphoryl-group donor in the synthesis of the adenylosuccinate intermediate from inosinate and aspartate. In accord with the use of GTP, the enzyme that promotes this conversion, adenylosuccinate synthetase, is structurally related to the G-protein family and does not contain an ATP-grasp domain.

Guanylate is synthesized by the oxidation of inosinate to xanthylate (XMP), followed by the incorporation of an amino group at C-2. NAD$^+$ is the hydrogen acceptor in the oxidation of inosinate, and xanthylate is activated by the transfer of an AMP group (rather than a phosphoryl group) from ATP to the oxygen atom in the newly formed carbonyl group. Ammonia, generated by the hydrolysis of glutamine, then displaces the AMP group to form guanylate, in a reaction catalyzed by GMP synthetase. Note that the synthesis of adenylate requires GTP, whereas the synthesis of guanylate requires ATP. This reciprocal use of nucleotides by the pathways creates an important regulatory opportunity.

Enzymes of the purine biosynthesis pathway associate with one another

Many of the intermediates in the de novo purine biosynthesis pathway degrade rapidly in water, suggesting that the product of one active site might be channeled directly to the next enzyme along the pathway in a manner similar to that described for carbamoyl phosphate synthetase II. In some bacteria, enzymes catalyzing consecutive steps in the purine biosynthetic pathway are fused into single polypeptide chains, likely promoting such channeling.

A more intricate process occurs in human cells. In the presence of sufficient levels of purines, the enzymes of the purine biosynthetic pathway do not appear to be associated with one another. However, when grown under conditions with low concentrations of purines, granules termed **purinosomes** form (**Figure 26.8**). The core of the purinosome comprises the enzymes catalyzing the first six steps of the pathway, with the remaining enzymes more loosely associated. The reversible formation of these large assemblies provides a powerful mode of regulation of purine biosynthesis. The assembly and disassembly of the purinosome is controlled by phosphorylation, enabling the linkage of these processes to the cell cycle and to other events through signal transduction pathways. Thus, the biosynthesis of purines, so central to DNA replication but also to other processes, can be regulated in response to a range of signals other than the concentrations of purines present.

Salvage pathways economize intracellular resource consumption

As we have seen, the de novo synthesis of purines requires a substantial investment of ATP as well as starting materials that play roles in many other pathways. Purine salvage pathways provide a more economical means of generating purines. Free purine bases, derived from the turnover of nucleotides or from the diet, can be attached to PRPP to form purine nucleoside monophosphates, in a reaction analogous to the formation of orotidylate. Two salvage enzymes with different specificities recover purine bases. Adenine phosphoribosyltransferase catalyzes the formation of adenylate (AMP):

$$\text{Adenine} + \text{PRPP} \longrightarrow \text{adenylate} + \text{PP}_i$$

whereas hypoxanthine-guanine phosphoribosyltransferase (HGPRT) catalyzes the formation of guanylate (GMP) as well as inosinate (inosine monophosphate, IMP), a precursor of guanylate and adenylate.

$$\text{Guanine} + \text{PRPP} \longrightarrow \text{guanylate} + \text{PP}_i$$
$$\text{Hypoxanthine} + \text{PRPP} \longrightarrow \text{inosinate} + \text{PP}_i$$

$^{2-}$O$_3$POH$_2$C

Hypoxanthine

HO   OH

Inosinate

An alternative to adenine is used by some viruses

When we first encountered the DNA double helix, you may have noticed that G–C base pairs are held together by three hydrogen bonds, whereas A–T base pairs have only two. However, if one of the hydrogens on adenine were replaced with an amino group, then this alternative could form three hydrogen bonds with T.

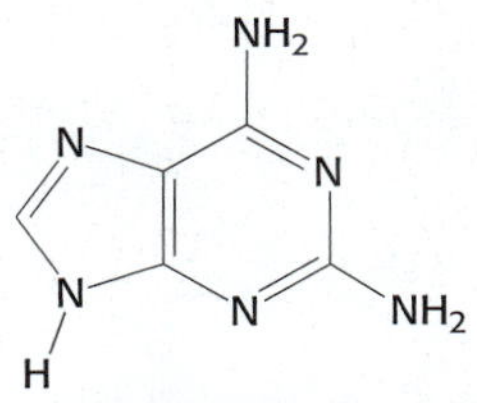

The genomes of certain bacterial viruses contain this nucleobase, termed *2-aminoadenine*, in place of adenine. Nucleotides containing this base are biosynthesized by a pathway that is analogous to that for the conversion of inosinate to adenylate (Figure 26.7) but differs in important details (**Figure 26.9**). The pathway begins with deoxyguanylate. In a first step, aspartate replaces the carbonyl oxygen in a process coupled to the hydrolysis of ATP in reaction catalyzed by the enzyme PurZ. Fumarate is then released to form deoxy 2-aminoadenylate in a subsequent step catalyzed by adenylosuccinate lyase, the same enzyme that catalyzes the removal to fumarate to form adenylate.

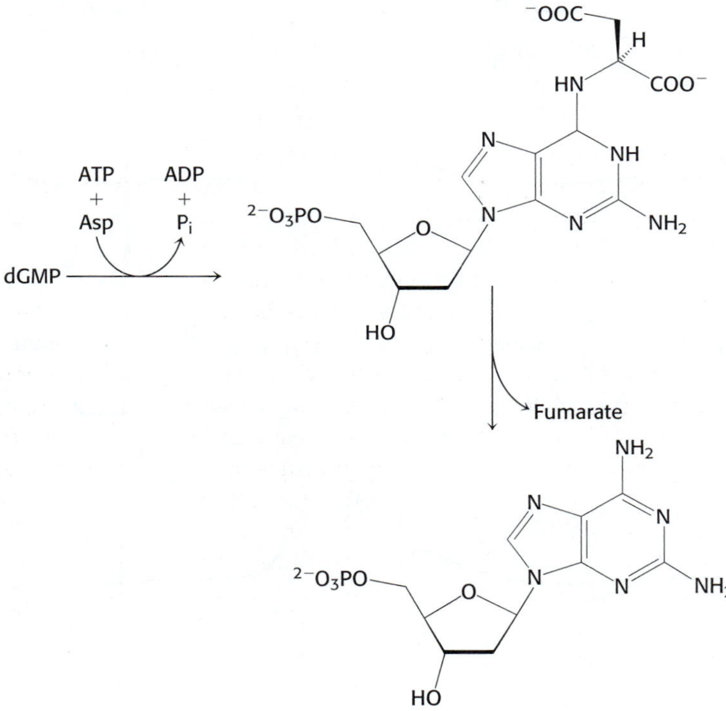

2-Aminoadenine

FIGURE 26.9 The nucleotide containing 2-aminoadenine is synthesized from deoxyguanylate. PurZ catalyzes the ATP-coupled replacement of the carbonyl oxygen on dGMP with aspartate. In the second step, catalyzed by adenylosuccinate lyase, fumarate is released, yielding the 2-aminoadenine-containing deoxyribonucleotide.

The identification of the gene encoding PurZ in other viral genomes indicates the likely presence of this pathway, without the need for characterizing the base composition of purified viral DNA. Upon the discovery of this bioinformatic approach, the number of bacterial viruses with evidence of 2-aminoadenine in their genomes increased from less than five to more than 50. As anticipated, DNA containing this base is more thermally stable, consistent with the formation of the additional hydrogen bonds. More important, the unusual base protects the viral genomes from bacterial endonucleases.

26.4 Deoxyribonucleotides Are Synthesized by the Reduction of Ribonucleotides

We turn now to the synthesis of deoxyribonucleotides. These precursors of DNA are formed by the reduction of ribonucleotides; specifically, the 2′-hydroxyl group on the ribose moiety is replaced by a hydrogen atom. The substrates are ribonucleoside diphosphates, and the ultimate reductant is NADPH. The enzyme **ribonucleotide reductase** is responsible for the reduction reaction for all four ribonucleotides.

The ribonucleotide reductases of different organisms are a remarkably diverse set of enzymes. Yet detailed studies have revealed that they have a common reaction mechanism, and their three-dimensional structural features indicate that these enzymes are homologous.

Ribonucleotide reduction occurs via a radical mechanism

The active sites of ribonucleotide reductases include three conserved cysteine residues and a glutamate residue, all four of which participate in the reduction of ribose to deoxyribose. In the synthesis of a deoxyribonucleotide, the OH bonded to C-2′ of the ribose ring is replaced by H, with retention of the configuration at the C-2′ carbon atom (**Figure 26.10**).

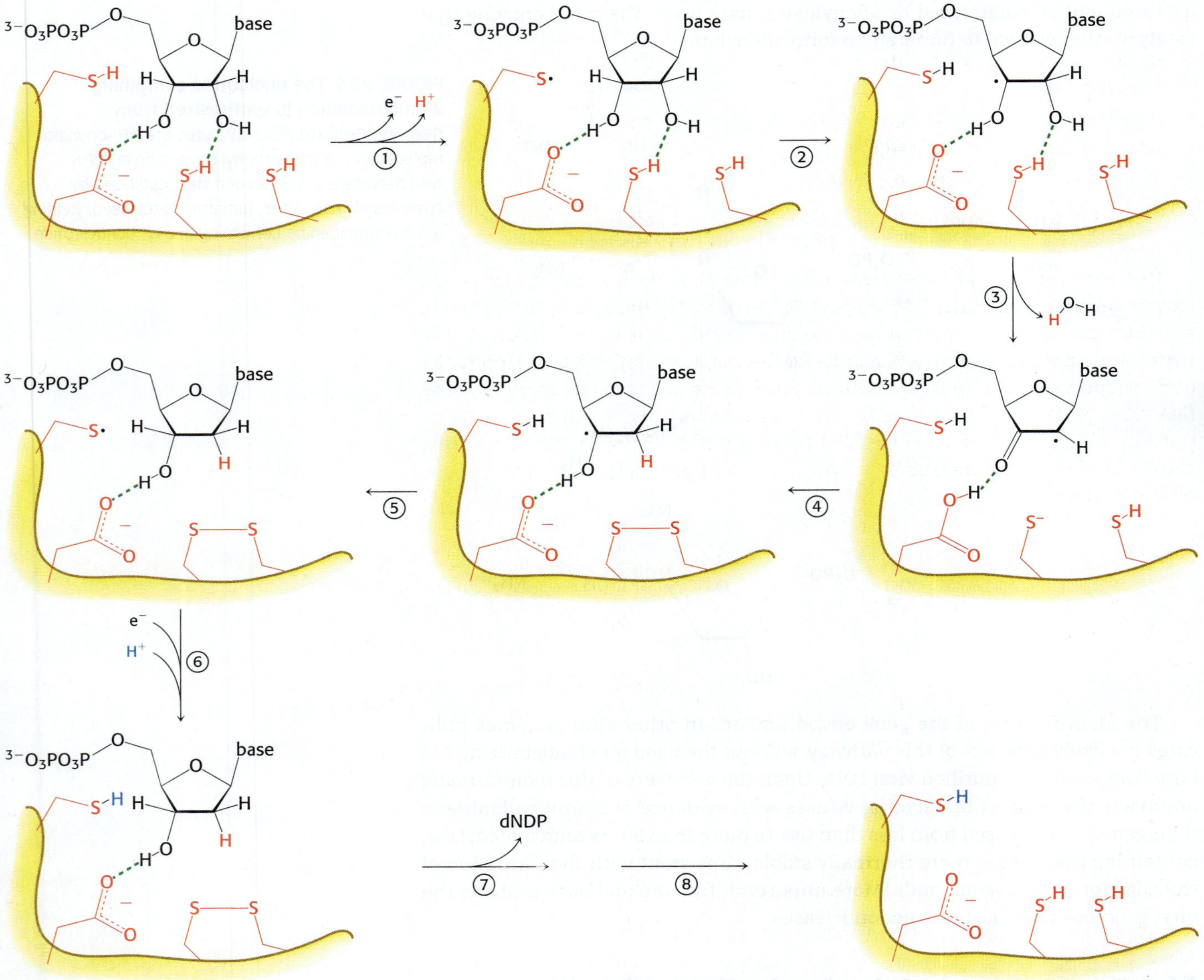

FIGURE 26.10 Ribonucleotide reductase uses a radical mechanism to produce deoxyribonucleotides. (1) An electron is transferred from a cysteine residue on R1 to a tyrosine radical on R2, generating a highly reactive cysteine thiyl radical. (2) This radical abstracts a hydrogen atom from C-3′ of the ribose unit. (3) The radical at C-3′ releases OH⁻ from the C-2′ carbon atom. Combined with a proton from a second cysteine residue, the OH⁻ is eliminated as water. (4) A hydride ion is transferred from a third cysteine residue, with the concomitant formation of a disulfide bond. (5) The C-3′ radical recaptures the originally abstracted hydrogen atom. (6) An electron is transferred from R2 to reduce the thiyl radical, which also accepts a proton. (7) The deoxyribonucleotide leaves R1. (8) The disulfide formed in the active site is reduced to begin another cycle.

1. The reaction begins with the transfer of an electron from a cysteine residue in the active site of ribonucleotide reductase to an electron acceptor on a second subunit of the enzyme. The loss of an electron generates a highly reactive cysteine thiyl radical.

2. This radical then abstracts a hydrogen atom from C-3′ of the ribose unit, generating a radical at that carbon atom.

3. The radical at C-3′ promotes the release of the OH⁻ from the C-2′ carbon atom. Protonated by a second cysteine residue, the departing OH⁻ leaves as a water molecule.

4. A hydride ion (a proton with two electrons) is then transferred from a third cysteine residue to complete the reduction of the position, form a disulfide bond, and regenerate the radical at the C-3′ carbon.

5. This C-3′ radical recaptures the same hydrogen atom originally abstracted by the first cysteine residue, and the deoxyribonucleotide is free to leave the enzyme.

6. An electron from the second subunit of ribonucleotide reductase is transferred to reduce the thiyl radical.

7. The newly synthesized deoxyribonucleotide exits the active site.

8. The disulfide bond generated in the enzyme's active site is reduced to regenerate the active enzyme.

The electrons for this reduction come from NADPH, but not directly. One carrier of reducing power linking NADPH with the reductase is thioredoxin, a 12-kDa protein with two exposed cysteine residues near each other. These sulfhydryls are oxidized to a disulfide in the reaction catalyzed by ribonucleotide reductase itself. In turn, reduced thioredoxin is regenerated by electron flow from NADPH. This reaction is catalyzed by thioredoxin reductase, a flavoprotein. Electrons flow from NADPH to bound FAD of the reductase, to the disulfide of oxidized thioredoxin, and then to ribonucleotide reductase, and finally to the ribose unit.

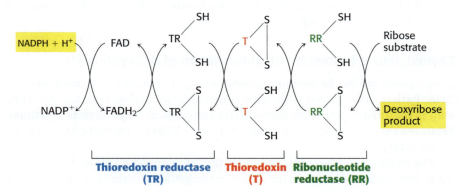

Stable radicals are present in ribonucleotide reductases

The most well-studied ribonucleotide reductase is that from *E. coli* grown under aerobic conditions. This enzyme consists of two subunits: R1 (an 87-kDa dimer) and R2 (a 43-kDa dimer). The R1 subunit contains the active site as well as two allosteric control sites (Section 26.5). Each R2 chain contains a tyrosyl radical with an unpaired electron delocalized onto its aromatic ring, generated by a nearby iron center consisting of two ferric (Fe^{3+}) ions bridged by an oxide (O^{2-}) ion (**Figure 26.11**). This very unusual free radical is remarkably stable, with a half-life of 4 days at 4°C. By contrast, free tyrosine radicals in solution have microsecond lifetimes.

The ribonucleotide reductases from many other organisms, including human beings, are similar to the *E. coli* enzyme, in that they contain the tyrosine radical

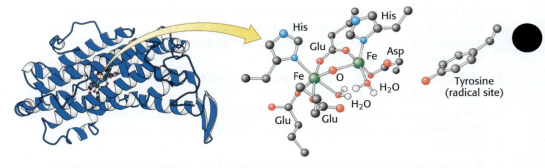

FIGURE 26.11 The ribonucleotide reductase R2 subunit contains a stable free radical on a tyrosine residue. This radical is generated by the reaction of oxygen (not shown) at a nearby site containing two iron atoms. Two R2 subunits come together to form a dimer. [Drawn from 1RIB.pdb.]

JOANNE STUBBE Trained in chemistry at the University of California, Berkeley, JoAnne Stubbe developed a deep interest in biochemistry and enzyme reaction mechanisms. Dr. Stubbe began her independent career on the faculty of Williams College, teaching and performing research with undergraduate student coworkers. As her interest and accomplishments in research grew, she took a leave of absence to work with a leading enzymologist at Brandeis University. She did not return to Williams but moved to more research-intensive departments, eventually ending up in the Chemistry Department at the Massachusetts Institute of Technology. There, with PhD students and postdoctoral fellows, she elucidated the mechanisms of several important biochemical processes, more notably the radical-based mechanism for ribonucleotide reductases. Her accomplishments have been recognized by important honors, including the National Medal of Science and the Priestley Medal from the American Chemical Society.

and the iron center. However, ribonucleotide reductases that do not contain tyrosyl radicals have been characterized in other prokaryotes. Instead, these enzymes contain distinct stable radicals that are generated by other processes. For example, in one class of reductases, the coenzyme adenosylcobalamin (from vitamin B_{12}) is the radical source (Section 22.3). Despite differences in the stable radical employed, the active sites of these enzymes are similar to that of the *E. coli* ribonucleotide reductase, and they appear to act by the same mechanism, based on the exceptional reactivity of cysteine radicals. Thus, these enzymes have a common ancestor but evolved a range of mechanisms for generating stable radical species that function well under different growth conditions. The primordial ribonucleotide reductases appear to have been inactivated by oxygen, whereas the one found in *E. coli* makes use of oxygen to generate the initial tyrosyl radical. Note that the reduction of ribonucleotides to deoxyribonucleotides is a difficult reaction, likely to require a sophisticated catalyst. The existence of a common protein enzyme framework for this process strongly suggests that proteins joined the RNA world before the evolution of DNA as a stable storage form for genetic information.

Thymidylate is formed by the methylation of deoxyuridylate

Uracil, produced by the pyrimidine synthesis pathway, is not a component of DNA. Rather, DNA contains *thymine*, a methylated analog of uracil. Another step is required to generate thymidylate from uracil. **Thymidylate synthase** catalyzes this finishing touch: deoxyuridylate (dUMP) is methylated to thymidylate (TMP).

The methyl group becomes attached to the C-5 atom within the aromatic ring of dUMP, but this carbon atom is not a good nucleophile and cannot itself attack the appropriate group on the methyl donor. Thymidylate synthase promotes methylation by adding a thiolate from a cysteine side chain to this ring to generate a nucleophilic species that can attack the methylene group of N^5,N^{10}-methylenetetrahydrofolate (**Figure 26.12**). This methylene group, in turn, is activated by distortions imposed by the enzyme that favor opening the five-membered ring. The attack of the activated dUMP on the methylene group forms the new carbon–carbon bond. The intermediate formed is then converted into product. First a hydride ion is transferred from the tetrahydrofolate ring to transform the methylene group into a methyl group, and then a proton is abstracted from the carbon atom bearing the methyl group to eliminate the cysteine and regenerate the aromatic ring. The tetrahydrofolate derivative loses both its methylene group and a hydride ion and, hence, is oxidized to dihydrofolate.

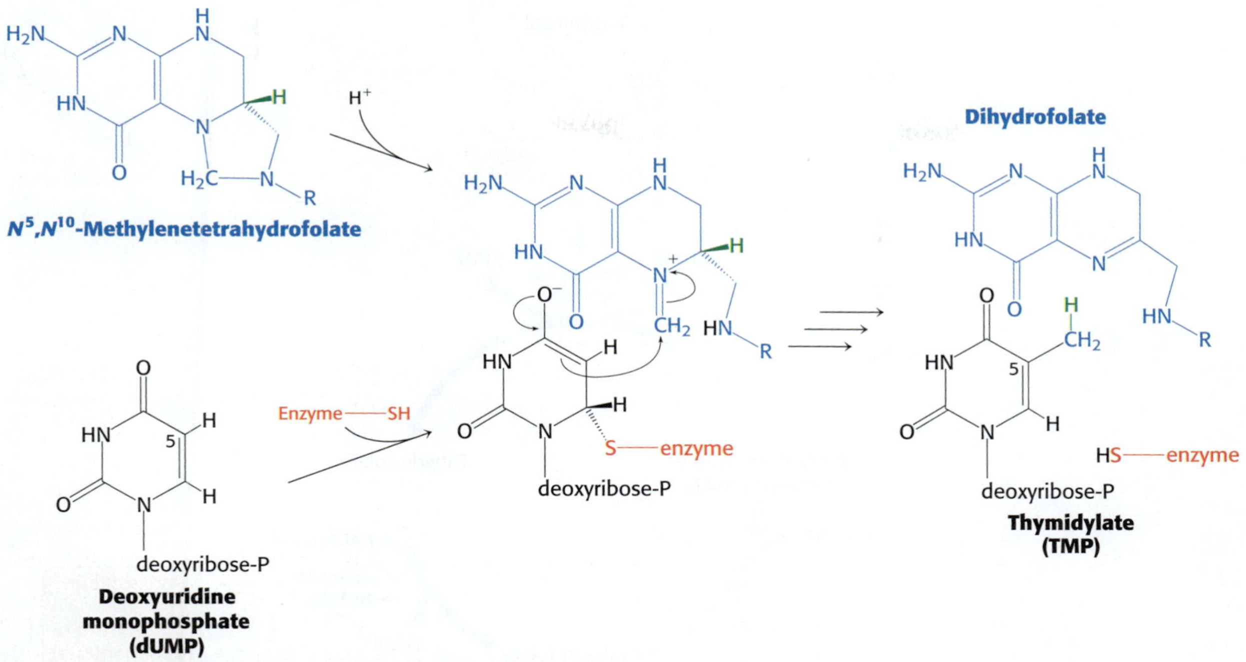

FIGURE 26.12 Thymidylate synthase catalyzes the addition of a methyl group to dUMP to form TMP. The methyl group is derived from N^5,N^{10}-methylenetetrahydrofolate. The addition of a thiolate from the enzyme activates dUMP. Opening the five-membered ring of the THF derivative prepares the methylene group for nucleophilic attack by the activated dUMP. The reaction is completed by the transfer of a hydride ion to form dihydrofolate.

For the synthesis of more thymidylate, tetrahydrofolate must be regenerated. This is achieved by **dihydrofolate reductase** with the use of NADPH as the reductant.

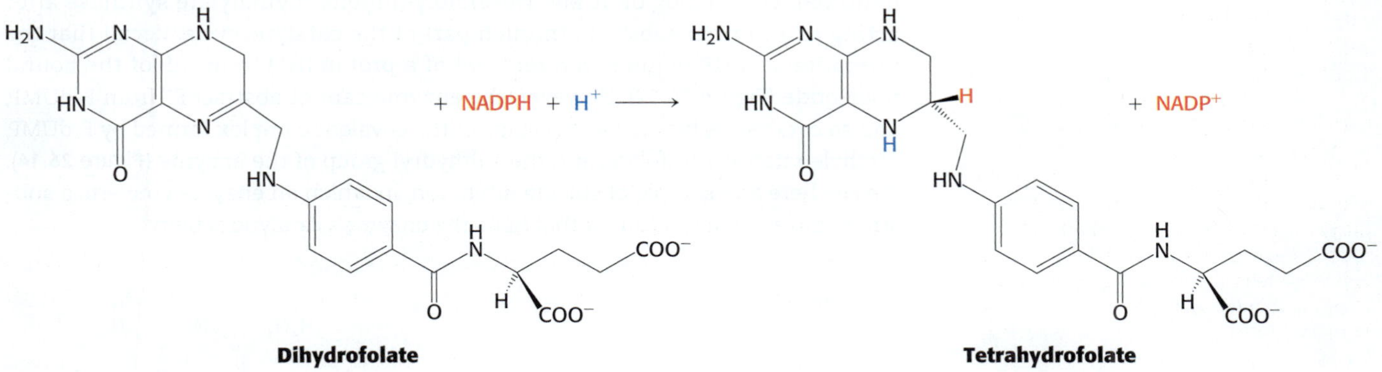

A hydride ion is directly transferred from the nicotinamide ring of NADPH to the pteridine ring of dihydrofolate. The bound dihydrofolate and NADPH are held in close proximity to facilitate the hydride transfer.

❖ **SELF-CHECK QUESTION**

Describe the path for the synthesis of TTP from UTP.

⚕ **Several valuable anticancer drugs block the synthesis of thymidylate**

Rapidly dividing cells require an abundant supply of thymidylate for the synthesis of DNA. The vulnerability of these cells to the inhibition of TMP synthesis has

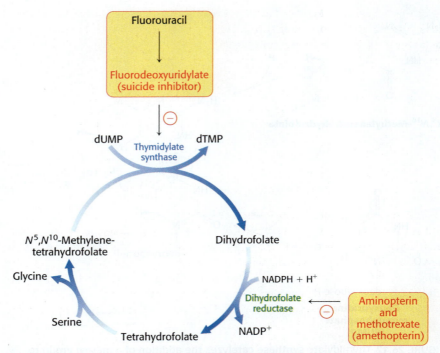

FIGURE 26.13 Enzymes crucial for nucleotide synthesis are anticancer drug targets. Thymidylate synthase and dihydrofolate reductase are choice targets in cancer chemotherapy because the generation of large quantities of precursors for DNA synthesis is required for rapidly dividing cancer cells.

been exploited in the treatment of cancer. Thymidylate synthase and dihydrofolate reductase are choice targets of chemotherapy (**Figure 26.13**).

Fluorouracil, an anticancer drug, is converted in vivo into fluorodeoxyuridylate (F-dUMP). This analog of dUMP irreversibly inhibits thymidylate synthase after acting as a normal substrate through part of the catalytic cycle. Recall that the formation of TMP requires the removal of a proton (H^+) from C-5 of the bound nucleotide (Figure 26.12). However, the enzyme cannot abstract F^+ from F-dUMP, and so catalysis is blocked at the stage of the covalent complex formed by F-dUMP, methylenetetrahydrofolate, and the sulfhydryl group of the enzyme (**Figure 26.14**). We see here an example of suicide inhibition, in which an enzyme converts a substrate into a reactive inhibitor that halts the enzyme's catalytic activity.

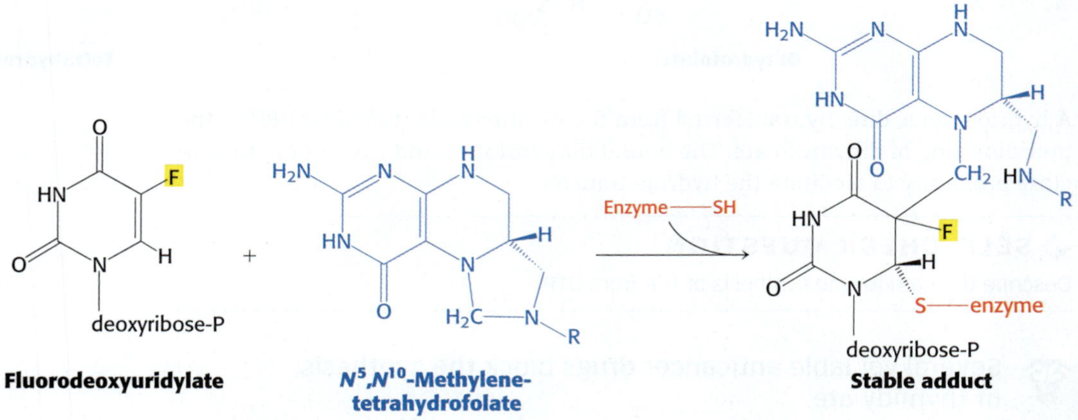

Fluorodeoxyuridylate N^5,N^{10}-**Methylene-tetrahydrofolate**   **Stable adduct**

FIGURE 26.14 Fluoruracil acts by suicide inhibition. Fluorodeoxyuridylate (generated from fluorouracil) forms an adduct with tetrahydrofolate that traps thymidylate synthase in a form that cannot proceed down the reaction pathway.

The synthesis of TMP can also be blocked by inhibiting the regeneration of tetrahydrofolate. Analogs of dihydrofolate, such as aminopterin and methotrexate (amethopterin), are potent competitive inhibitors ($K_i < 1$ nM) of dihydrofolate reductase.

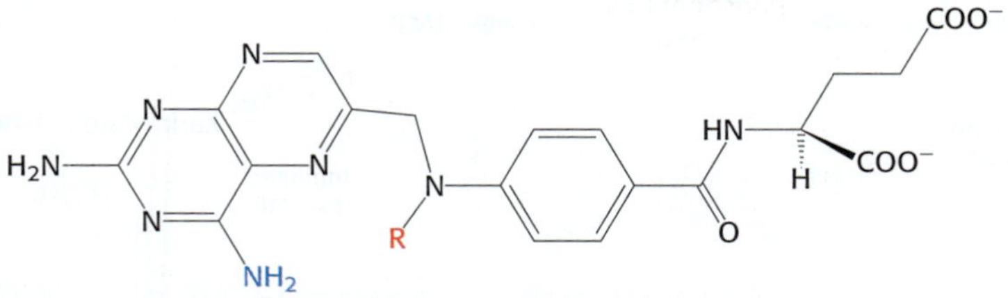

Aminopterin (R = H) or methotrexate (R = CH₃)

Methotrexate is a valuable drug in the treatment of many rapidly growing tumors, such as those in acute leukemia and choriocarcinoma, a cancer derived from placental cells. However, methotrexate kills rapidly replicating cells whether they are malignant or not. Stem cells in bone marrow, epithelial cells of the intestinal tract, and hair follicles are vulnerable to the action of this folate antagonist, accounting for its toxic side effects, which include weakening of the immune system, nausea, and hair loss.

Folate analogs such as trimethoprim have potent antibacterial and antiprotozoal activity. Trimethoprim binds 10^5-fold less tightly to mammalian dihydrofolate reductase than it does to reductases of susceptible microorganisms. Small differences in the active-site clefts of these enzymes account for the highly selective antimicrobial action. The combination of trimethoprim and sulfamethoxazole (an inhibitor of folate synthesis) is widely used to treat infections such as bronchitis, traveler's diarrhea, and urinary tract infections.

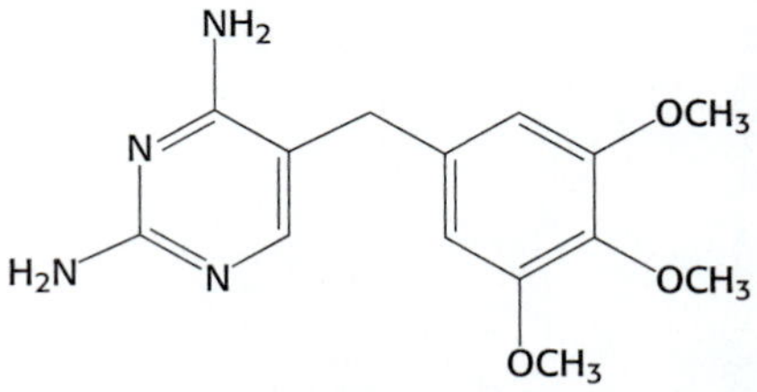

Trimethoprim

26.5 Key Steps in Nucleotide Biosynthesis Are Regulated by Feedback Inhibition

Nucleotide biosynthesis is regulated by feedback inhibition in a manner similar to the regulation of amino acid biosynthesis. These regulatory pathways ensure that the various nucleotides are produced in the required quantities.

Pyrimidine biosynthesis is regulated by aspartate transcarbamoylase

Aspartate transcarbamoylase (ATCase), one of the key enzymes for the regulation of pyrimidine biosynthesis in bacteria, was described in detail in Chapter 7. Recall that ATCase is inhibited by CTP, the final product of pyrimidine biosynthesis, and stimulated by ATP.

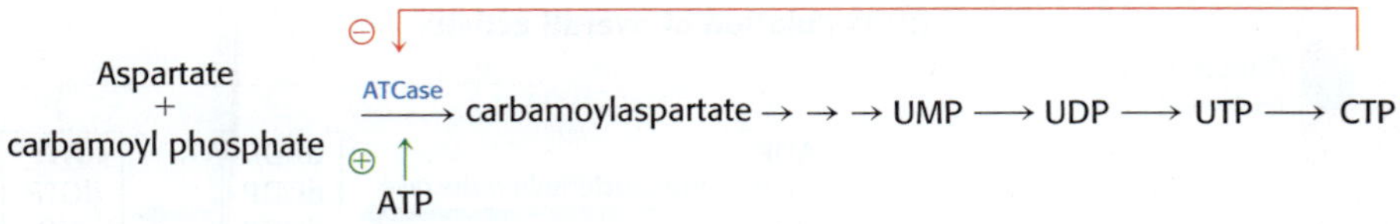

$$\text{Aspartate} + \text{carbamoyl phosphate} \xrightarrow[\text{ATCase}]{} \text{carbamoylaspartate} \rightarrow \rightarrow \rightarrow \text{UMP} \longrightarrow \text{UDP} \longrightarrow \text{UTP} \longrightarrow \text{CTP}$$

The synthesis of purine nucleotides is controlled by feedback inhibition at several sites

The regulatory scheme for purine nucleotides is more complex than that for pyrimidine nucleotides (**Figure 26.15**).

1. The committed step in purine nucleotide biosynthesis is the conversion of PRPP into phosphoribosylamine by glutamine phosphoribosyl amidotransferase. This important enzyme is feedback-inhibited by many purine ribonucleotides. It is noteworthy that AMP and GMP, the final products of the pathway, synergistically inhibit the amidotransferase.

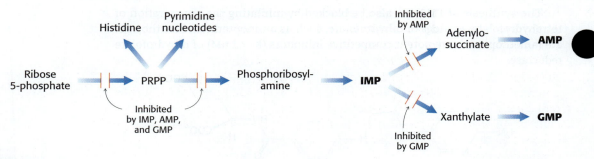

FIGURE 26.15 Purine biosynthesis is regulated at many steps. Feedback inhibition controls both the overall rate of purine biosynthesis and the balance between AMP and GMP production.

2. Inosinate is the branch point in the synthesis of AMP and GMP. The reactions leading away from inosinate are sites of feedback inhibition. AMP inhibits the conversion of inosinate into adenylosuccinate, its immediate precursor. Similarly, GMP inhibits the conversion of inosinate into xanthylate, its immediate precursor.

3. As already noted, GTP is a substrate in the synthesis of AMP, whereas ATP is a substrate in the synthesis of GMP (Figure 26.7). This reciprocal substrate relation tends to balance the synthesis of adenine and guanine ribonucleotides.

Note that the synthesis of PRPP by PRPP synthetase is highly regulated even though it is not the committed step in purine synthesis. Mutations have been identified in PRPP synthetase that result in a loss of allosteric response to nucleotides without any effect on catalytic activity of the enzyme. A consequence of this mutation is an overabundance of purine nucleotides that can result in gout, a pathological condition discussed in Section 26.5.

❖ SELF–CHECK QUESTION

Which enzyme catalyzes the committed step in the synthesis of pyrimidines? Which enzyme catalyzes the committed step in the synthesis of purines?

The synthesis of deoxyribonucleotides is controlled by the regulation of ribonucleotide reductase

The reduction of ribonucleotides to deoxyribonucleotides is precisely controlled by allosteric interactions. Each polypeptide of the R1 subunit of the aerobic *E. coli* ribonucleotide reductase contains two allosteric sites (**Figure 26.16**). One of them controls the overall activity of the enzyme, and the other regulates

FIGURE 26.16 Ribonucleotide reductase is highly regulated to balance the pool of deoxyribonucleotides. (A) Each subunit in the R1 dimer contains two allosteric sites in addition to the active site. One site regulates the overall activity, and the other site regulates substrate specificity. (B) The patterns of regulation with regard to different nucleoside diphosphates are demonstrated by ribonucleotide reductase.

(A)

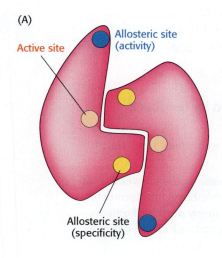

(B) **Regulation of overall activity**

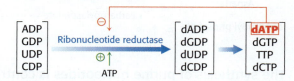

Regulation of substrate specificity

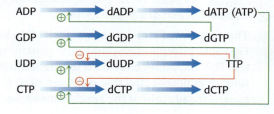

substrate specificity. The overall catalytic activity of ribonucleotide reductase is diminished by the binding of dATP, which signals an abundance of deoxyribonucleotides. The binding of ATP reverses this feedback inhibition.

The binding of dATP or ATP to the substrate-specificity control site enhances the reduction of UDP and CDP, the pyrimidine nucleotides. The binding of thymidine triphosphate (TTP) promotes the reduction of GDP and inhibits the further reduction of pyrimidine ribonucleotides. The subsequent increase in the level of dGTP stimulates the reduction of ATP to dATP. This complex pattern of regulation supplies the appropriate balance of the four deoxyribonucleotides needed for the synthesis of DNA.

26.6 Disruptions in Nucleotide Metabolism Can Cause Pathological Conditions

The nucleotides of a cell undergo continual turnover. This turnover facilitates the proper balance between ribo- and deoxyribonucleotides as well as the removal of damaged bases (**Figure 26.17**). This turnover begins with the hydrolytic conversion of nucleotides to nucleosides by nucleotidases. The sugars and bases are separated from one another by nucleoside phosphorylases in a reversible process. The free bases are then either reused to form nucleotides by salvage pathways or degraded to products that are excreted.

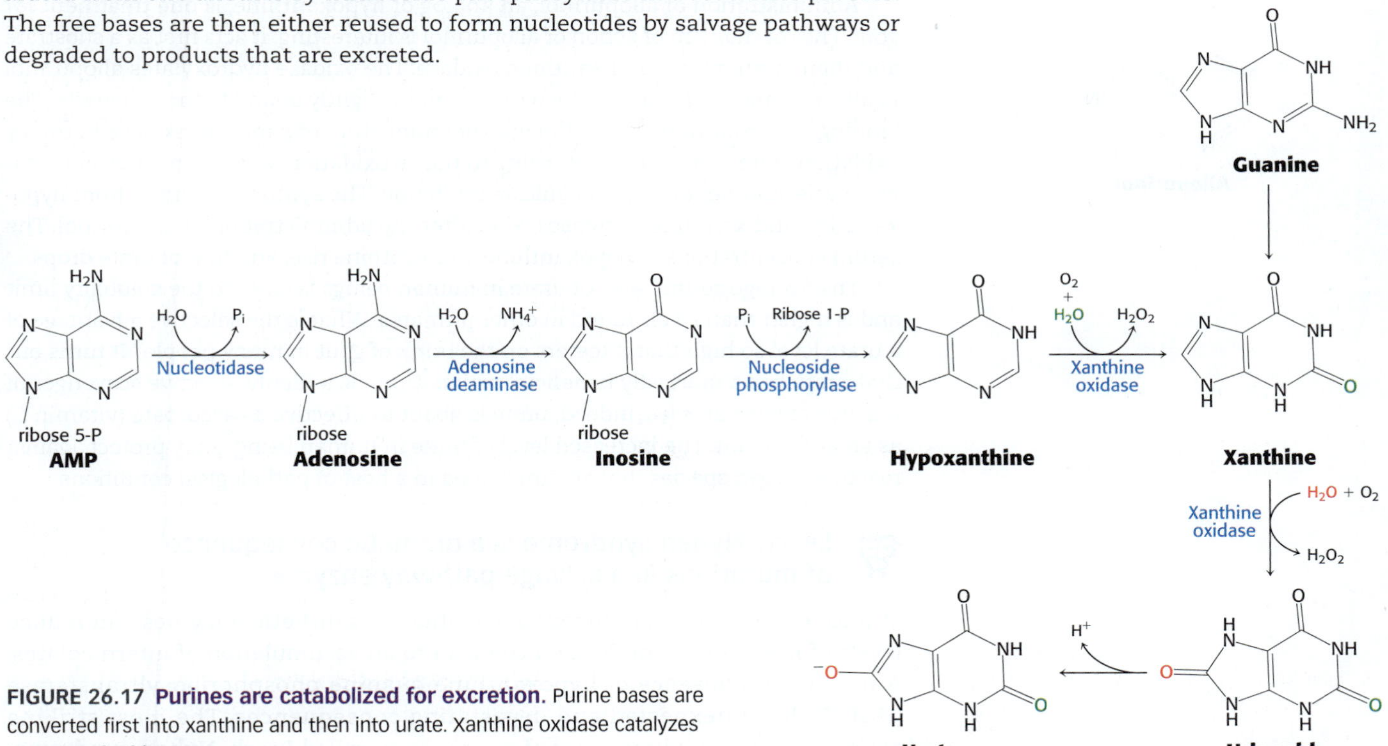

FIGURE 26.17 Purines are catabolized for excretion. Purine bases are converted first into xanthine and then into urate. Xanthine oxidase catalyzes two steps in this process.

The loss of adenosine deaminase activity results in severe combined immunodeficiency

Adenosine is not a substrate for nucleoside phosphorylases in eukaryotes, so that adenosine turnover requires an additional step compared to the turnover of guanosine. In the extra step, adenosine is deaminated by adenosine deaminase to form inosine (Figure 26.17).

A deficiency in adenosine deaminase activity is associated with some forms of **severe combined immunodeficiency (SCID)**, an immunological disorder. Persons with the disorder have acute recurring infections, often leading to death at an early age. SCID is characterized by a loss of T cells, which are crucial to the immune response. A lack of adenosine deaminase results in an accumulation of 50 to 100 times the normal level of dATP, which is toxic to many cells through a variety of mechanisms. Adenosine deaminase deficiency is one of the first conditions to be treated by gene therapy, and recent progress is encouraging.

Gout is induced by high serum levels of urate

Inosine generated by adenosine deaminase is subsequently metabolized by nucleoside phosphorylase to hypoxanthine, which is oxidized by xanthine oxidase—a molybdenum- and iron-containing flavoprotein—to xanthine, and then to uric acid. Uric acid loses a proton at physiological pH to form urate which, in human beings, is the final product of purine degradation and is excreted in the urine.

High serum levels of urate induce the painful joint disease **gout** as the sodium salt of urate crystallizes in the fluid and lining of the joints. The small joint at the base of the big toe is a common site for sodium urate buildup, although the salt also accumulates at other joints. Painful inflammation results when cells of the immune system engulf the sodium urate crystals. The kidneys, too, may be damaged by the deposition of urate crystals. Gout is a common medical problem, affecting 1% of the population of Western countries.

Administration of allopurinol, an analog of hypoxanthine, is one treatment for gout. The mechanism of action of allopurinol is interesting: it acts first as a substrate and then as an inhibitor of xanthine oxidase. The oxidase hydroxylates allopurinol to alloxanthine (oxipurinol), which then remains tightly bound to the active site. The binding of alloxanthine keeps the molybdenum atom of xanthine oxidase in the +4 oxidation state instead of it returning to the +6 oxidation state as in a normal catalytic cycle, another example of suicide inhibition. The synthesis of urate from hypoxanthine and xanthine decreases soon after the administration of allopurinol. The serum concentrations of hypoxanthine and xanthine rise, and that of urate drops.

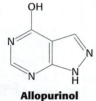

Allopurinol

The average serum level of urate in human beings is close to the solubility limit and is higher than levels found in other primates. What is the selective advantage of a urate level so high that it teeters on the brink of gout in many people? It turns out that urate has a markedly beneficial action. Urate is a highly effective scavenger of reactive oxygen species. Indeed, urate is about as effective as ascorbate (vitamin C) as an antioxidant. The increased level of urate in human beings may protect against reactive oxygen species that are implicated in a host of pathological conditions.

Lesch–Nyhan syndrome is a dramatic consequence of mutations in a salvage-pathway enzyme

Mutations in genes that encode nucleotide biosynthetic enzymes can reduce levels of needed nucleotides and can lead to an accumulation of intermediates. A nearly total absence of hypoxanthine-guanine phosphoribosyltransferase (HGPRT) has unexpected and devastating consequences. The most striking expression of this inborn error of metabolism, called **Lesch–Nyhan syndrome**, is compulsive, self-destructive behavior. At age 2 or 3, children with this disease begin to bite their fingers and lips and will chew them off if unrestrained. These children also behave aggressively toward others and often show cognitive disabilities and spasticity. Elevated levels of urate in the serum lead to the formation of kidney stones early in life, followed by the symptoms of gout years later. The disease is inherited as a sex-linked recessive disorder.

What is the connection between the absence of HGPRT activity and the behavioral characteristics of Lesch–Nyhan syndrome? The answer has not been established, but several explanations may contribute. The brain has limited capacity for de novo purine synthesis. Consequently, lack of HGPRT results in a deficiency of purine nucleotides. ATP and ADP, formed from inosinate, are important in the brain as signal molecules. These nucleotides bind to and activate G-protein coupled receptors, which require guanine nucleotides to function. Finally, a cofactor required for dopamine biosynthesis is biosynthesized from GTP. Thus, the lack of HGPRT may result in an imbalance of key neurotransmitters and the pathways they stimulate. Lesch–Nyhan syndrome demonstrates that the salvage pathway for the synthesis of IMP and GMP is not of minor importance. Moreover, Lesch–Nyhan syndrome reveals that abnormal behavior such as self-mutilation and extreme hostility can be caused by the absence of a single enzyme. Psychiatry will no doubt benefit from the unraveling of the molecular basis of such disorders.

Summary

26.1 Nucleotides Can Be Synthesized by de Novo or Salvage Pathways

- Nucleotide biosynthesis can be divided into either de novo pathways or salvage pathways.
- In de novo pathways, the nucleobases are synthesized from simpler compounds.
- In salvage pathways, pre-formed bases are recovered and reconnected to a ribose unit.

26.2 The Pyrimidine Ring Is Assembled from CO_2, Ammonia, and Aspartate

- In de novo synthesis, pyrimidine rings are first assembled, then attached to a ribose phosphate.
- 5-Phosphoribosyl-1-pyrophosphate is the donor of the ribose phosphate moiety.
- The synthesis of the pyrimidine ring starts with the formation of carbamoylaspartate from carbamoyl phosphate and aspartate, catalyzed by aspartate transcarbamoylase.
- Dehydration, cyclization, and oxidation yield orotate, which reacts with PRPP to give orotidylate, which serves as the precursor to the pyrimidine nucleotides.
- Nucleotide mono-, di-, and triphosphates are interconvertible by the action of nucleoside monophosphate kinases and nucleoside diphosphate kinase.

26.3 Purine Bases Can Be Synthesized from Glycine, Aspartate, and Other Components

- In the de novo pathway, the purine ring is assembled on ribose phosphate from glutamine, glycine, aspartate, N^{10}-formyltetrahydrofolate, and CO_2.
- The committed step in the de novo synthesis of purine nucleotides is the formation of 5-phosphoribosylamine from PRPP and glutamine.
- The purine ring is assembled over nine synthetic steps, predominantly using the activation of a carbon-bound oxygen atom by phosphorylation and subsequent displacement of an amino group.
- AMP and GMP are formed from IMP.

26.4 Deoxyribonucleotides Are Synthesized by the Reduction of Ribonucleotides

- Deoxyribonucleotides, the precursors of DNA, are formed by the reduction of ribonucleoside diphosphates.
- These reactions are catalyzed by ribonucleotide reductases through a radical mechanism.
- TMP is formed by the methylation of dUMP by thymidylate synthase. This reaction oxidizes N^5,N^{10}-methylenetetrahydrofolate to dihydrofolate, which is regenerated by dihydrofolate reductase.
- Thymidylate synthase and dihydrofolate reductase are targets of various anticancer drugs.

26.5 Key Steps in Nucleotide Biosynthesis Are Regulated by Feedback Inhibition

- Pyrimidine biosynthesis is regulated by the feedback inhibition of aspartate transcarbamoylase.
- Regulation of purine biosynthesis is more complex and includes the feedback inhibition of glutamine-PRPP amidotransferase by purine nucleotides.
- The synthesis of deoxyribonucleotides is controlled by the regulation of ribonucleotide reductase at two allosteric sites: one controls overall enzyme activity, and another regulates substrate specificity.

26.6 Disruptions in Nucleotide Metabolism Can Cause Pathological Conditions

- Severe combined immunodeficiency results from the absence of adenosine deaminase, an enzyme in the purine degradation pathway.
- Purines are degraded to urate in human beings. Gout, a disease that affects joints and leads to arthritis, is associated with an excessive accumulation of urate.
- Lesch–Nyhan syndrome, a genetic disease characterized by self-mutilation, cognitive disabilities, and gout, is caused by the absence of hypoxanthine-guanine phosphoribosyltransferase. This enzyme is essential for the synthesis of purine nucleotides by the salvage pathway.

Key Terms

de novo pathway (p. 792)

salvage pathway (p. 792)

ATP-grasp fold (p. 793)

5-phosphoribosyl-1-pyrophosphate (PRPP) (p. 795)

orotidylate (p. 795)

purinosome (p. 800)

ribonucleotide reductase (p. 801)

thymidylate synthase (p. 804)

dihydrofolate reductase (p. 805)

severe combined immunodeficiency (SCID) (p. 809)

gout (p. 810)

Lesch–Nyhan syndrome (p. 810)

Problems

1. Differentiate between the de novo synthesis of nucleotides and salvage-pathway synthesis. ❖ 1

2. How many molecules of ATP are required to synthesize one molecule of CTP from scratch? ❖ 1

3. Write a balanced equation for the synthesis of orotate from glutamine, CO_2, and aspartate. ❖ 1

4. What is the activated reactant in the biosynthesis of each of the following compounds?

(a) Phosphoribosylamine

(b) Carbamoylaspartate

(c) Orotidylate (from orotate)

(d) Phosphoribosylanthranilate

5. Amidotransferases are inhibited by the antibiotic azaserine (O-diazoacetyl-L-serine), which is an analog of glutamine. ❖ 2

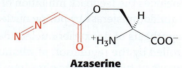

Azaserine

Which intermediates in purine biosynthesis would accumulate in cells treated with azaserine?

6. What intermediate in purine synthesis will accumulate if a strain of bacteria is lacking each of the following biochemicals? ❖ 2

(a) Aspartate

(b) Tetrahydrofolate

(c) Glycine

(d) Glutamine

7. What major biosynthetic reactions utilize PRPP?

8. Purine biosynthesis is allowed to take place in the presence of [^{15}N] aspartate, and the newly synthesized GTP and ATP are isolated. What positions are labeled in the two nucleotides? ❖ 2

9. Write a balanced equation for the synthesis of TMP from dUMP that is coupled to the conversion of serine into glycine. ❖ 3

10. Arrange the steps to outline the mechanism by which ribonucleotide reductase converts a ribonucleotide to a deoxyribonucleotide. ❖ 3

(a) C-2′ of the substrate releases an OH⁻.

(b) The tyrosyl radical on R2 accepts an electron from a cysteine on R1.

(c) C-2′ accepts a hydride ion from a cysteine residue.

(d) A hydrogen atom binds to C-3′ of the substrate.

(e) The cysteine radical abstracts a hydrogen atom from C-3′ of the substrate, forming a radical.

11. Which of the following molecules are substrates of the ribonucleotide reductase reaction? Which are products? ❖ 3

(a) ADP

(b) CDP

(c) dADP

(d) dCTP

12. Both side-chain oxygen atoms of aspartate 27 at the active site of dihydrofolate reductase form hydrogen bonds with the pteridine ring of folates. The importance of this interaction was assessed by studying two mutants at this position, Asn 27 and Ser 27. The dissociation constant of methotrexate was 0.07 nM for the wild type, 1.9 nM for the Asn 27 mutant, and 210 nM for the Ser 27 mutant, at 25°C. Calculate the standard free energy of the binding of methotrexate by these three proteins. What is the decrease in binding energy resulting from each mutation? ❖ 3

13. Which nucleoside triphosphate regulates aspartate transcarbamoylase (ATCase) by feedback inhibition? Which one acts as an allosteric activator of ATCase? ❖ 4

14. Which of the following statements correctly describe the regulatory steps of purine synthesis? ❖ 4

(a) The committed step in purine synthesis is the conversion of PRPP into phosphoribosylamine.

(b) Inosinate is the branch point in the synthesis of inhibitors AMP and GMP.

(c) The final reaction in purine synthesis, the generation of either AMP or IMP, is the rate-limiting step.

(d) AMP is a feedback inhibitor, regulating the conversion of inosinate into xanthylate, its immediate precursor.

(e) ATP is not required for the synthesis of GMP.

15. Which two nucleoside triphosphates bind to the allosteric site of the ribonucleotide reductase R1 subunit that controls overall enzyme activity? ❖ 4

16. Identify the correct statements about Lesch–Nyhan syndrome. ❖ 5

(a) This deficiency does not allow purine nucleotide production via the salvage pathway.

(b) This syndrome is caused by a deficiency in PRPP.

(c) Some symptoms of Lesch–Nyhan syndrome include involuntary movements and self-mutilation.

(d) Gout causes Lesch–Nyhan syndrome.

17. In humans, the multifunctional enzyme uridine 5′-monophosphate (UMP) synthase has orotate phosphoribosyltransferase and orotidylate decarboxylase activity.

Both enzymatic functions are impaired in patients with orotic aciduria. Which substance is present in high levels in the urine of these patients? ❖ 5

(a) Carbamate

(b) Carbamoyl phosphate

(c) Uridylate (UMP)

(d) Orotate

(e) Orotidylate

18. Suppose that a person is found who is deficient in an enzyme required for IMP synthesis. How might this person be treated? ❖ 5

19. Mutant cells unable to synthesize nucleotides by salvage pathways are very useful tools in molecular and cell biology. Suppose that cell A lacks thymidine kinase, the enzyme catalyzing the phosphorylation of thymidine to thymidylate, and that cell B lacks hypoxanthine-guanine phosphoribosyl transferase.

(a) Cell A and cell B do not proliferate in a HAT medium containing hypoxanthine, aminopterin or amethopterin (methotrexate), and thymine. However, cell C, formed by the fusion of cells A and B, grows in this medium. Why?

(b) Suppose that you want to introduce foreign genes into cell A. Devise a simple means of distinguishing between cells that have taken up foreign DNA and those that have not.

Biosynthesis of Membrane Lipids and Steroids

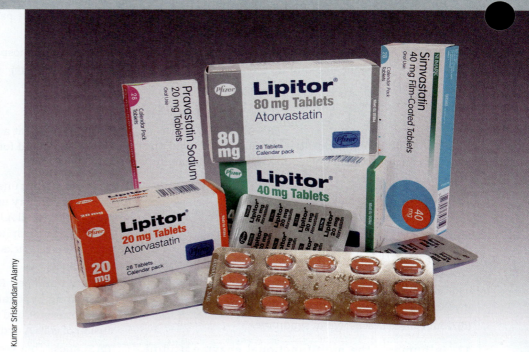

Statins ✗

Lower cholesterol levels

Enzymes in biosynthetic pathways involving lipids are important drug targets. For example, statins block the committed step in the biosynthesis of cholesterol, thus helping to control cardiovascular disease.

Kumar Sriskandan/Alamy

OUTLINE

27.1 Phosphatidate Is a Common Intermediate in the Synthesis of Phospholipids and Triacylglycerols

27.2 Cholesterol Is Synthesized from Acetyl Coenzyme A in Three Stages

27.3 The Regulation of Cholesterol Biosynthesis Takes Place at Several Levels

27.4 Important Biochemicals Are Synthesized from Cholesterol and Isoprene

◈ LEARNING GOALS

By the end of this chapter, you should be able to:

1. Describe the relationship between triacylglycerol synthesis and phospholipid synthesis.

2. Describe the three stages of cholesterol biosynthesis.

3. Identify the regulatory processes controlling cholesterol synthesis.

4. Describe some important molecules synthesized from cholesterol precursors and from cholesterol.

Living cells must replenish their cell membranes and enlarge them when growing and dividing. Furthermore, even seemingly minor variations in lipid structure are associated with participation in different functions, including intracellular and intercellular signaling. Thus, the pathways by which lipids are made and interconverted are of considerable interest.

This chapter examines the biosynthesis of three important components of biological membranes—phospholipids, sphingolipids, and cholesterol. Triacylglycerols also are considered here because the pathway for their synthesis overlaps that of phospholipids. Cholesterol is of interest both as a membrane component and as a precursor of potent signaling molecules, such as the steroid hormones estradiol, testosterone, progesterone, and cortisol. Knowledge of these biosynthetic pathways, particularly that for cholesterol, has been of great importance for drug development and understanding the mechanisms of drug action.

27.1 Phosphatidate Is a Common Intermediate in the Synthesis of Phospholipids and Triacylglycerols

Lipid synthesis requires the coordinated action of gluconeogenesis and fatty acid metabolism (**Figure 27.1**). The common step in the synthesis of both phospholipids for membranes and triacylglycerols for energy storage is the formation of **phosphatidate** (diacylglycerol 3-phosphate). In mammalian cells, phosphatidate is synthesized in the endoplasmic reticulum and the outer mitochondrial membrane.

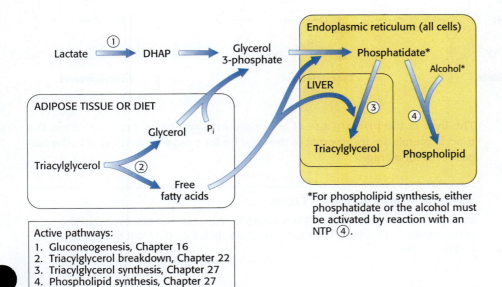

Active pathways:
1. Gluconeogenesis, Chapter 16
2. Triacylglycerol breakdown, Chapter 22
3. Triacylglycerol synthesis, Chapter 27
4. Phospholipid synthesis, Chapter 27

FIGURE 27.1 Phosphatidate is a key intermediate in lipid biosynthesis. Phosphatidate, synthesized from dihydroxyacetone phosphate (DHAP) produced in gluconeogenesis and fatty acids, can be further processed to produce triacylglycerol or phospholipids. Phospholipids and other membrane lipids are continuously produced in all cells.

EXPLORE this pathway further in the Metabolic Map in Achieve

The pathway for the biosynthesis of phosphatidate begins with glycerol 3-phosphate, which is formed primarily by the reduction of dihydroxyacetone phosphate (DHAP) synthesized by the gluconeogenic pathway, and to a lesser extent by the phosphorylation of glycerol. The addition of two fatty acids to glycerol-3-phosphate yields phosphatidate. First, acyl coenzyme A contributes a fatty acid chain to form lysophosphatidate, and then a second acyl CoA contributes a fatty acid chain to yield phosphatidate.

Glycerol 3-phosphate   **Lysophosphatidate**   **Phosphatidate**

These acylations are catalyzed by glycerol phosphate acyltransferase. In most phosphatidates, the fatty acid chain attached to the C-1 atom is saturated, whereas the one attached to the C-2 atom is unsaturated. Phosphatidate can also be synthesized from diacylglycerol (DAG), in what is essentially a salvage pathway, by the action of diacylglycerol kinase:

$$\text{Diacylglycerol} + \text{ATP} \longrightarrow \text{phosphatidate} + \text{ADP}$$

The phospholipid and triacylglycerol pathways diverge at phosphatidate. In the synthesis of triacylglycerols, a key enzyme in the regulation of lipid synthesis, phosphatidic acid phosphatase, hydrolyzes phosphatidate to give a diacylglycerol. This intermediate is acylated to a **triacylglycerol** through the addition of a third fatty acid chain in a reaction that is catalyzed by diglyceride acyltransferase. Both enzymes are associated in a triacylglycerol synthetase complex that is bound to the endoplasmic reticulum membrane.

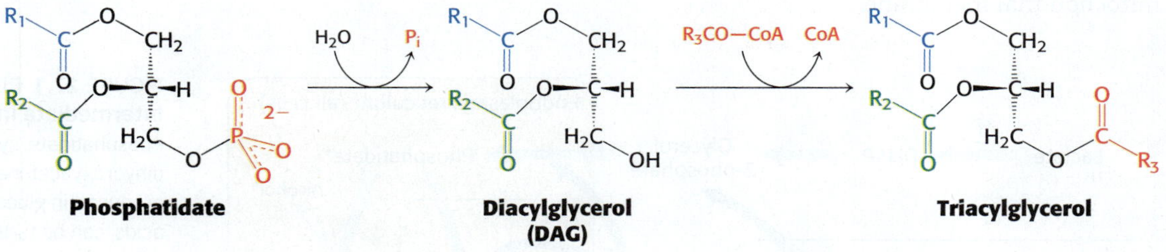

Phosphatidate   **Diacylglycerol**   **Triacylglycerol**
                              (DAG)

The liver is the primary site of triacylglycerol synthesis. From the liver, the triacylglycerols are transported to the muscles for energy conversion or to the adipose cells for storage.

❖ SELF–CHECK QUESTION

You find two samples labeled "Pure phosphatidic acid" but discover that the two samples have slightly different properties including different molecular weights. Explain.

The synthesis of phospholipids requires an activated intermediate

Membrane-lipid synthesis continues in the endoplasmic reticulum and in the Golgi apparatus. Phospholipid synthesis requires the combination of a diacylglycerol with an alcohol. As in most anabolic reactions, one of the components must be activated. In this case, either the diacylglycerol or the alcohol may be activated, depending on the source of the reactants.

The synthesis of some phospholipids begins with the reaction of phosphatidate with cytidine triphosphate (CTP) to form the activated diacylglycerol, **cytidine diphosphodiacylglycerol (CDP-diacylglycerol)**. This reaction, like those of many biosyntheses, is driven forward by the hydrolysis of pyrophosphate.

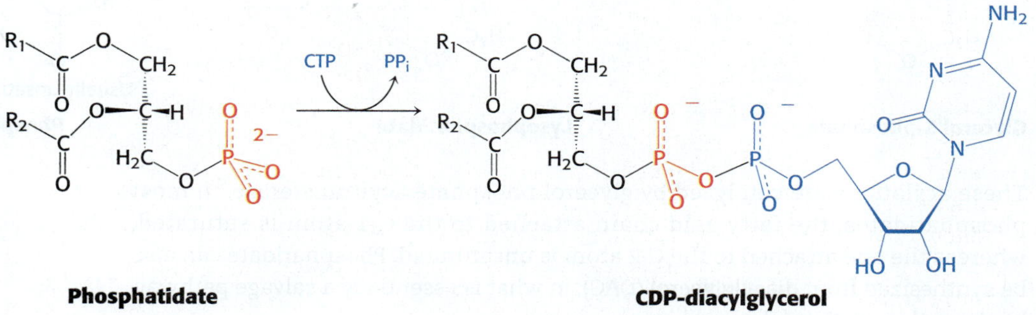

Phosphatidate   **CDP-diacylglycerol**

The activated phosphatidyl unit then reacts with the hydroxyl group of an alcohol to form a phosphodiester linkage. If the alcohol is inositol, the products are phosphatidylinositol and cytidine monophosphate (CMP).

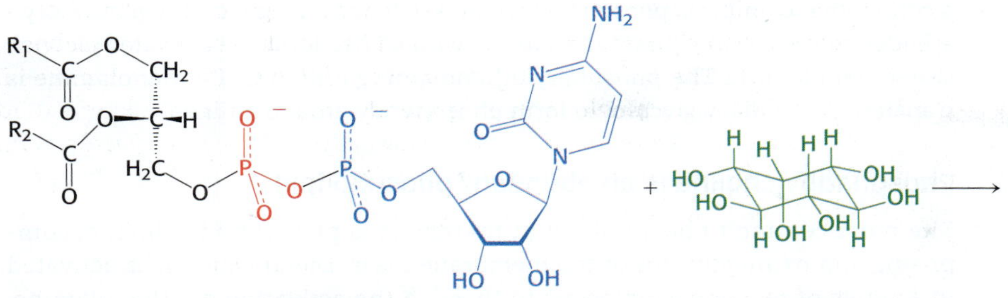

CDP-diacylglycerol   +   **Inositol**   →

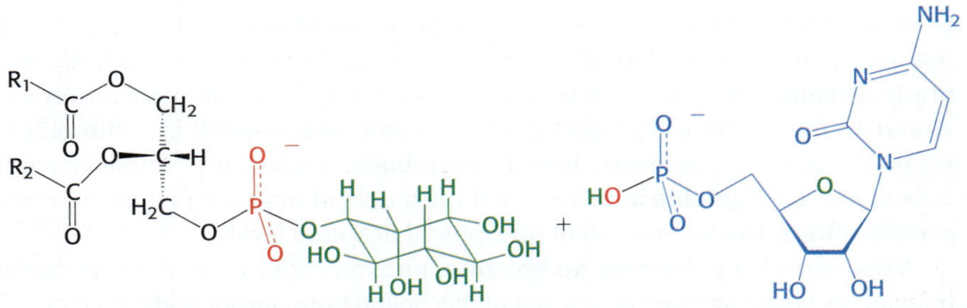

Phosphatidylinositol   +   **CMP**

Subsequent phosphorylations catalyzed by specific kinases lead to the synthesis of phosphatidylinositol 4,5-bisphosphate, the precursor of two intracellular messengers—diacylglycerol and inositol 1,4,5-trisphosphate. If the alcohol is phosphatidylglycerol, the products are diphosphatidylglycerol (cardiolipin) and CMP. In eukaryotes, cardiolipin is synthesized in the mitochondria and is located exclusively in inner mitochondrial membranes.

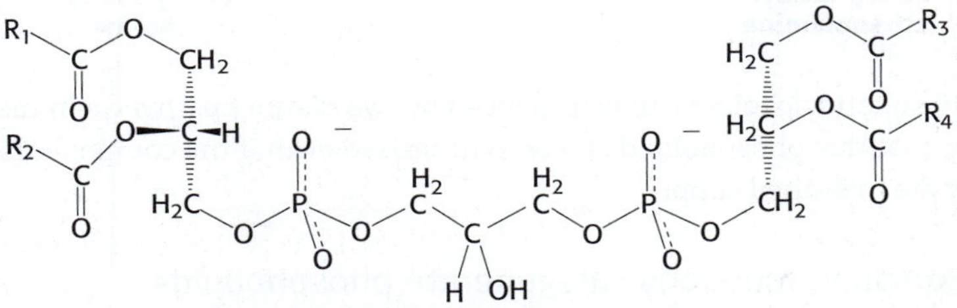

Diphosphatidylglycerol (Cardiolipin)

Cardiolipin is abundant in the inner membrane of mitochondria and plays an important role in the organization of the protein components participating in oxidative phosphorylation. Scientists have also discovered that cardiolipin facilitates the proper folding of α-synuclein in nerve cells. Misfolding of α-synuclein causes the formation of aggregates that are toxic. These α-synuclein deposits are believed to play a role in the development of Parkinson's disease.

The fatty acid components of phospholipids may vary, and thus cardiolipin, as well as most other phospholipids, represents a class of molecules rather than a single species. As a result, a single mammalian cell may contain thousands of distinct phospholipids. Phosphatidylinositol is unusual in that it has a nearly fixed fatty acid composition. Stearic acid usually occupies the C-1 position and arachidonic acid the C-2 position.

Some phospholipids are synthesized from an activated alcohol

Phosphatidylethanolamine, the major phospholipid of the inner leaflet of cell membranes, is synthesized from the alcohol ethanolamine. To activate the alcohol, ethanolamine is phosphorylated by ATP to form the precursor, phosphoryl-ethanolamine. This precursor then reacts with CTP to form the activated alcohol, CDP-ethanolamine. The phosphorylethanolamine unit of CDP-ethanolamine is transferred to a diacylglycerol to form phosphatidylethanolamine.

Phosphatidylcholine is an abundant phospholipid

The most common phospholipid in mammals is phosphatidylcholine, comprising approximately 50% of the membrane mass. Dietary choline is activated in a series of reactions analogous to those in the activation of ethanolamine. CTP-phosphocholine cytidylyltransferase (CCT) catalyzes the formation of CDP-choline, the rate-limiting step in phosphatidylcholine synthesis. CCT is termed an *amphitropic enzyme*—a class of enzymes whose regulator ligand is the membrane itself. A portion of the enzyme, normally associated with the membrane, is sensitive to a decrease in phosphatidylcholine concentration as an alteration in the physical properties of the membrane. When this occurs, another portion of the enzyme inserts into the membrane, leading to enzyme activation. Indeed, the k_{cat}/K_M value increases by three orders of magnitude upon activation, contributing to the restoration of phosphatidylcholine levels.

The liver also possesses an enzyme, phosphatidylethanolamine methyltransferase, which synthesizes phosphatidylcholine from phosphatidylethanolamine when dietary choline is insufficient, a testament to the importance of this vital phospholipid. The amino group of this phosphatidylethanolamine is methylated three times to form phosphatidylcholine. S-adenosylmethionine is the methyl donor.

**Phosphatidyl-
ethanolamine** → **Phosphatidylcholine**

Thus, phosphatidylcholine can be produced by two distinct pathways in mammals, ensuring that this phospholipid can be synthesized even if the components for one pathway are in limited supply.

Base-exchange reactions can generate phospholipids

Phosphatidylserine makes up 10% of the phospholipids in mammals. This phospholipid is synthesized in a base-exchange reaction of serine with phosphatidylcholine or phosphatidylethanolamine. In the reaction, serine replaces choline or ethanolamine.

$$\text{Phosphatidylcholine} + \text{serine} \longrightarrow \text{choline} + \text{phosphatidylserine}$$

$$\text{Phosphatidylethanolamine} + \text{serine} \longrightarrow \text{ethanolamine} + \text{phosphatidylserine}$$

Phosphatidylserine is normally located in the inner leaflet of the plasma membrane bilayer but is moved to the outer leaflet in apoptosis. There, it serves to attract phagocytes to consume the cell remnants after apoptosis is complete. Phosphatidylserine is translocated from one side of the membrane to the other by an ATP-binding cassette translocase.

Note that a cytidine nucleotide plays the same role in the synthesis of these phosphoglycerides as a uridine nucleotide does in the formation of glycogen (Section 21.4). In all of these biosyntheses, an activated intermediate (UDP-glucose, CDP-diacylglycerol, or CDP-alcohol) is formed from a phosphorylated substrate (glucose 1-phosphate, phosphatidate, or a phosphorylalcohol) and a nucleoside triphosphate (UTP or CTP). The activated intermediate then reacts with a hydroxyl group (the terminus of glycogen, an alcohol, or a diacylglycerol).

Ethanolamine
↓ ATP
↓ ADP

Phosphorylethanolamine
↓ CTP
↓ PP$_i$

CDP-ethanolamine
↓ Diacylglycerol
↓ CMP

Phosphatidylethanolamine

Sphingolipids are synthesized from ceramide

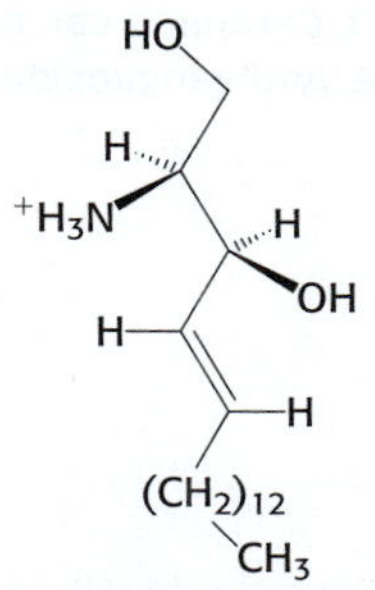

We now turn from glycerol-based phospholipids to another class of membrane lipid—the **sphingolipids**. These lipids are found in the plasma membranes of all eukaryotic cells, although the concentration is highest in the cells of the central nervous system. The backbone of a sphingolipid is sphingosine rather than glycerol. Palmitoyl CoA and serine condense to form 3-ketosphinganine. The serine–palmitoyl transferase catalyzing this reaction is the rate-limiting step in the pathway and requires pyridoxal phosphate; this reveals again the dominant role of this cofactor in transformations that include amino acids. Ketosphinganine is then reduced to dihydrosphingosine before conversion into **ceramide**, a lipid consisting of a fatty acid chain attached to the amino group of a sphingosine backbone (**Figure 27.2**).

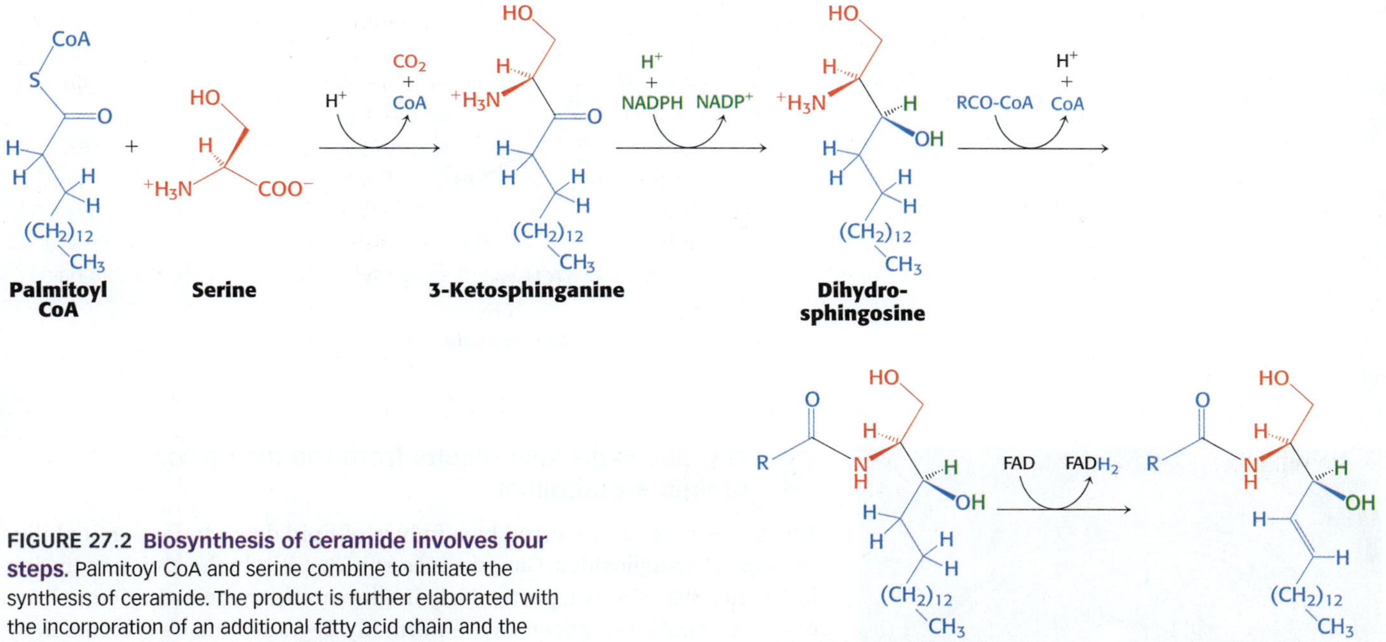

Sphingosine

FIGURE 27.2 Biosynthesis of ceramide involves four steps. Palmitoyl CoA and serine combine to initiate the synthesis of ceramide. The product is further elaborated with the incorporation of an additional fatty acid chain and the introduction of a carbon–carbon double bond.

Palmitoyl CoA + **Serine** → **3-Ketosphinganine** → **Dihydro-sphingosine**

Dihydroceramide → **Ceramide**

In all sphingolipids, the amino group of ceramide is acylated. The terminal hydroxyl group also is substituted. In sphingomyelin, a component of the myelin sheath covering many nerve fibers, the substituent is phosphorylcholine, which comes from phosphatidylcholine. In a cerebroside, the substituent is glucose or galactose. UDP-glucose or UDP-galactose is the sugar donor.

Gangliosides are the most complex sphingolipids. In a ganglioside, an oligosaccharide chain is linked to the terminal hydroxyl group of ceramide by a glucose residue (**Figure 27.3**). This oligosaccharide chain contains at least one acidic sugar, either N-acetylneuraminate or N-glycolylneuraminate. These acidic sugars are called *sialic acids*. Their nine-carbon backbones are synthesized from phosphoenolpyruvate (a three-carbon unit) and N-acetylmannosamine 6-phosphate (a six-carbon unit).

Gangliosides are synthesized by the ordered, step-by-step addition of sugar residues to ceramide (**Figure 27.4**). The synthesis of these complex lipids requires the activated sugars UDP-glucose, UDP-galactose, and UDP-N-acetylgalactosamine, as well as the CMP derivative of N-acetylneuraminate. CMP-N-acetylneuraminate is synthesized from CTP and N-acetylneuraminate. The sugar composition of the resulting ganglioside is determined by the specificity of the glycosyltransferases in the cell. Almost 200 different gangliosides have been characterized.

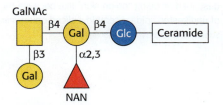

FIGURE 27.3 The ganglioside G$_{M1}$ consists of five monosaccharides linked to ceramide. These sugars include one glucose (Glc) molecule, two galactose (Gal) molecules, one N-acetylgalactosamine (GalNAc) molecule, and one N-acetylneuraminate (NAN) molecule.

FIGURE 27.4 Ceramide can be converted into sphingomyelin, cerebroside, and gangliosides.

Ceramide

Phosphatidyl-choline → DAG

UDP-glucose → UDP

Sphingomyelin

Cerebroside

Activated sugars → Gangliosides

FIGURE 27.5 Lysosome sometimes will fill up with lipids. An electron micrograph shows a lysosome engorged with lipids. Such lysosomes are sometimes described as being "onion-skin" like because the layers of undigested lipids resemble a sliced onion. [Photo Credit: Science Photo Library/Science Source]

Tay–Sachs disease results from the disruption of lipid metabolism

Tay–Sachs disease is caused by a failure of lipid degradation: an inability to degrade gangliosides. Gangliosides are found in highest concentration in the nervous system, particularly in gray matter, where they constitute 6% of the lipids. Gangliosides are normally degraded inside lysosomes by the sequential removal of their terminal sugars, but in Tay–Sachs disease this degradation does not take place. As a consequence, neurons become significantly swollen with lipid-filled lysosomes (**Figure 27.5**). An affected infant displays weakness and limited psychomotor skills before 1 year of age and usually dies before age 3.

The ganglioside content of the brain of an infant with Tay–Sachs disease is greatly elevated. The concentration of ganglioside G_{M2} is many times higher than normal because its terminal N-acetylgalactosamine residue is removed very slowly or not at all. The missing or deficient enzyme is a specific β-N-acetylhexosaminidase.

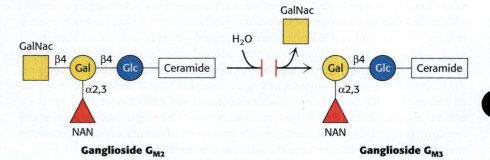

Ganglioside G_{M2}

H_2O

Ganglioside G_{M3}

Carriers of Tay–Sachs can be identified by genetic testing. Moreover, Tay–Sachs disease can be diagnosed in the course of fetal development. Cells obtained from either the villi of the placenta (chorionic villus sampling) or from the amniotic fluid (amniocentesis) are tested for the presence of the defective gene. Tay–Sachs disease was especially prominent among Ashkenazi Jews (descendants of Jews from central and eastern Europe). Genetic testing programs, initiated in the early 1970s upon the development of a simple blood test to identify carriers, have dramatically reduced the incidence of the disease in this population.

Phosphatidic acid phosphatase is a key regulatory enzyme in lipid metabolism

The enzyme phosphatidic acid phosphatase (PAP), working in concert with diacylglycerol kinase, plays a key role in the regulation of lipid synthesis. PAP, also called *lipin* 1 in mammals, controls the extent to which triacylglycerols are synthesized relative to phospholipids and regulates the type of phospholipid synthesized (**Figure 27.6**). For instance, when PAP activity is high, phosphatidate is dephosphorylated and diacylglycerol is produced, which can react with the appropriate activated alcohols to yield phosphatidylethanolamine, phosphatidylserine, or phosphatidylcholine. Diacylglycerol can also be converted into triacylglycerols. Evidence suggests that the formation of triacylglycerols may act as a fatty acid buffer. This buffering helps to regulate the levels of diacylglycerol and sphingolipids, which serve signaling functions.

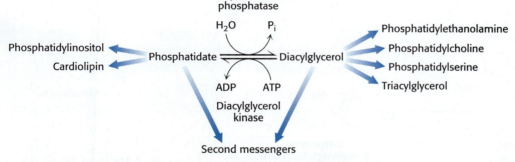

FIGURE 27.6 Phosphatidic acid phosphatase is a key regulatory enzyme in lipid synthesis. When active, PAP generates diacylglycerol (DAG), which can react with activated alcohols to form phospholipids or with fatty acyl CoA to form triacylglycerols. When PAP is inactive, phosphatidate is converted into CMP-DAG for the synthesis of different phospholipids. PAP also controls the amount of DAG and phosphatidate, both of which function as second messengers.

EXPLORE this pathway further in the Metabolic Map in ⚡ Achieve

When PAP activity is lower, phosphatidate is used as a precursor for different phospholipids, such as phosphatidylinositol and cardiolipin. Moreover, phosphatidate is a signal molecule itself. Phosphatidate regulates the growth of endoplasmic reticulum and nuclear membranes and acts as a cofactor that stimulates the expression of genes in phospholipid synthesis.

27.2 Cholesterol Is Synthesized from Acetyl Coenzyme A in Three Stages

We now turn our attention to the synthesis of the fundamental lipid **cholesterol**. This steroid modulates the fluidity of animal cell membranes and is the precursor of steroid hormones such as progesterone, testosterone, estradiol, and cortisol.

All 27 carbon atoms of cholesterol are derived from acetyl CoA in a three-stage synthetic process (**Figure 27.7**). The first stage takes place in the cytoplasm, and the next two in the endoplasmic reticulum.

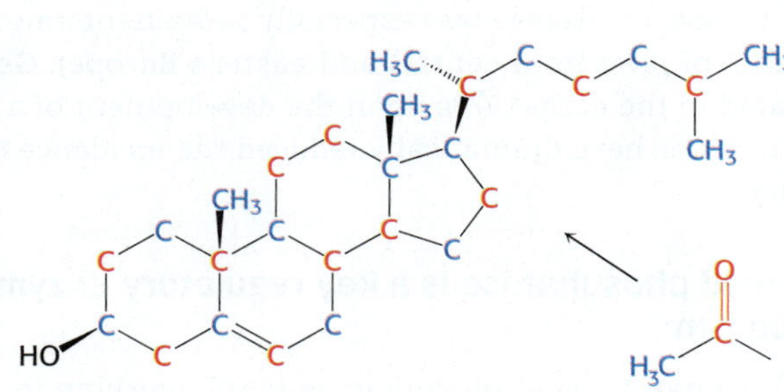

FIGURE 27.7 Isotope-labeling experiments have revealed the source of carbon atoms in cholesterol synthesized from acetate. In this depiction, the source of carbons from acetate's methyl group are shown in blue; carbons from acetate's carboxylate group are shown in red.

Stage 1: The synthesis of mevalonate initiates the synthesis of cholesterol

The first stage in the synthesis of cholesterol is the formation of isopentenyl pyrophosphate from acetyl CoA. This set of reactions starts with the formation of 3-hydroxy-3-methylglutaryl CoA (HMG CoA) from acetyl CoA and acetoacetyl CoA. This intermediate is reduced to **mevalonate** for the synthesis of cholesterol (**Figure 27.8**). Recall that, alternatively, 3-hydroxy-3-methylglutaryl CoA may be generated in the mitochondria and processed to form ketone bodies, which are subsequently secreted to provide fuel for other tissues, notably the brain under starvation conditions.

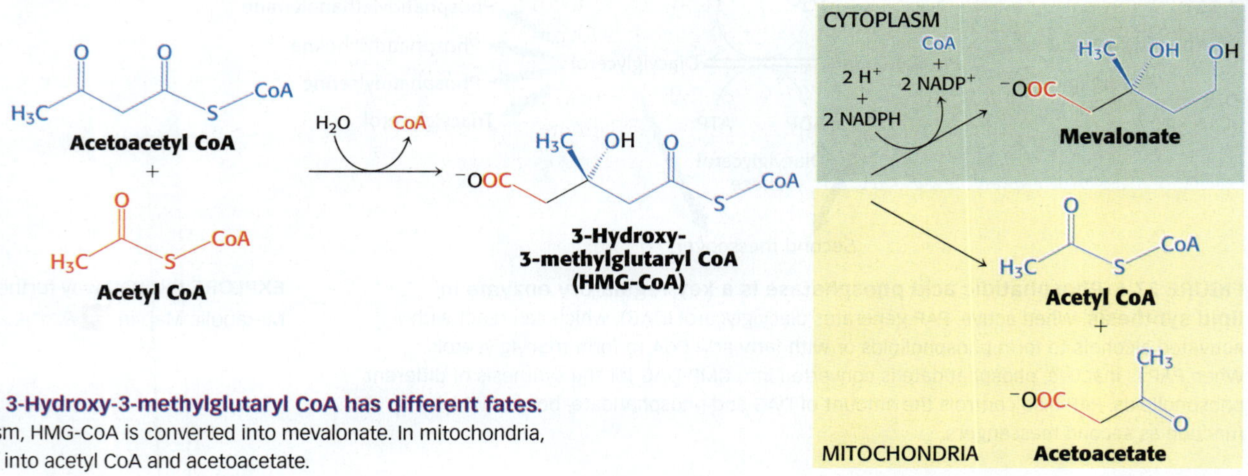

FIGURE 27.8 3-Hydroxy-3-methylglutaryl CoA has different fates. In the cytoplasm, HMG-CoA is converted into mevalonate. In mitochondria, it is converted into acetyl CoA and acetoacetate.

The synthesis of mevalonate is the committed step in cholesterol formation. The enzyme catalyzing this irreversible step, **3-hydroxy-3-methylglutaryl CoA reductase (HMG-CoA reductase)**, is an important control site in cholesterol biosynthesis, as will be discussed shortly.

3-Hydroxy-3-methylglutaryl CoA + 2 NADPH + 2 H$^+$ $\longrightarrow$

mevalonate + 2 NADP$^+$ + CoA

HMG-CoA reductase is an integral membrane protein in the endoplasmic reticulum. Mevalonate is converted into 3-isopentenyl pyrophosphate in three consecutive reactions requiring ATP (**Figure 27.9**). In the last step, the release of CO$_2$ yields isopentenyl pyrophosphate, an activated isoprene unit that is a key building block for many important biomolecules throughout the kingdoms of life.

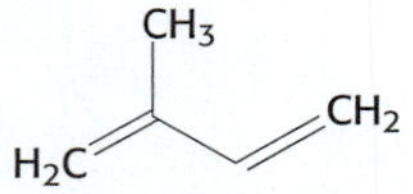

Isoprene

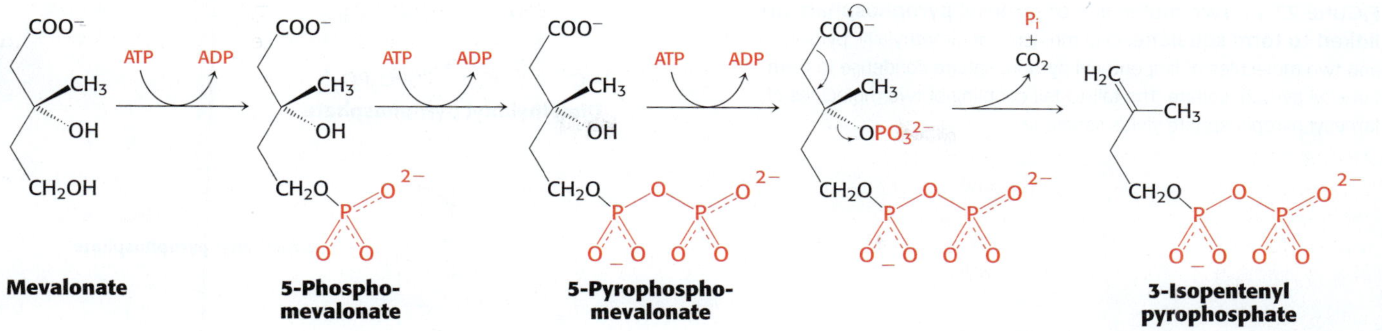

Mevalonate   **5-Phospho-mevalonate**   **5-Pyrophospho-mevalonate**   **3-Isopentenyl pyrophosphate**

FIGURE 27.9 Mevalonate is converted into isopentenyl pyrophosphate. This activated intermediate is formed from mevalonate in three steps requiring ATP, followed by a decarboxylation.

Stage 2: Squalene (C_{30}) is synthesized from six molecules of isopentenyl pyrophosphate (C_5)

Squalene is synthesized from isopentenyl pyrophosphate by the reaction sequence

$$C_5 \longrightarrow C_{10} \longrightarrow C_{15} \longrightarrow C_{30}$$

This stage in the synthesis of cholesterol starts with the isomerization of isopentenyl pyrophosphate to dimethylallyl pyrophosphate.

Isopentenyl pyrophosphate   **Dimethylallyl pyrophosphate**

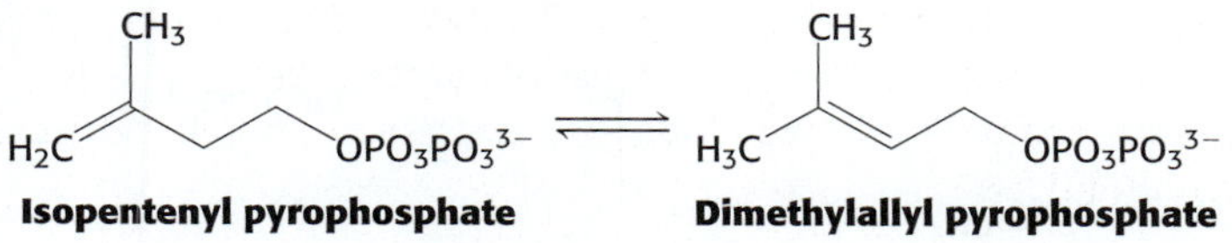

These two isomeric C_5 units (one of each type) condense to form a C_{10} compound: isopentenyl pyrophosphate attacks an allylic carbocation formed from dimethylallyl pyrophosphate to yield geranyl pyrophosphate (**Figure 27.10**). The same kind of reaction takes place again: geranyl pyrophosphate is converted into an allylic carbonium ion, which is attacked by isopentenyl pyrophosphate. The resulting C_{15} compound is called farnesyl pyrophosphate. The same enzyme, geranyl transferase, catalyzes each of these condensations.

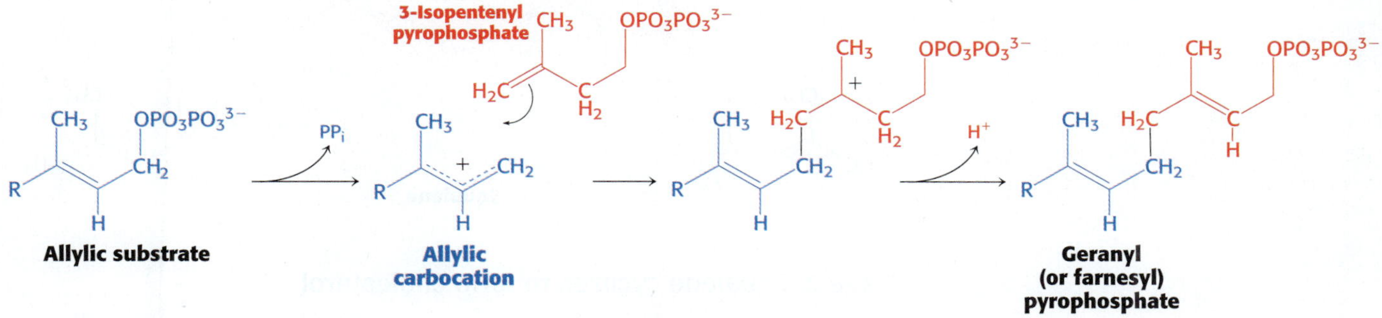

Allylic substrate   **Allylic carbocation**   **Geranyl (or farnesyl) pyrophosphate**

FIGURE 27.10 Condensation reactions build carbon–carbon bonds that eventually form the cholesterol skeleton. The isomeric C_5 units dimethylallyl pyrophosphate and isopentenyl pyrophosphate join to form geranyl pyrophosphate. The same mechanism is used to add an additional isopentenyl pyrophosphate to form farnesyl pyrophosphate.

The last step in the synthesis of squalene is a reductive tail-to-tail condensation of two molecules of farnesyl pyrophosphate catalyzed by the endoplasmic reticulum enzyme squalene synthase.

2 Farnesyl pyrophosphate (C_{15}) + NADPH $\longrightarrow$ squalene (C_{30}) + 2 PP_i + NADP$^+$ + H$^+$

The reactions leading from C_5 units to squalene, a C_{30} isoprenoid, are summarized in **Figure 27.11**.

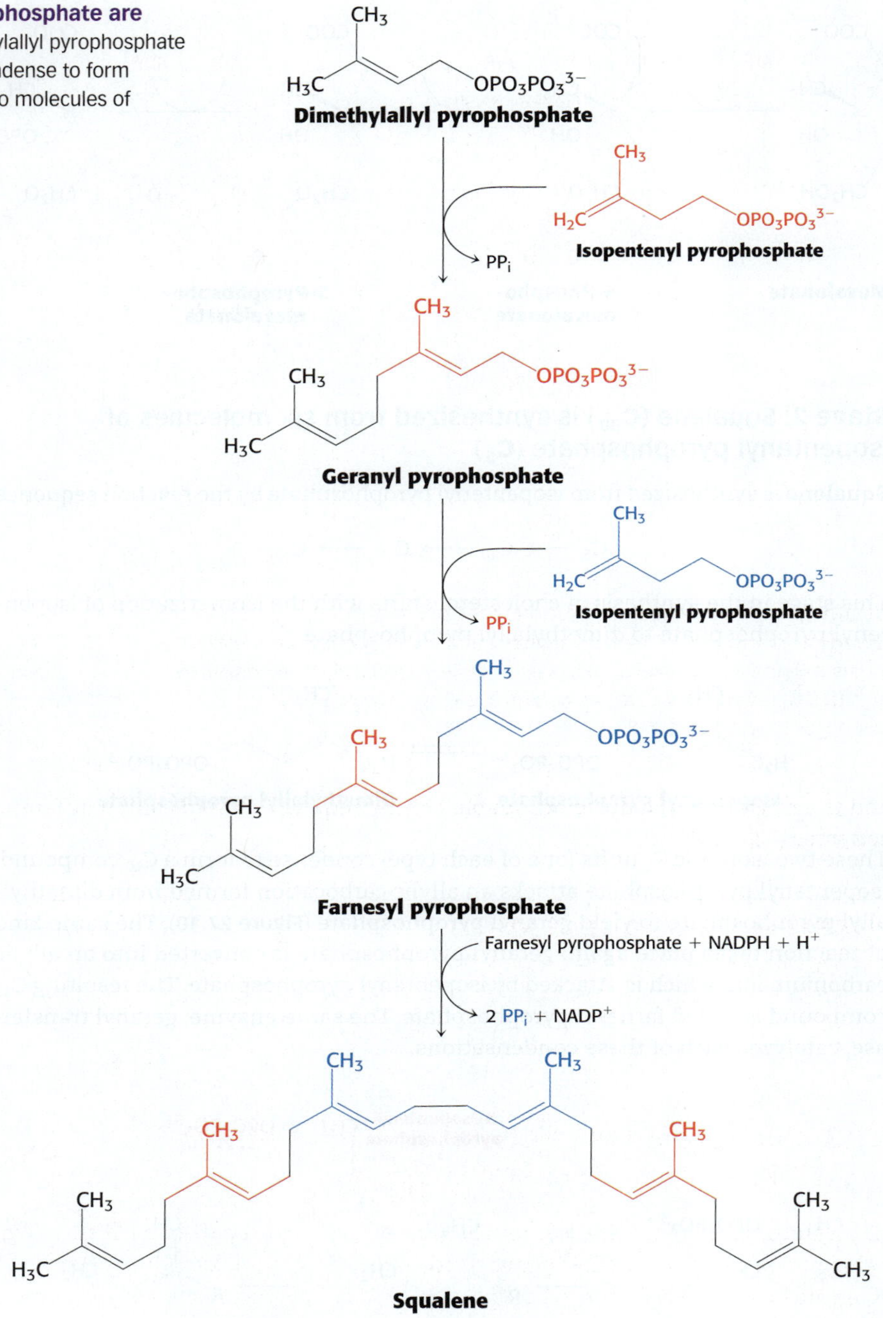

FIGURE 27.11 Two molecules of farnesyl pyrophosphate are linked to form squalene. One molecule of dimethylallyl pyrophosphate and two molecules of isopentenyl pyrophosphate condense to form farnesyl pyrophosphate. The tail-to-tail coupling of two molecules of farnesyl pyrophosphate yields squalene.

Dimethylallyl pyrophosphate

Isopentenyl pyrophosphate

PP$_i$

Geranyl pyrophosphate

Isopentenyl pyrophosphate

PP$_i$

Farnesyl pyrophosphate

Farnesyl pyrophosphate + NADPH + H$^+$

2 PP$_i$ + NADP$^+$

Squalene

Stage 3: Squalene cyclizes to form cholesterol

The final stage of cholesterol biosynthesis starts with the cyclization of squalene (**Figure 27.12**). Squalene is first activated by conversion into squalene epoxide (2,3-oxidosqualene) in a reaction that uses O$_2$ and NADPH. Squalene epoxide is then cyclized to lanosterol by oxidosqualene cyclase. The enzyme holds squalene epoxide in an appropriate conformation and initiates the reaction by protonating the epoxide oxygen. The carbocation formed spontaneously rearranges to produce lanosterol. This is a remarkable chemical transformation, forming the four rings that are characteristic of steroids through four new carbon–carbon bonds. In addition, six new chiral centers are generated. In contrast to this remarkably efficient step, the conversion of lanosterol to cholesterol—which involves the removal of three methyl groups, the reduction of a double bond,

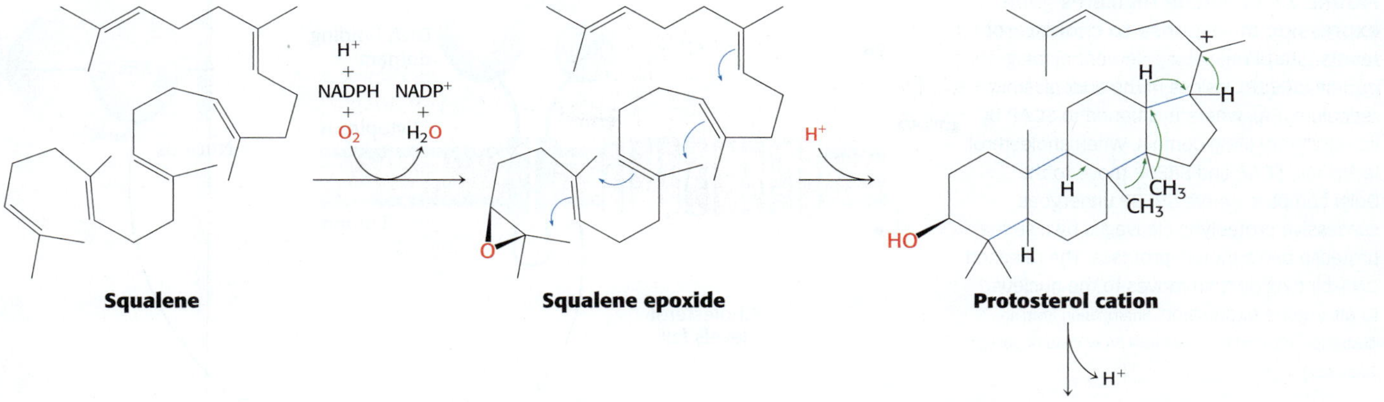

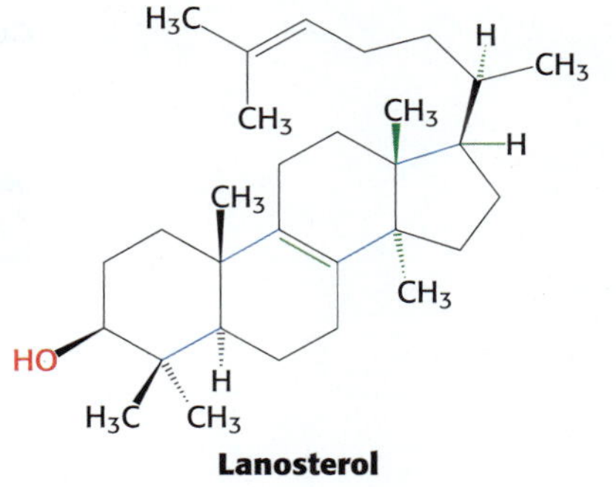

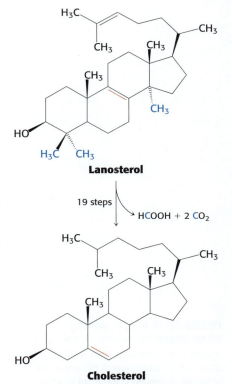

FIGURE 27.12 Squalene cyclization leads to the steroid skeleton in the form of lanosterol. The formation of the steroid nucleus from squalene begins with the formation of squalene epoxide. This intermediate is protonated to form a carbocation that cyclizes to form a tetracyclic structure, which rearranges to form lanosterol.

Squalene   Squalene epoxide   Protosterol cation   Lanosterol

and the migration of a different double bond—requires 19 steps and 9 different enzymes (**Figure 27.13**).

❖ SELF–CHECK QUESTION

The biosynthesis of cholesterol requires 20 steps and 10 different enzymes. Which enzyme is responsible for the formation of the five-membered ring in the cholesterol structure?

27.3 The Regulation of Cholesterol Biosynthesis Takes Place at Several Levels

Cholesterol can be obtained from the diet or cells can synthesize it de novo, using one of the most highly regulated metabolic pathways known. Biosynthetic rates may vary several hundred-fold, depending on how much cholesterol is consumed in the diet. An adult on a low-cholesterol diet typically synthesizes about 800 mg of cholesterol per day. The liver is the major site of cholesterol synthesis in mammals, although the intestine also forms significant amounts.

The rate of cholesterol formation by the liver and intestine is highly responsive to the cellular level of cholesterol. This feedback regulation is mediated primarily by changes in the amount and activity of 3-hydroxy-3-methylglutaryl CoA reductase. As described earlier, this enzyme catalyzes the formation of mevalonate, the committed step in cholesterol biosynthesis. HMG CoA reductase is controlled in multiple ways:

1. *The rate of synthesis of reductase mRNA is controlled by a transcription factor called SREBP.* **Sterol regulatory element binding protein (SREBP)** is a member of a family of transcription factors that regulate the proteins required for lipid synthesis. SREBP binds to a short DNA sequence called the sterol regulatory element (SRE) on the 5′ side of the reductase gene when cholesterol levels

Lanosterol

19 steps → HCOOH + 2 CO₂

Cholesterol

FIGURE 27.13 Cholesterol is formed from lanosterol. This complex transformation involves 19 steps and 9 different enzymes.

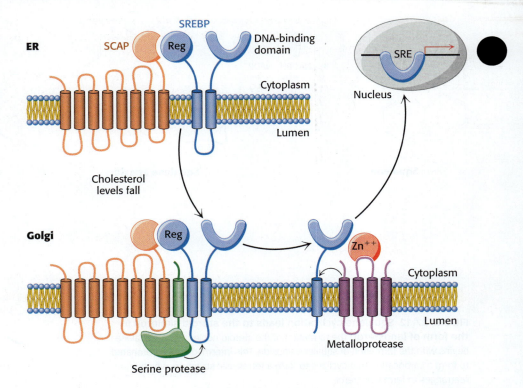

FIGURE 27.14 SREBP regulates gene expression in response to cholesterol levels. Sterol regulatory element binding protein (SREBP) resides in the endoplasmic reticulum (ER), where it is bound to SCAP by its regulatory (Reg) domain. When cholesterol levels fall, SCAP and SREBP move to the Golgi complex, where SREBP undergoes successive proteolytic cleavages by a serine protease and a metalloprotease. The released DNA-binding domain moves to the nucleus to alter gene expression. [Information from an illustration provided by Dr. Michael Brown and Dr. Joseph Goldstein.]

are low, enhancing transcription of the gene. In its inactive state, SREBP resides in the endoplasmic reticulum membrane, where it is associated with the <u>S</u>REBP <u>c</u>leavage <u>a</u>ctivating <u>p</u>rotein (SCAP), an integral membrane protein. SCAP is the cholesterol sensor.

When cholesterol levels fall, SCAP escorts SREBP in small membrane vesicles to the Golgi complex, where it is released from the membrane by two specific proteolytic cleavages (**Figure 27.14**). The first cleavage frees a fragment of SREBP from SCAP, whereas the second cleavage releases the regulatory domain from the membrane. The released protein migrates to the nucleus and binds the SRE of the HMG-CoA reductase gene, as well as several other genes in the cholesterol biosynthetic pathway, to enhance transcription. When cholesterol levels rise, the proteolytic release of the SREBP is blocked, and the SREBP in the nucleus is rapidly degraded by proteasomes located in the nucleus. These two events halt the transcription of genes of the cholesterol biosynthetic pathways.

What is the molecular mechanism that retains SCAP–SREBP in the ER when cholesterol is present but allows movement to the Golgi complex when cholesterol concentration is low? When cholesterol is low, SCAP binds to vesicular proteins that facilitate the transport of SCAP–SREBP to the Golgi apparatus. When cholesterol is present, SCAP binds cholesterol, which causes a structural change in SCAP so that it binds to another endoplasmic reticulum protein called Insig (<u>ins</u>ulin-<u>i</u>nduced <u>g</u>ene) (**Figure 27.15**). Insig is the anchor that retains SCAP and thus SREBP in the endoplasmic reticulum in the presence of cholesterol. The interactions between SCAP and Insig can also be

FIGURE 27.15 Insig regulates SCAP–SREBP movement. In the presence of cholesterol, Insig interacts with SCAP–SREBP and prevents the activation of SREBP. Cholesterol binding to SCAP or 25-hydroxycholesterol binding to Insig facilitates the interaction of Insig and SCAP, retaining SCAP–SREBP in the endoplasmic reticulum.

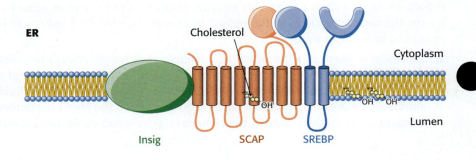

forged when Insig binds 25-hydroxycholesterol, a metabolite of cholesterol. Thus, two distinct steroid–protein interactions prevent the inappropriate movement of SCAP–SREBP to the Golgi complex.

2. *The rate of translation of reductase mRNA is inhibited by nonsterol metabolites derived from mevalonate.* This results in an 80% reduction in the rate of reductase protein production.

3. *The degradation of the reductase is stringently controlled.* The enzyme is bipartite: its cytoplasmic domain carries out catalysis, and its membrane domain senses signals that lead to its degradation. The membrane domain may undergo structural changes in response to increasing concentrations of sterols such as lanosterol and 25-hydroxycholesterol. Under these conditions, the reductase appears to bind to a subset of Insigs that are also associated with the ubiquitinating enzymes (**Figure 27.16**). The reductase is polyubiquitinated and subsequently extracted from the membrane in a process that requires geranylgeraniol. The extracted reductase is then degraded by the proteasome. The combined regulation at the levels of transcription, translation, and degradation can alter the amount of enzyme in the cell more than 200-fold.

4. *Phosphorylation decreases the activity of the reductase.* This enzyme, like acetyl CoA carboxylase, which catalyzes the committed step in fatty acid synthesis, is switched off by an AMP-activated protein kinase. Thus, cholesterol synthesis ceases when the ATP level is low.

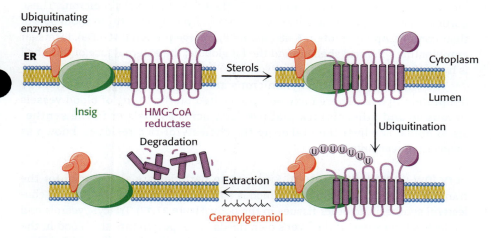

FIGURE 27.16 Insig facilitates the degradation of HMG-CoA reductase. In the presence of sterols, a subclass of Insig associated with ubiquitinating enzymes binds HMG-CoA reductase. This interaction results in the ubiquitination of the enzyme. This modification and the presence of geranylgeraniol results in extraction of the enzyme from the membrane and degradation by the proteasome.

Lipoproteins transport cholesterol and triacylglycerols throughout the organism

Cholesterol and triacylglycerols are transported in body fluids in the form of **lipoprotein particles**, which are important for a number of reasons. First, lipoprotein particles are the means by which triacylglycerols are delivered to tissues, from the intestine or liver, for use as fuel or for storage. Second, the fatty acid constituents of the triacylglycerol components of the lipoprotein particles are incorporated into phospholipids for membrane synthesis. Likewise, cholesterol is a vital component of membranes and is a precursor to the powerful signal molecules—the steroid hormones. Finally, cells are not able to degrade the steroid nucleus. Consequently, the cholesterol must be used biochemically or excreted by the liver. Excess cholesterol plays a role in the development of atherosclerosis. Lipoprotein particles function in cholesterol homeostasis, transporting the molecule from sites of synthesis to sites of use, and finally to the liver for excretion.

Each lipoprotein particle consists of a core of hydrophobic lipids surrounded by a shell of more-polar lipids and proteins. The protein components of these macromolecular aggregates, called apoproteins, have two roles: they solubilize hydrophobic lipids and contain cell-targeting signals. Apolipoproteins are synthesized and secreted by the liver and the intestine.

TABLE 27.1 Properties of plasma lipoproteins

Plasma lipoproteins	Density (g ml⁻¹)	Diameter (nm)	Apolipoprotein	Physiological role	TAG	CE	C	PL	P
					Composition (%)				
Chylomicron	< 0.95	75–1200	B-48, C, E	Dietary fat transport	86	3	1	8	2
Very low-density lipoprotein	0.95–1.006	30–80	B-100, C, E	Endogenous fat transport	52	14	7	18	8
Intermediate-density lipoprotein	1.006–1.019	15–35	B-100, E	LDL precursor	38	30	8	23	11
Low-density lipoprotein	1.019–1.063	18–25	B-100	Cholesterol transport	10	38	8	22	21
High-density lipoprotein	1.063–1.21	7.5–20	A	Reverse cholesterol transport	5–10	14–21	3–7	19–29	33–57

Abbreviations: TAG, triacylglycerol; CE, cholesteryl ester; C, free cholesterol; PL, phospholipid; P, protein.

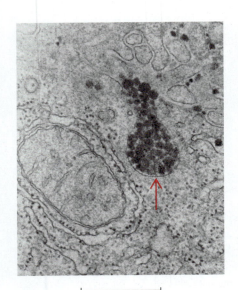

├─────────────┤
        500 nm

FIGURE 27.17 The liver is the major site of cholesterol synthesis. An electron micrograph shows a part of a liver cell actively engaged in the synthesis and secretion of very low-density lipoprotein (VLDL). The arrow points to a vesicle that is releasing its content of VLDL particles. [Photo Credit: Courtesy of Dr. George Palade/Yale University, Harvey Cushing/John Hay Whitney Medical Library.]

Lipoprotein particles are classified according to increasing density (Table 27.1). Lipoprotein particles can shift between classes as they release or pick up cargo, thereby changing their density.

Chylomicrons. Triacylglycerols, cholesterol, and other lipids obtained from the diet are carried away from the intestine in the form of large chylomicrons. These particles have a very low density because triacylglycerols constitute about 90% of their content. Apolipoprotein B-48 (apo B-48), a large protein (240 kDa), forms an amphipathic spherical shell around the fat globule; the external face of this shell is hydrophilic.

The triacylglycerols in chylomicrons are released through hydrolysis by lipoprotein lipases. These enzymes are located on the lining of blood vessels in muscle and other tissues that use fatty acids as fuels or in the synthesis of lipids. The liver then takes up the cholesterol-rich residues, known as *chylomicron remnants*.

Very low-density lipoprotein. Lipoprotein particles are also crucial for the transport of lipids from the liver, which is a major site of triacylglycerol and cholesterol synthesis, to other tissues in the body (Figure 27.17). Triacylglycerols and cholesterol in excess of the liver's own needs are exported into the blood in the form of very low-density lipoproteins (VLDL). These particles are stabilized by two apolipoproteins—apo B-100 (513 kDa) and apo E (34 kDa).

Intermediate-density lipoprotein. Triacylglycerols in very low-density lipoproteins, as in chylomicrons, are hydrolyzed by lipases on capillary surfaces, with the released fatty acids being taken up by the muscle and other tissues. The resulting remnants, which are rich in cholesteryl esters, are intermediate-density lipoproteins (IDL). These particles have two fates. Half of them are taken up by the liver for processing, and half are converted into low-density lipoprotein by the removal of more triacylglycerol by tissue lipases that absorb the released fatty acids.

Low-density lipoprotein. The major carrier of cholesterol in blood is **low-density lipoprotein (LDL)** (Figure 27.18). It contains a core of some 1500 cholesterol molecules esterified to fatty acids. The most common fatty acid chain in these esters is linoleate, a polyunsaturated fatty acid. A shell of phospholipids and unesterified cholesterol molecules surrounds this highly hydrophobic core. The shell also contains a single copy of apo B-100, which is recognized by target cells. In addition to transporting cholesterol to peripheral tissues, LDL regulates de novo cholesterol synthesis at these sites.

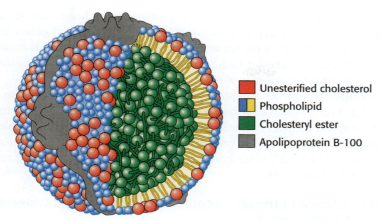

FIGURE 27.18 LDL consists of a core of cholesterol esters surrounded by a single molecule of apolipoprotein B-100. A low-density lipoprotein (LDL) particle is approximately 22 nm (220 Å) in diameter.

- 🔴 Unesterified cholesterol
- 🟡🔵 Phospholipid
- 🟢 Cholesteryl ester
- ⬛ Apolipoprotein B-100

High-density lipoprotein. A different purpose is served by **high-density lipoprotein (HDL)**, which picks up cholesterol released into the plasma from dying cells and from membranes undergoing turnover and delivers the cholesterol to the liver for excretion. An acyltransferase in HDL esterifies these cholesterols, which are then returned by HDL to the liver.

Low-density lipoproteins play a central role in cholesterol metabolism

Cholesterol metabolism must be precisely regulated to prevent atherosclerosis. The mode of control in the liver, the primary site of cholesterol synthesis, has already been considered: dietary cholesterol reduces the activity and amount of 3-hydroxy-3-methylglutaryl CoA reductase, the enzyme catalyzing the committed step. In general, cells outside the liver and intestine obtain cholesterol from the plasma rather than synthesizing it de novo. Specifically, their primary source of cholesterol is the low-density lipoprotein. The process of LDL uptake, called **receptor-mediated endocytosis**, serves as a paradigm for the uptake of many molecules (**Figure 27.19**).

Endocytosis begins when apolipoprotein B-100 on the surface of an LDL particle binds to a specific receptor protein on the plasma membrane of nonhepatic cells. The receptors for LDL are localized in specific regions called *coated pits*, which contain a protein called *clathrin*. The receptor–LDL complex is then internalized by endocytosis; that is, the plasma membrane in the vicinity of the complex invaginates and then fuses to form an endocytic vesicle called an *endosome*. The endosome is acidified by an ATP-dependent proton pump homologous to the Na^+/K^+ ATPase, which causes the receptor to release its cargo. Some of the receptor is returned to the cell membrane in a recycling vesicle, while a portion is degraded along with its cargo. The round-trip time for a receptor is about 10 minutes; in its lifetime of about a day, each receptor may bring hundreds of LDL particles into the cell.

The vesicles containing LDL, and in some cases the receptor, subsequently fuse with *lysosomes*, acidic vesicles that carry a wide array of degradative enzymes. The protein component of LDL is hydrolyzed to free amino acids, while the cholesteryl esters in LDL are hydrolyzed by a lysosomal acid lipase. The released unesterified cholesterol can then be used for membrane biosynthesis. Alternatively, it can be reesterified for storage inside the cell. In fact, free cholesterol activates a̲cyl C̲oA:cholesterol a̲cyltransferase (ACAT), the enzyme catalyzing this reaction.

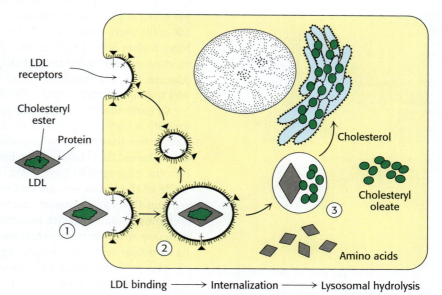

FIGURE 27.19 LDL enters cells through receptor-mediated endocytosis. (1) LDL binds to a specific receptor, the LDL receptor. (2) This complex invaginates to form an endosome. (3) After separation from its receptor, the LDL-containing vesicle fuses with a lysosome, leading to the degradation of the LDL and the release of the cholesterol.

Reesterified cholesterol contains mainly oleate and palmitoleate, which are monounsaturated fatty acids, in contrast with the cholesteryl esters in LDL, which are rich in linoleate, a polyunsaturated fatty acid (Table 12.1). It is imperative that the cholesterol be reesterified, because high concentrations of unesterified cholesterol disrupt the integrity of cell membranes.

The synthesis of the LDL receptor is itself subject to feedback regulation. Studies of cultured fibroblasts show that, when cholesterol is abundant inside the cell, new LDL receptors are not synthesized, and so the uptake of additional cholesterol from plasma LDL is blocked. The gene for the LDL receptor, like that for the reductase, is regulated by SREBP, which binds to a sterol regulatory element that controls the rate of mRNA synthesis.

The absence of the LDL receptor leads to hypercholesterolemia and atherosclerosis

The pioneering studies of familial hypercholesterolemia by Michael Brown and Joseph Goldstein revealed the physiological importance of the LDL receptor. The total concentration of cholesterol and LDL in the blood plasma is markedly elevated in this genetic disorder, which results from a mutation at a single autosomal locus. The cholesterol level in the plasma of homozygotes is typically 680 mg dl^{-1}, compared with 300 mg dl^{-1} in heterozygotes (clinical assay results are often expressed in milligrams per deciliter, which is equal to milligrams per 100 milliliters). A value of < 200 mg dl^{-1} is regarded as desirable, but many people have higher levels. In familial hypercholesterolemia, cholesterol is deposited in various tissues because of the high concentration of LDL cholesterol in the plasma. LDL accumulates under the endothelial cells lining the blood vessels, and nodules of cholesterol called *xanthomas* are prominent in skin and tendons.

Of particular concern is the oxidation of the excess LDL to form oxidized LDL (oxLDL), which can instigate the inflammatory response by the immune system, a response that has been implicated in the development of cardiovascular disease. The oxLDL is taken up by immune system cells called macrophages, which become engorged to form foam cells. These foam cells become trapped in the walls of the blood vessels and contribute to the formation of atherosclerotic plaques that cause arterial narrowing and lead to heart attacks (**Figure 27.20**). In fact, most people who are homozygous for the hypercholesterolemia allele die of coronary artery disease in childhood. The disease in heterozygotes (1 in 500 people) has a milder and more variable clinical course.

The molecular defect in most cases of familial hypercholesterolemia is an absence or deficiency of functional receptors for LDL. Receptor mutations

FIGURE 27.20 Arterial plaques are one of the effects of excess cholesterol. Cross sections of (A) a normal artery and (B) an artery blocked by a cholesterol-rich plaque are shown. [Photo Credit: SPL/Science Source.]

(A)

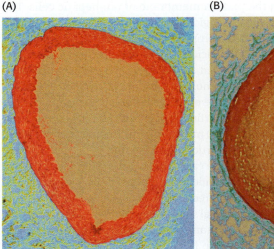

(B)

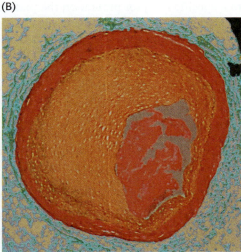

that disrupt each of the stages in the endocytotic pathway have been identified. Homozygotes have almost no functional receptors for LDL, whereas heterozygotes have about half the normal number. The entry of LDL into liver and other cells is impaired, leading to an increased level of LDL in the blood plasma. Furthermore, less IDL enters liver cells because IDL entry, too, is mediated by the LDL receptor. Consequently, IDL stays in the blood longer in familial hypercholesterolemia, and more of it is converted into LDL than in normal people. All deleterious consequences of an absence or deficiency of the LDL receptor can be attributed to the ensuing elevated level of LDL cholesterol in the blood.

Mutations in the LDL receptor prevent LDL release and result in receptor destruction

One class of mutations that results in familial hypercholesterolemia generates receptors that are reluctant to give up the LDL cargo. Let us explore why this is by examining the makeup of the receptor. The human LDL receptor is a 160-kDa glycoprotein with a complicated modular structure (**Figure 27.21**). The receptor exists in two interconvertible states: an extended or open state, capable of binding LDL, and a closed state that results in release of the LDL in the endosome. The receptor maintains the open state while in the plasma membrane, on binding LDL, and throughout its journey to the endosome. The conversion from the open state into the closed state takes place on exposure to the acidic environment of the endosome.

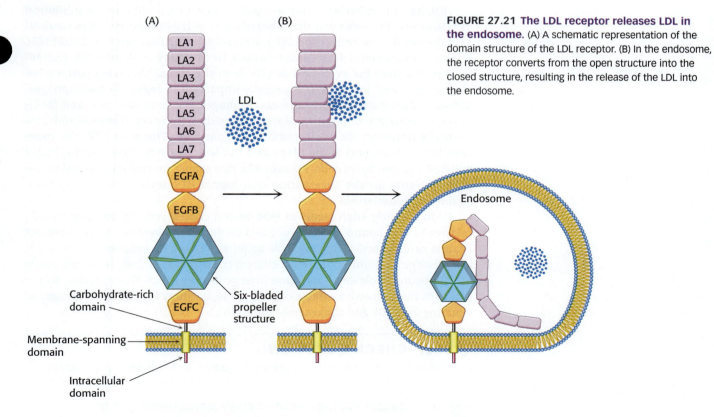

FIGURE 27.21 The LDL receptor releases LDL in the endosome. (A) A schematic representation of the domain structure of the LDL receptor. (B) In the endosome, the receptor converts from the open structure into the closed structure, resulting in the release of the LDL into the endosome.

The importance of this process is highlighted by the fact that more than half of the point mutations that result in familial hypercholesterolemia are due to disruptions in the interconversion between the open state and the closed state. These mutations result in a failure to release the LDL cargo and loss of the receptor by degradation.

Cycling of the LDL receptor is regulated

PCSK9 (proprotein convertase subtilisin/kexin type) is a protease, secreted by the liver, that plays a crucial role in the regulation of cycling of the LDL receptor. However, despite the fact that PCSK9 is a protease, enzymatic activity of the protein is not required for cycling regulation. PCSK9 in the blood binds to the LDL receptor and locks it in the open conformation even under the acidic conditions of the endosome. Failure to adopt the closed conformation prevents the receptor from returning to the plasma membrane, and it is degraded in the lysosome along with its cargo.

Individuals having a mutation that reduces the amount of PCSK9 in the blood have greatly reduced levels of LDL in the blood and display an almost 90% reduction in the rate of cardiovascular disease. Presumably, reduced levels of PCSK9 allow more receptor cycling and more efficient removal of LDL from the blood. Several monoclonal antibodies to PCSK9 have been approved for treatment for high LDL levels, but their use has been limited by their high cost.

HDL appears to protect against atherosclerosis

Although the events that result in atherosclerosis take place rapidly in familial hypercholesterolemia, a similar sequence of events takes place in people who develop atherosclerosis over decades. In particular, the formation of foam cells and plaques are especially hazardous occurrences. HDL and its role in returning cholesterol to the liver appear to be important in mitigating these life-threatening circumstances.

HDL has a number of antiatherogenic properties, including the inhibition of LDL oxidation. However, HDL's best-characterized property is the removal of cholesterol from cells, especially macrophages. Earlier, we noted that HDL retrieves cholesterol from other tissues in the body to return the cholesterol to the liver for excretion as bile or in the feces. This transport, called *reverse cholesterol transport*, is especially important in regard to macrophages. Indeed, when the transport fails, macrophages become foam cells and facilitate the formation of plaques. Macrophages that collect cholesterol from LDL normally transport the cholesterol to HDL particles. The more HDL, the more readily this transport takes place and the less likely that the macrophages will develop into foam cells. Presumably, this robust reverse cholesterol transport accounts for the observation that higher HDL levels confer protection against atherosclerosis.

Until recently, high levels of HDL-bound cholesterol ("good cholesterol") relative to LDL-bound cholesterol ("bad cholesterol") were believed to protect against cardiovascular disease. This belief was based on epidemiological studies. However, a number of recent clinical trials revealed that increased levels of HDL-bound cholesterol may have little protective effect. These studies do not discount the protective effects of HDL alone, but they illustrate the danger of equating free HDL and cholesterol-bound HDL.

❖ SELF–CHECK QUESTION

List three differences between low-density lipoprotein and high-density lipoprotein.

 ### The clinical management of cholesterol levels can be understood at the biochemical level

Homozygous familial hypercholesterolemia can be treated only by a liver transplant. A more generally applicable therapy is available for heterozygotes and others with high levels of cholesterol. The goal is to reduce the amount of cholesterol in the blood by stimulating the synthesis of more than the customary number of LDL receptors. We have already observed that the production of LDL receptors is controlled by the cell's need for cholesterol. The therapeutic strategy is to deprive the cell of ready sources of cholesterol. When cholesterol is required, the amount of mRNA for the LDL receptor rises, and more receptors are found on the cell surface. This state can be induced by a two-pronged approach. First, the reabsorption of **bile salts**—cholesterol derivatives that promote the absorption of dietary cholesterol and dietary fats—from the intestine is inhibited. Second, de novo synthesis of cholesterol is blocked.

The reabsorption of bile is impeded by oral administration of positively charged polymers, such as cholestyramine, that bind negatively charged bile salts and are not themselves absorbed. Cholesterol synthesis can be effectively blocked by a class of compounds called *statins*. A well-known example of such a compound is lovastatin, which is also called mevacor (**Figure 27.22**). Statins are potent competitive inhibitors ($K_i = 1$ nM) of HMG-CoA reductase, the essential control point in the biosynthetic pathway. Plasma cholesterol levels decrease by 50% in many patients given both lovastatin and inhibitors of bile-salt reabsorption. Lovastatin and other inhibitors of HMG-CoA reductase are widely used to lower the plasma-cholesterol level in people who have atherosclerosis, which is the leading cause of death in industrialized societies.

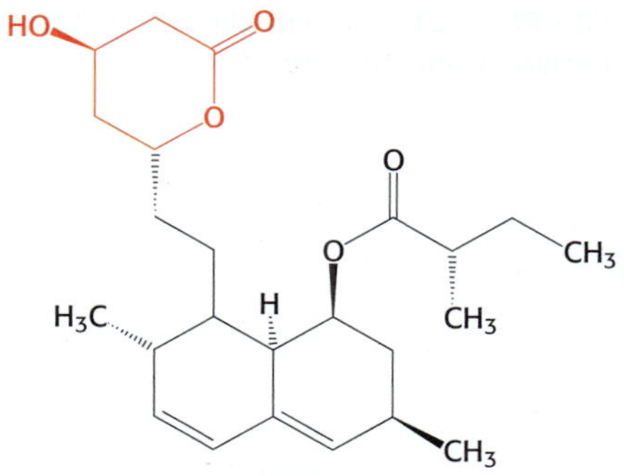

Lovastatin

FIGURE 27.22 The statin lovastatin is a competitive inhibitor of HMG-CoA reductase. The part of the structure that resembles the 3-hydroxy-3-methylglutaryl moiety is shown in red.

27.4 Important Biochemicals Are Synthesized from Cholesterol and Isoprene

Cholesterol is the precursor of several other important classes of compounds. Bile salts, the major constituent of bile, are synthesized in the liver, stored and concentrated in the gallbladder, and then released into the small intestine. Bile salts are highly effective at solubilizing dietary lipids because they contain both polar and nonpolar regions. Solubilization increases the effective surface area of lipids with two consequences: (1) more surface area is exposed to the digestive action of lipases and (2) lipids are more readily absorbed by the intestine. Bile salts are also the major breakdown products of cholesterol. The bile salts glycocholate, the primary bile salt, and taurocholate are shown in **Figure 27.23**.

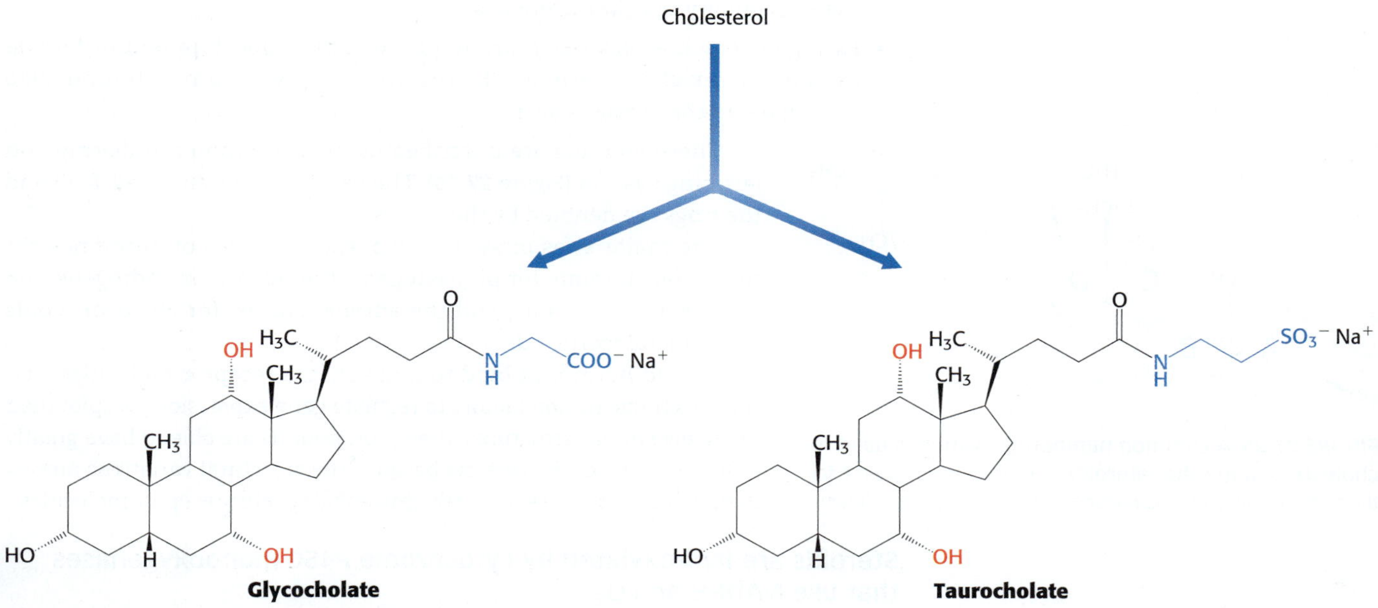

FIGURE 27.23 Bile salts are synthesized from cholesterol. The OH groups in red are added to cholesterol, as are the groups shown in blue. These polar groups make bile salt effective in detergents.

FIGURE 27.24 Different classes of steroid hormones are derived from cholesterol.

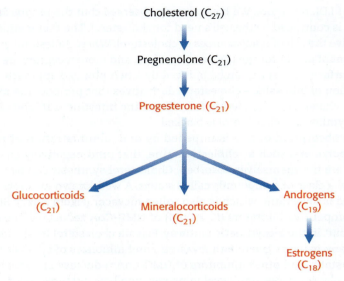

Cholesterol is also the precursor of the five major classes of steroid hormones—progestogens, glucocorticoids, mineralocorticoids, androgens, and estrogens (**Figure 27.24**). These hormones are powerful signal molecules that regulate a host of organismal functions.

- Progesterone, a progestogen, prepares the lining of the uterus for the implantation of an ovum. Progesterone is also essential for the maintenance of pregnancy by preventing premature uterine contractions.

- Glucocorticoids (such as cortisol) promote gluconeogenesis and glycogen synthesis, enhance the degradation of fat and protein, and inhibit the inflammatory response. They enable animals to respond to stress; indeed, the absence of glucocorticoids can be fatal.

- Mineralocorticoids (primarily aldosterone) act on the distal tubules of the kidney to increase the reabsorption of Na^+ and the excretion of K^+ and H^+, which leads to an increase in blood volume and blood pressure.

- Androgens (such as testosterone) are responsible for the development of male secondary sex characteristics.

- Estrogens (such as estradiol) are required for the development of female secondary sex characteristics. Estrogens, along with progesterone, also participate in the ovarian cycle.

These steroids are described using a common numbering and lettering system (**Figure 27.25**). The carbons are numbered 1–27 and the rings are denoted by the letters A–D.

The major sites of synthesis of these classes of hormones are the corpus luteum, for progestogens; the testes, for androgens; the ovaries, for estrogens; and the adrenal cortex, for glucocorticoids and mineralocorticoids.

Steroid hormones bind to and activate receptor molecules that serve as transcription factors to regulate gene expression. Despite their relatively similar structures, these compounds are able to have greatly differing physiological effects because the structural variations among them allow each class to be recognized by specific receptor molecules.

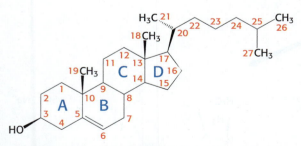

FIGURE 27.25 A common numbering system is used for cholesterol and other steroids. The numbering scheme for the carbon atoms and the naming system for rings are shown.

Steroids are hydroxylated by cytochrome P450 monooxygenases that use NADPH and O_2

The addition of oxygen atoms plays an important role in the synthesis of cholesterol from squalene and in the conversion of cholesterol into steroid hormones.

These oxidation reactions require O_2 but, also, the reducing agent NADPH. One oxygen atom of the O_2 molecule goes into the substrate, while the other is reduced to water. The enzymes catalyzing these reactions are called *monooxygenases* (or *mixed-function oxygenases*). The general form of the most common type of reaction catalyzed by these enzymes is

$$RH + O_2 + NADPH + H^+ \longrightarrow ROH + H_2O + NADP^+$$

In the synthesis of steroid hormones and bile salts, these reactions are catalyzed by members of the cytochrome P450 family, a family of heme-containing enzymes that absorb light maximally at 450 nm when complexed in vitro with exogenous carbon monoxide. Oxygen is activated through its binding to the iron atom in the heme group.

The catalytic mechanism for P450 enzymes is shown in **Figure 27.26**. NADPH transfers its high-potential electrons to a flavoprotein, which transfers them, one at a time, to adrenodoxin, a nonheme iron protein. Adrenodoxin transfers one electron to reduce the ferric (Fe^{3+}) form of P450 to the ferrous (Fe^{2+}) form. Without the addition of this electron, P450 will not bind oxygen. Recall that only the ferrous form (Fe^{2+}) of myoglobin and hemoglobin binds oxygen. The binding of O_2 to the heme is followed by the acceptance of a second electron from adrenodoxin. This second electron leads to cleavage of the O—O bond. One of the oxygen atoms is then protonated and released as water. The remaining oxygen atom forms a highly reactive ferryl (Fe^{4+})Fe=O intermediate with an additional electron removed from

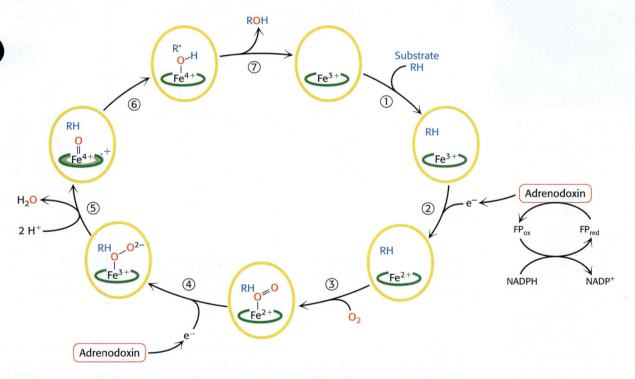

FIGURE 27.26 The reaction mechanism for cytochrome P450 enzymes reveals how oxygen is inserted into R-H bonds. (1) Substrate binds to the enzyme. (2) Adrenodoxin donates an electron, reducing the heme iron. (3) Oxygen binds to Fe^{2+}. (4) Adrenodoxin donates a second electron. (5) The bond between the oxygen atoms is cleaved, a molecule of water is released, and an intermediate with an $Fe^{4+} = O$ group and one electron removed from the porphyrin ring (indicated by shaded ring) is formed. (6) The $Fe^{4+} = O$ intermediate abstracts a hydrogen atom from the substrate to form a substrate radical. (7) The $Fe^{4+} -OH$ intermediate combines with the radical through the oxygen atom to form the hydroxylated product and return the iron to the Fe^{3+} state, ready for another reaction cycle.

the porphyrin ring of the heme group. This intermediate removes a hydrogen atom from the substrate RH to form R•. This transient free radical captures the OH group from the iron atom to form ROH, the hydroxylated product, returning the iron atom to the ferric state.

Cytochrome P450s are widespread and perform many functions

The human genome encodes 57 members of the cytochrome P450 family, many of which have relatively broad substrate preferences. In addition to their roles in lipid and other biosynthetic pathways, members of the cytochrome P450 family are also important in the metabolism of foreign substances (xenobiotic compounds) including most drugs. For example, the hydroxylation of phenobarbital, a barbiturate, increases its solubility and facilitates its excretion. The duration of action of many drugs depends on their rate of metabolism by cytochrome P450 enzymes.

Some compounds are not active as drugs until they are modified by cytochrome P450 enzymes. These are referred to as *prodrugs*. For example, the anticoagulant clopidogrel (Plavix®) is inactive until it is oxidized by a cytochrome P450 family named cytochrome P450 2C19 (**Figure 27.27**). This requirement is particularly important because the genes encoding this enzyme are variable in the population, with more than 20% of human beings having variants that lead to reduced activity. Clopidogrel does not function well in these individuals. In some settings, a simple genetic is used prior to prescribing clopidogrel, and alternative drugs are used if clopidogrel activation is likely to be inefficient. The appreciation and use of the roles of genomic variations on drug action, a field referred to as *pharmacogenomics*, is of increasing importance.

FIGURE 27.27 Cytochrome P450 2C19 converts the prodrug clopidogrel to an active form. The substance taken as clopidogrel is not active. The molecule is hydroxylated by cytochrome P450 2C19, and the product spontaneously hydrolyzes to produce a compound with a reactive thiol group (yellow).

Despite its general protective role in the removal of foreign chemicals, the action of the cytochrome P450 system can also do harm. Some of the most powerful carcinogens are generated from harmless compounds by the cytochrome P450 system in vivo in the process of metabolic activation. In plants, the cytochrome P450 system plays a role in the synthesis of toxic compounds as well as the pigments of flowers.

Pregnenolone is a precursor of many other steroids

Steroid hormones contain 21 or fewer carbon atoms, whereas cholesterol contains 27. Thus, the first stage in the synthesis of steroid hormones is the removal

NAMANDJÉ BUMPUS After developing an interest in science from tinkering with chemistry sets as a child, Namandjé Bumpus (shown seated in the photo above) now pursues research at the interface between chemistry and medicine. After majoring in biology and becoming the first person in her family to graduate from college in 2003, she went on to earn a PhD in pharmacology that focused on a member of the cytochrome P450 family, a group of enzymes that has remained a primary interest throughout her career. She has also pursued research on the metabolism of drugs used for the treatment of HIV infections, including the importance of variations in genes for drug-metabolizing genes in determining individual responses to drugs. In 2020, Dr. Bumpus was appointed director of the Department of Pharmacology and Molecular Sciences at Johns Hopkins University School of Medicine, becoming the first Black woman to chair a department in the 128-year history of this prestigious institution. She has received numerous awards including, appropriately enough, the John J. Abel Award in Pharmacology, named for the biochemist and pharmacologist who, in 1893, founded the department she now leads. In 2022, she was named Chief Scientist of the U.S. Food and Drug Administration (FDA).

of a six-carbon unit from the side chain of cholesterol to form the steroid pregnenolone. The side chain of cholesterol is hydroxylated at C-20 and at C-22, and then a third time resulting in cleavage of the bond between these atoms. Each of these steps is catalyzed by cytochrome P450$_{scc}$ (for side-chain cleavage), also known as desmolase and cytochrome P450 11A1.

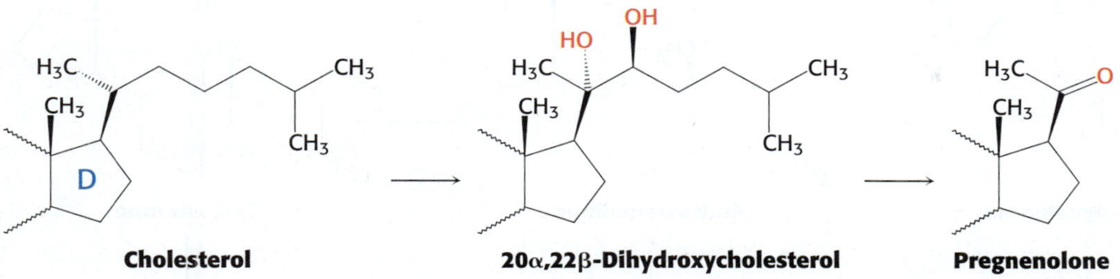

Progesterone and corticosteroids. Progesterone is synthesized from pregnenolone in two steps. The 3-hydroxyl group of pregnenolone is oxidized to a 3-keto group, and the Δ^5 double bond is isomerized to a Δ^4 double bond (**Figure 27.28**). Cortisol, the major glucocorticoid, is synthesized from progesterone by hydroxylations at C-11, C-17, and C-21; C-17 must be hydroxylated before C-21 is hydroxylated, whereas C-11 can be hydroxylated at any stage.

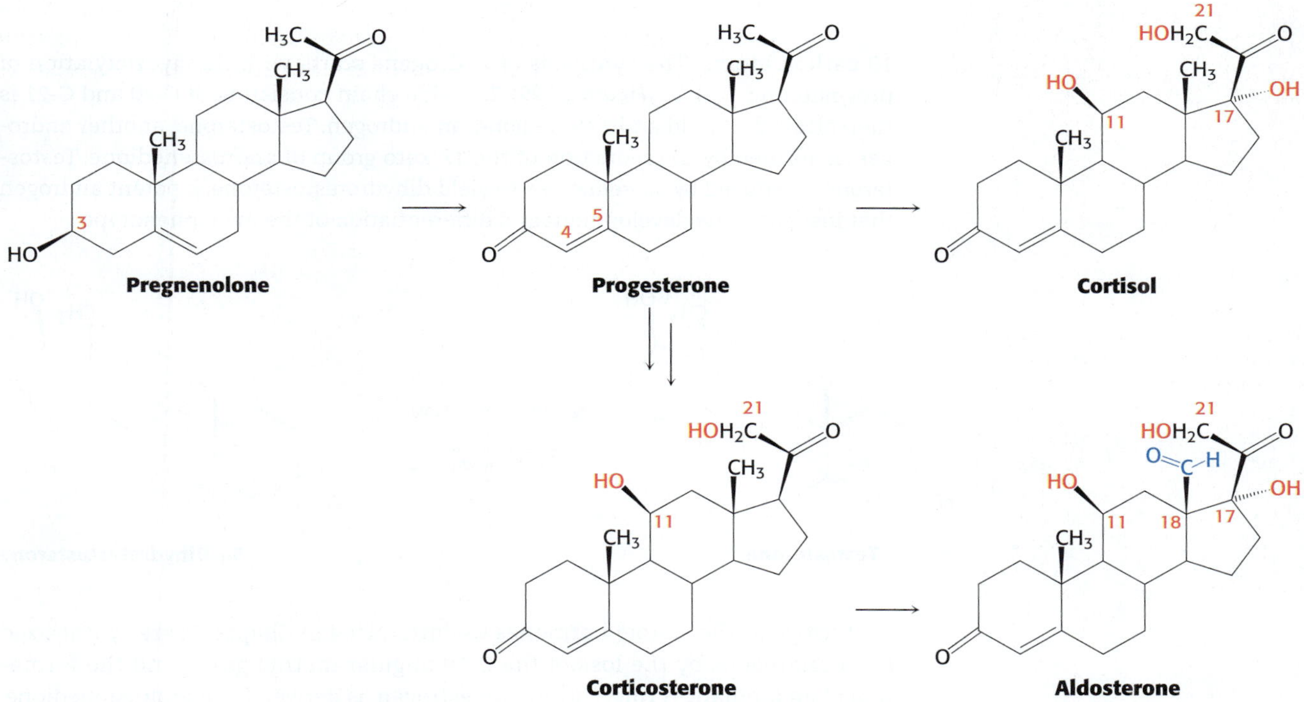

FIGURE 27.28 Pregnenolone is converted into progesterone, cortisol, and aldosterone.

The enzymes catalyzing these hydroxylations are highly specific. The initial step in the synthesis of aldosterone, the major mineralocorticoid, is the hydroxylation of progesterone at C-21. The resulting deoxycorticosterone is hydroxylated at C-11. The oxidation of the C-18 angular methyl group to an aldehyde then yields aldosterone.

Androgens and estrogens. Androgens and estrogens also are synthesized from pregnenolone through the intermediate progesterone. Androgens contain

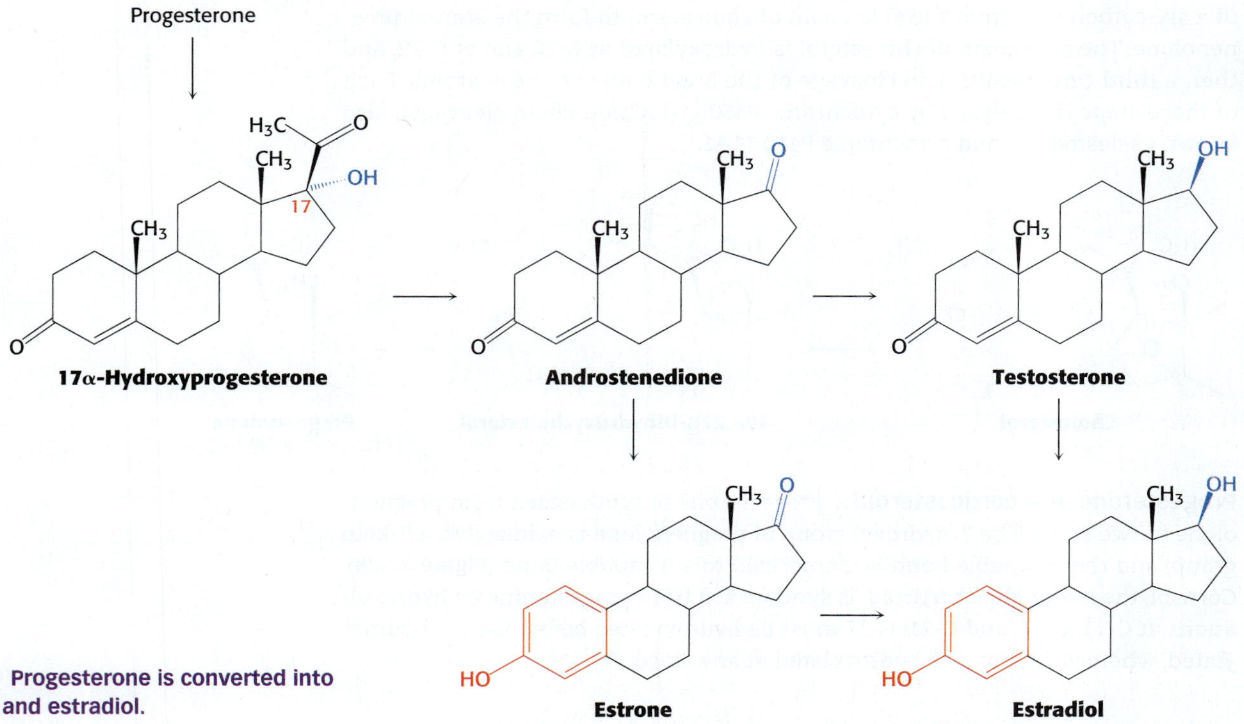

Progesterone

17α-Hydroxyprogesterone

Androstenedione

Testosterone

Estrone

Estradiol

FIGURE 27.29 Progesterone is converted into testosterone and estradiol.

19 carbon atoms. The synthesis of androgens starts with the hydroxylation of progesterone at C-17 (**Figure 27.29**). The side chain consisting of C-20 and C-21 is then cleaved to yield androstenedione, an androgen. Testosterone, another androgen, is formed by the reduction of the 17-keto group of androstenedione. Testosterone is reduced by 5α-reductase to yield dihydrotestosterone, a potent androgen that instigates the development and differentiation of the male phenotype.

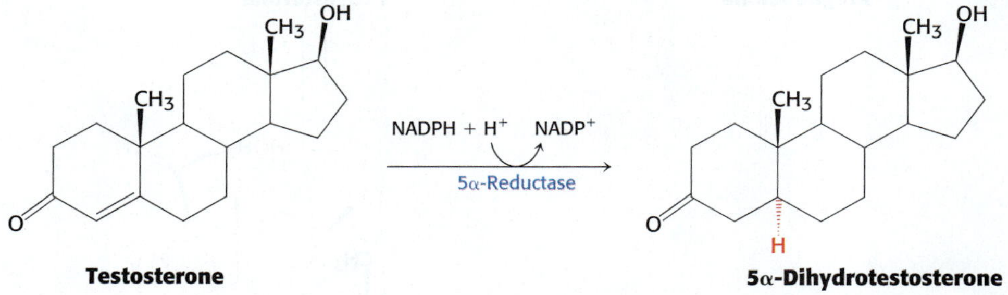

NADPH + H⁺ → NADP⁺

5α-Reductase

Testosterone

5α-Dihydrotestosterone

Estrogens, the steroid hormones we introduced in Chapter 2, are synthesized from androgens by the loss of the C-19 angular methyl group and the formation of an aromatic A ring. Estrone, an estrogen, is derived from androstenedione, whereas estradiol, the biologically most potent estrogen, is formed from testosterone. Estradiol can also be formed from estrone. The formation of the aromatic A ring is catalyzed by the cytochrome P450 enzyme aromatase.

Because breast and ovarian cancers frequently depend on estrogens for growth, aromatase inhibitors are often used as a treatment for these cancers. Anastrozole (Arimidex) is a competitive inhibitor of the enzyme, whereas exemestane (Aromasin) is a suicide inhibitor that covalently modifies and inactivates the enzyme (**Figure 27.30**). The development of these drugs again reveals how the understanding of biosynthetic pathways can yield important medical and other applications.

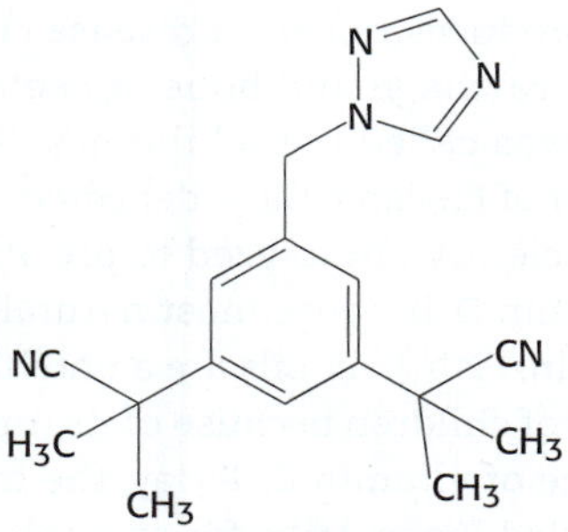

FIGURE 27.30 Drugs such as these for breast cancer can be developed based on knowledge of biosynthetic pathways.

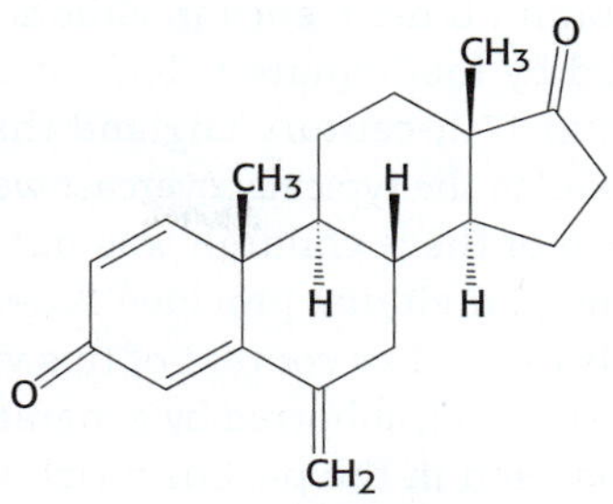

Anastrozole

Exemestane

Vitamin D is derived from cholesterol by the ring-splitting activity of light

Cholesterol is also the precursor of vitamin D, which plays an essential role in the control of calcium and phosphorus metabolism. 7-Dehydrocholesterol (provitamin D_3) is photolyzed by the ultraviolet light of sunlight to previtamin D_3, which spontaneously isomerizes to vitamin D_3 (**Figure 27.31**). Vitamin D_3 (cholecalciferol) is converted into the hormone calcitriol (1,25-dihydroxycholecalciferol), the active form of vitamin D, by hydroxylation reactions in the liver and kidneys. Although not a steroid hormone, vitamin D acts in an analogous fashion. It binds to receptors, homologous to the steroid receptors, to form complexes that function as transcription factors, regulating gene expression.

❖ **SELF-CHECK QUESTION**

Label the carbon atoms from 1 to 27 in calcitriol using the numbering scheme from Figure 27.25.

FIGURE 27.31 Vitamin D is synthesized from 7-dehydrocholesterol in a process initiated by absorption of ultraviolet light. The pathway for the conversion of 7-dehydrocholesterol into vitamin D_3 and then into calcitriol (vitamin D) is shown.

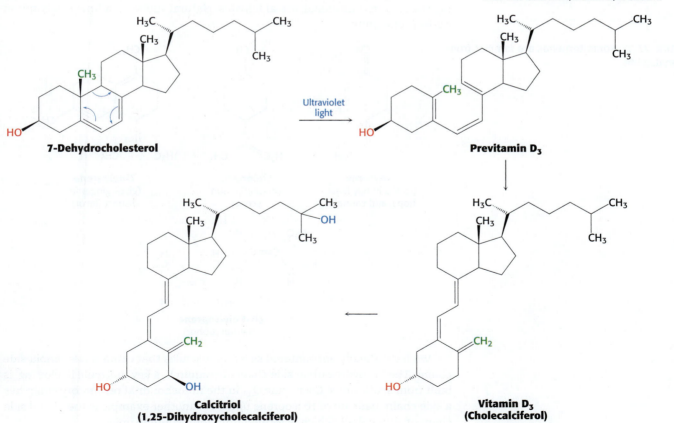

7-Dehydrocholesterol

Ultraviolet light

Previtamin D₃

Vitamin D₃
(Cholecalciferol)

Calcitriol
(1,25-Dihydroxycholecalciferol)

Vitamin D deficiency in childhood produces rickets, a disease character-ized by inadequate calcification of cartilage and bone. Rickets was so common in 17th-century England that it was called the "children's disease of the English." In the typically overcast weather of England, the 7-dehydrocholesterol in the skin of these children was not sufficiently photolyzed to previtamin D_3. Furthermore, their diet provided little vitamin D, because most naturally occur-ring foods have a low content of this vitamin. Fish-liver oils are a notable excep-tion. Cod-liver oil, abhorred by generations of children because of its unpleasant taste, was used in the past as a rich source of vitamin D. Today, the most reli-able dietary sources of vitamin D are fortified foods. Milk, for example, is forti-fied to a level of 400 international units per quart (10 µg per quart). The recommended daily intake of vitamin D is 200 international units until age 50, after which it increases with age as the skin becomes less able to manufacture previtamin D_3 in response to sunlight; this is a striking reminder that vitamin D is needed by adults as well as by children.

Five-carbon units are joined to form a wide variety of biomolecules

We have seen that a remarkable array of compounds are formed from isopentenyl pyrophosphate, the basic five-carbon building block. The first step, the synthesis of squalene (C_{30}) from isopentenyl pyrophosphate (C_5), exemplifies a fundamen-tal mechanism for the assembly of carbon skeletons of biomolecules. We will end this chapter by looking at a few more biologically important compounds produced from C_5 units.

The fragrances of many plants arise from volatile C_{10} and C_{15} compounds, which are called *terpenes*. For example, myrcene ($C_{10}H_{16}$) from bay leaves, hops, and cannabis, among other plants, consists of two isoprene units, as does limonene ($C_{10}H_{15}$) from lemon oil (**Figure 27.32**). Zingiberene ($C_{15}H_{24}$), from the oil of ginger, is made up of three isoprene units. Some terpenes, such as geraniol from geraniums and menthol from peppermint oil, are alcohols, and others, such as citronellal, are aldehydes. Natural rubber is a linear polymer of cis-isoprene units.

FIGURE 27.32 Some terpenes are familiar from everyday life.

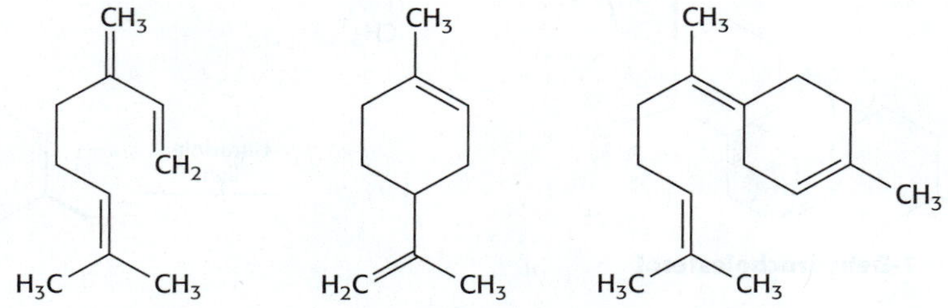

Myrcene
(found in bay leaves, hops, and cannabis)

Limonene
(natural lemon scent)

Zingiberene
(gives ginger its distinct flavor)

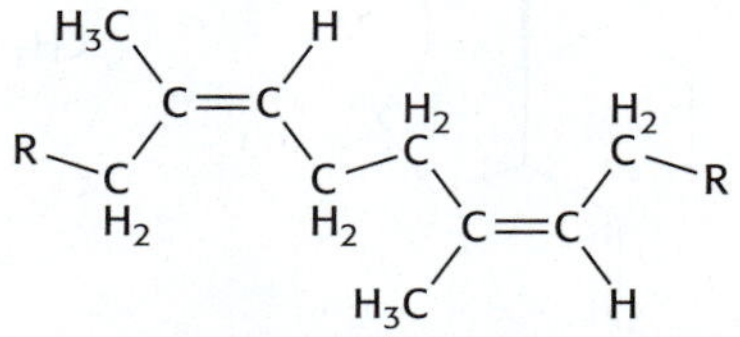

***cis*-Polyisoprene**
(natural rubber)

We have already encountered several molecules that contain isoprenoid side chains. The C_{30} hydrocarbon side chain of vitamin K, a key molecule in clotting, is built from six C_5 units. Coenzyme Q_{10} in the mitochondrial respiratory chain has a side chain made up of 10 isoprene units. Yet another example is the phytol side chain of chlorophyll, which is formed from four isoprene units.

Isoprenoids can delight by their color as well as by their fragrance. The color of tomatoes and carrots comes from *carotenoids*, specifically from lycopene and β-carotene, respectively. These compounds absorb light because they contain extended networks of single and double bonds—that is, they are *polyenes*. Their C_{40} carbon skeletons are built by the successive addition of C_5 units to form geranylgeranyl pyrophosphate, a C_{20} intermediate, which then condenses tail-to-tail with another molecule of geranylgeranyl pyrophosphate. This biosynthetic pathway is like that of squalene, except that C_{20} rather than C_{15} units are assembled and condensed.

$$C_5 \longrightarrow C_{10} \longrightarrow C_{15} \longrightarrow C_{30} \quad \text{(Squalene)}$$

$$C_5 \longrightarrow C_{10} \longrightarrow C_{15} \longrightarrow C_{20} \longrightarrow C_{40} \quad \text{(Phytoene)}$$

Phytoene, the C_{40} condensation product, is dehydrogenated to yield lycopene. Cyclization of both ends of lycopene gives β-carotene (**Figure 27.33**). Carotenoids serve as light-harvesting molecules in photosynthetic assemblies and also play a role in protecting bacteria from the deleterious effects of light. Carotenoids are also essential for vision. β-Carotene is the precursor of retinal, the chromophore in visual pigments.

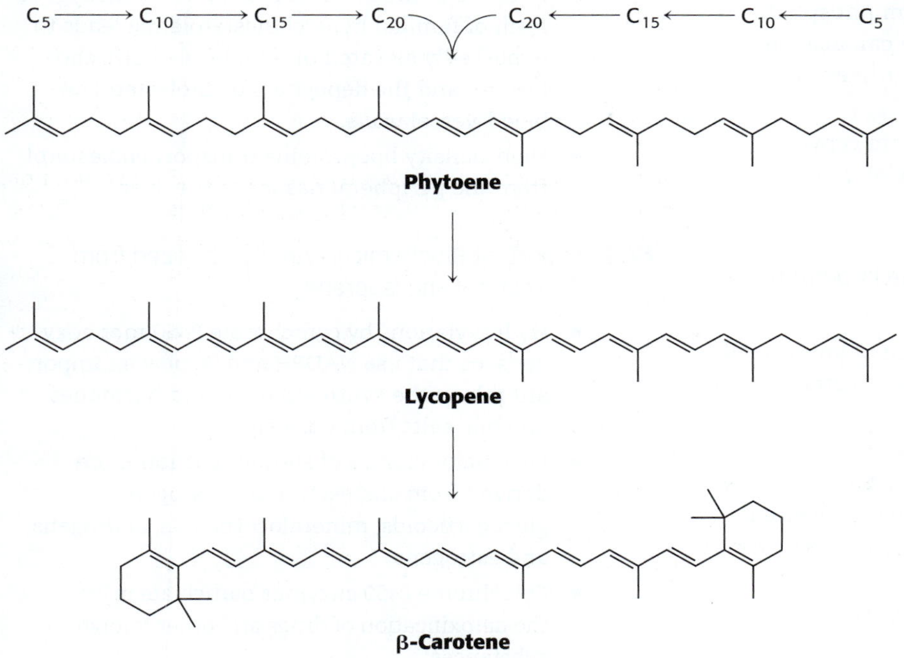

$$C_5 \longrightarrow C_{10} \longrightarrow C_{15} \longrightarrow C_{20} \qquad C_{20} \longleftarrow C_{15} \longleftarrow C_{10} \longleftarrow C_5$$

Phytoene

Lycopene

β-Carotene

FIGURE 27.33 Two carotenoids responsible for the color of tomatoes and carrots are large isotrenes. Lycopenes give ripe tomatoes their red color, and β-carotene gives carrots their orange hue.

Some isoprenoids have industrial applications

Farnesene is an isoprenoid that has potential to meet a wide variety of practical needs (**Table 27.2**).

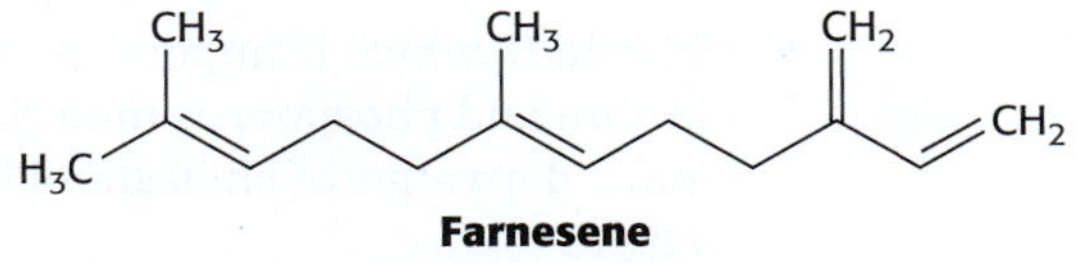

Farnesene

However, to meet these potential uses in an economical manner, large amounts of farnesene are produced in modified baker's yeast, with genes for four enzymes that were not native to yeast introduced. These genes enable the yeast to ferment sugar cane syrup to farnesene, and then to secrete the farnesene. The secreted farnesene is 93% pure, and distillation enhances the purity to 98%. This process can be scaled up to industrial levels, thus generating a versatile, renewable biomolecule with a variety of applications.

TABLE 27.2 Uses of farnesene and its derivatives

As a high-energy density biofuel
As sealants and adhesives
As solvents and lubricants
To make automobile tires that yield improved gas mileage and wet road grip
As a component of cosmetics
To make flavors and fragrances

Summary

27.1 Phosphatidate Is a Common Intermediate in the Synthesis of Phospholipids and Triacylglycerols

- Phosphatidate is formed by successive acylations of glycerol 3-phosphate by acyl CoA.

- Hydrolysis of the phosphoryl group from phosphatidate followed by acylation yields triacylglycerol.

- CDP-diacylglycerol is formed from phosphatidate and CTP. The activated phosphatidyl unit is then transferred to the hydroxyl group of a polar alcohol to form a phospholipid.

- In mammals, phosphatidylethanolamine is formed by CDP-ethanolamine and diacylglycerol. Phosphatidylethanolamine is methylated by S-adenosylmethionine to form phosphatidylcholine. Phosphatidylcholine can also be synthesized by a pathway that utilizes dietary choline.

- Sphingolipids are synthesized from ceramide, which is formed by the acylation of sphingosine.

27.2 Cholesterol Is Synthesized from Acetyl Coenzyme A in Three Stages

- Cholesterol is a steroid component of animal cell membranes and is also a precursor of steroid hormones.

- The committed step in its synthesis is the formation of mevalonate from 3-hydroxy-3-methylglutaryl CoA. Mevalonate is converted into farnesyl pyrophosphate, which condenses with itself to form squalene.

- Squalene cyclizes to lanosterol, which is modified to form cholesterol.

27.3 The Regulation of Cholesterol Biosynthesis Takes Place at Several Levels

- In the liver, cholesterol synthesis is regulated by changes in the amount and activity of 3-hydroxy-3-methylglutaryl CoA reductase. Transcription of the gene, translation of the mRNA, degradation of the enzyme, and enzyme activity are stringently controlled.

- Triacylglycerols exported by the intestine are carried by chylomicrons and then hydrolyzed by lipases lining the capillaries of target tissues. Cholesterol and other lipids in excess of those needed by the liver are exported in the form of very low-density lipoprotein.

- After delivering its content of triacylglycerols to adipose tissue and other peripheral tissue, very-low density lipoprotein is converted into intermediate-density lipoprotein (IDL) and then into low-density lipoprotein (LDL). IDL and LDL carry cholesteryl esters, primarily cholesteryl linoleate.

- Liver and peripheral tissue cells take up LDL by receptor-mediated endocytosis. The LDL receptor, a protein spanning the plasma membrane of the target cell, binds LDL and mediates its entry into the cell.

- Absence of the LDL receptor in the homozygous form of familial hypercholesterolemia leads to a markedly elevated plasma level of LDL cholesterol and the deposition of cholesterol on blood-vessel walls.

- High-density lipoproteins transport cholesterol from the peripheral tissues to the liver.

27.4 Important Biochemicals Are Synthesized from Cholesterol and Isoprene

- Hydroxylations by cytochrome P450 monooxygenases that use NADPH and O_2 play an important role in the synthesis of steroid hormones and bile salts from cholesterol.

- Five major classes of steroid hormones are derived from cholesterol: progestogens, glucocorticoids, mineralocorticoids, androgens, and estrogens.

- Cytochrome P450 enzymes participate in the detoxification of drugs and other foreign substances.

- Pregnenolone is an essential intermediate in the synthesis of steroids.

- Estrogens are synthesized from androgens by the loss of an angular methyl group and the formation of an aromatic A ring.

- Vitamin D, which is important in the control of calcium and phosphorus metabolism, is formed from a derivative of cholesterol by the action of light.

- A remarkable array of biomolecules in addition to cholesterol and its derivatives are synthesized from isopentenyl pyrophosphate, the basic five-carbon building block.

Key Terms

phosphatidate (p. 815)

triacylglycerol (p. 816)

cytidine diphosphodiacylglycerol (CDP-diacylglycerol) (p. 816)

sphingolipid (p. 819)

ceramide (p. 819)

cholesterol (p. 821)

mevalonate (p. 822)

3-hydroxy-3-methylglutaryl CoA reductase (HMG-CoA reductase) (p. 822)

squalene (p. 823)

sterol regulatory element binding protein (SREBP) (p. 825)

lipoprotein particles (p. 827)

low-density lipoprotein (LDL) (p. 828)

high-density lipoprotein (HDL) (p. 829)

receptor-mediated endocytosis (p. 829)

bile salt (p. 833)

Problems

1. How many high-phosphoryl-transfer-potential molecules are required to synthesize phosphatidylethanolamine from ethanolamine and diacylglycerol? Assume that the ethanolamine is the activated component. ❖ **1**

2. What is the activated reactant in each of the following biosyntheses? ❖ **1**

(a) Phosphatidylinositol from inositol

(b) Phosphatidylethanolamine from ethanolamine

(c) Ceramide from sphingosine

(d) Sphingomyelin from ceramide

(e) Cerebroside from ceramide

(f) Ganglioside G_{M1} from ganglioside G_{M2}

(g) Farnesyl pyrophosphate from geranyl pyrophosphate

3. Write a balanced equation for the synthesis of a triacylglycerol, starting from glycerol and fatty acids. ❖ **1**

4. Phosphatidic acid phosphatase (PAP) converts phosphatidate to diacylglycerol (DAG). Diacylglycerol kinase converts DAG to phosphatidate.

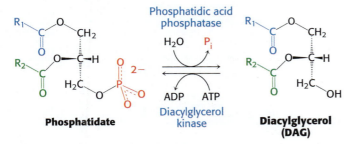

Phosphatidate / Phosphatidic acid phosphatase / H_2O / P_i / ADP / ATP / Diacylglycerol kinase / **Diacylglycerol (DAG)**

Consider the effect of an inactivating mutation in the gene encoding PAP on triacylglycerol and phospholipid synthesis. Classify each lipid based on whether its concentration will increase or decrease when the PAP carries this mutation.

(a) Phospholipids generated from DAG and activated head groups

(b) Triacylglycerols

(c) Phospholipids generated from CDP-DAG and a polar head group

(d) Phosphatidate

(e) Phosphatidylinositol

(f) Phosphatidylethanolamine

5. Below is the characteristic fused-ring core structure of a steroid. Add atoms as necessary to show the entire structure of cholesterol. ❖ **2**

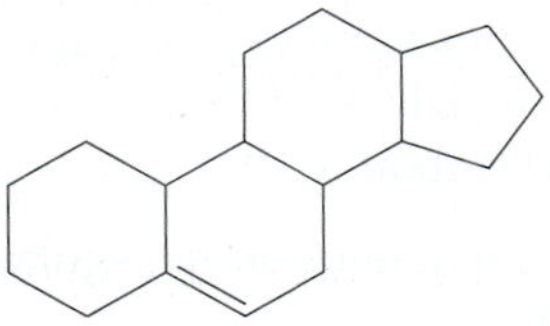

6. What are the three stages required for the synthesis of cholesterol? ❖ **2**

7. Arrange the molecules in the order they appear in cholesterol synthesis. Note that there may be intermediate steps between these molecules. ❖ **2**

(a)

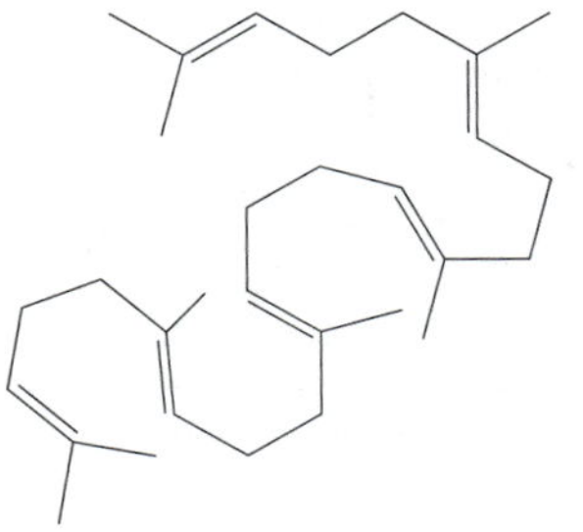

(b)

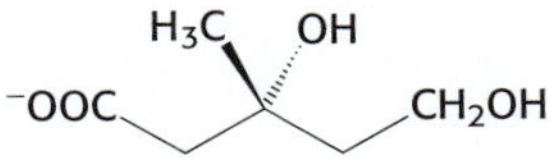

(c)

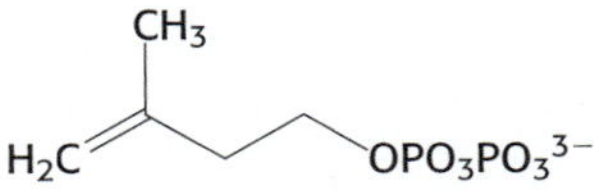

(d)

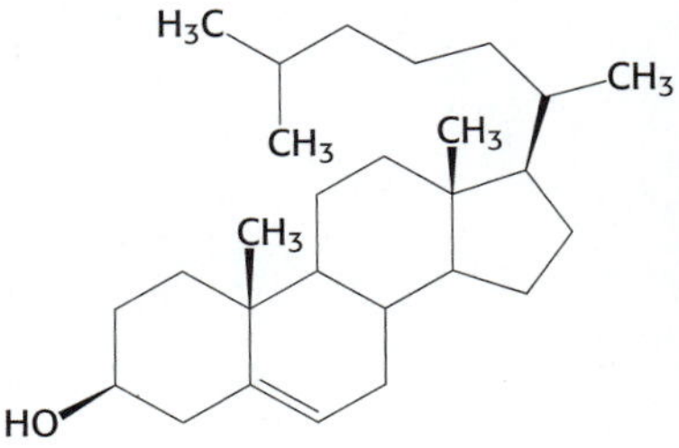

8. A key intermediate in the cyclization of squalene epoxide (in which the epoxide oxygen atom has been protonated) is shown below. Draw arrows indicating the movement of electrons to form the protosterol cation.

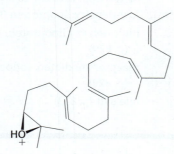

9. Arrange the following intermediates in the biosynthesis of cholesterol into their correct order: ❖ 2

(a) Geranyl pyrophosphate   (f) Squalene epoxide

(b) Lanosterol   (g) HMG-CoA

(c) Acetyl CoA   (h) Isopentenyl pyrophosphate

(d) Mevalonate   (i) Farnesyl pyrophosphate

(e) Cholesterol   (j) Squalene

10. Identify the atoms in mevalonate and isopentenyl pyrophosphate that will be labeled from acetyl CoA labeled with ^{14}C in the carbonyl carbon. ❖ 2

11. Outline the mechanisms of the regulation of cholesterol biosynthesis. ❖ 3

12. The synthesis of cholesterol is a process that involves over 30 different steps. Which step is the rate-determining step of cholesterol synthesis? ❖ 3

13. What are statins? What is their pharmacological function? ❖ 3

14. Describe the process of receptor-mediated endocytosis by using LDL as an example. ❖ 3

15. (a) Which lipoprotein is the least dense?

(b) Which lipoprotein is the most soluble? (Hint: Use Table 27.1 for this question.) ❖ 3

16. Match each description to either low-density lipoproteins (LDL) or high-density lipoproteins (HDL): ❖ 3

(a) Are known as the "bad" cholesterol

(b) Have levels that increase with regular exercise

(c) Help to remove cholesterol from arteries

(d) Can invade artery walls and trigger an inflammatory response

17. At a biochemical level, vitamin D functions like a steroid hormone. Therefore, it is sometimes referred to as an honorary steroid. Why is vitamin D not an actual steroid? ❖ 4

18. Match the steroid or steroid class with its respective function ❖ 4

(a) Bile salts   1. Regulation of electrolytes (e.g., Na$^+$, K$^+$)

(b) Androgens

(c) Glucocorticoids   2. Development of female secondary sex characteristics

(d) Mineralocorticoids

(e) Estrogens   3. Solubilizing dietary lipids

4. Development of male secondary sex characteristics

5. Regulation of blood glucose

19. 3-Hydroxy-3-methylglutaryl CoA is on the pathway for cholesterol biosynthesis. It is also a component of another pathway. Name the pathway. What determines which pathway 3-hydroxy-3-methylglutaryl CoA follows?

DNA Replication, Repair, and Recombination

Lou Linwei/Alamy

Newly synthesized DNA

Template DNA

Faithful copying is essential to the storage of genetic information. With the precision of a careful artisan copying an exemplar, a DNA polymerase (above) copies DNA strands, preserving the precise sequence of bases with very few errors.

◆ LEARNING GOALS

By the end of this chapter, you should be able to:

1. Describe the reactions catalyzed by DNA polymerase, DNA primase, and DNA ligase and the roles of the template, the primer, and Okazaki fragments.

2. Discuss the reaction catalyzed by helicases and key aspects of their mechanisms.

3. Define topoisomers of circular DNA molecules and explain the relationship between twist and writhe in determining the linking number.

4. Describe how different activities of DNA replication are coordinated to faithfully copy and maintain the genome of an organism.

5. Compare and contrast prokaryotic and eukaryotic replication.

6. Define telomeres and describe how they are synthesized.

7. Discuss types of DNA damage and how such lesions can be repaired.

8. Discuss the mechanism of DNA recombination and its role in a variety of biological processes.

OUTLINE

28.1 DNA Replication Proceeds by the Polymerization of Deoxyribonucleoside Triphosphates Along a Template

28.2 DNA Unwinding and Supercoiling Are Controlled by Topoisomerases

28.3 DNA Replication Is Highly Coordinated

28.4 Many Types of DNA Damage Can Be Repaired

28.5 DNA Recombination Plays Important Roles in Replication, Repair, and Other Processes

Perhaps the most exciting aspect of Watson and Crick's deduction of the structure of DNA was their realization "that the specific pairing we have postulated immediately suggests a possible copying mechanism for the genetic material." A double helix separated into two single strands can be replicated because each strand serves as a template on which its complementary sequence can be assembled. To preserve the information encoded in DNA through many cell divisions, copying of the genetic information must be extremely faithful. Although DNA is remarkably robust, ultraviolet light and a range of chemical species can damage DNA, introducing changes in the sequence (mutations) or lesions that can block further DNA replication. All organisms contain DNA-repair systems that detect DNA damage and act to preserve the original sequence.

28.1 DNA Replication Proceeds by the Polymerization of Deoxyribonucleoside Triphosphates Along a Template

To preserve the information encoded in DNA through many cell divisions, the base sequences of newly synthesized DNA must faithfully match the sequences of parent DNA. To replicate the human genome without mistakes, an error rate of less than 1 bp per 6×10^9 bp must be achieved. Such remarkable precision is achieved through a multilayered system of accurate DNA synthesis (which has an error rate of 1 per 10^3–10^4 bases inserted), proofreading during DNA synthesis (which reduces that error rate to approximately 1 per 10^6–10^7 bp), and postreplication mismatch repair (which reduces the error rate to approximately 1 per 10^9–10^{10} bp).

To achieve faithful replication, each strand within the parent double helix acts as a **template** for the synthesis of a new DNA strand with a complementary sequence, as discussed in Section 8.3. The building blocks for the synthesis of the new strands are deoxyribonucleoside triphosphates. They are added, one at a time, to the 3′ end of an existing strand of DNA.

Although this reaction is in principle quite simple, it is significantly complicated by specific features of the DNA double helix. First, the two strands of the double helix run in opposite directions. Because DNA strand synthesis always proceeds in the 5′-to-3′ direction, the DNA replication process must have specific mechanisms to accommodate the oppositely directed strands. Second, the two strands of the double helix interact with one another in such a way that the bases, key templates for replication, are on the inside of the helix. Thus, the two strands must be separated from each other to generate appropriate templates. Finally, the two strands of the double helix wrap around each other. Thus, strand separation also entails the unwinding of the double helix. This unwinding creates supercoils that must themselves be resolved as replication continues. We begin by considering the chemistry that underlies the formation of the phosphodiester backbone of newly synthesized DNA.

DNA polymerases require a template and a primer

Polynucleotide chains are catalyzed by enzymes called **DNA polymerases**. Each incoming nucleoside triphosphate first forms an appropriate base pair with a base in the template. Only then does the DNA polymerase link the incoming base with the predecessor in the chain. Thus, DNA polymerases are template-directed enzymes.

DNA polymerases cannot start synthesis from scratch by adding nucleotides to a single-stranded DNA template—they can only add nucleotides to an existing hydroxyl group at the 3′ end of a polynucleotide chain. This free hydroxyl group is presented by a **primer**—an initial polynucleotide chain that is already base-paired to the template. The polymerase catalyzes the nucleophilic attack by the 3′-hydroxyl-group terminus of the polynucleotide chain on the α-phosphoryl group of the nucleoside triphosphate to be added (Figure 8.24). Primers are made by RNA polymerases, which, in contrast to DNA polymerases, can initiate RNA synthesis from scratch (as we shall see in Chapter 29).

DNA polymerases have common structural features

The three-dimensional structures of a number of DNA polymerase enzymes are known. The first such structure was elucidated by Thomas Steitz and coworkers, who determined the structure of the Klenow fragment of DNA polymerase I from E. coli (**Figure 28.1**). This fragment comprises two main parts of the full enzyme, including the polymerase unit. This unit approximates the shape of a right hand with domains that are referred to as the fingers, the palm, and the thumb. In addition to the polymerase unit, the Klenow fragment includes a domain with 3′ → 5′ **exonuclease** activity that participates in proofreading and correcting the polynucleotide product.

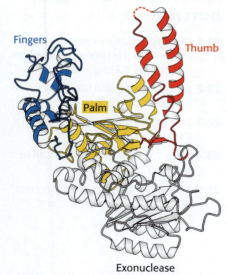

FIGURE 28.1 The structure of the *E.coli* DNA polymerase I Klenow fragment resembles a hand. This was the first DNA polymerase structure to be determined; like the structures of other DNA polymerases determined later, it resembles a right hand with fingers (blue), palm (yellow), and thumb (red). The Klenow fragment also includes an exonuclease domain that removes incorrect nucleotide bases. [Drawn from 1DPI.pdb.]

INTERACT with this model in
 Achieve

DNA polymerases are remarkably similar in overall shape, although they differ substantially in detail. At least five structural classes have been identified; some of them are clearly homologous, whereas others appear to be the products of convergent evolution. In all cases, the finger and thumb domains wrap around DNA and hold it across the enzyme's active site, which comprises residues primarily from the palm domain. Furthermore, all DNA polymerases use similar strategies to catalyze the polymerase reaction, making use of a mechanism in which three metal ions take part.

Bound metal ions participate in the polymerase reaction

Like all enzymes with nucleoside triphosphate substrates, DNA polymerases require metal ions for activity. Examination of the structures of DNA polymerases with bound substrates and substrate analogs initially revealed the presence of two metal ions in the active site, but recent data supports a three-metal-ion model. One metal ion binds both the deoxyribonucleoside triphosphate (dNTP) and the 3′-hydroxyl group of the primer, while the other two interact only with the dNTP (**Figure 28.2**). Two of the three metal ions are bridged by the carboxylate groups of two aspartate residues in the palm domain of the polymerase. These side chains hold the metal ions in the proper positions and orientations. The metal ion bound to the primer activates the 3′-hydroxyl group of the primer, facilitating its attack on the α-phosphoryl group of the dNTP substrate in the active site. The three metal ions together help stabilize the negative charge that accumulates on the pentacoordinate transition state. The two metal ions initially bound to the dNTP stabilize the negative charge on the pyrophosphate product.

❖ **SELF–CHECK QUESTION**

Provide a chemical explanation of why DNA synthesis proceeds in a 5′-to-3′ direction.

The specificity of replication is dictated by complementarity of shape between bases

Since DNA must be replicated with high fidelity, each base added to the growing chain should, with high probability, be the Watson–Crick complement of the base in the corresponding position in the template strand. The binding of the dNTP containing the proper base is favored by the formation of a base pair with its partner on the template strand. Although hydrogen bonding contributes to the formation of this base pair, overall shape complementarity is crucial. Studies show that a nucleotide with a base that is very similar in shape to adenine but lacks the ability to form base-pairing hydrogen bonds can still direct the incorporation of thymidine, both in vitro and in vivo (**Figure 28.3**).

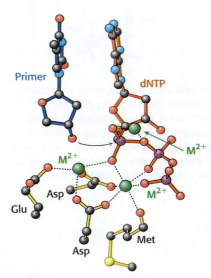

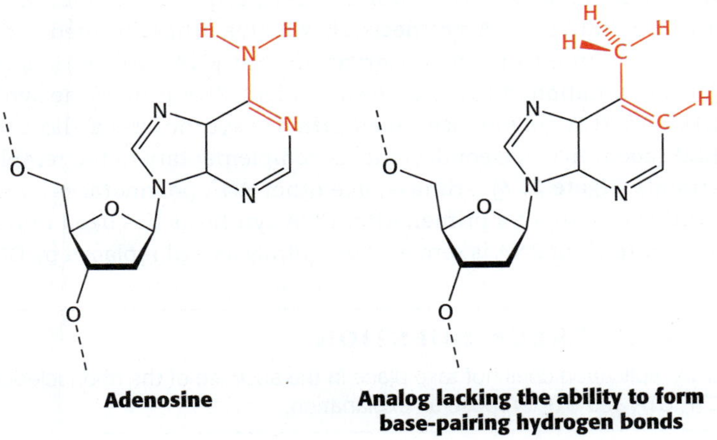

FIGURE 28.2 Three metal ions (typically, Mg²⁺) participate in the DNA polymerase reaction. One metal ion coordinates the 3′-hydroxyl group of the primer, whereas the other two metal ions interact only with the dNTP. The phosphoryl group of the nucleoside triphosphate bridges all three metal ions. Two of the three metal ions are also bridged by the carboxylate groups of two aspartate residues in the polymerase. The hydroxyl group of the primer attacks the phosphoryl group to form a new O—P bond.

FIGURE 28.3 Base analogs that have the same shape as normal bases can participate in DNA replication. The base analog on the right has the same shape as adenosine, but groups that form hydrogen bonds between base pairs have been replaced by groups not capable of hydrogen bonding (shown in red). Nonetheless, studies reveal that when this analog is incorporated into the template strand, it directs the insertion of thymidine in DNA replication.

Adenosine

Analog lacking the ability to form base-pairing hydrogen bonds

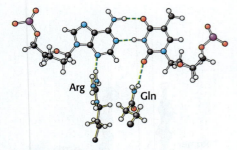

FIGURE 28.4 Minor-groove interactions promote the specificity of DNA replication. DNA polymerases donate two hydrogen bonds to base pairs in the minor groove. Hydrogen-bond acceptors are present in these two positions for all Watson–Crick base pairs, including the A–T base pair shown.

An examination of the crystal structures of various DNA polymerases reveals why shape complementarity is so important. First, residues of the enzyme form hydrogen bonds with the minor-groove side of the base pair in the active site (Figure 28.4). In the minor groove, hydrogen-bond acceptors are present in the same positions for all Watson–Crick base pairs. These interactions act as a "ruler" that measures whether a properly spaced base pair has formed in the active site.

Second, the binding of a dNTP in the active site of a DNA polymerase triggers a conformational change: the finger domain rotates to form a tight pocket into which only a properly shaped base pair will readily fit (Figure 28.5). Many of the residues lining this pocket are important to ensure the efficiency and fidelity of DNA synthesis. For example, mutation of a conserved tyrosine residue that forms part of the pocket results in a polymerase that is approximately 40 times more error prone as the parent polymerase.

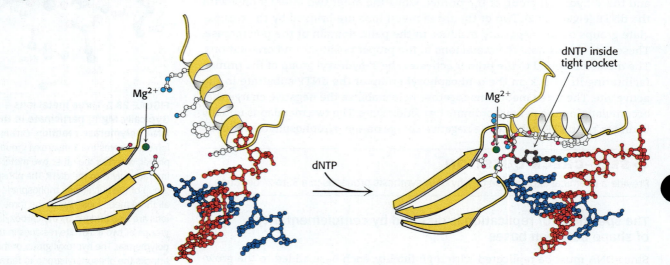

INTERACT with these models in Achieve

FIGURE 28.5 Shape selectivity promotes the specificity of DNA replication. The binding of a deoxyribonucleoside triphosphate (dNTP) to DNA polymerase induces a conformational change, generating a tight pocket for the base pair consisting of the dNTP and its partner on the template strand. Such a conformational change is possible only when the dNTP corresponds to the Watson–Crick partner of the template base. Only one of the three coordinating metal ions at the active site is shown. [Drawn from 2BDP.pdb and 1T7P.pdb.]

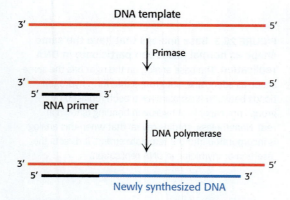

FIGURE 28.6 DNA polymerases require a primer. DNA replication is primed by a short stretch of RNA that is synthesized by primase, an RNA polymerase. The RNA primer is removed at a later stage of replication.

An RNA primer enables DNA synthesis to begin

As discussed above, DNA polymerases require a primer with a free 3′ end to initiate DNA synthesis. How is this primer formed? An important clue came from the observation that RNA synthesis is essential for the initiation of DNA synthesis. In fact, RNA primes the synthesis of DNA. An RNA polymerase called **primase** synthesizes a short stretch of RNA (about five nucleotides) that is complementary to the template DNA strands (Figure 28.6). Primase, like other RNA polymerases, can initiate synthesis without a primer. After DNA synthesis has been initiated, the short stretch of RNA is removed by hydrolysis and replaced by DNA.

❖ SELF–CHECK QUESTION

DNA replication does not take place in the absence of the ribonucleotides ATP, CTP, GTP, and UTP. Propose an explanation.

One strand of DNA is made continuously, whereas the other strand is synthesized in fragments

Both strands of parental DNA serve as templates for the synthesis of new DNA. The site of DNA synthesis is called the **replication fork** because the complex formed by the newly synthesized daughter helices arising from the parental duplex resembles a two-pronged fork. Recall that the two template strands are antiparallel; that is, they run in opposite directions. Since all known DNA polymerases synthesize DNA only in the 5′ → 3′ direction, the two daughter strands must be synthesized in opposite directions: one away from the fork, and one toward the fork. How exactly does this work?

An answer to this question was provided by Reiji Okazaki and his wife Tsuneko Okazaki, who found that a significant proportion of newly synthesized DNA exists as small fragments. These units of about a thousand nucleotides (called **Okazaki fragments**) are present briefly in the vicinity of the replication fork (**Figure 28.7**). As replication proceeds, these fragments become covalently joined, forming a continuous daughter strand. The other new strand is synthesized continuously. The strand formed from Okazaki fragments is termed the **lagging strand**, whereas the one synthesized without interruption is the **leading strand**.

Tsuneko Okazaki, University Professor, Nagoya University

TSUNEKO AND REIJI OKAZAKI

Growing up in Japan during World War II, Tsuneko and Reiji Okazaki persevered in their studies despite wartime hardships. Tsuneko's father wanted her to pursue medicine, but she was more interested in biology and became a researcher at a time when it was a very rare career path for women. Tsuneko met and married Reiji while they were both at Nagoya University, and in the early 1960s they briefly worked at Stanford University in the lab of Arthur Kornberg, who had recently extracted DNA polymerase activity from *E.coli*. Back at Nagoya University, they continued to investigate how DNA polymerase could be replicating antiparallel strands of DNA in only one direction, eventually leading to their discovery of the short-lived DNA fragments that now carry their name. In 1975 Reiji died of leukemia, a consequence of his exposure to intense radiation in Hiroshima as a teenager. Dr. Tsuneko Okazaki continued in research while also raising two children, and she went on to have a distinguished academic career.

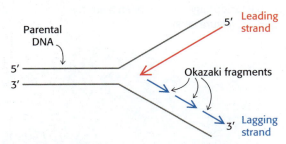

FIGURE 28.7 The lagging strand is made in short pieces called Okazaki fragments. At a replication fork, both strands are synthesized in the 5′ → 3′ direction, but only the leading strand is synthesized continuously

DNA ligase seals breaks in double-stranded DNA

The joining of Okazaki fragments requires an enzyme that can connect the ends of two DNA chains. The existence of circular DNA molecules also points to the existence of such an enzyme. In 1967, scientists in several laboratories simultaneously discovered **DNA ligase**, the enzyme that catalyzes the formation of a phosphodiester linkage between the 3′-hydroxyl group at the end of one DNA chain and the 5′-phosphoryl group at the end of the other (**Figure 28.8**). An energy source is required to drive this thermodynamically uphill reaction. In eukaryotes and archaea, ATP is the energy source. In bacteria, NAD^+ typically plays this role.

FIGURE 28.8 DNA ligase catalyzes the joining of two DNA strands. In this reaction, one DNA strand with a free 3′-hydroxyl group is joined to another with a free 5′-phosphoryl group. In eukaryotes and archaea, ATP is cleaved to AMP and PP_i to drive this reaction. In bacteria, NAD^+ is cleaved to AMP and nicotinamide mononucleotide (NMN).

DNA ligase cannot link two single-stranded DNA molecules or circularize single-stranded DNA. Rather, ligase seals breaks in double-stranded DNA molecules. The enzyme from *E. coli* ordinarily forms a phosphodiester bridge only if there are at least a few bases of single-stranded DNA on the end of a double-stranded fragment that can come together with those on another fragment to form base pairs. Ligase encoded by T4 bacteriophage can link two blunt-ended double-helical fragments, a capability that is exploited in recombinant DNA technology.

The separation of DNA strands requires specific helicases and ATP hydrolysis

For a double-stranded DNA molecule to replicate, the two strands of the double helix must be separated from each other, at least locally. This separation allows each strand to act as a template on which a new polynucleotide chain can be assembled. Specific enzymes, termed **helicases**, harness the energy of ATP hydrolysis to power strand separation.

While helicases are a large and diverse family of enzymes taking part in many biological processes, the helicases in DNA replication are typically oligomers containing six subunits that form a ring structure. The structure of one such helicase, that from bacteriophage T7, has been a source of considerable insight into the helicase mechanism (**Figure 28.9**).

FIGURE 28.9 Helicases are hexamers. The structure of the hexameric helicase from bacteriophage T7 is shown. The loops that participate in DNA binding are highlighted by a yellow oval in one of the subunits. Each subunit interacts closely with its neighbors, and the DNA-binding loops line the hole in the center of the structure. [Drawn from 1E0K.pdb.]

INTERACT with this model in
🅼 Achie/e

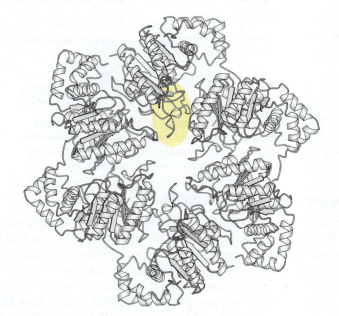

Each of the subunits within this hexameric structure has two loops that extend toward the center of the ring structure and interact with DNA. Each subunit interacts closely with its two neighbors within the ring structure. Closer examination of this structure reveals that the ring deviates significantly from sixfold symmetry, a deviation that is even more apparent when helicase is crystallized in the presence of the nonhydrolyzable ATP analog AMP-PNP.

AMP-PNP

The AMP-PNP binds to only four of the six subunits within the ring (**Figure 28.10**). Furthermore, the four nucleotide binding sites are not identical but fall into two classes. One class appears to be well positioned to bind ATP but not catalyze its hydrolysis, whereas the other class is more well-suited to catalyze the hydrolysis but not release the hydrolysis products. The classes are analogous to myosin's two different conformations—one for binding ATP and one for hydrolyzing it (Section 6.4). Finally, the six subunits fall into three classes with regard to their orientation within the overall ring structure, with differences in rotation around an axis in the plane of the ring of approximately 30°. These differences in orientation affect the positions of the two DNA-binding loops in each subunit.

These observations are consistent with the following mechanism for the helicase (**Figure 28.11**):

1. Only a single strand of DNA can fit through the center on the ring. This single strand binds to loops on two adjacent subunits, one of which has bound ATP and the other of which has bound ADP + P_i.

2. The binding of ATP to the domains that initially had no bound nucleotides causes a conformational change within the entire hexamer, leading to the release of ADP + P_i from two subunits and the binding of the single-stranded DNA by one of the domains that just bound ATP. This conformational change pulls the DNA through the center of the hexamer.

3. The protein acts as a wedge, forcing the two strands of the double helix apart.

4. This cycle then repeats itself, moving two bases along the DNA strand with each cycle.

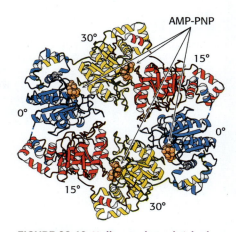

FIGURE 28.10 Helicases have intrinsic asymmetry. The structure of the T7 helicase complexed with the ATP analog AMP-PNP is shown. The three classes of helicase subunits are shown in blue, red, and yellow. The rotation relative to the plane of the hexamer is shown for each subunit. Notice that only four of the subunits, those shown in blue and yellow, bind AMP-PNP. [Drawn from 1E0K.pdb.]

INTERACT with this model in
Achieve

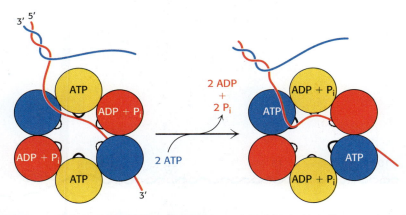

FIGURE 28.11 Helicases unwind DNA. One of the strands of the double helix passes through the hole in the center of the helicase, bound to the loops of two adjacent subunits. Two of the subunits do not contain bound nucleotides. On the binding of ATP to these two subunits and the release of ADP + P_i from two other subunits, the helicase hexamer undergoes a conformational change, pulling the DNA through the helicase. The helicase acts as a wedge to force separation of the two strands of DNA.

28.2 DNA Unwinding and Supercoiling Are Controlled by Topoisomerases

As a helicase moves along unwinding DNA, the DNA in front of the helicase will become overwound in the absence of other changes. DNA double helices that are torsionally stressed tend to fold up on themselves to form tertiary structures created by **supercoiling**. Supercoiling is most readily understood by considering circular DNA molecules, but it also applies to linear DNA molecules constrained to be in loops by other means. In fact, most DNA molecules inside cells are subject to supercoiling. We will first consider the supercoiling of DNA in quantitative terms and then turn to topoisomerases, enzymes that can directly alter DNA winding and supercoiling.

Consider a linear 260-bp DNA duplex in the B-DNA form (**Figure 28.12A**). Because the number of base pairs per turn in an unstressed DNA molecule averages 10.4, this linear DNA molecule has 25 (260/10.4) turns. The ends of this helix can be joined to produce what is called a *relaxed* circular DNA (**Figure 28.12B**). A different circular DNA can be formed by unwinding the linear duplex by two turns before joining its ends (**Figure 28.12C**). What is the structural consequence of unwinding before ligation? Two limiting conformations are possible. The DNA can fold into a structure containing 23 turns of B helix and an unwound loop (**Figure 28.12D**). Alternatively, the double helix can undergo supercoiling, in which it folds up to cross itself. In this example, a supercoiled structure with 25 turns of B helix and 2 turns of right-handed (termed *negative*) supercoils can be formed (**Figure 28.12E**). Unwinding will cause supercoiling in circular DNA molecules, whether covalently closed or constrained in closed configurations by other means.

(A)

(B) **Relaxed circle**

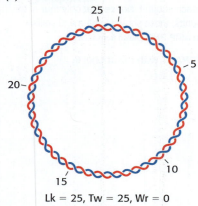

Lk = 25, Tw = 25, Wr = 0

(C) **Linear DNA unwound by two right-hand turns**

(D) **Unwound circle**

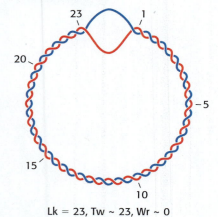

Lk = 23, Tw ~ 23, Wr ~ 0

(E) **Negative superhelix** (right-handed)

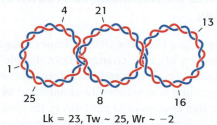

Lk = 23, Tw ~ 25, Wr ~ −2

FIGURE 28.12 The linking number (Lk), twist (Tw), and writhe (Wr) of circular DNA are related. [Information from W. Saenger, *Principles of Nucleic Acid Structure* (Springer Verlag, 1984), p. 452.]

Supercoiling markedly alters the overall form of DNA, because it makes the DNA more compact than a relaxed DNA molecule of the same length. Hence, supercoiled DNA moves faster than relaxed DNA when analyzed by centrifugation or electrophoresis (**Figure 28.13**).

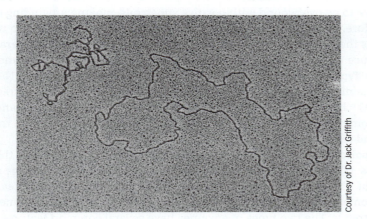

FIGURE 28.13 Supercoiling condenses DNA topoisomers. An electron micrograph showing negatively supercoiled and relaxed DNA.

Courtesy of Dr. Jack Griffith

The linking number, a topological property, determines the degree of supercoiling

Our understanding of the conformation of DNA is enriched by concepts drawn from topology, a branch of mathematics dealing with structural properties that are unchanged by deformations such as stretching and bending. A key topological property of a circular DNA molecule is its **linking number (Lk)**, which is equal to the number of times that a strand of DNA winds in the right-handed direction around the helix axis when the axis lies in a plane, as shown in Figure 28.12A. For the relaxed DNA shown in Figure 28.12B, Lk = 25. For the partly unwound molecule shown in part D and the supercoiled one shown in part E, Lk = 23 because the linear duplex was unwound two complete turns before closure. Molecules differing only in linking number are topological isomers, or **topoisomers**, of one another. Topoisomers of DNA can be interconverted only by cutting one or both DNA strands and then rejoining them.

The unwound DNA and supercoiled DNA shown in Figure 28.12D and E are topologically identical but geometrically different. They have the same value of Lk but differ in two measurements: twist and writhe. Although the rigorous definitions of twist and writhe are complex, **twist (Tw)** is a measure of the helical winding of the DNA strands around each other, whereas **writhe (Wr)** is a measure of the coiling of the axis of the double helix—that is, supercoiling. A right-handed coil is assigned a negative number (negative supercoiling), and a left-handed coil is assigned a positive number (positive supercoiling).

Is there a relation between Tw and Wr? Indeed, there is. Topology tells us that the sum of Tw and Wr is equal to Lk:

$$Lk = Tw + Wr$$

In Figure 28.12, the partly unwound circular DNA has Tw ~ 23, meaning the helix has 23 turns, and Wr ~ 0, meaning the helix has not crossed itself to create a supercoil. The supercoiled DNA, however, has Tw ~ 25 and Wr ~ −2. These forms can be interconverted without cleaving the DNA chain because they have the same value of Lk—namely, 23.

Topoisomerases prepare the double helix for unwinding

Most naturally occurring DNA molecules are negatively supercoiled. What is the basis for this? As already stated, negative supercoiling arises from the unwinding or underwinding of the DNA. In essence, negative supercoiling prepares DNA for processes requiring separation of the DNA strands, such as replication. Positive supercoiling condenses DNA as effectively, but it makes strand separation more difficult.

The action of the helicase at the replication fork causes the DNA ahead of the fork to be overwound with positive supercoils, making unwinding difficult. Therefore, negative supercoils must be continuously introduced to relax the DNA as the double helix unwinds. Specific enzymes called **topoisomerases** that introduce or eliminate supercoils were discovered by James Wang and Martin Gellert. Type I topoisomerases catalyze the relaxation of supercoiled DNA, a thermodynamically favorable process, while Type II topoisomerases use free energy from ATP hydrolysis to add negative supercoils to DNA. Both type I and type II topoisomerases play important roles in DNA replication as well as in transcription and recombination.

Topoisomerases alter the linking number of DNA by catalyzing a three-step process: (1) the cleavage of one or both strands of DNA, (2) the passage of a segment of DNA through this break, and (3) the resealing of the DNA break. Type I topoisomerases cleave just one strand of DNA, whereas type II enzymes cleave both strands. The two types of enzymes have several common features, including the use of key tyrosine residues to form covalent links to the polynucleotide backbone that is transiently broken.

Type I topoisomerases relax supercoiled structures

The three-dimensional structures determined for several type I topoisomerases reveal many features of the reaction mechanism (**Figure 28.14**). For example, human type I topoisomerase comprises four domains, which are arranged around a central cavity having a diameter of 20 Å, just the correct size to accommodate a double-stranded DNA molecule. This cavity also includes a tyrosine residue (Tyr 723), which acts as a nucleophile to cleave the DNA backbone in the course of catalysis.

From analyses of topoisomerases and the results of other studies, the relaxation of negatively supercoiled DNA molecules is known to proceed in the following manner (**Figure 28.15**):

FIGURE 28.14 Topoisomerase I binds to DNA. The structure of a complex between a fragment of human topoisomerase I and DNA is shown. The DNA occupies a central cavity within the enzyme. [Drawn from 1EJ9.pdb.]

INTERACT with this model in
 Achieve

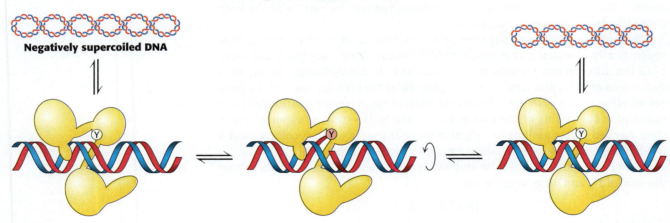

FIGURE 28.15 Topoisomerase I relaxes supercoiled DNA. On binding to DNA, topoisomerase I cleaves one strand of the DNA by means of a tyrosine (Y) residue attacking a phosphoryl group. When the strand has been cleaved, it rotates in a controlled manner around the other strand. The reaction is completed by re-ligation of the cleaved strand. This process results in partial or complete relaxation of a supercoiled plasmid.

1. First, the DNA molecule binds inside the cavity of the topoisomerase. The hydroxyl group of tyrosine 723 attacks a phosphoryl group on one strand of the DNA backbone to form a phosphodiester linkage between the enzyme and the DNA, cleaving the DNA and releasing a free 5′-hydroxyl group.

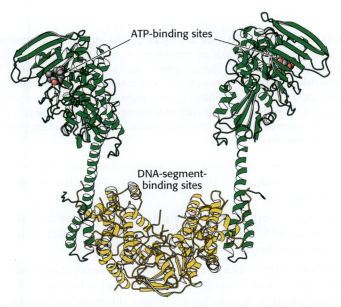

Tyr 723

2. With the backbone of one strand cleaved, the DNA can now rotate around the remaining strand, its movement driven by the release of energy stored because of the supercoiling. The rotation of the DNA unwinds the supercoils. The enzyme controls the rotation so that the unwinding is not rapid.

3. The free hydroxyl group of the DNA attacks the phosphotyrosine residue to reseal the backbone and release tyrosine. The DNA is then free to dissociate from the enzyme. Thus, reversible cleavage of one strand of supercoiled DNA allows controlled rotation to partly relax the supercoils.

Type II topoisomerases introduce negative supercoils through coupling to ATP hydrolysis

Supercoiling requires an input of energy because a supercoiled molecule, in contrast to its relaxed counterpart, is torsionally stressed. The introduction of an additional supercoil into a 3000-bp plasmid typically requires about 30 kJ mol^{-1} (7 kcal mol^{-1}).

Supercoiling is catalyzed by type II topoisomerases. These elegant molecular machines couple the binding and hydrolysis of ATP to the directed passage of one DNA double helix through another, temporarily cleaved DNA double helix. These enzymes have several mechanistic features in common with the type I topoisomerases.

Topoisomerase II molecules are dimeric with a large internal cavity (**Figure 28.16**).

ATP-binding sites

DNA-segment-binding sites

FIGURE 28.16 Topoisomerase II is a dimer. The structure of a typical topoisomerase II, this one from the archaeon *Sulfolobus shibatae*, is shown. Each half of the enzyme has one domain (shown in yellow) that contains a region for binding a DNA double helix and another domain (shown in green) that contains ATP-binding sites. [Drawn from 2ZBK.pdb.]

INTERACT with this model in
 Achie√e

The large cavity has gates at both the top and the bottom that are crucial to topoisomerase action:

1. The reaction begins with the binding of one double helix (hereafter referred to as the G, for gate, segment) to the enzyme (**Figure 28.17**). Each strand is positioned next to a tyrosine residue, one from each monomer, capable of forming a covalent linkage with the DNA backbone.

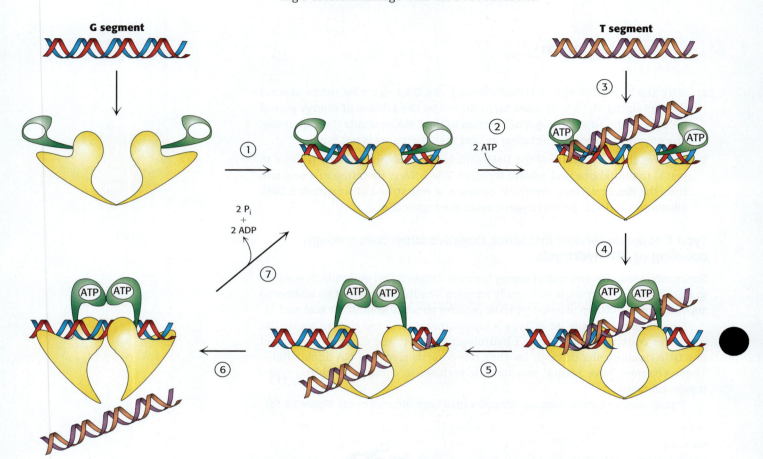

FIGURE 28.17 Topoisomerase II uses ATP hydrolysis to introduce negative supercoils into DNA. Topoisomerase II first binds one DNA duplex termed the G (for gate) segment (1). The binding of ATP to the two N-terminal domains brings these two domains together (2). This conformational change leads to the cleavage of both strands of the G segment and the binding of an additional DNA duplex, the T segment (3). This T segment then moves through the break in the G segment and out the bottom of the enzyme (4 through 6). The hydrolysis of ATP resets the enzyme with the G segment still bound (7). The orientation of the entering T segment results in the introduction of negative supercoils.

2. Each monomer of the enzyme has a domain that binds ATP, eventually leading to a conformational change that strongly favors the coming together of the two domains.

3. This complex then loosely binds a second DNA double helix (hereafter referred to as the T, for transported, segment). As the ATP-binding domains come closer together, they trap the bound T segment.

4. This conformational change also forces the separation and cleavage of the two strands of the G segment. Each strand is linked to the enzyme by a tyrosine–phosphodiester bond. Unlike the type I enzymes, the type II topoisomerases hold the DNA tightly so that it cannot rotate.

5. The T segment then passes through the cleaved G segment and into the large central cavity.

6. The ligation of the G segment leads to the release of the T segment through the gate at the bottom of the enzyme.

7. The hydrolysis of ATP and the release of ADP and orthophosphate allow the ATP-binding domains to separate, preparing the enzyme to bind another T segment. The overall process leads to a decrease in the linking number by two.

The bacterial topoisomerase II (often called DNA gyrase) is the target of several antibiotics that inhibit the prokaryotic enzyme much more than the eukaryotic one. Novobiocin blocks the binding of ATP to gyrase. Nalidixic acid and ciprofloxacin, in contrast, interfere with the breakage and rejoining of DNA chains. These two gyrase inhibitors are widely used to treat urinary tract and other infections including those due to *Bacillus anthracis* (anthrax). Camptothecin, an antitumor agent, inhibits human topoisomerase I by stabilizing the form of the enzyme covalently linked to DNA.

Nalidixic acid   **Ciprofloxacin**

28.3 DNA Replication Is Highly Coordinated

DNA replication must be very rapid, given the sizes of the genomes and the rates of cell division. The *E. coli* genome contains 4.6 million base pairs and is copied in less than 40 minutes. Thus, 2000 bases are incorporated per second. Enzyme activities must be highly coordinated to replicate entire genomes precisely and rapidly.

We begin our consideration of the coordination of DNA replication by looking at *E. coli*, which has been extensively studied. For this organism with a relatively small genome, replication begins at a single site and continues around the circular chromosome. The coordination of eukaryotic DNA replication is much more complex because there are many initiation sites throughout the genome and an additional enzyme is needed to replicate the ends of linear chromosomes.

DNA replication requires highly processive polymerases

Replicative polymerases are characterized by their very high catalytic potency, fidelity, and processivity. **Processivity** refers to the ability of an enzyme to catalyze many consecutive reactions without releasing its substrate. These polymerases are assemblies of many subunits that have evolved to grasp their templates and not let go until many nucleotides have been added.

The source of the processivity was revealed by the determination of the three-dimensional structure of the β_2 subunit of the *E. coli* replicative polymerase called DNA polymerase III (**Figure 28.18**). This unit, which has the form of a star-shaped ring, keeps the polymerase associated with the DNA double helix. A 35 Å-diameter hole in its center can readily accommodate a duplex DNA molecule (around 20 Å in diameter) and leaves enough space between the DNA and the protein to allow rapid sliding during replication. A polymerization rate of 1000 nucleotides per second requires that 100 turns of duplex DNA (a length of 3400 Å, or 0.34 μm) slide through the central hole of β_2 per second. Thus, β_2 plays a key role in replication by serving as a **sliding DNA clamp**.

How does DNA become entrapped inside the sliding clamp? Replicative polymerases also include assemblies of subunits that function as clamp loaders by

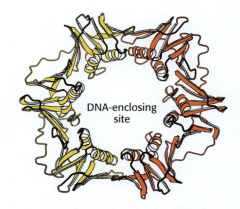

FIGURE 28.18 The sliding clamp subunit of DNA polymerase III is a dimer. The β subunit of DNA polymerase III forms a ring that surrounds the DNA duplex. The DNA template slides through the central cavity. Clasping the DNA molecule in the ring, the polymerase enzyme is able to move without falling off the DNA substrate. [Drawn from 2POL.pdb.]

INTERACT with this model in

 Achieve

grasping the sliding clamp and, using the energy of ATP binding, pulling apart one of the interfaces between the two subunits of the sliding clamp. DNA can move through the gap, inserting itself through the central hole. ATP hydrolysis then releases the clamp, which closes around the DNA.

The leading and lagging strands are synthesized in a coordinated fashion

Replicative polymerases such as DNA polymerase III synthesize the leading and lagging strands simultaneously at the replication fork (**Figure 28.19**). Let's first look at how DNA polymerase III begins the synthesis of the leading strand starting from the RNA primer formed by primase. While the duplex DNA ahead of the polymerase is unwound by a hexameric helicase called DnaB, copies of single-stranded-binding protein (SSB) bind to the unwound strands, keeping the strands separated so that both strands can serve as templates. The leading strand is synthesized continuously by polymerase III, as topoisomerase II concurrently introduces right-handed (negative) supercoils to avert a topological crisis.

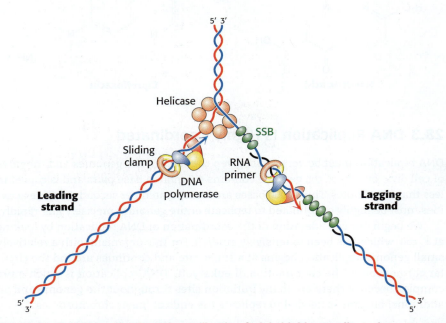

FIGURE 28.19 DNA synthesis at the replication fork is highly coordinated. A schematic view of the arrangement of DNA polymerase III and associated enzymes and proteins involved in the replication of DNA is shown. The helicase separates the two strands of the parent double helix, allowing DNA polymerases to use each strand as a template for DNA synthesis. Abbreviation: SSB, single-stranded-binding protein.

As mentioned earlier, the lagging strand is synthesized in fragments, so its synthesis is necessarily more complex. Yet the synthesis of the lagging strand is coordinated to be simultaneous with the synthesis of the leading strand. How is this coordination accomplished?

Examination of the subunit composition of the DNA polymerase III holoenzyme reveals an elegant solution for synthesis of the lagging strand (**Figure 28.20**). The holoenzyme includes two copies of the polymerase core enzyme, which consists of the DNA polymerase itself (the α subunit); the ε subunit, a 3′-to-5′ proofreading exonuclease; another subunit called θ; and two copies of the dimeric β-subunit sliding clamp. The core enzymes are linked to a central structure having the subunit composition $\gamma\tau_2\delta\delta'\chi\phi$. The $\gamma\tau_2\delta\delta'$ complex serves as the clamp loader, and the χ and ϕ subunits interact with the single-stranded-DNA–binding protein; meanwhile, the entire apparatus interacts with the hexameric helicase DnaB. Eukaryotic replicative polymerases have similar, albeit slightly more complicated, subunit compositions and structures.

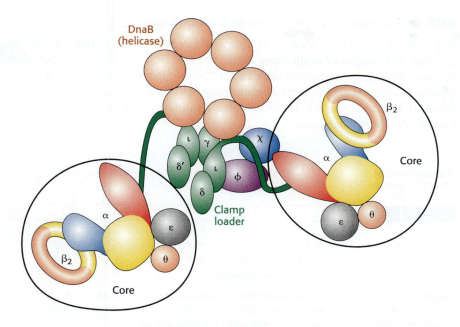

FIGURE 28.20 The DNA polymerase holoenzyme consists of multiple subunits. Each holoenzyme consists of two copies of the polymerase core enzyme, which comprises the α, ε, and θ subunits and two copies of the β subunit, linked to a central structure. The central structure includes the clamp-loader complex and the hexameric helicase DnaB.

The lagging-strand template is looped out so that it passes through the polymerase site in one subunit of a dimeric DNA polymerase III in the same direction as that of the leading-strand template in the other subunit. DNA polymerase III lets go of the lagging-strand template after adding about 1000 nucleotides by releasing the sliding clamp. A new loop is then formed, a sliding clamp is added, and primase again synthesizes a short stretch of RNA primer to initiate the formation of another Okazaki fragment. This mode of replication has been termed the **trombone model** because the size of the loop lengthens and shortens like the slide on a trombone (**Figure 28.21**).

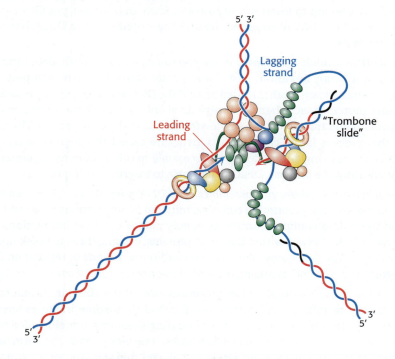

FIGURE 28.21 The trombone model describes the coordination of DNA synthesis at the replication fork. The replication of the leading and lagging strands is coordinated by the looping out of the lagging strand to form a structure that acts somewhat as a trombone slide, growing as the replication fork moves forward. When the polymerase on the lagging strand reaches a region that has been replicated, the sliding clamp is released and a new loop is formed.

The gaps between fragments of the nascent lagging strand are filled by DNA polymerase I. This essential enzyme also uses its $5' \rightarrow 3'$ exonuclease activity to remove the RNA primer lying ahead of the polymerase site. The primer cannot be erased by DNA polymerase III because the enzyme lacks $5' \rightarrow 3'$ editing capability. Finally, DNA ligase connects the fragments.

DNA replication in *E. coli* begins at a unique site

In *E. coli*, DNA replication starts at a unique site within the entire 4.6×10^6 bp genome. This **origin of replication**, called the *oriC locus*, is a 245-bp region that has several unusual features (**Figure 28.22**). The *oriC* locus contains five copies of a sequence that is the preferred binding site for the origin-recognition protein DnaA. In addition, the *oriC* locus contains a tandem array of 13-bp sequences that are rich in AT base pairs.

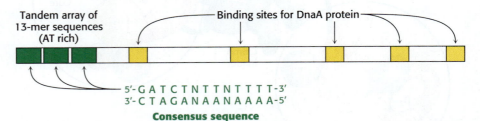

Tandem array of 13-mer sequences (AT rich)

Binding sites for DnaA protein

5′-G A T C T N T T N T T T T-3′
3′-C T A G A N A A N A A A A-5′

Consensus sequence

FIGURE 28.22 The *E. coli* chromosome has a single origin of replication. The *oriC* locus has a length of 245 bp and contains a tandem array of three nearly identical 13-nucleotide sequences (green) and five binding sites (yellow) for the DnaA protein.

Several steps are required to prepare for the start of replication:

1. *DnaA proteins bind to the* oriC. DnaA proteins are related to the hexameric helicases. Each DnaA monomer comprises an ATPase domain linked to a DNA-binding domain at its C-terminus. DnaA molecules are able to bind to each other through their ATPase domains; a group of bound DnaA molecules will break apart on the binding and hydrolysis of ATP. The DnaA proteins bind to the five high-affinity sites in *oriC* and then come together with other DnaA molecules bound to lower-affinity sites to form an oligomer, possibly a cyclic hexamer. The DNA is wrapped around the outside of the DnaA hexamer (**Figure 28.23**).

2. *Single DNA strands are exposed in the prepriming complex.* With DNA wrapped around a DnaA hexamer, additional proteins are brought into play. The hexameric helicase DnaB is loaded around the DNA with the help of the helicase loader protein DnaC. Then, local regions of *oriC*, including the AT regions, are unwound and trapped by the single-stranded-DNA–binding protein. The result of this process is the creation of a structure called the *prepriming complex*, which makes single-stranded DNA accessible to other proteins (**Figure 28.24**). Significantly, the primase, DnaG, is now able to insert the RNA primer.

3. *The polymerase holoenzyme assembles.* The DNA polymerase III holoenzyme assembles on the prepriming complex, initiated by interactions between DnaB and the sliding-clamp subunit of DNA polymerase III. These interactions also trigger ATP hydrolysis within the DnaA proteins, causing them to break up. The breakup of the DnaA assembly prevents additional rounds of replication from beginning at the *oriC*, signaling the end of the preparatory phase.

Once initiated, DNA replication continues around the circular chromosome. To efficiently terminate replication, the *E. coli* chromosome includes specific termination sites (or *Ter* sites) that act as binding sites for a protein called *termination utilization substance*, or *Tus* (**Figure 28.25**). The complexes that form when Tus binds to the Ter sites are not symmetrical, and this asymmetry is central for its function: when a replication fork, led by the helicase DnaB, reaches a Ter–Tus complex from one direction, it can move past it, but when it reaches it from the opposite direction, the Ter–Tus complex effectively blocks the replication fork. The positions and orientations of the Ter sites within the *E. coli* genome facilitate efficient termination when one cycle of termination is complete.

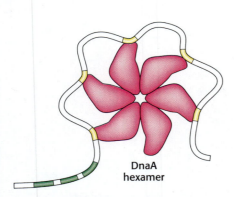

DnaA hexamer

FIGURE 28.23 The assembly of DnaA proteins initiates DNA replication. Monomers of DnaA bind to their binding sites (shown in yellow) in *oriC* and come together to form a complex structure, possibly the cyclic hexamer shown here. This structure marks the origin of replication and favors DNA strand separation in the AT-rich sites (green).

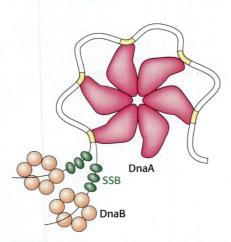

DnaA

SSB

DnaB

FIGURE 28.24 The prepriming complex prepares the template strands for DNA synthesis. The AT-rich regions are unwound and trapped by the single-stranded-binding protein (SSB). The hexameric DNA helicase DnaB is loaded on each strand. At this stage, the complex is ready for the synthesis of the RNA primers and assembly of the DNA polymerase III holoenzyme.

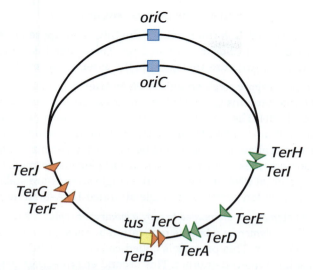

FIGURE 28.25 Replication termination in *E. coli* occurs at Ter sites. The *E. coli* genome includes 10 Ter sites that bind the termination utilization substance (Tus) protein. The Ter–Tus complexes block incoming replication forks in a unidirectional manner (indicated by the orientation of the triangles representing the Ter sites), facilitating termination of replication after a single cycle.

DNA replication in eukaryotes is initiated at multiple sites

Replication in eukaryotes is mechanistically similar to replication in prokaryotes but is more challenging for a number of reasons. One of them is sheer size: *E. coli* must replicate 4.6 million base pairs, whereas a human diploid cell must replicate more than 6 billion base pairs. Second, the genetic information for *E. coli* is contained on one chromosome, whereas, in human beings, 23 pairs of chromosomes must be replicated. Finally, whereas the *E. coli* chromosome is circular, human chromosomes are linear. Unless countermeasures are taken, linear chromosomes are subject to shortening with each round of replication.

The first two challenges are met by the use of multiple origins of replication. In human beings, replication requires about 30,000 origins of replication, with each chromosome containing several hundred. Each origin of replication is the starting site for a replication unit, or *replicon*. DNA replication can be monitored by single-molecule methods, revealing bidirectional synthesis from different sites (**Figure 28.26**). In contrast with *E. coli*, the origins of replication in human beings do not have sharply defined sequences. Instead, more broadly defined AT-rich sequences are the sites around which the **origin of replication complexes (ORCs)** are assembled.

FIGURE 28.26 Eukaryotic chromosomes have multiple origins of replication. The image shows a single molecule of DNA containing two origins of replication. The origins were identified by labeling newly replicated DNA in human cells first with one thymine analog (iodo-deoxyuridine, IdU) and then another (chloro-deoxyuridine, CldU). DNA molecules from these cells were then extended on a microscope slide and labeled with antibodies to IdU (green) and CldU (red) to visualize the DNA. This method allows the detection of replication origins as well as determination of the rate of DNA synthesis. [Courtesy Aaron Bensimon. Data from: Conti et al., Replication fork velocities at adjacent replication origins are coordinately modified during DNA replication in human cell, *Molecular Biology of the Cell* 18:3059–3067, 2007.]

Initiation of replication in eukaryotes proceeds as follows:

1. *The ORC is assembled.* In human beings, the ORC is composed of six different proteins, each homologous to DnaA in *E. coli*. These proteins come together to form a hexameric structure analogous to the assembly formed by DnaA.

2. *Licensing factors expose single strands of DNA.* After the ORC has been assembled, additional proteins are recruited, including Cdc6, a homolog of the ORC subunits, and Cdt1. These proteins, in turn, recruit a hexameric helicase with six distinct subunits called Mcm2-7. These proteins, including the helicase, are sometimes called *licensing factors* because they permit the formation of the initiation complex. After the initiation complex has formed, Mcm2-7 separates the parental DNA strands, and the single strands are stabilized by the binding of replication protein A, a single-stranded-DNA–binding protein.

3. *Two distinct polymerases are needed to copy a eukaryotic replicon.* An initiator polymerase called *polymerase α* begins replication but is soon replaced by a more processive enzyme. This process is called *polymerase switching* because one polymerase has replaced another. This second enzyme, *DNA polymerase δ*, is the principal replicative polymerase in eukaryotes (**Table 28.1**).

TABLE 28.1 Some types of DNA polymerases

Name	Function
Prokaryotic Polymerases	
DNA polymerase I	Removes primer and fills in gaps on lagging strand
DNA polymerase II (error-prone polymerase)	DNA repair
DNA polymerase III	Primary enzyme of DNA synthesis
Eukaryotic Polymerases	
DNA polymerase α	Initiator polymerase
Primase subunit	Synthesizes the RNA primer
DNA polymerase unit	Adds stretch of about 20 nucleotides to the primer
DNA polymerase β (error-prone polymerase)	DNA repair
DNA polymerase δ	Primary enzyme of DNA synthesis

Once initiated, replication proceeds with the binding of DNA polymerase α. This enzyme includes a primase subunit, used to synthesize the RNA primer, as well as an active DNA polymerase. After this polymerase has added a stretch of about 20 deoxynucleotides to the primer, another replication protein, called *replication factor C* (RFC), displaces DNA polymerase α. Replication factor C attracts a sliding clamp called *proliferating cell nuclear antigen* (PCNA), which is homologous to the β_2 subunit of *E. coli* DNA polymerase III. The binding of PCNA to DNA polymerase δ renders the enzyme highly processive and suitable for long stretches of replication. Replication continues in both directions from the origin of replication until adjacent replicons meet and fuse. At this stage the RNA primers are removed and the DNA fragments are ligated by DNA ligase.

The eukaryotic cell cycle ensures coordination of DNA replication and cell division

The use of multiple origins of replication requires mechanisms for ensuring that each sequence is replicated once and only once. Thus, the processes of DNA synthesis and cell division are coordinated in the eukaryotic **cell cycle** so that the replication of all DNA sequences (which occurs during S phase) is complete before the cell progresses into the next phase of the cycle (**Figure 28.27**). This coordination requires several so-called *checkpoints* that control the progression along the cycle. Checkpoints are managed by a family of small proteins termed *cyclins*, which are synthesized and degraded by proteasomal digestion in the course of the cell cycle. Cyclins act by binding to specific cyclin-dependent protein

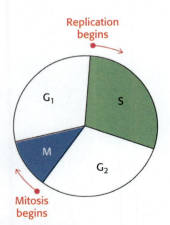

FIGURE 28.27 Eukaryotic DNA replication and cell division are highly coordinated. Mitosis (M) takes place only after DNA synthesis (S). Two gap phases (G₁ and G₂) separate the two processes.

kinases and activating them. One such kinase, cyclin-dependent kinase 2 (cdk2), binds to assemblies at origins of replication and regulates replication through a number of interlocking mechanisms.

The complexity of the eukaryotic genome includes the three-dimensional organization of chromosomes in specific compartments within the nucleus, a concept referred to as the *3D genome*. This level of organization is related to the fact that certain sections of the genome are replicated at distinct times during the S phase of the cell cycle. Moreover, regions of the genome that are replicated early in S phase also tend to be more highly expressed.

Telomeres are protective structures at the ends of linear chromosomes

Recall that whereas the genomes of essentially all prokaryotes are circular, the chromosomes of human beings and other eukaryotes are linear. The free ends of linear DNA molecules introduce several complications that must be resolved by special enzymes. In particular, complete replication of DNA ends is difficult because polymerases act only in the 5' → 3' direction. Thus, without some kind of intervention, the lagging strand would have an incomplete 5' end after the removal of the RNA primer, and each round of replication would further shorten the chromosome.

The first clue to how this problem is resolved came from sequence analyses of the ends of chromosomes, called **telomeres** (from the Greek *telos*, "an end"). Telomeric DNA contains hundreds of tandem repeats of a six-nucleotide sequence. One of the strands is G rich at the 3' end, and it is slightly longer than the other strand. In human beings, the repeating G-rich sequence is AGGGTT.

Extensive investigation of the structure adopted by telomeres suggests that they may form large duplex loops (**Figure 28.28**). The single-stranded region at the very end of the structure has been proposed to loop back to form a DNA duplex with another part of the repeated sequence, displacing a part of the original telomeric duplex. This loop-like structure is formed and stabilized by specific telomere-binding proteins. Such structures would nicely mask and protect the end of the chromosome.

FIGURE 28.28 Telomeres may form duplex loops. A single-stranded segment of the G-rich strand extends from the end of the telomere. In one model for telomeres, this single-stranded region invades the duplex to form a large duplex loop.

Telomeres are replicated by telomerase, a specialized polymerase that carries its own RNA template

How are these repeated sequences generated? An enzyme, termed **telomerase**, that executes this function has been purified and characterized. When a primer ending in GGTT is added to human telomerase in the presence of dNTPs, the sequences <u>GGTT</u>AGGGTT and <u>GGTT</u>AGGGTTAGGGTT, as well as longer products, are generated. Elizabeth Blackburn and Carol Greider (Chapter 1) discovered that the enzyme adding the repeats contains an RNA molecule that serves as the template for the elongation of the G-rich strand (**Figure 28.29**). Thus, telomerase carries the information necessary to generate the repeated telomeric sequences. Subsequently, a protein component of telomerases also was identified. This component is related to reverse transcriptases, enzymes first discovered in retroviruses that copy RNA into DNA (Section 8.4). Thus, telomerase is a specialized reverse transcriptase that carries its own template.

Telomerase is generally expressed at high levels only in rapidly growing cells, including cancer cells. Thus, telomerase is a potential target for anticancer

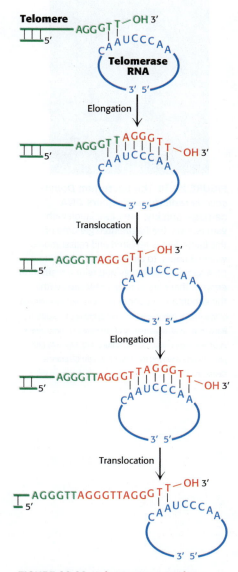

FIGURE 28.29 Telomerase uses its RNA template to synthesize the G-rich strand of telomeres. The RNA template of telomerase is shown in blue and the nucleotides added to the G-rich strand of the primer are shown in red. [Information from E. H. Blackburn, *Nature* 350:569–573, 1991.]

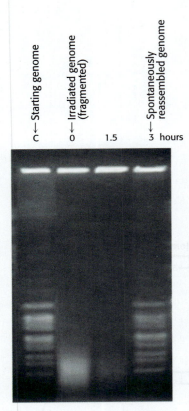

FIGURE 28.30 The bacterium *Deino-coccus radiodurans* repairs DNA damage quickly. After irradiation with gamma rays, the fragmented genome of this bacterium is repaired and reassembled within 3 hours. To aid analysis, the genomic DNA samples were digested with a restriction enzyme that cuts at only a few sites within the genome. C = control [Reprinted with permission of Macmillan Publishers Ltd, from Zahradka, K., Slade, D., Bailone, A. et al. Reassembly of shattered chromosomes in Deinococcus radiodurans. Nature 443, 569–573 (2006); permission conveyed through Copyright Clearance Cener, Inc.]

therapy. A variety of approaches for blocking telomerase expression or blocking its activity are under investigation for cancer treatment and prevention.

❖ SELF–CHECK QUESTION

Telomerase is not active in most human cells. Some cancer biologists have suggested that activation of the telomerase gene would be a requirement for a cell to become cancerous. Explain why this might be the case.

28.4 Many Types of DNA Damage Can Be Repaired

We have examined how even very large and complex genomes can, in principle, be replicated with considerable fidelity. However, DNA does become damaged, both in the course of replication and through other processes. Although DNA is remarkably robust, ultraviolet light as well as a range of chemical species can damage DNA, introducing changes in the DNA sequence (mutations) or lesions that can block further DNA replication. DNA damage can also take more complex forms such as the chemical modification of bases, chemical cross-links between the two strands of the double helix, or breaks in one or both of the phosphodiester backbones. The results may be cell death or cell transformation, changes in the DNA sequence that can be inherited by future generations, or blockage of the DNA replication process itself. While a variety of DNA-repair systems have evolved that can recognize DNA damage and, in many cases, restore the DNA molecule to its undamaged form, mutations in genes that encode components of DNA-repair systems are key factors in the development of cancer.

The bacterium *Deinococcus radiodurans* illustrates the extraordinary power of DNA-repair systems. This bacterium was discovered in 1956 when scientists were studying the use of high doses of gamma radiation to sterilize canned meat. In some cases, the meat still spoiled due to the growth of a bacterial species that withstood doses of gamma radiation more than 1000 times larger than those that would kill a human being. Each *D. radiodurans* cell contains between 4 and 10 copies of its genome. Even when these bacterial chromosomes are broken into many fragments by the ionizing radiation, they can reassemble and recombine to regenerate the intact genome with essentially no loss of information (Figure 28.30). These cells can also survive extreme desiccation—that is, drying out—much better than other organisms. This ability is believed to be the selective advantage that favored the evolution of this and related species.

Before we discuss various DNA-repair systems, let's begin with some of the sources of DNA damage.

Errors can arise in DNA replication

Errors introduced in the replication process are the simplest source of damage in the double helix. With the addition of each base, there is the possibility that an incorrect base might be incorporated, forming a non-Watson–Crick base pair. These non-Watson–Crick base pairs can locally distort the DNA double helix. Furthermore, such mismatches can be mutagenic; that is, they can result in permanent changes in the DNA sequence. When a double helix containing a non-Watson–Crick base pair is replicated, the two daughter double helices will have different sequences, because the mismatched base is very likely to pair with its Watson–Crick partner. Errors other than mismatches include insertions, deletions, and breaks in one or both strands. Furthermore, replicative polymerases can stall or even fall off a damaged template entirely, causing replication of the genome to halt before it is complete.

A variety of mechanisms have evolved to deal with such interruptions, including specialized DNA polymerases that can replicate DNA across many lesions. A drawback is that such polymerases are substantially more error-prone than are normal replicative polymerases. Nonetheless, these translesion or error-prone polymerases allow the completion of a draft sequence of the genome that can be at least partly repaired by DNA-repair processes. DNA recombination (Section 28.5) provides an additional mechanism for salvaging interruptions in DNA replication.

DNA can be damaged by oxidizing agents, alkylating agents, and light

A variety of chemical and physical agents, called **mutagens**, can alter specific bases within DNA after replication is complete. Let's look in depth at some of these mutagens.

Reactive oxygen species. For example, hydroxyl radical reacts with guanine to form 8-oxoguanine. 8-Oxoguanine is mutagenic because it often pairs with adenine rather than cytosine in DNA replication. Its pairing partner differs from that of guanine because it uses a different edge of the base to form base pairs (**Figure 28.31**).

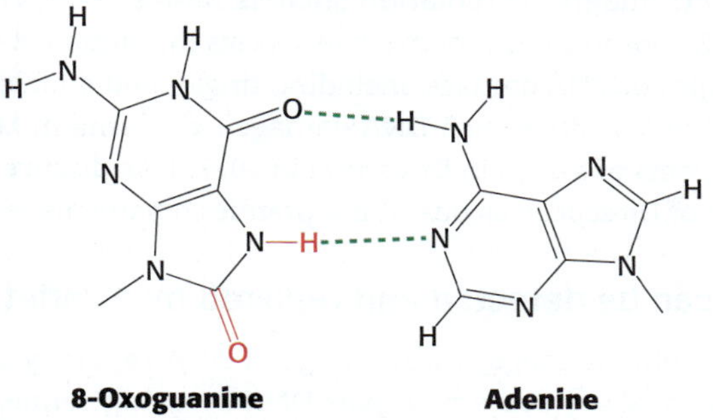

FIGURE 28.31 Damaged DNA bases can form base pairs. When guanine is oxidized to 8-oxoguanine, the damaged base can form a base pair with adenine through an edge of the base that does not normally participate in base-pair formation.

8-Oxoguanine   **Adenine**

Deaminating agents. Deamination is another potentially deleterious process. For example, nitrous acid will deaminate adenine to form hypoxanthine (**Figure 28.32**). This process is mutagenic because hypoxanthine pairs with cytosine rather than thymine. Guanine and cytosine can also be deaminated to yield bases that pair differently from the parent base.

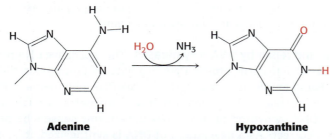

Adenine   **Hypoxanthine**

FIGURE 28.32 Adenine can be deaminated to form hypoxanthine. Hypoxanthine forms base pairs with cytosine in a manner similar to that of guanine, and so the deamination reaction can result in mutation.

Alkylating agents. In addition to oxidation and deamination, nucleotide bases are subject to alkylation, in which electrophilic centers are attacked by nucleophiles such as N-7 of guanine and adenine to form alkylated adducts. Some compounds are converted into highly active electrophiles through the action of enzymes that normally play a role in detoxification. A striking example is aflatoxin B_1, a compound produced by molds that grow on peanuts and other foods. A cytochrome P450 enzyme (Section 26.4) converts this compound into a highly reactive epoxide (**Figure 28.33**). This agent reacts with the N-7 atom of guanosine to form a mutagenic adduct that frequently leads to a G–C-to-T–A transversion.

FIGURE 28.33 Aflatoxin can be activated into a mutagen. Aflatoxin is activated by cytochrome P450 to form a highly reactive species that modifies bases such as guanine in DNA, leading to mutations.

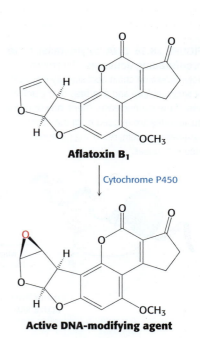

Aflatoxin B$_1$

Cytochrome P450

Active DNA-modifying agent

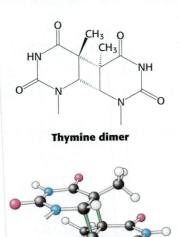

Thymine dimer

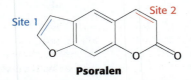

FIGURE 28.34 Ultraviolet light induces cross-links between adjacent pyrimidines along one strand of DNA.

Site 1   Site 2

Psoralen

FIGURE 28.35 Psoralen is a cross-linking agent. The compound psoralen and its derivatives can form interstrand cross-links through two reactive sites that can form adducts with nucleotide bases.

FIGURE 28.36 DNA polymerases have proofreading activities. The growing polynucleotide chain occasionally leaves the polymerase site and migrates to the active site of the exonuclease. There, one or more nucleotides are excised from the newly synthesized chain, removing potentially incorrect bases.

UV light and other cross-linking agents. The ultraviolet component of sunlight is a ubiquitous DNA-damaging agent. Its major effect is to covalently link adjacent pyrimidine residues along a DNA strand (**Figure 28.34**). Such a pyrimidine dimer cannot fit into a double helix, and so replication and gene expression are blocked until the lesion is removed. A thymine dimer is an example of an *intrastrand* cross-link because both participating bases are in the same strand of the double helix. In contrast, *interstrand* cross-links between bases on opposite strands disrupt replication because they prevent strand separation. Interstrand cross-links can be introduced by various agents, including compounds called *psoralens* (**Figure 28.35**) that are produced by a number of plants, including the common fig. When consumed in high doses and combined with exposure to UV light, psoralens can be mutagenic; but in lower doses, the combination of psoralens and UV light is an approved treatment method for a variety of skin conditions such as psoriasis and eczema.

High-energy electromagnetic radiation such as x-rays. X-rays can damage DNA by producing high concentrations of reactive species in solution. X-ray exposure can induce several types of DNA damage, including single- and double-stranded breaks in DNA. This ability to induce such DNA damage led Hermann Muller to discover the mutagenic effects of x-rays in *Drosophila* in 1927. This discovery contributed to the development of *Drosophila* as one of the premier organisms for genetic studies.

DNA damage can be detected and repaired by a variety of systems

To protect the genetic message, a wide range of DNA-repair systems are present in most organisms. Many systems repair DNA by using sequence information from the uncompromised strand. Such single-strand replication systems follow a similar mechanistic outline:

1. Recognize the offending base(s).
2. Remove the offending base(s).
3. Repair the resulting gap with a DNA polymerase and DNA ligase.

   We will briefly consider examples of several repair pathways. Although many of these examples are taken from *E. coli*, corresponding repair systems are present in most other organisms, including human beings.

Proofreading and repair by DNA polymerases. The replicative DNA polymerases themselves are able to correct many DNA mismatches produced in the course of replication. For example, the ε subunit of *E. coli* DNA polymerase III functions as a 3′-to-5′ exonuclease. This domain removes mismatched nucleotides from the 3′ end of DNA by hydrolysis. How does the enzyme sense whether a newly added base is correct? As a new strand of DNA is synthesized, it is, in a sense, proofread (**Figure 28.36**). If an incorrect base is inserted, then DNA synthesis slows down, owing to the difficulty of threading a non-Watson–Crick base pair into the polymerase. In addition, the mismatched base is weakly bound and therefore able to fluctuate in position. The delay from the slowdown allows time for these fluctuations to take the newly synthesized strand out of the polymerase active site and

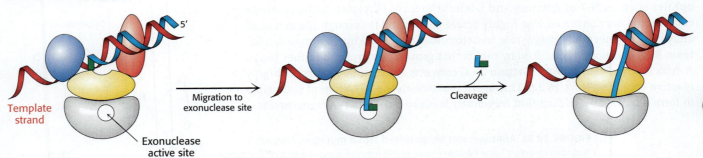

Template strand

Exonuclease active site

Migration to exonuclease site

Cleavage

into the exonuclease active site. There, the DNA is degraded, one nucleotide at a time, until it moves back into the polymerase active site and synthesis continues.

Mismatch repair. A second mechanism is present in essentially all cells to correct errors made in replication that are not corrected by proofreading (**Figure 28.37**). **Mismatch repair** is achieved by systems of at least two proteins, one for detecting the mismatch and the other for recruiting a restriction endonuclease that cleaves the newly synthesized DNA strand close to the lesion to facilitate repair. In *E. coli*, these proteins are MutS and MutL, and the endonuclease is MutH.

Direct repair. Another mechanism of DNA repair is **direct repair**, one example of which is the photochemical cleavage of pyrimidine dimers. Nearly all cells contain a photoreactivating enzyme called *DNA photolyase*. The *E. coli* enzyme, a 35-kDa protein that contains bound N^5, N^{10}-methenyltetrahydrofolate and flavin adenine dinucleotide (FAD) cofactors, binds to the distorted region of DNA. The enzyme uses light energy—specifically, the absorption of a photon by the N^5, N^{10}-methenyltetrahydrofolate coenzyme—to form an excited state that cleaves the dimer into its component bases.

Base-excision repair. Erroneous bases may also be removed. The excision of modified bases such as 3-methyladenine by the *E. coli* enzyme AlkA is an example of **base-excision repair**. The binding of this enzyme to damaged DNA flips the affected base out of the DNA double helix and into the active site of the enzyme (**Figure 28.38**). The enzyme then acts as a glycosylase, cleaving the glycosidic bond to release the damaged base. At this stage, the DNA backbone is intact, but a base is missing. This hole is called an *AP site* because it is <u>a</u>purinic (devoid of A or G) or <u>a</u>pyrimidinic (devoid of C or T). An AP endonuclease recognizes this defect and nicks the backbone adjacent to the missing base. Then, deoxyribose phosphodiesterase excises the residual deoxyribose phosphate unit, and DNA polymerase I inserts an undamaged nucleotide, as dictated by the base on the undamaged complementary strand. Finally, the repaired strand is sealed by DNA ligase.

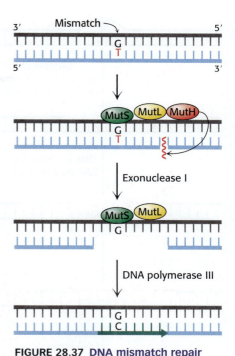

FIGURE 28.37 DNA mismatch repair in *E. coli* involves multiple proteins. First, a G–T mismatch is recognized by the MutS–MutL complex, and then MutH cleaves the backbone in the vicinity of the mismatch. Subsequently, a segment of the DNA strand containing the erroneous T is removed by exonuclease I and synthesized anew by DNA polymerase III. [Information from R. F. Service, *Science* 263:1559–1560, 1994.]

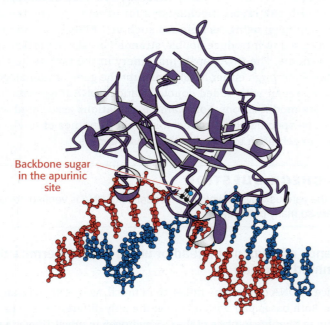

FIGURE 28.38 Base-excision repair enzymes remove damaged bases. A complex between the DNA-repair enzyme AlkA and an analog of a DNA molecule missing a purine base (an apurinic site) is shown. The backbone sugar in the apurinic site is flipped out of the double helix into the active site of the enzyme. [Drawn from 1BNK.pdb.]

INTERACT with this model in Achie⌾e

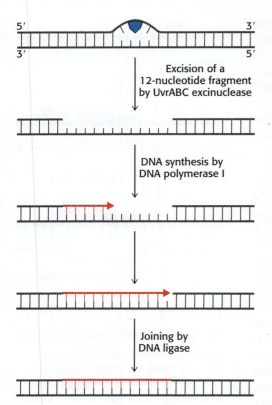

FIGURE 28.39 Nucleotide-excision repair involves the removal of a section of damaged DNA. Repair of a region of DNA containing a thymine dimer by the sequential action of a specific excinuclease, a DNA polymerase, and a DNA ligase. The thymine dimer is shown in blue and the new region of DNA is in red. [Information from P. C. Hanawalt, *Endeavour* 31:83, 1982.]

Nucleotide-excision repair. Sometimes, more than just the erroneous base is excised; the entire nucleotide may be removed. One of the best-understood examples of **nucleotide-excision repair** is utilized for the excision of a pyrimidine dimer. Three enzymatic activities are essential for this repair process in *E. coli* (**Figure 28.39**). First, an enzyme complex consisting of the proteins encoded by the *uvrABC* genes detects the distortion produced by the DNA damage. The UvrABC enzyme cuts the damaged DNA strand at two sites, 8 nucleotides away from the damaged site on the 5′ side and 4 nucleotides away on the 3′ side, and the 12-residue oligonucleotide excised by this highly specific *excinuclease* (from the Latin *exci*, "to cut out") then diffuses away. Second, DNA polymerase I enters the gap to carry out repair synthesis. The 3′ end of the nicked strand is the primer, and the intact complementary strand is the template. Third, the 3′ end of the newly synthesized stretch of DNA and the original part of the DNA chain are joined by DNA ligase.

Repair of breaks in both strands. While DNA ligase can seal simple breaks in one strand of the DNA backbone, alternative mechanisms are required to repair breaks on both strands that are close enough together that they would otherwise separate the DNA into two double helices. Several distinct mechanisms are able to repair such damage.

One mechanism, **non-homologous end joining (NHEJ)**, does not depend on other DNA molecules in the cell (see Section 9.4). In NHEJ, the free double-stranded ends are bound by a heterodimer of two proteins, Ku70 and Ku80. These proteins stabilize the ends and mark them for subsequent manipulations. Through mechanisms that are not yet well-understood, the Ku70/80 heterodimers act as handles used by other proteins to draw the two double-stranded ends close together so that enzymes can seal the break. Alternative mechanisms of double-stranded-break repair can operate if an intact stretch of double-stranded DNA with an identical or very similar sequence is present in the cell. These repair processes use homologous recombination, presented in Section 28.5.

As noted in Section 9.4, researchers can now take advantage of double-stranded-break repair mechanisms for executing targeted changes in eukaryotic genomes, using techniques such as CRISPR. Specific engineered nucleases are used to introduce double-stranded breaks in particular locations within the genome. The DNA-repair machinery in the targeted cells will then repair the break, through either NHEJ to disrupt the gene or homologous recombination-based repair with added double-stranded DNA fragments to introduce more elaborate modifications. For example, mutations associated with human diseases can be specifically introduced into the genomes of model organisms such as mice or zebrafish to examine the consequences.

❖ **SELF–CHECK QUESTION**

Summarize the various ways in which the sequence of DNA is verified or corrected. Why are there so many pathways?

The presence of thymine instead of uracil in DNA permits the repair of deaminated cytosine

The presence in DNA of thymine rather than uracil, as in RNA, was an enigma for many years. Both bases pair with adenine; the only difference between them is a methyl group in thymine in place of the C-5 hydrogen in uracil. Why is a methylated base employed in DNA and not in RNA? The existence of an active repair system to correct the deamination of cytosine provides a convincing solution to this puzzle.

Cytosine in DNA spontaneously deaminates at a perceptible rate to form uracil. The deamination of cytosine is potentially mutagenic because uracil pairs with adenine, and so one of the daughter strands will contain a U–A base pair rather than the original C–G base pair. This mutation is prevented by a repair system that recognizes uracil to be foreign to DNA (**Figure 28.40**). The repair enzyme, uracil DNA glycosylase, which is homologous to AlkA, hydrolyzes the glycosidic bond between the uracil and deoxyribose moieties but does not attack thymine-containing nucleotides. The AP site generated is repaired to reinsert cytosine.

Thus, we can conclude that the methyl group on thymine acts like a tag that distinguishes thymine from deaminated cytosine. If uracil instead of thymine were used in DNA, correctly placed uracil would be indistinguishable from uracil formed by deamination of cytosine. The defect would persist unnoticed, and so a C–G base pair would necessarily be mutated to U–A in one of the daughter DNA molecules. This mutation is prevented by a DNA-repair system that searches for uracil and leaves thymine alone. In summary, the use of thymine instead of uracil in DNA enhances the fidelity of the genetic message.

Some genetic diseases are caused by the expansion of repeats of three nucleotides

Some genetic diseases are caused by the presence of DNA sequences that are inherently prone to errors during replication and repair. A particularly important class of such diseases is characterized by the presence of long tandem arrays of multiple copies of a particular sequence of three nucleotides, called **trinucleotide repeats**. An example is Huntington's disease, an autosomal dominant neurological disorder with a variable age of onset. The mutated gene in this disease expresses a protein in the brain called *huntingtin*, which contains a stretch of consecutive glutamine residues encoded by a tandem array of CAG sequences within the gene. In unaffected persons, this array is between 6 and 31 repeats, whereas in those with the disease, the array is between 36 and 82 repeats or longer. Moreover, the array tends to become longer from one generation to the next. The consequence is a phenomenon called *anticipation* in which the children of an affected parent tend to show symptoms of the disease at an earlier age than did the parent.

The tendency of trinucleotide repeats to expand over time is explained by the formation of alternative structures in the course of DNA replication or repair. For example, part of the repeat can loop out without disrupting base-pairing outside this region. Then, in replication, DNA polymerase extends this strand through the remainder of the repeat, leading to an increase in the number of copies of the trinucleotide sequence (**Figure 28.41**).

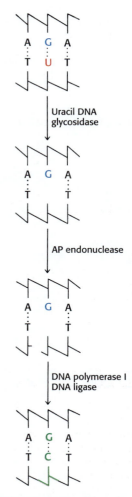

FIGURE 28.40 Uridine bases in DNA are excised and replaced by cytidine.

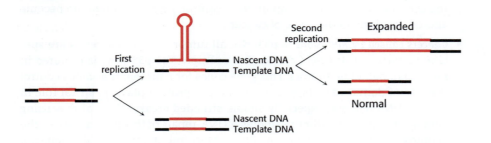

How do these long stretches of repeated amino acids cause disease? For huntingtin, it appears that the polyglutamine stretches become increasingly prone to aggregate as their length increases; the additional consequences of such aggregation are still under investigation.

Many cancers are initiated by the defective repair of DNA

As described in Chapter 14, cancers are caused by mutations in genes associated with growth control. Defects in DNA-repair systems increase the overall frequency of mutations and, hence, the likelihood of cancer-causing mutations.

FIGURE 28.41 Trinucleotide repeats can form structures that lead to their expansion during DNA replication. In this example, a hairpin structure forms that results in the expansion of the trinucleotide repeat segment (red) in one of the daughter strands after two rounds of replication.

Indeed, the synergy between studies of mutations that predispose people to cancer and studies of DNA repair in model organisms has been tremendous in revealing the biochemistry of DNA-repair pathways.

Genes for DNA-repair proteins are often **tumor-suppressor genes**; that is, they suppress tumor development when at least one copy of the gene is free of a deleterious mutation. When both copies of a gene are mutated, however, tumors develop at rates greater than those for the population at large. People who inherit defects in a single tumor-suppressor allele do not necessarily develop cancer but are susceptible to developing the disease because only the one remaining normal copy of the gene must develop a new defect to further the development of cancer. **Table 28.2** lists selected proteins involved in DNA-repair processes that, when mutated, can contribute to certain diseases, including cancers.

TABLE 28.2 Examples of diseases caused by mutations in repair proteins

Disease	Selected Proteins	Affected Repair Pathway
Xeroderma pigmentosum (skin)	XPA, XPB, XPC	Nucleotide-excision repair
Lynch syndrome (colorectal cancer)	MSH2, MLH1	Mismatch repair
Breast and ovarian cancer	BRCA1 and BRCA2	Double-strand break repair
Renal and lung cancer	OGG1	Base-excision repair

- *Xeroderma pigmentosum.* In the rare human skin disease xeroderma pigmentosum, the skin of an affected person is extremely sensitive to sunlight or ultraviolet light. Studies of xeroderma pigmentosum patients have revealed that mutations occur in genes for a number of proteins that are components of the human nucleotide-excision-repair pathway, including homologs of the UvrABC subunits. Severe changes in the skin start in infancy and worsen with time, usually resulting in the development of skin cancers at several sites. Many patients die before age 30 from metastases of these malignant skin tumors.

- *Hereditary nonpolyposis colorectal cancer.* As many as 1 in 200 people develop hereditary nonpolyposis colorectal cancer (HNPCC, or Lynch syndrome) resulting from defective DNA mismatch repair. Mutations in two genes, called *hMSH2* and *hMLH1*, account for most cases. The striking finding is that these genes encode the human counterparts of MutS and MutL of *E. coli*. Mutations in *hMSH2* and *hMLH1* likely allow other mutations to accumulate throughout the genome; in time, genes important in controlling cell proliferation become altered, resulting in the onset of cancer.

- *Cancers caused by mutations to p53.* Not all tumor-suppressor genes are specific to particular types of cancer. For example, the *p53* gene is mutated in more than half of all tumors. This gene codes for a protein that helps control the fate of damaged cells, p53. First, the p53 protein plays a central role in sensing DNA damage, especially double-stranded breaks. Then, after sensing damage, the protein either promotes a DNA-repair pathway or activates the apoptosis pathway, leading to cell death. Most mutations in the *p53* gene are sporadic; that is, they occur in somatic cells rather than being inherited, but the mutation can also be inherited. People who inherit a deleterious mutation in one copy of the *p53* gene have a high probability of developing several types of cancer.

Cancer cells often have two characteristics that make them especially vulnerable to agents that damage DNA molecules. First, they divide frequently, and so their DNA replication pathways are more active than they are in most cells. Second, as already noted, cancer cells often have defects in DNA-repair pathways. Several agents widely used in cancer chemotherapy, including cyclophosphamide and cisplatin, act by damaging DNA. Cancer cells are less able to avoid the effect of the induced damage than are normal cells, providing a therapeutic window for specifically killing cancer cells.

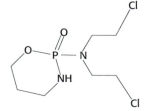

Cyclophosphamide

Cisplatin

Many potential carcinogens can be detected by their mutagenic action on bacteria

Many human cancers are caused by exposure to chemicals that are mutagens; therefore, it is important to identify such compounds and to ascertain their potency so that exposure to them can be minimized. Bruce Ames devised a simple and sensitive test, called the **Ames test**, for detecting chemical mutagens.

1. A thin layer of agar containing about 10^9 bacteria of a specially constructed tester strain of *Salmonella* is placed on an agar plate. These bacteria are unable to grow in the absence of histidine, because a mutation is present in one of the genes for the biosynthesis of this amino acid.

2. A chemical mutagen is added to the center of the plate, resulting in many new mutations. A small proportion of these mutations reverse the original mutation, allowing the bacteria to synthesize histidine.

3. These so-called *revertants* multiply in the absence of an external source of histidine and appear as discrete colonies after the plate has been incubated at 37°C for 2 days (**Figure 28.42**). For example, 0.5 µg of 2-aminoanthracene gives 11,000 revertant colonies, compared with only 30 spontaneous revertants in its absence.

4. A series of concentrations of the chemical mutagen that caused the reversion can be readily tested to generate a dose-response curve.

(A)   (B)

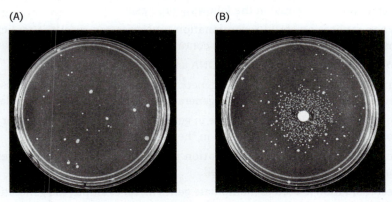

FIGURE 28.42 The Ames test determines the mutagenicity of various compounds. Two results are shown: (A) an agar plate containing about 10^9 *Salmonella* bacteria that cannot synthesize histidine, and (B) an agar plate containing a filter-paper disc with a mutagen, which produces a large number of revertants that can synthesize histidine. After 2 days, the revertants appear as rings of colonies around the disc. The small number of visible colonies in plate A are spontaneous revertants. [Republished with permission of Elsevier, from Mutation Resarch/Environmental Mutagenesis and Related Subjects, 31:6, Ames, B.N., et al. Methods for detecting carcinogens and mutagens with the salmonella/mammalian-microsome mutagenicity test, pp. 347–363, Copyright 1975; permission conveyed through Copyright Clearance Center, Ltd.]

Different tester strains of *Salmonella* used in the Ames test are sensitive to different kinds of mutations. For example, some of the tester strains are responsive to base-pair substitutions, whereas others detect deletions or additions of base pairs (frameshifts). These specially designed strains are particularly sensitive to mutations because the genes of their excision-repair systems have been deleted. In addition, potential mutagens enter the tester strains easily because the lipopolysaccharide barrier that normally coats the surface of *Salmonella* is incomplete in these strains. A key feature of this detection system is the inclusion of a mammalian liver homogenate, which promotes the conversion of some potential carcinogens into their active forms. Bacteria lack the enzymes required for this conversion.

The *Salmonella* test is extensively used to help evaluate the mutagenic and carcinogenic risks of a large number of chemicals. This rapid and inexpensive bacterial assay for mutagenicity complements epidemiological surveys and animal tests that are necessarily slower, more laborious, and far more expensive. The *Salmonella* test for mutagenicity is an outgrowth of studies of gene–protein relations in bacteria. It is a striking example of how fundamental research in molecular biology can lead directly to important advances in public health.

28.5 DNA Recombination Plays Important Roles in Replication, Repair, and Other Processes

Most processes associated with DNA replication function to copy the genetic message as faithfully as possible. However, several biochemical processes require **genetic recombination**, the exchange of genetic material between two DNA molecules that results in the reassortment of DNA sequences present on two different double helices (**Figure 28.43**).

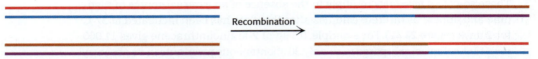

FIGURE 28.43 **Recombined DNA molecules have segments from both parent molecules.**

Recombination is essential in the following processes:

1. When replication stalls, recombination processes can reset the replication machinery so that replication can continue.

2. Some double-stranded breaks in DNA are repaired by recombination.

3. In meiosis, the limited exchange of genetic material between paired chromosomes provides a simple mechanism for generating genetic diversity in a population.

4. Recombination plays a crucial role in generating molecular diversity for antibodies and some other molecules in the immune system.

5. Some viruses employ recombination pathways to integrate their genetic material into the DNA of a host cell.

6. Recombination is used to manipulate genes in, for example, the generation of gene knockout mice and other targeted genome modifications (Section 9.4).

Recombination is most efficient between DNA sequences that are homologous. In homologous recombination, parent DNA duplexes align at regions of sequence similarity, and new DNA molecules are formed by the breakage and joining of homologous segments.

RecA can initiate recombination by promoting strand invasion

In many recombination pathways, a DNA molecule with a free end recombines with a DNA molecule having no free ends available for interaction. DNA molecules with free ends are the common result of double-stranded DNA breaks, but they may also be generated in DNA replication if the replication complex stalls. Double-stranded breaks are among the most potentially devastating types of DNA damage. With both strands of the double helix broken in a local region, neither strand is intact to act as a template for future DNA synthesis. Repairing such lesions relies on DNA recombination.

This type of recombination has been studied extensively in *E. coli*, but it also takes place in other organisms through the action of proteins homologous to those of *E. coli*. Often dozens of proteins participate in the complete recombination process. However, the key protein is RecA, the homolog of which is Rad51 in human cells. To accomplish the exchange, the single-stranded DNA first displaces one of the strands of the double helix (**Figure 28.44**). The resulting three-stranded structure is called a *displacement loop* or *D-loop*, and the process is often referred to as *strand invasion*. Because a free 3′ end is now base-paired to a contiguous strand of DNA, the 3′ end can act as a primer to initiate new DNA synthesis.

Strand invasion can initiate many processes, including the repair of double-stranded breaks and the reinitiation of replication after the replication apparatus has come off its template. In the repair of a break, the recombination partner is an intact DNA molecule with an overlapping sequence.

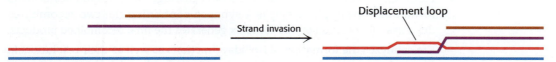

FIGURE 28.44 Strand invasion can initiate recombination. This process is promoted by proteins such as RecA.

Some recombination reactions proceed through Holliday-junction intermediates

In recombination pathways for meiosis and some other processes, intermediates form that are composed of four polynucleotide chains in a cross-like structure. Intermediates with these cross-like structures are referred to as **Holliday junctions**, after Robin Holliday, who proposed their role in recombination in 1964. Such intermediates have been characterized by a wide range of techniques, including x-ray crystallography.

Specific enzymes, termed **recombinases**, bind to these structures and resolve them into separated DNA duplexes. The Cre recombinase from bacteriophage P1 has been extensively studied. The mechanism begins with the recombinase binding to the DNA substrates (**Figure 28.45**).

FIGURE 28.45 Holliday junctions are formed and resolved by recombinases. Recombination begins as two DNA molecules and four recombinases come together to form a recombination synapse. One strand from each duplex is cleaved by transesterification with the 3′ end of each of the cleaved strands linked to a tyrosine residue on the recombinase enzyme. New phosphodiester bonds are formed when a 5′ end of the other cleaved strand in the complex attacks these tyrosine–DNA adducts. After isomerization, these steps are repeated to form the recombined products.

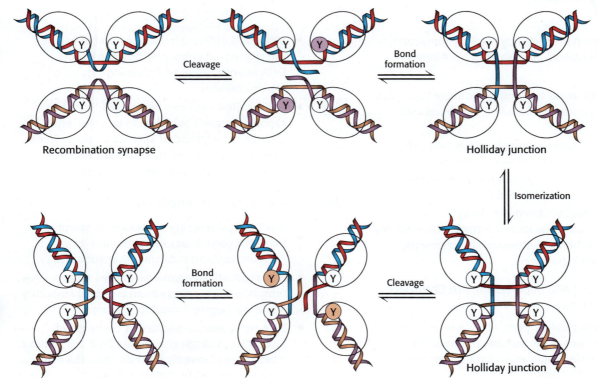

1. Four molecules of the enzyme and two DNA molecules come together to form a **recombination synapse**. The reaction begins with the cleavage of one strand from each duplex via transesterification reactions with the 3′-phosphoryl groups becoming linked to specific tyrosine residues in the recombinase while the 5′-hydroxyl group of each cleaved strand remains free.

2. The free 5′ ends invade the other duplex and participate in transesterification reactions to form new phosphodiester bonds and free the tyrosine residues. These reactions result in the formation of a Holliday junction.

3. This junction can then isomerize to form a structure in which the polynucleotide chains in the center of the structure are reoriented.

4. From this junction, the processes of strand cleavage and phosphodiester-bond formation repeat. The result is a synapse containing the two recombined duplexes. Dissociation of this complex generates the final recombined products.

Cre catalyzes the formation of Holliday junctions as well as their resolution. In contrast, other proteins bind to Holliday junctions that have already been formed by other processes and resolve them into separate duplexes. In many cases, these proteins also promote the process of branch migration whereby a Holliday junction is moved along the two component double helices. Branch migration can affect which segments of DNA are exchanged in a recombination process.

While most recombination events are between homologous sequences, recombination can also occur between different, non-homologous chromosomes. Nonhomologous recombination during cell division leads to genetic translocations, where parts of some chromosomes are moved to or exchanged with other chromosomes. Chromosomal translocations can be harmless or can result in diseases, depending on the chromosomes involved and the genes that are modified or disrupted. The formation of the *bcr-abl* gene, which leads to chronic myelogenous leukemia, was discussed in Section 14.5. During CRISPR manipulations of genomes, gene insertions occur primarily via homologous recombination, but, more rarely, sequences can also be inserted into non-homologous positions.

Summary

28.1 DNA Replication Proceeds by the Polymerization of Deoxyribonucleoside Triphosphates Along a Template

- DNA polymerases are template-directed enzymes that catalyze the formation of phosphodiester bonds.

- DNA polymerases require a primer with a free 3′-hydroxyl group to initiate strand synthesis.

- DNA polymerases always synthesize a DNA strand in the 5′-to-3′ direction.

- For both daughter strands to be synthesized simultaneously, one strand is made continuously while the other is synthesized in fragments, called *Okazaki fragments*.

28.2 DNA Unwinding and Supercoiling Are Controlled by Topoisomerases

- A key topological property of DNA is its linking number, which is the number of times one strand of DNA winds around the other in the right-hand direction.

- Molecules differing in linking number can be interconverted only by cutting one or both DNA strands; these reactions are catalyzed by topoisomerases.

- Supercoiled DNA can be relaxed by either topoisomerase I, which acts by transiently cleaving one strand of DNA in a double helix, or by topoisomerase II, which transiently cleaves both strands simultaneously.

28.3 DNA Replication Is Highly Coordinated

- The DNA polymerase holoenzyme comprises two DNA polymerase enzymes (one to act on each template strand) and other subunits.

- As a replicative polymerase moves along a DNA template, the leading strand is copied smoothly while the lagging strand forms loops.

- DNA replication is initiated at the origin of replication, a single site within the *E. coli* genome, where a set of specific proteins assembles the enzymes needed for DNA synthesis.

- The initiation of replication in eukaryotes is more complex than in prokaryotes; DNA synthesis is initiated at thousands of sites throughout the genome.
- Telomerase synthesizes specialized structures called telomeres at the ends of linear chromosomes.

28.4 Many Types of DNA Damage Can Be Repaired

- Different types of DNA damage include mismatched bases incorporated during DNA replication, individual bases damaged by oxidation or alkylation after replication, cross-linked bases, and single- or double-stranded breaks in the DNA backbone.
- Several different repair systems detect and repair DNA damage: the exonuclease activity of the DNA polymerase excises mismatches bases; specific enzymes reverse some DNA lesions such as thymine dimers; still other DNA-repair pathways act through the excision of single damaged bases or short segments of nucleotides.

- Double-stranded breaks in DNA can be repaired by homologous or non-homologous end-joining processes.
- Defects in DNA-repair components are associated with susceptibility to many different types of cancer.

28.5 DNA Recombination Plays Important Roles in Replication, Repair, and Other Processes

- Genetic recombination is important in some types of DNA repair.
- Some recombination pathways are initiated by strand invasion, where a single strand at the end of a DNA double helix forms base pairs with one strand of DNA in another double helix and displaces the other strand.
- In other recombination pathways a Holliday junction (a cross-like structure of four DNA strands) is formed.
- Recombinases promote recombination reactions through the introduction of specific DNA breaks and the formation and resolution of Holliday-junction intermediates.

Key Terms

template (p. 846)

DNA polymerases (p. 846)

primer (p. 846)

exonuclease (p. 846)

primase (p. 848)

replication fork (p. 849)

Okazaki fragments (p. 849)

lagging strand (p. 849)

leading strand (p. 849)

DNA ligase (p. 849)

helicase (p. 850)

supercoiling (p. 851)

linking number (Lk) (p. 853)

topoisomers (p. 853)

twist (Tw) (p. 853)

writhe (Wr) (p. 853)

topoisomerase (p. 854)

processivity (p. 857)

sliding DNA clamp (p. 857)

trombone model (p. 859)

origin of replication (p. 860)

origin of replication complex (ORC) (p. 861)

cell cycle (p. 862)

telomere (p. 863)

telomerase (p. 863)

mutagen (p. 865)

mismatch repair (p. 867)

direct repair (p. 867)

base-excision repair (p. 867)

nucleotide-excision repair (p. 868)

non-homologous end joining (NHEJ) (p. 868)

trinucleotide repeats (p. 869)

tumor-suppressor gene (p. 870)

Ames test (p. 871)

genetic recombination (p. 872)

Holliday junction (p. 873)

recombinases (p. 873)

recombination synapse (p. 874)

Problems

1. A nucleotide pairing between A (adenine) and T (thymine) in the original DNA strand is miscopied during DNA replication. On one strand of DNA, A correctly pairs with T. However, on the other strand, T pairs with G (guanine) instead of A. This mistake is not corrected, and the double strand with the mistake is incorporated into a new cell during cell division. The double strand with the G–T mistake is later replicated without additional mistakes. What nucleotide pairs will be found in the two new double strands at the site of the G–T mistake? ❖ 4

Original DNA sequence	Mistake during replication
A–T	A–T G–T

2. The oxidation of guanine bases in the context of triplet repeats such as CAGCAGCAG can lead to the expansion of the repeat. Explain. ❖ **4**

3. DNA polymerase I, DNA ligase, and topoisomerase I use an activated intermediate to catalyze the formation of phosphodiester bonds with the release of a leaving group. Match the activated intermediate and the leaving group to each enzyme. ❖ **1**

(a) DNA polymerase I
(b) DNA ligase
(c) DNA topoisomerase I

1. Tyrosine
2. AMP
3. DNA-adenylate
4. dNTPs
5. Pyrophosphate
6. DNA-tyrosyl

4. Draw a replication fork. Label the template strands and identify their 5′ and 3′ ends. Draw in and add labels for the leading strand, lagging strand, RNA primer(s), and Okazaki fragments. Add 5′ and 3′ labels to the daughter strands. Then place the following proteins on your diagram: SSBs, helicase (DnaB), DNA topoisomerase II (gyrase), primase. ❖ **1**

5. Under standard conditions, the free energy of ATP hydrolysis is about -30 kJ mol^{-1}. Assuming that the energy required to break an average base pair within double-stranded DNA is 10 kJ mol^{-1}, estimate the maximum number of base pairs that could be broken per ATP hydrolyzed by a helicase operating under standard conditions. ❖ **2**

6. Circular DNA from SV40 virus was isolated and subjected to gel electrophoresis. The results are shown in lane A (the control) of the adjoining gel patterns shown below. ❖ **3**

From W. Keller. PNAS 72(1975):2553.

(a) Why does the DNA separate in agarose gel electrophoresis? How does the DNA in each band differ?

(b) The DNA was then incubated with topoisomerase I for 5 minutes and again analyzed by gel electrophoresis, with the results shown in lane B. What types of DNA do the various bands represent?

(c) Another sample of DNA was incubated with topoisomerase I for 30 minutes and again analyzed as shown in lane C.

What is the significance of the fact that more of the DNA is in slower-moving forms?

7. Answer the following questions about supercoiling and linking number (Lk). ❖ **3**

(a) What is Lk for a relaxed, closed-circular DNA with 4784 base pairs?

(b) What is Lk for a negatively supercoiled 4784 bp DNA if it is underwound by 4 complete turns?

(c) How does Lk change when there is a break in one strand?

(d) What is Lk for a negatively supercoiled 4784 bp DNA after treatment with one molecule of topoisomerase II (DNA gyrase) and adequate ATP?

8. Which of the sequences below represents newly synthesized DNA that can be repaired by the proofreading activity of DNA polymerases? For the segments shown, the top strand represents the template strand; the bottom strand represents the newly synthesized strand. ❖ **4**

(a) 3′-GATCAGAATG-5′
    5′-CTAGGCTTAC-3′

(b) 3′-GATCAGAATG-5′
    5′-TTAGTCTTAC-3′

(c) 3′-GATCAGAATG-5′
    5′-CTAGTCTTAT-3′

(d) 3′-GATCAGAATG-5′
    5′-CTAGTCTTAC-3′

9. What of the following strategies do cells use to ensure that newly replicated DNA does not contain errors? Select all that apply. ❖ **4**

(a) Enzymes repair mistakes in the new DNA double helix after the new double helix separates from the original double helix.

(b) As DNA polymerase synthesizes new DNA, the DNA polymerase finds and corrects misplaced nucleotides.

(c) Enzymes proofread the DNA after the DNA has been replicated and replace any mismatched nucleotides.

(d) DNA polymerase replaces the newly replicated DNA on any chromosomes on which there are mistakes.

(e) Enzymes remove and resynthesize any misshapen sequences in the DNA prior to replication.

10. Place the following steps of prokaryotic DNA replication in order, starting with the bacterium initiating DNA replication, and ending with DNA replication being terminated. ❖ **5**

(a) The bacterium initiates DNA replication.

(b) DNA replaces RNA and lagging strands are joined.

(c) Tus protein stops replication forks from crossing each other and ends synthesis.

(d) The upstream region melts, and helicase binds and unwinds DNA.

(e) RNA primers are added to provide a 3′ end for elongation.

(f) DNA replication is terminated.

(g) 5′-to-3′ synthesis of the leading strand and short lagging strand is carried out by DNA polymerase.

(h) The DnaA protein binds to the single origin of replication.

11. Place the following steps of eukaryotic DNA replication in order, starting from when a germ cell enters gap 1 (G_1) phase and ending with the termination of one cell cycle. ❖ **5**

(a) The cell enters G_1.

(b) Each genomic origin of replication assembles a pre-replication complex.

(c) RNA primers are added to provide a 3′ end for elongation.

(d) Active telomerase can extend the lost telomere region.

(e) One round of the cell cycle terminates.

(f) 5′-to-3′ synthesis of the leading and lagging strands is carried out by DNA polymerase.

(g) The initiation complex creates an active replication fork as helicase unwinds DNA.

(h) RNA is replaced with DNA, and lagging strands are joined.

12. Place the events in which telomerase maintains chromosomal ends during replication in order. ❖ **6**

(a) Additional nucleotides are added to the 3′ end of the parental DNA strand.

(b) Telomerase moves along the newly elongated DNA strand toward its 3′ end.

(c) Nucleotides that are complementary to the RNA component of telomerase are added to the 3′ end of the DNA overhang.

(d) Telomerase is removed from the DNA strand entirely, making way for a DNA polymerase.

(e) The RNA component of telomerase binds to a complementary sequence on the 3′ overhang of the parental DNA strand.

(f) DNA polymerase adds bases complementary to the elongated overhang created by telomerase.

13. In contrast to prokaryotes, which have circular chromosomes, eukaryotes have linear chromosomes. A potential problem arises because DNA polymerase cannot replicate the 5′ ends of linear chromosomes. As a result, the lagging strands of eukaryotic chromosomes shorten with repeated rounds of replication.

Telomerase, an enzyme implicated in aging and cell lifespan, helps overcome this problem. Telomerase adds more deoxynucleotides to the end of the chromosome, lengthening it. This DNA sequence, or telomere, prevents the loss of essential DNA from the ends of chromosomes.

Which of the following statements accurately describe telomeres and the actions of telomerase? Select all that apply. ❖ **6**

(a) Telomerase contains a G-rich template sequence that adds a C-rich DNA sequence to the lagging strand of the chromosome.

(b) Inappropriate activation of telomerase can result in cellular immortality, one of the cellular changes implicated in the development of cancer.

(c) A telomere forms a loop with a section of the complementary DNA strand, protecting the chromosome from degradation by endonucleases.

(d) Telomerase activity is turned off in most human cells, causing the telomeres to gradually shorten as the individual ages.

(e) Telomerase consists of an RNA template and a protein component that adds complementary nucleotides to the 3′ end of DNA.

14. The following illustration shows four agar plates used for the Ames test. A piece of filter paper (white circle in the center of each plate) was soaked in one of four preparations and then placed on an agar plate. The four preparations contained (A) purified water (control), (B) a known mutagen, (C) a chemical whose mutagenicity is under investigation, and (D) the same chemical after treatment with liver homogenate. The number of revertants, visible as colonies on the plates, was determined in each case. ❖ **7**

(a) What was the purpose of the control plate, which was exposed only to water?

(b) Why was it wise to use a known mutagen in the experimental system?

(c) How would you interpret the results obtained with the experimental compound?

(d) What liver components would you think are responsible for the effects observed in preparation D?

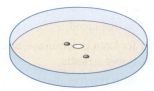

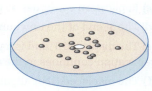

(A) Control: No mutagen   **(B)** + Known mutagen

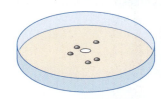

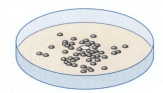

(C) + Experimental sample   **(D)** + Experimental sample after treatment with liver homogenate

15. Classify each statement on the right as describing DNA replication in prokaryotes, eukaryotes, both, or neither. ❖ **5**

(a) DNA replication in prokaryotes

(b) DNA replication in eukaryotes

(c) Both

(d) Neither

1. Several origins of replication per chromosome

2. Bidirectional replication

3. Several types of DNA polymerases

4. Occurs during M phase of cell cycle

5. One origin of replication per chromosome

16. Determine whether each phrase on the right describes recombination that enables replication with a single-stranded gap, a double-stranded break, or both. Each phrase may only be used once. ❖ 8

(a) Single-stranded gap

(b) Double-stranded break

(c) Both

1. Produces two DNA segments with no breaks
2. Occurs between arms of a replication fork
3. A free 3′ end invades another DNA segment
4. Occurs between homologous DNA segments
5. Involves removal of the 5′ end(s) of the broken DNA
6. Involves a gap in one strand of a DNA segment

17. DNA that contains a gap in a single strand can be replicated through recombination. Place the steps in order from when replication stops to when replication continues. ❖ 8

(a) The replication fork encounters a single-stranded gap.

(b) The crossed-over strands are resolved; a gap remains in one duplicated strand.

(c) Replication continues as the lesion is bypassed.

(d) The 3′ end of the unwound DNA is displaced.

(e) The displaced 3′ end is a primer for DNA polymerase, and replication continues in the crossed-over strands.

(f) The displaced DNA aligns itself with the complementary bases on the other arm of the replication fork.

(g) A helicase unwinds the newly synthesized DNA at the break.

18. Match each key term to its appropriate definition: ❖ 3

(a) Positive supercoiling

(b) Twist

(c) Writhe

(d) Negative supercoiling

(e) Supercoiling

1. The coiling, or looping, of the double helix around its axis
2. Results from the partial unwinding (right-handed, or underwinding) of the double helix; represents the topology of DNA in vivo; and permits easier strand separation
3. The winding of two DNA strands around each other
4. Permits DNA to maintain a relatively stable conformation when the helix is slightly underwound or overwound
5. Results from the overwinding of the DNA and double helix

19. The end-replication problem (telomere problem) exists in eukaryotic chromosomes and is characterized by the chromosomes shortening with each round of DNA replication. Select the statements that best explain why the end-replication problem exists in eukaryotic chromosomes. ❖ 6

(a) DNA ligase links the 5′-OH group of one fragment to the 3′ phosphate group of an adjacent fragment.

(b) DNA polymerase synthesizes DNA from the 5′ end to the 3′ end.

(c) The lagging strand is synthesized from the 3′ end to the 5′ end.

(d) DNA polymerase requires a primer for DNA synthesis.

(e) The RNA primer is removed in a 3′-to-5′ direction.

RNA Functions, Biosynthesis, and Processing

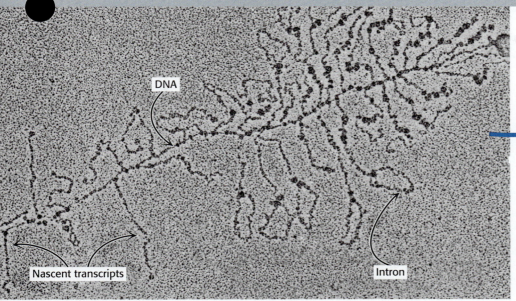

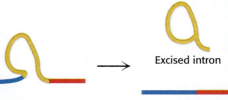

mRNA precursor · Excised intron · mRNA

Data from *Does the pachytene checkpoint, a feature of meiosis, filter out mistakes in double-strand DNA break repair and as a side-effect strongly promote adaptive speciation?* Foe, Victoria E, Integrative Organismal Biology, 10 Apr 2022, Vol. 4, Issue 1 Oxford University Press, Open Access article distributed under the terms of the Creative Commons Attribution License (https://creativecommons.org/licenses/by/4.0/)

RNA synthesis is a key step in the expression of genetic information. This electron micrograph shows a DNA template from a fruit fly in the genus *Drosophila* (the center "stem") being actively transcribed into RNAs (the "branches"). In eukaryotic organisms such as fruit flies, the initial RNA transcript often contains sections that do not encode protein sequence (introns). These sections are removed by splicing; proteins associated with RNA splicing can be seen as dark blobs on the mRNA transcripts in the image on the left.

❖ LEARNING GOALS

By the end of this chapter, you should be able to:

1. Discuss the primary function of RNA polymerases, the reaction they catalyze, and the chemical mechanism of that reaction.

2. Describe transcription, including the processes of initiation, elongation, and termination.

3. Compare the roles of eukaryotic RNA polymerases I, II, and III in producing ribosomal, transfer, and messenger RNAs.

4. Recognize the significance of transcription factors and enhancers in the regulation of transcription in eukaryotes.

5. Describe the process of RNA splicing, including the roles of the spliceosome and self-splicing RNA molecules.

6. Understand some of the differences between transcription in bacteria and in eukaryotes.

OUTLINE

29.1 RNA Molecules Play Different Roles, Primarily in Gene Expression

29.2 RNA Polymerases Catalyze Transcription

29.3 Transcription Is Highly Regulated

29.4 Some RNA Transcription Products Are Processed

29.5 The Discovery of Catalytic RNA Revealed a Unique Splicing Mechanism

DNA stores genetic information in a stable form that can be readily replicated. The expression of this genetic information requires its flow from DNA to RNA and, usually, to protein, as was introduced in Chapter 8. This chapter examines transcription, which, you will recall, is the process of synthesizing an RNA transcript from a DNA template, transferring the sequence information within the DNA to the new RNA molecule. We begin with a brief discussion of the diverse types of RNA molecules; then we will turn to RNA polymerases, the large and complex enzymes that carry out the synthetic process. This will lead into a discussion of transcription in bacteria and focus on the three stages of transcription: promoter binding and initiation, elongation of the nascent RNA transcript, and termination. We then examine transcription in eukaryotes, focusing on the distinctions between bacterial and eukaryotic transcription.

29.1 RNA Molecules Play Different Roles, Primarily in Gene Expression

While the function of some RNAs has been known for some time, other classes of RNAs have only recently been discovered. The investigation of some of these RNA molecules has been one of the most productive areas of biochemical research in recent years.

RNAs play key roles in protein biosynthesis

As we will explore in Chapter 30, the long-known ribosomal and transfer RNA molecules, along with messenger RNAs, are central to protein synthesis. Ribosomal RNAs are critical components of ribosomes (the sites of protein synthesis), and transfer RNAs play a role in delivering amino acids (the building blocks of proteins) to the ribosome. Messenger RNAs carry the information that ribosomes use for the production of specific protein sequences.

Some RNAs can guide modifications of themselves or other RNAs

In eukaryotes, one of the most striking examples of RNA modification is the splicing of mRNA precursors, a process that is catalyzed by large complexes composed of both proteins and small nuclear RNAs. These small nuclear RNAs play a crucial role in guiding the splicing of messenger RNAs.

Remarkably, some RNA molecules can splice themselves in the absence of other proteins and RNAs. This landmark discovery of self-splicing introns, which we will discuss later in this chapter, revealed that RNA molecules can serve as catalysts, which greatly influenced our view of molecular evolution. Many other types of RNAs, such as small regulatory RNAs and long noncoding RNAs, have been discovered more recently, and while their functions are still under active investigation, our understanding is rapidly expanding.

Some viruses have RNA genomes

While DNA is the genetic material in most organisms, some viruses have genomes made of RNA (Section 8.4). RNA viruses are responsible for several diseases, such as influenza, polio, mumps, Ebola, the common cold, and—of particular note—COVID-19. The coronavirus SARS-CoV-2, which causes COVID-19, belongs to a family of coronaviruses that have an unusually large, single-stranded RNA genome. Coronaviruses cause a variety of diseases in mammals and birds that have a wide range of symptoms, including respiratory distress in humans that can potentially be lethal.

RNA viruses vary in their genome organization. They have either single- or double-stranded RNA molecules arranged in single or multiple fragments. Inside their hosts, RNA viruses replicate their genomes with a virus-encoded RNA polymerase that uses RNA as a template. In some viruses, a complementary RNA molecule is made from the single-stranded viral RNA genome, while in other viruses, a double-stranded DNA copy is made that can integrate itself into the host's genome. Viral RNA polymerases do not have the same proofreading ability of other polymerases, which leads to high mutation rates of RNA viruses.

Messenger RNA vaccines provide protection against diseases

Messenger RNA (mRNA) vaccines take advantage of the fact that cells can be tricked into making proteins they don't usually make, even those from other organisms such as viruses. Unlike other vaccines that use parts of a weakened or inactivated pathogen to trigger an immune response, mRNA vaccines—including those available for the SARS-CoV-2 virus—use sections of pathogenic (usually viral) mRNA that have been generated in the laboratory. When human cells are injected with this mRNA, they will produce the corresponding protein from this set of instructions. This "foreign" protein will be presented on the surface of the cells, trigger an immune response, and provide some level of protection if and when the actual pathogen invades. Even though research on mRNA vaccines had been going on for many years, the COVID-19 pandemic accelerated the development, approval, and distribution of mRNA vaccines to the public.

❖ SELF–CHECK QUESTION

Compare the biological roles of RNA and DNA. What aspects of the structure and chemistry of RNA make it so versatile?

29.2 RNA Polymerases Catalyze Transcription

Transcription, the synthesis of RNA molecules from a DNA template, is catalyzed by large enzymes called **RNA polymerases**. The basic biochemistry of RNA synthesis is shared by all organisms, a commonality that has been beautifully illustrated by the three-dimensional structures of representative RNA polymerases from prokaryotes and eukaryotes (**Figure 29.1**). Despite substantial differences in size and number of polypeptide subunits, the overall structures of these enzymes are quite similar, revealing a common evolutionary origin.

FIGURE 29.1 RNA polymerase structures in prokaryotes and eukaryotes resemble each other. In these three-dimensional models of RNA polymerases from a prokaryote (*Thermus aquaticus*) and a eukaryote (*Saccharomyces cerevisiae*), the two largest subunits for each structure are shown in dark red and dark blue. Both structures contain a central metal ion (green) in their active sites, near a large cleft on the right. The similarity of these structures reveals that these enzymes have the same evolutionary origin and have many mechanistic features in common. [Drawn from 1I6V.pdb and 1I6H.pdb.]

INTERACT with these models in
≈ Achieve

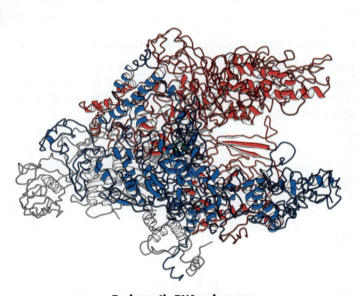

Prokaryotic RNA polymerase

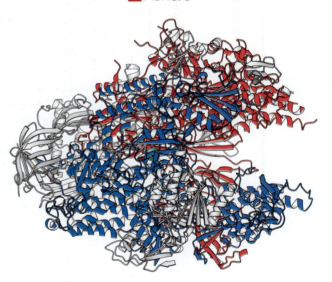

Eukaryotic RNA polymerase

RNA polymerases are very large, complex enzymes. For example, the core of the RNA polymerase of *E. coli* consists of five kinds of subunits with the composition $\alpha_2\beta\beta'\omega$ (**Table 29.1**). A typical eukaryotic RNA polymerase is larger and more complex, having 12 subunits and a total molecular mass of more than 500 kDa. Despite this complexity, the detailed structures of RNA polymerases have been determined by x-ray crystallography in work pioneered by Roger Kornberg and Seth Darst. The structures of many additional RNA polymerase complexes have been determined by cryo-electron microscopy.

TABLE 29.1 Subunits of RNA polymerase from *E. coli*

Subunit	Gene	Number	Mass (kDa)
α	*rpoA*	2	37
β	*rpoB*	1	151
β'	*rpoC*	1	155
ω	*rpoZ*	1	10
σ^{70}	*rpoD*	1	70

RNA synthesis comprises three stages: initiation, elongation, and termination

RNA synthesis, like all biological polymerization reactions, takes place in three stages: *initiation*, *elongation*, and *termination*. RNA polymerases perform multiple functions in this process:

1. They search DNA for initiation sites, also called *promoter sites* or simply **promoters**. For instance, *E. coli* DNA has about 2000 promoters in its 4.6×10^6 bp genome.

2. They unwind a short stretch of double-helical DNA to produce single-stranded DNA templates from which the sequence of bases can be easily read out.

3. They select the correct ribonucleoside triphosphate and catalyze the formation of a phosphodiester bond. This process is repeated many times as the enzyme moves along the DNA template. RNA polymerase is completely processive—a transcript is synthesized from start to end by a single RNA polymerase molecule.

4. They detect termination signals that specify where a transcript ends.

5. Their activity is regulated by activator and repressor proteins that interact with the promoter and modulate the ability of the RNA polymerase to initiate transcription. Gene expression is controlled substantially at the level of transcription, as will be discussed in detail in Chapter 31.

The chemistry of RNA synthesis is identical for all forms of RNA, including messenger RNAs, transfer RNAs, ribosomal RNAs, and small regulatory RNAs, so the basic steps just outlined apply to all forms. Their synthetic processes differ mainly in regulation, the specific RNA polymerase that creates them, and their posttranscriptional processing.

RNA polymerases catalyze the formation of a phosphodiester bond

The fundamental reaction of RNA synthesis, like that of DNA synthesis, is the formation of a phosphodiester bond. The 3′-hydroxyl group of the last nucleotide in the chain makes a nucleophilic attack on the α-phosphoryl group of the incoming nucleoside triphosphate, releasing a pyrophosphate.

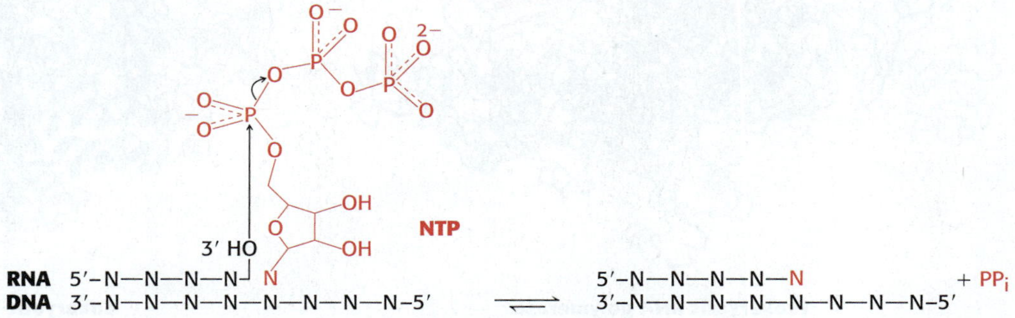

The catalytic sites of all RNA polymerases include two metal ions, normally magnesium ions (**Figure 29.2**). One ion remains tightly bound to the enzyme, whereas the other ion comes in with the nucleoside triphosphate and leaves with the pyrophosphate. Three conserved aspartate residues participate in binding these metal ions. Given the recent appreciation of the role of a third metal ion in the active site of DNA polymerases (see Section 28.1), it will be interesting to see if RNA polymerases show similarities to that model.

The polymerization reactions that are catalyzed by both prokaryotic and eukaryotic RNA polymerases take place within a complex in DNA termed a

FIGURE 29.2 The RNA polymerase active site contains metal ions. In this model of the transition state for phosphodiester-bond formation in the active site of RNA polymerase, the 3'-hydroxyl group of the growing RNA chain attacks the α-phosphoryl group of the incoming nucleoside triphosphate, resulting in the release of pyrophosphate. This transition state is structurally similar to that in the active site of DNA polymerase (see Figure 28.2).

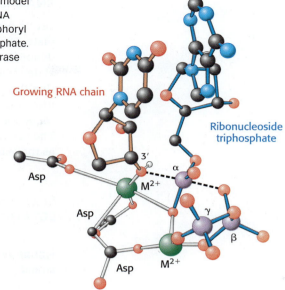

transcription bubble (**Figure 29.3**). This complex consists of double-stranded DNA that has been locally unwound in a region of approximately 17 base pairs. The edges of the bases that normally take part in Watson–Crick base pairs are exposed in the unwound region. We will begin with a detailed examination of the elongation process, including the role of the DNA template read by RNA polymerase and the reactions catalyzed by the polymerase, before returning to the more complex processes of initiation and termination.

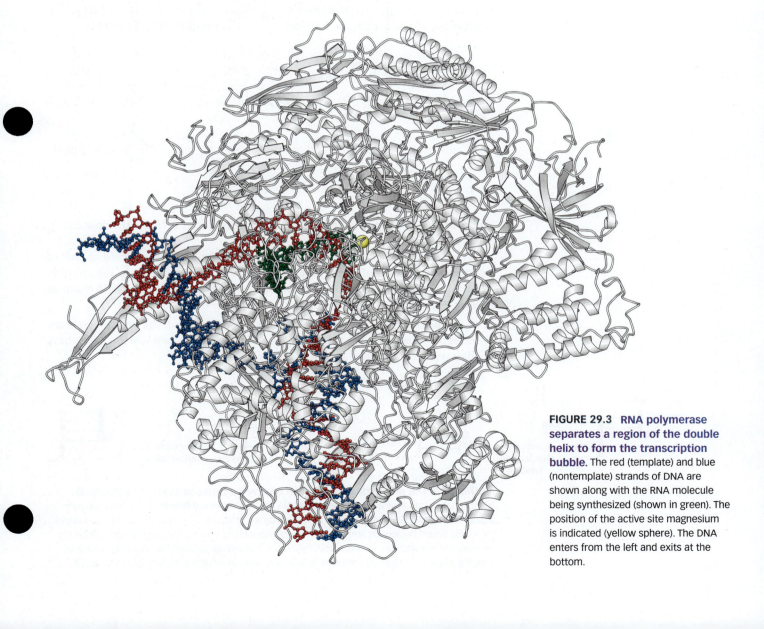

FIGURE 29.3 RNA polymerase separates a region of the double helix to form the transcription bubble. The red (template) and blue (nontemplate) strands of DNA are shown along with the RNA molecule being synthesized (shown in green). The position of the active site magnesium is indicated (yellow sphere). The DNA enters from the left and exits at the bottom.

RNA chains are formed de novo and grow in the 5′-to-3′ direction

Let us begin our examination of transcription by considering the DNA template. The first nucleotide (the start site) of a DNA sequence to be transcribed is denoted as +1 and the second one as +2; the nucleotide preceding the start site is denoted as −1. These designations refer to the coding strand of DNA. Recall that the sequence of the *template strand* of DNA is the complement of that of the RNA transcript (**Figure 29.4**). In contrast, the *coding strand* of DNA has the same sequence as that of the RNA transcript except for thymine (T) in place of uracil (U). The coding strand is also known as the *sense (+) strand*, and the template strand as the *antisense (−) strand*.

```
                        AACGUAGGGUCACAUC...   RNA transcript
...GCATACAACACACCTTGCATCCCAGTGTAG...   Template, or antisense (−), strand
...CGTATGTTGTGTGGAACGTAGGGTCACATC...   Coding, or sense (+), strand
                      −1 +1 +2
```

FIGURE 29.4 The RNA transcript is complementary to the template, or antisense (−), strand.

In contrast with DNA synthesis, RNA synthesis can start de novo, without the requirement for a primer. Most newly synthesized RNA chains carry a highly distinctive tag on the 5′ end: the first base at that end is either pppG or pppA.

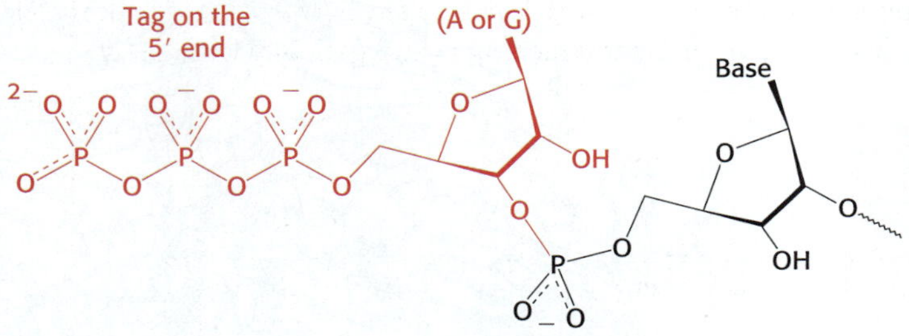

The presence of the triphosphate moiety confirms that RNA synthesis starts at the 5′ end.

The dinucleotide shown above is synthesized by RNA polymerase as part of the complex process of initiation, which will be discussed later in the chapter. After initiation takes place, RNA polymerase elongates the nucleic acid chain as follows (**Figure 29.5**).

1. A ribonucleoside triphosphate binds in the active site of the RNA polymerase directly adjacent to the growing RNA chain, and it forms a Watson–Crick base pair with the template strand.

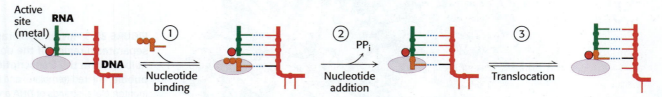

FIGURE 29.5 The elongation mechanism involves a series of steps. A ribonucleoside triphosphate binds adjacent to the growing RNA chain and forms a Watson–Crick base pair with a base on the DNA template strand (1). The 3′-hydroxyl group at the end of the RNA chain attacks the newly bound nucleotide and forms a new phosphodiester bond, releasing pyrophosphate (2). After nucleotide addition, the RNA–DNA hybrid can translocate through the RNA polymerase (3), bringing a new DNA base into position to base-pair with an incoming nucleoside triphosphate.

2. The 3′-hydroxyl group of the growing RNA chain, which is oriented and activated by the tightly bound metal ion, attacks the α-phosphoryl group to form a new phosphodiester bond, displacing pyrophosphate.

3. Next, the RNA–DNA hybrid must move relative to the polymerase to bring the 3′ end of the newly added nucleotide into proper position for the next nucleotide to be added. This translocation step does not include breaking any bonds between base pairs and is reversible; but, once it has taken place, the addition of the next nucleotide, favored by the triphosphate cleavage and pyrophosphate release and cleavage, drives the polymerization reaction forward.

The lengths of the RNA–DNA hybrid and of the unwound region of DNA stay rather constant as RNA polymerase moves along the DNA template. The length of the RNA–DNA hybrid is determined by a structure within the enzyme that forces the RNA–DNA hybrid to separate, allowing the RNA chain to exit from the enzyme and the DNA chain to rejoin its DNA partner (Figure 29.6).

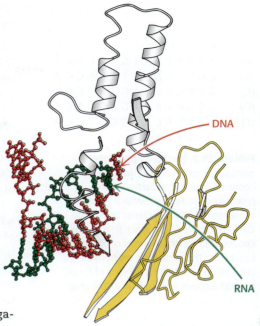

FIGURE 29.6 **A structure within RNA polymerase forces the separation of the RNA–DNA hybrid.** Notice that the DNA strand exits in one direction and the RNA product exits in another. [Drawn from 1I6H.pdb.]

INTERACT with this model in
📖 Achie⤵e

RNA polymerases backtrack and correct errors

The RNA–DNA hybrid can also move in the direction opposite that of elongation (Figure 29.7). This backtracking is energetically less favorable than moving forward because it breaks the bonds between a base pair. However, backtracking is very important for *proofreading*. The incorporation of an incorrect nucleotide introduces a non-Watson–Crick base pair. In this case, breaking the bonds between this base pair and backtracking is energetically less costly.

FIGURE 29.7 **The RNA–DNA hybrid can occasionally backtrack within the RNA polymerase.** In the backtracked position, hydrolysis can take place, producing a configuration equivalent to that after translocation. Backtracking is more likely if a mismatched base is added, facilitating proofreading.

After the polymerase has backtracked, the phosphodiester bond one base pair before the one that has just formed is adjacent to the metal ion in the active site. In this position, a hydrolysis reaction in which a water molecule attacks the phosphate can result in the cleavage of the phosphodiester bond and the release of a dinucleotide that includes the incorrect nucleotide.

Studies of single molecules of RNA polymerase have confirmed that the enzymes pause and backtrack to correct errors. Furthermore, these proofreading activities are often enhanced by accessory proteins.

The final error rate of the order of one mistake per 10^4 or 10^5 nucleotides is higher than that for DNA replication, including all error-correcting mechanisms. The lower fidelity of RNA synthesis can be tolerated because mistakes are not transmitted to progeny. For most genes, many RNA transcripts are synthesized; a few defective transcripts are unlikely to be harmful.

❖ **SELF–CHECK QUESTION**

List some ways in which RNA polymerases are similar to DNA polymerases, and how they are different.

Transcription
starts here
↓

	5′ −10 3′
(A)	G A T A T A A T G A G C A A A
(B)	C G T A T A A T T T G A C C A
(C)	G T T A A C T A G T A C G C A
(D)	G T G A T A C T G A G C A C A
(E)	G T T T T C A T G C C T C C A
	T A T A A T

FIGURE 29.8 Bacterial promoters contain conserved sequences. A comparison of five sequences from prokaryotic promoters reveals a recurring sequence of TATAAT centered on position −10. The −10 consensus sequence (in red) was deduced from a large number of promoter sequences. The sequences are from the (A) uvrD, (B) uncl, and (C) trp operons of E. coli, from (D) λ phage, and from (E) φX174 phage.

RNA polymerase binds to promoter sites on the DNA template in bacteria to initiate transcription

While the elongation process is common to all organisms, the processes of initiation and termination differ substantially in bacteria and eukaryotes. We begin with a discussion of these processes in bacteria, starting with initiation of transcription.

The bacterial RNA polymerase discussed earlier with the composition $\alpha_2\beta\beta'\omega$ is referred to as the *core enzyme*. The inclusion of an additional subunit (σ) produces the *holoenzyme* with composition $\alpha_2\beta\beta'\omega\sigma$. The σ **subunit** helps find the promoters, sites on DNA where transcription begins. At these sites, the σ subunit participates in the initiation of RNA synthesis and then dissociates from the rest of the enzyme.

Sequences upstream of the promoter site are important in determining where transcription begins. A striking pattern is evident when the sequences of bacterial promoters are compared: Two common motifs are present on the upstream side of the transcription start site. They are known as the −10 *sequence* and the −35 *sequence* because they are centered at about 10 and 35 nucleotides upstream of the start site. The region containing these sequences is called the *core promoter*. The −10 and −35 sequences are each 6 bp long. Their **consensus sequences**, deduced from analyses of many promoters (**Figure 29.8**), are:

−35   −10   +1
5′〰〰TTGACA〰〰〰〰〰〰〰TATAAT〰〰〰 Start
site

Promoters differ markedly in their efficacy. Some genes are transcribed frequently—as often as every 2 seconds in E. coli. The promoters for these genes are referred to as *strong promoters*. In contrast, other genes are transcribed much less frequently, about once in 10 minutes; the promoters for these genes are *weak promoters*. The −10 and −35 regions of most strong promoters have sequences that correspond closely to the consensus sequences, whereas weak promoters tend to have multiple substitutions at these sites. Indeed, mutation of a single base in either the −10 sequence or the −35 sequence can diminish promoter activity.

The distance between these conserved sequences also is important; a separation of 17 nucleotides is optimal. Thus, the efficiency or strength of a promoter sequence serves to regulate transcription. Regulatory proteins that bind to specific sequences near promoter sites and interact with RNA polymerase (Chapter 31) also markedly influence the frequency of transcription of many genes.

Outside the core promoter in a subset of highly expressed genes is the _upstream element_ (also called the UP element). This sequence is present from 40 to 60 nucleotides upstream of the transcription start site. The UP element is bound by the α subunit of RNA polymerase and serves to increase the efficiency of transcription by creating an additional interaction site for the polymerase.

Sigma subunits of RNA polymerase in bacteria recognize promoter sites

To initiate transcription, the $\alpha_2\beta\beta'\omega$ core of RNA polymerase must bind the promoter. However, it is the σ subunit that makes this binding possible by enabling RNA polymerase to recognize promoter sites. In the presence of the σ subunit, the RNA polymerase binds weakly to the DNA and slides along the double helix until it dissociates or encounters a promoter. The σ subunit recognizes the promoter through several interactions with the nucleotide bases of the promoter DNA. The structure of a bacterial RNA polymerase holoenzyme bound to a promoter site shows the σ subunit interacting with DNA at the −10 and −35 regions essential to promoter recognition (**Figure 29.9**). Therefore, the σ subunit is responsible for the specific binding of the RNA polymerase to a promoter site on the template DNA.

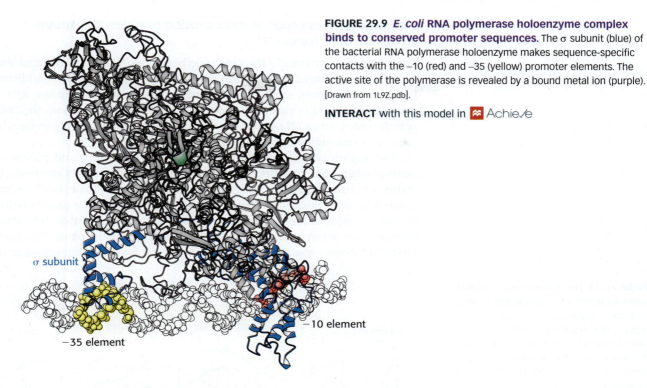

FIGURE 29.9 *E. coli* **RNA polymerase holoenzyme complex binds to conserved promoter sequences.** The σ subunit (blue) of the bacterial RNA polymerase holoenzyme makes sequence-specific contacts with the −10 (red) and −35 (yellow) promoter elements. The active site of the polymerase is revealed by a bound metal ion (purple). [Drawn from 1L9Z.pdb].

INTERACT with this model in ⮝ Achie√e

The σ subunit is generally released when the nascent RNA chain reaches 9 or 10 nucleotides in length. After its release, it can associate with another core enzyme and assist in a new round of initiation.

The template double helix must be unwound for transcription to take place

Although the bacterial RNA polymerase can search for promoter sites when bound to double-helical DNA, a segment of the DNA double helix must be unwound before synthesis can begin. The transition from the closed promoter complex (in which DNA is double helical) to the open promoter complex (in which a DNA segment is unwound) is an essential event in both bacterial and eukaryotic transcription. In bacteria it is the RNA polymerase itself that accomplishes this (**Figure 29.10**), while in eukaryotes additional proteins are required to unwind the DNA template.

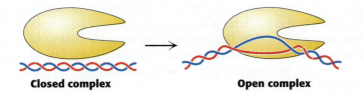

Closed complex   **Open complex**

FIGURE 29.10 *E. coli* **RNA polymerase binds to a promoter sequence.** The transition from the closed promoter complex to the open promoter complex requires the unwinding of approximately 17 base pairs of DNA.

We know that, in bacteria, the free energy necessary to break the bonds between approximately 17 base pairs in the double helix is derived from additional interactions between the template and the bacterial RNA polymerase. These interactions become possible when the DNA distorts to wrap around the RNA polymerase; they also occur between the single-stranded DNA regions and other parts of the enzyme. These interactions stabilize the open promoter complex and help pull the template strand into the active site. The −35 element remains in a double-helical state, whereas the −10 element is unwound. The stage is now set for the formation of the first phosphodiester bond of the new RNA chain.

Elongation takes place at transcription bubbles that move along the DNA template

The elongation phase of RNA synthesis begins with the formation of the first phosphodiester bond, after which repeated cycles of nucleotide addition can take place. However, until about 10 nucleotides have been added, RNA polymerase sometimes releases the short RNA, which dissociates from the DNA and gets degraded. Once RNA polymerase passes this point, the enzyme stays bound to its template until a termination signal is reached.

The region containing the unwound DNA template and nascent RNA corresponds to the transcription bubble (**Figure 29.11**). The newly synthesized RNA forms a hybrid helix with the template DNA strand. This RNA–DNA helix is about 8 bp long, which corresponds to nearly one turn of a double helix. The 3′-hydroxyl group of the RNA in this hybrid helix is positioned so that it can attack the α-phosphorus atom of an incoming ribonucleoside triphosphate. The core bacterial RNA polymerase also contains a binding site for the coding strand of DNA.

FIGURE 29.11 The transcription bubble moves with the RNA polymerase during elongation. In this schematic representation of a transcription bubble in the elongation of an RNA transcript, duplex DNA is unwound at the forward end of RNA polymerase and rewound at its rear end. The RNA–DNA hybrid rotates during elongation.

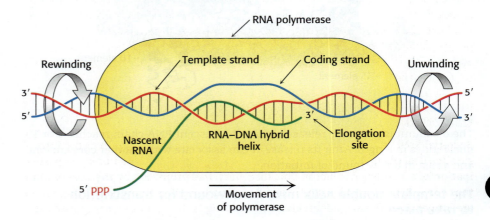

As in the initiation phase, about 17 bp of DNA are unwound throughout the elongation phase. The transcription bubble moves a distance of 170 Å (17 nm) in a second, which corresponds to a rate of elongation of about 50 nucleotides per second. Although rapid, it is much slower than the rate of DNA synthesis, which is 800 nucleotides per second.

Sequences within the newly transcribed RNA signal termination

How does the RNA polymerase know where to stop transcription? In the termination phase of transcription, the formation of phosphodiester bonds ceases, the RNA–DNA hybrid dissociates, the unwound region of DNA rewinds, and RNA polymerase releases the DNA. This process is as precisely controlled as initiation. So what determines where transcription is terminated? In both eukaryotes and bacteria, the transcribed regions of DNA templates contain so-called *intrinsic* termination signals.

In bacteria, the simplest intrinsic termination signal is a palindromic GC-rich region followed by an AT-rich region. The RNA transcript of this DNA palindrome is self-complementary (**Figure 29.12**). Hence, its bases can pair to form a hairpin structure with a stem and loop, a structure favored by its high content of G and C residues. Guanine–cytosine base pairs are more stable than adenine–thymine pairs, primarily because of the preferred base-stacking interactions in G–C base pairs (Section 1.3). This stable hairpin is followed by a sequence of four or more uracil residues, which also are crucial for termination. The RNA transcript ends within or just after them.

How does this combination hairpin–oligo(U) structure terminate transcription? First, RNA polymerase likely pauses immediately after it has synthesized

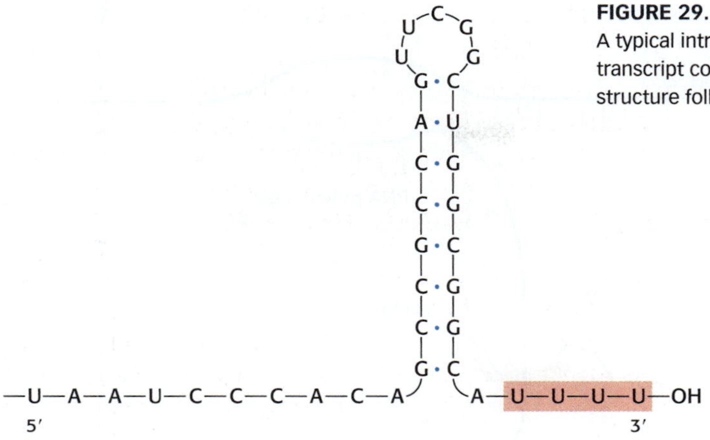

FIGURE 29.12 Some genes contain intrinsic termination signals. A typical intrinsic termination signal found at the 3′ end of an mRNA transcript consists of a series of bases that form a stable stem-loop structure followed by a string of U residues.

a stretch of RNA that folds into a hairpin. Furthermore, the RNA–DNA hybrid helix produced after the hairpin is unstable because its rU–dA base pairs are the weakest of the four kinds. Hence, the pause in transcription caused by the hairpin permits the weakly bound nascent RNA to dissociate from the DNA template and then from the enzyme. The solitary DNA template strand rejoins its partner to re-form the DNA duplex, and the transcription bubble closes.

In bacteria, the rho protein helps to terminate the transcription of some genes

Bacterial RNA polymerase needs no help to terminate transcription at the intrinsic sites described above. At other sites, however, termination requires the participation of an additional factor. This discovery was prompted by the observation that some RNA molecules synthesized in vitro by RNA polymerase acting alone are longer than those made in vivo. The missing factor, a protein that caused the correct termination, was isolated and named **rho (ρ)**.

Additional information about the action of ρ was obtained by adding this termination factor to an incubation mixture at various times after the initiation of RNA synthesis (**Figure 29.13**). RNAs with sedimentation coefficients of 10S, 13S, and 17S were obtained when ρ was added at initiation, a few seconds after initiation, and 2 minutes after initiation, respectively. If no ρ was added, transcription yielded a 23S RNA product. It is evident that the template contains at least three termination sites that respond to ρ (yielding 10S, 13S, and 17S RNA) and one termination site that does not (yielding 23S RNA). Thus, specific termination at a site producing 23S RNA can take place in the absence of ρ. However, ρ detects additional termination signals that are not recognized by RNA polymerase alone.

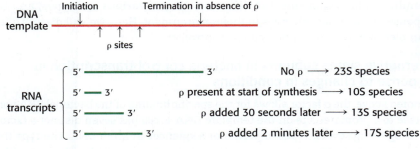

FIGURE 29.13 ρ protein promotes the termination of transcription. The DNA template shown here contains three ρ-dependent termination sites and one site that does not require the action of the ρ protein.

The ρ protein promotes about 20% of termination events in bacteria, but exactly how it selects its target termination signals is not clear. Unlike the hairpin–oligo(U) sequence of intrinsic termination sites, the identification of conserved patterns in ρ-dependent terminators has proven more difficult.

FIGURE 29.14 ρ protein is an ATP-dependent helicase. ρ promotes the termination of transcription by binding the nascent RNA chain and pulling it away from the RNA polymerase and the DNA template. The small arrows represent the movement of ρ protein along the transcript.

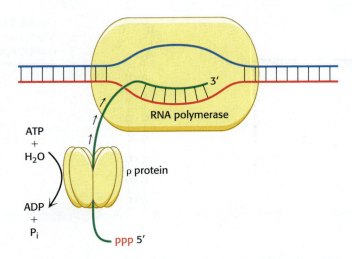

How does ρ promote the termination of RNA synthesis? A key clue is the finding that ρ is hexameric and hydrolyzes ATP in the presence of single-stranded RNA but not in the presence of DNA or duplex RNA. Thus ρ is a helicase, homologous to the hexameric helicases that we encountered in our discussion of DNA replication (Section 28.1). The role of ρ in the termination of transcription in bacteria is as follows (**Figure 29.14**):

- The ρ protein is brought into action by sequences located in the nascent RNA that are rich in cytosine and poor in guanine.
- A stretch of nucleotides is bound in such a way that the RNA passes through the center of the structure.
- The helicase activity of ρ enables the protein to pull the nascent RNA while pursuing RNA polymerase.
- When ρ catches RNA polymerase at the transcription bubble, it breaks the RNA–DNA hybrid by functioning as an RNA–DNA helicase.

Proteins in addition to ρ may promote termination. For example, the NusA protein enables RNA polymerase in E. coli to recognize a characteristic class of termination sites. A common feature of transcription termination, whether it relies on a protein or not, is that the functioning signals lie in newly synthesized RNA rather than in the DNA template.

29.3 Transcription Is Highly Regulated

As we will see in Chapter 31, the level at which different genes are transcribed is highly regulated. Regulated gene expression is critical for the development of multicellular organisms, the differentiation of various cell types, and the response of bacteria to changes in their environment. Here, we will discuss a few examples of how transcription can be controlled.

Alternative sigma subunits in bacteria control transcription in response to changes in conditions

As noted above, the σ factor allows for the specific binding of the bacterial RNA polymerase to a promoter site on the template DNA. E. coli has seven distinct σ factors for recognizing several types of promoter sequences in E. coli DNA. The type that recognizes the consensus sequences described earlier is called σ^{70} because it has a mass of 70 kDa. A different σ factor comes into play when the temperature is raised abruptly. E. coli responds by synthesizing σ^{32}, which recognizes the promoters of so-called *heat-shock genes*. These promoters exhibit –10 sequences that are somewhat different from the –10 sequence for standard promoters (**Figure 29.15**). The increased transcription of heat-shock genes leads to the coordinated synthesis

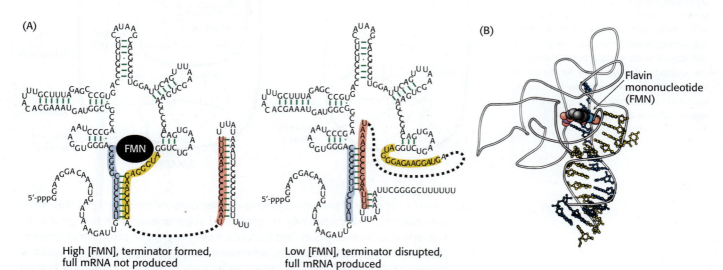

FIGURE 29.15 **Alternative promoters vary in their consensus sequences at conserved sites.** A comparison is shown of the consensus sequences of standard, heat-shock, and nitrogen-starvation promoters of *E. coli*. These promoters are recognized by σ^{70}, σ^{32}, and σ^{54}, respectively.

of a series of protective proteins. Other σ factors respond to environmental conditions, such as nitrogen starvation. These findings demonstrate that σ plays the key role in determining when and where RNA polymerase initiates transcription.

Some other bacteria contain a much larger number of σ factors. For example, the genome of the soil bacterium *Streptomyces coelicolor* encodes more than 60 σ factors recognized on the basis of their amino acid sequences. This repertoire allows these cells to adjust their gene-expression programs to the wide range of conditions, with regard to nutrients and competing organisms, that they may experience. Some messenger RNAs directly sense metabolite concentrations.

Some messenger RNAs directly sense metabolite concentrations

As we shall explore in Chapter 31, the expression of many genes is controlled in response to the concentrations of metabolites and signaling molecules within cells. One set of control mechanisms found in both prokaryotes and eukaryotes depends on the remarkable ability of some mRNA molecules to form secondary structures that are capable of directly binding small molecules. These structures are termed **riboswitches**.

Consider a riboswitch that controls the synthesis of genes that participate in the biosynthesis of riboflavin in the bacterium *Bacillus subtilis* (**Figure 29.16**). When flavin mononucleotide (FMN), a key intermediate in riboflavin biosynthesis, is present at high concentration, it binds to the RNA transcript. Binding of FMN to the transcript induces a hairpin structure that favors premature termination. By trapping the RNA transcript in this termination-favoring conformation,

High [FMN], terminator formed, full mRNA not produced

Low [FMN], terminator disrupted, full mRNA produced

Flavin mononucleotide (FMN)

FIGURE 29.16 **Riboswitches use metabolite concentrations to regulate transcription.** (A) The 5′ end of an mRNA that encodes proteins engaged in the production of flavin mononucleotide (FMN) folds to form a structure that is stabilized by binding FMN. This structure includes a terminator that leads to premature termination of the mRNA. At lower concentrations of FMN, an alternative structure that lacks the terminator is formed, leading to the production of full-length mRNA. (B) In the three-dimensional structure of a related FMN-binding riboswitch bound to FMN, the blue and yellow stretches correspond to regions highlighted in the same colors in part A. The yellow strand contacts the bound FMN, stabilizing the structure. [Drawn from 3F2Q.pdb].

INTERACT with this model in
≋ Achie√e

FMN prevents the production of functional full-length mRNA. However, when FMN is present at low concentration, it does not readily bind to the mRNA. Without FMN bound, the transcript adopts an alternative conformation without the terminator hairpin, allowing the production of the full-length mRNA. The occurrence of riboswitches serves as a vivid illustration of how RNAs are capable of forming elaborate, functional structures, though in the absence of specific information we tend to depict them as simple lines.

Control of transcription in eukaryotes is highly complex

We turn now to transcription in eukaryotes, a much more complex process than in bacteria. Eukaryotic cells have a remarkable ability to regulate precisely the time at which each gene is transcribed and how much RNA is produced. This ability led to the evolution of multicellular eukaryotes with distinct tissues. That is, multicellular eukaryotes use differential transcriptional regulation to create different cell types.

Gene expression is influenced by three important characteristics unique to eukaryotes: the nuclear membrane, complex transcriptional regulation, and RNA processing.

1. *The nuclear membrane allows transcription and translation to take place in differ-ent cellular compartments.* Transcription takes place in the membrane-bound nucleus, whereas translation takes place outside the nucleus in the cyto-plasm. In bacteria, the two processes are closely coupled (**Figure 29.17**). Indeed, the translation of bacterial mRNA begins while the transcript is still being synthesized. The spatial and temporal separation of transcription and translation enables eukaryotes to regulate gene expression in much more intricate ways, contributing to the richness of eukaryotic form and function.

FIGURE 29.17 Transcription and translation are closely coupled in prokaryotes, whereas they are spatially and temporally separate in eukaryotes. (A) In prokaryotes, the primary transcript serves as mRNA and is used immediately as the template for protein synthesis. (B) In eukaryotes, mRNA precursors are processed and spliced in the nucleus before being transported to the cytoplasm for translation into protein. [Information from J. Darnell, H. Lodish, and D. Baltimore. *Molecular Cell Biology*, 2nd ed. (Scientific American Books, 1990), p. 230.]

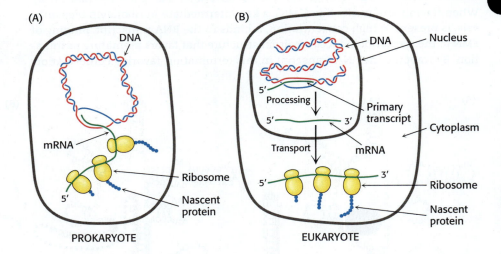

2. *A variety of types of promoter elements enables complex transcriptional regulation.* Like bacteria, eukaryotes rely on conserved sequences in DNA to regulate the initiation of transcription. But bacteria have only three promoter elements (the −10, −35, and UP elements), whereas eukaryotes use a variety of types of promoter elements, each identified by its own conserved sequence. Not all possible types will be present together in the same promoter. In eukaryotes, elements that regulate transcription can be found upstream or downstream of the start site and sometimes at distances much farther from the start site than in prokaryotes. For example, enhancer elements located on DNA far from the start site increase the promoter activity of specific genes.

3. *The degree of RNA processing is much greater in eukaryotes than in bacteria.* Although both bacteria and eukaryotes modify RNA, eukaryotes very extensively process

nascent RNA destined to become mRNA. This processing includes modifications to both ends and, most significantly, splicing out segments of the primary transcript. RNA processing is described in Section 29.4.

Eukaryotic DNA is organized into chromatin

Whereas bacterial genomic DNA is relatively accessible to the proteins involved in transcription, eukaryotic DNA is packaged into **chromatin**, a complex formed between the DNA and a particular set of proteins. Chromatin compacts and organizes eukaryotic DNA, and its presence has dramatic consequences for gene regulation. Although the principles for the construction of chromatin are relatively simple, the chromatin structure for a complete genome is quite complicated. Importantly, in any given eukaryotic cell, some genes and their associated regulatory regions are relatively accessible for transcription and regulation, whereas other genes are tightly packaged, less accessible, and therefore inactive. Eukaryotic gene regulation frequently requires the manipulation of chromatin structure.

Chromatin viewed with the electron microscope has the appearance of beads on a string (**Figure 29.18**). Partial digestion of chromatin with DNase exposes these particles, which consist of fragments of DNA (the "string") wrapped around octamers of proteins called *histones* (the "beads"). The complex formed by a histone octamer and a 145-bp DNA fragment is called the **nucleosome** (**Figure 29.19**).

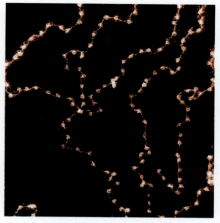

FIGURE 29.18 Eukaryotic chromatin structure resembles beads on a string. In this electron micrograph of chromatin, the "beads" correspond to DNA complexed with specific proteins into nucleosomes. Each bead has a diameter of approximately 100 Å. [Don. W. Fawcett/Science Source.]

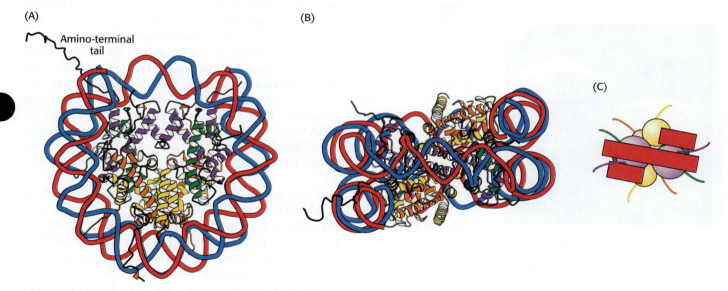

(A) Amino-terminal tail

(B)

(C)

FIGURE 29.19 The structure of the nucleosome consists of eight histone proteins surrounded by DNA. (A) A view showing the DNA wrapping around the histone core. (B) A 90-degree rotation of the view in part A. The DNA forms a left-handed superhelix as it wraps around the core. (C) A schematic view. [Drawn from 1AOI.pdb.]

The overall structure of the nucleosome was revealed through electron microscopic and x-ray crystallographic studies pioneered by Aaron Klug and his colleagues. More recently, the three-dimensional structures of reconstituted nucleosomes have been determined to higher resolution by x-ray diffraction methods. The histone octamer is a complex of four different types of histones (H2A, H2B, H3, and H4) that are homologous and similar in structure.

The eight histones in the core are arranged into a $(H3)_2(H4)_2$ tetramer and a pair of H2A–H2B dimers. The tetramer and dimers come together to form a left-handed superhelical ramp around which the DNA wraps. In addition, each histone has an amino-terminal tail that extends out from the core structure. These tails are flexible and contain many lysine and arginine residues. As we shall see in Chapter 31, covalent modifications of these tails play an essential role in regulating gene expression.

Three types of RNA polymerase synthesize RNA in eukaryotic cells

In bacteria, RNA is synthesized by a single kind of polymerase. In contrast, the nucleus of a typical eukaryotic cell contains three types of RNA polymerase differing in template specificity and location in the nucleus (**Table 29.2**). The three polymerases are named for the order in which they were discovered, which has no bearing on the relative importance of their function. We will discuss them in an order that reflects their similarities in localization, function, and regulation. We will emphasize RNA polymerase II, since it transcribes all of the protein-coding genes and has therefore been the focus of much research investigating transcriptional mechanisms.

TABLE 29.2 Eukaryotic RNA polymerases

Type	Location	Cellular transcripts	Effects of α-amanitin
I	Nucleolus	18S, 5.8S, and 28S rRNA	Insensitive
II	Nucleoplasm	mRNA precursors and snRNA	Strongly inhibited
III	Nucleoplasm	tRNA and 5S rRNA	Inhibited by high concentrations

 RNA polymerase I is located in specialized structures within the nucleus called nucleoli, where it transcribes the tandem array of genes for 18S, 5.8S, and 28S rRNA. The other rRNA molecule (5S rRNA) and all the tRNA molecules are synthesized by *RNA polymerase III*, which is located in the nucleoplasm rather than in nucleoli. *RNA polymerase II*, which also is located in the nucleoplasm, synthesizes the precursors of mRNA as well as several small RNA molecules, such as those of the splicing apparatus and many of the precursors to small regulatory RNAs. All three of the polymerases are large proteins, containing from 8 to 14 subunits and having total molecular masses greater than 500 kDa (or 0.5 MDa), and it is likely that they evolved from a single enzyme that was present in a common ancestor of eukaryotes, bacteria, and archaea. In fact, many components of the eukaryotic transcriptional machinery evolved from those in a common ancestor.

 Although all eukaryotic RNA polymerases are homologous to one another and to prokaryotic RNA polymerases, RNA polymerase II contains a unique **carboxyl-terminal domain** (CTD) on the 220-kDa subunit; this domain is unusual because it contains multiple repeats of a YSPTSPS consensus sequence. The activity of RNA polymerase II is regulated by phosphorylation, mainly on the serine residues of the CTD.

 The different polymerases were originally distinguished through their variable responses to the toxin α-amanitin, a cyclic octapeptide that contains several modified amino acids and is produced by a genus of poisonous mushroom (**Figure 29.20**). α-Amanitin binds very tightly ($K_d = 10$ nM) to RNA polymerase II and thereby blocks the elongation phase of RNA synthesis. Higher concentrations of α-amanitin (1 μM) inhibit RNA polymerase III, whereas RNA polymerase I is insensitive to this toxin. This pattern of sensitivity is highly conserved throughout the animal and plant kingdoms.

 Finally, eukaryotic polymerases differ from each other in the promoters to which they bind. Eukaryotic genes, like prokaryotic genes, require promoters for transcription initiation. Like prokaryotic promoters, eukaryotic promoters consist of conserved sequences that attract the

Tisha Razumovsky/Shutterstock

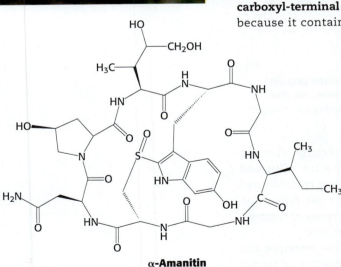

α-**Amanitin**

FIGURE 29.20 α-**Amanitin is produced by poisonous mushrooms in the genus *Amanita*.** Pictured is *Amanita phalloides*, also called the *death cap* or the *destroying angel*.

RNA polymerase I promoter

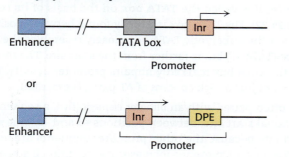

Upstream
promoter
element

Promoter

RNA polymerase II promoter

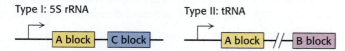

Enhancer

TATA box

Promoter

or

Enhancer

Inr

DPE

Promoter

RNA polymerase III promoter

Type I: 5S rRNA

A block — C block

Type II: tRNA

A block —//— B block

FIGURE 29.21 Eukaryotic promoters share common elements. Each eukaryotic RNA polymerase recognizes a set of promoter elements—sequences in DNA that promote transcription. The RNA polymerase I promoter consists of a ribosomal initiator (rInr) and an upstream promoter element (UPE). The RNA polymerase II promoter likewise includes an initiator element (Inr) and may also include either a TATA box or a downstream promoter element (DPE). Separate from the promoter region, enhancer elements bind specific transcription factors. RNA polymerase III promoters consist of conserved sequences that lie within the transcribed genes.

polymerase to the start site. However, eukaryotic promoters differ distinctly in sequence and position, depending on the type of RNA polymerase that binds to them (**Figure 29.21**).

- *The promoter sequences for RNA polymerase I are located in stretches of DNA separating the ribosomal DNA (rDNA) it transcribes.* These rRNA genes are arranged in several hundred tandem repeats, each containing a copy of each of three rRNA genes. At the transcriptional start site lies a TATA-like sequence called the *ribosomal initiator element* (rInr). Farther upstream, 150 to 200 bp from the start site, is the *upstream promoter element* (UPE). Both elements aid transcription by binding proteins that recruit RNA polymerase I.

- *Promoters for RNA polymerase II, like prokaryotic promoters, include a set of consensus sequences that define the start site and recruit the polymerase.* However, the promoter can contain any combination of a number of possible consensus sequences. Unique to eukaryotes, they also include enhancer elements that can be more than 1 kb from the start site.

- *Promoters for RNA polymerase III are within the transcribed sequence, downstream of the start site.* This is contrast to promoters for RNA polymerase I and II, which are *upstream* of the transcription start site. There are two types of intergenic promoters for RNA polymerase III. Type I promoters, found in the 5S rRNA gene, contain two short, conserved sequences, the A block and the C block. Type II promoters, found in tRNA genes, consist of two 11-bp sequences, the A block and the B block, situated about 15 bp from either end of the gene.

Three common elements can be found in the RNA polymerase II promoter region

RNA polymerase II transcribes all of the protein-coding genes in eukaryotic cells. Promoters for RNA polymerase II, like those for bacterial polymerases, are generally located upstream of the start site for transcription. Because these sequences are on the *same* molecule of DNA as the genes being transcribed, they are called *cis-acting elements*.

1. The most commonly recognized cis-acting element for genes transcribed by RNA polymerase II is called the **TATA box** on the basis of its consensus sequence (**Figure 29.22**). The TATA box is usually found between positions –30 and –100. Note that the eukaryotic TATA box closely resembles the prokaryotic –10 sequence (TATAAT) but is farther from the start site. The mutation of a single base in the TATA box markedly impairs promoter activity. Thus, the precise sequence, not just a high content of AT pairs, is essential.

2. The TATA box is often paired with an *initiator element* (Inr), a sequence found at the transcriptional start site, between positions –3 and +5. This sequence defines the start site because the other promoter elements are at variable distances from that site. Its presence increases transcriptional activity.

3. A third element, the *downstream core promoter element* (DPE), is commonly found in conjunction with the Inr in transcripts that lack the TATA box. In contrast with the TATA box, the DPE is found downstream of the start site, between positions +28 and +32.

$5'$ T_{82} A_{97} T_{93} A_{85} A_{63} A_{88} A_{50} $3'$

TATA box

FIGURE 29.22 The sequence of the TATA box is highly conserved. Comparisons of the sequences of more than 100 eukaryotic promoters led to the consensus sequence shown. The subscripts denote the frequency (%) of the base at that position.

Regulatory cis-acting elements are recognized by different mechanisms

Additional regulatory sequences are located between –40 and –150. Many promoters contain a *CAAT box*, and some contain a *GC box* (**Figure 29.23**). Constitutive genes (genes that are continuously expressed rather than regulated) tend to have GC boxes in their promoters. The positions of these upstream sequences vary from one promoter to another, in contrast with the quite constant location of the –35 region in prokaryotes. Another difference is that the CAAT box and the GC box can be effective when present on the template (antisense) strand, unlike the –35 region, which must be present on the coding (sense) strand.

These differences between prokaryotes and eukaryotes correspond to fundamentally different mechanisms for the recognition of cis-acting elements. The –10 and –35 sequences in prokaryotic promoters are binding sites for RNA polymerase and its associated σ factor. In contrast, the TATA, CAAT, and GC boxes and other cis-acting elements in eukaryotic promoters are recognized by proteins other than RNA polymerase itself.

$5'$ G G N C A A T C T $3'$

CAAT box

$5'$ G G G C G G $3'$

GC box

FIGURE 29.23 The sequences of the CAAT box and GC box are also highly conserved. Consensus sequences are shown for the CAAT and GC boxes of eukaryotic promoters for mRNA precursors.

The TFIID protein complex initiates the assembly of the active transcription complex in eukaryotes

Cis-acting elements constitute only part of the puzzle of eukaryotic gene expression. **Transcription factors** that bind to these elements also are required. For example, RNA polymerase II is guided to the start site by a set of transcription factors known collectively as *TFII* (TF stands for transcription factor, and *II* refers to RNA polymerase II). Individual TFII factors are called TFIIA, TFIIB, and so on.

In TATA-box promoters, the key initial event is the recognition of the TATA box by the TATA-box-binding protein (TBP), a 30-kDa component of the 700-kDa TFIID complex. In TATA-less promoters, other proteins in the TFIID complex bind the core promoter elements; however, because less is known about these interactions, we will consider only the TATA-box–TBP binding interaction. TBP binds 10^5 times as tightly to the TATA box as to nonconsensus sequences; the dissociation constant of the TBP–TATA-box complex is approximately 1 nM.

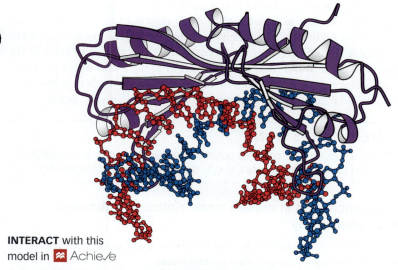

INTERACT with this
model in ✦ Achie√e

FIGURE 29.24 The TATA-box-binding protein binds to DNA. TBP is a saddle-shaped protein consisting of two similar domains. The saddle-like structure of the protein sits atop a DNA fragment, which is significantly unwound and bent. [Drawn from 1CDW.pdb.]

The TATA box of DNA binds to the concave surface of TBP, inducing large conformational changes in the bound DNA (**Figure 29.24**). The double helix is substantially unwound to widen its minor groove, enabling it to make extensive contact with the antiparallel β strands on the concave side of TBP. Hydrophobic interactions are prominent at this interface. Four phenylalanine residues, for example, are intercalated between base pairs of the TATA box. The flexibility of AT-rich sequences is generally exploited here in bending the DNA. Immediately outside the TATA box, classical B-DNA resumes. The TBP–TATA-box complex is distinctly asymmetric, a property that is crucial for specifying a unique start site and ensuring that transcription proceeds unidirectionally.

TBP bound to the TATA box is the heart of the initiation complex (**Figure 29.25**). The surface of the TBP saddle provides docking sites for the binding of other components, with additional transcription factors assembling on this nucleus in a defined sequence. TFIIA is recruited, followed by TFIIB; then TFIIF, RNA polymerase II, TFIIE, and TFIIH join the other factors to form a complex called the *pre-initiation complex* (PIC).

These additional transcription factors play specific roles in this complex. As we saw above, TFIID recognizes core promoter elements and is central to the assembly process. While TFIIA is not essential for the assembly or function of the PIC in vitro, it may aid in the binding of TFIID to the DNA. TFIIB is a DNA-binding protein that recognizes specific cis-acting promoter elements called *B recognition elements*, which are often found near the TATA box. TFIIF aids in the recruitment of polymerase II, while TFIIE brings TFIIH to the complex. TFIIH is a multisubunit complex with helicase and protein kinase activities, both of which are critical in the initiation of transcription. The helicase activity unwinds the DNA template, and the kinase activity phosphorylates specific amino acids in the CTD of polymerase II.

During the formation of the PIC, the carboxyl-terminal domain (CTD) is unphosphorylated and plays a role in transcription regulation through its binding to an enhancer-associated complex called *mediator* (Section 31.4). Phosphorylation of the CTD by TFIIH marks the transition from initiation to elongation. The phosphorylated CTD stabilizes transcription elongation by RNA polymerase II and recruits RNA-processing enzymes that act during the course of elongation. The importance of the carboxyl-terminal domain is highlighted by the finding that yeast cells containing mutant polymerase II with fewer than 10 repeats in the CTD are not viable.

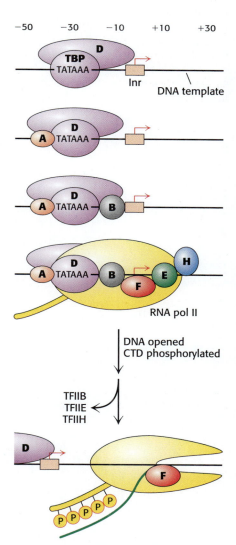

FIGURE 29.25 General transcription factors are essential in initiating transcription by RNA polymerase II. The step-by-step assembly of these general transcription factors begins with the binding of TFIID (purple) to the TATA box. [The TATA-box-binding protein (TBP), a component of TFIID, recognizes the TATA box.] After assembly, TFIIH opens the DNA double helix and phosphorylates the carboxyl-terminal domain (CTD), allowing the polymerase to leave the promoter and begin transcription. The red arrow marks the transcription start site.

The PIC described above initiates transcription at a low (basal) frequency, and the transcription factors associated with it are referred to as *basal* or *general* transcription factors. Additional transcription factors that bind to other sites are required to achieve a high rate of mRNA synthesis. Their role is to selectively stimulate *specific genes*. In summary, transcription factors and other proteins that bind to regulatory sites on DNA can be regarded as passwords that cooperatively open multiple locks, giving RNA polymerase access to specific genes.

❖ SELF-CHECK QUESTION

The function of the σ subunit of *E. coli* RNA polymerase is analogous to the function of general transcription factors in eukaryotes. Briefly describe their common function.

Enhancer sequences can stimulate transcription at start sites thousands of bases away

The activities of many promoters in higher eukaryotes are greatly increased by another type of cis-acting element called an **enhancer**.

- Enhancer sequences have no promoter activity of their own yet can exert their stimulatory actions over distances of several thousand base pairs.
- Enhancers can be upstream, downstream, or even in the middle of a transcribed gene.
- Enhancers are effective when present on either the coding or noncoding DNA strand.
- A particular enhancer is effective only in certain cells; for example, the immunoglobulin enhancer functions in B lymphocytes but not elsewhere.

Cancer can result if the relation between genes and enhancers is disrupted. In Burkitt lymphoma and B-cell leukemia, a chromosomal translocation brings the proto-oncogene *myc* (a transcription factor itself) under the control of a powerful immunoglobulin enhancer. The consequent dysregulation of the *myc* gene is hypothesized to play a role in the progression of the cancer.

The discovery of promoters and enhancers has allowed us to gain a better understanding of how genes are selectively expressed in eukaryotic cells. The regulation of eukaryotic gene transcription, discussed in Chapter 31, is the fundamental means of controlling gene expression.

29.4 Some RNA Transcription Products Are Processed

Virtually all the initial products of eukaryotic transcription are further processed, and even some prokaryotic transcripts are modified. As we will see next, the particular processing steps and the factors taking part vary according to the type of RNA precursor and the type of RNA polymerase that produced it.

Precursors of transfer and ribosomal RNA are cleaved and chemically modified after transcription

In bacteria, messenger RNA molecules undergo little or no modification after synthesis by RNA polymerase. Indeed, many mRNA molecules are translated while they are being transcribed. In contrast, transfer RNA (tRNA) and ribosomal RNA (rRNA) molecules are generated by modifications of nascent RNA chains.

- *The transcript can be cleaved at specific sites along its sequence.* For example, in *E. coli*, the three rRNAs and a tRNA are excised from a single primary RNA transcript that also contains spacer regions (**Figure 29.26**). Other transcripts contain arrays of several kinds of tRNA or several copies of the same tRNA. The nucleases that cleave and trim these precursors of rRNA and tRNA are

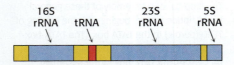

FIGURE 29.26 Cleavage of a primary RNA transcript in *E. coli* produces 5S, 16S, and 23S rRNA molecules and a tRNA molecule. Spacer regions are shown in yellow.

highly precise. *Ribonuclease P* (RNase P), for example, generates the correct 5′ terminus of all tRNA molecules in *E. coli*. Sidney Altman and his coworkers showed that this interesting enzyme contains a catalytically active RNA molecule. *Ribonuclease III* (Rnase III) excises 5S, 16S, and 23S rRNA precursors from the primary transcript by cleaving double-helical hairpin regions at specific sites.

- *Nucleotides can be added to the termini of some RNA chains.* For example, CCA, a terminal sequence required for the function of all tRNAs, is added to the 3′ ends of tRNA molecules for which this terminal sequence is not encoded in the DNA. The enzyme that catalyzes the addition of CCA is atypical for an RNA polymerase in that it does not use a DNA template.

- *Bases and ribose units of RNAs can be modified.* For example, some bases of rRNA are methylated. Furthermore, all tRNA molecules contain unusual bases formed by the enzymatic modification of a standard ribonucleotide in a tRNA precursor. For example, uridylate residues are modified after transcription to form ribothymidylate and pseudouridylate. These modifications generate diversity, allowing greater structural and functional versatility.

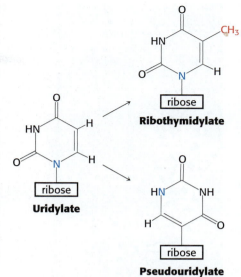

RNA polymerase I produces three ribosomal RNAs

Several RNA molecules are key components of ribosomes. In eukaryotes, RNA polymerase I transcription produces a single precursor (45S in mammals) that encodes three RNA components of the ribosome: the 18S rRNA, the 28S rRNA, and the 5.8S rRNA (**Figure 29.27**).

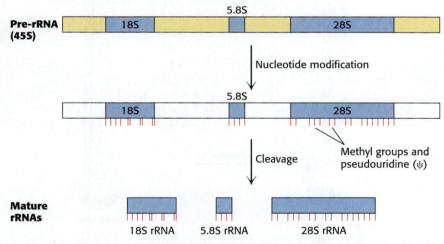

FIGURE 29.27 The processing of eukaryotic pre-rRNA produces mature rRNAs. The mammalian pre-rRNA transcript contains the RNA sequences destined to become the 18S, 5.8S, and 28S rRNAs of the small and large ribosomal subunits. First, nucleotides are modified: small nucleolar ribonucleoproteins methylate specific nucleoside groups and convert selected uridines into pseudouridines (indicated by red lines). Next, the pre-rRNA is cleaved and packaged to form mature ribosomes, in a highly regulated process in which more than 200 proteins take part.

The 18S rRNA is the RNA component of the small ribosomal subunit (40S), and the 28S and 5.8S rRNAs are two RNA components of the large ribosomal subunit (60S). The other RNA component of the large ribosomal subunit, the 5S rRNA, is transcribed by RNA polymerase III as a separate transcript. Processing of the precursor proceeds as follows:

- First, the nucleotides of the pre-rRNA sequences destined for the ribosome undergo extensive modification, on both ribose and base components, directed by many **small nucleolar ribonucleoproteins (snoRNPs)**, each of which consists of one snoRNA and several proteins.

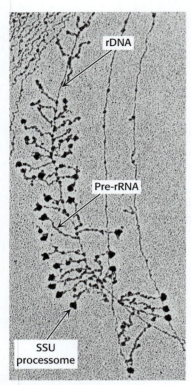

FIGURE 29.28 Electron microscopy gives a window into rRNA transcription and processing in eukaryotes. The assemblies of transcribed rRNA into precursor ribosomes resemble fir trees: the trunk is the rDNA, and each branch is a pre-rRNA transcript. Transcription starts at the top of the tree, where the shortest transcripts can be seen, and progresses down the rDNA to the end of the gene. The terminal knobs visible at the end of some pre-rRNA transcripts likely correspond to the SSU processome, a large ribonucleoprotein required for processing the pre-rRNA. [Reprinted by permission from Macmillan Publishers Ltd, from Dragon, F., Gallagher, J., Compagnone-Post, P. et al. A large nucleolar U3 ribonucleoprotein required for 18S ribosomal RNA biogenesis. Nature 417, 967–970 (2002); permission conveyed through Copyright Clearance Center, ltd.]

- The pre-rRNA is then assembled with ribosomal proteins, as guided by processing factors, to form a large ribonucleoprotein. For instance, the small-subunit (SSU) processome is required for 18S rRNA synthesis and can be visualized in electron micrographs as a terminal knob at the 5′ ends of the nascent rRNAs (**Figure 29.28**).

- Finally, rRNA cleavage (sometimes coupled with additional processing steps) releases the mature rRNAs assembled with ribosomal proteins as ribosomes. Like those of RNA polymerase I transcription itself, most of these processing steps take place in the cell's nucleolus.

RNA polymerase III produces transfer RNAs

Eukaryotic tRNA transcripts are among the most processed of all RNA polymerase III transcripts. Like those of prokaryotic tRNAs, the 5′ leader is cleaved by RNase P, the 3′ trailer is removed, and CCA is added by the CCA-adding enzyme (**Figure 29.29**). Eukaryotic tRNAs are also heavily modified on base and ribose moieties; these modifications are important for function. In contrast with prokaryotic tRNAs, many eukaryotic pre-tRNAs are also spliced by an endonuclease and a ligase to remove an intron.

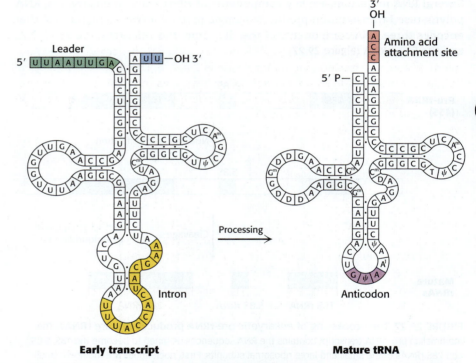

FIGURE 29.29 Transfer RNA precursors are processed. The conversion of a yeast tRNA precursor into a mature tRNA requires the removal of a 14-nucleotide intron (yellow), the cleavage of a 5′ leader (green), and the removal of UU and the attachment of CCA at the 3′ end (red). In addition, several bases are modified.

The product of RNA polymerase II, the pre-mRNA transcript, acquires a 5′ cap and a 3′ poly(A) tail

Perhaps the most extensively studied transcription product is the product of RNA polymerase II: most of this RNA will be processed to mRNA. The immediate product of RNA polymerase II is sometimes referred to as precursor-to-messenger RNA, or **pre-mRNA**. Most pre-mRNA molecules are spliced to remove the introns, which we will discuss in greater detail below. In addition, both the 5′ and the 3′ ends are modified, and both modifications are retained as the pre-mRNA is converted into mRNA.

As in prokaryotes, eukaryotic transcription usually begins with A or G. However, the 5′ triphosphate end of the nascent RNA chain is immediately modified:

- First, a phosphoryl group is released by hydrolysis.
- The diphosphate 5′ end then attacks the α-phosphorus atom of GTP to form a very unusual 5′ – 5′ triphosphate linkage. This distinctive terminus is called a **5′ cap** (Figure 29.30).
- The N-7 nitrogen of the terminal guanine is then methylated by S-adenosylmethionine to form cap 0. The adjacent riboses may be methylated to form cap 1 or cap 2.

Caps contribute to the stability of mRNAs by protecting their 5′ ends from phosphatases and nucleases. In addition, caps enhance the translation of mRNA by eukaryotic protein-synthesizing systems. Transfer RNA and ribosomal RNA molecules, in contrast with messenger RNAs and with small RNAs that participate in splicing, do not have caps.

As mentioned earlier, pre-mRNA is also modified at the 3′ end. Most eukaryotic mRNAs contain a string of adenine nucleotides—a **poly(A) tail**—at that end. This poly(A) tail is added *after* transcription has ended, since the DNA template does not encode this sequence. Indeed, the nucleotide preceding poly(A) is not the last nucleotide to be transcribed. Some primary transcripts contain hundreds of nucleotides beyond the 3′ end of the mature mRNA.

How is the 3′ end of the pre-mRNA given its final form? Eukaryotic primary transcripts are cleaved by a specific endonuclease that recognizes the sequence AAUAAA (Figure 29.31). Cleavage does not take place if this sequence or a segment of some 20 nucleotides on its 3′ side is deleted. The presence of internal AAUAAA sequences in some mature mRNAs indicates that AAUAAA is only part of the cleavage signal; its context also is important. After cleavage of the pre-RNA by the endonuclease, a *poly(A) polymerase* adds about 250 adenylate residues to the 3′ end of the transcript; ATP is the donor in this reaction.

The role of the poly(A) tail is still not firmly established despite much effort. However, evidence is accumulating that it enhances translation efficiency and the stability of mRNA. Blocking the synthesis of the poly(A) tail by exposure to 3′-deoxyadenosine (cordycepin) does not interfere with the synthesis of the primary transcript. Messenger RNA without a poly(A) tail can be transported out of the nucleus. However, an mRNA molecule without a poly(A) tail is usually much less effective as a template for protein synthesis than one with a poly(A) tail. Indeed, some mRNAs are stored in an unadenylated form and receive the poly(A) tail only when translation is imminent. The half-life of an mRNA molecule may be determined in part by the rate of degradation of its poly(A) tail.

Sequences at the ends of introns specify splice sites in mRNA precursors

Most genes in higher eukaryotes are composed of exons and introns (Section 8.7). The introns must be excised and the exons linked to form the final mRNA in a process called **RNA splicing**. This splicing must be exquisitely sensitive; splicing just one nucleotide upstream or downstream of the intended site would create a one-nucleotide shift, which would alter the reading frame on the 3′ side of the splice to give an entirely different amino acid sequence, likely including a premature stop codon. Thus, the correct splice site must be clearly marked.

Does a particular sequence denote the splice site? The sequences of thousands of intron–exon junctions within RNA transcripts are known. In eukaryotes from yeast to mammals, these sequences have a common structural motif: the

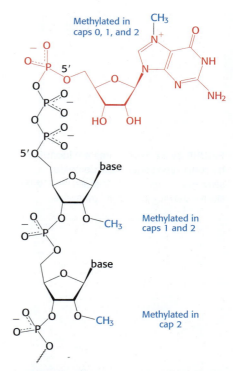

FIGURE 29.30 The 5′ ends of eukaryotic mRNAs are capped. Caps at the 5′ end of eukaryotic mRNA include 7-methylguanylate (red) attached by a triphosphate linkage to the ribose at the 5′ end. None of the riboses are methylated in cap 0, one is methylated in cap 1, and both are methylated in cap 2.

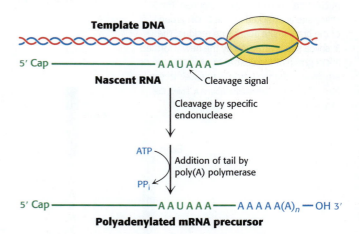

FIGURE 29.31 Primary transcripts in eukaryotes are polyadenylated. A specific endonuclease cleaves the RNA downstream of AAUAAA. Poly(A) polymerase then adds about 250 adenylate residues.

intron begins with GU and ends with AG. The consensus sequence at the 5′ splice in vertebrates is AGGUAAGU, where the GU is invariant (**Figure 29.32**). At the 3′ end of an intron, the consensus sequence is a stretch of 10 pyrimidines (U or C; termed the *polypyrimidine tract*), followed by any base, then by C, and ending with the invariant AG. Introns also have an important internal site located between 20 and 50 nucleotides upstream of the 3′ splice site; it is called the *branch* site for reasons that will be evident shortly. In yeast, the branch-site sequence is nearly always UACUAAC, whereas in mammals a variety of sequences are found.

FIGURE 29.32 Splice sites are identified by conserved sequences. Consensus sequences for the 5′ splice site and the 3′ splice site are shown. Py stands for pyrimidine.

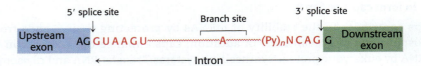

The 5′ and 3′ splice sites and the branch site are essential for determining where splicing takes place. Mutations in each of these three critical regions lead to aberrant splicing. Introns vary in length from 50 to 10,000 nucleotides, and so the splicing machinery may have to find the 3′ site several thousand nucleotides away. Specific sequences near the splice sites (in both the introns and the exons) play an important role in splicing regulation, particularly in designating splice sites when there are many alternatives. Researchers are currently attempting to determine the factors that contribute to splice-site selection for individual mRNAs. Despite our knowledge of splice-site sequences, predicting pre-mRNAs and their protein products from genomic DNA sequence information remains a challenge.

Splicing consists of two sequential transesterification reactions

The splicing of nascent mRNA molecules is a complicated process. It requires the cooperation of several small RNAs and proteins that form a large complex called a **spliceosome**. However, the chemistry of the splicing process is simple. Splicing begins with the cleavage of the phosphodiester bond between the upstream exon (exon 1) and the 5′ end of the intron (**Figure 29.33**). The attacking group in

FIGURE 29.33 The splicing mechanism for mRNA precursors involves a lariat intermediate. The upstream (5′) exon is shown in blue, the downstream (3′) exon in green, and the branch site in yellow. Y stands for a pyrimidine nucleotide, R for a purine nucleotide, and N for any nucleotide. The 5′ splice site is attacked by the 2′-OH group of the branch-site adenosine residue. The 3′ splice site is attacked by the newly formed 3′-OH group of the upstream exon. The exons are joined, and the intron is released in the form of a lariat. [Information from P. A. Sharp, *Cell* 42:397–408, 1985.]

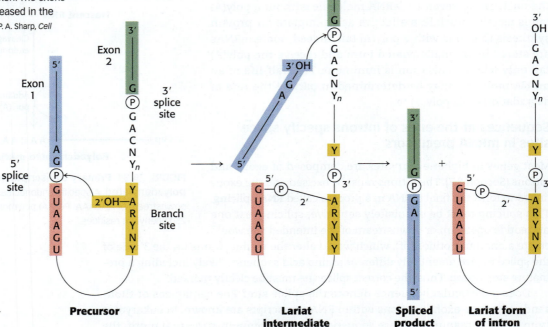

this reaction is the 2′-OH group of an adenylate residue in the branch site. A 2′–5′ phosphodiester bond is formed between this A residue and the 5′ terminal phosphate of the intron in a transesterification reaction.

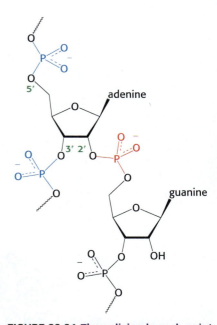

FIGURE 29.34 **The splicing branch point has a unique structure.** The structure of the branch point in the lariat intermediate in which the adenylate residue is joined to three nucleotides by phosphodiester bonds. The new 2′-to-5′ linkage is shown in red, and the usual 3′-to-5′ linkages are shown in blue.

Note that this adenylate residue is also joined to two other nucleotides by normal 3′–5′ phosphodiester bonds (**Figure 29.34**). Hence, a branch is generated at this site, and a lariat (loop) intermediate is formed.

The 3′-OH terminus of exon 1 then attacks the phosphodiester bond between the intron and exon 2. In another transesterification reaction, exons 1 and 2 become joined, and the intron is released in lariat form. Splicing is thus accomplished by two transesterification reactions rather than by hydrolysis followed by ligation.

Both transesterification reactions are promoted by the pair of bound magnesium ions, in reactions reminiscent of those for DNA and RNA polymerases. The first reaction generates a free 3′-OH group at the 3′ end of exon 1, and the second reaction links this group to the 5′-phosphate of exon 2. The number of phosphodiester bonds stays the same during these steps, which is crucial because it allows the splicing reaction itself to proceed without an energy source such as ATP or GTP.

Small nuclear RNAs in spliceosomes catalyze the splicing of mRNA precursors

The nucleus contains many types of small RNA molecules with fewer than 300 nucleotides, referred to as **small nuclear RNAs (snRNAs)**. A few of them—designated U1, U2, U4, U5, and U6—are essential for splicing mRNA precursors. The secondary structures of these RNAs are highly conserved in organisms ranging from yeast to human beings. snRNA molecules are associated with specific proteins to form complexes termed **small nuclear ribonucleoproteins (snRNPs)**; investigators often speak of them as "snurps" (**Table 29.3**). SnRNPs and their role in RNA splicing were discovered by Joan Steitz and Michael Lerner in 1980. One major piece of evidence suggesting a role for snRNPs in splicing was the base complementarity between portions of the U1 snRNA and the splice sites found in the unprocessed mRNAs.

TABLE 29.3 Roles of small nuclear ribonucleoproteins (snRNPs) in the splicing of mRNA precursors

snRNP	Size of snRNA (nucleotides)	Role
U1	165	Binds the 5′ splice site
U2	185	Binds the branch site
U5	116	Binds the 5′ splice site and then the 3′ splice site
U4	145	Masks the catalytic activity of U6
U6	106	Catalyzes splicing

SnRNPs associate with hundreds of other proteins (called *splicing factors*) and the mRNA precursors to form the large (60S) spliceosomes. The large and dynamic nature of the spliceosome made the determination of the detailed three-dimensional structure a great challenge. However, with the maturation

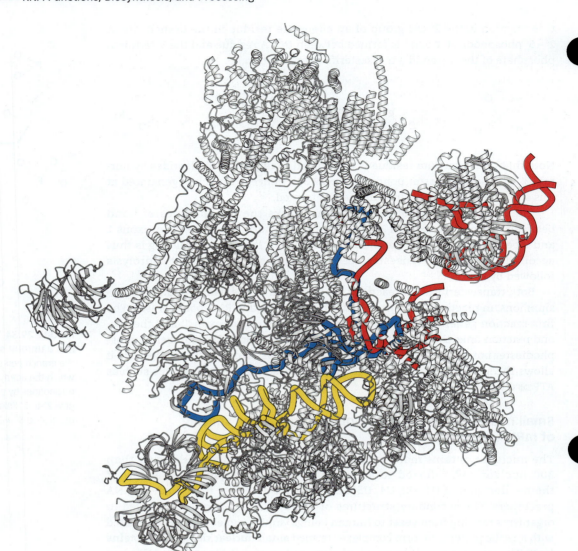

FIGURE 29.35 The spliceosome consists of proteins and RNAs. This three-dimensional structure displays one form of the spliceosome from a yeast determined by high-resolution cryo-electron microscopy. The key RNA components U2, U5, and U6 are shown in red, yellow, and blue, respectively. [Drawn from 3JB9.pdb.]

of cryo-electron microscopy (Section 4.5), the structures of spliceosomes from several species in a number of different stages of their function have been determined (**Figure 29.35**). These structures have added to our understanding of the splicing process (**Figure 29.36**).

1. Splicing begins with the recognition of the 5′ splice site by the U1 snRNP. U1 snRNA contains a highly conserved six-nucleotide sequence, not covered by protein in the snRNP, that base-pairs to the 5′ splice site of the pre-mRNA. This binding initiates spliceosome assembly on the pre-mRNA molecule.

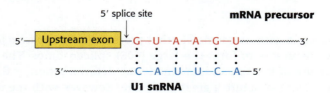

FIGURE 29.36 The spliceosome promotes the splicing of some genes. (1) U1 binds the 5′ splice site and (2) U2 binds to the branch point. (3) A preformed U4-U5-U6 complex then joins the assembly to form the complete spliceosome. (4) The U6 snRNA re-folds and binds the 5′ splice site, displacing U1. Extensive interactions between U6 and U2 displace U4. (5) Then, in the first transesterification step, the branch-site adenosine attacks the 5′ splice site, making a lariat intermediate. (6) U5 holds the two exons in close proximity, and the second transesterification takes place, with the 5′ splice-site hydroxyl group attacking the 3′ splice site. These reactions result in the mature spliced mRNA and a lariat form of the intron bound by U2, U5, and U6. [Information from T. Villa, J. A. Pleiss, and C. Guthrie, *Cell* 109:149–152, 2002.]

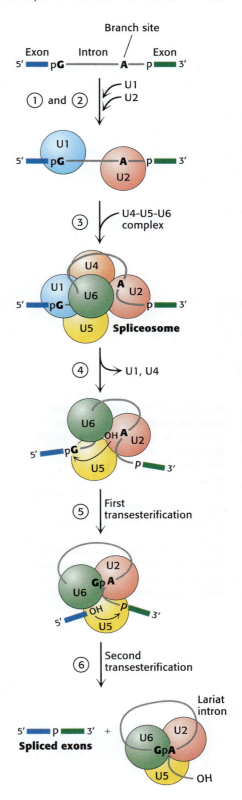

2. U2 snRNP then binds the branch site in the intron by base-pairing between a highly conserved sequence in U2 snRNA and the pre-mRNA. U2 snRNP binding requires ATP hydrolysis.

3. A preassembled U4-U5-U6 tri-snRNP joins this complex of U1, U2, and the mRNA precursor to form the spliceosome. This association also requires ATP hydrolysis. Experiments with a reagent that cross-links neighboring pyrimidines in base-paired regions revealed that in this assembly U5 interacts with exon sequences in the 5′ splice site and subsequently with the 3′ exon.

4. Next, U6 disengages from U4 and undergoes an intramolecular rearrangement that permits base-pairing with U2 as well as interaction with the 5′ end of the intron, displacing U1 and U4 from the spliceosome. U4 serves as an inhibitor that masks U6 until the specific splice sites are aligned. The catalytic center includes two bound magnesium ions bound primarily by phosphate groups from the U6 RNA (**Figure 29.37**).

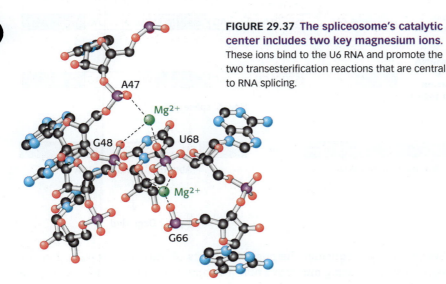

FIGURE 29.37 The spliceosome's catalytic center includes two key magnesium ions. These ions bind to the U6 RNA and promote the two transesterification reactions that are central to RNA splicing.

5. These rearrangements result in the first transesterification reaction, cleaving the 5′ exon and generating the lariat intermediate.

6. Further rearrangements of RNA in the spliceosome facilitate the second transesterification. In these rearrangements, U5 aligns the free 5′ exon with the 3′ exon such that the 3′-hydroxyl group of the 5′ exon is positioned to make a nucleophilic attack on the 3′ splice site to generate the spliced product. U2, U5, and U6 bound to the excised lariat intron are released, completing the splicing reaction.

Many of the steps in the splicing process require ATP hydrolysis. How is the free energy associated with ATP hydrolysis used to power splicing? To achieve the well-ordered rearrangements necessary for splicing, ATP-powered RNA helicases must unwind RNA helices and allow alternative base-pairing arrangements to form. Thus, two features of the splicing process are noteworthy. First, RNA molecules play key roles in directing the alignment of splice sites and in carrying out catalysis. Second, ATP-powered helicases unwind RNA duplex intermediates that facilitate catalysis and induce the release of snRNPs from the mRNA.

✚ Mutations that affect pre-mRNA splicing cause disease

Mutations in either the pre-mRNA (cis-acting) or the splicing factors (trans-acting) can cause defective pre-mRNA splicing that manifests in disease. In fact, mutations affecting splicing have been estimated to cause at least 15% of all genetic diseases. We will look at two examples here.

First, we will consider the possible effects of cis-acting mutations on hemoglobin function. Mutations in the pre-mRNA cause some forms of thalassemia, a group of hereditary anemias characterized by the defective synthesis of hemoglobin (Section 3.3). Cis-acting mutations that cause aberrant splicing can occur at the 5′ or 3′ splice sites in either of the two introns of the hemoglobin β chain or in its exons. Typically, mutations in the 5′ splice site alter that site such that the splicing machinery cannot recognize it, forcing the machinery to find another 5′ splice site in the intron and introducing the potential for a premature stop codon. The defective mRNA is normally degraded rather than translated. Alternatively, mutations in the intron itself may create a new 5′ splice site; in this case, either one of the two splice sites may be recognized (**Figure 29.38**). Consequently, some normal protein can be made, and so the disease is less severe.

FIGURE 29.38 Mutations in splice sites can cause diseases such as thalassemia. An A-to-G mutation within the first intron of the gene for the human hemoglobin β chain creates a new 5′ splice site (GU). Both 5′ splice sites are recognized by the U1 snRNP; thus, splicing may sometimes create a normal mature mRNA and an abnormal mature mRNA that contains intron sequences. The normal mature mRNA is translated into a hemoglobin β chain. Because it includes intron sequences, the abnormal mature mRNA now has a premature stop codon and is degraded.

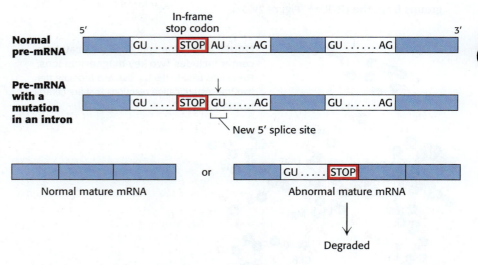

Second, we will consider the possible effects of trans-acting mutations on eyesight. Disease-causing mutations may also appear in splicing factors. Retinitis pigmentosa is a disease of acquired blindness, first described in 1857, with an incidence of 1/3500. About 5% of the autosomal dominant form of retinitis pigmentosa is likely due to mutations in the hPrp8 protein, a pre-mRNA splicing factor that is a component of the U4-U5-U6 tri-snRNP. How a mutation in a splicing factor that is present in all cells causes disease only in the retina is not clear; nevertheless, retinitis pigmentosa is a good example of how mutations that disrupt spliceosome function can cause disease.

Most human pre-mRNAs can be spliced in alternative ways to yield different proteins

As a result of **alternative splicing**, different combinations of exons from the same gene may be spliced into a mature RNA, producing distinct forms of a protein for specific tissues, developmental stages, or signaling pathways. What controls which splicing sites are selected? The selection is determined by the binding of trans-acting splicing factors to cis-acting sequences in the pre-mRNA. Most alternative splicing leads to changes in the coding sequence, resulting in proteins with different functions.

Alternative splicing provides a powerful mechanism for generating protein diversity. It expands the versatility of genomic sequences through combinatorial control. Consider a gene with five positions at which splicing can take place. With the assumption that these alternative splicing pathways can be regulated independently, a total of $2^5 = 32$ different mRNAs can be generated.

Sequencing of the human genome has revealed that most pre-mRNAs are alternatively spliced, leading to a much greater number of proteins than would be predicted from the number of genes. An example of alternative splicing leading to the expression of two different proteins, each in a different tissue, is provided by the gene encoding both calcitonin and calcitonin-gene-related peptide (CGRP; **Figure 29.39**). In the thyroid gland, the inclusion of exon 4 in one splicing pathway produces calcitonin, a peptide hormone that regulates calcium and phosphorus metabolism. In neuronal cells, the exclusion of exon 4 in another splicing pathway produces CGRP, a peptide hormone that acts as a vasodilator. A single pre-mRNA thus yields two very different peptide hormones, depending on cell type.

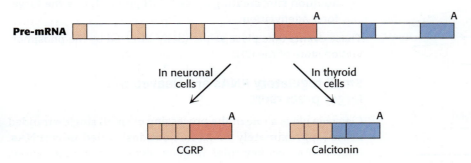

FIGURE 29.39 Alternative splicing produces different mRNAs. In human beings, two very different hormones are produced from a single calcitonin/CGRP pre-mRNA. Alternative splicing produces the mature mRNA for either calcitonin or CGRP (calcitonin-gene-related protein), depending on the cell type in which the gene is expressed. Each alternative transcript incorporates one of two alternative polyadenylation signals (A) present in the pre-mRNA.

In the above example, only two proteins result from alternative splicing; however, in other cases, many more can be produced. An extreme example is the *Drosophila* pre-mRNA that encodes DSCAM, a neuronal protein affecting axon connectivity. Alternative splicing of this pre-mRNA has the potential to produce 38,016 different combinations of exons, a greater number than the total number of genes in the *Drosophila* genome. However, only a fraction of these potential mRNAs appear to be produced, owing to regulatory mechanisms that are not yet well understood.

Several human diseases that can be attributed to defects in alternative splicing are listed in **Table 29.4**. Further understanding of alternative splicing and the mechanisms of splice-site selection will be crucial to understanding how the proteome represented by the human genome is expressed.

TABLE 29.4 Selected human disorders attributed to defects in alternative splicing

Disorder	Gene or its product
Acute intermittent porphyria	Porphobilinogen deaminase
Breast and ovarian cancer	*BRCA1*
Cystic fibrosis	*CFTR*
Frontotemporal dementia	τ protein
Hemophilia A	Factor VIII
HGPRT deficiency (Lesch–Nyhan syndrome)	Hypoxanthine-guanine phosphoribosyltransferase
Leigh encephalomyelopathy	Pyruvate dehydrogenase E1 α
Severe combined immunodeficiency	Adenosine deaminase
Spinal muscle atrophy	*SMN1* or *SMN2*

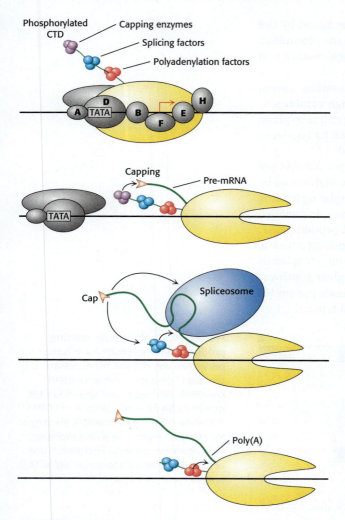

FIGURE 29.40 The CTD couples transcription to pre-mRNA processing. The transcription factor TFIIH phosphorylates the carboxyl-terminal domain (CTD) of RNA polymerase II, signaling the transition from transcription initiation to elongation. The phosphorylated CTD binds factors required for pre-mRNA capping, splicing, and polyadenylation. These proteins are brought in close proximity to their sites of action on the nascent pre-mRNA as it is transcribed during elongation. [Information from P. A. Sharp, *Trends in Biochemical Sciences* 30:279–281, 2005.]

Transcription and mRNA processing are coupled

Although we have described the transcription and processing of mRNAs as separate events in gene expression, experimental evidence suggests that the two steps are coordinated by the carboxyl-terminal domain of RNA polymerase II. We have seen that the CTD consists of a unique repeated seven-amino-acid sequence, YSPTSPS. Either S_2, S_5, or both may be phosphorylated in the various repeats. The phosphorylation state of the CTD is controlled by a number of kinases and phosphatases and leads the CTD to bind many of the proteins having roles in RNA transcription and processing. The CTD contributes to efficient transcription by recruiting certain proteins to the pre-mRNA (**Figure 29.40**). These proteins include:

1. Capping enzymes, which methylate the 5′ guanine on the pre-mRNA immediately after transcription begins

2. Components of the splicing machinery, which initiate the excision of each intron as it is synthesized

3. An endonuclease that cleaves the transcript at the poly(A) addition site, creating a free 3′-OH group that is the target for 3′ adenylation

These events take place sequentially, directed by the phosphorylation state of the CTD.

Small regulatory RNAs are cleaved from larger precursors

Cleavage plays a role in the processing of small single-stranded RNAs (approximately 20–23 nucleotides) called **microRNAs**. MicroRNAs play key roles in gene regulation in eukaryotes, as we shall see in Chapter 31. They are generated from initial transcripts produced by RNA polymerase II and, in some cases, RNA polymerase III. These transcripts fold into hairpin structures that are cleaved by specific nucleases at various stages (**Figure 29.41**). The final single-stranded RNAs are bound by regulatory proteins, where the RNAs help target the regulation of specific genes.

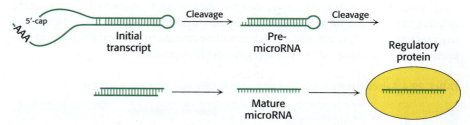

FIGURE 29.41 Small regulatory RNAs are involved in gene regulation. The initial transcription product is first cleaved to a small double-stranded RNA called a pre-microRNA. One of the strands of the pre-microRNA, the mature microRNA, is then bound by a regulatory protein.

RNA editing can lead to specific changes in mRNA

Remarkably, the amino acid sequence information encoded by some mRNAs is altered after transcription. This phenomenon is referred to as **RNA editing**, a posttranscriptional change in the nucleotide sequence of RNA that is caused by processes other than RNA splicing. RNA editing is prominent in some systems; next, we will consider three examples.

RNA editing is key to the process of lipid transport by apolipoprotein B (apo B). Apo B plays an important role in the transport of triacylglycerols and cholesterol by forming an amphipathic spherical shell around the lipids carried in lipoprotein particles (Section 27.3). Apo B exists in two forms, a 512-kDa *apo B-100* and a 240-kDa *apo B-48*. The larger form, synthesized by the liver, participates in the transport of lipids synthesized in the cell. The smaller form, synthesized by the small intestine, carries dietary fat in the form of chylomicrons. Apo B-48 contains the 2152 N-terminal residues of the 4536-residue apo B-100. This truncated molecule can form lipoprotein particles but cannot bind to the low-density-lipoprotein receptor on cell surfaces.

What is the relationship between these two forms of apo B? Experiments revealed that a totally unexpected mechanism for generating diversity is at work: the changing of the nucleotide sequence of mRNA *after* its synthesis (**Figure 29.42**). A specific cytidine residue of mRNA is deaminated to uridine, which changes the codon at residue 2153 from CAA (Gln) to UAA (stop). The deaminase that catalyzes this reaction is present in the small intestine, but not in the liver, and is expressed only at certain developmental stages.

RNA editing also plays a role in the regulation of postsynaptic receptors. Glutamate opens cation-specific channels in the vertebrate central nervous system by binding to receptors in postsynaptic membranes. RNA editing changes a single glutamine codon (CAG) in the mRNA for the glutamate receptor to the codon for arginine (CGG). The substitution of Arg for Gln in the receptor prevents Ca^{2+}, but not Na^+, from flowing through the channel.

In trypanosomes (parasitic protozoans), a different kind of RNA editing markedly changes several mitochondrial mRNAs. Nearly half the uridine residues in these mRNAs are inserted by RNA editing. A guide RNA molecule identifies the sequences to be modified, and a poly(U) tail on the guide RNA donates uridine residues to the mRNAs undergoing editing.

DNA sequences evidently do not always faithfully represent the sequence of encoded proteins; crucial functional changes to mRNA can take place. RNA editing is likely much more common than was formerly thought. The chemical reactivity of nucleotide bases—including the susceptibility to deamination that necessitates complex DNA-repair mechanisms—has been harnessed as an engine for generating molecular diversity at the RNA and, hence, protein levels.

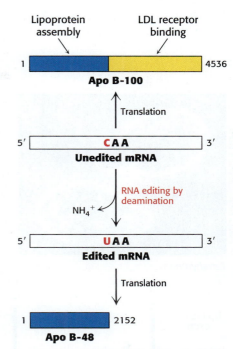

FIGURE 29.42 RNA editing alters the coding sequence of mRNAs. Enzyme-catalyzed deamination of a specific cytidine residue in the mRNA for apolipoprotein B-100 changes a codon for glutamine (CAA) to a stop codon (UAA). Apolipoprotein B-48, a truncated version of the protein lacking the LDL receptor-binding domain, is generated by this posttranscriptional change in the mRNA sequence. [Information from P. Hodges and J. Scott, *Trends in Biochemical Sciences* 17:77, 1992.]

29.5 The Discovery of Catalytic RNA Revealed a Unique Splicing Mechanism

RNAs form a surprisingly versatile class of molecules. As we have seen, splicing is catalyzed largely by RNA molecules, with proteins playing a secondary role. RNA is also a key component of ribonuclease P, which catalyzes the maturation of tRNA by endonucleolytic cleavage of nucleotides from the 5′ end of the precursor molecule. Finally, as we shall see in Chapter 30, the RNA component of ribosomes is the catalyst that carries out protein synthesis.

Some RNAs can promote their own splicing

The versatility of RNA first became clear from observations of the processing of ribosomal RNA in a single-cell eukaryote, a ciliated protozoan in the genus *Tetrahymena*. In *Tetrahymena*, a 414-nucleotide intron is removed from a 6.4-kb precursor to yield the mature 26S rRNA molecule.

In an elegant series of studies of this splicing reaction, Thomas Cech and his coworkers established that, in the absence of protein, the RNA spliced itself to precisely excise the intron. Indeed, the RNA alone is catalytic and, under certain conditions, is thus a **ribozyme**. More than 1500 similar introns have since been found in species as widely dispersed as bacteria and eukaryotes, though not in vertebrates. Collectively, they are referred to as group I **self-splicing introns**.

The self-splicing reaction in the group I intron requires an added guanosine nucleotide (**Figure 29.43**). Nucleotides were originally included in the reaction mixture because it was thought that ATP or GTP might be needed as an energy source. Instead, the nucleotides were found to be necessary as cofactors. The required cofactor proved to be a guanosine unit, in the form of guanosine, GMP, GDP, or GTP. G (denoting any one of these species) serves not as an energy source but as an attacking group that becomes transiently incorporated into the RNA. G binds to the RNA and then attacks the 5′ splice site to form a phosphodiester bond with the 5′ end of the intron. This transesterification reaction generates a 3′-OH group at the end of the upstream exon. This 3′-OH group then attacks the 3′ splice site in a second transesterification reaction that joins the two exons and leads to the release of the 414-nucleotide intron.

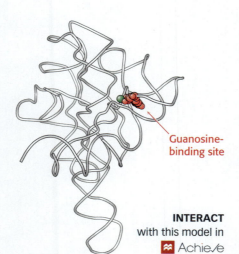

Tetrahymena

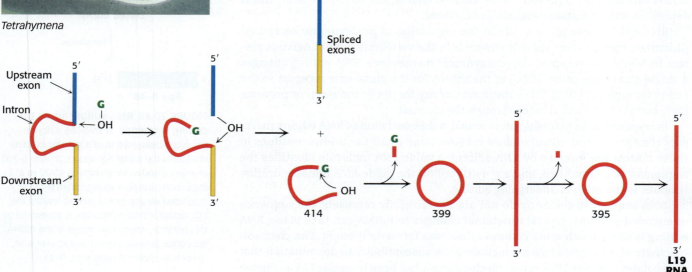

FIGURE 29.43 Some introns are capable of self-splicing. A ribosomal RNA precursor representative of the group I introns, from the protozoan *Tetrahymena*, splices itself in the presence of a guanosine cofactor (G, shown in green). A 414-nucleotide intron (red) is released in the first splicing reaction. This intron then splices itself twice again to produce a linear RNA that has lost a total of 19 nucleotides. This L19 RNA is catalytically active. [Information from T. Cech, RNA as an enzyme. Copyright © 1986 by Scientific American, Inc. All rights reserved.]

Self-splicing depends on the structural integrity of the RNA precursor. Much of the group I intron is needed for self-splicing. This molecule, like many RNAs, has a folded structure formed by many double-helical stems and loops (**Figure 29.44**), with a well-defined pocket for binding the guanosine. Examination of the three-dimensional structure of a catalytically active group I intron determined by x-ray crystallography reveals the coordination of magnesium ions in the active site analogous to that observed in protein enzymes such as DNA polymerase.

Analysis of the base sequence of the rRNA precursor suggested that the splice sites are aligned with the catalytic residues by base-pairing between the *internal guide sequence* (IGS) in the intron and the 5′ and 3′ exons (**Figure 29.45**). The IGS first brings together the guanosine cofactor and the 5′ splice site so that the 3′-OH group of G can make a nucleophilic attack on the phosphorus atom at this splice site. The IGS then holds the downstream exon in position for attack by the newly formed 3′-OH group of the upstream exon. A phosphodiester bond is formed between the two exons, and the intron is released as a linear molecule. Like catalysis by protein enzymes, self-catalysis of bond formation and breakage in this rRNA precursor is highly specific.

INTERACT
with this model in
Achieve

FIGURE 29.44 Self-splicing introns have a complex structure. The structure of a large fragment of the self-splicing intron from *Tetrahymena* reveals a complex folding pattern of helices and loops. The guanosine binding site is highlighted in red. [Drawn from 1GRZ.pdb].

FIGURE 29.45 The catalytic mechanism of the group I self-splicing intron includes a series of transesterification reactions. [Information from T. Cech, RNA as an enzyme. Copyright © 1986 by Scientific American, Inc. All rights reserved.]

The finding of enzymatic activity in the self-splicing intron and in the RNA component of RNase P has opened new areas of inquiry and changed the way in which we think about molecular evolution. As mentioned in an earlier chapter, the discovery that RNA can be a catalyst as well as an information carrier suggests that an RNA world may have existed early in the evolution of life, before the appearance of DNA and protein.

Messenger RNA precursors in the mitochondria of yeast and fungi also undergo self-splicing, as do some RNA precursors in the chloroplasts of unicellular organisms such as *Chlamydomonas*. Self-splicing reactions can be classified according to the nature of the unit that attacks the upstream splice site. Group I self-splicing is mediated by a guanosine cofactor, as in *Tetrahymena*. The attacking moiety in group II splicing is the 2′-OH group of a specific adenylate of the intron (**Figure 29.46**).

Group I and group II self-splicing resembles spliceosome-catalyzed splicing in two respects. First, in the initial step, a ribose hydroxyl group attacks the 5′ splice site. The newly formed 3′-OH terminus of the upstream exon then attacks the 3′ splice site to form a phosphodiester bond with the downstream exon. Second, both reactions are transesterifications in which the phosphate moieties at each splice site are retained in the products. The number of phosphodiester bonds stays constant.

Group II splicing is like the spliceosome-catalyzed splicing of mRNA precursors in several additional ways. First, the attack at the 5′ splice site is carried out by a part of the intron itself (the 2′-OH group of adenosine) rather than by an external cofactor (G). Second, the intron is released in the form of a lariat.

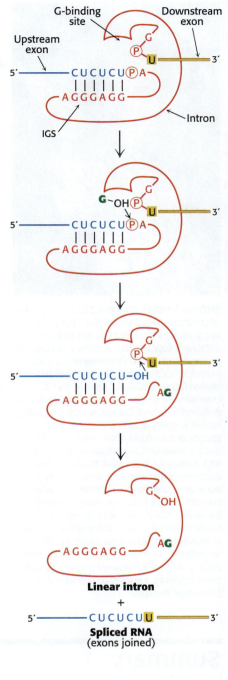

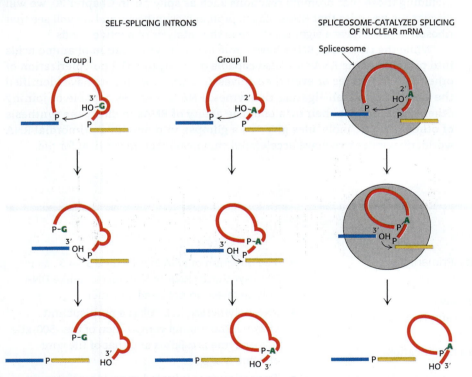

FIGURE 29.46 In self-splicing mechanisms, the catalytic site is formed by the intron itself. The exons being joined are shown in blue and yellow, and the attacking unit is shown in green. The catalytic site is formed by the intron itself (red) in group I and group II splicing. In contrast, the splicing of nuclear mRNA precursors is catalyzed by snRNAs and their associated proteins in the spliceosome. [Information from P. A. Sharp, *Science* 235:766–771, 1987.]

THOMAS CECH Fascinated by the structure of chromosomes, Thomas Cech pursued this topic as a graduate student and post-doctoral fellow. He then accompanied his wife, Dr. Carol Cech, to the University of Colorado in Boulder (UCB), where they had both been offered faculty positions. They had met in chemistry class while they were undergraduate students at Grinnell College. At UCB, Dr. Cech set out to purify enzymes involved in the splicing of a particular RNA molecule, but he and his coworkers soon discovered that the RNA molecule spliced itself. This fundamental discovery of RNA catalysis, which changed our view of both modern biochemistry and our evolutionary past, was recognized with the Nobel Prize in Chemistry in 1989. Dr. Cech has remained at UCB, teaching with infectious enthusiasm throughout his career. He also served as an investigator (and president from 2000 to 2009) of the Howard Hughes Medical Institute (HHMI), a leading private funder of biomedical research.

Third, in some instances, the group II intron is transcribed in pieces that assemble through hydrogen bonding to the catalytic intron, in a manner analogous to the assembly of the snRNAs in the spliceosome.

The similarities in mechanism have led to the suggestion that the spliceosome-catalyzed splicing of mRNA precursors evolved from RNA-catalyzed self-splicing. Group II splicing may well be an intermediate between group I splicing and the splicing in the nuclei of higher eukaryotes. A major step in this transition was the transfer of catalytic power from the intron itself to other molecules. The formation of spliceosomes gave genes a new freedom because introns were no longer constrained to provide the catalytic center for splicing. Another advantage of external catalysts for splicing is that they can be more readily regulated.

However, it is important to note that similarities do not establish ancestry. The similarities between group II introns and mRNA splicing may be a result of convergent evolution. Perhaps there are only a limited number of ways to carry out efficient, specific intron excision. The determination of whether these similarities stem from ancestry or from chemistry will require expanding our understanding of RNA biochemistry.

❖ SELF-CHECK QUESTION

Compare and contrast the three different splicing mechanisms that have been identified.

RNA enzymes can promote many reactions, including RNA polymerization

In this chapter, we have discussed the role of many different RNA molecules, including those that promote reactions such as splicing. In Chapter 30, we will discuss the key roles that RNAs play in protein biosynthesis, and we will see that ribosomal RNAs have a significant role in the catalysis of peptide bonds.

While the ribosomal RNAs have a role in the polymerization of amino acids into proteins, some RNAs are also capable of catalyzing the polymerization of other RNA molecules or even themselves. In vitro experiments have identified the possibility of self-ligating ribozymes: RNA molecules capable of joining other, short RNAs to their own end. The ability of RNAs to direct the synthesis of other catalytic molecules gives us a glimpse into how, in a primordial RNA world, ribozymes may have accelerated chemical reactions critical for life.

Summary

29.1 RNA Molecules Play Different Roles, Primarily in Gene Expression

- RNA molecules are integral to many different cellular functions; for example, protein biosynthesis.
- Some RNAs can direct their own modification.

29.2 RNA Polymerases Catalyze Transcription

- RNA polymerases synthesize all cellular RNA molecules according to instructions given by DNA templates.

- The direction of RNA synthesis is 5′ → 3′, as in DNA synthesis. RNA polymerases, unlike DNA polymerases, do not need a primer.
- RNA polymerase in *E. coli* is a multisubunit enzyme. The subunit composition of the ~500-kDa holoenzyme is $\alpha_2\beta\beta'\omega\sigma$ and that of the core enzyme is $\alpha_2\beta\beta'\omega$.
- Transcription is initiated at promoter sites.
- The σ subunit enables the holoenzyme to recognize promoter sites. The σ subunit usually dissociates from the holoenzyme after the initiation of the new chain.

- Elongation takes place at transcription bubbles that move along the DNA template at a rate of about 50 nucleotides per second.

- The nascent RNA chain contains stop signals that end transcription. One stop signal is an RNA hairpin, which is followed by several U residues. A different stop signal is read by the *rho* protein, an ATPase.

- In *E. coli*, precursors of transfer RNA and ribosomal RNA are cleaved and chemically modified after transcription, whereas messenger RNA is used unchanged as a template for protein synthesis.

29.3 Transcription Is Highly Regulated

- Bacteria use alternate σ factors to adjust the transcription levels of various genes in response to external stimuli.

- Some genes are regulated by riboswitches, structures that form in RNA transcripts and bind specific metabolites.

- Eukaryotic DNA is tightly bound to basic proteins called histones; the combination is called chromatin. DNA wraps around an octamer of core histones to form a nucleosome.

- There are three types of eukaryotic RNA polymerases in the nucleus, where transcription takes place: RNA polymerase I makes ribosomal RNA precursors, II makes messenger RNA precursors, and III makes transfer RNA precursors.

- Eukaryotic promoters are composed of several different elements. The activity of many promoters is greatly increased by enhancer sequences that have no promoter activity of their own.

29.4 Some RNA Transcription Products Are Processed

- The 5′ ends of mRNA precursors become capped and methylated during transcription.

- A 3′ poly(A) tail is added to most mRNA precursors after the nascent chain has been cleaved by an endonuclease.

- The splicing of mRNA precursors is carried out by spliceosomes, which consist of small nuclear ribonucleoproteins. Splice sites in mRNA precursors are specified by sequences at ends of introns and by branch sites near the 3′ ends of introns.

- RNA editing alters the nucleotide sequence of some mRNAs, such as the one for apolipoprotein B.

29.5 The Discovery of Catalytic RNA Revealed a Unique Splicing Mechanism

- Some RNA molecules undergo self-splicing in the absence of protein.

- Spliceosome-catalyzed splicing may have evolved from self-splicing.

- The discovery of catalytic RNA has opened new vistas in our exploration of early stages of molecular evolution and the origins of life.

Key Terms

transcription (p. 881)
RNA polymerase (p. 881)
promoter (p. 882)
transcription bubble (p. 883)
sigma (σ) subunit (p. 886)
consensus sequence (p. 886)
rho (ρ) protein (p. 889)
riboswitch (p. 891)
chromatin (p. 893)
nucleosome (p. 893)

carboxyl-terminal domain (CTD) (p. 894)
TATA box (p. 896)
transcription factor (p. 896)
enhancer (p. 898)
small nucleolar ribonucleoprotein (snoRNP) (p. 899)
pre-mRNA (p. 900)
5′ cap (p. 901)
poly(A) tail (p. 901)

RNA splicing (p. 901)
spliceosome (p. 902)
small nuclear RNA (snRNA) (p. 903)
small nuclear ribonucleoprotein (snRNP) (p. 903)
alternative splicing (p. 906)
microRNA (p. 908)
RNA editing (p. 908)
ribozyme (p. 909)
self-splicing introns (p. 909)

Problems

1. Why is RNA synthesis not as carefully monitored for errors as is DNA synthesis? ❖ 1, ❖ 2

2. What are the functions of RNA polymerases? Select all that apply. ❖ 1, ❖ 3
(a) Polymerization of polypeptides from RNA transcripts

(b) Elongation of ribosomal RNA (rRNA)
(c) Initiation of transcription at promoter sites
(d) Elongation of messenger RNA (mRNA) transcripts
(e) Initiation of translation from RNA transcripts

3. The overall structures of RNA polymerase and DNA polymerase are very different, yet their active sites show considerable similarities. What do the similarities suggest about the evolutionary relationship between these two important enzymes? ❖ 1

4. The sequence of part of an mRNA transcript is

5′-AUGGGGAACAGCAAGAGUGGGGCCCUGUCCAAGGAG-3′

What is the sequence of the DNA coding strand? Of the DNA template strand? ❖ 1

5. Sigma protein by itself does not bind to promoter sites. Predict the effect of a mutation enabling σ to bind to the –10 region in the absence of other subunits of RNA polymerase. ❖ 2

6. The molecular weight of an amino acid is approximately 110 Da, and *E. coli* RNA polymerase has a transcription rate of approximately 5050 nucleotides per second. What is the minimum length of time required by *E.coli* polymerase for the synthesis of an mRNA encoding a 100-kDa protein? Round your answer to the nearest whole number. ❖ 1

7. The autoradiograph below depicts several bacterial genes undergoing transcription. Identify the DNA. What are the strands of increasing length? Where is the beginning of transcription? The end of transcription? What can you conclude about the number of enzymes participating in RNA synthesis on a given gene? ❖ 2

Science Source

8. Fill in the blanks to complete the statements about the capping process: Capping occurs at the **(a)** _____ end of the mRNA. The nucleoside that is added is **(b)** _____. The nucleoside is connected to the mRNA through a **(c)** _____ linkage. The added nucleoside is modified by methylation at **(d)** _____ atom. The mRNA is further modified by methylation of ribose residues at the **(e)** _____ -hydroxyl position. ❖ 3

9. Suppose that RNA polymerase was transcribing a eukaryotic gene with several introns all contained within the coding region. In what order would the RNA polymerase encounter the elements in the DNA sequence of the gene, from earliest to latest? ❖ 2

(a) Splice donor site

(b) 3′ untranslated region (UTR)

(c) Transcription start site

(d) Translation start codon

(e) 5′ untranslated region (UTR)

(f) Translation termination codon

10. For the three genes shown below, transcription occurs from right to left for gene A and gene B, whereas transcription occurs from left to right for gene C. Identify which strand (top or bottom) is the template strand for each gene. ❖ 1

$5'$ ──────────────────── $3'$
| Gene A | Gene B | Gene C |
$3'$ ──────────────────── $5'$

11. Which would you expect to have more types of σ factors: a free–living prokaryote in a fluctuating environment, or a prokaryote in a relatively stable environment with adequate nutrients? Why? ❖ 2

12. The three different eukaryotic RNA polymerases perform different functions and can be distinguished experimentally by their sensitivity to the toxin α–amanitin. Match the appropriate function and α–amanitin sensitivity to each RNA polymerase. Each function and sensitivity will be used only once. ❖ 3

Type of polymerase	Function and sensitivity
(a) RNA polymerase I	1. Inhibited by high concentrations of α-amanitin
(b) RNA polymerase II	2. Not inhibited by α-amanitin
(c) RNA polymerase III	3. Synthesizes pre-ribosomal RNA
	4. Synthesizes primarily tRNA and small RNAs
	5. Synthesizes mostly pre-messenger RNA
	6. Inhibited by low concentrations of α-amanitin

13. Fill in the blanks to complete the statements about ribosomal RNA processing in eukaryotes: RNA polymerase I generates the **(a)** _____ kb rRNA transcript, which has a sedimentation coefficient of **(b)** _____. The RNA transcript is processed in a series of steps to yield three smaller rRNA molecules: **(c)** _____. The **(d)** _____ rRNA assembles with the € _____ rRNA, the **(f)** _____ rRNA, and proteins into **(g)** _____ ribosomal subunits. The 5S rRNA genes are transcribed by **(h)** _____. The 18S rRNA is associated with **(i)** _____ ribosomal subunits. ❖ 3

14. Which of the following statements regarding promoters is correct? Select all that apply. ❖ **1,** ❖ **4,** ❖ **6**

(a) In prokaryotes, the promoter is recognized by general transcription factors, which recruit the RNA polymerase holoenzyme.

(b) In both prokaryotes and eukaryotes, the promoter is located in the 5′ direction, upstream from the transcription start site.

(c) In eukaryotes, the promoter attracts the small and large ribosomal subunits with the help of initiation factors.

(d) In prokaryotes, the promoter contains a −35 and −10 region upstream of the transcription start site.

(e) In eukaryotes, the promoter recruits the preinitiation complex, which includes the TATA–binding protein.

15. Match each RNA type with its appropriate function. Each function will be used only once. ❖ **3**

Type of RNA	Function
(a) small nuclear RNA (snRNA)	1. Helps to regulate the expression of specific genes
(b) ribosomal RNA (rRNA)	2. Carries genetic information to ribosomes for protein synthesis
(c) transfer RNA (tRNA)	3. Combines with proteins to form ribosomes
(d) messenger RNA (mRNA)	4. Brings amino acids to sites of mRNA during protein synthesis
(e) microRNA (miRNA)	5. Involved in the splicing process when mRNA is formed

16. A gene contains eight sites where alternative splicing is possible. Assuming that the splicing pattern at each site is independent of that at all other sites, how many splicing products are possible? ❖ **5**

17. What processes considered in this chapter make the proteome more complex than the genome? What processes might further enhance this complexity? ❖ **5**

18. Nuclei were isolated from brain, liver, and muscle. The nuclei were then incubated with α-[^{32}P] UTP under conditions that allow RNA synthesis, except that an inhibitor of RNA initiation was present. The radioactive RNA was isolated and annealed to various DNA sequences that had been attached to a gene chip. In the adjoining graphs, the intensity of the shading indicates roughly how much mRNA was attached to each DNA sequence. ❖ **4**

  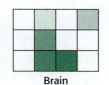

Liver   Muscle   Brain

(a) Why does the intensity of hybridization differ between genes?

(b) What is the significance of the fact that some of the RNA molecules display different hybridization patterns in different tissues?

(c) Some genes are expressed in all three tissues. What is the nature of these genes?

(d) Suggest a reason why an initiation inhibitor was included in the reaction mixture.

19. Eukaryotic messenger RNA can undergo postsynthetic processing after transcription and before translation. One of the processing steps is splicing, where portions of the RNA are removed and the remaining RNA are joined together. Classify the statements regarding mRNA splicing as true or false. ❖ **5**

(a) Splicing occurs while the mRNA is attached to the nucleosome.

(b) One mRNA can sometimes code for more than one protein by splicing at alternative sites.

(c) Splicing occurs while the mRNA is still in the nucleus.

(d) In splicing, intron sequences are removed from the mRNA in the form of lariats (loops) and are degraded.

(e) Splicing of mRNA does not involve any proteins.

Protein Biosynthesis

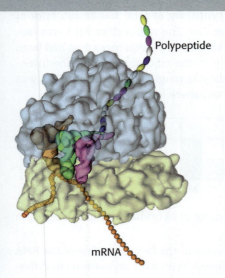

Polypeptide

mRNA

The ribosome, shown above, is a cellular machine that is not unlike a 3D printer. The ribosome follows specific instructions in mRNA to create a polypeptide from amino acids. Each amino acid is connected to a transfer tRNA molecule and brought to the ribosome, one at a time, to be joined to the growing polypeptide chain. Once complete, the polypeptide chain detaches from the ribosome and adopts a three-dimensional structure.

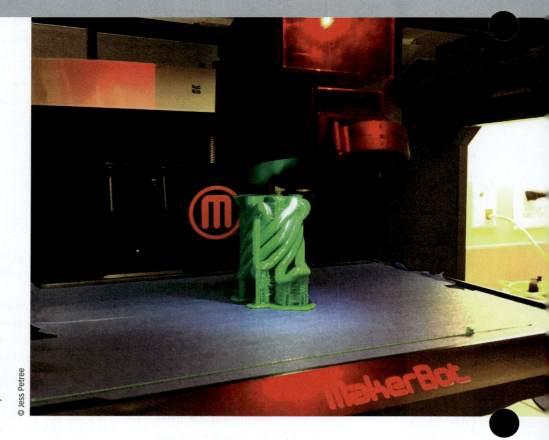

© Jess Petree

OUTLINE

30.1 Protein Biosynthesis Requires the Translation of Nucleotide Sequences into Amino Acid Sequences

30.2 Aminoacyl-tRNA Synthetases Establish the Genetic Code

30.3 The Ribosome Is the Site of Protein Biosynthesis

30.4 Ribosomes Bound to the Endoplasmic Reticulum Manufacture Secretory and Membrane Proteins

30.5 A Variety of Antibiotics and Toxins Inhibit Protein Biosynthesis

❖ LEARNING GOALS

By the end of this chapter, you should be able to:

1. Explain how nucleic acid information is translated into amino acid sequence and define the role of aminoacyl tRNA synthetases in this process.

2. Define the role of ribosomes in protein biosynthesis.

3. Compare and contrast bacterial and eukaryotic protein biosynthesis.

4. Describe the pathway for the biosynthesis of secretory and membrane proteins.

5. Describe how certain chemicals can inhibit protein biosynthesis.

Einstein is reported to have said, "The more I learn, the more I realize how much I don't know." He might as well have been talking about ribosomes, enormous molecular complexes containing several large RNA molecules and dozens of proteins. What makes ribosomes so fascinating? Ribosomes are responsible for protein biosynthesis, and since proteins play most of the functional roles in cells, ribosomes are therefore critical to cell survival. Any errors in this process can have detrimental effects on cells by interfering with normal cellular functions. In addition, toxins and antibiotics that bind to the ribosome and prevent its ability to make proteins can cause cell death. As we will see in this chapter, protein biosynthesis depends not just on the ribosome, but also on many additional nucleic acid and protein factors. There is still much to discover about this intricate process, and "ribosomology" is certain to keep biochemists busy for years to come.

30.1 Protein Biosynthesis Requires the Translation of Nucleotide Sequences into Amino Acid Sequences

In this chapter we discuss the mechanism of protein biosynthesis, a process called **translation** because the 4-letter alphabet of nucleic acids is translated into the entirely different 20-letter alphabet of proteins. Because of this conversion of one alphabet into another, translation is a conceptually more complex process than either replication or transcription, both of which take place within the framework of a common base-pairing language. The basics of protein biosynthesis are the same across all kingdoms of life—evidence that the protein-biosynthesis system arose very early in evolution. An mRNA is translated, or decoded, in the 5′-to-3′ direction one codon at a time, and the corresponding protein is synthesized in the amino-to-carboxyl direction by the sequential addition of amino acids to the carboxyl end of the growing peptide chain (**Figure 30.1**). Amino acids are delivered to the ribosome attached to specific tRNAs. Once the tRNA has delivered its amino acid, it is released from the ribosome and can be linked to another amino acid, and the cycle of peptide chain elongation continues. For each amino acid there is at least one kind of tRNA, and specific enzymes link amino acids to their corresponding tRNAs. This linking is essential to the accurate translation of the language of nucleic acids to that of proteins and thus to the accuracy of the genetic code.

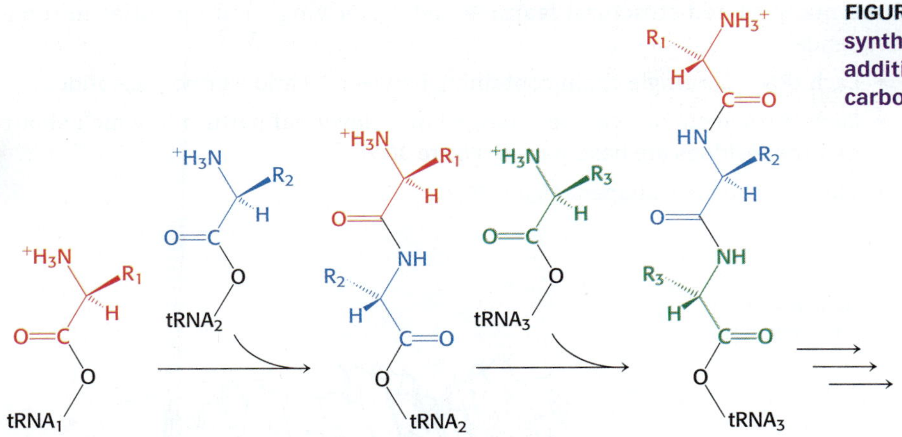

FIGURE 30.1 Polypeptides are synthesized by the successive addition of amino acids to the carboxyl terminus.

The biosynthesis of long proteins requires a low error frequency

The process of transcribing DNA into mRNA is analogous to copying, word for word, a page of a book. There is no change of alphabet or vocabulary, so the likelihood of a change in meaning is small. In contrast, translating the base sequence of an mRNA molecule into a sequence of amino acids is analogous to translating the page of a book into another language. Translation is a complex process, entailing many steps and dozens of molecules, and a potential for error exists at each step. The complexity of translation creates a conflict between two requirements: the process must be accurate yet fast enough to meet a cell's needs. In E. coli, translation can take place at a rate of 20 amino acids per second, a truly impressive speed considering the complexity of the process.

How accurate must protein biosynthesis be? Let us consider error rates. The probability of forming a protein with no errors depends on the number of amino acid residues and on the frequency (ε) of insertion of a wrong amino acid. As **Table 30.1** shows, an error frequency of 10^{-2} is intolerable, even for quite small proteins (~100 residues). An ε value of 10^{-3} usually leads to the error-free synthesis of an

TABLE 30.1 Accuracy of protein synthesis

Frequency of inserting an incorrect amino acid	Probability of synthesizing an error-free protein		
	Number of amino acid residues		
	100	300	1000
10^{-2}	0.366	0.049	0.000
10^{-3}	0.905	0.741	0.368
10^{-4}	0.990	0.970	0.905
10^{-5}	0.999	0.997	0.990

Note: The probability p of forming a protein with no errors depends on n, the number of amino acids, and ε, the frequency of insertion of a wrong amino acid: $p = (1 - \varepsilon)^n$.

average-sized 300-residue protein (~33 kDa) but not of a large 1000-residue protein (~110 kDa); the error frequency must not exceed approximately 10^{-4} to produce the larger proteins effectively. While lower error frequencies are conceivable, they will not dramatically increase the percentage of proteins with accurate sequences, except for really large proteins (>1000 residues). In addition, such lower error rates are likely to be possible only by a reduction in the rate of protein biosynthesis because additional time for proofreading is required. In fact, the observed values of ε are close to 10^{-4}, allowing for the accurate production of proteins consisting of as many as 1000 amino acids while maintaining a remarkably rapid rate for protein biosynthesis.

Transfer RNA (tRNA) molecules have a common design

The fidelity of protein biosynthesis requires accurate recognition of three-base **codons** on mRNA (Section 8.5). An amino acid cannot itself recognize a codon. Consequently, an amino acid is attached to a specific tRNA molecule that recognizes the codon by Watson–Crick base-pairing. **Transfer RNA** serves as the adapter molecule: not only does the tRNA contain an **anticodon** that binds to a specific codon, but it also brings with it an amino acid for incorporation into the polypeptide chain.

Thousands of tRNA sequences are known. All tRNA molecules must be able to interact in nearly the same way with the ribosomes, mRNAs, and protein factors that participate in translation. So, it's not surprising that tRNA molecules have many shared structural features and underlying commonalities in their sequences:

- Each tRNA is a single chain containing between 73 and 93 ribonucleotides.
- Each tRNA sequence can be arranged in a cloverleaf pattern in which about half the residues are base-paired (**Figure 30.2**).
- The molecule is L-shaped (**Figure 30.3**).

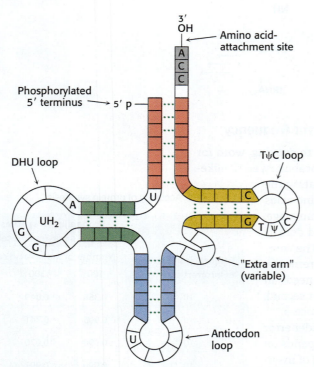

FIGURE 30.2 The secondary structure of all transfer RNAs resembles a cloverleaf. Comparison of the base sequences of many tRNAs reveals a number of conserved features.

FIGURE 30.3 Transfer RNAs have an L-shaped tertiary structure. The L-shaped structure is revealed by this skeletal model of yeast phenylalanyl-tRNA. The CCA region is at the end of one arm, and the anticodon loop is at the end of the other. [Drawn from 1EHZ.pdb.]

- They contain many unusual bases, typically between 7 and 15 per molecule. Some of these bases are methylated or dimethylated derivatives of A, U, C, and G formed by enzymatic modification of a precursor tRNA. Two examples of such modified bases are 5-methylcytidine (mC) and dihydrouridine (UH$_2$). Methylation imparts a hydrophobic character to some regions of tRNAs, which may be important for their interaction with synthetases and ribosomal proteins. Some methylations also prevent bases from forming proper base pairs, allowing them to interact with other bases. Other modifications alter codon recognition, as will be described shortly.

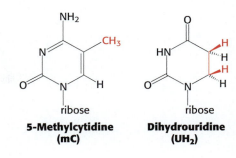

5-Methylcytidine (mC)   **Dihydrouridine (UH$_2$)**

- About half the nucleotides in tRNAs are base-paired to form double helices. The four helical regions are arranged to form two apparently continuous segments of double helix. These segments are like A-form DNA, as expected for an RNA helix (Section 4.2). One helix contains the 5′ and 3′ ends and runs horizontally in the model shown in **Figure 30.4**; it forms one arm of the L. The other helix, which contains the anticodon and runs vertically in Figure 30.4, forms the other arm of the L.

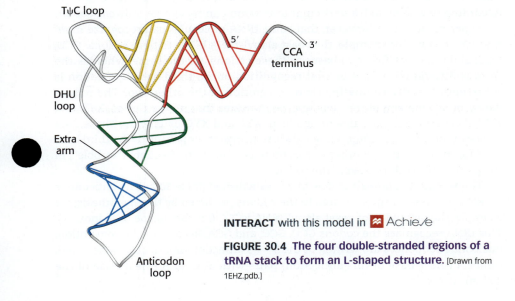

INTERACT with this model in ☕ Achieve

FIGURE 30.4 The four double-stranded regions of a tRNA stack to form an L-shaped structure. [Drawn from 1EHZ.pdb.]

Five groups of bases are not base-paired in this way: the 3′ CCA terminal region, which is part of a region called the *acceptor stem*; the TψC loop, which is named for its ribothymidine-pseudouracil-cytosine; the "*extra arm*," which contains a variable number of residues; the *DHU loop*, which contains several dihydrouracil residues; and the *anticodon loop*. Most of the bases in the nonhelical regions participate in hydrogen-bonding interactions, even if the interactions are not like those in Watson–Crick base pairs. Although they are structurally similar overall, the structural diversity generated by this combination of helices and loops containing modified bases ensures that the tRNAs can be uniquely distinguished.

- The 5′ end of a tRNA is phosphorylated. The 5′ terminal residue is usually pG.
- An activated amino acid is attached to a hydroxyl group of the adenosine residue in the amino-acid-attachment site, located at the end of the 3′ CCA component of the acceptor stem (**Figure 30.5**).
- The anticodon loop, which is present in a loop near the center of the sequence, is at the other end of the L, making the three bases that constitute the anticodon accessible for interaction with a codon.

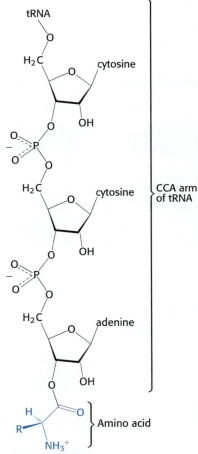

FIGURE 30.5 Ester linkages couple amino acids to tRNAs. While a linkage to the 3′-hydroxyl group is shown here, these bonds can be to either the 2′- or the 3′-hydroxyl group of the 3′-adenosine residue on the tRNA.

Thus, the architecture of the tRNA molecule is well suited to its role as adaptor: the anticodon is available to interact with an appropriate codon on mRNA, while the end that is linked to an amino acid is well positioned to participate in peptide-bond formation.

❖ SELF–CHECK QUESTION

What features are common to all tRNA molecules?

Some transfer RNA molecules recognize more than one codon because of wobble in base-pairing

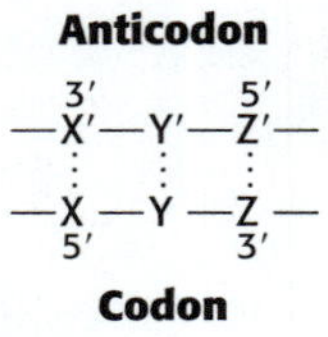

What are the rules that govern the recognition of a codon by the anticodon of a tRNA? A simple hypothesis is that each of the bases of the codon forms a Watson–Crick base pair with a complementary base on the anticodon of the tRNA. The codon and anticodon would then be lined up in an antiparallel fashion, where the prime denotes the complementary base in the anticodon. Thus X and X′ would be either A and U (or U and A) or G and C (or C and G). According to this model, a particular anticodon can recognize only *one* codon.

Experiments show, however, that some tRNA molecules can recognize more than one codon. For example, the yeast alanyl-tRNA binds to *three* codons: GCU, GCC, and GCA. The first two bases of these codons are the same, whereas the third is different. Could it be that recognition of the third base of a codon is sometimes less discriminating than recognition of the other two? The redundancy, or degeneracy, of the genetic code indicates that it might be so. XYU and XYC always encode the same amino acid; XYA and XYG usually do. These data suggest that the steric criteria might be less stringent for pairing of the third base than for the other two. In other words, there is some steric freedom ("wobble") in the pairing of the third base of the codon.

The **wobble hypothesis** is now firmly established (Table 30.2). The anticodons of tRNAs of known sequence bind to the codons predicted by this hypothesis; for example, the anticodon of yeast alanyl-tRNA is IGC (I is the purine base inosine). This tRNA recognizes the codons GCU, GCC, and GCA. Recall that, by convention, nucleotide sequences are written in the 5′ → 3′ direction unless otherwise noted. Hence, I (the 5′ base of this anticodon) pairs with U, C, or A (the 3′ base of the codon), as predicted.

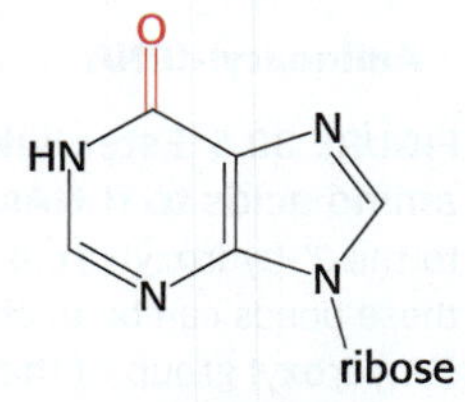

Anticodon

$$-\overset{3'}{X'}-\overset{}{Y'}-\overset{5'}{Z'}-$$
$$-\underset{5'}{X}-\underset{}{Y}-\underset{3'}{Z}-$$

Codon

TABLE 30.2 Allowed pairings at the third base of the codon according to the wobble hypothesis

First base of anticodon	Third base of codon
C	G
A	U
U	A or G
G	U or C
I	U, C, or A

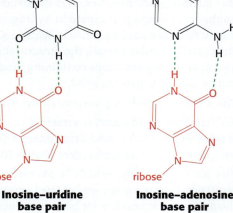

Inosine

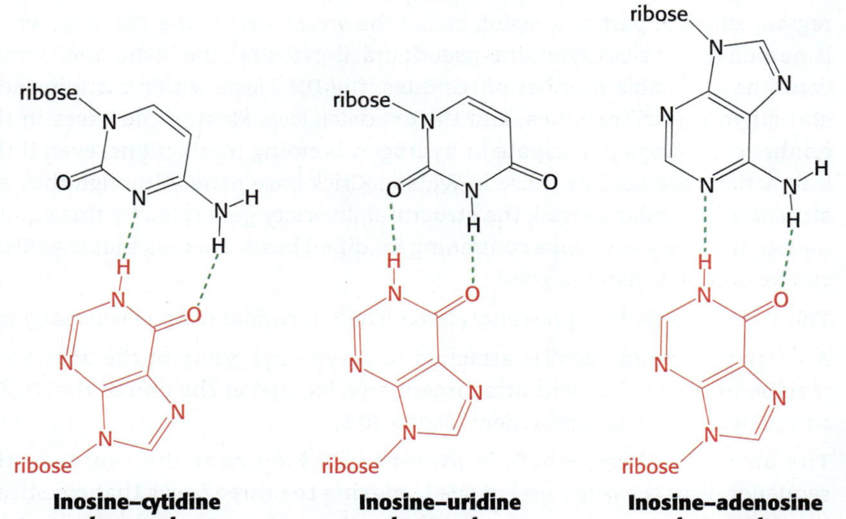

Inosine–cytidine base pair

Inosine–uridine base pair

Inosine–adenosine base pair

Two generalizations concerning the codon–anticodon interaction can be made:

1. *Codons that differ in either of their first two bases must be recognized by different tRNAs.* The first two bases of a codon pair in the standard way, and recognition is precise. For example, both UUA and CUA encode leucine but are read by different tRNAs.

2. *Part of the degeneracy of the genetic code arises from imprecision (wobble) in the pairing of the third base of the codon with the first base of the anticodon.* The first base of an anticodon determines whether a particular tRNA molecule reads one, two, or three kinds of codons: C or A (one codon), U or G (two codons), or I (three codons).

We see here a strong reason for the frequent appearance of inosine, one of the unusual nucleosides, in anticodons: inosine maximizes the number of codons that can be read by a particular tRNA molecule. The inosine bases in tRNA are formed by the deamination of adenosine after the synthesis of the primary transcript.

Why is wobble tolerated in the third position of the codon but not in the first two? This question is answered by considering the interaction of the tRNA with the ribosome. As we will soon examine in greater detail, **ribosomes** are huge RNA–protein complexes consisting of two subunits. One of these subunits has three universally conserved bases in its RNA molecule that form hydrogen bonds only with correctly formed base pairs in the codon–anticodon duplex (**Figure 30.6**). These interactions ensure that Watson–Crick base pairs are present in the first two positions of the codon–anticodon duplex, but not the third position, so more varied base pairs are tolerated. Thus, the ribosome plays an active role in decoding the codon–anticodon interactions.

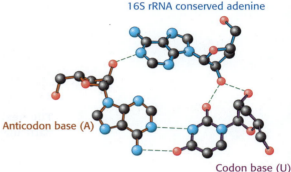

16S rRNA conserved adenine

Anticodon base (A)

Codon base (U)

FIGURE 30.6 16S rRNA monitors base-pairing between the codon and the anticodon. One of three universally conserved bases in 16S rRNA (an adenine) forms hydrogen bonds with the bases in both the codon and the anticodon only if the codon and anticodon are correctly paired. Only one anticodon base (adenine) and one codon base (uracil) are shown. Hydrogen atoms are omitted for clarity. [Information from J. M. Ogle and V. Ramakrishnan, *Annu. Rev. Biochem.* 74:129–177, 2005, Fig. 2a.]

❖ SELF–CHECK QUESTION

Explain how it is possible that some tRNA molecules recognize more than one codon.

30.2 Aminoacyl-tRNA Synthetases Establish the Genetic Code

Before codon and anticodon meet, the amino acids required for protein biosynthesis must first be attached to specific tRNA molecules, linkages that are crucial for two reasons:

- *The attachment of a given amino acid to a particular tRNA establishes the genetic code.* When an amino acid has been linked to a tRNA, it will be incorporated into a growing polypeptide chain at a position dictated by the anticodon of the tRNA.

- *Because the formation of a peptide bond between free amino acids is not thermodynamically favorable, the amino acid must first be activated for protein biosynthesis to proceed.* The activated intermediates in protein biosynthesis are amino acid esters, in which the carboxyl group of an amino acid is linked to either the 2′- or the 3′-hydroxyl group of the ribose unit at the 3′ end of tRNA. An amino acid ester of tRNA is called an **aminoacyl-tRNA** or sometimes a *charged* tRNA (Figure 30.6). For a specific amino acid attached to its cognate tRNA — for instance, threonine — the charged tRNA is designated Thr-tRNA^{Thr}.

Amino acids are first activated by adenylation

The activation reaction is catalyzed by specific **aminoacyl-tRNA synthetases**. These enzymes activate an amino acid and then link it to its corresponding tRNA in a two-step reaction. The first step is the formation of an aminoacyl adenylate from an amino acid and ATP.

$$\text{Amino acid} + \text{ATP} \rightleftharpoons \text{aminoacyl-AMP} + \text{PP}_i$$

This activated species is a mixed anhydride in which the carboxyl group of the amino acid is linked to the phosphoryl group of AMP; hence, it is also known as *aminoacyl-AMP*.

Aminoacyl-AMP

The next step is the transfer of the aminoacyl group of aminoacyl-AMP to a particular tRNA molecule to form aminoacyl-tRNA.

$$\text{Aminoacyl-AMP} + \text{tRNA} \rightleftharpoons \text{aminoacyl-tRNA} + \text{AMP}$$

The sum of these activation and transfer steps is

$$\text{Amino acid} + \text{ATP} + \text{tRNA} \rightleftharpoons \text{aminoacyl-tRNA} + \text{AMP} + \text{PP}_i$$

The $\Delta G^{\circ\prime}$ of this reaction is close to 0, because the free energy of hydrolysis of the ester bond of aminoacyl-tRNA is similar to that for the hydrolysis of ATP to AMP and PP_i. As we have seen many times, the reaction is driven by the hydrolysis of pyrophosphate. The sum of these three reactions is highly exergonic:

$$\text{Amino acid} + \text{ATP} + \text{tRNA} + \text{H}_2\text{O} \longrightarrow \text{aminoacyl-tRNA} + \text{AMP} + 2\,\text{P}_i$$

There are two key points to note about this reaction:

- *The equivalent of two molecules of ATP is consumed in the synthesis of each aminoacyl-tRNA.* One of them is consumed in forming the ester linkage of aminoacyl-tRNA, whereas the other is consumed in driving the reaction forward.

- *The aminoacyl-AMP intermediate does not dissociate from the synthetase.* Rather, it is tightly bound to the active site of the enzyme by noncovalent interactions. This means that the activation and transfer steps for a particular amino acid are catalyzed by the same aminoacyl-tRNA synthetase.

We have already encountered an acyl adenylate intermediate in fatty acid activation (Section 22.2). The major difference between these reactions is that the acceptor of the acyl group is CoA in fatty acid activation and tRNA in amino acid activation. The energetics of these biosyntheses are very similar: both are made irreversible by the hydrolysis of pyrophosphate.

Aminoacyl-tRNA synthetases have highly discriminating amino acid activation sites

Each aminoacyl-tRNA synthetase is highly specific for a given amino acid. Indeed, a synthetase will link an incorrect amino acid to a tRNA only once in 10^4 or 10^5 reactions. How is this level of specificity achieved? Each aminoacyl-tRNA synthetase takes advantage of the properties of its amino acid substrate. Consider the challenge faced by threonyl-tRNA synthetase. Threonine is especially similar to two other amino acids: valine and serine. Valine has almost exactly the same shape as that of threonine, except that valine has a methyl group in place of a hydroxyl group. Serine has a hydroxyl group, as does threonine, but lacks the methyl group. How can the threonyl-tRNA synthetase avoid coupling these incorrect amino acids to threonyl-tRNA (abbreviated tRNA^Thr)?

Acyl adenylate intermediate

Aminoacyl-tRNA

Fatty acyl CoA

Threonine

Valine

Serine

The structure of the amino acid-binding site of threonyl-tRNA synthetase reveals how valine is avoided (Figure 30.7). The synthetase contains a zinc ion bound to the enzyme by two histidine residues and one cysteine residue, with the remaining coordination sites available for substrate binding. The use of a zinc ion appears to be unique to threonyl-tRNA synthetase; other aminoacyl-tRNA synthetases have different strategies for recognizing their cognate amino acids.

Threonine binds to the zinc ion through its amino group and its side-chain hydroxyl group. The side-chain hydroxyl group is further recognized by an aspartate residue that hydrogen-bonds to it. The methyl group present in valine in place of this hydroxyl group cannot participate in these interactions; it is excluded from this active site and, hence, does not become adenylated and transferred to tRNA^{Thr}. The carboxylate group of the correctly positioned threonine is available to attack the α-phosphoryl group of ATP to form the aminoacyl adenylate.

The zinc site is less able to discriminate against serine because this amino acid does have a hydroxyl group that can bind to the zinc ion. Indeed, with only this mechanism available, threonyl-tRNA synthetase does mistakenly couple serine to tRNA^{Thr} at a rate 10^{-2} to 10^{-3} times that for threonine. This error rate is likely to lead to many translation errors, so how is a higher level of specificity achieved?

Proofreading by aminoacyl-tRNA synthetases increases the fidelity of protein biosynthesis

Threonyl-tRNA synthetase can be incubated with tRNA^{Thr} that has been covalently linked with serine (Ser-tRNA^{Thr}); the tRNA has been "mischarged." The reaction is immediate: a rapid hydrolysis of the aminoacyl-tRNA forms serine and free tRNA. In contrast, incubation with correctly charged Thr-tRNA^{Thr} results in no reaction. Thus, threonyl-tRNA synthetase contains an additional functional site that hydrolyzes Ser-tRNA^{Thr} but not Thr-tRNA^{Thr}. This editing site provides an opportunity for the synthetase to correct its mistakes and improve its fidelity to less than one mistake in 10^4.

The results of structural and mutagenesis studies revealed that the editing site for threonyl-tRNA synthase is more than 20 Å from the activation site (Figure 30.8). This editing site readily accepts and cleaves Ser-tRNA^{Thr} but does not cleave Thr-tRNA^{Thr}. The discrimination of serine from threonine is easy because threonine contains an extra methyl group; a site that conforms to the structure of serine will sterically exclude threonine. The structure of the complex between threonyl-tRNA synthetase and its substrate reveals that the aminoacylated CCA can swing out of the activation site and into the editing site (Figure 30.9). Thus, the aminoacyl-tRNA can be edited without dissociating from the synthetase. This proofreading depends on the conformational flexibility of a short stretch of polynucleotide sequence.

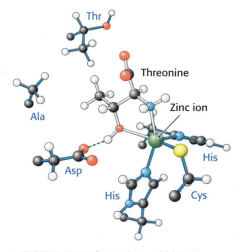

FIGURE 30.7 The amino acid-binding site of threonyl-tRNA synthetase contains a zinc ion. This zinc ion (green ball) binds to the amino and hydroxyl groups of threonine (white). Selected amino acids from the synthetase are shown in blue.

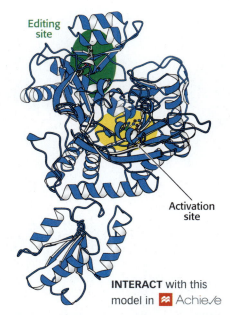

INTERACT with this model in ⧓ Achieve

FIGURE 30.8 Threonyl-tRNA synthetase also contains an editing site. Mutagenesis studies revealed the position of the editing site (shown in green) in threonyl-tRNA synthetase. The activation site is shown in yellow. Only one subunit of the dimeric enzyme is shown here and in subsequent illustrations. [Drawn from 1QF6.pdb.]

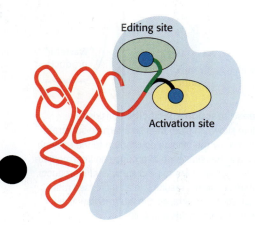

FIGURE 30.9 Amino acids move between the activation site and the editing site of an aminoacyl-tRNA synthetase. The flexible CCA arm of an aminoacyl-tRNA can swing the attached amino acid from the activation site (yellow) to the editing site (green). If the amino acid fits well into the editing site, the amino acid is removed by hydrolysis. Only one amino acid is attached to the tRNA at a time, and this amino acid can only occupy either the editing site or the activation site.

Most aminoacyl-tRNA synthetases contain editing sites in addition to activation sites. Together, these complementary sites ensure very high fidelity. In general, the acylation site rejects amino acids that are *larger* than the correct one because there is insufficient room for them, whereas the hydrolytic site cleaves activated species that are *smaller* than the correct one.

A few synthetases achieve high accuracy without editing. For example, tyrosyl-tRNA synthetase has no difficulty discriminating between tyrosine and phenylalanine; the hydroxyl group on the tyrosine ring enables tyrosine to bind to the enzyme 10^4 times as strongly as phenylalanine. In conclusion, proofreading has evolved only when fidelity must be enhanced beyond what can be obtained through an initial binding interaction.

Kinetic proofreading increases the fidelity of protein biosynthesis

As noted above, the integrity and validity of the genetic code depends on the fidelity with which tRNAs are linked to their cognate amino acids. The free energy difference of binding an incorrect versus a correct substrate is often too small to explain the accuracy of protein biosynthesis. So how is it possible? Aminoacyl-tRNA synthetases prevent many mistakes by introducing an irreversible step that facilitates their removal. This mechanism, called *kinetic proofreading*, locks in the equilibrium favoring the correct product, then selectively removes incorrect products to increase specificity further (**Figure 30.10**). For aminoacyl-tRNA synthetases, this results in additional hydrolysis of ATP, but it increases specificity.

Kinetic proofreading benefits other processes as well: aminoacyl-tRNA synthetases use it to discriminate between cognate and noncognate amino acids as well as tRNAs, and ribosomes use it during codon–anticodon recognition, thereby further adding to the integrity of the genetic code. Kinetic

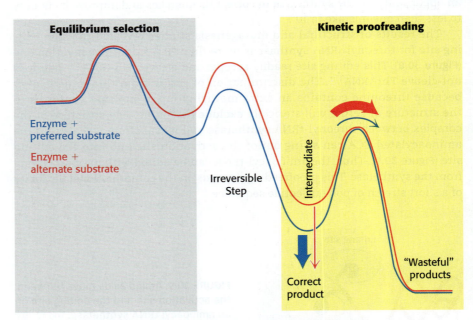

FIGURE 30.10 Kinetic proofreading can increase reaction specificity. Consider a substrate that binds to an enzyme more tightly (curve shown in blue) than an alternate substrate (curve shown in red). The preferred substrate is selected by the binding equilibrium to the enzyme. This selected equilibrium can be locked in place by an irreversible step to produce an intermediate that can either go on to the final product or undergo a side reaction to produce wasteful products. This side reaction is likely to be faster for the alternate substrate, resulting in an increased ratio of the preferred-to-alternate products. For aminoacyl-tRNA synthetases, the intermediate is an aminoacyl adenylate and the wasteful products are free amino acids and AMP.

proofreading is employed by many additional molecular processes (including DNA replication) to ensure the faithful selection of substrates.

Synthetases recognize various features of transfer RNA molecules

How do synthetases choose their tRNA partners? This enormously important step is the point at which the actual "translation" takes place: when the correlation between the amino acid and the nucleic acid worlds is made. In a sense, aminoacyl-tRNA synthetases are the only molecules in biology that "know" the genetic code. Their precise recognition of tRNAs is as important for high-fidelity protein biosynthesis as is the accurate selection of amino acids, and this recognition is sometimes referred to as the "second genetic code." In general, tRNA recognition by the synthetase is different for each synthetase and tRNA pair:

- *Some synthetases recognize their tRNA partners primarily on the basis of their anticodons.* The most direct evidence comes from crystallographic studies of complexes formed between synthetases and their cognate tRNAs. Consider, for example, the structure of the complex between threonyl-tRNA synthetase and tRNA^{Thr} (**Figure 30.11**). As expected, the CCA arm extends into the zinc-containing activation site, where it is well-positioned to accept threonine from threonyl adenylate. The enzyme interacts extensively not only with the acceptor stem of the tRNA, but also with the anticodon loop. The interactions with the anticodon loop are particularly revealing. Each base within the sequence 5'-CGU-3' of the anticodon participates in hydrogen bonds with the enzyme; those with the second two bases (G and U) appear to be more important because the synthetase interacts just as efficiently with the anticodons GGU and UGU.

- *Synthetases may also recognize other aspects of tRNA structure that vary among different tRNAs.* Although interactions between the enzyme and the anticodon are often crucial for correct recognition, **Figure 30.12** shows that many aspects of tRNA molecules are recognized by synthetases. Note that many of the recognition sites are loops rich in unusual bases that can provide structural identifiers.

FIGURE 30.11 Aminoacyl-tRNA synthetases bind to both the acceptor stem and the anticodon loop of a tRNA. This structure shows the complex between threonyl-tRNA synthetase (blue) and tRNA^{Thr} (red). [Drawn from 1QF6.pdb.]

INTERACT with this model in ≈ Achieve

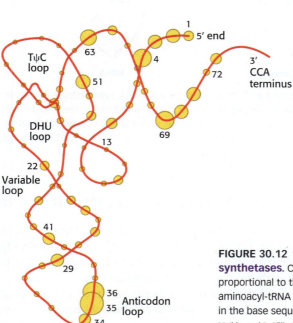

FIGURE 30.12 tRNAs have multiple recognition sites for aminoacyl-tRNA synthetases. Circles represent nucleotides, and the sizes of the circles are proportional to the frequency with which they are used as recognition sites by aminoacyl-tRNA synthetases. The numbers indicate the positions of the nucleotides in the base sequence, beginning from the 5' end of the tRNA molecule. [Information from M. Ibba and D. Söll, *Annu. Rev. Biochem.* 69:617–650, 1981, p. 636.]

Aminoacyl-tRNA synthetases are divided into two classes

The diverse sizes, subunit composition, and sequences of aminoacyl-tRNA synthetases were bewildering for many years. Could it be that essentially all synthetases evolved independently? The determination of the three-dimensional structures of several synthetases followed by more refined sequence comparisons revealed that different synthetases are, in fact, related.

Specifically, synthetases fall into two classes, termed *class I* and *class II*, each of which includes enzymes specific for 10 of the 20 amino acids (Table 30.3). Intriguingly, synthetases from the two classes bind to different faces of the tRNA molecule (Figure 30.13). The CCA arm of tRNA adopts different conformations to accommodate these interactions; the arm is in the helical conformation observed for free tRNA (Figures 30.3 and 30.4) for class II enzymes and in a hairpin conformation for class I enzymes. These two classes also differ in other ways:

- Class I enzymes acylate the 2′-hydroxyl group of the terminal adenosine of tRNA, whereas class II enzymes (except the enzyme for tRNA^Phe) acylate the 3′-hydroxyl group.

- The two classes bind ATP in different conformations.

- Most class I enzymes are monomeric, whereas most class II enzymes are dimeric.

Why did two distinct classes of aminoacyl-tRNA synthetases evolve? The observation that the two classes bind to distinct faces of tRNA suggests a possibility. Recognition sites on both faces of tRNA may have been required to allow the recognition of 20 different tRNAs.

TABLE 30.3 Classification and subunit structure of aminoacyl-tRNA synthetases in *E. coli*

Class I	Class II
Arg (α)	Ala (α4)
Cys (α)	Asn (α2)
Gln (α)	Asp (α2)
Glu (α)	Gly (α2β2)
Ile (α)	His (α2)
Leu (α)	Lys (α2)
Met (α)	Phe (α2β2)
Trp (α2)	Ser (α2)
Tyr (α2)	Pro (α2)
Val (α)	Thr (α2)

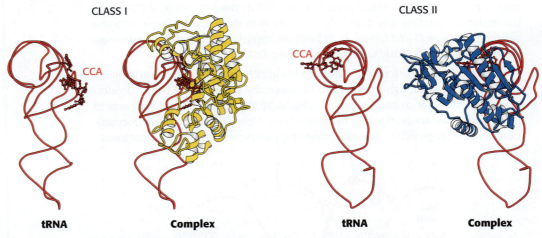

CLASS I   CLASS II

CCA   CCA

tRNA   Complex   tRNA   Complex

INTERACT with these models in ✿ Achieve

FIGURE 30.13 Class I and class II aminoacyl-tRNA synthetases recognize different faces of the tRNA molecule. The CCA arm of tRNA adopts different conformations in complexes with the two classes of synthetase. Note that the CCA arm of the tRNA is turned toward the viewer (as compared to Figures 30.3 and 30.4). Only the CCA-binding domains of the aminoacyl-tRNA synthetases are shown. [Drawn from 1EUY.pdb and 1QF6.pdb.]

30.3 The Ribosome Is the Site of Protein Biosynthesis

We turn now to ribosomes, the molecular machines that coordinate the interplay of aminoacyl-tRNAs, mRNA, and proteins that leads to protein biosynthesis. We will focus our discussion first on ribosomes and protein biosynthesis in bacteria, then discuss the differences observed in eukaryotes.

The 20,000 ribosomes in a bacterial cell constitute nearly a quarter of its mass. An *E. coli* ribosome is a ribonucleoprotein assembly with a mass of about 2300 kDa, a diameter of approximately 250 Å, and a sedimentation coefficient (Section 4.1) of 70S.

A 70S ribosome can dissociate into a **large subunit (50S)** and a **small subunit (30S)**, each of which can be further split into their constituent proteins and RNAs. The 30S subunit contains 21 different proteins (referred to as S1 through S21) and a 16S RNA molecule. The 50S subunit contains 34 different proteins (L1 through L34) and two RNA molecules, a 23S and a 5S species. A 70S ribosome contains one copy of each RNA molecule, two copies each of the L7 and L12 proteins, and one copy of each of the other proteins. The L7 protein is identical to L12 except that its amino terminus is acetylated (Section 7.3).

Both the 30S and the 50S subunits can be reconstituted in vitro from their constituent proteins and RNA. This reconstitution is an outstanding example of the principle that supramolecular complexes can form spontaneously from their macromolecular constituents.

The structures of both the 30S and the 50S subunits as well as the complete 70S ribosome have been determined (**Figure 30.14**). The features of these structures are in remarkable agreement with interpretations of less-direct experimental analyses of the structure and function of the ribosome. These structures provide an invaluable framework for examining the mechanism of protein biosynthesis.

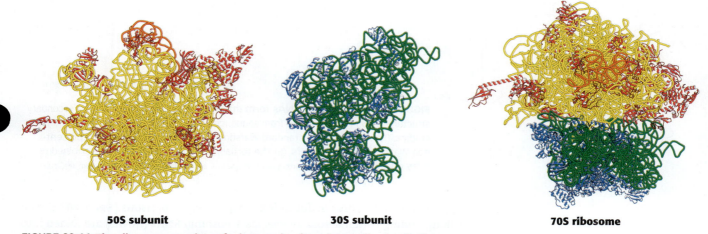

50S subunit   **30S subunit**   **70S ribosome**

FIGURE 30.14 The ribosome consists of a large subunit and a small subunit. These detailed models of a bacterial ribosome are based on high-resolution x-ray crystallographic data of the 70S ribosome and the 30S and 50S subunits. (Left) View of the part of the 50S subunit that interacts with the 30S subunit. (Center) View of the part of the 30S subunit that interacts with the 50S subunit. (Right) View of the 70S ribosome (note that this view is rotated with respect to the other two views). 23S rRNA is shown in yellow, 5S rRNA in orange, 16S rRNA in green, proteins of the 50S subunit in red, and proteins of the 30S subunit in blue. The interface between the 50S and the 30S subunits consists primarily of RNA. [Drawn from 1GIX.pdb and 1GIY.pdb.]

Ribosomal RNAs (5S, 16S, and 23S rRNA) play central roles in protein biosynthesis

The prefix *ribo* in the name *ribosome* is apt because RNA constitutes nearly two-thirds of the mass of these large molecular assemblies. The three RNAs present—5S, 16S, and 23S—are critical for ribosomal architecture and function. They are formed by the cleavage of primary 30S transcripts and further processing. These molecules fold into structures that allow them to form internal base pairs. Their base-pairing patterns were deduced by comparing the nucleotide sequences of many species to detect conserved sequences as well as conserved base pairings. For instance, the 16S RNA of one species may have a G–C base pair, whereas another may have an A–U base pair, but the location of the base pair is the same in both molecules. Chemical modification and digestion experiments

(A)

(B)

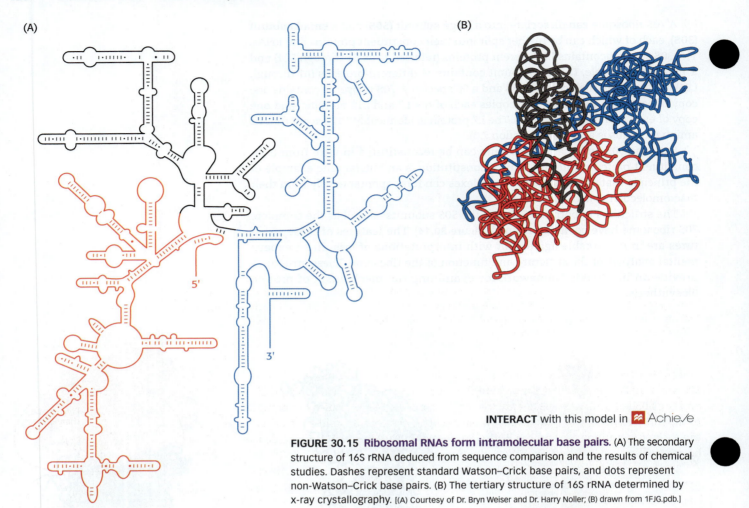

FIGURE 30.15 Ribosomal RNAs form intramolecular base pairs. (A) The secondary structure of 16S rRNA deduced from sequence comparison and the results of chemical studies. Dashes represent standard Watson–Crick base pairs, and dots represent non-Watson–Crick base pairs. (B) The tertiary structure of 16S rRNA determined by x-ray crystallography. [(A) Courtesy of Dr. Bryn Weiser and Dr. Harry Noller; (B) drawn from 1FJG.pdb.]

supported the structures deduced from sequence comparisons (**Figure 30.15**). The striking finding is that across all species, ribosomal RNAs (rRNAs) are folded into defined structures that have many short intramolecular duplex regions.

For many years, ribosomal proteins were presumed to orchestrate protein biosynthesis and ribosomal RNAs were presumed to serve primarily as structural scaffolding. The current view is almost the reverse. The discovery of catalytic RNA (Section 29.5) made biochemists receptive to the possibility that RNA plays a much more active role in ribosomal function. In fact, bacterial ribosomes can be stripped of most of their protein components in vitro and still retain their ability to catalyze peptide-bond formation. In addition, the detailed structures show that the key sites in the ribosome, such as those that catalyze the formation of the peptide bond and those that interact with mRNA and tRNA, are composed almost entirely of rRNA. Contributions from the proteins are minor. The almost inescapable conclusion is that the ribosome initially consisted only of RNA and that the proteins were added later to fine-tune its functional properties. This conclusion has the pleasing consequence of dodging a "chicken and egg" question: How can complex proteins be synthesized if complex proteins are required for protein biosynthesis?

Ribosomes have three transfer RNA-binding sites that bridge the 30S and 50S subunits

Three tRNA-binding sites in ribosomes are arranged to allow the formation of peptide bonds between amino acids encoded by the codons on mRNA (**Figure 30.16**).

(A)

(B)

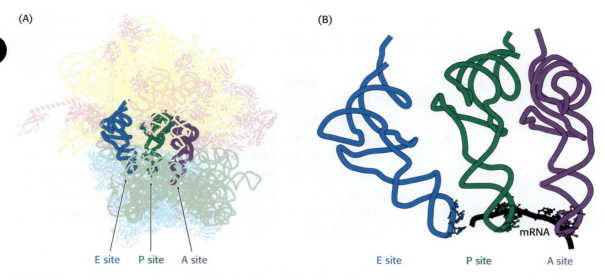

E site   P site   A site

E site   P site   A site

mRNA

FIGURE 30.16 The ribosome has three binding sites for transfer RNAs. (A) The three tRNA binding sites on the 70S ribosome are called the A (for aminoacyl), P (for peptidyl), and E (for exit) sites. Each tRNA molecule contacts both the 30S and the 50S subunit. (B) The tRNA molecules in sites A and P are base-paired with mRNA. [(B) Drawn from 1JGP.pdb.]

INTERACT with this model in ✷ Achieve

The mRNA fragment being translated at a given moment is bound within the 30S subunit. Each of the tRNA molecules is in contact with both the 30S subunit and the 50S subunit. At the 30S end, two of the three tRNA molecules are bound to the mRNA through anticodon–codon base pairs. These binding sites are called the A site (for aminoacyl) and the P site (for peptidyl). The third tRNA molecule is bound to an adjacent site called the E site (for exit).

The other end of each tRNA molecule, the end without the anticodon, interacts with the 50S subunit. The acceptor stems of the tRNA molecules occupying the A site and the P site converge at a site where a peptide bond is formed. A tunnel connects this site to the back of the ribosome, through which the polypeptide chain passes during synthesis (**Figure 30.17**).

The start signal is usually AUG preceded by several bases that pair with 16S rRNA

How does protein biosynthesis start? The simplest possibility would be for the first three nucleotides of each mRNA to serve as the first codon; no special start signal would then be needed. However, experiments show that translation in bacteria does not begin immediately at the 5′ terminus of mRNA. Indeed, the first translated codon is nearly always more than 25 nucleotides away from the 5′ end. Furthermore, in bacteria, many mRNA molecules encode two or more polypeptide chains. For example, a single mRNA molecule about 7000 nucleotides long specifies five enzymes in the biosynthetic pathway for tryptophan in E. coli. Each of these five proteins has its own start and stop signals on the mRNA. In fact, all known mRNA molecules contain signals that define the beginning and end of each encoded polypeptide chain.

A clue to the mechanism of initiation was the finding that nearly half the amino-terminal residues of proteins in E. coli are methionine. In fact, the initiating codon in mRNA is AUG (methionine) or, less frequently, GUG (valine) or, rarely, UUG (leucine). What additional signals are necessary to specify a translation start site? The first step toward answering this question was the isolation of initiator regions from a number of mRNAs. This isolation was accomplished by using ribonuclease to digest mRNA–ribosome complexes (formed under conditions in which protein biosynthesis could begin but elongation could not take place).

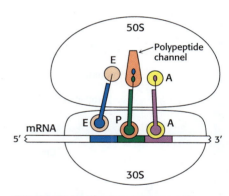

FIGURE 30.17 The key components of the translation machinery interact on an active ribosome. This schematic representation shows the relationship between the ribosomal subunits, tRNAs, and mRNA.

As expected, each initiator region usually displays an AUG codon (**Figure 30.18**). In addition, each initiator region contains a purine-rich sequence centered about 10 nucleotides on the 5′ side of the initiator codon.

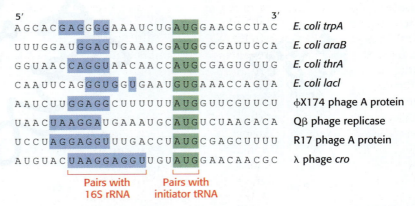

5′ 3′	
AGCACG**AGGGG**AAAUCUG**AUG**GAACGCUAC	*E. coli trpA*
UUUGG**AUGGAG**UGAAACG**AUG**GCGAUUGCA	*E. coli araB*
GGUAAC**CAGGU**AACAACC**AUG**CGAGUGUUG	*E. coli thrA*
CAAUUCAG**GGUGG**UGAAU**GUG**AAACCAGUA	*E. coli lacI*
AAUCUU**GGAGG**CUUUUUU**AUG**GUUCGUUCU	φX174 phage A protein
UAAC**UAAGG**AUGAAAUGC**AUG**UCUAAGACA	Qβ phage replicase
UCCU**AGGAGG**UUUGACCU**AUG**CGAGCUUUU	R17 phage A protein
AUGUAC**UAAGGAGGU**UGU**AUG**GAACAACGC	λ phage *cro*

Pairs with 16S rRNA   Pairs with initiator tRNA

FIGURE 30.18 Initiator regions in mRNA determine where protein biosynthesis starts. This sequence alignment of mRNA initiation sites in some bacterial and viral mRNA molecules reveals some recurring features. The purine-rich sequence on the 5′ side of the initiator codon pairs with a conserved sequence on the 16S rRNA. The first downstream AUG, GUG (less frequent), or UUG (rare) codon acts as the initiator codon for protein biosynthesis.

The role of this purine-rich region, called the **Shine–Dalgarno sequence** (after John Shine and Lynn Dalgarno, who first described the sequence), became evident when the sequence of 16S rRNA was determined. The 3′ end of this rRNA component of the 30S subunit contains a sequence of several bases that is complementary to the Shine–Dalgarno sequence in the mRNA. Mutagenesis of the CCUCC sequence near the 3′ end of 16S rRNA to ACACA markedly interferes with the recognition of start sites in mRNA. This result and other evidence show that the initiator region of mRNA binds very near the 3′ end of the 16S rRNA. The number of base pairs linking mRNA and 16S rRNA ranges from three to nine. This interaction between the 5′ end of the mRNA and the 3′ end of the 16S rRNA positions the initiator codon of the mRNA in the P site of the ribosome, ready for recognition by the anticodon of an initiator tRNA molecule.

Bacterial protein biosynthesis is initiated by *N*-formylmethionyl-transfer RNA

As stated earlier, methionine is the first amino acid in many *E. coli* proteins. However, the methionine residue found at the amino-terminal end of *E. coli* proteins is usually modified. In fact, protein biosynthesis in bacteria starts with the modified amino acid N-formylmethionine (fMet). A distinct tRNA brings N-formylmethionine to the ribosome to initiate protein biosynthesis. This tRNA, called *initiator* tRNA (abbreviated as tRNA^{fMet}), differs from the tRNA that inserts methionine in internal positions (abbreviated as tRNA^{Met}). The superscript "fMet" indicates that methionine attached to the initiator tRNA can be formylated, whereas it cannot be formylated when attached to tRNA^{Met}. Although virtually all proteins synthesized in *E. coli* begin with N-formylmethionine, in approximately one-half of the proteins, N-formylmethionine is removed when the nascent, or growing, chain is 10 amino acids long.

Methionine is linked to these two kinds of tRNAs by the same aminoacyl-tRNA synthetase. A specific enzyme then formylates the amino group of the methionine molecule that is attached to tRNA^{fMet} (**Figure 30.19**). The activated formyl donor in this reaction is N^{10}-formyltetrahydrofolate, a folate derivative that carries activated one-carbon units (Section 25.2). Free methionine and methionyl-tRNA^{Met} are not substrates for this transformylase.

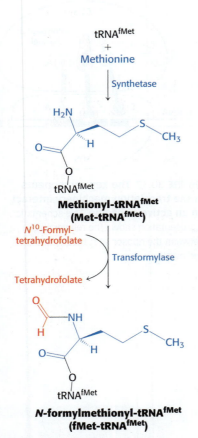

FIGURE 30.19 Methionyl-tRNA is formylated to create a modified amino acid. Initiator tRNA (tRNA^{fMet}) is first charged with methionine, and then a formyl group is transferred to the methionyl-tRNA^{fMet} from N^{10}-formyltetrahydrofolate.

N-Formylmethionyl-tRNA^{fMet} is placed in the P site of the ribosome in the formation of the 70S initiation complex

Messenger RNA and *N*-formylmethionyl-tRNA^{fMet} must be brought to the ribosome for protein biosynthesis to begin. How is this task accomplished? Three protein **initiation factors** (IF1, IF2, and IF3) are essential. The 30S ribosomal subunit first forms a complex with IF1 and IF3 (**Figure 30.20**). The binding of these factors to the 30S subunit prevents it from prematurely joining the 50S subunit to form a dead-end 70S complex, devoid of mRNA and fMet-tRNA^{fMet}. IF1 binds near the A site and directs the fMet-tRNA^{fMet} to the P site. IF2, a member of the G-protein family, binds GTP, and the concomitant conformational change enables IF2 to associate with fMet-tRNA^{fMet}. The IF2–GTP–initiator-tRNA complex binds with mRNA (correctly positioned by the interaction of the Shine–Dalgarno sequence with the 16S rRNA) and the 30S subunit to form the 30S initiation complex. Structural changes then lead to the ejection of IF1 and IF3. IF2 stimulates the association of the 50S subunit to the complex. The GTP bound to IF2 is hydrolyzed upon arrival of the 50S subunit, releasing IF2. The result is a 70S initiation complex. The formation of the 70S initiation complex is the rate-limiting step in protein biosynthesis.

When the 70S initiation complex has been formed, the ribosome is ready for the elongation phase of protein biosynthesis. The fMet-tRNA^{fMet} molecule occupies the P site on the ribosome, positioned so that its anticodon pairs with the initiating codon on mRNA. The other two sites for tRNA molecules, the A site and the E site, are empty. This interaction establishes what is termed the *reading frame* for the translation of the entire mRNA (Section 8.6). After the initiator codon has been located, groups of three nonoverlapping nucleotides are decoded by the ribosome to create the polypeptide chain.

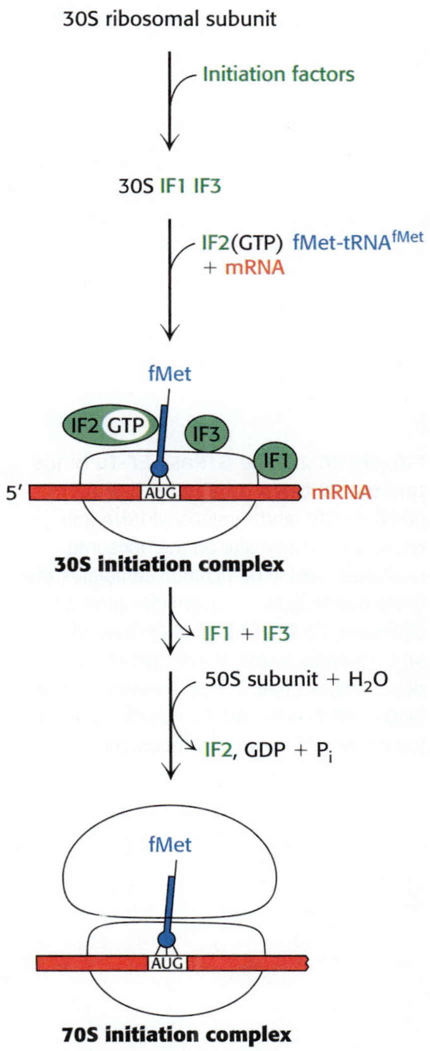

Elongation factors deliver aminoacyl-tRNAs to the ribosome

At this point, fMet-tRNA^{fMet} occupies the P site, and the A site is vacant. The particular aminoacyl-tRNA inserted into the empty A site depends on the mRNA codon in the A site. However, the appropriate aminoacyl-tRNA does not simply leave the synthetase and diffuse to the A site. Rather, it is delivered to the A site in association with a 43-kDa protein called **elongation factor Tu (EF-Tu)**, another member of the G-protein family. EF-Tu, the most abundant bacterial protein, binds aminoacyl-tRNA only in its GTP form (**Figure 30.21**).

The binding of EF-Tu to aminoacyl-tRNA serves two functions. First, EF-Tu protects the delicate ester linkage in aminoacyl-tRNA from hydrolysis. Second, EF-Tu contributes to the accuracy of protein biosynthesis because GTP hydrolysis and expulsion of the EF-Tu-GDP complex from the ribosome occurs only if the pairing between the anticodon and the codon is correct. EF-Tu interacts with the 16S rRNA, which monitors the accuracy of the base pairing at positions 1 and 2 of the codon (Figure 30.6).

Correct codon recognition induces structural changes in the 30S subunit that move the EF-Tu to a highly conserved site in the 23S rRNA in the 50S subunit called the sarcin-ricin loop (SRL). The interaction with the SRL of the 50S

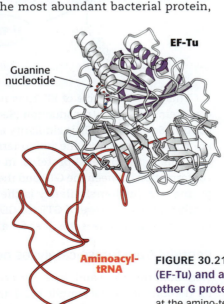

FIGURE 30.21 The structure of elongation factor Tu (EF-Tu) and an aminoacyl-tRNA resembles those of other G proteins. The P-loop NTPase domain (purple shading) at the amino-terminal end of EF-Tu is also found in other G proteins. [Drawn from 1B23.pdb.]

INTERACT with this model in 🅰 Achieve

30S ribosomal subunit
↓ Initiation factors
30S IF1 IF3
↓ IF2(GTP) fMet-tRNA^{fMet} + mRNA

fMet
IF2 GTP IF3 IF1
5′ ——AUG—— mRNA

30S initiation complex

↓ IF1 + IF3
  50S subunit + H₂O
  IF2, GDP + P_i

fMet
AUG

70S initiation complex

FIGURE 30.20 Prokaryotic initiation factors aid the assembly of initiation complexes. These factors first aid assembly of the 30S initiation complex, and then of the 70S initiation complex. IF1 interacts with the 30S near the A site and prevents any tRNAs from binding there. IF3 also binds to the 30S and prevents the 50S from prematurely interacting with the 30S. IF2 delivers the initiator tRNA (or *N*-formylmethionyl-tRNA^{fMet}) to the P site as a ternary complex with GTP.

subunit activates the GTPase activity of EF-Tu, releasing EF-Tu-GDP from the ribosome. These same structural changes also rotate the aminoacyl-tRNA in the A site so that the amino acid is brought into proximity with the aminoacyl-tRNA in the P site on the 50S subunit, a process called *accommodation*. Accommodation aligns the amino acids for peptide-bond formation.

Released EF-Tu is then reset to its GTP form by a second elongation factor, **elongation factor Ts** (Figure 30.22). EF-Ts induces the dissociation of GDP. GTP binds to EF-Tu, and EF-Ts concomitantly departs. It is noteworthy that EF-Tu does not interact with fMet-tRNA^{fMet}. Hence, this initiator tRNA is not delivered to the A site. In contrast, Met-tRNA^{Met}, like all other aminoacyl-tRNAs, does bind to EF-Tu. These findings account for the fact that internal AUG codons are not read by the initiator tRNA. Conversely, IF2 recognizes fMet-tRNA^{fMet} but no other tRNA.

FIGURE 30.22 The GTPase EF-Tu binds aminoacyl-tRNA only in its GTP form.
(1) EF-Tu-GTP binds aminoacyl-tRNA and delivers it to the A site on the ribosome.
(2) Correct codon recognition stimulates the GTPase activity of EF-Tu, which leaves it in its GDP form. (3) EF-Ts binds to EF-Tu-GDP. (4) EF-Ts induces release of GDP. EF-Ts departs as another GTP and aminoacyl-tRNA bind to EF-Tu-GTP, and the complex is ready for another delivery to the ribosome.

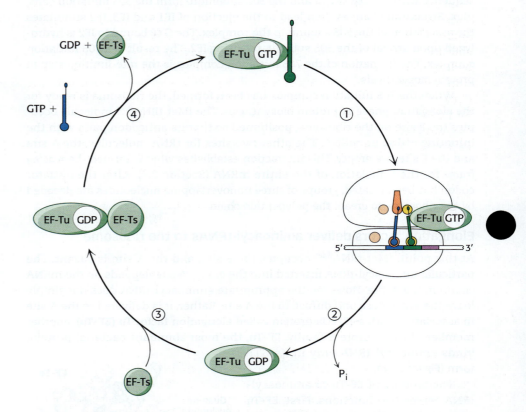

This GTP–GDP cycle of EF-Tu is reminiscent of those of the heterotrimeric G proteins in signal transduction (Section 14.2) and the Ras proteins in growth control (Section 14.4). This similarity is due to their shared evolutionary heritage, seen in the homology of the amino-terminal domain of EF-Tu to the P-loop NTPase domains in the other G proteins. In all these related enzymes, the change in conformation between the GTP and the GDP forms leads to a change in interaction partners. A further similarity is the requirement that an additional protein catalyzes the exchange of GTP for GDP; EF-Ts catalyzes the exchange for EF-Tu, just as an activated receptor does for a heterotrimeric G protein.

Peptidyl transferase catalyzes peptide-bond formation

With both the P site and the A site occupied by aminoacyl-tRNA, the stage is set for the formation of a peptide bond: the N-formylmethionine molecule linked to the initiator tRNA will be transferred to the amino group of the amino acid in the A site. The formation of the peptide bond, one of the most important reactions in life, is a thermodynamically spontaneous reaction catalyzed by a site on the

23S rRNA of the 50S subunit called the **peptidyl transferase center**. This catalytic center is located deep in the 50S subunit near the tunnel that allows the nascent peptide to leave the ribosome.

The ribosome, which enhances the rate of peptide-bond synthesis by a factor of 10^7 over the uncatalyzed reaction ($\sim 10^{-4}$ M^{-1} s^{-1}), derives much of its catalytic power from catalysis by proximity and orientation. The ribosome positions and orients the two substrates so that they are situated to take advantage of the inherent reactivity of an amine group (on the aminoacyl-tRNA in the A site) with an ester (on the initiator tRNA in the P site). The amino group of the aminoacyl-tRNA in the A site, in its unprotonated state, makes a nucleophilic attack on the ester linkage between the initiator tRNA and the N-formylmethionine molecule in the P site (**Figure 30.23A**). The nature of the transition state that follows the attack is not established, and several models are plausible. One model proposes roles for the 2′ OH of the adenosine of the tRNA in the P site and a molecule of water at the peptidyl transferase center. The nucleophilic attack of the α-amino group generates an eight-membered transition state in which three protons are shuttled about in a concerted manner. The proton of the attacking amino group hydrogen bonds to the 2′ oxygen of ribose of the tRNA. The hydrogen of 2′ OH in turn interacts with the oxygen of the water molecule at the center, which then donates a proton to the carbonyl oxygen. Collapse of the transition state with the formation of the peptide bond allows protonation of the 3′ OH of the now-empty tRNA in the P site (**Figure 30.23B**). The stage is now set for translocation and formation of the next peptide bond.

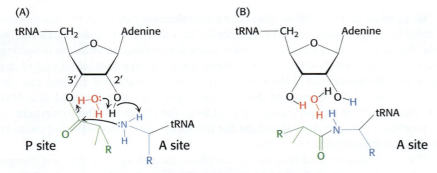

FIGURE 30.23 Peptide-bond formation links amino acids. (A) The amino group of the aminoacyl-tRNA in the A site attacks the carbonyl group of the ester linkage of the peptidyl-tRNA in the P site. An eight-membered transition state is formed with the addition of a molecule of water. Notice that not all atoms are shown, and some bond lengths are exaggerated for clarity. (B) This transition state collapses to form the peptide bond and release the deacylated tRNA.

❖ SELF-CHECK QUESTION

Why is the ribosome considered to be a ribozyme?

GTP hydrolysis-driven translocation of tRNAs and mRNA follows peptide-bond formation

With the formation of the peptide bond, the peptide chain is now attached to the tRNA whose anticodon is in the A site on the 30S subunit. The two ribosomal subunits rotate with respect to one another. This structural change places the acceptor end of the same tRNA and its peptide in the P site of the large subunit while its anticodon end is still in the A site (**Figure 30.24**). However, protein biosynthesis cannot continue unless the next codon is presented in the A site and the site is open for the next aminoacyl-tRNA to be delivered. This requires the translocation of the mRNA and the tRNAs within the ribosome.

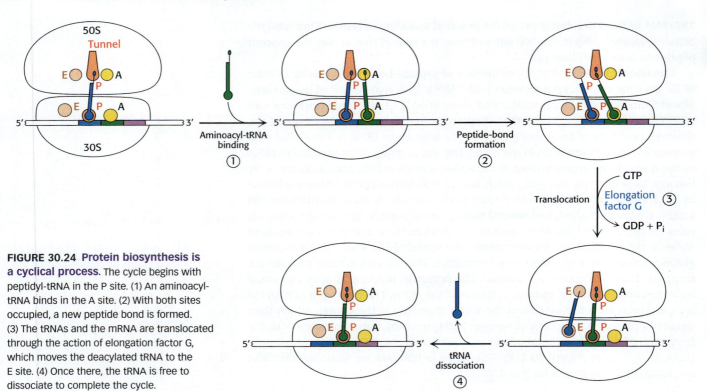

FIGURE 30.24 Protein biosynthesis is a cyclical process. The cycle begins with peptidyl-tRNA in the P site. (1) An aminoacyl-tRNA binds in the A site. (2) With both sites occupied, a new peptide bond is formed. (3) The tRNAs and the mRNA are translocated through the action of elongation factor G, which moves the deacylated tRNA to the E site. (4) Once there, the tRNA is free to dissociate to complete the cycle.

FIGURE 30.25 Translocation repositions tRNAs and mRNA with respect to the ribosome. In the GTP form, EF-G binds to the A site on the 50S subunit. This binding stimulates GTP hydrolysis, inducing a conformational change in EF-G that forces the tRNAs and mRNA to move through the ribosome by a distance corresponding to one codon.

Elongation factor G (EF-G, also called **translocase)** catalyzes the movement of mRNA, at the expense of GTP hydrolysis, by a distance of three nucleotides. Now, the next codon is positioned in the A site for interaction with the incoming aminoacyl-tRNA-EF-Tu-GTP complex. Simultaneously, the anticodon end of the peptidyl-tRNA moves out of the A site into the P site on the 30S subunit and the anticodon end of the deacylated tRNA moves out of the P site into the E site. This deacylated tRNA is subsequently released from the ribosome. The movement of the peptidyl-tRNA into the P site shifts the mRNA by one codon, exposing the next codon to be translated in the A site.

The three-dimensional structure of the ribosome undergoes significant change during translocation, and evidence suggests that translocation may result from properties of the ribosome itself. However, EF-G accelerates the process. A possible mechanism for accelerating the translocation process is shown in **Figure 30.25**. First, EF-G in the GTP form binds to the ribosome near the A site, interacting with the 23S rRNA of the 50S subunit. The binding of EF-G to the ribosome stimulates the GTPase activity of EF-G. On GTP hydrolysis, EF-G undergoes a conformational change that displaces the peptidyl-tRNA in the A site to the P site, which carries the mRNA and the deacylated tRNA with it. The dissociation of EF-G leaves the ribosome ready to accept the next aminoacyl-tRNA into the A site. This cycle of elongation is repeated, with mRNA translation taking place in the 5′ → 3′ direction.

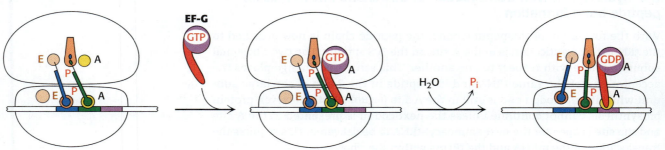

As new aminoacyl-tRNAs move into the A site, the polypeptide is elongated until a termination codon is met. Note that the peptide chain remains in the P site on the 50S subunit throughout this cycle, growing into the exit tunnel.

The direction of translation has important consequences. Recall that transcription is also in the $5' \rightarrow 3'$ direction (Section 29.2). If the direction of translation were opposite that of transcription, only fully synthesized mRNA could be translated. Since the directions of synthesis are the same, translation can occur while an mRNA is being synthesized.

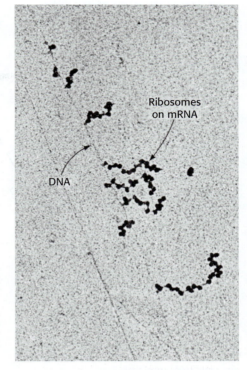

FIGURE 30.26 In prokaryotes, polysomes can simultaneously translate one mRNA. Transcription of a segment of DNA from *E. coli* generates mRNA molecules that are immediately translated by multiple ribosomes (polysomes). [Republished with permission of AAAS, from Visualization of Bacterial Genes in Action, O. L. Miller, Jr., B. A. Hamkalo, and C. A. Thomas, Jr. *Science* 169(1970): 392; permission conveyed through Copyright Clearance Center, Inc.]

❖ SELF–CHECK QUESTION

Summarize the role of G proteins and GTP hydrolysis in translation initiation and elongation in bacteria.

In bacteria, transcription and translation are coupled in space and time

In bacteria, almost no time is lost between transcription and translation. The 5′ end of mRNA interacts with ribosomes very soon after it is made, well before the 3′ end of the mRNA molecule is finished. Recent experiments with *E. coli* using cryo-electron microscopy (Section 4.5) have established the existence of an *expressome*, a transcribing and translating complex consisting of RNA polymerase and the 70S ribosome. The interaction of the polymerase with the ribosome occurs at the carboxyl-terminal domain of the polymerase.

Many ribosomes can be translating an mRNA molecule simultaneously. A group of ribosomes bound to an mRNA molecule is called a *polyribosome* or a **polysome** (**Figure 30.26**). Recent work shows that the ribosomes are arranged so as to protect the mRNA and to facilitate easy exchange of the substrates and products with the cytoplasm. The ribosomes in the polysome are in a helical array around the mRNA with the tRNA binding sites and peptide exit tunnel exposed to the cytoplasm.

Protein biosynthesis is terminated by release factors that read stop codons

The final phase of translation is termination. How does the synthesis of a polypeptide chain come to an end when a stop codon is encountered? No tRNAs with anticodons complementary to the stop codons—UAA, UGA, or UAG—exist in normal cells. Instead, these stop codons are recognized by proteins called **release factors (RFs)**. One of these release factors, RF1, recognizes UAA or UAG. A second factor, RF2, recognizes UAA or UGA. A third factor, RF3, another GTPase, catalyzes the removal of RF1 or RF2 from the ribosome upon release of the newly synthesized protein.

RF1 and RF2 are compact proteins that resemble a tRNA molecule. When bound to the ribosome, the proteins unfold to bridge the gap between the stop codon on the mRNA and the peptidyl transferase center on the 50S subunit (**Figure 30.27**). The RF interacts with the peptidyl transferase center using a loop

FIGURE 30.27 Release factors facilitate the termination of protein biosynthesis. A release factor recognizes a stop codon in the A site and stimulates the release of the completed protein from the tRNA in the P site.

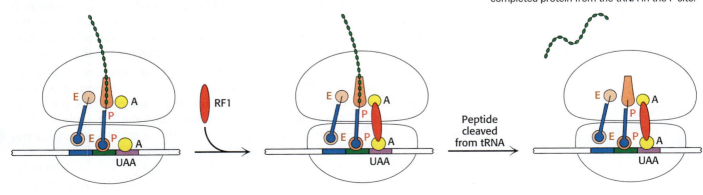

containing a highly conserved glycine-glycine-glutamine (GGQ) sequence, with the glutamine methylated on the amide nitrogen atom of the R group. This modified glutamine (assisted by the peptidyl transferase) is crucial in promoting a water-molecule attack on the ester linkage between the tRNA and the polypeptide chain, freeing the polypeptide chain.

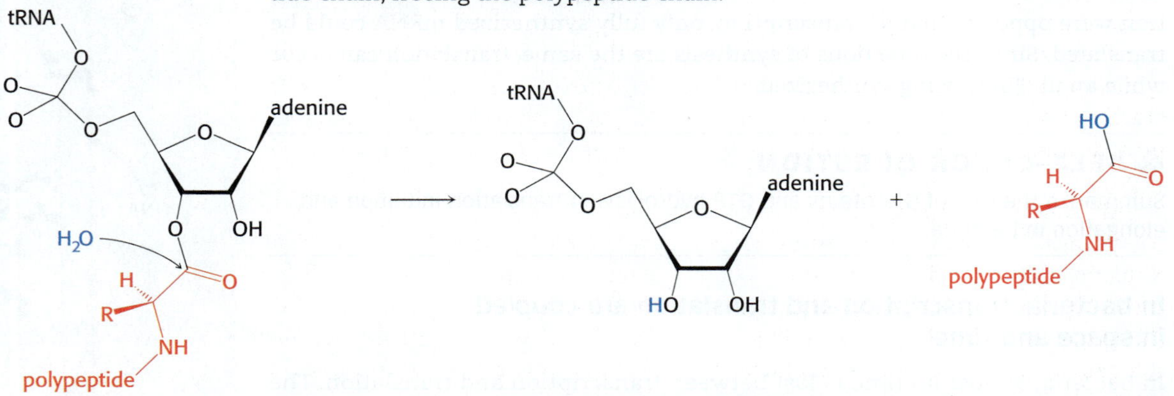

Next, the detached polypeptide leaves the ribosome. Transfer RNA and mRNA remain briefly attached to the 70S ribosome until the entire complex is dissociated through the hydrolysis of GTP in response to the binding of EF-G and another factor, called the *ribosome release factor* (RRF).

Eukaryotic protein biosynthesis differs from bacterial protein biosynthesis primarily in translation initiation

The basic plan of protein biosynthesis in eukaryotes and archaea is similar to that in bacteria. The major structural and mechanistic themes recur in all domains of life. However, eukaryotic protein biosynthesis entails more protein components than does bacterial protein biosynthesis, and some steps are more intricate. Some noteworthy similarities and differences include:

- *Eukaryotic ribosomes are larger than those of bacteria.* They consist of a 60S large subunit and a 40S small subunit, which come together to form an 80S particle having a mass of 4200 kDa, compared with 2300 kDa for the bacterial 70S ribosome. The 40S subunit contains an 18S rRNA that is homologous to the bacterial 16S rRNA. The 60S subunit contains three RNAs: the 5S rRNA, which is homologous to the bacterial 5S rRNA; the 28S rRNA, which is homologous to the bacterial 23S molecule; and the 5.8S rRNA, which is homologous to the 5′ end of the 23S rRNA of bacteria.

- *In eukaryotes, the initiating amino acid is methionine rather than N-formylmethionine.* However, as in bacteria, a special tRNA participates in initiation. This aminoacyl-tRNA is called Met-tRNAi^Met ("i" stands for *initiation*).

- *In eukaryotes, the initiating codon in eukaryotes is almost always AUG (sometimes it is CUG), while in bacteria it can infrequently be GUG or UUG in addition to AUG.* Eukaryotes, in contrast with bacteria, do not have a Shine–Dalgarno sequence on the 5′ side to distinguish initiator AUGs from internal ones. Instead, the AUG nearest the 5′ end of mRNA is usually selected as the start site. Eukaryotes use many more initiation factors than do bacteria, and the interplay of these factors is much more complex. The prefix *eIF* denotes a eukaryotic initiation factor.

Initiation begins with the formation of a ternary complex consisting of the 40S ribosome and Met-tRNAi^Met in association with eIF-2. This complex is called the *43S preinitiation complex* (PIC). The PIC binds to the 5′ end of mRNA and begins searching for an AUG codon by moving step-by-step in the 3′ direction. Meanwhile, initiation factor eIF-4E binds to the 5′ cap of the mRNA (Section 29.3) and facilitates binding of the PIC to the mRNA (**Figure 30.28**). This scanning process is catalyzed by helicases that move along the mRNA powered by ATP hydrolysis. When the anticodon of Met-tRNAi^Met finds and pairs with the AUG codon of mRNA, it signals that the target has been found. In almost all cases, eukaryotic mRNA has only one start site and hence is the template for only a single protein. In contrast, a bacterial mRNA can have multiple start sites and can serve as a template for the synthesis of several proteins.

The difference in initiation mechanism between bacteria and eukaryotes is, in part, a consequence of the difference in RNA processing. The 5′ end of mRNA is readily available to ribosomes immediately after transcription in bacteria. In contrast, in eukaryotes, pre-mRNA must be processed and transported to the cytoplasm before translation is initiated. The 5′ cap provides an easily recognizable starting point. In addition, the complexity of eukaryotic translation initiation provides another mechanism for regulation of gene expression that we shall explore further in Chapter 31.

Although most eukaryotic mRNA molecules rely on the 5′ cap to initiate protein biosynthesis, recent work has established that some mRNA molecules can recruit ribosomes for initiation without the use of a 5′ cap and cap-binding proteins. In these mRNAs, highly structured RNA sequences called *internal ribosome entry sites* (IRES) facilitate 40S ribosome binding to the mRNA. IRES were first discovered in the genomes of RNA viruses and have since been found in other viruses, as well as in a subset of cellular mRNA that appears to take part in development and stress responses (such as to a viral infection). The molecular mechanism by which IRES function to initiate protein biosynthesis remains to be determined.

- *Eukaryotic mRNAs adopt a unique structure.* Soon after the PIC binds the mRNA, eIF-4G links eIF-4E to a protein associated with the poly(A) tail, called *the poly(A)-binding protein 1* (PABP1; **Figure 30.29**). Cap and tail are thus brought together to form a circle of mRNA. One benefit of circularization of eukaryotic mRNAs is that it may prevent translation of mRNA molecules that have lost their poly(A) tails (Section 29.3).

- *Bacterial elongation and release factors have structural and functional counterparts in eukaryotes.* Eukaryotic elongation factors EF1α and EF1βγ are the counterparts of bacterial EF-Tu and EF-Ts. The GTP form of EF1α delivers aminoacyl-tRNA to the A site of the ribosome, and EF1βγ catalyzes the exchange of GTP for bound GDP. Eukaryotic EF2 mediates GTP-driven translocation in much the same way as does bacterial EF-G. Termination in eukaryotes is carried out by a single release factor, eRF1, compared with two in bacteria. Release factor eIF-3 accelerates the activity of eRF-1.

- *The components of the translation machinery in higher eukaryotes are organized into large complexes associated with the cytoskeleton.* This association is believed to facilitate the efficiency of protein biosynthesis. Recall that organization of elaborate biochemical processes into physical complexes is a recurring theme in biochemistry (for example, the ATP synthasome in Section 18.5 and purine biosynthesis in Section 26.2).

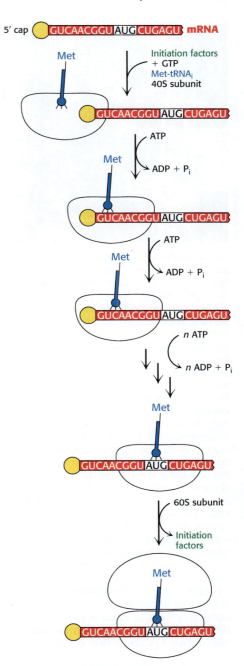

FIGURE 30.28 Translation initiation differs in eukaryotes as compared with prokaryotes. In eukaryotes, translation initiation starts with the assembly of a complex on the 5′ cap that includes the 40S subunit and Met-tRNAi^{Met}. Driven by ATP hydrolysis, this complex scans the mRNA until the first AUG is reached. The 60S subunit is then added to form the 80S initiation complex.

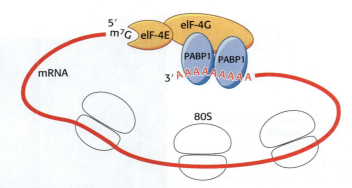

FIGURE 30.29 Protein interactions circularize eukaryotic mRNA. Eukaryotic IFs bound to the 5′ cap and PABP1 bound to the 3′ poly(A) tail come together to create a circular mRNA. This circularization may regulate translation of certain mRNAs. [Information from H. Lodish et al., *Molecular Cell Biology*, 5th ed. (W. H. Freeman and Company, 2004), Fig. 4.31.]

ADA YONATH Jerusalem-born Ada Yonath and her family experienced poverty and grief, but her parents always supported her desire for understanding the world. In 1970, after completing her postdoctoral work in the United States, Dr. Yonath returned to Israel and set up the country's first biological x-ray crystallography lab at the Weizmann Institute. Fascinated by genetic code translation, she began structural studies on ribosomes but, like many others, failed to arrive at a solution, owing to the ribosomes' complexity and marked instability. She continued only a few years later, after she read about how the ribosomes of hibernating polar bears are arranged in an ordered fashion within their cells, indicating to her the natural tendency to maintain stability. This inspired her to pursue the crystallization of ribosomes from bacteria that live under pressure. Then she confronted another problem: extreme damage to the ribosomal crystals by the x-rays used to visualize them, which prevented structure determination. Her novel solution—taking measurements of frozen crystals of ribosomes at sub-zero temperatures, called *cryo-crystallography*—became popular immediately and has allowed the determination of more than 100,000 structures of various biological materials. Dr. Yonath was awarded the Nobel Prize in Chemistry with Drs. Venkatraman Ramakrishnan and Thomas Steitz in 2009.

Ribosomes selectively control gene expression

Ribosomes were once thought to all be the same, regardless of cell type or physiological state. Recent evidence suggests otherwise and points to a more significant role for ribosomes in the regulation of gene expression.

First, the composition of ribosomes has been shown to change in a cell-specific manner. For instance, mutations in certain ribosomal proteins result in Diamond-Blackfan anemia, which is characterized by bone marrow failure in blood stem-cell differentiation. Although protein biosynthesis is impaired in the blood stem cells, protein biosynthesis in other tissues is apparently normal. If mutations in ribosomal proteins lead to loss of ribosomal function in a tissue-specific manner, might changes to ribosomal protein amounts or posttranslational modifications also regulate ribosomal function in a tissue-specific manner? This question remains to be answered.

Second, nonribosomal proteins can also control ribosome activity. One example is FMRP (fragile X mental retardation protein), which is essential for human cognitive development and also for reproductive function in women. FMRP, an RNA-binding protein, binds to the 80S ribosome to inhibit the translation of specific mRNAs. Mutations in FMRP result in cognitive disabilities, ovarian failure, and other pathologies. In another startling example, an isozyme of the glycolytic enzyme pyruvate kinase, pyruvate kinase M (Section 16.3), binds to the ribosome to control the translation of proteins destined for the endoplasmic reticulum.

How else might ribosomes regulate translation? Ribosome concentration, which varies 3- to 10-fold among different tissues, may play a role. For instance, a low ribosome concentration might discriminate against certain mRNAs that do not initiate well. This might be the case if mRNA has a complex three-dimensional structure in the 5′ untranslated region, the part of the mRNA upstream of the initiator codon. Indeed, an insufficient ribosome concentration that results in the impaired translation of a specific subset of mRNA is hypothesized to cause some cases of Diamond-Blackfan anemia. What determines ribosome concentration in different cell types remains unknown.

Scientists have manipulated protein biosynthesis pathways to incorporate unnatural amino acids in preselected positions

Knowledge of how aminoacyl-tRNA synthetases charge tRNAs with amino acids, and how the ribosome decodes the information in mRNA, has allowed biochemists to synthesize modified proteins with unnatural amino acids (amino acids that are not normally found in nature). These unnatural amino acids can be incorporated at specific positions within the polypeptide chain. Examples of unnatural amino acids include those that have been modified to contain a fluorescent moiety (**Figure 30.30**), heavy atoms, or other reactive groups,

(A)

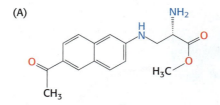

**(2S)-3-[(6-Acetylnaphthalen-2-yl)amino]-2-aminopropanoic acid
(L-ANAP)**

(B)

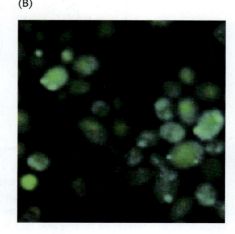

FIGURE 30.30 Unnatural amino acids can be incorporated into proteins. Unnatural amino acids give proteins novel characteristics that are useful for experimental studies. (A) The structure of L-ANAP ((2S)-3-[(6-Acetylnaphthalen-2-yl)amino]-2-aminopropanoic acid), a fluorescent unnatural amino acid, is shown. (B) Fluorescent proteins are easy to visualize and thus localize, as seen in this image of yeast cells that express a protein containing L-ANAP. [Image from Hsieh et al. 2014. Monitoring Protein Misfolding by Site-Specific Labeling of Proteins In Vivo. PLoS ONE 9:e99395, 10.1371/journal.pone.0099395.]

providing the protein that contains them with new and useful chemical and biological properties.

How are unnatural amino acids introduced into proteins? There are multiple strategies to accomplish this, but in all cases the three key players involved are a tRNA, the unnatural amino acid itself, and the aminoacyl-tRNA synthetase that charges the particular tRNA with the unnatural amino acid. In addition to these players, it is often necessary to usurp one of the three termination codons to introduce the unnatural amino acid into the polypeptide chain (since all other mRNA codons are already set to code for the 20 natural amino acids). Recall that termination codons are not recognized by any cellular tRNAs. To introduce an unnatural amino acid into a protein, the tRNA is engineered to recognize a termination codon, and the aminoacyl-tRNA synthetase is engineered to charge that tRNA with the unnatural amino acid. All of these components can be introduced into a living cell for in vivo protein production. More recent research focuses on creating unnatural proteins in vitro.

Modifying proteins through the incorporation of unnatural amino acids is likely less disruptive to overall protein function than posttranslational modification or labeling. These "designer" proteins can be used in experimental studies of protein folding, localization, and many other characteristics, both in vitro and in vivo. Consider, for example, a synthetic protein with fluorescent unnatural amino acids incorporated near its surface: the exact in vivo sub-cellular localization of this protein can be easily determined via fluorescent microscopy. If the functions of this protein are largely undetermined, knowing where it is located, and what other factors it colocalizes with, might give some insights into what its functions might be.

30.4 Ribosomes Bound to the Endoplasmic Reticulum Manufacture Secretory and Membrane Proteins

Not all newly synthesized proteins are destined to function in the cytoplasm. Eukaryotic cells can direct proteins to internal sites such as mitochondria, the nucleus, and the endoplasmic reticulum, a process called *protein targeting* or *protein sorting*. How is this sorting accomplished? There are two general mechanisms by which sorting takes place. In one mechanism, the protein is synthesized in the cytoplasm, and then the completed protein is delivered to its intracellular location posttranslationally. Proteins destined for the nucleus, chloroplasts, mitochondria, and peroxisomes are delivered by this general process. The other mechanism, termed the *secretory pathway*, directs proteins into the endoplasmic reticulum (ER), the extensive membrane system that comprises about half the total membrane of a cell. This mechanism delivers the proteins to the ER cotranslationally—that is, while the protein is being synthesized. Approximately 30% of all proteins are sorted by the secretory pathway, including secreted proteins, residents of the ER, the Golgi complex, lysosomes, and integral membrane proteins of these organelles as well as integral plasma membrane proteins. We will focus our attention on the secretory pathway only.

Protein biosynthesis begins on ribosomes that are free in the cytoplasm

In eukaryotic cells, a ribosome remains free in the cytoplasm unless it is directed to the ER. The region of the ER that ribosomes bind to is called the *rough ER* because of its studded appearance in contrast with the *smooth ER*, which is devoid of ribosomes (**Figure 30.31**). The synthesis of proteins sorted by the secretory pathway begins on free ribosomes that soon associate with the ER. Free ribosomes that are synthesizing proteins for use in the cell are apparently identical with those attached to the ER, so ribosomes synthesizing proteins destined to enter the ER need to bind to the ER. How is this accomplished?

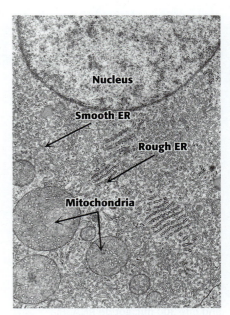

FIGURE 30.31 Ribosomes are bound to the endoplasmic reticulum. In this electron micrograph, ribosomes appear as small black dots binding to the cytoplasmic side of the endoplasmic reticulum to give a rough appearance. In contrast, the smooth endoplasmic reticulum is devoid of ribosomes. [Don W. Fawcett/Science Source.]

Signal sequences mark proteins for translocation across the endoplasmic reticulum membrane

The synthesis of proteins destined to leave the cell or become embedded in the plasma membrane begins on a free ribosome. However, shortly after synthesis begins, the process is halted until the ribosome is directed to the cytoplasmic side of the ER. When the ribosome docks with the ER membrane, protein biosynthesis begins again. As the newly forming peptide chain exits the ribosome, it is transported—while being translated—through the membrane into the lumen of the ER. The translocation consists of four components:

1. *The signal sequence.* The **signal sequence** is a sequence of 9 to 12 hydrophobic amino acid residues, sometimes containing positively charged amino acids (**Figure 30.32**). This sequence, which adopts an α-helical structure, is usually near the amino terminus of the nascent polypeptide chain. The presence of the signal sequence identifies the nascent peptide as one that must cross the ER membrane. Some signal sequences are maintained in the mature protein, whereas others are cleaved by a **signal peptidase** on the lumenal side of the ER membrane.

Cleavage site

Human growth hormone	M A T G S R T S L L L A F G L L C L P W L Q E G S A	F P T
Human proinsulin	M A L W M R L L P L L A L L A L W G P D P A A A	F V N
Bovine proalbumin	M K W V T F I S L L L F S S A Y S	R G V
Mouse antibody H chain	M K V L S L L Y L L T A I P H I M S	D V Q
Chicken lysozyme	M R S L L I L V L C F L P K L A A L G	K V F
Bee promellitin	M K F L V N V A L V F M V V Y I S Y I Y A	A P E
Drosophila glue protein	M K L L V V A V I A C M L I G F A D P A S G	C K D
Zea maize protein 19	M A A K I F C L I M L L G L S A S A A T A	S I F
Yeast invertase	M L L O A F L F L L A G F A A K I S A	S M T
Human influenza virus A	M K A K L L V L L Y A F V A G	D Q I

FIGURE 30.32 The signal sequences of eukaryotic secretory and plasma membrane proteins contain hydrophobic residues. This alignment of the amino-terminal signal sequences of some eukaryotic secretory and plasma membrane proteins highlights the hydrophobic core (yellow), which is preceded by basic residues (blue) and followed by a cleavage site (red) for signal peptidase.

2. *The signal-recognition particle.* The **signal-recognition particle (SRP)** recognizes the signal sequence and binds the sequence and the ribosome as soon as the signal sequence exits the ribosome. SRP then shepherds the ribosome and its nascent polypeptide chain to the ER membrane. SRP is a ribonucleoprotein consisting of a 7S RNA and six different proteins (**Figure 30.33**). One protein, SRP54, is a GTPase that is crucial for SRP function. SRP samples all ribosomes until it locates one exhibiting a signal sequence. After SRP binds to the signal sequence, interactions between the ribosome and the SRP block the elongation-factor-binding site, thereby halting protein biosynthesis.

3. *The SRP receptor.* The SRP–ribosome complex diffuses to the ER, where SRP binds the **SRP receptor (SR)**, an integral membrane protein consisting of two subunits, SRα and SRβ. SRα is, like SRP54, a GTPase.

4. *The translocon.* The SRP–SR complex delivers the ribosome to the translocation machinery, called the **translocon**, a multisubunit assembly of integral and peripheral membrane proteins. The translocon is a protein-conducting

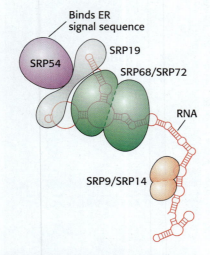

Binds ER
signal sequence

SRP54

SRP19

SRP68/SRP72

RNA

SRP9/SRP14

FIGURE 30.33 The signal-recognition particle (SRP) consists of six proteins and one 300-nucleotide RNA. The RNA has a complex structure with many double-helical stretches punctuated by single-stranded regions, shown as circles.

[Information from H. Lodish et al., *Molecular Cell Biology*, 5th ed. (W. H. Freeman and Company, 2004). Data from K. Strub et al., *Mol. Cell Biol.* 11:3949–3959, 1991, and S. High and B. Dobberstein, *J. Cell Biol.* 113:229–233, 1991.]

channel that opens when the translocon and ribosome bind to each other. Protein biosynthesis resumes with the growing polypeptide chain passing through the translocon channel into the lumen of the ER.

The interactions of the components of the translocation machinery are shown in **Figure 30.34**. Both the SRP54 and the SRα subunits of SR must bind GTP to facilitate the formation of the SRP–SR complex. For the SRP–SR complex to then deliver the ribosome to the translocon, the two GTP molecules—one in SRP and the other in SR—are aligned in what is essentially an active site shared by the two proteins. The formation of the alignment is catalyzed by the 7S RNA of the SRP. After the ribosome has been passed along to the translocon, the GTPs are hydrolyzed, SRP and SR dissociate, and SRP is free to search for another signal sequence to begin the cycle again. Thus, SRP acts catalytically. The signal peptidase, which is associated with the translocon in the lumen of the ER, removes the signal sequence from most proteins.

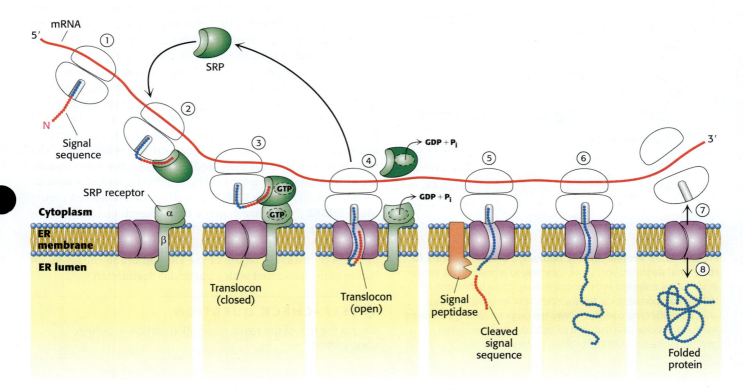

FIGURE 30.34 The SRP targeting cycle delivers nascent proteins containing a signal sequence to the ER. (1) Protein biosynthesis begins on free ribosomes. (2) After the signal sequence has exited the ribosome, it is bound by the SRP, and protein biosynthesis halts. (3) The SRP–ribosome complex docks with the SRP receptor in the ER membrane. (4) The SRP and the SRP receptor simultaneously hydrolyze bound GTPs. Protein biosynthesis resumes and the SRP is free to bind another signal sequence. (5) The signal peptidase may remove the signal sequence as it enters the lumen of the ER. (6) Protein biosynthesis continues as the protein is synthesized directly into the ER. (7) On completion of protein biosynthesis, the ribosome is released. (8) The protein tunnel in the translocon closes. [Information from H. Lodish et al., *Molecular Cell Biology*, 5th ed. (W. H. Freeman and Company, 2004), Fig. 16.6.]

Transport vesicles carry cargo proteins to their final destinations

As the proteins are synthesized, they fold into their three-dimensional structures in the lumen of the ER. Some proteins are modified by the attachment of N-linked carbohydrates (Section 11.3). Finally, the proteins must be sorted and transported to their final destinations.

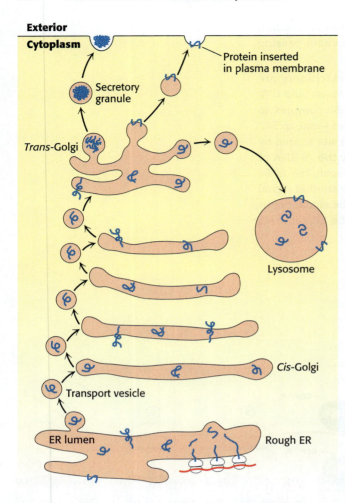

Exterior
Cytoplasm

Protein inserted in plasma membrane

Secretory granule

Trans-Golgi

Lysosome

Cis-Golgi

Transport vesicle

ER lumen   Rough ER

FIGURE 30.35 Protein-sorting pathways deliver proteins to their final destinations. Newly synthesized proteins in the lumen of the ER are collected into membrane buds. These buds pinch off to form transport vesicles. The transport vesicles carry the cargo proteins to the Golgi complex, where the cargo proteins are modified. Transport vesicles then carry the cargo to the final destination as directed by the v-SNARE and t-SNARE proteins.

Regardless of the destination, the principles of transport are the same. Transport is mediated by **transport vesicles** that bud off the ER (**Figure 30.35**). Transport vesicles from the ER carry their cargo (the proteins) to the Golgi complex, where the vesicles fuse and deposit the cargo inside the complex. There the cargo proteins are further modified—for instance, by the attachment of O-linked carbohydrates. From the Golgi complex, transport vesicles carry the cargo proteins to their final destinations, as shown in Figure 30.35.

How does a protein end up at the correct destination? A newly synthesized protein will float inside the ER lumen until it binds to an integral membrane protein called a *cargo receptor*. This binding sequesters the cargo protein into a small region of the membrane that can subsequently form a membrane bud. The bud will carry the protein to a specific destination—plasma membrane, lysosome, or cell exterior. The key to ensuring that the protein reaches the proper destination is that the protein must bind to a receptor in the ER region associated with the protein's destination. To ensure the proper match of protein with ER region, cargo receptors recognize various characteristics of the cargo protein, such as a particular amino acid sequence or an added carbohydrate.

The formation of buds is facilitated by the binding of coat proteins to the cytoplasmic side of the bud. The coat proteins associate with one another to pinch off the vesicle. After the transport vesicle has formed and is released, the coat proteins are shed to reveal another integral protein, called **v-SNARE** ("v" for vesicle) (Section 12.6). The binding of v-SNARE to a particular **t-SNARE** ("t" for target) in the target membrane leads to the fusion of the transport vesicle to the target membrane, and the cargo is delivered. Thus, the assignment of identical v-SNARE proteins to the same region of the ER membrane causes an ER region to be associated with a particular destination.

❖ **SELF–CHECK QUESTION**

What is the role of the signal-recognition particle in protein translocation?

30.5 A Variety of Antibiotics and Toxins Inhibit Protein Biosynthesis

Many chemicals that inhibit various aspects of protein biosynthesis have been identified. These chemicals are powerful experimental tools and clinically useful drugs.

Some antibiotics inhibit protein biosynthesis

The differences between eukaryotic and bacterial ribosomes can be exploited for the development of antibiotics (**Table 30.4**). For example, the antibiotic streptomycin, a highly basic trisaccharide, interferes with the binding of fMet-tRNA^{fMet} to ribosomes in bacteria and thereby prevents the correct initiation of protein biosynthesis. Other aminoglycoside antibiotics such as neomycin, kanamycin, and gentamycin interfere with the interaction between tRNA and the 16S rRNA of the 30S subunit of bacterial ribosomes. Chloramphenicol acts by inhibiting peptidyl transferase activity. Erythromycin binds to the 50S subunit and blocks

Streptomycin

TABLE 30.4 Antibiotic inhibitors of protein biosynthesis

Antibiotic	Action
Streptomycin and other aminoglycosides	Inhibit initiation and cause the misreading of mRNA (bacteria)
Tetracycline	Binds to the 30S subunit and inhibits the binding of aminoacyl-tRNAs (bacteria)
Chloramphenicol	Inhibits the peptidyl transferase activity of the 50S ribosomal subunit (bacteria)
Cycloheximide	Inhibits translocation (eukaryotes)
Erythromycin	Binds to the 50S subunit and inhibits translocation (bacteria)
Puromycin	Causes premature chain termination by acting as an analog of aminoacyl-tRNA (bacteria and eukaryotes)

translocation. Cycloheximide, another antibiotic, blocks translocation in eukaryotic ribosomes, making it a useful laboratory tool for blocking protein biosynthesis in eukaryotic cells.

The antibiotic puromycin inhibits protein biosynthesis in both bacteria and eukaryotes by causing nascent polypeptide chains to be released before their synthesis is completed. Puromycin is an analog of the terminal part of aminoacyl-tRNA (**Figure 30.36**) that binds to the A site on the ribosome and blocks the entry of aminoacyl-tRNA. Furthermore, puromycin contains an α-amino group. This amino group, like the one on aminoacyl-tRNA, forms a peptide bond with the carboxyl group of the growing peptide chain. The product, a peptide having a covalently attached puromycin residue at its carboxyl end, dissociates from the ribosome. While no longer used medicinally, puromycin remains an experimental tool for the investigation of protein biosynthesis.

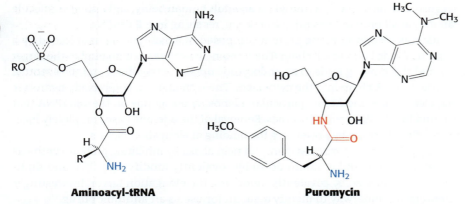

Aminoacyl-tRNA   **Puromycin**

FIGURE 30.36 The antibiotic puromycin resembles the aminoacyl terminus of an aminoacyl-tRNA. The amino group of puromycin joins the carboxyl group of the growing polypeptide chain to form peptidyl-puromycin that dissociates from the ribosome. Peptidyl-puromycin is stable because puromycin has an amide (shown in red) rather than an ester linkage.

Diphtheria toxin blocks protein biosynthesis in eukaryotes by inhibiting translocation

Many antibiotics, harvested from bacteria for medicinal purposes, are inhibitors of bacterial protein biosynthesis. However, some bacteria produce compounds that inhibit eukaryotic protein biosynthesis, leading to diseases such as diphtheria, which was a major cause of childhood death before the advent of effective immunization. Pathogenic strains of *Corynebacterium diphtheriae*, a bacterium that grows in the upper respiratory tract of an infected person, have a prophage inserted into their genomes that produces the diphtheria toxin. The toxin consists of a single polypeptide chain that is cleaved into a 21-kDa A fragment and a 40-kDa B fragment shortly after entering the target cell. The role of the B fragment in the intact protein is to bind to the cell, enabling the toxin to enter the

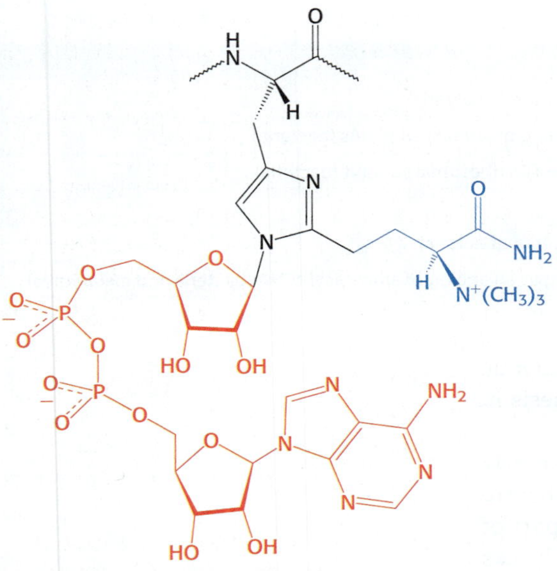

ADP-ribose

FIGURE 30.37 Diphtheria toxin inhibits protein biosynthesis in eukaryotes. Diphtheria toxin blocks translocation by catalyzing the transfer of an ADP-ribose unit from NAD$^+$ to diphthamide, a modified amino acid residue in EF2 (translocase). Diphthamide is formed by the posttranslational modification (blue) of a histidine residue.

cytoplasm of its target cell. The A fragment catalyzes the covalent modification of EF2, the elongation factor catalyzing translocation in eukaryotic protein biosynthesis.

A single A fragment of the toxin in the cytoplasm can kill a cell. Why is it so lethal? EF2 contains diphthamide, an unusual amino acid residue that enhances the fidelity of codon shifting during translocation. Diphthamide is formed by a highly conserved complicated pathway that posttranslationally modifies histidine. The A fragment of the diphtheria toxin catalyzes the transfer of the ADP ribose unit of NAD$^+$ to the diphthamide ring (**Figure 30.37**). This ADP ribosylation of a single side chain of EF2 blocks EF2's capacity to carry out the translocation of the growing polypeptide chain. Protein biosynthesis ceases, accounting for the remarkable toxicity of diphtheria toxin. A few micrograms of diphtheria toxin are usually lethal in an unimmunized person.

Some toxins modify 28S ribosomal RNA

α-Sarcin is a 17kDa ribonuclease secreted by filamentous fungi. It penetrates cells by interacting with membrane lipids and cleaves a specific single phosphodiester linkage in the 28S rRNA of the large ribosomal subunit. This cleavage completely inhibits protein biosynthesis.

Another toxin, ricin, also affects the 28S rRNA in the same region as α-sarcin (the sarcin-ricin loop; Section 30.3) but is a glycoside hydrolase, not a ribonuclease. Ricin is a small protein (65 kDa) found in the seeds of the castor oil plant, *Ricinus communis*. It is indeed a deadly molecule because as little as 500 μg is lethal for an adult human being, and a single molecule can inhibit all protein biosynthesis in a cell, resulting in cell death.

Ricin is a heterodimeric protein composed of a catalytic A chain joined by a single disulfide bond to a B chain. The B chain allows the toxin to bind to the target cell, and this binding leads to an endocytotic uptake of the dimer and the eventual release of the A chain into the cytoplasm. The A chain is an N-glycoside hydrolase that cleaves adenine from a particular adenosine nucleotide on the 28S rRNA that is found in all eukaryotic ribosomes. Removal of the adenine base completely inactivates the ribosome by preventing the binding of elongation factors.

Thus, diphtheria toxin, α-sarcin, and ricin all act by inhibiting protein-synthesis elongation; ricin and α-sarcin do so by covalently modifying rRNA, and diphtheria toxin does so by covalently modifying the elongation factor. Interestingly, attempts are underway to modify α-sarcin for use as an antitumor drug.

Summary

30.1 Protein Biosynthesis Requires the Translation of Nucleotide Sequences into Amino Acid Sequences

- Protein biosynthesis is called translation because information in a nucleic acid sequence is translated into a different language, the sequence of amino acids in a protein.

- This complex process is mediated by the coordinated interplay of more than a hundred macromolecules, including mRNA, rRNAs, tRNAs, aminoacyl-tRNA synthetases, and protein factors.

- Transfer RNAs are the adaptors that make the link between a nucleic acid and an amino acid.

30.2 Aminoacyl-tRNA Synthetases Establish the Genetic Code

- Each amino acid is activated and linked to a specific tRNA by an aminoacyl-tRNA synthetase. There is at least one specific aminoacyl-tRNA synthetase and at least one specific tRNA for each amino acid.

- By specifically recognizing both amino acids and tRNAs, aminoacyl-tRNA synthetases implement the instructions of the genetic code.

- The codons of mRNA recognize the anticodons of transfer RNAs rather than the amino acids attached to the tRNAs. A codon on mRNA forms base pairs with the anticodon of the tRNA.

- Some tRNAs are recognized by more than one codon because pairing of the third base of a codon is less crucial than that of the other two (the wobble mechanism).

30.3 The Ribosome Is the Site of Protein Biosynthesis

- Protein biosynthesis takes place on ribosomes, which are huge ribonucleoprotein particles consisting of large and small subunits.

- In *E. coli*, the 70S ribosome (2300 kDa) is made up of 30S and 50S subunits.

- The ribosome includes three sites for tRNA binding, called the A (aminoacyl) site, the P (peptidyl) site, and the E (exit) site.

- Protein biosynthesis takes place in three phases: initiation, elongation, and termination.

- In bacteria, mRNA, N-formylmethionyl-tRNA^{fMet} and a 30S ribosomal subunit come together with the assistance of initiation factors to form a 30S initiation complex. A 50S ribosomal subunit then joins this complex to form a 70S initiation complex, in which fMet-tRNA^{fMet} occupies the P site of the ribosome.

- During elongation, an EF-Tu-GTP complex delivers the appropriate aminoacyl-tRNA to the ribosome's A site as a ternary complex.

- A peptide bond is formed when the amino group of the aminoacyl-tRNA nucleophilically attacks the ester linkage of the peptidyl-tRNA.

- After peptide-bond formation, the tRNAs and mRNA are translocated for the next cycle to begin. The deacylated tRNA moves to the E site and then leaves the ribosome, and the peptidyl-tRNA moves from the A site into the P site.

- Protein biosynthesis is terminated by release factors, which recognize the termination codons UAA, UGA, and UAG and cause the hydrolysis of the ester bond between the polypeptide and tRNA.

- Eukaryotic ribosomes (80S) consist of a 40S small subunit and a 60S large subunit.

- The initiation of protein biosynthesis is more complex in eukaryotes than in bacteria.

- The regulation of translation in eukaryotes provides a means for regulating gene expression.

30.4 Ribosomes Bound to the Endoplasmic Reticulum Manufacture Secretory and Membrane Proteins

- Proteins contain signals that determine their ultimate destination. The synthesis of all proteins begins on free ribosomes in the cytoplasm.

- In eukaryotes, protein biosynthesis continues in the cytoplasm unless the nascent chain contains a signal sequence that directs the ribosome to the endoplasmic reticulum.

- The signal-recognition particle recognizes signal sequences and brings ribosomes bearing them to the ER. The nascent chain is then translocated across the ER membrane.

30.5 A Variety of Antibiotics and Toxins Inhibit Protein Biosynthesis

- Many clinically important antibiotics function by inhibiting protein biosynthesis.

- All steps of protein biosynthesis are susceptible to inhibition by antibiotics or toxins.

Key Terms

translation (p. 917)

codon (p. 918)

transfer RNA (tRNA) (p. 918)

anticodon (p. 918)

wobble hypothesis (p. 920)

ribosome (p. 921)

aminoacyl-tRNA (p. 921)

aminoacyl-tRNA synthetase (p. 921)

large subunit (50S) (p. 927)

small subunit (30S) (p. 927)

Shine–Dalgarno sequence (p. 930)

initiation factor (p. 931)

elongation factor Tu (EF-Tu) (p. 931)

elongation factor Ts (EF-Ts) (p. 932)

peptidyl transferase center (p. 933)

elongation factor G (EF-G) (translocase) (p. 934)

polysome (p. 935)

release factor (RF) (p. 935)

signal sequence (p. 940)

signal peptidase (p. 940)

signal-recognition particle (SRP) (p. 940)

SRP receptor (SR) (p. 940)

translocon (p. 940)

transport vesicle (p. 942)

v-SNARE (p. 942)

t-SNARE (p. 942)

Problems

1. Which of the following reaction steps are required to form an aminoacyl-tRNA? (Select all that apply.) ❖ 1

(a) Formation of an aminoacyl adenylate from an amino acid and ATP

(b) Transfer of an aminoacyl group to a tRNA molecule

(c) Formation of an acyl adenylate from a fatty acid and ATP

(d) Attack of the ester linkage of a peptidyl-tRNA by an amino group

2. Ribosomes were isolated from bacteria grown in a "heavy" medium (^{13}C and ^{15}N) and from bacteria grown in a "light" medium (^{12}C and ^{14}N). These 70S ribosomes were then added to an in vitro system engaged in protein synthesis. Several hours later, an aliquot was removed and analyzed by density-gradient centrifugation. How many bands of 70S ribosomes would you expect to see in the density gradient? ❖ 2

3. A tRNA with a ACA anticodon is enzymatically conjugated to ^{14}C-labeled cysteine. The cysteine unit is then chemically modified to alanine. The altered aminoacyl-tRNA is added to a protein-synthesizing system containing normal components except for this tRNA. The mRNA added to this mixture contains the following sequence:

5′-UUUUGCCAUGUUUGUGCU-3′

What is the sequence of the corresponding radiolabeled peptide? ❖ 1

4. Match each factor with its associated phase of translation in prokaryotes. Factors can be used more than once. ❖ 2

(a) Initiation
(b) Elongation
(c) Termination

1. GTP
2. AUG
3. fMet
4. RRF
5. IF2
6. Shine–Dalgarno
7. EF-Tu
8. Peptidyl transferase
9. UGA
10. Transformylase

5. The two basic mechanisms for the elongation of biomolecules are represented in the adjoining illustration. In type 1, the activating group (X) is released from the growing chain. In type 2, the activating group is released from the incoming unit as it is added to the growing chain. Indicate whether each of the following biosyntheses is by means of a type 1 or a type 2 mechanism: ❖ 1

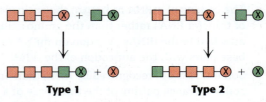

Type 1   **Type 2**

(a) Glycogen synthesis

(b) Fatty acid synthesis

(c) $C_5 \rightarrow C_{10} \rightarrow C_{15}$ in cholesterol synthesis

(d) DNA synthesis

(e) RNA synthesis

(f) Protein synthesis

6. What is the smallest number of molecules of ATP and GTP consumed in the synthesis of a 200-residue protein, starting from amino acids? Assume that the hydrolysis of PP_i is equivalent to the hydrolysis of ATP for this calculation. ❖ 1, ❖ 2

7. Compare and contrast protein biosynthesis by ribosomes with protein synthesis by the solid-phase method (Section 4.4). ❖ 1, ❖ 2

8. Suppose that you have a protein-synthesis system that is synthesizing a protein designated A. Furthermore, you know that protein A has four trypsin-sensitive sites, equally spaced in the protein, that, on digestion with trypsin, yield the peptides A_1, A_2, A_3, A_4, and A_5. Peptide A_1 is the amino-terminal peptide, and A_5 is the carboxyl-terminal peptide. Finally, you know that your system requires 4 minutes to synthesize a complete protein A. At t = 0, you add all 20 amino acids, each carrying a ^{14}C label. ❖ 2

(a) At t = 1 minute, you isolate intact protein A from the system, cleave it with trypsin, and isolate the five peptides. Which peptide is most heavily labeled?

(b) At t = 3 minutes, what will be the order of the labeling of peptides from greatest to least?

(c) What does this experiment tell you about the direction of protein synthesis?

9. Every ribosome has three tRNA binding sites: the A site, the P site, and the E site. For the given sentences, identify whether it refers to the P site, A site, or E site. ❖ 2

(a) This site binds the tRNA molecule that is attached to the growing peptide chain.

(b) This site binds the aminoacyl-tRNA.

(c) This site may bind deacylated (uncharged) tRNA.

(d) This site binds the peptidyl-tRNA.

10. EF-Tu, a member of the G-protein family, plays a crucial role in the elongation process of translation. Suppose

that a slowly hydrolyzable analog of GTP were added to an elongating system. What would be the effect on the rate of protein synthesis? ❖ **2**

11. List the differences between bacterial and eukaryotic protein biosynthesis. ❖ **3**

12. Match each label or component with its associated ribosomal subunit. Some labels can be used more than once. ❖ **3**

(a) Prokaryotic large ribosomal subunit

(b) Prokaryotic small ribosomal subunit

(c) Eukaryotic large ribosomal subunit

(d) Eukaryotic small ribosomal subunit

1. 30S subunit
2. 40S subunit
3. 50S subunit
4. 60S subunit
5. 5S rRNA
6. 5.8S rRNA
7. 16S rRNA
8. 18S rRNA
9. 23S rRNA
10. 28S rRNA

13. There are several key differences between prokaryotic and eukaryotic ribosomes which have been exploited by some antibiotics to selectively target bacterial ribosomes. Match each antibiotic with the best description of how it inhibits translation. Not all descriptions will be used. ❖ **5**

(a) Streptomycin is an aminoglycoside. It binds to the 16S rRNA, which interferes with the proofreading function of the 30S subunit.

(b) Tetracyclines bind to and distort the A site.

(c) Puromycin has a structure similar to the 3′ end of an aminoacyl-tRNA. It binds to the A site and reacts with the peptidyl-tRNA to form peptidylpuromycin, which cannot be translocated to the P site and has a weak affinity for the A site.

(d) Azithromycin is a macrolide. It binds to the 50S subunit and interferes with the exit of the growing polypeptide chain from the ribosome.

1. Prevents initiation
2. Causes dissociation of covalently modified polypeptide
3. Prevents binding of aminoacyl-tRNA at the A site
4. Causes misreading of mRNA codons
5. Prevents termination
6. Causes premature termination

14. Fill in the blanks in the following paragraph:
Transcription in prokaryotes begins as promoter sites bind RNA polymerase. Prokaryotic cells have two sites that function as promoters: the **(a)** _____, which has the consensus sequence TTGACA, and the **(b)** _____, which is centered at –10 and has the consensus sequence TATAAT. Transcription starts at position **(c)** _____. Transcription ends when RNA polymerase transcribes a **(d)** _____, which in E. coli is a hairpin loop. In translation, ribosomal RNA pairs with the **(e)** _____, a purine-rich region of the mRNA molecule upstream from the start codon. Translation starts as the first tRNA recognizes the start codon, **(f)** _____, on mRNA.

15. The protein-synthesizing machinery was isolated from eukaryotic cells and briefly treated with a low concentration of RNase. The sample was then subjected to sucrose density-gradient centrifugation. The gradient was fractionated, and the absorbance, or optical density (OD), at 254 nm was recorded for each fraction. The following plot was obtained. ❖ **3**

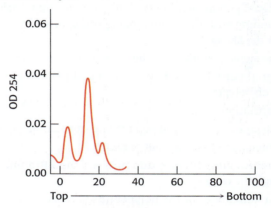

(a) What do the three peaks of absorbance represent?

The experiment was repeated except that, this time, the RNase treatment was omitted.

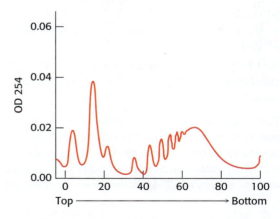

(b) Why is the centrifugation pattern now more complex? What does the series of peaks near the bottom of the centrifuge tube represent?

Before the isolation of the protein-synthesizing machinery, the cells were grown in low concentrations of oxygen (hypoxic conditions). Again the experiment was repeated without RNase treatment. The next graph shows the results.

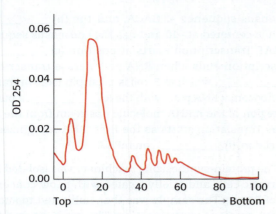

(c) What happens to cells grown in hypoxic conditions?

16. Place the following steps of the membrane translocation of eukaryotic secretory proteins in order, starting with the signal peptide exiting the ribosome during translation, and ending with the protein being transported by vesicles out of the ER. ❖ 4

(a) The signal peptide exits the ribosome during translation.

(b) Signal peptidase removes the signal peptide from the growing polypeptide.

(c) Translation resumes.

(d) SRP binds to the signal peptide and translation stops.

(e) Protein is transported out of the ER.

(f) The growing polypeptide crosses the ER membrane through the translocon.

(g) The remainder of the polypeptide is moved into the ER lumen.

(h) Ribosome docks to a translocon.

(i) SRP moves the ribosome to the ER membrane.

17. (a) What four components are required for the translocation of proteins across the endoplasmic reticulum membrane? ❖ 4

(b) What is the energy source that powers the cotranslational movement of proteins across the endoplasmic reticulum? ❖ 4

18. The initiation factor eIF-4A displays ATP-dependent RNA helicase activity. Another initiation factor, eIF-4H, has been proposed to assist the action of eIF-4A. The graph shows some of the experimental results from an assay that can measure the helicase activity of eIF-4A in the presence of eIF-4H. ❖ 3

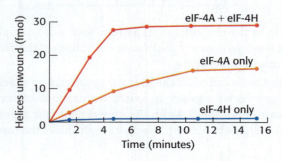

(a) What are the effects on eIF-4A helicase activity in the presence of eIF-4H?

(b) Why did measuring the helicase activity of eIF-4H alone serve as an important control?

(c) The initial rate of helicase activity of 0.2 μM of eIF-4A was then measured with varying amounts of eIF-4H. What ratio of eIF-4H to eIF-4A yielded optimal activity?

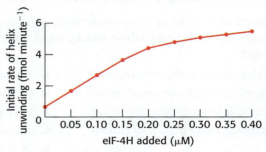

(d) Next, the effect of RNA–RNA helix stability on the initial rate of unwinding in the presence and absence of eIF-4H was tested. How does the effect of eIF-4H vary with helix stability?

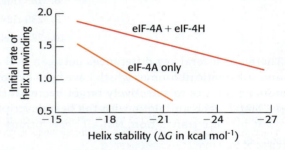

(e) How might eIF-4H affect the helicase activity of eIF-4A?

19. Consider the 5′-to-3′ DNA sequence of three consecutive codons below.

5′-TTAGAGCGC-3′

Determine the following sequences: ❖ 1

(a) The complementary DNA strand (3′ to 5′)

(b) The transcribed mRNA

(c) The tRNA anticodons

(d) The amino acids

20. Consider the following tRNA molecules. (a) Which tRNA is the activated version? (b) What is the name of the amino acid side chain? ❖ 1

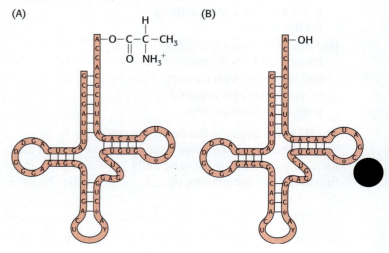

Control of Gene Expression

Complex biological processes often require coordinated control of the expression of many genes. The maturation of a tadpole into a frog is largely controlled by thyroid hormone. This hormone regulates gene expression by binding to a protein, the thyroid-hormone receptor, as shown at the right. In response to the hormone's binding, this protein binds to specific DNA sites in the genome and modulates the expression of nearby genes.

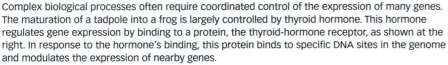 LEARNING GOALS

By the end of this chapter, you should be able to:

1. Evaluate the control of transcription by bacterial and eukaryotic DNA-binding proteins.
2. Analyze the common structural features of bacterial and eukaryotic DNA-binding proteins that allow them to recognize specific DNA sequences.
3. Understand the roles of the various components in the *E. coli lac* operon.
4. Describe how genetic switches work, using the lytic versus lysogenic path in bacteriophage lambda as an example.
5. Discuss the role of chromatin modifications in controlling eukaryotic gene expression.
6. Contrast the control of gene expression in bacteria and in eukaryotes.
7. Understand examples of mechanisms of posttranscriptional gene regulation.

OUTLINE

31.1 Bacterial DNA-Binding Proteins Bind to Specific Regulatory Sites

31.2 In Bacteria, Genes Are Often Arranged into Clusters Under the Control of a Single Regulatory Sequence

31.3 Regulatory Circuits Can Result in Switching Between Patterns of Gene Expression

31.4 Regulation of Gene Expression Is More Complex in Eukaryotes

31.5 The Control of Gene Expression in Eukaryotes Can Require Chromatin Remodeling

31.6 Gene Expression Can Be Controlled at the Posttranscriptional Level

The development of multicellular organisms; the changes that distinguish normal cells from cancer cells; bacteria responding to changes in their environment—all of these intriguing phenomena have one thing in common: they rely on the regulation of gene expression. A gene is expressed when it is transcribed into RNA and, for most genes, translated into proteins. While some of the thousands of genes that make up a genome are expressed all the time, many other genes are expressed only under some circumstances. For example, the level of expression of some genes in bacteria may vary more than a thousand-fold in response to the supply of nutrients or to environmental challenges. In this chapter, we will learn about several strategies that allow for the coordinated regulation of gene expression, both in bacteria and eukaryotes.

31.1 Bacterial DNA-Binding Proteins Bind to Specific Regulatory Sites

How do regulatory systems distinguish the genes that need to be activated or repressed from genes that are *constitutive*, or expressed all the time? The DNA sequences of genes themselves do not have any distinguishing features that would allow regulatory systems to recognize them. Instead, gene regulation depends on other sequences in the genome. In bacteria, these regulatory sites are close to the region of the DNA that is transcribed. Regulatory sites are usually binding sites for specific DNA-binding proteins, which can either stimulate or repress gene expression.

Many DNA-binding proteins match the symmetry in their target DNA sequences

Regulatory sites were first identified in *E. coli* through experiments that looked at changes in gene expression in response to nutrient availability. For example, if the usual energy source (the sugar glucose) is not available to the bacterium, but lactose is available, the bacterium starts to express a gene encoding β-**galactosidase**, an enzyme that can process lactose for use as a carbon and energy source. The sequence of the regulatory site for this gene is shown in **Figure 31.1**. The nucleotide sequence of this site shows a nearly perfect inverted repeat, indicating that the DNA in this region has an approximate twofold axis of symmetry. Recall that cleavage sites for restriction enzymes such as EcoRV have similar symmetry properties (Section 9.2). Symmetry in such regulatory sites usually corresponds to symmetry in the protein that binds the site. Symmetry matching is a recurring theme in protein–DNA interactions.

5′-...TGTGTGGAATTGTGAGCGGATAACAATTTCACACA...3′
3′-...ACACACCTTAACACTCGCCTATTGTTAAAGTGTGT...5′

FIGURE 31.1 The sequence of the *lac* regulatory site is a nearly perfect inverted repeat. This leads to twofold rotational symmetry in the DNA. Parts of the sequences that are related by this symmetry are shown in the same color.

To understand these protein–DNA interactions in detail, scientists determined the structure of the complex between the regulatory site for the gene encoding β-galactosidase and the DNA-binding unit from the protein that binds to it (**Figure 31.2**). The protein that recognizes this particular regulatory site is called the *lac repressor*, because it represses the expression of the lactose-processing gene when bound to the regulatory site.

As expected, the DNA-binding unit from the *lac* repressor binds as a dimer, and the twofold axis of symmetry of the dimer matches the symmetry of the DNA. An α helix from each monomer of the protein is inserted into the major

FIGURE 31.2 The *lac* repressor binds to operator DNA. The DNA-binding domain from the *lac* repressor binds to a DNA fragment containing its preferred binding site by inserting an α helix into the major groove of operator DNA. A specific contact forms between an arginine residue of the repressor and a G–C base pair in the binding site. [Drawn from 1EFA.pdb.]

INTERACT with this model in
Achieve

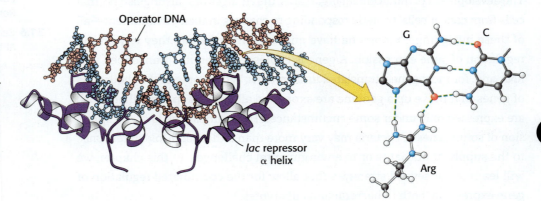

groove of the DNA, where amino acid side chains make specific contacts with exposed edges of the base pairs. For example, the side chain of an arginine residue of the protein forms a pair of hydrogen bonds with a guanine residue of the DNA, which would not be possible with any other base. This interaction and similar ones allow the *lac* repressor to bind more tightly to this site than to the wide range of other sites present in the *E. coli* genome.

The helix-turn-helix motif is common to many bacterial DNA-binding proteins

Do other bacterial DNA-binding proteins use similar strategies? The structures of many such proteins have now been determined, and amino acid sequences are known for many more. Strikingly, the DNA-binding surfaces of many, but not all, of these proteins consist of a pair of α helices separated by a tight turn, called a **helix-turn-helix motif** (Figure 31.3). In complexes with DNA, the second of these two helices (often called the *recognition helix*) lies in the major groove, where amino acid side chains make contact with the edges of base pairs. In contrast, residues of the first helix participate primarily in contacts with the DNA backbone. Helix-turn-helix motifs are present in many proteins that bind DNA as dimers, and thus two of the units will be present, one in each monomer.

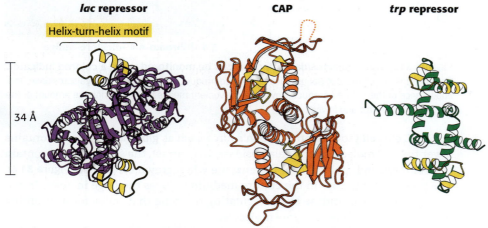

FIGURE 31.3 Sequence-specific DNA-binding proteins interact with DNA through a helix-turn-helix motif. In each case, the helix-turn-helix units (highlighted in yellow) within a protein dimer are approximately 34 Å apart, corresponding to one full turn of DNA. [Drawn from 1EFA.pdb, 1RUN.pdb, and 1TRO.pdb.]

INTERACT with the models shown in Figure 31.3 and Figure 31.4 in Achieve

Although the helix-turn-helix motif is the most commonly observed DNA-binding unit in bacteria, not all regulatory proteins bind DNA through such units. The *E. coli* methionine repressor (Figure 31.4), for example, binds DNA through the insertion of a pair of β strands into the major groove. A historically important example for gene regulation in *E. coli*, discussed next, reveals many common principles and elucidates the role of DNA-binding proteins in controlling transcription.

31.2 In Bacteria, Genes Are Often Arranged into Clusters Under the Control of a Single Regulatory Sequence

As noted above, bacteria such as *E. coli* usually rely on glucose as their source of carbon and energy, even when other sugars are available. However, when glucose is scarce, *E. coli* can use lactose as their carbon source, even though this disaccharide does not lie on any major metabolic pathways. β-Galactosidase is an essential enzyme in the metabolism of lactose because it hydrolyzes lactose

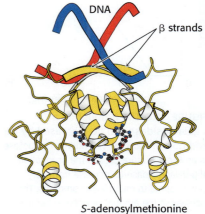

FIGURE 31.4 Some DNA-binding proteins recognize specific DNA sequences through β strands. The methionine repressor complexed with its co-repressor *S*-adenosylmethionine is shown bound to DNA (blue and red ribbon). In this example residues in β strands, rather than in α helices, participate in the crucial interactions between the protein and the DNA. [Drawn from 1CMA.pdb.]

into galactose and glucose. These products are then metabolized by pathways discussed in Chapter 16.

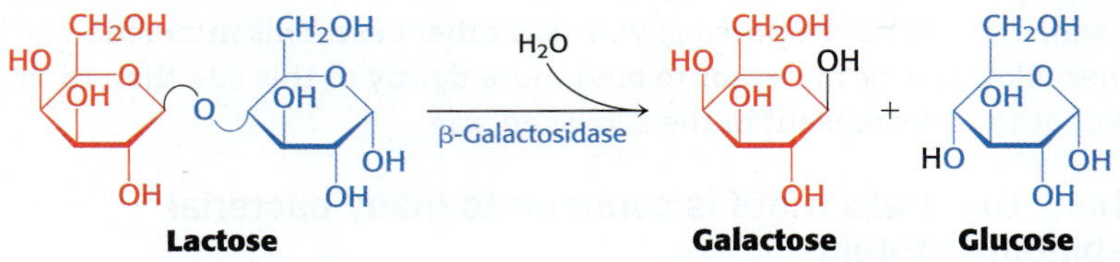

This reaction can be conveniently followed in the laboratory through the use of alternative galactoside substrates, such as X-Gal, that form colored products (**Figure 31.5**).

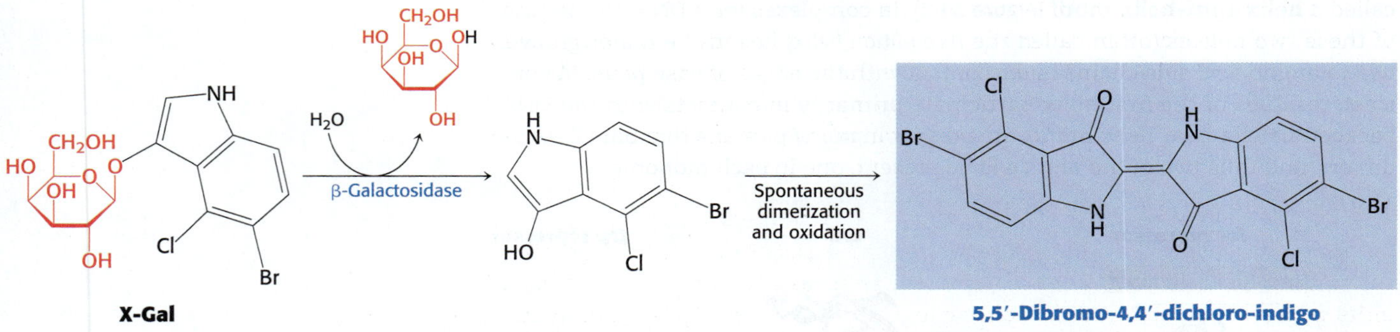

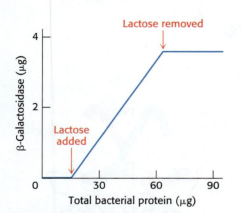

X-Gal

FIGURE 31.5 **The β-galactosidase reaction can be monitored using a colored indicator.** The galactoside substrate X-Gal produces a colored product when cleaved by β-galactosidase. The appearance of this colored product provides a convenient means for monitoring the amount of the enzyme both in vitro and in vivo.

5,5′-Dibromo-4,4′-dichloro-indigo

An E. coli cell growing on a carbon source such as glucose or glycerol contains fewer than 10 molecules of β-galactosidase. In contrast, the same cell will contain several thousand molecules of the enzyme when grown on lactose (**Figure 31.6**). The presence of lactose in the culture medium induces the large increase in the amount of β-galactosidase by up-regulating the gene that codes for it, resulting in the synthesis of new enzyme molecules.

A crucial clue to the mechanism of gene regulation was the observation that two other proteins are synthesized in concert with β-galactosidase—namely, galactoside permease and thiogalactoside transacetylase. The permease is required for the transport of lactose across the bacterial cell membrane (Section 13.3). The transacetylase is not essential for lactose metabolism but appears to play a role in the detoxification of compounds that also may be transported by the permease. Thus, the expression levels of a set of enzymes that all contribute to the adaptation to a given change in the environment are regulated together. Such a coordinated unit of gene expression is called an **operon**.

An operon consists of regulatory elements and protein-encoding genes

The parallel regulation of β-galactosidase, the permease, and the transacetylase suggested that the expression of genes encoding these enzymes is controlled by a common mechanism. François Jacob and Jacques Monod proposed the so-called *operon model* to account for this parallel regulation as well as the results of other genetic experiments. The genetic elements of the model are a regulator gene that encodes a regulatory protein (with its promoter), a regulatory DNA sequence called an **operator site**, and a set of structural genes with a separate promoter (**Figure 31.7**).

The regulator gene encodes a **repressor** protein that binds to the operator site. The binding of the repressor to the operator prevents transcription of the structural genes. The operator and its associated structural genes constitute

FIGURE 31.6 **The production of β-galactosidase can be increased by the addition of lactose.** The addition of lactose to an E. coli culture causes the production of β-galactosidase to increase from very low amounts to much larger amounts. The increase in the amount of enzyme parallels the increase in the number of cells in the growing culture. β-Galactosidase constitutes 6.6% of the total protein synthesized in the presence of lactose.

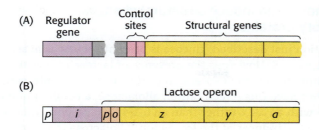

FIGURE 31.7 Operons are clusters of genes under a common control. (A) The general structure of an operon as conceived by Jacob and Monod. In some cases, the regulator gene may be some distance away from the structural genes, as indicated by the break. (B) The structure of the lactose operon. In addition to the promoter, *p*, in the operon, a second promoter is present in front of the regulator gene, *i*, to drive the synthesis of the regulator.

the operon. For the *lactose (lac) operon*, the i gene encodes the repressor, o is the operator site, and the z, y, and a genes are the structural genes for β-galactosidase, the permease, and the transacetylase, respectively. The operon also contains a promoter site (denoted by p), which directs the RNA polymerase to the correct transcription initiation site. The i gene is under the control of a separate promoter (also denoted p). The z, y, and a genes are transcribed to give a single mRNA molecule that encodes all three proteins. An mRNA molecule encoding more than one protein is known as a *polygenic* or *polycistronic* transcript.

The *lac* repressor protein can block transcription

In the absence of lactose, the lactose operon is repressed. How does the *lac* repressor mediate this repression? The *lac* repressor exists as a tetramer of 37-kDa subunits with two pairs of subunits coming together to form the DNA-binding unit previously discussed. In the absence of lactose, the repressor binds very tightly and rapidly to the operator, which is directly downstream of the promoter. When the *lac* repressor is bound to the operator DNA, it prevents transcription of the protein-coding genes because it blocks the progress of RNA polymerase along the DNA template.

How does the *lac* repressor locate the operator site in the E. coli chromosome? The *lac* repressor binds 4×10^6 times as strongly to operator DNA as it does to random sites in the genome. This high degree of selectivity allows the repressor to find the operator efficiently even with a large excess of other sites within the E. coli genome. The dissociation constant for the repressor–operator complex is approximately 0.1 pM (10^{-13} M). The rate constant for association ($\approx 10^{10}$ M^{-1} s^{-1}) is strikingly high, indicating that the repressor finds the operator primarily by diffusing along a DNA molecule (a one-dimensional search) rather than encountering it from the aqueous medium (a three-dimensional search). This diffusion has been confirmed by studies that monitored the behavior of fluorescently labeled single molecules of *lac* repressor inside living E. coli cells.

Inspection of the complete E. coli genome sequence reveals two sites within 500 bp of the primary operator site that approximate the sequence of the operator. When one dimeric DNA-binding unit binds to the operator site, the other DNA-binding unit of the *lac* repressor tetramer can bind to one of these sites with similar sequences. The DNA between the two bound sites forms a loop. No other sites that closely match the sequence of the *lac* operator site are present in the rest of the E. coli genome sequence. Therefore, the DNA-binding specificity of the *lac* repressor is sufficient to target two closely related sites within the E. coli genome.

The three-dimensional structure of the *lac* repressor has been determined in various forms. Each monomer consists of a small amino-terminal domain that binds DNA and a larger domain that mediates the formation of the dimeric DNA-binding unit and the tetramer (**Figure 31.8**). A pair of the amino-terminal domains come together to form the functional DNA-binding unit. Each monomer has a helix-turn-helix unit that interacts with the major groove of the bound DNA.

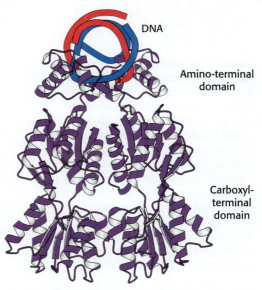

FIGURE 31.8 The *lac* repressor binds to DNA as a dimer. The amino-terminal domain binds to DNA, whereas the carboxyl-terminal domain forms a separate structure. A part of the structure that mediates the formation of *lac* repressor tetramers is not shown. [Drawn from 1EFA.pdb.]

INTERACT with this model in
 Achieve

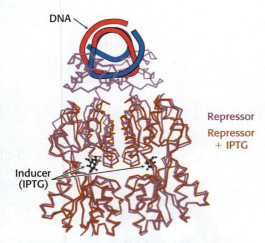

1,6-Allolactose

Isopropylthiogalactoside (IPTG)

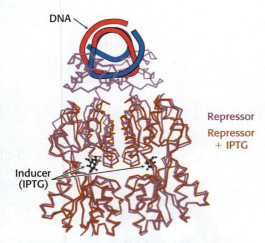

DNA

Repressor
Repressor
+ IPTG

Inducer
(IPTG)

FIGURE 31.9 Binding of IPTG alters the *lac* repressor structure and affects its ability to bind DNA. The structure of the *lac* repressor bound to the inducer isopropylthiogalactoside (IPTG), shown in orange, is superimposed on the structure of the *lac* repressor bound to DNA, shown in purple. The binding of IPTG induces only subtle structural changes, but these are large enough near the interface of the DNA-binding domains that the repressor cannot interact effectively with DNA. The DNA-binding domains of the *lac* repressor bound to IPTG are not shown, because these regions are not well ordered in the crystals studied.

Ligand binding can induce structural changes in regulatory proteins

In the situation just described, glucose is present, lactose is absent, and the *lac* operon is repressed. How does the presence of lactose trigger the relief of this repression and, hence, the expression of the *lac* operon? Interestingly, lactose itself does not have this effect; rather, allolactose, a combination of galactose and glucose with an α-1,6 rather than an α-1,4 linkage, does. Allolactose is thus referred to as the **inducer** of the *lac* operon. Allolactose is a side product of the β-galactosidase reaction and is produced at low levels by the few molecules of β-galactosidase that are present before induction. Some other β-galactosides such as isopropylthiogalactoside (IPTG) are potent inducers of β-galactosidase expression, although they are not substrates of the enzyme. IPTG is useful in the laboratory as a tool for inducing gene expression in engineered bacterial strains.

The inducer triggers gene expression by preventing the *lac* repressor from binding the operator. The inducer binds to the *lac* repressor and thereby greatly reduces the repressor's affinity for operator DNA. An inducer molecule binds in the center of the large domain within each monomer. This binding leads to subtle conformational changes that modify the relation between the two small DNA-binding domains (**Figure 31.9**). These domains can no longer easily contact DNA simultaneously, leading to a dramatic reduction in DNA-binding affinity.

How does this fit in with the regulation of gene expression in the lactose operon? Let's review the process (**Figure 31.10**). In the absence of inducer, the *lac* repressor is bound to DNA in a manner that blocks RNA polymerase from transcribing the z, y, and a genes. Thus, very little β-galactosidase, permease, or transacetylase is produced. The addition of lactose to the environment leads to the formation of allolactose. This inducer binds to the *lac* repressor, leading to conformational changes and the release of DNA by the *lac* repressor. With the operator site unoccupied, RNA polymerase can then transcribe the other *lac* genes and the bacterium will produce the proteins necessary for the efficient use of lactose.

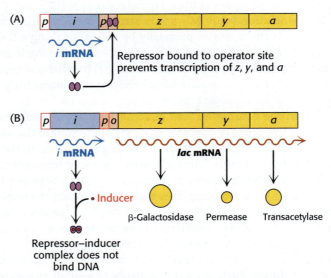

(A)   P   i   P   o o   z   y   a

i mRNA   Repressor bound to operator site
                   prevents transcription of z, y, and a

(B)   P   i   P   o   z   y   a

i mRNA   *lac* mRNA

                   • Inducer

Repressor–inducer   β-Galactosidase   Permease   Transacetylase
complex does not
bind DNA

FIGURE 31.10 The *lac* operon can be induced. (A) In the absence of lactose, the *lac* repressor binds DNA and represses transcription from the *lac* operon. (B) Allolactose or another inducer binds to the *lac* repressor, leading to its dissociation from DNA and to the production of *lac* mRNA.

The operon is a common regulatory unit in bacteria

The lac operon is just one well-studied example of an operon in *E. coli*. In fact, there are an estimated 600–700 operons in *E. coli*, all of which contain genes that are connected via a common metabolic function or pathway. Clustering genes in operons provides the clear advantage of a coordinated regulatory mechanism for enzymes that are functionally related.

Many other gene-regulatory networks are controlled in ways analogous to those of the *lac* operon. For example, genes taking part in purine and, to a lesser degree, pyrimidine biosynthesis are repressed by the *pur* repressor. This dimeric protein is 31% identical in sequence with the *lac* repressor and has a similar three-dimensional structure. However, the behavior of the *pur* repressor is opposite that of the *lac* repressor: whereas the *lac* repressor is released from DNA by binding to a small molecule, the *pur* repressor binds DNA specifically, blocking transcription, only when bound to a small molecule. Such a small molecule is called a **corepressor**. For the *pur* repressor, the corepressor can be either guanine or hypoxanthine.

The dimeric *pur* repressor binds to inverted-repeat DNA sites of the form 5′-ANGCAANCGNTTNCNT-3′, in which the bases shown in boldface type are particularly important. Examination of the *E. coli* genome sequence reveals the presence of more than 20 such sites, regulating 19 operons and including more than 25 genes (**Figure 31.11**).

Because the DNA-binding sites for these regulatory proteins are short, it is likely that they evolved independently and are not related by divergence from an ancestral regulatory site. Once a ligand-regulated DNA-binding protein is present in a cell, binding sites for the protein may arise by mutation adjacent to additional genes. Binding sites for the *pur* repressor have evolved in the regulatory regions of a wide range of genes taking part in nucleotide biosynthesis. All such genes can then be regulated in a concerted manner.

The organization of bacterial genes into operons is useful for the analysis of completed genome sequences. Sometimes a gene of unknown function is discovered to be part of an operon containing well-characterized genes. Such associations can provide powerful clues to the biochemical and physiological functions of the uncharacterized gene.

Some DNA-binding proteins stimulate transcription

All the DNA-binding proteins discussed thus far function by inhibiting transcription until some environmental condition, such as the presence of lactose, is met. There are also DNA-binding proteins that stimulate transcription. One particularly well-studied example is a protein in *E. coli* that stimulates the expression of catabolic enzymes.

E. coli grown on glucose, a preferred energy source, have very low levels of catabolic enzymes for metabolizing other sugars. Synthesizing these enzymes when glucose is abundant would be wasteful. In fact, glucose has an inhibitory effect on the genes encoding these enzymes, an effect called **catabolite repression**. How does this work? Glucose achieves this inhibitory effect by lowering the concentration of cyclic AMP in *E. coli*. To understand the consequences of this, let us first consider the role of cAMP in the regulation of transcription.

High concentrations of cAMP stimulate the concerted transcription of many catabolic enzymes by acting through a protein called the **catabolite activator protein (CAP)**, also known as the *cAMP receptor protein* (CRP). When bound to cAMP, CAP stimulates the transcription of lactose- and arabinose-catabolizing genes. CAP is a sequence-specific DNA-binding protein. Within the *lac* operon, CAP binds to an inverted repeat that is centered near position −61 relative to the start site for transcription (**Figure 31.12**). This site is approximately 70 base

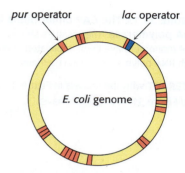

FIGURE 31.11 Operator sites are not distributed evenly across the *E. coli* genome. The *E. coli* genome contains only a single region that closely matches the sequence of the *lac* operator (shown in blue). In contrast, 19 sites match the sequence of the *pur* operator (shown in red). Thus, the *pur* repressor regulates the expression of many more genes than does the *lac* repressor.

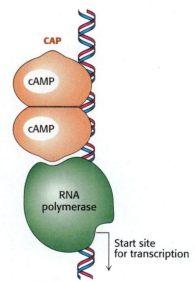

FIGURE 31.12 The catabolite activator protein (CAP) binds to DNA as a dimer. This protein binds to an inverted repeat that is at position −61 relative to the start site of transcription. The CAP-binding site on DNA is adjacent to the position at which RNA polymerase binds.

pairs from the operator site. As expected from the symmetry of the binding site, CAP functions as a dimer of identical subunits. CAP binding also bends DNA in a manner that favors interactions with RNA polymerase.

The CAP–cAMP complex stimulates the initiation of transcription by approximately a factor of 50. Energetically favorable contacts between CAP and RNA polymerase increase the likelihood that transcription will be initiated at sites to which the CAP–cAMP complex is bound (**Figure 31.13**). Thus, for the *lac* operon, gene expression is maximal when the binding of allolactose relieves the inhibition by the *lac* repressor and the CAP–cAMP complex stimulates the binding of RNA polymerase.

FIGURE 31.13 The CAP dimer binds to DNA and to RNA polymerase. The residues shown in yellow in each CAP monomer have been implicated in direct interactions with RNA polymerase. [Drawn from 1RUN.pdb.]

INTERACT with the models in Figure 31.13 and Figure 31.14 in Achieve

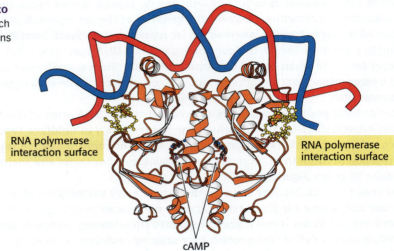

RNA polymerase interaction surface

RNA polymerase interaction surface

cAMP

The *E. coli* genome contains many CAP-binding sites in positions appropriate for interactions with RNA polymerase. Thus, an increase in the cAMP level inside an *E. coli* bacterium results in the formation of CAP–cAMP complexes that bind to many promoters and stimulate the transcription of genes encoding a variety of catabolic enzymes. Conversely, as noted above, when glucose levels are high, the concentration of cAMP is low. This results in fewer CAP–cAMP complexes and reduced expression of the genes encoding the catabolic enzymes.

❖ SELF–CHECK QUESTION

Under conditions of high glucose and high lactose, *E. coli* will first metabolize glucose before switching to lactose. How do you explain this?

31.3 Regulatory Circuits Can Result in Switching Between Patterns of Gene Expression

The study of viruses that infect bacteria has led to significant advances in our understanding of the processes that control gene expression. Again, sequence-specific DNA-binding proteins play key roles in these processes. Investigations of bacteriophage λ have been particularly revealing.

We examined the alternative infection modes of λ phage in Chapter 9. In the lytic pathway, most of the genes in the viral genome are transcribed, initiating the production of many virus particles and leading to the eventual lysis of the bacterial cell with the concomitant release of approximately 100 virus particles. In the lysogenic pathway, the viral genome is incorporated into the bacterial DNA where most of the viral genes remain unexpressed, allowing the viral genome to be carried along as the bacteria replicate. Two key proteins and a set of regulatory sequences in the viral genome are responsible for the switch that determines which of these two pathways is followed.

The λ repressor regulates its own expression

The first protein that we will discuss is the λ **repressor**, sometimes known as the λ cI protein. This protein is key because it blocks, either directly or indirectly, the transcription of almost all genes encoded by the virus. The one exception is the gene that encodes the λ repressor itself.

The λ repressor consists of an amino-terminal DNA-binding domain and a carboxyl-terminal domain that participates in protein oligomerization (**Figure 31.14**). This protein binds to a number of key sites in the λ phage genome. The sites of greatest interest for our present discussion are in the so-called *right operator* (**Figure 31.15**). This region includes three binding sites for the λ repressor dimer, as well as two promoters within a section of approximately 80 base pairs. One promoter drives the expression of the gene for the λ repressor itself, whereas the other drives the expression of several other viral genes.

The λ repressor does not have the same affinity for the three sites; it binds the site O_R1 with the highest affinity. In addition, the binding to adjacent sites is cooperative so that, after a λ repressor dimer has bound at O_R1, the likelihood that a protein will bind to the adjacent site O_R2 increases by approximately 25-fold. Thus, when the λ repressor is present in the cell at moderate concentrations, the most likely configuration has the λ repressor bound at O_R1 and O_R2, but not at O_R3. In this configuration, the λ repressor dimer bound at O_R1 blocks access to the promoter on the right side of the operator sites. This represses transcription of the adjacent gene, which encodes a protein termed **Cro** (controller of repressor and others), while the repressor dimer at O_R2 can be in contact with RNA polymerase and stimulate transcription of the promoter that controls the transcription of the gene that encodes the λ repressor itself.

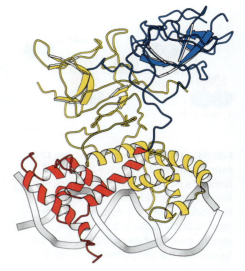

FIGURE 31.14 The λ repressor binds to DNA as a dimer. The amino-terminal domain of one subunit is shown in red and the carboxyl-terminal domain is shown in blue. In the other subunit, both domains are shown in yellow. Notice how α helices on the amino-terminal domains fit into the major groove of the DNA. [Drawn from 3BDN.pdb.]

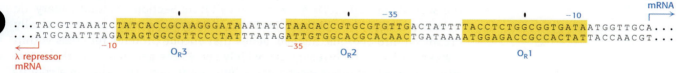

FIGURE 31.15 The sequence of the λ right operator is a binding target for regulatory proteins. The three operator sites (O_R1, O_R2, and O_R3) are shaded yellow with their centers indicated. The start sites for the λ repressor mRNA and the Cro mRNA are indicated, as are the –10 and –35 positions of their respective promoters.

Thus, the λ repressor stimulates its own production. As the concentration of the λ repressor increases further, an additional repressor dimer can bind to the O_R3 site, blocking the other promoter and repressing the production of additional repressor. Thus, the O_R3 site serves to maintain the λ repressor in a narrow, stable concentration range (**Figure 31.16**). The λ repressor also blocks other promoters in the λ phage genome so that the repressor is the only phage protein produced, maintaining the lysogenic state.

A circuit based on the λ repressor and Cro forms a genetic switch

What stimulates the switch to the lytic pathway? Changes such as DNA damage initiate the cleavage of the λ repressor at a specific bond between the DNA-binding and oligomerization domains. This process is mediated by the *E. coli* RecA protein (Section 28.5). After this cleavage has taken place, the affinity of the λ repressor for DNA is reduced. After the λ repressor is no longer bound to the O_R1 site, the Cro gene can be transcribed. Cro is a small protein that binds to the same sites as the λ repressor does, but with a different order of affinity for the three sites in the right operator. In particular, Cro has the highest affinity for O_R3. Cro bound at this site blocks the production of new λ repressor. The absence of λ repressor activates the expression of other phage

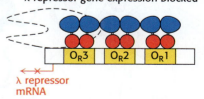

FIGURE 31.16 The λ repressor controls its own synthesis. (A) When λ repressor levels are relatively low, the repressor binds to sites O_R1 and O_R2 and stimulates the transcription of the gene that encodes the λ repressor itself. (B) When λ repressor levels are higher, the repressor also binds to site O_R3, blocking access to its promoter and repressing transcription from this gene.

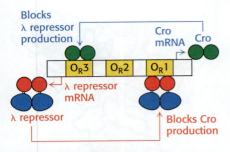

FIGURE 31.17 The λ repressor and Cro form a genetic circuit. The λ repressor blocks the production of Cro by binding most favorably to site O_R1, whereas Cro blocks the production of the λ repressor by binding most favorably to site O_R3. This circuit forms a switch that determines whether the lysogenic or the lytic pathway is followed.

genes, leading to the production of virus particles and the eventual lysis of the host cells.

Thus, this genetic circuit acts as a switch with two stable states: (1) λ repressor high, Cro low, corresponding to the lysogenic state; and (2) Cro high, λ repressor low, corresponding to the lytic state (**Figure 31.17**). Regulatory circuits with two stable states are often referred to as being *bistable*. In these bistable circuits, positive and negative transcriptional control are tightly linked, with different DNA-binding proteins controlling the expression of each other's genes. This is a common motif for controlling gene expression in all systems, from viruses to eukaryotes. In eukaryotes, for example, genetic switches can determine distinct cellular fates, analogous to the lysogenic and lytic infection modes of λ phage.

31.4 Regulation of Gene Expression Is More Complex in Eukaryotes

Gene regulation in eukaryotes is significantly more complex than in bacteria in several ways. First, the genomes being regulated are significantly larger. The *E. coli* genome consists of a single, circular chromosome containing 4.6 megabases (Mb). This genome encodes approximately 2000 proteins. In comparison, one of the simplest eukaryotes, *Saccharomyces cerevisiae* (baker's yeast), contains 16 chromosomes ranging in size from 0.2 to 2.2 Mb. The yeast genome totals 12 Mb and encodes approximately 6000 proteins. The genome within a human cell contains 23 pairs of chromosomes ranging in size from 50 to 250 Mb. Approximately 20,000 protein-coding genes are present within the 3000 Mb of human DNA. Second, recall from Section 29.3 that eukaryotic DNA is packaged into chromatin. Third, eukaryotic genes are not generally organized into operons. Instead, genes that encode proteins for steps within a given pathway are often spread widely across the genome. This characteristic requires that other mechanisms function to regulate genes in a coordinated way.

Despite these differences, some aspects of gene regulation in eukaryotes are quite similar to those in bacteria. In particular, proteins that recognize specific DNA sequences are central to many gene-regulatory processes. Eukaryotic transcription factors—DNA-binding proteins similar in many ways to the bacterial proteins that we encountered earlier—can act directly by interacting with the transcriptional machinery, or indirectly by influencing chromatin structure. Key features of eukaryotic promoters, transcription factors, and enhancer sequences that can act far away from transcription start sites were introduced in Section 29.3.

A range of DNA-binding motifs are employed by eukaryotic DNA-binding proteins

The structures of many eukaryotic DNA-binding proteins have been determined and a range of structural motifs have been observed, but we will focus on three that reveal the common features and the diversity of these motifs. The first type of eukaryotic DNA-binding motif that we will consider is the **homeodomain** (**Figure 31.18**). The structure of the homeodomain and its mode of recognition of DNA are very similar to those of the bacterial helix-turn-helix proteins. In eukaryotes, homeodomain proteins often form heterodimeric structures, sometimes with other homeodomain proteins, that recognize asymmetric DNA sequences.

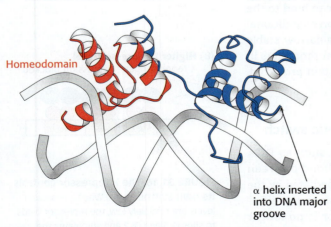

FIGURE 31.18 Two different DNA-binding homeodomains can form a heterodimer. Each homeodomain has a helix-turn-helix motif with one helix inserted into the major groove of DNA. [Drawn from 1AKH.pdb.]

Another class of eukaryotic DNA-binding motif comprises the **basic-leucine zipper (bZip) proteins** (Figure 31.19). This DNA-binding motif consists of a pair of long α helices. The first part of each α helix is a basic region that lies in the major groove of the DNA and makes contacts responsible for DNA-site recognition. The second part of each α helix forms a coiled-coil structure with its partner. When these units are stabilized by appropriately spaced leucine residues, they are referred to as *leucine zippers*.

The final class of eukaryotic DNA-binding motifs that we will consider here are the **Cys₂His₂ zinc-finger domains** (Figure 31.20). This motif comprises tandem sets of small domains, each of which binds a zinc ion through conserved sets of two cysteine and two histidine residues. These structures, often just called zinc-finger domains, form a string that follows the major groove of DNA. An α helix from each domain makes specific contacts with the edges of base pairs within the groove. Some proteins contain arrays of 10 or more zinc-finger domains, potentially enabling them to contact long stretches of DNA. The human genome encodes several hundred proteins that contain zinc-finger domains of this type. We will encounter another class of zinc-based DNA-binding domain when we consider nuclear hormone receptors in Section 31.5.

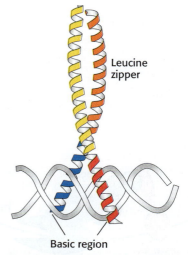

FIGURE 31.19 Two basic-leucine zipper proteins can form a heterodimer. The basic region lies in the major groove of DNA. The leucine zipper stabilizes the protein dimer. [Drawn from 1FOS.pdb.]

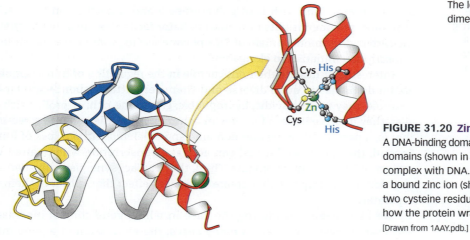

FIGURE 31.20 Zinc-finger domains bind to DNA. A DNA-binding domain comprising three Cys₂His₂ zinc-finger domains (shown in yellow, blue, and red) is shown in a complex with DNA. Each zinc-finger domain is stabilized by a bound zinc ion (shown in green) through interactions with two cysteine residues and two histidine residues. Notice how the protein wraps around the DNA in the major groove. [Drawn from 1AAY.pdb.]

INTERACT with the models in Figures 31.18–31.20 in Achieve

❖ SELF–CHECK QUESTION

The α helix is the most common element used by both bacterial and eukaryotic DNA-binding motifs. Why is this?

DNA-binding transcription factors are key to gene regulation in eukaryotes, just as they are in bacteria. However, the roles of eukaryotic transcription factors are different in several ways:

- Whereas the DNA-binding sites crucial for the control of gene expression in bacteria are usually quite close to promoters, those in eukaryotes can be farther away from promoters and can exert their action at a distance.

- Most bacterial genes are regulated by single transcription factors, and multiple genes in a pathway are expressed in a coordinated fashion because such genes are often transcribed as part of a polycistronic mRNA. In eukaryotes, the expression of each gene is typically controlled by multiple transcription factors, and the coordinated expression of different genes depends on having similar transcription-factor-binding sites in each gene in the set.

- In bacteria, transcription factors usually interact directly with RNA polymerase. In eukaryotes, while some transcription factors interact directly with RNA polymerase, many others act more indirectly, interacting with other proteins associated with RNA polymerase or modifying chromatin structure.

Eukaryotic transcription factors usually consist of several domains. As discussed above, the **DNA-binding domain** binds to regulatory sequences that can either be adjacent to the promoter or at some distance from it. Most commonly, transcription factors include additional domains that help activate transcription. When a transcription factor is bound to the DNA, its **activation domain** promotes transcription by interacting with RNA polymerase, by interacting with other associated proteins, or by modifying the local structure of chromatin.

Activation domains interact with other proteins

The activation domains of transcription factors generally recruit other proteins that promote transcription. In some cases, these activation domains interact directly with RNA polymerase. In other cases, the activation domains bind to closely associated proteins that bridge the transcription factors and the polymerase.

One example of an important bridge between transcription factors and promoter-bound RNA Polymerase II is **mediator**, a protein complex of 25 to 30 subunits that is conserved from yeast to human beings (**Figure 31.21**). Studies using cryo-electron microscopy are revealing the details of mediator structure and its interactions with RNA polymerase II and transcription factors. These structures provide insights into how mediator facilitates the phosphorylation of the carboxyl-terminal domain of RNA polymerase to promote the transition from transcription initiation to elongation.

Interestingly, despite their critical role in the regulation of transcription, activation domains are quite diverse and share very little sequence similarity. For example, they may be acidic, hydrophobic, glutamine rich, or proline rich. However, they do have certain features in common. First, they are often redundant; that is, a part of the activation domain can be deleted without loss of function. Second, they are modular and can activate transcription when paired with a variety of DNA-binding domains. Third, they can act synergistically: two activation domains acting together create a stronger effect than either of them acting separately.

We have been considering the case in which gene control increases the expression level of a gene. In many cases, the expression of a gene must be decreased by blocking transcription. The agents in such cases are *transcriptional repressors*. In many cases, transcriptional repressors act by altering chromatin structure.

Multiple transcription factors interact with eukaryotic regulatory regions

The pre-initiation complex described in Chapter 29 initiates transcription at a low frequency. Recall that several general transcription factors join with RNA polymerase II to form the pre-initiation complex (PIC). Additional transcription factors must bind to other sites that can be near the promoter or quite distant for a gene to achieve a higher rate of mRNA synthesis. In contrast with the regulators of bacterial transcription, few eukaryotic transcription factors have any effect on their own. Instead, each factor recruits other proteins to build up large complexes that interact with the transcriptional machinery to activate transcription.

A major advantage of this mode of regulation is that a given regulatory protein can have different effects, depending on what other proteins are present in the same cell. This phenomenon, called *combinatorial control*, is crucial to multicellular organisms that have many different cell types. Even in unicellular eukaryotes such as yeast, combinatorial control allows for the generation of distinct cell types.

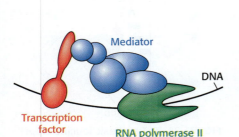

FIGURE 31.21 The mediator complex binds to transcription factors and to RNA polymerase II. Mediator acts as a bridge between transcription factors bearing activation domains and RNA polymerase II. These interactions help recruit and stabilize RNA polymerase II near specific genes that are then transcribed.

31.5 The Control of Gene Expression in Eukaryotes Can Require Chromatin Remodeling

Recall from Section 29.3 that to fit inside the nucleus, eukaryotic DNA is compacted into chromatin, a complex of DNA and a special set of proteins. One of the fundamental features of chromatin is the nucleosome, where DNA is wrapped around an octamer of histone proteins H2A, H2B, H3, and H4. In this section we will discuss how chromatin modifications affect gene expression.

Early studies investigating the effect of nucleases on chromatin suggested a correlation between chromatin structure and gene expression. DNA that is densely packaged into chromatin is less susceptible to cleavage by the nonspecific DNA-cleaving enzyme DNase I. Regions adjacent to genes that are being transcribed are more sensitive to cleavage by DNase I than are other sites in the genome, suggesting that the DNA in these regions is less compacted than it is elsewhere and is thus more accessible to regulatory proteins.

In addition, some sites, called *hypersensitive sites*, are extremely sensitive to DNase I and other nucleases. These sites are usually within 1 kb of the start site of an active gene and correspond to regions that have few nucleosomes or contain nucleosomes in an altered conformation. Hypersensitive sites are cell-type specific and developmentally regulated. For example, globin genes in the precursors of erythroid cells from 20-hour-old chicken embryos are insensitive to DNase I. However, when hemoglobin synthesis begins at 35 hours, regions adjacent to these genes become highly susceptible to digestion. In tissues such as the brain that produce no hemoglobin, the globin genes remain resistant to DNase I throughout development and into adulthood. These studies and additional lines of evidence suggest that chromatin structure is altered in active genes compared with inactive ones and that one prerequisite for gene expression is a relaxing of the chromatin structure.

Chromatin remodeling and DNA methylation regulate access to DNA-binding sites

If, as mentioned above, the relaxing of chromatin structure allows for gene expression, then it makes sense that the packaging of DNA into nucleosomes will inhibit gene expression. Another mechanism of regulating gene expression is provided by the degree of methylation of DNA, especially at cytosines. Specific methyltransferases methylate carbon 5 of cytosine, converting it into 5-methylcytosine.

About 70% of the 5′-CpG-3′ sequences (where "p" represents the phosphate residue in the DNA backbone) in mammalian genomes are methylated. This modification is associated with gene silencing because the methyl group of 5-methylcytosine protrudes into the major groove of DNA where it may interfere with the binding of proteins that stimulate transcription.

The distribution of these methylated CpG sequences in mammalian genomes is not uniform. Many CpG sequences have been converted into TpG through mutation by the deamination of 5-methylcytosine to thymine. In addition, some regions contain primarily *non-methylated* CpG dinucleotides, referred to as **CpG islands**. The promoters of ca. 70% of mammalian genes are associated with CpG islands, making them more accessible by transcription factors. The relative absence of 5-methylcytosines near gene promoters is referred to as *hypomethylation*.

The distribution of methylated CpG sequences may also depend on the cell type. Consider the β-globin gene. In cells that are actively expressing hemoglobin, the region from approximately 1 kb upstream of the start site to approximately

NH$_2$

CH$_3$

deoxyribose

5-Methylcytosine

PBH Images/Alamy

SARAH STEWART Dr. Sarah Stewart was born in Mexico in 1905 to an American father and a Mexican mother. Her family relocated to the United States in 1911, and she pursued her interests in bacteriology and chemistry, earning a PhD in microbiology in 1939. Dr. Stewart wanted to research the possible link between cancer and viruses at a time when that idea was considered preposterous, even though research as early as 1911 suggested such a link. Unable to obtain funding for her research, she instead pursued a medical degree. In 1949 she was the first woman to ever receive an MD from Georgetown University School of Medicine, having completed that degree in only two years. She was 43 at the time. Dr. Stewart returned to research, and in 1953 she isolated a small agent from cancerous tissue that is also capable of causing cancer in healthy tissue, thus pioneering the field of viral oncology. Sadly, Dr. Stewart succumbed to the same disease she was investigating; she died of cancer in 1976.

100 bp downstream of the start site is less methylated than the corresponding region in cells that do not express this gene.

These observations suggest that methylation of CpG sequences may block DNA recognition by transcription factors, thus influencing gene expression. Recent evidence suggests that non-methylated CpG sequences are specifically bound by certain transcription factors. These transcription factors may also recruit histone-modifying enzymes that alter the chromatin structure. We will discuss histone modifications in more detail below.

Epigenetic modifications influence gene expression

DNA methylation and histone modification (to be discussed below) are two mechanisms associated with **epigenetics**—modifications to the genome that affect gene expression but do not alter the DNA sequence. A third mechanism involves RNA-mediated silencing, which we will discuss in Section 31.6. Several human disorders, including Fragile X syndrome, Angelman's syndrome, and various forms of cancer, are thought to be caused by epigenetic modifications and influenced by diet, exercise, age, or exposure to environmental toxins.

Cancer was the first human disease to be linked to epigenetics when it was discovered that cancerous cells had abnormally high levels of methylation as compared to normal cells. Further investigations showed that in cancer cells, many CpG islands become methylated, silencing genes such as tumor-suppressor genes that are critical for the control of cell division and should not be turned off.

Enhancers stimulate transcription by recruiting activator proteins that alter chromatin structure

Let us now turn to the action of enhancers, which were introduced in Section 29.3. Recall that enhancers are DNA sequences that have no promoter activity of their own but greatly increase transcription, even when they are located several thousand base pairs from the gene being expressed.

Enhancers function by serving as binding sites for specific transcription factors. An enhancer is effective only in the specific cell types in which appropriate regulatory proteins are expressed. In many cases, these DNA-binding proteins influence transcription initiation by altering the local chromatin structure to expose a gene or its regulatory sites rather than by direct interactions with RNA polymerase. This mechanism accounts for the ability of enhancers to act at a distance.

The properties of enhancers are illustrated by studies of the enhancer controlling the gene Islet-1. This gene encodes a homeodomain-containing protein that plays important roles in the developing nervous system. DNA fragments were constructed by connecting the promoter and enhancer regions associated with the Islet-1 gene to a gene expressing green fluorescent protein (GFP; **Figure 31.22**). When this DNA construct was introduced into zebrafish, an organism useful for imaging studies because of its size and transparency, the expression of green fluorescent protein could be visualized by microscopy. The green fluorescence of GFP was only seen in motor neurons and only during one stage of zebrafish development (**Figure 31.23**), revealing the power of the enhancer in limiting gene expression to particular cells.

❖ **SELF–CHECK QUESTION**

Why is eukaryotic DNA compacted into chromatin? How does this impact the activation or repression of gene expression? What must happen to chromatin in order for genes to be transcribed or silenced?

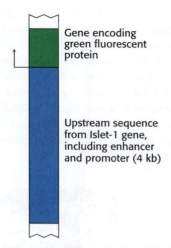

Gene encoding
green fluorescent
protein

Upstream sequence
from Islet-1 gene,
including enhancer
and promoter (4 kb)

FIGURE 31.22 Enhancer function can be assayed with a reporter construct. A DNA construct that includes a gene encoding green fluorescent protein adjacent to 4 kilobases of the DNA upstream of the start site for the Islet-1 gene. If this construct is introduced into cells in which Islet-1 is expressed, then the enhancer and promoter should lead to the production of green fluorescent protein.

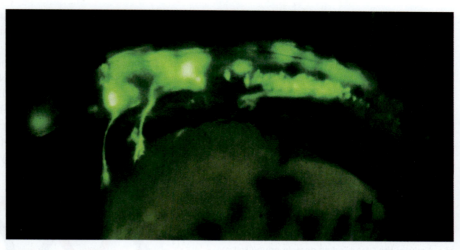

FIGURE 31.23 Enhancer function can be demonstrated in vivo. This microscopic image shows a developing zebrafish embryo into which the probe for enhancer function (Figure 31.22) has been introduced. The green fluorescence reveals that green fluorescent protein is expressed only in a subset of cells, shown to be cranial motor neurons. [Republished with permission of Society for Neuroscience, from Higashijima S., Hotta Y., Okamoto H., Visualization of Cranial Motor Neurons in Live Zebrafish. *J. Neurosci.*, January 1, 2000, 20(1):206–218; permission conveyed through Copyright Clearance Center, Inc.]

Nuclear hormone receptors are transcription factors that cause changes in chromatin structure

How do transcription factors stimulate changes in chromatin structure and thus influence gene expression? To illustrate this, let's discuss the system that detects and responds to estrogens, cholesterol-derived steroid hormones (Section 27.4). Synthesized and released by the ovaries, estrogens, such as estradiol, are required for the development of female secondary sex characteristics and, along with progesterone, participate in the ovarian cycle.

Because they are hydrophobic molecules, estrogens easily diffuse across cell membranes. When inside a cell, estrogens bind to highly specific, soluble receptor proteins. Estrogen receptors are members of a large family of proteins that act as receptors for a wide range of hydrophobic molecules, including other steroid hormones, thyroid hormones, and retinoids.

Estradiol
(an estrogen)

All-*trans*-retinoic acid
(a retinoid)

Thyroxine
(L-3,5,3′,5′-Tetraiodothyronine)
(a thyroid hormone)

The human genome encodes approximately 50 members of this family of receptors, often referred to as **nuclear hormone receptors**. The genomes of other multicellular eukaryotes encode similar numbers of nuclear hormone receptors, although they are absent in yeast.

All these receptors have a similar mode of action. On binding of the signal molecule (called, generically, a *ligand*), the ligand–receptor complex modifies the expression of specific genes by binding to control elements in the DNA. Estrogen receptors bind to specific DNA sites (referred to as estrogen response elements or EREs) that contain the consensus sequence 5′-**AGGTCANNNTGACCT**-3′. As expected from the symmetry of this sequence, an estrogen receptor binds to such sites as a dimer.

A comparison of the amino acid sequences of members of this family reveals two highly conserved domains: a DNA-binding domain and a ligand-binding domain (**Figure 31.24**). The DNA-binding domain lies toward the center of the molecule and consists of a set of zinc-based domains different from the Cys_2His_2 zinc-finger proteins introduced in Section 31.4. These zinc-based domains bind to specific DNA sequences by virtue of an α helix that lies in the major groove in the specific DNA complexes formed by estrogen receptors.

FIGURE 31.24 Nuclear hormone receptors contain two important conserved domains. These domains are: (A) a DNA-binding domain toward the center of the protein sequence, and (B) a ligand-binding domain toward the carboxyl terminus. The structure of a dimer of the DNA-binding domain bound to DNA is shown, as is one monomer of the normally dimeric ligand-binding domain. [Drawn from 1HCQ.pdb and 1LBD.pdb.]

INTERACT with these models in
 Achieve

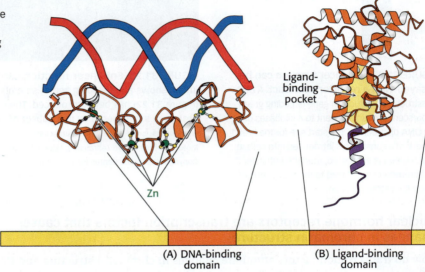

The second highly conserved domain of the nuclear receptor proteins is the ligand-binding site, which lies near the carboxyl terminus. This domain folds into a structure that consists almost entirely of α helices, arranged in three layers. The ligand binds in a hydrophobic pocket that lies in the center of this array of helices (**Figure 31.25**). This domain changes conformation when it binds its ligand, estrogen.

FIGURE 31.25 Nuclear hormone receptors interact with ligands. The ligand lies completely surrounded within a pocket in the ligand-binding domain. Notice that the last α helix, helix 12 (shown in purple), folds into a groove on the side of the structure once the ligand binds. [Drawn from 1LDB.pdb and 1ERE.pdb.]

INTERACT with this model in
Achieve

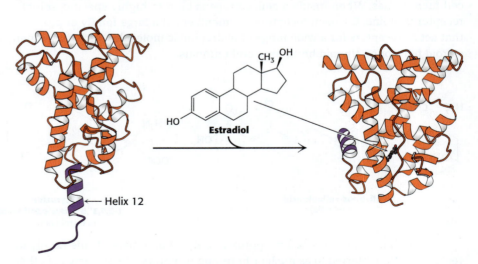

How does ligand binding lead to changes in gene expression? The simplest model would have the binding of ligand alter the DNA-binding properties of the receptor, much like the *lac* repressor in bacteria. However, experiments with purified nuclear hormone receptors revealed that ligand binding does *not* significantly alter DNA-binding affinity and specificity. The effect must be due to a different mechanism.

Nuclear hormone receptors regulate transcription by recruiting coactivators to the transcription complex

Is it possible that nuclear hormone receptors interact with specific proteins only in the presence of a ligand? Experiments to answer this question led to the identification of several related proteins called **coactivators**, such as SRC-1 (steroid receptor coactivator-1), GRIP-1 (glucocorticoid receptor interacting protein-1), and NcoA-1 (nuclear hormone receptor coactivator-1). These coactivators are referred to as the p160 family because of their size. The binding of ligand to the receptor induces a conformational change that allows the recruitment of a coactivator (**Figure 31.26**). In many cases, these coactivators are enzymes that catalyze reactions that lead to the modification of chromatin structure.

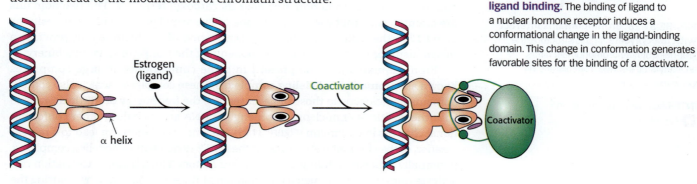

FIGURE 31.26 Nuclear hormone receptors recruit coactivators upon ligand binding. The binding of ligand to a nuclear hormone receptor induces a conformational change in the ligand-binding domain. This change in conformation generates favorable sites for the binding of a coactivator.

Chromatin structure is modulated through covalent modifications of histone tails

We have seen that nuclear receptors respond to signal molecules by recruiting coactivators. So now the question is: how do coactivators modulate transcriptional activity? The answer is that these proteins act to loosen the histone complex from the DNA, exposing additional DNA regions to the transcription machinery.

Much of the effectiveness of coactivators appears to result from their ability to covalently modify the amino-terminal tails of histones as well as regions on other proteins. Some of the p160 coactivators and the proteins that they recruit catalyze the transfer of acetyl groups from acetyl CoA to specific lysine residues in these amino-terminal tails.

Lysine in histone tail   **Acetyl CoA**

Enzymes that catalyze such reactions are called **histone acetyltransferases (HATs)**. The histone tails are readily extended so they can fit into the HAT active site and become acetylated (**Figure 31.27**).

What are the consequences of histone acetylation? Lysine bears a positively charged ammonium group at neutral pH. The addition of an acetyl group generates an uncharged amide group. This change dramatically reduces the affinity of the histone tail for DNA and modestly decreases the affinity of the entire histone complex for DNA, loosening the histone complex from the DNA and relaxing the chromatin structure.

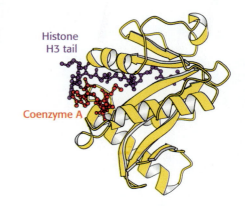

Histone H3 tail

Coenzyme A

FIGURE 31.27 Histone acetyltransferase modifies histone tails. In this structure, the amino-terminal tail of histone H3 extends into a pocket in the histone acetyltransferase enzyme (yellow). A lysine side chain on the histone tail can accept an acetyl group from acetyl Coenzyme A bound in an adjacent site on the enzyme. [Drawn from 1QSN.pdb.]

INTERACT with this model in

 Achieve

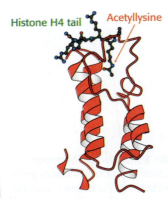

FIGURE 31.28 Bromodomains interact with acetyllysine. In this structure, the four-helix-bundle bromodomain binds an acetylated peptide from histone H4.

[Drawn from 1E6I.pdb.]

INTERACT with this model in

 Achieve

In addition, the acetylated lysine residues interact with a specific acetyllysine-binding domain that is present in many proteins that regulate eukaryotic transcription. This domain, also termed a **bromodomain**, comprises approximately 110 amino acids that form a four-helix bundle containing a peptide-binding site at one end (**Figure 31.28**).

Bromodomain-containing proteins are components of two large complexes essential for transcription. One is a complex of more than 10 polypeptides that binds to the TATA-box-binding protein (TBP). Recall that TBP is an essential transcription factor for many genes (Section 29.3). Proteins that bind to the TATA-box-binding protein are called **TAFs** (for **TATA-box-binding protein associated factors**), and together with TBP they form TFIID (Section 29.3). In particular, TAF1 contains a pair of bromodomains near its carboxyl terminus. The two domains are oriented such that each can bind one of two acetyllysine residues at positions 5 and 12 in the histone H4 tail. Thus, acetylation of the histone tails provides a mechanism for recruiting other components of the transcriptional machinery.

Bromodomains are also present in some components of large complexes known as **chromatin-remodeling complexes**. These complexes, which also contain domains homologous to those of helicases, use the free energy of ATP hydrolysis to shift the positions of nucleosomes along the DNA and to induce other conformational changes in chromatin (**Figure 31.29**). Histone acetylation can lead to a reorganization of the chromatin structure via these ATP-dependent remodeling complexes, potentially exposing binding sites for other factors. Thus, histone acetylation can activate transcription through a combination of three mechanisms: by reducing the affinity of the histones for DNA, by recruiting other components of the transcriptional machinery, and by initiating the remodeling of the chromatin structure.

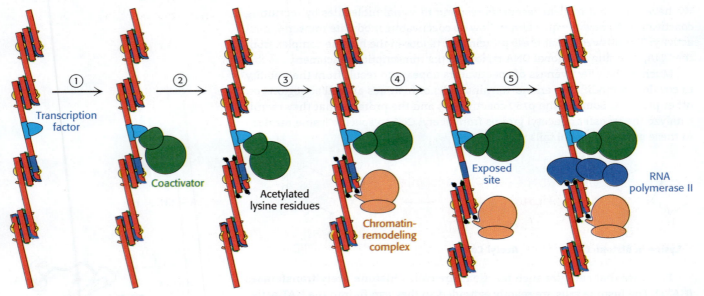

FIGURE 31.29 Gene regulation can involve chromatin remodeling. Eukaryotic gene regulation begins with an activated transcription factor bound to a specific site on DNA. One scheme for the initiation of transcription by RNA polymerase II requires five steps: (1) recruitment of a coactivator, (2) acetylation of lysine residues in the histone tails, (3) binding of a remodeling-engine complex to the acetylated lysine residues, (4) ATP-dependent remodeling of the chromatin structure to expose a binding site for RNA polymerase II or for other factors, and (5) recruitment of RNA polymerase II. Only two subunits are shown for each complex, although the actual complexes are much larger. Other schemes are possible.

Nuclear hormone receptors also include regions that interact with components of the mediator complex. Thus, two mechanisms of gene regulation can work in concert. Modification of histones and chromatin remodeling can open up regions of chromatin into which the transcription complex can be recruited through protein–protein interactions.

Transcriptional repression can be achieved through histone deacetylation and other modifications

Just as in bacteria, some changes in a cell's environment lead to the repression of genes that had been active. The modification of histone tails again plays an important role. However, in repression, a key reaction appears to be the *deacetylation* of acetylated lysine, catalyzed by specific **histone deacetylase** enzymes.

In many ways, the acetylation and deacetylation of lysine residues in histone tails (and, likely, in other proteins) is analogous to the phosphorylation and dephosphorylation of serine, threonine, and tyrosine residues in other stages of signaling processes. Like the addition of phosphoryl groups, the addition of acetyl groups can induce conformational changes and generate novel binding sites. Without a means of removal of these groups, however, these signaling switches will become stuck in one position and lose their effectiveness. Like phosphatases, deacetylases help reset the switches.

Acetylation is not the only modification of histones and other proteins in gene-regulation processes. The methylation of specific lysine and arginine residues also can be important. Some of the more common modifications are shown in **Table 31.1**.

The elucidation of the roles of these modifications is a very active area of research. The relation between various histone modifications and their roles in controlling gene expression is sometimes referred to as "the histone code," and important generalizations have been elucidated. For example, trimethylation of lysine 27 on histone H3 (H3 K27) is associated with repression of gene expression. This modification is promoted by a protein complex termed *polycomb repressive complex-2 (PRC-2)*. PRC-2 both binds to histones containing the trimethylated H3 K27 and catalyzes the addition of methyl groups to other H3 K27 substrates. This allows an initial modification to spread to adjacent histone octamers, facilitating gene repression.

The development and application of methods that allow probing of histone modifications on a genome-wide scale in a range of cell types has provided large data sets that can be examined to test and refine models of transcription regulation via chromatin modifications.

TABLE 31.1 Selected histone modifications

Modification	Effect
H3 K9 acetylation	Activation
H3 K27 acetylation	Activation
H4 K16 acetylation	Activation
H3 K4 monomethylation	Activation
H3 K9 trimethylation	Repression
H3 K27 trimethylation	Repression
H3 R17 methylation	Activation
H2B S14 phosphorylation	DNA repair
H2B K120 ubiquitination	Activation

❖ SELF–CHECK QUESTION

Summarize the various aspects of chromatin modifications discussed above and how they affect gene expression.

31.6 Gene Expression Can Be Controlled at the Posttranscriptional Level

The modulation of the rate of transcription initiation is the most common mechanism of gene regulation. However, other stages of transcription also can be targets for regulation. In addition, the process of translation provides other points of intervention for regulating the level of a protein produced in a cell. In Chapter 29, we considered riboswitches that control transcription termination (Section 29.3). Other riboswitches control gene expression by other mechanisms, such as the formation of structures that inhibit translation. Additional mechanisms for posttranscriptional gene regulation have been discovered in both bacteria and eukaryotes, some of which will be described here.

Attenuation regulates transcription in bacteria through the modulation of nascent RNA secondary structure

A different mechanism for regulating transcription in bacteria was discovered by Charles Yanofsky and his colleagues as a result of their studies of the tryptophan operon. This operon encodes five enzymes that convert chorismate into tryptophan. Analysis of the 5′ end of *trp* mRNA revealed the presence of a leader sequence of 162 nucleotides before the initiation codon of the first enzyme.

The next striking observation was that bacteria produced a transcript consisting of only the first 130 nucleotides when the tryptophan level was high, but they produced a 7000-nucleotide *trp* mRNA, including the entire leader sequence, when tryptophan was scarce. Thus, when tryptophan is plentiful and the biosynthetic enzymes are not needed, transcription is abruptly broken off before any mRNA coding for the enzymes is produced. The site of termination is called the attenuator, and this mode of regulation is called **attenuation**.

Attenuation depends on features at the 5′ end of the mRNA product (**Figure 31.30**). The first part of the leader sequence encodes a 14-amino-acid leader peptide. Following the open reading frame for the peptide is the attenuator, a region of RNA that is capable of forming several alternative structures. Recall that transcription and translation are tightly coupled in bacteria. Thus, the translation of the *trp* mRNA begins soon after the ribosome-binding site has been synthesized.

FIGURE 31.30 The leader region of *trp* mRNA includes tryptophan codons. (A) The nucleotide sequence of the 5′ end of *trp* mRNA includes a short open reading frame that encodes a peptide comprising 14 amino acids; the leader encodes two tryptophan residues and has an untranslated attenuator region that includes a region that can form a terminator structure (red and blue). (B and C) The attenuator region can adopt either of two distinct stem-loop structures.

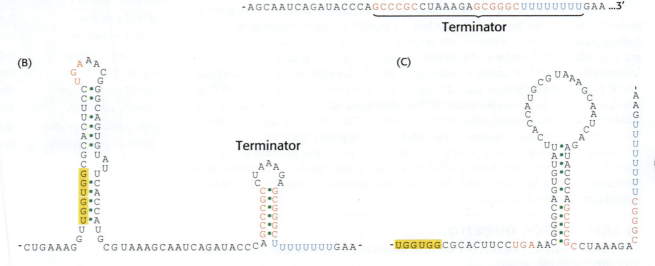

How does the level of tryptophan alter transcription of the *trp* operon? An important clue was the finding that the 14-amino-acid leader peptide includes two adjacent tryptophan residues. A ribosome is able to translate the leader region of the mRNA product only in the presence of adequate concentrations of tryptophan. When enough tryptophan is present, a stem-loop structure forms in the attenuator region, which leads to the release of RNA polymerase from the DNA (**Figure 31.31**). However, when tryptophan is scarce, transcription is

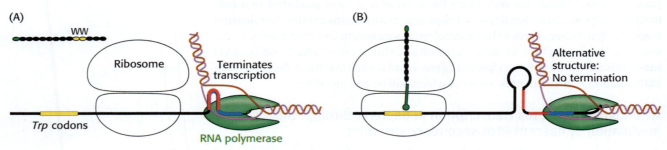

FIGURE 31.31 Attenuation regulates transcription. (A) In the presence of adequate concentrations of tryptophan (and, hence, Trp-tRNA), translation proceeds rapidly and an RNA structure forms that terminates transcription. (B) At low concentrations of tryptophan, translation stalls while awaiting Trp-tRNA, giving time for an alternative RNA structure to form that prevents the formation of the terminator and allows transcription to proceed.

terminated less frequently. Little tryptophanyl-tRNA is present, and so the ribosome stalls at the tandem UGG codons encoding tryptophan. This delay leaves the adjacent region of the mRNA exposed as transcription continues. An alternative RNA structure that does not function as a terminator is formed, and transcription continues into and through the coding regions for the enzymes. Thus, attenuation provides an elegant means of sensing the supply of tryptophan required for protein synthesis.

Several other operons for the biosynthesis of amino acids in E. coli also are regulated by attenuator sites. The leader peptide of each contains an abundance of the amino acid residues of the type synthesized by the operon (**Figure 31.32**). For example, the leader peptide for the phenylalanine operon includes 7 phenylalanine residues among 15 residues. The threonine operon encodes enzymes required for the synthesis of both threonine and isoleucine; the leader peptide contains 8 threonine and 4 isoleucine residues in a 16-residue sequence. The leader peptide for the histidine operon includes 7 histidine residues in a row. In each case, low levels of the corresponding charged tRNA cause the ribosome to stall, trapping the nascent mRNA in a state that can form a structure that allows RNA polymerase to read through the attenuator site. Evolution has apparently converged on this strategy repeatedly as a mechanism for controlling amino acid biosynthesis.

(A)

Met - Lys - Arg - Ile - Ser - Thr - Thr - Ile - Thr - Thr - Thr - Ile - Thr - Ile - Thr - Thr -

5′   AUG AAA CGC AUU AGC ACC ACC AUU ACC ACC ACC AUC ACC AUU ACC ACA   3′

(B)

Met - Lys - His - Ile - Pro - Phe - Phe - Phe - Ala - Phe - Phe - Phe - Thr - Phe - Pro - Stop

5′   AUG AAA CAC AUA CCG UUU UUC UUC GCA UUC UUU UUU ACC UUC CCC UGA   3′

(C)

Met - Thr - Arg - Val - Gln - Phe - Lys - His - His - His - His - His - His - His - Pro - Asp -

5′   AUG ACA CGC GUU CAA UUU AAA CAC CAC CAU CAU CAC CAU CAU CCU GAC   3′

FIGURE 31.32 The leader sequences of several operons contain codons for the specific amino acid that they regulate. Shown here are the amino acid sequences and the corresponding mRNA nucleotide sequences of the (A) threonine operon, (B) phenylalanine operon, and (C) histidine operon. In each case, an abundance of one amino acid in the leader peptide sequence leads to attenuation.

This example of posttranscriptional gene regulation that we have been describing comes from bacteria, as opposed to archaea. The transcriptional apparatus present in archaea shares many features with that found in eukaryotes. This commonality is frequently interpreted to suggest that eukaryotes evolved after a cell fusion event in which a bacterial cell was engulfed by an archaeal cell. Nonetheless, the key principles of gene regulation such as the occurrence of operons and the roles of DNA-binding proteins in directly blocking or stimulating RNA polymerase are the same in bacteria and archaea, perhaps because of similar genome sizes. As we shall see next, eukaryotic cells with larger genomes and, often, many distinct cell types, use quite different strategies.

Eukaryotes use different mechanisms to control gene expression at the posttranscriptional level

Just as in bacteria, gene expression in eukaryotes can be regulated subsequent to transcription. We shall consider two examples. The first is the regulation of genes taking part in iron metabolism through key features in RNA secondary structure, similar in many ways to attenuation in bacteria. The second entails an entirely new mechanism, in which certain small regulatory RNA molecules allow the regulation of gene expression through interaction with a range of target mRNAs. Remarkably, this mechanism, discovered relatively recently, affects the expression of approximately 60% of all human genes.

(A)

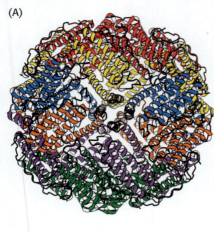

(B)

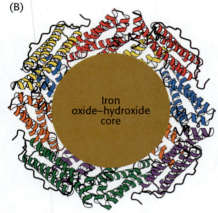

Iron oxide–hydroxide core

FIGURE 31.33 The structure of ferritin forms a sphere. (A) Twenty-four ferritin polypeptides form a nearly spherical shell. (B) A cutaway view reveals the core that stores iron as an iron oxide–hydroxide complex. [Drawn from 1IES.pdb.]

INTERACT with the above model in Achie√e

Genes associated with iron metabolism are translationally regulated in animals

RNA secondary structure plays a role in the regulation of iron metabolism in eukaryotes. Iron is an essential nutrient, required for the synthesis of hemoglobin, cytochromes, and many other proteins. However, excess iron can be quite harmful because, untamed by a suitable protein environment, iron can initiate a range of free-radical reactions that damage proteins, lipids, and nucleic acids.

Animals have evolved sophisticated systems for the accumulation of iron in times of scarcity and for the safe storage of excess iron for later use. Key proteins include **transferrin**, a transport protein that carries iron in the serum, **transferrin receptor**, a membrane protein that binds iron-loaded transferrin and initiates its entry into cells, and **ferritin**, an impressively efficient iron-storage protein found primarily in the liver and kidneys. Twenty-four ferritin polypeptides form a nearly spherical shell that encloses as many as 2400 iron atoms, a ratio of one iron atom per amino acid (**Figure 31.33**).

Ferritin and transferrin receptor expression levels are reciprocally related in their responses to changes in iron levels. When iron is scarce, the amount of transferrin receptor increases and little or no new ferritin is synthesized. Interestingly, the extent of mRNA synthesis for these proteins does not change correspondingly. Instead, regulation takes place at the level of translation.

Consider ferritin first. Ferritin mRNA includes a stem-loop structure termed an *iron-response element* (IRE) in its 5′ untranslated region (**Figure 31.34**). A 90 kDa protein, the IRE-binding protein (IRP), binds to this stem-loop and blocks the initiation of translation at the downstream coding region. When the iron level increases, the IRP binds iron as a 4Fe–4S cluster. The IRP bound to iron cannot bind RNA, because the binding sites for iron and RNA substantially overlap. Thus, in the presence of iron, ferritin mRNA is released from the IRP and translated to produce ferritin, which sequesters the excess iron.

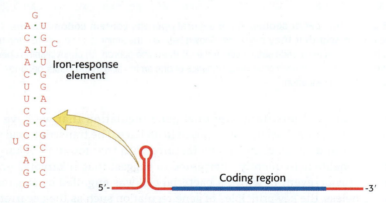

FIGURE 31.34 Ferritin mRNA includes an iron-response element in its 5′ untranslated region. A specific protein binds to the IRE stem-loop structure in the 5′ untranslated region of the mRNA and blocks the translation of this mRNA under low-iron conditions.

An examination of the nucleotide sequence of transferrin receptor mRNA reveals the presence of several IRE-like regions. However, these regions are located in the 3′ untranslated region rather than in the 5′ untranslated region (**Figure 31.35**).

FIGURE 31.35 Transferrin receptor mRNA has a set of iron-response elements (IREs) in its 3′ untranslated region. The binding of the IRE-binding protein to these elements stabilizes the mRNA but does not interfere with translation.

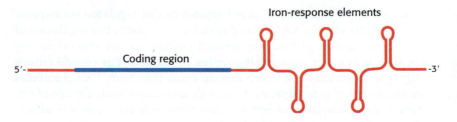

Under low-iron conditions, IRP binds to these IREs. However, given the location of these binding sites, the transferrin receptor mRNA can still be translated.

What happens when the iron level increases and the IRP no longer binds transferrin receptor mRNA? Freed from the IRP, transferrin receptor mRNA is rapidly degraded. Thus, an increase in the cellular iron level leads to the destruction of transferrin receptor mRNA and, hence, a reduction in the production of transferrin receptor protein.

The purification of the IRP and the cloning of its cDNA were sources of truly remarkable insight into evolution. The IRP was found to be approximately 30% identical in amino acid sequence with the citric acid cycle enzyme aconitase from mitochondria. Further analysis revealed that the IRP is, in fact, an active aconitase enzyme; it is a cytoplasmic aconitase that had been known for a long time, but its function was not well understood (**Figure 31.36**). The iron–sulfur center at the active site of the IRP is rather unstable, and loss of the iron triggers significant changes in protein conformation. Thus, this protein can serve as an iron-sensing factor.

Other mRNAs, including those taking part in heme synthesis, have been found to contain IREs. Thus, genes encoding proteins required for iron metabolism acquired sequences that, when transcribed, provided binding sites for the iron-sensing protein. An environmental signal—the concentration of iron—controls the translation of proteins required for the metabolism of this metal. Thus, mutations in the untranslated region of mRNAs have been selected for beneficial regulation by iron levels.

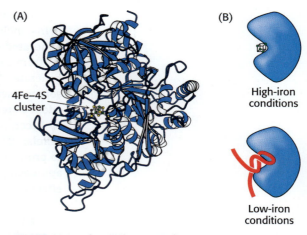

FIGURE 31.36 The IRP is an aconitase. (A) Aconitase contains an unstable 4Fe–4S cluster at its center. (B) Under conditions of low iron, the 4Fe–4S cluster dissociates and appropriate RNA molecules can bind in its place. [Drawn from 1C96.pdb.]

INTERACT with this model in Achieve

Small RNAs are involved in posttranscriptional gene regulation in eukaryotes

Other mechanisms of posttranscriptional gene regulation, particularly silencing, involve small regulatory RNAs, most of which were only recently discovered. These small RNAs represent an important class of gene-regulatory molecules.

One prominent mechanism of gene silencing is known as **RNA interference (RNAi)**, in which small RNAs interact with a protein complex and guide it to a target mRNA. Subsequently, the target mRNA will either be degraded or its translation will somehow be inhibited.

How was RNAi discovered? Genetic studies of development in *C. elegans* revealed that a gene called *lin-4* encodes an RNA molecule 61 nucleotides long that can regulate the expression of certain other genes. The 61-nucleotide *lin-4* RNA does not encode a protein, but rather is cleaved into a 22-nucleotide RNA that possesses the regulatory activity. This discovery was the first view of a large class of regulatory RNA molecules known as microRNAs (miRNAs), which were briefly introduced in Section 29.4. The key to the activity of microRNAs and their specificity for particular genes is their ability to form Watson–Crick base-pair-stabilized complexes with the mRNAs of those genes.

MicroRNAs do not function on their own, however. They bind to members of a highly conserved family of proteins called **Argonaute** (**Figure 31.37**). Argonaute–miRNA complexes bind target mRNAs due to sequence complementarity between the miRNA and the mRNA. Thus, the miRNAs serve as guide RNAs that determine the specificity of the Argonaute protein, which will cleave the target mRNA through its intrinsic RNase activity (**Figure 31.38**). Argonaute proteins form part of a larger complex called the *RNA-induced silencing complex* (RISC), which is central to the process of RNAi.

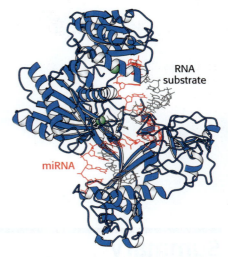

FIGURE 31.37 MicroRNAs bind to the Argonaute complex. The miRNA (shown in red) is bound by the Argonaute protein. The miRNA serves as a guide that binds the RNA substrate (shown in gray) through the formation of a double helix. Two magnesium ions are shown in green. [Drawn from 3HK2.pdb.]

INTERACT with the above model in Achieve

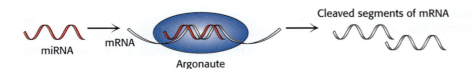

FIGURE 31.38 MicroRNA–Argonaute complexes target specific mRNA molecules for cleavage.

Interestingly, Argonaute proteins can bind to other types of small noncoding RNAs, including small interfering RNAs (siRNAs). Both siRNAs and miRNAs are generated through cleavage by an RNA endonuclease called *Dicer*, but they differ from each other in that they originate from different sources. MicroRNAs are generated from large, genetically encoded, single-stranded pri-microRNA precursors transcribed in the cell, as described in Figure 29.41. These miRNA precursors adopt a stem-loop structure which is cleaved and processed into mature miRNAs by Dicer. In contrast, siRNAs are derived from longer double-stranded extracellular RNAs, many of which might be invading viral sequences. It is also possible to produce siRNAs in vitro and introduce them into cells experimentally. This approach provides researchers with the opportunity to examine the phenotypic effects of controlled targeted gene silencing.

The occurrence of gene regulation by small noncoding RNAs was originally thought to be limited to a relatively small number of species. However, subsequent studies have revealed that this mode of gene regulation is nearly ubiquitous in eukaryotes. Indeed, more than 1900 miRNAs encoded by the human genome have been identified. Each miRNA can regulate many different genes because many different target sequences are present in each mRNA. An estimated 60% of all human genes are regulated by one or more miRNAs.

As one example, consider the human miRNA called miR-206. This miRNA down-regulates the expression of one isoform of the estrogen receptor. In addition, it appears to down-regulate the expression of several different coactivators that interact with the estrogen receptor. Thus, this miRNA can mute the influence of estrogen by blocking the estrogen-initiated signaling pathway at several different steps.

The microRNA pathway has tremendous implications for the evolution of gene-regulatory pathways. Most of the target sites for miRNAs are present in the 3′-untranslated regions of mRNAs. These sequences are quite free to mutate because they do not encode proteins and are not required to fold into any specific structures. Thus, in the context of a set of expressed miRNAs, mutations in this region of any particular gene could, in principle, increase or decrease the affinity for one or more miRNAs and, hence, alter the regulation of the gene.

❖ **SELF–CHECK QUESTION**

Gene expression in all systems involves the conversion of information present in a genomic DNA sequence into a functional protein. Give general examples of the control of gene expression at different steps along this pathway.

Summary

31.1 Bacterial DNA-Binding Proteins Bind to Specific Regulatory Sites

- In bacteria, specific DNA-binding proteins recognize regulatory sites that usually lie adjacent to the genes whose transcription is regulated by these proteins.
- The helix-turn-helix motif is a common DNA-binding motif.

31.2 In Bacteria, Genes Are Often Arranged into Clusters Under the Control of a Single Regulatory Sequence

- In bacteria, many genes are clustered into operons, which consist of control sites (an operator and a promoter) and a set of structural genes.

- Regulator genes encode proteins that interact with the operator and promoter sites in an operon to stimulate or inhibit transcription.
- The treatment of E. coli with lactose induces an increase in the production of β-galactosidase and two additional proteins that are encoded in the lactose operon.
- In the absence of lactose, the *lac* repressor protein binds to an operator site on the DNA and blocks transcription.
- The binding of allolactose, a derivative of lactose, to the *lac* repressor induces a conformational change that leads to dissociation from DNA. RNA polymerase can then move through the operator to transcribe the *lac* operon.

- The binding of the cAMP–CAP complex to a specific site in the promoter region of an inducible catabolic operon enhances the binding of RNA polymerase and the initiation of transcription.

31.3 Regulatory Circuits Can Result in Switching Between Patterns of Gene Expression

- Bacteriophage λ can maintain either a lytic or a lysogenic life cycle.
- A key regulatory protein, the λ repressor, regulates its own expression, promoting transcription of the gene that encodes the repressor when repressor levels are low and blocking transcription of the gene when levels are high.
- The Cro protein binds to the same sites as does the λ repressor, but with reversed affinities. When Cro is present at sufficient concentrations, it blocks the transcription of the gene for the λ repressor while allowing the transcription of its own gene.

31.4 Regulation of Gene Expression Is More Complex in Eukaryotes

- In eukaryotes, transcription factors are required to activate the expression of most genes. Transcription factors bind DNA and interact directly or indirectly with RNA polymerases.
- Eukaryotic transcription factors are modular: they consist of separate DNA-binding and activation domains.
- Important classes of DNA-binding domains include the homeodomains, the basic-leucine zipper proteins, and Cys_2His_2 zinc-finger proteins.
- Activation domains interact with RNA polymerases or their associated factors, or with other protein complexes such as mediator.
- Enhancers are DNA elements that can modulate gene expression from more than 1000 bp away from the start site of transcription.

31.5 The Control of Gene Expression in Eukaryotes Can Require Chromatin Remodeling

- Chromatin structure is crucial to the control of gene expression; it is more open and the DNA is less methylated near the transcription start sites of actively transcribed genes.
- Steroids such as estrogens bind to eukaryotic transcription factors called nuclear hormone receptors. The binding of ligands to the receptors induces a conformational change that allows the recruitment of additional proteins called coactivators.
- Coactivators can catalyze the addition of acetyl groups to lysine residues in the tails of histone proteins. Histone acetylation decreases the affinity of the histones for DNA, making additional genes accessible for transcription.
- In addition, acetylated histones are targets for proteins containing specific binding units called bromodomains. Bromodomains are components of two classes of large complexes that open up sites on chromatin and initiate transcription.

31.6 Gene Expression Can Be Controlled at the Posttranscriptional Level

- In bacteria, many operons important in amino acid biosynthesis are regulated by attenuation, a process that depends on the formation of alternative structures in mRNA.
- In eukaryotes, genes encoding proteins that transport and store iron are regulated at the translational level.
- RNA interference (RNAi) is a common gene-silencing mechanism that involves small non-coding RNAs. These small regulatory RNAs function as guides to identify specific target mRNA molecules for cleavage by nucleases.

Key Terms

β-galactosidase (p. 950)
helix-turn-helix motif (p. 951)
operon (p. 952)
operator site (p. 952)
repressor (p. 952)
inducer (p. 954)
corepressor (p. 955)
catabolite repression (p. 955)
catabolite activator protein (CAP) (p. 955)
λ repressor (p. 957)
Cro (p. 957)

homeodomain (p. 958)
basic-leucine zipper (bZip) protein (p. 959)
Cys_2His_2 zinc-finger domain (p. 959)
DNA-binding domain (p. 960)
activation domain (p. 960)
mediator (p. 960)
CpG island (p. 961)
epigenetics (p. 962)
nuclear hormone receptor (p. 963)
coactivator (p. 965)
histone acetyltransferase (HAT) (p. 965)

bromodomain (p. 966)
TATA-box-binding protein associated factor (TAF) (p. 966)
chromatin-remodeling complex (p. 966)
histone deacetylase (p. 967)
attenuation (p. 968)
transferrin (p. 970)
transferrin receptor (p. 970)
ferritin (p. 970)
RNA interference (RNAi) (p. 971)
Argonaute (p. 971)

Problems

1. Predict the effects of deleting the following regions of DNA:

(a) The gene encoding *lac* repressor

(b) The *lac* operator

(c) The gene encoding CAP ❖ **1,** ❖ **3**

2. The *lac* repressor and the *pur* repressor are homologous proteins with very similar three-dimensional structures, yet they have different effects on gene expression. Describe two important ways in which the gene-regulatory properties of these proteins differ. ❖ **2**

3. Which types of mutation in the *lac* operon stop *E. coli* from using lactose as a carbon source? Select all that apply. ❖ **3**

(a) Operator deletion

(b) Promoter deletion

(c) Lactose-binding site mutation

(d) Repressor DNA-binding site mutation

4. Consider a hypothetical mutation in O_R2 that blocks both λ repressor and Cro binding. How would this mutation affect the likelihood of bacteriophage λ entering the lytic phase? ❖ **4**

5. Place the events of gene regulation by the *lac* operon in order of their occurrence, starting with the introduction of lactose into the environment, and ending with the cell digesting lactose. ❖ **3**

(a) The repressor is inactivated by allolactose.

(b) RNA polymerase binds to the promoter region.

(c) The repressor is removed from the operator.

(d) The cell begins to digest lactose.

(e) Lactose is introduced into the environment.

(f) Lactose enzyme genes are expressed by the operon.

6. Growth of mammalian cells in the presence of 5-azacytidine results in the activation of some normally inactive genes. Propose an explanation. ❖ **5**

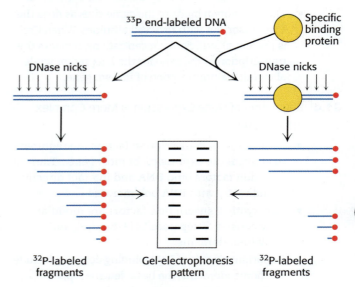

deoxyribose
5-Azacytidine

7. What is the effect of an increased Cro concentration on the expression of the gene for the λ repressor? Of an increased concentration of λ repressor on the expression of the Cro gene? Of an increased concentration of λ repressor on the expression of the λ repressor gene? ❖ **1,** ❖ **4**

8. A powerful method for examining protein–DNA interactions is called *DNA footprinting*. In this method, a DNA fragment containing a potential binding site is radiolabeled on one end. The labeled DNA is then treated with a DNA-cleaving agent such as DNase I such that each DNA molecule within the population is cut only once. The same cleavage process is carried out in the presence of the DNA-binding protein. The bound protein protects some sites within the DNA from cleavage. The patterns of DNA fragments in the cleaved pool of DNA molecules are then examined by electrophoresis followed by autoradiography.

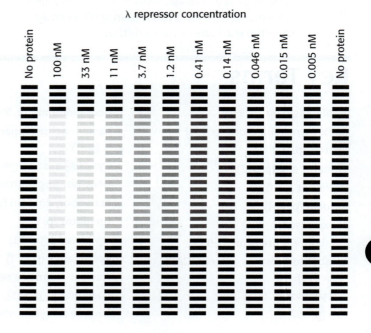

This method is applied to a DNA fragment containing a single binding site for the λ repressor in the presence of different concentrations of the λ repressor. The results are shown below:

Estimate the dissociation constant for the λ repressor–DNA complex and the standard free energy of binding. ❖ **4**

9. Which of the following statements about miRNAs and RNA interference are true? ❖ **7**

(a) miRNAs suppress gene expression by interfering with transcription.

(b) Small RNAs interfere with gene expression by interacting with mRNA, bringing about mRNA degradation.

(c) siRNAs and miRNAs are produced by the Dicer enzyme complex from various precursors.

(d) miRNAs are about 45 nucleotides long.

10. The restriction enzyme HpaII is a powerful tool for analyzing DNA methylation. This enzyme cleaves sites of the form 5′-CCGG-3′ but will not cleave such sites if the DNA is methylated on any of the cytosine residues. Genomic DNA from different organisms is treated with HpaII, and the results are analyzed by gel electrophoresis (see below). Provide an explanation for the observed patterns. ❖ **5**

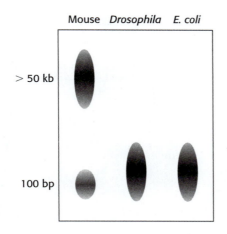

11. Which of the following statements are true? (Select all that apply.) ❖ **6**

(a) Chromatin packaging occurs in eukaryotes.

(b) The structure of adenine is different in eukaryotic DNA compared with prokaryotic DNA.

(c) Eukaryotic DNA contains enhancers.

(d) Multiple general transcription factors are found in eukaryotes.

(e) Nuclear export of RNA occurs in bacteria.

12. A protein domain that recognizes 5-methylcytosine in the context of double-stranded DNA has been characterized. What role might proteins containing such a domain play in regulating gene expression? Where on a double-stranded DNA molecule would you expect such a domain to bind? ❖ **1**, ❖ **2**

13. What is the effect of acetylation of a lysine residue on the charge of a histone protein? Of lysine methylation? ❖ **5**

14. Place the events in the order necessary for an epigenetic modification to be inherited in the next generation, starting with methylated CpG sites present on one inherited DNA strand, and ending with the epigenetic silencing being passed to the offspring. ❖ **5**

(a) A methyl group is added to the cytosine residue of the DNA sequence.

(b) Certain CpG methylation sites are not erased during gametogenesis or embryogenesis.

(c) Epigenetic silencing is passed to the offspring.

(d) Methylated CpG sites are present on one inherited DNA strand.

(e) DNA methyltransferase maintains the methylation pattern on both DNA strands.

(f) DNA methyltransferase recognizes and binds CpG sites on DNA.

15. Through recombinant DNA methods, a modified steroid hormone receptor was prepared that consists of an estrogen receptor with its ligand-binding domain replaced by the ligand-binding domain from the progesterone receptor. Predict the expected responsiveness of gene expression for cells treated with estrogen or with progesterone. ❖ **1**

16. What effect would you expect from the addition of an IRE to the 5′ end of a gene that is not normally regulated by iron levels? To the 3′ end? ❖ **7**

17. The following is the amino acid sequence of a eukaryotic transcription factor that belongs to one of the three structural classes discussed in Section 31.4. Identify the class. ❖ **2**

HTCDYAGCGKTYTKSSHLKAHLRTHTGEKPYHCDWDGC GWKFARSDELTRHYRKHTGHRPFQCQKCDRAFSRSDHL ALHMKRHF

18. Modifications to chromatin can affect transcriptional activity by changing the accessibility of DNA to the transcription machinery. The given descriptions are examples of various processes that cause remodeling of chromatin. Match each description to the effect it has on transcriptional activity. ❖ **5**

Chromatin modification description	Effect on transcriptional activity via chromatin remodeling
(a) Histone methylation occurs at different amino acids.	**1.** Activates
(b) Histone acetyltransferases attach acetyl groups to the N-terminus of histones.	**2.** Inactivates
(c) Methylated CpG regions are found close to transcription start sites.	**3.** Activates and inactivates

19. Nuclear hormone receptors form a complex with their ligands, other proteins, and DNA control elements in order to regulate the expression of specific genes. Match each nuclear hormone receptor domain to the characteristic(s) with which it is associated. ❖ 2

**Nuclear hormone
receptor domain**

(a) Activation domain

(b) DNA-binding domain

(c) Ligand-binding domain

Characteristic

1. Zinc-finger domain

2. Binds coactivators

3. Center of the receptor

4. Carboxyl-terminal end

5. Steroid-binding hydrophobic pocket

20. Classify the following examples of gene expression as positive or negative regulation. ❖ 6

(a) In the presence of lactose and low glucose, the *lac* operon is expressed 20-fold higher than in the absence of lactose.

(b) In the absence of lactose, the *lac* repressor binds to the *lac* operator.

(c) Low levels of tryptophan in E. coli result in the production of the full-length *trp* mRNA.

(d) Elevated iron levels in eukaryotes prevent the IRP from binding to the transferrin receptor mRNA, allowing it to be translated.

(e) miRNAs direct the Argonaute nuclease to a target mRNA.

Principles of Drug Discovery and Development

Inga Spence/Science Source

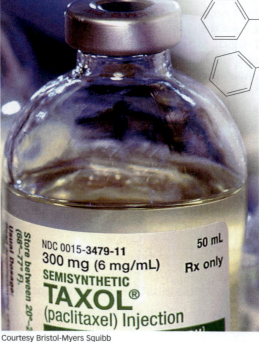

Courtesy Bristol-Myers Squibb

Paclitaxel
(Taxol)

Many drugs are based on natural products. Paclitaxel (marketed as Taxol) is a molecule isolated from the bark of the Pacific yew tree (*Taxus brevifolia*) that potently inhibits cell division and is used to treat various forms of cancer. Recently, methods have been developed to synthesize paclitaxel without destruction of the Pacific yew, significantly reducing the ecological impact of drug preparation.

❖ LEARNING GOALS

By the end of this chapter, you should be able to:

1. Describe the importance of target validation and tractability in the early drug discovery process.

2. List some ways in which lead molecules are identified.

3. Explain the criteria the compounds must meet in order to be developed into drugs.

4. Describe the processes by which the body interacts with drug molecules.

5. Describe how structural information can be used to assist in drug development.

6. Differentiate small molecule drugs and biologics.

7. Distinguish the clinical phases of drug development.

OUTLINE

32.1 Drug Discovery Begins with Target Identification and Validation

32.2 Lead Molecules Can Be Discovered in Many Ways

32.3 Compounds Must Meet Stringent Criteria to Be Developed into Drugs

32.4 Biologics are a Growing Family of Drugs

32.5 The Clinical Development of Medicines Proceeds Through Several Phases

The development of drugs is among the most important interfaces between biochemistry and medicine. In most cases, drugs act by binding to proteins and inhibiting or otherwise modulating their activities. An effective drug is much more than a potent modulator of its target, however. Drugs must be readily administered to patients and must persist within the body long enough to reach their targets. Furthermore, to prevent unwanted physiological effects, drugs must not modulate the properties of biomolecules other than the intended targets. These requirements tremendously limit the number of compounds that have the potential to be clinically useful drugs.

32.1 Drug Discovery Begins with Target Identification and Validation

FIGURE 32.1 The journey to an approved drug is divided into a discovery phase and a development phase. Each phase is subdivided into stages. At the end of each stage, the data obtained up to that point are evaluated, and a decision is made on whether to progress to the next step.

In this chapter, we will explore the science of **pharmacology**, which encompasses the discovery, chemistry, composition, identification, biological and physiological effects, uses, and manufacture of drugs. The journey from knowledge of a biochemical pathway to a clinically approved medicine is long, laborious, and expensive. This process is typically divided into two phases (**Figure 32.1**). In the **discovery phase**, molecules with the desired physiological effects are identified and optimized such that they can be introduced into humans; in the **development phase**, clinical trials are conducted with the optimal candidate molecule. Each of these phases is subdivided into stages. At the transition point from one stage to the next, the data are evaluated and a decision is made to approve the investment required to progress into the next stage.

FIGURE 32.1 The journey to an approved drug is divided into a discovery phase and a development phase. Each phase is subdivided into stages. At the end of each stage, the data obtained up to that point are evaluated, and a decision is made on whether to progress to the next step.

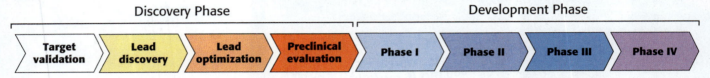

Discovery Phase | Development Phase

Target validation | Lead discovery | Lead optimization | Preclinical evaluation | Phase I | Phase II | Phase III | Phase IV

Drug targets must be validated and tractable

The initial steps of the discovery phase first require the identification of a **target**—a protein or an oligonucleotide whose activity, when modulated by a drug, will alter the progression of a human disease process. Throughout the previous chapters in this book, we have encountered a number of types of proteins—enzymes, receptors, and transporters—that serve as targets for a majority of the drugs in clinical use (**Figure 32.2**). Before embarking on the process of developing a drug against a particular target, researchers must first establish confidence that the target exhibits two critical properties.

There should be sufficient evidence that the target plays an important role in disease progression in humans. The process of obtaining this evidence is referred to as **target validation**. Researchers achieve validation of a target from a variety of experiments, such as human genetic analyses, RNA interference, or transgenic or knockout animal models (Section 9.4); they may also conduct studies with *tool compounds*—molecules that are usable in cellular or animal experiments but not necessarily suitable as drugs for humans.

There should be sufficient evidence that a molecule can be designed to interact with the target in a potent and specific manner. This property is referred to as a drug's **tractability**, or **druggability**. Assessment of target tractability is made significantly easier if the structure of the target, or a closely related protein, is known. A tractable target contains structural features, such as ligand-binding pockets, with which drug-like molecules can interact with a high degree of potency and specificity.

FIGURE 32.2 Drug targets span a variety of protein classes. In this pie chart, currently used drugs that modulate human proteins are sorted by target type: enzymes in green, receptors in blue, transporters in red, and other targets in gray.
[Data from M. Rask-Andersen et al., *Nat. Rev. Drug Discov.* 10:579–590, 2011.]

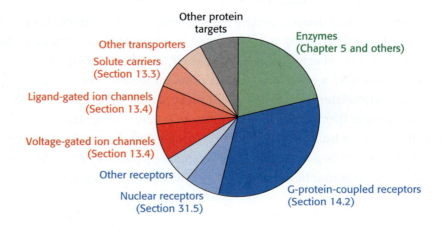

Other protein targets

Other transporters

Solute carriers (Section 13.3)

Ligand-gated ion channels (Section 13.4)

Voltage-gated ion channels (Section 13.4)

Other receptors

Nuclear receptors (Section 31.5)

Enzymes (Chapter 5 and others)

G-protein-coupled receptors (Section 14.2)

In practice, very few targets are known to be well-validated and highly tractable during the early stages of the discovery phase. As the target proceeds through the further investigation, additional data are obtained that either support or refute initial projections of the target's validation and tractability. One of the primary goals of the pharmaceutical scientist is to design experiments which can provide this information as early as possible. If the data do not lend support for a particular target, it can be abandoned in favor of more attractive targets.

Serendipitous observations can drive drug development

One cannot recount the history of drug discovery without acknowledging the role played by serendipity, or chance observations. Perhaps the most well-known example is Alexander Fleming's observation in 1928 that colonies of the bacterium *Staphylococcus aureus* died when they were adjacent to colonies of the mold *Penicillium notatum*, after spores of the mold had landed accidentally on plates growing the bacteria. Fleming determined that the mold produced a substance that could kill disease-causing bacteria. This discovery led to a fundamentally new approach to the treatment of bacterial infections. Howard Florey and Ernest Chain developed a powdered form of the substance, termed penicillin, that became a widely used antibiotic in the 1940s. When the structure of penicillin was elucidated in 1945, it was found to contain a four-membered β-lactam ring. This unusual feature is key to the antibacterial function of penicillin, as noted in Section 5.6.

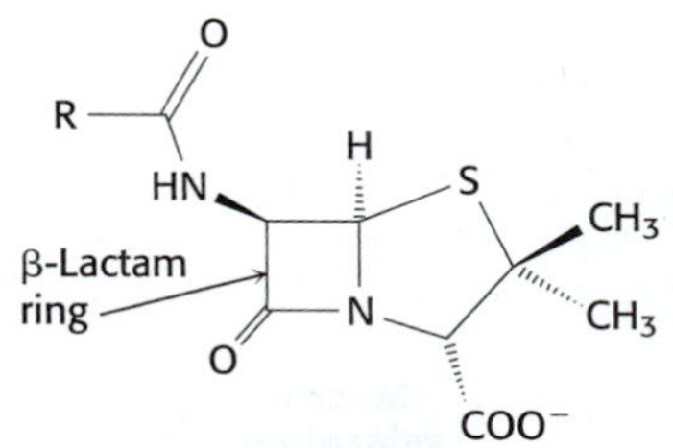

Three steps were crucial to fully capitalize on Fleming's discovery. First, an industrial process was developed for the production of penicillin from *Penicillium* mold on a large scale. Second, penicillin and its derivatives were chemically synthesized. The availability of synthetic penicillin derivatives opened the way for scientists to explore the relationship between structure and function. Finally, in 1965, Jack Strominger and James Park independently determined that penicillin exerts its antibiotic activity by blocking a critical transpeptidase reaction in bacterial cell-wall biosynthesis (**Figure 32.3**). Many penicillin derivatives remain widely used today.

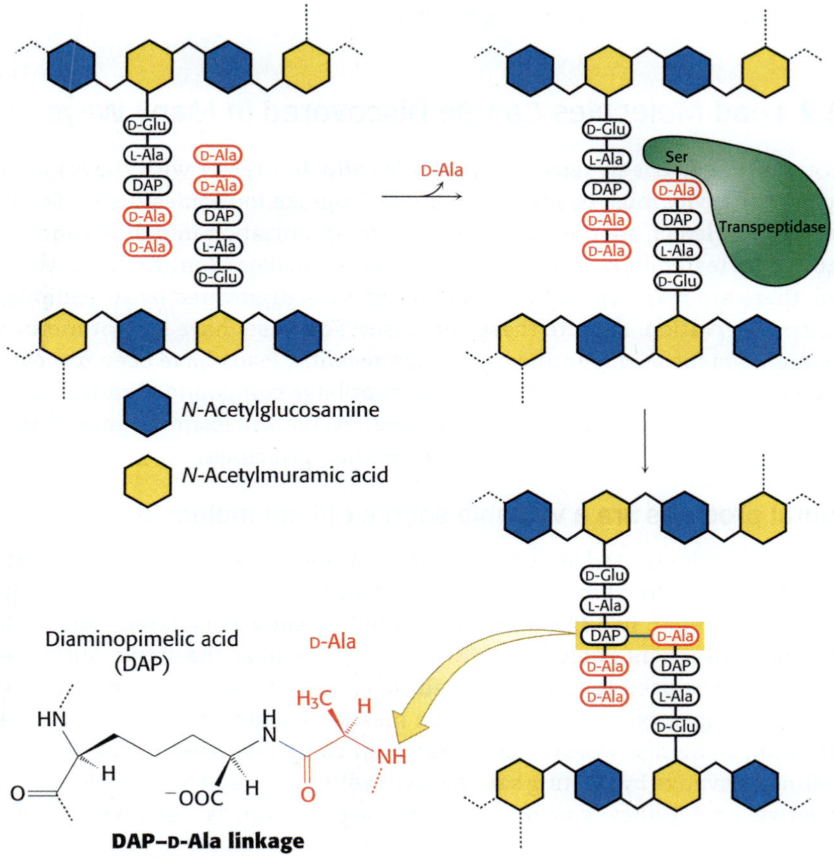

Penicillin

FIGURE 32.3 Penicillin disrupts the biosynthesis of the bacterial cell wall. A transpeptidase enzyme catalyzes the formation of cross-links between peptidoglycan groups. In the case shown, the transpeptidase catalyzes the linkage of D-alanine at the end of one peptide chain to the amino acid diaminopimelic acid (DAP) on another peptide chain. The diaminopimelic acid linkage (bottom left) is found in Gram-negative bacteria such as *E. coli*. Linkages of glycine-rich peptides are found in Gram-positive bacteria. Penicillin inhibits the action of the transpeptidase, so bacteria exposed to the drug have weak cell walls that are susceptible to lysis.

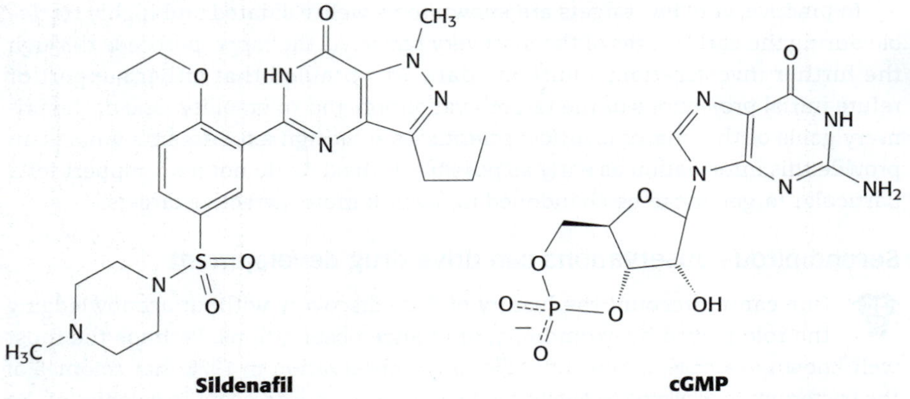

Sildenafil   **cGMP**

A more recent example of a drug discovered by serendipity is sildenafil. This compound was developed as an inhibitor of phosphodiesterase 5 (PDE5), an enzyme that catalyzes the hydrolysis of cGMP to GMP (**Figure 32.4**). The compound was intended as a treatment for hypertension and angina because cGMP plays a central role in the relaxation of smooth-muscle cells in blood vessels (**Figure 32.5**). The inhibition of PDE5 was expected to increase the concentration of cGMP by blocking the pathway for its degradation. In the course of early clinical trials in Wales, some men reported unusual penile erections, but whether this chance observation by a few men was due to the compound or to other effects was unclear. However, the observation made some biochemical sense because smooth-muscle relaxation due to increased cGMP levels had been previously discovered to play a role in penile erection.

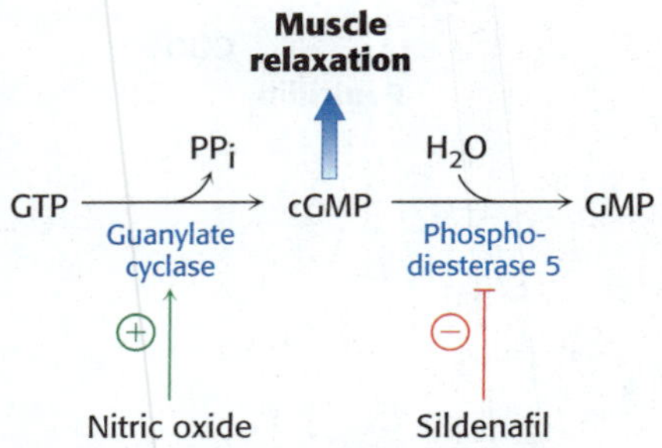

Subsequent clinical trials directed toward the evaluation of sildenafil for erectile dysfunction were successful. This account testifies to the importance of collecting comprehensive information from clinical-trial participants. In this case, incidental observations led to a new treatment for erectile dysfunction and a multibillion-dollar-per-year drug market.

Muscle relaxation

$$GTP \xrightarrow[\text{Guanylate cyclase}]{PP_i} cGMP \xrightarrow[\text{Phospho-diesterase 5}]{H_2O} GMP$$

Nitric oxide   Sildenafil

FIGURE 32.5 cGMP stimulates muscle relaxation. Increases in nitric oxide levels stimulate guanylate cyclase, which produces cGMP. PDE5 hydrolyzes cGMP, which lowers the cGMP concentration. The inhibition of PDE5 by sildenafil maintains elevated levels of cGMP.

32.2 Lead Molecules Can Be Discovered in Many Ways

Once a target has been selected, molecules which interact with the target and impact its activity must be identified. Initial drug-like molecules, also called **lead molecules** or **leads**, are the starting points for optimization into drug candidates that can be tested in humans. How are lead molecules identified? As we have seen, there are many examples where drugs were discovered by serendipity, or chance observation. Despite these successes, scientists have sought more reliable methods for lead identification. More recently, leads have been discovered by screening collections of natural products or large compound libraries for molecules that have desired biochemical properties. We will examine several examples of each of these pathways to reveal common principles.

Natural products are a valuable source of lead molecules

No drug is as widely used as aspirin. Observers at least as far back as Hippocrates (~400 B.C.E.) have noted the use of extracts from the bark and leaves of the willow tree for pain relief. In 1829, a mixture called *salicin* was isolated from willow bark, and subsequent analysis identified salicylic acid as the active component. Although it was used to treat pain, salicyclic acid often irritated the stomach. Several investigators attempted to find a means to neutralize salicylic acid. Felix Hoffmann, a chemist working at the German company Bayer, developed a less irritating derivative by treating salicylic acid with a base and acetyl chloride. This derivative, acetylsalicylic acid, was named *aspirin* from "a" for acetyl chloride,

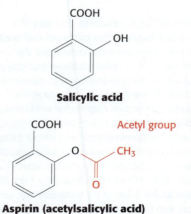

Salicylic acid

Acetyl group

Aspirin (acetylsalicylic acid)

"spir" for *Spiraea ulmaria* (meadowsweet, a flowering plant that also contains salicylic acid), and "in" (a common ending for drugs). Each year, approximately 35,000 tons of aspirin (nearly the weight of the *Titanic*) are taken worldwide.

As discussed in Section 12.4, the acetyl group in aspirin is transferred to the side chain of a serine residue that lines the hydrophobic channel from the membrane to the active site of the cyclooxygenase component of prostaglandin H_2 synthase (Figure 12.24). In this position, the acetyl group blocks access to the active site. Thus, even though aspirin binds in the same pocket on the enzyme as salicylic acid, the acetyl group of aspirin dramatically increases its effectiveness as a drug. This account illustrates the value of screening extracts from plants and other materials that are believed to have medicinal properties for active compounds. The large number of herbal and folk medicines are a treasure trove of new drug leads.

Let us consider another example, which has also had a significant impact on the clinical management of individuals with cardiovascular disease. More than 100 years ago, a fatty, yellowish material was discovered on the arterial walls of patients who had died of vascular disease. The presence of the material was termed *atheroma* from the Greek word for porridge. This material proved to be cholesterol. The Framingham Heart Study, initiated in 1948, documented a correlation between high blood-cholesterol levels and high mortality rates from heart disease. This observation led to the notion that blocking cholesterol synthesis might lower blood-cholesterol levels and, in turn, lower the risk of heart disease.

Initial attempts at blocking cholesterol synthesis focused on targets near the end of the pathway. However, these efforts were abandoned because the accumulation of the insoluble substrate for the inhibited enzyme led to the development of cataracts and other side effects. Investigators eventually identified a more favorable target—namely, the enzyme HMG-CoA reductase (Section 27.2). This enzyme acts on a substrate, HMG-CoA (3-hydroxy-3-methylglutaryl coenzyme A), that does not accumulate, because it is water-soluble and can be utilized by other pathways.

A promising natural product, called *compactin*, was discovered in a screen of compounds from a fermentation broth from *Penicillium citrinum* in a search for antibacterial agents. In some animal studies, compactin was found to inhibit HMG-CoA reductase and to lower serum cholesterol levels. In 1982, a new HMG-CoA reductase inhibitor was discovered in a fermentation broth from another mold, *Aspergillus terreus*. This compound, now called *lovastatin*, was found to be structurally very similar to compactin, bearing one additional methyl group. After many studies, the Food and Drug Administration (FDA) approved lovastatin for treating high serum cholesterol levels. A structurally related HMG-CoA reductase inhibitor was later shown to cause a statistically significant decrease in deaths due to coronary heart disease, validating the clinical benefits of lowering serum cholesterol levels.

Compactin

Lovastatin

High-throughput screening expands the opportunity for lead identification

Drug developers can screen large libraries of either natural products or purely synthetic compounds to identify lead molecules with desired activities. Under favorable circumstances, millions of compounds can be tested in this process, termed **high-throughput screening (HTS)**.

The success of a high-throughput screen depends on the development of a robust and sensitive assay. Earlier, we introduced the necessity of a specific assay to assess the progress of a protein purification scheme (Section 4.1). A similar assay is needed to rapidly and sensitively detect molecules that have a desired activity such as the ability to inhibit an enzyme or block an ion channel. Since a high-throughput screen will assess a large number of compounds, screening

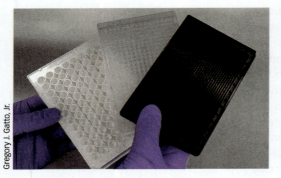

FIGURE 32.6 High-throughput screens are conducted in microtiter plates. Shown here are plates that contain 96, 384, and 1536 wells. As the number of wells increases, the maximum volume that each well can hold is reduced.

assays are typically run in very small volumes (usually between 1 and 100 µl) in microtiter plates that contain hundreds, if not thousands, of wells (**Figure 32.6**). Additionally, screening is often conducted using automation equipment to handle the repetitive workload and ensure precise timing.

To achieve the required level of assay sensitivity, methods that rely on fluorescence or luminescence detection are often employed. While numerous methods are available, let us consider one approach to introduce a sensitive detection method into an enzyme assay. Recall the reaction of a protein kinase (Section 7.3), an enzyme which uses ATP to attach phosphoryl groups to the side chains of serine, threonine, or tyrosine residues.

$$\text{Protein-OH} + \text{ATP} \underset{}{\overset{\text{Protein kinase}}{\rightleftharpoons}} \text{Protein-OPO}_3{}^{2-} + \text{ADP}$$

Neither the substrates (ATP or the unphosphorylated protein) nor the products (ADP or the phosphorylated protein) of this reaction are inherently fluorescent. However, by including additional enzymes (pyruvate kinase, pyruvate oxidase, and peroxidase) and the necessary substrates in the reaction, the production of ADP by the kinase can be coupled to the generation of the fluorescent dye resorufin (**Figure 32.7**). The successive actions of multiple enzymes to detect a reaction product is referred to as a **coupled assay**. Such assays can be quite effective for use in high-throughput screening. Follow-up assays, also called *secondary assays*, must be run on any hits from a coupled assay screen to ensure that the hits are not blocking the activity of the coupling enzymes, rather than the target kinase.

FIGURE 32.7 Coupled assays link the activity of an enzyme to the production of a fluorescent product. In this example, the hydrolysis of ATP to ADP by a protein kinase is linked via the successive action of three enzymes — pyruvate kinase, pyruvate oxidase, and peroxidase — to the production of the fluorescent dye resorufin. [Information from N. W. Charter et al., *J. Biomol. Screen.* 11:390–399, 2006.]

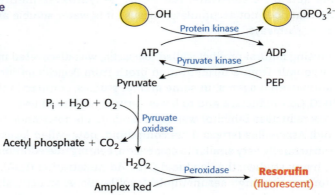

❖ **SELF–CHECK QUESTION**

Some compounds in a screening library are themselves fluorescent. Why might they be problematic in a high-throughput screening experiment?

Screening libraries can be prepared using combinatorial chemistry

The final component needed for a high-throughput screen is a diverse chemical library. While compounds in these libraries can be synthesized one at a time for testing, an alternative approach is to synthesize all at once a large number of structurally related compounds that differ from one another at only one or a few positions. This approach is termed **combinatorial chemistry**. Here, compounds are synthesized with the use of the same chemical reactions but a variable set of reactants, enabling the generation of thousands of possible compounds over a few chemical steps.

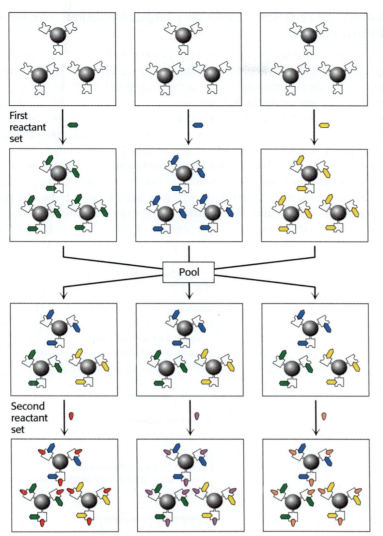

FIGURE 32.8 Compound libraries can be generated by split-pool synthesis. Reactions are performed on beads, shown in gray, that have the same initial structural scaffold, shown in white. Each of the reactions with the first set of reactants is performed on a separate set of beads. The beads are then pooled, mixed, and split into sets. The second set of reactants is then added. Many different compounds will be produced, but all of the compounds on a single bead will be identical.

A key method in combinatorial chemistry is **split-pool synthesis (Figure 32.8)**. The method depends on solid-phase synthetic methods, first developed for the synthesis of peptides (Section 4.4). Compounds are synthesized on small beads. Beads containing an appropriate starting scaffold are produced and divided (split) into n sets, with n corresponding to the number of building blocks to be used at one site. Reactions adding the reactants at the first site are run, and the beads are isolated by filtration. The n sets of beads are then combined (pooled), mixed, and split again into m sets, with m corresponding to the number of reactants to be used at the second site. Reactions adding these m reactants are run, and the beads are again isolated. The important result is that each bead contains only one compound, even though the entire library of beads contains many. Furthermore, although only $n + m$ reactions were run, $n \times m$ compounds are produced. Suppose that a molecular scaffold is constructed with two reactive sites and that $n = 20$ reactants can be used in the first site and $m = 40$ reactants can be used in the second site. Over two synthetic steps, $20 + 40 = 60$ reactions produce $20 \times 40 = 800$ compounds.

In some cases, assays can be performed directly with the compounds still attached to the bead to find compounds with desired properties (**Figure 32.9**). Alternatively, each bead can be isolated and the compound cleaved from the bead to produce free compounds for analysis. After an interesting compound has been identified, analytical methods of various types must be used to identify which of the $n \times m$ compounds are present.

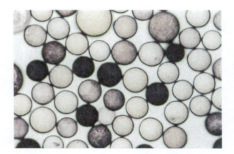

FIGURE 32.9 A library of synthesized carbohydrates can be screened using an "on-bead" assay. A small combinatorial library of carbohydrates synthesized on the surface of 130-mm beads is screened for carbohydrates that are bound tightly by a lectin from peanuts. Beads that have such carbohydrates are darkly stained through the action of an enzyme linked to the lectin. [From Figure 3 in Liang et al., "Polyvalent binding to carbohydrates immobilized on an insoluble resin," Proceedings of the National Academy of Sciences, USA, vol. 94, pp. 10554–10559. Copyright 1997 National Academy of Sciences, USA.]

DNA-encoded libraries provide very large compound libraries for lead identification

The "universe" of drug-like compounds is vast. More than an estimated 10^{40} compounds are possible with molecular weights less than 750. So, even with a library of millions of compounds, only a tiny fraction of the chemical possibilities, referred to as **chemical space**, is represented in a particular compound collection.

In an effort to develop libraries that encompass even greater regions of chemical space, scientists have developed **DNA-encoded libraries (DELs)**. In a DEL, compounds are synthesized using split-pool methods directly on a short sequence of DNA. At each synthetic step, a new synthetic oligonucleotide is also ligated to the DNA portion of the molecule. The sequence of this oligonucleotide is unique to the identity of the building block added. Essentially, the synthetic DNA tag acts as a bar code. In this manner, a vast array of compounds can be generated, each with its own uniquely identifying DNA sequence (**Figure 32.10**).

FIGURE 32.10 Diverse DNA-encoded libraries are generated by split pool synthesis. In the construction of a DNA-encoded library (DEL), compounds are synthesized directly on a short strand of double-stranded DNA. At each step, the pool of compounds is split, and different building blocks are added. Before pooling, a short oligonucleotide is ligated to the DNA strand. The sequence of this oligonucleotide is unique to the particular chemical building block that was added. In this example, after four steps, a library containing over 800 million unique molecules is constructed, each linked to a DNA sequence that specifically encodes the reaction history of that molecule. [Information from P. A. Harris et al., *Nature Chem. Biol.* 5:647–654, 2009, Fig. 1.]

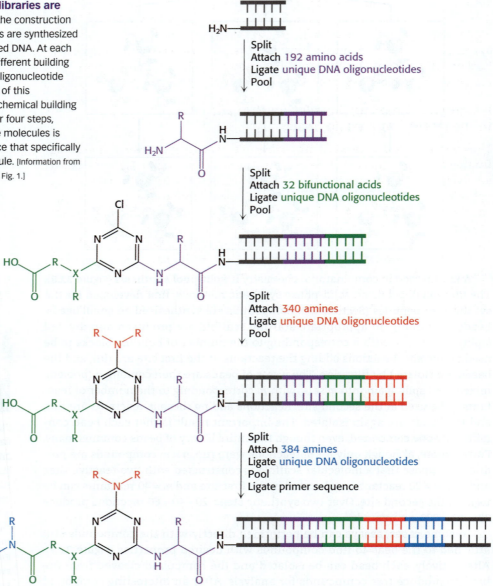

DNA-encoded library containing 802,160,640 unique molecules

How are DELs then screened? In one approach reminiscent of affinity chromatography, the target of interest is attached to a solid resin support, usually through a recombinantly added affinity tag (Section 4.1). The DEL is then passed over the resin; molecules with high affinity for the target will remain bound to the target, while the others pass through unbound. The protein is then denatured, and the bound molecules are eluted. The DNA tags are then amplified by PCR and sequenced using next-generation methods (Section 9.3), revealing the identity of the hits. Since DELs are tested in a single pool, as opposed to individual compounds each dispensed into separate wells in a standard HTS, much larger libraries can be screened over a shorter period of time. The PCR amplification step enables identification of very small amounts of the hit compounds.

A number of follow-up studies must be performed to confirm the hits and explore their properties. For example, once the structures of the positive hits are inferred by DNA sequencing, the compounds must be prepared "off-DNA" to ensure that their affinity to the target was driven solely by the compound, and not influenced by any region of the attached DNA. Furthermore, this screening approach identifies only compounds that *bind* to the target; separate experiments must be conducted to determine if these molecules have *functional impact* on the activity of the target. Despite these challenges, leads have been identified by DEL screening for a number of protein targets, and several of these have been further optimized into highly potent candidates that have entered clinical testing, as we will explore in Section 32.3.

Phenotypic screening provides an alternative to the target-centered approach

Thus far, we have discussed a path to drug discovery that starts with the identification and validation of desired targets. Then, assays specific for these targets are used to screen compound libraries for chemical starting points for further optimization. This approach, called **target-based screening** (**Figure 32.11A**), is widely applied throughout the pharmaceutical industry. However, target-based screening presents significant challenges that must be overcome: compelling validation evidence for the target must be accumulated; assays specific to the target must be developed; and, as we shall see, compounds must be chemically optimized to be functional in physiological environments.

Alternatively, leads may be identified by **phenotypic screening** (**Figure 32.11B**). In this approach, libraries of compounds are screened in an assay designed to

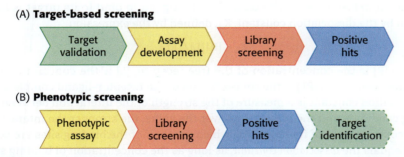

(A) Target-based screening

Target validation → Assay development → Library screening → Positive hits

(B) Phenotypic screening

Phenotypic assay → Library screening → Positive hits → Target identification

FIGURE 32.11 Target-based screening and phenotypic screening are two paths to early drug discovery. (A) In target-based screening, a molecular target is identified and validated. An assay for this target's activity is used to screen a compound library. Hits can then be tested for their physiological effects and optimized into drug candidates. (B) In phenotypic screening, an assay for a disease-relevant phenotype is developed in a cellular or animal model system. This assay is used to screen a compound library, from which hits can be optimized into drug candidates. These compounds can also be used to identify the molecular target.

detect a desired phenotype in a cellular system or animal model. For example, compounds may be tested for their ability to promote cellular survival or enhance resistance to specific toxic stimuli. In phenotypic screening, a biological effect is known before the target is identified. One advantage to this method is that any positive screening hits are already known to have the desired pharmacological effect in a physiological setting.

A significant challenge to phenotypic screening is that the target is not always obvious. As we have seen in our discussions on signal transduction (Chapter 14), a physiological phenotype could be modulated at a wide variety of potential points. Hence, the mode of action of the compound, including the identity of its target or targets, must be identified later. These target identification experiments typically require a substantial amount of effort.

❖ SELF–CHECK QUESTION

Propose an experiment that could enable identification of a target, using the knowledge of a phenotypic screening hit.

32.3 Compounds Must Meet Stringent Criteria to Be Developed into Drugs

In the previous section, we described the process by which leads are identified. During the next stage, **lead optimization**, numerous molecules related to the leads are synthesized to identify those which meet the strict criteria for further development into a **candidate**, a molecule suitable for testing in humans. Many compounds have significant effects when taken into the body, but only a very small fraction of them have the potential to be useful drugs. A foreign compound, not adapted to its role in the cell over the course of evolution, must have a range of special properties to function effectively without causing serious harm. Let us consider some of the challenges faced by drug developers.

Drugs must be potent and selective

Most drugs bind to specific proteins, usually receptors or enzymes, within the body. To be effective, a drug should bind a sufficient number of its target proteins when taken at a dose that can be reasonably administered to patients. One factor in determining drug effectiveness is the strength of the interaction between the drug and its target. Recall from Section 3.1 that a molecule that binds to some target protein is often referred to as a *ligand*. A prototypical ligand-binding curve is shown in **Figure 32.12**. The tendency of a ligand to bind to its target is measured by the dissociation constant, K_d, defined by the expression

$$K_d = [R][L]/[RL]$$

where [R] is the concentration of the free receptor, [L] is the concentration of the free ligand, and [RL] is the concentration of the receptor–ligand complex. The dissociation constant is a measure of the strength of the interaction between the drug candidate and the target; the lower the value, the stronger the interaction. The concentration of free ligand at which one-half of the binding sites are occupied equals the *dissociation constant*, as long as the concentration of binding sites is substantially less than the dissociation constant.

In many cases, biological assays using living cells or tissues (rather than direct enzyme or binding assays) are employed to examine the potency of drug candidates. More potent molecules exert their effects at a lower concentration than less potent ones. For example, the fraction of bacteria killed by a drug might indicate the potency of a potential antibiotic. In these cases, values such as EC_{50} are used. EC_{50}, where EC stands for *effective concentration*, is

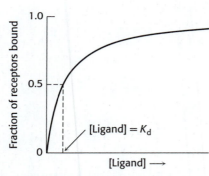

FIGURE 32.12 Simple ligand binding results in a hyperbolic curve. The titration of a receptor, R, with a ligand, L, results in the formation of the complex RL. In uncomplicated cases, the binding reaction follows a simple saturation curve. Half of the receptors are bound to ligand when the ligand concentration equals the dissociation constant, K_d, for the RL complex.

the concentration of the drug candidate required to elicit 50% of the maximal biological response (**Figure 32.13**). Similarly, EC_{90} is the concentration required to achieve 90% of the maximal response. In the example of an antibiotic, EC_{90} would be the concentration required to kill 90% of bacteria exposed to the drug. For drug candidates that are inhibitors, the corresponding terms IC_{50} and IC_{90}, where IC stands for *inhibitory concentration*, are often used to describe the concentrations of the inhibitor required to reduce a response by 50% or 90%, respectively.

Values such as the EC_{50} and IC_{50} are measures of the potency of a drug candidate in modulating the activity of the desired biological target. Additionally, to prevent unwanted effects, often called **side effects**, ideal drug candidates should be selective. That is, they should not bind biomolecules other than the target to any significant extent. The degree of selectivity can be described in terms of the ratio of the K_d values for the binding of the drug candidate to any other molecules to the K_d value for the binding to the desired target.

Under physiological conditions, achieving potent and selective binding of a target site by a drug can be quite challenging. Most drug targets also bind ligands normally present in tissues; often, the drug and these ligands will compete for binding sites on the target. We encountered this scenario when we considered competitive inhibitors in Section 5.6. Suppose that the drug target is an enzyme and the drug candidate is competitive with respect to the enzyme's natural substrate. The concentration of the drug candidate necessary to inhibit the enzyme will depend on the physiological concentration of the normal substrate (**Figure 32.14**). Biochemists Yung-Chi Cheng and William Prusoff described the relationship between the IC_{50} of an enzyme inhibitor and its inhibition constant K_i (analogous to the dissociation constant, K_d, of a ligand). For a competitive inhibitor,

$$IC_{50} = K_i \left(1 + [S]/K_M\right)$$

This relationship, referred to as the Cheng–Prusoff equation, demonstrates that the IC_{50} of a competitive inhibitor will depend on the concentration and the Michaelis constant (K_M) for the substrate S. The higher the concentration of the natural substrate, the higher the concentration of drug needed to inhibit the enzyme.

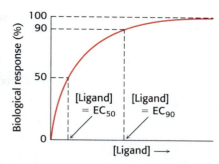

FIGURE 32.13 Effective concentration (EC) describes the amount of compound required to elicit a particular biological response. The EC_{50} value describes the concentration required to elicit 50% of the maximum response, while the EC_{90} describes that required to elicit 90% of the maximum response.

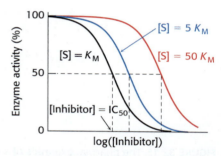

FIGURE 32.14 Inhibitors compete with substrates for enzyme active sites. The measured IC_{50} of a competitive inhibitor to its target enzyme depends on the concentration of substrate present.

EXAMPLE Determining the IC_{50} for an Inhibitor

PROBLEM: The activity of an enzyme was measured in the presence of varying concentrations of Compound A and reported in the table shown. Estimate the IC_{50} for this compound. If Compound A is known to be a competitive inhibitor with respect to the enzyme substrate and the assay was run at a concentration of substrate equal to its K_M, what is the K_i of Compound A?

Compound A (nM)	Percent enzyme activity
0	100
45	90
140	84
410	65
800	51
3700	22
11,200	8
33,100	4

GETTING STARTED: The IC_{50} is defined as the concentration at which the activity of an enzyme is inhibited by 50%. Reassuringly, the first row of the table confirms that the activity of the enzyme in the absence of Compound A is 100%. Hence, a good first approximation for the IC_{50} would be close to the concentration where the enzyme activity is 50%. From the table, this is about 800 nM.

CALCULATE: Since we know that Compound A is competitive with respect to substrate, and the assay was run at a substrate concentration equal to its K_M, we can use the Cheng–Prusoff equation to estimate the K_i:

$$IC_{50} = K_i\left(1 + \frac{[S]}{K_M}\right)$$

$$IC_{50} = K_i\left(1 + \frac{K_M}{K_M}\right) = 2K_i$$

$$K_i = \frac{IC_{50}}{2} = \frac{800 \text{ nM}}{2} = 400 \text{ nM}$$

REFLECT: Pharmacologists describe inhibitor potency in a variety of ways. Note that while IC_{50} values are commonly used, they are dependent upon assay conditions, whereas K_i values are reflective of the intrinsic affinity of the inhibitor for its target. Also, be careful when applying the Cheng–Prusoff equation! The equation provided in the text is specific for competitive inhibitors with respect to the substrate S. More complicated equations are needed to describe this relationship for other inhibitory modalities.

Drugs must have suitable properties to reach their targets

Thus far, we have focused on the ability of molecules to act on specific target molecules. However, an effective drug also has other requisite characteristics. It must be easily administered and must reach its target at sufficient concentration to be effective. After a drug molecule has entered the body, it is acted upon by several processes—absorption, distribution, metabolism, and excretion—that will determine its effective concentration over time (**Figure 32.15**). A drug's response to these processes is referred to as its **ADME** (pronounced "add-me") **properties**.

FIGURE 32.15 The pharmacokinetics of a compound in a living organism is described by its absorption, distribution, metabolism, and excretion (ADME) properties. The concentration of a compound at its target site (yellow) is affected by the magnitudes and rates of absorption, distribution, metabolism, and excretion.

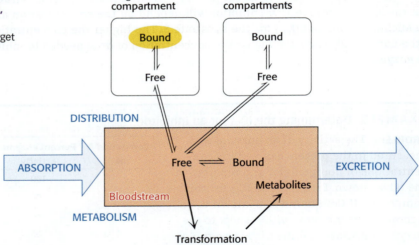

Administration and absorption. Ideally, a drug can be taken orally as a small tablet. An orally administered active compound must be able to survive the acidic conditions in the gut and then be absorbed through the intestinal epithelium. Thus, the compound must be able to pass through cell membranes at a significant rate. Larger molecules such as proteins cannot be administered orally, because they often cannot survive the acidic conditions in the stomach and, if they do, they are not readily absorbed. Even many small molecules are not absorbed well; they may be too polar and unable to pass through cell membranes easily, for example. The ability to be absorbed is often quantified in terms of the **oral bioavailability**. This quantity is defined as the ratio of the peak concentration of a compound given orally to the peak concentration of the same dose injected directly into the bloodstream. Bioavailability can vary considerably from species to species, so results from animal studies may be difficult to apply to human beings. Despite this variability, some useful generalizations have been made. One powerful set consists of Lipinski's rules, developed by the pharmaceutical chemist Christopher A. Lipinski.

Lipinski's rules tell us that poor absorption is likely when

1. The molecular weight is greater than 500 daltons.
2. The number of hydrogen-bond donors is greater than 5.
3. The number of hydrogen-bond acceptors is greater than 10.
4. The partition coefficient [measured as log(*P*)] is greater than 5.

The partition coefficient is a way to measure the tendency of a molecule to enter lipid membranes, which correlates with its ability to dissolve in organic solvents. It is determined by allowing a compound to equilibrate between water and an organic phase, *n*-octanol. The log(*P*) value is defined as $\log_{10}$ of the ratio of the concentration of a compound in *n*-octanol to the concentration of the compound in water:

$$\log(P) = \log\left(\frac{[\text{compound}]_{\text{n-octanol}}}{[\text{compound}]_{\text{water}}}\right)$$

For example, if the concentration of the compound in the *n*-octanol phase is 100 times that in the aqueous phase, then log(*P*) is 2. Although the ability of a drug to partition in organic solvents is ideal, as it implies that the compound can penetrate membranes, a log(*P*) value that is too high suggests that the molecule may be poorly soluble in an aqueous environment.

Morphine, for example, satisfies all of Lipinski's rules and has moderate bioavailability (**Figure 32.16**). Still, a drug that violates one or more of these rules may also have satisfactory bioavailability. Nonetheless, these rules, and other similar guidelines that have since been developed, serve as useful principles for evaluating new drug candidates.

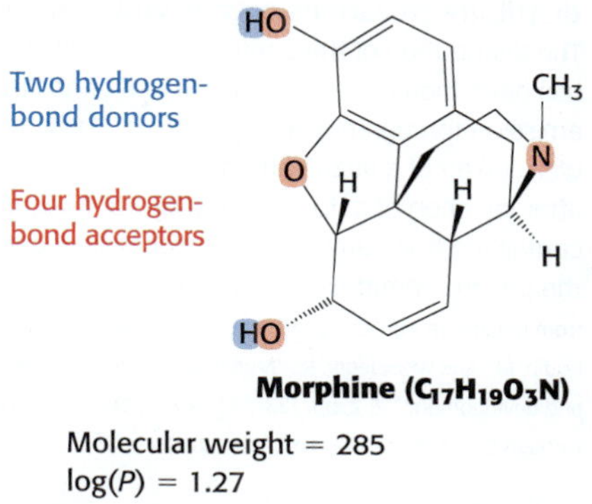

Distribution. Compounds taken up by intestinal epithelial cells can pass into the bloodstream. However, hydrophobic compounds and many others do not freely dissolve in the bloodstream. These compounds bind to proteins, such as albumin (**Figure 32.17**), that are abundant in the blood serum. Because albumin is present in the circulation at high concentrations (nearly 0.7 mM) and can bind a wide variety of compounds, it can dramatically affect the free concentration of drug in the bloodstream. Nevertheless, the binding capacity of albumin enables transport of the drug throughout the circulatory system.

Two hydrogen-bond donors

Four hydrogen-bond acceptors

Morphine (C$_{17}$H$_{19}$O$_3$N)

Molecular weight = 285
log(*P*) = 1.27

FIGURE 32.16 Lipinski's rules serve as a guide to predict oral bioavailability. Morphine satisfies all of Lipinski's rules and has an oral bioavailability in human beings of 33%.

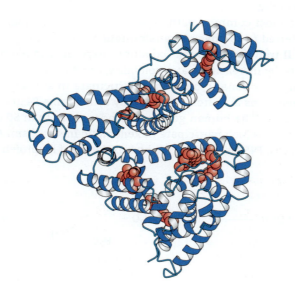

FIGURE 32.17 Albumin is abundant in the bloodstream and binds hydrophobic compounds. Seven hydrophobic molecules (in red) are shown bound to a single molecule of serum albumin. [Drawn from 1BKE.pdb.]

When a compound has reached the bloodstream, it is distributed to different fluids and tissues, which are often referred to as *compartments*. Some compounds are highly concentrated in their target compartments, either by binding to the target molecules themselves or by other mechanisms. Other compounds are distributed more widely (**Figure 32.18**). An effective drug will reach the target compartment in sufficient quantity; the concentration of the compound in the target compartment is reduced whenever the compound is distributed into other compartments.

Some target compartments are particularly hard to reach. For example, many compounds are excluded from the central nervous system by the tight junctions between endothelial cells that line blood vessels within the brain and spinal cord, referred to as the *blood-brain barrier*.

FIGURE 32.18 Once absorbed, compounds distribute to various organs within the body. The distribution of the antifungal agent fluconazole has been monitored through the use of positron emission tomography (PET) scanning. These images were taken of a healthy human volunteer 90 minutes after injection of a dose of 5 mg kg^{-1} of fluconazole containing trace amounts of fluconazole labeled with the positron-emitting isotope ^{18}F. [Reprinted by permission from Macmillan Publishers Ltd: Nature Reviews Drug Discovery, Rudin, M., and Weissleder, R., "Molecular imaging in drug discovery and development," 2: 2, pp. 123–131, (2003); permission conveyed through Copyright Clearance Center, Inc.]

Metabolism. A potential drug molecule must evade the body's defenses against foreign compounds, also called **xenobiotics**. Many compounds are released from the body in the urine or stool, often after having been metabolized to aid in excretion. This **drug metabolism** poses a considerable threat to drug effectiveness, because the concentration of the desired compound decreases as it is metabolized. Thus, a rapidly metabolized compound must be administered more frequently or at higher doses.

Drug metabolism takes place primarily in the liver. Because blood flows from the intestine directly to the liver through the portal vein, xenobiotic metabolism often alters drug compounds before they ever reach full circulation. This process, referred to as *first-pass metabolism*, can substantially limit the availability of compounds taken orally.

Two of the most common pathways in xenobiotic metabolism are oxidation, often referred to as **phase I transformations**, and conjugation (which are called **phase II transformations**). Oxidation reactions can aid excretion in at least two ways: by increasing water solubility, and thus ease of transport, and by introducing functional groups that participate in subsequent metabolic steps. These reactions are often promoted by cytochrome P450 enzymes in the liver (Section 27.4). The human genome encodes more than 50 different P450 isozymes, many of which participate in xenobiotic metabolism. A typical reaction catalyzed by a P450 isozyme is the hydroxylation of ibuprofen (**Figure 32.19**).

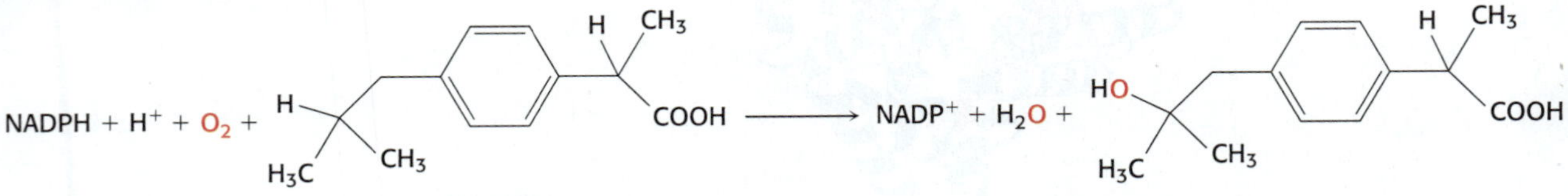

FIGURE 32.19 Cytochrome P450 isozymes catalyze xenobiotic metabolic reactions such as hydroxylation. In this example, one P450 isozyme introduces an oxygen atom derived from molecular oxygen into the anti-inflammatory drug ibuprofen.

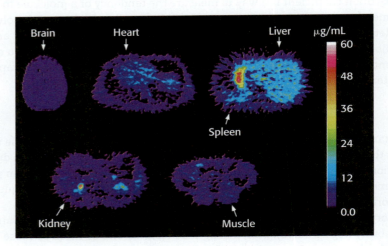

Conjugation is the addition of particular groups to the xenobiotic compound. Common groups added are glutathione (Section 20.5), glucuronic acid, and sulfate (**Figure 32.20**). These additions often increase water solubility and promote excretion. Note that an oxidation reaction often precedes conjugation because the oxidation reaction can generate modifications, such as the hydroxyl group, to which groups such as glucuronic acid can be attached.

FIGURE 32.20 Compounds that have certain functional groups are often modified by conjugation reactions. Such reactions include the addition of glutathione (top), glucuronic acid (middle), or sulfate (bottom). The conjugated product is shown boxed.

Glutathione

UDP-α-ᴅ-glucuronic acid

3′-Phosphoadenosine-5′-phosphosulfate (PAPS)

Examples of conjugation include the addition of glutathione to the anticancer drug cyclophosphamide, the addition of glucuronidate to the analgesic morphine, and the addition of a sulfate group to the hair-growth stimulator minoxidil. While the metabolic products of a drug are usually less active than the drug itself, there are notable exceptions. For example, the sulfation of minoxidil produces a compound that is more active in stimulating hair growth than is the unmodified compound.

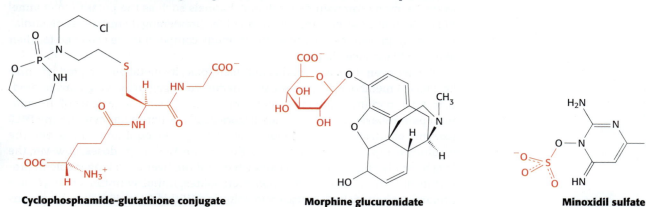

Cyclophosphamide-glutathione conjugate   **Morphine glucuronidate**   **Minoxidil sulfate**

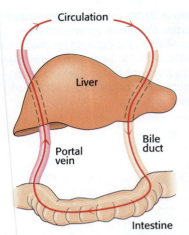

FIGURE 32.21 Drugs can move between the intestine and the circulation via enterohepatic cycling. In this pathway, some drugs can move from the blood circulation to the liver, into the bile, into the intestine, to the liver, and back into circulation. This cycling decreases the rate of drug excretion.

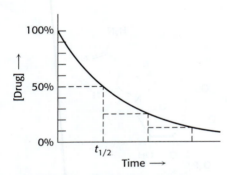

FIGURE 32.22 Drug excretion may be measured by its half-life. The concentration of a drug in the bloodstream decreases to one-half of its value in a period of time, $t_{1/2}$, referred to as its half-life.

Excretion. After compounds have entered the bloodstream, they can be removed from circulation and excreted from the body by two primary pathways. First, they can be absorbed through the kidneys and excreted in the urine. In this process, the blood passes through networks of fine capillaries, called *glomeruli*, in the kidney that act as filters. Compounds with molecular weights less than approximately 60,000 pass though the glomeruli. Many of the water molecules, glucose molecules, nucleotides, and other low-molecular-weight compounds that pass through the glomeruli are reabsorbed into the bloodstream, either by transporters that have broad specificities or by the passive transfer of hydrophobic molecules through membranes. Drugs and metabolites that pass through the first filtration step and are not reabsorbed are subsequently excreted.

Second, compounds can be actively transported into bile by a process that takes place in the liver. After concentration in the gallbladder, bile flows into the intestine. In the intestine, the drugs and metabolites can be reabsorbed into the bloodstream, further degraded by digestive enzymes, or excreted through the stool. Sometimes, compounds are recycled from the bloodstream into the intestine and back into the bloodstream, a process referred to as *enterohepatic cycling* (**Figure 32.21**). This process can significantly decrease the rate of excretion of some compounds, because they escape from an excretory pathway and reenter the circulation.

The kinetics of compound excretion is often complex. In some cases, a fixed percentage of the remaining compound is excreted over a given period of time (**Figure 32.22**). This pattern of excretion results in exponential loss of the compound from the bloodstream that can be characterized by a half-life ($t_{1/2}$). The half-life is the fixed period of time required to eliminate 50% of the remaining compound. It is a measure of how long an effective concentration of the compound remains in the system after administration. As such, the half-life is a major factor in determining how often a drug must be taken. A drug with a long half-life might need to be taken only once per day, whereas a drug with a short half-life might need to be taken three or four times per day.

Toxicity can limit drug effectiveness

An effective drug must not be so toxic that it seriously harms the person who takes it. A drug may be toxic for any of several reasons. First, it may modulate the target molecule itself too effectively, referred to as **mechanism-based**, or **on-target**, **toxicity**. For example, the presence of too much of the anticoagulant drug coumadin can result in dangerous, uncontrolled bleeding and death.

Second, the compound may modulate the properties of proteins that are distinct from the target molecule itself, referred to as **off-target toxicity**. Compounds that are directed to one member of a family of enzymes or receptors often bind to other family members. For example, an antiviral drug directed against viral proteases may be toxic if it also inhibits proteases that perform critical functions in the body, such as those that regulate blood pressure. A compound may also be toxic if it modulates the activity of a protein unrelated to its intended target. For example, many compounds block ion channels such as the potassium channel hERG (Section 13.4), causing potentially life-threatening disturbances of cardiac rhythm. To prevent cardiac side effects, many compounds are screened for their ability to block such channels.

Finally, even if a compound is not itself toxic, its metabolic by-products may be. Phase I metabolic processes can generate damaging reactive groups in products. An important example is liver toxicity observed with large doses of the common pain reliever acetaminophen (**Figure 32.23**). A particular cytochrome P450 isozyme oxidizes acetaminophen to N-acetyl-p-benzoquinone imine, and the resulting compound is conjugated to glutathione. With large doses, however, the liver's supply of glutathione is exhausted, and the liver is no longer able to protect itself from the imine and other metabolites. Initial symptoms of excessive acetaminophen include nausea and vomiting. Within 24 to 48 hours, symptoms

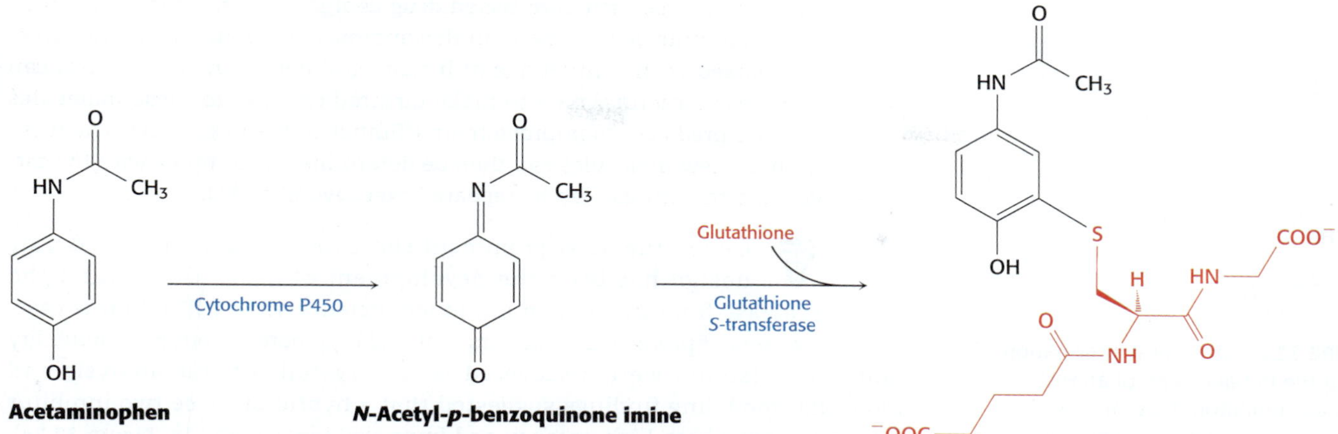

FIGURE 32.23 Acetaminophen metabolism results in its toxicity. A minor metabolic product of acetaminophen is *N*-acetyl-*p*-benzoquinone imine, which is conjugated to glutathione. Large doses of acetaminophen can thus deplete liver glutathione stores.

of liver failure may appear. Acetaminophen poisoning accounts for about 35% of cases of severe acute liver failure in the United States. A liver transplant is often the only effective treatment.

The toxicity of a drug candidate can be described in terms of the **therapeutic index**. This measure of toxicity is determined through animal tests, usually with mice or rats. The therapeutic index is defined as the ratio of the dose of a compound that is required to kill one-half of the animals (referred to as the LD_{50} for "lethal dose") to a comparable measure of the effective dose, usually the EC_{50}. So, if the therapeutic index is 1000, then lethality is significant only when 1000 times the effective dose is administered. Analogous indices can provide measures of toxicity less severe than lethality.

Many compounds have favorable properties in vitro, yet they fail when administered to a living organism because of difficulties with ADME and toxicity. Expensive and time-consuming animal studies are required to verify that a drug candidate is not toxic, yet differences between animal species in their response can confound decisions about moving forward with a compound toward human studies. One hope is that, with more understanding of the biochemistry of these processes, scientists can develop computer-based models to replace or augment animal tests. Such models would need to accurately predict the fate of a compound inside a living organism from its molecular structure or other properties that are easily measured in the laboratory without the use of animals.

Lead molecules can be optimized on the basis of three-dimensional structural information about their targets

Many drugs bind to their targets in a manner reminiscent of Emil Fischer's lock-and-key model (Figure 5.8). In principle, we should be able to design a key, given enough knowledge about the shape and chemical composition of the lock. In the idealized case, we could design a small molecule that is complementary in shape and electronic structure to a target protein so that it binds effectively to the targeted site. Yet, despite our ability to determine three-dimensional structures rapidly, the achievement of this goal remains elusive. Designing stable compounds from scratch that have the correct shape and properties to fit precisely into a binding site is difficult because predicting the structure that will best fit into a binding site is difficult. Prediction of binding affinity requires a detailed understanding of the interactions between a compound and its binding partner and of the interactions between the compound and the solvent when the compound is free in solution.

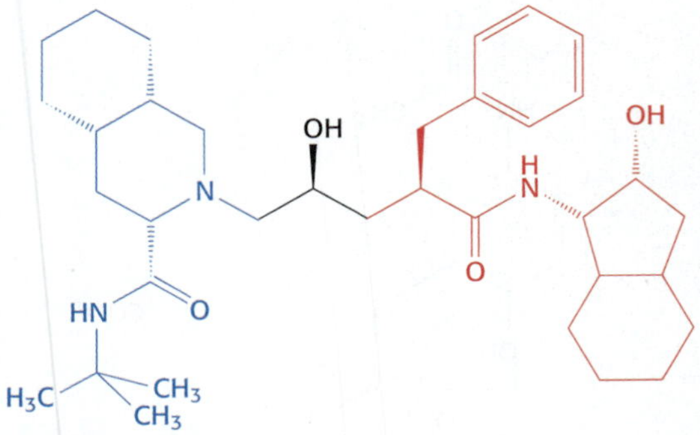

FIGURE 32.24 Structural information led to the initial design of an HIV protease inhibitor. This compound was designed by combining part of one compound with good inhibition activity but poor solubility (shown in red) with part of another compound with better solubility (shown in blue).

Nonetheless, **structure-based drug design** has proved to be a powerful tool in drug development. In this approach, the structure of the target complexed with compounds of interest are determined. Chemists can use these structural data to make directed changes to these molecules that are predicted to improve their affinities to the target. The structures of these new molecules can then be determined in complex with the target, and the process can be repeated over several cycles.

One of the most prominent successes of structure-based drug design has been the development of drugs that inhibit the protease from the human immunodeficiency virus (HIV). In this case, two sets of promising inhibitors that had high potency but poor solubility and bioavailability were discovered. X-ray crystallographic analysis and molecular-modeling findings suggested that a hybrid of these two inhibitor sets might have both high potency and improved bioavailability (**Figure 32.24**). The synthesized hybrid compound did show improvements but required further optimization.

Guided by structural data, a series of compounds were prepared and tested for their ability to inhibit the protease (**Figure 32.25**). These data demonstrate a **structure–activity relationship (SAR)**; they provide an opportunity to correlate each compound's structure with its function to guide the design of further molecules. The most active compound showed poor bioavailability, but one of the other compounds (highlighted in yellow in Figure 32.25) showed good bioavailability and acceptable activity. The maximum serum concentration available through oral administration was significantly higher than the levels required to suppress replication of the virus. Ultimately, this molecule became the marketed HIV protease inhibitor indinavir (Section 6.2). This drug, as well as other protease inhibitors developed at about the same time, has been used in combination with other drugs to treat HIV infection and AIDS with much more encouraging results than had been obtained previously.

FIGURE 32.25 Optimizing lead molecules requires the balancing of multiple compound properties. Four HIV protease inhibitors were evaluated for characteristics, including the IC_{50}, $\log(P)$, and c_{max} (the maximal concentration of compound present) measured in the serum of dogs. The compound shown at the bottom (highlighted in yellow) has the weakest inhibitory power (measured by IC_{50}) but by far the best bioavailability (measured by c_{max}). This compound was selected for further development, ultimately leading to the drug indinavir.

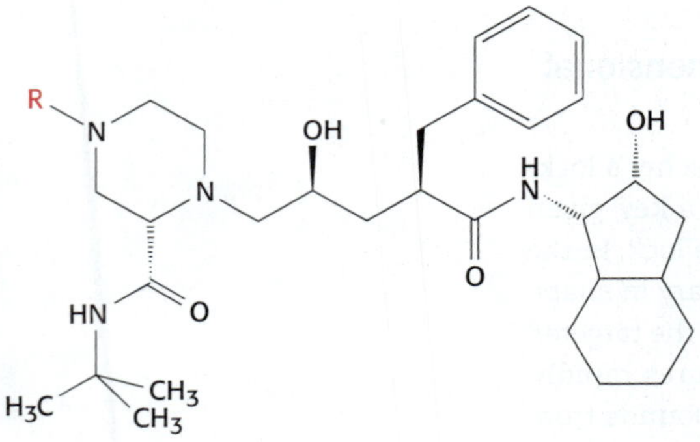

R group	IC_{50} (nM)	$\log(P)$	c_{max} (μM)
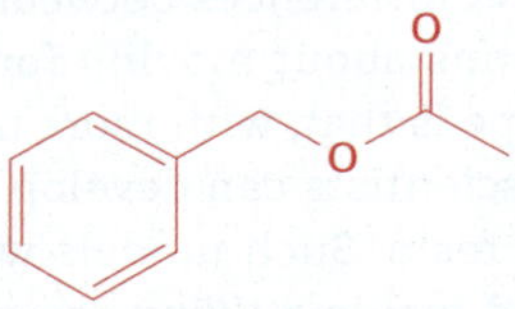	0.4	4.67	< 0.1
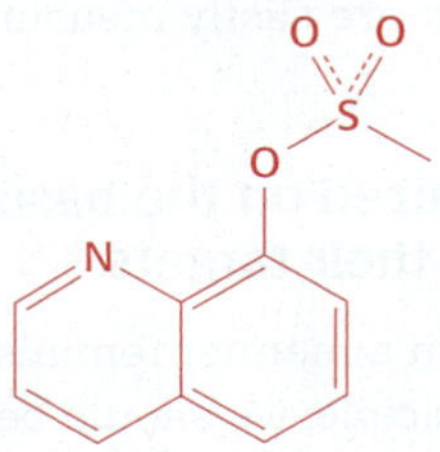	0.01	3.70	< 0.1
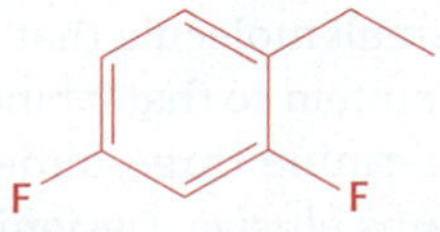	0.3	3.69	0.7
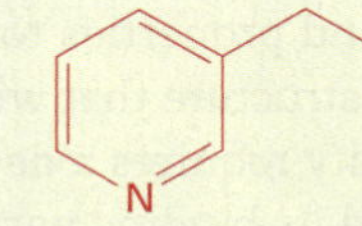	0.6	2.92	11

The development of indinavir represents an example of the challenges faced by pharmaceutical scientists during the process of lead optimization. Potency is only one of multiple properties that must be optimized during this stage. During the course of a discovery program, thousands of compounds may be synthesized and tested. Improvements in one property may be to the detriment of other properties. The collaboration of many scientists—biochemists, pharmacologists, chemists, and biologists, just to name a few—is required to attain a molecule suitable for testing in humans.

32.4 Biologics Are a Growing Family of Drugs

In this chapter, we have primarily focused on small molecules as drugs. These molecules are typically less than 500 Da in molecular weight (as described by Lipinski's rules), may be orally bioavailable, and bind to specific structural pockets on their targets. Many small molecule drugs are membrane-permeable, enabling them to access intracellular targets. However, they also exhibit some limitations, including short half-lives and a high susceptibility to off-target toxicity.

Over the past several decades, a new class of therapeutics has emerged. Broadly defined, **biologics** are medicines derived from living cells or through biological processes. With each scientific advancement, the number and different types of biologics have grown significantly (**Table 32.1**). As this is a highly diverse group of medicines, we will consider some of the more prominent examples here.

TABLE 32.1 Examples of biologics

Recombinant proteins
Monoclonal antibodies
Vaccines
Blood and blood components
Allergenics (molecules that elicit an allergic response)
Cells and tissues

The majority of biologics are recombinant proteins

Earlier, we discussed the first approved biologic: recombinant human insulin produced in *E. coli* (Section 9.2). Approved by the U.S. Food and Drug Administration (FDA) in 1982, this groundbreaking drug enabled the treatment of millions of diabetic patients without the risk of allergic reactions caused by insulins produced by other animal species.

Since the approval of insulin, a number of recombinant proteins have been approved for clinical use. For example, granulocyte colony-stimulating factor (G-CSF) is a 174-amino-acid glycoprotein that stimulates the production of granulocytes in the bone marrow. A recombinant form of G-CSF (filgrastim) was approved by the FDA in 1991 for patients suffering from low neutrophil counts due to HIV infection or some forms of chemotherapy. Similarly, other signaling proteins have been developed as biologics.

Recombinant protein biologics may be chemically modified to improve their pharmacokinetic properties. For example, filgrastim has a short half-life of around 3.5 hours, requiring once- or twice-daily dosing. Through the covalent attachment of a 20-kD polyethylene glycol (PEG) moiety to filgrastim, a process referred to as *pegylation*, its half-life can be extended to 40 hours. A single injection of the pegylated form of filgrastim, pegfilgrastim, exhibits the same biological effect as 11 daily doses of the nonpegylated protein.

Monoclonal antibodies are highly specific and potent recombinant protein biologics

As we discussed in Section 3.4, antibody-producing cells have evolved remarkable mechanisms for generating a broad range of recognition diversity. Scientists can take advantage of nature's own combinatorial chemistry to generate numerous distinct antibodies that can bind to a specific ligand. In addition, the hybridoma method (Section 4.2) can be used to prepare large quantities of monoclonal antibodies that exhibit high affinities for their targets.

As drugs, monoclonal antibodies have unique advantages over small molecules. For example, they can recognize a broader surface of the target; they are not

GERTRUDE ELION At the age of 15, Gertrude Elion was inspired to pursue a career in science and medicine after her grandfather's death from stomach cancer. Her family was bankrupted by the stock market crash in 1929, but she was able to attend Hunter College for free, where she achieved great academic success. At the age of 19, she took jobs as a secretary, a chemistry teacher, and an unpaid laboratory assistant, after which she had saved enough money to enroll in the graduate program at New York University, where she obtained her Master of Science degree in chemistry in 1941. During World War II, with many male chemists serving overseas, Elion was able to secure a research position at Burroughs Wellcome, where she began a lifelong collaboration with George Hitchings (both scientists are shown above). Together, they eschewed the "trial-and-error" approach to drug discovery, exploiting differences in metabolism between human cells and pathogens to target infectious organisms and tumors. Her work with Hitchings led to the discovery of numerous antibacterial, antiviral, immunosuppressive, and chemotherapeutic medicines, and their efforts were recognized with the Nobel Prize in Physiology or Medicine in 1988.

limited to the smaller pockets typically bound by small molecule drugs. This larger interaction surface enables the development of drugs with very high potency and specificity. Furthermore, the Fc region of monoclonal antibodies (Figure 3.40) binds to an endogenous salvage pathway receptor, the *neonatal Fc receptor* (FcRn), permitting the drug to persist in an active form for a very long half-life, on the order of weeks. Hence, monoclonal antibodies can be dosed much less frequently (e.g., weekly to monthly) than small-molecule drugs (e.g., daily).

There are some limitations to using monoclonal antibodies as drugs. First, as they are proteins, they are unable to withstand the harsh environment of the gastrointestinal tract. Hence, antibodies cannot be dosed orally and are usually administered under the skin (subcutaneously) or directly into blood vessels (intravenously). Additionally, antibodies do not cross membranes readily, so they generally are usable only against secreted or cell-surface targets. Despite these limitations, however, over 60 monoclonal antibodies are currently in clinical use as therapeutics, with hundreds more in clinical development.

How do monoclonal antibodies exert their therapeutic effects? In many cases, they serve as antagonists, blocking receptors from interacting with their corresponding ligands. This was the case for cetuximab, the chemotherapeutic monoclonal antibody that blocks the epidermal growth factor (EGF) receptor from binding EGF (Section 14.5). Alternatively, some monoclonal antibodies bind specific circulating ligands, preventing them from exerting their cellular effects. In other cases, monoclonal antibodies can be used to target specific cell-surface proteins—such as markers specific for tumor cells—for degradation by the host's own immune system.

Another unique example is the *antibody-drug conjugate* (ADC). In this case, a small molecule drug is conjugated to an antibody specific for a particular cell type. Once the antibody-target complex is internalized by the cell, the linker between the antibody and the drug is cleaved, and the free drug can exert its effects. In this manner, the off-target toxicity of a particular drug can be limited by directing its activity to a small subset of cells within the body.

Monoclonal antibodies represent a rapidly growing and exciting drug class. These versatile molecules have been exploited in numerous creative ways to provide clinical benefit for a variety of diseases besides cancer, including inflammatory, infectious, cardiovascular, and ophthalmologic conditions.

❖ SELF–CHECK QUESTION

A number of Fc domain mutations have been discovered that enhance the affinity of an antibody for FcRn. What effect would those mutations have on the half-life of a monoclonal antibody drug?

32.5 The Clinical Development of Medicines Proceeds Through Several Phases

In the United States, the FDA requires demonstration that drug candidates be effective and safe before they may be used in human beings on a large scale. This requirement is particularly true for drug candidates that are to be taken by people who are relatively healthy. More side effects are acceptable for drug candidates intended to treat significantly ill patients such as those with serious forms of cancer, where there are clear, unfavorable consequences for not having an effective treatment.

Clinical trials are time-consuming and expensive

Clinical trials test the effectiveness and potential side effects of a candidate drug before it is approved by the FDA for general use. These trials proceed in at least

FIGURE 32.26 **Clinical trials proceed in phases examining safety and efficacy in increasingly large groups.**

three phases (**Figure 32.26**). In Phase I, a small number (typically from 10 to 100) of usually healthy volunteers take the drug for an initial assessment of safety. These volunteers are given a range of doses and are monitored for signs of toxicity. The efficacy of the drug candidate is not specifically evaluated.

In Phase II, the efficacy of the drug candidate is tested in a small number of persons who might benefit from the drug, and further data regarding the drug's safety are obtained. Such trials are often controlled and double-blinded. In a **controlled study**, subjects are divided randomly into two groups. Subjects in the treatment group are given the treatment under investigation. Subjects in the control group are given either a **placebo**—that is, a treatment such as sugar pills known to not have intrinsic value—or the best standard treatment available, if withholding treatment altogether would be unethical. In a **double-blinded study**, neither the subjects nor the researchers know which subjects are in the treatment group and which are in the control group. A double-blinded study prevents bias in the course of the trial. Only when the trial has been completed are the assignments of the subjects into treatment and control groups unsealed and the results for the two groups compared. A variety of doses are often investigated in Phase II trials to determine which doses appear to be free of serious side effects and which doses appear to be effective.

The power of the **placebo effect**—that is, the tendency to perceive improvement in subjects who believe that they are receiving a potentially beneficial treatment—should not be underestimated. In a study of arthroscopic surgical treatment for knee pain, for example, subjects who were led to believe that they had received surgery through the use of videotapes and other means showed the same level of improvement, on average, as subjects who actually received the procedure.

In Phase III, similar studies are performed on a larger and more diverse population. This phase is intended to more firmly establish the efficacy of the drug candidate and to detect side effects that may develop in a small percentage of the subjects who receive treatment. Thousands of subjects may participate in a typical Phase III study.

Clinical trials can be extremely costly and time-consuming. Hundreds or thousands of patients must be recruited and monitored for the duration of the trial. Many physicians, nurses, clinical pharmacologists, statisticians, and others participate in the design and execution of the trial. Costs can run from tens to hundreds of millions of dollars. Extensive records must be kept, including documentation of any adverse reactions. These data are compiled and submitted to the FDA. Furthermore, many drugs fail to show efficacy in these larger studies where the patient population is more diverse. The full cost of developing a drug is currently estimated to be more than $800 million.

Even after a drug has been approved and is in use, difficulties can arise. Clinical trials that are run after a drug has entered the market, referred to as Phase IV or *postmarketing surveillance* studies, are designed to identify low-frequency side effects that may emerge only after widespread or long-term use. For example, the cyclooxygenase-2 inhibitor rofecoxib was withdrawn from the market after significant cardiac side effects were detected in Phase IV clinical trials. Such events highlight the necessity for doctors and patients to balance beneficial effects against potential risks for any medicine.

The evolution of drug resistance can limit the utility of drugs for infectious agents and cancer

Many drugs are used for long periods of time without any loss of effectiveness. However, in some cases, particularly for the treatment of cancer or infectious diseases, drug treatments that were initially effective become less so. In other words, the disease becomes resistant to the drug therapy. Why does this resistance develop? Infectious diseases and cancer have a common feature—namely, that an affected person contains many cells (or viruses) that can mutate and reproduce, meeting the two conditions necessary for evolution to take place. Thus, an individual virus, microorganism, or cancer cell may, by chance, have a genetic variation that makes it more suitable for growth and reproduction in the presence of the drug. These microorganisms or cells are more fit than others in their population, and they will tend to take over the population. As the selective pressure due to the drug is continually applied, the population of microorganisms or cancer cells will tend to become more and more resistant to the presence of the drug. Note that resistance can develop by a number of mechanisms.

The HIV protease inhibitors discussed in Section 32.3 provide an important example of the evolution of drug resistance. Retroviruses such as HIV are very well suited to this sort of evolution because reverse transcriptase carries out replication without a proofreading mechanism. In a genome of approximately 9750 bases, each possible single point mutation is estimated to appear in a virus particle more than 1000 times per day in each infected person. At this mutation rate, multiple point mutations occur. While most of these mutations either have no effect or are detrimental to the virus, a few of the mutant virus particles encode proteases that are less susceptible to inhibition by the drug. In the presence of an HIV protease inhibitor, these virus particles will tend to replicate more effectively than does the population at large. With the passage of time, the less susceptible viruses will come to dominate the population and the virus population will become resistant to the drug.

Pathogens may become resistant to antibiotics by completely different mechanisms. Some pathogens contain enzymes that inactivate or degrade specific antibiotics. For example, many bacteria are resistant to β-lactams such as penicillin because they contain β-lactamase enzymes. These enzymes hydrolyze the β-lactam ring and render the drugs inactive.

Penicillin

Many of these enzymes are encoded in plasmids, small circular pieces of DNA often carried by bacteria (Section 9.2). Many plasmids are readily transferred from one bacterial cell to another, transmitting the capability for antibiotic resistance. Plasmid transfer thus contributes to the spread of antibiotic resistance, a major health-care challenge.

Drug resistance commonly emerges in the course of cancer treatment. Cancer cells are characterized by their ability to grow rapidly without the constraints that apply to normal cells. Many drugs used for cancer chemotherapy inhibit processes that are necessary for this rapid cell growth. However, individual cancer cells may accumulate genetic changes that mitigate the effects of such drugs. These altered cancer cells will tend to grow more rapidly than others and will become dominant within the cancer-cell population. This ability of cancer cells to mutate quickly has been a challenge to one of the major breakthroughs in cancer treatment: the development of inhibitors for proteins specific to cancer cells present in certain leukemias (Section 14.5). For example, tumors become undetectable in patients treated with imatinib, a Bcr-Abl protein kinase inhibitor, but in many patients the tumors recur after a period of years. In many of these cases, mutations have altered the Bcr-Abl protein so that it is no longer inhibited by the concentrations of imatinib used in therapy.

Finally, cancer patients often take multiple drugs concurrently in the course of chemotherapy, and, in many cases, cancer cells become simultaneously resistant to many or all of them. This multiple-drug resistance can be due to the proliferation of cancer cells that overexpress a number of ABC transporter proteins that pump drugs out of the cell (Section 13.2). Thus, cancer cells can evolve drug resistance by overexpressing normal human proteins or by modifying proteins responsible for the cancer phenotype.

Summary

32.1 Drug Discovery Begins with Target Identification and Validation

- The process of drug discovery is divided into two phases: the discovery phase, where molecules are optimized for testing in humans, and the development phase, where clinical trials are conducted.

- A drug target is a protein or oligonucleotide whose activity, when modulated by a drug, will alter the progression of a disease process.

- Targets should be validated, in that sufficient evidence suggests that it plays an important role in disease progression.

- Targets should also be tractable, in that sufficient evidence suggests that a molecule can be designed to interact with a target.

- Serendipity has played a significant role in the history of drug discovery.

32.2 Lead Molecules Can Be Discovered in Many Ways

- Lead molecules are starting points for optimization into drug candidates.

- Natural products are often a robust source of lead molecules. For example, the first HMG-CoA reductase inhibitors were derived from compounds isolated from the broths of bacteria and fungi.

- High-throughput screening (HTS), in which up to millions of compounds can be tested for their activity against a target, is a valuable source of lead molecules.

- HTS depends on the construction of a large chemical library. Combinatorial chemistry can be used to prepare large populations of compounds using only a few synthetic steps.

- In phenotypic screening, compounds are tested for their ability to affect a disease-relevant phenotype. The identity of the targets of hits from such screens are determined in subsequent steps.

32.3 Compounds Must Meet Stringent Criteria to Be Developed into Drugs

- Lead optimization is the process by which leads are modified into drug candidates suitable for testing in humans.

- To be effective, drugs must bind to their targets with high affinity and specificity.

- Most compounds are poorly absorbed and rapidly excreted from the body, or they are modified by metabolic pathways that target foreign compounds. Consequently, a drug's ADME properties, related to its absorption, distribution, metabolism, and excretion, must be optimized.

- A compound may also not be a useful drug because it is toxic; it may either modulate the target molecule too effectively (mechanism-based toxicity) or bind to proteins other than the target (off-target toxicity).

- Structural information on the interaction of a compound and its target can be used to drive further optimization of potency and selectivity.

32.4 Biologics Are a Growing Family of Drugs

- Biologics are a diverse class of drugs that are derived from living cells or through biological processes.

- The majority of biologics are recombinant proteins.

- Monoclonal antibodies represent a large proportion of recombinant protein biologics, as they are capable of a high degree of potency and selectivity.
- Because of their interaction with an endogenous salvage pathway, monoclonal antibodies typically exhibit very long half-lives, allowing them to be dosed weekly or monthly.

32.5 The Clinical Development of Medicines Proceeds Through Several Phases

- Before compounds can be given to human beings as drugs, they must be extensively tested for safety and efficacy.

- Clinical trials are performed in stages: first testing safety, then safety and efficacy in a small population; finally, safety and efficacy in a larger population to detect rarer adverse effects.
- With regard to infectious diseases and cancer, patients often develop resistance to a drug after it has been administered for a period of time because variants of the disease agent that are less susceptible to the drug have a selective advantage over more susceptible variants when the drug is present.

Key Terms

pharmacology (p. 978)

discovery phase (p. 978)

development phase (p. 978)

target (p. 978)

target validation (p. 978)

tractability (druggability) (p. 978)

lead molecules (leads) (p. 980)

high-throughput screening (HTS) (p. 981)

coupled assay (p. 982)

combinatorial chemistry (p. 982)

split-pool synthesis (p. 983)

chemical space (p. 984)

DNA-encoded libraries (DELs) (p. 984)

target-based screening (p. 985)

phenotypic screening (p. 985)

lead optimization (p. 986)

candidate (p. 986)

side effect (p. 987)

ADME properties (p. 988)

oral bioavailability (p. 988)

xenobiotic compounds (p. 990)

drug metabolism (p. 990)

phase I transformation (p. 990)

phase II transformation (p. 990)

mechanism-based (on-target) toxicity (p. 992)

off-target toxicity (p. 992)

therapeutic index (p. 993)

structure-based drug design (p. 994)

structure–activity relationship (SAR) (p. 994)

biologics (p. 995)

controlled study (p. 997)

placebo (p. 997)

double-blinded study (p. 997)

placebo effect (p. 997)

Problems

1. Sildenafil induces its physiological effects by increasing the intracellular concentrations of cGMP, leading to muscle relaxation. On the basis of the scheme shown in Figure 32.5, identify another approach for increasing cGMP levels with a small molecule. ❖ 1

2. For each of the following drugs, indicate whether the drug discovery was accidental, a result of natural product screening, or the result of targeted design. ❖ 2

(a) Penicillin

(b) Lovastatin

(c) Sildenafil

(d) Aspirin

(e) Indinavir

3. Which of the following compounds satisfy all of Lipinski's rules? Note that log(P) values are given in parentheses. ❖ 3

(A)

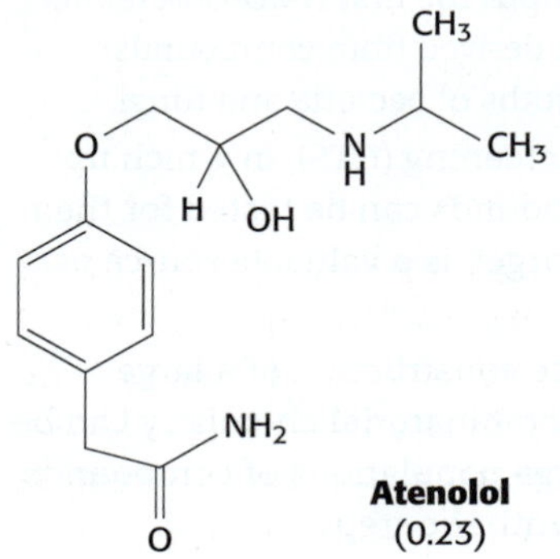

Atenolol
(0.23)

(B)

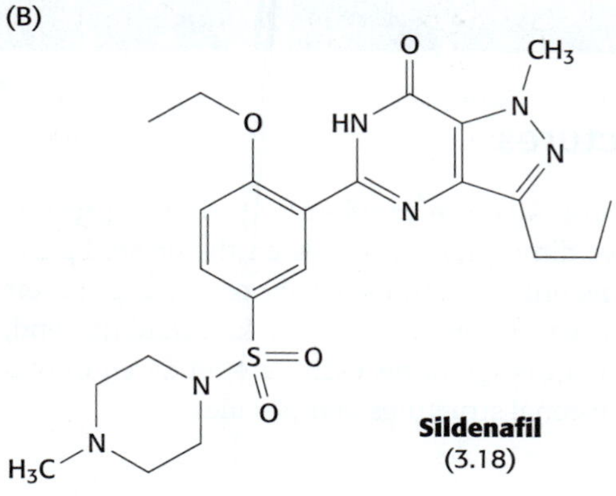

Sildenafil
(3.18)

(C)

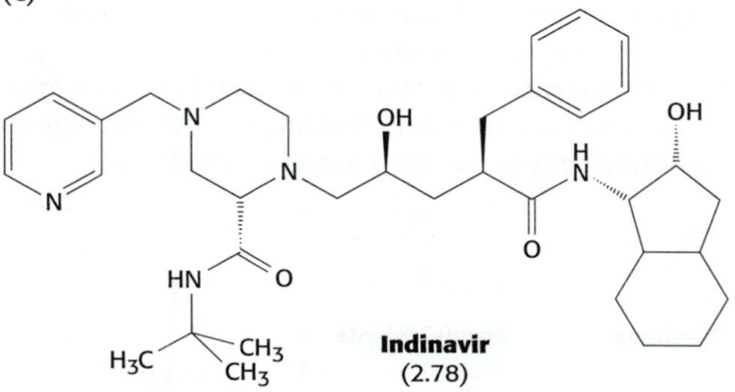

Indinavir
(2.78)

4. Name one advantage of a noncompetitive inhibitor as a potential drug compared with a competitive inhibitor. ❖ 3

5. Explain why drugs that inhibit P450 enzymes may be particularly dangerous when used in combination with other medications. ❖ 4

6. Coumadin can be a very dangerous drug because too much can cause uncontrolled bleeding. Persons taking coumadin must be careful about taking other drugs, particularly those that bind to albumin. Propose a mechanism for this drug–drug interaction.

7. Compound A is one of a series that were designed to be potent inhibitors of HIV protease.

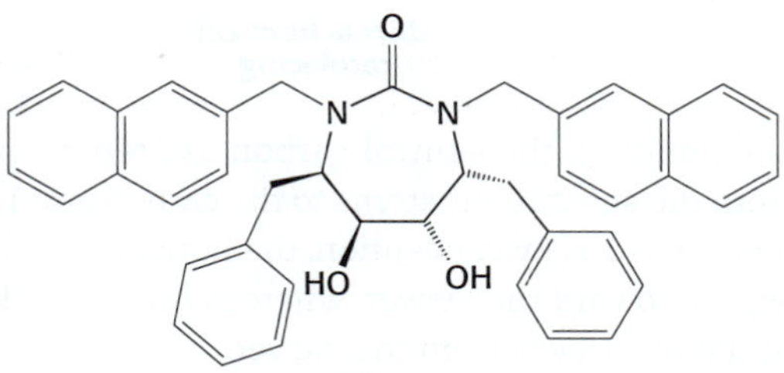

Compound A

Compound A was tested by using two assays: (1) direct inhibition of HIV protease in vitro and (2) inhibition of viral RNA production in HIV-infected cells, a measure of viral replication; results are shown in the following table. The HIV protease activity is measured with a substrate peptide present at a concentration equal to its K_M value.

Compound A (nM)	HIV protease activity (arbitrary units)	Compound A (nM)	Viral RNA production (arbitrary units)
0	11.2	1	4.0
0.2	9.9	2	2.2
0.4	7.4	10	0.9
0.6	5.6	100	0.2
0.8	4.8	0	760
		1.0	740
		2.0	380
		3.0	280
		4.0	180
		5.0	100
		10	30
		50	20

(a) Estimate the values for the K_I of compound A in the protease-activity assay and for its IC_{50} in the viral-RNA-production assay.

(b) Treating rats with the relatively high oral dose of 20 mg kg^{-1} results in a maximum concentration of compound A of 0.4 μM. On the basis of this value, do you expect compound A to be effective in preventing HIV replication when taken orally? Explain.

8. Considerable effort has been expended to develop computer programs that can estimate log(P) values entirely on the basis of chemical structure. Why would such programs be useful? ❖ 3

9. Distinguish between Phase I and Phase II clinical trials in regard to number of persons enrolled, the state of health of the subjects, and the goals of the study. ❖ 7

10. The metabolism of amphetamine by cytochrome P450 enzymes results in the conversion shown here. Propose a mechanism, and indicate any additional products.

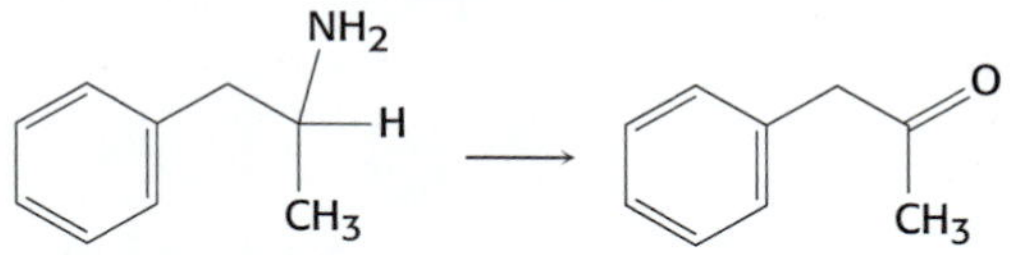

Amphetamine

11. You are studying a collection of competitive inhibitors of an exciting new enzyme. The K_M value for the natural substrate is 1 μM. One member of your team presents to you a list of IC_{50} values for these compounds using an assay that employs a substrate concentration of 20 μM. A week later, another member of your team tells you about her assay that requires a substrate concentration of only 1 μM. She presents you a list of IC_{50} values for the same set of compounds, which are consistently about 10-fold lower than those previously determined. Explain why this drop in IC_{50} is not at all surprising.

12. What are some advantages of using a monoclonal antibody as a drug, as opposed to using a small molecule? ❖ 6

Visualizing Molecular Structures

The authors of a biochemistry textbook face the problem of trying to present three-dimensional molecules in the two dimensions available on the printed page. The interplay between the three-dimensional structures of biomolecules and their biological functions are discussed extensively throughout this book. Toward this end, we frequently use representations that, although of necessity are rendered in two dimensions, emphasize the three-dimensional structures of molecules.

Depicting Structural Formulas

Structural formulas in this book are often shown in two different ways.

Stereochemical Renderings. Most of the chemical formulas in this book are drawn to depict the geometric arrangement of atoms—the stereochemistry—crucial to chemical bonding and reactivity, as accurately as possible. For example, the carbon atom of methane is tetrahedral, with H–C–H angles of 109.5 degrees, whereas the carbon atom in formaldehyde has bond angles of 120 degrees.

Methane   **Formaldehyde**

To illustrate the correct stereochemistry about tetrahedral carbon atoms, wedges are used to depict the direction of a bond into or out of the plane of the page. A solid wedge with the broad end away from the carbon atom denotes a bond coming toward the viewer out of the plane. A dashed wedge, with its broad end at the carbon atom, represents a bond going away from the viewer behind the plane of the page. The remaining two bonds are depicted as straight lines.

Fischer Projections. Although representative of the actual structure of a compound, stereochemical structures are often difficult to draw quickly. An alternative, less-representative method of depicting structures with tetrahedral carbon centers relies on the use of Fischer projections.

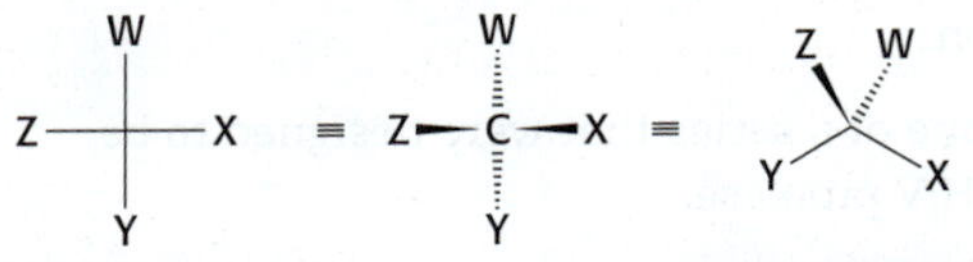

Fischer projection   **Stereochemical rendering**

In a Fischer projection, the bonds to the central carbon are represented by horizontal and vertical lines from the substituent atoms to the carbon atom, which is assumed to be at the center of the cross. By convention, the horizontal bonds are assumed to project out of the page toward the viewer, whereas the vertical bonds are assumed to project behind the page away from the viewer.

Depicting Small Molecules

For depicting the molecular architecture of small molecules in more detail, two types of models are often used in this book: space filling and ball-and-stick. These models show structures at the atomic level.

Space-Filling Models. The space-filling models are the most realistic. The size and position of an atom in a space-filling model are determined by its bonding properties

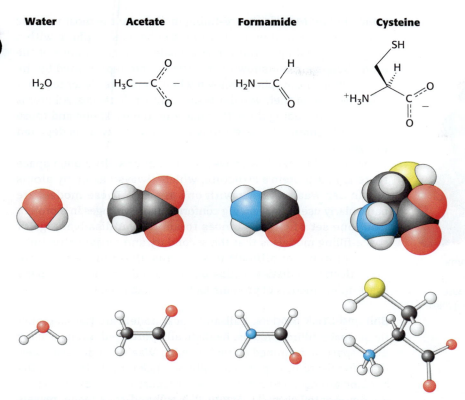

FIGURE 1 Molecules are depicted in various ways. Structural formulas (top), space-filling representations (middle), and ball-and-stick models (bottom) of several molecules are shown.

and van der Waals contact distance, or the van der Waals radius. The van der Waals radius describes how closely two atoms can approach each other when they are not linked by a covalent bond. The colors of the model are set by convention.

Carbon, black	Hydrogen, white	Nitrogen, blue
Oxygen, red	Sulfur, yellow	Phosphorus, purple

Space-filling models of several simple molecules are shown in **Figure 1**.

Ball-and-Stick Models. Ball-and-stick models are not as realistic as space-filling models, because the atoms are depicted as spheres of radii smaller than their van der Waals radii. However, the bonding arrangement is easier to see because the bonds are explicitly represented as sticks. The taper of a stick helps to depict perspective and indicates which of a pair of bonded atoms is closer to the reader. Thus, a ball-and-stick model reveals a complex structure more clearly than a space-filling model does. Ball-and-stick models of several simple molecules are shown in Figure 1.

Depicting Protein Structures

Many of the powerful techniques for the determination of protein structure discussed in Chapter 4 allow the positions of the thousands of atoms within a protein to be determined. The final results include the x, y, and z coordinates for each atom in the structure. These coordinate files are compiled in the Protein Data Bank (https://www.wwpdb.org), from which they can be readily downloaded. These structures comprise thousands or even tens of thousands of atoms.

The complexity of proteins with thousands of atoms presents a challenge for the depiction of their structure. Several different types of representations are used to portray proteins, each with its own strengths and weaknesses. The types that you will see most often in this book are space-filling models, ball-and-stick models, backbone models, and ribbon diagrams. Where appropriate, structural features of particular importance or relevance are noted in an illustration's legend.

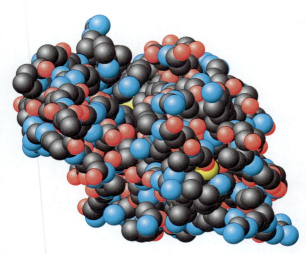

FIGURE 2 A space-filling model shows the atoms tightly packed. Hydrogen atoms are often omitted because their positions are not often readily determined and because excluding them aids visualization of the other atoms. The protein lysozyme is depicted here and in the next four figures.

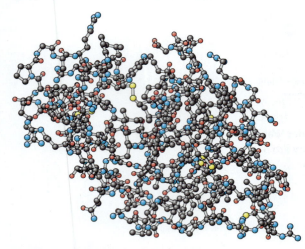

FIGURE 3 In a ball-and-stick model, bonds are represented by sticks. Again, hydrogen atoms are omitted.

FIGURE 4 In a backbone model, only the backbone polypeptide chain is represented. In this model, the α carbons of the adjacent amino acids are shown in gray and lines relating their positions in black. Many backbone models omit the atoms and include only the lines.

Space-Filling Models. Space-filling models are the most realistic type of representation. Each atom is shown as a sphere with a size corresponding to the van der Waals contact distance of the atom. Bonds are not shown explicitly but are represented by the intersection of the spheres shown when atoms are closer together than the sum of their van der Waals contact distances. All atoms are shown, including those that make up the backbone and those in the side chains. A space-filling model of lysozyme is depicted in **Figure 2**.

Space-filling models convey a sense of how little open space there is in a protein's structure, which always has many atoms in van der Waals contact with one another. These models are particularly useful in showing conformational changes in a protein from one set of circumstances to another. A disadvantage of space-filling models is that the secondary and tertiary structures of the protein are difficult to see. Thus, these models are not very effective in distinguishing one protein from another—many space-filling models of proteins look very much alike.

Ball-and-Stick Models. Ball-and-stick models are not as realistic as space-filling models. Realistically portrayed atoms occupy more space, determined by their van der Waals contact distances, than do the atoms depicted in ball-and-stick models. However, the bonding arrangement is easier to see because the bonds are explicitly represented as sticks (**Figure 3**). A ball-and-stick model reveals a complex structure more clearly than a space-filling model does. Yet, the depiction is so complicated that structural features such as α helices or potential binding sites are difficult to discern.

Because space-filling and ball-and-stick models depict protein structures at the atomic level, the large number of atoms in a complex structure makes it difficult to distinguish the relevant structural features. Thus, representations that are more schematic—such as backbone models and ribbon diagrams—have been developed for the depiction of macromolecular structures. In these representations, most or all atoms are not shown explicitly.

Backbone Models. Backbone models show only the backbone atoms of a polypeptide chain, or sometimes only the α-carbon atom of each amino acid. Atoms are linked by lines representing bonds; if only α-carbon atoms are depicted, lines connect α-carbon atoms of amino acids that are adjacent in the amino acid sequence (**Figure 4**). While in Figure 4, the α-carbon atoms are shown, elsewhere in this book backbone models show only the lines connecting the α-carbon atom positions; bonds to other carbon atoms are not depicted.

A backbone model shows the overall course of the polypeptide chain much better than a space-filling or ball-and-stick model does. However, secondary structural elements are still difficult to see.

Ribbon Diagrams. Ribbon diagrams are highly schematic and most commonly used to accent a few dramatic aspects of protein structure, such as the α helix, depicted as a coiled ribbon or a cylinder. The β strand is depicted as a broad arrow, and loops, turns, and random coil structure are all shown with thin tubes. These structural depictions provide the clearest views of the folding patterns of proteins (**Figure 5**). The ribbon diagram allows the course of a polypeptide chain to be traced and readily shows

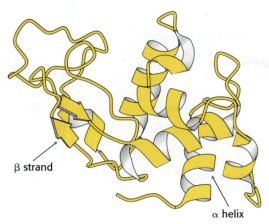

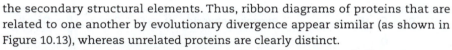

FIGURE 5 In a simple ribbon diagram, the α helices are shown as coiled ribbons and β strands are depicted as arrows. More irregular structures are shown as thin tubes.

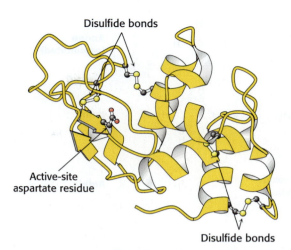

FIGURE 6 An enhanced ribbon diagram may show functionally important areas using ball-and-stick. Four disulfide bonds and a functionally important aspartate residue of lysozyme are shown in ball-and-stick form.

the secondary structural elements. Thus, ribbon diagrams of proteins that are related to one another by evolutionary divergence appear similar (as shown in Figure 10.13), whereas unrelated proteins are clearly distinct.

In this book, coiled ribbons are generally used to depict α helices. However, for membrane proteins, which are often quite complex, cylinders are used rather than coiled ribbons. This convention also makes membrane proteins with their membrane-spanning α helices easy to recognize (as shown in Figure 12.18).

Bear in mind that the open appearance of ribbon diagrams is deceptive. As noted earlier, protein structures are tightly packed and have little open space. However, the openness of ribbon diagrams makes them particularly useful as frameworks in which to highlight additional aspects of protein structure. Active sites, substrates, bonds, and other structural fragments can be included in ball-and-stick or space-filling form within a ribbon diagram (**Figure 6**).

Functional Groups

Biochemistry depends on many concepts and principles from organic chemistry. The properties of individual atoms are substantially affected by the other atoms to which they are bonded. Functional groups are specific combinations of a small number of atoms that often have properties that are relatively independent of other aspects of the molecules in which they are found. There are many functional groups in organic chemistry, but a relatively small number of these are central to biochemistry:

- *Hydroxyl group,* –OH A hydroxyl group comprises an oxygen atom bonded to a hydrogen atom. The hydroxyl group can both donate and accept hydrogen bonds. The hydroxyl group is both a very weak acid and a very weak base.

- *Amino group,* –NH$_2$ An amino group comprises a nitrogen atom bonded to two hydrogen atoms. The amino group can both donate and accept hydrogen bonds. The amino group often behaves as a base, accepting a hydrogen ion to form an ammonium group, –NH$_3^+$.

- *Carbonyl group,* C=O A carbonyl group comprises a carbon atom double bonded to an oxygen atom. The carbonyl group can act as a hydrogen bond acceptor but not a donor. The carbonyl group is found in many different classes of compounds, depending on the other groups that are bonded to the carbon atom. The most important carbonyl-containing functional group in biochemistry is the amide, in which the carbonyl carbon is bonded to one carbon and one nitrogen.

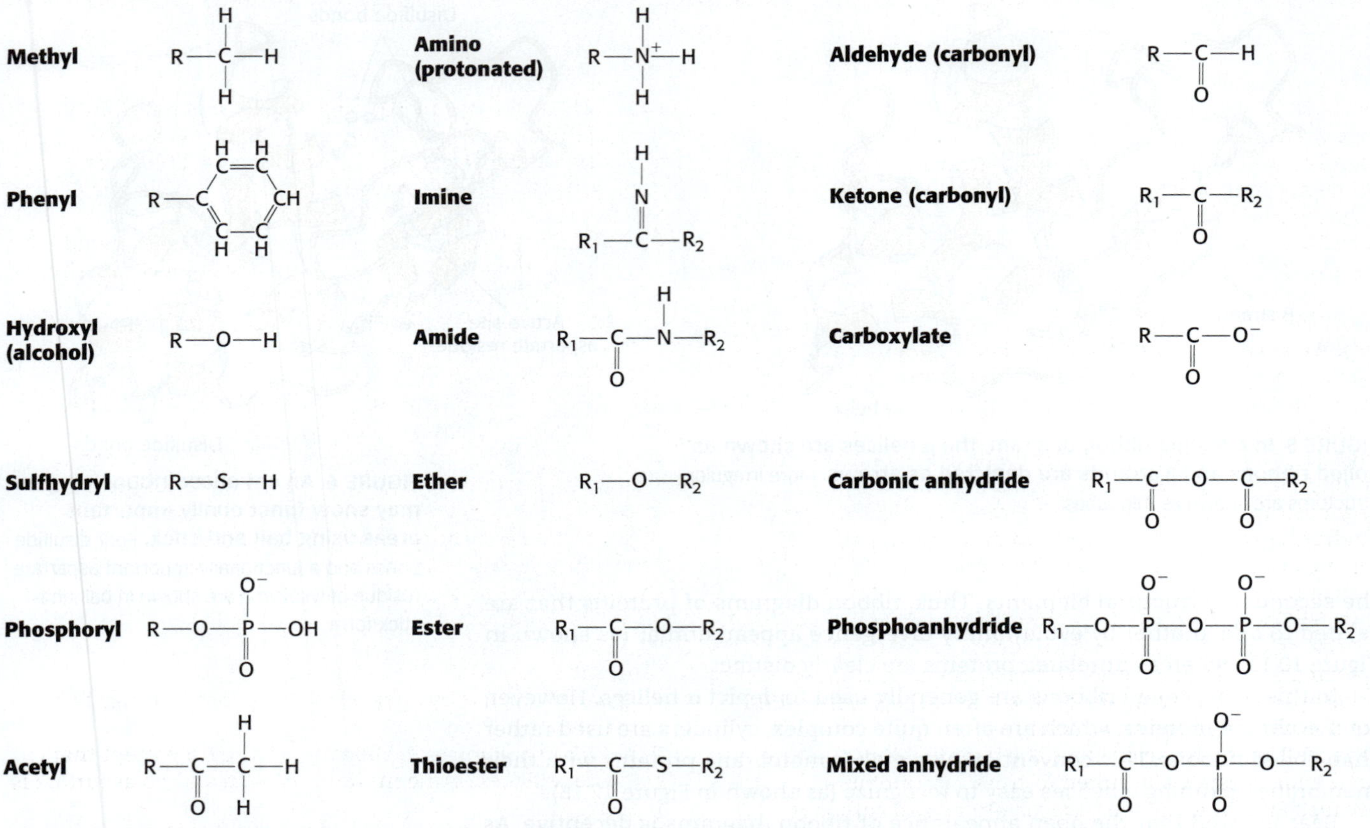

Methyl	$R-\underset{\underset{H}{\textstyle\vert}}{\overset{\overset{H}{\textstyle\vert}}{C}}-H$	**Amino (protonated)**	$R-\underset{\underset{H}{\textstyle\vert}}{\overset{\overset{H}{\textstyle\vert}}{\overset{+}{N}}}-H$	**Aldehyde (carbonyl)**	$R-\underset{\underset{O}{\textstyle\Vert}}{C}-H$
Phenyl	phenyl ring	**Imine**	$R_1-\underset{\underset{R_2}{\textstyle\vert}}{\overset{\overset{H}{\textstyle\vert}}{\overset{\textstyle N}{C}}}$	**Ketone (carbonyl)**	$R_1-\underset{\underset{O}{\textstyle\Vert}}{C}-R_2$
Hydroxyl (alcohol)	$R-O-H$	**Amide**	$R_1-\underset{\underset{O}{\textstyle\Vert}}{C}-\overset{\overset{H}{\textstyle\vert}}{N}-R_2$	**Carboxylate**	$R-\underset{\underset{O}{\textstyle\Vert}}{C}-O^-$
Sulfhydryl	$R-S-H$	**Ether**	R_1-O-R_2	**Carbonic anhydride**	$R_1-\underset{\underset{O}{\textstyle\Vert}}{C}-O-\underset{\underset{O}{\textstyle\Vert}}{C}-R_2$
Phosphoryl	$R-O-\underset{\underset{O}{\textstyle\Vert}}{\overset{\overset{O^-}{\textstyle\vert}}{P}}-OH$	**Ester**	$R_1-\underset{\underset{O}{\textstyle\Vert}}{C}-O-R_2$	**Phosphoanhydride**	$R_1-O-\underset{\underset{O}{\textstyle\Vert}}{\overset{\overset{O^-}{\textstyle\vert}}{P}}-O-\underset{\underset{O}{\textstyle\Vert}}{\overset{\overset{O^-}{\textstyle\vert}}{P}}-O-R_2$
Acetyl	$R-\underset{\underset{O}{\textstyle\Vert}}{C}-\overset{\overset{H}{\textstyle\vert}}{\underset{\underset{H}{\textstyle\vert}}{C}}-H$	**Thioester**	$R_1-\underset{\underset{O}{\textstyle\Vert}}{C}-S-R_2$	**Mixed anhydride**	$R_1-\underset{\underset{O}{\textstyle\Vert}}{C}-O-\underset{\underset{O}{\textstyle\Vert}}{\overset{\overset{O^-}{\textstyle\vert}}{P}}-O-R_2$

FIGURE 7 Functional groups define many important properties of macromolecules.

It may be beneficial to review the name and structure of some common functional groups so that you can recognize them readily during your study of biochemistry. **Figure 7** lists some commonly encountered groups.

Reaction Mechanisms: "Arrow Pushing"

Many aspects of biochemistry depend on chemical reactions in which covalent bonds are broken and formed. These reactions involve the flow of electrons out of bonds and into spaces between atoms in which new bonds are formed. It is often useful to depict such reactions with arrows to represent this electron flow. The process of analyzing reactions in this way is sometimes referred to informally as "arrow pushing" or "electron pushing."

Consider the reaction between ammonia, NH_3, and methyl iodide, H_3C-I. NH_3 has a lone pair of electrons on the nitrogen atom that can participate in bond formation. The iodine atom in H_3C-I can accept electrons to form the relatively stable iodide ion, I^-. The reaction and its mechanism are shown below:

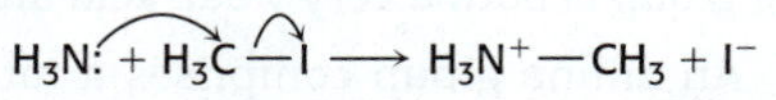

$$H_3N\colon + H_3C-I \longrightarrow H_3N^+-CH_3 + I^-$$

The first arrow depicts the flow of the electron pair from the nitrogen to the space between the nitrogen and the carbon to form the new nitrogen–carbon bond. The second arrow shows the flow of electrons from the carbon–iodine bond to the iodine to form the iodide ion. The initial product of this reaction is methylammonium ion, methylamine in its protonated form. In this reaction, ammonia is the nucleophile, meaning the molecule or group that donates an electron pair during a reaction.

In this book, we encounter many examples of reaction mechanisms, particularly when we discuss the actions of enzymes—the essential catalysts that facilitate so much of biochemistry.

Chapter 1
Page 7 TATCGGACCT
Page 14 The enthalpy change (heat released or absorbed by the system) accounts for entropy changes in the surroundings.
Page 20 15.14
Page 26 No

Chapter 2
Page 41 Isoleucine and threonine. Both these atoms feature four unique groups emerging from the first carbon atom in their side chains. These atoms are indicated by an asterisk in Figures 2.6 and 2.7.
Page 46 Proline and glycine. The cyclic side chain of proline linking the nitrogen and α-carbon atoms limits φ to a very narrow range (around –60 degrees). The lack of steric hindrance exhibited by the side chain hydrogen atom of glycine enables this amino acid to access a much greater area of the Ramachandran plot.
Page 50 There are about 3.6 amino acids per turn of an α helix, so residues three or four amino acids apart will appear on the same face. By contrast, adjacent amino acids appear on opposite faces of a β strand, so every other residue will lie on the same face.
Page 53 The amino acids would be hydrophobic in nature. An α helix is especially suited to crossing a membrane because all of the amide hydrogen atoms and carbonyl oxygen atoms of the peptide backbone take part in intrachain hydrogen bonds, thus stabilizing these polar atoms in a hydrophobic environment.
Page 57 If the protein concentration was high upon denaturation, groups from different molecules might interact to form a large, insoluble aggregate.

Chapter 3
Page 72 50 nanograms/ml = 2.3×10^{-9} M = 2.3 nM. The $L_{1/2}$ value would be expected to be in the same range.
Page 72 The dissociation rate is equal to the association rate at equilibrium. 5 s^{-1}.
Page 79 b.
Page 80 Since the proximal histidine is not attached to the protein, it is free to move into the plane of the heme when oxygen binds; so, the oxygen affinity is high, similar to that of myoglobin.
Page 88 Total number, 12. Total not counting duplicates, 6; 3 from the light chain and 3 from the heavy chain.
Page 94 The association rate constant is $k_1 = 10^4 s^{-1}/10^{-3} M = 10^7 M^{-1}s^{-1}$. The dissociation rate constant is $k_{-1} = 50$ s^{-1}. The dissociation constant is $K_d = 50 s^{-1}/10^7 M^{-1}s^{-1} = 5 \times 10^6$ M.

Chapter 4
Page 107 Note in Figure 4.10 that the relative electrophoretic mobility of proteins is proportional to the *logarithm* of their individual masses. The mass of the protein is 50 kDa.
Page 116 A sandwich ELISA experiment requires both antibodies to be able to simultaneously bind the antigen. If the capture antibody on the plate and the detection antibody both recognize the same epitope, then each antibody will displace the other. Either the antigen will remain bound to the capture antibody and will not be detected, or the antigen will bind to the soluble detection antibody and will be washed away before the detection step.

Page 121 Isoleucine and leucine are isomers and, hence, have identical masses. Peptide sequencing by mass spectrometry as described in this chapter is incapable of distinguishing these two residues. Further analytical techniques are required to differentiate them.
Page 132 Protein crystal formation requires the ordered arrangement of identically positioned molecules. Proteins with flexible linkers can introduce disorder into this arrangement and prevent the formation of suitable crystals.

Chapter 5
Page 144 All three terms apply, but prosthetic group is the most specific.
Page 158 K_M is an intrinsic property of an enzyme because it is independent of the quantity of enzyme measured. In contrast, V_{max} is an extrinsic property because it does depend on the quantity of enzyme observed.
Page 159 The mutation made the enzyme a more efficient catalyst as the catalytic efficiency (k_{cat}/K_M) increased by 50%.
Page 169 Competitive.

Chapter 6
Page 180 Catalysis by approximation (since it brings two substrates together).
Page 189 The same as is shown in Figure 6.8, except that cysteine replaces serine in the active site and no aspartate is present.
Page 193 10^4 s^{-1}. The maximum expected buffer-assisted rate is 10^9 $M^{-1}s^{-1} \times 10^{-6}$ M = 10^3 s^{-1}, but the unassisted maximum rate is 10^4 s^{-1}.
Page 199 EcoRV binds all sequences with approximately equal affinity but discriminates at the cleavage step.
Page 204 ATP is reversible hydrolyzed in the myosin active site with incorporation of ^{18}O. Some of this ATP is released as ATP rather than ADP + P_i, and this ATP will include ^{18}O.

Chapter 7
Page 216 All of the enzyme would be in the R form all of the time. There would be no cooperativity. The kinetics would look like that of a Michaelis–Menten enzyme.
Page 221 The net outcome of the two reactions is the hydrolysis of ATP to ADP and P_i, which has a ΔG of –50 kJ mol^{-1} (–12 kcal mol^{-1}) of under cellular conditions.
Page 226 Although quite rare, cases of enteropeptidase deficiency have been reported. The affected person has diarrhea and fails to thrive because digestion is inadequate. In particular, protein digestion is impaired.
Page 232 Antithrombin III is a very slowly hydrolyzed substrate of thrombin. Hence, its interaction with thrombin requires a fully formed active site on the enzyme.

Chapter 8
Page 242 There would be too much charge repulsion from the negative charges on the phosphoryl groups. These charges must be countered by the addition of cations.
Page 248 (a) Hydrogen bonds between base pairs and van der Waals interactions among the bases. The van der Waals interactions come into play because of the hydrophobic effect, which forces the bases to the interior of the helix.
(b) One possible explanation is that GC base pairs have three hydrogen bonds compared with two for AT base pairs; the higher content of GC would mean more hydrogen bonds

and greater helix stability. Another possibility is that the base stacking interactions are stronger for GC base pairs than for AT base pairs. Careful measurements support the second explanation: the favorable base-stacking energy of GC base pairs largely contributes to the increases in melting temperature.

Page 256 The genetic code is degenerate. Of the 20 amino acids, 18 are specified by more than one codon. Hence, many nucleotide changes (especially in the third base of a codon) do not alter the nature of the encoded amino acid. Mutations leading to an altered amino acid are usually more deleterious than those that do not and hence are subject to more stringent selection.

Chapter 9

Page 266 The presence of the AluI sequence would, on average, be $(1/4)^4$, or 1/256, because the likelihood of any base being at any position is one-fourth and there are four positions. By the same reasoning, the presence of the NotI sequence would be $(1/4)^8$, or 1/65,536. Thus, the average product of digestion by AluI would be 250 base pairs (0.25 kb) in length, whereas that by NotI would be 66,000 base pairs (66 kb) in length.

Page 270 At high temperatures of hybridization, only very close matches between primer and target would be stable because all (or most) of the bases would need to find partners to stabilize the primer–target helix. As the temperature is lowered, more mismatches would be tolerated; so the amplification is likely to yield genes with less sequence similarity. In regard to the yeast gene, synthesize primers corresponding to the ends of the gene, and then use these primers and human DNA as the target. If nothing is amplified at 54°C, the human gene differs from the yeast gene, but a counterpart may still be present. Repeat the experiment at a lower temperature of hybridization.

Page 278 The average frequency of EcoRI sites in the genome is $(1/4)^6 = 1/4,096$, or about once in every 4 kb. Because most human genes are much longer than 4 kb, this would not be a suitable cloning strategy, as most fragments would contain only a small part of a complete gene.

Page 294 Since PAM sequences are required downstream of a particular target site, variants of Cas9 that recognize different PAMs would increase the likelihood that a sgRNA could be designed against a particular target sequence.

Chapter 10

Page 305 Placement of too many gaps in an alignment could artificially increase the number of identities.

Page 309 44% identity.

Page 314 To detect pairs of residues with correlated mutations, there must be variability in these sequences. If the alignment is overrepresented by closely related organisms, there may not be enough changes in their sequences to allow the identification of potential base-pairing patterns.

Chapter 11

Page 330 D-Fructopyranose; the chair conformation of D-fructopyranose; bulky groups in equatorial positions about a ring structure; β-anomer of D-glucopyranose

Page 333 All three are reducing sugars.

Page 337 Maltose, cellulose, starch, amylose, and glycogen; cellulose and chitin.

Page 346 The genome is the least complex; the glycome is likely the most complex.

Chapter 12

Page 358 Phospholipids, glycolipids (which are a subclass of sphingolipids), and cholesterol. Examples of polar head groups include serine, ethanolamine, choline, and inositol.

Page 360 Essentially an "inside-out" membrane. The hydrophilic groups would come together on the interior of the structure, away from the solvent, whereas the hydrocarbon chains would interact with the solvent.

Page 369 In a hydrophobic environment, the formation of intrachain hydrogen bonds stabilizes the amide hydrogen atoms and carbonyl oxygen atoms of the polypeptide chain, and so an α helix forms. In an aqueous environment, these groups are stabilized by interaction with water, and so there is no energetic reason to form an α helix. Thus, the α helix would be more likely to form in a hydrophobic environment.

Page 372 The presence of a cis double bond introduces a kink in the fatty acid chain that prevents tight packing and reduces the number of atoms in van der Waals contact. The kink lowers the melting point compared with that of a saturated fatty acid. Trans fatty acids do not have the kink, and so their melting temperatures are higher, more similar to those of saturated fatty acids. Because trans fatty acids have no structural effect, they are rarely observed.

Chapter 13

Page 390 SERCA, a P-type ATPase, uses a mechanism by which a covalent phosphorylated intermediate (at an aspartate residue) is formed. At steady state, a subset of the SERCA molecules is trapped in the E_2-P state and, as a result, radiolabeled. The MDR protein is an ABC transporter and does not operate through a phosphorylated intermediate. Hence, a radiolabeled band would not be observed for MDR.

Page 392 FCCP effectively creates a pore in the bacterial membrane through which protons can pass rapidly. Protons that are pumped out of the bacteria will pass through this pore preferentially (the "path of least resistance"), rather than participate in H^+/lactose symport.

Page 398 An ion channel must transport ions in either direction at the same rate. The net flow of ions is determined only by the composition of the solutions on either side of the membrane.

Page 407 Cardiac muscle must contract in a highly coordinated manner in order to pump blood effectively. Numerous gap junctions, abundant in connexin proteins, mediate the orderly cell-to-cell propagation of the action potential through the heart during each beat.

Chapter 14

Page 424 In cardiac muscle, increased cAMP levels lead to increased protein kinase A activity, which leads to increased phosphorylation of troponin, resulting in muscle contraction. In smooth muscle, decreased cAMP levels lead to decreased protein kinase A activity, which decreases the level of phosphorylation of myosin light chains, thus decreasing the extent of muscle contraction.

Page 430 No, the substrate preference and the activation loop sequence are quite different.

Page 433 Both Ras and a heterotrimeric G protein are activated by GTP for GDP exchange. Ras is a small monomeric protein, whereas heterotrimeric G proteins are larger and exist as trimers in their inactive states. (Other answers are possible.)

Page 437 One would expect the loss of sensitivity to the odor of one or a small number of compounds detectable by most individuals.

Page 441 Olfaction: odorant molecules; vision: light in the visible range; hearing: pressure waves (vibrations) in the air. Odorant molecules bind to odorant receptors; light absorption results in isomerization of 11-cis-retinal in visual pigments; air vibrations result in relative displacement of stereocilia in hair bundles.

Chapter 15

Page 449 A can still be converted to B despite a positive $\Delta G^{\circ\prime}$ value if the concentrations of A and B are sufficiently imbalanced such that ΔG is negative. Alternatively, the reaction that converts A into B can be coupled to another reaction with a negative $\Delta G^{\circ\prime}$ value by an enzyme such that the overall reaction now has a negative $\Delta G^{\circ\prime}$ value.

Page 459 Statement 3 is the only true statement. Statement 1 is false because energy is required to break covalent bonds and is released upon bond formation with water, and Statement 2 is false because ATP has an intermediate phosphoryl transfer potential.

Page 468 Coenzyme A; $NADP^+$; Vitamins.

Chapter 16

Page 484 The 1-arseno-3-phosphoglycerate that will be formed will be rapidly hydrolyzed so no ATP can be generated by the next reaction of glycolysis by phosphoglycerate kinase. Thus, arsenate eliminates one of the two ATP-synthesizing steps in glycolysis. In fact, all enzymes that use phosphate could use arsenate, severely compromising the energy production of the cell, leading to cell damage. If the arsenate concentration were high enough, death would result.

Page 486 Two molecules of NADH and four molecules of ATP.

Page 488 Two molecules of ATP and zero molecules of NADH.

Page 492 Classic galactosemia is much more detrimental to human health. If a person was galactosemic, it would be better to be lactose intolerant so that less galactose would be absorbed when lactose is consumed in the diet.

Page 499 ATP initially stimulates PFK activity, as would be expected for a substrate, but at higher concentrations, it inhibits the enzyme. Although this effect seems counterintuitive for a substrate, recall that the function of glycolysis in muscle is to generate ATP. Consequently, high concentrations of ATP signal that the ATP needs are met and glycolysis should stop. In addition to being a substrate, ATP is an allosteric inhibitor of PFK, with the allosteric site having a lower affinity for ATP than the active site.

Page 508 The energy needs of a muscle cell vary widely, from rest to intense exercise, so the rapid regulation of phosphofructokinase by ATP and AMP is vital. In other tissues, such as the liver, ATP concentration is less likely to fluctuate and will not be a key regulator of phosphofructokinase, but rather the enzyme, and its counterpart in gluconeogenesis, fructose 1,6-bisphosphatase, are subject to hormonal control through the action of fructose 2,6-bisphosphate instead.

Page 509 Forming lactate allows tissues that are anaerobic, either temporarily, like skeletal muscle, or permanently, like some parts of the skin, to continue to produce ATP without oxygen. The burden of resynthesizing glucose from lactate at the expense of ATP can be borne by other tissues, like the liver, that are replete with oxygen and energy stores. Thus, the Cori cycle allows an important division of labor in the body.

Page 511 Fructose 2,6-bisphosphate, present at high concentration when glucose is abundant, normally slows gluconeogenesis by inhibiting fructose 1,6-bisphosphatase. In this genetic disorder, the phosphatase is active irrespective of the glucose level. Hence, substrate cycling is increased. The level of fructose 1,6-bisphosphate is consequently lower than normal. Less pyruvate is formed, and thus less ATP is generated.

Chapter 17

Page 521 The remaining steps regenerate oxidized lipoamide, which is required to begin the next reaction cycle. Moreover, this regeneration results in the production of high-energy electrons in the form of NADH.

Page 529 The fact that fluorocitrate is formed tells us something about the immediate metabolic fate of fluoroacetate: It must first be converted to fluoroacetyl CoA and enter the citric acid cycle. Fluoroacetyl CoA must react with oxaloacetate to form fluorocitrate, but while citrate is normally converted into isocitrate by the enzyme aconitase, it appears that fluorocitrate is not a substrate for aconitase, and so it accumulates.

Page 531 The enzyme nucleoside diphosphokinase transfers a phosphoryl group from GTP (or any nucleoside triphosphate) to ADP according to the reversible reaction:
$$GTP + ADP \rightleftharpoons GDP + ATP$$

Page 534 The CAC depends on a steady supply of NAD^+ as an oxidant, generating NADH. O_2 is never directly utilized in the cycle. However, NAD^+ is regenerated via donation of electrons to O_2 by way of the electron transport chain, so eventually a lack of O_2 will cause the cycle to cease due to a lack of NAD^+. The same is true of FAD and $FADH_2$.

Chapter 18

Page 545 From the cytosol it would pass through the mitochondrial outer membrane, into the intermembrane space (IM space), then through the mitochondrial inner membrane, and finally into the mitochondrial matrix. The infoldings are called cristae, and they are formed by the mitochondrial inner membrane.

Page 549 Pyruvate accepts electrons and is thus the oxidant. NADH gives up electrons and is the reductant.

Page 552 The 10 isoprene units render coenzyme Q soluble in the hydrophobic environment of the inner mitochondrial membrane. The two oxygen atoms can reversibly bind two electrons and two protons as the molecule transitions between the quinone form and quinol form.

Page 561 The answer to all three questions is Succinate-Q reductase (Complex II).

Page 562 Complex I would be reduced, whereas Complexes II, III, and IV would be oxidized. The citric acid cycle would halt because it has no way to oxidize NADH.

Page 567 Recall from the discussion of enzyme-catalyzed reactions that the direction of a reaction is determined by the ΔG difference between substrate and products and that enzymes accelerate the rate of both the forward and the backward reactions. The hydrolysis of ATP is exergonic, and so ATP synthase will enhance the hydrolytic reaction.

Page 570 3 ATP per full rotation; 14 protons/3 ATP = 4.7 protons per ATP if there are 14 **c** subunits.

Page 573 The extra negative charge on ATP relative to that on ADP accounts for ATP's translocation out of the mitochondrial matrix toward the more positively charged side of the inner mitochondrial membrane. Mg^{2+} lessens the charge differences between ATP and ADP, reducing the impact of charge on the directionality of transport.

Additionally, by decreasing the difference between the two ligands, the metal may make ADP more readily compete with ATP for transport to the cytoplasm.

Page 579 DNOC must be an uncoupler, similar to DNP. Recall that uncouplers deplete the proton gradient, allowing the electron-transport chain to operate in the absence of ATP synthesis. In the presence of DNOC, the energy released by the electron-transport chain is released as heat instead of being used to synthesize ATP. Oxygen consumption will increase as more fuel molecules are metabolized to try to make up for the lack of ATP produced by ATP synthase.

Page 579 In the presence of poorly functioning mitochondria, the only means of generating ATP is by anaerobic glycolysis, which will lead to an accumulation of lactic acid in blood.

Page 581 A proton-motive force across the plasma membrane is necessary to drive the flagellar motor. Under conditions of starvation, this proton-motive force is depleted. In acidic solution, the pH difference across the membrane is sufficient to power the motor.

Chapter 19

Page 587 Phototrophy and autotrophy.

Page 589 Chloroplasts are descended from an ancient bacterium that was engulfed and became codependent with another organism, a phenomenon called endosymbiosis. Within chloroplasts are membrane-enclosed compartments called thylakoids. The internal compartment of the thylakoid is called the stroma.

Page 593 The special pair is a specific pair of chlorophyll molecules in the reaction center that absorb light energy and transfer an electron away, creating photoinduced charge separation, the key to photosynthesis.

Page 599 PS I generates reduced ferredoxin, a powerful reductant that reduces $NADP^+$ to NADPH. PS II activates the WOC, a powerful oxidant capable of oxidizing water, generating O_2. The WOC releases the electrons that will ultimately reduce $NADP^+$ and generates protons to form the proton gradient used to drive the synthesis of ATP.

Page 602 The two components are the chemical gradient and the charge gradient. The chemical gradient is nearly entirely responsible for ATP production in chloroplasts because the thylakoid membrane is quite permeable to Cl^- and Mg^{2+}. Thus, as H^+ is pumped into the thylakoid lumen, Mg^{2+} moves out of the lumen while Cl^- move in, dissipating any charge gradient that would otherwise form.

Chapter 20

Page 615 When ribulose 1,5-bisphosphate reacts with CO_2, it forms a six-carbon compound that splits into two molecules of 3-phosphoglycerate. Only one will have ^{14}C.

Page 618 In the dark reactions, magnesium serves to bind ribulose 1,5-bisphosphate and activate it for reaction with CO_2. In the light reactions, magnesium is present in chlorophyll.

Page 634 In the pentose phosphate pathway, erythrose 4-phosphate reacts with xylulose 5-phosphate to produce fructose 6-phosphate and glyceraldehyde 3-phosphate in a reaction catalyzed by transketolase. In the Calvin–Benson cycle, erythrose 4-phosphate reacts with dihydroxyacetone phosphate to form sedoheptulose 1,7-bisphosphate in a reaction catalyzed by aldolase.

Chapter 21

Page 646 Phosphorylase, debranching enzyme, phosphoglucomutase, and glucose 6-phosphatase.

Page 647 Free glucose must be phosphorylated at the expense of one ATP, whereas glucose 6-phosphate derived from glycogen is formed by phosphorolytic cleavage, sparing one molecule of ATP. Thus, the net yield of ATP when glycogen-derived glucose is processed to pyruvate is three molecules of ATP compared with two molecules of ATP from free glucose.

Page 649 Glucose is an allosteric inhibitor of phosphorylase a. Hence, crystals grown in its presence are in the T state. The addition of glucose 1-phosphate, a substrate, causes a conformational change in the enzyme toward the R state which is sufficiently large that the crystal shatters.

Page 660 In the liver, glucagon stimulates the cAMP-dependent pathway that activates protein kinase A. Epinephrine binds to a 7TM α-adrenergic receptor in the liver plasma membrane, which activates phospholipase C and the phosphoinositide cascade. This activation causes calcium ions to be released from the endoplasmic reticulum, which bind to calmodulin, and further stimulates phosphorylase kinase and glycogen breakdown.

Chapter 22

Page 668 Two times each. The triacylglycerol will be hydrolyzed in the intestinal lumen, resynthesized in the intestinal cell, hydrolyzed again in the blood, and resynthesized again in the adipocyte.

Page 671 Fat mobilization in adipocytes is activated by phosphorylation. Hence, overproduction of the cAMP-activated kinase will lead to an accelerated breakdown of triacylglycerols and a depletion of fat stores.

Page 674 Fatty acids cannot be transported into mitochondria for oxidation. The muscles cannot use fats as a fuel. Muscles could use glucose derived from glycogen. However, when glycogen stores are depleted, as after a fast, the effect of the deficiency is especially apparent.

Page 683 Eight molecules of acetyl CoA combine to form four molecules of acetoacetate for release into the blood, and so they do not contribute to the energy yield in the liver. However, the $FADH_2$ and NADH generated in the preparation of acetyl CoA can be processed by oxidative phosphorylation to yield ATP.

$$1.5\, ATP/FADH_2 \times 7 = 10.5\, ATP$$

$$2.5\, ATP/NADH \times 7 = 17.5\, ATP$$

The equivalent of 2 ATP were used to form palmitoyl CoA. Thus, 26 ATP were generated for use by the liver.

Page 687 HCO_3^- is attached to acetyl CoA to form malonyl CoA. When malonyl CoA condenses with acetyl CoA to form the four-carbon ketoacyl CoA, the HCO_3^- is lost as CO_2.

Page 696 Phosphofructokinase controls the flux down the glycolytic pathway which generates ATP or building blocks for biosynthesis, depending on the tissue. The presence of citrate in the cytoplasm means that the cell favors fatty acid synthesis over degradation, indicating that those needs are met, and there is no need to catabolize glucose.

Chapter 23

Page 703 Dietary amino acids are absorbed in the upper part of the small intestine (the duodenum). The primary fate of both dietary and cellular-derived amino acids is as building blocks for cellular proteins and other nitrogen-containing molecules.

Page 707 Exposure of such a region suggests that a component of a multiprotein complex has failed to form properly or that one component has been synthesized in excess. This exposure leads to rapid degradation and the restoration of appropriate stoichiometries.

Page 709 Aminotransferases transfer the α-amino group to α-ketoglutarate to form glutamate. Glutamate is oxidatively deaminated to form an ammonium ion in the liver by glutamate dehydrogenase.

Page 713 Ser and Thr. The presence of a hydroxyl group on the β-carbon.

Page 718 Carbamoyl phosphate and aspartate.

Page 728 As shown in Figure 23.27, alanine, a gluconeogenic amino acid, is released during the metabolism of tryptophan to acetyl CoA and acetoacetyl CoA.

Chapter 24

Page 736 Cholecystokinin (CCK) and glucagon-like peptide 1 (GLP-1) are short-term and anorexigenic. Ghrelin is short-term and orexigenic.

Page 741 Phosphorylation of dietary glucose after it enters the liver; gluconeogenesis; glycogen breakdown.

Page 742 When glycogen stores are filled, the excess carbohydrates are metabolized to acetyl CoA, which is then converted into fats. Human beings cannot convert fats into carbohydrates, but they can certainly convert carbohydrates into fats.

Page 743 Such a mutation would increase the phosphorylation of the insulin receptor and IRS in muscle and would improve insulin sensitivity. Indeed, PTP1B is an attractive therapeutic target for type 2 diabetes.

Page 745 The inability of muscle mitochondria to process all of the fatty acids leads to excessive levels of diacylglycerol (DAG) and ceramide in the muscle cytoplasm. DAG is a second messenger that activates protein kinase C, which phosphorylates IRS, reducing its ability to propagate the insulin signal. Ceramide inhibits Akt.

Page 746 In the absence of insulin, lipid mobilization will take place to an extent that it overwhelms the ability of the liver to convert the lipids into ketone bodies. The excess is re-esterified and released into the blood.

Page 753 A role of the liver is to provide glucose for other tissues. In the liver, glycolysis is used not for energy production but for biosynthetic purposes. Consequently, in the presence of glucagon, liver glycolysis stops so the glucose can be released into the blood.

Page 754 Adipose tissue. Other tissues are the liver and muscle which store glycogen, and the skeletal muscle which technically acts as a storage of energy if necessary in the form of muscle proteins.

Page 757 Ethanol is oxidized to yield acetaldehyde by alcohol dehydrogenase, which is subsequently oxidized to acetate and acetaldehyde. Ethanol is also metabolized to acetaldehyde by the P450 enzymes, with the subsequent depletion of NADPH.

Chapter 25

Page 767 Glutamine synthetase: Glutamate + ATP + NH₃ → glutamine + ADP + P$_i$

Glutamate synthase: α-Ketoglutarate + glutamine + NADPH + H⁺ → 2 glutamate + NADP⁺

The sum of these reactions is:

NH₄⁺ + 2 α-ketoglutarate + NADPH + ATP → glutamate + NADP⁺ + ADP + P$_i$

Page 771 Both conversions are amidation reactions that require ATP hydrolysis. In the conversion of glutamate to glutamine, the nitrogen donor is ammonia and the side chain carboxylate is activated by phosphorylation. In the conversion of aspartate to asparagine in human beings, the

nitrogen donor is glutamine and the side chain carboxylate is activated by adenylation.

Page 775 (a) N^5, N^{10}-Methylenetetrahydrofolate; (b) N^5-methyltetrahydrofolate.

Page 788 Succinyl CoA is formed in the mitochondrial matrix as part of the citric acid cycle.

Chapter 26

Page 795 Aspartate and carbamoyl phosphate.

Page 799 Carbon dioxide, glycine, aspartate, glutamine, and N^{10}-formyltetrahydrofolate.

Page 805 UTP is first converted into UDP. Ribonucleotide reductase generates dUDP. DeoxyUDP is converted to dUMP. Thymidylate synthase generates TMP from dUMP. Monophosphate and diphosphate kinases subsequently form TTP.

Page 808 The committed step in pyrimidine biosynthesis is catalyzed by aspartate transcarbamoylase (ATCase). The committed step in purine biosynthesis is catalyzed by glutamine phosphoribosyl amidotransferase.

Chapter 27

Page 816 Phosphatidic acid includes two acyl groups that can vary in structure, leading to molecular weight differences.

Page 825 Oxidosqualene cyclase.

Page 832 Density, protein composition, lipid composition. See Table 27.1.

Page 839

Chapter 28

Page 847 The nucleotides used for DNA synthesis have the triphosphate attached to the 5'-hydroxyl group with free 3'-hydroxyl groups. Such nucleotides can be used only for 5'-to-3' DNA synthesis.

Page 848 DNA replication requires RNA primers. Without appropriate ribonucleotides, such primers cannot be synthesized.

Page 864 A hallmark of most cancer cells is prolific cell division, which requires DNA replication. If the telomerase were not activated, the chromosomes would shorten until they became nonfunctional, leading to cell death.

Page 868 DNA polymerase has proofreading activity, so it will immediately fix any bases that were wrongly incorporated, much like using a backspace key to correct typos. Any errors that are not fixed during replication can be repaired afterwards by mismatch repair. Damaged DNA is fixed via base-excision repair or nucleotide-excision repair. More significant damage, such as double-stranded breaks, can be repaired via recombination pathways. There are so many systems because the integrity of the DNA sequence is critically important, and it must be duplicated with fidelity. Each system provides added levels of accuracy.

Chapter 29

Page 881 RNA is mostly single-stranded, although it can form short duplexed sections, such as hairpins. The sugar in RNA is ribose, which has an extra hydroxyl group on the 2′ carbon. This extra 2′-OH gives RNA its catalytic characteristics.

Page 885 RNA polymerases are similar to DNA polymerases in that they catalyze the formation of a phosphodiester bond in an active site that contains metal ions. Unlike DNA polymerases, RNA polymerases do not require an existing 3′-OH for this reaction and can therefore build RNA strands de novo. RNA polymerases have a greater error rate than DNA polymerases.

Page 898 The σ subunit as well as the eukaryotic transcription factors play a significant role in recruiting the RNA polymerase (bacterial or eukaryotic, respectively) to the promoter in order to facilitate the initiation of transcription at the appropriate site(s) and not randomly within the genome.

Page 912 All three reaction mechanisms involve two sequential transesterification reactions. Group I and group II self-splicing introns do not require the spliceosome. In group 1 splicing, the first reaction is an intermolecular nucleophilic attack (the 3′-OH group from a G). In group II and spliceosomal splicing, the first reaction is an intramolecular nucleophilic attack by the 2′-OH group from an adenylate within the intron.

Chapter 30

Page 920 (i) Each is a single chain. (ii) They contain unusual bases. (iii) Approximately half of the bases are base-paired to form double helices. (iv) The 5′ end is phosphorylated and is usually pG. (v) The amino acid is attached to the hydroxyl group of the A residue of the CCA sequence at the 3′ end of the tRNA. (vi) The anticodon is located in a loop near the center of the tRNA sequence. (vii) The molecules are l-shaped.

Page 921 The first two bases in a codon form Watson–Crick base pairs with their corresponding bases in the anticodon. These interactions are checked for fidelity by bases of the 16S rRNA. The interaction of the third base with its corresponding base in the anticodon is not inspected for accuracy, and so some variation is tolerated. This leads to the redundancy of the genetic code.

Page 933 The ribosome is a ribozyme because the peptide bond is catalyzed by rRNA. The peptidyl transferase center in the large subunit consists entirely of rRNA, with no ribosomal proteins near that active site. In addition, a ribosome stripped of its protein components is still able to catalyze peptide bond formation.

Page 935 During translation initiation in bacteria, the IF-2-GTP-initiator-tRNA complex forms part of the 30S initiation complex. Hydrolysis of GTP occurs once the 50S subunit joins this complex and will result in the release of IF2. During translation elongation, the EF-Tu-GTP complex binds the aminoacyl-tRNA and delivers it to the ribosome. Hydrolysis of GTP will release the EF-Tu-GDP complex from the ribosome.

Page 942 The SRP binds to the signal sequence and inhibits further translation. The SRP ushers the inhibited ribosome to the ER, where it interacts with the SRP receptor (SR). The SRP–SR complex binds the translocon and simultaneously hydrolyzes GTP. On GTP hydrolysis, SRP and SR dissociate from each other and from the ribosome. Protein synthesis resumes, and the nascent protein is channeled through the translocon.

Chapter 31

Page 956 High concentrations of lactose cause the inducer (allolactose) to prevent the repressor from binding to the operator DNA, allowing for transcription of the structural genes in the operon. High concentrations of glucose, however, result in low levels of cAMP, meaning there would be no positive regulation (activation) of the *lac* operon by CAP. Under these conditions, the lack of positive regulation results in little to no expression of the structural genes, despite the negative control (repression) having been removed. In the absence of β-galactosidase, glucose is metabolized first. Once glucose levels have been depleted, cAMP levels are high, and cAMP–CAP complexes stimulate the transcription of the operon.

Page 959 Alpha helices fit into the major groove of DNA and are able to make specific contacts with the bases presented there.

Page 962 Eukaryotic DNA must be compacted and organized into chromatin in order to fit into the nucleus. This compaction makes it difficult for RNA polymerase and other regulatory proteins to gain access to the DNA. Chromatin structure must be relaxed in order for transcription to be activated. Gene expression can also be silenced by chromatin remodeling.

Page 967 The various aspects of chromatin modification include DNA methylation, histone methylation, histone acetylation, and histone deacetylation. DNA methylation, histone methylation, and histone deacetylation typically silence gene expression by making the DNA less accessible to transcription factors. Histone acetylation activates transcription by relaxing the chromatin structure and recruiting transcription factors and other components of the transcriptional machinery.

Page 972 During transcription, regulatory proteins determine the amount of mRNA produced from certain genes. Posttranscriptionally, regulatory mechanisms can affect the stability and translatability of mRNAs. Once a protein is made, its functionality can be regulated via covalent modifications.

Chapter 32

Page 982 If an enzyme assay used in a high-throughput screen relies on a fluorescence readout, compounds that are themselves fluorescent may interfere with the assay and appear to be hits (i.e., false positives).

Page 986 In a manner reminiscent of affinity chromatography, the compound identified as a phenotypic screening hit could be attached to a solid support, and the extract of the type of cell used in the initial screen could be passed over the resin. Proteins that bind to the resin could be potential targets for the compound. If numerous proteins are identified, then this experiment could be repeated on other screening hits, and one could look for proteins that were common to both.

Page 996 A monoclonal antibody with higher affinity for FcRn would be more likely to be salvaged and retained in the body, thus prolonging its half-life.

Chapter 1

1. The hydrogen-bond donors are the NH and NH_2 groups. The hydrogen-bond acceptors are the carbonyl oxygen atoms and those ring nitrogen atoms that are not bonded to hydrogen or to deoxyribose.

2.

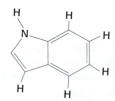

3. (a) Ionic interactions; (b) van der Waals interactions.
4. a and b.
5. $\Delta S_{system} = -661\ J\ mol^{-1}\ K^{-1}\ (-158\ kcal\ mol^{-1}\ K^{-1})$
   $\Delta S_{surroundings} = +842\ J\ mol^{-1}\ K^{-1}\ (+201\ cal\ mol^{-1}\ K^{-1})$
6. 0.815
7. 55.6 M
8. 447; 0.00050
9. 6.0; 5.5
10. 7.8
11. 1.86
12. (a) 0.45; (b) 1.00; (c) 1.62
13.

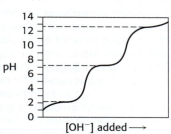

14. (a) No; (b) Around pH 2.
15. 90 mM acetic acid; 160 mM sodium acetate, 0.18 moles acetic acid; 0.32 moles sodium acetate; 10.81 g acetic acid; 26.25 g sodium acetate.
16. (a) MOPS; (b) MES
17. Buffer, because the sodium ions will shield the electrostatic repulsion between the phosphate groups.
18. $+1.45\ kJ\ mol^{-1}\ (+0.35\ kcal\ mol^{-1}); +57.9\ kJ\ mol^{-1}$ $(+13.8\ kcal\ mol^{-1})$
19. The hydrophobic effect
20. 100%. No, it would not be a useful clock because it changes too slowly.
21. 7.9%

Chapter 2

1. (A) Proline, Pro, P; (B) tyrosine, Tyr, Y; (C) leucine, Leu, L; (D) lysine, Lys, K.
2. (a) D; (b) none; (c) A, C; (d) B.
3. Ser, Glu, Tyr, Thr.
4. (a) Alanine-glycine-serine; (b) alanine; (c) 2.

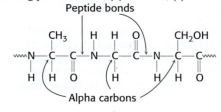

5. At pH 5.5, the net charge is +1:

At pH 7.5, the net charge is 0:

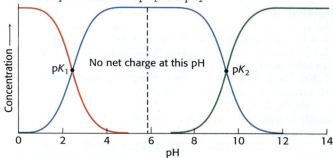

6.

7. The solution would have no net charge at the pH in the midpoint of the zwitterionic species curve (blue), effectively at the midpoint between pK_1 and pK_2:

8. (a) pH 1.5; (b) pH 3.0 and pH 4.9
9. (a) Each strand is 35 kDa and hence has about 318 residues (the mean residue mass is 110 daltons). Because the rise per residue in an α helix is 1.5 Å, the length is 477 Å. More precisely, for an α-helical coiled coil, the rise per residue is 1.46 Å; so the length is 464 Å. (b) Eighteen residues in each strand (40 minus 4 divided by 2) are in a β-sheet conformation. Because the rise per residue is 3.5 Å, the length is 63 Å.
10. The methyl group attached to the β-carbon atom of isoleucine sterically interferes with α-helix formation. In leucine, this methyl group is attached to the γ-carbon atom, which is farther from the main chain and hence does not interfere.
11. The native conformation of insulin is not the thermodynamically most stable form, because it contains two separate chains linked by disulfide bonds. Insulin is formed from proinsulin, a single-chain precursor, that is cleaved to form insulin, a 51-residue molecule, after the disulfide bonds have formed.
12. A 12 kDa single-stranded α helix contains about 109 residues:

$$12\ kDa\ \times \frac{1000\ Da}{1\ kDa} \times \frac{1\ residue}{110\ Da} = 109\ residues$$

A single turn of the helix (each repeating unit) extends about 5.4 Å. There are 3.6 residues per turn. Therefore, the rise per residue is 1.5 Å. The length of the segment is 164 Å

$$109 \text{ residues} \times \frac{1.5 \text{ Å}}{1 \text{ residue}} = 164 \text{ Å}$$

13. (a). This peptide has a large number of aromatic and beta-branched amino acids, which tend to be disfavored in α helices and accommodated well by β sheets.

14. (a) a; (b) a, b, c.

15. (a) All; (b) β sheets; (c) α helices; (d) reverse turn; (e) α helices.

16. Recall that hemoglobin exists as a tetramer while myoglobin is a monomer. Consequently, the hydrophobic residues on the surface of hemoglobin subunits are probably involved in van der Waals interactions with similar regions on the other subunits, and will be shielded from the aqueous environment by this interaction.

17. Using the Henderson–Hasselbalch equation, we find the ratio of alanine-COOH to alanine-COO$^-$ at pH 7 to be 10^{-4}. The ratio of alanine-NH$_2$ to alanine-NH$_3^+$, determined in the same fashion, is 10^{-1}. Thus, the ratio of neutral alanine to the zwitterionic species is $10^{-4} \times 10^{-1} = 10^{-5}$.

18. ELVISISLIVINGINLASVEGAS

19. No, Pro—X would have the characteristics of any other peptide bond. The steric hindrance in X—Pro arises because the R group of Pro is bonded to the amino group. Hence, in X—Pro, the proline R group is near the R group of X, which would not be the case in Pro—X.

20. A, c; B, e; C, d; D, a; E, b.

21. Disulfide pairings are not positioned correctly unless weak bonding interactions are present.

Chapter 3

1. ½

2. 1.44×10^{-2} grams; 1.44×10^{-1} grams; 128

3. No, because oxygen binding requires partial electron transfer to oxygen.

4. No. The Bohr effect requires changes in quaternary structure and myoglobin in monomeric.

5. Less tight, because n-hexyl isocyanide is likely too large to easily fit into the oxygen-binding cavity.

6. lowered; raised; raised

7. (a) 1.0, expected for a monomeric protein; (b) 2.8. This suggests substantial cooperativity but is less that 4.0 expected for full cooperativity

8. (a) 61.7%; (b) Her hemoglobin P$_{50}$ increases, promoting oxygen delivery to tissues.

9. The two copper ions move together upon binding oxygen, and this motion could promote cooperativity.

10. (c) because of its negative charge.

11. Uracil is found in RNA, characteristic of potential viral infection.

12. Uracil to avoid the potential innate immune response.

13. Each IgG has two arms and each hemoglobin has two copies of each potential binding site. These could come together to form a large insoluble polymer. This is not possible for myoglobin because it is monomeric.

14. (a) LLQATYSAV

15. 0.17; 0.67, 0.95, 0.995

16. (a) R$_{free}$ = R = 1.0 − 0.14 − 0.71 = 0.15R$_{total}$.
(b) Estradiol$_{free}$ = 1 nM − 0.14R$_{total}$.
(c) BPA$_{free}$ = 1 μM − 0.71R$_{total}$.
(d) Estradiol K_d = (R$_{free}$ × Estradiol$_{free}$)/[R-Estradiol] = (0.15 R$_{total}$) × (1 nM − 0.14 R$_{total}$)/0.14R$_{total}$ = (0.15/0.14) × (1 nM − 0.14R$_{total}$) ≈ 1 nM assuming R$_{total}$ is much less than 1 nM, BPA K_d = (R$_{free}$ × BPA$_{free}$)/[R-BPA] = (0.15 R$_{total}$) × (1 μM − 0.71R$_{total}$)/0.71R$_{total}$ = (0.15/0.71) × (1 μM − 0.71R$_{total}$) ≈ 200 nM.

17. Rate = k_1 [L] = 5×10^{-3} s^{-1} when [L] = 10 μM so that $k_1 = 5 \times 10^8$ M^{-1}s^{-1}.
$k_{-1} = K_d \times k_1 = (20 \text{ nM}) \times (5 \times 10^8$ M^{-1}s$^{-1}) = 10$ s^{-1}.

18. 0.5 nM

19. 10^{-6} M

20. Essentially all (100%) of the irons will be bound to carbon dioxide.

Chapter 4

1. (a), (c), (d)

2.

Purification procedure	Total protein (mg)	Total activity (units)	Specific activity (units mg^{-1})	Purification level	Yield (%)
Crude extract	20,000	4,000,000	200	1	100
(NH$_4$)$_2$SO$_4$ precipitation	5000	3,000,000	600	3	75
DEAE-cellulose chromatography	1500	1,000,000	667	3.3	25
Gel-filtration chromatography	500	750,000	1500	7.5	19
Affinity chromatography	45	675,000	15,000	75	17

3. (a) 0.4 mM; (b) 0.02 mM

4. A fluorescence-labeled derivative of a bacterial degradation product (e.g., a formylmethionyl peptide) would bind to cells containing the receptor of interest.

5. (a) Trypsin cleaves after arginine (R) and lysine (K), generating AVGWR, VK, and S. Because they differ in size, these products could be separated by molecular exclusion chromatography. (b) Chymotrypsin, which cleaves after large aliphatic or aromatic R groups, generates two peptides of equal size (AVGW) and (RVKS). Separation based on size would not be effective. The peptide RVKS has two positive charges (R and K), whereas the other peptide is neutral. Therefore, the two products could be separated by ion-exchange chromatography.

6. (a) Ion exchange chromatography will remove Proteins A and D, which have a substantially lower isoelectric point; then gel filtration chromatography will remove Protein C, which has a lower molecular weight. (b) If Protein B carries a His tag, a single affinity chromatography step with an immobilized nickel(II) column may be sufficient to isolate the desired protein from the others.

7. Mass spectrometry is highly sensitive and capable of detecting the mass difference between a protein and its deuterated counterpart. Fragmentation techniques can be used to identify the amino acids that retained the isotope label. Alternatively, NMR spectroscopy can be used to detect the isotopically labeled atoms because the deuteron and the proton have very different nuclear-spin properties.

8. (c), (e), (b), (a), (d).

9. The difference between the predicted and the observed masses for this fragment equals 28.0, exactly the mass shift that would be expected in a formylated peptide. This peptide is likely formylated at its amino terminus and corresponds to the most N-terminal fragment of the protein.

10. (b)

11. (a) Three fragments; the smallest is Ser-Arg. (b) One "fragment"; this peptide does not contain any chymotrypsin cleavage sites (no Phe, Tyr, Trp, Leu, or Met). The sequence would remain Gly-Ser-Lys-Ala-Gly-Arg-Ser-Arg.

12. (a) d and e; (b) b.

13. TRMKPAMLICWMPD

14. Met-Val-Lys-Tyr-Thr-Trp-Ala-Phe-Gly-Arg

15. (a) (A): Native PAGE; (B): Non-reducing SDS-PAGE; (C) Reducing SDS-PAGE
(b) Protein X consists of two 50 kDa polypeptide chains, resulting in a total mass of 100 kDa. According to the nonreducing SDS-PAGE gel, intact protein X has a total molecular weight of 100 kDa. Therefore, the molecular weights of the protein X polypeptide chains must add up to 100 kDa. The only peptides indicated by the reducing SDS-PAGE gel that combine to yield a 100-kDa protein are two 50-kDa chains. (c) Protein Y consists of two 60-kDa chains and one 30-kDa chain, for a total mass of 150 kDa. According to the nonreducing SDS-PAGE gel, the intact protein Y has a total molecular weight of 150 kDa. Therefore, the molecular weights of the protein Y polypeptides must add up to 150 kDa. Although three 50 kDa could make up a 150-kDa polypeptide, the reducing SDS-PAGE method shows bands of 30 kDa and 60 kDa. Therefore, protein Y must include chains of 30 kDa and 60 kDa. The association of two 60-kDa chains and one 30-kDa chain would yield the 150-kDa protein Y.

16. If the protein did not contain any disulfide bonds, then the electrophoretic mobility of the trypsin fragments would be the same before and after performic acid treatment: all the fragments would lie along the diagonal of the paper. If one disulfide bond were present, the disulfide-linked trypsin fragments would run as a single peak in the first direction, and then as two separate peaks after performic acid treatment. The result would be two peaks appearing off the diagonal:

No disulfides present

First direction of electrophoresis →

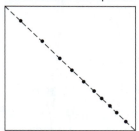

One disulfide present

First direction of electrophoresis →

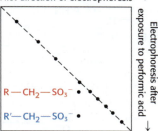

$R\!-\!CH_2\!-\!SO_3^-$

$R'\!-\!CH_2\!-\!SO_3^-$

Electrophoresis after exposure to performic acid ↓

These fragments could then be isolated from the chromatography paper and analyzed by mass spectrometry to determine their amino acid composition and thus identify the cysteines participating in the disulfide bond.

17. From top to bottom: (d), (e), (f), (a), (c), (b)

18. From low pH to high pH: (d), (e), (f), (c), (b), (a)

Chapter 5

1. A cofactor.

2. d, e, and f.

3. The energy required to reach the transition state (the activation energy) is returned when the transition state proceeds to product.

4. The product is more stable than the substrate in graph A; so ΔG is negative and the reaction is exergonic. In graph B, the product has more energy than the substrate has; ΔG is positive, meaning that the reaction is endergonic.

5. (a) $K = \dfrac{[P]}{[S]} = \dfrac{k_F}{k_R} = \dfrac{10^{-4}}{10^{-6}} = 100$. Using equation 5 in the text, $\Delta G^{\circ\prime} = -11.42\ \text{kJ mol}^{-1}\ (-2.73\ \text{kcal mol}^{-1})$. (b) $k_F = 10^{-2}\ \text{s}^{-1}$ and $k_R = 10^{-4}\ \text{s}^{-1}$. The equilibrium constant and $\Delta G^{\circ\prime}$ values are the same for both the uncatalyzed and catalyzed reactions.

6. There would be no catalytic activity. If the enzyme–substrate complex is more stable than the enzyme–transition-state complex, the transition state would not form and catalysis would not take place.

7. (a) 0; (b) +28.53; (c) –22.84; (d) –11.42; (e) +5.69.

8. (a) $\Delta G^{\circ\prime} = -RT \ln K'_{eq}$
$+1.8 = -(1.98 \times 10^{-3}\ \text{kcal}^{-1}\ \text{K}^{-1}\ \text{mol}^{-1})(298\,\text{K})$
$\qquad (\ln[\text{G1P}]/[\text{G6P}])$
$-3.05 = \ln\,[\text{G1P}]/[\text{G6P}]$
$+3.05 = \ln\,[\text{G6P}]/[\text{G1P}]$
$K'^{-1}_{eq} = 21$ or $K'_{eq} = 4.8 \times 10^{-2}$

Because $[\text{G6P}]/[\text{G1P}] = 21$, there is 1 molecule of G1P for every 21 molecules of G6P. Because we started with 0.1 M, the [G1P] is 1/22 (0.1 M) = 0.0045 M and [G6P] must be 21/22 (0.1 M) or 0.096 M. Consequently, the reaction does not proceed as written to a significant extent. (b) Supply G6P at a high rate and remove G1P at a high rate by other reactions. In other words, make sure that the [G6P]/[G1P] is kept large.

9. $K_{eq} = 19$, $\Delta G^{\circ\prime} = -7.3\ \text{kJ mol}^{-1}\ (-1.77\ \text{kcal mol}^{-1})$

10. No, K_M is not equal to the dissociation constant and thus does not measure affinity because the numerator also contains k_2, the rate constant for the conversion of the enzyme substrate complex into enzyme and product. If, however, k_2 is much smaller than k_{-1}, $K_M \approx K_d$.

11. Competitive inhibition: 2, 3, 9; uncompetitive: 4, 5, 6; noncompetitive: 1, 7, 8.

12. When $[S] = 10\,K_M$, $V_0 = 0.91\,V_{max}$. When $[S] = 20\,K_M$, $V_0 = 0.95\,V_{max}$. So, any Michaelis–Menten curves showing that the enzyme actually attains V_{max} are pernicious lies.

13. (a) 31.1 µmol; (b) 0.05 µmol; (c) 622 s^{-1}, a midrange value for enzymes (Table 8.5).

14. (a) Yes, $K_M = 5.2 \times 10^{-6}$ M; (b) $V_{max} = 6.8 \times 10^{-10}$ mol minute^{-1}; (c) 337 s^{-1}.

15. (a) V_{max} is 8.9 µmol minute^{-1} K_M is 9.9×10^{-5} M, the same as without inhibitor. (b) Noncompetitive. (c) 2.5×10^{-5} M (d) $f_{ES} = 0.73$, in the presence or absence of this noncompetitive inhibitor.

16. (a) $V_0 = V_{max} - (V_0/[S])\,K_M$; (b) Slope = $-K_M$, y-intercept = V_{max}, x-intercept = V_{max}/K_M; (c) An Eadie Hofstee plot.

17. Sequential reactions are characterized by the formation of a ternary complex consisting of the enzyme and both substrates. Double-displacement reactions always require the formation of a temporarily substituted enzyme intermediate.

18. (a) K_M is a measure of affinity only if k_2 is rate limiting, which is the case here. Therefore, the lower K_M means higher affinity. The mutant enzyme has a higher affinity. (b) 50 µmol minute^{-1}. 10 mM is K_M, and K_M yields $\frac{1}{2}$ V_{max}. V_{max} is 100 µmol minute^{-1}, and so (c) Enzymes do not alter the equilibrium of the reaction.

19. If the total amount of enzyme $[E]_T$ is increased, V_{max} will increase, because $V_{max} = k_2[E]_T$. But $K_M = (k_{-1} + k_2)/k_1$; that is, it is independent of enzyme concentration. The middle graph describes this situation.

20. Enzyme 2. Despite the fact that enzyme 1 has a higher V_{max} than enzyme 2, enzyme 2 shows greater activity at the concentration of the substrate in the environment because enzyme 2 has a lower K_M for the substrate.

21.

Experimental condition	V_{max}	K_M
(a) Twice as much enzyme is used	Doubles	No change
(b) Half as much enzyme is used	Half as large	No change
(c) A competitive inhibitor is present	No change	Increases
(d) An uncompetitive inhibitor is present	Decreases	Decreases
(e) A pure noncompetitive is present	Decreases	No change

22. The first step will be the rate-limiting step. Enzyme E_B is operating at $1/2\ V_{max}$, and E_C is very close to maximum velocity, whereas the K_M for enzyme E_A is greater than the substrate concentration. E_A would be operating at approximately $10^{-2}\ V_{max}$.

23. At substrate concentrations near the K_M, the enzyme is elastic, meaning that it displays significant catalysis yet is still sensitive to changes in substrate concentration.

24. (a) This piece of information is necessary for determining the correct dosage of succinylcholine to administer. (b) The duration of the paralysis depends on the ability of the serum cholinesterase to clear the drug. If there were one-eighth the amount of enzyme activity, paralysis could last eight times as long. (c) K_M is the concentration needed by the enzyme to reach $1/2\ V_{max}$. Consequently, for a given concentration of substrate, the reaction catalyzed by the enzyme with the lower K_M will have the higher rate. The patient with the higher K_M will clear the drug at a much lower rate.

Chapter 6

1. For the amide substrate, the formation of the acyl-enzyme intermediate is slower than the hydrolysis of the acyl-enzyme intermediate, and so no burst is observed. A burst is observed for ester substrates, because the formation of the acyl-enzyme intermediate is faster, leading to the observed burst.

2. The histidine residue in the substrate can substitute to some extent for the missing histidine residue of the catalytic triad of the mutant enzyme.

3. No. The catalytic triad works as a unit. After this unit has been made ineffective by the mutation of histidine to alanine, the further mutation of serine to alanine should have only a small effect.

4. The substitution corresponds to one of the key differences between trypsin and chymotrypsin, and so trypsin-like specificity (cleavage after lysine and arginine) might be predicted. In fact, additional changes are required to affect this specificity change.

5. Imidazole is apparently small enough to reach the active site of carbonic anhydrase and compensate for the missing histidine. Buffers with large molecular components cannot do so, and the effects of the mutation are more evident.

6. No, because the enzyme would destroy the host DNA before protective methylation could take place.

7. EDTA will bind to Zn^{2+} and remove the ion, which is required for enzyme activity, from the enzyme.

8. (a) The aldehyde reacts with the active-site serine.
(b) A hemiacetal is formed.

9. lysine; trypsin; aspartate

10. $k_{cat} = 60,000\ s^{-1}$. The reaction is expected to be slower by a factor of 10 because the rate depends on the pK_a of the zinc-bound water.

11. EDTA binds the Mg^{2+} necessary for the enzymatic reaction to proceed.

12. If the aspartate is mutated, the protease is inactive and the virus will not be viable.

13. Water substitutes for the hydroxyl group of serine 236 in mediating proton transfer from the attacking water and the γ-phosphoryl group.

14. For subtilisin, the catalytic power is approximately $30\ s^{-1}/10^{-8}\ s^{-1} = 3 \times 10^9$. For carbonic anhydrase (at pH 7), the catalytic power is approximately $500,000\ s^{-1}/0.15\ s^{-1} = 3.3 \times 10^6$. Subtilisin is the more powerful enzyme by this criterion.

15. The triple mutant still catalyzes the reaction by a factor of approximately 1000-fold compared to the uncatalyzed reaction. This could be due to the enzyme binding the substrate and holding it in a conformation that is susceptible to attack by water.

16. (a) Cysteine protease. The same as Figure 6.8 except that cysteine replaces serine and no aspartate is present.
(b) Aspartyl protease.

(c) Metalloprotease.

17. From a biological perspective, highly specific enzymes allow control of which reactions occur at appreciable rates. From a chemical perspective, enzymes generally require a tight fit between enzyme and substrate, likely leading to high specificity. An enzyme with high catalytic activity but low specificity would require the ability to bind substrates with substantial parts of the substrate not in close contact with the enzyme.

18. Myosin will bind and hydrolyze ATP but will release the products ADP and P_i very slowly, limiting the overall rate of hydrolysis.

Chapter 7

1. a, b, and d are true.
2. a, c, and d.
3. Homotropic effectors are the substrates of allosteric enzymes. Heterotropic effectors are the regulators of allosteric enzymes. Homotropic effectors account for the sigmoidal

nature of the velocity versus substrate concentration curve, whereas heterotropic effectors alter the K_M of the curve. Ultimately, both types of effectors work by altering the T/R ratio.
4. CTP is formed by the addition of an amino group to UTP. Evidence indicates the UTP is also capable of inhibiting ATCase in the presence of CTP.
5. Kinases: (a), (b), (f); Phosphatases: (c), (g); Neither: (d), (h); Both: (e).
6. (b), (d), (f).
7. (a) CTP and UTP; (b) PALA; (c) ATP.
8. (c), (d), (b), (a).
9. (a) Decrease; (b) Decrease; (c) Decrease; (d) The activity of enzyme 1 increases.
10. Replacing methionine with leucine would be a good choice. Leucine is resistant to oxidation and has nearly the same volume and degree of hydrophobicity as methionine has.
11. (b).
12. (a) Enteropeptidase and trypsin; (b) All four zymogens are directly activated by trypsin!
13. (b), (d), (f), (a), (e), (c).
14. The simple sequential model predicts that the fraction of catalytic chains in the R state, f_R, is equal to the fraction containing bound substrate, Y. The concerted model, in contrast, predicts that f_R increases more rapidly than Y as the substrate concentration is increased. The change in f_R leads to the change in Y on addition of substrate, as predicted by the concerted model.
15. (a) e_1, e_4, e_5, and e_7; (b) The product of e_1, B, has two possible fates, conversion to C or B′. Thus, the product is not committed to the synthesis of G. The first enzyme that catalyzes a reaction that commits the pathway to the synthesis of G is enzyme e_4. This enzyme is most likely the allosteric enzyme.
16. (a) When crowded, the control group displays gregarious behavior. (b) Inhibition of PKA appears to prevent gregarious behavior, while inhibition of PKG has no effect on the behavior. (c) The effect of PKG inhibition was investigated to establish that the effect seen with PKA inhibition is specific and not just due to the inhibition of any kinase. (d) PKA plays a role in altering behavior of insects.
17. The binding of PALA switches ATCase from the T to the R state because PALA acts as a substrate analog. PALA disrupts the equilibrium in favor of R. Thus, the limited amount of substrate can more readily bind the R state, and the ATCase activity initially increases with increasing PALA. Once the inhibitor occupies more and more active sites, the real substrate will be unable to bind the enzyme. When PALA occupies all of the active sites, the catalytic activity falls to zero.
18. (a) 100. If the binding of the substrate is 100 times tighter for the R form, then c = 1/100 (recall that tighter binding means a lower dissociation constant). Thus, $\dfrac{[T_1]}{[R_1]} = c^1 L = \dfrac{1}{100} L$
and the change in [R]/[T] will 100-fold. Note that the change in the [R]/[T] ratio on binding one substrate molecule must be the same as the ratio of the substrate affinities of the two forms.
(b) 10. With four substrates bound, we can use the equation to determine:
$$\dfrac{[T_4]}{[R_4]} = c^4 L = \left(\dfrac{1}{100}\right)^4 L = 10^{-8} \cdot 10^7 = \dfrac{1}{10}$$
Hence, the [R]/[T] ratio in the fully liganded molecule is 10.

Chapter 8
1. Nucleosides: (c), (f); Nucleotides: (a), (e), (g); Both: (b), (d), (h)
2. (a) Guanosine 5′-diphosphate (GDP); (b) Deoxyuridine; (c) Adenosine

3. T is always equal to A, and so these two nucleotides constitute 40% of the bases. G is always equal to C, and so the remaining 60% must be 30% G and 30% C.
4. (a) TTGATC; (b) GTTCGA; (c) ACGCGT; (d) ATGGTA.
5. DNA: (b), (c), (f); RNA: (a), (d), (e), (g)
6. (a) purine, ribose; (b) pyrimidine, ribose; (c) purine, deoxyribose
7. B-DNA: (b), (f); A-DNA: (a), (c), (i); Z-DNA: (d), (g); All three: (e), (h)
8. (a) Tritiated thymine or tritiated thymidine. (b) dATP, dGTP, dCTP, and TTP labeled with ^{32}P in the innermost (α) phosphorus atom.
9. Molecules in parts (a) and (b) would not lead to DNA synthesis, because they lack a 3′-OH group (a primer). The molecule in part (d) has a free 3′-OH group at one end of each strand but no template strand beyond. Only the molecule in part (c) would lead to DNA synthesis.
10. (a) 50%; (b) 0%
11. The diameter of DNA is 20 Å, and 1 Å = 0.1 nm, so the diameter is 2 nm. Because 1 μm = 10^3 nm, the length is 2×10^4 nm. Thus, the axial ratio is 1×10^4.
12. There are 2 μm missing in the deletion mutant, or 20,000 Å of DNA. Given a base separation of 3.4 Å in B-DNA, this corresponds to 20000/3.4, or about 5.88×10^3 base pairs.
13. (a) 2, 4, 8; (b) 1, 6, 10; (c) 3, 5, 7, 9.
14. (a) 5′-UAACGGUACGAU-3′
(b) Leu-Pro-Ser-Asp-Trp-Met
(c) Poly(Leu-Leu-Thr-Tyr)
15. (a) Deoxyribonucleoside triphosphates versus ribonucleoside triphosphates. (b) 5′ → 3′ for both. (c) Semiconserved for DNA polymerase I; conserved for RNA polymerase. (d) DNA polymerase I needs a primer, whereas RNA polymerase does not.
16. Three nucleotides encode an amino acid; the code is nonoverlapping; the code has no punctuation; the code exhibits directionality; the code is degenerate.
17. tRNA: (c), (e); mRNA: (a), (d); rRNA: (b)
18. Prokaryotic promoters: (a), (d); Eukaryotic promoters: (c), (e); Both: (b)
19. It shows that the genetic code and the biochemical means of interpreting the code are common to even very distantly related life forms. It also testifies to the unity of life: that all life arose from a common ancestor.
20. (a) 3; (b) 6; (c) 2; (d) 5; (e) 7; (f) 1; (g) 4.

Chapter 9
1.

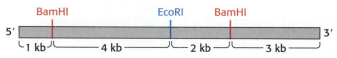

2. A mutation in person B has altered one of the alleles for gene X, leaving the other intact. The fact that the mutated allele is smaller suggests that a deletion has occurred in one copy of the gene. The one functioning copy is transcribed and translated and apparently produces enough protein to render the person asymptomatic.

Person C has only the smaller version of the gene. This gene is neither transcribed (negative northern blot) nor translated (negative western blot).

Person D has a normal-size copy of the gene but no corresponding RNA or protein. There may be a mutation in the promoter region of the gene that prevents transcription.

Person E has a normal-size copy of the gene that is transcribed, but no protein is made, which suggests that a mutation prevents translation. There are a number of possible explanations, including a mutation that introduced a premature stop codon in the mRNA.

Person F has a normal amount of protein but still displays the metabolic problem. This finding suggests that the mutation affects the activity of the protein—for instance, a mutation that compromises the active site of enzyme Y.

3. Chongqing: residue 2, L → R, CTG → CGG
   Karachi: residue 5, A → P, GCC → CCC
   Swan River: residue 6, D → G, GAC → GGC

4. Careful comparison of the sequences reveals that there is a 7-bp region of complementarity at the 3′ ends of these two primers:

5′-GGATCGATG**CTCGCGA**-3′
          3′-**GAGCGCT**GGGCTAGGA-5′

In a PCR experiment, these primers would likely anneal to one another, preventing their interaction with the template DNA. During DNA synthesis by the polymerase, each primer would act as a template for the other primer, leading to the amplification of a 25-bp sequence corresponding to the overlapped primers.

5. The encoded protein contains four repeats of a specific sequence.

6. (b), (d), (a), (c)

7. EcoRI and HindIII. Cleavage with PstI would disrupt the ampicillin resistance gene in pSAP, while cleavage with EcoRI and SmaI would remove the origin of replication from pSAP. Based on the information given, the other remaining restriction enzymes do not have cleavage sites on the fragment.

8. On the basis of the comparative genome map shown in Figure 9.29, the region of greatest overlap with human chromosome 20 can be found on mouse chromosome 2.

9. The codon(s) for each amino acid can be used to determine the number of possible nucleotide sequences that encode each peptide sequence (Table 8.5):

Ala–Met–Ser–Leu–Pro–Trp:

$4 \times 1 \times 6 \times 6 \times 4 \times 1 = 576$ total sequences

Gly–Trp–Asp–Met–His–Lys:

$4 \times 1 \times 2 \times 1 \times 2 \times 2 = 32$ total sequences

Cys–Val–Trp–Asn–Lys–Ile:

$2 \times 4 \times 1 \times 2 \times 2 \times 3 = 96$ total sequences

Arg–Ser–Met–Leu–Gln–Asn:

$6 \times 6 \times 1 \times 6 \times 2 \times 2 = 864$ total sequences

The set of DNA sequences encoding the peptide Gly-Trp-Asp-Met-His-Lys would be most ideal for probe design because it encompasses only 32 total oligonucleotides.

10. In this sequence, the GTG codon encodes valine. To change this to an alanine codon, one should use the primer sequence: ATG–CGC–GAA–CTG–G**C**G–AAC–TAA

11. Since the N could be any one of the four bases, the presence of the NGG trinucleotide would occur with a frequency of $\dfrac{4}{4} \times \dfrac{1}{4} \times \dfrac{1}{4}$, or once every 16 bases, on average.

12. (a) In order to confirm the success of a knockdown experiment, one should use a method that can rapidly detect the loss of mRNA transcript. The most convenient method would be qPCR (real-time PCR) to compare the transcript levels in a knockdown study versus a control sample. One could also look for changes in the expressed protein level by western blotting if an appropriate antibody is available. (b) Whole-genome complementary DNA (cDNA)

includes fully processed mRNA transcripts that have already had their introns spliced out. Therefore, cDNA represents the best source for an RNAi library because the mature mRNA is what researchers want to target and degrade in order to knock down the gene products.

13. 1: (b); 2: (c); 3: (a)

14. (a) Individual II; (b) Individual IV; (c) Cannot be determined with the current probe.

15. This particular person is heterozygous for this particular mutation: one allele is wild type, whereas the other carries a point mutation at this position. Both alleles are PCR amplified in this experiment, yielding the "dual peak" appearance on the sequencing chromatogram.

16. Although the two enzymes cleave the same recognition site, they each break different bonds within the 6-bp sequence. Cleavage by KpnI yields an overhang on the 3′ strand, whereas cleavage by Acc65I produces an overhang on the 5′ strand. These sticky ends do not overlap.

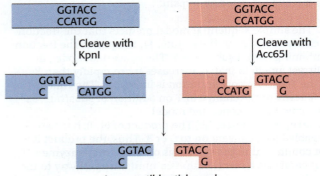

Incompatible sticky ends

17. The amount of dsDNA doubles with each cycle, so the amount of dsDNA present is equal to 2^x, where x is the number of cycles. Hence, after 1 cycle, there will be 2 molecules of dsDNA; after 3 cycles, 8 molecules of dsDNA; and after 30 cycles, ~1 billion molecules of dsDNA.

18. (c); (d); (e); (g).

19. (a): 2, 8; (b): 1, 6; (c): 3, 7; (d): 4, 5.

20. RNA-seq detects transcripts whether or not a gene is annotated in the genome, while in microarray analysis, researchers quantify expression levels based on binding to an array of oligonucleotides known to represent annotated genes, so transcripts of unannotated genes are not measured. RNA-seq has a wider dynamic range than microarray analysis, meaning it detects changes in high and low expression level genes. Another advantage of RNA-seq is its sensitivity to splice variants among mRNAs produced from the same gene. In fact, microarray analysis cannot identify splice variants unless each variant is represented and distinguishable in the oligonucleotides on the microarray.

Chapter 10

1. Paralogs are homologs that are present within one species and often differ in their detailed biochemical functions. Orthologs are homologs that are present within different species and have very similar or identical functions.

2. Orthologs: (b), (e), (f); Paralogs: (a), (c), (d)

3. There are 27 identities (highlighted in yellow) and two gaps, for a score of 220. The two sequences are approximately 27% identical. For an alignment of 27% identity over almost 100 residues, it is likely these two proteins are evolutionarily related and structurally similar.

```
WYLGKITRMDAEVLLKKPTVRDGHFLVTQCESSPGEF
WYFGKITRRESERLLLNPENPRGTFLVRESETTKGAY

SISVRFGDSVQ-----HFKVLRDQNGKYYLWAVK-FN
CLSVSDFDNAKGLNVKHYKIRKLDSGGFYITSRTQFS

SLNELVAYHRTASVSRTHTILLSDMNV
SSLQQLVAYYSKHADGLCHRLTNV
```

4. Protein shuffling involves randomly rearranging one of the protein sequences and comparing the alignment between the two sequences. Researchers repeat this process multiple times to produce a distribution plot that displays alignment scores of the randomly aligned sequences. If the alignment score for the two sequences is significantly higher than the randomly generated plot, then the sequences are likely significantly homologous.

5. Alignment score of sequences (1) and (2) is $6 \times 10 = 60$. Many answers are possible, depending on the randomly reordered sequence. A possible result is

Shuffled sequence: (2)LYEGADKAKT

Alignment: (1)ASNFLDKAGK
                        (2)LYEGADKAKT

Alignment score is $3 \times 10 = 30$

6. Replacement of cysteine, glycine, and proline never yields a positive score. Each of these residues exhibits features unlike those of its other 19 counterparts: cysteine is the only amino acid capable of forming disulfide bonds, glycine is the only amino acid without a side chain and is highly flexible, and proline is the only amino acid that is highly constrained through the bonding of its side chain to its amine nitrogen.

7. According to the Blosum-62 matrix, substitution of an aspartate residue with a glutamate residue yields a score of +2.

8. (1) Identity score = −25; Blosum score = 14; (2) Identity score = 15; Blosum score = 4.

9. (a) The proteins almost certainly diverged from a common ancestor. (b) The proteins almost certainly diverged from a common ancestor. (c) The proteins may have diverged from a common ancestor, but the sequence alignment may not provide supporting evidence. (d) The proteins may have diverged from a common ancestor, but the sequence alignment is unlikely to provide supporting evidence.

10. There are 4^{40}, or 1.2×10^{24}, different molecules. Each molecule has a mass of 2.2×10^{-20} g, because 1 mol of polymer has a mass of 330 g mol$^{-1} \times 40$, and there are 6.02×10^{23} molecules per mole. Therefore, 26.4 kg of RNA would be required.

11. #1, Solvin A; #2, Solvin D; #3, Solvin C; #4, Solvin B. Solvin B and Solvin E are most similar, having diverged from each other 100 million years ago.

12. Convergent evolution: (a), (g); Divergent evolution: (d), (e), (f); Both: (b), (c)

13. (a) Divergent evolution; (b) Convergent evolution; (c) Divergent evolution.

14. Protein A is clearly homologous to protein B, given 65% sequence identity, and so A and B are expected to have quite similar three-dimensional structures. Likewise, proteins B and C are clearly homologous, given 55% sequence identity, and so B and C are expected to have quite similar three-dimensional structures. Thus, proteins A and C are likely to have similar three-dimensional structures, even though they are only 15% identical in sequence.

15. After RNA molecules have been selected and reverse transcribed, PCR is performed to introduce additional mutations into these strands. The use of this error-prone, thermostable polymerase in the amplification step would enhance the efficiency of this random mutagenesis.

Chapter 11

1. a, b, and c
2. b, c, d, and e
3. d
4. Three amino acids can be linked by peptide bonds in only six different ways. However, three different monosaccharides can be linked in a plethora of ways. The monosaccharides can be linked in a linear or branched manner, with α or β linkages, with bonds between C-1 and C-3, between C-1 and C-4, between C-1 and C-6, and so forth. In fact, the three monosaccharides can form 12,288 different trisaccharides.
5. (a) 10; (b) 6; (c) 8; (d) 9; (e) 2; (f) 4; (g) 1; (h) 5; (i) 7; (j) 3.
6. (a) aldose-ketose; (b) epimers; (c) aldose-ketose; (d) anomers; (e) aldose-ketose; (f) epimers.
7. b and c
8. Erythrose: tetrose aldose; ribose: pentose aldose; glyceraldehyde: triose aldose; dihydroxyacetone: triose ketose; erythrulose: tetrose ketose; ribulose: pentose ketose; fructose: hexose ketose.
9. The proportion of the α anomer is 0.36, and that of the β anomer is 0.64.
10. (a) Not a reducing sugar; no open-chain forms are possible; (b) D-galactose, D-glucose, D-fructose; (c) D-galactose and sucrose (glucose + fructose).
11. (a) Each glycogen molecule has one reducing end, whereas the number of nonreducing ends is determined by the number of branches, or α-1,6 linkages. (b) Because the number of nonreducing ends greatly exceeds the number of reducing ends in a collection of glycogen molecules, all of the degradation and synthesis of glycogen takes place at the nonreducing ends, thus maximizing the rate of degradation and synthesis.
12. b, c, and e
13. (a) Cellulose; (b) chitin; (c) glycogen; (d) chitin, glycogen, cellulose, and starch; (e) glycogen and starch.
14. Simple glycoproteins are often secreted proteins and thus play a variety of roles. For example, the hormone EPO is a glycoprotein. Usually, the protein component constitutes the bulk of the glycoprotein by mass. In contrast, proteoglycans and mucoproteins are predominantly carbohydrates. Proteoglycans have glycosaminoglycans attached and play structural roles as in cartilage and the extracellular matrix. Mucoproteins often serve as lubricants and have multiple carbohydrates attached through an N-acetylgalactosamine moiety.
15. b and c
16. Asparagine, serine, and threonine.
17. A glycoprotein is a protein that is decorated with carbohydrates. A lectin is a protein that specifically recognizes carbohydrates. A lectin can also be a glycoprotein.
18. Each site either is or is not glycosylated, and so there are $2^6 = 64$ possible proteins.
19. If the carbohydrate specificity of the lectin is known, an affinity column with the appropriate carbohydrate attached could be prepared. The protein preparation containing the lectin of interest could be passed over the column. Indeed, the use of this method was how the glucose-binding lectin concanavalin A was purified.
20. (a) Aggrecan is heavily decorated with glycosaminoglycans. If glycosaminoglycans are released into the media, aggrecan must be undergoing degradation. (b) Another enzyme might be present that cleaves glycosaminoglycans from aggrecan without degrading

aggrecan. Other experiments not shown established that glycosaminoglycan release is an accurate measure of aggrecan destruction. (c) The control provides a baseline of "background" degradation inherent in the assay. (d) Aggrecan degradation is greatly enhanced. (e) Aggrecan degradation is reduced to the background system. (f) It is an in vitro system in which not all the factors contributing to cartilage stabilization in vivo are present.

Chapter 12

1. This fatty acid contains 18 carbons and two cis double bonds, which are located at carbons 9 and 12. Hence, its designation is 18:2(cis, cis-$\Delta^{9,12}$). Its common name is linoleate. Counting from the distal, or ω carbon, the two double bonds are located at positions ω-6 and ω-9.

2.

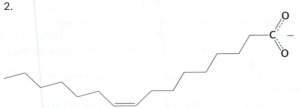

The common name for this fatty acid is palmitoleate.

3. The C_{16} alkyl chain is attached by an ether linkage. The C-2 carbon atom of glycerol has only an acetyl group attached by an ester linkage instead of a fatty acid, as is the case with most phospholipids.

4. (a) Phospholipids; (b) Sphingolipids, glycolipids, cholesterol; (c) Cholesterol; (d) Sphingomyelin, glycolipids; (e) Glycolipids; (f) Archaeal lipids.

5. 2×10^{-7} cm, 6×10^{-6} cm, and 2×10^{-4} cm.

6. Phospholipids and sphingolipids. Lipid molecules capable of aligning side by side to form a lipid bilayer must be amphipathic as well as have an appropriate geometry. Cholesterol molecules are almost completely hydrophobic and are therefore not capable of forming a lipid bilayer. Fatty acids, triacylglycerols, glycerophospholipids, and sphingomyelins are all amphipathic lipid molecules; however, the geometries of fatty acids and triacylglycerols are not favorable for aligning side by side to form a lipid bilayer.

7. Integral membrane proteins: (a); Peripheral membrane proteins: (b), (d); Lipid-anchored membrane protein: (c).

8. Membrane-spanning α helix: (c); Membrane-spanning β strands: (e); Lipid-anchored membrane protein: (a); Peripheral membrane protein: (b), (d).

9. The protein plotted in part c is a transmembrane protein from C. elegans. It spans the membrane with four α helices that are prominently displayed as hydrophobic peaks in the hydropathy plot. Interestingly, the protein plotted in part a also is a membrane protein, a porin. This protein is made primarily of β strands, which lack the prominent hydrophobic window of membrane helices. This example shows that, although hydropathy plots are useful, they are not infallible.

10. The protein may contain an α helix that passes through the hydrophobic core of the protein. This helix is likely to feature a stretch of hydrophobic amino acids similar to those observed in transmembrane helices.

11. The radius of this molecule is 3.1×10^{-7} cm, and its diffusion coefficient is 7.4×10^{-9} cm^{2}s^{-1}. The average distances traversed are 1.7×10^{-7} cm in 1 μs, 5.4×10^{-6} cm in 1 ms, and 1.7×10^{-4} cm in 1 s.

12. (a) The graph shows that, as temperature increases, the phospholipid bilayer becomes more fluid. T_m is the temperature of the transition from the predominantly less-fluid state to the predominantly more-fluid state. Cholesterol broadens the transition from the less-fluid to the more-fluid state. In essence, cholesterol makes membrane fluidity less sensitive to temperature changes. (b) This effect is important because the presence of cholesterol tends to stabilize membrane fluidity by preventing sharp transitions. Because protein function depends on the proper fluidity of the membrane, cholesterol maintains the proper environment for membrane-protein function.

13. S = 7.2 μm.

14. From lowest melting point to highest: d, c, b, e, a. The fatty acid with the lowest melting point is the one with the highest number of double bonds, α-linoleic acid (18:3). Palmitoleic acid (16:1) has a higher melting point than α-linoleic acid (18:3) because it only contains one double bond. Oleic acid (18:1) also contains one double bond, but it has a longer fatty acid chain than palmitoleic acid, and subsequently, a higher melting point. Fatty acids without double bonds have the highest melting points, and those that are longer have higher melting points. Thus, palmitic acid (16:0) has a higher melting point than oleic acid (18:1). Stearic acid (18:0) has the highest melting point.

15. Each of the 21 v-SNARE proteins could interact with each of 7 t-SNARE partners. Multiplication gives the total number of different interacting pairs: $7 \times 21 = 147$ different v-SNARE–t-SNARE pairs.

16. The membrane underwent a phase transition from a highly fluid to a nearly frozen state when the temperature was lowered. A carrier can shuttle ions across a membrane only when the bilayer is highly fluid. A channel, in contrast, allows ions to traverse its pore even when the bilayer is quite rigid.

17. (a) H_2O; (b) Indole; (c) H_2O. Permeability is most directly correlated with the solubility in a nonpolar solvent.

18. Hibernators selectively feed on plants that have a high proportion of polyunsaturated fatty acids with lower melting temperature.

19. Saturated phospholipids: c; Unsaturated phospholipids: a, d; Both: b.

20. (a) Prostaglandin H_2 synthase-1 recovers its activity immediately after removal of ibuprofen, suggesting that this inhibitor dissociates rapidly from the enzyme. In contrast, the enzyme remains significantly inhibited 30 minutes after removal of indomethacin, suggesting that this inhibitor dissociates slowly from its active site. (b) Aspirin covalently modifies prostaglandin H_2 synthase-1, indicating that it would dissociate very slowly (if at all). Hence, one would anticipate that very low activity would be evident in all conditions where inhibitor has been added (columns 2, 3, and 4).

Chapter 13

1. In simple diffusion, the substance in question can diffuse down its concentration gradient through the membrane. In facilitated diffusion, the substance is not lipophilic and cannot directly diffuse through the membrane. A channel or carrier is required to facilitate movement down the gradient.

2. Uniporters act as enzymes do; their transport cycles include large conformational changes, and only a few molecules interact with the protein per transport cycle. In contrast, channels, after having opened, provide a pore in the membrane through which many ions may pass. As such, channels mediate transport at a much higher rate than do uniporters.

3. Passive transport: (b), (d); Active transport: (a), (c).

4. For chloride, $z = -1$. At the concentrations given, the equilibrium potential for chloride is -97 mV.

5. (a) Using $\Delta G = RT \ln(c_2/c_1) + ZF\Delta V$, where the inside of the cell is side 2 and the outside is side 1, the free energy of sodium transport is:

$$\Delta G_{Na} = (8.315 \times 10^{-3} \, kJ \, mol^{-1} \, deg^{-1}) \times (310 \, deg) \times \ln(17/155)$$
$$+ (1) \times (96.5 \, kJ \, mol^{-1} \, V^{-1}) \times (-0.055 \, V)$$

$$\Delta G_{Na} = -11.0 \, kJ \, mol^{-1}$$

Because two Na^+ ions are transported for each glucose molecule, the maximum energy available for establishing the glucose gradient is -22.0 kJ mol^{-1}.

(b) For an uncharged solute like glucose, the minimum energy required to pump glucose against a concentration gradient is given by:

$$\Delta G_{gluc} = RT \ln([glucose]_{in}/[glucose]_{out})$$

Rearranging this equation, we obtain:

$$[glucose]_{in}/[glucose]_{out} = e^{\frac{\Delta G_{gluc}}{RT}} = e^{\frac{(22.0 \, kJ \, mol^{-1})}{(8.315 \times 10^{-3} \, kJ \, mol^{-1} \, K^{-1})(310 \, K)}} = 5100$$

6. At pH 7.4, $[H^+] = 3.98 \times 10^{-8}$ M, and at pH 2.0, $[H^+] = 0.01$ M. Using $\Delta G = RT \ln(c_2/c_1) + ZF\Delta V$:

$$\Delta G = (8.315 \times 10^{-3} \, kJ \, mol^{-1} \, deg^{-1}) \times (310 \, deg) \times \ln\left(\frac{0.01}{3.98 \times 10^{-8}}\right)$$

$$+ (1) \times (96.5 \, kJ \, mol^{-1} \, V^{-1}) \times (0.070 \, V)$$

$$\Delta G = 38.8 \, kJ \, mol^{-1}$$

7. ABC transporters: (b), (f); P-type ATPases: (c), (d); Both: (a), (e).

8. (b), (c), (f), (g).

9. After the addition of ATP and calcium, SERCA will pump Ca^{2+} ions into the vesicle. However, the accumulation of Ca^{2+} ions inside the vesicle will rapidly lead to the formation of an electrical gradient that cannot be overcome by ATP hydrolysis. The addition of calcimycin will allow the pumped Ca^{2+} ions to flow back out of the vesicle, dissipating the charge buildup, and enabling the pump to operate continuously.

10. The three types of carriers are symporters, antiporters, and uniporters. Symporters and antiporters can mediate secondary active transport.

11. Facilitated diffusion: (e), (i); Primary active transport: (b), (g); Secondary active transport: (c), (d), (h); Both primary and secondary active transport: (a), (f).

12. When a potassium ion passes through the channel, it remains hydrated for most of the distance. However, the channel narrows after the ion passes about 22 Å into the membrane, and K^+ must relinquish its hydration shell. This is energetically costly, but it is compensated by the interaction of K^+ with the carbonyl oxygen atoms in the interior of the channel at its narrow point. Since Na^+ is small compared to K^+, it cannot interact optimally with these oxygen atoms, and can thus not adequately compensate for the high energy cost of Na^+ desolvation.

13. A mutation that impairs the ability of the sodium channel to inactivate would prolong the duration of the depolarizing sodium current, thus lengthening the cardiac action potential.

14. Batrachotoxin blocks the transition from the open to the closed state.

15. (a) Only ASIC1a is inhibited by the toxin. (b) Yes; when the toxin was removed, the activity of the acid-sensing channel began to be restored. (c) 0.9 nM.

16. The blockage of ion channels inhibits action potentials, leading to loss of nervous function. Like tetrodotoxin, these toxin molecules are useful for isolating and specifically inhibiting particular ion channels.

17. Ligand-gated ion channels: (c), (e); Voltage-gated ion channels: (a); Both: (b), (d).

18. After repolarization, the ball domains of the ion channels engage the channel pore, rendering them inactive for a short period of time. During this time, the channels cannot be reopened until the ball domains disengage and the channel returns to the "closed" state.

19. (b), (d), (e), (c), (a), (f).

20. Compound A is TEA, Compound B is TTX.

Chapter 14

1. c, d, a, f, g, e, b

2. c, b, d, a, e, f

3. The rate of cAMP production will be 10^5 per second; in 10 seconds, 10^6 molecules of cAMP will be produced.

4. The mutated α subunit will always be in the GTP (that is, active) form; thus, it would stimulate its signaling pathway.

5. $G_{\alpha s}$ stimulates adenylate cyclase, leading to the generation of cAMP. This signal then leads to glucose mobilization (Chapter 21). If cAMP phosphodiesterase were inhibited, then cAMP levels would remain high even after the termination of the adrenaline signal, and glucose mobilization would continue.

6. (a)

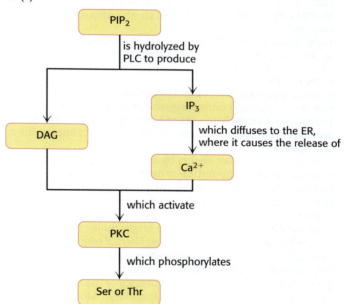

(b) IP_3, DAG, and Ca^{2+} are considered second messengers.

7. b, c, f

8. G protein-coupled receptors: c, d, g, i
Receptor tyrosine kinases: a, e, f
Both: f
Neither: b

9. Imatinib (Gleevec) is a small molecule that targets the active site of a tyrosine kinase inside the cell; cetuximab (Erbitux) is an antibody that binds to the extracellular domain of the receptor, preventing ligand binding.

10. a, d, e

11. In the reaction catalyzed by adenylate cyclase, the 3′-OH group undergoes a nucleophilic attack on the α-phosphorus atom attached to the 5′-OH group, leading to displacement of pyrophosphate. The reaction catalyzed by DNA polymerase is similar except that the 3′-OH group is on a different nucleotide.

12. GTPγS is not hydrolyzed by the G_α subunit, leading to prolonged activation.

13. (a) Use ATP labeled at the γ (gamma) phosphate; (b) Use ATP labeled at the α (alpha) phosphate.

14. (a) X ≈ 10^{-7} M; Y ≈ 5×10^{-6} M; Z ≈ 10^{-3} M. (b) Because much less X is required to fill half of the sites, X displays the highest affinity. (c) The binding affinity almost perfectly matches the ability to stimulate adenylate cyclase, suggesting that the hormone receptor complex leads to the stimulation of adenylate cyclase. (d) Try performing the experiment in the presence of antibodies to $G_{\alpha s}$.

15. The transgenic nematode would avoid the compound. The identity of the ligand is determined by the receptor, whereas the behavioral response is dictated by the neuron in which the receptor is expressed.

16. Only a mixture of compounds C_5–COOH and HOOC–C_7–COOH is predicted to yield this pattern.

17. Sound travels 0.15 m in 428 μs. The human hearing system is capable of sensing time differences of close to a microsecond, and so the difference in arrival times at the two ears is substantial. A system based on G proteins is unlikely to be able to reliably distinguish between signals arriving at the two ears, because G proteins typically respond in milliseconds.

18. (a) 380 (one for each receptor); (b) there are $(380 \times 379)/2! = 72,010$ combinations of two receptors; (c) $(380 \times 379 \times 378)/3! = 9,073,260$ combinations of three receptors.

19. The absorption of light converts 11-*cis*-retinal into all-*trans*-retinal.

Chapter 15

1. Anabolism is the set of biochemical reactions that build new molecules and ultimately new cells and generally use energy. Catabolism is the set of biochemical reactions that break down biomolecules and generally extract energy from fuel sources.

2. Phototrophs derive energy by transforming light into chemical energy. Chemotrophs derive energy from the oxidation of chemicals, either organic such as glucose and fatty acids, or inorganic sources like iron, sulfur, or ammonia.

3. Stage 1 is digestion, where the larger molecules in food are broken into smaller molecules. No ATP is generated. In stage 2, these smaller molecules are converted to key molecules of metabolism. A small amount of ATP is generated in this stage. Stage 3 consists of the complete oxidation of fuel molecules to CO_2 and H_2O. Most of the ATP is produced in stage 3.

4. Cellular movements and the performance of mechanical work; active transport; biosynthetic reactions.

5. Charge repulsion, resonance stabilization, increase in entropy, and stabilization by hydration.

6. Increasing the concentration of ATP or decreasing the concentration of cellular ADP or P_i (by rapid removal by other reactions, for instance) would make the reaction more exergonic. Likewise, altering the Mg^{2+} concentration could raise or lower the ΔG of the reaction (see problem 20).

7. Reactions in parts (a) and (c), to the left; reactions in parts (b) and (d), to the right.

8. (a) $\Delta G^{\circ\prime} = +31.4$ kJ mol^{-1} (+7.5 kcal mol^{-1}) and K'_{eq} = 3.06×10^{-6}; (b) 3.28×10^4.

9. $\Delta G^{\circ\prime} = +7.1$ kJ mol^{-1} (+1.7 kcal mol^{-1}). The equilibrium ratio is 17.5.

10. (a) Acetate + CoA + H^+ goes to acetyl CoA + H_2O, $\Delta G^{\circ\prime} = -31.4$ kJ mol^{-1} (–7.5 kcal mol^{-1}). ATP hydrolysis to AMP

and PP_i, $\Delta G^{\circ\prime} = -45.6$ kJ mol^{-1} (–10.9 kcal mol^{-1}). Overall reaction, $\Delta G^{\circ\prime} = -14.2$ kJ mol^{-1} (–3.4 kcal mol^{-1}). (b) With pyrophosphate hydrolysis, $\Delta G^{\circ\prime} = -33.4$ kJ mol^{-1} (–7.98 kcal mol^{-1}). Pyrophosphate hydrolysis makes the overall reaction even more exergonic.

11. Arginine phosphate in invertebrate muscle, like creatine phosphate in vertebrate muscle, serves as a reservoir of high-potential phosphoryl groups. Arginine phosphate maintains a high level of ATP in muscular exertion.

12. An ADP unit.

13. (a) The rationale behind creatine supplementation is that it would be converted into creatine phosphate and thus serve as a rapid means of replenishing ATP after muscle contraction. (b) If creatine supplementation is beneficial, it would affect activities that depend on short bursts of activity; any sustained activity would require ATP generation by fuel metabolism, which, as Figure 15.7 shows, requires more time.

14. Under standard conditions, $\Delta G^{\circ\prime} = -RT \ln$ [products]/[reactants]. Substituting +23.8 kJ mol^{-1} (+5.7 kcal mol^{-1})q for $\Delta G^{\circ\prime}$ and solving for [products]/[reactants] yields 9.9×10^{-5}. In other words, the forward reaction does not take place to a significant extent. Under intracellular conditions, ΔG is –1.3 kJ mol^{-1} (–0.3 kcal mol^{-1}). Using the equation $\Delta G = \Delta G^{\circ\prime} + RT \ln$ [products]/[reactants] and solving for [products]/[reactants] gives a ratio of 5.96×10^{-5}. Thus, a reaction that is endergonic under standard conditions can be converted into an exergonic reaction by maintaining the [products]/[reactants] ratio below the equilibrium value. This conversion is usually attained by using the products in another coupled reaction as soon as they are formed, thus artificially decreasing their concentration and driving the proceeding reaction forward (Le Chatelier's Principle).

15. Liver: –45.2 kJ mol^{-1} (–10.8 kcal mol^{-1}); muscle: –48.1 kJ mol^{-1} (–11.5 kcal mol^{-1}); brain: –48.5 kJ mol^{-1} (–11.6 kcal mol^{-1}). The ΔG is most negative in brain cells.

16. (a) Ethanol; (b) lactate; (c) succinate; (d) isocitrate; (e) malate.

17. NADH and $FADH_2$ are electron carriers for catabolism; NADPH is the carrier for anabolism.

18. Oxidation–reduction reactions; ligation reactions; isomerization reactions; group-transfer reactions; hydrolytic reactions; cleavage of bonds by means other than hydrolysis or oxidation.

19. Controlling the amount of enzymes; controlling enzyme activity; controlling the availability of substrates.

20. (a) As the Mg^{2+} concentration falls, the ΔG of hydrolysis becomes more negative. Note that pMg is a logarithmic plot, and so each number on the x-axis represents a 10-fold change in $[Mg^{2+}]$. (b) Mg^{2+} would bind to the phosphates of ATP and help to mitigate charge repulsion. As the $[Mg^{2+}]$ falls, charge stabilization of ATP would be less, leading to greater charge repulsion and ΔG of hydrolysis becoming more negative.

Chapter 16

1. a and e.

2. b and d.

3. (a) 4; (b) 3; (c) 1; (d) 6; (e) 8; (f) 2; (g) 10; (h) 9; (i) 7; (j) 5.

4. In both cases, the electron donor is glyceraldehyde 3-phosphate. In lactic acid fermentation, the electron acceptor is pyruvate, converting it into lactate. In ethanol fermentation, acetaldehyde is the electron acceptor, forming ethanol.

5. Glucokinase enables the liver to remove glucose from the blood when hexokinase is saturated, ensuring that glucose is captured for later use.

6. The GAP formed is immediately removed by subsequent reactions, resulting in the conversion of DHAP into GAP by the enzyme.

7. Glucose 6-phosphate must have other fates. Indeed, it can be converted into glycogen (Chapter 21) or processed to yield reducing power for biosynthesis (Chapter 20).

8. The conversion of glucose into glucose 6-phosphate by hexokinase; the conversion of fructose 6-phosphate into fructose 1,6-bisphosphate by phosphofructokinase; the formation of pyruvate from phosphoenolpyruvate by pyruvate kinase.

9. GLUT2 transports glucose only when the blood concentration of glucose is high, which is precisely the condition in which the β cells of the pancreas secrete insulin.

10. Without triose isomerase, only one of the two three-carbon molecules generated by aldolase could be used to generate ATP. Only two molecules of ATP would result from the metabolism of each glucose. But two molecules of ATP would still be required to form fructose 1,6-bisphosphate, the substrate for aldolase. The net yield of ATP would be zero, a yield incompatible with life.

11. The equilibrium concentrations of fructose 1,6-bisphosphate, dihydroxyacetone phosphate, and glyceraldehyde 3-phosphate are 7.8×10^{-4} M, 2.2×10^{-4} M, and 2.2×10^{-4} M, respectively.

12. (a) 2, 3, 6, 9; (b) 1, 4, 5, 7, 8.

13. Glucose is an important energy source for both tissues and is essentially the only energy source for the brain. Consequently, these tissues should never release glucose. Glucose release is prevented by the absence of glucose 6-phosphatase.

14. 6 NTP (4 ATP and 2 GTP); 2 NADH.

15. 50 pyruvate, 100 ATP, and 50 NADH.

16. (a) Increased; (b) increased; (c) increased; (d) decreased.

17. b and e.

18. This example illustrates the difference between the *stoichiometric* and the *catalytic* use of a molecule. If cells used NAD$^+$ stoichiometrically, a new molecule of NAD$^+$ would be required each time a molecule of lactate was produced. The synthesis of NAD$^+$ requires ATP. On the other hand, if the NAD$^+$ that is converted into NADH could be recycled and reused, a small amount of the molecule could regenerate a vast amount of lactate, which is the case in the cell. NAD$^+$ is regenerated by the oxidation of NADH and reused. NAD$^+$ is thus used catalytically.

19. Using the Michaelis–Menten equation to solve for [S] when $K_M = 50$ μM and $V_o = 0.9V_{max}$ shows that a substrate concentration of 0.45 mM yields 90% of V_{max}. Under normal conditions, the enzyme is essentially working at V_{max}.

20. (a) If both enzymes operated simultaneously, the net result would be simply:
$$ATP + H_2O \rightarrow ADP + P_i$$
The energy of ATP hydrolysis would be released as heat.
(b) Not really. For the cycle to generate heat, both enzymes must be functional at the same time in the same cell. (c) The species *B. terrestris* and *B. rufocinctus* might show some futile cycling because both enzymes are active to a substantial degree. (d) No. These results simply suggest that simultaneous activity of phosphofructokinase and fructose 1,6-bisphosphatase is unlikely to be employed to generate heat in the species shown.

Chapter 17

1. The pyruvate dehydrogenase complex catalyzes the following reaction, linking glycolysis and the citric acid cycle:
$$Pyruvate + CoA + NAD^+ \rightarrow acetyl\ CoA + NADH + H^+ + CO_2$$

2. a

3. b

4. Thiamine pyrophosphate plays a role in the decarboxylation of pyruvate. Lipoic acid (as lipoamide) transfers the acetyl group. Coenzyme A accepts the acetyl group from lipoic acid to form acetyl CoA. FAD accepts the electrons and hydrogen ions when reduced lipoic acid is oxidized. NAD$^+$ accepts electrons from FADH$_2$.

5. Catalytic coenzymes (TPP, lipoic acid, and FAD) are modified but regenerated in each reaction cycle. Thus, they can play a role in the processing of many molecules of pyruvate. Stoichiometric coenzymes (coenzyme A and NAD$^+$) are used in only one reaction because they are the components of products of the reaction.

6. $-41.0\ kJ\ mol^{-1}$ ($-9.8\ kcal\ mol^{-1}$)

7. (a) DCA inhibits pyruvate dehydrogenase kinase. (b) The fact that inhibiting the kinase results in more dehydrogenase activity suggests that there must be some residual activity that is being inhibited by the kinase.

8. (a) Enhanced kinase activity will result in a decrease in the activity of the PDH complex because phosphorylation by the kinase inhibits the complex. (b) Phosphatase activates the complex by removing a phosphate. If the phosphatase activity is diminished, the activity of the PDH complex also will decrease.

9. (a) 6; (b) 10; (c) 1; (d) 7; (e) 2; (f) 8; (g) 3; (h) 4; (i) 5; (j) 9.

10. (a) 5; (b) 7; (c) 1; (d) 10; (e) 2; (f) 4; (g) 9; (h) 3; (i) 8; (j) 6.

11. Succinate dehydrogenase is the only enzyme in the citric acid cycle that is embedded in the mitochondrial membrane, which makes it associated with the electron-transport chain.

12. (a) The steady-state concentrations of the products are low compared with those of the substrates. (b) The ratio of malate to oxaloacetate must be greater than 1.57×10^4 for oxaloacetate to be formed.

13.

$$Pyruvate + CoA + NAD^+ \xrightarrow{\text{Pyruvate dehydrogenase complex}} acetyl\ CoA + CO_2 + NADH$$

$$Pyruvate + CO_2 + ATP + H_2O \xrightarrow{\text{Pyruvate carboxylase}} oxaloacetate + ADP + P_i + H^+$$

$$Oxaloacetate + acetyl\ CoA + H_2O \xrightarrow{\text{Citrate synthase}} citrate + CoA + H^+$$

$$Citrate \xrightarrow{\text{Aconitase}} isocitrate$$

$$Isocitrate + NAD^+ \xrightarrow{\text{Isocitrate dehydrogenase}} \alpha\text{-ketoglutarate} + CO_2 + NADH$$

Net: $2\ Pyruvate + 2\ NAD^+ + ATP + H_2O \longrightarrow \alpha\text{-ketoglutarate} + CO_2 + ADP + P_i + 2\ NADH + 3\ H^+$

14. Succinate will increase in concentration, followed by α-ketoglutarate and the other intermediates "upstream" of the site of inhibition. Succinate has two methylene groups that are required for the dehydrogenation, whereas malonate has but one.

15. It enables organisms such as plants and bacteria to convert fats, through acetyl CoA, into glucose.

16. We cannot get the net conversion of fats into glucose, because the only means to get the carbon atoms from fats into oxaloacetate, the precursor of glucose, is through the citric acid cycle. However, although two carbon atoms enter the cycle as acetyl CoA, two carbon atoms are lost as CO_2 before oxaloacetate is formed. Thus, although some carbon atoms from fats may end up as carbon atoms in glucose, we cannot obtain a *net* synthesis of glucose from fats.

17. Acetyl CoA will inhibit the complex. Glucose metabolism to pyruvate will be slowed because acetyl CoA is being derived from an alternative source.

18. (a) Citrate is a symmetric molecule. Consequently, the investigators assumed that the two —CH_2COO^- groups in it would react identically. Thus, for every citrate molecule undergoing the reactions shown in path 1, they thought that another citrate molecule would react as shown in path 2. If so, then only *half* the label should have emerged in the CO_2.

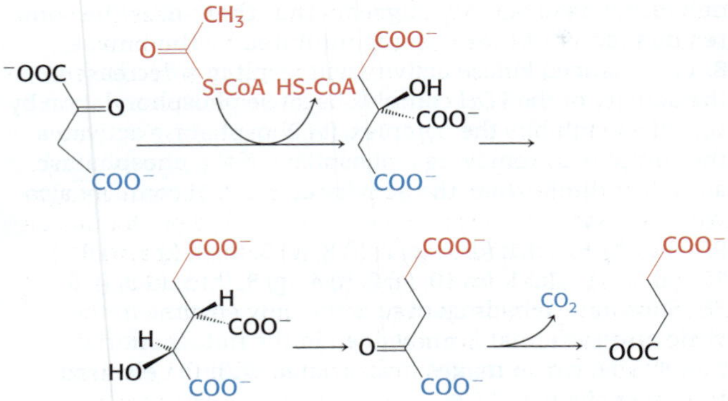

Path 1

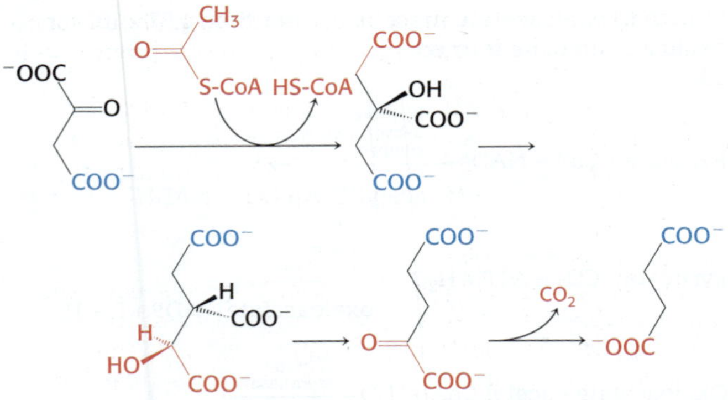

Path 2 (does not occur)

(b) Call one hydrogen atom A and the other B. Now suppose that an enzyme binds three groups of this substrate—X, Y, and H—at three complementary sites. The adjoining diagram shows X, Y, and H_A bound to three points on the enzyme. In contrast, X, Y, and H_B cannot be bound to this active site; two of these three groups can be bound, but not all three. Thus, H_A and H_B will have different fates.

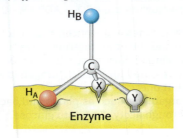

Sterically nonequivalent groups such as H_A and H_B will almost always be distinguished in enzymatic reactions. The essence of the differentiation of these groups is that the enzyme holds the substrate in a specific orientation. Attachment at three points, as depicted in the diagram, is a readily visualized way of achieving a particular orientation of the substrate, but it is not the only means of doing so.

19. (a) The complete oxidation of citrate requires 4.5 μmol of O_2 for every micromole of citrate.

$$C_6H_8O_7 + 4.5\ O_2 \rightarrow 6\ CO_2 + 4\ H_2O$$

Thus, 13.5 μmol of O_2 would be consumed by 3 μmol of citrate. (b) Citrate led to the consumption of far more O_2 than can be accounted for simply by the oxidation of citrate itself. Citrate thus facilitated O_2 consumption.

20. (a) The initial infection is unaffected by the absence of isocitrate lyase, but the absence of this enzyme inhibits the latent phase of the infection. (b) Yes. (c) A critic could say that, in the process of deleting the isocitrate lyase gene, some other gene was damaged, and it is the absence of this other gene that prevents latent infection. Reinserting the isocitrate lyase gene into the bacteria from which it had been removed renders the criticism less valid. (d) Isocitrate lyase enables the bacteria to synthesize carbohydrates that are necessary for survival, including carbohydrate components of the cell membrane.

Chapter 18

1. b

2. a

3. b

4. d

5. Biochemists use E_0', the value at pH 7, whereas chemists use E_0, the value in 1 M H^+. The prime denotes that pH 7 is the standard state.

6. The reduction potential of $FADH_2$ is less than that of NADH (Table 18.1). Consequently, when those electrons are passed along to oxygen, less energy is released. The consequence of the difference is that electron flow from $FADH_2$ to O_2 pumps fewer protons than do the electrons from NADH. The $\Delta G^{\circ\prime}$ for the reduction of oxygen by $FADH_2$ is –200 kJ mol^{-1} (–48 kcal mol^{-1}).

7. $\Delta G^{\circ\prime}$ is +67 kJ mol^{-1} (+16.1 kcal mol^{-1}) for oxidation by NAD^+ and –3.8 kJ mol^{-1} (–0.92 kcal mol^{-1}) for oxidation by FAD. The oxidation of succinate by NAD^+ is not thermodynamically feasible.

8. c, e, b, a, d.

9. (a) 4; (b) 3; (c) 1; (d) 5; (e) 2.

10. Rotenone: NADH, NADH-Q oxidoreductase will be reduced. The remainder will be oxidized. Antimycin A: NADH, NADH-Q oxidoreductase and coenzyme Q will be reduced. The remainder will be oxidized. Cyanide: All will be reduced.

11. (a) 12.5; (b) 14; (c) 32; (d) 13.5; (e) 30; (f) 16.

12. (a) It blocks electron transport and proton pumping at Complex IV. (b) It blocks electron transport and ATP synthesis by inhibiting the exchange of ATP and ADP across the inner mitochondrial membrane. (c) It blocks electron transport and proton pumping at Complex I. (d) It blocks ATP synthesis without inhibiting electron transport by dissipating the proton gradient. (e) It blocks electron transport and proton pumping at Complex IV. (f) It blocks electron transport and proton pumping at Complex III.

13. ATP synthase will from an arsenate (AsO_4^{3-}) anhydride with ADP. Such compounds are unstable and rapidly hydrolyzed. Under these conditions, no ATP synthesis can occur, but the synthase will continue catalyzing the useless reaction. Consequently, the electron-transport system will

keep running, most likely at a higher than normal rate. In essence, arsenate is acting as an uncoupler.

14. (a) No effect; mitochondria cannot metabolize glucose. (b) No effect; no fuel is present to power the synthesis of ATP. (c) [O_2] falls because citrate is a fuel and ATP can be formed from ADP and P_i. (d) Oxygen consumption stops because oligomycin inhibits ATP synthesis, which is coupled to the activity of the electron-transport chain. (e) No effect, for the reasons given in part (d). (f) [O_2] falls rapidly because the system is uncoupled and does not require ATP synthesis to lower the proton-motive force. (g) [O_2] falls, though at a lower rate. Rotenone inhibits Complex I, but the presence of succinate will enable electrons to enter at Complex II. (h) Oxygen consumption ceases because Complex IV is inhibited and the entire chain backs up.

15. (a) The P:O ratio is equal to the product of ($H^+/2\,e^-$) and (P/H^+). Note that the P:O ratio is identical with the P:2 e^- ratio. (b) 2.5 and 1.5, respectively.

16. This inhibitor (like antimycin A) blocks the reduction of cytochrome c_1 by QH_2.

17. If oxidative phosphorylation were uncoupled, no ATP could be produced. In a futile attempt to generate ATP, much fuel would be consumed. The danger lies in the dose. Too much uncoupling would lead to tissue damage in highly aerobic organs such as the brain and heart, which would have severe consequences for the organism as a whole. The energy that is normally transformed into ATP would be released as heat. To maintain body temperature, sweating might increase, although the very process of sweating itself depends on ATP.

18. In humans, the proton motive force is also used to drive the export of ATP from the matrix and the import of phosphate into the matrix. In bacteria, it drives flagellar motion.

19. (a) DNP is an uncoupler that prevents the use of the proton-motive force for ATP synthesis. Consequently, the rate of oxygen consumption rises (reflecting the rate of the electron transport chain) in a futile attempt to synthesize ATP. Because mitochondrial ATP synthesis is inhibited, the rate of glycolysis, as measured by ECAR, increases in an attempt meet the cells ATP needs. (b) Because glycolysis is now inhibited, no lactic acid will be produced, and the rate of extracellular acidification will fall. Because DNP is still present, oxygen consumption will still occur at a high rate. (c) A key step in glycolysis is the isomerization of glucose 6-phosphate to fructose 6-phosphate. 2-Deoxyglucose is incapable of undergoing the isomerization. (d) Rotenone inhibits electron flow through Complex I, the electron transport chain is inhibited, and oxygen consumption ceases.

20. (a) Succinate is oxidized by Complex II, and the electrons are used to establish a proton-motive force that powers ATP synthesis. (b) The ability to synthesize ATP is greatly reduced. (c) The goal was to measure ATP hydrolysis. If succinate had been added in the presence of ATP, no reaction would have taken place, because of respiratory control. (d) The mutation has little effect on the ability of the enzyme to catalyze the hydrolysis of ATP. (e) They suggest two things: (1) the mutation did not affect the catalytic site on the enzyme, because ATP synthase is still capable of catalyzing the reverse reaction, and (2) the mutation did not affect the amount of enzyme present, given that the controls and patients had similar amounts of activity.

Chapter 19

1. d
2. a
3. a
4. c

5. The oxygen we require is produced by photosynthesis. Moreover, most of the carbon atoms of which we are made, not just carbohydrates, enter the biosphere through the process of photosynthesis.

6. $2\ NADP^+ + 3\ ADP^{3-} + 3\ P_i^{2-} + H^+ \rightarrow O_2 + 2\ NADPH + 3\ ATP^{4-} + H_2O$

7. (a) PS I; (b) PS II; (c) PS II; (d) PS II; (e) PS I.

8. Photosystem II, in conjunction with the water-oxidizing complex, powers oxygen release. The reaction center of photosystem II absorbs light maximally at 680 nm. Oxygen consumption will be maximal when photosystems I and II are operating cooperatively. Oxygen will be efficiently generated when electrons from photosystem II fill the electron holes in photosystem I, which were generated when the reaction center of photosystem I was illuminated by light of 700 nm.

9. The light reactions take place on thylakoid membranes. Increasing the membrane surface increases the number of ATP- and NADPH-generating sites.

10. These complexes absorb more light than a reaction center alone can absorb by using additional pigments. The light-harvesting complexes funnel light energy to the reaction centers using resonance energy transfer.

11. $NADP^+$ is the acceptor. H_2O is the donor. Light energy.

12. Protons released by the oxidation of water; protons pumped into the lumen by the cytochrome bf complex; protons removed from the stroma by the reduction of $NADP^+$ and plastoquinone.

13. $\Delta E_0' = 10.11$ V, and $\Delta G^{\circ\prime} = -21.3$ kJ mol^{-1} (-5.1 kcal mol^{-1}).

14. Not at all. Chemicals other than water can donate electrons and protons as they do in anoxygenic photosynthesis on Earth today.

15. DCMU inhibits electron transfer in the link between photosystems II and I. O_2 can evolve in the presence of DCMU if an artificial electron acceptor such as ferricyanide can accept electrons from Q.

16. DCMU will have no effect, because it blocks photosystem II, and cyclic photophosphorylation uses photosystem I and the cytochrome bf complex.

17. (a) +120 kJ einstein^{-1}(+28.7 kcal einstein^{-1}); (b) 1.24 V; (c) One 1000-nm photon has the free-energy content of 2.4 molecules of ATP. A minimum of 0.42 photon is needed to drive the synthesis of a molecule of ATP.

18. The cristae, formed by the inner mitochondrial membrane, and the matrix.

19. In eukaryotes, both processes take place in specialized organelles. Both depend on high-energy electrons to generate ATP. In oxidative phosphorylation, the high-energy electrons originate in fuels and are extracted as reducing power in the form of NADH. In photosynthesis, the high-energy electrons are generated by light and are captured as reducing power in the form of NADPH. Both processes use redox reactions to generate a proton gradient, and the enzymes that convert the proton gradient into ATP are very similar in both processes. In both systems, electron transport takes place in membranes inside organelles.

Chapter 20

1. b.
2. c, d, a, b.
3. Calvin–Benson cycle: b, c, f, g, h. Krebs cycle: a, d, e, i, j.
4. c.
5. For CO_2, $k_{cat}/K_M = 3 \times 10^5$ M^{-1} s^{-1}; for O_2, $k_{cat}/K_M = 4 \times 10^3$ M^{-1} s^{-1}
6. a, c, e.
7. a.
8. [Ru5P]/[R5P] = 3.8; d.
9. Light reactions: a, e, f; Calvin–Benson cycle: b, **c**, g, h; Both: d.

10. Brought in and used in six turns of the Calvin–Benson cycle: a, d, f; Produced during six turns of the Calvin–Benson cycle that leave the cycle: b, e, g; Regenerated within the Calvin–Benson cycle during six turns: c, h.

11. Increases; NADPH, ribose; increase.

12. a, b, c.

13. a, b, c.

14. CO_2; a.

15. a.

16. a.

17. Falciparum malaria, glucose 6-phosphate dehydrogenase; sickle-cell anemia.

18. Fava beans can be dangerous for individuals with glucose 6-phosphate deficiency.

19. (a) 5 Glucose 6-phosphate + ATP → 6 ribose 5-phosphate + ADP + H^+. (b) Glucose 6-phosphate + 12 $NADP^+$ + 7 H_2O → 6 CO_2 + 12 NADPH + 12 H^+ + P_i.

20. The C_4 pathway allows the CO_2 concentration to increase at the site of carbon fixation. High concentrations of CO_2 inhibit the oxygenase reaction of Rubisco. This inhibition is important for tropical plants because the oxygenase activity increases more rapidly with temperature than does the carboxylase activity.

21. d.

Chapter 21

1. Step 1 is the release of glucose 1-phosphate from glycogen by glycogen phosphorylase. Step 2 is the formation of glucose 6-phosphate from glucose 1-phosphate, a reaction catalyzed by phosphoglucomutase. Step 3 is the remodeling of the glycogen by the transferase and the α-1,6-glucosidase.

2. b.

3. d.

4. As an unbranched polymer, α-amylose has only one nonreducing end. Therefore, only one glycogen phosphorylase molecule could degrade each α-amylose molecule. Because glycogen is highly branched, there are many nonreducing ends per molecule. Consequently, many phosphorylase molecules can release many glucose molecules per glycogen molecule, providing a rapid source of energy.

5. In muscle, the *b* form of phosphorylase is activated by AMP. In the liver, the *a* form is inhibited by glucose. The difference corresponds to the difference in the metabolic role of glycogen in each tissue. Muscle uses glycogen as a fuel for contraction, whereas the liver uses glycogen to maintain blood-glucose concentration.

6. Water is excluded from the active site to prevent hydrolysis. The entry of water could lead to the formation of glucose rather than glucose 1-phosphate. A site-specific mutagenesis experiment is revealing in this regard. In phosphorylase, Tyr 573 is hydrogen bonded to the 2'-OH group of a glucose residue. The ratio of glucose 1-phosphate to glucose product is 9,000:1 for the wild-type enzyme and 500:1 for the Phe 573 mutant. Model building suggests that a water molecule occupies the site normally filled by the phenolic OH group of tyrosine and occasionally attacks the carbocation intermediate to form glucose.

7. Epinephrine binds to its G-protein-coupled receptor. The resulting structural changes activate a G_α protein, which in turn activates adenylate cyclase. Adenylate cyclase synthesizes cAMP, which activates protein kinase A. Protein kinase A partly activates phosphorylase kinase, which phosphorylates and activates glycogen phosphorylase. The calcium released during muscle contraction also activates the phosphorylase kinase, leading to further stimulation of glycogen phosphorylase.

8. First, the signal-transduction pathway is shut down when the initiating hormone is no longer present. Second, the inherent GTPase activity of the G protein converts the bound GTP into inactive GDP. Third, phosphodiesterases convert cyclic AMP into AMP. Fourth, PP1 removes the phosphoryl group from glycogen phosphorylase, converting the enzyme into the usually inactive *b* form.

9. It prevents both from operating simultaneously, which would lead to a useless expenditure of energy.

10. Phosphoglucomutase, UDP-glucose pyrophosphorylase, pyrophosphatase, glycogenin, glycogen synthase, and branching enzyme.

11. The enzyme pyrophosphatase converts the pyrophosphate into two molecules of inorganic phosphate. This conversion renders the overall reaction irreversible.

$$\text{Glucose 1-phosphate} + \text{UTP} \rightleftharpoons \text{UDP-glucose} + PP_i$$

$$PP_i + H_2O \longrightarrow 2\ P_i$$

$$\overline{\text{Glucose 1-phosphate} + \text{UTP} + H_2O \longrightarrow \text{UDP-glucose} + 2\ P_i}$$

12. The presence of high concentrations of glucose 6-phosphate indicates that glucose is abundant and that it is not being used by glycolysis. Therefore, this valuable resource is saved by incorporation into glycogen.

13. $\text{Glycogen}_n + P_i$ → glycogen_{n-1} + glucose 6-phosphate

Glucose 6-phosphate → glucose 1-phosphate

UTP + glucose 1-phosphate → UDP-glucose + $2\ P_i$

Glycogen_{n-1} + UDP-glucose → glycogen_n + UDP

Sum: Glycogen_n + UTP → glycogen_n + UDP + P_i

14. Insulin binds to its receptor and activates the tyrosine kinase activity of the receptor, which in turn triggers a pathway that activates protein kinases. The kinases phosphorylate and inactivate glycogen synthase kinase. Protein phosphatase 1 then removes the phosphate from glycogen synthase and thereby activates the synthase.

15. Galactose + ATP + UTP + H_2O + glycogen_n ⟶ glycogen_{n+1} + ADP + UDP + $2\ P_i$ + H^+.

16. a, b, c, and d.

17. This disease can also be produced by a mutation in the gene that encodes the glucose 6-phosphate transporter. Recall that glucose 6-phosphate must be transported into the lumen of the endoplasmic reticulum to be hydrolyzed by phosphatase. Mutations in the other three essential proteins of this system can likewise lead to von Gierke disease.

18. (a) Apparently, the glutamate, with its negatively charged R group, can mimic to some extent the presence of a phosphoryl group on serine. That the stimulation is not as great is not surprising in that the carboxyl group is smaller and not as charged as the phosphate. (b) Substitution of aspartate would give some stimulation, but because it is smaller than the glutamate, the stimulation would be smaller.

19. (a) Glycogen was too large to enter the gel and, because analysis was by western blot with the use of an antibody specific to glycogenin, we would not expect to see background proteins. (b) α-Amylase degrades glycogen, releasing the protein glycogenin, which can be visualized by a western blot. (c) Glycogen phosphorylase, glycogen synthase, and protein phosphatase 1. These proteins might be visible if the gel were stained for protein, but a western analysis using only an anti-glycogenin antibody reveals the presence of glycogenin only.

20. (a) The smear was due to molecules of glycogenin with increasingly large amounts of glycogen attached to them. (b) In the absence of glucose in the medium, glycogen is

metabolized, resulting in a loss of the high-molecular-weight material. (c) Glycogen could have been resynthesized and added to the glycogenin when the cells were fed glucose again. (d) No difference between lanes 3 and 4 suggests that, by 1 hour, the glycogen molecules had attained maximum size in this cell line. Prolonged incubation does not apparently increase the amount of glycogen. (e) α-Amylase removes essentially all of the glycogen, and so only the glycogenin remains.

Chapter 22

1. Glycerol + 2 NAD$^+$ + P$_i$ + ADP $\longrightarrow$ pyruvate + ATP + H$_2$O + 2 NADH + H$^+$
Glycerol kinase and glycerol phosphate dehydrogenase
2. b, c, a, g, h, d, e, f.
3. The citric acid cycle. The reactions that take succinate to oxaloacetate, or the reverse, are similar to those of fatty acid metabolism (Section 17.3).
4. Palmitic acid yields 106 molecules of ATP. Palmitoleic acid has a double bond between carbons C-9 and C-10. When palmitoleic acid is processed in β-oxidation, one of the oxidation steps (to introduce a double bond before the addition of water) will not take place, because a double bond already exists. Thus, FADH$_2$ will not be generated, and palmitoleic acid will yield 1.5 fewer molecules of ATP than palmitic acid, for a total of 104.5 molecules of ATP.
5. To form stearoyl CoA requires the equivalent of 2 molecules of ATP.
Stearoyl CoA + 8 FAD + 8 NAD$^+$ + 8 CoA + 8 H$_2$O $\longrightarrow$
9 acetyl CoA + 8 FADH$_2$ + 8 NADH + 8 H$^+$

9 acetyl CoA at 10 ATP/acetyl CoA	+90 ATP
8 NADH at 2.5 ATP/NADH	+20 ATP
8 FADH$_2$ at 1.5 ATP/FADH$_2$	+12 ATP
Activation fee	−2.0
Total	**122 ATP**

6. Stearate + ATP + 13.5 H$_2$O + 8 FAD + 8 NAD$^+$ $\longrightarrow$
4.5 acetoacetate + 14.5 H$^+$ + 8 FADH$_2$ + 8 NADH + AMP + 2 P$_i$.
7. a, b, and d.
8. a, b, and c.
9. The formation of malonyl CoA from acetyl CoA by acetyl CoA carboxylase 1.
10. (a) Oxidation in mitochondria; synthesis in the cytoplasm. (b) Coenzyme A in oxidation; acyl carrier protein for synthesis. (c) FAD and NAD$^+$ in oxidation; NADPH for synthesis. (d) The l isomer of 3-hydroxyacyl CoA in oxidation; the d isomer in synthesis. (e) From carboxyl to methyl in oxidation; from methyl to carboxyl in synthesis. (f) The enzymes of fatty acid synthesis, but not those of oxidation, are organized in a multienzyme complex.
11. We will need six acetyl CoA units. One acetyl CoA unit will be used directly to become the two carbon atoms farthest from the acid end. The other five units must be converted into malonyl CoA. The synthesis of each malonyl CoA molecule costs 1 molecule of ATP; so 5 molecules of ATP will be required. Each round of elongation requires 2 molecules of NADPH, 1 molecule to reduce the keto group to an alcohol, and 1 molecule to reduce the double bond. As a result, 10 molecules of NADPH will be required. Therefore, 5 molecules of ATP and 10 molecules of NADPH are required to synthesize lauric acid.
12. e, b, d, a, c.
13. (a) False. Biotin is required for acetyl CoA carboxylase activity. (b) True. (c) False. ATP is required to synthesize

malonyl CoA. (d) True. (e) True. (f) False. Fatty acid synthase is a dimer. (g) True. (h) False. Acetyl CoA carboxylase is stimulated by citrate, which is cleaved to yield its substrate acetyl CoA.
14. All of the labeled carbon atoms will be retained. Because we need 8 acetyl CoA molecules and only 1 carbon atom is labeled in the acetyl group, we will have 8 labeled carbon atoms. The only acetyl CoA used directly will retain 3 tritium atoms. The 7 acetyl CoA molecules used to make malonyl CoA will lose 1 tritium atom on addition of the CO$_2$ and another one at the dehydration step. Each of the 7 malonyl CoA molecules will retain 1 tritium atom. Therefore, the total retained tritium is 10 atoms. The ratio of tritium to carbon is 1.25.
15. The mutant enzyme will be persistently active because it cannot be inhibited by phosphorylation. Fatty acid synthesis will be abnormally active. Such a mutation might lead to obesity.
16. (a) Palmitoleate; (b) linoleate; (c) linoleate; (d) oleate; (e) oleate; (f) linolenate.
17. Citrate works by facilitating, in cooperation with the protein MIG12, the formation of active filaments from inactive monomers. In essence, it increases the number of active sites available, or the concentration of enzyme. Consequently, its effect is visible as an increase in the value of V_{max}. Allosteric enzymes that alter their V_{max} values in response to regulators are sometimes called V-class enzymes. The more common type of allosteric enzyme, in which K_M is altered, comprises K-class enzymes. Palmitoyl CoA causes depolymerization and thus inactivation.
18. Regulation by insulin: a, b, e; regulation by glucagon or epinephrine: c, d, f.
19. The first reaction

$$\text{Oxaloacetate} + \text{NADH} + \text{H}^+ \rightleftharpoons \text{malate} + \text{NAD}^+$$

is catalyzed by cytoplasmic malate dehydrogenase. The next reaction is catalyzed by malic enzyme:

$$\text{Malate} + \text{NADP}^+ \longrightarrow \text{pyruvate} + \text{CO}_2 + \text{NADPH}$$

Finally, oxaloacetate is regenerated from pyruvate by pyruvate carboxylase.

$$\text{Pyruvate} + \text{CO}_2 + \text{ATP} + \text{H}_2\text{O} \longrightarrow$$
$$\text{oxaloacetate} + \text{ADP} + \text{P}_i + 2\,\text{H}^+$$

The sum of the reactions is

$$\text{NADP}^+ + \text{NADH} + \text{ATP} + \text{H}_2\text{O} \longrightarrow$$
$$\text{NADPH} + \text{NAD}^+ + \text{ADP} + \text{P}_i + \text{H}^+$$

20. (a) The V_{max} is decreased and the K_M is increased. V_{max}(wild type) = 13 nmol minute^{-1} mg^{-1}; K_M (wild type) = 45 μM; V_{max} (mutant) = 8.3 nmol minute^{-1} mg^{-1}; K_M (mutant) = 74 μM.
(b) Both the V_{max} and the K_M are decreased. V_{max} (wild type) = 41 nmol minute^{-1} mg^{-1}; K_M wild type = 104 μM; V_{max} (mutant) = 23 nmol minute^{-1} mg^{-1}; K_M (mutant) = 69 μM.
(c) The wild type is significantly more sensitive to malonyl CoA.
(d) With respect to carnitine, the mutant displays approximately 65% of the activity of the wild type; with respect to palmitoyl CoA, approximately 50% activity. On the other hand, 10 μM of malonyl CoA inhibits approximately 80% of the wild type but has essentially no effect on the mutant enzyme.
(e) The glutamate appears to play a more prominent role in regulation by malonyl CoA than in catalysis.
21. (a) Soraphen A inhibits fatty acid synthesis in a dose-dependent manner. (b) Fatty acid oxidation is increased in the presence of soraphen A. (c) Recall that acetyl CoA carboxylase 2 synthesizes malonyl CoA to inhibit the transport of fatty

acids into the mitochondria, thereby preventing fatty acid oxidation. Soraphen A apparently inhibits both forms of the carboxylase. (d) Phospholipid synthesis was inhibited in a dose-dependent manner. (e) Phospholipids are required for membrane synthesis. (f) Soraphen A inhibits cell proliferation, especially at higher concentrations.

Chapter 23

1. First, the ubiquitin-activating enzyme (E1) links ubiquitin to a sulfhydryl group on E1 itself. Next, the ubiquitin is transferred to a cysteine residue on the ubiquitin-conjugating enzyme (E2) by E2. The ubiquitin-protein ligase (E3), using the ubiquitinated E2 as a substrate, transfers the ubiquitin to the target protein.
2. (a) The ATPase activity of the 26S proteasome resides in the 19S subunit. The energy of ATP hydrolysis is used to unfold the substrate, which is too large to enter the catalytic barrel. ATP may also be required for translocation of the substrate into the barrel. (b) Substantiates the answer in part (a). Because they are small, the peptides do not need to be unfolded. Moreover, small peptides could probably enter all at once and not require translocation.
3. b
4. d
5. c
6. b
7. (a) Pyruvate; (b) oxaloacetate; (c) α-ketoglutarate; (d) α-ketoisocaproate; (e) phenylpyruvate; (f) hydroxyphenylpyruvate.
8. (a) Aspartate + α-ketoglutarate + GTP + ATP + $2 H_2O + NADH + H^+ \longrightarrow$ ½ glucose + glutamate + CO_2 + $ADP + GDP + NAD^+ + 2 P_i$.
The required coenzymes are pyridoxal phosphate in the transamination reaction and $NAD^+/NADH$ in the redox reactions.
(b) Aspartate + $CO_2 + NH_4^+ + 3 ATP + NAD^+ + 4 H_2O \longrightarrow$ oxaloacetate + urea + $2 ADP + 4 P_i + AMP + NADH + H^+$.
9. Aspartate (oxaloacetate), glutamate (α-ketoglutarate), alanine (pyruvate).
10. (a) 4; (b) 5; (c) 1; (d) 6; (e) 7; (f) 3; (g) 2.
11. A, arginine; B, citrulline; C, ornithine; D, argininosuccinate. The order of appearance: C, B, D, A.
12. $CO_2 + NH_4^+ + 3 ATP + NAD^+ + aspartate + 3 H_2O \longrightarrow$ urea + $2 ADP + 2 P_i + AMP + PP_i + NADH + H^+$ + oxaloacetate.
Four high-transfer-potential phosphoryl groups are spent. Note, however, that an NADH is generated if fumarate is converted into oxaloacetate. NADH can generate 2.5 ATP in the electron-transport chain. Taking these ATP into account, only 1.5 high-transfer-potential phosphoryl groups are spent.
13. The synthesis of fumarate by the urea cycle is important because it links the urea cycle and the citric acid cycle. Fumarate is hydrated to malate, which, in turn, is oxidized to oxaloacetate. Oxaloacetate has several possible fates: (1) transamination to aspartate, (2) conversion into glucose by the gluconeogenic pathway, (3) condensation with acetyl CoA to form citrate, or (4) conversion into pyruvate.
14. Aspartame, a dipeptide ester (L-aspartyl-L-phenylalanine methyl ester), is hydrolyzed to L-aspartate and L-phenylalanine. High levels of phenylalanine are harmful in phenylketonurics.
15. The carbon skeletons of ketogenic amino acids can be converted into ketone bodies or fatty acids. Only leucine and lysine are purely ketogenic. Glucogenic amino acids are those with carbon skeletons that can be converted into glucose.

16. The branched-chain amino acids leucine, isoleucine, valine, and threonine. The required enzyme is the branched-chain α-ketoacid dehydrogenase complex.
17. Pyruvate (glycolysis and gluconeogenesis), acetyl CoA (citric acid cycle and fatty acid synthesis), acetoacetyl CoA (ketone-body formation), α-ketoglutarate (citric acid cycle), succinyl CoA (citric acid cycle), fumarate (citric acid cycle), and oxaloacetate (citric acid cycle and gluconeogenesis).
18. Double-displacement. A substituted enzyme intermediate is formed.
19. (a) Depletion of glycogen stores. When they are gone, proteins must be degraded to meet the glucose needs of the brain. The resulting amino acids are deaminated, and the nitrogen atoms are excreted as urea. (b) The brain has adapted to the use of ketone bodies, which are derived from fatty acid catabolism. In other words, the brain is being powered by fatty acid breakdown. (c) When the glycogen and lipid stores are gone, the only available energy source is protein.
20. (a) Virtually no digestion in the absence of nucleotides. (b) Protein digestion is greatly stimulated by the presence of ATP. (c) AMP-PNP, a nonhydrolyzable analog of ATP, is no more effective than ADP. (d) The proteasome requires neither ATP nor PAN to digest small substrates. (e) PAN and ATP hydrolysis may be required to unfold the peptide and translocate it into the proteasome. (f) Although *Thermoplasma* PAN is not as effective with the other proteasomes, it nonetheless results in threefold to fourfold stimulation of digestion. (g) In light of the fact that the archaea and eukarya diverged several billion years ago, the fact that *Thermoplasma* PAN can stimulate the rabbit muscle proteasome suggests homology not only between the proteasomes, but also between PAN and the 19S subunit (most likely the ATPases) of the mammalian 26S proteasome.

Chapter 24

1. Normal circumstances: b; after fasting: d.
2. c.
3. Over the 40 years under consideration, our test subject will have consumed

40 years × 365 days year^{-1} × 8400 KJ(2000 kcal) day^{-1}
$= 1.2 \times 10^8$ kJ $(2.9 \times 10^7$ kcals) in 40 years.

Thus, over the 40-year span, our subject has ingested

$1.2 \times 108(2.9 \times 10^7$ kcal)/38 kJ(9 kcalg^{-1})
$= 3.2 \times 10^6$ g = 3200 kg of food,

which is equivalent to more than 6 tons of food!
4. 2.55 pounds = 25 kg = 25,000 g = total weight gain

40 years × 365 days year^{-1} = 14,600 days
25,000 g/14,600 days = 1.7 g day^{-1}

which is equivalent to an extra pat of butter per day. Her BMI is 26.5, and she would be considered overweight but not obese.
5. Adipose tissue is now known to be an active endocrine organ, secreting signal molecules called adipokines.
6. Leptin and insulin.
7. CCK produces a feeling of satiety and stimulates the secretion of digestive enzymes by the pancreas and the secretion of bile salts by the gallbladder. GLP-1 also produces a feeling of satiety; in addition, it potentiates the glucose-induced secretion of insulin by the β cells of the pancreas.
8. (a) 1, 2; (b) 6; (c) 3, 4; (d) 3, 4; (e) 3; (f) 6; (g) 5; (h) 5; (i) 5.
9. Type 1 diabetes is due to autoimmune destruction of the insulin-producing cells of the pancreas. Type 1 diabetes is

also called insulin-dependent diabetes because affected people require insulin to survive. Type 2 diabetes is characterized by insulin resistance. Insulin is produced, but the tissues that should respond to insulin, such as muscle, do not.

10. a, b, and c.

11. (a) A watt is equal to 1 joule (J) per second (0.239 calorie per second). Hence, 70 W is equivalent to 0.07 kJ s^{-1} ($0.017 \text{ kcal s}^{-1}$). (b) A watt is a current of 1 ampere (A) across a potential of 1 volt (V). For simplicity, let us assume that all the electron flow is from NADH to O_2 (a potential drop of 1.14 V). Hence, the current is 61.4 A, which corresponds to 3.86×10^{20} electrons per second ($1 \text{ A} = 1 \text{ coulomb s}^{-1} = 6.28 \times 10^{18} \text{ charge s}^{-1}$). (c) About 2.5 molecules of ATP are formed per molecule of NADH oxidized (two electrons). Hence, 1 molecule of ATP is formed per 0.8 electron transferred. A flow of 3.86×10^{20} electrons per second therefore leads to the generation of 4.83×10^{20} molecules of ATP per second, or 0.80 mmols^{-1}. (d) The molecular weight of ATP is 507 g/mol. The total body content of ATP of 50 g is equal to 0.099 mol. Hence, ATP turns over about once in 125 seconds when the body is at rest.

12. (a) The stoichiometry of the complete oxidation of glucose is

$$C_6H_{12}O_6 + 6\ O_2 \longrightarrow 6\ CO_2 + 6\ H_2O$$

and that of tripalmitoylglycerol is

$$C_{51}H_{98}O_2 + 72.5\ O_2 \longrightarrow 51\ CO_2 + 49\ H_2O$$

Hence, the RQ values are 1.0 and 0.703, respectively. (b) An RQ value reveals the relative use of carbohydrates and fats as fuels. The RQ of an amateur marathon runner typically decreases from 0.97 to 0.77 in the course of a race. The lowering of the RQ indicates the shift in fuel from carbohydrates to fat.

13. A high blood-glucose level triggers the secretion of insulin, which stimulates removal of glucose from the blood for the synthesis of glycogen and triacylglycerols. A high insulin level would impede the mobilization of fuel reserves during the marathon.

14. (a) Insulin inhibits lipid utilization. Insulin stimulates protein synthesis, but there are no amino acids in the children's diet. Moreover, insulin inhibits protein breakdown. Consequently, muscle proteins cannot be broken down and used for the synthesis of essential proteins. (b) Because proteins cannot be synthesized, blood osmolarity is too low. Consequently, fluid leaves the blood. An especially important protein for maintaining blood osmolarity is albumin.

15. During strenuous exercise, muscle converts glucose into pyruvate through glycolysis. Some of the pyruvate is processed by cellular respiration. However, some of it is converted into lactate and released into the blood. The liver takes up the lactate and converts it into glucose through gluconeogenesis. Muscle may process the carbon skeletons of branched-chain amino acids aerobically. The nitrogens of these amino acids are transferred to pyruvate to form alanine, which is released into the blood and taken up by the liver. After the transamination of the amino group to α-ketoglutarate, the resulting pyruvate is converted into glucose. Finally, muscle glycogen may be mobilized, and the released glucose can be used by muscle. The conversion of pyruvate to lactate allows muscle to function anaerobically. NAD^+ is regenerated when pyruvate is reduced to lactate, and so energy can continue to be extracted from glucose during strenuous exercise. The liver converts the lactate into glucose.

16. Fatty acids and glucose, respectively.

17. (a) The oxygen consumption at the end of exercise is used to replenish ATP and creatine phosphate and to oxidize any lactate produced. (b) After strenuous exercise, oxygen is used in oxidative phosphorylation to resynthesize ATP and creatine phosphate. The liver converts lactate released by the muscle into glucose. Blood must be circulated to return the body temperature to normal, and so the heart cannot return to its resting rate immediately. Hemoglobin must be reoxygenated to replace the oxygen used in exercise. The muscles that power breathing must continue working at the same time as the exercised muscles are returning to resting states. In essence, all the biochemical systems activated in intense exercise need increased oxygen to return to the resting state.

18. (a) The increase in all of the ratios is due to the NADH glut caused by the metabolism of alcohol. (b) The increased amounts of lactate and d-3-hydroxybutyrate are released into the blood, accounting for the acidosis. (c) Drinking on an empty stomach suggests that glycogen stores are low. Because of the NADH glut, gluconeogenesis cannot occur. Consequently, hypoglycemia results. A well-fed drinker will have glucose in the blood from the meal and thus will not experience hypoglycemia.

19. The precise cause of all of the symptoms is not firmly established, but a likely explanation depends on the centrality of oxaloacetate to metabolism. A lack of pyruvate carboxylase would reduce the amount of oxaloacetate. The lack of oxaloacetate would reduce the activity of the citric acid cycle and so ATP would be generated by lactic acid formation. If the concentration of oxaloacetate is low, aspartate cannot be formed and the urea cycle would be compromised. Oxaloacetate is also required to form citrate, which transports acetyl CoA to the cytoplasm for fatty acid synthesis. Finally, oxaloacetate is required for gluconeogenesis.

20. (a) Red blood cells always produce lactate, and fast-twitch or type II muscle fibers also produce a large amount of lactate. (b) At that point, the athlete is beginning to move into anaerobic exercise, in which most energy is produced by anaerobic glycolysis. (c) The lactate threshold is essentially the point at which the athlete switches from aerobic exercise, which can be done for extended periods, to anaerobic exercise, essentially sprinting, which can be done for only short periods. The idea is to race at the extreme of his or her aerobic capacity until the finish line is in sight and then to switch to anaerobic. (d) Training increases the number of blood vessels and of muscle mitochondria. Together, they increase the ability to process glucose aerobically. Consequently, a greater effort can be expended before the switch to anaerobic energy production.

Chapter 25

1. The reductase provides electrons with high reducing power, whereas the nitrogenase, which requires ATP hydrolysis, uses the electrons to reduce N_2 to NH_3.

2. a, b, d, f.

3. (a) The α-keto acid precursor to glutamate is α-ketoglutarate:

α-Ketoglutarate

(b) $NAD(P)^+$ and H_2O

4. 16

5. The reductase protein provides electrons with high reducing power. The nitrogenase uses these electrons to reduce N_2 to NH_3.

6. Oxaloacetate, pyruvate, ribose-5-phosphate, phosphoenol-pyruvate, erythrose-4-phosphate, α-ketoglutarate, and 3-phosphoglycerate.

7. Glucose + 2 ADP + 2 P_i + 2 NAD^+ + 2 glutamate $\longrightarrow$ 2 alanine + 2 α-ketoglutrate + 2 ATP + 2 NADH + 2 H_2O + 2 H^+

8. Alanine from pyruvate; aspartate from oxaloacetate; glutamate from α-ketoglutarate.

9. Serine, glycine, and cysteine.

10. (b). As an essential amino acid, methionine is not synthesized de novo in human beings, but it can, however, be regenerated via methylation of homocysteine to yield methionine. Homocysteine is also an intermediate in cysteine synthesis. Homocysteine condenses with serine to form cystathionine, which is deaminated and cleaved to yield α-ketobutyrate and cysteine.

11. The nitrogen atom shaded red is derived from glutamine. The carbon atom shaded blue is derived from serine.

12. Lysine cyclodeaminase converts L-lysine into the six-membered ring analog of proline, also referred to as L-homoproline or L-pipecolate:

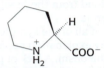

Pipecolate

13. Y could inhibit the C → D step, Z could inhibit the C → F step, and C could inhibit A → B. This scheme is an example of sequential feedback inhibition. Alternatively, Y could inhibit the C → D step, Z could inhibit the C → F step, and the A → B step would be inhibited only in the presence of both Y and Z. This scheme is called concerted feedback inhibition.

14. The rate of the A → B step in the presence of high levels of Y and Z would be 24 s^{-1} (0.6 × 0.4 × 100s^{-1}).

15. N_2 $\longrightarrow$ NH_4^+ $\longrightarrow$ glutamate $\longrightarrow$ serine $\longrightarrow$ glycine $\longrightarrow$ δ-aminolevulinate $\longrightarrow$ porphobilinogen $\longrightarrow$ heme

16. Acetate $\longrightarrow$ acetyl-CoA $\longrightarrow$ citrate $\longrightarrow$ isocitrate $\longrightarrow$ α-ketoglutarate $\longrightarrow$ succinyl-CoA

17. c, e, f.

18. S-Adenosylmethionine, tetrahydrofolate, and methylcobalamin.

19. In human beings, tyrosine is synthesized from phenylalanine by phenylalanine hydroxylase. If this enzyme is defective, the biosynthetic route to tyrosine is blocked, and tyrosine must be obtained from the diet. Tyrosine is sometimes said to be "conditionally essential" because its synthesis is dependent on the essential amino acid phenylalanine.

Chapter 26

1. In de novo synthesis, the nucleotides are synthesized from simpler precursor compounds, in essence from scratch. In salvage pathways, preformed bases are recovered and attached to riboses.

2.

The synthesis of carbamoyl phosphate requires 2 ATP	2 ATP
The formation of PRPP from ribose 5-phosphate yields an AMP*	2 ATP
The conversion of UMP to UTP requires 2 ATP	2 ATP
The conversion of UTP to CTP requires 1 ATP	1 ATP
Total	**7 ATP**

*Remember that AMP is the equivalent of 2 ATP because an ATP must be expended to generate ADP, the substrate for ATP synthesis.

3. Glutamine + aspartate + CO_2 + 2 ATP + NAD^+ $\longrightarrow$ orotate + 2 ADP + 2 P_i + glutamate + NADH + H^+.

4. (a, c, and d) PRPP; (b) carbamoyl phosphate.

5. PRPP and formylglycinamide ribonucleotide.

6. (a) Carboxyaminoimidazole ribonucleotide; (b) glycinamide ribonucleotide; (c) phosphoribosyl amine; (d) formylglycinamide ribonucleotide.

7. PRPP is the activated intermediate in the synthesis of phosphoribosylamine in the de novo pathway of purine formation; of purine nucleotides from free bases by the salvage pathway; of orotidylate in the formation of pyrimidines; of nicotinate ribonucleotide; and of phosphoribosylanthranilate in the pathway leading to tryptophan.

8. N-1 in both cases, and the amine group linked to C-6 in ATP.

9. dUMP + serine + NADPH + H^+ $\longrightarrow$ TMP + $NADP^+$ + glycine.

10. b, e, a, c, d.

11. Substrates: a, b; Product: c.

12. The free energies of binding are −57.7 (wild type), −49.8 (Asn 27), and −38.1 (Ser27) kJ mol^{-1} (−13.8, −11.9, and −9.1 kcal mol^{-1}, respectively). The loss in binding energy is +7.9 kJ mol^{-1} (+1.9 kcal mol^{-1}) and +19.7 kJmol^{-1} (+4.7 kcal mol^{-1}).

13. CTP is a feedback inhibitor of ATCase, while ATP is an allosteric activator.

14. a and b.

15. ATP and dATP.

16. a and c.

17. d.

18. Inosine or hypoxanthine could be administered.

19. (a) Cell A cannot grow in a HAT medium, because it cannot synthesize TMP either from thymidine or from dUMP. Cell B cannot grow in this medium, because it cannot synthesize purines by either the de novo pathway or the salvage pathway. Cell C can grow in a HAT medium because it contains active thymidine kinase from cell B (enabling it to phosphorylate thymidine to TMP) and hypoxanthine guanine phosphoribosyltransferase from cell A (enabling it to synthesize purines from hypoxanthine by the salvage pathway). (b) Transform cell A with a plasmid containing foreign genes of interest and a functional thymidine kinase gene. The only cells that will grow in a HAT medium are those that have acquired a thymidylate kinase gene; nearly all of these transformed cells will also contain the other genes on the plasmid.

Chapter 27

1. Three. One molecule of ATP to form phosphorylethanolamine and two molecules of ATP to regenerate CTP from CMP.

2. (a) CDP-diacylglycerol; (b) CDP-ethanolamine; (c) acyl CoA; (d) phosphatidylcholine; (e) UDP-glucose or UDP-galactose; (f) UDP-galactose; (g) geranyl pyrophosphate.

3. Glycerol + 4 ATP + 3 fatty acids + 4 H_2O $\longrightarrow$ triacylglycerol + ADP + 3 AMP + 7 P_i + 4 H^+.

4. If PAP activity were to decline due to a mutation, phosphatidate conversion to DAG would decrease. Therefore, the concentration of DAG would decline and the concentration of phosphatidate would be high. Hence, the concentrations of lipids derived from activated phosphatidate will likely increase, and the concentrations of lipids derived from DAG and an activated head group would decrease.

The following would decrease: a, b, f. The following would increase: c, d, e.

5.

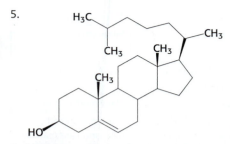

6. (1) The synthesis of activated isoprene units (isopentyl pyrophosphate), (2) the condensation of six of the activated isoprene units to form squalene, and (3) cyclization of the squalene to form cholesterol.

7. b, c, a, d

8.

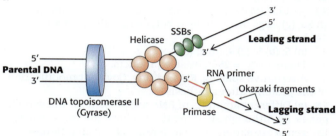

Protosterol cation

9. c, g, d, h, a, i, j, f, b, e

10.

Labeled mevalonate   **Labeled isopentenyl pyrophosphate**

11. The amount of reductase and its activity control the regulation of cholesterol biosynthesis. Transcriptional control is mediated by SREBP. Translation of the reductase mRNA also is controlled. The reductase itself may undergo regulated proteolysis. Finally, the activity of the reductase is inhibited by phosphorylation by AMP kinase when ATP levels are low.

12. The conversion of HMG-CoA to mevalonate by HMG-CoA reductase is the rate-determining step.

13. Statins are competitive inhibitors of HMG-CoA reductase. They are used as drugs to inhibit cholesterol synthesis in patients with high levels of cholesterol.

14. The LDL contains apolipoprotein B-100, which binds to an LDL receptor on the cell surface in a region known as a coated pit. On binding, the complex is internalized by endocytosis to form an internal vesicle. The vesicle is separated into two components. One, with the receptor, is transported back to the cell surface and fuses with the membrane, allowing continued use of the receptor. The other vesicle fuses with lysosomes inside the cell. The cholesteryl esters are hydrolyzed, and free cholesterol is made available for cellular use. The LDL protein is hydrolyzed to free amino acids.

15. (a) Lipids are less dense than proteins. Chylomicrons, which have the highest percentage by weight of lipids, are the least dense lipoproteins. (b) In a lipoprotein, the protein components are generally soluble and the lipid components are not. Lipoproteins with more proteins are more soluble. High-density lipoproteins (HDLs) have more protein than other lipoproteins and are the most soluble.

16. LDL: a and d; HDL: b and c

17. The core structure of a steroid is four fused rings: three cyclohexane rings and one cyclopentane ring. In vitamin D, the B ring is split by ultraviolet light.

18. (a) 3; (b) 4; (c) 5; (d) 1; (e) 2.

19. 3-Hydroxy-3-methylglutaryl CoA is also a precursor for ketone-body synthesis. If fuel is needed elsewhere in the body, as might be the case during a fast, 3-hydroxy-3-methylglutaryl CoA is converted into the ketone body acetoacetate. If energy needs are met, the liver will synthesize cholesterol. Remember also that cholesterol synthesis occurs in the cytoplasm while ketone bodies are synthesized in the mitochondria.

Chapter 28

1. One new strand will have an A–T pair, and the other strand will have a G–C pair.

2. The oxidation of guanine could lead to DNA repair: DNA strand cleavage could allow looping out of the triplet repeat regions and triplet expansion.

3. (a) 4, 5; (b) 3, 2; (c) 6, 1

4.

5. In principle, it could be used to break three base pairs.

6. (a) Size; the top is relaxed and the bottom is supercoiled DNA. (b) Topoisomers. (c) The DNA is becoming progressively more unwound, or relaxed, and thus slower-moving.

7. (a) Lk = 460; (b) Lk = 456; (c) Lk is undefined; (d) Lk = 458

8. c

9. b, c, e

10. a, h, d, e, g, b, c, f

11. a, b, g, c, f, h, d, e

12. e, c, b, a, d, f

13. b, d, e

14. (a) It was used to determine the number of spontaneous revertants—that is, the background mutation rate. (b) The known mutagen was used to firmly establish that the system was working. A known mutagen's failure to produce revertants would indicate that something was wrong with the experimental system. (c) The chemical itself has little mutagenic ability but is apparently activated into a mutagen by the liver homogenate. (d) Components of the cytochrome P450 system.

15. (a) 5; (b) 1; (c) 2, 3; (d) 4

16. (a) 6, 4; (b) 1, 5; (c) 3, 4

17. a, g, d, f, e, b, c

18. (a) 5; (b) 3; (c) 1; (d) 2; (e) 4

19. b, d

Chapter 29

1. An error will affect only one molecule of mRNA of many synthesized from a gene. In addition, the errors do not become a permanent part of the genomic information.

2. b, c, d.

3. The active sites are related by convergent evolution.

4. The sequence of the coding (+, sense) strand is
5'-ATGGGGAACAGCAAGAGTGGGGCCCTGTCCAAGGAG-3'
and the sequence of the template (–, antisense) strand is
3'-TACCCCTTGTCGTTCTCACCCCGGGACAGGTTCCTC-5'

5. This mutant σ will competitively inhibit the binding of holoenzyme and prevent the specific initiation of RNA chains at promoter sites.

6. A 100-kDa protein contains about 910 residues, which are encoded by 2730 nucleotides. At a maximal transcription rate of 50 nucleotides per second, the mRNA would be synthesized in 55 s.

7. DNA is the single strand that forms the trunk of the tree. Strands of increasing length are RNA molecules; the beginning of transcription is where growing chains are the smallest; the end of transcription is where chain growth stops. Many enzymes are actively transcribing each gene.

8. (a) 5′; (b) guanosine; (c) triphosphate; (d) a nitrogen; (e) 2′.

9. c, e, d, a, f, b.

10. Gene A, top; gene B, top; gene C, bottom.

11. The organism in the fluctuating environment, because it is exposed to more extremes.

12. (a) 2, 3; (b) 5, 6; (c) 1, 4

13. (a) 13.7; (b) 45S; (c) 18S, 5.8S, and 28S; (d) 28S; (e) 5.8S; (f) 5S; (g) large (60S); (h) RNA polymerase III; (i) small (40S).

14. b, d, e.

15. (a) 5; (b) 3; (c) 4; (d) 2; (e) 1.

16. There are $2^8 = 256$ possible products.

17. Alternative splicing, RNA editing. Covalent modification of the proteins after synthesis.

18. (a) Different amounts of RNA are present for the various genes. (b) Although all of the tissues have the same genes, the genes are expressed to different extents in different tissues. (c) These genes are called housekeeping genes—genes that most tissues express all the time. They might include genes for glycolysis or citric acid cycle enzymes. (d) The point of the experiment is to determine which genes are initiated in vivo. The initiation inhibitor is added to prevent initiation at start sites that may have been activated during the isolation of the nuclei.

19. True: b, c, d; False: a, e.

Chapter 30

1. (a) and (b) only.

2. Four bands: light, heavy, a hybrid of light 30S and heavy 50S, and a hybrid of heavy 30S and light 50S.

3. The peptide would be Phe-Cys-His-Val-Ala-Ala. The codons UGC and UGU encode cysteine but, because the cysteine was modified to alanine, alanine is incorporated in place of cysteine.

4. (a) 1, 2, 3, 5, 6, 10; (b) 1, 2, 7, 8; (c) 1, 4, 8, 9.

5. (a, d, and e) Type 2; (b, c, and f) Type 1.

6. Two hundred molecules of ATP are converted into 200 AMP + 400 P_i to activate the 200 amino acids, which is equivalent to 400 molecules of ATP. One molecule of GTP is required for initiation, and 398 molecules of GTP are needed to form 199 peptide bonds, for a total of 399 molecules of GTP.

7. Proteins are synthesized from the amino to the carboxyl end on ribosomes, whereas they are synthesized in the reverse direction in the solid-phase method. The activated intermediate in ribosomal synthesis is an aminoacyl-tRNA; in the solid-phase method, it is the adduct of the amino acid and dicyclohexylcarbodiimide.

8. (a) A_5. (b) $A_5 > A_4 > A_3 > A_2$. (c) Synthesis is from the amino terminus to the carboxyl terminus.

9. (a) P site; (b) A site; (c) E site; (d) P site.

10. The rate would fall because the elongation step requires that the GTP be hydrolyzed before any further elongation can take place.

11.

	Bacteria	Eukaryotes
Ribosome size	70S	80S
Initiation	Shine–Dalgarno is required	First AUG is used
Protein factors	Required	Many more required
Relation to transcription	Translation can start before transcription is completed	Transcription and translation are spatially separated
First amino acid	fMet	Met

12. (a) 3, 5, 9; (b) 1, 7; (c) 4, 5, 6, 10; (d) 2, 8.

13. (a) 4; (b) 3; (c) 2; (d) 6.

14. (a) −35 sequence; (b) Pribnow box; (c) +1; (d) terminator sequence; (e) Shine–Dalgarno sequence; (f) AUG.

15. (a) The three peaks represent, from left to right, the 40S ribosomal subunit, the 60S ribosomal subunit, and the 80S ribosome. (b) Not only are ribosomal subunits and the 80S ribosome present, but polysomes of various lengths also are apparent. The individual peaks in the polysome region represent polysomes of discrete length. (c) The treatment significantly inhibited the number of polysomes while increasing the number of free ribosomal subunits. This outcome could be due to inhibited protein-synthesis initiation or inhibited transcription.

16. a, d, i, h, c, f, b, g, e.

17. (a) The signal sequence, signal-recognition particle (SRP), the SRP receptor, and the translocon. (b) The formation of peptide bonds, which in turn are powered by the hydrolysis of the aminoacyl-tRNAs.

18. (a) eIF-4H has two effects: (i) the extent of unwinding is increased; and (ii) the rate of unwinding is increased, as indicated by the increased rise in activity at early reaction times. (b) To firmly establish that the effect of eIF-H4 was not due to any inherent helicase activity. (c) Half-maximal activity was achieved at 0.11 μM of eIF-4H. Therefore, maximal stimulation would be achieved at a ratio of 1:1. (d) eIF-4H enhances the rate of unwinding of all helices, but the effect is greater as the helices increase in stability. (e) The results in graph C suggest that eIF-4H increases the processivity.

19. (a) 3′-AATCTCGCG-5′; (b) UUAGAGCGC; (c) AAU, CUC, GCG; (d) leucine, glutamic acid, arginine.

20. (a) A; (b) alanine.

Chapter 31

1. (a) and (b) Cells will express β-galactosidase, *lac* permease, and thiogalactoside transacetylase even in the absence of lactose. (c) The levels of catabolic enzymes such as β-galactosidase and arabinose isomerase will remain low even at low levels of glucose.

2. The *lac* repressor does not bind DNA when the repressor is bound to a small molecule (the inducer), whereas the *pur* repressor binds DNA only when the repressor is bound to a small molecule (the corepressor). The *E. coli* genome contains only a single *lac* repressor-binding region, whereas it has many sites for the *pur* repressor.

3. b, c.

4. Bacteriophage λ would be more likely to enter the lytic phase because the cooperative binding of the λ repressor to O_R2 and O_R1, which supports the lysogenic pathway, would be disrupted.

5. e, a, c, b, f, d.

6. 5-Azacytidine cannot be methylated. Some genes, normally repressed by methylation, will be active. 5-Azacytidine is also an inhibitor of DNA methyltransferase, resulting in lower levels of methylation and, hence, lower levels of gene repression.

7. Increased Cro concentration reduces the expression of the λ repressor gene. Increased λ repressor concentration reduces the expression of the Cro gene. At low λ repressor concentrations, increased λ repressor concentration increases the expression of the λ repressor gene. At higher λ repressor concentrations, increased λ repressor concentration decreases the expression of the λ repressor gene.

8. The footprint appears to have approximately 50% of its intensity near 3.7 nM so that the dissociation constant is approximately 3.7 nM, corresponding to a standard free energy of binding of –48 kJ/mol (–11 kcal/mol) at T = 298K.

9. b, c.

10. In mouse DNA, most of the HpaII sites are methylated and therefore not cut by the enzyme, resulting in large fragments. Some small fragments are produced from CpG islands that are unmethylated. For *Drosophila* and *E. coli* DNA, there is no methylation, and all sites are cut.

11. a, c, d.

12. Proteins containing these domains will be targeted to methylated DNA in repressed promoter regions. They would likely bind in the major groove, because that is where the methyl group is located.

13. The acetylation of lysine will reduce the charge from +1 to 0. The methylation of lysine will not reduce the charge.

14. d, f, a, e, b, c.

15. Gene expression is not expected to respond to the presence of estrogen. However, genes for which expression normally responds to estrogen will respond to the presence of progesterone.

16. The addition of an IRE to the 5′ end of the mRNA is expected to block translation in the absence of iron. The addition of an IRE to the 3′ end of the mRNA is not expected to block translation, but it might affect mRNA stability.

17. On the basis of the pattern of cysteine and histidine residues, this region appears to contain three zinc-finger domains.

18. (a) 3; (b) 1; (c) 2.

19. (a) 2; (b) 3, 1; (c) 4, 5.

20. Positive: a, c, d; negative: b, e.

Chapter 32

1. Sildenafil increases cGMP levels by inhibiting the phosphodiesterase-mediated breakdown of cGMP to GMP. Intracellular cGMP levels can also be increased by activating its synthesis. This activation can be achieved with the use of NO donors (such as sodium nitroprusside and nitroglycerin) or compounds that activate guanylate cyclase activity. Drugs that act by the latter mechanism are currently in clinical trials.

2. (a) Accidental; (b) natural product screening; (c) accidental; (d) natural product screening; (e) targeted design.

3. (A) Yes; (B) yes; (C) no (MW > 600).

4. Unlike competitive inhibition, noncompetitive inhibition cannot be overcome with additional substrate. Hence, a drug that acts by a noncompetitive mechanism will be unaffected by changing levels of the physiological substrate.

5. A drug that inhibits a P450 enzyme may dramatically affect the disposition of another drug that is metabolized by that same enzyme. If this inhibited metabolism is not accounted for when dosing, the second drug may reach very high, and sometimes toxic, levels in the blood.

6. The binding of other drugs to albumin could cause extra coumadin to be released. (Albumin is a general carrier for hydrophobic molecules.)

7. (a) $K_I \approx 0.3$ nM. $IC_{50} \approx 2.0$ nM. (b) Yes, compound A should be effective when taken orally because 400 nM is much greater than the estimated values of K_I and IC_{50}.

8. The computational prediction of log(P) values on the basis of chemical structure enables a more rapid assessment of drug-like properties of a molecule, and it avoids the need for experimental phase partitioning studies. Many such algorithms are now available to pharmacologists.

9. In Phase I clinical trials, approximately 10 to 100 usually healthy volunteers are typically enrolled in a study designed to assess safety. In contrast, a larger number of subjects are enrolled in a typical Phase II trial. Moreover, these persons may benefit from the drug administered. In a Phase II trial, efficacy, dosage, and safety can be assessed.

10. A reasonable mechanism would be an oxidative deamination following an overall mechanism similar to that in Figure 32.19, with release of ammonia.

$$NADPH + H^+ + O_2 + \text{[structure]} \longrightarrow$$

$$NADP^+ + H_2O + NH_3 + \text{[structure]}$$

11. Recall that for a competitive inhibitor, the IC_{50} will shift depending on the substrate concentration. According to the Cheng–Prusoff equation, for an enzyme with a K_M of 1 μM the IC_{50} of a competitive inhibitor at $[S] = 20$ μM will be $21 \times K_I$, while at $[S] = K_M$, the IC_{50} will be $2 \cdot K_I$, a 10-fold shift.

12. Because monoclonal antibodies interact with a broader surface on the target, they can achieve greater potency than small molecules, as well as achieve a higher degree of selectivity versus proteins similar to the target. Also, because they are retained in the body via the FcRn receptor-mediated salvage pathway, they can maintain much longer half-lives than small molecules.

Note: Page numbers followed by f, t, and b refer to figures, tables, and boxed material. **Boldface** page numbers indicate structural formulas and ribbon diagrams.

A site, ribosomal, 929, 929f, 933, 934, 935
AAA ATPases, 706
Ab initio prediction, 61
ABC transporters
  ATP hydrolysis reaction and, 390
  defined, 388
  domain arrangement of, 388, 388f
  in drug resistance, 999
  eukaryotic, 390
  mechanism, 388–390, 389f
  prokaryotic, 390
  structure of, 388, **389f**
ABO blood groups, 343–344, 343f, 344f
Abortive complex, 709
Absorption
  of dietary proteins, 702, 702f
  drug, 988–989
  light, 439–440, 590, 590f
Acarbose, 334, **334**
Acceptor control. *See* Respiratory control
Acceptor stem, 919, 925, 925f
Accessory pigments in photosynthesis,
  602–605, 603f, 604f
Accommodation process, 932
Acetaldehyde, 486f
Acetaminophen, 992–993, 993f, **993f**
Acetate, 822f
Acetoacetate, **681**. *See also* Ketone bodies
  from branched chain amino acids,
    722–723
  conversion into acetyl CoA, 681–682,
    681f
  formation of, 680, 724f
  in gluconeogenesis, 753–754
  from phenylalanine, 724, **724**
  in respiration, 680–681
  from tryptophan, 724, 724f
  utilization as fuel, 680–681, 681f
Acetoacetyl CoA, 718
Acetyl ACP, 686–687
Acetyl CoA (acetyl coenzyme A)
  acetoacetate conversion into, 681–682,
    681f
  acetyl-group-transfer potential, **462**
  in amino acid degradation, 718,
    722–723
  cholesterol synthesis and, 821–825
  in citric acid cycle, 502, 516, 518–520,
    522–524, 524f, 535
  defined, 462, 516
  in fatty acid metabolism, 532, 670–671,
    672–673, 690, 690f, 691
  in fatty acid oxidation, 672–673
  in fatty acid synthesis, 532
  formation from ketone bodies, 679–680
  formation of, 488, 520
  formation of ketone bodies, 680f
  in glyoxylate cycle, 536–537, 537f
  oxidation of, 459
  pyruvate and, 486f

pyruvate dehydrogenase complex and,
  520
  structure of, **462**, **516**, **726b**
  synthesis of, 518–520
  transfer to cytoplasm, 690, 690f
Acetyl CoA acetyltransferase, 758
Acetyl CoA carboxylase, 502
  in citrate concentration, 695, 695f
  in fatty acid metabolism, 685, 694–696,
    694f, 722
  filaments of, 695f
  regulation of, 694f, 695–696
Acetyl transacylase, 686
Acetylated lysine, **218**
Acetylation
  defined, 219t
  in gene regulation, 965–966, 967
  glycogen phosphorylase and, 658
  in metabolism regulation, 218
  protein, 219–220
Acetylcholine, 400–402
Acetylcholine receptor, 404
  defined, 400
  as ligand-gated channel, 400–402
  as nonspecific cation channel, 403
  structure of, **400**, **401f**
Acetylgalactosamine, **332f**

N-acetylglucosamine phosphotransferase,
  345
N-acetylglutamate, 714–715
N-acetylglutamate synthase, 714
N-acetyl-l-phenylalanine p-nitrophenyl
  ester, 182, 182f
Acetyllysine-binding domain, 966, 966f
N-acetylneuraminate, 819
Acetylsalicylic acid, 141, 980–981, **980**
Achiral molecules, 36
Acid-base reactions, 17–18
Acids. *See also* Amino acids; Fatty acids;
    Nucleic acids
  bile, 667, 667f
  protonation/deprotonation of, 19–20
  sialic, 819
Aconitase, 524–525, 525f, 971, 971f
Acquired immune deficiency
    syndrome (AIDS). *See* Human
    immunodeficiency virus (HIV)
    infection
Actin/actin filaments, 207f
  myosin and, 204–206, 205f, 206f
  structure of, 310, **311f**
Action potentials, 405
  defined, 392, 400
  integration, 402–404
  ion channels and, 392, 392f
  mechanism, 403–404, 404f
Activated carriers
  defined, 460
  as derived from vitamins, 463–464

  of electrons for fuel oxidation, 460–461
  of electrons for reductive biosynthesis,
    461
  in metabolism, 460–463, 463t
  recurring set of, 463
  of two-carbon fragments, 462–463
Activated glycoaldehyde unit, 629
Activated methyl cycle, 775, 775f
Activation domains, 960
Activation energy
  defined, 148
  enzyme decrease of, 148–149, 148f
  product or substrate accumulation, 151
  symbol, 148
Activation loop, 426, 427f
Active membrane transport, 382–383, 384
Active sites
  ATCase, 213–214, **213f**, **214f**
  chymotrypsin, 183f
  concentration of, 156
  as crevice, 150
  defined, 149
  enzyme and substrate interaction,
    150–151
  mapping with irreversible inhibitors,
    169–172
  as microenvironments, 150
  myosin, 204, 204f
  proteases, 188f
  residues, 150, 150f, 486f
  RNA polymerases, 882, 883f
  volume of enzyme, 150
Active transport, 382–383
Acute intermittent porphyria, 788
Acyclovir monophosphate, 797
Acyclovir triphosphate, 797
Acyl adenylate, 670–671
Acyl carnitine, 671–672, 671f
Acyl carrier protein (ACP), 685, 686f, 689,
  689f
Acyl CoA, 462
Acyl CoA synthetase, 670–671
Acyl CoA: cholesterol acyltransferase
  (ACAT), 829
Acyl intermediates, 173
Acyl-enzyme intermediates, 183, 183f, 706
Adaptive control, 696
Adaptive immune system, 86–88
Adaptor proteins, 428
ADC. *See* Antibody-drug conjugate (ADC)
Adenine, 4, 5
  deamination of, 865, 865f
  methylation of, 200, 200f
  resonance structures, **8**
  structure of, **4**, **5f**
  synthesis of, **783f**
Adenine nucleotide translocase (ANT).
    See ATP-ADP translocase
Adenine phosphoribosyltransferase, 800
Adenosine, **424**

Adenosine 5′-triphosphate. *See* ATP (adenosine triphosphate)
Adenosine deaminase, 809–810
Adenosine diphosphate. *See* ADP (adenosine diphosphate)
Adenosine monophosphate, **332f**
Adenosine nucleotides, 505–506
Adenosine triphosphate. *See* ATP (adenosine triphosphate)

S-adenosylhomocysteine, 774, **775**
S-adenosylmethionine (SAM)
   in activated methyl cycle, 774–775, 775f
   in adenylation of methionine, 721
   in amino acid degradation, 721, 721f
   in amino acid synthesis, 774–775
   in phospholipid synthesis, 818
Adenylate, 799
Adenylate cyclase, 418, 418f, 423–424, 424f
Adenylated intermediates, 770–771
Adenylation
   in amino acid activation, 921–922
   in amino acid synthesis, 782–783, 782f
   in translation, 921–922
Adenylosuccinate synthetase, 800
Adenylyl transferase, 782, 782f
Adipocytes, 666, 666f, 737
Adipokines, 736, 737f
Adiponectin, 736–737, 737f
Adipose cells (fat cells), 666
Adipose tissue
   brown, 575, 576f
   fat in, 748
   glycerol released from, 753
   leptin secretion, 736–737
   triacylglycerols in, 753
   white, 575
Adipose triglyceride lipase (ATGL), 669, 669f
Adler, Julius, 580
ADME properties of drugs, 988–989, 988f
ADP (adenosine diphosphate)
   ATP hydrolyzed to, 450–451
   formation of, 200
   in oxidative phosphorylation, 572–573, 572f, 575f
   in random sequential reaction, 160
   structure of, **200**, **219**, **450f**
Adrenaline, 221, 412, 414, **414**, 650. *See also* Epinephrine
Adrenergic receptors, 415
Adrenocorticotropic hormone, 63
Advanced glycation end product (AGE), 331, 480, 510
Aequorea victoria, 63, 117
Aerobic exercise, 747
Aerobic glycolysis, 497–498, 531
Affinity chromatography, 104–105, 104f, 110
Affinity labels, 170, 170f
Affinity tags, 112, 170f
Aflatoxin activation, 865, 865f
Aggrecan, 340–341
Agonists, 415, 416f
Agouti-related peptide (AgRP), 736, 736f
Agre, Peter, 1, 27–28, 407

Agrobacterium tumefaciens, 295–296, 295f
Akt. *See* Protein kinase B (PKB)
Alanine
   defined, 36
   formation of, 769–770
   in gluconeogenesis, 509
   in muscle contraction, 509
   pyruvate formation from, 719, 719f
   site-directed mutagenesis and, 187–188, 187f
   structure of, **37f**, 38
   from tryptophan, 724f
   tryptophan in, 719
Alanine aminotransferase, 708
Albinism, 727t
Albumin, **989f**
Alcaptonuria, 726
Alcohol dehydrogenase, 755
Alcoholic fermentation, 472, 486–487
Alcoholic hepatitis, 756
Alcoholic scurvy, 756–757, 757f
Alcohols
   fermentation of, 472, 486–487
   formation from pyruvate, 486–487
   metabolism of, 755–759
   monosaccharides and, 331–332
   phospholipids, synthesis of, 817–818
   toxicity of, 756–757
Aldehyde dehydrogenase, 755
Aldehydes, oxidation of energy, 480–481
Alder, Julius, 580
Aldimine, 710, **710**
Aldol cleavages, 711
Aldolase, 478
   in Calvin–Benson cycle, 616
   in transaldolase reaction, 629–630
Aldoses
   conversion to ketoses, 477
   defined, 325
   structure of, **615**
Aldosterone, 834, 837, **837f**
Alkali cations, properties of, 396t
Alkali digestion, 278
Alkylating agents, 865
Allergenics, 995t
Allolactose, 954, 956
Allopurinol, 810, **810**
Allosteric coefficient (L), 215
Allosteric control, 210–217
Allosteric effector, 81
Allosteric enzymes, 161f
   concerted model, 215
   control of, 468
   defined, 161
   homotropic effects on, 215
   sequential model, 215
   sigmoidal kinetics, 212, 212f
   substrate concentration and, 215
   threshold effects, 215f, 216
   T-to-R equilibrium in, 216–217, 216f, 217f
   T-to-R state transition, 215, 216f
Allosteric sites, 211
Alloxanthine, 810
All-*trans*-retinal, 439f
All-*trans*-retinoic acid, **963**

α₁-antiproteinase, 227, 228f
AlphaFold, 61, 62f
α carbon, 35
α chains, 75
α helix, 46–48
   formation of, 57–59
   hydrogen-bonding scheme, 47f
   of membrane proteins, 368–369, 368f, 368t
   pitch of, 47
   as right-handed, 47, 47f
   schematic view of, 48f
   screw sense, 47
   structure of, **47f**, 58f
α-1,4-glycosidic linkage, 333
α-1,6 linkages, 655, 655f
α-amino acids, 35, 41
α-glucosidase, 473–474
α-glucosidase inhibitors, 334
α-helical coiled coil, 53
α-hemoglobin
   alignment with gap insertion, 304f
   amino acid sequences, 303–305, 303f
α-keratin, 53
α-sarcin, 944
αβ dimers, 75
α-synuclein, 817
Alternative splicing
   defects in, 907t
   defined, 906
   example of, 907f
   pre-mRNA, 906–907
Altman, Sidney, 899
Alu sequences, 284
Alzheimer's disease, 62–63, 63f
Amanita phalloides, 894f
α-amanitin, 894, **894f**
American College of Sports Medicine, 751
American Medical Association, 751
Ames, Bruce, 871
Ames test, 871, 871f
Amide bonds. *See* Peptide bonds
Amines, monosaccharides and, 331–332
Amino acid degradation, 701–728
   acetoacetyl CoA in, 718
   acetyl CoA in, 718, 722–723
   amino acid use through, 701
   aminotransferases in, 708–711
   of aromatic amino acids, 723–724
   of branched-chain amino acids, 708–713, 722–723
   carbon atoms of as metabolic intermediates, 718–726
   digestive enzymes in, 702
   fumarate in, 718, 719f
   glucose-alanine cycle in, 713, 713f
   glutamate dehydrogenase in, 708–709
   a-ketobutyrate in, 720
   α-ketoglutarate in, 718, 719, 719f, 720, 720f
   liver in, 708–709, 712–713
   metabolic products of, 725b–726b
   metabolism errors and, 723–728, 727t
   overview of, 702–703
   oxaloacetate in, 718, 719–720, 719f
   oxygenases in, 723–724

propionyl CoA in, 722–723
proteasomes in, 705–707, 706f, 707f
in protein turnover, 703–705
pyridoxal phosphate (PLP) in, 709–710, 711–712
pyruvate in, 712–713, 718, 719, 719f
regulation function of, 707, 707t
Schiff-base intermediates in, 709–711, 711f
serine dehydratase in, 712
starvation and, 754, 754f
succinyl CoA in, 718, 719f, 721, 721f
threonine dehydratase in, 712
transaminases in, 708–713
ubiquitin in, 703–705, 704f
ubiquitination in, 703–705
urea cycle in, 713–718
Amino acid sequences, 33–34
alignment of, 302–309, 303f, 304f
collagen, 53–54, 54f
comparison of, 123–124
determination of, 100
directionality of, 42, 42f
in DNA probe creation, 124
DNA sequences and, 123f
evolution of, 44
evolutionary pathways and, 123
gap insertion, 303, 304f
of hemoglobin, 303–305, 303f
homologous, 308–309
identities, 303
importance of, 43–44
information provided by, 123–124
of insulin, 43, 43f
of myoglobin, 303–305, 303f
porin, 364–365, 364f, 365f
probes generated from, 277, 277f
in protein identification, 123–124
in protein structure determination, 55–64
protein structures and, 43, 43f
recombinant DNA technology and, 123
ribonuclease, 56–57, 56f, 57f
searching for internal repeats, 124, 124f
shuffled, 305, 305f
as signals, 124
statistical comparison of alignment, 305, 305f
substitution in, 305–308, 306f, 307f, 309f
transmembrane helices from, 367–369, 368f
Amino acid side chains
aliphatic, 36
aromatic, 37
charge of, 39–40, 39f, 41f
covalent modifications, 63f
defined, 35
distribution of, 51
hydrocarbon, 36
hydrophilic, 38
hydrophobic, 38
hydroxyl-containing, 38, 38f
ionizable, 40, 40t
sulfhydryl-containing, 39
thiol-containing, 39
types of, 36–40

Amino acid synthesis, 763–788
S-adenosylmethionine in, 774–775
adenylation in, 782–783, 782f
ammonia in, 764–767
aromatic amino acids, 776–779
branched pathways in, 780–782
chorismate in, 776–779, 777f
enzyme multiplicity in, 781
essential amino acids, 768–769, 769f
feedback inhibition/activation in, 780–783, 780f
glutamine synthetase in, 782–783, 782f
human, 768–769
α-ketoglutarate in, 766–767, 769
nitrogen fixation in, 764–767, 764f
nonessential amino acids, 768, 768f
overview of, 763
regulation of, 780–783
regulatory domains in, 780–782, 781f
Schiff bases in, 766, 767f
shikimate in, 776–779
tetrahydrofolate in, 772–774, 773f
threonine deaminase in, 780–781, 780f, 781f
by transamination, 767, 769–770, 769f, 770f
Amino acids. *See also* Proteins
abbreviations for, 40, 41t
activation by adenylation, 921–922
activation sites, 922–923, 923f
aromatic, 38, 723–724, 776–779
basic set of, 768t
biosynthetic families of, 768f
branched-chain, 708–713, 722–723
as building block, 25
carbon skeletons, fates of, 719f
chirality of, 35, 769f, 770
conservation across species, 40–41
D isomer, 35, 35f
defined, 25, 35
dipolar form of, 35
essential, 702, 702t, 768–769, 768f, 768t
glucogenic, 718, 719f
in gluconeogenesis, 499
hydrophilic, 38, 52
hydrophobic, 36–38, 37f, 51f, 52f
hydroxyl groups of, 38–39, 63
"inside out" distribution of, 52f
ionization state of, 36, 36f
ketogenic, 718, 719f
L isomer, 35, 35f
metabolism, 726–728, 727t
negatively charged, 39–40, 39f
nitrogen in, 764
nonessential, 768, 768f, 768t
nonproteinogenic, 126
peptide bonds, 42, 42f
polar, 38–39, 38f
positively charged, 39, 39f
as precursors, 783–788
as precursors for neurotransmitters, 785
protein folding and, 25, 25f
proteinogenic, 126
pyruvate formation from, 719, 719f
reactive, 38, 41, 41f

residue, 42, 51
in secondary structures, 58t
undesirable reactivity in, 41f
unnatural, 938–939, 938f
Amino sugar, 332
Amino-acid signal sequences, 940–941, 940f
Aminoacrylate, Schiff bases of, 779, **779**
Aminoacyl adenylate, **922**
Aminoacyl-AMP, 922
Aminoacyl-tRNA
defined, 921
editing of, 923–924, 923f
elongation factors and, 931–932, 931f
Aminoacyl-tRNA synthetases
accuracy of, 931
activation sites of, 922–923, 923f
amino acid activation sites, 922–923
binding to acceptor stem and anticodon loop, 925, 925f
classes of, 926, 926f, 926t
defined, 921
editing sites, 923–924, 923f
evolution of, 926
genetic code and, 921–926
proofreading by, 923–925, 924f
subunit structure, 926, 926t
2-aminoadenine, 801, 801f, **801**
Aminoglycoside antibiotics, 942
δ-aminolevulinate, 786–787, 787f
δ-aminolevulinate synthase, 786–787
Aminopeptidase N, 702
Aminopterin, 806f, 807
Aminotransferases
in amino acid degradation, 708–711
blood levels of, 711
defined, 708
Schiff-base intermediates formation in, 709–711
Ammonia
in amino acid degradation, 708–709
in amino acid synthesis, 764–767
conversion to urea, 713–718
formation of, 708–709
in glutamate synthesis, 766–767
in glutamine synthesis, 766–767
in purine synthesis, 800
in pyrimidine synthesis, 793
Ammoniotelic organisms, 718
AMP (adenosine monophosphate)
ATP hydrolyzed to, 450–451
energy charge and, 505–506
phosphofructokinase (PFK), regulation of, 493–494
in purine synthesis, 799–800, 799f
structure of, **450f**
synthesis of, 807–808, 808f
AMP-activated protein kinase (AMPK), 695, 696, 737, 737f, 745
Amphibolic pathways, 448
Amphipathic (amphiphilic) molecules, 358
Amphitropic enzyme, 818
AMP-PNP, 850–851, **850**, **851f**
α-amylase, 335, 473
Amylin, 735t

Amyloid fibrils, 62–63, 62f, 63f
Amyloid plaques, 62, 63f
Amylopectin, 335
Amylose, 335
Amyotrophic lateral sclerosis (ALS), 271, 290
Amytal, 577, 577f
Anabolic reactions, 451–452
Anabolism, 448
Anaerobic exercise, 747
Anaerobic glycolysis, 531
Analytes, 118
Anaplerotic reaction, 535
Anastrozole, 838, **839f**
Anchor residues, 90
Ancient DNA, 318–319
Anderson disease, 661t
Androgens
    functions of, 834
    pathways for formation of, 838f
    synthesis of, 837–838, 838f
Androstenedione, 838, **838f**
Anemia, 21, 83–84, 339
Anfinsen, Christian, 56–57
Angiogenin, 302, **302f**
Animals
    fatty acid synthesis in, 687–689
    fatty acids and, 683
    as models in drug target testing, 993
    vision in, 442
Anion exchange, 104
Anomeric carbon atom, 328
Anomers, 326f, 328
Anoxygenic photosynthesis, 586
Antagonists, 424
Anthranilate branch, 778–779
Anthrax, 857
Antibiotics
    aminoglycoside, 942
    protein synthesis inhibition by, 942–944, 943f, 943t, 944f
    resistance to, 998
    translocation inhibition by, 943–944, 944f
Antibodies. See also Immunoglobulins
    in adaptive immune system, 89
    amino acid sequence data in making, 124
    antibody binding of, 88–89
    antigen binding of, 88–89
    antigen interaction, 113f
    defined, 69, 86
    diversity of, 89
    generation of, 112–113
    monoclonal, 113–114, 113f, 114f, 434, 434f, 995–996, 995t
    polyclonal, 113, 113f
    primary, 116
    secondary, 116
    structures of, 87–88
Antibody-drug conjugate (ADC), 996
Antibody-protein interactions, 87–88
Anticancer drugs
    in blocking thymidylate synthesis, 805–807, 806f

resistance to, 998–999
    targets, 806f
Anticipation, 869
Anticoagulants, 231, 231f
Anticodon loop, 919, 925, 925f
Anticodons
    base/base pair, 920–921, 921f
    defined, 918
    in translation, 919–921, 921f
Antidiuretic hormone, 126, 127f
Antigenic determinants, 113
Antigens
    ABO blood group, 343–344, 343f, 344f
    antibody interaction, 113, 113f
    defined, 86, 112–113
    synthetic peptides as, 126
Antimycin A, 577, 577f
Antiporters, 390, 390f
Antisense (-) strand, 884, 884f
Antiserum, 113
Antithrombin III, 232
AP endonuclease, 867
Apoenzyme, 144
Apolipoprotein A-IV, 735t
Apolipoprotein B (apo B), 909, 909f
Apolipoproteins, 827
Apoproteins, 827
Apoptosis, 224, 579
Apoptosome, 579
Apoptotic peptidase-activating factor 1 (APAF-1), 579
Approximation, catalysis by, 180
Aptamers, 320–321
Apyrase, 285
Aquaporins, 407, 408f
Arabidopsis thaliana, 2f, 283
Arachidonate, 693, 693f
Arachidonic acid, 366, 366f
Archaea
    defined, 3, 3f
    membranes of, 358, 358f, 373
    proteasomes of, 707
Archaeal lipid, 359f
Arcuate nucleus, 735
Arginase, 716
Arginine
    argininosuccinase and, 715, 716
    degradation of, 720, 720f
    nitric oxide from, 784–785, 784f
    structure of, 39, **39f, 40t**
    supplementation of, 717
    synthesis of, 771
Argininosuccinase, 715, 716
Argininosuccinase deficiency, 717, 717f
Argininosuccinate, 715, **715**, 717
Argininosuccinate synthetase, 715
Argonaute complex, 971–972, 971f
Aromatase, 838
Aromatase inhibitors, 838
Aromatic amino acids
    degradation of, 723–724
    side chains, 38
    synthesis of, 776–779
Arrestin, 420
Arsenate, 459
Arsenite poisoning, 535–536, 536f

Arthropods, polysaccharides of, 336–337, 337f
Artificial intelligence, 61
Artificial photosynthetic systems, 606–607, 607f, 620, 620f
Ascorbic acid (vitamin C)
    deficiency, alcohol related, 756–757, 757f
    forms of, 757, 757f
    function of, 464t
Asparaginase, 719
Asparagine
    conversion into oxaloacetate, 719–720
    structure of, **38f**, 39
    synthesis of, 770
Aspartate, **160**
    conversion into oxaloacetate, 719–720
    deamination of, 535
    formation of, 769–770
    phosphorylation of, 781
    structure of, 39, **39f, 160**
    synthesis of, 771
Aspartate aminotransferase, 708, 711, 711f
Aspartate transcarbamoylase. See ATCase
Aspartic acid, 39, **40t**
Aspartokinases, 781, 781f
Aspartyl proteases, 188, 188f, **188f**
Aspergillus terreus, 981
Aspirin, 141
    discovery and development of, 980–981
    effect on prostaglandin H2 synthase-1, 366, 366f
    as prostaglandin inhibitor, 694, 981
Assays
    "on-bead," 983f
    coupled, 982, 982f
    defined, 101
    enzyme-linked immunosorbent (ELISA), 114–115, 115f
    in protein purification, 101
    secondary, 982
    sensitivity of, 982
ATCase (aspartate transcarbamoylase)
    active site of, 213–214, **213f, 214f**
    allosteric interactions in, 212–216
    catalytic subunit, 212, 212f, 213f
    concerted model, 215
    cooperativity, 215
    defined, 211
    feedback inhibition, 211
    impact of CTP vs. ATP, 217f
    inhibition by CTP, 807
    reaction, 211, 211f
    regulatory subunit, 212, 212f
    separable subunits, 212
    sigmoidal curve, 215, 215f
    sigmoidal kinetics, 212, 212f, 215, 215f
    structure of, 212–216, **213f, 214f**
    tense and relaxed states, 214f, 215
    T-to-R transition, 214f, 215
Atheroma, 981
Atherosclerosis, 679, 827, 830–831, 832
ATP (adenosine triphosphate)
    as activated carrier, 460
    binding of, 320–321, 320f, 321f

in Calvin–Benson cycle, 618
catabolism generation, 456
in catabolism stages, 458–459, 459f
as cellular energy currency, 221
in citric acid cycle, 517, 526–531, 533, 534–535
defined, 448, 449–450
effect on ATCase kinetics, 216f, 217, 217f
electrostatic repulsion, 453
as energy source, 792
as energy-coupling agent, 452
as energy-rich molecule, 450
entropy increase in, 453
enzymes and, 145
in exercise, 454–455, 455f, 498–499, 747–749, 747t
export, inhibition of, 578
formation of, 448, 601, 601f, 602
free energy and, 449–456
generation by NADH, 461
glucose oxidation and, 573–574
hydration stabilization in, 453
hydrolysis of, 179
as immediate donor of free energy, 456
in insulin release, 744, 745f
ion gradients in, 458–459, 458f
membrane proteins and, 384–390
in mitochondria, 570–571, 572f
in muscle, 747, 749
in muscle contraction, 493, 747–749
in nitrogen fixation, 764–769
in oxidative phosphorylation, 573–574, 574t, 575f
phosphoryl-transfer potential, 453–455, 454f
in random sequential reaction, 160
resonance structures, 453, 453f
structure of, **200**, **219**, 450–451, **450f**
synthesis, 456, 543–545, 543f
ATP formation
1,3-bisphosphoglycerate (1,3-BPG) and, 482–484
citric acid cycle and, 529–531
by cyclic photophosphorylation, 601, 601f, 602
enzyme-bound, 565
fermentation in, 486
from glucose to pyruvate conversion, 484–485
from glycolysis, 485–486
from phosphoryl transfer, 482–484
proton-motive force and, 565–567, 565f
pyruvate in, 485–486
ATP hydrolysis
ABC transporters and, 388
ATP as base to promote, 202, 202f
in DNA replication, 850–851, 850f
in DNA supercoiling, 855–857, 856f
in driving metabolism, 451–452
enzyme use of, 201
as exergonic, 450–451
free energy from, 449–456
γ subunit in, 567
hydrolysis, 201
magnesium ions in, 201–202

myosin and, 200–207
myosin conformational changes, 202f, 203
reversible, within myosin active site, 204, 204f
rotational motion, 567
in transcription, 905
transition state formation for, 202–203
ATP synthase
α subunit of, 564
a subunit of, 564, 567–569, 567f
ATP synthesis and, 565–566, 565f
ATP-driven rotation in, 566, 567f
b subunit of, 564
β subunit of, 564, 566, 567, 600
bind-change mechanism for, 565–567, 566f
c ring of, 567–569, 568f
c subunit of, 566, 567–569
of chloroplasts, 600–601
cristae and, 565f
defined, 562, 563
δ subunit of, 564
dimer of, 564f
discovery, 563
efficiency of, 570f
F_0 subunit of, 564
F_1 subunit of, 564
G proteins and, 569
γ subunit of, 564, 566, 569
γε stalk, 564, 566
inhibition of, 577
in mitochondria, 564f, 570–571
nucleotide-binding sites, 566, 566f
proton flow through, 563, 565–567
proton motion across membrane in, 568f
proton path through membrane in, 568f
proton-conducting unit, 564, 567f
regulation of, 575
rotational catalysis in, 567, 567f
structure of, 564, 564f
ATP synthasome, 572
ATP synthesis
ATP synthase and, 565–566, 565f
in chloroplasts, 599–602
efficiency of, 567
as endergonic process, 562–563
mechanism, 564–567, 565f
oxidative phosphorylation uncoupling from, 575–576, 576f
proton flow around c ring in, 567–569
proton gradient across thylakoid membrane, 599–602
proton gradients in, 562–569
proton motion across membrane in, 567–569
proton path through membrane in, 567–569
uncoupling electron transport from, 577–578
ATP yield, 573–574, 574t
ATP–ADP cycle, 451, 456f
ATP-ADP translocase, 572–573, 572f, 573f, 578

ATPases, 201
  AAA, 706
  domains of myosin, 201, 201f, 202, 202f
  mitochondrial. See ATP synthase
  Na⁺–K⁺, 384, 387
  P-type, 384–387, 388
  SERCA, 384–387, **385f**, 386f
ATP-binding cassettes (ABCs). See also ABC transporters
  ATP binding with, 388, 388f
  defined, 388
  distance between, 388
ATP-citrate lyase, 690
ATP-driven pump, 384
ATP-grasp fold, 793–794, 793f
Atractyloside, 578
Atrazine, **605**
Attenuation, 967–969, 968f
Autism, 705
Autophagy, 707
Autotrophs, 586, 611
Avian influenza, 348
Avogadro's number, 15
Axel, Richard, 436
Axial bonds, 329–330, 329f
Azide, 577, 577f

B recognition elements, 897
B. stearothermophilus, 520f
B vitamins. See Vitamins
Bacillus amyloliquefaciens, 186
Bacillus anthracis, 857
Bacillus subtilis, 891
Backbones, in polypeptide chains, 42, 42f
Bacteria, 3f. See also *Escherichia coli (E. coli)*
  attenuation in, 967–969, 968f
  carcinogen action on, 871–872, 871f
  cell membranes of, 371, 373
  chromosomes, 864, 864f
  defined, 3
  flagella movement of, 579–580, 579f
  gene expression in, 935, 950–951
  glyoxylate cycle in, 536–538, 537f
  gram staining of, 373, 374f
  intrinsic termination signals in, 888
  nitrogen fixation in, 764–767
  photosynthetic, 606–607, 606t, 607f
  proinsulin synthesis by, 279, 279f
  promoters in, 886–887, 886f, 887f
  reaction centers, 591f, 592–593, 592f, 593f
  restriction enzymes in, 194, 265–266
  ribosomes, 936–937
  RNA polymerases in, 886–887
  sigma factors in, 890–892
  transcription and translation in, 935, 935f
  transcription factors in, 958
  transcription in, 886, 935, 935f
  translation in, 930, 931f, 935, 935f, 936–937
  upstream promoter elements in, 886
Bacterial reaction center, 592f, 593f
  cyclic electron flow in, 592–593
  defined, 591
  electron chain in, 591f

Bacteriochlorophylls, 591, **591**
Bacteriophage λ, 55, 55f, 272, 275–276
Bacteriopheophytins, 591, **591**, 592, 592f, 593f
Bacteriorhodopsin, 364, 364f, 365, 586, 586f
Baculovirus, 290
Ball-and-chain model, 399–400, 400f
Basal transcription factors, 898
Base-exchange reactions, 818
Base-excision repair, 867, 867f
Base-pair substitutions, tests for, 871
Bases/base pairs
    adenine, 5f, **5**
    anticodon, 920–921, 921f
    cytosine, 5f, **5**
    damage to, 864–865
    double helix, 11
    guanine, 5f, **5**
    hydrogen bonds, 5
    minor-groove side of, 848
    mutagenic, 864
    nomenclature of, 792t
    non-Watson–Crick, 864, 866, 885
    rRNA in, 927–928, 928f
    stacking of, 12, 12f
    thymine, 5f, **5**
    in translation, 919
    tRNA, 919
    Watson–Crick, 920–921
Basic-leucine zipper (bZip), 959, 959f
Bayer, 980
B-cell leukemia, 898
Bcr-abl gene, 435, 435f, 874
Benson, Andrew, 610, 611b
Benzaldehyde, **436**
Benzoate, 717
Benzoyl CoA, 717
Berg, Howard, 580
Berg, Jeremy, 28
Beriberi, 535–536, 756
Bertozzi, Carolyn, 347b
β chains, 82f, 83
β oxidation, 748, 753
β pleated sheets, 46, 48–50
    amino acid residues in, 57–59, 58t
    antiparallel, 48, 49f
    defined, 48
    formation of, 57–59
    mixed, structure of, 49f
    parallel, 48, 49f
    protein rich, 50f
    twisted, schematic, 50f
β strands
    defined, 48
    DNA recognition through, 951, 951f
    of membrane proteins, 364–365, 364f
    Ramachandran plot for, 48f
    structure of, 48f, 58f
β turns, 46, 50, 57–59
β-adrenergic receptor, 415, 415f, 416f, 419f
β-carotene, 841, 841f, **841f**
β-galactosidase, 950
β-lactam ring, 172, 172f
β-oxidation pathway, 672–673
Bicarbonate, 793–794
Bifunctional enzyme, 506–507, 506f

Bile, 832, 833
Bile acids, 667, 667f
Bile salts, 833, 833f
Bilirubin, 787, 788f
Biliverdin, 787, 788f
Biliverdin reductase, 787
Bimolecular reactions, 152
Bimolecular sheets, 359–362
Binding
    association and dissociation, 72, 72f
    as fundamental process, 69–72
    immune system and, 84–91
    myoglobin and, 73–75
    oxygen and, 75–84
    partners, concentrations of, 69–70, 70f
    proteins and, 70–72
Binding energy, 151, 180, 199
Binding partners of protein, 116
Binding propensity, 91–95
    dissociation constant for, 91–93, 92f
    kinetic parameters, 94–95
    specificity, 93–94
Binding-change mechanism, 565–567, 566f
Biochemical timescale, 7
Biochemistry
    biological diversity and, 2–4
    biological molecules, properties of, 6–21
    careers in, 27f
    chemistry concepts in, 6–21
    defined, 1
    DNA, form and function of, 4–6
    DNA sequencing, 21–27
    evolution timeline, 3f
    genome sequencing and, 21–22
    genomic revolution and, 21–27
    research in, 27–29, 27f
    unifying concepts, 2–4
Bioenergetics, 580–581, 581f
Bioinformatics, 265
Biological diversity
    biochemical unity underlying, 2–4
    similarity and, 2f
Biological macromolecules, 2
Biological membranes. *See* Membranes
Biologics, 995–996, 995t
Biomarkers, A1C as, 331
Biomolecules, 840–841
Biopterin, 723
Biotechnology, 264, 265
Biotin, 463t, 501, 501f
Biotin carboxyl carrier protein (BCCP), 502, 502f
Bisphenol A (BPA), 93–94, 93f, **93**
1,3-bisphosphoglycerate (1,3-BPG)
    as acyl phosphate, 480–481
    ATP formation and, 482–484
    in glycolysis, 457–458, 480–481, 483–484
    oxidation of energy, 457–458
    phosphoryl-transfer potential, 453–454, 454f
    structure of, **457**, **458**
2,3-bisphosphoglycerate (2,3-BPG)
    allosteric effector, 81
    binding of, 80–81, 80f
    defined, **80**

in oxygen affinity determination, 80–81, 80f
Bistable circuits, 958
Bisubstrate analog, 213, 213f
Bjorkman, Pamela, 90, 91b
Blackburn, Elizabeth, 863
BLAST (Basic Local Alignment Search Tool) search, 309, 309f
The Blind Watchmaker (Dawkins), 60
Blood clotting, 223
    Ca^{2+}, 229
    clots, 231–232
    enzymatic cascades, 228–229, 228f
    extrinsic pathway of, 228
    intrinsic pathway of, 228
    regulation of, 231–232
    zymogen activation in, 228–229, 228f
Blood glucose homeostasis, 334
Blood groups, 343–344, 343f, 344f
Blood sugar, 330. *See also* Glucose homeostasis
Blood-brain barrier, 990
Blood-glucose concentration. *See* Glucose homeostasis
Blosum-62
    alignment of identities only versus, 307f
    defined, 306
    graphic view of, 306f
    myoglobin and leghemoglobin, 308f
Blotting techniques, 265, 266–267, 267f
Boat form, 329–330, 329f
Body mass index (BMI), 733f, 734
Body weight, regulation of, 733–738. *See also* Obesity
Bohr, Christian, 81
Bohr effect, 81–83
Bombardment-mediated transformation, 296
Bombesin, 735t
Bonds
    axial, 329–330, 329f
    covalent, 5, 7–8
    disulfide, 43, 43f
    electrostatic interactions, 8
    equatorial, 329–330, 329f
    glycosidic, 331–332, 332f, 336, 336f, 338
    hydrogen. *See* hydrogen bonds
    hydrophobic effect and, 10–11, 11f
    hydrophobic interactions and, 11
    ionic interactions, 8–9, 11f
    isopeptide, 703, 704f
    noncovalent, 7–11
    peptide. *See* peptide bonds
    van der Waals interactions, 9–10, 10f, 12, 52, 360
Bortezomib, **707**
Bound receptors, fraction of, 92b–93b
Bovine ribonuclease, **302f**
Brain
    glucose in, 740
    ketone bodies and, 753–754
    leptin effects in, 736, 736f
    role in caloric homeostasis, 734–735
Branch points, 902, 903f
Branched-chain amino acids, 708–713, 722–723

Branched-chain ketoaciduria, 727
Branching enzyme, 655, 655f
Branson, Herman, 46, 46b
Brittle bone disease, 54
Bromodomains, 966, **966f**
Bronchitis, 341
Brown, Michael, 830
Brown adipose tissue (BAT), 575, 576f
Brown fat mitochondria, 576
Buchner, Edward, 472
Buchner, Hans, 472
Buck, Linda, 436
Buffers
    action, 19f
    in deprotonation, 193
    pH and, 19–20, 19f
    protonation, 19f
Bumpus, Namandjé, 836b
Burkitt lymphoma, 898
Butyryl CoA, **725b**

C_3 plants, 623, 624f
C_4 pathway, 622–623, 622f
C_4 plants, 623, 624f
CAAT box, 896, 896f
Caenorhabditis elegans, 125, 283, 295
Caffeine, **424**
Calcitriol, 464t, 839, 839f, 840
Calcium, 749, 750f
Calcium ion channels, 397, 397f
Calcium ion pump, 384–387, 386f
Calcium ions
    calmodulin activation, 423
    coordination to multiple ligands, 422,
        422f
    imaging, 422, 422f
    in phosphorylase activation, 650–651,
        650f
    pump structure, 386f
    pumping, 384–387, 386f
    as second messenger, 413–414,
        421–422
    in signal transduction, 420–422, 422f
Calmodulin (CaM), 124, 124f
    defined, 420, 423, 651
    in mitochondrial biogenesis, 749
    in signal transduction, 420, 423, 423f
Calmodulin-dependent protein kinases
        (CaM kinases), 423, 423f
Calnexin, 347
Caloric homeostasis, 733–738
    in body weight regulation, 733–738
    brain role in, 734–735
    defined, 733
    in diabetes, 741–746
    evolution and, 734
    exercise and, 752b
    impact of athletic activity on, 752b
    insulin in, 736
    leptin in, 736–737
    obesity and, 734
    satiation signals in, 735–736, 735f
    signaling in, 735–736
    suppressors of cytokine signaling
        (SOCS) in, 737–738, 738f
Calreticulin, 347
Calvin, Melvin, 610, 611b

Calvin–Benson cycle, 610–624
    ATP (adenosine triphosphate) in, 618
    C_4 pathway and, 622–623, 622f
    carbon dioxide in, 611–620, 622–623
    defined, 587, 610
    environmental factors in, 620–624
    five-carbon sugar formation in, 615,
        616f
    fructose 6-phosphate in, 615, 615f, 616f,
        618, 619f
    glycine in, 614f, 615
    glycolate in, 614f, 615
    hexose sugar formation in, 615–618, 616f
    illustrated, 612f
    light regulation of, 620–621, 620f
    magnesium ion and, 612–613, 613f, 621
    NADPH in, 610, 618, 622
    overview of, 610
    oxygenase reaction in, 614–615, 614f
    pentose phosphate pathway and,
        633–634
    3-phosphoglycerate formation in,
        611–612, 612f, 613f, 614f, 615, 615f
    photorespiration in, 614f, 615
    reactions in, 611–620, 617f
    regulation of, 620–622
    ribulose 1,5-bisphosphate in, 611–612,
        613f
    rubisco in, 611–615, 612f, 613f, 614f, 621
    salvage pathway in, 614–615
    stages, 611–612, 612f, 613f, 615–618,
        615f
    thioredoxin in, 621–622, 621t, **621f**
cAMP. See Cyclic AMP (cAMP)
cAMP receptor protein (CRP). See
        Catabolite activator protein (CAP)
Camptothecin, 857
Cancer
    adenocarcinomas, 341
    aerobic glycolysis in, 497–498
    breast, 838, 839f, 870t
    colorectal, 870
    defective DNA repair in, 869–871, 870t
    drug resistance in, 998–999
    epigenetics and, 962
    ethanol metabolism in, 758–759
    fatty acid metabolism in, 691–692
    glycolysis in, 497–498
    lung, 870t
    ovarian, 838, 870t
    renal, 870t
    signal transduction abnormalities in,
        433, 434, 434f, 435f
    skin, in xeroderma pigmentosum, 870
    treatments for, 870–871
    tumor hypoxia in, 498
    tumor suppressor genes and, 870
    tumor visualization, 497, 498f
    Warburg effect in, 497
Cancer therapy, 805–807
Candidates, in drug development, 986.
        See also Drug candidates
Capillary electrophoresis, 268
Captopril, 189
Carbamate
    in Calvin–Benson cycle, 613, 614
    formation of, 82, **82f**

Carbamate groups, 82
Carbamoyl phosphate
    in pyrimidine synthesis, 793
    reaction, 714
    in urea cycle, 714–716
Carbamoyl phosphate synthetase,
        deficiency of, 717, 717f
Carbamoyl phosphate synthetase I,
        714–715
Carbamoyl phosphate synthetase II
    in pyrimidine synthesis, 793, 793f, 794,
        794f
    structure of, 793, **793f**
Carbanion, 629
Carbanion intermediates, 628–630, 631f
Carbocation, 644f
Carbohydrate–asparagine adduct, **63f**
Carbohydrates, 324–348
    disaccharides, 333–334
    as fuel, 457f
    glucose generation from, 473–474
    isomeric forms of, 326f
    lectins and, 346–348, 347f
    monosaccharide, 325–333
    N-linked, 338, 338f
    oligosaccharides, 333, 342–343, 343f
    O-linked, 338
    overview of, 324
    polysaccharides, 335, 336f
    proteins linkage, 337–346
Carbon, oxidation of, 456–459
Carbon dioxide
    binding site, 192, 192f
    in Calvin–Benson cycle, 611–612,
        622–623
    defined, 82
    fixation of, 619–620, 620f
    hydration acceleration, 190
    hydration of, 193, 194f
    oxygen release and, 81–83
    pH and, 82, 82f
    transport from tissue to lungs, 82, 82f
Carbon fuels, free energy of, 456–459, 457f
Carbon monoxide, 74
    defined, 82
    myoglobin and, 74–75, **75f**
    in oxygen transport disruption, 81,
        82–83
    poisoning, 82–83
Carbonic anhydrases, 141
    bound zinc in, 190–191
    in carbon dioxide hydration
        acceleration, 190
    catalytic strategies of, 179, 190–194
    defined, 190
    evolution of, 193
    mechanism of, 192, 192f
    pH effect on, 190–191, 191f
    proton availability and, 192–194
    proton shuttle, 193, 194f
    regeneration of, 192–194
    structure of, **190**
Carbon–nitrogen bond, 181
Carboxamides, 38f, 39
Carboxybiotin, 501f
γ-carboxyglutamate, **63f**, 231, 231f
Carboxyhemoglobin, 83, 83f

Carboxylation of pyruvate, 501–502, 535
Carboxyl-terminal domain (CTD)
  in coupling transcription to pre-mRNA,
    908, 908f
  defined, 894
    phosphorylation of, 897, 897f
    serine residues of, 894
Carcinogens, tests for, 871–872, 871f
Cardiac muscles, 416, 416f, 419, 419f
Cardiotonic steroids, 387, 387f
Cargo receptor, 942
Carnitine, 671–672, 671f
Carnitine acyltransferase I, 671
Carnitine acyltransferase II, 672
Carotenoids, 602–603, 841, 841f
Carriers, 390
Cartilage, 340–341, 340f
Caspases, 579
Cassette mutagenesis, 281, 281f
Catabolism
  ATP generation by, 456
  defined, 448
  energy from, 456–459
  free energy of, 456–459, 459f
  roles of, 456
  stages of, 459, 459f
Catabolite activator protein (CAP),
    955–956, 955f, 956f
Catabolite repression, 955
Catalase, 561
Catalysis
  by approximation, 180
  by ATCase, 211
  Circe effect in, 159
  covalent, 180, 181, 181f, 183f
  efficiency of, 158–159
  enzyme–substrate binding in, 151
  enzyme–substrate complex formation
    in, 149, 149f
  general acid–base, 180
  maximal rate, 154–155, 156
  metal ion, 180
  specificity constant, 158
  transition state formation and, 148
  velocity of, 158
  zinc activation of water molecule,
    191–192, 192f
Catalysts, 142, 181
Catalytic RNA, 909–912
Catalytic strategies, 180
  of carbonic anhydrases, 179, 190–194
  in chymotrypsin, 181–182, 181f, 182f
  of myosins, 179, 200–207
  overview of, 179
  of restriction endonucleases, 179,
    194–200
  of serine proteases, 179, 183–185
Catalytic triads
  defined, 183
  in elastase, 186–187
  in hydrolytic enzymes, 184f, 186–187,
    186f
  serine in, 183–185, 183f
  site-directed mutagenesis and,
    187–188, 187f
  of subtilisin, 186–187, 187f
  in trypsin, 186–187, 186f

Cataracts, 491, 491f
Catecholamines, 785, 785f
Cation exchange, 104
cDNA library, 279
CDP-ethanolamine, 818, **818**
Cech, Carol, 912b
Cech, Thomas, 912b
Cell cycle, 862
Cell-mediated immunity, 89
Cells, 2, 114
Cell-to-cell channels, 406–407, 406f, 407f
Cellular immune response, 86
Cellular respiration
  citric acid cycle in, 516–517, 517f
  defined, 543
  enzymes in, 144
  in oxidative phosphorylation, 543–545
  vs. photosynthesis, 587
  regulation of, 573–579
Cellulose, 335–336, **336f**
Centrifugation
  differential, 102, 102f
  ultracentrifugation, 111
Cephalopods, 337
Ceramide
  insulin resistance and, 743–744, 744f
  as precursor, 820f
  synthesis of, 819, 819f
Cerebroside, 357, **357**, 819, **820f**
CETCH cycle, 619–620
Cetuximab (Erbitux), 434, 434f, 996
cGMP, 980, **980f**
cGMP phosphodiesterase, 439, 439f
Chain, Ernest, 979
Chair form, 329–330, 329f
Changeux, Jean-Pierre, 78
Channelrhodopsin, 607, 607f
Channels, 382
Chaperones, 57
Charged tRNA, 921
Charpentier, Emmanuelle, 294b
Checkpoints, 862
Chemical modification reaction, 182
Chemical shifts, 133
Chemical space, 984
Chemiosmotic hypothesis, 562–563, 563f
Chemotroph, 447, 586
Cheng, Yung-Chi, 987
Cheng–Prusoff equation, 987
Chirality
  of amino acids, 35, 769f, 770
  establishment of, 767f
Chitin, 336–337, 337f
Chitosan, 337
Chlamydomonas, 911
Chloramphenicol, 942, 943t
Chlorobium thiosulfatophilum, 606
Chlorophyll *a*, 589–590, 602, 603f
  absorbance spectrum, 590f
  structure of, **589f**
Chlorophyll *b*, 602–603, **602**, 603f
Chlorophylls, 786
  defined, 589
  in photosynthesis, 589–593
  light absorption by, 590, 590f
  phytol, 589
  structure of, 589

Chloroplasts
  ATP synthase, 600–601
  ATP synthesis in, 599–602
  defined, 588
  electron micrograph of, 588f
  evolution of, 589
  genome of, 589
  in photosynthesis, 588–589, 588f
  reaction center, 590–591, 591f
  redox conditions, 601
  structure of, 588, 588f
Cholecystokinin (CCK), 735, 735f, 735t
Cholera toxin, 344, 344f
Cholesterol
  "bad," 832
  carbon numbering, 834, 834f
  defined, 355, 821
  discovery of, 981
  excess, effects of, 830, 830f
  fatty acid chains and, 371–372, 372f
  formation of, 825f
  "good," 832
  from lanosterol, **825f**
  levels, clinical management of,
    832–833
  lipid rafts, 372–373, 372f
  as membrane fluidity regulator,
    371–372
  in membrane lipid, 357–358, 821–823
  as precursor, 833–841, 833f
  as precursor for steroid hormones, 827,
    834, 834f
  as precursor for vitamin D, 839–840
  structure of, 357–358, **357**
  transport of, 827–829
Cholesterol homeostasis, 827
Cholesterol synthesis, 821–825, 822f
  blocking of, 833, 981
  condensation mechanism of, 823,
    823f
  defined, 822
  isopentenyl pyrophosphate in, 822–823,
    823f
  liver in, 825, 828f
  low-density (LDLs) in, 829–830
  metabolism, 829–830
  mevalonate in, 822, 822f, 823f, 827
  as precursor for steroid hormones, 834,
    834f
  rate of, 825–827
  regulation of, 825–833
  site of, 827, 828f, 829
  squalene in, 823, 824–825, 824f, 825f
  stages of, 822–825
  sterol regulatory element binding
    protein (SREBP) in, 825–827, 826f
Choline, 818
Chorismate in amino acid synthesis,
    776–779, 777f
Chromatin, 961–967
  defined, 893
  in gene expression, 893, 961–967
  higher-order, structure of, 958
  histone tails and, 965–966, 965f
  remodeling of, 961–967, 966f
  structure of, **893f**
Chromatin-remodeling complexes, 966

Chromatography
  affinity, 104–105, 104f, 110
  gel-filtration, 103, 103f, 110
  high-performance liquid (HPLC), 105, 105f
  ion-exchange, 103–104, 104f
Chromogenic substrate, 182, 182f
Chromosomes
  accessory, 272, 274
  bacterial, 864, 864f
  eukaryotic, 861, 861f, 863
  in human genome, 283, 283f, 286f
  yeast, 958
Chronic myelogenous leukemia (CML), 435, 435f
Chylomicron remnants, 828
Chylomicrons, 667–668, 668f, 828, 828t
Chymotrypsin
  active site, 183f
  catalysis kinetics, 182, 182f
  catalytic mechanism, 181–182, 181f
  catalytic triad in, 183–185
  conformations of, 225, 225f
  covalent catalysis in, 181, 181f, 183f
  diisopropylphosphofluoridate (DIPF) and, 181–182, 182f
  elastase and, 186, 186f
  mammalian, 313
  natural substrate for, 170, **170f**
  oxyanion hole, 185, 185f
  peptide hydrolysis by, 183, 184f
  S₁ pockets of, 186f
  serine residue, 181–182, 182f
  specificity of, 181f
  specificity pocket of, 185, 185f
  structure of, 183–185, **313f**
  substrate preferences of, 158–159, 158t
  trypsin and, 186, 186f
Chymotrypsinogen
  conformations of, 225, 225f
  defined, 224
  proteolytic activation of, 224f, 225–226
Ciproflaxin, 857, **857**
Circe effects, 159
Cirrhosis, 756
Cis configuration, peptide bonds, 44, 44f, 45f
Cis-acting elements, 896
Cisplatin, **871**
cis-Polyisoprene, **840f**
Citrate
  in fatty acid metabolism, 690, 695f
  in glyoxylate cycle, 537, 537f
  isomerization of, 524–525, 525f
  in isomerization reactions, 467
  vs. phosphate, 459
  phosphofructokinase inhibition by, 495
  in polymerization, 695
  synthesis of, 533
  transport to cytoplasm, 533
Citrate synthase
  in citric acid cycle, 522–523, 523f
  conformational changes on binding, 523, 523f
  defined, 523
  mechanism of, 523–524, 524f
Citric acid cycle

acetyl CoA in, 502, 516, 518–520, 522–524, 524f, 535
aconitase in, 524–525, 525f
ADP in, 503
ATP in, 517, 526–531, 533, 534–535
biosynthetic roles of, 534–536, 534f
carbon oxidation in, 522–531
in cellular respiration, 516–517, 517f
citrate isomerization in, 524–525, 525f
citrate synthase in, 522–523, 523f
citryl CoA in, 522–524, 524f
components of, 530t
control sites of, 533, 533f
decarboxylation in, 518–519, 519f
defined, 516
entry to, 531–533
evolution of, 536
in exercise, 532
glycolysis and, 517–521, 518f
high-energy electron harvesting in, 516
illustrated, 529f
intermediates in, 534
isocitrate dehydrogenase in, 525, 533, 533f
isocitrate in, 524–525
α-ketoglutarate dehydrogenase complex in, 526, 533, 533f
α-ketoglutarate in, 525
ketone bodies in, 753–754, 754f
lipoamide in, 520–521, 522f
in mitochondria, 516, 516f
overview of, 459, 516–517, 517f
oxaloacetate in, 517, 522–523, 523f, 528–529, 534
oxidation by, 488
pathway integration, 535
pattern of, 517, 517f
phosphorylation in, 531–532, 532f
in photosynthesis. See dark reactions
pyruvate dehydrogenase complex in, 516, 517–521, 518f, 520f, 522f, 531–533, 531f, 532f
pyruvate dehydrogenase in, 531–533
pyruvate dehydrogenase phosphatase (PDP) in, 531
rate of the cycle, 533, 533f
reactions of, 528
regulation of, 531–532, 533, 533f
replenishment, 534–535
substrate channeling in, 530
succinate dehydrogenase in, 528–529
succinyl CoA in, 526–527, 527f
succinyl CoA synthetase in, 527
thiamine pyrophosphate (TPP) in, 518–519, 519f
Citrulline, 715–716, **715**
Citrullinemia, 727t
Citryl CoA (citryl coenzyme A), 522–524, 524f
Clamp loaders, 857–858, 859f
Clapham, David, 425b
Class I MHC proteins, 90, 90f
Classic galactosemia, 491
Clathrin, 375, 829
Clathrin-coated pit, 375
Cleavage
  DNA. See DNA, cleavage of
  protein, 121–123, 122f

Cleland, W. Wallace, 160
Clinical trials, 996–997, 997f
Cloned genes, 104–105
Clones, 113
Cloning
  of cDNA, 394
  expression, 279, 279f
  plasmid vectors in, 274–275
  in recombinant DNA technology, 277–278
Cloning vectors, 274
  mutant λ phage as, 276, 276f
Clopidogrel (Plavix), 836, 836f
Closed promoter complex, 887, 887f
Clostridium perfringens, 488
Clotting. See Blood clotting
Clustered regularly interspaced palindromic repeats See CRISPRs
Coactivators, 965, 965f
Coat proteins (COPs), 942
Coated pits, 829
Cobalamin, 676–677, 676f, 677f, 678f. See also Vitamin B₁₂
Cobalt atom, 676–677
Cochlea, 440
Coding strand, 884
Codon-anticodon interactions, 921
Codons
  defined, 918
  maximizing number of, 921
  stop, 935–936, 935f
  in translation, 920–921, 921f
Coenzyme A, 462–463, **462f**, 486f
Coenzyme B₁₂, 676–677, **676f**, 677f, 678
Coenzyme Q, 528, 550, 551–552
Coenzyme Q₁₀, 840
Coenzymes
  defined, 144
  prosthetic groups, 144
  stoichiometric, 144
  types of, 144t
  vitamin, 463–464, 463t
Cofactors, 144
Cognate DNA, 194, 197–199, **198f**, 199f
Cohesive ends, DNA, 272–273, 273f
Coiled-coil proteins, 53, **53f**
Co-immunoprecipitation, 116, 117f
Colipase, 667
Collagen
  amino acid sequence of, 53–54, 54f
  ascorbate and, 757
  in cartilage, 340–341, 340f
  defined, 53
  structure of, **54f**
  triple helix, 54
Color blindness, 442
Color vision, 440f
  in animals, 442, 442f
  cone receptors, 440
  defective, 442
  evolution of, 442, 442f
Combinatorial chemistry, 319, 982–983
Combinatorial control, 960
Committed step, 211, 494, 780, 780f
Common ancestor, 3, 3f, 25
Compactin, 981, **981**
Comparative genomics, 22–23, 22f
Compartmentalization, 468

Compartments, drug target, 990
Competitive inhibition. *See also* Enzyme inhibition
 defined, 164–165
 dissociation constant, 166–167
 double-reciprocal plot, 156, 156f
 kinetics of, 166–167, 167f
 use of, 166
Complementarity-determining regions (CDRs), 86
Complementary DNA (cDNA)
 cloning and sequencing of, 394
 defined, 278
 formation of, 278–279, 278f, 279f
 screening of, 279, 279f
 synthesis of, 289
Complementary sequences, 6, 6f
Complementary single-stranded ends, DNA, 272
Complex I, 549, 552, 553f, 559. *See also* NADH-Q oxidoreductase
Complex I mutations, 578
Complex II, 549, 553. *See also* Succinate-Q reductase
Complex III, 549, 554, 555f, 559. *See also* Q-cytochrome *c* oxidoreductase
Complex IV, 549, 556, 559. *See also* Cytochrome *c* oxidase
Complex V, 562
Concerted model
 allosteric enzymes, 215
 defined, 78, 215
 T and R states, 78f
 tetramers, 78–79
Condensing enzyme, 687
Cones, 438, 440, 442, 442f
Conformation selection, 151
Congenital disorders of glycosylation, 344–345
Congenital erythropoietic porphyria, 788
Congenital hemolytic anemia, 509
Congestive heart failure, 387
Conjugation in drug metabolism, 990–991, 991f
Connexin, 406
Connexons, 406
Consensus sequence, 220, 886
Conservative substitutions, 305, 307f, 308f
Conserved sequences, 886f
Constitutional isomers, 325, 326f
Controlled studies, 997
Controlled termination of replication, 267–268
Convergent evolution, 313, 313f
Cooperative binding
 concerted model, 78–79, 78f
 of oxygen, 76–77
 oxygen delivery and, 76–77, 77f
 sequential model, 78–79, 79f
Cooperative transition, 59
Copper ions in cytochrome c oxidase, 556
Coproporphyrin I, 788
Coproporphyrinogen III, 786, 787f
Core enzyme, 886
Corepressors, 955

Corey, Robert, 46
Cori, Carl, 509b, 661
Cori, Gerty, 509b, 661
Cori cycle, 509, 509f
Cori disease, 661t
Coronary artery disease, 830
Corrin ring, 676–677, 676f
Corticosteroids, 837, 837f
Corticosterone, **837f**
Cortisol, 222, 837, **837f**
Corynebacterium diptheriae, 943
Cotransporters, 390. *See also* Secondary transporters
Coulomb energy, 8
Coupled assays, 982, 982f
Coupling, 268
Covalent bonds
 in biological molecules, 7–8
 defined, 5
 multiple, 7–8
 as strongest bonds, 7–8
Covalent catalysis
 chymotrypsin as example, 181, 181f, 183f
 defined, 180
Covalent enzyme-bound intermediate, 481
Covalent modifications
 acetylation, 218, 219t
 common, 219t
 dephosphorylation, 218
 in enzyme regulation, 218–223
 irreversible, 218
 phosphorylation, 219t
COVID-19, 26, 89, 437, 881. *See also* SARS-CoV-2
CpG island, 961
Crassulacean acid metabolism (CAM), 624, 624f
Creatine, **160**, **454**, **747**
Creatine kinase, 454
Creatine phosphate, 453–455, 454f, **454**, 747, 747t, **747**
Crick, Francis, 5, 6, 845
CRISPR-Cas system, 293, 294b, 294f, 868, 874
CRISPRs (clustered regularly interspaced palindromic repeats), 293, 294f
Cristae, 543, 565f
Critical Assessment of Structure Prediction (CASP), 61
Cro in genetic circuit formation, 957–958, 958f
Cross talk, 414
Cross-links, 43, 43f
Cross-peaks, 134
Crotonyl-CoA, **619**
Crown gall, 295
Cryo-crystallography, 938b
Cryo-electron microscopy (cryo-EM), 100, 135–136, 136f
Crystal violet dye, 373
CTP. *See* Cytidine triphosphate (CTP)
CTP-phosphocholine cytidylyltransferase (CCT), 818
C-type lectins, 347, 347f

Cumulative feedback inhibition, 781–782
Cumulative selection, 60
Cupriavidus necator, 606–607
Cushing syndrome, 222–223
Cushing's disease, 222–223
Cut and display, 89–90
Cyanide, 577, 577f
Cyanidiales, 317
Cyanobacteria, 589, 589f, 593
Cyanocobalamin, 676f, 677
Cyclins, 862–863
Cyclic AMP (cAMP)
 activation of, 740
 binding to regulatory subunit, 221
 defined, 792
 in eukaryotic cells, 221, 418
 generation by adenylate cyclase, 418
 in glycogen metabolism, 652–653, 657f
 phosphorylation stimulation, 418
 as second messenger, **413**, 424
 in signal transduction, 412, 418–419
 structure of, **221**
Cyclic electron flow, 592–593
Cyclic GMP (cGMP), **413**, 792
Cyclic nucleotides, 792
Cyclic photophosphorylation, 601, 601f
Cyclin-dependent protein kinases, 863
Cycloheximide, 943, 943t
Cyclooxygenase inhibitors, 366
Cyclooxygenases, 141
Cyclophosphamide, **871**, 991
Cys$_2$His$_2$ zinc-finger domains, 959
Cystathionase, 776
Cystathionine, 776
Cystathionine β-synthase, 776
Cysteine, **43f**, 719, 719f
 in glycolysis, 483f
 linked, 42
 pyruvate formation from, 719f
 structure of, **38f**, 39, **40t**, 41
 synthesis of, 772, 776
Cysteine proteases, 188, 188f, **188f**
Cysteine thiyl radical, 803
Cystic fibrosis, 44, 341
Cystine, 43
Cytidine, 869f
Cytidine diphosphodiacylglycerol (CDP-diacylglycerol), **816**, 817, **817**
Cytidine monophosphate (CMP), 816, **817**
Cytidine triphosphate (CTP), 216f, 217f, 451
 ATCase inhibition by, 211, 211f, 807
 defined, 210, 211
 effect on ATCase kinetics, 216–217, 216f, 217f
 in pyrimidine synthesis, 796
 T state stabilization, 216, 216f, 217f
Cytidine triphosphate synthetase, 796
Cytochrome, in photosynthetic reaction center, 591f, 592–593, 593f
Cytochrome *b5*, 692–693, 693f
Cytochrome *bf* complex, 596, 596f, 605
Cytochrome *c*, in oxidative phosphorylation, 551, 551f

Cytochrome c oxidase, 549, 549f, 550t
  defined, 554
  electron flow in, 577
  heme, 557
  mechanism, 558, 558f
  peroxide bridge, 558, 558f
  proton transport by, 559, 559f
  in respiratory chain, 554–555, 556–559
  structure of, 557, 557f
  subunits, 556
Cytochrome P450, 865, 990, 990f
  defined, 835
  hydroxylation by, 834–836, 835f
  mechanism, 835
  protective function, 836, 836f
  research in, 836b
Cytochromes, 551
Cytoplasm, 502–503, 502f, 624, 641, 641f, 690, 690f
Cytoplasmic side, 544
Cytosine, 4, 5, 5f, 783f, 868–869
Cytoskeleton, 35, 206
Cytotoxic (killer) T cells, 86

D amino acids, 35, 35f
Dalgarno, Lynn, 930
Daltons, 43
Daly, Marie M., 357b
Dark reactions, 587
Darst, Seth, 881
Dawkins, Richard, 60
De novo pathways. See also Nucleotide synthesis
  defined, 792, 792f
  purine, 796f, 797, 797f, 798–799, 798f, 799t
  pyrimidine, 792, 792f, 796
Deamination
  agents, 865
  in amino acid degradation, 712
  example of, 865, 865f
  pyridoxal phosphate (PLP) in, 711
Death cap, 894f
Debranching enzyme, 646, 661
Decarboxylation
  in citric acid cycle, 518–519, 519f
  in gluconeogenesis, 502–503
  in pentose phosphate pathway, 624
  pyridoxal phosphate (PLP) in, 711
DeepMind, 61
Deficiency disorders, 63–64, 63f
Degrons, 704, 707
Dehydratases, 712
7-dehydrocholesterol (pro-vitamin D$_3$), 839–840, 839f
Deinococcus radiodurans, 864, 864f
Deleterious reactions, 480
Denaturation
  of DNA, 17–18, 18f
  of proteins, 56, 59, 59f, 162
Denisovans, 318, 319f
5'-deoxyadenosyl radical, 677, 677f
Deoxyguanylate, 801f
Deoxyhemoglobin
  αβ dimers, 75
  2,3-BPG binding to, 80f

carbon dioxide and, 82
  defined, 78
  quaternary structure of, 77f, 78
  structure of, 79f
Deoxymyoglobin, 73, 74f
Deoxyribonuclease (DNase), 289
Deoxyribonucleic acid (DNA). See DNA
Deoxyribonucleoside 3'-phosphoramidites, 268, 268f, 269
Deoxyribonucleoside triphosphate (dNTP), 847, 847f, 848, 848f
Deoxyribonucleotide synthesis
  control of, 808–809
  deoxyuridylate in, 804–805
  dihydrofolate reductase in, 805
  overview of, 801–802
  ribonucleotide reductase in, 801–803, 802f, 804f, 808–809, 808f
  thymidylate, 804–807, 805f, 806f
Deoxyribose, 4, 326f, 327
Deoxyribose phosphodiesterase, 867
Deoxyuridylate (dUMP), 804–805, 805f
Dephosphorylation, 218, 220, 221
Deprotonation
  effect of buffer on, 193, 193f
  water, kinetics of, 192f, 193
Designer genes, 281
Desmolase, 837
Desmopressin, 127f
Destroying angel, 894f
Development phase, of drug development, 978, 978f
α-dextrinase, 473
DHU loop, 919
Diabetes insipidus, 741
Diabetes mellitus
  defined, 741
  exercise and, 745
  insulin in, 742–743
  insulin resistance and, 741, 742–743, 743f
  ketosis in, 682, 682f
  monitoring of, 331
  obesity and, 741–746
  trans fats and, 679
  type 1, 741, 746
  type 2, 741, 743
Diabetic ketoacidosis, 682–683
Diabetic neuropathy (DN), 532–533
Diacylglycerol (DAG), 667f, 669, 743, 744f
  in lipid synthesis, 815–816, 821
  in signal transduction, 420–421, 420f
  structure of, 413
Diacylglycerol kinase, 815
Dialysis, 57
Dialysis in protein purification, 103, 103f
Diamond-Blackfan anemia, 938
Diastereoisomers, 325–326, 326f
Diazotrophic microorganisms, 764
Dicer, 972
Dictyostelium discoideum, 201, 201f
Dicyclohexylcarbodiimide (DCC), 128f, 577

Diels-Alder reaction, 778
2,4-dienoyl CoA, 675, 675f
Diet
  fatty acid metabolism and, 696
  ketogenic, 682
Differential centrifugation, 102, 102f
Diffusion
  facilitated, 382
  lateral, 369, 370, 370f
  lipid, 369
  membrane protein, 370
  simple, 382
  transverse, 370, 370f
Diffusion coefficient, 369, 370
Digestion
  of dietary proteins, 702, 702f
  enzymes in, 223, 224, 224f
  of lipids, 667–668, 667f
  proteolytic enzymes in, 223–228
Digestive enzymes, 63, 702
  chymotrypsinogen, 224–225, 224f
  synthesis of, 223
Digitalis, 387
Digitoxigenin, 387
Diglyceride acyltransferase, 816
Digoxin, 387, 387f
Dihedral angles, 45
Dihydrobiopterin, 723
Dihydrofolate reductase, 805
Dihydrolipoyl transacetylase, 518t, 520
Dihydropteridine reductase, 723
Dihydrotestosterone (DHT), 838
Dihydroxyacetone, 325, 325, 326f
Dihydroxyacetone phosphate (DHAP), 332
  in Calvin–Benson cycle, 616
  in glycolysis, 466, 476f, 478, 479f, 489
  in lipid synthesis, 815
  structure of, 332
Dihydroxyphenylalanine, 162
Diisopropylphosphofluoridate (DIPF), 170, 170f, 170f, 181–182, 182f
Dimerization arm, 431, 431f, 432f, 434
Dimethylallyl pyrophosphate, 823
Dimethylbenzimidazole, 677
2,4-dinitrophenol (DNP), 577–578, 578
2,4-dinitrophenylhydrazine, 727
Dioxygenases, 724, 724f
Diphosphatidylglycerol, 356f
Diphosphatidylglycerol (cardiolipin), 817, 817
2,3-diphosphoglycerate (2,3-DPG). See 2,3-bisphosphoglycerate (2,3-BPG)
Diphthamide, 944, 944f
Diphtheria toxin, 943–944, 944f
Diplococcus pneumoniae, 280
Dipolar ions, 35
Dipoles, 9
Direct DNA repair, 867
Disaccharides
  defined, 333
  structure of, 333–334, 333f
  types of, 333–334, 333f
Discovery phase, of drug development, 978, 978f

Diseases and disorders
  albinism, 727t
  alcaptonuria, 726
  alcohol-related, 755–757, 757f
  Alzheimer's disease, 62–63, 63f
  amino acid sequences and, 44
  amyotrophic lateral sclerosis (ALS),
      271, 290
  Anderson disease, 661t
  anemia, 21, 83–84, 339
  Angelman syndrome, 705
  anthrax, 857
  argininosuccinase deficiency, 717, 717f
  arsenite poisoning, 535–536, 536f
  atherosclerosis, 679, 827, 830–831, 832
  autism, 705
  B-cell leukemia, 898
  beriberi, 535–536, 756
  bronchitis, 341
  Burkitt lymphoma, 898
  cancer. See Cancer
  carbamoyl phosphate synthetase
      deficiency, 717, 717f
  carnitine deficiency, 672
  cataracts, 491, 491f
  cholera, 344, 344f
  citrullinemia, 727t
  congenital disorders of glycosylation,
      344–345
  congestive heart failure, 387
  Cori disease, 661t
  coronary artery disease, 830
  Cushing syndrome, 222–223
  cystic fibrosis, 44, 341
  diabetes. See Diabetes mellitus
  drug-resistant, 998–999
  emphysema, 228
  epigenetics and, 962
  familial hypercholesterolemia, 830
  galactosemia, 491
  glucose 6-phosphate dehydrogenase
      deficiency, 634–636
  glycogen-storage, 660–661, 661t, 749
  gout, 810
  Hartnup disease, 702
  heart disease, 833, 981
  hemolytic anemia, 634–635
  Hers disease, 661t
  HIV infection. See Human
      immunodeficiency virus (HIV)
      infection
  homocystinuria, 727t
  Huntington disease, 62, 703, 869
  hyperammonemia, 717
  hyperlysinemia, 727t
  I-cell disease, 344–345
  lactose intolerance, 492
  Leber hereditary optic neuropathy
      (LHON), 578
  Lesch-Nyhan syndrome, 810–811
  long QT syndrome (LQTS), 404–405
  Lynch syndrome, 870, 870t
  maple syrup urine disease, 727
  McArdle disease, 661, 661t, 749
  mercury poisoning, 535–536
  mitochondrial, 578–579
  multidrug resistance in, 388–390
  multiple myeloma, 113–114
  mutations causing, 271–272
  neurological, protein misfolding in,
      62–63
  ornithine transcarbamoylase
      deficiency, 717, 717f
  osteoarthritis, 341
  osteogenesis imperfecta, 54
  Parkinson disease, 62, 703, 705, 817
  phenylketonuria, 727–728
  Pompe disease, 661, 661f, 661t
  predisposition to, 23
  protein aggregates in, 62–63
  pyruvate carboxylase deficiency (PCD),
      510–511
  retinitis pigmentosa, 906
  ricin poisoning, 944
  rickets, 840
  scurvy, 63, 756–757
  severe combined immunodeficiency
      (SCID), 809–810
  sickle-cell anemia, 21, 23, 44, 83, 83f
  Tay-Sachs disease, 820–821
  thalassemia, 906, 906f
  triose phosphate isomerase deficiency
      (TPID), 509–510
  tuberculosis, 538, 707
  tyrosinemia, 727t
  vitamin D deficiency, 464t, 840
  von Gierke disease, 660–661, 661t
  Wernicke-Korsakoff syndrome, 756
  xeroderma pigmentosum, 870, 870t
  Zellweger syndrome, 679
Dispensation sequence, 285
Displacement loop (D-loop), 873
Dissociation constant (K_d), 91–93, 92f, 95f,
      157, 166–167, 986
Distal histidine, 74f, 79f, 82–83
Distant evolutionary relationships,
      305–308
Distribution, drug, 989–990, 989f, 990f
Disulfide bonds
  formation of, 43, 43f
  reduction, 122–123, 123f
Dithiothreitol, 106
Diuron, **605**
Divergent evolution, 313
DNA
  ancient, amplification and sequencing
      of, 318–319, 319f
  bases in, 4, **4**, 11–12, 12f. See also Bases/
      base pairs
  building blocks, 4–5
  chemically synthesized linker, 274
  coding strand of, 884
  cognate, 194, 197–199, **198f**
  cohesive ends, 272–273, 273f
  complementary (cDNA), 278–279, 279f,
      289, 394, 401
  complementary single-stranded ends,
      272
  covalent structure, 4, 5f
  damage to, 864–872, 864f
  denaturation, 17–18, 18f
  distortion of, 198, 199f
  double helix of, 5, 6, 11–12, 17–18.
      See also Double helix
  double-stranded, 282
  electroporation and, 296, 296f
  exchange between species, 317, 317f
  fingerprint, 266
  forensics and, 271, 271f
  functions of, 2
  heredity and, 6
  ionic interactions, 11f
  as linear polymer, 5f
  linking number of, 852f, 853
  methylation, 200
  microinjection of, 288f, 289–290
  promoter sites in, 882
  proofreading of, 866–867, 866f
  properties of, 4–6
  proviral, 290
  recombinant. See recombinant DNA
      technology
  restriction fragment, 265–266
  in storage of genetic information, 4, 6
  structure of, 4–5, 6
  sugars, 4
  supercoiled, 851–857, 856f
  synthesis of, 268–269, 268f
  topoisomers of, 853, 853f
  transferred (T-DNA), 296
  unwinding of, 851–857, 882, 887, 887f, 888f
DNA, cleavage of, 272, 273f
  by CRISPR-Cas system, 293, 294f
  EcoRV endonuclease, 194
  metal ion catalysis in, 196–197, 197f
  restriction enzymes and, 194–200,
      265–266
  stereochemistry, 196, 196f
DNA fragments
  covalent insertion of, 272
  insertion and replication of, 274
  insertional inactivation of, 274, 275f
  joined by DNA ligases, 272
  library of, 277–278, 277f
DNA gyrase, 857
DNA ligase, 867
  defined, 849
  DNA fragments joined by, 272
  in DNA replication, 850, 862
  in forming recombinant DNA
      molecules, 272–273
  reaction, 849–850, 849f
DNA methylase, 200
DNA microarrays
  defined, 288
  gene-expression analysis with,
      288–289, 288f, 289f
  measuring gene expression using, 287f,
      288
DNA photolyase, 867
DNA polymerase α, 862, 862t
DNA polymerase β, 862t
DNA polymerase δ, 862, 862t
DNA polymerase I, 846, **846f**, 859, 862t
DNA polymerase II, 862t
DNA polymerase III, 857–859, 857f, 858f,
      862t
DNA polymerase switching, 862

DNA polymerases
  common structural features, 846–847
  defined, 846
  DNA repair, 866–867
  error-prone, 864–865
  eukaryotic, 861–862, 862t
  holoenzyme, 858, 859f, 860
  mechanism, 846, 846f
  metal ion catalysis in, 847, 847f
  primer, 846, 848f
  prokaryotic, 862t
  proofreading activities, 866–867, 866f
  specificity of, 847–848
  structure of, 846–847, **846f**
  template, 846
  as template-directed enzymes, 846
  types of, 862t
DNA probes
  amino acid sequences in making, 124
  defined, 266–267
  generated from protein sequences, 277,
      277f
  as primer, 269
  synthesis by automated solid-phase
      methods, 268–269
DNA recombination
  in cell division, 874
  defined, 872
  functions of, 872–874
  Holliday junctions in, 873–874, 873f
  illustrated, 872, 872f
  initiation of, 872–873, 873f
  mechanism, 873, 873f
  RecA in, 872–873
  recombinases in, 873
  strand invasion in, 872–873, 873f
  synapse of, 874
DNA repair
  base-excision, 867, 867f
  direct, 867
  double-strand, 868
  in E. coli, 866–867, 867f
  enzyme complexes in, 867f, 868
  example of, 864, 864f
  ligase in, 867
  mismatch, 867, 867f
  nonhomologous end joining (NHEJ) in,
      868
  nucleotide-excision, 868, 868f
  overview of, 864, 864f, 866
  proofreading in, 866–867, 866f
  single-strand, 858–859
  thymine in, 868–869
  trinucleotide repeats and, 869, 869f
  uracil DNA glycosylase in, 869
DNA replication, 6f, 846–851
  ATP hydrolysis in, 850–851, 850f
  cell cycle and, 861f, 862–863
  clamp loaders in, 857–858, 859f
  controlled termination of, 267–268, 267f
  coordinated process in, 857–864
  cross-linking in, 866, 866f
  defective, in cancer, 869–871, 870t
  DNA polymerase III in, 857–859, 859f
  DNA polymerases in, 847, 860–861, 862t
  in E. coli, 857, 860–861, 860f

  errors in, 864–865
  in eukaryotes, 861–862, 861f, 862f
  in eukaryotic cell cycle, 862–863, 862t
  helicases in, 850–851, 851f
  lagging strand in, 849, 849f, 858–859,
      858f
  leading strand in, 849, 849f, 858–859,
      858f
  licensing factors in, 862
  ligase in, 849–850, 849f, 862
  minor-groove interactions in, 848, 848f
  Okazaki fragments in, 849, 849f, 859
  origin of, 860–861, 860f, 861f
  origin of replication complexes (ORCs),
      861–862, 861f
  overview of, 846
  polymerase switching in, 862
  prepriming complex in, 860, 860f
  primer in, 848, 848f
  processivity in, 857–858
  replication fork in, 849, 858, 858f, 859f
  replicon, 861
  RNA polymerases in, 848, 848f
  shape complementarity in, 847–848,
      848f
  shape selectivity in, 848, 848f
  sites of, 860–861
  sliding DNA clamp in, 857–858, 857f
  specificity of, 847–848
  telomeres in, 863, 863f
  template in, 846
  termination of, 860, 861f
  trombone model in, 859, 859f
  tumor-suppressor genes in, 870
DNA sequencing, 123
  amino acid sequences and, 123f
  amplification by PCR, 269–270
  ancient DNA, 318–319, 319f
  by controlled termination of
      replication, 267–268, 267f
  decreasing costs of, 22f
  defined, 265
  human migrations and, 23f
  methods for, 21–22
  next-generation (NGS), 284–286, 285f,
      286f
  pyrosequencing, 284
  reversible terminator method, 285
  semiconductor, 285, 286f
DNA vectors, 272, 274–276
DnaA, 860, 860f
DnaB, 858
DNA-binding domains, 960
DNA-binding proteins
  basic-leucine zipper (bZip) in, 959, 959f
  binding to regulatory sites in operons,
      951–956
  in eukaryotes, 958–960
  helix-turn-helix motif in, 951, 951f
  homeodomains in, 958, 958f
  in prokaryotes, 950–956
  symmetry matching in, 951
  transcription inhibition by, 953
  transcription simulation by, 955–956
  zinc-finger domains in, 959, 959f
DNA-binding sites, 954f, 955, 961

DNA-encoded libraries, 984–985, 984f
Dolichol phosphate, **342**
Domains
  acetyllysine-binding, 966, 966f
  defined, 3
  DNA-binding, 950–956, 958–960
  homeodomains, 958, 958f
  immunoglobulin, 86–87
  nuclear-hormone receptor, 963–964, 964f
  protein, 52, 52f
Doolittle, Russell, 318b
Dopamine, 785, 785f
Dopaquinone, **162**
Double helix, 5–6, 5f, 845
  Acid–base reactions and, 17–18
  base pairs, 11
  destabilization, 17–18
  as expression of rules of chemistry,
      11–12
  features of, 846
  formation of, 6, 15–17, 15f
  minor groove, 897
  right-handed, 853
  single-molecule methods of
      monitoring, 15–17, 15f, 16f
  structure, 5, **5**
  topoisomerases and, 853–854, 854f
  in transcription, 897, 897f
  unwinding of, 853–854
Double helix formation
  from component strands, 5f, 6–7
  entropy and, 14f
  heat release in, 14–15
  illustrated, 5f
  principles of, 12
Double-blind studies, 997
Double-displacement reactions, 160–161
Double-reciprocal plot, 156, 156f, 167f
Double-stranded DNA, 282
Double-stranded RNA, 293, 294f
Doudna, Jennifer, 294b
Downstream core promoter element
      (DPE), 896
Doxycycline, 166
Drosophila melanogaster
  bases and genes, 125
  fluorescence micrograph of, 117f
  in genetic studies, 866
  genome of, 84, 283
  Toll receptor in, 84
Drug candidates
  absorption of, 988–989, 989f
  ADME properties of, 988–989, 988f
  administration routes for, 988–992
  characteristics of, 986–995
  distribution of, 989–990, 989f, 990f
  effective concentrations of, 986–987,
      987f
  excretion of, 992
  ligand binding in, 986, 986f
  metabolism of, 990–991
  oral bioavailability of, 988–989, 989f
  potency of, 986–987
  selectivity of, 986–987
  side effects of, 987
  therapeutic index of, 993

Drug development, 977–999
  animal testing in, 993
  clinical trials in, 996–997, 997f
  combinatorial chemistry in, 982–983
  criteria, 986–995
  DNA-encoded libraries in, 984–985, 984f
  high-throughput screening in, 981–982, 982f
  lead molecules, 980–986, 982f, 984f, 985f, 994f
  natural products in, 980–981
  overview of, 978–980, 978f
  phases of, 996–999
  phenotypic screening approach, 985–986, 985f
  screening libraries in, 982–983, 983f
  serendipitous observations in, 979–980
  split-pool synthesis in, 982–983, 983f, 984f
  structure-based, 993–995, 994f
  target-based screening approach, 985–986, 985f
Drug targets, 978–979, 978f
  animal-model testing of, 993
  compartments, 990
  validation of, 978
Druggability, 978
Drugs
  absorption of, 988–989, 989f
  administration routes for, 988–992
  delivery with liposomes, 361
  discovery approaches, 978–986, 978f
  distribution of, 989–990, 989f, 990f
  excretion of, 992, 992f
  half-life of, 992, 992f
  metabolism of, 990–991, 991f
  natural products in, 980–981
  resistance to, 998–999
  side effects of, 987
  targets of, 978–979, 978f
  therapeutic index of, 993
  toxicity of, 992–993, 993f
Duodenum, 702
Dynamin, 375

E. coli. See Escherichia coli (E. coli)
E site, ribosomal, 929, 929f, 934
Ear, hair cells of, 440–441, 440f
EcoRV endonuclease
  cleavage of DNA, 194, 196, 196f
  magnesium ion-binding site in, 196–197, 197f
  nonspecific and cognate DNA with, 199f
  recognition site structure, 197f
  twofold rotational symmetry, 197, 197f
Ectotherms, 162
Edman degradation, 120, 120f
Effective concentration, 986–987
EF-hand protein family, 423, 423f
EGF signaling, 431–433
Eicosanoids, 693–694, 693f, 694f
Elastase
  blocking of, 228
  catalytic triad in, 186, 187f
  S_1 pockets of, 186f
Elasticity, 157
Electric dipoles, 9, 9f

Electrochemical potential, 383. See also Membrane potential
Electromagnetic radiation, and DNA damage, 866
Electron transfer
  half-reactions in, 548
  inhibitors, 577–578, 577f
  in oxidative phosphorylation, 545–548, 561, 562f
  in photosynthesis, 589–593
  rate of, 561, 562f
  uncoupling, 575–576
Electron-density maps, 131, 131f, 132f
Electron-proton transfer reactions, 553f
Electron-transferring flavoprotein (ETF), 672
Electron-transport chain, 517, 559, 559f
  components of, 549f, 550t
  defined, 543
  in desaturation of fatty acids, 692–693, 693f
  electron flow through, 543, 543f
  FAD (flavin adenine dinucleotide) and, 517, 531
  $FADH_2$ (flavin adenine dinucleotide reduced) and, 517, 531, 545, 553–554
  inhibition of, 577
  iron–sulfur clusters in, 551–552, 551f
  NADH (nicotinamide adenine dinucleotide reduced) and, 517, 530, 545, 547, 548–549
  oxidation by, 488
  in oxidative phosphorylation, 543, 548–549, 569f
Electron-transport potential, 545
Electrophilic catalyst, 181
Electrophoresis
  capillary, 268
  gel, 105–107, 105f, 106f, 107f, 266–267, 266f, 363
  SDS, 105f, 106–107
  two-dimensional, 108–109, 108f, 109f
Electroporation, 296, 296f
Electrospray ionization (ESI), 118
Electrostatic repulsion, 453
Elion, Gertrude, 996b
Elongation factors
  bacterial versus eukaryotic, 937
  G (EF-G), 934, 934f
  Ts (EF-Ts), 932
  Tu (EF-Tu), 931–932, 931f
Elongation reactions, 692
Embden-Meyerhof-Parnas pathway, 473
Emphysema, 228
Enantiomers, 325, 326f
Endocytosis, 829
Endoplasmic reticulum (ER), 374, 374f
  defined, 341
  membrane, 692
  protein glycosylation in, 341–342, 341f
  protein synthesis in, 939–942
  protein targeting from, 939–941
  ribosome binding to, 939–942, 939f
  rough, 939
  smooth, 667, 668f, 939
  stress in, 745
Endosomes, 376f
  defined, 376, 829
  LDL release in, 831, 831f

Endosymbiosis
  mitochondria evolution from, 374
  receptor-mediated, 375–376, 375f, 376f
Endosymbiotic event, 544–545
Endosymbiotic theory, 545
Endotherms, 162
Enediol intermediate in glycolysis, 479, 479f
Energy. See also Thermodynamics
  activation, 148–149
  binding, 151, 180
  free. See Free energy
  kinetic, 12
  light, 606–607, 607f
  potential, 12–13
  status, 468
  total, 12
Energy charge, 468f
  defined, 468
  in regulation, 575–576
Energy coupling agent, 481
Energy homeostasis. See Caloric homeostasis
Energy metabolism, 732–759
  body weight regulation, 733–738
  diabetes and obesity, 741–746
  ethanol and, 755–759
  exercise and, 746–752
  fasted-fed cycle, 738–741
  starvation, 752–754, 753t
Energy transfer
  from accessory pigments to reaction centers, 603–604, 604f
  resonance, 603–604, 603f, 604f
Enhancers
  defined, 898, 962
  experimental demonstration of, 963f
  in gene expression, 962, 963f
  probe for, 963f
  in transcription, 898
Enol, 484
Enol intermediates, 524
Enol phosphate, 484
Enolase, 484
Enoyl CoA hydratase, 672–673
Enoyl reductase, 687
Ensemble studies, 163
Enterohepatic cycling, 992, 992f
Enteropeptidase, 226, 226f
Enterostatin, 735t
Enthalpy, 13
Entropy
  change during chemical reaction, 13
  double helix formation and, 14f
  overall, 14
  total, 13
Envelope form, 330, 330f
Enzymatic cascades, 228, 228f
Enzymatic velocity, 158
Enzyme inhibition
  affinity labels, 170, 170f
  competitive, 164–165, 167f
  by diisopropylphosphofluoridate (DIPF), 170, 170f
  illustrated, 170f
  irreversible, 164, 169–172
  mechanism-based (suicide), 171, 173f
  mixed noncompetitive, 166

noncompetitive, 168
pure noncompetitive, 166, 168f
reversible, 164, 166–168, 166f, 167f
by specific molecules, 164
by transition-state analogs, 139f, 173f
transition-state analogs in, 169
type of, determining, 171b–172b
uncompetitive, 166, 167f
Enzyme kinetics
for allosteric enzyme, 161, 161f
chymotrypsin, 182–183, 182f
of competitive inhibitors, 166–167, 167f
defined, 151
in first-order reactions, 152
Lineweaver-Burk plot, 156, 156f
maximal rate, 154–155, 156
Michaelis constant in, 154–157, 157t
Michaelis–Menten model for, 151–162, 167, 183
of noncompetitive inhibitors, 168, 168f
in pseudo-first-order reactions, 152
in second-order reactions, 152
sigmoidal, 212, 212f
specificity constant, 158–159
steady-state assumption and, 154–155
of uncompetitive inhibitors, 167–168, 167f, 168f
of water deprotonation, 192f, 193
Enzyme multiplicity, 781
Enzyme regulation, 210–232
by covalent modification, 218–223
of protein kinase A (PKA), 221–222, 222f
strategies, 210
Enzyme-linked immunosorbent assay (ELISA)
defined, 114
indirect, 115, 115f
sandwich, 115, 115f
Enzyme-pyridoxal phosphate complex (E-PLP), 710
Enzyme-pyridoxamine phosphate complex (E-PMP), 710
Enzymes, 34, 141–173
acceleration of reactions, 148–151
accessory enzyme systems, 692–694
activation energy and, 148, 148f, 149
active sites, 149, 150–151, 150f, 156
activity of, 101
allosteric, 161, 161f, 211. *See also* Allosteric enzymes
amount control, 210
amphitropic, 818
assay measurement of, 101
ATP hydrolysis and, 201
binding energy between substrate and, 151
branching, 655, 655f
catalytic activity, control of, 468
catalytic efficiency, 158–159, 158t
catalytic reaction, 149, 149f
in cellular respiration, 144
classes of, 143–144, 143t
cofactors for, 144, 144t
condensing, 687
debranching, 646, 661
defined, 142
digestive, 63, 223, 224–225, 224f, 702
energy-transducing, 14–15

ensemble studies, 163
free energy change, 145–147
Gibbs free energy and, 145–148
in gluconeogenesis, 535
of glycolysis, 474, 497
inhibition by molecules, 164–173
isozymes, 217–218, 218f
kinetic properties of, 151–162
kinetically perfect, 480
metabolism control of, 467
molecular dynamics, 165, 165f
multiple forms of, 210
overview of, 416
peptide-cleaving, 188–189
in photosynthesis, 144
as proteins, 142
proteolytic, 143, 210, 223–228
of purine nucleotide synthesis, 800
pyridoxal phosphate, 711–712, 712f
rate enhancement by, 142, 142t
reaction rate and, 148, 148f
regulatory, 821
restriction, 194–200, 265. *See also* Restriction enzymes (endonucleases)
reversible covalent modification, 210
from RNA, 912
single-molecule studies, 162–165, 163f, 164f
specificity, 142, 143f
spectroscopic characteristics of, 149
speed, 142
substituted intermediate, 161
substrate interaction, 150–151
temperature and, 161–162
ternary complex of, 159
transition state formation and, 148–151
turnover number of, 157, 157t
urea cycle, 716
Enzyme–substrate (ES) binding
in catalysis, 150–151
conformation selection, 151
induced-fit model of, 151, 151f
lock-and-key model of, 150–151, 151f
multiple weak attractions, 150
Enzyme–substrate (ES) complex, 168
defined, 149
diffusion-controlled encounter, 159
dissociation constant for, 157
evidence of existence, 149
fates, 153
formation in catalysis, 149, 149f
hydrogen bonds in, 150, 151f
reformed, 153
structure of, **150f**
Enzyme–substrate-inhibitor (ESI) complex, 167–168
Epidermal growth factor (EGF)
defined, 431
Ras activation and, 432, 432f
in signal transduction, 414f, 431–433
signaling pathway, 414f, 431–433, 432f, 433f
structure of, **430f**, 431
Epidermal growth factor receptor (EGFR)
defined, 432
dimerization of, 431f, 432
modulator structure of, 431f

overexpression of, 433
phosphorylation of carboxyl-terminal tail, 432
unactivated, structure of, 431f, 434
Epigenetics, 962
Epimers, 326f, 327
Epinephrine
defined, 414, 650
effects of, 221
in fatty acid metabolism, 695
in glycogen metabolism, 651–653, 651f, 656, 657f, 658, 658f
in signal transduction, 414–425, 414f, **414**, 416f, 419f
structure of, **650**
synthesis of, 783, 783f, 785f
Epitopes, 113
Equatorial bonds, 329–330, 329f
Equilibrium constant, 17
dissociation of water, 17
between enzyme-bound reactants and products, 204
metabolism and, 451
rate constants, 148
of reactions, 91–92, 146–147
standard free-energy change and, 147t
Equilibrium point, 148f
Equilibrium potential, 402, 402f, 403, 403b
Equilibrium reaction, 148
Erbitux, 996
Error-prone polymerases, 864–865
Erythromycin, 942–943, 943t
Erythropoietin (EPO), 339, 339f
Erythrose 4-phosphate, **627**, **628**
Escherichia coli (E. coli), 3
DNA polymerase I, 846, 846f
DNA recombination in, 872
DNA repair in, 866–867, 867f
DNA replication in, 857, 860–861, 860f
fatty acid synthesis in, 687
flagella movement of, 579–580
gene expression in, 950–956
genome of, 958
lac operon of, 953, 953f, 954f
membranes, 371
methionine repressor, 951, 951f
operons in, 955
primary RNA transcript, 898f
promoter sequences, 887f
pyruvate dehydrogenase complex in, 518t
RecA protein, 957
recombinant systems and, 112
restriction enzymes in, 194
ribonucleotide reductase of, 803
ribosomes in, 926
RNA polymerases in, 881, 881t, 882, 887f, 890
RNA sequence alignments, 314, 314f
sigma factors in, 890–892
translation in, 917
two-dimensional electrophoresis and, 108
Essential amino acids, 768f, 768t
defined, 702, 768
pathway steps, 768–769, 768f
table of, 702t

Estradiol
  binding of, 70–71, 93–94, 93f
  precursors, 838, 838f
  receptors for, 71f
  structure of, **70f**, 71, **838f**, **963**
Estrogen response elements (EREs), 963–964
Estrogens
  defined, 834
  nuclear hormone receptors and, 963–964
  pathways for formation of, 838f
  receptors for, 70–72, 71f, 72f, 93–94, 93f,
    95f
  synthesis of, 837–838, 838f, **838f**
Estrone, 70, **70f**, 838, **838f**
Ethanol
  cancer and, 758–759
  fermentation of, 473f, 486–487
  formation of, 486f
  in glycolysis, 473
  metabolism, 755–759
  vitamin metabolism and, 756–757
Ethanolamine, 817–818, **818**
Eukarya, 3, 3f
Eukaryotes
  ABC transporters in, 390
  chromatin and, 893, 893f
  chromosomes, 861, 861f, 863
  defining characteristics, 3
  DNA replication in, 861–862, 861f, 862f
  DNA-binding proteins in, 958–960
  fatty acid synthesis in, 687
  gene expression in. *See* Gene
    expression (eukaryotes)
  genomes of, 282–283
  glycolysis in, 473, 476–477
  membranes, 373–376
  new genes, expression of, 289–290
  nuclear envelope, 374, 374f
  oxidative phosphorylation in, 543–544
  photosynthesis in, 588–589, 593
  primary transcripts in, 901f
  promoters in, 895f
  protein synthesis inhibition in, 943–944
  ribosomes, 936–937
  RNA polymerases in, 881, 881f, 894–895,
    894t, 896f
  RNA processing, 892–893, 908, 908f
  transcription factors in, 896–898, 897f,
    958–960
  transcription in. *See* Transcription in
    eukaryotes
  translation in, 892, 892f, 936–937, 937f
  translation initiation in, 936–937, 937f
Eukaryotic cell cycle, 861f, 862–863
Eukaryotic genes
  organization of, 958
  quantitation and manipulation of,
    287–296
Eukaryotic RNA polymerases, 881, **881f**
Eversion, 386
Evolution
  of amino acid sequences, 44
  of aminoacyl-tRNA synthetases, 926
  ancient DNA amplification and
    sequencing in, 318–319, 319f
  of blood types, 344
  caloric homeostasis and, 734

  of carbonic anhydrases, 193
  of chloroplasts, 589
  of citric acid cycle, 536
  of color vision, 442, 442f
  comparative genomics and, 25–27
  convergent, 313, 313f
  divergent, 313
  of DNA-binding sites, 955
  of drug resistance, 998–999
  of globins, 316, 316f
  of glycolysis, 511
  of immune system, 84–86
  in laboratory, 319, 320f
  of metabolism, 468
  molecular, 271, 318, 319–321
  obesity and, 734
  of proteasomes, 707, 707f
  of sensory systems, 441–442, 441f
  of vitamins, 464
Evolutionary relationships, 301, 302
  distant, 305–308
  three-dimensional structures and,
    310–315
Evolutionary trees, 316–318, 316f
Excinuclease, 868
Excretion, drug, 992, 992f
Exemestane, 838, **839f**
Exercise
  aerobic, 747
  anaerobic, 747
  ATP in, 454–455, 455f, 498–499,
    747–749, 747t
  caloric homeostasis and, 752b
  citric acid cycle in, 532
  creatine phosphate in, 747, 747t
  diabetes and, 745
  effects on muscle and whole-body
    metabolism, 750–751, 751t
  energy metabolism and, 746–752
  fatty acid metabolism in, 748, 750
  fuel sources for, 746–750, 747t
  gene expression in, 750, 751
  glycolysis during, 499
  long-distance running, fuel for, 748–749
  middle-distance running, fuel for,
    747–748
  mitochondrial biogenesis and,
    749–750, 750f
  molecular adaptation to, 750–751, 750f
  oxidative phosphorylation in, 747–749
  oxygen concentration and, 77f
  in proteins phosphorylation, 223
  reactive oxygen species and, 561
  running velocity, 748, 748f
  sprinting, fuel for, 747
Exercise is Medicine initiative, 751
Exons, splicing, 901, 902f
Exonuclease, 846
Expression cloning, 279, 279f
Expression vectors, 275, 275f
Expressome, 935
Extracellular signal-regulated kinases
  (ERKs), 433
Extrinsic pathway, clotting, 228

2-^{18}F-2-D-deoxyglucose (FDG), 497, 498f
F_{ab} fragments, 87–88, 87f, 88f

Facilitated diffusion, 382
Facultative anaerobes, 488
FAD (flavin adenine dinucleotide)
  as catalytic cofactor, 518
  defined, 460
  electron transfer potential of, 520
  electron-transport chain and, 517, 531
  structure of oxidized form, **461f**
  structure of reactive components of,
    461, **461f**
$FADH_2$ (flavin adenine dinucleotide
  reduced), 672–673
  defined, 460
  electron-transport chain and, 517, 531,
    545, 553–554
  structure of reactive components of,
    461, **461f**
Familial hypercholesterolemia, 830–831,
  832
Faraday constant, 383, 546
Farnesene, 841, 841t, **841**
Farnesyl pyrophosphate, 823, 824f, **824f**
Fast twitch muscle fibers, 487, 493
Fasted state, 739–740
Fasted-fed cycle, 738–741
  defined, 738
  postabsorptive state, 739–740
  postprandial state, 739
  refed state, 741
Fasting, 753
Fast-twitch muscle fibers, 649, 747
Fat
  in adipose tissue, 748
  body, energy storage in, 666
  brown, 576
  as fuel, 457f
  neutral. *See* triacylglycerols
Fat cells, 666
Fatty acid chains
  cholesterol and, 371–372, 372f
  melting temperature, 371, 371t
  packing of, 371, 371f, 372f
Fatty acid metabolism, 665–696
  acetyl carnitine in, 671–672
  acetyl CoA carboxylase in, 685, 694–696,
    694f, 722
  acetyl CoA in, 532, 670–671, 672–673,
    690, 690f, 691
  activation in, 670–671
  acyl carrier protein (ACP) in, 685, 686f,
    689, 689f
  acyl CoA synthase in, 672
  β-oxidation pathway in, 672f
  chylomicrons in, 667–668, 668f
  citrate in, 690, 695f
  coenzyme B_{12} in, 676, 677f
  degradation in, 667, 667f, 672f, 679f,
    683–692
  diet and, 696
  enoyl CoA hydratase in, 672–673, 674f
  epinephrine in, 695
  in exercise, 748, 750
  FADH2 in, 672–673
  glucagon in, 695
  hormones in, 693–694
  inhibition of by NADH, 755
  insulin in, 696

ketone bodies in, 679–683
lipases in, 668–669
lipid mobilization and transport in, 667–668, 668f
lipolysis in, 669, 670f
malonyl CoA in, 685, 695
methylmalonyl CoA in, 677–678, 678f
NADH in, 672–673
NADPH sources for, 691
overview of, 665
oxidation in, 669–670, 671–674, 673t, 678–679
palmitoyl CoA in, 674, 674f, 695
pancreatic lipases in, 667, 667f
pathway integration, 691, 691f
peroxisomes in, 678–679, 679f
phosphatidate and, 815–821, 815f
steps in, 667f
triacylglycerols in, 666–668, 669f
in tumor cells, 691–692
Fatty acid synthase
catalytic cycle of, 689, 689f
defined, 683, 685, 686
structure of, 688, 688f
Fatty acid synthesis
in animals, 687–689
in cytoplasm, 690, 690f
degradation versus, 684–685, 684f
by fatty acid synthase, 683–692
glucagon and, 740
malonyl CoA in, 685
maximal, 695
reactions in, 684–689, 686t, 687f
steps in, 667f, 686–687, 686t, 687f
Fatty acid thiokinase, 673t. See also Acyl CoA synthetase
Fatty acids
animals and, 683
β oxidation of, 753
chain length in, 354, 688
cholesterol and, 371–372, 372f
defined, 353
degree of unsaturation, 354, 371
derivatives, 665
desaturation of, 692–693, 693f
as fuel, 665, 668–674
generation by lipolysis, 670, 670f
in lipids, 353–354, 353f, 354t
in membrane lipids, 353–354, 353f, 354t
in muscle, 743–744, 748
naturally occurring, 354t
in obesity, 743–744, 744f
odd-chain, 675–676, 677f
pathological conditions by, 679
physiological roles of, 665
polyunsaturated, 674–675, 675f, 679, 693–694
saturated, 679
structures of, 353f, 354
unsaturated, 674–679, 674f, 675f, 692–694
Fatty liver, 755–756
Fatty-acid binding proteins (FABPs), 667
Fatty-acid transport proteins (FATPs), 667
Favism, 634
F_c fragment, 87f, 88
Feedback inhibition, 211, 468

in amino acid synthesis, 780–783, 780f
cumulative, 781–782
enzyme multiplicity and, 781
in nucleotide synthesis, 807–809
in purine synthesis control, 807–808, 808f
Feedforward activation, 494
Feedforward stimulation, 494
Fehling's solution, 331
FeMo cofactor, 765–766, 765f
Fermentation
of alcohol, 472, 486–487
as anaerobic process, 488
defined, 486
in glycolysis, 486–487
to lactic acid, 473, 487, 487f
starting and ending points, 488t
Ferredoxin, 597–598, 597f, 602, 765
Ferredoxin-NADP$^+$, 597f, 598
Ferredoxin-NADP$^+$ reductase, 598, 598f
Ferredoxin–thioredoxin reductase, 601
Ferritin, 48, 48f, 786, 970, 970f
Ferrochelatase, 786
Fetal hemoglobin, 81, 81f
Fiber types, muscle, 649, 649t
Fibrils, 336
Fibrin, 229–230, 230f
Fibrin clots, 229–230, 230f
Fibrin monomer, 230
Fibrinogen
conversion by thrombin, 229–230
structure of, 229, 229f
Fibrinopeptides, 230
Fibronectin, 370
Fibrous proteins, 53–54
Fidelis, Krzysztof, 61
50S subunit of ribosomes, 927, 927f, 929f, 931–932, 933, 935
Fight-or-flight response, 412, 414, 785
Filgrastim, 995
Fingerprint, DNA, 266
First Law of Thermodynamics, 12–13, 734
First-order reactions, 152
First-pass metabolism, 990
Fischer, Emil, 150, 993
5′ cap, 900–901, 902f
Five-carbon units, 840–841
Flagella
bacterial, 579–580, 579f
rotary movement of, 579–580
structure of, 580, 580f
Flagellar motor
components of, 580, 580f
rotation of, 579–580
schematic view of, 580f
structure of, 580f
Flagellin, 579f
Flavin adenine dinucleotide (FAD), 867
Flavin mononucleotide (FMN), 461, 461f, 552, 552f, 891–892, 891f
Flavoproteins, 520, 553–554, 679, 679f
Fleming, Alexander, 979
FliG protein, 580
FliM protein, 580
FliN protein, 580
Flip-flop, 370, 370f

Flippases, 388
Florey, Howard, 979
Fluconazole, 990f
Fluid mosaic model, 370
Fluorescence microscopy, 117–118, 117f, 207, 207f
Fluorescence recovery after photobleaching (FRAP), 369, 370f
Fluorouracil, 806, 806f
Folding funnel, 60, 61f
Folic acid, 463t
Food production, 604
Foodstuffs, energy extraction from, 459, 459f
Formylglycinamidine ribonucleotide, 798
N-formylkynurenine, 724f, **724**
N-formylmethionine (fMet), 930–931
Formylmethionyl (fMet) peptide, **126**
Formylmethionyl-tRNA, 930, 930f
Formyltetrahydrofolate, 798, 798f, 799
43S preinitiation complex (PIC), 936
Fossil record, 316
Fourier transform, 131
Foxglove, 387, 387f
Fractional saturation, 73
Fragile X mental retardation protein (FMRP), 938
Framington Heart Study, 981
Franklin, Rosalind, 5, 129b
Frataxin, 552
Free energy
ATP and, 449–456
change during chemical reaction, 145–147, 455b–456b
defined, 13
Gibbs, 13, 145–148
in membrane transport, 382–383, 382f
negative, 14
of oxidation of single-carbon compounds, 456, 457f
in oxidation–reduction reactions, 546, 547–548
of phosphorylated compounds, 454t
of phosphorylation, 221
standard change, 146–147
standard free-energy change and, 147b
thermodynamics of metabolism and, 449
Free radicals. See Reactive oxygen species (ROS)
Friedreich's ataxia, 552
Fructofuranose, 327–328, 328f
Fructokinase, 489–490
Fructopyranose, 328f, 329
Fructose, 325, 326f, 327
excessive consumption of, 489
in glycolysis, 489–490, 489f
metabolism, 489–490, 489f
ring structures, 327, 328f
Fructose 1,6-bisphosphate, 740, 740f
in gluconeogenesis, 500f, 503
in glycolysis, 466, 476f, 477, **477**
Fructose 1-phosphate in glycolysis, 489, 489f

Fructose 2,6-bisphosphate (F-2,6-BP), 740
  activation of phosphofructokinase by, 495, 496f
  in gluconeogenesis, 505f, 506–507, 506f, 507f, 508
  in glycolysis, 495, 496f
  regulation of phosphofructokinase by, 495, 495f
Fructose 6-phosphate, 619f, 740f
  in Calvin–Benson cycle, 615, 615f, **616f**, 618, 619f
  in gluconeogenesis, 500f, 503
  in glycolysis, 477, **477**, 495
  in pentose phosphate pathway, **627**, **628**
  phosphorylation of, 477
Fructose bisphosphatase 2 (FBPase2), 506
Fucopyranose, **332f**
Fuel reserves in human body, 752, 752t
Fumarase, 528
Fumarate
  in amino acid degradation, 718, 719f
  in oxidation-reduction reactions, **465**
  from phenylalanine, 724, 724f, **724**
  in purine synthesis, 799
  in urea cycle, 715, **716**
4-fumarylacetoacetate, 724, **724**
Functional groups, of proteins, 34
Fungi, polysaccharides of, 336–337, 337f
Funny current (I_f), 405
Furan, 327
Furanose, 327, 328, 328f, 329–330
Futile cycles, 508

G elongation factor (EF-G), 934, 934f
G proteins
  activation of, 415–418, 417f, 439f
  adenylate cyclase and, 423–424
  ATP synthase and, 569
  defined, 416
  in glycogen metabolism, 651–653
  heterotrimeric, 415–418, 417f
  resetting of, 419–420, 419f
  role in signaling pathways, 416
  seven-transmembrane-helix (7TM) receptor and, 424–425, 425f
  in signal transduction, 415–418, 417f
  small, 432
  subunits, 416, 417f, 424, 424f
Gain, 653
Galactitol, 491
Galactokinase, 490, 491
Galactolipids, 588
Galactose, 164f, 326f, 327
  defined, 490
  in glycolysis, 489f, 490–491
  missing transferase and, 491
  ring form, 329f
  structure of, **164**
Galactose 1-phosphate uridyl transferase, 491
Galactosemia, 491
β-galactosidase, 164, 165f, 274, 274f, 334, 951–952, 952f
β-galactosides, 954
Galacturonic acid, **336**
Galdieria sulphuraria, 317, 317f

Gamma helix, 46
Gamma phosphoryl group, 201–202
Gangliosides, 820f
  disorders of, 820–821
  structure of, 819f, **820**
  synthesis of, 819, 819f
Gap junctions, 406–407, 406f, 407f
Gas-phase ions, 118
Gastric H+–K+ ATPase, 384
Gastric inhibitory peptide, 735t
Gastrointestinal peptides, 735t
Gate opening, 706
Gatto, Gregory, 29
GC box, 896, 896f
GDP (guanosine diphosphate)
  in citric acid cycle, 526–527
  hydrolysis of, 451
  in olfaction, 436
  in signal transduction, 416, 432, 432f
  in translation, 932
  in vision, 439
Gel electrophoresis
  defined, 106
  isoelectric focusing and, 107–108, 108f
  polyacrylamide, 105f, 106–107, 106f
  protein separation by, 105–107, 105f, 106f, 107f
  restriction fragments separation by, 266–267
  SDS, 105f, 106–107
  SDS-PAGE, 106–107, 108
  two-dimensional electrophoresis and, 108–109, 108f, 109f
Gel-filtration chromatography, 103, 103f, 110
Gellert, Martin, 854
Gene disruption, 290–294, 292f, 294f
  consequences of, 290–291, 291f
  defined, 290
  by homologous recombination, 290–291, 291f
Gene expression
  analysis with microarrays, 288–289, 288f, 289f
  bacterial, 935, 950–951
  chromatin in, 893, 961–967
  constitutive, 950
  control of, 938, 949–972
  epigenetics and, 962
  exercise and, 750, 751
  levels of, 287–289
  methylation in, 961–962
  nuclear hormone receptors in, 963–965
  regulated, 950
  regulatory circuits, 956–958
  ribosomes and, 938
  transcription factors in, 958–960
  transcriptional, 935, 950
  transcriptional, in eukaryotes, 967–972
  transcriptional, in prokaryotes, 967–969
  translational, 935, 970–971
Gene expression (eukaryotes), 958–960
  activation domains in, 960
  basic-leucine zipper (bZip) in, 959, 959f
  chromatin remodeling in, 961–967
  coactivators in, 965, 965f
  combinatorial control in, 960

DNA methylation and, 961–962
  DNA-binding structures in, 958–960
  enhancers in, 962, 963f
  homeodomain in, 958, 958f
  hypomethylation in, 961
  mediator in, 960, 960f
  nuclear hormone receptors in, 963–965, 964f
  overview of, 958
  posttranscriptional, 970–971
  at posttranscriptional level, 969
  versus in prokaryotes, 958
  regulation of, 958–960
  transcription factors in, 958–960
  zinc-finger domains in, 959, 959f
Gene expression (prokaryotes), 950–951
  attenuation in, 967–969, 968f
  catabolite activator protein (CAP) in, 955–956, 955f, 956f
  catabolite repression in, 955
  chromatin remodeling in, 966f
  corepressors in, 955
  DNA-binding proteins in, 950–951
  DNA-binding sites in, 955, 955f, 956f
  versus in eukaryotes, 958
  helix-turn-helix motif, 951, 951f
  lac operon in, 953, 953f
  lac repressor in, 950, 953, 953f, 954f
  λ repressor in, 957–958, 957f, 958f
  ligand binding in, 954
  operon model, 953
  overview of, 949
  posttranscriptional, 967–972
  pur repressor in, 955, 955f
  regulatory sites in, 950–951
  RNA in, 880–881
  symmetry matching in, 950–951
  transcriptional, 950–951
"Gene guns," 296
Gene knockdown, 295
Gene knockout. *See* Gene disruption
Gene therapy, 296
Gene transfer, horizontal, 317–318, 317f
General acid–base catalysis, 180
General transcription factors, 898
Genes
  cloned, 104–105
  comparative analysis of, 286–287
  designer, 281
  horizontal transfer, 317–318, 317f
  number of, in human genome, 283
  protein-coding, 25
  regulator, 952
  reporter, 274
  structural, 952
  synthesis by automated solid-phase methods, 268–269
  tumor suppressor, 870
Genetic code, aminoacyl-tRNA synthetases and, 921–926
Genetic recombination, 872. *See also* DNA recombination
Genetically modified organisms (GMOs), 296
Genomes
  analysis of, 284–287
  of *Arabidopsis thaliana*, 283

of *Caenorhabditis elegans*, 283
of chloroplasts, 589
comparative, 25–27, 286–287, 286f
complete, 282f
defined, 125
of *Drosophila melanogaster*, 84, 283
editing, 291–294, 292f
eukaryotic, 282–283
of *Haemophilus influenzae*, 282, 282f, 309
human, 283–284, 283f, 286–287, 286f
long interspersed elements, 284
of mice, 286f
of mitochondria, 544–545
next-generation sequencing, 284–286, 285f, 286f
noncoding DNA in, 284
number of genes in, 283
protein encoding, 25
proteome as functional representation, 125–126
of puffer fish, 286–287
of *Reclinomonas americana*, 545
of *Rickettsia prowazekii*, 545
of *Saccharomyces cerevisiae*, 283, 958
sequencing of, 21–22, 282–287
Genomic library, 277–278, 277f, 278f
Genomic variation, 22–24, 23f
Genomics
comparison, 22–23, 22f
revolution, 21–27
Gentamycin, 942
Geraniol, **436**
Geranyl pyrophosphate, 823, 823f
Geranyl transferase, 823
Ghrelin, 735t, 736
Gibbs, Josiah Willard, 13
Gibbs free energy, 13, 145–148
Gibbs free energy of activation, 148
Gigaseal, 393
Gilman, Alfred, 425b
Gla domain in prothrombin, 229, 229f
GlcNAcase, 339
Gleevec (imatinib mesylate), 435, 435f
Globin fold, 75
Globins. *See also* Hemoglobin; Myoglobin
defined, 303
evolution of, 316, 316f
sequence alignment in, 303–305
Globular proteins, 51–52
Glomeruli, 992
Glucagon
defined, 650
in essential enzyme regulation, 507–508, 507f
excess in type 1 diabetes, 746
in fatty acid metabolism, 695
in glycogen metabolism, 651–653, 656, 657f, 658, 658f
liver and, 739–740
in postabsorptive state, 739–740
secretion of, 735
Glucagon-like peptide 1 (GLP-1), 735, 735f, 735t
Glucocorticoids, 834
Glucogenic amino acids, 718, 719f
Glucokinase, 495–496, 739

Glucokinase regulatory protein (GKRP), 496
Gluconeogenesis, 499–504
acetoacetate in, 753–754
decarboxylation in, 502–503
defined, 473, 499
energy change and, 501
energy charge and, 505–506
evolution of, 511
fasting and, 753
fructose 1,6-bisphosphate in, 500f, 503
fructose 2,6-bisphosphate in, 505f, 506–507, 506f, 507f, 508
fructose 6-phosphate in, 500f, 503
glucagon and, 740
glucose 6-phosphate in, 500f, 503–504, 504f, 646, 648
glucose generation and, 503–504, 504f
glycolysis and, 499–501, 505–511, 505f
hormones in, 507–508
in liver, 499, 503, 506–508, 753
in muscle contraction, 508–509, 509f, 510f
overview of, 473–474, 499
oxaloacetate in, 499–503, 500f, 502f
pathway integration, 509, 510f
pathway of, 499, 500f
phosphoenolpyruvate (PEP) in, 500f
phosphorylation in, 502–503, 507
protein metabolism and, 725
pyruvate conversion in, 501–502, 502f
reactions of, 504t
reciprocal regulation in, 505–510, 505f
regulation of, 505–510, 505f, 740f, 746
stoichiometry of, 504
substrate cycles in, 508, 508f
urea cycle in, 716
Gluconeogenic pathway, 680
Glucophage, 745
Glucopryanose, 327–328, **327**
Glucose, 326f, 327
in brain, 740
as cellular fuel, 473
characteristics of, 473
complete oxidation of, 573–574, 574t
conversion into pyruvate, 485–486
as essential energy source, 325, 330
formation of, 473, 740
generation from glucose 6-phosphate, 503–504, 504f
generation of, 473–474
in glycogen, 335, 335f, 640, 641–642
homopolymers of, 335
levels of, 425f
in liver-glycogen metabolism, 659–660
metabolism, 447f, 473f
in postprandial period, 739
as reducing sugar, 330–331
in refed state, 741
ring form, 329f
storage forms of, 335, 335f
structure of, **2**, **327**
synthesis from noncarbohydrate precursors, 499–504
transporters, 474, 474t
Glucose 1-phosphate in glycogen metabolism, 642–643, **643**, 646

Glucose 6-phosphate dehydrogenase (G6PD), 626
deficiency, 635–636
in pentose phosphate pathway, 634–636
in reactive oxygen species protection, 634–636
Glucose 6-phosphate (G-6P), 332, **332**, 739, 740
galactose conversion into, 490–491
in gluconeogenesis, 500f, 503–504, 504f, 646, 648
in glycolysis, 476, **477**, 494, 495f
isomerization of, 477
in liver, 646–647, 661
metabolism of, 631–634, 641–642, 642f
in pentose phosphate pathway, 624, 631–634
transporter, 661
Glucose homeostasis
blood, 334
defined, 739
in diabetes, 741–746
starvation and, 752–754, 753t
Glucose transporters (GLUTs), 498, 498t, 742
Glucose-alanine cycle, 713, 713f
α-1,6-glucosidase, 661
α-1,6-glucosidase in glycogen metabolism, 645, 645f
Glutamate, **160**
chirality of, 767f
formation of, 708, 713, 720, 720f, 769–770
in nitrogen fixation, 766–767
oxidative deamination of, 708–709
structure of, 39, **39f**, **708**
synthesis of, 766–767, 771, 773f
Glutamate dehydrogenase, 708–709, 766–767
Glutamate synthase, 767
Glutamic acid, 39, **40t**
Glutamine, 766–767
carbon monoxide poisoning and, 83
in citric acid cycle, 535
degradation of, 720
in nitrogen transport, 713
in pyrimidine synthesis, 794, 794f
structure of, **38f**, 39
synthesis of, 766–767, 771
Glutamine phosphoribosyl amidotransferase, 797
Glutamine synthetase, 713, 767
in amino acid synthesis, 782–783, 782f
regulation of, 782–783, 782f
structure of, 782–783
Glutathione, 634
conjugation reaction, 991, **991f**
defined, 783–784
structure of, **784f**
Glutathione peroxidase, 635, 784, **784f**
Glutathione reductase, 635, 784
Glycan-binding proteins, 346
Glycated hemoglobin, 331
Glycation, 331
Glyceraldehyde, 325, **325**, 326f, 489
Glyceraldehyde 3-phosphate dehydrogenase (GAPDH), 480–481, 482f, **482f**, 483f, 622

Glyceraldehyde 3-phosphate (GAP), 332, **332**
  in Calvin–Benson cycle, 615
  catalytic mechanism of, 482
  free-energy profiles for, 481, 481f
  in fructose metabolism, 489
  in glycolysis, 466, 476f, 478, 479f, 480–481, 480f, 481f
  NADH generation, 487, 487f
  oxidation of, 457–458, 482
  in pentose phosphate pathway, 626, **627**, **628**
  structure of, **457**
Glycerol
  generation by lipolysis, 670, 670f
  in gluconeogenesis, 499
  in liver, 669–670
  in phosphoglycerides, 356
  release of, 740
  structure of, **2**
Glycerol 3-phosphate, 739
  in lipid synthesis, 815
  reoxidation of, 571
Glycerol 3-phosphate shuttle, 570–571, 570f
Glycerol phosphate acyltransferase, 815
Glycine, 719, 719f
  in Calvin–Benson cycle, 614f, 615
  collagen chain and, 54, 54f
  conversion into serine, 719, 719f
  defined, 36
  in liposomes, 360, 360f
  porphyrins from, 786–787
  pyruvate formation from, 719, 719f
  structure of, **37f**
  synthesis of, 772
Glycine cleavage enzyme, 774
Glycine synthase, 774
Glycine-glycine-glutamine (GGQ) sequence, 936
Glycobiology, 347b
Glycocholate, 667f, **833f**
Glycoforms, 338
Glycogen
  branch point in, 335f
  defined, 335, 640
  glucose storage in, 335, 335f, 640, 641–642
  in glycogen metabolism, 344–345, 651f, 652f
  in liver, 641, 641f, 651, 740, 741
  molecules, 640, 641f, 655f
  in muscle, 641, 641f
  phosphorolytic cleavage of, 643–645
  remodeling, 645, 645f
  storage efficiency, 656
  storage sites of, 641, 641f
  structure of, **336f, 641f, 643**
Glycogen lakes, 661f
Glycogen metabolism, 640–661
  branching enzyme in, 655, 655f
  cAMP in, 652–653, 657f
  coordinate control of, 656, 657f
  degradation of, 641–642
  enzyme breakdown, 642–647
  epinephrine in, 651–653, 651f, 656, 657f, 658, 658f

G proteins in, 651–653
glucagon in, 651–653, 656, 657f, 658, 658f
α-1,6-glucosidase in, 645f
glycogen in, 344–345, 651f, 652f
glycogen phosphorylase in, 642–643
glycogen synthase in, 654–655, 655f
glycogen synthase kinase (GSK) in, 655–656, 657f
glycogenin in, 654
hormones in, 651–653, 651f, 656–661, 657f
insulin in, 658–659, 659f
in liver, 646–647, 659–660, 659f, 660f
overview of, 640–642
pathway integration, 652f
phosphoglucomutase in, 646, 646f, 647f
phosphoglycerate mutase in, 646
phosphorolysis in, 642–645
phosphorylase in, 642–647, 645f, 659, 659f, 660f
protein kinase A (PKA) in, 651, 652f, 656, 657f, 658f
protein phosphatase 1 (PP1) in, 653, 656–658, 657f, 659
pyridoxal phosphate (PLP) in, 643–645, 644f
regulation of, 642, 656–661, 657f
regulatory cascade for, 651–653, 652f
7TM receptors in, 652, 652f
termination of, 653
transferase in, 645–646, 645f
UDP-glucose and, 654–655
Glycogen phosphorylase
  a, 647–649, **647f**, 650, 650f
  allosteric regulation of, 647–651, 648f
  b, 647–649, **647f**, 650, 650f
  carboxyl-terminal domain of, 643, 644f
  defined, 642, 646
  in glycogen metabolism, 642–647
  glycogen-binding site of, 643–645, 644f
  in liver, 647–649, 647f, 648f, 657f
  in muscle, 648, 648f, 650f
  regulation of, 647–651, 653
Glycogen synthase
  α-1,6 linkages, 655, 655f
  defined, 654
  in glycogen synthesis, 655–656
  isozymes of, 654–655
Glycogen synthase kinase (GSK), 742
  defined, 655–656
  in glycogen metabolism, 655–656, 657f
Glycogenin, 641, 654
Glycogen-storage diseases, 661f, 661t
Glycolate, 614f, 615
Glycolipids, 355, 357, 359
N-glycolylneuraminate, 819
Glycolysis, 447f
  aerobic, 497–498, 531
  aldehyde oxidation in, 480–481
  anaerobic mode, 531, 747, 749
  ATP formation in, 482–484
  1,3-bisphosphoglycerate (1,3-BPG) in, 457–458, 480–481, 483–484
  in cancer, 497–498
  citric acid cycle and, 517–521, 518f
  control of, 492–499
  covalent enzyme-bound intermediate in, 481

defined, 473
dihydroxyacetone phosphate (DHAP) in, 466, 476f, 478, 479f, 489
as Embden-Meyerhof-Parnas pathway, 473
enediol intermediate in, 479, 479f
energy charge and, 505–506
as energy-conversion pathway, 474–492
enzymes of, 474, 497
in eukaryotic cells, 473, 476–477
evolution and, 511
fermentation in, 486–487
first stage of, 475f, 476f
fructose 1,6-bisphosphate in, 466, 476f, 477
fructose 1-phosphate in, 489, 489f
fructose 2,6-bisphosphate in, 495, 496f
fructose 6-phosphate in, 477, 495
fructose in, 489–490, 489f
galactose in, 489f, 490–491
gluconeogenesis and, 499–501, 505–511, 505f
glucose 6-phosphate (G-6P) in, 476, 477, 494, 495f
glucose transporters in, 498, 498t
glyceraldehyde 3-phosphate (GAP) in, 476f, 478, 479f, 480–481, 480f
hexokinase in, 476–477, 476f, 494, 495–496, 495f
historical perspective on, 473
isomerization of carbon sugars in, 477, 478
in liver, 494–497, 495f, 496f, 497f, 506–508
in muscle, 493–494, 495f
NAD+ in, 482, 482f, 486–488, 487f
NADH in, 482, 486–487, 486f
overview of, 473f, 475f, 476f
pathway integration, 509, 510f
pentose phosphate pathway and, 624–628
phosphofructokinase (PFK) in, 493–495, 493f, 496f
3-phosphoglycerate in, 484–485
2-phosphoglycerate in, 484–485
3-phosphoglycerate in, 485–486
phosphoglycerate mutase in, 484
pyruvate in, 473, 473f, 476f, 485–486, 486f
pyruvate kinase in, 494, 495f, 496–497, 497f
reactions of, 485t
reciprocal regulation in, 505–511, 505f
regulation of, 740f, 746
second stage of, 475f, 480–481, 480f
stages of, 474–476, 475f, 480f
thioester intermediate in, 481, 481f, 482, 483f
triose phosphate isomerase (TPI) in, 478–480, 479f
Glycolytic intermediates, 633
Glycolytic pathway
  glucokinase in, 495–496
  hexokinase in, 495–496
  phosphofructokinase (PFK) in, 493–495, 496f
  pyruvate kinase in, 496–497, 497f

Glycopeptide transpeptidase, 173
Glycophorin, 348f, 369
Glycoproteins, 337–346
Glycosaminoglycans, 338, 339–340, 340f
Glycosidic bonds, 331–332, 332f, 336, 336f, 338
Glycosidic linkage, 331–332, 332f, 335f, 336f, 338f
O-glycosidic linkage, 332, 332f
N-glycosidic linkage, 332, 332f
Glycosylation, 331, 341–342
    congenital disorders of, 344–345
    errors in, 344–345
    location of, 341f
    in nutrient sensing, 339, 339f
Glycosyltransferases, 341–342, 343f
Glyoxylate, 537, 537f, 615
Glyoxylate cycle, 536–538, 537f
Glyoxysomes, 537
Glyphosate, 777
Glyset, 334
GMP (guanosine monophosphate)
    in purine synthesis, 799–800, 799f
    synthesis of, 807–808, 808f
GMP synthetase, 800
Golden rice, 296
Goldstein, Joseph, 830
Golgi complex, 341, 341f, 342f, 344, 942
Gout, 810
G-protein-coupled receptors (GPCRs), 418, 735–736
Gram, Hans Christian, 373
Gram negative bacteria, 373
Gram positive bacteria, 373
Gram staining technique, 373, 374f
Granulocyte colony-stimulating factor (G-CSF), 995
Granum, 588
Grb2, 432, 432f
Green fluorescent protein (GFP), 63, 64f, 117, 117f
Greider, Carol, 1, 27–28, 863
Group-specific reagents, 169–170, 170f
Group-transfer reactions, 465–466, 465t
GTP (guanosine triphosphate)
    as energy source, 792
    hydrolysis, 419–420
    in olfaction, 436
    in resetting G proteins, 419–420
    in signal transduction, 416
    in translation, 932
    in vision, 439
GTPase-activating protein (GAP), 434
GTPases
    defined, 433
    Ras family of, 432
    in signal transduction, 433
GTP-GDP cycle, 932, 932f
Guanidinium chloride, 59
Guanidinium group of arginine, 39, 39f
Guanine, **4, 5, 5f, 18**
Guanine-nucleotide-exchange factor (GEF), 432
Guanosine triphosphate (GTP), 451
Guanylate, 800
Guide strand, 295

Guirard, Beverly, 770b
Gustation. See Taste
G$_{\beta\gamma}$ dimer, 424, 424f

H5N1 flu, 348
Haemophilus influenzae, 282, 282f, 309
Hair cells, 440–441, 440f, 441f
Hairpin turns. See Reverse turns
Half-life
    of drugs, 992, 992f
    of proteins, 703, 704, 705t
Half-reactions
    defined, 546
    in electron transport, 548
Halobacterium salinarum, 586, 586f
Hartnup disease, 702
Hatch-Slack pathway, 622
Haworth projections, 327–328
HCN channels, 405
HCN4 gene, 405–406, 405f
Hearing
    hair cells in, 440–441, 440f, 441f
    ion channels in, 440–441
    signal transduction in, 440–441, 441f
Heart disease, 833, 981. See also Atherosclerosis
Heat
    defined, 12
    oxidative phosphorylation uncoupling in generation of, 575–576, 576f
    release in double helix formation, 14–15
    thermodynamics and, 12
Heat shock protein 70 (Hsp70), 310, **311f**
Heat-shock genes, 890–891
Heavy (H) chains
    in antibody diversity, 89
    defined, 86–87
    in immunoglobulin classes, 87f
Heinz bodies, 635, 635f
Helicase DnaB, 860
Helicases
    asymmetry, 851f
    defined, 850
    in DNA replication, 850–851
    mechanism, 850, 850f, 851f
    structure of, 850, **850f**
    symmetry, 850, 850f
Helix-turn-helix motif, 52, 52f, 951, 951f
Heller, Jutta, 29
Hemagglutinin, 347–348, 348f
Heme, 69, **73f**
    biosynthetic pathway, 786f
    in cytochrome c oxidase, 557
    degradation of, 788f
    labeling of, 786, 786f
    in oxygen-binding, 73–74
    structural changes, 79–80
    structure of, **557, 589f**
    synthesis of, 787f
Heme prosthetic group, 554–555
Hemiacetal intramolecular, **327**
Hemichannels, 406
Hemiketal intramolecular, 327, **327**
Hemoglobin, 21
    alignment comparison, 305, 305f
    alignment with gap insertion, 303, 304f

α chains, 75
α$_2$β$_2$ tetramer of, 55, **55f**
αβ dimers, 75
amino acid sequences of, 303–305, 303f
β chains, 75, 82f, 83
Bohr effect, 81–83
2,3-BPG binding to, 81
comparative genomics and, 26, 26f
conformational changes in, 78, 78f
cooperative oxygen binding by, 76–77
cooperativity, 78–79
defined, 69, 303
fetal, 81, 81f
globin fold, 75
glucose and, 331
glutathione and, 635
glycated, 331
mutations and, 83–84
myoglobin comparison, 75
oxygen affinity of, 80–81, 80f, 81f, 82f
oxygen delivery by, 77f
oxygen transport by, 81
oxygen-binding, 73–74
oxygen-binding curve, 76, 76f, 77f
polypeptide chains, 75
purified, 80, 80f
quaternary structure of, 75–76, **76f,** 77–78, 77f
R state, 78, 78f
receptor-mediated endocytosis and, 375–376
sequence alignment of, 307f, 308f
sickle-cell, 83, 83f
space-filling model, 76f
subunits, 75–76, 76f
T state, 78, 78f, 82
tertiary structure of, 310–311, **310f**
Hemoglobin A (HbA), 75
Hemoglobin β-chain gene (HbB), 83
Hemoglobin S (HbS), 83–84, 84f
Hemolytic anemia, G6PD deficiency and, 634–635
Hemostasis, 228
Henderson–Hasselbalch equation, 19, 20b–21b
Heparin, 232
Hepatic steatosis, 743
Hepatocytes, 669
Heptad repeat, 53, **53f**
Heptoses, 325
HER2 receptor, 434, 434f
Herbicides, photosynthesis and, 605
Hereditary nonpolyposis colorectal cancer (HNPCC), 870
Herpes simplex infections, 797
Hers disease, 661t
Heterotrimeric G proteins, 415–418, 417f
Heterotrophs, 586
Heterotropic effects, 217
Hexokinase
    defined, 476
    in glycolysis, 476–477, 476f, 494, 495–496, 495f
    inhibition of, 494
    liver and glycolysis and, 495–496
    metal ion requirement, 476
    muscle and glycolysis and, 494

Hexose monophosphate pathway.
      See Pentose phosphate pathway
Hexose monophosphate pool, 615
Hexose phosphates, 615, 615f
Hexose sugars formation in Calvin–
      Benson cycle, 615–618, 616f
Hexoses, 325, 326f, 476, 616f
High-density lipoproteins (HDLs),
      828t, 829
   defined, 829
   properties of, 832
   protective effects and, 832
   reverse cholesterol transport in, 832
High-performance liquid chromatography
      (HPLC), 105, 105f
High-throughput screening, 981–982, 982f
Hind, Geoffrey, 599b
Hines, Justin, 29
Hippocrates, 980
His 64 (distal histidine), **313f**
His tag, 105
Histamine, 783, **783f**
Histidine
   degradation, 720, 720f
   distal, 74, 79f, 82–83
   ionization of, 39, 39f
   proton shuttle, 193, 194f
   proximal, 73, 79, 79f
   residue, 183
   structure of, **38f**, 39, 39f, **39f**, **40t**
Histidine operon, 969, 969f
Histone acetyltransferases (HATs),
      965–966, **965f**
Histone code, 967
Histone deacetylases, 967
Histones
   acetylation of, 965–966
   amino-terminal tails of, 965–966
   in gene regulation, 967t
   homologous, 893
   modifications, 967, 967t
   modifications of, 967t
   tails, 965–966, 965f
Hitchings, George, 996b
HIV. *See* Human immunodeficiency virus
      (HIV) infection
HMG-CoA reductase, 981
HMG-CoA reductase inhibitor, 981
Hodgkin, Alan, 392
Hodgkin, Dorothy, 677b
Hoffmann, Felix, 980
Holliday, Robin, 873
Holliday junctions, 873–874, 873f
Holoenzyme, 144, 886, 887f
Homeodomain, 958, 958f
Homeostasis, 467. *See also* Caloric
      homeostasis; Glucose homeostasis
Homo sapiens, 2f
Homocysteine
   in activated methyl cycle, 774, 775f
   in amino acid synthesis, 775
   in cysteine synthesis, 776
   levels, reducing, 776
   structure of, 41
   in vascular disease, 776
Homocystinuria, 727t
Homodimer, 55

Homogenate, 101–102
Homogenization in protein purification,
      101–102
Homogentisate, 724, 726, **726**
Homogentisate oxidase, 724
Homologies
   databases for, 308–309
   defined, 302
   detection of, 302–309
   repeated motifs and, 311–312
   sequence alignment in, 305, 305f, 311
   shuffling in, 305, 305f
   statistical analysis in, 305
   substitution matrices in, 305–308
   three-dimensional structures and, 311
Homologous recombination, 290–291,
      291f
Homologous sequences, 308–309
Homologs, 302, 302f
Homology directed repair (HDR), 292, 292f
Homolytic cleavage reaction, 677, 677f
Homopolymers, 335
Homoserine, 41, **41f**
Homotropic effects, 215
Horizontal gene transfer, 317–318, 317f
Hormone receptors, 963–964, 964f
Hormone-receptor complex, 417
Hormones
   in acetyl CoA carboxylase regulation,
      694–696
   androgens, 834, 837–838, 838f
   antidiuretic, 126, 127f
   eicosanoid, 693–694, 693f, 694f
   in fatty acid metabolism, 693–694
   in gluconeogenesis, 507–508
   in glucose metabolism, 507
   in glycogen metabolism, 651–653, 651f,
      656–661, 657f
   local, 694
   metabolic functions of, 468
   in signal transduction, 415–416
   steroid, 827, 834–836, 834f
HT1171, **707**
Human genome, 283–284, 283f, 286–287,
      286f
Human Genome Project, 22f
Human history, genome sequencing and,
      22–23, 23f
Human immunodeficiency virus (HIV)
      infection
   drug development effects, 994f
   protease, 189, 189f
   protease inhibitors, 189, 994, 994f, 998
   zinc-finger domains and, 294
Human milk oligosaccharides, 335
Human papillomavirus (HPV), 705
Human ribonuclease, **302f**
Humoral immune response, 86
Huntington disease, 62, 703, 869
Huxley, Andrew, 392
Hybrid-biological-inorganic (HBI) system,
      606–607
Hybridoma cells, 114, 114f
Hydrogen bonds, 9f
   acceptors, 9, 18
   in base pairs, 5
   breaking with membranes, 367

   covalent bonds, compared to, 9
   donors, 9
   in enzyme-substrate complex, 150,
      150f, 151f
   interaction, 9
   as noncovalent bond, 5, 9, 9f, 10, 11
   strength of, 9
   water, 10, 11
Hydrogen ions
   in acid–base reactions, 17
   concentration in solutions, 17
   oxygen release and, 81–82
Hydrogen-bond donors, 9
Hydrolases, 143t
Hydrolysis. *See also* ATP hydrolysis
   GTP (guanosine triphosphate), 419–420
   of lipids, 668–669
   in metabolism, 451–452, 466
   of peptide bonds, 180–181, 183, 184f
   of phosphodiester bond, 194–195, 195f
Hydrolytic enzymes, 184f, 186–187, 186f
Hydrolytic reaction, 465t, 466
Hydronium ions, 17
Hydropathy index, 368
Hydropathy plots, 368, 368f
Hydrophilic amino acids, 38, 52
Hydrophilic moiety, 358, 359f
Hydrophobic amino acids, 51f, 52f
   structures of, 37f
   types of, 36–38
Hydrophobic effect, 10–11, 11f, 13, 38, 52
Hydrophobic interactions
   defined, 11
   as noncovalent bond, 9
Hydrophobic moiety, 358, 359f
3-hydroxy-3-methylglutaryl CoA, 722
3-hydroxy-3-methylglutaryl CoA
      reductase (HMG-CoA reductase),
      822, 822f, 826, 827f, 833, 833f
3-hydroxyanthranilate, 724, **724**
D-3-Hydroxybutyrate, 753
3-hydroxykynurenine, 724f
Hydroxyl groups, 38–39, 63
Hydroxylation, 63
   cytochrome P450 in, 835–836, 835f
   steroid hormones, 834–836
P-hydroxymercuribenzoate, 212–213, 212f
P-hydroxyphenylpyruvate, 724, **724**, 778
P-hydroxyphenylpyruvate hydroxylase,
      724
Hydroxyproline, 54, **63f**, 757, 757f
Hyperammonemia, 717
Hypercholesterolemia, 830–831. *See also*
      Familial hypercholesterolemia
Hyperglycemia, 533
Hyperlysinemia, 727t
Hyperpolarization-activated cyclic
      nucleotide-gated (HCN) channels,
      405
Hyperpolarization-activated ion
      channels, 405–406, 405f
Hypersensitive sites, 961
Hypervariable loops, 86–88, 87f, 88f
Hypervariable regions, 86, 87f
Hypoglycemia, 510
Hypomethylation, 961
Hypoxanthine, 810, 865, 865f

Hypoxanthine-guanine phosphoribosyltransferase (HGPRT), 800, 810
Hypoxia
    defined, 498, 758
    tumor, 498, 499f
Hypoxia-inducible factor 1 (HIF-1), 498, 499, 499f, 751, 758

Ibuprofen, 166, 990, **990f**
Ice, 10, 10f
I-cell disease, 344–345
Identity, 303
IF2-GTP-initiator-tRNA complex, 931
Imatinib mesylate (Gleevec), 999
Imidazoles, 38f
Immune response, 86
Immune system
    adaptive, 86–88
    binding proteins and, 84–91
    evolution of, 84–86
    innate, 84–86
    MHC proteins in, 89–91
    overview of, 84
    T cells. See T cells, killer
    T-cell receptors. See T-cell receptors (TCRs)
Immunity, cell-mediated, 89
Immunoelectron microscopy, 118, 118f
Immunoglobulin A (IgA), 88, 88f
Immunoglobulin D (IgD), 88, 88f
Immunoglobulin domains, 86–87. See also Immunoglobulin fold
Immunoglobulin E (IgE), 88, 88f
Immunoglobulin fold, 86, **86f**
Immunoglobulin G (IgG)
    antigen binding of, 88
    cleavage of, 87, 87f
    defined, 86
    structure of, 86, **87f**
Immunoglobulin M (IgM), 88, 88f, 89
Immunoglobulins. See also Antibodies
    classes of, 88–89, 88f
    defined, 86
    heavy chains, 89
    light chains, 89
    myeloma and, 114
    sequences of, 86f
Immunologic techniques
    enzyme-linked immunosorbent assay (ELISA), 114–115, 115f
    fluorescence microscopy, 117–118, 117f
    hybridoma cells in, 114
    monoclonal antibodies in, 113–114, 113f, 114f
    polyclonal antibodies in, 113, 113f
    in protein studies, 112–118
    western blotting, 116, 116f, 265
Immunotoxins, 281
Indinavir (Crixivan), 189, 189f, 994–995
Indinavir complex, 189, 189f
Indirect ELISA, 115, 115f
Indole, 779
Indole groups, 37–38
Induced-fit model, 151, 151f, 180
Inducer, 954
Influenza virus, 347–348, 348f

Information science, 21
Ingram, Vernon, 83
Inhibition
    competitive, 166–167, 167f
    noncompetitive, 166, 168f
    reversible, 166–168
    uncompetitive, 167–168, 168f
Inhibition constant (K_i), 987, 987b–988b
Inhibitory concentration, 987
Inhibitory factor 1 (IF1), 575
Initiation factors, 931, 931f
Initiator element (Inr), 896
Initiator tRNA, 930
In-line displacement, 195
Innate immune system, 84–86
Inosinate, 799, 799f, 808
Inosine, 810, 921
Inositol, **817**
Inositol 1,4,5-trisphosphate (IP_3), **413**, 420–421, 420f
Insertional inactivation, 274, 275f
Insig, 826–827, 826f, 827f
Insulin, 735
    adiponectin and, 737
    amino acid sequence of, 43, 43f
    binding of, 426–427, 427f, 430f
    defined, 33, 425
    in essential enzyme regulation, 507–508, 507f
    in fatty acid metabolism, 696
    formation of, 224
    glucokinase and, 496
    in glycogen metabolism, 658–659, 659f
    levels of, 425f
    MALDI-TOF mass spectrum of, 119f
    pancreatic failure and, 744–745
    in postprandial period, 739
    recombinant human, 995
    release of, 744, 745f
    research in, 677b
    secretion of, 735, 739
    sensitivity to in athletes, 750
    in signal transduction, 414f, 425–431, 742–743, 742f, 744f
Insulin receptor
    activation of, 426–427, 427f
    binding sites for, 742
    defined, 426
    structure of, 426, 427f
    subunits, 426f, 427
Insulin resistance, 737
    biochemical basis of, 742–743, 744f
    ceramide and, 743–744, 744f
    defined, 741
    in muscle, 743–744
    in pancreatic failure, 744–745, 745f
Insulin signaling, 425–431
    illustrated, 427f
    lipid kinase in, 428, 428f, 429f
    pathway, 428, 430f
    termination of, 430–431
Insulin-dependent diabetes. See Type 1 diabetes
Insulin-receptor kinase, 428–430, 429f
Insulin-receptor substrates (IRS), 428, 428f, 658, 742, 742f, 743
Insulin-signaling pathway, 742–743, 742f

Integral membrane proteins, 363, 363f, 364f
Intermediary metabolism. See Metabolism
Intermediate-density lipoproteins, 828, 828t
Intermembrane space (IM space), 543
Internal guide sequence (IGS), 910
Internal ribosome entry sites (IRES), 937
International Union of Biochemistry, 143
Intracellular pathogen, 89
Intramolecular hemiacetal, 327
Intramolecular hemiketal, 327
Intramolecular rearrangements, 677, 678
Intrinsic pathway, clotting, 228
Intrinsic termination signals, 888, 889f
Introns
    self-splicing, 909–910, 910f
    splicing, 901–902, 902f
Inverse PCR, 281, 281f
Invertase, 334
Inverted repeats, 197, 197f
Ion channels, 392–393
    acetylcholine receptor, 400–401, 401f, 402f
    action potentials and, 392, 392f, 402–404
    ball-and-chain model for inactivation, 399–400, 400f
    calcium, 396, 397, 397f
    cell-to-cell, 406–407, 406f, 407f
    defined, 392, 402
    disruption of, 404–405
    energetic basis of selectivity, 396, 397f
    equilibrium potential and, 402–404, 402f
    gap junction, 406–407, 406f, 407f
    in hearing, 440–441
    hyperpolarization-activated, 405–406, 405f
    inactivation of, 399–400, 400f
    ligand-gated, 400–401
    membrane permeability and, 399–400
    nerve impulses and, 392
    patch-clamp technique, 393, 393f, 394f
    potassium, 394–400, 395f, 396f, 397f, 398f, 399f, 405
    selectivity filter, 396, 397f
    sequence relations of, 394
    shaker, 394
    sodium, 392, 394, 396–397, 397f, 399–400, 404
    structure of, 394f
    in vision, 439, 439f, 440
    voltage-gated, 398–399, 399f
Ion gradients, 458, 458f
Ion semiconductor sequencing, 285, 286f
Ion source, 118
Ion-exchange chromatography, 103–104, 104f
Ionic interactions, 8–9, 11f
IRE-binding protein (IRP), 970–971, 970f
Iron
    metabolism of, 970–971
    oxygen-binding and, 73–74, 79f
    receptor-mediated endocytosis and, 375–376, 376f
    structural transition in subunit, 79
Iron center of ribonucleotide reductase, 803, 804

Iron-response element (IRE), 970–971, 970f
Iron–sulfur clusters, 551–552, 551f
Iron–sulfur proteins, 525f
    defined, 525, 551
    in nitrogen fixation, 764–765, 765f
    in oxidative phosphorylation, 552
    structure of, **765f**
Irreversible inhibition. *See also* Enzyme
        inhibition
    defined, 164
    in mapping active site, 169–172
    mechanism-based (suicide), 171, 173f
Islet-1 gene, 962
Isocitrate
    in citric acid cycle, 525
    in glyoxylate cycle, 537, 537f
    in isomerization reactions, 467
Isocitrate dehydrogenase, in citric acid
        cycle, 525, 533, 533f
Isocitrate lyase, 537
Isoelectric focusing, 107–108, 108f
Isoelectric point, 107, 108f
Isoenzymes. *See* Isozymes
Isoform, 217
Isoleucine, 36, **37f**, 718
    degradation of, 722–723, 725b–726b
    in maple syrup urine disease, 727
    structure of, **37f**
Isomerases, 143t
Isomerization, 439–440, 439f
Isomerization reactions in metabolism,
        465t, 467
Isomers
    constitutional, 325, 326f
    diastereoisomers, 325–326, 326f
    stereoisomers, 325, 326f
Isopentenyl pyrophosphate, 822–823,
        822f, **823**
Isopeptide bonds, 703, 704f
Isoprene, **822**, 833–841
Isoprenoids, 840–841
Isopropylthiogalactoside (IPTG), 954, 954f
Isotope labeling studies, 530, 786, 786f,
        822f
Isotrenes, 841f
Isovaleryl CoA, **722**
Isovaleryl CoA dehydrogenase, 722
Isozymes
    defined, 217
    distinguishing, 217–218
    of lactate dehydrogenase (LDH), 217,
        218f
Itano, Harvey, 83

Jacob, François, 952
Jagendorf, André, 599, 599b
Jagendorf's demonstration, 599

K⁺ ions. *See* Potassium ions
Kanamycin, 942
k_{cat} (rate constant), 157
k_{cat}/K_M (specificity constant), 158–159,
        159f
K_d (dissociation constant), 157, 166
Kendrew, John, 51
Ketimine, **708**
α-ketoacids, 722–723, 769–770

β-ketoacyl synthase, 687
α-ketobutyrate, 721
Ketogenic amino acids, 718, 719f
Ketogenic diets, 682
Ketoglutarate, **160**
α-ketoglutarate, **160**
    in amino acid degradation, 718, 719,
        719f, 720, 720f
    in amino acid synthesis, 766–767, 769
    in citric acid cycle, 525, **525**
    in double-displacement reactions,
        160–161
    in glutamate synthesis, 766, 767
    in nitrogen fixation, 766
    oxidative decarboxylation of, 526
α-ketoglutarate dehydrogenase, 535
α-ketoglutarate dehydrogenase complex
    in citric acid cycle, 526, 533, 533f
    defined, 526
α-ketoisocaproate, 722
Ketone, **327**
Ketone bodies
    acetyl CoA formation from, 679–680
    brain and, 753–754
    in citric acid cycle, 753–754, 754f
    defined, 680
    in diabetes, 682, 682f
    in fatty acid metabolism, 679–683
    formation of, 680, 680f, 752–754
    as fuel source, 679–683
    high levels of, 682–683
    in liver, 681, 681f, 753, 754f
    starvation and, 752–754, 753f
    as water soluble, 682
Ketoses
    aldoses conversion to, 477, 615
    defined, 325
Ketosis, 682, 682f
3-ketosphinganine, 819, **819**
β-ketothiolase, 673
K_i (inhibition constant), 987,
        987b–988b
Kidney
    drug excretion in, 992
    gluconeogenesis in, 499
Killer T cells, 86
Kilocalories (kcal), 146
Kilodaltons (kDa), 43
Kilojoules (kJ), 146
Kinase fold, 222
Kinases, 476
Kinetic energy, 12
Kinetic perfection, 159
Kinetic proofreading, 924–925, 924f
Kinetics, 141
    defined, 151
    enzyme. *See* enzyme kinetics
    as study of reaction rates, 152
    of water deprotonation, 192f, 193
Klenow fragment, 846, 846f
Klug, Aaron, 893
K_M. *See* Michaelis constant
Knowledge-based methods, 61
Köhler, Georges, 113
Kornberg, Roger, 881
Krebs, Hans, 459, 528b
Krebs cycle. *See* Citric acid cycle

Kringle domain, 229
Kynurenine, 724f

ʟ amino acids, 35, 35f, 43
Lac operon, 953, 953f, 954, 954f
Lac regulatory site, 950f
Lac repressor
    binding of, 953
    components of, 954
    defined, 950
    effects of IPTG on, 954f
    inhibition relief, 956
    structure of, 953, **953f**, 955
Lac repressor-DNA complex, 950–951, 950f
Lactase, 334, 474, 492
Lactate, **101**, **159**, 533
    in Cori cycle, 509, 509f
    formation from pyruvate, 487
    formation of, 486f, 487
    in gluconeogenesis, 499
    in glucose metabolism, 473f
    in glycolysis, 473
    in muscle contraction, 508–509, 509f
Lactate dehydrogenase (LDH), 217, 218f,
        487f, 509
Lactic acid fermentation, 473, 487, 487f
Lactic acidosis, 510
Lactobacillus, 492f
Lactobacillus acidophilus, 488
Lactoferrin, 35f
β-lactoglobulin, 119f
Lactonases, 626
Lactose, 952f
    defined, 334
    in human milk, 335
    structure of, 333–334, **333f**
Lactose intolerance, 492
Lactose operon. *See Lac* operon;
        Lac repressor
Lactose permease, 390–392, 391f
LacZ gene, 274, 274f
Lagging strand
    defined, 849
    synthesis of, 849f, 858–859, 858f
    template, 859
λ phages
    alternative infection modes for,
        275–276, 276f
    for DNA cloning, 272, 275–276
    mutant, as cloning vector, 276, 276f
λ repressor
    defined, 957
    in gene expression, 957–958
    in genetic circuit formation, 957f,
        958, 958f
    structure of, **957f**
Lamprey, 316, 316f
Lanosterol, 824, **825f**, 827
Large subunit (50S) of ribosomes, 927,
        927f, 929f, 931–932, 933, 935
Lateral diffusion, 369, 370, 370f
Lavoisier, Antoine, 764
Laws of thermodynamics, 12–14
Lead molecules, 980–986
    DNA-encoded libraries, 984–985, 984f
    high-throughput screening, 981–982,
        982f

from natural products, 980–981
phenotypic screening, 985–986, 985f
screening libraries, 982–983, 983f
structure-based design, 993–995, 994f
target-based screening, 985–986, 985f
Lead optimization, 986
Leader peptide sequences, 969f
Leader sequences, 967–968, 969f
Leading strand
    defined, 849
    synthesis of, 849f, 858–859, 858f
Leaflets, 359
Leber hereditary optic neuropathy
    (LHON), 578
Lectins
    as carbohydrate-binding proteins,
        346–348
    classes, 346–347
    C-type, 347, 347f
    defined, 346
    in interactions between cells, 346
    L-type, 347
Leghemoglobin
    sequence alignment of, 307, 307f, 308f
    tertiary structure of, 310, **310f**
Leptin
    in caloric homeostasis, 736–737
    defined, 736
    effects in brain, 736, 736f
    obesity and, 736–737, 736f
    secretion of, 736
Leptin resistance, 737–738
Lerner, Michael, 903
Lesch–Nyhan syndrome, 810–811
Leucine, 36, 718
    degradation of, 723
    in maple syrup urine disease, 727
    structure of, **37f**
Leucine-rich repeats (LRRs), 84, 85f
Leuconostoc citrovorum, 488
Leukemia, 497
Leukocyte antigen A2 (HLA-A2), 90
Leukotrienes, 693
Levinthal, Cyrus, 59
Levinthal's paradox, 59
Licensing factors, 862
Ligand binding
    concerted model, 78–79, 78f
    in drug development, 986, 986f
    sequential model, 78–79, 79f
    7TM receptor and, 415–418
Ligand-gated channels, 400–401
Ligands, 70f
    defined, 69, 191, 963, 986
    features of, 72f
    receptors and, 413, 964f
Ligases, 143t
Ligation reactions in metabolism, 465t,
    467
Light
    absorption, 439–440, 589–590, 590f
    electromagnetic spectrum of, 438f
Light (L) chains, 87, 87f, 89
Light absorption, 439–440, 589–590, 590f
Light energy, 606–607, 607f
Light independent reactions, 587
Light reactions. See also Photosynthesis

in Calvin–Benson cycle, 620, 620f
    defined, 587
    herbicides and, 605
    illustrated, 587f
    stoichiometry for, 602
Light-harvesting complexes, 604
Light-independent reactions. See Dark
    reactions
Limit dextrin, 473
Limonene, 840, **840f**
lin-4 gene, 971
Linear polymers, 5f, 41
Lineweaver–Burk plot, 156, 156f, 164, 164f,
    167, 167f
α-1,6 linkages, 655, 655f
Linking number, 852f, 853
Linoleate, 675
Linoleoyl CoA, 675f
Lipases, 667, 667f
Lipid bilayers. See also Membrane lipids
    as cooperative structures, 360
    defined, 355, 359
    diagram, 359f
    formation of, 359–360
    as permeability barrier, 361–362, 361f
    self-assembly process, 359
    as self-sealing, 360
    stabilization of, 360
Lipid droplet, 666
Lipid metabolism
    diacylglycerol in, 821
    disruption of, 820–821
    liver in, 669
    phosphatidic acid phosphatase in, 821,
        821f
Lipid phosphatases, 431
Lipid rafts, 372–373, 372f
Lipid vesicles, 360–361, 360f
Lipids
    defined, 353
    digestion of, 667–668, 667f
    energy storage in, 666–668
    fatty acids in, 353–354, 353f, 354t
    hydrolysis of, 668–669
    membrane. See membrane lipids
    metabolism of. See fatty acid
        metabolism
    synthesis of, 815–816, 821
    transport of, 667, 668f
Lipin 1, 821
Lipinski, Christopher A., 989
Lipinski's rules, 989, 989f
Lipmann, Fritz, 770b
Lipoamide in citric acid cycle, 520–521,
    522f
Lipoic acid, 518, **518**
Lipolysis, 669–670, 670f
Lipophilic molecules, 382
Lipoprotein particles, 827–828
Lipoproteins
    chylomicrons, 828, 828t
    components of, 827
    high-density (HDLs), 828t, 829, 832
    intermediate-density (IDLs), 828, 828t
    low-density (LDLs), 828, 828t, 829–832,
        829f, 831f
    metabolism, 829f

properties of, 828t
    as transporters, 827–829
    very-low-density (VLDLs), 828, 828f,
        828t
Liposomes
    defined, 360
    formation of, 360–361, 360f
    therapeutic applications of, 361
Lipoxygenase, 693, 693f
Liver
    alcoholic injury of, 755–756
    in amino acid degradation, 708–709,
        712–713
    in cholesterol synthesis, 825, 828f
    cirrhosis of, 756
    drug metabolism in, 990
    ethanol metabolism in, 755–759
    fatty, 755–756
    glucagon and, 739–740
    gluconeogenesis in, 499, 503, 506–508,
        753
    glucose 6-phosphate (G-6P) in, 646–647,
        661
    glucose and, 739
    glucose levels and, 646–647
    glucose sensor in, 659
    glycerol in, 669–670
    glycogen in, 641, 641f, 651, 740, 741
    glycogen metabolism in, 646–647,
        659–660, 659f, 660f
    glycogen phosphorylase in, 647–649,
        647f, 648f, 657f
    glycolysis in, 494–497, 495f, 496f, 497f,
        506–508
    ketone bodies formed in, 680f
    ketone bodies in, 681, 681f, 753, 754f
    ketone body synthesis in, 753–754, 754f
    in lipid metabolism, 669
    peroxisomes in, 679f
    in starvation, 753
    transplant of, 832
    in triacylglycerol synthesis, 816
Local hormones, 694
Lock-and-key model, 150–151, 151f
Long interspersed elements (LINES), 284
Long QT syndrome (LQTS), 404–405
Loops, 50, 50f
Lovastatin, 833, 833f, 981, **981**
Low-density lipoprotein receptor
    absence of, 830–831
    destruction of, 831
    LDL release, 831, 831f
    mutations in, 831
Low-density lipoproteins (LDLs), 828t,
    829f
    in cholesterol metabolism, 829–830,
        829f
    cycling of, 832
    defined, 828
    in familial hypercholesterolemia,
        830–831
    oxidized, 830
    receptor-mediated endocytosis of, 829,
        829f
    schematic model of, 829f
L-type lectins, 347
Lupine leghemoglobin, 307–308

Lyases, 143t
Lycopenes, 841, 841f, **841f**
Lynch syndrome, 870, 870t
Lysine, 644f, 718
　acetylated, **218**, 966
　structure of, 39, 39f, **39f**, **40t**
　synthesis of, 781
Lysogenic pathway, 275
Lysosomes
　cholesterol synthesis and, 829
　defined, 344–345
　function of, 344–345, 345f
　in I-cell disease, 344–345
　with lipids, 820, 820f
　in Pompe disease, 661f
Lysozyme, antibodies against, 87–88
Lytic pathway, 275

MacKinnon, Roderick, 395, 399
Macrophages, 830, 832
Macugen, 321
Magnesium ions
　in ATP hydrolysis, 201–202
　in Calvin–Benson cycle, 612–613, 613f, 621
　in DNA cleavage, 194–196
　RNA polymerases and, 882
Magnetic moment, 132
Main chains. See Backbones, in polypeptide chains
Major histocompatibility complex (MHC), 84
Major histocompatibility complex proteins, 89–91
　anchor residues, 90
　class I, 90, 90f
　peptide presentation by, 89–91, 90f
　structure of, 89–91
　T-cell receptors and, 89–91, 91f
Malaria, 635–636
Malate, 537, 537f
　in Calvin–Benson cycle, 622f, 623, 624
　in citric acid cycle, 528
　in gluconeogenesis, 502, 502f
　in oxidation-reduction reactions, **465**
Malate dehydrogenase, 528, 691
Malate synthase, 537
Malate–aspartate shuttle, 571, 571f
Malathion, **182**
MALDI-TOF mass spectroscopy, 118–119, 119f, 345, 345f
4-maleylacetoacetate, **724**
Malic enzyme, 691
Malignant hyperthermia, 508
Malnutrition, 756
Malonyl ACP, 685, 686–687
Malonyl CoA, 685, 695
Malonyl transacylase, 686
Malonyl/acetyl transacylase (MAT), 688f, 689
Maltase, 334, 473
Maltase inhibitors, 334
Maltose, 333–334, **333f**
Mammalian fatty acid synthase, 688, **688f**, 689f
Mammalian liver homogenate, 871

Manganese, 595, 595f
Manganese center, 595, 595f
Mannose, 326f, 327
Mannose 6-phosphate, 345, 345f
Mannose 6-phosphate receptor, 346
Maple-syrup urine disease, 727
Margulis, Lynn, 545, 545b
Mass spectroscopy
　defined, 118
　MALDI-TOF, 118–119, 119f
　peptide sequencing by, 120–121, 121f
　in protein and peptide identification, 118–126
　proteomic analysis by, 125f, 126
　tandem, 121, 121f
Mass-to-charge ratio, 118
Mast cells, 232
Matrix, 118, 543
Matrix side, 544
Matrix-assisted laser desorption/ ionization (MALDI), 118–119, 345
Maximal rate, 154–155, 156
McArdle disease, 661, 661t, 749
Mechanism-based (suicide) inhibition, 171
Mechanism-based toxicity, 992
Mediator, 960, 960f
Meiosis, 872
Melanin, 162, 783
Melanocyte-stimulating hormone (MSH), 736, 736f
Melanoma, 64f
Melatonin, 785, 785f
Melting temperature
　of fatty acid chains, 371, 371t
　of phospholipid membrane, 371, 371f
Membrane anchors, 367, 367f
Membrane channels
　ion, 382, 392–406
　membrane permeability to, 407–408, 408f
　water, 407–408, 408f
Membrane diffusion
　facilitated, 382
　lateral, 369, 370, 370f
　lipid, 369
　protein, 370
Membrane lipids. See also Lipid bilayers
　as amphipathic molecule, 358
　archaeal, 358, 358f
　carbohydrate moieties, 357
　cholesterol, 357–358, 821–833
　defined, 355
　diffusion coefficient, 369
　fatty acids in, 353–354, 353f, 354t
　glycolipids, 357, 372
　hydrophilicity of, 358, 359f
　hydrophobicity of, 358, 359f, 360
　lateral diffusion, 369, 370, 370f
　metabolism, 821
　movement in membranes, 370, 370f
　permeability barrier, 361
　phospholipids, 355–357
　phospholipids, synthesis of, 815–821
　rate of diffusion, 369
　representations of, 359f

sphingolipids, synthesis of, 819, 819f
synthesis of, 814–841
triacylglycerols, synthesis of, 815–821
　as two-dimensional solutions, 355
Membrane potential, 383
Membrane proteins. See also specific proteins
　α helix of, 364, 364f, 365f, 367–369, 368f, 368t
　ATP hydrolysis, 384–390
　β strands of, 365f
　carriers, 390
　channels. See membrane channels
　content variation, 362
　defined, 355
　diffusion, 370
　function of, 362
　gel patterns of, 363, 363f
　hydropathy plots for, 368, 370f
　hydrophobicity of, 367, 367f
　integral, 363, 363f, 364f
　membrane interaction, 363–367
　peripheral, 363, 363f
　pumps. See pumps
　as two-dimensional solutions, 355
Membrane transport
　ABC transporters in, 388–390, 388f, 389f
　active, 382–383, 384
　energetic cost of, 383b
　free energy in, 382–383, 382f
　passive, 382–383
　potassium ion channels and, 397, 397f
　primary active, 390
　secondary active, 390
　transporters. See transporters
Membranes, 352–376
　archaeal, 358, 358f
　as asymmetric, 355, 373, 373f
　budding and fusion of, 375–376, 375f
　common features of, 355
　defined, 363
　diffusion, 369–373
　as electrically polarized, 355
　endoplasmic reticulum (ER), 692
　enzymatic activities, 373
　fluidity of, 370–372
　functions of, 352
　ion gradients across, 458, 458f
　lipid bilayer, 355, 359–362
　mitochondria, 543–544
　as noncovalent assemblies, 355
　overview of, 363–364
　permeability of, 360–361, 361f, 382
　planar bilayer, 360, 361f
　polarity scale for transmembrane helices, 368t
　processes of, 362–369
　protein interaction, 363–367
　as sheetlike structures, 355
　synthesis of, 373
Membrane-spanning α helices, 364, 364f, 365f, 368f, 369
Menten, Maud, 153, 154b
β-mercaptoethanol, 56–57, **56**, **56f**, 57f, 106

Mercury poisoning, 535–536
Merrifield, R. Bruce, 127
Metabolic pathways. *See also specific pathways*
 amphibolic, 448
 biosynthetic, 448
 classes of, 448
 defined, 448
 degradative, 448
 formation of, 449
 illustrated, 447f
 motifs, 460–468
 regulation of, 448
 thermodynamically favorable, 449
 thermodynamically unfavorable, 449
Metabolic syndrome, 743, 743f
Metabolism, 446–468
 activated carriers in, 460–463
 AMPK in, 696
 anabolic reactions in, 451–452
 carbon bonds cleavage, 466
 catabolic reactions in, 466
 catalytic activity control and, 468
 cholesterol synthesis, 829–830
 common motifs in, 448, 460–468
 defined, 447
 drug, 990–991
 enzyme control and, 467
 evolution of, 468
 first-pass, 990
 free energy in, 449–450
 fructose, 489–490, 489f
 glucose, 447f
 of glucose 6-phosphate, 631–634
 group-transfer reactions in, 465–466
 hormones and, 468
 hydrolytic reactions in, 466
 inborn error of, 509–511
 integration of, 732–759
 interconnected interactions in, 447–449, 447f
 iron, 970–971
 isomerization reactions in, 467
 kinetic stability of, 462
 ligation reactions in, 467
 lipid, 821
 lipoproteins, 828f
 overview of, 447–448
 oxidation-reduction reactions in, 465
 of pyruvate, 486–488
 regulation of, 467–468
 substrate accessibility control and, 467–468
 substrate cycles in, 508, 508f
 thermodynamics of, 448–449
 types of reactions in, 465t
Metabolites, 2, 672
Metabolomics, 447
Metal ion catalysis
 in ATP hydrolysis, 201–202
 in carbon dioxide hydrolysis, 191–192
 defined, 180
 in DNA cleavage, 196–197, 197f
Metal ions, 847, 847f, 882, 883f
Metalloproteases, 188f, 189
Metals, cofactor, 144t

Metformin, 745
Methionine
 degradation, 721, 721f
 from homocysteine, 774
 metabolism, 721, 721f
 oxidation of, 228, 228f
 regeneration of, 775
 structure of, 36, 37f, **37f**
 synthesis of, 781
 in translation, 929, 930, 930f
Methionine sulfoxide, 228, 228f
Methionine synthase, 775
Methotrexate, 806f, 807
Methylases, 200
Methylation
 of adenine, 200, 200f
 in amino acid synthesis, 774–775
 DNA protection by, 199f, 200
 in gene expression, 961–962
 of phospholipids, 775
 reaction, 676
3-methylbutane-1-thiol, **436**
2-methylbutyryl CoA, **725b**
Methylcobalamin, 676f, 775
β-methylcrotonyl CoA, 722
Methylglucopyranose, **332f**
β-methylglutaconyl CoA, 722
Methylglyoxal, 510, **510**
Methylmalonyl CoA, 677–678, 677f, 678f, **678f**
Metmyoglobin, 74
Mevalonate, 822, 822f, 823f, 827
Mice, genome of, 286f
Micelles, 359, 359f
Michaelis, Leonor, 153
Michaelis constant (K_M)
 in Cheng–Prusoff equation, 987
 defined, 154
 determination of, 156
 as enzyme characteristic, 156–157
 physiological consequences of variations, 156
 values of enzymes, 157t
Michaelis–Menten equation, 154–155, 155b, 161
Michaelis–Menten kinetics, 151–162, 153f, 167, 167f, 183, 212, 215, 215f
Microbiomes, 22f, 23–24, 24f, 682
MicroRNAs, 908, 909f, 971–972, 971f
Microsomal ethanol-oxidizing system (MEOS), 755
Miglitol, 334, **334**
Migrations, genome sequencing and, 23, 23f
Migratory birds, protein metabolism in, 725
Milstein, César, 113
Mineralocorticoids, 834
Minoxidil, 991
Mismatch DNA repair, 867, 867f
Mitchell, Peter, 562–563, 581, 599, 599b
Mitochondria
 in apoptosis, 579
 ATP synthase in, 564f, 570–571
 brown fat, 576
 in citric acid cycle, 516, 516f

 electron entry via shuttles, 570–571
 electron micrograph and diagram of, 544f
 endosymbiotic origin of, 544–545
 genome of, 544–545
 matrix, 543
 membrane, 543
 membranes and, 374
 oxidative phosphorylation in, 543–544, 562–564, 563f, 564f
 properties of, 543–544
 structure of, 543–544, 544f
 transporters, 572–573, 572f, 573f
Mitochondrial ATP-ADP translocase, 572–573, 572f, 573f, 578
Mitochondrial ATPase, 562. *See also* ATP synthase
Mitochondrial biogenesis, 749–750, 750f
Mitochondrial diseases, 578–579
Mitochondrial mutants, 578
Mitochondrial outer membrane permeabilization (MOMP), 579
Mitochondrial porin, 544
Mitochondrial transporters
 defined, 573f
 illustrated, 573f
 for metabolites, 572–573, 573f
 structure of, 572, 572f
Mitosis, 862f
Mixed inhibition, 166. *See also* Enzyme inhibition
Mixed noncompetitive inhibition, 166
Mixed-function oxygenase, 723, 835
Mobile genetic elements, 284
Molecular clock hypothesis, 25–27, 26f, 316
Molecular evolution, 318, 319–321
Molecular imaging agents, 422
Molecular motors, flagella, 579–580
Molecular oxygen, toxic derivatives of, 560–561
Molecular recognition, 69, 71f
 antibodies and, 88, 89
 complementarity-determining regions and, 88
 example of, 82
 immune system and, 84
 intracellular pathogens and, 91
 utility of, 81
Molecules
 amphipathic nature of, 359
 homologous, 302
 permeability of, 361, 361f
Moloney murine leukemia virus, 290
Monoacylglycerols, 667f, 668f
Monoclonal antibodies, 995–996, 995t
 defined, 113
 generation for proteins, 113–114
 illustrated, 113f
 preparation of, 114f
 in signal transduction inhibition, 434, 434f
 as therapeutic agents, 114
Monod, Jacques, 78, 952
Monomers, 5f
Monooxygenase, 723, 724f, 835

Monosaccharides. *See also* Carbohydrates
  alcohols and, 331–332
  aldoses, 325
  amines and, 331–332
  boat form of, 329–330, 329f
  chair form of, 329–330, 329f
  common, 326–327, 326f
  defined, 325
  diastereoisomers, 325–326, 326f
  envelope form of, 330, 330f
  glycosidic bonds, 331–332, 332f
  isomers of, 325–326, 326f
  ketoses, 325
  linkage of, 333–337
  modified, 331–332, 332f
  nomenclature of, 325
  reducing sugars, 330–331
  ring forms of, 327–329, 327f, 329f
  stereoisomers, 325, 326f
  structure of, 325–327
Morphine, 989, **989f**
MotA–MotB pairs, 580, 580f
Motifs
  defined, 52
  helix-turn-helix, 52, 52f, 951, 951f
  metabolism, 448
  repeating, 124, 124f
Moult, John, 61
mRNA (messenger RNA)
  cDNA prepared from, 278–279
  5′ cap of, 900–901, 901f
  FMN and, 891–892, 891f
  poly(A) tail of, 900–901
  precursors, 898–899, 903–906
  RNA editing and, 908–909, 909f
  structure of, 937, 937f
  transcription and, 908, 908f
  transferrin-receptor, 970–971, 970f
  vaccines, 880–881
Mucins, 338, 341, 341f
Mucoprotein, 338
Mucus, 341
Muller, Hermann, 866
Mullis, Kary, 269
Multidrug resistance, 388–390
Multidrug resistance protein, 388
Multiple myeloma, 113–114
Multiple-drug resistance, 999
Multiple-substrate reactions
  classes of, 159
  defined, 159
  double-displacement, 160–161
  sequential, 159–160
Muscle
  ATP in contraction, 493
  biochemical adaptations to exercise,
    750–751, 751t
  excess fatty acids in, 743–744
  fast twitch, 487, 493
  fatty acids in, 743–744, 748
  fiber types, 649, 649t
  glycogen in, 641, 641f
  glycogen phosphorylase in, 648, 648f,
    650f
  glycolysis in, 493–494, 495f
  insulin resistance in, 743–744
  PP1 regulation in, 657f, 658, 658f

signal transduction in, 742–743
Muscle contraction
  alanine in, 509
  ATP formation in, 747–749
  fuel sources for, 746–750, 747t
  gluconeogenesis in, 508–509, 509f, 510f
  lactate in, 508–509, 509f
Muscle-relaxation pathway, 980, 980f
Mutagenesis
  cassette, 281, 281f
  by PCR, 281, 281f
  site-directed, 187–188, 187f, 280, 280f
Mutagens, 865–866, 865f
Mutant λ phage, 276f
Mutarotation, 328
Mutase, 484
Mutations
  defined, 25
  deletion, 279, 281, 281f
  disease-causing, functional effects of,
    271–272
  in genes encoding hemoglobin
    subunits, 83–84
  hemoglobin and, 83–84
  insertion, 279, 281
  at level of function, 310
  at level of sequence, 310
  point, 280
  in protein kinase A, 222–223
  in recombinant DNA technology,
    279–281
  substitution, 279
MWC model, 78
myc gene, 898
Mycobacterium tuberculosis, 271, 538, 707
Myelin, 362, 362f
Myelin sheath, 352f
Myogenin, 291, 291f
Myoglobin, 69
  alignment comparison, 305, 305f
  alignment with gap insertion, 303, 304f
  amino acid sequences of, 303–305, 303f
  carbon monoxide and, 74–75, 75f
  as compact molecule, 51, 51f
  defined, 50–51, 69, 73, 303
  distribution of amino acids in, 51f
  oxygen and, 73–75
  oxygen-binding curve, 73f, 76
  receptor-mediated endocytosis and,
    375–376
  sequence alignment of, 307
  structure of, 73, **73f**
  tertiary structure of, 310, **310f**
  three-dimensional structure of, **51f**
Myosin, 145
  actin filaments and, 204–206, 205f, 206f
  active sites, 204, 204f
  altered conformation persistence,
    203–207
  ATP complex structure, 201, 201f
  ATP hydrolysis and, 200–207
  ATPase transition-state analog, 202,
    202f
  catalytic strategies of, 179, 200–207
  conformational changes, 202f, 203, 203f
  defined, 200
  kinetic studies of, 201

as slow enzymes, 203
  structure of, **201f**
  two-headed structure, 206f
Myosin V, 205, 205f
Myosteatosis, 743
Myrcene, 840, **840f**

Na⁺ ions. *See* Sodium ions
Na⁺–K⁺ ATPase, 384
Na⁺–K⁺ pumps, 384, 387, 387f
NAD⁺ (nicotinamide adenine dinucleotide
    oxidized)
  in glycolysis, 482, 482f, 486–488, 487f
  light absorption and, 101
  from metabolism of pyruvate, 486–488
  reduction potential of, 547
  regeneration of, 487, 487f, 488
  structures of oxidized form of, 460–461,
    460f
  synthesis of, 783, 783f
NADH (nicotinamide adenine
    dinucleotide reduced), 487f
  in ATP generation, 461
  electrons from, 570–571
  electron-transport chain and, 517, 530,
    545, 547, 548–549
  ethanol metabolism in, 755
  in fatty acid oxidation, 672–673
  in glycolysis, 482, 486–487, 486f
  light absorption, 101
  in oxidative phosphorylation, 545,
    547–548
  pyruvate reduction by, 487
  in respiratory chain, 552–553, 553f
  structures of oxidized form of, 460
  transport of, 570–571
NADH-cytochrome *b*5 reductase, 692, 693f
NADH-Q oxidoreductase, 549, 549f, 550t,
    552–553, 553f
NADP⁺ (nicotinamide adenine
    dinucleotide phosphate), 626, 631
NADPH (nicotinamide adenine
    dinucleotide phosphate reduced), 835
  in Calvin–Benson cycle, 610, 618, 622
  dihydrofolate reductase and, 805
  in fatty acid synthesis, 691
  formation by photosystem I, 601
  formation of, 624–630
  in glucose 6-phosphate conversion, 624
  oxidative generation of, 624
  pathways requiring, 624t
  in pentose phosphate pathway, 631–633
  in reductive biosynthesis, 461
Nalidixic acid, 857, **857**
National Center for Biotechnology
    Information, 309
Native chemical ligation, 129, 129f
Native protein, 57
N-degron, 704
Neanderthals, 318, 319f
Neer, Eva, 425b
Negative nitrogen balance, 768
Negatively charged amino acids, 39–40,
    39f
N-end rule, 704
Neomycin, 942
Neonatal Fc receptor (FcRn), 996

Nernst equation, 403
Nerve impulses, 392, 400
Neuraminidase, 348
Neuroglobin, 82–83
Neurological diseases. *See* Diseases and disorders
Neuron, myelination of, 362, 362f
Neuropeptide Y (NPY), 736, 736f
Neurotransmitter release, 375, 375f
Neurotransmitters, 400, 785
Neutral fat. *See* Triacylglycerols
Next-generation sequencing (NGS)
    defined, 284
    detection methods in, 284–286, 285f, 286f
    as highly parallel, 284
    ion semiconductor, 285, 286f
    pyrosequencing, 284
    reversible terminator method, 285
    RNA-seq, 289
NF-κB, 707
Niacin (vitamin B₃), 463t, 464f
Nicotinamide, 783, **783f**
Nitric oxide (NO)
    functions of, 784t
    synthesis of, 784–785, 784f
Nitrogen
    in amino acids, 764
    excess, disposal of, 713–718
    metabolism, 701, 716f, 783
    peripheral tissue transport of, 712–713
Nitrogen fixation
    in amino acid synthesis, 764–767, 764f
    ATP in, 765–769
    in bacteria, 764–767
    defined, 764
    FeMo cofactor in, 765, 765f
    glutamate in, 766–767
    glutamine in, 766–767
    iron–sulfur proteins in, 765–766, 765f
    P clusters in, 766
Nitrogenase, 764, 765–766, **765f**
Nitrogenase complex, 764–765
Nitrogenous bases. *See* Bases/base pairs
Nitrous acid, 865
N-linkage, 338, 338f, 341
N-linked oligosaccharides, 338, 338f, 345
Nonalcoholic fatty liver disease (NAFLD), 695
Noncompetitive inhibition. *See also* Enzyme inhibition
    defined, 166
    example of, 168
    inhibitor binding, 168
    kinetics of, 168, 168f
Nonconservative substitutions, 305, 307, 307f
Noncovalent bonds
    in biological molecules, 7–11, 9f
    hydrogen bonds, 5, 9, 9f, 10, 11
    hydrophobic effect and, 10–11, 11f
    types of, 8
    van der Waals interactions, 9–10, 10f, 12
Nonessential amino acids, 768, 768f, 768t
Nonheme iron proteins, 525, 551–552
Non-homologous end joining (NHEJ), 292, 292f, 868

Nonphotochemical quenching (NPQ), 604
Nonpolar molecules, 11
Nonproteinogenic amino acids, 126
Nonshivering thermogenesis, 575, 577, 580–581
Norepinephrine, 785, 785f
Northern blotting, 267
Novobiocin, 857
Nuclear envelope, 374, 374f
Nuclear hormone receptors
    defined, 963
    domain structures, 964, 964f
    in gene expression, 963–965
    ligand-binding to, 963, 964f, 965f
Nuclear magnetic resonance (NMR) spectroscopy, 132–135, 132t
    basis of, 133, 133f
    chemical shifts, 133
    defined, 132
    NOESY and, 133–134, 134f, 135f
    one-dimensional, 133, 133f
    RNA structure, 320, 321f
    signals, 133f
    two-dimensional, 133–135
Nuclear membrane, gene expression and, 892
Nuclear Overhauser effect (NOE), 133–134
Nuclear Overhauser enhancement spectroscopy (NOESY), 134–135, 134f, 135f
Nuclear pores, 374
Nucleation-condensation model, 60, 60f
Nucleic acids, 265. *See also* DNA; RNA
Nucleophilic catalyst, 181
Nucleoside diphosphate kinase, 451, 796
Nucleoside diphosphokinase, 527
Nucleoside monophosphate, 451
Nucleoside monophosphate kinases, 451, 796
Nucleoside phosphorylases, 809
Nucleosides
    defined, 792
    nomenclature of, 792, 792t
Nucleosome, 893, **893f**
Nucleotidases, 809
Nucleotide synthesis, 791–812
    de novo pathways in, 792, 798
    deoxyribonucleotide, 801–807, 802f, 808–809, 808f
    disorders of, 809–811
    inosinate in, 808
    overview of, 791–792
    of purine, 783, 796–801, 797f, 807–808, 808f
    of pyrimidine, 783, 792f, 793–797, 793f, 807
    regulation of, 807–809
    salvage pathways in, 792, 792f
    substrate channeling in, 794
Nucleotide-excision repair, 868, 868f
Nucleotides
    cyclic, 792
    defined, 792
    metabolism disruptions, 809–811
    nomenclature of, 792, 792t

sequence of, 43
in signal transduction pathways, 792
NusA protein, 890

Obesity, 734
    causes of, 734
    diabetes and, 741–746
    evolution and, 734
    health consequences of, 734t
    insulin in, 736
    insulin resistance and, 743, 743f
    leptin resistance in, 736f, 737–738
    storage capacity of fat tissue, 743f
Obligate anaerobes, 488, 488t
Octadecanoic acid, 353
Odd-chain fatty acids, 675–676, 676f
Odorant receptors (ORs)
    activation of, 436–437, 437f
    evolution of, 441, 441f
    in olfaction, 436–437
Off-diagonal peaks, 134
Off-target toxicity, 992
O-GlcNAc transferase, 339
Okazaki, Reiji, 849, 849b
Okazaki, Tsuneko, 849, 849b
Okazaki fragments, 849, 849f, 859
Oleate, 693
Olfaction. *See also* Sensory systems
    7TM receptors in, 436–437
    COVID-19 and, 437
    odorant receptors (ORs) in, 436–437, 437f, 441, 441f
    signal transduction in, 436–437, 436f, 437f
Olfactory neurons, 436
Oligomycin, 577
Oligopeptides. *See* Peptides
Oligosaccharide chain, 819
Oligosaccharides, 345f
    antigen structures, 343, 343f
    attached to erythropoietin, 339, 339f
    defined, 333
    enzymes responsible for assembly, 342–343
    human milk, 335
    mass spectroscopy and, 345, 345f
    N-linked, 338, 338f
    O-linked, 338, 338f
    sequencing of, 338, 339f
O-linkage, 338, 338f, 341
Omega (Ω) loop, 46
ω-carbon atom, 354
"On-bead" assays, 983f
Oncogenes, 433
One-dimensional NMR, 133, 133f
On-target toxicity, 992
Open promoter complex, 887, 887f
Open reading frames (ORFs), 309
Operator site, 952
Operon model, 952
Operons, 953f
    in bacteria, 955
    defined, 952
    histidine, 969, 969f
    lac, 953, 953f
    structure of, 952, 953f
    threonine, 969, 969f
    trp, 968, 968f

Opsin, 438

Optogenetics, 607, 607f

Oral bioavailability, 988–989, 989f

Origin of replication complexes, 861–862

Origin of replication (oriC locus), 860, 861f

Ornithine, 715–716

Ornithine transcarbamoylase, 715, 717

Ornithine transcarbamoylase deficiency, 717f, 718–719

Orotate, 794–795

Orotate phosphoribosyltransferase, 795

Orotidylate, 795

Orotidylate decarboxylase, 795

Orthologs, 302, 302f

Orthophosphate, 453, **453f**, 482, 483f, 503, 642–643

Oseltamivir, 348

Osteoarthritis, 341

Osteogenesis imperfecta, 54

Overlap peptides, 122, 122f

Oxaloacetate, **160**
  in amino acid degradation, 718, 719–720, 719f
  in amino acid synthesis, 769
  in Calvin–Benson cycle, 622, 622f
  in citric acid cycle, 517, 522–523, 523f, 528–529, 534
  conversion into phosphoenolpyruvate, 502–503
  in fasting or diabetes, 680
  formation of, 501–502
  in gluconeogenesis, 499–503, 500f, 502f
  in glyoxylate cycle, 537f
  in ligation reactions, 467
  in oxidation-reduction reactions, **465**
  in starvation, 753–754

Oxidant, 546

Oxidation
  activated carriers of electrons for, 460–461
  of aldehydes, 480–481
  carbon, 456–459
  by citric acid cycle, 488
  in drug metabolism, 990
  by electron transport chain, 488
  energy, 457–458
  in fatty acid metabolism, 669–670, 671–674, 673t, 678–679
  of fatty acids, 678–679
  of glyceraldehyde 3-phosphate (GAP), 482
  of membrane proteins, 674f
  phase I transformations in, 990
  pyruvate dehydrogenase complex and, 518–519
  of succinate, 528–529
  of unsaturated fatty acids, 674–675, 674f

Oxidation-reduction potential. *See* Reduction potential

Oxidation-reduction reactions
  free-energy change in, 546, 547–548
  in metabolism, 465, 465t

Oxidative phosphorylation, 542–581
  ADP in, 572–573, 572f, 575f
  ATP in, 573–574, 575f
  ATP yield in, 573–574, 574t
  ATP-ADP translocase in, 572–573, 572f, 573f, 578
  cellular respiration in, 543–545
  chemiosmotic hypothesis in, 562–563, 563f
  defined, 458, 543
  electron transfer in, 545–548, 561, 562f
  electron-transport chain in, 543, 548–549, 569f
  in eukaryotes, 543–544
  in exercise, 747–749
  inhibition of, 577–578
  iron–sulfur clusters in, 551–552, 551f
  in mitochondria, 543–544, 562–564, 563f, 564f
  mitochondrial transporters in, 572–573, 572f
  NADH in, 545, 547–548
  overview of, 542, 543f
  photosynthesis comparison, 600f
  proton gradients in, 458, 458f, 543, 548–549, 562–569
  proton-motive force in, 543, 562–563, 563f, 565–567, 565f
  rate of, 574
  reduction potential in, 545–548, 546f
  respiratory chain in, 549–562
  respiratory control in, 573–574, 575f
  shuttles in, 570–573, 570f, 571f
  transporters in, 570–573, 572f
  uncoupling of, 575–576, 576f

Oxidized LDL (oxLDL), 830

Oxidized lipoamide, 520

Oxidizing agent, 546

Oxidoreductases, 143t

Oxidosqualene cyclase, 824

Oxoguanine-adenine base pair, 865, 865f

Oxyanion hole
  of chymotrypsin, 185, 185f
  defined, 185
  of subtilisin, 186–187, 187f

Oxygen
  affinity in red blood cells, 80–81, 80f, 81f, 82f
  carbon monoxide and, 81, 82–83
  concentration in tissues, 77, 77f
  cooperative binding of, 76–77
  cooperative release of, 76
  hemoglobin quaternary structure and, 77–78, 77f
  in hydroxylation function, 834–836
  myoglobin and, 73–75
  partial pressure, 76
  pyruvate and, 486
  reactive, 74
  toxic derivatives of, 560–561

Oxygen affinity of hemoglobin, 80–81, 80f, 81f, 82f

Oxygen binding
  cooperative, 76–77
  by heme iron, 73–74
  hemoglobin quaternary structure and, 75–76, 77f
  iron and, 79f
  pure hemoglobin, 80, 80f
  sites, 73

Oxygenase reaction, 614–615, 614f

Oxygenases
  in aromatic amino acid degradation, 723–724
  mixed-function, 723

Oxygen-binding curve
  defined, 73
  fractional saturation, 73
  for hemoglobin, 76, 76f, 77f
  for myoglobin, 73f, 76
  sigmoid, 76
  T-to-R transition, 78, 78f, 80–81

Oxygenic photosynthesis, 586, 587f. *See also* Photosynthesis

Oxyhemoglobin, **77f**, 78, 79f

Oxymyoglobin, 73–74, **74f**

P clusters, 765f, 766

P site, ribosomal, 929, 929f, 931, 934, 935

p53, 705, 870

P450, 756

P680, 594, 595f

P700, 597–598, 597f

P960, 591, 592, 592f, 593f

Pacemakers, 405–406

Paclitaxel, 977

Paddles, 398–399

PALA, 213–214, **213f**, 214f, **214f**

Palindromes, 265

Palmitate, 673–674, 690

Palmitoleate, 674, 674f

Palmitoyl CoA, 674, 674f, 695

Pamaquine, 634

p-aminobenzoic acid (PABA), 166, **166**

Pancreas, β cells of, 735, 744

Pancreatic failure, 744–745

Pancreatic lipases, 667, 667f

Pancreatic trypsin inhibitor, 226–228, 227f

Pancreatic zymogens, 225f, 226, 226f

Pantothenic acid (vitamin B_5), 463t, 464f

Papain, 87, 143

Paracoccus denitrificans, 551

Paralogs, 302, 302f

Paraquat, 605

Park, James, 979

Parkinson disease, 62, 703, 705, 817

Partial pressure, oxygen, 76

Partition coefficient, 989

Passenger strand, 295

Passive membrane transport, 382–383

Pasteur, Louis, 472

Patch-clamp technique, 393, 393f, 394f

Pathogen-associated molecular patterns (PAMPs), 84–85, 85f

Pauling, Linus, 25, 46, 83, 169

PCSK9 (proprotein convertase subtilisin/kexin type), 832

Pectin, 296, 336

Pegfilgrastim, 995

Pegylation, 995

Penicillin, 979f
  conformations of, 173, **173f**
  mechanism of action, 172–173, 173f
  reactive site of, 172–173, 172f
  research in, 677b
  structure of, **979**, **998**

Penicillium citrinum, 981

Penicilloyl-enzyme derivative, 173f

Pentose phosphate pathway
  Calvin–Benson cycle and, 633–634
  cell growth and, 633
  decarboxylation in, 624

defined, 610, 624
erythrose 4-phosphate in, 627–628
fructose 6-phosphate in, 627, 628
glucose 6-phosphate dehydrogenase in, 634–636
glucose 6-phosphate in, 624, 631–634
glycolysis and, 624–628
illustrated, 625f
modes, 631–633, 632f
$NADP^+$ in, 631
NADPH in, 631–633
nonoxidative phase of, 624–630, 628t
overview of, 610
oxidative phase of, 624, 626f, 628t, 631
rate of, 631
reactions, 627–630, 629f, 630f
ribulose 5-phosphate in, 624, 626, 627, 631–633
sedoheptulose 7-phosphate in, 627
tissues with, 632t
transaldolase in, 626–628, 629–630, 630f
transketolase in, 626–628, 629f
Pentose shunt. See Pentose phosphate pathway
Pentoses, 325, 326–327, 326f
Peptide bonds
chemical nature of, 180
cis configuration, 44, 44f, 45f
cleavage of, 63–64, 185
defined, 41
formation in translation, 933–935, 934f
formation of, 42f
hydrolysis of, 180–181, 183, 184f
lengths, 44, 44f
as planar, 44, 44f
rigidity of, 46
rotation of, 45, 45f
stability of, 42
torsion angles, 45
trans configuration, 44–45, 44f, 45d
Peptide mass fingerprinting, 124, 125f
Peptide sequences, 59f
Peptide-ligation methods, 129
Peptides
mass spectroscopy for sequencing, 120–121, 121f
overlap, 122, 122f
presentation by MHC proteins, 89–91, 90f
sequencing of, 277
synthesis of, 126–129, 128f, 129f
Peptidoglycan, 173, 173f
Peptidyl transferase center, 932–933
Perilipin, 669
Peripheral membrane proteins, 363, 363f
Periplasm, 373
Periplasmic side, 591
Permeability
lipid bilayers, 360–361, 361f
membrane, 382, 406–407
of potassium ion channels, 396
Permeability coefficients, 361, 361f
Peroxide bridge in cytochrome c oxidase, 558, 558f
Peroxisome proliferator-activated receptor (PPARα), 756
Peroxisomes
in Calvin–Benson cycle, 615, 615f

defined, 615
electron micrograph of, 614f
in fatty acid oxidation, 678–679, 679f
Perutz, Max, 75
PEST sequences, 704
P-glycoprotein, 388
pH
buffers and, 19–20, 19f
carbon dioxide and, 82, 82f
defined, 17
effect on carbonic anhydrase activity, 190–191, 191f
effect on oxygen affinity of hemoglobin, 81f, 82f
ionization state as function of, 36f
Pharmacogenomics, 836
Pharmacology, 978. See also Drug development
Phase I transformations in oxidation, 990
Phase II transformations in conjugation, 990
Phenobarbital, 836
Phenotypic screening approach, 985–986, 985f
Phenyl isothiocyanate, 120
Phenylacetate, 717
Phenylacetyl CoA, 718
Phenylacetylglutamine, 718
Phenylalanine, 718
degradation of, 723–724, 724f, **724**, 768, 769
phenylketonuria and, 727–728
structure of, 37, 37f, **37f**, 38, **728**
synthesis of, 776–777, 778f
Phenylalanine hydroxylase, 724
Phenylalanine operon, 969f
Phenylketonuria, 727–728
Phenylpyruvate, **728**, 778
Phi (φ) angle, 45, 45f, 46f
Phosphatases
defined, 220, 503
in gluconeogenesis, 503
insulin signaling and, 430–431, 743
in phosphorylation, 219–220
PP2A, 220
Phosphate carrier, 572, 573, 573f
Phosphate esters, 458
Phosphates, 20, 458–459
Phosphatidate, 356
defined, 815
as intermediate, 815–821, 815f
structure of, **356f, 816**
Phosphatidic acid phosphatase (PAP), 816, 821, 821f
Phosphatidylcholine, **356f**
Phosphatidylcholine, synthesis of, 818
Phosphatidylethanolamine, **356f**, 817–818, **818**
Phosphatidylethanolamine methyltransferase, 818
Phosphatidylinositol, **356f**, 816, **817**
Phosphatidylinositol 3,4,5-trisphosphate (PIP$_3$), 428, **428f**, 429f, 742, 742f
Phosphatidylinositol 4,5-bisphosphate (PIP$_2$)
in insulin signaling, 428, **428f**, 429f
in phosphoinositide cascade, 420, 420f

in signal transduction pathway, 742, 742f
synthesis of, 817
Phosphatidylinositol-dependent protein kinase (PDK), 742
Phosphatidylserine, **356f**, 818
Phosphite triester method, 268f, 269
Phosphocreatine, **160**
Phosphodiester bonds, 194–195, 195f, 882–884
Phosphoenolpyruvate (PEP)
formation of, 484, 501–503, 502f
in gluconeogenesis, 500f
in glycolysis, 466, 484–485
phosphoryl-transfer potential, 453–454, 454f, 484–485
Phosphoenolpyruvate carboxykinase (PEPCK), 502, 503
Phosphoenolpyruvate carboxylase, 623
Phosphofructokinase (PFK), 477, 740, 740f
activation of, 495
in liver, 494–495, 495f
in muscle, 493–494, 493f
regulation of, 494–495, 495f
Phosphofructokinase 2 (PFK2), 506–507, 507f
Phosphoglucomutase, 646, 646f, 647f
6-phosphogluconate dehydrogenase, 626
Phosphogluconate pathway. See Pentose phosphate pathway
6-phosphoglucono-δ-lactone, 626, 626f
Phosphoglucose isomerase, 477
2-phosphoglycerate, 466
3-phosphoglycerate
in Calvin–Benson cycle, 611–612, **611**, 612f, 613f, 615, 615f
in cysteine synthesis, 772
in glycine synthesis, 772
in glycolysis, 484–486
in oxygenase reaction, 614, 614f
in serine synthesis, 772
3-phosphoglycerate dehydrogenase, 780–781, **780f, 781f**
2-phosphoglycerate in glycolysis, 482–484
Phosphoglycerate kinase, 482
Phosphoglycerate mutase, 484, 646
3-phosphoglyceric acid, **457, 458**
Phosphoglycerides, 356, 356f, 359f
Phosphoglycolate, 614, 614f
Phosphoinositide 3-kinases (PI3Ks), 428, 742
Phosphoinositide cascade, 420–421, 420f, 421f, 653
Phosphoinositide-dependent kinase-1 (PDK1), 429–430, 430f
Phospholipase C, 420, 420f
Phospholipid synthesis
from activated alcohol, 817–818
activated intermediate, 816–817
phosphatidate and, 815–821, 815f
phosphatidylcholine, 818
sources of intermediates in, 815–821

Phospholipids, 355–357
  alcohols and, 817–818
  base-exchange reactions in generation of, 818
  in bimolecular sheet formation, 359–362
  components of, 355f
  defined, 355
  flip-flop, 370, 370f
  lipid vesicles formed from, 360–361
  melting temperature, 371, 371f
  membrane, 355–357
  platform for, 356
  synthesis of, 815–821
Phosphopantetheine, 686f
Phosphopentose epimerase, 617, 617f
Phosphopentose isomerase, 309f, 617, 617f
Phosphoproteome, 223
Phosphoproteomics, 223
5-phosphoribosyl-1-pyrophosphate (PRPP)
  in purine synthesis, 797, 807–808
  in pyrimidine synthesis, 779, 795
Phosphoribulokinase, 622
Phosphorolysis in glycogen metabolism, 642–647
Phosphorothioates, 196, 196f
Phosphoryl group, 220
Phosphorylase. *See* Glycogen phosphorylase
Phosphorylase *a*, 647–649, **647f**, 650, 650f, 659, 660f, 739, 740
Phosphorylase *b*, 647–649, **647f**, 650, 650f
Phosphorylase kinase
  activation of, 650–651, 650f
  defined, 650
  subunits, 650, 650f
Phosphorylaspartate, **384**
Phosphorylation
  amplified effects of, 221
  ATP formation by, 601, 601f, 602
  cAMP stimulation of, 418
  of carboxyl-terminal domain (CTD), 897, 897f
  of carboxyl-terminal tail, 432
  in citric acid cycle, 531–532, 532f
  control of, 219–220
  in covalent modification, 218, 219t
  dephosphorylation and, 220
  EGF receptor, 432
  in enzyme regulation, 218, 219t
  free energy of, 221
  of fructose 6-phosphate, 477
  in gluconeogenesis, 502–503, 507
  in glycogen metabolism, 642–643
  in glycolysis, 476–477
  insulin binding in, 427, 428f
  in insulin signaling, 425–431
  in metabolism, 451, 458
  of nucleoside monophosphates, 451
  oxidative. *See* oxidative phosphorylation
  phosphoinositide-dependent kinase 1 (PDK1), 430f
  in phosphorylase activation, 650–651, 650f
  in phosphorylase conversion, 650

of proteins, 223
in purine nucleotide synthesis, 797–799, 798f
of pyruvate dehydrogenase, 531–532, 532f
reactive intermediates, 333
of serine residue, 507
substrate-level, 482–483
sugars, 332–333
Phosphorylethanolamine, 818, **818**
Phosphoryl-transfer potential, 453–455, 454f, 457–458
Phosphoserine, **63f**
Phosphotriesters, 268f, 269
Photoinduced charge separation, 590, 590f, 592f
Photophosphorylation, cyclic, 601, 601f, 602
Photorespiration
  in C$_4$ pathway, 622–623, 622f
  in Calvin–Benson cycle, 614f, 615
Photosynthesis, 585–607
  accessory pigments in, 602–605, 603f, 604f
  artificial photosynthetic systems and, 606–607, 607f, 620, 620f
  in autotrophs, 611
  bacterial reaction center, 591f, 592–593, 592f, 593f
  basic equation of, 586
  biochemical principles of, 587
  chlorophylls in, 589–593
  in chloroplasts, 588–589, 588f
  components of, location of, 605, 605f
  in cyanobacteria, 593
  cyclic photophosphorylation in, 601, 601f, 602
  dark reactions, 587, 610
  defined, 586
  efficiency of, 604
  electron transfer in, 589–593
  energy conversion in, 586–588
  enzymes in, 144
  in eukaryotes, 588–589, 593
  herbicide inhibition of, 605
  in heterotrophs, 586
  light absorption in, 590, 590f
  light energy conversion in, 586–588
  light reactions in, 587, 587f, 602, 605, 610
  light-harvesting complexes in, 604, 604f
  overview of, 586–588
  oxidative phosphorylation comparison, 600f
  photoinduced charge separation in, 590, 591f
  photosystem I. *See* photosystem I
  photosystem II. *See* photosystem II
  in prokaryotes, 606t
  proton-motive force in, 599
  reaction centers in, 591f, 592–593, 593f, 596, 596f, 602–605, 604f
  resonance energy transfer in, 603–604, 603f
  vs. respiration, 587
  in thylakoid membranes, 588–589
  Z scheme of photosynthesis, 598, 599f

Photosynthetic systems, 606–607, 606t, 607f
Photosystem I. *See also* Photosynthesis
  defined, 588, 594f
  electron flow through, 601
  ferredoxin generated by, 597–598, 597f
  illustrated, 593f
  inhibitors of, 605
  link to photosystem II, 596, 596f
  NADPH formation by, 601
  reaction center, 597–598, 597f
  structure of, 597–598, 597f
  Z scheme of photosynthesis, 598, 599f
Photosystem II. *See also* Photosynthesis
  defined, 588
  electron flow through, 593–594, 595f
  electron source, 593–594
  illustrated, 593f
  inhibitors of, 605
  link to photosystem I, 596, 597f
  location of, 596
  photochemistry of, 594
  proton release, 596f
  reaction center, 594–596, 594f
  structure of, 594f
  Z scheme of photosynthesis, 598, 599f
Photosystems, 588
  proton gradients in, 593–599
Phototroph, 447, 586
Phototrophic halobacteria, 585
Phototrophy
  defined, 586
  light energy conversion, 586–588
  overview of, 586f
Photovoltaic cells, 606, 607f
Phytoene, 841, **841f**
Phytol, 589
Pichia pastoris, 112
Pickart, Cecile, 708b
Pig embryos, UCP-1 into, 577
Ping-pong reactions, 160–161
Pinnix, Chelsea, 29, 29f
pK_a values, 18, 40t
Placebo effect, 997
Planar bilayer membrane, 361, 361f
Plants
  C$_3$, 623, 624f
  C$_4$, 623, 624f
  C$_4$ pathway and, 623
  genetically engineered, 295–296, 295f
  glyoxylate cycle in, 536–538, 537f
  oxaloacetate synthesis, 683
  photosynthesis in. *See* photosynthesis
  polysaccharides of, 335–336
  recombinant DNA technology in, 295–296, 295f
  starch in, 618
  sucrose synthesis in, 618, 619f
  thylakoid membranes of, 605, 605f
Plaques in genomic library screening, 277, 278f
Plasma lipoproteins, 828t
Plasma-membrane proteins, 940, 940f
Plasmids
  antibiotic resistance and, 998
  defined, 272

engineered, 274
expression vectors, 275
genes, 274
microinjection of, 290, 290f, 291f
polylinker region, 274, 274f
Ti, 295–296, 295f
as vectors for DNA cloning, 274–275
Plastocyanin, 605
Plastoquinol, **594**, 596
Plastoquinone, 593–594, **594**, 595f, 596, 605
Pleckstrin homology domain, 428, 430
PLP-Schiff-base linkage, 643–644, 644f
Point mutation, 280
Poisoning
arsenite, 535–536, 536f
mercury, 535–536
ricin, 944
Polar amino acids, 38–39, 38f
Polar head group, 358
Polarity, 4
Poly(A) polymerase, 901, 901f
Poly(A) tail, 901
Poly(A) tail-binding protein I (PABPI), 937, 937f
Polyacrylamide gel electrophoresis (PAGE), 105f, 106–107, 106f
Polyacrylamide gels, 266
Polycistronic transcript, 953
Polyclonal antibodies, 113, 113f
Polycomb repressive complex-2 (PRC-2), 967
Polyenes, 841
Polygenic transcript, 953
Polylinker region, plasmid, 274, 274f
Polymerase chain reaction (PCR)
cycle steps, 269–270, 269f, 270f
defined, 265, 269
DNA sequence amplification by, 269–270, 269f, 270f
first cycle of, 269f
as gene exploration tool, 265
inverse, 281, 281f
multiple cycles of, 270f
mutagenesis by, 281, 281f
quantitative (qPCR), 287, 287f
uses of, 271
Polymerase switching, 862
Polymerases, processive, 857–858
Polymorphisms, 271
Polypeptide chains
backbone, 42, 42f
bond rotation in, 45, 45f
cleavage of, 121, 122t
components of, 42f
cross-linked, 43, 43f
directionality of, 42, 42f
disulfide bonds of, 43, 43f
flexibility of, 44–46
formation of, 42
loops, 50, 50f
random-coil conformation, 56
residue, 42
reverse turns, 50, 50f
side chains, 42, 42f, 52
subunit structures, 55
Polypyrimidine tract, 902

Polysaccharides
defined, 335
of fungi and arthropods, 336–337, 337f
glycosidic bonds, 335f, 336f
glycosidic linkages, 336f
of plants, 335–336
Polysomes, 935, 935f
Polyunsaturated fatty acids, 674–675, 675f, 679, 693–694
Pompe disease, 661, 661f, 661t
Porin, 52f
amino acid distribution in, 52, 52f
amino acid sequence of, 364–365, 364f, 365f
defined, 364
hydropathy plot for, 369f, 370f
Porphobilinogen, 788
Porphobilinogen deaminase, 786
Porphyrias
disorders of, 788
synthesis of, 786–787
Positively charged amino acids, 39, 39f
Positron emission topography (PET) scanning, 63, 63f
Postabsorptive state, 739–740
Postmarketing surveillance, 997
Postprandial period, 735
Postprandial state, 739
Posttranscriptional gene expression
in eukaryotes, 970–971
in prokaryotes, 967–972
Posttranslational modifications, 63–64, 63f, 123
Potassium ion channels
as archetypical structure, 394–395
hERG, 405
inactivation of, 399–400, 400f
path through, 396f
permeability of, 400
purification of, 394
selectivity filter, 396, 396f, 397f
structure of, 394–397, 395f, 397f, 398
transport model, 397, 397f
voltage-gated, 398–399, 399f
Potassium ions
action potentials and, 393
dehydration of, 396, 397f
hydration of, 397f
Potential energy, 12–13
PP2A, 220
PPARα (peroxisome proliferator-activated receptor), 756
Precose, 334
Precursor ions, 120
Predisposition, 23
Pregnenolone, **837f**
androgen synthesis by, 837–838, 838f
corticosteroids from, 837, 837f
defined, 837
estrogen synthesis by, 837–838, 838f
progesterone from, 837, 837f
structure of, **837f**
synthesis of, 836–838
Pre-initiation complex (PIC), 897–898
Pre-mRNA processing, 900–901, 901f, 908, 908f

Pre-N-degrons, 704
Prephenate branch, 777–778
Prepriming complex, 860, 860f
Pre-rRNA, 900f
Primaquine, 634, 636
Primary active transport, 390
Primary antibody, 116
Primary messenger, 413
Primary protein structure, 34, 41–46
Primary transcripts, 901f
Primase, 848, 848f
Primers
annealing, 269
defined, 846
DNA probes as, 269
in DNA replication, 848, 848f
in polymerase chain reaction (PCR), 269–270, 269f
RNA, 859
Probes. *See* DNA probes
Procaspases, 224
Processivity
defined, 645, 857
in DNA replication, 857–858
Procollagen, 224
Prodrugs, 836
Product ions, 121
Proenzyme, 223
Progesterone, 834, 837, **837f**
Progestogen, 834
Programmed cell death. *See* Apoptosis
Progressive neuromuscular disorder, 509
Proinsulin, synthesis of, 224, 279, 279f, 744
Prokaryotes. *See also* Bacteria
ABC transporters in, 390
defined, 3
DNA-binding proteins in, 950–956
gene expression in. *See* Gene expression (prokaryotes)
glycolysis in, 473
photosynthetic, 606t
proteasomes, 707, 707f
RNA polymerases in, 881, 881f
transcription and translation in, 892f
transcription factors in, 958
Prokaryotic RNA polymerases, 881, **881f**
Proliferating cell nuclear antigen (PCNA), 862
Proline
degradation of, 720, 720f
structure of, 37, **37f**
synthesis of, 771
Prolyl hydroxylase, 758
Promoters
alternative sequences, 890–892, 890f
bacterial, 886–887, 886f, 887f
closed complex, 887, 887f
common elements, 895f
core, 886
defined, 882
open complex, 887, 887f
strong, 886
upstream of, 886
weak, 886
Pro-N-degrons, 704

Proofreading
  in DNA repair, 866–867, 866f
  kinetic, 924–925, 924f
  in transcription, 885
  in translation, 923–925, 924f, 970–971
Proopiomelanocortin (POMC), 736
Propionyl CoA, 675–676, 676f, 721, 722–723, **726b**
Prostacyclin, 693, 693f
Prostacyclin synthase, 693, 693f
Prostaglandin, 693, 693f
Prostaglandin H_2
  defined, 365
  formation of, 365, 366f
  synthase-1, 365–366, 365f, 366f
Prostate-specific antigen (PSA), 186
Prosthetic groups, 144
Protease inhibitors, 189
  α_1-antiproteinase, 227, 228f
  drugs, 189, 189f
  in HIV infection treatment, 189, 189f, 994, 994f, 998
  initial design of, 994, 994f
  pancreatic trypsin inhibitor, 226–228, 227f
  serine, 228
Proteases, 346
  active sites, 188f
  applied to glycoproteins, 346
  aspartyl, 188, **188f**
  catalytic triad in, 186–187
  cysteine, 188, **188f**
  defined, 142, 181
  metalloproteases, **188f**, 189
  reaction facilitation, 180–189
  serine, 179
Protease-substrate interactions, 185, 185f
Proteasomes. See also Amino acid degradation
  19S regulatory unit, 705–706, 705f
  20S, 705–706, 705f
  26S, 705, 705f
  in amino acid degradation, 705–707, 706f, 707f
  defined, 705
  evolution of, 707, 707f
  free amino acid generation, 706f
  prokaryotic, 707, 707f
Protein Data Bank (pdb), 136
Protein domains, 52, 52f
Protein folding, 25, 939
  essence of, 60
  funnel, 60, 61f
  as highly cooperative process, 59
  illustrated, 25f
  Levinthal's paradox in, 59
  misfolding and, 745
  nucleation-condensation model, 60, 60f
  pathway of chymotrypsin inhibitor, 60f
  by progressive stabilization, 59–60
  transition from folded to unfolded, 59f
  typing-monkey analogy, 60, 60f

Protein identification
  cleavage in, 121–123, 122f
  co-immunoprecipitation in, 116, 117f
  genomic and proteomic methods as complementary, 123
  MALDI-TOF mass spectroscopy in, 118–119, 119f
  mass spectroscopy in, 118–126, 119f, 121f, 125f
  peptide mass fingerprinting, 124, 125f
  in protein studies, 101
Protein kinase A (PKA)
  cAMP activation of, 221–222, 222f
  chains, 418
  consensus sequence, 220
  Cushing syndrome, 222–223
  defined, 219
  gene stimulation, 418
  in glycogen metabolism, 651, 652f, 656, 657f, 658f
  regulation of, 221–222, 222f
  in signal transduction, 418–419, 424
Protein kinase B (PKB), 742, 742f
Protein kinase C (PKC), 421, 421f, 743
Protein kinase inhibitors, as anticancer drugs, 435
Protein kinases
  defined, 219
  in phosphorylation, 219–220, 220t
  serine, 219–220, 220t
  in signal transduction, 433
  specificity, 220
  threonine, 219–220, 220t
Protein phosphatase, 220
Protein phosphatase 1 (PP1)
  functions of, 653, 656–658
  in glycogen metabolism, 653, 656–658, 657f, 659
  phosphorylase a and b and, 659–660, 660f
  regulation in muscle, 657f, 658, 658f
  regulation of glycogen synthesis by, 656–658, 657f
  substrate for, 659
  subunits, 658
Protein purification
  affinity chromatography in, 104–105, 104f
  assays in, 101
  binding affinity and, 102–105
  cell release and, 101–102
  charge and, 102–105
  dialysis in, 103, 103f
  effectiveness, calculating, 110b–111b
  electrophoretic analysis of, 109f
  gel electrophoresis in, 105–107, 105f, 106f, 107f
  gel-filtration chromatography in, 103, 103f
  high-performance liquid chromatography (HPLC) in, 105, 105f
  homogenization in, 101–102
  ion-exchange chromatography in, 103–104, 104f
  isoelectric focusing in, 107–108, 108f
  in protein studies, 101

quantitative evaluation of, 109–111, 109t
  recombinant DNA technology in, 112
  salting out in, 102–103
  SDS-PAGE in, 106–107, 107f, 108
  sedimentation coefficient and, 111
  sedimentation-equilibrium technique in, 111
  separation of proteins in, 105–109
  size and, 102–105
  solubility and, 102–105
  two-dimensional electrophoresis in, 108–109, 108f, 109f
  ultracentrifugation and, 111
Protein sorting. See also Protein targeting
  defined, 939
  pathways, 941–942, 942f
Protein structures, 2–3
  amino acid sequence as determinant, 55–64
  amino acid sequences and, 39f, 42f
  complex assembly, 34, 34f
  determination by cryo-EM, 135–136, 136f
  dictating function, 34f
  elucidation of, 136
  family of, 135, 135f
  primary, 34, 41–46, 310–311
  quaternary, 34, 55, 55f
  secondary, 34, 46–50, 52, 58t
  synthetic peptides and, 126
  tertiary, 34, 50–54, 310–311, 310f
  three-dimensional prediction from sequence, 61–62, 62f
Protein studies
  enzyme-linked immunosorbent assay (ELISA) in, 114–115, 115f
  fluorescence microscopy in, 117–118, 117f
  immunology in, 112–118
  mass spectroscopy in, 118–126, 119f, 121f, 125f
  peptide synthesis, 126–129
  purification methods for, 101–112
  steps in, 100
  western blotting in, 116, 116f
  X-ray crystallography in, 129–132
Protein synthesis, 916–944. See also Translation
  accuracy of, 917–918, 931
  antibiotic inhibitors of, 942–944, 943f, 943t, 944f
  endoplasmic reticulum (ER) in, 939–942
  eukaryotic versus bacterial, 936–937
  insulin and, 739
  manipulation of, 938–939, 938f
  peptide-ligation methods, 129
  polypeptide-chain growth in, 917, 917f
  ribosomes as site of, 917, 926–939
  RNA in, 880
Protein targeting, 939–942
Protein turnover, regulation of, 703–707. See also Amino acid degradation
Protein wasting, 752–754, 754f
Proteinogenic amino acids, 126
Proteins, 33–64

acetylation of, 219–220
adaptor, 428
affinity tags and, 112, 170f
aggregated, in neurological diseases, 62–63, 63f
aggregation of, 703
allosteric, 210, 211–217
alpha helix of, 46–48, 47f
amino acids. *See* Amino acids
antibody generation to, 112–113
β pleated sheets of, 48–50, 49f, 50f
binding and, 70–72
binding partners of, 116
building blocks of, 2
cargo, 941–942
cellular, degradation of, 703
chaperone, 57
cleavage of, 121–123, 122f
coat proteins (COPs), 942
coiled-coil, 53, **53f**
covalent modification of, 218–223, 219t
crystallization of, 130
defined, 2, 43
degradation of, 754, 754f. *See also* Amino acid degradation
denatured, 56, 59, 59f, 162
dietary, digestion and absorption of, 702–703, 702f
disulfide bonds, 43
DNA-binding, 950–956
encoding, 25
enzymes as, 142
fibrous, 53–54
flexibility of, 35, 35f
folding of, 25, 25f
functional groups of, 34
functions of, 33–35
genetically engineered, 281
glycan-binding, 346
glycosylation of, 339
half-life of, 703, 704, 705t
interaction of, 34
internal repeats, 124, 124f
as linear polymers, 34, 41
loops, 50, 50f
membrane. *See* membrane proteins
misfolding of, 62–63
molecular weights of, 111t
native, 57
overview of, 100
phosphorylation of, 223
plasma-membrane, 940, 940f
posttranslational modifications, 63–64
properties of, 33–35
pure, 101
purification of, 101–112
refolding of, 57
regulatory, 954
repressor, 953
reverse turns, 50, 50f
ribosomal. *See* ribosomes
rigidity of, 35
S values of, 111t
secretory, 940, 940f
secretory pathway, 939
separation of, 100, 105–109

subunits of, 55
tagging for destruction, 703–705
translocation of, 940–941
unfolding of, 59, 59f
Proteoglycans
in cartilage, 340–341, 340f
defined, 338
properties of, 339–340
structural roles, 339–340
Proteolysis, 142, 753
Proteolytic activation, 223–228
apoptosis by, 224
of chymotrypsinogen, 224–225, 224f
zymogens, 225f, 226
Proteolytic enzymes
activation of, 223–228
in blood clotting, 228–229
chymotrypsinogen as, 224, 224f
in digestion, 224–225
function of, 143
inhibitors of, 226–228
Proteomes, 125–126
Proteomics, 125
Prothrombin, 228–229, 229f
Proton abstraction, 193
Proton dissociation, 18
Proton gradients
across thylakoid membrane, 599–602
in ATP synthesis, 562–569
biochemical processes and, 579–581
cytochrome *bf* contribution to, 596, 596f
formation of, 548–549
in oxidative phosphorylation, 458, 458f, 543, 548–549, 562–569
in photosynthesis, 592
photosystems and, 593–599
power transmission by, 579–581, 581f
Proton shuttle, 193, 194f
Proton transport, 558–559, 559f
Proton-motive force
ATP forms without, 565–566, 565f
energy in, 599
in oxidative phosphorylation, 543, 562–563, 563f, 565–567, 565f
in photosynthesis, 599
Proto-oncogenes, 433
Protoplasts, 296
Protoporphyrin IX, 786, 787f
Protospacer-adjacent motif (PAM), 294
Proviral DNA, 290
Proximal histidine, 73, 79, 79f
Prusoff, William, 987
Pseudo-first-order reactions, 152
Pseudogenes, 283–284
Pseudosubstrate sequence, 222
Psi (ψ) angle, 45, 45f, 46f
Psoralens, 866, 866f
P-type ATPases
defined, 384
as evolutionarily conserved, 387–388
roles of, 387–388
SERCA, 384–387
Puffer fish, genome of, 286–287
Pumps
ATP-driven, 384
calcium ion, 384–387, 385f

defined, 384
ion gradients, 382
MDR, 388
Na$^+$–K$^+$, 384, 387, 387f
purification of, 384
P*ur* repressor, 954f, 955
Pure noncompetitive inhibition, 166, 168f
Purine catabolism, 809f
Purine nucleotide synthesis, 783, 797–801
AMP in, 799–800, 799f
control of, 807–808, 808f
de novo pathway for, 796f, 797, 797f, 798–799, 798f, 799t
enzymes of, 800
GMP in, 799–800, 799f
5-phosphoribosyl-1-pyrophosphate (PRPP) in, 797, 807–808
phosphorylation in, 797–799, 798f
ribose phosphate in, 797
salvage pathway for, 797, 800
Purine nucleotides, 792, 797
Purine synthesis, 799–800, 799f
Purinosomes, 800, 800f
Puromycin, 943, 943t, **943f**
PurZ, 801, 801f
Pyran, 327
Pyranose, 327, 327f, 329–330
Pyridoxal phosphate enzymes
bond cleavage in, 712, 712f
reaction choice, 711–712, 711f, 712f
stereoelectronic effects, 712, 712f
Pyridoxal phosphate (PLP)
in amino acid degradation, 709–710, 711–712
in glycogen metabolism, 643–645, 644f
in Schiff-base intermediates formation, 709–710
in Schiff-base linkage, 769
structure of, **709**, **711f**
Pyridoxamine phosphate (PMP), 710, 710f, **710**, 769f, 770
Pyridoxine (vitamin B$_6$), 463t, 464f, 709–710, **709**
Pyrimidine nucleotide synthesis, 783, 793–797
carbamoyl phosphate in, 793f, 794
control of, 807
cytidine triphosphate (CTP) in, 796
de novo pathway for, 792, 792f, 796
glutamine in, 796
mono-, di-, and triphosphates in, 796
orotate in, 794–795
5-phosphoribosyl-1-pyrophosphate (PRPP) in, 779, 795
regulation of, 807
salvage pathway for, 796–797
substrate channeling in, 794
Pyrimidine nucleotides
defined, 792
recycling of, 796–797
Pyrimidines, 866, 866f
Pyrosequencing, 284, 285f
Pyrrolidine rings, 37, 54, 54f

Pyruvate
alcohols, formation from, 486–487
in amino acid degradation, 712–713, 718, 719, 719f
in amino acid synthesis, 769
in ATP formation, 484–485
carboxylation of, 501–502, 535
conversion into phosphoenolpyruvate, 484, 501–503, 502f
fates of, 486, 486f
formation of, 484–485, 719, 719f
glucose conversion into, 485–486
in glycolysis, 473, 473f, 476f, 485–486, 486f
in ligation reactions, 467
metabolism disruption, 535–536
oxaloacetate synthesis from, 467
reduction of, 487, 487f
structure of, **101**, **159**, **712**
Pyruvate carboxylase
in citric acid cycle, 534–535
in gluconeogenesis, 500f, 501–502, 501f, 502f
Pyruvate carboxylase deficiency (PCD), 510–511
Pyruvate decarboxylase, 486
Pyruvate dehydrogenase, 531–533, 532f, 535
diabetic neuropathy from inhibition of, 532–533
Pyruvate dehydrogenase complex, 515, 516
of *B. stearothermophilus*, 520f
carbon dioxide production, 518–521
in citric acid cycle, 516, 517–521, 518f, 520f, 522f, 531–533, 531f, 532f
components of, 518–521, 520f
of *E. coli*, 518t
mechanism, 518–520, 519f
reactions of, 520–521, 522f
regulation of, 531–532, 531f
response to energy change, 531, 532f
Pyruvate dehydrogenase, in citric acid cycle, 531–533
Pyruvate dehydrogenase phosphatase, 531
Pyruvate kinase
in glycolysis, 494, 495f, 496–497, 497f
liver and glycolysis and, 496–497, 496f
muscle and glycolysis and, 494
Pyruvate kinase isozyme (PKM), 633
Pyruvate kinase M, 498

Q cycle
defined, 555
in respiratory chain, 555–556, 555f
Q pool, 553
Q-cytochrome c oxidoreductase, 549, 549f, 550t
defined, 554–555
in respiratory chain, 554–555
structure of, 554f, 555
Quantitative PCR (qPCR), 287, 287f
Quaternary protein structure, 34, 55, 55f
Quinones
oxidative states of, 550–551, 550f
in photosynthesis, 592, 592f
ubiquitous, 561
Quinonoid intermediate, 710

R groups, 35, 37f. *See also* Amino acid side chains

R state of hemoglobin, 78, 78f, 79
Racemizations, 711
Raf, 432
Ramachandran, Gopalasamudram, 45
Ramachandran plot
for angles of rotation, 46f
β strands, 48f
defined, 45–46
for helices, 47, 47f
Random-coil conformation, 56
Ras
activation of, 432, 432f
affixation to cytoplasmic face, 218
GTPase activity of, 432
in protein kinase cascade initiation, 432–433
Ras genes, 271
Rate constants, 152
RBP4, 737
R-carvone, **436**
Reaction centers
accessory pigments and, 602–605, 603f, 604f
bacterial, 591f, 592–593, 592f, 593f
chloroplasts and, 590–591, 591f
cyclic electron flow in, 592–593
defined, 590
photosystem I, 597–598, 597f
photosystem II, 594–596, 594f
Reaction rates
defined, 152
enzymes and, 148, 148f
kinetics as study of, 152
rate constants, 152
Reactions
acid–base, 17–18
ATP hydrolysis, 204, 204f
bimolecular, 152
in Calvin–Benson cycle, 611–620, 617f
carbonic anhydrases and, 190–194
of citric acid cycle, 528
deleterious, 480
DNA ligase, 849–850, 849f
double-displacement, 160–161
elongation, 692
enzyme acceleration of, 148–151
equilibrium, 146–147
equilibrium constant of, 146
in fatty acid oxidation, 673t
in fatty acid synthesis, 684–689, 686t, 687f
first-order, 152
free-energy difference of, 145
of gluconeogenesis, 504t
group-transfer, 465–466
hydrolytic, 466
isomerization, 467
ligation, 467
multiple-substrate, 159–161
oxidation-reduction, 465
oxygenase, 614–615, 614f
pentose phosphate pathway, 627–630, 629f, 630f
pseudo-first-order, 152
reduction potentials of, 548
second-order, 152
sequential, 159–160
standard reduction potentials of, 547t
transaldolase, 629–630, 630f, 631f

transketolase, 627–628, 629f, 631f
velocity versus substrate concentration, 149, 149f
Reactive oxygen species (ROS)
accessory pigments and, 604
defined, 560
DNA damage and, 865
exercise and, 561
glucose 6-phosphate dehydrogenase in protection against, 634–635
pathological conditions that may entail injury, 560t
release prevention, 74
source of, 677
Reactive substrate analogs, 170, 170f
Real-time PCR, 287
RecA, 872–873, 957
Receptor tyrosine kinase, 426
Receptor-mediated endocytosis, 375–376, 375f, 376f, 829, 829f
Receptors, 413
Reclinomonas americana, 545
Recognition helix, 951
Recognition sequence, 194
Recognition sites (sequences)
in cognate and noncognate DNA, 197–199, 198f, 199f
defined, 194
distortion of, 198, 198f
EcoRV endonuclease, 197f, 198
structure of, 197, 197f
Recombinant DNA
formation of, 272–273
manipulation of, 272
Recombinant DNA technology
amino acid sequence information, 123
in biology, 272–281
blotting techniques in, 265, 266–267, 267f
cloning in, 277–278
cohesive-end method in, 272, 273f
complementary DNA (cDNA) in, 278–279, 278f, 279f
complete genome, 282–287, 282f
disease-causing mutations, functional effects of, 271–272
in disease-causing mutations identification, 271–272
DNA probes in, 266–267, 268–269
DNA sequencing in, 265, 267–268
DNA synthesis in, 268–269, 268f
electroporation in, 296, 296f
eukaryotic genes, 287–296
functional effects of disease-causing mutations and, 271–272
gel electrophoresis in, 266–267, 266f
gene disruption in, 290–294, 291f, 292f, 294f
gene expression analysis in, 287–289, 287f, 288f
gene insertion into eukaryotic cells, 289–290
gene therapy in, 296
genome comparison, 286–287, 286f
genome sequencing, 282–283
genomic libraries in, 277–278, 277f, 278f
human genome, 283–284, 283f
λ phage in, 275–276, 276f

mutation creation, 279–281
in mutation identification, 271–272
next-generation sequencing methods, 284–286, 285f, 286f
in plants, 295–296, 295f
plasmids in, 274–275
polymerase chain reaction (PCR) in, 265, 269–270, 269f, 270f
in protein purification, 112
restriction-enzyme analysis in, 265–266, 272–273
RNA interference in, 294–295, 295f
solid-phase approach in, 268–269, 268f
tools of, 265–272
transformed into animal cells, 289–290
transgenic animals in, 290, 291f
vectors in, 274–276
Recombinant human insulin, 995. See also Insulin
Recombinant proteins, 995, 995t
Recombinases, 873, 873f
Recombination, DNA. See DNA recombination
Recombination synapse, 874
Recovery, 369, 400
Red blood cells, 80–81, 80f, 81f, 82f
Red fluorescent protein (RFP), 64f
Redox balance, 486
Redox couples, 546
Redox potential. See Reduction potential
Reducing agent, 546
Reducing end, 333
Reducing sugars, 330–331
Reductant, 546
Reductase, 764, **765f**
5α-reductase, 838
Reductase mRNA in cholesterol regulation, 827
Reduction potential
measurement of, 546, 546f
of NAD+, 547
in oxidative phosphorylation, 545–548, 546f
of reactions, 548
Refed state, 741
Refractory period, 404
Regulator genes, 952
Regulatory domains
in amino acid synthesis, 780–782, 781f
recurring, 781f
structures of, 780–781
Regulatory protein P_II, 782–783, 782f
Regulatory proteins, 954
Regulatory sites, 211
Release factors
bacterial, 937
ribosome (RRF), 935
translation termination by, 935–936, 935f
Relenza, 348
Renewable energy, 606–607, 607f
Repeating motifs, 124, 124f, 311–312, 312f
Replication factor C (RFC), 862
Replication fork, 859f. See also DNA replication
defined, 849
schematic view of, 858f
Replication protein A, 862

Replicon, 861
Reporter genes, 274
Repressors
corepressors, 955
lac, 953, 953f, 954f
λ, 957–958, 957f
protein, 952
pur, 955, 955f
Residue, 42, 53, 54
Resistin, 737
Resonance energy transfer, 603–604, 603f
Resonance structures
ATP (adenosine triphosphate), 453, 453f
defined, 8
depiction of, 8
of orthophosphate, 453f
Respirasome, 559, 559f
Respiration. See Cellular respiration
Respiratory chain
coenzyme Q in, 551–552
components of, 549f, 550t
cytochrome c oxidase in, 554–555, 556–559
defined, 543
electron transfer in, 561, 562f
NADH-Q oxidoreductase in, 552–553, 553f
in oxidative phosphorylation, 549–562
Q cycle in, 555–556, 555f
Q-cytochrome c oxidoreductase in, 554–555, 554f
succinate-Q reductase in, 553–554
Respiratory control, 573–574, 575f
Respiratory quotient (RQ), 748, 748f
Restriction endonucleases. See Restriction enzymes (endonucleases)
Restriction enzymes (endonucleases), 179
analysis, 265
catalytic strategies of, 179, 194–200
defined, 194, 265
in DNA cleavage, 194–200, 265–266
in E. coli, 194
in forming recombinant DNA molecules, 272–273
hydrolysis of phosphodiester bond, 194–195, 195f
inverted repeats, 197, 197f
recognition sites, 197, 197f, 198f
restriction-modification systems, 200
specificity, 197, 200, 266, 266f
type II, 194
Restriction fragments
separation by gel electrophoresis, 266–267, 266f
in Southern blotting, 266
Restriction-fragment-length polymorphisms (RFLPs), 271
Restriction-modification systems, 200
11-cis-retinal, 438–440, **438**, 439f
Retinitis pigmentosa, 906
Retinol (vitamin A), 464t, 756
Retrovir, 189
Retroviruses
in drug resistance evolution, 998
in gene introduction, 290
transduction of, 290
Reverse cholesterol transport, 832
Reverse transcriptase, 863

Reverse turns, 50, 57–59
Reversible covalent modification, 468, 782–783, 782f
Reversible inhibition, 164, 166–168, 166f, 167f. See also Enzyme inhibition
Reversible terminator method, 285
Revertants, 871
Rhinovirus, 55
Rhizobium bacteria, 764
Rho, 889–890, 889f, 890f
Rhodobacter capsulatus, 364
Rhodopseudomonas blastica, 364f, 365f
Rhodopseudomonas viridis, 590, 591f, 606
Rhodopsin, 370, 438–440, 439f
Rhodospirillum rubrum, 551
Riboflavin (vitamin B_2), 463t, 464f
Ribonuclease
amino acid sequences in, 56–57, 56f, 57f
denatured, 56
reduction and denaturation of, 56f
sequence comparison of, 302, **302f**
structure of, 56–57, 56f
Ribonuclease III (RNase III), 899
Ribonuclease P (RNase P), 899
Ribonucleic acid (RNA). See RNA
Ribonucleotide reductase
defined, 801
in deoxyribonucleotide synthesis, 801–803, 802f, 804f, 808–809, 808f
elements of, 802, 802f
mechanism, 803
regulation of, 808–809, 808f
stable radicals, 803–804, 804f
tyrosyl radicals, 802f, 803
Ribose, 326–327, 326f, 329f
Ribose phosphate, 797
Ribosomal initiator element (rInr), 895
Ribosome release factor (RRF), 936
Ribosomes
bacterial, 936–937
catalysis by proximity and orientation, 933
defined, 916, 921
endoplasmic reticulum bound, 939–942, 939f, 940f
eukaryotic, 936–937
gene expression and, 938
large subunit (50S) of, 927, 927f, 929f, 931–932, 933, 935
polysomal, 935, 935f
in protein synthesis, 916–917, 926–939
schematic representation, 930f
70S subunit of, 927, 927f, 929f, 931
sites of, 928–929, 929f
small subunit (30S) of, 927, 927f, 929f, 931
structure of, 927, 927f
subunits of, 927, 927f
in translation, 921, 926–939
tRNA-binding sites, 928–929, 929f
Riboswitches, 891, 891f
Ribozyme, 142, 909
Ribulose 1,5-bisphosphate, 611–612, **611**, 613f, 615–618, 617f
Ribulose 5-phosphate
in pentose phosphate pathway, 624, 626, **627**, 631–633
structure of, **626**

Ricin, 944
Ricinus communis, 944
Rickets, 840
Rickettsia prowazekii, 545
Rieske center, 555
Rifampicin, 538
Right operator, 957, 957f
RNA, 6. *See also* Transcription
    ATP binding of, 320–321, 320f, 321f
    compared to DNA, 85
    double-stranded, 294–295, 295f
    enzymes from, 912
    guide strand, 295
    messenger. *See* mRNA (messenger RNA)
    micro, 908, 909f, 971–972, 971f
    passenger strand, 295
    primer, 859
    ribosomal. *See* rRNA (ribosomal RNA)
    roles of, 880–881
    single guide (sgRNA), 293, 293f
    small interference (siRNA), 295, 295f, 972
    small nuclear (snRNA), 903–906
    small regulatory, 908, 908f
    splicing of, 902f, 903–906, 905f
    structure of, 320, 321f
    synthesis of, 882
    transfer. *See* tRNA (transfer RNA)
RNA editing, 908–909, 909f
RNA interference (RNAi), 294–295, 295f, 971
RNA polymerase holoenzyme complex, 886, **887f**
RNA polymerase I, 894, 895f
    promoter sequences, 895
    in RNA synthesis, 894, 895, 895f
    rRNA production, 899–900
RNA polymerase II, 894, 895f
    promoter region, 896
    promoter sequences, 895
    in RNA synthesis, 894, 895f, 896
RNA polymerase III, 894, 895f, 897f, 899, 899f
    promoter sequences, 895
    tRNA (transfer RNA), 900
RNA polymerases, 896f
    active site of, 882
    backtracking and, 885, 885f
    in bacteria, 886–887
    catalytic action of, 881–890
    defined, 881
    in DNA replication, 848, 848f
    DNA templates and, 882
    in *E. coli*, 881, 881t, 882, 887f, 890
    eukaryotic, 881, 881f, 894–895, 894t, 896f
    eukaryotic promoter elements, 895–896, 896f
    functions of, 881
    metal ions and, 882, 883f
    phosphodiester bonds, formation of, 882–884
    prokaryotic, 881, 881f
    promoter sites in, 886–887
    in proofreading, 885
    structure of, 881, **881f**
    subunits of, 886
    in transcription, 881–890

types of, 894–895, 894t
RNA processing
    eukaryotic pre-rRNA, 899f, 900
    of pre-mRNA, 899f, 900–901, 901f
    in transcription of eukaryotes, 892–893, 908, 908f
    tRNA, 898–899, 901f
RNA sequences, comparison of, 314, 314f, 315b, 316f
RNA splicing, 901–902, 902f. *See also* Splicing
RNA-DNA hybrid
    backtracking, 885, 885f
    during elongation, 888, 888f
    lengths of, 885
    separation, 885, 885f
    translocation, 885, 885f
RNA-induced silencing complex (RISC), 295, 295f
RNA-seq, 289
Rodbell, Martin, 425b
Rods, 438, 438f
Rofecoxib (Vioxx), 997
Rotational catalysis in ATP synthase, 567, 567f
Rotenone, 577, 577f
rRNA (ribosomal RNA)
    in base pairing, 927
    folding of, 928
    in structural scaffolding, 928
    transcription of, 898–899, 899f, 900f
    in translation, 927–928, 928f
Rubisco
    in Calvin–Benson cycle, 611–615, 612f, 613f, 614f, 621
    defined, 611
    magnesium ion in, 612–613, 613f
    oxygenase reaction, 613f, 614–615, 614f
    structure of, 611–612, 612f

S_1 pockets, 185, 185f, 186
Saccharomyces cerevisiae, 283, 488, 881f, 958
Salicin, 980
Salicylic acid, 980, **980**
Salmonella test, 871–872, 871f
Salmonella typhimurium, 579–580, 579f
Salt fractionation, 110
Salting out, 102–103
Salvage pathways. *See also* Nucleotide synthesis
    defined, 792, 792f
    for IMP and GMP synthesis, 811
    purine, 797, 800
    pyrimidine, 796–797
Sample half-cell, 546
Sandwich ELISA, 115, 115f
Sanger, Frederick, 43, 267, 282
Sanger dideoxy method, 267, 267f
Sarcin-ricin loop (SRL), 931
Sarcoplasmic reticulum Ca^{2+} ATPase. *See* SERCA
Sarin, **182**
SARS-CoV-2, 89, 89f, 100, 880, 881. *See also* COVID-19
Saturated fatty acids, 679
S-carvone, **436**
Schiff bases

in amino acid degradation, 709–711, 711f
    in amino acid synthesis, 766, 767f
    in aminoacrylate, 779, **779**
    linkage with PLP, 644f
    in retinal, 438f
    in transaldolase reaction, 629–630, 630f, 631f
Scissile bond, 185, 185f
Scoring system, 307, 307f
Screening libraries, 982–983, 983f
Screw sense, 47, 47f
scRNA-seq, 289
Scurvy, 63, 756–757
SDS (sodium dodecyl sulfate), 106
SDS–polyacrylamide gel electrophoresis (SDS-PAGE), 106–107, 107f, 109, 110, 363, 363f
Second Law of Thermodynamics, 13–15
Second messengers, 413–414, 413f, 421–422, 424–425
Secondary active transport, 390
Secondary antibody, 116
Secondary assays, 982
Secondary protein structure, 34, 46–50, 52, 58t
Secondary transporters, 390–392, 390f, 391f
Second-order reactions, 152
Secretory pathway, 939
Secretory proteins, 940, 940f
Sedimentation coefficients, 111
Sedimentation equilibrium, 111
Sedoheptulose 7-phosphate, **627**
Selectins, 347
Selectivity filter, 396, 397f
Self-splicing, 909–912
    defined, 909
    example of, 910, 910f
    intron structure, 910f
    mechanism, 910, 911f
Sense (+) strand, 884
Sensory systems
    evolution of, 441–442, 441f
    7TM receptor and, 436–437, 436f
    signal transduction and, 435–442, 436f
Sequence alignment
    with conservative substitution, 307f
    defined, 303
    distant evolutionary relationships, 305–308
    with gap insertion, 303, 304f
    of hemoglobin, 307f, 308f
    homologous sequences, 308–309
    of identities only versus Blosum-62, 307f
    of leghemoglobin, 307, 307f, 308f
    of repeated motifs, 311–312, 312f
    scoring system, 307f
    shuffling and, 304f, 305, 305f
    statistical analysis of, 302–305
    statistical comparison of, 305, 305f
    three-dimensional structures in evaluation of, 311
Sequence comparison methods, 302, 310
Sequence identities, 303

Sequence template, 311
Sequential model
    allosteric enzymes, 215
    of hemoglobin-oxygen binding, 78–79, 78f
Sequential reactions, 159–160
SERCA
    defined, 384
    P-type ATPase, 384–387
    pumping by, 385–386, 385f
    structure of, 385–386
Serine, **41f**, 719, 719f
    in catalytic triads, 183–185, 183f
    in chymotrypsin, 181–182, 182f
    in cysteine synthesis, 772, 776
    defined, 38
    from glycine, 719, 719f
    in glycine synthesis, 772
    pyruvate formation from, 719, 719f
    in sphingolipid synthesis, 819, 819f, **819**
    sphingosine from, 783
    structure of, **38f**, 41, 41f
    synthesis of, 772, 776
Serine dehydratase, 712
Serine hydroxymethyltransferase, 772
Serine kinases, 219–220, 220t
Serine phosphatases, 431
Serine protease inhibitors, 226–227
Serine proteases
    catalytic strategies of, 179, 183–185
    convergent evolution and, 313, 313f
Serotonin, synthesis of, 783, **783f**, 785, 785f
Serpins, 226–227
Seven-transmembrane-helix (7TM)
        receptor
    defined, 415
    G proteins and, 424–425, 425f
    in glycogen metabolism, 652, 652f
    ligand binding to, 415–418
    in olfaction, 436–437
    phosphoinositide cascade activation, 420–421, 420f, 421f
    sensory systems and, 436–437, 436f
    in signal termination, 419f
    in signal transduction, 415f, 419f, 420–421
70S initiation complex, 931, 931f
70S ribosomes, 927, 927f, 929f, 931
Severe Acute Respiratory Syndrome (SARS), 26
Severe combined immunodeficiency (SCID), 809–810
SgRNA (single guide RNA), 293, 293f
Shaker channel, 394, 399, 399f
Shaker gene, 394
Shape complementarity, 847–848, 847f
Shemin, David, 786
Shikimate in amino acid synthesis, 776–779
Shine, John, 930
Shine–Dalgarno sequences, 930
Short interspersed elements (SINES), 284
Shuffled sequences, 304f, 305, 305f
Sialic acids, **332f**, 819
Sick sinus syndrome, 405
Sickle-cell anemia, 21, 23, 44, 83, 83f

Sickled red blood cells, 83f
Side chains
    amino acid. *See* amino acid side chains
    hydrophilic, 52
    hydrophobic, 52
    polypeptide chains, 42, 42f, 52
Side effect, 987, 997
Sigma (σ) subunit, 886–887, 887f, 890–892
Sigmoid oxygen-binding curve, 76
Sigmoidal curve, 215, 215f
Sigmoidal kinetics, 212, 212f, 215
Signal peptidase, 940
Signal sequence, 940
Signal transduction, 412–442
    abnormalities in cancer, 433, 434, 434f, 435f
    adaptor proteins in, 428
    β-adrenergic receptor in, 415, 415f, 416f, 419f
    calcium ions in, 420–422, 422f
    calmodulin in, 420, 423, 423f
    cAMP in, 412, 418–419
    cross talk, 414
    defects, 433–435
    defined, 412
    EGF signaling, 431–433
    epidermal growth factor (EGF) in, 414f, 431–433, 432f, 433f
    epinephrine in, 414–425, 414f, 416f, 419f
    function of, 413, 414f
    G proteins in, 415–418, 417f
    in glycogen metabolism, 651–653
    GTPases and, 433
    in hearing, 440–441, 441f
    insulin in, 414f, 425–431, 742–743, 742f, 744f
    molecular circuits, 413–414
    monoclonal antibodies as inhibitors, 434, 434f
    nucleotides in, 792
    in olfaction, 436–437, 436f, 437f
    overview of, 413–414, 413f
    primary messenger, 413
    principles of, 413–414, 414f
    protein kinase A in, 418–419, 424
    protein kinases in, 433
    Ras in, 432, 432f
    second messengers, 413–414, 413f, 421–422, 424
    sensory systems and, 435–442, 436f
    7TM receptors, 415f, 419f, 420–421
    in vision, 437–439, 439f
Signal-recognition-particle (SRP), 940, 940f
Sildenafil, 980, **980f**
Simple diffusion, 382
Single guide RNA (sgRNA), 293, 293f
Single-cell RNA-seq, 289
Single-molecule methods, 15–17, 16f
Single-molecule Michaelis–Menten equation, 164
Single-molecule studies, enzyme, 162–165, 163f
Single-particle analysis, 135
Single-stranded DNA, 266
Single-stranded-binding protein (SSB), 858, 858f

Sinoatrial (SA) node, 405
Site-directed mutagenesis, 187–188, 187f, 280, 280f
Sliding DNA clamp, 857, 857f
Slow-twitch muscle fibers, 649, 747
Small G proteins, 432
Small interference RNA (siRNA), 295, 295f, 972
Small nuclear ribonucleoprotein particles (snRNPs), 903–906, 903f
Small nuclear RNA (snRNA), 903–906
Small nucleolar ribonucleoprotein (snoRNPs), 899
Small subunit (30S) of ribosomes, 927, 927f, 929f, 931
Smell. *See* Olfaction
Smooth endoplasmic reticulum (SER), 667, 668f, 939
SNARE proteins, 376, 376f
Snell, Esmond, 770b
SOD1 gene, 272
Sodium dodecyl sulfate (SDS)
    in gel electrophoresis, 105f, 106–107
    PAGE, 106–107, 107f
Sodium ion channels, 396–397
    inactivation of, 399–400
    paddles, 404
    purification of, 394
    selectivity filter, 397f
Sodium ions, 393
Solid-phase DNA synthesis, 268–269, 268f
Solid-phase peptide synthesis, 127–129, 128f
Somatostatin, 735t
Sonicating, 360
Sos, 432, 432f
Southern, Edwin, 266
Southern blotting, 266, 267f, 290
Space-filling models, 76f
Special pair, 591, 592, 592f
Specific activity, 101
Specificity, dissociation constants and, 93–94
Specificity constant, 158–159
Sperm whale myoglobin, 73
Sphingolipids
    ceramide and, 819, 819f
    defined, 819
    diversity and, 821
    gangliosides, 819
    synthesis of, 819, 819f
Sphingomyelin, 359f
    defined, 357, 819
    structure of, **357f, 820f**
Sphingosine, 356
    defined, 357
    structure of, **357f, 819**
    synthesis of, 783, **783f**
Spike protein, 89, 89f
Spliceosomes, 902
    assembly and action, 903–906, 905f
    catalytic center, 905, 905f
    defined, 902
    mRNA precursors and, 903–904
    structure of, **904f**

Splicing
  alternative, 906–907, 907f, 907t
  branch points in, 902, 903f
  catalytic center, 905, 905f
  defined, 902
  group I, 910, 911f
  group II, 911, 911f, 912
  mechanism, 902f, 905f
  mRNA precursors, 903–906
  mutations affecting, 906, 906f
  pathways comparison, 911f
  self-splicing and, 909–912, 910f, 911f
  sites of, 902f, 903–906
  snRNPs in, 903–906
  spliceosomes in, 902–906, 905f
  transesterification in, 902–903, 903f, 905f
Splicing factors, 903
Split-pool synthesis, 982–983, 983f, 984f
Squalene
  cyclization of, 824–825, 825f
  synthesis of, 823, 824–825, 824f, 825f
Squalene synthase, 823
SREBP (sterol regulatory element binding protein), 756, 825–827, 826f, 830
SREBP cleavage activating protein (SCAP), 826–827, 826f
SRP receptor, 940, 940f, 941f
Stable radicals, 803–804, 804f
Standard free-energy change, 146–147, 147b, 147t
Standard reference half-cell, 546
Staphylococcus aureus, 172–173, 173f, 979
Starch, **336f**, 618
Starvation, metabolic adaptations in, 752–754, 753t
Statins, 166, 833
Stator, 566
Steady-state assumption
  defined, 152–153
  in enzyme kinetics, 152–153
Steady-state system, 152–153
Stearoyl CoA desaturase, 692, 693f
Steatorrhea, 667
Steitz, Joan, 903
Steitz, Tom, 846
Stereochemistry
  of cleaved DNA, 196, 196f
  observation of, 196
  of proton addition, 769f, 770f
Stereocilia, 440–441, 440f, 441f
Stereoelectronic control, 712, 712f
Stereoisomers, carbohydrate, 325, 326f
Steric exclusion, 46
Steric repulsion, 44
Steroid hormones
  binding and activation, 834
  cholesterol as precursor, 827, 834, 834f
  classes of, 834
  defined, 834
  hydroxylation of, 834–836
  identification of, 834
  synthesis of, 834–836, 834f
Sterol regulatory element binding protein (SREBP), 756, 825–827, 826f, 830
Sterol regulatory element (SRE), 825–826
Stewart, Sarah, 962b

Sticky ends, DNA, 272–273, 273f
Stochastic processes, 16
Stoichiometric coenzymes, 144
Stoichiometry
  of gluconeogenesis, 504
    for light reactions, 602
  of palmitate synthesis, 690
Stoma, 623f, 624
Strand invasion, 872–873, 873f
Strand separation, 269, 269f
Streptococcus lactis, 488
Streptococcus pyogenes, 293, 294b
Streptococcus thermophilus, 488
Streptomyces achromogenes, 265
Streptomyces coelicolor, 891
Streptomyces lividans, 395
Streptomycin, 942, **942**, 943t
Stroma, 588, 594f, 596f
Stroma lamellae, 588
Strominger, Jack, 91, 979
Structural genes, 952
Structure-activity relationship (SAR), 994
Structure-based drug development, 993–995, 994f
Stryer, Lubert, 28
Stubbe, JoAnne, 804b
Substituted enzyme intermediates, 161
Substitution matrix, 305–308, 306f, 307f, 309f
Substitutions
  in amino acid sequences, 305–308, 306f, 307f, 309f
  conservative, 305, 307f, 308f
  nonconservative, 305, 307f
Substrate binding
  in catalysis, 151
  conformation selection, 151
  induced-fit model, 151, 151f, 180
  lock-and-key model, 150–151, 151f
Substrate channeling
  in amino acid synthesis, 779, 779f
  defined, 497, 530
  in pyrimidine synthesis, 794
Substrate cycles, 508, 508f
Substrate-induced cleft closing, 477
Substrate-level phosphorylation, 482–483
Substrates. *See also* Enzyme-substrate (ES) complex
  accessibility of, controlling, 467–468
  in biochemical reactions, 159–161
  chromogenic, 182, 182f
  concentration, 152f, 153–155, 153f, 156
  defined, 142
  enzyme interaction, 150–151
  homotropic effect on allosteric enzymes, 215
  multiple, in reactions, 159–161
  reciprocal relation, 808
  spectroscopic characteristics of, 149
Subtilisin
  catalytic triad of, 186–187, 187f
  oxyanion hole of, 186–187, 187f
  site-directed mutagenesis of, 187, 187f
  structure of, 313, **313f**
Subunit, polypeptide chain, 55

Succinate
  in glyoxylate cycle, 537, 537f
  oxidation of, 528–529
  in oxidation-reduction reactions, **465**
Succinate dehydrogenase in citric acid cycle, 528–529
Succinate-Q reductase, 549, 549f, 550t, 553–554
Succinyl CoA
  in amino acid degradation, 718, 719f, 721, 721f
  in citric acid cycle, 526–527, 527f
  formation of, 526, 677–678, 678f, **678f**
  porphyrins from, 786–787
Succinyl CoA synthetase
  in biochemical transformation, 527
  reaction mechanism of, 527, 527f
Sucrase, 334, 474
Sucrose
  defined, 334
  structure of, 333–334, **333f**
  synthesis of, 618, 619f
Sucrose 6-phosphate, 618, 619f
Sugars
  in cyclic forms, 327–329
  DNA, 4
  five-carbon, 615–616, 616f, 624–630
  monosaccharides. *See* Monosaccharides
  N-linked, 341–342
  nonoxidative interconversion of, 624
  O-linked, 341–342
  phosphorylation, 333
  reducing, 330–331
Suicide inhibition, 171, 806, 806f, 810
Suicide inhibitors, 171
Sulfamethoxazole, 807
Sulfanilamide, 166, **166**
Sulfhydryl groups, 39
Sulfolipids, 588–589
Sulfolobus archaea, 2f
Sulfolobus shibatae, 855f
Summers, Michael, 28, 28f
Supercoiling
  ATP hydrolysis in, 855–857, 856f
  catalyzation of, 855
  defined, 851
  degree of, 851–853
  DNA, condenses, 853f
  linking number and, 852f, 853
  negative, 853f, 854, 855–857, 856f
  positive, 854
  relaxation of, 854–855, 854f
  right-handed, 852
  topoisomerases and, 854–857, 854f, 856f
  twist in, 852f, 853
  writhe in, 852f, 853
Superhelical cable, 54
Superoxide anion, 74
Superoxide dismutase, 560–561, 560f
Superoxide radicals, 560–561
Supersecondary structures, 52, 52f
Suppressors of cytokine signaling (SOCS), 737–738, 738f
Surface complementarity, 12
Surroundings, 12
Sutherland, Earl, 425b
Svedberg units, 111

Symmetry matching, 951
Symporters, 390, 390f
Synaptic cleft, 400, 400f
Synaptic vesicles, 375f
Synchrotron radiation, 130
Syngas, 692
Synthetic biology, 620, 620f
Synthetic oligonucleotides, 320–321
Synthetic peptides
　as antigens, 126
　construction of, 126–129, 128f
　as drugs, 126, 127f
　linking of, 129
　in receptor isolation, 126
　in three-dimensional structure of
　　proteins, 126
Synthetic vaccines, 281
Systems, 12

T cells, killer, 86
T state of hemoglobin, 78, 78f, 82
Tamiflu, 348
Tamoxifen, **93**, 95f
Tandem mass spectroscopy, 121, 121f
Taq DNA polymerase, 270
Target, in drug development, 978
Target validation, in drug development,
　978
Target-based screening approach,
　985–986, 985f
Taste. *See* Sensory systems
TATA box, in transcription of eukaryotes,
　896, 896f
TATA-box-binding protein associated
　factor (TAF), 966
TATA-box-binding protein (TBP), 312, 312f,
　896–897, 897f, 966
Taurocholate, **833f**
Taxol, 977, **977**
Tay–Sachs disease, 820–821
T-cell receptors (TCRs)
　defined, 91
　MHC proteins and, 89–91
　structure of, **91f**
Telomerase, 863–864, 863f
Telomeres, 863, 863f
Temperature, 13, 161–162
Template, DNA
　coding strands and, 884, 884f
　defined, 846
　DNA polymerases, 846
　lagging strand, 859
　in replication, 846
　RNA polymerases and, 882
　transcribed regions of, 888–889
　transcription bubbles and, 888
Template strand, 884, 884f
Termination utilization substance (Tus),
　860
Ternary complex, 159
Terpenes, 840, 840f
Tertiary protein structure, 34, 50–54,
　310–311, 310f
Testosterone, 70–71, 71f, **71**, 834,
　838, **838f**
Tetracycline, 943t
Tetrahydrobiopterin, 723

Tetrahydrofolate, 773f
　in amino acid synthesis, 772–774, 773f
　defined, 772
　one-carbon groups carried by, 773f, 773t
Tetrahymena, 909, 910f, 911
Tetrapyrrole, 786
Tetraubiquitin chain, **704f**
Tetrodotoxin, **394**
Tetroses, 325
TFIID protein complex, 896–898, 897f
Thalassemia, 906, 906f
Therapeutic index, 993
Thermodynamics
　of coupled reactions, 449, 451–452
　laws of, 12–14, 734
　of metabolism, 448–449
Thermogenin, 576
Thermus aquaticus, 270, 881f
Thiamine (vitamin B$_1$), 463t, 535, 756
Thiamine pyrophosphate (TPP), 615
　in citric acid cycle, 518–519, 519f
　deficiency, 535
　defined, 535
　structure of, **518**
　in transketolase reaction, 629, 629f
Thioester intermediate, 481, 481f, 482,
　483f
Thioesterase, 689
Thioesters, 462
Thioether groups, 36
Thiol groups, 39
Thioredoxin, 601
　in Calvin–Benson cycle, 621–622, 621t,
　　621f
　enzyme activation by, 622f
　enzymes regulated by, 621t
Thioredoxin reductase, 803
30S initiation complex, 931, 931f
3D genome, 863
Three-dimensional structure
　conservation of, 310–311, 310f
　convergent evolution, 313
　repeated motifs, 312
　RNA sequences, 314, 314f, 315b
　in sequence alignment evaluation, 311
　tertiary structure and, 310–311
　in understanding evolutionary
　　relationships, 308
Threonine
　defined, 38
　degradation of, 721–722
　pyruvate and, 719
　structure of, **38f**
　synthesis of, 781
Threonine deaminase, 721–722, 780–781,
　780f, **781f**
Threonine dehydratase, 712, 721
Threonine kinases, 219–220, 220t
Threonine operon, 969, 969f
Threonyl-tRNA synthetase, 922–923,
　923f
Threshold effects, 215f, 216
Thrombin, 143, **143f**
　antithrombin and, 232
　in blood clotting, 228–229, 231
　dual function of, 231
　inhibitors of, 232

Thromboxane synthase, 693
Thromboxanes, 693, 693f
Thunnus alalunga, 551
Thylakoid membranes, 594f
　defined, 588
　pH gradient across, 599
　photosynthesis in, 588–589
　photosystem II, 596f
　photosystem II and, 594
　proton gradient across, 599–602
　stacked, 605, 605f
　unstacked, 605, 605f
Thylakoid spaces, 588–589, 599
Thylakoids, 588
Thymidine, 796
Thymidine kinase, 796–797
Thymidine monophosphate (TMP), 804, 805f
Thymidine phosphorylase, 796
Thymidylate
　blocking synthesis of, 805–807, 806f
　defined, 804
　synthesis of, 804–805, 805f
Thymidylate synthase, 804, 805f
Thymine, 796
　in DNA repair, 868–869
　structure of, **4**, **5**, **5f**
　synthesis of, 774
Thyroxine, **963**
Thyroxine, synthesis of, 783, **783f**
Time-of-flight (TOF) mass analyzer, 119,
　119f
Tip links, 441, 441f
Tissue factor pathway inhibitor (TFPI),
　232
Tissue factor (TF), 228
Titin, 43
Titration, 19
Toll-like receptors (TLRs)
　defined, 84
　extracellular domain of, 85
　illustrated, 85f
　PAMPs and, 85f
Topoisomerases
　bacterial, 855–857
　defined, 853–854
　type I, 854–855, 854f
　type II, 854, 855–857, **855f**, 856f
Topoisomers, 853, 853f
Torpedo marmorata, 401, 401f
Torsion angles, 45
Tosyl-l-phenylalanine chloromethyl
　ketone (TPCK), 170, **170**, **170f**, 171
Toxicity, drug, 992–993, 993f
Tractability, 978
Trans configuration, peptide bonds,
　44–45, 45f
Trans fat, 679
Trans unsaturated fatty acids, 679
Transacetylase, 520–521, 521f
Transaldolase
　in pentose phosphate pathway,
　　626–628, 629–630, 630f
　reactions, 629–630, 630f, 631f
Transaminases
　in amino acid degradation, 708–713
　defined, 708
　mechanism, 710, 710f

Transamination
in amino acid degradation, 709–711
amino acid synthesis by, 767, 769–770, 769f, 770f
Schiff-base intermediates in, 710
Transcription
ATP hydrolysis in, 905
in bacteria, 886, 935, 935f
chemistry of, 882
defined, 881
elongation, 882, 884, 884f, 888, 888f
in eukaryotes, 892–893, 892f
in gene expression, 935, 950
initiation of, 882
mRNA processing and, 908, 908f
nuclear hormone receptors in, 963–965
overview of, 881
in prokaryotes, 892f
promoters in, 882, 886, 886f
proofreading in, 885
regulation of, 890–898, 892f
repression of, 967
rho in, 889–890, 889f, 890f
riboswitches in, 891, 891f
RNA polymerases in, 881–890
of rRNA, 898–899, 899f, 900f
self-splicing in, 909–912, 910f, 911f
splicing in, 880, 902f, 903–906, 903f, 905f
stages of, 882
start site, 886
termination of, 882, 888–890, 889f, 890f
upstream promoter element (UPE) in, 886, 895
Transcription activator-like effector nucleases (TALENs), 292–293, 292f, 293f
Transcription bubbles
defined, 882
elongation at, 888, 888f
schematic representation, 888f
structure of, **883f**
Transcription factors
activation domain, 960
defined, 291, 896, 959
DNA-binding domains of, 958–960
ethanol metabolism and, 756
in eukaryotes, 896–898, 897f, 958–960
in gene expression, 958–960
nuclear hormone receptor, 963–964, 964f
in prokaryotes, 958
regulatory domains of, 960
Transcription in eukaryotes, 890–898
in bacteria versus, 892, 892f
CAAT box in, 896, 896f
carboxyl-terminal domain (CTD) in, 894
control of, 892–893
downstream core promoter element (DPE) in, 896
enhancers in, 898
eukaryotic promoter elements, 896f
GC box in, 896, 896f
initiation of, 892, 896–898, 897f
initiator element (Inr) in, 896
microRNAs in, 902f, 908
nuclear membrane in, 892

pre-mRNA processing and, 899–900, 901f, 908, 908f
products of, 898–909
RNA editing in, 908–909, 909f
RNA polymerases in, 896f
RNA processing and, 892–893, 908, 908f
splicing in, 905f
TATA box in, 896, 896f
transcription factors in, 896–898, 897f
translation and, 892, 892f
upstream promoter element (UPE) in, 895
Transcription initiation, 882
in bacteria, 886
de novo, 884–885
in eukaryotes, 892, 896–898, 897f
Transcription repressors, 960
Transcriptome, 288
Transducin, rhodopsin and, 439
Transduction, 290, 413
Transesterification reactions, 902–903
Transfection, 289
Transferases, 143t
Transferred DNA (T-DNA), 296
Transferrin, 376, 970
Transferrin receptor, 376, 376f, 970
Transferrin-receptor mRNA, 970–971, 970f
Transgenic mice, 290
Transglutaminase, 230
Transition state
ATP hydrolysis, 202–203
in catalysis, 149
collapse of, 151
defined, 148
formation facilitation of, 148–151
symbol, 148
Transition-state analogs, 169, 169f, 173f, 202, 202f
Transketolase, 535, 615
in Calvin–Benson cycle, 616, 616f, 618
in pentose phosphate pathway, 626–628, 629f
reaction, 628–629, 629f, 631f
Translation, 47, 917–921
accuracy of, 917–918, 917t
activation sites, 922–923, 923f
adenylation in, 921–922
aminoacyl-tRNA synthetase in, 921–926
anticodons in, 919–921, 921f
in bacteria, 930, 931f, 935, 935f, 936–937
base pairing in, 919–920, 927–928, 928f
codon-anticodon interactions in, 921
codons in, 920–921, 921f
defined, 917
direction of, 935
elongation factors in, 931–932, 931f, 937
in eukaryotes, 892, 892f, 936–937, 937f
eukaryotic initiation of, 936–937, 937f
formylmethionyl-tRNA in, 930, 930f
in gene expression, 935
initiation factors, 931, 931f
initiation of, 929–930, 930f, 931f, 936–937
initiation sites, 929–930, 930f
mRNA in, 937
organization in, 937
overview of, 916–917

peptide bond formation in, 932–933, 933f
proofreading in, 923–925, 924f, 970–971
reading frame, 931
release factors in, 935–936, 935f
ribosomes in, 921, 926–939
rRNA in, 927–928, 928f
Shine–Dalgarno sequences in, 930
signaling in, 940–941
termination of, 935–936, 935f, 937
translocation in, 933–935, 934f
tRNA in, 918–920, 929, 930f, 936–937
wobble hypothesis and, 920–921, 920t
Translesion polymerases, 864–865
Translocase, 934
Translocation
inhibition by antibiotics, 943–944, 944f
mechanism of action, 934
of proteins, 940–941
RNA-DNA hybrid, 885, 885f
signal sequences in, 940–941, 940f, 941f
in translation, 933–935, 934f
Translocon, 940–941
Transmembrane helices, 367–369
Transport vesicles, 941–942
Transporters
ABC, 388–390, 388f, 389f, 999
antiporters, 390, 390f
ATP-ADP, 572–573, 572f, 573f
dicarboxylate, 573, 573f
glucose, 390, 474, 474t, 498, 498t
glucose 6-phosphate, 661
mitochondrial, 572–573, 572f, 573f
in oxidative phosphorylation, 570–573
phosphate, 572, 573f
in photosynthesis, 589–593, 589f
secondary, 390–392
symporters, 390, 390f
tricarboxylate, 573, 573f
uniporters, 390, 390f
Transverse diffusion, 370, 370f
Trastuzumab (Herceptin), 114, 434
Tree of life, 3f
Triacylglycerol synthesis, 815–821
Triacylglycerol synthetase complex, 816
Triacylglycerols
in adipose tissue, 753
defined, 666–668
electron micrograph of, 666f
as energy source, 692
in fatty acid metabolism, 668–669, 669f
hydrolysis of, 499
pancreatic lipases and, 667f
structure of, **666**
synthesis of. *See* Triacylglycerol synthesis
transport of, 827–829
Tricarboxylic acid (TCA) cycle. *See* Citric acid cycle
Triglycerides. *See* Triacylglycerols
Trimethoprim, 807, **807**
Trinucleotide repeats, 869, 869f
Triose phosphate isomerase deficiency (TPID), 509–510
Triose phosphate isomerase (TPI)
catalytic mechanism of, 479f, 480
in glycolysis, 478–480, 479f
structure of, **479f**

Trioses, 325
Triple helix of collagen, 54f
tRNA (transfer RNA)
 acceptor stem, 919
 anticodons of, 918, 919–920
 bases, 919–920
 CCA terminal region of, 919, 919f
 charged, 921
 codons in, 920–921
 common features of, 918–920
 defined, 918
 extra arm, 919
 helix stacking in, 919, 919f
 initiator, 936–937
 inosine in, 921
 molecule design, 918–920, 919f
 precursors of, 900f, 919
 recognition sites on, 925, 925f
 RNA polymerase III and, 900
 RNA processing, 898–899, 901f
 structure of, **918f**, 919, 919f, **919f**
 synthetase recognition of, 925, 925f
 in translation, 918–920, 929, 930f,
  936–937
 translocation of, 933–935, 934f
 wobble hypothesis and, 920–921, 920t
Trombone model, 859, 859f
Troponin complex, 419
Trp operon, 968, 968f
TRPV1, 136
Trypanosomes, 909
Trypsin, 143, **143f**
 catalytic triad in, 186–187, 186f
 chymotrypsin and, 186, 186f
 defined, 226
 generation of, 226
 S$_1$ pockets, 186f
 S$_1$ pockets of, 186f
 structure of, 186f
Trypsin inhibitor, 226–228, 227f
Trypsinogen, 226
Tryptophan, 718, **724**
 in alanine, 719
 degradation of, 724, 724f
 nicotinamide from, 783
 as precursor for neurotransmitters,
  785, 785f
 serotonin from, 783
 structure of, 37–38, 37f, **37f**
 synthesis of, 778–779, 778f
Tryptophan synthase, 779, **779f**
Tryptophan synthetase, 794
Ts elongation factor (EF-Ts), 932
t-SNARE, 942
T-to-R equilibrium, 216–217, 216f, 217f
T-to-R state transition, ATCase, 214f, 215
T-to-R transition, 78f, 80–81
Tu elongation factor (EF-Tu), 931–932, 931f
Tuberculosis (TB), 538, 707
Tumor hypoxia, 498
Tumor-inducing (Ti) plasmids, 295–296,
  295f
Tumor-suppressor genes, 433, 434, 870
Turnover number, 157, 157t
Twin studies, 23, 24t
Twist in supercoiling, 852f, 853
2,′ 3′-dideoxy analog, 267, **267**

Two-dimensional electrophoresis, 108–
  109, 108f, 109f
Two-dimensional NMR, 133–135
Twofold rotational symmetry, 197, 197f
Type 1 diabetes
 defined, 741
 glucagon excess in, 746
 insulin insufficiency in, 746
 metabolic derangements in, 746
 treatment for, 746
Type 2 diabetes
 defined, 741
 exercise and, 746
 metabolic syndrome and, 743, 743f
 treatments for, 660, 745
Type I muscle fibers, 649
Type I topoisomerases, 854–855, 854f
Type II topoisomerases, 854, 855–857,
  855f, 856f
Type IIb muscle fibers, 649
Typing-monkey analogy, 60, 60f
Tyrosine phosphatases, 431
Tyrosine, 718
 defined, 38
 degradation of, 723–724, 724f
 from phenylalanine, 768, 769
 phenylketonuria and, 728
 as precursor for neurotransmitters,
  785f
 structure of, **38f**, **40t**, **162**, **724**
 synthesis of, 776–777, 778f
 thyroxine from, 783
Tyrosine kinases, 219–220, 426, 427f, 428f
Tyrosine phosphatase 1B, 743
Tyrosinemia, 727t
Tyrosyl radical, 802f, 803, 804
Tyrosyl-tRNA synthetase, 924

Ubiquinol, 553–554, 672
Ubiquinone, 550, 672
Ubiquinone reductase, 672
Ubiquitin
 conjugation, 703–704, 704f
 defined, 703
 pathway, 707
 in protein tagging, 703–705
 research in, 708b
 structure of, **703f**
Ubiquitination, 704–705
Ubiquitin-conjugating enzyme (E2), 703,
  704f
Ubiquitin-protein ligase (E3), 703–705,
  704f
Ubiquitin-activating enzyme (E1),
  703, 704f
UDP (uridine diphosphate), 656
UDP-galactose 4-epimerase, 491
UDP-glucose
 defined, 490, 653, 792
 in galactose conversion, 491
 in glycogen synthesis, 654–655
 synthesis of, 653–654
UDP-glucose pyrophosphorylase, 653
Ultracentrifugation, 111
Ultraviolet light, and DNA damage, 866,
  866f
UMP kinase, 796

Uncompetitive inhibition, 166, 167f, 168f.
 See also Enzyme inhibition
Uncoupling proteins, 576, 576f
Unfolded protein response (UPR), 745
Unicellular red algae, 317, 317f
Uniporters, 390, 390f
Unnatural amino acids, 938–939, 938f
Unsaturated fatty acids, 674–679, 674f,
  675f, 692–694
Upstream promoter element (UPE), 886,
  895
Uracil DNA glycosylase, 869, 869f
Urate, 810
Urea, 56–57, **56**, 712
Urea cycle
 in amino acid degradation, 713–718
 carbamoyl phosphate in, 714–716
 disorders of, 717–718, 717f
 fumarate in, 715, **716**
 in gluconeogenesis, 716, 716f
 illustrated, 714f
Ureotelic, 718
Ureotelic organisms, 718
Uric acid, 718, **718**
Uridine diphosphate glucose (UDP-
  glucose), 653–654, **653**
Uridine monophosphate synthetase, 795
Uridine triphosphate (UTP), 211, 451, 653
Uridylate, 795
Uridylyl transferase (UT), 782f
Urine, disease diagnosis and, 726, 726f
Uroporphyrinogen III, 787f, 788

Vaccines
 adaptive immune system and, 89, 89f
 as biologics, 995t
 mRNA (messenger RNA), 880–881
 synthetic, 281
Vaccinia virus, 290
Valine
 degradation of, 722–723
 in maple syrup urine disease, 727
 structure of, 36, **37f**, 38
van der Waals forces, 150
van der Waals interactions
 contact distance, 10, 10f, 12
 defined, 9
 energy of, 10, 10f
 hydrocarbon tails, 360
 maximization of, 12
 as noncovalent bond, 9–10, 10f, 12
 protein stability and, 52
Variable number of tandem repeats
  (VNTR) region, 341, 341f
Vascular endothelial growth factor
  (VEGF), 321, 498
Vasopressin, 126, 127f
VDAC (voltage-dependent anion channel),
  544
Vectors
 cloning, 274, 275, 276f
 defined, 272
 expression, 275
 plasmid, 274–275, 274f
 viral, 290
Very-low-density lipoproteins (VLDLs),
  828, 828f, 828t

Vibrio cholerae, 344
Vicine, 634, **635f**
Vioxx (rofecoxib), 997
Viral hepatitis, 711
Viral receptors, 348f
Viral vectors, 290
Virions, 275
Viruses
    coats, 55, 55f
    defined, 26
    in gene introduction, 290
    infectious mechanisms of, 347–348
    influenza, 347–348, 348f
    progeny, 275
    retroviruses, 290, 998
    RNA genomes in, 880
    transduction of, 290
Visible light, 438f
Vision. *See also* Sensory systems
    in animals, 442
    color, 440, 440f, 442, 442f
    color blindness and, 442
    cones in, 440
    evolution of, 442, 442f
    ion channels in, 439, 439f, 440
    light absorption in, 439–440
    7TM receptors in, 437–439
    signal transduction in, 437–439, 439f
Visual pigments, 438, 440, 440f
Vitamin A (retinol), 464t, 756
Vitamin B_1 (thiamine), 463t, 535, 756
Vitamin B_2 (riboflavin), 463t, 464f
Vitamin B_3 (niacin), 463t, 464f
Vitamin B_5 (pantothenic acid), 463t, 464f
Vitamin B_6 (pyridoxine), 463t, 464f, 709–710, **709**
Vitamin B_7 (biotin), 463t, 501, 501f
Vitamin B_9 (folic acid), 463t
Vitamin B_{12}, 463t, 677b
    as coenzyme, 676–677, 676f, 678
    corrin ring and cobalt atom, 676–677
    in fatty acid metabolism, 676–677
Vitamin C (ascorbic acid)
    deficiency, 464t
    deficiency of, alcohol related, 756–757, 757f
    forms of, 757, 757f
    function of, 464t
Vitamin D (calcitriol)
    biochemical role, 839, 840
    cholesterol as precursor, 839–840

deficiency, 464t, 840
function of, 464t
synthesis of, 839–840, 839f
Vitamin E (α-tocopherol), 464t
Vitamin K
    in blood clotting, 231, 231f, 464t
    deficiency, 464t
    structure of, **231f**
    in γ-carboxyglutamate formation, 231, 231f
Vitamins
    coenzyme, 463–464, 463t
    defined, 463
    evolution of, 464
    metabolism of and ethanol, 756–757
    noncoenzyme, 464t
    in reducing homocysteine levels, 776
V_{max}, 156. *See also* Maximal rate
Voltage-gated ion channels, 398–399, 399f
von Gierke disease, 660–661, 661t
von Gierke, Edgar, 660
v-SNARE, 942

Wald, George, 439
Wang, James, 854
Warburg, Otto, 497
Warburg effect, 497, 575, 758
Warfarin, 231, **231f**
Water
    attack, facilitating, 201–202, 202f
    concentration of, 17
    dissociation, equilibrium constant of, 17
    as highly cohesive, 10
    hydrogen bonds, 10, 11
    as polar molecule, 10
    properties of, 10
Water-oxidizing complex (WOC), 594f, 595, 595f, 596f
Watson, James, 5, 6, 845
Watson-Crick base pairs. *See* Bases/base pairs
Wernicke–Korsakoff syndrome, 756
Wernicke's encephalopathy, 535
Western blotting, 116, 116f, 265
Wet beriberi, 535
White adipose tissue (WAT), 575
Wiley, Don, 90, 91
Williams, Roger, 770b
Windows, 368

Withering, William, 387
Wobble hypothesis, 920–921, 920t
Woese, Carl, 3
Wood-Ljungdahl pathway, 692
Wool, 53
Wooly mammoth, 301f
Writhe in supercoiling, 852f, 853
Wyman, Jeffries, 78

Xanthine oxidase, 810
Xanthohumol, **578**
Xanthomas, 830
Xanthylate (XMP), 800
Xenobiotic compounds, 990–991, 990f
Xeroderma pigmentosum, 870, 870t
X-ray crystallography
    defined, 129
    diffraction patterns, 130–131, 130f
    electron-density map, 131, 131f, 132f
    experiment, 129–132, 130f
    reflections, 130
    resolution, 131, 132f
    synchrotron radiation and, 130
X-rays, 866
Xylulose 5-phosphate, **627**, **628**

Yanofsky, Charles, 967
Yeast chromosomes, 958
Yonath, Ada, 938b

Z scheme of photosynthesis, 598, 599f
Zanamivir, 348
Zellweger syndrome, 679
Zinc
    activation of water molecule, 191–192, 191f, 192f
    in biological systems, 190
    in carbonic anhydrase, 190–191, 191f
    in translation, 923
Zinc-finger domains, 959, 959f
Zinc-finger nucleases (ZFNs), 292, 293f, 294
Zingiberene, **436**, 840, **840f**
Zwitterions, 35, 36f
Zymogens
    cascade of activations, 228–229, 228f
    conversion into proteases, 226
    defined, 223
    developmental process control by, 224
    proteolytic activation, 226, 226f
    secretion of, 224, 224f